ASM Handbook®
Volume 10
Materials Characterization

Contents

See page xi for a more detailed table of contents.

ASM Handbook®

Volume 10
Materials Characterization

Prepared under the direction of the
ASM International Handbook Committee

Division Editors

Thomas J. Bruno, National Institute of Standards and Technology
Ryan Deacon, United Technologies Research Center
Jeffrey A. Jansen, The Madison Group
Neal Magdefrau, Electron Microscopy Innovative Technologies
Erik Mueller, National Transportation Safety Board
George F. Vander Voort, Vander Voort Consulting L.L.C
Dehua Yang, Ebatco

ASM International Staff

Victoria Burt, Content Developer
Steve Lampman, Senior Content Developer
Amy Nolan, Content Developer
Susan Sellers, Content Developer
Madrid Tramble, Manager of Production
Vincent Katona, Production Coordinator
Jennifer Kelly, Production Coordinator
Karen Marken, Senior Managing Editor
Scott D. Henry, Senior Content Engineer

Editorial Assistance

Elizabeth Marquard
Lilla Ryan

ASM INTERNATIONAL

ASM International®
Materials Park, Ohio 44073-0002
www.asminternational.org

First printing, December 2019

Library of Congress Cataloging-in-Publication Data

ASM International

ASM Handbook
Includes bibliographical references and indexes
Contents: v.1. Properties and selection—irons, steels, and high-performance alloys—v.2. Properties and selection—nonferrous alloys and special-purpose materials—[etc.]—v.23. Materials for medical devices

1. Metals—Handbooks, manuals, etc. 2. Metal-work—Handbooks, manuals, etc. I. ASM International. Handbook Committee. II. Metals Handbook.
TA459.M43 1990 620.1'6 90-115
SAN: 204-7586

ISBN-13: 978-1-62708-211-2 (print)
ISBN-13: 978-1-62708-212-9 (pdf)
ISBN-13: 978-1-62708-213-6 (electronic)

ASM International®
Materials Park, OH 44073-0002
www.asminternational.org

Printed in the United States of America

Foreword

The 2019 edition of *ASM Handbook*, Volume 10, *Materials Characterization* is a revised and updated work written and reviewed by leading experts in the field. Like the previous *ASM Handbook* volume on materials characterization published in 1986, the updated Volume 10 provides detailed technical information that can enable readers to select and use analytical techniques that could be appropriate for their problem.

ASM International and the Board of Trustees is grateful for the work and dedication of volunteer editors, authors, and reviewers who devoted their time and expertise to develop a reference publication of high technical and editorial quality. A special note of thanks is offered to the division editors who put forth extraordinary efforts to keep this project focused and completed on schedule. The result is a comprehensive reference work that can help readers solve problems and gain insights.

Dr. David U. Furrer
President
ASM International

William T. Mahoney
Chief Executive Officer
ASM International

Preface

This iteration of *ASM Handbook* Volume 10 is the culmination of years of effort and planning. The division editors for Volume 10 did a tremendous job finding authors to write about new techniques, and reviewers to update articles on established techniques. Without their efforts the book would not be what it is today, a useful tool for engineers, who likely are not themselves material characterization experts.

In keeping with the format of the 1986 edition, the articles describing each characterization technique begin with an overview. This summary describes the method in simplified terms and lists common applications as well as limitations. Sample size, form, and special preparation requirements are listed upfront to help readers quickly decide if the techniques are appropriate to solve their problem.

Another useful tool kept and expanded on from the previous edition are the tables and charts in the articles at the beginning of the book. For different classes of materials, the most common methods are listed in table form, showing, for example, whether the technique is suitable for elemental analysis, qualitative analysis, surface analysis, or alloy verification. The articles also describe in general terms material characterization according to material type and serve as a jumping-off point to the more specific technique articles.

Victoria Burt
Content Developer
ASM International

List of Contributors

Rachel Anderson
Navy Research Lab

Juan Asensio-Lozano
Universidad de Oviedo

Nabil Bassim
McMaster University

Arne Bengtson
SWERIM AB

Bharat Bhushan
The Ohio State University

Thomas N. Blanton
International Centre for Diffraction Data

Nerea Bordel
University of Oviedo

Olaf J. Borkiewicz
Argonne National Laboratory

David Brown
St. John's University

Thomas J. Bruno
Applied Chemicals and Materials Division
National Institute of Standards and Technology

John Cammett
Consultant

Bob Chaney
Service Physics Corporation

Patrick Chapon
Horiba Jobin Yvon S.A.S

Sumit Chaudhary
Exponent Inc.

Yu-Sheng Chen
The University of Chicago

Kenton Childs
Sandia National Laboratories

Richard E. Chinn
National Energy Technology Laboratory

Robert Chirico
Thermodynamics Research Center
National Institute of Standards and Technology

Wei-Kan Chu
University of Houston

Ryan Deacon
United Technologies Research Center

Tom Depover
Ghent University

Gary Doll
The University of Akron

Gregory P. Dooley
Colorado State University

António M. dos Santos
Oak Ridge National Laboratory

Michael Dudley
Stony Brook University

Giovanni Ferraris
Politecnico di Torino

Wesley Frey
UC Davis

Raymond S. Fuentes
Sandia National Laboratories

Oumaïma Gharbi
Sorbonne Université

Jim Gibson
EAG

Franz Giessibl
Universitat Regensburg

Cristina Gonzalez-Gago
University of Oviedo

Mark Goorsky
UCLA

Hugh E. Gotts
Air Liquide Electronics - Balazs NanoAnalysis

Vladislav V. Gurzhiy
St. Petersburg State University

BoYing Han
EAG Laboratories

Larry Hanke
Materials Evaluation and Engineering, Inc.

Megan Harries
National Institute of Standards and Technology

Andrew Anthony Havics
pH2 LLC

April A. Hill
Metropolitan State University of Denver

Odile Hirsch
Horiba Jobin Yvon S.A.S

Christina Hoffmann
Oak Ridge National Laboratory

Douglas J. Hornbach
Lambda Research, Inc.

Bo Hu
North American Hoganas, Inc.

Michael Janicke
Los Alamos National Laboratory

Jeffrey A. Jansen
The Madison Group

Russell L. Jarek
Sandia National Laboratories

N. Jayaraman
Lambda Technolgoies

Kavita M. Jeerage
Applied Chemicals and Materials Division
National Institute of Standards and Technology

A. Joshi
Advanced Technology Center, Lockheed Martin Space

Michael Kangas
Doane University

Vassili Karanassios
University of Waterloo

Martin Kasik
MK2 Technologies, Inc.

Yongchang Kim
Argonne National Laboratory

Tiffany L. Kinnibrugh
DeNovX

Melanie Kirkham
Oak Ridge National Laboratory

Petr Klapetek
Czech Metrology Institute

Takane Kobayashi
RIKEN

Mary Kosarzycki
Element Materials Technology

S. Lampman
ASM International

Saul H. Lapidus
Argonne National Laboratory

Matteo Leoni
University of Trento
Saudi Aramco

Fuhe Li
Air Liquide Electronics - Balazs NanoAnalysis

Tara M. Lovestead
Applied Chemicals and Materials Division
National Institute of Standards and Technology

Sydney Luk
North American Hoganas, Inc.

Neil Magdefrau
Electron Microscopy Innovative Technologies

Curtis Marcott
Light Light Solutions, LLC

Patrick McKeown
EAG Laboratories

Dana Medlin
EAG Laboratories

Todd J. Menna
Element Materials Technology

Joseph R. Michael
Sandia National Laboratories

Curtis D. Mowry
Sandia National Laboratories

Erik Mueller
National Transportation Safety Board

Meredith S. Nevius
Exponent Inc.

Dale E. Newbury
National Institute of Standards and Technology

Reinhard Noll
Fraunhofer Institute for Laser Technology ILT

Gert Nolze
Bundesanstalt für Materialforschung und –prüfung, BAM

John Notte
Carl Zeiss SMT, Inc.

Lisa S. Ott
California State University, Chico

Binayak Panda
Marshall Space Flight Center / NASA

D.F. Paul
Physical Electronics USA

Yoosuf N. Picard
Carnegie Mellon University

Adam S. Pimentel
Sandia National Laboratories

Jorge Pisonero
University of Oviedo

Christopher A. Pohl
Thermo Fisher Scientific

Paul S. Prevey
Lambda Research, Inc.

Balaji Raghothamachar
Stony Brook University

Gary D. Rayson
New Mexico State University

Nicholas W. M. Ritchie
National Institute of Standards and Technology

Jessica Román-Kustas
Sandia National Laboratories

Bibhudutta Rout
University of North Texas

Richard Rucker
Exponent Inc.

Gregory Schmidt
Thermo Fisher Scientific

David T. Schoen
Exponent Inc.

Craig J. Schroeder
Element Materials Technology

Sven Schroeder
University of Leeds

Lin Shao
Texas A&M University

Paul Shiller
The University of Akron

Vince Smentkowski
GE Global Research

Deane K. Smith
The Pennsylvania State University

Alice Stankova
Horiba Jobin Yvon S.A.S

Fred Stevie
North Carolina State University

Kevin Stone
SLAC National Accelerator Laboratory

Paris Svoronos
Queensborough Community College-CUNY

Ian Swainsson
International Atomic Energy Agency

Jim Sydnor
JWS, Inc.

Agnieszka Szczotok
Silesian University of Technology

John Tartaglia
Element Materials Technology

Amber C. Telles
Sandia National Laboratories

Agnes Tempez
Horiba Jobin Yvon S.A.S.

Cristian Teodorescu
National Institute of Materials Physics

Brian Toby
Argonne National Laboratory, APS

Bryan E. Tomlin
Texas A&M University

Kenji Umezawa
Osaka Prefecture University

Kimberly N. Urness
Applied Chemicals and Materials Division
National Institute of Standards and Technology

George F. Vander Voort
Vander Voort Consulting L.L.C.

Cornel Venzago
Evonik Industries

Farooq Wahab
University of Texas at Arlington

Masashi Watanabe
Lehigh University

Paul West
AFMWorkshop

Kamila M. Wiaderek
Argonne National Laboratory

Leszek Wojnar
Politechnika Krakowska im Tadeusza Kosciuszki

Wenqian Xu
Argonne National Laboratory

Andrey Y. Yakovenko
Argonne National Laboratory

Dehua Yang
Ebatco

Randall E. Youngman
Corning Incorporated

Amirali Zangiabadi
Columbia University

Officers and Trustees of ASM International (2018–2019)

Members of the ASM Handbook Committee (2018–2019)

Chairs of the ASM Handbook Committee

Policy on Units of Measure

By a resolution of its Board of Trustees, ASM International has adopted the practice of publishing data in both metric and customary U.S. units of measure. In preparing this Handbook, the editors have attempted to present data in metric units based primarily on Système International d'Unités (SI), with secondary mention of the corresponding values in customary U.S. units. The decision to use SI as the primary system of units was based on the aforementioned resolution of the Board of Trustees and the widespread use of metric units throughout the world.

For the most part, numerical engineering data in the text and in tables are presented in SI-based units with the customary U.S. equivalents in parentheses (text) or adjoining columns (tables). For example, pressure, stress, and strength are shown both in SI units, which are pascals (Pa) with a suitable prefix, and in customary U.S. units, which are pounds per square inch (psi). To save space, large values of psi have been converted to kips per square inch (ksi), where 1 ksi = 1000 psi. The metric tonne (kg $\times$ 10^3) has sometimes been shown in megagrams (Mg). Some strictly scientific data are presented in SI units only.

To clarify some illustrations, only one set of units is presented on artwork. References in the accompanying text to data in the illustrations are presented in both SI-based and customary U.S. units. On graphs and charts, grids corresponding to SI-based units usually appear along the left and bottom edges. Where appropriate, corresponding customary U.S. units appear along the top and right edges.

Data pertaining to a specification published by a specification-writing group may be given in only the units used in that specification or in dual units, depending on the nature of the data. For example, the typical yield strength of steel sheet made to a specification written in customary U.S. units would be presented in dual units, but the sheet thickness specified in that specification might be presented only in inches.

Data obtained according to standardized test methods for which the standard recommends a particular system of units are presented in the units of that system. Wherever feasible, equivalent units are also presented. Some statistical data may also be presented in only the original units used in the analysis.

Conversions and rounding have been done in accordance with IEEE/ASTM SI-10, with attention given to the number of significant digits in the original data. For example, an annealing temperature of 1570 °F contains three significant digits. In this case, the equivalent temperature would be given as 855 °C; the exact conversion to 854.44 °C would not be appropriate. For an invariant physical phenomenon that occurs at a precise temperature (such as the melting of pure silver), it would be appropriate to report the temperature as 961.93 °C or 1763.5 °F. In some instances (especially in tables and data compilations), temperature values in °C and °F are alternatives rather than conversions.

The policy of units of measure in this Handbook contains several exceptions to strict conformance to IEEE/ASTM SI-10; in each instance, the exception has been made in an effort to improve the clarity of the Handbook. The most notable exception is the use of g/cm3 rather than kg/m3 as the unit of measure for density (mass per unit volume).

SI practice requires that only one virgule (diagonal) appear in units formed by combination of several basic units. Therefore, all of the units preceding the virgule are in the numerator and all units following the virgule are in the denominator of the expression; no parentheses are required to prevent ambiguity.

Contents

Introduction to Materials Analysis Methods

ASM Handbook, Volume 10, *Materials Characterization*
ASM Handbook Committee
DOI 10.31399/asm.hb.v10.a0006669

Introduction to Materials Characterization*

The macro, micro, and surface composition and structure of materials are determined by their processing and service histories. These conditions, in turn, control material properties and performance. Characterization of a material's composition and structure at these levels provides important information that can be used for several purposes:

- *Quality assurance:* to determine whether processing is done properly
- *Failure analysis:* to determine why properties and performance are different from expected
- *Research and product improvement:* to develop an understanding of how processing and service history influence properties and performance, and use this understanding to develop improved materials

Two useful ways of classifying materials characterization techniques are by the type of information they obtain and the characteristic sample dimensions from which this information is obtained.

According to the types of information they obtain, techniques are often classified as:

- *Elemental:* What elements are present (qualitative elemental analysis)? In what concentration is each element present (quantitative elemental analysis)?
- *Structural:* How are the atoms crystallographically arranged and spaced? In the case of organic materials, what is the molecular structure?
- *Morphological:* What are the sizes, shapes, arrangements, and appearances of key features, such as grains, cracks, welds, and interfaces?

According to the dimensions of material sampled, techniques are frequently categorized as:

- *Bulk:* information typically obtained and integrated over surface dimensions on the order of 10^{-2} m (0.39 in.) or more and depths of at least 50×10^{-6} m (1.9×10^{-3} in.), presumably characteristic of the bulk material
- *Micro:* able to obtain information from individual microstructural features; typically able to resolve and characterize features with dimensions of 10^{-6} m (3.93×10^{-5} in.) or less
- *Surface:* information obtained only from the first atomic layer or so of the sample, such as sampling depths on the order of 10^{-9} m (3.93×10^{-8} in.)

Selection of Materials Characterization Methods

It is very important to keep the previously described categories in mind when selecting appropriate analytical techniques. There is no universal characterization method that provides all information an analyst might want to know about a sample. For each problem, a decision must be made regarding the type of information needed. On the basis of this decision, one or more analytical techniques that will provide this information must be selected.

Selection and use of inappropriate methods can result in data that are analytically correct but inappropriate to the problem at hand. This misuse usually confounds the problem rather than helps solve it. For example, it would be inappropriate to use a bulk analysis technique in an effort to identify a thin layer of surface contamination on a material. The results would reflect the composition of the underlying material, not the contamination on its surface. Conversely, it would be equally inappropriate to use a surface analytical technique to determine the bulk composition of an alloy. The results would reflect whatever contaminants happened to be present on the surface, rather than the composition of the underlying alloy. In selecting analytical techniques, it is critical to think through the current problem, determine what type of information is needed, and select one or more techniques that will provide appropriate types of information.

In addition, in cases where several different types of analyses may be needed to solve a problem, it is important to consider the order in which these analyses are performed, as some types of analysis are destructive or alter the sample, making subsequent types of analysis impossible.

Scope of this Volume

Materials Characterization has been developed with the goal of providing the engineer or scientist who has little background in materials analysis with an easily understood reference book on analytical methods. Although there is an abundance of excellent in-depth texts and manuals on specific characterization methods, they frequently are too detailed or theoretical to serve as useful guides for the average engineer who is primarily concerned with solving a problem rather than becoming an analytical specialist. This Volume describes analytical methods in simplified terms and emphasizes the most common applications and limitations of each method. The intent is to familiarize the reader with the techniques that may be applied to a problem, help identify the most appropriate technique or techniques, and provide sufficient knowledge to interact with the appropriate analytical specialists, thereby enabling materials characterization and troubleshooting to be conducted effectively and efficiently. The intent of this Volume is *not* to make an engineer a materials characterization specialist.

Materials characterization represents many different disciplines depending upon the background of the user. These concepts range from that of the scientist, who thinks of it in atomic terms, to that of the process engineer, who thinks of it in terms of properties, procedures, and quality assurance, to that of the mechanical engineer, who thinks of it in terms of stress distributions and heat transfer. The definition selected for this book states, "Characterization describes those features of composition and structure (including defects) of a material that are significant for a particular preparation, study of properties, or use,

*Adapted from K.H. Eckelmeyer, Introduction and Overview of Materials Characterization, *Metals Handbook Desk Edition*, ASM International, 1998 and R.E. Whan, Introduction to Materials Characterization, *Materials Characterization*, Vol 10, *ASM Handbook*, ASM International, 1986

and suffice for reproduction of the material." This definition limits the characterization methods included herein to those that provide information about composition, structure, and defects and excludes those methods that yield information primarily related to materials properties, such as thermal, electrical, and mechanical properties.

Most characterization techniques (as defined previously) that are in general use in well-equipped materials analysis laboratories are described in this Volume. These include methods used to characterize materials such as alloys, glasses, ceramics, organics, gases, and inorganics. Techniques used primarily for biological or medical analysis are not included. Some methods that are not widely used but that provide unique or critical information are also described. Several techniques may be applicable for solving a particular problem, providing the engineer, materials scientist, or analyst with a choice or with the possibility of using complementary methods.

Organization

This Volume has been organized for ease of reference by the user. The articles at the beginning of the book include tables, flow charts, and descriptions that can be used to quickly identify techniques applicable to a given problem. In these tables, the most common methods (not necessarily all-inclusive) for analyzing a particular class of materials are listed on the left. The types of information available are listed as column headings. When a particular technique is applicable, an entry appears in the appropriate column. It should be emphasized that lack of an entry for a given technique does not mean that it cannot be adapted to perform the desired analysis; it means simply that that technique is not usually used and others are generally more suitable.

The subdivisions within the Volume address a set of related techniques, for example, spectroscopy. Within each section are several articles, each describing a separate analytical technique. For example, in the Section "Spectroscopy" are the articles "Optical Emission Spectroscopy," "Atomic Absorption Spectroscopy," and "Infrared Spectroscopy," to name a few. Each article begins with a summary of general uses, applications, limitations, sample requirements, and capabilities of related techniques, which is designed to give the reader a quick overview of the technique, and to help him or her decide whether the technique might be applicable to the problem. This summary is followed by text that describes in simplified terms how the technique works, how the analyses are performed, what types of information can be obtained, and what types of materials problems can be addressed. Included are several brief examples that illustrate how the technique has been used to solve typical problems. A list of references at the end of each article directs the reader to more detailed information on the technique.

ASM Handbook, Volume 10, *Materials Characterization*
ASM Handbook Committee
DOI 10.31399/asm.hb.v10.a0006678

Introduction to Characterization of Metals

MATERIALS IDENTIFICATION, characterization, and verification are essential for companies to know that their materials or products have been manufactured to correct alloy grade and conform to standards. Metal chemical analysis is also used in reverse engineering to determine the alloys used to produce a component and in failure investigations to establish whether the correct alloy was used.

Two ways to classify materials characterization techniques are by the type of information they obtain and the characteristic sample dimensions from which this information is obtained.

When classified by the types of information they obtain, techniques are often classified as:

- *Elemental:* What elements are present (qualitative elemental analysis)? In what concentration is each element present (quantitative elemental analysis)?
- *Structural:* How are the atoms crytallographically arranged and spaced? In the case of organic materials, what is the molecular structure?
- *Morphological:* What are the sizes, shapes, arrangements, and appearances of key features, such as grains, cracks, welds, and interfaces?

When classified by the dimensions of material sampled, techniques are frequently categorized as:

- *Bulk:* Information typically obtained and integrated over surface dimensions on the order of 10^{-2} m or more and depths of at least 50×10^{-6} m, presumably characteristic of the bulk material
- *Micro:* Able to obtain information from individual microstructural features, typically able to resolve and characterize features with dimensions of 10^{-6} m or less
- *Surface:* Information obtained only from the first atomic layer or so of the sample, such as sampling depths on the order of 10^{-9} m

Popular techniques for metals characterization are briefly discussed here, with references to other articles in this Volume. These articles include detailed discussion as well as practical tips for use. An overview of techniques is listed Table 1 and Fig. 1.

Bulk Elemental Analysis

Chemical analysis is often performed by one or more complimentary techniques. The specific technique chosen depends on the type of sample, quantity of material available for analysis, desired result, and cost constraints. In most cases, the applicable analysis techniques can detect parts per million concentrations or better.

Table 1 Analytical methods used for characterization of metals and alloys

Technique	Elemental analysis	Speci-ation	Compound	Isotopic or mass analysis	Qualitative analysis (identification of constituents)	Semiquantitative analysis (order of magnitude)	Quantitative analysis (precision of +20% relative standard deviation)	Macroanalysis or bulk analysis	Microanalysis (≤10 μm)	Surface analysis	Major component (>10 wt%)	Minor component (0.1–10 wt%)	Trace component (1–1000 ppm, or 0.0001–0.1 wt%)	Struc-ture	Morph-ology
Atomic force microscopy	N	...	N	...	•	N	...	...	•	•	•	•	...	...	•
Auger electron spectroscopy	•	...	...	...	•	•	...	...	•	•	•	•	...	...	...
Electron probe x-ray microanalysis	N	...	...	...	N	N	N	...	N	...	N	N	N	...	...
Electron spin resonance	N	N	...	...	N	N	N	N	...	...	...	N	N	N	...
Elemental and functional group analysis	•	...	•	...	•	•	•	•	...	...	•	•	...	...	...
Energy-dispersive x-ray spectroscopy	N	...	...	...	N	N	...	...	...	N	...	...	...	...	...
Fourier transform infrared spectroscopy	N	D, •	•	...	•	N	N	•	•	•	•	•	...	•	...
Gas chromatography	...	...	V	...	V	V	V	V	V	...	V	V	V	N,S	...
Gas chromatography/ mass spectrometry	N	N	V	•	V	V	V	V	...	...	V	V	V	V	...
High-temperature combustion	N	...	...	...	...	...	N	N	...	...	N	N	N	...	...

(continued)

Key: • = generally usable; N = limited number of elements or functional groups; D = after dissolution; V = volatile liquids; S = under special conditions (i.e., with tandem mass spectrometer detection); C = crystalline solids

Table 1 (Continued)

Technique	Elemental analysis	Speci -ation	Compound	Isotopic or mass analysis	Qualitative analysis (identification of constituents)	Semiquantitative analysis (order of magnitude)	Quantitative analysis (precision of +20% relative standard deviation)	Macroanalysis or bulk analysis	Microanalysis (≤10 μm)	Surface analysis	Major component (>10 wt%)	Minor component (0.1–10 wt%)	Trace component (1–1000 ppm, or 0.0001–0.1 wt%)	Struc- ture	Morph- ology
Image analysis	...	...	...	...	...	...	...	•	•	...	...	...	...	...	•
Inductively coupled plasma mass spectrometry	•	...	...	•	•	•	•	•	•	•	•	•	•	...	...
Ion chromatography	D,N	...	D,N	D,N	D,N	D,N	D,N	D,N	...	N	D,N	D,N	D,N	...	...
Liquid chromatography	...	...	D	...	D	D	D	D	...	...	D	D	D	...	...
Liquid chromatography mass spectrometry	D, •	D,N	D,•	D,•	D	D	D	D	...	...	D	D	D	...	...
Low-energy ion-scattering spectroscopy	•	...	...	...	•	•	...	...	...	•	•	•	S	...	...
Molecular fluorescence spectroscopy	D,N	D,N	D,N	...	D,N	D,N	D,N	D,N	...	...	D,N	D,N	D,N	...	...
Neutron activation analysis	N	...	...	N	N	N	N	N	...	...	N	N	N	...	...
Nuclear magnetic resonance	N	D,N	•	D,N,S	N	N	N	N	...	...	N	N	S	N	S
Optical metallography	...	...	...	...	...	...	...	•	•	...	...	...	...	...	•
Raman spectroscopy	N	D, •	D,•	...	D,•	D,•	D,•	•	•	S	•	•	...	•	...
Scanning electron microscopy	...	...	...	...	...	...	...	...	•	...	N	N	...	...	•
Secondary ion mass spectroscopy	•	...	S	•	•	N	...	...	•	•	•	•	•	N	N,S
Small-angle x-ray scattering	•	...	•	...	•	...	...	•	...	...	...	...	...	•	...
Transmission electron microscopy	S	...	C	...	N	N	...	...	•	...	N	N	...	C	•
Ultraviolet/visible absorption spectroscopy	D, •	D, •	D,•	...	D,•	D,•	D,•	D,•	...	...	D,•	D,•	D,•	...	...
X-ray diffraction	...	...	C	...	C	C	C,S	C	...	...	C	C	...	C	...
X-ray photoelectron spectroscopy	•	N	S	...	•	•	S	...	...	•	•	•	...	...	...
X-ray spectrometry	N	...	...	...	N	N	N	N	...	...	N	N	N	...	...

Key: • = generally usable; N = limited number of elements or functional groups; D = after dissolution; V = volatile liquids; S = under special conditions (i.e., with tandem mass spectrometer detection); C = crystalline solids

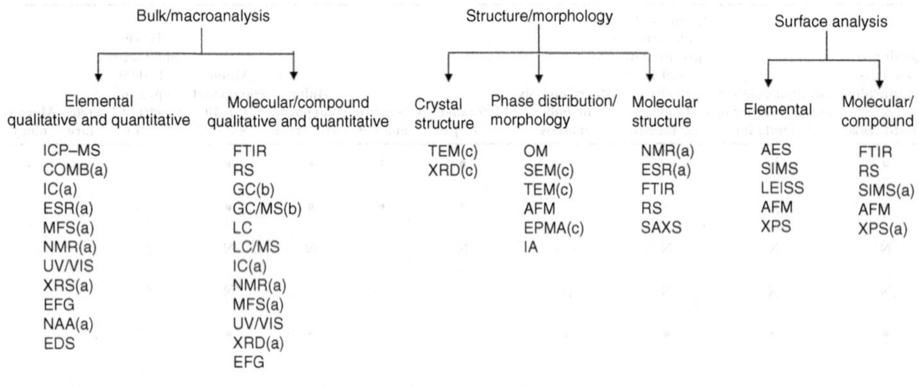

Fig. 1 Flow charts of common techniques for characterization of metals and alloys. AES: Auger electron spectroscopy; AFM: atomic force microscopy; COMB: high-temperature combustion; EDS: energy-dispersive x-ray spectroscopy; EFG: elemental and functional group analysis; EPMA: electron probe x-ray microanalysis; ESR: electron spin resonance; FTIR: Fourier transform infrared spectroscopy; GC, gas chromatography; GC/MS: gas chromatography/mass spectrometry; IA: image analysis; IC: ion chromatography; ICP-MS, inductively coupled plasma mass spectrometry; LC: liquid chromatography; LC/MS: liquid chromatography/mass spectrometry; LEISS: low-energy ion-scattering spectroscopy; MFS: molecular fluorescence spectroscopy; NAA: neutron activation analysis; NMR: nuclear magnetic resonance; OM: optical metallography; RS: Raman spectroscopy; SAXS: small-angle x-ray scattering; SEM: scanning electron microscopy; SIMS: secondary ion mass spectroscopy; TEM: transmission electron microscopy; UV/VIS: ultraviolet/visible absorption spectroscopy; XPS: x-ray photoelectron spectroscopy; XRD: x-ray diffraction; XRS: x-ray spectrometry. (a) Limited number of elements or groups. (b) Volatile liquids, solids, or components. (c) Under special conditions

The most common techniques for determining chemical composition of metals and alloys include x-ray fluorescence, optical emission spectroscopy, inductively coupled plasma optical emission spectroscopy, and wet chemistry. While these methods work well for most elements, they are not useful for dissolved gases and some nonmetallic elements that can be present in metals as alloying or impurity elements. High-temperature combustion and inert gas fusion methods are typically used to analyze for dissolved gases (oxygen, nitrogen, hydrogen) and, in some cases, carbon and sulfur in metals.

X-Ray Fluorescence

X-ray spectrometry, or x-ray fluorescence (XRF) spectrometry, is an emission spectroscopic technique that has found wide application in elemental identification and determination (see the article "X-Ray Spectroscopy" in this Volume). The technique depends on the emission of characteristic x-radiation, usually in the 1 to 60 keV energy range, following excitation of atomic electron energy levels by an external energy source, such as an electron beam, a charged particle beam, or an x-ray beam. In most sample matrices, laboratory x-ray spectrometry can detect elements at concentrations of less than 1 μg/g of sample (1 ppm); in a thin-film sample, it can detect total amounts of a few tenths of one microgram.

X-ray spectrometry is one of the few techniques that can be applied to solid samples of various forms. Examples of specialized applications have included the determination of sulfur and wear elements in petroleum products, analysis of forensic samples, and in measurements of electronic and computer-related materials.

Portable XRF instruments are increasingly used for in situ nondestructive characterization, including measurement of lead in painted surfaces in housing. Continuing advances in computer and detector technology have led to further advances in the capabilities and sensitivity of portable XRF instruments for applications such as soil analysis and nondestructive qualitative and quantitative analysis of inorganic elements in archaeological objects. Portable XRF instruments cannot generally provide the low detection limits attained by laboratory methods, but the sensitivity of portable XRF instruments has increased and made possible their use for measurement of lead and other metal in dusts and soils (Ref 1). For example, field XRF can easily provide lead-in-soil detection limits of less than 100 ppm, well below typical regulatory levels of 300 to 1500 ppm.

Optical Emission Spectroscopy

Optical emission spectroscopic (OES) methods originated in experiments performed in the mid-1800s, yet they remain some of the most useful and flexible means of performing elemental analysis. See the article "Optical Emission Spectroscopy" in this Volume for detailed discussion. Free atoms, when placed in an energetic environment, emit light at a series of narrow wavelength intervals. These intervals, termed emission lines, form a pattern—the emission spectrum—that is characteristic of the atom generating it. The energetic environment typically is a type of plasma, and the device used to produce such a plasma is commonly termed an emission source.

The presence of an element in a sample is indicated by the presence in light from the source (plasma) of one or more of its characteristic emission lines. The concentration (mass fraction) of that element can be determined by measuring line intensities. Thus, the characteristic emission spectrum forms the basis for qualitative elemental analysis, and the measurement of intensities of the emission lines forms the basis of quantitative elemental analysis.

The spark discharge remains the most commonly used source for bulk analysis of metal alloys. The current trend is mainly toward complete automation. Spark OES systems are increasingly incorporated into automatic laboratories with facilities for sample preparation and sample handling by robots. There also is continued development of more sophisticated electronics to control the waveform of the spark, for increased stability and flexibility.

Inductively Coupled Plasma-Optical Emission Spectroscopy

Inductively coupled plasma optical emission spectrometry (ICP-OES) is a technique for analyzing the concentration of metallic elements in solid and liquid samples. Similar to spark OES, ICP-OES uses the optical emission principles of exited atoms to determine the elemental concentration. However for ICP-OES, solid samples are dissolved (digested) in an appropriate solvent (typically acid) to produce a solution for analysis. The resulting sample solution (or an original liquid solution for analysis) is often diluted in water to obtain a final specimen suitable for analysis (Ref 2).

The ICP-OES technique exhibits the unique properties of providing simultaneous multielement (up to 70) analysis within a few minutes, as well as ensuring over 6 orders of magnitude dynamic range, applicable for most of the elements on the periodic table. This specificity renders ICP-OES the preferred technique to determine both trace, major, and minor elements in a single experiment with minimum chemical interferences.

Development of a similar technique, for isotope and trace analysis—namely inductively coupled plasma mass spectrometry (ICP-MS)—began in the 1980s and is currently considered as equally indispensable as ICP-OES (Ref 3).

The ICP-MS technique is recommended for trace analysis, because it provides a significantly lower detection limit (parts per trillion for the majority of the elements). In addition, the technique is not subject to spectral overlap or interferences originated from the optical spectra. In fact, the ICP-MS technique is recommended when elements, such as the rare earth and actinide elements, produce complex optical spectra. Nevertheless, this technique is not suitable for concentrated matrices or high dissolved solid contents, which could saturate the detector. Some minor mass interferences are observed, particularly in the low- to mid-mass range, due to doubly charged ions, metal oxide ions, and molecular fragments that are originally present in the plasma or formed at the orifice and interface.

Common applications of the ICP-MS technique include isotope ratio analysis and trace impurities for pharmaceutical or biological applications. However, ICP-MS instruments are less robust than ICP-OES, and consumables as well as maintenance can be rather expensive.

High-Temperature Combustion (Ref 2)

High-temperature combustion is used to determine carbon and sulfur content in a variety of materials. The sample is accurately weighed and placed in a ceramic crucible or combustion boat, often along with combustion accelerators. The crucible is placed in a high-temperature furnace, which is then flooded with oxygen. The furnace is heated to 1370 to 1425 °C (2500 to 2600 °F), causing combustion of the carbon and sulfur in the sample to form CO, CO_2, and SO_2. The gases are separated and analyzed by infrared absorption or thermal conductivity detectors. A heated catalyst is used to convert the CO to CO_2 prior to detection.

The infrared absorption detector measures the absorption of the infrared wavelengths characteristic to CO_2 and SO_2.The amount of infrared absorption at these wavelengths is correlated to a quantitative content based on standards and the weight of the original specimen.

The thermal conductivity detectors monitor the thermal conductivity of the carrier gas. As the evolved gases pass the detector, changes in the thermal conductivity correspond to a change in the gas (e.g., from the inert carrier gas to hydrogen) and the amount of evolved gas present. These changes correspond to the amount of CO_2 and SO_2 generated and indicate the amount of sulfur or carbon in the original specimen.

Inert Gas Fusion (Ref 2)

Inert gas fusion is a quantitative analytical technique for determining the concentrations of nitrogen, oxygen, and hydrogen in ferrous and nonferrous materials. The sample is accurately weighed and placed in a pure graphite crucible in a fusion furnace with an inert gas atmosphere. The crucible is heated to 2000 to 3000 °C (3630 to 5430 °F), resulting in the sample fusing to a molten state. The hydrogen and nitrogen gases dissociate from the molten material and are carried away from the fusion chamber as H_2 and N_2. The oxygen released from the material bonds with carbon (from the graphite crucible) to form CO or CO_2 and is carried away.

An inert carrier gas flushes the gases evolved from the sample out of the fusion chamber. The fusion gases are separated and carried to the detector. The individual concentrations for the evolved gases are detected by infrared absorption (for CO and CO_2 only) or thermal conductive techniques (N_2, H_2, CO, and CO_2).

Bulk Structural Analysis

Diffraction techniques are used to gain insight to the atomic structure of the sample surface. Depending on the problem, x-rays, electrons, or neutrons are used, which provide the electron density, electrostatic potential density, and nuclear density (also magnetic spin density) distribution, respectively, in a crystal (Ref 4). Most x-ray diffraction techniques are rapid and nondestructive; some instruments are portable and can be transported to the sample.

The great utility of x-rays for determining the structure and composition of materials is due to the differential absorption of x-rays by materials of different density, composition, and homogeneity. The higher the atomic number of a material, the more x-rays are absorbed.

X-Ray Diffraction

X-ray diffraction (XRD) is the most extensively used method for identifying and characterizing various aspects of metals related to the arrangements and spacings of their atoms. The XRD techniques are equally applicable to other crystalline materials, such as ceramics, geologic materials, and most inorganic chemical compounds. The article "Single-Crystal X-Ray Diffraction" in this Volume explains how single-crystal x-ray diffraction is used to characterize the crystal structure, including determining symmetry, unit cell parameters, atomic coordinates and thermal displacement parameters, bond lengths and angles between the atoms, and structural motive (or topology).

The article "X-Ray Powder Diffraction" (XRPD) in this Volume describes techniques used to characterize samples in the form of loose powders, aggregates of finely divided material, or polycrystalline specimens. The powder method, as it is referred to, is a good phase-characterization tool, partly because it can routinely differentiate between phases having the same chemical composition but different crystal structures (polymorphs). Although chemical analysis can indicate that the empirical formula for a given sample is $FeTiO_3$, it cannot determine whether the sample is a mixture of two phases (FeO and one of the three polymorphic forms of TiO_2) or whether the sample is the single-phase mineral $FeTiO_3$ or ilmenite. The ability of XRPD to perform such identifications more simply, conveniently, and routinely than any other analytical method explains its importance in many industrial applications as well as its wide availability and prevalence.

Neutron Diffraction

Neutron diffraction techniques can be used to determine the atomic structure of crystals. Advantages include higher sensitivity to atoms with low atomic numbers (especially hydrogen), ability to distinguish isotopes, deep penetration, and information about magnetic structure. Disadvantages include the requirement for larger crystals (approximately 1 mm^3), longer data-collection times, and a nuclear reactor. Despite sharing the scattering mechanism, the interaction of neutrons with matter is fundamentally different than that of x-rays: Photons strongly interact with the electron shell of the atom. Because all electrons interact with radiation, independently of which atoms they are part of, the scattering for x-rays is related only to the electronic density being proportional to Z^2, where Z is the atomic number. Nonetheless, both fission and spallation remain quite energy intensive, and thus, usable neutron beams for materials science are only available in large-scale research facilities that house a nuclear reactor or a spallation neutron source. While this requirement has arguably been the biggest hurdle to the growth of neutron sciences, the importance—and uniqueness—of neutron diffraction is demonstrated by the construction of high-flux spallation neutron facilities and reactors as well as the technical upgrades and expansions to existing facilities across the world. Improved access has significantly grown the user community and moved the technique from esoteric measurements that once required several cubic centimeter samples measured over days or weeks to time-resolved measurements at subsecond time scales on samples of a few hundred milligrams or less. This technique is described in detail in the article "Neutron Diffraction" in this Volume.

Electron Diffraction

Electron diffraction is similar to x-ray and neutron diffraction and is often used in transmission electron microscopy (see the article "Transmission Electron Microscopy" in this Volume) or scanning electron microscopy (see the article "Scanning Electron Microscopy" in this Volume) as well as electron backscatter diffraction (see the article "Crystallographic Analysis by Electron Backscatter Diffraction in the Scanning Electron Microscope" in this Volume). Electron diffraction is usually used for symmetry determination and raw determination of unit cell parameters and atomic positions; however, recent techniques can determine the structures from these data. Advantages include a nanometer size of single crystals. Disadvantages of this method include the necessity for very careful sample preparation, the requirement for the sample to be electron transparent, and the possibility of damaging the sample by the incident electrons.

Microstructural Analysis

Microstructural analysis is the combined characterization of the morphology, elemental composition, and crystallography of microstructural features through the use of a microscope. Light microscopes have been used to characterize the microstructures of metals for over 100 years. The methods and applications of light microscopy are covered in the article "Light Optical Metallography" in this Volume. While light microscopy is an important metallurgical tool, it has a number of limitations:

- *Spatial resolution:* Conventional light microscopes cannot resolve features smaller than ~1 μm.
- *Depth of field:* Light microscopes cannot image rough surfaces; samples must be flat to be in focus.
- *Type of information provided:* Light microscopes provide images and morphological information, but they do not provide any direct chemical or crystallographic information about the microstructural features observed.

Over the years, several types of electron microscopy have been developed and refined to greatly extend the ability to resolve and characterize the morphologies of much smaller microstructural features, to image rough surfaces, and to obtain direct chemical and crystallographic information about microstructural features.

The following sections briefly review three types of electron microscopies most commonly used in metallurgical studies: scanning electron microscopy, electron probe microanalysis, and transmission electron microscopy.

Scanning Electron Microscopy

The scanning electron microscope (SEM) provides a valuable combination of high-resolution imaging, elemental analysis, and, recently, crystallographic analysis. The primary use of an SEM is to produce high-resolution images at magnifications unattainable by optical microscopy, while also providing direct topographic and compositional information. Magnifications up to 100,000× or more are possible by modern instruments. The SEMs produce nanoscale-resolution imaging and mapping because the electron beam is focused to a ~1 to 10 nm sized probe. Two major advantages of the SEM over the optical microscope as a tool for examining surfaces are improvements in resolution and depth of field.

The components of a typical SEM include an electron-optics column, specimen chamber, support system, and control and imaging system. The electron beam is generated at an electron gun and accelerated toward the sample housed inside a specimen chamber, typically below the electron-optics column. Electromagnetic lenses below the electron gun focus the electron beam to a small probe at the sample surface. Scanning coils deflect the electron probe across the sample surface, and detectors housed either in the specimen chamber or in the electron-optics column collect and report resultant signal intensity as a function of beam position. The SEM operator uses a computer to control the electron optics, detectors, and a motorized stage for sample positioning. During SEM operation, a support system provides cooling water for the electromagnetic lenses and maintains a low pressure within the electron-optics column and the specimen chamber by using vacuum pumps. See the article "Scanning Electron Microscopy" in this Volume for details on the technique and its variations.

Developments in SEM instrumentation during the 1990s and 2000s produced compact

SEMs, smaller and more robust instruments as well as some that are portable for remote analysis. More powerful computers, automated stages, and automated SEM control enabled significant advances in high-throughput analysis. Focused ion beams were combined with the SEM to produce dual-beam instruments. Specially designed SEMs, called variable-pressure/environmental SEMs, are equipped with differential pumping so the specimen chamber can operate under higher background pressure environments, a capability that accommodates high-vapor-pressure specimens inside the SEM, such as biological samples, polymers, and liquids.

Electron Probe Microanalysis

Electron-excited x-ray microanalysis is a technique that enables spatially resolved elemental composition measurement at the micrometer to submicrometer lateral and depth resolution. With the exception of hydrogen and helium, which do not produce characteristic x-rays, all elements of the periodic table can be measured. Generally referred to as electron probe microanalysis, the technique is typically performed in an SEM equipped with an energy-dispersive x-ray spectrometer (EDS). The SEM-EDS is capable of quantitative measurements of constituents at the major (concentration $C > 10$ wt%), minor ($1 \text{ wt\%} \leq C \leq 10$ wt%), and trace ($0.05 \text{ wt\%} < C < 1$ wt%) levels. When tested on known materials, demonstrated accuracy is generally within ±5% relative in 95% of analyses for major and minor constituents. Low-atomic-number elements, for example, fluorine, oxygen, nitrogen, carbon, and boron, can be measured with useful accuracy as major constituents in fluorides, oxides, nitrides, carbides, and borides. Quantitative compositional mapping can be performed on x-ray spectrum image databases, resulting in elemental images in which the gray or color scale is related to elemental concentration. See the article "Electron Probe X-Ray Microanalysis" in this Volume for detailed explanations and applications.

Transmission Electron Microscopy

A major benefit of the transmission electron microscope (TEM) is its superior resolution over optical microscopy and the SEM. The TEM provides the highest-resolution imaging, elemental analysis, and crystallographic analysis of all the techniques described. It is capable of:

- Imaging of features as small as several tenths of nanometers, down to the scale of individual crystallographic planes and atoms
- Qualitative and semiqualitative elemental analyses with spatial resolution approaching 10 nm, roughly 100 times better than SEMs or electron microprobes
- Identifying crystalline compounds and determining crystallographic orientations of microstructural features as small as 30 nm

The TEM approach is unique among materials characterization techniques in that it enables essentially simultaneous examination of microstructural features through imaging from lower magnifications to atomic resolution and the acquisition of chemical and crystallographic information from small (submicrometer down to the atomic level) regions of the specimen. Specifications and best practices for using the TEM are covered in depth in the article "Transmission Electron Microscopy" in this Volume.

Surface Analysis

A number of methods can be used to obtain information about the chemistry of the first one to several atomic layers of samples of metals, as well as of other materials, such as semiconductors and various types of thin films. A summary of surface analysis techniques is listed in Table 2, showing the topics covered in the Surface Analysis Section of

Table 2 Quick reference summary of surface analysis methods

Analysis method	Analysis probe	Detection signal	Analysis information	Lateral resolution	Depth resolution	Typical applications
Atomic force microscopy	Coated or noncoated cantilever probes made of various materials	Laser light	Images of surface topographical features or other near-field and far-field interactions between probe and sample surface	0.2–10 nm	10–80 pm	Biological molecules, biomaterials, cells, crystallography, electrochemistry, polymer chemistry, thin-film studies, nanomaterials, nanotechnology, failure analysis, process development, process control, surface metrology
Auger electron spectroscopy	Electrons	Auger electrons	Elemental composition analysis for all elements except H and He	20 nm	0.5–5 nm	Adhesion, catalysis, corrosion, oxidation, surface chemical reaction, surface contamination, wear, depth profile of each element with ion gun sputtering
Low-energy electron diffraction	Electrons	Diffraction electrons	Surface crystallography and microstructure	10 μm	0.4–2 nm	Adsorption, catalysis, chemical reactions, crystallography, crystal structure in epitaxial growth, film-growth kinetics, grain size and boundary, microstructure, reconstruction, segregation
Scanning probe microscopy	Probes made of various materials	Laser light, electrical current, or other probe-sample interactions	Three-dimensional image with atomic resolution of surface properties such as height, electron tunneling current, electrostatic force, magnetic force, etc.	0.2–10 nm	10–80 pm	Broad usage in research and development of nanomaterials and applications of nanotechnology and micromanufacturing that involves understanding, characterization, and manipulating surfaces at atomic or nanometer scale
Secondary ion mass spectroscopy	Ion beam	Secondary ions	Elemental, isotopic, or molecular composition of the surface through detection of the species with different mass-to-charge ratios	50 nm–10 μm	0.5–10 nm	Concentration depth profiling, identification of inorganic or organic surface layers, isotopic abundances in geological and lunar samples, trace element detection including hydrogen
Thermal desorption spectroscopy	Thermal energy	Desorbed atoms and/or molecules	Desorption rate of desorbing gases from surfaces as a function of temperature	Not applicable	Not applicable	Adsorption, desorption, and reaction of adsorbed atoms and molecules on surfaces; catalysis; personal exposure to toxic chemicals; indoor air-quality monitoring; identification of volatiles in soil and water; analysis of environmental pollutants; quantification of hydrogen in metals (especially steels); corrosion mechanisms; electrochemistry; tribology
X-ray photoelectron spectroscopy	X-ray	Photoelectrons	Chemical state and composition	20–500 μm for monochromatic x-ray; 10–30 mm (0.4–1.2 in.) for nonmonochromatic x-ray	10 nm	Elemental analysis of all elements with an atomic number of lithium and higher, chemical state identification of surface elements, composition and chemical state depth profiles with ion gun sputtering, determination of oxidation states of metal atoms in metal compounds, identification of surface contaminations, measurement of surface film thicknesses, identification and degradation of polymers

this Volume. These techniques are capable of providing elemental composition, chemical state, and other important properties of the outermost atomic layers of metals, semiconductors, ceramics, organic materials, and biomaterials in either bulk, thin-film and coating, powder, or particulate format. This Section of the Volume includes articles on recently developed techniques, such as "Introduction to Scanning Probe Microscopy" and "Atomic Force Microscopy," as well as established methods, such as "Auger Electron Spectroscopy," "Low-Energy Electron Diffraction," and "Secondary Ion Mass Spectroscopy."

Metallography

Metallography is the scientific discipline of examining and determining the constitution and the underlying structure of the constituents in metals and alloys. The objective of metallography is to accurately reveal material structure at the surface of a sample and/or from a cross-sectional specimen. Examination may be at the macroscopic, mesoscopic, and/or microscopic levels. For example, cross sections cut from a component or sample may be macroscopically examined by light illumination to reveal various important macrostructural features (on the order of 1 mm to 1 m, or 0.04 in. to 3 ft), such as the ones shown in Fig. 2 and listed here:

- Flow lines in wrought products
- Solidification structures in cast products
- Weld characteristics, including depth of penetration, fusion zone size and number of passes, size of heat-affected zone, and type and density of weld imperfections
- General size and distribution of large inclusions and stringers
- Fabrication imperfections, such as laps, cold welds, folds, and seams, in wrought products
- Gas and shrinkage porosity in cast products
- Depth and uniformity of a hardened layer in a case-hardened product

Macroscopic examination of a component surface is also essential in evaluating the condition of a material or the cause of failure. This may include:

- Characterization of the macrostructural features of a fracture surface to identify the fracture initiation site and changes in the crack-propagation process
- Estimations of surface roughness, grinding patterns, and honing angles
- Evaluation of coating integrity and uniformity
- Determination of extent and location of wear
- Estimation of plastic deformation associated with various mechanical processes
- Determination of the extent and form of corrosive attack; readily distinguishable types of attack include pitting, uniform, crevice, and erosion-corrosion
- Evaluation of tendency for oxidation

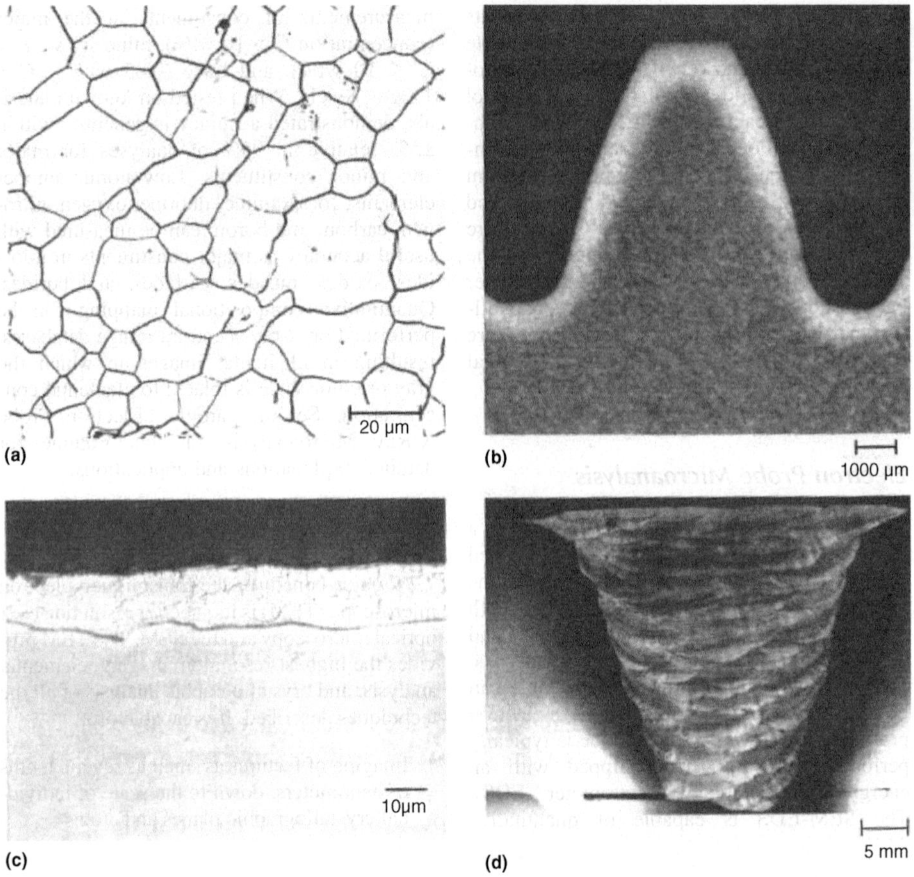

Fig. 2 Examples of uses for metallography. (a) Equiaxed ferrite grain size in plain carbon steel. (b) Ion-carburized gear tooth showing case depth. (c) Microstructure of galvanized coating on steel—thickness and quality. (d) Multipass weld quality in type 304 stainless steel plate. Source: Ref 5–7

- Association of failure with welds, solders, and other processing operations

This listing of macrostructural features in the characterization of metals, although incomplete, represents the wide variety of features that can be evaluated by light macroscopy.

See the articles "Light Optical Metallography" and Quantitative Metallography" in this Volume for further examples and discussion.

ACKNOWLEDGMENT

Parts of this article were adapted from K.H. Eckelmeyer, Introduction and Overview of Materials Characterization, *Metals Handbook Desk Edition*, 2nd ed., J.R. Davis, Ed., ASM International, 1998; and from articles to be published in *Materials Characterization*, Volume 10, *ASM Handbook*, ASM International, 2019.

REFERENCES

1. S. Clark, W. Menrath, M. Chen, S. Roda, and P. Succop, Use of a Field Portable X-Ray Fluorescence Analyzer to Determine the Concentration of Lead and Other Metals in Soil Samples, *Ann. Agricul. Environ. Med.*, Vol 6 (No. 1), 1999, p 27–32
2. *Handbook of Analytical Methods for Materials*, Materials Evaluation and Engineering, Inc., 2016
3. C. Boss and K. Fredeen, *Concepts, Instrumentation and Techniques in Inductively Coupled Plasma Optical Emission Spectrometry*, 3rd ed., PerkinElmer, 2004, http://www.perkinelmer.co.uk/CMSResources/Images/44-159043GDE_Concepts-of-ICP-OES-Booklet.pdf
4. A. Guinier et al., Diffraction Methods and Crystal Structure Analysis, *Methods and Instrumentations: Results and Recent Developments. Advanced Mineralogy*, Vol 2, A.S. Marfunin, Ed., Springer, Berlin, Heidelberg, 1995
5. A.O. Benscoter and B.L. Bramfitt, Metallography and Microstructures of Low-Carbon and Coated Steels, *Metallography and Microstructures*, Vol 9, *ASM Handbook*, ASM International, 2004, p 588–607
6. Metallography and Microstructures of Case-Hardening Steel, *Metallography and Microstructures*, Vol 9, *ASM Handbook*, ASM International, 2004, p 627–643
7. Metallography and Microstructures of Weldments, *Metallography and Microstructures*, Vol 9, *ASM Handbook*, ASM International, 2004, p 1047–1056

ASM Handbook, Volume 10, *Materials Characterization*
ASM Handbook Committee
DOI 10.31399/asm.hb.v10.a0006670

Semiconductor Characterization

David T. Schoen, Meredith S. Nevius, and Sumit Chaudhary, Exponent Inc.

Introduction

This article introduces various techniques commonly used in the characterization of semiconductors and semiconductor devices. Many of the common materials characterization techniques have a variety of applications to semiconductors. Research into semiconductor materials has also led to the development of a large variety of highly specialized characterization approaches that do not possess demonstrated applications outside of the semiconductor field. This article focuses on the first set of techniques, but where possible, references to outside works in the second set are made. Because of space limitations, it is not possible to offer a comprehensive overview of all the various methods and approaches that can be used in characterizing semiconductor materials; consequently, the focus is on the most commonly available techniques and those that are covered in the later sections of this Volume.

As a class of materials, semiconductors are defined not by their elemental composition, position on the periodic table, chemical arrangement, processing history, or microstructure, but instead by a specific property: electrical resistance. Semiconductors are materials with an electrical resistance intermediate between metals and insulators (Ref 1). From a functional perspective, semiconductors form the active material components of structures whose electrical behavior, such as resistance, can be controlled through external stimulation, such as the gate voltage applied to a metal-oxide-semiconductor field-effect transistor (MOSFET) or visible light illumination on a photovoltaic. The ability for this coupling to occur arises from interactions between the outer shell electrons and the potential energy landscape created by the atomic cores within semiconducting materials.

This class of materials is characterized by an energy gap between an electronic conduction band and valence band, both of which are partially occupied at room temperature with electronic carriers. These carriers are electrons in the valence band and holes in the conduction band. The population of these carriers can be manipulated, most frequently through addition or removal of specific atomic constituents called dopants, through application of an external electric field, and by exposure to radiation such as visible light. The carriers generated by these various stimuli move either in an electric field via drift or due to concentration gradients via diffusion. This fundamental relationship between the density of electronic carriers, the currents and voltages generated by their motion, and these external stimuli form the basis for a variety of solid-state transducers that can convert information or energy from one form to another. The modern world contains many examples of these devices, including transistors, integrated circuits, diodes, photovoltaics, digital cameras, and displays.

The large family of semiconducting materials that has been identified and applied to problems in science and technology spans a wide range of material classes. To guide the interested reader most quickly to the relevant material, this article is divided into sections based on classes of semiconductors that share a common set of analytic techniques. A number of techniques have uses across these categories, such as scanning electron microscopy for imaging surfaces at high spatial resolution, while other techniques may have a more specific and specialized application, such as Raman spectroscopy for assessing the material quality of graphene sheets.

Single-crystal semiconductors include some of the most common semiconducting materials, particularly single-crystal silicon, which is the basis for the majority of the microelectronics industry. This section also considers germanium, gallium arsenide (GaAs), and the various epitaxial thin film semiconductors that can be grown on single-crystal wafer substrates. These materials are characterized first by their single-crystal nature and lack of transverse phase domains or grain boundaries, and often are fabricated at very high purity levels. Despite their high purity, the behavior of devices fabricated from these materials can still be quite sensitive to trace elemental constituents. These trace elements may be added intentionally, such as in the case of dopants, which are added to manipulate the relative concentration of positive to negative charge carriers in the material. Trace elements may also be unintentional contaminants. In either case, techniques with extremely low detection limits are often required for effective analysis in these systems. In many cases, it may not be practical to directly detect these trace constituents, but it may be simpler and more cost-effective to indirectly measure them through electrical testing.

Polycrystalline and amorphous semiconductors have wide application in technologies that require large-area devices such as electronic displays and photovoltaics. These materials by definition contain structural defects. In the case of polycrystalline materials, these consist of grain or phase boundaries, dislocations, and various point crystal defects. One way to conceive of amorphous materials is that they lie at the end of a continuum, where the density of defects becomes so high the material no longer exhibits the long-range structural order required to produce discrete features in x-ray, electron, or neutron diffraction. Organic semiconductors essentially lie between the amorphous and polycrystalline regimes, and are discussed in a separate section in this article.

The most common element in industrial application for amorphous and polycrystalline materials is also silicon, and amorphous silicon and polycrystalline silicon each have many applications where some electronic performance can be sacrificed in order to reduce cost. However, a number of other materials are also used in thin film form for various applications including silicon-germanium alloy and chalcogenides such as copper indium gallium selenide ($Cu[In,Ga]Se_2$) and cadmium telluride (CdTe). As a class, these materials tend to lack the extremely high purity characteristic of single-crystal semiconductors; therefore, a larger variety of chemical analysis techniques can be used to adequately determine their composition. A wider variety of microstructural questions are also relevant to polycrystalline materials, whose properties are often strongly dependent on the density and distribution of crystalline defects.

While many oxides are insulating, a subset of oxide materials exhibit resistances and band gaps within the semiconducting range. These semiconductors are commonly transition metal oxides that are typically polycrystalline or polyphasic. Some oxides form naturally on metallic surfaces, but fabrication by other methods is typically utilized for commercial quality and processing requirements. Given the diverse nature of the oxides that fall into the category of semiconductors, it may not be surprising that the routes to fabricating these materials are equally diverse. Techniques related to the confirmation of fabrication processing parameters varies depending on the route employed. Independent of the route, both the grain structure and composition can dominate the final oxide material properties. Consequently, techniques that are sensitive to these properties can be necessary for analysis of these systems.

Organic semiconductors are small molecules or polymers possessing a backbone of alternating single and double bonds between carbon atoms. Primary applications of organic semiconductors are in displays and photovoltaics. Organic semiconductors are synthesized via wet chemistry techniques; thus, structural and optoelectronic properties can be easily tuned. Thin films of semiconducting polymers are fabricated using coating or printing techniques, and vacuum-based thermal evaporation is typically utilized for high-performance small-molecule-based thin films. Organic semiconductor thin films typically consist of a mix of amorphous and polycrystalline domains; properties of these thin films are commonly characterized using electrical, optical, and surface scanning probe-based techniques.

Low-dimensional semiconductors are materials and structures with at least one critical dimension that is within the nanoscale. This can include monolayers, quantum wells, nanowires/nanoribbons/nanorods, and quantum dots. While some of these nanoscale materials are composed of materials that are also semiconductors in their bulk form, others are composed of materials whose bulk and nanoscale properties greatly differ. In these cases, sufficiently reducing the size of the material's volume in one or more dimensions can lead to quantum confinement and opening of a band gap. To create these low-dimensional materials, two general fabrication schemes are commonly used: a top-down approach, which generally involves lithographic patterning and etching of a bulk or thin film material; and a bottom-up approach, which generally entails building nanostructures from atomic or molecular precursors. These length scales, usually less than optical wavelengths, can limit the usefulness of diffraction-limited characterization techniques (e.g., optical microscopy), because they may not be able to achieve the necessary spatial resolution. That being the case, techniques that are commonly utilized to characterize low-dimensional materials are usually real-space electron- or probe-based microscopy techniques.

As an overview, Figure 1 and Table 1 list some of the most common tools used for the characterization of the various classes of semiconducting materials considered in this article.

Single-Crystal Semiconductors

The starting point for many microelectronic devices is a single-crystal wafer of an elemental or compound semiconductor, most frequently silicon or GaAs, although a limited number of other materials are also available in wafer form. Insulating or metal substrates may also be used in some applications; however, their characterization is not the focus here. In some cases the wafer itself is the basis for a device, such for single-crystal silicon photovoltaics, but in many other cases devices are patterned onto the outer surface of the wafer by means of lithographic patterning and a variety of growth, deposition, etching, diffusion, and implantation processes.

Several semiconductor materials, particularly silicon, are routinely prepared with specialized processing that generates material of very high purity with virtually no incorporated crystalline defects or other inhomogeneities. Material of this high quality is often required for the repeatable fabrication of dense integrated circuitry. Nonetheless, a number of materials questions may arise regarding the compositional purity, crystallographic structure, surface morphology, and chemistry of these materials during research and development, prototyping, manufacturing, and end use of these materials.

There is an even larger set of semiconducting materials that are available only in the form of thin films grown on the surface of single-crystal wafers. For a few specific classes of materials and deposition techniques, these thin films can be grown epitaxially on the surface of a single-crystal substrate, meaning that the atoms align in a second single crystal of the new material with a well-defined crystallographic

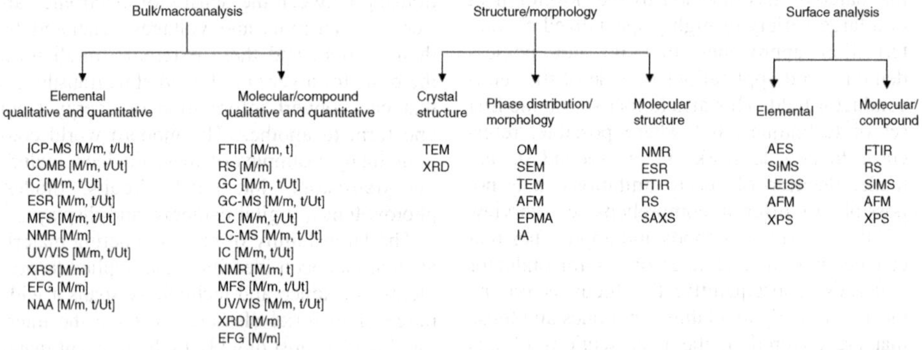

Fig. 1 Flow charts of common techniques for characterization of semiconductors. ICP-MS: inductively coupled mass spectrometry; ESR: electron spin resonance; NMR: nanomagnetic resonance; UV-vis: ultraviolet-visible spectroscopy; XRS: x-ray Raman spectroscopy; NAA: neutron activation analysis; FTIR: Fourier-transform infrared spectrometry; RS: Raman spectroscopy; GC: gas chromatography; GC-MS: gas chromatography-mass spectrometry; LC: liquid chromatography; LC-MS: liquid chromatography-mass spectrometry; XRD: x-ray diffraction; TEM: transmission electron microscopy; OM: optical microscopy; SEM: scanning electron microscopy; AFM: atomic force microscopy; EPMA: electron probe microanalysis; SAXS: x-ray solution scattering; AES: Auger electron spectroscopy; SIMS: secondary ion mass spectroscopy; LEISS: low-energy ion scattering spectroscopy

Table 1 Selected techniques

	Bulk/elemental	Microstructural	Surface analysis
Single crystal	EPMA, IR/FTIR, PL/tr:PL	XRD, XRF, TEM, EELS, RBS, EBIC, CL	SIMS, RBS, XPS, AFM, LEED, RHEED, SEM
Polycrystalline/ amorphous	EPMA, IR/FTIR, PL/tr:PL	ND, TEM, EELS, EBIC/ LBIC, SEM, SKPM	SIMS, XPS, AFM, SEM, RBS
Oxides	EPMA, IR/FTIR, PL/tr:PL	XRD, EBSD, TEM, SAED, AES, RS	SEM, XPS, LEED, RHEED, OM, EDS, STM, AFM, NIM, RS
Organic	UV-vis, PL/tr:PL	X-ray (GIWAXS)	AFM
Low-dimensional	(See surface analysis)	SAED, TEM	LEED, RHEED, SEM, SPM, AFM, STM, SCM, EFM, SKPM, EDS, EELS, XPS, SIMS, LEAP, ARPES

EPMA: electron microprobe analysis; IR/FTIR: infrared spectrometry, Fourier-transform infrared spectrometry; PL/tr:PL: photoluminescence/ time-resolved photoluminescence; XRD: x-ray diffraction; XRF: x-ray fluorescence; TEM: transmission electron microscopy; EELS: electron energy loss spectroscopy; EBIC/LBIC: electron-beam-induced current/light-beam-induced current; CL: cathodoluminescence; SEM: scanning electron microscopy; SKPM: scanning Kelvin probe microscopy; SIMS: secondary ion mass spectroscopy; RBS: Rutherford backscattering spectrometry; XPS: x-ray photoelectron spectroscopy; AFM: atomic force microscopy; EFM: electrostatic force microscopy; SCM: scanning capacitance microscopy; NIM: nanoimpedance microscopy; LEED: low-energy electron diffraction; RHEED: reflection high-energy electron diffraction; ND: neutron diffraction; LEAP: local electron atom probe; ARPES: angle-resolved photoemission spectroscopy; UV-vis: ultraviolet-visible spectroscopy; SAED: selected area electron diffraction; AES: Auger electron spectroscopy; GIWAXS: grazing-incidence wide angle x-ray scattering; OM: optical microscopy; energy dispersive x-ray spectrometry; STM: scanning tunneling microscopy; SPM: scanning probe microscopy; RS: Raman spectroscopy relationship to the substrate. These materials have many of the same characteristics as the single-crystal substrates themselves: they typically have very high purities and low densities of crystallographic defects. For this reason, there is a large degree of overlap in the characterization tools and strategies used for epitaxial films.

Single-crystal wafer substrates also play a role in a variety of devices that may use other classes of semiconductor for their active layers. Other examples of thin films utilizing these single-crystal wafer substrates span several classes of materials discussed in this article, including amorphous and polycrystalline semiconductors, and oxide semiconductors. There are also many examples of devices featuring organic semiconductors and low-dimensional materials that are built on single-crystal wafers, especially silicon. For this reason, an understanding of the behavior of single-crystal semiconductors in the various common characterization tools and techniques is an important place to begin a discussion of semiconductor characterization in general.

Material Classification

The main characteristic that defines a *single crystal* is a lack of internal crystallographic boundaries, such as grain boundaries or phase boundaries. Of course, every object in the world has an external surface, so in this sense every single crystal has one continuous grain boundary that defines its external surface. This definition captures both single-crystal wafers as well as epitaxial thin films, which can be viewed as a separate single crystal stacked on top of a single-crystal wafer substrate, or a series of such crystals for a multilayer structure. Such multilayer structures may contain a high spatial density of planar interfaces perpendicular to the stacking direction, but in order to fall in the single crystal category, they must lack internal grain boundaries.

Another level of classification that is frequently used is the division of these materials into elemental semiconductors and compound semiconductors. Elemental semiconductors are single crystals of a single element, primarily the group IV elements carbon, silicon, germanium, and tin, of which silicon and germanium are by far the most common in application. Compound semiconductors may be binary combinations of rows 3 and 5 of the periodic table (e.g., GaAs), rows 2 and 6 of the periodic table (e.g., CdTe), or more complex compounds arising from three or more elements, such as aluminum gallium arsenide and $Cu(In,Ga)Se_2$ (Ref 1, 2).

Fabrication Methods

Bulk single crystals suitable for creating wafers can be fabricated by a variety of melt-solidification techniques. The most common technique is the Czochralski (CZ) process, in which a single-crystal boule is drawn out of a molten pool of the source material. This is accomplished by placing a single-crystal seed in contact with the liquid surface. At the interface of the colder solid in contact with the molten liquid, epitaxial solidification occurs, and the growing crystal is slowly drawn away from the liquid surface where additional solidification occurs. The orientation of the growing crystal is then determined by the orientation of the single-crystal seed. Modifications to the process are required to grow compound semiconductors, in particular the addition of a background pressure of the more volatile component or the use of a boron oxide liquid encapsulant on the surface of the liquid melt (Ref 1, 2).

Compositional purity for intrinsic materials, as well as accurate homogeneous control of dopant atoms, is accomplished through a number of steps in the process. Raw materials are extracted from mineral sources via a large variety of processing steps. For the most common semiconductor, silicon, the raw material is SiO_2 sand. A lower-grade metallurgical silicon is separated from the oxygen through high-temperature reaction with a carbon source, such as coal. Metallurgical-grade silicon contains a number of undesirable atomic constituents, particularly iron. A secondary purification step is performed in a chemical reaction producing trichlorosilane or silane gas, which is subsequently purified by fractional distillation and converted back to silicon through chemical decomposition (Ref 1, 2).

The crystal growth process itself provides another degree of refinement due to segregation of minor constituents preferentially into the solid phase rather than the liquid phase. Table 2, adapted from Ref 1, shows that this segregation can provide a large degree of additional purification for some elements such as iron, which is an undesirable contaminant in silicon. Other elements, such as boron, do not segregate as strongly, which provides an advantage when those elements can be used to provide a background doping level, if the goal is to produce an electron majority or hole majority conducting material of uniform composition. Other techniques exist for additional purification of semiconductor materials, such as zone refinement (Ref 1).

Table 2 Segregation ratios of impurities in silicon. A smaller ratio indicates that a solid solidifying from a melt will contain less of the indicated species; a higher ratio indicates that it will contain more. Adapted from Ref 1

Impurity	C_{solid}/C_{liquid}
B	0.8
Al	0.002
P	0.35
As	0.3
C	0.07
Fe	8×10^{-6}
O	1.25

The CZ process is capable of producing high-purity single-crystal samples of silicon that are nearly defect free. Dislocations may be generated during the initial stages of growth due to thermal shock in the single-crystal seed when it touches the molten silicon pool, but their propagation into the bulk of the boule can be prevented by inclusion of a thin neck region between the seed and the primary bulk of the boule. However, the CZ process is not capable of producing defect-free materials for all semiconductors. Twin defects are difficult to suppress in indium phosphide (InP), and GaAs wafers cut from CZ boules may have a dislocation density of several thousands per squared centimeter. Other processes including the Bridgeman technique and the vertical gradient freeze method, are employed to produce GaAs materials with lower crystal defect density (Ref 2).

Thin film epitaxy can be conducted by a variety of strategies in which precursors of the desired elements are transported to the surface of a single-crystal wafer substrate. Conditions at the sample surface are controlled to encourage condensation of the desired solid via several mechanisms, including precipitation from a supersaturated solution during liquid phase epitaxy, and high-temperature chemical reaction between a metal chloride gas or a metal organic liquid precursor vapor and group V hydride gases in the case of vapor phase epitaxy and metal organic chemical vapor deposition.

A discussion of semiconductor processing would be incomplete without reference to how these materials are used to form electronic devices. The first requirement for many electronic devices is the formation of electronic junctions between volumes of material that have either electrons or holes as the majority electronic charge carriers. Elemental semiconductors can be fabricated as intrinsic semiconductors, where the shorthand is *i*, meaning the number of electrons and holes is equal at room temperature. These materials can also be doped by adding small concentrations of other elements to produce either an electron majority, called an *n*-type material, or a hole majority, called a *p*-type material. Doping is accomplished through the addition of dilute concentrations of other elements, such as boron and phosphorous in silicon.

Doping is one of the key features facilitating standard silicon electronics, because a silicon wafer can be doped in some locations as *n*-type, and in other locations as *p*-type via ion implantation or diffusion of appropriate dopant atoms. This allows the formation of a variety of lateral combinations of *p* and *n* regions, forming the basic structure required for diodes, transistors, and other electronic elements which are the constituents of integrated circuits. A number of compound materials can also be doped; however, many of them can be fabricated in only either n or p form, requiring the use of multiple materials to form active electronic structures. The joining of a single material with regions of n-type and p-type conduction is called a *homojunction*, and the formation of an n-type to p-type pair between dissimilar materials is called a *heterojunction*.

Depending on the device architecture, the volumes of p and n material may be arranged laterally, in the x and y directions of the wafer, or they may be arranged vertically, in the z direction (Fig. 2). Integrated circuits are often arranged primarily laterally, with every transistor or diode structure having its own p and n regions. Photovoltaics are often arranged vertically, with several layers of doping, starting with a heavily doped p^+- or n^+-type material, followed by planar layers that may include lightly doped p or n materials or intrinsic regions, and ending with a layer of the opposite type, so that the entire structure forms a single diode, p-n or p-i-n. Exceptions exist in both cases, with lateral photovoltaics and vertical transistors. Light-emitting diodes may be formed vertically or laterally. In characterization of the various parts of the semiconductor structure, understanding the physical arrangement of the p and n regions is critical for selecting the proper analysis tools, because each technique has a characteristic lateral resolution and depth resolution. While some techniques can be used to extract depth-dependent information from a sample, such as Rutherford backscattering spectrometry (RBS), other techniques may be more suited for lateral mapping, such as electron microprobe microanalysis, and some techniques are capable of either or both functions, such as secondary ion mass spectrometry (SIMS). For many microelectronic applications, the majority of the electrical functionality is confined to a surface region no deeper than 5–10 μm below the wafer surface, and the bulk of the wafer provides a primarily mechanical function as an inert substrate supporting the thin layer of active electronics.

Once the p and n regions have been defined, it is necessary to generate electrical conduction pathways to carry signals or power to the electronics, between the electronics, and away from the electronics. For microelectronics, this is accomplished by the deposition of a variety of insulating and conducting layers that are laterally patterned to define the various interconnects that form a series of layers on the surface of the semiconductor (Fig. 3). For optoelectronics, it is common that only a single connection is required on each side of the semiconductor, one on the p side and one on the n side. Often one of these contacts needs to be optically transparent to allow the semiconductor to absorb or emit light, depending on the function.

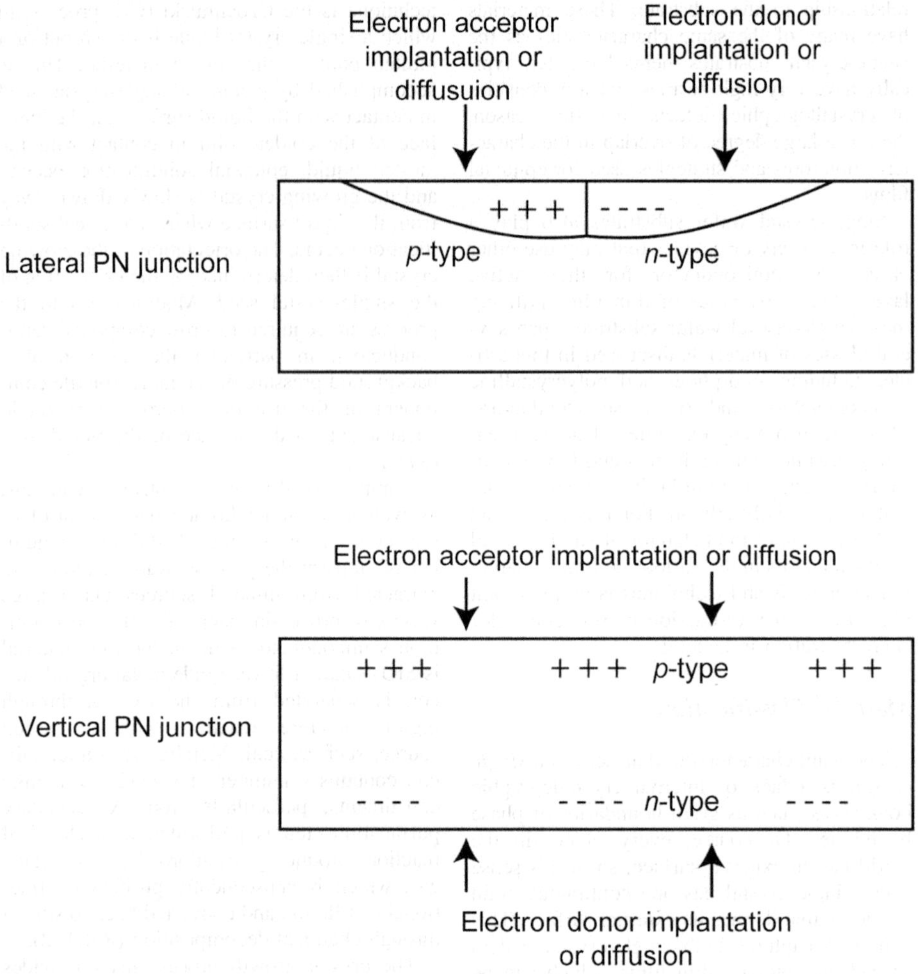

Fig. 2 Lateral and vertical *pn* junction geometry

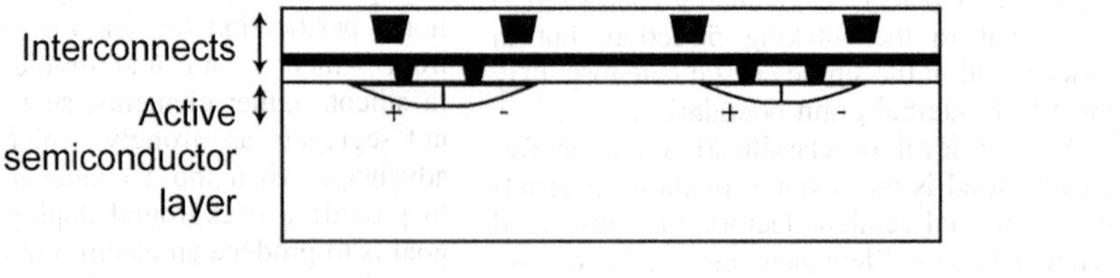

Fig. 3 Example microelectronics interconnect geometry, utilizing laterally patterned stacks of various insulating and conducting layers on the surface of a semiconductor

Sample Preparation

The region of interest in the semiconductor that the engineer desires to obtain information about is often buried within a matrix of other materials that may or may not be of interest. Some analytic tools are capable of gathering information through certain thicknesses of other layers that may be on the semiconductor surface, while several require the region of interest to be directly exposed in order to collect information from the semiconductor. The process of exposing the semiconductor for analysis is an intrinsically destructive one, since it requires removal of material. This can be done in cross section or in plane view by a variety of techniques.

Cross sectioning by mechanical or ion beam techniques (Fig. 4) can be useful if the desired analysis tool has a sufficiently small spatial resolution to image the semiconducting layer on edge. The simplest cross-sectioning approach is to fracture the substrate, which is often brittle. If the fracture proceeds at a low stress, it often propagates through the thin film stack in a relatively flat fracture plane, forming a usable cross section. A low-stress fracture can be encouraged by creating a large defect on the opposite surface of the waver via mechanical scribing (see Fig. 4, Cleavage).

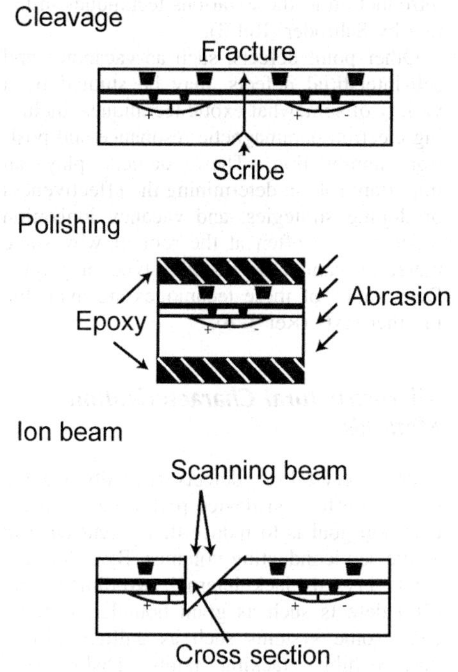

Fig. 4 A few methods for cross-sectioning microelectronics device architecture for analysis of buried layers

This approach can be difficult to control and perform repeatably. Mechanical cross sectioning similar to that used for metallographic analysis (see Fig. 4, Polishing) is frequently employed for imaging the macroscopic details of a semiconductor device such as the solder bonds to the printed circuit board or the overall dimensions of various components. Frequently the edges of brittle ceramic components exhibit chipping damage during mechanical sectioning that precludes use of this technique for analysis of the interconnect or functional layers, but if appropriate polishing media and material removal rates are used, it is possible to generate high-quality cross sections via this approach. It is often possible to target structures with dimensions as small as 30 to 50 μm by hand, or even smaller with automated polishing systems after conducting the proper instrument calibration.

Ion beams provide an alternative approach to mechanical sectioning (see Fig. 4, Ion Beam). Ion beam systems include argon ion mills, which produce high-current diffuse beams appropriate for larger-scale sectioning; more precise tools such as focused ion beam (FIB) systems appropriate for 5 to 50 μm scale cross sections; and plasma FIB systems, which lie between these two approaches. With the right ion beam, it is possible to make larger cross sections that may fully transect parts, with vertical dimensions up to 1 to 2 mm, as well as thin film lamella appropriate for TEM imaging, which are typically 5– to 10 μm in lateral dimensions with thicknesses of 10 to 200 nm. Ion beam systems can often produce high-quality cross sections and are particularly useful when there is a close and unavoidable collection of soft and hard materials in close proximity, which can be problematic for mechanical cross-sectioning approaches. One advantage of ion beam cross sectioning is that the section location can easily be placed with submicron precision, allowing the targeting of very small structures for analysis, including individual transistors within an integrated circuit (see article "Focused Ion Beam Instruments" in this Volume) (Ref 3).

Delayering, or removing the insulating and conducting layers in the interconnect structure from the surface of the semiconductor, can be accomplished with mechanical, ion, or chemical strategies. Each of these has distinct advantages. Mechanical abrasion similar to cross sectioning is a materials agnostic approach that can be employed on any substrate without prior knowledge of the material composition, though depending on the material, it can be very difficult to uniformly and completely remove layers from the surface of the semiconductor. If the chemistry of the various layers is well known, it is sometimes possible to use chemically selective etchants to target specific conducting or insulating layers for removal, thus gradually removing layers until the layer of interest is revealed. For an unknown substrate, identifying appropriate etchants and developing etching recipes may be very time consuming. Ion beams are commonly used for depth profiling in a variety of analysis tools, such as x-ray photoelectron spectroscopy (XPS) and SIMS, and these can gradually remove material from the surface of a part while continually collecting elemental or chemical information. Typically, the depth dimension is determined by assuming a constant material removal rate and then by recording the time for each data collection event. The removal rate is calculated by measuring the final depth of the ion-etched pit at the end of the etching process. This technique works most effectively on flat and homogenous regions of a sample, as differential material etching rates can introduce a variety of artifacts in the data and uncertainty as to the real depth of a given data point.

An important consideration for all cross-sectioning strategies is the possibility of the cross-sectioning process itself changing the analysis volume. Ion beam approaches commonly implant the ion used for ablation into the surface of the cross section. This implantation may lead to a significant change in the composition of the material of interest. For instance, gallium implantation in an FIB may produce materials with >10% gallium content after sectioning, and the degree to which this is an issue is highly material dependent. Ion radiation also induces a variety of other changes in material, including accelerated diffusion, which leads to material mixing at the section surface, and radiation-induced phase transformations such as amorphization and recrystallization. Many of these effects can be minimized by tuning the equipment parameters, such as current and accelerating voltage, but in all cases the optimum parameters are highly dependent on the material system. Another common issue in ion beam cross sectioning is curtaining, which is introduced due to the geometry of the technique. A region of slow sputtering at a vertical location between the ion source and the region of interest generates an ion shadow, leading to topology in and out of the plane of the cross section. This presents itself in the final cross-section images as vertical streaking parallel to the ion beam direction. In some cases, this may preclude measurement of the features of interest (Ref 4).

Sample damage can also occur during mechanical cross sectioning. Mechanical abrasion creates a deformed zone within a certain distance from the abraded surface, which depends on the material. For softer materials, this may extend 1 to 5 μm into the material and is a zone characterized by heavy plastic deformation. Most semiconductors are relatively hard and brittle, but surface cracking, edge chipping, and foreign species from the polishing media and any lubricating fluid (typically water) all may be introduced into the mechanically sectioned surface. Material that has been worn from other regions of the samples may also be transported via the polishing media and fluid to areas it was not originally present.

Depth profiling by ion ablation carries with it a variety of risks for sample damage, which are discussed in detail in the relevant technique articles of the Handbook. See, for instance, "Secondary Ion Mass Spectroscopy" in this Volume.

Bulk/Elemental Characterization Methods

By definition, many of the important bulk material properties of a single-crystal semiconductor are already determined. It must be a single crystal, so the composition must be uniform enough to avoid internal phase boundaries, and it does not contain internal grain boundaries. In practice, most fabrication and processing strategies to produce these materials also produce materials of exceptionally high purity with a minimum number of crystalline defects. Materials that are available in bulk crystal form include silicon, germanium, silicon carbide, GaAs, InP, gallium antimonide, gallium phosphide, zinc oxide (ZnO), zinc sulfide, zinc selenide, cadmium sulfide, cadmium selenide, and CdTe. Other materials including aluminum oxide, silicon dioxide, magnesium oxide, and strontium titanate ($SrTiO_3$) are also available as single-crystal substrates, though they are insulators.

Many important parameters of single-crystal semiconductors can be determined and controlled during initial processing; therefore, in many practical situations in design and manufacturing, it is not necessary to characterize them. The bulk atomic arrangement of a single-crystal wafer is fully defined if the following information is determined: composition, trace constituent content, density and distribution of crystalline defects, and finally crystallographic orientation. Some of these quantities, such as the material composition, background doping level (if any), and degree of final refinement, are specified during initial wafer fabrication to determine the purity and background crystal defect concentration. Crystallographic orientation for the wafer surface is determined during growth and wafer dicing, and frequently another crystal plane, such as <100> or <110>, is indicated by cutting a small portion from the side of the wafer, referred to as a flat, which can be used for wafer alignment into automated wafer handling systems. However, not all characterization of materials and devices is conducted with firsthand knowledge of the design and raw materials, so any of these quantities may need to be determined on an unknown sample. Interestingly, for semiconductors, even full knowledge of the atomic arrangement may not provide sufficient information. For example, the fraction of dopant atoms that are electronically active is a critical factor for even basic electronic structures. Measurement of this property still requires electrical testing, and it is not directly measurable by materials analysis techniques.

The bulk composition of many compound semiconductors is tightly confined to well-defined stoichiometries at specific line compounds in the binary and ternary phase diagrams. The case of the common elemental wafers, silicon and germanium, is of course trivial. These materials are miscible, and alloys of silicon and germanium can be fabricated in any ratio (Ref 1). For binary compounds, such as GaAs, the stoichiometry of a single-phase material is typically a simple whole number ratio, in most cases 1:1 (Ref 2). Deviations from these line compounds result in secondary phases and thus must be avoided during single-crystal growth. Such materials would be unusual in commercially available raw materials. Many binary compounds can be alloyed in pseudobinary, pseudoternary, pseudoquartenary, etc. mixtures. Frequently, the minor component composition is in the 1 to 50% range, allowing for determination by a variety of techniques. Common approaches to assessing unknown compositions include energy dispersive x-ray spectrometry (EDS) performed in a scanning or transmission electron microscope, electron energy loss spectroscopy (EELS) in a transmission electron microscope for light elements, electron probe microanalysis, XPS, and Auger electron spectroscopy (AES). The lattice constant of pseudobinary compounds in many cases obeys Vegard's law, where the lattice constant a_0 of an alloy with formula $A_{1-x}B_x$ can be determined by (Ref 1):

$$a_0(A_{1-x}B_x) = (1-x)a_0(A) + xa_0(B)$$

For these systems, diffraction-based techniques such as x-ray diffraction (XRD) would also be able to determine the overall alloy stoichiometry.

Point defects have received a significant amount of study due to their importance for the doping required to produce electronic junctions. The direct measurement of bulk impurities in single-crystal semiconductors can be very difficult because these species may have measurable and problematic electronic effects at levels below the parts per billion (ppb) concentration level, which remains a distinct experimental challenge. The technique able to directly detect trace atomic constituents at the lowest concentrations is SIMS. The lower-bound detectability limit varies depending on the impinging ion, the analyte, and the matrix material, and may vary over many orders of magnitude from the parts per million (ppm) to ppb level (Ref 3). Quantification of concentrations may also be a challenge except in a handful of materials systems where careful calibration work with known samples has been performed. Within silicon, infrared frequency adsorption features may occur at specific known wavelengths that have been tabulated and can be used for impurity identification. A variety of species may be identified via this technique (Ref 3, 5). By using the channeling effect, RBS can be used in some situations to determine the location of an impurity point defect within the atomic lattice.

In many cases, the most sensitive approaches for understanding trace constituents in semiconductors are electrical and optical in nature. What is detected are not the atomic constituents themselves, but the carriers or electronic energy states created by or associated with these atoms. Often these techniques may not be able to identify the nature of the defect, but they focus on the electronic effects of the defect, which may or may not be specific enough to use for identification. These approaches vary from somewhat simple, such as direct-current (dc)-based resistance measurements or current-voltage characteristics, to more complex, such as direct current and alternating current measurements on two-terminal and three-terminal device configurations to extract device-relevant performance parameters. Such measurements can be performed at different frequencies and temperatures, and assessed against device physics models to more deeply understand charge dynamics for given structural variations. A detailed discussion of these techniques is outside the scope of this article since they are more in the area of property measurement than in physical characterization, but frequently these approaches offer much greater sensitivity and lower detection floors than is currently possible with physical characterization. The interested reader is directed to an excellent introduction to these various techniques in the text by Schroder (Ref 3).

Other point defects, such as vacancies and self-interstitial defects, may be studied by a variety of somewhat exotic techniques, including electron paramagnetic resonance and positron annihilation. These defects play an important role in determining the effectiveness of doping strategies, and vacancy ionization processes are often at the root of why some materials cannot be doped *n*-type or *p*-type. Discussions of these techniques are available in other texts (Ref 3, 6).

Microstructural Characterization Methods

Other crystalline defects typically have a negative effect on device performance, and a common goal is to reduce their occurrence in active semiconducting volumes. By definition, single crystals lack internal two-dimensional (2D) defects such as grain boundaries; however, some systems such as gallium nitride may exhibit stacking faults. Dislocations, one-dimensional (1D) defects, may occur within the bulk of a material or at interfaces. Dislocations and stacking faults contain a variety of distinct electronic states that may fall in the band gap of the semiconductor, and are often highly detrimental to device performance. They also place the surrounding material in a state of mechanical stress, which leads to the segregation of solute atoms. This can have further detrimental effects on the material due to small-scale rearrangement of impurities or dopants. For several materials systems, including silicon, it is possible to produce single crystals with no 2D defects or dislocations. However, even in initially dislocation-free materials, thin film growth processes can generate stresses sufficient to nucleate dislocations due to thermal expansion mismatch and lattice strains. The avoidance of dislocations is often one of the central concerns for the design of multilayer epitaxial structures.

A variety of techniques are capable of locating, counting, and imaging 1D and 2D crystalline defects. As with the case of impurities, sometimes the easiest way to detect these defects is through electrical characterization, but such approaches may not allow definitive identification of the defect. The task of locating defects can be made easier in some cases by mapping the local optical or electronic performance of a material in order to identify areas with electronic inhomogeneities. Photoluminescence and time-resolved photoluminescence can be used to measure the internal quantum efficiency and minority carrier lifetime of an optoelectronic device and can be performed using confocal optical microscopy at high lateral resolution to permit 2D mapping. This lifetime is typically shorter near 1D or 2D crystalline defects. Light-beam-induced current is a relatively simple approach that can be used in photovoltaics and other sensors to map the conversion efficiency of incident light to electrical current in a device and may reveal the location of crystal defects when they are present in sufficiently low concentrations (Ref 7). For higher-resolution imaging, scanning electron microscopy (SEM) can be used to produce a local source of hot electrons. This can generate valence band electrons or conduction band holes by impact ionization, and for certain structures such as diodes, the current induced by the electron beam can be recorded in a 2D map during scanning and reveals local electronic traps such as dislocations. This technique is referred to as electron-beam-induced current. There are a large variety of similar techniques that generally function by connecting the electronic device under test in a number of electrical characterization systems and measuring voltages, capacitances, and currents dynamically while the electron beam is being scanned over the sample, and these approaches in some cases are sensitive to the presence of dislocations and stacking faults (Ref 8).

When the crystallographic features of the defects are of interest, they can be imaged through transmission electron microscopy (TEM) with diffraction-based contrast techniques such as bright field/dark field imaging, weak beam diffraction contrast, and high-resolution TEM. These techniques require imaging the sample in multiple angular orientations with respect to the incident electron beam, so a two-angle tilt stage is typically required. Transmission electron microscopy imaging can allow for the determination of the orientation, type, and Burgers vector of dislocations in a sample (see "Transmission Electron Microscopy" in this Volume). Sample preparation is often the first challenge to successful TEM work, but thin foil sample preparation by FIB liftout is becoming increasingly automated and routine. The high resolution of TEM imaging can be used to count dislocations at higher densities than other approaches. Sample sizes for TEM are typically constrained to the 5 to 50 μm level, so in cases where the sample contains dislocations at low densities, it may be necessary to first locate and target a dislocation by another technique before performing TEM sample preparation (Ref 9).

The crystallographic orientation of a substrate or a film on a substrate is typically investigated by diffraction. The most common approaches for single crystals are XRD and electron diffraction, though neutron diffraction can also be used for this purpose. Single crystals have a three-dimensional (3D) diffraction pattern, so the appearance of a Bragg peak depends on both the θ-2θ angle as in powder diffraction as well as the χ and Φ angles. The operation is similar to a pole figure collection for a textured metal film. A specific Bragg reflection is selected, such as <111>, which determines the θ-2θ angle. Then χ and Φ are varied in order to locate the reflection, for example, by forming a 2D intensity map and selecting the corresponding χ and Φ that produce the maximum intensity. Various other scattering angles are selected and located until a pole diagram of the sample can be developed. This technique can also be used to determine offcut angles. For epitaxial films, a similar procedure is used to determine the relative alignment of the crystal planes of the film to those of the substrate (Ref 3). Selected area electron diffraction (SAED) can also be used in a similar fashion; however, the angular resolution of this technique is typically inferior to that of XRD techniques, so small differences lower than 1° may be difficult to resolve. Surface-based diffraction techniques such as reflection high-energy electron diffraction (RHEED) and low-energy electron diffraction (LEED) are also used during the growth of epitaxial thin films to characterize the alignment of the growing film on the substrate in real time.

Surface Characterization Methods

Surface characterization is very important in single-crystal semiconductors. In most cases, the functional electronic layer of material is very close to the outer surface of the device. The typical thickness of a wafer is 300 to 500 μm, while the active material and interconnect structure is usually confined to the outer 5 to 30 μm of the final device. There are exceptions: single-crystal silicon solar cells may use the entire wafer volume for light absorption and electronic transport, for instance, but in many electronic and optoelectronic applications, the *p* and *n* regions are defined laterally through selective area doping or vertically through a stack of epitaxial films. Most device architectures feature a large number of interfaces. Even laterally defined *p* and *n* devices still require dielectric and metal interconnect layers. Interfaces can have a large influence on the mechanical, electrical, chemical, and optical properties of a device, and the control of interfacial structure and composition is critical for repeatable manufacturing and consistent device performance.

Questions relating to general surface structure and composition may often be first approached through the use of SEM (see "Scanning Electron Microscopy" in this Volume). This technique has a lateral imaging resolution that can be lower than 10 nm in certain circumstances, but can frequently fall in the 10 to 50 nm range. For wafer-based single-crystal semiconductors specifically, SEM is a powerful and versatile tool. Typically, the background conductivity of the wafer is sufficient to discharge the surface of the sample, so semiconductors can be imaged in high vacuum with no surface metal coating. Even when the outer surface of a device may consist of insulating films containing metal interconnect structures, it is possible to choose an accelerating voltage where the bulk of the secondary electrons generated by the incident beam can be transported away from the imaging area. Scanning electron microscopy can provide topographic information, particularly at lower accelerating voltages, while forming an image using the secondary electron current. Composition information can be collected via imaging with backscattered electrons and x-ray spectroscopy, since many scanning electron microscopes are also equipped with an energy dispersive x-ray spectrometer. The accelerating voltage can be chosen in order to sample volumes at different depths below the surface. A high voltage can be used to visualize the arrangement of buried metal structures, sometimes many microns below the outer surface, while a lower accelerating voltage permits imaging of the outer surface of the device. For this reason, an awareness of the accelerating voltage is critical for the interpretation of SEM images. Care should always be taken with image interpretation. While backscattered electrons provide more contrast due to atomic composition, and secondary electrons provide more contrast due to surface topography, both techniques contain contrast from each source to a different degree. Other effects such as local work function variation, surface charge, crystal orientation, and sample tilt can also have effects on the electron signal and the contrast between different areas on the sample surface. Quantified topographic measurements with SEM can be performed by means of stereoscopic reconstruction or local cross sectioning by FIB.

For the quantified measurement of step heights and surface roughness, other tools are more common than SEM. Optical profilometry and optical scatterometry are fast measurement techniques that can provide high vertical resolution and roughness characterization down to the single nanometer level, although often with limited lateral resolution. Typically, 1 to 5 μm is a practical limit. They are equally capable of imaging large areas because the sampled area can be controlled through the use of focusing or dilating optics. More sophisticated optical techniques, such as scanning confocal microscopy, can achieve 100 nm level lateral resolution and so may be comparable with SEM in many situations, but are slower compared with other optical techniques since they rely on serial measurements across a 3D voxel grid. The highest imaging resolution for topography is available with scanning probe techniques. Stylus profilometry is a simple and fast technique appropriate for larger features. Atomic force microscopy (AFM) is the most common tool; it can be performed on a variety of surfaces and is often employed for the measurement of step heights even in the sub-nanometer range with a typical lateral resolution of 30 to 50 nm for relatively flat surfaces. Given the very low roughness of most wafer surfaces, AFM is a common technique for roughness measurements. The expected lateral length scale of roughness should be considered when selecting a roughness measurement strategy. Typically, the roughness can be calculated for lateral wavelengths greater than the width of 3 pixels in each line or image and up to ~25% of the total line or image width. An 80 μm wide AFM scan with 1024 pixels is sensitive to roughness with a lateral wavelength from ~80 nm to 20 μm (Ref 10). This technique would not be appropriate for assessing the overall flatness of a 10 cm diameter wafer, because a large-scale variation in the surface height would not be captured with a local measurement. A technique such as optical profilometry would be more appropriate.

Techniques commonly used for surface chemistry overlap to some extent with those used for bulk analysis. The definition of what specific volume of the wafer counts as "surface" as opposed to "bulk" is dependent on the context and purpose of the information, and this context often determines the most appropriate technique. During assessment of the difference in water contact angle measurements or adhesive debonding, often the outer 1 to 3 nm of the solid provides the most important information. When studying the distribution of dopant atoms defining *n* and *p* regions, it may be necessary to gather information to depths of several microns. The various surface chemistry techniques available for semiconductor characterization also provide slightly different but often overlapping information.

For problems relating to surface energy and surface contamination, XPS, AES, and SIMS are frequently the optimal tools. They provide very near surface information, within the outer 1 to 5 nm of the sample when operated without depth profiling. Secondary ion mass spectrometry provides elemental identification, and can also be used to distinguish between different organic molecules on the basis of analysis of the fragmentation pattern. Quantitative analysis of surface species can be quite difficult, though it can be accomplished in some special circumstances discussed later. XPS and AES both also provide elemental information, particularly for elements with atomic numbers greater than 11, in addition to the oxidation state. XPS can also be used for studying electronic band offsets and band gaps. These techniques are much less sensitive than SIMS, with lower-bound detection limits in the 1% range, whereas SIMS can detect species to the ppm or even ppb level. This high sensitivity with a lack of quantifiability can produce a practical issue during use of SIMS in that many species may be observable at levels far below what would be required to produce a degradation in performance, and often a large variety of species are observed. Distinguishing between species that have a functional impact on the device as opposed to those that do not can be difficult. This issue can be particularly pronounced in cases where semiconductor devices have been through packaging and assembly, and in end-use cases where the high cleanliness levels typical of a cleanroom have not been maintained.

For electronic dopant measurements from implantation and diffusion processes, dynamic SIMS is often the preferred technique (Ref 3). Dynamic SIMS is destructive. The ion beam used to generate the secondary ions is continually scanned over the surface while the SIMS spectra are collected. The surface material is thus gradually removed, and the spectra change over time as the concentration at the milled interface descends into the material. For some common electronic materials, SIMS data on substrates with known dopant concentrations have been collected for common dopants, and highly quantitative depth profiles with detection floors as low as 10^{14} or 10^{15} per cm^3 are achievable. Technologically relevant doping concentrations can be measured. There is an inherent tradeoff between detection floor and sampling volume. In order to detect reasonable counting statistics of very trace elements, it may be necessary to sample a large volume of the matrix material. This limitation can make it difficult to map such lateral concentration gradients as might be present in transistors and diodes with small critical dimensions. Another technique that can be employed for higher-concentration dopants is RBS, which provides depth information without the need for destructive ion milling. However, RBS often uses an ion beam at very high energies of 1 MeV, which may cause radiation damage in the material.

When it is necessary to measure both interfacial chemistry and interface structure, TEM is a common approach. As discussed previously, thin foil samples of semiconductors can be produced using the FIB liftout approach and then imaged in cross section. In ideal conditions, TEM and scanning transmission electron microscopy (STEM) both have a subatomic resolution and are commonly used for imaging transistor structures and epitaxial multilayers used in optoelectronic devices such as light-emitting diodes (LEDs) and quantum confined lasers. A variety of spectroscopic tools are available in TEM, including EDS and EELS. Electron energy loss spectroscopy in particular can give enhanced sensitivity to light elements compared with EDS for chemical composition and can be used to study chemical bonding. Interfacial and volume defects such as dislocations and stacking faults can be imaged, identified, and placed in a geometric context in a way that is often not possible with optical and electronic measurements.

The surface structure of thin films and thin film multilayers can also be studied without cross sectioning by a variety of diffraction-based approaches. These include the electron diffraction techniques of LEED and RHEED, which use the electron energy and incident electron angle respectively to confine the collected information to the surface of the sample. An x-ray analog for RHEED is grazing incidence XRD, which can be used to gather similar information.

Electronic Characterization

As briefly mentioned in the section "Surface Characterization Methods" in this article, in many cases the most sensitive measurements that are available for determining the number and density of defects in semiconductor materials are electronic in nature. A detailed discussion of all the various electronic characterization methods that can be employed is outside the scope of this article, but in brief, the main properties that are typically of interest are the electrical resistivity of the material, carrier density, carrier type, electron and hole effective mass, electronic mobility, band gap magnitude, type (direct or indirect), band edge energies, and recombination rates for the various radiative and nonradiative processes. To this list could be added the various performance characteristics of diodes and transistors fabricated from the semiconducting materials (Ref 3).

Electrical resistance typically must be measured with a four-terminal geometry in order to remove the influence of contact resistance. Conducting this testing properly requires consideration of the overall sample geometry and the electrode placement, and is easiest for a few simple standard coupon shapes. Carrier type, carrier density, and carrier mobility can be measured through a combination of resistivity measurement and Hall measurements. Quantitative determination of the band gap and band alignment is most effectively completed through measurement of the IV characteristics of diodes and transistors made of the materials of interest. Optical transmission spectrometry can be used to estimate the band gap, particularly for direct band gap materials due to their abrupt adsorption edge, and XPS can be used to estimate both the band gap as well as the energies of the valence band and conduction band edges (Ref 3).

Polycrystalline and Amorphous Semiconductors

Many applications do not require the high mobilities and minority carrier lifetimes characteristic of high-purity single-crystal semiconductors. In these applications, it is often possible to lower cost by using less pure material with more crystalline defects, because these materials can typically be fabricated with faster processes and require fewer steps with lower processing temperatures. There is a corresponding trade-off in the form of less desirable electronic properties, but in applications that require large total areas such as high-volume components, transistor arrays for displays based on organic light emitting diodes (OLEDs) and liquid crystal displays (LCDs), and photovoltaics, the benefit of the cost savings can outweigh the loss of performance.

Material Classification

Polycrystalline and amorphous semiconductors are composed of the same basic elements used in single-crystal semiconductors, but differ primarily in their processing. From a definitional standpoint, polycrystalline semiconductors contain internal grain boundaries. In most applications, materials should still be single phase, so the grain boundaries separate regions of differing local crystallographic orientation but not composition. Amorphous materials can be defined in a few different ways. They lack long-range crystalline order; consequently, they do not produce discrete diffraction peaks via Bragg diffraction in x-ray, electron, or neutron diffraction, unlike crystalline materials. Amorphous materials are always metastable, so they exhibit recrystallization when heated to a temperature lower than the melting point of the bulk material. When cooled from a liquid state at a sufficiently high cooling rate, they undergo a second-order phase transition termed the glass transition. Many amorphous semiconductors are condensed from the vapor phase or grown via a low-pressure surface chemical reaction and so may not actually undergo a glass transition on formation. They may also recrystallize at a temperature lower than the glass transition; therefore, this common thermodynamic perspective is less relevant than the diffraction-based definition (Ref 11).

Fabrication Methods

Bulk polycrystalline materials can be solidified from a melt but do not require a seed crystal or the slow drawing or slow cooling processes used for the formation of bulk single-crystal material. In many cases, control of the grain size is very important for the final material properties. Bulk samples of polycrystalline semiconductors can be solidified in arbitrary shapes through casting and subsequently diced into wafers in a process similar to that used for single-crystal materials. Square wafers that fill space better for many applications are often produced, particularly for silicon. In many applications, polycrystalline material can be deposited onto other substrates by chemical or physical vapor deposition. Because the need for single-crystal growth is relaxed, the processes selected are typically performed at much higher rates. Common examples include plasma enhanced chemical vapor deposition (PECVD) and radio frequency or direct current sputtering.

A wide variety of materials are used as polycrystalline semiconductors, with polycrystalline silicon being the most common material. Grain boundaries significantly reduce the performance of electronic devices for most materials, particularly by serving as scattering and trap sites for minority carriers and thus reducing electronic mobility and minority carrier diffusion lengths. One approach for avoiding this issue is to significantly increase the grain size. An example of this approach is multicrystalline silicon, which typically features grain diameters on the order of several centimeters in the plane of the wafer. This material is common in photovoltaics. The collection of photocurrent from the volume of material within approximately one bulk minority carrier diffusion length is heavily suppressed, but the overall density of grain boundaries in a photocell is low enough that the vast majority of the available area is unaffected. There are also a few materials in which grain boundaries are not as deleterious as in silicon. Common examples of this set of materials include $Cu(In,Ga)Se_2$ and CdTe, and a variety of other chalcogenide compounds.

Similar approaches, particularly sputtering, PECVD, and hot-wire CVD, are used to fabricate amorphous semiconducting thin films (Ref 1, 11). The most common materials used for this purpose are silicon and germanium. Some compound semiconductors, particularly among the chalcogenide family, can also be fabricated in amorphous phases; however, they are less common in application. Amorphous materials are produced by a combination of lower substrate temperatures and faster deposition rates compared with polycrystalline materials, both of which are also compatible with low-cost deposition on a variety of substrates, including plastics. An important factor in the fabrication of amorphous materials is the incorporation of hydrogen into amorphous silicon and germanium, which can passivate dangling bonds that occur due to undercoordinated host atoms. This improves device performance and permits the material to be doped via chemistries similar to those used in single-crystal and polycrystalline silicon.

Bulk/Elemental Characterization Methods

There is a significant degree of overlap between the characterization approaches used for single-crystal and polycrystalline materials, with a greater difference between either of these classes and amorphous semiconductors. The purity levels of polycrystalline materials are lower; thus, compositional analysis by less sensitive techniques, including EDS and electron microprobe microanalysis, is more common than it is in single-crystal semiconductors. Many polycrystalline materials cannot be doped, but are already p-type or n-type as fabricated due to intrinsic defects. In these materials, transistors and diodes are formed by junctions of dissimilar materials (heterojunctions). For this reason, the need for very-high-sensitivity techniques such as SIMS is reduced; however, the ability to perform high-vertical-resolution depth profiling with this technique can allow it to offer other advantages over the lower-sensitivity approaches. X-ray photoelectron spectroscopy is also capable of depth profiling, typically using an argon ion beam. X-ray photoelectron spectroscopy has a significantly higher detection floor than SIMS, but quantification of XPS data is easier than that of SIMS data, and XPS also permits characterization of the chemical bonding in the material. Preferential sputtering of the substrate atoms should be carefully considered during XPS depth profiling, however, because it can both influence the measured concentration and provide misleading chemical bonding information if the ion beam causes chemical rearrangement of the atoms during sputtering. The vapor pressure contrast between the various constituents can provide a simple rule: during sputtering of a material with one high vapor pressure component, such as selenium, and a lower vapor pressure component, such as copper, this issue may be particularly pronounced.

Polycrystalline Material Microstructural Characterization Methods

The most important difference from a characterization standpoint between polycrystalline and single-crystal materials is the presence of grain boundaries. Grain boundaries are themselves features that have physical locations, distributions, and orientations. The nature of a grain boundary also depends on the relative orientation between the grains it separates, which introduces the need to characterize the grain orientation distribution, or texture, of a polycrystalline thin film. The first step for characterizing polycrystalline films is typically XRD. The presence of grains introduces Scherrer broadening in the XRD peaks, and XRD permits direct estimation of the grain size for grains smaller than approximately 100 nm on a direction parallel to the scattering vector for that reflection. Scans in the Φ, χ, and ω angles can permit the development of pole figures that provide information about the general texture of a film. In many circumstances, a set of roughly aligned grains connected by low-angle grain boundaries perpendicular to the film surface may be desirable from a functional perspective.

Physical vapor deposition commonly produces a columnar grain morphology due to nucleation on the substrate surface followed by vertical growth of those initial grains. When this is the dominant growth morphology, it is possible to visualize the grain structure on the outer surface of the film after growth by simple optical microscopy or SEM, depending on the grain size. Frequently, there is a different growth rate at the grain interiors compared with the boundaries, so the grain boundaries are correlated with topographic features on the free surface that are visible to these techniques. If the surface is very flat, the grains may still be visualized via channeling contrast in SEM or in FIB imaging, or by selective chemical etching of the grain boundaries. Electron backscatter diffraction (EBSD) allows for grain orientation determination via Kikuchi pattern analysis of the backscatter diffraction pattern. This allows for grain counting and grain orientation mapping, which permits the determination of the angle of each grain boundary.

Imaging of grain boundaries that are not vertically oriented typically requires a variety of cross sectioning methods to be employed. For the case of a thin film on a brittle substrate such as glass or silicon, introducing a brittle fracture in the substrate via scribing and bending can be a fast and inexpensive strategy. The fracture of the thin film often proceeds in an intergranular manner, thus providing an easily visible contrast mechanism in SEM. This approach has a few drawbacks, because the roughness created may or may not directly correlate with the grain structure, the film may spall from the surface rather than cleave in a plane parallel to the substrate fracture, and the roughness on the cross section can interfere with compositional analysis.

When a higher-quality section is desired, or the goal is to study segregation of lower-concentration constituents to the grain boundaries, the most common cross-sectioning approach is FIB, which permits a high-quality flat section to be made in any location on the sample surface. The FIB liftout technique previously discussed in the context of epitaxial thin films is often also employed, permitting visualization of the grain boundary by TEM and STEM. Scanning transmission electron microscopy EDS has significantly higher resolution than SEM EDS due to the reduced interaction volume size, and reduced secondary x-ray production makes quantification easier, although good standards are still required to calculate accurate correction factors. Electron energy loss spectroscopy is also possible in STEM mode, permitting visualization and measurement of the distribution of light elements. High-resolution TEM imaging coupled with selected area electron diffraction (SAED) is another approach for determining grain orientation and grain boundary type, though the small sample size limitation of the liftout approach means that such studies are necessarily conducted over many fewer grains than a lower-magnification approach such as XRD or EBSD (Ref 9).

The electronic properties of grain boundaries can be studied by a variety of scanning probe, optical, and electron techniques. Conductive AFM and scanning Kelvin probe microscopy have been used to characterize the effective photocurrent and electric potential surrounding grain boundaries in biased and illuminated conditions (Ref 12). Electron-beam-induced current and the related electron beam imaging techniques in which the electron current is generated by the sample can be used for a similar purpose.

Amorphous Material Microstructural Characterization Methods

Amorphous materials present some unique challenges from a characterization standpoint. The discussion here focuses on amorphous silicon and silicon germanium, both of which are typically grown with a significant hydrogen content. Hydrogen is a very important constituent in these materials, and many analysis techniques based on electron or x-ray spectroscopy are not capable of detecting hydrogen due to the limited electronic transitions available for hydrogen's single electron (ruling out EDS, XPS, and AES) and the small interaction cross section of hydrogen with most energetic particles (making EELS analysis typically impractical). Hydrogen can be detected with ion-based approaches, such as RBS in a forward scattering configuration and SIMS. Hydrogen content can also be estimated by use of infrared spectroscopy (Ref 11).

X-ray, electron, and neutron diffraction have all been used to study the distribution of atoms in amorphous silicon (Ref 11). Even from a theoretical perspective, it is not yet clear how to fully describe the structure of an amorphous material. Significant work in this area has focused on study of the pair and radial distribution functions, which can be extracted via analysis of diffraction data. Neutron diffraction can provide both analysis deeper into a material, due to lower scattering cross sections than x-rays, and enhanced contrast for light elements. Amorphous silicon and germanium still exhibit a relatively uniform tetrahedral coordination that produces significant atomic order over short ranges. Despite the lack of long-range order, a large variety of structural motifs including bonds of various stretching geometries and dangling bonds may produce specific electronic states that may be benign or detrimental, and the processing details of the amorphous material can increase or decrease the frequency of these structures. Study of these states in amorphous materials can be conducted with a variety of approaches, including photoluminescence (PL) spectroscopy, frequency-resolved spectroscopy, and electron magnetic resonance (Ref 11).

Surface Characterization Methods

Surface characterization of amorphous and polycrystalline materials is performed by a set of techniques similar to that used for single-crystal semiconductors. The external surface topography can be imaged via SEM, and surface roughness may be characterized by optical and stylus profilometry, or AFM at higher resolutions. Common tools for surface chemical analysis include XPS, SIMS, and AES. The various approaches to image grain size and grain boundaries discussed previously are often also conducted on the surface of an as-grown polycrystalline thin film.

Electronic Characterization Methods

The main techniques relevant for single-crystal semiconductors are equally applicable to polycrystalline and amorphous semiconductors. For polycrystalline semiconductors, the main influence on the material is the introduction of localized electronic states on the grain boundaries and the potential for small-scale reorganization of the various impurities in the material toward the grain boundaries. This acts to create states, some of which are in the band gap, at and near the grain boundary, which cannot participate in band conduction. Carriers that enter these trap states become trapped in that location, greatly increasing their probability of recombining before exiting the semiconductor volume, or participating in a radiative recombination pathway for the case of an LED. One would expect to see both a decrease in the minority carrier lifetime and a decrease in the electronic mobility for higher grain boundary concentrations (Ref 3).

For amorphous semiconductors, there is a continuum of electronic states near the band edges and deep into the band gap. Many of these states contain carriers that are not mobile, but quantum mechanical tunneling between similar states near the band edges permits a certain degree of bulk conductivity. The main difference from an electronic point of view between amorphous semiconductors and single-crystal materials of the same composition is a greatly reduced carrier mobility. Interestingly, due to the increased symmetry of amorphous materials, it is not possible for amorphous semiconductors to have indirect band gaps, so all amorphous semiconductors are direct band gap materials. Amorphous materials also exhibit a much more gradual absorption edge in optical transmission than is observed for single-crystal materials, because an increased number of electronic states inside the band gap near the band edges are able to participate in optical absorption (Ref 11).

Semiconducting Oxides

There are a number of applications for which semiconducting oxides are chosen over elemental or compound single-crystal semiconductors, including transparent conducting films. Transparent conducting films are heavily used in consumer electronics for LCDs, OLEDs, touch screens, and photovoltaic devices. While indium tin oxide (ITO) is one of the most common materials for these purposes, its cost has driven other materials such as aluminum-doped zinc oxide (AZO) to be researched and utilized. Most thin films for these purposes are polycrystalline or amorphous. Oxides such as indium tin zinc oxide have also been used within the thin film transistor structures for displays, including large-format OLEDs such as those in televisions, because they are less expensive than their amorphous silicon counterparts.

Another common application for semiconducting oxides are varistors, which are electrical devices used for protection against large voltage surges. In semiconducting oxide varistors, grain boundaries can be understood to capture charge and lead to effective Schottky barriers. Thus, the steady-state current-voltage behavior can be highly nonlinear "through the interplay between the applied bias and the occupation of the defect states at the interface and in the depletion region" (Ref 13). When there are grain boundaries with large effective potential barriers along with high doping levels and elevated bias, large electric fields build up within the depletion regions, leading to minority carrier generation and strongly enhancing the nonlinearities in charge transport in these materials. ZnO is a commonly used oxide material for this application.

Other common uses for semiconducting oxides include PTC (positive temperature coefficient) thermistors, capacitors with a high dielectric constant, gas sensors, and electro-optic modulators (Ref 14). Copper(I) oxide has been used in rectifier diodes for decades. While ZnO has been found useful in transducers, AZO has been used in solar cells.

Material Classification

Many well-known oxide materials are insulating. Silicon dioxide, for example, is a native oxide layer widely used in microelectronics for its convenient insulating nature. Semiconducting oxides, also known as semiconducting ceramics, are commonly transition metal oxides that, instead of being insulating, possess band gaps between ~1.2 and ~3.8 eV. ZnO, for example, is a semiconducting material with a wide band gap of 3 to 3.3 eV that is typically n-type doped (Ref 15). Copper(I) and copper (II) oxides have long been known and in use as semiconductors. Other well-known semiconducting oxide materials include bismuth (III) oxide, $SrTiO_3$, and lithium niobate and oxides for transparent conductive films such as ITO and AZO.

Common Semiconducting Oxides

Copper(I) oxide has been in use as a semiconductor in industrial applications since as early as the 1920s, being utilized in rectifier diodes. It has a direct band gap of 2.1 eV. While stable in dry air, copper(I) oxide can oxidize into copper(II) oxide or convert to a carbonate after exposure to humid air. Copper (II) oxide is also a semiconducting material, with a band gap of 1.2 eV. Copper(II) oxide can also be formed by heating copper metal in air, though other fabrication methods may be more suited for laboratory or industrial purposes.

Zinc oxide (ZnO) is a wide-band-gap semiconductor, with a direct band gap of 3.3 eV. It forms naturally as the mineral zincite; however, due to its rarity, the majority of ZnO is synthetic. It can be fabricated in numerous ways that can be considered broadly either chemical or metallurgical (Ref 16). A few such methods include controlled precipitation, sol-gel, and solvothermal or hydrothermal methods (Ref 16). The exact band gap of ZnO can be tunable, if the material is doped with other elements (such as aluminum) or alloyed with other metal oxides (such as cadmium or magnesium oxide) (Ref 17). While it has been studied since the beginnings of semiconductor electronics (Ref 18) and used in varistors since the 1980s (Ref 19), ZnO also enjoys a wide range of other applications, including ultraviolet (UV) lasers, gas sensors, and solar cells (Ref 16).

Strontium titanate has both a direct band gap of 3.75 eV and an indirect band gap of 3.25 eV (Ref 20). It is an extremely rare mineral to find naturally synthesized; however, research in laboratory synthesis led to patents for fabrication methods in the 1940s and 1950s. Today, thin films can be grown using various methods including molecular beam epitaxy, atomic layer deposition, and pulsed laser deposition (PLD). Strontium titanate has a perovskite structure and a large dielectric constant, and is also ferroelectric (Ref 21).

One of the semiconducting oxide materials more widely in use commercially today is ITO. It has a band gap of ~4 eV, which makes it primarily transparent to optical light (though opaque to UV) and useful as an optoelectronic material; it is often utilized in transparent conductive coatings for electronic displays, and it has also found use in organic LEDs, solar cells, and other such applications. Physical vapor deposition is the typical fabrication method; due to the expense of indium, fabrication of ITO can be costly compared with other semiconducting oxides. Indium tin oxide is not affected by moisture and can survive in a solar cell for multiple decades. However, ITO fabricated for these applications is typically quite brittle and can fracture under impact or unintended tensile loading.

Aluminum-doped zinc oxide is under investigation for an alternative material to ITO in transparent conducting films, in large part due to the decreased cost for production. The band gap of AZO can vary depending on the doping concentration of aluminum and final microstructure of the film, but typically ranges from ~3.2 to 3.6 eV (Ref 22, 23). Aluminum-doped zinc oxide films have a hexagonal wurtzite structure (Ref 21). Other materials similarly being considered as alternatives to ITO include gallium- and indium-doped zinc oxides, as well as indium-doped cadmium oxides (which use a smaller concentration of indium than ITO).

Structure of Semiconducting Oxides

Semiconducting oxides are commonly polycrystalline and polyphasic materials (Ref 15). In particular, many such semiconducting ceramics contain grain boundaries with residual glass phase material (Ref 24). As such, the properties of the material after production and the stability of the phases can be greatly affected by the properties of the grains and grain boundaries. Specifically, trapping of charge at grain boundaries can lead to the creation of electrostatic potential barriers (Ref 14). Grain boundaries can also provide a pathway for the rapid diffusion of impurities (Ref 15). If processed properly, these grain boundary properties can be leveraged to lead to successful use of semiconducting oxide materials in various applications.

Additional Properties of Interest

Some semiconducting oxides can exhibit superconducting behavior under certain conditions. Lanthanum copper oxide, for example, typically has a band gap of 2 eV (Ref 25). Substitutions of barium or strontium for lanthanum within the structure in the presence of oxygen can lead to the formation of a semiconducting metal. These so-called high-T_c (critical temperature) superconductors exhibit superconductive behavior above the ~30 K temperature limit typically exhibited by metallic superconductors. Copper oxide or cuprate superconductors are typically based on a layered perovskite structure with oxygen deficiencies, leading to a conductive anisotropy.

Fabrication

Unlike the majority of elemental and compound semiconductors, which can typically be created as bulk crystals or thin films utilizing widely employed techniques, the fabrication methods needed to create a semiconducting oxide for commercial electronic devices can vary greatly by material of choice. For example, copper(II) oxide can be created by heating certain copper compounds such as copper(II) nitrate (Ref 26), or by sol-gel spin coating (Ref 27). Copper(I) oxide, on the other hand, can be fabricated by electrochemical deposition (Ref 28), by oxidation of copper sheets under controlled conditions (Ref 29), or from copper(II) acetate aqueous solution (Ref 30). Because of the difficulties arising from the fabrication method, some semiconducting oxides have found limited use in industrial-level applications as yet (Ref 24).

Metal oxides for transparent conductive film applications, such as photovoltaic devices and electronic displays, are typically deposited onto glass substrates using metal organic chemical vapor deposition, solution deposition, PLD, or magnetron sputtering (Ref 31). Changing thin film deposition parameters such as the target or initial material content, power, substrate temperature, and working pressure (Ref 21) can greatly affect the structural and electrical properties of the final film.

Microstructure Characterization Methods

Postfabrication characterization of oxides largely involves the confirmation of ordering and grain size, composition, and electrical properties. Changes in processing parameters can affect any number of structural factors as well as final properties of the material. Understanding the effect of processing parameters on the final material product is important for reproducibility and knowing that the material for use in devices is consistent with the properties needed, and any changes in processing parameters should be duly characterized before proceeding to device fabrication.

The characterization techniques useful for semiconducting oxides can vary depending on the fabrication method and final form of the material. Given the typical polycrystalline and polyphasic structure of these oxide materials, particular attention must be paid to the processing variables as well as characterizing the final composition and microstructure of the grains (Ref 15). For example, grain boundaries and defects can introduce allowed states within the band gap of the semiconducting material, greatly affecting the final material's conductive behavior. Thus, the final grain structure of a semiconducting oxide should likely be characterized for each set of processing variables to assess how they affect the structure and final properties of the material.

One versatile technique useful for characterizing grain structure is SEM, which has a number of available imaging modes. For polycrystalline materials, differences in the crystal orientation of grains within a sample can be leveraged to image grain structure by utilizing various forms of diffraction. Electron backscatter diffraction, discussed previously in the polycrystalline section, functions by using diffraction of electrons within the electron interaction volume to form a Kikuchi diffraction pattern. Analysis of the pattern provides both phase information and local crystallographic orientation of the material within the interaction volume. With this technique, as with many diffraction imaging techniques, a flat material is typically required to minimize the effects of topography on diffraction angle and intensity.

For obtaining the average crystallographic properties of many grains within a film, XRD is a common choice. Many of the common defects and structural features of oxide thin films that influence their electronic properties can be characterized by XRD. The lattice parameters, defect density, and crystallographic texture of the oxide films can be assessed, as well as thin film strains. For thin films with sufficiently small grains, analysis similar to that performed during powder diffraction can be used. X-ray diffraction is also capable of highly quantitative phase composition determination averaged over large areas.

Additional diffraction techniques useful for understanding the ordering of a material are low-energy electron diffraction (LEED) and reflection high-energy electron diffraction (RHEED). Both techniques use an electron beam incident on the surface of a thin film or bulk material to characterize structure and ordering. For LEED, beam energies are typically 10 to 1000 eV. In this energy range, penetration into the material is typically limited; thus, information obtained from diffracted electrons is particularly sensitive to the surface material. If too many crystallites with varying crystallographic orientation are present, the diffraction patterns for each orientation become superimposed and may not be easily analyzed. Thus, for small grain sizes, decreasing the beam spot size to less than the typical sample grain size can enable investigation of individual grains.

While LEED utilizes an electron beam that is perpendicular to the surface, RHEED measurement geometry utilizes a glancing angle setup. Due to the grazing incidence angle, RHEED is commonly used to monitor crystal structure during growth of thin films without obstructing the sample sightline to source targets during growth.

If an exemplar sample from a particular batch of known processing conditions can be locally thinned until electrons can successfully pass through (typically <200 nm), then imaging with TEM enables atomic and microstructural analysis of the oxide material, with similar imaging strategies used for the general class of polycrystalline material. For the case of polyphasic materials, TEM SAED can provide a powerful tool for local phase analysis and can allow the identification in cross section of phase domains using dark field imaging conditions, in which a series of images is formed with diffracted electrons characteristic of each phase, if their crystal structures differ sufficiently. Additional techniques for understanding the microstructure of a fabricated oxide material include optical microscopy and AES.

Raman spectroscopy is a technique that utilizes inelastic light scattering resulting from the excitation of vibrations in crystalline materials. For metal oxides such as ZnO or other semiconductors, Raman spectroscopy can inform users on structural properties of the material such as order, phase, composition, and orientation (Ref 32, 33). Lateral resolutions as low as 1 μm can be achieved (Ref 34).

Composition Characterization Methods

To confirm composition either macroscopically or locally with mapping capabilities, techniques such as EDS, electron probe microanalysis, and XPS can help assess spatial distributions of elements. While XPS is not a trace element detection method, the technique has an additional benefit of chemical binding information that could be useful for the assessment of different charge states within the oxide if a variation in chemical composition is suspected. The influence of the sputtering process during depth profiling on the chemical bonding in the analysis volume should be carefully considered when interpreting data gathered from this method. Additionally, peak fitting can be utilized to assess ordering within the material, which may be helpful for comparison between samples prepared under different fabrication and processing parameters.

One other technique particularly helpful for semiconductors, PL spectroscopy, can provide information concerning the composition of semiconducting compounds; however, a calibration curve is necessary to relate the band gap as a function of composition to make the connection between composition and the measured photoemission spectra (Ref 15). Given a calibration curve of this type, PL could be helpful for characterizing materials such as aluminum-, gallium-, and indium-doped zinc oxides, which can vary in final composition depending on processing parameters such as composition of the initial target, power, pressure, and sample temperature during sputtering or deposition (Ref 21).

Surface Characterization

Other surface-level techniques of interest for oxide materials include scanning probe microscopy (SPM) techniques such as scanning tunneling microscopy, AFM, and other related and more advanced SPM techniques such as nanoimpedance microscopy.

Electrical Characterization

Because the end application for semiconductors is generally an electronic or optoelectronic device, the relationship between processing parameters, subsequent material structure, and final electrical properties should be well characterized. Common electrical measurements for semiconducting materials aim to characterize properties such as electrical resistivity (or, conversely, conductivity), the band gap of the material, the relative levels of any impurity states, charge carrier mobility, lifetime and diffusion length characteristics of minority carriers, and majority carrier concentration (Ref 15). Once a material is incorporated into a device, device-level characteristics of interest include contact resistance, channel length and width, and barrier height (Ref 15).

Several common methods exist for measuring the resistivity of a semiconductor, including four-point probe and van der Pauw measurements. While material-level characterization is typically necessary before device fabrication, additional device-level characterization can provide information on the properties of the material after it has undergone any additional fabrication steps.

Other techniques of use specifically for semiconductors include techniques that assess the state of the band gap of the final material. Such techniques include PL spectroscopy, modulation spectroscopy, and angle-resolved photoemission spectroscopy. In PL spectroscopy, a light source at a known wavelength is utilized to excite electrons within a material. In relaxation down to the ground state, electrons can then emit a photon characteristic of the change in energy from their excited state to their final ground state. These photons can be detected, and are characteristic of the material from which they originated. Photoluminescence spectroscopy is a sensitive technique, and can identify impurities and defect centers in semiconducting materials as well as assess dopant levels. Photoluminescence spectroscopy can also provide information concerning the relative composition of compounds, if the relationship between the composition and subsequent band gap of the material is known. Conversely, if samples are prepared with varying but known doping concentration, the effect of change in doping concentration as a function of the band structure can be observed (Ref 35).

Organic Semiconductors

Introduction to organic semiconducting materials perhaps must begin with a definition of *organic*, which means carbon based. However, most carbon-based materials we encounter are not interesting with regard to electronic properties or applications. These electrically uninteresting materials include plastics such as polyethylene (plastic bags) and acrylonitrile butadiene styrene (plastic products such as toys). There are also some carbon-based materials that are electrically interesting but not typically mentioned in the discussion of organic semiconductors. These include carbon nanotubes and graphene; they are discussed separately in the family of low-dimensional semiconductors. They are omitted from discussion on organic semiconductors because they are purely carbon and are too conductive. Organic semiconductors are less conductive by definition, and not purely carbon.

The structure of organic semiconductors consists of a carbon backbone of alternating single and double bonds. Such a backbone may also have side chains consisting of carbon or noncarbon elements such as hydrogen, nitrogen, and sulfur. In insulating plastics, also called saturated polymers, all four valence electrons of carbon are used in covalent bonds. However, in semiconducting polymers, chemical bonding leads to one unpaired electron (the π electron) per carbon atom. Pi bonding, in which the carbon orbitals are in an sp^2p_z configuration and in which the orbitals of neighboring carbon atoms overlap, leads to electronic delocalization or an electron cloud. This delocalization responds to an external electric field, provides charge mobility along the carbon backbone, and results in the semiconducting nature of the material. Regarding nomenclature, semiconducting polymers are sometimes simply referred to as conductive polymers. Semiconducting polymers are also called conjugated polymers, which implies that they have a backbone chain of alternating double and single bonds.

In comparison with inorganic semiconductors, typical advantages of organic semiconductors include facile band gap or electrical/optical property tunability utilizing wet chemistry techniques, simpler or cheaper fabrication methods, the possibility of device fabrication on flexible substrates, solution processability, and moderate processing temperatures. Typical disadvantages include environmental stability and electronic properties such as reduced charge carrier mobility compared with crystalline inorganic materials.

Common Semiconducting Organic Materials

Semiconducting organic materials can be classified in several ways. On a size basis, there are three types of materials: small molecules, oligomers, and polymers. Small molecules are one monomer unit, oligomers are a few monomer units, and polymers are a larger (in principle, infinite) number of monomer units. Generally, the optoelectronic properties of small molecules are superior to those of polymers, but small molecules typically require vacuum processing and are difficult to process as thin films from solution. Polymers can be processed from solution using simple methods such as printing or coating, but they are difficult to purify and not every polymer chain in a sample is identical. Organic semiconductors also can be classified in other ways such as their ability to accept or donate electrons, or in other words, *n*-type or *p*-type nature.

Examples of Organic Semiconducting Materials

There are a large number of examples of chemical structures that have been synthesized by the organic semiconductor chemistry community over the years. Not all can be mentioned here; some notable examples are shown in Fig. 5 (Ref 36).

Band Gap Engineering

Band gap engineering or tuning for organic semiconductors is performed through chemical synthesis techniques or wet chemistry, informed by intuition as well as computational methods. Band gaps are typically in the 1 to 3 eV range. Generally speaking, the more delocalized the electrons are, the lower the band gap is. It is also worthwhile to briefly mention what doping means for organic semiconductors. Doping in organic semiconductors is not substitutional, as in the case of silicon, where substituting a silicon atom with phosphorus makes the material *n*-type and with boron makes it *p*-type. Doping in organics is interstitial, meaning that the dopants are present within the matrix of the host material. Doping must be substitutional in crystalline inorganic materials; otherwise, the dopant acts as a defect. Organic semiconductor thin films, on the other hand, are already amorphous or defective to an extent; they can tolerate defects such as interstitial dopants.

Common Applications

In decreasing order of commercial success to date, applications of organic semiconductors include light emitting diodes (OLEDs) or displays, solar photovoltaics (OPVs), field effect transistors (OFETs), and lasers. Commercially successful OLEDs emerged in the 1990s and are ubiquitous today in displays for consumer electronics devices such as phones and televisions. Organic LEDs are typically multiple layers (thin films) of small molecules sandwiched between two electrodes, one electrode being metallic and the other transparent, typically ITO. Polymers have not been the preferred choice for OLEDs due to purity issues. On the other hand, OPVs are based on both polymers and small molecules. The most common OPV material combination is a polymer blended with fullerenes, wherein light is absorbed by the polymer material, which then donates electrons to fullerene, and separated charges are transported to respective electrodes. There have been attempts to commercialize organic photovoltaics, but they have achieved limited success for two main reasons: First, the proof of concept that reasonably efficient devices obtain in a lab setting is often not as efficient when scaled to larger areas. Second, the environmental stability of organic semiconductors is typically inferior to that of their inorganic counterparts. Transistors and lasers based on organic semiconductors have been demonstrated in lab settings but are behind OLEDs and photovoltaics in their commercial viability or promise.

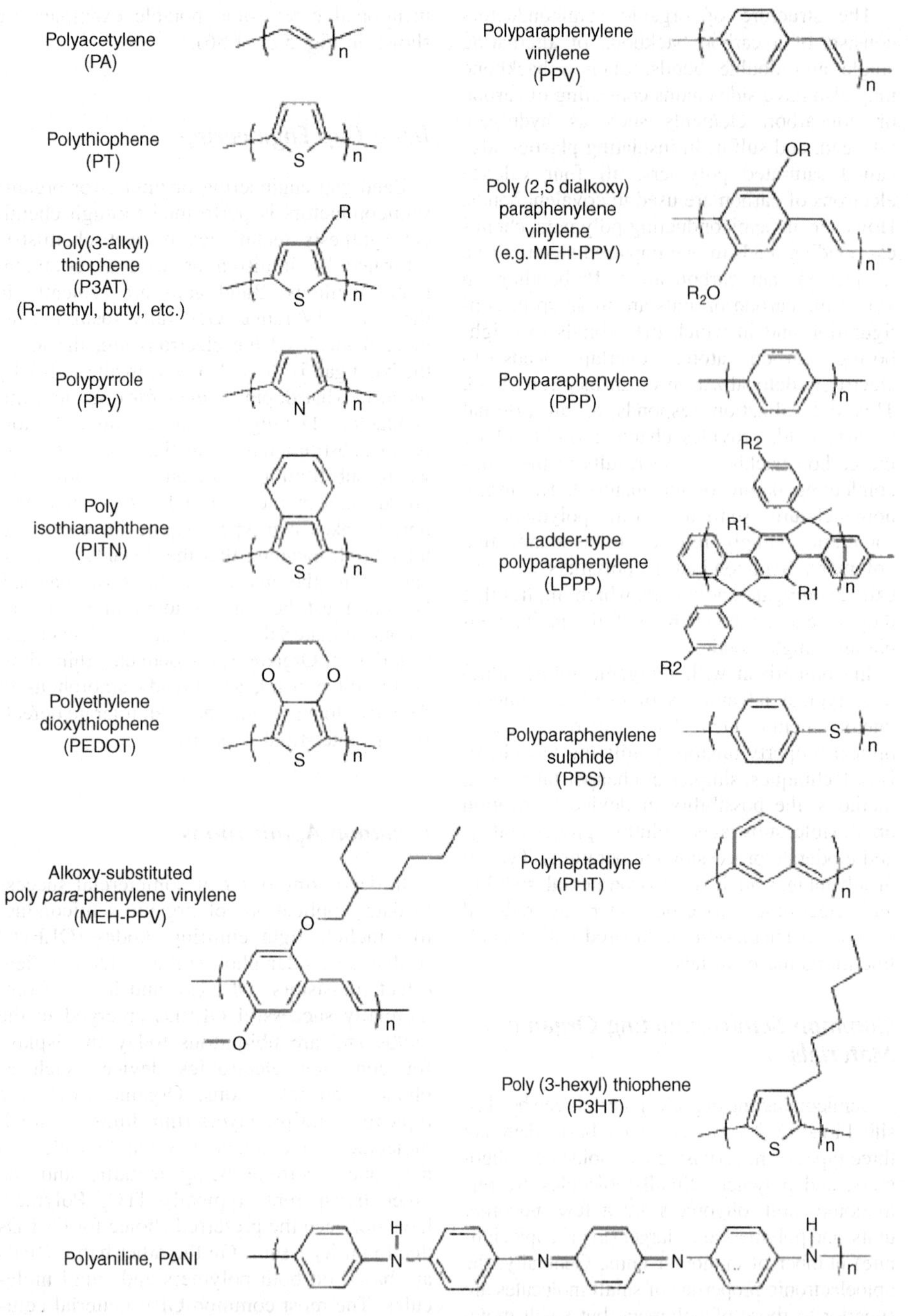

Fig. 5 Molecular structures of some conjugated polymers; a commonality among all structures is alternating single and double bonds between carbon atoms. Adapted from Ref 35

Fabrication

Though single crystals of organic semiconducting small molecules (e.g., pentacene) may have electronic properties rivaling inorganics, typical application device structures do not employ single crystals because single crystals do not have a thin film form factor that devices such as OLEDs or OPVs are designed around. These thin films are realized using fabrication methods that differ between small-molecule-based devices and polymer-based devices. Small-molecule thin films are typically fabricated using thermal (vacuum) evaporation or vapor phase deposition. Vapor phase and vacuum deposition are similar in the sense that the organic material is heated and it then evaporates; however, in vapor phase deposition, the vapors are carried toward a substrate using a carrier gas. By contrast, polymers are typically solution processable, and thin films can be realized using techniques such as spin coating, screen printing, drop casting, and inkjet printing. It should be noted that a lot of effort in the last two decades has been focused on developing solution-processable small molecules; this is done by attaching long aliphatic side chains to the conjugated backbone. Such side chains interact with organic solvents and enable dissolution and thin film processing from solution. Typically, OLED or OPV devices are fabricated on ITO-coated glass on which organic layers are deposited and finally capped by a thermally evaporated metal electrode. Commercial applications may employ flexible polymeric substrates or stacked fabrication on top of other electronics. For both OLEDs and OPVs, other than the active organic layer, there are also buffer layers employed at the interface of the active layer and electrodes to balance the charge injection process for OLEDs, or the charge extraction process for OPVs, and the charge transport process for both types of devices. Due to environmental stability issues, device fabrication is typically performed in nitrogen or argon-filled glove boxes; encapsulation can further improve device lifetime.

Characterization of Composition, Band Gap, and Energy Levels

The compositional characterization of these materials is typically performed at the synthesis stage by utilizing nuclear magnetic resonance, gel permeation chromatography, optical methods, and electrochemical methods. Nuclear magnetic resonance and GPC are used to assess the size and size distribution in polymers and are typically performed on liquid samples. Ultraviolet (UV)-visible absorption in solution or in a thin film state is utilized to ascertain the optical band gap. Cyclic voltammetry is performed to determine the energy level of the highest occupied molecular orbital (HOMO) and the lowest unoccupied molecular orbital (LUMO); HOMO and LUMO levels are analogous to the valence and conduction bands of inorganic semiconductors. Because the polymer chain packing is different in solution versus in thin films, and even in solution it changes with the choice of the solvent and concentration, the band gap and energy level properties measured in solution are often different from those measured in thin films. Properties measured in thin films are more relevant for applications as applications also employ these materials in thin film configuration.

Structure and Surface Characterization

Once a material is synthesized, thin films are typically fabricated and various characterizations are performed to understand process-structure-property relationships. Atomic force microscopy is the most common technique employed to evaluate thin film morphology. See Ref 37 for sample AFM images of organic semiconductor thin films. The noncontact AFM mode is typically employed to yield topographical and phase scans; the topographical image provides information about thin film surface roughness with a resolution of approximately 1 nm. Phase image contrast provides information about heterogeneity; it entails mapping the distribution of crystalline domains versus amorphous domains, or assessing phase separation when multiple materials are mixed together, such as in the classic bulk-heterojunction architecture of OPVs, in which a polymer is blended with fullerenes or another organic material. Some research groups and AFM companies have also been working on more progressive AFM-based techniques (Ref 38) for OPV application wherein the sample is illuminated through a transparent substrate (such as ITO glass) from the bottom, ITO acts as one electrode, the AFM tip acts as another electrode, and thus topography and phase information can be simultaneously measured along with optoelectronic properties; the resultant data provide rich process-structure-property correlation at the nanoscale.

Though some information about crystallinity or an amorphous nature can be ascertained from AFM imaging, there are other techniques such as grazing-incidence wide angle x-ray scattering (Ref 39), which is suitable for probing molecular length scales in thin films and yields information on grain sizes, whether the conjugated carbon backbone is oriented edge-on on a substrate or face-on, and other structural information. Though TEM is also occasionally employed by researchers on a cross-sectional slice of thin films to image phase separation in a blend of materials, SEM is rarely used and not very useful in the organic electronics community because of the charging effect that electron beams have on organic films. Electron beam damage is also an important concern for organic materials in general. This damage can arise from a variety of mechanisms, but melting due to Joule heating from absorbed electrons and electrolysis are both potential damage mechanisms (Ref 9).

Electrical Characterization

Regardless of application (OLEDs, OPVs, or OFETs), current-voltage (IV) characteristics are routinely measured on thin films between two electrodes. For OLEDs, IV characteristics are coupled with measuring light emission. Both the intensity of the light emission and the spectrum of the emitted light are typically of interest. For OPVs, natural sunlight (or light from a solar simulator) is incident on the device while the IV characteristics are measured to determine the photovoltage, photocurrent, and fill factor produced by the device. For OFETs, three terminal (source, drain, gate) IV characteristics are measured to assess properties such as mobility, threshold voltages, and switching speed.

While direct current IV characterization assesses overall performance characteristics, small-signal alternating current characterization is also useful for identification and characterization of defect states within the band gap of a material, or at the interface of two materials (Ref 40). Alternating current characterization at different frequencies and temperatures may be performed for this purpose, and data are analyzed with respect to models originally developed for inorganic semiconductor solid-state physics.

Optical Characterization

Optical characterization methods such as PL and UV-visible absorption, as mentioned previously, are utilized both by chemists in assessing the band gaps and optical properties of their materials and by device engineers to probe properties of thin films. Thin films with high crystallinity exhibit UV-visible absorption curves with more pronounced peaks and shoulders than curves of amorphous films. Photoluminescence helps to assess light emission properties of materials relevant for OLEDs, as well as exciton (electron-hole pair) quenching properties of blends relevant for OPV applications. Time-resolved versions of these techniques, called transient absorption or PL lifetime, are also utilized for detailed understanding of electron-hole pair and charge dynamics on nano-, pico-, or femtosecond time scales.

The mentioned techniques are the most common characterization techniques utilized in the community. In addition to these, nearly every materials science technique that is utilized for inorganics can be utilized for organics as well; however, a number of the most common inorganic analysis spectroscopies such as EDS and electron microprobe analysis are less sensitive to the light elements characteristic of organic semiconductors.

Low-Dimensional Semiconductors

With an industrial trend toward smaller, denser electronic components and optoelectronics (Ref 41), a significant amount of research has been devoted to novel materials that are meant to function on the nanoscale. Many materials exhibit properties at these length scales that differ from the properties of their bulk material counterparts. For a material to be considered low dimensional, its size is typically constrained in at least one dimension to less than 100 nm, causing charge carriers to experience only one or two dimensions rather than the full three dimensions of a bulk semiconducting material. In these systems, electronic and optical properties are often best described by quantum mechanical paradigms, and final properties of the material are largely affected by their structural boundary conditions. As such, understanding the size, structure, ordering, and properties of the final semiconducting material after production is critical to obtaining a device that functions as intended.

Common Applications

Low-dimensional semiconductor research is multifaceted in its motivation for end-use applications, and the interested reader is directed to other sources for a more detailed discussion beyond the scope of this brief summary (Ref 41, 42). Some examples include the continuing miniaturization of electronic devices, advanced optical waveguides, nanowire lasers, and alternatives to silicon-based technology. A number of ambitious goals could potentially be addressed through novel applications of low-dimensional materials. Currently, many of these proposals have not yet found successful commercial application and remain in the realm of fundamental research. However, some applications of low-dimensional semiconductors are in use at the time of this publication, including quantum dot display technology (Ref 43).

Examples of Low-Dimensional Semiconducting Materials

The last two decades have seen a large volume of research in the fields of materials science, chemistry, physics, and mechanical and electrical engineering devoted to the fabrication, characterization, and integration of novel low-dimensional semiconducting materials for electronics and optoelectronics. Thus, a wide array of nanomaterials of many chemistries, sizes, and shapes can be found in recent literature, from ZnO nanowires and cadmium selenide nanorods to semiconducting polymer nanoparticles. Initial reports of low-dimensional semiconductor fabrication were published earlier. Carbon nanotubes and other fullerenes, for example, were discovered in the late 1900s and have been researched for several decades. Carbon nanotubes can be either single walled or multiwalled, as well as semiconducting or metallic, depending on the size and orientation of the carbon lattice (Ref 44). There also exist layered semiconductor systems that can be considered low-dimensional due to the constraint of charge carriers in very thin epitaxial films, where electronic states created at the interfaces of certain material pairs act as a 2D electron gas (Ref 45).

Structure of Low-Dimensional Semiconductors

The main attribute of a low-dimensional semiconductor is its confinement in at least one dimension. For these materials, the equivalent bulk material may not be semiconducting; if that is the case, the creation of a band gap in the low-dimensional material is typically attributable to quantum confinement. In the case of materials that are already semiconducting in their macroscopic state (for example, bulk ZnO as well as ZnO nanorods), the low-dimensional structure of the fabricated material may be desirable for the decrease in scale of the overall device, for an increase in surface-to-volume ratio, or due to size-related changes in the material's optical or electronic properties.

In broad terms, a number of similar geometries can be observed to fall under the header of low-dimensional semiconducting materials, with the number of constrained material dimensions being the primary differentiating factor: (1) 2D systems, where the material extends largely in two directions but exists only on the order of nanometers in the third dimension (e.g., monolayers, and some layered semiconductor systems), and (2) nanostructures that are effectively 1D or zero-dimensional (0D), constrained in two or three dimensions respectively. Some examples of both of these types of structures are quantum dots (0D); nanotubes, nanowires, and nanorods (1D); and superlattices, monolayers, quantum wells, and some layered semiconductor systems (2D) (Ref 41).

Fabrication

Two broad categories of fabrication of low-dimensional materials exist: top-down approaches, which typically utilize lithography to pattern structures from macroscopic materials down to the nanoscale; and bottom-up approaches, which leverage growth or self-assembly to build nanostructures from atomic or molecular precursor materials (Ref 46). Efforts have also been made to develop combined top-down/bottom-up approaches to leverage the strengths of each (Ref 47).

Top-Down Techniques

Top-down approaches typically involve a substrate or thin film that is lithographically patterned in a desirable way to create the final material. A number of forms of lithography are available, including photolithography, which uses ultraviolet or x-ray light to expose a photoresist; electron beam lithography, which utilizes a well-focused beam of electrons to expose an electron resist; and a large variety of imprint lithographies, which mechanically replicate a self-organized structure from a mold into a resist. After exposure and development of either resist, a combination of mask deposition and selective etching can be performed using a variety of gases, plasmas, and wet chemistries chosen on the basis of the substrate or thin film material.

Using this approach, there is a limitation on the smallest size of a feature that can be reproducibly patterned, dependent on the mask quality and lithography technique. For radiation-based lithographies, which use a scanning beam or a mask for pattern transfer, the fundamental limit to resolution arises from diffraction. In these cases, a reduction in wavelength may allow for patterning of smaller features, which motivates the move from UV light, to x-rays, and finally to electron beams. For electron-beam lithography, the resolution is currently limited more by the interaction volume and the sidewall roughness achievable for the available electron beam resists than the electron wavelength at typical accelerating voltages. State-of-the-art lithographic techniques implemented by the integrated circuit industry are capable of producing critical lateral dimensions for some classes of devices lower than the 10 nm range. For silicon integrated circuitry, the main challenges to further scaling are no longer the lithographic patterning, but a variety of materials issues including solubility limits of dopants in silicon, leakage current through gate oxides, and the need to dissipate the large quantities of heat generated by highly dense arrays of transistors (Ref 1). Bottom-up techniques can fabricate high-quality single-crystal structures with critical dimensions down to the several nanometer range, but the deterministic arrangement of these elements into useful patterns over technologically relevant areas remains a distinct challenge (Ref 45).

Bottom-Up Techniques

Bottom-up techniques are diverse and can be used to produce everything from thin films to quantum wells and quantum dots. For these techniques, a substrate that is compatible with the necessary processing and structure of the final desired material is generally utilized. The thin epitaxial films discussed in the section "Single-Crystal Semiconductors" in this article can also be considered 2D structures. A number of common thin film fabrication techniques can be used to form nanostructures when they are used to deposit coatings with sub-monolayer coverage. Techniques that are suitable for bottom-up fabrication of nanostructures include chemical vapor deposition, metal-organic chemical vapor deposition, and molecular beam epitaxy. Condensing atoms may aggregate into nanometer-scale particles, or along step edges or other substrate features into wires or other shapes. Deposition of material onto templates fabricated by other methods, such as anodic etching of alumina (Ref 48) or on cleaved multilayer sidewalls (Ref 49), is another approach. A variety of liquid phase and vapor phase approaches that make use of surfactants, anisotropic crystal growth, and metal-catalyst-particle-mediated growth have also been demonstrated (Ref 50). However, nanostructures are often formed in a random distribution on the wafer or carrier material. Unless the structures can be grown or assembled repeatably in a desired distribution on a suitable electronic substrate, the material requires transfer to the substrate to be used for device fabrication, and no general scalable approach for deterministic arrangement of materials during such transfer has yet been demonstrated. Current commercially successful applications of bottom-up nanomaterials rely on creating random arrays or networks of sufficient density of the constituent nanomaterials to achieve the desired electronic or optoelectronic properties without reaching such a high density that the quantum confined nature of the nanomaterials is lost.

Band Gap Engineering

Because low-dimensional semiconducting materials are often best described quantum mechanically, it might not be surprising that the allowed electron states within the low-dimensional structures can be determined by the final size and shape of the material. Analogous to a quantum mechanical "particle in a box" or square well potential, the size of the well affects the allowed energy states. Similarly, it is possible to affect the electronic properties of a low-dimensional material by changing its final size and structure (Ref 51). On these length scales, strains in the material lattice can also influence the band gap (Ref 52) and have also been predicted to open a band gap in materials such as graphene (Ref 53).

Structure and Surface Characterization

Due to the size of low-dimensional semiconductors, there are inherent limitations in the techniques that can be utilized to characterize them. "Bulk"-type techniques that require large volumes of uniform material are generally not well suited for characterizing nanostructures. X-ray- and neutron-diffraction-based techniques, for example, are limited in their utility of characterizing nanoscale materials for fundamental reasons: while the material may be periodic, the small size interrupts the periodicity, which can broaden diffraction peaks, and these peaks are often weak compared with the background signal arising from the substrate supporting the nanomaterial due to mismatch between the total size of the structure being analyzed and the x-ray or neutron penetration depth (Ref 45). Electron diffraction techniques, such as selected area electron diffraction, low-energy electron diffraction, and reflection high-energy electron diffraction, may provide more information. In general, techniques that can provide localized characterization of structure, composition, and surfaces are required to understand the final properties and structure of low-dimensional semiconducting materials. These types of techniques generally include real-space or diffraction imaging by probe-based, electron-based, or optical microscopy.

To characterize the final structure of a fabricated low-dimensional material, electron and optical microscopy techniques offer real-space characterization of dimensions and surfaces. Scanning electron microscopy can be used on most types of low-dimensional samples and is useful for visualizing the size and shape of 0D and 1D nanomaterials, as well as surfaces, steps, and some defects. Transmission electron microscopy can provide information on structure at the atomic level but depends on the ability to prepare samples such that electrons can penetrate the thickness. A variety of commercially available electron-transparent supports are available, such as thin amorphous carbon and thin supported silicon nitride membranes. When nanomaterials can be grown or transferred onto these support structures, TEM may be the fastest and most information-rich characterization approach, as the composition, size, shape, and internal microstructure of the nanomaterials can all be assessed at the same time. Collection of TEM images from a variety of viewing angles can also allow for 3D tomographic reconstruction of nanomaterials to obtain information about their 3D shape. In certain special conditions, it may be possible to fabricate freestanding membranes of a 2D nanomaterial, such as suspended graphene. In these systems, chemical identification of single atoms has been performed (Ref 54). If a low-dimensional structure can be probed only while on a solid substrate, imaging through transmission may not be possible.

The lateral resolution of optical microscopy is limited to one quarter of the optical wavelength by the diffraction limit, in the 100 nm range for visible light. Thus, optical microscopy may not be sufficient for imaging nanostructures that are in the deep submicron range. However, in cases where individual nanoparticles are well separated, optical approaches may allow direct excitation of electronic or optical transitions in materials that can allow physical dimensions to be indirectly inferred, for instance, on the basis of the particle color (Ref 55).

Another group of techniques that can be immensely useful for low-dimensional materials is scanning probe microscopy and its various subbranches. In all such techniques, a small probe is brought near the surface of interest and scanned within a controlled area. Two such techniques that are particularly useful for low-dimensional semiconductors are AFM and scanning tunneling microscopy (STM). Both provide structural information, with AFM resolutions being as low as a few nanometers and STM being able to measure individual atoms under certain conditions. Other offshoots of AFM (such as electrostatic force microscopy, or EFM) can simultaneously measure surface structure and surface-related material or electronic properties.

Composition Characterization

Due to their small dimensions, the full volume of a nanomaterial can often be considered to be on the "surface." For this reason, many of the surface chemistry approaches used in other semiconductor systems can be applied to nanomaterials to characterize their average composition. In many cases, it can be difficult to cleanly separate the nanomaterials from some type of supporting structure, such as a solid substrate, and care should be taken to select such a substrate whose own signal in the selected analysis technique does not overlap or interfere with that of the nanomaterial of interest. Techniques discussed in previous sections, such as EDS, EELS, XPS, and SIMS can all be used to probe the composition of nanomaterials, and for EELS and XPS, also their chemical bonding. Where possible, EDS should be performed on an electron transparent substrate, such as in TEM, to increase the signal coming from the nanomaterial compared with the unavoidable background generated by the support structure. Raman spectroscopy has a special application for the case of carbon nanomaterials and is often used to assess the material structure and properties, including disorder, grain boundaries, thickness, and strain (Ref 56).

A specialized form of ion spectroscopy, local electron atom probe (LEAP), enables performing mass spectroscopy to identify atomic constituents, similar to SIMS, as well as mapping in three dimensions the original location of the atom before sputtering and ionization. Challenges with detection efficiency and accurate spatial reconstruction remain, but this technique may enable the full 3D reconstruction of the atoms within an individual nanoparticle, for instance, the counting and physical location of dopant atoms within silicon nanowires (Ref 57). Because the detection system relies on sputtering the host atoms, the technique is inherently destructive.

Electronic Property Characterization

Before fabrication of any devices utilizing low-dimensional semiconductors, the electronic properties of the low-dimensional material should be well understood. For these material properties, a few probe-based techniques can be particularly helpful, including STM and the related scanning tunneling spectroscopy, EFM, scanning capacitance microscopy, and scanning Kelvin probe microscopy. The band gap of the materials can be assessed by PL or angle-resolved photoemission spectroscopy, if the material of interest is a thin film or can be fabricated with a dense array of nanostructures.

REFERENCES

1. A. Rockett, *The Materials Science of Semiconductors*, Springer, 2007
2. F. Scholz, *Compound Semiconductors*, Pan Stanford Publishing, 2018, p 1–307
3. D.K. Schroder, *Semiconductor Material and Device Characterization*, John Wiley & Sons, 2006
4. C.A. Volkert and A.M. Minor, Focused Ion Beam Microscopy and Micromachining, *MRS Bulletin*, Vol 32 (No. 5), May 2007, p 389–399
5. G. Davies, Optical Measurements of Point Defects, *Identification of Defects in Semiconductors*, Vol 51, M. Stavola, Ed., Academic Press, 1999, p 1–92
6. K. Saarinen, P. Hautojärvi, and C. Corbel, Positron Annihilation Spectroscopy of Defects in Semiconductors, *Identification of Defects in Semiconductors*, Vol 51, M. Stavola, Ed., Academic Press, 1998, p 209–285
7. J. Marek, Light-Beam-Induced Current Characterization of Grain Boundaries, *Journal of Applied Physics*, Vol 55 (No. 2), Jan 1984, p 318–326
8. R.J. Ross, *Microelectronics Failure Analysis*, 6th ed., ASM International, 2011
9. D.B. Williams and C.B. Carter, *Transmission Electron Microscopy*, 2nd ed., Springer, 2009
10. J. Stover, *Optical Scattering: Measurement and Analysis*, 3rd ed., SPIE, 2012
11. K. Morigaki, S. Kugler, and K. Shimakawa, *Amorphous Semiconductors: Structural, Optical, and Electronic Properties*, John Wiley & Sons, 2017
12. Y. Yan, C.S. Jiang, R. Noufi, S.H. Wei, H.R. Moutinho, and M.M. Al-Jassim, Electrically Benign Behavior of Grain Boundaries in Polycrystalline CuInSe2-Films, *Phys. Rev. Lett.*, Vol 99 (No. 23), Dec 2007, p 235504–4
13. F. Greuter, Electrical Properties of Grain Boundaries in Polycrystalline Compound Semiconductors, *Semi. Sci. Tech.*, Vol 5 (No. 2), 1990, p 111
14. B.G. Yacobi, Semiconductor Materials: An Introduction to Basic Principles, 2003, p 153–186
15. A. Janotti, Fundamentals of Zinc Oxide as a Semiconductor, *Rep. Prog. Phys.*, Vol 72, 2009, p 126501
16. A. Kołodziejczak-Radzimska, Zinc Oxide—From Synthesis to Application: A Review, *Materials*, Vol 7, April 2014, p 2833–2881
17. Ü. Özgür, A Comprehensive Review of ZnO Materials and Devices, *J. Appl. Phys.*, Vol 98 (No. 4), 2005, p 1118.
18. H.E. Brown, *Zinc Oxide Rediscovered*, New Jersey Zinc Company, 1957
19. K. Eda, Zinc Oxide Varistors, *IEEE Electrical Insulation Magazine*, Vol 5 (No. 6), 1989, p 28–30.
20. K. van Benthem, Bulk Electronic Structure of SrTiO3: Experiment and Theory, *J. Appl. Phys.*, 2001, Vol 90, p 6156
21. X.X. Xi, Dielectric Properties and Applications of Strontium Titanate Thin Films for Tunable Electronics, *Nano-Crystalline and Thin Film Magnetic Oxides*, Springer, 1999, p 195–208

22. K.H. Kim, Structural, Electrical and Optical Properties of Aluminum Doped Zinc Oxide Films Prepared by Radio Frequency Magnetron Sputtering, *J. Appl. Phys.*, Vol 81 (No. 12), 1997, p 7764–7772
23. H.L. Shen, Preparation and Properties of AZO Thin Films on Different Substrates, *Prog. Nat. Sci.*, Vol 20, Nov 2010, p 44–48
24. D.R. Clarke, Grain Boundaries in Polyphase Ceramics, *J. Phys. Colloq.*, Vol 46 (No. C4), 1985, p C4–51
25. P.Y. Yu and M. Cardona, *Fundamentals of Semiconductors: Physics and Material Properties*, 4th ed., H.E. Stanley and W. T. Rhodes, Ed., 2010, p 3
26. C.R. Crick, CVD of Copper and Copper Oxide Thin Films via the in situ Reduction of Copper (II) Nitrate: A Route to Conformal Superhydrophobic Coatings, *J. Mat. Chem.*, Vol 21 (No. 38), 2011, p 14712–14716
27. J.H. Lee, Electrical and Optical Properties of ZnO Transparent Conducting Films by the Sol–gel Method, *J. Cryst. Growth*, Vol 247 (No. 1-2), 2003, p 119–125
28. L.C.K. Liau, Fabrication Pathways of p–n Cu2O Homojunction Films by Electrochemical Deposition Processing, *J. Phys. Chem. C*, Vol 117 (No. 50), 2013, p 26426–26431
29. T. Minami, High-Efficiency Oxide Solar Cells with ZnO/Cu2O Heterojunction Fabricated on Thermally Oxidized Cu2O Sheets, *Appl. Phys. Expr.*, Vol 4 (No. 6), 2011, p 062301
30. R.V. Kumar, Sonochemical Synthesis of Amorphous Cu and Nanocrystalline Cu2O Embedded in a Polyaniline Matrix, *J. Mat. Chem.*, Vol 11 (No. 4), 2001, p 1209–1213
31. T. Minami, Present Status of Transparent Conducting Oxide Thin-Film Development for Indium-Tin-Oxide (ITO) Substitutes, *Thin Solid Films*, Vol 516 (No. 17), 2008, p 5822–5828
32. J. Richter, The One Phonon Raman Spectrum in Microcrystalline Silicon, *Solid State Commun.*, Vol 39 (No. 5), 1981, p 625
33. I.H. Campbell, The Effects of Microcrystal Size and Shape on the One Phonon Raman Spectra of Crystalline Semiconductors, *Solid State Comm.*, Vol 58 (No. 10), 1986, p 739–741
34. C.R. Brundle, C.A. Evans, Jr., and S. Wilson, *Encyclopedia of Materials Characterization*, 1992, p 33
35. M. Wang, Optical and Photoluminescent Properties of Sol-gel Al-doped ZnO Thin Films, *Mat. Lett.*, Vol 61 (No. 4-5), 2007, p 1118–1121
36. A.J. Heeger, "Semiconducting and Metallic Polymers: The Fourth Generation of Polymeric Materials," Nobel Lecture, Dec 2000, https://www.nobelprize.org/uploads/2018/06/heeger-lecture.pdf
37. J.A. Carr, Controlling Nanomorphology in Plastic Solar Cells, *Nanomat. and Energy*, Vol. 1 (No. 1), 2012, p 18–26
38. D.C. Coffey, Mapping Local Photocurrents in Polymer/Fullerene Solar Cells with Photoconductive Atomic Force Microscopy, *Nano Lett.*, Vol 7 (No. 3), 2007, p 738–744
39. N.C. Miller, Use of X-Ray Diffraction, Molecular Simulations, and Spectroscopy to Determine the Molecular Packing in a Polymer-Fullerene Mimolecular Crystal, *Adv. Mat.*, Vol 24 (No. 45), 2012, p 6071–6079
40. J.A. Carr, The Identification, Characterization and Mitigation of Defects in Organic Photovoltaic Devices: A Review and Outlook, *Energy Environ. Sci.*, Vol 6 (No. 12), 2013, p 3414–3438
41. R.J. Martin-Palma and J.M. Martinez-Duart, *Nanotechnology for Microelectronics and Photonics*, 2nd ed., 2017, p 1–15
42. O. Manasreh, *Introduction to Nanomaterials and Devices*, 1st ed., 2012, p 1–46
43. "Quantum Dots: Solution for a Wider Color Gamut," *Samsung Display*, https://pid.samsungdisplay.com/en/learning-center/white-papers/quantum-dot-technology
44. M.S. Dresselhaus, G. Dresselhaus, and P. C. Eklund, *Science of Fullerenes and Carbon Nanotubes*, 1996, p 756–869
45. D. Weiss, K.V. Klitzing, K. Ploog, and G. Weimann, Magnetoresistance Oscillations in a Two-Dimensional Electron Gas Induced by a Submicrometer Periodic Potential, *Europhysics Letters*, Vol 8 (No. 2), 1989, p 179
46. C. Kittel, *Introduction to Solid State Physics*, 8th ed., 2005, p 515–562
47. J.D. Hicks, "A Combined Top-Down/Bottom-Up Route to Fabricating Graphene Devices," thesis, Georgia Institute of Technology Department of Electrical and Computer Engineering, Aug 2013
48. D. Routkevitch, Electrochemical Fabrication of CdS Nanowire Arrays in Porous Anodic Aluminum Oxide Templates, *J. Phys. Chem.*, Vol 100 (No. 33), p 14037–14047
49. N.A. Melosh, Ultrahigh-Density Nanowire Lattices and Circuits, *Science*, Vol 300 (No. 5616), Apr 2003, p 112–115
50. A.M. Morales, A Laser Ablation Method for the Synthesis of Crystalline Semiconductor Nanowires, *Science*, Vol. 279 (No. 5348), Jan 1998, p 208–211
51. N.M. Park, Band Gap Engineering of Amorphous Silicon Quantum Dots for Light-Emitting Diodes, *Appl. Phys. Lett.*, Vol 78 (No. 17), 2001, p 2575–2577
52. E.D. Minot, Tuning Carbon Nanotube Band Gaps with Strain, *Phys. Rev. Lett.*, Vol 90 (No. 15), 2003, p 156401
53. G. Gui, Band Structure Engineering of Graphene by Strain: First-Principles Calculations, *Phys. Rev. B*, Vol 78 (No. 7), 2008, p 075435
54. O.L. Krivanek, Atom-by-Atom Structural and Chemical Analysis by Annular Dark-Field Electron Microscopy, *Nature*, Vol 464, Mar 2010, p 571–574
55. C.B. Murray, Synthesis and Characterization of Nearly Monodisperse CdE (E = Sulfur, Selenium, Tellurium) Semiconductor Nanocrystallites, *J. Am. Chem. Soc.*, Vol 115 (No. 19), 1993, p 8706–8715
56. I. Childres, Raman Spectroscopy of Graphene and Related Materials, *New Developments in Photon and Materials Research*, J.I. Jang, Ed., 2013
57. D.E. Perea, Direct Measurement of Dopant Distribution in an Individual Vapour–Liquid–Solid Nanowire, *Nature Nano*, Vol 4, 2009, p 315–319

ASM Handbook, Volume 10, *Materials Characterization*
ASM Handbook Committee
DOI 10.31399/asm.hb.v10.a0006671

Characterization of Ceramics and Glasses*

THE CHARACTERIZATION, testing, and nondestructive evaluation of ceramics and glasses are vital to manufacturing control, property improvement, failure prevention, and quality assurance. A fundamental understanding of ceramic and glass materials is requisite as well. This article provides a broad overview of characterization methods and their relationship to property control, both in the production and use of ceramics and glasses.

Important aspects that are covered in this overview include:

- The means for characterizing ceramics and glasses and the corresponding rationale behind them
- The relationship of chemistry, phases, and microconstituents to engineering properties
- The effects that the structure of raw ceramic materials and green (unfired) products and processing parameters have on the ultimate structure and properties of the processed piece
- The effects that trace chemistry and processing parameters have on glass properties, which are more significant than the effects of raw glass material microstructure

Those who have only an informal background in this technology should be aware of several sources of potential frustration. First, slight variations in chemistry of the ionic-covalent chemical bond that is typical of ceramics and glasses, and subtle differences in phase distribution in a ceramic are often difficult to observe but can cause major changes to properties during production or use. Second, experience based on metals and plastics can be misleading because these materials have wider ranges of composition and microstructure, which often lead to indistinguishable properties. Third, preparation procedures associated with metals and plastics are not totally appropriate to the characterization activities associated with ceramics and glasses. For example, the techniques used to polish hard metals prior to microstructural analysis can often lead to artifacts when applied to ceramics, which, in turn, can lead to misinterpretation of key features. Careful sample preparation and application of analytical techniques are very important to a correct correlation between the structure and properties of ceramics and glasses. Fourth, there are neither comparable standards of preparation or heat treatment nor compendia of compositions and structures for ceramics and glasses, as there are for metals.

Chemical Analysis

The chemistry of a ceramic or glass is an important starting point in the characterization process because it is almost always vital to specification and property control. Very small quantities of additive can have a major effect on both processing and ultimate properties. Sintering aids are often added at the level of tenths of one percent to promote rapid consolidation of ceramic powders and prevent grain (crystal) growth. A similar level of iron impurity gives a noticeable brown tone to alumina, clay-based whitewares, and silicate glasses. This effect is significantly enhanced by the presence of titanium at trace levels. Small percentages of alkaline elements in alumina can drastically alter its dielectric constant and loss values. Similar levels of alkaline or alkaline earths may significantly deteriorate the chemical durability of alumina.

As an overview guide, Fig. 1 and Table 1 list some of the most common tools used for the characterization of the ceramics and glasses discussed in this article.

Bulk Chemistry

The general methods of spectrochemical and wet chemical analysis can be applied to ceramic materials and glasses. Methods of chemical analysis are applicable to both ceramics and glasses, but often characterization techniques are specific to either ceramics or glasses. It is generally necessary to dissolve the material because melting and/or vaporization of the ceramic or glass is difficult. This task requires some experience because powerful acids or bases are usually needed to perform the digestion. The whole material must be reduced without leaving either a residue or colloidal material. Interaction effects between species in the ceramic and/or reagent can have major effects on observed responses and sensitivities.

The determination of oxygen, nitrogen, or carbon in oxide, nitride, carbide, and mixed ceramics is frequently calculated by compound balance from the anion determination. This can lead to error when variable valences, complex compounds, or vacancy structures occur. It is important that the analytical facility have experience with ceramics in general and with the specific ceramic and species to be measured. This cannot be stressed strongly enough for the novice who may rely, based on previous good experiences, on an excellent analytical laboratory that is actually inexperienced in ceramic analysis.

Methods

Optical emission spectroscopy is perhaps the most common method used for major and minor constituent analyses, whereas inductively coupled plasma and atomic absorption are usually chosen for trace element work. The bulk chemistry of solid samples and powders can be determined by x-ray spectroscopy, down to 0.1% for elements heavier than fluorine, in most cases. Organics in unfired ceramics and important chemical bonds in glass or other transparent/translucent ceramics may be analyzed by infrared spectroscopy or Fourier transform infrared spectroscopy. In some cases, gases given off during drying and firing are analyzed by a gas chromatograph/mass spectrometer. Other chemical analysis methods that yield specific information, are sensitive to light elements, are more suitable for the specific sample, and/or are the most readily available can also be used.

Microchemical Analysis

It is often important to know the local chemical distributions in a ceramic, glass, or composite. This distribution may reveal the homogeneity of

*Adapted from V.A. Greenhut, Characterization of Ceramics and Glasses: An Overview, *Ceramics and Glasses*, Volume 4, *Engineered Materials Handbook*, ASM International, 1991

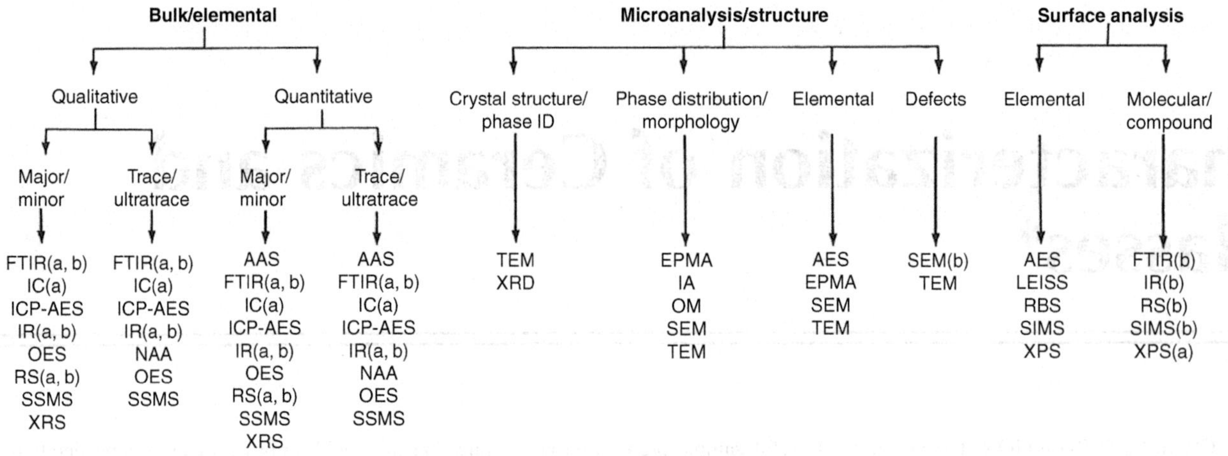

Fig. 1 Flow charts of common techniques for characterization of glasses and ceramics. AAS, atomic absorption spectrometry; AES, Auger electron spectroscopy; EPMA, electron probe x-ray microanalysis; FTIR, Fourier transform infrared spectroscopy; IA, image analysis; IC, ion chromatography; ICP-AES, inductively coupled plasma atomic emission spectroscopy; IR, infrared spectroscopy; LEISS, low-energy ion-scattering spectroscopy; NAA, neutron activation analysis; OES, optical emission spectroscopy; OM, optical metallography; RBS, Rutherford backscattering spectrometry; RS, Raman spectroscopy; SEM, scanning electron microscopy; SIMS, secondary ion mass spectroscopy, SSMS, spark source mass spectrometry; TEM, transmission electron microscopy; XPS, x-ray photoelectron spectroscopy; XRD, x-ray diffraction; XRS, x-ray spectrometry. (a) Limited number of elements or groups. (b) Under special conditions

raw material powders or the distribution of phases in a fired piece. Such an analysis is most commonly performed down to micrometer levels by using x-ray fluorescent analysis or x-ray spectroscopy (XRS), coupled with a scanning electron microscope (SEM) to perform electron probe microanalysis (EPMA). However, this analysis can also be done at submicrometer levels when coupled with a transmission electron microscope (TEM) or analytical electron microscope (AEM). The electron beam can be localized to an area, line, or spot to perform local qualitative or quantitative chemical analyses. Both TEM/XRS and AEM can be used to analyze cross-sectioned interfaces down to nearly atomic levels. Elemental maps can be produced that relate directly to observed microstructure. The emitted intensities of various elements can show the location of phases. Often, heterogeneous microstructure, such as a grain-boundary phase, may strongly influence properties.

The surfaces, or interfaces, of a material often have a very different chemistry than the bulk, which can affect processing characteristics such as rheology and sintering rate. The performance of fired ceramics may be dictated by surface resistance to chemical attack and ability to bond to other materials. The SEM/XRS reveals near-surface effects at a depth of approximately 1 lxm (40 pin.). Auger electron spectroscopy, scanning Auger microscopy, x-ray photoelectron spectroscopy, secondary ion mass spectroscopy, and ion-scattering spectroscopy can be used to show what is on or within the first few atomic layers of the surface. These techniques yield specific chemical bonding information and identify the elements present and, in contrast to x-ray fluorescent analysis, are sensitive to the lighter elements. Ion sputtering can be used to remove surface layers and thereby profile in depth.

Phase Analysis

The specific phases present in a ceramic can strongly influence properties. For example, alumina containing up to approximately 15 wt% metastable zirconia in an optimum distribution can show a several-fold increase in toughness and strength. The metastable phase in zirconia can be controlled by trace chemical additives, particle size, forming methodology, and thermal treatment. It is increasingly being realized that minor oxide glassy phases, even just a few atoms thick, on grain boundaries of nonoxide and oxide ceramics can promote creep and plastic stress relief at elevated temperature and affect ambient strength as well.

High-temperature ceramic superconductors require close control of the formation of specific phases. Many commercial ceramic compositions show a wide multiplicity of phases, depending on thermal history, trace element content, or crystal growth conditions. Zirconia shows three phases that are stable in different temperature domains. Alumina has approximately five common forms and as many as two dozen minor variants, when trace impurities are considered. Silicon carbide has two major phases and a large set of polytype structures.

Although chemical analysis can provide an important indication of the phases present in a ceramic, when a number of phases are possible, the analysis must be augmented by other techniques. For instance, optical microscopy of thin sections gives optical indexes and extinction angles. Crystal shape and color can be used to identify phases in either transmission or reflection modes, providing additional knowledge of the system. An example of a common application is the identification of stones (inclusions) in glass. Either SEM/XRS or EPMA may also be used to infer phases. The coordinated observation of local, microstructural-level chemistry and microstructural features such as grain shape, size, or orientation may help to show a particular phase. Other localized spectral or chemical analysis methods that may reveal specific bonding information useful for indicating phases can also be used.

Diffraction

The most common analytical tool for phase identification is crystal diffraction, which employs x-rays, high-energy electrons, or neutrons that undergo wave interference with the regular arrangement of atoms in a crystalline lattice. Crystal diffraction techniques can be used to give both qualitative and quantitative data on the phases present.

By far the most common technique used is x-ray powder diffraction. The Joint Committee on Powder Diffraction Standards (International Center for Diffraction Data, Swarthmore, PA) publishes a file listing the characteristic patterns of several hundred thousand crystalline phases in terms of interatomic (*d*) spacings and relative diffraction peak intensities. When an x-ray beam interacts with a material, the diffraction pattern that results is recorded with a spectrometer-like arrangement, called a diffractometer. The *d*-spacings and relative intensities are used as a fingerprint of the crystalline phase(s).

Modern systems are highly computer automated and may analyze as many as 40 samples automatically. Film methods can also be used to record the interference pattern, paradoxically, either when very superior data are required or when a modem system cannot be afforded. Special attachments allow analysis of solid samples at elevated temperature and for films, coatings, liquids, or unconsolidated

Table 1 Characterization techniques for ceramics and glasses

Wet analytical chemistry, ultraviolet/visible absorption spectroscopy, and molecular fluorescence spectroscopy can generally be adapted to perform many of the bulk analyses listed.

Method	Elemental analysis	Speciation	Isotopic or mass analysis	Qualitative analysis (identification of constituents)	Semiquantitative analysis (order of magnitude)	Quantitative analysis (precision of +20% relative standard deviation)	Macroanalysis or bulk analysis	Microanalysis (≤10 μm)	Surface analysis	Major component (>10 wt%)	Minor component (0.1–10 wt%)	Trace component (1–1000 ppm or 0.0001–0.1 wt%)	Phase identification	Structure	Morphology
Atomic absorption spectrometry	D	...	...	...	...	D	D	...	...	D	D	D	...	...	...
Auger electron spectroscopy	●	...	...	●	●	...	...	●	●	●	●	S	...	...	S
Electron probe x-ray microanalysis	●	...	...	●	●	●	...	●	...	●	●	S	S	...	●
Image analysis	...	...	...	...	...	...	●	●	...	...	...	...	...	...	●
Ion chromatography	D, N	...	...	D, N	D, N	D, N	D, N	...	...	D, N	D, N	D, N	...	...	...
Inductively coupled plasma atomic emission spectroscopy	D	...	...	D	D	D	D	...	...	D	D	D	...	...	...
Infrared spectroscopy/ Fourier transform infrared spectroscopy	S	S	...	S	S	S	S	...	...	S	S	S	S	S	...
Low-energy ion-scattering spectroscopy	●	...	...	●	●	...	...	S	●	●	●	●	...	...	...
Neutron activation analysis	●	...	N	●	●	●	●	...	...	S	●	●	...	...	...
Optical emission spectroscopy	●	...	...	●	●	●	●	...	...	●	●	●	...	...	...
Optical metallography	...	...	...	...	...	...	●	●	...	...	...	...	...	...	●
Rutherford backscattering spectrometry	●	...	...	●	●	●	...	...	●	●	S	S	...	...	...
Raman spectroscopy	S	S	...	S	S	S	S	S	...	S	S	S	S	...	...
Scanning electron microscopy	●	...	...	●	●	...	...	●	...	●	●	...	S	...	●
Secondary ion mass spectroscopy	●	...	●	●	...	...	...	...	●	●	●	S	...	...	...
Spark source mass spectrometry	●	...	●	●	●	●	●	...	...	●	●	●	...	...	...
Transmission electron microscopy	●	...	...	●	●	S	...	●	...	●	●	...	●	●	●
X-ray photoelectron spectroscopy	●	N	...	●	●	...	...	...	●	●	●	...	...	...	...
X-ray diffraction	...	...	...	●	●	S	●	...	...	●	●	...	●	●	...
X-ray spectrometry	●	...	...	●	●	●	●	...	...	●	●	N	...	...	...

● = generally usable; D = after dissolution; S = under special conditions; N = limited number of elements or groups

powders. Some error-causing factors to which ceramic samples can be particularly prone include composition differences between standards and samples, preferred orientation of plate- or needle-shaped grains, inadequate number of crystals sampled, matrix or microabsorption by some phases in a mixture, interaction with the environment, and sample-preparation effects, such as phase segregation during reduction to a fine powder.

Electron diffraction is usually used to identify very fine (<0.1 μm, or 4 μin.) phases and determine their distribution. Because this technique employs a very small sample, care must be taken to ensure that it is typical. Neutron diffraction can enhance information by accentuating differences for phases containing light elements and elements that are close in atomic number.

X-ray diffraction is also the principal method used to determine the atomic arrangement, that is, the crystal structure. Those who have experience with this analytical tool use it to determine preferred crystal orientation (texture), amorphous (glass) content, residual stresses, particle and crystallite size, impurity level, atomic defect structure, and the structure of amorphous material.

Structure and Properties of Glasses

The chemistry of common commercial glasses is the chief feature that requires characterization, because it usually has the greatest correlation with all properties. Often, spectrochemical absorption methods are used to characterize the effects of specific impurities or additives that affect optical properties, such as color. In optical fibers, trace elements must be determined and controlled for species as common as water and iron to either minimize optical transmission losses or to produce an active fiber.

Defects in glasses, such as seeds (small bubbles) and stones (small inclusions), can be analyzed by using microstructural analysis and microchemical analysis techniques. Similar methods are used to determine the structure and local chemistry of glasses that either separate into two or more glass phases or partially crystallize to form a glass-ceramic. A polariscope, consisting of crossed polarizers, can be used to determine residual stresses by stress birefringence. It may be important to determine temperature versus viscosity in order to fabricate a glass and allow residual stresses to be either relieved during annealing or incorporated during tempering. Electrical, mechanical, chemical durability, and optical properties can be characterized for specific applications.

Microstructural Analysis

Ceramography, the study of microstructural features at magnifications greater than 50×, is a primary analytical tool that is used to understand the nature of ceramic raw materials, green (unfired) pieces, and fired samples. Important features are the particle size and shape distribution in raw materials, as well as similar features in green and fired ceramics. Other important aspects are the distribution of various phases in a ceramic, the size and spatial distribution of porosity, the nature of surface and volume flaws, and the degree of microstructural uniformity.

Although there is no compendium of microstructures available for ceramics, as there is for metals, someone who is familiar with expected microstructural features can develop a considerable understanding of both structure-property relationships and the types of production changes that may be necessary to obtain desired behavior characteristics.

Very fine-scale features, such as pores and large grains, can act as critical flaws in a ceramic or glass, because of their brittle nature. A typical desirable ceramic microstructure consists of a minimum volume of small pores and a porosity with fine grain size. Phase heterogeneity and orientation may have a significant effect on all properties. For example, the honeycomb monolithic catalytic converter substrate used in automotive applications must have a carefully controlled porosity to achieve optimum catalytic performance, coupled with controlled grain size and orientation to achieve the desired directional thermal expansion and strength properties. Proper grain orientation is the key to the prevention of thermal shock failure during use.

The preparation of a ceramic sample for microstructural analysis requires great care and attention. Powders can easily separate due to gravity; size and surface chemistry effects can arise during mounting for optical or scanning electron microscopy; and green ceramics can be altered by the drying or embedding procedure used to prepare them for analysis.

Although fired samples are extremely hard, polishing cross-sectioned samples can easily cause pull-out of individual grains, leading to a significant error in the interpretation of pore distributions. It is best to polish with diamond using an extremely flat polishing wheel, such as hardwood, with a minimum of polish and a great deal of extender. Light pressure is preferred. This is usually accomplished by polishing sequentially on flat wheels with successively finer polishing compound. At each polishing level, the abrasion scratches of the prior step are removed by polishing normal to the prior direction. Care must be taken to thoroughly remove embedded polishing compound from prior steps. Coarser polishing can introduce pull-out damage and chipping.

Because many ceramics are attacked or altered by common reagents, such as water, preparation artifacts are possible. Etching a ceramic sample to reveal microstructure usually involves the use of strong acids and bases, often at elevated temperatures. Controlling the etch is difficult, and it is complicated by the profound effect that slight differences in chemistry and/or chemical distribution can have. Preparation procedures are not standardized and are frequently undocumented.

Optical microscopy is a primary analytical technique used to examine microstructure. Thick sections and fracture surfaces can be examined with reflected light to characterize microstructure. Thin sections can be more revealing and provide a better opportunity for phase analysis but with increased effort in sample preparation. Frequently, phases and structures that are difficult to show using other microstructural techniques can be distinguished in thin sections. A number of different contrast techniques can reveal structure. Often, slight differences in chemistry can cause a significant difference in color. When using reflected light, care must be taken to focus on the surface of interest; subsurface features may make visualization difficult.

Optical microscopy is often well suited to the characterization of grain structure, glassy bonding phase, and porosity in traditional ceramics. It is also useful for identifying flaws and defects in ceramics and glasses, as well as relatively planar fracture surfaces. However, it is limited in that many ceramic microstructures, particularly those of advanced ceramics, are either at or beneath the scale that an optical microscope can resolve.

Scanning electron microscopy can be a useful tool for looking at finer structures and powders. Secondary electron contrast is provided mainly by topography, atomic number, and conductivity resulting from bombardment with a high-energy electron beam. The most common images, employing secondary electrons, show light and shadow illumination as if the sample were illuminated by an oblique light source. This image shows three-dimensional topographic features. Higher-atomic-number areas and less electrically conductive microstructural features will appear brighter on polished and etched section samples and, to a lesser extent, topographic samples.

Compositional contrast and information can be obtained with a backscatter electron detector and by x-ray fluorescent spectroscopy, which permits analysis of both structure and microchemistry.

The scale of the SEM is particularly suitable for many ceramic parts and raw materials. The depth of focus provided by this instrument makes it very well suited for rough failure surfaces. Indeed, many samples are prepared for SEM by fracturing to reveal microstructure. However, caution must be exercised in this case, because failure may follow a specific path in the material that is not typical of the general structure.

Transmission electron microscopy is limited to very detailed, usually scientific, studies at high magnification. It is powerful enough to reveal and identify ultrastructure, submicrometer chemistry, and very fine crystalline or amorphous phases. Sample preparation is usually difficult for ceramics, involving polishing to a very thin 3 mm (1/8 in.) disk and dimpling the center. Then, a sample is generally ion milled for an extended time to yield a small electron-transparent area that may be suitable for TEM study. The transmission electron microscope is particularly useful for elucidating grain-boundary phases, interfaces, ultrafine particles, nanoscale microstructures, and crystal defect chemical analysis techniques to give micro-micro chemistry.

Image analysis and feature analysis permit the quantification of morphological aspects of images obtained by optical microscopy, SEM, and TEM. This has become rather convenient to do with the advent of computer processing of images and the application of quantitative stereology to the computer. Major limitations are the contrast that must be obtained for the features of interest and the operator skill that is required to analyze real features. The size and shape distribution of powder particles, grain sizes in a green or fired ceramic, and porosity can all be obtained. Volume fractions of various phases and pores can also be obtained. Because two-dimensional features are analyzed, either a separate normal section must be made or the three-dimensional structure must be inferred.

Ceramic Powder Characterization

The characteristics of a ceramic powder are important to final properties. For example, particle size and surface chemistry affect forming and firing response as well as ultimate properties. In addition, the degree of powder agglomeration may affect forming and lead to larger-scale porosity.

The microscopic methods described in the section "Microstructural Analysis" in this article are quite suitable for determining particle size. However, more rapid and industrially applicable methods have been developed. Screens of various sizes can be used to measure intervals of particle size in terms of the weight fraction that will pass through a mesh of one size, while remaining on a finer mesh. Settling of a suspension can be used to show particle size distribution in terms of the relative settling rate of different-sized particles in a medium. Usually, a simplifying assumption is made that the particles are spherical, a condition infrequently met by ceramics. A particular application is the sedigraph, an instrument in which the turbidity of a constantly falling column is measured by continuously scanning the column. Because these techniques miss the fine particles that have a profound effect on forming behavior, newer techniques have been developed to measure submicrometer-sized particles by time-of-flight or interparticle interference.

Surface properties of powders are quite important. Surface chemistry can be obtained by using the methods defined in the section "Chemical Analysis" in this article. Surface charge, a property that affects agglomeration, can be determined with a zeta potential meter, whereas the dispersion of a powder can be determined by measuring its rheology. Metal oxide and other ceramics or glasses form an electrical double (Stem) layer in water and other polar liquids. As the extent of the double layer increases, so does the repulsive force between similarly charged particles. As the concentration of surface potential determining is changed (pH), the layer thickness changes. When the activity of positive and negative sites on the sample is equal, there is no electrical double layer. This zero point of charge gives particle agglomeration and high viscosity. The extent of the double layer can be determined by electrokinetic measurements that give the potential at the shear plane between liquid bound to the particles and free liquid—the zeta potential. The surface charge of the ceramic particles will go through a zero point and particles will agglomerate as a function of additives or pH. When different materials are used, the system must be treated for all components to obtain good forming properties and to heterodisperse the multiple phases for a homogeneous microstructure. Under other conditions, a best dispersion can be developed, which, together with particle size, gives best flow characteristics for forming a ceramic and obtaining best unfired density.

Rheometry is used to measure flow, or viscosity, behavior of a ceramic suspension by shear rate and time. Generally, satisfactory ceramic forming occurs when the ceramic suspension is pseudoplastic and yields and flows readily above a certain shear level. This means that flow will occur when the slurry is forced into a mold, but the material will remain in the mold when the force is removed. For slips, thixotropy is often desired. In this case, agitation thins the material progressively with time so that it will pour into a mold. Upon standing for a time, the material sets and no longer flows readily.

Porosity can be determined from microstructural analysis, as previously indicated. Alternatively, instruments such as a mercury intrusion porosimeter can be used to determine porosity. Mercury or another nonwetting fluid is forced into open pores under pressure. The pressure required to penetrate with a volume of mercury is measured. This can be related to the diameter of cylindrical capillary pores to yield a pore size distribution based on the assumption of cylindrically shaped pores. Neither closed pores nor small pores (<0.05 μm, or 2 μin.) are determined by this method. Water penetration (boiling) and resulting weight gain can also be used to determine porosity. Ultrasonic attenuation can also furnish some information about porosity, particularly when scanning in frequency.

Density is determined by either pycnometry or the Archimedes experiment (liquid displacement and weight change). Because open porosity may be penetrated, it is not considered in the density measurement. This can be circumvented by coating a sample with wax or lacquer to seal the open pores.

Surface area measurements are frequently made by a technique employing the Brunauer-Emmett-Teller equation, which is used for high-specific-surface-area materials, such as very fine powders, catalytic substrates, and sol-gel glasses. It determines the adsorption / desorption of a gas in order to calculate the open surface area.

For a simple pore shape, the measurements of pore size, frequency and distribution, density, and surface area should be interrelated, particularly if open and closed porosity is well understood. When discrepancies arise, it is useful to examine whether the microstructure and phase distribution are properly understood or whether a significant experimental error has been made.

Surface roughness can have a significant effect on the mechanical properties of a ceramic or glass. Surface scratches act as flaws that promote failure. The larger and sharper the crack emanating from the bottom of the scratch, the lower the stress at which failure occurs. Chemical exposure, even to atmospheric water, under an applied stress can cause slow crack growth, historically termed static fatigue, and, ultimately, catastrophic failure in a process analogous to stress-corrosion cracking in metals and craze cracking in polymers. The finer the surface finish, the more resistant to failure a ceramic will be. Usually, a fine finish also gives lower friction forces and lower material wear. Surface roughness is often determined by profilometry, optical microscopy, or SEM.

Testing

A wide variety of mechanical strength tests are used for ceramics, but the most common tests involve three- and four-point loading of bar samples. The typical "dog bone" tensile sample commonly used for metals and plastics is expensive to machine in the case of ceramics, because diamond grinding is required to produce a fine surface finish. Special precautions and fixtures are required to prevent failure in clamping grips.

Recent advances in testing procedures dictate that two lower hardened steel rollers support rectangular bar samples. One central roller on the upper surface transfers load from a universal testing machine in three-point loading. Two symmetrical rollers are used in four-point loading. These are termed three- and four-point modulus of rupture tests. Care must be taken to make the span-to-thickness ratio approximately 10:1 or more.

Strength is not an innate property of a ceramic. Surface finish has a profound effect on strength. Large flaws will cause brittle failure at a low stress. For this reason, the surface is often machined to a uniform finish (600 grit is common) and edges are chamfered to establish a consistency that will show material differences. A sandblasted (damaged) ceramic or glass rod may show a strength value that is several orders of magnitude lower than a polished and lacquered surface. If a typical production surface is to be tested for strength, it should be used as the tensile side of the bar, with carefully chamfered edges.

The results of a three-point test will yield a higher modulus of rupture than a four-point test. This is because the tensile load drops off in the three-point test as one moves away from the center. There is a greater probability that the largest flaw will be in a region of high stress in the four-point case. The two strength values can be related by a statistics-of-failure argument. The statistical nature of failure also requires a large sample population.

One way of treating strength is to report the probability of failure as a function of stress. Such a procedure gives rise to the so-called Weibull plots. This can be viewed as a technique for minimizing flaws by increasing the modulus with refined production methods or surface finish. Failure reliability can be improved by increasing the Weibull modulus (the slope of such a plot), or the strength can be increased by moving the whole probability curve to higher stresses.

Proof testing subjects a part to stresses in order to eliminate parts having large flaws. Understressing uses a relatively low stress to remove most flawed parts and improve the statistics of failure for a moderate load. Overstressing applies loads greater than the design stress to obtain an additional margin of safety. If the proof test enlarges flaws, it could lessen performance and reliability. Recent studies suggest that proof testing can be used successfully.

If a flaw of known size and sharpness is put into a test sample and either the tensile stress to cause failure or the work required to fail the material is determined, a fracture toughness value, K_{Ic}, can be attained. This value shows the resistance of the ceramic to catastrophic failure.

A ceramic can be indented with either a Knoop or Vickers shaped diamond indenter and the indentation size measured to give a hardness value. Hardness is related to properties such as wear resistance, which is often measured as the material-removal rate that occurs as a pin of similar or differing material is run against a rotating plate. A single-stroke device can also be used. The friction coefficient can also be determined and, along with wear rate, is important for applications such as cutting tools.

The reversible expansion, as well as the irreversible shrinkage, that occurs during sintering consolidation of a ceramic is tracked by a dilatometer. The irreversible behavior can be used to monitor the rate and extent of consolidation. Resistance to thermal shock is determined by repeatedly heating and cooling a ceramic at a controlled rate and either determining deterioration of strength or onset of failure after a number of repeated cycles. Thermal analysis techniques measure relative or time differential weight change (thermogravimetry), temperature difference relative to a reference (differential thermal analysis), and scanning calorimetry relative energy difference (differential scanning calorimetry). These techniques can explain important phase transformations and reactions that occur during either the firing or use of a ceramic. This information can be used as input to improve firing procedures or to attain a more thermally stable system for use under particular conditions. These systems can be coupled to a gas chromatograph/mass spectrometer to determine gaseous reaction products. The mechanical behavior of ceramics at elevated temperature in terms of creep resistance, mechanical strength, and toughness is an area of much interest because of the potential of turbine engines and other high-temperature structural applications.

Nondestructive evaluation (NDE) techniques include optical examination, liquid penetrant inspection, ultrasonic testing, acoustic emission, microwave inspection, radiographic inspection, computed tomography, magnetic spin resonance imaging, neutron radiography, thermal inspection, optical holography, and acoustic holography.

As already noted, flaws and cracks in ceramics represent locations where mechanical failure initiates and therefore dictate the mechanical properties. Identifying these flaw sites is necessary to eliminate parts with flaws and disqualify parts for which failure is likely. Because flaws and microstructure are frequently of a submicrometer scale for high-performance ceramics, NDE techniques must progress somewhat to be successfully applied to ceramics.

Failure Analysis

Fractography is the study of fracture surfaces to determine the manner and origin of failure. Based on this determination, a change in design and/or elimination of a class of defects may be undertaken. Ceramic materials fracture in a brittle mode. Several fracture surface features are observable in the optical microscope. Cracks branch as they propagate away from the failure origin, giving rise to a distinctive river pattern. Tracing these hackle and rib patterns back to their origins can reveal the imperfection or feature responsible for failure. Initial failure zones that propagate by slow crack growth can often be discerned and their size related to the critical failure load. Products on the surface may give evidence of environmental interaction. The procedures are similar to those employed in the fractography of brittle metals.

The SEM also provides topographic and compositional information from rough fracture surfaces and is therefore used for investigating failure surfaces after optical analysis. The surface analysis techniques can be useful for identifying environmental products that result during slow crack growth. Often, the failure path is important because it can propagate through a weak phase or interface. A modification of such a material to strengthen the weak structure may improve performance.

SELECTED REFERENCES

- L.V. Azaroff, *Elements of X-Ray Crystallography*, McGraw-Hill, 1968
- F.D. Bloss, *An Introduction to the Methods of Optical Crystallography*, Holt Rhinehart and Winston, 1961
- D. Briggs and M.P. Seah, *Practical Surface Analysis*, John Wiley & Sons, 1983
- B.D. Cullity, *Elements of X-Ray Diffraction*, Addison-Wesley, 1978
- J.I. Goldstein et al., *Scanning Electron Microscopy and X-Ray Microanalysis*, Plenum Press, 1981
- P.B. Hirsch et al., *Electron Microscopy of Thin Crystals*, Plenum Press, 1965
- *Materials Characterization*, Vol 10, *Metals Handbook*, 9th ed., American Society for Metals, 1986
- *Mechanical Testing*, Vol 8, *Metals Handbook*, 9th ed., American Society for Metals, 1985
- *Metallography and Microstructures*, Vol 9, *Metals Handbook*, 9th ed., American Society for Metals, 1985
- *Metals Handbook Desk Edition*, American Society for Metals, 1985
- L.E. Murr, *Electron Optical Applications in Materials Science*, McGraw-Hill, 1970
- *Nondestructive Evaluation and Quality Control*, Vol 17, *Metals Handbook*, 9th ed., ASM International, 1989
- G.W. Phelps et al., *Rheology and Rheometry of Clay-Water Systems*, Cyprus Industrial Minerals, 1981
- H.H. Willard et al., *Instrumental Methods of Analysis*, 6th ed., Wadsworth, 1981

ASM Handbook, Volume 10, *Materials Characterization*
ASM Handbook Committee
DOI 10.31399/asm.hb.v10.a0006652

Introduction to Characterization of Organic Solids and Organic Liquids

Richard Rucker, Exponent, Inc.

THE ART AND SCIENCE of organic and materials synthesis has progressed a great deal since the publication of the 1986 *Materials Characterization*, Volume 10 of *ASM Handbook*. Today (2019), the materials scientist has access to metrological tools with vastly greater analytical power than were commonly available then. While this is no doubt enabled by the continual progress in computational power so beneficial to all enterprises and industries, such progress is also a product of society's desire to advance technology and satisfy curiosity in the pursuit, application, and expansion of scientific knowledge.

Characterization of Organic Solids

The field of organic solids is still constituted by encompassing categories such as polymers, plastics, epoxies, long-chain hydrocarbons, esters, foams, resins, detergents, dyes, organic composites, coal and coal derivatives, wood products, chemical reagents, and organometallics; however, many materials incorporate components from multiple of the aforementioned categories as well as from inorganic components, such as semiconductors and glasses. As technology has progressed, the number and applications of these organic/inorganic composites, nanomaterials, and related substances, which are complex and rely on multiple methods for thorough characterization, have only increased. Consequentially, a great number of the techniques available for the characterization of inorganic solids, liquids, effluents, and other materials are also applicable to the characterization of organic solids. Due to their ever-increasing complexity, there are tremendous opportunities to contribute to the development of useful methods relying on application of several analytical techniques for their successful characterization.

These challenges are also opportunities for interdisciplinary collaborations to apply techniques from different fields of study to characterize single systems. As was highlighted in the 1986 introduction and still very relevant today, the collaborative approach of team-building and integrating scientists and engineers of different fields of research to solve a technical challenge, such as the characterization of an organic material, is more effective than what would be expected if they were working alone.

As an example, consider the process involved in characterizing a commercially available multiwalled carbon nanotube (MWCNT)-filled epoxy resin (a polymer nanocomposite), which may be under consideration for use due to its excellent mechanical or thermally conductive properties or its reduced weight compared to conventionally filled epoxies. To ensure that the epoxy is compatible with the system, it is necessary to evaluate its chemical and physical properties and behavior. Due to the complexity of the material, several characterization techniques are likely necessary. These epoxies are typically provided in a two-part formulation consisting of the epoxy prepolymer and a curative, or hardener. The two components undergo reaction after mixing to provide the hardened final product. In this example, the type of chemistry involved in the epoxy curing (the reaction between the two components) can be established by using Fourier transform infrared (FTIR) spectroscopic analysis of the uncured epoxy components as well as the cured material. The type and concentration of curative (hardener) used in the epoxy resin formulation can be assessed by using gas chromatography/mass spectrometry (GC/MS). In consultation with a polymer chemist, one would learn that the epoxide-containing component is often oligomeric; therefore, the average molecular weight of that component should be tracked by using liquid chromatography/mass spectrometry or perhaps gel permeation chromatography. A materials scientist would explain that the type of filler, including its dimensions, can greatly influence the final material properties. From this conversation, it is apparent that determining the physical parameters of the carbon nanotubes used is important, which can be done by dissolving the uncured prepolymer in an organic solvent, followed by treatment with ultracentrifugation to concentrate it. The solid mass can then be dispersed and characterized by using transmission electron microscopy (TEM) with image analysis to obtain the average diameter and length of the nanotubes. The carbon nanotube loading is also important and affects the mechanical properties as well; therefore, the residual mass of the nanotubes left after combustion or thermogravimetric analysis can be measured as a useful quality-control parameter of the material.

Once cured, the mechanical and thermally conductive properties of the epoxy depend on the distribution of the MWCNTs throughout the epoxy as well as the extent of cure of the epoxy resin. The MWCNT distribution may be studied through a variety of techniques, such as secondary ion mass spectrometry (SIMS) (time-of-flight SIMS, for example) or microscopic imaging, such as atomic force microscopy or TEM. The extent of cure can be established through chemical analysis, such as by correlating the disappearance or appearance of signals related to the starting materials, intermediates, or final cured product by spectroscopic techniques, such as FTIR analysis. This parameter, along with information about the MWCNT distribution, can then be correlated to the mechanical or thermal performance of the resin.

The preceding example readily illustrates the complexity of organic solids, which commonly consist of formulations of multiple components added to meet a particular requirement, and the need for multiple experts working together using multiple analytical techniques for their successful characterization. Although it is specific to epoxy resin, the same approach can be used for the characterization of virtually any organic solid by using combinations of techniques described in this Volume. Table 1 and Fig. 1 list techniques used for the analysis of organic solids.

Some general advice in approaching a material analysis includes:

- Consider the strengths and limitations of the proposed characterization approach. There are many instrumental techniques listed in this section that can be used to accomplish a particular analytical task, such as elemental

Table 1 Analytical methods used for characterization of organic solids

Technique (a)	Elemental analysis	Speciation	Compound	Isotopic or mass analysis	Qualitative analysis (identification of constituents)	Semiquantitative analysis (order of magnitude)	Quantitative analysis (precision of +20% relative standard deviation)	Macroanalysis or bulk analysis	Microanalysis (≤10 μm)	Surface analysis	Major component (>10 wt%)	Minor component (0.1–10 wt%)	Trace component (1–1000 ppm, or 0.0001–0.1 wt%)	Structure	Morphology
AES	•				•	•			•	•	•	•			
AFM	N		N		•	N			•	•	•	•			•
COMB	N						N	N			N	N	N		
EFG	•		•		•	•	•	•			•	•			
EPMA	N				N	N	N		N		N	N	N		
ESR	N	N			N	N	N	N				N	N	N	
FTIR	N	D, •	•		•	N	N	•	•	•	•	•		•	
GC			V		V	V	V	V	V		V	V	V	N,S	
GC/MS	N	N	V	•	V	V	V	V			V	V	V	V	
IA								•	•						•
IC	D,N		D,N	D,N	D,N	D,N	D,N	D,N		N	D,N	D,N	D,N		
ICP-MS	•			•	•	•	•	•	•	•	•	•	•		
LC			D		D	D	D	D			D	D	D		
LC/MS	D, •	D,N	D,•	D,•	D	D	D	D			D	D	D		
LEISS	•				•	•				•	•	•	S		
MFS	D,N	D,N	D,N		D,N	D,N	D,N	D,N			D,N	D,N	D,N		
NAA	N			N	N	N	N	N			N	N	N		
NMR	N	D,N	•	D,N,S	N	N	N	N			N	N	S	N	S
OM								•	•						•
RS	N	D, •	D,•		D,•	D,•	D,•	•	•	S	•	•		•	
SAXS	•		•		•			•						•	
SEM	N				N	N			•		N	N			•
SIMS	•		S	•	•	N			•	•	•	•	•	N	N,S
TEM	S		C		N	N			•		N	N		C	•
UV/VIS	D, •	D, •	D,•		D,•	D,•	D,•	D,•			D,•	D,•	D,•		
XPS	•	N	S		•	•	S			•	•	•			
XRD			C		C	C	C,S	C			C	C		C	
XRS	N				N	N	N	N			N	N	N		

Key: • = generally usable; N = limited number of elements or functional groups; S = under special conditions (i.e., with tandem mass spectrometer detection); D = after dissolution/extraction; V = volatile solids or components, pyrolyzed solids; C = crystalline solids. (a) AES, Auger electron spectroscopy; AFM, atomic force microscopy; COMB, high-temperature combustion; EFG, elemental and functional group analysis; EPMA, electron probe x-ray microanalysis; ESR, electron spin resonance; FTIR; Fourier transform infrared spectroscopy; GC, gas chromatography; GC/MS, gas chromatography/mass spectrometry; IA, image analysis; IC, ion chromatography; ICP-MS, inductively coupled plasma mass spectroscopy; LC, liquid chromatography; LC/MS: liquid chromatography/mass spectrometry; LEISS, low-energy ion-scattering spectroscopy; MFS, molecular fluorescence spectroscopy; NAA, neutron activation analysis; NMR, nuclear magnetic resonance; OM, optical metallography; RS, Raman spectroscopy; SAXS, small-angle x-ray scattering; SEM, scanning electron microscopy; SIMS, secondary ion mass spectroscopy; TEM, transmission electron microscopy; UV/VIS, ultraviolet/visible absorption spectroscopy; XPS, x-ray photoelectron spectroscopy; XRD, x-ray diffraction; XRS, x-ray spectrometry

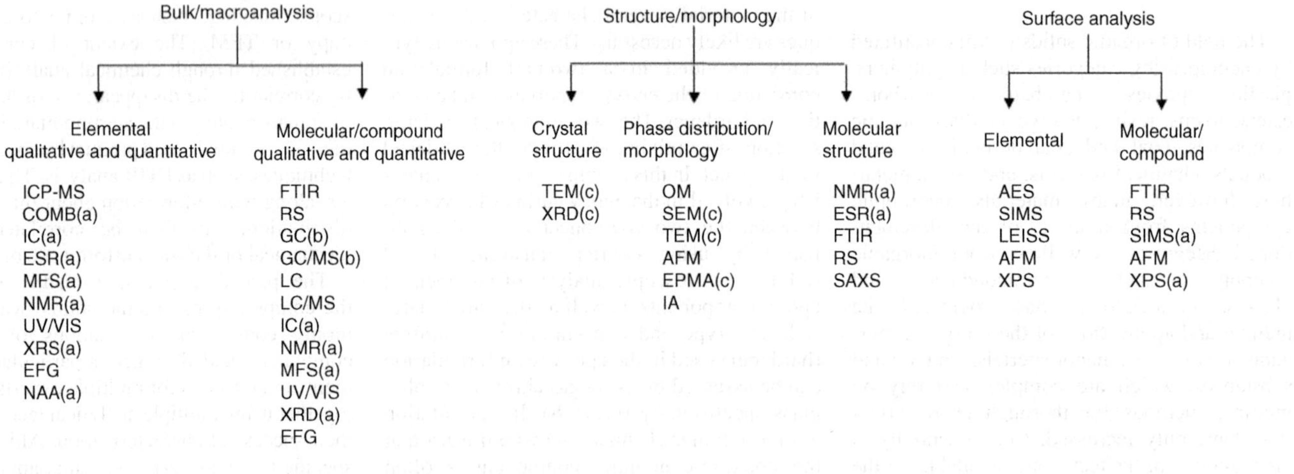

Fig. 1 Flow charts of common techniques for characterization of organic solids. AES: Auger electron spectroscopy; AFM: atomic force microscopy; COMB: high-temperature combustion; EFG: elemental and functional group analysis; EPMA: electron probe x-ray microanalysis; ESR: electron spin resonance; FTIR: Fourier transform infrared spectroscopy; GC: gas chromatography; GC/MS: gas chromatography/mass spectrometry; IA: image analysis; IC: ion chromatography; ICP-MS: inductively coupled plasma mass spectrometry; LC: liquid chromatography; LC/MS: liquid chromatography/mass spectrometry; LEISS: low-energy ion-scattering spectroscopy; MFS: molecular fluorescence spectroscopy; NAA: neutron activation analysis; NMR: nuclear magnetic resonance; OM: optical metallography; RS: Raman spectroscopy; SAXS: small-angle x-ray scattering; SEM: scanning electron microscopy; SIMS: secondary ion mass spectroscopy; TEM: transmission electron microscopy; UV/VIS: ultraviolet/visible absorption spectroscopy; XPS: x-ray photoelectron spectroscopy; XRD: x-ray diffraction; XRS: x-ray spectrometry. (a) Limited number of elements or groups. (b) Volatile liquids, solids, or components. (c) Under special conditions

analysis of a solid. Although many techniques are available, some will be more applicable to the particular analytical requirement than others. Incorporate those techniques that will be selective enough to provide usable data for the analysis, and appreciate that multiple techniques may need to be employed to obtain satisfactory data about the sample.

- Always clean and dry instruments used for sample handling and preparation. Many scalpels and razor blades contain finishing oils on the surface. These should be cleaned with isopropanol and then dried in an oven at an elevated temperature to evaporate residual solvent; this is especially important when performing a trace or ultratrace analysis.
- Always acquire instrument and reagent blanks. Verify the cleanliness of the instrument used for analysis by acquiring data of a blank, which is an acquisition of the data from the instrument under the same experimental and instrumental parameters as will be used for the analysis of an actual sample, with the exception that no sample is analyzed. For example, if extracting an organic solid into a solvent for analysis by GC/MS, it is good practice to prepare a separate sample consisting of the solvent used for extraction. Analysis and comparison of the signals from both the actual sample dissolved in the solvent and the solvent itself enables the determination of which signals are due to the sample and which ones are introduced by the solvent. This is especially important for the analysis of trace/ultratrace components.
- Be aware of points and sources of contamination. The packaging used for storing or preparing samples for analysis may contain components that can transfer onto the sample. Care should be taken in selecting the appropriate storage medium for the analysis. For example, an analysis using SIMS of a plastic surface for the presence of contaminants that prevent adhesion would be greatly confounded if the plastic sample was placed without protective coverings into a plastic bag or a cardboard box, because these containers would likely transfer additional components onto the surface of the sample. In this example, it is good practice to wrap the sample in an inert covering, such as aluminum foil, which will not contribute extraneous signals and will allow for more selective acquisition of data to diagnose the problem.

Characterization of Organic Liquids

The characterization of organic liquids, solutions, solvents, and related media uses many of the same tools that are available for the characterization of organic solids. This is in large part due to the fact that many dispersions, solutions, and related liquids consist of one or more solid organic solids dissolved or otherwise dispersed in a matrix consisting of organic solvent(s). As a result, the team of scientists practiced in performing many of the characterization tools for organic solids that first require dissolution or extraction will be familiar with many of these same tools for the characterization of organic liquids, such as inductively coupled plasma mass spectrometry (ICP-MS), nuclear magnetic resonance (NMR), electron paramagnetic resonance, ultraviolet/visible spectrophotometry, liquid chromatography, and many other techniques.

Due to the myriad applications that use these materials, organic liquids can be relatively simple, consisting of just a solvent or chemical reagent, for instance. Alternatively, they can be very complex, consisting of a number of solvents and organic and inorganic materials present in the formulation to meet certain performance requirements. In some solutions, the species and concentrations of dissolved gas can affect the behavior or performance and must be evaluated.

An additional challenge in performing quantitative analysis with solutions/mixtures composed of volatile solvents or solutes is evaporation of the solvent or components during sample handling, including initial preparation and any dilution steps. Any unaccounted losses in solvent or components will directly affect the calculated concentrations in a semiquantitative or quantitative analysis. Consult an analytical chemist, who is likely to be familiar with the appropriate and practical steps that can be taken during solution preparation and analysis to minimize such errors.

Many higher-boiling liquid or solid components of an organic liquid can be analyzed while either dissolved or dispersed in the liquid medium or after evaporation of the solvent. For example, the characterization of a dispersion of 3-aminopropyl-functionalized silica nanoparticles in an organic solvent may proceed in the following manner: First, the hydrodynamic radii of the particles, which are related to particle diameters, could be measured by dynamic light scattering. An end-functional group analysis could be performed by titration of the pendant amine group functionality present on the nanoparticle surface with an appropriate titrant. The solvent identity could be determined by analysis of a small aliquot using NMR, FTIR, or GC/MS techniques. On the other hand, many of these same properties could be assessed in the dried state. For instance, the size of the nanoparticles could be assessed by drying an aliquot of the solution onto a stub for TEM or scanning electron microscopy characterization, followed by image analysis. The nitrogen content could be determined after drying with combustion analysis (carbon-hydrogen-nitrogen analysis). The silica nanoparticle loading could be determined in several ways: by ICP-MS analysis of the silicon content, followed by determination of the silica nanoparticle size by image analysis, or by evaporation of the solvent from a known mass of the solution to obtain the residual mass of the nanoparticles.

An example of a more complex organic solution is a lithium ion battery electrolyte formulation, which is often composed of a concentrated lithium salt, such as lithium hexafluorophosphate, in mixtures of volatile organic solvents, such as carbonate solvents, together with small amounts of organic and inorganic additives designed to improve the performance of the battery system. To complicate matters, these mixtures are often air and moisture sensitive. Fortunately, the materials scientist who is consulted has familiarity with these formulations. After consulting with the analytical chemist on the team, it is learned that the volatile and semivolatile organic components, such as solvent and certain additives, can be separated and quantitated by dilution of the electrolyte with a solvent such as dichloromethane, followed by GC/MS analysis of the resulting solution after separation of the precipitated inorganic salts by filtration or centrifugation. The concentration of lithium hexafluorophosphate can be measured by 31P-NMR spectroscopic analysis, which will provide the molar concentration of phosphate in solution, which is proportional to the lithium molar concentration. With the same solution, other nuclei can be probed (e.g., 1H, 13C, 19F, 3Li) to obtain elemental and molecular information about the electrolyte composition. The concentrations of halides, such as chloride, or pseudohalides dissolved in the electrolyte could be assessed by using ion chromatography, and ICP-MS analysis could be performed to provide information about the concentrations of most other elements in the electrolyte.

The preceding examples were selected due to their complex compositions that require input, analysis, and collaboration by various scientists and engineers to perform a successful characterization. The same approach can be applied to other complex or simple organic liquids or solutions. Table 2 and Fig. 2, show the methods applicable for the characterization of organic liquids. If the organic liquid contains dissolved or otherwise dispersed solid components, then analysis of those solid components can be performed after removal of the solvent—either by evaporation, distillation, or solvent partitioning—using any of the techniques applicable for the analysis of inorganic or organic solids.

Table 2 Analytical methods used for characterization of organic liquids

Technique (a)	Elemental analysis	Speciation	Compound	Isotopic or mass analysis	Qualitative analysis (identification of constituents)	Semiquantitative analysis (order of magnitude)	Quantitative analysis (precision of +20% relative standard deviation)	Macroanalysis or bulk analysis	Major component (>10 wt%)	Minor component (0.1–10 wt%)	Trace component (1–1000 ppm, or 0.0001–0.1 wt%)	Structure
EFG	•		•		•	•	•	•	•	•	•	
ESR	N	N			N	N	N	N		N	N	N
FTIR	S	S	•		•	•	•	•	•	•		•
GC		N	V		V	V	V	V	V	V	V	
GC/MS	N	N	V	•	V	V	V	V	V	V	V	V,S
IC	N		N	N,S	N	N	N	N	N	N	N	
ICP-MS	•			•	•	•	•	•	•	•	•	
LC			•		•	•	•	•	•	•	•	
LC/MS	•	•	•	•	•	•	•	•	•	•	•	
MFS	N	N	N		N	N	N	N	N	N	N	
NAA	N			N	N	N	N	N	N	N	N	
NMR	N	N	•	N,S	N	N	N	N	N	N	S	N
RS	S	S	•		•	•	•	•	•	•		•
SIMS	•		S	•	•	N			•	•	•	N
UV/VIS			•		•	•	•	•	•	•	•	
XRS	N				N	N	N	N	N	N	N	

Key: • = generally usable; N = limited number of elements or functional groups; S = under special conditions (i.e., with tandem mass spectrometer detection); V = volatile liquids. (a) EFG: elemental and functional group analysis; ESR: electron spin resonance; FTIR: Fourier transform infrared spectroscopy; GC: gas chromatography; GC/MS: gas chromatography/mass spectrometry; IC: ion chromatography; ICP-MS: inductively coupled plasma mass spectrometry; LC: liquid chromatography; LC/MS: liquid chromatography/mass spectrometry; MFS: molecular fluorescence spectroscopy; NAA: neutron activation analysis; NMR: nuclear magnetic resonance; RS: Raman spectroscopy; SIMS: secondary ion mass spectroscopy; UV/VIS: ultraviolet/visible absorption spectroscopy; XRS: x-ray spectrometry

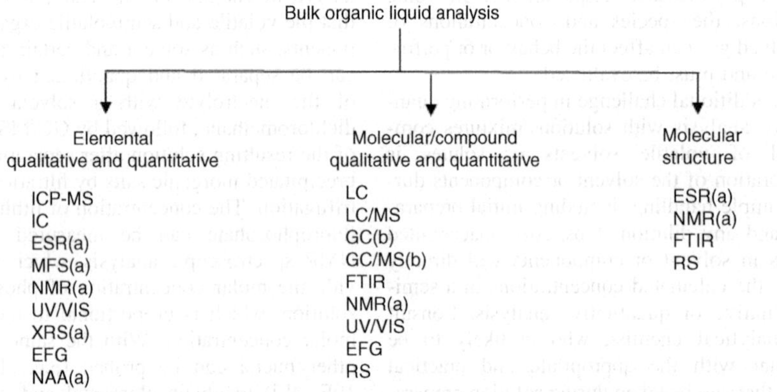

Fig. 2 Flow charts of common techniques for characterization of organic liquids. EFG: elemental and functional group analysis; ESR: electron spin resonance; FTIR: Fourier transform infrared spectroscopy; GC: gas chromatography; GC/MS: gas chromatography/mass spectrometry; IC: ion chromatography; ICP-MS: inductively coupled plasma mass spectrometry; LC: liquid chromatography; LC/MS: liquid chromatography/mass spectrometry; MFS: molecular fluorescence spectroscopy; NAA: neutron activation analysis; NMR: nuclear magnetic resonance; RS: Raman spectroscopy; UV/VIS: ultraviolet/visible absorption spectroscopy; XRS: x-ray spectrometry. (a) Limited number of elements or groups. (b) Volatile liquids, solids, or components

ASM Handbook, Volume 10, *Materials Characterization*
ASM Handbook Committee
DOI 10.31399/asm.hb.v10.a0006649

Introduction to Characterization of Powders

Bo Hu, North American Hoganas, Co.

POWDER is a dry assembly of a large number of small particles (solid and/or porous) with either a uniform particle size or a range of particle size distribution from submicrometer to millimeters. Materials in powder form are widely applied in science and engineering fields such as food, pharmaceutical, chemical, mining, metallurgical industries, and so on. Traditionally, powder can be produced by crushing and grinding from large pieces of substances such as rocks, spray drying of liquid such as milk, or atomization of molten metals. The types of powder that are commonly used for commercial applications can be divided into two large groups: inorganic powder and organic powder. Examples of inorganic powder include metals and alloys, ceramics, cements, minerals, ores, pigments, inorganic compounds, chemical reagents, and so on. Examples of organic powder include polymers, plastics, esters, resins, detergents, dyes, organic composites, coal and coal derivatives, chemical reagents, organometallics, and so on. This article uses metal and alloy powders as examples to briefly discuss how to perform the characterization of powders.

The characteristics of a powder are determined by its material compositions and manufacturing methods. For large-scale production, for example, commercial metal and alloy powders are typically produced in atomization, chemical-reduction, electrodeposition, and thermal decomposition processes. Figure 1 summarizes conventional manufacturing methods for plain elemental iron powders.

The atomized iron powder is produced with a disintegrating molten iron stream into fine droplets by high-velocity water jets so that it has a solid and irregular particle structure.

The reduced iron powders are mainly produced by two major chemical-reduction processes. One uses carbon monoxide and the other uses hydrogen as the reducing agent. The raw material used in the reduction method is iron oxide, Fe_2O_3 or Fe_3O_4, either high-grade iron ore or selected mill scale, which is reduced continuously at 1000 to 1200 °C (1830 to 2190 °F) for 5 to 10 h. The particles of reduced iron powder are irregularly porous shaped, and therefore, it is referred to as sponge iron powder. Generally, the hydrogen-reduced iron powder has a more porous structure with very fine pores compared to carbon-reduced iron powder.

The electrolytic iron powder is produced by electrolytic deposition from ferrous sulfate solution at 50 °C (120 °F) with a 5 day batch process. The obtained electrolytic deposits (usually 45 by 45 by 0.5 cm, or 18 by 18 by 0.20 in., plates) are then ground into the required particle size, so that the final milling techniques often determine the particle morphology and powder quality that relate to certain specific performances. For example, one of the electrolytic iron powders is produced into layered solid particles with a flake shape.

Carbonyl iron powder is produced by thermal decomposition of iron carbonyl molecules [$Fe(CO)_5$], which are formed at >130 atm (13 MPa) from a reaction between the reduced iron powder and CO gas. The carbonyl iron powder consists of small, solid particles (1 to 10 μm) with near-true spherical shape.

Therefore, these manufacturing processes generate the featured particle shape and morphology. Figure 2 illustrates the particle shape and morphology of iron powders produced by different manufacturing processes.

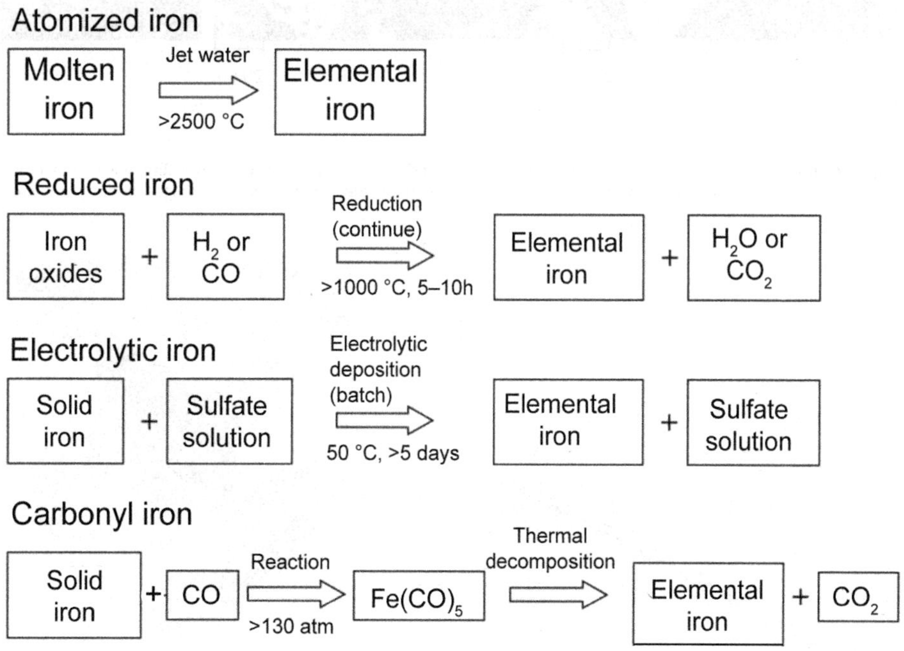

Fig. 1 Manufacturing methods for elemental iron powder. Source: Ref 1

For characterization of powders, both individual particles and bulk powders are used to evaluate their physical and chemical properties. To ensure accurate characterization processes, it is extremely important to use adequate sampling techniques to collect representative samples for analysis.

Sampling of Powder

Generally, powder is a bulk material with particle size distributions (PSDs). Segregation happens easier in powder with coarser PSD compared to powder with finer PSD, so caution should be paid in the sampling of coarse powder. On the other hand, powder with less irregular particle shape is easier to flow; thus, it segregates easily during handling and storage compared to powder with more irregular particle shape.

Powder can be sampled from stored material and flowing streams. For stored material such as powders already packaged in containers, a sampling spear such as a Keystone sampler (Fig. 3) can be used to withdraw samples from containers such as drums, bags, trucks, railway wagons, and so on. The number of samples to be taken is dependent on the total number of containers, but sampling should be performed in a systematic way from the selected containers. When the samples are collected, they are combined together to form a composite sample after blending. For flowing streams, that is, powders in the process of being packaged from a blender or storage tank, a rectangular receptacle can be used to pass completely through the stream of flowing powder at a constant speed. The number of samples and the amount of each sample to be taken is dependent on the process of the powder to be packaged, and the collected samples are combined to form a composite sample after blending. If the composite sample is too large in quantity as a test sample, then sample reduction is required by using a sample splitter or spinning rifflers (Fig. 4) to obtain an adequate amount of

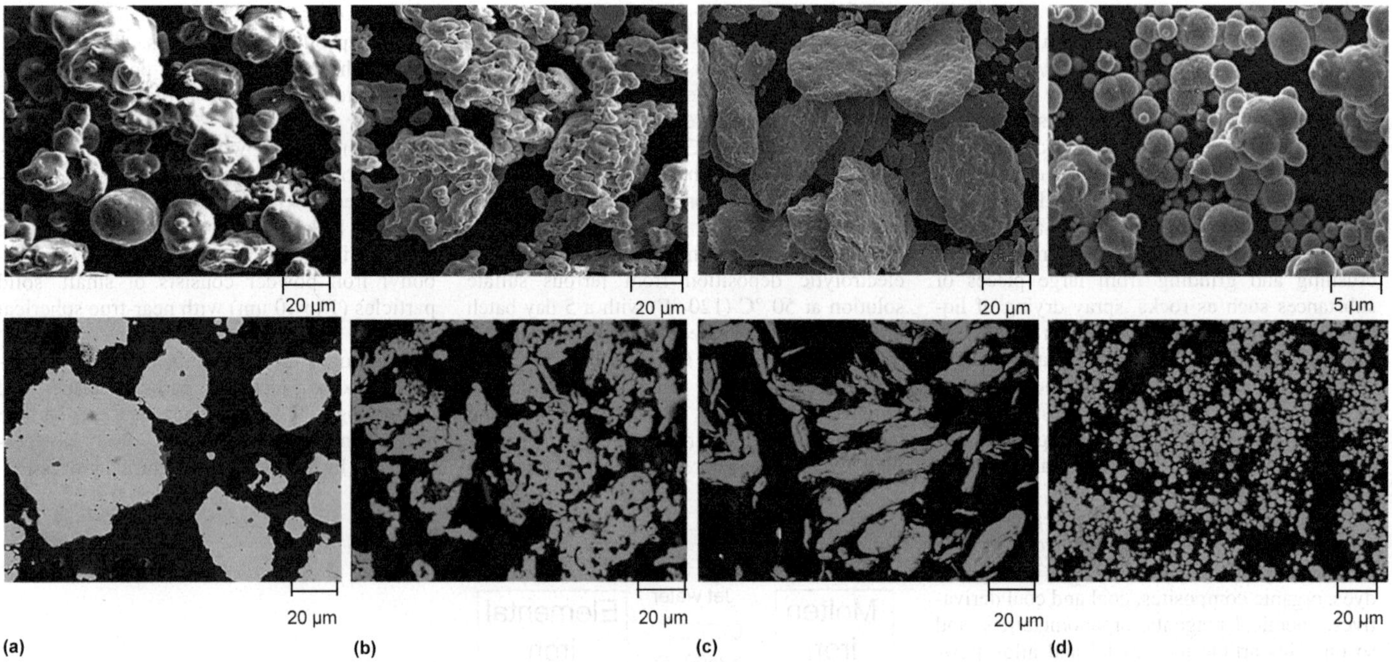

Fig. 2 Images of elemental iron powders (95% <325 mesh, or 45 μm) produced from different manufacturing processes. (a) Atomized iron. (b) Hydrogen-reduced iron. (c) Electrolytic iron. (d) Carbonyl iron. Top row: scanning electron microscopy images; bottom row: cross-sectional optical microscopy images

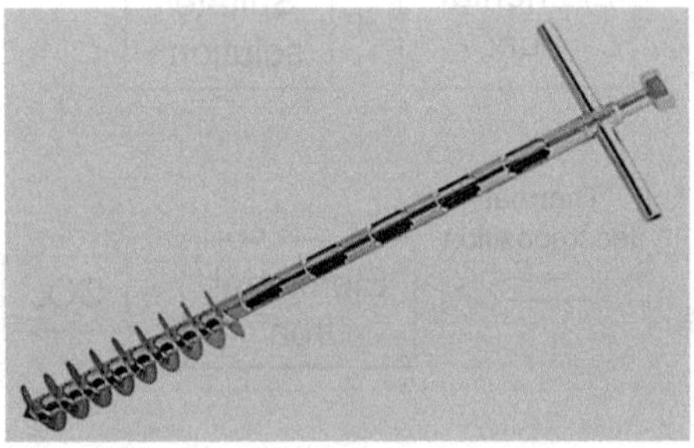

Fig. 3 Keystone sampler

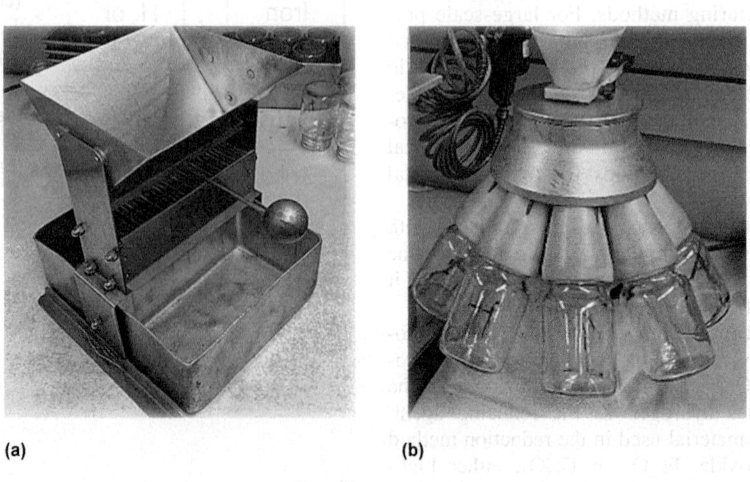

Fig. 4 (a) Sample splitter. (b) Spinning rifflers

powder as the test sample. Figure 5 shows a sampling scheme for taking samples from containers and combining them for preparing test samples.

Bulk and Particle Properties of Powders

Particle Size and Particle Size Distribution

In most cases, powder consists of a large number of small particles with different particle sizes. Therefore, the particle size of a powder is commonly described in the form of PSDs. Depending on the analysis method, expression of the PSD is different. Sieving analysis is the most common and convenient analysis method for PSDs of a powder when the majority of the particles are more than 20 μm. It uses a set of sieves with different opening sizes, called mesh size (Table 1), on the screen to generate narrowly classified fractions of the powder (Fig. 6). Sieving can be run in dry (air) or wet (water) manually or by machine. For 45 μm or coarser powder, sieving in dry is sufficient, while when sieving 45 μm or finer powder, wet sieving in liquids is effective. If the powder is easy to flocculate, a dispersing agent is required to add into the liquids. After sieving, the fractions obtained from each sieve screen are weighed, and the weight percentage between each opening size (mesh size) is calculated. It is important to follow a proper sieving procedure designated for particulate powder in order to minimize analysis error. Table 2 shows an example of sieve analysis on common metal powders. The results obtained from sieve analysis are given as weight percentages in each fraction of sieve set.

Coulter counter and laser diffraction are two effective methods for particle size analysis of fine powders (Ref 3). The former is called a stream scanning method, where particle volumes are measured through the changes in voltage pulse when particles are suspended in an electrolyte solution and pass through an orifice. The latter is called a field scanning method, where particle volumes are measured when particles pass through a laser beam to cause light scattering, which is collected and calculated into the PSD. Both analysis methods present the PSDs in volume percentage under the particulate size. Table 3 presents laser diffraction analysis on common industrial powders. The laser diffraction analysis normally presents the PSDs as volume percentages of a given particle size in micrometers (Fig. 7). It can also convert the results into volume size distribution, similar to the sieve analysis. Table 4 gives a comparison of analysis results from sieve and laser analyses. The results obtained from the laser analysis were very close to those of the sieve analysis.

In addition, the powder particle size can be measured by a visual method using optical and electron microscopy to examine individual particles. Figure 8 shows optical and scanning electron microscopy image analysis on a nickel-base alloy powder.

The sedimentation method is another particle size analysis technique. It is based on the settling of solid particles under gravitational

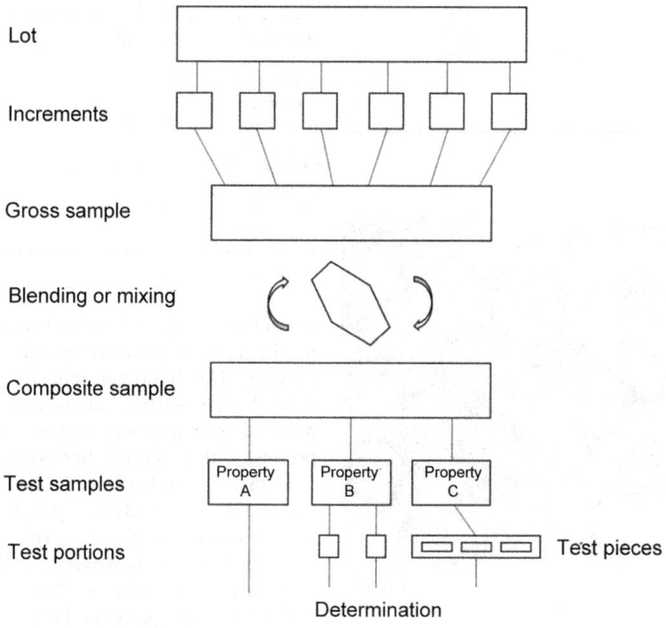

Fig. 5 Sampling scheme. Source: Ref 2

Fig. 6 Set of sieves used for analysis of particle size distributions

Table 1 Sieve openings

U.S. standard mesh size	Opening size, μm
20	850
40	425
60	250
70	212
80	180
100	150
140	105
200	75
325	45
500	25
625	20

Note: Mesh size is the number of openings per inch on the sieve screen.

Table 2 Sieve analysis of common metal powders

	Mesh fraction, wt%					
Type of powder	+80	−80 + 100	−100 + 140	−140 + 200	−200 + 325	−325
Iron powder (atomized)	1.0	7.7	16.9	23.7	27.1	23.6
Iron powder (CO-reduced)	0.1	1.6	25.7	32.1	23.0	17.5
Iron powder (H-reduced)	. . .	. . .	0	0.1	1.5	98.4
Copper powder (atomized)	. . .	0	0.4	11.9	28.8	58.9
Aluminum powder	0	0.4	26.7	40.6	29.6	2.7
Zinc powder	. . .	. . .	. . .	0	12.3	87.7
Stainless steel powder	0	1.2	8.1	14.2	28.4	48.1

Table 3 Laser diffraction particle size analysis of common industrial powders

	Particle size, μm			
Type of powder	D10	D50	D90	D99
Iron powder (atomized)	36.2	88.4	177.8	254.7
Iron powder (CO-reduced)	38.9	97.6	161.4	219.0
Copper powder (atomized)	21.8	51.9	102.4	152.1
Natural graphite powder	3.2	8.4	17.6	28.4
Ethylene bis stearamide wax (lubricant) powder	1.6	7.5	23.4	40.9
Manganese sulfur (MnS) powder	1.5	5.3	9.9	14.7
h-boron nitride (hBN) powder	0.7	2.7	20.9	42.7

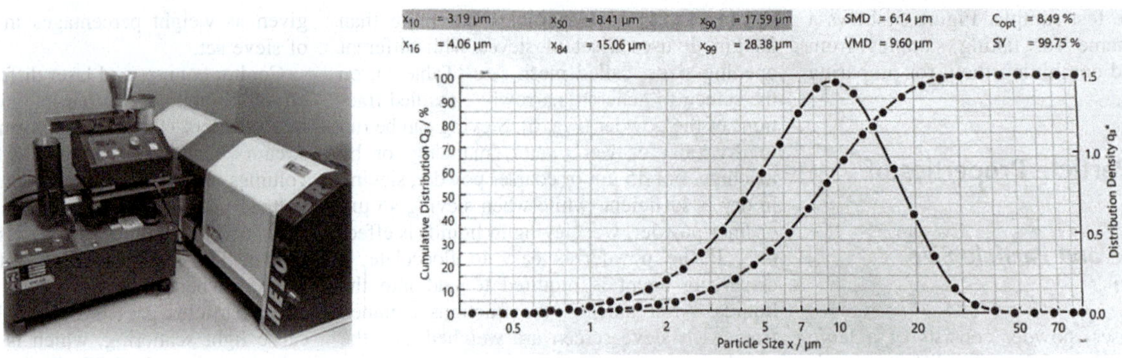

Fig. 7 Laser diffraction particle size analyzer and example of analysis report

Table 4 Comparison of sieve analysis and laser diffraction particle size analysis

Type of powder	Analysis(a)	+80	−80 + 100	−100 + 140	−140 + 200	−200 + 325	−325
Iron powder (atomized)	Sieve	1.0	7.7	16.9	23.7	27.1	23.6
	Laser	0.1	6.4	18.0	22.5	28.4	24.6
Iron powder (CO-reduced)	Sieve	0.1	1.6	25.7	32.1	23.0	17.5
	Laser	0	2.7	19.6	31.5	27.2	19.0
Copper powder (atomized)	Sieve	…	0	0.4	11.9	28.8	58.9
	Laser	…	0.2	2.4	11.9	31.9	53.6

(a) Sieve analysis, wt%; laser analysis, vol%

Table 5 Surface area of various iron powders

Type of iron powder	Particle size, µm D50	Particle size, µm D95	Brunauer-Emmett-Teller specific surface area, m²/kg
H-reduced (fine)	20	40	320
H-reduced (coarse)	230	425	225
H-reduced (regular)	75	150	200
CO-reduced (regular)	75	150	100
Atomized (regular)	75	180	50

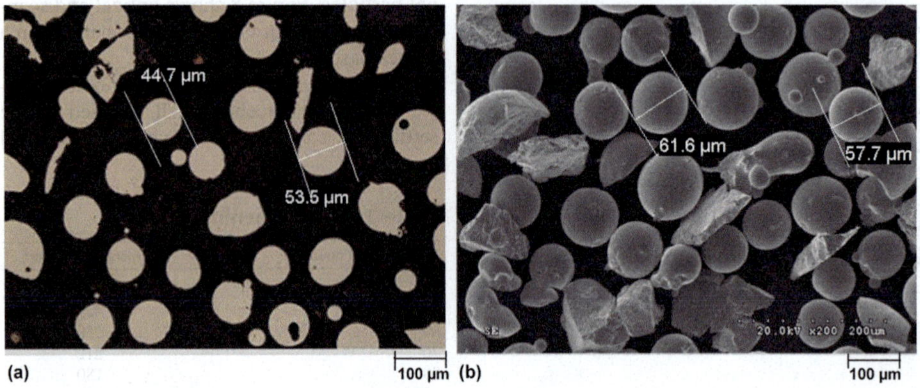

Fig. 8 Particle size analysis of nickel-base alloy powder. (a) Optical microscopy. (b) Scanning electron microscopy

or centrifugal forces in liquids. The particle size is expressed in the weight distribution in the particulate size, calculated by Stock's law.

Surface Area, Density, and Porosity

Because powder is composed of a large number of particles, surface area is an important property used to characterize the powder. The surface area of total particles measured in a unit mass of powder is defined as the specific surface area (SSA) (m^2/g or m^2/kg). In a general rule, the finer the powder particle size is, the larger the SSA. Powder with an irregular shape and porous structure presents a larger surface area than powder consisting of regular and dense particles. Therefore, depending on the particle structures of a powder, finer powder does not always have a larger SSA than a coarser powder.

To determine the surface area of a powder, a physical adsorption method based on the Brunauer-Emmett-Teller (BET) equation is used, commonly through gas adsorption. The adsorptive gas, such as nitrogen gas, forms a monolayer molecule on all accessible particle surfaces in the powder, including pores, cavities, cracks, and so on. The SSA value is calculated by the amount of adsorbed gas molecules; in this case, it is called the BET specific surface area. Table 5 presents the BET specific surface area of elemental iron powders with various particle size and morphology. Even though they have similar PSDs, reduced iron powder has more than two times the surface area of atomized iron powder, due to the porous structure of the reduced iron powder. The hydrogen-reduced iron powders have higher surface areas than the carbon-reduced iron powder, because they are more porous, with much smaller pores.

Density is another basic property of a powder. Generally, the density of a powder is defined as the ratio of mass to volume (g/cm^3). If there is internal porosity, however, the density is not necessarily the same as that of the material which forms the particles. Four types of density can be used to describe the density of a powder:

- True density
- Particle density
- Apparent density (or bulk density)
- Tap density

True density (or solid density) is defined as the ratio of particle mass to its actual volume, excluding all porosity. It has the same value as the specific gravity of the material. Particle density is defined as the particle mass divided by the particle volume, including the closed pores. Apparent density is defined for powder and particle beds (a mass-to-volume ratio of the bed in a vessel, including the voids between the particles). Tap density is defined for a powder and particle bed to be tapped or vibrated under specified conditions. Therefore, the apparent density and tap density are greatly affected by PSD, particle shape, and particle porosity. For a given powder, the value of these densities follows this order: bulk density < tap density < particle density < true density (or specific gravity).

Bulk density and tap density are normally measured according to standardized procedures (Ref 4). Particle density can be measured from the liquid volume increase found upon adding the particles into a liquid, or, in the case of a gaseous medium, it is determined from the increased amount of the volume or the pressure (Ref 3). The pycnometer method uses liquid as a medium and is the most representative method for measuring particle density. The liquid can be water or solvent, but it must be a nonvolatile liquid in which the particles are both wettable and insoluble. Table 6 shows the difference in particle density, tap density, and apparent density between the atomized iron powder and the hydrogen- and carbon-reduced iron powders. The hydrogen-reduced iron powder has the lowest apparent (bulk) density, tap density, and particle density, due to its irregular particle shape and high internal porosity. The particle density of atomized iron powder is close to its true density, because it has a solid particle structure almost without internal pores.

Particle Hardness, Compressibility, and Green Strength

Particle hardness is the resistance to plastic deformation of a powder. For metal powders, for example, particle hardness directly relates to powder compressibility, that is, the harder the particle, the more difficult to compact the powder into a component. Higher particle hardness also results in more abrasive wear and difficulty in size reduction.

The particle hardness of metal powders is dependent on the chemical composition, especially the carbon content and the amount of alloying. The diffusion of carbon in a small amount can cause a significant increase in hardness. Therefore, particle hardness can be adjusted by controlling the carbon content. In addition, the manufacturing process also influences the particle hardness of metal powders. Iron powder made by water atomization of molten iron becomes hard due to residual stresses. Hydrogen annealing the iron powder releases the stresses and minimizes the carbon and oxygen content, so that the iron powder softens.

To measure particle hardness, indentation hardness tests, such as the Vickers microhardness (MHV) test, can be used. Under an optical measurement system, a pyramidal-pointed diamond is impressed into the particle surface by a controlled loading force ranging from 0.10 to 1.0 N (0.01 to 0.1 kgf) to form a permanent indentation that is measured and converted to a hardness value, called the MHV (Ref 5).

Compressibility and green strength are two properties for evaluating how easily a powder is compacted to form a component. The compressibility of powder is the ability of a powder to be compacted into green-state compacts under pressure. It is determined by measuring the green density of a test specimen, which results from subjecting the powder to uniaxial compressive loading in a confining die (Ref 6).

Green strength is the strength of green-state compacts with a certain green density or pressed at a certain pressure, such as 414 MPa (30 tsi) (Ref 7). Sufficient green strength is required to prevent compacts from cracking or being damaged during processing, handling, and transporting. Increasing the compact density generally provides increased green strength, but the powder particle shape and morphology greatly affect the green strength. Table 7 presents the microhardness values, compressibility, and green strength of various elemental iron powders, including the MHV of iron oxides for comparison.

Flowability of Powders

Flowability is an important characteristic for powders that must be transferred to bins, hoppers, and conveyors or filled into dies for making compacts. Powder followability is dependent on particle shape and PSD. Powder with more irregular particle shape is more difficult to flow than powder with more spherical particle shape. Powder with less fine particles is easier to flow compared to powder with more fine particles (commonly, particles less than 45 μm). The flowability of a powder can be determined by measuring the powder flow rate with a flowmeter funnel. There are two types of flowmeters: the Hall funnel and the Carney funnel (Fig. 9) (Ref 8). The former has a smaller orifice size, suitable for measurement of free-flowing powder, while the latter has a large orifice size that can be

Table 6 Density and porosity of various iron powders

Type of iron powder	Apparent density, g/cm^3	Tap density, g/cm^3	Particle density, g/cm^3	True density, g/cm^3	Porosity(a), %
H-reduced	1.3	1.8	6.48	7.87	17.7
CO-reduced	2.4	3.3	7.23	7.87	8.3
Atomized	3.0	3.9	7.83	7.87	0.5

(a) Internal porosity of particles calculated from particle density and true density

Table 7 Microhardness, compressibility, and green strength of various iron powders

Type of iron powder	Vickers microhardness value	Compressibility(a), g/cm^3	Green strength(a) MPa	Green strength(a) psi
H-reduced (coarse)	124	5.9	87	12,650
CO-reduced (regular)	126	6.5	18	2,540
Atomized (regular, annealed)	136	6.9	11	1,520
Atomized (regular, unannealed)	324	5.9	6	900
Iron oxide (Fe_3O_4, or magnetite)	598	...	...	...
Iron oxide (Fe_2O_3, or hematite)	1,186	...	...	...

(a) Compacted at 414 MPa (30 tsi)

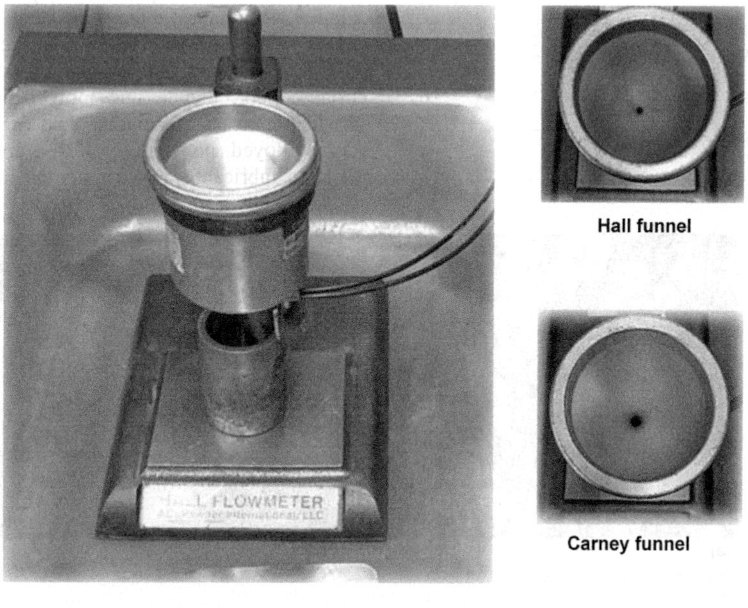

Fig. 9 Powder flow-rate tester with Hall and Carney funnels

used when a powder is not free-flowing and cannot be measured with the Hall funnel. Table 8 shows the flow rate of various metal powders with different bulk densities and the amount of fine particles less than 325 mesh or 45 μm.

Bulk and Surface Characterization of Powder

Particle Shape and Morphology

The particle shape and morphology, together with particle size and PSD, are important physical characteristics of a powder. A powder particle can be described by four main morphological features:

- Shape or form, defined by axial dimensions
- Roundness, defined by the curvature of corners
- Surface texture, defined by the irregularity of the surface between the corners
- Internal porosity with various structures

All of these features or descriptions determine the apparent density, flowability, compressibility, green strength, sintered properties, and dimensional changes of a powder and a component.

Table 8 Flowability of metal powders

Type of powder	Apparent density, g/cm^3	Amount of fines < 45 μm, %	Hall flow rate, s/50 g
H-reduced iron (coarse)	1.3	2.6	No flow
H-reduced iron (fine)	2.0	98.4	No flow
CO-reduced iron	2.4	19.8	32.6
Atomized iron	2.9	25.1	25.8
Atomized copper	2.7	58.9	30.0
Stainless steel	2.9	48.1	31.0
Fe-base high alloy	3.2	42.7	25.9
Ni-base high alloy	3.9	1.2	19.0
Ni-base high alloy (fine)	4.0	97.6	No flow

To characterize the particle shape and morphology of a powder, particle image analysis can be performed under optical macro- and microscopes. A macroscope is used to analyze particle shape and morphologies with a typical magnification from 1 to 10×, while a microscope is used to analyze particle cross sections for internal porosity with a typical magnification from 25 to 500×. Both types of scopes can be used to analyze particle size. However, a scanning electron microscope (SEM) can cover all works run by optical macro- and microscopes (Fig. 10). In addition, as shown in Fig. 11, it can perform elemental analysis on the particles if it is equipped with an energy-dispersive x-ray spectroscope (EDS). Compared to optical scopes, the SEM provides high-resolution images and can present the details of particle morphology.

In particle image analysis, no special treatment is normally required for the powder samples, except for the analysis of internal porosity. In this case, the powder sample must be mounted, ground, and polished to achieve a cross section of the particles (Fig. 2). For analysis of particle shape and morphologies, a powder sample can be used as-is under the optical macroscope and the SEM. Figure 12 presents SEM particle images of metal powders and additives that are commonly used for manufacturing powdered metal components. Figure 13 shows the image analysis of a powder mix made with these metal powders and additives.

Powder Compositions, Impurities, Solubility, and Moisture

Powder can be composed with single elemental particles, multiple elemental particles, or a mixture of such particles. In powder metallurgy, for example, there are plain powders made with pure iron, copper, and nickel; partially bonded and fully alloyed powders with iron, copper, nickel, and molybdenum; and powder mixes that contain plain metal powders, alloyed powders, and/or graphite and organic lubricants. Powder compositions and the contained impurities directly affect the powder properties and the performances of materials made with the powder. The solubility of a powder, which is how easy it is to dissolve in a solution, is one important characteristic for some applications. Ignition loss and moisture content also are two useful factors to characterize a powder.

Powder compositions can be determined by dry and wet analysis methods. The former is a nondestructive technique, while the latter is required to dissolve the powder in an aggressive solution. Today (2019), an SEM equipped with an EDS is a quick and convenient tool as a dry analysis method for determining the elemental compositions of a powder. Although it is not an accurate analysis for chemical compositions, it can provide relative elemental analysis of the analyzed area. X-ray diffraction is another dry analysis method that can provide phase and compound information for a powder. X-ray fluorescence provides a dry analysis method that can quantitatively determine the elemental compositions of powders. Other dry analysis methods, such as high-temperature combustion in inert gas fusion, can be used to analyze the carbon, sulfur, oxygen, and nitrogen content of a powder.

In wet analysis, atomic absorption, inductively coupled plasma, and ion-coupled plasma mass spectrometry are three common techniques used for elemental analysis to identify and quantify the individual elements presented in powders, especially their capability in determining trace elements in parts per million and parts per billion levels. For these methods, however, the powder samples must be completely dissolved in a solution such as a strong acidic solution. Impurities such as metallic oxides that cannot be dissolved in a solution are counted as acid-insoluble in the powder.

The solubility of a powder is its ability to dissolve in a solution, normally an acidic solution. The solubility varies depending on the pH of the solution, the surface area, and the PSD of the powder. Ignition loss (or loss on ignition) is a weight loss when the powder is heated at a specified temperature to burn off volatile components in the powder. The

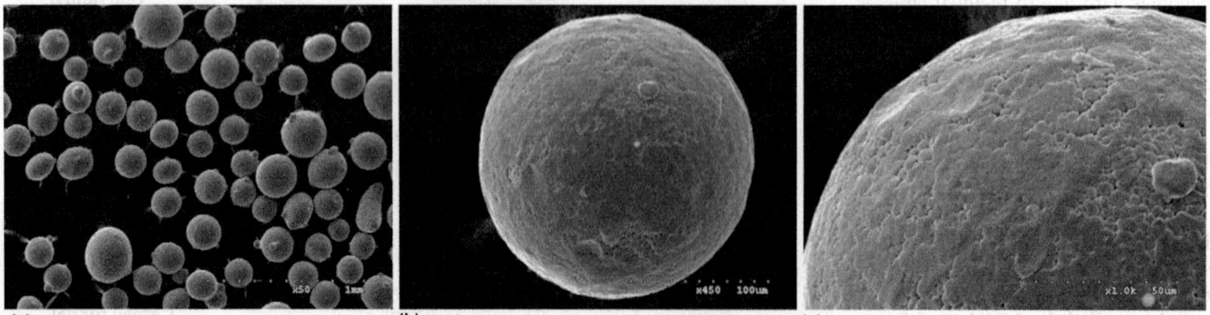

(a) (b) (c)

Fig. 10 Scanning electron microscope images of Fe-Cr-Mo-Mn alloyed powder in (a) 50×, (b) 450×, and (c) 1000× magnifications test is commonly performed at 1000 °C (1832 °F) in air. For powders such as metal powders that can be oxidized, the test should be done in a reactive or inert atmosphere. The moisture content of a powder can be measured simply by drying the powder in an oven at 100 to 105 °C (212 to 221 °F) in air. In addition, advanced moisture analyzers are available for automatic and quick analysis. Table 9 presents the chemical analysis results of various elemental iron powders.

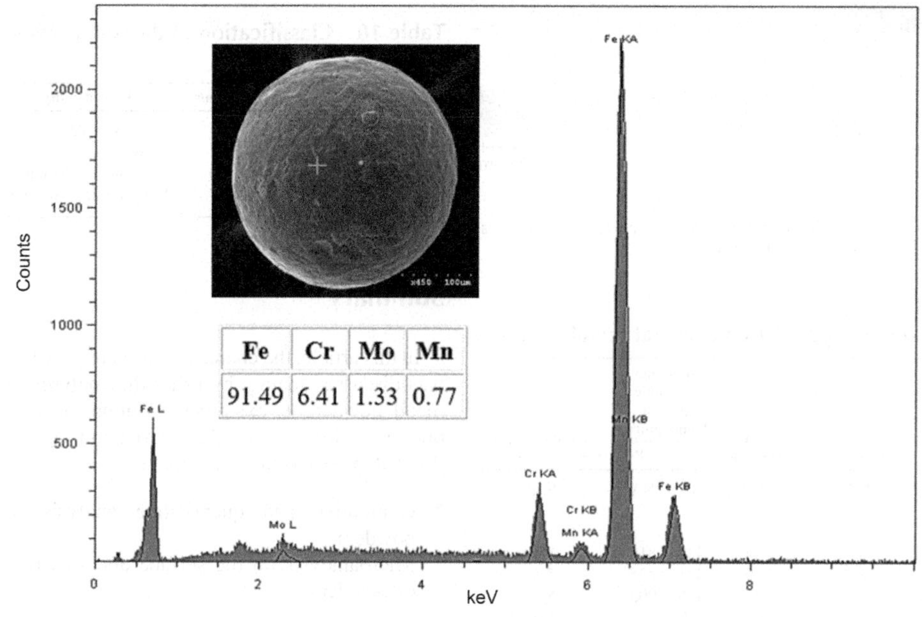

Fig. 11 Energy-dispersive x-ray spectroscope analysis of the surface of Fe-Cr-Mo-Mn alloyed powder

Particle Ignition and Dust Explosion

When handling powders in manufacturing processes, great care must be taken to prevent particle ignition and dust explosion. It is important to understand the ignitability and the explodability of a powder by measuring its ignition and dust-explosion characteristics, such as minimum automatic ignition temperature (MAIT), minimum ignition energy, limiting oxygen concentration (LOC), maximum deflagration pressure, maximum rate of pressure, and Kst value. As shown in Table 10, the dust generated from powders or materials processing is ranked into four classes of explodability based on its Kst value measured in an explosion severity test. Table 11 presents the ignition and dust-explosion characteristics of an organic powder and several metal powders. The Kst value of the organic powder, ethylene bis stearamide lubricant, was 301 bar · m/s, so it was classified as a very strong explosion powder, while the elemental iron powders were classified as weak explosion powders because their Kst values were less than 200 bar · m/s. The hydrogen-reduced iron powder is easier to ignite than the electrolytic and carbonyl iron powders, because it has lower MAIT and LOC values. For atomized iron powder, which has a much coarser PSD compared to other tested iron powders, no Kst value could be generated, so it was classified as a no-explosion powder.

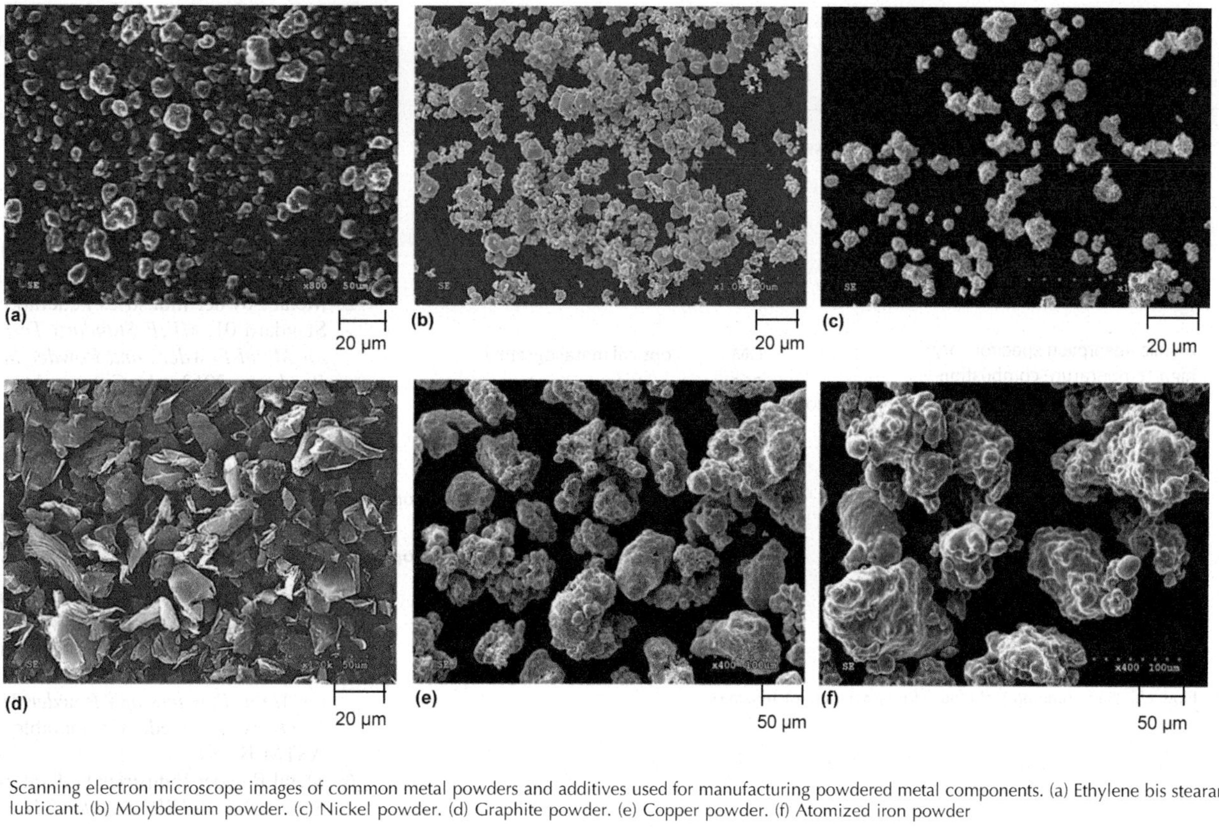

Fig. 12 Scanning electron microscope images of common metal powders and additives used for manufacturing powdered metal components. (a) Ethylene bis stearamide wax lubricant. (b) Molybdenum powder. (c) Nickel powder. (d) Graphite powder. (e) Copper powder. (f) Atomized iron powder

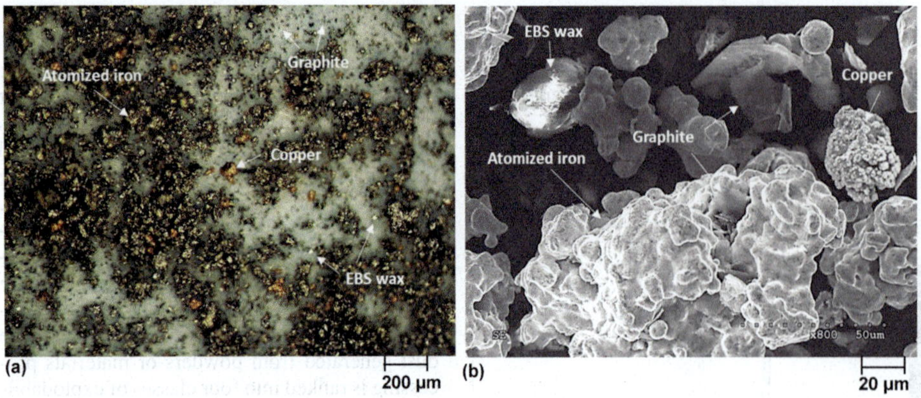

Fig. 13 Image analysis of powder mix commonly used for manufacturing powdered metal components. (a) Optical macroscope. (b) Scanning electron microscope. EBS, ethylene bis stearamide

Table 9 Chemical analysis of elemental iron powders

Type of powder	Fe, %	O, %	C, %	Acid insoluble, %
Atomized iron	99.6	0.07	0.003	0.04
CO-reduced iron	98.7	0.47	0.004	0.15
H-reduced iron	98.1	0.81	0.02	0.22

Table 10 Classification of dusts according to their explodability

Dust explosion class	Kst, bar · m/s	Explodability
St 0	0	No explosion
St 1	0–200	Weak explosion
St 2	200–300	Strong explosion
St 3	>300	Very strong explosion

Table 11 Ignitability and explodability of an organic powder and several metal powders

Powder type	Particle size distribution, μm D50	Particle size distribution, μm D90	Maximum deflagration pressure, bar/g	Maximum rate of pressure, bar/s	Kst, bar · m/s	Minimum automatic ignition temperature °C	Minimum automatic ignition temperature °F	Limiting oxygen concentration, %
Ethylene bis stearamide wax (lubricant)	7.5	23.4	8.6	1111	301	NA	NA	NA
H-reduced iron	24.9	47.7	3.45	163	44	432	810	5.5
Electrolytic iron	24.7	51.1	3.76	185	50	441	825	7.5
Carbonyl iron	7.5	18.4	4.61	403	109	485	905	8.5
Atomized iron	88.4	177.8	(a)	(a)	(a)	NA	NA	NA

(a) No ignition happened in the designated screen test, which classified the powder as not explodable, and no maximum deflagration pressure and Kst tests could be performed. NA, data not available

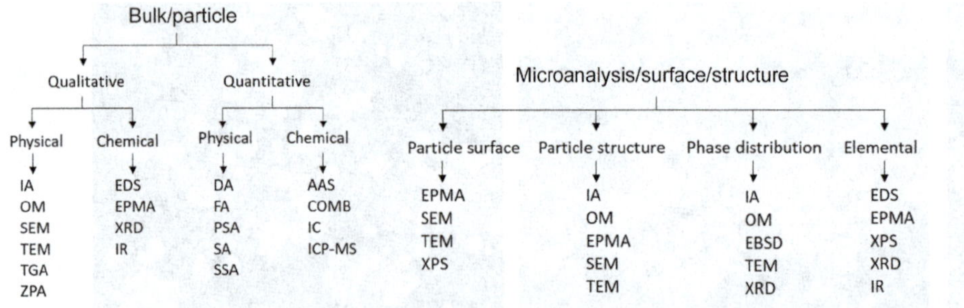

AAS	atomic absorption spectrometry	OM	optical metallography
COMB	high-temperature combustion	PSA	particle size analyzer
DA	density analysis (liquid and gas volume methods)	SA	sieve analysis
EBSD	electron backscatter diffraction	SEM	scanning electron microscopy
EDS	energy-dispersive x-ray spectroscopy	SSA	specific surface area analyzer
EPMA	electron probe x-ray microanalysis	TEM	transmission electron microscopy
FA	flowability analyzer	TGA	thermogravimetric analysis
IA	image analysis	XPS	x-ray photoelectron spectroscopy
IC	ion chromatography	XRD	x-ray diffraction
ICP-MS	ion-coupled plasma mass spectrometry	ZPA	zeta-potential analyzer
IR	infrared spectroscopy		

Fig. 14 Flow charts of common techniques for powder characterization

Summary

In this article, the characterization of powders was discussed mainly by using the analyses of metal powders as examples. Common techniques used for powder characterization can be divided into two major categories:

- Qualitative and quantitative analysis of powders
- Microanalysis on the surface and structure of powders

Figure 14 presents flow charts for these two categories and how to select the most commonly used techniques to achieve the desired information for the powders to be characterized. The capability of each analytical method used for typical powder characterization is summarized in Table 12.

REFERENCES

1. B. Hu, Developing a PM Answer for a Dietary Problem, Special Feature Article, *Met. Powder Rep.*, June 2007, p 28–31, https://doi.org/10.1016/S0026-0657(07)70127-3
2. Metal Powder Industries Federation (MPIF) Standard 01, *MPIF Standard Test Methods for Metal Powders and Powder Metallurgy Products*, 2012 ed.; Comparable standard: ASTM B 215
3. K. Gotoh, H. Masuda, and K. Higashitani, Ed., *Powder Technology Handbook*, 2nd ed., Marcel Dekker, Inc., 1997, p 28–36, 57–60
4. Metal Powder Industries Federation (MPIF) Standard 04/28/46, *MPIF Standard Test Methods for Metal Powders and Powder Metallurgy Products*, 2012 ed.; Comparable standards: ASTM B 212/417/527
5. Metal Powder Industries Federation (MPIF) Standard 51, *MPIF Standard Test Methods for Metal Powders and Powder Metallurgy Products*, 2012 ed.; Comparable standard: ASTM B 384
6. Metal Powder Industries Federation (MPIF) Standard 45, *MPIF Standard Test Methods*

Table 12 Capability of analytical methods used for characterization of powders

Method	Elemental analysis	Alloy verification	Qualitative analysis (identification of constituents)	Semiquantitative analysis (order of magnitude)	Quantitative analysis (precision of +20% relative standard deviation)	Macroanalysis or bulk analysis	Microanalysis (≤10 μm)	Surface analysis	Major component (>10 wt%)	Minor component (0.1–10 wt%)	Trace component (1–1000 ppm or 0.0001–0.1 wt%)	Compound/ phase	Structure	Morphology
Atomic absorption spectrometry	D	D	...	...	D	...	...	...	D	D	D	...	...	...
Auger electron spectroscopy	●	●	●	●	...	...	S	●	●	●	S	...	...	S
High-temperature combustion	G	...	...	G	G	G	...	...	G	G	G	...	...	...
Density analysis (liquid and gas volume methods)	...	...	...	●	●	●	...	...	●	...	...	...	...	...
Electron backscatter diffraction	...	●	●	●	...	●	●	...	●	●	S	●	S	●
Energy-dispersive x-ray spectroscopy	●	●	●	●	...	●	●	●	●	●	...	...	...	...
Electron probe x-ray microanalysis	●	●	●	●	●	...	●	●	●	●	...	S	...	...
Flowability analyzer	...	...	...	...	●	●	...	...	...	...	...	...	...	...
Image analysis	...	...	...	...	...	●	●	●	...	...	...	S	...	●
Ion chromatography	D	...	D	D	D	D	...	...	D	D	D	...	...	...
Ion-coupled plasma mass spectrometry	D	...	D	D	D	D	...	...	D	D	D	...	...	...
Infrared spectroscopy	S, D	...	S, D	S, D	S, D	S, D	...	S	S, D	S, D	S, D	S, D	S, D	...
Optical metallography	...	...	...	...	...	●	●	...	...	...	...	S	...	●
Particle size analyzer	...	...	...	...	●	●	●	...	...	...	...	...	...	...
Sieve analysis	...	...	...	...	●	●	...	...	...	...	...	...	...	...
Scanning electron microscopy	●	...	●	●	...	...	●	●	●	●	...	...	...	●
Specific surface area analyzer	...	...	...	...	●	●	●	...	...	...	...	...	...	...
Transmission electron microscopy	●	...	●	●	S	...	●	●	●	●	...	●	●	●
Thermogravimetric analysis	...	...	...	...	●	●	...	...	...	...	...	...	...	...
X-ray photoelectron spectroscopy	●	...	●	●	...	...	...	●	●	●	...	S	...	...
X-ray diffraction	...	...	●	●	S	●	...	...	●	●	...	●	●	...
Zeta-potential analyzer	...	...	...	...	...	●	...	●	●	...	...	...	...	...

Powder: metals, alloys, ceramics, minerals, ores, pigments, inorganic compounds, chemical reagents, composites, and catalysts. Wet analytical chemistry, electrochemistry, ultraviolet/visible absorption spectroscopy, and molecular fluorescence spectroscopy can generally be adapted to perform many of the bulk analyses listed. ● = generally usable; S = under special conditions; D = after dissolution; G = carbon, nitrogen, hydrogen, sulfur, or oxygen

for Metal Powders and Powder Metallurgy Products, 2012 ed.; Comparable standard: ASTM B 331

7. Metal Powder Industries Federation (MPIF) Standard 15, *MPIF Standard Test Methods for Metal Powders and Powder Metallurgy Products*, 2012 ed.; Comparable standard: ASTM B 312
8. Metal Powder Industries Federation (MPIF) Standard 03, *MPIF Standard Test Methods for Metal Powders and Powder Metallurgy Products*, 2012 ed.; Comparable standard: ASTM B 213

Spectroscopy

ASM Handbook, Volume 10, *Materials Characterization*
ASM Handbook Committee
DOI 10.31399/asm.hb.v10.a0006640

Optical Emission Spectroscopy

Arne Bengtson, SWERIM AB

Overview

General Uses

- Quantitative determination of major and trace elemental constituents in various sample types
- Qualitative elemental analysis

Examples of Applications

- Rapid determination of mass fractions of alloying elements in steels and other alloys
- Fast elemental depth profiling of technical coatings
- Determination of trace impurity concentrations in semiconductor materials
- Wear metals analysis in oils
- Rapid classification of scrap pieces for sorting
- Elemental analysis of geological materials

Samples

- *Form:* Conducting solids (sparks, glow discharges, laser-induced breakdown spectroscopy), liquids (sparks), and solutions (flames)
- *Size:* Depends on specific technique; from approximately 10^{-6} to 100 g
- *Preparation:* Machining or grinding (metals), dissolution (for flames), laser ablation of nonrepresentative surface layers for subsequent bulk analysis by laser-induced breakdown spectroscopy

Limitations

- Some elements are difficult to determine, such as nitrogen, oxygen, hydrogen, halogens, and noble gases.
- Sample form must be compatible with specific technique.
- All methods provide matrix-dependent responses.

Estimated Analysis Time (Including Sample Preparation)

- 30 s to several hours, depending on sample-preparation requirements

Capabilities of Related Techniques

- *X-ray fluorescence:* Bulk and minor constituent elemental analysis; requires sophisticated data reduction for quantitative analysis; not useful for light elements (atomic number ≤ 9)
- *Inductively coupled plasma optical emission spectroscopy:* Rapid quantitative elemental analysis with parts per billion detection limits; samples must be in solution; not useful for hydrogen, nitrogen, oxygen, halides, and noble gases
- *Atomic absorption spectroscopy:* Favorable sensitivity and precision for most elements; single-channel technique; inefficient for multielement analysis

Introduction

History

Optical emission spectroscopic methods originated in experiments performed in the mid-1800s (Ref 1), yet they remain some of the most useful and flexible means of performing elemental analysis. Free atoms, when placed in an energetic environment, emit light at a series of narrow wavelength intervals. These intervals, termed emission lines, form a pattern—the emission spectrum—that is characteristic of the atom generating it. The energetic environment typically is a type of plasma, which is defined as an ionized gas consisting of positive ions and free electrons in proportions resulting in more or less no overall electric charge. A device used to produce such a plasma is commonly termed an emission source, and this article describes a few of those most commonly used in optical emission spectroscopy (OES).

The presence of an element in a sample is indicated by the presence in light from the source (plasma) of one or more of its characteristic emission lines. The concentration (mass fraction) of that element can be determined by measuring line intensities. Thus, the characteristic emission spectrum forms the basis for qualitative elemental analysis, and the measurement of intensities of the emission lines forms the basis of quantitative elemental analysis.

Development Trends

The spark discharge remains the most commonly used source for bulk analysis of metal alloys. The current trend is mainly toward complete automation. Spark OES systems are increasingly incorporated into automatic laboratories with facilities for sample preparation and sample handling by robots. There also is continued development of more sophisticated electronics to control the waveform of the spark, for increased stability and flexibility.

Glow discharge spectrometer systems are undergoing strong development for improved stability and capability for depth profile analysis. The introduction of the radio-frequency-powered source—allowing analysis of nonconducting coatings—has been a major breakthrough. Another important development trend is special sample holders for small and nonflat samples.

The OES technique undergoing the fastest development is laser-induced breakdown spectroscopy (LIBS). Driven by the development of new laser types, for example, fiber lasers, the number of applications and types of LIBS systems is steadily increasing. Similar to the spark OES system, LIBS can be used for routine analysis of metallurgical samples, but the main strength of LIBS is to enable measurements not possible by other methods, for example, in terms of distance; size, shape, and composition of the sample; and speed and spatial resolution. These characteristics have made LIBS an outstanding technique for applications involving inline analysis of materials in various industrial processes.

Another general trend for all OES techniques is that the photomultiplier tubes are gradually being replaced by solid-state detectors of the charge-coupled- and charge-injection-type devices. With these detectors, development of compact miniature spectrometers for special applications is another strong trend.

The previously rather common sources of direct current arcs and combustion flames are becoming obsolete, and very few commercial instruments are currently available.

General Principles

The characteristic spectrum that an atom produces reflects the electronic structure of the atom. Changes in the energy of the valence or outer-shell electrons result in the atomic lines used in emission spectroscopy. Each atom has a ground state in which all of its electrons occupy positions of minimum potential energy. As an atom absorbs energy, one or more of the outer electrons may be promoted to higher energies, producing an excited energy state. The energy of an atomic state is a function of the energies of the individual electrons and of energy changes resulting from interactions among the electrons. Each possible combination of electron configurations produces what is known as a spectroscopic term that describes the electronic state of the atom.

Electronic Energy Levels

The simplest atoms, such as hydrogen and the alkali metals, have only one electron outside a core, with the net charge +1. In hydrogen, this core is the atomic nucleus, consisting of a single proton (or proton + neutron for the heavier isotope deuterium); in the alkali metals, the core is the nucleus surrounded by filled electron shells, with a net charge of +1. The simple electron configurations of these hydrogen-like atoms produce several possible terms, as illustrated by the energy-level diagram for lithium in Fig. 1. Atomic emission lines result when the atom undergoes a spontaneous transition from one excited state to another lower-energy state. Not all possible combinations of states produce emission lines. Only transitions obeying quantum mechanically derived selection rules occur spontaneously. Diverse factors control the relative intensities of the lines. Those transitions between a low excited state and the ground state, termed resonance transitions, generally yield the most intense emission. The energy of the excited electron increases with decreasing spacing between excited states until it reaches an ionization limit. At this point, the electron is no longer bound to the atom and may assume a continuous range of energies. Such unbound electrons may undergo transitions to bound states. Because the upper state of the transition is not limited to discrete values, the light from such transitions is spread continuously over a range of wavelengths. However, this type of emission is normally very weak and of no interest for chemical analysis.

The ionization limit for the atom corresponds to the ground state of the singly charged ion of the same element. Excitation of the remaining bound electrons yields a different term system and a different set of emission lines. Ionization and excitation may continue until an atom is completely stripped of its electrons. In practical emission sources, ionization rarely proceeds beyond removal of two electrons, and, in most cases, only the first stage of ionization needs to be considered. Therefore, lines from the neutral atom and the first ion spectrum are commonly used in chemical analysis.

Spectral Overlap

The use of optical emission for elemental analysis requires measurability of the emission intensity from a line of a specific element independent of overlapping emission from other species in the sample. The probability of undesired overlap depends on the number of lines in the spectrum and on the wavelength spread (linewidth) of each transition. If all atomic term systems were as simple as that shown for lithium in Fig. 1, the probability of spectral overlap would be low. However, lithium is one of the simplest atoms.

Atoms with more complex electronic structures produce correspondingly complex emission spectra. The iron spectrum, a small section of which is shown in Fig. 2, exemplifies such spectral complexity. The spectrum

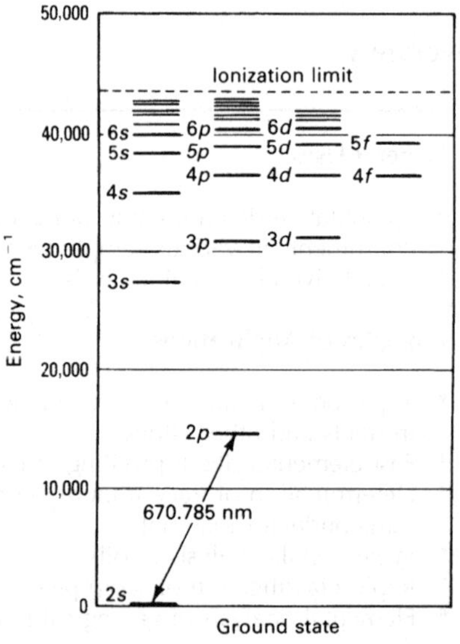

Fig. 1 Energy-level diagram for lithium. With the exception of the *s* states, each horizontal line corresponds to two closely spaced energy levels. The numbers and letters to the left of the lines are designations for the orbitals that the single electron can occupy. Transitions from the two 2*p* states to the ground 2*s* state produce a pair of closely spaced resonance lines at 670.785 nm.

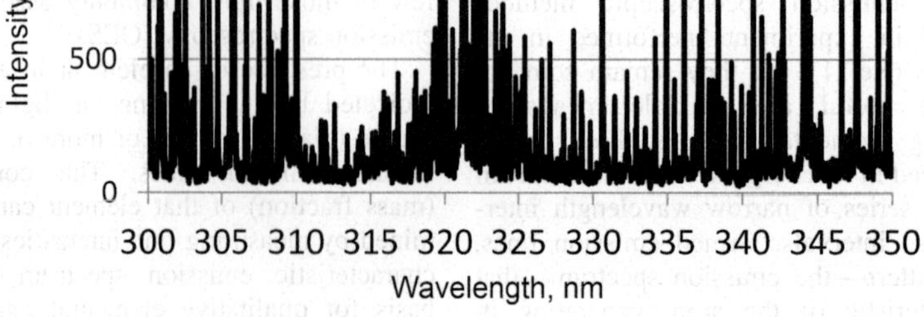

Fig. 2 Example of spectral complexity exhibited in a small section of a steel sample

from one ionization stage of a single element may, given sufficient excitation energy, consist of hundreds or even thousands of emission lines. In the range 300 to 350 nm shown in Fig. 2, there are 1057 known iron atomic lines and 538 ion lines, all according to the spectral database provided by the National Institute of Standards and Technology. A large number of these lines are so weak that they are not observed in this spectrum, but at least a few hundred are present. The complexity is compounded when several elements are present in a sample, each generating neutral (atomic) and ionic spectra. For methods to deal with spectral overlap, see the section "Calibration and Quantification of OES for Direct Solids Analysis—The Ratio Method" in this article.

Line Broadening

Spectral complexity would not be a problem if emission lines were strictly monochromatic and instruments were available with infinite spectral resolution. Several factors contribute to line broadening. The normally dominant type is Doppler broadening, resulting from motion of the emitting species relative to the device detecting the emission. For fixed transition energy, the emission recorded from an atom moving toward the detector is at shorter wavelengths than that recorded from an atom at rest. The emission from an atom moving away from the detector is at longer wavelengths.

Wavelength broadening of the emission lines also can result from uncertainty in the energy levels due to frequent collisions of the emitting atom or ion with other species in the plasma, and placement of the emitting atom in an electric field. The first type of line broadening is termed collisional broadening; the second is Stark broadening.

The relative magnitude of these three line-broadening contributions depends strongly on the type of emission source. The collisional contribution to linewidth is primarily a function of source pressure. The Doppler contribution for a given element depends on source temperature. The magnitude of the Stark contribution depends on the density of charged species near the emitter.

However, the broadening of spectral lines resulting from the previously described mechanisms usually is considerably smaller than the limitation in spectral resolution of the optical spectrometer used. The line broadening usually is < 10 pm (10^{-12} m), while the optical spectral resolution seldom is below 20 pm.

Self-Absorption

Atomic line profiles produced by any of the aforementioned effects also can be altered by self-absorption. At high concentrations of atoms in the spectroscopic source, the probability that the radiation an atom emits will be absorbed by another atom of the same element can be relatively high. The probability of absorption is greater at wavelengths near the center of the line profile than at wavelengths near the wings. The emission profiles observed under such conditions are flatter and broader than those observed in the absence of self-absorption. In practical analytical work, this leads to nonlinear calibration functions; that is, the analytical sensitivity is reduced at high mass fraction of the analyte.

Molecular Emission

The energetic emitting volume of a spectroscopic source may contain small molecules in addition to free atoms. Like the atoms, the molecules produce optical emission that reflects change in the energies of the outer electrons of the molecule. Unlike the atoms, the molecules have numerous vibrational and rotational levels associated with each electronic state. Each electronic transition in the molecule therefore produces an emission band composed of a large number of individual lines reflecting the vibrational and rotational structure of the electronic states involved in the transition.

Molecular bands appear in a recorded spectrum as intense edges, out of which develop at higher or lower wavelengths less intense lines with a spacing that increases with distance from the edge. The edge is termed a band head. Composed of many closely spaced lines, molecular bands may dominate a region of the spectrum, complicating detection of emission from other species in that region. An example of a molecular band is shown in Fig. 3. Emission sources are often designed to minimize molecular emission. Less frequently, band intensities are used in place of atomic line intensities to measure mass fractions of elements making up the molecules.

Optical Systems

Atomic emission is analytically useful only to the extent that the emission from one atomic species can be measured and its intensity recorded independent of emission from other sources. This detection and quantification demands high-resolution wavelength-sorting (dispersing) instrumentation. Further, before the light can be sorted, it must be collected efficiently, sometimes only from an isolated region in a spatially heterogeneous emission source. Background information on optical systems is cited in Ref 3 and 4.

Wavelength Sorters

The key element in modern wavelength-sorting instruments is the diffraction grating, a precisely shaped reflective surface having many closely spaced parallel grooves. Figure 4 shows the principle of a diffraction grating. Parallel rays of light strike adjacent grooves on the grating. The incident rays are in phase with each other. The rays scattered from the grating have traversed different paths. At angles producing a path difference that is an integral number m of wavelengths, the exiting light waves are in phase, and light is diffracted at that angle. At other angles, the exiting waves are out of phase, and destructive interference occurs. The diffraction condition is:

$$m\lambda = d(\sin\alpha \pm \sin\beta) \quad \text{(Eq 1)}$$

where d is the diffraction grating groove spacing, α is the angle of incidence, β is the angle of diffraction, and m is an integer called the spectral order. The minus sign enters when the incident and diffracted beams are on opposite sides of the grating normal. Note that light

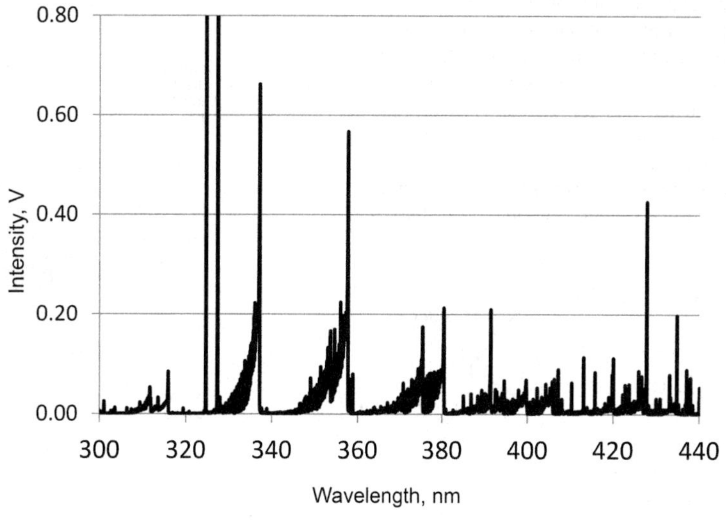

Fig. 3 Spectrum of N_2 recorded with nitrogen added to the argon gas flow of a glow discharge lamp, run with a pure copper sample. Source: Ref 2. Reprinted with permission of Elsevier

waves with wavelength $\lambda/2$ in spectral order 2 are diffracted at the same angle as λ in the first order, as illustrated in Fig. 4.

Two types of wavelength-sorting devices are most commonly used for emission spectroscopy. The first, the grating monochromator, is used for single-channel detection of radiation. Figure 5 shows the light path through a Czerny-Turner monochromator, a commonly used configuration. Light enters the monochromator through the entrance slit and passes to the collimating mirror. The collimated light strikes the plane diffraction grating and is diffracted at an angle dependent on its wavelength. Some of the light is diffracted at angles such that it strikes the focusing mirror. It then is focused to form an array of entrance slit images in the focal plane of the monochromator. The position in the array of a slit image depends on the angle at which the light that forms it exited the grating. The wavelength of the image centered on the exit slit is:

$$m\lambda = 2d \sin\theta \cos\varphi \quad \text{(Eq 2)}$$

where θ is the angle through which the grating is rotated, and φ, the instrument angle, is the angle that a line through the center of the grating and the center of either mirror makes with the centerline into a spectrometer. The relationships between θ and φ and the angles α and β used in Eq 1 are shown in Fig. 5. As the grating is rotated, images from different wavelengths pass sequentially through the exit slit and are detected by a photomultiplier tube.

The second general type of wavelength sorter is the polychromator. Most modern polychromators are variations on the Rowland spectrometer invented in 1873 (Fig. 6). The diffraction grating is concave (focusing), with a radius of curvature R. If an entrance slit is located on a circle of radius $R/2$ tangent to the grating face, the diffracted images of the slit are focused around the circle. Exit slits and photomultiplier tubes may be placed at positions on the focal curve corresponding to wavelengths of lines from various elements. Line intensities from 20 to more than 60 elements, depending on instrument capability, can be determined simultaneously with this type of instrument.

Alternatively, one or several array detectors of the charge-coupled device or charge-injection-type device can be placed on the circle, converting the polychromator into a spectrograph. An entire emission spectrum can be recorded in a short time on these detectors, which are the same type as those used in electronic cameras, including modern smartphones. Array detectors allow more flexibility in line selection and provide more information than the combination of fixed slits and photomultiplier tubes. These types of instruments are becoming increasingly common, in particular, very compact (miniature) spectrometers of limited spectral resolution.

Before the introduction of electronic array detectors, photographic plates were commonly used in spectrographs to record complete spectra. However, such instruments are rarely used anymore.

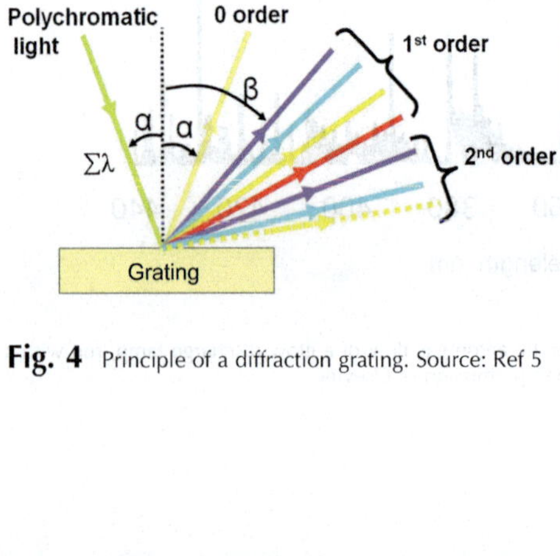

Fig. 4 Principle of a diffraction grating. Source: Ref 5

Collection Optics

Collection optics for a spectroscopic instrument must transfer light from the source to the detector with maximum efficiency and resolve or, in some cases, scramble spatial heterogeneities in the emission from the source. The first requirement is met if radiation from the source fills the entrance slit and the collimating optic of the spectrometer. A simple lens of appropriate size is commonly used to image the source on the entrance slit with sufficient magnification to fill it.

The size of the lens is selected such that radiation passing through the slit just fills the collimating optic. The entrance slit then defines the area of the source viewed by the system, and any source nonuniformity within

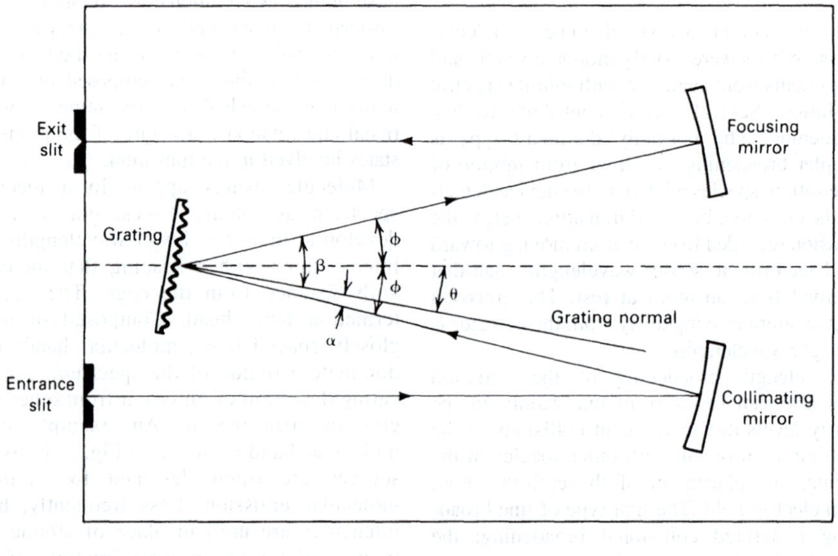

Fig. 5 Light path through a Czerny-Turner monochromator

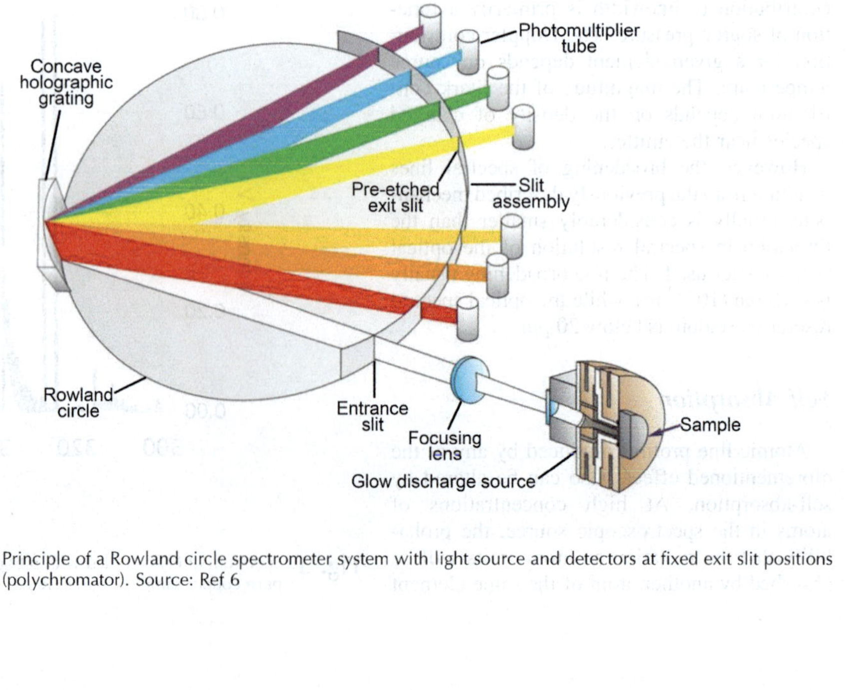

Fig. 6 Principle of a Rowland circle spectrometer system with light source and detectors at fixed exit slit positions (polychromator). Source: Ref 6

that area is transferred to the detector. Photographic detection often requires spatial uniformity of the slit images. The desired uniformity is obtained if the source is imaged onto the collimating optic by a lens near the slit. Other lenses then are used to generate an intermediate image of the source at an aperture to provide spatial resolution.

Emission Sources

An emission light source must decompose the sample from some easily prepared form into an atomic vapor and then excite that vapor with sufficient efficiency to produce a measurable emission signal from the sample components of interest. Each of the types of emission sources— high-voltage sparks, glow discharges, laser-induced plasma, electrical arcs, and combustion flames—has a set of physical characteristics with accompanying analytical assets and liabilities. They all can be classified as plasma sources. Background information on excitation mechanisms is cited in Ref 7.

Excitation Mechanisms

The property of an emission source that is most closely linked to its excitation characteristics is temperature. Temperature indicates the amount of accessible energy in the source. Because energy can be partitioned variously among different species, different temperatures may reflect that partitioning.

Gas kinetic temperature and electron temperature indicate the kinetic energies of heavy particles (atoms and ions) and electrons, respectively. Excitation and ionization temperatures reflect the electronic energy content of atomic, ionic, and molecular species. In addition, molecules store energy in rotational and vibrational modes, which is expressed as vibrational and rotational temperatures. In many source environments, excess energy in one mode is rapidly exchanged, or transferred, to another. In such cases, all the aforementioned temperatures are equal, and the source is in local thermodynamic equilibrium (LTE). When LTE exists, excitation conditions may be described without an understanding of the microscopic mechanisms of energy transfer. The population distribution among the possible excited states for a given species is given by the Boltzmann equation:

$$\frac{n_q}{n} = \frac{g_q \exp\frac{-E_q}{kT}}{\sum_{m=0}^{j} g_m \exp\frac{-E_m}{kT}} \qquad \text{(Eq 3)}$$

Equation 3 gives the fraction of the total population of a given species that is in excited state q. If, for example, the species is a neutral atom, n_q is the number density of atoms in state q, and n is the number density of atoms in all states. E_q is the energy above the ground state of state q, g values are statistical weights reflecting the fact that different combinations of quantum numbers can give rise to states with the same energy, j (the limit for the summation in the denominator) is the number of possible excited states of the atom, k is the Boltzmann constant, and T is the absolute temperature.

Examination of Eq 3 illuminates several important characteristics of thermal excitation. At the high-temperature limit, the population n is distributed nearly equally among all of the possible excited states. The populations of the individual states depend only on their statistical weights. At practically attainable temperatures, the distribution favors states with low excitation energies. Therefore, the quantum mechanically allowed resonance transitions from low-lying excited states almost always produce the most intense emission. The higher the energy of such transitions, the hotter the source required for efficient excitation. The denominator on the right side of Eq 3 is the partition function for the species. The value of the partition function increases with the number of excited states. This implies that thermal excitation of a given level is more efficient for a simple atom with few excited states than for a complex atom with many excited states.

Equation 3 alone does not characterize excitation completely, because it describes only the distribution among excited states of the atom or ion in question, omitting the equilibrium among the possible stages of ionization. That equilibrium is described by the Saha equation:

$$S = \frac{n_i n_e}{n_a} = 4.83 \times 10^{15} T^{3/2} \frac{Z_i}{Z_a} \times 10^{-(5040/T)V_i} \qquad \text{(Eq 4)}$$

where n_i, n_e, and n_a are the number densities of the ions, electrons, and atoms, respectively; Z_i and Z_a are the partition functions of the ion and atom; and V_i is the ionization potential in volts. Please note that this is a simplified description, because the lowering of the ionization energy in a plasma is not considered. Furthermore, it is a tailored equation for numbers only, lacking units for several of the physical quantities. However, it is generally accepted as a good description of the degree of ionization in emission sources for OES. This function reaches large values at temperatures routinely generated by emission sources; therefore, ionization of some atomic species is virtually complete. The emission spectra from the ionic forms of an element are useful in such cases, but emission from neutral atomic species is weak or absent.

When LTE does not exist, Eq 3 and 4 are not rigorously applicable. A complete description of excitation in such cases must account for microscopic collisional processes that may excite or de-excite a given energy level with an efficiency far different from that predicted using LTE. For example, in low-pressure discharges, a small portion of the electron population usually has a temperature far higher than the gas temperature in the discharge. These fast electrons may produce highly excited atoms or ions in much greater numbers than would be generated under LTE conditions. The excitation efficiency in non-LTE sources often depends on close matches in the kinetic or internal energy of colliding species and therefore displays sharp variations as the chemical composition of the excitation region changes.

Spark Sources

The high-voltage spark is by far the most commonly used source for analysis of solid metal alloy samples, and it is also one of the oldest source types, dating back to the mid-19th century. Several instrument companies produce spark OES instruments for primarily the metallurgical industry, where it has been the "workhorse" for chemical analysis for approximately half a century.

The spark stand essentially is a very simple device: a small enclosure where the sample (cathode) is placed on a grounded flat plate with a hole opposite to a pin-shaped electrode. A diagram of a typical spark stand (electrode and sample geometry) is shown in Fig. 7. It illustrates the conventional point-to-plane configuration (electrode to sample). The electrode typically is made of tungsten. The spark stand in modern instrumentation is flushed with argon or (rarely) nitrogen to prevent oxidation effects and allow transmission of short ultraviolet wavelengths absorbed by oxygen. A very basic electrical circuit diagram used to provide the discharge is shown in Fig. 8, where V is a voltage source; S, a switch; C, a capacitor; L, an inductor; R, a resistor; and D, a diode. When the switch is closed, the capacitor is charged to the voltage, V. When the switch is opened, the electrical energy stored in the capacitor is released through the inductor and resistor, producing a discharge across the spark gap of a few millimeters. To cause dielectric breakdown of the gas and thereby trigger the discharge, a much higher voltage, on the order of 10 kV or more, is required. This is provided by an ignitor circuit (similar to the ignition coil of an older automobile motor), whose sole function is to bridge the gap. Once this short path has been ionized, the lower voltage is

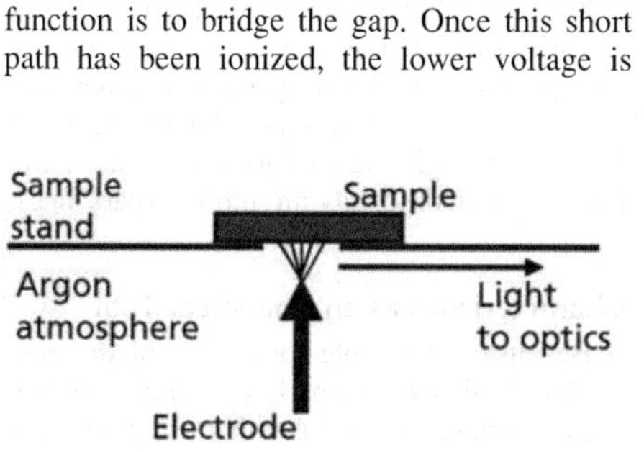

Fig. 7 Schematic of a typical spark stand design. Source: Ref 8. Reprinted with permission of Elsevier

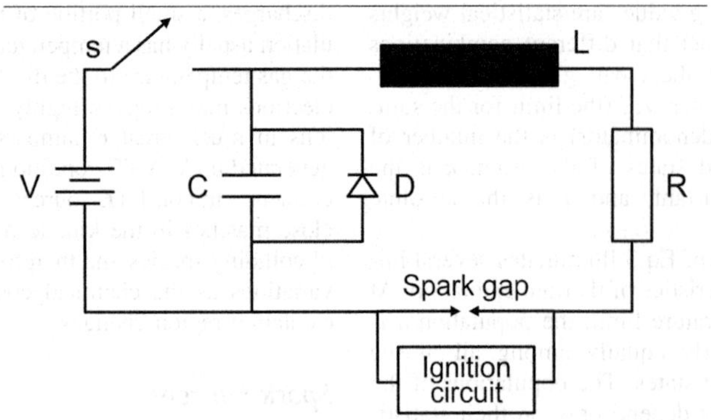

Fig. 8 Basic electrical circuit diagram of a spark source. V, voltage source; S, switch; C, capacitor; L, inductor; R, resistor; and D, diode. Here, the ignition circuit is connected in parallel to the spark gap; there also are types connected in series. Source: Ref 8. Reprinted with permission of Elsevier

sufficient for a continued discharge. The discharge is sustained at typically 20 to 100 μs with a voltage of 400 to 1000 V and peak currents of 50 to 150 A. The current is oscillatory in nature at frequencies of typically a few hundred megahertz, and the diode, D, ensures that the current is unidirectional. Material from the sample (cathode) is vaporized and atomized by a combination of the heat generated and sputtering by energetic argon ions.

The excitation and ionization of atomized elements in the spark discharge column is caused by a wide variety of mechanisms, as is the case of all analytical plasma sources. Inelastic collisions with slow (thermalized) electrons account for a large fraction of both excitation and ionization. Charge-transfer reactions with ionic species of the plasma gas play an important role in the excitation of certain levels of ions from the sample (Ref 9). Emission lines from neutral as well as singly ionized atoms make up a large fraction of the emission spectra, but lines from multiply charged ions also are observed. Ions formed also will recombine to neutral atoms by electron capture, creating atoms in excited states, giving rise to atomic emission spectra. The spark discharge is characterized by considerable temporal and spatial inhomogeneity. However, the degree of excitation of atomized material can be described by an average excitation temperature, T_e, typically in the range of 5,000 to 15,000 K (4,700 to 14,700 °C). At the higher end of this range, efficient excitation of higher energy levels generates rather line-rich spectra. There also is a large fraction of lines from singly charged ions, the reason why these lines traditionally are termed spark lines.

Characteristics as an Analytical Tool

Because the sample forms one of the electrodes, analysis by spark emission spectroscopy is limited to samples that are conductive or can be made so. The individual sparks in a train strike different locations on the sample electrode in a random fashion similar to lightning, generating a several-millimeter-wide burn pattern on a planar sample. For an average analysis of a sample with inhomogeneity at the microscopic level, this is in fact an analytical advantage. To further reduce the effect of inhomogeneity, as well as effects of the sample surface preparation, a sparking-off (preburn) period of typically several hundred high-energy sparks normally precedes the spark train employed for the actual analysis.

Modern spark sources are very fast, with spark frequencies of 200 to 800 Hz. A complete analysis normally takes less than 20 s, including the preburn to condition the sample surface. Sophisticated solid-state digital electronics allow precise control of the duration and electrical parameters of the discharge. Therefore, the discharge parameters in an analytical sequence can be changed several times to optimize the analytical performance for different sets of elements. For major elements, a relatively short pulse of high voltage typically is used; for trace elements, a longer pulse of lower voltage and higher current (arclike) often is used. The spark is a pulsed source, and high-end modern spark spectrometers are equipped with time-gated detectors to obtain the best analytical figures of merit for each element. In particular, the best detection limits for several trace elements are obtained in the "afterglow" when the current is turned off.

The spark is operated at atmospheric pressure, and the narrow spark column forms a rather dense plasma. This has the effect that calibration curves, to a large extent, are nonlinear due to self-absorption. However, this is not a problem for analytical accuracy. By wise selection of analytical emission lines and spark conditions, modern spark OES instruments are extremely accurate and stable analytical tools. For most elements of interest, the detection limits are in the few parts per million range, in most cases sufficient also for analysis of trace elements.

Sample Preparation

Metallurgical samples are prepared by milling or dry grinding with 80-mesh or finer abrasive paper. In an automatic lab system, milling is the predominant method.

Glow Discharges

Glow discharges (GDs) are direct current low-pressure discharges that are characterized primarily by the current density in the discharge and by the mechanism of electron production at the cathodic electrode. Current conduction in an arc or spark is confined to a well-defined channel that tends to anchor itself to the cathode, causing intense local heating. Electrons are thermally ejected from the cathode. In contrast, current conduction in a GD is diffuse, and the cathode is maintained at relatively low temperatures. Thermionic emission of electrons from the cathode is unimportant. Rather, bombardment of the cathode by high-energy photons and ions ejects electrons. Glow discharge light sources can be designed in almost any conceivable form and size; well-known examples are neon lights and fluorescent tubes (low-energy lamps). The sputtering of the cathode is a well-controlled process of atomization, making GDs good candidates for spectrochemical analysis. The GDs used as spectroscopic emission sources are operated in the current-voltage domain characteristic of abnormal GDs. The conducting region in such discharges envelopes the entire cathode. Increases in current necessarily increase current density and thus discharge voltage. The increase in voltage with increasing current continues until the cathode becomes sufficiently hot to emit electrons thermally, and an arc forms. Decreases in the current below a threshold value contract the conducting region; current density and voltage remain constant. This constant voltage discharge is a normal GD. The values of the current at which the transitions from normal to abnormal GD and from abnormal GD to arc occur depend on electrode geometry, gas pressure, and electrode cooling efficiency.

At typical operating pressures of 0.5 to 1 kPa (0.07 to 0.15 psi), 10 to 500 mA currents produce an abnormal GD. For a given electrode configuration, current, voltage, and fill-gas pressure regulate operation of a GD. The values of two of these may be varied independently, fixing that of the third. At a fixed current, an increase in pressure decreases voltage. The current in a GD is not self-regulating; a ballast resistor must be included in the discharge circuit to limit the current.

Glow discharges are spatially inhomogeneous, particularly in the region near the cathode. Most of the voltage applied between the electrodes is dropped across a layer near the cathode surface, because electrons are repelled from the cathode faster than the heavier positive ions are attracted to it, and a positive space charge develops. Electrons exiting the cathode are accelerated through this region of high electric field, termed the cathode fall, and acquire sufficient energy to ionize the buffer gas, normally argon. The buffer gas ions are in turn accelerated toward the cathode, gaining sufficient energy to eject atoms of the cathode material when they strike its surface. The sputtered atoms diffuse from the cathode surface and then may be excited in the plasma. The excitation and ionization mechanisms of atomized elements in the GD are, to a large extent, the same as those of the electrical spark, with electron impact the dominating type. The difference is mainly in the relative importance of the different type of mechanisms. One important excitation mechanism in the GD is Penning ionization, caused by collisions with gas atoms (normally argon) in metastable states, that is, atoms excited to a high energy level with a relatively long lifetime. The transfer of energy in the collision results in the ionization of the target atom, resulting in optical emission from both the ion and atomic species in cascade reactions. Penning ionization occurs when the target atom has an ionization potential lower than the excited energy of the metastable atom or molecule. Another important excitation mechanism is asymmetric charge transfer (ACT), which occurs when an ionized atom collides with a neutral atom of a different element. During the collision, one electron is transferred from the neutral atom to the ion. As a result, the atom that was initially neutral has been ionized, most likely in an excited ionic state. The ACT has increased probability (cross section) when there is a close match between the energy carried by the impacting atom/ion and the energy needed for the reaction leaving the target atom in an excited ionic state. The result is that certain emission lines are very intense due to such resonance conditions. Compared with the spark discharge, the GD plasma operates at low pressure (and thereby lower atom number density) as well as a much lower current and electron density. Therefore, it is a cooler plasma with generally less probability to excite higher energy levels, giving rise to somewhat less line-rich spectra.

The Grimm Emission Source

In 1967, a glow discharge lamp (GDL) was developed by Grimm for easy analysis of metallurgical samples (Ref 10). Similar to the point-to-plane spark source, the sample is a flat piece of metal forming the cathode of the electrical discharge. This source has become very popular, and several commercial spectrometer systems with this type of source are being produced. A Grimm-type GD consists of a small vacuum chamber with a flat front plate (cathode) and a cylindrical hole, where a tubular-shaped anode extends to approximately 0.2 mm (0.008 in.) from the plane defined by the cathode plate (Fig. 9). A flat sample is placed on the cathode plate and sealed from the surrounding air by an O-ring. When operated, a vacuum pump first evacuates the lamp house before the discharge gas (normally argon) is admitted at a small flow rate, adjusted to a pressure of typically 0.5 to 1 kPa (0.07 to 0.15 psi). Applying a voltage of 500 to 1200 V causes a dielectric breakdown of the low-pressure gas to form a constricted discharge in the opening of the tubular anode, with the sample as the cathode. The inner diameter of the anode is in the range of 1 to 8 mm (0.04 to 0.32 in.), with 4 mm (0.16 in.) as the standard in current commercial instruments. The discharge current with a 4 mm anode is in the range of 10 to 50 mA. This means that the average power dissipation in the GD plasma is just a few percent of the typical spark plasma, generating much less heat at the sample surface. As a consequence, sample removal and atomization is normally entirely due to sputtering by energetic ions and atoms striking the sample surface. Due to the short mean free path of atoms and ions at typical operating pressures, ions generated in the negative glow of the plasma undergo several collisions before reaching the surface. Therefore, the average kinetic energy of gas ions striking the cathode is < 100 eV, in spite of the accelerating voltage of 500 to 1200 V. Furthermore, a substantial fraction of the primary ions become neutralized while maintaining direction and momentum toward the cathode, with the result that neutral atoms also contribute significantly to the sputtering. It also has been shown that self-sputtering by ions and atoms originating from the sample itself contribute significantly to the total mass-removal rate.

The sputtering in a GDL normally is very even over the sample surface, resulting in sputter craters with very flat bottoms (Fig. 10). This makes the GDL a very useful tool for surface and depth profile analysis; by recording the emission signals as a function of time, an elemental depth profile is obtained. This is a unique feature of the GDL, which has contributed immensely to the popularity of GD-OES and the development of sophisticated commercial instruments.

An important development of the GDL was the introduction of the modified radiofrequency (RF)-powered Grimm lamp in 1987 (Ref 11). This allows measurement of nonconducting materials, mainly metal sheet with nonconducting oxides and coatings. The development of RF lamps for improved performance is still ongoing.

Characteristics as an Analytical Tool

The GDL can be used for bulk analysis of the same types of metallurgical samples commonly analyzed by spark OES. Compared with

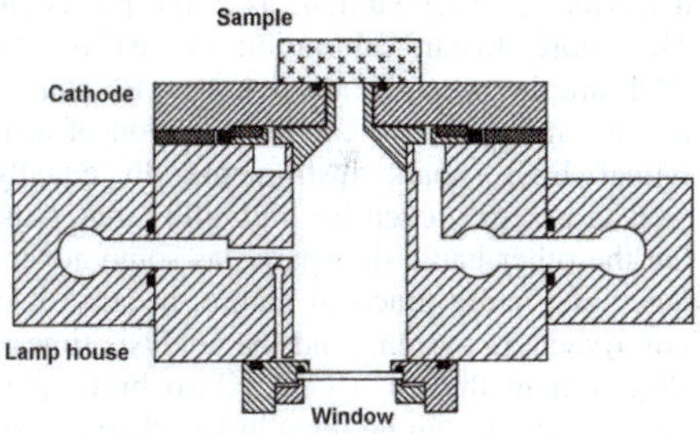

Fig. 9 Schematic of a Grimm-type glow discharge lamp. Source: Ref 8. Reprinted with permission of Elsevier

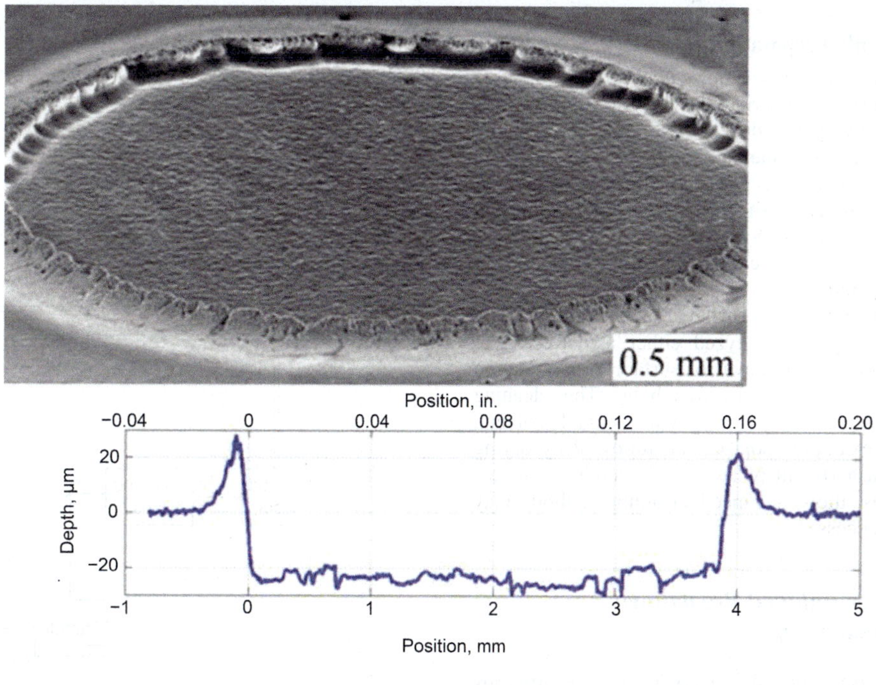

Fig. 10 Sputtering crater from a Grimm-type glow discharge lamp

the spark, a number of advantages from the analytical point of view have been put forth, all based on the fact that the GD plasma is less dense and has lower ionization and excitation temperatures than that of the spark:

- Less self-absorption, thus more linear calibration curves
- Less ionic lines in the spectra, thus fewer line interferences
- More narrow emission lines, allowing more effective separation of analytical and interfering lines

While all of these claims are essentially true, it also is true that no significant advantage over the spark in terms of analytical figures of merit has been shown. The last point is almost irrelevant, because the intrinsic linewidths even of a spark source rarely exceed 10 pm, which is approximately the best resolution high-end analytical spectrometers can offer. The spectral backgrounds and lower limits of detection are very similar, as is the precision. The more linear calibration curves of the GDL are, in principle, an advantage for analytical accuracy, but by careful selection of analytical lines, spark instruments do equally well in practice, even for high-alloy materials. On the other hand, the spark has some advantages of a more practical nature, because it is not a vacuum system, and the analysis time is shorter than that of a GDL. Two burn spots on one sample can be done in less than a minute. On the other hand, the sputtering process for sample removal has been shown to be an advantage of the GDL for certain complex alloys, in which the spark source suffers from more severe matrix effects due to the complex metallurgical structure.

Sample Preparation

Metallurgical samples normally are prepared by dry or wet grinding with 180-mesh or finer abrasive paper, depending on the type of alloy. The reason a finer surface is required compared with the spark is that the GDL is a vacuum system; therefore, the sample must form a gastight seal against an O-ring. For the same reason, milling normally cannot be used for surface preparation.

Samples for surface analysis and depth profiling are analyzed either as-delivered or cleaned for removal of surface contamination in the form of hydrocarbons. The cleaning often is done by rinsing with a suitable solvent, for example, isopropanol, followed by drying with a stream of air. For demanding applications, more advanced cleaning methods may be necessary.

Laser-Induced Breakdown Spectroscopy

Laser-induced breakdown spectroscopy (LIBS) is a technique based on the optical emission from a laser-induced plasma (laser spark), generated in the very intense electromagnetic field of a laser pulse. As such, it cannot be described as a certain type of emission source in the same way as the other types described in this article. The first mention of an LIBS plasma as a spectral emission source occurred in 1962 (Ref 12), only two years after the first ruby laser was demonstrated. Since then, LIBS has progressed to become a major technique in analytical spectroscopy; it is incredibly versatile, with a steadily increasing array of applications. As a result, it has become one of the most widely used types of OES today (2019). Further reading on LIBS is available in Ref 13.

Laser-induced plasmas are very complex, due to the dynamics of the very short-lived and rapidly expanding plasma. There is little resemblance to continuous GD plasmas at low pressure but more similarities to an electrical spark, which also is short-lived and rapidly expanding after ignition. When laser-induced breakdown occurs, the initial plasma is very dense and hot, emitting intense radiation with a more or less continuous spectrum without element-specific features. If the laser pulse duration is a few nanoseconds (typical of most LIBS applications), the sample ablation is mainly due to vaporization and melting of the sample material. However, particle ejection of target material also occurs. As the plasma expands with an initial radial velocity of typically several thousand meters per second, it rapidly cools and begins to emit element-specific spectra similar to the electrical spark. With this rapid expansion, the plasma mass compresses the surrounding medium (typically air or a noble gas) and produces shockwaves. The continuous radiation decays within a few hundred nanoseconds as the plasma expands to millimeter size; after this time, mainly the element-specific spectra with relatively low backgrounds are observed. Therefore, it is common to apply a time delay after the laser pulse of a few hundred nanoseconds to several microseconds in LIBS measurements to increase signal-to-background of the line emission. However, when using high-frequency lasers operating in the kilohertz regime or higher, the charge-coupled device (CCD) detectors commonly used in compact spectrometers cannot be gated at these high frequencies. In such cases, the LIBS spectra are averaged over longer times, including typically a few hundred shots, and therefore have a substantial continuum background overlaid on the line spectra. On the other hand, spectrometers with gated photomultipliers or intensified CCD detectors do not have this limitation.

Characteristics as an Analytical Tool

Laser-induced breakdown spectroscopy is not like any of the other techniques described in this article, due to the fact that the sample does not have to be placed inside a source; instead, the laser beam comes to the sample. Virtually any type of matter can be analyzed by LIBS: solids, liquids, gases, aerosols, and so on. In most cases, no sample preparation is needed, unless a surface layer must be removed prior to analysis. The analysis can be anything from microscale with a tightly focused laser beam at short distance from the sample, to tele-LIBS applications up to 150 m (500 ft) from the instrument, using powerful lasers and telescopes to collect the plasma light. An example of an LIBS system for analysis of samples at some distance is shown in Fig. 11. Due to this capability to analyze samples at considerable distance, new inline applications in industrial processes, not possible with any other technique, are steadily increasing.

The spectra generated by LIBS often are similar to spark spectra; with good-quality spectrometers equipped with time-gated detectors, the analytical figures of merit therefore are similar to, but not better than, spark OES.

Sample Preparation

In most applications, there is no need for surface preparation at all. In fact, this is one of the attractive features of LIBS. However, if it is

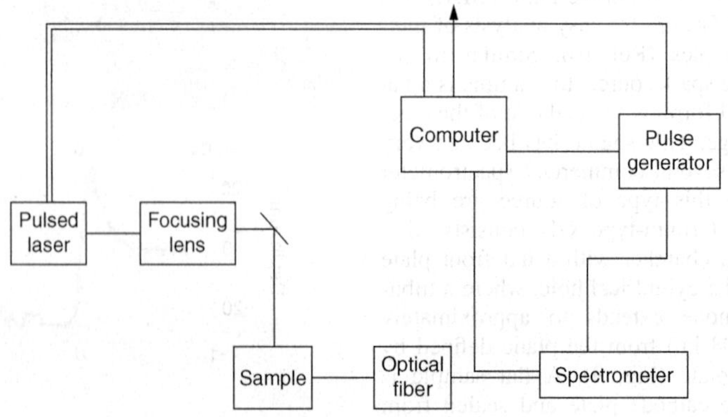

Fig. 11 Laser-induced breakdown spectroscopy system for analysis of samples at a distance

necessary to, for example, remove dirt or a surface oxide prior to analysis, this often can be done with the laser itself. By subjecting the sample to a series of high-intensity laser pulses, the underlying material can be exposed to the next burst of laser pulses for the actual analysis.

Flame Sources

Flames were the main emission sources for emission spectroscopy as it evolved into an analytical science in the last half of the 19th century, with some of the most important work done by Kirchhoff and Bunsen (Ref 1). Flame emission sources differ significantly from the other sources described in this article in that the energy required for the decomposition and excitation of the sample derives from chemical combustion rather than an electrical discharge, or breakdown to plasma in an intense electromagnetic field. Analytical flames are generated by the burning of a mixture of two gases—a fuel and an oxidant—in a burner arrangement that allows introduction of the sample into the flame as a fine aerosol. However, the use of combustion flames as emission sources is very rarely used nowadays and therefore is not described in detail in this article. In the last several decades, these sources have almost completely been replaced by inductively coupled plasma. The article "Inductively Coupled Plasma Optical Emission Spectroscopy" in this Volume is devoted to this widely used, very powerful technique.

Arc Sources

The term *arc* describes various electrical discharges; this description is limited to the direct current (dc) arc. Similar to the spark discharge, dc arcs are operated at voltages below that required to break down the gap between the electrodes; therefore, they must be started by touching the electrodes together or by ionizing the gas in the gap with a transient high-voltage discharge. Once current begins to flow, the motion of the electrons heats the electrodes and the gas in the gap. This heating increases the number of charged species in the gap and thus reduces gap resistance.

This reduction in resistance with increased current necessitates the resistor in the circuit. For a given value of ballast resistance, the gap resistance depends on the composition of the arc plasma. To minimize the effect of changes in plasma composition on arc current, the ballast resistor should be selected as the major source of resistance in the discharge circuit. Typical operating currents for the dc arc are 5 to 20 A. This current range can be obtained from sources operating at 250 V, with ballast resistances from 10 to 50 Ω.

Electrodes

Electrodes for a dc arc are arranged coaxially on a vertical axis with a separation of a few millimeters. They are most commonly machined from graphite or amorphous carbon. The lower electrode, usually the anode, has a cup machined in it to hold the sample. Heating of the anode vaporizes the material in the cup into the interelectrode region, where it is heated to 5000 to 7000 K (4700 to 6700 °C) in a generally LTE environment. At these temperatures, ionization of most elements is minimal, and neutral atomic lines dominate the spectrum. Therefore, neutral atom emission lines traditionally are designated as arc lines. The dc arc is used for its ability to sample and excite diverse samples with favorable sensitivity, providing low detection limits for trace analysis. Careful sample preparation, electrode design, and selection of arcing atmosphere minimize the impact of some of the undesirable characteristics of the arc.

Direct current arc emission spectroscopy is not precise. Its primary value is in qualitative or semiquantitative analysis of samples not easily dissolved in preparation for analysis using a more precise emission source, such as inductively coupled plasma (ICP), or when a more precise source is not available (see the article "Inductively Coupled Plasma Optical Emission Spectroscopy" in this Volume).

Similar to combustion flames, dc arc sources rarely are used nowadays. For trace analysis of very low contents, techniques based on mass spectrometry (MS) have almost completely replaced the dc arcs. The sources used for OES also are the most common types used in MS instruments for elemental analysis: ICP-MS, laser ablation MS, and GD-MS.

Calibration and Quantification of OES for Direct Solids Analysis—The Ratio Method

To perform quantitative analysis, the measured emission intensities from each element must be converted to mass fractions. For analysis of solid samples, this requires a calibration using reference samples of known composition. There are several suppliers of such samples on the market. If the procedures to determine the mass fractions are in accordance with internationally agreed standard methods, such samples are termed Certified Reference Materials (CRM) or Standard Reference Materials (SRM), which are trademarks of the National Institute of Standards and Technology. In the metallurgical industry in particular, it also is common to use in-house reference materials analyzed by well-established wet chemical methods.

Depending on the type of source and the application, a number of different calibration methods exist, and it is not possible to describe all of them in this article. However, for bulk elemental analysis of solids, the ratio method has become the most common and is described here.

The ratio method is based on the assumption that the emission intensities of all elements in the sample are proportional not just to the mass fraction of the element in the sample but also to the total amount of sample material in the plasma. This assumption has been shown to function well for spark, GDL, and LIBS spectrometry. This fact makes it possible to use an emission line of the major (matrix) element in the sample material as an internal standard, meaning the calibration is based on the ratio of the analyte line intensity to the line intensity of the matrix element. This procedure both reduces "flicker" noise in the plasma and compensates for variations in the atomization rate of material from different sample types (alloys). For each element i, the analysis function then is defined as the ratioed mass fractions versus the ratioed intensities obtained from the calibration reference materials (RMs):

$$C_i/C_M = f(I_i/I_M) \quad \text{(Eq 5)}$$

where C_i and C_M are the mass fractions of element i and the matrix element M (e.g., iron in steel), and I_i and I_M are the signal intensities of element i and the matrix element M. The function f typically is a linear or quadratic polynomial with a small background term. Polynomials of higher order are allowed and often necessary in spark OES and LIBS but are less common in GD-OES. Note that here the mass fraction is expressed as a function of intensity rather than the other way around. This is for practical reasons; a regression calculation to fit the calibration points to a curve then directly provides the analysis function used when analyzing an unknown. It can be argued that this is a backward way to perform a regression calculation to fit calibration points to a function, but it has been used for several decades with excellent results in OES and has become the standard method. Examples of calibration curves from spark OES using Eq 5 are shown in Fig. 12.

The ratio method also rests on the assumption that the sum of the determined elements and the matrix make up nearly 100% (mass fraction 1) of the samples. This can be expressed as:

$$1 = C_M + \sum C_i \quad \text{(Eq 6)}$$

Divide Eq 6 by C_M:

$$1/C_M = 1 + \sum C_i/C_M \quad \text{(Eq 7)}$$

Solve for C_M:

$$C_M = 1/\left(1 + \sum C_i/C_M\right) \quad \text{(Eq 8)}$$

When measuring an unknown, the intensity ratios I_i/I_M are first calculated, and the mass fraction ratios C_i/C_M then are obtained directly from the analysis functions. After summing all mass fraction ratios, C_M is calculated

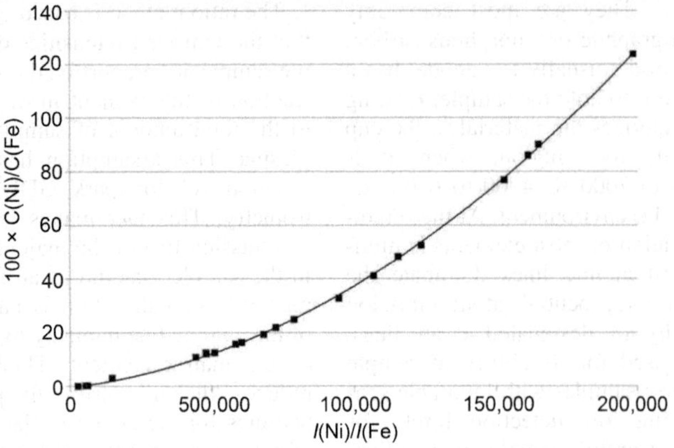

Fig. 12 Examples of calibration curves in spark optical emission spectroscopy using the ratio method. Source: Ref 8. Reprinted with permission of Elsevier

according to Eq 8. The mass fractions of the analyte elements then are calculated as:

$$C_i = C_M \times C_i/C_M \quad \text{(Eq 9)}$$

The mass fractions can be expressed as percentages (multiply by 100) or any other unit the analyst desires.

Note that this calculation includes sum normalization to 100%; it therefore will give incorrect results for all elements if one or more of the major elements of an alloy are not included in the quantification. However, this is very unlikely to be the case for steel and other metallurgical samples of well-known alloy types.

The ratio method automatically compensates even for rather large variations in both sample ablation and mass fraction of the matrix element. In cases where the spectral background is dominated by stray light and/or weak line interference from the matrix element, which often is the case, an automatic compensation of variations in the spectral backgrounds also is obtained. It is only in cases in which the spectral background is dominated by emission from the plasma gas (normally argon) and thereby matrix independent that the ratio method can have a detrimental effect on the accuracy of trace elements. This is because the nearly constant background signal then undergoes shifts in the ratio calculation due to variations in the signal from the matrix element. In such cases, it is common to omit the ratio to the matrix element and set up an analysis function in the form of mass fraction versus direct intensity.

While the ratio method allows a single calibration to be used for a rather wide range of different alloys, it is common practice to set up separate calibrations for at least each class of alloy to be analyzed. This is simply to increase analytical accuracy.

With the increased use of solid-state array detectors, providing complete spectra with a lot more analytical information than polychromators with photomultiplier tube detectors, it is likely that there will be an increased use of methods for calibration and quantification based on more advanced chemometric techniques. However, at the time this article was written, the ratio method still dominated quantitative OES analysis for solid materials.

In GD-OES, it was pointed out that probably the most common applications involve depth profiling of technical layers, for example, zinc alloys on steel. In such applications, there often is no common matrix element throughout the depth profile; therefore, the ratio method cannot be used. Instead, a quantification method based on the concept of emission yield (EY) has been developed. The EY is defined as the amount of emitted light (number of photons) per unit sputtered weight of an element given for a specific wavelength (emission line). It is well documented that the EY is matrix-independent in a GDL, at least to a first approximation.

The intensity of element i at wavelength λ then can be expressed as:

$$I_{i\lambda} = \mathrm{EY}_{i\lambda} \times c_i \times Sr_M \quad \text{(Eq 10)}$$

where c_i is the mass fraction of element i, and Sr_M is the sputtering rate of the sample M.

Equation 10 also contains information on the sample sputtering rate, a fact that can be used to determine sputtering depth by summing over all elements of the sample. For the interested reader, more information on quantitative GD-OES depth profiling is provided in Ref 14.

Interelement Corrections

In practice, it often is necessary to correct for line interference, that is, spectral overlap from elements other than the analyte i. This problem is handled by adding terms from additional elements to the basic calibration function (additive corrections):

$$C_i/C_M = f(I_i/I_M) + \Sigma k_{ij} \times I_j/I_M \quad \text{(Eq 11)}$$

Provided there is a sufficient number of calibration samples with varying mass fractions of both the analyte i and interfering elements j, a regression calculation can determine the constants needed to subtract the line interference contribution to the signal in elemental channel i. All constants k_{ij} must have a negative sign, because the interfering elements add to the signal in the I_j spectral channel. Caution must be taken not to include an excessive number of line interference corrections. It is tempting to do so, because the regression fit to a function tends to improve with each additional free parameter, but with too few calibration samples, the results can be incorrect or even nonsensical. What happens is the calibration algorithm begins to fit noise into the spectra rather than elemental information. To alleviate this problem, most OES regression software also provides the possibility of adding line interference corrections manually.

In spark OES, it also is common to provide provisions for multiplicative corrections to the calibration function. Such corrections are employed if, for example, the presence of an element suppresses or enhances the emission of other elements due to more complex reactions in the plasma. Properly used, such corrections can be very effective in improving the accuracy of the analysis, but the same caution as for the line interference holds: an excessive number of parameters to fit can cause the regression calculation to give an incorrect analysis function.

Drift Correction

All analytical instruments have drift of the analytical signals with time; in OES spectrometers the most common cause is contamination and degradation of optical components (windows, lenses, mirrors). The alignment of optical components also may shift slightly with ambient temperature changes, and detector sensitivity may change over time. For an instrument used several times daily, correction for drift may be necessary on a daily basis to maintain high analytical accuracy. Because the calibration normally is based on a very large number of RMs (in many cases >100), it would be very impractical and time-consuming to perform a complete new calibration every time a drift correction is necessary. Therefore, each set of calibration RMs contains a limited set of samples called drift standards. These samples are selected in such a way that for each element, there is a high and low point on the calibration curve. (Actually, these samples need not be part of the samples that define the calibration function.) When performing a drift correction, the drift standards are measured and the intensities recorded. Combined with the original intensities from the first calibration, a linear polynomial is calculated that defines the changes in sensitivity (slope) and background (offset). When evaluating data from unknown samples, these polynomials first are applied to the intensities of each spectral channel, to restore the signal response to the condition at the time of calibration. Obviously, there is a limit to the degree of signal drift that can be corrected in this way. If the signal of an analyte decreases to nearly zero, drift correction will no longer be effective, and the instrument must be serviced to restore the signal.

In the fully automated spectrometer labs common in the metallurgical industry today (2019), another smaller set of check standards often is employed to determine when a drift correction is needed. The system is programmed to run the check standards with regular intervals, and the measured mass fractions are checked for accuracy. If one or more elements are out of a specified range, the system will automatically run the drift standards and calculate a new set of drift-correction constants.

Table 1 Short-term precision of the analysis of a low-alloy steel, based on six burns with a Thermo iSpark spectrometer

Element	Average, mass%	Standard deviation, mass%	Relative standard deviation, %
C	1.040	0.004	0.41
Mn	0.253	0.002	0.71
Si	0.426	0.005	1.1
P	0.031	0.001	3.2
S	0.0072	0.0002	2.5
Ni	0.0915	0.0003	0.29
Cr	1.778	0.002	0.11
Cu	0.119	0.0005	0.40
Mo	0.305	0.002	0.72
V	0.097	0.0009	0.91
Ti	0.056	0.0006	1.0
Al	0.0202	0.0003	1.6
Nb	0.176	0.0015	0.86
W	0.0106	0.0003	2.4
As	0.0106	0.0003	2.8
Sn	0.021	0.0002	1.0
Co	0.0127	0.0005	3.8
Pb	0.0136	0.0004	3.0
B	0.0035	6×10^{-5}	1.7
Ca	0.0013	8×10^{-5}	5.2
Mg	0.0004	1×10^{-5}	2.5

Table 2 Short-term precision of the analysis of a high-alloy steel, based on six burns with a Thermo iSpark spectrometer

Element	Average, mass%	Standard deviation, mass%	Relative standard deviation, %
C	0.0173	0.0007	4.0
Mn	1.452	0.008	0.57
Si	0.466	0.005	0.92
P	0.0287	0.0007	2.4
S	0.0256	0.0009	3.4
Ni	10.75	0.05	0.44
Cr	17.77	0.03	0.15
Cu	0.417	0.003	0.7
Mo	2.039	0.008	0.41
V	0.0594	0.0003	0.44
Ti	0.0408	0.0017	4.1
Al	0.0068	0.0004	6.5
Nb	0.0120	0.0003	2.7
W	0.0548	0.0003	0.56
As	0.0116	0.0003	2.7
Sn	0.0150	0.0002	2.3
Co	0.235	0.001	0.39
Pb	0.0009	4.00×10^{-5}	4.6
B	0.0005	2×10^{-5}	4.1
Ca	0.006	0.0005	9.0
N	0.0448	0.0002	1.7

Data Analysis and Reliability

Because OES is frequently used in industry for process control and quality control of the final products, method validation often is crucial. In most cases, the validation is done by regular measurement of check samples of well-known composition. Preferably, these samples should be CRM, with certificates traceable to primary pure materials. When this is not possible, production samples analyzed with methods used for the production of CRM are used. Such standard methods often are classical wet chemical methods.

As mentioned in the previous section, regular measurement of check samples for verification often is automated, particularly in spark OES.

The short-term precision of modern OES instrumentation often is outstanding, with relative standard deviation figures < 1%. Tables 1 and 2 provide examples of analytical performance of a state-of-the-art spark OES system.

Applications

Spark Sources

The spark source is used mainly in the metallurgical industry for routine analysis in production, where it is the most widely used instrument for fast multielemental analysis. The reason is that it is accurate, stable, and extremely fast. Furthermore, the simple design of the spark stand lends itself very well to automation, and spark instruments in steel industries are increasingly being completely automated, operated by robots.

Special spark sources exist for applications other than solid metals, for example, elemental analysis of wear oils. In this case, the cathode is a rotating disc electrode. The disc is immersed in the oil sample on one side, forming a thin layer that is evaporated and atomized in the spark gap. Liquids other than oil also can be analyzed with this device, for example, coolants and fuel.

The spark has another unique capability: it can provide rather detailed information about microscopic nonmetallic inclusions simultaneously with the average (bulk) analysis. When a spark hits an area with a nonmetallic inclusion, for example, an oxide particle, that particle is vaporized and the elements that make up the particle give higher intensities than when steel only is vaporized. The volume of steel vaporized by a single spark is typically ~10,000 μm^3. An average-sized inclusion (2 to 5 μm diameter) occupies a volume of only 10 to 100 μm^3, that is, less than 1% of the total vaporized volume. Therefore, it appears as a contamination of the steel and is completely vaporized by a single spark. However, larger inclusions (>10 μm) normally are only partially vaporized. This must be taken into account when interpreting the data. By recording the signals of each individual spark, individual inclusions can be detected, identified, and even quantified in terms of mass and size. This technique has become known as pulse distribution analysis, and it is increasingly used in the steel industry for fast quality control and even process control. Examples of pulsograms of the intensities of spark trains with hits on inclusions are shown in Fig. 13.

Glow Discharges

The GDL can be used for bulk analysis of metallurgical samples just like the spark, which was the original intended main application of the Grimm source. It has taken some market share of bulk analysis from the spark but not to the extent originally anticipated. The main metallurgical applications are for certain special alloys with complex metallurgical structures, for example, gray and nodular cast irons. Another important application is silicon wafers, which cannot be run in a spark stand. However, the most important applications probably have turned out to be surface and depth profile analysis of technical coatings. An example of the oxide of a heat treated steel is shown in Fig. 14. Among the optical emission sources, this is a unique capability of the Grimm-type GD source. The high sample throughput and relatively low cost compared with competing (high-vacuum) techniques has made routine quality control and troubleshooting of coating processes in industry possible and affordable. In addition to the metallurgical industry, the automotive and other mechanical industries have become users of GD-OES systems for this reason. The addition of the RF source with the capability to profile nonconducting coatings also has increased the range of such applications to a large extent.

Laser-Induced Breakdown Spectroscopy

A large variety of lasers exist that are useful for LIBS applications; therefore, a large number of different applications also exist. A few are listed here:

- Direct solids analysis in laboratory instruments. In such instruments, the sample is put into a chamber that can be flushed with, for example, nitrogen or argon, allowing analysis of light elements (carbon, sulfur, phosphorus, nitrogen, oxygen) with strong emission lines < 200 nm. It is common to include a translation stage in such instruments to move the sample, allowing elemental surface mapping.

- Inline analysis of moving industrial material at moderate distances (0.5 to 5 m, or 1.6 to 16 ft). One of the major types of such applications is scrap sorting/identification, in which a fast qualitative analysis is used to classify scrap pieces in a limited set of alloy types. Often a single laser burst (one pulse or a sequence of pulses in rapid succession) is sufficient to do a correct classification, and therefore, up to 40 scrap pieces per second can be processed. The same type of system also has been used to check for material mixup to avoid delivery of, for example, the wrong steel grade to customers.
- Direct analysis of molten metal for process control. This type of application so far is restricted to metals of relatively low melting point, primarily zinc baths in hot dip galvanization processes.
- Analysis of minerals with field-portable instruments. The most spectacular application of this type is the ChemCam instrument (Fig. 15) on the Mars rover Curiosity, which has sent thousands of spectra from different rocks on Mars.
- Hand-held LIBS instruments for field applications, for example, identification of alloys in construction elements with corrosion problems and identification of metal pieces in the recycling industry
- Monitoring the exhaust from combustion furnaces by analysis of dust particles in the smokestack

Arc Sources

Direct current arc excitation is primarily a technique for determining trace metal impurities in various materials. It used to be a widely used technique for trace analysis of high-purity metals and very special materials, such as uranium. However, as mentioned earlier, dc arc OES has largely been replaced by various MS techniques.

Flame Sources

Flames are among the most efficient sources for excitation of spectra from alkali metals and have their greatest utility in the trace determination of these elements. A still rather frequent application of emission measurement with the flame is the regulation of alkali metals for pharmaceutical analytics.

ACKNOWLEDGMENT

Revised from Paul B. Farnsworth, Optical Emission Spectroscopy, *Materials Characterization*, Vol 10, *ASM Handbook*, ASM International, 1986, p 21–30.

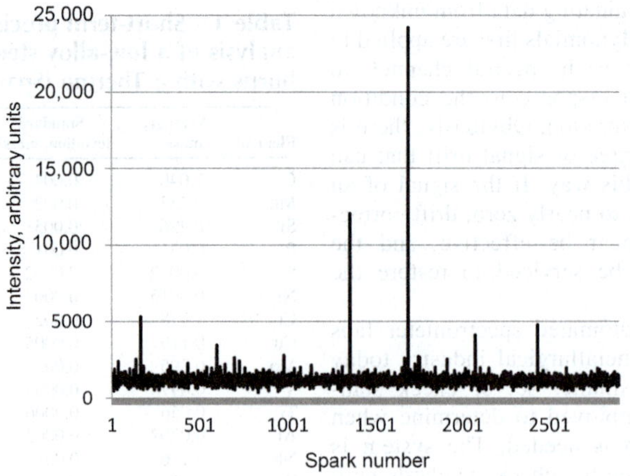

Fig. 13 Spark pulse distribution analysis pulsogram of the Al 396, 15 nm line in a high-alloy steel. The high-intensity peaks correspond to spark hits on areas with aluminum-rich nonmetallic inclusions. Source: Ref 8. Reprinted with permission of Elsevier

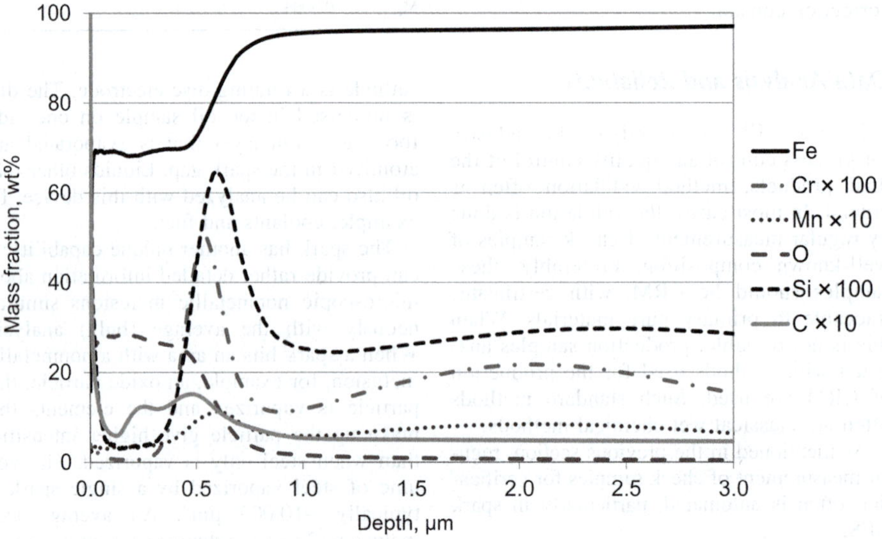

Fig. 14 Glow discharge-optical emission spectroscopy quantitative depth profile of a ~0.5 μm oxide layer on a low-alloy steel. Note that the scale factors for chromium, manganese, silicon, and carbon are expanded. Source: Ref 8. Reprinted with permission of Elsevier

Fig. 15 Laser-induced breakdown spectroscopy system ChemCam on the Mars rover Curiosity. Courtesy of NASA/JPL-Caltech

REFERENCES

1. G.R. Kirchhoff and R.W. Bunsen, Chemische Analyse durch Spectralbeobachtungen, *Ann. Phys.*, Vol 110, 1860, p 161
2. A. Bengtson, The Impact of Molecular Emission in Compositional Depth Profiling Using Glow Discharge-Optical Emission Spectroscopy, *Spectrochim. Acta B, At. Spectrosc.*, Vol 63, 2008, p 917–928
3. R.M. Barnes and R.F. Jarrell, in *Analytical Emission Spectroscopy*, Vol 1, Part I, E.L. Grove, Ed., Marcel Dekker, 1971
4. H.T. Betz and G.L. Johnson, in *Analytical Emission Spectroscopy*, Vol 1, Part I, E.L. Grove, Ed., Marcel Dekker, 1971
5. "Diffraction Gratings Ruled and Holographic," *Horiba*, https://www.horiba.com/en_en/diffraction-gratings-ruled-holographic/
6. "GDS850 Brochure," LECO Corporation, https://www.leco.com/index.php/component/edocman/?view=document&id=49&Itemid=49
7. P.W.J.M. Boumans, in *Analytical Emission Spectroscopy*, Vol 1, Part II, E.L. Grove, Ed., Marcel Dekker, 1971
8. A. Bengtson, Laser Induced Breakdown Spectroscopy Compared with Conventional Plasma Optical Emission Techniques for the Analysis of Metals—A Review of Applications and Analytical Performance, *Spectrochim. Acta B: At. Spectrosc.*, Vol 134, 2017, p 123–132
9. S.A. Goldstein and J.P. Walter, Charge Transfer Excitation in Atmospheric Spark Discharges, *Tenth Natl. Mtg. Soc. Appl. Spectros.* (Abstracts only), Oct 18–22, 1971 (St. Louis MO), p 53
10. W. Grimm, *Spectrochim. Acta B*, Vol 23, 1968, p 443
11. M. Chevrier and R. Passetemps, European Patent 0 296 920 A1, 1988
12. F. Brech and L. Cross, Optical Microemission Stimulated by a Ruby Laser, *Appl. Spectrosc.*, Vol 16, 1962, p 59
13. R. Noll, *Laser-Induced Breakdown Spectroscopy; Fundamentals and Applications*, Springer Verlag, 2012
14. Z. Weiss, Calibration Methods in Glow Discharge Optical Emission Spectroscopy: A Tutorial Review, *J. Anal. At. Spectrom.*, Vol 30, 2015, p 1038

SELECTED REFERENCES

- C.T.J. Alkemade, T.J. Hollander, W. Snelleman, and P.J.T. Zeegers, *Metal Vapours in Flames*, Pergamon Press, 1982
- P.W.J.M. Boumans, *Theory of Spectrochemical Excitation*, Plenum Press, 1966
- G. Herzberg, *Atomic Spectra and Atomic Structure*, Dover, 1944
- R. Payling, D. Jones, and A. Bengtson, Ed., *Glow Discharge Optical Emission Spectrometry*, J. Wiley & Sons, 1997
- J.P. Singh and S.N. Takhur, *Laser-Induced Breakdown Spectroscopy*, Elsevier B.V., 2007
- M. Slavin, *Emission Spectrochemical Analysis*, Wiley-Interscience, 1971

ASM Handbook, Volume 10, *Materials Characterization*
ASM Handbook Committee
DOI 10.31399/asm.hb.v10.a0006653

Atomic Absorption Spectroscopy

Gary D. Rayson, New Mexico State University

Overview

General Use

- Used for the quantitative determination of 68 metals and metalloids

Examples of Applications

- Determination of parts per million and parts per billion concentrations in environmental samples
- Water purity verification for regulatory compliance
- Waste water treatment quality assurance and quality control
- Determination of metal content in foods

Samples

- *Form:* Liquids, solutions, and digests (limited direct solids)
- *Size:* Milligrams of solid using direct solids and graphite furnace atomic absorption spectroscopy (GFAAS) to hundreds of microliters using GFAAS to milliliters using flame atomic absorption spectroscopy (AAS)
- *Preparation:* Solid typically requires digestion, but aqueous samples require only minimal preparation. Depends on the type of atomizer used; usually a solution must be prepared

Limitations

- Dynamic range typically limited to 1 to 2 orders of magnitude above the limit of quantitation (approximately a factor of 50)
- Typically requires a line source constructed using the element (s) of interest (i.e., a hollow cathode lamp or an electrodeless discharge lamp)
- Poorer sensitivity for elements forming refractory compounds (e.g., oxides)
- Limited to single-element detection (other than a few notable exceptions)
- Susceptible to sample-matrix interferences
- Use of nebulizer for sample introduction with flame AAS results in loss of 95% of sample solution

Estimated Analysis Time

- Sample preparation is typically the most time-consuming:
  - Can require minutes to hours
  - Significant source of contamination and interferences
- Typical sample-analysis times range from less than 1 min (flame AAS) to several minutes (GFAAS)

Capabilities of Related Techniques

- Inductively coupled plasma atomic emission spectrometry and inductively coupled plasma mass spectrometry are capable of simultaneous multielement determinations but require large volumes of sample solution (1 to 10 mL). While they enable determinations over large dynamic ranges (typically 6 orders of magnitude), they often incorporate a nebulizer/spray chamber introduction system with efficiencies of only 1%. Additionally, they are considerably more expensive to acquire and operate (with argon consumption rates of nearly 15 L/min^{-1}).

Introduction

Since the initial landmark papers by Walsh and Alkemade (Ref 1, 2), atomic absorption has steadily evolved into a frequently used analytical tool for the determination of metals (Ref 3). From drinking water samples to complex biological samples, atomic absorption spectroscopy (AAS) has become the workhorse analytical technique for metal-concentration determination (Ref 4, 5). While the procedures have evolved to meet ever-changing analytical challenges, the fundamental processes described in initial papers have remained. Therefore, any thorough discussion of AAS must begin with the basics.

Inherent to all forms of analytical atomic spectroscopy is the necessity to produce atomic vapors from condensed phase samples (i.e., solids or solutions). This is accomplished through a series of processes depicted in Fig. 1:

1. Sample solutions must have the solvent (typically water) removed by desolvation.
2. The resulting solid material is vaporized.
3. The molecular species are decomposed to yield the atomic vapor.
4. Because of the high temperatures required to accomplish these processes, the first ionization potential of many elements results in gaseous ion formation.

Therefore, while the first three processes are required for AAS determinations, the fourth must be considered as a potentially interfering process (Ref 3).

Initial efforts in AAS development used high-temperature flames as the preferred energetic environment (Ref 1). Conversion of aqueous metal ion solutions to gaseous atoms involved generation of an aerosol, with subsequent injection of those droplets into a flame. Early efforts used a direct-injection nebulizer-burner configuration in which the solution aerosol was generated by using the oxidant gas immediately prior to the flame. While this enabled quantitative introduction of the sample solution to a high-temperature flame (e.g., oxy-acetylene), a broad distribution of droplet sizes containing varied amounts of the metal analyte resulted in excessive measurement variations (Ref 3). This negatively impacted achievable limits of detection. Additionally, the geometry of these flames yielded effective path lengths for absorption measurements to only a few millimeters.

The measured absorbance is proportional to the number of absorbing species (i.e., gaseous atoms) within the optical viewing volume (Ref 1, 6). Therefore, the absorbance is related to both the path length of the incident light within the metal atom source (e.g., a flame or graphite furnace) and the residence time of each metal atom within the probed volume. The longer a metal atom resides within the incident light beam and the greater the number of atoms, the higher the probability of absorbing the incident radiation. The Beer-Lambert relationship is written as:

$$\text{Abs} = a\,b\,n\,t_{\text{res}} = -\log\,(I/I_0) \qquad \text{(Eq 1)}$$

where the absorbance (Abs) is proportional to the optical path length (b), average residence time of each atom within the optical volume (t_{res}), and the number of metal atoms in that volume (n). This is related to the sample solution metal concentration. The factor a includes parameters such as the efficiency of solution aerosol introduction to the flame, the probability of a metal atom absorbing light at that wavelength, and atom production (Fig. 1) (Ref 1). As measured, absorbance is logarithmically related to the ratio of the incident (I_0) and transmitted (I) light intensities, which limits the range of usable measurements (Ref 3).

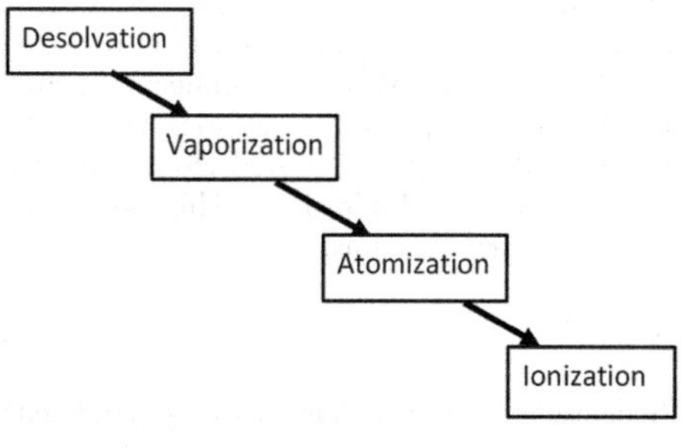

Fig. 1 Fundamental processes occurring within atomizers for atomic absorption measurements

Obtaining the maximum sensitivity requires optimization of each parameter, that is, maximum production of atoms for observation with the optical volume. Narrowing the analyte atom distribution within each aerosol droplet was required to provide a more uniform time-dependent atomic absorption signal. To achieve this result, an alternate approach to sample introduction than the direct-insertion technique was developed, that is, a spray chamber. This enabled introduction of only the smallest droplets to the flame. By reducing the distribution of droplet volumes introduced to the flame, variation in the measured signal resulting from their desolvation, vaporization, and atomization (i.e., noise) was greatly reduced. This configuration also enabled the use of a flame with a longer optical path length (i.e., several centimeters). Figure 2 shows an example of this configuration using the oxidant gas (e.g., air or NO_2) for aerosol generation and mixed with the fuel (e.g., acetylene) within the spray chamber. Unfortunately, this configuration enabled propagation of the flame into the chamber containing a mixture of both fuel and oxidant, with explosive results. This limited the choices of fuels and oxidant that could be used, to avoid an explosive flashback into the spray chamber. This resulted in cooler flames with lower vaporization and atomization efficiencies, that is, lower absorbance measurements (Ref 3).

Flame Atomic Absorption

Use of a nebulizer/spray chamber configuration with a slot burner (Fig. 2) yielded increased signal-to-noise by more uniform droplet size introduction and a longer path length (Ref 3). However, it exhibited very poor sampling efficiencies (Ref 7). Only 5% of the nebulized sample solution is introduced to the flame and atomic vapor generation (Ref 7). Because sufficiently small droplets are able to traverse the tortuous path with the spray chamber without collision with a surface, the remaining sample solution is removed as waste.

In addition to an atomic vapor, critical components of atomic absorption instrumentation are the sources of incident radiation, isolation of used wavelengths of light, and detection of the incident and transmitted intensities of that light (Fig. 3). Because gas-phase atoms exhibit absorption line spectra, it was critical for the development of AAS that the effective spectral bandwidth of the instrumentation be comparable (Ref 7). This was accomplished by using atomic line emission sources such as hollow cathode lamps (Ref 1, 8) for metals and electrodeless discharge lamps (Ref 9) for more volatile elements (e.g., metalloids and some nonmetals). The construction of these lamps using a single element results in limitation of their respective emission spectra to only that of the element of interest and the associated fill gas (i.e., neon or argon). Consequently, subsequent wavelength-isolation components (monochromator or interference filter) were only required to isolate the selected resonance transition wavelength. This is contrasted with high-resolution instrumentation required to select specific emission wavelengths by using atomic emission spectroscopy (Ref 1).

Early atomic absorption spectrometers were single-beam designs that required separate

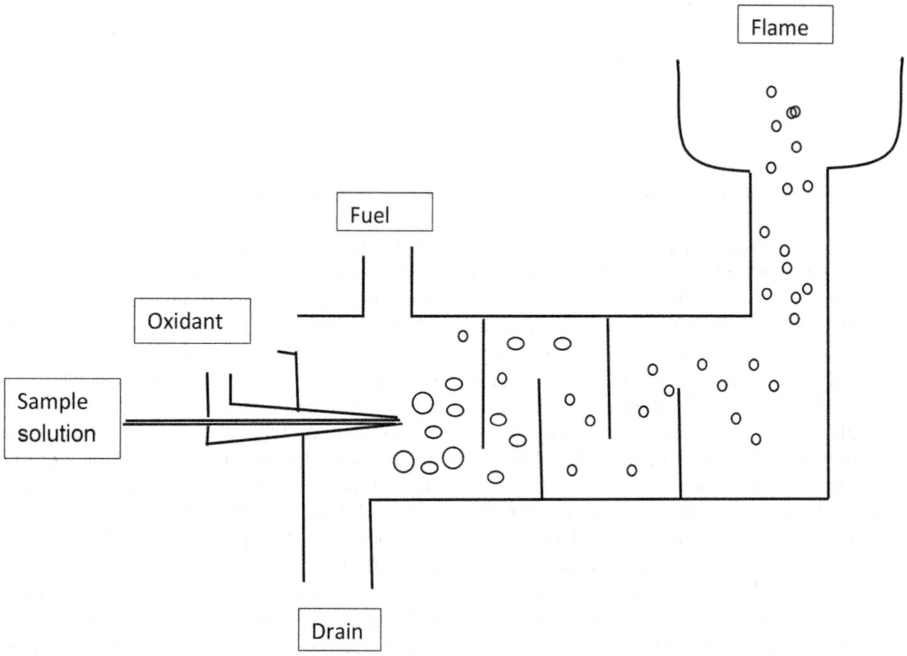

Fig. 2 Schematic representation of a nebulizer/spray chamber assembly used in flame atomic absorption

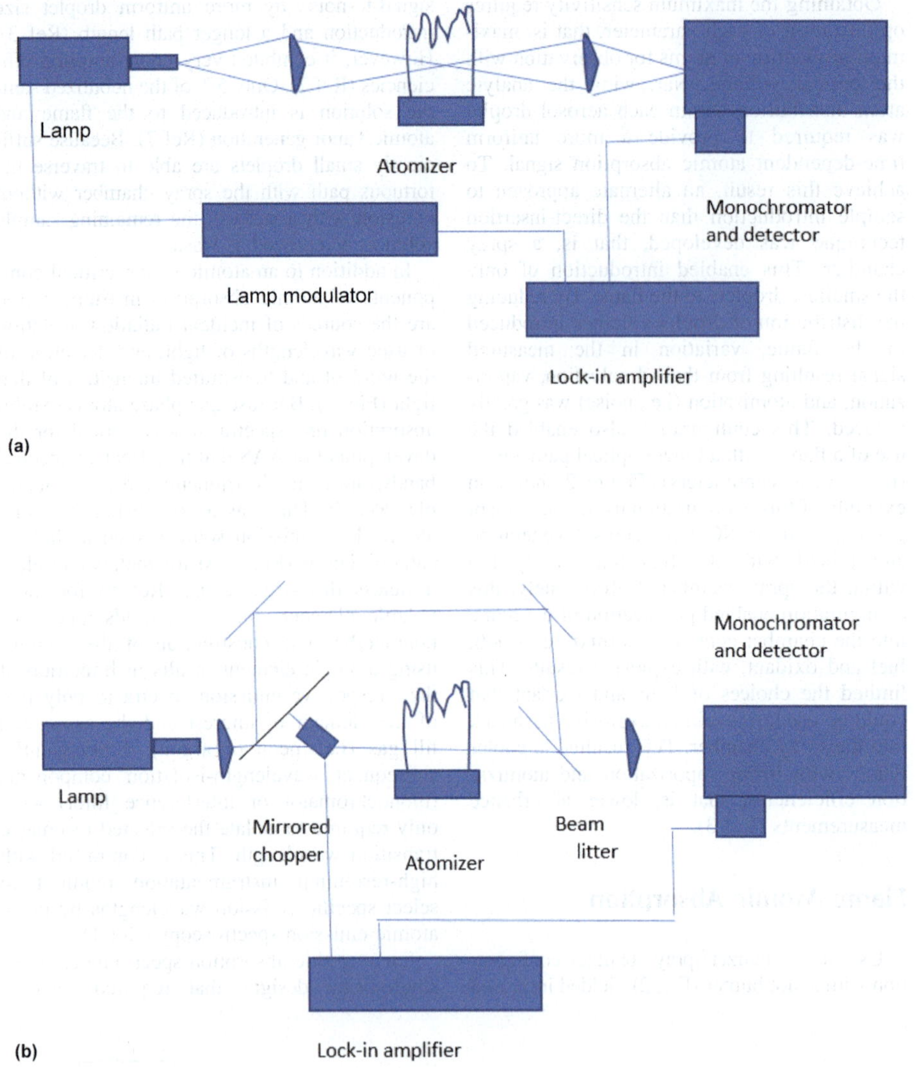

Fig. 3 Schematic representations of (a) single-beam and (b) double-beam atomic absorption spectrometers. Source: Ref 7

measurements of the incident light intensity, typically during aspiration of a blank solution, and the transmitted radiation intensity for each sample or standard solution (Fig. 3a). This configuration was sufficient but required a modulated light source and detection electronics (Ref 10). Incorporation of a mirrored chopper enabled a reference beam to be diverted around the flame and alternatingly directed through it (Fig. 3b). This also enabled the incorporation of a lock-in amplifier for signal-to-noise enhancement (Ref 3).

As indicated previously, AAS requires the analyte to be converted to gas-phase atoms (Fig. 1). Fundamentally, this involves each of three sequential processes with a potentially interfering fourth. Specifically, a sample solution must undergo desolvation, vaporization of any molecular species, and generation of the atomic vapor, that is, atomization (Ref 11). In the presence of sufficient energy to enable these processes, production of the singly-charged ion is possible. Successful conversion of an analyte to free atoms involves the realization of each of the first three processes while minimizing the fourth. Any conditions or sample-matrix concomitants that interfere with any of these processes will result in an erroneous determination.

Desolvation typically occurs within the flame upon introduction of the sample aerosol (Ref 1). In a classic paper by Hieftje and Malmstad (Ref 11), the dependence on droplet size was measured through controlled introduction of uniform droplets into a laminar flow air-acetylene flame (Ref 12). Using this same technique, the impact of flame parameters and sample composition on each of these processes was determined (Ref 12). Through this increased understanding of these fundamental processes, significant improvements in analytical performance of metal determination by using flame AAS were realized (Ref 12).

Vaporization Interferences

Determination of metal concentrations in sample matrices that are prone to the formation of more refractory metal salts is an example of the application of vaporization and atomization mechanisms within a high-temperature flame. Specifically, the determination of calcium in sample solutions containing large amounts of sulfates, phosphates, or aluminates can be problematic for flame AAS. On desolvation of the individual droplets, solubility equilibria will favor the formation of the respective metal salts (i.e., $CaSO_4$, $Ca_3(PO_4)_2$, or $CaAl_2O_4$). Because of the relatively strong ionic bonding in each salt, decomposition within the flame is incomplete within the typical viewing region (Ref 13). The diminished atomization efficiency then results in an erroneous low-calcium determination within the sample solution (Ref 14). Knowing the potential formation of these problematic species, an analytical method can be devised to minimize or eliminate these interfering interactions during the desolvation process. One approach is the addition of another metal ion, such as lanthanum(III) at sufficiently high concentration (e.g., 1% by mass) (Ref 15). This enables formation of the respective lanthanum salts, thus allowing the calcium to form either the hydroxide or oxide, which is more efficiently atomized to yield gaseous calcium atoms. Another approach is to add a strong chelator (e.g., ethylene diamine tetra acetic acid, or EDTA) to the solution. The chelator then binds to the calcium ions in solution (Ref 16), preventing subsequent formation of the refractory salts upon desolvation. The resulting CaEDTA containing salt is then easily decomposed within the high-temperature flame, which yields the calcium atomic vapor (Ref 16).

Ionization Interferences

As indicated in Fig. 1, the presence of sufficient energy to generate an atomic vapor may also promote the generation of the singly-charged ion of that species (i.e., the energy available exceeds the first ionization potential for the element). When a significant fraction of the metal vapor exists as ions within the optical volume, the measured atomic absorption is greatly diminished. This ionization reaction is described as:

$$M_{(g)} \rightleftharpoons M^{+}_{(g)} + e^{-} \quad \text{(Eq 2)}$$

Because the number densities of ground-state metal atoms ($M_{(g)}$), the corresponding ion, ($M^{+}_{(g)}$), and free electrons (e^{-}) are governed by an equilibrium condition, the relative values can be controlled by using Le Châtelier's principle. By introducing more free electrons, this relation can be shifted to favor the gas-phase atoms, yielding a larger atomic absorbance signal. This can be accomplished through the concurrent introduction of larger amounts of another element that has a low first ionization potential, for example, an alkali metal. An example of this approach is in the determination of magnesium in water. Generation of a calibration curve using standard magnesium solutions in pure water will not enable the accurate determination of magnesium in samples containing varied mounts of sodium and potassium, both easily ionized elements, because of this ionization interference. If, however, the standards and all samples were prepared with 1000 ppm cesium, the number density of free electrons from the cesium ionization would dominate all other, sample-dependent, sources of electrons; this would maximize the number density of atomic magnesium within the flame.

Flame AAS can achieve low limits of detection for many metals in aqueous samples (Ref 3). Unfortunately, the poor efficiency of nebulizer/spray chamber configurations (i.e., ~5%) becomes problematic for detection of low concentrations in limited sample volumes (e.g., sub parts per million metals in biological fluids) (Ref 14). This limits attainable detection limits for many metals (Ref 3). Both of these challenges were addressed in landmark papers by L'vov (Ref 17, 18) describing metal atom vapor generation using a graphite tube furnace atomizer and microliter volumes of a sample solution (Ref 19). This innovation enabled detection of extremely small amounts of metals and the attainment of lower detection limits (Table 1) (Ref 3).

While the incorporation of the graphite tube furnace for metal atom vapor generation offers several advantages compared to flame AAS, increased sample-dependent matrix interferences have been observed with its use (Ref 20). This is made possible by its ability to segregate many of the required processes shown in Fig. 1. By application of controlled heating of the graphite furnace atomizer (GFA), total desolvation (drying) of the sample can be achieved prior to vaporization of many sample-matrix concomitants during a thermal pretreatment cycle. Finally, the furnace can be rapidly heated at >1500 °C/s (>2730 °F/s) atomization; this is schematically represented in Fig. 4.

Deposition of discrete sample volumes into the GFA enables the analysis of microliter-sized samples with 100% transfer efficiency (i.e., no losses to waste, as realized by using a nebulizer/spray chamber configuration). Additionally, the GFA can be slowly heated for total solvent removal (i.e., desolvation) and then heated to a controlled temperature to enable, ideally, vaporization of the sample matrix (Ref 21). All of this is accomplished prior to the rapid heating of the GFA

Table 1 Reported detection limits for selected elements using flame and graphite furnace atomization

Element	Flame atomic absorption spectroscopy	Graphite furnace atomic absorption spectroscopy(a)
Ag	3	0.01
Al	30	0.1
As	200	0.5
Au	20	0.5
B	2000	...
Ba	20	0.25
Be	2	0.03
Bi	30	0.1
Ca	1	0.25
Cd	1	0.01
Co	4	0.3
Cr	4	0.03
Cu	2	0.05
Fe	6	0.25
Hg	500	5
Mg	0.2	0.002
Mn	2	0.01
Mo	5	0.5
Ni	3	0.5
Pb	8	0.1
Sn	15	5
V	25	1
Zn	1	0.005

Note: All concentrations in units of ng/mL^{-1} (ppb). (a) Sample volume = 10 μL. Source: Ref 3

(~1500 °C/s) to yield the atomic vapor cloud that can be retained within the optical path for 1 to 3 s rather than the 10 to 20 ms realized with a flame (Ref 19). This enables the realization of very low limits of detection for most metals (Table 1) (Ref 3).

Unfortunately, this sample-increased residence time of the atomic vapor within the GFA also enables the increased opportunity for gas-phase reactions with covaporized species, resulting in formation of molecular species, which are not detected by using AAS (Ref 21, 22). An example of such a situation is the determination of tin in a sample with significant amounts of sodium sulfate (Ref 23). During the atomization of the metal, the sulfate salt decomposes to yield SO_2 (Ref 24). Within the GFA, the metal atoms will react with the SO_2 to yield SnO vapor. Because the incident radiation from a tin hollow cathode lamp is wavelength-specific for atomic tin, it is not absorbed to the same extent by the metal oxide and is effectively removed from further detection. This results in an erroneously low value for the measured tin concentration in the sample solution (Ref 23). By introduction of a graphite platform within the GFA, which is heated more slowly than the tube furnace, the sample is vaporized into a higher-temperature environment (Fig. 5), thus inhibiting formation of the metal oxide (Ref 25).

With increased understanding of the mechanisms of atom vapor generation and potentially interfering subsequent reactions, the design of the GFA evolved from L'vov's initial graphite tube. Commercialization of the GFA as an alternate atomization source inspired each of two designs. These were the Massmann-

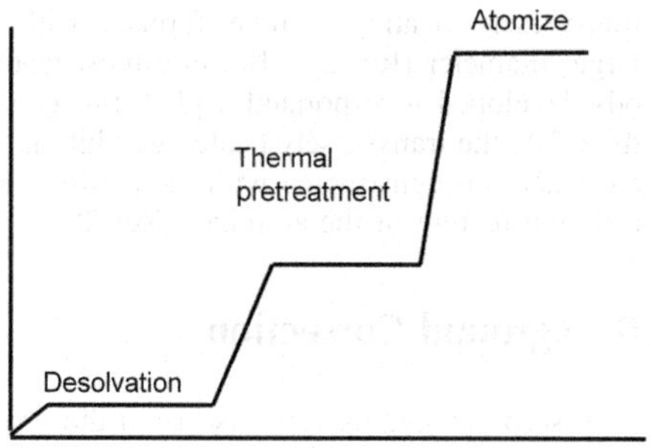

Fig. 4 Schematic representation of graphite furnace atomizer temperature-heating program

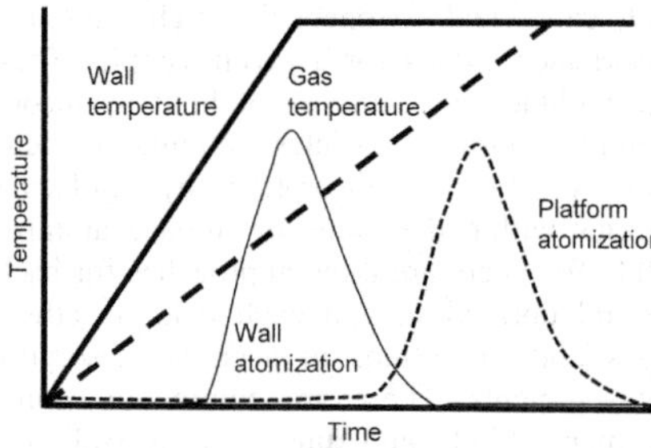

Fig. 5 Schematic representation of graphite furnace temperature and atomic absorption signal when the sample is deposited on the wall of the furnace (no platform) and when it is placed on a platform within the graphite furnace atomizer

type design and the carbon rod atomizer (CRA) (Ref 26, 27). Each of these involved the resistive heating of a tube of graphite. While the Massmann design involved passing current along the length of a 2.5 cm (1 in.) tube (i.e., parallel to the optical axis and the path of the incident light) with a 6 mm (0.2 in.) inside diameter, the CRA design incorporated a smaller tube furnace (1.0 cm, or 0.4 in., in length with a 3 mm, or 0.1 in., diameter) with current flow directed perpendicular to the light path. Because of the high temperatures involved during the atomization stage of the furnace-heating program, connecting electrodes were water cooled. This resulted in a temperature gradient along the Massmann furnace and cooler ends. Consequently, the atomic vapor could condense at the ends of the furnace and generate error in that and subsequent measurements (Ref 27). While the CRA did not exhibit as dramatic a temperature gradient by heating with electrodes orthogonal to the optical axis, its shorter length and smaller volume yielded shorter transient absorption signals (i.e., higher limits of detection and lower sensitivities). Additionally, the smaller inside diameter of the CRA prohibited the insertion of a platform. Therefore, the clear majority of GFA instruments initially used the Massmann furnace design.

To minimize the negative impact of the cooler ends with a longitudinally heated tube furnace, an alternate design provided for transversely heating a tube furnace with a larger diameter (Ref 28). Because most methods developed incorporated a platform within the GFA, the transversely heated graphite tube atomizer was constructed with the platform as a design feature of the atomizer (Ref 28).

Background Correction

Absorption signals can arise from the presence of anything within the optical path that attenuates the incident light. While the assumption is that this will be dominated by its absorption by gas-phase atoms, there can be numerous non-analyte-related processes that can yield erroneously high measured absorbance values for a given sample. These can include the scattering of light by incompletely vaporized particles or remaining droplets or by its absorption by molecular species within the flame or furnace atomizer (Ref 29). There are three approaches for background correction, and each exhibits advantages and limitations to accurate absorption measurements. These are continuum source, Zeeman effect, and line self-reversal (i.e., Smith-Hieftje background correction).

Continuum-Source Correction

The continuum-source correction method was first described by Koirtyohann and Pickett (Fig. 6) (Ref 30) and involves an alternate light source emitting over a large range of wavelengths (Ref 30). These are typically low-pressure arc discharges (H_2 or D_2) for ultraviolet wavelengths and a tungsten-filament source for visible (or near-infrared) wavelengths. Light from a line source is alternatingly directed through the flame or furnace with that from the continuum source. The bandpass of the monochromator is adjusted to either 0.5 or 1.0 nm. Consequently, radiation across this bandwidth illuminates the detector during the continuum-source portion of the measurement cycle, while only the very narrow wavelength range corresponding to a single atomic transition (<0.005 nm, Ref 1) reaches the detector during the remaining times. During this latter half of a measurement cycle, the indecent light can be attenuated by both gas-phase metal atoms and any other scattering or absorbing species (i.e., background) at only the atomic transition wavelength. Conversely, the continuum radiation is attenuated by background species across the entire bandpass and diminished by the analyte at only a small fraction of the detected range. Because the contribution of the analyte atoms to the measured signal is small, the calculated absorbance during the continuum-source measurements can be considered as only resulting from background species. Simple subtraction of the two measurements results in a background-corrected AAS signal.

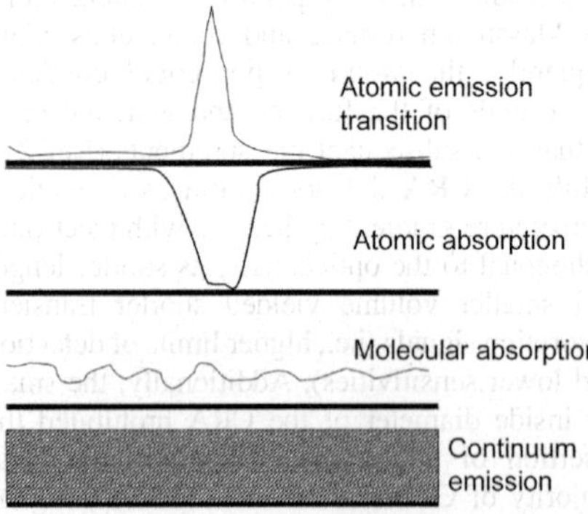

Fig. 6 Diagrammatic representation of signals associated with continuum background correction (spectral features are not to scale)

Because the light sources are alternatingly directed through the flame or furnace using a mirrored optical chopper, corrected signals can be recorded at frequencies dictated by chopper rotation rate divided by the number of blades. Hence, background and analyte/background measurements can be made nearly simultaneously at moderately high frequencies (Ref 30). A limitation of this approach is the assumption that background species attenuate the light uniformly across the spectral bandpass of the monochromator (Ref 19). While scattering species (e.g., soot particles) meet this requirement, some molecular concomitants do not and can result in overcorrection of the absorbance measurement (Ref 3). Another limitation of this is the need for a separate power supply for the continuum source and the need for separate sources for the ultraviolet and visible regions of the spectrum as well as the added complication required for its implementation (Ref 29, 30).

Zeeman Correction

When metal atoms are placed within a strong magnetic field, normally energy-degenerate transitions can be differentiated (Ref 31). Specifically, when the applied field is orthogonal to the optical axis, it is split into three components: σ_-, π, and σ_+, which differ in both energy and the plane of light polarization (the Voigt effect) (Ref 3, 9). While the π component interacts with light polarized parallel to the applied magnetic field at the normal transition wavelength, the two σ components absorb light that is circularly polarized in opposite directions at wavelengths symmetrically shifted from the normal transition wavelength to higher and lower energies, respectively. Therefore, by selecting only light perpendicularly polarized relative to an applied alternating current magnetic field, measurements can be made at both maximum field strength (i.e., greatest separation of the σ components) and zero field (zero splitting). Under conditions of maximum field, the measured absorbance results from only background species. Conversely, with no applied field, the measurement results from both the metal atoms and background (Fig. 7). Again, simple subtraction of these two measurements provides the corrected signal.

A limitation of Zeeman-effect background correction using either effect is the inability to modulate the required high magnetic fields (~1 to 1.5 T, Ref 19) at greater than 60 Hz, corresponding to typical alternating current line frequency applied to an electromagnet (Ref 3, 9). Although each of two sets of field on/field off measurements can be made during each line current cycle, this limits the time difference between when the signal and background measurements are made (Ref 3). This becomes especially problematic when applied to the transient species generated within a GFA (Ref 19). Additionally, Zeeman correction can yield lower corrected absorbances at higher analytic concentration, resulting in double-valued calibration curves (i.e., two different concentrations yield the same corrected absorbance value) (Ref 19). Another limitation is the need for an additional power supply for the magnet (Ref 9).

Line Self-Reversal Correction (Smith-Hieftje)

To decrease the time difference between analyte and background measurements (as with the Zeeman correction), to simplify the optical systems to a single light source, and to provide background correction at the analytical wavelength, the line self-reversal background-correction system was developed by Smith and Hieftje (Ref 32). In this approach, a single hollow cathode lamp is operated in each of two modes. The first is by using the typically low operating currents, providing narrow emission lines for the corresponding metal (Fig. 8). The second is by using much higher sputtering currents. This results in both Doppler broadening of the emission lines and generation of a cloud of the metal atoms in front of the hollow cathode. Because these ejected atoms will be a significantly lower temperature than those within the cathode, their absorption line width is considerably less (Fig. 8a). This results in attenuation of those central wavelengths from the emitted radiation and thus line self-reversal (Ref 32).

As this broadened emission line passes through the atomizer (flame or furnace), the central portion can be absorbed by the analyte atoms, while most of the light is only attenuated by background species. The difference in these measurements (i.e., using lower or "normal" lamp current and applying higher operating sputtering current) provides a background-corrected measurement "near" the analytical wavelength (Fig. 8b). This also enables background measurements to be made only a few milliseconds after the analyte-plus-

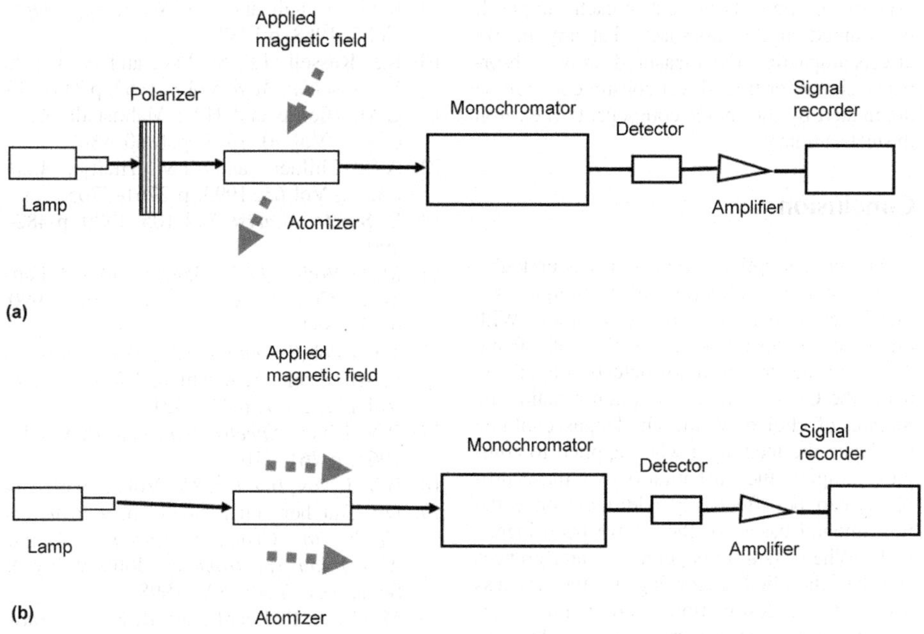

Fig. 7 Schematic representation of Zeeman background configurations using the (a) Voigt and (b) Faraday effects

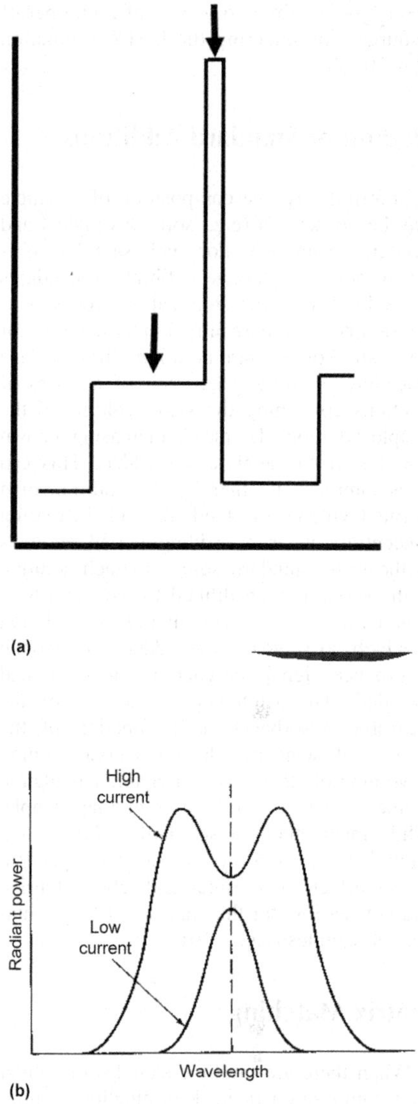

Fig. 8 (a) Schematic representation of the lamp current as a function of time used with the line self-reversal background-correction scheme. The arrows indicate approximate sampling times for the low- and high-current signals. (b) Representation of the corresponding lamp emission signals

background measurement. The limitation of this technique results from the time required for the lamp to recover from its operation at high current. This reduces the frequency of measurements, which can be problematic for GFAs that generate transient analyte and background species within the optically sampled volume. Advantages include the need for a single light source and no separate power supply. Additionally, background measurement more closely approximates those species present at the time of the analyte absorption measurement.

Methods of Analysis

As with any spectrochemical method of analysis, the amount of an analyte in a sample is the objective. The measured response, for example, absorbance, must be related to the concentration of analyte within a sample solution and ultimately within the initial sample. To accomplish the conversion of measured absorbance to a concentration requires the generation of a calibration or growth curve. As indicted previously, the proportionality constant relating absorbance to concentration depends on several parameters that can be impacted by the sample matrix, that is, the total composition of the sample solution. To minimize the impact of sample concomitants on the analytical signal, that is, the attenuation of light by analyte gas-phase atoms, several methods can be employed.

For sample solutions in which the diluent is the primary sample-matrix component, a calibration curve can be generated from the analyte metal salt, typically as the nitrate, dissolved in water or a diluted acid solution. As solutions of known and increasing concentration are each introduced to the atomizer and the resulting absorbance is recorded, the relationship between metal concentration and absorbance can be determined through the application of simple linear regression of those standards within the linear portion of the curve. For flame AAS, the steady-state absorbance is measured over a fixed period of time, and its average or integrated value is recorded. When a GFA is employed, the integrated area of the measured transient absorbance signal is recorded by using a constant volume of sample solution deposited into the furnace.

It is imperative that operating parameters for the absorbance measurement remain constant. Variations that could impact the number of gas-phase analyte metal atoms can significantly impact the validity of any calibration. The flame AAS parameters can include the fuel-to-oxidant ratio for the flame, the nebulizer efficiency, the rate of sample solution introduction to the nebulizer, the fuel and oxidant gas-flow rates, the observation height within the flame, and the flame composition. Similarly, parameters using graphite furnace atomic absorption spectroscopy include the volume of sample deposited, the temperature and duration of each stage of the furnace program (Fig. 4), the use of a platform, and the absorbance integration time. Each of these operating parameters can impact the efficiency of atomization or the residence time of the metal atoms within the observation volume. Additionally, any reagents added to address potential chemical interferences (e.g., the 1% La or EDTA solutions discussed previously) must be included for each standard solution used to generate the calibration curve.

Matrix Modifiers

Because of the potential myriad of chemical interactions that can occur within the GFA, it is typically necessary to adjust those reactions to maximize production of the gas-phase analyte atoms. While the permutations and combinations of added chemicals to address these sample-matrix-dependent interactions are large, each has the explicit purpose of enhancing the atomization efficiency of the analyte during the atomization stage.

The reader is directed to any of a number of resources for selecting the best combination (Ref 20, 33).

Method of Standard Additions

Unfortunately, the composition of a sample may be not known (e.g., soils, geological and biological samples). For such samples, it is not possible to generate calibration standards that exhibit the same atomization processes as the sample. These require implementation of the method of standard additions (Ref 9). This procedure involves generation of a series of solutions containing the same volume of the sample solution. To these, increasing known amounts of the analyte are added. This can be accomplished either by the addition of a constant volume of standards with increasing concentrations or by adding varied amounts of the same standard solution. Each solution in the series is then diluted to the same total volume. In each approach, there is one solution to which no standard was added. The atomic absorbance signal for each is measured and recorded. The x-intercept of a plot of the measured absorbance as a function of the amount of standard added (as concentration or volume of standard) will enable calculation of the analyte concentration in that sample. While the method of standard additions will minimize the impact of sample-dependent matrix effects, it is time- and labor-intensive and not feasible for the analysis of large numbers of samples (e.g., >10).

Matrix Matching

When there are multiple samples of similar total composition (e.g., determination of trace metals in an aluminum alloy), an alternate approach can be employed in which a series of standard solutions are generated containing the major components of the sample matrix. An example is the determination of trace metals within alloy samples. While the characteristics of an alloy can vary significantly dependent upon the presence and relative concentrations of trace elements (Ref 34), the major components within each sample are relatively constant (e.g., aluminum in an aluminum alloy or iron in steel samples). Therefore, standard solutions can be prepared that contain the quantities of primary components present in each sample. It is assumed in this approach that any matrix effects impacting the measured atomic absorbance signal for trace-level constituents will be dominated by the major components (i.e., iron in steel samples).

Conclusion

Atomic absorption spectroscopy is undoubtedly a mature collection of techniques for metal determination in many samples. With the ease of operation available with flame AAS and the low limits of detection available using the GFA, atomic absorption remains the method of choice. While challenges continue to plague the technique when applied to complex samples, the vast majority of these problems can be effectively addressed once the fundamental bases for them have been identified. When sample-dependent interferences can be classified according to the process involved (i.e., desolvation, vaporization, atomization, or ionization), methods can be efficiently modified to yield accurate results.

ACKNOWLEDGMENTS

This article was revised from Darryl D. Siemer, Atomic Absorption Spectrometry, *Materials Characterization*, Volume 10, *ASM Handbook*, ASM International, 1986, p 43–59.

REFERENCES

1. A. Walsh, *Spectrochim. Acta*, Vol 7, 1955, p 108
2. C.T.J. Alkemade and J.M.W. Milatz, *J. Opt. Soc. Am.*, Vol 45, 1955, p 583
3. J.D. Ingle and S.R. Crouch, *Spectrochemica Analysis*, Prentice-Hall, Englewood Cliffs, NJ, 1988
4. S.R. Koirtyohann, *Anal. Chem. A*, Vol 52, Vol 736, 1980, p 5
5. G.M. Hieftje, *J. Anal. At. Spectrom.*, Vol 4, 1989, p 117–122
6. K. Fuwa and B.L. Vallee, *Anal. Chem.*, Vol 35, 1963, p 942–946
7. D.A. Skoog, F.J. Holler, and S.R. Crouch, *Principles of Instrumental Analysis*, 6th ed., Thomson Brooks Cole, Australia, 2007
8. W.W. Harrison and N.J. Prakash, *Anal. Chim. Acta*, Vol 49, 1970, p 151
9. R.M. Dagnall and T.S. West, *Appl. Opt.*, Vol 7, 1968, p 1287
10. B.J. Russell, J.P. Shelton, and A. Walsh, *Spectrochim. Acta*, Vol 8, 1957, p 317–318
11. G.M. Hieftje and H.V. Malmstadt, *Anal. Chem.*, Vol 40, 1968, p 1860–1867
12. A.G. Childers and G.M. Hieftje, *Anal. Chem.*, Vol 65, 1993, p 2761–2765
13. B. Smets, *Analyst*, Vol 105, 1980, p 482–490
14. M.W. Welch, D.W. Hamar, and M.J. Fettman, *Clinical Chem.*, Vol 36, 1990, p 351–354
15. A.P. Udoh, *Talanta*, Vol 52, 2000, p 749–754
16. L. Alierf and A. Mashlah, *Microchem. J.*, Vol 132, 2017, p 411–421
17. B.V. L'vov, *Spectrochim. Acta B*, Vol 17, 1961, p 761–770
18. B.V. L'vov, *Ing. Fiz. Zh.*, Vol 2, 1959, p 44
19. D.J. Butcher and J. Sneddon, *A Practical Guide to Graphite Furnace Atomic Absorption Spectrometry*, John Wiley & Sons, New York, NY, 1998
20. M. Grotti, R. Leardi, and R. Frache, *Anal. Chim. Acta*, Vol 376, 1998, p 293–304
21. G.N. Brown and D.A. Styris, *J. Anal. At. Spectrom.*, Vol 8, 1993, p 211
22. S. Ratlift, *Anal. Chim. Acta*, Vol 333, 1996, p 285–293
23. G.D. Rayson and J.A. Holcombe, *Anal. Chim. Acta*, Vol 136, 1982, p 319–330
24. J.G. Jackson, R.W. Fonseca, and J.A. Holcombe, *Spectrochim. Acta B*, Vol 50, 1978, p 1449–1457
25. B.V. L'vov, *Spectrochim. Acta B*, Vol 33, 1978, p 153–193
26. H. Massmann, *Spectrochim. Acta B*, Vol 23, 1968, p 215
27. F.J.M.J. Maessen and F.D. Posma, *Anal. Chem.*, Vol 46, 1974, p 1439–1444
28. Z. Li, G. Carnrick, and W. Slavin, *Spectrochim. Acta B*, Vol 48, 1993, p 1435–1443
29. I. Wegsheider, G. Knapp, and H. Spitzy, *Z. Anal. Chem.*, Vol 283, 1977, p 9–14
30. S.R. Koirtyohann and E.E. Pickett, *Anal. Chem.*, Vol 37, 1965, p 601–603
31. H. Koizumi and K. Yasuda, *Spectrochim. Acta B*, Vol 31, 1976, p 237–255
32. S.B. Smith, Jr. and G.M. Hieftje, *Appl. Spectrosc.*, Vol 37, 1983, p 419–424
33. A.B. Volynskii, *J. Anal. Chem.*, Vol 58, 2003, p 905–921
34. K. Charleton, T. Buffie, and D.M. Goltz, *Talanta*, Vol 74, 2007, p 7–13

ASM Handbook, Volume 10, *Materials Characterization*
ASM Handbook Committee
DOI 10.31399/asm.hb.v10.a0006647

Inductively Coupled Plasma Optical Emission Spectroscopy

Oumaïma Gharbi, Sorbonne Université
Odile Hirsch, Patrick Chapon, and Alice Stankova, Horiba Jobin Yvon S.A.S.

Overview

General Uses

- Simultaneous multielement analysis
- Quantitative and qualitative measurements for more than 70 elements, with detection limits in the parts per billion (ppb, μg/L) range
- Determination of major, minor, and trace elementals

Examples of Applications

- Mineralogy
- Metals and alloys analysis
- Biology and medical applications
- Water quality control
- Process control
- Mass loss for corrosion-rate analysis

Samples

- *Form:* Liquids, gases, and solids; however, liquids are most adapted
- *Size:* Only 1 to 5 mL of solution is necessary, 10 to 500 mg of solids
- *Preparation:* Most samples are analyzed as solutions. Samples can be analyzed as-received, diluted, or preconcentrated if required. Ideally, solids should be dissolved prior to analysis, but gases may be analyzed directly. In general, samples should be in liquid form; however, modern inductively coupled plasma optical emission spectroscopy can analyze micrometer-sized solids suspended in slurries with the appropriate sample-introduction system.

Advantages

- Robust equipment
- Low matrix effect
- Excellent detection limits (ppb level)
- Fast analysis (one sample in < 1 min)
- Large dynamic range
- Can analyze up to 70 elements of the periodic table
- Only small amount of sample needed for analysis (1–3 mL)

Limitations

- Not suitable for noble gases analysis
- Detection limit does not go below the ppb level (inductively coupled plasma mass spectrometry would be more appropriate for parts per trillion, or ppt, level)

Estimated Analysis Time

- According to the sample nature, dissolution of solids during preparation may require several hours.
- Full-sequence calibration and analysis may require minutes to several hours, according to the number of samples to be run.

Capabilities of Related Techniques

- *Atomic absorption spectroscopy:* Suitable for single-element analysis; higher sensitivity for most elements, especially if using electrothermal atomization, but not as good for refractory elements. More limited dynamic range
- *Inductively coupled plasma mass spectrometry:* Multielement technique providing higher detection limits (ppt level). Ideal for samples with dilute matrices (water-based samples) and widely used in pharmaceutical or biological applications. More complex samples must be diluted to minimize the matrix effect.
- *Glow discharge optical emission spectroscopy:* Multielement technique. The main advantage is direct solid analysis without sample preparation prior to experiment. Can be used to analyze conductive and nonconductive samples

Introduction

In the context of materials science, elemental analysis is routinely used and offers considerable support in understanding the chemical nature of organic and inorganic materials. In fact, elemental analysis is used in a diverse range of applications, from research and development to quality-control laboratories. Although a multitude of elemental surface and bulk characterization techniques are available in the market (i.e., x-ray diffraction or energy-dispersive spectroscopy), the accurate and quantitative analysis of samples in solid or liquid form can be more challenging.

The development and use of spectroscopic methods for elemental detection was first introduced by Bunsen and Kirchoff in 1860 with the discovery of cesium and rubidium by using a spectroscope invented in 1859 (Ref 1). In 1934, the work of Lundergårdh demonstrated the use of a flame as an excitation source, combined with a prism as a dispersive system, to separate spectral lines, giving rise to the flame spectroscopy technique (Ref 2). However, it was only in the 1960s that the first inductively coupled plasma optical emission spectrometer (ICP-OES), operating at room temperature, was reported (Ref 3–6). More than a decade later, in 1974, the ICP-OES became commercially available (Ref 7). Thereafter, the potential was exploited by many researchers, and over the years, the performances (detection limits, spectral range) improved tremendously. Constant development of the ICP-OES technique resulted in the expansion of what is now considered an operational, high-performance analytical tool (Ref 8). Although multiple excitation sources (such as combustion flame or electrical arc) were used in past decades, the ICP technique is the most recent technology used in commercially available instruments.

This relatively cost-effective technique permits the quantitative analysis of approximately all the elements of the periodic table, except argon (the plasma source is argon based), with a parts per billion (ppb) detection-limit level (Ref 4).

This article aims to provide a clear but nonexhaustive description of the general principle of atomic emission, with a particular focus on instrumentation, and summarize the main characteristics of the ICP-OES technique. The general principle of atomic emission is presented as well as the instrument characteristics and their influence on the instrument performances. The advantages and drawbacks of this technique are discussed, and, finally, alternative techniques and examples of recent applications are presented.

Basic Atomic Theory

The modus operandi of atomic emission spectroscopy rests on the excitation-radiative/relaxing process of atoms, whereby the emission of characteristic light of excited atoms is produced and collected by an optical system (Ref 8–10). This light (photon) is in the form of energy packets and exhibits two important characteristics: a specific wavelength and an intensity, specific to the atom and its concentration (Ref 10).

Each atom is comprised of a nucleus surrounded by electrons. The nucleus is composed of protons (positive charge) and neutrons (neutral charge) and for neutral atoms, the total charge of the surrounding electrons (negative charge) is equal to the nucleus charge.

The excitation of atoms, caused by collisions with another particle (Ref 11), results in the promotion of an electron from the lowest (ground) energy level to a higher energy level. However as this energy state is not stable, the electron will rapidly go back to its ground level by emitting a photon, with an energy corresponding to the difference between the two energy levels. According to the laws of quantum mechanics, the electrons surrounding an atom can only attain a limited number of discrete energy levels. Therefore, each element is associated with a characteristic emission spectrum.

This energy difference between the ground state and the higher energy state is linked to the frequency of radiation according to Max Planck's equation:

$$\Delta E = h\nu \quad \text{(Eq 1)}$$

where ΔE is the difference between the higher and the lower energy level ($\Delta E = E_{\text{excited state}} - E_{\text{ground state}}$), h is Planck's constant (h = 3.336×10^{-11} s/cm), and ν is the radiation frequency. The characteristic wavelength can be determined through the relationship in Eq 2:

$$E = hc/\lambda \quad \text{(Eq 2)}$$

where c is the velocity of light. ($c = 2.997 \times 10^8$ m/s)

The energies of the photons emitted by excited atoms in spectroscopic sources, such as the ICP-OES, correspond to wavelengths in the ultraviolet and visible region of the electromagnetic spectrum.

The radiation is collected and dispersed by a spectrometer into characteristic component wavelengths, allowing the element by element analysis of the sample analyte. The intensity of the light at a specific wavelength (spectral line) increases linearly with the concentration of the atom.

Background Continuum and Electron Recombination

The radiation collected by the spectrometer stems from the atoms of interest in the analyte but also, in the case of sample analysis, the matrix (the electrolyte) and the species (gas) present in the ICP source. The ICP-OES background, also known as background continuum, corresponds to the spectra of all the species present in the chamber (argon, oxygen, etc.). The background continuum is caused by the recombination of ions and electrons in the plasma.

Interference Effects

Interference effects arise from the presence of multiple components in the sample and are defined as phenomena that interfere with the linear intensity versus concentration relationship. Interference effects, sometimes referred to as matrix effects, may be classified as:

- Chemical interferences from low-atomization-efficiency compounds
- Physical interferences (change in electrolyte viscosity)
- Spectral interferences or line interference (superposition of spectral lines)
- Ionization interferences (very common for alkali metals)

Spectral/Line Interference

Spectral/line interferences are the consequence of an incomplete isolation of spectral lines emitted by the analyte from spectral lines emitted by other present species. In reality, analyzed samples often contain more than one element, resulting in complex spectra (Ref 12). In addition, several transitions from multiple excited states of a single atom can occur during a single analysis, increasing the probability of spectral line overlap (partial or complete). In addition to sample complexity, if the detector resolution is too low, two lines can be too close or even partially overlap. Spectral interference will stem from the presence of such close or coincident lines not resolved by the detector, resulting in an erroneous concentration level for a given element. Spectral interferences remain a fundamental problem in emission spectroscopy and will always occur during ICP-OES analysis. Although partial overlaps may be reduced with the use of high-resolution spectrometers, direct spectral overlap necessitates the selection of alternative analytical wavelengths (most elements have more than one analytical wavelength) (Ref 8). For this reason, careful wavelength selection (using line coincidence tables available in the literature, Ref 13–15) for each element is important.

Fortunately, current software packages are equipped with predetermined interelement correction factors, which are able to compensate for partial spectral overlap.

In addition, repetitive scans of the analyte and suspected interferents at the wavelength of interest can be graphically overlaid and inspected for potential problems. Another type of spectral interference is continuum overlap caused by electron recombination continuum that is part of the ICP-OES background emission. This also can be corrected by background subtraction from the analytical signal (Ref 16).

Ionization Interference

Ionization interferences are related to an increase of the emission intensity of an element with a low ionization potential (Ref 17). Typically, alkali metals, sodium or potassium, are affected by ionization interferences (Ref 18), particularly when another readily ionizable element is present in the sample as well. This problem is observed in most ICP-OES systems, particularly when the emission observation is realized in axial view, and instrument manufacturers still are investigating methods to reduce such interferences during analysis. Nonetheless, several studies have presented promising results whereby the addition of a low-ionization-potential element at high concentration in the sample significantly reduced such interference (Ref 17, 19).

Self-Absorption Effects

In some cases, when high concentration of one atom is present in the plasma, the reabsorption (by another atom of the same element) of the emitted radiation may occur. The emission line profiles observed under such conditions are flatter and broader than those observed in the absence of self-absorption. In practical analytical work, this leads to nonlinear calibration functions; that is, the analytical sensitivity is reduced at high mass fraction of the analyte (see the section "Analytical Procedure" in this article for information about calibration).

Principles of Operation

The ICP-OES instrument generally is comprised of an electrolyte-introduction system (nebulizer and spray chamber), a plasma chamber containing a torch and argon gas supplies, a radio-frequency generator and the associated electronics, and the spectrometer involving the detectors for the optical system (i.e., polychromator and monochromator equipped with phototubes) (Fig. 1).

Plasma Generation and Radio Frequency

The plasma consists of an internal gas, essentially argon, operating at room temperature. The plasma is generated in a so-called torch, composed of two concentric quartz tubes—one large and a second, called the auxiliary tube, inserted inside—and an internal injector. Most injectors are composed of quartz/alumina; however, for hydrofluoric-acid-containing electrolytes, quartz injectors are replaced by alumina injectors. Between the outer tube and the auxiliary tube, the walls are cooled by an auxiliary argon gas flow, because the high plasma temperature (7000–8000 K) can melt the quartz tubes. This tangential flow with a rate of approximately 15 mL/min of gas is used to center the plasma, stabilize it, and isolate it from the torch (Ref 8).

To ignite the plasma, a short pulse of a very high voltage is required. This pulse is provided by a Tesla coil, which produces high-voltage, low-current, and high-frequency alternating current density. The brief Tesla discharge must be generated to supply the argon gas with electrons. Once the plasma is ignited, it is maintained by a water-cooled induction coil at a radio-frequency (RF) electromagnetic field (27 or 40 MHz). The high-frequency current of up to 100 A flows in the water-cooled copper induction coils, generates oscillating magnetic fields oriented axially inside the quartz tube, and follows closed elliptical paths outside the tubes. The magnetic field forces the electrons in the argon gas to flow in oscillating closed annular paths within the internal quartz tube. The coil alternates the current, modifying the magnetic and electric field, and therefore increasing the collision between the argon gas and the accelerated electrons (Ref 20). These electrical and magnetic fields are responsible for the electron flow (often termed the eddy current) and the plasma, as represented in Fig. 2 and 3.

As the analyte is introduced in the center of the plasma, the atoms will undergo desolvation, atomization, ionization, and finally excitation, resulting in the optical emission of element-specific wavelengths varying from the

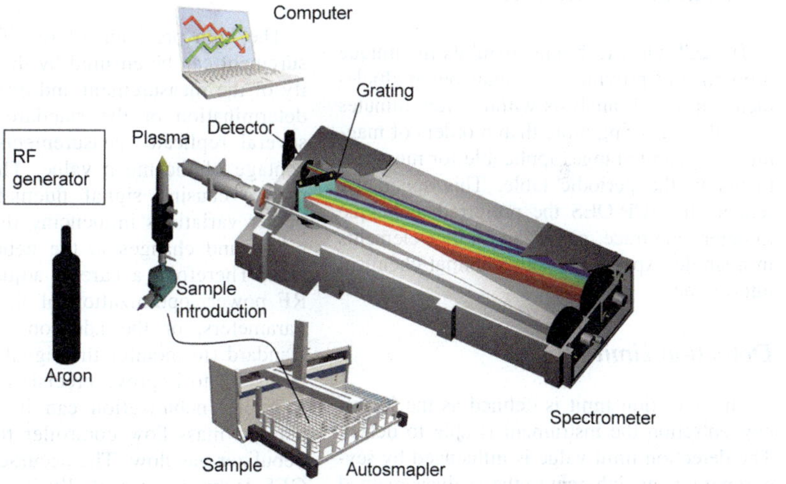

Fig. 1 Schematic diagram illustrating the various components constituting the modern inductively coupled plasma optical emission spectrometer (ICP-OES) system. The system includes the sample-introduction system, the ICP-OES torch and argon gas supplies, the radio-frequency (RF) generator and associated electronics, the spectrometer with the detection electronics, and the system computer with appropriate hardware and software.

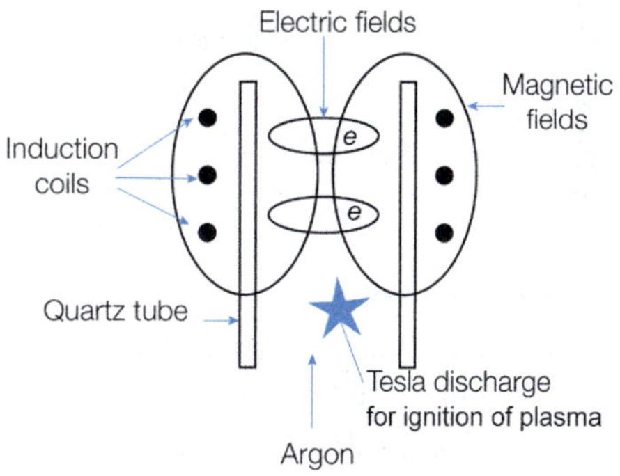

Fig. 2 Schematic diagram illustrating the plasma-generation process. Radio-frequency power is applied to the induction coils surrounding the torch (or quartz tube), generating electric and magnetic fields. An argon flow is carried at the center of the torch, where a Tesla discharge will ignite the plasma.

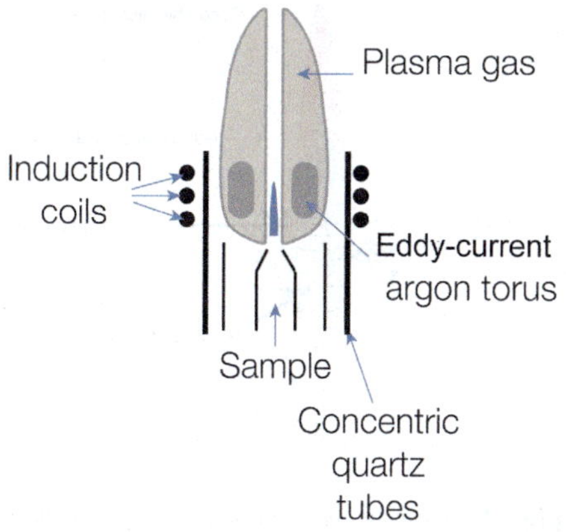

Fig. 3 General structure of the inductively coupled plasma optical emission spectrometer plasma torch during analysis. The electron flow (eddy current) is maintained by the electric and magnetic fields initiated by the induction coils. The plasma is retained within two concentric quartz tubes, and the sample is directly injected at the core of the plasma flame.

ultraviolet to the visible as the electrons return to a lower energy level (Ref 8, 9). Plasma observation can be realized via radial or axial observations, the latter providing a higher sensitivity. The atomic emission is present mostly at the lower section of the plasma. The ionic emission is largely present at the upper (vertical) section, usually called the analytical zone (Ref 21). The analytical zone is used for spectroscopic measurements. Finally, in the upper part of the plasma, above the induction coils, the atomic and some molecular emissions are observed (Fig. 4) (Ref 22).

Nevertheless, spectral interferences can be more important in axial observation due to an increase in background and analyte emission detection (Ref 21). Modern ICP-OES systems also can provide dual view.

Analytical Characteristics

The ICP-OES technique exhibits the unique properties of providing simultaneous multielement (up to 70) analysis within a few minutes as well as ensuring more than 6 orders of magnitude dynamic range, applicable for most elements in the periodic table. This specificity renders the ICP-OES the preferred technique to determine trace, major, and minor elements in a single experiment with minimal chemical interferences.

Detection Limit

The detection limit is defined as the lowest concentration the instrument is able to detect. The detection limit value is influenced by several parameters inherent to the analysis method and varies between instruments. The detection limit is determined by calculating the standard deviation the standard deviation of the noise in the background of each element (σ) and is defined as 2σ, expressed in units of concentration (Ref 23). To determine this value, a calibration curve must be established for each element considered. In practice, the concentration of the elements to be analyzed must be within the calibration range, and a standard concentration should be chosen accordingly, as illustrated in Fig. 5. Some sample-preparation procedures may diminish the achievable detection limits of the elements in the original sample material (such as dissolution and dilution) and must be taken into consideration. The ICP-OES technology offers a 6 orders of magnitude dynamic range (from µg/L to g/L) and relatively good detection limits for the majority of the elements, with a precision and accuracy of approximately 1% (approximately the parts per billion level) (Ref 8, 9, 24).

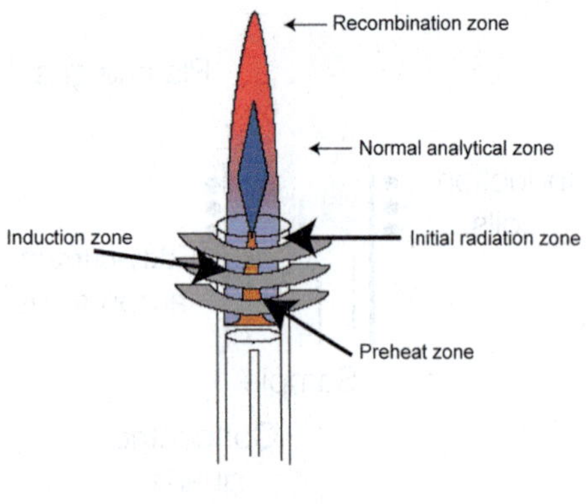

Fig. 4 Schematic representation of the plasma flame structure showing the various analytical areas. The atomization process mainly occurs in the preheating zone. The initial radiation zone covers predominantly the excitation and ionization, and the normal analytical zone comprises the emission collected by the spectrometer. The upper part of the plasma corresponds with the recombination zone, where electrons are recombined with the surrounding argon ions.

Precision and Accuracy

The high precision of the ICP-OES measurement can be ensured by the reproducibility of the measurement and evaluated by the determination of the standard deviation of several replicate measurements as the percentage of the mean value. The main parameters causing signal fluctuations are RF power variations influencing the plasma stability, and changes in the nebulization process. Therefore, a careful adjustment of the RF power, optimization of the nebulization parameters, or the addition of an internal standard (to monitor the signal stability) can be useful to improve precision (Ref 25). For example, nebulization can be stabilized by using a mass flow controller to regulate the nebulizer gas flow. The accuracy of the ICP-OES technique essentially is limited by the precision and systematic errors, such as interference effects (Ref 12, 17, 18).

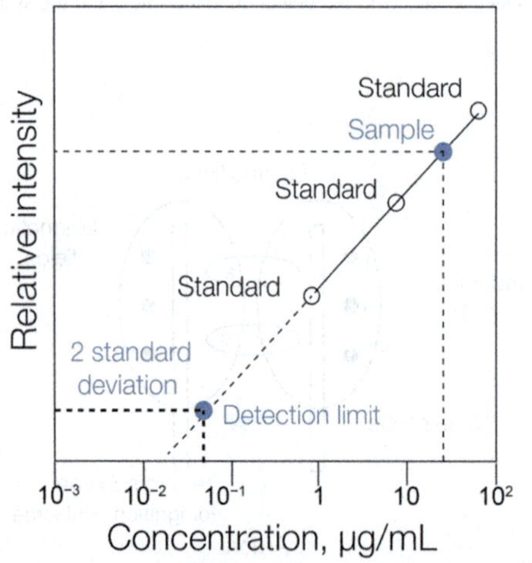

Fig. 5 Typical calibration curve representing the concentration of the analyzed element (µg/L) as a function of the relative intensity. The detection limit is defined as the standard deviation of the lowest measurable concentration. As illustrated, the standard concentration should be within the calibration concentration range.

Analytical Procedure

Almost all analytical procedures for ICPs-OES involve the use of solutions. If the sample is originally in solid form, dissolution prior to analysis should be carried out, and appropriate preparation procedures must be chosen accordingly.

The volume of solution required for analysis varies, as it mostly depends on the type of spectrometer and the number of elements to be determined. For simultaneous spectrometer systems (polychromators), as many as 70 elements can be determined with less than 5 mL of solution, although replicate measurements or longer integration times will require larger volumes. In contrast, the monochromator calibration only requires approximately 0.2 to 1 mL of solution for a single analysis (Ref 26); higher volumes also are required for replicates or longer integration times. Sample dilution may be necessary for smaller sample volumes; however, if dilution is not appropriate, an alternative sample-introduction technique can be used (see the section "Electrothermal Vaporization" in this article).

Instrument calibration involves standard solution preparation (with a minimum of three different concentrations and one blank corresponding to the solvent) for all the elements to be analyzed. The calibration procedure can be performed with a multielement standard solution if no interference effects are expected. In practice, from the linear calibration curve ($I = f(C)$), the slope, corresponding to the sensitivity factor k, is used to calculate the concentrations of the samples (Ref 27). The sensitivity factor value varies with the element and the solution matrix and should be determined at each measurement.

Calibration curves for ICP-OES generally are linear from approximately 10 ppb (µg/L) to 1000 ppm (mg/L), although self-absorption effects can cause curvature of the calibration curves, particularly for alkali elements and some alkaline earths at 100 ppm and greater (Ref 28). These calibration curves remain analytically useful if enough standards were used to sufficiently define the curves. To ensure high precision and accuracy, the calibration procedure must be integrated as a daily routine, alternating between samples and standards for good results.

Each element has a characteristic emission intensity maximum position in the plasma. In addition, every element responds differently to experimental parameters (i.e., plasma parameters, nebulizer type, and electrolyte), which therefore must be carefully selected. The alkali elements in particular require special conditions for optimal results.

System Components

All modern ICP-OES systems include a sample-introduction system, an ICP-OES torch and argon gas supplies, an RF generator and associated electronics, a spectrometer with detection electronics, and a system computer with appropriate hardware and software, as shown in Fig. 1.

Sample Introduction

Over the years, sample-introduction development has become one of the primary focuses for ICP-OES and is considered to be a critical step, requiring careful adjustments to obtain optimal results. In general, the sample is transported to the plasma by means of a peristaltic pump coupled with a nebulizer/spray chamber system. Optimization of the peristaltic pump flow rate is important as the quality of the analysis is intrinsically linked to this step. Normally, samples are in the liquid form, and appropriate preparation is needed (not discussed in this section), particularly if the samples are originally in a gas or solid form. The sample is transported through specific pump tubing (the internal diameter size and composition can be adapted for multiple applications) at a fixed flow rate (from 0.1 to 3 mL/min). The sample subsequently is introduced to a nebulizer, where an argon gas is mixed with the sample, breaking the liquid into small droplets to generate an aerosol. In fact, only a small amount of the electrolyte can be injected in the plasma, as large electrolyte volumes can disturb the plasma and generate an excessive amount of noise in the signal. As the electrolyte is transported to the ICP-OES system, it will enter the introduction system part, which includes the nebulizer and spray chamber.

Use of Peristaltic Pumps

Peristaltic pumps are conventionally used for ICP-OES analysis. However, historically, peristaltic pumps were not provided with the ICP-OES system and were only recently integrated in every ICP-OES instrument. A peristaltic pump is commonly used to supply sample solution to the nebulizer, where the aerosol is created. The peristaltic pump consists of a series of rollers on a rotating head that passes the solution to the nebulizer or evacuates it to the drain (Ref 29). Typical rates lie between 0.1 and 3.0 mL/min solution uptake with a 2 Hz pump pulsation. Without the use of a pump, the uptake rate of the sample solution will depend on the viscosity of the solution and the nebulizer gas flow rate. However, minimal changes in the flow rate can significantly alter the amount of analyte reaching the plasma and cause errors in the analytical measurement. As a consequence, the use of a peristaltic pump that delivers solution to the nebulizer at a fixed volume rate is necessary to eliminate such experimental errors.

Nebulizers (Ref 30)

The peristaltic pump injects the analyte into a micrometer-sized capillary inlet (the nebulizer inlet), whereby a high-velocity argon gas flow is injected to atomize the electrolyte into fine droplets ranging from 0.1 to 100 μm in diameter. This aerosol is subsequently introduced to the spray chamber, where only droplets with a diameter < 10 μm can be transported to the injector and introduced into the plasma.

However, when electrolytes with high salt concentrations are used, the capillary system can be clogged with solids. The crystallization of salts at the output of the nebulizer inner capillary tube substantially reduces the quality of the analysis. To ensure good analysis, an additional argon humidifier system can be added to prevent salt accumulation and crystallization at the nebulizer output.

Conventionally, nebulizers can be categorized into two families: pneumatic and non-pneumatic. Their design has an impact on the aerosol and the detection sensitivity (Ref 29, 31, 32). Over the years, nebulizer performances were expanded to more aggressive and specific environments (high salt concentrations, solvent-containing samples, or hydrofluoric-acid-containing electrolyte) to suit a large portfolio of applications. The most common nebulizers and their specificities are summarized in Table 1.

Concentric Nebulizer

The concentric nebulizer, or Meinhard nebulizer (Fig. 6), displays a characteristic T-form and can be constructed from borosilicate glass

Table 1 List of various commercially available nebulizers for the inductively coupled plasma spectroscopy technique

Nebulizer	Sensitivity	Stability	Suitability to salt-containing solution	Risk of clogging	Adapted to hydrofluoric acid
Concentric glass nebulizer	++	++	Up to 10 g/L with argon humidifier	High	No
Parallel-flow nebulizer	+	++	Up to 40 g/L	Low	Yes
Cross-flow nebulizer	+	+	No	High	No
V-groove nebulizer	+	+	Up to 35%	Low	Yes, if made from polytetrafluoroethylene
Ultrasonic nebulizer	+++	+	No	High	No

Note: The symbols "+" to "+++" refer to the applicability of the design to the considered application.

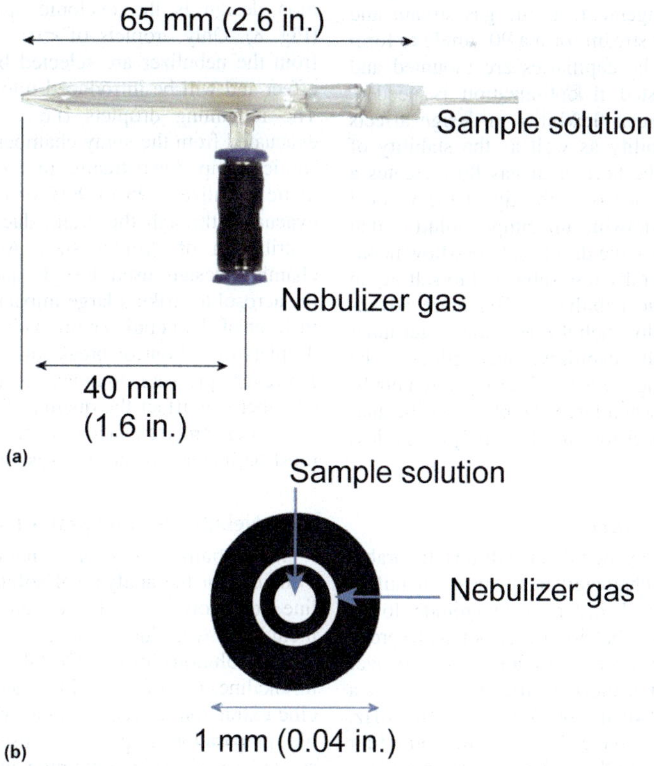

Fig. 6 Meinhard nebulizer displaying (a) the sample and nebulizer gas inlets and (b) the structure of the internal capillaries that allow generation of an aerosol

or polymer (perfluoroalkoxy alkane) (Ref 32). Two inlets allow electrolyte transport (left or right side) and argon flow, respectively (bottom). In general, the argon flow is fixed at 1.0 L/min at a line pressure of approximately 275 kPa (40 psi) through an outer annulus, as low pressure draws the sample solution through the inner capillary tube at a rate between 0.01 and 3 mL/min. Multiple technical issues may be encountered with this design, because the thin capillaries may be obstructed by small particles. This phenomenon can alter the signal and the quality of the analysis. In addition, high salt contents can induce accumulation of undesired dry deposits at the nebulizer tip. Such technical constraints will significantly slow the gas and electrolyte flow rate and ultimately lead to instrument breakdown. In such conditions, a parallel-flow nebulizer (one-piece synthetic fluorine-containing resin nebulizer) can be used (Fig. 7). The parallel-flow nebulizer configuration is ideal for high dissolved solids or electrolytes with particle suspensions, particularly for hydrofluoric-acid-containing samples.

The addition of an argon humidifier is an alternative option that can prevent residual salt accumulation by incorporating water in the system, promoting solid dissolution. Nonetheless, the use of only a parallel-flow nebulizer is not recommended for the study of suspended particles.

Cross-Flow Nebulizer

The cross-flow nebulizer displays a slightly different arrangement, as the gas stream and the electrolyte stream form a 90° angle to form the aerosol. The capillaries are mounted and may be adjusted if optimization is needed. Nonetheless, such flexibility in design affects the reproducibility as well as the stability of the system. The horizontal gas flow creates a low-pressure zone over the tip of the vertical sample tube, drawing up sample solution that is shattered into fine droplets. Cross-flow nebulizers are generally less subjected to salting-up than concentric nebulizers. The main advantage of cross-flow nebulizers is the larger inner diameter of the capillary, thus reducing the risk of clogging. In general, cross-flow nebulizers are used when large samples must be analyzed and precision in the analysis is less important.

Ultrasonic Nebulizer

The ultrasonic nebulizer exhibits favorable performance characteristics (most manufacturers claim a 1 order of magnitude lower detection limit), but its use is not widespread (Ref 30). The sample solution is passed over the surface of a piezoelectric transducer at a frequency between 200 kHz and 10 MHz. The pressure produced breaks the liquid-air interface into small droplets, producing a fine aerosol. In theory, the formation of smaller droplets (less than 10 μm in size), the narrow droplet size distribution, and the high-rate electrolyte introduction to the plasma promote a higher efficiency. Conversely, greater quantities of water are also delivered to the plasma, causing a cooling effect. Therefore, aerosol desolvation prior to plasma introduction is necessary and can be performed by using a heated spray chamber, which promotes water evaporation, along with a cooled condensing chamber. Another problem is a memory effect (persistence of a signal after removing the sample) caused by inadequate cleaning of the nebulizer between solutions. These disadvantages and the relatively high price of ultrasonic nebulizers can hinder their use.

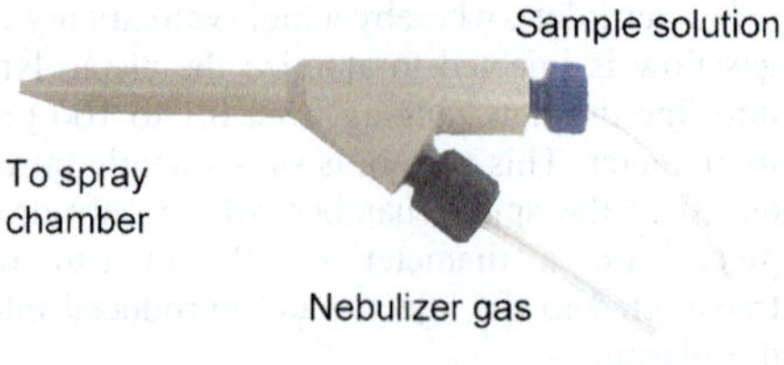

Fig. 7 Configuration of synthetic-fluorine-containing-resin-based parallel-flow nebulizer, which is highly recommended for hydrofluoric-acid-based electrolytes

Spray Chambers

The spray chamber is the last step prior to introducing the electrolyte to the plasma. Once generated by the nebulizer, the aerosol passes into a spray chamber (Ref 33). The most common design is the cyclonic spray chamber (Fig. 8). Only droplets of <10 μm generated from the nebulizer are selected by the inertia effect and will be introduced into the injector. The remaining droplets (i.e., >10 μm) are evacuated from the spray chamber by the peristaltic pump downstream. In fact, for pneumatic nebulizers, 95 to 99% of the aerosol is evacuated through the drain, due to the large distribution of droplet size. Another spray chamber design used less frequently causes the aerosol to strike a large impact bead placed in front of the nebulizer tip. This causes large droplets to fall out or break into smaller ones. However, pressure fluctuations in the spray chamber can affect the quality of the signal; a smooth electrolyte evacuation and a liquid trap must be integrated into the system.

Inert Nebulizers and Spray Chambers

Inert nebulizers and spray chambers also are available for the analysis of solutions in alkaline electrolytes and concentrated acids, including hydrofluoric acid. Although most sample solutions for the ICP-OES are prepared in alkaline, neutral, or acidic media, few specific extraction procedure applications in sample preparation require the use of organic solvents or fluoride-containing electrolyte for metallurgy applications. In the case of organic solvents, several modifications of the operating parameters (higher power and the use of auxiliary plasma gas) and the use of solvent-resistant pump tubing are recommended. Conversely, fluoride-containing medium (often used for the dissolution of silicon-containing alloys) requires the use of synthetic fluorine-containing resin-based equipment (torch, nebulizer, spray chamber, and alumina-base injector). This will prevent further silicon contamination from the glass ICP-OES consumables and their degradation during analysis.

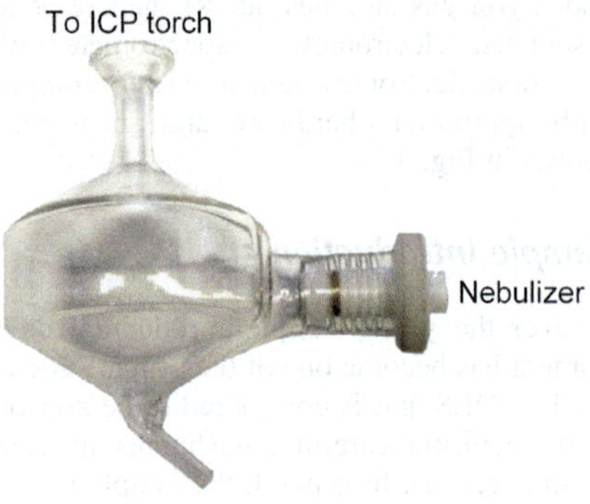

Fig. 8 General configuration of a cyclonic spray chamber. The nebulizer is connected to the spray chamber on the right side, where the aerosol droplets will either condense on the spray chamber walls and be evacuated through the drain or be transported to the plasma chamber.

Alternate Designs

The ICP-OES field has not encountered significant developments over the last decades in terms of nebulizer designs; the nebulizers currently available on the market are summarized in Table 1.

Hydride-Generation Systems

The principle of using a gas phase instead of liquids would short-circuit the sample-introduction process, as nebulization would no longer be required. In theory, this would allow a 100% sample-introduction efficiency; nevertheless, most samples are either in solid or liquid phase and cannot be readily volatilized. For some elements (arsenic, antimony, bismuth, lead, tin, tellurium, and selenium), the generation of gaseous hydrides from acidic solutions was attempted. The goal was to improve the detection limit and this methodology was applied to atomic absorption spectroscopy.

Electrothermal Vaporization

Electrothermal vaporization (ETV) is used for small-volume (microliter-sized liquids, solid residues) analysis (Ref *34)*. The development of ETV was originally initiated to avoid issues created by the pneumatic nebulization process. Essentially, a small current is applied to the sample to remove the solvent. Following this step, the elements undergo vaporization from an electrothermal device and are transported by an argon flow to the center of the plasma. The main advantage of this technique is the significant increase in measurement efficiency (from <5% for pneumatic nebulizers to >60% for ETV/ICP-OES).

However, ETV is not suitable for experiments requiring continuous or simultaneous multielement analysis. In addition, graphite material is mostly used for ETV, and carbide generation can be problematic for refractory element analysis (Ref 35). Although several designs have been developed in the past decades (use of tungsten coil instead of graphite), they have not yet reached a commercial scale.

Torch and Gas Supplies

Favorable analytical performances also rely on the ICP-OES torch design. In fact, the design should ensure that the nebulizer gas punches through the base of the plasma to carry the sample up the central channel. Conventionally, two main designs are available in the market: fixed or partially demountable, as shown in Fig. 9 (Ref 36). The argon gas is supplied between the outer and middle quartz tube at 10 to 20 mL/min flow rate to maintain the primary plasma flow. During plasma ignition, an auxiliary gas flow is added at approximately 1 L/min to contain the plasma in the inner tube. Once the aerosol is produced, the analyte is transported by the argon gas flow to the central quartz tube at 1 L/min. At this stage, this environment should not contain traces of air, as the plasma in operating conditions cannot sustain more than a few minutes in oxygen-containing environments.

Modern ICP-OES systems are inherently flexible and have the ability to analyze a large range of matrices. For example, when volatile samples (i.e., organic solvents) are analyzed, the use of an auxiliary gas is highly recommended to ensure plasma stability. In addition, electrolytes containing high amounts of dissolved solids should be analyzed with the addition of sheath gas, because an argon gas flow will isolate the laminar flux from the spray chamber and minimize salt deposits in the injector and memory effects.

Nonetheless, maintenance of ICP-OES equipment is extremely critical, because the torch will progressively deteriorate. In fact, a typical sign of usage includes a brownish coloration of the central tube and the quartz tubes, which could extend downward a few millimeters. The discolorations may not systematically diminish the analytical performances; however, undesired solid accumulation at the injector tip can significantly alter the quality of the analysis, or worse, stop sample introduction into the plasma. Maintenance recommendations include periodic nitric acid cleaning of the torch and nebulizer to ensure long service life.

The stability of gas flows in the torch is also critical for analytical measurement precision. The nebulizer gas flow rate is particularly important as it controls the amount of analyte introduced into the plasma. The use of mass flow controllers to regulate gas flows provides the most reliable performance of the system. Low-flow, low-power torches (minitorches) for ICP-OES also are available, to reduce operating costs by consuming less argon. These devices, which may operate with only a few liters per minute of total argon flow, require less power, and solid-state RF generators providing approximately 1 kW of power are sufficient.

Radio Frequency

As mentioned previously, the plasma is initiated and sustained by a high-frequency alternating current supplied by an RF generator. The operating conditions of most available instruments include a 100 A alternating current at 27 or 40 MHz frequency.

The plasma is ignited by the spark from a Tesla coil, which provides electrons and ions in the argon gas. The choice of RF is closely related to its incidence on sample introduction and the shape of the plasma flame (Ref 37). In fact, the choice of higher frequencies for solids suspended in a slurry sample is recommended, because their residence time increases and the plasma stability improves (Ref 38).

More importantly, power variations can significantly affect the plasma stability, and minimal power fluctuations can considerably modify emission intensities and the analysis quality. Essentially, the generator power is considered optimized when the highest signal-to-noise ratio is obtained. Higher powers are required when organic solutions are nebulized into the plasma or if mixed-gas or diatomic-gas plasmas, such as argon, helium, and nitrogen mixtures, are used for improved precision and better sensitivity (Ref 39).

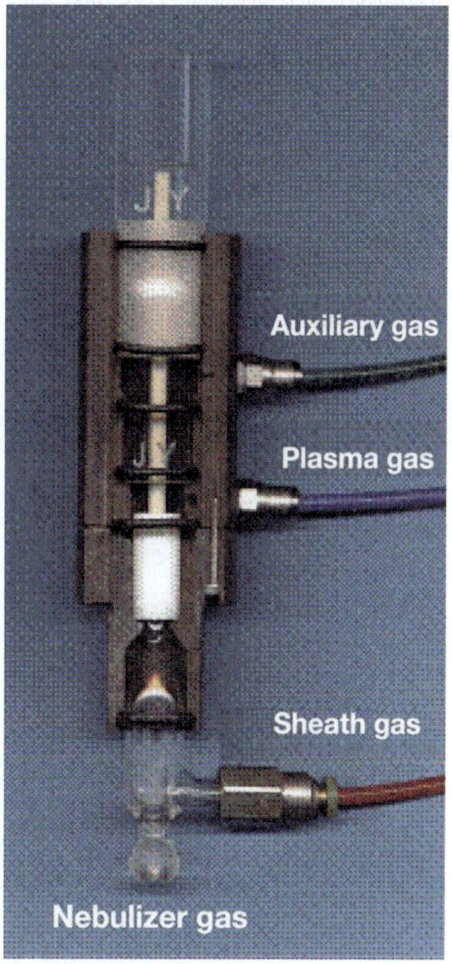

Fig. 9 Cross section of a partially demountable plasma torch displaying its general structure. The lower part is connected to the cyclonic spray chamber. The upper part is composed of one injector (the internal white tube) and two concentric quartz tubes.

Detection Instrumentation

Inductively coupled plasma atomic emission spectrometry is inherently a multielement technique, and different spectrometer configurations are used to detect the emission of interest: polychromators for simultaneous multielement analysis and monochromators for sequential multielement analysis.

Grating Spectrometers

The primary function of the ICP-OES technique is to separate the wavelength and allow element identification. For this purpose, gratings are commonly used as dispersive devices for ICP-OES systems. The diffraction grating is a reflecting material (usually aluminum-coated glass or quartz) composed of multiple, equally spaced, and parallel grooves (e.g., 1200 grooves per millimeter), as shown in Fig. 10. Once the incident beam arrives on the grating, some of it is diffracted. The position, particularly the angle at which the exit slit and the detector are placed, will define the

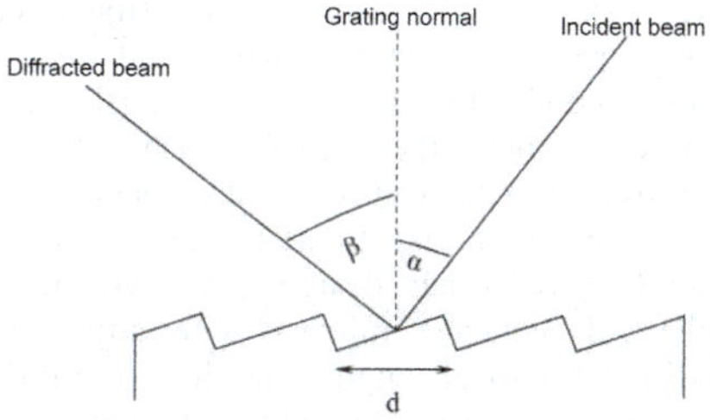

Fig. 10 Schematic diagram illustrating the diffraction grating design. The incident beam arrives at an angle α and is diffracted at an angle β. The presence of grooves and multiple exit slit diffractions at predetermined angles enables light dispersion and element-specific wavelength monitoring with high spectral range.

measured wavelength. The finer the ruling spacing (defined as "d" in Fig. 10) of the grating, the higher the light-dispersing ability or resolution (Ref 11, 40, 41).

Grating spectrometers equipped with an additional cross-dispersive system (also known as echelle spectrometers) have improved spectral resolution. The general principle consists of using higher angles (for incidence and diffraction) than typical grating spectrometers. The cross-dispersive system usually is a prism or a second grating, allowing separation of multiple spectral orders from the primary grating.

Grating Spectrometer Designs

A large variety of grating designs are described in the literature, and they can be classified into two categories: plane and concave grating mounts. In general, plane grating mounts are found in single-wavelength analysis (scanning monochromators), and concave grating mounts are used for simultaneous multielement analysis (polychromators). This section briefly covers the most commonly used gratings in commercially available instruments.

Concave Gratings

Concave gratings are used as both a dispersive and a focusing element. The Rowland design was the first concave grating mount used for spectrometers (Ref 40), and several other mounting designs, such as the Abney, Eagle, and Paschen-Runge, were developed. The Paschen-Runge is the most common concave grating mount found in modern polychromator designs, because it offers a wide wavelength coverage (Ref 40).

Plane Gratings

Plane gratings are less susceptible to aberration (compared to concave gratings), because the focusing and dispersive optics are separate. The plane grating is comprised of a focusing lens, a concave mirror to collimate the incident light onto a plane grating, and a second concave mirror to focus the dispersed light onto the secondary slit. Several designs were developed to address the technical constraints of the original Ebert mounting (Ref 42), such as the mirror size (the Ebert design requires a large mirror) and aberrations. The Czerny-Turner mounting (Ref 43), derived from the Ebert-Fastie design, improved the aberration issues encountered with the Ebert and Ebert-Fastie mounting. The Czerny-Turner design uses two small, spherical, concave mirrors with adjustable position for focal length selection, diminishing aberration risks. In addition to aberration correction, the Czerny-Turner design is a cost-effective alternative to the original Ebert mounting (the mirrors used are less expensive than the large mirror used for the Ebert-Fastie design) and is found in most commercially available monochromators.

Polychromator

The polychromator is equipped with a concave grating mount, a focusing lens, and a primary and secondary slit (namely, the entrance and exit slits). Individual electronic detection channels (phototubes) to detect and measure the intensity of the spectral lines are integrated (Fig. 11) (Ref 8, 9).

The entrance slit, diffraction grating, and secondary slits are placed on a 0.5 to 1 m (1.6 to 3.3 ft) diameter Rowland circle. The diffracted spectral lines are focused on the Rowland circle, and the secondary slits (also called exit slits) are positioned according to the wavelengths of interest. The spectrometer components are mounted in a lighttight box that must be carefully temperature controlled for instrument stability. For vacuum spectrometers, the box must be airtight as well.

Modern polychromators cover a 200 to 800 nm wavelength range. Vacuum or inert-gas-purged spectrometers (e.g., nitrogen purge) are required to detect analytical wavelengths from 170 to 200 nm (Ref 44). Important lines in this region include phosphorus (178.2 nm), sulfur (180.7 nm), carbon (193.1 nm), and aluminum (167.08 nm) (Ref 45).

The lens, usually quartz, is mounted between the ICP-OES source and the spectrometer primary slit and focuses the emitted radiation onto the slit. In a vacuum spectrometer, the lens also will be part of the airtight seal of the spectrometer box, and the optical path between the lens and the plasma also must be purged.

Several polychromator designs also provide a quartz refractor plate, sometimes termed a spectrum shifter, situated just behind the primary or entrance slit. Slight rotation of this refractory plate displaces the incident beam and enables scanning of the spectrum at each analytical wavelength over a small range of approximately 1 nm. This option is used to investigate the spectral vicinity of each line for performing background corrections or identifying spectral interferences.

The secondary slits that isolate the spectral lines may be permanently fixed on a single metal mask along the Rowland circle or may be individual adjustable assemblies. The space required for each slit and associated electronics limits the number of spectral lines that can be simultaneously detected; 20 to 60 lines usually are observed. A spectral bandpass of approximately 0.05 nm typically will pass through the 50 μm width of the secondary slit. This light then is focused, sometimes using a small secondary mirror, onto the cathode of a photomultiplier tube (PMT). The PMT is an excellent detector device for ICP-OES equipment, because it can provide linear responses over a light-intensity range as great as 10^8. The PMT generates a current proportional to the light intensity. This photocurrent is converted into a voltage, amplified, and recorded over preselected integration times by the system computer for data handling of the analytical information.

The polychromator enables the simultaneous detection of up to 70 different wavelengths. However, the polychromator configuration consists of an array of preselected spectral line channels, making changes or additions in the spectral range difficult. Nevertheless, some manufacturers provide ICP-OES instruments with an auxiliary monochromator, to be used

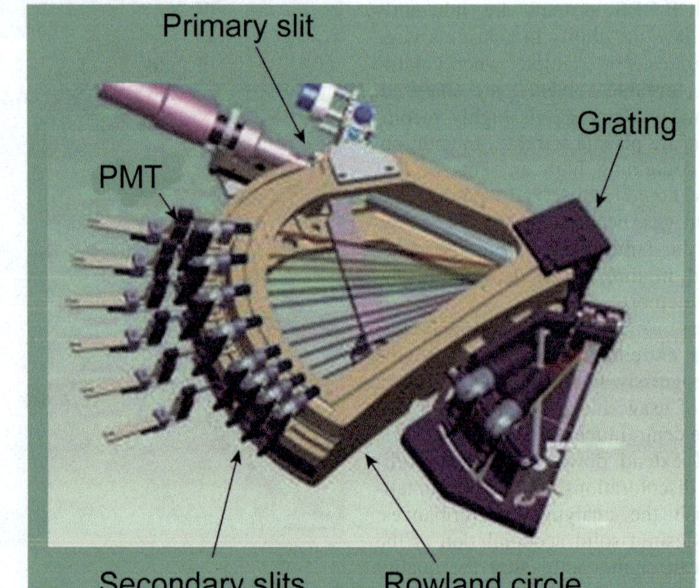

Fig. 11 General structure of the polychromator, comprised of primary and secondary slits and a grating positioned on the Rowland circle. Each exit slit is equipped with an individual photomultiplier tube (PMT) to collect and monitor the emission of each element at its specific wavelength.

in case an additional analytical wavelength (not available in the polychromator) is needed.

Scanning Monochromator

The scanning monochromator is the alternate spectrometer configuration for detecting ICP-OES radiation. The Czerny-Turner optical design is the most common (Fig. 12). Only a single wavelength can be detected at a time, and wavelength selection at the detector is accomplished by rotating the plane grating by a computer-controlled stepper motor. In sequential multielement analysis, the grating is driven at a fast slew rate to a position near each of the preselected analytical wavelengths. The grating drive then must scan more slowly across the estimated position of the desired wavelength, recording integrated light intensities at each step, for example, nine steps across a 0.05 nm scan.

Because of unavoidable mechanical and thermal instabilities in the instrument, a peak-seeking routine must be performed at each analytical wavelength to ensure that the analytical measurement is made at the proper wavelength. The peak is located by fitting the intensities recorded during the stepwise scan of the estimated position of the analytical wavelength to a mathematical model of the spectral line. The peak intensity is calculated as the maximum value in the mathematically fitted curve through the three to five highest measurements in the scan. Background intensities can be measured on either side of the peak for simple background correction by subtraction from the peak intensity.

An alternative to the Czerny-Turner design is a scanning monochromator that uses a fixed grating and a focal curve along which a mask of equally spaced secondary slits is placed (Ref 46, 47). The photodetector moves behind this mask on a computer-controlled carriage. Selection of a given wavelength requires rapid movement of the carriage to the secondary slit closest to the selected wavelength on the focal curve, and fine-tuning of the selection is accomplished by precise and small movement of the primary slit. This provides rapid and precise wavelength selection.

The scanning monochromator offers freedom of choice in analytical wavelengths for each analysis. This is valuable in research and development or in an analytical laboratory where the elements to be determined and the sample types are variable, requiring a high level of flexibility. The major disadvantage of this type of spectrometer is the increased amount of time necessary for each analysis. Because each measurement at each wavelength will require approximately 10 s or longer, the determination of several elements will require several minutes and more samples. Therefore, the choice between instrument configurations involves a compromise between flexibility and speed. As mentioned earlier, some instrument manufacturers offer a combination of simultaneous and sequential spectrometers in one system.

Modern ICP-OES Systems

Modern ICP-OES systems generally are equipped with simultaneous systems. In addition to the polychromator with PMT detectors, an increasingly common type consists of a semiconductor detector, such as a charge-coupled device, charge-injection device, or segmented charge-coupled device, and an echelle spectrometer. The main benefits are simultaneous measurement of the sample with, if needed, the analysis of an internal standard and the background intensities. These systems are more compact compared to polychromator or monochromator systems.

Alternate Designs

One approach to simultaneously detecting all of the wavelengths of the emitted radiation involves replacing the PMTs of conventional spectrometers with photodiode arrays (PDAs) (Ref 48–50). In fact, these multichannel light-sensing detectors consist of up to 2048 detectors, covering a large spectral range in a single measurement.

The PDAs show excellent linearity and an overall good spectral response (190 to 1100 nm for silicon-base PDAs), with a relatively low noise compared to the photomultiplier. In addition, they can be of interest because they are relatively low cost compared to photomultipliers and have a longer lifetime.

Depending on the compromise between resolution and spectral range, a spectral window of approximately 20 nm can be observed at moderate resolution on one PDA, and all the emission lines and spectral background can be simultaneously monitored. One arrangement is to use a single PDA at the exit plane of a monochromator to observe such a spectral window; the wavelengths under observation then are selectable by rotating the grating. Another arrangement is to mount several PDAs along the focal curve of a polychromator to cover several such spectral windows. However, the use of PDAs involves a loss of sensitivity by a factor of 10 compared to that of PMTs, because the light detectors have a slower response time.

An associated computer must perform a mathematical Fourier transform to convert the observed interferogram of signal versus mirror movement into an analytically useful spectrum of intensity versus wavelength. The resolution of the spectrum is determined by the extent of mirror movement in the interferometer. The use of Fourier transform spectroscopy (FTS) as a detection system for ICP-OES provides simultaneous and comprehensive detection for an entire selected spectral region (Ref 51–53). Although FTS has found widespread application in infrared spectroscopy, this approach is experimental for atomic spectroscopy in the ultraviolet and visible light wavelength range, and no commercial instrumentation is yet available.

Detection Electronics and Interface

In summary, the analytical information consists of the emitted light and its intensity of excited ions and atoms. The light is subsequently collected and separated according to its wavelengths in the spectrometer. The photomultipliers convert the light intensity into electrical signals transmitted to the system computer. To reduce statistical noise in the signal, the photocurrent is integrated over 5 to 20 s in each measurement and stored on a small capacitor.

At the end of the integration period, an analog-to-digital converter (ADC) reads and digitizes the charge on the capacitor. The digital value is transferred to the computer for data management. In the polychromator, all of the wavelength channels integrate their signals simultaneously. At the end of the integration period, the ADC reads the integrators in sequence. In the monochromator, however, integration of the signal followed by conversion to the digital value is accomplished in sequence at each selected wavelength.

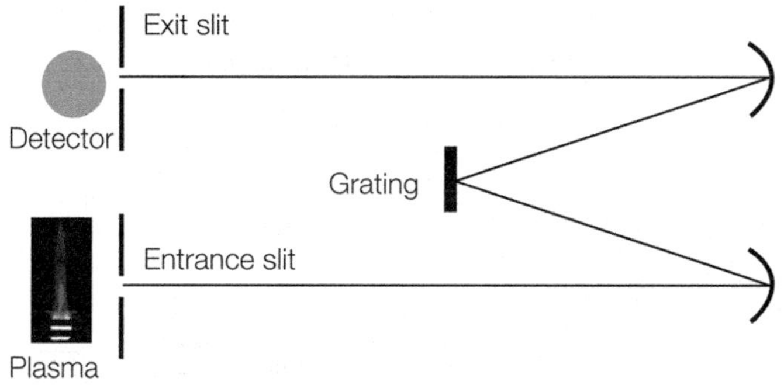

Fig. 12 Schematic diagram illustrating the general configuration of a scanning monochromator. The monochromator is comprised of a concave mirror to collimate the incident light onto a plane grating, a second concave mirror to focus the dispersed light onto the secondary slit, and a photomultiplier detector at the exit slit.

System Computer

Modern ICP-OES systems include dedicated software used to monitor the analysis and the instrument state as well as to correct, store, and process the analytical data. The ICP-OES software packages include:

- An instrument configuration program that stores information such as the individual channel numbers, the wavelengths and corresponding elements, and the predefined detection limits
- A computer program, enabling the design of task file programs, tailored to specific procedures that include the elements to be determined, wavelengths to be used, signal integration times for each wavelength, concentrations of the standard solutions to determine the calibration curves, and interelement or interference corrections to be made on the analytical data
- A calibration program that establishes calibration curves for concentration determinations. The programs can be designed to collect intensities measured from a set of two or more standard solutions of known concentration for each preselected element and to calculate the detection limits from the measured data.
- A background-correction option (automatic or manual option), allowing background emission subtraction from the peak of interest
- The analysis routine in which the intensities of selected wavelengths are measured over the predefined integration times and compared to the calibration curves to determine the concentrations of elements in the sample
- Data storage and reporting of analytical results in a variety of formats

Other Developments

To date, ICP-OES is considered a fundamental technique for elemental analysis of sample materials. Development of a similar technique for isotope and trace analysis—namely, inductively coupled plasma mass spectrometry (ICP-MS)—began in 1980. The ICP-MS technique currently is considered as equally indispensable as ICP-OES (Ref 9).

The ICP-MS source is built horizontally and is connected to a sample-introduction system, similar to the ICP-OES system described previously. The mass spectrum can be recorded in approximately 1 min, enabling sample analysis at a much faster rate than with conventional mass spectrometric techniques.

The free ions produced in the plasma subsequently are introduced to the mass spectrometer, where they are separated according to their mass-to-charge ratio. The quadrupole mass filter, composed of four rods, allows the mass range of 3 to 2400 to be scanned sequentially in under 1 s with high resolution and sensitivity. Typical quadrupole mass filters exhibit a 0.7 to 1.0 atomic mass per unit resolutions, adequate for most applications. Final data management is performed by an associated computer system, as illustrated in Fig. 13.

The ICP-MS technique is recommended for trace analysis, because it provides a significantly lower detection limit (parts per trillion for the majority of the elements). In addition, the technique is not subject to spectral overlap or interferences originated from the optical spectra. In fact, the ICP-MS technique is recommended when elements produce complex optical spectra, such as rare earth and actinide elements. Nevertheless, this technique is not suitable for concentrated matrices or high-dissolved-solid contents, which could saturate the detector. Some minor mass interferences are observed, particularly in the low-to-mid-mass range, due to doubly charged ions, metal oxide ions, and molecular fragments that are originally present in the plasma or formed at the orifice and interface.

Common applications of the ICP-MS technique include isotope ratio analysis and trace impurities for pharmaceutical or biological applications. However, ICP-MS instruments are less robust than those of ICP-OES, and consumables as well as maintenance can be rather expensive.

Direct Current Plasma

An alternative plasma source for analysis is the direct current plasma (DCP), which was developed in the 1960s to 1970s and is now a well-established technology. This technique uses a direct current for the emission source flowing between a carbon anode and a tungsten cathode. The main advantage of this technique is the relatively low argon consumption compared to the ICP-OES technique.

In addition, the DCP is adapted to complex samples, including solids and slurries, and provides the same detection range as the ICP-OES (up to 70 elements between the 200 and 700 nm range) with a similar detection limit range (parts per million to parts per billion) (Ref 54). However, the DCP has a more limited dynamic range (only 3 orders of magnitude linearity for most elements) and also suffers from matrix interference effects, particularly for alkali and alkaline earth elements. The argon gas flow introduces the aerosol from the bottom into the center of the plasma jet, where the excitation and characteristic emission will occur. Photomultiplier tubes monitor line intensities, and spectra are produced with an echelle grating spectrometer. Cross dispersion by the grating and the prism achieves resolution of the spectrum in two dimensions at the focal plane of the spectrometer. Spectral orders are separated in one direction, and wavelengths within an order are dispersed in the other direction (Ref 55).

The DCP provides good-quality multielement analysis at a more affordable price and could be of interest, for example, in the metallurgy, oil, and photography industries as well as for biological and environmental analysis.

Microwave-Induced Plasma

The microwave-induced plasma technique is derived from ICP-OES but with the use of a low-temperature plasma, and it can be used as a source for atomic emission spectrometry (Ref 56). A magnetron-microwave generator generates electron oscillations. The collision of these oscillating electrons in the flowing gas with other atoms maintains the plasma during the analysis.

New Developments

Since the development of the ICP-MS technique, the field has not witnessed, per se, any significant change in technique instrumentation. However, the emergence of elaborated experimental setups for engineering applications (Ref 57–59) (i.e., ICP-OES/ICP-MS

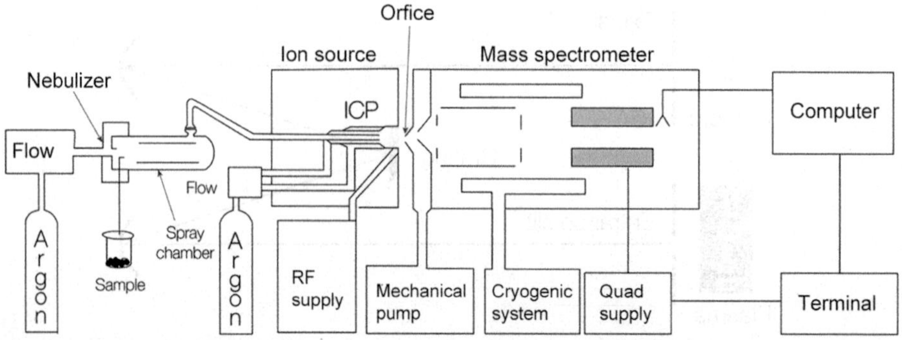

Fig. 13 General structure of modern inductively coupled plasma (ICP) mass spectrometry system consisting of (from left to right): a sample-introduction system, an ion source with a radio-frequency (RF) supply, the mass spectrometer equipped with a quadrupole mass filter, and a computer system and associated electronics

coupling with electrochemical analysis) or for materials science applications has become rather widespread, and one example is presented in the following section "Applications" in this article (Ref 58, 60, 61).

Applications

Example 1: Online Corrosion-Rate Measurement via ICP-OES—Electrochemical Flow Cell Coupling

In the context of materials development and durability, the degradation of alloys needs to be monitored to predict lifetime performances of engineered structure as well as ensure user safety. The corrosion of metals and alloys often is monitored through mass-loss analysis, whereby the sample undergoes an accelerated corrosion test, simulating the exposure to an aggressive environment. Typically, aliquots are regularly collected to analyze the concentration of released elements by using ICP-OES. Although this methodology is quite useful, recent developments in this field resulted in the coupling of an electrochemical flow cell and an ICP-OES, allowing online measurement of the dissolution rate element by element (Ref 58). The equipment consists of a three-electrode electrochemical flow cell comprised of two compartments. The first compartment allows the electrolyte to flow and react with the sample, and the second contains the reference and counterelectrode for electrochemical measurements.

A typical sample analysis can involove an Al-Cu-Mg alloy sample exposed to 0.5 M NaCl solution to monitor the dissolution rate of aluminum, copper, and magnesium as a function of time. The dissolution profiles of aluminum, copper, magnesium, iron, silicon, and manganese were monitored as a function of time. The dissolution rates, v_M (in g/s/cm^2) can be expressed as:

$$v_M = C_M f/A \qquad \text{(Eq 3)}$$

where f corresponds to the electrolyte flow rate in mL/min, A is the exposed aluminum alloy surface area in cm^2, and C_M is the concentration of the element M in g/L.

For example, when an electrochemical experiment is conducted (potentiodynamic, potentiostatic/galvanostatic polarization), the dissolution rate can be expressed as an equivalent current density, assuming Faraday's law:

$$j_M = zFv_M \qquad \text{(Eq 4)}$$

where j_M is the current density, F is Faraday's constant, and v_M is the dissolution rate. In this case, the individual current densities can be superposed to the total current density measured by the potentiostat (after accounting for time offset due to flow all hydrodynamics and signal broadening of concentration transients) and can simultaneously monitor electrochemical and chemical processes.

Example 2: Analysis of Stainless Steel Sample

Solid samples always require dissolution (digestion) prior to analysis. Different procedures are prescribed on how to digest samples, such as microwave digestion or open systems using digestion blocks. The advantage of microwave digestion is the use of pressure to obtain complete digestion of a math element sample. In the case of a stainless steel sample, 0.5 g of the sample is weighed, and a mixture of nitric acid and hydrochloric acid is used for digestion. Once the digestion is complete, the sample is diluted to 50 mL volume. Reference standard solutions are always used for the preparation of calibration standards.

The sample analysis involves:

- Design of an analytical method containing elements and wavelengths used for measurement
- Calibration of the spectrometer by measuring the emission intensities of each of the standard solutions and a blank for each of the desired elements. Calibration curves are calculated and stored by the computer using a linear least-squares program. It is highly recommended to use at least three different concentrations spanning the concentration range of the elements present in the sample.
- Measurement of each emission intensity corresponing to the desired elements for each sample. Measurements at each wavelength should be made in triplicate.
- Calculation of elemental concentrations from the intensities, sample weight and dilution factors, and printing of analytical results as the average concentration of each element from triplicate and a standard deviation of concentration from triplicate. All modern computer systems are equipped with calibration programs.

Example 3: Analysis of Black Carbon Using ETV/ICP-OES

When complex samples are analyzed, direct sample introduction is the most suitable method, especially when it involves low concentrations analysis. For example, impurity levels of a black carbon sample can be determined using ETV/CP-OES analysis. Approximately 0.1 g of the sample is weighed and, using a graphite tube, and introduced into the ETV.

The sample analysis involves:

- Determination of the exact weight of the sample by using a precise analytical balance
- Calibration of the spectrometer by measuring the emission intensities of each of the standard solutions and a blank for each of the desired elements. Calibration curves are calculated by using a linear least-squares program. Commercial software is able to calculate the surface area under the peak.
- Analysis of the black carbon sample without further treatment. Emission intensities of each of the desired elements in each sample are recorded and stored by the computer.
- Computer calculation of elemental concentrations from the recorded emission intensities and printing of the analytical results

ACKNOWLEDGMENT

Revised from Paul B. Farnsworth, Optical Emission Spectroscopy, *Materials Characterization*, Vol 10, *ASM Handbook*, ASM International, 1986.

REFERENCES

1. G. Kirchhoff and R. Bunsen, Analyse Chimique Fondée sur les Observations du Spectre, *Ann. Chim. Phys.*, Vol 3 (No. LXII), 1861, p 452–486
2. W.G. Schrenk, Flame Emission Spectroscopy, *Analytical Atomic Spectroscopy*, Plenum Press, New York, 1975, p 212–242
3. K. Ohls and B. Bogdain, History of Inductively Coupled Plasma Atomic Emission Spectral Analysis: From the Beginning up to Its Coupling with Mass Spectrometry, *J. Anal. At. Spectrom.*, Vol 31, 2016, p 22–31, doi:10.1039/c5ja90043c
4. R.H. Wendt and V.A. Fassel, Induction-Coupled Plasma Spectrometric Excitation Source, *Anal. Chem.*, Vol 37, 1965, p 920–922, doi:10.1021/ac60226a003
5. S.R. Koirtyohann, J.S. Jones, C.P. Jester, and D.A. Yates, Use of Spatial Emission Profiles and a Nomenclature System as Aids in Interpreting Matrix Effects in the Low-Power Argon Inductively Coupled Plasma, *Spectrochim. Acta B, At. Spectrosc.*, Vol 36, 1981, p 49–59, doi:10.1016/0584-8547(81)80007-9
6. J.M. Mermet, Optical Atomic Emission Spectrometry-Inductively Coupled Plasma, *Encyclopedia of Analytical Science*, Elsevier, 2005, p 2122–2129
7. R.H. Scott, V.A. Fassel, R.N. Kniseley, and D.E. Nixon, Inductively Coupled Plasma-Optical Emission Analytical Spectrometry: A Compact Facility for Trace Analysis of Solutions, *Anal. Chem.*, Vol 46, 1974, p 75–80, doi:10.1021/ac60337a031
8. X. Hou, R.S. Amais, B.T. Jones, and G.L. Donati, Inductively Coupled Plasma Optical Emission Spectrometry, *Encyclopedia of Analytical Chemistry*, R.A. Meyer, Ed., John Wiley & Sons, Ltd., 2000, p 9468–9485
9. C. Boss and K. Fredeen, *Concepts, Instrumentation and Techniques in Inductively*

Coupled Plasma Optical Emission Spectrometry, 3rd ed., PerkinElmer, 2004, www.perkinelmer.co.uk/CMSResources/Images/44-159043GDE_Concepts-of-ICP-OES-Booklet.pdf
10. L.H.J. Lajunen and P. Perämäki, *Spectrochemical Analysis by Atomic Absorption and Emission,* 2nd ed., Royal, Cambridge, 2004
11. P.W.J.M. Boumans, *Inductively Coupled Plasma Emission Spectroscopy, Part 1: Methodology, , Instrumentation and Performance,* John Wiley and Sons, Ltd., New York, 1987
12. J.M. Mermet and C. Trassy, A Spectrometric Study of a 40 MHz Inductively Coupled Plasma, Part V: Discussion of Spectral Interferences and Line Intensities, *Spectrochim. Acta B, At. Spectrosc.,* Vol 36, 1981, p 269–292
13. C.C. Wohlers, ICP-AES Wavelength Table, *ICP Inf. Newsl.,* Vol 10, 1985
14. P.W.J.M. Boumans, *Line Coincidence Tables for Inductively Coupled Plasma Atomic Emission Spectrometry,* Pergamon Press, Oxford, 1984
15. P.W.J.M. Boumans and F.J. Boerde, Studies of an Inductively-Coupled High-Frequency Argon Plasma for Optical Emisson Spectrometry, Part II: Compromise Conditions for Simultaneous Multielement Analysis (extension of a paper presented at the 17th Colloquium Spectroscopicum Internationale [Florence], 1973), *Spectrochim. Acta B, At. Spectrosc.,* Vol 30, 1975, p 309–334, doi:10.1016/0584-8547(75)80030-9; for Part I, see *Spectrochim. Acta B, At. Spectrosc.,* Vol 27, 1972, p 391
16. G.J. Schmidt and W. Slavin, Inductively Coupled Plasma Emission Spectrometry with Internal Standardization and Subtraction of Plasma Background Fluctuations, *Anal. Chem.,* Vol 54, 1982, p 2491–2495, doi:10.1021/ac00251a020
17. Y. Morishige and A. Kimura, Ionization Interference in Inductively Coupled Plasma-Optical Emission Spectroscopy, *Ind. Mater.,* 2008, p 106–111
18. M.W. Blades and G. Horlick, Interference from Easily Ionizable Element Matrices in Inductively Coupled Plasma Emission Spectrometry—A Spatial Study, *Spectrochim. Acta B, At. Spectrosc.,* Vol 36, 1981, p 881–900, doi:10.1016/0584-8547(81)80080-8
19. H. Sanui and N. Pace, Chemical and Ionization Interferences in the Atomic Absorption Spectrophotometric Measurement of Sodium, Potassium, Rubidium, and Cesium, *Anal. Biochem.,* Vol 25, 1968, p 330–346
20. A. Goldwasser and J.M. Mermet, Contribution of the Charge-Transfer Process to the Excitation Mechanisms in Inductively Coupled Plasma Atomic Emission Spectroscopy, *Spectrochim. Acta B, At. Spectrosc.,* Vol 41, 1986, p 725–739, doi:10.1016/0584-8547(86)80087-8
21. J. Dennaud, A. Howes, E. Poussel, and J. M. Mermet, Study of Ionic-to-Atomic Line Intensity Ratios for Two Axial Viewing-Based Inductively Coupled Plasma Atomic Emission Spectrometers, *Spectrochim. Acta B, At. Spectrosc.,* Vol 56, 2001, p 101–112, doi:10.1016/S0584-8547 (00)00299-8
22. N. Furuta and G. Horlick, Spatial Characterization of Analyte Emission and Excitation Temperature in an Inductively Coupled Plasma, *Spectrochim. Acta B, At. Spectrosc.,* Vol 37, 1982, p 53–64, doi:10.1016/0584-8547(82)80008-6
23. M. Carré, S. Excoffier, and J.M. Mermet, A Study of the Relation between the Limit of Detection and the Limit of Quantitation in Inductively Coupled Plasma Spectrochemistry, *Spectrochim. Acta B, At. Spectrosc.,* Vol 52, 1997, p 2043–2049, doi:10.1016/S0584-8547(97)00114-6
24. M. Thompson, *Handbook of Inductively Coupled Plasma Spectrometry,* 2nd ed., Springer, 1989
25. S.A. Myers and D.H. Tracy, Improved Performance Using Internal Standardization in Inductively-Coupled Plasma Emission Spectroscopy, *Spectrochim. Acta B, At. Spectrosc.,* Vol 38, 1983, p 1227–1253, doi:10.1016/0584-8547(83)80066-4
26. J.L. Todoli and J.M. Mermet, Elemental Analysis of Liquid Microsamples through Inductively Coupled Plasma Spectrochemistry, *Trends Anal. Chem.,* Vol 24, 2005, p 107–116, doi:10.1016/j.trac.2004.11.005
27. J.M. Mermet, Quality of Calibration in Inductively Coupled Plasma Atomic Emission Spectrometry, *Spectrochim. Acta B, At. Spectrosc.,* Vol 49, 1994, p 1313–1324, doi:10.1016/0584-8547(94)80111-8
28. J.M. Mermet, Calibration in Atomic Spectrometry: A Tutorial Review Dealing with Quality Criteria, Weighting Procedures and Possible Curvatures, *Spectrochim. Acta B, At. Spectrosc.,* Vol 65, 2010, p 509–523, doi:10.1016/j.sab.2010.05.007
29. L.R. Layman, J.G. Crock, and F.E. Lichte, Multichannel Peristaltic Pump with a Pneumatic Nebulizer for Atomic Absorption or Emission Spectrometry, *Anal. Chem.,* Vol 53, 1981, p 747–748, doi:10.1021/ac00227a046
30. K.W. Olson, W.J. Haas, and V.A. Fassel, Multielement Detection Limits and Sample Nebulization Efficiencies of an Improved Ultrasonic Nebulizer and a Conventional Pneumatic Nebulizer in Inductively Coupled Plasma-Atomic Emission Spectrometry, *Anal. Chem.,* Vol 49, 1977, p 632–637, doi:10.1021/ac50012a032
31. E. Paredes, J. Bosque, J.M. Mermet, and J. L. Todoli, Influence of Nebulizer Design and Aerosol Impact Bead on Analytical Sensitivities of Inductively Coupled Plasma Mass Spectrometry, *Spectrochim. Acta B, At. Spectrosc.,* Vol 65, 2010, p 908–917, doi:10.1016/j.sab.2010.08.006
32. J.L. Todoli and J.M. Mermet, Pneumatic Nebulizer Design, *J. Liq. Sample Introd. ICP Spectrom.,* 2008, p 17–76, doi:10.1016/B978-0-444-53142-1.00003-0
33. J.M. Todoli and J.L. Mermet, Spray Chamber Design, *J. Liq. Sample Introd. ICP Spectrom.,* 2008, p 77–118, doi:10.1016/B978-0-444-53142-1.00004-2
34. J. Wang, J.M. Carey, and J.A. Caruso, Direct Analysis of Solid Samples by Electrothermal Vaporization Inductively Coupled Plasma Mass Spectrometry, *Spectrochim. Acta B, At. Spectrosc.,* Vol 49, 1994, p 193–203, doi:10.1016/0584-8547(94)80018-9
35. D.R. Hull and G. Horlick, Electrothermal Vaporization Sample Introduction for Inductively Coupled Plasma-Mass Spectrometry, *Spectrochim. Acta B, At. Spectrosc.,* Vol 39, 1984, p 843–850
36. H.A. Phillips, T.R. Smith, B.D. Webb, and M.B. Denton, Simplified Construction of Demountable ICP Torches, *Appl. Spectrosc.,* Vol 43, 1989, p 1488–1489, doi:10.1366/0003702894204218
37. E. Michaud-Poussel and J.M. Mermet, Influence of the Generator Frequency and the Plasma Gas Inlet Area on Torch Design in Inductively Coupled Plasma Atomic Emission Spectrometry, *Spectrochim. Acta B, At. Spectrosc.,* Vol 41, 1986, p 125–132, doi:10.1016/0584-8547(86)80144-6
38. P.W.J.M. Boumans, Direct Element Analysis of Solids by ICP-AES, *Inductively Coupled Plasma Emission Spectroscopy, Part II: Applications and Fundamentals,* Wiley, 1987, p 217–243
39. J.A.C. Broekaert, F. Leis, and K. Laqua, The Application of an Argon/Nitrogen Inductively-Coupled Plasma to the Analysis of Organic Solutions, *Talanta,* Vol 28, 1981, p 745–752, doi:10.1016/0039-9140(81)80116-6
40. H.A. Rowland, XXIX, On Concave Gratings for Optical Purposes, *London, Edinburgh, Dublin Philos. Mag. J. Sci.,* Vol 16, 1883, p 197–210, doi:10.1080/14786448308627419
41. P.N. Keliher and C.C. Wohlers, Echelle Grating Spectrometers in Analytical Spectrometry, *Anal. Chem.,* Vol 48, 1976, p 333A–340A, doi:10.1021/ac60367a051
42. H. Ebert, Zwei Formen von Spectrographen, *Ann. Phys.,* Vol 38, 1889, p 489–493
43. M. Czerny and A.F. Turner, Über den Astigmatismus bei Spiegelspektrometern, *Z. Phys.,* Vol 61, 1930, p 792–797, doi:10.1007/BF01340206
44. T. Hayakawa, F. Kikui, and S. Ikeda, The Determination of I, P, B, S, As and Sn by Inductively Coupled Plasma Emission Spectroscopy Using Lines at Vacuum Ultra-Violet Wavelengths, *Spectrochim. Acta B, At. Spectrosc.,* Vol 37, 1982, p 1069–1073, doi:10.1016/0584-8547(82)80036-0

45. P. Wu, X. Wu, X. Hou, and C.G. Young, Inductively Coupled Plasma/Optical Emission Spectrometry (ICP/OES), *Appl. Spectrosc. Rev.,* Vol 44, 2009, p 507–533
46. G. Tondello and F. Zanini, High-Resolution Czerny-Turner Monochromator for Application to Undulators, *Rev. Sci. Instrum.,* Vol 60, 1989, p 2116–2119, doi:10.1063/1.1140840
47. G. Wunsch, A. Wennemer, and J.W. McLaren, On the Design and Performance of the Czerny-Turner Monochromator in ICP-AES, *Spectrochim. Acta B, At. Spectrosc.,* Vol 46, 1991, p 1517–1531
48. S.W. McGeorge and E.D. Salin, Enhancement of Wavelength Measurement Accuracy Using a Linear Photodiode Array as a Detector for Inductively Coupled Plasma Atomic Emission Spectrometry, *Anal. Chem.,* Vol 57, 1985, p 2740–2743, doi:10.1021/ac00290a069
49. G.M. Levy, A. Quaglia, R.E. Lazure, and S.W. McGeorge, A Photodiode Array Based Spectrometer System for Inductively Coupled Plasma-Atomic Emission Spectrometry, *Spectrochim. Acta B, At. Spectrosc.,* Vol 42, 1987, p 341–351, doi:10.1016/0584-8547(87)80075-7
50. R.K. Winge, V.A. Fassel, and M.C. Edelson, Capabilities of a Photodiode Array Based System for Cataloguing the Atomic Emission Spectra of the Inductively Coupled Plasma, *Spectrochim. Acta B, At. Spectrosc.,* Vol 43, 1988, p 85–91, doi:10.1016/0584-8547(88)80043-0
51. L.M. Faires, B.A. Palmer, and J.W. Brault, Line Width and Line Shape Analysis in the Inductively Coupled Plasma by High Resolution Fourier Transform Spectrometry, *Spectrochim. Acta B, At. Spectrosc.,* Vol 40, 1985, p 135–143, doi: http://dx.doi.org/10.1016/0584-8547(85)80017-3
52. L.M. Faires, Fourier Transforms for Analytical Atmoic Spectroscopy, *Anal. Chem.,* Vol 58, 1986, p 1023A–1034A, doi:10.1021/ac00122a779
53. L.M. Faires, B.A. Palmer, R. Engleman, and T.M. Niemczyk, Temperature Determinations in the Inductively Coupled Plasma Using a Fourier Transform Spectrometer, *Spectrochim. Acta B, At. Spectrosc.,* Vol 39, 1984, p 819–828, doi:10.1016/0584-8547(84)80090-7
54. W.C. Grogan, Applications of a dc Plasma Emission Spectrometer (DCP) to the Analysis of Envrionmental Samples, *Spectrochim. Acta B, At. Spectrosc.,* Vol 38, 1983, p 357–367, doi:10.1016/0584-8547(83)80134-7
55. M.S. Epstein and R.E. Jenkins, Automation and Application of a Direct-Current Plasma Emission Spectrometer, *J. Res. Natl. Bur. Stand.,* Vol 93, 1988, p 458–462
56. J.A.C. Broekaert, Trends in Optical Spectrochemical Trace Analysis with Plasma Sources, *Anal. Chim. Acta,* Vol 196, 1987, p 1–21, doi:10.1016/S0003-2670(00)83065-2
57. C. Gabrielli, M. Keddam, F. Minouflet-Laurent, K. Ogle, and H. Perrot, Investigation of Zinc Chromatation, Part I: Application of QCM-ICP Coupling, *Electrochim. Acta,* Vol 48, 2003, p 965–976, doi:10.1016/S0013-4686(02)00809-5
58. K. Ogle, Atomic Emission Spectroelectrochemistry: A New Look at the Corrosion, Dissolution and Passivation of Complex Materials, *Corros. Mater.,* Vol 37, 2012, p 58–65
59. K. Ogle and S. Weber, Anodic Dissolution of 304 Stainless Steel Using Atomic Emission Spectroelectrochemistry, *J. Electrochem. Soc.,* Vol 147, 2000, p 1770–1780, doi:10.1149/1.1393433
60. L. Rossrucker, A. Samaniego, J.-P. Grote, A.M. Mingers, C.A. Laska, N. Birbilis, G.S. Frankel, and K.J.J. Mayrhofer, The pH Dependence of Magnesium Dissolution and Hydrogen Evolution during Anodic Polarization, *J. Electrochem. Soc.,* Vol 162, 2015, p C333–C339, doi:10.1149/2.0621507jes
61. O. Gharbi, N. Birbilis, and K. Ogle, Li Reactivity Measurement during the Surface Pretreatment of Al-Li Alloy AA2050-T3, *Electrochim. Acta,* 2017, doi:10.1016/j.electacta.2017.05.038

ASM Handbook, Volume 10, *Materials Characterization*
ASM Handbook Committee
DOI 10.31399/asm.hb.v10.a0006662

Infrared Spectroscopy

Curtis Marcott, Light Light Solutions

Overview

General Uses

- Identification and structure determination of organic and inorganic materials
- Quantitative determination of molecular components in mixtures
- Identification of molecular species adsorbed on surfaces
- Identification of chromatographic effluents
- Determination of molecular orientation
- Determination of molecular conformation and stereochemistry

Examples of Applications

- Identification of chemical reaction species; reaction kinetics
- Quantitative determination of nontrace components in complex matrices
- Determination of molecular orientation in stretched polymer films
- Identification of flavor and aroma components
- Determination of molecular structure and orientation of thin films deposited on metal substrates (oxidation and corrosion products, soils, adsorbed surfactants, etc.)
- Depth profiling of solid samples (granules, powders, fibers, etc.)
- Characterization and identification of different phases in solids or liquids

Samples

- *Form:* Almost any solid, liquid, or gas sample
- *Size (minimum):* Solids—10 ng if it can be ground in a transparent matrix, such as potassium bromide; 10 μm diameter for a single particle; 1 to 10 ng if soluble in a volatile solvent (methanol, methylene chloride, chloroform, etc.). Flat metal surfaces—1 by 1 cm (0.4 by 0.4 in.) or larger. Liquids—10 μL if neat, considerably less if soluble in a transparent solvent. Gases—1 to 10 ng
- *Preparation:* Minimal or none; may have to grind in a potassium bromide matrix or dissolve in a volatile or infrared-transparent solvent

Limitations

- Little elemental information
- Molecule must exhibit a change in dipole moment in one of its vibrational modes upon exposure to infrared radiation
- Background solvent or matrix must be relatively transparent in the spectral region of interest

Estimated Analysis Time

- 1 to 10 min per sample

Capabilities of Related Techniques

- *Raman spectroscopy:* Complementary molecular vibrational information
- *X-ray fluorescence:* Elemental information on bulk samples
- *X-ray photoelectron spectroscopy:* Elemental information on adsorbed species
- *High-resolution electron energy loss spectroscopy:* Molecular vibrational surface information
- *Mass spectrometry:* Molecular weight information
- *Nuclear magnetic resonance:* Additional molecular structure information

Introduction

Infrared (IR) spectroscopy is a useful technique for characterizing materials and providing information on the molecular structure, dynamics, and environment of a compound. When irradiated with infrared light (photons), a sample can transmit, scatter, or absorb the incident radiation. Absorbed infrared radiation usually excites molecules into higher-energy vibrational states. This can occur when the energy (frequency) of the light matches the energy difference between two vibrational states (or the frequency of the corresponding molecular vibration).

Infrared spectroscopy is particularly useful for determining functional groups present in a molecule. Many functional groups vibrate at nearly the same frequencies independent of their molecular environment. This makes IR spectroscopy useful in materials characterization. Further, many subtle structural details can be gleaned from frequency shifts and intensity changes arising from the coupling of vibrations of different chemical bonds and functional groups.

Recent advances in computerized IR spectroscopy, particularly Fourier transform infrared (FTIR) spectroscopy, have made it possible to obtain infrared spectra using various sampling techniques. Infrared spectra have traditionally been produced by transmission, that is, transmitting light through the sample, measuring the light intensity at the detector, and comparing it to the intensity obtained with no sample in the beam, all as a function of the infrared wavelength. Techniques such as attenuated total reflectance, diffuse reflectance, specular reflectance, reflection-absorption spectroscopy, and photoacoustic spectroscopy have recently become more common. This article discusses the sampling techniques, applications, and the molecular structure information the resulting infrared spectra can provide.

Basic Principles

Infrared spectra are typically presented as plots of intensity versus energy (in ergs), frequency (in s^{-1}), wavelength (in micrometers), or wavenumber (in cm^{-1}). Wavenumber units are preferred, but several books of reference spectra are plotted in wavelength. Units can be converted easily by using:

$$E = h\nu = \frac{hc}{\lambda} \qquad \text{(Eq 1)}$$

where E is energy in ergs, h is Planck's constant (6.63×10^{-27} erg · s), ν is the frequency in s^{-1}, λ is the wavelength in centimeters, and c is the velocity (speed) of light (3×10^{10} cm/s); and:

$$\bar{\nu} = \frac{1}{\lambda} \qquad \text{(Eq 2)}$$

where $\bar{\nu}$ is the wavenumber in cm^{-1}. When λ is expressed in micrometers (the unit normally used for wavelength), Eq 2 becomes:

$$\bar{\nu} = \frac{10,000}{\lambda} \qquad \text{(Eq 3)}$$

In practice, wavenumber is often called frequency, and the symbol ν is used instead of $\bar{\nu}$. Although, formally, the infrared region of the electromagnetic spectrum is between the wavelengths of 0.78 and 1000 μm (12,820 to 10 cm^{-1}), this article considers the midinfrared region from 2.5 to 25 μm (4000 to 400 cm^{-1}). This region, where most fundamental vibrational modes occur, is the most useful for materials characterization.

Intensity can be expressed as percent transmittance (%T) or absorbance (A). If I_0 is the energy, or radiant power, reaching the infrared detector with no sample in the beam, and I is the energy detected with a sample present, transmittance is:

$$T = \frac{I}{I_0} \qquad \text{(Eq 4)}$$

and percent transmittance is:

$$\%T = \frac{100I}{I_0} \qquad \text{(Eq 5)}$$

Absorbance is:

$$A = \log\left(\frac{1}{T}\right) = \log\left(\frac{I_0}{I}\right) \qquad \text{(Eq 6)}$$

Strong and weak bands are more easily visualized simultaneously without changing scale when spectra are plotted in transmittance, because the absorbance scale ranges from zero to infinity, while transmittance ranges from 0 to 100% T (0% T corresponds to an absorbance of infinity). For quantitative analyses (including spectral subtraction), absorbance or some other intensity scale proportional to concentration must be used.

The molecular geometry, atomic masses, and a complete description of the forces between the atoms (force field) are required to calculate the vibrational frequencies and describe the fundamental vibrations of a molecule. This is normal-coordinate analysis (Ref 1, 2). In practice, the molecular structure is usually not known, and the infrared spectra are used to assist in determining it. However, accomplishing this requires an understanding of how molecular vibrations contribute to the observed infrared spectrum. Molecular vibrations are complicated, because individual bond stretches or bond angle bends are often highly coupled to each other. Progress in understanding the nature of molecular vibrations has derived mainly from empirical observation. Although of little practical use in materials characterization, normal-coordinate analysis reveals the nature of molecular vibrations, which can be useful in analyzing complicated infrared spectra.

Normal-Coordinate Analysis

Normal-coordinate analysis begins by describing a molecular structure as a collection of balls and massless springs. This is a reasonable model based on chemical intuition. The balls represent atoms, and the springs are the forces or chemical bonds between atoms. The problem would be simple if bonds stretched or bond angles bent independently, without affecting the motion of other atoms or bonds. Because this is seldom the case, there is the mechanical problem of solving for the normal modes of vibrations. The number of normal modes in a given molecule equals the total number of degrees of freedom minus the number of rotational and translational degrees of freedom. For a nonlinear molecule, this number is $3N - 6$, where N is the number of atoms in the molecule. It is assumed that there is no coupling of the vibrational degrees of freedom with the rotations and translations.

Of the parameters necessary to begin a normal-coordinate analysis, only the atomic masses are always easily obtained. The molecular geometry can often be acquired from an x-ray structure or by using typical bond distances and angles from the literature. Problems can arise when several conformations are possible. Unfortunately, accurate force fields are usually unavailable, except for very simple molecules. Obtaining a force field requires knowing the potential energy of the molecule as a function of atomic displacements from the equilibrium geometry. These displacements are usually described relative to internal coordinates consistent with chemical intuition, such as bond stretches and angle bends. The potential energy (V) is usually expanded in a Taylor's series as a function of the internal coordinates (R_a):

$$V = V_O + \sum_a \left(\frac{\partial V}{\partial R_a}\right)_O R_a + \left(\frac{1}{2}\right)\sum_{a,b}\left(\frac{\partial^2 V}{\partial R_a \partial R_b}\right)_O R_a R_b + \ldots \qquad \text{(Eq 7)}$$

The derivatives of the potential energy with respect to the internal coordinates evaluated at the equilibrium geometry are the force constants. Force fields are typically derived empirically from the infrared frequencies they were designed to predict, and the number of parameters (force constants) often significantly exceeds the number of experimental frequencies. Therefore, a unique set of force constants that will predict the observed frequencies does not exist.

Most force-field (normal-coordinate) calculations of the vibrational frequencies are performed in the harmonic-oscillator approximation. Because the atomic displacements are considered to be small, the motions are approximated as harmonic. That is, the potential-energy function along an internal-displacement coordinate can be approximated as a parabola near its lowest potential energy.

In the harmonic-oscillator approximation, the only nonzero force constants are from the quadratic terms in Eq 7. The quadratic force constants are the second derivatives of the potential-energy function with respect to the internal coordinates. The harmonic-oscillator approximation reduces dramatically the number of force constants. The number of parameters can be further lessened by assuming that force constants coupling internal coordinates more than one or two atoms apart are zero. That many force constants do not differ significantly from one molecule to another also helps in determining an initial set of quadratic force constants. Therefore, force constants can be transferred from simpler molecules containing the same functional groups.

The net result of a normal-coordinate analysis is a complete description of the atomic displacements for each normal mode, the vibrational frequency of that normal mode, and the percentage contribution of each internal displacement coordinate to each vibration (potential-energy distribution). That is, a normal-coordinate analysis provides a complete picture (within the constraints of the initial model) of the molecular vibration responsible for each fundamental absorption band in an infrared spectrum of a pure compound.

Largely as a result of massive improvements in computational power, it now is possible to calculate normal modes of molecular vibrations for reasonably sized molecules using ab initio theory without the need to transfer force fields from simpler molecules containing the same functional groups. Density functional theory (DFT) is a computational quantum mechanical modeling methodology for calculating molecular geometries and intermolecular forces based on atomic masses and electronic structure (Ref 3, 4). Thus, instead of starting from an initial molecular geometry and determining force constants by iteratively fitting them to reproduce observed vibrational frequencies, DFT can be used to determine the initial equilibrium geometry and normal modes of vibration by displacing the atomic positions and recalculating the molecular energy. Both vibrational frequencies and infrared intensities now can be calculated for many molecules by using DFT approaches.

Molecular Vibrations

Even with the approximations and assumptions discussed previously, IR spectroscopy would be of limited practical use if it were necessary to rely completely on a normal-coordinate calculation for every structure identification or materials characterization. Fortunately, certain functional groups consistently produce absorption bands in the same spectral regions independent of the remainder of the molecular structure. These molecular vibrations are known as group frequencies. For example, methylene stretching vibrations always occur from 3000 to 2800 cm^{-1}, methylene deformations from 1500 to 1300 cm^{-1}, and methylene rocking motions from 800 to 700 cm^{-1}. These relatively broad regions can be made more specific if the molecular environment of the CH_2 group is considered in more detail. Similar sets of rules for many other functional groups have been derived empirically and with the assistance of normal-coordinate calculations on simple molecules.

A few general rules are useful when applying the mechanical ball-and-spring model to molecular vibrations:

- Stretching vibrations generally have a higher frequency than bending vibrations.
- The higher the bond order, the higher the stretching frequency.
- The lighter the atoms involved in the vibration, the higher the vibrational frequency.

The last two rules are apparent in the simple solution for the only molecular vibration of a diatomic molecule:

$$v = \left(\frac{1}{2}\pi c\right)\sqrt{k\left(\frac{1}{m_1}+\frac{1}{m_2}\right)} \qquad \text{(Eq 8)}$$

where v is the frequency in cm^{-1}, m_1 and m_2 are the atomic masses in grams, and k is the Hooke's law force constant in dynes/cm. This is the solution that would result from a normal-coordinate calculation for a diatomic molecule. Vibrational frequency increases with the magnitude of the force constant, that is, bond order. As masses increase, vibrational frequency decreases.

For a molecular vibration to absorb infrared radiation, dipole moment must change during the vibration (Ref 2). The infrared photon frequency must resonate with the vibrational frequency to excite the molecule into the higher vibrational state. In addition, the electric dipole-transition moment associated with the molecular vibration being excited must have a component parallel to the polarization direction of the incident infrared photon. The dipole strength of the transition from vibrational state $|n>$ to vibrational state $|k>$, $D_{n,k}$, equals the square of the electric dipole-transition moment:

$$D_{n,k} = |<n|\vec{\mu}|k>|^2 \qquad \text{(Eq 9)}$$

where the electric dipole-moment operator, $\vec{\mu}$, is the sum over i of the charge multiplied by the position, $\sum_i q_i r_i$, for each charged particle, i, relative to an arbitrarily selected origin fixed in the molecular frame. The electric dipole-transition moment is proportional to the first derivative of the electric dipole-moment operator with respect to the normal coordinate, that is, the motion of all atoms in the vibrational normal mode. Dipole strength can also be experimentally measured and is proportional to the frequency-weighted area under the absorption band for the transition:

$$D_{n,k} = K\int_{\text{Band}}\left[\frac{A(v)}{v}\right]dv \qquad \text{(Eq 10)}$$

where $A(v)$ is the absorbance at wavenumber v, and K is a known constant. The observed infrared intensity contains contributions from the motions of all the atoms and electronic charges during molecular vibration. As mentioned previously, motions of the atoms and electronic charges that occur during a particular molecular vibration can be described by using DFT. This means both vibrational frequencies and infrared intensities now can be calculated for many reasonably sized molecules.

Instrumentation

Obtaining an infrared spectrum necessitates detection of intensity changes as a function of wavenumber. Commercial infrared instruments separate light into single infrared wavenumber intervals using a dispersive spectrometer (prism, grating, or variable filter), a Fourier transform spectrometer, or a tunable infrared laser source.

Dispersive Infrared Spectroscopy

In dispersive infrared spectroscopy, a monochromator separates light from a broadband source into individual wavenumber intervals. The monochromator contains a dispersing element, such as a prism, diffraction grating, or variable filter, that rotates to enable a mechanical slit to select individual wavenumber regions sequentially. The spatial distance separating the individual wavelengths after dispersion and the mechanical slit width determine spectral resolution and optical throughput. For a given instrument, the higher the spectral resolution, the lower the signal-to-noise ratio of the spectrum. Figure 1 shows a typical double-beam grating spectrometer.

Fourier Transform Infrared Spectroscopy

Fourier transform infrared spectroscopy uses an interferometer to modulate the intensity of each wavelength of light at a different audio frequency (Ref 5). A beam splitter divides the light from a broadband infrared source into two optical paths. Recombination of the beams at the beam splitter generates a modulated optical path difference. The recombined beam undergoes constructive and destructive interference as a function of the instantaneous optical path difference. Several methods can be used to generate the modulated optical path difference. The most common is the Michelson interferometer (Ref 6, 7), as shown in Fig. 2.

After being split into two optical paths by the beam splitter, the light from each path strikes two flat mirrors, one fixed and one moving, that return it to the beam splitter. When the moving mirror is positioned so that the two optical paths are equal (zero path-difference position), all wavelengths of light simultaneously undergo constructive interference. At any other position of the moving mirror, a given wavelength of light may interfere constructively or destructively, depending on the phase relationship between the light rays in the two paths. For example, if the optical path difference is an integral multiple of the wavelength, constructive interference will result.

Spectrometer resolution equals the reciprocal of the distance the moving mirror travels. The modulation frequency for a particular wavenumber depends on mirror velocity. Modulation frequencies are typically from 50 Hz to 10 kHz. The detector signal, which is digitally sampled in fixed increments defined by interference fringes from a helium-neon laser, displays signal intensity as a function of moving mirror position. This signal is the interferogram. It must be Fourier transformed by a computer into the single-beam infrared spectrum.

Fourier transform infrared spectroscopy affords several advantages over dispersive IR spectroscopy. Information on the entire infrared spectrum is contained in an interferogram, which requires 1 s or less to collect.

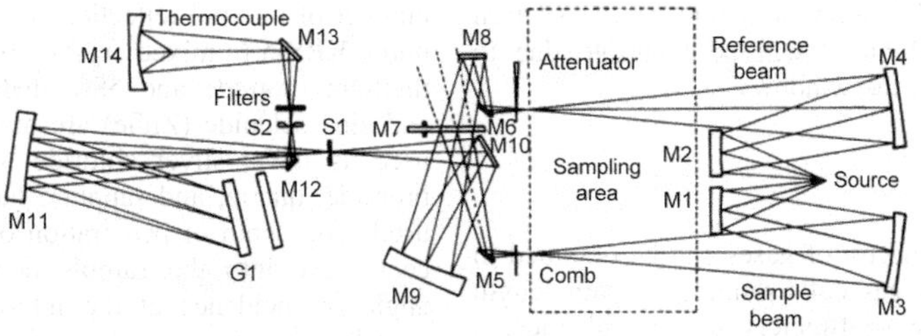

Fig. 1 Optical diagram of a double-beam grating spectrometer. M, mirror; G, grating; S, slit. Courtesy of Perkin-Elmer

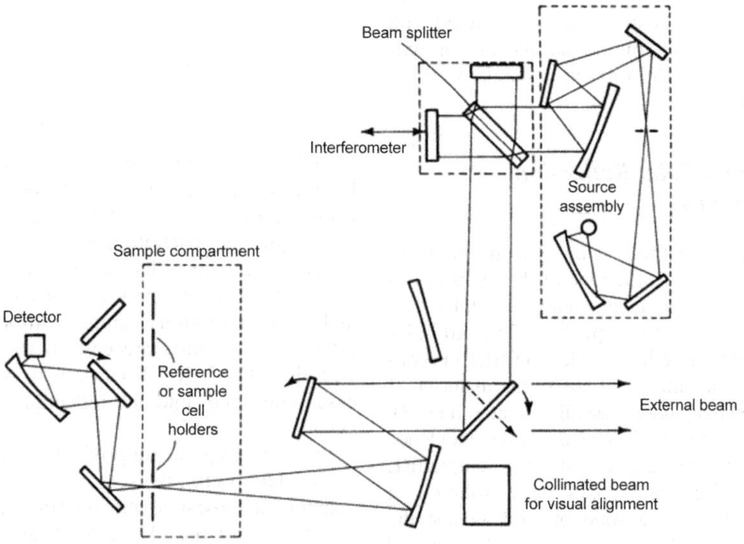

Fig. 2 Optical diagram of a Fourier transform infrared spectrometer. Courtesy of Digilab

A spectrum with N resolution elements can be collected in the same amount of time as a single-resolution element. This advantage is the multiplex, or Felgett's, advantage (Ref 8). Because the signal-to-noise ratio is proportional to the square root of the number of scans, the FTIR time advantage can become a signal-to-noise ratio advantage of $N^{1/2}$ for the same data-collection time as with a dispersive spectrometer. The greater the total number of resolution elements in the spectrum, the more important becomes Felgett's advantage.

Because no slit is required, FTIR spectrometers have a light throughput advantage over dispersive spectrometers (Jacquinot's advantage) (Ref 9, 10). The helium-neon laser that controls sampling of the interferogram also provides precise calibration for the wavenumber position (Connes's advantage) (Ref 11). Finally, because a digital computer is used to perform Fourier transforms, it is available for data processing. A computerized dispersive spectrometer can also handle data processing.

Interferometers other than the Michelson design are also available (Ref 12–15). The Genzel interferometer is similar, except the light beam is focused at the beam splitter, and the moving mirror modulates the optical path length in both arms of the interferometer (Ref 12). In other interferometers, optical path length is changed by moving a refractive element in one of the arms of the interferometer (Ref 13–15).

Tunable Infrared Lasers

Tunable infrared lasers can be used as an alternate method of obtaining single infrared wavenumber intervals (Ref 16). Although lasers can provide a tremendous amount of light intensity at each wavenumber compared to conventional broadband infrared sources, their stability is generally not as favorable, and they are usually tunable over only short wavenumber ranges, limiting their general usefulness as a tool for broadband spectroscopic measurements. Until recently, infrared lasers have been applied mainly for specific applications in which high resolution or only a limited portion of the spectrum is required. For example, infrared diode laser systems have been used in many process-monitoring situations in which the material of interest transmits almost no light. More broadly tunable infrared laser sources have been developed and are being used increasingly often to perform broadband spectroscopic measurements (Ref 17–20). Examples include pulsed tunable optical parametric oscillator (OPO) lasers, covering the spectral range from 4000 to 900 cm^{-1}, and quantum cascade lasers (QCLs). Broadband external cavity QCLs can be operated in either pulsed or continuous wave and typically cover the spectral range from 1900 to 800 cm^{-1}. The OPO, QCL, and broadband femtosecond pulsed laser sources are all starting to be used in commercial IR spectroscopic instrumentation. In particular, their use enables microspectroscopic measurements that far exceed the diffraction limit of conventional IR spectroscopy (3 to 10 μm) (Ref 21).

Sample Preparation, Sampling Techniques, and Accessories

Sample preparation is usually necessary in obtaining the infrared spectrum of a material. Most materials are almost totally opaque in infrared light and must be dissolved or diluted in a transparent matrix before the transmittance spectrum can be obtained. Sampling techniques and accessories are discussed in this section (Ref 22).

Nonvolatile Liquid Samples

Nonvolatile liquid samples can often be deposited as a thin film spread between two infrared-transmitting windows. A drop or two of sample is placed on the face of one of the clean, polished windows, and the second window placed on top. All the air bubbles must be squeezed out, and the sample must cover the entire window area. The windows are clamped using a screw-down holder, and the path length adjusted, if necessary, so the absorbance maximum of the strongest band is approximately 5 to 10% *T*.

Solvent Evaporation

Solvent evaporation can be used to deposit as a film a nonvolatile liquid or solid sample that is soluble in a relatively volatile solvent whose absorption bands would mask or interfere with those of the sample. A few drops (depending on the sample concentration) of the sample solution are transferred to the face of a cleaned, polished window, and the solvent allowed to evaporate. When the solvent appears to have evaporated, the spectrum is obtained. If solvent bands are apparent, it may be necessary to continue drying the sample until they disappear.

Meltable Solids

Meltable solids, that is, samples with melting points under approximately 60 °C (140 °F) and that will not decompose upon heating, can be melted and pressed between two infrared-transmitting windows. Approximately 10 to 30 mg of sample are usually transferred to the face of one window and placed on or under a heating source. When the sample has melted, the top window is pressed down and the air bubbles squeezed out to distribute the melted sample over the entire area of the windows. The windows can then be clamped together, and the sample thickness adjusted as with the neat film.

Potassium Bromide Pellets

Potassium bromide (KBr) pellets can often be prepared from solid samples that are difficult to melt or dissolve. The sample is dispersed in a KBr matrix and pressed into a transparent pellet. Approximately 0.4 to 1.0 mg of sample is usually ground in 200 to 400 mg of KBr or other infrared-transparent pressing powder. The sample and KBr must be ground so that the particle size is less than the wavelength of light, minimizing band distortion due to scattering effects. The matrix material must be pure and dry.

Preparing a Mull

Preparing a mull is also an alternative for a grindable solid sample. The sample is ground with the mulling agent to yield a paste that is examined as a thin film between infrared-transmitting windows. As opposed to the pellet technique, mulling agents are not as hygroscopic as KBr and are less likely to react with some samples. However, mulling agents have some infrared absorption bands that can interfere with absorptions in the sample. This can be overcome by preparing a split mull. Half the spectrum is obtained in hydrocarbon oil, and the other half in fluorocarbon oil. The region of the spectrum below 1365 cm^{-1} is taken from the hydrocarbon oil spectrum and combined with the 4000 to 1365 cm^{-1} portion of the fluorocarbon oil spectrum, yielding a spectrum free of mulling-agent bands.

Diamond-Anvil Cell

The diamond-anvil cell, a transmission accessory, is useful for very small single-particle samples or for obtaining infrared spectra under extremely high pressures (Ref 23). Beam-condensing optics are often necessary to focus the light beam on the sample, which is pressed between two small diamond windows. A screw and spring mechanism provides high pressures. The diamond cell is perhaps the best way to obtain a transmittance spectrum of samples such as a single hair. The sample size is well suited, and the high pressures can be used to compress the sample. This reduces the optical path length and prevents excessive intensity in the absorption bands. If thin type II diamonds are used, only a small region of the infrared spectrum around 2000 cm^{-1} will be unusable due to absorption of the windows.

Gas Cells

Infrared spectra of gases can be obtained in vacuumtight gas cells ranging in path length from a few centimeters to several meters. These glass or metal cells require valves to facilitate filling from an external vacuum system. Gas pressures necessary to obtain reasonable infrared spectra depend on the sample absorbance and the path length of the cell. Favorable spectra can usually be obtained with a partial pressure for the sample of approximately 6500 Pa (50 torr) in a standard 10 cm (4 in.) gas cell.

Attenuated Total Reflection Spectroscopy

Reflection methods can also be used to obtain infrared spectra. One of the more useful reflectance techniques is attenuated total reflectance (ATR) spectroscopy (Ref 24). An ATR plate or internal-reflection element (IRE) is constructed of an infrared-transparent material of high refractive index (usually greater than 2). Light enters the IRE at near-normal incidence and is internally reflected at each sample/IRE interface. The sample can be a solution, a deposited film, or a solid pressed against the IRE. Figure 3 shows a solid sample clamped against a multiple-internal-reflection element. If the angle of incidence at the interface exceeds the critical angle, θ_c, the light will be totally internally reflected. At each internal reflectance point, a standing, or evanescent, wave is generated that penetrates a short distance (on the order of the wavelength of light) into the sample. The detector senses intensity loss due to absorption of the evanescent wave by the sample. Multiple internal reflections can be used to build up signal. Monolayer quantities of adsorbed material can be detected by using this technique.

Internal-reflection elements are available in various shapes and sizes, depending on the geometry of the reflection optics and the number of internal reflections desired. Germanium, KRS-5 (a mixed crystal containing 42% thallium bromide and 58% thallium iodide), and zinc selenide (ZnSe) are most commonly used as IREs; silver chloride, silicon, silver bromide, quartz, and sapphire have also been used. The depth of penetration of the evanescent wave into the sample depends on the angle of incidence at the interface (θ_i), the wavelength of light (λ), and the refractive indexes of the sample and IRE (n_{sam} and n_{IRE}). Defining d_p as the distance required for the electric field amplitude to fall to e^{-1} of its value at the surface, the depth of penetration is (Ref 24):

$$d_p = \frac{(\lambda/n_{IRE})}{2\pi\left[\sin^2\theta_i - (n_{sam}/n_{IRE})^2\right]^{1/2}} \quad \text{(Eq 11)}$$

The depth of penetration can be decreased by increasing the angle of incidence or by using an IRE with a higher refractive index. The depth of penetration will be greater at lower wavenumber than at higher wavenumber. Table 1 shows how changes in λ, n_{IRE}, and θ_i for an arbitrary sample with $n_{sam} = 1.5$ affect d_p. No values are entered for $\theta_i = 30°$ with the KRS-5 IRE, because $\theta_c = 40°$ in this case, and total internal reflection does not occur.

In extracting depth-profiling information from ATR spectra, the spectrum of the sample component closest to the interface is always weighted most heavily in ATR spectroscopy, regardless of the value of d_p. Contributions to the spectrum from beyond d_p are also present, although less heavily weighted. In addition, establishing favorable contact between the IRE and the sample is important, especially with solid samples. Further, the refractive index of the sample is not constant as a function of wavenumber and can change dramatically in the region of an absorption band. This can distort the ATR spectrum, particularly in regions in which the refractive index of the sample is close to that of the IRE. Moreover, in changing the depth of penetration by varying the angle of incidence, an IRE cut at the desired angle should be used at normal incidence, or the effective angle of incidence at the IRE/sample interface will vary only slightly.

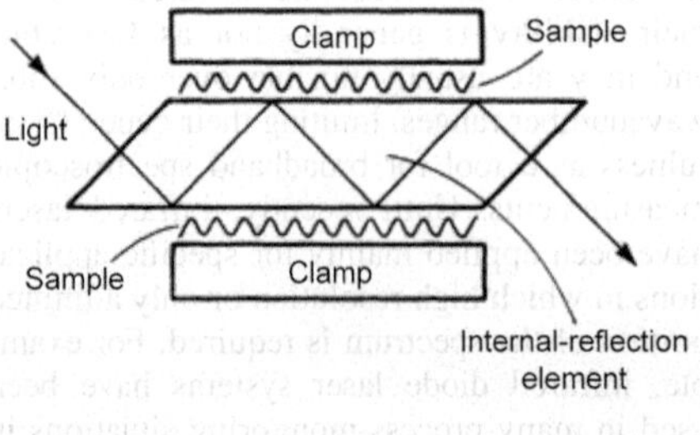

Fig. 3 Top view of micro-KRS-5 (thallium bromide/thallium iodide) internal-reflection element against which a solid is clamped

Table 1 Depth of penetration, d_p, for a sample (n_{sam} = 1.5) as a function of attenuated total reflectance plate material, angle of incidence, and wavenumber

Material	Wavenumber, cm[1]	d_p, µm 30°	d_p, µm 45°	d_p, µm 60°
KRS-5(a)	3000	···	0.74	0.33
KRS-5(a)	1000	···	2.23	1.16
Germanium(b)	3000	0.40	0.22	0.17
Germanium(b)	1000	1.20	0.66	0.51

(a) Refractive index, n = 2.35. (b) Refractive index, n = 4.0

When light enters a medium of high refractive index from air (n_{air} = 1.0) at nonnormal incidence ($\theta_i \neq 0$), it is refracted according to Snell's law:

$$n_{air} \sin \theta_i = n_{IRE} \sin \theta_r \quad \text{(Eq 12)}$$

where θ_r is the angle of refraction. Because $(n_{air}/n_{IRE}) < 1$, the angle of incidence at the IRE/sample interface will not differ greatly from the $\theta_i = 0$ (normal incidence) case. Table 2 shows the angle of incidence achieved at the IRE/sample interface using 45° KRS-5 and germanium IREs at 30°, 45°, and 60°, along with the corresponding values of d_p (for n_{sam} = 1.5). This is not an effective technique for depth profiling a sample using IR spectroscopy. Perhaps the best overall method of depth profiling using only ATR is to alternate between two 45° IREs of germanium and KRS-5 (Table 1). The refractive indexes of ZnSe and diamond IREs are essentially the same as KRS-5, so the values of d_p and the optical band distortion effects noted earlier apply equivalently to these more commonly used materials. In addition to providing the most surface-sensitive ATR spectra, germanium IREs, with their high refractive index of 4.0, better reproduce the proper peak positions of strong IR-absorbing bands, which can shift to lower wavenumber by as much as 10 cm^{-1} or more compared to transmission spectrum values when using lower-refractive-index (2.5) IREs, such as KRS-5, ZnSe, and diamond.

Attenuated total reflection measurements can be performed on dispersive or Fourier transform instrumentation. When attempting ATR on a double-beam dispersive spectrometer, a matched pair of IREs and reflection optics are usually used to improve the baseline. A clean IRE is used as a reference in one beam, and the sample is placed on the IRE in the other beam path. With FTIR spectrometers, the reference and sample are usually analyzed sequentially in the same beam using the same optics. Attenuated total reflection spectra recorded on FTIR instruments are generally superior to dispersive ATR spectra.

Sensitivity improvements in FTIR spectrophotometers over the years have enabled ATR spectra to be measured with single-bounce accessories. This makes sample preparation much easier, and many low-cost FTIR instruments now are typically fitted with such single-bounce accessories. For this reason, ATR has become the approach of choice for obtaining infrared spectra in many analytical laboratories.

Diffuse Reflectance Spectroscopy

Diffuse reflectance spectroscopy (DRS) is another reflectance technique that has application in the infrared region of the spectrum (Ref 25). Until the advent of FTIR spectroscopy, the technique had been used almost exclusively in ultraviolet (UV), visible (VIS), and near-infrared (NIR) regions of the spectrum, where brighter sources and more sensitive detectors exist.

Infrared radiation is focused on a cup filled with the sample; the resulting diffusely scattered light is collected and refocused on the detector. Although used in most UV-VIS-NIR applications to collect scattered radiation from the sample, integrating spheres are not very efficient in the midinfrared region. Commercial attachments for FTIR spectrometers typically incorporate large ellipsoidal mirrors for focusing and collecting the light. The technique can be used on bulk samples or samples ground in a KBr or potassium chloride (KCl) matrix. Potassium bromide and KCl powders are excellent diffuse reflectors and are also used as reference standards. Spectra of matrixed samples are similar in appearance to KBr-pellet absorbance spectra when plotted in units proportional to concentration. Such units are log $(1/R)$ or $(1 - R)^2/2R$ (Kubelka Munk units) (Ref 26, 27), where $R = R$(sample)/R(reference) is the sample reflectance measured relative to the reference standard. In DRS, the sample matrix need not be pressed into a transparent pellet but simply packed loosely in a sample cup. Scattering artifacts that often occur in cloudy KBr pellets are not such a problem with matrixed diffuse reflectance samples.

Diffuse reflectance spectra collected on bulk, unmatrixed samples should be interpreted with extreme caution. Unless the diffuse reflectance accessory is designed and aligned so that little or none of the light reflected directly from the surface of the sample (specular component) reaches the detector, the observed spectrum will be a complicated combination of diffuse and specular components. Diffuse reflectance signals will appear as a decrease in reflected intensity versus a KBr or KCl reference standard, due to absorbance by the sample. However, specular reflectance signals can have a positive or negative sign, depending on the optical constants of the sample.

Diffuse reflectance spectroscopy can also be used to study adsorbed species on catalyst surfaces. This approach is suitable for high-surface-area infrared-transparent substrates. Evacuable cells that can be heated have been designed for observing catalysts under reaction conditions.

Table 2 Actual angle of incidence (at the interface between the attenuated total reflectance plate and sample) and depth of penetration at 1000 cm^{-1} as a function of attenuated total reflectance plate material and apparent angle of incidence

Material	Apparent angle	Actual angle	d_p, µm
KRS-5	3°	38.7°	...
KRS-5	45°	45°	2.23
KRS-5	60°	51.3°	1.51
Germanium	30°	41.3°	0.73
Germanium	45°	45°	0.66
Germanium	60°	48.7°	0.61

Infrared Reflection-Absorption Spectroscopy

Infrared reflection-absorption spectroscopy (IRRAS) is a useful technique for studying material adsorbed on flat metal surfaces (Ref 28). Unlike many other surface techniques, IRRAS does not require ultrahigh vacuum and provides more information on molecular structure and functional groups. Using a single external reflection at near-grazing angle of incidence, single monolayers adsorbed on low-area surfaces (flat metal plates) can be detected. Figure 4 illustrates a typical external reflectance attachment. Because the component of the incident light polarized parallel to the surface (perpendicular to the plane of incidence) has a node at the surface, only those dipole-transition moments of adsorbed molecules with a component perpendicular to the surface are observed. Therefore, information on the orientation of the adsorbed material can be obtained if the dipole-transition moment directions are known.

Infrared reflection-absorption spectroscopy experiments can be performed on dispersive and FTIR spectrometers. When double-beam dispersive instruments are used, identical reflection devices are usually aligned in each beam path, one path containing a clean metal surface (reference) and the other the metal surface with the adsorbed film (sample). The difference signal due to the adsorbed film is then recorded directly. Because commercial FTIR spectrometers are single-beam instruments, the reference and sample metal surfaces

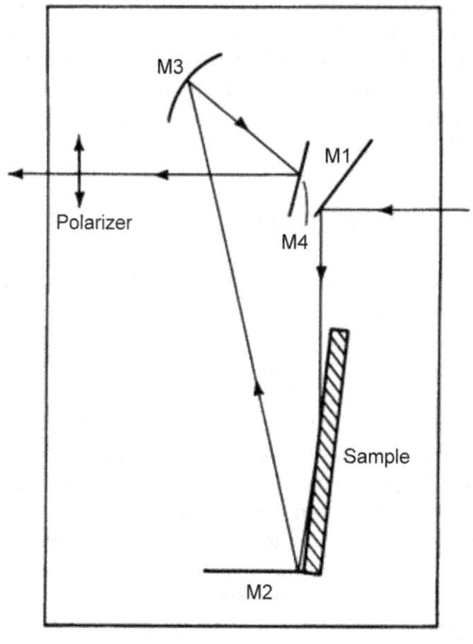

Fig. 4 Top view of a sample compartment containing a reflection-absorption attachment. M, mirror

must be analyzed sequentially, and the difference or ratio taken later. The signal difference of interest is usually extremely small compared to the size of the individual single-beam signals. Favorable signal-to-noise ratios and a stable interferometer are required to detect adsorbed monolayers on metals. Inserting a linear polarizer oriented to pass light polarized perpendicular to the surface (p-polarized component) in the beam path may improve sensitivity. Adsorbed films significantly thinner than the wavelength of light will not absorb the component of polarization oriented parallel to the surface (s-polarized component).

Polarization Modulation

Polarization modulation can be used to measure infrared reflection-absorption spectra of adsorbed monolayers in a single-beam experiment (Ref 29, 30). A linear polarizer is oriented to pass p-polarized light and is followed by a photo-elastic modulator (PEM) oriented with the stress axis at 45° to the plane of polarization. The optical element of a PEM is a highly isotropic infrared-transmitting material, the most common being ZnSe and calcium fluoride (CaF_2). Two piezoelectric drivers cemented to either side of the optical element induce a stress birefringence that rotates the plane of polarization 90° when the retardation is 180° or one-half the wavelength of the incident light.

Photo-elastic modulators typically modulate at approximately 50 kHz and flip the plane of polarization between the two perpendicular states at exactly twice that modulation frequency. Because monolayers adsorbed on metal substrates "see" only the p-polarized component, a spectrum attributable to the difference in reflectivity of the p and s components can be detected directly at twice the PEM modulation frequency. This polarization-modulation approach can also be used to measure linear dichroism of oriented samples and vibrational circular dichroism of optically active compounds on dispersive or FTIR spectrometers.

Specular Reflectance

Specular reflectance refers to the component of incident light that bounces off the surface of the sample (Ref 31). Unlike reflection-absorption spectroscopy, in which the films are thin and the light reflects off a metal substrate, specular reflectance can be performed without a metal substrate on thick samples or films. Under these conditions, the incident electric field vectors do not have a node at the surface, and angles closer to normal incidence can be used. However, the sensitivity does not equal that of reflection-absorption spectroscopy, and spectra often contain a significant contribution from the refractive index of the sample. Extracting useful information from these spectra may be difficult.

Emission Spectroscopy

Emission spectroscopy is another technique for obtaining infrared spectra of difficult samples (Ref 32). The principal applications are for remote samples, such as stars or smokestack emissions, and thin films adsorbed on metals. The sample becomes the source in emission spectroscopy. For weak emission signals, the spectrometer temperature should be less than the sample temperature, or infrared emission from the background and spectrometer optics may be larger than the source signal of interest. Heating the sample is usually more convenient than cooling the spectrometer. The sample is often placed on a heating element in a location optically equivalent to the normal source position. For remote emission studies, suitable optics direct the light into the interferometer or monochromator. A reference signal is obtained by positioning a blackbody, such as a metal plate painted black, in place of the sample and holding it at the same temperature as the sample to be measured. The emission spectrum of the sample is the ratio of the emission of the sample and blackbody reference.

Emission signals can result when molecules in an excited vibrational state return to the ground state. Spectral emissivity equals spectral absorptivity. However, the observed emission spectrum of thick samples can become complicated when light emitted from the interior of the sample is self-absorbed by the outer part of the sample before detection.

Photoacoustic Spectroscopy

While in an excited vibrational state, a molecule can return to its ground vibrational state through the mechanism of emission, fluorescence, or nonradiative transfer of the vibrational energy to the lattice as kinetic energy. Fluorescence in the infrared is rare. Most molecules return to the ground state by the third mechanism. If the light that initially excited a molecule is modulated by a light chopper or interferometer, the kinetic energy in the lattice as a result of the nonradiative relaxation to the ground vibrational state is also modulated. This is the photoacoustic effect.

In photoacoustic spectroscopy (PAS), a microphone or piezoelectric device detects the lattice modulations, and the signal amplitude is proportional to the amount of light the sample absorbs (Ref 33, 34). Photoacoustic spectroscopy is suitable for highly absorbing samples that are difficult to measure by transmission. In addition, because the PAS signal is proportional to the total amount of light absorbed, bulk samples can usually be analyzed without dilution or any other sample preparation. Modulation signals from solid samples are usually detected after the acoustic waves in the solid lattice are transferred to gas molecules. The microphone detects the acoustic wave in the gas. Photoacoustic spectroscopy is more sensitive to infrared absorbances by gases than solids, because the acoustic wave need not be transferred from the solid. The PAS sample cell must be carefully sealed, free of absorbing gases, and isolated vibrationally and acoustically from the environment.

Although PAS has been performed using dispersive spectrometers, it is far easier using FTIR spectroscopy. On a FTIR spectrometer, a modulated interferogram signal at the microphone is Fourier transformed in the usual manner. Each wavenumber is modulated at an independent frequency. The lower the modulation frequency, the deeper in the sample is the origin of the overall PAS signal. Thus, lower-wavenumber absorbances come from deeper in the sample than higher-wavenumber absorbances, and the resulting spectrum tends to be skewed in a manner similar to ATR spectra. Depth profiling can be achieved by varying mirror velocity. The depth of penetration depends on the modulation frequency, the thermal diffusion length of the sample, and the extinction coefficient of the sample. This value can vary dramatically from one sample to another but is typically between 10 and 100 μm.

Although suitable PAS spectra can be obtained within minutes on many FTIR spectrometers, other techniques, such as ATR or diffuse reflectance, often yield higher-quality spectra in less time. Photoacoustic spectroscopy is seldom the infrared technique of first choice, although it can be useful in specific applications such as analyses of optically thick samples (Ref 33, 34).

Other Photothermal Infrared Methods

Two photothermal infrared approaches analogous to PAS have recently emerged for directly detecting the infrared absorbance of a sample. Instead of measuring the photothermal expansion of a solid sample by using a microphone to detect the sound generated in a sealed cell filled with helium gas, the absorbance can be detected directly by contacting the tip of an atomic force microscope (AFM) to the sample surface (Ref 21, 35, 36). The AFM cantilever oscillates at its resonant frequency when perturbed by the photothermal expansion of the sample. The amplitude of the cantilever ring-down oscillation is directly proportional to the amount of infrared radiation actually absorbed by the sample at a particular infrared wavenumber. The infrared source in this AFM-IR experiment typically is a pulsed, tunable laser (such as an OPO laser or QCL) that is scanned one wavenumber at a time through the infrared spectral region. The spatial resolution of an AFM-IR photothermal measurement ultimately is limited by the diameter of the AFM tip, typically between 10 and 25 nm.

Alternatively, a visible laser can be used to detect the photothermal expansion of the sample following absorbance of infrared radiation into a molecular vibration (Ref 37, 38). The measurement of the visible laser can be made in either transmission. In this case, the spatial resolution of the infrared measurement is on the same order of the visible laser wavelength selected and is typically a factor of approximately 20 better than the diffraction-limited spatial resolution of a conventional infrared measurement (3 to 10 μm). Such an instrument can also be configured to simultaneously collect the Raman scattered signal. Thus, complementary infrared and Raman spectral information can be collected simultaneously on the same sample spot at exactly the same spatial resolution (Ref 39).

Chromatographic Techniques

The ability of Fourier transform spectrometers to obtain infrared spectra within 1 s allows interfacing FTIR spectroscopy with various chromatographic techniques. The most advanced and useful of these techniques is gas chromatography-infrared (GC-IR) spectroscopy (Ref 40, 41). Although not as sensitive as the more widely used technique of gas chromatography/mass spectrometry (GC/MS), GC-IR spectroscopy can provide complementary information. Gas chromatography/mass spectrometry provides molecular-weight information but cannot satisfactorily distinguish isomers. Gas chromatography-infrared spectroscopy can usually distinguish isomers effectively. Using capillary columns and flow-through infrared cells (light pipes) consisting of gold-coated glass tubes 10 to 30 cm (4 to 12 in.) long with inside diameters of 1 to 2 mm (0.04 to 0.08 in.), 10 to 20 ng of typical infrared absorbers can be detected in real-time ("on the fly") during gas chromatography. With some FTIR systems, spectra may be stored every second during a GC-IR spectroscopy operation lasting 1 h or more. The infrared spectra of each of the chromatographic peaks, which can number 100 or more, can be searched against libraries of vapor-phase spectra or can be interpreted manually. Much of the computer software developed for GC-IR spectroscopy can be used to study the kinetics of such processes as chemical reactions, polymer curing, and adsorption on surfaces from solution.

Gas chromatography-infrared spectroscopy usage remains well below that of GC/MS primarily because of its relative lack of sensitivity. Although some small incremental improvements in GC-IR using light-pipe technology still occur, an approach that uses a matrix-isolation interface and is approximately 2 orders of magnitude more sensitive has been developed (Ref 42). Approximately 1.5% of an argon-matrix gas is mixed with helium carrier gas in the gas chromatograph. The gas chromatograph effluent is frozen onto a rotating cryogenic (12 to 13 K) gold-coated disk surface. A gas chromatography peak can be concentrated in the argon matrix to an area less than 0.2 mm^2 (0.0003 $in.^2$) while the helium carrier gas is pumped away by the vacuum system. The infrared beam is focused onto the matrix band, reflected off the gold-coated disk, and refocused on an infrared detector after passing out of the vacuum system. The improved sensitivity of this approach derives from the reduced cross-sectional area of the sample, which results in a longer path length for the same amount of sample. In addition, matrix-isolation spectra generally have sharper bands with higher peak absorbances, the system has higher light throughput, and gas chromatography peaks can be held in the beam and scanned for much longer times.

High-performance liquid chromatography (HPLC) (Ref 43) and supercritical fluid chromatography (SFC) (Ref 44) are examples in which chromatographs have been interfaced with FTIR spectrometers. Liquid flow cells are available for HPLC-IR interfaces. Unlike GC-IR spectroscopy, in which a nonabsorbing carrier gas (helium) is used, all liquid chromatography solvents absorb somewhere in the infrared region of the spectrum. Overlapping bands from the solvent often complicate interpretation of the spectra. One advantage of HPLC-IR over GC-IR spectroscopy is that effluent peaks of interest may often be isolated and analyzed later after elimination of the solvent. It is not as crucial that the spectra be taken in real-time ("on the fly"). Although SFC-IR has potential, it is in an early stage of development. Carbon dioxide, a common SFC solvent, is transparent through the most critical portions of the infrared spectrum. It absorbs strongly only near 2350 and 667 cm^{-1}.

Infrared Microsampling

Infrared microsampling can be performed by using many of the techniques discussed. Spectra of approximately 1 ng of sample are obtainable with diffuse reflectance and ATR, assuming the sample can be dissolved in a volatile solvent. The diamond-anvil cell can be used to obtain infrared spectra of single small particles. Infrared microsampling accessories are available (Ref 45). One design uses 32× Cassegrainian optics to focus the light on the sample. Spectra of samples 15 μm in diameter can be obtained. The stage can be translated, and the exact area to be sampled can be viewed by using normal microscope optics. Single small particles or a high-resolution infrared-spectral map can be measured.

Infrared Microscopes

Transmission, specular reflection, and ATR versions of the infrared microscope are available. The reflectance scope is designed for use when the sample of interest is optically too thick. However, spectra recorded using the reflectance microscope exhibit potential optical artifact complications of regular specular reflectance spectra.

The ultimate spatial resolution of the infrared microscope is determined by the diffraction limit. When the aperture size is nearly the same as the wavelength of light, band shape distortions due to diffraction begin to occur. These distortions appear first in the lower-wavenumber regions of the spectrum.

Qualitative Analysis

Qualitative identification is an important use of IR spectroscopy (Ref 46). The infrared spectrum contains much information related to molecular structure. The most positive form of spectral identification is to locate a reference spectrum that matches that of the unknown material. Many collections of reference spectra are published, and many of these volumes are being computerized. When an exact reference spectrum match cannot be found, a band-by-band assignment is necessary to determine the structure. In this case, the infrared spectrum alone will usually not be sufficient for positive identification. Nuclear magnetic resonance and mass spectrometry measurements may often be necessary to confirm a molecular structure.

The interpretation of infrared spectra has recently become computerized. Vapor-phase spectral databases have been used to assist identification of GC-IR spectra. Libraries of condensed-phase infrared spectra are also gradually being enlarged, although the number of compounds in the available databases is only a small fraction of the total number of known compounds.

Infrared spectra databases can be searched in several ways. The important factors are peak locations, band intensities, and band shapes. Because a typical infrared spectrum contains approximately 2000 data points, many databases and their corresponding search strategies are designed to minimize the amount of data needed for searching without sacrificing a great deal of spectral selectivity. This can shorten search times and reduce the storage space required for the spectral libraries. It is more practical to search for unknown infrared spectra in laboratories in which large volumes of samples are analyzed daily and the compounds of interest are likely to be in existing databases. Even when an exact match to an unknown spectrum cannot be found in a database, search routines usually list the closest matches located. This can be useful in identifying at least the molecular type.

Absorbance-Subtraction Techniques

Absorbance-subtraction techniques, or spectral-stripping techniques, can be useful in interpreting spectra of mixtures or in removing solvent bands. Spectral subtraction should be attempted only when the spectrum is plotted in absorbance or some other units proportional to sample concentration. In addition, absorbance subtractions should be interpreted with extreme caution in any region in which sample absorbance exceeds approximately 0.6 to 0.8 absorbance units. Subtractions can be performed with or without scaling. An isolated band in the spectrum of the component to be subtracted out can often be minimized in intensity by interactively observing the difference between the two spectra as a function of scale factor. Subtraction of solid ATR spectra should be avoided because of the wavelength dependence of the depth of penetration and the difficulty in obtaining reproducible contact of the solid with the IRE. Absorbance subtractions are most effective when the optics are not perturbed between scans of the two spectra involved in the subtraction, for example, kinetic studies or flow cell systems.

Factor Analysis

Factor analysis is a mathematical procedure for determining the number of components in a set of mixtures (Ref 47). Knowledge of spectra of the pure components is unnecessary. A matrix of mixture spectra (**A**) is constructed with each column vector representing the infrared spectrum of one mixture. This matrix is multiplied by its transpose ($\mathbf{A}^T$) to yield an $\mathbf{m} \times \mathbf{m}$ square matrix (**C**), where **m** is the number of mixture spectra. The **C** matrix is then diagonalized, and the number of nonzero eigenvalues represents the number of independent components. Noise in the spectra can cause small eigenvalues, making it difficult to distinguish nonzero and zero eigenvalues. The presence of numerous components, some having similar infrared spectra, complicates factor analysis. Factor-analysis Fortran programs are available with many infrared software packages.

Resolution-Enhancement Methods

Infrared spectra of condensed phases often contain many overlapping bands that cannot be resolved even by obtaining the spectra at high resolution, because the natural linewidths of the spectra limit resolution. However, spectral lineshapes contain information on these broad overlapping bands. Several methods exist for enhancing the resolution of infrared spectra. Derivative spectroscopy can be used to determine the exact peak location of broad profiles. Second and fourth derivative spectra are much sharper, and more bands may appear; however, the signal-to-noise ratio deteriorates with each additional derivative.

Fourier self-deconvolution is another method of resolution enhancement (Ref 48). A region of an infrared absorption spectrum containing overlapping bands is Fourier transformed back to the time domain using the fast Fourier transform algorithm. The resulting damped ringing pattern is multiplied by a function (usually containing an exponential term) that weights the tail portion of the pattern more heavily. The net effect is an interferogram whose tail extends farther in the direction of increasing time. This ringing pattern must be truncated before the noise becomes significant. The interferogram is Fourier transformed back to the frequency domain, where the resulting absorbance bands are now narrower.

Two parameters are varied during this procedure: the bandwidth and the number of points in the time domain spectrum retained before truncation. If the selected bandwidth is too broad, negative lobes will appear in the deconvolved spectrum. If too many points in the time domain spectrum are retained, noise will be deconvolved. Proper bandwidth selection is important. It is impossible to deconvolve bands of significantly different bandwidth simultaneously and completely.

Quantitative Analysis

The basis for quantitative analysis in infrared spectroscopy is Beer's law, which for a single compound at a single wavenumber is:

$$A = abc \qquad \text{(Eq 13)}$$

where A is the sample absorbance at a specific wavenumber, a is the absorptivity of the sample at that wavenumber, b is the pathlength, and c is the concentration. In practice, Beer's law may not hold due to matrix or concentration effects that change absorptivity. Calibration curves must be obtained to confirm linearity of the relationship between A and c.

As in spectral subtraction, derivative methods, and Fourier self-deconvolution, quantitative infrared analysis should not be attempted unless the spectra are plotted in absorbance units. Infrared intensity can be determined with favorable results by measuring peak height or the area under the absorption band. In either case, intensity measurements must be made relative to some baseline. Baseline determination can be subjective, particularly when there are overlapping bands, and can be a major source of error in intensity measurements. Consistent procedures for measuring intensities provide optimum results.

Before the concentration of a component in a mixture can be determined from the absorbance spectrum, the absorptivities of bands sensitive to the presence of that component must be known. This is usually accomplished by obtaining spectra of calibration standards for which the concentrations are known. Infrared bands sensitive to the presence of the component of interest are then determined. Linear plots of absorbance versus concentration indicate the validity of Beer's law over the concentration range of the calibration set for the bands selected. The path length is usually held constant or accurately measured. When the absorbance of a sample component of unknown concentration is measured, concentration is determined using the Beer's law plot.

Matrix Methods

Matrix methods are needed in analyzing complex mixtures. Computerized infrared spectrometers and commercial software packages facilitate multicomponent analysis on many instruments. A basic understanding of multicomponent analysis can help in avoiding errors. The **K**-matrix method is one approach to multicomponent analysis (Ref 49).

In the **K**-matrix method, absorptivity multiplied by the path length is defined as a single constant, k, and Beer's law becomes:

$$A = kc \qquad \text{(Eq 14)}$$

Assuming Beer's law is additive, the absorbance at frequency i of sample j is:

$$a_{ij} = \sum_{l=1}^{n} k_{il} C_{lj} \qquad \text{(Eq 15)}$$

where the summation is over all components, l, from 1 to n. Equation 14 can be written in matrix form as:

$$A = \mathbf{KC} \qquad \text{(Eq 16)}$$

The **K** matrix is determined from the calibration spectra. The number of calibration samples and frequencies used should each exceed or equal the number of components, n. Solving for the **K** matrix yields:

$$\mathbf{K} = A\mathbf{C}^T(\mathbf{CC}^T)^{-1} \qquad \text{(Eq 17)}$$

where the superscript T is the transform of the matrix, and the superscript −1 is the inverse of the matrix. The unknown concentrations of the components in the mixture can then be obtained from the absorbance spectrum of the mixture by using:

$$\mathbf{C} = (\mathbf{KK}^T)^{-1}\mathbf{K}^T A \qquad \text{(Eq 18)}$$

Sources of Error

Several potential sources of error may arise in quantitative infrared analysis. The matrix ($\mathbf{CC}^T$) must be nonsingular to be inverted. Therefore, none of the rows or columns of C should be linear combinations of other rows or columns, that is, determinant ($\mathbf{CC}^T$) $\neq 0$. The **K**-matrix method assumes that a linear relationship exists between the absorbances and the concentrations and that Beer's law is additive in the multicomponent case. The theory does not rigorously account for band shifts or absorptivity changes due to interaction of the components. The calibration data are always least squares fitted to the linear expression (Eq 16).

Other sources of error may occur during experimental measurement. The photometric accuracy of the instrument is important in quantitative analysis. Apparent breakdowns of Beer's law may result from spectrometer nonlinearities rather than sample-component interactions. Many commercial infrared instruments become nonlinear when sample absorbances approach l. Other instrument manufacturers claim ability to measure absorbances linearly to values as high as 3. Regardless of the photometric accuracy specification of the instrument, detection of stray light signals that do not pass through the sample can cause serious errors in absorbance measurements. For example, if during a transmittance measurement the sample solution contains an air bubble that allows passage of 1% of the total light, the largest absorbance that can be recorded at any frequency is 2. This will cause errors in the measured absorbance values that worsen progressively as sample absorbance increases.

Scattering artifacts, such as those described earlier, can severely affect quantitative accuracy and precision. Spectra that contain significant contributions from the refractive index can result in inaccurate peak heights or peak areas. Scattering effects can also cause sloped baselines that complicate intensity measurement. Noncontinuous samples can also lead to errors in quantification. Films that are not uniformly thick or inhomogeneous powders for which a representative aliquot is not collected can lead to absorbance spectra that do not reflect the average composition of the entire sample. Overlapping bands from atmospheric absorptions, such as water vapor and carbon dioxide, can also affect intensity measurements. Finally, sample temperature can affect band shapes and intensities due to phase transitions or changes in sample emission.

Curve Fitting

Curve fitting is a final alternate method of infrared quantitative analysis (Ref 50). When the spectrum of a mixture is known to consist of specific pure components, and spectra of these components are available, it should be possible to generate a linear combination of the pure component spectra that reproduces the spectrum of the mixture. The coefficients then represent the amount of each pure component in the mixture. This is a specific case of the **K**-matrix approach in which the **C** matrix is a square matrix with ones along the diagonal and zeros elsewhere, and every frequency in the spectrum is used in the **K** and **A** matrices. Each mixture spectrum to be analyzed must contain only the pure components anticipated, and bands must not occur due to interaction of the components. Curve fitting usually is effective only with fewer than five components and with significantly differing pure component spectra.

Applications

Example 1: Factor Analysis and Curve Fitting Applied to a Polymer Blend System

Application of factor analysis and curve fitting to a polymer blend system of polystyrene (PS) and poly-2,6-dimethyl-1,4-phenylene oxide (2MPPO) has been documented (Ref 47). Five polymer blend (polyblend) films of PS and 2MPPO as well as the two pure-component films were prepared, and factor analysis was applied to the 3200 to 2700 cm^{-1} region of their infrared spectra. Figure 5 shows a plot of the log of the eigenvalue versus possible number of components. This plot indicates that the spectra of each of the several samples can be expressed as a linear combination of two spectra. By least-squares curve fitting the pure-component infrared spectra of PS and 2MPPO in the 3200 to 2700 cm^{-1} region, the composition of each polyblend film can be accurately determined (Table 3). Figure 6 illustrates the quality of the fit of the pure-component PS and 2MPPO spectra to the 1:3 PS/2MPPO polyblend.

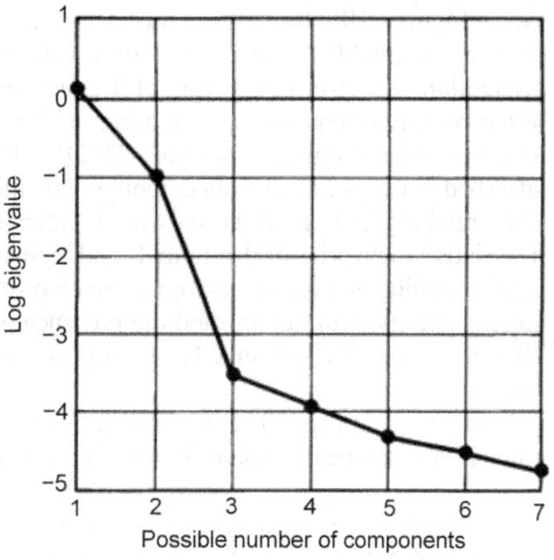

Fig. 5 Factor analysis from 3200 to 2700 cm^{-1} of polystyrene/poly-2,6-dimethyl-1,4-phenylene oxide polymer blend system. Source: Ref 47

Table 3 Compositional analysis of polystyrene (PS)/ poly-2,6-dimethyl-1,4-phenylene oxide (2MPPO) polyblend films by least-squares curve fitting 3200 to 2700 cm^{-1}

Known		Calculated	
wt% PS	wt% 2MPPO	wt% PS	wt% PPO
50.0 ± 0.5	50.0 ± 0.5	50.00 ± 0.38	50.00 ± 0.54(a)
75.0 ± 0.5	25.0 ± 0.5	76.33 ± 0.52	23.67 ± 0.73
25.0 ± 0.5	75.0 ± 0.5	25.18 ± 0.50	74.82 ± 0.71
90.0 ± 0.5	10.0 ± 0.5	91.27 ± 0.30	8.73 ± 0.42
10.0 ± 0.5	90.0 ± 0.5	9.95 ± 0.63	90.05 ± 0.89

(a) Calibration. Source: Ref 47

Factor analysis applied to the 1800 to 1100 cm^{-1} region of the PS/2MPPO polyblend spectra indicated the presence of the independent components. This region of the infrared spectrum is evidently more sensitive to conformation effects, suggesting the occurrence of a conformation transition in one of the blend components.

Example 2: Examination of Structural Changes in Surfactant Molecules in Water

Infrared spectroscopy can be used to study molecular aggregation in dilute aqueous solutions (Ref 51). Figure 7 shows a transmittance spectrum in the CH-stretching region of a 10 m*M* solution of $C_{12}H_{25}N(CH_3)_3Cl$ in deuterium oxide (D_2O). A 50-μm CaF_2 cell was used for the sample and reference (D_2O). As the surfactant concentration is increased, the bands shift to lower wavenumber and become narrower. The concentration at which the shift begins coincides with the known critical micelle concentration of 21 m*M* (Fig. 8). Similar results were obtained in water. Problems with window

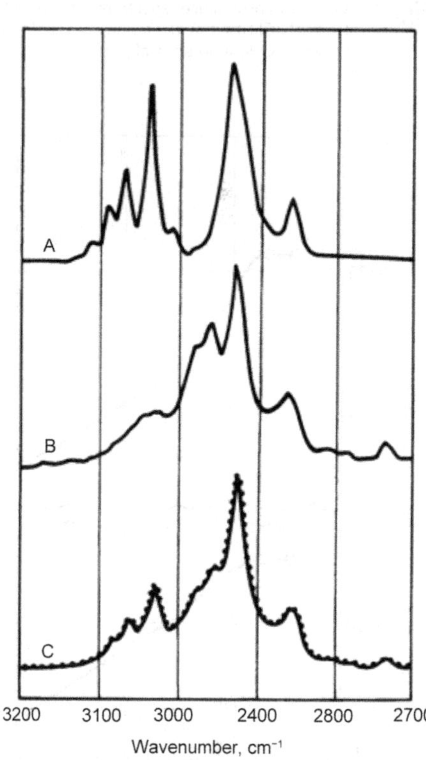

Fig. 6 Accuracy of fit of polystyrene (PS) and poly-2,6-dimethyl-1,4-phenylene oxide (2MPPO) spectra to a polymer-blend spectrum by least-squares curve fitting from 3200 to 2700 cm^{-1}. A, PS; B, 2MPPO; C, experimental PS/2MPPO polyblend spectrum (solid line). Best least-squares fit of PS plus 2MPPO (dotted). Source: Ref 47

solubility and/or adsorption onto the CaF_2 windows prevented obtaining data below 10 m*M*.

Similar, more recent experiments have been conducted with other surfactants by using a cylindrical internal reflection cell with a ZnSe IRE (Ref 52, 53). Adsorption and solubility seem to be less of a problem with this device, and the lowest concentration for detection of the spectrum with reasonable signal-to-noise ratios has been improved by an order of magnitude. The infrared transmission method has also been used to study structural changes in aqueous solution as a function of temperature and to characterize bilayer-to-nonbilayer transitions (Ref 54).

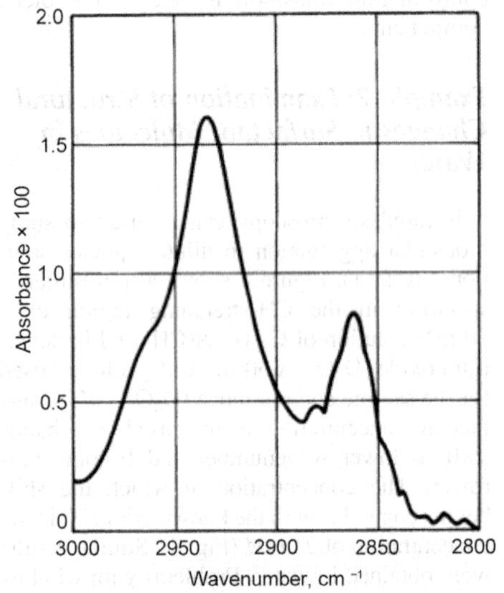

Fig. 7 Infrared-transmittance spectrum of 10 m*M* $C_{12}H_{25}N(CH_3)_3Cl$ in deuterium oxide (D_2O) versus D_2O in a 50 μm CaF_2 cell

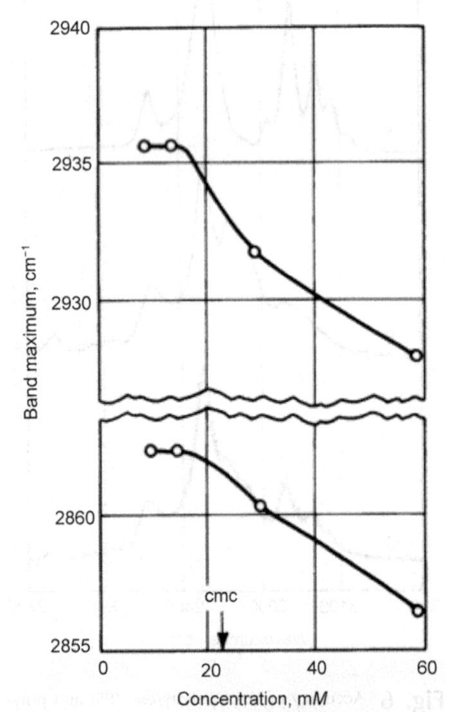

Fig. 8 Frequency of the CH_2 antisymmetric (top) and symmetric (bottom) stretching modes as a function of $C_{12}H_{25}N(CH_3)_3Cl$ concentration in deuterium oxide. cmc, critical micelle concentration

Example 3: Examination of Monolayers Adsorbed on Metal Surfaces

Structure transformations similar to those observed in aqueous solution are apparent in the infrared reflection-absorption spectra of monolayer films adsorbed on metal surfaces. Figure 9 illustrates the CH-stretching region of $C_{14}H_{29}$dimethylammonium hexanoate ($C_{14}AH$) adsorbed on polished carbon steel coupons. Spectrum A, of a dry coupon that had been soaked 5 days in a 0.1% aqueous solution of $C_{14}AH$, exhibits broad CH_2-stretching bands centered at 2925 and 2854 cm^{-1}. Spectrum B, of a dry coupon soaked 5 days in a 1.0% $C_{14}AH$ aqueous solution, has much narrower bands that are shifted to 2920 and 2850 cm^{-1}. Spectrum B suggests a more ordered structure on the surface than spectrum A.

The IRRAS spectra in Fig. 9 do not clearly reveal any molecular orientation on the surface. Figure 10, however, shows spectral changes assignable to differences in molecular orientation. Spectrum A is an ATR spectrum of a randomly oriented evaporated film of ditallowdimethylammonium chloride (DTDMAC) adsorbed on a polished copper coupon in the CH-stretching region. It is shown to indicate the relative intensities of the methyl- and methylene-stretching bands in a randomly oriented situation. The spectrum confirmed the presence of 10 times more CH_2 than CH_3 groups in the molecule.

Spectrum C in Fig. 10 is of a polished copper coupon that had been soaked 30 min in a dilute aqueous solution of DTDMAC. The relative intensities of the methyl- and methylene-stretching modes suggest an attenuation of the CH_2 bands and a slight enhancement of the CH_3 bands. This result can be explained in terms of an orientation effect. A tendency of the hydrocarbon tails to orient vertically from the surface in an all-trans configuration would lead to significant attenuation of the CH_2 antisymmetric and symmetric stretching motions at 2920 and 2854 cm^{-1}. The dipole-transition moments for these molecular vibrations would lie in a plane parallel to the surface, making them inactive according to the surface selection rule. However, the methyl-stretching vibrations would have significant components of their dipole-transition moments perpendicular to the surface for vertically oriented hydrocarbon chains.

Determining the orientation of DTDMAC on nonmetallic surfaces using IR spectroscopy presents two problems. First, nonmetallic substrates absorb infrared radiation, and the bulk substrate spectrum usually severely overlaps the desired surface spectrum of interest. Second, because the surface selection rule holds only for metallic substrates, if an infrared spectrum is obtainable, the orientation information is lost. One solution is to form a thin-film model substrate adsorbed on the metal surface and deposit the monolayer film of interest on top of this model surface. If the model substrate film is thin enough, its spectrum can often be completely subtracted without affecting the surface selection rule and orientation information about the top monolayer film. Spectrum B in Fig. 10 shows the spectrum of DTDMAC deposited in the

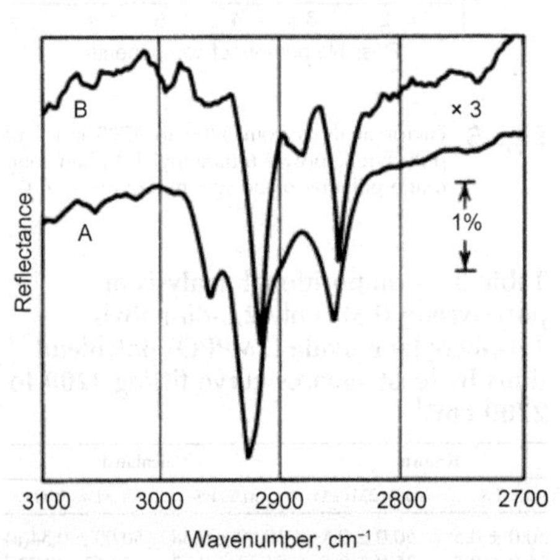

Fig. 9 Infrared reflection-absorption spectra of $C_{14}H_{29}$dimethylammonium hexanoate adsorbed on steel from 1.0 (A) and 0.1% (B) aqueous solutions (soaked 50 days and dried)

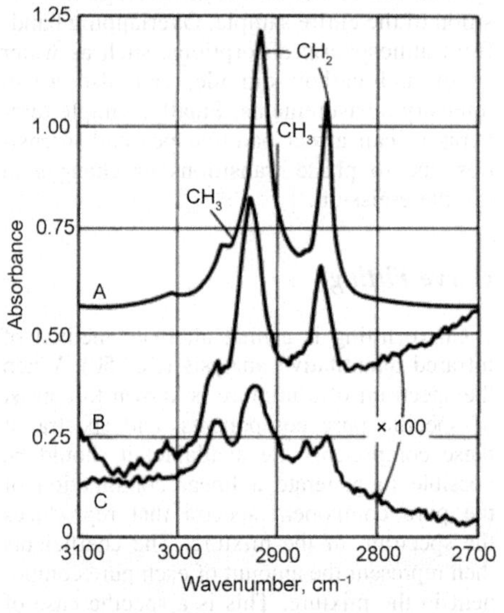

Fig. 10 Spectrum of bulk ditallowdimethylammonium chloride (DTDMAC) and DTDMAC adsorbed on metallic and nonmetallic substrates. A, attenuated total reflectance spectrum of bulk DTDMAC; B, infrared reflection-absorption spectroscopy spectrum of DTDMAC adsorbed on a 2 nm thick film of cellulose acetate on copper; C, infrared reflection-absorption spectrum of DTDMAC adsorbed on copper

same manner as in spectrum C on a 2 nm thick model surface of cellulose acetate adsorbed on copper. The cellulose acetate spectrum has been subtracted out. The relative intensities of the CH-stretching bands now match those observed in the randomly oriented case shown in spectrum A, suggesting that DTDMAC orients differently on cellulose acetate than on bare copper.

Figure 11 shows the orientation of a long-chain molecule on a model surface. The compound is dioctadecyldimethylammonium bromide, and the model surface is an 8 nm thick film of keratin adsorbed on a polished copper coupon. The film represents three monolayers deposited using the Langmuir-Blodgett technique (Ref 55–57). The relative intensities clearly indicated evidence of vertical orientation even on top of the 8 nm model protein surface.

Example 4: Depth Profiling a Granular Sample Using ATR, Diffuse Reflectance, and Photoacoustic Spectroscopy

Although ATR and PAS can be used to depth profile a sample, neither approach provides more than approximately one decade of dynamic range. The spectra shown in Fig. 12 illustrate how ATR, PAS, and DRS can be combined to depth profile a granular sample. The spectra show the carbonyl-stretching region of a granule containing diperoxydodecanedioic acid, monoperoxydodecanedioic acid, and dodecanedioic acid. The bands at 1753 and 1735 cm^{-1} are due to peracid groups, and the bands at 1695 cm^{-1} (labeled with an asterisk) are assigned to the carbonyl stretch of the carboxylic acid component. The estimated depth of penetration of the infrared beam into the sample for each experiment is shown in parentheses. Combining three infrared techniques enables acquisition of sample spectra from 0.4 μm to 1 mm (the diameter of the granule). Much more peracid is on the surface of the granule than in the interior. This result is more dramatic than that achieved using only one approach, in which dynamic range would have been narrower by at least a factor of 25.

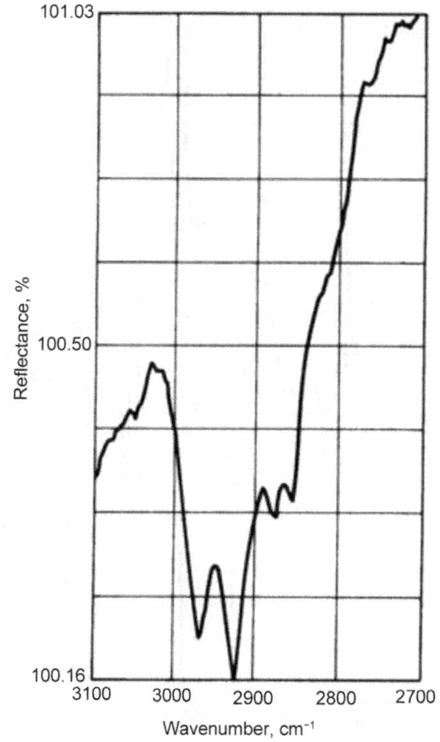

Fig. 11 Three Langmuir-Blodgett monolayers of dioctadecyldimethylammonium bromide adsorbed on an 8 nm thick film of keratin on copper

Example 5: Determination of Molecular Orientation in Drawn Polymer Films

Infrared linear dichroism spectroscopy is useful for studying the molecular orientation in polymeric materials (Ref 58, 59). The ultimate properties of a polymer depend on the conditions under which it was formed. This study illustrates the use of infrared linear dichroism to determine molecular orientation in a polymer film as a function of processing conditions.

When a polymer film is drawn, the macromolecular chains tend to align in a specific direction. The oriented film may then absorb, to different extents, incident infrared radiation polarized parallel and perpendicular to a reference direction usually defined as the drawing direction. The dichroic ratio, $R = A_{\parallel}/A_{\perp}$, associated with a specific absorbance band in the infrared spectrum can be used to assist determination of molecular chain orientation (Ref 58, 59). Absorbances of the components parallel and perpendicular to the reference direction are given by $A_{\parallel}$ and $A_{\perp}$, respectively. The farther from 1.00 that R is (greater or less), the greater the degree of orientation suggested.

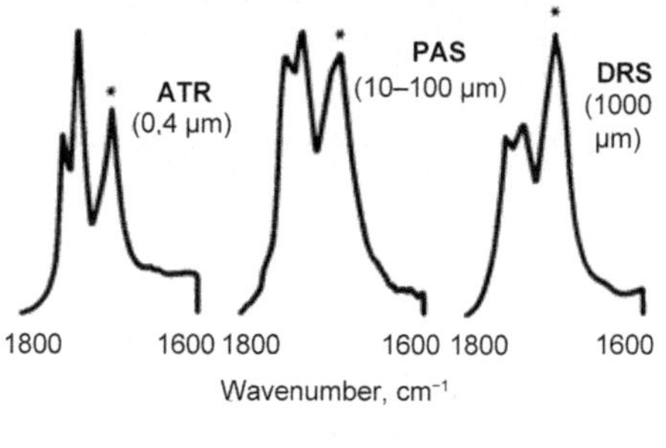

Fig. 12 Attenuated total reflectance (ATR) spectroscopy, photoacoustic spectroscopy (PAS), and diffuse reflectance spectroscopy (DRS) spectra of a granule containing diperoxydodecanedioic acid, monoperoxydodecanedioic acid, and dodecanedioic acid. The band at 1695 cm^{-1} (indicated by an asterisk) is due to carboxylic acid. The estimated depth of the infrared beam into the sample is shown in parentheses.

Infrared spectra of eight samples of isotactic polypropylene were obtained with the sample oriented parallel and perpendicular to a linear polarizer placed in the beam of the FTIR spectrometer. The areas of the peaks at 528, 941, and 1104 cm^{-1} were measured, and the dichroic ratios calculated using the integrated intensities in place of the absorbance values. The results are shown in Table 4. The draw ratio and temperature at which the sample was drawn are indicated. Peak absorbances varied from near zero for parallel polarization of the 528 cm^{-1} band in the highly oriented samples to near 1.0 for the 1104 cm^{-1} band in the unstretched sample. Nevertheless, the same relative degree of orientation is predicted for the sample using each dichroic ratio. The 941 cm^{-1} band has a profile similar to band 2 in Fig. 13, and the baseline drawn from C to D was used to determine the peak area (Ref 59). The absorbance value A_2 could also be used in place of the peak area. The results in Table 4 indicate that the lower the temperature when stretching, the greater the orientation induced at a given draw ratio. At a given temperature, the greater the draw ratio, the greater the orientation induced.

Example 6: Monitoring Polymer-Curing Reactions Using ATR

Fourier transform infrared spectroscopy can be used to follow changes in polymer films as they cure. In this example, the polymer sample is a paint applied to a KRS-5 IRE. The method used is similar to a documented technique in which the paint samples were analyzed inside the instrument with a dry air purge infrared transmission (Ref 60). Use of a Wilks' ATR attachment (described subsequently) enabled exposing the paint to ambient

Table 4 Order of samples according to degree of orientation ($R = A_{\parallel}/A_{\perp}$) for sample polypropylene bands at three wavenumbers

	Polypropylene sample	R at 528 cm^{-1}	R at 941 cm^{-1}	R at 1104 cm^{-1}
Most oriented	7 × at 105 °C (225 °F)	0.017	0.049	0.161
	7 × at 135 °C (275 °F)	0.027	0.070	0.188
	4 × at 105 °C (225 °F)	0.030	0.070	0.190
	4 × at 135 °C (275 °F)	0.046	0.075	0.205
	7 × at 160 °C (325 °F)	0.130	0.147	0.263
	4 × at 160 °C (325 °F)	0.250	0.244	0.340
Least oriented	Unknown at 150 °C (300 °F)	0.590	0.425	0.548
	Unstretched	1.020	1.059	1.092

atmosphere. Absorbance subtraction was used to aid understanding of the chemical reactions. Subtraction is optimized, because the sample is not touched between analyses, and the alignment of the ATR device remains constant.

The Wilks' ATR attachment used was designed as a skin analyzer for a dispersive spectrometer. When fitted into the front beam of the spectrometer, the attachment allows the sample to be exposed to the atmosphere during purging of the sample compartment with dry nitrogen. This avoids water vapor and carbon dioxide interference in the optical path while allowing oxygen to reach the sample. A paint sample was spread on the KRS-5 IRE, and spectra were obtained every 15 min for the first 90 min and every hour thereafter for 5 h. A final measurement was taken after 24 h. Each spectrum was ratioed with a blank IRE reference spectrum that was converted to absorbance. Difference spectra were generated by subtraction of the most recent absorbance spectrum from the one immediately preceding it. The subtraction reveals the types of bonds forming or breaking. Bands that are positive in the difference spectra indicate functional groups forming, and those negative suggest disappearing functional groups.

Only the layer of paint next to the surface of the IRE is sampled in the ATR experiment. The depth of penetration of the infrared beam at 1000 cm^{-1} through a KRS-5 IRE is approximately 2 μm. Therefore, the spectra represent the innermost surface, away from the atmosphere.

These results are for paint dissolved in a methylene chloride solvent. The first part of the curing process involved loss of C=C and solvent with a gain of C=O and OH. Between 6 and 24 h, the OH, in the form of hydroperoxide, had disappeared. Subtraction of the 24 h spectrum from the time-zero spectrum did not reveal any hydroperoxide bands. Thus, the hydroperoxide is formed as an intermediate species. Figure 14 shows the result of subtracting a 15 min spectrum from a 30 min spectrum. Table 5 summarizes band frequencies and their assignments.

Example 7: Quantitative Analysis of Hydroxyl and Boron Content in Glass

The properties of a seven-component glass were found to vary over a wide range of values. Careful study of the composition of the glass after preparation revealed that two components, boron oxide (B_2O_3) and hydroxyl groups, caused most of the variation in glass properties. These components were not well controlled because boron oxide is volatile, and some boron is lost during the melt preparation of the glass, and water in the atmosphere of the furnace can react with the glass to increase hydroxyl content. Because consistency of glass properties was required, the boron and OH content of each glass batch was monitored to identify proper melt operating conditions and to achieve better quality control of the final product. However, because of the large number of samples generated, a rapid method of analysis of the glass was desired. Infrared spectroscopy was selected for its rapidity and accuracy for these two components in the glass.

Samples of 6 mm (0.24 in.) diameter glass rod were poured from the original melt and cut into 5 mm (0.2 in.) thick samples for the infrared analysis. The samples were cleaned ultrasonically, but the faces of the samples were not polished. Spectra were taken using a FTIR spectrometer at 4 cm^{-1} resolution (full width at half maximum.). Figure 15 shows the infrared spectrum of a typical glass. The band at approximately 3560 cm^{-1} is due to the stretching vibration of OH groups. The band at approximately 2685 cm^{-1} has been assigned to the overtone band of the B–O stretching vibration. The peak intensities of these bands were found to follow Beer's law if the intensity was measured at the peak band position. The OH band remained at relatively constant frequency over the concentration range studied. The B–O band was found to shift to higher energy with higher boron content. The OH content of the glass was determined on a relative basis, because there was no convenient method of absolute determination of OH in this glass. The boron oxide content of a set of glass samples used for infrared calibration was measured independently using ion chromatography exclusion (ICE). These samples covered the concentration range of B_2O_3 and OH expected in normal glass processing.

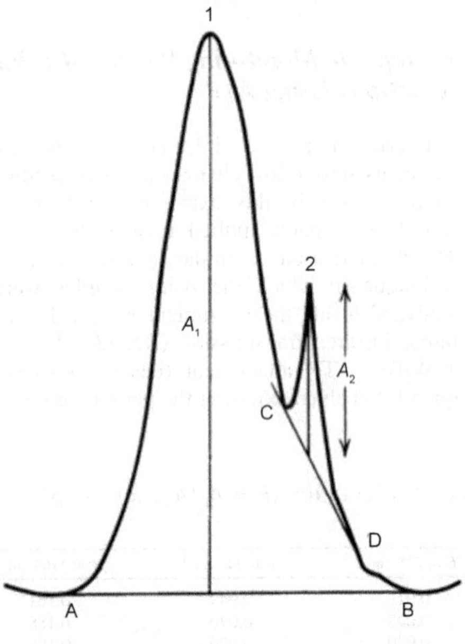

Fig. 13 Synthetic spectrum showing baseline choices for two overlapping bands. The baseline from C to D is acceptable for band 2. The baseline drawn from A to B is correct for band 1 but if used for band 2 would give an incorrect value.

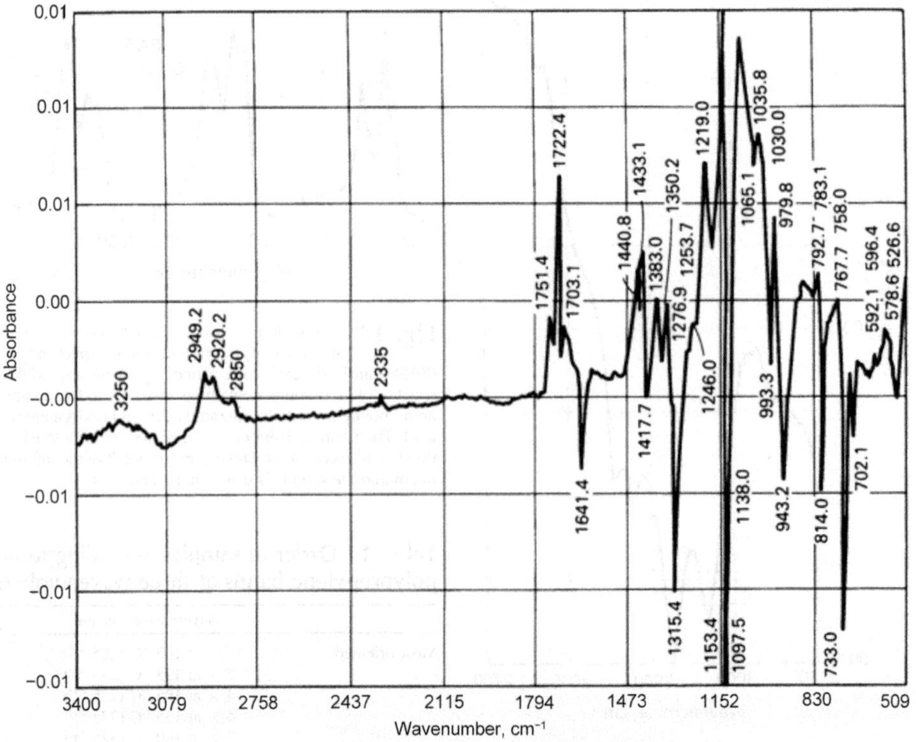

Fig. 14 KRS-5 attenuated total reflectance spectrum of paint. Spectrum after 15 min cure subtracted from spectrum after 30 min

Table 5 Bands and assignments for difference spectrum
30 min spectrum minus 15 min spectrum

Band, cm^{-1}	Forming (F) or disappearing (D)	Assignment
3250	F	Hydroperoxide (O–H stretching)
1751	F	6-membered lactone (C=O stretching)
1722	F	Ketone or conjugated ester (C=O stretching)
1641	D	*cis*-vinyl or vinylidine (C=C stretching)
1437	F	Methylene adjacent to carbonyl (CH_2 deformation)
1417	D	Methylene adjacent to C=C (CH_2 deformation)
1315	D	Not assigned
1153	F	Ester (C–O stretching)
1138	D	Not assigned
993	D	Vinyl C=C (=CH_2 wagging)
943	D	Vinyl adjacent to ester (O – C(=O) – C=CCH$_2$ wagging)
814	D	R–O–CH=CH_2 (=CH_2 wagging)
733	D	Methylene chloride
702	D	Methylene chloride

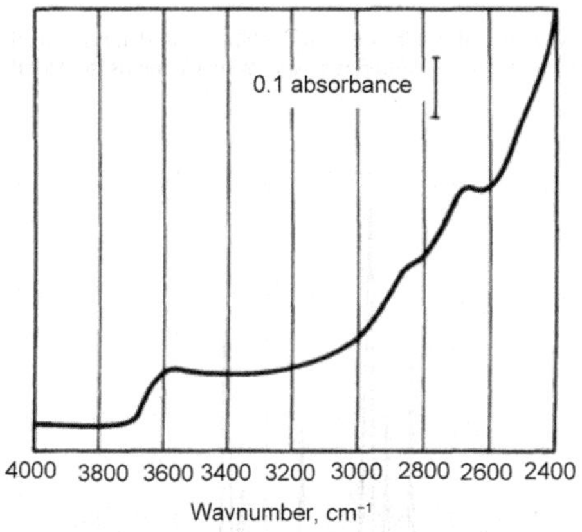

Fig. 15 Typical Fourier transform infrared spectrum of a glass sample

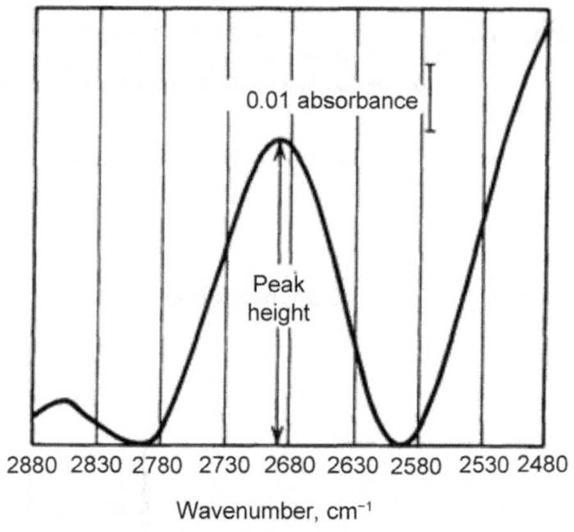

Fig. 16 Spectrum of a glass sample after pathlength normalization and linear baseline correction

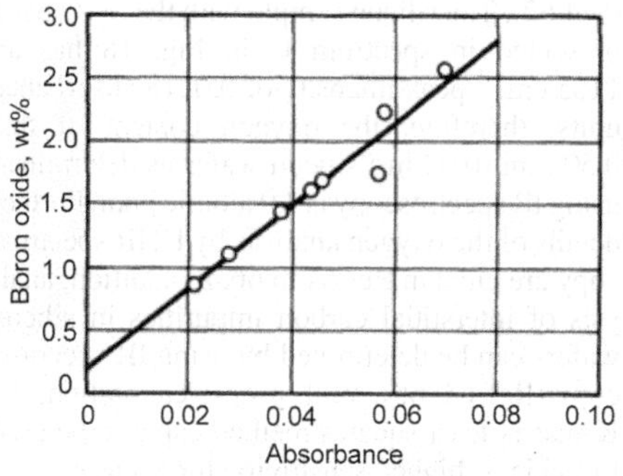

Fig. 17 Calibration of B_2O_3 concentration from ion chromatography exclusion versus height of the B–O overtone band. The line drawn through the data is from the linear least-squares fit of the data.

Because samples were not polished, spectra of the glass samples were sloped slightly. The sloping baseline, a result of scattered light from the rough glass surfaces, was removed using standard software to correct for linear baseline offsets. Sample thickness was measured accurately using a micrometer, and the spectrum was multiplied by the factor required to yield spectra in units of absorbance/5.0 mm (0.2 in.). Figure 15 shows a baseline-corrected and scaled spectrum. The OH content was then determined from the height of the 3560 cm^{-1} peak measured from the high-energy baseline.

For B_2O_3 determination, the spectra were further baseline corrected to facilitate measurement of the B–O overtone band (Fig. 16). The peak height of the baseline-corrected band was used to quantify the B_2O_3 content of the glass. Based on eight glass samples analyzed independently using ICE and covering the concentration range of 0.88 to 2.58 wt% B_2O_3, the infrared calibration for B_2O_3 was found to be linear (Fig. 17). The average relative percent error in determining boron content of the eight calibration samples using IR spectroscopy was 4.0%, which compares favorably with the 3% average relative error of the ICE analysis. The boron content of unknown glass samples was determined directly from the calibration shown in Fig. 17.

Following analysis of several glass batches, glass samples exhibiting the appropriate properties were found to have an OH content corresponding to a 3560 cm^{-1} band intensity less than 0.048 absorbance/mm and a B_2O_3 content from 1.0 to 1.4 wt%. Starting concentrations were then adjusted and the humidity controlled for reproducibility of these ideal concentrations. Infrared spectroscopy was then used to rapidly monitor the quality of the final glass batches. Further details of the analysis are cited in Ref 61.

More accurate and automated analysis of the glass can be obtained with application of the **K**-matrix least-squares approach. However, because the B–O band shifts with frequency, a nonlinear model (rather than the linear Beer's law model) must be used to achieve the higher accuracy and to model the frequency shift with boron concentration. Use of the **K**-matrix method for multicomponent quantitative analysis is discussed in the section "Matrix Methods" in this article. Additional information is cited in Ref 49.

Example 8: Quantitative Analysis of Oxygen Contained in Silicon Wafers

The oxygen content of silicon wafers can affect the properties, rejection rates, and long-term reliability of integrated circuits produced on the wafers. Therefore, rapid measurement of oxygen in silicon wafers is desired for quality control. Infrared spectroscopy is well suited to rapid and nondestructive analysis of oxygen in silicon wafers. The proper methods for infrared determination of interstitial oxygen in silicon are cited in Ref 62. However, because the methods were designed for dispersive spectrometers with silicon wafers 2 to 4 mm (0.08 to 0.16 in.) thick and polished on both sides, modifications are required when using FTIR spectrometers applied to commercial silicon wafers approximately 0.5 mm (0.02 in.) thick that are normally polished on one side.

One possible procedure involves taking a spectrum of a float-zoned silicon wafer with negligible oxygen content as a reference. The spectrum of the sample to be analyzed is also obtained at the same spectral resolution. If interference fringes due to multiple reflections in the sample are observed, spectra taken at lower resolution may be obtained that exhibit reduced fringe intensity. Alternatively, the fringes can be removed with software processing (Ref 63). A scaled subtraction of the float-zoned and sample spectra is performed to yield the spectrum of the Si–O band in the absence of any silicon phonon band; the float-zoned silicon spectrum is scaled according to its thickness relative to the sample spectrum.

Figure 18 shows the spectra of the float-zoned material and the sample to be analyzed as well as the result of the scaled subtraction. Conversion of absorbance to concentration in atomic parts per million for the 1105 cm^{-1} band of Si–O uses the function 11.3 atomic ppm per cm^{-1} at 300 K

(Ref 62). The silicon sample with the Si–O band presented in spectrum C in Fig. 18 has an 1105 cm^{-1} peak intensity of 0.0218 absorbance units; therefore, the oxygen content of this 0.50 mm (0.02 in.) silicon wafer as determined using IR spectroscopy is 4.9 atomic ppm. Further details of the oxygen analysis by FTIR spectroscopy are cited in Ref 64 to 66. In addition, analysis of interstitial carbon impurities in silicon wafers can be determined by using IR spectroscopy (Ref 64–66). With proper calibration, the **K**-matrix least-squares method can also be used to achieve higher sensitivity for measurement of oxygen in the thin silicon wafers.

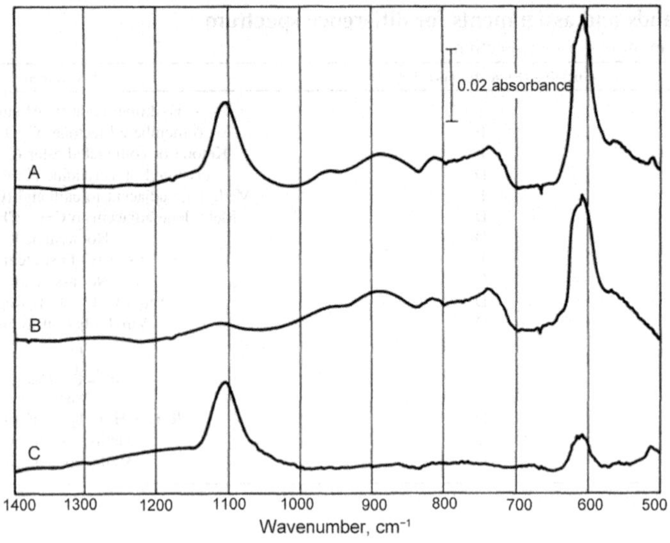

Fig. 18 Fourier transform infrared spectra of silicon. A, silicon wafer with oxygen content to be determined; B, float-zoned silicon wafer with negligible oxygen content; C, difference spectrum (A minus scaled B) showing the Si–O stretching band at 1105 cm^{-1}

Example 9: Identification of Polymer and Plasticizer Materials in a Vinyl Film

A sheet of vinyl film was submitted for analysis to determine the identity of the polymer and any other information that may be available by using IR spectroscopy.

A piece of the polymer approximately 25 by 10 mm (1 by 0.4 in.) was cut from the film, rinsed in methanol to remove surface contamination (fingerprints and so on), then allowed to air dry. This sample was then soaked in acetone to extract the plasticizer from the sample. The acetone solution was retained to recover the plasticizer later. Being too thick for analysis, the film was prepared as a cast film on a KBr window. This was accomplished by dissolving the sample in methyl ethyl ketone (MEK) and adding several drops of the MEK solution to a KBr window. The MEK evaporated, leaving a cast film of the sample on the window. The KBr window was then placed in an appropriate sample holder and the infrared spectrum taken. Because this spectrum exhibited MEK retention, a reference spectrum of a MEK capillary film was taken to subtract the spectrum of the MEK from the cast film spectrum (Fig. 19). Using a flow chart (Ref 67), the spectrum was identified as that of polyvinylchloride. This assignment was confirmed by comparison with published polyvinylchloride spectra.

The plasticizer material was recovered from the acetone extraction solution by evaporating the solution until only the viscous plasticizer remained. A small amount of the plasticizer was pressed between two KBr windows to form a thin film. Figure 20 shows the plasticizer spectrum. A computerized spectral search routine was used to identify the plasticizer. Although an exact match was not found, the compound was identified as a di-alkyl phthalate, a class of compounds commonly used as plasticizers in polymers.

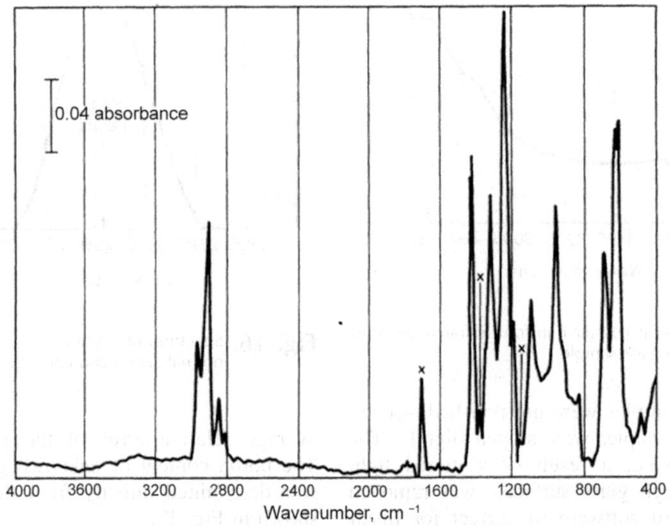

Fig. 19 Infrared spectrum of polymer material after subtraction of methyl ethyl ketone spectrum. Residual artifact peaks are denoted by the "x's."

ACKNOWLEDGMENTS

Revised from C. Marcott, Infrared Spectroscopy, *Materials Characterization*, Vol 10, *ASM Handbook*, ASM International, 1986, p 109–125

Examples 7 and 8 were supplied by D.M. Haaland, Sandia National Laboratories. Example 9 was supplied by M.C. Oborny, Sandia National Laboratories.

REFERENCES

1. E.B. Wilson, J.C. Decius, and P.C. Cross, *Molecular Vibrations*, McGraw-Hill, 1955
2. P.C. Painter, M.M. Coleman, and J.L. Koenig, *The Theory of Vibrational Spectroscopy and Its Application to Polymeric Materials*, John Wiley & Sons, 1982
3. R.G. Parr and W. Yang, *Density-Functional Theory of Atoms and Molecules*, Oxford University Press, New York, 1989, ISBN 978-0-19-504279-5
4. D. Yang and A. Rauk, in *Reviews in Computational Chemistry*, Vol 7, K.B. Lipkowitz and D.B. Boyd, Ed., VCH Publishers, New York, 1996, p 261
5. P.R. Griffiths, *Chemical Infrared Fourier Transform Spectroscopy*, John Wiley & Sons, 1975
6. A.A. Michelson, *Philos. Mag.*, Vol 31 (Ser. 5), 1891, p 256
7. A.A. Michelson, *Philos. Mag.*, Vol 34 (Ser. 5), 1892, p 280
8. P. Fellgett, *J. Phys. Radium*, Vol 19, 1958, p 187
9. P. Jacquinot and J.C. Dufour, *J. Rech. C. N.R.S.*, Vol 6, 1948, p 91
10. P. Jacquinot, *Rep. Prog. Phys.*, Vol 23, 1960, p 267

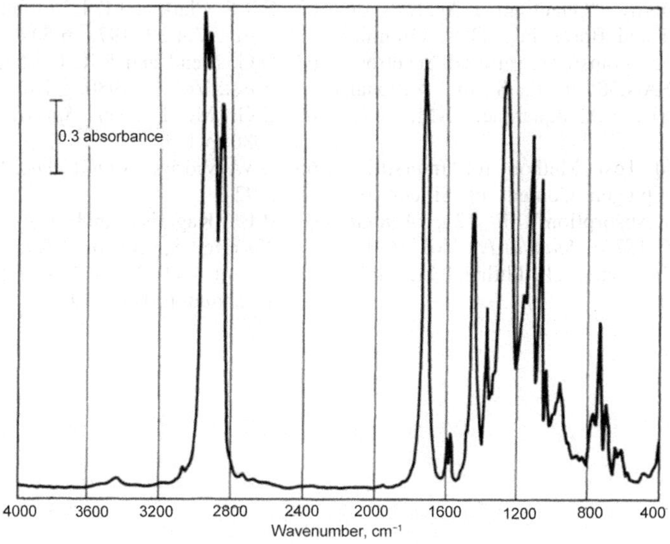

Fig. 20 Infrared spectrum of plasticizer material

11. J. Connes and P. Connes, *J. Opt. Soc. Am.*, Vol 56, 1966, p 896
12. L. Genzel and J. Kuhl, *Appl. Opt.*, Vol 17, 1978, p 3304
13. W.M. Doyle, B.C. McIntosh, and W.L. Clark, *Appl. Spectrosc.*, Vol 34, 1980, p 599
14. R.P. Walker and J.D. Rex, in *Proceedings of the Society of Photo-Optical Instrumentation Engineers,* Vol 191, G.A. Vannasse, Ed., Society of Photo-Optical Instrumentation Engineers, Bellingham, WA, 1979
15. P.R. Griffiths, in *Advances in Infrared and Raman Spectroscopy*, Vol 10, R.J.H. Clark and R.E. Hester, Ed., Heyden, 1983
16. R.S. McDowell, in *Advances in Infrared and Raman Spectroscopy,* Vol 5, R.J.H. Clark and R.E. Hester, Ed., Heyden, 1980,
17. K. Vodopyanov, G. Hill, J. Rice, S. Meech, D. Craig, M. Reading, A. Dazzi, K. Kjoller, and C. Prater, in *Nano-Spectroscopy in the 2.5–10 Micron Wavelength Range Using Atomic Force Microscope*, Optical Society of America, 2009
18. K.L. Vodopyanov and P.G. Schunemann, Broadly Tunable Noncritically Phase-Matched $ZnGeP_2$ Optical Parametric Oscillator with a 2-μJ Pump Threshold, *Opt. Lett.*, Vol 28, 2003, p 441–443
19. F. Lu and M.A. Belkin, Infrared Absorption Nano-Spectroscopy Using Sample Photoexpansion Induced by Tunable Quantum Cascade Lasers, *Opt. Express*, Vol 19, 2011, p 19942–19947
20. F. Lu, M. Jin, and M.A. Belkin, Tip-Enhanced Infrared Nanospectroscopy via Molecular Expansion Force Detection, *Nat. Photon*, Vol 8 (No. 4), 2014, p 307–312
21. A. Dazzi and C.B. Prater, AFM-IR: Technology and Applications in Nanoscale Infrared Spectroscopy and Chemical Imaging, *Chem. Rev.*, Vol 117, 2016, p 5146–5173
22. W.J. Potts, *Chemical Infrared Spectroscopy*, Vol I, John Wiley & Sons, 1963
23. J.L. Lauer, in *Fourier Transform Infrared Spectroscopy, Applications to Chemical Systems*, Vol I, J.R. Ferraro and L.J. Basile, Ed., Academic Press, 1978
24. N.J. Harrick, *Internal Reflection Spectroscopy*, John Wiley & Sons, 1967
25. M.P. Fuller and P.R. Griffiths, *Appl. Spectrosc.*, Vol 34, 1978, p 1906
26. P. Kubelka and F. Munk, *Z. Tech. Phys.*, Vol 12, 1931, p 593
27. P. Kubelka, *J. Opt. Soc. Am.*, Vol 38, 1948, p 448
28. D.L. Allara, in *Characterization of Metal and Polymer Surfaces*, Vol II, Academic Press, 1977
29. L.A. Nafie and D.W. Vidrine, in *Fourier Transform Infrared Spectroscopy, Techniques Using Fourier Transform Interferometry*, Vol III, J.R. Ferraro and L.J. Basile, Ed., Academic Press, 1982
30. A.E. Dowrey and C. Marcott, *Appl. Spectrosc.*, Vol 36, 1982, p 414
31. W.W. Wendlandt and H.G. Hecht, *Reflectance Spectroscopy*, Interscience, 1966
32. J.B. Bates, in *Fourier Transform Infrared Spectroscopy, Applications to Chemical Systems*, Vol I, J.R. Ferraro and L.J. Basile, Ed., Academic Press, 1978
33. D.W. Vidrine, in *Fourier Transform Infrared Spectroscopy, Techniques Using Fourier Transform Interferometry*, Vol III, J.R. Ferraro and L.J. Basile, Ed., Academic Press, 1982
34. J.A. Graham, W.M. Grim III, and W.G. Fateley, in *Fourier Transform Infrared Spectroscopy, Applications to Chemical Systems*, Vol IV, J.R. Ferraro and L.J. Basile, Ed., Academic Press, 1985
35. A. Dazzi, R. Prazeres, F. Glotin, and J.M. Ortega, Local Infrared Microspectroscopy with Subwavelength Spatial Resolution with an Atomic Force Microscope Tip Used as a Photothermal Sensor, *Opt. Lett.*, Vol 30 (No. 18), 2005, p 2388–2390
36. A. Dazzi, C.B. Prater, Q. Hu, D.B. Chase, J. F. Rabolt, and C. Marcott, AFM-IR: Combining Atomic Force Microscopy and Infrared Spectroscopy for Nanoscale Chemical Characterization, *Appl. Spectrosc.*, Vol 66 (No. 12), 2012, p 1365–1384
37. J.A. Reffner, Advances in Infrared Microspectroscopy and Mapping Molecular Chemical Composition at Submicrometer Spatial Resolution, *Spectroscopy*, Vol 33 (No. 9), 2018, p 12–17
38. D. Zhang, C. Li, C. Zhang, M.N. Slipchenko, G. Eakins, and J.-X. Cheng, *Sci. Adv.*, Vol 2 (No. 9), 2016, p e1600521
39. "Simultaneous Submicron IR and Raman Microscopy," Photothermal Spectroscopy Corp., www.photothermal.com/iraman, unpublished results
40. P.R. Griffiths, in *Fourier Transform Infrared Spectroscopy, Applications to Chemical Systems*, Vol I, J.R. Ferraro and L.J. Basile, Ed., Academic Press, 1978
41. P.R. Griffiths, J.A. Hasethde, and L.V. Azarraga, *Anal. Chem.*, Vol 55, 1983, p 1361A
42. G.T. Reedy, S. Bourne, and P.T. Cunningham, *Anal. Chem.*, Vol 51, 1979, p 1535
43. D.W. Vidrine, in *Fourier Transform Infrared Spectroscopy, Applications to Chemical Systems*, Vol II, J.R. Ferraro and L.J. Basile, Ed., Academic Press, 1979
44. K.H. Shafer and P.R. Griffiths, *Anal. Chem.*, Vol 55, 1983, p 1939
45. K. Krishnan, *Polym. Prepr.*, Vol 25 (No. 2), 1984, p 182
46. N.B. Colthup, L.H. Daley, and S.E. Wiberley, *Introduction to Infrared and Raman Spectroscopy*, 2nd ed., Academic Press, 1975
47. M.K. Antoon, L. D'Esposito, and J.L. Koenig, *Appl. Spectrosc.*, Vol 33, 1979, p 351
48. J.K. Kauppinen, D.J. Moffatt, H.H. Mantsch, and D.G. Cameron, *Appl. Spectrosc.*, Vol 35, 1981, p 271
49. C.W. Brown, P.F. Lynch, R.J. Obremski, and D.S. Lavery, *Anal. Chem.*, Vol 54, 1982, p 1472
50. M.K. Antoon, J.H. Koenig, and J.L. Koenig, *Appl. Spectrosc.*, Vol 31, 1977, p 518
51. J. Umemur, H.H. Mantsch, and D.G. Cameron, *J. Colloid Interface Sci.*, Vol 83, 1981, p 558
52. A. Rein and P.A. Wilks, *Am. Lab.*, Oct 1982
53. P.A. Wilks, *Ind. Res. Dev.*, Sept 1982
54. H.H. Mantsch, A. Martin, and D.G. Cameron, *Biochemistry*, Vol 20, 1981, p 3138

55. I. Langmuir, *J. Am. Chem. Soc.*, Vol 39, 1917, p 1848
56. K.B. Blodgett, *Phys. Rev.*, Vol 55, 1939, p 391
57. K.B. Blodgett, *J. Chem. Rev.*, Vol 57, 1953, p 1007
58. R.J. Samuels, *Structured Polymer Properties*, John Wiley & Sons, 1974
59. B. Jasse and J.L. Koenig, *J. Macromol. Sci.-Rev. Macromol. Chem.*, Vol C17 (No. 2), 1979, p 61
60. J.H. Hartshorn, *J. Coatings Technol.*, Vol 54, 1982, p 53
61. M.C. Oborny, "Quantitative Analysis of Hydroxyl and Boron in S Glass-Ceramic by Fourier Transform Infrared Spectroscopy," SAND85-0738, Sandia National Laboratories, Albuquerque, NM, July 1985
62. "Standard Test Method for Interstitial Atomic Oxygen Content of Silicon by Infrared Absorption," F 121, *Annual Book of ASTM Standards*, Vol 10.05, ASTM International, Philadelphia, PA, p 240–242
63. F.R.S. Clark and D.J. Moffatt, *Appl. Spectrosc.*, Vol 32, 1978, p 547
64. D.G. Mead and S.R. Lowry, *Appl. Spectrosc.*, Vol 34, 1980, p 167
65. D.G. Mead, *Appl. Spectrosc.*, Vol 34, 1980, p 171
66. D.W. Vidrine, *Anal. Chem.*, Vol 52, 1980, p 92
67. R.E. Kagarise and L.A. Weinberger, "Infrared Spectra of Plastics and Resins," Report 4369, Naval Research Laboratory, Washington, D.C., 1954

ASM Handbook, Volume 10, *Materials Characterization*
ASM Handbook Committee
DOI 10.31399/asm.hb.v10.a0006685

Raman Spectroscopy*

Overview

General Uses

- Molecular analysis of bulk samples and surface or near-surface species as identified by their characteristic vibrational frequencies
- Low-frequency vibrational information on solids for metal-ligand vibrations and lattice vibrations
- Determination of phase composition of solids

Examples of Applications

- Identification of effects of preparation on glass structure
- Structural analysis of polymers
- Determination of structural disorder in graphites
- Determination of surface structure of metal oxide catalysts
- Identification of corrosion products on metals
- Identification of surface adsorbates on metal electrodes

Samples

- *Form:* Solid, liquid, or gas
- *Size:* Single crystal of material to virtually any size the Raman spectrometer can accommodate

Limitations

- *Sensitivity:* Poor to fair without enhancement
- Raman spectroscopy requires concentrations greater than approximately 1 to 5%
- Analysis of surface or near-surface species difficult but possible
- Sample fluorescence or impurity fluorescence may prohibit Raman characterization

Estimated Analysis Time

- 30 min to 8 h per sample

Capabilities of Related Techniques

- *Infrared spectroscopy and Fourier-transform infrared spectroscopy:* Molecular vibrational identification of materials; lacks sensitivity to surface species; difficult on aqueous systems
- *High-resolution electron energy loss spectroscopy:* Vibrational analysis of surface species in ultrahigh-vacuum environment; extremely sensitive; requires ultrahigh-vacuum setup; low resolution compared to Raman spectroscopy; cannot be used for *in situ* studies

Introduction

Raman spectroscopy is a valuable tool for the characterization of materials due to its extreme sensitivity to the molecular environment of the species of interest. Information on molecular vibrations can provide much structural, orientational, and chemical information that can assist in defining the environment of the molecule of interest to a high degree of specificity. The materials applications for which Raman spectroscopy can be used continue to expand with improvements in requisite instrumentation and methodology.

This article will introduce principles of Raman spectroscopy and the representative materials characterization applications to which Raman spectroscopy has been applied. The section "The Raman Effect" includes a discussion of light-scattering fundamentals and a description of the experimental aspects of the technique. Emphasis has been placed on the different instrument approaches that have been developed for performing Raman analyses on various materials. The applications presented reflect the breadth of materials characterization uses for Raman spectroscopy and highlight the analysis of bulk material and of surface and near-surface species.

The Raman Effect

Fundamentals

Raman spectroscopy is one of many light-scattering phenomena. All these phenomena originate from the principle that the intensity of a beam of light decreases measurably when it passes through a nonabsorbing medium. The energy lost is not significantly degraded to heat. Rather, some of the light energy is scattered into the space surrounding the sample.

The Raman effect is named after C.V. Raman, who, with K.S. Krishnan, first observed this phenomenon in 1928 (Ref 1). It belongs to the class of molecular-scattering phenomena. The molecular-scattering phenomena that must be considered are Rayleigh scattering, Stokes scattering (the normal Raman effect), and anti-Stokes scattering. The nature of this scattered radiation is predicted by quantum theory and classical electromagnetic theory.

The quantum theory of Raman scattering involves consideration of radiation of frequency ν_0 as consisting of photons that have energy $h\nu_0$. Scattering of this radiation occurs

*Reprinted from J.E. Pemberton and A.L. Guy, Raman Spectroscopy, *Materials Characterization*, Vol 10, *ASM Handbook*, ASM International, 1986, p 126–138

when these photons undergo two types of collisions with the molecules of a medium. These collisions are shown as energy-level diagrams in Fig. 1. Elastic collisions are those in which the energy of the scattered photon, $h\nu_s$, is unchanged relative to the initial energy of the incident photon; that is, $h\nu_s = h\nu_0$. This is known as Rayleigh scattering and is the most probable scattering that will occur in a molecular system.

Much less probable is the inelastic collision of a photon with a molecule. In this case, energy is exchanged between the photon and the molecule such that the scattered photon is of higher or lower energy than the incident photon. The energy of the scattered photons in these types of scattering events is $h(\nu_0 \pm \nu_n)$. Because the energy levels of the molecule are discrete and well defined, energy can be lost or gained by the molecule only in quantized or discrete amounts. Therefore, two types of scattered radiation result from these inelastic scattering events.

Stokes radiation, the first type, is observed for molecules that gain a vibrational or rotational quantum of energy from the incident photon. When this occurs, the scattered photon is lower in energy than the incident photon by an amount $h\nu_n$ that equals the amount of energy required to excite a vibration or rotation in the molecule. The energy of the Stokes-scattered photon, $h\nu_S$, is $h(\nu_0 - \nu_n)$. Anti-Stokes radiation, the second type, is observed for molecules that lose a vibrational or rotational quantum of energy to the incident photon. The energy of the anti-Stokes scattered photon, $h\nu_{AS}$, is $h(\nu_0 + \nu_n)$. All these scattering events occur within 10^{-12} to 10^{-13} s.

Because anti-Stokes scattering can occur only for molecules that are in an excited vibrational or rotational state before scattering, the intensity of anti-Stokes radiation is significantly less than that of Stokes radiation at room temperature. Therefore, Raman spectroscopy generally uses Stokes radiation. Overall, however, the total amount of inelastically scattered Stokes and anti-Stokes radiation is small compared to the elastically scattered Rayleigh radiation. This feature of molecular scattering makes the detection of Stokes radiation a serious problem.

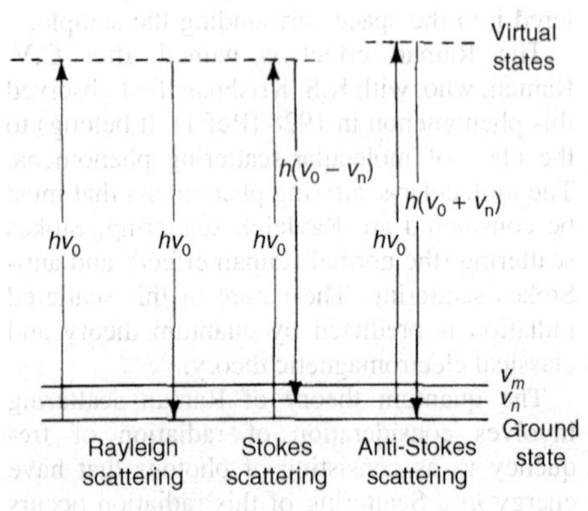

Fig. 1 Energy-level diagram of molecular light-scattering processes.

When a molecule is in an electromagnetic field, it is distorted by the attraction of the electrons to the positive pole of the electric field and the attraction of the nuclei to the negative pole of the electric field. The extent to which this distortion occurs is a characteristic of the molecule known as its polarizability. The resulting separation of charge produces a momentary induced electric dipole moment that is usually expressed as the dipole moment per unit volume and is known as the polarization, P. Under these circumstances the molecule is considered to be polarized.

The magnitude of polarization of a molecule depends on the magnitude of the electric field, E, and on the characteristics of the molecule describing the ease with which the molecule can be distorted, its polarizability (α). Therefore:

$$P = \alpha E \qquad \text{(Eq 1)}$$

The oscillating electric field in an electromagnetic wave is:

$$E = E_0 \cos(2\pi\nu_0) \qquad \text{(Eq 2)}$$

The induced dipole also oscillates at frequency ν_0. Therefore:

$$P = \alpha E_0 \cos(2\pi\nu_0 t) \qquad \text{(Eq 3)}$$

According to classical electromagnetic theory, such an oscillating dipole moment can act as a source of radiation. Rayleigh scattering arises from radiation that the oscillating dipole emits at its own frequency, ν_0. If the molecule also undergoes some internal motion, such as vibration or rotation, that periodically changes the polarizability, the oscillating dipole will have superimposed on it the vibrational or rotational frequency. This effect is mathematically based on the equation describing the polarizability of the molecule. Polarizability, α, is:

$$\alpha = \alpha_0 + \sum_n \alpha_n \cos(2\pi\nu_n t) \qquad \text{(Eq 4)}$$

where α_0 is the static polarizability of the molecule, which in part produces Rayleigh scattering. The second term in the polarizability expression is a sum of terms having the periodic time dependence of the normal frequencies of the internal motions of the molecule. Substituting this expression for the polarizability into Eq 3 yields:

$$P = E_0\alpha_0 \cos(2\pi\nu_0 t) + E_0 \sum_n \alpha_n \cos(2\pi\nu_0 t)\cos(2\pi\nu_n t) \qquad \text{(Eq 5)}$$

This can be expanded to provide:

$$P = E_0\alpha_0 \cos(2\pi\nu_0 t) + \tfrac{1}{2}E_0 \sum_n \alpha_n[\cos 2\pi(\nu_0 - \nu_n)t + \cos 2\pi(\nu_0 + \nu_n)t] \qquad \text{(Eq 6)}$$

This equation predicts three components of the scattered radiation. The first term predicts scattering of radiation at the incident frequency, ν_0, or Rayleigh scattering. The second term predicts scattering at frequencies lower than the incident frequency by amounts corresponding to the normal frequencies of the molecule, $(\nu_0 - \nu_n)$. This is Stokes scattering. The third term predicts scattering at frequencies higher than that of the incident frequency by amounts corresponding to the normal frequencies of the molecule, $(\nu_0 + \nu_n)$. This is anti-Stokes scattering.

Polarizability is a tensor that leads to important consequences in the angular dependence and polarization of the scattered radiation. Therefore, the relationship between polarization and the electric field vector is more accurately written in matrix notation:

$$\begin{vmatrix} P_x \\ P_y \\ P_z \end{vmatrix} = \begin{vmatrix} \alpha_{xx} & \alpha_{xy} & \alpha_{xz} \\ \alpha_{yx} & \alpha_{yy} & \alpha_{yz} \\ \alpha_{zx} & \alpha_{zy} & \alpha_{zz} \end{vmatrix} \begin{vmatrix} E_x \\ E_y \\ E_z \end{vmatrix} \qquad \text{(Eq 7)}$$

This relationship also has consequences regarding selection rules in Raman spectroscopy. If a vibrational mode is to be Raman active, the vibration must alter the polarizability of the molecule; that is, α_n must not equal zero. This selection rule is best put into context by contrasting it with the selection rule of another vibrational spectroscopy, infrared spectroscopy (the article "Infrared Spectroscopy" in this Volume supplies additional information on this technique). Infrared active modes of a molecule produce a change in true electric dipole moment existing in the molecule. This fundamental difference between these two vibrational spectroscopies leads to the complementary and sometimes mutually exclusive nature of the vibrational modes measured by infrared and Raman spectroscopies.

In terms of the experimental utility of Raman spectroscopy, the Raman intensity of a particular vibrational mode is proportional to the intensity of the incident radiation and proportional to the fourth power of the scattered-light frequency:

$$I_n = \frac{2^3}{3^3 c^4} I_0 (\nu_0 - \nu_n)^4_{i,j} (\alpha_{ij})^2_n \qquad \text{(Eq 8)}$$

This relationship indicates that the sensitivity of a Raman analysis can be improved by using higher excitation powers or by increasing the energy (frequency) of the excitation. Additional information on Raman intensities is cited in Ref 2 and 3.

Experimental Considerations

The inherent weakness of Raman scattering that produces the poor sensitivity of the technique precluded the widespread use of Raman spectroscopy for materials characterization until recently. The advent of the laser as an intense, monochromatic light source revived interest in use of the Raman effect for the acquisition of molecularly specific information about materials.

Two types of Raman spectrometers are commonly used to analyze materials. A conventional scanning monochromator system is shown in Fig. 2. Figure 3 illustrates a Raman system developed around one of several multichannel detectors. The difference between these Raman spectrometers is the method of obtaining Raman intensity as a function of frequency information.

Lasers are used almost exclusively as excitation sources in Raman spectroscopy. Laser radiation, possessing intensity, monochromaticity, and collimation, is well suited as a Raman excitation source. The most commonly used lasers for Raman spectroscopy are continuous-wave gas lasers. The most prevalent of these include the argon, krypton, and helium-neon lasers. Broadband tunable dye lasers are also commonly used to extend excitation capabilities further into the red region of the spectrum. Information on the fundamentals of lasers is cited in Ref 4.

Typical laser powers used in Raman analyses range from several milliwatts to several watts. The laser beam is usually focused on the sample using a series of mirrors and lenses. Focusing of the beam results in luminous power densities of several watts to thousands of watts per square centimeter. For absorbing samples, these power densities can cause significant heating. One means of reducing the extent of heating is to focus the laser beam to a line on the sample using a cylindrical lens. This approach also produces scattered radiation in a lineshape such that the entrance slit of the monochromator can be completely filled with the slit-shaped image.

The output of either type of Raman system is a plot of scattered-light intensity as a function of frequency shift in which the shift is calculated relative to the laser line frequency that is assigned as zero. Presentation of the spectra in this way facilitates comparison with infrared spectra, because both spectra are on equivalent frequency scales.

Raman spectra can be plotted on a recording device in real time in the above-mentioned fashion. However, conventional recording devices are being replaced by microcomputer-based data systems that provide for data storage and subsequent manipulation. A data system is required when using multichannel detection systems.

The heart of any Raman system is the monochromator-detector assembly. Conventional scanning monochromators are usually based on the use of two dispersion stages (a double monochromator) or three dispersion stages (a triple monochromator). Multiple dispersion stages are essential in obtaining Raman spectra to reduce the amount of stray radiation reaching the detector. Figure 4 shows a plot of the intensity ratio of the grating scatter to the

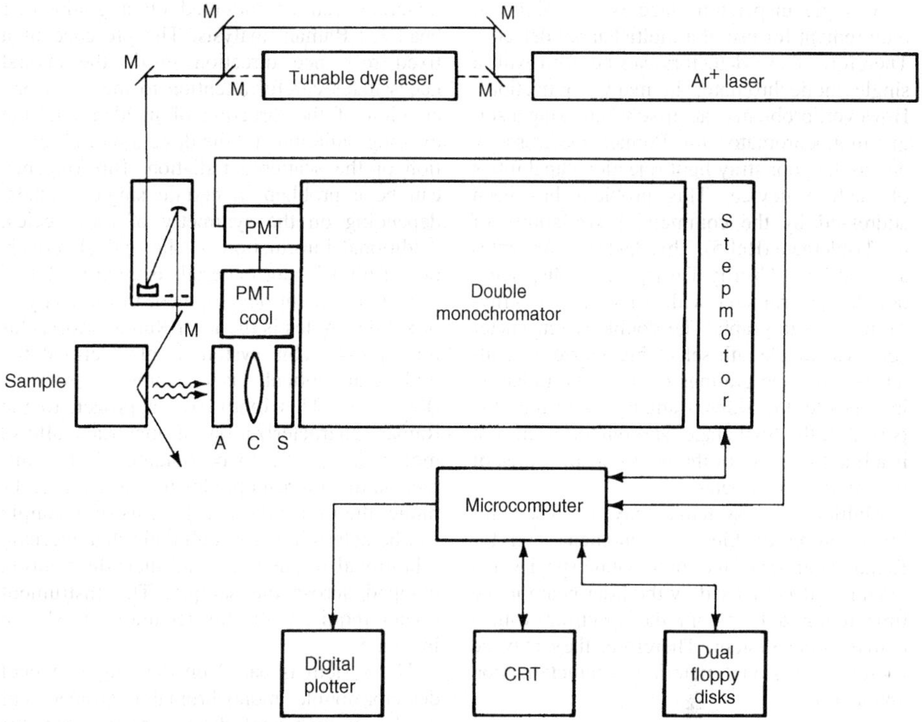

Fig. 2 Conventional Raman spectrometer. M, mirror; A, polarization analyzer; C, collection optics; S, polarization scrambler

Rayleigh scattering as a function of the Raman shift. This plot demonstrates that without multiple dispersion stages, the intensity of stray radiation can overshadow the much less intense Stokes-scattered radiation. In these devices, the dispersion elements are ruled gratings. Commercial scanning systems generally incorporate gratings that are ruled holographically to reduce the effects of optical artifacts in the observed spectra.

Raman spectra are acquired using scanning monochromators by mechanical movement of the dispersion elements such that the single-element detector sequentially detects the frequencies of interest. High-sensitivity photomultiplier tubes (PMTs) cooled to −20 to −40 °C (−4 to −40 °F) to reduce the dark current are typically used.

An alternative for the acquisition of Raman spectra is use of multichannel detectors in conjunction with a dispersion stage. Vidicon and diode array detectors may be used. To meet the requirements of Raman spectroscopy, these detectors usually incorporate image intensifiers that increase sensitivity. The benefit of the multichannel detector is known as Fellgett's advantage or the multiplex advantage. This signal-to-noise ratio advantage or time advantage relative to the performance of a single-channel detector is realized, because many frequencies are detected simultaneously. Relative to a single-channel detector, a multichannel detector can increase the signal-to-noise ratio proportional to the square root of the number of individual spectral resolution

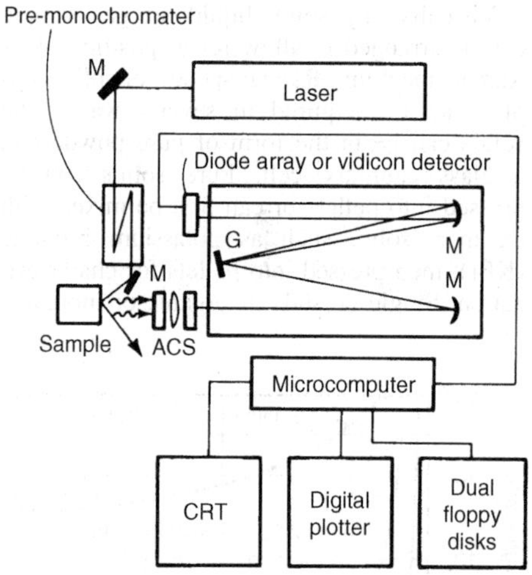

Fig. 3 Raman spectrometer with multichannel detector. M, mirror; G, grating; A, polarization analyzer; C, collection optics; S, polarization scrambler

elements simultaneously monitored by the multichannel detector.

Alternatively, Fellgett's advantage can be viewed as a time-saving benefit proportional to the square root of the number of spectral resolution elements. This is because a signal-to-noise ratio equivalent to that measured with a single-channel detector can be obtained in less time using a multichannel detector, assuming such factors as sensitivity and resolution equal those of a single-channel detector.

A single dispersion stage is the minimum requirement for use of a multichannel detector. Therefore, these detectors can be used with a single monochromator in many applications. However, problems can arise when using a single monochromator for Raman spectroscopy due to the poor stray light rejection capabilities of such a device. This problem has been addressed by the commercial availability of the Triplemate (Ref 5). This device incorporates a modified Czerny-Turner, zero-dispersion double spectrometer with a modified Czerny-Turner spectrograph. The double spectrometer acts as a wavelength-selectable interference filter, because the gratings disperse the radiation in opposite directions. Radiation is further dispersed in the final stage, and output at the exit is a line the width of the photosensitive area of the multichannel detector.

Multichannel systems may be used for investigation of kinetic phenomena and for Raman analysis of thermally labile species that would be decomposed by the laser beam in the time required to obtain the spectrum with a conventional system. Therefore, they may be useful in various materials characterization applications.

Sampling

Virtually any solid, liquid, or gas sample can be arranged to allow for acquisition of its Raman spectrum. Raman spectra of solid samples can be acquired in several ways. The solids can be in the form of pure powders in a glass capillary cell. Pure solids can be pressed into pellets or can first be mixed with an inert solid, such as potassium bromide (KBr), then pressed into pellets for characterization. Single crystals of organic or inorganic materials can be mounted on a goniometer head for Raman analysis. The presence of a fixed reference direction inside the crystal necessitates careful attention to the exact orientation of the direction of incidence of the exciting radiation and the direction of observation of the scattered radiation. Birefringence can be a problem in certain single crystals, depending on the symmetry of the species. Additional information on the optical properties of birefringent materials is cited in Ref 6.

A more recent development in the analysis of solid samples is the laser Raman molecular microprobe. This system is also termed the molecular optical laser examiner (MOLE) (Ref 7, 8). This innovative approach to the Raman characterization of materials allows molecular spectra to be obtained from samples on the microscopic level. Using this technique, the molecular components of a sample can be determined through their characteristic vibrational frequencies, and their distribution mapped across the sample. The instrument layout required for this technique is shown in Fig. 5.

The system is based on the single-channel detector/double monochromator arrangement or the multichannel detector/monochromator arrangement described above. A conventional optical microscope with bright- and dark-field illumination is the imaging system. The sample is placed on the microscope stage and can be analyzed in air, liquid, or a transparent medium. Two detectors are used in the system. The first is the PMT or multichannel detector for the acquisition of the actual spectra. The second is a TV detector that permits observation of the microstructure.

The MOLE can be operated in the punctual illumination or global illumination mode (Fig. 6). Punctual illumination allows recording of the Raman spectrum of one spot on the surface of the sample. This is also known as the spectral mode. Global illumination allows for obtaining the distribution or map of one component across the sample. Operation in this manner is also known as the imaging mode. The primary advantage of the laser Raman microprobe (MOLE) is that it provides molecular information on the microscale in a nondestructive analysis.

Raman spectroscopy may be used to analyze surface species, but its lack of sensitivity complicates these types of analyses. Several approaches are available to overcome the constraints imposed by the inherent weakness of the technique.

Raman spectra of surface species on high-surface-area solids, such as powders, are acquired easily in a glass capillary tube. Alternatively, powders can be pressed into pellets and analyzed. Careful analysis of species on metal surfaces also provides useful information. Raman analyses of these samples are usually performed by reflecting the laser beam off the metal surface and collecting the scattered radiation in the specular direction. Several mechanisms that enhance the intensity of the Raman-scattered radiation at metal surfaces under certain conditions may be used.

Several problems can arise in the Raman analysis of solids or surface species. Many solids frequently exhibit weak fluorescence, due to their inherent fluorescence or the presence of small amounts of fluorescent surface impurities. This fluorescence, even if weak,

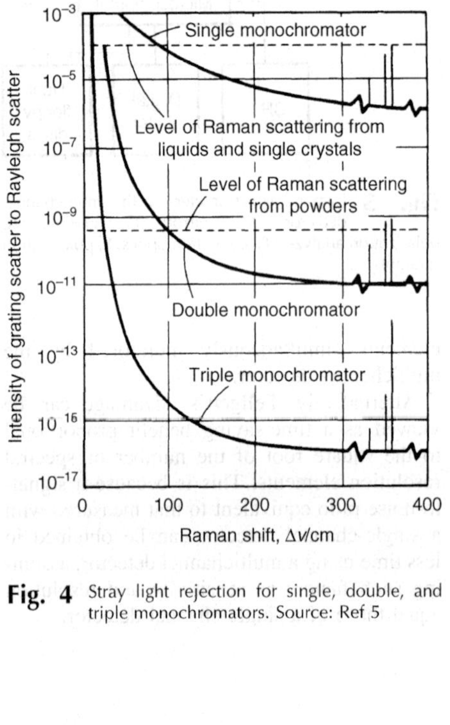

Fig. 4 Stray light rejection for single, double, and triple monochromators. Source: Ref 5

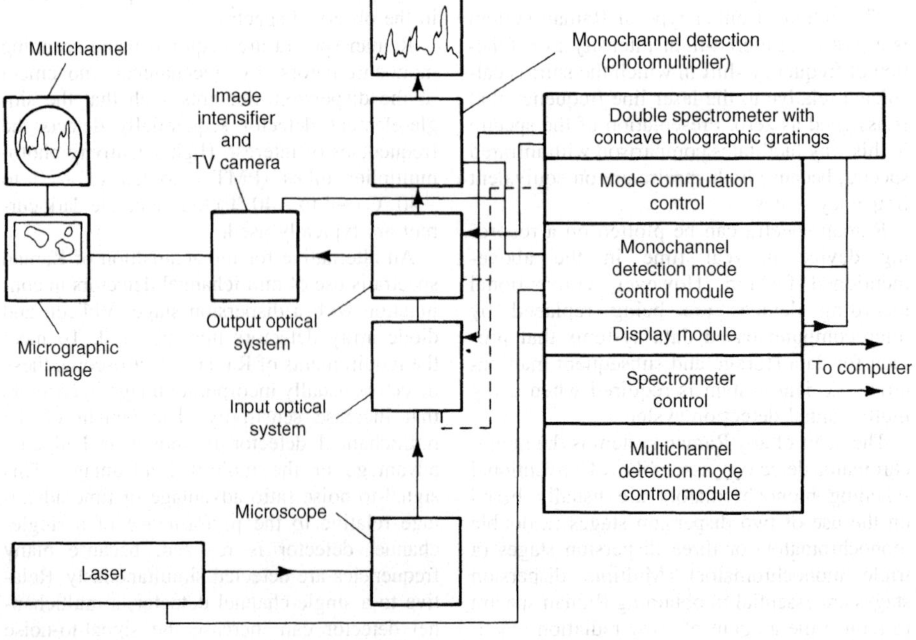

Fig. 5 Laser Raman microprobe. Source: Ref 8

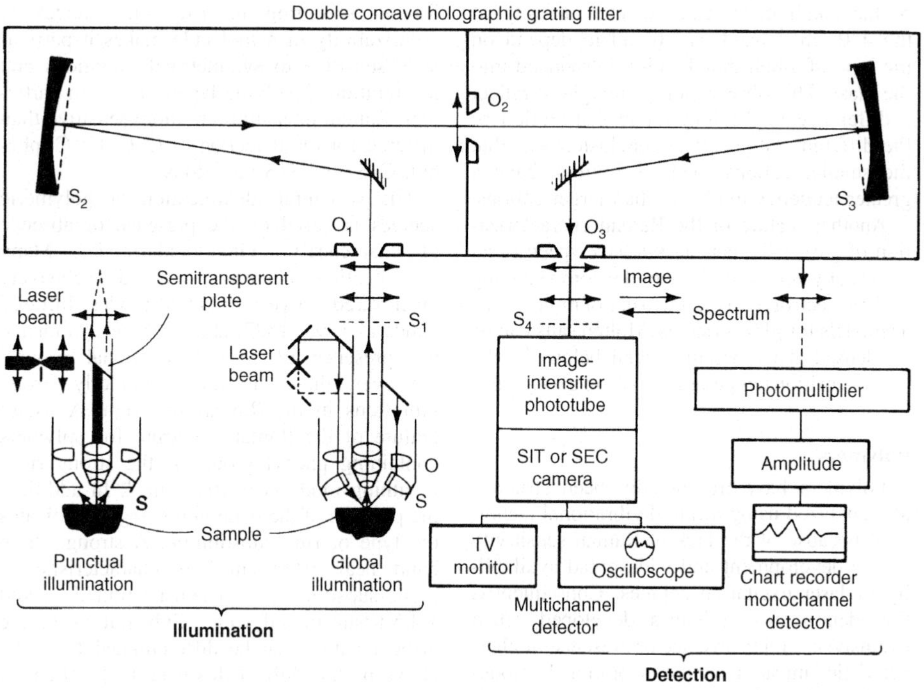

Fig. 6 Optical scheme of MOLE instrument. Source: Ref 8

will be more intense than the scattered radiation and will have noise associated with it. The Raman bands, which are superimposed on the fluorescence background, are often difficult to locate in the noise associated with fluorescence. The background fluorescence for silicas is fairly weak, but that for the alumina and silica-aluminas is considerably more intense (Ref 9). Various fluorescent impurities on solid surfaces have been identified. Hydrocarbons are commonly found on metal oxide surfaces (Ref 10-12). Trace amounts of transition metals have also been identified as sources of background fluorescence on aluminas and zeolites (Ref 13).

Few options are available in the analysis of inherently fluorescent solids. Extensive signal averaging to minimize the effect of the fluorescence noise may allow obtainment of partial Raman spectra. Significantly lowering the excitation energy may reduce the overall intensity of the fluorescence.

Removal of fluorescing impurities from the surface of solids to be analyzed requires treatment of these solids under rigorously controlled conditions. The most common surface impurities encountered are hydrocarbons (Ref 10-12). These can usually be removed by holding the sample several hours at elevated temperature in an oxygen or air atmosphere. Once cleaned, extreme care must be taken to avoid recontamination of sample surfaces through exposure to the ambient environment. Another approach to eliminating or reducing background fluorescence from impurities is preactivation of the sample by exposure to the laser beam for several hours before acquisition of the Raman spectrum (Ref 9).

A second problem frequently encountered in the Raman characterization of solids and surfaces is decomposition of the sample in the laser beam. One method of minimizing or eliminating this problem is to alter the laser conditions under which the sample is analyzed. Decreasing the laser power or changing the excitation frequency to a more suitable energy can help to eliminate the decomposition problem. A second approach is to rotate the sample such that the laser beam does not remain on any one spot on the sample long enough to cause extensive local heating and decomposition.

Sample decomposition is further exacerbated when analyzing samples that are highly absorbing at the excitation frequency used. The most effective method for handling highly absorbing samples is sample rotation (Ref 14). This approach has been frequently used for absorbing solids, such as graphites.

Information Obtainable from Raman Analyses

The type of molecular vibrations that produce Raman scattering must alter the polarizability of the molecule. Therefore, those vibrations that originate in relatively nonpolar bonds with symmetrical charge distributions produce the greatest polarizability changes and thus yield the most intense Raman scattering. Organic functional groups that fit these criteria include such moieties as C=C, C≡C, C≡N, C–N, N=N, C–S, S–S, and S–H.

However, functional group information is not the only type of vibrational information present in a Raman spectrum. Raman spectra of solids and crystals also contain contributions from lattice vibrations at low frequencies. These vibrations are due to the vibration of the molecules around their centers of mass or the restricted translation of molecules relative to each other. Lattice vibrations can provide a wealth of information on crystal forces (Ref 2).

Analysis of Bulk Materials

Metal Oxide Systems

Raman spectroscopy has been used with considerable success in the analysis of metal oxide systems. Metal oxide glasses provide particularly illustrative examples. Raman spectroscopy was initially applied to the investigation of glasses to overcome the problems of infrared analysis of these materials. Metal oxides exhibit strong absorption in the infrared region, making analysis of bulk metal oxides virtually impossible. Therefore, alkali halide pellets, for example, KBr, of the metal oxides usually must be prepared.

One possible consequence of this type of preparation—ion exchange of the metal oxide with the alkali halide—is a serious limitation to the infrared analysis of these materials. The chemical interaction between the metal oxide and the matrix changes the composition of the metal oxide under investigation. However, Raman scattering from metal oxides is usually only of weak to medium intensity. Therefore, bulk metal oxides can be analyzed easily without the chemical complications of infrared analysis.

An early Raman study of the influence of various cations on the bonding in the phosphate skeleton in binary phosphate glasses has been reported (Ref 15). Binary phosphate glasses containing sodium oxide (Na_2O), beryllium oxide (BeO), magnesium oxide (MgO), calcium oxide (CaO), strontium oxide (SrO), barium oxide (BaO), zinc oxide (ZnO), cadmium oxide (CdO), aluminum oxide (Al_2O_3), gallium oxide (Ga_2O_3), lead oxide (PbO), and bismuth oxide (Bi_2O_3) were used at metaphosphate stoichiometry. Several important vibrational bands were observed in the spectra. A band at 700 cm^{-1} was assigned to the symmetrical vibration of the –P–O–P– group. Bands observed from 1155 to 1230 cm^{-1} represent the symmetric and antisymmetric vibrations of the $-PO_2$ group.

Effects on the frequency and the intensity of these bands were observed upon addition of the above-mentioned cations to these glasses. A quadratic increase in frequency of the $-PO_2$ feature occurred with an increase in the ionic potential (charge-to-radius ratio) of the cation. However, a linear decrease in intensity of this band was observed with an increase in ionic potential of the cation. These cationic dependencies were rationalized in terms of an increase in the ionic character of the oxygen-phosphorus bonds, P ··· O, as a result of the donor-acceptor interaction between oxygen and the metal, and oxygen and phosphorus.

Inclusions in glasses that contain gases associated with various chemical reactions and processing stages of glass formation have been studied (Ref 16). Because these inclusions generally degrade the appearance and mechanical strength of the glass, it is desirable to identify and eliminate their causes. Test samples were prepared for Raman analysis by bubbling carbon dioxide (CO_2) or sulfur dioxide (SO_2) through molten glass ($68SiO_2$-$14NaO_2$-$12BaO$-$6ZnO$). Raman bands associated with CO_2 and SO_2 were monitored, and their appearance correlated with the preparation conditions.

This approach was later expanded to the Raman analysis of glasses by using the Raman microprobe (MOLE) to characterize deposits and gaseous contents of bubbles in the glass (Ref 17). The MOLE technique enabled sampling of only the bubbles in the glass to the exclusion of the bulk glass matrix. In these studies, clear soda-lime-silica glass was prepared by the float glass process. Carbon dioxide and SO_2 gaseous inclusions were identified by their Raman bands at 1389 and 1286 cm^{-1} for CO_2 and 1151 cm^{-1} for SO_2. The ratio of the CO_2 to SO_2 concentrations was quantitatively determined by the relative band intensities to be 11:1. The Raman microprobe analysis further indicated that no nitrogen, oxygen, sulfite, or water vapor was present in the glass bubbles. However, solid deposits in the bubbles showed polymeric sulfur in the $S_\infty + S_8$ structure, as indicated by the Raman bands at 152, 216, and 470 cm^{-1}. Monitoring the 1151 cm^{-1} band of SO_2 revealed that the concentration of SO_2 in these bubbles can be decreased by heating to 450 °C (840 °F). The loss of SO_2 was attributed to the reaction:

$$2Na_2O + 3SO_2 \rightleftharpoons 2Na_2SO_4 + S$$

which also indicates the source of the sulfur deposits. Finally, the presence of sulfur or the absence of SO_2 in the bubbles was concluded to be a function of the cooling to which the glass is subjected during fabrication.

In a recent Raman characterization of SiO_2 glasses, the effect of alkali cations on the spectral response of SiO_2 glasses was monitored as a function of weight percent of lithium, sodium, potassium, rubidium, and cesium cations (Ref 18). High-frequency bands greater than 800 cm^{-1} have been assigned to the local distribution of silicon tetroxide (SiO_4) tetrahedra in these glasses. These features were found to be insensitive to large-scale clustering of alkali metal species. A Raman feature at 440 cm^{-1} is also characteristic of silica glass and indicates that regions of the three-dimensional SiO_4 network remain after introduction of the alkali metal cations. The intensity of the 440 cm^{-1} band was found to depend on the type of alkali metal cation introduced into the glass. This observation is thought to reflect a difference in the long-range distribution of the different cations. The conclusion was that the smaller cations, such as lithium, have a greater tendency to cluster than larger cations.

Another feature of the Raman characterization of glasses is that glasses have spectra that are usually similar to those of the corresponding molten electrolyte. This led to use of molten salts as models for glass systems. Molten salts can be condensed at temperatures well below T_g, the glass transition temperature, to form glassy salts.

Polymers

Polymers have traditionally been structurally analyzed using infrared vibrational techniques because of the lack of Raman sensitivity before development and widespread availability of laser excitation sources. Consequently, the identification schemes developed relied exclusively on the absence or presence of characteristic infrared active vibrational modes (Ref 19, 20). Nevertheless, because Raman analysis of polymer systems offers several advantages, many Raman investigations have appeared in the literature.

The weak Raman scattering of water makes the Raman analysis of polymers in aqueous media particularly attractive. Little sample preparation is required to obtain at least a survey spectrum of a polymer, and sample size and thickness present no problem as in infrared spectroscopy. One of the major advantages of Raman spectroscopy is the availability of the entire vibrational spectrum using one instrument. Vibrations that occur at frequencies lower than 50 cm^{-1} can be observed with little difficulty. The main problem that remains in handling polymeric samples involves sample fluorescence. The solution to this problem is similar to that used for other Raman samples and has been discussed above.

Infrared and Raman spectroscopy are complementary techniques. Neither can give all the information that a combination of the two techniques can provide because of differences in selection rules. However, depending on the type of information required, Raman spectroscopy can prove to be better suited for polymer characterization in terms of required sensitivity, simplicity of sampling methods, or vibrational region of interest.

One area of interest in the Raman characterization of polymers is the identification of components that may be present in only 5 to 10% concentrations. Studies of various types of degradation and polymerization have involved observing changes in vibrational features that affect only a small part of the polymer. In such cases, it is advantageous to monitor vibrational bands that are sensitive to the particular functional group that will reflect the change of interest. The dependence of Raman scattering intensity on changes in polarizability of a molecule makes it particularly sensitive to symmetrical vibrations and to vibrations involving larger atoms. In particular, Raman intensities are more sensitive than infrared for the detection of C=C, C≡C, phenyl, C–S, and S–S vibrations.

The structural identification of polymeric species is based on the presence or absence of characteristic vibrational modes. Many identification schemes are based exclusively on infrared active vibrational modes. Infrared bands at 1493, 1587, and 1605 cm^{-1} indicate the presence of phenyl functional groups. However, these are not particularly strong vibrations in the Raman spectrum. A useful feature of the Raman spectrum for polymers containing phenyl groups is the strong ring-breathing mode near 1000 cm^{-1}. In addition, the position of the band in this region indicates the type of ring substitution. A strong, sharp band near 1000 cm^{-1} is characteristic of mono-substituted, metadisubstituted, and 1,3,5-trisubstituted rings. Substitution in the ortho position can be differentiated from the above by the shift of this band to 1050 cm^{-1}. However, a band in this region is absent for para-substituted compounds. The presence of phenyl groups can also be characterized by a strong aromatic C–H stretch from 3000 to 3100 cm^{-1}.

The strong Raman vibration associated with the C=C symmetric stretch facilitates identification of trans-substituted alkene polymers and investigation of several polymerization reactions. In particular, Raman spectroscopic studies have been used to follow the polymerization of butadiene (Ref 21) and styrene (Ref 22). Raman results indicate that polymerization of butadiene proceeds by several competing mechanisms (Ref 21). Because the polymer products all contain an unsaturated skeletal backbone, the Raman active C=C stretching mode can be conveniently monitored during polymerization due to its enhanced sensitivity. In this study, the *trans*-1,4-butadiene product was identified by its Raman spectrum.

The thermal polymerization of styrene has also been investigated using Raman spectroscopy (Ref 22). The decrease in the intensity of the C=C stretch at 1632 cm^{-1}, relative to that of an internal standard, was used to obtain kinetic information on the styrene polymerization reaction. Values obtained for the activation energy and percent styrene conversion were found to be in reasonable agreement with results from other methods.

The intensity of the asymmetric C–H stretching vibration has been used to determine quantitatively the percent vinyl chloride in the vinyl chloride/vinylidene chloride copolymer (Ref 23). Calibration curves showed a linear relationship between the ratio of the intensity of the C–H stretch at 2906 cm^{-1} to the scattering reference up to 100% vinyl chloride with an accuracy of ±2%. By comparison, an analysis based on a characteristic infrared absorption at 1205 cm^{-1} showed a correlation between vinyl chloride copolymer content and infrared intensity for concentrations only up to 25%.

Strong Raman vibrations of the S–S and C–S stretching modes in the regions from 400 to 500 cm^{-1} and 600 to 700 cm^{-1} are particularly useful for identification of polysulfides, because these modes are infrared inactive. These vibrational features have been used to investigate the structural changes accompanying vulcanization of *cis*-1,4-polybutadiene (Ref 24, 25). Several spectral features in these regions were assigned tentatively for vibrations corresponding to disulfide, polysulfide, and five- and six-membered thioalkane and thioalkene structures (Ref 24). The structures found after vulcanization varied with the length of time and temperature of the process (Ref 25). Corresponding information obtained from the C=C region provides information on skeletal chain modifications. The vibrational information from this spectral region suggests the formation of *cis*-1,4-*trans*-1,4, vinyl, and conjugated triene groups. The number of terminal mercapto groups in polythioethers has been quantitatively determined using the S–H stretching vibration at 2570 cm^{-1} (Ref 26). The method, based on a comparison of the peak area of the S–H band with that of an internal standard, has been shown to be effective to 0.5% mercapto group content with a precision of ±1%.

The Raman spectrum of aromatic silicones is characterized by a strong band near 500 cm^{-1} that is attributable to the symmetric Si–O–Si mode. The infrared-active asymmetric stretch is a strong, broad absorption from 1000 to 1100 cm^{-1} that overlaps characteristic bands due to vinyl and phenyl groups. Therefore, the determination of small amounts of vinyl and phenyl groups copolymerized in silicone systems is particularly well suited to Raman analysis. In addition, a Raman active C–Si stretch can be observed at 710 cm^{-1}. These Raman vibrational features have been used to investigate helix formation in polydimethylsiloxane (Ref 27). Depolarization measurements were obtained for the methyl C–H stretching mode at 2907 cm^{-1} and the Si–O–Si mode at 491 cm^{-1} as a function of temperature. Estimated values were obtained for the enthalpy of helix formation, the change in entropy, and the lower limit for the fraction of polymers existing in helix conformation.

A method has been devised to determine the percent conversion of polyacrylamide to poly (N-dimethylaminomethylacrylamide) using intensity ratios of the characteristic C=N stretching bands for the reactant at 1112 cm^{-1} and the product at 1212 cm^{-1} (Ref 28). The percent conversion results were in excellent agreement with results obtained using ^{13}C nuclear magnetic resonance (NMR).

Raman spectroscopic methods have been used to investigate formation of phenolformaldehyde resins (Ref 29). The polymerization reaction is carried out in aqueous media that severely limit applicability of infrared analysis. The intensity of four characteristic Raman bands were measured as a function of time during the early stages of the condensation reaction. Raman results were consistent with results obtained by the analysis of the reaction mixture using paper chromatography. However, the Raman results provide more detailed information on structural changes occurring during the early stages of the reaction.

The extent of crystallinity of polyethylene has been monitored using Raman spectroscopy (Ref 30). Crystallization of polyethylene was found to produce a marked narrowing of the Raman band corresponding to the C=O stretch at 1096 cm^{-1}. This change in bandwidth was correlated with density changes in the polymer and found to be a reliable indicator of the degree of crystallinity.

Graphites

Raman spectroscopy has been used extensively to characterize the extent of surface structural disorder in graphites. More recently, Raman spectroscopy has been developed to investigate intercalated graphites. The analysis of graphites is experimentally complicated by strong absorption of laser radiation, which has been observed to damage the surface significantly. To avoid significant surface decomposition during analysis, low laser powers (20 to 40 mW) on a stationary graphite sample are used. An alternate, more prevalent technique is the use of a rotating sample cell (Ref 14, 31, 32) with which higher incident powers (300 to 400 mW) can be used.

The utility of Raman spectroscopy for graphite characterization derives from the various vibrational behaviors observed for different graphites. Group theory predicts two Raman active modes for smooth single-crystalline graphite. These vibrations are an in-plane mode at 1581 cm^{-1} and a low-frequency plane rigid shear mode at 42 cm^{-1}. Single-crystalline and highly oriented pyrolytic graphite (HOPG) exhibits a single sharp vibrational feature near 1580 cm^{-1} (Ref 33). Less highly ordered graphites, such as activated charcoal, vitreous carbon, and stress-annealed pyrolytic graphite, show additional vibrational modes. A vibrational feature near 1355 cm^{-1} is associated with surface structural defects in the graphite lattice. The relative intensity of the 1355 cm^{-1} band to that of the 1580 cm^{-1} band increases with the degree of surface disorder (Ref 34). The behavior of the 1355 cm^{-1} feature has been used extensively to characterize the effects of annealing (Ref 35), grinding (Ref 36), mechanical polishing (Ref 37), and ion implantation (Ref 38, 39) on the structural integrity of various graphites.

A similar increase in the intensity of the 1355 cm^{-1} band occurs after electrochemical oxidation and reduction of HOPG in 0.05 *M* sulfuric acid (H_2SO_4) solution (Ref 31). An additional vibrational feature at 1620 cm^{-1} appears after oxidation at 800 °C (1470 °F) (Ref 37), mechanical polishing (Ref 37), and grinding (Ref 36) of various graphites. This feature has been associated with a hexagonal ring stretching mode that has been modified by formation of carbon-oxygen complexes near the graphite surface or crystallite edges (Ref 37).

Raman microprobe analysis has been used with transmission electron microscopy (TEM) to characterize vapor deposition of carbon films on alkali halide cleavages (Ref 40). These results indicate that the graphite films graphitize in five distinct stages characterized by the release of a given structural defect at each stage.

Raman spectroscopy has also been a valuable tool for the analysis of intercalated graphite species. These studies have focused on the vibrational behavior of donor intercalates, such as K^+ (Ref 41), Li^+ (Ref 42), Rb^+ (Ref 43), and Cs^+ (Ref 44), and acceptor intercalates, such as ferric chloride ($FeCl_3$) (Ref 45), bromine (Br_2) (Ref 46-48), iodine chloride (ICl) (Ref 48), and iodine bromide (IBr) (Ref 48) of various stage numbers. The stage number refers to the number of carbon layers between any pair of intercalant layers. Therefore, the number of carbon layers between intercalant layers increases with the stage number. Although there is general agreement of the Raman data for stage two and greater species, some controversy remains regarding stage one compounds (Ref 49).

Several models have been proposed to explain the Raman vibrational behavior observed for various intercalated species (Ref 44, 50, 51). The nearest layer model (Ref 44) is based on a perturbation of the pristine graphite modes by the presence of intercalate nearest neighbor layers. This model predicts that pure stage one and stage two donor or acceptor compounds should exhibit a single Raman band in the vicinity of but displaced from that of HOPG, near 1580 cm^{-1}. For stage three and higher compounds, an additional vibration near 1580 cm^{-1} is anticipated due to the presence of graphite layers that do not have intercalates as nearest neighbors. The intensity of the pristine graphite mode at 1580 cm^{-1} is expected to increase with the stage number relative to the displaced mode.

Several intercalate systems have been examined to test the nearest layer model. The Raman behavior of potassium (Ref 41) and rubidium (Ref 43) intercalated HOPG followed that predicted by the model. Further data analysis of the ratio of intensities of the 1581 cm^{-1} band to that of the displaced band as a function of stage number provides valuable information on charge transfer and charge localization in the various layers. The similarity in slope for potassium and rubidium cations indicates that both systems are electrically similar and that charge exchange to the carbon layers is not completely localized (Ref 49). Qualitative spectral agreement has been observed for intercalated $FeCl_3$ (Ref 45), an acceptor intercalate.

Raman spectroscopy has also provided evidence for the intercalation of Br_2, ICl, and IBr as molecular entities (Ref 48). The high-frequency bands were found to be insensitive to the intercalant species. The position of the low-frequency band was found to depend on the nature of the intercalant. These low-frequency bands were assigned to intramolecular modes of the intercalant species. The strong wavelength dependence of the Br_2 intramolecular mode has been interpreted as a resonance-enhancement effect due to an electronic Br_2 excitation (Ref 46).

Analysis of Surfaces

Use of the laser as an intense monochromatic light source exploits the advantages of Raman spectroscopy in the vibrational analysis of surfaces and surface species while surmounting the inherent sensitivity limitations of the technique that would preclude its applicability to surfaces. Moreover, Raman spectroscopy adequately overcomes several limitations of infrared spectroscopy, which has been used extensively over the past several decades to provide vibrational information on surfaces and surface species.

An advantage of Raman spectroscopy is its ready accessibility to the low-frequency region of the spectrum. Vibrational behavior can be characterized as close to the Rayleigh line as 10 cm^{-1} with conventional instrumentation. This frequency region, particularly below 200 cm^{-1}, is not accessible with infrared detectors. Such low-frequency data are important in the complete vibrational analysis of surface species, especially for investigations of the nature of the chemical interaction of a surface species with the underlying surface.

Raman spectroscopy is also valuable in probing surface processes in aqueous environments due to the extreme weakness of the Raman scattering of water. This advantage has made feasible the vibrational characterization of such materials problems in aqueous environments as corrosion. A third problem with infrared analysis of certain surface systems that Raman overcomes is interference from absorption of the radiation by the underlying bulk material. In particular, this advantage has been realized in the study of metal oxide systems. These species are strong infrared absorbers, but only weak to moderate Raman scatters.

Surface Structure of Materials

The study of metal oxide and supported metal oxide catalysts best illustrates application of Raman spectroscopy to analysis of the surface structure of materials. An example of the Raman analysis of metal oxide surface structure is a study on the effect of specific chemical treatments on the surface structure of molybdenum oxide catalysts used for coal hydrosulfurization (Ref 52). Raman surface spectra characteristic of molybdenum disulfide (MoS_2) after sulfidation of the catalyst by H_2/H_2S were observed. After these catalysts were used for coal hydrosulfurization, the Raman spectra were dominated by intense scattering characteristic of carbon. Furthermore, the Raman analysis of used catalysts subjected to regeneration showed that all the original features of the unused catalysts are not recovered in regeneration (Ref 52).

Raman spectroscopy has been used extensively to characterize the surface structure of supported metal oxides. The Raman investigation of chemical species formed during the calcination and activation of tungsten trioxide (WO_3) catalysts supported on silica (SiO_2) and alumina (Al_2O_3) has been documented (Ref 53). These Raman results indicate that crystalline and polymeric forms of WO_3 are present on SiO_2-supported surfaces. However, only polymeric forms of WO_3 are present on γ-Al_2O_3 supports.

Vanadium oxide catalysts supported on γ-Al_2O_3, cerium dioxide (CeO_2), chromium oxide (Cr_2O_3), SiO_2, titanium dioxide (TiO_2), and zirconium dioxide (ZrO_2) have been characterized using Raman spectroscopy (Ref 54). These catalysts are of industrial importance for the oxidation Of SO_2, CO, and other hydrocarbons. In this study, the effects of catalyst preparation and catalyst surface coverage on the Raman vibrational behavior were investigated. For low surface coverages of vanadium oxide (1 to 40 wt%) on any support, the Raman spectra were found to be characteristic of a two-dimensional surface vanadate phase. This behavior was found to be independent of catalyst preparation. For medium to high surface coverages of vanadium oxide on any support, the wet-impregnated preparation method resulted in a crystalline V_2O_5 surface phase.

Many Raman characterization studies have been performed on the effects of the nature of the support, the presence of other metals, impregnation order, molybdenum oxide loading, pH, and calcination and regeneration conditions on the resulting surface structure of molybdena catalysts (Ref 55-61). Results indicate the presence of MoO_3 or MoO_4^{2-} surface species. An example of these studies is the Raman investigation of molybdena catalyst supported on γ-Al_2O_3 and n-Al_2O_3 (Ref 61). Various catalysts were prepared by impregnation from aqueous molybdate solutions of pH 6 and 11, and the development of the final catalytic moiety was followed using Raman spectroscopy and ultraviolet photoelectron spectroscopy (UPS). Vibrational modes of surface species were assigned on the basis of solution spectra of various isopolymolybdates. Results show that the initial species undergoes ion exchange with surface hydroxides to form MoO_4^{2-} regardless of the solution pH, and depending on the surface coverage of MoO_4^{2-}, the formation of Mo–O–Mo bridging species can occur during subsequent preparation to give a final polymeric surface species.

Surface Species on Nonmetals

One of the most prevalent chemical probes of the surface chemical environment in adsorption studies is pyridine. Its utility as a surface probe stems from the extreme sensitivity of the ring-breathing vibrational modes to chemical environment. Furthermore, its π-electron system, which is responsible for the large Raman scattering cross section for this molecule, makes pyridine useful for Raman studies in terms of detectability. Assignments of the ring-breathing vibrational modes of adsorbed pyridine species are usually made by comparison to a series of model environments of the pyridine molecule. Table 1 lists the accepted model environments for pyridine. The various interactions produce substantial shifts in the peak frequency, ν_1, of the symmetrical ring-breathing vibration of pyridine. In general, the more strongly interacting the lone pair of electrons on the pyridine nitrogen, the larger the shift in ν_1 to higher frequencies.

Similar behavior in the ν_1, ring-breathing mode of pyridine is observed when pyridine adsorbs to various metal oxide and related solid surfaces. Table 2 summarizes the Raman studies performed on pyridine adsorbed on diverse adsorbents. In general, for high coverages of pyridine on any solid adsorbent, the Raman spectrum closely resembles that of liquid pyridine. In these cases, the interaction of the pyridine with the underlying adsorbent is thought to be weak. Therefore, the pyridine is considered physisorbed.

Two types of strong interaction of pyridine with the underlying adsorbent are possible for low coverages of pyridine on metal oxide and related surfaces. Hydrogen-bonded pyridinium surface species can be formed at Brönsted sites on the surface. These species give rise to the ν_1 pyridine band near 1010 cm^{-1}. Strong chemisorption of the pyridine species can also occur at Lewis acid sites, such as Al^{3+}, on these surfaces. This type of interaction produces the higher frequency ν_1 band near 1020 cm^{-1}.

In the Raman spectroscopy of pyridine on X and Y zeolites, the frequency of the ν_1 symmetric ring-breathing mode can be linearly correlated with the electrostatic potential (charge-to-radius ratio) of the balancing cation within the cation-exchange zeolite (Ref 13, 67, 69, 70). This correlation has been interpreted as indicating pyridinecation interactions of varying strength in these systems. The strength of interaction of pyridine with the cation is thought to increase with the electrostatic potential of the cation, as indicated by the corresponding shift of ν_1 to higher frequencies.

Table 1 Model environments for pyridine vibrational behavior

Model compound	ν_1/cm	Nature of interaction	Ref
Pyr	991	Neat liquid	62
Pyr in $CHCl_3$	998	H-bond	12, 13, 63
Pyr in CH_2Cl_2	992	No interaction	12, 13, 63
Pyr in CCl_4	991	No interaction	12, 13, 63
Pyr in H_2O	1003	H-bond	12, 13, 63–65
$PyrH^+ BF_4^-$	1012	Pyridinium	64, 65
Pyr:$ZnCl_2$	1025	Coordinately bound	12, 64, 65
Pyr N-Oxide	1016	Coordinately bound	63
Pyr:$GaCl_3$ (benzene solution)	1021	Coordinately bound	12

Table 2 Vibrational behavior of adsorbed pyridine

Adsorption environment	ν_1/cm	Nature of interaction	Ref
Pyr	991	Neat liquid	62
Pyr/chromatographic grade silica	1010	Lewis acid site	12, 13, 63
Pyr/Cab-O-Sil HS5 (high coverage)	991	Physisorption	63
Pyr/Cab-O-Sil HS5 (low coverage)	1010	H-bonded	63
Pyr/aerosil	1006	H-bonded	63
Pyr/silica with excess Al^{3+}	1020	Lewis acid site	63
Pyr/porous vycor glass	1006	H-bonded	11
Pyr/γ-Al_2O_3	1019	Lewis acid site	13, 63, 66
Pyr/n-Al_2O_3	1019	Lewis acid site	63
Pyr/chlorided γ-Al_2O_3	1022	Lewis acid site	67, 68
Pyr/chlorided n-Al_2O_3	1022	Lewis acid site	63
Pyr/13% Al_2O_3-silica(a)	1007, 1020	Lewis acid site and H-bonded	66, 68
Pyr/13% Al_2O_3-silica(b)	999	Physisorption	66, 68
Pyr/TiO_2	1016	Lewis acid site	63
Pyr/NH_4^+ -mordenite	1004	H-bonded	63
Pyr/magnesium oxide	991	Physisorption	63
Pyr/zeolites (X and Y)	998–1020	Physisorption and Lewis acid site	13, 69, 70

(a) Low temperature pretreament. (b) High temperature pretreatment

Another feature of the pyridine surface studies that has important implications is the linear increase in pyridine band intensity with coverage (Ref 71). This observation suggests the utility of Raman spectroscopy as a probe of adsorption isotherms for surface species. Such studies may be significant in understanding catalytic systems that are important in industry.

Raman microprobe characterization of pyridine adsorbed on metal oxide surfaces has been reported. Pyridine adsorbed on Ni-Mo-γ-Al_2O_3 has been studied using the Raman microprobe (MOLE) (Ref 72). The catalyst used was 3 wt% nickel oxide (NiO) and 14 wt% molybdenum trioxide MoO_3) with γ-Al_2O_3. The ν_1 feature of pyridine at 1014 cm^{-1} was attributed to pyridine chemisorbed at Lewis acid sites. It was determined that the physisorbed pyridine can be removed by heating to 100 °C (212 °F) as monitored by the disappearance of the bands at 991 and 1031 cm^{-1}. The chemisorbed pyridine remained on the surface even after heating to 200 °C (390 °F). The similarity between these results and those for pyridine on Co-Mo-γ-Al_2O_3 (Ref 73) was noted.

Corrosion on Metals

Raman spectroscopy is finding widespread use in the characterization of corrosion processes on metal surfaces. Corrosion can be easily monitored under gas phase or liquid conditions. The principal advantage of using Raman spectroscopy for corrosion is in corrosive aqueous environments, such as acidic or alkaline solutions, in which the Raman scattering of the aqueous medium is weak and does not interfere with detection of the metal corrosion products. Most of the corrosion products of interest involve metal oxide species. Due to the relatively weak scattering of metal oxides, Raman spectroscopy was not successfully applied to the *in situ* characterization of corrosion until the use of lasers as excitation sources had become commonplace. However, advances in this materials characterization area since the late 1970s suggest that the Raman investigation of corrosive environments has the potential to provide much molecularly specific information about metal surface corrosion products, such as chemical composition, stoichiometry, and crystallographic phase.

In situ Raman spectroscopy has been used to study the surface oxides formed on common alloys during oxidation at elevated temperatures in air (Ref 74). By comparison of the surface spectra with those of mixed pure oxides, metal oxides, such as ferric oxide (Fe_2O_3), chromic oxide (Cr_2O_3), nickel oxide (NiO), and manganese chromite ($MnCr_2O_4$), were identified.

In a more recent gas-phase corrosion study, the chemical composition of iron oxide films formed on iron by air oxidation at 400 °C (750 °F) for 2 h were identified using Raman spectroscopy (Ref 75). The characteristic lattice vibrations of the different iron oxides enabled differentiation between Fe_2O_3 and iron oxide (Fe_3O_4) films. Further analysis using Ar^+ sputtering and Raman spectroscopy to depth-profile the oxide layer formed revealed the presence of two zones of different oxides. Raman spectra obtained after various sputtering times indicated that the composition of this two-zone layer was 200 nm Fe_2O_3 on 800 nm Fe_3O_4.

Electrochemically based corrosion systems in aqueous environments have been studied using Raman spectroscopy. The Raman characterization of the galvanostatic reduction of different crystallographic forms of FeOOH on weathering steel surfaces has been reported (Ref 76). Atmospheric corrosion of metals has been explained relative to the electrochemical response of the metal in which different regions of the metal act as anode and cathode of the electrochemical cell. This study was motivated by a previous claim that an inner layer of α-FeOOH is formed on weathering steels under atmospheric corrosion conditions. This layer of α-FeOOH presumably resists electrochemical reduction to Fe_3O_4 such that, upon formation of a layer of α-FeOOH of sufficient thickness, further corrosion is inhibited. The results of this study confirmed the previous claims. The Raman intensity of the 300 and 380 cm^{-1} bands of α-FeOOH were monitored in an electrochemical cell under reducing conditions. No change in intensity of these bands was observed, suggesting that no significant reduction of α-FeOOH occurs after 9 h.

Similar studies indicated that Fe_3 also is not reduced after 9 h under these conditions. In contrast, γ-FeOOH is reduced to Fe_3O_4, as shown by the disappearance of the band at 258 cm^{-1} and the appearance of the Fe_3O_4 band at 675 cm^{-1}. Furthermore, amorphous FeOOH can also be reduced to Fe_3O_4. Amorphous FeOOH is reduced more easily than γ-FeOOH. The overall conclusions of this study were that three of the four polymorphs of FeOOH present on weathering steels can be reduced to Fe_3O_4. The only form of FeOOH that resists reduction is γ-FeOOH.

Several reports of the Raman characterization of the corrosion of lead surfaces have appeared in the literature. Early research on lead in 0.1 M sulfate solutions showed the presence of surface films of compositions not in complete agreement with the predictions of the Pourbaix (potential-pH) diagram (Ref 77). The Pourbaix diagram does not predict the formation of lead oxide (PbO) under any conditions. However, the recorded spectra indicated the presence of PbO at certain potentials in acid and neutral solutions and at all potentials above the immunity region in basic solutions. Despite the lack of agreement of the Raman spectra with the Pourbaix diagrams, the Raman spectra were in agreement with the potentiodynamic polarization curves for these systems.

A later study of this system helped to resolve the above-mentioned anomalies (Ref 78). The objective of this study was to monitor the surface phases formed on lead during potentiodynamic cycling to obtain information on the cycle life and failure mechanisms in lead-acid batteries. The approach used was to anodize lead foils and acquire Raman spectra of the resulting lead surface films formed under potential control and after removal from the electrochemical cell. The surface spectra were then compared with spectra of the corresponding pure lead oxide for assignment. Lead surfaces anodized at −0.45 V versus a mercury/mercurous sulfate (Hg/Hg_2SO_4) reference electrode for 12 to 72 h were covered by a film of lead sulfate ($PbSO_4$), as indicated by bands at 436, 450, and 980 cm^{-1}. The surface expected to exist for lead anodized at +1.34 V versus Hg/Hg_2SO_4 is β-PbO_2, according to the Pourbaix diagram. This surface phase for *in situ* or *ex situ* analysis cannot be assigned unequivocally to this oxide in agreement with the previous study. Evidence for damage of the original phase due to laser irradiation was noted, however. Bands were observed that suggested that the β-PbO_2 film is converted to PbO during irradiation.

Raman spectroscopy has been used to study the oxidation of silver electrodes in alkaline environments (Ref 79). Earlier studies on this system suggested a two-step oxidation process of silver in which silver oxide (Ag_2O) is formed followed by further oxidation to AgO. It was also known that Ag_2O could be photoelectrochemically oxidized to AgO. However, the mechanism of this process was controversial. Therefore, this study was undertaken to monitor this process *in situ*.

Ex situ Raman analysis of silver electrodes anodized at +0.6 V versus a mercury/mercuric oxide (Hg/HgO) reference electrode showed no distinct vibrational features, although the Pourbaix diagram for this system predicts the formation of Ag_2O. Therefore, the Ag_2O was concluded to be a weak Raman scatterer or decomposed in the laser beam. However, *ex situ* Raman analysis of silver electrodes anodized at +0.8 V showed a strong peak at 430 cm^{-1}, with weaker features at 221 and 480 cm^{-1}. The surface phase formed at this potential was assigned to AgO by comparison with the Raman spectrum obtained on a sample of pure AgO.

When these analyses were performed *in situ* under potential control, the spectrum of AgO was always observed regardless of the applied potential. This observation was explained as evidence for the photoelectrochemical conversion of Ag_2O to AgO. The kinetics of this conversion were followed with Raman spectroscopy in potential step experiments in which the growth of the 430 cm^{-1} AgO band was monitored as a function of time.

Raman spectroscopy has been used to characterize the corrosion of nickel and cobalt in aqueous alkaline media (Ref 80). The metals were anodized in 0.05 *M* sodium hydroxide (NaOH), and the Raman spectra of the surface phases were acquired. Comparison of the surface phase formed on nickel with various pure oxides of nickel indicated the presence of $Ni_2O_{3.4}{\cdot}2H_2O$, with vibrational bands at 477 and 555 cm^{-1}. The surface phase formed during anodization of cobalt was determined to be a mixture of cobaltous oxide (CoO), with Raman bands a 515 and 690 cm^{-1}, and cobalt oxide (Co_3O_4), with bands at 475 and 587 cm^{-1}. These assignments were also confirmed by comparison of the surface spectra with those of the pure cobalt oxides.

Surface-Enhanced Raman Scattering

The sensitivity constraints imposed by the normal Raman-scattering effect severely limits applicability of this technique to the study of species on smooth and low-surface-area area surfaces. Therefore, Raman characterization of monolayer amounts of materials on metals was not feasible for some time. A significant advance in this field that prompted surface Raman spectroscopy was the 1973 Raman study of mercurous chloride (Hg_2Cl_2), mercurous bromide (Hg_2Br_2), and mercuric oxide (HgO) on a thin mercury film electrode in an operating electrochemical environment (Ref 81). Pyridine adsorbed at silver and copper electrodes was also studied (Ref 82, 83). Spectra of good quality were presented and attributed to a monolayer of adsorbed pyridine on high-surface-area silver and copper electrodes produced by anodization.

In 1977, it was recognized that the pyridine/silver spectra were anomalously intense (Ref 84, 85). The intensity enhancement of the pyridine surface species was estimated at approximately 10^5 to $10^6\times$ that which would be expected for an equivalent amount of pyridine in solution. This began the extensive investigation of the phenomenon known appropriately as surface-enhanced Raman scattering (SERS). Since the early efforts in this field, a variety of adsorbates at metal surfaces have been studied. An extensive list has been compiled of atomic and molecular species whose surface vibrational behavior has been characterized using SERS (Ref 86). An in-depth review of the field through 1981 has also been published (Ref 87).

In addition, SERS can be observed in diverse materials environments. Along with the metal/solution studies performed as indicated above, SERS investigations have been readily performed at metal/gas and metal/vacuum interfaces (Ref 87-90) as well as metal/solid interfaces in tunnel junction structures (Ref 87, 91, 92).

The major limitation of SERS as a materials characterization tool is that surface enhancement is not supported by all surfaces. Only a limited number of metals can support surface enhancement. The list of metals for which SERS has been documented remains controversial. Although the three most prevalent SERS metals are silver, copper, and gold (Ref 87), other metals have also been previously demonstrated or claimed to exhibit SERS.

The alkali metals lithium and potassium exhibit SERS in a vacuum environment for adsorbed benzene (Ref 93). Several reports of surface enhancement at platinum in a vacuum, sol (a suspension of metal colloids), and electrochemical environments have appeared (Ref 94-96). A brief report of the SERS of pyridine adsorbed at a cadmium electrode appeared in 1981 (Ref 97). Palladium (Ref 98) and nickel (Ref 99, 100) have been claimed to support SERS in vacuum environments. Beta-PdH electrodes are capable of surface enhancement of adsorbed pyridine and CO (Ref 101). Several recent reports of surface enhancement on semiconductor surfaces have also appeared. The SERS spectra of pyridine on NiO and TiO_2 surfaces in the gas phase have been reported (Ref 102, 103).

The general use of SERS as a surface characterization tool involves serious limitations. However, the potential for chemical modification of nonenhancing surfaces to allow for surface enhancement is under investigation. Surface-enhanced Raman scattering from pyridine adsorbed at a platinum electrode modified by small amounts of electrochemically deposited silver has been reported (Ref 104).

A similar approach has been used to investigate species adsorbed onto GaAs semiconductor surfaces (Ref 105, 106). Surface-enhanced resonant Raman scattering from $Ru(bipyridine)_3^{2+}$ has been observed on *n*-GaAs modified with small islands of electrochemically deposited silver (Ref 105). Normal SERS from $Ru(bipyridine)_3^{2+}$ has been documented on a silver-modified *p*-GaAs[100] electrode (Ref 106), and SERS has been reported from molecules adsorbed on thin gold overlayers on silver island films (Ref 107). These studies suggest that the potential exists in many systems for suitable modification of the surface to exploit the increase in sensitivity provided by SERS.

Relative to the restrictions imposed by the limited metals that can support SERS, it is necessary to determine the surface properties required for surface enhancement (Ref 87, 108). Proposed theoretical contributions to SERS can be classified as electromagnetic and chemical. The surface properties required to activate the chemical contributions fully are subtle and have eluded systematic investigation. The requisite surface properties necessary for electromagnetic effects are better understood and have begun to yield to systematic investigation. The latter include surface roughness and surface dielectric properties. These properties are related, because surface roughness dictates the resulting surface electronic structure.

Electromagnetic enhancement effects are based on the enhanced electric field found at roughness features on metal surfaces having the appropriate dielectric properties. The role of surface roughness has been recognized in electrochemical SERS (Ref 84). However, the anodization procedure used in these systems has not been completely investigated. Research is underway to understand systematically the chemistry, electrochemistry, and resulting surface morphology of the electrochemically generated surface roughness in SERS (Ref 109).

Systematic studies of the functional relationship between the extent of surface enhancement and surface dielectric properties are in their infancy. One approach in electrochemical systems is to alter the surface dielectric properties of an electrode by electrochemically depositing submonolayer and monolayer amounts of a foreign, that is, different metal (Ref 104, 110-113). The ability of that surface to support SERS for an adsorbate can then be correlated with some parameter describing electronic properties of the surface (Ref 113). This approach may yield a level of predictability about whether or not the surface of a new material can support SERS.

Despite the lack of general applicability of SERS, the wealth of information this technique can yield warrants further study. The ease of acquiring Raman spectral data from surface-enhancing systems is unsurpassed due to the remarkable intensities observed. Therefore, SERS should continue to receive consideration as a tool capable of providing molecular vibrational information about surfaces and interfaces. Although SERS will probably never gain acceptance as a general surface analytical tool, it can and should be used with other vibrational surface probes, such as infrared spectroscopy, to help provide a complete molecular picture of a given surface or interface.

Exploration of SERS for the study of relevant materials systems has only begun. Surface-enhanced Raman scattering has been used to study catalytic oxidation of nitric oxide (NO), nitrogen dioxide (NO_2), nitrogen peroxide (N_2O_4), and sulfur dioxide (SO_2) on silver powders. Surface SO_3^{2-} was detected on the surface of silver powder exposed to SO_2 gas in a helium atmosphere. Further, thermal desorption as SO_2 and oxidation to SO_4^{2-} were followed spectroscopically as the temperature was slowly raised to 108 °C (225 °F) in an oxygen-containing atmosphere (Ref 114). In a later study, brief exposure of oxygenated silver powder to NO and NO_2/N_2O_4 gases was found to result in SERS spectra of NO_2^- and NO_3^- (Ref 115).

The electropolymerization of phenol on silver electrodes in the presence and absence of amines has been studied using SERS (Ref 116). Results show the polymerization to be similar to that observed on iron. Surface-enhanced Raman scattering elucidation of the role of amines in polymerization indicates that amines displace phenoxide ions, which are adsorbed flat at the silver surface, and allow formation of thick protective films of the polymer (Ref 116). Further SERS studies on this system involving the role of the surfactant Triton in improving the adhesion characteristics of the polymer on the silver substrate have revealed that Triton is found at the silver/polymer interface and is dispersed throughout the polymer, chemically bonding to the polymer after curing in air (Ref 117).

The interest in SERS has signified the desirability of vibrational information about species at metal surfaces. This has led to the development of the technology for performing surface Raman measurements on metals without enhancement. The availability of sensitive multichannel detectors and appropriate optical components compatible with such systems has enabled obtainment of surface Raman spectra of molecules adsorbed on smooth metal surfaces. The first successful demonstration of surface Raman spectroscopy without enhancement was published in 1982 (Ref 118). High-quality Raman spectra from molecules adsorbed on well-characterized surfaces at low coverage were reported. The unenhanced Raman approach has since been used for the Raman spectroscopic investigation of molecules in tunnel junction structures (Ref 119) and metal/gas environments (Ref 120-122).

REFERENCES

1. C.V. Raman and K.S. Krishnan , *Nature*, Vol 122, 1928, p 501
2. D.A. Long, *Raman Spectroscopy*, McGraw-Hill, 1977
3. M.C. Tobin, *Laser Raman Spectroscopy*, John Wiley & Sons, 1971
4. D.C. O'Shea, W.R. Callen, and W.T. Rhodes, *Introduction to Lasers and Their Applications*, Addison-Wesley, 1977
5. Spex Industries, Metuchen, NJ, 1981
6. E.E. Wahlstrom, *Optical Crystallography*, 4th ed., John Wiley & Sons, 1969
7. M. Delhaye and P. Dhemalincourt, *J. Raman Spectrosc.*, Vol 3, 1975, p 33
8. P. Dhamelincourt, F. Wallart, M. Leclercq, A.T. NGuyen, and D.O. Landon, *Anal. Chem.*, Vol 51, 1979, p 414A
9. P.J. Hendra and E.J. Loader, *Trans. Faraday Soc.*, Vol 67, 1971, p 828
10. E. Buechler and J. Turkevich, *J. Phys. Chem.*, Vol 76, 1977, p 2325
11. T.A. Egerton, A. Hardin, Y. Kozirovski, and N. Sheppard, *Chem. Commun.*, 1971, p 887
12. R.O. Kagel, *J. Phys. Chem.*, Vol 74, 1970, p 4518
13. T.A. Egerton, A.H. Hardin, Y. Kozirovski, and N. Sheppard, *J. Catal.*, Vol 32, 1974, p 343
14. W. Kiefer and H.J. Bernstein, *Appl. Spectrosc.*, Vol 25, 1971, p 609
15. Y.S. Bobovich, *Opt. Spectrosc.*, Vol 13, 1962, p 274
16. G.J. Rosasco and J.H. Simmons, *Am. Cer. Soc. Bull.*, Vol 53, 1974, p 626
17. G.J. Rosasco and J.H. Simmons, *Am. Cer. Soc. Bull.*, Vol 54, 1975, p 590
18. D.W. Matson and S.K. Sharma, *J. Noncryst. Solids*, Vol 58, 1983, p 323
19. H.J. Sloane, in *Polymer Characterization: Interdisciplinary Approaches*, C. Craver, Ed., Plenum Press, 1971, p 15–36
20. R.E. Kagarise and L.A. Weinberger, "Infrared Spectra of Plastics and Resins," Report 4369, Naval Research Laboratory, Washington, DC, 26 May 1954
21. J.L. Koenig, *Chem. Technol.*, 1972, p 411
22. B. Chu and G. Fytas, *Macromolecules*, Vol 14, 1981, p 395
23. J.L. Koenig and M. Meeks, *J. Polymer Sci.*, Vol 9, 1971, p 717
24. J.L. Koenig, M.M. Coleman, J.R. Shelton, and P.H. Stramer, *Rubber Chem. Technol.*, Vol 44, 1971, p 71
25. J.R. Shelton, J.L. Koenig and M.M. Coleman, *Rubber Chem. Technol.*, Vol 44, 1971, p 904
26. S. K. Mukherjee, G.D. Guenther, and A. K. Battacharya, *Anal. Chem.*, Vol 50, 1978, p 1591
27. A.J. Hartley and I.W. Sheppard, *J. Polymer Sci.*, Vol 14, 1976, p 64B
28. B.R. Loy, R.W. Chrisman, R.A. Nyquist, and C.L. Putzig, *Appl. Spectrosc.*, Vol 33, 1979, p 174
29. S. Chow and Y.L. Chow, *J. Appl. Polymer Sci.*, Vol 18, 1974, p 735
30. A.J. Melveger, *J. Polymer Sci.*, A2, Vol 10, 1972, p 317
31. A.J. McQuillan and R.E. Hester, *J. Raman Spectrosc.*, Vol 15, 1984, p 17
32. T.P. Mernagh, R.P. Cooney, and R.A. Johnson, *Carbon*, Vol 22, 1984, p 1
33. F. Tunistra and J. L. Koenig, *J. Chem. Phys.*, Vol 53, 1970, p 1126
34. M. Nakamizo, R. Kammerzck, and P.L. Walker, *Carbon*, Vol 12, 1974, p 259
35. M. Nakamizo, *Carbon*, Vol 15, 1977, p 295
36. M. Nakamizo, *Carbon*, Vol 16, 1978, p 281
37. M. Nakamizo and K. Tami, *Carbon*, Vol 22, 1984, p 197
38. B.S. Elman, M.S. Dresselhaus, G. Dresselhaus, E.W. Maby, and H. Mazurek, *Phys. Rev. B.*, Vol 24, 1981, p 1027
39. B.S. Elman, M. Shayegan, M.S. Dresselhaus, H. Mazurek, and G. Dresselhaus, *Phys. Rev. B.*, Vol 25, 1982, p 4142
40. J.M. Rouzand, A. Oberlin, and C. Beny-Bassez, *Thin Solid Films*, Vol 105, 1983, p 75
41. N. Caswell and S.A. Solin, *Bull. Am. Phys. Soc.*, Vol 23, 1978, p 218
42. P.C. Eklund, G. Dresselhaus, M.S. Dresselhaus, and J.E. Fischer, *Phys. Rev. B.*, Vol 21, 1980, p 4705
43. S.A. Solin, *Mater. Sci. Eng.*, Vol 31, 1977, p 153
44. R.J. Nemanich, S.A. Solin, and D. Guerard, *Phys. Rev. B.*, Vol 16, 1977, p 2965

45. N. Caswell and S.A. Solin, *Solid State Commun.*, Vol 27, 1978, p 961
46. P.C. Eklund, N. Kambe, G. Dresselhaus, and M.S. Dresselhaus, *Phys. Rev. B.*, Vol 18, 1978, p 7068
47. A. Erbil, G. Dresselhaus, and M.S. Dresselhaus, *Phys. Rev. B.*, Vol 25, 1982, p 5451
48. J.J. Song, D.D.L. Chung, P.C. Eklund, and M.S. Dresselhaus, *Solid State Commun.*, Vol 20, 1976, p 1111
49. S.A. Solin, *Physica B & C*, Vol 99, 1980, p 443
50. S.Y. Leng, M.S. Dresselhaus, and G. Dresselhaus, *Physica B & C*, Vol 105, 1981, p 375
51. H. Miyazaki, T. Hatana, T. Kusunaki, T. Watanabe, and C. Horie, *Physica B & C*, Vol 105, 1981, p 381
52. J. Medema, C. van Stam, V.H.J. deBear, A.J.A. Konings, and D.C. Konigsberger, *J. Catal.*, Vol 53, 1978, p 385
53. R. Thomas, J.A. Moulijn, and F.R.J. Kerkof, *Recl. Trav. Chim. Pays-Bas.*, Vol 96, 1977, p m134
54. F. Roozeboom and M.C. Metelmeijer-Hazeleger, *J. Phys. Chem.*, Vol 84, 1980, p 2783
55. B.A. Morrow, Vibrational Spectroscopies for Adsorbed Species, in *ACS Symposium Series No. 137*, A.T. Bell and M.L. Hair, Ed., American Chemical Society, Washington, 1980
56. F.R. Brown, L.E. Makovsky, and K.H. Ree, *J. Catal.*, Vol 50, 1977, p 385
57. C.P. Cheng and G.L. Schrader, *J. Catal.*, Vol 60, 1979, p 276
58. H. Knozinger and H. Jeziorowski, *J. Phys. Chem.*, Vol 82, 1978, p 2002
59. D.S. Znigg, L.E. Makovsky, R.E. Tischer, F.R. Brown, and D.M. Hercules, *J. Phys. Chem.*, Vol 84, 1980, p 2898
60. F.R. Brown, L.E. Makovsky, and K.H. Rhee, *J. Catal.*, Vol 50, 1977, p 162
61. H. Jeziorowski and H. Knozinger, *J. Phys. Chem.*, Vol 83, 1979, p 1166
62. J.K. Wilhurst and H.J. Bemstein, *Can. J. Chem.*, Vol 35, 1957, p 1183
63. P.J. Hendra, J.R. Horder, and E.J. Loader, *J. Chem. Soc. A*, 1971, p 1766
64. R.P. Cooney and T.T. Nguyen, *Aust. J. Chem.*, Vol 29, 1976, p 507
65. R.P. Cooney, T.T. Nguyen, and G.C. Curthoys, *Adv. Catal.*, Vol 24, 1975, p 293
66. P.J. Hendra, I.D.M. Turner, E.J. Loader, and M. Stacey, *J. Phys. Chem.*, Vol 78, 1974, p 300
67. T.A. Egerton and A.H. Hardin, *Catal. Rev. Sci. Eng.*, Vol 11, 1975, p 1
68. P.J. Hendra, A.J. McQuillan, and I.D.M. Turner, *Dechema*, Vol 78, 1975, p 271
69. T.A. Egerton, A.H. Hardin, and N. Sheppard, *Can. J. Chem.*, Vol 54, 1976, p 586
70. A.H. Hardin, M. Klemes, and B.A. Morrow, *J. Catal.*, Vol 62, 1980, p 316
71. R.P. Cooney and N.T. Tam, *Aust. J. Chem.*, Vol 29, 1976, p 507
72. E. Payen, M.C. Dhamelincourt, P. Dhamelincourt, J. Grimblot, and J.P. Bonnelle, *Appl. Spectrosc.*, Vol 36, 1982, p 30
73. C.P. Cheng, and G.L. Schrader, *Spectrosc. Lett.*, Vol 12, 1979, p 857
74. R.L. Farrow, R.E. Benner, A.S. Nagelberg, and P.L. Mattern, *Thin Solid Films*, Vol 73, 1980, p 353
75. J.C. Hamilton, B.E. Mills, and R.E. Benner, *Appl. Phys. Lett.*, Vol 40, 1982, p 499
76. J.T. Keiser, C.W. Brown, and R.H. Heidersbach, *J. Electrochem. Soc.*, Vol 129, 1982, p 2686
77. R. Thibeau, C.W. Brown, G. Goldfarb, and R.H. Heidersbach, *J. Electrochem. Soc.*, Vol 127, 1980, p 37
78. R. Varma, C.A. Melendres, and N.P. Yao, *J. Electrochem. Soc.*, Vol 127, 1980, p 1416
79. R. Kötz and E. Yeager, *J Electroanal. Chem.*, Vol 111, 1980, p 105
80. C. A. Melendres and S. Xu, *J. Electrochem. Soc.*, Vol 131, 1984, p 2239
81. M. Fleischmann, P.J. Hendra, and A.J. McQuillan, *J. Chem. Soc. Chem. Comm.*, Vol 3, 1973, p 80
82. M. Fleischmann, P.J. Hendra, and A.J. McQuillan, *Chem. Phys. Lett.*, Vol 26, 1974, p 163
83. R.L. Paul, A.J. McQuillan, P.J. Hendra, and M. Fleischmann, *J. Electroanal. Chem.*, Vol 66, 1975, p 248
84. D.L. Jeanmaire and R.P. VanDuyne, *J. Electroanal. Chem.*, Vol 84, 1977, p 1
85. M.G. Albrecht and J.A. Creighton, *J. Am. Chem. Soc.*, Vol 99, 1977, p 5215
86. H. Seki, *J. Electron Spectrosc, Related Phenom.*, Vol 30, 1983, p 287
87. R.K. Chang and T.E. Furtak, Ed., *Surface Enhanced Raman Scattering*, Plenum Press, 1982
88. T.H. Wood and M.V. Klein, *J. Vac. Sci. Technol.*, Vol 16, 1979, p 459
89. J.E. Rowe, C.V. Shank, D.A. Zwemet, and C.A. Murray, *Phys. Rev. Lett.*, Vol 44, 1980, p 1770
90. H. Seki and M.R. Philpott, *J. Chem. Phys.*, Vol 73, 1980, p 5376
91. J.C. Tsang and J.R. Kirtley, Light Scattering in Solids, in *Proceedings of Second Joint USA-USSR Symposium*, J.L. Birman, H.Z. Cummins, and K.K. Rebane, Ed., Plenum Press, 1979, p 499
92. J.C. Tsang and J.R. Kirtley, *Solid State Commun.*, Vol 30, 1979, p 617
93. M. Moskovits and D.P. DiLella, in *Surface Enhanced Raman Scattering*, R.K. Chang and T.E. Furtak, Ed., Plenum Press, 1982, p 243
94. H. Yamada, Y. Yamamoto, and N. Tani, *Chem. Phys. Lett.*, Vol 86, 1982, p 397
95. R.E. Benner, K.U. van Raben, K.C. Lee, J.F. Owen, R.K. Chang, and B.L. Laube, *Chem. Phys. Lett.*, Vol 96, 1983, p 65
96. B.H. Loo, *J. Phys. Chem.*, Vol 87, 1983, p 3003
97. B.H. Loo, *J. Chem. Phys.*, Vol 75, 1981, p 5955
98. B.H. Loo, *J. Electron Spectrosc. Related Phenom.*, Vol 29, 1983, p 407
99. C.C. Chou, C.E. Reed, J.C. Hemminger, and S. Ushioda, *J. Electron Spectrosc. Related Phenom.*, Vol 29, 1983, p 401
100. H. Yamada, N. Tani, and Y. Yamamoto, *J. Electron Spectrosc. Related Phenom.*, Vol 30, 1983, p 13
101. M. Fleischmann, P.R. Graves, J.R. Hill, and J. Robinson, *Chem. Phys. Lett.*, Vol 95, 1983, p 322
102. H. Yamada, N. Tani, and Y. Yamamoto, *J. Electron Spectrosc. Related Phenom.*, Vol 30, 1983, p 13
103. H. Yamada and Y. Yamamoto, *Surf. Sci.*, Vol 134, 1983, p 71
104. J. E. Pemberton, *J. Electroanal. Chem.*, Vol 167, 1984, p 317
105. R.P. Van Duyne and J.P. Haushalter, *J. Phys. Chem.*, Vol 87, 1983, p 2999
106. Y. Mo, H. van Känel, and P. Wachter, *Solid State Commun.*, Vol 52, 1984, p 213
107. C.A. Murray, *J. Electron Spectrosc. Related Phenom.*, Vol 29, 1983, p 371
108. G.C. Schatz, *Acc. Chem. Res.*, Vol 17, 1984, p 370
109. D.D. Tuschel, J.E. Pemberton, and J.E. Cook, *Langmuir*, Vol 2, 1986
110. B. Pettinger and L. Moerl, *J. Electron Spectrosc. Related Phenom.*, Vol 29, 1983, p 383
111. L. Moerl and B. Pettinger, *Solid State Commun.*, Vol 43, 1982, p 315
112. T. Watanabe, N. Yanigaraha, K. Honda, B. Pettinger, and L. Moerl, *Chem. Phys. Lett.*, Vol 96, 1983, p 649
113. A.L. Guy, B. Bergami, and J.E. Pemberton, *Surf. Sci.*, Vol 150, 1985, p 226
114. P.B. Dorain, K.U. Von Raben, R.K. Chang, and B.L. Laube, *Chem. Phys. Lett.*, Vol 84, 1981, p 405
115. K.U. Von Raben, P.B. Dorain, T.T. Chen, and R.K. Chang, *Chem. Phys. Lett.*, Vol 95, 1983, p 269
116. M. Fleischmann, I.R. Hill, G. Mengoli, and M.M. Musiani, *Electrochim. Acta*, Vol 28, 1983, p 1545
117. G. Mengoli, M.M. Musiani, B. Pelli, M. Fleischmann, and I.R. Hill, *Electrochim. Acta*, Vol 28, 1983, p 1733
118. A. Campion, J.K. Brown, and V.M. Grizzle, *Surf Sci.*, Vol 115, 1982, p L153
119. J.C. Tsang, Ph. Avouris, and J.R. Kirtley, *Chem. Phys. Lett.*, Vol 94, 1983, p 172
120. A. Campion and D.R. Mullins, *Chem. Phys. Lett.*, Vol 94, 1983, p 576
121. V.M. Hallmark and A. Campion, *Chem. Phys. Lett.*, Vol 110, 1984, p 561
122. D. R. Mullins and A. Campion, *Chem. Phys. Lett.*, Vol 110, 1984, p 565

ASM Handbook, Volume 10, *Materials Characterization*
ASM Handbook Committee
DOI 10.31399/asm.hb.v10.a0006650

Nuclear Magnetic Resonance

Randall E. Youngman, Corning Incorporated

Overview

General Uses

- Molecular structure of organic compounds
- Element coordination numbers
- Short- and intermediate-range structure in disordered solids
- Phase analysis
- Polymer chain dynamics

Examples of Applications

- Determination of atomic structure in inorganic glasses
- Identification of organic polymer components
- Phase identification in crystalline solids
- Identification and quantification of amorphous and crystalline content in glass-ceramic materials

Samples

- *Form:* Crystalline, semicrystalline, or amorphous solids in the form of powders, films, and crystals; also liquids and sometimes gases
- *Size:* Milligram to several grams, depending on experiment
- *Preparation:* Analysis of solid materials typically requires powdering or reducing the specimen size to fit into the sample holder (rotor); dissolution of organic materials in a solvent for liquid nuclear magnetic resonance (NMR) is common.

Limitations

- Sample generally cannot be magnetic or contain more than ~1 wt% of paramagnetic species.
- Concentration of nuclei of interest usually must be on the order of several percent or higher.
- Study of metals is limited to thin or small-diameter samples due to radio-frequency penetration effects.
- Isotopic enrichment may be necessary for certain experiments or for analysis of certain isotopes.

Estimated Analysis Time

- 5 min to several days per sample, depending on type of experiment and material being studied

Capabilities of Related Techniques

- *X-ray diffraction:* Crystalline phase identification; much more sensitive and accurate than NMR, and generally able to identify phases regardless of elemental content; sometimes limited in study of solid solutions or in materials with very small-sized or low concentration of crystalline phases
- *Infrared spectroscopy:* Organic compound identification; short-range glass structure information; excellent for surface and thin-film characterization, as well as analysis of gas-phase species; interpretation can be complicated in amorphous materials
- *Raman spectroscopy:* Intermediate-range structure in disordered solids; organic compound identification; limited quantitation and interpretation can be challenging
- *Electron paramagnetic resonance:* Very sensitive technique for analysis of paramagnetic species in all manner of materials; good technique for determination of optical defects in materials and oxidation states of certain polyvalent atoms; limited resolution of species with g-factor of approximately 2
- *Mössbauer spectroscopy:* Very sensitive method to measure quadrupole splittings, isomer shifts, and magnetic hyperfine splittings; gamma radiation source must contain same isotopes as being absorbed in sample, limiting application and availability
- *Mass spectrometry:* Accurate molecular weight determination and identification of molecular species in organic materials; end-group analysis in organic polymers; excellent analysis capabilities when coupled to separation (chromatographic) technologies; limited application in inorganic solids

Introduction

Nuclear magnetic resonance (NMR) spectroscopy is an analytical method based on interaction of a nuclear magnetic moment with an external magnetic field, resulting in resonances that can be measured by using radio-frequency (RF) methods. The interaction of the nuclear magnetic moment with magnetic fields, including those internal to the sample, determines the resonance frequency of the nucleus of interest and, more specifically, the shift in frequency from some known standard reference material. There are a number of interactions that alter these resonance frequencies and provide highly detailed structural and dynamical information. Applications of NMR in materials science have grown exponentially since the discovery of this methodology in the 1940s. Other disciplines, including especially chemistry and biology, have significantly benefitted from this spectroscopy, taking advantage of the nuclear specificity of NMR and the ability to apply these techniques to a large variety of sample types. In this overview, emphasis is placed on the application of solid-state NMR spectroscopy in materials science, especially for inorganic and organic polymer solids. There are a large number of reviews on biomolecular solid-state NMR and liquid NMR available to the reader if interested in topics not covered herein.

Brief History

Nuclear magnetic resonance was first described by Rabi in 1938 (Ref 1) and then extended to liquids and solids by Bloch and Purcell in 1945 to 1946 (Ref 2, 3). Much of the early research into the phenomenon was conducted by physicists interested in measuring nuclear properties of various isotopes. With commercialization of the first instrument in the 1950s by Varian and Associates, the technique rapidly expanded into chemistry and a variety of related applications. Chief among these was the realization that NMR spectra could provide a molecular fingerprint of organic molecules in solution, leading to adoption of this analytical technique in organic and eventually biomolecular chemistry.

Nuclear magnetic resonance of solids was instrumental in some of the earliest investigations of this technique, but adoption of NMR in materials science occurred slowly. Even with the introduction of magic-angle spinning (MAS) to narrow resonances (Ref 4, 5), wide deployment was hampered by the specialization of the technique. Led by the Bray group at Brown, as well as a few others around the world, NMR applications in glass science were demonstrated in the late 1950s (Ref 6). Other materials applications followed, but it was not until the invention of cross-polarization with magic-angle spinning in the early 1970s (Ref 7), leading to a substantial improvement in spectral resolution and sensitivity, that widespread adoption of solid-state NMR occurred. Since these seminal discoveries during the early development of NMR, advances in sample spinning, magnetic field homogeneity and size, as well as other methods, including multidimensional NMR, have made this an indispensable characterization tool in most fields of materials science, chemistry, and biology. The technique continues to grow in application and capability, led by recent advancements in computational methods (Ref 8) and the desire to expand this technique into new research areas (Ref 9).

General Principles

This section discusses nuclear spin descriptions, the impact of magnetic field on nuclear spins, factors determining resonance frequency, and line narrowing and spectral resolution.

Nuclear Spin Descriptions

The NMR technique is based on measuring resonance frequencies for nuclei with magnetic moments. A nucleus with both an even number of protons and neutrons has zero nuclear spin and thus zero magnetic moment. The ^{16}O isotope of oxygen and the ^{12}C isotope of carbon are two of the more prevalent examples of nuclei with zero nuclear spin. Fortunately, most nuclei have nonzero magnetic moments and thus are amenable for NMR investigation. Integer spins, like that of ^{2}H and ^{10}B, have odd numbers of both protons and neutrons. Half-integer spins, which are most common, arise from a nucleus having either an even number of protons and odd number of neutrons (e.g., ^{13}C) or an odd number of protons and even number of neutrons, as in the ^{15}N isotope of nitrogen.

Certain elements have multiple isotopes with different magnetic properties, which can be quite beneficial. For example, the inability to study the most common isotopes of carbon and oxygen (both ^{12}C and ^{16}O have zero nuclear spin) can be circumvented by performing NMR on the ^{13}C and ^{17}O isotopes, both of which have nonzero spin. For some other elements, multiple isotopes having nonzero spin enable NMR of more than a single isotope (e.g., ^{1}H and ^{2}H) or the ability to select the isotope with more favorable nuclear properties, such as natural abundance or nuclear spin characteristics. For example, ^{11}B is much easier to study with NMR than the ^{10}B isotope of boron.

The distribution of charges in the nucleus determines the magnitude and shape of the magnetic moment. A spherical distribution of charges leads to a magnetic dipole moment, while higher-order moments can be realized for nonspherical charge distributions. The easiest nuclei for NMR study are those with only dipole moments, because their interaction with surrounding charges is simplest. Quadrupolar nuclei, with $I > 1/2$ and thus possessing an electric quadrupole moment, interact strongly with nearby electric field gradients, significantly complicating their resonance frequency and absorption lineshape.

Other properties of nuclei that control the NMR response and either facilitate or negatively impact NMR spectroscopy measurements include the natural abundance of the isotope, the gyromagnetic ratio (γ), and the magnitude of the quadrupole moment (Q). Table 1 lists these properties for all of the elements, demonstrating a large range and level of complexity in their nuclear properties.

Impact of Magnetic Field on Nuclear Spins

Nuclear magnetic resonance works because an externally applied magnetic field removes the degeneracy of the nuclear spin energy ground state, splitting this into $2I + 1$ levels, where I represents the nuclear spin quantum number. This effect, termed the Zeeman effect, allows for measurement of transitions between nuclear energy states. The diagram in Fig. 1 demonstrates the Zeeman effect, in which increasing magnitude of the magnetic field results in a larger separation of the spin states. Once this splitting occurs, transitions between the $+1/2$ and $-1/2$ levels (central transition), which are those typically measured by NMR, can be detected using RF-based methods. Other transitions, called satellite transitions, also are made possible through the Zeeman effect, including those for higher-order spin states such as 3/2 and 5/2, depending on the spin quantum number described previously, but these are largely unused in routine solid-state NMR.

For most nuclei with nonzero spin, the Zeeman effect represents the largest change in spin states, in which the transition energies are on the order of tens to hundreds of megahertz, depending on the gyromagnetic ratio of the nucleus. This constant, specific to each nucleus (Table 1), determines the Larmor frequency in a given applied magnetic field:

$$\omega_L = \gamma B_0 \qquad \text{(Eq 1)}$$

where γ is the gyromagnetic ratio of the nucleus, B_0 is the strength of the externally applied magnetic field, and ω_L is the Larmor frequency, also called the precession frequency. A Boltzmann distribution describes a very small difference in each spin state population, which then allows one to probe the resonance, or transition energy between levels, by manipulating the bulk magnetization of the spins with RF energy proportional to the Larmor frequency.

Factors Determining Resonance Frequency

As mentioned earlier, the main factors in determining the resonance frequency of a particular nucleus are the gyromagnetic ratio and the size of the externally applied magnetic field. However, if that were the only consideration, then NMR would be useful only in determining which nuclei are present in a material, because the resonance frequencies of a given nucleus would be identical, regardless of their structural or chemical environment. What makes NMR spectroscopy useful are several different perturbations to the Zeeman splitting due to local structure (symmetry, coordination number, etc.), and also local dynamics of different elements. Essentially any magnetic or electronic interaction with the observed nuclear spin affecting the energy levels in Fig. 1 results in the ability to distinguish different environments of the nuclear spin, which is ultimately the reason for considering NMR spectroscopy in solving materials

Table 1 Important nuclear properties of most nuclear magnetic resonance (NMR)-active nuclei: gyromagnetic ratio (γ), nuclear spin (*I*), and quadrupole moment (*Q*)

Several atoms are shown with more than one NMR-active isotope.

Element	Isotope	γ/2π, MHz/T	*I*	Natural abundance, %	*Q*, barns
Hydrogen	^{1}H	42.577	1/2	99.9885	···
	^{2}H	6.536	1	0.0115	0.00286
Helium	^{3}He	−32.434	1/2	1.37×10^{-4}	···
Lithium	^{6}Li	6.27	1	7.59	−0.00082
	^{7}Li	16.546	3/2	92.41	−0.0406
Beryllium	^{9}Be	5.98	3/2	100	0.0529
Boron	^{10}B	4.58	3	19.9	0.0847
	^{11}B	13.66	3.2	80.1	0.0407
Carbon	^{13}C	10.7084	1/2	1.07	···
Nitrogen	^{14}N	3.077	1	99.632	0.02001
	^{15}N	−4.316	1/2	0.368	···
Oxygen	^{17}O	−5.772	5/2	0.038	−0.02578
Fluorine	^{19}F	40.052	1/2	100	···
Neon	^{21}Ne	−3.36	3/2	0.27	0.103
Sodium	^{23}Na	11.262	3/2	100	0.1045
Magnesium	^{25}Mg	−2.61	5/2	10.00	0.199
Aluminum	^{27}Al	11.103	5/2	100	0.1466
Silicon	^{29}Si	−8.465	1/2	4.6832	···
Phosphorus	^{31}P	17.235	1/2	100	···
Sulfur	^{33}S	3.27	3/2	0.76	??
Chlorine	^{35}Cl	4.18	3/2	75.78	−0.08249
	^{37}Cl	3.48	3/2	24.22	−0.06493
Potassium	^{39}K	1.99	3/2	93.2581	0.585
Calcium	^{43}Ca	−2.87	7/2	0.135	−0.0408
Scandium	^{45}Sc	10.36	7/2	100	0.046
Titanium	^{47}Ti	−2.40	5/2	7.44	0.30
	^{49}Ti	−2.40	7/2	5.41	0.247
Vanadium	^{51}V	11.21	7/2	99.750	−0.043
Chromium	^{53}Cr	−2.41	3/2	9.501	−0.15
Manganese	^{55}Mn	10.58	5/2	100	0.33
Iron	^{57}Fe	1.382	1/2	2.119	···
Cobalt	^{59}Co	10.08	7/2	100	0.41
Nickel	^{61}Ni	−3.81	3/2	1.1399	0.162
Copper	^{63}Cu	11.319	3/2	69.17	??
	^{65}Cu	12.10	3/2	30.83	−0.195
Zinc	^{67}Zn	2.669	5/2	4.10	0.150
Gallium	^{71}Ga	13.02	3/2	39.892	0.1040
Germanium	^{73}Ge	−1.49	9/2	7.73	−0.17
Arsenic	^{75}As	7.32	3/2	100	0.314
Selenium	^{77}Se	8.16	1/2	7.63	···
Bromine	^{79}Br	10.70	3/2	50.69	0.318
Krypton	^{83}Kr	−1.64	9/2	11.49	0.259
Rubidium	^{87}Rb	13.98	3/2	27.83	0.19
Strontium	^{87}Sr	−1.85	9/2	7.00	0.33
Yttrium	^{89}Y	−2.09	1/2	100	···
Zirconium	^{91}Zr	−3.97	5/2	11.22	−0.176
Niobium	^{93}Nb	10.45	9/2	100	−0.32
Molybdenum	^{95}Mo	−2.79	5/2	15.92	−0.022
Ruthenium	^{99}Ru	−1.96	5/2	12.76	0.079
Rhodium	^{103}Rh	−1.35	1/2	100	···
Palladium	^{105}Pd	−1.96	5/2	22.33	0.660
Silver	^{109}Ag	−1.99	1/2	48.161	···
Cadmium	^{113}Cd	−9.49	1/2	12.22	···
Indium	^{115}In	9.39	9/2	95.71	0.81
Tin	^{119}Sn	−15.97	1/2	8.59	···
Antimony	^{121}Sb	10.26	5/2	57.21	−0.36
Tellurium	^{125}Te	−13.55	1/2	7.07	···
Iodine	^{127}I	8.58	5/2	100	−0.710
Xenon	^{129}Xe	−11.777	1/2	25.44	···
	^{131}Xe	3.52	3/2	21.18	−0.116
Cesium	^{133}Cs	5.62	7/2	100	−0.00355
Barium	^{137}Ba	4.76	3/2	11.232	0.245
Lanthanum	^{139}La	6.06	7/2	99.910	0.20
Hafnium	^{177}Hf	1.73	7/2	18.60	3.37
	^{179}Hf	−1.09	9/2	13.62	3.79
Tantalum	^{181}Ta	5.16	7/2	99.988	3.17
Tungsten	^{183}W	1.80	1/2	14.31	···
Rhenium	^{187}Re	9.82	5/2	62.60	2.07
Osmium	^{187}Os	0.99	1/2	1.96	···
	^{189}Os	3.35	3/2	16.15	0.86
Iridium	^{193}Ir	0.83	3/2	62.7	0.751
Platinum	^{195}Pt	9.29	1/2	33.832	···
Gold	^{197}Au	0.75	3/2	100	0.547
Mercury	^{199}Hg	7.71	1/2	16.87	···
	^{201}Hg	−2.85	3/2	13.18	0.38
Thallium	^{205}Tl	24.97	1/2	70.476	···
Lead	^{207}Pb	8.88	1/2	22.1	···
Bismuth	^{209}Bi	7.23	9/2	100	−0.516

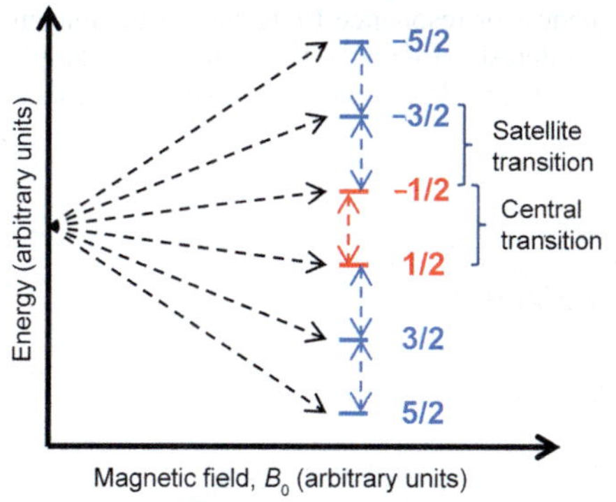

Fig. 1 Schematic showing the nuclear spin energy levels as a function of spin quantum number, *I*, and externally applied magnetic field, B_0. The central transition (+1/2 to −1/2, in red) is denoted, and satellite transitions (in blue) also are shown. The Larmor frequency (splitting), ω_L, increases with magnetic field strength.

science problems. Although quite complex, the position, shape, and intensity of measured NMR signals enable quantitative study of the structure of molecules and materials.

One of the most important factors is the chemical shielding perturbation of the spins, leading to very small changes in resonance frequency, depending on the shielding of the nuclear spin. This shielding is dominated by the manner in which an atom is bonded into the material (i.e., local electronic environment), so different chemical environments exhibit different chemical shielding and therefore different resonance frequencies, thus the most useful attribute of NMR spectroscopy. This is the foundation for the many decades of NMR spectroscopy of organic molecules—a chemical fingerprint from NMR can aid in identification of molecules.

Another important consideration, applicable only for nuclei with quadrupolar moments (spin > 1/2) is the quadrupolar coupling interaction. The impact of this interaction on energy levels is dependent on both the magnitude of the nuclear quadrupole moment and the surrounding electric field gradient (EFG) of the nucleus. The frequency of this quadrupolar coupling is approximated by:

$$C_Q = e^2qQ/\hbar \qquad \text{(Eq 2)}$$

where Q is the nuclear quadrupole moment (shown for each $I > 1/2$ nucleus in Table 1), q is the EFG, $\hbar$ is the reduced Planck's constant ($h/2\pi$), and C_Q is the quadrupolar coupling constant, which often can be measured in solid-state NMR spectra of quadrupolar nuclei. In many cases, C_Q can be as large as, or larger than, the Zeeman splitting (i.e., tens to hundreds of megahertz), making these NMR measurements very difficult. Fortunately in many materials of interest, measurement of C_Q can be highly informative about the local environment of that particular quadrupolar nucleus, as shown in some of the following examples. The development of extremely high magnetic fields can facilitate some measurement of spins having large C_Q values (Ref 10), and nuclear quadrupole resonance often can provide estimates for C_Q when traditional NMR is ineffective (Ref 11).

Other perturbations to the resonance frequency of a nucleus include dipolar coupling between spins (like and unlike) and interaction of nuclear spins with occupied conduction bands, the latter leading to what is called the Knight shift. Both of these interactions can be beneficial in application of NMR but are less commonly used in routine solid-state NMR experiments. Dipolar coupling through space is determined by the natural abundance of the isotopes (Table 1) and the distance between spins. In solid-state NMR, this often is restricted to coupling of ^{1}H spins with other nuclei of interest, for example, ^{13}C in organic polymers. J-coupling, which is prevalent in liquid NMR, involves spin-spin coupling through bonds, and because this typically is small in magnitude, it is not often observed in solid-state NMR. There are, however, exceptions where utilization of J-coupling in NMR of zeolites has proven quite useful (Ref 12), as well as extensive studies of phosphate group connectivity in polycrystalline and amorphous phosphates using J-resolved two-dimensional NMR (Ref 13).

Line Narrowing and Spectral Resolution

The interactions described previously, which ultimately determine the resonance frequency of a nucleus, are extremely beneficial and ultimately determine the value of NMR spectroscopy, but in solid materials, these interactions also are sources of substantial line broadening and loss of resolution. Unlike liquids, which typically allow for isotropic averaging of the molecule through rapid tumbling, the atoms of a solid are frozen in place/orientation, and only certain environments show any motional averaging. These may include methyl groups in organic molecules and polymers, as well as ions that are highly mobile at the measurement temperature (e.g., those leading to ionic conduction in solids). The resulting NMR lineshape from a solid sample therefore contains information from a distribution of crystallite or molecular orientations relative to the magnetic field, giving rise to what is termed a powder pattern, and thus is much broader than a signal from a molecule in solution.

Loss of resolution makes interpretation of solid-state NMR spectra much more difficult, especially in materials with multiple atomic environments or those with substantial structural disorder. Fortunately, one of the most important advances in NMR spectroscopy has been the discovery and implementation of line-narrowing methods for solids. Magic-angle spinning, mentioned previously, is by far the most common technique for high-resolution solid-state NMR. The MAS NMR works by averaging the spatial dependence of many NMR interactions to their isotropic values through rapid sample spinning at an angle of 54.74° from the external magnetic field, eliminating, for example, chemical shift anisotropy and most dipolar coupling. The result for spin 1/2 nuclei, including especially ^{13}C, ^{29}Si, and ^{31}P, is improved spectral resolution and the ability to distinguish unique chemical environments in the material. However, this comes at a cost, because MAS NMR then requires that one discard the information that originally led to this line broadening. Most common applications of NMR spectroscopy make use of MAS to obtain highly resolved spectra and the chemical information contained in identifying and quantifying these peaks, but there are methods that have been developed for recovering interactions such as chemical shift anisotropy, which can be useful in certain situations. These details and a few select examples are described later.

Quadrupolar coupling, the other key perturbation to the Zeeman splitting in Fig. 1, is an even larger issue in determining the potential resolution of an NMR spectrum. The MAS NMR technique only partially narrows the resonance of a spin experiencing quadrupolar coupling; as can be seen in Table 1, this includes most of the elements. The spatial dependence of the quadrupolar coupling, which has a different angle relative to the external field where the interaction can be averaged to zero, still is narrowed under MAS NMR, and hence, the technique retains its value. Because the interaction is only partially removed with MAS (54.74°), the resulting peaks exhibit some line broadening and thus the potential for less-than-ideal spectral resolution. More recent developments in solid-state NMR, beginning in the 1980s, fortunately have provided other experimental means by which to eliminate both quadrupolar and other line-broadening mechanisms and thus have been useful in solid-state NMR spectroscopy of quadrupolar nuclei. The current method of choice is called multiple-quantum magic-angle spinning (MQMAS) NMR, and it is a two-dimensional NMR experiment conducted with standard instrumentation and sample spinning at the 54.74° magic angle (Ref 14). This technique provides high-resolution spectra from quadrupolar nuclei, succeeding in the elimination of all line broadening due to the nuclear spin interactions. The only remaining broadening is that from the material itself, either the random distribution of spins in a polycrystalline solid, or the disorder inherent to glasses and polymers.

Even with the easy availability of the MQMAS NMR method, one also is able to use knowledge of quadrupolar lineshapes in simple MAS NMR spectra to deconvolute and describe multiple resonances, gaining useful information and spectral resolution without the extra time and complication of a two-dimensional NMR experiment. This is by far the most common approach to making NMR measurements on solids and is the focus of the examples in this article.

Systems and Equipment

Modern NMR instruments are based on RF electronics, an NMR probe, and some type of stable magnetic field, as shown schematically in Fig. 2. The magnetic field splits the nuclear energies into multiple states (Zeeman effect), providing the ability to use RF to probe resonant transitions between different energy levels. Magnets typically are based on superconducting solenoids, with cryogenic stabilization of the field. Advances in magnet technology have resulted in commercially available magnetic field strengths as high as 25 T, although these are more commonly found in the range of 4 to 16 T. Weaker fields, based on permanent magnets, are being used in a variety of benchtop and in-line/near-line process NMR instruments, and applications in materials science are limited but growing.

The NMR consoles are comprised of RF generators, transmitters, mixers, and receivers and are necessary to generate the Larmor frequency, or resonance frequency of the nucleus of interest; efficiently transmit to the sample; and then detect a very weak resonance

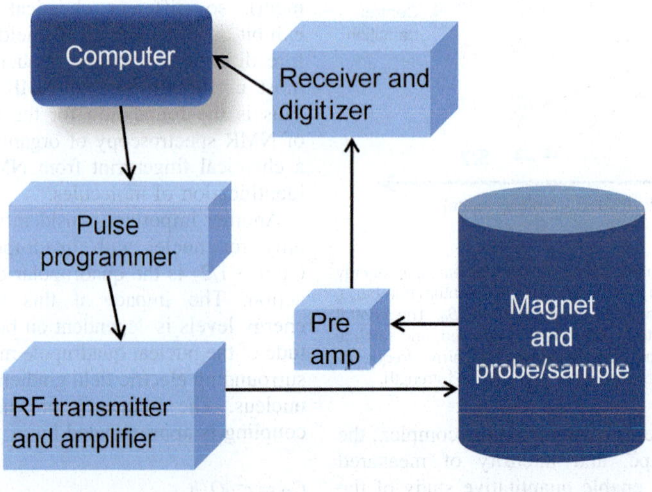

Fig. 2 Simple block diagram of a modern nuclear magnetic resonance (NMR) instrument. Key components involved in generation of radio frequency (RF) and detection of NMR signal are shown. Other important components, as well as second and third RF channels, are omitted for clarity.

response from the sample. Solid-state NMR instruments are distinct from common liquid or solution NMR instrumentation due to the fact that much more powerful RF pulses are needed to excite a full solid-state NMR spectrum. This is accomplished with high-power amplifiers, on the order of 1 kW, for each of the RF channels. This specialized equipment can provide high-power, uniform excitation pulses that cover a large frequency range, inversely proportional to the length of the RF pulse, and also can deliver strong decoupling fields to eliminate line broadening from dipolar coupling to ^{1}H. The other special requirement for solid-state NMR console electronics is a high-speed digitizer. Pulse Fourier transform NMR involves digitization of the time domain signal, and the sampling rate defines the maximum frequency range of the experiment. Because solid-state NMR spectra typically are very broad, at least relative to liquid NMR data, the instrumentation for solid-state NMR is capable of much faster digitization of the data.

The other unique and critical component for performing solid-state NMR is the NMR probe. A standard MAS NMR probe houses the sample and provides the local antenna (coil) for excitation and detection of the resonance signal, as well as air handling for sample spinning and management of temperature. A picture of a standard MAS NMR probe is given in Fig. 3. Most of the critical components are housed in the upper 10% of the probe, which can be easily situated in the homogeneous center of a superconducting magnet. All of these components occupy space, so these probes typically are designed for wide-bore (89 mm, or 3.5 in.) superconducting magnets. However, most new instrumentation allows for MAS NMR probes to be designed for narrow-bore (54 mm, or 2.1 in.) magnets used for high-resolution liquid NMR. The only constraint is that MAS NMR probes designed for the smaller-magnet bores cannot be too large, because sample holders (rotors) and the associated housing must still fit in a small space. Large-volume MAS NMR rotors and probes still require larger-bore magnets due to these limitations.

Fig. 3 Photo of a standard magic-angle spinning (MAS) nuclear magnetic resonance (NMR) probe (Chemagnetics/Varian wide-bore design, 89 mm, or 3.5 in.). The sample rotor is inserted into the MAS housing through the opening near the top of the probe. Air attachments, for both drive and bearing gas supplies, and radio-frequency cable attachment(s) are at the bottom of the probe. Frequency-tuning circuitry is located within the probe, usually near the MAS housing, and consists of both fixed and variable capacitors. The entire probe assembly is mounted to the bottom of the magnet, allowing the sample to be easily centered in the magnetic field for NMR measurements.

Specimen Preparation

This section includes discussion on general sampling for solid-state NMR, sample-spinning requirements, and extraneous signals.

General Sampling for Solid-State NMR

Nuclear magnetic resonance is a bulk measurement in which the nuclei measured are those contained in the sample holder. For something like ^{13}C NMR of polymers, this means that all ^{13}C nuclei in the sample are generally detected, regardless of location in the material. Surfaces and bulk sites are detected alike, without any specificity. This means that making NMR measurements with any type of spatial resolution is difficult or even impossible. If a sample can be prepared before NMR measurement to isolate regions of interest, then spatially separated data can be collected. However, this means isolating a region of interest from other, noninteresting parts of the specimen and being certain that the isolated sample is what was intended. If regions of a specimen are easy to separate, for example, large, single crystals, then this is quite simple. However, in the case of glasses and polymers, even those exhibiting large phase separation, the ability to separate one region from another, for example, surface from bulk, is quite challenging.

Sample-Spinning Requirements

Sampling in solid-state NMR is relatively simple, because this is a bulk technique and thus the sample contained in the rotor is what gets measured. The critical factor is getting from specimen to packed sample rotor. The NMR rotor, shown in Fig. 4, is designed to contain the sample and to spin at fairly high rates. Choice of rotor size is dictated either by what is available or, more desirable, by the MAS NMR spinning-frequency requirements for the sample of interest or the specific experiment. For example, most solid-state NMR studies of dipolar nuclei, from ^{13}C in organic polymers to ^{29}Si and ^{31}P in ceramics and glasses, can be conducted with modest spinning rates, because the MAS NMR linewidths can be fairly narrow and thus the sample spinning required to move sidebands away from the resonances of interest can be pretty modest. So, if sample-spinning speeds for these types of samples are not very high, then larger rotors and their slower spinning rates can be used. Larger rotors mean more sample, which also is generally quite favorable for natural abundance studies of ^{13}C and ^{29}Si. However, some nuclei require much faster sample spinning, either to effectively eliminate dipolar broadening, as is often the case for ^{1}H and ^{19}F in solids, or to move spinning sidebands and satellite transition peaks from quadrupolar nuclei to spectral regions without signals of interest. Significant quadrupolar broadening of resonances for ^{71}Ga and ^{93}Nb, to name a couple of common examples, requires much faster sample spinning, which thus requires smaller rotors and inherently small sample volumes.

Therefore, sample spinning to perform MAS NMR and related experiments is a big factor in determining how much sample is required. The other main considerations are the natural abundance of the isotope of interest (Table 1) and the absolute concentration of that element in the specimen. Solid-state NMR inherently is not very sensitive, so elemental concentrations of at least a few atomic weight percent often are needed, and this depends on the isotope. Proton (^{1}H) NMR is benefitted from large natural abundance and large gyromagnetic ratio, so it exhibits high sensitivity and thus can be measured in a solid specimen to low concentrations, perhaps as low as a few hundred parts per million (ppm). Other elements, especially those with low natural abundance such as ^{29}Si (4.7%), must be present in fairly high concentrations to obtain data in reasonable times.

All of these factors dictate the size of the sample holder (rotor) and thus the form in which a specimen may be studied with NMR. Rotors generally are small and experience rapid spinning at the magic angle, so uniform and high packing densities are beneficial. This increases the amount of material in the rotor, contributing to improved sensitivity, and is easier to spin at high rates without spinning instability or rotor crashes. Thus, the most common sample form is a relatively fine powder. Glasses and ceramics can be crushed and powdered with a mortar and pestle, and organic polymers can be cut into small pieces or even squished into the rotor to achieve high packing density. In the early days of MAS NMR, samples often would be prespun on benchtop devices to uniformly distribute the material, but this does not seem to be required with modern commercial NMR probes.

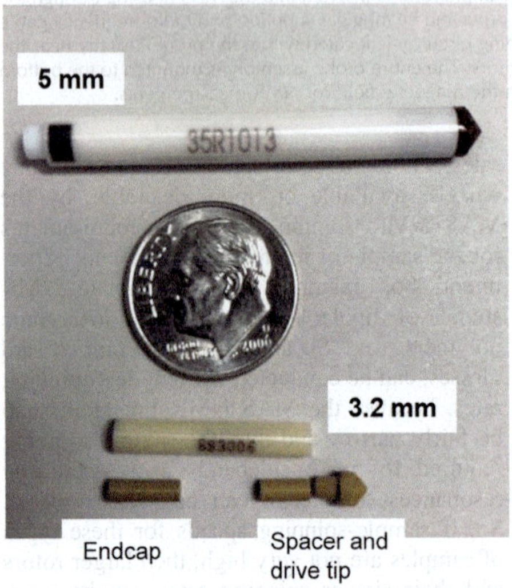

Fig. 4 Examples of two widely available standard magic-angle spinning nuclear magnetic resonance rotor (sample holder) sizes. The cylindrical sleeve typically is zirconia, although silicon nitride also is used in some applications. The endcap, spacer, and drive tip shown for the 3.2 mm (0.13 in.) rotor are machined from a polyamide-imide polymer and can be spun routinely at 25 kHz. The sample volume for the Agilent 3.2 mm rotor shown here is 22 μL, roughly centered in the rotor assembly. The 5 mm (0.20 in.) rotor assembly contains a synthetic fluorine-containing resin spacer and endcap, with a high-performance polyimide-based plastic drive tip, capable of spinning up to 12 kHz. This particular rotor holds 160 μL of sample. Rotors generally cover the range of ~0.9 to 14 mm (0.04 to 0.55 in.) outer diameter, with sample spinning up to ~100 kHz for the smallest rotors.

Another critical consideration for sample spinning is the density of the specimen. Spinning speeds for all ceramic rotors are subject to limitations from the sample density. Due to the large forces developed during sample spinning, highly dense materials can create instabilities or simply be difficult to safely spin. Partially packed rotors can mitigate this issue, and mixing the powdered specimen with an inert, low-density material works well to reduce these effects.

Other sample types can be studied with solid-state NMR. For example, liquid crystals and gels (e.g., semisolids) can be packed into MAS rotors and then fully contained by using special endcaps to keep any liquid from exiting the sample during the measurement. Another common approach, especially in the organic polymer and sol-gel applications of NMR, is to seal the liquid samples into rotor inserts, which are specially designed glass containers that can be tightly fitted in a ceramic rotor but also sealed with epoxy to provide a contained sample for spinning and NMR measurement. Figure 5 shows a typical glass insert for a 9.5 mm (0.37 in.) rotor, which can be cut and sealed at the neck to effectively hold a nonsolid sample for MAS NMR.

Extraneous Signals

One of the issues with solid-state NMR is that extra signals can be detected and confused with the sample of interest. Care must be taken to eliminate or at least understand the source of potential sample contamination, especially if solids are ground into powders by using a mortar and pestle or if the material of interest requires significant handling. Grinding materials, necessary for powdering solids, commonly are made from ceramics, and under the right circumstances, some material can be incorporated into the sample of interest. Oils from skin can easily contribute ^{1}H NMR signals if samples are handled carelessly. If using inert material to sufficiently pack a rotor, either because of limited sample quantity or the need to spin odd-shaped samples, be certain that the inert filler is indeed free of nuclei of interest. Clean sand works well unless the measurement involves ^{29}Si, and a variety of organic powders can be useful in studies of ceramics or glasses. Polycrystalline KBr also is an excellent choice of filler, because the salt can be washed away to recover samples, and this compound is widely used to calibrate the magic angle, as is further described later.

Fig. 5 A 9.5 mm (0.37 in.) magic-angle spinning nuclear magnetic resonance (MAS NMR) rotor (drive tip not fully inserted for clarity) along with a typical glass insert that can contain nonsolid samples and fit tightly into the zirconia rotor for standard MAS NMR experiments. A synthetic fluorine-containing resin spacer and endcap then are used to properly position the sample insert.

Another source of background signal in an NMR experiment is the NMR probe. These are made using a variety of polymeric or ceramic materials, and any such material in close proximity to the NMR sample will potentially contribute an NMR signal. This is especially true for ^{1}H NMR, because most polymers and adhesives contain hydrogen. Similarly, even the fluorinated polymers such as synthetic fluorine-containing resin will give rise to a ^{13}C (and ^{19}F) NMR peak. In addition to probe components, the NMR rotor containing the sample of interest also will contain some elements that give background signals, as shown in Fig. 4 and 5. Some of the common occurrences of extraneous background signal in NMR spectra published in the literature include ^{11}B or ^{27}Al signal from zirconia and other rotor materials, and ^{13}C and ^{1}H signal from polymeric spacers, drive tips, and endcaps.

There are multiple solutions for identifying and even removing unwanted NMR signal from these nonspecimen sources. The brute-force approach involves making an identical NMR measurement on an empty rotor in order to identify peaks that are due to sources other than the actual sample. If done properly, with exactly the same measurement conditions, this signal often can be subtracted from data to yield spectra that are mostly free of the background. This works very well in the case of rotor-derived background signals, as in the studies of ^{27}Al or ^{11}B using different commercial rotors. Other background, including polymeric material in the rotor or near the coil, can be substantially suppressed by using spin echo pulse sequences. In these types of experiments, refocusing of the NMR signal with RF pulses can be highly selective, and only spins experiencing precise tip angles ($\pi/2$ and π pulses) will be detected. These types of experiments are readily available with commercial instruments and often are denoted as background-suppression experiments.

Calibration and Accuracy

The MAS NMR Technique

Magic-angle spinning NMR often is necessary for obtaining high-resolution solid-state NMR. This technique, as described earlier, is the most common experimental method in modern NMR of solids and is routinely available with high-field NMR instruments. The sampling descriptions discussed earlier provided details on packing rotors for MAS NMR, but additional care is needed to properly execute and interpret MAS NMR spectra. First and foremost, the identification of spinning sideband artifacts is necessary to interpret a MAS NMR spectrum. The underlying mechanism of MAS NMR is line narrowing to reduce or eliminate chemical shielding anisotropy and quadrupolar broadening. What happens is that MAS NMR breaks up the broad, static NMR signal into a pattern of peaks, with spinning sidebands spaced around the isotropic resonance by the sample-spinning speed. If samples are spun sufficiently fast, that is, faster than the static linewidth of the resonance, sidebands can be eliminated. Spin 1/2 nuclei, such as ^{13}C, ^{29}Si, and ^{31}P, can be studied without the complication of spinning sidebands, because the combination of rotor size (MAS rate) and magnetic field strength often overcomes the static linewidth of these nuclei.

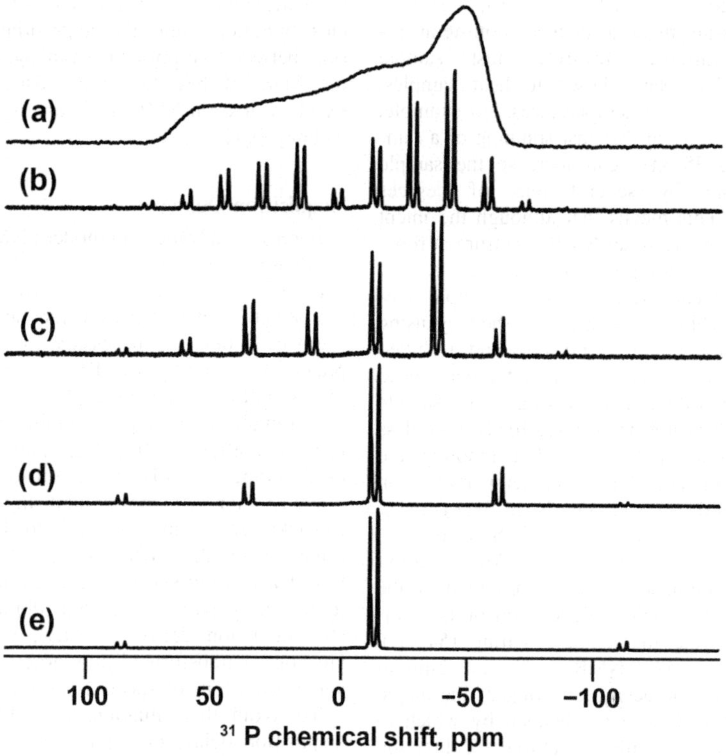

Fig. 6 The ^{31}P nuclear magnetic resonance (NMR) spectra of crystalline $Sn_2P_2O_7$ at 11.7 T (ω_L = 202 MHz) using a 3.2 mm (0.13 in.) magic-angle spinning (MAS) NMR probe. The spectrum in (a) was obtained under static conditions (i.e., no sample spinning), while the other spectra were collected with different spinning rates: (b) 2 kHz, (c) 5 kHz, (d) 10 kHz, and (e) 20 kHz. This compound contains pyrophosphate dimers with inequivalent Q^1 phosphate polyhedra (Ref 15), giving rise to the two peaks at chemical shifts of −11.7 and −14.6 ppm. All other resonances are spinning sideband artifacts, which change position and intensity with MAS NMR rotation rate.

Figure 6 shows how the static spectrum of ^{31}P in crystalline tin (II) pyrophosphate ($Sn_2P_2O_7$) can be broken up into much more narrow peaks with MAS NMR. This example also shows how the position of spinning sidebands can be altered with spinning speed, which is an important approach to separating sidebands from peaks of interest. In many solid-state NMR studies of materials, the isotropic (real) peaks are unknown, and the presence of multiple sidebands can complicate or even invalidate the interpretation. The shape of the sideband pattern can lead to further uncertainty in the peak assignment, because the strongest signal is not always the isotropic peak, as can be seen in Fig. 6. Spin 1/2 nuclei with asymmetric chemical shift anisotropies (CSAs) will give asymmetric sidebands, and additional care is needed to properly identify the real peak. The best way to approach this problem, especially for unknown samples or when trying a new nucleus or experiment, is to spin as fast as possible and to try to reduce or eliminate sidebands, or to acquire NMR data with different MAS NMR rates, as in Fig. 6. The former may be limited by the choice of MAS NMR probes and rotors available for the experiment of interest or simply by the practical aspects of sample size and signal level. Checking data with more than one MAS NMR rate also is a good practice, to be sure signals of interest are not obscured by spinning sidebands from another peak. This is quite common in ^{13}C NMR of polymers, where different functional groups have different chemical shifts and CSAs and where chemical complexity of the material increases the likelihood of interfering signals from isotropic peaks and spinning sidebands. The practical downside to this approach is that it increases the total experiment time.

There is an alternative solution to alleviating spinning sideband complications in MAS NMR spectroscopy, which is to use sideband-suppression techniques. Total sideband-suppression sequences, based on multiple RF pulses, have enabled measurements of MAS NMR data without the complication of sidebands (Ref 16). This approach may be necessary when sample spinning cannot be sufficiently fast to move sidebands away from peaks of interest or when more sample is desired in order to detect more NMR signal, thus requiring larger rotors and slower sample-spinning rates.

Another consideration when performing MAS NMR is the nature of signal quantification. Many experiments may be conducted to qualitatively determine the composition of a polymer or the number of unique sites in a glass or ceramic. However, there are many times in which quantitative information is sought, and, given the nature of NMR in general, site populations or chemical composition can be obtained by careful quantitative analysis of the NMR peaks. The signal is split into spinning sidebands unless the spinning speed is sufficiently high. If sidebands are present, these also must be included in quantitative analysis of the peak areas. The exception to this rule is when MAS NMR of quadrupolar nuclei generates satellite transition peaks, which themselves serve to complicate the analysis and are described later. If spinning sidebands are present in a spectrum, then their peak areas also are representative of the overall peak areas or site populations. If only one site is present, and thus only one isotropic peak and its related spinning sidebands comprise the NMR spectrum, then sidebands can be ignored. However, this generally is unlikely to be the case, and more than one isotropic peak and related sidebands are possible. The analysis of spinning sidebands is further discussed in the examples for both spin 1/2 and quadrupolar nuclei.

One other general consideration when performing MAS NMR experiments concerns sample temperature. While standard MAS NMR probes are designed to cover a range of temperatures, usually from approximately

−80 to +200 °C (−110 to +390 °F), most measurements are made at or near room-temperature conditions. However, fast sample spinning has been shown to heat samples, often to fairly high temperatures. For example, one study showed that fast spinning of a sample up to 35 kHz can increase the sample temperature by several tens of degrees (Ref 17). This means that although the intent may be to measure an NMR spectrum at room temperature, extra precautions are necessary to understand and properly control sample temperature. This can be accomplished by using variable temperature controllers and a chilled nitrogen gas supply to regulate temperature to a desired value. The heating from sample spinning may not be an issue for certain glass and ceramic applications, but ignoring the potential impact of elevated temperatures on samples and corresponding NMR data is risky.

Finally, when performing MAS NMR spectroscopy, the accuracy of the NMR spectrum and the optimized resolution provided by the MAS NMR technique depend on proper calibration of the magic-angle setting (54.74°). Commercial MAS NMR probes are built to reliably provide accurate setting of the magic angle but include a mechanism by which to adjust this angle with respect to the Z-direction of the external magnetic field. While the range of motion is quite limited, regular handling of the probes, constant insertion and removal of rotors, and overall use tend to jostle the rotor housing in the probe and thus lead to changes in the angle setting of the sample housing. Accurate calibration of the magic angle is quite simple, making use of the spinning sideband signal of ^{79}Br in polycrystalline KBr, for example (Ref 18). Sideband intensity in this data is directly proportional to the rotation angle of the sample, being maximized at the magic angle of 54.74°. By monitoring the spinning sideband intensities in the frequency domain data or by maximizing the intensity and number of sideband spikes in the time domain (free-induction decay), the angle can be adjusted until suitably close to 54.74°. Small angle offsets, which sometimes cannot be avoided, generally are acceptable for routine MAS NMR experiments, but significant deviation from 54.74° will result in lineshape distortions, because the dipolar, CSA, and quadrupolar coupling interactions are no longer properly averaged. It is advisable to check and calibrate the magic angle whenever reattaching a probe to a magnet or periodically as one makes measurements over a series of samples.

Data-Acquisition Parameters

The accuracy with which an NMR spectrum is collected and reported depends on other experimental parameters, including proper shift referencing, calibration of the RF pulses, consideration of how RF pulses excite resonances with different quadrupolar frequencies (nutation), and the appropriate recycle time between RF pulses used in signal averaging. Many of these factors are discussed in an excellent text on NMR hardware and practical techniques (Ref 19).

Acquisition Delays

Even with advances in modern NMR instrumentation, there always is a short delay between the application of an RF pulse and the detection of signal from the sample. This acquisition delay is necessary to protect the preamp from high-power RF generated during the excitation pulse (Fig. 2), but it also is beneficial in allowing the probe-tuning circuitry to ring down after the RF pulse. Many capacitors used to tune the probe RF circuit to the appropriate Larmor frequency exhibit ringing, which can take tens of microseconds to decay. One often can see the probe ringing if the acquisition delay is too short, where artifacts appear in the early part of the time domain signal (free-induction decay, or FID), which then are Fourier transformed into large and bothersome distortions of spectral baselines.

To avoid this situation, a slightly longer acquisition delay, or a pulse echo sequence, can be used, similar to how one can suppress broad background signals from the probe or rotor. Unfortunately, a nonzero acquisition delay, which always is required, means that some of the early part of the FID is not measured. For many applications involving narrow resonances, which have long time domain signals, this is not an issue; however, for spectra containing very broad signals, especially those involving quadrupolar nuclei in disordered solids, the loss of even a few data points in the time domain signal can be devastating. For example, if an NMR peak has a linewidth of 20 kHz or larger (not impossible), then the time domain signal will decay within 50 μs, and thus, a short acquisition delay of 10 μs will eliminate a significant portion of the desired NMR signal. It also eliminates the broadest part of a signal. This can lead to phasing issues, loss of quantitative accuracy in peak areas, and even complete elimination of very broad signals.

In addition to echo sequences, which may be necessary for very broad signals, computational prediction of the missing FID points is available in commercial spectrometers, but prediction of the missing signal may be inaccurate and certainly depends on the quality of the time domain signal collected in the experiment. It is advisable to consider the impact of such timing delays and to consider all of the possible ways in which the measured NMR spectrum is potentially incomplete or incorrect.

Shift Referencing

Shift referencing is necessary to understand how the measured NMR resonance and its frequency are correlated to known structures or chemical environments. The most common NMR shift reference is tetramethylsilane (TMS), which is used to calibrate the 0 ppm position of ^{1}H, ^{13}C, and ^{29}Si for both liquid and solid-state NMR. Established and agreed-upon shift references are readily available for all nuclei of interest.

The International Union of Pure and Applied Chemistry (IUPAC) has issued a recommendation on standardizing how chemical shifts are to be reported for NMR data. Based on the ^{1}H chemical shift of TMS, any other nucleus can be properly referenced at the same magnetic field by the simple ratio of ^{1}H to the nucleus of interest frequencies, and this ratio was published in the original and follow-up recommendations (Ref 20, 21). This means that once the proper ^{1}H chemical shift of TMS has been established, one can change nuclei and retain proper shift referencing. This approach has been incorporated into commercial NMR instrumentation, for example, by setting the lock frequency (^{2}H) to match the proper shift reference (ideally ^{1}H of TMS), and then all other nuclei will be referenced using the IUPAC recommendation. However, given the nature of solid-state NMR studies in materials science, and especially due to the large variety of nuclei studied, this particular reference approach may not be as useful as using agreed-upon shift references already established in the scientific literature. Even small offsets of just a part per million or two could lead to differences in interpretation of measured NMR spectra.

Another consideration is that the standard shift reference compound of a given nucleus may be unavailable, inconvenient, or even pose safety concerns due to toxicity. In these cases, secondary shift references often are used. For example, the standard shift reference for ^{11}B, with a shift of 0 ppm, is boron trifluoride etherate ($BF_3{\cdot}O(C_2H_5)_2$), but knowing that solid BPO_4 has a shift of −4.5 ppm from the standard or that 1.0 M aqueous boric acid has a shift of +19.6 ppm relative to the standard allows one to select other materials for proper frequency referencing. It is imperative that the frequency axis used to plot NMR spectra has been properly referenced, even when using a secondary standard. Proper frequency axis calibration allows for the acquired NMR spectrum to be compared with those measured by other researchers, provided they also used the same reference.

Radio-Frequency Pulse Calibration and Nutation

The MAS NMR signals are generated through irradiation of the samples with high-power, short RF pulses. These RF pulses typically are less than 10 μs in length, and the actual pulse width is inversely proportional to the frequency window that can be covered by that pulse. For example, a perfectly square

5 μs RF pulse will span a frequency window of 200 kHz, sufficiently broad when performing NMR on most routine samples. It is important to keep in mind the frequency range covered by RF pulses and, for the most accurate and uniform excitation of a resonance, to be sure the pulse widths are adequate. This is more of a concern when making NMR measurements on quadrupolar nuclei with very large linewidths (e.g., larger than that defined by even the shortest RF pulse). In these instances, the NMR signal may be detected but may not be quantitatively accurate, although this often is sufficient for establishing the number of unique structural environments for a given element.

Radio-frequency pulses generally are calibrated to provide a 90° or $\pi/2$ tip angle, which means that maximum spin magnetization has been shifted away from equilibrium, giving rise to the maximum signal intensity. For a typical spin 1/2 nucleus, such as ^{31}P or ^{13}C, the $\pi/2$ tip angle can be calibrated on the shift reference by measuring the signal versus the pulse width to find the $\pi/2$ maximum and other key features of the sine-like behavior, as shown in Fig. 7. It is convenient to look for the 2π null in signal and divide that pulse width by 4 to obtain a calibrated $\pi/2$ pulse. Accurate RF pulse calibrations are not always necessary for single-pulse experiments that are most common in solid-state NMR, but in other cases, especially for echo-type sequences, in which RF pulses are used to refocus magnetization, the accuracy of $\pi/2$ and π pulses is critical. Many multiple-pulse experiments, including the background- or sideband-suppression techniques described earlier, also require accurate pulse-width calibrations to maintain their effectiveness.

Pulse widths for quadrupolar nuclei are much more complex, and it is highly unusual to find a quadrupolar spin system that mimics the response in Fig. 7. Generally, the $\pi/2$ pulse width for a quadrupolar nucleus in a solid is the liquid $\pi/2$ pulse width divided by $(I + ½)$. This means that the solid $\pi/2$ pulse width for ^{27}Al ($I = 3/2$) is approximately ½ the measured liquid $\pi/2$. However, this works to generate quantitative NMR spectra only if all of the ^{27}Al spins in the sample have the same quadrupolar coupling, and, because this is not common, the aforementioned relation between solid and liquid $\pi/2$ pulse widths is only an initial starting point. Instead, due to differing quadrupolar frequencies, the tip angle necessary to make quantitative measurements is chosen to be very short. One typically would calibrate a $\pi/2$ pulse width by using a liquid sample, usually the shift reference for that quadrupolar nucleus, and then divide this pulse length by at least a factor of 6 or larger to obtain a very short RF pulse width. By using these $\pi/12$ or shorter pulses, one can be assured that all of the signals in the spectrum are uniformly excited and that differences in C_Q (Eq 2) do not affect the response of those resonances to the RF pulses. The classic demonstration of this effect in solid-state MAS NMR is for ^{11}B spins in crystalline or amorphous borates. In these materials, boron can be found in both three- and fourfold coordination, where the EFG for these two environments is very different, and thus, the measured C_Q values are also quite distinct. The C_Q for threefold coordinated boron typically is in the range of 2.5 to 2.7 MHz, while that of an average fourfold coordinated boron is on the order of 0.5 MHz. If one were to measure the optimal RF pulse length, as in Fig. 7, for each of these resonances, the maximum signal would be reached at two different pulse widths, with the $\pi/2$ value for fourfold coordinated boron being much longer and more typical of a spin 1/2 nucleus due to the much smaller C_Q value. To uniformly excite and detect accurate NMR signals when a material contains these very different environments, the solution is to work with very short pulse widths that equally excite all of the spins. The $\pi/12$ pulse widths commonly are used for MAS NMR of ^{11}B, ^{27}Al, ^{17}O, and other quadrupolar nuclei.

Recycle Delays and T_1 Relaxation

One of the most critical experimental parameters to be understood before making NMR measurements is the spin-lattice relaxation time (T_1) of the nucleus of interest. This time constant describes the time necessary for the return to equilibrium of the spin system after an RF pulse. It generally is important to understand how long the spins require for full relaxation before trying to excite them again with RF, as is routinely the case when performing signal averaging. If each subsequent RF pulse were to excite a diminishing amount of magnetization due to lack of relaxation, the signal can be saturated and defeat the purpose of signal averaging. Instead, with an understanding of the T_1 relaxation of the spin system, one can optimize the excitation and signal averaging to achieve the desired signal-to-noise levels.

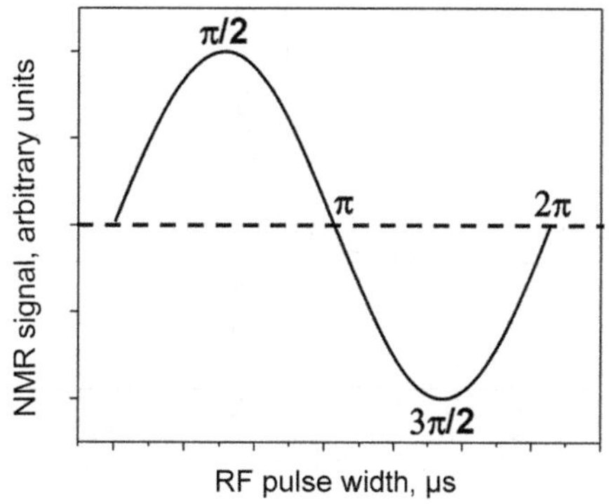

Fig. 7 Idealized diagram of nuclear magnetic resonance (NMR) signal as a function of radio-frequency (RF) pulse width. Maximum signal can be achieved by applying a 90° or $\pi/2$ pulse width.

The potential for negative impact of T_1 is more important when considering a sample in which the nuclei being measured exhibit different T_1 values. If a portion of the NMR spectrum were saturated, as described previously, then the signals are no longer quantitative. This can easily occur in organic polymers, where backbone carbon atoms without large numbers of attached hydrogen have T_1 values that can be much longer than CH_3 and CH_2 groups. If the recycle time (delay between acquisitions) is too short, the fast-relaxing spins (those with attached hydrogen) would be preferentially detected over carbonyls and other less-protonated environments (e.g., aromatic groups), and the resulting MAS NMR spectrum would be misleading in terms of peak intensities. The same situation occurs in any material in which some fraction of the spins of interest relaxes differently than the others. Materials with partial ionic or molecular motion may fall into this category, and certainly, a mixed disordered/ordered sample such as a glass-ceramic would create these types of complications.

The other factor when considering T_1 relaxation and overall length of an NMR experiment is the optimization of spin excitation and signal averaging to generate sufficient quality data in a reasonable time. The standard recommendation is that if one fully excites the spins with a $\pi/2$ pulse, then one must wait for a time of $5 \times T_1$ before pulsing again, because this delay provides recovery of greater than 99% of the bulk magnetization. If one considers a shorter tip angle, less than $\pi/2$, then the amount of time necessary for equilibration of the spin system is shorter, because less of the initial magnetization has been affected. One practical suggestion then is to use a $\pi/6$ pulse width, allowing for full relaxation in $0.14 \times T_1$, and thus much more rapid signal averaging (Ref 19). Of course, the pulse width is less than optimal for full signal, but considering the shorter time between scans, a comparable NMR spectrum may be measured in a shorter time.

One manner in which to address T_1 relaxation and impact on peak intensities is to make multiple NMR measurements with different recycle delays. If the same spectrum, or peak ratios, is measured with one time and some longer time, for example, tens and hundreds of seconds, then it generally is safe to consider that the shorter recycle time is sufficient. One often sees general statements like this in the literature to account for T_1. However, the most accurate and robust approach to making any NMR measurement is to measure T_1 relaxation and either employ $5 \times T_1$ delays or the combination of shorter pulse width and more rapid cycling.

The T_1 relaxation times can be measured using a variety of techniques, but one of the most common and fastest ways to measure T_1 is to use a saturation-recovery pulse sequence (Ref 19). The spins are saturated using a series of $\pi/2$ pulses spaced closely together and then a variable delay before applying another $\pi/2$ pulse to measure the signal. Signal averaging and total experimental time can be short, as very short recycling delays are used because the signal is always going to be saturated. Determining T_1 is very important for unknown materials or new experiments (i.e., spin systems), especially because material purity and crystallinity are two factors that significantly impact T_1. There also are ways to induce relaxation and effectively shorten T_1, including doping with paramagnetic centers such as Cr^{3+}. If material synthesis allows for incorporation of a dopant, then a few hundred to a few thousand parts per million can greatly reduce T_1 and lead to much shorter NMR experiments. One must be certain that all spins of interest are similarly affected by the dopant and that any longer T_1 environments are addressed when selecting the pulse widths and recycle delays.

Cross-Polarization MAS NMR

A favorite technique for rapid NMR of materials containing ^1H or ^{19}F is cross-polarization MAS (CPMAS), in which magnetization from the abundant spin (e.g., ^1H) is transferred to a less abundant spin (e.g., ^{13}C or ^{29}Si). This approach results in a much greater NMR signal but also is faster due to the fact that T_1 relaxation and recycle delays (see previous discussion) are determined by the abundant spin system. Typically, the T_1 of ^1H is much shorter than ^{13}C or ^{29}Si, so signal averaging can be done much more rapidly than if measuring ^{29}Si or ^{13}C MAS NMR directly (i.e., direct-polarization NMR). Because the experiment also relies on spatial proximity of the two spin systems, CPMAS also is quite common for organic polymers and sol-gel materials. An example of the latter is included toward the end of this article. It also is one way to specifically probe surface chemistry in materials.

Calibration of the CPMAS NMR experiment is quite simple, especially if one uses a good calibration sample, for example, a polymer with substantial aliphatic carbons (i.e., many attached protons) or a sol-gel material having substantial organic groups attached to the silicon atoms. The experiment depends on excitation of the abundant spin, so proper calibration of the $\pi/2$ pulse width of ^1H or ^{19}F follows that described earlier. The transfer of this magnetization is facilitated by applying RF pulses simultaneously to both spin systems, described as the Hartmann-Hahn match (Ref 22). The proper power levels for this RF radiation can be calculated based on $\pi/2$ pulse widths of the two spins but, in many instances, can be further optimized by sweeping through a series of power levels on one of the spins to find where the signal is highest. The Hartmann-Hahn matching condition also is sensitive to MAS spinning rate, so a bit of calibration is almost always necessary to ensure ideal measurement conditions when performing CPMAS NMR.

Data Processing

Data processing is almost always performed by using vendor software that operates the NMR instrumentation. There are a number of factors to consider when converting time domain NMR signal (e.g., FID) into frequency domain NMR spectra. Among the most critical are apodization, back and forward prediction of signal, and phasing. Multidimensional NMR experiments require even more data processing but are outside the scope of this discussion.

Apodization, or line broadening, can be quite useful in improving the signal-to-noise quality of NMR data. One key drawback is that application of excessive line broadening, which can be visually desirable for plotting purposes, also can decrease resolution and eliminate separation of closely spaced peaks. For example, the ^{31}P MAS NMR spectrum of crystalline $Sn_2P_2O_7$, which was used to demonstrate the impact of sample-spinning speeds in Fig. 6, contains two closely spaced resonances from the two inequivalent Q^1 polyhedra. The ^{31}P MAS NMR spectrum in Fig. 8(a) shows the data processed without any line broadening, so linewidths are naturally defined by the time domain NMR signal. However, as can be seen in Fig. 8(b), Gaussian line broadening of only 200 Hz, which is 1 ppm for ^{31}P at 11.7 T, leads to a significant increase in linewidth (0.5 to 1.3 ppm). Eventually, the line broadening could be large enough to merge these peaks into one resonance, losing valuable structural information. The best approach when considering signal apodization, regardless of the experiment, is to use the minimum necessary to obtain the desired quality of spectra. If sufficient signal averaging of a fairly abundant spin is possible, then apodization may not even be necessary. However, some amount certainly can clean up the data and allow for better detection of weak signals. A general recommendation is that apodization should be no larger than the full width at half maximum of the resonances of interest. Anything larger certainly will impact spectral resolution.

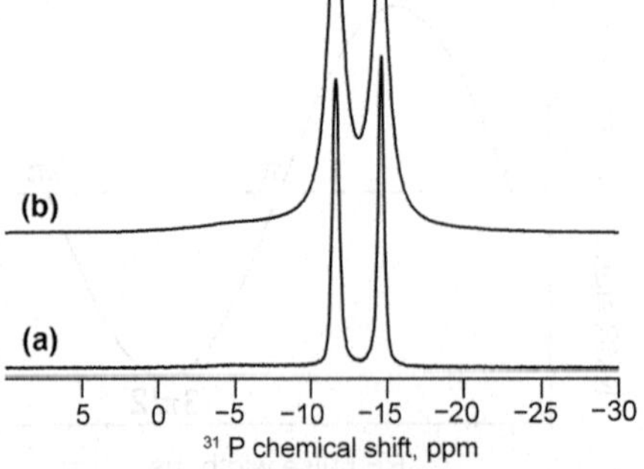

Fig. 8 The ^{31}P magic-angle spinning nuclear magnetic resonance spectra of crystalline $Sn_2P_2O_7$ with (a) no apodization and (b) 200 Hz Gaussian line broadening. Only the isotropic peaks due to the two Q^1 polyhedra are shown.

Another processing parameter that can be both beneficial and problematic is linear prediction of data points in the time domain. As mentioned earlier, when considering the acquisition delay and missing data points, back prediction of these points can be useful and sometimes is needed to properly phase the data. However, the accuracy of this back prediction depends on the overall quality of signal being used in the calculation. Thus, poor signal levels may lead to unreasonable results. It also is possible to delete time points in the FID and then include these in back-prediction routines. Again, this can be useful if the first few points contain probe ringing or background signal, but too much manipulation of the NMR data can lead to erroneous and highly unreliable results. Forward prediction of data points, also in the time domain, can help with Fourier transformation of data that is truncated or cut off due to insufficient acquisition times. The best situation is to always look at the FID before processing, to be sure the signal decays to zero within the acquisition window (time), and if it appears to be cut off, then increase the acquisition time to collect the full FID. Sometimes this cannot be done after performing a long, complex experiment, so zero value points can be added at the end (zero-filling) and then predicted based on the existing NMR signal. Again, any artificial changes to the data should be avoided or at least understood when interpreting the results.

Properly phased spectra mean that each resonance is displayed as a positive peak (absorptive lineshape) with a relatively flat baseline. Phase corrections are routine and necessary to generate these types of spectra, and baseline-correction procedures also are readily available to improve the appearance of NMR spectra. With practice and adequate choice of data-acquisition parameters, this type of spectral manipulation can be minimized. Not only is this important for the visual quality of the NMR data, but properly phased data are easier to fit and thus better for accurate quantification of peak areas.

Applications and Interpretation

Applications in Glass Science

The most common use of NMR in glass science is simple MAS NMR of different network-forming cations, including ^{31}P, ^{29}Si, ^{11}B, and ^{27}Al (Ref 23). These are common to most commercial glass compositions and generally are included in experimental compositions for many different research purposes. ^{31}P and ^{29}Si are both spin 1/2 and have relatively favorable properties, including γ and natural abundance (Table 1). Unfortunately, both also have rather lengthy T_1 relaxation, so these seemingly simple experiments can take several hours to days for each measurement. ^{31}P is 100% abundant, so receptivity is high; ^{29}Si is only 4.7% abundant, so much more signal averaging, or even isotopic enrichment, is necessary to obtain data of sufficient quality. ^{11}B and ^{27}Al are very common quadrupolar nuclei and have been studied in glasses for many years. Their abundances and γ are both high, so measurements can be made using readily available commercial instrumentation. The magnitude of Q (Table 1) is relatively small, so even though these exhibit complex quadrupolar coupling effects, their lineshapes are not excessively broad.

As an example of the typical application of spin 1/2 MAS NMR data from glasses, consider the ^{31}P MAS NMR spectra in Fig. 9. Here, for a ternary soda-lime phosphate glass, ^{31}P NMR data contain two partially resolved isotropic peaks (Fig. 9a) and associated spinning sidebands (Fig. 9b). As discussed earlier, spinning sidebands appear at intervals of the MAS rate, so these data show that sample-spinning speed was 20 kHz, which is approximately 99 ppm for ^{31}P at this magnetic field (ω_L = 202 MHz). The MAS NMR at 20 kHz, which is typical for 3.2 mm (0.13 in.) rotors similar to that shown in Fig. 4, is sufficient for moving these sidebands away from all isotropic peaks but insufficient for complete removal of the sidebands. The two isotropic peaks, because of their chemical shift values (−8.9 and −24.8 ppm) as well as consideration of the glass composition, are assigned to Q^1 and Q^2 phosphate groups. The ability to resolve and identify structural units in glasses, for example, in this simple phosphate glass, is the main benefit to conducting solid-state NMR studies of glasses. Once this type of insight is gained, then properties and compositional trends can be connected and rationalized by using the underlying network structure.

In addition to identifying the types of network polyhedra, which also is the main application of ^{29}Si NMR in glasses, the other critical determination is their populations. If NMR data similar to those in Fig. 9 are collected properly, then the peak areas correspond directly to Q^1 and Q^2 site occupancy or population. Proper intensity determination for the two ^{31}P resonances requires fitting of all peaks, including any detectable sidebands. If one simply fits the two isotropic peaks, ignoring all spinning sideband intensities, one measures 24.8 and 75.2% Q^1 and Q^2, respectively. If one includes all of their sidebands, as shown in Fig. 9(b), then these populations are properly determined to be 22.1 and 77.9%, respectively, indicating that the number of Q^2 sites in this glass would be underestimated if one only fits the isotropic peaks. This difference in peak area determination is a reflection of how the local symmetry of the nucleus and its impact on CSA determines the signal-intensity distribution in spinning sidebands, much like that discussed previously for crystalline $Sn_2P_2O_7$. Thus, different Q^n sites, having different local symmetries, can be further identified by estimating CSAs using the spinning sideband-intensity distribution or by making static rather than MAS NMR measurements (compare with Fig. 6a).

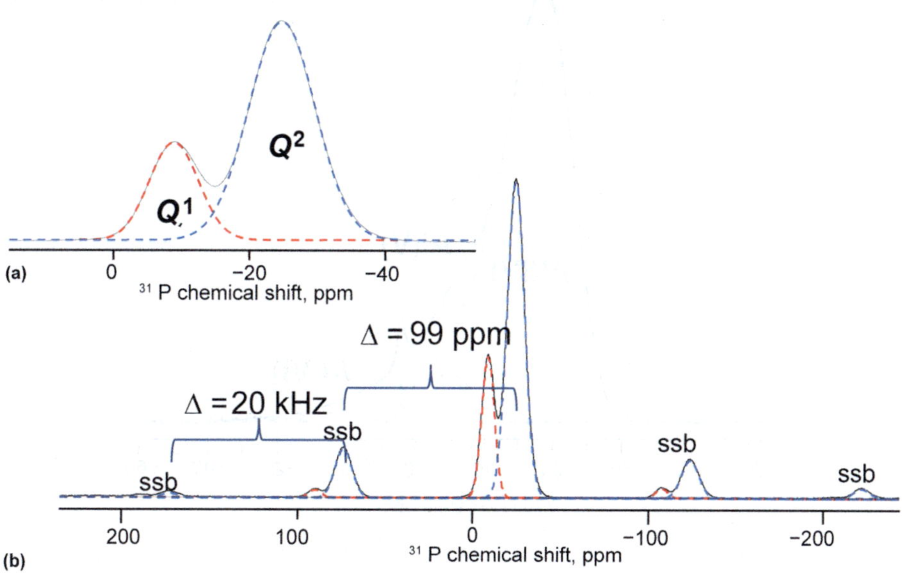

Fig. 9 The ^{31}P magic-angle spinning nuclear magnetic resonance spectra of glassy $10Na_2O{\cdot}45CaO{\cdot}45P_2O_5$. (a) Spectrum showing only the isotropic peaks. (b) Full spectrum with all spinning sidebands (ssb). Gaussian fits to the isotropic peaks and associated spinning sidebands are shown with the dashed lines. Spacing of the sidebands is shown in frequency units of both kHz and ppm.

Because ^{11}B and ^{27}Al NMR studies of glasses are so common and useful, one example of each is discussed. First, consider the ^{11}B MAS NMR spectrum of a simple sodium borosilicate glass, plotted in Fig. 10(a). The spectrum consists of two groups of resonances, centered roughly at 15 and 1 ppm, originating from boron in three- and fourfold coordination, respectively. The symmetry around fourfold coordinated boron generally is very high, so the NMR lineshape from this environment is narrow and can be approximated with a Gaussian or mixed Gaussian/Lorentzian peak. This signal in borosilicate glasses typically can include more than one resonance, due to fourfold coordinated boron with different arrangements of next-nearest neighbor cations. However, as shown in Fig. 10(a), the presence of two tetrahedral boron resonances does not necessarily mean two distinct environments. The weaker resonance, shown as a filled curve, actually is a peak due to the satellite transitions of the fourfold coordinated boron, which can be identified by the spinning sideband pattern. The satellite transition for threefold coordinated boron is the very small peak marked by the arrow. These satellite peaks are a direct result of exciting the noncentral transition of spin >1/2 nuclei, as depicted in Fig. 1.

The threefold coordinated boron sites, which are planar units having three bonded oxygen atoms, have lower symmetry, and thus, the EFG around this site is higher and effectively interacts with the quadrupole moment of the ^{11}B nucleus to yield a complex resonance. Not only does this alter the lineshape (i.e., not a simple Gaussian peak such as in spin 1/2 nuclei or quadrupolar nuclei in high-symmetry sites) but the measured shift is not a chemical shift but rather a combination of chemical shift and quadrupolar shift. Thus, one must consider the contribution of the quadrupolar shift in determining the true isotropic chemical shift. This can be done with measurements at multiple magnetic field strengths, at which only the quadrupolar shift changes with field, or by fitting the complex lineshape to separate out the various contributions. This fitting, which is fairly standard in modern solid-state NMR, results in the dashed lines in Fig. 10(a), which allows for deconvolution of the threefold coordinated boron peak into two distinct resonances, both with characteristic lineshapes due to partial averaging of the ^{11}B quadrupolar interactions. The fits provide estimates for C_Q, isotropic chemical shift, and also the site populations as derived from the integrated peak areas, the latter of which is critical for determining the coordination number of boron and its impact on glass properties.

The ^{27}Al MAS NMR data from aluminum-containing oxide glasses exhibit similar information and spectral complexity. As with ^{11}B, the ^{27}Al nucleus is quadrupolar (spin 5/2) and thus in MAS NMR data exhibits non-Gaussian lineshapes. Here, the ^{27}Al MAS NMR spectrum (Fig. 10b) of a peraluminous calcium aluminosilicate glass, having nominal composition $20CaO\text{-}30Al_2O_3\text{-}50SiO_2$, shows the presence of a main peak at approximately 66 ppm, reflecting fourfold coordinated aluminum that is charge-balanced by the calcium cations. Given that this glass composition is peraluminous, there is insufficient Ca^{2+} for full charge-balancing of all aluminum polyhedra, so some of the aluminum goes into higher-coordination environments. The partially resolved shoulder in Fig. 10(b), with a shift near 38 ppm, corresponds to fivefold coordinated aluminum. Proper deconvolution of these data, which requires the use of peak shapes that incorporate quadrupolar coupling and distributions in chemical shift and quadrupolar coupling, as shown by the dashed lines, allows for quantification of the different aluminum peaks. This particular result gives 88% Al in fourfold coordination, 11% Al in fivefold coordination, and a very weak (1%) signature of sixfold coordinated aluminum. Also of note is the relatively sharp peak at 16 ppm, which is due to aluminum in the zirconia rotor, which can be a major source of background signal in ^{27}Al MAS NMR spectroscopy. This peak can be fit and then ignored when calculating the aluminum coordination number for these glasses or, for clarity in publications or reports, can be subtracted by acquiring the same ^{27}Al MAS NMR data for the empty rotor.

Applications of NMR in glass science continue to advance, with new techniques and new elements being characterized to further establish the correlations between glass structure and material properties.

Applications of NMR in Ceramics

The ^{27}Al and ^{29}Si NMR of ceramics also are well understood and represent common applications of the method (Ref 24). The resolution in ceramic NMR can be quite high, especially for well-ordered crystalline systems such as alumina and zeolites. With high-resolution MAS NMR data, significant structural insight can be gained, which is highly complementary with x-ray diffraction and other characterization methods. For example, the ^{29}Si MAS NMR spectrum of a cordierite (magnesium aluminosilicate) ceramic, plotted in Fig. 11, shows the two tetrahedral sites and their occupancy by silicon. In addition to this general categorization of the silicon tetrahedra, there is clear evidence for additional resonances within the regions defined for the T_1 and T_2 sites. Each of these peaks corresponds to a T_1 or T_2 tetrahedral site with different numbers of next-nearest neighbor aluminum atoms. Each of the peak areas accurately reflects the number of distinct silicon atoms that occupy these sites, and with changes in composition or processing, the impact of these variables on local atomic structure can be easily ascertained.

Another example of MAS NMR for a ceramic is the ^{27}Al MAS NMR spectrum of γ-alumina. This material contains both tetrahedral and octahedral aluminum sites, both of which are resolved in the ^{27}Al NMR data (Fig. 12a) and can be quantified by proper fitting of the two resonances.

In a completely different application, the ^1H MAS NMR spectrum of H-ZSM-5 zeolite in Fig. 12(b) shows how different hydrogen environments can be detected and understood. In this case, the data consist of a distinct resonance from SiOH groups (silanols) and also a very different resonance due to protons that are attached to oxygen atoms bridging between the zeolite framework silicon and aluminum. The latter is highly reactive, and by understanding their abundance and characteristics as a function of composition, much can be learned about these materials.

Applications of NMR in Glass-Ceramics

One of the more recent applications of solid-state NMR in materials science is the study of glass-ceramics, which are materials formed as glasses but then, under controlled heat treatment, are partially converted to a crystalline material. Such hybrid solids are interesting for their mechanical and optical properties,

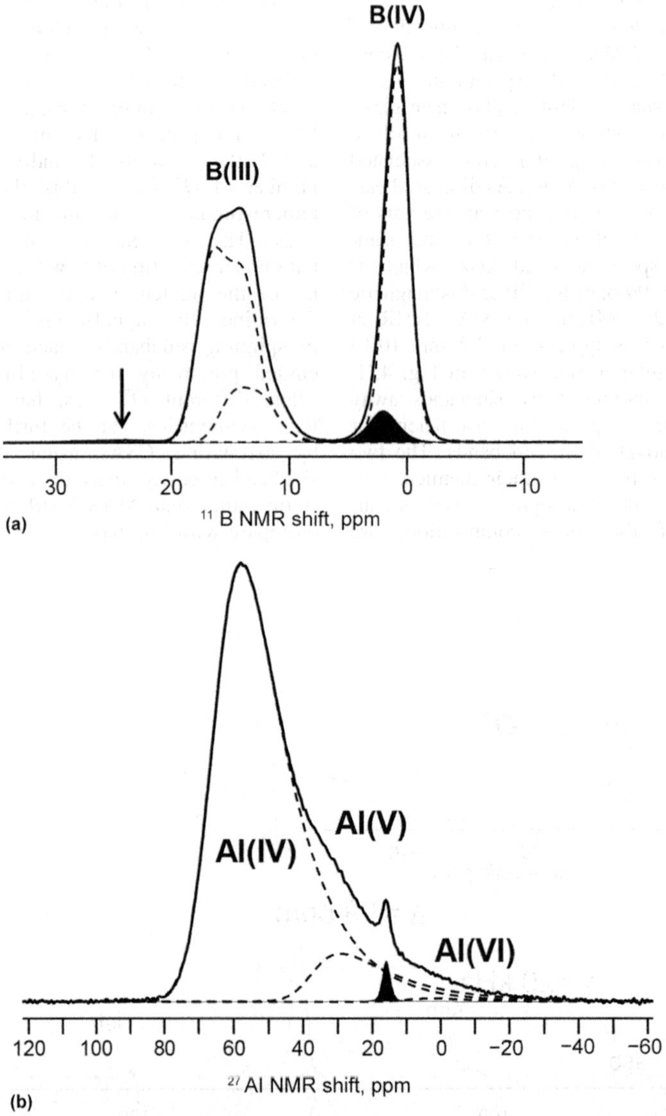

Fig. 10 Examples of (a) ^{11}B magic-angle spinning nuclear magnetic resonance (MAS NMR) spectrum (16.4 T) of a sodium borosilicate glass with composition of $33Na_2O\text{-}53B_2O_3\text{-}14SiO_2$ and (b) ^{27}Al MAS NMR spectrum (16.4 T) of a glass with composition of $20CaO\text{-}30Al_2O_3\text{-}50SiO_2$. Deconvoluted peaks are shown with dashed lines in both spectra. The filled curve in (a) is the satellite transition of fourfold coordinated boron, and the arrow in (a) marks the threefold coordinated boron satellite transition. The filled curve in (b) is the background sixfold coordinated aluminum signal from the zirconia rotor.

but the presence of both glass and crystalline phase(s) presents an enormous challenge for structural characterization. X-ray diffraction, for example, is very useful in identifying and quantifying the crystalline fraction but cannot readily describe the glassy material. Nuclear magnetic resonance, on the other hand, as shown earlier for the different glass and ceramic applications, is uniquely positioned to be able to detect certain elements in both the glassy and crystalline fractions of a glass-ceramic. The only real limitation is that these types of NMR measurements still are limited to isotopes with favorable nuclear properties (Table 1), and additional consideration of differential relaxation must be incorporated into the design of an NMR experiment. Crystalline solids, especially those with high purities, have much longer T_1 values than their analogous glasses, so to properly quantify the elemental partitioning between glass and ceramic, the longer T_1 is the determining factor in signal averaging and overall experimental time.

Figure 13 contains a series of NMR spectra for different elements in a β-spodumene glass-ceramic, both before and after the crystallization heat treatment. The material contains a number of NMR-active nuclei, including ^{29}Si and ^{27}Al, and thus, much can be learned about the structure and how this evolves with the crystallization process. The ^{29}Si MAS NMR of the glass and final glass-ceramic (Fig. 13a) demonstrates clear changes in resonances and linewidths due to a change from fully disordered aluminosilicate glass to a material in which most of the silicon has been converted to an aluminosilicate crystal phase. The data are highly quantitative, and thus, peaks from crystalline phases and any residual glass in the glass-ceramic can be analyzed and the amount of silicon in each phase determined from these peak areas. Similarly, one can follow the evolution of aluminum with ^{27}Al MAS NMR, as shown in Fig. 13(b). Here, the same qualitative changes occur upon crystallization, with most of the aluminum ending up in the crystalline β-spodumene phase. The main ^{27}Al NMR peak after heat treatment is from the ceramic material, but upon close examination of the baseline and with additional NMR experiments, it can be shown that some of the aluminum remains in a glassy phase, and in this case, partitioning of aluminum is approximately 90:10 crystal to glass. This type of structural information can be obtained for other nuclei that may form crystalline phases in glass-ceramics, for example, ^{19}F, ^{23}Na, ^{31}P, and so on. For each of these systems, NMR can be very useful in determining how glass-ceramics evolve as a function of initial glass composition, heat treatment, and other materials processing (Ref 25).

Applications of NMR in Organic Solids and Polymers

Nuclear magnetic resonance has been used in organic chemistry for many decades, taking advantage of the chemical specificity of ^{13}C NMR and the complementary nature of NMR with Fourier transform infrared and mass spectrometry. Thus, only a couple of examples are included here, although the interested reader can find many outstanding review papers and textbooks on this particular application of solid-state NMR (Ref 26, 27).

One common example in polymer science is the identification and quantification of structural groups in complex polymers. Often, these data are measured by using CPMAS NMR, in which the signal from abundant protons (^1H) is excited and transferred to the less abundant ^{13}C spins. This works exceptionally well to detect weak ^{13}C signals, but because the data are based on ^1H-to-^{13}C transfer of magnetization, the spectra are not quantitative. Similar to the previous discussion of T_1 relaxation,

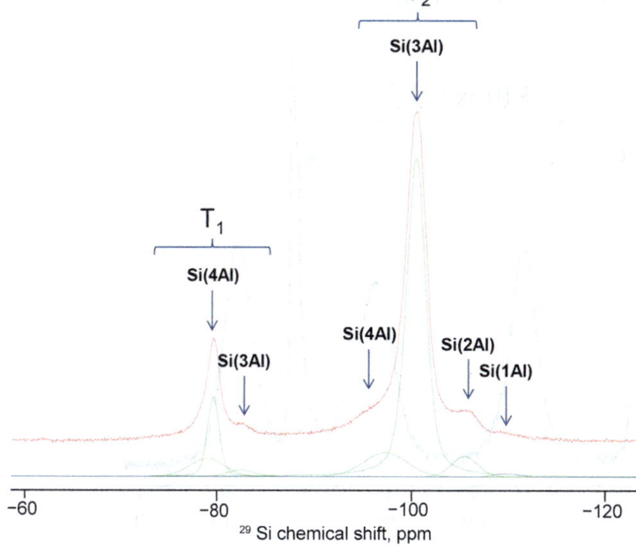

Fig. 11 The ^{29}Si magic-angle spinning nuclear magnetic resonance spectrum of a cordierite (aluminosilicate) ceramic. This ceramic contains two different tetrahedral sites, denoted T_1 and T_2, each of which is occupied by silicate tetrahedra having different numbers of next-nearest neighbor aluminum.

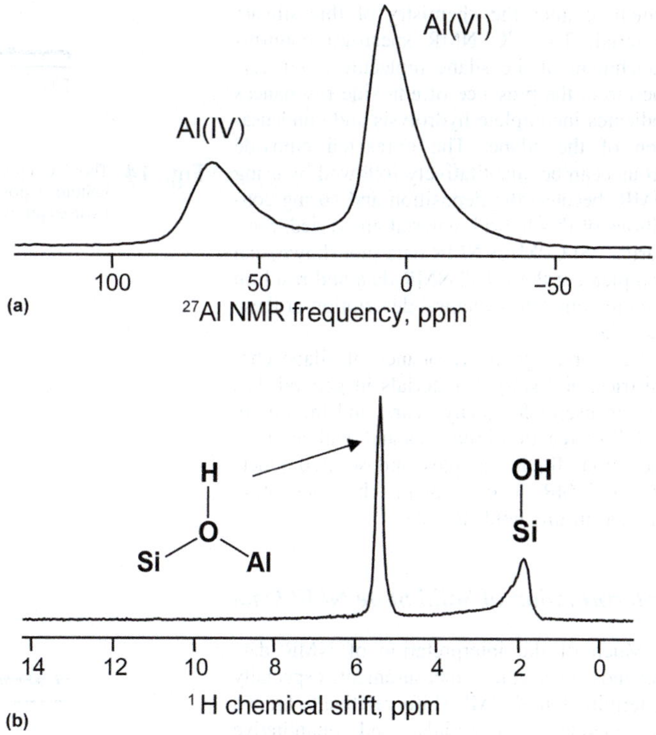

Fig. 12 Additional examples of solid-state nuclear magnetic resonance studies of ceramics. (a) ^{27}Al magic-angle spinning nuclear magnetic resonance (MAS NMR) spectrum of γ-alumina, showing resolution of aluminum in tetrahedral and octahedral sites. (b) ^1H MAS NMR spectrum of H-ZSM-5 zeolite, with identification of silanol and protonated linkages between the framework silicon and aluminum atoms

the detection of CH_3 and CH_2 groups is much more enhanced than carbon environments with few to no attached protons, such as carbonyls and aromatic groups. Thus, much care is needed when interpreting the peak intensities in CPMAS NMR spectroscopy. However, qualitative data, which are highly useful for structural confirmation, can be obtained in relatively short experimental times.

The ^{13}C CPMAS NMR spectrum of chemically modified cellulose is shown in Fig. 14, where the carbon environments from the cellulose and various functional groups are detected. This chemical fingerprint provides confirmation of the chemical derivitization (i. e., methoxy and hydroxypropyl groups). Similar data from complex mixtures of organic molecules can aid identification of components, including the nature of monomers in complex polymeric solids.

Another example, which spans the organic-inorganic material space, is that of silane chemistry on glass surfaces. Organometallic compounds often contain hydrocarbon groups and, in many cases, alkoxide chemistries for reaction and attachment to substrates. The ^{13}C CPMAS NMR data in Fig. 15 gives an example of the type of data and corresponding information from a γ-aminopropyl silane molecule reacted with high-surface-area silica gel. The ethoxide groups on this triethoxysilane react with hydroxyl (silanol) species on the silica gel surfaces, enabling a means by which to alter the chemistry of this silicate material. The ^{13}C NMR spectrum confirms attachment of the silane molecules, and, furthermore, the presence of ethoxide resonances indicates incomplete hydrolysis and condensation of the silane. The unreacted ethoxide groups can be quantitatively followed by using NMR, because the deposition and curing conditions of this hybrid material are varied. Similarly, ^{29}Si CPMAS NMR data (not shown) can complement these ^{13}C NMR data and result in a more complete structural description of these materials.

Nuclear magnetic resonance of silane chemistries, and sol-gel materials in general, has proven useful for many years, and the ability to follow reaction processes and evaluate both the initial liquid samples and solid products makes NMR a convenient characterization method in this field (Ref 28).

Interpretation of Solid-State NMR Data

Much of the interpretation of NMR data depends on accurate measurement, especially determination of NMR shifts relative to known compounds or materials, and quantitative accuracy. The former is critical when making measurements on unknown materials or when investigating new nuclei with NMR. The scientific literature is continually growing in the number of materials science studies using NMR, and each of these provides valuable

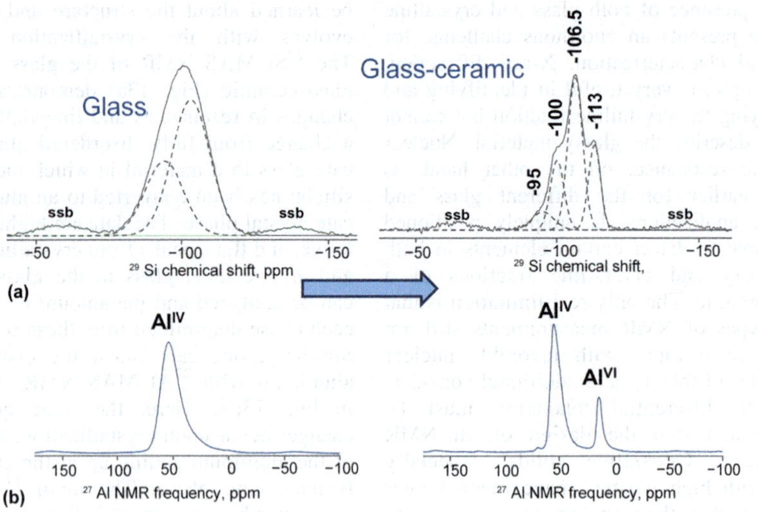

Fig. 13 Solid-state magic-angle spinning nuclear magnetic resonance (NMR) spectra of (a) ^{29}Si and (b) ^{27}Al in a β-spodumene glass-ceramic, before and after heat treatment. Spinning sidebands are marked as "ssb," and dashed curves denote fitted peaks.

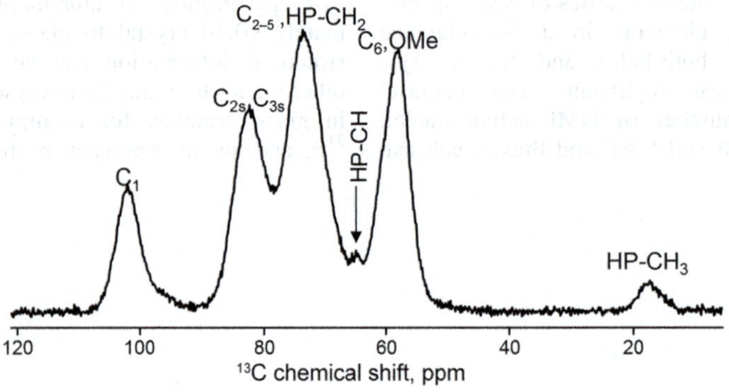

Fig. 14 The ^{13}C cross-polarization magic-angle spinning nuclear magnetic resonance spectrum of modified cellulose powder. Different peaks are labeled with their position in the cellulose backbone: the hydroxypropyl (HP) and methoxy (Me) modifications.

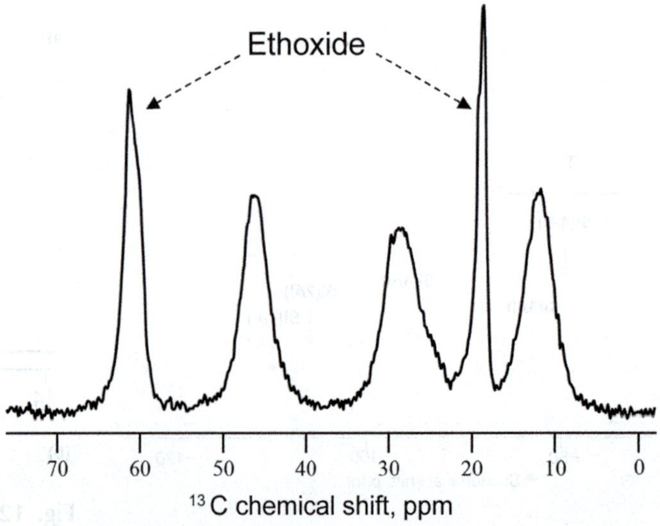

Fig. 15 The ^{13}C cross-polarization magic-angle spinning nuclear magnetic resonance spectrum of silica gel treated with γ-aminopropyl triethoxysilane. Two resonances from unreacted ethoxide functionality are identified, and the other three resonances are from the aminopropyl group on the silane molecules.

insight into interpretation of a measured peak position, whether something routine, such as ^{29}Si or ^{11}B, or perhaps something more exotic, such as some of the quadrupolar transition metals (^{93}Nb, ^{53}Cr, etc.) (Ref 29). The other critical aspect to data interpretation is having a good understanding of the material being studied and making sure the NMR results make physical sense. If one determines that silicate groups are highly cross-linked (e.g., Q^4 tetrahedra) in a glass or ceramic, but this material is made to have large amounts of nonbridging oxygen, then there is serious inconsistency between the NMR interpretation and the compositional reality of the sample. Similarly, fitting NMR data to extract peak areas and more quantitative descriptions of cation coordination environments, such as shown earlier, must be guided by the knowledge of similar studies from the literature and common sense.

Nuclear magnetic resonance is incredibly powerful for research in materials science and many other fields, and the applications continue to grow exponentially. Much more complex experiments are regularly applied to these types of materials, and access to NMR expertise can provide additional guidance on what is possible and what may be unavailable by using solid-state NMR methods. Finally, this short summary of simple MAS NMR in materials science is but one example of a large number of review papers and book chapters detailing the many ways to leverage NMR in materials science research and development.

ACKNOWLEDGMENTS

The author would like to thank Professor Josef Zwanziger (Dalhousie) and my many colleagues at Corning Incorporated for their encouragement, interest, and constant desire to apply and find new applications for NMR spectroscopy in materials research and development.

REFERENCES

1. I.I. Rabi, J.R. Zacharias, S. Millman, and P. Kusch, A New Method of Measuring Nuclear Magnetic Moment, *Phys. Rev.*, Vol 53, 1938, p 318–327, https://doi.org/10.1103/PhysRev.53.318
2. F. Bloch, Nuclear Induction, *Phys. Rev.*, Vol 70, 1946, p 460–474, https://doi.org/10.1103/PhysRev.70.460
3. E.M. Purcell, H.C. Torrey, and R.V. Pound, Resonance Absorption by Nuclear Magnetic Moments in a Solid, *Phys. Rev.*, Vol 69, 1945, p 37–38, https://doi.org/10.1103/PhysRev.69.37
4. E.R. Andrew, A. Bradbury, and R.G. Eades, Nuclear Magnetic Resonance Spectra from a Crystal Rotated at High Speed, *Nature*, Vol 182, 1958, p 1659, https://doi.org/10.1038/1821659a0
5. I.J. Lowe, Free Induction Decays of Rotating Solids, *Phys. Rev. Lett.*, Vol 2, April 1959, p 285, https://doi.org/10.1103/PhysRevLett.2.285
6. A.H. Silver and P.J. Bray, Nuclear Magnetic Resonance Absorption in Glass, Part I: Nuclear Quadrupole Effects in Boro Oxide, Soda-Boric Oxide, and Borosilicate Glasses, *J. Chem. Phys.*, Vol 29, Nov 1958, p 984–990, https://doi.org/10.1063/1.1744697
7. J. Schaefer and E.O. Stejskal, Carbon-13 Nuclear Magnetic Resonance of Polymers Spinning at the Magic Angle, *J. Am. Chem. Soc.*, Vol 98, Feb 1976, p 1031, https://pubs.acs.org/doi/abs/10.1021/ja00420a036
8. T. Charpentier, M.C. Menziani, and A. Pedone, Computational Simulations of Solid State NMR Spectra: A New Era in Structure Determination of Oxide Glasses, *RSC Adv.*, Vol 3, March 2013, p 10550–10578, https://doi.org/10.1039/C3RA40627J
9. S.E. Ashbrook and D. McKay, Combining Solid-State NMR Spectroscopy with First-Principles Calculations—A Guide to NMR Crystallography, *Chem. Commun.*, Vol 52, April 2016, p 7186–7204, https://doi.org/10.1039/C6CC02542K
10. C. Berthier, M. Horvatic, M.-H. Julien, H. Mayaffre, and S. Kramer, Nuclear Magnetic Resonance in High Magnetic Field: Application to Condensed Matter Physics, *C.R. Physique*, Vol 18, Oct 2017, p 331–348, http://doi.org/10.1016/j.crhy.2017.09.009
11. A.D. Bain, Quadrupole Interactions: NMR, NQR, and in between from a Single Viewpoint, *Magn. Resonance Chem.*, Vol 55, 2017, p 198–205, https://doi.org/10.1002/mrc.4418
12. R.E. Morris, S.J. Weigel, N.J. Henson, L.M. Bull, M.T. Janicke, B.F. Chmelka, and A.K. Cheetham, A Synchrotron X-Ray Diffraction, Neutron Diffraction, 29Si MAS-NMR, and Computational Study of the Siliceous Form of Zeolite Ferrierite, *J. Am. Chem. Soc.*, Vol 116, 1994, p 11849–11855, https://pubs.acs.org/doi/10.1021/ja00105a027
13. F. Fayon, I.J. King, R.K. Harris, J.S.O. Evans, and D. Massiot, Application of the Through-Bond Correlation NMR Experiment to the Characterization of Crystalline and Disordered Phosphates, *C.R. Chimie*, Vol 7, March 2004, p 351–361
14. J.-P. Amoureux and M. Pruski, *MQMAS NMR: Experimental Strategies and Applications, eMagRes*, Dec 2008, https://doi.org/10.1002/9780470034590.emrstm0319.pub2
15. V. Chernaya et al., Synthesis and Investigation of Tin(II) Pyrophosphate $Sn_2P_2O_7$, *Chem. Mater.*, Vol 17, 2005, p 284–290, https://doi.org/10.1021/cm048463h
16. D.P. Raleigh, E.T. Olejniczak, S. Vega, and R.G. Griffin, An Analysis of Sideband Suppression Techniques in Magic-Angle Sample Spinning NMR, *J. Magn. Resonance*, Vol 72, April 1987, p 238–250, https://doi.org/10.1016/0022-2364(87)90286-1
17. X. Guan and R.E. Stark, A General Protocol for Temperature Calibration of MAS NMR Probes at Arbitrary Spinning Speeds, *Solid State NMR*, Vol 38, Oct 2010, p 74–76, https://doi.org/10.1016/j.ssnmr.2010.10.001
18. S. Penzel, A.A. Smith, M. Ernst, and B.H. Meier, Setting the Magic Angle for Fast Magic-Angle Spinning Probes, *J. Magn. Resonance*, Vol 293, Aug 2018, p 115–122, https://doi.org/10.1016/j.jmr.2018.06.002
19. E. Fukushima and S.B.W. Roeder, *Experimental Pulse NMR: A Nuts and Bolts Approach*, Addison-Wesley Publishing Company, Reading, MA, 1981
20. R.K. Harris, E.D. Becker, S. Cabral de Menezes, R. Goodfellow, and P. Granger, *Pure Appl. Chem.*, Vol 73, 2001, p 1795
21. R.K. Harris, E.D. Becker, S.M. Cabral de Menezes, P. Granger, R.E. Hoffman, and K.W. Zilm, Further Conventions for NMR Shielding and Chemical Shifts, *Pure Appl. Chem.*, Vol 80, 2008, p 59–84, https://doi.org/10.1371/pac200880010059
22. R.E. Taylor, Setting up 13C CP/MAS Experiments, *Concepts Magn. Resonance A*, Vol 22, May 2004, p 37–49, https://doi.org/10.1002/cmr.a.20008
23. M. Edén, NMR Studies of Oxide-Based Glasses, *Annu. Rep. Prog. Chem., Sect. C: Phys. Chem.*, Vol 108, 2012, p 177–221, https://doi.org/10.1039/c2pc90006h
24. G. Engelhardt and D. Michel, *High-Resolution Solid-State NMR of Silicates and Zeolites*, John Wiley & Sons, New York, NY, 1987
25. A. Ananthanarayanan, G. Tricot, G.P. Kothiyal, and L. Montagne, A Comparative Overview of Glass-Ceramic Characterization by MAS-NMR and XRD, *Crit. Rev. Solid State Mater. Sci.*, Vol 36, Dec 2011, p 229–241, http://doi.org/10.1080/10408436.2011.593643
26. K. Schmidt-Rohr and H.W. Spiess, *Multidimensional Solid-State NMR and Polymers*, Academic Press, London, 1994
27. H. Saitô, I. Ando, and A. Naito, *Solid State NMR Spectroscopy for Biopolymers: Principles and Applications*, Springer, Dordrecht, The Netherlands, 2006
28. R.A. Assink and B.D. Kay, Study of Sol-Gel Chemical Reaction Kinetics by NMR, *Annu. Rev. Mater. Sci.*, Vol 21, 1991, p 491–513, https://doi.org/10.1146/annurev.ms.21.080191.002423
29. R.W. Schurko, Ultra-Wideline Solid-State NMR Spectroscopy, *Acc. Chem. Res.*, Vol 46, June 2013, p 1985–1995, https://doi.org/10.1021/ar400045t

Mass and Ion Spectrometry

ASM Handbook, Volume 10, *Materials Characterization*
ASM Handbook Committee
DOI 10.31399/asm.hb.v10.a0006651

Solid Analysis by Mass Spectrometry

Jessica Román-Kustas, Raymond S. Fuentes, and Curtis D. Mowry, Sandia National Laboratories

Overview

General Uses

- Qualitative and quantitative elemental and molecular analysis of inorganic and organic materials
- Bulk material analysis
- Measurement of trace impurities in materials
- Elemental and molecular profiling/mapping in materials

Examples of Applications

- Verification of alloy and ceramic compositions
- Determination of molecular weight and structural/compositional information of synthetic polymers
- Trace impurity analysis of epoxies, plastics, and metal alloys
- Material profiling of natural and synthetic materials

Samples

- *Form:* Solid
- *Preparation:* Varies with ionization/introduction method, ranging from no sample preparation to some requiring digestion prior to analysis

Limitations

- Multiple ionization methods require dissolution of solid samples prior to analysis.
- Often complicated data analysis of unknown materials and impurities
- Analytes of interest must be ionizable.

Estimated Analysis Time

- Sample preparation varies between ionization/introduction methods from no sample preparation to dissolution of materials, which can be extensive.
- Analysis requires up to 1 h.
- Data-reduction time varies depending on the type of analysis, that is, bulk, trace impurities, isotope ratio, or profiling analysis.

Capabilities of Related Techniques

- *Glow discharge mass spectrometry:* Exhibits little to no ion interferences but is not as fast, precise, or sensitive as inductively coupled plasma mass spectrometry for solutions and easily dissolvable materials
- *Secondary ion mass spectrometry:* Capable of profiling elemental, isotopic, and molecular composition at high depth resolution (~50 nm), with quantitation possible through appropriate standards
- *Laser ionization mass spectrometry:* Little sample preparation with fast analysis but not typically quantitative

Introduction

Since its development, mass spectrometry (MS) has become a research tool used throughout scientific fields. With roots in physical and chemical sciences, MS opened the door to multiple avenues of research, such as quantitative gas analysis, characterization of new elements, and fast identification of trace elements and contaminants. Mass spectrometry was even used in the Manhattan Project for the separation of uranium isotopes. With advancements in, and expansion of, the technique, a range of MS methods are used to analyze solids, including metals, ceramics, plastics, polymers, semiconductors, and biological materials.

Novel ionization techniques are continually being developed, with ambient ionization greatly advancing the capabilities of solid analysis by MS. Ambient ionization techniques ionize analyte material outside of the mass spectrometer mostly through thermal desorption, laser desorption, and impact by charged droplets or ions. This article endeavors to familiarize the reader with a selection of different ionization designs and instrument components to provide knowledge for sorting the various analytical strategies in the large field of solid analysis by MS.

General Principles of Mass Spectrometry

Mass spectrometry is widely used for material analysis, providing information ranging from elemental composition to molecular identification, with options for surface analysis and depth profiling of solid materials. However, the term *mass spectrometry* is a misnomer, with MS techniques determining the mass-to-charge (m/z) ratio instead of mass. Generally, the principle consists of ionizing neutral analytes (elements or molecules) to generate charged gas species, which travel differently based on the instrument environment, including magnetic and electric fields. By discerning the way in which the charged species move through different environments, the m/z of the species can be determined. However, data analysis can be complicated and is specific to the type of MS technique and the information desired. For example, many mass spectrometers can operate in both positive ion and negative ion mode, that is, observing positively or negatively charged species, while many analytes can only ionize in one mode. Various ionization mechanisms are discussed further in the article "Gas Analysis by Mass Spectrometry" in this Volume.

Ionization sources, which serve as the introduction method to MS, can also result in different types of adducts, including different salts and ammonia. For instance, an analyte (A) can often be ionized to $[A + H]^+$, where H is hydrogen. However, the same analyte in many systems is detected as $[A + Na]^+$, where Na is sodium salt (Ref 1). These species have the same charge but different mass. Ionization sources can also be "hard," resulting in species fragmentation, or "soft," resulting in little fragmentation. Understanding and discerning differences in the various techniques within the wide field of MS can inform experimental design and lead to improved analysis.

Depending on the sample type (i.e., solid, liquid, or gas), the introduction of the analyte can vary. Driven in large part by materials characterization and interest in the analysis of solid samples, ambient ionization methodology expanded and diversified, with many ionization sources requiring little or no sample preparation and/or little sample consumption. Table 1 summarizes select solid analysis MS techniques.

However, for all mass spectrometers, the mass analyzer separates ions based on m/z values. The process is typically electrically driven, while many traditional analyzers use magnetic fields for the separation. Various mass analyzer options include, but are not limited to, quadrupole, triple quadrupole, magnetic sectors, time of flight, ion trap, and orbitrap. Two of the main analyzers, triple quadrupole and time of flight, are discussed here, while many are examined in other articles in this Volume. As with ionization sources, there are multiple mass-analyzer systems that can be used based on experimental design.

Combinations of different ionization sources and mass spectrometer techniques enable analysts to customize the material analysis to suit their sample and the desired information to be drawn from the analysis. Understanding fundamental principles, instrumentation, and operating requirements related to solid sample analysis by MS aids in identifying appropriate characterization techniques for various materials. Instrumentation discussed here includes the triple quadrupole mass spectrometer and the time-of-flight mass spectrometer.

Triple Quadrupole Mass Spectrometer

A quadrupole analyzer consists of four metal parallel electrical rods around a central axis. Through applying direct current (dc) and oscillating radio-frequency potentials across the quadrupole rods (two rods of each potential), an oscillating electric field is created within the quadrupole. Ionized species traveling through the electric field follow oscillating paths around the central axis. However, the oscillation path varies species to species based on the species weight and charge. Therefore, some species are destabilized traveling through the quadrupole (nonresonant ions), with only ions of a certain m/z following a stable path, which enables passing fully through the system (resonant ions) for a certain applied frequency. By changing and tuning the potentials applied to the rods, the oscillating electric field is changed, and therefore, the range of m/z that can move through the quadrupole is adjusted, effectively filtering those ions.

While there are single quadrupole mass analyzers (see the article "Gas Analysis by Mass Spectrometry" in this Volume), many MS systems capable of solid analysis use a triple quadrupole mass spectrometer (TQMS) composed of three sets of quadrupole cells in succession. In these systems, while full m/z scans (i.e., scanning a range of m/z) can be collected, the additional quadrupole cells provide distinct advantages. By adjusting the potentials applied, the first quadrupole cell can serve as a focuser, allowing only a select m/z species to pass to the next quadrupole. This quadrupole, termed the collision cell, contains neutral or reactive gas molecules, which "smash" into the analyte species, forming fragments. The fragments pass to the final quadrupole, which scans an m/z range to analyze the fragments.

The design of the triple quadrupole is complex and elegant. Figure 1 shows a schematic of the overall design layout. A TQMS consists of an ionization source (discussed in later sections), three quadrupoles (one of which is a collision cell), and an ion detector.

Quadrupole

Each quadrupole chamber consists of four rods: two opposite rods with applied potential of $(U + V\cos(\omega t))$, and two opposite rods with $-(U + V\cos(\omega t))$, where U is the dc voltage, $V\cos(\omega t)$ is the radio-frequency alternating current voltage, and ω is the angular frequency. The different applied voltages determine the oscillating trajectory of ions traveling through

Table 1 Summary of select ionization techniques

Ionization technique name	Acronym	Sample state	Typical mass analyzer types	Type of ionization	Typical analytes	Typical uses (a)	Spatial resolution
Laser-ablation inductively coupled plasma	LA-ICP-MS	Solid	Quadrupole	Hard	Elemental ions	B, P	100–10 μm
Inductively coupled plasma	ICP	Solid, liquid	Quadrupole	Hard	Elemental ions	B	...
Direct analysis in real-time	DART	Solid, liquid, gas	Triple quadrupole, time of flight, ion trap	Soft	Small molecules, polymers	B, P	NA
Glow discharge	GDMS	Solid, liquid	Sector, quadrupole, time of flight	Hard	Elemental ions	B	...
Secondary ion	SIMS	Solid	Time of flight	Hard	Elemental ions, small molecules, polymers	P	~100 nm
Laser-ablation electrospray ionization	LAESI	Solid, liquid	Triple quadrupole, time of flight, orbitrap	Soft	Small molecules, proteins	B, P	350–15 μm
Electrospray laser-desorption ionization	ELDI	Solid, liquid	Quadrupole, ion trap, orbitrap	Soft	Small molecules, polymers, proteins	B, P	~5 μm
Laser ionization	LIMS	Solid	Triple quadrupole, time of flight, orbitrap	Hard	Elemental ions	B, P	~10 μm
Matrix-assisted laser-desorption ionization	MALDI	Solid	Triple quadrupole, time of flight	Soft	Small molecules, polymers, proteins	B, P	~10 μm
Desorption electrospray ionization	DESI	Solid, liquid	Triple quadrupole, ion trap, orbitrap	Soft	Small molecules	B, P	50–20 μm
Thermal ionization	TIMS	Solid, liquid	Sector	Hard	Elemental ions, isotope	B	...

NA, not applicable. (a) B, bulk; P, profiling

the center of the system, allowing only certain m/z ions to remain on the correct path. All other ions are ejected from the flight path required to pass fully through the quadrupole, encounter one of the rods, and are discharged. Essentially, the quadrupole functions as a mass filter. Varying the potentials and angular frequency enables not just a change in the range of m/z that can pass through the filter but also enables isolation of single m/z ions for fragmentation in the collision cell quadrupole.

Collision Cell

The advantage of using three quadrupoles as opposed to a single quadrupole is the ability to fragment a particular ionized species. By selecting an m/z species in the first quadrupole chamber and introducing the species to a collision-inducing quadrupole chamber, a targeted MS/MS spectrum can be generated, facilitating species identification. In collision-induced dissociation, excited ions from the first quadrupole energetically collide with an inert gas (typically helium or argon) pumped into the collision cell. Collisions result in an increase in internal energy, which induces the ionized species (precursor ions) to dissociate into fragments. The degree of fragmentation can be increased or decreased by adjusting the collision energy (1 to 100 eV). The collision quadrupole is also only subjected to radio-frequency potentials, allowing all fragmented product ions to pass through the cell to the final quadrupole, which functions in the same manner as the initial quadrupole.

Time-of-Flight Mass Spectrometer

A time-of-flight mass spectrometer (TOF-MS) measures the m/z of an analyte by recording the ion species flight time over a specific distance. Imagine a group of ions with varying masses moving in the same direction with constant kinetic energy. With acceleration from an external electric field toward an ion detector, the ions have different velocities based inversely on the square root of their m/z, resulting in lighter ions traveling faster and reaching the detector earlier than heavier ions. This simplified description assumes that all the ions have a similar initial kinetic energy and start at the same position, with both affected by the nature of the ionization source. However, a significant advantage of TOF-MS compared with other MS systems is its ability to perform parallel ion detection. Unlike the TQMS and many mass spectrometers that function as "mass filters," scanning a narrow range of masses of interest, TOF-MS systems do not require scanning as the ions arrive at the detector at different times. The TOF-MS not only detects all ions present in an analyte but also has an essentially unlimited mass range compared with other MS systems, which is of interest for the analysis of biological macromolecules.

Basic TOF-MS systems consist of four parts: an ionization source (discussed in subsequent sections), an acceleration chamber, a field-free drift region, and an ion detector.

Acceleration Chamber

Following sample ionization, the resulting ions are exposed to a strong electrical field within the acceleration chamber, which typically consists of a stack of plates with center holes, except for the back plate. Ions enter the stack from the side, and a high-voltage pulse is applied to the back plate, causing the ions to accelerate through the stack of plates and into the field-free drift region. The potential energy of the ion is related to both the strength of the voltage pulse and the ion charge. When accelerated into the drift region, the energy is converted into kinetic energy (i.e., velocity), and ions of different charge and mass travel at different speeds.

Field-Free Drift Region

Ions entering the drift region have a kinetic energy (K) that is proportional to the species charge (z). With energy (U) applied within the acceleration chamber, the amount of kinetic energy and the velocity (v) of an ion with mass (m) are given by:

$$K = Uz \qquad \text{(Eq 1)}$$

and

$$v = \sqrt{\frac{2Uz}{m}} \qquad \text{(Eq 2)}$$

The time (t) it takes for an ion to travel the length of the drift region (L) and reach the detector is given by:

$$t = L/\sqrt{\frac{2Uz}{m}} \times \sqrt{\frac{m}{z}} \qquad \text{(Eq 3)}$$

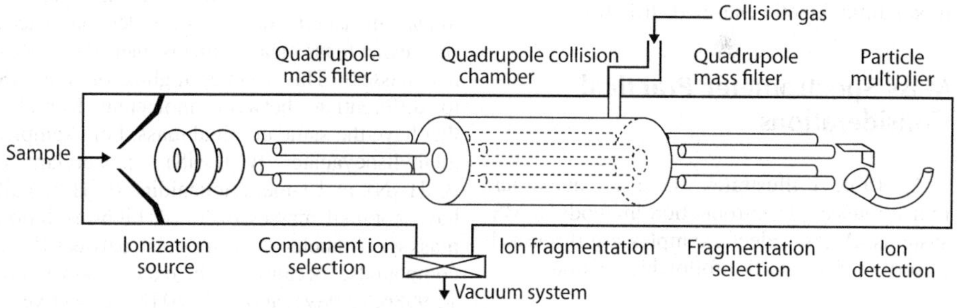

Fig. 1 Schematic of triple quadrupole mass spectrometer

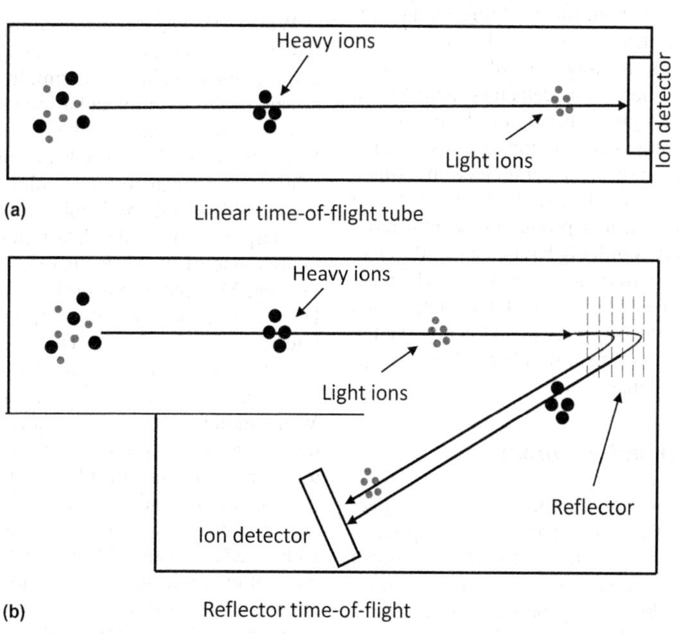

Fig. 2 Schematics of time-of-flight mass spectrometer with (a) linear flight path and (b) reflected flight path, also called a reflectron

Therefore, the time an ion takes to reach the detector is related to the square root of the species m/z (Ref 2). The TOF-MS uses a pulsed-energy technique, meaning that ions are accelerated as a group into the drift region, where their various velocities cause them to reach the detector at different times.

Linear versus Reflectron Flight Tube

The TOF-MS systems are often combined with other MS technologies. For instance, quadrupole TOF systems are common dual MS systems, with the mass filter and collision cell increasing the utility of the TOF-MS. Many TOF-MS systems use a reflected flight path (also called a reflectron) rather than the linear flight path explained previously (Fig. 2). Reflectron instruments incorporate an electrostatic ion mirror at the end of the linear flight tube, which uses potential gradients to reverse the direction of the ions (Ref 2). Off axis by design, ions are reflected toward the detector rather the ionization source. This configuration provides better m/z resolution by correcting for broad initial kinetic energy distributions.

Mass Spectrometer Practical Considerations

As Table 1 illustrates, there are numerous ionization/sample-introduction methods to MS systems. A few select examples are discussed in further depth in the following sections.

Ion Detector

As with the types of ionization/sample-introduction methods, there are various types of ion detectors, with most relying on amplification of the detected signal. The most common detectors in modern MS are electron multipliers. These ion detectors amplify an ion signal through a cascade of electron emissions, beginning with an ion species striking a dynode and emitting several electrons, which are drawn by an electric field to a second dynode, hitting it and producing several more electrons each. Similar behavior causes the electrons to hit a third dynode and so on. The electrons are collected by a metal anode, which generates an electrical signal and is observed in the spectra. Electron multipliers show exceptionally high signal gain.

Calibration and Accuracy

To obtain quality m/z spectra, the mass spectrometer must be tuned and calibrated prior to use. While this varies between instrument manufacturers, it typically consists of a mixture of ionizable, highly purified molecules of known m/z or known atomic composition. Using a calibration standard enables accounting for instrument variations over time, and mass accuracy is determined by using analyte peaks. One method to evaluate instrument mass accuracy measurement is by calculating the root mean square (rms) error using the expression:

$$\text{rms} = \sqrt{\frac{\sum (E_{\text{ppm}})^2}{n}} \qquad \text{(Eq 4)}$$

where E_{ppm} is the parts per million (ppm) error, and n is the number of masses considered in the calibration. Typically, the error should be below 5 ppm rms. Adjusting instrument parameters and mass calibration ensures good sensitivity and accurate mass analysis. Frequency of instrument tuning and mass calibration depends on the desired use and ionization method. Some ionization sources are interchangeable on a single instrument, which, when switched, require retuning.

Relating the separation of two m/z values to the width of their peaks is measured as mass resolution, and mass-resolving power is defined as full width at half the maximum height. High-resolution MS instruments can provide ion m/z measurements to several decimal places (i.e., exact mass measurements instead of nominal masses). This enables high-resolution MS to differentiate between molecular formulas that have the same nominal mass. For example, methylisocyanate (C_2H_3NS), isobutylamine ($C_4H_{11}N$), and 1-methylguanidine ($C_2H_7N_3$) all have nominal masses of 73. A high-resolution mass spectrometer can distinguish between these compounds, because their exact masses are 72.9986, 73.0891, and 73.0640 Da, respectively.

Data Analysis and Reliability

While MS spectra appear complicated, analytical chemists rarely try to assign species to every single peak in a spectrum. Instead, characteristic peaks and molecular ions are identified. This reveals the m/z of the molecule present. When tandem MS is used, precursor ions are selected and fragmented, creating product ions that provide further molecular information.

Interpretation of MS data relies on multiple factors, one of which is charge. As stated previously, MS techniques measure m/z not mass. Determining the mass of a singly charged analyte ion (A^+) is simple. However, molecules can often form multiple charged ions, complicating the determination of the molecule mass. Many ambient ionization techniques discussed here or in Table 1 are soft ionization techniques, meaning that in addition to little or no fragmentation, multiple charged ions are far less likely. However, with these sources, adduct effects can still influence the assignment of the molecular ion, and there are differences in the relative abundance of these ions compared with that of the molecular ion. Often, isotope effects facilitate the identification of atoms present in a molecule. For instance, molecules containing chlorine (Cl) illustrate two molecular ion peaks, one at ⅓ the height of the other and separated by two atomic mass units. This is due to both 37-Cl and 35-Cl isotopes within the population of molecules, with 37-Cl being 25% the natural abundance of chlorine compared to 75% 35-Cl.

Important considerations in reliability and tuning are the peak shape and the need for calibration. Mass spectrometry reliability relies heavily on the upkeep of the instrument. Timely maintenance, tuning, and mass calibration are crucial to acquiring accurate MS spectra.

Ionization Sources

As discussed previously, ion sources create atomic and molecular ions, typically through desorption by a laser or high heat. Developments in ambient ionization have led to numerous ionization techniques, and select methods are shown in Table 1. Two atomic and two molecular ionization sources are discussed here. Inductively coupled plasma and thermal ionization MS provide atomic information, and direct analysis in real-time and matrix-assisted laser-desorption ionization MS are used to analyze molecular compositions.

Inductively Coupled Plasma Mass Spectrometry

Inductively coupled plasma mass spectrometry (ICP-MS) (Fig. 3) is a common atomic analysis technique used to determine the metallic and nonmetallic compositions of a variety of solids and liquids. Due to improved technology, ICP-MS now offers low detection limits with high throughput. Some instrument manufacturers claim detection limits in the parts per quadrillion range, but such capability depends on a variety of factors, including sample matrix, sample load, introduction system, age of instrument, contamination/cleanliness, and so on. Although solids can be analyzed by using this technique, they must either first be dissolved into liquid form or, with optional add-on instrumentation, be analyzed directly. In liquid form, samples can be analyzed to determine metals and nonmetals both semi-quantitatively and quantitatively, with the ability to scan most elements in the periodic table.

A typical ICP-MS instrument includes a peristaltic pump, nebulizer, spray chamber, torch/injector, RF coil, ion source (plasma), cones, and collision/reaction cells (Fig. 3).

Peristaltic Pump

There are several pump sizes and rates (revolutions per minute) that various manufacturers use to get the sample liquid from the sample vial to the instrument. Most common is the peristaltic pump, which uses two pieces of pump tubing, one for sample intake and the other for waste.

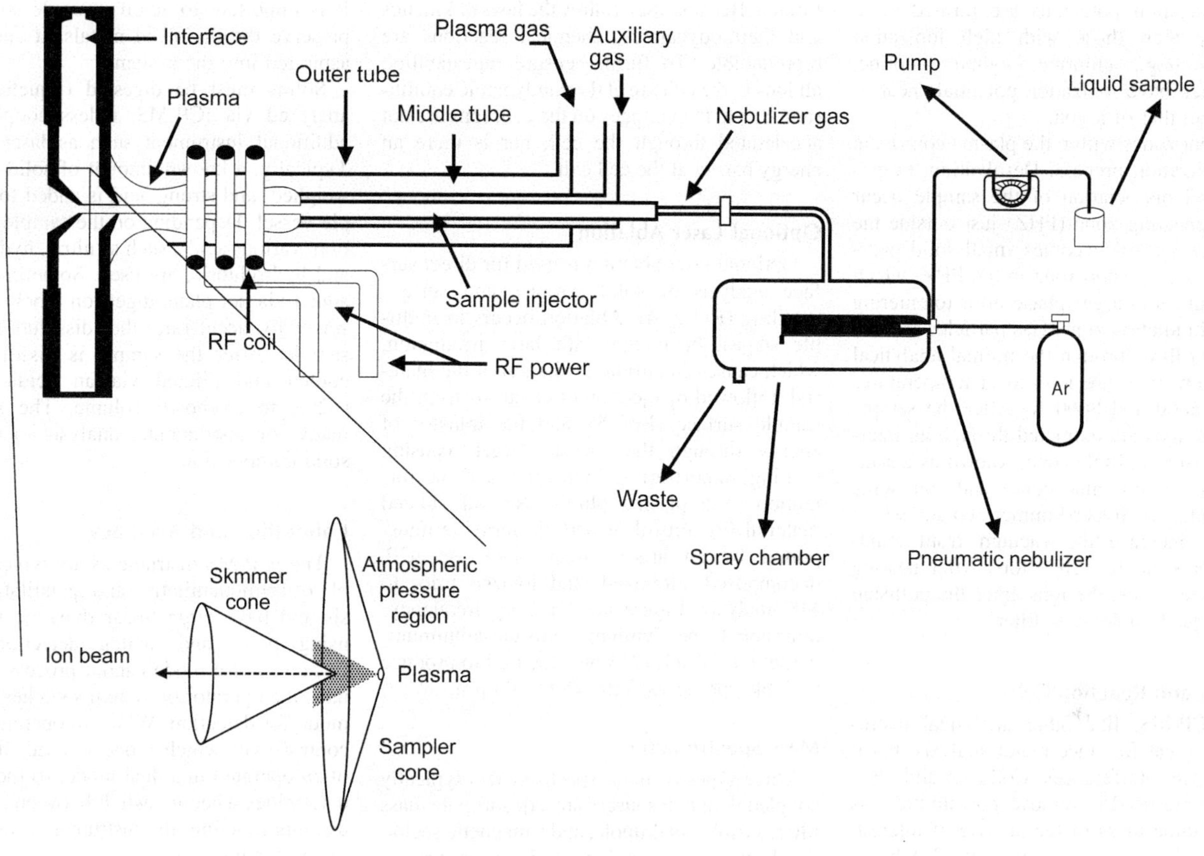

Fig. 3 Schematic of inductively coupled plasma ionization source (components are not scaled in relation to each other). RF, radio frequency

Nebulizer

The function of the nebulizer is to convert the liquid sample to an aerosol. The most common of a wide variety of different types and materials for nebulizers are pneumatic nebulizers usually made of glass or, for highly corrosive sample liquids, various polymers. The nebulizer uses a gas, typically argon, to force the liquid sample through a small orifice, creating a fine aerosol. Both sample matrix and oxide ratios can cause low signal intensities and create spectral interferences, including those caused by polyatomic species. Depending on the sample matrix and oxide ratio, the nebulizer flow can be changed to maximize signal intensity and/or minimize interferences. Nebulization plays a vital role with instrument optimization and could cause erroneous results if not optimized.

Spray Chamber

Within the spray chamber, the aerosol collides with chamber walls, which sends the larger sample droplets to waste and allows only the smallest droplets to be introduced to the plasma. The spray chamber also stabilizes the nebulization stream, enabling a more constant flow pattern by removing residual pulse flow from the peristaltic pump. Spray chambers are made of glass, quartz, and a variety of different polymers, with cyclonic, barrel, and conical configurations. The correct spray chamber can help improve precision between measurements and decrease system contamination. For example, barrel spray chambers have a larger surface area that the sample encounters and therefore require more rinses of the system to prevent carryover and memory effects, resulting in increased analysis time.

Spray chambers can be coupled with a cooling device, such as a Pelletier cooler or chiller, to provide thermal stability and control over the amount of sample reaching the plasma, in addition to decreasing oxide species. Generally, proper instrument optimization suppresses oxide species to levels less than 3%, although newer technology has enabled achieving less than 1.5%.

Injector and Torch

The injector is part of the introduction system that guides the fine aerosol from the spray chamber through the torch and directly into the plasma. The injector is available in different sizes, depending on the analysis. Quartz and alumina are most common, although other materials are available, depending on the type of sample material. The torch, usually made of quartz, contains and sustains the plasma. It consists of three concentric channels that enable argon to flow freely to the torch, with one end of the torch centered within an RF coil, and both working together to create and sustain the plasma. Energy from the RF generator to the coil produces the plasma power, typically 750 to 1500 W.

Ion Source

The plasma is used to create atomic ions. An ICP uses high-purity gas, usually argon, which flows inside the concentric channels of the torch, while the RF load coil creates an oscillating electric and magnetic field. A spark from the ignitor ionizes argon atoms by removing an electron from the outer orbital of an argon atom, creating positively charged argon ions and electrons, which begin to interact with the electric and magnetic field. This creates a chain reaction of collisions, resulting in additional argon ions and electrons. The process generates heat, and the reactions are sustained until the flow of argon gas is stopped. The RF coil sits above the torch, and a bullet-shaped plasma (6000 to 10,000 K) lies at the tip of the torch, serving as an excellent ion source. Ion species formed by the plasma typically are singly charged species because the ICP is very effective at removing a single electron from the outer orbital, although doubly charged species (i.e., ions with a +2 charge) also form. Ions with lower ionization potentials are formed more efficiently than those with high ionization potentials (e.g., chlorine, iodine, fluorine, etc.), which have ionization potentials near or higher than that of argon.

Different zones within the plasma contribute to the ionization process. Desolvation, evaporation, and dissociation of the sample occur in the preheating zone (PHZ) just outside the torch. The aerosol becomes small solid particles after a very short time in the PHZ, which quickly turn into a gas phase prior to entering the initial radiation zone. Gas particles are atomized and flow through the normal analytical zone, where they are ionized at temperatures between 5000 and 8000 K. After the sample is ionized, ions are extracted through an interface by using multiple cones known as a sampler cone, a skimmer cone, and, on some instruments, a hyperskimmer cone, which gradually increase the vacuum from atmospheric pressure to ~10^{-6} torr. After passing through the cones, the ions enter the collision cell and quadrupole mass filter.

Collision and Reaction Cells

The ICP-MS, like other analytical instruments, is great for trace metal analysis, but it suffers from interferences. Collision and reaction cells are used to remove polyatomic ions with the same mass of the analyte of interest. For example, argon (40)-chloride (35) has a mass of 75 Da, similar to that of arsenic (As), thus becoming an interference if arsenic is the analyte of interest. Collision cells use a nonreactive gas (such as helium) to fill a small chamber, and, as the ion of interest and polyatomic ion (interferent) pass through the cell, collisions with helium atoms result in a decrease in the ion kinetic energy. The larger polyatomic ion $ArCl^+$ collides with more helium atoms than the smaller arsenic ion, thus losing more energy than the As^+. An energy barrier is established at the exit of the cell, so ions with less energy (i.e., $ArCl^+$) cannot exit the cell, while ions with more energy (i.e., As^+) can exit the cell and pass to the analytical quadrupole and detector. This is known as kinetic energy discrimination.

Reaction cells differ from collision cells in several ways. A reactive gas is used to remove interferences via chemical reactions instead of nonreactive collisions. Because chemical reactions are used, the laws of kinetics and thermodynamics are followed, meaning that interferences can be removed efficiently. Reactions with large rate constants occur very quickly, enabling a higher degree of interference reduction than in collision cells. Any reactive gas can be used in reaction cell technology, with some of the most common being hydrogen, oxygen, ammonia, and methane. The choice of gas depends on the analyte and interference. Reactions are controlled by using a low-mass bandpass so that new interferences are not formed. Because they follow the laws of kinetics and thermodynamics, chemical reactions are reproducible. To further ensure repeatability, all ions in the cell are at thermodynamic equilibrium due to the voltages on the cell; ions are not accelerated through the cell, nor is there an energy barrier at the cell exit.

Optional Laser Ablation

Optional laser ablation is used for direct surface analysis of solid samples and surface profiling (Ref 3, 4). Ablation occurs in multiple stages, beginning with laser irradiation, which causes electronic excitation of the material, followed by ejection of electrons from the sample surface (Ref 5), and the transfer of energy through the sample target, causing melting, vaporization, ionization, and the formation of a plasma plume. Neutral ablated material (in aerosol or particle form) is transported by an argon stream into the ICP, decomposed, atomized, and ionized prior to MS analysis. Lasers are typically frequency-quintupled neodymium: yttrium-aluminum-garnet (Nd:YAG) (213 nm) focused to produce variable spot sizes from <5 to 300 μm.

Mass Spectrometer

Three types of mass spectrometers typically coupled with ICP sources are a quadrupole mass filter, a triple quadrupole, and a magnetic sector. Single quadrupoles and magnetic sectors are discussed in depth in the article "Gas Analysis by Mass Spectrometry" in this Volume. A single quadrupole is similar to the triple quadrupole but uses only one quadrupole mass filter. Varying the applied potential across the metal rods of the quadrupole affects the trajectory of ions through the quadrupole, allowing only ions of small m/z ranges to pass through to the ion detector. In a similar way, magnetic sectors alter the trajectory of ions through a flight tube by using an applied electric and/or magnetic field.

Additional Instrumentation

The ICP-MS is also coupled with other instrumentation to aid in the analysis process, depending on the analytical goals. Solid sampling is typically accomplished with spark and laser-ablation systems, while gas and liquid chromatography systems are used for sample introduction when required to separate different forms of the same element prior to measurement. Sample preparation varies and is highly dependent on the sample type and analytical goals.

Specimen Preparation

The ICP-MS can analyze liquids and solids. While gases can theoretically be analyzed, calibration is difficult. For liquids, the solution must be less than or equal to 0.3% total dissolved solids entering the plasma, although techniques and accessories are available to accommodate higher levels of dissolved solids. It is important to acidify sample solutions to preserve the dissolved metals. It can then be aspirated into the system.

Solids must be digested (liquefied) to be analyzed via ICP-MS, unless coupled to an additional instrument such as laser ablation. Typically, a known amount of solid sample is weighed, and strong acid is added to the sample vessel. Depending on the sample composition, various acids, such as nitric, hydrochloric, and hydrofluoric, are used. Sometimes heat is added via hot plate, digestion block, or microwave to accelerate the dissolution of the sample. After the sample is dissolved, it is cooled and diluted via an acidic solution (~2%) to a known volume. The sample is ready for instrumental analysis to determine solid composition.

Calibration and Accuracy

The ICP-MS instruments are typically capable of semiquantitative and quantitative analysis and have large linear dynamic ranges by using pulse and analog detection modes. Because pulse mode cannot process large signals, the operator or system switches to analog mode for detection. While an operator has full control over which mode is used, the unit is often operated in a dual mode, so the software determines when to switch between modes. To use this function, the instrument must be calibrated for both modes.

Data Analysis and Reliability

Quantitative data analysis is carried out by using a set of known solutions to construct a calibration curve. Calibration standards are prepared by using certified multielement stock standards and diluted by using acidified blank solution. The calibration curve is usually checked with quality checks prepared in the same fashion. Quality checks are of known values to target a specific concentration within the calibration curve. However, quality checks are typically prepared by using a different certified multielement stock standard than those from which calibration standards were prepared. Calibration standards, quality checks, and samples are analyzed on the instrument by using the same instrument setting and parameters. The percent recovery of the quality checks provides a quantitative analysis to measure the reliability of the instrument/calibration and thereby the data provided in the sample analysis.

Examples 1 and 2 presented later in this article discuss ICP-MS analyses of glass and aluminum, respectively. More information on ICP-MS can be found in Ref 3 and 4.

Thermal Ionization Mass Spectrometry

Thermal ionization sources operate based on a simple theory to atomically analyze materials. Ionization is achieved through deposition of material, either liquid or solid, on a conducting metal filament, which is heated to a very high temperature. Ions are formed by electron transfers between the analyte species and the filament. The predominant application of thermal ionization mass spectrometry (TIMS) has long been bulk elemental analysis, including isotope ratio analysis in areas ranging from geochemistry to nuclear forensics and from cosmochemistry to urine analysis.

Typically, TIMS is composed of the filament(s), accelerator, sector, and detector. Unlike many of the ionization techniques described here, TIMS systems typically use magnetic sector mass spectrometers. Sectors are briefly discussed here and in more detail in the article "Gas Analysis by Mass Spectrometry" in this Volume.

Filament Ionization Chamber

The ionization source of TIMS systems uses heat to thermally ionize solid and dried liquid samples for introduction into a mass spectrometer. Following sample deposition onto a conducting metal filament (typically tungsten, platinum, or rhenium), electric current heats the metal and sample to a high temperature. At temperatures often exceeding 1000 °C (1830 °F), element ions are formed through electron transfer with the filament. Positive species are formed from electron transfer from the atom to the filaments, while negative species are formed from electron transfer from the filament to the atom. Ionization efficiency depends heavily on an atom or molecule ionization potential and electron affinity. Typically, metals can be analyzed in positive ion mode, while nonmetals and transition metals can be observed in negative ion mode.

The TIMS systems use single-, double-, and triple-filament methodologies to ionize sample atoms. Their use depends largely on sample characteristics and experiment methodology. Single-filament sources evaporate and ionize the sample by using the same filament surface. In double-filament systems, the filament holding the sample is used for evaporation, while the second filament ionizes the evaporated sample. Double-filament sources enable more flexibility of sample evaporation rates and ionization temperatures than single-filament sources where the settings are tied to each other. This is often used with samples that evaporate at lower temperatures than that at which they are ionized. Triple-filament systems are often used where direct comparison of two different samples under identical source conditions is required.

Accelerator and Sector

Similar to TOF-MS systems, TIMS accelerates ions toward the mass spectrometer by exposing species to an electrical potential gradient (up to 10 kV), which is focused into a beam through slits and electrostatically charged plates. Typically, TIMS systems separate ions by their m/z by passing them through an electromagnetic sector analyzer. Sector mass spectrometers use electric and/or magnetic fields to alter the trajectory of ions based on how they move through the applied field, which is dependent on the size and charge of the ion. As described in the TOF-MS theory, ions of different masses have different velocities (v). When applied to an electric field (E) and a magnetic field (B), the force (F) applied to a particle with charge q is given by:

$$F = q(E + v \times B) \quad \text{(Eq 5)}$$

With velocity after acceleration dependent on mass, lighter ions are deflected by the applied fields more than heavier ions, resulting in separate beams of ions based on the ions m/z, which are directed to ion detectors and converted into voltage. Examples of various TIMS analyses can be found in the literature (Ref 6, 7).

Specimen Preparation

The TIMS sample preparation varies from sample to sample. Often, liquid samples are deposited onto the conducting metal filament and allowed to dry fully, creating a thin film. Many solid samples require complete dissolution prior to analysis, similar to that described for ICP-MS sample preparation.

Calibration and Accuracy

Like other mass spectrometers, TIMS systems require calibration and tuning by using standard elemental reference materials (National Institute of Standards and Technology) or internal calibration techniques. Instrumental biases such as mass fractionation (isotope ratio changes during evaporation) must be corrected for in measured isotope ratios. Thermal ionization mass spectrometry provides excellent sensitivity, precision, and accuracy for precise isotope ratio measurements. However, progress in ICP-MS is increasingly replacing these systems due to less restriction on element ionization potential.

Analysis of TIMS data can be complicated compared with many other MS systems. Isotope count rates typically must be adjusted by subtracting background and isobar signal contributions, that is, species that have the same mass number but different exact masses.

The TIMS reliability is dependent on mass calibration and tuning, with isotope identification requiring especially accurate m/z measurements.

Direct Analysis in Real-Time

Ambient MS was first introduced with desorption electrospray ionization (DESI) and direct analysis in real-time (DART), simplifying the molecular analysis of a large variety of samples. By creating ions outside of the instrument at atmospheric pressure (in the open laboratory environment), ambient MS is a soft-ionization technique requiring little to no sample preparation or preseparation (Ref 8, 9). Since the introduction(s) of DESI and DART, another ~50 different atmospheric pressure desorption/ionization techniques have been developed, all with similar underlying methodology. The methods form ions by "wiping off" sample analyte molecules using energetically charged particles, species, or laser beams. Of these methods, DART is one of the most established, with a vast range of applications ranging from pesticide monitoring to detection of explosives and warfare agents and to forensic and environmental analysis, to name just a few.

The DART mass spectrometry is based on atmospheric pressure interactions of electronically or vibronically excited-state species from gases such as helium, argon, and nitrogen with the sample and atmospheric molecules. The interaction ionizes species directly from solid surfaces, liquids, and gases. A typical DART instrument includes an enclosed ionization source, a reaction zone (extending from the source exit to the sample), and the interface (extending from the sample to the entrance of the mass spectrometer), as shown in Fig. 4.

Ion Source

Formation of ions that will interact with atmospheric gases and sample material occurs within three compartments in the ionization source. A cutaway view in Fig. 4 shows the general principle of the DART ionization source. Within the first compartment, helium (typically) gas (~3.5 L/min; 50 to 550 °C, or 120 to 1020 °F) flows through the chamber, where a corona discharge (~2 mA) between a needle electrode and a perforated, grounded disk electrode produces ions, electrons, and excited-state (metastable) atoms and molecules. Passing through perforated electrode grids results in cations, anions, and electrons being extracted from the gas stream. The perforated electrodes are set to positive or negative potentials, depending on the ionization mode needed for sample analysis. Excited neutral-gas molecules, including metastable species, remain to exit the ionization source into the reaction zone. Three main roles of the perforated exit-grid electrode are preventing ion-ion and ion-electron recombination, promoting ion drift toward the mass spectrometer interface by exerting a repelling force to push the excited neutral-gas molecules, and serving as a source of electrons for some ionization mechanisms.

Reaction Zone and Interface

Ions formed in the DART source are released into the reaction zone and immediately interact with and ionize the surrounding gas (laboratory air) and sample. While the excited ionizing species released by the ion source are typically helium species, ionized air created through ionization with the helium species is the primary source for sample analyte ionization. Following excitation in the ion source by the reaction:

$$He + Energy \rightarrow He^{*} \quad \text{(Eq 6)}$$

the energy stored in helium is greater than what is possible for other noble gas atoms (He*>Ne*>Ar*), with an excited electronic state energy of 19.8 eV, above that needed to ionize most molecules. Several sample (M) ionization mechanisms can occur, depending on the mode of analysis. However, the dominant positive ion-formation mechanism is through ionized water (H_2O) clusters as:

$$He^{*} + H_2O \rightarrow H_2O^{+} + He + e^{-} \quad \text{(Eq 7)}$$

$$H_2O^{+} + H_2O \rightarrow H_3O^{+} + OH^{\cdot} \quad \text{(Eq 8)}$$

$$H_3O^{+} + nH_2O \rightarrow [(H_2O)_n + H]^{+} \quad \text{(Eq 9)}$$

$$[(H_2O)_n + H]^{+} + M \rightarrow [M + H]^{+} + nH_2O \quad \text{(Eq 10)}$$

Negative ion formation has multiple ionization methods as well, including deprotonation by dissociation, direct electron capture, and anion attachment. In addition, surface Penning ionization, where electrons are produced by the reaction of excited atoms with the surface of the exit electrode ($N_{surface}$) or neutral gas species (N_{gas}), is also a possible ionization route. The negative ion-formation mechanism is:

$$He^{*} + N_{gas} \rightarrow N_{gas}^{+\cdot} + He + e^{-} \text{ or}$$
$$He^{*} + N_{surface} \rightarrow N_{surface}^{+\cdot} + He + e^{-} \quad \text{(Eq 11)}$$

$$O_2 + e^{-} \rightarrow O_2^{-\cdot} \quad \text{(Eq 12)}$$

$$O_2^{-\cdot} + M \rightarrow M^{-\cdot} + O_2 \quad \text{(Eq 13)}$$

Between the reaction zone and interface, helium flow may be insufficient to maintain proper mass spectrometer vacuum conditions. Therefore, a pump interface can be inserted to provide additional suction of the gas-containing ions toward the inlet of the mass spectrometer.

Specimen Preparation

The main advantage of a DART introduction method is the requirement for little to no sample preparation. A solid, liquid, or gas, and even living organisms, can be analyzed by using DART-MS in an open atmosphere without pretreatment. However, DART can be coupled with various separation techniques with typically no additional DART instrument manipulation. This includes gas chromatography-DART (GC-DART), liquid chromatography-DART (LC-DART), and thin-layer chromatography (TLC). Spot analysis of TLC plates has even been used to track nanogram levels of pharmaceuticals. The GC- and LC-DART coupling only requires that the GC or LC gas or liquid be eluted in the reaction zone of the DART source.

Calibration and Accuracy

Analysis by a typical DART-MS system is qualitative, but DART-MS is also capable of semiquantitative analysis. As a possible analyte-introduction method to a variety of MS types, including TOF, TQMS, ion trap, and orbitrap, the sensitivity, resolution, and mass accuracy are more dependent on these instruments than the DART ion source. However, calibration of the interface distance and angle can have dramatic effects on sensitivity. For example, with the DART source at a distance farther than optimal from the MS inlet (a large reaction zone and interface section), signal intensity can decrease significantly due to poor transfer of gas and ionized analyte from the interface to the MS inlet. With the DART source too close to the inlet, the signal-to-noise ratio can decrease due to overload of excited or ionized material in the MS.

Data Analysis and Reliability

Analysis of DART-MS data is identical to that discussed in the various MS detectors. The reliability of the detector used in addition to the reliability of the DART ionization source must be considered for data analysis. These systems, as with any MS method, rely heavily on the analyte material, and, while the energy stored in the excited gas stream is above that required to ionize most molecules, there can be difficulty in how efficiently the species are ionized.

Example 3 presented later in this article discusses DART-MS analysis of polymers. Additional information on DART-MS analysis can be found in Ref 10.

Matrix-Assisted Laser-Desorption Ionization

Desorption ionization sources are considered soft-ionization techniques, meaning that molecular ion formation occurs without breaking chemical bonds within the species. Matrix-assisted laser-desorption ionization (MALDI) MS is a powerful tool for the analysis of large, nonvolatile, and thermally labile compounds, including proteins, large inorganic compounds, synthetic polymers, and various other materials often difficult for other ionization sources. The MALDI ionization occurs via two steps (Fig. 5). The sample material is first dissolved in solvent containing small organic-matrix molecules. The mixture is dried, resulting in a thin film of analyte-doped matrix crystals. The second step relies on the matrix molecules having a strong absorption in the laser wavelength. Through short pulses of irradiation, the laser rapidly heats the matrix crystals and ablates portions of the film. Ablation results in material being expelled in the gas phase concurrent with ionization, and the ions are accelerated into the mass spectrometer. Most sample preparation in MALDI involves dissolving a sample in solution, which is mainly due to the extensive use of MALDI in biological studies. However, MALDI imaging has used ionic-matrix solutions to open the door for MALDI in the analysis of solid samples.

The MALDI ionization systems mainly consist of the matrix, acceleration grid, focusing lens, laser, and mass spectrometer.

Matrix

Choosing a matrix is the most important step in MALDI analysis, with different matrix molecules being ideal for different analytes. Criteria for picking a suitable matrix compound include having a strong laser absorbance, being stable under vacuum, and being acidic to function as a proton donor. Matrix molecules tend to be small organic molecules, such as sinapinic acid (used for proteins) and 2,5-dihydroxybenzoic acid (used for polar synthetic polymers). Picking a matrix compound

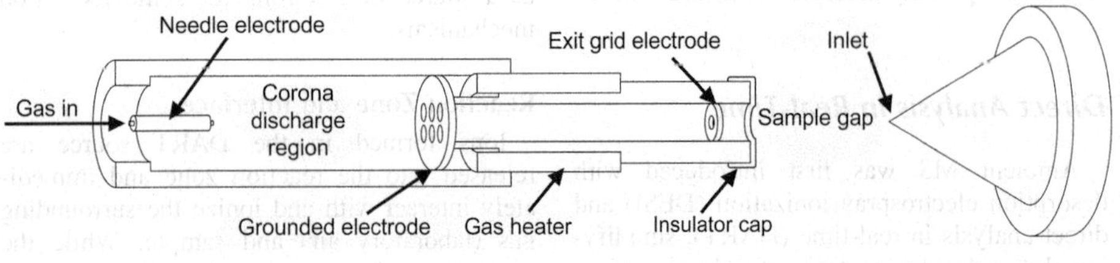

Fig. 4 Schematic of direct analysis in real-time ionization source. MS, mass spectrometer

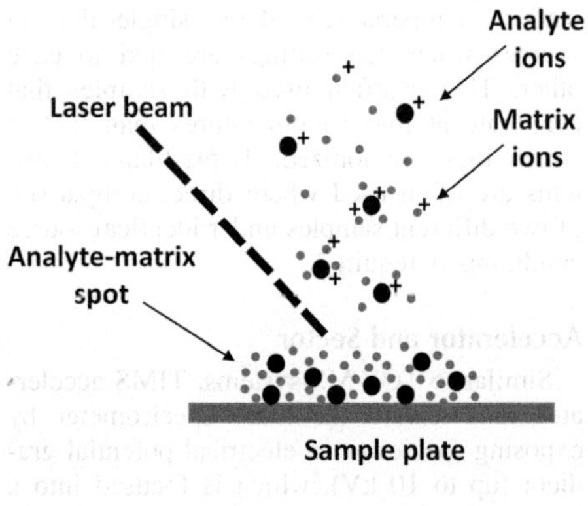

Fig. 5 Schematic of matrix-assisted laser-desorption ionization

is often based on trial and error or experience with similar analyte materials. This is a result of the still poorly understood ionization mechanism, with multiple mechanisms likely at play. Prevailing mechanisms are discussed in more detail as follows.

Acceleration Grid and Focusing Lens

The role of these components is to extract and focus ions from the source and efficiently pass them through the mass spectrometer. The acceleration grid uses voltage variations to accelerate ions toward the mass spectrometer. More information for this component can be found in the previous TOF-MS section. The focusing lens typically uses electric potentials spread along parallel plates to create an electric field, which bends the ion beam toward a focal point to optimally be transferred to the mass spectrometer. The focusing lens also separates the acceleration region from the field-free drift region in the mass spectrometer, typically a TOF-MS.

Laser and Ionization

Typical lasers used in MALDI systems are nitrogen gas lasers (337.7 nm) and frequency-tripled (355 nm) and -quadrupled (266 nm) Nd:YAG lasers. The lasers irradiate a sample-matrix film, which causes matrix molecules to absorb the laser energy. The energy transforms into thermal energy and causes sublimation of the matrix, driving analytes into the gas phase. Within the hot plume produced, the analyte species are ionized, but the exact mechanisms for ionization are debated. One hypothesis is multiphoton absorption, where the matrix (M_{at}) absorbs one photon (*, excited state), then another (charged radical), and then interacts with the analyte (A) species:

$$M_{at}M_{at} \xrightarrow{2hv} M^*_{at}M^*_{at} \rightarrow M_{at} + M^{+\cdot}_{at} + e^- \qquad \text{(Eq 14)}$$

$$M^*_{at}M^*_{at} + A \rightarrow M_{at}M_{at} + A^{+\cdot} + e^- \qquad \text{(Eq 15)}$$

where hv is the laser energy. This results in a deprotonated matrix and protonated analyte.

However, the prevailing mechanism is believed to be the transfer of protons between matrix molecules and analyte molecules, creating singly charged ($[M + H]^+$ or $[M - H]^-$) ions. This mechanism presumes that a matrix molecule in the excited state is more acidic or basic than the analyte, and therefore, it can give up or accept a proton from the analyte as:

$$M_{at} + hv \rightarrow M^*_{at} \qquad \text{(Eq 16)}$$

$$M^*_{at} + A \rightarrow (M_{at} - H)^- + [A + H]^+ \qquad \text{(Eq 17)}$$

While the exact mechanisms are not fully understood, analyte ion signals depend not only on matrix compounds but also on laser intensity.

Mass Spectrometer and Imaging System

Due to the analysis of large molecules, MALDI systems are typically coupled with a TOF-MS system.

The utility of MALDI ionization sources increased significantly with the development of imaging mass spectrometry. The technique typically involves a calibrated sample stage, which can move while the mass spectrum is recorded, resulting in spatial distribution analysis of molecular species. Images are constructed by correlating ion intensity to the relative position from which the data were acquired and are similar to heatmaps in Example 3 for DART-MS analysis presented later in this article.

Specimen Preparation

The MALDI sample preparation can be more extensive than many other ionization sources, with three potential methods discussed here: dried droplet, thin layer, and sandwich method (similar to thin layer). Dried droplet is the easiest, most used method. Solid samples are dissolved in solvent and mixed with the matrix solution. Sample solutions should be acidic to prevent neutralization of the matrix. The mixture is dispensed on a metal plate and dried at ambient temperature. The thin-layer method involves dissolving matrix molecules in acetone and dispensing the solution on the metal target. Acetone is used due to how quickly it spreads and dries. The sample solution is deposited on top of the target and allowed to dry. This method provides more homogeneous matrix crystals and can yield high-resolution spectra and low detection limits. The sandwich method is identical to the thin layer but applies a final layer of matrix on top of the sample layer.

Preparation of multiple matrix-analyte spots on the metal plate enables faster analysis of multiple different samples and matrix-analyte mixtures. Advances in MALDI techniques have made solid analysis more straightforward, with sample integrity maintained. This is typically achieved by attaching a solid sample to the MALDI metal plate, followed by deposition of ionic matrix solution on top, which is allowed to dry prior to analysis.

Calibration and Accuracy

Homogeneous dispersion of the analyte and matrix improves signal intensity, consistency, and mass resolution. Therefore, MALDI analysis often requires optimization of both matrix constituents and laser intensity, depending on sample type. The MALDI systems are as accurate and precise as the MS system to which they are coupled, typically TOF-MS. However, advances in MALDI matrices greatly expanded the range of samples that can be ionized using this methodology.

Data Analysis and Reliability

The MALDI spectra are typically simple to analyze due to the majority of displayed ions being singly charged. Coupled with a TOF analyzer, MALDI systems are capable of fast, precise analysis of molecules ranging from 100 Da to 100 kDa.

Example 4 presented later in this article discusses MALDI-MS analysis of polymers. Other information about MALDI-MS analysis can be found in Ref 11.

Related Ionization Source Techniques

The vast range of ionization methods prohibits full coverage of every technique. Two other common ionization methods, glow discharge and secondary ion, are discussed briefly here and in greater detail in the articles "Glow Discharge Mass Spectrometry" and "Secondary Ion Mass Spectroscopy" in this Volume.

Glow Discharge Mass Spectrometry

Glow discharge mass spectrometry (GDMS) ionizes compounds from solid surfaces via a gas discharge between a cathode and anode in a low-pressure gas (0.1 to 10 torr). The sample is introduced as the cathode in these systems, followed by the application of an electric current across the electrodes, which causes the breakdown of the gas, typically argon, and acceleration of ions toward the electrodes. The bombardment of argon ions on the sample surface results in kinetic energy transfer and surface atom ejection. Atoms are ionized via multiple ionization mechanisms.

The main application for GDMS is in bulk metal analysis (Ref 12). As opposed to inductively coupled plasma mass spectrometry, GDMS can analyze solid samples using very little material (i.e., nondestructive), while it is limited to positive ion analysis and requires solid standards. Additional advantages of GDMS compared with ICP-MS include:

- Decreased possibility of the loss of volatile elements and false positives related to contamination during sample preparation
- High-resolution MS can provide superior sensitivity and better resolution.
- Greater ability to measure insoluble and difficult-to-digest materials (e.g., silicon carbide, aluminum oxide, graphite, etc.)

Secondary Ion Mass Spectrometry

Secondary ion mass spectrometry (SIMS) is a technique that uses an internally generated beam of ions, either positive or negative, to generate ions from a sample surface, which are then accelerated and analyzed by a mass spectrometer. The ion beams bombard a sample, ejecting charged particles (secondary ions) from the surface. Generally, beams of positive ions (e.g., Cs^+, O_2^+, Ar^+) generate negative ions from the surface, while negative ion beams (e.g., O^-) produce positive ions from the surface. Because SIMS is performed under vacuum, samples must be solid materials that are stable in a vacuum system.

Most SIMS systems are used for elemental and isotopic analysis, due to high sensitivity and depth profiling. The SIMS analysis also consumes very little sample, making it useful for samples of limited size (Ref 13–15), for example, quantification of ion implants or contaminants in conductors and semiconductors, with small niches in analysis of meteorites and nuclear materials. Disadvantages of the technique include requiring suitable standards to obtain quantitative information, and the sensitivity is dependent on the matrix and ion beam selected.

Applications and Interpretation

The large number of ionization methods can make choosing a technique for specific samples daunting, requiring consideration of sample type, desired information (i.e., molecular versus elemental versus isotopic, etc.), ionization advantages and limitations, cost, and availability. Analysis requirements vary between sample materials, which requires an experienced analyst to choose efficient ionization techniques. Specifics on the characterization of various materials are covered in more depth throughout this Volume. Information can be found on semiconductors (Ref 16), microelectronics (Ref 15), photovoltaics (Ref 14), battery technology (Ref 13), and proteomics (Ref 11). Examples 1 and 2 that follow present ICP-MS analyses of glass and aluminum, respectively; Example 3 presents DART-MS analysis of polymer; and Example 4 presents MALDI-MS analysis of polymer.

Example 1: Glass Filters

The ICP-MS is used to determine the concentration of residue content on a glass filter, in this case, lead and calcium. The ICP-MS can measure most of the elements in the periodic table except light elements, such as hydrogen, helium, carbon, nitrogen, oxygen, fluorine, and neon, and many of the actinides.

Sample Preparation

Three 46 mm (1.8 in.) glass filters were collected from different locations within the sampling area. Samples were weighed to approximately 0.1 g, and 10 mL nitric acid, 3 mL hydrochloric acid, and 2 mL hydrofluoric acid were transferred to a synthetic fluorine-containing resin microwave digestion vessel. The mixture was heated in the digester at 180 °C (355 °F) for 60 min, and, after completely dissolving, samples were normalized to 50 mL with acidified water and weighed.

Data Reduction

Requirements to obtain reliable quantitative measurements for this sample included:

- Analyzing instrument performance check to ensure the instrument was optimized, which required meeting the instrument manufacturer's target specifications
- Analyzing calibration blank and calibration standards for each analyte
- Calibration curves were plotted by known analyte concentration versus signal intensity. A correlation coefficient close to 1 was ideal for a more accurate analysis.
- Analyzing quality-control checks following calibration to verify calibration curves were valid, then applying accurate measurements
- Using internal standards to monitor matrix interference and instrument drift in the 80 to 125% recovery range. Quantitative measurements were suspect if values were outside this range.

Data Analysis

Signal intensities of analyzed samples were compared against the calibration curve to yield a concentration. Concentration information together with intensities and internal standard recoveries of each analyzed sample were generated, with concentration given:

Sample identification	Calcium, wt%	Lead, wt%
Filter blank	4.38	0.30
Filter 1	6.50	1.12
Filter 2	6.45	1.13
Filter 3	6.04	1.15

When raw data values were within the working calibration range, calculations were assessed by using the initial weight (sample weight), final weight (50 mL), and dilution factor(s). This yielded a final concentration in parts per million, and weight percent in the original material was calculated.

Example 2: Aluminum Wire and Sheet Verification

The ICP-MS is used to verify the composition of aluminum wire and aluminum sheet, analyzing for magnesium, vanadium, copper, zinc, titanium, gallium, and iron.

Sample Preparation

Five types of aluminum wire and sheet were provided for analysis. Triplicate replicates of sample were weighed to 0.1 g, and acids (10 mL nitric acid and 5 mL hydrochloric acid) were added into synthetic fluorine-containing resin microwave digestion vessels for digestion. Samples were heated for 10 min at 95 °C (200 °F), and the temperature was increased to 120 °C (250 °F) for an additional 20 min, but an additional 10 mL nitric acid and 5 mL hydrochloric acid was required to dissolve samples completely. Samples were normalized to 100 mL, and subsequent dilutions were prepared to obtain a signal in the working range of the instrument. For data reduction and analysis, see Example 1 plus Table 2.

Example 3: Nonpolar Poly(Dimethyl Siloxane) Polymers

Direct analysis in real-time mass spectrometry is used to determine the extent of spreading of mold-release formulations consisting of poly(dimethyl siloxane) (PDMS) and solvent.

Sample Preparation

Materials received were liquid (trimethylsilyl (TMS)-terminated PDMS), and the sample required deposition and drying on a surface prior to analysis to observe variations in spreading behavior in a mold release. Standards of TMS-PDMS ranging in viscosity from 5 to 50 cSt were diluted to 20 mg/mL in hexane to facilitate spreading, similar to how the polymers would be used as a mold release. Approximately 2 μL are spotted on two locations on a clean glass slide, spaced evenly from the edge of the slide as well as from each other.

Data Reduction

Obtaining a reliable-quality spectrum in this sample required:

- Tuning and mass calibrating the MS to ensure good sensitivity, peak shape, and accurate *m/z* values. This can vary from instrument to instrument and between ionization modes but consists of a mixture of highly purified and ionizable molecules of known *m/z* to adjust and optimize instrument parameters. Mass calibration should span that of the instrument mass range.
- Collecting a background spectrum of the glass slide prior to spotting and drying the TMS-PDMS solution, ensuring no

Table 2 Inductively coupled plasma mass spectrometry analysis of aluminum wire and sheet

	Concentration, ppm								
Sample	Magnesium	Vanadium	Copper	Zinc	Titanium	Gallium	Iron	Carbon	Sulfur
Wire	0.92 ± 0.18	0.16 ± 0.02	0.34 ± 0.05	2.03 ± 0.32	11.95 ± 4.38	0.24 ± 0.03	ND	85.87 ± 12.13	5.95 ± 5.07
Sheet	0.58 ± 0.16	1.23 ± 1.89	1.08 ± 0.25	0.78 ± 0.47	0.22 ± 0.092	0.08 ± 0.03	19.71 ± 4.41	107.37 ± 13.27	3.80 ± 0.68

ND, not determined

contaminants were present on the slide or within the instrument

- Adjusting the DART spray according to the position of the spots on the glass slide to ensure that the linear rail moved the samples within the optimal reaction-zone positions. This was completed by randomly spotting standard solutions on a glass slide and observing the signal-to-location correlations, adjusting to identify ideal width placement on the slide.

Data Acquisition

Following air drying, the glass slide was mounted in the linear rail of a DART ion-trap MS system. Data was acquired by setting the gas (helium) temperature of the DART ion source to 500 °C (930 °F) with an MS scan range of *m*/*z* 50 to 2000. The linear rail then moved the glass slide across the reaction zone of the DART gas at a raster speed of 0.2 mm/min (0.008 in./min). The *m*/*z* signals versus retention time (related to the *y*-axis location on the glass slide) were plotted via heatmap (Fig. 6). With this method, the TMS-PDMS thin-film polymer profile was not only observable, but also, spatial analysis revealed that with changes in the heat treatment temperature during drying, the degree of TMS-PDMS spreading on the slide increased for the lower-average-molecular-weight 5 cSt polymer (Fig. 7).

Example 4: Parachute Nylon Polymer Material

Matrix-assisted laser-desorption ionization mass spectrometry is used to directly measure the molecular mass of high-mass polymeric materials such as nylon. While gel permeation chromatography can provide a rough mass distribution, MALDI can perform the analysis directly from the solid material.

Sample Preparation

Materials were received as microcalorimetry devices with a nylon 6,6 coating, the chemical structure shown in Fig. 8. The MALDI matrix 2-(4′-hydroxybenzeneazo)benzoic acid was deposited on the surface and allowed to dry under ambient conditions. Figure 9 shows a scanning electron microscopy image of the matrix crystals following drying. A poly(methyl methacrylate) calibration standard was prepared by mixing with the same matrix compound and dispensing on an uncoated microcalorimetry device. The standard was allowed to dry under ambient conditions prior to analysis.

Data Reduction

Requirements for accurate, reliable spectra in these materials include:

- Tuning and mass calibrating the TOF-MS by using a mixture of highly purified, ionizable molecules, which ensures accurate *m*/*z* values and good sensitivity
- Collecting a background spectrum of the matrix compound on an uncoated microcalorimetry device prior to the analysis of nylon 6,6-coated devices to ensure matrix effects were accounted for and understood
- Adjusting the plate location to ensure efficient irradiation of the sample spots on the device
- Adjusting laser intensity according to signal intensity and the amount of energy required for matrix excitation and ablation

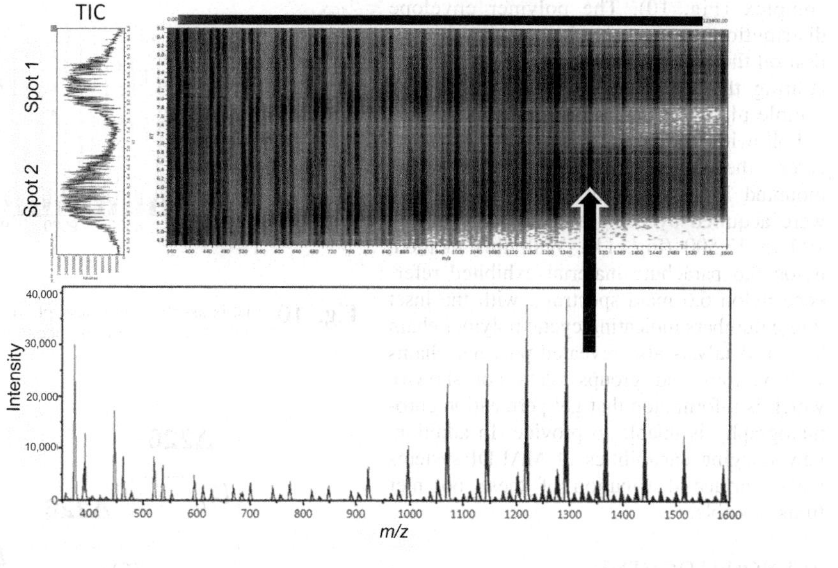

Fig. 6 Total ion chromatogram (TIC), heatmap, and sample spectra of trimethylsilyl-terminated poly(dimethyl siloxane) (50 cS) mold-release formula on a glass slide. *m*/*z*, mass-to-charge ratio

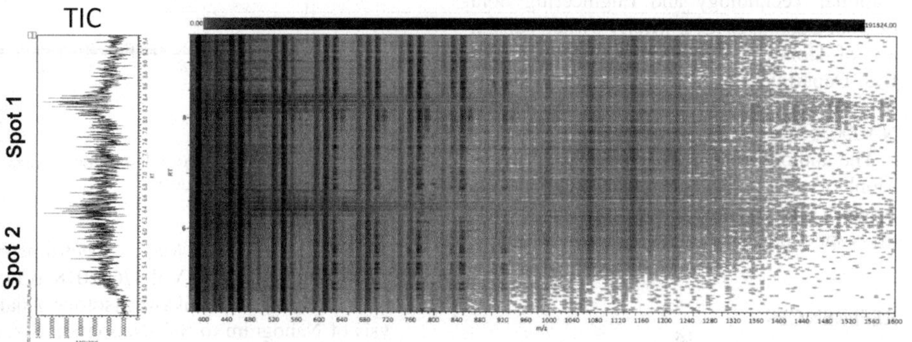

Fig. 7 Total ion chromatogram (TIC) and heatmap of heat treated trimethylsilyl-terminated poly(dimethyl siloxane) mold-release formula on a glass slide

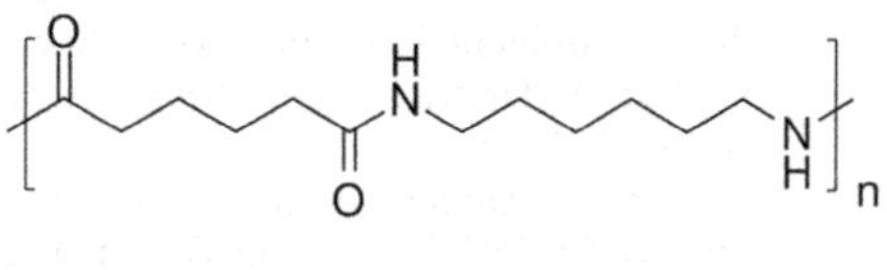

$(C_{12}H_{22}N_2O_2)_n$

Base molar mass: 226.32 g/mol.

Fig. 8 Chemical structure of nylon 6,6

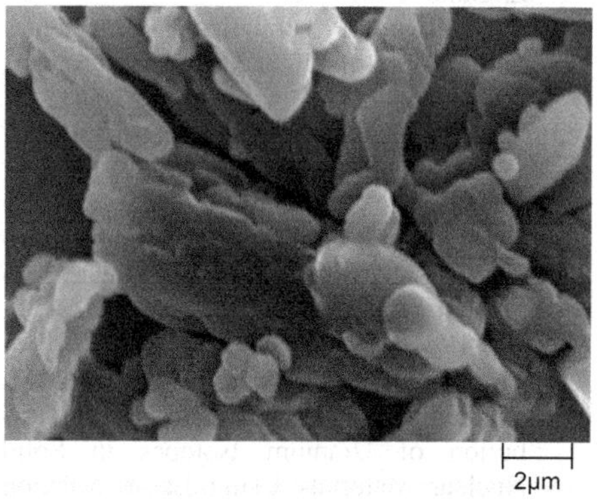

Fig. 9 Scanning electron microscopy image of 2-(4′-hydroxybenzeneazo)benzoic acid matrix crystals formed on a microcalorimetry device

Data Acquisition

Prior to the analysis of the nylon 6,6 coating, a poly(methyl methacrylate) calibration standard was analyzed directly from the same type of device and same matrix complex (Fig. 10). The polymer envelope distribution was preserved even after deposition on the microcalorimetry device, demonstrating the viability of these devices as a sample platform.

Following air drying of the matrix molecules, the nylon 6,6-coated devices were mounted in the MALDI chamber, and data were acquired by using an MS range of *m/z* 100 to 13,500 (Fig. 11). With this method, nylon 6,6 parachute material exhibited reference nylon 6,6 mass spectrum, with the inset figure numbers indicating cyclic polymer chain length. Analysis also revealed polymer chains with various end groups (data not shown), which is information that gel permeation chromatography is unable to provide. In addition, new imaging capabilities of MALDI systems make surface distribution of these polymer films possible.

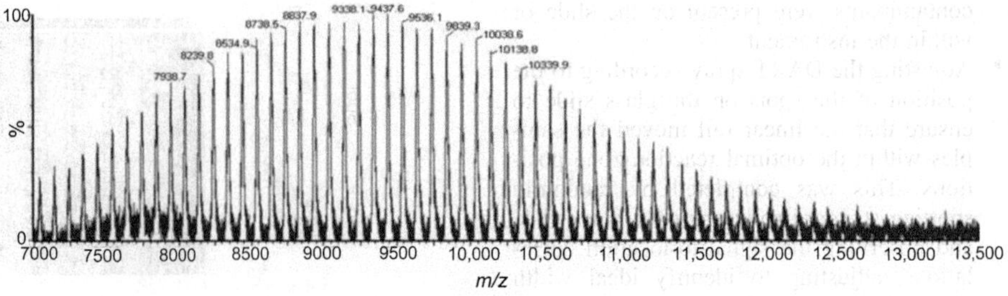

Fig. 10 Matrix-assisted laser-desorption ionization mass spectrometry spectrum of poly(methyl methacrylate) calibration standard on uncoated microcalorimetry device

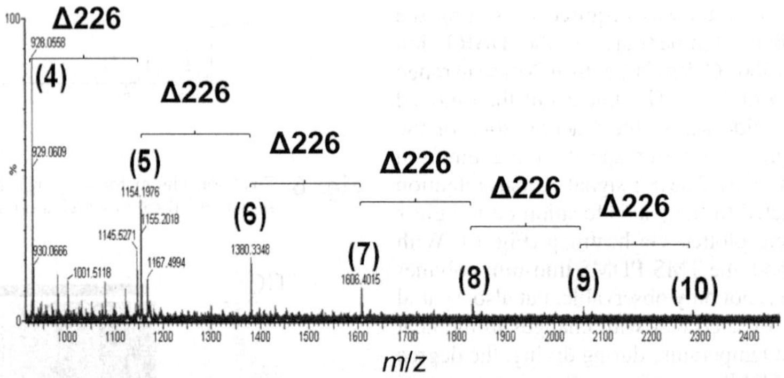

Fig. 11 Matrix-assisted laser-desorption ionization mass spectrometry spectrum of nylon 6,6-coated microcalorimetry device

ACKNOWLEDGMENTS

Sandia National Laboratories is a multimission laboratory managed and operated by National Technology and Engineering Solutions of Sandia, LLC, a wholly owned subsidiary of Honeywell International Inc., for the U.S. Department of Energy's National Nuclear Security Administration under contract DE-NA0003525.

The authors thank Anthony Esparsen, GlowMonkeyStudio, for schematic creations and Amber C. Telles for inductively coupled plasma mass spectrometry examples.

REFERENCES

1. E. de Hoffmann and V. Stroobant, *Mass Spectrometry Principles and Applications*, John Wiley & Sons Ltd., West Sussex, England, 2007, p 65
2. M. Guilhaus, Principles and Instrumentation in Time-of-Flight Mass Spectrometry, *J. Mass Spectrom.*, Vol 30, 1995, p 1519
3. K. Drost, D. Chew, J. Petrus, F. Scholze, J. Woodhead, J. Schneider, and D. Harper, An Image Mapping Approach to U-Pb LA-ICP-MS Carbonate Dating and Applications to Direct Dating of Carbonate Sedimentation, *Geochem., Geophys., Geosys.*, Vol 19, 2018, p 4631
4. M. Krachler, Z. Varga, A. Nicholl, M. Wallenius, and K. Mayer, Spatial Distribution of Uranium Isotopes in Solid Nuclear Materials Using Laser Ablation Multi-Collector ICP-MS, *Microchem. J.*, Vol 140, 2018, p 24
5. A. Ammann, Inductively Coupled Plasma Mass Spectrometry (ICP MS): A Versatile Tool, *J. Mass Spectrom.*, Vol 42, 2007, p 419
6. M. Willig and A. Stracke, Accurate and Precise Measurement of Ce Isotope Ratios by Thermal Ionization Mass Spectrometry (TIMS), *Chem. Geol.*, Vol 476, 2018, p 119
7. S. Wakaki and T. Ishikawa, Isotope Analysis of Nanogram to Sub-Nanogram Sized Nd Sample by Total Evaporation Normalization Thermal Ionization Mass Spectrometry, *Int. J. Mass Spectrom.*, Vol 424, 2018, p 40
8. M.-Z. Huang, S.-C. Cheng, Y.-T. Cho, and J. Shiea, Ambient Ionization Mass Spectrometry: A Tutorial, *Anal. Chim. Acta*, Vol 702, 2011, p 1
9. J.H. Gross, Direct Analysis in Real Time—A Critical Review on DART-MS, *Anal. Bioanal. Chem.*, Vol 406, 2014, p 63
10. M. Marić, J. Marano, R. Cody, and C. Bridge, DART-MS: A New Analytical Technique for Forensic Paint Analysis, *Anal. Chem.*, Vol 90, 2018, p 6877
11. S. Kabaria, I. Mangion, A. Makarov, and G. Pirrone, Use of MALDI-MS with Solid-State Hydrogen Deuterium Exchange for Semi-Automated Assessment of Peptide and Protein Physical Stability in Lyophilized Solids, *Anal. Chim. Acta*, Vol 1054, 2019, p 114
12. V. Bodnar, A. Ganeev, A. Gubal, N. Solovyev, O. Glumov, V. Yakobson, and I. Murin, Pulsed Glow Discharge Enables Direct Mass Spectrometric Measurement of Fluorine in Crystal Materials—Fluorine Quantification and Depth Profiling in Fluorine Doped Potassium Titanyl Phosphate, *Spectrochim. Acta B, At. Spectrosc.*, Vol 145, 2018, p 20
13. C.-K. Chiuhuang, C. Zhou, and H.-Y. Shadow Huang, In Situ Imaging of Lithium-Ion Batteries via the Secondary Ion Mass Spectrometry, *J. Nanotechnol. Eng. Med.*, Vol 5, 2014, p 21002
14. S. Harvey, Z. Li, J. Christians, K. Zhu, J. Luther, and J. Berry, Probing Perovskite Inhomogeneity beyond the Surface: TOF-SIMS Analysis of Halide Perovskite Photovoltaic Devices, *ACS Appl. Mater. Inter.*, Vol 10, 2018, p 28541
15. C. Parks, Comparative Ion Yields by Secondary Ion Mass Spectrometry from Microelectronic Films, *J. Vac. Sci. Technol. A*, Vol 19, 2001, p 1134
16. J.-J. Gaumet and G. Strouse, Electrospray Mass Spectrometry of Semiconductor Nanoclusters: Comparative Analysis of Positive and Negative Ion Mode, *J. Am. Soc. Mass Spectrom.*, Vol 11, 2000, p 338

ASM Handbook, Volume 10, *Materials Characterization*
ASM Handbook Committee
DOI 10.31399/asm.hb.v10.a0006663

Gas Analysis by Mass Spectrometry

Curtis D. Mowry, Russell L. Jarek, Jessica Román-Kustas, Amber C. Telles, and Adam S. Pimentel
Sandia National Laboratories

Overview

General Uses

- Qualitative and quantitative analysis of bulk and trace inorganic and organic compounds and simple mixtures, field measurements, process gases

Examples of Applications

- Gas purity, contaminants, and isotope ratios
- Process gas monitoring (industrial, petroleum, steel)
- Environmental gases (air, SO_2, CO_2, underwater, volcanic)
- Volatile organics (fermentation, pharmaceutical)
- Rapid analysis
- Continuous monitoring
- Internal atmospheres of sealed components
- Gases in inorganic and geologic materials

Samples

- Form: Primarily gases, but volatile liquid from headspace sample can be analyzed
- Size: 1×10^{-4} mL (STP) or larger
- Sample can be collected in a compatible vessel or directly sampled by the instrument from a flowing gas stream or sample environment

Limitations

- Target detection limits, measurement speed, and mass resolution must be matched with instrument capabilities
- Reactive gases are difficult to introduce intact and detect
- Complex mixtures might not be resolved
- Species must have some volatility and, for gas analysis, be less than ~500 amu

Estimated Analysis Time

- Milliseconds to minutes, not including sample preparation and calibration

Capabilities of Related Techniques

- Gas chromatography (GC): may require larger sample size
- Gas chromatography-mass spectrometry (GC-MS): greater capability for compound identification in complex organic mixtures
- Liquid chromatography-mass spectrometry (LC-MS): can measure larger molecular weight and less volatile species
- Infrared/Raman spectroscopy: requires larger sample size; infrared spectroscopy does not detect some inorganic gases

Introduction

Gas analysis by mass spectrometry, or gas mass spectrometry, is a general technique using a family of instrumentation that creates a charged ion from a gas phase chemical species (atomic or molecular) and measures the mass-to-charge ratio (*m*/*z*). Typical ionization and measurement occur in a vacuum, and, therefore, samples must comprise permanent gases or volatile species, where larger molecules must use alternative vaporization-ionization methods. Because each empirical molecular formula has a different mass, gas mass spectrometry has the potential to both identify and quantitate the components within a sample. This article covers gas analysis applications that do not use chromatographic separation to physically isolate components of the sample prior to analysis.

Typical applications of gas mass spectrometry fall into two categories: static and continuous. In static analysis, a measurement is made on a fixed or static gas volume, and a plot of signal intensity versus *m*/*z* value for each species in the gas is assembled by the instrument. In some cases, fragmentation occurs, which can complicate data interpretation. Static analysis is used for both qualitative and quantitative compositional analysis of individually captured samples. In the category of continuous analysis, specific *m*/*z* values are monitored (when analytes are known) or the instrument is scanned repeatedly across an *m*/*z* range. Output is typically a plot of analyte *m*/*z* versus time or total ion signal (per scan) versus time. Continuous analysis is used for applications such as reactions, industrial processes, and environmental monitoring and can be qualitative or quantitative.

This article is intended to provide an understanding of gas analysis instrumentation and terminology so informed decisions can be made in choosing an instrument and methodology appropriate for the data needed.

General Principles

Every gas mass spectrometer requires the same common functions to perform an analysis: sample introduction, ionization, mass separation, and detection (Fig. 1). Each function can be achieved using a variety of methods and techniques based on the target data goal and sample particulars. In addition to these common functions, general principles that should be considered in matching instrumentation to an analysis goal include, but are not limited to:

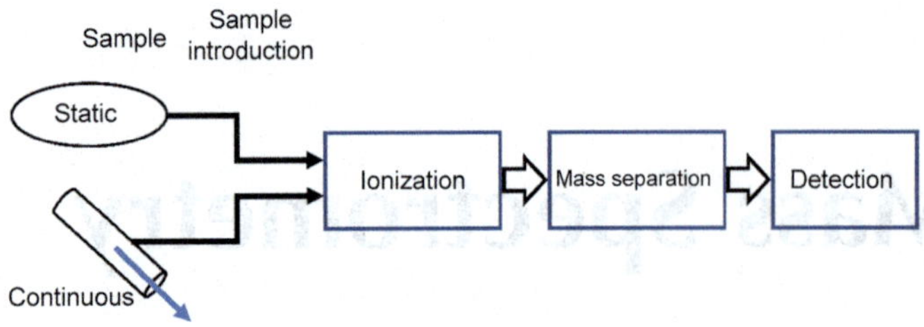

Fig. 1 Schematic of common instrument functions for a gas mass spectrometer

- It is important that the gas volume measured by the instrument is representative of the desired sample, or that the relationship is understood. Sampling bias can be an advantage or disadvantage.
- The complexity of the sample must be considered. A complex gas stream might require advanced sampling techniques to provide adequate data output.
- The mass range of the gases or potential targets should be properly estimated. This can affect the cost of the instrumentation needed.
- The measurement goal should be considered, in conjunction with complexity and mass range, in relation to the sampling frequency. For static measurements sampling frequency relates to the quantity of gas required, while for continuous measurements, sampling frequency relates to the time resolution needed. For example, a sampling frequency of one minute would likely be adequate for environmental monitoring, but not for a reaction monitoring application.

Considerations common to all instrument types are mass resolution, cost (and associated size, weight, and power), and robustness. These factors have a tremendous range in the instrumentation for gas mass spectrometry. Mass resolution is the numeric factor that defines the ability to differentiate between different *m*/*z* and, thus, different analytes apart. Sample complexity and data goals can determine whether an inexpensive, small, low-resolution instrument is suitable.

These general principles are overarching and apply regardless of system details. The following sections discuss some instrument options in the context of these principles to choose appropriate methodology to achieve the data goals.

Sample Introduction

The introduction system serves a common function among instruments; i.e., to bring sample gas into the ionization region of the mass spectrometer in a controlled quantity and controlled pressure. A wide variety of sample pressures ranging from vacuum to >100 psi (> 5000 torr) can usually be accommodated. These goals are accomplished differently for static and continuous sample configurations. In all cases, the mass analyzers require low ($<10^{-4}$ torr) pressures; thus, the fraction of sample gas being analyzed is often small.

Static gas introduction systems vary from a simple valve and effusive leak orifice to complex automated transfer systems that perform evacuation, dilution, and transfer. The primary consideration for choosing an introduction system is to preserve sample gas mixture ratios. Moisture and reactive gases can interact with tubing and sample vessels, therefore vessels and tubing are coated with an inert material ("passivated"). If transferring gas samples in a system below about 0.1 torr, note that as gases transition into a molecular flow regime, the gas mixtures can fractionate, with lighter/faster molecules moving ahead of heavier/slower ones; which requires some extra time to allow equilibrium to be established. Molecular leaks can also bias gas transfer due to mass-dependent flow and pumping efficiency, so care is taken with gas mixtures of diverse gases to characterize this effect if applicable. For quantitative work, it is usually necessary to know both pressure (P) and volume (V) of the sample gas; therefore, typical introduction systems often measure P and use a controlled V for the gas that enters the ionization region. Static gas measurements are typically used in situations where the sample gas volume is small (microliters to liter), or where pressure is low (few torr or lower), and where gas is emitted from a fixed size material (mineral, sealed component, etc.).

Controlling the quantity and pressure of sample gases are primary considerations for continuous gas introduction systems, which can vary from simple pulsed, fixed-volume valves to multistream samplers, capillary transfer lines, molecular leaks, of membrane inlets. Sampling frequency of continuous introduction must balance source/analyzer pressure requirements with the flow rate entering the instrument. For example, a long capillary inlet enables continuous measurement, but the capillary transit time delays the measurement with respect to the flowing sample. Membrane barriers between the sample gas and the ionization region can be beneficial, reducing the pressure load and/or providing chemical selectivity, such as in the monitoring of hydrocarbons within seawater. The membrane rejects the liquid while allowing hydrocarbons to pass. Therefore, the tradeoff consideration is that the membrane must be permeable to the analytes of interest. Membranes can also create a delay time due to analyte diffusion, which may be an important consideration in reaction and process gas monitoring.

Ionization Source

The ionization source converts the neutral analyte species into charged species, which can be manipulated using magnetic or electric fields for mass analysis. In research environments, ionization techniques use plasmas, lasers, and surfaces under vacuum or atmospheric pressure. Methods to create negative ions are mature, but are needed only in special applications like explosives detection and other niche areas in which the analytes have physical or electronic properties favoring negative ionization; this is outside the discussion of this article. Two methods in widespread use in the context of gas analysis, electron ionization (EI) and chemical ionization (CI), are discussed here. A specific variant of CI called proton-transfer reaction (PTR) is enjoying increased use (Ref 1).

Electron Ionization

Electron ionization is the most common method for creating positive ions of analyte species within the ionization region. Electrons are generated by passing current through a tungsten or rhenium filament and are accelerated and focused into a beam that intersects the cloud of neutral gas brought into the vacuum by the sample introduction system. A kinetic energy (ionization peak) of 70 eV, which has been adopted as the standard condition, creates consistent fragmentation of the analyte species, enabling spectral comparison across systems and generation of standard libraries for identification. In some cases, a lower energy can be used to reduce the observed fragmentation, but results in reduced ionization efficiency and, thus, reduced signal. The electron energy is sufficient to interact with neutral atoms and molecules, stripping an electron and resulting in a positive analyte ion as in the reactions:

$$\text{analyte} + e^- \rightarrow [\text{analyte}]^+ + 2e^- \quad \text{(Eq 1)}$$

and

$$\text{analyte} + e^- \rightarrow [\text{analyte}]^{2+} + 3e^- \quad \text{(Eq 2)}$$

Analyte fragmentation results in a charged fragment ion species (fragment) and neutral product (neutral) in the reaction:

$$\text{analyte} + e^- \rightarrow [\text{fragment}_1]^+ + \text{neutral}_1 + 2e^- \quad \text{(Eq 3)}$$

Because the electron cloud has a distribution of energies, other fragmentation pathways also occur (generating fragment_2, fragment_3, etc.). While there are some species that have relatively low ionization probabilities, EI is considered a universal ionization method.

Chemical Ionization and Proton-Transfer Reaction Ionization

The goal of CI is to reduce the probability of fragmentation by reducing the excess energy of the ionization process. The result can be increased signal as the analyte signal is no longer distributed among $[\text{analyte}]^+$ and some number of fragments $[\text{fragment}_n]^+$. Implementation of CI typically uses EI (sometimes other methods) to create a reagent ion, which interacts with the analytes to produce a charged analyte. Proton-transfer reaction (PTR) ionization is a specific version of CI that limits the reaction, as the name implies, to exclusively proton transfer as:

$$\text{analyte} + H_3O^+ \rightarrow [\text{analyte} + H]^+ + H_2O \quad \text{(Eq 4)}$$

Both CI and PTR provide the advantage of tuning this reaction using different reagents to provide selectivity. Data is simplified and sensitivity can be increased by ionizing a single species in a mixture. To capitalize on these advantages, CI and PTR are usually used for known targets. PTR instruments also use alternative reagent ions such as NO^+ and O^{2+} to achieve reaction selectivity for particular analyte targets.

Mass Analyzer

The mass analyzer manipulates the charged analyte species to determine their *m*/*z*. For applications and instrumentation described in this article, the objective typically is to generate a single positive charge. Mass-analyzer technologies commonly used for gas mass spectrometry include quadrupole mass filters (also called quad-MS), magnetic sector mass filters (also called sector-MS), and time-of-flight mass analyzers (also called TOF-MS). The first two are technically filters; i.e. they allow only one *m*/*z* value to be detected at a given moment and, therefore, are scanned across an *m*/*z* range to obtain a spectrum. Common factors to consider in choosing an analyzer for static or continuous gas measurement are mass range; resolution (*m*/*z*) and sensitivity; analysis frequency/scanning speed; and size, weight, and power (SWAP).

Mass Range

General-use systems generate primarily singly charged analytes (or fragments), so the mass analyzer needs to span only the atomic or molecular weight range of the targets. For example, an air (O_2, N_2, Ar) or CO_2 measurement only requires a range containing *m*/*z* from 28 to 44, a Xe isotope measurement requires a range of *m*/*z* up to 136, and a volatile organic compound analysis might need to analyze a range of *m*/*z* up to 250 or more.

Resolution (*m*/*z*) and Sensitivity

Resolution of a mass analyzer is the effective separation between two *m*/*z* values in relation to the width of the peaks (Fig. 2). The separation can be defined with respect to the height of the overlap between the peaks (valley) and is often related to the full width at half of the maximum height (FWHM) (Ref 2). The mass-resolving power can be defined for a single peak using its *m*/*z* value divided by width at FWHM, expressed as $m/\Delta m_{50\%}$.

Sensitivity refers to the concept of the number of ions detected relative to the number of atoms or molecules within a volume of the sample gas. Many factors contribute to the sensitivity of a given combination of sample introduction, ionization source, mass analyzer, and ion-detection components including gas transport, ionization efficiency, ion extraction efficiency, and ion transmission through the system. Because of the complexity of these factors, a clear understanding of the sample stream (or gas volume) and measurement requirements is required to evaluate whether a system has appropriate sensitivity.

Analysis Frequency/Scanning Speed

The frequency of measuring a particular *m*/*z* value relates directly to the mass range scanned (quads and sectors) or the pulse frequency of a times-of-flight (TOF) analyzer. This is important in static measurements, because measurements are desired at intervals faster than gas can change due to evacuation. Continuous gas measurements can be slow-changing process gases where one data point per minute might be suitable, but a reaction-monitoring process might require a data point per second. Instrument scan modes can vary by application, so it is important to resolve the desired data rate with the effective rate provided by the analyzer.

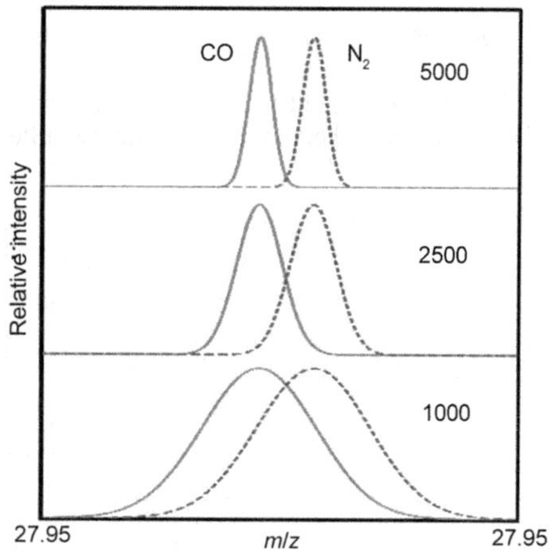

Fig. 2 Simulated peaks for nitrogen (N_2) and carbon monoxide (CO) at instrument mass resolving powers of 1000, 2500, and 5000. *m*/*z*, mass-to-charge ratio

SWAP (Size, Weight, and Power)

Analyzers used for gas analysis vary widely in size, weight, and power requirements, which are proportional to analyzer vacuum requirements, inlet system and somewhat proportional to mass resolution. Simple quadrupole analyzers can be benchtop (~0.02 m^3) and even small enough to use in underwater probes and unmanned aerial vehicles. Process gas sector instruments can be large (~0.2 m^3) due to industrial protections (temperature, vibration, fire); dual-sector instruments are typically even larger due to their electromagnet and vacuum systems. It is important to balance analyzer size, weight, and power restrictions with the data requirements. Types of analyzers discussed here include quadrupole, time-of-flight, and magnetic sector and double focusing.

Quadrupole Analyzer

The quadrupole mass analyzer consists of typically four (or greater even number) parallel rods shown schematically in the center of Fig 3. The rods are axially precision aligned, with opposing rods electrically coupled. An electric field is created by applying a dc potential bias and an RF oscillating potential between the pairs. The dynamic electrical field creates a stable trajectory for a particular *m*/*z* value such that it travels through the length of the rods from ion source to detector. At that potential, the trajectory of all other *m*/*z* values prevents them from reaching the exit of the rods (i.e., unstable trajectories). For this reason, the quadrupole mass analyzer is really an *m*/*z* filter. The dc and RF amplitudes are scanned so different *m*/*z* values exit at different times. "Jump" scans can also be performed to

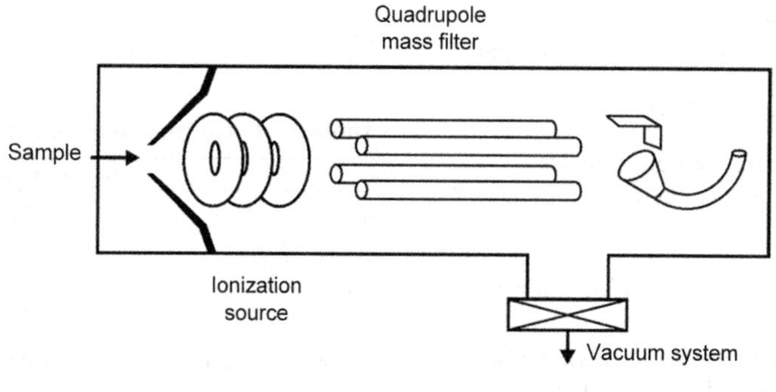

Fig. 3 Schematic of a quadrupole mass spectrometer with four parallel rods

reduce the time required for an analysis. Mass resolution is related to the amplitude of applied potentials, rod lengths, and rod radii and diameters. Ideal fields require hyperbolic rods, so there is a practical compromise in using simple rods between manufacturability and efficiency losses. Typical performance parameters are listed in the analyzer comparison table at the end of this section.

Advantages of quadrupole mass filters compared with other analyzers include:

- Relatively small size and cost
- Wide range of performance available
- Possibility to be configured with EI and CI ion sources

Disadvantages compared with other analyzers include:

- Slower scan speed
- Lower mass resolution

Time-of-Flight (TOF) Analyzer

Time-of-flight (TOF) analyzers consist of a field-free region into which all ions produced in the ionization source are introduced with constant kinetic energy by a pulsed voltage. Because kinetic energy is defined as KE = $\frac{1}{2}mv^2$, ions having different masses have different velocities. Therefore, each value of m/z analyte ion requires a different time to travel the distance of the flight tube (time-of-flight) (Ref 2). The difference can be in the nanosecond or larger time regime. Therefore, the TOF analyzer is a pulsed system (rather than scanning like the quadrupole), so the repetition rate depends on the quality of the electronics and the upper m/z to be analyzed. The repetition rate should not allow the slowest ion of pulse n be overtaken by the fastest ion of pulse $n + 1$. In practice, this still enables kHz acquisition rates; a rate where data analysis and data storage might be a consideration. Additional details and a schematic of this type of analyzer can be found in the article "Solid Analysis by Mass Spectrometry" in this volume.

Typical performance parameters are listed in Table 1. Because mass resolution is related to flight time, TOF analyzers are typically larger than quadrupole analyzers.

Advantages of TOF analyzers include:

- Nonscanning, rapid (kHz) acquisition
- Higher m/z resolution than quadrupoles

Disadvantages include:

- Larger in size than quadrupoles
- Typically not configured for permanent gas analysis

Magnetic Sector and Double Focusing Analyzers

The magnetic sector analyzer also functions as a mass filter. After ions are formed in the ionization source, they are equally accelerated and focused into a magnetic field. Because of their charge, the magnetic field exerts a force of magnitude $qv\mathbf{B}$ (the force direction being orthogonal to the velocity and magnetic field vectors); the product of the magnitude of the charge (q), the velocity (v), and the strength of the magnetic field (**B**). The force produces a curved flight path, which has a radius defined as:

$$r_m = \frac{mv}{q\mathbf{B}} \qquad \text{(Eq 5)}$$

This radius is created only in the x-y plane, but clever design of the magnet edges provides fringe focusing, which creates focusing in the z plane as well. At a given magnetic field, only one m/z value travels fully to the detector, functioning as a filter, sending other m/z values to the vacuum chamber walls (Ref 2).

A double-focusing magnetic sector analyzer adds a curved electrostatic analyzer as shown in Fig. 4. While there are other geometries and other tandem analyzer arrangements, the EB double-focusing configuration is the most common double-focusing type available for gas analysis.

The curved electrostatic analyzer applies an electric force (qE) that defines another radius of curvature according to the expression:

$$r_e = mv^2/qE \qquad \text{(Eq 6)}$$

This radius is related to the kinetic energy of the ions, thus becoming an energy filter, because only those ions satisfying Eq 6 travel fully to the magnetic sector for m/z separation.

The electrostatic sector produces a narrower ion energy distribution, which has the effect of removing ions that might have been indirectly generated near (but not within) the source's intended ionization region. Examples of indirect ions are those created by charge transfer, bimolecular ion-molecule reaction, and delayed ion dissociation/fragmentation. The energy-resolved nature of the ion beam entering the magnetic sector generates a cleaner dispersion, thus resulting in a higher mass-resolving power. In practice, both electrostatic and magnetic sectors are scanned together across the m/z range desired. Because both sectors create a physical dispersion for their filtering, physical size options are limited. Therefore, the double-focusing class are typically the largest form factor of the systems considered in this article.

Advantages of a double-focusing analyzer include:

- Quantitative results (1% relative or better)
- Much higher m/z resolution than typical quads

Disadvantages include:

- Large and expensive
- Scanning speeds can be slow

Typical performance parameters are listed in Table 1.

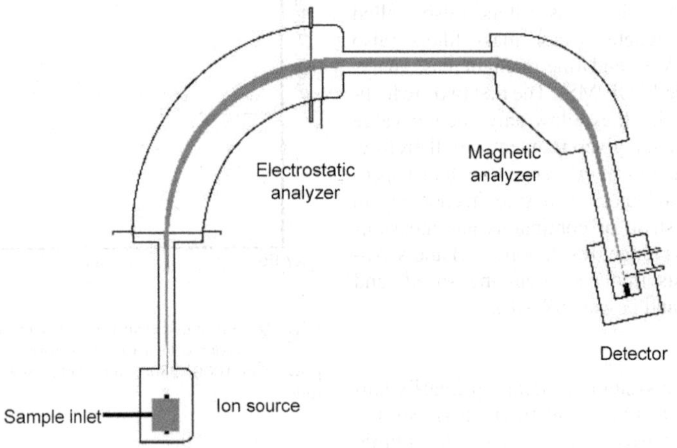

Fig. 4 Schematic of a double-focusing-sector mass spectrometer

Table 1 Performance parameters for analyzer types used in gas analysis

	Quadrupole mass filter	Time-of-flight	Magnetic sector	Double-focusing sector
Ionization mode(a)	EI, PTR	PTR	EI	EI, CI
Mass range	2–300	1–500	1–300	1–200
Mass resolution	100–3000	~12,000	60–5000	1000–60,000
Mass accuracy, ppm	10^2–10^5	10	<10	<5
Scan rate, amu/s	1–200	Nonscanning instrument	0.05–1	0.1–10,000
Detection limit	ppm–ppb	ppm–ppb	1 ppm	1 ppm
Dimensions (volume), m^3	0.02	0.44	0.2	2.7

(a) EI = electron ionization; PTR = proton transfer reaction ionization; CI = chemical ionization.

Ion Detection

The final stage of any mass spectrometer is a detector that can provide a current proportional to the number of ions entering. Often, some level of amplification is required to obtain the necessary sensitivity to enable useful detection limits (optimal signal to noise) for trace analysis. Even strong ion signals on the order of high pA to nA range use basic current amplifiers to put the signals into an easily measurable range to be digitized with accuracy. Ion detectors can have as much variation between each other as the mass analyzers discussed previously.

Faraday Cup

The Faraday cup is possibly one of the simplest ion detectors used in mass spectrometers. The Faraday cup, consisting of a conductive metal cup (element of the circuit) residing in the vacuum chamber, is designed to catch charged ions guided to the surface through the magnetic and/or electric fields produced by the mass spectrometer. The charged particles are neutralized due to electron flow within the metal. Upon striking the metal cup, electric current flow through the metal circuit can be measured and amplified. It is designed with an entrance such that nearly any incoming ion is neutralized within its boundary by wall collisions, and any secondary electrons produced from high-energy collisions are contained by a negative voltage bias at the Faraday cup entrance; loss of these electrons would falsely induce a larger ion current. The signal produced by wall collisions is directly proportional to the number of ions hitting the Faraday cup, where one charge is approximately 1.6×10^{-19} C (coulombs).

Faraday cups can also measure electrons in the vacuum. In those cases, a current is produced in the circuit when electrons hit the metal cup. One of the main advantages of Faraday cup ion detectors is its quantitative accuracy, where the number of ions can be directly calculated from the measured current, with a 1 nA current corresponding to six billion singly charged ions hitting the Faraday cup each second. However, Faraday cups have relatively low sensitivity, providing no direct signal gain like other ion detectors, such as electron multipliers.

Electron Multiplier

While more complex, electron multipliers have become one of the most common detectors in mass spectrometers due to high signal gain and relatively low additional noise compared with other ion detectors. Electron multipliers amplify incident charges to increase the amount of signal from one charged ion. The basic principle focuses on secondary electron emission, which can occur when an electron or charged ion strikes an electrode surface (dynode), typically an active layer on a metal plate, with enough kinetic energy to cause the release of secondary electrons. An avalanche of this process occurs when an electric potential is applied that accelerates the electrons toward another plate, secondary emission occurs, those electrons are accelerated to the next plate, and so on. The final metal plate is an anode that collects the cascade of electrons released, resulting in a current correlating to the electron or ion that triggered the cascade.

Compared with the Faraday cup, electron multipliers have significant advantages, including fast response and increased sensitivity due to the high signal gain and significantly higher signal-to-noise ratio for low intensity signals, thereby lowering some detection limits by several orders of magnitude. Utilizing gated pulse detecting electronics one can even perform discrete ion counting using such detectors. Disadvantages include potential amplification changes due to contamination of the dynode's active layer, analyte saturation of the anode, and mass discrimination caused by variation of the ion velocity. Mass discrimination often results in less signal gain for higher m/z. Channeltrons, or channel electron multipliers, are a type of electron multiplier with a funnel-shaped input aperture which captures the primary particle that impacts it and emits secondary electrons which further produce an electron cascade in a continuous avalanche down the channel.

Specimen Preparation

Gas sample preparation, transport, and storage can have a large impact on the quality of gas analysis, especially for determining low concentration level components (ppm or less). In general, there is more perturbation of the gas sample with static gas sampling than with continuous sampling. Continuously flowing sample gas of a consistent nature can passivate a transfer line such that the gas makeup is not affected in a measurable way while in transit. However, material of the transfer line could present some background signals, especially when heated (e.g., hydrogen from steel). Some of the best practices and pitfalls of gas sampling are presented here.

Samples from static gas sampling, because they are of limited quantity, are susceptible to contamination that is sufficient enough to affect the results of analysis. They also require containment within a vessel that can range from a classic glass bulb with stopcock, to a steel cylinder with valve welded or threaded on, to specially cleaned and inert-coated metal containers.

The most important consideration with any sample vessel (also referred to as a bottle) is material compatibility with the sample. Also of importance is the level of vessel cleanliness including all internal "wetted" surfaces, particularly the sealing material, which might not be metal but a polymer. Stainless-steel vessels, while common, have a varying effect on the contained gas composition. Certain reactive gases including acidic gases (HCl, HNO_3) and strongly oxidizing or reducing agents (O_3, H_2S) will not remain long in stainless steel cylinders. Other species could have surface-mediated reactions affecting composition distribution over time (isotopic mixing of labile hydrogens, and even of H_2 and D_2 forming HD). There also is the ever-present outgassing of hydrogen from the steel itself (as well as adsorption if the gas sample contains hydrogen), together with lesser amounts of CO and CO_2 depending on the level of cleanliness. Both reactions can be reduced by using a cylinder that has been either internally electropolished to reduce surface area, or a passivated steel (typically by formation of a chromium-rich oxide film) if warranted to reduce reactivity. Containers that have inert coatings can reduce or eliminate these effects. Regardless of the container used, the use of a blank sample bottle can provide insight on background gas levels building up within the bottle and possibly other components of the sampling system (vacuum chamber, transfer lines, and vacuum pump).

There are also common contamination issues that can affect a sample vessel including gas leaks (typically air) into a sample or a slow leak out of a pressurized sample and the presence of internal "virtual" leaks (tight internal volumes that only slowly transfer gas to/from the bulk sample). Cross contamination should also be considered if reusing sample containers. If a sample container is suspect or if trace-level analysis is desired, pretreatment of the sample vessel prior to use is recommended. Such treatment would typically consist of evacuation to high vacuum ($<10^{-6}$ torr, typical of a turbomolecular pumped system) preferably while being heated. If an enclosure or wrapping is not available, the use of a heat gun on metal surfaces helps to remove residual adsorbed water.

For continuous gas monitoring there are various equipment combinations available to bring the sample stream into the ionization source of the mass spectrometer. Various configurations can address common considerations to:

- Step down sample pressure to vacuum levels (sometimes with a membrane)
- Provide pressure adjustment and control
- Measure flow so the signal can be traced to a time-point in the process gas
- Provide the ability to transfer gas volumes from many different gas streams
- Reduce cross contamination with gas flushing and transfer line heating

Because most applications of continuous gas monitoring are in industrial settings, robustness toward the environment and sample stream should also be considered (temperature, reactivity of gases, etc.).

M/Z and Sensitivity Calibration

Quality mass spectrometer data relies on calibration and tuning with appropriate primary standards. Tuning enables determining response factors (sensitivity), mass accuracy, and closure ratio (calculated pressure to measured pressure), indicating overall instrument suitability. Tuning a quadrupole ion source and mass spectrometer is relatively simple and automated, typically using two mass peaks (e.g., nitrogen and argon, substituting hydrogen if low mass is desired). It optimizes maximum sensitivity, resolution, and peak shape with minimum mass discrimination. The TOF mass analyzer can have peak shape tuned in the middle of its relevant mass range, but a broader mixture of the gas mass range should be used. Detailed tuning of sector instruments is rarely needed as it primarily involves adjusting the ion source acceleration, extraction, and focusing voltages to attain good peak shape and intensity. Sector instruments and quadrupole analyzers have very good mass stability and might only require mass calibration once per week, while the TOF could require more frequent tuning as it is based on the accuracy of multiple electronic timing signals. Spiked internal mass calibration is used for high-resolution instruments if the highest mass accuracy is required for the analysis. An accurate mass is helpful in determining the identity of the component(s) when working with unknown compositions.

It is necessary to calibrate a mass spectrometer's response factors (or sensitivity factors) and confirm the linearity of that response to obtain quantitative analytical results. Signal responses are typically based on a maximum peak height, and in sector instruments run at lower resolutions, peak shape is ideally an isosceles trapezoid (a flat peak top), which results in a high-precision response. A set of gaseous compounds representative of permanent gas components analyzed often typically includes H_2, N_2, O_2, CO, CO_2, and He. Use standards that vary gas partial pressures within the mixture, preferably ranging from near the lower detection limit to the upper limit of the detector.

Knowing the gas concentration in a standard mixture often is the limiting factor to quantitative accuracy for sector instruments. If a mixed gas standard is used, it is highly recommended to obtain more than one mixture so they can be cross checked against one another (supplier reliability varies greatly regardless of the certificate). A better alternative, while less convenient, is to perform calibrations (or occasional cross checks) from pure gases, where the gas purity can be initially confirmed by simple absence of any significant other species.

A problem with using pure gas standards is that the pressure required that will not saturate the detector is usually quite low and below the discrimination of the inlet pressure gauge (or one can use a spinning rotor gauge). One method to overcome this situation is to use a pressure-volume division, or divider, within the inlet system, which produces a constant fractional pressure drop with every repeatable cycle. Multiple cycles are used to extend from where the pressure is measurable (to obtain the divider factor itself) to below gauge resolution, but at desired signal levels and pressure is calculated using the measured divider factor. This same method can be extended to determine/confirm response factor to gas pressure linearity.

Note that the mass calibration mixture need not be a well-quantified mixture as this tuning is distinct from sensitivity calibration to be discussed.

Data Analysis and Reliability

Interpretation of gas analysis mass spectrometer data can be complicated. As indicated previously and in the article "Solid Analysis by Mass Spectrometry" in this volume, ionization sources can introduce multiple types of species to the mass spectrometer from a single gas species, including fragments, multicharged species, and reaction products if the source pressures are high. Stemming from the ideal gas law:

$$\rho = p/kT, \text{or density} = \text{partial pressure}/(\text{Boltzmann constant})(\text{temperature}) \quad \text{(Eq 7)}$$

signal peak heights (and sometimes areas) are proportional to the concentration of gas within the sample.

Mass spectrometry reliability depends heavily on tuning, mass calibration, and gas introduction. Peak shape, resolution, and sensitivity are important in selecting the appropriate mass spectrometer technique for a given application.

To achieve precise quantitative measurements in gas analysis by MS, it is important to consider both sample- and instrument-related factors. Sample-related factors include ionization efficiencies, concentration, preparation, and detection interferences. Instrument-related factors include transmission and detection efficiency, contamination, source parameter tuning, and calibration.

A common difficulty is the presence of contaminant species, and a concern of whether such potential isobaric (same unit mass) signals can be resolved; i.e., mass resolution must be sufficient to discriminate the potential interference peak if present (Fig. 2). Use of the naturally occurring isotopes can sometimes provide further insight on the presence or absence of a convoluted spectrum (Ref 3). For example, it is possible to discriminate the mass difference between a molecule containing sulfur with one containing two oxygen atoms (e.g., CS and CO_2 with a Δm = 0.018 amu) using a resolution of 2500. It can easily be achieved using most sector and TOF MS if mass stability is on the order of 0.001 *m/z*. However, there is doubt as to its precise mass using a relatively low-resolution quadrupole or if only one peak is present; finding the ~4% isotopic peak for $C^{34}S$ can confirm its identity (CO_2 has a much smaller isotopic signal of ~0.4% for $CO^{18}O$). In similar ways, the fragmentation of a parent species can help elucidate the composition of that parent ion.

Applications and Interpretation

Common applications relate to the factors of analysis speed, *m/z* resolution, size, and cost discussed previously. For a particular application, these factors are not equally weighted, and there is significant overlap among the capabilities of the different analyzer types. The systems covered here do not use separation schemes, and therefore can require more complicated data interpretation when multiple species are ionized at the same time. The more complex the sample stream, the more complex the data generated. This can be exacerbated with EI sources, which create potentially many fragment signals for each species in the sample stream; yet simplified in CI/PTR sources where a single *m/z* can be produced per species.

Quadrupole-based systems can be small, rapid, and relatively inexpensive, and so are used in portable applications such as environmental monitoring (e.g., pollution, volcanic emissions, and in-situ seawater contamination).

Sector-based systems are stable but slower, and are used for many process measurement scenarios including gas monitoring (O_2, CO_2, H_2, NH_3, and CH_4) in fermentation, iron and steel production, and natural gas; and monitoring larger molecules (hydrocarbons and volatile organics and solvents) in pharmaceutical and polymer production. In combination with array detectors, they are ideal for precise, simultaneous measurement of isotopic ratio distributions.

TOF-based systems in the context of gas analysis (typically PTR ionization source-based) are similar in speed and more sensitive, but much larger compared with quadrupole systems. Higher resolution and a simplified (one *m/z* per species) data stream makes these systems ideal in more complex sample streams including complex environmental applications, breath monitoring, and fermentation.

Dual-focusing sector systems are typically used in the highest precision applications where analysis speed is not the main constraint. Small gas volume and high precision requirements (quantitation and *m/z* resolution) make dual-focusing sector systems the best choice. However, the precision necessitates higher costs and instrument complexity. Applications include medical and geological isotopic measurements.

The capabilities of the different analyzer types are illustrated in the following examples.

Example 1: Quantification In An Atmospheric-Pressure Plasma Jet Using A Quadrupole MS (Ref 4)

A molecular beam quadrupole mass spectrometer and EI was used to monitor ambient air transport through an atmospheric-pressure plasma jet. The application required the rapid measurement of multiple analyte species and low detection limit capabilities of a quadrupole mass spectrometer.

Sample Preparation

The effect on ambient air species around the plasma jet required no sample preparation. However, it required optimization of the plasma jet and mass spectrometer orifice positions in relation to each other. Due to visible contact between the orifice, plasma emission, and effluent (active plasma flame), interferences were avoided by turning off the plasma jet at separation distances less than 10.5 mm.

Data Reduction

Several factors had to be considered to obtain accurate, reliable analysis of ambient air species in the active plasma region of the atmospheric-pressure plasma jet:

- The technique had to be able to measure multiple analyte species rapidly. A molecular-beam mass spectrometer (MBMS), quadrupole, was used to measure multiple analyte species inside and around the plasma jet. Rapid analysis was possible by scanning a small range for ambient gases.
- It was necessary to calibrate neutral MBMS intensities to correlate them with the different partial pressures of each analyzed particle due to composition distortion effects; i.e., radial diffusion effect of lighter particles versus heavier particles. This was achieved using an airtight, well-defined chamber in front of the MBMS systems with dry compressed air passed through to calibrate ambient air species.
- It was necessary to avoid contact between the plasma jet and mass spectrometer to prevent interferences, as described previously in sample preparation. Therefore, the plasma jet was switched off when closer than 10.5 mm to the orifice of the mass spectrometer.

Data Acquisition

It was determined through calibration of ambient air species (argon, nitrogen, and oxygen) that composition distortion effects were minimal; therefore, intensities and absolute densities were obtained through Eq 7. On-axis measurements of the ambient air species transported through the plasma jet showed different compositions at mass spectrometer distances closer to the jet nozzle versus farther from the nozzle; i.e., argon was higher near the nozzle, but nitrogen and oxygen were higher at greater distances. Figure 5 shows the spectrum observed, with many species being produced inside the plasma jet due to either impurities in the feed gas or formation in the active effluent of the plasma jet through molecular collisions and/or recombination of air species. This includes air fragments, such as O^+, which illustrates the fragmentation occurring in the plasma jet and/or the EI.

Example 2: In Situ Volcanic Plume Analysis Using A Quadrupole MS (Ref 5)

Previously inaccessible, unobtainable data became possible with the development of unmanned aerial systems (UAS), such as miniature mass spectrometers. In this example, both species detection and abundance were determined from in situ volcanic plume analysis. The application required the capability of quadrupole mass spectrometers, including the ability for miniaturization (small and light), large dynamic ranges, and sensitivity.

Sample Preparation and Data Reduction

In situ analysis is the analysis of a material in its original form and position, which requires no sample preparation. However, extensive development of the UAS platforms was required including:

- Integration of a mass spectrometer into small UAS platforms capable of obtaining a 3D chemical map of the volcanic plume. Prior to this development, ground-based, fixed sensors were used to monitor volcanic activity. The UAS-MS systems consisted of a mass analyzer capable of bulk and trace gas analysis and a low-power, miniature scroll vacuum pump.
- A multigas sensor capable of determining chemical species and a large dynamic range. Mass spectrometers can be both qualitative and quantitative, being able to detect components at concentrations less than one ppm. In this case, the appropriate mass analyzer required sensitivity ranging from ppm to 100% concentration, resolution of less than 1 amu between the mass range of 1 to 65 amu, and scan speeds below 2 s/scan. In addition, the small mass range enabled faster scan rates.

Data Acquisition

Prior to analysis of volcano plumes (2014), UAS-mass spectrometry platforms were designed using a small quadrupole mass spectrometer with a direct gas inlet and a sub-ppm limit of detection. Targeted molecular gas species included He, H_2, O_2, CO_2, SO_2, H_2S, and H_2O. Mass spectrometer calibration was accomplished in a laboratory prior to field deployment using calibration mixtures consisting of different compositions of gases, as well as different ppmv (parts per million by volume) and pressure control. Demonstrating sub-ppm limits of detection, good linearity, and high dynamic range, the UAS-mass spectrometry platforms were deployed to multiple volcanic events for plume characterization. Spectra of in situ, direct MS sampling (Fig. 6) of the Las Hornillas, Bocca Nuova, and Bocca Grande volcanic plumes following calibration in multiple-ion mode, showed CO_2 concentrations well above 50%, while H_2S and SO_2 varied more dramatically between each other, as well as the different volcanoes. Advances in mass spectrometer sizes and sensitivity have made the in situ, unmanned analysis of harsh environments possible, even potentially satellite-based remote sensing data.

Example 3: TOF-MS For Fermentation Monitoring (Ref 6)

A PTR ionization source and TOF analyzer was used to monitor the VOC (volatile organic compound) profile during fermentation of beer.

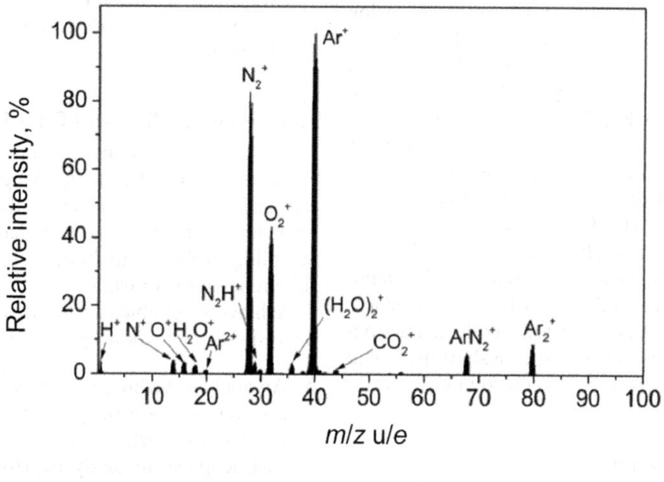

Fig. 5 Positive ion measurements from a quadrupole mass spectrometer illustrate the variety of transported species in ambient air. *m/z*, mass-to-charge ratio. Source: Ref 4

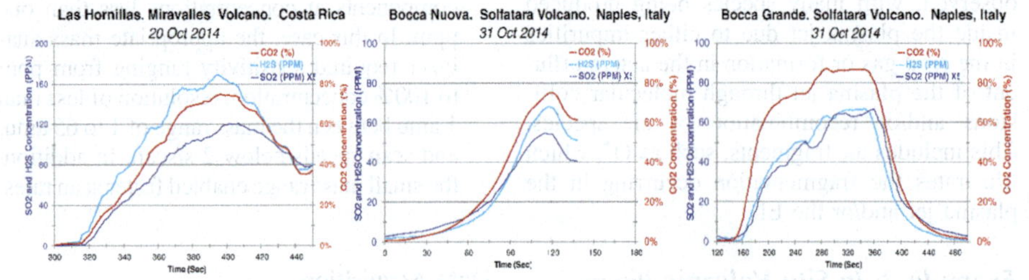

Fig. 6 Volcanic plume analysis of CO_2, H_2S, and SO_2 from in situ direct mass spectrometry sampling of three different volcanic plumes. Source: Ref 5

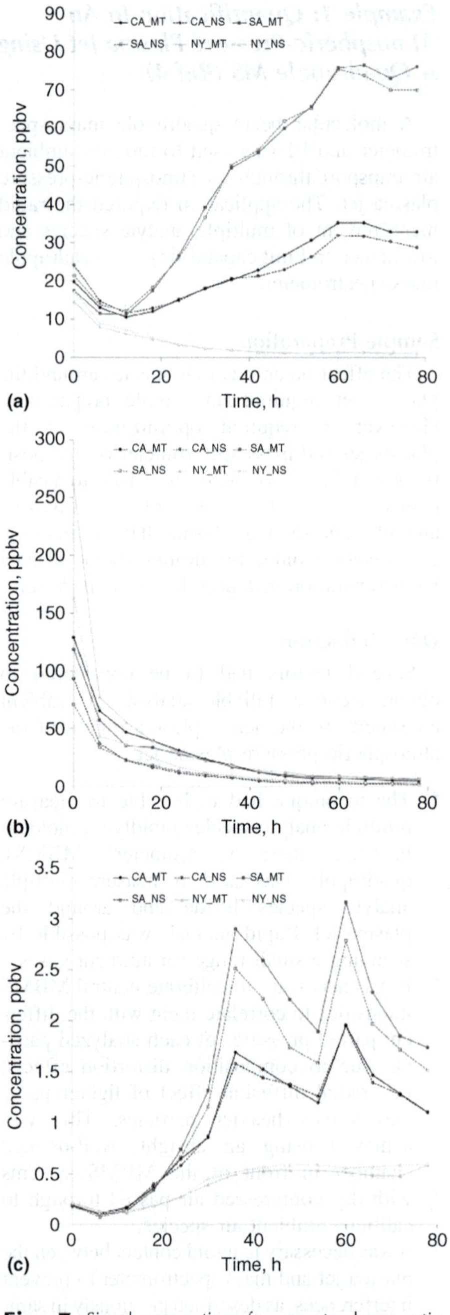

Fig. 7 Signal versus time for mass-to-charge ratio (*m/z*) values of (a) 89.057, (b) 137.131, and (c) 173.153 show different trends based on yeast (CA, California ale; SA, Scottish ale; NY, no yeast) and hops (MT, Motueka; NS, Nelson Sauvin) used in the fermentation of beer. ppbv, parts per billion by volume. Source: Ref 6

The range of concentrations and number of species in the vapor phase above the fermentation broth is complex, and PTR (through soft ionization) provides simpler data than would EI. The high mass resolution of the TOF enables potential identification of any volatiles detected.

Sample Preparation

The goal of the study was to measure volatiles produced under different conditions of yeast strain and hop cultivar, and to monitor the volatile profile over time (four days). A wort mixture was generated through brewing processes involving heating, cooling, and adding ingredients at specified intervals. Each variant treatment was transferred to a headspace vial (20 mL glass with septum cap) for long-term fermentation at 20 °C (68 °F). An autosampler robot transferred a volume of gas from the vial directly into the instrument.

Data Reduction

The signal is from protonated analytes, $[A + H]^+$, which would most likely be polar species such as alcohols, ketones, etc. Principal component analysis and advanced processing of the signal was required to determine correlated and statistically significant differences among the sample set. Even with high mass resolution, specific signals were compared with other fermentation literature (and methods using separation techniques such as GC/MS) to attempt identification or at least classification. For example, three *m/z* values are plotted versus time in Fig 7. Tentative identification for these ions, based on isotope signals and prior fermentation literature, was assigned as: *m/z* 89.057 ($C_4H_8O_2$: acetoin or ethyl acetate), *m/z* 137.131 ($C_{10}H_{16}$: β-pinene, myrcene, α/γ-terpinene, or limonene), *m/z* 173.153 ($C_{10}H_{20}O_2$: isoamyl isovalerate, ethyl octanoate, 2-methylbutyl-2-methyl-butyrate, decanoic acid, or pentyl pentanoate). While it may not seem to be very specific, such a short list would not be possible using a lower resolution analyzer or EI ionization source.

Data Acquisition

At set intervals of six hours, the headspace was flushed with nitrogen, then sampled and 20 spectra averaged. The autosampler allowed for the same time intervals for each variant, and rapid data collection would allow for only minutes or less time between variant measurements. The variants shown are combinations of yeast (CA = California ale, SA = Scottish ale, NY = no yeast) and hops (NS = Nelson Sauvin, MT = Motueka). The signals in Fig. 7 show three types of VOC emission during fermentation: increase over time, decrease over time, and lifecycle variation. Note that the concentration axis is in parts per billion and the time axis is hours. Concentration was calculated using a PTR rate coefficient rather than individual response factors, which is not possible due to the complexity and lack of absolute identification. The data is still very useful for purposes of comparison and contributing to the understanding of the fermentation process.

Example 4: Sector MS For Steel Process Gas Monitoring (Ref 7)

Performed under vacuum, many steel production processes need continuous gas analysis to monitor the composition of the exhaust gas. Detecting variations during the process identifies materials out of specification prior to completion of the process. Using a magnetic sector mass spectrometer illustrates the increased accuracy and precision of magnetic sectors compared with most quadrupoles.

Sample Preparation and Data Reduction

Due to high amounts of dust produced in the materials for the steelmaking process, the integration system connecting the mass spectrometer to the process included two heated sampling probes with built-in filters to clean the probes prior to use.

Achieving reliable, accurate spectra for the analysis of steelmaking processes requires:

- An analysis technique capable of analyzing a system under a range of pressures. Previous methodologies were confined to atmospheric pressure analysis. However, many modern steelmaking processes, including basic oxygen furnaces and electric arc furnaces, have dramatic pressure fluctuations, including down to <0.75 torr. Mass spectrometers operate under vacuum and, therefore, are ideal for the analysis of vacuum processes, but it is important that the pressure within the mass spectrometer remains constant. Therefore, in this case, a variable pressure inlet was used that has two valves, which work in opposition to control the analyzer pressure.

- An ability to analyze inert gases with signal stability. The ability of mass spectrometers, such as magnetic sectors, to separate ions based on *m*/*z* enables monitoring of multiple inert gases rapidly. In addition to having slow response, previous experimental methodology required the use of at least three types of non-mass spectrometry analyzers to monitor inert gases and their concentrations.
- Integration of the mass spectrometer to the steelmaking processes, including multiple components discussed previously in the sample preparation section.

Data Acquisition

Analyzing furnace exhaust gas compositions provides information on the development process and defects can be detected prior to process completion. In the case of decarburization, the species of interest were N_2, CO, CO_2, O_2, and Ar. All of these species are below 50 amu. Figure 8 shows an example of a vacuum oxygen decarburization (VOD) real-time monitoring spectrum where rapid changes in composition can be seen. This illustrates the importance of the fast analysis provided by the mass spectrometer. The process pressure also illustrates the variation in pressure, for which many other analyzers would be unable to compensate. With the precision of the magnetic sector (between 2 and 10 times that of a quadrupole) combined with the speed of analysis (only needing to be 200 amu/s) and the inlet pressure control, this work illustrated the utility of mass spectrometers in monitoring the steelmaking process.

Example 5: Double-Focusing Sector MS For Isotopic Xe Gas Measurement (Ref 8)

Double-focusing magnetic sector instrumentation has the largest physical footprint of the gas-analysis mass spectrometers discussed in this article. However, this type of instrument satisfies data requirements in precision applications such as isotope analysis. The abundance of different isotopes of an element can vary by orders of magnitude, and the overall abundance of some isotopes can be very small (<0.1% relative to the highest abundant isotope). Small changes in relative abundance can be meaningful, so the high stability, precision, and resolution provided by dual-focusing instrumentation and multicollector configurations is desirable. This example illustrates precise quantitative comparison between xenon isotopes in the study of earth's formation, using static gas samples of limited volume.

Sample Preparation

Gases and radioactive elements are trapped as molten rock cools and solidifies, with subsequently generated gases created by radioactive decay also being trapped. Xenon has nine stable isotopes, five of which are produced from radioactive decay of other elements. Therefore, the ratio of radiogenic ^{129}Xe and non-radiogenic ^{130}Xe isotopes changes with the age of the rock and how much of the parent atom ^{129}I (iodine) it originally contained. Because Xe is within the rock, a "step-crush" method is used inside the vacuum of the source region of the mass spectrometer. As the rock is crushed, additional gas is released, and another measurement is made. In some cases, the gas is filtered (for reactive gases) and cryo-trapped for controlled introduction into the source region. Care is taken to minimize air contamination of the sample and source.

Data Reduction

To account for potential mass discrimination issues or sensitivity variations that could affect the precision needed, ratios between signals are often used rather than absolute signal intensity. Also, calibration or reference gases can be introduced alternately with sample gas. It is important to use controls and standards to separate instrumental artifacts and detection drift from the variations present in samples. Sample preparation method, sample inlet, ionization methods, and detector sensitivity can all affect the final ratio of isotopes measured, together referred to as mass discrimination, so care is taken to determine instrument and method correction factors. The $^{129}Xe/^{130}Xe$ ratio change plotted for the Icelandic sample represents only a 6% change in the fractional abundance ratio for $^{129}Xe/^{130}Xe$ where ^{130}Xe has only 4% natural abundance, thus highlighting the importance of precision potentially needed in isotope ratio measurements.

Data Acquisition

In this example, a multicollector detector is used where multiple xenon isotopes are measured simultaneously. This type of detector eliminates some of the systematic instrumentation mass discrimination effects mentioned above. Measurements are made on many replicates of gas escaping from the rock matrix and span a range of isotopic ratios (partial pressures). Figure 9 demonstrates that mixing of mantle-gas sources with air is a linear relation. The slope of the line created by the ratios of radiogenic to non-radiogenic isotopes provides information about the primordial gas source and early mantle mixing. The two slopes observed show that the molten rocks were geochemically distinct and remained separated 4.45 Gyr when mantle solidification was complete and had not undergone theorized whole mantle convective mixing.

ACKNOWLEDGMENTS

Sandia National Laboratories is a multimission laboratory managed and operated by National Technology and Engineering Solutions of Sandia, LLC, a wholly owned subsidiary of Honeywell International, Inc., for the U.S. Department of Energy's National Nuclear Security Administration under contract DE-NA0003525.

The authors thank Anthony Esparsen, GlowMonkeyStudio, for schematic creations.

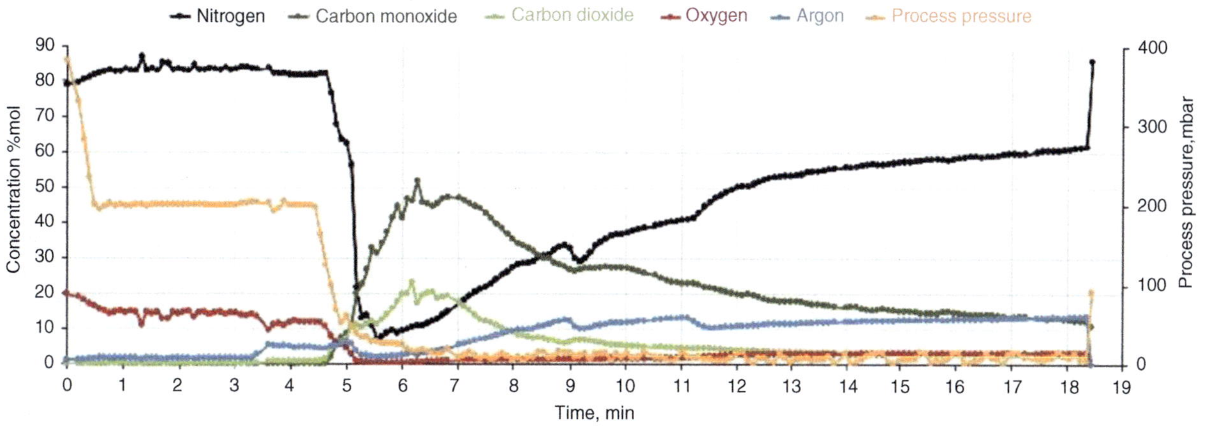

Fig. 8 Typical real-time monitoring spectrum for the vacuum oxygen decarburization steelmaking process illustrates the capability of the mass spectrometer to show the rapid changes in composition. Source: Ref 7

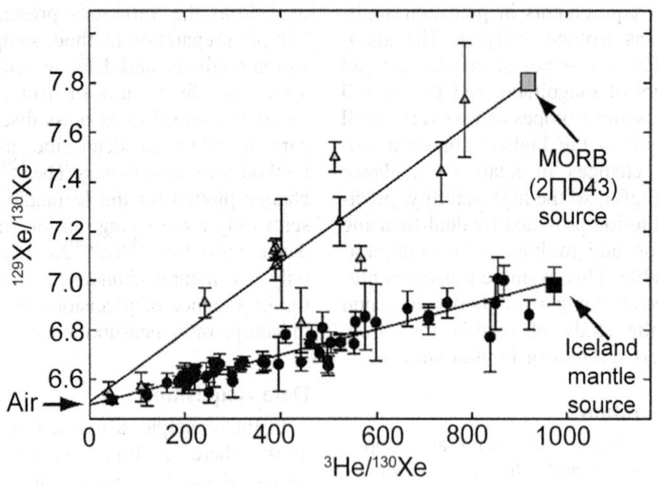

Fig. 9 Plot of isotope ratios for two basalts in geochronology study. MORB, midocean ridge basalt. Source: Ref 8

Revised from Q.G. Grindstaff, J.C. Franklin, and J.L. Marshall, Gas Analysis by Mass Spectrometry, *Materials Characterization*, Vol 10, *ASM Handbook*, ASM International, 1986.

REFERENCES

1. R.S. Blake, P.S. Monks, A.M. Ellis, Proton-Transfer Reaction Mass Spectrometry, in Chem Rev, vol. 109, pp. 861-96, 2009.
2. J.T. Watson, O.D. Sparkman, Introduction to Mass Spectrometry: Instrumentation, Applications, and Strategies for Data Interpretation, Wiley, 2013.
3. T.R. Ireland, Invited Review Article: Recent Developments in Isotope-Ratio Mass Spectrometry for Geochemistry and Cosmochemistry, in The Review of Scientific Instruments, vol. 84, pp. 011101, 2013.
4. M. Dünnbier, A. Schmidt-Bleker, J. Winter, M. Wolfram, R. Hippler, K.D. Weltmann, S. Reuter, Ambient Air Particle Transport into the Effluent of a Cold Atmospheric-Pressure Argon Plasma Jet Investigated by Molecular Beam Mass Spectrometry, in Journal of Physics D: Applied Physics, vol. 46, pp. 435203, 2013.
5. J.A. Diaz, D. Pieri, K. Wright, P. Sorensen, R. Kline-Shoder, C.R. Arkin, M. Fladeland, G. Bland, M.F. Buongiorno, C. Ramirez, E. Corrales, A. Alan, O. Alegria, D. Diaz, J. Linick, Unmanned Aerial Mass Spectrometer Systems for in-Situ Volcanic Plume Analysis, in J. Am. Soc. Mass Spectrom., vol. 26, pp. 292-304, 2015.
6. T.M. Richter, P. Silcock, A. Algarra, G.T. Eyres, V. Capozzi, P.J. Bremer, F. Biasioli, Evaluation of Ptr-Tof-Ms as a Tool to Track the Behavior of Hop-Derived Compounds During the Fermentation of Beer, in Food Res Int, vol. 111, pp. 582-89, 2018.
7. G. Lewis, Prima PRO Process Mass Spectrometer: Improving Low Carbon Steel Production in Specialty Steel Processes, Thermo Fisher Scientific Inc., Application Note AN_Steel_0618, 2018.
8. S.J.N. Mukhopadhyay, Early Differentiation and Volatile Accretion Recorded in Deep-Mantle Neon and Xenon, in Nature, vol. 486, pp. 101, 2012.

ASM Handbook, Volume 10, *Materials Characterization*
ASM Handbook Committee
DOI 10.31399/asm.hb.v10.a0006648

Glow Discharge Mass Spectrometry

Cristina Gonzalez-Gago, Nerea Bordel, and Jorge Pisonero, University of Oviedo

Overview

General Uses

Direct solid analysis:

- Bulk analysis
- Depth profiling

Examples of Applications

- Determination of impurities in precious metals
- Quantification of trace elements in high-purity metals, alloys, and semiconductors
- Bulk and depth profiling of high-purity materials (e.g., solar-grade silicon)
- Depth profiling of thin and ultrathin layers (e.g., hard disk)

Samples

- Bulk homogeneous samples
- Multilayered conductive and nonconductive materials
- Pressed powder

Limitations

- No or poor lateral resolution (approximately millimeters)

Estimated Analysis Time

- Sample-preparation time of a few minutes (no need for thin-film depth profiling):
  - Polishing for bulk samples
  - Pressing for powders
- Typical analysis time from seconds to tens of minutes

Capabilities of Related Techniques

- *Spark source mass spectrometry (SSMS):* Allows quantitative multielemental analysis with high sensitivity. It is limited by poor precision.
- *Glow discharge optical emission spectroscopy (GD-OES):* Used for bulk and depth-profile analysis. It provides lower sensitivity than glow discharge mass spectrometry (GDMS).
- *Secondary ion mass spectroscopy (SIMS):* Provides structural chemical information with high spatial resolution (down to nanometer resolution using nano-SIMS). It requires ultrahigh-vacuum conditions, and the analysis time is long (hours).
- *Laser ablation inductively coupled plasma mass spectrometry (LA-ICP-MS):* Operates at atmospheric pressure, providing high elemental sensitivity and lateral resolution (approximately micrometers). Determination of light elements such as hydrogen, carbon, nitrogen, and oxygen is restricted. Depth resolution is conditioned by the characteristics of the laser used.

Introduction

Glow discharge (GD) ion sources have been largely investigated since the early days of mass spectrometry at the beginning of the 20th century (Ref 1–3). Moreover, they have been coupled to various mass analyzers, including quadrupole, sector field, and time of flight (Ref 4, 5). Originally, the rapid development of glow discharge mass spectrometry (GDMS) was partly motivated by the need to improve the handling capabilities and relatively poor precision (>10%) of spark source mass spectrometry. Currently, the advantageous features of GDMS include high stability and sensitivity, simplicity of operation, low matrix effects, low gas consumption, and short analysis times, which have contributed to establish GDMS as a reference technique for direct determination of trace elements in solid samples and for fast depth profiling in a great variety of innovative materials.

General Principles

A low-pressure GD is produced in an inert gas, normally argon, by the application of a potential difference (above a threshold value) between two electrodes: the cathode (the sample to be analyzed) and the anode (the GD cell body). After breakdown of the plasma gas, ions are accelerated toward the sample surface, producing the sputtering of the sample atoms (e.g., removal of analyte species) and the emission of secondary electrons that sustain the plasma. For analytical applications, GDs are operated in the abnormal current-voltage mode, where applied voltage is proportional to electrical current. After the sputtering process, neutrals from the sample diffuse into the negative glow region of the GD, where they undergo various excitation/ionization collisions. Here, it should be highlighted that the temporal and spatial separation of the sputtering and ionization/excitation processes results in low matrix effects in GD spectroscopies. Three main ionization processes occur in

GDMS, including electron ionization related to collisions with high energetic electrons (considered to be the main process responsible for the ionization of the discharge gas atoms); asymmetric charge transfer between ions of the discharge gas and sputtered species, which is a selective ionization mechanism only possible if the sputtered atoms have ionic excited levels with similar energy to the ionization potential of the discharge gas (Ref 6); and Penning ionization, which results from the interaction between plasma gas metastable species and the sputtered atoms. (This process is considered to be a nonselective mechanism because it is possible to ionize elements with ionization potentials lower than the energy of the plasma gas metastable levels, e.g., 11.55 and 11.72 eV in the case of argon metastable species, respectively.)

Glow discharge can be generated by application of direct current (dc), but its use is limited to the analysis of conducting and semiconductive samples. Nevertheless, nonconductive samples may be analyzed by using various approaches, including the use of secondary cathodes, sample mixing with conductive powders, or deposition of thin conductive coatings on the sample (Ref 7, 8). The use of radio-frequency-powered GD sources opens the possibility of directly analyzing nonconductive samples, because the electrodes are polarized by the fluctuating voltage. The analytical applications of GD could be further extended by employing the pulsed-power mode. This method allows the application of higher instantaneous power in short periods (milliseconds or microseconds), which results in higher ionization efficiencies without increasing the thermal stress. Moreover, pulsed GD results in the formation of a dynamic plasma with various temporal regions (e.g., prepeak, plateau, afterglow), which are characterized by the predominance of different excitation/ionization mechanisms.

Systems and Equipment

A typical GDMS system consists of a GD ion source coupled with a mass analyzer and a detector(s).

Ion Source

At present, there are two fundamental types of ion sources used for the GDMS: the low-pressure or static source, and the fast-flow sources. Both types can accept pin samples or samples with a flat surface (Ref 9). Typical pin samples have a length of approximately 10 to 20 mm (0.4 to 0.8 in.) and a diameter of approximately 3 mm (0.12 in.), while flat samples usually must have at least 20 mm (0.8 in.) of diameter. Smaller or wider samples, as well as curved surfaces, also may be analyzed by using special sources or adaptors.

In the static source, the argon flow rate is approximately 1 mL/min, employing argon of very high purity (e.g., 99.9999%). In this case, the GD cell is a sealed unit held within a vacuum chamber (samples are introduced using an insertion probe) and with a small hole or slit, allowing the ions to exit the GD cell and to be subsequently accelerated toward the mass analyzer. The typical potential difference between sample (cathode) and GD cell (anode) is approximately 1 kV, while current is approximately 2 to 3 mA. Therefore, plasma power is relatively low (e.g., approximately 2 to 3 W). In addition, the static source usually is cooled by contact with a heat-exchanger unit through which liquid nitrogen (LN_2) flows. This cooling reduces the formation of molecular species (e.g., hydrides, carbides, nitrides, and oxides) and also allows the analysis of low-melting-point and volatile materials (e.g., gallium, arsenic).

In the fast-flow source (modified-Grimm hollow anode source containing a flow tube, Ref 5–11), the argon flow rate is in the range of 100 to 500 mL/min, employing argon of high purity (e.g., at least 99.999%). The potential difference between anode and cathode is typically between 800 and 1200 V, and the current ranges from 30 to 60 mA. In this source, cooling and reduction of temperature variations, which may affect ion signals, are achieved by Peltier cooling or circulating fluid from a chiller. Fast-flow design does not exhibit as many molecular species, due to the higher power (50 to 100 W) in the plasma; therefore, cooling is less critical than in the static GD source.

Hollow cathode GD ion sources were employed in the first GDMS prototypes (Ref 12). This configuration is widely used as the photon source in optical emission spectrometry because it provides enhanced sensitivity related to high sputtering rates and excitation efficiencies; however, it is scarcely used in mass spectrometry because it hampers ion extraction, and samples are difficult to machine (Ref 13). Nevertheless, a pulsed dc combined hollow cathode ion source, which consists of a cylindrical cathode and a flat sample in the bottom of the cylinder, was introduced recently as a new ion source for GDMS and was applied for the analysis of semiconductors and nonconductors (Ref 14). In this ion source, the potential difference between anode and cathode typically is between 1000 and 2000 V, while the current can increase up to 3 A. Figure 1 shows the typical designs of a fast-flow source and a combined hollow cathode.

Mass Analyzer

The interface between the GD source and the mass analyzer comprises the extraction cones (one or two) that help to compensate the pressure differences between the GD chamber (~10 to 10^2 Pa) and the mass spectrometer (below 10^{-5} Pa), and the ion optics (extraction lenses) that help to transfer the ions to the entrance of the mass analyzer. For example, in some mass spectrometers, ions are accelerated in the extraction region by a high potential (e.g., 6 to 8 kV for sector field mass spectrometry) and are focused and steered by a series of plates or cylinders at different potentials.

Ions of the constituent elements of the sample are separated as a function of their mass-to-charge ratio in the mass analyzer. All mass analyzers operate under vacuum to ensure that the original ions reach the detector. Usually, all GDMS systems incorporate a valve to isolate the mass spectrometer from the source, which must be exposed to atmosphere frequently. Different types of mass analyzers—including quadrupole, sector field (also denoted as high resolution), and time of flight—have been combined with low-pressure GD ion sources.

Glow discharge sector field mass spectrometry (GD-SFMS), or high-resolution glow discharge mass spectrometry, is the most popular system in the market because it offers the best performance for bulk analysis with high sensitivity and high resolution (e.g., GD-SFMS is able to resolve multiple polyatomic interferences from the peak/isotope of interest). Spectral interferences always are present in mass spectrometry as a result of the existence of isobaric atomic ions, molecular ions, or the formation of multiply charged ions. In GDMS spectra, the most intense ion signals are those from matrix isotopes and discharge gas species ($^{40}Ar^+$, $^{40}Ar^1H^+$, $^{40}Ar_2{}^+$). In addition, the presence of argide ions, which are especially noticeable in the case of low-flow conditions, harms the detection of some elements (e.g., $^{40}Ar^{12}C^+$ and $^{52}Cr^+$, or $^{40}Ar^{16}O^+$ and $^{56}Fe^+$), mainly if their concentration is low. In some cases, the interferences can be avoided by considering other isotopes of the element. In general, many of the polyatomic interferences (e.g., also related to the presence of hydrogen, oxygen, nitrogen, or carbon in the plasma) can be resolved with a medium mass resolution (R = 4000). However, for some particular applications, the separation of some analyte signals from spectral interferences may require higher resolution. Most polyatomic interferences can be resolved with R < 10,000, which can be accomplished by most sector-field-based mass analyzers (e.g., allowing the detection of $^{72,74}Ge$, ^{75}As, or ^{78}Se) (Ref 16). Selected polyatomic interferences and required mass resolution to separate the isotopes of interest from the interfering polyatomic ions are listed in Table 1.

On the other hand, the combination of a pulsed GD with a time-gated detection system, such as glow discharge time-of-flight mass spectrometry (GD-TOFMS), offers the possibility of achieving low average sputtering rates

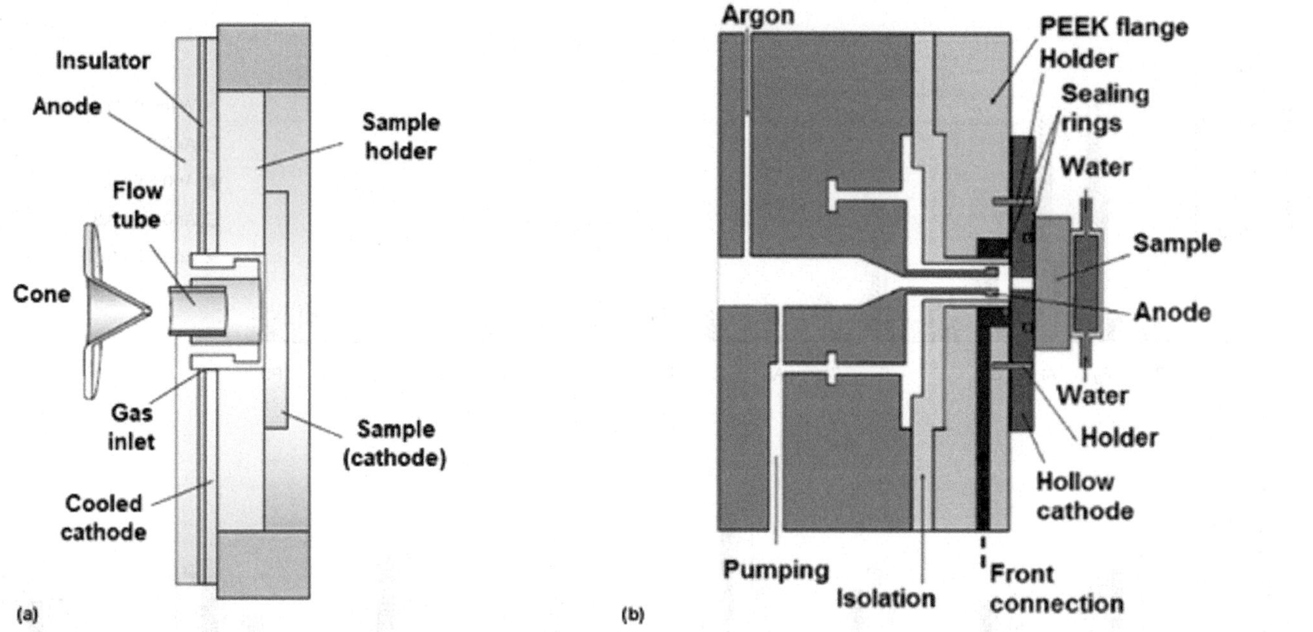

Fig. 1 (a) Schematic of fast-flow source in the Element GD. Courtesy of Thermo Fisher Scientific. Source: Ref 15. (b) Grimm cell with hollow cathode construction. PEEK, polyetheretherketone. Reproduced from Ref 14 with permission from the Royal Society of Chemistry

Table 1 List of selected polyatomic interferences and required mass resolution to separate the isotope of interest from the interfering polyatomic ions

Isotope	Polyatomic ion	Resolution
$^{28}Si^+$	$^{14}N_2^+$	958
$^{31}P^+$	$^{14}N^{16}O^1H^+$	968
$^{16}O^+$	$^{15}N^1H^+$	1230
$^{31}P^+$	$^{15}N^{16}O^+$	1458
$^{28}Si^+$	$^{12}C^{16}O^+$	1557
$^{32}S^+$	$^{16}O_2^+$	1801
$^{64}Zn^+$	$^{32}S^{16}O_2^+$	1952
$^{54}Cr^+$	$^{40}Ar^{14}N^+$	2031
$^{54}Fe^+$	$^{40}Ar^{14}N^+$	2088
$^{52}Cr^+$	$^{12}C^{40}Ar^+$	2376
$^{56}Fe^+$	$^{40}Ar^{16}O^+$	2502
$^{48}Ti^+$	$^{32}S^{16}O^+$	2519
$^{51}V^+$	$^{35}Cl^{16}O^+$	2572
$^{44}Ca^+$	$^{28}Si^{16}O^+$	2688
$^{64}Zn^+$	$^{32}S_2^+$	4261
$^{75}As^+$	$^{40}Ar^{35}Cl^+$	7775

and fast multielemental information (e.g., complete mass spectra, up to a mass/charge number of ions, or m/z, of 300, typically is recorded every 30 μs). Figure 2 shows the mass spectrum of an iron-base sample obtained by radio-frequency GD-TOFMS. These features are of great interest for depth-profile analysis of thin and ultrathin layers and for the identification of unknown samples. The increasing interest for depth-profiling analysis with high in-depth resolution (down to nanometer resolution) and high sensitivity motivated the development of commercial GD-TOFMS instruments.

Major drawbacks of GD-TOFMS are the reduced linear dynamic range (typically 10^5 to 10^7) in comparison to that of GD-SFMS (up to 10^{12}) and the high levels of abundance sensitivity. Intense ion signals in GD-TOFMS, typically coming from the matrix and the argon species, result in long signal tails that affect neighbor isotopes. This effect can be avoided with the use of mass filters that allow the attenuation of ion signals from matrix and plasma-gas-related species (Ref 17). Figure 3 shows that enhanced sensitivity is achieved for the detection of isotopes that are in the neighborhood of major elements/species when appropriate blanking is applied before entering the TOFMS (Ref 18).

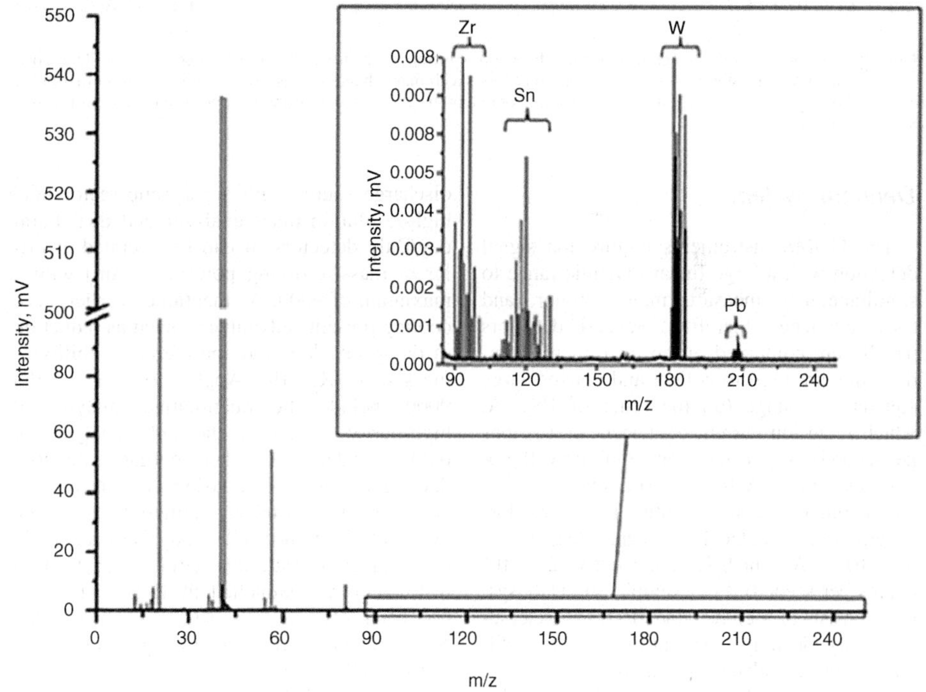

Fig. 2 Mass spectrum of an iron-base sample obtained by radio-frequency glow discharge time-of-flight mass spectrometry. Ion signals from plasma gas species and major, minor, and trace elements in the sample are highlighted. m/z, mass/charge number of ions

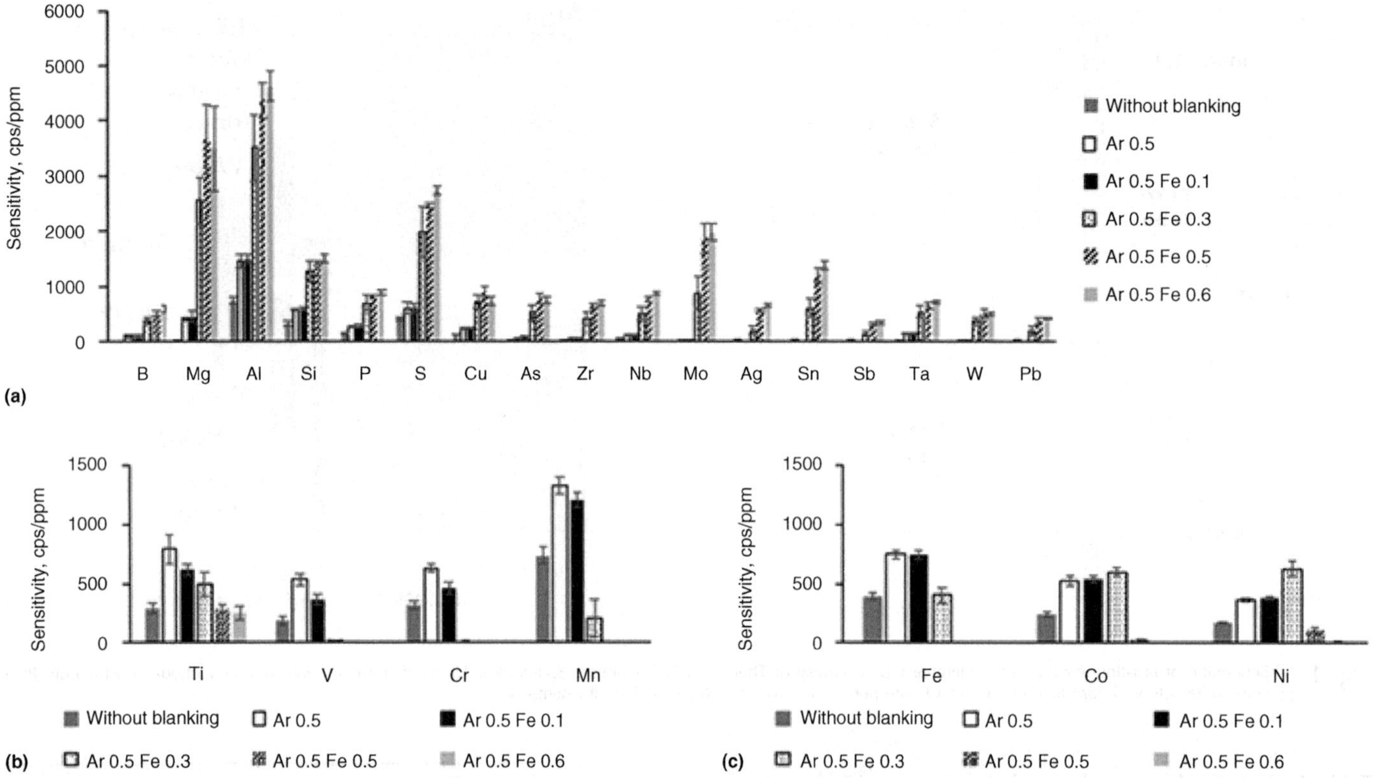

Fig. 3 Enhanced sensitivity achieved for the detection of isotopes that are in the neighborhood of major elements/species when appropriate blanking is applied before entering the time-of-flight mass spectrometer. (a) Isotopes with mass/charge number of ions (m/z) far from that of the attenuated ions. (b) Isotopes with m/z in between that of the two blanked ions. (c) Isotopes with m/z close to that of one of the attenuated ions (the isotope collected for iron was ^{57}Fe). Reproduced from Ref 18 with permission from the Royal Society of Chemistry

Detector System

The GDMS instruments require ion signal detection with a large linear dynamic range to simultaneously measure major, minor, and trace elements; therefore, several detectors usually are combined on an instrument. Faraday cups are used to collect and convert large signals to voltage (on the order of 10^{-9} A, which is an ion current equivalent of 10^9 ions per second). A typical current amplifier will produce a signal of 1 V for an ion current of 10^{-9} A. On the other hand, ion-counting-based detectors are employed to detect lower signals (e.g., below 3×10^{-13} A, which is equivalent to 2×10^6 counts per second). For example, the Daly system consists of a conversion electrode (releases secondary electrons when hit by ions), a scintillator (converts electrons into photons), and a photomultiplier tube that can be placed outside the vacuum region of the mass analyzer. Other types of ion-counting detectors include electron multipliers and microchannel plates. In all of these cases, cross calibration of the detectors (e. g., analog and digital) is required as well as a protection switch to avoid overexposure and damage of the more sensitive multipliers.

Commercial GDMS Instruments

The first commercial GDMS system, the VG 9000, combined a static cryo-cooled dc glow discharge source with a magnetic sector field high-resolution mass analyzer and dual Faraday/Daly detectors. It can be operated at different mass-resolving powers (m/Δm) with a maximum of 9000. As mentioned earlier, cryo cooling presents advantages, such as reduction of the gas background and thermal stability of the source (Ref 19). Applications of the VG 9000 include the quantitative analysis of high-purity materials, the bulk analysis of oxide powders and other inorganic nonconducting matrices (e.g., using conductive binders or a secondary cathode), and the quantitative depth-profile analysis of relatively thick layered materials (Ref 20–22). This instrument was launched in 1985, but it has not been offered commercially since 2005. Nevertheless, it still is being significantly utilized by GDMS users.

Currently (2019), there are five GDMS instruments available on the market: Element GD (Thermo Fisher Scientific), Autoconcept GD 90 (Mass Spectrometry Instruments), Astrum (Nu Instruments), Plasma Profiling TOFMS (Horiba Scientific), and Lumas 30 (Lumex). The main features of these instruments are summarized in Table 2. Three of these instruments (Element GD, Astrum, and Autoconcept GD 90) consist of GD-SFMS systems specially designed to provide outstanding performance in the analysis of high-purity materials. They include a dual-detection mode (Faraday cup and electron multiplier), providing more than 12 orders of magnitude dynamic range, which is essential to simultaneously determine matrix and ultratrace elements. Particular features of these instruments are described here in more detail.

Element GD

The Element GD was introduced on the market in 2005. It consists of a fast-flow direct current GD source with an 8 mm (0.3 in.) diameter hollow anode that includes an internal flow tube. The source is cooled by Peltier elements and is typically operated at flow rates of several hundreds of milliliters per minute, with discharge powers of up to 100 W (Ref 23). The flow tube and sampler cone (components of the GD source and interface) can be easily exchanged to reduce cross contamination. In addition, different sets of these components are available for each specific application (e.g., graphite parts for trace-metal analysis, stainless steel parts for more accurate carbon determination). Moreover, the GD source can be operated in continuous or microsecond pulsed-power mode.

Autoconcept GD 90

The Autoconcept GD 90 has been commercially available since 2008. It consists of a static GD source prepared to analyze both pin

Table 2 Glow discharge mass spectrometry commercial instruments

Instrument	Samples	Power supply	Mass analyzer
Element GD	Pin and flat	Direct current and micropulsed direct current	Sector field
Autoconcept GD 90	Pin and flat	Direct current and radio frequency	Sector field
Astrum	Pin and flat	Direct current	Sector field
Plasma Profiling TOFMS	Flat	Radio frequency and pulsed radio frequency	Time of flight
Lumas 30	Combined hollow cathode	Pulsed direct current	Time of flight

and flat samples. In the standard mode, the power is supplied by a dc source, but a radio-frequency (RF) system also is available. In routine analysis, it usually works at a mass resolving power of 4000.

Astrum

The Astrum, with similar characteristics to the VG 9000, has been commercially available since 2010. It is equipped with a tantalum GD cell cooled by a cryogenic circuit that permits rapid sample change. The analysis of low-melting-point samples is possible thanks to cryogenic cooling, which also contributes to the reduction of background gases. The resolution is improved in comparison to that of the VG 9000, reaching a maximum of 10,000.

Plasma Profiling TOFMS

The Plasma Profiling TOFMS system has been commercially available since 2012. It consists of a fast-flow Grimm-type GD source (e.g., with flow tube) coupled to a TOFMS. This instrument is mainly focused on depth-profiling applications and can be operated in continuous and pulsed RF modes. Under typical operating conditions, the mass resolving power is approximately 3500 at mass 208, but it can be increased to 5200. The interface comprises the extraction cones and the electrostatic ion optics. In addition, a blanking stage is included to attenuate the high signals from argon-ion plasma gas and matrix ions and thus avoid saturation of the detector. The polarity of the TOFMS can be inversed for negative-ion detection of electronegative elements (e.g., halogens) and negatively charged molecular fragments (Ref 24–26).

Lumas 30

The Lumas 30 consists of a pulsed dc combined hollow cathode ion source coupled to an orthogonal TOFMS, which provides a mass resolution of approximately 800. This instrument allows for the analysis of conducting and nonconducting solid samples under different approaches as well as depth-profile analysis capabilities (Ref 7, 27, 28).

Specimen Preparation

Sample preparation in GDMS generally is quite simple and fast, which results in a reduction of the contamination risk as well as the overall analytical cycle/time (compared to analysis techniques requiring the dissolution of the solid sample). For the analysis of bulk homogeneous solid materials, including metals, alloys, and semiconductors, samples may be formed into the required geometry by many methods, for example, casting, rolling, drawing, cutting, machining, or etching. Moreover, the sample surface is finished by machining, polishing, etching, solvent cleaning, and/or drying, to obtain a mirrorlike surface that favors the sealing of the source and the stability of the discharge. Once the sample is located in the analysis position, presputtering of the sample for a specified time enables any surface contaminants to be removed and ensures that representative bulk analysis is performed. The time taken for presputtering depends on various parameters, including the chemical and mechanical sample preparation prior to analysis, the material being analyzed, the types of analyte, and/or the required sensitivity of the analysis.

The analysis of powder samples also is possible, but the powder must be conformed to an adequate shape for the GD source design (pin or pellet). Powder is first homogenized and then compacted by pressing or sintering processes. For the analysis of nonconductive samples using a direct current GD, a secondary cathode (e.g., metal mask containing an aperture) is located in contact with the surface of the nonconductive sample. The secondary cathode provides a stable discharge to sputter and analyze the nonconductive sample. Nevertheless, the material of the secondary cathode should be chosen to minimize interferences with the analyte(s) to be determined. Analysis of nonconductive materials using a pulsed-dc combined hollow cathode also is possible, facilitating the ignition of the plasma by the deposition of a thin conductive layer on top of the surface to be analyzed.

For depth-profile analysis of coated samples, direct analysis without any presputtering or pretreatment is desired, to avoid the destruction of the external layers. Additionally, it is important to begin the data acquisition before switching on the discharge.

Cleaning

Any components that fit within the vacuum system should be handled with powder-free gloves, to avoid chemical residues such as silicon, calcium, sodium, and potassium. Some GDMS systems operate with a GD cell made from tantalum, which can be thoroughly cleaned (mechanically and with most concentrated acids, such as aqua regia) to remove sputter-deposited coating (e.g., reduced memory effects). Moreover, tantalum produces low background interferences. On the other hand, stainless steel or copper components of the GD source should be treated more carefully (e.g., use of diluted acids) than the tantalum ones.

Calibration and Accuracy

Calibrations are needed to find the proportionality between the measured intensity and the element concentration in bulk analysis. This proportionality can be directly obtained by using matrix-matched certified reference materials (CRMs) whenever both sample and calibration standards are measured under the same discharge conditions. Moreover, CRMs and samples must be consecutively measured to avoid any instrument drift. However, this quantification procedure is, in some cases, limited by the availability of CRMs.

In addition, various approaches based on the concept of relative sensitivity are commonly used for quantification. These procedures are of special interest for routine analysis because they do not require calibrations, thus reducing the effort and analysis time. Studies from Vieth et al. (Ref 29), based on measurements with the VG 9000, demonstrated that intensity ratios of any two elements in a given sample are proportional to the ratios of their concentrations. Indeed, this proportionality is given by the so-called relative sensitivity factor (RSF):

$$RSF_X = \frac{I_{IS}/c_{IS}}{I_X/c_X} = \frac{I_{IS}\, c_X}{I_X\, c_{IS}}$$

The RSF of an element "X" is given by the ratio between the sensitivity of an element used as internal standard (I_{IS}/c_{IS}) and the sensitivity of the element "X" (I_X/c_X). This procedure compensates the differences related to the sputtering rate and the ionization and transport processes. Nevertheless, the RSF must be determined for a select set of working conditions. In high-purity samples, the matrix typically is considered as the internal standard, and the concentration of the matrix element can be approximated to 100%. Then, $c_{IS} = 1$, and c_X can be directly determined as:

$$c_X = \frac{I_X}{I_{IS}} RSF_X = IBR_X \cdot RSF_X$$

where IBR is the ion beam ratio.

The RSFs can be measured on different days by obtaining reproducible values, and they can be carefully transferred between the same types of instruments if the discharge conditions are comparable. Nevertheless, they must be determined by using standards similar to the sample of interest, because accurate quantification is possible only when matrix-matched RSFs are considered. Therefore, the limiting factor for this procedure again is the lack of adequate calibration samples.

The lack of matrix-matched reference materials can be partly overcome with the use of self-prepared standards, in which it is possible to select the elemental concentrations that cover a desired range. For example, the quantification of almost 50 elements with concentrations in the ng · g^{-1} level was reported by Gusarova et al. based on a matrix-adapted calibration with standards prepared by pressing high-purity matrix powder doped with standard solutions of analytes (Ref 30). This quantification approach provided uncertainties lower than 30% for most elements (Ref 31). Self-prepared standards are especially relevant for multielemental calibrations in GDMS employed to certify high-purity metals as primary standards; however, these samples must be carefully prepared to avoid losses and/or contamination. The determination of elements such as hydrogen, carbon, nitrogen, and oxygen is essential for adequate characterization of primary standards in metrology; however, there is a lack of calibration materials for these elements. To overcome this problem, the use of samples made of pressed or sintered powder mixtures containing oxides for oxygen calibration standards was investigated (Ref 32).

Another approach is based on the use of standard relative sensitivity factors (StdRSFs), which are calculated based on calibration with reference materials of different matrices that include iron as a matrix or constituent element (Ref 13, 33). In particular, a StdRSF usually is defined as the RSF of an analyte normalized to the RSF of iron (as one of the most common chemical elements present in analyzed samples). For example, the StdRSFs of aluminum and copper could be calculated as:

$$\text{StdRSF}_{\text{Al}} = \frac{\text{RSF}_{\text{Al/matrix1}}}{\text{RSF}_{\text{Fe/matrix1}}}$$

$$\text{StdRSF}_{\text{Cu}} = \frac{\text{RSF}_{\text{Cu/matrix2}}}{\text{RSF}_{\text{Fe/matrix2}}}$$

The StdRSFs then are applied for the quantification of an analyte in a sample of any matrix. For example, the determination of element aluminum in copper is calculated as:

$$C_{\text{Al/Cu}} = \frac{I_{\text{Al}}}{I_{\text{Cu}}} \frac{\text{StdRSF}_{\text{Al}}}{\text{StdRSF}_{\text{Cu}}} = \frac{I_{\text{Al}}}{I_{\text{Cu}}} \text{RSF}_{\text{Al/Cu}}$$

The StdRSFs, widely used for quantification in GDMS industrial applications, usually are implemented by the manufacturer in commercial instruments. This concept works well for survey analyses in various matrices, and it delivers accuracy of results within a factor of 2 from true values for many elements.

Quantification based on multimatrix calibration also has been demonstrated to offer good features for accurate quantification in depth-profile analysis by GDMS. In this case, sputtering rate must be considered, to compensate for the differences in the sensitivities related to the various sample matrices (Ref 34).

Analytical Setup

Usually, mass spectrometry acquisition parameters must be preselected, including selection of the isotopes of interest (e.g., most abundant isotope of a given element free of spectral interferences), selection of mass resolving power (e.g., in SFMS), selection of integration times, and so on. In the case of TOFMS, isotope selection can be done after mass spectra acquisition. On the other hand, operating conditions of the GD ion source also must be optimized for the analysis of various types of samples. Parameters to be considered include applied voltage, current, pressure, applied power, gas flow rates, type of plasma gas, and so on. Due to interrelationships, not all of these parameters can be independently fixed. For example, if the applied voltage and pressure are fixed, the current will take a specific value depending on the sample being analyzed.

Alternative gases and gas mixtures were investigated in GD spectroscopy, pursuing the improvement of ionization and excitation efficiencies as well as the reduction of spectral interferences (Ref 35, 36). Most elements can be ionized by Penning collisions in an argon GD, except for some nonmetals, such as hydrogen, nitrogen, oxygen, fluorine, chlorine, and bromine (Ref 37). Therefore, the ionization efficiency of these elements is low in an argon plasma. Helium appears to be a strong candidate to improve the ionization of the aforementioned elements, because it has a first-ionization potential of 24.5 eV and two metastable levels at 19.82 and 20.62 eV. However, due to the low mass of the helium atoms, it produces significantly lower intensities due to the low sputtering rate. The addition of small amounts of H_2 into an argon GD also was evaluated to enhance sensitivity by using GDMS systems; however, the presence of polyatomic interferences is increased due to the formation of hydride species (Ref 38, 39).

Data Analysis and Reliability

Once the data acquisition is performed, appropriate algorithms are used to select the peaks of interest and to integrate their area or measure their height. Corrections such as the integration time used for each isotope or normalization factor(s) between detectors are taken into account by the instrument software.

Then, ion currents are corrected for the abundance of each particular isotope, and the ratio of each individual elemental ion current to that of the matrix element (single-element matrix) or all matrix elements (multielement matrix) is calculated. As mentioned previously, this is the so-called ion beam ratio. By making use of a library of RSFs stored in the instrument software, it is possible to achieve more accurate quantitative results, especially when using RSFs derived from the analysis of matrix-matched standard samples.

Application and Interpretation

Glow discharge mass spectrometry is a relatively simple method that has a wide range of applications, including the determination of major, minor, and trace elements in bulk samples or the depth-profile analysis of thin and ultrathin coatings. Moreover, it allows trace analysis of all metal and nonmetal elements.

Bulk Analysis

Most GDMS instruments nowadays are applied in the industry for quality control of high-purity metals and alloys. Several measurement standards are listed in Table 3.

In particular, multiple applications of GDMS are in the nonferrous metals industry,

Table 3 Selected standards produced for industrial applications

ASTM F 1593-08	"Standard Test Method for Trace Metallic Impurities in Electronic Grade Aluminum by High Mass-Resolution Glow-Discharge Mass Spectrometer"
ASTM F 1710-08	"Standard Test Method for Trace Metallic Impurities in Electronic Grade Titanium by High Mass-Resolution Glow-Discharge Mass Spectrometer"
ASTM F 1845-08	"Standard Test Method for Trace Metallic Impurities in Electronic Grade Aluminum-Copper, Aluminum-Silicon, and Aluminum-Copper-Silicon Alloys by High Mass-Resolution Glow-Discharge Mass Spectrometer"
ASTM F 2405-04	"Standard Test Method for Trace Metallic Impurities in High Purity Copper by High Mass-Resolution Glow-Discharge Mass Spectrometer"
SEMI PV1-0211	"Test Method for Measuring Trace Elements in Silicon Feedstock for Silicon Solar Cells by High Mass-Resolution Glow-Discharge Mass Spectrometer"
ISO-TS 15338	"Glow Discharge Mass Spectrometry (GDMS)—Introduction to Use"

Source: Ref 5

where metals and alloys containing copper, aluminum, silicon, titanium, gallium, nickel, and chromium as well as refractories and precious metals are analyzed for their purity. Additional application may include the steel industry, nuclear materials, and environmental samples. References related to various industrial and material research applications of GDMS are listed in Table 4.

Table 4 Industrial and research applications of glow discharge mass spectrometry in bulk analysis

Bulk analysis applications	References
High-purity metals for primary standard reference materials in metrology	40–44
Quantification of trace elements in steel	45–49
Analysis of high-purity aluminum and aluminum alloys	50–54
Round-robin using copper-matrix BCR® standards	55, 56
Quality control of pure titanium	57–59
Determination of impurity elements in precious metals such as platinum, palladium, silver, and gold	55, 60–63
Analysis of bulk high-purity silicon for photovoltaic applications	64, 65
Analysis of semiconductors related to specific materials such as ZnO, LiF:Mg,Cu,P, La/B composition of LaB_6 chemical-vapor-deposited layers, purity of Al_2O_3/sapphire, impurity analysis of CdZnTe, impurities in rare earth elements, and AlCu, Ti, InP, In, GaAs	56, 66–72

Depth-Profiling Analysis

The relatively soft and fast sputtering processes achieved by GD sources allow for fast depth-profile analysis with high depth resolution. The synergy between pulsed GD sources and time-gated detection of ions via TOFMS is responsible for the recent increasing number of applications in depth-profile analysis of thin and ultrathin coatings. References related to various applications of GDMS on depth profiling are listed in Table 5.

Table 5 Applications of glow discharge mass spectrometry on depth-profile analysis

Depth-profile applications	References
Nanometer metallic bilayers (Al/Nb) on silicon wafers	73
Solar-grade silicon in photovoltaics	74
CdTe solar cells	75
Single (Ni) and multilayered (Au/Ni(Fe)/Au) metallic nanowires	76
Coated glasses	77, 78
Anodic alumina films enriched in ^{18}O	79
Polymers deposited on silicon	80
PBrS/Cu-polymethyl methacrylate/polyethylene terephthalate multilayer	81
Oxidized thin-film composite	61
$CuInS_2$/Mo layers on glass (using high-resolution glow discharge mass spectrometry)	82, 83
Au (6 nm)/Ni/brass (using high-resolution glow discharge mass spectrometry)	84

As an illustrative example, GD-TOFMS was recently used for the characterization of the oxidation degree of porous and flexible thin-film composite (TFC) membranes (Ref 29). These TFC membranes, which are used for water purification or desalinization, consist of polyamide on top of a porous polysulfone deposited on a polyester substrate. During purification/desalinization, the membranes are irreversibly degraded by oxidation. Depth profiles of anion and cation species provided a detailed evaluation of the membrane degradation as a function of the exposure time and the oxidant employed (NaClO or ClO_2). Depth profiles were obtained by using a pulsed-RF GD-TOFMS prototype consisting of a modified Grimm-type ion source (EMPA, Switzerland) with a 4 mm (0.16 in.) diameter anode and a 2.5 mm (0.10 in.) inner diameter flow tube coupled to the interface of a fast orthogonal TOFMS (Tofwerk, Switzerland). The analyses were carried out under optimized conditions in the pulsed operation mode (e.g., 1 ms pulse width and 25% duty cycle) at 200 Pa of plasma pressure and a constant forward power of 30 W, using high-purity argon as the discharge gas. Figure 4 shows the depth profiles of a virgin (nonoxidized) sample (Fig. 4a) and oxidized samples with NaClO for 1 h (Fig. 4b) and 3 h (Fig. 4c). These profiles have been built by using positive ion detection for all the isotopes except the one at m/z = 127, which was acquired in negative-mode detection. The main results show that an intense $^{79}Br^+$ signal was linked to the polyamide layer in the oxidized membranes. In addition, very interestingly and due to the capability of the technique to provide complete mass spectra, an intense signal was registered in the negative mode at m/z 127 for the virgin membrane. This result showed the presence of a very thin external layer on the polyamide. In addition, it was observed that this signal decreased in the depth profiles obtained for the oxidized sample, showing that this external thin layer progressively was degraded by oxidation.

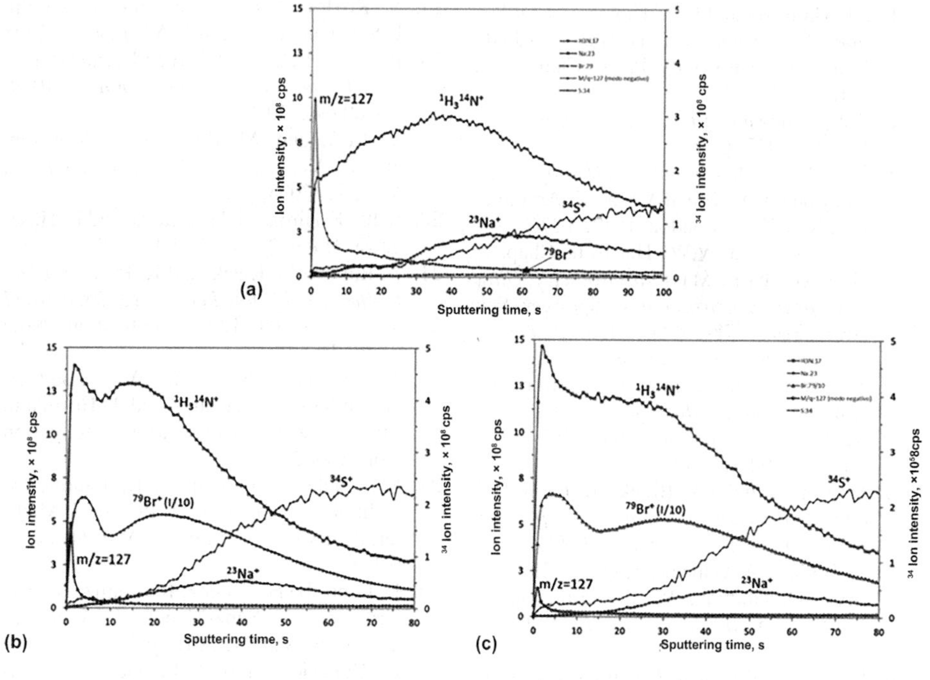

Fig. 4 Qualitative depth profiles of thin-film composite membranes. (a) Virgin membrane. (b) Sample oxidized with NaClO for 1 h. (c) Sample oxidized with NaClO for 3 h. All the isotopes were detected in positive mode except the one at m/z = 127, which was detected in negative mode. Reproduced from Ref 26 with permission from the Royal Society of Chemistry

The analytical performance of GD-TOFMS for depth-profile analysis of thin-coated samples was recently compared in terms of depth resolution, analysis time, and capability to detect light elements to that of other complementary reference techniques, such as secondary ion mass spectrometry (SIMS) and laser ablation inductively coupled plasma mass spectrometry (LA-ICP-MS) (Ref 85). Figure 5 shows the qualitative depth profiles of CdTe photovoltaic cells obtained by using pulsed-RF GD-TOFMS, TOF-SIMS, and LA-ICP-MS, respectively.

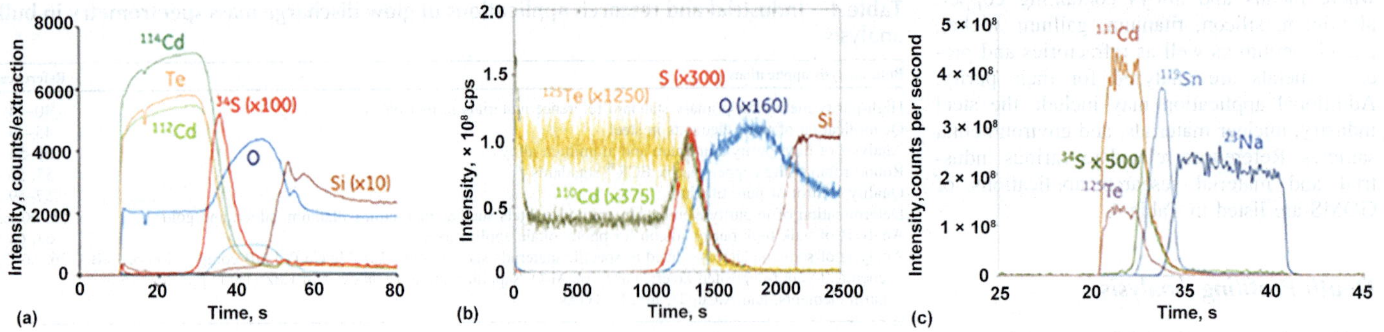

Fig. 5 Qualitative depth profiles of CdTe photovoltaic cells obtained by using (a) pulsed radio-frequency glow discharge time-of-flight mass spectrometry, (b) time-of-flight secondary ion mass spectrometry, and (c) laser ablation inductively coupled plasma mass spectrometry. Source: Ref 85, 86

ACKNOWLEDGMENTS

The authors would like to acknowledge the financial support from the Government of the Principality of Asturias through project IDI/2018/000186 and from the Ministerio de Economia (Spain) through national project MINECO-17-CTQ2016-77887-C2-1-R.

REFERENCES

1. J.J. Thomson and G.P. Thomson, *Conduction of Electricity through Gases*, 3rd ed., Cambridge University Press, Cambridge, U.K., 1928, p 1
2. A.J. Dempster, *Proc. Am. Philos. Soc.*, Vol 75, 1935, p 755
3. F.W. Aston, *Mass Spectra and Isotopes*, 2nd ed., Longmans, Green & Co.; E. Arnold & Co., New York, London, 1942, p 276
4. F.L. King and W.W. Harrison, Chap. 5, Glow Discharge Mass Spectrometry, *Glow Discharge Spectroscopies*, Springer Science, ISBN: 978-1-4899-2396-7, 1993
5. C. Venzago and J. Pisonero, Chap. 13, New Developments in Mass Spectrometry, *Glow Discharge Mass Spectrometry*, The Royal Society of Chemistry, ISBN: 978-1-84973-392-2, 2015
6. K. Marcus and J.A. Broekaert, *Glow Discharge Plasmas in Analytical Spectroscopy*, Wiley, ISBN: 9780471606994, 2003
7. V. Bodnar, A. Ganeev, A. Gubal, N. Solovyev, O. Glumov, V. Yakobson, and I. Murina, *Spectrochim. Acta B*, Vol 145, 2018, p 20
8. S. Jung, S. Kim, and J. Hinrichs, *Spectrochim. Acta B*, Vol 122, 2016, p 46
9. M. Hohl, A. Kanzari, J. Michler, T. Nelis, K. Fuhrer, and M. Gonin, *Surf. Interface Anal.*, Vol 38, 2006, p 292
10. W. Grimm, *Spectrochim. Acta B*, Vol 23, 1968, p 443
11. P.W.J.M. Boumans, *Anal. Chem.*, Vol 44, 1972, p 1219
12. D.L. Donohue and W.W. Harrison, *Anal. Chem.*, Vol 47, 1975, p 1528
13. R.K. Marcus, F.L. King, and W.W. Harrison, *Anal. Chem.*, Vol 58, 1986, p 972
14. A. Gubal, A. Ganeev, V. Hoffmann, M. Voronov, V. Brackmann, and S. Oswald, *J. Anal. At. Spectrom.*, Vol 32, 2017, p 354
15. "Defining Quality Standards for the Analysis of Solid Samples," *Thermo Fisher Scientific*, https://assets.thermofisher.com/TFS-Assets/CMD/brochures/BR-30066-GD-MS-ELEMENT-GD-PLUS-BR30066-EN.pdf
16. N. Jakubowski, T. Prohaska, L. Rottmann, and F. Vanhaecke, *J. Anal. At. Spectrom.*, Vol 26, 2011, p 693,
17. V. Kozlovski, K. Fuhrer, and M. Gonin, U.S. Patent 2008/149825 A1, June 26, 2008
18. R. Muñiz, L. Lobo, B. Fernández, and R. Pereiro, *J. Anal. At. Spectrom.*, DOI: 10.1039/c8ja00334c, 2019
19. V. Hoffmann, M. Kasik, P.K. Robinson, and C. Venzago, *Anal. Bioanal. Chem.*, Vol 381, 2005, p 173
20. S. Ito, F. Hirose, S. Hasegawa, and R. Hasegawa, *Mater. Trans.*, Vol 36, 1995, p 664
21. A. Efimov, M. Kasik, and K. Putyera, *Electrochem. Solid State Lett.*, Vol 3, 2000, p 477
22. VG Elemental Ltd., Application Note APP/GD9/006
23. J. Hinrichs, L. Rottmann, W. Schoettker, N. Frerichs, C. Venzago, and T. Hofmann, Thermo Electron Corporation, Application Note 30082, 2005
24. S. Canulescu, J. Whitby, K. Fuhrer, M. Hohl, M. Goning, P. Horvath, and J. Michler, *J. Anal. At. Spectrom.*, Vol 24, 2009, p 178
25. L. Lobo, B. Fernandez, R. Muñiz, R. Pereiro, and A. Sanz-Medel, *J. Anal. At. Spectrom.*, Vol 31, 2016, p 212
26. C. Gonzalez-Gago, J. Pisonero, R. Sandrin, J.F. Fuertes, A. Sanz-Medel, and N. Bordel, *J. Anal. At. Spectrom.*, Vol 31, 2016, p 288
27. A.A. Ganeev, M.A. Kuz'menkov, V.A. Lyubimtsev, S.V. Potapov, A.I. Drobyshev, S.S. Potemin, and M.V. Voronov, *J. Anal. Chem.*, Vol 62, 2007, p 52007
28. A. Ganeev, A. Titova, B. Korotetski, A. Gubal, N. Solovyev, A. Vyacheslavov, E. Iakovleva, and M. Sillanpää, *Anal. Lett.*, https://doi.org/10.1080/00032719.2018.1485025, 2018
29. W. Vieth and J.C. Huneke, *Spectrochim. Acta B*, Vol 46, 1991, p 137
30. T. Gusarova, T. Hofmann, H. Kipphardt, C. Venzago, R. Matschat, and U. Panne, *J. Anal. At. Spectrom.*, Vol 25, 2010, p 314
31. T. Gusarova, "Wege zur genauen Charakterisierung hochreiner Materialien mit der Glimmentladungs-Massenspektrometrie (GDMS)," Ph.D. thesis, BAM Bundesanstalt für Materialforschung und -prüfung, ISBN: 978-3-9813346-1-6, 2010
32. C. González-Gago, P. Smid, T. Hofmann, C. Venzago, V. Hoffmann, and W. Gruner, *Anal. Bioanal. Chem.*, Vol 406, 2014, p 7473
33. F.L. King, J. Teng, and R.E. Steiner, *J. Mass Spectrom.*, Vol 30, 1995, p 1061
34. M. Bouza, R. Pereiro, N. Bordel, A. Sanz-Medel, and B. Fernandez, *J. Anal. At. Spectrom.*, Vol 30, 2015, p 1108
35. J.J. Giglio and J.A. Caruso, *Appl. Spectrosc.*, Vol 49, 1995, p 900
36. B. Lange, R. Matschat, and H. Kipphardt, *Anal. Bioanal. Chem.*, Vol 389, 2007, p 2287
37. R.L. Smith, D. Serxner, and K.R. Hess, *Anal. Chem.*, Vol 61, 1989, p 1103
38. M. Saito, *Fresenius' J. Anal. Chem.*, Vol 357, 1997, p 18
39. A. Menendez, R. Pereiro, N. Bordel, and A. Sanz-Medel, *J. Anal. At. Spectrom.*, Vol 21, 2006, p 531
40. R. Matschat, J. Hinrichs, and H. Kipphardt, *Anal. Bioanal. Chem.*, Vol 386, 2006, p 125
41. H. Kipphardt, R. Matschat, J. Vogl, T. Gusarova, M. Czerwensky, H.-J. Heinrich, A. Hioki, L.A. Konopelko, B. Methven, T. Miura, O. Petersen, G. Riebe, R. Sturgeon, G.C. Turk, and L.L. Yu, *Accred. Qual. Assur.*, Vol 15, 2010, p 29
42. S. Rudtsch, M. Fahr, J. Fischer, T. Gusarova, H. Kipphardt, and R. Matschat, *Int. J. Thermophys.*, Vol 29, 2008, p 139
43. M. Lucic and V. Krivan, *J. Radioanal. Nucl. Chem.*, Vol 207, 1996, p 444
44. B. Beer and K.G. Heumann, *Fresenius' J. Anal. Chem.*, Vol 343, 1992, p 741
45. Y. Ishikawa, K. Mimura, and M. Isshiki, *Mater. Trans., Jpn. Inst. Met. Mater.*, Vol 41, 2000, p 420

46. S. Itoh, H. Yamaguchi, I. Hamano, T. Hobo, and T. Kobayashi, *Tetsu-to-Hagane (J. Iron Steel Inst. Jpn.)*, Vol 89, 2003, p 962
47. V.D. Kurochkin, *Powder Metall. Met. Ceram.*, Vol 47, 2008, p 248
48. T. Takahashi and T. Shimamura, *Anal. Chem.*, Vol 66, 1994, p 3274
49. S. Itoh, F. Hirose, and R. Hasegawa, *Spectrochim. Acta B*, Vol 47, 1992, p 1241
50. C. Venzago, L. Ohanessian-Pierrard, M. Kasik, U. Collisi, and S. Baude, *J. Anal. Atom. Spectrom.*, Vol 13, 1998, p 189
51. C. Venzago and M. Weigert, *Fresenius' J. Anal. Chem.*, Vol 350, 1994, p 303
52. A.P. Mykytiuk, P. Semeniuk, and S. Berman, *Spectrochim. Acta Rev.*, Vol 13, 1990, p 1
53. G. Kudermann, K.H. Blaufuss, C. Luhrs, W. Vielhaber, and U. Collisi, *Fresenius' J. Anal. Chem.*, Vol 343, 1992, p 734
54. J. Hinrichs and M. Hamester, Application Note 30142, Thermo Fisher Scientific, 2011
55. M. Kasik, C. Venzago, and R. Dorka, *J. Anal. Atom. Spectrom.*, Vol 18, 2003, p 603
56. T.C. Chen and T.G. Stoebe, *Radiat. Protect. Dosim.*, Vol 100, 2002, p 243
57. H.W. Rosenberg and J.E. Green, Analyzing High Purity Titanium, *Titanium '92, The Science and Technology*, Minerals, Metals and Materials Society, Warrendale, PA, 1993, p 2371
58. H.M. Dong and V. Krivan, *J. Anal. Atom. Spectrom.*, Vol 18, 2003, p 367
59. A. Held, P. Taylor, C. Ingelbrecht, P. de Bièvre, J.A.C. Broekaert, M. van Straaten, and R. Gijbels, *J. Anal. Atom. Spectrom.*, Vol 10, 1995, p 849
60. Y. Nakamura, S. Maeda, I. Nagai, H. Inoue, M. Ohtaki, M. Yamazaki, M. Hosoi, K. Shinzawa, Y. Sayama, and T. Kawabata, *Bunseki Kagaku*, Vol 40, 1991, p 209
61. M.V. Straaten, K. Swenters, R. Gijbels, J. Verlinden, and E. Adriaenssens, *J. Anal. Atom. Spectrom.*, Vol 9, 1994, p 1389
62. M. Resano, E. Garcia-Ruiz, K.S. McIntosh, J. Hinrichs, I. Deconinck, and F. Vanhaecke, *J. Anal. Atom. Spectrom.*, Vol 21, 2006, p 899
63. D.M. Wayne, T.M. Yoshida, and D.E. Vance, *J. Anal. Atom. Spectrom.*, Vol 11, 1996, p 861
64. M. Di Sabatino, A.L. Dons, J. Hinrichs, and L. Arnberg, *Spectrochim. Acta B*, Vol 66, 2011, p 144
65. C. Venzago, H. von Campe, and W. Warzawa, in *11th European Photovoltaic Solar Energy Conference* (Montreux), 1992, p 484
66. B. Wang, M.J. Callahan, C. Xu, L.O. Bouthillette, N.C. Giles, and D.F. Bliss, *J. Cryst. Growth*, Vol 304, 2007, p 73
67. S.S. Kher and J.T. Spencer, *J. Phys. Chem. Solids*, Vol 59, 1998, p 1343
68. X. Jianwei, Z. Yongzong, Z. Guoqing, K. Xu, D. Peizhen, and J. Xu, *J. Cryst. Growth*, Vol 193, 1998, p 123
69. J.P. Tower, S.P. Tobin, M. Kestigian, P. W. Norton, A.B. Bollong, H.F. Schaake, and C.K. Ard, *J. Electron. Mater.*, Vol 24, 1995, p 497
70. A.B. Bollong, G. Feldewerth, J.P. Tower, S.P. Tobin, M. Kestigian, P.W. Norton, H.F. Schaake, and C.K. Ard, *Adv. Mater. Optics Electron.*, Vol 5, 1995, p 87
71. W. Gao, P.R. Berger, M.H. Ervin, J. Pamulapati, R.T. Lareau, and S. Schauer, *J. Appl. Phys.*, Vol 80, 1996, p 7094
72. R.J. Guidoboni and F.D. Leipziger, *J. Cryst. Growth*, Vol 89, 1988, p 16
73. R. Valledor, J. Pisonero, N. Bordel, J.I. Martín, C. Quirós, A. Tempez, and A. Sanz-Medel, *Anal. Bioanal. Chem.*, Vol 396, 2010, p 2881
74. J. Pisonero, L. Lobo, N. Bordel, A. Tempez, A. Bensaoula, N. Badi, and A. Sanz-Medel, *Sol. Energy Mater. Sol. Cells*, Vol 94, 2010, p 1352
75. C. Gonzalez-Gago, J. Pisonero, N. Bordel, A. Sanz-Medel, N.J. Tibbetts, and V.S. Smentkowski, *J. Vac. Sci. Technol. A*, Vol 31 (No. 6), Nov/Dec 2013, p 06F106-1
76. M. Bustelo, B. Fernandez, J. Pisonero, R. Pereiro, N. Bordel, V. Vega, V.M. Prida, and A. Sanz-Medel, *Anal. Chem.*, Vol 83, 2011, p 329
77. J. Pisonero, J.M. Costa, R. Pereiro, N. Bordel, and A. Sanz-Medel, *Anal. Bioanal. Chem.*, Vol 370, 2004, p 667
78. A.C. Muñiz, J. Pisonero, L. Lobo, C. González, N. Bordel, R. Pereiro, A. Tempez, P. Chapon, N. Tuccitto, A. Licciardello, and A. Sanz-Medel, *J. Anal. At. Spectrom.*, Vol 23, 2008, p 1239
79. A. Tempez, S. Canulescu, I.S. Molchan, M. Döbeli, J.A. Whitby, L. Lobo, J. Michler, G.E. Thompson, N. Bordel, P. Chapon, P. Skeldon, I. Delfanti, N. Tuccitto, and A. Licciardello, *Surf. Interface Anal.*, Vol 41, 2009, p 966
80. N. Tuccitto, L. Lobo, A. Tempez, I. Delfanti, P. Chapon, S. Canulescu, N. Bordel, J. Michler, and A. Licciardello, *Rapid Commun. Mass Spectrom.*, Vol 23, 2009, p 549
81. L. Lobo, N. Tuccitto, N. Bordel, R. Pereiro, J. Pisonero, A. Licciardello, A. Tempez, P. Chapon, and A. Sanz-Medel, *Anal. Bioanal. Chem.*, Vol 396, 2010, p 2863
82. M. Voronov, P. Smíd, V. Hoffmann, T. Hofmann, and C. Venzago, *J. Anal. At. Spectrom.*, Vol 25, 2010, p 511
83. V. Hoffmann, D. Klemm, V. Efimova, C. Venzago, A.A. Rockett, T. Wirth, T. Nunney, C.A. Kaufmann, and R. Caballero, Chap. 16, Advanced Characterization Techniques for Thin Film Solar Cells, *Elemental Distribution Profiling of Thin Films for Solar Cells*, Wiley-VCH, ISBN:9783527410033, 2011, p 411
84. J. Pisonero, I. Feldmann, N. Bordel, A. Sanz-Medel, and N. Jakubowski, *Anal. Bioanal. Chem.*, Vol 382, 2005, p 1965
85. A. Gutierrez-Gonzalez, C. Gonzalez-Gago, J. Pisonero, N. Tibbetts, A. Menendez, M. Velez, and N. Bordel, *J. Anal. At. Spectrom.*, Vol 30, 2015, p 191
86. *J. Vac. Sci. Technol. A*, Vol 31 (No. 6), Nov/Dec 2013

ASM Handbook, Volume 10, *Materials Characterization*
ASM Handbook Committee
DOI 10.31399/asm.hb.v10.a0006641

Inductively Coupled Plasma Mass Spectrometry

Fuhe Li and Hugh E. Gotts, Air Liquide Electronics—Balazs NanoAnalysis

Overview

General Uses

- Quantitative analysis of inorganic and organic materials and mixtures
- Determination of elemental concentration, isotopic ratio, and nanoparticle concentration and distribution information, depending on the type of samples being analyzed and the instrument configuration

Examples of Applications

- Analysis of blood, hair, serum, tissue, urine, and other bio and clinical materials
- Analysis of drinking water, groundwater, sludge, and soils
- Analysis of bagels, chocolate bars, salmon fish, seaweed snacks, wine, and food products
- Analysis of ultrapure water, acids, bases, organic solvents, cleanroom air, high-purity silicon, wafers, and ceramic parts, as well as atomic-layer-deposition precursor molecules for semiconductor manufacturing
- Analysis of nuclear hot waste, uranium fuel, and so on
- Analysis of brick, clay, metal, and various other archaeological burial materials
- Analysis of igneous rocks, sediment, seawater, plants, and geological ores

Samples

- *Form:* Gases, liquids, and solids
- *Size:* Varies depending on the application
- *Preparation:* Sample can be analyzed directly or after preparation

Limitations

- Not sensitive enough to measure nonmetals such as carbon, oxygen, chlorine, and sulfur
- Cannot detect fluorine, nitrogen, and hydrogen

Estimated Analysis Time

- 1 to 30 min per sample depending on the application, type of samples, and number of elements of interest; time does not include sample preparation, instrument calibration, quality control, and data-reduction steps

Capabilities of Related Techniques

- *Glow discharge mass spectrometry:* Good survey analysis technique but is less quantitative, less sensitive, requires sample preparation, and is difficult to analyze thin films
- *Inductively coupled plasma optical emission spectrometry:* A simultaneous multielement quantitative technique but less sensitive
- *Neutron activation analysis:* More expensive nondestructive method requiring a nuclear reactor and disposal of radioactive samples after analysis; very long analysis time
- *Spark source mass spectrometry:* Less quantitative and less sensitive

Introduction

Inductively coupled plasma mass spectrometry (ICP-MS) was developed in the 1970s, and the first commercial instrument was introduced into the market in 1983. Its unmatched analytical capabilities led to rapid acceptance by analytical chemists, materials scientists, and spectroscopists in academia and industry, with the number of applications in various fields increasing significantly. ICP-MS is more sensitive than the established atomic optical emission and atomic absorption spectrometers. ICP-MS analysis is also a more quantitative method compared with spark (or arc) source mass spectrometry, glow discharge mass spectrometry, secondary ion mass spectrometry, and other classical techniques despite their usefulness for many applications. Today (2019), ICP-MS is recognized as the most widely used and most quantitative technique for trace elemental analysis.

ICP-MS provides quantitative elemental concentration, isotopic ratio, and nanoparticle concentration and distribution information, depending on the type of samples being analyzed and the instrument configuration. The technique enables determination of elements with atomic mass ranging from 7 (lithium) to 238 (uranium). Its superior detection limit (parts per trillion to parts per quadrillion), wide linear dynamic range (9 to 12 orders of magnitudes), and simultaneous multielement analysis capability (70+ elements per run), together with the availability of National Institute of Standards and Technology (NIST)- and International Organization for Standardization (ISO)-certified calibration standards, enable the successful application of ICP-MS to a wide range of gas, liquid, and solid samples. Samples include, but are not limited to, archaeological, biological, clinical, environmental, water, geological, medical, metallurgical, nuclear, and semiconductor and other electronic materials.

Basics

An ICP-MS instrument is a synergistic coupling of an atmospheric-pressure argon inductively coupled plasma and a mass spectrometer under vacuum. A schematic diagram of a typical ICP-MS instrument is shown in Fig. 1. In a typical ICP-MS analysis, a liquid sample is introduced as an aerosol into the ICP via a nebulizer. The aerosol is vaporized, atomized, excited, and ionized in the ICP, producing predominantly singly charged atomic ions. The analyte ions produced are accelerated through interface cones (a sampler and skimmer) into the vacuum system of a mass spectrometer and pass sequentially through ion optics, a reaction/collision cell, and a mass analyzer, where they are sorted and separated according to their mass/charge (m/z) ratio. Separated analyte ions emerging from a mass analyzer eventually reach a detector. As an analyte ion impinges the detector surface, an electron is ejected, the number of which is directly proportional to the number of analyte ions impinging on the detector. The detector essentially converts the number of ions striking the detector into an electrical signal, which is quantitatively measured and related to the number of atoms of that element in the sample via a standard calibration.

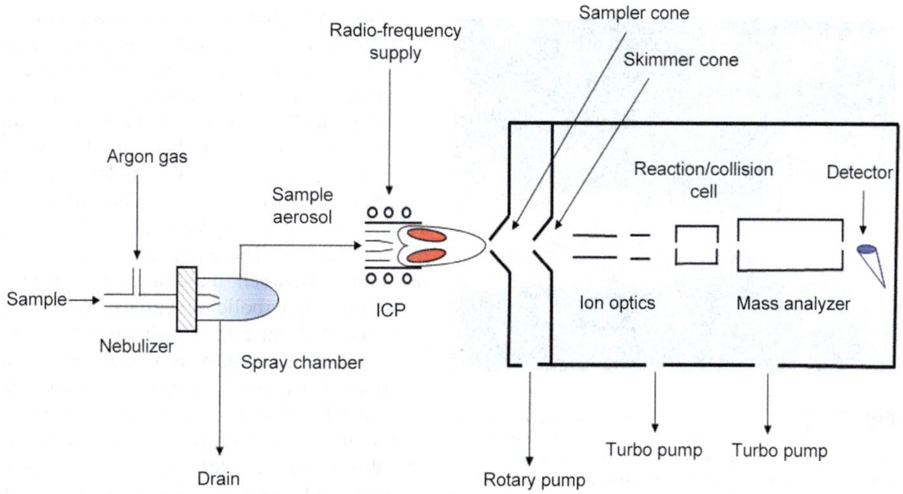

Fig. 1 Schematic diagram of an inductively coupled plasma (ICP) mass spectrometry instrument

Instrumentation

Sample-Introduction System

Sample introduction into the ICP is crucial in ICP-MS analysis. To achieve the best detection limit and highest possible analytical accuracy, analyte ions should be produced efficiently and in sufficient quantity in the ICP. A sample must be introduced into the plasma continuously and constantly in gaseous form, an aerosol, or an atomic vapor that can be efficiently excited and ionized by the ICP. Depending on sample type and preparation, various sample-introduction systems were developed to transport a gas, liquid, or solid sample into the ICP (Ref 1).

Liquid Sample Introduction

The majority of ICP analyses are performed on liquid samples, whether or not the original sample is a liquid. The most convenient and effective way to introduce a liquid sample into the ICP is as an aerosol from a pneumatic nebulizer/spray chamber assembly. The nebulizer converts the liquid into an aerosol, defined as a finely dispersed liquid mist or spray suspended in a gas. An aerosol facilitates uniform sample introduction for both reproducible signal output and stable ICP operation. The spray chamber separates large aerosol droplets from smaller ones and serves as an ICP-MS system noise dampener. Smaller droplets are introduced into the ICP for ionization, and large ones condense on the inner surface of the spray chamber and flow down to the drain. Of the various nebulizers, the perfluoroalkoxy alkane (PFA) nebulizer is the most commonly one used in ICP-MS analysis. A PFA nebulizer is made entirely of high-purity, chemically resistant fluoropolymers. It is an ideal liquid sample-introduction device for ultratrace analysis of ultrapure water, inorganic corrosive chemicals, reactive organic compounds, and even high-salt solutions such as environmental and geochemical samples. Previously, liquid was often introduced into the nebulizer/spray chamber assembly via a peristaltic pump. Potential contamination from the various types of pump tubing and flow pulses from the peristaltic pump presented a challenge to trace and ultratrace analysis. Currently, the PFA nebulizer enables a liquid to be self-aspirated directly and smoothly into the spray chamber, providing exceptionally low analytical background and long-term ICP-MS signal stability necessary for performing single parts per trillion (ppt) and sub-ppt-level analysis.

Gaseous Sample Introduction

A gas can be directly introduced into the ICP through the gas phase as a part of an argon carrier flow. This is carried out by adding a gaseous sample directly into the spray chamber via an additional gas port. Gaseous samples such as industrial gases (air, argon, helium, N_2, O_2, xenon) are regulated by using a needle valve, and gas flow is controlled by using a conventional flow meter. Depending on the type of gas, the concentration of the gas sample (especially organic gaseous compounds) in the argon carrier gas must be low (e.g., <0.1%) so it will not alter the fundamental nature of the ICP. The challenge with this approach is the standard calibration, due to the limited availability of gas standards. In addition, the gaseous sample is heavily diluted prior to ICP-MS analysis. As a result, the detection limits obtained are high and not optimal. Elemental impurities in industrial gases are typically in particle form. Trapping particles in a gaseous sample using a synthetic fluorine-containing resin membrane filter, followed by acid digestion and ICP-MS analysis of a liquid sample, is a practical alternative. A gas sample such as cleanroom air can also be passed through an acid solution for a long period of time (Ref 2). Gaseous and nongaseous elemental impurities in the sample are trapped and dissolved in the acid. The resultant solution is digested to ensure complete dissolution of all particles and nonparticulate contaminants. These two approaches are widely used because the methods permit accurate quantitative analysis attributed to the availability of NIST- and ISO-certified liquid elemental standards, and the long trapping time concentrates elemental impurities, which, in turn, leads to low detection limits of these elements. For a select group of elements (e.g., arsenic, bismuth, germanium, lead, antimony, selenium, tin, tellurium) that can form volatile hydrides at ambient temperature, hydride-generation ICP-MS has been used. In this case, a sample is either liquid or solid. Elements of interest in the sample are converted into volatile hydrides, which are introduced into the ICP in the form of gas (Ref 1).

Solid Sample Introduction

Many solids can be analyzed in the liquid form. Solid materials can be digested using high-purity chemicals (usually concentrated mineral acids) prior to ICP-MS analysis (Ref 2). Digestion can be carried out via either an open system heated on a hotplate or a closed microwave system. Fully digested solutions are introduced into the ICP via a PFA nebulizer and a spray chamber. Such a sample preparation prior to ICP-MS is analytically advantageous because the unwanted sample matrix is removed, and analytes of interest are preconcentrated, resulting in exceptionally low detection limits (Ref 2). Unfortunately, not all solids can be digested. Some solid materials, including, but not limited to, carbides, ceramics, high-strength oxides, graphite, nitrides, and precious metals, are chemically inert and insoluble in concentrated chemicals. In addition, sample digestion is a time-consuming process and does not provide spatially resolved information of a heterogeneous solid sample.

A refractory solid sample can be introduced directly into the ICP by direct insertion (Ref 3, 4), electrothermal vaporization (Ref 5), high-voltage spark (Ref 6), laser ablation (Ref 7), and many other approaches. Laser ablation is by far the most versatile, widely used direct solid-introduction technique. Laser ablation involves the conversion of a solid material into a plume of atomic vapor and microparticles by focusing a short-pulsed, high-power laser beam onto a solid sample surface. The plume of atomic vapor and microparticles is transported in an argon carrier gas to the steady-state ICP for atomization and ionization (Ref 8, 9). Using laser ablation as the solid-introduction system coupled to ICP-MS enables the direct analysis of a solid material not only for average bulk elements but also for spatially resolved information via either line scan or depth profiling. Direct solid sample introduction by laser ablation offers several advantages over liquid sample introduction for analysis of a solid sample. Sample preparation and handling time are significantly reduced or eliminated. There is no chemical solution involved in the process. A solid material is essentially introduced as a plume of atomic vapor and microparticles into a dry ICP. Therefore, the level of mass interferences derived from liquid sample introduction (e.g., ArO^+ and ArH^+) is significantly reduced. In addition, spatial distribution of elements in a heterogeneous solid can be studied. Such analysis is critical in the electronics and other high-technology industries.

Inductively Coupled Plasma Ion Source

The ICP is an annular flamelike electric discharge formed inductively by radio-frequency (RF) energy in a stream of inert gas. It consists of an assembly of three concentric quartz tubes (plasma torch) surrounded by a copper induction coil connected to a free-running or crystal-controlled RF generator, typically operated at 27 or 40 MHz with an output power of 1 to 2 kW. An inert gas (usually argon) is used to form the ICP. Two argon gases (plasma gas, or coolant gas, and auxiliary gases) flow tangentially through the outer two quartz tubes. The RF power applied through the induction coil forms an oscillating magnetic field, inducing a current in the argon gas stream inside the quartz tubes within the induction coil. A spark from a high-voltage Tesla coil is used to generate "seed" electrons and ions in the argon stream. Argon gas seeded with energetic electrons from the Tesla discharge becomes conductive, and a stable, self-sustaining ICP forms spontaneously and expands to its full dimension. Figure 2 shows a typical ICP in operation (Ref 10).

Typical plasma gas flow of 15 to 18 L/min (0.009 to 0.01 ft^3/s) is used to support the plasma, to thermally isolate the plasma from the outer quartz tube, and to prevent overheating. Typical auxiliary gas flow of 0.8 to 1.2 L/min (0.0005 to 0.0007 ft^3/s) is used to adjust the horizontal position of the plasma in the quartz tube. The third argon gas used in the ICP operation is called carrier gas (or sample gas). Typical carrier gas flow of 0.8 to 1.2 L/min is used to transport a liquid sample from the sample-introduction system to the center core of the plasma. As viewed from the top, the plasma has a circular "doughnut" shape.

Fig. 2 Photograph of inductively coupled plasma. The plasma flows from left to right. In inductively coupled plasma mass spectrometry, the sampler cone tip would be just to the right of the end of the first red region. Source: Ref 10. Reproduced with permission of Royal Society of Chemistry in the format Book via Copyright Clearance Center

The sample is introduced horizontally through the center of the doughnut, where the sample is excited and ionized. The argon ICP has a relatively high electron density (on the order of 10^{15} cm^{-3}) and a very high electron temperature (11,000 K, or 10,700 °C). Additionally, argon has a first-ionization potential of 15.8 V, which is higher than that of all other elements in the periodic table except helium, fluorine, and neon. Therefore, as an ionization source, argon ICP is capable of efficiently ionizing a wide range of elements, particularly metals. It produces predominantly singly charged atomic ions (M^+) that are directly proportional to the number of atoms of that element in the sample. Argon plasma gas creates an inert environment in the plasma and significantly reduces the number of molecular ions formed. As a result, ICP-MS atomic mass spectra are rather simple with much less mass interference, and consequently, they are very easy to interpret.

Mass Analyzer

After ions produced exit the ICP (the ion source), they are accelerated through sampler/skimmer cones and ion lenses, entering a mass analyzer for separation. The function of a mass analyzer is to separate the ions according to their different mass/charge (m/z) ratios. The most crucial parameter of a mass analyzer is its resolution, or resolving power. Depending on the application, different mass analyzers can be selected, and a variety of analyzers have successfully been used for ICP-MS analysis. Five major types of ICP-MS instruments used today (2019) are quadrupole; dynamic reaction cell, or collision cell, quadrupole; triple quadrupole; high-resolution magnetic sector; and time-of-flight, with quadrupole being the most commonly and widely used.

Quadrupole Mass Analyzer

In most early ICP-MS instruments, mass separation was performed using a conventional quadrupole mass analyzer. A quadrupole (Q) field is formed by four parallel electrically conducting rods, with opposite pairs of electrodes electrically connected. One diagonally opposite pair of rods is held at $+U_{dc}$ V and the other pair at $-U_{dc}$ V. An RF oscillator supplies a signal to the first pair of rods at $+U_{dc}$ V and an RF signal retarded by 180° to the second pair. The equipotential surfaces in the region between the four rods appear as oscillating hyperbolic potentials. When ions from the ICP enter the quadrupole mass analyzer along the axis into one end at the velocity determined by their energy and mass, the applied RF voltages deflect all the ions into oscillatory paths through the rods. If the RF and direct current (dc) voltages are selected properly, only ions of a given m/z ratio have stable paths through the rods and emerge from the other end. Other ions are deflected too much, striking the rods, and are nebulized and lost there. The quadrupole mass analyzer is compact and convenient to use with a high-pressure ICP ion source.

The main challenge with a quadrupole system is its one-unit mass separation (mass resolution), which is insufficient for many analysis applications. Although ICP-MS has relatively few mass interferences compared with other mass spectrometers, it has some polyatomic and even isobaric interferences. For example, the argon ICP generates argon-base atomic and polyatomic ions (e.g., $^{40}Ar^+$, $^{40}Ar^{12}C^+$, $^{40}Ar^{16}O^+$, $^{38}Ar^{1}H^+$, $^{40}Ar^{35}Cl^+$, and $^{40}Ar^{40}Ar^+$). The signals of these ions and several important analyte ions, such as $^{40}Ca^+$, $^{52}Cr^+$, $^{56}Fe^+$, $^{39}K^+$, $^{75}As^+$, and $^{80}Se^+$, are superimposed in an ICP-MS spectrum, making the low-level determination of these important elements very difficult. The polyatomic mass interferences limit ICP-MS analytical accuracy and degrade the achievable detection limit for many elements, especially those below m/z 80. A cold-plasma analysis was often used to mitigate the effect of polyatomic mass interference from the argon species in a conventional quadrupole ICP-MS. *Cold plasma* refers to a lower-temperature ICP obtained by using a reduced forward RF power, typically at 600 to 700 W, and an increased nebulizer flow. The cold-plasma condition reduces the background caused by the mass interference from the argon species and significantly improves the signal-to-background ratios for the elemental analytes that were formerly interfered with.

Unfortunately, cold-plasma analysis only works for low-mass elements that have relatively lower first-ionization potential. Elements with relatively high first-ionization potential cannot be ionized by such a weak plasma. As a result, a sample must be analyzed at least two times under both hot- and cold-plasma conditions to obtain the concentration of all elements. This analysis arrangement reduces analysis throughput and lowers instrument productivity. Furthermore, the reduced RF power makes the plasma susceptible to matrix effects when analyzing a complicated sample.

Dynamic Reaction Cell, or Collision Cell, Mass Analyzer

The dynamic reaction cell (DRC), or collision cell, mass analyzer is a quadrupole-based mass spectrometer, but the addition of the reaction/collision cell in the system enables chemical elimination of the most challenging mass interference encountered in ICP-MS by using controlled ion-molecular chemistry. Sample analysis can be completed under a single hot-plasma condition at 1.2 kW or higher, enabling single-ppt- and sub-ppt-level detection for calcium, iron, potassium, chromium, arsenic, and selenium, along with all other elements of interest. Typically, the reaction/collision cell is placed between the interface-ion optics and the quadrupole mass analyzer of the ICP-MS system. The reaction/collision cell consists of another quadrupole that operates as an ion-transfer device rather than a mass analyzer. The ion-transfer quadruple is located inside the cell filled with a selected reaction/collision gas (ammonia, methane, oxygen, or hydrogen). The purpose of the reaction/collision gas is to create reactions or collisions with atomic and molecular ions coming from the ICP. The collisions enable the conversion of argon ions and argon-base polyatomic ions (e.g., $^{40}Ar^+$, $^{40}Ar^{16}O$, and $^{38}Ar^{1}H$) into neutral species through a so-called charge-transfer process, while allowing atomic ions ($^{40}Ca^+$, $^{56}Fe^+$, and $^{39}K^+$) to pass through the reaction/collision cell without losing the charge. The collision process significantly reduces mass interferences from the polyatomic species present in the ion beam, improves the signal-to-background ratios and the detection limit for critical but challenging elements, and ultimately increases the analysis throughput by eliminating the need for the cold-plasma run.

Time-of-Flight Mass Analyzer

The time-of-flight (TOF) mass analyzer operates in a pulsed rather than continuous mode. Ions from the ICP are extracted via sampling and skimmer cones exactly as with other types of mass analyzers. The ions are guided in a straight path by lenses to an acceleration region where a large voltage pulse is imposed, pushing all the ions equally and directing them into the field-free flight tube. The essential principle of a TOF mass analyzer is that if ions with different masses are all accelerated to the same kinetic energy, each ion acquires a characteristic velocity that depends on its m/z ratio. The light ions with low m/z ratios travel faster than heavy ions with high m/z ratios. Ions with different m/z ratios are thereby spatially separated as they travel down the field-free flight tube. Ions with high velocity speed on ahead and arrive at the detector before lower-velocity heavier ions. The balance of acceleration voltage, flight time, and spectral cycles is selected so that the slowest and heaviest ion (e.g., uranium-hydroxide ion) reaches the detector before the fastest and lightest ion (e.g., lithium) from the next pulse arrives at the detector. Acceleration can be repeated at a frequency of up to 30 kHz or 3000 ion pulses per second. Because each short transient signal can produce a simultaneous full mass range spectrum, TOF ICP-MS is unique in its ability to analyze a microvolume of solution, a plume from laser ablation of solids, an eluent from a gas chromatography, and a vapor from an electrothermal vaporization sample-introduction system. It is also suitable for analyzing a small sample with high total dissolved solids and for overcoming the physical effects imposed by the sample matrix. The fast data-acquisition speed of TOF ICP-MS eliminates continuous sample introduction of a highly concentrated sample, which can plug a sample-introduction system and sampling and skimmer cones. However, TOF ICP-MS has only a unit mass resolution due largely to the kinetic energy spread. As a result, it cannot resolve many polyatomic and isobaric mass interferences expected in a unit mass ICP-MS, such as a conventional quadrupole system. Mass resolution can be improved by increasing the length of the field-free flight tube, but this results in degradation of the ion transmission efficiency and overall TOF ICP-MS sensitivity.

Triple-Quadrupole Mass Analyzer

The conventional quadrupole ICP-MS, the DRC- (or collision cell-) based ICP-MS, and the TOF ICP-MS are useful in real-world analyses. However, their mass resolution is 0.7 to 1 atomic mass unit (a unit mass resolution) and is inadequate in some challenging applications. Limitations in their high resolving power led to the development and commercialization of two major types of high-resolution ICP-MS instruments. An example of one such high-resolution instrument is the triple-quadrapole (QQQ) ICP-MS, defined by the International Union of Pure and Applied Chemistry as a tandem mass spectrometer comprising two transmission quadrupole mass spectrometers in series, with a (nonselecting) RF-only quadrupole (or other multipole) between them to serve as a collision cell (Ref 11).

In this arrangement, ions produced from the ICP are separated by mass dispersion in the first high-frequency and hyperbolic quadrupole, functioning as a mass filter. The first quadrupole mass filter (Q1) is used to reject all masses except atomic analyte ions of interest (or target mass). Ions from the Q1 enter an octopole reaction system, where atomic analyte ions and polyatomic interference ions are further separated by consistent, predictable reaction chemistry. Polyatomic interference ions that emerge from the octopole reaction system are filtered by the second quadrupole (Q2) system. The atomic analyte ions pass through Q2 and reach the detector. The high resolution provided by a QQQ ICP-MS enables separation of analyte signals from many polyatomic and isobaric interferences, which cannot be resolved chemically by a DRC-, or collision cell-, based quadrupole system.

High-Resolution Magnetic Sector Mass Analyzer

Significantly high mass resolution and very high resolving power are achieved using a double-focusing magnetic-sector-based mass analyzer by passing ions from the ICP through an electric field prior to a magnetic field. The analyzer uses ion velocity focusing with the electric field and ion deflection with an electromagnet. The high resolution enables a baseline physical separation of analyte signals from many polyatomic and isobaric interferences, which cannot be resolved chemically by a DRC-, or collision cell-, based quadrupole system. Many different geometries are used for the magnetic sector arrangements. The most popular commercial magnetic-sector-based high-resolution ICP-MS instrument uses a reverse Nier-Johnson geometry, which is appropriate when an argon ICP is used as the ion source. The Nier-Johnson geometry consists of a 90° electric sector, or electrostatic analyzer; a long intermediate drift length; and a 60° electromagnetic sector of the same curvature direction. Unlike the standard design, the electrostatic analyzer is positioned after the electromagnet. The electromagnet has a curved ion path located in the gap between the poles of the magnet. It deflects ions based on their momentum-to-charge ratio (not m/z) when the ions from the ICP are accelerated into the magnetic field, which is perpendicular to the flight direction of the ions. Ions of the same mass do not necessarily have the same energy, while ions with different masses can have the same momentum. Therefore, the energy spread of the ion beam produces a spread in the ion radii of curvature in the magnetic field and limits the ultimate resolving power achievable with a deflecting magnetic sector. The use of an electrostatic analyzer after the magnet permits translational energy focusing. The combination of an electromagnet and an electrostatic analyzer enables the dispersion of ions according to their momentum and translational energy, resulting in a resolving power up to 10,000, which is sufficient to resolve the majority of polyatomic and isobaric mass interferences encountered in ICP-MS.

To achieve high mass resolution, two mechanical slits are also used in conjunction with the electromagnet and the electrostatic analyzer—one at the entrance to the mass analyzer (entrance slit) and another at the exit (exit slit), each with three different slit widths. Low resolution is achieved with wide slits, and high resolution is achieved with narrow slits. Changing the instrument operating resolution is essentially accomplished by varying the width of both entrance and exit slits. When mass interferences are not involved, a low-resolution setting is selected. Both entrance and exit slits are wide open to provide maximum ion transmission and the highest analytical signals. A medium- or high-resolution setting (or a combination) must be used when analyzing difficult matrices or trying to resolve challenging mass interferences that a unit mass resolution ICP-MS cannot resolve. Although absolute ion transmission and signal intensity from both settings are less than those from a low-resolution setting, a better detection limit and more reliable analytical data are often obtained because mass interference is eliminated.

In addition to the major mass analyzers described earlier, other mass analyzers have been coupled to the ICP for ICP-MS analysis, including but limited to direct reader, ion cyclotron resonance, ion trap, multicollector magnetic sector, and twin quadrupole.

Ion Detector

An ion detector is a structure or device that multiplies incident charges and converts an incident or impinging ion beam emerged from a mass analyzer into an amplified electric current. The electric currents are quantitatively related to the number of atoms of that element (element concentration) in the sample through a standard calibration. A variety of detectors have been used in ICP-MS instruments. Previously, a channel electron multiplier (channeltron) and Faraday cup collector were commonly used. The channeltron detector has a reasonable sensitivity, but its linear dynamic range is limited. Faraday cup technology was only used in some applications (e.g., isotopic ratio measurement) where ultratrace detection limits are not required due to insufficient sensitivity. To improve the linear dynamic range, a dual-detector system was used in some of the early quadrupole ICP-MS instrumentation. In this arrangement, a channel electron multiplier measured low-current signals, and a Faraday cup measured high-ion currents. The process worked reasonably well but presented challenges in some applications, because it required rapid switching between the two detectors. Today (2019), a dual-mode, discrete dynode electron multiplier, often called an active film multiplier, is used on the majority of high-sensitivity ICP-MS instruments, working in a similar manner to the channeltron. However, it uses a series of discrete dynodes to carry out the electron multiplication, unlike a channeltron, which uses a continuous dynode. The dual-mode, discrete dynode electron multiplier enables the detector to operate in both the pulse-counting and analog modes; therefore, high and low concentrations can be determined in the same sample. The detector also provides up to 11 orders of magnitude of linear dynamic range and an acquisition time (or dwell time) as short as 0.1 ms, which facilitates many fast-transient analyses such as laser ablation ICP-MS and single-nanoparticle ICP-MS.

Select Applications

As a premier, reliable analytical technique, ICP-MS is routinely used in archaeological (Ref 12), environmental (Ref 13), food, geological, and earth sciences (Ref 14) as well as metallurgical, nuclear (Ref 15), and semiconductor applications. ICP-MS is also ideally suited to the analysis of a wide variety of biological, clinical, and pharmaceutical matrices, including toxicological, nutritional, and biomedical materials (Ref 16). Many applications have already been discussed in detail elsewhere, and typical sample types for each application are listed in Table 1. This article emphasizes ICP-MS applications in the semiconductor, photovoltaic, materials science, and other electronics and high-technology areas.

Water

ICP-MS is an ideal analytical technique for analysis of deionized water, drinking water, groundwater, natural fresh water, source water, surface water, wastewater, recycled/reclaimed water, and ultrapure water (UPW) for trace and ultratrace elements. With its wide linear dynamic range, superior sensitivity, and simultaneous multielement analysis capability, ICP-MS can determine 70+ elements at high and low concentrations in a single run. A detection limit of <1 ppt (pg/g or pg/mL) can be readily and routinely achieved for most elements. To minimize contamination levels and reduce quality fluctuation, analysis of UPW for trace and ultratrace elements has been used routinely in semiconductor and photovoltaic (PV) wafer production. ICP-MS analysis has greatly benefited the semiconductor and PV industries, which require low production costs and high manufacturing yield. The technique enables both industries to maintain a competitive edge.

Chemicals

The semiconductor industry has some of the most demanding analysis applications for ICP-MS, including a diverse range of highly corrosive deposition and processing chemicals (Ref 2), such as concentrated acids, bases, organic solvents, and oxides. The most common chemicals routinely analyzed by ICP-MS in the semiconductor industry include:

- Hydrogen peroxide (H_2O_2)
- Hydrofluoric acid (HF)
- Hydrochloric acid (HCl)
- Nitric acid (HNO_3)
- Sulfuric acid (H_2SO_4)
- Phosphoric acid (H_3PO_4)

Table 1 Inductively coupled plasma mass spectrometry applications and sample types

Application	Typical sample types
Archaeological	Alloy, brick, clay, metal, various burial materials
Biological	Blood, hair, serum, tissue, urine
Environmental	Drinking water, groundwater, sludges, soils
Food	Bagels, chocolate bars, salmon fish, seaweed snacks, wine
Geological	Igneous rocks, sediments, seawater, plants
Metallurgical	High-temperature alloys, high-purity metals
Nuclear	Hot waste, uranium, nuclear fuel
Semiconductor	Acids, bases, organic solvents, precursors, ion implants

- Isopropanol (IPA)
- Ammonium hydroxide (NH_4OH)
- n-butyl acetate (NBA)
- n-methyl pyrrolidone (NMP)
- Propylene glycol monomethyl ether acetate (PGMEA)
- Tetramethylammonium hydroxide (TMAH)

The elemental impurities in these process chemicals must be strictly controlled before an acceptable manufacturing yield can be obtained.

Advances in semiconductor manufacturing led to the introduction of unconventional materials (reactive organometallics synthesized as a precursor compound) into the marketplace for use in the atomic-layer-deposition (ALD) process (Ref 17). Atomic layer deposition is an organometallic thin-film-deposition method based on sequential, self-saturating surface reactions involving two or more precursor chemicals, which are sequentially condensed on a surface at ~1 mbar pressure. Each cycle consists of a single monolayer surface-saturation step and a pump cycle to remove residual chemistry, after which the second chemical is introduced, pumped, and reacted at temperature with the prior layer to produce a uniform, integrated thin film. Typical cycle times, ranging from one to several seconds, have been developed for metals, oxides, nitrides, carbides, fluorides, chlorides, and organic layers. The ALD process was first suggested in 1952. It was commercially introduced for making thin-film electroluminescent displays and was used in other industrial applications, including semiconductor device manufacturing. Considerable research is conducted in universities, research centers, and industrial laboratories (Ref 18).

Analysis of these high-matrix materials by ICP-MS is difficult due to interferences caused by the presence of the predominant constituents (matrix element) in the sample (Ref 17). Two types of interference are space charge effect (Ref 19) and isobaric.

Space charge effect is caused by a buildup of positive charge in the ion path due to the presence of high concentrations of the matrix element, which hinders the progress of minor constituents through the cones and lenses by electrostatic repulsion, resulting in fewer ions making it into the detector because they are forced off course. The analytical result is that the actual concentration of trace metals is underestimated. This signal suppression is a serious problem with quadrupole ICP-MS instruments but is more easily overcome in magnetic-sector ICP-MS instruments, which have a much higher accelerating voltage. It can be minimized by diluting the sample to reduce the concentration of the matrix element, but dilution also raises the detection limit for trace metals. This requires a compromise between minimizing the detection limit and reducing signal suppression caused by the matrix element.

Isobaric interference must be taken into account to obtain accurate data. It occurs when two different species have the same or nearly the same mass-to-charge ratio (*m*/*z*), for example, $^{40}Ca^+$ and $^{40}Ar^+$, $^{50}Ti^+$ and $^{50}Cr^+$, and $^{182}Ta^+$ and $^{182}W^+$. Molecular and polyatomic interferences are similar and can be caused by chemical reactions in the ICP itself, including a combination of sample matrix, argon plasma gas, and solvent constituents. Examples include hydrides ($^{39}K^+$ with $^{38}Ar^1H^+$), doubly charged species ($^{48}Ti^+$ with $^{96}Zr^{++}$), and oxides ($^{48}Ti^{16}O^+$ with $^{64}Zn^+$). In some cases, it is possible to select an isotope for measurement that is free of the interference, while in other cases, analytical techniques other than ICP-MS must be used. The resolution requires the separation of two ions with the same nominal mass; for example, Cu^+ and TiO^+ can be calculated, because the exact masses are well known: $^{65}Cu^+$ has mass 64.927793, and TiO^+ has mass 64.942786. A resolution of 4300 is required to baseline separate these two ions.

The resolution of the common quadrupole ICP-MS is less than 300. An instrument of this design is not adequate to resolve two ions with the same nominal mass, such as Cu^+ and TiO^+. However, a high-resolution magnetic-sector-based ICP-MS with up to 10,000 resolution is more than capable of separating them. Other interference challenges, such as the interference of ZrO^+ species on $^{107}Ag^+$ and $^{109}Ag^+$ in analyzing zirconium-base precursor compounds, require resolutions of 133,000 and 2,178,100, respectively. Such resolutions are not available even with high-resolution magnetic-sector-based ICP-MS. One approach used to deal with matrix interference problems is to selectively remove the matrix. In some cases, investigations of several compounds were successful in removing the matrix without losing analytes of interest. For example, sample preparation and analysis of chlorosilanes illustrates this approach. Metal contaminants in dichlorosilane (DCS), which is a gas at ambient temperature, can be quantified by decomposing the DCS in an aqueous solution of HF. After collection, a portion of the HF solution is evaporated under an inert atmosphere using heat, and the silicon matrix is removed as volatile silicon fluorides. Spike studies demonstrate that analytes of interest are not lost during the evaporation process, and detection limits as low as 25 ppt W were achieved. The method has also been used for hexachlorodisilane (HCDS), although direct hydrolysis must be used rather than using an impinger, because HCDS is a liquid at room temperature.

Unfortunately, applying this method to the processing of amidosilanes such as tris(dimethylamido)silane, tetrakis(dimethylamido)silane, and bis(diethylamido)silane was not successful. It is believed that the HF solution does not hydrolyze the nitrogen-silicon bond in these compounds, and silicon is not converted to volatile silicon fluorides, which can be removed upon sample evaporation. The molecule can be hydrolyzed through chemical modification. Subsequent evaporation of the excess of an alcohol such as methanol, followed by the addition of concentrated HF and the evaporation of HF using heat under an inert gas, results in the removal of the silicon matrix. Spike studies show that analytes of interest are not lost. Upon adding concentrated HF and heating under inert gas, a reaction occurs, releasing SiF_4 (Ref 4):

$$SiH_x(OiPr)_{4-x} + HF(4-x)iPrOH$$
$$+ H_2SiF_6(4-x)iPrOH\uparrow\ + HF\uparrow\ SiF_4\uparrow$$

Data obtained using this procedure show that a detection limit of 1 part per billion (ppb) is easily obtained, along with good spike recoveries, as shown in Table 2.

Gases

High-purity gases are used throughout the semiconductor manufacturing process in a wide range of applications, from providing inert atmospheres (e.g., N_2, helium, and argon) to serving as raw materials (e.g., arsine and phosphine) (Ref 20). Gases can be introduced directly into the ICP-MS, but they are commonly bubbled through an acidic solution. ICP-MS is used to analyze impurities in various industrial and specialty gases by bubbling the gases through an acid solution while trapping metallic impurities. The resultant solution is quantitatively analyzed by ICP-MS against NIST or ISO Guide 34 elemental standards. Specially designed perfluoroalkoxy alkane vessels together with proprietary solvents can be used, depending on the gas and analytical needs. Filter membranes can also be used to trap particulate material, followed by acid digestion and ICP-MS analysis (Ref 20).

Table 2 Inductively coupled plasma mass spectrometry analysis and spike recovery results from a bis(diethylamido)silane precursor

Element	Bis(diethylamido)silane, ppb	Spike recovery, %
Li	<1	91
Be	<1	97
B	<1	87
Na	16	84
Mg	5	83
Al	4	95
K	2	89
Ca	4	84
Ti	2	84
V	1	90
Cr	2	99
Mn	<1	96
Fe	4	96
Co	<1	73
Ni	3	83
Cu	<1	93
Zn	9	78

Cleanroom Air Quality

ICP-MS is useful in monitoring cleanroom air quality. A serious problem in semiconductor manufacturing is hazy wafers. Reasons for the occurrence of hazy wafers include particles, roughness, adsorbed organics and metal oxides, and salt formation. Cleanroom air monitoring for airborne molecular contaminants is commonly performed as part of an overall fabrication-monitoring program. Process chemicals and materials used in wafer fabrication can be present in cleanroom air and can contaminate wafers, chemical baths, and other fabrication areas. Acids and bases (e.g., NH_4^+, F^-, and NO^{3-}), organic compounds (e.g., amides, esters, and siloxanes), dopants (e.g., boron, phosphorus, antimony, and arsenic), and trace metals (e.g., residuals from chemical vapor deposition chemicals and chemical-mechanical processes) can result in surface contamination, negatively affecting yield.

Many approaches to cleanroom air analysis are employed to address these concerns, using ion chromatography (acids and bases), thermal desorption gas chromatography-mass spectrometry (TD-GC-MS) (organic compounds), and ICP-MS. While real-time monitoring has the advantage of identifying spikes, grab sampling after several hours or days of collection time enables extremely low detection limits by concentrating the analyte on filters. Commonly used sample-collection methods for ICP-MS analysis are the witness wafer and bubbling technique described previously. Although their trapping efficiency of elemental impurities is difficult to assess, these sampling-collection techniques produce consistent results and are widely used to evaluate cleanroom and facility environments. These types of sampling and offline analysis are also more amenable to ICP-MS.

Semiconductor cleanroom air is monitored to detect toxic phosphine (PH_3) gas through filter sampling, followed by high-resolution ICP-MS analysis. Spectral interferences from $^{19}F^{12}C^+$ on $^{31}P^+$ are minimized by operating high-resolution ICP-MS in medium-resolution (R = 4000) mode. Low blank and high sensitivity of the technique enable low detection limits of 8.0 parts per trillion volume over 120 min sampling time at 2.0 L/min (0.001 ft^3/s). This was sufficient for use to monitor phosphine contaminants in the manufacturing environment. Similarly, filter collection for boron particulates and a solution-absorption method for both particulate and gaseous boron are used to monitor total boron content in cleanroom air. The difference between the two collection methods was used to determine whether contamination was mostly gaseous or particulate based. Total boron concentrations of <0.1 ng/m^3 were measured in well-controlled cleanrooms, which met cleanliness requirements for the manufacturing process. The use of TD-GC-MS to screen materials for organo-phosphorus, siloxanes, and other compounds prior to bringing them into fabrication plants is also effective in detecting possible contaminants before they impact manufacturing processes (Ref 20).

Silicon Materials

Silicon is the most important semiconductor material for the semiconductor industry, with increasingly sophisticated integrated circuits built almost entirely on a silicon substrate. Silicon is also one of the most widely used semiconductor materials in PV technology to manufacture solar cells. Crystalline silicon accounts for more than 80% of total PV market revenue, owing partially to its desirable energy-conversion efficiency. The use of silicon as a semiconductor in a microchip or solar cell requires a much higher purity. Impurities, especially metallic impurities, in silicon have a serious effect on electronic devices, creating defects and lowering their manufacturing yield. To control the amount of impurities in high-purity silicon and facilitate optimization of the silicon material-purification process, ICP-MS is used to analyze various crystalline and amorphous silicon materials for 70+ trace elements. Sample shapes include ingots, powders, and crystals.

A detection limit of $\leq$0.1 ppb (ng/g) is routinely and reliably achieved (Ref 21). Compared with classical spark (or arc) source mass spectrometry and glow discharge mass spectrometry, ICP-MS analysis is > 100 times more sensitive. Importantly, ICP-MS is always calibrated using either NIST- or ISO-certified elemental standards prior to analysis. Therefore, ICP-MS results are more quantitative and consistent. Further, ICP-MS uses a much larger sample size (~1 g or larger) and analyzes the sample entirely. ICP-MS analysis results are statistically more representative than other mass spectrometers that analyze only the surface and subsurface of the silicon materials, because trace and ultratrace impurities are not evenly distributed in the bulk material. In contrast to sophisticated neutron activation analysis (NAA), the analysis cost of ICP-MS is much lower and the analysis time is much shorter. Also, it does not require a nuclear reactor, available only in a few university and industrial laboratories. The NAA sensitivity varies considerably for different elements, and some key elements do not have signals. Samples irradiated in NAA remain radioactive over time (often years) following analysis, and the samples require special handling and disposal protocols.

Silicon Wafers

In semiconductor manufacturing, the ideal single-crystal silicon wafer surface is modeled as a perfect silicon crystalline lattice terminated with silicon-silicon or silicon-hydrogen bonds. Real-world silicon wafer surfaces are often covered with an oxide layer trapped with various particles (metallic and ionic) and contaminants, which must be removed through various surface-cleaning processes prior to semiconductor device fabrication. Cleaning effectiveness and quality control of wafer surface cleanliness are largely determined by monitoring elemental impurities on wafer surfaces by the vapor-phase decomposition (VPD) ICP-MS technique (Ref 22). With VPD, the sample is placed in a closed container where it is exposed to isothermally distilled HF vapor, which etches the surface oxide layer, either native or thermally grown dielectric silicon oxide. After the surface layer is decomposed by VPD, a droplet-collection step is performed using a small droplet typically containing ultrapure water and high-purity mineral acids. The droplet is deposited on the wafer and scanned across the surface, collecting metallic impurities released from the decomposed layer. The droplet is collected and analyzed for metal contamination by ICP-MS. With VPD ICP-MS analysis, most metallic elements on a 300 mm (12 in.) wafer can be detected at or below 1×10^7 atoms/cm^2.

Silicon wafer, silicon electrode, and other silicon-base material surfaces can also be extracted directly with an acid without involving the VPD process. The acid extraction solution is analyzed by ICP-MS for surface metallic contaminants and thickness of native oxide film (Si_xO_y) (Ref 23). With advances in ultralarge-scale integration miniaturization, native oxide formed on a silicon surface is rapidly becoming a new interfacial contaminant that must be removed and controlled. The presence of interfacial native oxide is recognized as an impediment to the formation of high-quality ultrathin-gate atomic-layer epitaxy and small metal contacts on the surfaces. Using a high-resolution magnetic sector ICP-MS, a rapid, sensitive method for ultrathin native-oxide thickness measurement was developed to facilitate the removal of interfacial native oxide for interface control (Ref 23). Coupling a rapid acid-etching process with low-level extractable silicon determination by ICP-MS achieves a thickness-detection limit of 0.1 Å, enabling precise, reliable measurements of ultrathin native oxides within a monolayer range. Achieving such a low detection limit is largely attributed to the ability of the magnetic sector ICP-MS to spatially separate $^{28}Si^+$ signal at *m/z* 27.97693 from the two common polyatomic mass interferences, namely $^{12}C^{16}O^+$ at *m/z* 27.99491 and $^{14}N^{14}N^+$ at *m/z* 28.00614. The separation reduces the silicon background equivalent concentration and significantly improves the detectability of silicon (Ref 23).

Semiconductor tool manufacturers and integrated circuit fabrication facilities adopted a similar technique called drop scan etch

ICP-MS (DSE ICP-MS), or acid extraction ICP-MS (AE ICP-MS). The technique is surface sensitive, which enables monitoring surface cleanliness of various original equipment manufacturer (OEM) parts and tool components (new, used, and refurbished), optimizing the wet and dry cleaning processes, and troubleshooting process-related issues. Data obtained from DSE-ICP-MS, or AE ICP-MS, are used to correlate postcleaned parts with wafer surface contamination (Ref 20).

Laser ablation ICP-MS is used for quantitative analysis of the total dopant dose (total number of implanted ions per unit wafer area) implanted in crystalline silicon wafers (Ref 8, 24). Quantification of dopant doses was traditionally the domain of secondary ion mass spectrometry, sheet resistance, capacitance voltage, and thermal wave techniques. However, coupling laser ablation sampling with ICP-MS analysis offers advantages and functionality not previously available, including extremely fast wafer analysis and simultaneous availability of all metallic contamination concentrations without ion source modification. Additionally, quantification of dopant doses by laser ablation does not depend on achievable depth profiling resolution, thereby ensuring measurement accuracy for shallow and even ultrashallow ion implants performed by ultralow-energy (ULE) implantation (Ref 24).

Table 3 shows laser ablation ICP-MS total dose results and theoretically calculated (expected) doses from $^{11}B^+$ implanted at 0.5 keV and $^{75}As^+$ implanted at 2 to 5 keV. Laser ablation ICP-MS results from the ULE implants are in excellent agreement with nominal doses calculated based on ion implanter beam currents and the length of time for implant. Load-to-load repeatability and precision is approximately 2 to 3%. In addition to dopant dose measurement for ULE ion implantation, laser ablation ICP-MS is also used to simultaneously monitor all metallic contaminants on and in crystalline silicon. Inadvertent metal contaminants commonly found during ion implantation (e.g., aluminum, iron, chromium, nickel, and titanium) due to sputter erosion and implanter construction material outgassing can adversely affect the electrical properties of devices.

Other Solid Materials

In addition to silicon wafers, laser ablation ICP-MS is used for quantitative analyses of semiconductor, PV, and other electronics materials (Ref 25, 26). Materials analyzed include particles, raw silicon materials, quartz, aluminum nitride, ceramic wafers, graphite, low-k fluoropolymer, copper interconnects, bonding wires, metal films, solder joints, and silicon carbide. Both conductive and insulating materials are analyzed in their natural state without any sample pretreatment, such as gold coating or charge neutralization. Both compositional analysis and trace-element determination are performed using laser ablation ICP-MS (Ref 27).

Unlike glow discharge mass spectrometry (GD-MS) and secondary ion mass spectrometry (SIMS) that use one source for both sampling and ionization, laser ablation ICP-MS spatially and temporally separates its ionization from the sampling process. The laser in this arrangement is solely used for sampling. The steady-state ICP, with a typical electron temperature of approximately 11,000 K (10,700 °C), is used for ionization due to its constant high thermal energy and relatively long sample-plasma interaction time. Sampling and ionization are two fundamentally different processes requiring different optimization conditions. Spatial and temporal separation of sampling and ionization enables separate operation and independent optimization, resulting in much more efficient ionization and more quantitative measurements with less matrix effects than GD-MS and SIMS analysis (Ref 9).

By properly tuning laser parameters, laser ablation ICP-MS can be used to quantitatively analyze some microscopic features for elemental compositions. In this analysis mode, the laser is focused on one spot, and a single pulse or brief burst of pulses is applied. Laser spot sizes as small as 5 μm are possible, providing good spatial resolution. Examples of the materials analyzed by laser ablation ICP-MS range from inclusions on the quartz used as the inner-wall coating for a chemical vapor deposition chamber to the lead-tin solder bumps (~100 μm in diameter) on a finished chip used for bonding/packaging. Quantitative results obtained for lead-tin bumps were confirmed and verified using wet chemical analysis. Laser ablation ICP-MS was also used to study both vertical and horizontal diffusion of a refill material deposited via focused ion beam (FIB) in bulk silicon. Although it is beneficial to use a FIB to refill some pits left on a wafer, the refill material that contains high percentages of dopant such as gallium can become a source of contamination, especially during a high-temperature annealing step. The worst case showed that gallium spreading could be as large as 1 cm (0.4 in.) from the center of the refilled material on a wafer. Laser ablation ICP-MS was used to measure vertical and lateral distributions of gallium around an FIB cut to help better understand and control the diffusion (Ref 9).

Laser ablation ICP-MS analysis has expanded into many semiconductor and PV-related application areas, from failure analysis to monitoring ion migration in various substrates (Ref 20). In these applications, there is not one analytical tool capable of solving the problem by itself in many cases; this requires a combination of analytical tools, including scanning electron microscopy/energy-dispersive spectroscopy (SEM-EDS), SIMS, laser ablation ICP-MS, electron spectroscopy for chemical analysis/x-ray photoelectron spectroscopy, glow discharge optical emission spectroscopy, GD-MS, and others, with the trade-offs of each technique being complemented by another (Ref 28, 29). For example, the deep depth profiles possible with laser ablation ICP-MS are difficult to achieve with SIMS and x-ray photoelectron spectroscopy, but those techniques are better suited for differentiating subtle differences in the top 1 to 2 μm at the surface. The use of laser ablation ICP-MS in conjunction with SIMS, FIB, and SEM-EDS reportedly can evaluate failures in microelectronic devices. For example, in analyzing a functional and a defective sample, laser ablation ICP-MS identified slightly higher levels (0.1 to 12 versus <0.1 μg/g) of nine metals (sodium, vanadium, chromium, manganese, iron, cobalt, nickel, copper, and zinc) in the defective sample. Follow-up analysis using SIMS identified depth profiles of the metals in the top 2 μm of the sample, showing that all metals except sodium resided solely on the device surface. The migration of sodium into the sample toward the Si-SiO_2 interface suggested that surface inversion was degrading oxide isolation and causing reduction of the breakdown voltage. Separately, 200 nm thick metal jumpers were sectioned using FIB and analyzed using SEM-EDS and laser ablation ICP-MS. Analysis of the defect areas by laser ablation ICP-MS showed elevated levels of chlorine and phosphorus, indicating insufficient cleaning of etchants (HCl, H_3PO_4, and KCl) used in the process as a possible root cause. The levels of chlorine and phosphorus found were

Table 3 Total dose results obtained using laser ablation inductively coupled plasma mass spectrometry

	Ultralow-energy B^+ implants, 0.5 keV			Ultralow-energy As^+ implants, 2.5 keV		
Wafer No.	Expected, ions/cm²	Found, ions/cm²	Difference, %	Expected, ions/cm²	Found, ions/cm²	Difference, %
1	8.0×10^{14}	8.25×10^{14}	3.1	1.0×10^{14}	1.04×10^{14}	4.0
2	2.0×10^{15}	2.10×10^{15}	5.0	1.0×10^{15}	0.99×10^{15}	1.0
3-1	1.5×10^{15}	1.58×10^{15}	5.1	3.0×10^{14}	3.00×10^{14}	0.0
3-2	1.5×10^{15}	1.51×10^{15}	0.7	3.0×10^{14}	3.12×10^{14}	4.0
3-3	1.5×10^{15}	1.55×10^{15}	3.6	3.0×10^{14}	2.90×10^{14}	3.3
3-4	1.5×10^{15}	1.51×10^{15}	0.7	3.0×10^{14}	3.00×10^{14}	0.0
3-5	1.5×10^{15}	1.51×10^{15}	0.7	3.0×10^{14}	2.90×10^{14}	3.3
Average		1.53×10^{15}	2.2	**Average**	2.98×10^{14}	0.7
Standard deviation		0.03×10^{15}	...	**Standard deviation**	0.09×10^{14}	...
Relative standard deviation		1.99%	...	**Relative standard deviation**	2.95%	...

below the limit of detection for SEM-EDS (Ref 20).

The feasibility of using laser ablation ICP-MS was investigated to evaluate a thin, 1 μm titanium nitride (TiN) layer on silicon wafers, commonly used as a diffusion barrier material in copper- and aluminum-base metallization (Ref 20). The ablated particles and crater morphology were characterized by transmission electron microscopy, FIB-SEM, and a surface-mapping microscope to evaluate laser ablation repetition rate, laser power, and beam diameter. Under optimized conditions, the average ablation rate was 40 ± 4 μm/shot, but ablation craters had an inverse triangular shape, with severe thermal degradation at the brink. While it was possible to fully penetrate through the TiN layer after ~50 shots, the physical properties of the two materials are sufficiently different to require optimizing laser ablation parameters independently for each layer. Ion migration from the bulk surface glass of solar cells through the encapsulant and into the cell due to potential-induced degradation (PID) was examined (Ref 30). Modules were exposed to damp heat and humidity (60 °C, or 145 °F, and 85% relative humidity) under negative 1000 V bias in an environmental chamber for increasing lengths of time. Cross sections of the modules were examined using laser ablation ICP-MS to track sodium migration into the cell, with increasing sodium levels correlating to exposure time and decreasing power retention. No sodium accumulation was observed at the interface at $t = 0$ h, but after 425 h exposure, there was a significant increase in sodium at the interface and subsequently into the cell. There was no sodium accumulation or power loss observed in an ionomer/ethylene vinyl acetate bilayer encapsulant developed to prevent PID (Ref 20, 30).

Thin Films and Coatings

ICP-MS is also useful for measuring major, minor, and trace elements in thin films and coatings used in the electronics and semiconductor industries. Table 4 shows ICP-MS detection limits and spike recoveries for determination of trace-elemental impurities in a copper thin film deposited on a 200 mm (8 in.) silicon wafer. The copper film was dissolved chemically with mineral acids. The copper matrix and trace-elemental impurities in the resultant solution were electrochemically separated. The matrix was removed, and trace elements were enriched. Enriched trace elements were quantitatively determined using ICP-MS against NIST and ISO Guide 34 elemental standards. Copper is used in ultralarge-scale integrated circuits as an interconnect, which requires exceptionally high purity. High-purity copper (a better conductor) replaces aluminum in semiconductor manufacturing, significantly improving the electromigration resistance of the interconnect and reducing propagation delays and power consumption with significantly narrower dimensions.

Table 4 Detection limits and spike recoveries of trace impurities in copper film

Element	Detection limit, ng/g	Spike recovery, %
Aluminum (Al)	4	98
Arsenic (As)	10	84
Antimony (Sb)	3	95
Barium (Ba)	1	94
Beryllium (Be)	4	90
Boron (B)	70	90
Calcium (Ca)	30	95
Cerium (Ce)	1	84
Chromium (Cr)	5	93
Cobalt (Co)	1	88
Gallium (Ga)	3	88
Germanium (Ge)	4	84
Lanthanum (La)	1	99
Lead (Pb)	4	104
Lithium (Li)	3	88
Iron (Fe)	30	102
Magnesium (Mg)	3	86
Manganese (Mn)	3	92
Molybdenum (Mo)	6	81
Nickel (Ni)	5	89
Potassium (K)	50	113
Phosphorus (P)	50	82
Sodium (Na)	10	94
Strontium (Sr)	1	114
Tin (Sn)	10	94
Titanium (Ti)	5	100
Tungsten (W)	10	89
Vanadium (V)	4	94
Zinc (Zn)	10	95
Zirconium (Zr)	10	95

Table 5 shows laser ablation ICP-MS analysis results from two rare earth yttrium oxide (Y_2O_3) films and a ceramic coating. One yttrium oxide film was deposited on a ceramic substrate and the other on a silicon substrate. Duplicate analyses were performed using laser ablation ICP-MS to examine film reproducibility. Yttrium oxide exhibits exceptional plasma, heat, and corrosion resistance. Therefore, it is often used in semiconductor etching and deposition processes as a film on plasma chamber parts and components to stabilize substrate compositions such as alumina, quartz glass, and zirconia. Due to the corrosive nature of the process gases and plasma in such processing chambers used for plasma etching, chemical vapor deposition, and resist stripping, chamber parts and components can be corroded and eroded quickly after being exposed. Corroded and eroded parts produce metallic contamination and particles in the semiconductor wafer processes, adversely impacting the manufacturing yield. Yttrium oxide film is erosion and corrosion resistant to such process gases and plasma. The use of yttrium oxide as a protection film on top of the ceramic components and parts can significantly reduce metal contamination from the substrate material, minimize particle shredding, and prolong the life of process chambers. The film must be high purity to avoid metal contamination from the coating itself. An ICP-MS purity analysis of such a film and yttrium oxide target materials has become essential. ICP-MS results are routinely used to select raw materials, optimize the coating process, and examine surface-cleaning effectiveness. The yttrium oxide film and target are directly analyzed in their natural state by using laser ablation ICP-MS, unlike GD-MS, which requires a test pin to be made via machining and some other sample preparation. Therefore, potential sample contamination during preparation is avoided. Laser ablation ICP-MS is more sensitive than GD-MS in analyzing films and coatings, and it is always calibrated prior to analysis. No minimum film thickness is required for analysis. Furthermore, indium, tantalum, and tungsten content can be reliably determined at a very low detection limit.

Table 5 Laser ablation inductively coupled plasma mass spectrometry analysis results from yttrium oxide films and ceramic coating

	Y_2O_3 film on Al_2O_3		Y_2O_3 film on Si		Ceramic coating	
	Film 1	Film 2	Film 1	Film 2	Coating 1	Coating 2
Element	ppm (μg/g)	ppm (μg/g)	ppm (μg/g)	ppm (μg/g)	ppm (μg/g)	ppm (μg/g)
Li	5.2	4.8	<0.05	<0.05	0.67	0.69
Be	<0.05	<0.05	<0.05	<0.05	<0.05	<0.05
B	<0.07	<0.07	54	49	<0.07	<0.07
Na	33	31	<0.05	<0.05	210	230
Mg	11	12	0.49	0.52	4100	4300
Al	750	780	<0.05	<0.05	Matrix	Matrix
K	98	92	<0.05	<0.05	62	69
Ca	23	21	<0.05	<0.05	170	140
Ti	4.5	4.7	0.89	0.82	2.7	2.3
V	1.7	2.1	0.71	0.65	1.3	1.5
Cr	71	78	<0.05	<0.05	140	130
Mn	7.5	6.9	<0.05	<0.05	2.1	2.3
Fe	190	180	0.92	0.89	510	560
Co	<0.01	<0.01	<0.01	<0.01	0.21	0.28
Ni	32	37	<0.05	<0.05	18	21
Cu	420	430	<0.05	<0.05	88	93
Zn	2.1	1.7	<0.05	<0.05	30	26
Ga	<0.05	<0.05	<0.05	<0.05	42	46
Ge	<0.05	<0.05	<0.05	<0.05	<0.05	<0.05
Sr	<0.01	<0.01	<0.01	<0.01	1.1	0.99
Y	Matrix	Matrix	Matrix	Matrix	<0.05	<0.05
Zr	23	27	<0.01	<0.01	3.2	3.6
Mo	0.78	0.84	<0.01	<0.01	2.1	2.30
Cd	<0.01	<0.01	<0.01	<0.01	<0.01	<0.01
In	<0.05	<0.05	<0.05	<0.05	<0.05	<0.05
Sn	0.95	0.89	<0.01	<0.01	7.6	8.0
Sb	<0.01	<0.01	<0.01	<0.01	0.051	0.052
Ba	53	52	0.67	0.61	1.7	1.5
Ta	<0.01	<0.01	<0.01	<0.01	<0.01	<0.01
W	<0.05	<0.05	<0.05	<0.05	0.29	0.34
Pb	7.8	7.9	<0.01	<0.01	<0.01	<0.01
Bi	<0.01	<0.01	<0.01	<0.01	<0.01	<0.01

The applicability of laser ablation ICP-MS as a depth-profiling tool was also studied for dielectric thin films, copper interconnects, ceramic coating, and many other thick films used in the optical-communication and electronics manufacturing fields. Although depth resolution obtained with laser ablation ICP-MS (0.05 to 0.1μm, depending on the material) is moderate compared with SIMS, the simultaneous multielement depth-profiling capability of laser ablation ICP-MS has shown great applicability to some problems encountered in ultralarge-scale integration technology processes, especially in monitoring the consistency of major constituents in metal films. A qualitative screening for trace impurities in the film can also be performed. Because nonuniformity does occur in some films, the depth profile obtained using simultaneous multielement measurement at a single spot is more representative than overlapping the depth profiles obtained from separate analyses, especially those done at different sampling spots using more than one ionization source under various instrumental conditions.

Laser ablation ICP-MS is routinely used to profile the depth of thick (typically >50 μm) anodized coatings widely used in semiconductor, defense, aerospace, automotive, architectural, medical, marine, sporting goods, home appliances, and recreation applications. Using a laser for material sputtering and an ICP for excitation and ionization avoids many intrinsic limitations associated with traditional electron, ion, and x-ray techniques used in characterizing surface, interface, and bulk anodized coatings. Laser ablation ICP-MS can profile the depth of a >50 μm anodized coating (e.g., type III hard coatings) throughout its entire thickness in real-time. Deep profiles obtained are used to examine coating uniformity, interfacial contamination, and surface stoichiometry. Laser ablation ICP-MS depth profiles are useful in optimizing the anodization process, facilitating base materials (aluminum alloys) selection, controlling surface and interfacial contamination, and preventing premature failure of an anodized coating in the process. Another successful application of the laser ablation technique is to profile the depth of a hard NiP coating used in the disk-drive industry (Ref 9).

Nanoparticles

Nanoparticles are ultrafine particles ranging in size from 1 to 100 nm. Naturally occurring nanoparticles are often present as a contaminant in all aspects of semiconductor chip-fabrication processes, including wafer surface cleaning and preparation, deposition, etching, chemical-mechanical planarization (CMP), implantation, and lithography (Ref 31, 32). Many nanoparticles are found in the process chemicals and ultrapure water (UPW) used in wafer surface cleaning and preparation, as well as on the surface of various OEM parts on the plasma etch and deposition equipment (Ref 33). These nanoparticle contaminants are very difficult to remove from the process, can cause random device defects, and have a detrimental impact on current and next-generation semiconductor manufacturing. Microprocessors using 10 nm processes were being made in 2017, and 7 nm processes were used in 2018, so nanoparticle control of process chemicals, UPW, and equipment parts used in the processes is crucial to the success and profitability of semiconductor manufacturing (Ref 31–33).

Monitoring nanoparticles in a liquid chemical and on a part surface is challenging, and there is no suitable technique yet. Continued advancement in the scanning speed of ICP-MS coupled with ultrafast electronics is facilitating the development of single-particle ICP-MS analysis (Ref 31–33).

Single-particle ICP-MS enables differentiation of a solid particle passing through the ICP versus the signal from dissolved ionic species when operating in a time-resolved data-acquisition mode. This differentiation is enhanced if the scanning speed of the mass analyzer is high and the detector used can integrate the signal at a very short dwell time (Ref 20). This unique feature enables the ICP-MS to count and characterize metal-containing nanoparticles down to single-nanometer size, producing results in the form of particle concentration, particle size, and size distribution (Ref 31). Single-particle ICP-MS is used to characterize both naturally occurring (e.g., unwanted particle contaminants) and engineered nanoparticles (e.g., CMP slurries) (Ref 31–33). Analytical results are used by chipmakers, OEMs, and raw-material suppliers to monitor nanoparticle concentration and particle-size distribution in UPW and process chemicals such as nitric acid, sulfuric acid, peroxide, and organic solvents. The surfaces of various OEM parts used in the etch and deposition chambers, such as quartz, silicon carbide, and ceramics, have been studied (Ref 33). The consistency of analytical results enabled development of nanoparticle specifications to control the quality of incoming UPW and process chemicals, as well as the part-cleaning process. Furthermore, material suppliers use the results to guide efforts to reduce and remove nanoparticles from various chemicals through, for example, filtration.

The smallest particle that can be detected using single-particle ICP-MS for most elements is single-nanometer range. With such single-nanometer background equivalent diameter, single-particle ICP-MS is more sensitive than the commonly used laser-scattering nanoparticle counter, which allows for only 40 nm resolution of particle sizes. Single-particle ICP-MS can be used to study the behavior of nanoparticles (e.g., agglomeration) in conjunction with nanoparticle shape and size information provided by transmission electron microscopy and high-resolution field-emission scanning electron microscopy (Ref 20). A combination of these techniques could be useful in exploring and better understanding the impact of nanoparticles on the fabrication process of current and next-generation semiconductor devices. The single-particle ICP-MS baseline background signal must be continually reduced before this technique is widely applied (Ref 20). The background signal is generated by dissolved analytes in the sample solution, instrument noise, and mass interference, all of which can hinder the detectability of single-particle ICP-MS and impede its resolution of the particle signal from the background (Ref 33, 34). Additional standard reference materials must be developed to accurately discriminate and quantify nanoparticles using elements other than just gold and silver (Ref 20, 31–33).

REFERENCES

1. A. Montaser, M.G. Minnich, J.A. McLean, H. Liu, J.A. Caruso, and C.W. McLeod, Sample Introduction in ICP-MS, *Inductively Coupled Plasma Mass Spectrometry*, A. Montaser, Ed., Wiley-VCH Inc., 1998
2. K. Kawabata, Y. Kishi, F. Li, and S. Anderson, Sample Preparation for Semiconductor Materials, Chap. 29, *Comprehensive Analytical Chemistry XLI*, Z. Mester, Ed., Elsevier Science B.V., New York, 2003
3. R. Rattray and E.D. Salin, *J. Anal. At. Spectrom.*, Vol 10, 1995, p 829
4. V. Karanassios and T.J. Wood, *Appl. Spectrosc.*, Vol 53, 1999, p 197
5. A. Rowland, T.B. Housh, and J.A. Holcombe, *J. Anal. At. Spectrom.*, Vol 23, 2008, p 167
6. J.M. Goldberg and D.S. Robinson, *Spectrochim. Acta B*, Vol 50, 1995, p 885
7. R.E. Russo, *Appl. Spectrosc. A*, Vol 49 (No. 9), 1995 p 14
8. F. Li, M.K. Balazs, and R. Pong, *J. Anal. At. Spectrom.*, Vol 15, 2000, p 1139
9. F. Li and S. Anderson, in *Characterization and Metrology for ULSI Technology*, D.G. Seiler et al., Ed., American Institute of Physics Press, New York, 2003, p 715
10. D.B. Aeschliman, S.J. Bajic, D.P. Baldwin, and R.S. Houk, *J. Anal. At. Spectrom.*, Vol 18, 2003, p 872
11. K.K. Murray, R.K. Boyd, M.N. Eberlin, G.J. Langley, L. Li, and Y. Naito, IUPAC 2013 Recommendation, Term 538, *Pure Appl. Chem.*, Vol 85, 2013, p 1515
12. M.E. Hall, S.P. Brimmer, F. Li, and L. Yablonsky, *J. Arch. Sci.*, Vol 25, 1998, p 545
13. S.N. Willie, The Determination of Trace Elements in Water, Chap. 29, *Comprehensive Analytical Chemistry XLI*, Z. Mester, Ed., Elsevier Science B.V., New York, 2003
14. P.J. Potts and P. Robinson, Sample Preparation of Geological Samples, Soils, and Sediments, Chap. 29, *Comprehensive Analytical Chemistry XLI*, Z. Mester, Ed., Elsevier Science B.V., New York, 2003
15. S.F. Wolf, D.L. Bowers, and J.C. Cunnane, *J. Radioanal. Nucl. Chem.*, Vol 263, 2005, p 581
16. R.M. Barnes, *Fresenius' J. Anal. Chem.*, Vol 355, 1996, p 433

17. P. Clancy, L.S. Milstein, H.E. Gotts, D. Cowles, P. Chitrathorn, Z. Wan, L. Vanatta, and Q. Bales, *ECS Trans.*, Vol 28, 2010, p 349
18. R.L. Puurunen, *J. Appl. Phys.*, Vol 97, 2005, p 121301
19. K. Busch, *Spectroscopy*, Vol 19, 2004, p 35
20. C. Westphal and F. Li, ICP-MS Applications in the Semiconductor and Electronics Industry, *Inductively Coupled Plasma Mass Spectrometry*, A. Montaser, Ed., Wiley-VCH Inc., 2019
21. H.E. Gotts, "SEMI PV Si Material Analysis Guideline," SEMI PV49-0613, June 2013
22. J. Fucsko, S.S. Tan, and M.K. Balazs, *J. Electrochem. Soc.*, Vol 140, 1993, p 1105
23. F. Li, M.K. Balazs, and S. Anderson, *J. Electrochem. Soc.*, Vol 152, 2005, p G669
24. F. Li and S. Anderson, *Solid State Technol.*, Vol 49, 2006, p 65
25. F. Li and S. Anderson, *Proc. IEEE/SEMI Int. Electron. Manuf. Technol. Symp.*, Vol 29, 2004, p 147
26. F. Li and S. Anderson, *PV Int. J.*, Vol 7, 2010, p 111
27. F. Li and S. Anderson, in *Frontiers of Characterization and Metrology for Nanoelectronics*, Vol 1173, D.G. Seiler et al., Ed., American Institute of Physics Press, New York, 2009, p 62
28. Z. Pan, W. Wei, and F. Li, *Proc. 36th Int. Symp. Test. Fail. Anal.* (*ISTFA*), ASM International, 2010, p 285
29. Z. Pan, W. Wei, and F. Li, *J. Mater. Sci.: Mater. Electron.*, Vol 22, 2011, p 1594
30. J. Kapur, K.M. Stika, C.S. Westphal, and J.L. Norwood, *IEEE J. Photovolt.*, Vol 5, 2015, p 219
31. L.M. Mey-Ami, J. Wang, H. Gotts, and F. Li, *Proc. Int. Surface Preparation and Cleaning Conference* (*SPCC*) (Cambridge, MA), April 2018
32. J. Wang, L.M. Mey-Ami, H. Gotts, and F. Li, "Nanoparticle," UP Micro Conference (Austin, TX), May 2018
33. L.M. Mey-Ami, J. Wang, and F. Li, *Proc. Int. Surface Preparation and Cleaning Conference* (*SPCC*) (Portland, OR), April 2019
34. D.M. Schwertfeger, J.R. Velicogna, A.H. Jesmer, R.P. Scroggins, and J.I. Princz, *Anal. Chem.*, Vol 88, 2016, p 9908

ASM Handbook, Volume 10, *Materials Characterization*
ASM Handbook Committee
DOI 10.31399/asm.hb.v10.a0006637

Rutherford Backscattering Spectrometry

Lin Shao, Texas A&M University
Wei-Kan Chu, University of Houston

Overview

General Uses

- Quantitative compositional analysis of thin films, layered structures, or bulks
- Quantitative measurements of surface impurities of heavy elements on substrates of lighter elements
- Defect distribution depth profile in single-crystal sample
- Surface atom relaxation in single-crystal sample
- Interfacial studies on heteroepitaxy layers
- Lattice location of impurities in single-crystal sample
- Crystal orientation mapping

Examples of Applications

- Analysis of silicide or alloy formation; identification of reaction products; obtaining reaction kinetics, activation energy, and moving species
- Composition analysis of bulk garnets
- Depth distribution of heavy ion implantation and/or diffusion in a light substrate
- Surface damage and contamination on reactive-ion-etched samples
- Providing calibration samples for other instrumentation, such as secondary ion mass spectroscopy and Auger electron spectroscopy
- Defect depth distribution due to ion implantation damage or residue damage from improper annealing
- Lattice location of impurities in single-crystal sample
- Surface atom relaxation of single-crystal sample
- Lattice strain measurement of heteroepitaxy layers or superlattices

Samples

- *Form:* Solid samples with smooth surfaces, thin films on smooth substrates, self-supporting thin foils, and so on
- *Size:* Typically 1 cm by 1 cm by 1 mm (0.40 by 0.40 by 0.04 in.); can accept sample as small as 2 by 2 mm (0.08 by 0.08 in.) for traditional analysis and as small as 1 by1 μm for helium ion microscope
- *Preparation:* No special preparation required other than the surface must be smooth

Limitations

- Composition information may be obtained but not chemical bonding information
- Poor lateral resolution. Typical beam spot is 1 by 1 mm (0.04 by 0.04 in.). With attachment, beam spot may be reduced to 1 by 1 μm for traditional analysis. The beam spot size can be reduced down to a few nanometers for the recently developed helium ion microscope.
- Poor mass resolution for heavy (high-*Z*) elements; it cannot distinguish surface impurities of gold from platinum, tantalum, tungsten, and so on. Mass resolution is better for low- and mid-*Z* elements; for example, it can distinguish ^{37}Cl from ^{35}Cl.
- Poor sensitivity for low-*Z* elements on substrates with elements heavier than the impurity
- Depth resolution is generally approximately 20 nm. Glancing-angle Rutherford backscattering spectrometry provides 1 to 2 nm. The depth resolution can be reduced down to approximately 1 Å for heavy metal film by using a magnet-bending-based high-resolution detection system instead of traditional semiconductor detectors.

Estimated Analysis Time

- For routine Rutherford backscattering spectrometry, up to four samples per hour; for channeling Rutherford backscattering spectrometry, approximately 1 h per sample

Capabilities of Related Techniques

- *Particle-induced x-ray emission:* Mass distinction
- *Nuclear reaction:* Low-*Z* element sensitivity
- *Low-energy ion-scattering spectroscopy:* Low-energy Rutherford backscattering spectrometry using kiloelectron volt ions rather than typical Rutherford backscattering spectrometry, which uses megaelectron volt ions
- *Secondary ion mass spectrometry:* Impurity depth profiling at nanometer resolution
- *Cross-section transmission electron microscopy:* Depth and types of defect
- *Atom probe tomography:* Composition analysis at atomic scales

Introduction

Rutherford backscattering spectrometry (RBS) has evolved in the past few decades into a major materials characterization technique primarily due to its versatility and the amount of information it can provide in a short analysis time. Because quantitative information may be obtained without standard samples, RBS analyses often serve as standards for other techniques that are much more sensitive but require calibration. Detailed information on RBS is available in Ref 1 to 3.

Rutherford's first scattering experiment established the analytical utility of the ion beam. Scattering techniques often are used in atomic and nuclear physics to check targets for impurities, thickness, and composition, but only since the late 1960s has ion beam analysis taken hold. This has been due to the need for rapid growth of planar technology in semiconductors and the availability of compatible data processing systems. In the mid-1960s, the discovery of channeling phenomena and the recognition of ion implantation in material doping and alteration provided additional motivation for RBS.

Backscattering is a two-body elastic collision process. The first basic concept of backscattering spectrometry is that in an elastic collision, such as backscattering, energy transfer from a projectile to a target atom can be calculated from collision kinematics. The mass of the target atom can be determined by measuring the energy transfer due to collision or by measuring the scattered energy of the backscattered particle. The second basic concept is that the probability of elastic collision between the projectile and target atoms is highly predictable, and the scattering probability or scattering cross section enables quantitative analysis of atomic composition. Lastly, slowing of the projectile in a medium through inelastic energy loss may be treated as a continuous process, leading to the perception of depth. The combination and interplay of these three concepts allow RBS analysis to perceive depth distribution of masses.

Principles of the Technique

Collision Kinematics

An ion backscattered from a relatively heavy target atom carries a higher backscattered energy and provides smaller energy to the target atom than an ion backscattered from a lighter target atom. Figure 1 defines a backscattering situation. The projectile of atomic mass m and energy E_0 collides with a stationary target atom of atomic mass M_2 ($M_2 > m$). The projectile then backscatters with a scattering angle θ, and its energy becomes E_1. The target atom recoils forward.

A scattering kinematic factor (K) that is defined as the ratio of E_1 to E_0 can be derived from conservation of energy and conservation of momentum along a longitudinal direction and an incident direction for cases before and after scattering:

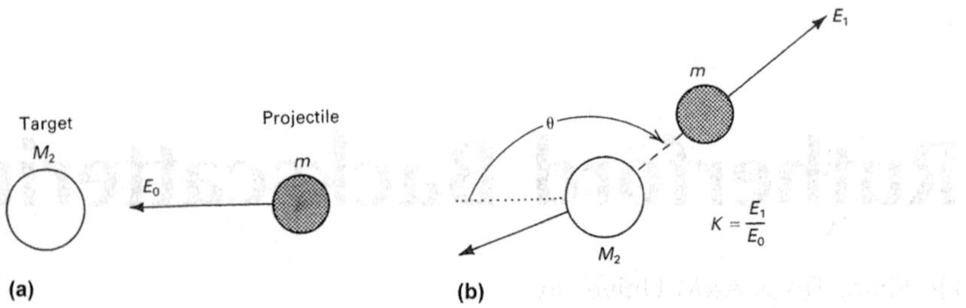

Fig. 1 Collision kinematics between a projectile atom m and a target atom M^2. (a) Before collision. (b) After collision

$$K \equiv \frac{E_1}{E_0} = \left[\frac{(M_2^2 - m^2\sin^2\theta)^{1/2} + m\cos\theta}{M_2 + m}\right]^2 \quad \text{(Eq 1)}$$

Equation 1 reduces to $(M_2 - m)/(M_2 + m)$ at $\theta = 90°$ and to $(M_2 - m/M_2 + m)^2$ at $\theta = 180°$. The kinematic energy loss from a collision is greatest for full backward scattering (180°). Because this type of scattering yields optimum mass resolution, the detector should be placed at the most backward angle possible.

According to Eq 1, when the incident energy E_0 is known, a measurement of the scattered energy E_1 is a measurement of the mass of the target atom at which scattering occurs. That is, the energy scale for a given backscattering spectrum can be translated into a mass scale. Such a translation is expressed in Fig. 2 when ^{4}He is used as the probing beam for RBS. The mass resolution for RBS is directly related to the energy resolution of the detecting system. Helium ion scattering of 2 MeV typically has energy resolution of approximately 15 keV (Fig. 2). This is adequate to distinguish scattering from different chlorine isotopes but not sufficient to separate tantalum from gold and platinum.

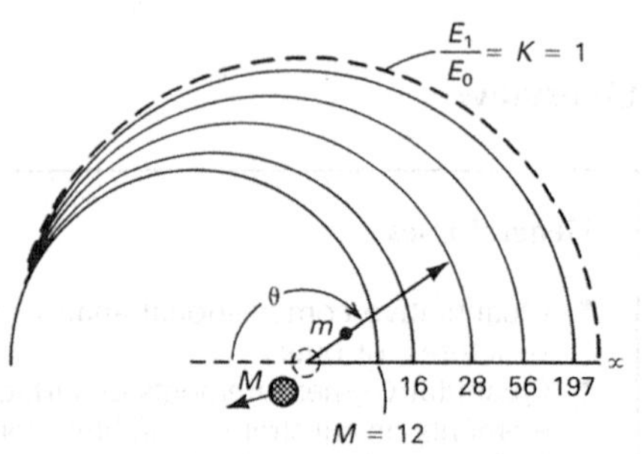

Fig. 2 The energy E of a ^{4}He atom scattered elastically at an angle θ on collision with an atom of mass M is given by the length of the arrow at that angle. The incident energy E_0 is given by the radius of the outer semicircle.

Scattering Cross Section

Scattering events occur rarely. Most of the projectiles are moving forward and eventually stop inside the solid, as in ion implantation. Only a small portion of the ion beam is backscattered and detected for RBS. The total number of detected particles, A, is:

$$A = \sigma\Omega Q(Nt) \quad \text{(Eq 2)}$$

where σ has the dimension of area (cm^2) and is termed the scattering cross section, Ω is the solid detection angle, Q is the total number of incident projectiles, and Nt is the number of target atoms per unit area of the sample.

The scattering cross section for an elastic collision where Coulomb repulsion is the force between the two nuclei can be calculated using:

$$\sigma = \left(\frac{Z_1Z_2e^2}{4E}\right)^2 \frac{4}{\sin^4\theta} \times \frac{\left\{\left[1 - \left(\left(\frac{m}{M_2}\right)\sin\theta\right)^2\right]^{1/2} + \cos\theta\right\}^2}{\left[1 - \left(\left(\frac{m}{M_2}\right)\sin\theta\right)^2\right]^{1/2}} \quad \text{(Eq 3)}$$

where Z_1 is the atomic number of the projectile of mass m, Z_2 is the atomic number of the target atom of mass M_2, and e is the electron charge; the other terms are defined in Eq 1. Equation 3 (Ref 4) reveals that the significant functional dependence of the Rutherford differential scattering cross section is proportional to Z_1^2 (the heavier the projectile, the larger the scattering cross section; for helium ions, $Z_1 = 2$), Z_2^2 (for a given projectile, it is more sensitive to heavy elements than light elements), and E^{-2} (scattering yield increases with decreasing ion beam energy). In thin-film analysis, ratios of scattering cross sections between two elements Z_2 and Z'_2 often are used to determine stoichiometry, and σ/σ' reduces to Z_2^2/Z'^2_2 as an approximation, because Eq 3 is not sensitive to the change of M_2.

Equation 2 demonstrates that calculation of σ and the measurement of the number of detected particles provide a direct measure of the number of target atoms per unit area, (Nt), where N is the number of target atoms per unit volume, and t is the thickness of the target. The concept of scattering cross section leads to the capability of quantitative analysis of atomic composition. Figure 3 shows the approximated scattering cross section and approximated K scale for helium backscattering.

Rutherford scattering cross section is based on the assumption that the projectile and target are point charged and their interaction follows the Coulomb law. The factor involving mass ratio in Eq 3 derives from the laboratory/center-of-mass correction. Correction due to electron screening, more pronounced for high-Z target atoms, has been confirmed (Ref 6), revealing that the empirical cross-section formula:

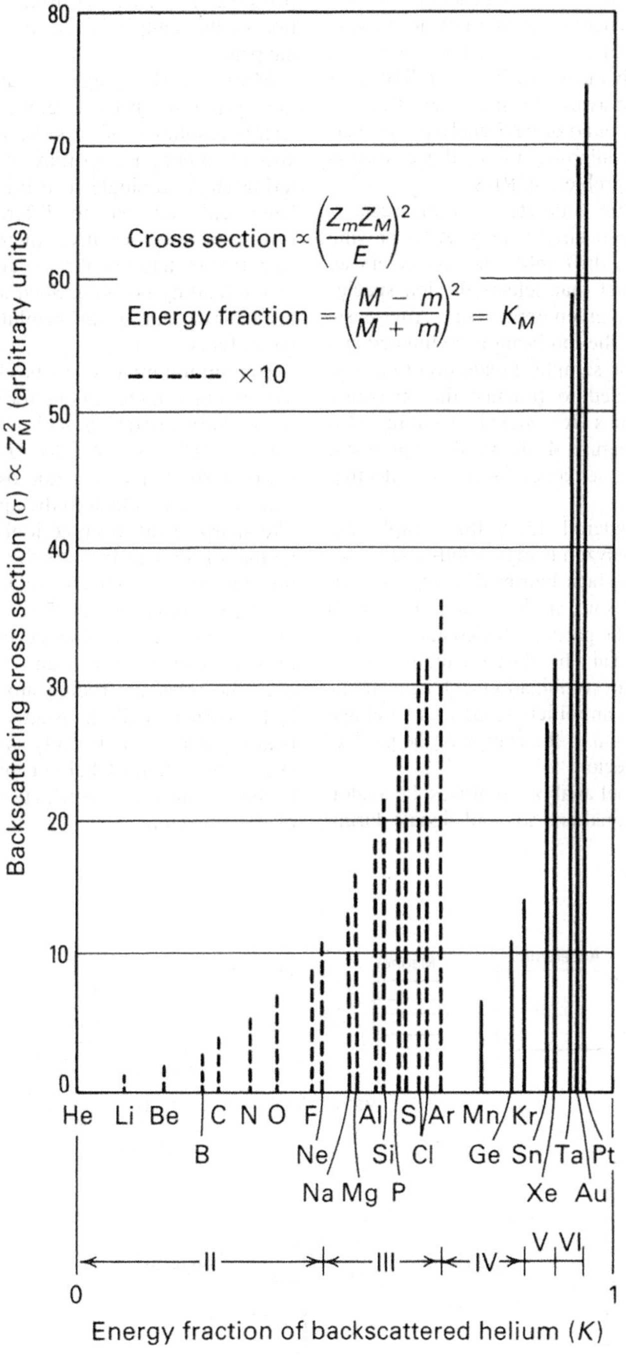

Fig. 3 The sensitivity of Rutherford backscattering spectrometry to the various elements is proportional to the backscattering cross section σ, which varies as the square of the charge Z_m contained in the nucleus of an atom of mass m (Eq 4). The ordinate provides Z_m^2 for a selection of elements. The kinetic energy left in a projectile of mass m after an elastic backscattering collision with an atom of mass m is a fraction k_m of the incident energy (Eq 1 and 2). The abscissa yields K_m for a selection of elements, assuming that the projectiles are helium ions (m = 4). Source: Ref 5

$$\sigma(\text{corrected}) = \sigma\left(1 - \frac{0.049 Z_1 Z_2^{4/3}}{E}\right) \qquad \text{(Eq 4)}$$

agrees with a measurement of helium ion 150 to 180° scattering within 1 to 2% throughout the energy range 0.6 to 2.3 MeV for various Z_2. At a scattering angle within a few tenths of a degree from 180°, an unusual enhancement of scattering yield has been observed (Ref 7). This enhancement has a factor of 2 to 3, a consequence of a correlation between the incoming and outgoing paths of the backscattered ions (Ref 8).

Energy Loss

An energetic ion impinging on a target penetrates it, but large-angle scattering from a target atom is highly unlikely. The Coulomb interaction between the moving ions and electrons in the target material largely determines the fate of all of the incident ions. The moving ions lose their energies by ionization and excitation of target electrons. The numerous discrete interactions with electrons slow the moving ions by transferring the ion energy to the target electrons in an almost continuous frictionlike manner. This type of interaction does not alter significantly the direction of the ion beam. The energy loss per unit path length (dE/dx) can be measured experimentally by passing the ion beam through a thin foil of thickness Δx and measuring the energy loss ΔE; it also can be calculated theoretically. Energy loss produces the depth perception for RBS.

The value of dE/dx is a function of projectile, target, their atomic numbers, mass, and projectile energy. Another definition, used more frequently in RBS, is stopping cross section (ε), which is:

$$\varepsilon \equiv \left[\frac{1}{N}\frac{dE}{dx}\right]$$

where N is the atomic density of the target material, that is, the number of target atoms per cubic centimeter. The stopping cross section ε carries the unit of electron volts times square centimeters (eV · cm^2), hence the term *cross section*. This is similar in name but different in meaning from the scattering cross section, which carries the unit of square centimeters. The term ε is understandable as the energy loss on the atomic level. For example, if a thin layer contains $N\Delta x$ atoms per square centimeter, the energy loss of the ion beam passing through this thin layer becomes ΔE (eV) and:

$$\Delta E = \varepsilon(N\Delta x)$$

where ε is the stopping cross section of the given ion in the given elements. The concept of stopping cross section can be generalized for a molecule or for a mixture by the principle of additivity of stopping cross sections:

$$\varepsilon^{A_m B_n} = m\varepsilon^A + n\varepsilon^B$$

where ε^A and ε^B are stopping cross sections of elements A and B, constituents of molecule $A_m B_n$ (a mixture of A and B with atomic ratio of m to n). For RBS using megaelectron volt helium ions, tabulations of ε for all elements are given in Ref 9 and 10.

The Channeling Effect

Charged particles that penetrate a single crystal along or near a major axis experience a collective string potential produced by the rows of atoms along that axis. If the incident direction of the ion beam is nearly parallel with the string of atoms, the string potential steers (channels) the charged particle forward. A critical angle is defined as the largest angle between the incident direction of the ion beam and the row of atoms such that the steering effect exists. When charged particles are incident in a direction exceeding the critical angle, those particles have transverse kinetic energy exceeding the collective string potential; the collective steering effect subsequently disappears.

Channeling was first predicted in 1912 (Ref 11). In 1963, channeling was accidently observed through Monte Carlo simulations (Ref 12, 13), which immediately triggered intensive experimental and modeling studies (Ref 14, 15). In 1965, the systematic channeling theory was developed (Ref 16).

Ion channeling has been used extensively in conjunction with ion backscattering measurements to study single crystals near the surface. One such application of RBS is discussed in the section "Applications" in this article. Information on the theory of channeling and the experimental study of channeling effects, with emphasis on materials analysis, is given in Ref 17 to 19.

Energy Loss under Channeling Conditions

The precise description of stopping powers of channeled ions is complicated, because the energy loss is a function of impact parameter for different projectile charge states (Ref 20). In the majority of channeling RBS analyses, random stopping powers often are used for data analysis, although it is well known that channeling stopping powers are systematically lower than random stopping powers. Care must be taken in channeling RBS analysis if high accuracy in depth conversion is required. Parameters of fitted stopping powers of helium and lithium ions under various channeling conditions in silicon are available in Ref 21 and 22. However, in general, such data are difficult to obtain and are missing for other important substrates. It was proposed that a certain energy window exists in silicon at which stopping powers between channeled and nonchanneled helium ions are very close to each other. So, the error in depth conversion can be minimized if random stopping powers are used for channeling RBS analysis (Ref 23).

Rutherford Backscattering Spectrometry Equipment

A basic piece of equipment in RBS analysis is a small accelerator capable of generating helium ions of typical energies of 400 keV to 5 MeV. Such units, which may be housed in a 6 by 9 m (20 by 30 ft) lab and are dedicated to RBS analyses, have been available since 1980. Because of the fast turnaround of RBS analyses, accelerator time and availability of small accelerators are not a major problem in RBS.

The accelerator generates ions and accelerates them to the desired energy. After passing through a short drift tube, the ion enters an analyzing magnet that selects the ion species and energy for a given experiment. After passing the magnet, the ion beam is collimated and directed onto the sample. In addition to an ion accelerator needed to produce the analyzing beam, electronics for signal handling also are required. Figure 4 shows the equipment involved and the sequence for signal collecting and processing.

Ions backscattered from the sample are detected and analyzed using a solid-state detector. A silicon surface barrier detector approximately 2 cm (0.8 in.) in diameter commonly is used to collect the particles backscattered into a small solid angle at a fixed backscattering angle. This detector—in combination with a preamplifier and a linear amplifier—generates a voltage signal proportional to the energy of the particle entering the detector.

A multichannel analyzer digitizes the analog input voltage signal and sorts all signals during an experiment. Therefore, at the conclusion of a measurement, the surface barrier detectors receive approximately 10^6 particles with various energies. A measurement typically takes 10 to 30 min. The multichannel analyzer plots backscattering energy spectrum counts versus channel number. That is, it provides the number of scattering events per unit energy interval as a function of energy of the backscattered particles. The RBS spectrum carries information of the sample in which scattering events happen.

Most recently, a high-resolution RBS detection system has been developed to achieve the energy resolution of ~1 keV. The system consists of an entrance aperture to define the scattering angle, a single focusing 90° magnet to bend scattered ions to different trajectories, and a position-sensitive microchannel plate detection system to collect counts as a function of ion landing position. Such a system is able to achieve Angstrom resolution for heavy metal films.

Helium ion microscopy has been developed as a commercial technique for localized structural characterization with a probe size as small as 0.25 nm (Ref 24). The technique is evolved from field ion microscopy in which a gas atom can attach to the apex of a needle-like sharp metal, become ionized due to tunneling of its electron to the metal, and be injected away from the apex under a large positive voltage. Figure 5 shows a schematic of a helium ion microscope in which helium pressure controls the beam current. Because only the most protruding atoms of the apex can produce helium beams, the resulting beam spot size is extremely small. In a typical emission pattern of helium ions, three atoms at the needle apex contribute more than 90% of the emission.

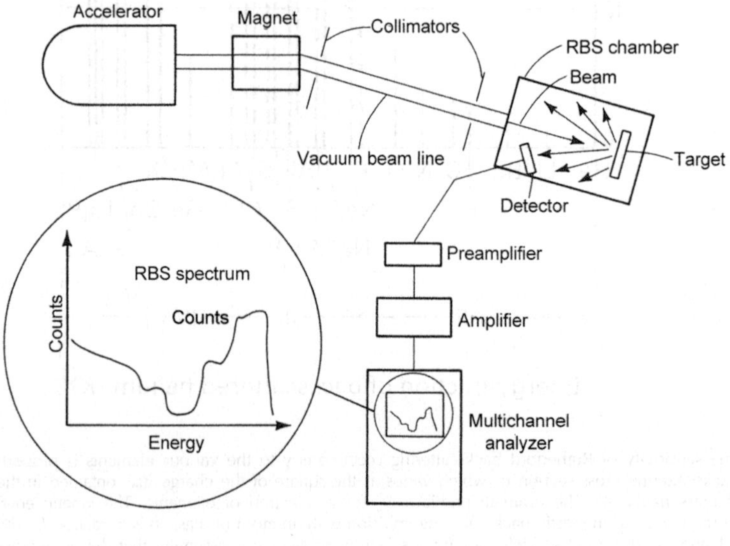

Fig. 4 System for backscattering analysis and signal processing. RBS, Rutherford backscattering spectrometry

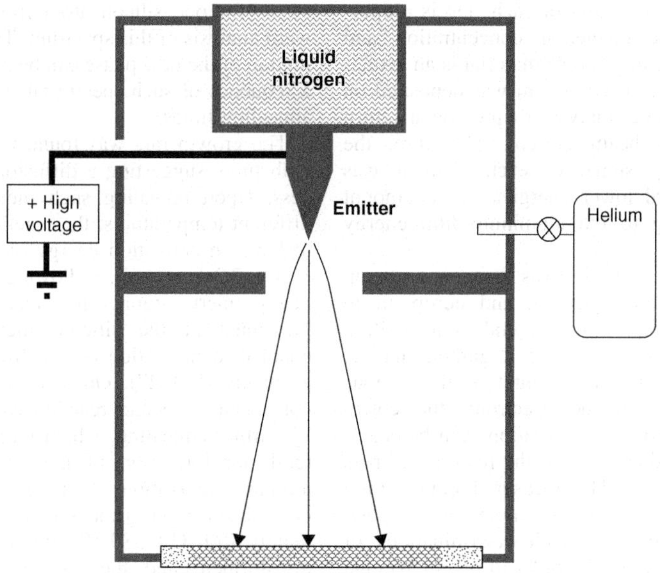

Fig. 5 Schematic of a helium ion microscope. Source: Ref 24

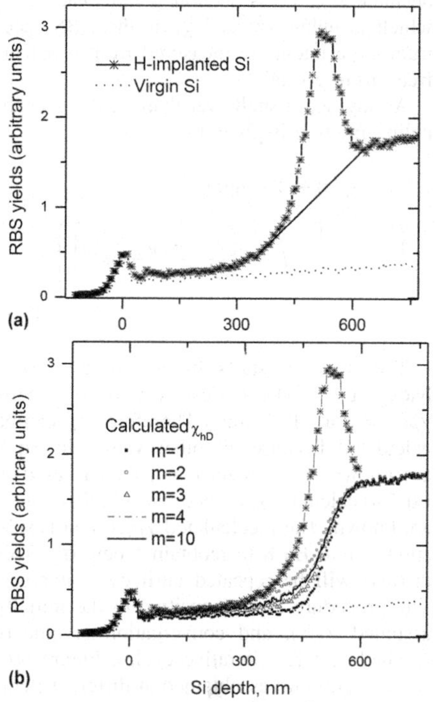

Fig. 6 Channeling Rutherford backscattering spectrometry (RBS) spectrum and calculated dechanneling component for hydrogen-implanted silicon. (a) Line approximation. (b) Double-iteration procedure after different cycles (*m*) of the iterative process. Source: Ref 25

Major Steps of RBS Data Analysis

This section discusses channel-energy conversion, energy-depth conversion, and separation of the dechanneling background as the main steps of RBS data analysis.

Channel-Energy Conversion

Channel numbers in a backscattering spectrum are related to the energies of the backscattered particles, with their conversions to energies adjustable by instrument setups including detector bias, preamplifier, main amplifier, analog-to-digital convertor, and beam-sample-detector configurations. The channel-energy conversion can be easily determined by using calibration samples, which contain several elements of big mass differences on the sample surface. The corresponding energies of the backscattered ions from each element on the surface are calculated using collision kinematics. A linear fitting of the plot of the channel numbers versus the energies produces the channel-energy conversion for experiments performed under the same instrument setups.

Energy-Depth Conversion

After channel-energy conversion, the energy-depth conversion is performed by using:

$$t = \frac{kE_0 - E(t)}{\frac{k}{\cos\phi_1}\left|\left(\frac{dE}{dx}\right)_{E_0}\right| + \frac{1}{\cos\phi_2}\left|\left(\frac{dE}{dx}\right)_{E}\right|} \quad \text{(Eq 5)}$$

where $E(t)$ is the energy recorded by a detector for particles backscattered at the depth t. ϕ_1 and ϕ_2 are the beam incident angle and the particle exit angles, with respect to sample normal, respectively. $(dE/dx)_{E_0}$ is the energy loss for particles in its inward path, and $(dE/dx)_E$ is the energy loss for particles in its outward path.

Equation 5 is valid for thin-layer characterization, assuming the energy loss is constant within the layer. For a thick layer in which energy loss changes along the projectile penetration, depth conversion requires an iteration approach.

Separation of Dechanneling Background

Yields in a channeling RBS spectrum are contributed by backscattering of channeled particles from displacements within the channel and backscattering of dechanneled particles from target lattice atoms. The yield separation between these two contributions is necessary to extract defect depth profiles in a crystalline target. The normalized yield (normalized to the random spectrum yield) at depth z is given by:

$$\chi_D(z) = \chi_R(z) + [1 - \chi_R(z)]n_D(z)/n \quad \text{(Eq 6)}$$

where n_D is the density of displaced atoms, χ_R is the fraction of the particles dechanneled, and n is the total atomic density.

Line Approximation

Line approximation is the easiest way to separate dechanneling contributions. At depths immediately before and after the damage peak, a straight line is plotted. The yield above the line is approximated to be $\chi_D(z)$. Figure 6(a) shows one example of line approximation.

Iterative Procedure

In a more accurate approach, the substrate is divided into thin layers. Starting from the first layer, assuming $\chi_R(z = 0) = 0$, $\chi_D(z = 0)$ is calculated. In the second layer, the χ_R contributed from the first layer χ_D is calculated and used to obtain the second layer χ_D. This procedure is repeated for all subsequent layers. In each layer, dechanneling background from accumulated dechanneling by χ_D in all shallower layers is considered. This procedure, however, requires the knowledge of the dechanneling cross sections, which can be calculated from classical scattering theory.

Double-Iterative Procedure

The double-iteration procedure does not require the knowledge of dechanneling cross sections (Ref 25). For a depth beyond a disorder region (denoted as z_1), there is no displacement at this depth, and the yield is given by:

$$\chi_D(z_1) = \chi_V(z_1) + [1 - \chi_V(z_1)] \left\{1 - \exp\left[-\int_0^{z_1} \sigma_D \frac{\chi_D(z') - \chi_R(z')}{1 - \chi_R(z')} n dz'\right]\right\} \quad \text{(Eq 7)}$$

where χ_D is the yield from a defective solid; χ_R is the dechanneled fraction; z is the depth; n is the crystal atom density; σ_D is the dechanneling cross section per displacement, which is unknown; and χ_V is the RBS spectrum experimentally measured from a defect-free virgin crystal.

At any depth shallower than z_1, the dechanneling fraction is given by:

$$\chi_R(z) = \chi_V(z) + [1 - \chi_V(z)] \left\{ 1 - \exp\left[- \int_0^z \sigma_D \frac{\chi_D(z') - \chi_R(z')}{1 - \chi_R(z')} n dz' \right] \right\} \quad \text{(Eq 8)}$$

The iteration starts by assuming a $\chi_R(z)$ background. (The easiest way is to assume $\chi_R(z) = 0$ at all depths.) Therefore, σ_D can be calculated because all other values in Eq 8 are known. The obtained σ_D then is used in Eq 7 to calculate $\chi_R(z)$ because all other values are known. The recalculated $\chi_R(z)$ then is substituted into Eq 8 to reobtain a new σ_D. This process will be repeated until σ_D converges. The procedure is not sensitive to the initially assumed $\chi_R(z)$, and convergation is quickly reached in a few iterative cycles. Figure 6(b) shows $\chi_R(z)$ curves obtained at different iterative cycles (Ref 25).

Applications

Composition of Bulk Samples

An example of the composition analysis of bulk material is shown in Fig. 7. The sample consists of a magnetic bubble material grown on a gallium gadolinium garnet. Although the bubble material actually is a film, its thickness is much greater than the measurable depth; thus, the film appears as a bulk material.

The spectrum consists of a series of steps, each generated by one of the elements in the material. The position of the step indicates the mass of the element (arrows) as obtained from Eq 1. The height of each step is proportional to the elemental concentration and to σ. Because this bubble material is an insulator, a thin film of aluminum was deposited on top of it before analysis to provide a return path for the beam current. Therefore, the backscattering signal of each element was shifted toward lower energies by an amount corresponding to the aluminum film energy attenuation.

The bubble material was known to contain iron, gadolinium, yttrium, and europium in nominal amounts; the overall composition was known to be that of a garnet, that is, X_8O_{12}. From the step height of the various metal signals in the spectrum, the corresponding relative concentrations can be calculated by dividing through the respective cross sections (Eq 3). The measured composition agrees favorably with the nominal composition. The errors for such determinations of composition are of the order of 5% in average cases and 1 or 2% in the optimum cases.

Thin-Film Composition and Layer Thickness

Thin-film composition and layer thickness can be obtained routinely using RBS, which has become a powerful technique in thin-film analysis. Figure 8 illustrates an example of thin-film study, showing the changes in a nickel film on silicon as-deposited and after heat treatment at 250 °C (480 °F) for 1 and 4 h, respectively (Ref 26). A new phase develops at the interface with an atomic composition that may be identified as two nickel atoms for one silicon atom from straightforward analysis of this spectrum. The rate of formation of the new phase can be measured from a sequence of such spectra taken after various annealing times.

The growth rate was found to be parabolic with time, suggesting a diffusion-limited process. Upon repeating such measurements at different temperatures, the reaction was found to have an activation energy of 1.5 ± 0.1 eV (3.4 ± 2.3 kcal/mol). Following the signal of heavy inert atoms, for example, xenon, implanted at the silicon/ nickel interface, enabled identification of the diffusing species as nickel (Ref 27). Once the nickel is fully consumed, a second reaction can be initiated by further annealing, which terminates in the total transformation of the layer to a new atomic composition of one silicon atom for one nickel atom (not shown). At approximately 750 °C (1380 °F), this layer in turn is transformed, and the silicon-to-nickel ratio increases 2:1. This last layer grows epitaxially, that is, under preservation of the crystalline orientation of the substrate—a fact established using RBS by channeling, which is present only when the target is a single crystal.

Impurity Profiles

Depth profiles of heavy element impurities in a light substrate can be obtained easily by using RBS. Figure 9 depicts RBS profiles of arsenic in silicon on three ion-implanted silicon samples, showing the energy spectrum of 2.0 MeV ^4He ions backscattered from a silicon

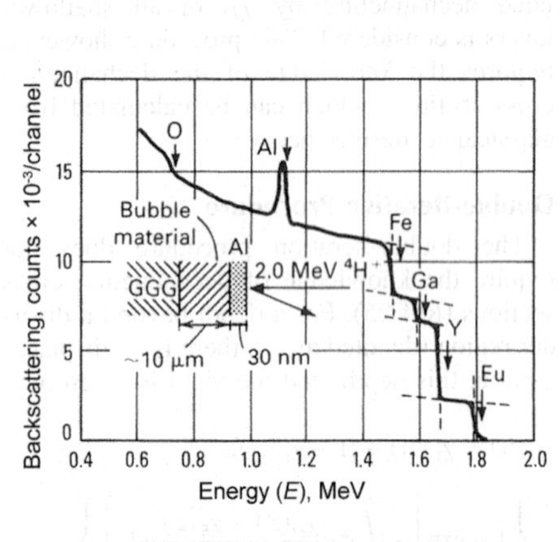

Fig. 7 Backscattering spectrum of a thick target consisting of magnetic bubble material. The material was known to have the garnet composition X_8O_{12}. The spectrum yields a similar composition ratio. The measured composition was $Y_{2.57}Eu_{0.48}Ga_{1.2}Fe_{3.75}O_{12}$; the nominal composition, $Y_{2.45}Eu_{0.55}Ga_{1.2}Fe_{3.8}O_{12}$. GGG, gallium gadolinium garnet

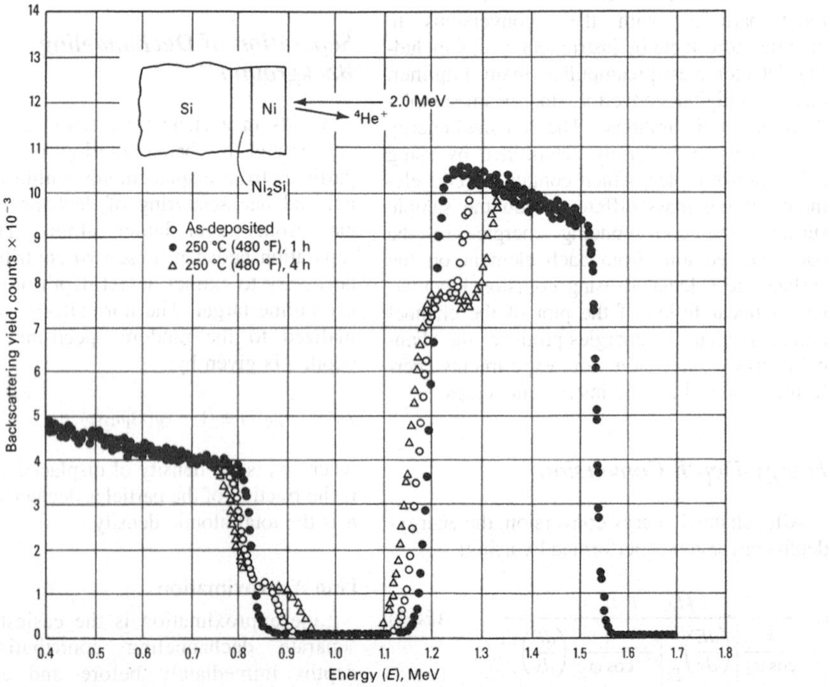

Fig. 8 Energy spectra of 2.0 MeV $^4He^+$ ions backscattered from a sample of 200 nm nickel on a surface of silicon before and after annealing at 250 °C (480 °F) for 1 and 4 h. Ni_2Si formation is visible for the Rutherford backscattering spectrometry spectra on the two annealed samples. Source: Ref 26

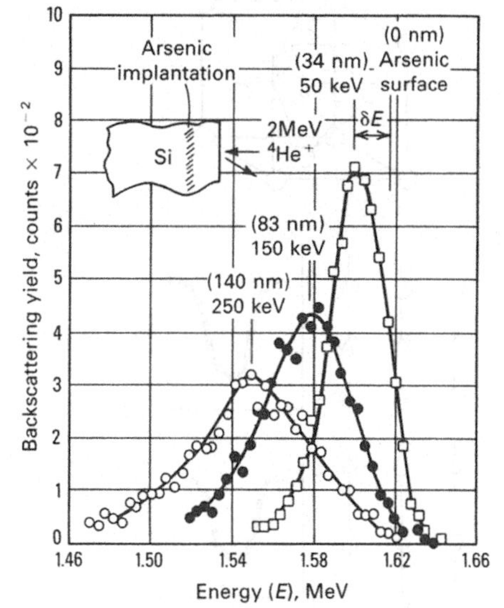

Fig. 9 2.0 MeV ^{4}He Rutherford backscattering spectrometry spectra of arsenic-implanted silicon samples

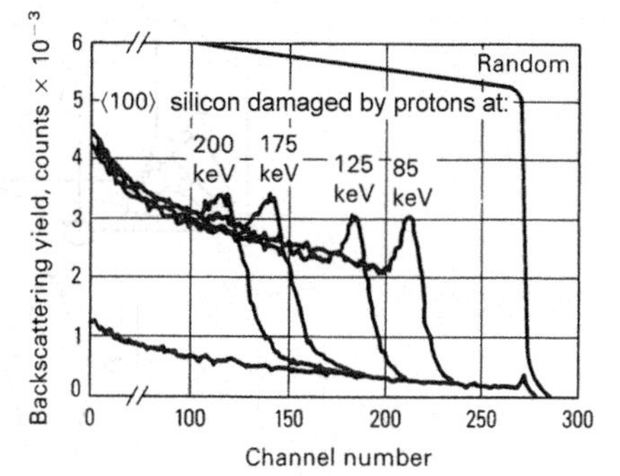

Fig. 10 Channeled backscattering spectra of 2.4 MeV ^{4}He ions from silicon bombarded with 4 × 10^{16} protons/cm^2 at various energies. Channeled and random spectra for unbombarded silicon are shown as the background. Source: Ref 29

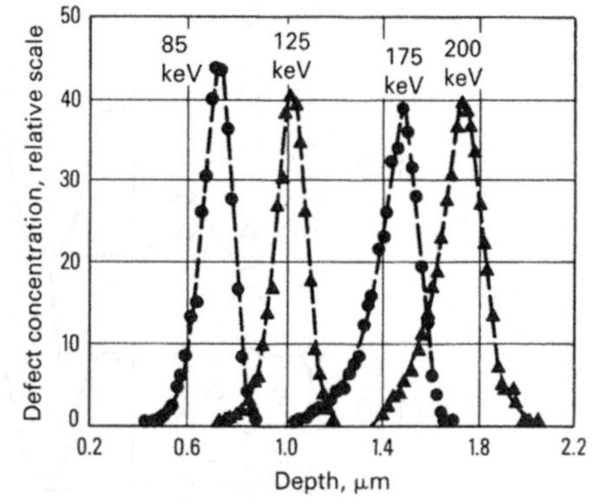

Fig. 11 Defect distribution extracted from Fig. 10. Source: Ref 29

sample implanted with 2 × 10^{15} As/cm^2 at 50, 150, and 250 keV. The peak positions are below the surface by an energy ΔE that corresponds to the projected range R_p of the arsenic distribution. The surface position of arsenic is defined by calculation from Eq 1 or by scattering from a GaAs calibration sample. The full width at half maximum (FWHM) of the arsenic spectrum corresponds to a FWHM of a depth distribution after subtraction from detector resolution and energy straggling. A detailed analysis of this example is given in Ref 28. A substitutional portion of arsenic can be obtained by channeling. Peak concentration can be obtained from the scattering cross section ratio of arsenic to silicon and from comparing peak highs of arsenic to silicon.

Damage Depth Profile

Ion channeling, in conjunction with ion backscattering measurements, has been used extensively to study near-surface defects in single crystals. Most defect studies by channeling involve backscattering of light particles, such as protons or helium ions, with energies of a few hundred kiloelectron volts to a few megaelectron volts.

Figure 10 illustrates the conversion from channeling spectrum to depth distribution of the defect, representing channeled backscattering spectra of 2.4 MeV He^+ ions from single-crystal silicon prebombarded with high-dose H^+ at various energies (Ref 29). Figure 11 shows a direct translation from energy spectra to the buried defect distribution. The defect concentration given in Fig. 11 is on a relative scale. Assuming the peaks in Fig. 10 are produced by the scattering of silicon atoms randomly distributed in the defect region, the relative concentration scale may be considered to be the percentage of silicon atoms randomly displaced from lattice sites. Unless there is clear evidence that the irradiation has produced a totally amorphous surface layer, the term *displaced atoms* should be replaced by *scattering* and *dechanneling* to describe the degree of disordered atoms or defects in the crystal (Ref 30). Scattering and dechanneling should not be used to measure the number of disordered atoms or defects in the crystal; rather, they are useful indicators of these quantities.

Backscattering is useful for determining the depth distribution of the disorder, but transmission electron microscopy is better for defining the nature of the defect. The energy dependence of dechanneling can be observed to obtain information on the nature of the defects. Additional information is given in Ref 17 to 19.

Surface Peak

The use of ion scattering to study surface structure by channeling and by blocking is a well-developed technique. Detailed methods and examples are available in Ref 17 to19. Rutherford backscattering spectrometry channeling of silicon sputtered by low-energy noble gas can provide information on sputter-induced crystal damage. The quantitative number of displaced silicon atoms is obtained by subtracting the surface peak contributions from the native oxide and the normal surface peak for crystalline silicon. Noble gas atoms heavier than silicon, such as argon, krypton, and xenon, also can be incorporated. Similar studies have been extended to reactive ion etching involving fluor- or chlorine-containing gases in combination with argon (Ref 31).

Another example using channeling on surfaces is the study of melting of single-crystal lead (Ref 32). This example shows that a simple analysis of surface peaks yields fundamental knowledge of melting phenomena.

Lattice Location of Dopant Atoms

Channeling can reveal the lattice location of the solute atoms in single-crystal solid solutions. In general, a foreign atom species dissolved in a single crystal could take the position of a substitutional site, a well-defined interstitial site, a well-defined displacement from a substitutional site, or a random distribution within the lattice.

Channeling can show the foreign atom location in a lighter matrix. Foreign atoms sometimes may take more than one of these positions. Superposition of the different locations produces superposition of various channeling results, making it difficult to provide a unique answer on the lattice locations.

Substitutional Case

From the analytical point of view, detectable foreign atoms located at the substitutional site are the simplest case in channeling characterization. The yield attenuation of the foreign atoms along all axial and planar channeling directions should be identical to that of the host lattice.

Gold atoms in single-crystal copper assume the substitutional site perfectly (Ref 33). Figure 12 shows the angular yield profiles of the normalized backscattering yield of 1.2 MeV helium ions from gold and from copper atoms in a single-crystal copper sample containing 2% Au. The identical half-angle on the angular scan for impurity and first atoms indicates that the gold is substitutional and that the gold atoms are completely shadowed by the copper atoms. This 100% substitutional case is rarely observed. A small percentage of impurity usually occupies a nonsubstitutional site, producing a higher minimum yield for the backscattering signals from the impurity atoms.

An angular scan along more than one channeling direction often is required to deduce the lattice location. Planar and axial channeling are useful in the triangulation analysis, especially when foreign atoms take interstitial locations.

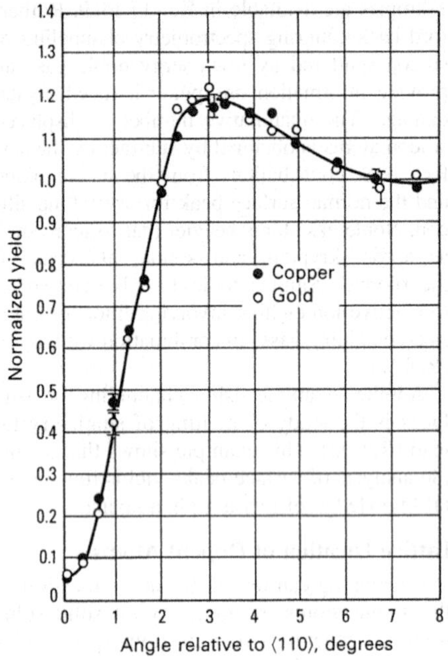

Fig. 12 Angular yield profiles of 1.2 MeV ^{4}He ions backscattering from gold and copper atoms in single-crystal copper containing 2 at.% Au. Source: Ref 33

Interstitial Case

For a single cubic crystal, an interstitial site is located at the center of the cube. Body-centered cubic (bcc), face-centered cubic (fcc), or diamond structures have more than one type of interstitial site. Figure 13 illustrates the tetrahedral interstitial and octahedral interstitial site of an fcc single crystal. For the (100) planar direction, the tetrahedral interstitials are located between the (100) planes, and the angular scans indicate a peak for the foreign atom at the well-channeled direction. This is not the case for octahedral interstitials, which are hidden behind the host atoms in (100) and (110) directions.

For planar channeling cases, the ion beam is incident parallel to the lattice planes, which are defined by the atoms on the plane. For axial channeling cases, the ion beam is incident parallel to the major axial direction. Flux peak is shown in ⟨100⟩ tetrahedral sites and ⟨110⟩ octahedral sites, and good shadowing in ⟨110⟩ tetrahedral interstitials. The foreign atoms are off-center of the channel at a well-defined distance, not on center. A double peak can be expected. Through careful triangulation in various channeling directions, a well-defined lattice location can be determined. Examples of hydrogen locations in an fcc crystal (Ref 34) and a bcc crystal (Ref 35, 36) are the classic demonstrations of this type.

Small Displacement from Substitutional Site

Many cases indicate that favorable channeling conditions for impurity atoms are observed. However, the width of the angular scan from the impurities sometimes is much narrower than that of the host atoms. Angular width narrowing indicates small displacement from a substitutional site. Foreign atoms displace from a lattice site for many reasons. For example, foreign atom and vacancy pairs could move the foreign atom off the lattice site (toward the vacancy). Small foreign atom clusters could have slightly different bond lengths between them that force the foreign atom clusters to exit the lattice site. The cause of the displacement is unimportant. The amount of the displacement can be measured by observing the amount of half-angle narrowing versus that of the angular scan of the host atoms.

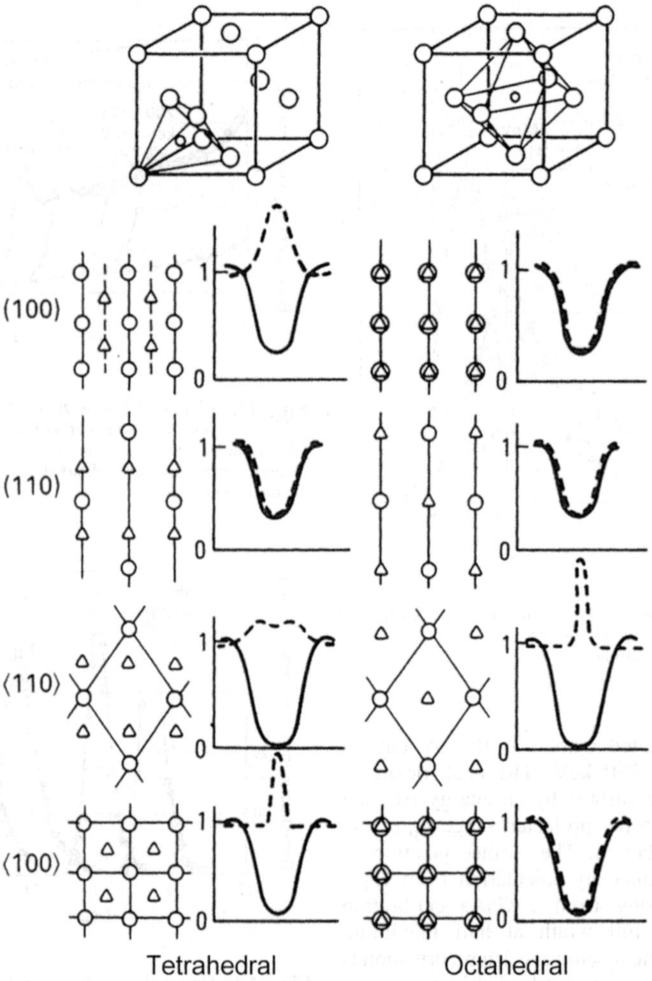

Fig. 13 Impurity atom taking tetrahedral and octahedral interstitial sites for face-centered cubic crystals. The schematics show the lattice planes and strings and their corresponding form of the angular yield profiles. Source: Ref 34

This concept can be illustrated by noting that when silicon crystals are heavily doped with arsenic, arsenic atoms tend to form small clusters and are displaced from the lattice by approximately 0.015 nm. When the sample is subjected to laser irradiation, arsenic clusters dissolve in the silicon solution, and arsenic atoms take an exact substitutional site. The channeling study of atom displacement has been discussed previously (Ref 37, 38).

Superlattice/Interface Studies

Channeling can be applied to interfacial problems, especially the investigation of strained layered superlattices. Superlattices are alternating layered structures of fundamental interest and importance for potential application in electronic and optical devices. Lattice matching of heteroepitaxy has been emphasized in the materials selection for superlattice layers. However, high-quality superlattices also can be grown from lattice-mismatched materials; these are termed strained-layer superlattices (SLSs).

In SLS structures, the lattice mismatch is accommodated by uniform layer strains to prevent generation of misfit defects for sufficiently thin layers (Ref 39, 40). This built-in strain provides additional flexibility in tailoring superlattice properties, including the band gap and transport parameters. These both depend strongly on the amount of strain in the SLS. Therefore, strain measurements in

SLSs are important for the characterization of these materials. Relative to channeling, strain can be measured in SLSs by the dechanneling method, by angular scan, and by a resonance effect.

Lattice Strain Measurement by Dechanneling

Early studies of InAs/GaSb superlattices found anomalously large dechanneling along the inclined [110] axis relative to the [100] growth direction. These results were interpreted in terms of a change in lattice spacing at the interfaces. However, computer simulation calculations show that the expected lattice spacing differences provide significantly less dechanneling than was observed, but the dechanneling is consistent with that expected due to the strain in the layers.

Recent studies have shown that dechanneling provides a depth-sensitive monitor of the strain in SLS structures. For a given layer thickness, this technique can be used to determine the maximum amount of strain that could be incorporated into SLS structures through different growth techniques. Quantitative analysis of the magnitude of the strain is more difficult. The only general quantitative analysis available is the use of computer simulation calculations for comparison to experiment. More such calculations are needed to establish general scaling rules for the tilt and thickness dependence of the dechanneling.

Lattice Strain Measurement by Channeling Angular Scan

A second approach to the measurement of strain in SLS structures is to examine the channeling angular scans along inclined directions where the crystal rows are tilted. Figures 14 and 15 show an example.

Studies (Ref 41) indicate that backscattering and channeling measurements of the relative changes in the channel direction in combination with computer simulation are sensitive to relative changes in the channel direction; the strains result in various crystal distortions that alter symmetry directions slightly. Angular scan measurements along ⟨110⟩ and ⟨100⟩ directions with an error of ±0.03° allow bond angle distortion as small as 0.1° to be measured in superlattices. In a two-layered system, pseudomorphic growth of Ge_xSi_{1-x} on silicon by molecular beam epitaxy has been studied by the ion channeling technique (Ref 42); channeling studies indicated that for thin epitaxial layers, lattice mismatch may be accommodated by strain rather than dislocation formation. This pseudomorphic growth condition may be maintained at alloy compositions or layer thicknesses substantially greater than those predicted by equilibrium theory.

Resonance Effect

When an ion beam enters a single crystal in a direction parallel to a set of planes, the collective atomic potential steers the ions back and forth between the planes. This oscillation during channeling can be observed through the oscillations in the backscattering spectra (Fig. 16). A resonance effect in planar dechanneling when the wavelength of the channeled-particle oscillations is matched to the period of an SLS has been observed (Fig. 17). A simple phase-rotation model is developed to calculate the dechanneling on each layer. The angular dependence of resonance channeling has been studied; its utility in superlattice studies has been extended using more sophisticated potential and detail treatment (Ref 44). This type of analysis provides a means of measuring small lattice strain and establishes a basis for using planar-channeled focusing for the structural study of interface phenomena, such as impurity location, interface reconstruction, and other structural effects.

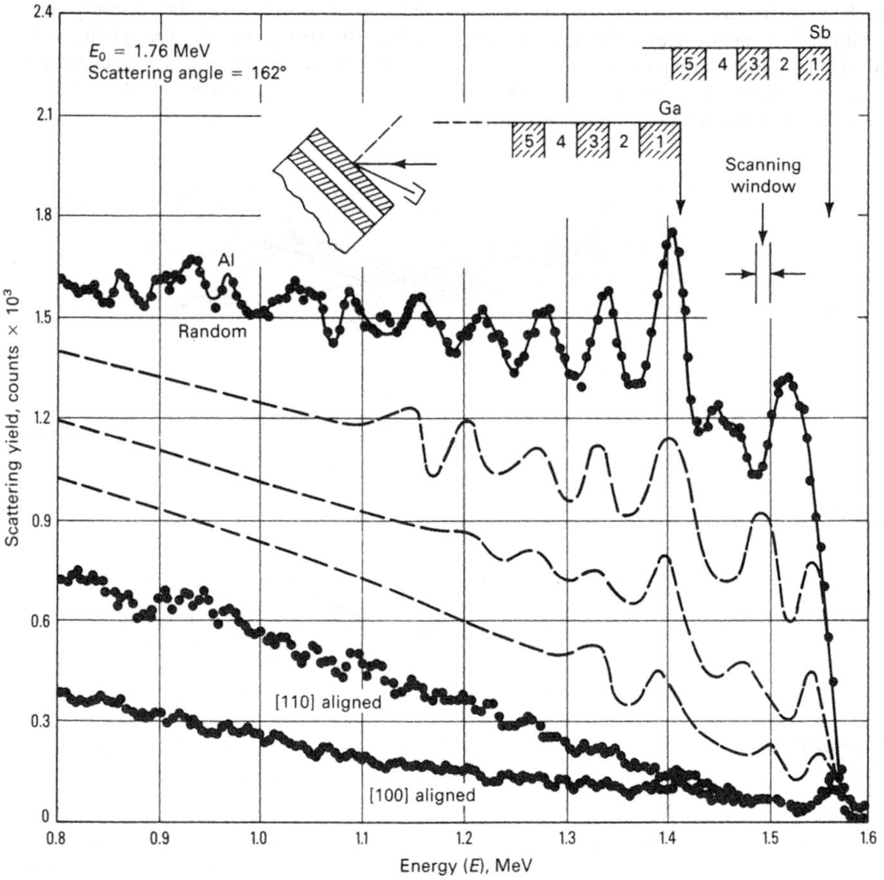

Fig. 14 Energy spectra of 1.76 MeV He⁺ ions backscattered from [100] GaSb/AlSb superlattices. Depth scales based on antimony and gallium signals are marked in units of the number of layers (30 nm per layer). [100]- and [110]-aligned spectra, a random spectrum taken at an angle of 3° relative to the [110] direction, and three more spectra between the [110] and random spectra are given. Source: Ref 40

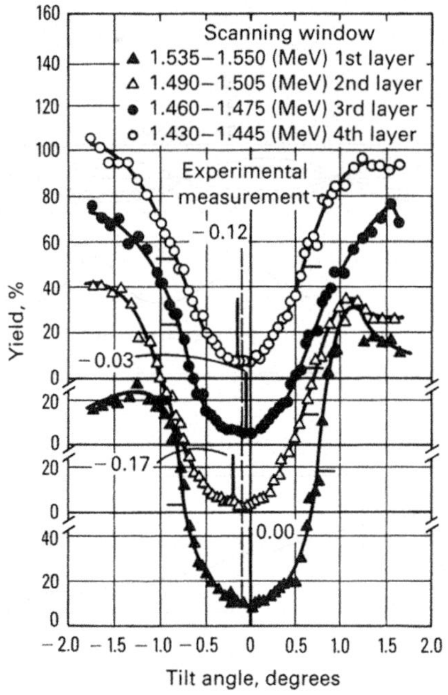

Fig. 15 Angular scan performed by setting an energy interval (window) from the first to fourth layer from 52 spectra run at 52 different angles. The center position of the angular scan changes from layer to layer, indicating that the ⟨110⟩ direction varies. The vertical dashed line is the limiting center position for layers much deeper in the sample. Source: Ref 40

Characterization of Nanostructures

By using the helium ion microscope, a beam spot of subnanometer size can be scanned over a sample surface for various characterizations. Secondary electrons, photons, and backscattered helium ions can be collected to obtain topographic (due to secondary electron yield sensitivity on the surface slope), composition information (due to helium backscattering energy dependence on target atomic numbers), electrical information (due to beam-induced voltage contrast due to capacitive/resistive effects), and crystallographic effects (due to channeling). Under channeling conditions, the helium backscattering yield is reduced. In addition, secondary electron yields are reduced as well due to reduced dE/dX. Hence, both yields can be collected to show crystallographic information.

Rutherford Backscattering Spectrometry Simulation Codes

Modeling of Random RBS

Various codes have been developed to simulate RBS and other ion beam analysis spectra. The codes can be roughly divided into two major approaches: the Monte Carlo approach and the analytical approach. Regardless of their detail approaches, codes must consider energy-loss straggling, geometrical straggling, multiple small-angle scattering, plural large-angle scattering, and surface roughness. All of them have data-fitting capabilities to extract composition information. A comparison of various codes has been organized by the International Atomic Energy Agency (Ref 45). As shown in Fig. 18, all participant codes show good agreement. For the interest of the readers, technical contact information of various codes is listed in Table 1.

The popular Monte Carlo codes include MCERD (Ref 46) and CORTEO (Ref 47). In these codes, trajectories of projectiles are recorded through binary collision approximation. The computation cost is relatively high due to the need to store lattice locations of projectiles and crystallography information. Advantages in Monte Carlo approaches are the accuracy to consider multiple small- and large-angle scattering effects, and the flexibility to consider complicated detector geometries.

The popular analytical codes include RUMP (Ref 48), SIMNRA (Ref 49), DEPTH (Ref 50), GISA (Ref 51), DataFurnace (NDF) (Ref 52), and RBX (Ref 53). In these codes, particle trajectories are approximated as straight lines, with multiple small-angle scattering effects and geometrical effects due to finite detector sizes added as additional corrections to the beam energies. Such approximation is valid if particle energies are high. The analytic approaches have the main advantage of low computation costs.

Using SIMNRA as one example, SIMNRA code can simulate RBS, elastic backscattering spectrometry with non-Rutherford cross sections, nuclear reaction analysis, elastic recoil detection analysis, and particle-induced gamma-ray emission. The program can simulate arbitrary ion-target combinations with SigmaCalc included for non-Rutherford backscattering and nuclear reaction analysis (Ref 54).

SIMNRA allows the use of arbitrary layer thickness distribution functions as input. The spectra are calculated by using a linear superposition of subspectra. Figure 19 compares the simulated and experimental RBS spectra (Ref 55).

Modeling of Channeling RBS

Principles of Continuum Approximation

For a series of correlated collisions, it was shown by Lindhard that the ion can be considered to move in a transverse potential, V_T, that results from an averaging of the potentials of each atom in a string (Ref 16). This average (continuum) potential thus is expressed as:

$$V_T(\rho) = \frac{1}{d}\int_{-\infty}^{\infty} V\left[\left(\rho^2 + x^2\right)^{1/2}\right] dx \qquad \text{(Eq 9)}$$

where ρ is the distance of the ion from the atom string, x is the distance traveled along the string, and d is the spacing between atoms in the string. An average potential is used if the string consists of different atomic species.

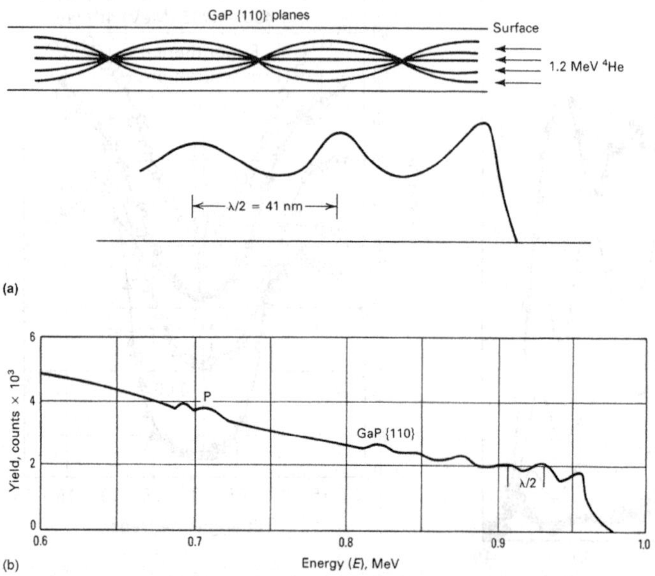

Fig. 16 Resonance effect (oscillations of the ion beam) in Rutherford backscattering spectrometry channeling. (a) Focusing and defocusing that produces oscillations. (b) Backscattering spectrum of 1.2 MeV ^{4}He ions along the {110} plane in a single GaP crystal. The oscillations in the planar-channeled spectrum provide a measure of wavelength. Source: Ref 43

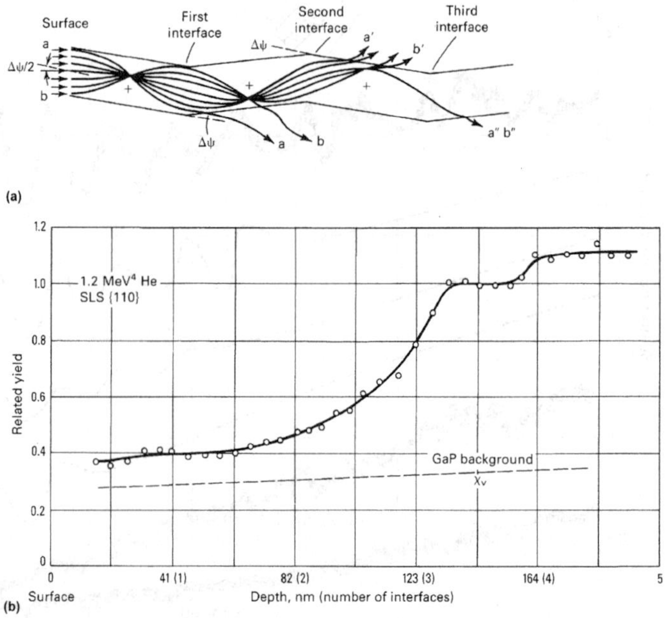

Fig. 17 Resonance effect in planar dechanneling, with the wavelength of the channeled-particle oscillations matched to the period of a strained-layer superlattice (SLS). (a) Catastrophic planar dechanneling condition for matching the wavelength of the beam (82 nm) to the effective pair layer thickness of an SLS. Tracing of the position and direction of the beam is based on the phase rotation analysis. (b) Aligned {110} yield versus depth for 1.2 MeV ^{4}He backscattering from a GaP crystal (background) and from an SLS with a $GaAs_{0.12}P_{0.88}/GaP$ path length per layer of 41 nm. Source: Ref 43

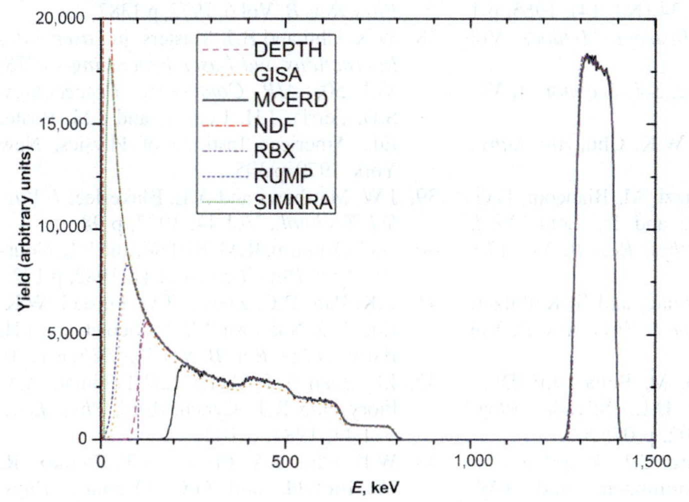

Fig. 18 Simulated Rutherford backscattering spectrometry spectrum from silicon bulk/SiO_2 200 nm/Au 50 nm. Source: Ref 45

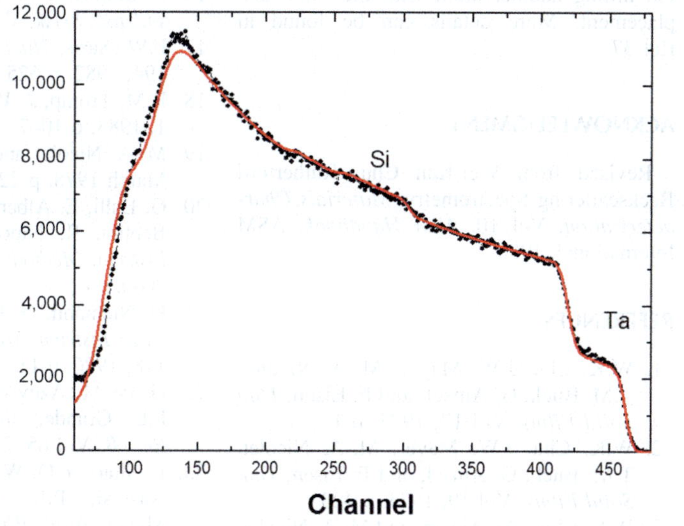

Fig. 19 Experimental and simulation spectra. Source: Ref 55

Due to steering under a channeling condition, ions move toward the channel center, resulting in a higher flux there. Ignoring the energy loss and assuming that the transverse component, $E_\perp$, of the ion energy is conserved:

$$E_\perp = E\psi^2 + V_T(\rho_{in}) \quad \text{(Eq 10)}$$

where ψ is the initial incident angle of the analysis beam, and ρ_{in} is the distance between the atomic row and the initial striking position.

Upon reaching statistical equilibrium, ions are uniformly distributed within an accessible area specified by:

$$E_\perp = V_T(\rho_A) \quad \text{(Eq 11)}$$

Under these conditions, a simple estimate of the flux distribution within a channel can be made for a given value of ψ.

Angular Distribution

When ions are incident with angle ψ, the yield of close encounters is given by:

$$\chi(\psi) = \int F(r)\, dP(r) \quad \text{(Eq 12)}$$

where P(r) is the displacement probability. Selecting different ψ values, the angular distribution is obtained by repeating the aforementioned calculation procedure.

Half-Way-Plane Approach

In the half-way-plane approach, a displacement is assumed to emit particles isotropically, and the distribution in the transverse energy of these particles is calculated at a plane halfway between the atoms in the row. Changes in the transverse energy distribution of these particles during their exits toward the surface are neglected. The calculated angular distribution is directly related to the probability of one

Table 1 General information on various Rutherford backscattering spectrometry simulation codes

Analysis program	Technical contact	Operating systems	Distribution mode	Status of source code
DEPTH	Edit Szilágyi KFKI Research Institute for Particle and Nuclear Physics Budapest, Hungary szilagyi@rmki.kfki.hu; www.kfki.hu/~ionhp/	DOS (or emulators), Windows	No charge. Downloadable from the web	Restricted to author—not available
GISA	Eero Rauhala University of Helsinki Helsinki, Finland Eero.rauhala@helsinki.fi and Jaakko Saarilahti Technical Research Center of Finland Jaakko.Saarilahti@vtt.fi	DOS (or emulators)	No charge. Write to author for copy	Restricted to author—not available
MCERD	Kai Arstila University of Helsinki Helsinki, Finland and IMEC Leuven, Belgium Kai.Arstila@iki.fi	Linux, UNIX	No charge. Write to author for copy	Source code available
NDF: DataFurnace	Nuno Barradas Technological and Nuclear Institute Sacavem, Portugal nunoni@itn.pt; https://www.surrey.ac.uk/ion-beam-centre/research-areas/ion-beam-analysis	Windows, UNIX	Commercial through University of Surrey. Evaluation copies available by request	Restricted to author—not available
RBX	Endre Kótai KFKI Research Institute for Particle and Nuclear Physics Budapest, Hungary kotai@rmki.kfki.hu	Windows	No charge. Write to author for copy	Restricted to author—not available
RUMP	Mike Thompson Department of Materials Science Cornell University Ithaca, NY motl@cornell.edu; www.genplot.com	Windows, Linux, UNIX, OS2	Commercial through Computer Graphics Service. Evaluation copies available on web	Source code available
SIMNRA	Matej Mayer MPI for Plasma Physics Garching, Germany Matej.Mayer@ipp.mpg.de; www.rzg.mpg.de	Windows	Commercial through MPI for Plasma Physics. Evaluation copies available on web	Restricted to author—not available

Source: Ref 45

ion hitting another atom with the same displacement. More details can be found in Ref 37.

ACKNOWLEDGMENT

Revised from Wei-Kan Chu, Rutherford Backscattering Spectrometry, *Materials Characterization*, Vol 10, *ASM Handbook*, ASM International, 1986

REFERENCES

1. W.K. Chu, J.W. Mayer, M.-A. Nicolet, T.M. Buck, G. Amsel, and F. Eisen, *Thin Solid Films*, Vol 17, 1973, p 1
2. W.K. Chu, J.W. Mayer, M.-A. Nicolet, T.M. Buck, G. Amsel, and F. Eisen, *Thin Solid Films*, Vol 19, 1973, p 423
3. W.K. Chu, J.W. Mayer, and M.-A. Nicolet, *Backscattering Spectrometry*, Academic Press, 1978
4. J.F. Ziegler and R.F. Lever, *Thin Solid Films*, Vol 19, 1973, p 291
5. J.R. Macdonald, J.A. Davies, T.E. Jackman, and L.C. Feldman, *J. Appl. Phys.*, Vol 54, 1983, p 1800
6. P.P. Pronko, B.R. Appleton, O.W. Holland, and S.R. Wilson, *Phys. Rev. Lett.*, Vol 43, 1979, p 779
7. J.H. Barrett, B.R. Appleton, and O.W. Holland, *Phys. Rev. B*, Vol 22, 1980, p 4180
8. J.F. Ziegler and W.K. Chu, *At. Data Nucl. Data Tables*, Vol 13, 1974, p 463
9. J.F. Ziegler, *Helium: Stopping Powers and Ranges in All Elemental Matter*, Pergamon Press, 1978
10. L.C. Feldman, J.W. Mayer, and S.T. Picraux, *Materials Analysis by Ion Channeling*, Academic Press, 1982
11. J. Stark, *Phys. Z.*, Vol 13, 1912, p 973
12. M.T. Robinson and O.S. Oen, *Appl. Phys. Lett.*, Vol 2, 1963, p 30
13. M.T. Robinson and O.S. Oen, *Phys. Rev.*, Vol 132, 1963, p 2385
14. G.R. Piercy, F. Brown, J.A. Davies, and M. McCargo, *Phys. Rev. Lett.*, Vol 10, 1963, p 399
15. R.S. Nelson and M.W. Thompson, *Philos. Mag.*, Vol 8, 1963, p 1677
16. J. Lindhard, *Mat. Fys. Medd. K. Dan. Vidensk. Selsk*, Vol 34 (No. 14), 1965, p 1
17. F.W. Saris, *Nucl. Instrum. Methods*, Vol 194, 1982, p 625
18. R.M. Tromp, *J. Vac. Sci. Technol. A*, Vol 1, 1983, p 1047
19. M.-A. Nicolet and W.K. Chu, *Am. Labs.*, March 1978, p 22
20. G. Lulli, E. Albertazzi, M. Bianconi, G.G. Bentini, R. Nipoti, and R. Lotti, *Nucl. Instrum. Methods Phys. Res. B*, Vol 170, 2000, p 1
21. D. Niemann, G. Konac, and S. Kalbitzer, *Nucl. Instrum. Methods Phys. Res. B*, Vol 118, 1996, p 11
22. G. de M. Azevedo, M. Behar, J.F. Dias, P.L. Grande, and D.L. Silvada, *Phys. Rev. B*, Vol 65, 2002, p 075203
23. L. Shao, Y.Q. Wang, C.J. Wetteland, M. Nastasi, P.E. Thompson, and J.W. Mayer, *Appl. Phys. Lett.*, Vol 86, 2005, p 221913
24. J. Notte, B. Ward, N. Economou, R. Hill, R. Percival, L. Farkas, and S. McVey, *AIP Conf. Proc.*, Vol 931, 2007, p 489
25. L. Shao and L.M. Nastasi, *Appl. Phys. Lett.*, Vol 87, 2005, p 064103
26. K.N. Tu, W.K. Chu, and J.W. Mayer, *Thin Solid Films*, Vol 25, 1975, p 403
27. W.K. Chu, H. Krautle, J.W. Mayer, H. Muller, M.-A. Nicolet, and K.N. Tu, *Appl. Phys. Lett.*, Vol 25, 1974, p 454
28. T.W. Sigmon, W.K. Chu, H. Muller, and J. W. Mayer, *Appl. Phys.*, Vol 5, 1975, p 347
29. W.K. Chu, R.H. Kastl, R.F. Lever, S. Mader, and B.J. Masters, *Phys. Rev. B*, Vol 16, 1977, p 3851
30. Y. Quéré, *Phys. Status. Solidi*, Vol 30, 1968, p 713
31. T. Mizutani, C.J. Dale, W.K. Chu, and T.M. Mayer, *Nucl. Instrum. Methods B*, Vol 7/8, 1985, p 825
32. J.W.M. Frenken and J.F. Veenvan der, *Phys. Rev. Lett.*, Vol 54, 1985, p 134
33. R.B. Alexander and J.M. Poate, *Radiat. Eff.*, Vol 12, 1972, p 211
34. J.P. Bugeat, A.C. Chami, and E. Ligeon, *Phys. Lett. A*, Vol 58, 1976, p 127
35. H.D. Carstangen and R. Sizmann, *Phys. Lett. A*, Vol 40, 1972, p 93
36. S.T. Picraux and F.L. Vook, *Phys. Rev. Lett.*, Vol 33, 1974, p 1216
37. S.T. Picraux, W.L. Brown, and W.M. Gibson, *Phys. Rev. B*, Vol 6, 1972, p 1382
38. W.K. Chu and B.J. Masters, in *Laser-Solid Interactions and Laser Processing—1978*, Vol 50, *AIP Conference Proceedings*, S.D. Ferris, J.H. Leamy, and J.M. Poate, Ed., American Institute of Physics, New York, 1979, p 305
39. J.W. Matthews and A.E. Blakeslee, *J. Vac. Sci. Technol.*, Vol 44, 1977, p 98
40. G.C. Osbourn, R.M. Biefeld, and P.L. Gourley, *Appl. Phys. Lett.*, Vol 41, 1982, p 172
41. C.K. Pan, D.C. Zheng, T.G. Finstad, W.K. Chu, V.S. Speriosu, M.-A. Nicolet, and J.H. Barrett, *Phys. Rev. B*, Vol 31, 1985, p 1270
42. J.C. Bean, T.T. Sheng, L.C. Feldman, A.T. Fiory, and R.T. Lynch, *Appl. Phys. Lett.*, Vol 44, 1984, p 102
43. W.K. Chu, J.A. Ellison, S.T. Picraux, R. M. Biefeld, and G.C. Osbourn, *Phys. Rev. Lett.*, Vol 52, 1984, p 125
44. S.T. Picraux, W.R. Allen, R.M. Biefeld, J. A. Ellison, and W.K. Chu, *Phys. Rev. Lett.*, Vol 54, 1985, p 2355
45. N.P. Barradas, K. Arstila, G. Battistig, M. Bianconi, N. Dytlewski, C. Jeynes, E. Kótai, G. Lulli, M. Mayer, E. Rauhala, E. Szilágyi, and M. Thompson, *Nucl. Instrum. Methods Phys. Res. B*, Vol 262, 2007, p 281
46. T. Sajavaara, K. Arstila, A. Laakso, and J. Keinonen, *Nucl. Instrum. Methods B*, Vol 161–163, 2000, p 235
47. F. Schiettekatte, *Nucl. Instrum. Methods Phys. Res. B*, Vol 266, 2008, p 1880
48. L.R. Doolittle, *Nucl. Instrum. Methods B*, Vol 9, 1985, p 344
49. M. Mayer, *Technical Report IPP9/113, Max-Planck-Institut für Plasmaphysik*, Garching, Germany, 1997
50. E. Szilágyi, F. Pászti, and G. Amsel, *Nucl. Instrum. Methods B*, Vol 100, 1995, p 103
51. J. Saarilahti and E. Rauhala, *Nucl. Instrum. Methods B*, Vol 64, 1992, p 734
52. N.P. Barradas, C. Jeynes, and R.P. Webb, *Appl. Phys. Lett.*, Vol 71, 1997, p 291
53. E. Kótai, *Nucl. Instrum. Methods B*, Vol 85, 1994, p 588
54. A.F. Gurbich, *Nucl. Instrum. Methods B*, Vol 136–138, 1998, p 60
55. H. Langhuth, M. Mayer, and S. Lindig, *Nucl. Instrum. Methods B*, Vol 269, 2011, p 1811

ASM Handbook, Volume 10, *Materials Characterization*
ASM Handbook Committee
DOI 10.31399/asm.hb.v10.a0006628

Low-Energy Ion-Scattering Spectroscopy

Kenji Umezawa, Osaka Prefecture University

Overview

General Uses

- Identification of elements present on solid surfaces
- Surface atomic structure

Examples of Applications

- Study of surface atomic structure of Pd/Pt(111)
- Study of metal epitaxy
- Determination of evaporation rate on a substrate
- Low-energy atom-scattering spectroscopy for insulator surfaces

Samples

- Form: Solids (metals, semiconductors, thin films)
- Size: φ5 × 0.5 mm (minimum size); φ25 × 0.5 mm (maximum size)
  Minimum size is determined by the probing beam size
  Maximum size is determined by the stage of the *xyz* manipulator

Limitations

- Space resolution: Less than 0.1 Å (depends on radius of shadow cone)
- Adjacent elements are difficult to separate due to insufficient time-of-flight mass resolution
- Samples must be set in an ultrahigh vacuum chamber

Estimated Analysis Time

- 30 s for one time-of-flight spectrum
- 45 min for polar or azimuth scans
- 60 min for scattering image of atoms on a crystal surface

Capabilities of Related Techniques

- Auger electron spectroscopy (elements present in the first 0.1 to 1 nm)
- Low-energy electron diffraction (surface atomic structure in the first 0.1 to 1 nm)
- Secondary ion mass spectroscopy (mass/charge identification of elements in the first 0.2 to 2 nm)
- Scanning tunneling microscopy (surface atomic structure in the first 0.1 to 0.6 nm)
- X-ray photoelectron spectroscopy (elements present in the first 0.1 to 1 nm)

Introduction

Low-energy ion-scattering spectroscopy (LEIS), known simply as ion-scattering spectroscopy (ISS), can be used to determine the atomic structure of solid surfaces (Ref 1). Turkevich began a series of investigations into using large angle scattering energetic charged particles for the chemical analysis of surfaces in 1961 (Ref 2). In 1963, Robinson and Oen performed computer simulations that showed that ions traveling through a crystal scatter depending on the direction of travel (Ref 3). In 1967, Smith built an analytical instrument in which noble gas ions (He^+, Ne^+, or Ar^+) of lower primary energies (0.5 to 3 keV) were directed at a solid target, and an energy spectrum of ions that scattered from the surface was collected with a 127° electrostatic analyzer (Ref 4). By the late 1970s, scientists had realized that the primary ion-scattering techniques were Rutherford backscattering spectroscopy (RBS) in the mega-electron volt range, medium-energy ion scattering spectroscopy (MEIS) in the 100 keV range, and LEIS in the several kiloelectron volt range. The velocities of 1 MeV $^4He^+$ and 1 keV $^4He^+$ are 6.9×10^6 m/s and 2.2×10^5 m/s, respectively. LEIS has two major features: (a) larger cross sections for the ion-atom interaction and (b) a higher neutralization rate for noble-gas ion scattering compared to RBS and MEIS. The larger cross section makes LEIS a very surface-sensitive technique. Most of the noble gas ions scattered from the topmost layers are neutralized. The neutralization is sensitive to the local electron density. In the outermost region where the ion-surface distance is large (>5 Å), resonance neutralization and ionization with the valence-band electrons of the solid target can take place. In the region where the ion-surface separation is approximately 2 Å, the ion core orbitals overlap appreciably with the valence electron cloud of the target and Auger neutralization is more dominant. Neutralized particles cannot be detected by an electrical energy analyzer. In order to overcome the neutralization problem, both noble-gas ions and neutrals have been detected by a time-of-flight (TOF) method (Ref 5, 6). Furthermore, low-energy atom-scattering spectroscopy can be used for analyzing insulator surfaces.

General Principles

Binary Elastic Collision Model

A simple model of binary elastic collision is applied to interpret experimental spectra (Ref 7). Figure 1 shows atomic collisions illustrating scattering and recoiling from a solid surface. There are direct relationships between the initial energy (E_0) of the primary ions before scattering and their kinetic energy after scattering (E_1) and between E_0 and the kinetic energy transferred to the target atoms (E_2). The kinematic relations between incident particles and target elements are given in Eq 1 and 2. The projectile mass and target mass are denoted by M_1 and M_2, respectively. For example, incident particles are 3 keV $^4He^+$ ions and target atoms are ^{195}Pt. E_0 is 3 keV (3000 eV $\times$ 1.6 $\times$ 10^{-19} J/eV). M_1 is 4 (4 $\times$ 1.67 $\times$ 10^{-27} kg) and M_2 is 195 (195 $\times$ 1.67 $\times$ 10^{-27} kg). E_1 is the scattering energy of $^4He^+$ ions after collision with ^{195}Pt atoms.

$$\frac{E_1}{E_0} = \left[\frac{\left(M_2{}^2 - M_1{}^2 sin^2\theta\right)^{1/2} + M_1 cos\,\theta}{M_1 + M_2}\right]^2 = \left[\frac{\left((M_2/M_1)^2 - sin^2\theta\right)^2 + cos\,\theta}{1 + M_2/M_1}\right]^2 = \left[\frac{\left(A^2 - sin^2\theta\right)^2 + cos\,\theta}{1 + A}\right]^2 \quad \text{(Eq 1)}$$

$$\frac{E_2}{E_0} = \frac{4M_1M_2}{(M_1+M_2)^2} cos^2\varnothing = \frac{4M_1M_2/M_1{}^2}{(1+(M_2/M_1))^2} cos^2\varnothing = \frac{4A}{(1+A)^2} cos^2\varnothing \quad \text{(Eq2)}$$

Here, $A = M_2/M_1$ is the ratio between the target mass and projectile mass. Figure 2 shows the scattered energy ratio E_1/E_0 as a function of the scattering angle for different values of the mass ratio $A = M_2/M_1$. Figure 3 shows the recoiled energy ratio E_2/E_1 as a function of the recoiling angle for different values of the mass ratio $A = M_2/M_1$.

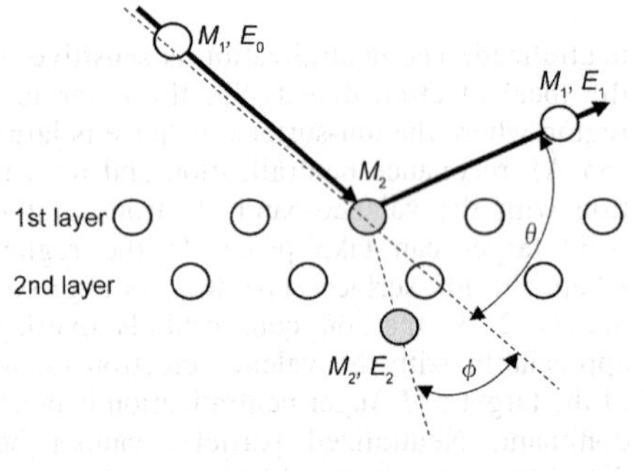

Fig. 1 Atomic collisions illustrating scattering and recoiling from a solid surface. The symbols θ and Ø denote the scattering angle and recoiling angle, respectively. M_1 and M_2 denote the masses of the projectile and target atoms, respectively. E_0 is the initial energy of the projectile atom with M_1. E_1 and E_2 are the energies of scattered particles with M_1 and the recoiled target atom with M_2.

For the calculation of the classical trajectory between projectile and target atoms, the modified universal potential is needed. In the energy region under 10 keV, the screening effect produced by the electron clouds, which partially mask the Coulomb potential, has to be taken into account. There are two representative potentials for low-energy ion-scattering study, namely the Ziegler, Biersack, and Littmark (ZBL) and Thomas Fermi-Molière (TFM) potential models (Ref 8–11).

Both potential models are commonly used in the energy region under 10 keV (Ref 10, 11). The ZBL potential function depends on the distance r between the nuclei. The ZBL potential is shown by Eq 3, and 4 (Ref 10):

$$V(r) = \frac{Z_1Z_2}{r} e^2 \left[0.1818\, exp\left(-3.2\frac{r}{a}\right) + 0.5099\, exp\left(-0.9423\frac{r}{a}\right) + 0.2802\, exp\left(-0.4029\frac{r}{a}\right) + 0.02817\, exp\left(-0.2016\frac{r}{a}\right)\right] \quad \text{(Eq 3)}$$

$$a = 0.8854\,\left(Z_1{}^{0.23} + Z_2{}^{0.23}\right)^{-1} a_0 \quad \text{(Eq 4)}$$

where Z_1 is the atomic number of the projectile, Z_2 is the atomic number of the target atom, e is the unit electrical charge, a is the ZBL screening length, and a_0 is the first Bohr radius.

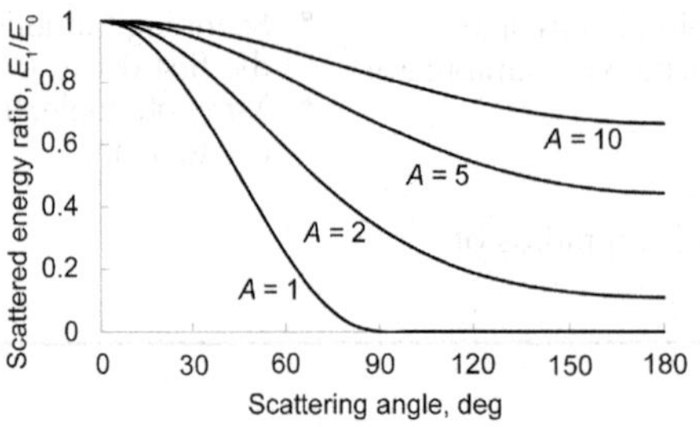

Fig. 2 Scattered energy ratio E_1/E_0 as a function of the scattering angle for different values of the mass ratio $A = M_2/M_1$

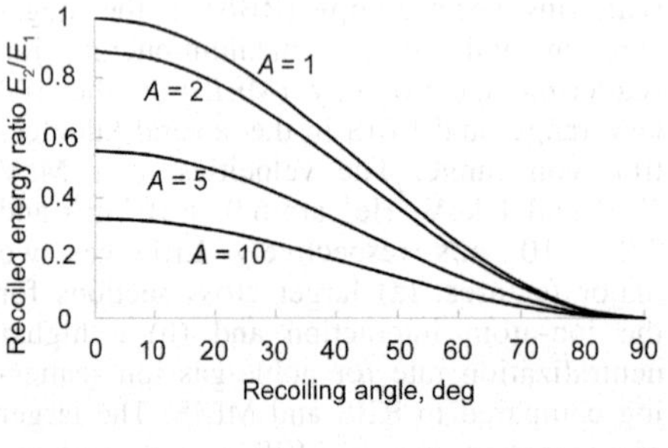

Fig. 3 Recoiled energy ratio E_2/E_1 as a function of the recoiling angle for different values of the mass ratio $A = M_2/M_1$

The TFM potential is expressed as (Ref 11):

$$V(r) = \frac{Z_1Z_2}{r} e^2 \left[0.35\, exp\left(-0.3\frac{r}{a}\right) + 0.55\, exp\left(-1.2\frac{r}{a}\right) + 0.1\, exp\left(-6.0\frac{r}{a}\right)\right] \quad \text{(Eq 5)}$$

$$a_F = 0.8854\,\left(Z_1{}^{0.5} + Z_2{}^{0.5}\right)^{-2/3} a_0 \quad \text{(Eq 6)}$$

$$a = C \cdot a_F (C \leq 1.0) \quad \text{(Eq 7)}$$

A constant factor C is adapted to fit the experimental data.
The force F is obtained from Eq 8 using Eq 3 or 5.

$$F = -\nabla V(r) \quad \text{(Eq 8)}$$

To vary the potential strength, the screening length is changed via a multiplicative correction factor C. An experimental study in combination with extensive simulations has shown that TOF-LEIS spectra are reproduced very well by a TFM potential with a screening correction of between C = 0.75 and 1.0 (Ref 11). In the case of binary systems, the choice of the potential has a minor influence. The difference between these interatomic potentials is small (Ref 11). These two potentials are mostly used in LEIS.

The rate at which scattered ions (I_i) arrive at the detector depends on the number of incident ions that strike the sample surface, that is, the current of incident ion beams I_0, the scattering cross section σ (scattering cross sections for ions colliding sequentially with two target atoms are described in a previous paper (Ref 12)), a solid angle of the detector $d\Omega$, and the detection efficiency η. K is a constant factor. The rate is shown in Eq 9.

$$Ii = K\, I_0 \sigma\, d\Omega \eta \quad \text{(Eq 9)}$$

A microchannel plate (MCP) is used as a detector. The ion-survival probability P^+ does not need to be considered, because an MCP detects both ion and neutral particles that are produced when a low-energy ion interacts with a solid surface.

Shadowing and Blocking Phenomena

Shadowing and blocking are specific phenomena characteristic of scattering by repulsive potentials (Ref 1). If a single atom is placed in a parallel flux of incoming particles that are lighter than the atom, the scattering angles of incoming particles within a certain range of small impact parameters (<0.1 Å) are large enough to prevent them from penetrating the region immediately behind the atom. The region behind the atom that the incoming particles cannot penetrate is called a shadow cone. An area free of incoming particles is thus formed.

Figures 4 and 5 show the shadow cones for 3 keV He^+ and 3 keV Ne^+ ions, respectively, scattering from an Au atom calculated using the ZBL potential. The x-axis shows the distance behind the Au atom in angstroms. The y-axis shows the shadow cone radius and impact parameters of incident 3 keV $^4He^+$ and 3 keV $^{20}Ne^+$ ion beams in angstroms. The edges of the cones are areas of increased flux density arising from the focusing of the ion trajectories. The flux of incident ion beams flows toward nuclei, and then scattered flux is formed according to the repulsive potential. The shadow cone (excluded region behind the scattering center inside which no primary ions can penetrate) takes the form of a paraboloid with radius R_{sh} as a function of distance from the target atom. Using a small-angle approximation and an unscreened Coulomb potential, the shadow cone radius R_{sh} is obtained:

$$R_{sh} = 2\left(Z_1 Z_2 e^2 d/E\right)^{1/2} \quad \text{(Eq 10)}$$

where d is the distance behind the target atom, E is the primary energy, e is the charge of an electron, and Z_1 and Z_2 are the atomic numbers of the projectile and target atoms, respectively.

Parallel incident ions impinging on a single atom A form a shadow cone, as shown in Fig. 6. An atom B in a shadow cone cannot be scattered by incident ions.

Blocking is another phenomenon involving a region free of scattered particles behind an atom. Blocking is the interaction between ions scattered from one atom (B) and a second atom (C), as shown in Fig. 7. The presence of at least two atoms is needed for blocking. As shown in Fig. 7, when a beam of kiloelectron volt ions with parallel trajectories interacts with the first atom A or B, the latter acts as a source of scattered particles with an angular distribution. Some of these trajectories are deflected by the repulsive potential of the second atom C (blocking atom). Figure 8 shows focusing effects. First, a shadow cone forms due to the interaction between incident ions and the target atom A. Then, at a particular angle called a critical angle (αcr) of incidence, the focused particles at the edge of the shadow cone collide with the second atom B at an impact parameter close to zero, and the focused particles are backscattered. This leads to an increase in the detected intensity at a critical angle (αcr). The shadow cone, blocking cone, and focusing concepts are used for the determination of the surface structure in 180° backscattering configurations (Ref 5).

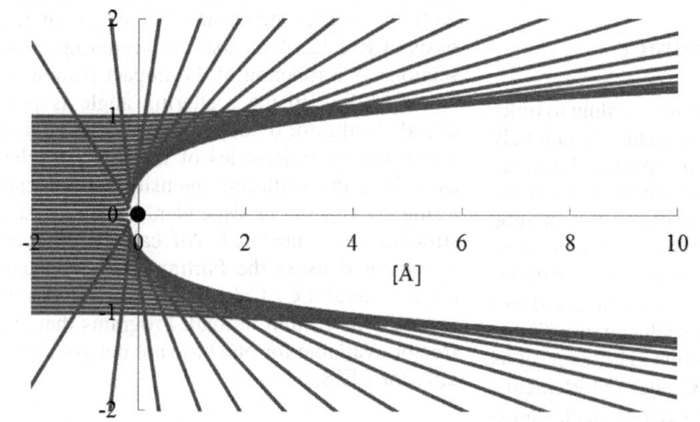

Fig. 4 Simulation of a parallel beam of 3 keV $^4He^+$ ions scattering from a single Au atom (●) illustrating a shadow cone. Calculation was performed using the ZBL potential. The unit for the x- and y-axes is the angstrom.

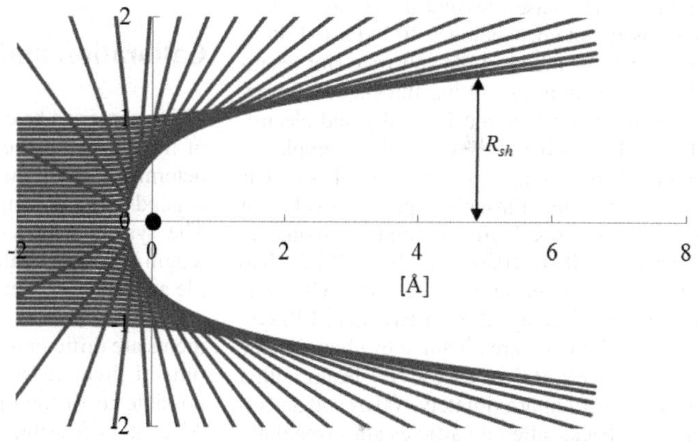

Fig. 5 Simulation of a parallel beam of 3 keV $^{20}Ne^+$ ions scattering from a single Au atom (●) illustrating a shadow cone. Calculation was performed using the ZBL potential. The unit for the x- and y-axes is the angstrom.

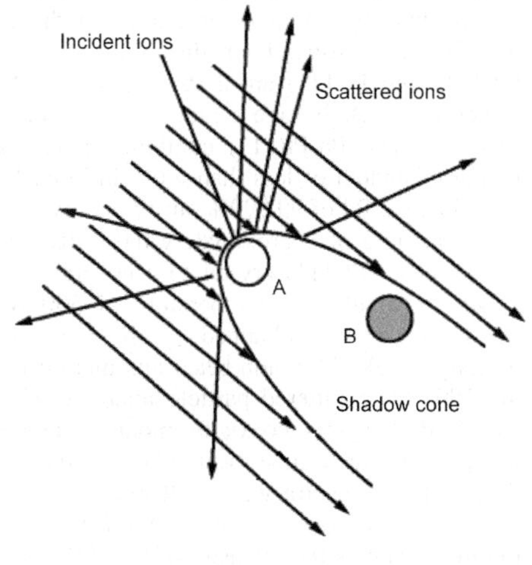

Fig. 6 Schematic illustration of shadowing effects. Atom B is within a shadow cone formed by atom A. Incident ions cannot penetrate the shadow cone area.

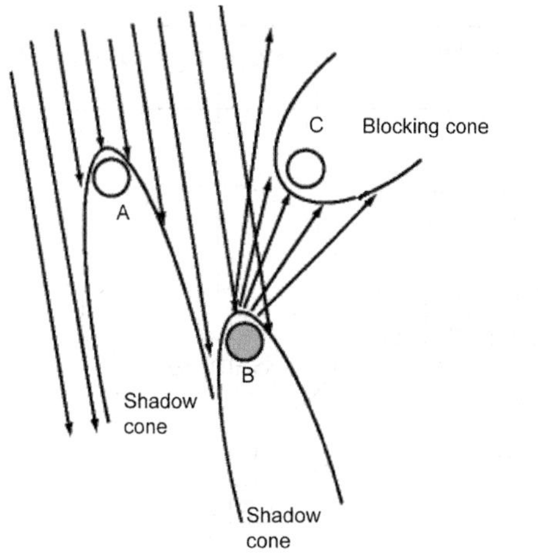

Fig. 7 Schematic illustration of a blocking cone created by scattering ions

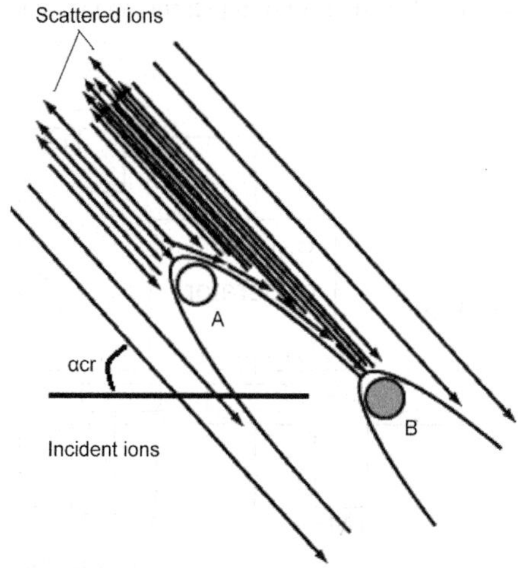

Fig. 8 Schematic illustration of focusing effects. The edge of the shadow cone created by the first atom A is on the center of the second atom B. The number of backscattered ions increases due to the focusing effects at a critical angle of αcr.

Systems and Equipment

A schematic view of the experimental equipment for LEIS is shown in Fig. 9 (Ref 11). An LEIS system comprises an ultrahigh vacuum (UHV) chamber with the primary atom beamline: (1) ion beam source, (2) MCP detector, (3) precision sample manipulator with stepping motors (custom made), (4) preamplifier, (5) amplifier, (6) four-channel multiple-stop time-to-digital converter (time resolution 10 ns), (7) pulse generator (100 kHz, 50 V), (8) low-energy electron diffraction (LEED) optics and Auger electron spectroscopy (AES) equipment, and (9) software to collect data and control the stepping motor. A neutralizer is shown in Fig. 9. It is not needed for LEIS. A neutralizer is useful for low-energy atom scattering for the analysis of metal oxide or insulator surfaces. It is described in the section Applications and Interpretation. The base pressure during the experiment is maintained below ~10^{-8} Pa (~1.45 × 10^{-12} psi). A solid sample substrate (for example, φ = 10 mm, or .39 in.; thickness 0.5 mm, or .02 in.; crystal) is mechanically and electrochemically polished. After the sample is mounted on a standard *xyz* manipulator, it is cleaned in situ through repeated cycles of 500 eV Ar^+ bombardment and subsequently annealed at 700 to 1000 °C (1292 to 1832 °F) to remove the surface damage. The state of the surface is monitored by AES, LEED, and LEIS systems LEIS spectra are obtained by chopping the primary 3 keV He^+ or 3 keV Ne^+ beam with a pulse width of approximately 100 ns and detecting 180° backscattered particles after free flight through a drift tube of 60 cm (23.62 in.) by means of an MCP, which is coaxially mounted along the primary tube. A flight time for a spectrum is less than 20 μs. Polar angle scans are performed from −85° to 85° in 1° increments with an average beam current of ~10 Pa (0.0015 psi). The incident angle of 0° represents an incidence perpendicular to the sample surface. The time required for a complete sample scan is usually 1.5 h at a dose of ~3 × 10^{13} cm^2, which provides sufficiently low coverage to avoid radiation damage from the primary beam. The metal deposition rate on a sample surface is approximately 1.0 monolayer (ML) per 10 to 20 min using an evaporator.

Specimen Preparation

A single crystal is mounted on a standard *xyz* manipulator after mechanical and electrochemical polishing. The crystal is cleaned in situ through repeated cycles of 500 eV Ar^+ bombardment followed by annealing at 700 °C to remove the surface damage. After this treatment, no carbon, oxygen, or sulfur surface impurities are detectable by LEIS and AES.

Calibration and Accuracy

Elements can be calibrated according to time-of-flight spectra, because elements are uniquely determined by flight time on spectra. Accuracy depends on tube length and accelerated energy. The system that is described here has a tube length of approximately 60 cm between the sample and the MCP detector. Accelerated particles are 2 to 3 keV $^4He^+$ or $^{20}Ne^+$ ions. Adjacent elements are difficult to separate due to insufficient time-of-flight mass resolution. However, it is possible to distinguish ^{12}C and ^{16}O elements. Also, it is possible to detect hydrogen atoms using forward scattering.

Data Analysis and Reliability

Data analysis is carried out using the intensity of each element as a function of polar and azimuth scans. Polar scans are carried out every 2°. Azimuth scans are carried out every 3°. Because azimuth scans are not as sensitive as polar scans, 3° is sufficient for the analysis of surfaces. The reliability of the experiments depends on the stability of the incident particles. The intensity of the incident particles should be stable for at least 1.5 h. Accordingly, the gas pressure (He or Ne) in the ion gun should be stable for more than 1.5 h.

Computer simulations are based on in-plane scattering only within planes of atoms that represent discrete slices of the three-dimensional structure. The computer program for polar scans or azimuth scans consists of three separate subprograms. In the first subprogram, a database of scattering angle and scattering energy as a function of impact parameters for binary atomic collisions is produced on the basis of work by M.T. Robinson and I.M. Torrens (Ref 7). In the second subprogram, scattering cross sections for ions colliding sequentially with two target atoms are calculated on the basis of Ref 12. A database of scattering cross sections as a function of the impact parameter corresponding to the scattering angle is produced. In the third subprogram, the incoming and outgoing trajectories of incident particles as well as the scattering intensity due to scattering cross sections dependent on the crystal structure are calculated. All calculations can be obtained using the Fortran or C language on a general PC. Calculation takes approximately 5 to 10 min in total. Programs that are free or available for purchase are not yet available for LEIS.

Applications and Interpretation

Surface Structural Analysis

An example is presented to illustrate analysis of surface structure using LEIS. The bimetallic Pd-Pt catalyst has been used in industry for the hydrogenation of aromatics in diesel oil. The nearest-neighbor distances of bulk Pd(111) and Pt(111) are 2.75 Å and 2.77 Å, respectively. Both have a face-centered cubic structure. The Pd-Pt(111) combination has a lattice mismatch of less than 1.0% in bulk. In experiments, Pd of 99.999% purity was evaporated at a rate of approximately 0.05 ML/min onto a clean Pt(111) crystal to a coverage of 3 ML at a substrate temperature of 300 K. LEIS spectra were obtained by chopping the primary 2 keV $^{20}Ne^+$ ion beam and measuring the 180° backscattered particle intensity with an MCP that was coaxially mounted along the primary drift tube (Ref 13). Pd and Pt time-of-flight spectra were well resolved as a function of Pd coverage, as shown in Fig. 10. Figure 11 shows the intensity of the ISS spectra of the Pt(111) clean surface as a function of the polar angle along the $[\bar{1}10]$ azimuth. Solid circles denote the experimental data of the backscattering intensity based on a clean Pt

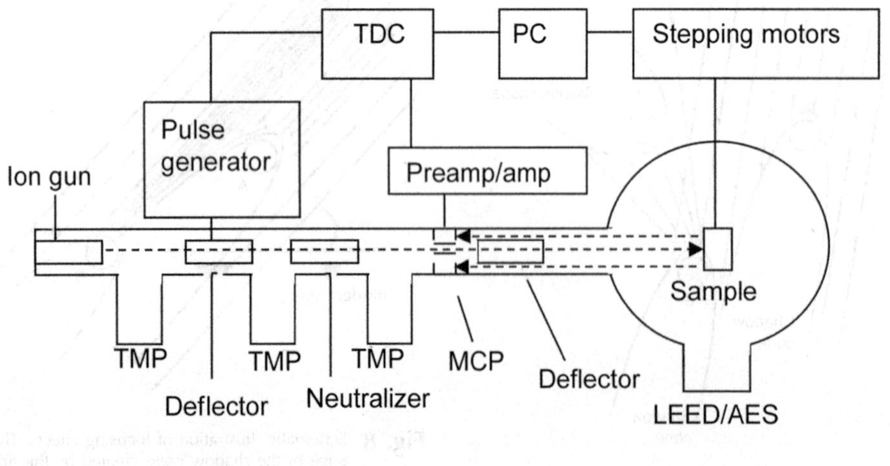

Fig. 9 Schematic view of low-energy ion-scattering spectroscopy combined with an ultrahigh vacuum chamber. AES, Auger electron spectroscopy; LEED, low-energy electron diffraction; MCP, microchannel plate; TDC, time-to-digital converter; TMP, turbo molecular pump

(111) surface structure. Figure 12 shows simulations based on the three-dimensional cross section for ions that scatter sequentially and classically from two atoms. The solid line labeled "Sum" represents the total calculated intensity from the first through the third layers of Pt atoms. The peaks at ±50° are attributable to multiple focusing effects due to out-of-plane scattering by the second- and third-layer Pt atoms. The peak at 0° is attributable mainly to focusing effects from the third-layer Pt atoms. Figure 13 shows the intensity of the ISS spectra of the Pt(111) clean surface as a function of the polar angle along the [11$\bar{2}$] azimuth. In Fig. 13, the contribution of backscattering from the first layer is seen at ±70° ("sp"). The peaks at −60° ("a") and 46° ("c + d") show signals coming from the second and third layers of Pt atoms, respectively. Other peaks at 0° ("b") come from the third layer. The peak at −10° probably comes from the fourth layer. Figure 14 shows the side view of Pt(111) [11$\bar{2}$]. Arrows indicate the incident beams striking Pt atoms. These arrows correspond to labels shown in Fig. 13. Figure 15 shows the Pt intensity obtained from computer simulations as a function of the polar angle along the [11$\bar{2}$] azimuth. The symbol "Sum" denotes the total calculated intensity from the first through the third layers. The polar angle was fixed at −18° from the sample surface during azimuth scans. Along the [$\bar{1}$10] azimuth, the Pt-Pt spacing thus obtained was 2.77 Å for a clean surface structure. This obtained value is in good agreement with the Pt-Pt spacing in the bulk. This means that surface reconstruction is not seen for clean Pt(111) surfaces. Figure 16 shows the variation of the intensity of 2 keV $^{20}Ne^+$ ions scattered from Pt atoms as a function of the azimuth angle. The solid line and dashed line show the experimental and simulation results, respectively. The figure clearly shows a periodicity of 60° for the azimuth angle (sixfold symmetry for the first layer). Figures 17 and 18 show the variation of the intensity of 2 keV $^{20}Ne^+$ ions scattered from Pd atoms at a coverage of 3 ML along the [$\bar{1}$10] and [11$\bar{2}$] azimuths, respectively. Figure 19 shows the variation of the intensity of 2 keV $^{20}Ne^+$ ions scattered from Pd atoms at a coverage of 3 ML as a function of the azimuth angle. Figures 17 to 19 illustrate that 3 ML Pd atoms have a parallel orientation with respect to the Pt substrate. There were no Pd atoms that had an antiparallel orientation with respect to the Pt substrate because the lattice mismatch between Pd and Pt is less than 1.0%. In a large lattice mismatch (16%) such as that between Au(111) and Ni(111), there are two growth orientations of Au(111).

Layer-by-Layer (Frank van der Merwe) Growth

Figure 20 shows the intensity from Sb atoms as a function of Sb deposition time. Sb deposition was carried out on a Cu(111) sample at room temperature. The Sb intensity was measured using 3 keV $^{20}Ne^+$ ions backscattered from Sb atoms along the Cu(111) [11$\bar{2}$] azimuth at a polar angle of 67°. Figure 20 was used to determine the ML coverage. One ML is defined to be 1.86×10^{15} atoms/cm^2 from the ideal Cu density in a (111) plane of the bulk crystal. The slope of the Sb curve changes at about the same coverage (the same deposition time). This indicates that 10 min of Sb deposition time corresponds to the completion of first layer coverage. The Sb evaporation rate was 0.1 ML/min. The second layer and third layer of Sb atoms were completed at the deposition times of 20 min and 30 min, respectively. The Sb atoms show layer-by-layer (Frank van der Merwe) growth on Cu(111) surfaces.

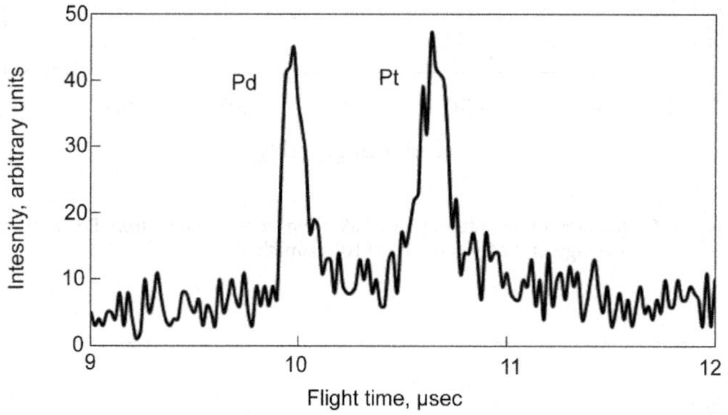

Fig. 10 Time-flight-spectra at Pt(111) substrate temperature of 300 K during Pd deposition

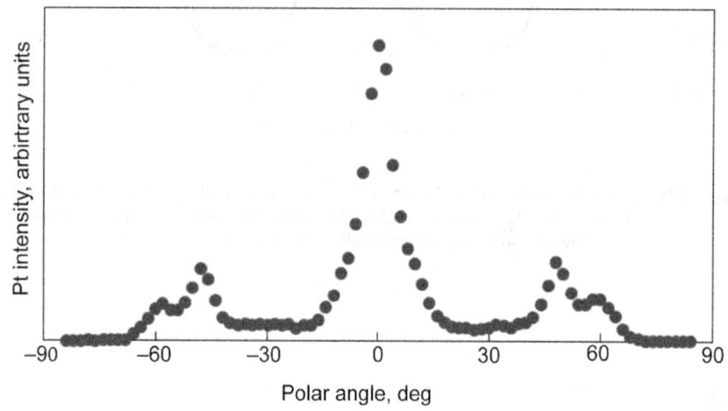

Fig. 11 A series of ion-scattering spectroscopy polar angle scans for 2 keV $^{20}Ne^+$ ions backscattered from a Pt(111) surface along the [$\bar{1}$10] azimuth

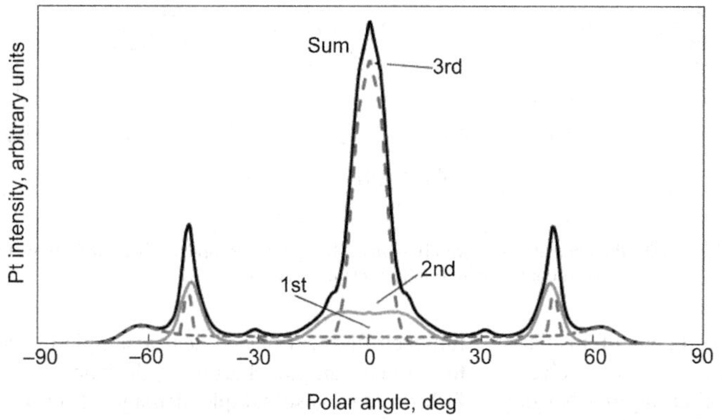

Fig. 12 Computer simulations along the [$\bar{1}$10] azimuth performed using the TFM scattering potential with an adjustable parameter $C = 0.9$ for Pt atoms

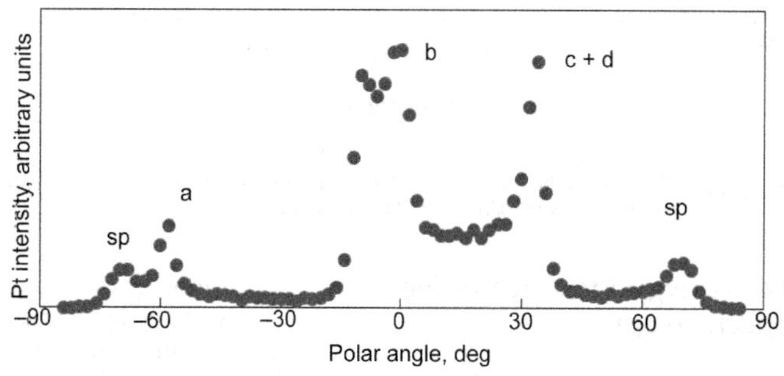

Fig. 13 A series of ion-scattering spectoscopy polar angle scans for 2 keV $^{20}Ne^+$ ions backscattered from a Pt(111) surface along the [11$\bar{2}$] azimuth

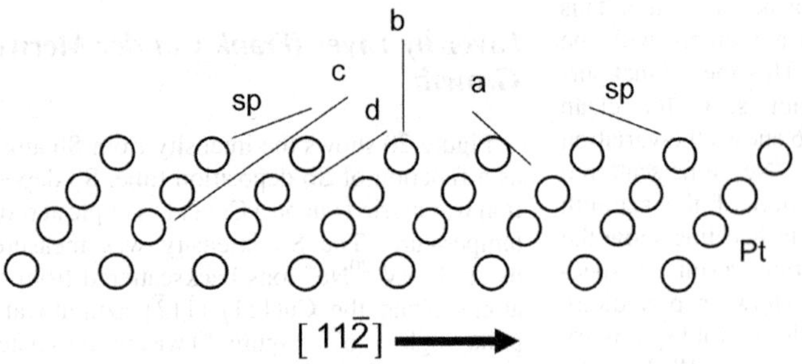

Fig. 14 Side view of Pt(111) [11$\bar{2}$]

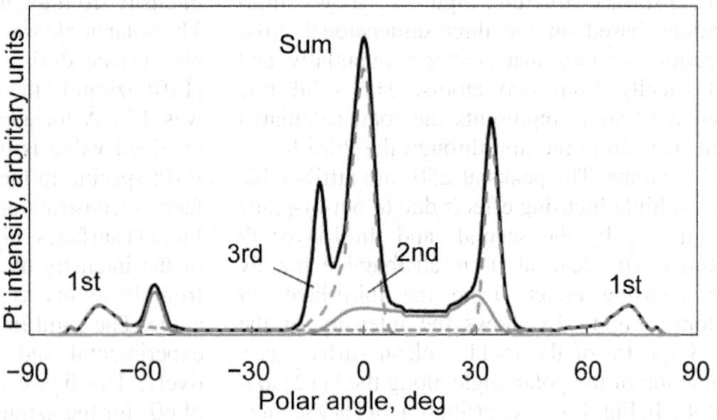

Fig. 15 Computer simulations along the [11$\bar{2}$] azimuth using the TFM scattering potential with an adjustable parameter $C = 0.9$ for Pt atoms

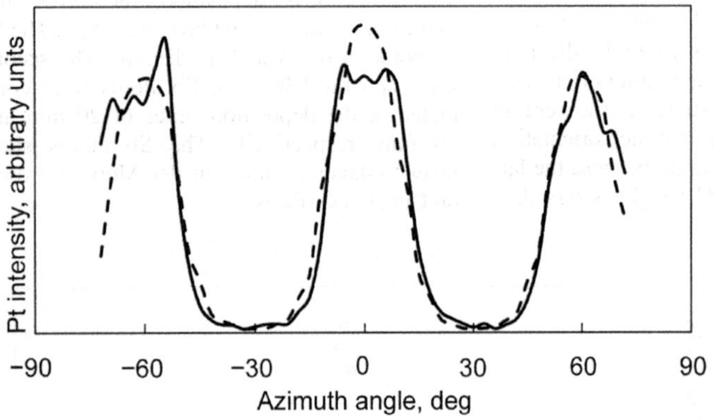

Fig. 16 Variation of the intensity of 2 keV $^{20}Ne^+$ ions scattered from Pt atoms as a function of the azimuth angle. The solid line and dotted line show the experimental and simulation results, respectively.

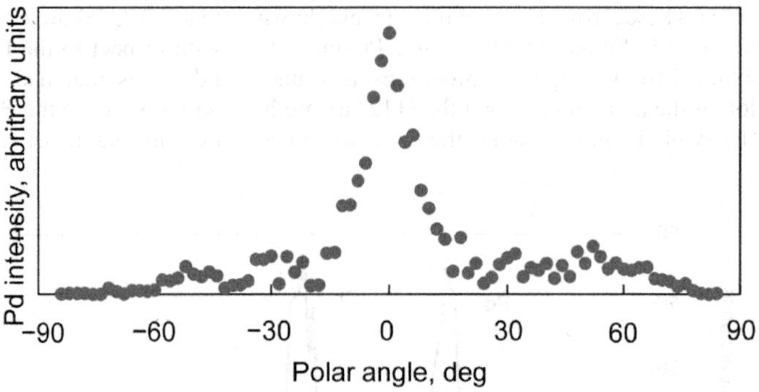

Fig. 17 Variation of the intensity of 2 keV $^{20}Ne^+$ ions scattered from Pd atoms at a coverage of 3 ML along the [$\bar{1}$10] azimuth

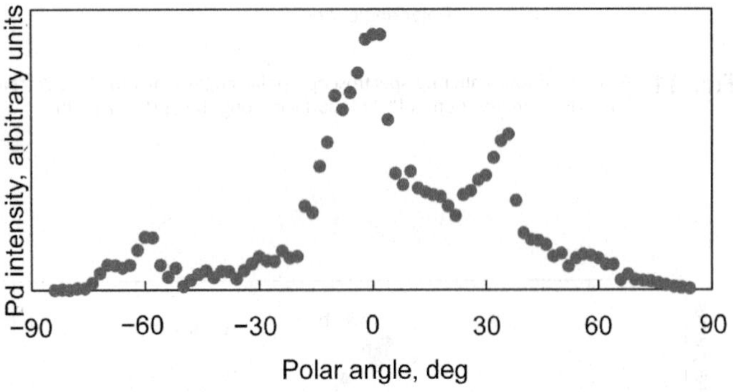

Fig. 18 Variation of the intensity of 2 keV $^{20}Ne^+$ ions scattered from Pd atoms at a coverage of 3 ML along the [11$\bar{2}$] azimuth

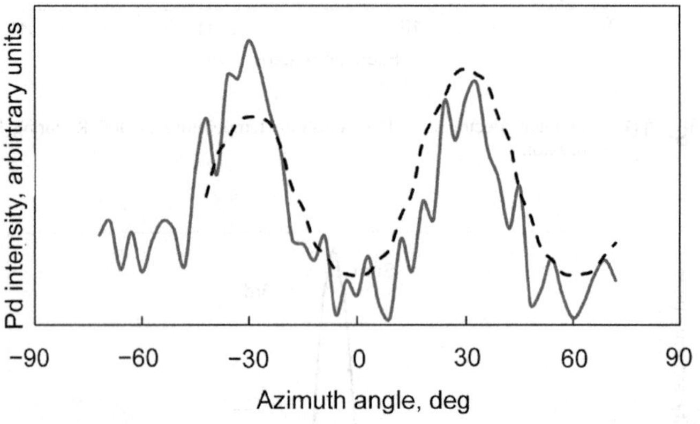

Fig. 19 Pd intensity of azimuth scans. The solid line and dashed line show the experimental and simulation results, respectively.

Low-Energy Atom-Scattering Spectroscopy

Low-energy atom-scattering spectroscopy is described as follows (Ref 14, 15). LEIS is useful for the analysis of metal and semiconductor surfaces. However, bombardment of insulator surfaces by charged ions can induce a charge on the surfaces. The charging/discharging dynamics of the insulating material are evident during this ion-beam bombardment. Sometimes, an electron shower emitted from a tungsten filament placed near a sample is used to reduce sample charging. Electron shower failure can cause sample damage. Therefore, we developed a low-energy atom-scattering spectroscopy system for the analysis of insulator surfaces (Ref 14). Low-energy atom beams

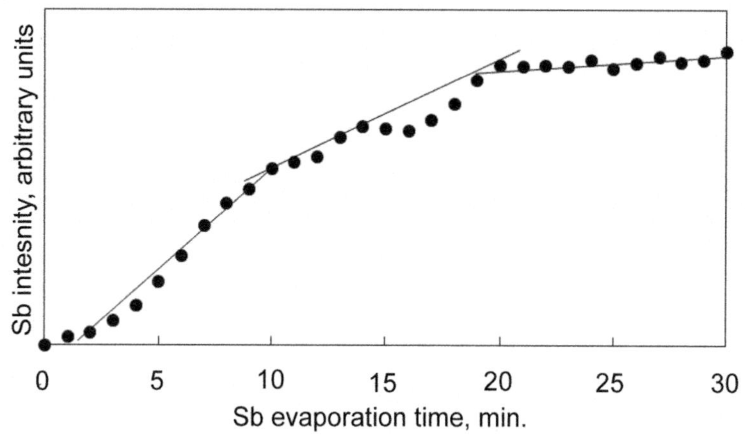

Fig. 20 Sb intensity on a clean Cu(111) surface as a function of Sb deposition time

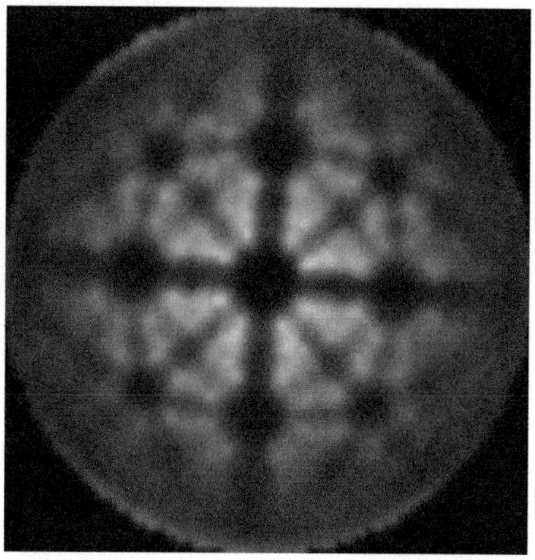

Fig. 21 Image of magnesium atoms on a clean MgO (100) surface obtained using low-energy atom-scattering spectroscopy (TOF-LAS3000, Pascal Co., Japan)

are produced using ion beams. Atom beams are converted from ion beams through charge exchange while the ion beams pass through a small gas chamber, which is labeled as the "Neutralizer" in Fig. 9. The gas is the same element as the ion beams, and the pressure of the gas is in a range of approximately 10^{-1} Pa (1.45×10^{-5} psi). The conversion efficiency from a 3 keV $^4He^+$ beam to a 3 keV $^4He^0$ beam is approximately 34%. Residual $^4He^+$ ions are removed by the deflector at the entrance of the UHV chamber. $^4He^0$ particles impinge on a sample, and the scattered particles are detected by an MCP that is located at 180°. Fig. 21 shows Mg atoms on a clean MgO(100) surface. This figure was obtained by rotating a clean MgO(100) sample against the incident 3 keV $^4He^0$ beams (Ref 16) and recording Mg signals on time-of-flight spectra. The (100) crystal is clearly shown because of a fourfold rotation axis.

ACKNOWLEDGMENT

Revised from G.C. Nelson, Low-Energy Ion-Scattering Spectroscopy, *Materials Characterization*, Vol 10, *ASM Handbook*, ASM International, 1986

REFERENCES

1. J.W. Rabalais, *Low Energy Ion-Surface Interactions*, Wiley, 1994
2. A. Turkevich, Chemical Analysis of Surfaces by Use of Large-Angle Scattering of Heavy Charged Particles, *Science*, Vol 134, 1961, p 672
3. M.T. Robinson, O.S. Oen, Computer Studies of the Slowing Down of Energetic Atoms in Crystals, *Phys. Rev.*, Vol 132, Issue 6, 1963, p 2385
4. D.P. Smith, Scattering of Low-Energy Noble Gas Ions from Metal Surfaces, *J. Appl. Phys.*, Vol 38 (No. 1), 1967, p 340
5. M. Katayama, E. Nomura, H. Soejima, S. Hayashi, and M. Aono, Real-Time Monitoring of Molecular-Beam Epitaxy Processes with Coaxial Impact-Collision Ion Scattering Spectroscopy (CAICISS), *Nucl. Instrum. Meth. B*, Vol 45 (No. 1–4), 1990, p 408
6. K. Umezawa, S. Nakanishi, M. Yoshimura, K. Ojima, K. Ueda, and W. M. Gibson, Ag/Cu(111) Surface Structure and Metal Epitaxy by Impact-Collision Ion-Scattering Spectroscopy and Scanning Tunneling Microscopy, *Phys. Rev. B*, Vol 63, 2000, p 035402
7. M.T. Robinson and I.M. Torrens, Computer Simulation of Atomic-Displacement Cascades in Solids in the Binary-Collision Approximation, *Phys. Rev. B*, Vol 9 (No. 12), 1974, p 5008
8. J.F. Ziegler, J.P. Biesrsack, and U. Littmark, *The Stopping and Range of Ions in Solids*, Pergamon Press, New York, 1985
9. C.S. Chang, U. Knipping and I.S.T. Tsong, Shadow Cones Formed by Target Atoms Bombarded by 1 to 3 keV He^+, Li^+, Ne^+ and Na^+ ions, *Nucl. Instrum. Meth. B*, Vol 18 (No. 1–6), 1986, p 11
10. D. O'Connor and J. Biersack, Comparison of Theoretical and Empirical Interatomic Potentials, *Nucl. Instr. Meth. Phys. Res. B*, Vol 15 (No. 1–6), 1986, p 14
11. D. Goebl, B. Bruckner, D. Roth, C. Ahamer, and P. Bauer, Low-Energy Ion Scattering: A Quantitative Method?, *Nucl. Instr. Meth. Phys. Res. B*, Vol 354, 2015, p 3
12. R.S. Williams, M. Kato, R.S. Daley, and M. Aono, Scattering Cross Sections for Ions Colliding Sequentially with Two Target Atoms, *Surf. Sci.*, Vol 225 (No. 3), 1990, p 355
13. W. Eckstein, *Computer Simulation of Ion-Solid Interactions*, Springer-Verlag, 1991
14. K. Umezawa, E. Narihiro, Y. Ohira, and M. Yohimura, Pd/Pt(111) Surface Structure and Metal Epitaxy by Time-of-Flight Impact-Collision Ion Scattering Spectroscopy and Scanning Tunneling Microscopy: Does Lattice Mismatch Really Determine the Growth Mode?, *Nucl. Instrum. Meth. B*, Vol 266 (No. 8), 2008, p 1903
15. T. Suzuki and R. Souda, Structure Analysis of CsCl Deposited on the MgO(001) Surface by Coaxial Impact Collision Atom Scattering Spectroscopy (CAICASS), *Surf. Sci.*, Vol 442 (No. 2), 1999, p 283
16. K. Umezawa, Low Energy Ion and Atom Scattering Spectroscopy for Surface Structural Analysis of Single Metal and Insulator Crystals, *Nucl. Instrum. Meth. B*, Vol 266 (No. 8), 2008, p 1892–1896

Chemical Analysis and Separation Techniques

Thomas J. Bruno, National Institute of Standards and Technology

ASM Handbook, Volume 10, *Materials Characterization*
ASM Handbook Committee
DOI 10.31399/asm.hb.v10.a0006634

Calibration and Experimental Uncertainty*

Thomas J. Bruno, National Institute of Standards and Technology

Overview

General Uses

- Consideration of experimental uncertainty must be part of any reported measurement.
- Calibration must be applied to any device or instrument that is not absolute.
- Validation in measurement is confirmation, through objective evidence, that the requirement for a specific intended use or application is fulfilled.

Examples of Applications

- A gas chromatographic analysis of diesel fuel requires calibration of peak responses and subsequent analysis of the uncertainty of quantitative measurements.
- Measurement of the density of a food product requires calibration of the densimeter, uncertainty analysis of the result, and validation with any accepted standards for the food product to be sold.
- The custody transfer of natural gas across countries requires multiple measurements that depend on calibration, the uncertainty of which directly affect economics, and the validation of which directly impacts consumer confidence.

Samples

- Samples subject to measurement calibration, uncertainty analysis, and validation span all phases of matter under a broad range of conditions.

Limitations

- A normal distribution is often ascribed to a population, which may not be valid, but which may in many cases is useful.
- During the development of uncertainty analyses, an uncertainty budget (a listing of all contributors to uncertainty) is useful. The uncertainty budget may contain items that are poorly understood or not quantifiable.
- Although an uncertainty statement may be given (in sound scientific language), this is no guarantee of validity. The basis for the assertion must be provided.
- In developing calibration, one must observe the cautions outlined below regarding data coverage and curvature.

Estimated Time

- Uncertainty analysis in the past was often a very time-consuming process, but tools are available to make this much easier.
- Calibration in the past was also often a very time-consuming process, but tools are readily available to make this much easier.

Capabilities of Related Techniques

- Note that uncertainty analysis, calibration, and validation are very broad topics, and progress is being made to expand applicability, especially under the umbrella of "decision making."
- Among the many techniques that are not discussed in this article are sensitivity analysis, principal component analysis, and the Bayesian approach. These should be considered where appropriate.

Introduction

Most modern instrumental techniques produce an output or signal that is not absolute; the signal (or peak) is, for example, not a direct quantitative measure of concentration or target analyte quantity. This is true of all the instrumental techniques discussed in Division 4, "Chemical Analysis and Separation Techniques," in this Volume, including gas, liquid, and ion chromatography, and electrochemical methods. Thus, to obtain quantitative information, the raw output from an instrument (information) must be converted into a physical quantity (knowledge). This is done by standardizing or calibrating the raw response from an instrument and subsequently analyzing the uncertainty from both the calibration process

*Contribution of the United States Government. No copyright in the United States. This material was adapted from entries prepared by the author for the CRC Handbook of Chemistry and Physics, 94th through 99th editions, from 2013 to 2017.

and the measurement process (Ref 1–4). Here, we briefly summarize the most common calibration and uncertainty analysis methods, recognizing that this is a very large field. Note that the common use of the term "standardization" is equivalent to calibration and is not to be confused with the application of standard methods as specified by regulatory or consensus standard organizations.

In this article, it is assumed that the sample has been properly drawn from the parent population material, and properly prepared. The term "sampling," which describes the process of obtaining the sample from the population material, implies the existence of a sampling uncertainty, arising mainly from population material heterogeneity (Ref 5–7). Thus, the measurement result is an estimate of what would be obtained from the parent population material. Sampling uncertainty is that part of the total uncertainty in a measurement procedure that results from using only a fraction of the population material. In this respect, sampling by any method is an extrapolation process. Since the sampling uncertainty is usually ignored for an individual measurement on an individual test portion, the sampling uncertainty is considered as being due entirely to the variability of the test portion. It is therefore assessed, when necessary, by replication of the sampling from the parent population material and statistically isolating the uncertainty thus introduced by analysis of variance. Sampling uncertainty is often minimized by field and laboratory processing, with procedures that can include mixing, reduction, coning, quartering, riffling, milling, and grinding.

Another aspect that must be considered subsequent to sampling is sample preservation and handling. The integrity of the sample must be preserved during the inevitable delay between sampling and analysis. Sample preservation may include the addition of preservatives or buffer solutions, pH adjustment, use of an inert gas "blanket," and cold storage or freezing.

Calibration and Standardization

External Standard Methods

The external standard method can be applied to nearly all instrumental techniques, within the general limits discussed in this article, and the specific limitations that may be applicable with individual techniques. This method is conceptually simple; the user constructs a calibration curve of instrumental response with prepared samples containing known amounts of the substances of interest (the analytes) at a range of concentrations, an example of which is shown in Fig. 1. Thus, the curve represents the raw instrumental response of the analyte as a function of analyte concentration or amount. Each point on the plot must be measured several times so that the repeatability can be assessed. Only random

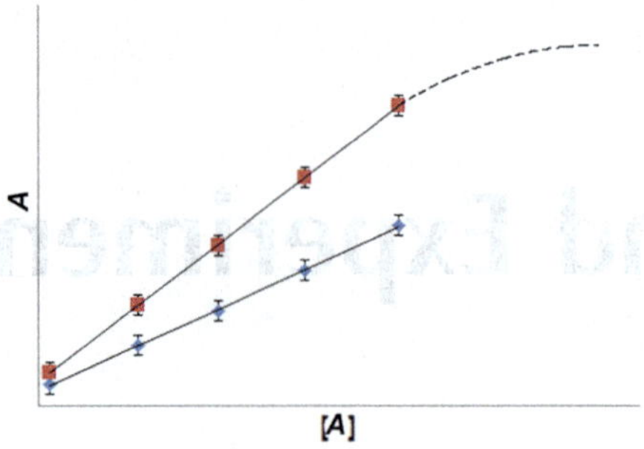

Fig. 1 An example calibration curve prepared by use of the external standard method. The instrument response is represented by *A*, and the concentration resulting in that response is [*A*]. While curves for two analytes are shown, in principle as many analytes as desired can be plotted. While five points per analyte are shown, as many as required can be measured. Note that a region of nonlinearity is shown in the latter part of the curve for one of the components. This requires a larger number of points to adequately represent and fit any nonlinear areas.

uncertainty should be observed in the replicates; trends of increasing or decreasing response (hysteresis) must be remedied by identifying the source and adjusting the method accordingly. The calibration samples should be randomized; that is, measured in random order. Although called a calibration "curve," ideally the signal versus concentration plot is linear, or substantially linear. In other words, areas of nonlinearity are unimportant; otherwise, they are localized, minor, and properly treated by the measurement technique. In some cases, the response may be linearizable, for example, by calculating the logarithm of the raw response. If a curve shows nonlinearity in an area that is important for the analysis, one must measure more concentrations (data points) in the region of curvature.

In practice, the line that results from the calibration is fit with an appropriate software program, and the desired value for the unknown concentration is calculated. The curve can be used graphically if approximation suffices. Samples prepared for external standard calibration can contain one or many analytes. Once a calibration curve is prepared, it can often be used for some time period, provided such a procedure has been previously validated; that is, the stability of the standards and the instrument over the time of use has been assessed. Otherwise, it is best to measure the unknown and the standards within a short period of time. Moreover, if any major change is made to the instrumentation, such as changing a detector or detector parameters, or changing a chromatographic column, the standards must be remeasured.

To successfully use the method, the standard samples must be in a concentration range that is comparable to that of the unknown analyte, and ideally should bracket the unknown. Multiple measurements of each standard mixture should be made to establish repeatability of points on the curve. Many instrumental methods have operation ranges (e.g., frequency, temperature) in which the uncertainty is minimized, so components and concentrations for standard samples must respect this. The standard mixtures should be in the same matrix as the unknown, and the matrix must not interfere with the unknown or other standard mixture components. Any pretreatment of the unknown must be reflected in the standard mixtures. As with any calibration method, components in the standard mixtures must be available at a high (or at least known) purity, must be stable during preparation, and must be soluble in the required matrix. Unless the physical phenomenon of a measurement is well understood, extrapolation beyond the calibration data is not recommended, and indeed is usually strongly discouraged; nevertheless, extrapolation is occasionally performed in practice. In those cases, one must be cautious, report exactly how the extrapolation was performed, and assess any increase in uncertainty that may result. Note that the curve might not extrapolate through the origin. This is usually the result of adsorption (of components on container walls), carryover, hysteresis, absorption (of components in seals or septa), or component degradation, oxidation, or evaporation.

A major consideration with external standardization is that, typically, the sample size, such as the injection volume in chromatography, must be maintained constant for standard mixtures and the unknowns. If the sample size varies slightly, it is often possible to apply a correction to the raw signal. Generating a calibration curve by varying the sample size should not be attempted (for example, injecting increasing volumes into a gas chromatograph). This caution does not preclude serial dilution methods (see the section "Serial Dilution" in this article), in which multiple solutions are generated for separate measurement. Other issues that can hinder successful application of the external standards method include instrumental aspects that might not be readily apparent, such as overload. In older instruments, settings of signal attenuation were typically made manually, while in newer instruments, this may occur through software, sometimes without operator interaction or knowledge.

In Fig. 1 and all the examples presented in this article, uncertainty is indicated only for the variable on the *y*-axis. In reality, there is uncertainty for the values plotted on the *x*-axis as well, but the largest uncertainty is often the only treated, or the uncertainty that is most important for the application. It is critical to maintain the integrity of standards; decomposition, degradation, oxidation, and moisture uptake adversely affect the validity of the calibration. See the section "Uncertainty" in this article for more detail, as well as several international standards (see Ref 8). The standard or expanded uncertainty is the preferred quantity in the figures.

Abbreviated External Standard Methods

In many situations, a full calibration curve is not prepared because of the complexity, time, or cost. In such situations, abbreviated external standard methods are often used. Under no circumstances can an abbreviated method be used if the raw signal response is nonlinear. Moreover, these methods generally are not appropriate for analyses in regulatory, forensic, or health care environments where the consequences can be far-reaching.

Discussion of these methods concerns "determinate error" and "bias," which are related terms that describe uncertainty that arises from a fixed cause, and that can, in principle, be eliminated if recognized. Determinate error (or systematic error) is most often associated with a measurement, while bias can be associated with either a measurement or with the sampling procedure.

Single Standard

This method uses a simple proportion approach to standardize an instrument response. It can be used only when the system has no constant, determinate error, or bias, and when the reagents used give a zero blank response, that is, the instrument response from the matrix and measurement system only, without the analyte. A standard should be prepared such that the concentration is close to that of the unknown. To calculate the concentration of the unknown, $[X]$, use:

$$[X] = (A_x/A_s)\,[S] \qquad \text{(Eq 1)}$$

where A_x is the instrument response of the unknown, A_s is the instrument response of the standard, and $[S]$ is the concentration of the standard.

Single Standard plus Assumed Zero

This method, illustrated in Fig. 2, assumes that the blank reading will be zero. A two-point calibration in which the origin is included as the first point is used. It is important to ensure, by experiment or experience, that such a method is adequate to the task.

Single Standard plus Blank

If the analytical method has no determinate error or bias, but does produce a finite blank value, then a blank measurement must be performed, which is subtracted from the instrument response of the standard and the unknown. The same procedure (Eq. 1) is used for the single standard. If multiple samples are to be measured, it is important to measure the blank between each measurement.

Two Standards plus Blank

When the analytical method has both a determinate error or bias and a finite blank value, at least three calibrations must be made: two standards and one blank. The standard concentrations are typically prepared widely spaced in concentration, and the higher concentration should be chosen to represent the limit of linearity of the instrument or method. If this is not practical, the higher concentration should simply be the highest expected concentration of the analyte (unknown). This method is illustrated in Fig. 3. Figure 3 shows the instrument reading with the blank value subtracted. This usually gives a small but finite departure from the origin. If multiple samples are to be measured, it is important to measure the blank between each measurement.

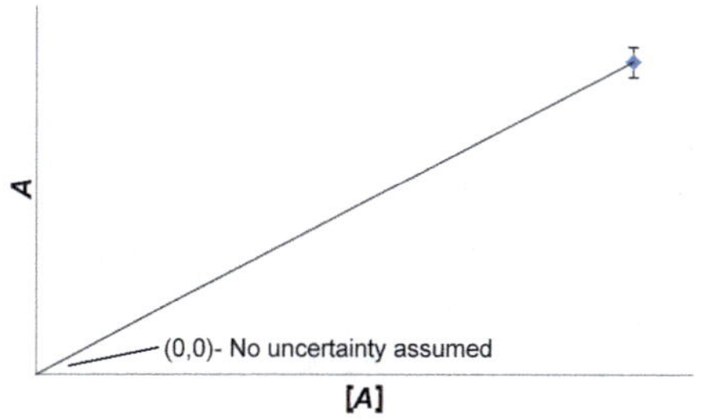

Fig. 2 An example of a single-point calibration curve. The instrument response is represented by *A*, and the concentration resulting in that response is [*A*]. The origin (0,0) is assigned as part of the curve, and is assumed to have no uncertainty.

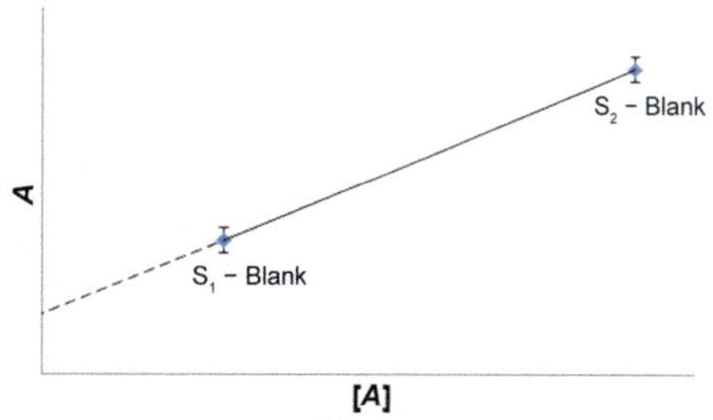

Fig. 3 An example of two standards plus a blank calibration curve. The blank is subtracted from each of the standards. The instrument response is represented by *A*, and the concentration resulting in that response is [*A*].

Internal Normalization Method

As mentioned in the section "External Standard Methods" in this article, the raw signal from an analytical instrument is typically not an absolute measure of concentration of the analytes, because the instrument may respond differently to each component. In some cases, such as with chromatographic methods, it is possible to apply response factors, determined from a standard mixture containing all constituents of the unknown sample, for standardization (Ref 9). The standard mixture is gravimetrically prepared (with known mass percentages for each component), and the instrument response is measured, for example, as chromatographic areas. The total mass percentage and the total area percentage each sum to 100. The ratio of each mass percentage to each area percentage is calculated, with *one* component chosen as the reference, which is assigned a response factor of unity. To obtain the response factors of all the other components, the mass percentage to area ratio with that of the reference is divided. This is done for all components, producing a response factor for all components, except for the reference, defined as unity. When the unknown sample is measured, the response factor is multiplied by each raw area, and the resulting area percentage provides the normalized mass percentages of each component in the unknown.

This method corrects minor variations in sample size (defined as the test portion in the section "Introduction" in this article), although large differences in sample size must be avoided so that consistent instrument performance can be ensured. Although the method corrects for the different responses of samples, large differences must be avoided. This also means that the detector must respond linearly to the concentrations of each component, even if the concentrations are very different. This may require dilution or concentration of the sample in some situations. In chromatographic applications, all components of a mixture must be analyzed and standardized, since normalization must be performed on the entire sample.

Some techniques, such as gas chromatography with flame ionization detection and thermal conductivity detection, have well-defined physical phenomena associated with output signals. With these techniques, there are some limited, published response factor data that can be used in an approximate way to standardize the response from these devices (Ref 10).

In Situ Standardization

While it is rare that an analytical method can be calibrated by use of a single solution, some instances of spectrophotometry and electroanalytical methods can qualify. To use this method, known masses of standard analytes are sequentially and incrementally added to a solution, and an instrument response is measured after each addition. This procedure can be used only if the analytical method itself does not change the analyte concentration (nondestructive) and does not lead to a loss of solution volume. A solid crystalline analyte is an example. Changes in solution volume over the course of the standardization must be minimized.

Standard Addition Methods

Samples presented for analysis often are contained in complex matrices with many impurities that may interact with the analyte, potentially enhancing or diminishing a signal from an instrumental technique. In such cases, the preparation of an external standard calibration curve is impossible, because it is difficult to reproduce the matrix. In these cases, the standard addition method may be used. A standard solution containing the target analyte is prepared and added to the sample, thus accounting for the unknown impurities and their effects. While the quantity of target analyte in the target sample is unknown, the added quantity is known, and its incremental additive effect on the instrument signal can be measured. Then, the quantity of the unknown analyte is determined by what is effectively an extrapolation. In practice, the volume of standard solution added is kept small to avoid dilution of the unknown impurities by no more than 1% of the total signal. This method can be used only if there is a verified linear relationship between the signal and quantity of the analyte. If a determinate error is present, then the slope of the line must be known. Moreover, the sample cannot contain any components that can respond as the analyte (that is, masquerade).

Single Standard Addition

In the simplest case, one addition of analyte is made after first measuring the response of the analyte in the unknown sample. Thus, two measurements are required:

$$A_{xo} = m[X_O] \quad \text{(Eq 2)}$$

$$A_{xi} = m([X_O] + [S]) \quad \text{(Eq 3)}$$

where A_{xo} is the instrument response of the analyte in the unknown sample, $[X_O]$ is the concentration in the unknown sample, and A_{xi} is the instrument response upon the addition of the standard, $[S]$ (additive in the equation because X and S are the same compound). The assumed slope is the proportionality constant, m. The two equations are solved simultaneously for $[X_O]$. This technique is very rapid and economical, but there are serious drawbacks. There is no built-in check for mistakes on the part of the analyst, there is no means to average random uncertainties, and there is no way to detect interference (mentioned previously as masquerade).

Multiple Standard Addition

This standard addition method alleviates some of the problems inherent in single standard addition. Here, the unknown sample is first measured in the instrument. Then that sample is "spiked" with incrementally increasing concentrations of the analyte, generating a curve such as that shown in Fig. 4. The curve should extrapolate to zero signal at zero concentration. The concentration of the analyte in the unknown is read or calculated from the abscissa (x-axis).

Internal Standard Methods

An internal standard is a compound added to a sample at a known concentration, the purpose of which is to exhibit a similar signal when measured in an instrument, but to be distinguishable from the signal of the desired analyte. It typically provides the highest level of reliability in quantitation by many methods and is not affected by large differences in sample size (Ref 8). Unlike the internal normalization method, it is not necessary to elute or measure all the components of the sample; the focus is only on the components of interest. In atomic spectrometry of metals, this method is not affected by changes in gas flow rates, sample aspiration rates, and flame suppression or enhancement. Another situation in which this method is valuable is when the sample matrix is either unknown or very complex, precluding the preparation of external standards.

Multiple Internal Standards

A set of calibration solutions is prepared by mass, containing the target analyte, X, and a standard that is not present in the unknown sample, A. The instrument response is measured for each calibration solution, and a plot is made to establish linearity, as in Fig. 1. The ordinate axis is the ratio of the response of the unknown analyte component, A_x, to the response of the chosen standard, A_s. The abscissa is the ratio of the mass of X to the mass of S for that standard mixture. Once the linearity is confirmed in the concentration range of interest, the unknown is spiked with a known mass of S, the instrument response is measured, and the area ratio A_x/A_s is calculated. Either the graph or a fit of the data on Fig. 5 is then used to determine the corresponding mass fraction, from which the mass of X can be determined. Note that the calculations can be simplified if the same mass per volume of internal standard is added to both the unknown samples and the calibration standards.

Single Internal Standard

In practice, once the linearity is established for a given mixture, it is no longer necessary to use multiple standards, although this is the most precise method. Subsequent to the verification of linearity, one standard solution can be used to fix the slope, provided it is close in concentration to that of the target analyte. In this case, the mass of the unknown can be found from:

$$X/S = (A_x/A_S)(1/R) \quad \text{(Eq 4)}$$

where X is the mass of the unknown analyte in the sample, S is the mass of the added internal standard in the sample, and A_x and A_s are the instrument responses (areas) of the unknown and internal standard, respectively. R is a ratio determined from the standard solution prepared with both X and S:

$$\frac{\left(\dfrac{\text{Mass, Unknown Analyte}}{\text{Mass, Internal Standard}}\right)}{\left(\dfrac{\text{Signal, Unknown Analyte}}{\text{Signal, Internal Standard}}\right)} = R \quad \text{(Eq 5)}$$

Since R is the slope of the calibration curve, once linearity is established, one solution suffices. There are many conditions that must be fulfilled in order to use the internal standard method, and it is rare that all of them can actually be met (mandatory conditions are italicized here). The compound chosen *must not be present* already in the unknown. The response of the compound chosen *must be separable from the analyte* present in the unknown. An exception occurs when an isotopically labeled standard is used, in conjunction with mass discrimination or radioactive counting detection. In a chromatographic measurement, this is typically at least baseline resolution, although this would be a minimally acceptable degree of separation. On the other hand, the unknown analyte peak and the internal standard peak should be close to each other (temporally) on the chromatogram. The compound chosen *must be miscible* with the solvent at the temperature of reagent preparation and measurement. The compound chosen must not react chemically with the sample or solvent,

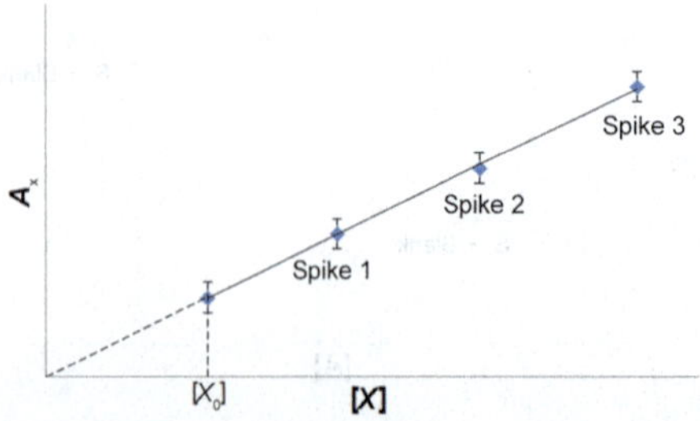

Fig. 4 An example of calibration by multiple standard addition. Three additions (spikes) of the analyte X are shown, as is the extrapolation to the unknown concentration, X_O.

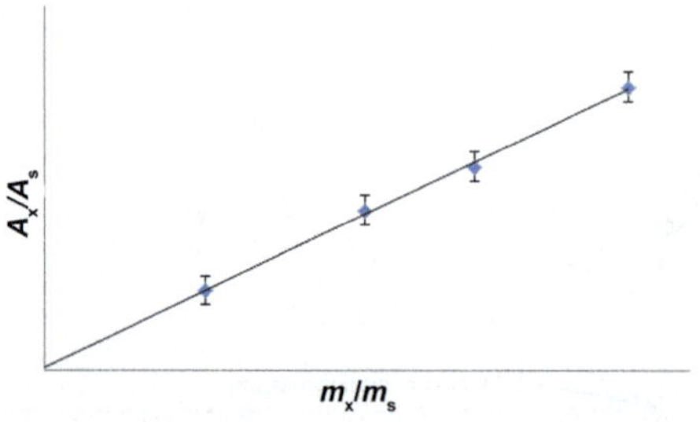

Fig. 5 An example of the multiple internal standard method. The ordinate (y) axis is the ratio of the response of the unknown analyte component, A_x, to the response of the chosen standard, A_s. The abscissa (x) axis is the ratio of the mass of X to the mass of S for that standard mixture.

or interfere in any way with the analysis. It is critical to maintain the integrity of standards; decomposition, degradation, and moisture uptake will adversely affect the validity of the calibration. In the case of a chromatographic measurement, the same applies to interactions with the stationary phase. The compound chosen must be chemically similar (for example, in functional groups, thermophysical properties) to the analyte. If such a compound is not available (for example, in a chromatographic measurement), an appropriate hydrocarbon should be chosen as a surrogate. The standard solution should be prepared at a similar concentration as in the unknown matrix; ratio correction of large differences is no substitute for an appropriate concentration. In a chromatographic measurement, the compound chosen must be eluted as closely as possible in retention time to the analyte, and should not be the last peak to be eluted; the final peak often shows different geometry such as tailing. The compound chosen must be sufficiently nonvolatile to allow for storage as needed. When there is the potential for the unknown analyte to be lost by adsorption, absorption, or some other interaction with the matrix or container, a compound called a carrier is sometimes added in large excess. The carrier is similar, chemically and physically, to the unknown analyte, but easily separated from it. Its purpose is to saturate or season the matrix and prevent analyte loss.

Serial Dilution

Serial dilution is less a standardization method as it is a method of generating solutions to be used for standardizations. Nevertheless, its importance and utility, as well as the popularity of its application, warrant mention. A serial dilution is the stepwise dilution of a substance, observant of a specified, constant progression, usually geometric (or logarithmic). A known volume of stock solution of a known concentration is prepared, and then some small fraction of it is withdrawn to another container or vial. This subsequent container is then filled to the same volume as the stock solution with the same solvent or buffer. The process is repeated for as many standard solutions as are desired. A 10-fold serial dilution can be 1 M, 0.1 M, 0.01 M, 0.001 M, etc. A 10-fold dilution for each step is called a logarithmic dilution or log-dilution, a 3.16-fold ($10^{0.5}$-fold) dilution is called a half-logarithmic dilution or half-log dilution, and a 1.78-fold ($10^{0.25}$-fold) dilution is called a quarter-logarithmic dilution or quarter-log dilution. In practice, the 10-fold dilution is the most common. The serial dilution procedure is used not only in chemical analysis but also in serological preparations in which cellular materials such as bacteria are diluted. A critical aspect of serial dilution is that the initial solution concentration must be prepared and determined with great care because any mistake will be propagated into all resulting solutions.

Traceability

Analytical measurements and certifications often contain a statement of traceability. Traceability describes the "property of a result of a measurement whereby it can be related to appropriate standards, generally international or national standards, through an unbroken chain of comparisons" (Ref 11). Traceability typically includes the application of reference materials or National Institute of Standards and Technology (NIST) Standard Reference Materials for instrument calibration before standardization for the analytes of interest. The true value of a measured quantity (τ) cannot typically be determined. The true value is defined as characterizing a quantity that is perfectly defined. It is an ideal value that can be arrived at only if all causes of measurement uncertainty are eliminated and the entire population is sampled.

Uncertainty

The result of a measurement is only an approximation or estimate of the true value of the measurand (quantity subject to measurement). In the determination of the combined standard uncertainty and ultimately the expanded uncertainty, it is critical to include the uncertainty of calibration in the process, as discussed in this section. The term "combined" standard uncertainty is often truncated to standard uncertainty. The process of arriving at the uncertainty, U_y, of a quantity y that is based upon measured quantities $x_i, \cdots, x_z$ is called the propagation of uncertainty. A full discussion of propagation of uncertainty is beyond the scope of this entry; a simplified prescription, in the form of general and specific formulas, is provided here. In general, the propagated random uncertainty in y can be determined from:

$$U_y = \sqrt{\left(\frac{\partial y}{\partial x_i} U_{x_i}\right)^2 + \left(\ldots \frac{\partial y}{\partial x_z} U_{x_z}\right)^2} \quad \text{(Eq 6)}$$

This approach can be used when the uncertainties are random (not systematic), are relatively small, and are independent or uncorrelated (that is, in the absence of covariance). Relatively large uncertainties, such as those approaching the magnitude of the measurand itself, cannot be treated with this approach, especially if the measurand is a nonlinear function of a measured quantity. Note that, by convention, the use of uppercase U_y denotes the expanded uncertainty, which is the uncertainty multiplied by a coverage factor k (for the 95% confidence level, the coverage factor is 2, assuming a Gaussian distribution). The effect of this multiplier on the confidence level is shown in Fig. 6. The assumption of a Gaussian distribution is justified in that it covers many analytical problems. There are some situations in which so much data are collected that the assumption of a Gaussian distribution is actually demonstrable. This is rarely the case in analytical measurements.

It is possible to reduce this general formulation to more specific formulas in the cases of common mathematical operations. These are provided in Table 1.

REFERENCES

1. R.A. Chalmers, Standards and Standardization in Chemical Analysis, *Comprehensive Analytical Chemistry*, Vol 3, Elsevier, 1975
2. K. Danzer and L.A. Currie, Guidelines for Calibration in Analytical Chemistry, Part I: Fundamentals and Single Component Calibration, *Pure Appl. Chem.*, Vol 70, 1998, p 993
3. K. Danzer, M. Otto, and L.A. Currie, Guidelines for Calibration in Analytical Chemistry, Part 2: Multicomponent Calibration, *Pure Appl. Chem.*, Vol 76, 2004, p 1215
4. B.W. Woodget and D. Cooper, *Samples and Standards (Analytical Chemistry by*

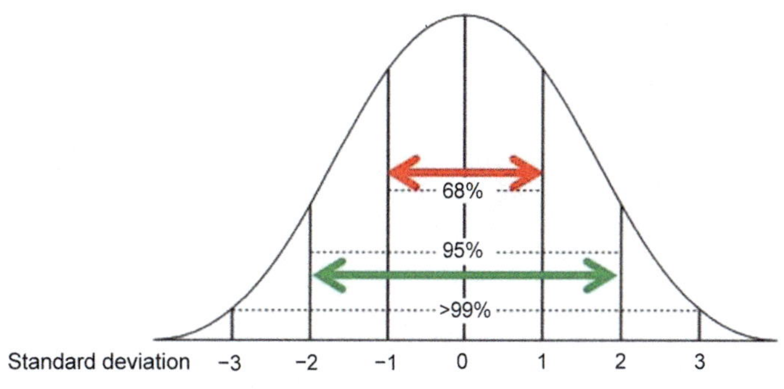

Fig. 6 A graphical depiction of the relationship between the coverage factor k and the confidence level on a hypothetical normal distribution.

Table 1 Specific formulas for the propagation of random, independent uncertainty in the absence of covariance

Measurand argument	Arithmetic uncertainty formula
y (where y is a counted random event over a time interval)	$U_y = \sqrt{y}$
$y = A \times x$ (where A is a constant with no uncertainty)	$U_y = \lvert A \rvert \times U_x$
$y = x_1 + x_2$	$U_y = \sqrt{U_{x_1}^2 + U_{x_2}^2}$
$y = x_1/x_2$ $y = x_1 \times x_2$	$\frac{U_y}{\lvert y \rvert} = \sqrt{\left(\frac{U_{x_1}}{x_1}\right)^2 + \left(\frac{U_{x_2}}{x_2}\right)^2}$
$y = (x_1 \times x_2)/x_3$	$\frac{U_y}{\lvert y \rvert} = \sqrt{\left(\frac{U_{x_1}}{x_1}\right)^2 + \left(\frac{U_{x_2}}{x_2}\right)^2 + \left(\frac{U_{x_3}}{x_3}\right)^2}$
$y = \log(x)$	$U_y = \left(\frac{1}{2.303}\right)\left(\frac{U_x}{x}\right)$
$y = \ln(x)$	$U_y = \frac{U_x}{x}$
$y = e^x$	$U_y = y \times U_x$
$y = x^a$	$\frac{U_y}{\lvert y \rvert} = \lvert a \rvert \times \left(\frac{U_x}{x}\right)$
$y = 10^x$	$\frac{U_y}{\lvert y \rvert} = 2.303 \times U_x$

Open Learning), John Wiley and Sons, 1987
5. P. Gy, *Sampling for Analytical Purposes*, John Wiley and Sons, 1998
6. J.E. Vitt and R.C. Engstrom, Effect of Sample Size on Sampling Error, *J. Chem. Educ.*, Vol 76, **1999**, p 99
7. W. Horowitz, Nomenclature for Sampling in Analytical Chemistry, *Pure Appl. Chem.*, Vol 62, **1990**, p 1193
8. Evaluation of Measurement Data—Guide to the Expression of Uncertainty in Measurement, Joint Committee for Guides in Metrology, 2008
9. R.L. Grob, *Modern Practice of Gas Chromatography*, Wiley Interscience, 1995
10. H.M. McNair, *Basic Gas Chromatography*, John Wiley and Sons, 2009
11. J. Inczedy, T. Lengyel, and A.M. Ure, *Compendium of Analytical Nomenclature*, 3rd ed., International Union of Pure and Applied Chemistry, 1997

SELECTED REFERENCES

Selected References are provided to assist the reader in the evaluation of uncertainty in general. These two NIST documents are very readable and, while written for use by NIST scientists and engineers, they are generally applicable and useful. They are also downloadable.

- B.N. Taylor and C.E. Kuyatt, "Guidelines for Evaluating and Expressing the Uncertainty of NIST Measurement Results," NIST Technical Note 1297, 1994 Edition, Washington, D.C., 1994
- Possolo, "Simple Guide for Expressing Uncertainty of NIST Measurement Results," NIST Technical Note 1900, Washington, D.C., 2015

ASM Handbook, Volume 10, *Materials Characterization*
ASM Handbook Committee
DOI 10.31399/asm.hb.v10.a0006635

Chemical Spot Tests and Presumptive Tests

April A. Hill, Metropolitan State University of Denver

Overview

General Uses

- Rapid classification and, in some cases, presumptive identification of unknown substances
- Rapid screening to determine appropriate chromatographic and/or instrumental methods for confirmation and quantitation

Examples of Applications

- Water-quality monitoring in home, laboratory, and field settings
- Presumptive identification of suspected controlled substances in forensic laboratories
- Metal alloy testing (e.g., kits for detecting the presence of nickel in metal jewelry)

Samples

- *Form:* Organic and inorganic substances can be tested as pure liquids, powdered solids, or solutions.
- *Size:* Samples can be as small as a few nanograms of solid or one drop (~50 μL) of liquid or dissolved solid unknown.
- *Preparation:* Bulk solids must be powdered or dissolved; complex liquids (e.g., body fluids) often require clean-up steps, such as filtration, prior to analysis.

Limitations

- The tests are generally considered nonspecific, because they indicate functional groups or characteristic structures/properties rather than specific molecules.
- Similar functional groups (e.g., alcohols or metal ions) often give similar color test results.

Estimated Time

- A single spot test can be performed in seconds.
- A series of several spot tests is usually required to arrive at a classification or presumptive identification.
- Most reagents must be prepared shortly before use, which adds to the analysis time.

Capabilities of Related Techniques

- *Biochemical spot tests (e.g., immunoassays):* These can provide greater specificity than chemical spot tests, but they require the development of an antibody specific to the analyte of interest.
- *Thin-layer chromatography:* Compounds are separated on a thin layer of adsorbent material bound to a glass or plastic plate; a spot test reagent can then be sprayed directly onto the adsorbent material, where it can react with the compound of interest.

Introduction

Chemical spot tests, also known as presumptive tests, provide a rapid, simple method for screening samples for the presence of certain functional groups or classes of elements. Spot tests are typically colorimetric (based on the formation of colored compounds), making color interpretation a significant analytical tool. Currently, there are colorimetric methods for detecting most of the elements, functional groups of organic compounds, oxidation states of inorganic compounds, and even specific molecules (Ref 1–3). When performed in systematic combinations, these chemical spot tests can provide a presumptive (qualitative) identification of nearly any substance.

There is evidence that chemical spot tests have been used in materials testing for at least two millennia. In his book *Historia Naturalis*, published in 77 A.D., Pliny the Elder described a test used by ancient Romans to detect iron adulteration in the popular medicine known as verdigris (Ref 4). Perhaps the first use of a solid-phase, color-based sensor was by Lewis, who developed litmus paper for the detection of acids and bases in 1767 (Ref 5). In 1834, Runge developed a test for chlorine using paper impregnated with iodine and starch, and in 1859, Schiff published a method for determining uric acid using filter paper impregnated with silver carbonate (Ref 3, 6).

Perhaps the most comprehensive inventory of chemical spot tests can be found in Fritz Feigl's books *Spot Tests in Organic Analysis* and *Spot Tests in Inorganic Analysis* (Ref 2, 3). The two volumes, last published in 1960 and 1972, respectively, discuss the theory and application of over 1000 spot tests. A more recent inventory is provided in the 9th edition (1980) of *Colorimetric Chemical Analytical Methods* by Thomas and Chamberlin (Ref 1).

This article presents a summary of the chemical fundamentals, techniques, and applications of chemical spot testing as well as a brief overview of recent innovations and specialized applications.

Chemical Fundamentals

Basis of Color

White light is composed of a mixture of wavelengths that correspond to different colors. White light can be thought of in terms of a color wheel, with complementary colors "canceling" each other out. The appearance of visible color is based on the absorption of particular wavelengths of visible light and the transmission, scattering, reflection, or refraction of other wavelengths. When an atom or molecule absorbs a certain wavelength (λ) of light, the human eye perceives the sample as the complementary color to that absorbed. For example, a molecule with a maximum absorbance at 450 nm (blue light) will appear orange to the human eye.

The mechanisms of absorbance vary across the electromagnetic spectrum. Spot tests involve transitions in the visible region (~380 to 780 nm), which correspond to electronic transitions within the atom or molecule. Atoms and molecules have unique energy levels that dictate their chemical properties (e.g., molecular structure, electron arrangement, etc.) as well as the wavelengths of light that can be absorbed. That is, visible light can only be absorbed by a molecule if the energy of the photon (which is inversely proportional to its wavelength) directly corresponds to an energy level gap within the molecule, as shown in Fig. 1. Therefore, the wavelength absorbed (or the color observed) can be used to identify a compound. Moreover, using the Beer-Lambert law, the intensity of the absorbance can be directly correlated to the concentration of the colored species. (See the section "Color Interpretation and Quantitation" in this article.) If the color is observed by the human eye rather than an instrument, the identification will be more subjective.

Chemistry and Color

Presumptive tests involve reactions in which the energy levels of an atom or molecule are shifted, causing the color of the substance to change. This provides a visual clue to the nature of the reaction; that is, observing a color change, or lack thereof, upon addition of a spot test reagent can indicate the presence or absence of certain elements and/or functional groups in a sample. Most spot tests use one of three general reaction mechanisms to produce a color change:

- The screening reagent adds an electron to or removes an electron from the target molecule (analyte). The addition or removal of an electron from a colorless structure can change the relative positions of its energy levels, as shown in Fig. 1. If the new gap in energy levels corresponds to absorption of a photon with a wavelength in the visible region of the electromagnetic spectrum, the compound will have a visible color.
- The screening reagent complexes with the analyte. This creates a new structure, often with highly conjugated pi-bonding systems (cyclic structures with alternating double bonds, allowing for delocalization of electrons), which tend to absorb photons in the visible region.
- The screening reagent serves as a bridge to couple two or more of the analyte molecules, again resulting in highly conjugated systems.

It is important to note that the color of the resulting products is often pH dependent. The presence or absence of a proton (H^+ ion) on a structure affects the number of electrons that can participate in the conjugation, which therefore affects color (this is the basis of acid-base indicator function). As such, many chemical spot tests involve the use of strong acids or bases to ensure protonation or deprotonation as desired.

There are, of course, other types of reactions that result in a color change, such as immunoassays that employ dye-labeled antibodies to indicate the presence of anything from drugs to explosives to hormones (Ref 7). However, immunoassay techniques are beyond the scope of this article, which is focused on chemical spot tests that use the three reaction mechanisms listed earlier.

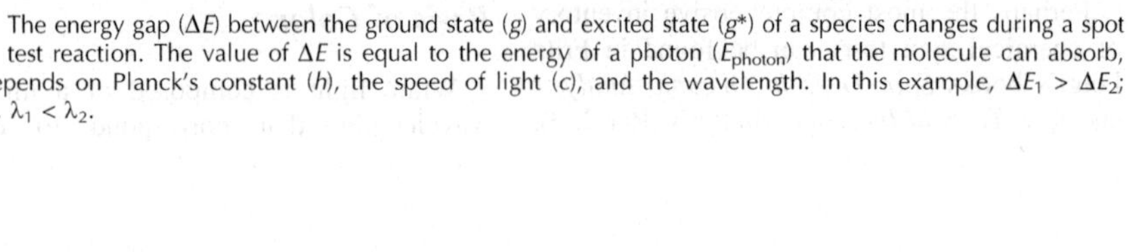

Fig. 1 The energy gap (ΔE) between the ground state (g) and excited state (g^*) of a species changes during a spot test reaction. The value of ΔE is equal to the energy of a photon (E_{photon}) that the molecule can absorb, which depends on Planck's constant (h), the speed of light (c), and the wavelength. In this example, $\Delta E_1 > \Delta E_2$; therefore, $\lambda_1 < \lambda_2$.

General Techniques

Chemical spot tests can be applied to nearly any sample type, ranging from pure powdered solids to complex body fluids. Sample preparation and testing techniques vary depending on the sample and the conditions required for the colorimetric reaction. Hence, this section includes only a brief overview of the many types of sample preparation that can be employed and is focused on the two most common modes of colorimetric testing: liquid-phase testing using dissolved reagents, and solid-phase testing using impregnated reagents.

Reagent Preparation

Reagents can be prepared as liquid solutions or as dry mixtures or by immobilizing on paper, membranes, fibers, and silicates. Dry or immobilized reagents are typically preferred for field applications. Dry reagents are most often prepared as powders, pellets, or "pillows" that contain the additional reagents necessary to adjust the reaction conditions on dissolution (e.g., buffers, oxidizing agents, masking agents, etc.). Immobilized reagents are embedded in matrices such as absorbent cellulose paper or polymer membranes, which are then exposed to the test solution, causing a color change within the matrix. Commercial test strips for determining a wide range of analytes both qualitatively and semiquantitatively are readily available from many manufacturers.

Reagent solutions are typically used for laboratory-based qualitative analysis. Typically, different tests are used for identifying organic compounds than are used for identifying inorganic (often ionic) compounds/elements. Therefore, a list of selected common spot tests, including instructions for preparing reagents and/or performing the test, is included in the relevant sections of this article.

Sample Preparation

Sample preparation will vary greatly depending on the nature of the sample being tested. For samples in powdered solid form or in the solution phase, colorimetric spot testing can be as simple as mixing a drop (~50 μL) of the reagent with a drop of the sample solution or a small amount (as little as a few nanograms, or enough to cover the head of a pin) of powdered sample. For more complex samples, such as body fluids and solid metals, additional preparation steps are often required (Ref 8).

In complex liquid samples, the goal of sample preparation is typically to separate the analyte from interfering substances. Removal of interfering substances can be accomplished by using a variety of techniques, including centrifugation, column chromatography, extraction with activated charcoal or ion-exchange resins, and precipitation reactions. Alternately, the analyte can be selectively extracted from the sample matrix using similar techniques as well as liquid-liquid extraction or solid-phase extraction. In the case of the latter, the colorimetric reagent can often be added directly to the solid matrix either before (i.e., immobilized reagent) or after the extraction of the analyte, with the color change occurring directly on the solid surface.

For solid samples, preparation may involve grinding, mixing, solid-liquid extraction, or other techniques. A coarse solid sample should be ground and mixed so that the aliquot to be tested has the same composition as the bulk. Many inorganic materials can be dissolved in strong acids with heating. Glass vessels are often useful but are not recommended for trace metals analysis due to their tendency to leach metal ions. Moreover, synthetic fluorine-containing resin, platinum, or silver vessels are required when using hydrofluoric acid (HF), which dissolves silicates. Wet ashing is an extension of acid dissolution that is useful in the analysis of trace metals in the presence of organic materials. In this technique, the sample is digested by heating in a mixture of oxidizing acids (such as nitric and sulfuric), which decomposes the organic material.

If acid digestion fails, fusion in a molten salt (flux) is generally possible, but the large quantity of flux involved adds trace impurities. Finely powdered unknown is mixed with 2 to 20 times its mass of solid flux, and fusion (melting) is carried out in a platinum-gold alloy crucible at 300 to 1200 °C (570 to 2190 °F) in a furnace or over a burner. The molten flux is then poured into beakers containing strong acid (e.g., 10% by mass aqueous nitric acid) to dissolve the product. Most fusions use lithium tetraborate ($Li_2B_4O_7$) or lithium metaborate ($LiBO_2$). Fusion with sodium metal is often used as a preliminary step in colorimetric spot testing protocols. (See the section "Sodium Fusion (Lassaigne's) Test" in this article.)

Color Interpretation and Quantitation

Many spot tests rely on a simple visual observation of a positive (color change) or negative (no color change) result. However, spot tests can be combined with comparison charts or spectrophotometric techniques to provide semiquantitative or even quantitative results. Although color comparison charts can be used to evaluate chemical spot test results, it is generally recommended to run positive controls whenever conducting presumptive tests of an unknown. A positive control contains the functional group or element that is indicated by the spot test reagent and therefore provides confirmation of a positive test result (formation of the same color as the positive control) as well as an indication of faulty test solutions (failure to produce the expected color for the control sample).

The quantitative interpretation of color results falls into two main categories: assigning numerical values to a color (often to facilitate the differentiation of similar colors) and assigning a concentration to a colored species based on the amount of light absorbed.

Numerical Descriptions of Color

Color measurement involves associating numerical values with the perception of color (Ref 9). There are three components that contribute to a perceived color: the light source, the object, and the color sensitivity of the observer. The spectral characteristics of a light source are represented by a plot of relative energy as a function of wavelength. Objects contribute to the perception of color by modifying the light from the light source, that is, by reflecting, transmitting, or absorbing certain wavelengths, as described in the section "Chemistry and Color" in this article. The third component is the sensitivity of the observer (or detector). The human eye contains three types of color receptors: one sensitive to red, one to green, and the other to blue.

The Commission Internationale de l'Éclairage (CIE) conducted a series of experiments to quantify the ability of the human eye to perceive color. The resulting experimentally derived functions, x_λ, y_λ, and z_λ, respectively quantify the red, green, and blue sensitivity of the average human observer. The CIE tristimulus color values X_λ, Y_λ, and Z_λ of any object are obtained by combining the values for the light source intensity, the reflectance of the object, and the standard observer color sensitivity functions at each wavelength. The resulting product is then summed over all wavelengths of the visible spectrum to give the X, Y, and Z tristimulus values.

The CIE tristimulus values do not define a uniform color space when plotted as three-dimensional coordinates. That is, distances between plotted points representing different colors do not correspond to differences in perceived color. To address this issue, the $L^*a^*b^*$ color space was developed by CIE. The $L^*a^*b^*$ values are derived from tristimulus values, as shown in Eq 1 to 3, where X_n, Y_n, and Z_n are tristimulus values for a selected white object (ideally, a perfect reflecting diffuser), and X, Y, and Z are tristimulus values for the colored object:

$$L^* = 116\sqrt[3]{\frac{Y}{Y_n}} - 16 \quad \text{(Eq 1)}$$

$$a^* = 500\left(\sqrt[3]{\frac{X}{X_n}} - \sqrt[3]{\frac{Y}{Y_n}}\right) \quad \text{(Eq 2)}$$

$$b^* = 200\left(\sqrt[3]{\frac{Y}{Y_n}} - \sqrt[3]{\frac{Z}{Z_n}}\right) \quad \text{(Eq 3)}$$

The CIE $L^*a^*b^*$ color space is defined by the three mutually perpendicular L^*, a^*, and b^* axes. The L^* axis runs from top to bottom. The maximum for L^* is 100, which defines a perfect diffuse reflector (white), while the minimum for L^* is zero, which represents black. The positive a^* axis is red, and negative a^* is green. Positive b^* is yellow, and negative b^* is blue. Differences in color between, for example, a sample and a standard can be described by differences in the individual color coordinates. That is, $\Delta L^* = L^*_{\text{sample}} - L^*_{\text{standard}}$ and similarly for the other axes.

For evaluating color differences, it is more convenient to work with a single parameter, the value of which indicates how close two colors are to one another. For this, the total color difference, ΔE^*, is determined by using Eq 4:

$$\Delta E^* = \sqrt{\Delta L^2 + \Delta a^2 + \Delta b^2} \quad \text{(Eq 4)}$$

The value of ΔE^* is therefore the geometric distance between color coordinates of the sample and the standard in $L^*a^*b^*$ space. Because color measurement is ultimately based on human perception, a limitation of this approach is that it cannot be directly applied to ultraviolet (<400 nm) or infrared (>700 nm) measurements. A comprehensive explanation of the CIE $L^*a^*b^*$ color space can be found in the book *Colorimetry: Understanding the CIE System* (Ref 9).

Quantitative Analysis via Absorbance/ Reflectance

For samples that are transparent, concentration is most often determined from absorbance measurements. Typically, a liquid sample (containing the colored product) in a cuvette is placed in a spectrophotometer (Fig. 2a). The sample is exposed to a known intensity of monochromatic light (a narrow range of wavelengths) selected to correspond to an energy gap in the colored product, and the intensity of the transmitted radiation is recorded. Transmittance (T) is defined as the ratio of incident light that passes through a sample, giving a range of values from 0 to 1 (0% transmittance to 100% transmittance). T decreases as analyte concentration increases, making it a counterintuitive means of calibration. Therefore, T values are converted to absorbance (A) values, where $A = -\log T$. Absorbance values are proportional to analyte (colored product) concentration. This relationship is given by the Beer-Lambert law (also known as Beer's law), which is expressed as:

$$A_\lambda = \varepsilon_\lambda bC \quad \text{(Eq 5)}$$

Equation 5 shows that the amount of light absorbed by a colored solution at a given wavelength (A_λ) depends on the path length (width of sample the light traverses, b), the molar absorptivity (amount of light attenuated by the substance at the selected wavelength, ε_λ), and the concentration of the colored compound in the sample (C). Beer's law demonstrates that the concentration of the analyte is directly proportional to its absorbance. Therefore, if the path length is constant, a plot of absorbance as a function of concentration (often called a Beer's law plot) for a set of standard solutions can be generated. The line of best fit for such a plot yields an equation that can be used to solve for the concentration of an unknown sample of the same analyte based on its measured absorbance.

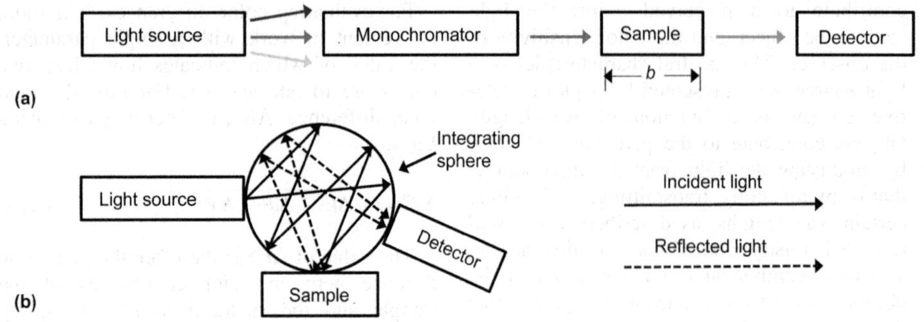

Fig. 2 Schematics of instrumentation used for measuring (a) absorbance and (b) diffuse reflectance

For samples that are opaque, concentration can be determined by measuring the light reflected rather than that transmitted. When light is shined on a surface, some of the incident radiation will penetrate into the sample, where it undergoes a combination of scattering (i.e., reflection, refraction, and diffraction) and absorption within the sample. As a result of these interactions, some of the radiation is redirected toward the surface of the sample, where it exits in all directions with equal intensity (i.e., it is diffuse), and the surface appears uniformly colored (Ref 10). The Kubelka-Munk function ($F(R)$) is calculated from the diffuse reflectance (R) of the sample at the analytical wavelength measured (Fig. 2b) relative to a nonabsorbing (colorless) standard. This function, the reflectance analog of Beer's law for absorption measurements, relates the diffuse reflectance of an opaque sample to its absorption coefficient, k, and scattering coefficient, s, as given by (Ref 11):

$$F(R) = \frac{(1-R)^2}{2R} = \frac{k}{s} \quad \text{(Eq 6)}$$

When measuring the reflectance of an absorbing species diluted in a nonabsorbing matrix (e.g., a colorimetric product embedded in a test strip or a membrane) relative to the reflectance of the pure matrix, the absorption coefficient may be replaced by the product $2.303\varepsilon_\lambda C$, where ε_λ is the molar absorptivity of the colorimetric product at the analytical wavelength, and C is the concentration of the colored product (e.g., concentration of the analyte). This gives:

$$F(R) = \frac{2.303\varepsilon_\lambda}{s} C \quad \text{(Eq 7)}$$

For measurements of varying concentrations of a single colored compound on a given matrix (e.g., a paper strip or a membrane), ε_λ and s can be reasonably assumed to be constant throughout an analysis. To determine analyte concentrations, reflectance measurements are converted to Kubelka-Munk values at the appropriate analytical wavelength. These $F(R)$ values are used to quantify the analyte via a calibration curve analogous to that used in a Beer's law analysis.

Limitations of Spot Tests

Given that chemical spot tests are typically responsive to functional groups (specific groups of atoms that form the building blocks of molecules) or characteristic structures/properties, they are generally nonspecific and therefore provide only presumptive identifications. That is, most colorimetric reagents can indicate what building blocks are present in a molecule, but they cannot reveal how those building blocks are assembled. (They cannot tell the precise structure, which is required for a confirmed identification.) Also, if a reagent gives a color change with one functional group (e.g., a halogen atom such as chlorine), it will usually give the same or very similar result with other halogens (e.g., iodine). As is evident in the section "Spot Tests for Inorganic Analysis" in this article, this presents a particular challenge when using colorimetric spot tests to identify metal ions, many of which give similar color changes when tested with a particular colorimetric reagent.

Although the vast majority of spot tests cannot provide a definitive identification of a compound, they can indicate the presence or absence of classes of compounds in a sample, which helps to narrow down the possibilities and determine what confirmatory tests will be effective.

Spot Tests for Organic Analysis

The qualitative identification of most organic compounds can be accomplished by completing a series of spot tests to determine the presence of functional groups (atoms or groups of atoms that give a molecule predictable properties). A catalog of selected organic functional groups is shown in Table 1. Table 2 presents a list of spot tests that are commonly used for detecting each of these functional groups, and Fig. 3 gives a suggested protocol (flow chart) for performing a full functional

Table 1 Structures of selected organic functional groups, where "R" indicates a variable structure attached to the relevant group (i.e., the rest of the molecule), and "X" indicates a halogen atom

Functional group		Formula
Alcohol		HO — R
Aldehyde		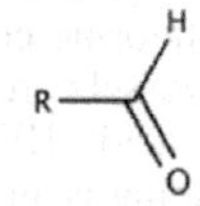
Alkane		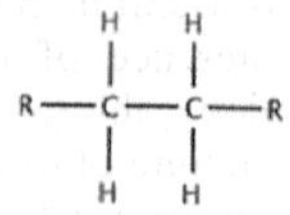
Alkene		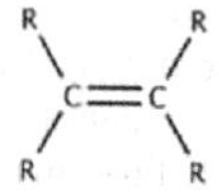
Alkyne		$R-C\equiv C-R$
Amide		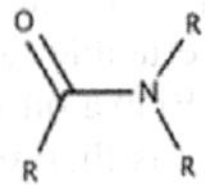
Amine	Primary	$R-NH_2$
	Secondary	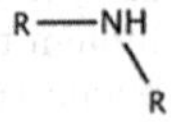
	Tertiary	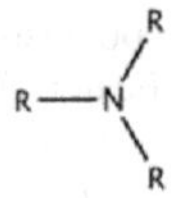
Arene (aromatic ring)		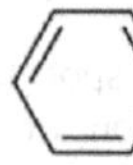
Aryl halide		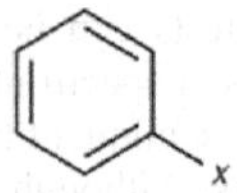
Carboxylic acid		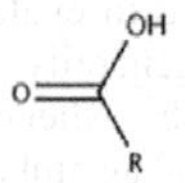
Ketone		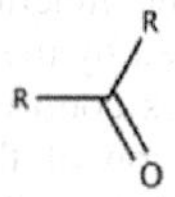
Nitrile		R — CN
Phenol		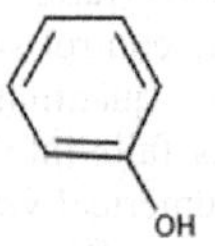
Sulfonamide		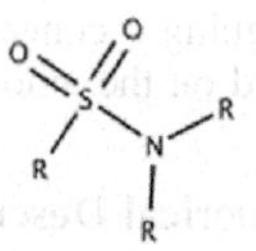
Sulfonic acid		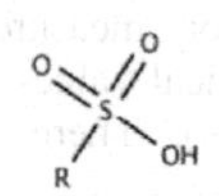

Table 2 Presumptive tests for organic functional groups

Family	Test	Notes
Alcohols	Ceric ammonium nitrate	Positive for all alcohols
	Dichromate test	Positive for 1° and 2° alcohols; negative for 3° alcohols
	Iodoform test	Positive for all alcohols of the general formula $CH_3CH(OH)R$
	Lucas test	Immediate reaction for 3°, allylic, or benzylic alcohols; slower reaction (2–5 min) for 2°; no reaction tor 1° alcohols
Aldehydes	Benedict's test	Positive for all aldehydes
	Dichromate test	Positive for all aldehydes
	2,4-dinitrophenylhydrazine (2,4-DNP)	Positive for all aldehydes (and ketones)
	Fehling's test	Positive for all aldehydes
	Iodoform test	Positive only for acetaldehyde
	Oxime	Positive for all aldehydes (and ketones)
	Permanganate test	Positive for all aldehydes
	Semicarbazone	Positive for all aldehydes (and ketones)
	Tollen's test	Positive for all aldehydes
Alkanes	No test	
Alkenes	Bromine test	Positive for all alkenes
	Permanganate test	Positive for all alkenes
	Solubility in conc, sulfuric acid	All alkenes dissolve
Alkynes	Bromine test	Positive for all alkynes
	Permanganate test	Positive for all alkynes
	Silver nitrate	Positive for all terminal alkynes only
	Sodium metal addition	Positive for all terminal alkynes only
	Sulfuric acid	Positive for all alkynes
Amides	Basic (reflux) hydrolysis	All amides yield ammonia or the corresponding amine detected by odor or by placing wet blue litmus paper on top of the condenser.
Amines	Diazotization	All 1° amines give red azodyes with β-naphthol.
	Hinsberg test	Distinguishes between 1°, 2°, or 3°
	Solubility In dilute HCl	All amines are soluble.
Arenes	Aluminum chloride-chloroform	Positive for all arenes
Aryl halides	Aluminum chloride-chloroform	Positive for all aryl halides
Carboxylic acids	Solubility in dilute sodium bicarbonate	All carboxylic acids are soluble
	Solubility in dilute sodium hydroxide	All carboxylic acids are soluble
Ketones	2,4-dinitrophenylhydrazine (2,4-DNP)	Positive for all ketones (and aldehydes)
	Hydrazine	Positive for all ketones (and aldehydes)
	Iodoform	Positive for methyl ketones
	Oxime	Positive for all ketones (and aldehydes)
	Semicarbazone	Positive for all ketones (and aldehydes)
Nitriles	Basic hydrolysis	Positive for all nitriles
Phenols	Acetylation	Precipitate of a characteristic melting point
	Benzoylation	Precipitate of a characteristic melting point
	Sulfonation	Precipitate of a characteristic melting point
	Ferric chloride test	Variety of colors characteristic of the individual phenol
	Solubility in aqueous base	Most phenols are soluble to dilute sodium hydroxide but insoluble in dilute sodium bicarbonate. Phenols with strong electron withdrawing groups (e.g.. picric acid) are soluble to sodium bicarbonate.
Sulfonamides	Basic (reflux) hydrolysis	Positive for all sulfonamides
	Sodium fusion test	Presence of sulfur and nitrogen
Sulfonic acids	Sodium fusion test	Presence of sulfur
	Solubility in aqueous base	Most sulfonic acids are soluble to dilute sodium hydroxide and generate carbon dioxide with sodium bicarbonate.

Source: Ref 12

group analysis of an unknown (Ref 12). What follows is a list of selected reagents from Table 2, including abbreviated instructions for preparing the reagent solution(s), for performing the test, and for interpreting the results (Ref 1–3, 12–14). Most of the tests in use today (2019) have been in use for well over 50 years, and many variations of reagent preparations and protocols exist. Moreover, specialized applications, such as the forensic analysis of illicit drugs, often use a different set of reagents and protocols. (See the section "Example Applications and Innovations" in this article.)

Common Tests and Protocols

All solutions are aqueous unless otherwise stated. All reagents should be considered toxic and handled only in a fume hood with gloves.

Aluminum Chloride/Chloroform Test

Prepare a solution (in a fume hood) that contains three drops of liquid unknown (or 0.1 g if solid) in 2.0 mL of dry chloroform ($CHCl_3$). Heat 0.2 g of anhydrous aluminum chloride ($AlCl_3$) in a test tube held at an angle so that gaseous $AlCl_3$ sublimes onto the tube walls. Once cool enough to touch, pipette a few drops of the chloroform solution into the tube. Formation of an intense color (shades can range from orange to purple) indicates a compound with an aromatic ring.

Basic Hydrolysis Test

Reflux 0.1 g (or three drops if liquid) of the unknown in 5 mL of a 10% (by mass) sodium hydroxide solution for 30 min (or until mixture is homogeneous). Amides are detected based on odor (ammonia) or by placing wet red litmus paper (positive tests turn the paper blue) on top of the test tube.

Benedict's Test

With the aid of heat, dissolve 173 g of sodium citrate ($Na_3C_6H_5O_7$) and 100 g of sodium carbonate (Na_2CO_3) in 800 mL of water. Filter if necessary. Dissolve 17.3 g of copper sulfate pentahydrate ($CuSO_4 \cdot 5H_2O$) in 100 mL of water. To make Benedict's reagent, pour (with constant stirring) the copper solution into the citrate solution and dilute to 1 L.

Add a small amount of unknown (approximately two drops of liquid or 80 mg solid) to 1 to 2 mL of water in a test tube. Add 1 to 2 mL of the Benedict's reagent and heat to boiling. A positive test for aldehydes involves both a color change from blue copper(II) to red copper (I) and formation of a copper(I) oxide precipitate.

Bromine Test

The unknown is treated with a few drops of a 1 to 5% solution of bromine (Br_2) in dichloromethane (CH_2Cl_2). The decolorization of the orange-brown bromine indicates unsaturation in the unknown (C=C or C≡C).

Ceric Ammonium Nitrate Test

Add ten drops of the compound to be tested to 1 to 2 mL of 5% (by mass) ceric ammonium nitrate [$(NH_4)_2Ce(NO_3)_6$]. The appearance of an orange/red color is indicative of the presence of an alcohol.

Dichromate (Jones, Chromic Acid) Test

To ten drops of the unknown (or 0.4 g if solid) add 1 mL of 1% sodium dichromate ($Na_2Cr_2O_7$) and five drops of concentrated H_2SO_4. Conversion of the orange reagent to a blue-green solution in a few seconds is indicative of a primary (1°) or secondary (2°) alcohol. Aldehydes will also give this color change, but more slowly (~10 to 90 s). Tertiary (3°) alcohols do not react. Slight heating may be necessary for water-immiscible alcohols. Extensive heat may give a positive test for tertiary alcohols, due to the water elimination of the alcohol and oxidation of the formed alkene.

2,4-dinitrophenylhydrazine (2,4-DNP) Test

The 2,4-DNP reagent is prepared by dissolving 1 g of 2,4-DNP in 5 mL concentrated

H_2SO_4, then adding 8 mL of water and 20 mL of 95% ethanol. Add ten drops of the compound to be tested to 1 mL of the 2,4-DNP reagent. A yellow to orange-red precipitate is considered a positive test for aldehydes and ketones.

Ferric Chloride Test

Dissolve one drop (or 40 mg of solid) of unknown in 1.0 mL of water (or chloroform if insoluble in water). Add five drops of 3% aqueous $FeCl_3$ solution. Phenols give red, blue, purple, or green colorations.

Hinsberg Test

To three drops of the unknown (or 0.1 g of solid) in a test tube, add 6 mL of 10% sodium hydroxide (NaOH). Add 0.2 g of p-toluenesulfonyl chloride, stopper the tube, and shake for 3 to 5 min. Remove the stopper and warm the tube with shaking in a hot water bath (70 °C, or 160 °F) for approximately 1 min. The lack of a reaction indicates a 3° amine; the amine becomes soluble upon acidification (pH 2 to 4) with 10% HCl. If precipitate forms in the alkaline solution, dilute with 5 to 8 mL of deionized water and shake. If the precipitate does not dissolve, this indicates a 2° amine. If the solution is clear, acidify (pH ~ 4) with 10% HCl. The formation of a precipitate at this point is indicative of a 1° amine.

Iodoform Test

The reagent is prepared by dissolving 10 g of iodine (I_2) and 20 g of potassium iodide (KI) in 100 mL of water. The unknown (three drops of liquid or 0.1 g of solid) is dissolved in 2 mL of water (or methanol if insoluble in water) and 1 mL of 10% aqueous NaOH. The reagent is then added dropwise to this mixture until a persistent brown color remains (even when heating in a hot water bath at 60 °C, or 140 °F). A yellow precipitate is indicative of iodoform (CHI_3) formation and is characteristic of a methyl ketone, acetaldehyde, or an alcohol.

Molisch Test

The reagent is a solution of 95% (vol/vol) of 1-naphthol in ethanol. This reagent is added to the unknown solution, which is then acidified with sulfuric acid. The development of a purple color at the interface of the test solution with the reagent mixture is indicative of a carbohydrate.

Lucas Test

The reagent is made by dissolving 16 g anhydrous zinc chloride ($ZnCl_2$) in 10 mL of concentrated hydrochloric acid and cooling to avoid HCl loss. Add five drops of the unknown alcohol to 2 mL of the reagent in a test tube. Stopper the tube immediately and shake vigorously. Tertiary alcohols form an emulsion that appears as two layers (due to the formation of a water-insoluble alkyl halide) almost immediately. Secondary alcohols form this emulsion

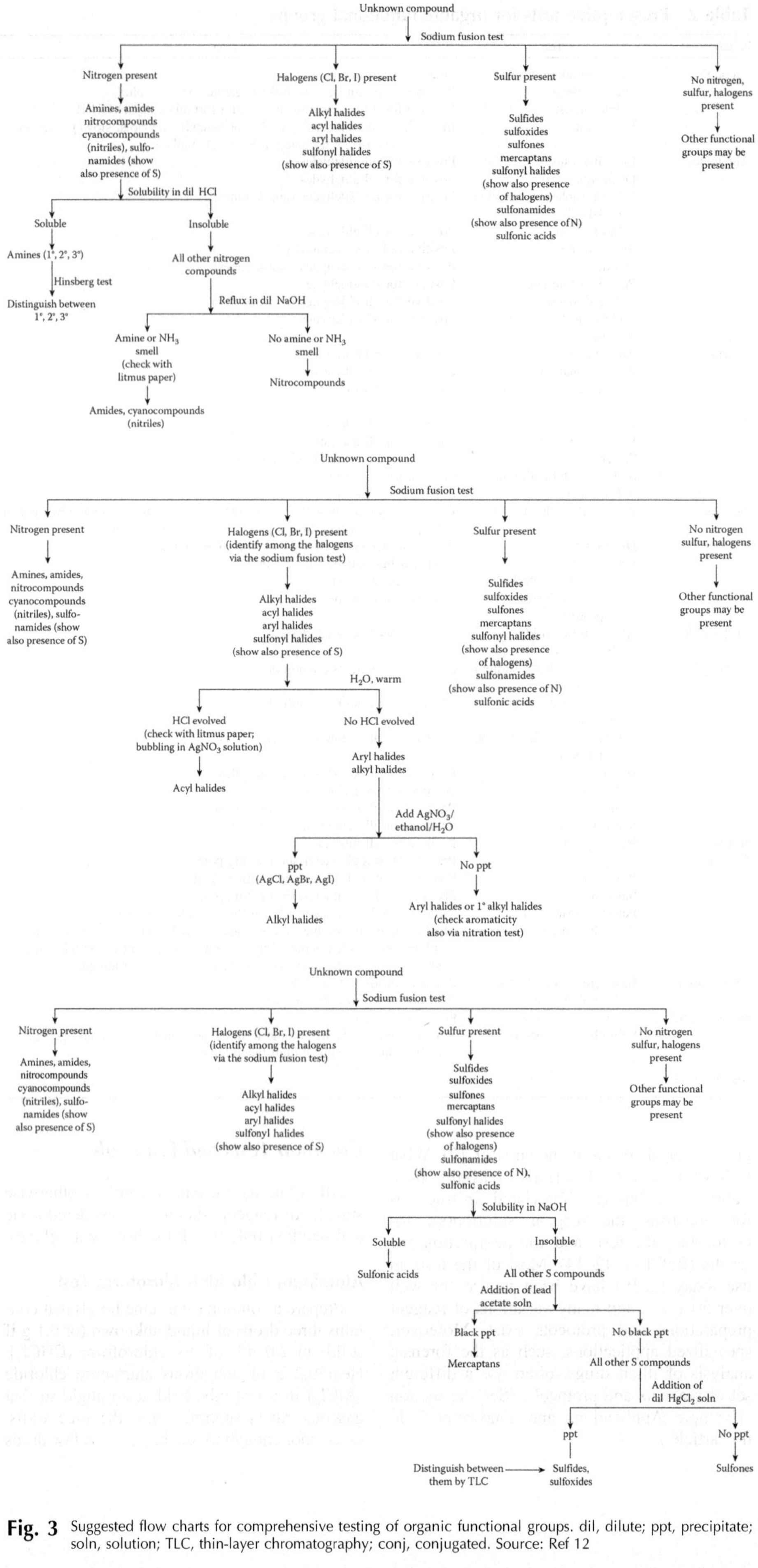

Fig. 3 Suggested flow charts for comprehensive testing of organic functional groups. dil, dilute; ppt, precipitate; soln, solution; TLC, thin-layer chromatography; conj, conjugated. Source: Ref 12

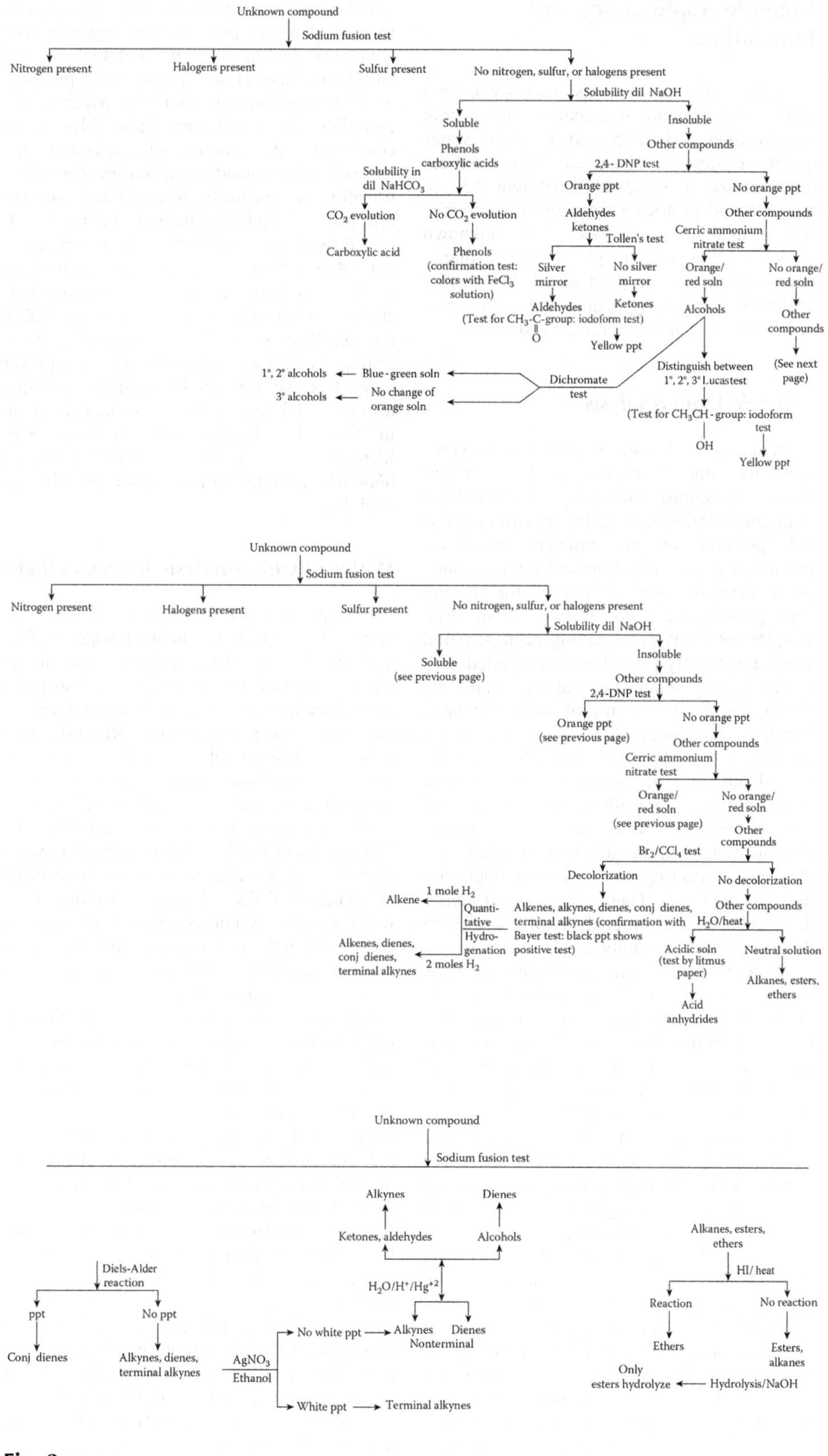

Fig. 3 (continued)

after 5 to 10 min, while primary alcohols do not react. Some secondary alcohols (e.g., isopropyl) may not form a separate layer, because the low-boiling alkyl halide may evaporate.

Permanganate Test

The compound to be tested (one drop of liquid or 40 mg of solid) is dissolved in 2.0 mL of water or 95% ethanol in a test tube. This solution is treated with 10 to 15 drops of 1% potassium permanganate ($KMnO_4$) solution. A positive test for aldehydes is indicated by the decolorization of the solution and subsequent formation of a reddish-brown manganese dioxide (MnO_2) precipitate.

Silver Nitrate Test

The unknown compound (one drop of liquid or 50 mg of solid) dissolved in a small amount of ethanol is added to 2.0 mL of 1% silver nitrate ($AgNO_3$) in ethanol. A precipitate indicates the presence of a halide on the unknown. A white precipitate is indicative of silver chloride (chloride present in the unknown as an alkyl halide), silver bromide is pale yellow or cream color, and silver iodide is yellow. Some carboxylic acids and alkynes will yield precipitates, but these should redissolve upon addition of acid.

Sodium Fusion (Lassaigne's) Test

Place a piece of sodium metal the size of a small pea in a 100 mm (4 in.) test tube. The test tube is warmed gently to melt the sodium metal. Slowly add 100 mg (or three to five drops of liquid) unknown and observe that decomposition (indicated by charring) of the compound occurs. When it appears that all the volatile material has been decomposed, the test tube is strongly heated until the residue acquires a red color. The red-hot test tube is plunged into a small beaker containing 10 to 15 mL distilled water and covered with a watch glass or a wire gauze. The test tube should shatter, allowing the sodium salts to dissolve in the water. This solution is then filtered and analyzed according to the following three procedures:

- *Test for nitrogen:* To 1 mL of the fusion solution add four drops of 6 M NaOH (pH adjusted to 13), two drops of a saturated solution of $Fe(NH_4)_2(SO_4)_2$ (ferrous ammonium sulfate), and two drops of 30% KF (potassium fluoride). The mixture is then boiled for 30 s, followed by immediate addition of 3 M sulfuric acid, dropwise, until the precipitate of iron hydroxide just dissolves. The appearance of the deep-blue color of potassium ferrocyanide (Prussian blue) indicates the presence of nitrogen in the original organic compound.
- *Test for sulfur:* To a 1 mL aliquot of the sodium fusion solution, add ten drops of 6 M acetic acid and two to three drops of 5% lead(II) acetate solution. A black precipitate (lead sulfide) is indicative of sulfur.

- *Test for halogens:* Acidify a 2 mL aliquot of the fusion solution by dropwise addition of 6 M nitric acid (verify acidification with blue litmus paper). Boil the solution gently for 2 to 3 min to expel any hydrogen sulfide or cyanide that may be present, because these may interfere with the test for the halogens. Cool the solution and add several drops of 10% aqueous silver nitrate solution. A precipitate of silver halide indicates the presence of chlorine (white precipitate), bromine (pale yellow), or iodine (canary yellow) in the original organic compound. A faint turbidity should not be interpreted as a positive test.

Solubility Tests

Some qualitative analysis schemes recommend starting with solubility tests to provide a basic classification. Solubility in water, acids (concentrated H_2SO_4 or dilute HCl), or bases (dilute sodium bicarbonate or dilute NaOH) can indicate a solubility class to further aid in identification. Useful solubility tests and their interpretation are included in Table 2.

Tollen's Test

The reagent should be freshly prepared by mixing two solutions (A and B). Solution A is 10% aqueous silver nitrate, and solution B is 10% aqueous sodium hydroxide. The reagent mixture (A + B) must be prepared immediately prior to use, because explosive silver fulminate can form in a matter of minutes. The test is performed by mixing 1 mL of solution A and 1 mL of solution B and dissolving the silver oxide thus formed by dropwise addition of 10% aqueous ammonia. To the clear solution, one drop of the unknown (or 40 mg of solid unknown) is added. The formation of a silver mirror on the walls of the test tube is indicative of the presence of an aldehyde. The silver mirror is usually deposited either immediately or after a short warming period in a hot water bath. Residual mixed reagent (A + B) should be disposed of immediately by dissolving it in dilute HNO_3.

Spot Tests for Inorganic Analysis

Presumptive tests for inorganic species typically involve color changes or formation of precipitates or both. Inorganic species tend to exist as charged ions (anions are negatively charged, cations are positively charged) in aqueous solutions. Common spot test for detecting anions are presented in Table 3, while Table 4 lists common tests for detecting cations, which are organized into groups based on common behaviors. The presumptive testing reagents discussed in this section are in aqueous solution unless otherwise stated. For details on reagent preparation, see *Vogel's Qualitative Inorganic Analysis* by Vogel, Svehla, and Suehla (Ref 15).

Example Applications and Innovations

In the following sections, two specialized applications of qualitative analysis (illicit drug identification and spacecraft drinking water quality testing) are discussed. These applications utilize a somewhat different set of reagents and protocols from those used in a typical qualitative analysis of an unknown organic or inorganic sample. A brief explanation of these new tests is given, but further details on reagent preparation and testing procedures are given in the referenced texts.

Forensic Drug Analysis

The analysis of suspected illicit drugs comprises the bulk of casework in the chemistry unit of any forensic laboratory. While a structural confirmation (i.e., mass spectrum, infrared spectrum, nuclear magnetic resonance spectrum) is generally required for the courtroom, there are several reasons for starting with presumptive color testing. First, drug samples are rarely pure, having been prepared using rudimentary procedures and diluted with cutting agents along the supply chain. To obtain reasonable spectra of such mixtures, lengthy chromatographic steps are often required. Chromatographic procedures can be carried out more efficiently if a presumptive identification of the illicit drug present (if any) is obtained. Moreover, a presumptive identification is often sufficient to result in a plea deal, rendering full structural confirmation unnecessary. Therefore, most forensic drug chemists start an analysis of a suspected illicit drug with presumptive testing.

As with any organic qualitative analysis scheme, the reagents chosen and the sequence in which they are used can vary significantly. However, there are some fairly common tests and protocols across the discipline. One of the most common first-pass reagents used is the Marquis reagent (formaldehyde in concentrated sulfuric acid). This reagent gives a color result with a wide variety of illicit drugs and can therefore help to categorize unknowns into broad classes. This reagent turns purple with opiates (e.g., morphine and heroin), red-orange to brown with amphetamines, faint pink with phencyclidine (PCP, a synthetic hallucinogen), and orange with mescaline and psilocybin (natural hallucinogens). Another reagent that gives a variety of colors with different classes of drugs is the Mandelin reagent (ammonium vanadate in concentrated sulfuric acid). This reagent turns blue-gray to brown with morphine and heroin, green with amphetamines and psilocybin, and orange with PCP.

There are also a few tests that are considered presumptive positive for a specific drug. Cobalt thiocyanate (cobalt chloride and ammonium thiocyanate in water), for example, reacts with cocaine to form a vibrant blue precipitate in the otherwise pink reagent solution. The Duquenois-Levine reagent (acetaldehyde and vanillin in ethanol) turns purple in the presence of Δ^9-tetrahydrocannabinol (the psychoactive ingredient in marijuana). This color result combined with a microscopic analysis of the leaves is often considered a confirmatory identification of marijuana. Simon's test (sodium nitroprusside and acetaldehyde in water, followed by alkalization with sodium carbonate) is used to distinguish amphetamine (a primary amine, which turns red) from methamphetamine (a secondary amine, which turns blue). An excellent overview of presumptive drug-testing techniques, including instructions for reagent preparation (and associated hazards), testing, and interpretation of results, is given in "Color Test Reagents/Kits for Preliminary Identification of Drugs of Abuse" (standard 0604.01) from the National Institute of Justice (Ref 16).

Water Quality Analysis for Spaceflight

Perhaps the most well-known use of colorimetric spot tests is the field of water quality analysis. The use of test strips to evaluate pH and contaminant levels, whether in fish tanks or in drinking water, is a fairly common practice even among nonscientists. Recently, spot tests have been employed in the analysis of spacecraft drinking water supplies by using an innovative technique called colorimetric solid-phase extraction (CSPE) (Ref 17). The CSPE is a spectrophotometric technique based on measuring the change in diffuse reflectance of indicator disks following exposure to a water sample. A known amount of water is passed through a membrane that has been impregnated with a colorimetric reagent. The analyte is simultaneously extracted, concentrated, and converted to a colored product. Quantification is achieved by using a commercially available, handheld spectrophotometer to measure the diffuse reflectance ($F(R)$) of the membrane. The overall process (using impregnated disks prepared preflight) requires only 45 to 90 s. The technique is now a part of the water-reclamation system aboard the International Space Station (ISS).

Schematics depicting the procedures for analyzing both silver(I) and total silver by CSPE are shown in Fig. 4. The results obtained from a CSPE analysis of total silver in potable water from the ISS were shown to be consistent with those obtained from a more cumbersome and expensive analysis via inductively coupled plasma mass spectrometry (Ref 18). These data suggest CSPE-based methods could provide quantitative interpretation of presumptive test results at a fraction of the cost of most quantitative instrumentation, allowing CSPE to serve as a versatile platform for the development of rapid, quantitative screening methodologies for use in many settings.

Table 3 Presumptive tests for anions

Acetates, CH_3COO^-	
1. Sulfuric acid, dilute	Evolution of acetic acid (vinegar-like odor); concentrated sulfuric acid also evolves sulfur dioxide under mild heating
2. Dry ethanol and concentrated sulfuric acid	Evolves ethyl acetate (fruity odor) upon heating; dry isoamyl alcohol may be substituted for ethanol
3. Silver nitrate	Formation of white precipitate of silver acetate that is soluble in dilute ammonia solution
4. Iron(III) chloride	Deep-red coloration (coagulates on boiling, forming a brownish-red precipitate)
Benzoates, $C_6H_5COO^-$ (or $C_7H_5O_2^-$)	
1. Dilute sulfuric acid	Formation of white precipitate of benzoic acid
2. Dilute hydrochloric acid	Formation of a crystalline precipitate melting between 121 and 123 °C (250 and 253 °F)
3. Silver nitrate	White precipitate of silver benzoate from cold solutions, soluble in hot water and also in dilute ammonia solution
4. Iron(III) chloride	Buff-colored (light yellow-red) precipitate of iron (III) benzoate from neutral solution, soluble in hydrochloric acid
Borates, BO_3^{3-}, $B_4O_7^{2-}$, BO_2^-	
1. Concentrated sulfuric acid	Upon heating solution, white fumes of boric acid are evolved
2. Silver nitrate	White precipitate of silver metaborate, soluble in dilute ammonia solution and in acetic acid
3. Barium chloride	White precipitate of barium metaborate that is soluble in excess reagent, dilute acids, as well as ammonium salt solutions
Bromates, BrO_3^-	
1. Concentrated sulfuric acid	Evolution of red bromine vapors even when cold
2. Silver nitrate	White precipitate of silver bromate that is soluble in dilute ammonia
3. Sodium nitrite	Brown color develops after addition and subsequent acidification with dilure nitric acid.
Note: Bromates are reduced to bromides by sulfur dioxide, hydrogen sulfide, or sodium nitrite solution.	
Bromides, Br^-	
1. Concentrated sulfuric acid	Reddish-brown coloration, followed by reddish-brown vapors (hydrogen bromide + bromine) evolution
2. Manganese dioxide + sulfuric	Reddish-brown bromine vapors evolve upon mild heating
3. Silver nitrate	Pale-yellow, curdy precipitate of silver bromide, slightly soluble in ammonia solution; insoluble in nitric acid
4. Lead acetate	White crystalline precipitate of lead bromide that is soluble in hot water
Special tests: The addition of an aqueous solution of chlorine (or sodium hypochlorite) will liberate free bromine, which may be isolated in a layer of carbon tetrachloride or carbon disulfide.	
Carbonates, CO_3^{2-}	
1. Hydrochloric acid	Decomposition with effervescence and evolution of carbon dioxide (odorless)
2. Barium chloride	White precipitate of barium carbonate that is soluble in HCl (calcium chloride may be substituted for barium chloride)
3. Silver nitrate	Gray precipitate of silver carbonate
4. Magnesium sulfate	White precipitate is formed, which can be dissolved by the addition of dilute acetic acid.
Special tests: Effervescence with all acids, producing carbon dioxide that makes limewater cloudy	
Chlorates, ClO_3^-	
1. Concentrated sulfuric acid	Liberates chlorine dioxide gas (green); solids decrepitate (crackle explosively) when warmed; large quantities may result in a violent explosion
2. Concentrated HCl	Chlorine dioxide gas evolved, imparts yellow color to acid
3. Manganes(II) sulfate + phosphoric acid	Violet coloration due to diphosphatomanganate formation; peroxydisulfate nitrates, bromates, iodates, periodates react similarly
4. Heat of neat sample	Decomposition and formation of gaseous oxygen
Chlorides, Cl^-	
1. Concentrated sulfuric acid	Evolution of hydrogen chloride gas (pungent odor)
2. Silver nitrate	White precipitate of silver chloride that is soluble in ammonia solution (reprecipitate with HNO_3)
3. Lead acetate	White precipitate of lead bromide, soluble in boiling water
Special tests: 1. MnO_2 + H_2SO_4 evolves Cl_2 gas. 2. An aqueous solution of chlorine + carbon disulfide produces no coloration.	
Chromates, CrO_4^{2-}, Dichromates, $Cr_2O_7^{2-}$	
1. Barium chloride	Pale-yellow precipitate of barium chromate, soluble in dilute mineral acids; insoluble in water and in acetic acid
2. Silver nitrate	Brownish-red precipitate of silver chromate, soluble in dilute nitric acid and in ammonia solution; insoluble in acetic acid
3. Lead acetate	Yellow precipitate of lead chromate, soluble in dilute nitric acid; insoluble in acetic acid
4. Hydrogen peroxide	Deep-blue coloration in acidic solution, which quickly turns green, with the subsequent liberation of oxygen
5. Hydrogen sulfide	Dirty-yellow deposit of sulfur is produced in acidic solutions.
Citrates, $C_6H_5O_7^{3-}$	
1. Concentrated sulfuric acid	Evolution of carbon dioxide and carbon monoxide (highly poisonous)
2. Silver nitrate	White precipitate of silver citrate that is soluble in dilute ammonia solution
3. Cadmium acetate	White gelatinous precipitate of cadmium citrate, practically insoluble in boiling water; soluble in warm acetic acid
4. Pyridine + acetic anhydride, 3:1 (vol/vol)	A red-brown color develops upon addition to the reagent mixture.
Cyanates, OCN^-	
1. Sulfuric acid, concentrated and dilute	Vigorous effervescence, due largely to evolution of carbon dioxide, with concentrated acid producing a more dramatic effect
2. Silver nitrate	Curdy white precipitate of silver cyanate
3. Copper sulfate-pyridine	Lilac-blue precipitate (interference by thiocyanates) Reagent is prepared by adding 2–3 drops of pyridine to 0.25 M $CuSO_4$ solution

(continued)

Source: Ref 12

Table 3 (Continued)

Cyanides, CN^-	
1. Cold dilute HCl	Liberation of hydrogen cyanide (odor of bitter almond; Caution: highly toxic)
2. Silver nitrate	White precipitate of silver cyanide
3. Concentrated sulfuric acid, hot	Liberation of carbon monoxide (caution)
4. Mercury(I) nitrate	Gray precipitate of mercury
5. Copper sulfide	Formation of colorless tetracyanocuprate (I) ions. This test can be done on a section of filter paper.
Dithionites, $S_2O_4^{2-}$	
1. Dilute sulfuric acid	Orange coloration that disappears quickly, accompanied by evolution of sulfur dioxide gas and deposition of pale-yellow sulfur
2. Concentrated sulfuric acid	Fast evolution of sulfur dioxide and precipitation of pale-yellow sulfur
3. Silver nitrate	Black precipitate of silver
4. Copper sulfate	Red precipitate of copper
5. Mercury(II) chloride	Gray precipitate of mercury
6. Methylene blue	Decolorization in cold solution
7. Potassium hexacyanoferrate(II) and iron(II) sulfate	White precipitate of dipotassium iron(II) hexacyanoferrate(II); turns from white to Prussian blue
Fluorides, F^-	
1. Concentrated sulfuric acid	Evolution of hydrogen fluoride dimer
2. Calcium chloride	White slimy precipitate of calcium fluoride, slightly soluble in dilute hydrochloric acid
3. Iron(III) chloride	White precipitate
Special tests: HF etches glass (only visible after drying)	
Formates, $HCOO^-$	
1. Dilute sulfuric acid	Formic acid is evolved (pungent odor).
2. Concentrated sulfuric acid	Carbon monoxide (highly poisonous, odorless, colorless) is evolved on warming.
3. Ethanol and concentrated H_2SO_4, heat	Ethyl formate evolved (pleasant odor)
4. Silver nitrate	White precipitate of silver formate in neutral solutions, forming a black deposit of elemental silver upon mild heating
5. Iron(III) chloride	Red coloration due to complex formation
6. Mercury(II) chloride	White precipitate of calomel produced on warming; upon boiling, a black deposit of elemental mercury is produced
Hexacyanoferrate(II) ions, $[Fe(CN)_6]^{4-}$	
1. Silver nitrate	White precipitate of silver hexacyanoferrate(II)
2. Iron(III) chloride	Prussian blue is formed in neutral or acid conditions, which is decomposed by alkali bases
3. Iron(II) sulfate (aq)	White precipitate of potassium iron(II) hexacyanoferrate, which turns blue by oxidation
4. Copper sulfate	Brown precipitate of copper hexacyanoferrate(II)
5. Thorium nitrate	White precipitate of thorium hexacyanoferrate(III)
Hexacyanoferrate(III) ions $[Fe(CN)_6]^{3-}$	
1. Silver nitrate	Orange-red precipitate of silver hexacyanoferrate(III), which is soluble in ammonia solution but not in nitric acid
2. Iron(II) sulfate	Dark-blue precipitate in neutral or acid solution (Prussian or Turnbull's blue)
3. Iron(III) chloride	Brown coloration
4. Copper sulfate	Green precipitate of copper(II) hexacyanoferrate(III)
5. Concentrated hydrochloric acid	Brown precipitate of hexacyanoferric acid
Hexafluorosilicates (silicofluorides), $[SiF_6]^{2-}$	
1. Barium chloride	White crystalline precipitate of barium hexafluorosilicate, insoluble in dilute HCl; slightly soluble in water
2. Potassium chloride	White gelatinous precipitate of potassium hexafluorosilicate, slightly soluble in water
3. Ammonia solution	Gelatinous precipitate of silica acid
Hydrogen peroxide, H_2O_2	
1. Potassium iodide and starch	If sample is previously acidified by dilute sulfuric acid, a deep-blue coloration occurs due to the production of iodine complexation with starch
2. Potassium permanganate	Decolorization, evolution of oxygen
3. Titanium(IV) chloride	Orange-red coloration; very sensitive test
Special tests: 1. $4H_2O_2 + PbS \rightarrow PbSO_4\downarrow + 4H_2O$; black lead sulfide reacts to produce white lead sulfate. 2. A reagent prepared from p-hydroxyphenylacetic acid (HPPA), 7.6 mg, hematin (typically from pig albumin), 1.0 mg, in 100 mL of 0.1 M KOH (aq) will produce a fluorescent dimer (6,6'-dihydroxy-3,3'-biphenyl acetic acid) with hydrogen peroxide. This test is extremely sensitive.	
Hypochlorites, OCl^-	
1. Dilute hydrochloric acid	Yellow coloration, followed by chlorine gas evolution
2. Lead(II) acetate or nitrate	Brown lead(IV) oxide forms upon heating
3. Cobalt nitrate	Black precipitate of cobalt(II) hydroxide
4. Mercury	On shaking slightly acidified solution of a hypochlorite with Hg, a brown precipitate of mercury(II) chloride is formed.
Hypophosphites, $H_2PO_2^-$	
1. Silver nitrate	White precipitate of silver hypophosphite
2. Mercury(II) chloride	White precipitate of calomel in cold solution that darkens upon warming
3. Copper(II) sulfate	Red precipitate of copper(I) oxide forms upon warming
4. Potassium permanganate	Immediate decolorization under cold conditions
Iodates, IO_3^-	
1. Silver nitrate	White curdy precipitate of silver iodate, soluble in dilute ammonia solution
2. Barium chloride	White precipitate of barium iodate, sparingly soluble in hot water or dilute nitric acid; insoluble in ethanol and methanol
3. Mercury(II) nitrate	White precipitate of mercury(II) iodate

(continued)

Source: Ref 12

Table 3 (Continued)

Iodides, I^-	
1. Concentrated sulfuric acid	Produces hydrogen iodide and iodine
2. Silver(I) nitrate	Yellow precipitate of silver(I) iodide that is slightly soluble in ammonia solution and insoluble in dilute nitric acid
3. Lead acetate	Yellow precipitate of lead iodide, soluble in excess hot water
4. Potassium dichromate and concentrated sulfuric acid	Liberation of iodine
5. Sodium nitrite	Liberation of iodine
6. Copper sulfate	Brown precipitate
7. Mercury(II) chloride	Scarlet precipitate of mercury(II) iodide
Special tests: 1. MnO_2 + H_2SO_4 produces I_2 2. Cl_2 (aq)/CS_2 produces I_2 in CS_2 (purple) 3. Starch paste + Cl_2 (aq), deep-blue coloration	
Lactates, $CH_3CH(OH)COO^-$	
1. Potassium permanganate solution	The odor of acetaldehyde is observed upon the addition of dilute potassium permanganate solution, followed by acidification with dilute sulfuric acid and heating.
Metaphosphates, PO_3^-	
1. Silver nitrate	White precipitate, soluble in dilute nitric acid, in dilute ammonia solution, and in dilute acetic acid
2. Albumin and dilute acetic acid	Coagulation
3. Zinc sulfate solution	White precipitate on warming; soluble in dilute acetic acid
Nitrates, NO_3^-	
1. Concentrated sulfuric acid	Solid nitrate with concentrated sulfuric acid evolves reddish-brown vapors of nitrogen dioxide + nitric acid vapors when heated
Special tests: 1. Add iron(II) sulfate, shake, then add concentrated sulfuric acid; produces brown ring. 2. White precipitate is formed upon addition of nitron reagent ($C_{20}H_{16}N_4$); test is not specific to only nitrates.	
Nitrites, NO_2^-	
1. Dilute hydrochloric acid	Cautious addition of acid to a solid nitrite in cold gives a transient pale-blue (of nitrous acid or the anhydride) liquid and consequent evolution of brown fumes of nitrogen dioxide
2. Silver nitrate	White precipitate of silver nitrite
3. Iron(II) sulfate solution (25%, acidified with either acetic or sulfuric acid)	A brown ring forms at the junction of the two liquids due to the formation of a complex
4. Acidified potassium permanganate	Decolorization with no gas evolution
5. Ammonium chloride (solid)	Boiling with excess of solid reagent causes nitrogen to be evolved
6. Concentrated sulfuric acid	Liberates brown nitrogen dioxide gas
Special tests: Acidified solutions of nitrites liberate iodine from potassium iodide	
Orthophosphates, PO_4^{3}	
1. Silver nitrate	Yellow precipitate of silver orthophosphate, soluble in dilute ammonia and in dilute nitric acid
2. Barium chloride	White precipitate of barium hydrogen phosphate, soluble in dilute mineral acids and acetic acid
3. Magnesium nitrate reagent or magnesia mixture	White crystalline precipitate of magnesium ammonium phosphate, soluble in acetic acid and mineral acids; practically insoluble in 2.5% ammonia solution
4. Ammonium molybdate	Addition of 2–3 mL excess reagent to approximately 0.5 mL sample gives yellow precipitate of ammonium phosphomolybdate that is soluble in ammonia solution and in solutions of caustic alkalis. Large quantities of hydrochloric acid interfere.
5. Iron(III) chloride	Yellowish-white precipitate of iron(III) phosphate, soluble in mineral acids; insoluble in dilute acidic acid
6. Ammonium molybdate-quinine	Yellow precipitate of unknown composition; reducing agents interfere
Note: The orthophosphates are salts of orthophosphoric acid, H_3PO_4 and are simply referred to as phosphates.	
Oxalates, $(COO)_2^{2-}$	
1. Silver nitrate	White precipitate of silver oxalate that is soluble in ammonia solution and dilute nitric acid
2. Calcium chloride	White precipitate of calcium oxalate that is insoluble in dilute acetic acid, oxalic acid, and in ammonium oxalate solution; soluble in dilute hydrochloric acid and in dilute nitric acid
3. Potassium permanganate	Decolorization upon warming to 60–70 °C (140–160 °F) in acidified solution
Perchlorates, ClO_4^-	
1. Potassium chloride	White precipitate of potassium perchlorate, insoluble in alcohol
Special tests: 1. Neutral ClO_4^- + cadmium sulfate in concentrated ammonia produces $[Cd(NH_3)_4](ClO_4)_2$ (white precipitate). 2. Cautious heating of solids evolves oxygen.	
Peroxydisulfates, $S_2O_8^{2-}$	
1. Water	On boiling, decomposes into the sulfate, free sulfuric acid, and oxygen
2. Silver nitrate	Black precipitate of silver peroxide
3. Barium chloride	On boiling or standing for some time, a precipitate of barium sulfate is formed.
4. Manganese(II) sulfate	Brown precipitate of hydrate complex in neutral or alkaline test solution
Phosphites, HPO_3^{2-}	
1. Silver nitrate	White precipitate of silver phosphite that yields black metallic silver on standing
2. Barium chloride	White precipitate of barium phosphite, soluble in dilute acids
3. Mercury(II) chloride	White precipitate in cold solutions that yields gray metallic mercury on warming
4. Copper sulfate	Light-blue precipitate that dissolves in hot acetic acid
5. Lead(II) acetate	White precipitate of lead(II) hydrogen phosphite
Pyrophosphates, $P_2O_7^{4-}$	
1. Silver nitrate	White precipitate, soluble in dilute nitric acid and in dilute acetic acid
2. Copper sulfate	Plate-blue precipitate
3. Magnesia mixture or magnesium reagent	White precipitate, soluble in excess reagent but reprecipitated on boiling
4. Cadmium acetate and dilute acetic acid	White precipitate
5. Zinc sulfate	White precipitate, insoluble in dilute acetic acid; soluble in dilute ammonia solution, yielding a white precipitate on boiling

(continued)

Source: Ref 12

Table 3 (Continued)

Salicylates, $C_6H_4(OH)COO^-$ (or $C_7H_5O_3^-$)	
1. Concentrated sulfuric acid	Evolution of carbon monoxide and sulfur dioxide (poisonous)
2. Concentrated sulfuric acid and methanol	0.5 g sample + 3 mL reagent + heat evolves methyl salicylate (odor of wintergreen).
3. Dilute hydrochloric acid	Crystalline precipitate of salicylic acid
4. Silver nitrate	Heavy crystalline precipitate of silver salicylate (that is soluble in boiling water and recrystallizes upon cooling)
5. Iron(III) chloride	Violet-red coloration that clears upon the addition of dilute mineral acids
Silicates, SiO_3^{2-}	
1. Dilute hydrochloric acid	Gelatinous precipitate of metasilicic acid, insoluble in concentrated acids; soluble in water and dilute acids
2. Ammonium chloride or ammonium carbonate	Gelatinous precipitate
3. Silver nitrate	Yellow precipitate of silver silicate, soluble in dilute acids as well as ammonia solution
4. Barium chloride	White precipitate of barium silicate that is soluble in dilute nitric acid
Succinates, $C_4H_4O_4^{2-}$	
1. Silver nitrate	White precipitate of silver succinate, soluble in dilute ammonia solution
2. Iron(III) chloride	Light-brown precipitate of iron(III) succinate
3. Barium chloride	White precipitate of barium succinate
4. Calcium chloride	Slow precipitation of calcium succinate
Sulfates, SO_4^{2-}	
1. Barium chloride	White precipitate of barium sulfate, insoluble in warm dilute hydrochloric acid and in dilute nitric acid; slightly soluble in boiling hydrochloric acid
2. Lead acetate	White precipitate of lead sulfate, soluble in hot concentrated sulfuric acid, ammonium acetate, ammonium tartrate, and sodium hydroxide
3. Silver nitrate	White precipitate of silver sulfate
4. Mercury(II) nitrate	Yellow precipitate of mercury(II) sulfate
Sulfides, S^{2-}	
1. Dilute hydrochloric acid or sulfuric acid	Hydrogen sulfide gas is evolved and detected by odor or lead acetate paper
2. Silver nitrate	Black precipitate of silver sulfide, soluble in hot, dilute nitric acid
3. Lead acetate	Black precipitate of lead sulfide
4. Sodium nitroprusside solution, ($Na_2[Fe(CN)_5NO]$)	Transient purple color in the presence of solutions of alkalis
Special tests: Catalysis of iodine: azide reaction: Solution of sodium azide (NaN_3) and iodine reacts with a trace of a sulfide to evolve nitrogen. Thiosulfates and thiocyanates act similarly and therfore must be absent.	
Sulfites, SO_3^{2-}	
1. Dilute hydrochloric acid	Decomposition (which becomes more rapid on warming) and evolution of sulfuric dioxide (odor of burning sulfur)
2. Barium chloride strontium chloride	White precipitate of the respective sulfite, the precipitate being soluble in dilute hydrochloric acid
3. Silver nitrate	At first no change; upon addition of more reagent, white crystalline precipitate at silver sulfite forms that darkens to metallic silver upon heating
4. Potassium permanganate solution acidified with dilute sulfuric acid	Decolorization (Fuchsin test)
5. Potassium dichromate in with dilute sulfuric acid	Green color formation
6. Lead acetate or lead nitrate solution	White precipitate of lead sulfite
7. Zinc and sulfuric acid	Hydrogen sulfide gas evolved, detected by holding lead acetate paper to mouth of test tube
8. Concentrated sulfuric acid	Evolution of sulfur dioxide gas
9. Sodium nitroprusside-zinc sulfate	Red compound of unknown composition
Tartrates, $C_4H_4O_6^{2-}$	
1. Concentrated sulfuric acid	When sample is heated, the evolution of carbon monoxide carbon dioxide, and sulfur dioxide (burned sugar odor) results
2. Silver nitrate	White precipitate of silver tartrate
3. Calcium chloride	White precipitate of calcium tartrate, soluble in dilute acetic acid, dilute mineral acids, and cold alkali solutions
4. Potassium chloride	White precipitate, the reaction is: $C_4H_4O_6^{2-} + K + +CH_3COOH \rightarrow C_4H_5O_6K \downarrow +CH_3COO^-$
Special texts: One drop 25% iron(II) sulfate, 2–3 drops hydrogen peroxide: produces deep violet-blue color (Fenton's test).	
Thiocyanates, SCN^-	
1. Sulfuric acid	In cold solution, yellow coloration is produced; upon warming, violent reaction occurs and carbonyl sulfide is released. In basic solution, carbonyl sulfide is hydrolyzed to hydrogen sulfide.
2. Silver nitrate	White precipitate of silver thiocyanate
3. Copper sulfate	First a green coloration, then black precipitate of copper(II) thiocyanate is formed
4. Mercury(II) nitrate	White precipitate of mercury(II) thiocyanate
5. Iron(III) chloride	Blood-red coloration due to complex formation
6. Dilute nitric acid	Upon warming, red coloration is observed, with nitrogen oxide and hydrogen cyanide (poisonous) being evolved.
7. Cobalt nitrate	Blue coloration due to complex ion formation
Thiosulfates, $S_2O_3^{2-}$	
1. Iodine solution	Decolorized; a colorless solution of tetrathionate irons is formed
2. Barium chloride	White precipitate of barium thiosulfate
3. Silver nitrate	White precipitate of silver thiosulfate
4. Lead(II) acetate or nitrate solution	First no change; no further addition of reagent, a white precipitate of lead thiosulfate forms
5. Iron(III) chloride solution	Dark-violet coloration due to complex formation
6. Nickel ethylenediamine nitrate, $[Ni(NH_2(CH_2)_2NH_2)_3](NO_3)_2$	Violet complex precipitate forms; hydrogen sulfide and ammonium sulfide interfere
Special test: Blue ring test: When solution of thiosulfate mixed with ammonium molybdate solution is poured slowly down the side of a test tube that contains concentrated sulfuric acid, a blue ring is formed temporarily at the contact zone.	

Source: Ref 12

Table 4 Presumptive tests for cations

This table provides summary of the common tests for cations, primarily in aqueous solution. The captions are grouped according to the usual convention of reactivity to a set of common reagents.

Group I: Pb(II), Ag(I), Hg(I)

All members are precipitated by dilute HCl to give lead chloride ($PbCl_2$), silver chloride (AgCl), or mercury(I) chloride (Hg_2Cl_2).

Lead(II), Ph^{2+}

1. Potassium chromate	Yellow precipitate of lead(II) chromate
2. Potassium iodide	Yellow precipitate of lead(II) iodide
3. Sulfuric acid, dilute	White precipitate of lead(II) sulfate
4. Hydrogen sulfide gas	Black precipitate of lead(II) sulfide
5. Potassium cyanide	White precipitate of lead(II) cyanide
6. Tetramethyldiaminodiphenyl-methane	Blue oxidation product (presence of Bi, Ce, Mn, Th, Co, Ni, Fe, Cu may interfere)
7. Gallocyanine	Deep-violet precipitate, unknown composition (Bi, Cd, Cu, Ag may interfere)
8. Diphenylthiocarbazone	Brick-red complex in neutral or ammoniacal solution

Silver(I), Ag^+

1. Potassium chromate	Reddish-brown precipitate of silver chromate
2. Potassium iodide	Yellow precipitate of silver iodide
3. Hydrogen sulfide gas	Black precipitate of silver sulfide
4. Disodium hydrogen phosphate	Yellow precipitate of silver phosphate
5. Sodium carbonate	Yellow-white precipitate of silver carbonate, forming the brown oxide upon heating
6. p-dimethylaminobenzylidene-rhodanine	Reddish-violet precipitate in acidic solution
7. Ammonia solution	Brown precipitate of silver oxide, dissolving in excess to form Ag_3N, which is explosive

Mercury(I), Hg_2^{2+}

1. Potassium carbonate	Red precipitate of mercury(I) chromate
2. Potassium iodide	Green precipitate of mercury(I) iodide
3. Dilute sulfuric acid	White precipitate of mercury(I) sulfate
4. Elemental copper, aluminum, or zinc	Amalgamation occurs
5. Hydrogen sulfide	Black precipitate (in neutral or acid medium) of mercury(I) sulfide and mercury
6. Ammonia solution	Black precipitate of $HgO \times Hg(NH_2)(NO_3)$
7. Diphenylcarbizide (1% in ethanol, with 0.2 M nitric acid	Violet-colored complex results (high sensitivity and selectivity)
8. Potassium cyanide	Mercury(I) cyanide solution, with precipitation of elemental mercury (mercury(II) interferes)

Group II: Hg(II), Cu(II), Bi(III), Cd(II), As(III), As(V), Sb(III), Sb(V), Sn(II), Sn(IV)

All members show no reaction with HCl; all form a precipitate with H_2S.

Bismuth(III), Bi^{3+}

1. Potassium iodide	Black precipitate of bismuth(III) iodide
2. Potassium chromate	Yellow precipitate of bismuth(III) chromate
3. Ammonia solution	White precipitate of variable composition, approximate formula: $Bi(OH)_2NO_3$
4. Pyrogallol (10%)	Yellow precipitate of bismuth(III) pyrogallate
5. 8-hydroxyquinoline (5%) + potassium iodide (6 M)	Red precipitate of the tetraiodobismuthate salt (characteristic in the absence of Cl^-, F^-, Br^-)
6. Sodium hydroxide	White precipitate of bismuth(III) hydroxide

Copper(II), Cu^{2+}

1. Potassium iodide	Brown precipitate of copper(I) iodide, colored brown due to I_3^-
2. Potassium cyanide	Yellow precipitate of copper(II) cyanide, which then decomposes
3. Potassium thiocyanate	Black precipitate of copper(II) thiocyanate, which then decomposes
4. α-benzoin oxime (or cupfon), 5% in ethanol	Green precipitate of the α-benzoin oxime salt derivative
5. Salicylaldoxime (1%)	Greenish-yellow precipitate of the copper complex
6. Rubeanic acid (0.5%) (dithio-oxamide)	Black precipitate of the rubeanate salt

Cadmium(II), Cd^{2+}

1. Ammonia solution	White precipitate of cadmium hydroxide that dissolves in excess ammonia
2. Potassium cyanide	White precipitate of cadmium cyanide that dissolves in excess potassium cyanide
3. Sodium hydroxide	White precipitate of cadmium hydroxide that is insoluble in excess sodium hydroxide
4. Dinitro-p-diphenyl carbizide	Brown precipitate with cadmium hydroxide

Arsenic(III), As^{3+}

1. Silver nitrate	Yellow precipitate of silver arsenite in neutral solution
2. Copper(II) sulfate	Green precipitate of copper(II) arsenite (or $Cu_3(AsO_3)_2 \cdot xH_2O$)
3. Potassium triiodide ($KI + I_2$)	Decolorization due to oxidation
4. Tin(II) chloride + concentrated hydrochloric acid	Black precipitate forms in the presence of excess reagent.

Arsenic(V), As^{5+}

1. Silver nitrate	Brownish-red precipitate of silver arsenate from neutral solutions
2. Ammonium molybdate	Yellow precipitate (in presence of excess reagent) of ammonium arsenomolybdate, $(NH_4)_3AsMo_{12}O_4$
3. Potassium iodide + concentrated hydrochloric acid	Iodine formation

Small amounts of As(III) or As(V) can be identified by the response to the Marsh, Gutzeit, or Fleitmann tests.

Antimony(III), Sb^{3+}

1. Sodium hydroxide	White precipitate of the hydrated oxide $Sb_2O_3 \cdot xH_2O$
2. Elemental zinc or tin	Black precipitate of antimony
3. Potassium iodide	Yellow color of $[SbI_6]^{3-}$ ion
4. Phosphomolybdic acid, $H_3[PMo_{12}O_{40}]$	Blue color produced; Sn(II) interferes; sensitivity: 0.2 μg

(continued)

conc, concentrated; dil, dilute; g, gaseous; EtOH, ethanol. Source: Ref 12

Table 4 (Continued)

Group II: Hg(II), Cu(II), Bi(III), Cd(II), As(III), As(V), Sb(III), Sb(V), Sn(II), Sn(IV) (Continued)	
Antimony(V), Sb^{5+}	
1. Water	White precipitate of basic salts and ultimately antimonic acid, H_3SbO_4
2. Potassium iodide	Formation of iodine as a floating precipitate
3. Elemental zinc or tin	Black precipitate of antimony (in the presence of hydrochloric acid)
Small amounts of antimony can be identified using Marsh's test and/or Gutzeit's test.	
Tin(II), Sn^{2+}	
1. Mercury(II) chloride	White precipitate of mercury(I) chloride (in an excess of tin ions, precipitate turns gray)
2. Bismuth nitrate	Black precipitate of bismuth metal
3. Cacotheline (nitro-derivative of brucine, $C_{21}H_{21}O_7N_3$)	Violet coloration with stannous salts. The following interfere: strong reducing agents (hydrogen sulfide, dithionites, sulfites, and selenites); also, U, V, Te, Hg, Bi, Au, Pd, Se, Sb
4. Diazine green (dyestuff formed by coupling diazotized safranine with N,N-dimethylaniline)	Color change blue → violet → red
Tin(IV), Sn^{4+}	
1. Iron powder	Reduces Sn(IV) to Sn(II)
Group III: Fe(II), Fe(III), Al(III), Cr(III), Cr(VI), Ni(II), Co(II), Mn(II), Mn(VIII), Zn(II)	
All members are precipitated by H_2S in the presence of ammonia and ammonium chloride, or ammonium sulfide solutions.	
Iron(II), Fe^{2+}	
1. Ammonia solution	Precipitation of iron(II) hydroxide. If large amounts of ammonium ion are present, precipitation does not occur.
2. Ammonium sulfide	Black precipitate of iron(II) sulfide
3. Potassium cyanide (poison)	Yellowish-brown precipitate of iron(II) cyanide, soluble in excess reagent, forming the hexacyanoferrate(II) ion
4. Potassium hexacyanoferrate(II) solution	In complete absence of air, white precipitate of potassium iron(II) hexacyanoferrate If air is present, a pale-blue precipitate is formed.
5. Potassium hexacyanoferrate(III) solution	Dark-blue precipitate, called Turnbull's blue
6. α,α'-dipyridyl	Deep-red bivalent cation $[Fe(C_5H_4N)_2]^{2+}$formed with iron(II) salts in mineral acid solution; sensitivity: 0.3 μg
7. Dimethylglyoxime (DMG)	Red, iron(II) dimethylglyoxime; nickel, cobalt, and large quantities of copper salts interfere; sensitivity: 0.04 μg
8. o-phenanthroline (0.1 wt% in water)	Red coloration due to the complex cation $[Fe(C_{12}H_8N_2)_3]^{2+}$ in slightly acidic conditions
Iron(III), Fe^{3+}	
1. Ammonia solution	Reddish-brown gelatinous precipitate of iron(III) hydroxide
2. Ammonium sulfide	Black precipitate mixture of iron(II) sulfide and sulfur
3. Potassium cyanide	When added slowly, reddish-brown precipitate of iron(III) cyanide is formed, which dissolves in excess potassium cyanide to yield a yellow solution
4. Potassium hexacyanoferrate(III)	A brown coloration is produced due to the formation of iron(III) hexacyanoferate(III)
5. Disodium hydrogen phosphate	Yellowish-white precipitate of iron(III) phosphate
6. Sodium acetate solution	Reddish-brown coloration caused by complex formation
7. Cupferron ($C_6H_5N(NO)ONH_4$) aqueous solution, freshly prepared	Reddish-brown precipitate formed in the presence of hydrochloric acid
8. Ammonium thiocyanate + dilute acid	Deep-red coloration of iron(III) thiocyanate complex
9. 7-iodo-8-hydroxyquinoline-5-sulfonic acid (ferron)	Green or greenish-blue coloration in slightly acidic solutions; sensitivity: 0.5 μg
Cobalt(III), Co^{2+}	
1. Ammonia solution	In the absence of ammonium salts, small amounts of $Co(OH)NO_3$ precipitate that are soluble in excess aqueous ammonia
2. Ammonium sulfide	Black precipitate of cobalt(II) sulfide from neutral or alkaline solutions
3. Potassium cyanide (poison)	Reddish-brown precipitate of cobalt(II) cyanide that dissolves in excess potassium cyanide
4. Potassium nitrite	Yellow precipitate of potassium hexacyanocobaltate(III), $K_3[Co(NO_2)_6]$
5. Ammonium thiocyanate (crystals)	Gives blue coloration when added to neutral or acid solution of cobalt, due to a complex formation (Vogel's reaction); sensitivity: 0.5 μg
6. α-nitroso-β-naphthol (1% in 50% acetic acid)	Red-brown (chelate) precipitate, extractable using carbon tetrachloride; sensitivity: 0.05 μg
Nickel(II), Ni^{2+}	
1. Ammonia solution	Green precipitate of nickel(II) hydroxide that dissolves in excess ammonia
2. Potassium cyanide (poison)	Green precipitate of nickel(II) cyanide that dissolves in excess potassium cyanide
3. Dimethylglyoxime (DMG), ($C_4H_8O_2N_2$)	Red precipitate of nickel-DMG chelate complex in ammoniacal solution; sensitivity: 0.16 μg
Manganese(II), Mn^{2+}	
1. Ammonia solution	Partial precipitation of white manganese(II) hydroxide
2. Ammonium sulfide	Pink precipitate of manganese(II) sulfide, which is soluble in mineral acids
3. Sodium phosphate (in the presence of ammonia or ammonium ions)	Pink precipitate of manganese ammonium phosphate, $Mn(NH_4)$ PO4 · $7H_2O$, which is soluble in acids
Aluminum(III), Al^{3+}	
1. Ammonia	White gelatinous precipitate of aluminum hydroxide
2. Sodium hydroxide	White gelatinous precipitate of aluminum hydroxide, which is soluble in excess sodium hydroxide
3. Ammonium sulfide	White precipitate of aluminum sulfide
4. Sodium acetate	Upon boiling with excess reagent, a precipitate of basic aluminum acetate, $Al(OH)_2CH_3COO$, is formed
5. Sodium phosphate	White gelatinous precipitate of aluminum phosphate
6. Aluminon (a solution of the ammonium salt of aurine tricarboxylic acid)	Bright-red solution
7. Quinalizarin, alizarin-S, alizarin	Red precipitate or "lake"
Chromium(III), Cr^{3+}	
1. Ammonia solution	Gray-green to gray-blue gelatinous precipitate of chromium(III)
2. Sodium carbonate	Precipitate of chromium(III) hydroxide

(continued)

conc, concentrated; dil, dilute; g, gaseous; EtOH, ethanol. Source: Ref 12

Table 4 (Continued)

Group III: Fe(II), Fe(III), Al(III), Cr(III), Cr(VI), Ni(II), Co(II), Mn(II), Mn(VIII), Zn(II) (Continued)	
Zinc(II), Zn^{2+}	
1. Ammonia solution	White precipitate of zinc hydroxide, which is soluble in excess ammonia
2. Disodium hydrogen phosphate	White precipitate of zinc phosphate, which is soluble in dilute acids
3. Potassium hexacyanoferrate(II)	White precipitate of variable composition, which is soluble in sodium hydroxide
4. Ammonium tetrathiocyanato-mercurate(II): copper sulfate, slightly acidic	Solution is treated with 5 drops 0.25 M copper(II) sulfate solution followed by 2 ml ammonium tetrathiocyanato-mercurate to give a violet precipitate.
Group IV: Ba^{2+}(II), Sr^{2+}(II), Ca^{2+}(II)	
All members of this group react with ammonium carbonate.	
Barium(II), Ba^{2+}	
1. Ammonium carbonate	White precipitate of barium carbonate, which is soluble in dilute acids
2. Ammonium oxalate	White precipitate of barium oxalate, which is soluble in dilute acids
3. Dilute sulfuric acid	Heavy, white, finely divided precipitate of barium sulfate
4. Saturated calcium sulfate (or strontium sulfate)	White precipitate of barium sulfate
5. Sodium rhodizonate	Red-brown precipitate; sensitivity: 0.25 μg
Strontium(II), Sr^{2+}	
1. Ammonium carbonate	White precipitate of strontium carbonate
2. Dilute sulfuric acid	White precipitate of strontium sulfate
3. Saturated calcium sulfate	White precipitate of strontium sulfate
4. Potassium chromate	Yellow precipitate of strontium chromate
5. Ammonium oxalate	White precipitate of strontium oxalate that is soluble in mineral acids
Calcium(II), Ca^{2+}	
1. Ammonium carbonate	White precipitate of calcium carbonate
2. Dilute sulfuric acid	White precipitate of calcium sulfate
3. Ammonium oxalate	White precipitate of calcium oxalate that is soluble in mineral acids
4. Potassium chromate	Yellow precipitate of strontium chromate that is soluble in mineral acids
5. Sodium rhodizonate	Red-brown precipitate; sensitivity: 4 μg
Group V: Mg^{2+}(II), Na^+(I), K^+(I), NH_4^+(I)	
No common reaction or reagent	
Magnesium(II), Mg^{2+}	
1. Ammonia solution, sodium hydroxide	Partial precipitation of white magnesium hydroxide
2. Ammonium carbonate	White precipitate of magnesium carbonate, only in the absence of ammonia salts
3. Oxine + ammoniacal ammonium chloride solution	Yellow precipitate of $Mg(C_9H_6NO)_2 \cdot 4H_2O$
4. Quinalizarin	Blue precipitate, or blue-colored solution that can be cleared by a few drops of bromine-water
Sodium(I), Na^+	
1. Uranyl magnesium acetate solution (in 30 vol/vol ethanol)	Yellow precipitate of sodium magnesium uranyl acetate
2. Uranyl zinc acetate solution	Yellow precipitate of sodium zinc uranyl acetate; sensitivity: 12.5 μg Na
Potassium(I), K^+	
1. Sodium hexanitrocobaltate(III) ($Na_3[Co(NO_2)_6]$)	Yellow precipitate of potassium hexanitrocobaltate(III); insoluble in acetic acid
2. Tartaric acid solution (sodium acetate buffered)	White precipitate of potassium hydrogen tartrate
3. Perchloric acid	White precipitate of potassium perchlorate
Note: Perchloric acid is a powerful oxidizing agent that must be handled carefully.	
4. Dipicrylamine	Orange-red complex precipitate (NH_4^+ interferes); sensitivity: 3 μg K
5. Sodium tetraphenylboron + acetic acid	White precipitate of potassium tetraphenylboron
Ammonium, NH_4^+	
1. Sodium hydroxide	Evolution of ammonia gas upon heating
2. Potassium tetraiodomercurate (Nessler's reagent)	Brown-yellow color, or brown precipitate of mercury(II) amidoiodide; high sensitivity; all other metals (except Na and K) interfere
3. Tannic acid: silver nitrate	Precipitate of black elemental silver from neutral solution; very sensitive
4. p-nitrobenzene-diazonium chloride	Red-colored solution results in the presence of sodium hydroxide; sensitivity: 0.7 μg NH_4^+

Note: Ammonium ions will cause a similar reaction to that of potassium in the presence of sodium hexanitrocobaltate(III) sodium hydrogen tartrate.

conc, concentrated; dil, dilute; g, gaseous; EtOH, ethanol. Source: Ref 12

Summary

Presumptive tests, or spot tests, provide a rapid, simple method for screening an unknown for the presence of certain functional groups, ions, or elements. Although they do not typically provide a definitive identification, spot tests can provide a classification of the unknown, which both narrows down the list of possible substances and provides guidance on carrying out confirmatory tests (such as solubility of the unknown in various solvents). Spot tests have been in use for thousands of years, and their utility continues to expand, covering fields as varied as forensic drug analysis and water quality monitoring for human space exploration. Presumptive tests are a staple of the analytical toolbox and will continue to be so for the foreseeable future.

ACKNOWLEDGMENT

The author gratefully acknowledges the assistance of Dr. Thomas J. Bruno and Dr. Joshua P. Martin in the preparation of this manuscript.

REFERENCES

1. L.C. Thomas and G.J. Chamberlin, *Colorimetric Chemical Analytical Methods,* 9th ed., John Wiley and Sons, New York, 1980
2. F. Feigl, *Spot Tests in Organic Analysis,* 6th ed., Elsevier, New York, 1960
3. F. Feigl and V. Anger, *Spot Tests in Inorganic Analysis,* Elsevier Publishing Co., 1972
4. L.S. Ettre, Was Moses the First Chromatographer?: Chromatography in the Ancient

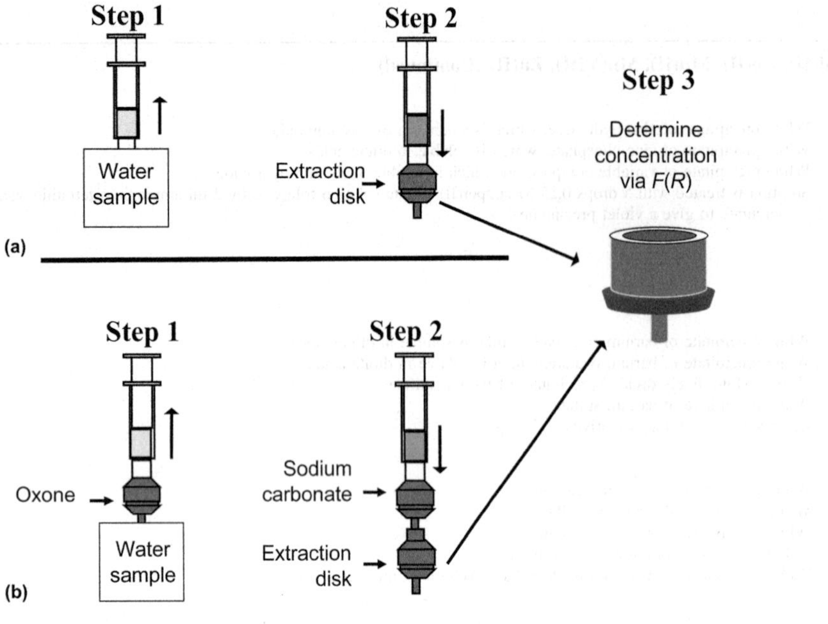

Fig. 4 Schematics depicting the analysis of (a) silver(I) and (b) total silver by colorimetric solid-phase extraction. Source: Ref 18

World, *LCGC North Am.,* Vol 24 (No. 12), 2006, p 1280–1283
5. V.G. Amelin, Chemical Test Methods for Determining Components of Liquids, *J. Anal. Chem.,* Vol 55 (No. 9), 2000, p 808–836
6. L.S. Ettre, The Predawn of Paper Chromatography, *Chromatograph.,* Vol 54 (No. 5/6), 2001, p 409–414
7. E. Jungreis, *Spot Test Analysis: Clinical, Environmental, Forensic, and Geochemical Applications,* 2nd ed., Wiley, New York, 1997
8. D.C. Harris and C.A. Lucy, *Quantitative Chemical Analysis,* 9th ed., W.H. Freeman, New York, 2016, p 777–787
9. J. Schanda, Ed., *Colorimetry: Understanding the CIE System,* Wiley-Interscience, Hoboken, NJ, 2007
10. J.P. Blitz, Diffuse Reflectance Spectroscopy, *Modern Techniques in Applied Molecular Spectroscopy,* F.M. Mirabella, Ed., John Wiley and Sons, Inc., New York, 1998, p 185–217
11. G. Kortüm, *Reflectance Spectroscopy: Principles, Methods, Applications,* Springer, 1969
12. T.J. Bruno and P.D.N. Svoronos, Qualitative Tests, *Handbook of Basic Tables for Chemical Analysis,* 3rd ed., CRC Press, 2011, p 659–691
13. R.P. Gower and I.P. Rhodes, A Review of Techniques in the Lassaigne Sodium-Fusion, *J. Chem. Educ.,* Vol 46 (No. 9), 1969, p 606–607
14. J.W. Lehman, *Multiscale Operational Organic Chemistry: A Problem-Solving Approach to the Laboratory Course,* 2nd ed., Pearson Prentice Hall, Upper Saddle River, NJ, 2009
15. A.I. Vogel, G. Svehla, and G. Suehla, *Vogel's Qualitative Inorganic Analysis,* 7th ed., Prentice Hall, New York, 1996
16. "Color Test Reagents/Kits for Preliminary Identification of Drugs of Abuse," Standard 0604.01, Law Enforcement and Corrections Standards and Testing Program, National Institute of Justice, Washington, D.C., 2000
17. D. Gazda et al., In-Flight Water Quality Monitoring on the International Space Station (ISS): Measuring Biocide Concentrations with Colorimetric Solid Phase Extraction (CSPE), *41st International Conference on Environmental Systems* (Portland, OR), July 2011, American Institute of Aeronautics and Astronautics, p 1–8
18. A.A. Hill, R.J. Lipert, and M.D. Porter, Determination of Colloidal and Dissolved Silver in Water Samples Using Colorimetric Solid-Phase Extraction, *Talanta,* Vol 80 (No. 5), 2010, p 1606–1610

SELECTED REFERENCES

- T.J. Bruno and P.D.N. Svoronos, Qualitative Tests, *Handbook of Basic Tables for Chemical Analysis,* 3rd ed., CRC Press, 2011, p 659–691
- F. Feigl, *Spot Tests in Organic Analysis,* 6th ed., Elsevier, New York, 1972
- F. Feigl and V. Anger, *Spot Tests in Inorganic Analysis,* Elsevier Publishing Co., 1972
- I. Vogel, G. Svehla, and G. Suehla, *Vogel's Qualitative Inorganic Analysis,* 7th ed., Prentice Hall, New York, 1996

ASM Handbook, Volume 10, *Materials Characterization*
ASM Handbook Committee
DOI 10.31399/asm.hb.v10.a0006629

Classical Wet Analytical Chemistry

Lisa S. Ott, California State University, Chico

Overview

General Uses

- Quantitative elemental analysis
- Qualitative identification of material type
- Isolation and characterization of inclusions and phases
- Determination of solution concentration
- Determination of oxidation state

Examples of Applications

- Characterization of a homogeneous sample for instrument calibration
- Quantitative determination of a sample that is too small or not stable enough for instrumental analysis
- Gross composition estimation for heterogeneous samples
- Identification of metal ions in solution using simple spot tests
- Isolation of a single species from a solid or liquid solution
- Investigation of compositional differences in surface versus core of a solid material

Samples

- *Form:* Solids (crystalline or amorphous, in any particle size) and liquids
- *Size:* Strongly dependent on the chosen method of analysis, but generally less than 2 g for solid samples and less than 20 mL for liquids
- *Preparation:* Large solid samples are milled, drilled, crushed, or ground into small particles (approximately 2 mm, or 0.08 in., in diameter); heterogeneous samples require careful sampling design to ensure representative samples and meaningful results.

Limitations

- Some methods are slow in comparison to instrumental methods.
- In many cases, relatively large sample sizes are required.

Capabilities of Related Techniques

- *Optical emission spectroscopy:* Multielement detection is possible; can be more rapid than wet chemical techniques.
- *Atomic absorption spectrophotometry:* Detection limits down to parts per billion range
- *Electrochemical analysis:* Accurate quantitation of low concentrations
- *Electron and ion microprobes:* Suitable for very small samples
- *Secondary ion mass spectroscopy, x-ray photoelectron spectroscopy, and Auger spectroscopy:* More sensitive studies of surface layers are possible.
- *Mössbauer spectroscopy:* Solid-state technique, very sensitive for a limited number of oxidation states

Introduction

Classical wet analytical chemistry ranges from qualitative identification of a chemical substance to quantitative determination of how much of that substance is present. While instrumental methods, including spectrometric methods, are useful in some situations, they are purposefully omitted from the present article. Instead, this article focuses on quintessential methods that have stood the test of time in laboratories around the world. In the most general terms, wet analytical chemistry is comprised of gravimetry, titrimetry, and separations.

To be appropriate for any of these three methods, samples require extensive pretreatment. The pretreatment may be as simple as adjusting the pH, or it may be as complicated as isolating the analyte from a complicated matrix without changing the composition of the analyte.

The methods discussed in this article are not intended for functional group analysis of organic compounds; these methods are addressed separately in this Handbook. Additional methods that are considered to be wet chemical methods but are not discussed herein include environmental/industrial hygiene, forensic studies, and engineering tests specific to individual products.

Appropriateness of Classical Wet Methods

Although advances in chemical instrumentation have made a wide range of analyses possible with reasonable cost and time considerations, there are still a number of situations in which classical wet chemical techniques are the preferred analysis method. There are a variety of reasons an analyst may select a wet chemical method. For example, many wet chemical methods can be less costly than instrumental methods and can be done in-house. Additionally, the analyst does not need to be trained on specialized instrumentation, rendering the cost to employers significantly less. Ion chromatography, for example, can be employed as a means to quantitate halogens in a sample. However, it is much more straightforward and less expensive to carry out this quantitation via precipitation and subsequent gravimetric analysis.

Sampling

Obtaining a sample that is representative of the whole is of utmost importance in wet chemical analysis (Ref 1). It is essential to properly represent the entire sample in order to convey correct composition (and other) information. In fact, the International Union for Pure and Applied Chemistry states that use of the term *sample* implies the existence of a sampling error, because any measurement is just an estimate based on a portion of the parent sample (Ref 2). A sampling plan should be developed that includes selection, withdrawal, precautions for preservation and/or transportation, and preparation of the portions for analysis. This sampling plan should carefully delineate the range of samples for which the method is appropriate, and the limits of the method's relevance (in terms of concentration and sample matrices) must also be defined. After analysis of the material, an analysis of variance statistical test is recommended to assess possible biases (Ref 3).

Larger samples of solid material can have compositional variation within the sample itself (Ref 4). Metals can be sampled appropriately by drilling the piece to be sampled at regular, predefined intervals along all sides of the sample. The drilled pieces are then analyzed separately or mixed thoroughly before analysis to provide an average composition measurement. Alternatively, the sample can be melted in a platinum crucible to yield a homogeneous sample.

Basic Chemical Equilibria and Analytical Chemistry

Chemical Equilibria

Entire tomes of work have been devoted to the topic of chemical equilibria. Indeed, the position of equilibrium of chemical reactions can have far-reaching effects, including safety and economic. In the most straightforward presentation of the concept, the position of equilibrium is evaluated in aqueous solution based on a ratio of the concentration of the products to the concentration of the reactants. This ratio is referred to as the equilibrium constant, K:

$$\text{Reactants}_{(aq)} \leftrightharpoons \text{Products}_{(aq)} \quad K = \frac{[\text{products}]}{[\text{Reactants}]}$$

Large values of K favor the product of the chemical reaction, small values favor the reactants, and intermediate values yield a mixture of reactants and products. An assessment of the position of equilibrium is important in all of the analysis discussed herein, with the magnitude of K controlling everything from the strength of an acid solution to the ability of a molecule to bind a metal cation. The equilibrium constant can also be assessed for gas-phase reactions. In gas-phase reactions, the pressure of the component is used in place of the concentration.

Acid/Base Reactions

Central to wet chemical methods is the examination of the acidic or basic properties of an analyte. There are a number of ways to define acids and bases; here, the Brønsted-Lowry definition is used. In this definition, acids (HA) are species that donate protons to yield the conjugate base A^-, and bases (B) are species that accept protons to yield the conjugate acid BH^+ (Ref 5). Their generic hydrolysis reactions (reactions with water) are shown, along with a presentation of their equilibrium constant expressions:

$$HA_{(aq)} + H_2O_{(l)} \leftrightharpoons A^-_{(aq)} + H_3O^+_{(aq)}$$

$$K_a = \frac{[A^-][H_3O^+]}{[HA]}$$

$$B_{(aq)} + H_2O_{(l)} \leftrightharpoons HB^+_{(aq)} + OH^-_{(aq)}$$

$$K_b = \frac{[BH^+][OH^-]}{[B]}$$

Notable from the equations is that water can act as a Brønsted-Lowry acid and as a Brønsted-Lowry base. As such, water is termed amphiprotic. This amphiprotic nature leads a well-known autoprotolysis reaction that generates to a measureable quantity of both H_3O^+ and OH^- in water:

$$H_2O_{(l)} + H_2O_{(l)} \leftrightharpoons H_3O^+_{(aq)} + OH^-_{(aq)}$$

$$K_w = [H_3O^+][OH^-] = 1.0 \times 10^{-14}$$

In dilute solutions of strong acids and bases, the autoprotolysis reaction is significant enough to measurably change the pH of the solution.

In wet chemical analysis, the acidic or basic nature of the solution is most commonly probed by an acid/base titration. See the section "Titrimetry" in this article for a discussion about titrations.

Oxidation-Reduction Reactions

For analytes that can change oxidation states, analysis by oxidation-reduction reactions (more commonly referred to as redox reactions) is a useful method. In reduction reactions, the species of interest gains electrons; in oxidation reactions, the species of interest loses electrons. In the following exemplary reactions, Fe^{3+} is reduced while tin (Sn) is oxidized:

$$\text{Reduction half-reaction: } Fe^{3+}_{(aq)} + e^- \leftrightharpoons Fe^{2+}_{(aq)}$$

$$\text{Oxidation half-reaction: } Sn_{(s)} \leftrightharpoons Sn^{2+}_{(aq)} + 2e^-$$

$$\text{Balanced redox reaction: } Sn_{(s)} + 2\,Fe^{3+}_{(aq)} \leftrightharpoons Sn^{2+}_{(aq)} + 2\,Fe^{2+}_{(aq)}$$

Redox reactions must occur in pairs; that is, there cannot be a reduction without a corresponding oxidation. Additionally, the number of electrons transferred between the elements participating in the reaction must be the same. This is the reason that the Fe^{3+} half-reaction is multiplied by two in the balanced redox reaction.

For a redox reaction to occur without an external driving force (i.e., a spontaneous chemical reaction), the redox pair must be selected with consideration of their standard reduction potentials. These standard reduction potentials are essentially a measure of the tendency of a species to gain an electron or electrons. A spontaneous chemical reaction occurs if subtracting the standard reduction potential of the oxidation half-reaction (written as a reduction per convention) from the standard reduction potential of the reduction half-reaction yields a positive number. Tables of standard reduction potentials are found in every general and analytical chemistry textbook and will not be reproduced in full here; selected standard reduction potentials are presented later in this article in the context of redox titrations.

Complexometric Reactions

Complexometric reactions are generally used to quantify the amount of a free metal cation in aqueous solution. The metal cation (M^{a+}) is complexed with a ligand (L^{b-}) through a Lewis acid/base interaction where lone pairs on the ligand are donated to the metal cation to form a stable complex ($M\text{-}L^{a+b}$). This complexation is described as:

$$M^{a+}_{(aq)} + L^{b-}_{(aq)} \leftrightharpoons M\text{-}L^{(a+b)}_{(aq)}$$

The strength of this interaction between the metal cation and the ligand is measured by the formation constant (K_f), where K_f is a ratio of the reaction products to its reactants:

$$K_f = \frac{[M\text{-}L^{a+b}]}{[M^{a+}][L^{b-}]}$$

Larger values of the formation constant correspond to a stronger interaction between the metal and the ligand. To be analytically useful, the magnitude of the formation constant should be greater than 10^3. Because the extent of the interaction depends on the availability of lone electron pairs on the ligand molecule, the pH of the solution must be carefully controlled. Buffer solutions are generally selected for this purpose.

The single most analytically important complexometric reaction uses ethylenediamine tetraacetic acid (EDTA) as the ligand (Ref 6). In strongly basic solutions, EDTA is fully deprotonated and has up to six electron pairs available to donate to a metal cation. The physical structure of EDTA, along with multiple electron pairs available for coordination to a metal cation, leads to formation constants averaging on the order of 10^{13} to 10^{18}. Other suitable ligands for complexometric reactions include triethylenetetramine (four electron pairs) and ethylenediamine (two electron pairs).

Complexometric interactions can also be exploited to selectively bind an interfering ion in solution. A ligand is selected that has a large K_f value for the interfering ion and a negligible K_f value for the analyte ion. This ligand is added in excess, and the interfering ion is bound in solution. The interfering ion then will not be bound in a subsequent complexometric titration of the analyte ion.

Gravimetry

Also referred to as gravimetric analysis, the purpose of gravimetry is to quantitate an analyte by measuring mass as the analytical signal. This mass measurement is made after the addition of a reagent that reacts completely with the analyte to form an isolable precipitate. This reagent is commonly referred to as the precipitating agent or simply the precipitant. The suitability of the precipitant is determined by the magnitude of its solubility product constant (K_{sp}) with the analyte. A sample K_{sp} equation is shown as follows for the dissolution of the hypothetical sparingly soluble compound AB; by convention, K_{sp} equations are shown as the dissolution of the sparingly soluble solid. Smaller values of K_{sp} indicate decreased solubility of the precipitate. In general, K_{sp} values of 10^{-9} and smaller are desirable for quantitative analysis:

$$AB_{(s)} \leftrightharpoons A^{+}{}_{(aq)} + B^{-}{}_{(aq)}$$

A number of wet chemical techniques can be used to maximize the yield of the precipitate. Good control of the initial nucleation step of precipitation is achieved by employing elevated temperature, a slow addition of the precipitate, and a large solution volume. To maintain the solubility equilibrium in favor of the precipitate over its dissolved constituents, the precipitation reaction is generally performed with excess precipitant. As the desired solid precipitates from solution, the equilibrium is shifted in favor of the production of more solid.

For the nucleated precipitant particles to grow to a filterable and isolable size, the solution is often heated near the boiling point of the solution. The growth of the particles can also be promoted by adding an electrolyte. Generally, the goal is to have a clear and colorless electrolyte solution above a suspension of large, pure, easily filterable products. The product is gathered on a sintered glass filter and washed with an electrolyte solution to remove any adsorbed impurities. After a suitable drying step, the gravimetric product is weighed on an analytical balance.

Electrogravimetry, or the precipitation of a solid by means of electrolysis, was first described in 1864 by Wolcott Gibbs (Ref 7). In this technique, an electroactive analyte is deposited on the surface of the working electrode in a coulometric cell. The working electrode is first weighed. Then, a controlled potential is applied to change the oxidation state of the analyte and deposit that analyte on the working electrode surface. The analyte deposition must be quantitative so the solution is stirred during electrolysis. After deposition, the working electrode is removed from the analyte solution, rinsed thoroughly, dried, and weighed. The difference in masses corresponds to the mass of the analyte. Information on the initial oxidation state can also be gained by measuring the current applied during the electrogravimetric process (Ref 8).

Solvent Extraction

In classic wet chemistry, solvent extraction is employed when the desired analyte is present in a mixture. The solubility characteristics of the desired analyte can be exploited to separate the analyte from the mixture using a two-phase system. The two phases, or solvents, chosen for extraction must be mutually immiscible. The analyte exists in equilibrium between the two phases:

$$[\text{Analyte}]_{\text{phase 1}} \leftrightharpoons [\text{Analyte}]_{\text{phase 2}}$$

The extent of transfer of the analyte from phase 1 to phase 2 is controlled by the magnitude of the equilibrium constant for the phase transfer, referred to as the partition coefficient (K_c). The partition coefficient is strongly dependent on the identity of the solvents being used. Commonly, one of the phases is aqueous and the second phase is organic. Common choices for the organic solvent include diethyl ether, dichloromethane, and hexanes. Large values for K_c lead to more complete transfer of the analyte to the extracting phase. Additionally, phase transfer of the analyte is more efficient when multiple extractions are carried out. This is quantitatively described in the following equation:

$$q = \left(\frac{V_1}{V_1 + K_c V_2}\right)^n$$

where q is the fraction of analyte remaining in phase 1 after extraction, V_1 is the volume of phase 1, V_2 is the volume of phase 2, K_c is the partition coefficient, and n is the number of extractions. Hence, it is more efficient to extract an analyte by doing three extractions using 10 mL of phase 2 than by doing one extraction with 30 mL of phase 2.

Ideally, K_c for any other components in the mixture that are not the desired analyte will be small, leading to an effective separation between the desired analyte and the remaining components of the mixture. Once isolated in phase 2, the desired analyte can be quantified by any appropriate wet chemical method.

In solvent extraction with metal ion species, the metal ion is suspended in an aqueous phase and extracted into an immiscible organic phase. The organic phase is commonly loaded with an extractant reagent designed to complex the metal ion and make it preferentially soluble in the organic phase (Ref 9). Ideally, the extractant reagent is selected for the ion of interest, making separation of a mixture possible at this stage. Extractants are generally categorized as cation exchangers, anion exchangers, or solvating extractants. A list of common organic phases, extractants, and metal ion targets is provided in Table 1.

Ion Exchange

Ion exchange is used for a method of quantification when the analyte of interest has a charge. Both anions and cations can be analyzed via ion exchange after proper selection of an ion exchange material. These materials are usually polymer resin beads with ionizable surface functional groups. If the analyte is a cation, the ionizable surface groups are maintained in an anionic form with a balancing countercation present; if the analyte is an anion, the ionizable surface groups are maintained in cationic form with a balancing counteranion present. The resin material is held stationary, usually in a glass or plastic column, and the analyte solution is passed over the resin material. The analyte material is then

Table 1 Common solvent extraction separations

Type of extractant	Reagent	Examples
Cation exchangers	Carboxylic acids, mono- or bi-alkylated phosphoric acids, sulfonic acid	Acorga M5640 salicylaldoxime in kerosene selectively extracts Cu from sulfuric acid solutions.
	Ketoximes, hydroxyquinolines, organophosphonic acid, mono- and dithiophosphonic acid	Two equivalents of hydroxyoxime chelate 1 Cu ion, phosphinic acid forms a 2:1 chelate with metal ions; ammoniacal solutions of Cu are extracted with betadiketone.
Anion exchangers	Alkylammonium salts	Co^{2+} can be separated from Ni^{2+} using tri-isooctyl amine.
Solvating extractants	Tributyl phosphate, tri(n-octyl)phosphine oxide	Tributyl phosphate in a hydrocarbon organic phase is used to extract Fe^{3+} from hydrochloric leaching solutions of Ni mattes.

Source: Ref 9

adsorbed on the surface of the resin material, displacing the corresponding counterion.

There are two methods for quantifying the analyte after adsorption. The first method requires the analyst to know the charge on the analyte ion. For example, an analyst may be interested in quantifying a cationic analyte. This analyte solution would be passed over a stationary bed of anionically functionalized resin beads, with the analyte adsorbing the anionic surface groups and displacing the balancing countercations. For example, the countercation may be hydronium ions, which are easily quantified by acid/base titration. Then, with knowledge of the stoichiometry that relates the countercation to the analyte ion, the initial quantity of analyte ions can be calculated. The second method involves chemically stripping the analyte off the resin bead and analyzing the stripped material directly.

Karl Fischer Titrations

The quantification of water presents a particular challenge in analytical chemistry. In most academic, government, and industrial labs, this challenge is met by Karl Fischer titration (KFT) (Ref 10). The KFTs have been studied extensively since the initial report in 1935, and now commercial titrators are available for purchase with preprepared mixtures of the required reagents. Optimum conditions have been determined to be a methanolic solution of weak nitrogenous base (RN) in the pH range of 5 to 7. Initially, SO_2 reacts with the base in methanol as:

$$CH_3OH + SO_2 + RN \leftrightharpoons [RNH]^+SO_3CH_3^-$$

The sulfite ion then reacts with any water present:

$$H_2O + I_2 + [RNH]^+SO_3CH_3^- + 2\,RN \leftrightharpoons [RNH]^+SO_4CH_3^- + 2\,[RNH]^+I^-$$

The net overall reaction is the reduction of $I_{2(aq)}$ and the corresponding oxidation of $SO_{2(aq)}$:

$$SO_{2(aq)} + I_{2(aq)} + 3\,H_2O_{(l)} \leftrightharpoons SO_{3(aq)} + 2\,H_3O^+_{(aq)} + 2I^-_{(aq)}$$

The stoichiometry of the net overall reaction describes a 1:1 relationship between I_2 and H_2O. The redox and/or colorimetric characteristics of I_2 are then used to quantitate the water present in the sample. In the redox (or coulometric) method, the KFT materials are placed in an electrochemical cell. A carefully measured current is applied to the analyte solution until the moisture has been consumed and the I_2 is no longer reduced to I^-. An abrupt drop in voltage signals the end of this reduction. Alternately, the colorimetric properties of iodine can be exploited to determine the endpoint. Because the complex formed between I_2 and RN is dark brown, this is a self-indicating reaction. The first drop of excess titrant changes the color of the solution, and a visual determination of the endpoint is possible.

Sample Dissolution

Many samples for wet chemical analysis are in the aqueous phase, such as effluent from a chemical plant, runoff from a parking lot, or stream water. However, many samples that require analysis by a wet chemical method are solids. The analyst must select a preparative method for transforming the solid-phase sample into an aqueous-phase sample (sample dissolution). The entire sample must be dissolved so that crucial sample information is not lost. Ideally, the analyst will select the mildest conditions possible for dissolution, to minimize any potential chemical changes to the analyte.

Mechanical Methods

Large-particle-sized samples can be challenging to dissolve. To expedite the process and ensure complete dissolution, large-particle-sized samples are first crushed or ground. While these two terms are often used interchangeably, crushing a sample generally produces coarser sample sizes, and grinding a sample generally produces finer sample sizes (Ref 4).

Crushing or grinding is simply done with a mortar and pestle. The composition of the mortar and pestle themselves must be selected carefully to avoid porosity (and thus sample loss) and any possible chemical interference. Additionally, the hardness of the mortar and pestle material must be greater than the hardness of the sample to successfully grind the sample into smaller particles (for example, an agate mortar and pestle may be required) (Ref 11).

Large-particle-sized samples can also be ground with a ball mill, jaw crusher, pan crusher, or hammer mill (Ref 4). In a ball mill, a rotating chamber is filled with the sample and stainless steel or porcelain balls. The chamber is rotated, and the balls crush the sample into smaller particle sizes. Jaw crushers are commonly used for rocks or minerals. A hardened steel jaw moves back and forth across a plane supporting the sample, crushing the sample in the process. Pan crushers are used for crushing solid samples as well as mixing liquids and solids. Two heavy wheels, often called mullers, roll around on the inside of a large pan. The material loaded in the pan is crushed and mixed by the motion of the mullers. Finally, hammer mills have a large rotor with freely rotating hammers affixed to the rotor. Circular rotation of the rotor causes the hammers to swing out and crush the loaded material against the side of the mill. Hammer mills are often used to obtain very fine sample powders.

If a particle size is specified, the sample can be sieved through a series of screens with defined mesh sizes. Mesh size refers to the number of standard wires per inch of screen. For example, a 50-mesh powder will pass through a screen having 50 wires per inch of screen (see Table 2 for more examples).

Nonoxidizing Acids and Acid Mixtures

If a mechanical approach is insufficient for analysis, solid samples can be dissolved using nonoxidizing acids and/or acid mixtures. For this purpose, any of the hydrohalic acids can be employed; hydrochloric acid (HCl) is the most common choice. HCl is capable of dissolving a wide range of metals, ranging from alkali metals (sodium, potassium, etc.) to transition metals (cobalt, nickel, etc.) (Ref 12). These dissolutions, which oxidize the metal, also yield hydrogen gas. The evolution of this gas can present safety hazards. In the case of Ge(IV), As(III), Sb(III), Sn(IV), Se(IV), and Hg(II), volatile chlorides can also be formed (Ref 13). Nonoxidizing acids can be used in a high-pressure reaction vessel, which eliminates the loss of gases and may extend the range of samples that can be dissolved by the acid. Carbonate salts are particularly soluble in HCl, as evolution of $CO_{2(g)}$ drives the dissolution reaction forward.

Hydrobromic acid (HBr) can be used for many similar dissolutions but must be kept protected from sunlight to avoid the formation of bromine in solution. Hydroiodic acid (HI) is the least stable of these three hydrohalic species. It is susceptible to oxidation under mild conditions by atmospheric oxygen, and sunlight can promote this decomposition. Therefore, the use of HI is often limited.

Hydrofluoric acid (HF) can dissolve glasses, silicates, quartz, and other siliceous materials but must be used with great caution (Ref 14). HF causes severe skin burns, sometimes leading to gangrene, and may cause ulcers of the upper respiratory tract. HF must be contained in non-silicon-base materials, such as synthetic fluorine-containing resin or polyethylene, which

Table 2 Mesh sieve sizes and approximate particle size diameter

	Particle size	
Mesh	μm	mils
5	4000	158
10	2000	79
25	707	28
50	297	12
70	210	8
100	149	6
200	74	3
400	37	1.5

are not susceptible to etching or dissolution by the acid. Additionally, it will react violently with solids such as calcium oxide, sodium, and sodium hydroxide.

Concentrated sulfuric acid is a very powerful acidic agent that can ignite or cause autoignition on contact with many common laboratory materials. Additionally, some alcohols can polymerize in concentrated acids, yielding tarry oxidation products. Sulfuric acid is miscible with water and alcohols, although the dissolution of sulfuric acid in either of those media is extremely exothermic. Sulfuric acid dissolution of a solid analyte material can result in the emission of toxic SO_x fumes, so care must be taken to undertake the reaction either with appropriate ventilation or in an inert high-pressure vessel. With appropriate safety consideration, however, sulfuric acid is an effective dissolution medium for iron-base materials, mineral halides, and rare earth phosphates. The concentrated acid may need to be heated to be effective. In its diluted aqueous form, sulfuric acid can also act as a strong oxidizer.

Phosphoric acid is another option for the dissolution of solid analyte materials. Although it has the advantage of a higher boiling point compared to other mineral acids, it has the tendency to cause the precipitation of phosphate salts during the dissolution process (Ref 13). It is a weaker acid than sulfuric acid and as such is frequently used in combination with other acids. For the dissolution of mineral samples, phosphoric acid can be mixed with sulfuric acid, hydrogen peroxide, nitric acid, and perchloric acid.

Oxidizing Acids and Acid Mixtures

Nitric acid is the most commonly used oxidizing acid. It is capable of dissolving most metals, with the exceptions of gold, chromium, group 3A and 8B elements, and a few others. It is particularly useful for dissolving sulfides and phosphates. When mixed in a 1:3 volume ratio with hydrochloric acid, it is referred to as aqua regia. Aqua regia is an aggressive oxidizing medium for sample dissolution but also can liberate toxic HCl, HNO_3, Cl_2, and NO_x fumes. Despite this risk, aqua regia is a frequently used medium for dissolving gold, silver, platinum, stainless steel, other high-temperature alloys, and a wide variety of minerals/ores (including most sulfides, selenides, tellurides, and arsenides). If a 3:1 volume ratio of nitric to hydrochloric acid is used, this mixture is referred to as LaFort aqua regia. LaFort aqua regia is used extensively in the geosciences to dissolve sulfide-containing materials (Ref 15). These sulfides are oxidized to sulfates during the process and then are suitable for gravimetric analysis after precipitation.

Perchloric acid is similar in many qualities to sulfuric acid but is a strong oxidizer at high temperatures. With only a few exceptions, the perchlorates generated during sample dissolution with perchloric acid are soluble in water, alcohols, ethyl acetate, and acetone. The solubility of the products presents an advantage in situations where halogen salts form precipitates. Perchloric acid is a severe irritant to the eyes, skin, and mucous membranes and is classified as a poison by inhalation, ingestion, and subcutaneous incorporation. Great caution should be taken with anhydrous perchloric acid, which can explode spontaneously. Additionally, perchloric acid reacts violently with a number of common laboratory items, including cellulose, dehydrating agents, steel, and wood. Consequently, many laboratories are equipped with dedicated fume hoods for the use of perchloric acid.

Hydrogen peroxide, the simplest of all peroxides, can be a useful dissolution agent by virtue of its reactive oxygen-oxygen bond. It is generally used as an oxidizing agent, often in combination with a strong acid such as hydrochloric or sulfuric acid. In combination with hydrochloric acid, hydrogen peroxide is useful for dissolving iron, nickel, and cobalt metals and alloys.

Common acids and acid combinations for sample dissolution are shown in Table 3.

Fusions

If a solid analyte cannot be dissolved by any acid solution, another option is fusion. Fusion is the high-temperature reaction of a finely powdered analyte in the presence of an auxiliary substance or flux. Fusion reactions are carried out in the molten state and often require up to 10 times as much flux as sample. The products of a fusion reaction may be soluble in either water or in an acid solution (Ref 13). Common fluxes include carbonates, tetraborates, and hydroxides. Because of the high temperatures required to bring solid analytes and the added fluxes to a molten state, the fusion reactions are generally carried out in metal crucibles. Care must be taken to select a crucible that is not susceptible to attack by the flux or the resultant fusion product(s). There is also some danger in losing the analyte due to volatilization under the high-temperature conditions required for fusion. Some commonly used fluxes are shown in Table 4.

Miscellaneous Techniques

Frequently, an inorganic analyte is dissolved in an organic matrix. To remove the matrix, the sample is combusted with oxygen, resulting in the loss of carbonaceous compounds as CO_2. However, this method can produce the possibility of loss of volatile analytes or compounds. Some solutions to this problem include bomb calorimetry, where the sample is ignited and combusted in an O_2-atmosphere high-pressure vessel; Schöniger flask combustion, where the sample is combusted under atmospheric pressure with an aqueous solution

Table 3 Commonly used acid media for digesting solid samples

Acid medium	Suitable for:	Comments
HCl	Metal oxides, metals that lie above H_2 in electromotive series	Possible loss of volatile chlorides
HNO_3	Most common metals and common metal alloys	Al and Cr show surface passivation; Sn, Sb, and W form insoluble acids.
H_2SO_4	Most metals, many alloys	Hot solutions of the high-boiling (340 °C, or 645 °F) solutions are frequently employed.
$HClO_4$	Ferrous alloys, stainless steels	Hot concentrated solutions are excellent oxidizers; explosion hazard with organic substances
HCl/HNO_3 (aqua regia)	Stainless steels, Au, Pt	Widely employed as 3:1 ratio
Mineral acid + bromine or H_2O_2	See mineral acids above	Increases solvent action; speeds oxidation of organics
HF	Glasses, quartz	Synthetic fluorine-containing resin or Pt labware required; any Si lost as SiF_4; significant health hazard

Source: Ref 4

Table 4 Commonly used fluxes for fusing powder samples

Medium (boiling point, °C)	Crucible material(s)	Uses	Notes
Li_2CO_3 (723), Na_2CO_3 (851), K_2CO_3 (891)	Pt	Silicates, some organics	High temperatures and lids required
Na_2O_2 (675)	Zr, Ni, Fe	Minerals, slags, refractory compounds	Dangerous with carbonaceous materials
$NaHSO_4$ (59), $K_2S_2O_7$ (300)	Pt; quartz; high-silica, high-temperature glass	Aluminas, metal alloys	H_2SO_4, SO_3 evolved
B_2O_3 (577), $Li_2B_4O_7$ (917), $Na_2B_4O_7$ (743), $LiBO_2$ (849)	Pt, C (graphite)	Minerals, slags, glasses	Can be used in combination with carbonates
NaOH (318), KOH (360)	Ni, Ag, Au	Sand, glass, quartz	Initial drying step to remove H_2O from flux, especially for NaOH

Source: Ref 13

available to absorb any volatiles; and oxygen plasma dry ashing, which uses radiofrequency-generated oxygen plasma to remove carbon at relatively low temperatures. Organic matrices can also be removed by reaction with alkali metals or their salts. The analyst can use fusion with an alkali metal, sodium peroxide decomposition in a pressure vessel, or reaction with a sodium metal dissolved in liquid ammonia.

Alternatively, the analysis can be designed to remove the analyte as a volatile compound. For example, many silicon halides are gaseous compounds. Reaction of a homogeneous or heterogeneous sample containing silicon with the appropriate ammonium halide or dihalide gas at high temperatures generates the volatile product. These compounds can then be collected for analysis on a cold finger.

Qualitative Methods

Many of the wet chemical methods discussed in this article are quantitative in nature. However, in many cases, a simple qualitative analysis of a sample or a sample component is sufficient. A qualitative analysis relies on the chemical or physical properties of the sample, such as solubility, melting point, or reactivity with certain reagents. Chemical "spot tests" are designed to be a sensitive and selective reaction of the analyte with a drop of a reagent solution. These spot tests are used to identify an analyte based on a color change or precipitation (Ref 16). A wide variety of free elements, anions, cations, alloys, and even organic compounds can be identified by chemical spot tests with simple reagents. After qualitative identification, an appropriate instrumental method is selected for quantitative analysis. Table 5 contains some commonly used reagents for qualitative spot tests.

Separation of Chemical Mixtures

Separation by Precipitation

When two or more ions are present and soluble in an analyte solution, precipitation can sometimes be used to separate the ions from each other. A selective precipitation of one ion is accomplished by selecting a precipitating agent that has a small solubility product (K_{sp}) with one ion and a large K_{sp} for the other ion(s) in solution. The desired result is the precipitation of one species while the other species remains in solution. For example, consider a solution containing equal concentrations of Ca^{2+} and Cu^{2+}. Addition of sodium hydroxide will cause the precipitation of each ion as the hydroxide salt at very different concentrations:

$$Ca(OH)_{2(s)} \leftrightharpoons Ca^{2+}{}_{(aq)} + 2\ OH^{-}{}_{(aq)}$$
$$K_{sp} = 6.5 \times 10^{-6}$$

$$Cu(OH)_{2(s)} \leftrightharpoons Cu^{2+}{}_{(aq)} + 2\ OH^{-}{}_{(aq)}$$
$$K_{sp} = 4.8 \times 10^{-20}$$

Because the stoichiometry of the two reactions is the same, it is apparent that $Cu(OH)_{2(s)}$ will precipitate from solution, leaving soluble Ca^{2+}. With knowledge of initial concentrations, it is straightforward to calculate whether or not all Cu^{2+} can be precipitated from solution without co-precipitating Ca^{2+}.

Distillation

Distillation is the most widely employed method for separating mixtures of miscible species (Ref 11). In the process of distillation, a boiling kettle containing the mixture is gradually heated. Once the boiling point of a component in the mixture is met, the component is converted from the liquid phase to the gas phase and rises out of the kettle. At the top of the kettle, a glass column takes the gas phase away from the boiling kettle. A temperature measurement of the gas phase is made at the top of the column, and the gas is condensed back into a liquid for collection. If the boiling points of the components in a mixture are sufficiently different, each component can be separated and collected by this simple distillation method. If the components have similar boiling points, an additional stage of separation known as fractional distillation may be required. In fractional distillation, the glass column is replaced with a fractioning column that has a much higher surface area. This fractioning column allows for a series of equilibria between gas and liquid phases on the internal surface of the column. Small differences in volatility are magnified by the number of equilibration processes the sample undergoes. In this manner, components that have similar boiling points can be separated.

Particulate Gravimetry

In particulate gravimetry, the analyte is a particulate suspended in some gaseous or liquid matrix. The particulate analyte is separated from the matrix by using filtration or extraction; no chemical reaction is required (Ref 17). For example, particulate gravimetry is commonly used for the analysis of suspended solids in water samples. If the particulate is a solid suspended in a liquid matrix, the separation can be as simple as gravity filtration through a paper filter. If the matrix is viscous or the sample size is large, the filtration process can be expedited by applying suction with an aspirator. Other filters for liquid matrices are composed of cellulose, glass, or polytetrafluoroethylene. The type of filter is selected based on the size of the pores in the filter compared to the size of the particulate material.

Gravimetry

In its simplest terms, gravimetry is a determination by weight. As previously discussed, a soluble analyte reacts with a reagent, causing the precipitation of the analyte as an insoluble product. The mass of the product coupled with the stoichiometry of the reaction will yield quantitative information on the analyte. The scale for analysis by gravimetry is limited only by the sensitivity of the balance and the quantity of the sample available. There are a wide variety of reagents that cause the precipitation of analytes (Ref 18). Table 6 lists a selection

Table 5 Qualitative chemical tests used in materials identification

Spot test for:	Reagent	Positive result
Cr	Diphenylcarbazide	Violet color
Co	Ammonium thiocyanate/ acetone	Blue color
Cu	Dithizone	Purple color
Fe	Potassium ferricyanide	Blue precipitate
Pb	Sulfuric acid	White precipitate
Mo	Potassium thiocyanate/ stannous chloride	Pink color
Ni	Dimethylglyoxime	Red precipitate

Table 6 Widely applicable precipitation separations

Precipitant	Conditions	Elements precipitated	Comments
NH_4OH	...	Fe^{3+}, Cr^{3+}, Al, Ga, Be, Sn, In, Nb, Ta, Ti, Zr, Hf, U, rare earths	NH_4^+ can be difficult to remove.
NaOH	...	Mn, Ni, Cu, Co, Fe, Cr, Sn, In, Nb, Ta, Ti, Zr, Hf, U, rare earths	Oxides and carbonates are possible with Cr, V, and U.
H_2S	HCl solution (0.25–12 M)	Cu, Ag, Hg, Pb, Bi, Cd, Ru, Rh, Pd, Os, As, Au, Pt, Sn, Sb, Ir, Ge, Se, Te, Mo	Co-precipitation of Tl, In, Ga is possible; tartaric acid prevents V and W from co-precipitation.
	pH 2–3	Zn (incomplete with Tl, In, Ga)	...
H_2S or $(NH_4)_2S$	pH >7	Mn, Fe, Ni, Co	Tartrate is sometimes required to prevent interference.
Cupferron	Dilute H_2SO_4, cold HCl	Fe, Ga, Sn, Nb, Ta, Ti, Zr, Hf, V, Mo, W, Pd, Sn, Bi	...
8-hydroxyquinoline	Weak acid or base	Most metals except Re, Pt; in weak acid, Pb, Be, and alkali earth metals do not precipitate.	HCl can dissolve residue.
F^-	Strong acid	Th, U^{4+}, Mg, Ca, Sr, Ba, rare earths	Al precipitates from basic solution.
$C_2O_4^{2-}$	Strong acid	Th, Mg, Ca, Sr, Ba, Al, rare earths, Co, Cu, Ni, Fe^{2+}, Pb, Bi, Hg, Ag, Zn	Preliminary separation is frequently necessary.

of widely applicable precipitants; Table 7 shows some narrowly applicable but selective precipitants.

Weighing as the Oxide

More than 40 elements can be determined gravimetrically as the oxide. The analyte cation of interest is frequently first isolated as the hydroxide or hydrated oxide. Ignition of the isolated species to the highest stable oxide yields the analytically useful product. Alternately, an organic precipitant can be used to remove the analyte from solution. This precipitate can then be ashed to the metal oxide form. In either of these two processes, the temperature must be carefully maintained to both ensure complete loss of the carbonaceous material and avoid the loss of volatile analyte. The latter concern is particularly important when the sublimation temperature of the product oxide is low (for example, MoO_3 begins to sublime at approximately 500 °C, or 930 °F). Finding this intermediate temperature for each product can be challenging; melting, splattering, and volatilization can occur before reaching the temperature required to form the oxide. Some examples of conversion to the oxide for weighing and other similar transformations are collected in Table 8.

Weighing as the Metal

Relatively few elements can be isolated and weighed in their free metallic state. The formation of metal oxide surface layers is the most frequent issue encountered when attempting to weigh the metal. In most cases, surface oxide layers can be eliminated by ignition of the sample in the presence of hydrogen. Oxides of semimetals, such as arsenic and tellurium, can be chemically reduced to their elemental forms. Perhaps the most general form of weighing the metal is through electrogravimetry. Electrogravimetry is a very common method for the determination of Cu^{2+} solutions.

Weighing as the Sulfate

At least seven metals can be gravimetrically investigated by precipitation as the sulfate salt, principally, the alkali earth metal. Of the alkali earth metals, barium is the least soluble as the sulfate salt. As such, this is likely the most common method for barium determination. The $BaSO_4$ precipitate is finely divided, however, so care must be taken in the filtration, isolation, and drying of this solid.

Weighing as the Phosphate or Pyrophosphate

A wide variety of metals can be gravimetrically determined as their phosphate (PO_4^{3-}) or pyrophosphate ($P_2O_7^{4-}$) salt. Lead, silver, iron, and zinc are particularly insoluble as their phosphate salts. The alkali earth metals can be precipitated as the hydrated and protonated HPO_4^{2-} salts as well. The two most analytically useful pyrophosphate salts are formed with Zr^{4+} and Mg^{2+}.

Weighing as the Sulfide

First-row transition metals are particularly well suited to gravimetric analysis as their sulfide salts. Manganese, iron, cobalt, nickel, and zinc all form sparingly soluble salts, with K_{sp} values ranging from 10^{-11} to 10^{-49}. Additionally, heavier metals, such as mercury and lead, have very low solubility as the sulfide salt, with K_{sp} values on the order of 10^{-53} for the mercury salts.

Weighing as the Chloride

The principal gravimetric analysis of a chloride salt is carried out in the analysis of silver. While AgCl precipitates readily from aqueous solution, it is also light sensitive. Exposure to light causes reduction of the compound to metallic silver and introduces a significant error in gravimetric determinations. Hg_2Cl_2 has a similarly small K_{sp} value, but the toxicity rating of this compound indicates that other methods for the quantification of mercury should be preferentially selected if possible.

Weighing as the Chromate

Precipitation of the CrO_4^{2-} salt is a feasible gravimetric assay for Ba^{2+}, Cu^{2+}, Ag^+, Hg_2^{2+}, and Tl^+. In cases were Ba^{2+} is to be quantified but is present in solution with either Ca^{2+} or Sr^{2+}, precipitation with chromate avoids the co-precipitation that is possible with sulfate.

Weighing as the Dimethylglyoxime Complex

As opposed to the previously mentioned precipitating agents, which are all charged species, dimethylglyoxime is a neutral compound. Dimethylglyoxime has two nitrogens with lone pairs, making it a strong chelating ligand for metal cations. It is particularly useful for the gravimetric determination of Ni^{2+}. Addition of dimethylglyoxime to an aqueous solution of Ni^{2+} results in the rapid precipitation of the bright-red solid Ni(dimethylglyoxime)$_2$. Pd^{2+} reacts similarly to form a yellow crystalline solid.

Miscellaneous Compounds

A wide variety of other precipitants for metals exist in chemical literature. The primary requirement for selecting a precipitant is that the precipitant has a well-defined final composition. In many cases, the composition is simply the anion/cation pair in their appropriate stoichiometric ratio. In some cases, the precipitant is generated in situ from another reagent. For example, hydrolysis of sulfamic acid generates SO_4^{2-}, which can subsequently be used to precipitate Pb^{2+} or Sr^{2+}

$$NH_2HSO_{3(aq)} + 2\,H_2O_{(l)} \leftrightharpoons NH_4^+{}_{(aq)} + H_3O^+{}_{(aq)} + SO_4^{2-}{}_{(aq)}$$

$$Pb^{2+}{}_{(aq)} + SO_4^{2-}{}_{(aq)} \leftrightharpoons PbSO_{4(s)}$$

Table 7 Common narrow-range precipitants

Precipitant	Elements	Comments
Dimethylglyoxime	Ni, Pd	Co can co-precipitate if present in high concentrations.
Cl^-	Ag, Hg(I)	AgCl is photosensitive; $HgCl$ has slight solubility.
Mandelic acid	Zr, Hf	Precipitate from hot solution; p-bromo- and p-chloro-mandelic acid is also useable.
$Na[B(C_6H_5)_4]$	K, Rb, Cs, NH_4^+	Other monovalent cations, including quaternary ammoniums, may interfere.
SO_4^{2-}	Ba, Sr, Ca, Pb	At pH = 4.0, Ba selectively precipitates.
Alphabenzoinoxime	Mo	Co-precipitation by Nb, Si, Pd, W, Ta
Cinchonine	W	Co-precipitation by Si, Sn, Sb, Nb, Ta; interference by Mo, P, As

Table 8 Common gravimetric finishes

Analyte	Precipitant	Precipitate formed	Precipitate weighed
Ba	$(NH_4)_2CrO_4$	$BaCrO_4$	$BaCrO_4$
Pb	K_2CrO_4	$PbCrO_4$	$PbCrO_4$
Ag	HCl	AgCl	AgCl
Al	NH_3	$Al(OH)_3$	Al_2O_3
Be	NH_3	$Be(OH)_2$	BeO
Ca	$Na_2C_2O_4$	CaC_2O_4	CaO
Hg	H_2S	HgS	HgS
Be	$(NH_4)_2HPO_4$	NH_4BePO_4	$Be_2P_2O_7$
Sr	KH_2PO_4	$SrHPO_4$	$Sr_2P_2O_7$
Ni	Dimethylglyoxime $C_4H_8O_2N_2$	$Ni(C_4H_7O_2N_2)_2$	$Ni(C_4H_7O_2N_2)_2$
Fe	Cupferron $C_6H_5N(NO)ONH_4$	$Fe(C_6H_5N(NO)O)_3$	Fe_2O_3
Co	1-nitroso-2-napthol $C_{10}H_6(NO)OH$	$Co(C_{10}H_6(NO)O)_2$	Co or $CoSO_4$

Additionally, for some products, deliquescence or instability requires the sample to be dried at temperatures that remove volatiles. In some cases, the elevated temperature causes decomposition of the anion to a more stable form. Some exemplars for gravimetric analyses are collected in Table 9. More extensive lists can be found in most general and analytical chemistry texts (Ref 12, 18).

Titrimetry

In general terms, titration is the process of determining the quantity of an analyte by adding measured increments of a titrant. The analyte and titrant must participate in a chemical reaction, and there must be some means for indicating the endpoint of the reaction. At the endpoint of the reaction (also called the equivalence point), essentially all of the analyte has reacted with the titrant. To gain useful information on the concentration of the analyte, a well-standardized titrant must be employed. Table 10 lists common titrants and methods for their standardization.

A wide variety of titrations are useful as analytical tools. The most common are acid/base, precipitation, complexometric, redox, and iodimetric or iodometric titrations.

Table 9 Additional species commonly weighed in gravimetry

Analyte	Weighed as:
Li	Li_2SO_4
F	CaF_2, PbClF, ThF_4, LaF_3
Mg	$Mg_2P_2O_7$, $Mg(C_9H_6NO)_2$ [Mg(8-hydroxyquinolate)$_2$]
Al	$AlPO_4$, $Al(C_9H_6NO)_3$ [Al(8-hydroxyquinolate)$_3$]
Si	SiO_2
P	$Mg_2P_2O_7$
S	$BaSO_4$
Cl	AgCl
K	$KB(C_6H_5)_4$
Ca	CaF_2, $CaSO_4$
Ti	TiO_2
Mn	$Mn_2P_2O_7$
Fe	Fe_2O_3
Co	Co, Co_3O_4, $CoSO_4$, CoS
Ni	NiS
Cu	CuO, CuS
Zn	Zn, $ZnHPO_4$, $Zn(CN)_2$, ZnS
Ga	Ga_2O_3
Br	AgBr
Sr	$SrSO_4$, SrO
Y	Y_2CO_3
Zr	ZrO_2
Mo	$PbMoO_4$
Ag	Ag, AgCl, $Ag_3Co(CN)_6$, $Ag_4Fe(CN)_6$
In	In_2O_3
Sn	SnO_2
Sb	Sb_2S_3, Sb_2O_4
I	AgI
Cs	$CsB(C_6H_5)_4$
Ba	$BaCrO_4$
In	In_2S_3
Pb	PbO_2
Bi	$BiPO_4$
Lanthanides	Oxides

Acid/Base Titrations

Acid/base titrations are perhaps the most fundamental wet chemical analysis. During an acid/base titration, the pH of an unknown solution is continuously monitored while a known concentration of acid or base is added via a buret. Using acid/base titrations, an analyst can determine the concentration of an unknown species as well as infer identity information (Ref 19).

Consider a known concentration of base in a buret suspended above a solution of acid that has an unknown concentration and identity. An initial pH measurement will yield information on the activity of hydronium ion in solution present from autoionization of the unknown acid. As base is slowly added from the buret, the pH of the solution will slowly rise as the acid/base pair present in solution generates a buffer that resists change in pH. The pH of the solution begins to rise more quickly when approaching the equivalence point, or the point at which the number of moles of added base is equal to the number of moles of acid in the initial solution.

The equivalence point is often determined using a visual indicator. The indicator is selected so that the color change of the indicator takes place over the same pH range as the equivalence point. Table 11 shows several commonly used acid/base indicators. For example, Fig. 1 shows two titration curves of acids with 0.200 M NaOH (a strong base). One curve shows a strong acid being titrated. This curve shows a very pronounced "S" shape with almost no positive slope before the equivalence point. At the equivalence point, pH = 7. An appropriate visual indicator for this titration would be bromothymol blue. The second curve shows a weak acid being titrated. This curve has an appreciable positive slope before the equivalence point, and the pH at equivalence is slightly basic (in this case, the pH at equivalence is approximately 8.8). An appropriate indicator for this titration would be phenolphthalein.

Correct identification of the equivalence point is of utmost importance in quantifying the amount of acid present and subsequently gaining some identity information. Qualitatively, the equivalence point is estimated by plotting pH versus volume of base added (V_{base}) and identifying the steepest point on the plot. This qualitative estimate can be improved by taking the first derivative of this plot, which should lead to an apex at the equivalence point. However, if a pH measurement was not recorded exactly at the equivalence point, this method could lead to an erroneous identification of the equivalence point. To avoid this erroneous identification, the second derivative of the data may prove more accurate. Using the second derivative, the equivalence point is determined where a plot of the data crosses the *x*-axis. Thus, having a point recorded at the equivalence point is not necessary. Once the equivalence point has been identified, the initial molarity of the acid can easily be calculated with knowledge of V_{base} and the molarity of the base. Then, the identity of the acid can be inferred using the half-equivalence point. At the half-equivalence point, the pH of the solution is equal to the pK_a of the acid ($pK_a = -\log K_a$, where K_a is the acid dissociation equilibrium constant shown in the section "Acid/Base Reactions" in this article). Comparing the experimental pK_a to a table of known pK_a values can lead to the identification of the unknown acid.

Precipitation Titrations

In a precipitation titration, the analyte and titrant react to form an insoluble product. At the endpoint, or equivalence point, exactly enough titrant has been added to precipitate all of the analyte initially present. The concentrations of the analyte and titrant, as well as the magnitude of the solubility product (K_{sp}), affect the sharpness of the endpoint. The analyte concentration can be monitored during titration by using an appropriate ion-selective electrode to generate the familiar shape of a titration curve. In some cases, adding titrant past the equivalence point can lead to the formation of soluble complex ions, which diverts analyte ions from the insoluble product and is undesirable (Ref 17). The gravimetric product can be isolated by filtration, dried, and weighed to yield quantitative information about the analyte.

Precipitation titrations can be followed with a suitable indicator (Ref 20). There are three types of precipitation indicators. For the first type, the first drop of excess titrant forms a

Table 10 Common standardization methods for titrants

Titrant	Primary standard	Remarks
Dilute strong base	Potassium hydrogen phthalate	CO_2 must be removed from solution to avoid acidification; endpoint is pH 8.6.
Dilute strong acid	Na_2CO_3; standardized NaOH solution	See above
$KMnO_4$ solution	$Na_2C_2O_4$	Reaction is carried out in warm dilute H_2SO_4.
Ferrous ammonium sulfate	$K_2Cr_2O_7$	Reaction is carried out in dilute H_2SO_4; diphenylamine indicator
Iodine solution	Standardized $Na_2S_2O_3$ solution	Starch indicator added near endpoint; As_2O_3 also possible but has toxicity issues
$Na_2S_2O_3$	Standardized iodine solution	Starch indicator added near endpoint

precipitate with the indicator. Frequently, the indicator is chosen such that the indicator-titrant complex is brightly colored and thus easy to identify. For the second type, the first drop of excess titrant forms a colored, soluble complex with the indicator. Thus, the solution changes color at the equivalence point. For the third type, the endpoint is detected when the indicator adsorbs to the precipitate. The indicator, such as the dichlorofluoroscein dye used in the Fajans method, is initially repelled by the growing precipitate due to adsorbed negative surface charges. At the equivalence point, the surface charge changes to positive, and the dye is adsorbed to the precipitate surface. With the adsorption of the dye comes a color change and visual detection of the equivalence point.

Complexometric Titrations

In a complexometric titration, the complexing agent is added via buret to a solution of an analyte cation. The complexometric titration is followed with an appropriate visual indicator, often called a metallochromic indicator. In a solution of the free metal ion to be titrated, the indicator binds to the metal ion. The indicator binds the metal ion less strongly than the complexometric ligand, so as the ligand is titrated in, the indicator is displaced from the metal ion. At the equivalence point, all indicator molecules are displaced from the metal ion, and the solution changes color. Metallochromic indicators are selected based on both the pH range of the analyte solution and the metal ion of interest; common indicators are shown in Table 12.

The shape of complexometric titrations resembles that of a strong acid/strong base titration curve. The concentration of the analyte ion gradually decreases until just before the equivalence point. At approximately the equivalence point, the concentration changes rapidly, resulting in a sharp drop in the curve. The steepest point in the curve (the inflection point) is at the equivalence point, when the moles of complexing agent equal the moles of analyte ion originally in solution. Following the equivalence point, the concentration of the analyte ion is very low, because the ion is only present in solution as a slight back reaction from the precipitated solid. The analyte ion concentration is then dictated by the magnitude of K_{sp} and the amount of excess complexing agent.

Redox Titrations

To determine the quantity of a redox active analyte in an unknown solution, a useful wet chemical approach is a redox titration. In a redox titration, the unknown analyte is probed by adding a known concentration of redox-active titrant by buret. The redox reaction between the analyte and titrant must be rapid and stoichiometric. Additionally, the redox reaction between the two must be spontaneous. To ensure spontaneity, the difference between the standard reduction potential of the oxidizing agent and the reducing agent must be a positive value.

To be analytically useful, the endpoint of the redox titration involving the analyte must be detectable by use of a redox indicator. Redox indicators are compounds added in small concentrations that change color when the indicator itself undergoes a redox process. Redox indictors are selected such that a slight excess of the titrant induces the redox reaction of the indicator (and subsequent color change). Some frequently selected redox indicators are shown in Table 13.

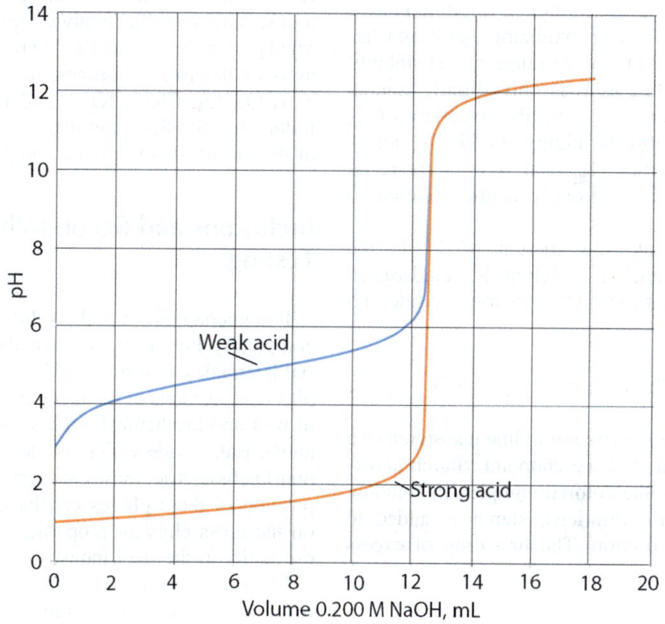

Fig. 1 Titration curves of 0.100 M strong and weak acids, titrated with 0.200 M NaOH

Table 11 Frequently used acid-base indicators

Name	Acid color	Approximate pH range for color change	Base color
Methyl violet	Yellow	0.0–1.6	Violet
Cresol purple	Red	1.2–2.8	Yellow
Bromophenol blue	Yellow	3.0–4.6	Blue
Bromocresol green	Yellow	3.8–5.4	Blue
Methyl red	Red	4.8–6.0	Yellow
Bromothymol blue	Yellow	6.0–7.6	Blue
Cresol red	Yellow	7.2–8.8	Red
Phenolphthalein	Colorless	8.0–9.6	Pink
Alizarin yellow	Yellow	10.1–12.0	Red

Table 12 Common complexometric (metallochromic) indicators

Name	Color of free indicator	Appropriate metal ions	pH range	Color of complexed indicator
Eriochrome Black T	Blue	Mg^{2+}, Ca^{2+}, Sr^{2+}, Mn^{2+}	6.3–11.6	Red/violet
Napthyl azoxine S	Red/violet	Cu^{2+}, Zn^{2+}, Pb^{2+}	3–9	Yellow
Xylenol orange	Yellow	Bi^{3+}, Th^{4+}	1.5–3.0	Violet
Pyridylazonaphthol	Yellow	Ca^{2+}, Cu^{2+}, Co^{2+}, Ni^{2+}	4–6	Red/violet

Source: Ref 6

Table 13 Frequently used redox indicators

Indicator	Color		Standard reduction potential, V
	Oxidized form	Reduced form	
Phenosafranine	Red	Colorless	0.28
Methylene blue	Blue	Colorless	0.53
Sodium diphenylamine sulfonate	Red-violet	Colorless	0.80
n-ethoxychrysoidine	Red	Yellow	1.00
Iron (III) phenanthroline, $Fe(phen)_3^{3+}$	Pale blue	Red	1.06
Tris(5-nitro-1,10-phenanthroline) iron	Pale blue	Red-violet	1.25

The two oxidizing reagents first developed for use in redox titrations were MnO_4^- (permanganate ion) and $Cr_2O_7^{2-}$ (dichromate) (Ref 19). These reagents are still in common use for quantitating redox-active species such as titanium, chromium, manganese, molybdenum, and cerium (all using MnO_4^-) and iron (using $Cr_2O_7^{2-}$). Another common strong oxidizer used as a titrant in redox titrations is Ce^{4+}. Ce^{4+} is strong enough to oxidize water to O_2 gas (but slowly) and has been used to quantify analytes such as Fe^{2+}, As^{3+}, I^-, Sb^{3+}, Mo^{5+}, Pu^{3+}, Sn^{2+}, Tl^{1+}, U^{4+}, and V^{4+}. Common oxidizing and reducing agents used in redox titrations are shown in Table 14 (Ref 18).

Iodimetric and Iodometric Titrations

Also referred to as iodimetry, iodine is used as the titrant and an oxidizing agent and has the advantage of a wide range of pH stability (Ref 21). Iodine as I_2 is only slightly soluble in water, and as such, the reaction of I_2 with I^- to form the highly soluble I_3^- ion is common practice. I_3^- can also be prepared using KIO_3 in a strongly acidic solution to produce I_3^-.

To quantitate the amount of I^- in the titrant solution, the solution is standardized with thiosulfate ($S_2O_3^{2-}$) to form iodide and tetrathionate:

$$I_{2(aq)} + 2\,S_2O_3^{2-}{}_{(aq)} \leftrightharpoons 2\,I^-{}_{(aq)} + S_4O_6^{2-}{}_{(aq)}$$

Most titrations involving iodine use starch as a visual indicator of the endpoint. Starch and I_2 form a dark-blue-colored complex in solution. For iodimetric titrations, starch is added to the analyte solution. The first drop of excess iodine then causes the analyte solution to turn dark blue.

A related method is iodometric titrations. In iodometric methods, a chemical reaction first produces iodine, which is then titrated. First, an excess of I^- is added to the analyte solution. The analyte and I^- react to form I_2, but excess I^- remains in solution. The excess I^- is titrated with a standardized solution of $S_2O_3^{2-}$. When all of the I^- has been consumed, the starch indicator reacts with I_2 to form a dark-blue solution. The quantity of I_2 generated by chemical reaction is then determined by difference. Generally, the starch is added close to the endpoint, which can be visually estimated by the initial yellow color of the iodine solution fading (Ref 22).

Iodimetric and iodometric titrations can be used for a wide variety of redox-active analytes. Iodimetric titrations are commonly used to measure metallic analytes including Fe^{3+}, MnO_4^-, $Cr_2O_7^{2-}$, and Ce^{4+}. They can also be used to determine nonmetallic analytes such as H_2O_2, O_3, ClO^-, IO_3^-, and Br^- (Ref 21). Iodometric titrations can determine the number of double bonds in organic compounds.

Table 14 Oxidizing and reducing agents for redox titrations

Oxidizing agents		Reducing agents(a)	
Species	$E°$, V	Species	$E°$, V
BiO_3^-	1.6	Ascorbic acid	0.390
BrO_3^-	1.50	BH_4^-	−2.25
Br_2	1.098	Cr^{2+}	−0.89
Ce^{4+}	1.44–1.72(b)	$S_2O_4^{2-}$	1.130
Cl_2	1.3604	Fe^{2+}	0.68–0.771(b)
ClO_2	1.068	N_2H_4	−1.160
$Cr_2O_7^{2-}$	1.36	Hydroquinone	0.70
FeO_4^{2-}	0.80	NH_2OH	−3.04
H_2O_2	1.763	H_3PO_2	−0.48
OCl^-	0.81	Sn^{2+}	−0.141
IO_3^-	0.269	SO_3^{2-}	−0.936
I_2	1.154	SO_2	0.450
HNO_3	0.960	$S_2O_3^{2-}$	0.40
O_2	1.229	...	...
O_3	2.075	...	...
IO_4^-	1.589	...	...
MnO_4^-	1.692	...	...
$S_2O_8^{2-}$	2.01	...	...

(a) The reducing agent is oxidized in solution but, by convention, is always written as the reduction. Therefore, these are reduction potentials for the species that is oxidized. (b) $E°$ is dependent on the nature of the acid in which the reagent is dissolved.

Inclusions and Second-Phase Testing

It is uncommon to find a solid sample that is completely homogeneous, whether that material is an alloy or a mineral. Based on the complex compositions of the materials and the stress of heat and mechanical work during fabrication and/or use, a wide variety of stoichiometric and nonstoichiometric phases are formed. The composition of these phases can have implications on the gross physical properties of the sample, especially in the steel industry.

Inclusions are nonmetallic compounds that form or are trapped within an alloy during the manufacturing process. Some common inclusions in stainless steel manufacturing are oxides, nitrides, carbides, sulfides, and carbonitrides. The stability and reactivity of these compounds vary widely, making their removal difficult. Strategies for inclusion removal include digestion in dilute acid, ionic displacement, dissolution in bromine-methanol, dissolution in bromine-methyl acetate, and chlorination. Some of these methods and their common applications are shown in Table 15. The most general approach may be electrolysis. In electrolysis, a current is applied that engenders a redox reaction and solubilizes the inclusion(s). The mass of the inclusion is determined by difference.

Table 15 Other dissolution media

Medium	Target
Aqueous strong base (NaOH, KOH, Na_2CO_3)	Al, Al alloys
$NaOH + H_2O_2$	Rh, U
$NH_4OH + H_2O_2$	Mo, Cu
$CH_3CO_2H + H_2O_2$	Pb, Pb alloys
CH_3OH	Mg
Cl_2 + NaCl + Heat	Ni, Pd, Pt

Once the inclusion has been removed, its composition can be determined using a wide variety of wet chemical or instrumental techniques. Commonly, the sample is submitted to an external company for elemental analysis of the entire inclusion. X-ray fluorescence (XRF) and x-ray photoelectron spectroscopy (XPS) can also give elemental analysis on the surface of the inclusion, should the removed inclusion be a solid. For powders, x-ray diffraction (XRD) is used to determine crystal structures. Infrared spectrophotometry is used to identify functional groups of some compounds. Electron microscopy, such as scanning electron microscopy and transmission electron microscopy, can also be helpful for compound identification. Finally, thermal methods such as differential thermal analysis and thermogravimetric analysis can be employed to aid in compositional understanding.

Chemical Surface Studies

Especially with recent advances in instrumentation, the majority of chemical surface studies are carried out instrumentally. Techniques such as XRD, XRF, and XPS give detailed atomic-level information on surface composition. However, these techniques are costly and may not be available at the analyst's location. In those cases, there are some wet chemical techniques that may provide some insight in a rapid, cost-effective, accessible manner.

Solid samples can be milled or turned to remove turnings (thin slices) of the sample. Then, the sample slices can be analyzed to gain insight on discrepancies between surface slices and core slices. This approach is used to probe compositional differences due to case hardening (also known as surface hardening) and decarburization. In both of these cases, the slices are analyzed for carbon content. Similar work is carried out to determine the difference in nitrogen and boron content in surface slices versus the bulk material. The compilation of slice-by-slice data from these analyses gives valuable engineering data to surface depths inaccessible by XRD or XPS.

Wet chemical techniques can also be useful for determining the mass of surface coatings. For example, a metal coating can be chemically stripped from the surface it was deposited on. Then, the now-soluble metal can be analyzed by a number of titrimetric methods. Organic coatings on metal surfaces can be removed by using an appropriate solvent. After removal of the solvent, the mass of the organic coating can be determined by difference with gravimetry. Alternately, the coating in solvent can be probed using a technique such as gas chromatography or nuclear magnetic resonance spectrometry.

In addition, contaminants that are present by simple surface adsorption are suitable for analysis through wet chemical techniques. These surface-adsorbed impurities can be removed by boiling in water or dilute acid; then, these aqueous solutions can be analyzed by any number of the techniques discussed herein. Surface impurities on large-surface-area samples can be removed by using an absorbent material moistened with a reagent in which the impurity is soluble. Then, the impurity is extracted from the material and subjected to the analysis of choice.

Partitioning Oxidation States

Wet chemical analysis is well suited to the quantitative determination of an analyte that is present in different oxidation states. This speciation is valuable for predicting future product properties and optimizing process parameters.

Iron

Iron can adopt a variety of oxidation states, ranging from −2 to +7. In compounds, +2 and +3 are the most common oxidation states. In steelmaking, iron is often found as FeO, Fe_2O_3, and Fe. The total iron content is measured by taking a small sample and subjecting it to redox titration with $K_2Cr_2O_7$. To specify individual oxidation states, a series of reactions are performed on a small portion of the sample (see the following for the net ionic equations). First, $CuSO_4$ is added to selectively dissolve Fe. The solid copper is filtered off, and then excess Cu^{2+} in solution is precipitated out using a redox reaction with solid aluminum. The soluble Fe^{2+} is then titrated to a diphenylamine endpoint in acidic solution using $K_2Cr_2O_7$. This procedure quantitatively determines the metallic Fe content in a solid sample:

$$Fe_{(s)} + Cu^{2+}{}_{(aq)} \leftrightharpoons Fe^{2+}{}_{(aq)} + Cu_{(s)}$$

$$3\ Cu^{2+}{}_{(aq)} + 2\ Al_{(s)} \leftrightharpoons 3\ Cu_{(s)} + 2\ Al^{3+}{}_{(aq)}$$

$$14\ H_3O^{+}{}_{(l)} + 6\ Fe^{2+}{}_{(aq)} + Cr_2O_7{}^{2-}{}_{(aq)} \leftrightharpoons 2\ Cr^{3+}{}_{(aq)} + 6\ Fe^{3+}{}_{(aq)} + 7\ H_2O_{(l)}$$

On another portion of the sample, the FeO content of the sample is measured by using a separate titration. This sample is placed in dilute HCl and brought to reflux under a carbon dioxide (CO_2) atmosphere. The CO_2 atmosphere is maintained as the solution is cooled and titrated with $K_2Cr_2O_7$ as before; under these conditions, the iron content is a combination of Fe and FeO. The FeO content is determined by difference, using the metallic Fe content previously determined. If the sample resists dissolution by HCl, it may be feasible to dissolve the sample via reaction with a strong oxidant such as $Ce(SO_4)_2$ in an inert atmosphere. The excess Ce^{4+} can be determined by titration with $(NH_4)_2Fe(SO_4)_2$, yielding the Fe^{2+} present in the sample by difference. Finally, the Fe^{3+} present in the original sample is determined by the difference in total iron and all other oxidation states present in solution.

Other Metallic and Semimetallic Elements

The many oxidation states of chromium have widely disparate properties and chemical health effects. For example, Cr^{6+} is a well-known carcinogen, while other oxidation states are recommended dietary supplements. It is also important to assess the concentration and oxidation state of chromium when formulating plating baths. Quantitative determination of these oxidation states requires solubilizing any solid samples. Dissolution of Cr, CrO, and CrO_3 can be accomplished by subjecting the samples to a bromine-methanol solution, an acidic $FeCl_3$ solution, and a boiling $NaCO_3$ solution, respectively. Care must be taken with bromine-methanol solutions, which are used as etchants for semiconducting applications, because their preparation is highly exothermic and can lead to sudden expansions and release of vapor (Ref 23).

In some steel samples, silicon is present as either FeSi or SiO_2. To distinguish between these types of silicon, the total silicon is first determined gravimetrically. Then, a suitable source of chlorine is provided at high temperatures to react away Fe as $FeCl_3$. These reaction conditions also convert FeSi to $SiCl_4$, which sublimes away. The remaining silicon must come from SiO_2 and can be determined gravimetrically. This analysis is complicated by the presence of SiC and/or Si_3N_4.

Nonmetallic Elements

Interstitial nitrogen is frequently found in steel, and nitrides are particularly important in the strength of steel samples. To determine the total quantity of interstitial nitrogen, fine millings of the sample are heated to 550 °C (1020 °F) under an atmosphere of H_2. Under these conditions, the N_2 reacts with H_2 to form NH_3. The formed NH_3 is quantified by using one of a variety of different methods. After this quantification, the Kjeldahl method is used to determine total nitrogen. Following the total nitrogen determination, the nitride nitrogens, which are strongly bound to the lattice of the metal, are determined by the difference of the previous two analyses. The nitride content of a steel has a strong correlation to the strain-aging behavior of the steel.

Any mobile carbon contained in steel samples is also susceptible to conversion to methane by using the heat and H_2 atmosphere conditions previously described. There are a variety of options for quantifying methane as well. Graphitic carbon or graphite can be liberated from steel samples by dilute nitric acid and then subsequently analyzed by combustion.

Applications

Example 1: Detection of Nickel Leached from Stainless Steel

Chemical engineers studied the pressure, volume, and temperature behavior of compressed fluids in stainless steel cells (Ref 24). The engineers found that after a fairly acidic fluid had been brought to high temperature and pressure, a greenish fluid was recovered from the cell. Because the expected fluid was clear and colorless, the engineers suspected that metal was leaching from the stainless steel cells during the harsh experimental conditions. A series of instrumental tests were carried out to evaluate the organic compounds, and then a series of spot tests were employed to check for metal content. Spot testing with dimethylglyoxime yielded a bright-red precipitate, indicating that nickel had leached into the fluid from the reaction cell.

Example 2: Analysis of Chloride in Concrete Samples

Professors and students in a concrete industry management program were designing a method to determine the filtering capabilities of porous concrete. One of the important parameters was the chloride concentration of solutions that had been passed through different samples of porous concrete. To determine the chloride concentration, the solutions were titrated with $AgNO_3$ using a K_2CrO_4 indicator. As the $AgNO_3$ was delivered with the buret, the analyte solution became cloudy due to suspended $AgCl_{(s)}$. The endpoint was determined when the bright-yellow chromate indicator turned red/brown. This method is pH sensitive; the pH in the example was maintained between 6.5 and 9.0 to enable optimum behavior of the indicator without precipitation of silver hydroxides.

Example 3: Determination of Calcium and Total Acidity in Olive Oil Brine

A commercial olive oil manufacturer required the determination of calcium and total acidity in samples of olive brine for quality-control and consistency measurements. The total acidity was determined by titrating the solution with standardized NaOH, using a phenolphthalein indicator to determine the endpoint. The total calcium concentration was determined by EDTA titration. Samples of olive brine were buffered to pH 10 using an ammonia/ammonium chloride buffer. This pH maintains EDTA in a favorably deprotonated state to ensure chelation of the calcium ions. Additionally, this pH is optimum for the Eriochrome Black T indicator. The initially red/violet calcium solution was titrated with EDTA to the blue endpoint.

ACKNOWLEDGMENT

This article was revised from Thomas R. Dulski, Classical Wet Analytical Chemistry, *Materials Characterization*, Volume 10, *ASM Handbook*, ASM International, 1986, p 161–180.

REFERENCES

1. C. Burgess and J.J. Wilson, Samples and Sampling, *Valid Analytical Methods and Procedures: A Best Practice Approach to Method Selection, Development, and Evaluation*, 2000, p 15
2. A.D. McNaught and A. Wilkinson, *Compendium of Chemical Terminology*, 2nd ed., Blackwell Science Ltd., Malden, MA, 1997
3. J.N. Miller and J.C. Miller, *Statistics and Chemometrics for Analytical Chemistry*, 6th ed., Pearson Education Ltd., Gosport, U.K., 2010
4. J.T. Ballinger and G.J. Shugar, *Chemical Technicians' Ready Reference Handbook*, 5th ed., B.A. Ballinger and C.T. Ballinger, Ed., McGraw-Hill, New York, 2011
5. P. Atkins, L. Jones, and L. Laverman, *Chemical Principles: The Quest for Insight*, 6th ed., W.H. Freeman, New York, 2013
6. R. Pribil, *Analytical Applications of EDTA and Related Compounds*, 1st ed., Pergamon Press, New York, 1972
7. W. Gibbs, *Z. Anal. Chem.*, Vol 3, 1864, p 334
8. J.W. Robinson, E.M. Skelly Frame, and G.M. FrameII, *Undergraduate Instrumental Analysis*, 7th ed., CRC Press, Boca Raton, FL, 2014
9. A. Vignes, *Extractive Metallurgy 2: Metallurgical Reaction Processes*, 1st ed., John Wiley & Sons, Inc., Hoboken, NJ, 2013
10. K. Fischer, A New Method for the Volumetric Determination of the Water Content of Liquids and Solids, *Angew. Chem.*, Vol 48, 1935, p 394–396
11. D.A. Aikens, R.A. Bailey, J.A. Moore, G. G. Giachino, and R.P. Tomkins, *Principles and Techniques for an Integrated Chemistry Laboratory*, Waveland Press, Prospect Heights, IL, 1984
12. J.E. McMurry, R.C. Fay, and J.K. Robinson, The Activity Series of the Elements, *Chemistry*, Pearson, 2015, p 135–138
13. J. Dolezal, P. Povondra, and Z. Sulcek, *Decomposition Techniques in Inorganic Chemistry*, Iliffe Books Ltd., London, 1968
14. N.I. Sax and R.J. Lewis, Sr., *Hazardous Chemicals Desk Reference*, New York, 1987
15. J. Azain, "USGS Method 36: Aqua Regia Digestion," https://minerals.usgs.gov/science/analytical-chemistry/method36.html, accessed May 14, 2018
16. F. Fiegl and V. Anger, *Spot Tests in Inorganic Analysis*, 6th ed., Elsevier, New York, 1972
17. D.T. Harvey, "Analytical Chemistry 2.1," http://dpuadweb.depauw.edu/harvey_web/eTextProject/version_2.1.html, accessed May 9, 2018
18. D.C. Harris, *Quantitative Chemical Analysis*, W.H. Freeman, New York, 2016
19. C.G. Enke, *The Art and Science of Chemical Analysis*, John Wiley & Sons, Inc., New York, 2001
20. L. Meites, Ed., *Handbook of Analytical Chemistry*, 1st ed., McGraw-Hill, New York, 1963
21. D.S. Hage and J.D. Carr, *Analytical Chemistry and Quantitative Analysis*, 1st ed., Pearson Prentice Hall, Upper Saddle River, NJ, 2011
22. J.H. Nelson and K.C. Kemp, Rates of Chemical Reactions I: A Clock Reaction, *Laboratory Experiments*, Pearson, Upper Saddle River, NJ, 2009, p 331–344
23. P.T. Bowman, E.I. Ko, and P.J. Sides, A Potential Hazard in Preparing Bromine-Methanol Solutions, *J. Electrochem. Soc.*, Vol 137 (No. 4), 1990, p 1309–1311
24. T.J. Bruno and G.C. Straty, Thermophysical Property Measurement on Chemically Reacting Systems—A Case Study, *J. Res. Natl. Bur. Stand. (1934)*, Vol 91 (No. 3), 1986, p 135–138

ASM Handbook, Volume 10, *Materials Characterization*
ASM Handbook Committee
DOI 10.31399/asm.hb.v10.a0006659

Gas Chromatography*

Tara M. Lovestead, National Institute of Standards and Technology

Overview

General Uses

- Analysis of complex mixtures made up of anything that can be heated to form stable vapors

Examples of Applications

- Mixtures of volatile compounds in petroleum oil, coal gasification and liquefaction products, oil shale, and tar sands
- Pollutants and impurities in air, water, soil, and other solids
- Drugs and metabolites
- Pesticides
- Additives, such as antioxidants and plasticizers in plastics

Samples

- *Form:* solids, liquids, and gases; all organics and some inorganics
- *Size:* For liquid samples, a 0.1 to 10 μL injection depending on the sample concentration, inlet liner, and split ratio. For gaseous samples, a gastight syringe can be used to inject 100 to 500 μL volumes.
- *Preparation:* Samples should be prepared to conform to the sample size restrictions given above and to the limitations of the detector response to concentration.

Limitations

- Compound(s) must form stable vapors upon heating.
- Detection limit is dependent on the detector. A thermal conductivity detector is ~1 ng, whereas an electron capture detector can be ~1 pg. A gas chromatograph/mass spectrometer operated in selected ion monitoring can achieve detection limits as low as 0.5 pg.

Estimated Analysis Time

- When analyzing one compound, direct introduction of the sample takes 1 to 20 min per analysis. For gas chromatography/mass spectrometry, analysis of one to two compounds takes approximately 15 min, while analysis of 20 or more compounds takes 180 min or more.
- Analysis and interpretation of data is variable (15 min to days, depending on the number of compounds analyzed).

Capabilities of Related Techniques

- *Gas chromatography/Fourier transform infrared spectroscopy:* functional group analysis, but at least an order of magnitude less sensitive
- *Nuclear magnetic resonance:* at least 2 orders of magnitude less sensitive
- *Secondary ion mass spectroscopy:* a mass spectrometry method for looking only at surfaces of materials

CHROMATOGRAPHY is about separating (often) closely related components (such as plant pigments or components of fuel). Modern chromatography is used to purify drugs, determine the level of pollutants in water or soil for environmental analyses, compare samples from a crime scene, determine the identity of unknown substances, and for process engineering quality control.

A basic chromatographic separation requires:

- Sample
- Column/stationary phase
- Mobile phase
- Detector

To begin, the sample is introduced into the column. In gas chromatography, the column is typically long (from 10 to 60 m, or 33 to 200 ft). The column is wound into a coil and housed in a temperature-controlled oven. The column is coated with a stationary phase that is a solid (or a liquid supported on a solid). Chromatography also requires a mobile phase, which can be a liquid or a gas that flows past the stationary phase and carries the components of the mixture. The movement of components through the column can be facilitated by heating the column or by temperature ramping. The separation of the components in the sample depends on their interactions with the column stationary and mobile phases. The sample components ultimately arrive at a detector at different times (retention times), based on interactions with the column stationary and mobile phases (Ref 1).

One way to classify a chromatographic technique is by the choice of mobile phase. The mobile phase can be either a liquid (liquid chromatography, or LC) or a gas (gas

*Contribution of the United States government; not subject to copyright in the United States

chromatography, or GC). This article is dedicated to gas chromatography. Gas chromatography is the method of choice for analyzing mixtures made up of anything that can be heated to form stable vapors. Gas chromatography is robust (can be used with minimal preparation), is rapid (minutes), and is efficient at separating complex mixtures of both organic and inorganic components.

The principles that govern separations are based on the affinity of a molecule for the stationary phase. A component with no interaction with the stationary phase is simply carried by the gas and is said to be nonretained. If a component strongly interacts with the stationary phase, it will take longer to travel the entire length of the column. The affinity of a component for the stationary phase is governed by adsorption (adhesion) onto a solid substrate or by absorption or partitioning (dissolving and vaporizing) into and out of the liquid phase. Enthalpy of adsorption, ΔH_{ads}, describes the amount of energy required to move a molecule out of the solid substrate. Molecules with different ΔH_{ads} will move at different rates through a column, determined by the instrumentation and adjustable parameters, for example, the temperature-ramping rate.

A chromatogram, a graph of the detector response as a function of time, is a way to visualize the molecule separations. A chromatogram shows the peaks that are representative of the response of a detector to a molecule; the area under the peak relates to the concentration of the molecule in the mixture. Chromatograms are used to determine the performance of a column or the resolving power (ability to separate two peaks). Resolving power is determined by the retention factor, the selectivity factor, and the efficiency. These parameters can be optimized by changing pressure, temperature, amount of sample introduced, and heating profile and are dependent on the column length, stationary phase (chemistry), and the molecule of interest.

The instrument parts and the relevant adjustable parameters (temperature, pressure, phase chemistry, and solvent choice) for obtaining the best possible chromatographic separation are discussed in detail and include carrier gas cylinder, flow controller and pressure regulator, sample inlet and injection port, column, column oven, detector (for example, a flame ionization detector), and computer.

The mobile phase or carrier gas is supplied by a high-pressure gas cylinder through regulators and tubing. The sample is introduced into the GC instrument at the inlet or sample injection port. The molecules in the sample are vaporized in the injection port, and the sample vapors (along with the carrier gas) travel through the column. The column is housed in a temperature-programmed oven. The gas stream is transported through tubing from the tank to the column exit and eventually into a detector. Gas chromatography relies on having an appropriate detector to measure and translate a signal response to useful data in the form of a chromatogram.

Common GC detectors are discussed, including the thermal conductivity cell detector, the flame ionization detector, the electron capture detector, and the sulfur chemiluminescence detector. Mass spectrometry for detection is discussed in more detail in the following article of this Volume, because it is a stand-alone experimental technique (does not require GC separation).

Gas Chromatography Method

Gas chromatography is the chromatographic method in which gas is the mobile phase, and the column stationary phase is a solid or a nonvolatile liquid attached to a solid support. Gas chromatography can be used for analyzing mixtures made up of molecules (gases, liquids, and solids) that can be heated to form vapors (from solid or liquid into their gaseous form) and that do not break apart or decompose when heated. Gas chromatography is different from liquid chromatography because the mobile phase is essentially inert and thus is only used to carry the vapors; the vapors do not interact with the carrier gas. In LC, the liquid phase also interacts with molecules. Gas chromatography is advantageous because very small sample sizes can be used with minimal preparation, the technique is rapid (1 to 100 min), and it is efficient at separating complex mixtures into the individual components. Both organic and inorganic molecules (that can be vaporized in the inlet and are not immobilized in the column phase) can be separated. Also, GC can be automated to analyze hundreds of samples per day, and all for a very reasonable cost (Ref 1).

Details on Molecule Separation

In gas chromatography, the principles that govern separations are based on the affinity of a molecule for the stationary phase. A component with no interaction with the stationary phase is simply carried by the gas and is said to be nonretained. If a component strongly interacts with the stationary phase, it will take longer to travel the entire length of the column. If the stationary phase is a solid, the affinity of a component for the stationary phase is governed by adsorption (adhesion) onto the solid substrate; if the stationary phase is a liquid adhered to a solid support, the affinity is described by the absorption or partitioning (dissolving and vaporizing) of the molecule into and out of the liquid phase (Ref 1, 2).

The stronger the interaction between the stationary phase and the molecule, the larger the quantity of heat required to remove the molecule (sample molecule and carrier gas) from the surface of the stationary phase. A quantity can be assigned to this amount of heat and is referred to as the enthalpy of adsorption, ΔH_{ads}. A stationary phase must be selected so that the types of molecules to be separated have different adsorption or partitioning/absorption behaviors; additionally, the molecules cannot adhere too well, or they will not leave the column. A GC separation is the process of molecules adsorbing (or partitioning) and releasing many times as they move through the column and separate out of the mixture. This process is often temperature dependent.

Separation is visualized in a chromatogram, which is a graph of the detector response as a function of time. Figure 1 presents an example of a chromatogram. A peak is the response of a detector to a molecule separated from the mixture. The area under the peak can be related to how much of a molecule is in the sample mixture. By using standards, the identity of the peak can be determined by the time it takes after sample injection for the peak to exit the column (the retention time, t).

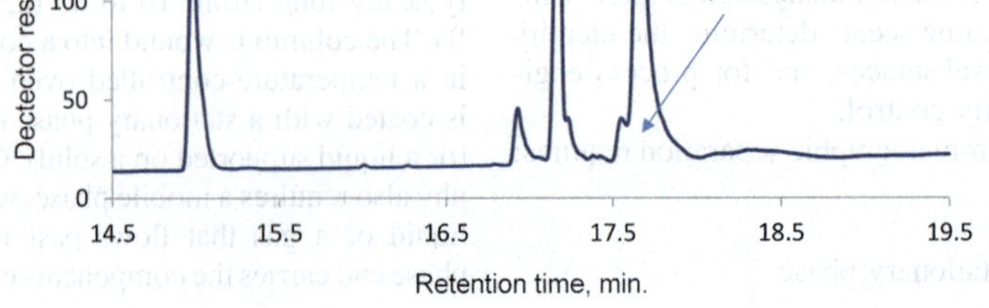

Fig. 1 Example of a chromatogram

The chromatogram can be examined to determine the column performance or resolving power, which is the ability of a column to separate two peaks. Three chromatographic parameters that are used to evaluate the resolving power of a column are the retention factor, the selectivity factor, and the efficiency. These parameters can be optimized by changing pressure, temperature, amount of sample introduced, and heating profile and are dependent on the column length, stationary phase (chemistry), and molecule of interest. Both the retention factor and selectivity factor describe the migration of solutes on the column. Equation 1 describes the retention factor for molecule A:

$$k_A = \frac{t_A - t_D}{t_D} \quad \text{(Eq 1)}$$

Here, t_A is the retention time for molecule A, and t_D is the retention time for a nonretained solute to exit the column. The selectivity factor, α, is used to compare two eluting molecules and describes peak separation or isolation. It is the ratio of the retention factors of molecules A and B:

$$\alpha = \frac{k_B}{k_A} \quad \text{(Eq 2)}$$

Efficiency can be visualized on a chromatogram as a measure of the width or sharpness of a peak. Efficiency can be measured by calculating the plate number, N, a dimensionless measurement based on the peak width at half the peak height and the retention time for a given molecule. A good column has a large N (Ref 3).

Gas Chromatography Instrumentation

A schematic of a typical modern GC apparatus is shown in Fig. 2, which consists of primary components such as:

- Carrier gas cylinder
- Flow controller and pressure regulator
- Sample inlet and injection port
- Column
- Column oven
- Detector (for example, a flame ionization detector)
- Computer

In summary, a high-pressure gas cylinder supplies the mobile phase (the carrier gas) to the GC instrument through regulators and tubing. The sample is introduced into the GC instrument at the inlet or sample injection port. The molecules in the sample are vaporized in the injection port, and the sample vapors (along with the carrier gas) travel through the column. It is important to note that the injector port is typically maintained at a temperature 20 to 40 °C (35 to 70 °F) above the boiling point of the sample. The column phase is selected to optimize molecule separation. The forced-air oven that houses the column is programmed to either remain constant at a chosen temperature (isothermal) or, more commonly, to change the temperature throughout the separation to control, speed up, or improve separation efficiency. Usually, the temperature is increased or plateaued during the separation. The gas stream is transported through tubing from the tank to the column exit and eventually into a detector. In Fig. 2, a flame ionization detector (FID) is shown; however, there are many detector options that are discussed subsequently. Finally, GC software is used to control all of the adjustable parameters: flow rate, sample injection volume, vaporization temperature, oven temperature ramping program, and detector temperature. The computer software is also used to translate the detector response (electrical signal) into interpretable results in the form of a chromatogram (Fig. 1).

Safety is always a concern in the analytical lab. The potential hazards to be aware of in GC are the use of compressed gases, the possibility of exposure to heated surfaces, and the potential exposure to the samples injected into the instrument. Good laboratory practice will minimize any of these hazards.

Carrier Gas

The purpose of the carrier gas (mobile phase) is to carry the sample from the inlet, through the tubing, column, and, ultimately, to the detector. The carrier gas must be inert, meaning that it should not interact and/or react with the sample (analytes or solvent) that is being injected onto the column. The carrier gas also must be dry, nonreactive, safe, and suitable for the detector used (this is discussed in the section "Detectors" in this article). Among all the gas choices, there are advantages and disadvantages. For example, hydrogen gas, the gas that filled the Hindenburg airship that exploded in 1937, is sometimes used as a carrier gas. However, because hydrogen has the potential to explode, necessary safety controls must be put in place before it can be used as the carrier gas. Helium is more common.

Impurities in the gas stream (oxygen, water, and particulates) must be removed before entering the GC instrument. These can attack the stationary phase or cause high background noise and, therefore, this is typically accomplished by having the appropriate filter (adsorbent trap) inline between the gas cylinder and the GC instrument or by using an ultra-high-purity compressed gas source.

In addition to choosing an appropriate gas, the gas flow rate is also important. The higher the flow rate of the carrier gas, the shorter the run time of the experiment and the lower the oven temperature must be to move a molecule through the column. However, lower flow rate may be necessary for hard-to-separate compounds, for example, samples with similar retention times. An optimum carrier gas flow rate will depend on the column diameter, the mixture molecules to be separated, and the type of detector.

Flow Control and Measurement

The carrier gas flow rate should be measured and controlled to identify molecules and to achieve an efficient separation. Often, identification is determined by the amount of time required to elute a given molecule from the column (retention time). If the flow rate is inconsistent between experiments, the retention times will be inconsistent. The appropriate gas flow rate will depend on the size and type of column (packed or capillary) and the film thickness of the stationary phase. Packed columns are made from glass or metal (copper or stainless steel) with 1 to 4 mm (0.04 to 0.16 in.) internal diameters, are 1 to 5 m (3 to 16 ft) long, and are typically operated with 20 to 30 mL/min flow rates, whereas capillary columns are made from fused silica with 0.1 to 0.5 mm (0.004 to 0.02 in.) internal diameters, are 10 to 150 m (33 to 490 ft) long, and are typically operated with 1 to 10 mL/min flow rates.

The flow rate is determined by the desired separation, the size of the column, and the choice of detector. It is selected to not only optimize separation but also to be sufficient

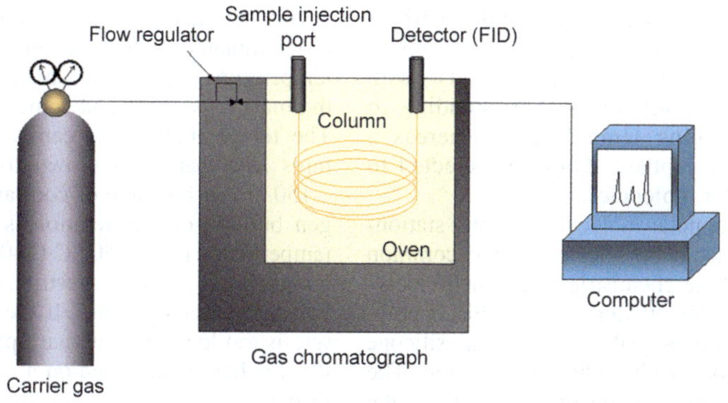

Fig. 2 Schematic of a typical gas chromatograph apparatus. FID, flame ionization detector

for the detector operation. For example, gas chromatography/mass spectrometry (GC/MS) requires the flow rate to be approximately 1 mL/min, while GC-FID can handle flow rates up to 20 mL/min.

Sample Introduction

There are several methods by which sample (both liquid and gas) can be introduced into the gas chromatograph. A liquid sample can be manually introduced into the injector by pushing a known volume of solution directly into the inlet with a small-volume syringe (1 to 10 μL). The syringe needle pierces a gastight rubber seal (septum). A gaseous sample can be introduced by directly flowing a gas into the inlet, a gas sampling valve, or by using a gastight syringe (100 to 500 μL). Liquid samples amenable to syringe injection can be injected manually or with an automatic liquid sampler (ALS). An ALS is often employed because it is much more accurate at injecting consistent solution volumes into the injection port with repeat injections. Reproducible volumes are necessary for quantitative results, because the area under the peak on the chromatogram relates to the concentration of the molecule in the sample solution.

The main function of the injection port is to properly place the sample onto the beginning of the column. Typically, the injection port is heated to above the boiling point of the molecules of interest, so that they can be applied to the column in vapor form. When choosing a proper GC inlet, there are several aspects of the GC setup to consider, including the diameter of the column that follows the sample inlet, the sample concentration, and the injection volume. The most common inlet encountered today (2019) is the split/splitless. This inlet allows some of the vaporized sample volume to be deposited onto the column and the remainder of the vaporized sample to be vented. The splitting of the sample is achieved by selecting an appropriate inlet liner, with sufficient volume to contain the vapor volume (upon injection of the sample into the heated chamber), and the correct gold seal. The vapor is formed in the inlet liner. Slits in the gold seal (at the base of the inlet liner) allow a path for some of the vapor volume to be diverted between the liner and the inlet body and out the split vent. Splitting the vapor volume injected onto the column avoids column overload and improves reproducibility. One disadvantage to splitting the injection volume is that higher-boiling-point molecules may not be accurately represented in the injection volume. Splitless injection is also possible with this inlet. In this case, the split vent is closed, and all the vapor volume is injected onto the capillary column.

Some additional aspects of sample introduction that must be mentioned are the proper choice of sample solvent, injection volume, and inlet temperature. These three parameters must be optimized so that the solvent vapor volume does not exceed the total volume that the inlet liner can contain.

Columns for Gas Chromatography

The stationary phase is held in a tube called the column. Typically, there are two types of columns that are used: either a packed column with small particles coated with a stationary phase, or a capillary column that is hollow and contains the stationary phase on the inner wall. Packed columns are used in the separation of permanent gases and when standard methods (such as ASTM International or Environmental Protection Agency) call for their use. Most modern GC methods use capillary columns because they are far more efficient and sensitive than similar applications of packed columns. The remainder of this article focuses on instrumentation for capillary columns with liquid stationary phases.

Capillary columns can be further classified into porous-layer open tubular columns or wall-coated open tubular (WCOT) columns. The WCOT columns are so ubiquitous that they are often referred to as simply capillary columns. Capillary columns are made from fused silica. The columns are usually coated on the outside with polyimide (a strong, flexible, and durable polymer or plastic that protects the more brittle fused silica from scratches) and coated on the inside with a stationary phase. The column is housed in the temperature-controlled oven. A stationary phase is selected to achieve the desired separation of mixtures. Stationary phases should reasonably solubilize the sample molecules and should have a negligible vapor pressure at the GC operating temperatures (Ref 2).

All the molecules eventually exit the column to be analyzed by the detector. The separation of the molecules is dependent on many factors, for example, the chemical properties of the molecules being separated (molecular weights, boiling temperatures, volatility, and polarities) as well as the physical and chemical properties of the stationary phase (density, surface area, and polarity). For example, a nonpolar stationary phase will separate nonpolar compounds with increasing retention times corresponding to increasing boiling temperatures, whereas a polar stationary phase should be selected to separate polar molecules.

To separate polar molecules, a polar stationary phase should be selected. Some common stationary phase chemistries are dimethylpolysiloxane and polyethylene glycol (PEG) polymers. Dimethylpolysiloxane has a silicone oxide backbone with methyl substitution. The least polar stationary phase occurs when the backbone is completely substituted with methyl groups. The methyl groups can be replaced with either phenyl or cyanopropyl groups to increase the polarity of the stationary phase. The stationary phase can also be entirely made up of PEG polymer to obtain the most polar stationary phase.

In addition to stationary phase chemistry, separation can be controlled by the stationary phase film (or particle film) layer thickness (0.1 to 1 μm), capillary column length, and tube internal diameter. For example, a typical specification would be 30 m (98 ft) (coiled into a 150 mm, or 6 in., diameter configuration) with a 250 μm internal diameter and a film thickness of 0.25 μm.

Chromatography configurations are not limited to only one column. A guard column, a short column placed between the inlet and the analytical column to remove impurities and particulate matter, is sometimes used to protect the column. Additionally, two capillary columns can be used in series. This practice was originally developed for the identification of hundreds of hydrocarbons in fuel mixtures in the 1990s. Today (2019), it is referred to as comprehensive two-dimensional chromatography and is used to analyze very complex sample mixtures for fuels, environmental analyses, and food and fragrance applications. It works by pairing two GC columns, connected in series through a modulator, which is sometimes cooled. The carrier gas and sample molecules are carried from the first column to the modulator and are held there for a fixed period of time before being focused and injected into the second column. The first column typically is nonpolar and thus separates nonpolar compounds essentially by boiling point. The second column is typically polar (to separate by polarity) and shorter, and thus, separation is more rapid. The chromatograms obtained through repeated trapping and injecting are rendered in two dimensions, with boiling point on one axis and polarity on the other. Understanding and interpreting the resulting chromatogram requires complex software and a great deal of training.

Column Oven

Gas chromatography ovens are temperature programmable, although operation at room temperature and/or constant temperature (isothermal) is possible and sometimes necessary. The temperature range can include temperatures less than zero (down to −100 °C, or −150 °F), if a source of cool air (liquid nitrogen boiloff) or refrigeration is available, and temperatures up to 450 °C (840 °F). The oven temperatures are often restricted based on the column stationary phase. If the oven temperature is too low, the stationary phase will solidify (for liquid supported on a solid), making it difficult for absorption to take place. If the oven temperature is too high, the stationary phase will vaporize, potentially increasing the background signal and possibly fouling the detector.

In general, the oven is programmed to have a low initial temperature to resolve the low-boiling-point molecules, and the temperature is increased during separation to resolve the less volatile molecules.

Detectors

The detector indicates what and how much is in the carrier gas that exits the column. Detectors must have characteristics such as:

- High sensitivity
- Low noise
- Linear response to concentration
- Response to the types of chemicals of interest to the user
- Cost-effectiveness

A detector should be selected with sufficient sensitivity for the intended analysis. Sensitivity is a measure of the smallest detectable quantity that can be measured above the noise, divided by the width of the peak base in seconds. In general, sensitivity increases as the noise (the background electrical response inherent in the detector and electronics) decreases. Detectors are further defined by a range of compound concentrations for which the response of the detector can be fit with a straight line when plotted on a graph (Fig. 3). In the linear response range, the detector response can be converted to the compound concentration, providing a quantitative analysis of a sample with a previously unknown analyte concentration. Different detectors also respond differently to the types of chemicals of interest to the user.

Common GC detectors include the thermal conductivity cell detector, FID, electron capture detector, sulfur chemiluminescence detector, and MS detectors. Descriptions of each of these detectors as well as the operating principles, limits of detection, linear range, advantages, and disadvantages are presented next, while GC/MS is treated in more detail in the next article of this Volume.

Thermal Conductivity Cell Detector

The first gas chromatographs were built in the mid-1950s. Typically, the thermal conductivity cell detector (TCD) was the detector used for these instruments because it is rugged, reliable, and easy to use (Ref 4).

The TCD works on the principle that a hot body will lose heat at a rate depending on the composition of the surrounding gas. For a hot wire-based detector, the flow of carrier gas (typically helium or hydrogen) through the wires (across the filament) removes heat at a specific rate, depending on the thermal conductivity of the reference gas. When a sample is present, there is a change in the thermal conductivity of the gas mixture (a decrease in the rate of heat removal), which raises the temperature of the wire. The change in temperature causes an electrical resistance change, which can be measured and translated into a signal response in the form of a chromatogram (Ref 5, 6).

An advantage of the TCD is that it uses a nondestructive process, so it is often used for prep-scale chromatography application. A disadvantage is that it is not as sensitive as other GC detectors. The TCD is very useful for analyzing inorganic gases (argon, nitrogen, hydrogen, carbon monoxide, carbon dioxide, ammonia, carbon disulfide, etc.) and small hydrocarbon molecules (that may be found in natural gas, for example). The TCD is not very useful if the sample has water or oxygen impurities in it. These impurities can interact with the stationary phase and cause high baseline noise, which is problematic for the TCD because of its low sensitivity.

Flame Ionization Detection

The limitations of the TCD (its lack of sensitivity and use of helium gas, which was difficult to procure outside of the United States in the 1950s) led researchers to develop other detectors. In 1958, two detectors were introduced: the FID and the electron capture detector, which is discussed later. With high sensitivity, an extended linear range, and lack of a radioactive source, the FID became the most widely used detector within a few years (Ref 4).

The FID is based on the principle that the electrical resistance of a flame changes with the composition of the gases being combusted (Ref 4). The FID is highly sensitive to carbon-containing molecules and, in fact, has been referred to as a carbon counter. The FID typically uses a hydrogen and air flame, which has a high electrical resistance. (It is very difficult to pass an electrical current through the flame.) When a substance with carbon-hydrogen bonds exits the column and enters the flame, carbon atoms produce ions, which decreases the resistance (increases the current across two electrodes). The corresponding chromatographic response is proportional to the change in resistance. Since hydrogen gas is used, there is an explosion hazard when operating a FID.

The FID responds well to organic and combustible gases. It is not sensitive to the noble gases (helium, neon, argon, krypton, xenon, and radon), carbonyls (compounds with C=O groups), alcohols (compounds with C–OH groups), halogens (compounds with fluorine, chlorine, bromine, or iodine), nitrogen-containing compounds, and noncombustible gases such as water, carbon dioxide, sulfur dioxide, and oxides of nitrogen. The combination of sensitivity toward organics and insensitivity toward water make the FID particularly well suited for detection of pollutants in water samples (Ref 3).

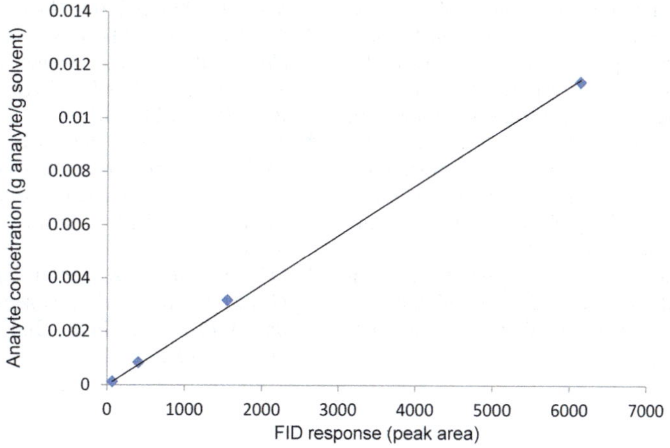

Fig. 3 Graph of analyte concentration in solvent versus flame ionization detector (FID) response. A linear increase in FID response with increased analyte concentration is observed.

Electron Capture Detector

At the same time that the FID was introduced, the argon ionization detector was also introduced, which was later modified to become the electron capture detector (ECD). This detector is most sensitive when measuring molecules that contain halogens (for example, pesticides) and polychlorinated biphenyls (toxic substances that were used widely as heat-transfer fluids). The ECD uses a radioactive energy source to charge, or ionize, the carrier gas. The radioactive energy source is a β-emitter. (β-rays or particles are electrons.) The emitted electrons ionize the carrier gas, generating a standard current between a pair of electrodes. Molecules in the gas stream that capture or absorb electrons, such as halogens, cause the current to decrease sharply. The response signal is proportional to this decrease.

Sulfur Chemiluminescence Detector

Sulfur molecules are ubiquitous in refining and petrochemical products. Sulfur has a negative impact on product quality, causes corrosion, poisons catalysts, and pollutes the air. For example, higher sulfur content in diesel fuels leads to higher emission of particulates from diesel engines. A sulfur chemiluminescence detector (SCD) can be used to determine the sulfur concentration of diesel fuel and thus determine the impact of diesel fuel emissions on the environment.

In the SCD, the gas mixture flowing out of the end of the column is mixed with hydrogen and air and is burned, just as in the FID. The resulting gases are then mixed with ozone, trioxygen (O_3). It is a pale-blue gas with a distinctively pungent smell. It is present in the Earth's atmosphere in low concentrations and prevents damaging ultraviolet light from reaching the Earth's surface. It is also a powerful oxidant that reacts with sulfur molecules. When sulfur molecules encounter ozone, they react and luminesce (emit light). The intensity of the luminescence is proportional to the sulfur concentration. In an SCD, the luminescence is due to a chemical reaction and thus is termed chemiluminescence. Some disadvantages of the SCD are that it requires stabilization (on the order of days) and can be tricky to operate. Also, the ceramic reaction tubes are costly and are prone to contamination from hydrocarbon residue.

Nitrogen-Phosphorus Detector

Another detector for GC is the nitrogen-phosphorus detector. This detector is basically an FID with a bead of metal (rubidium or cesium) just above the flame. The hydrogen-to-air ratio is optimized to minimize hydrocarbon ionization, while the alkali ions on the bead surface facilitate ionization of nitrogen- and phosphorus-containing molecules, so it can detect the things that the FID cannot. This detector is well suited for environmental and forensic applications, for example, detecting pesticide residues in food and water.

Conclusion

Gas chromatography is the method of choice for analyzing mixtures made up of anything that can be heated to form stable vapors. Gas chromatography is robust (can be used with minimal preparation), rapid (minutes), and efficient at separating complex mixtures of both organic and inorganic components. The principles that govern separations are based on the affinity of a molecule for the stationary phase. A component with no interaction with the stationary phase is simply carried by the gas and is said to be nonretained. If a component strongly interacts with the stationary phase, it will take longer to travel the entire length of the column. The affinity of a component for the stationary phase is governed by adsorption (adhesion) onto a solid substrate or by absorption or partitioning (dissolving and vaporizing) into and out of the liquid phase. Enthalpy of adsorption, ΔH_{ads}, describes the amount of energy required to move a molecule out of the solid substrate. Molecules with different ΔH_{ads} will move at different rates through a column determined by the instrumentation and adjustable parameters, for example, temperature ramping rate.

Separations are visualized by a chromatogram, a graph of the detector response as a function of time. A peak is the response of a detector to a molecule separated from the mixture, and the area under the peak provides quantitative information. The chromatogram can be examined to determine the column performance or resolving power (ability to separate two peaks). Resolving power is determined by the retention factor, the selectivity factor, and the efficiency. These parameters can be optimized by changing pressure, temperature, amount of sample introduced, and heating profile and are dependent on the column length, stationary phase (chemistry), and molecule of interest.

The instrument parts and the relevant adjustable parameters (temperature, pressure, phase chemistry, and solvent choice) for obtaining the best possible chromatographic separation are discussed in detail. The mobile phase or carrier gas is supplied by a high-pressure gas cylinder through regulators and tubing. The sample is introduced into the GC instrument at the inlet or sample injection port. The molecules in the sample are vaporized in the injection port, and the sample vapors (along with the carrier gas) travel through the column. The column is housed in a temperature-programmed oven. The gas stream is transported through tubing from the tank to the column exit and eventually into a detector. Gas chromatography relies on having an appropriate detector to measure and translate a signal response to useful data in the form of a chromatogram.

Common GC detectors include the TCD, FID, ECD, and SCD. Mass spectrometry is another extremely powerful detector commonly available for GC analyses. Gas chromatography/mass spectrometry is a much more complicated experimental technique than GC with the aforementioned detectors and thus is treated in more detail in the next article of this Volume.

REFERENCES

1. H.M. McNair and J.M. Miller, *Basic Gas Chromatography: Techniques in Analytical Chemistry*, John Wiley & Sons, Inc., New York, NY, 1998
2. H.M. McNair and E.J. Bonelli, *Basic Gas Chromatography*, Consolidated Printers, Berkeley, CA, 1969
3. D.A. Skoog, F.J. Holler, and T.A. Nieman, *Principles of Instrumental Analysis*, 5th ed., Thomson Learning, Inc., 1998
4. L.S. Ettre, The Invention, Development and Triumph of the Flame Ionization Detector, *LCGC Europe*, June 2002
5. R. Trautner et al., "Visual Encyclopedia of Chemical Engineering," College of Engineering, Chemical Engineering, University of Michigan, http://encyclopedia.che.engin.umich.edu/Pages/ProcessParameters/ChromatographyColumns/ChromatographyColumns.html.
6. T.J. Bruno, *Chromatographic and Electrophoretic Methods*, Prentice Hall, Inc., New Jersey, 1991

SELECTED REFERENCES

- T.J. Bruno, Method and Apparatus for Precision In-Line Sampling of Distillate, *Sep. Sci. Technol.*, Vol 41 (No. 2), 2006, p 309–314
- T.J. Bruno and B.L. Smith, Enthalpy of Combustion of Fuels as a Function of Distillate Cut: Application of an Advanced Distillation Curve Method, *Energy Fuels*, Vol 20, 2006, p 2109–2116
- T.J. Bruno and P.D.N. Svoronos, Ed., *CRC Handbook of Basic Tables for Chemical Analysis*, 3rd ed., CRC Taylor and Francis, Boca Raton, FL, 2011
- Y.M. Moustafa and R.E. Morsi, Ion Exchange Chromatography—An Overview, Chap. 01, *Column Chromatography*, D.F. Martin and B.B. Martin, Ed., InTech, Rijeka, 2013
- O.D. Sparkman, Z. Penton, and F. Kitson, *Gas Chromatography and Mass Spectrometry: A Practical Guide*, 2nd ed., Elsevier, Oxford, U.K., 2011

ASM Handbook, Volume 10, *Materials Characterization*
ASM Handbook Committee
DOI 10.31399/asm.hb.v10.a0006664

Gas Chromatography/Mass Spectrometry*

Tara M. Lovestead and Kimberly N. Urness, National Institute of Standards and Technology

Overview

General Use

- Analysis of complex mixtures of volatile compounds and analysis of nonvolatile compounds

Examples of Applications

- *Gas chromatography/mass spectrometry (GC/MS):* mixtures of volatile compounds in petroleum oil, coal gasification and liquefaction products, oil shale, and tar sands; pollutants in air, water, and solids; drugs and metabolites; pesticides; and additives, such as antioxidants and plasticizers in plastics
- *Mass spectrometry/mass spectrometry:* mixtures of nonvolatile and high-molecular-weight solids
- *Pyrolysis gas chromatography/mass spectrometry:* analysis of polymers and their additives

Samples

- *Form:* solids, liquids, and gases; all organics and some inorganics
- *Size:* For liquid samples, 0.1 to 10 μL injection in which each compound of interest is in the 5 to 20 ng range. For selected ion GC/MS, samples in which each compound of interest is in the 100 to 500 pg range and, in some cases, down to 0.5 pg. For gaseous samples, a gastight syringe can be used to inject 100 to 500 μL volumes.
- For mass spectrometry/mass spectrometry, sample sizes are 10 to 500 ng per compound of interest.
- *Preparation:* Samples should be prepared to conform to the sample size restrictions given above.

Limitations

- Compound(s) must be ionizable.
- Detection limit is from 5 to 20 ng, depending on the compound. In selected ion-monitoring GC/MS, the detection limit can be as low as 0.5 pg.

Estimated Analysis Time

- When analyzing one compound, direct introduction of sample takes 2 to 20 min per analysis. For GC/MS, analysis of one to two compounds takes approximately 15 min, while analysis of 20 or more compounds takes 180 min or more.
- Analysis and interpretation of data is variable (15 min to days, depending on the number of compounds analyzed).

Capabilities of Related Techniques

- *Gas chromatography/Fourier transform infrared spectroscopy:* functional group analysis but at least an order of magnitude less sensitive
- *Liquid chromatography/mass spectrometry:* analysis of heat-sensitive and degradable compounds, such as biological materials
- *Nuclear magnetic resonance:* at least 2 orders of magnitude less sensitive
- *Secondary ion mass spectroscopy:* a mass spectrometry method for looking only at surfaces of materials

MASS SPECTROMETRY (MS) is a technique that is used to analyze molecular mass and molecular structure for qualitative compound identification and/or quantitative analysis. While MS can be done without gas chromatography (GC) separation, interpretation of the data becomes increasingly more difficult when analyzing mixtures.

Gas chromatographic separation prior to MS detection is a very powerful technique that improves analysis. The preceding article in this Volume, "Gas Chromatography," discusses GC fundamentals as well as common GC detectors: the flame ionization detector, thermal conductivity cell detector, electron capture detector, and sulfur chemiluminescence detector. The mass spectrometer is another common GC detector and requires a detailed article of its own.

*Contribution of the United States government; not subject to copyright in the United States

Gas chromatography/mass spectrometry (GC/MS) is used to qualitatively and quantitatively determine organic (and some inorganic) compound purity and stability and to identify components in a mixture. Gas chromatography/mass spectrometry is commonly used in scientific fields, including environmental chemistry for atmospheric, soil, and water research; forensic science for detection of drugs of abuse (or metabolites) and in arson fire debris analysis; food science for determination of food or beverage quality and authenticity; and in developing renewable fuels.

In brief, a GC/MS experiment begins with sample preparation, sample injection, and molecule separation on a GC column (as discussed in the previous article in this Volume). After GC separation, the molecules flow into a mass spectrometer. Because the operation of a mass spectrometer requires a high-vacuum system, an interface is necessary. When the molecules leave the GC column, they enter an ionization chamber. This article focuses primarily on electron ionization, where the molecules are bombarded in the ionization chamber with a stream of energetic electrons, which ionize and fragment some of the molecules. The ions can include unfragmented molecular ions and/or ions generated due to fragmentation or rearrangement reactions. Ions are accelerated and rapidly sorted according to their mass-to-charge ratio (m/z, where m is the mass and z is the charge) in a mass analyzer by use of a magnetic or electric field. Mass analyzers can typically sort thousands of different ion masses (m/z) per second. Then, detectors count ion abundance by measuring the current of electrons generated when the ions strike the detector.

Detectors generate chromatograms, which indicate the quantity of each compound as a function of retention time. The underlying dimension of data specific to MS is called a mass spectrum, which is a histogram of the abundance of each ion according to its m/z and serves as the fingerprint to identify the compound represented by a peak on the chromatogram (Ref 1–3). In some cases, the structure of an unknown compound can be deduced from the molecular ion, fragmentation or rearrangement ions, or the isotopic abundance of some ions. In other cases, known analytical standards or comparisons of mass spectra with compounds found in library databases can be used.

This article also discusses sample preparation, which is very important in GC/MS to avoid erroneous data and to minimize maintenance and troubleshooting of the instrument. Samples must be dilute, and molecules that can negatively affect the GC/MS instrument or interfere with analyte detection must be removed prior to injection. It is beneficial to obtain a good understanding of chromatography separation before performing GC/MS experiments.

This article also highlights the current state of the art in MS detector technology, which now makes it possible to record data for ions up to m/z = 1050. Data can also be collected continuously in scan mode and selected ion-monitoring mode. Additionally, high-resolution mass spectrometers can provide data with accurate mass of 5 ppm or less. Structural information and trace analysis of complex mixtures can also be deciphered with tandem mass spectrometry (MS/MS), a technique in which two mass analyzers are separated by a collision chamber. Mass spectrometry/mass spectrometry has been used successfully for protein sequencing and toxicology. The following sections of this article discuss in more detail GC/MS instrumentation, interpreting mass spectra, GC/MS methodology, and GC/MS advances.

GC/MS Instrumentation

Gas chromatography/mass spectrometry is used to qualitatively and quantitatively determine organic (and some inorganic) compound purity and stability and to identify components in a mixture. Gas chromatography/mass spectrometry is more difficult than GC coupled with other detectors because the mass spectrometer requires a high-vacuum system. Thus, when conducting maintenance and troubleshooting GC/MS, the instrument requires cooling and venting the MS, a very time-consuming process. Mass spectrometry detectors are also more easily fouled than other GC detectors, and thus, careful sample preparation and good chromatography are more necessary.

GC/MS Interface

The vapor stream that exits a gas chromatograph is under pressure, sometimes more than 5 times greater than atmospheric pressure. The mass spectrometer analyzer operates at a high vacuum to facilitate the ions traveling through the analyzer. High vacuum is achieved by use of a fore pump and a turbomolecular pump. For typical conditions (capillary column, 1 to 2 mL/min flow rate, and an internal diameter of 250 to 320 μm), the turbomolecular pump can handle the gas flow out of the GC. In this case, the GC/MS interface is a heated metal tube. For larger columns or higher flow rates, or both, some of the vapor flow must be separated by another pump that draws most of the gas out of the interface tube prior to the sample and carrier gas entering the mass spectrometer. Care must be taken to ensure the analyte molecules do not condense or decompose in the GC/MS interface tube. The interface temperature is usually set to 10 °C (18 °F) above the final oven temperature.

GC/MS Ionization Techniques

The gas exiting the GC enters an ionization chamber, where the carrier gas and analytes are bombarded with high-energy electrons. The most common method is electron impact or electron ionization, a hard ionization technique (that is, there is an excess of energy applied to the analytes beyond what is required for ionization, thus causing ion fragmentation). In this technique, the high-energy electron beam is created by applying a voltage to a heated filament. The neutral molecules that enter the ionization chamber are bombarded by the high-energy electron beam, which removes valence shell electrons, converting some of the molecules to molecular ions. Here, radical cations, M+· (abbreviated hereafter as M+), are formed, where M is the analyte molecule, the + sign indicates that the ions have a positive charge, and the dot symbolizes the unpaired electron. The ionized molecules may include the molecular ion; however, the molecular ion may also have sufficient energy to undergo rearrangement, bond cleavage, or fragmentation into mass fragment ions (product ions). Most of the sample, as well as the other products formed during electron ionization, is removed by the vacuum system (Ref 4).

Typical operation of electron ionization MS uses a beam of electrons with a potential of 70 eV. This has been established as the energy required to efficiently create ions in a reproducible way for organic molecules with sizes < 1000 atomic mass units (amu, or Daltons). An example of the ions produced from electron ionization MS of ethanol (Chemical Abstracts Service, or CAS, registry number 64-17-5) is presented in Table 1, which was adapted from Raphaelian's previously published edition of this article (Ref 5). When universal electron ionization GC/MS conditions are used, mass spectral databases can be

Table 1 Molecular or parent ion and characteristic fragment ions produced during electron ionization mass spectrometry of ethanol

Molecular/parent ion	$CH_3CH_2OH \xrightarrow{-e} (CH_3CH_2OH)^+$		m/z = 46
Fragment ion	$CH_3CH_2OH \xrightarrow{-H} (CH_3CH_2O)^+$		m/z = 45
Fragment ion	$CH_3CH_2OH \xrightarrow{-H} (CH_2CH_2OH)^+$		m/z = 45
Fragment ion	$CH_3CH_2OH \xrightarrow{-CH_3} (CH_2OH)^+$		m/z = 31
Fragment ion	$CH_3CH_2OH \xrightarrow{-OH} (CH_3CH_2)^+$		m/z = 29
Fragment ion	$CH_3CH_2OH \xrightarrow{CH_2OH} (CH_3)^+$		m/z = 15

m/z, mass-to-charge ratio. Adapted from Ref 5

employed to identify the compound from inspection of the mass spectrum (Ref 6, 7).

Chemical ionization is a common soft ionization technique (Ref 2). This technique produces an adduct ion (MH+), where a proton (H+) is added to the molecule. The main difference between these ionization techniques is in the amount of energy that is transferred to the molecule and the type of ions generated. There is typically not enough energy to fragment the molecular ions, thus increasing the strength of signal produced by these ions. Chemical ionization and electron ionization are often complementary techniques. Electron ionization is the most commonly used ionization technique for GC/MS; thus, for the remainder of this article, only electron ionization MS is discussed (Ref 2–4).

Mass Analyzers

The direction and velocity of the positively charged ions are controlled by a repeller and accelerator plates that steer the ions out of the ionization chamber. Single-focusing, magnetic-sector mass spectrometers use a magnetic field to focus the trajectory of ions of the same *m*/*z* value on the analyzer exit slit (Ref 2). The velocity and trajectory of the ions is controlled by continuously varying the accelerating voltage and the magnetic field. The interested reader is referred to Ref 2 for more details on magnetic-sector mass analyzers. Neutral molecules are removed by vacuum pumps, and molecules with negative charges are drawn to the repeller plate (Ref 3).

The most commonly used mass analyzer is the quadrupole mass filter. The quadrupole consists of four solid, circular, or hyperbolic in cross section rods aligned equidistant and parallel to the ion path. The quadrupole uses electric fields to separate the ions by their *m*/*z* instead of the magnetic fields used in the magnetic-sector mass spectrometer. Separation is achieved by application of a combination of radio-frequency and direct current voltages to each rod, with opposite polarities on the pairs of opposing rods (Ref 3, 5). The field strengths are varied to allow ions with stable oscillations to continue to the detector and ions with unstable oscillations to strike a rod, become neutral, and be pumped out of the analyzer by the vacuum system (Ref 2).

Mass analyzers have also been built based on separation of ions due to their time of flight (TOF) or flight time. Here, ions of similar initial kinetic energy travel down a long field-free flight tube. The smaller ions travel faster and reach the detector before the heavier (slower) ions. Time of flight is simple and fast (more than 100 spectra per second) and is theoretically boundless in mass range (Ref 5, 8).

Mass analyzer performance is often characterized by mass resolution, Δm. The International Union of Pure and Applied Chemistry definition states that this is the smallest mass difference between two equal-magnitude peaks so that the valley between them is a specified fraction of the peak height. Two common specifications are for the 10% valley definition or the 50% full peak width at half maximum definition. Quadrupole mass analyzers are well known for their unit mass resolution that is constant across the entire mass range. Magnetic-sector and TOF mass analyzers are different than the previous mass analyzers. Their mass resolution is not constant across the entire mass range; it is smaller for smaller masses. To describe the performance of these mass analyzers, the resolving power, R, is defined as $m/\Delta m$.

Interpreting Mass Spectra

Data from GC/MS experiments are two-dimensional and are displayed first by a chromatogram that indicates the quantity of each compound as a function of retention time, and second by a mass spectrum, which is a histogram of the abundance of each ion according to its *m*/*z*. Several examples of how to interpret GC/MS data are presented.

The Mass Spectrum

The mass spectrometer acquires a mass spectrum and displays these data as a histogram of the abundance of the ions that reach the detector according to their *m*/*z*; the spectrum is often plotted on a relative-abundance scale. An example electron ionization MS spectrum is presented in Fig. 1. All peak intensities are normalized to the highest peak, in this case, the *m*/*z* = 31 peak. Peaks for the molecular/parent ion (*m*/*z* = 46) and the fragmentation ions (also listed in Table 1) *m*/*z* = 45, 31, 29, and 15 are all present (Ref 5, 9).

In some cases, the structure of an unknown compound can be deduced from the molecular ion and fragmentation ions. In other cases, known analytical standards or comparisons of mass spectra with compounds found in library databases can be used. As of 2018, the National Institute of Standards and Technology/Environmental Protection Agency/National Institutes of Health Mass Spectral Library has electron ionization MS spectra for 267,376 chemical compounds, with new species constantly being added (Ref 7).

Isotope Abundance

In addition to the molecular ion (M^+) and the fragmentation ions, another useful tool for structural identification is the isotopic abundance of select ions. For example, in addition to the molecular ion, a small quantity of a species one mass unit larger, written as $(M+1)^+$, is often observed. In the case of carbon, which is predominately ^{12}C, there is a small natural abundance of ^{13}C atoms. In fact, approximately 1.1% of all carbon atoms in an organic molecule are ^{13}C. Figure 2 shows the electron ionization MS spectrum for 1,2,3,4-tetrahydro naphthalene (CAS registry number 119-64-2), with the molecular formula $C_{10}H_{12}$. This molecule contains approximately 11% ^{13}C atoms. The result of this, in terms of ion abundance, is that the abundance of $(M+1)^+$ (*m*/*z* = 133)

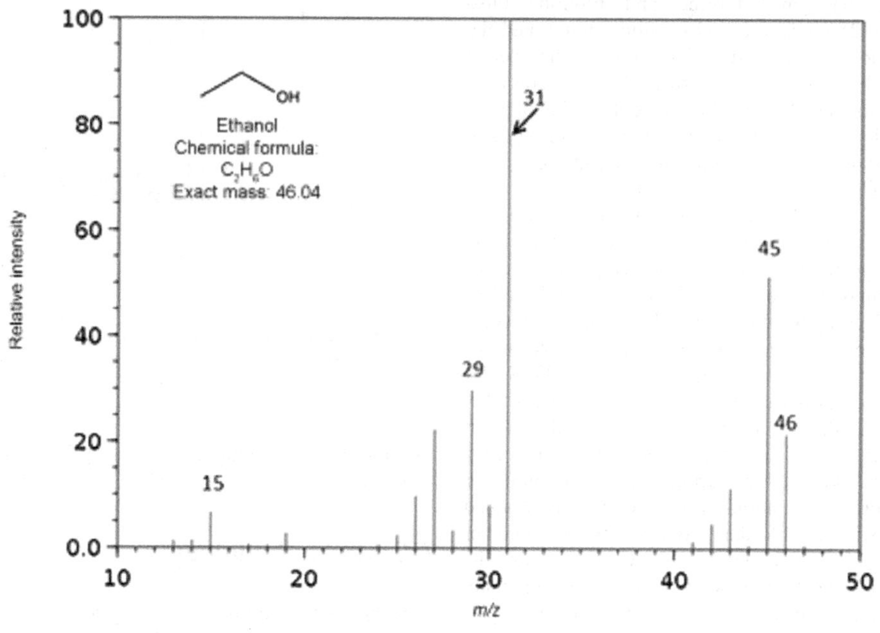

Fig. 1 Electron ionization mass spectrometry spectrum for ethanol. The ions (mass-to-charge ratio, *m*/*z*) are presented as a function of relative intensity normalized to the most abundant at *m*/*z* = 31. The molecular ion at *m*/*z* = 46 is also observed. Source: Ref 9

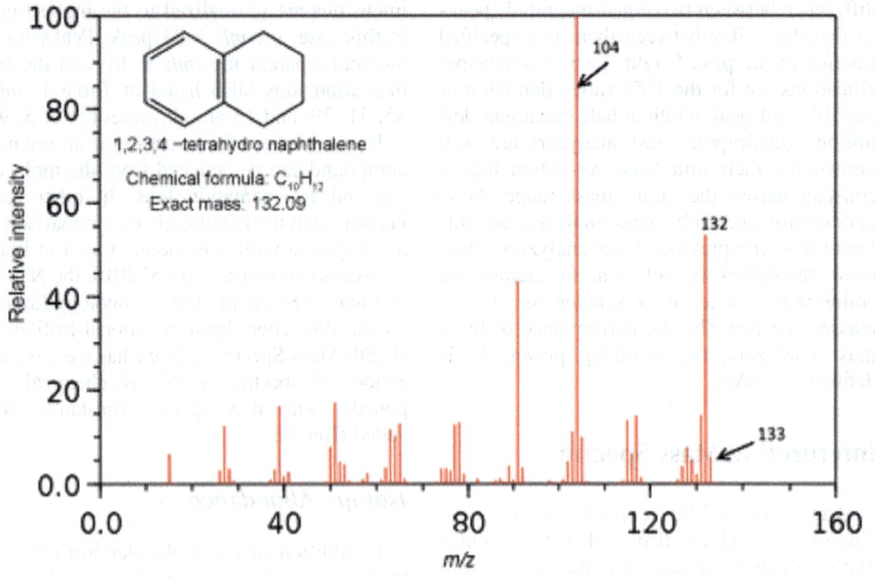

Fig. 2 Electron ionization mass spectrometry spectrum for 1,2,3,4-tetrahydro naphthalene (tetralin). The ions (mass-to-charge ratio, *m/z*) are presented as a function of relative intensity normalized to the most abundant at m/z = 104. The molecular ion at m/z = 132 is observed, and the molecular ion containing one ^{13}C atom is observed at m/z = 133. Source: Ref 9

Table 2 Relative natural isotopic abundances of common elements

	$(M)^+$		$(M+1)^+$		$(M+2)^+$	
Element	Mass	%	Mass	%	Mass	%
H	1	99.98	2	0.015	...	...
B	10	19.7	11	80.22	...	...
C	12	98.89	13	1.11	...	...
N	14	99.63	15	0.37	...	...
O	16	99.76	17	0.04	18	0.2
F	19	100	...	...	...	...
Mg	24	78.7	25	10.13	26	11.17
Si	28	92.21	29	4.7	30	3.09
P	31	100	...	...	...	...
S	32	95.02	33	0.76	34	4.22
Cl	35	75.53	...	...	37	24.47
Br	79	50.5	...	...	81	49.5
I	127	100	...	...	...	...

Adapted from Ref 5, 8

relative to $(M)^+$ (m/z = 132) should be approximately 11%.

Carbon is only one of several elements with a significant natural abundance of multiple isotopes. Isotopic abundances of other common elements are presented in Table 2. Oxygen, for example, has an isotope two mass units larger than the predominant ^{16}O; the presence of an ^{18}O isotope in a molecule makes it a (M+2) species. The isotopic contributions of the halogens chlorine and bromine (also (M+2) species) are also quite unique, and the ratios can be used to determine the number of each element in an unknown. These patterns have historically been useful for identification of polychlorinated biphenyls, pesticides, refrigerants, and other commonly halogenated species.

Currently, a prevalent chemical identification need that relies on accurate measurement of isotopic ratios is the proper identification of pesticides. There are many pesticide compounds that are banned from use under all or select applications. To determine if a compound has been illegally applied, analysis of soil samples by GC/MS (and GC/MS/MS, which is discussed later) is common practice, in addition to identification with known standards with an electron capture detector. (See the previous article in this Volume for more information.) For example, the banned pesticide chlordane (CAS registry number 57-74-9) has the formula $C_{10}H_6Cl_8$, and its mass spectrum is shown in Fig. 3. For many banned substances, their ion fragmentation patterns have been previously documented in searchable databases. For additional information on how to identify halogenated species by their isotopic ratios, the reader is referred

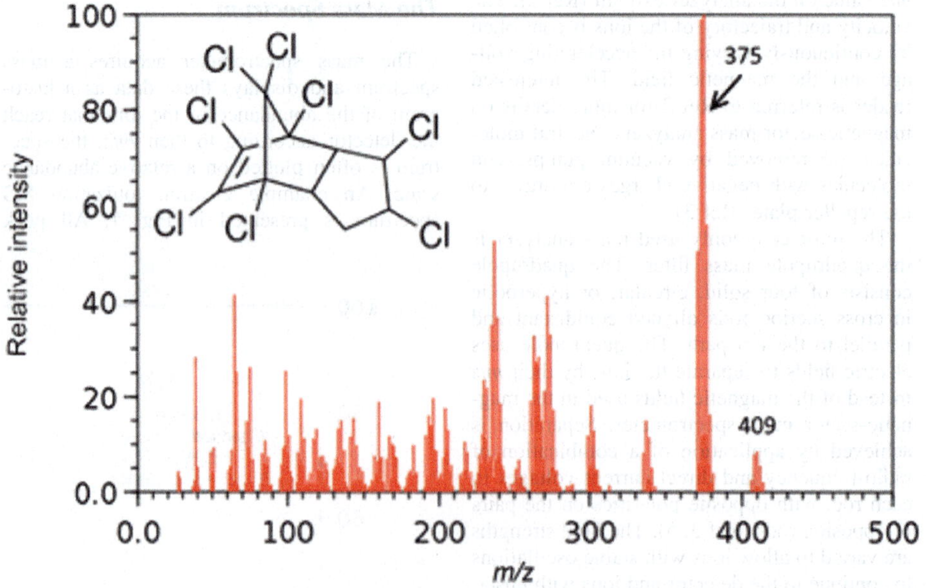

Fig. 3 Electron ionization mass spectrometry spectrum for chlordane. The ions (mass-to-charge ratio, *m/z*) are presented as a function of relative intensity normalized to the most abundant at m/z = 375. The molecular ion at m/z = 409 is observed. Source: Ref 9

to a more comprehensive analysis by McLafferty (Ref 2).

GC/MS Methodology

Sample preparation is necessary to optimize analyte concentration and to remove compounds that can negatively affect the GC/MS instrument or interfere with analyte detection. Sample preparation depends on the parameters selected for the GC/MS instrument, column phase and dimensions, sample composition, and detector limitations. Some sample preparations are done inline, such as headspace or thermal desorption. Samples can also be diluted in an appropriate solvent. The solvent purity is important, and the solvent must dissolve and not coelute with the analytes of interest. Hubschmann provides an excellent review and detailed discussion of sample-preparation techniques (Ref 8).

Operational Considerations

For injection of the prepared sample onto the column, the injector can be cooled or heated to optimize the vaporization of the entire sample. For a GC/MS instrument equipped with a split/splitless injector, the vapor volume injected onto the capillary column is controlled by operating the inlet in either splitless or split mode. High-purity GC carrier gas, free from oxygen, moisture, and hydrocarbons, is necessary to avoid fouling the detector. The GC flow rate can be operated with either constant head pressure or constant flow, with the flow rate through the GC column optimized for the detector. For a typical MS detector in high vacuum, less than 2.0 mL/min flow out of the GC is necessary. A low-bleed GC column must be used to avoid high background noise. More detail about the chromatography can be found in the previous article of this Volume.

Data Acquisition

Mass spectra can be collected for each peak from m/z = 1.6 to 800 and up to m/z = 1050 for the newest mass spectrometers. Data can be collected continuously in scan mode, selected ion-monitoring (SIM) mode, or simultaneous scan/SIM mode.

In scan mode, a range of ions is selected (e.g., m/z = 15 to 500), and the voltages on the quadrupole are ramped continuously to scan over this predetermined range. In GC/MS analyses, a total ion chromatogram (TIC) is constructed, which is a plot of the sum of the intensity for all ions in a predetermined range as a function of time. The TIC has two dimensions of data, with each point on the TIC containing a distribution of the relative intensity of each ion scanned. For example, Fig. 4 shows a TIC for the jet fuel JP-8. Examining specifically the peak at 4.66 min, Fig. 5 shows the mass spectrum with a molecular ion at m/z = 142 and fragmentation ions of those typically observed for electron ionization MS analysis of n-decane. By comparing the mass spectra across a peak observed in the TIC, the peak purity (resolution of separation) can be assessed. Additionally, an average of the mass spectra collected across a peak is often useful for identifying an unknown analyte. Background ions can also be subtracted to improve compound identification with library searching.

The SIM mode enhances the sensitivity for predefined ions that the analyst selects. This is done by setting the analyzer voltages to scan for only a single ion or a group of ions. The SIM mode results in improved signal to noise because the analyzer can spend more time collecting specific ions and noise is reduced. Selecting an appropriate ion to analyze for with SIM mode requires a priori knowledge of the mass spectrum for that compound. This can be obtained by analyzing a reference standard of the analyte of interest in scan mode. Figure 6 shows the mass spectrum for cannabidiol (CBD), the nonpsychoactive component of cannabis that is of interest due to its potential for medical applications.

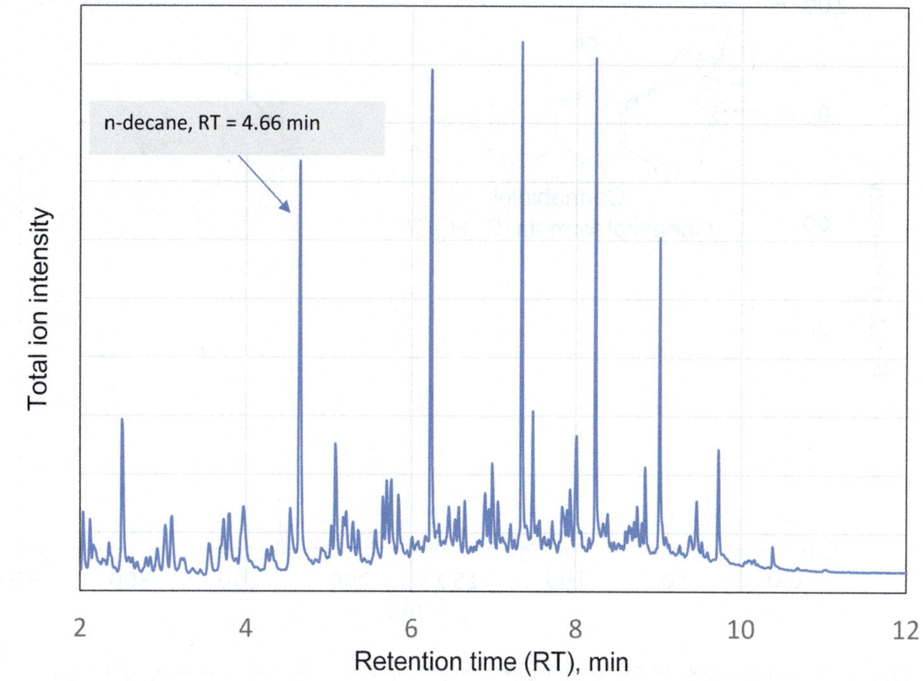

Fig. 4 Total ion chromatogram (plot of total ion intensity versus time) obtained with electron ionization gas chromatography/mass spectrometry for the kerosene-based jet fuel JP-8

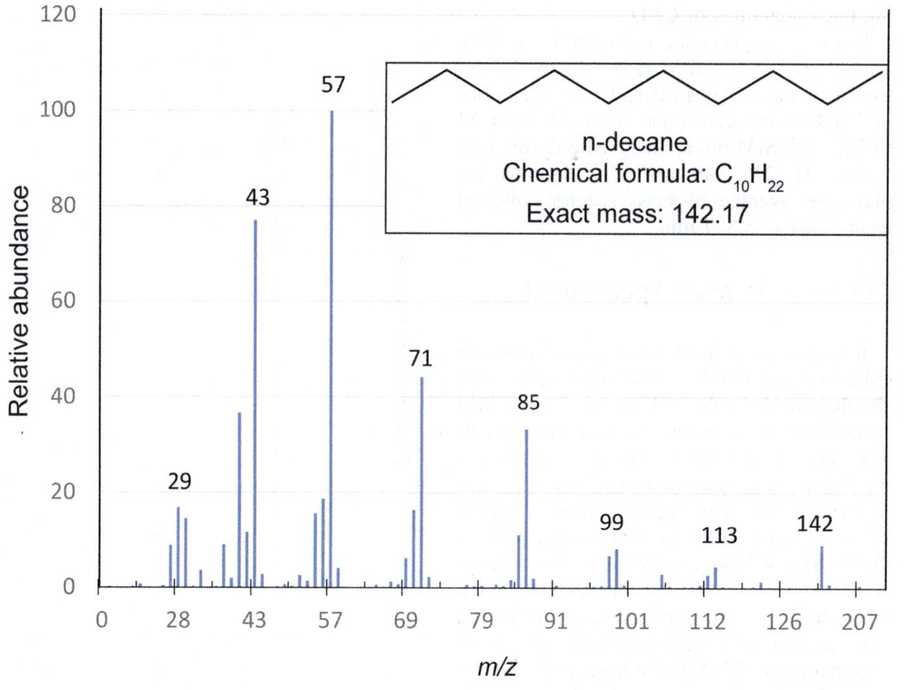

Fig. 5 Electron ionization mass spectrometry spectrum of the peak at 4.66 min from the total ion chromatogram in Fig. 4. This spectrum displays the molecular ion at mass-to-charge ratio m/z = 142 and the fragmentation ions typically observed for n-decane.

Inspection of Fig. 6 shows that the molecular ion for CBD is observed (m/z = 314); additionally, the most abundant ions observed are m/z = 231 and 246. Selecting these three ions for SIM-mode analysis would be a good

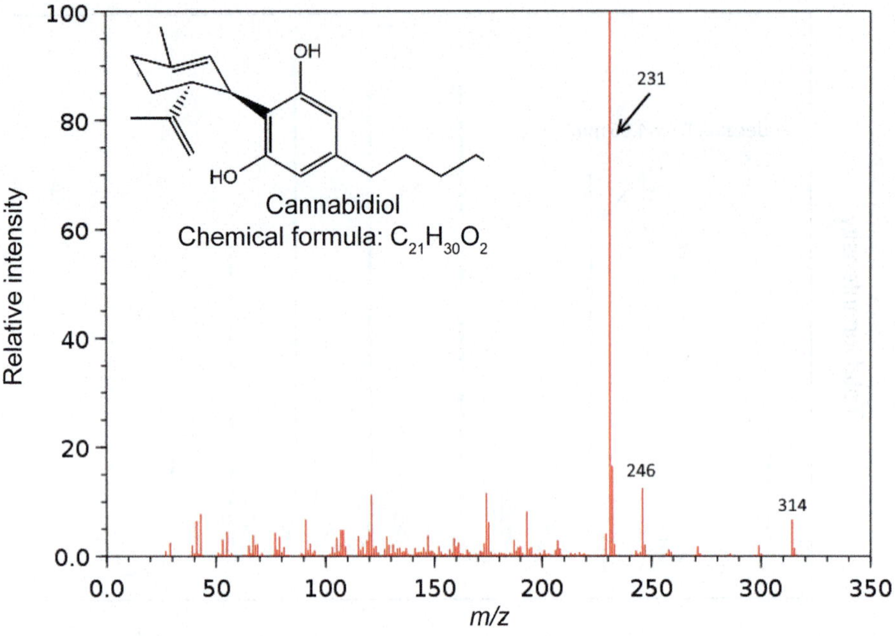

Fig. 6 Electron ionization mass spectrometry spectrum for cannabidiol. The ions (mass-to-charge ratio, *m/z*) are presented as a function of relative intensity normalized to the most abundant at *m/z* = 231. The molecular ion at *m/z* = 314 is observed. Source: Ref 9

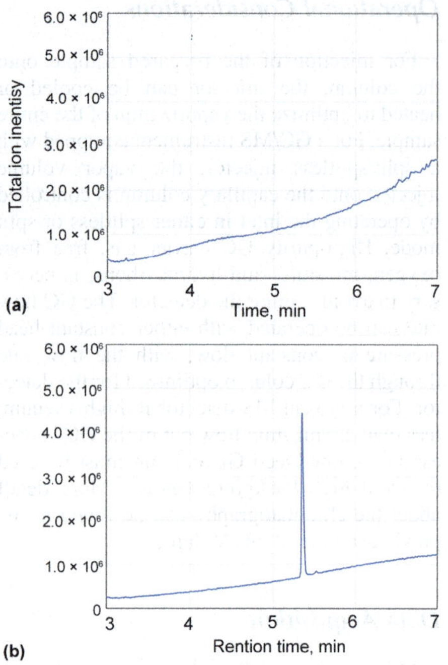

Fig. 7 Total ion chromatograms for the molecule cannabidiol, acquired simultaneously with electron ionization gas chromatography/mass spectrometry operated in (a) scan mode or (b) selected ion-monitoring mode

starting point to develop a method for detecting trace quantities of CBD.

Figure 7 displays the resulting TICs when scan (Fig. 7a) and SIM (Fig. 7b) modes are used simultaneously to detect trace quantities of CBD. Here, scan mode analyzed *m/z* = 33 to 550, and SIM mode analyzed only the ions *m/z* = 231, 246, 299, and 314. Figure 8 displays the average of mass spectra collected from 5.387 to 5.537 min.

Advances in Mass Spectrometry

It is now possible for mass spectrometers to collect data up to *m/z* = 1050 and for data to be simultaneously collected in scan and SIM mode. Recent advances also make unambiguous ion identification possible with high-resolution mass spectrometers, which are able to provide data with accurate mass of 5 ppm or less. Last, tandem mass spectrometry (MS/MS), a technique in which two mass analyzers are separated by a collision chamber, makes it possible to decipher structural information and trace analysis of complex mixtures. For more information on GC/MS, the interested reader is referred to the references herein.

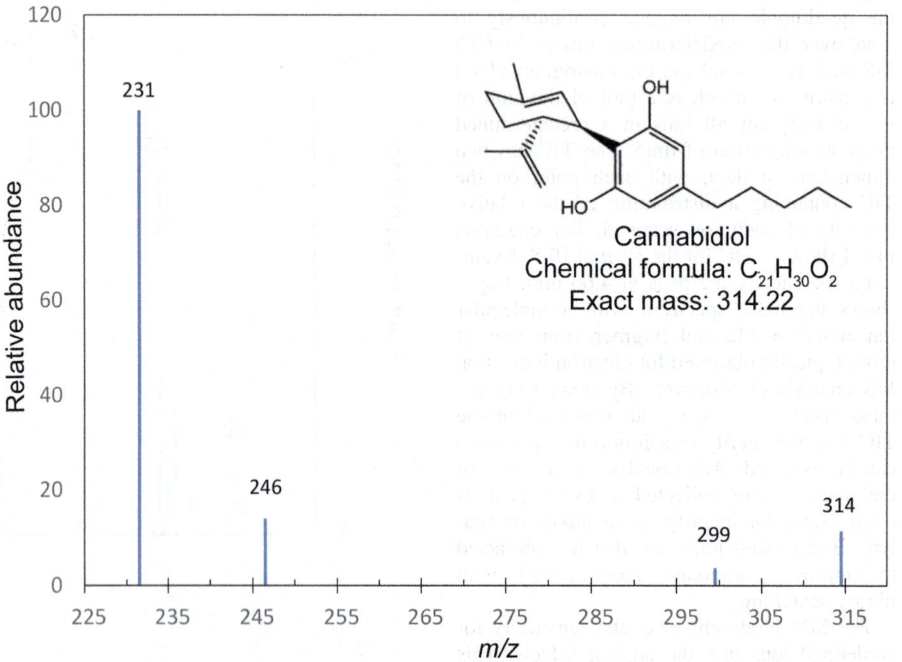

Fig. 8 Average of mass spectra acquired in selected ion-monitoring mode for cannabidiol obtained with electron ionization gas chromatography/mass spectrometry with the ion peaks as a function of relative intensity

Exact Mass Measurements

Low-resolution analyzers (e.g., rounded quadrupoles) are limited to producing data with an integer *m/z* value. This does not allow for unambiguous ion identification. In many cases, mixtures can still be analyzed with the use of known analytical standards or comparisons of mass spectra with compounds found in library databases (Ref 7). Certain analyzers (e.g., TOF; Fourier transform, or FT, orbitrap; magnetic sector; hyperbolic triple quadrupole; and FT ion cyclotron resonance) provide data with accurate mass of 5 ppm or less (Ref 8).

Recent advances in optical instrumentation, high-speed electronic recording, and signal processing have made it possible for TOF analyzers to be a viable addition to the GC/MS system, providing high-resolution detection of compounds without a priori knowledge of chromatographic retention time. The maximum resolving power (full width at half maximum) of the TOF is up to 50,000 ($m/\Delta m$), which equates to a mass accuracy of up to 5 ppm. For example, both perhydrofluorene ($C_{13}H_{22}$) and anthracene ($C_{14}H_{10}$) have the same nominal molecular mass of 178 Da; however, their exact masses differ by 0.094 Da, as shown in Fig. 9. Therefore, these two species could easily be resolved by quadrupole-TOF-MS. Exact mass, however, cannot be used to distinguish between isomers (Ref 10). It is possible that MS/MS could be used to distinguish these isomers. Tandem mass spectrometry is discussed next.

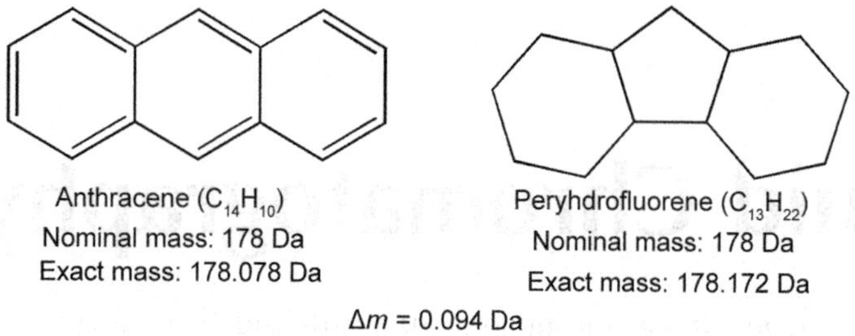

Fig. 9 Example of two hydrocarbons with the same nominal mass but different exact masses that can be distinguished by using the time-of-flight analytical technique

Tandem Mass Spectrometry

In addition to high-resolution mass spectra, structural information and trace analysis of complex mixtures can be deciphered with MS/MS. Tandem MS is a technique in which two mass analyzers are separated by a collision chamber. Tandem MS can be used for determining the exact mass of organic molecules, to determine complex structures (e.g., protein sequencing), or for confirmation of a drug identification (Ref 8).

Tandem MS uses the first mass analyzer to select only the precursor ion of interest (in the example considered earlier, m/z = 178) to enter a separate cell, where it is fragmented by collision-induced dissociation with an inert gas that is used to quench or fragment the parent (or precursor) ion into additional product ions called daughter ions. The product ions are then analyzed in the second mass analyzer, which produces a mass spectrum (Ref 6).

The resulting product ions correlate to a molecule, making it possible to differentiate between isomers and provide an additional identification tool.

The analytical capabilities of high-resolution MS and MS/MS are even further enhanced by the option to perform two-dimensional GC. This technique serves the purpose of separating complex mixtures into families of polarity prior to detection by TOF-MS. Cryogenic modulation is used to improve the separation by taking the output of the first GC column (e.g., a polar material) and injecting it into a column of differing material (nonpolar).

REFERENCES

1. H.M. McNair and J.M. Miller, *Basic Gas Chromatography: Techniques in Analytical Chemistry*, John Wiley & Sons, Inc., New York, NY, 1998
2. F.W. McLafferty, Ed., *Interpretation of Mass Spectra*, 3rd ed., *Organic Chemistry Series*, N.J. Turro, Ed., University Science Books, Mill Valley, CA, 1980
3. D.L. Pavia, G.M. Lampman, and G.S. Kriz, *Introduction to Spectroscopy: A Guide for Students of Organic Chemistry*, 3rd ed., Harcourt College Publishers, Philadelphia, PA, 2001
4. S.E.V. Bramer, *An Introduction to Mass Spectrometry*, Widener University, Chester, PA, 1997
5. L.A. Raphaelian, Gas Chromatography/Mass Spectrometry, *Materials Characterization*, Vol 10, *ASM Handbook*, R.E. Whan, Ed., ASM International, 1986, p 639–648
6. O.D. Sparkman, Z.W. Penton, and F.G. Kitson, Ed., *Gas Chromatography and Mass Spectrometry: A Practical Guide*, 2nd ed., Elsevier, Oxford, U.K., 2011
7. "NIST/EPA/NIH Mass Spectral Database," SRD Program, National Institute of Standards and Technology, Gaithersburg, MD, 2011
8. H.-J. Hubschmann, *Handbook of GC-MS: Fundamentals and Applications*, 3rd ed., Wiley-VCH Verlag GmbH & Co. KGaA, Weinheim, Germany, 2015
9. P.J. Linstrom and W.G. Mallard, Ed., NIST Standard Reference Database Number 69, *NIST Chemistry WebBook*, National Institute of Standards and Technology, Gaithersburg, MD
10. D.P. Joachim and K.I. Kristin, High-Resolution Mass Spectrometry: Basic Principles for Using Exact Mass and Mass Defect for Discovery Analysis of Organic Molecules in Blood, Breath, Urine and Environmental Media, *J. Breath Res.*, Vol 10 (No. 1), 2016, p 012001

ASM Handbook, Volume 10, *Materials Characterization*
ASM Handbook Committee
DOI 10.31399/asm.hb.v10.a0006636

Liquid Chromatography

Kavita M. Jeerage, National Institute of Standards and Technology
Gregory P. Dooley, Colorado State University

Overview

General Uses

- Separation of chemical components within a complex liquid mixture. Purification is possible on analytical, preparative, or industrial scales.
- Identification of chemical components requires infrared spectroscopy, nuclear magnetic resonance spectroscopy, or mass spectrometry.
- Quantitation of chemical components requires ultraviolet-visible, fluorescence, electrochemical, mass spectrometry, or refractive index detectors.

Examples of Applications

- Isolation of synthesized chemicals prior to identification or further synthesis
- Purification of peptides, oligonucleotides, proteins, or small-molecule drugs. Isolation of impurities in pharmaceutical products
- Identification of illegal drugs or quantitation of therapeutic drugs in biofluids
- Isolation of phytochemicals such as terpenes and flavonoids. Quantitation of additives, vitamins, sugars, and lipids in foods and supplements
- Quantitation of fragrance allergens in perfumes and personal-care products. Quantitation of pesticides in water and toxic residues on plant material

Samples

- Samples must be dissolved in pure solvents free of particles and organic impurities.
- Complex sample matrices require solid-phase extraction or liquid-liquid extraction. Dilute samples may require preconcentration.

Limitations

- Cannot identify chemical components unambiguously, other techniques are needed
- Solids must be soluble in an appropriate solvent or solvent mixture; liquids must be miscible with it.

Estimated Analysis Time

- Development and validation of a new method can be a lengthy process.
- Time requirements for a single analysis vary from less than a minute to an hour, depending on chemical components, elution strategy, and instrument capabilities.

Capabilities of Related Techniques

- *Size-exclusion chromatography, gel-permeation chromatography:* Liquid chromatography modes in which analytes are separated by size
- *Ion-exchange chromatography, ion-pair chromatography:* Liquid chromatography modes in which analytes are separated by electric charge
- *Gas chromatography:* Separates chemical components within a complex gas mixture; restricted to volatile and thermally stable analytes

HIGH-PERFORMANCE LIQUID CHROMATOGRAPHY (HPLC) and ultrahigh-performance liquid chromatography (UHPLC) can be used to separate and quantitate the chemical components in any sample that can be dissolved in a liquid. This includes pharmaceutical drugs, medicinal plant extracts, food constituents, flavors, fragrances, industrial chemicals, pesticides, and pollutants. HPLC/UHPLC may be used for identification (e.g., of illicit drugs in urine), for quantitation (e.g., of sugars and sugar alcohols in food), or for purification (e.g., of an active pharmaceutical ingredient). In the original implementation of liquid chromatography, the botanist Mikhail Tswett used glass cylinders packed with powdered chalk and saturated with solvent (Ref 1, 2). He added sample to the top of the column and then poured more solvent into the column (Fig. 1a). As solvent flowed through the column, sample components separated into different bands based on their relative affinities for the chalk (stationary phase) versus the solvent (mobile phase) (Fig. 1b). Individual sample components were collected by physically cutting the chromatogram, named for the colored bands that resulted from the separation of plant pigments such as chlorophylls and carotenoids. Tswett identified key factors driving the quality of the separation, including the need for an inert adsorbent material, the importance of tight packings and fine particles, the influence of the solvent, and the advantages of solvent mixtures; however, his work was largely ignored for several decades.

Modern HPLC/UHPLC separations are still based on differential affinities between the components in the sample, the stationary phase, and the mobile phase. Three characteristics are mainly used to drive a separation: size, electrical charge, and polarity. Separations based on size differences use particles with pores that are tuned to allow some, but not all, dissolved analytes to diffuse into the pores. Therefore, low-molecular-weight analytes are retained more than high-molecular-weight analytes. Size-exclusion chromatography is for biomolecule separations, whereas gel-permeation chromatography is for polymers. In both these modes, the largest molecules elute first. Ion-exchange chromatography is based on differences in electrical charge and can be used for protein separations. To retain anionic analytes, resins with positively charged functional groups are employed as stationary phases. Similarly, cationic analytes are retained by resins with negatively charged functional groups. In both cases, retention is due to the formation of ionic bonds. Ion-pair chromatography is also based on differences in electrical charge.

Separations based on polarity can employ a polar stationary phase and a nonpolar mobile phase (normal phase) or a nonpolar stationary phase and a polar mobile phase (reverse phase). Normal-phase chromatography goes all the way back to the original packed column, which used a polar stationary phase (calcium carbonate) with a nonpolar mobile phase (aliphatic hydrocarbons). Typically, for normal-phase chromatography on silica, the mobile phase is 100% organic solvent; no water is used. Analytes are eluted in order of increasing hydrophilicity. Reverse-phase chromatography is just the opposite—analytes are eluted in order of increasing hydrophobicity.

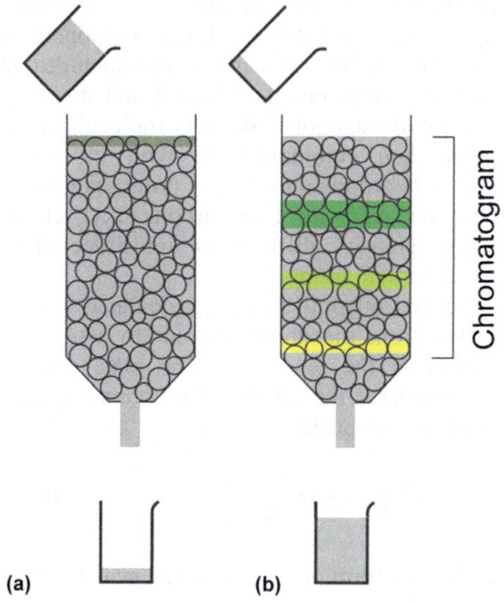

Fig. 1 The original implementation of liquid chromatography employed single-use columns packed with natural powders and gravity-driven flow.

Hydrophilic-interaction chromatography is a variant of normal-phase chromatography that adds a small amount of water (<20%) to the mobile phase to elute polar analytes that are too strongly retained in normal-phase mode. Water actually competes with polar analytes for stationary-phase sites. Hydrophobic-interaction chromatography is a variant of reverse-phase chromatography for large biomolecules (e.g., proteins) that need to stay in aqueous solution with no organic solvents that may denature them. A moderate-polarity stationary phase is used. Under high salt concentrations, the proteins are "salted out" on the column, and then the gradient decreases in salt concentration.

This introduction to liquid chromatography cannot provide detailed information on each of these separation modes or instrument design and detectors. A comprehensive introduction can be found in the book by Snyder, Kirkland, and Dolan that was published in 2010 (Ref 3). This article is intended for researchers and technicians who may want to use liquid chromatography but are not familiar with its terminology or options beyond the conceptual picture in Fig. 1. Readers are introduced to the most commonly employed mode, reverse-phase chromatography, with examples and an exclusive focus on commercially available instruments and consumables. Recent developments reported in the literature are included when the material is relevant to new chromatographers.

Instrument Modules

The HPLC/UHPLC instruments can be configured in different ways to meet user requirements regarding separation efficiency, solvent consumption, detection mode(s), detection sensitivity, flexibility, ease of use, and cost. Understanding instrument capabilities is important when developing and troubleshooting analytical methods. Figure 2 shows the flow path from solvent reservoirs to waste. A single solvent reservoir may contain a premixed mobile phase, but generally, multiple reservoirs are used for pure solvents or buffers that are mixed within the system. Air can be less soluble in a solvent mixture compared to the individual solvents, causing bubbles that interfere with detection. Helium sparging is the gold standard for degassing, but most instruments conveniently include online degassers ahead of the pump (Fig. 2). Such degassers pass each solvent through gas-permeable polymeric tubes that are held under vacuum to pull out dissolved gases. Degassed solvents can then be mixed and pumped. Water, methanol, and acetonitrile are common mobile-phase solvents for reverse-phase chromatography; relevant physical properties for these and other solvents are available (Ref 4).

Pumps must deliver sufficient pressure to drive the mobile phase through the tightly packed column stationary phase. Pumps for HPLC systems deliver up to 40 MPa (equivalent to 400 bar, or 6000 psi), although many separations are designed to operate at lower pressures. Pumps for UHPLC systems deliver more than 40 MPa; the maximum pressure depends on the specific instrument. High-pressure mixing systems use a separate pump for each solvent, and the solvent ratio is controlled by the relative flow rates from each pump. These systems are usually limited to two solvents, because each solvent requires its own pump. Low-pressure mixing systems (Fig. 2) combine the solvents before they reach the (single) pump and are therefore more economical. The solvent ratio is controlled by a valve that opens for a time proportional to the concentration of each solvent; such valves can mix up to four solvents. In low-pressure mixing systems, any flow restrictions will reduce the volume of solvent delivered to the valve, creating proportioning errors.

Autosamplers deliver precise volumes of sample into the mobile phase with a rotary valve that switches between loading into and injecting from a sample loop. In load position,

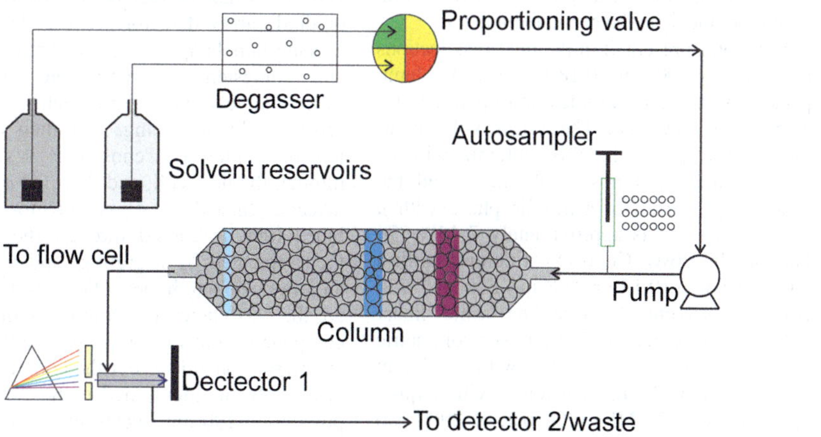

Fig. 2 Modern implementations of liquid chromatography employ reusable columns packed with engineered particles and high-pressure flow.

a mechanically controlled syringe transfers a small volume of sample (e.g., 10 μL) from a vial or well plate to the sample loop. The needle can be automatically rinsed to prevent vial-to-vial contamination. When the valve switches to inject position, the pump pushes liquid from the sample loop into the high-pressure mobile phase ahead of the column. Although most autosamplers allow a variable sample volume to be injected, several factors can contribute to inaccuracy in the injected volume. Therefore, it is best to design methods for constant-volume injections, such that samples and calibration standards are injected with high precision, and any inaccuracy is canceled in the analysis. Note that the maximum sample volume depends on the column dimensions. Furthermore, if the mass of one or more sample components overloads the column, the peaks appear distorted due to saturation of the stationary phase.

The actual separation occurs in the column, which is housed in an oven to provide a stable temperature. Columns are designed for high pressure and to resist chemical attack by the mobile phase within specified pH ranges. Stainless steel is the most common column material. For analytical applications, commercially available packed columns have lengths between 30 and 250 mm (1.2 and 10 in.). Standard analytical columns with a 4.6 mm (0.18 in.) diameter can be packed with fully porous particles from 5 to 3 μm in diameter. These columns are easy to use, do not demand high pressures, and are suitable for quantifying major and trace components. Equation 1 describes the pressure drop during laminar flow through a packed column (Darcy's law) as a function of column and experimental parameters:

$$\Delta P = \phi\mu\frac{4FL}{\pi d_c^2 d_p^2} \quad \text{(Eq 1)}$$

In this equation, Φ is a dimensionless parameter that characterizes resistance to flow ($\Phi \approx 900$), μ is the mobile-phase viscosity, F is the flow rate, L is the column length, d_c is the column diameter, and d_p is the particle diameter.

Examples presented in the next two sections use a 4.6 by 100 mm (0.18 by 4.0 in.) column packed with 5 μm particles and operated at a 2 mL/min flow rate. The linear velocity of the mobile phase in this column will be approximately 2 mm/s (0.08 in./s), and the required pressure (for a mobile phase with $\mu \approx 0.001$ Pa · s) is approximately 7 MPa (70 bar, or 1000 psi). Choosing a column with a 3.0 or 2.1 mm (0.12 or 0.08 in.) diameter will conserve solvent, because the same linear velocity can be achieved at a lower volumetric flow rate. Choosing a column with a 2.1 mm diameter and 3 μm particles will require approximately 20 MPa (200 bar, or 2900 psi) to achieve the same linear velocity, which is within the capability of a conventional system. Fully porous particles with diameters larger than 2 μm push the limits of conventional instruments; diameters smaller than 2 μm require high-pressure instruments. However, narrow columns and/or smaller particles place additional demands on the chromatographer and the instrument, as discussed later.

HPLC/UHPLC is a separation technique; identifying or quantifying the (ideally) well-resolved sample components requires one or more detectors. Detectors are positioned directly after the column to minimize spreading of the analyte bands. The mobile phase passes through the detector flow cell and from there continues to a waste container, fraction collector, or a second detector. Sample-specific detectors require that the analytes of interest have a specific characteristic, for example, absorption of ultraviolet or visible (UV-VIS) light, which is likely for analytes containing unsaturated bonds, aromatic rings, or heteroatoms. Other sample-specific detectors are based on the ability of an analyte to fluoresce, conduct electricity, or participate in oxidation or reduction reactions. Refractive index (RI) detectors are universal detectors. These detectors compare the refraction of visible light by the mobile phase plus analyte to the refraction by the mobile phase alone. The RI detectors are extremely sensitive to changes in temperature and mobile-phase composition and cannot be used with online mixing systems. Mass spectrometry (MS) detectors require that the analyte is ionizable at atmospheric pressure. The MS detectors provide structural information and are frequently employed for identification and quantitation of structurally similar molecules. Liquid chromatography-mass spectrometry (LC-MS) is commonly applied to the analysis of natural products, pharmaceuticals, proteins and peptides, and clinical and forensic toxicology. PubMed searches indicate that LC-MS publications have doubled in the last decade.

Even when MS detection is implemented, UV-VIS detection is very commonly included as a first detector and is described in more detail. The UV-VIS detectors can be configured as variable-wavelength detectors (Fig. 2) or diode array detectors. These detectors use a deuterium lamp for ultraviolet wavelengths and a tungsten lamp for visible wavelengths. In a variable-wavelength detector, the light is separated by a grating, and then the wavelength of interest is chosen by a slit, passed through the flow cell, and detected by a photodetector. In a diode array detector, the combined light is passed through the flow cell, then separated by a grating, and detected by an array of photodiodes. Diode array detectors are the best choice for samples with unknown components, but variable-wavelength detectors are more economical. For the common reverse-phase solvents mentioned earlier, detection is possible to 190 nm (water and acetonitrile) or 205 nm (methanol); below these wavelengths, the solvent also absorbs light (Ref 4). Traditional flow cells (e.g., volume = 10 μL; path length = 1 cm, or 0.4 in.) provide high sensitivity due to the long path length and are used for standard analytical columns. For narrow columns, a large flow cell volume leads to mixing of chromatographically resolved components. It is not possible to maintain the path length while reducing the cell volume, because light scattering increases as the diameter of the flow cell decreases (because of imperfectly collimated beams). Therefore, with a traditional design, a smaller flow cell volume means a shorter path length (e.g., volume = 2 μL; path length = 0.3 mm, or 0.01 in.), so sensitivity decreases. Newer light pipe designs permit flow cell volumes smaller than 0.5 μL while maintaining long path lengths. The UV-VIS quantitation is based on the well-known relationship between absorbance and concentration (Beer-Lambert law) and is not discussed further.

Assessing a Separation

In the simplest implementation of reverse-phase chromatography, the column is eluted with a mobile phase of fixed composition (isocratic elution). As the injected sample travels through the column, the sample components interact with the stationary phase. Analytes that are hydrophilic tend to remain in the mobile phase and therefore travel through the column rapidly. Analytes that are hydrophobic tend to be retained by the stationary phase and therefore travel through the column slowly. If the column is sufficiently long, this process leads to the separation (in space) of the sample components. Separation in space leads to separation in time as each analyte reaches the detector and is detected in the chromatogram. Figure 3 shows several chromatograms for a three-component mixture of eugenol, cuminaldehyde, and butylbenzene. Each component in this mixture has ten carbons and includes an aromatic ring, but the alcohol and methoxy groups of eugenol make this molecule more polar than cuminaldehyde with its single aldehyde group and far more polar than butylbenzene with its four-carbon alkyl chain. These polarity differences are reflected in the chromatograms.

Several quantities can be calculated or estimated from the chromatogram to assess the quality of the separation. The plate number (N) indicates the ability of a column to produce narrow peaks:

$$N = 16\left(\frac{t_R}{W}\right)^2 = 5.54\left(\frac{t_R}{W_{1/2}}\right)^2 \quad \text{(Eq 2)}$$

In the first form of Eq 2, W is the peak width at base, which can be measured with tangent lines. In the second form of this equation, $W_{1/2}$ is the peak width at half-height, which is often easier to measure from the chromatogram and may be measured automatically by chromatography

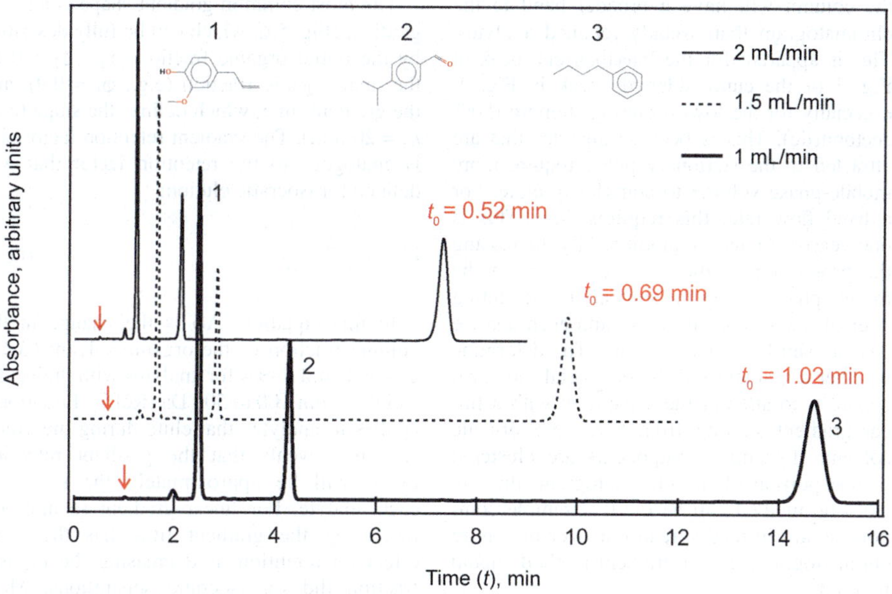

Fig. 3 Chromatograms for the separation of eugenol (1), cuminaldehyde (2), and butylbenzene (3) by a C_{18} stationary phase with fully porous particles and a mobile phase consisting of 60% acetonitrile/40% water. Column conditions: particle diameter, 5 μm; column diameter, 4.6 mm (0.18 in.); column length, 100 mm (4.0 in.); temperature, 30 °C (85 °F). Detection at 200 nm. Chemical structures are adapted from Ref 5.

software. For both equations, t_R is the retention time. Columns with larger plate numbers are more efficient than columns with smaller plate numbers, because the more efficient column will produce a narrower peak at any given retention time. For the cuminaldehyde peak in Fig. 3, N is approximately 8500 when F = 2 mL/min and 10,000 when F = 1 mL/min. If the greater column efficiency achieved at 1 mL/min is not needed, the slightly lower column efficiency achieved at 2 mL/min cuts the run time in half. Plate number is a function of plate height (H), which is a measure of efficiency per unit length of column ($N = L/H$). Column efficiency can be increased by increasing the column length, because H is constant for a given particle diameter and flow rate. Of course, increasing the column length increases the run time and the required pressure, which is apparent from Eq 1.

The retention factor (k) is an important characteristic of each sample component. It is defined as the quantity of an analyte in the stationary phase divided by its quantity in the mobile phase. An analyte must be in either the stationary phase or the mobile phase, so if the fraction of analyte A in the mobile phase is R_A, then the fraction in the stationary phase is $1 - R_A$, meaning $k_A = (1 - R_A)/R_A$. For the analyte of interest, the retention time is defined by the column length divided by the band velocity: $t_A = L/u_A$. Dead time (t_0) is the time required for an unretained sample component to reach the detector: $t_0 = L/u$. Combining these equations with $R_A = u_A/u$ gives:

$$k_A = \frac{t_A - t_0}{t_0} \quad \text{(Eq 3)}$$

Equation 3 can be used to estimate the retention factor for each peak in the chromatogram once the dead time is determined. If the sample solvent is different than the mobile phase, the unretained solvent may cause a small baseline disturbance in the chromatogram due to a change in refractive index. For the chromatograms in Fig. 3, the sample components were dissolved in methanol, which is detectable at 200 nm; therefore, small solvent peaks are visible in the chromatogram, and the dead times are indicated by red arrows. For situations where the sample is dissolved in the mobile phase, an unretained component such as thiourea or uracil can be added to the sample. The dead time can also be estimated by dividing the column volume by the flow rate, incorporating the fact that a portion of the column volume is occupied by the particles. For fully porous particles, this estimate is:

$$t_0 \sim 0.5\frac{L d_c^2}{F} \quad \text{(Eq 4)}$$

Values estimated by Eq 4 differed by up to 10% from measured values for several hundred reverse-phase columns (Ref 6).

For the chromatograms in Fig. 3, the dead time is estimated to be 0.53 min (for 2 mL/min), 0.71 min (for 1.5 mL/min), or 1.1 min (for 1 mL/min) by Eq 4, within 10% of measured values. Although methanol was used here to generate a solvent peak, this is not a good practice in general, because analytes dissolved in a different solvent spend a portion of the column length with a mobile phase that does not have the intended composition. If the sample is dissolved in a weaker solvent (e.g., 100% water), then analytes will remain in the column inlet. Solvent focusing is sometimes used intentionally to concentrate analytes. If the sample is dissolved in a stronger solvent (e.g., 100% acetonitrile), then analytes rapidly move through the column, preventing interaction with the stationary phase and defeating the purpose of chromatography. For the chromatograms in Fig. 3, the retention factors are approximately 1.4 for eugenol, 3.0 for cuminaldehyde, and 12.8 for butylbenzene for the separation at 2 mL/min. Selectivity (α) represents the time or distance between peak maxima ($\alpha = k_B/k_A$), where A is the less-retained component. The critical pair of analytes are the ones that are most difficult to separate. If $\alpha = 1$, the retention times are identical, meaning the two analytes co-elute and there is no separation, whereas $\alpha >> 1$ indicates a good separation. For the chromatograms in Fig. 3, $\alpha = 2.2$ for eugenol/cuminaldehyde for the separation at 2 mL/min, but the critical pair is actually eugenol plus an impurity, for which $\alpha = 1.5$.

The resolution (R_s) quantifies the ability to separate (resolve) peaks of interest. For two analytes, A and B, resolution is the difference in retention times divided by the average peak width. For the chromatograms in Fig. 3, all the peaks are well resolved with $R_s > 3$. Resolution can be calculated by:

$$R_s = \frac{t_B - t_A}{0.5(W_A + W_B)} = 1.18\frac{t_B - t_A}{(W_{1/2,A} + W_{1/2,B})} \quad \text{(Eq 5)}$$

For peaks of approximately equal size, $R_s = 0.6$ is required to discern a valley, and $R_s = 1$ is considered the minimum value for a measurable separation. $R_s = 1.6$ indicates baseline separation; however, $R_s > 2$ is desirable because the safety factor accounts for minor peak tailing and allows for peaks of dissimilar size. For example, for preparative separations, baseline resolution with an additional safety factor allows each analyte to be 100% recovered with 100% purity. Although Eq 5 defines resolution, an approximate expression can be useful for method development. This expression assumes that the two peaks have equal widths:

$$R_s = \frac{1}{4}\left(\frac{k}{k+1}\right)(\alpha - 1)N^{0.5} \quad \text{(Eq 6)}$$

In this equation, k is the retention factor of the first peak (defined by Eq 3), α is the selectivity, and N is the plate number (defined by Eq 2). Equation 6 shows that while resolution can be improved by increasing the retention factor, by increasing the selectivity, or by increasing the plate number, improving the selectivity is the most powerful approach.

Adjusting the Mobile Phase

When an initial separation does not yield well-separated peaks in the chromatogram, several strategies may be employed to improve the selectivity or relative retention of the sample components. One simple approach is to decrease the solvent strength or the percentage of the mobile phase that is organic. Retention varies with mobile-phase composition according to:

$$\log k = \log k_w - S\varphi \qquad \text{(Eq 7)}$$

In this equation k_w is the extrapolated value of k for a hypothetical separation with water as the mobile phase, S is a constant that depends on the analyte, and ϕ is the volume fraction of the organic solvent. For a high percentage of organic solvent (high eluting strength), the analytes travel through the column rapidly, and some analytes may not be fully resolved. For a low percentage of organic solvent (low eluting strength), the analytes spend more time in the column, and, ideally, all analytes will be fully resolved. An intermediate mobile-phase composition that keeps $1 < k < 10$ for all analytes is ideal. To achieve this, one procedure is to attempt the separation with 80% organic solvent and then reduce the fraction of organic solvent in 10% increments. For analytes that form a homologous series (e.g., linear alkylbenzenes), the resolution of all peaks will improve, but their relative spacing will stay the same. For analytes with greater molecular diversity, their relative spacing may change, meaning that peak order may change.

Figure 4 shows several chromatograms for a three-component mixture containing eugenol and cuminaldehyde, but with linalool instead of butylbenzene. For the strongest mobile phase, 80% acetonitrile, linalool and cuminaldehyde are not resolved, but each reduction in the eluting strength improves their resolution. Choosing a mobile-phase composition between 60 and 70% acetonitrile will yield a good separation of all sample components, including the impurity that elutes ahead of eugenol. When changes in the fraction of organic solvent cannot adequately resolve all peaks, another strategy is to change the organic solvent. Methanol has less eluting power than acetonitrile, so when changing between these solvents, eluting strength should be considered. For example, a mobile phase with 40% acetonitrile has similar eluting strength to a mobile phase with 50% methanol. Mobile-phase mixtures of equivalent strength are available (Ref 7). If optimized concentrations of water/acetonitrile or water/methanol do not yield adequate resolution, water, acetonitrile, and methanol can be combined to fine-tune the mobile phase.

For isocratic elution, the mobile phase remains constant in composition and therefore eluting strength. What this means in practice is that analytes that are strongly retained by the column will have a broader band in the chromatogram than weakly retained analytes. This is apparent for the butylbenzene peak in Fig. 3 or the cuminaldehyde peak in Fig. 4, especially for the lowest eluting strength (50% acetonitrile). This is because analytes that are attracted to the stationary phase require more mobile-phase volume to completely elute. For a fixed flow rate, this requires time. This is one reason to use a gradient. By increasing the proportion of the organic solvent in the mobile phase during the separation, the eluting strength increases with time, and each analyte has a similar peak width. To determine whether a gradient will be beneficial, one procedure is to attempt the separation with a linear gradient starting from 5 to 95% organic solvent. If sample components are clustered in one portion of the chromatogram, then an isocratic method can be used. If sample components are instead spread out over the entire chromatogram, then a gradient method should be used.

The most common gradient shape is a linear gradient (Fig. 5a), which can be fully described by the initial organic fraction (e.g., $\phi_i = 0.1$), the final organic fraction (e.g., $\phi_f = 0.9$), and the gradient time, which defines the slope (e.g., $t_G = 20$ min). The gradient retention factor (k^*) is analogous to the retention factor that was defined for isocratic elution:

$$k^* = \frac{t_G}{1.15 t_0 \Delta\varphi S} \qquad \text{(Eq 8)}$$

In this equation, $\Delta\phi$ is the change in the volume fraction of the organic solvent ($\Delta\phi = \phi_f - \phi_i$), and $S \approx 4$ for analytes with molecular weights from 100 to 500 Da (Ref 7). Equation 8 applies to analytes that elute during the gradient and reveals that the gradient retention factor will be approximately the same for each analyte. For linear gradient separations, increasing the gradient time has the same effect on retention as decreasing the organic fraction did for isocratic separations. More

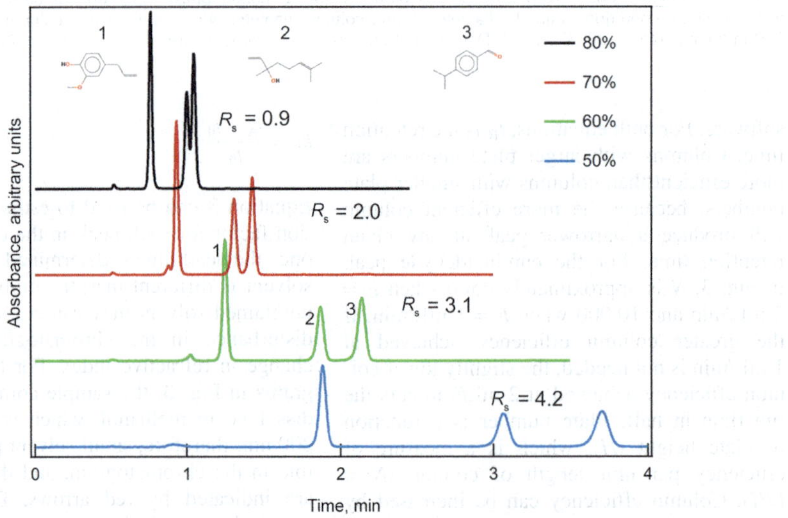

Fig. 4 Chromatograms for the separation of eugenol (1), linalool (2), and cuminaldehyde (3) by a C_{18} stationary phase with fully porous particles and mobile-phase compositions ranging from 80 to 50% acetonitrile. Column conditions: particle diameter, 5 μm; column diameter, 4.6 mm (0.18 in.); column length, 100 mm (4.0 in.); flow rate, 2 mL/min; temperature, 30 °C (85 °F). Detection at 200 nm. Chemical structures are adapted from Ref 5.

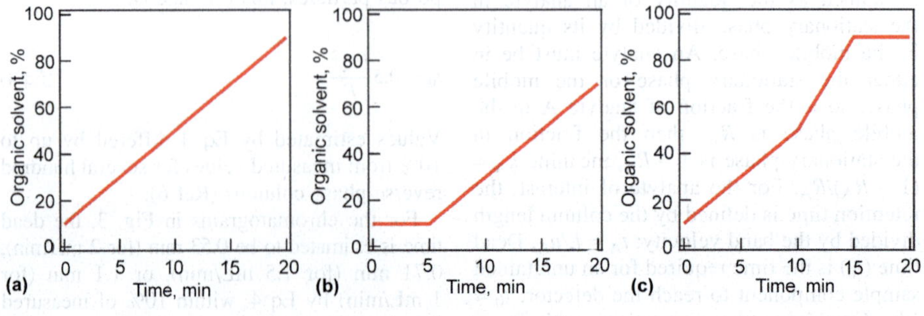

Fig. 5 (a) Linear gradient profile, (b) gradient delay profile, and (c) segmented gradient profile adjust eluting strength for different sample components.

specifically, increasing the gradient time increases the gradient retention factor, spreading out the peaks and improving resolution. If a gradient time can be found that adequately resolves each peak, the initial and final organic fractions (ϕ_i and ϕ_f) can be adjusted to eliminate blank space in the chromatogram. The gradient time should be simultaneously reduced to maintain the same gradient retention time (Eq 8) or equivalently the same slope (Fig. 5a). If early peaks are not well resolved, perhaps because of impurities, one can incorporate an isocratic hold prior to the start of the gradient (Fig. 5b). This gradient delay will spread out peaks at the beginning of the chromatogram, helping to separate sample impurities from the components of interest. Improvements in the resolution of early peaks can also be achieved by decreasing ϕ_i if the stationary phase can tolerate a nearly pure aqueous mobile phase. Gradients that incorporate an isocratic hold at the end (Fig. 5c) will flush highly retained components from the column.

Gradient methods improve separation speed in addition to peak quality. However, gradient methods can still require substantial analysis time, because additional time is required to re-equilibrate the column back to the initial mobile-phase composition. Schellinger et al. demonstrated that run-to-run repeatability can be achieved after re-equilibration with 2 column volumes, whereas full equilibration may require 20 column volumes or more for a standard analytical column, depending on mobile-phase choices (Ref 8). The authors provided quantitative evidence that incorporating a small fraction of propanol or butanol in the mobile phase dramatically reduces the number of column volumes required for full equilibration, because these alcohols effectively wet the stationary phase. They suggest designing gradients that increase in eluting strength from 1% butanol/89% water/10% acetonitrile to a maximum of 1% butanol/99% acetonitrile.

Choosing the Stationary Phase

Up to this point, the column has been treated as a black box that retains hydrophobic sample components more than hydrophilic components. Analytical columns are filled with rigid, fully porous or superficially porous support structures that permit mobile-phase flow through the column and surfaces for analyte retention. Early packed columns contained large (10 μm), fully porous silica particles with irregular shape and wide size distribution. By improving both the sphericity of the particles and their size distribution, packed columns were created with higher efficiency that required lower pressure to operate. Improvements in the purity of silica minimized undesirable and poorly understood interactions with sample components. Finally, the diameter of fully porous particles steadily decreased, creating columns with higher efficiency that also required higher pressure to operate. One recent innovation is the development of superficially porous particles, which are discussed later. The silica support structure influences operating parameters and instrument requirements, while stationary-phase chemistry determines analyte retention. Modifying the stationary-phase chemistry requires more effort and expense than modifying the mobile-phase composition but is the most powerful way to modify selectivity.

The stationary phase provides surfaces where chemical species adsorb or partition from the mobile phase, and its differential selectivity toward sample components is the basis for achieving separation as the sample flows through the column. In early implementations of liquid chromatography that used organic mobile phases which are nonpolar, the stationary phase was simply the surface of the porous silica support, which is hydrophilic and possesses polar silanol (Si–OH) groups analogous to alcohol (C–OH) groups. This stationary phase interacts strongly with polar analytes and weakly with nonpolar analytes, such that nonpolar analytes elute ahead of polar analytes. The polarity of the silica surface can be modified by chemically bonding it to less polar or nonpolar functional groups (bonded phase). For example, C_8, C_{18}, and C_{30} dimethylchlorosilane ligands form bonded silica phases with increasing hydrophobicity. These stationary phases interact strongly with nonpolar analytes and weakly with polar analytes (which tend to remain in the polar water/methanol or water/acetonitrile mobile phases), such that polar analytes elute ahead of nonpolar analytes. Normal-phase chromatography works well for analytes that are not water soluble. However, any water or protic organic solvents (hydrogen bound to oxygen or nitrogen) in the mobile phase changes the hydration state of silica, causing poor reproducibility of retention times. For this reason, reverse-phase chromatography has come to dominate separations based primarily on polarity.

In the manufacture of a monomeric bonded phase, a single silanol group reacts with a single dimethylchlorosilane ligand (Fig. 6a). An unbonded silica has a surface silanol concentration of approximately 8 μmol/m^2 (Ref 9). The fraction of sites that react depends on the ligand bulkiness but is not greater than 50%. As a result, most commercial columns are

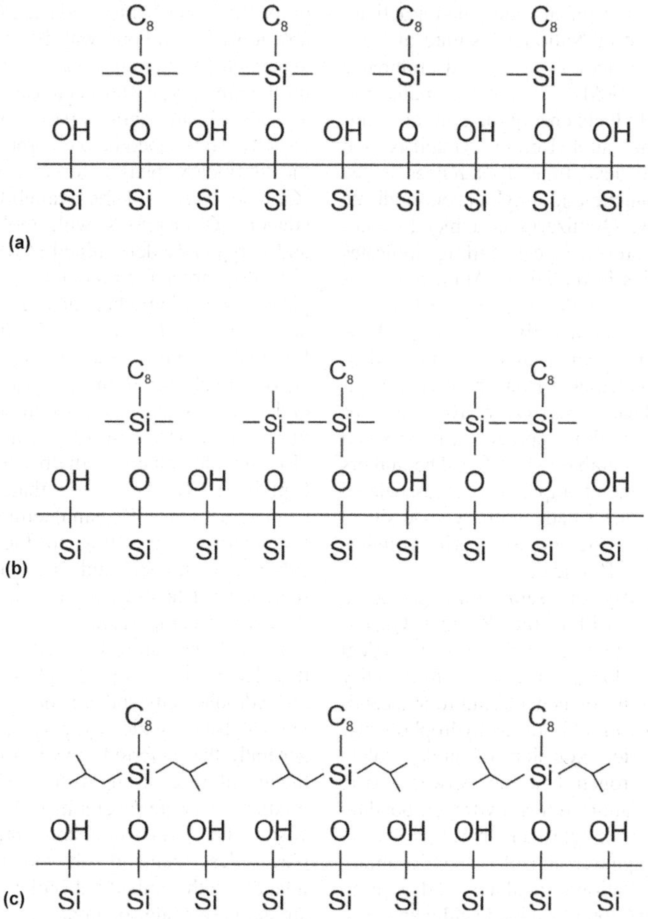

Fig. 6 Monomeric C_8-bonded phases (a, b, c) can be end-capped (b) or sterically protected (c).

end-capped with a small ligand such as trimethylchlorosilane (Fig. 6b), which increases the fraction of reacted silanol groups. Sterically protected bonded phases substitute bulkier isopropyl or isobutyl groups for the methyl groups (Fig. 6c) to protect the Si–O bond from hydrolysis in acidic environments. The bulkiness of this ligand also reduces the fraction of reacted silanol groups. Alkyl ligands available in commercial columns include C_1, C_3, C_8, C_{18}, and C_{30}. Difunctional (or trifunctional) silanes react with two (or three) silanol groups to form cross-linked or polymeric bonded phases that are more rigid than monomeric phases. Polymeric phases have a smaller fraction of unreacted silanol groups and are more stable in acidic environments.

It may seem counterintuitive, but the C_{18} phase from one manufacturer is unlikely to be interchangeable with the C_{18} phase from another manufacturer, meaning their selectivity will be different. In response to the proliferation of C_{18}-bonded phases (over 200 were commercially available in 2005), the United States Pharmacopeia, which sets standards for chemical and biological drugs including chromatographic procedures, created a working group to identify traits that distinguish C_{18} phases (Ref 10). This group developed a procedure to quantify hydrophobicity, the presence of interfering metals, and silanol activity by measurement of National Institute of Standards and Technology standard reference material (NIST SRM) 870, which contains uracil, toluene, ethylbenzene, quinizarin, and amitriptyline in methanol (Ref 11). Uracil is used to indicate the dead time. Differences in the retention of toluene and ethylbenzene indicate hydrophobicity. Quinizarin is a metal-chelating agent; therefore, peak tailing indicates metal impurities in the silica. Amitriptyline is basic; therefore, for this analyte, peak tailing indicates high silanol activity. Finally, shape selectivity is inferred from the bonding density, which is calculated from the carbon loading and surface area. Over 100 columns characterized by this approach are provided in a searchable database (Ref 12). The authors point out that metal impurities are associated with old columns (made with type-A silica) and are not an issue for high-purity columns (made with type-B silica).

Within roughly the same time period, a working group within the Product Quality Research Institute Drug Substance Technical Committee developed a procedure to quantify column selectivity by considering four secondary interactions in addition to hydrophobicity. These are steric exclusion of bulky solute molecules, hydrogen bonding between basic analytes and silanol groups, hydrogen bonding between acidic analytes and basic groups in the stationary phase, and electrostatic attraction between cationic analytes and ionized silanols or repulsion of ionized acidic analytes. The hydrophobic subtraction model, developed in a series of publications in the early 2000s (Ref 13), defines stationary-phase selectivity by the sum of these contributions:

$$\log\alpha = \log\frac{k}{k_{EB}} = (\eta'\boldsymbol{H}) - (\sigma'\boldsymbol{S}^*) + (\beta'\boldsymbol{A}) + (\alpha'\boldsymbol{B}) + (\kappa'\boldsymbol{C}) \quad \text{(Eq 9)}$$

In this equation, α is the selectivity for an analyte relative to ethylbenzene. The analyte is described by its hydrophobicity (η'), bulkiness (σ'), hydrogen bond basicity (β'), hydrogen bond acidity (α'), and effective ionic charge (κ'). The remaining quantities are properties of the stationary phase, and each term in Eq 9 is the product of a solute parameter and a stationary-phase parameter.

Hydrophobicity (***H***) increases when the ligand is longer (e.g., C_{30} versus C_{18} versus C_8) or at a higher concentration on the silica surface. ***H*** increases slightly when the stationary phase is end-capped, because this process decreases the fraction of unreacted silanol groups. Steric resistance (***S****) describes the difficulty that bulky analytes experience when interacting with stationary-phase ligands; therefore, this term is negative. ***S**** increases with ligand length and ligand concentration and is not the same as shape selectivity (discussed later). Hydrogen bond acidity (***A***) depends on the fraction of unreacted silanol groups available to hydrogen bond with basic analytes, such as unprotonated amines and amides. ***A*** depends on the silica type and decreases when stationary phases are end-capped. Hydrogen bond basicity (***B***) depends on sorbed water from the stationary phase. Ion-exchange capacity (***C***) is a measure of the availability of ionized silanols. ***C*** increases with mobile-phase pH and is typically determined at pH 2.8 and 7.0. ***C*** is also larger for non-end-capped stationary phases. For a hypothetical C_{18} column, $\boldsymbol{H} = 1$ and $\boldsymbol{S}^* = \boldsymbol{A} = \boldsymbol{B} = \boldsymbol{C} = 0$. To find the values for real columns, the hydrophobicity term ($\eta'\boldsymbol{H}$) is subtracted from log α, then the secondary contributions are simultaneously determined for a series of 18 test analytes (which also include toluene, ethylbenzene, and amitriptyline) (Ref 14). More than 700 columns from more than 30 manufacturers have been characterized according to the hydrophobic subtraction model and are provided in a searchable database (Ref 15), along with tools for column comparison.

For the separation of small molecules (less than 1000 Da), C_8 and C_{18} phases have moderate hydrophobicity and are commonly employed. For alkylsilica stationary phases, acetonitrile is generally the preferred organic solvent. It has a higher eluting strength than methanol, and mixtures of water/acetonitrile have lower viscosity than water/methanol mixtures. (This places less demand on the pump.) Alkyl ligands with embedded polar groups (e.g., amide, carbamate, or urea) have lower hydrophobicity and higher hydrogen bond basicity; the polar group also suppresses silanol activity (Ref 13). Alkyl chains terminated with polar groups are compatible with highly aqueous mobile phases, but their selectivity parameters do not differ greatly from equivalent stationary phases without the polar group (Ref 13). Phenyl phases (e.g., phenyl-propyl, phenyl-hexyl, and diphenyl) contain a spacer chain that binds to the silica surface and aromatic ring (s). Having an aromatic ring in the stationary phase causes greater retention of analytes with aromatic rings due to π–π interactions. For these stationary phases, methanol is the preferred organic solvent, because acetonitrile contains π electrons that can interact with the aromatic analyte or the aromatic stationary phase, reducing the impact of the aromatic groups in the stationary phase (Ref 16). Cyano-propyl phases (propyl chain terminated by a nitrile) retain analytes by π–π and dipole–dipole interactions.

The following examples illustrate how the choice of stationary phase can tune selectivity. During opioid replacement therapy, simultaneously determining the plasma levels of multiple long-lasting opioid drugs may be required. Employing a C_8 phase, Somaini et al. attempted to separate methadone (an opioid receptor agonist), buprenorphine (a partial agonist), norbuprenorphine (an active metabolite), and naloxone (an opioid antagonist) by modifying the fractions of phosphate buffer, acetonitrile, and methanol in isocratic mobile phases (Ref 17). The C_8 phase did not retain naloxone, which is more hydrophilic than the other opioids, and it eluted too close to components in the plasma matrix. Switching to a cyano-propyl phase provided greater retention of naloxone and allowed elution of the other opioids in a reasonable time. In another example, Ali et al. needed to quantitate curcumin, demethoxycurcumin, and bis-demethoxycurcumin, each of which contain two substituted phenyl rings, to evaluate the medicinal properties of turmeric rhizomes. Although curcuminoid pigments can be separated by C_{18} phases, a phenyl phase provides π–π interactions and resulted in robust separations with lower detection limits (Ref 18). In a final example, a completely different stationary phase (not alkylsilica) was employed. Terpenes and terpenoids are major components of essential oils, which are used as flavors, fragrances, and in traditional medicine. Stevenson and Guiochon examined whether a porous graphitized-carbon (PGC) stationary phase, which contains delocalized π electrons, could resolve closely related terpenes/terpenoids (Ref 19). The PGC has a rigid planar surface that retains large planar molecules, with reduced retention of highly branched molecules due to steric hindrance that limits analyte-surface contact. The PGC resolved isomers such as α-pinene, β-pinene, camphene, and d-limonene ($C_{10}H_{16}$) with similar physical structures (six-carbon rings, one or two π bonds); C_{18} phase did not.

Shape selectivity describes the ability of stationary phases to separate sample components based on their molecular structure. This trait is relevant to molecules with rigid structures, such as polycyclic aromatic hydrocarbons, polycyclic aromatic sulfur heterocycles, polychlorinated biphenyls, steroids, and carotenoids. Isomer separations are usually motivated by different biological activities, whether toxic or beneficial. Shape selectivity increases with bonding density and alkyl chain length; monomeric and polymeric C_{30} phases are shape selective, whereas C_8 phases are not. For example, a polymeric C_{30} phase was used to resolve four tocopherol isomers with different antioxidant capacities (Ref 20). For intermediate alkyl chain lengths, the polymeric phase has greater shape selectivity than the corresponding monomeric phase. NIST SRM 869a contains three polycyclic aromatic hydrocarbons whose elution order changes with the degree of shape selectivity (Ref 21); the relative selectivity for two of the components identifies stationary phases with high or low shape selectivity (Ref 22). In contrast to other contributors to selectivity, shape selectivity is only weakly influenced by mobile-phase composition but is enhanced at low temperatures.

Optimizing a Separation

If a separation will be repeated hundreds of times, then after the mobile-phase composition and stationary-phase chemistry are identified, column parameters should be optimized. Optimization can mean achieving a specific plate number to resolve the critical analytes in a separation (time is not optimized) or maximizing the plate number achieved in a specific amount of time. Essentially, the compromise is between efficiency (plate number) and time. Mobile-phase velocity, column length, particle size, and temperature can be adjusted to increase efficiency or decrease time. Each of these choices has different implications in terms of instrument demands (Ref 23). Many improvements are linked to the specific structure of the silica support material in the column. Small, fully porous particles and superficially porous particles are described, because these support materials are available in commercial columns. A brief introduction to optimization is provided by discussing the parameters identified previously. This first requires a discussion of factors that contribute to peak broadening and therefore influence column efficiency.

Sample injection results in a narrow band of molecules. This article has already described how analytes separate into individual bands based on their differing affinities for the stationary phase. Simultaneously, there are multiple processes that cause each analyte band to spread as it flows through the column in the mobile phase. As a reminder, the plate height (H) describes the efficiency of a column per unit length. Columns with small plate heights have more resolving power per unit length than columns with large plate heights. The plate height or the reduced plate height (h), which normalizes plate height by the particle diameter ($h = H/d_p$), is influenced by the design of the silica support material and by operating parameters. The reduced velocity ($\nu = u_e d_p / D_m$) normalizes the interstitial velocity (u_e), which is related to the linear velocity by the porosity of the column. Reduced plate height can be related to reduced velocity by the Knox equation ($n = 1/3$) (Ref 24) or by the simpler van Deemter equation ($n = 0$) (Ref 25):

$$h = A\nu^n + \frac{B}{\nu} + C\nu \qquad \text{(Eq 10)}$$

In Eq 10, each term describes one or more physical processes that contribute to peak broadening. The A-term describes the contribution of eddy diffusion and mobile-phase mass transfer. Eddy diffusion is caused by differences in flow streams between particles (solvent flow is slower through constricted paths and faster through wide paths). Mobile-phase mass transfer refers to broadening caused by mobile phase flowing faster in the center of streams relative to the edges. Together, these phenomena are largely independent of reduced velocity; broadening instead depends on particle size and packing quality. The B-term describes the contribution of longitudinal diffusion, which is simply ordinary diffusion along the column axis that would occur without any mobile-phase movement. Broadening depends on time and is therefore inversely proportional to reduced velocity. The C-term describes the contribution of stationary-phase mass transfer, which affects broadening because some analytes diffuse further into particle pores than others. During this time, analytes that do not diffuse as far into the pores travel along the column. Broadening is therefore directly proportional to reduced velocity. It is apparent from Eq 10 that at low reduced velocities, the B-term dominates; at high reduced velocities, the C-term dominates.

Although Eq 10 is used by chromatographers to develop a theoretical understanding of different silica support structures, it is not particularly useful for chromatographers designing a separation method. Following the recommendation of Desmet et al., who point out that plots that relate efficiency to operating parameters can have great practical utility (Ref 26), this article examines efficiency as a function of pressure, flow rate, and length for standard analytical columns packed with 5 μm particles (Fig. 7). Consider a situation in which both column length and particle size are fixed (e.g., there is only one column available with the correct stationary phase). If this single

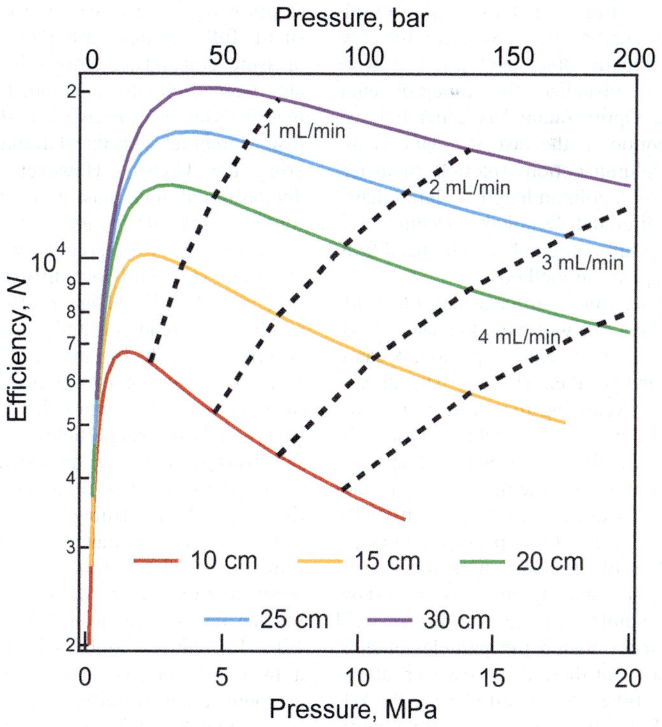

Fig. 7 Approximate relationship between efficiency, pressure, and flow rate for analytical columns with particle diameter = 5 μm, column diameter = 4.6 mm (0.18 in.), and variable lengths. Each curve was developed starting from Eq 10, with A = 1, B = 5, and C = 0.05, which are representative of commercial materials (Ref 23). The analyte is assumed to have a diffusivity (D_m) of 10^{-5} cm²/s. Typical porosities for commercial materials (Ref 23) yield u_e = 1.8 u. To use Eq 1, the mobile phase is assumed to have a viscosity (μ) of 0.00065 Pa · s (40% water/60% acetonitrile at 30 °C, or 85 °F). The flow resistance (Φ) is assumed to be 900.

available column is 15 cm (6 in.) in length, the maximum plate number (N_{max}) will be achieved at a pressure below 5 MPa (50 bar, or 725 psi) and a flow rate below 1 mL/min. This does not imply that a separation with N_{max} is optimal. The flow rate should be increased, staying below the maximum desired operating pressure (e.g., P_{max} = 20 MPa, or 200 bar, or 2900 psi), until the plate number decreases to the minimum acceptable efficiency for the separation (e.g., N = 8000). This approach decreases the analysis time. This was seen in the discussion of Fig. 3, where doubling the flow rate reduced the plate number, but all peaks were resolved in half the time. Note that when high flow rates are employed, the dimension of the capillary tubes and the viscosity of the mobile phase must be chosen to avoid a Reynolds number that exceeds a critical value, leading to turbulent flow conditions and a dramatic increase in detector noise (Ref 27).

For a situation in which multiple column lengths are available, the mobile-phase velocity and column length should be optimized together. Column length does not vary continuously, so only lengths that are easily created by joining columns together must be considered. The minimum acceptable efficiency for the separation (e.g., N = 12,000) can be achieved with a variety of flow rate and length combinations. One problem with plotting efficiency against operating parameters (Fig. 7) is that analysis time is not explicitly considered. Poppe curves (Ref 28) were the first example of kinetic plots (Ref 26) and were developed to visualize the time/efficiency compromise. Optimization has generated significant attention in the last 15 years as the number of column options (particle diameter, particle porosity, column length, column diameter) has proliferated. Detailed procedures will be helpful to chromatographers who need fully optimized separation methods.

Fully porous silica particles may have different diameters but will have the same A, B, and C constants in the Knox equation, because they are scaled by their characteristic dimension (Ref 29). Reducing the particle size proportionally reduces the plate height (H), which proportionally increases the plate number (N). Therefore, it is tempting to conclude that columns packed with smaller particles are always better than those packed with larger particles. Plots of efficiency as a function of dead time (which indicates analysis time) show that, for example, 1.8 μm particles yield greater efficiency than 5 μm particles at short analysis times, but there is a crossover, and at long analysis times, 5 μm particles are the better choice (Ref 30). Small, fully porous particles (d_p < 2 μm) are therefore often used to improve speed rather than efficiency, because a reasonable plate number can be achieved with a shorter column. The HPLC instruments cannot operate columns packed with small, fully porous particles. The UHPLC instruments were introduced commercially in the early 2000s to address their additional challenges. Small, fully porous particles lead to heating caused by viscous friction of the mobile phase, which becomes a significant concern when the column pressure exceeds 40 MPa (400 bar, or 5800 psi) (Ref 23). When heat dissipates from the walls of the column, the center of the column will be at a higher temperature, causing a radial temperature gradient. This has two negative consequences. Higher temperature leads to lower retention. Therefore, analytes move faster at the center of the column compared to the wall. Higher temperature also reduces the viscosity of the mobile phase, increasing the linear velocity at the center of the column compared to the wall. Together, these phenomena cause peak broadening and therefore a decrease in separation efficiency. To minimize radial gradients, small, fully porous particles require narrow columns ($d_c \leq 2.1$ mm, or 0.08 in.). The narrow columns require redesigning the instrument to minimize extra-column contributions to peak broadening.

Superficially porous particles or core-shell particles were introduced commercially in the 2000s. These particles consist of a solid core surrounded by a thin shell of porous material bonded to the stationary phase. For example, a 1.7 μm diameter solid core may be surrounded by a 0.5 μm thick porous shell for an overall particle diameter of 2.7 μm. Core-shell particles with intermediate dimensions (2 μm < d_p < 3 μm) can perform similarly to small, fully porous particles but have lower pressure demands. Superficially porous particles were originally developed with the idea that analytes would have less distance to diffuse to interact with the stationary phase, lowering the C-term. However, Gritti et al. demonstrated that mass transfer was similar for low-molecular-weight analytes and that the C-term was lower only when separating large analytes with low diffusivities, such as proteins (Ref 31). Higher efficiency instead results from reducing eddy dispersion (lower A-term) and reducing the dead volume of the column (lower B-term) (Ref 32). The HPLC instruments can operate columns packed with superficially porous particles with intermediate dimensions; however, the instrument must be modified to minimize extra-column contributions to peak broadening.

Finally, a parameter is discussed that improves analysis time and requires only minor modification to the instrument. Increasing the temperature from 20 to 50 °C (70 to 120 °F) reduces mobile-phase viscosity by a factor of approximately 2 for both water/acetonitrile and water/methanol mixtures (Ref 33), and analyte diffusivity increases with temperature. Gritti and Guiochon measured the effect of temperature on the relationship between plate height and linear velocity for a column with 5 μm particles, with phenol as a representative analyte (Ref 34). The temperature increase flattened the high-velocity region (C-term), indicating that a temperature increase to 55 °C (130 °F) allows linear velocity to be doubled without a loss in efficiency. The authors point out that a temperature increase could also be used to (approximately) double the efficiency if a column with double the length is operated at the same linear velocity. Increasing the temperature to improve efficiency or reduce analysis time is obviously only possible if all sample components and the stationary phase remain stable at elevated temperatures. The HPLC instruments can be used for high-temperature operation with a solvent preheater to heat the mobile phase to within 5 °C (9 °F) of the required temperature as it enters the column (Ref 23).

Preparing Real Samples

Analytes of interest are often found within matrices with many components, or components that are not compatible with packed columns. Examples include whole blood, urine, food, plant material, and soil. Blood and urine samples are often used to detect illegal drug use. Blood samples are also used to monitor the concentration of drugs with a narrow therapeutic index. Solid samples such as food and plant material are analyzed to develop nutrient profiles. Various approaches can be implemented to extract the analytes prior to liquid chromatography and avoid compromising the analytical column. This article describes solid-phase extraction and liquid-liquid extraction. It is noted that analytes in matrices such as rain or seawater may require concentration prior to analysis; this can also be accomplished with extraction methods.

Solid-phase extraction (SPE) is a useful technique to extract analytes from complex matrices prior to HPLC analysis. If an analyte in a liquid sample is passed over a solid surface, it can be retained on the solid surface by means of mechanisms such as hydrophobic interactions, hydrophilic interactions, or electrostatic interactions. Solid-phase extraction is based on the principle that this retention is reversible; therefore, the analyte can be eluted from the solid surface and collected for analysis. Matrix components from the original sample that are not retained on the solid surface are washed away and do not end up in the final purified sample. Solid-phase extraction is typically used to extract analytes from liquid samples such as blood, urine, or water, but solid samples such as animal or plant tissues, foods, or soils can be treated after a liquid-extraction technique.

The solid-phase material is typically silica particles with a particular functional group bonded to the surface that dictates the chemical interaction with the analyte. These particles can be packed into a column or incorporated into a disk, depending on preference. Solid-phase materials can be generally classified into the same categories as HPLC packing materials: reverse phase, normal phase, and ion exchange. Reverse-phase functional groups are hydrophobic (C_{18}, C_8, and phenyl), and the retention mechanism is van der Waals and other nonpolar interactions. Nonpolar analytes in a polar sample matrix will be retained on these surfaces, while other (polar) matrix components will pass through to waste. The retained analytes can then be eluted with a nonpolar solvent. Normal-phase functional groups are hydrophilic (amino, cyano, hydroxyl), and the retention mechanism is hydrogen bonding and other polar interactions. Polar analytes in a nonpolar sample matrix will be retained on these surfaces, while other (nonpolar) matrix components will pass through to waste. The retained analytes can then be eluted with a polar solvent.

Ion-exchange materials use positively or negatively charged functional groups to retain analytes containing oppositely charged functional groups by means of electrostatic interactions. The pH of the sample matrix and the elution solvent become very important in this scenario. Cation-exchange SPE can be used to retain positively charged analytes (basic compounds) by negatively charged functional groups on the solid phase. The pH of the sample matrix must be below the pKa of the basic analyte for it to be in a charged state and retained. A solvent with a pH above the pKa of the basic analyte will cause the analyte to become neutral and be released from the electrostatic interaction. Anion-exchange SPE can be used to retain negatively charged analytes (acidic compounds) by positively charged functional groups on the solid phase. The pH of the sample matrix must be above the pKa of the acidic analyte for it to be in a charged state and retained. A solvent with a pH below the pKa of the acidic analyte will cause it to become neutral and be released from the electrostatic interaction.

Solid-phase extraction typically follows a three-step procedure: loading, washing, and elution. The analytes are loaded by filling a reservoir above the solid phase and allowing the liquid sample to percolate through the solid phase at a flow rate that is slow enough to allow chemical interactions to develop. A vacuum can be applied by means of a manifold to assist viscous samples through the sorbent, but it should not pull the sample through at a rate quicker than a drop per second. Washing removes weakly retained compounds from the solid phase, while leaving analytes of interest still retained. The strength of the washing solvent should be as strong as possible but must not elute analytes of interest; it must be determined empirically during method development. Finally, the analytes are eluted from the solid phase by passing a solvent through that is strong enough to disrupt or neutralize the retention mechanism, thereby releasing the analytes.

Cation-exchange SPE can be used to prepare blood samples for opioid quantitation by removing proteins, lipids, and other interferences from the blood matrix. When working with blood samples, protein precipitation is a necessary first step to prevent clogging the solid phase. This can be accomplished by the addition of zinc sulfate followed by cold acetonitrile and centrifugation. The supernatant that remains will contain the opioids in a liquid form suitable for loading onto the solid phase. Most of the commonly encountered opioids have a basic nitrogen group that will be ionized at low pH (morphine's pKa ≈ 7.9 for the tertiary amine). Because basic functional groups such as the tertiary amine found in opioids will be 100% ionized at pH < 3, adjusting the pH of the sample is critical to maximize analyte retention. Diluting the supernatant 4X with 0.1% formic acid in water prior to loading will ensure the opioids are ionized. Once the sample has been loaded, washing the solid phase with methanol will remove unwanted lipophilic sample components that may be weakly retained. Because the dominant retention interaction is between the positively charged amino groups on the opioids and the negatively charged functional groups on the solid phase, using a solvent such as methanol will not disrupt this interaction as it would with a nonpolar interaction. Opioids can then be eluted with acetonitrile/methanol/ammonium hydroxide (80:10:10), because the pH of this solvent will be greater than the pKa of the opioids, causing deprotonation and neutralizing the amino group involved in the ionic retention.

Liquid-liquid extraction (LLE) is another useful technique to extract analytes from complex matrices prior to HPLC analysis. Two immiscible solvents such as water (aqueous/polar) and hexane (organic/nonpolar) are mixed and allowed to separate into two layers. Liquid-liquid extraction is based on the principle that "like attracts like," and therefore, polar analytes will preferentially distribute into the aqueous solvent, whereas nonpolar analytes will distribute into the organic solvent. If a nonpolar analyte such as Δ-9-THC (tetrahydrocannabinol, the psychoactive component of cannabis) is present in blood (aqueous), it can be extracted by mixing the blood with hexane. The Δ-9-THC will be drawn into the hexane, which can be separated from the water by centrifugation. The preferential distribution of a neutral analyte between two solvents depends on its partition coefficient (K). High K-values indicate that a higher fraction of the analyte dissolves in the organic solvent; analytes with low K-values will remain in the aqueous solvent.

For ionizable analytes, the pH of the solvent and the pKa of the analyte determine the analyte distribution and will therefore greatly affect the efficiency of extraction. Analytes in an ionized form tend to be more water soluble, while neutral molecules are more organic soluble. Methamphetamine contains an ionizable amine group (pKa ≈ 9.9) that will be primarily protonated and ionized in urine samples at pH 6. Extracting the methamphetamine with chlorobutane would not succeed, because methamphetamine would remain in the aqueous urine layer. To extract the methamphetamine, the pH of the urine must be increased to greater than pH 10 with the addition of a base such as borate. At a pH higher than its pKa, methamphetamine will deprotonate, becoming more organic soluble and easily extracted with chlorobutane. After the analyte equilibrates between the immiscible solvents, they must be separated. It is important to recognize that density differences of the two solvents will dictate which one is on top and bottom, so determine which solvent is which prior to mixing to prevent collecting the wrong solvent for analysis.

The solvent (eluent) from LLE (SPE) should be collected and dried, which allows the extracted analytes to be resuspended in an HPLC-compatible solvent and to be concentrated if that volume is less than the original sample. Resuspending a dried extract from LLE or SPE can result in insoluble particulate matter being suspended in the new sample. If not removed, the particles can clog the analytical column, resulting in increased system pressure and changes in chromatographic performance. Therefore, all samples should be either centrifuged or filtered prior to HPLC injections.

Analyzing Complex Samples

Complex samples may have structurally similar components that are difficult to resolve, trace components that co-elute with major components, or many components. Two-dimensional chromatography has been applied to these challenges. Two-dimensional methods involve collecting (at a minimum) one fraction of the effluent from a first column (^{1}D) and injecting it into a second column (^{2}D) with different stationary-phase selectivity. The goal of the second column is to separate components that were poorly resolved by the first column. Two-dimensional chromatography modes are introduced here; a recent tutorial review explains key decisions that are unique to two-dimensional experiments (Ref 35).

Two-dimensional chromatography should be considered when the analysis time for a simple sample is controlled by one pair of components that is difficult to resolve. Adjusting the mobile-phase composition proves insufficient. Changing the stationary-phase chemistry resolves one pair of components but creates a different unresolved pair. If the ^{1}D effluent containing the unresolved components is collected and injected as the ^{2}D sample, this is referred to as a "heart-cutting" method. Heart-cutting methods can be implemented offline with a fraction collector. To maintain accurate quantitation, the fraction volume must be much larger than the peak volume, such that all the analyte is transferred. For example, after ^{1}D separation with a C_{18} stationary phase, five fractions containing warfarin and hydroxywarfarin isomers were collected for ^{2}D separation with a different stationary phase (Ref 36). Although offline methods are simple to implement in principle, the manipulation of multiple fractions can lead to sample contamination or sample loss and requires a lot of analysis time. For online implementation, the ^{2}D column must allow a high-speed separation such that it is ready for subsequent injections from the ^{1}D column.

For pharmaceuticals, identification and quantitation of impurities and degradation products is critical because of their potential toxicity. Components present at concentrations greater than 0.05% relative to the active pharmaceutical ingredient must be quantified; however, in a ^{1}D separation, components that co-elute with the active pharmaceutical ingredient may not even be identified. Stoll et al. investigated the influence of transfer volume and solvent strength on the sensitivity of UV-VIS detection for strongly retained components such as butylbenzene and naproxen (Ref 37). The authors determined that collecting a large volume of effluent from the ^{1}D column, diluting it with weak solvent (water), and injecting it improved the limit of quantitation by focusing the analytes at the inlet of the ^{2}D column. Other strategies include cooling the temperature to trap the analytes, partial evaporation of the mobile-phase solvents, and choosing chromatography modes that inherently refocus analytes.

Two-dimensional chromatography has many advantages for samples that simply contain too many components. The peak capacity is an estimate of the maximum number of components that can be resolved into single peaks with $R_s \geq 1$; for instruments with a single dimension (one column), peak capacities of 200 can be achieved (Ref 35). Comprehensive methods involve continually collecting ^{1}D effluent fractions and injecting them as ^{2}D samples, providing maximum information about the original sample from a single experiment. This process requires a valve or modulator that switches between collecting (from the ^{1}D column) and injecting (into the ^{2}D column). Online comprehensive methods have been applied to the analysis of food components, such as triacylglycerols, phospholipids, carotenoids, and polyphenols (Ref 38), biochemicals, and pharmaceuticals. Despite the long analysis times required, offline methods may be chosen when the ^{2}D column cannot keep up with the sampling frequency from the ^{1}D column.

Two-dimensional methods are frequently designed with two reverse-phase columns, because stationary phases used with other chromatography modes may be incompatible with reverse-phase mobile phases (Ref 39). One final example illustrates a systematic strategy for choosing two stationary phases. Pyrolysis oils formed from lignocellulosic biomass contain components with high polarity, low volatility, poor thermal stability, and/or high molecular weight that are not suitable for gas chromatography. Optimizing the upgrading process (which reduces the proportion of oxygenated molecules to stabilize the mixture and prepare it for conventional refining processes) requires detailed composition information. Le Masle et al. chose 38 representative chemicals and injected these into 28 columns with different stationary phases (including phenyl hexyl, cyanopropyl, pentafluorophenyl, graphitic porous carbon, zirconia, and polymer phases) under linear gradient conditions (Ref 40). The authors found that combining a silica-based stationary phase with a nonsilica phase yielded the best overall separation.

This article has introduced theoretical and practical aspects of reverse-phase liquid chromatography without assuming any prior background in chromatography. Commercial liquid chromatography instruments have been engineered into easy-to-use instruments for identification, quantitation, and/or purification, and it is possible to develop separations without extensive theoretical knowledge. Even so, fine-tuning the interactions between sample components, stationary phase, and mobile phase for complex samples requires an understanding of the chemistry involved. Column technology and instrument design also require attention to avoid attempting separations with incompatible elements. The authors encourage new chromatographers to develop their expertise so they can take full advantage of the many options to solve their analytical problems.

REFERENCES

1. L.S. Ettre and K.I. Sakodynskii, M.S. Tswett and the Discovery of Chromatography I: Early Work (1899–1903), *Chromatogr.*, Vol 35, 1993, p 223–231
2. L.S. Ettre and K.I. Sakodynskii, M.S. Tswett and the Discovery of Chromatography II: Completion of the Development of Chromatography (1903–1910), *Chromatogr.*, Vol 35, 1993, p 329–338
3. L.R. Snyder, J.J. Kirkland, and J.W. Dolan, *Introduction to Modern Liquid Chromatography*, 3rd ed., Wiley, 2010
4. T.J. Bruno and P.D.N. Svoronos, High Performance Liquid Chromatography, *Handbook of Basic Tables for Chemical Analysis*, 3rd ed., CRC Press, 2011, p 137–208
5. P.J. Linstrom and W.G. Mallard, *NIST Chemistry WebBook, NIST Standard Reference Database Number 69*, National Institute of Standards and Technology, Gaithersburg, MD
6. L.R. Snyder, J.J. Kirkland, and J.W. Dolan, Basic Concepts and the Control of Separation, *Introduction to Modern Liquid Chromatography*, John Wiley Sons, Inc., 2010, p 19–86
7. L.R. Snyder, J.J. Kirkland, and J.W. Dolan, Gradient Elution, *Introduction to Modern Liquid Chromatography*, John Wiley Sons, Inc., 2010, p 403–473
8. A.P. Schellinger, D.R. Stoll, and P.W. Carr, High Speed Gradient Elution Reversed-Phase Liquid Chromatography, *J. Chromatogr. A*, Vol 1064, 2005, p 143–156
9. L.R. Snyder, J.J. Kirkland, and J.W. Dolan, The Column, *Introduction to Modern Liquid Chromatography*, John Wiley Sons, Inc., 2010, p 199–252
10. B. Bidlingmeyer, C.C. Chan, P. Fastino, R. Henry, P. Koerner, A.T. Maule, M.R. C. Marque, U. Neue, L. Ng, H. Pappa, L. C. Sander, C. Santasania, L.R. Snyder, and T. Wozniak, HPLC Column Classification, *Pharm. Forum*, Vol 31, 2005, p 637–645
11. "Certificate of Analysis for Standard Reference Material 870: Column Performance Test Mixture for Liquid Chromatography," National Institute of Standards and Technology, 2016
12. "The USP Approach for Selecting Columns of Equivalent Selectivity," The United States Pharmacopeial Convention, www.usp.org/resources/usp-approach-column-equiv-tool
13. L.R. Snyder, J.W. Dolan, and P.W. Carr, A New Look at the Selectivity of RPC Columns, *Anal. Chem.*, Vol 79, 2007, p 3254–3262
14. L.R. Snyder, J.W. Dolan, and P.W. Carr, The Hydrophobic-Subtraction Model of Reversed-Phase Column Selectivity, *J. Chromatogr. A*, Vol 1060, 2004, p 77–116
15. "Common Selectivity Database," HPLC Columns, www.hplccolumns.org/database/index.php
16. S. Bocian, P. Vajda, A. Felinger, and B. Buszewski, Solvent Excess Adsorption on the Stationary Phases for Reversed-Phase Liquid Chromatography with Polar Functional Groups, *J. Chromatogr. A*, Vol 1204, 2008, p 35–41
17. L. Somaini, M.A. Saracino, C. Marcheselli, S. Zanchini, G. Gerra, and M.A. Raggi, Combined Liquid Chromatography-Coulometric Detection and Microextraction by Packed Sorbent for the Plasma Analysis of Long Acting Opioids in Heroin Addicted Patients, *Anal. Chim. Acta*, Vol 702, 2011, p 280–287
18. I. Ali, A. Haque, and K. Saleem, Separation and Identification of Curcuminoids in Turmeric Powder by HPLC Using Phenyl Column, *Anal. Methods*, Vol 6, 2014, p 2526–2536
19. P.G. Stevenson and G. Guiochon, Retention Divergence of Terpenes with Porous Graphitized Carbon and C_{18} Stationary Phases, *J. Chromatogr. A*, Vol 1247, 2012, p 57–62
20. S. Strohschein, M. Pursch, D. Lubda, and K. Albert, Shape Selectivity for C_{30} Phases for RP-HPLC Separation of Tocopherol Isomers and Correlation with

MAS NMR Data from Suspended Stationary Phases, *Anal. Chem.*, Vol 70, 1998, p 13–18
21. "Certificate of Analysis for Standard Reference Material 869a: Column Selectivity Test Mixture for Liquid Chromatography," National Institute of Standards and Technology, 2007
22. L.C. Sander, M. Pursch, and S.A. Wise, Shape-Selectivity for Constrained Solutes in Reversed-Phase Liquid Chromatography, *Anal. Chem.*, Vol 71, 1999, p 4821–4830
23. P.W. Carr, D.R. Stoll, and X. Wang, Perspectives on Recent Advances in the Speed of High-Performance Liquid Chromatography, *Anal. Chem.*, Vol 83, 2011, p 1890–1900
24. P.A. Bristow and J.H. Knox, Standardization of Test Conditions for High-Performance Liquid-Chromatography Columns, *Chromatogr.*, Vol 10, 1977, p 279–289
25. J.J. Deemtervan, F.J. Zuiderweg, and A. Klinkenberg, Longitudinal Diffusion and Resistance to Mass Transfer as Causes of Nonideality in Chromatography, *Chem. Eng. Sci.*, Vol 5, 1956, p 271–289
26. G. Desmet, D. Cabooter, and K. Broeckhoven, Graphical Data Representation Methods to Assess the Quality of LC Columns, *Anal. Chem.*, Vol 87, 2015, p 8593–8602
27. D. Cabooter, F. Lynen, P. Sandra, and G. Desmet, Turbulence as a Source of Excessive Baseline Noise during High-Speed Isocratic and Gradient Separations Using Absorption Detection, *Anal. Chem.*, Vol 80, 2008, p 1679–1688
28. H. Poppe, Some Reflections on Speed and Efficiency of Modern Chromatographic Methods, *J. Chromatogr. A*, Vol 778, 1997, p 3–21
29. G. Desmet, D. Clicq, and P. Gzil, Geometry-Independent Plate Height Representation Methods for the Direct Comparison of the Kinetic Performance of LC Supports with a Different Size or Morphology, *Anal. Chem.*, Vol 77, 2005, p 4058–4070
30. P.W. Carr, X. Wang, and D.R. Stoll, Effect of Pressure, Particle Size, and Time on Optimizing Performance in Liquid Chromatography, *Anal. Chem.*, Vol 81, 2009, p 5342–5353
31. F. Gritti, A. Cavazzini, N. Marchetti, and G. Guiochon, Comparison between the Efficiencies of Columns Packed with Fully and Partially Porous C_{18}-Bonded Silica Materials, *J. Chromatogr. A*, Vol 1157, 2007, p 289–303
32. R. Hayes, A. Ahmed, T. Edge, and H. Zhang, Core-Shell Particles: Preparation, Fundamentals, and Applications in High Performance Liquid Chromatography, *J. Chromatogr. A*, Vol 1357, 2014, p 36–52
33. P.W. Carr and J. Li, Accuracy of Empirical Correlations for Estimating Diffusion Coefficients in Aqueous Organic Mixtures, *Anal. Chem.*, Vol 69, 1997, p 2530–2536
34. F. Gritti and G. Guiochon, The Current Revolution in Column Technology: How It Began, Where Is It Going? *J. Chromatogr. A*, Vol 1228, 2012, p 2–19
35. D.R. Stoll and P.W. Carr, Two-Dimensional Liquid Chromatography: A State of the Art Tutorial, *Anal. Chem.*, Vol 89, 2016, p 519–531
36. E.L. Regalado, J.A. Schariter, and C.J. Welch, Investigation of Two-Dimensional High Performance Liquid Chromatography Approaches for Reversed Phase Resolution of Warfarin and Hydroxy Warfarin Isomers, *J. Chromatogr. A*, Vol 1363, 2014, p 200–206
37. D.R. Stoll, E.S. Talus, D.C. Harmes, and K. Zhang, Evaluation of Detection Sensitivity in Comprehensive Two-Dimensional Liquid Chromatography Separations of an Active Pharmaceutical Ingredient and Its Degradants, *Anal. Bioanal. Chem.*, Vol 407, 2015, p 265–277
38. F. Cacciola, P. Donato, D. Sciarrone, P. Dugo, and L. Mondello, Comprehensive Liquid Chromatography and Other Liquid-Based Comprehensive Techniques Coupled to Mass Spectrometry in Food Analysis, *Anal. Chem.*, Vol 89, 2017, p 414–429
39. D. Li, C. Jakob, and O. Schmitz, Practical Considerations in Comprehensive Two-Dimensional Liquid Chromatography Systems (LC × LC) with Reversed Phases in Both Dimensions, *Anal. Bioanal. Chem.*, Vol 407, 2015, p 153–167
40. A. Le Masle, D. Angot, C. Gouin, A. D'Attoma, J. Ponthus, A. Quignard, and S. Heinisch, Development of On-Line Comprehensive Two-Dimensional Liquid Chromatography Method for the Separation of Biomass Compounds, *J. Chromatogr. A*, Vol 1340, 2014, p 90–98

ASM Handbook, Volume 10, *Materials Characterization*
ASM Handbook Committee
DOI 10.31399/asm.hb.v10.a0006630

Ion Chromatography

Christopher A. Pohl, Thermo Fisher Scientific

Overview

General Uses

- Qualitative and quantitative analysis of a wide range of inorganic and organic anions as well as organic cations and certain inorganic cations in aqueous and organic solutions

Examples of Applications

- Analysis of aqueous solutions, such as surface and groundwater, leachates, brines, condensates, and high-purity water
- Evaluation of organically bound halides and sulfur following combustion utilizing a Schöniger combustion flask, an oxygen-pressurized vessel, or an instrument designed for combustion analysis via ion chromatography
- Determination of anions and cations on contaminated surfaces
- Plating bath solution analysis
- Determination of anions, cations, and carbohydrates in foods and beverages
- Determination of pharmaceutical counterions
- Analysis of ionic liquids
- Analysis of ions in air and particulates suspended in air by utilizing impingers, impactors, or denuders

Samples

- *Form:* solids, solutions, or gases
- *Size:* minimum of 0.1 to 0.5 mg for solids, minimum of 200 μL for solutions, 0.5 μg can be detected on surfaces
- *Materials:* inorganic and organic materials, geological samples, environmental samples, industrial process liquids, glasses, ceramics, leachates, explosives, alloys, ionic liquids, and pyrotechnics
- *Preparation:* aqueous samples can be analyzed as received or after dilution; organic solutions can be analyzed as received if the injection volume is minimal or by utilizing matrix elimination techniques in trace analysis applications; water-soluble solids can be analyzed upon dissolution in water; analysis of water-insoluble solids must follow a sample preparation and dissolution procedure; gases can be analyzed using impingers, impactors, or denuders

Limitations

- Detection limits at the single-digit parts per billion level or lower for many ions; sub–parts per billion detection limits are achievable using concentrator columns or large volume injections
- Cations: Applications that involve utilizing conductivity detection under aqueous conditions are limited to alkali metals and alkaline earth cations, ammonia, moderate- to low-molecular-weight amines, quaternary ammonium compounds, phosphonium compounds, and sulfonium compounds. Most other inorganic cations require alternative detection methods such as spectrophotometric detection, atomic absorption detection spectroscopy, inductively coupled plasma detection, or inductively coupled plasma mass spectrometry detection.
- Analytes must be ionic in solution
- Water solubility is preferred
- Ions with limited water solubility require the addition of solvent to the mobile phase and generally exhibit lower sensitivity than ions detected under 100% aqueous conditions
- Not all columns are compatible with organic solvents
- Electrolytic suppressors have limited solvent compatibility

Estimated Analysis Time

- Requires 3 to 45 min per sample if the sample is already in aqueous solution
- Requires 60 min per sample for organically bound elements
- Times for other sample matrices are highly variable depending upon the complexity of the sample and the difficulty of preparing the sample solution for injection

Capabilities of Related Techniques

- Wet analytical chemistry: much slower and less sensitive when mixtures of ions are to be analyzed
- Discrete analyzers: generally limited to one analyte per analysis, but throughput is generally less than 1 min per sample. Subject to interferences depending upon the specificity of the chemistry. Advantageous for applications where only a limited number of analytes are of interest
- Atomic absorption spectroscopy: generally, not capable of analyzing anions; more versatile for analysis of cations
- Inductively coupled plasma spectroscopy: generally, not capable of analyzing anions; faster and more versatile for the analysis of cations except in the case of very complex matrices
- Inductively coupled plasma mass spectrometry: generally, not capable of analyzing anions; faster, more sensitive, and more versatile for the analysis of cations except in the case of very complex matrices

Introduction

Ion chromatography (IC) originated from the work of Hamish Small at Dow Chemical Company in the early 1970s (Ref 1). Although the technique has broadened considerably in scope, the majority of ion chromatography utilizes an ion-exchange column, a suppressor device, and a conductivity detector. A high-pressure pump is used to pump the mobile phase through the ion-exchange column. An injection valve capable of switching a sample loop in and out of the flow path is generally used to introduce samples into the column.

Separation Principles

The most common separation mode in IC is ion exchange. While the use of natural materials for ion-exchange processes dates to biblical times, development of modern ion-exchange materials took place in the early 1940s in support of the Manhattan Project (Ref 2). The high-capacity ion-exchange materials initially developed were based on styrenic monomers providing a combination of high capacity and excellent chemical stability. Such polymers are composed of an ionic polymer backbone that has been cross-linked to form an immobilized ionic gel, rendering them highly hydrated but water insoluble. The concentration of ion-exchange groups in such a structure is generally in the 1–5 molar range. The affinity of ions for ion-exchange materials is highly correlated with the hydration enthalpy and entropy of the ion. Ions with the highest hydration energy tend to have the largest hydration sphere in solution. Because ion-exchange materials are generally cross-linked, this restricts the hydration of the ion-exchange polymer. Highly hydrated ions (e.g., fluoride or lithium) have the greatest difficulty entering the ion-exchange polymer and hence generally exhibit the lowest affinity for the ion-exchange material. Conversely, poorly hydrated ions tend to have easy access to the ion-exchange polymer and therefore exhibit the highest affinity for the ion-exchange material.

Ion-exchange materials are available in four basic types (Ref 3): strong base anion-exchange materials with quaternary ion-exchange sites; weak base anion-exchange materials with primary, secondary, or tertiary amine ion-exchange sites; strong acid cation-exchange materials with sulfonate ion-exchange sites; and weak acid cation-exchange materials with carboxylate or phosphonate ion-exchange sites. Nearly all anion-exchange materials used in IC are strong base anion-exchange materials because they provide the highest capacity and good selectivity for most anions of interest. By contrast, strong acid cation-exchange materials are now rarely used in IC because such materials exhibit excessively high affinities for divalent cations (Ref 4). Although such materials were used in the early years of IC, the vast majority of all cation-exchange materials used in IC today incorporate carboxylic acid ion-exchange sites. Such materials have a much lower affinity for divalent cations, enabling the efficient elution of both monovalent cations and divalent cations with mobile phases suitable for use in combination with the suppressor.

Because ion-exchange materials are composed of fixed ionic sites attached to a polymer backbone with a mobile counterion of opposite charge, counterions remain fixed within the ion-exchange material when the solution surrounding the ion-exchange material is free of ions, as in the case of deionized water. If the solution surrounding the ion-exchange material contains ions of the same type as the counterions present in the ion-exchange material, the ions in both phases will be in dynamic equilibrium with individual ions, rapidly moving from the aqueous phase surrounding the ion-exchange phase to the ion-exchange phase. The rapid equilibration process can easily be demonstrated by immersing an ion-exchange resin containing counterions with one isotope (e.g., sodium 23) in a solution of an electrolyte containing counterions with a radioactive isotope that can be detected using suitable instrumentation (e.g., sodium 22) (Ref 5). After a matter of minutes, the sodium 23 and sodium 22 will be distributed in the same ratio throughout both the ion-exchange resin and the surrounding solution. Similarly, an ion-exchange resin containing only one type of counterion will rapidly reach equilibrium with the surrounding solution containing a different counterion. However, in the latter case, the ratio of the concentrations of the two ions in the resin phase will generally not be identical to the concentration ratio in the surrounding solution. Such a disparity is because, in most cases, two different ions with identical net charge will be distributed in the ion-exchange phase such that the ion with the highest affinity will be preferentially concentrated in the ion-exchange material while the ion with the lowest affinity will be present in higher relative concentrations in the surrounding aqueous phase.

Using the process described above, it is generally not possible to isolate one ion from a mixture of two ions. Ion-exchange chromatography allows quantitative separation of mixtures of ions by taking advantage of the differences in ion affinity for the ion-exchange phase under flowing conditions. For example, if a cation-exchange material in the hydronium form is packed into a length of tubing with porous supports on each end that are designed to prevent the ion-exchange material from exiting the tube (this assembly is commonly referred to as the "column" in chromatography), it can be used for the separation of cations. A solution of hydrogen chloride passed through the column will not affect the composition of the column because the cation in the solution passing through the column (commonly referred to as the mobile phase or eluent) is identical to the cation in the ion-exchange material. If a sample containing sodium chloride is injected into the column, the sodium from the sample will first bind to the top of the cation-exchange column, displacing hydronium into the surrounding solution. The sodium will travel the length of the cation-exchange column as the solution of hydrogen chloride passes over the column at a rate that is related to the relative affinity of the cation-exchange material for sodium and hydronium. If sodium has a higher affinity for the cation-exchange phase than hydronium, it will travel through the column at a rate that is slower than the migration rate of hydronium. Conversely, if sodium has a lower affinity for the cation-exchange phase than hydronium, it will travel through the column at a rate that is higher than the migration rate of hydronium. The sodium will move through the column in a relatively narrow zone, uniformly distributed across the cross section of the column bed until the sodium-rich zone exits the column and travels to the detector. Separation of two cations, for example, sodium and potassium, occurs similarly, except that the separation of sodium from potassium is dictated by the relative affinity of the stationary phase for sodium, potassium, and hydronium. Upon injection of both sodium and potassium onto the top of the column, both cations will bind to the cation-exchange resin, displacing hydronium with the two cations mixed in the cation-exchange resin. Provided that the affinity of sodium is different from the affinity for potassium, the mixture at the inlet of the column will begin to separate into two separate bands. These bands will move at a rate indicative of the relative affinity of each cation in comparison to hydronium with the lowest affinity ion (typically sodium) moving at a faster rate than the highest affinity ion (typically potassium).

Selectivity in Ion Chromatography

In ion exchange, selectivity is expressed by (Ref 6):

$$K\frac{A^x}{E^y} = \frac{[A]_r^y[E]_e^x}{[E]_r^x[A]_e^y}$$

where K is defined as the distribution coefficient of analyte A in the presence of eluent species E. The term in the equation with the subscript r represents the concentration of the specified ion in the resin phase. The term in the equation with the subscript e represents the concentration of the specified ion in the eluent. The superscript x represents the net charge of the analyte species A, while the superscript y represents a net charge of the eluent species E. In the most common case in IC where the eluent species is monovalent (i.e., hydronium or hydroxide eluent), the equation can be simplified to:

$$K\frac{A^x}{E}=\frac{[A]_r[E]_e^x}{[E]_r^x[A]_e}$$

Under ideal conditions, the slope of the log of K versus the log of eluent concentration (E) is a straight line with a slope that is equal to the charge of the analyte (x). Under such conditions, the slope for all monovalent species is -1, the slope for all divalent species is -2, the slope for all trivalent species is -3, and so on. This relationship presents the opportunity to manipulate the gap between ions of different charge or even to change the elution order of such ions depending upon the mobile phase concentration. Likewise, although the mathematics is beyond the scope of this discussion, under gradient conditions the slope of the gradient can be used to manipulate the relative location of ions of different charge. This relationship provides a convenient vehicle for optimizing analytical conditions in ion exchange.

A concrete example of the application of this concept can be understood by examining the consequences of doubling the concentration of a monovalent eluent species on the retention of a monovalent, a divalent, and a trivalent analyte ion. Doubling the eluent concentration results in a 50% reduction in the retention of the monovalent species, a 4-fold reduction in the retention of a divalent species, and a 9-fold reduction in the retention of a trivalent species. Given this relationship, the eluent concentration can be adjusted to manipulate the relative position of the monovalent, the divalent, and the trivalent species.

Alternative Separation Modes

In addition to ion exchange, which comprises greater than 90% of all separations used in IC, additional separation modes used in IC include ion exclusion chromatography, ion pair chromatography, and reversed phase chromatography (Ref 7). Ion exclusion chromatography is a chromatographic technique with limited resolving power but is suitable for analysis of organic acids in the presence of much higher concentrations of inorganic anions. Although ion exclusion chromatography is, in principle, applicable to both anions and cations, in practice the technique is exclusively employed in the analysis of organic acids. Ion pair chromatography is applied in cases where analytes have excessively high retention during operation in the ion-exchange mode. Ion pair chromatography generally has inferior selectivity for polyvalent species and is usually employed only in cases where target analytes are monovalent. Reversed-phase chromatography is generally employed for ions that are substantially hydrophobic, such as ionic surfactants and long-chain fatty acids.

Modes of Detection

The vast majority of IC applications employ conductivity detection. This is unsurprising given the fact that ions are inherently conductive when dissolved in a suitable solvent. Ions can be detected in both direct current and alternating current modes, but conductivity detection is invariably performed using alternating current. Direct current can result in oxidation or reduction of analyte ions during conductivity measurement. Oxidation and reduction reactions result in anomalous conductivity measurements. By contrast, alternating current avoids such issues by rapidly altering the direction of the electric field. Even if ions are oxidized when the applied field is suitably polarized, this oxidation is reversed when the field is reversed, thus canceling any errors associated with oxidation and reduction. As a result, alternating current results in a clean signal dependent only upon the conductivity of the species passing between the detector electrodes.

Although measurement of conductivity was well known long before the invention of IC, its application to ion-exchange chromatography was uncommon before the introduction of IC as an analytical technique. The general problem with applying conductivity detection to ion-exchange chromatography stems from the fact that both the eluent species and the analyte species are by definition conductive and the concentration of the eluent species is generally 100 to 100,000 times higher than the concentration of the analyte species in a typical IC application. Under such conditions, the eluent ion produces a high conductivity background, making it difficult to detect the small signal associated with the analyte species. Ion exchange is the ideal separation mode for ions while conductivity is the ideal detector for ions, but the high background associated with the eluent ion largely precluded its application to ion exchange. It was Hamish Small who developed the idea of eluent "suppression" as a means of eliminating this background associated with the eluent species (Ref 1).

Suppressed Conductivity Detection

A suppressor is an added element in IC that enhances conductivity performance. An example layout of a suppressed IC system is shown in Fig. 1. Suppression in IC can be utilized with a wide variety of eluent species. Nonetheless, it is easiest to understand the concept of suppression by considering its simplest form: the application of the suppressor for anion analysis when the eluent species is the hydroxide anion. For example, if the eluent species is sodium hydroxide, the stationary phase is an anion-exchange material, and the analyte is sodium chloride, the chloride will be retained by the anion-exchange phase and eluted in the presence of a large concentration excess of sodium hydroxide. As this solution of sodium hydroxide containing a small fraction of sodium chloride enters the conductivity detector, it will be difficult to detect traces of chloride in the presence of this large, highly conductive excess of sodium hydroxide. If instead a column filled with cation-exchange resin in the hydronium form is placed after the anion-exchange column, the cation-exchange resin will remove any sodium from the eluent and replace it with hydronium. Hydronium will react with a hydroxide previously associated with the sodium to produce water. In the process, the background conductivity associated with the eluent species is reduced to negligible levels. Because detector noise is proportional to the background signal associated with the eluent, the detector noise is greatly reduced as a consequence of the suppression process. At the same time, the sodium previously associated with the chloride will also be removed and replaced with hydronium. Because hydronium is substantially more conductive than any other cation, the suppression process not only reduces the background noise but also enhances the analyte signal because the conductivity of both the anion and the cation are measured by the detector. The consequence of these two factors, reduced background noise and an enhanced analyte signal, is a substantial enhancement of the signal-to-noise ratio of the analytical system.

The very first IC systems commercialized in 1975 used packed-bed suppressors of the type described above. While such devices accomplished the suppressor function and some IC systems still utilize this suppressor format, packed-bed suppressors suffer from several disadvantages. First, such a suppressor can neutralize only a finite volume of eluent before its capacity is exhausted. Upon exhaustion, the suppressor must be taken off-line and chemically regenerated with an acid solution in the case of the cation-exchange suppressor used for anion analysis or a hydroxide solution in the case of the anion-exchange suppressor used for cation analysis.

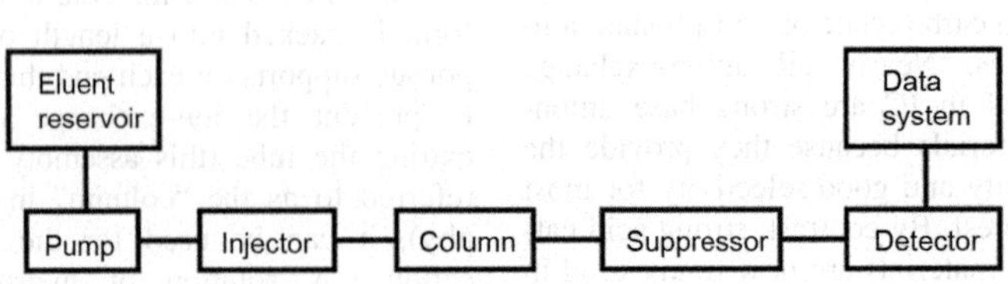

Fig. 1 Schematic representation of a suppressed ion chromatography system. Courtesy of Thermo Fisher Scientific

Two considerations apply to the design of the suppressor when packed-bed suppressors are employed. The larger the volume of the suppressor, the less frequently the suppressor must be regenerated. On the other hand, the larger the volume, the greater the degradation of the chromatographic signal due to dilution. As a consequence, relatively small suppressors that are regenerated after every analysis are preferred. However, such suppressors tend to be limited to applications where operation under isocratic conditions is feasible. Operation under gradient conditions generally requires suppressor capacities not practical for such suppressors. In addition, early IC columns had very low ion-exchange capacity in order to facilitate the use of eluents compatible with low-capacity suppressors. While low-capacity columns improve the life of the suppressor column, low-capacity separation columns significantly limit the concentration of sample that can be injected without overloading the column, resulting in distorted peaks and degraded separations.

The limitations of packed-bed suppressors led to a series of new developments that substantially improved the performance of suppressors. First, fiber suppressors were introduced in the early 1980s (Ref 8). These early devices allowed, for the first time, continuous operation with regenerant solution bathing the exterior of the fiber while the eluent flowed through the interior of the fiber performing the suppression process in the manner described above. However, fiber suppressors had relatively limited suppression capacity, limiting the suppression flux to 20 millimoles of eluent ion per minute. The limited capacity of fiber suppressors constrained column design and necessitated the use of low-capacity columns to accommodate the limitations of the fiber suppressor.

Fiber suppressors were subsequently supplanted by membrane suppressors (Ref 9), providing suppression capacity in excess of 150 millimoles of eluent ion per minute. These devices were designed with three flow chambers sandwiched between two ion-exchange membranes. Eluent flowed between the two membranes while chemical regenerant flowed across the outer surfaces of the two membranes, bathing the eluent flow channel with chemical regenerant on both sides. Substantially increased suppression capacity of second-generation continuous suppressors opened the possibility of two new innovations in IC: 5- to 10-fold increases in column capacity and application of concentration gradients, expanding the range of ions that could be eluted in a single analysis.

Finally, the early 1990s saw the introduction of electrolytic suppressors (Ref 10). These devices utilize the architecture of the earlier membrane suppressors but incorporate an electrode above and below the regenerant chambers (Fig. 2). Making use of electrolytic water splitting, electrolytic suppressors produce hydronium at the positive electrode and hydroxide at the negative electrode. Instead of making use of acid or base regenerant, electrolytic suppressors allow the use of water as the regenerant, greatly simplifying the operation of the suppressor. Furthermore, because the effluent from the conductivity cell in suppressed IC is essentially deionized water, the waste stream from the conductivity cell can be redirected through the regenerant chambers of the electrolytic suppressor, eliminating the need for any regenerant solution whatsoever (Ref 11). The use of an electric field not only allows the electrolytic production of the regenerant ion but also enhances the suppression capacity of the device. As a consequence, electrolytic suppressors generally have higher suppression capacity than membrane suppressors. The electrolytic suppressor is quite popular given its considerable advantages, and most suppressors in use today are of this type.

Nonsuppressed Conductivity Detection

As noted previously, suppression is widely used as a means of reducing the background conductivity of the mobile phase while simultaneously enhancing analyte response. In the early days of IC, operation without a suppressor was relatively common. An example layout of a nonsuppressed IC system is shown in Fig. 3. It might seem preferable to avoid the use of a suppressor, given the added complexity and band broadening consequences of the packed-bed suppressor used in the earliest commercial implementation of IC. However, as noted already, there are significant detection limit advantages to operating with a suppressor. Because the suppressor column was covered by a Dow Chemical Company patent licensed to Dionex Corporation, other companies wishing to participate in this analytical space found it necessary to work without a suppressor. This drove considerable development effort toward the use of nonsuppressed conductivity detection. Thus, in the late 1970s and early 1980s, there were numerous papers published using this approach (Ref 12). Several methods were employed to render this approach feasible, including use of columns with very low ion-exchange capacity, high-elution-power ions, and electronic noise filters. Low-ion-exchange capacity columns minimize the concentration of the eluent necessary to elute ions of interest and thus minimize the background conductivity of the eluent and the associated detector noise. The disadvantage of this approach, however, is that it reduces the sample load that the column can withstand without severe peak distortion. Likewise, using a high-elution-power ion allows the use of lower eluent concentrations, reducing the

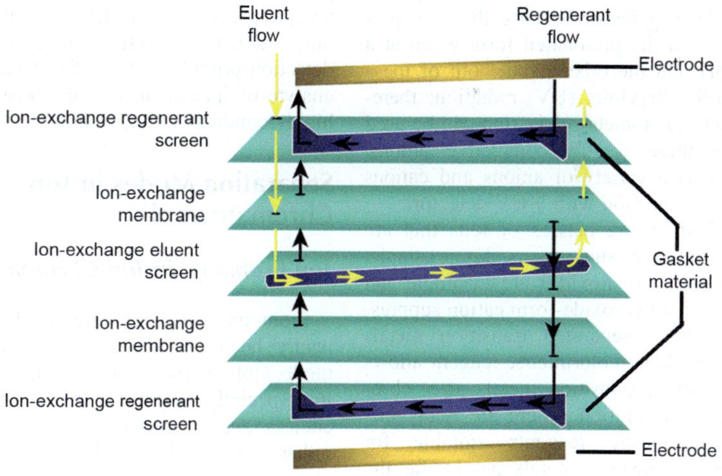

Fig. 2 Schematic representation of the fluid flow paths in an SRS suppressor with the yellow lines representing the eluent flow path and the black lines representing the separate and isolated regenerant flow path. The membranes are cation-exchange membranes in anion analysis and anion-exchange membranes in cation analysis. The outer electrodes serve to electrolyze water, producing regenerant ions from pure water. Courtesy of Thermo Fisher Scientific

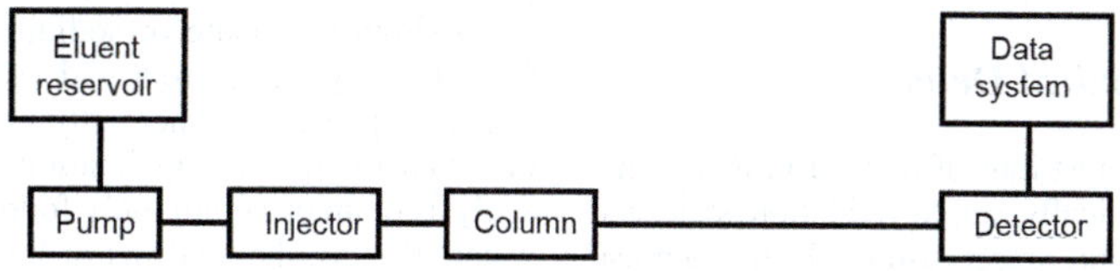

Fig. 3 Schematic representation of a nonsuppressed ion chromatography system. Courtesy of Thermo Fisher Scientific

magnitude of the eluent conductivity background. This approach also reduces the dynamic loading capacity of the column because high affinity of the eluent ion for the stationary phase makes it more difficult for low-affinity ions to gain access to stationary-phase ion-exchange sites. Filtering can indeed reduce noise, but generally it also affects peak shape; consequently, extreme filtering is generally not desirable. Considering all these factors, it is probably unsurprising that relatively few analytical chemists still utilize nonsuppressed IC after the expiration of the original Dow Chemical Company patents. Today, use of IC for the analysis of ions is almost universally performed using a suppressor. In the case of cations, operation in the nonsuppressed mode is somewhat more common, although most practitioners use the suppressed mode, given the sensitivity advantages of the suppressor.

Spectrophotometric Detection

Although most IC is performed using conductivity detection, there are several special cases where spectrophotometric detection is advantageously used in IC. First, there are ions that exhibit relatively low conductivity response due to a low level of acidity or basicity. Sulfide and arsenite are good examples of this situation. They are such weak acids that they are virtually undetectable once they pass through the suppressor because they are predominantly in the protonated form when at a neutral pH. On the other hand, both of these ions absorb ultraviolet (UV) radiation; therefore, spectrophotometric detection can be used to detect these weakly conductive anions. Table 1 lists a variety of anions and cations suitable for detection with a UV detector.

Second, there are classes of ions that are incompatible with suppressors. For example, transition metal hydroxides are insoluble and precipitate in a hydroxide-form cation suppressor. Combining separated transition metals with a postcolumn colorimetric reagent allows for detection of transition metals after chromatographic separation. There are many different colorimetric reagents suitable for detection of transition metals but by far the most common is 4-(2-pyridylazo)resorcinol (PAR) (Ref 13). The postcolumn addition of PAR to chromatographically operated transition metals allows for detection of transition metals using a visible wavelength detector at 520 nm.

Electrochemical Detection

Although most ions of interest in IC are in a thermodynamically stable oxidation state and are difficult to detect using electrochemical detection, there are a number of ions amenable to electrochemical detection (Ref 14, 15). Table 2 lists a variety of ions suitable for electrochemical detection. Included in this list are ions such as sulfide, arsenite, and cyanide that have poor sensitivity due to their relatively high pKa. Other listed ions have good suppressed conductivity sensitivity. For these ions, electrochemical detection is employed only when the selectivity of electrochemical detection provides a basis for detection of the analyte of interest in the presence of another highly conductive matrix ion.

Table 1 Ultraviolet activity of common ion chromatography analytes

Anion	UV active	Anion	UV active	Cation	UV active
Fluoride	No	Arsenate	Yes	Lithium	No
Sulfide	Yes	Thiosulfate	Yes	Sodium	No
Iodate	Yes	Bromide	Yes	Ammonium	No
Chlorite	Yes	Arsenite	Yes	Potassium	No
Hypophosphite	No	Sulfite	Yes	Rubidium	No
Sulfamate	No	Azide	Yes	Cesium	No
Bromate	Yes	Nitrate	Yes	Magnesium	No
Chloride	Yes	Chlorate	Yes	Calcium	No
Nitrite	Yes	Selenocyanate	Yes	Strontium	No
Tetrafluoroborate	No	Perchlorate	No	Barium	No
Phosphite	No	Sulfate	No	Methylamine	No
Iodide	Yes	Selenite	Yes	Ethanolamine	No
Phosphate	No	Hydroxide	Yes	Morpholine	No
Thiocyanate	Yes	Periodate	Yes	Benzylamine	Yes

UV, ultraviolet. Courtesy of Thermo Fisher Scientific

Table 2 Preferred electrode material for electrochemical detection of select ion chromatography analytes

Ion	Detection electrode	Ion	Detection electrode	Ion	Detection electrode
Cyanide	Silver	Iodide	Silver	Thiosulfate	Platinum
Sulfide	Silver	Arsenite	Platinum	Selenite	Platinum
Chloride	Silver	Sulfite	Platinum	Hydroxylamine	Platinum
Bromide	Silver	Hypochlorite	Platinum	Hydrazine	Platinum

Courtesy of Thermo Fisher Scientific

Separation Modes in Ion Chromatography

Ion Exchange in Ion Chromatography

From its very beginning, IC has predominantly been performed in the ion-exchange mode. Unlike other modes of chromatography, IC, especially suppressed IC, constrains eluent choices and concentrations. For that reason, rather than freely manipulating the mobile phase to achieve different selectivities for different applications, it is more common to see a wide variety of stationary phases designed to operate with just a few different mobile phases based on electrolytes that provide advantageous suppression products such as potassium hydroxide, methanesulfonic acid, and sodium carbonate or sodium bicarbonate. Consequently, there are a wide variety of ion-exchange phases commercially available today designed for specific applications, such as the analysis of anions commonly found in drinking water (Fig. 4), the analysis of bromate in bottled water, and the analysis of cations commonly found in drinking water (Fig. 5). Because most eluents used in suppressed IC are either highly acidic or highly basic, ion-exchange phases used in IC are typically based on polymeric materials not susceptible to attack by acids or bases such as styrenic or ether-based polymers. Ester- or amide-based polymers are not sufficiently stable in acids or bases to be used with hydroxide or hydronium mobile phases. Inorganic substrates such as silica are not sufficiently pH stable to be used at high pH, and the stability of bonded

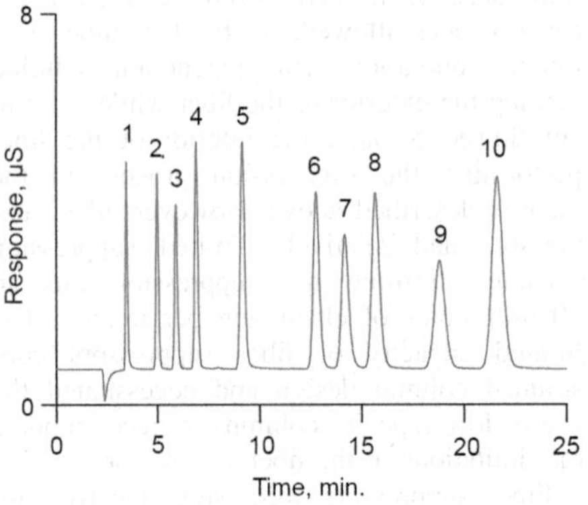

Fig. 4 Isocratic separation of common anions and disinfection byproduct anions commonly found in drinking water with the IonPac (Thermo Fisher Scientific) AS9-HC column and the AG9-HC column. Column dimensions: 4 mm (0.16 in.) internal diameter (ID) by 50 mm (2 in.) and 4 mm ID by 250 mm (10 in.); eluent: 9 mM sodium carbonate; flow rate: 1 mL/min; injection volume: 25 μL; temperature: 30 °C (85 °F); suppressor: ASRS; conductivity detection. Peaks: (1) fluoride, 3 mg/L; (2) chlorite, 10 mg/L; (3) bromate, 20 mg/L; (4), chloride 6 mg/L; (5), nitrite 15 mg/L; (6) bromide, 25 mg/L; (7) chlorate, 25 mg/L; (8) nitrate, 25 mg/L; (9) phosphate, 40 mg/L; (10) sulfate, 40 mg/L. Courtesy of Thermo Fisher Scientific

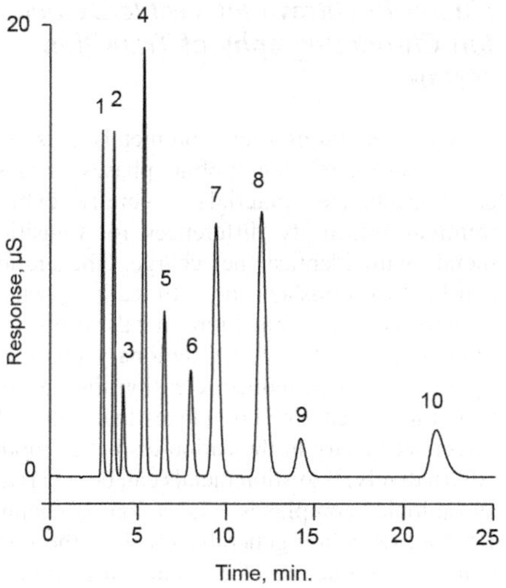

Fig. 5 Isocratic separation of the ammonium ion, alkali metal cations, and alkaline earth cations with the IonPac (Thermo Fisher Scientific) CS12A column. Column dimensions: 4 mm ID by 250 mm; eluent: 18 mM methanesulfonic acid; flow rate: 1 mL/min; injection volume: 25 μL; temperature: 24 °C (75 °F); suppressor: CSRS; conductivity detection. Peaks: (1) lithium, 1 mg/L; (2) sodium, 4 mg/L; (3) ammonium, 5 mg/L; (4) potassium, 10 mg/L; (5) rubidium, 10 mg/L; (6) cesium, 10 mg/L; (7), magnesium, 5 mg/L; (8) calcium, 10 mg/L; (9) strontium, 10 mg/L; (10) barium, 10 mg/L. Courtesy of Thermo Fisher Scientific

phases is poor at either high pH or low pH. Many inorganic substrates such as silicate, alumina, and zirconia are reactive to ions of interest and not suitable for use in suppressed IC. The pH constraints during operation in the nonsuppressed mode are not as extreme, so there are a few examples of ion-exchange phases designed for operation in the nonsuppressed mode utilizing silica-based substrates and ester-based polymeric substrates.

Ion Exclusion in Ion Chromatography

While the vast majority of all IC is practiced using the ion-exchange mode of separation, ion exclusion is also used in IC (Ref 16). Ion exclusion chromatography generally utilizes a fully sulfonated styrenic cation-exchange resin in the hydrogen form. The separation mode is based on electrostatic repulsion preventing permeation of anions into the stationary phase as it passes through the column. Only uncharged or partially ionized analytes can partition into the stationary phase. Retention is controlled by adjusting the mobile phase pH. Partition into the stationary phase occurs when the mobile phase pH is near the pKa of the analyte. Ion exclusion thus provides a convenient way to separate high-pKa analytes such as acetate and formate from low-pKa analytes such as chloride and sulfate, which tend to be eluted in the void under normal conditions. Ion exclusion mode retention is particularly advantageous in samples containing large excesses of common anions such as chloride and sulfate in the presence of weak acids such as acetate and propionate. While the cation mode of ion exclusion chromatography is possible, there are currently no commercial products targeting this mode of operation.

Ion Pairing in Ion Chromatography

Ion pair chromatography makes use of ion pair reagents to produce in situ ion-exchange sites through adsorption of ion pair reagents onto the surface of reversed phase hydrophobic chromatographic media (Ref 17). While C18 bonded phase silica is the most common reversed phase material used in high-performance liquid chromatography (HPLC), the extreme pH conditions of IC make such materials unsuitable for use in IC. Instead, the most common chromatographic material used in ion pair chromatography is highly cross-linked macroporous particles based on an ethylvinylbenzene/divinylbenzene copolymer. Such materials tend to have somewhat lower chromatographic performance than bonded phase silica with an equivalent particle size, but these materials can withstand exposure to extreme pH virtually indefinitely and thus are ideal for use in IC. Ion pair chromatography finds use in applications where analytes are difficult to elute from ion-exchange materials. Typical examples are ions with high retention in IC due to a high degree of polarizability, such as hexacyanoferrate, hexacyanocobaltate, and polythionate, or highly charged, high-retention ions such as polystyrene sulfonate and polyvinyl phosphonate. Ion pair chromatography generally provides reasonably good selectivity for monovalent ions but selectivity for higher valence ions is generally superior in the ion-exchange mode.

Eluents in Ion Chromatography

Anion-Exchange Eluents in Suppressed Ion Chromatography

As previously mentioned, suppressed IC provides the optimum detection limits when hydroxide is the eluent. Because the suppression product of hydroxide is water, it provides the lowest possible background conductivity upon suppression. In addition, hydroxide-based anion-exchange mobile phases are the best option when it comes to operation in the concentration gradient mode. However, a carbonate/bicarbonate buffer system is frequently employed in anion-exchange IC. Carbonate provides inferior detection limits and nonlinear calibration plots when compared with hydroxide, but it can offer operational simplicity compared with hydroxide-based mobile phases. Because many samples contain substantial amounts of carbonate, using a carbonate mobile phase helps masks the presence of additional carbonate in the sample. Carbonate in the sample has a concentration similar to that of carbonate in the mobile phase; therefore, the presence of carbonate in the sample produces a minimal baseline disturbance during chromatography. By contrast, when hydroxide-based mobile phases are used, the response for carbonate is enhanced, resulting in a significant baseline disturbance, which sometimes compromises quantitation of adjacent analytes (Ref 18). This problem can be mitigated by incorporating a carbon dioxide removal device after the suppressor (Ref 19) (the suppression product of carbonate is carbonic acid that is in equilibrium with carbon dioxide), but in many cases, it is simpler to use a carbonate-based mobile phase for samples containing large amounts of carbonate. Another eluent species frequently used in anion-exchange IC is sodium tetraborate. Sodium tetraborate produces a lower background conductivity than carbonate but is a relatively weak eluent species; therefore, it is generally not widely used except in cases where the sample matrix contains significant amounts of borate. One application area where borate-containing samples are frequently found is in pressurized water reactors in the nuclear power industry. In this application, boron, a neutron-flux-moderating additive, is incorporated into the high-purity water that recirculates through the reactor core.

Anion-Exchange Eluents in Nonsuppressed Ion Chromatography

To minimize background conductivity using nonsuppressed IC, eluents are chosen on the basis of conductivity and elution power. An ideal eluent anion is an anion with high elution power and low equivalent conductance. A wide variety of anions have been described in the literature for this application. The most common eluents used for analysis of anions in nonsuppressed IC are phthalate and p-hydroxybenzoate. They exhibit high elution strength due to their aromatic structure (aromatic anions tend to have high affinity for anion-exchange sites in styrenic ion-exchange materials) and because they can be divalent when operated at a suitable mobile phase pH. Because such ions can interact with the stationary phase through both electrostatic and van der Waals forces, it is important to control the temperature of the column when using such mobile phases. Otherwise, substantial deviation in the baseline conductivity is observed, compromising the ability to detect analytes of interest.

Cation-Exchange Eluents in Suppressed Ion Chromatography

The optimum eluent species from the point of view of suppressed conductivity detection in cation-exchange is the hydronium cation.

When suppressed, it provides the lowest possible background conductivity: the conductivity of deionized water. However, styrene sulfonate cation-exchange materials tend to have a very high affinity for divalent cations relative to the affinity of monovalent cations. As such, hydronium tends to provide inadequate elution power for styrene sulfonate cation-exchange materials, necessitating the use of an additive to boost the elution strength of hydronium. A number of cations have been utilized for this purpose, but the most widely used is diaminopropionic acid. This cation acts as a divalent cation under low pH conditions typical of IC, and it has an isoelectric point close to pH 7, thus contributing very little background conductivity upon suppression. A typical eluent utilizes 40 mM methanesulfonic acid containing 4 mM diaminopropionic acid (Fig. 6).

Most cation-exchange phases designed over the last 30 years have been designed for use with hydronium-based mobile phases. They circumvent the issues associated with styrene sulfonate by making use of carboxylate stationary phases. Because carboxylate ion-exchange sites are much weaker acids than styrene sulfonate ion-exchange sites, they have a much higher affinity for hydronium. Consequently, it is possible to elute both monovalent cations and divalent cations using simple hydronium-based mobile phases with columns employing carboxylate ion-exchange sites. In principle any strong acid can be used as a hydronium source, but in practice there are several considerations that dictate the choice of the acid used. Several acids including hydrogen chloride and nitric acid cause damage to the suppressor when operated in the electrolytic mode due to the formation of highly oxidizing byproducts in the anode chamber of the suppressor. Therefore, the most commonly used acids are sulfuric acid and methanesulfonic acid, with the latter being the most widely used in suppressed cation-exchange IC.

Cation-Exchange Eluents in Nonsuppressed Ion Chromatography

To minimize background conductivity for cation-exchange separations during operation in the nonsuppressed mode, the optimum eluent species must have high elution power with relatively low conductivity compared with other cations. Common eluent species are weak acid chelating agents such as citric acid, tartaric acid, and dipicolinic acid used alone or in combination with additives such as acetone, which further reduces the conductivity of the eluent acids, or 18-crown-6, which beneficially modifies the selectivity of monovalent cations. Although such eluent systems exhibit relatively high background conductivities and thus adversely affect method detection limits, they advantageously affect calibration linearity for weakly basic amines such as ammonia, ethanolamine, diethanolamine, and triethanolamine. Such bases tend to have nonlinear calibration characteristics during operation in the suppressed mode.

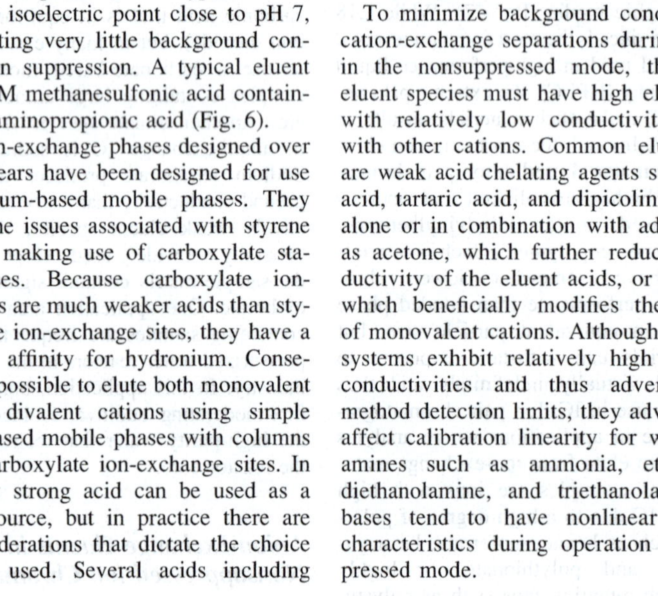

Fig. 6 Comparison of the isocratic separation of common cations with the OmniPac PCX-100 (A) and the IonPac CS10 column (B). Column dimensions: 4 mm ID by 250 mm; eluent: 40 mM hydrogen chloride and 4 mM diaminopropionic acid; flow rate: 1 mL/min; injection volume: 25 µL; temperature: 24 °C; suppressor: CMMS; conductivity detection. Peaks: (1) lithium, 0.5 mg/L; (2) sodium, 2 mg/L; (3) ammonium, 2.5 mg/L; (4) potassium, 5 mg/L; (5) magnesium, 15 mg/L; (6) calcium, 25 mg/L. Courtesy of Thermo Fisher Scientific

Eluents in Ultraviolet-Visible Detection Ion Chromatography of Transition Metals

When separating transition metals, it is critical to use chelating mobile phases because cation-exchange materials generally exhibit minimal selectivity differences for transition metals with identical net charge. The elution can be accomplished either by cations, which selectively elute transition metals from cation-exchange sites in the stationary phase, or by anions, which reduce retention through the formation of anionic complexes that indirectly affect selectivity in the cation-exchange mode. Alternatively, transition metals can be separated as anionic complexes via anion exchange. Because detection generally involves the post-column introduction of a complexing chromophore, it is important to choose an eluent species that does not compete with the complexing chromophore. Otherwise, the sensitivity of the detection method will be compromised. Common eluent species for operation in the cation-exchange mode include tartaric acid, oxalic acid, citric acid, and hydroxybutyric acid. The most common eluent species for operation in the anion-exchange mode is dipicolinic acid.

Ion Exclusion Eluents in Ion Chromatography

During operation in the ion exclusion mode, retention control is achieved by manipulating the mobile phase pH. Retention is increased by decreasing the mobile phase pH. Common eluent species for operation in the suppressed mode include hexanesulfonic acid, perfluorobutyric acid, and sulfuric acid. Ion exclusion suppressors replace hydronium with a low-conductivity cation such as the tetrabutylammonium ion to minimize the conductivity of the eluent species while increasing calibration linearity for the organic acids separated via ion exclusion. As such, larger anions such as hexanesulfonate and perfluorobutyrate provide the lowest equivalent conductivity, beneficially reducing background conductivity. Because the suppression byproduct is a salt (e.g., tetrabutylammonium hexanesulfonate), the background conductivity in suppressed ion exclusion chromatography and the corresponding detection limits are significantly higher when organic acids are detected using this mode of suppression. During operation in the UV detection mode, ion exclusion chromatography normally utilizes sulfuric acid as the mobile phase because it has negligible UV absorbance at the normal operating wavelength (200 nm). Because the extinction coefficient for carboxylic acids is quite low even at 200 nm, detection limits during operation in the ion exclusion mode are generally superior when conductivity detection is used, except for organic acids that contain a UV chromophore. Reversed phase separation is an additional separation mode that utilizes the aforementioned ion exclusion suppression mechanism. In this case, low-millimolar strong acid eluent is employed to enable separation of organic acids in a fully protonated state using a reversed phase column. Subsequent detection utilizes ion exclusion suppression to convert carboxylic acids to a more highly conducting salt form prior to detection.

Ion Pair Eluents in Ion Chromatography

During operation in the suppressed mode, the choice of the mobile phase is based on multiple factors, including the suppressibility of the ion pair reagent, the affinity of the ion pair reagent for the reversed phase column, and the background conductivity of the associated counterion of the ion pair reagent. As with conventional suppressed IC, the lowest background conductivity is achieved when the ion pair reagent counterion is either hydronium or hydroxide. Commonly used ion pair reagents for the analysis of anions include tetrapropylammonium hydroxide, tetrabutylammonium hydroxide, hexanesulfonic acid, octanesulfonic acid, and perfluorobutyric acid. In general, the ion pair reagent with the highest number of carbon atoms provides the highest retention at a given concentration as well as the greatest challenge with respect to suppression capacity. Because of the relatively low suppression capacity for large ions, the concentration of such reagents in the mobile phase is typically relatively low, in the 2 to 5 mM range. During operation in the nonsuppressed mode, counterions are chosen to minimize the conductivity of the ion pair. Since in the nonsuppressed mode suppressibility is not an issue, operation with higher-affinity ion pair reagents such as dodecyltrimethyl ammonium chloride and sodium laurylsulfate is a suitable option. In this case as well, the concentrations are kept to a minimum, generally less than 5 mM.

Gradients in Ion Chromatography

As with most modes of chromatography, it is frequently useful in IC to utilize concentration gradients to solve the "general elution problem," where resolution of weakly retained analytes excessively extends the retention of strongly retained analytes unless a gradient is applied after the elution of the weakly retained analytes. Although there were early attempts at gradient elution in nonsuppressed IC, gradients are virtually never used in nonsuppressed IC due to excessive baseline drift as the concentration of the mobile phase increases. By contrast, at least when hydroxide or hydronium mobile phases are used, gradients are commonly employed in suppressed IC. Because the background conductivity of suppressed hydroxide or suppressed hydronium is nearly zero, the effect of a concentration gradient is negligible under most conditions, resulting in a baseline with minimal shift in conductivity even when the concentration of the mobile phase increases by 2 orders of magnitude. A side benefit of using the gradient mode is that the concentration gradient across each peak as the analyte is eluted results in peak focusing and corresponding increases in peak height. The major disadvantage of gradient chromatography is the time required for re-equilibration at the completion of each gradient. Given this trade-off between time lost due to re-equilibration and time gained by reduced retention, gradient elution is typically used only in cases where the analysis time is already rather long and the sample is rather complex.

Eluent Generators as an Ion Source

Just as an electrolytic suppressor can be used to remove ions from the mobile phase, the reverse process can be used to add ions to the mobile phase. For example, if a destination chamber containing a negatively charged electrode is separated by a cation-exchange membrane from a solution of concentrated potassium hydroxide containing a positively charged electrode, applying a voltage between the two electrodes sufficient to drive potassium ions through the cation-exchange membrane produces potassium hydroxide in the receiver chamber (Fig. 7). The concentration produced by this process is a function of the current of ions to the membrane and the flow rate of the receiver solution. Because both parameters can be controlled with high precision, it is possible to produce eluent with much greater accuracy than is possible by mechanical means using proportioning valves as is common in HPLC. Furthermore, the concentration of the source solution can be 50 to 100 times greater than the solution concentration in the receiver chamber; consequently, such an eluent generation device can easily last 6 to 12 months before it needs to be replaced. Because the source solution is deionized water, the pump is exposed only to deionized water, which extends the life of the pump and the pump seals. Furthermore, especially in the case of hydroxide, contamination from environmental sources such as carbon dioxide or ammonia affects the background conductivity of manually prepared hydroxide solutions. Thus, electrolytically produced mobile phases result in background conductivities that are significantly lower than those associated with manually prepared hydroxide or hydronium solutions (Fig. 8). This is especially true when an electrolytic eluent generator is used in combination with a continuously regenerated trap column (CR-TC), a device that can remove trace contaminants originally present in the feed deionized water. The combination of an electrolytic eluent generator, a CR-TC, and a suppressor after the column can produce background conductivities of 0.1 μS or lower, less than 1/10 of what is achievable with manually prepared eluents. An elevated background conductivity not only increases the noise of the system but also increases baseline drift when a concentration gradient is used to reduce analysis time and increase sensitivity. Thus, this combination of components provides the convenience of high precision, low maintenance, and high sensitivity, which is especially useful during operation in the gradient mode.

Analyte Range in Ion Chromatography

While the original scope of IC was limited to the small set of ions commonly found in

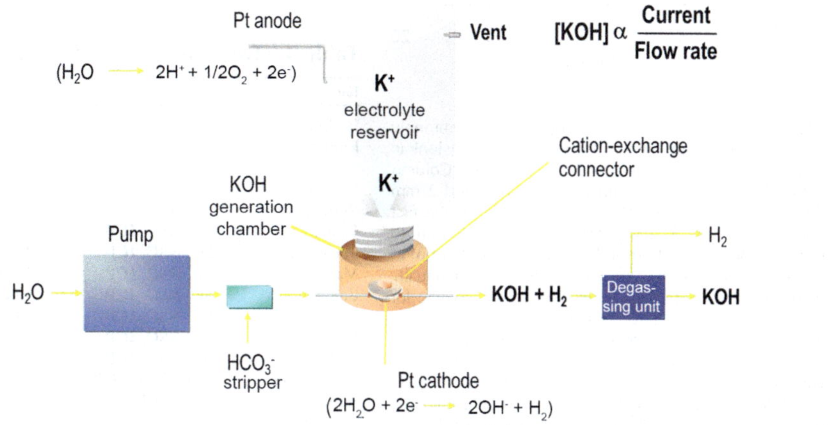

Fig. 7 Schematic representation of an eluent generator configured for production of potassium hydroxide. Courtesy of Thermo Fisher Scientific

drinking water along with a few common organic acids and amines, the full scope of IC has been vastly expanded over the past 40 years. In the case of anions, analytes suitable for analysis via IC range from the most weakly retained anions such as fluoride and methyl phosphonate, to highly polarizable anions such as perchlorate, perrhenate, and hexafluorophosphate, to highly charged ions such as tripolyphosphate and phytic acid. Likewise, cations suitable for analysis via IC range from the most weakly retained cations such as lithium, to simple amines such as ammonia and triethanolamine, to highly polarized cations such as cesium and barium, and to highly charged cations such as spermine and spermidine. Generally, if both the salt form and the acid or base form are at least marginally soluble in water, they can be analyzed by IC using conductivity detection. Solvent can be added to the mobile phase to enhance the solubility of marginal analytes. The nature of suppressed conductivity detection results in relatively limited sensitivity for very weak acids and very weak bases since they tend to have limited ionization in the free acid or free base form, respectively. Such analytes are more effectively analyzed using nonsuppressed IC or by adding trace levels of base in the case of weak acids or acid in the case of weak bases after the suppressor but before the conductivity detector to enhance ionization of the analyte. Tables 3 to 6 detail a wide variety of analytes with the detection modes generally employed for each ion or class of ions.

Sample Preparation in Ion Chromatography

Liquid samples are generally the easiest to handle in IC. If the sample is an aqueous solution, it is generally sufficient to directly inject the sample. High concentrations of analyte can overload the column, so it is generally advisable to inject aqueous samples containing less than 0.1% total solids. Higher-concentration samples should be diluted in deionized water as required to reduce the concentration to a useful range. If the sample is dissolved in an aqueous miscible fluid, it is generally still acceptable to directly inject the sample, although it is advisable to start with a relatively small injection volume to minimize matrix effects on early eluting peaks. If the sample solvent is an aqueous immiscible solvent, there are three basic options: perform a liquid-liquid extraction using water as the second solvent and analyze the aqueous fraction; add solvent to the mobile phase to improve the solubility of the sample solvent; or use a concentrator column in the matrix elimination mode, passing the sample solution over the concentrator, rinsing any remaining solvent from the concentrator using a water solvent mixture, and then switching the concentrator in line with the separation column for subsequent analysis.

Handling of solid samples depends upon the nature of the sample. If analytes of interest are mixed with an insoluble matrix such as a soil sample, samples can be analyzed by slurring in deionized water, stirring or sonicating the sample to improve the dissolution kinetics, and then filtering the sample to remove any particulate. If the solid sample includes

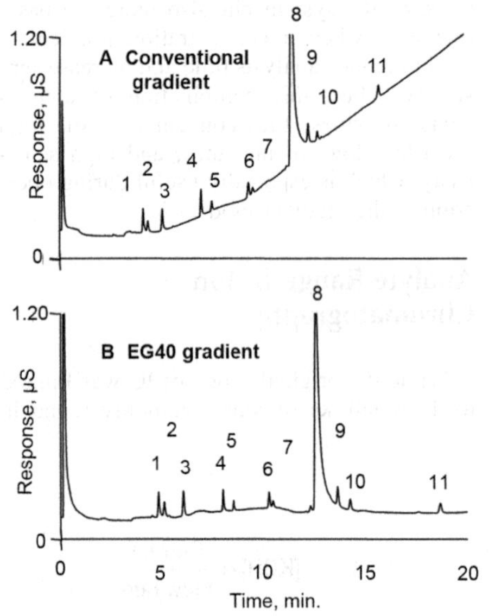

Fig. 8 Gradient separation of anions commonly found in high-purity water with the IonPac AS11 column and the AG11 column. Column dimensions: 2 mm (0.08 in.) ID by 50 mm and 2 mm ID by 250 mm; eluent: 0.5 mM hydroxide for the first 2.5 min, 0.5 to 5 mM hydroxide from 2.5 min to 6 min, and 5 to 26 mM hydroxide from 6 min to 20 min; eluent source for chromatogram A, manually prepared sodium hydroxide eluent; eluent source for chromatogram B, EG40 eluent generator, potassium hydroxide eluent; flow rate: 0.5 mL/min; injection volume: 1 mL; temperature: 30 °C (85 °F); suppressor: ASRS; conductivity detection. Peaks: (1) fluoride, 0.37 μg/L; (2) acetate, 1 μg/L; (3) formate, 0.93 μg/L; (4), chloride, 0.44 μg/L; (5), nitrite, 0.27 μg/L; (6) bromide, 1 μg/L; (7) nitrate, 33 μg/L; (8) carbonate; (9) sulfate, 0.64 μg/L; (10) oxalate, 0.39 μg/L; (11) phosphate, 1.1 μg/L. Courtesy of Thermo Fisher Scientific

Table 3 Detection modes for inorganic anions

Ion	Detection mode	Ion	Detection mode	Ion	Detection mode
Fluoride	**SC**, NSC	Hypophosphite	R	Sulfide	NSC, **EC,** UV
Hypochlorite	R	Phosphite	**SC**, NSC	Sulfite	**SC**, NSC, EC, UV
Chlorite	**SC**, NSC, PCR, UV	Phosphate	**SC**, NSC	Sulfate	**SC**, NSC
Chloride	**SC**, NSC, EC, UV	Pyrophosphate	**SC**, NSC, PCR	Thiosulfate	**SC**, NSC, EC, UV
Chlorate	**SC**, NSC, UV	Tripolyphosphate	**SC**, NSC, PCR	Persulfate	**SC**, NSC
Perchlorate	**SC**, NSC	Trimetaphosphate	**SC**, NSC, PCR	Dithionate	**SC**, NSC, UV
Bromate	**SC**, NSC, PCR, UV	Tetrapolyphosphate	**SC**, NSC, PCR	Trithionate	**SC**, NSC, UV
Bromide	**SC**, NSC, EC, UV	Tetrametaphosphate	**SC**, NSC, PCR	Tetrathionate	**SC**, NSC, UV
Iodate	**SC**, NSC, PCR	Hexametaphosphate	SC, NSC, **PCR**	Thiocyanate	SC, NSC, **UV**
Iodide	**SC**, NSC, EC, UV	Polyphosphate	SC, NSC, **PCR**	Selenite	**SC**, NSC, EC, UV
Nitrite	**SC**, NSC, EC, UV	Hexafluorophosphate	**SC**, NSC	Selenate	**SC**, NSC, UV
Nitrate	**SC**, NSC, UV	Silicate	NSC, **PCR**	Selenocyanate	SC, NSC, **UV**
Azide	**SC**, NSC, UV	Hexafluorosilicate	R	Tungstate	**SC**, NSC
Carbonate	SC, **NSC**	Borate	NSC, **PCR**	Molybdate	**SC**, NSC
Cyanate	**SC**, NSC	Tetrafluorborate	**SC**, NSC	Chromate	SC, NSC, **PCR**, UV
Cyanide	SC, NSC, **EC**	Permanganate	R	Perrhenate	**SC**, NSC
Hydroxide	**NSC**, UV	Hydrogen peroxide	NSC, **EC**	Pertechnetate	**SC**, NSC

Bold indicates preferred detection mode. SC, suppressed conductivity detection; NSC, nonsuppressed conductivity detection; UV, ultraviolet detection; EC, electrochemical detection; PCR, postcolumn reaction with visible or UV detection; R, reacts with ion exchange material or eluent, not suitable for ion exchange. Courtesy of Thermo Fisher Scientific

Table 4 Detection modes for inorganic cations

Ion	Detection mode	Ion	Detection mode	Ion	Detection mode
Hydronium	**NSC**	Scandium	NSC, PCR, **ICP**	Copper	NSC, PCR, **ICP**
Lithium	SC, NSC, **ICP**	Yttrium	NSC, PCR, **ICP**	Silver	NSC, PCR, **ICP**
Sodium	SC, NSC, **ICP**	Titanium	NSC, PCR, **ICP**	Zinc	NSC, PCR, **ICP**
Potassium	SC, NSC, **ICP**	Zirconium	NSC, PCR, **ICP**	Cadmium	NSC, PCR, **ICP**
Rubidium	SC, NSC, **ICP**	Hafnium	NSC, PCR, **ICP**	Mercury	NSC, PCR, **ICP**
Cesium	SC, NSC, **ICP**	Vanadium	NSC, PCR, **ICP**	Aluminum	NSC, PCR, **ICP**
Francium	SC, NSC, **ICP**	Niobium	NSC, PCR, **ICP**	Gallium	NSC, PCR, **ICP**
Magnesium	SC, NSC, **ICP**	Tantalum	NSC, PCR, **ICP**	Indium	NSC, PCR, **ICP**
Calcium	SC, NSC, **ICP**	Chromium	NSC, PCR, **ICP**	Thallium	NSC, PCR, **ICP**
Strontium	SC, NSC, **ICP**	Manganese	NSC, PCR, **ICP**	Tin	NSC, PCR, **ICP**
Barium	SC, NSC, **ICP**	Iron	NSC, PCR, **ICP**	Lead	NSC, PCR, **ICP**
Radium	SC, NSC, **ICP**	Cobalt	NSC, PCR, **ICP**	Antimony	NSC, PCR, **ICP**
Lanthanides	NSC, PCR, **ICP**	Nickel	NSC, PCR, **ICP**	Bismuth	NSC, PCR, **ICP**
Actinides	NSC, PCR, **ICP**				

Bold indicates preferred detection mode. SC, suppressed conductivity detection; NSC, nonsuppressed conductivity detection; UV, ultraviolet detection; EC, electrochemical detection; PCR, postcolumn reaction with visible or UV detection; ICP, inductively coupled plasma detection or inductively coupled plasma mass spectrometry. Courtesy of Thermo Fisher Scientific

Table 5 Detection modes for organic anions

Ion	Detection mode	Ion	Detection mode
Monovalent aliphatic carboxylic acids, C1–C8	**SC**, NSC, UV	Monovalent aliphatic carboxylic acids, C9–C20	**ISC**, NSC, UV
Monovalent olefinic carboxylic acids, C1–C8	**SC**, NSC, UV	Monovalent olefinic carboxylic acids, C9–C20	**ISC**, NSC, UV
Monovalent hydroxycarboxylic acids, C1–C10	**SC**, NSC, UV	Monovalent hydroxycarboxylic acids, C11–C24	**ISC**, NSC, UV
Monovalent ketocarboxylic acids, C1–C8	**SC**, NSC, UV	Monovalent ketocarboxylic acids, C9–C20	**ISC**, NSC, UV
Divalent aliphatic carboxylic acids, C1–C12	**SC**, NSC, UV	Divalent aliphatic carboxylic acids, C13–C24	**ISC**, NSC, UV
Divalent olefinic carboxylic acids, C1–C12	**SC**, NSC, UV	Divalent olefinic carboxylic acids, C13–C24	**ISC**, NSC, UV
Divalent hydroxycarboxylic acids, C1–C14	**SC**, NSC, UV	Divalent hydroxycarboxylic acids, C15–C24	**ISC**, NSC, UV
Divalent ketocarboxylic acids, C1–C12	**SC**, NSC, UV	Divalent ketocarboxylic acids, C13–C24	**ISC**, NSC, UV
Haloacetic acids	**SC**, NSC, UV	Monovalent halocarboxylic acids C3–6	**SC**, NSC, UV
Uronic acids	**SC**, NSC, EC	Phosphated carbohydrates	**SC**, NSC, EC
Sulfated carbohydrates	**SC**, NSC, EC	Phytic acid	**SC**, NSC
Phenates	NSC, **UV**, EC	Arylcarboxylates	SC, NSC, **UV**
Alkylphosphonates	SC, NSC, **UV**	Alkylsulfates	SC, NSC, **UV**
Alkylsulfonates	SC, NSC, **UV**	Arylsulfonates	SC, NSC, **UV**

Bold indicates preferred detection mode. SC, suppressed conductivity detection; NSC, nonsuppressed conductivity detection; UV, ultraviolet detection; EC, electrochemical detection. Courtesy of Thermo Fisher Scientific

Table 6 Detection modes for organic cations

Ion	Detection mode	Ion	Detection mode
Monovalent aliphatic amines, C1–C8	**SC**, NSC	Monovalent aliphatic carboxylic acids, C9–C20	SC, **NSC**
Hydroxylamines	**SC**, NSC, EC	Hydrazines	SC, NSC, **EC**
Alkanolamines, C1–C8	**SC**, NSC, UV	Alkanolamines, C9–C20	SC, **NSC**
Diamines, C1–C12	**SC**, NSC, UV	Diamines, C13–C24	SC, **NSC**, UV
Polyamines	**SC**, NSC, UV	Spermine	**SC**, NSC
Spermidine	**SC**, NSC, UV	Cadaverine	**SC**, NSC
Histamine	SC, NSC, **UV**	Agmatine	**SC**, NSC
Pyridinium compounds	SC, NSC, **UV**	Anilinium compounds	SC, NSC, **UV**
Quaternary ammonium compounds	**SC**, NSC	Sulfonium compounds	**SC**, NSC
Phosphonium compounds	**SC**, NSC		

Bold indicates preferred detection mode. SC, suppressed conductivity detection; NSC, nonsuppressed conductivity detection; UV, ultraviolet detection; EC, electrochemical detection. Courtesy of Thermo Fisher Scientific

components that have ion-exchange properties, it might be necessary to add an electrolyte to quantitatively release analytes of interest. Classical soil analysis protocols often used concentrated electrolyte solutions for this purpose, but such solutions can easily overwhelm the IC column and should generally be avoided. A better option is to use low-ionic-strength analogs of the original method since often the concentration specified is much greater than the minimum concentration required. Better still is to use the mobile phase as the extraction solvent since this will be transparent to the final analysis. Of course, in this case, recoveries must be validated against established extraction electrolyte methods. Caustic fusion can be used to take glass and mineral samples into aqueous solution. The powdered sample is mixed with a flux, such as sodium carbonate or sodium metaborate, in a porcelain or platinum crucible and heated to the molten state for a suitable period. The cooled melt is then taken into solution with hot deionized water. These solutions have a high ionic background that must be considered when selecting a separation and detection mode for analysis.

Ion chromatography has been used to analyze organic solids for such inorganic elements as the halogens, sulfur, and phosphorus. The sample is prepared by burning it in a closed container and dissolving the combustion products into aqueous solution. Schöniger or oxygen-flask combustion is the technique most commonly used to burn organics under controlled conditions. The weighed sample is wrapped in a special paper and placed in a platinum sample carrier to be suspended inside the combustion flask. Immediately before combustion, the flask, containing a small volume of suitable absorbing solution, is swept thoroughly with oxygen. Parr oxygen bombs have been used to determine trace levels of inorganics in organic materials. Although more cumbersome, the apparatus can handle larger sample sizes than a Schöniger flask, and high-purity combustion aids are available to minimize sample blanks.

Removal of Particulates

For samples containing particulates, it is generally advisable to remove them prior to injection into the ion chromatograph to avoid clogging of the chromatographic column, possibly necessitating its replacement. Generally, the most convenient method for removing particulates from samples is to use a disposable single-use filter cartridge designed for use with a disposable syringe. It is advisable to select a filter certified for use in IC, since filter cartridges contain substantial levels of ion contaminants as received from the supplier. Likewise, it is useful to verify that disposable syringes are free from any ionic contaminants or are suitably cleaned prior to use to avoid blank problems. An alternative option is to use a centrifuge to remove suspended particulates prior to analysis. Some samples, for example, orange juice, clog cartridge syringe filters easily, making them unsuitable for such applications. Although slower than using a syringe filter, a centrifuge can quantitatively remove suspended particulates for such difficult samples. In extreme cases, such as the analysis of ions in toothpaste, it may be necessary to first dilute and centrifuge the sample and then use a small particle syringe filter to remove any remaining particles.

Standardization

Ion chromatography can be used to identify ions present in solution and to quantitatively determine the concentrations of ions in solution. In both cases, it is common to use commercially available stock solutions containing analytes of interest for calibration. It is best to avoid use of stock solutions designed for elemental spectroscopy, as these solutions tend to be diluted in mineral acids. Look for suppliers providing standards designed specifically for IC to avoid this issue. If calibration solutions are not commercially available, it is often convenient to make 1000 ppm stock solutions for each of the analytes of interest and then make a calibration stock mixture of the analytes of interest, choosing concentrations that are representative of the concentrations in samples. In most cases, a single-point calibration is sufficiently accurate, assuming analyte concentrations are within a factor of 5 of the concentration of the same ion in the standard. When working with samples where ions of interest vary over a wide range, it is useful to utilize a multipoint calibration. To create a calibration plot, prepare several standard dilutions covering typical concentration ranges. It is generally preferable not to calibrate using standards with concentrations more than 100 ppm, as such concentrations generally overload most commercially available IC columns using standard injection volumes. While data systems can handle data derived from overloaded columns, generally the accuracy and precision of the quantitation is degraded under such conditions.

For the most part, inorganic ion stock solutions and standards are quite stable and can be stored at room temperature, assuming the sample container is clean and free from ionic contaminants. However, organic-acid- or amine-containing standards are subject to degradation by microbial contaminants and should be stored under refrigerated conditions to minimize sample degradation. Alternatively, adding a drop of chloroform to a 100 mL standard solution generally prevents microbial degradation of standards containing organic acids or amines without compromising the analytical performance of the standard.

Preliminary confirmation of the identity of unknown analytes can be accomplished by comparing the retention time of an unknown analyte with that of analytes in the calibration standard. If in doubt, the sample can be spiked with the suspect ion at a similar concentration. If the efficiency of the peak does not degrade significantly upon spike addition, this indicates a possible match. Further confirmation can be achieved by repeating the analysis on a different column with significantly different selectivity. Alternatively, coupling the ion chromatograph to a mass spectrometer allows for confirmation of the molecular weight of the unknown analyte in comparison with the putative assignment of ion identity.

Instrumentation

There are a number of manufacturers of IC instrumentation. Most provide not only pumping systems optimized for IC eluents but also temperature-controlled chromatography compartments and temperature-controlled conductivity detectors to minimize baseline drift and detector noise. Most manufacturers also provide suppressors for anion applications, and several provide suppressors for cation applications. While on the surface IC instrumentation appears to be quite like HPLC instrumentation, it is desirable to use nonmetallic hardware for IC applications due to the extreme pH conditions typical of IC. While stainless steel hardware might seem suitable for relatively noncorrosive mobile phases such as potassium hydroxide or methanesulfonic acid, alternating between acidic and basic mobile phases can remove the passivation layer from stainless steel, resulting in considerable corrosion, contaminating columns, suppressors, and detectors with corrosion byproducts. Thus, it is generally advisable to use an instrument designed specifically for IC rather than adapting an HPLC instrument for IC. All suppliers offer data analysis software and autosamplers for automation of sample introduction, data analysis, and reporting.

Applications

By far the most common application of IC is the analysis of drinking water (Fig. 9). Modern columns allow for the analysis of anions commonly found in drinking water as well as a variety of disinfectant byproducts formed during the disinfection process.

Analysis of ions in high-purity water is another common application of IC. Two approaches are commonly employed for high-purity water analysis: large loop injection and use of a concentrator column. Large loop injection is relatively easy to implement, but injection of volumes greater than 2 mL is uncommon due to the added time associated with this mode of analysis. For example, when a 4 mm column with a 1 mL/min flow rate is used, injecting a 2 mL sample will add 2 min to the analysis time. Large loop injection also tends to compromise retention time reproducibility, as it takes a significant amount of time to take the loop from atmospheric pressure to column pressure, which can affect the flow rate downstream. A concentrator column method allows for unlimited sample volumes to be concentrated without directly affecting the analysis time. The concentration process can potentially take as long as the analysis process, but modern software allows for overlap of the two processes such that the concentration process generally does not compromise the analysis speed.

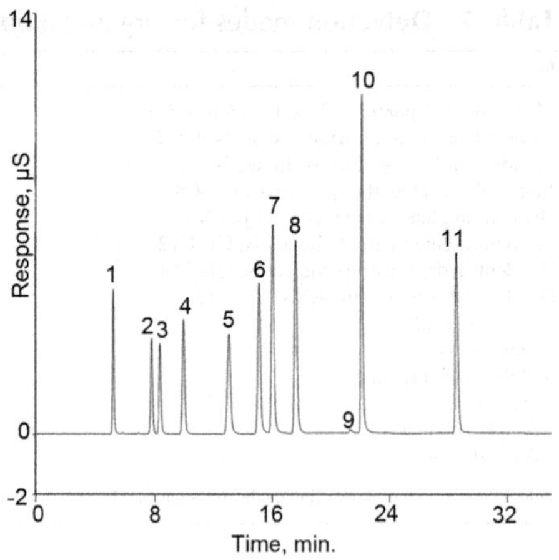

Fig. 9 Separation of anions on 4 mm ID IonPac AS19-4µm and AG19 columns; eluent: 10 mM potassium hydroxide from 0 to 10 min, 10 to 45 mM potassium hydroxide from 10 to 25 min; flow rate: 1 mL/min; injection volume: 10 µL; temperature: 30 °C; suppressed conductivity detection. Peaks: (1) fluoride, 3 mg/L; (2) chlorite, 10 mg/L; (3) bromate, 20 mg/L; (4) chloride, 6 mg/L; (5) nitrite, 15 mg/L; (6) bromide, 25 mg/L; (7) chlorate, 25 mg/L; (8) nitrate, 25 mg/L; (9) carbonate; (10) sulfate, 25 mg/L; (11), phosphate, 40 mg/L. Courtesy of Thermo Fisher Scientific

Future Directions in Ion Chromatography

Although IC has evolved considerably since its initial introduction in 1975, the technique continues to evolve. As in the case of HPLC, there has been a steady decline in the particle size of the chromatographic materials used in IC, resulting in improved sensitivity and reduced analysis time. As engineering science tackles the challenge of developing materials with sufficient strength and inertness to allow the application of such particles and the associated increases in pressure to IC, further reductions in particle size can be expected. Column capacities have increased substantially since the original introduction of IC as suppressors with increased capacity have become available. Higher column capacities provide a number of benefits including compatibility with samples containing high concentrations of matrix ions and longer column life when such columns are exposed to challenging samples. As IC continues to evolve, further increases in suppression and column capacity can be expected. Continued development of new column chemistries expanding the range of analytes that can be eluted with hydronium and hydroxide eluent systems can also be expected. A perfect example of this trend is the recent introduction of the IonPac AS32 column, which for the first time allows isocratic elution of tetrathiocyanate and persulfate with hydroxide eluent systems. Finally, improvements in instrumentation technology and software make the application of two-dimensional (2D) IC increasingly practical. Two-dimensional applications provide a number of advantages including the ability to quantitate trace components in a matrix containing much more concentrated matrix ions, the ability to simplify and automate sample prep, and the ability to enhance detection sensitivity. The latter feature is enabled when 2D applications are combined in such a way that the first dimension is significantly larger in internal diameter than the second dimension. Assuming a suppressor is used after the first dimension to remove the eluent prior to injection into the second dimension, sensitivity enhancement should be directly proportional to the ratio of the cross-sectional areas of the two dimensions. For example, a 4 mm (0.16 in.) internal diameter (ID) column in the first dimension coupled to a 0.4 mm ID column in the second dimension should result in a 100-fold enhancement of concentration sensitivity. Figure 10 shows an illustrative example of such a 2D application to the determination of trace levels of perchlorate in a simulated sample. In the future, specialized instrumentation developed explicitly to take full advantage of the power of 2D IC can be expected.

ACKNOWLEDGMENTS

This article was revised from Raymond M. Merrill, Ion Chromatography, *Materials Characterization*, Volume 10, *ASM Handbook*, ASM International, 1986, p 658–667.

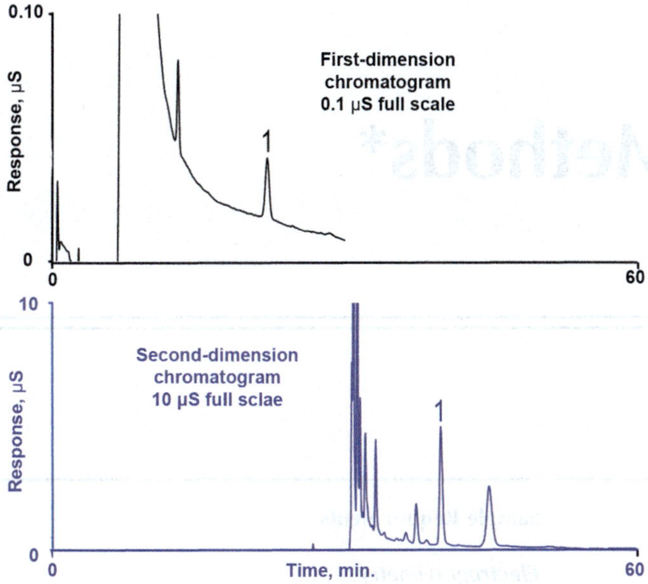

Fig. 10 Separation of perchlorate in the first dimension on 4 mm ID IonPac AS16 and AG16 columns and in the second dimension on a 0.4 mm (0.016 in.) ID IonPac AS20 column. First dimension: eluent, 65 mM potassium hydroxide; flow rate, 1 mL/min; injection volume, 4 mL; temperature, 30 °C; suppressed conductivity detection. Second dimension: eluent, 35 mM potassium hydroxide, flow rate: 10 μL/min, temperature, 30 °C; suppressed conductivity detection. Five milliliters of the suppressed first-dimension eluent (19 to 24 min) was diverted to a capillary concentrator and subsequently brought in line with the capillary IonPac AS20. Peak: (1) perchlorate, 1 μg/L. First-dimension perchlorate peak area: 0.0115 μS · min. Second-dimension perchlorate peak area: 1.75 μS · min. Courtesy of Thermo Fisher Scientific

REFERENCES

1. H. Small, T.S. Stevens, and W.C. Bauman, Novel Ion Exchange Chromatographic Method Using Conductimetric Detection, *Anal. Chem.*, Vol 47, 1975, 1801
2. C.A. Lucy, Evolution of Ion-Exchange: From Moses to the Manhattan Project to Modern Times, *J. Chromatogr. A*, Vol 1000, 2004, p 711–724
3. O. Samuelson, *Ion Exchange Separations in Analytical Chemistry*, Almqvist and Wiksell/Wiley, 1963
4. D. Jensen, J. Weiss, M. Rey, and C.J. Pohl, Novel Weak Acid Cation-Exchange Column, *J. Chromatogr. A*, Vol 640, 1993, p 65
5. J. Plicka, J. Cabicar, and A. Gosman, and K. Štamberg, Kinetics, Equilibrium and Isotope Exchange in Ion Exchange Systems, *J. Radioanal. Nucl. Chem.*, Vol 88, 1985, p 325
6. H. Small, *Ion Chromatography*, Plenum, 1989
7. J. Weiss, *Ion Chromatography*, 4th ed., Wiley, 2016
8. T.S. Stevens, J. C. Davis, and H. Small, Hollow Fiber Ion-Exchange Suppressor for Ion Chromatography, *Anal. Chem.*, Vol 53, 1981, p 1488
9. J. Stillian, An Improved Suppressor for Ion Chromatography, *LC*, Vol 3, 1985, p 802
10. C. Pohl, R. Slingsby, J. Stillian, and R. Gajek, Modified Membrane Suppressor and Method for Use, U.S. Patent 4,999,098, 1991
11. J. Stillian, V. Barreto, K. Friedman, S. Rabin, and M. Toofan, Ion Chromatography System Using Electrochemical Suppression and Detection Effluent Recycle, U.S. Patent 5,352,360, 1994
12. C.A Pohl, Mechanisms: Ion Chromatography, *Elsevier Reference Module in Chemistry, Molecular Sciences and Chemical Engineering*, J. Reedijk, Ed., Elsevier, 2016
13. E. Ohyoshi, Relative Stabilities of Metal Complexes of 4-(2-pyridylazo)resorcinol and 4-(2-thiazolylazo)resorcinol, *Polyhedron*, Vol 5, 1986, p 1165–1170
14. W. Wu, Q. Xiao, P. Zhang, M. Ye, Y. Wan, and H. Liang, Rapid Measurement of Free Cyanide in Liquor by Ion Chromatography with Pulsed Amperometric Detection, *Food Chem.*, Vol 172, 2015, p 681–684
15. Z.I. Li, S.F. Mou, Z.M. Ni, and J.M. Riviello, Sequential Determination of Arsenite and Arsenate by Ion Chromatography, *Anal. Chim. Acta*, Vol 307, 1995, p 79–87
16. K. Tanaka and P.R. Haddad, Ion Exclusion Chromatography: Liquid Chromatography, *Encyclopedia of Separation Science*, I.D. Wilson, Ed., Academic Press, 2000, p 3193–3201
17. J. Ståhlberg, Liquid Chromatography: Ion Pair Liquid Chromatography, *Encyclopedia of Separation Science*, I.D. Wilson, Ed., Academic Press, 2000, 676–684
18. M. Doury-Berthod, P. Giampaoli, H. Pitsch, C. Sella, and C. Poitrenaud, Theoretical Approach of Dual-Column Ion Chromatography, *Anal. Chem.*, Vol 57, 1985, p 2257
19. S.M.R. Ullah, S.L. Adams, K. Srinivasan, and P.K. Dasgupta, Asymmetric Membrane Fiber-Based Carbon Dioxide Removal Devices for Ion Chromatography, *Anal. Chem.*, Vol 76, 2004, p 7084–7093

ASM Handbook, Volume 10, *Materials Characterization*
ASM Handbook Committee
DOI 10.31399/asm.hb.v10.a0006679

Electrochemical Methods*

Paris Svoronos, Queensborough Community College

Overview

General Uses

Electrogravimetry

- Removal of easily reduced ions before analysis by another method
- Quantitative determination of ions after removal of interfering ions
- Quantitative determination of metal ions in the presence of other metal ions
- Quantitative determination of metal ions using electrogravimetry in conjunction with other techniques

Controlled-potential coulometry

- Quantitative chemical analysis; major constituent assay of solutions, alloys, nonmetallic materials, and compounds
- Accuracy and precision are typically 0.1%.
- Primarily applicable to the transition and heavier elements
- Studies of electrochemical reaction pathways and mechanisms

Voltammetry and polarography

- Qualitative and quantitative analysis of metals and nonmetals in solutions of concentration 10^{-2} to 10^{-9} mol/L
- Multicomponent, effectively nondestructive repeatable analysis
- Elucidation of solute-solute and solute-solvent equilibria
- Kinetic investigations
- Structure determination in solution

Electrometric titration

- Widely applicable as a branch of volumetric analysis
- Automated determinations
- High-precision determinations
- Potentiometric continuous monitoring and process control

Potentiometric membrane electrodes

- Quantification of cationic and anionic substances
- Quantification of gaseous species in aqueous solutions
- Detector for analytical titrations

Sample Requirements

Electrogravimetry

- *Form:* Solution
- *Size:* Down to decigram amounts of solid
- *Preparation:* Solutions of analyte

Controlled-potential coulometry

- *Form:* Samples must be dissolved in a solvent suitable for electrolysis—usually aqueous solutions.
- *Size:* Quantity sufficient for 1 to 10 mg analyte per determination

Voltammetry and polarography

- *Form:* Solution, mostly aqueous
- *Size:* Cell capacities from 10 to 100 mL are common, but cells less than 1 mL have been devised for special purposes.
- *Preparation:* Bulk samples (solids or liquids) must be pretreated to obtain required species in an acceptable and manageable concentration range together with an excess concentration of electroinactive background electrolyte.

Electrometric titration

- *Form:* Any
- *Size:* Small, unless lacking homogeneity or of very low analyte concentration
- *Preparation:* If a solution, frequently none. Solids must be made into a suitable solution without loss of analyte; interfering substances must be masked or removed.

Potentiometric membrane electrodes

- *Form:* The measurement must ultimately be made in a solution; aqueous solutions are generally used.
- *Size:* Several milliliters of sample are typically needed, but samples of less than a microliter can be analyzed using a microelectrode. Samples as small as 0.5 mL can be measured using commercial gas sensors.

(continued)

*Adapted from Ref 1 to 4

- *Preparation:* Depending on the system, sample preparation can be extensive to remove interferents and to release the ion of interest from binding agents in the sample but is sometimes unnecessary.

Application Examples

Electrogravimetry

- Quantitative analysis of metals in alloys
- Precision analysis of metallurgical products and samples
- Quantitative measurement of microgram amounts of metals
- Quantitative determination of metals in the presence of other ions, such as chloride

Controlled-potential coulometry

- High-accuracy assays of nuclear fuels for uranium and plutonium
- Assays of electroplating solutions and alloys for gold, silver, palladium, or iridium
- Measurement of Fe^{3+}/Fe^{2+} ratios in ceramics
- Verification of standards for other analytical techniques, for example, titanium in titanium dioxide
- Determination of molybdenum in molybdenum-tungsten alloys
- Assays of organic compounds containing nitro groups

Voltammetry and polarography

- Analysis and characterization of metals in commercial chemicals, pharmaceuticals, high-purity metals, and alloys
- Monitoring of pollutant metals and nonmetals in foodstuffs, water, effluents, herbage, biological/medical systems, and petroleum
- Detection of herbicide and pesticide residues in plant and animal tissue
- Continuous monitoring of major and minor metallic and nonmetallic compounds in commercial electroplating baths

Electrometric titration

- Determinations in colored, turbid, or very dilute solutions that preclude use of chemical indication
- By electrogeneration, use of titrants that are unstable if stored or have no real lifetime
- Assay of primary standards
- Analyses that must be controlled remotely, for example, of radioactive samples
- Maintenance of constant conditions, for example, of pH or component concentrations, during fermentation, sludge treatment, and so on

Potentiometric membrane electrodes

- Activity or concentration determination of selected cationic, anionic, or gaseous species in a variety of materials, including supply waters, waste waters, plating baths, mineral ores, biological fluids, soils, food products
- Elemental analysis of organic compounds, especially for nitrogen and halide content
- Titrimetric determination of major components in alloys
- Detector for chromatographic processes
- Detector to follow chemical reaction kinetics

Advantages

Electrogravimetry

- More accurate (0.1%) than controlled-potential coulometry (0.2 to 5%) or polarography (2%)

Controlled-potential coulometry

- Shorter analysis time than electrogravimetry

Voltammetry and polarography

- Provides direct electrochemical measurement of metals and nonmetals and is the basis of amperometric titration

Electrometric titration

- Submicrogram amounts of analyte can be determined, because a single drop of solution can be titrated.
- Electrical quantities need not be measured absolutely, except in coulometric titration. Because results are determined from changes in these quantities, instrumental requirements and conditions are usually less rigid than indirect-measurement determinations.

Potentiometric membrane electrodes

- Insensitive to sample turbidity
- Short analysis times
- Small sample volume requirement
- Simple to operate
- Inexpensive
- Portable
- Easy to automate
- Measures activity

Limitations

Electrogravimetry

- Limited to the analysis of ionizable species; complete deposition (99.5%) is essential for precise results
- Time-consuming and more skill required to eliminate interferences from codeposition

Controlled-potential coulometry

- Not recommended for trace or minor impurity analysis
- Requires dissolution of samples—destructive analysis technique
- Not useful for determination of alkali and alkaline earth metals
- Qualitative knowledge of sample composition is required
- For inorganic constituent analysis, organic constituents must be completely destroyed.
- Requires good electrolysis cell design and potential control
- May require elimination of oxygen from sample solutions

(continued)

Voltammetry and polarography

- Sample preparation may sometimes be time-consuming relative to the usually short probe time.
- Interference from electrochemical signals of species other than those whose analysis is required
- Restricted anodic or cathodic ranges of some electrode materials
- Perfect renewal of electrode surface between analyses is not always feasible.
- Complexities in electrochemical or chemical behavior of required species in solution may prevent straightforward detection.

Electrometric titration

- Precision and accuracy are usually better when the titration determines a single analyte, although the successive titration of several analytes may be possible.
- Concentration and reaction speed must be high enough to permit rapid equilibration and to yield reproducible end points. Approximately 10^{-5} *M* may be taken as the lowest practical limit, although titration at lower concentrations is sometimes possible.

Potentiometric membrane electrodes

- Potential drift
- Interferences
- Often requires aqueous solutions

Capabilities and Related Techniques

Electrogravimetry

- *Estimated analysis time*: After sample preparation, electrolysis requires 15 min to well over an hour, depending on conditions.
- *Coulometric titrations:* Applicable for all volumetric reactions
- *Atomic absorption spectrophotometry:* Quantitative and qualitative determination of metal ions in the presence of other ions
- *X-ray fluorescence:* Qualitative and quantitative determination of elements

Controlled-potential coulometry

- *Estimated analysis time:* 5 to 30 min per measurement, after solution preparation
- Determination of major constituents, metals, and certain organics; study of electrochemical reactions. Suitable for determinations requiring greater selectivity than obtainable by constant-current techniques. Small-scale preparations
- Quantitative determination of metal ions
- *Controlled-potential electrolysis:* Can be used as a separation technique before other measurement techniques
- *Polarography and solid-electrode voltammetry:* These techniques measure electrolysis current-potential characteristics of solution samples, provide more global qualitative information on sample composition, and are more suitable for minor constituent and trace analysis but not as accurate or precise.

Voltammetry and polarography

- *Estimated analysis time*: 15 min to 3 h per sample, depending on sample-preparation time. In batch analysis, the probing of individual solutions for several components may require only minutes.
- *Coulometry:* Small amounts of reagents may be electrogenerated without the need for standardization and storage of dilute solutions. Analysis possible in same lower regions as voltammetry but without the same versatility of multielement analysis
- *Potentiometry:* The exceptional sensitivity of many ion-sensitive systems makes them widely and often simply used as direct probes and in titration methods. Some redox couples may be slow in establishing equilibrium at indicator electrodes.
- *Amperometry:* Offers more flexibility in selection of convenient solid electrode materials, because electrode history is less significant in titration techniques. May be used at low concentrations at which other titration methods are inaccurate (~10^{-4} mol/L), but cannot reach the low levels attained by voltammetry for direct analysis
- *Conductometry:* May not be used in the presence of high concentrations of electrolyte species other than that required
- *Classical wet chemistry:* Generally more accurate, but electrochemical methods offer better detection limits for trace analyses

Electrometric titration

- *Estimated analysis time:* At least several minutes after sample preparation
- *Catalytic techniques:* Determinations by effect of a substance upon the speed of a normally slow reaction. The effect can be monitored electrochemically, or electrochemical titration can be used to maintain constancy of the reacting system.
- *Stripping analysis:* Quantitative determination of analytes, usually metals in very dilute solutions
- *Electrographic analysis:* Qualitative or, at best, semiquantitative analysis of metallic samples

Potentiometric membrane electrodes

- *Estimated analysis time:* Several minutes per sample after dissolution or other sample preparation
- *Amperometric gas sensors:* Quantification of oxygen and hydrogen peroxide
- *Ultraviolet/visible spectroscopy:* Direct or indirect determination of cation, anion, gaseous, and molecular species
- *Ion chromatography:* Determination of cation and anion species concentrations
- *Atomic spectroscopies:* Quantification of sample elemental components
- *Voltammetry:* Quantification of cations, anions, and certain organics

ELECTROCHEMISTRY is a branch of physical chemistry that deals with the chemical changes produced by electricity and the production of electricity by chemical changes. The voltaic pile discovered by Volta in 1799 was the first electrochemical device, which lead to the production of batteries for reliable production of electrical current. Conversely, electricity can bring about chemical changes in a process known as electrolysis (meaning "electric-splitting"), whereby substances are separated by the application an electric current in an electrochemical cell. This process of chemical splitting was quickly discovered in 1800 by William Nicholson, who copied Volta's battery and observed the decomposition (electrolysis) of water into oxygen and hydrogen. Soon thereafter in 1839, Christian Schonbein and William Grove demonstrated the reverse reaction of a fuel cell, whereby an electric current is generated by the recombination of hydrogen and oxygen.

The process of electrolysis has a number of general uses in separating elements from naturally occurring sources. Applications are too numerous to list, but some examples include:

- Reduction of salts to yield pure metals
- Reduction of neutral organic molecules
- Production of oxygen for spacecraft and nuclear submarines via electrolysis of acidified water
- Production of pure hydrogen to serve as a fuel through electrolysis of acidified water (even possible with wind-power-generated electrolysis). Further uses include hydrocracking, reduction of unsaturated hydrocarbons, as well as a reactant in the Haber procedure for industrial ammonia synthesis.
- Isolation and purification of pure metals by reduction of the corresponding salts (aluminum, calcium, magnesium, sodium, potassium, copper)
- Industrial production of chlorine and chlorate ions
- Cleaning of metallic artifacts (coins, statues, and ancient weapons) (Ref 5, 6) and restoration of parts (Ref 7)
- Synthesis of polyfluorinated organic compounds (electrofluorination)
- Reduction of CO_2 to produce useful fuel hydrocarbons (methane, ethane, ethylene) (Ref 8)
- Growth of conductive crystals on an electrode (electrocrystallization) (Ref 9, 10)

Various electrochemical methods, based on the process of electrolysis, have been developed for chemical analysis. The oldest method is electrogravimetry or electrodeposition (Ref 11). Various electrochemical methods of chemical analysis also have been developed based on the measurement of an electrical signal generated by the process of electrolysis in an electrochemical cell. Electrochemical techniques are of three types: potentiometric, voltammetric (amperometric), and coulometric. Potentiometry is the most commonly used of the electrochemical techniques and involves the measurement of a potential (voltage) generated by a cell under essentially equilibrium conditions. Coulometry involves the determination of the weight of metal deposited or released at the cathode by measuring the electrical charge (electrical current over a determined time frame) produced by an electrochemical cell.

The basic foundation of these methods is Faraday's law(s) of electrolysis, which state that the amount of any substance deposited or liberated during electrolysis is proportional to the quantity of electric charge passed and to the equivalent weight of the substance. Faraday's law of electrolysis is involved in several electroanalytical techniques. The mathematical formulation is:

$$Q = n \times F \times N$$

where F is the Faraday constant (96,495 C/mol), and Q is the minimum number of coulombs (C) required to alter the charge of N moles of a given species by n units. The reaction may involve deposition (of silver, for example) on an electrode; dissolution from an attackable anode, such as of silver or tin; or alteration of the charge of a species in the solution, for example, $Fe^{3+} + e \rightarrow Fe^{2+}$. If the current is constant at I amperes and flows for t seconds, the number of moles, N, of the substance electrolyzed can be calculated by using Faraday's law:

$$N = \frac{1}{nF} \int_0^t I dt$$

Faraday's laws also can be summarized in terms of mass by the equation:

$$m = (Q/F) \times (M/z)$$

where m is the mass (in grams, g) of the isolated substance, Q is the total electric charge (in coulombs, C) passed used in the isolation of the mass at the electrode, M is the molar mass (g/mol) of the isolated substance, and z is the valence (charge) of the substance ions involved in the process (number of electrons transferred per ion). If the reaction proceeds with 100% current efficiency, the total electric charge (Q) is calculated from the integrated equation:

$$Q = \int_0^t I dt$$

This article describes various methods of electrochemical analysis. The oldest electroanalytical technique is electrogravimetry, where the element of interest is deposited electrolytically onto an electrode and weighed. A more common method is coulometry, which is comprised of several techniques that use an instrument known as a coulometer (or electric charge counter). Coulometers may be a chemical device (chemical coulometer) or an electronic device (integrator). In either case, coulometers can be used to determine the amount of an analyte by measuring the electrical current and time (i. e., total charge) required for complete oxidation or reduction of the analyte. Coulometers also are used in a technique known as voltammetry, where quantitative information about the analyte is obtained by measuring electrical current as a function of applied voltage. The methods of coulometry do not require weighing and allow for shorter analysis times. Coulometry is used in quantitative chemical analysis and the study of electrochemical mechanisms. Example applications of coulometry include:

- Measurement and determination of inorganic substances and metals of different oxidation state in the presence of each other (Ref 12)
- Examination of corroded materials (Ref 13)
- Determination of uranium and plutonium assays in nuclear fuel (Ref 14, 15)

Electrochemical Cells

The basic process of an electrochemical reaction requires an electrochemical cell comprised of two half cells with an electrode (cathode and anode) that are in a medium (electrolyte) that can conduct electricity. The term *anode* or *cathode* is applied to one or the other of the electrodes depending on whether the electrochemical reaction is an oxidation or a reduction reaction, regardless of the voltage potential between the electrodes. The oxidized species (O) of metal (M) occurs at a site when an atom dislodges from the solid and enters the environment (or solution) as a positive ion (or cation):

$$M \rightarrow M^{n+} + ne^- \quad \text{(Eq 1)}$$

This formation of a cation (M^{n+}) occurs with n number of electrons (with a charge of ne) remaining in the solid. A conducting medium (electrolyte) between the electrodes is necessary, because the oxidation reaction can only be sustained if the number of electrons (n) left over from the oxidized metal (O) can be consumed (reduced) at the cathode of the cell. Hence, the cathodic reaction is referred to as reduction. With stoichiometric coefficients (b and c) that depend on the chemical specie, the reduction reaction at the cathode is expressed as:

$$b \times O + ne^- \rightarrow c \times R \qquad \text{(Eq 2)}$$

where O denotes a soluble oxidant, and R is the reduced form of the oxidant that may or may not be a soluble species. It was decided in 1953 at the International Union of Pure and Applied Chemistry (IUPAC) meeting that the half-cell reactions should be conventionally written in the reduction direction (Eq 2).

Equilibrium Potential

Either the anodic (Eq 1) or the cathodic (Eq 2) reaction can control the overall rate of the reaction. The equilibrium potential, also called the reversible potential, is the electrode potential measured under zero current (rest potential) when the electrochemical equilibrium occurs. If the resistance between the electrodes of the galvanic cell is made very high, so that very little current flows, the extent of reaction being small enough not to change the activities of reactants and products, the potential difference between the two electrodes remains constant and is the maximum cell voltage, called the electromotive force of the cell.

The potential of an electrochemical cell under equilibrium conditions is defined as the Nernst equation:

$$U_0 = U^* + \frac{RT}{nF} \ln\left[\frac{a_O^b}{a_R^c}\right] = U^* - \frac{RT}{nF}\frac{c}{b} \ln\left[\frac{a_R}{a_O}\right] \qquad \text{(Eq 3)}$$

where U_0 is the equilibrium potential relative to a reference potential (U^*), and where a_O and a_R are the activities (concentrations) of the oxidized and reduced species. The reference potential (U^*) established by the IUPAC is based on a half cell with standard hydrogen electrode (SHE), where reaction is based on electron activity from hydrogen in an aqueous solution with reduction into hydrogen gas at 1 bar ($H^+ + e^- \leftrightarrow$ ½ H_2[gas]). The equilibrium potential also can be expressed in terms of the chemical potential (or Gibbs free energy), which depends on the concentration (activity, a) and chemical formula of the species involved. At equilibrium, the change in Gibbs free energy (G) is balanced for oxidation (Ox) and reduction (Red):

$$-\frac{\Delta G_{Red}}{nF} = +\frac{\Delta G_{Ox}}{nF} \qquad \text{(Eq 4)}$$

With the SHE potential as a reference, the equilibrium potential expressed in terms of chemical potentials (μ) is:

$$U_0 = \frac{(b\mu_O + n\mu_e - c\mu_R)}{nF} \qquad \text{(Eq 5)}$$

where μ_e is the standard chemical potential or Gibbs free energy of an electron in the SHE half cell, where $\mu_e = \mu\,(H_2, \text{gas}) - \mu\,(H^+, \text{aqueous})$.

Electrolytic Cells

The external circuit connection between the cathode and anode and the voltage/current relationship in an electrochemical cell determines whether the cell operates as a battery, an electrolytic cell, or perhaps as a fuel cell. Either the anodic reaction or the cathodic reaction can control the overall rate of the reaction, and electrochemical methods can drive the reactions away from the equilibrium potential. The relationship between current (I) and potential difference (ΔU, from equilibrium potential U_0) in a cell is described by the Tafel equation:

$$I = nFk \exp\left[\pm\alpha\, nF\frac{\Delta U}{RT}\right] \qquad \text{(Eq 6)}$$

where the plus sign under the exponent refers to an anodic reaction, a minus sign refers to a cathodic reaction, n is the number of electrons involved in the electrode reaction, and k is the rate constant for the electrode reaction. The α term is an exponential factor related to the specific electrochemical conditions. Even with the same reagents and process, the voltage-current behavior can be modified by changing things such as additives in the electrolyte or surface conditions of the electrode. The α term, also referred to as the transfer or symmetry coefficient, is generally taken to be 0.5.

At equilibrium, no measurable current flows, such that the rate of metal dissolution ($M \rightarrow M^{n+} + ne^-$) results in an anodic current (I_A) that is equal to the cathodic current (I_C) from the rate of metal cation deposition ($M^{n+} + ne^- \rightarrow M$). Both reactions occur simultaneously, and at equilibrium they are balanced with an equal exchange of electrons back and forth. This inherent current at equilibrium is termed the exchange current (I_0), such that $I_0 = -I_C = I_A$.

The current-voltage relationship is not linear but changes exponentially with changes in the potential from the equilibrium point. As an example, Fig. 1 is an example of the current-voltage relationships for metal dissolution/deposition reactions. This curve follows the Butler-Volmer equation:

$$I = I_0\left\{\exp\left(\frac{\alpha F}{RT}\eta\right) - \exp(1-\alpha)\frac{F}{RT}\eta\right\} \qquad \text{(Eq 7)}$$

where I is the current, I_0 is the exchange current, F is the Faraday constant, R is the gas constant, T is the absolute temperature, and α is the transfer or symmetry coefficient. The term η is the overpotential (ΔU) defined by:

$$\eta = V - U_0 \qquad \text{(Eq 8)}$$

where V is the experimentally applied voltage, and U_0 is the equilibrium potential given by the Nernst equation. The overpotential is a measure of how far the reaction is from equilibrium.

Electrolysis leads to the separation and isolation of metals originally in a molten or a solution (ionic) mixture on an electrode by using a direct current and a voltage called the decomposition potential. In an electrolytic cell (Fig. 2), the anions (negative ions) go to the anode (positive electrode) and the cations (positive ions) go to the cathode (negative

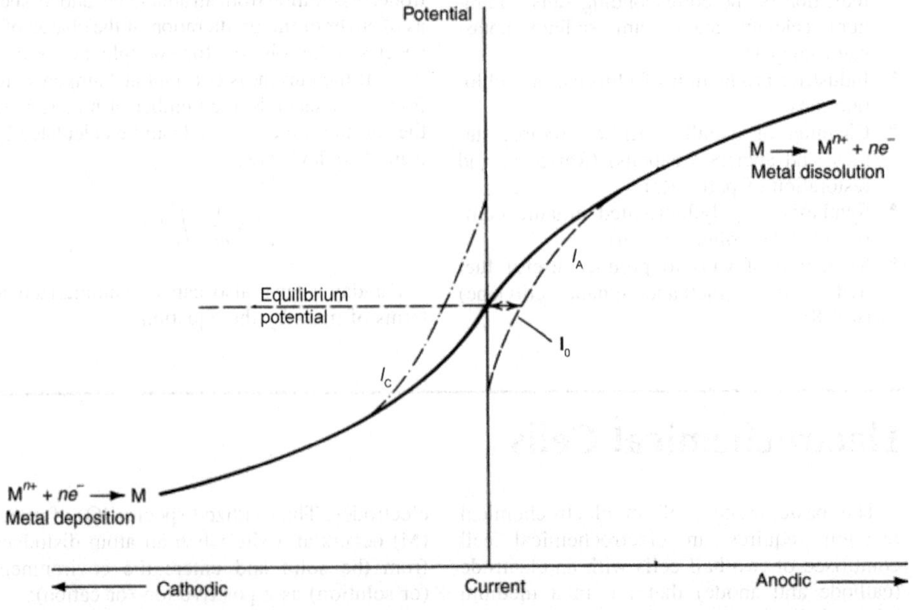

Fig. 1 Current-potential relationships for a metal dissolution or a deposition reaction in an electrochemical cell. The exchange current (I_0) at the equilibrium potential (U_0) has two components, $I_A = -I_C$, which results in zero net current at equilibrium. The solid line shows the measurable current; the dashed lines (I_C, I_A) show the partial dissolution/deposition currents.

electrode). So, for example, oxygen would go to the anode and iron would go to the cathode. The electrolyte serves as the source of the mobile ions that will eventually deposit the metal at the cathode. The electric circuit provides the energy for the ions to deposit due to electron transfer.

The electrolytic cell is fundamental in electrochemical analysis. The two classic types of cells used in the major techniques are the constant-current cell and the controlled-potential cell (Fig. 3). The difference between the two basic types of cells is the addition of a reference electrode in controlled-potential electrolysis (Fig. 3a, bottom). Both types of cells typically contain large platinum gauze electrodes (Fig. 3a) or a mercury cathode and a flat spiral wire anode (Fig. 3b). The mercury cathode has the advantages of a high hydrogen overpotential and the ability to form an amalgam with most metals. In both cases, efficient stirring is used to prevent concentration polarization.

In a constant-current (galvanostatic) cell, the total charge is defined as the product of the current and the exactly measured time of the reaction. In controlled-potential electrolysis, however, the current decreases exponentially in time as the analyte is consumed (Fig. 4). As the analyte is consumed during electrolysis, the current decays more or less exponentially toward zero, and the electrolysis is terminated when the current reaches a suitably low value, indicating complete electrolysis. In controlled-potential cells, where currents are changing continuously in time, it is convenient to use coulometers that measure charge that results from the oxidation or reduction of an analyte. Coulometers may be electrochemical devices based on the laws of Faraday or electronic charge integrators (see the section "Coulometry" in this article).

Determining the voltage of the cell necessary to achieve the required separation requires knowing the reactions that occur at each electrode and the potential of each electrode by using the Nernst equation. If the potential is made more positive than the equilibrium potential, then $I_A > |I_C|$ (i.e., the first exponential term in Eq 7 increases in value, and the second one decreases), and the metal dissolution reaction will proceed. Similarly, if E is made more negative than the equilibrium potential, $I_A > |I_C|$, then metal cation deposition proceeds. Over a short potential range, the two reactions oppose each other, but for sufficiently large overpotentials (η_A, anodic; or η_C, cathodic), one reaction occurs at a negligible rate; that is, one or the other of the exponential terms in the Butler-Volmer equation (Eq 7) becomes negligible.

Controlled-Potential Electrolysis

In controlled-potential electrolysis (Fig. 3a, bottom), the cell includes the addition of a reference electrode, which acts as a sensor of the

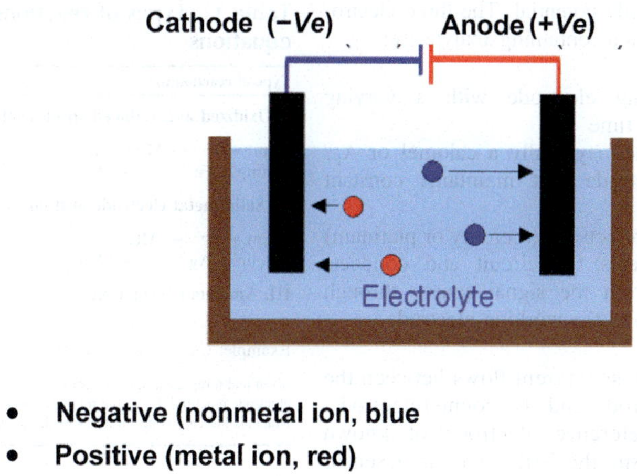

Fig. 2 A simple electrolysis setup. The electrodes (black bars) are often made out of graphite.

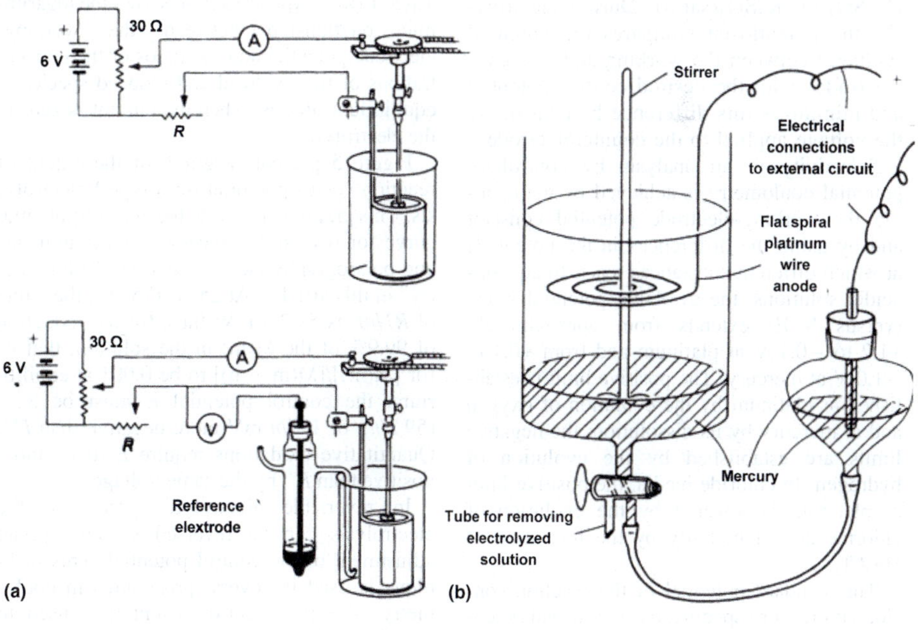

Fig. 3 Classical cell types. (a) Constant-current (top) and controlled-potential (bottom) cells. (b) Cell for constant-current electrolysis with mercury cathode. Source: Ref 16

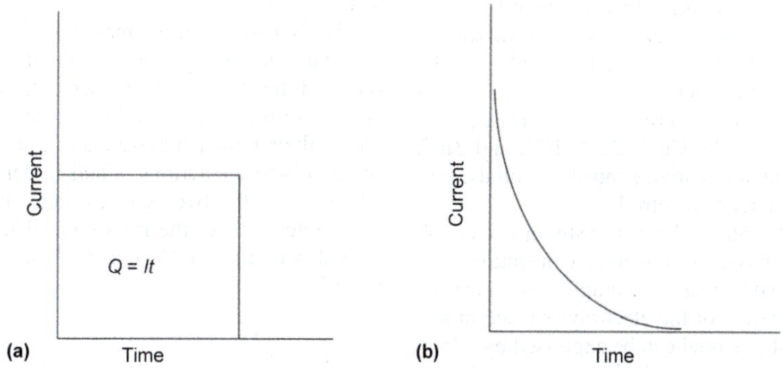

Fig. 4 Current-time curves of electrolytic cells with (a) constant-current circuit and (b) controlled-potential circuit

working-electrode potential. The three electrodes in the solution containing analyte are:

- The working electrode with a varying potential in time
- The reference (typically a calomel or Ag/AgCl) electrode that maintains constant potential
- The counterelectrode (mercury or platinum) that completes the circuit and conducts electrons from the signal source through the solution to the working electrode

The electrolysis current flows between the working electrode and the counterelectrode. The stable reference electrode of known potential (versus the SHE) acts as a sensor of the working-electrode potential. Typical reference electrodes are the types used for measuring pH, such as $Hg\text{-}Hg_2Cl_2$, KCl (sat'd) (the saturated calomel electrode, SCE); Ag-AgCl(s), KCl(sat'd); or $Hg\text{-}Hg_2SO_4(s)$, K_2SO_4(sat'd). During electrolysis, the potentiostat compares the potential (voltage) between the working and reference electrodes with the desired control potential and minimizes this difference by controlling the voltage applied to the counterelectrode.

Selectivity in an analysis by controlled-potential coulometry is achieved by maintaining the working-electrode potential constant and by using the differences in the potentials at which different substances react. In aqueous acidic solutions, the available potential range (versus SCE) extends from approximately +1.2 to −0.1 V at platinum and from +0.1 to −1.0 V at mercury. The positive limit is established at platinum by the evolution of oxygen and at mercury by its dissolution; the negative limits are established by the evolution of hydrogen. In chloride media, the positive limit at platinum is lowered by the evolution of chlorine and at mercury by the formation of Hg_2Cl_2.

The standard potential of the reaction considered or, more specifically, the formal potential in the particular medium of the electrolysis determines the potential at which a substance is electroactive. Metal ion species having reactions that have positive formal reduction potentials, for example, Fe^{3+}, Ce^{4+}, and Au^{3+} in certain media, and which are therefore considered oxidizing agents, are electrolytically reduced at positive control potentials, usually at a platinum electrode. The ions of metals with negative or slightly positive formal potentials, for example, Cu^{2+}, Pb^{2+}, Ti^{4+}, and Zn^{2+}, are reduced at negative control potentials, usually at a mercury electrode.

When the rate of charge transfer in an electrolysis reaction is large compared to the rate of mass transport and there are no complicating side reactions, the extent of the electrolysis reaction as a function of potential can be expressed by a form of the Nernst equation. Such processes are generally known in electrochemistry as reversible. The exact form of the equation depends on the type of reaction considered. Table 1 summarizes the three most common types of reactions encountered in controlled-potential electrolysis, together with the relevant Nernst equations. Equation A in Table 1 (for a type I reaction) shows the logarithmic relationship between the working-electrode potential and the ratios of the concentrations of the oxidized and reduced species at equilibrium, that is, when the current is zero in the electrolysis.

Table 1 Types of reactions in controlled-potential electrolysis and applicable Nernst equations

Type of reaction(a)	Nernst equation(b)
I. Oxidized and reduced species soluble in solution	
$M(m) + ne^- \rightarrow M(m-n)$ Example: $Fe^{3+} + e^- \rightarrow Fe^{2+}$	$E = E^{\circ\prime} + RT/nF \cdot \log \{[M(m)]/[M(m-n)]\}$ (Eq A)
II. Solid metal electrodeposition	
$M(m) + me^- \rightarrow M(s)$ Example: $Ag^+ + e^- \rightarrow Ag(s)$	$E = E^{\circ\prime} + RT/nF \cdot \log [M(m)]$ (Eq B)
III. Amalgam formation	
$M(m) + me^- \rightarrow M(Hg)$ Example: $Cu^{2+} + 2e^- \rightarrow Cu(Hg)$	$E = E^{\circ\prime} + RT/nF \cdot \log \{[M(m)]/[M(Hg)]\}$ (Eq C)

(a) m and n represent the number of electrons in the reaction or, in M(m), the oxidation state of the metal. (b) In the equations, E is the working-electrode potential, $E^{\circ\prime}$ is the reference potential of the reaction, R is the gas constant, T is the absolute temperature, and F is the Faraday constant. Brackets indicate concentrations of the oxidized and reduced species of the metal at equilibrium.

Figure 5 presents a graph of the extent of reaction versus potential for a type I electrolysis. This graph is similar to the current-potential curve for reversible processes in voltammetry and polarography (see the section "Voltammetry" in this article). At 25 °C (75 °F), the value of RT/nF is $59.2/n$ mV; thus, for the reduction of 99.9% of the M(m) in the solution, that is, for $[M(m)]/[M(m-n)]$ to be 0.001 at equilibrium, the control potential E must be $3 \times (59.2/n)$, or $177/n$ mV more negative than $E^{\circ\prime}$. Quantitative oxidations require E to be more positive than $E^{\circ\prime}$ by the same voltage.

In reversible electrolytic processes, the electrolysis can be reversed by appropriate adjustment of the control potential. This technique is used in several procedures in coulometry. The first control potential is used to transform all the substance determined in the sample to one oxidation state. The potential is then changed to reverse the electrolysis. The current is integrated during the second electrolysis to measure the quantity of the substance present.

The Nernst equation may also be used to calculate the difference in formal potentials required for the complete separation of two species in the same sample solution or to calculate their mutual interference. For example, for two type I reactions, equal molar concentrations of the two species, and less than 0.1% interference, the minimum difference in formal potentials at 25 °C (75 °F) in millivolts must be:

$$E_1^{\circ\prime} - E_2^{\circ\prime} = \frac{177}{n_1} + \frac{177}{n_2} \quad \text{(Eq 9)}$$

For a type II reaction—the electrodeposition of metal on a substrate of the same metal—the

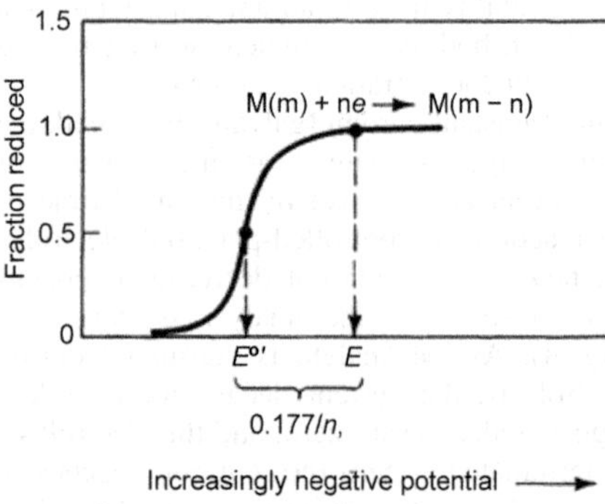

Fig. 5 Completeness of reaction as a function of potential for the reversible reduction of a metal ion species to another soluble species. E is the control potential required for 99.9% conversion of M(m) to M(m − n).

metal film is assumed (for greater than monolayer deposits) to have an activity of 1; thus, only the concentration of the solution species appears in Eq B in Table 1. Therefore, unlike type I reactions, the fraction of metal deposited is not independent of potential. A decreasing initial concentration of the metal requires an increasingly negative control potential to deposit a given fraction. Very low concentrations of metals and depositions on foreign substrates exhibit different characteristics.

Type III processes, in which the metal ion is reduced at a mercury electrode to form an amalgam, are generally reversible if the metal is soluble in mercury to the extent required by electrolysis. Otherwise, a metal film will form on the electrode, and the process will be similar to a type II reaction. Equation C in Table 1 for amalgam formation is similar to that for type I, except that the concentrations of M(m) and M in the logarithmic term are their concentrations in the solution and mercury, respectively. Because of their solubility in mercury, metals that participate in electrolytic amalgam formation, for example, copper, lead, indium, and thallium, have formal potentials for the reactions that are significantly more positive than those for the formation of the corresponding solid metal.

Many other electrolytic processes exhibit lower rates of charge transfer (termed irreversible or non-Nernstian) or involve chemical reactions, such as a follow-up reaction, coupled to the electron transfer. Many of the analytical methods of controlled-potential coulometry are based on such reactions; examples are the procedures for vanadium and uranium. Irreversible processes require a greater difference between the control potential and the $E^{\circ\prime}$ to effect rapid electrolysis, and, depending on the degree of irreversibility, a reverse electrolysis may not be feasible. In addition, separation factors for systems involving irreversible couples must be determined empirically.

Coulometry

Coulometers measure charge and can be used to determine the amount of an analyte involved in an oxidation or reduction reaction. In experiments with currents changing continuously in time (e.g., controlled-potential electrolysis), it is more convenient to use counters for the quantity of charge passed. The electrolysis current can be integrated by an electronic integrator (coulometer) or with an electrochemical coulometer. Electrochemical coulometers are based on the laws of Faraday. The basic components consist of an electrolytic cell and a chemical coulometer placed in series with a working electrode (Fig. 6). In operation of the cell, the electrolysis current flows between the working electrode and counterelectrode. The working electrode, where the desired reaction occurs, is usually a mercury pool or a platinum screen, although other inert materials have been used. The area of the working electrode is as large as possible, and the solution is vigorously stirred to maximize the rate of electrolysis. The counterelectrode is also an inert material, usually platinum. The volume of gas or mercury liberated, which is proportional to charge, is measured.

There are basically three types of coulometric techniques:

- *Potentiostatic coulometry:* (also known as controlled-potential coulometry) where the electric potential is held constant via a potentiostat
- *Coulometric titration:* (also known as amperostatic coulometry) where the current is kept constant via an amperostat or galvanostat
- *Voltammetry:* where quantitative information about the analyte is obtained by measuring electrical current as a function of applied voltage

The latter two methods are described in separate sections of this article.

Controlled-Potential Coulometry (Ref 1)

Controlled-potential coulometry is similar in applicability to the classical solution techniques of gravimetry and titrimetry. Like the classical methods, it is used primarily for major constituent analysis and is characterized by high precision and accuracy (typically 0.1%). Coulometry routinely requires only approximately 1 to 10 mg of analyte substance. This relatively small sample size is advantageous when the quantity of material available for analysis is small or must be limited because of its value, toxicity, or radioactivity. Controlled-potential coulometry is especially applicable to assays of alloys, compounds, and nonmetallic materials for the transition and heavier elements. Organic compounds and nonmetals—for example, halides and nitrogen-oxygen compounds—are also suitable for determination.

Accurate coulometry requires that electrolysis be performed as rapidly as possible by controlling the potential at a point where the rate of electrolysis is limited only by mass transport of the reacting species. For reversible processes, E is not far from the value shown in Fig. 5, which will also ensure complete electrolysis. When E is set at this value, the electrolysis current as a function of time is:

$$I = I_o \exp(-pt) \qquad \text{(Eq 10)}$$

where I_o is a constant (a theoretical initial current), and p is an overall rate constant of the electrolysis. These constants incorporate parameters that are determined by the mass transport characteristics of the cell and the diffusion coefficient of the species electrolyzed.

Figure 7 shows the current-time curves observed in the coulometric determinations of gold and silver. Because the current scale is logarithmic, the curves should theoretically be straight lines. These curves deviate slightly from Eq 10 at the beginning of electrolysis due to imperfect compensation of the solution resistance by the potentiostat and at the end of electrolysis because the current tends to level off at the background current.

Table 2 lists the metals for which accurate methods have been developed and the basic electrochemistry of the procedures. A double arrow in the reaction indicates that the method involves a multistep or reverse electrolysis. In reversible electrolytic processes, the electrolysis can be reversed by appropriate adjustment of the control potential. This technique is used in several procedures in coulometry (Table 2). The first control potential is used to transform all the substance determined in the sample to one oxidation state. The potential is then changed to reverse the electrolysis. The current is integrated during the second electrolysis to measure the quantity of the substance present.

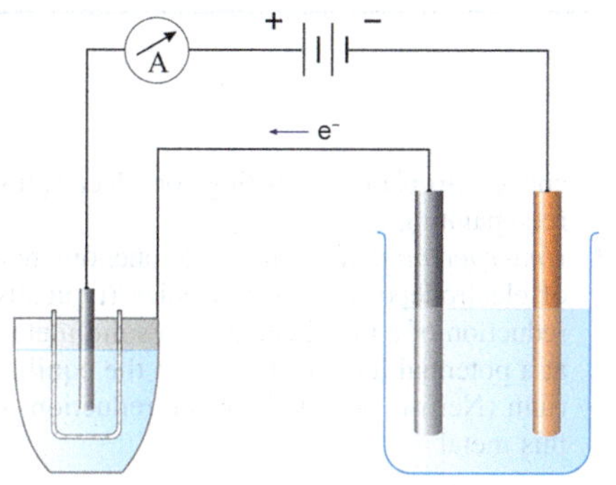

Fig. 6 The coulometer-electrochemical cell setup. Source: Ref 17

Determination of Gold

Because the quantities of metal available for analysis are usually small and high accuracy (0.1%) is frequently desired, the assay for gold in various materials is a useful example of the application of coulometry. The current-time curve for this electrolysis was discussed and shown in Fig. 7. If the sample contains none of the few interfering elements, such as iridium or ruthenium, the gold can usually be determined immediately after dissolution without further pretreatment. Each determination requires approximately 10 min. Aqua regia—3 parts hydrochloric acid (HCl) and 1 part nitric acid (HNO_3)—typically is used for the dissolution, and sulfamic acid (NH_2SO_3H) is included in the supporting electrolyte to remove the interfering nitrogen oxides and nitrite ion produced during the dissolution. If

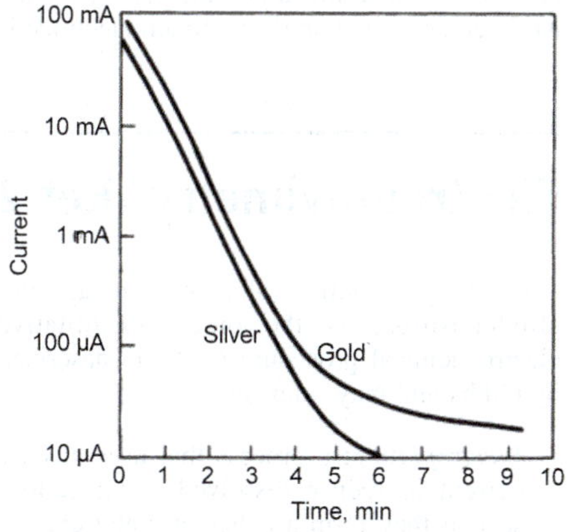

Fig. 7 Current-time curves for the reduction of Ag^+ to Ag(s) and Au^{3+} to Au(s) on a platinum electrode. Electrolysis conditions: silver, 0.1 *M* H_2SO_4, E = +0.16 V versus saturated calomel electrode (SCE); gold, 0.5 *M* HCl, E = +0.48 V versus SCE

Table 2 Examples of metals determined by controlled-potential coulometry

Metal	Working electrode	Overall reaction	Supporting electrolyte
Antimony	Hg	$Sb^{3+} \rightarrow Sb(Hg)$	0.4 *M* tartaric acid, 1 M HCl
Arsenic	Pt	$As^{3+} \rightarrow As^{5+}$	1 *M* H_2SO_4
Bismuth	Hg	$Bi^{3+} \rightarrow Bi(Hg)$	0.5 *M* tartrate, 0.2 *M* HCl
Cadmium	Hg	$Cd^{2+} \rightarrow Cd(Hg)$	1 *M* HCl
Chromium	Hg	$Cr^{3+} \rightleftarrows Cr^{2+}$	6 *M* HCl
	Au	$Cr^{6+} \rightarrow Cr^{3+}$	1 *M* H_2SO_4
Cobalt	Pt	$Co^{2+} \rightarrow Co^{3+}$	1,10-phenanthroline, acetic acid
Copper	Hg	$Cu^{2+} \rightarrow Cu(Hg)$	1 *M* H_2SO_4
Europium	Hg	$Eu^{3+} \rightleftarrows Eu^{2+}$	0.1 *M* HCl or $HClO_4$
Gold	Pt	$Au^{3+} \rightarrow Au(s)$	0.5 *M* HCl
Indium	Hg	$In^{3+} \rightarrow In(Hg)$	1 *M* KCl, 0.25 *M* HCl
Iridium	Pt	$Ir^{4+} \rightleftarrows Ir^{3+}$	0.5 *M* HCl
Iron	Pt	$Fe^{3+} \rightleftarrows Fe^{2+}$	0.5 *M* H_2SO_4
Lead	Hg	$Pb^{2+} \rightarrow Pb(Hg)$	1 *M* $HClO_4$
Manganese	Pt	$Mn^{2+} \rightarrow Mn^{3+}$	0.25 *M* $Na_4P_2O_7$
Mercury	Hg	$Hg^{2+} \rightarrow Hg(liq)$	1.5 *M* $HClO_4$
Molybdenum	Hg	$Mo^{6+} \rightarrow Mo^{5+}$	0.2 *M* $(NH_4)_2C_2O_4$, 1.3 *M* H_2SO_4
Neptunium	Pt	$Np^{6+} \rightleftarrows Np^{5+}$	1 *M* H_2SO_4
Nickel	Hg	$Ni^{2+} \rightarrow Ni(Hg)$	1 *M* pyridine, 0.5 *M* HCl
Palladium	Pt	$Pd^{2+} \rightleftarrows Pd^{4+}$	0.2 *M* NaN_3, 0.2 *M* Na_2HPO_4
Plutonium	Pt	$Pu^{4+} \rightleftarrows Pu^{3+}$	1 *M* $HClO_4$ or, to avoid iron interference, 5.5 *M* HCl, 0.015 *M* sulfamic acid, NaH_2PO_4
Rhodium	Hg	$Rh^{3+} \rightarrow Rh(Hg)$	0.2 *M* HCl
Ruthenium	Pt	$Ru^{4+} \rightarrow Ru^{3+}$	5 *M* HCl
Silver	Pt	$Ag^{+} \rightarrow Ag(s)$	0.1 *M* H_2SO_4
Technetium	Hg	$Tc^{7+} \rightarrow Tc^{3+}$	Acetate-tripolyphosphate
Thallium	Pt	$Tl^{+} \rightarrow Tl^{3+}$	1 *M* HCl
Tin	Hg	$Sn^{4+} \rightleftarrows Sn(Hg)$	3 *M* KBr, 0.2 *M* HBr
Titanium	Hg	$Ti^{4+} \rightarrow Ti^{3+}$	6–9 *M* H_2SO_4
Uranium	Hg	$U^{6+} \rightarrow U^{4+}$	0.5 *M* H_2SO_4
Vanadium	Pt	$V^{5+} \rightarrow V^{4+}$	1.5 *M* H_3PO_4
		$V^{4+} \rightarrow V^{5+}$	
Zinc	Hg	$Zn^{2+} \rightleftarrows Zn(Hg)$	1 *M* NH_4H citrate, 3 *M* NH_4OH

Source: Ref 1

large quantities of iron are present in the sample, phosphoric acid (H_3PO_4) can also be added to remove its interference.

Gold electroplating solutions that contain cyanide or sulfite ions and scrap materials containing organic compounds must be subjected to a more severe pretreatment before measurement of the gold. This typically involves a sulfuric acid-nitric acid (H_2SO_4-HNO_3) boildown or fuming with perchloric acid ($HClO_4$), which removes or destroys the complexing species and organics. This treatment precipitates most of the gold as the metal; it is then redissolved in aqua regia for coulometric measurement. After several determinations, the gold metal that has been deposited on the platinum electrode is stripped electrolytically by controlling the potential at +1.00 V in 2 *M* HCl.

Determination of Uranium

Uranium is another element that has been determined frequently in various sample types by controlled-potential coulometry. The adaptability of coulometry to automated and remote operation makes it useful for nuclear materials that contain uranium, plutonium, or both. Highly radioactive samples can be handled if there are no chemical interferences. Methods have been developed for separating the uranium from fission products in nuclear-reactor and waste samples before coulometric determination. A procedure is also available for coulometrically measuring the ratio of uranium oxidation states in uranium oxide fuels.

As indicated in Table 2, coulometric determination involves conversion of U^{6+} to U^{4+}. However, electrolysis proceeds in two steps. First, the U^{6+} is reduced directly at the electrode by the reaction:

$$U^{6+} + e^{-} \rightarrow U^{5+}$$

The U^{5+} then disproportionates:

$$2U^{5+} \rightarrow U^{6+} + U^{4+}$$

Thus, an exhaustive electrolysis results in the reduction of U^{6+} to U^{4+} and an overall two-electron change in valence.

The reduction is carried out at a mercury-pool electrode, usually in H_2SO_4 (also containing sulfamic acid if HNO_3 was used to dissolve the sample), at a control potential of −0.325 V versus SCE. A preliminary electrolysis performed at +0.075 V pre-reduces any oxides of mercury and substances such as Fe^{3+}, Cr^{6+}, and Pu^{4+} that would interfere in the subsequent uranium electrolysis. Copper, if present in the sample, can also be determined, because the reduction of uranium is irreversible at the mercury electrode. The Cu^{2+} and U^{6+} are reduced together at −0.325 V. The Cu(Hg) is then oxidized back to Cu^{2+} at +0.100 V. Integration of the current during the second electrolysis yields the amount of copper, and subtraction of this quantity of electricity from the total in the first electrolysis provides the amount of uranium.

Coulometric measurement requires approximately 30 min, including removal of oxygen from the solution (which also would be reduced at −0.325 V), pre-electrolysis, and uranium electrolysis. Because of the homogeneous disproportionation reaction, the uranium electrolysis is somewhat slower than most, and its electrolysis current-time curve differs slightly from that predicted by Eq 10 for simple reactions.

Electrogravimetry (Ref 2)

Electrogravimetry (also known as electrodeposition) is the oldest quantitative electrochemical procedure (Ref 11) described by Gibbs and may refer to:

- *Electroplating*: a process that uses electric current to reduce dissolved metal cations so that they form a coherent metal coating on an electrode
- *Electrophoretic deposition:* a term for a broad range of industrial processes that includes electrocoating, e-coating, cathodic electrodeposition, anodic electrodeposition, and electrophoretic coating, or electrophoretic painting
- *Underpotential deposition:* a phenomenon of electrodeposition of a species (typically reduction of a metal cation to a solid metal) at a potential less negative than the equilibrium (Nernst) potential for the reduction of this metal

Unlike most electrochemical techniques, the reaction in electrogravimetry often must be allowed to go to completion efficiently, prolonging analysis times. However, the technique is more accurate (0.1%) than controlled-potential coulometry (0.2 to 5%) or polarography (2%). Use of efficient stirring lessens analysis time. Overall analysis time depends on the technique used.

To achieve separation and complete deposition requires knowing the potential of each electrode by using the Nernst equation (Eq 3). For a cell reaction ($M^{n+} + ne^{-} \rightarrow M$) occurring at 25 °C (75 °F), the required potential (*E*) is:

$$E = E^{*} + \frac{0.0591}{n} \log\ [M^{+}] \qquad \text{(Eq 11)}$$

where E^* is the potential of the reference electrode. Therefore, the voltage becomes more negative by $0.0591/n$ V or $59/n$ mV for each tenfold decrease in concentration.

Electrogravimetry normally requires complete or nearly complete deposition. The electrode must be totally covered by the metal to ensure formation of a monolayer. If not, the activity of the solid metal cannot be considered to be unity, and Eq 11 will not hold. This presents problems in the deposition of small amounts of metal.

Potential for Separation

The applied potential (E_{APP}) for separation may be determined from the calculated (E_{CAL}, Eq 11). This result, coupled with knowledge of the IR drop and the overpotentials of the cathode and anode (η_C and η_A, respectively), yields:

$$E_{APP} = E_{CAL} + IR + \eta_C \text{ and } \eta_A \quad \text{(Eq 12)}$$

Application of an electromotive force (emf) to the cell initiates current flow and thus the desired reaction. The overpotential developed at an electrode usually results from the difference in concentration of the ions in the bulk solution and at the electrode surface and provides one source of resistance to current flow. For an anodic process exhibiting an overpotential effect, the applied emf must be more positive than the calculated potential; for a cathodic process with associated overpotential, it must be more negative than the calculated value. In a dilute solution, depletion in concentration near the electrode increases resistance (R), yielding a change (IR) in potential. This IR drop may lead to serious disadvantages in instrumental conditions, because it is not always constant and is sometimes unknown.

Such factors as the applied voltage, the electrode potential at the electrodes of interest, the current flowing, the amount of electricity used, and the nature of the deposit at the electrodes affect electrolysis of a solution. The deposition of copper on a platinum cathode best exemplifies the principles of electrogravimetry. The cell may be represented as:

$$\text{Pt, Cu} \rightarrow \text{Cu}^{2+}(0.100\ M)$$

$$\text{H}^+(1.000\ M) \rightarrow \text{O}_2(0.2\ \text{atm}), \text{Pt}$$

Using the Nernst equation (at 25 °C, or 75 °F) for the electrode oxidation reaction of $Cu \rightarrow Cu^{2+} + 2e^-$:

$$E_{Cu} = 0.337 + \frac{0.0591}{2}\log\left[\text{Cu}^{2+}\right]$$

$$E_{Cu} = 0.337 + 0.0296 \log[0.100] = 0.307\ \text{V}$$

For the reduction electrode ($O_2 + 4H^+\ 4e^- \rightarrow 2H_2O$):

$$E_{O_2} = 1.229 + \frac{0.0591}{4}\log\left[\frac{[O_2][H^{+4}]}{|H_2O|}\right]$$

Because $|H_2O|$ concentration is approximately constant and $|O_2|$ partial pressure (p_{O_2}) is approximately 0.20 atm, then:

$$E_{O_2} = 1.229 + 0.0148 \log\left\{\left[H^{+4}\right] p_{O_2}\right\}$$

$$E_{O_2} = 1.229 + 0.0148 \log\left\{\left[1.000^4\right] 0.20\right\} = 1.219\ \text{V}$$

Therefore, the cell potential for the reaction $2Cu + O_2 + 4H^+ \rightarrow 2Cu^{2+} + 2H_2O$ is:

$$E_{cell} = E_{O_2} - E_{Cu} = 1.219\ \text{V} - 0.307\ \text{V} = 0.912\ \text{V}$$

To deposit copper metal, the reaction must be reversed by applying an external voltage at least equal to the cell potential together with the overpotentials and any IR drop. This value of voltage is the decomposition potential. The anodic overvoltage of oxygen in this cell may be taken as 0.40 V. The decomposition potentials vary with concentration, as does the IR drop; this must be considered. Current will flow only when:

$$E_{APP} = 0.912\ \text{V} + 0.40\ \text{V} + IR\ \text{drop}$$

Therefore, E must be increased until it exceeds this sum to sustain electrolysis. The limiting current, proportional to concentration, decreases with concentration. In this case, the decomposition potential increases, and IR decreases.

A major problem is that voltages frequently must be maintained within close limits to ensure efficient separation of metals. Other factors affect separation. For example, the smoother the deposited metal, the higher the overpotential. In addition, increases in temperature affect the result. As temperature is raised, overpotential is decreased. In some cases, particularly in strong acid solution, overpotential may become pH dependent. Determination of a working potential may be difficult, and often a rough value is obtained by application of the Nernst equation (Eq 3 or 11) and modified by experiment.

Various parameters are occasionally advantageous in determination of the voltage conditions required by two metals to enable deposition of one in the presence of the other. The use of complexing agents may provide solutions with decomposition potentials very different from those in the noncomplexed state. This sometimes allows reversal of the order of deposition.

Physical Properties of Deposits

To ensure purity of the metal and quantitative deposition, electrolysis must exhibit the following characteristics. The ability of metals to adhere to the cathode is most important, because any loss from the cathode causes an error in the weight of the metal that cannot be accommodated. Smoother deposits generally adhere more favorably to the electrode. Therefore, effects that cause the metal deposit to be flaky or spongelike must be reduced or eliminated. The most common cause of this failing—the evolution of a gas coincident with metal deposition on the same electrode—must be prevented if possible by arranging electrolysis conditions.

The chemical nature of the medium may affect the smoothness of the deposit. For example, copper is deposited readily from a solution in nitric acid, but silver yields a better deposit when precipitated from the silver cyanide complex. In samples of cobalt, nickel, or copper containing chloride ions, the chloride ions must be removed to deposit the metals quantitatively (Ref 18).

Temperature increases hasten the rate of diffusion and thus the current density, which decreases deposition time. Hydrogen overpotential is also decreased; this may affect the degree of separation, particularly when metal complexes are involved.

The rate of flow of ions to an electrode is hastened by increasing their rate of movement in the solution. This is achieved by efficient stirring using a stirrer or a rotating or vibrating electrode (Ref 19). Again, the increased rate of diffusion decreases deposition time but lessens the concentration potential. This increases current density, which further decreases deposition time, as outlined previously.

Use of Incomplete Deposition

Small mass changes arising from electrolytic deposition may be monitored by using the piezoelectric effect. For example, in the determination of cadmium, metallic electrodes of a piezoelectric crystal are used as cathodes. The weight of cadmium deposited is related linearly to the fundamental frequency of the crystal; this is used to determine incomplete deposition of cadmium from a solution of 0.1 M $NaClO_4$. Analysis takes 15 min.

Methods and Instrumentation

The cell is fundamental to electrogravimetry equipment. The circuit requirements for electrogravimetry are a direct current source, a variable resistance, an ammeter for measuring current, and a voltmeter for voltage measurement. Direct current may be provided by batteries or a power supply that converts alternating current to direct current. Separation is normally carried out under conditions of constant current or constant or controlled voltage.

As noted, the two primary classical types of cells are constant-current and controlled-potential circuits (Fig. 3) with large platinum gauze electrodes or a mercury cathode and a flat spiral wire anode. In the case of the mercury cathode, solution may be removed without interfering with the electrolysis current. Thus, there is no oxidation of the elements deposited at the solution-amalgam interface. In the simple cell shown in Fig. 3(b), this may be accomplished by using the stop cock and siphon. The cell constant and specific resistance of the solution are low, allowing flow of a current of several amperes—normally sufficient to reduce all the solution components. This can be prevented by efficient stirring and by the heating effects of the current. New and modified cells and electrodes are available to improve precision or to be used at the microgram level.

Microelectrogravimetry is conducted easily in the following simply constructed cell using controlled potential (Ref 20). The cell contains a rapidly rotating helical platinum wire, 10 to 15 cm (4 to 6 in.) long and 0.05 cm (0.02 in.) in diameter. From 3 to 10 mg copper in copper foil have been recovered (99.5%) in 90 min by using this cell. Recovery from brass was lower (97.4 ± 0.8%). Lead and tin have also been deposited from brass. Cadmium was analyzed in micromolar concentration by using a piezoelectric electrode, as described earlier.

An electrolytic cell fitted with a mechanically homogenized system for decreasing concentration polarization yields dense, adhering, smooth, brilliant deposits that provide high-precision analysis of metallurgical samples and that may be used in quality control of metallurgical products. Use of vibrating electrodes improves the structure of the deposit and reduces the analysis time of electrogravimetry. Table 3 shows typical results. Table 4 lists results obtained by using internal electrolysis to deposit copper.

Table 3 Electrogravimetric determination of some metals by using vibrating electrodes

Metal	Mass, g	Relative error, %	Analysis time(a), min
Copper	0.1994	0.06	10
Lead	0.0972–0.6635	0.20	12
Bismuth	0.2013	0.05	15
Antimony	0.1022	0.11	14

(a) Note reduction in analysis time compared to that shown in Table 4. Source: Ref 2

Table 4 Determination of copper by using internal electrolysis

Mass, g	Relative error, %	Analysis time, min
0.02019	0.23	20
0.05029	0.90	35
0.10056	1.62	45

Source: Ref 2

Method Selection

Constant-current methods are the oldest in electrogravimetry. Continual increases in voltage maintain the constancy of current. Two problems arise in the separation of two ions in aqueous solution. First, the overpotential of hydrogen, if reached, causes deterioration in the smoothness of the deposit, with the effects considered earlier. Second, any second metal to be deposited must have a decomposition potential more negative than that of the hydrogen overpotential or the first deposited metal.

Considerable increases in voltage maintain analysis speed and ensure complete deposition. For example, when separating copper from other metals, such as zinc, copper is reduced to approximately $\frac{1}{100}$ of its concentration at 1.43 V. The current may be increased to give a final voltage of 2.2 V before hydrogen is evolved. For example, using 0.5 *M* sulfuric acid as the electrolyte, the hydrogen ion remains effectively constant while hydrogen is evolved. Codeposition will not occur if the metal from which copper is to be separated has a more negative potential than that for hydrogen.

Constant-Voltage Electrogravimetry

In the earlier example, hydrogen is not evolved below 2.2 V. Therefore, if electrolysis is conducted at a constant voltage from 1.43 to 2.2 V, copper may be deposited without hydrogen. However, the method, involving small currents and long separation times, is rarely used.

Internal Electrolysis

In this method, current is obtained from a secondary reaction usually consisting of attack of the anode by the electrolyte. The anode and cathode are connected directly to one another, creating a short-circuited galvanic cell (Fig. 8). Analysis of copper in the presence of lead necessitates use of the cell:

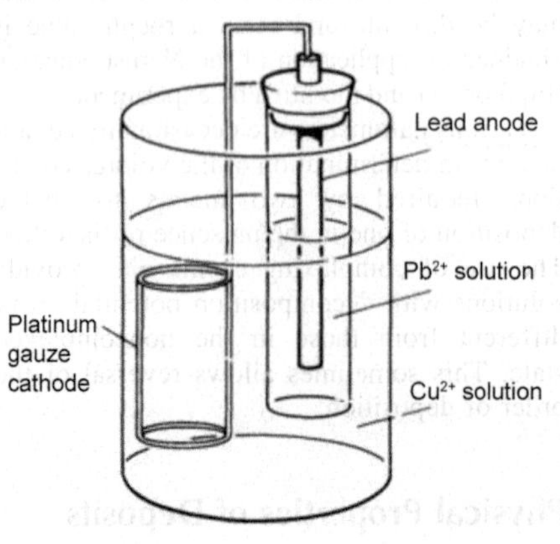

Fig. 8 Typical internal electrolysis cell

$$-\text{Pb} \mid \text{Pb}^{2+} \mid \mid \text{Cu}^{2+} \mid \text{Cu}+$$

where lead is the attackable anode. Energy is lost only by ohmic resistance, which controls the maximum current flowing through the cell:

$$IR = 0.22 + 0.0296 \log \left[\text{Cu}^{2+}\right]$$

for a 1 *M* Pb^{2+} solution at 25 °C (75 °F). Metal ions with decomposition potentials below −0.12 V, for example, silver and copper but not cadmium, may be deposited using this technique. To make *I* as large as possible, *R* must be kept as low as possible by using a high electrolyte concentration.

Electrolysis will take too long if large quantities of deposit are handled. In addition, the deposit becomes spongy, and metal ions may diffuse to the anode during prolonged electrolysis. Deposits are usually restricted to approximately 25 mg.

Considerations of High Precision and Automation

Advances in instrumentation and computer techniques have furthered high-precision analysis, automated techniques of analysis, and process on-line control. Electroanalytical probes are gaining acceptance in process control.

Metals may be separated using automated techniques. As an example, one such device consists of a stabilized operational amplifier that acts as a power potentiostat/galvanostat with voltages ≤10 V and a current ≤2 A. At the beginning of electrolysis, the apparatus maintains constant current until the working electrode reaches a preselected value of potential. This value is then constant throughout electrolysis. The apparatus was used to separate 0.1 g of Ag^{+} from 0.3 g Cu^{2+} using platinum-rhodium working and auxiliary electrodes and a 1.0 *M* Hg_2SO_4 solution as the reference electrode. The instrument may be used as a potentiostat for polarography as well as a voltage amplifier and current/voltage converter.

The anode may be selected not to affect the potential of the cell. In addition, dual anodes may be used to enlarge the surface area (Fig. 9) and may be protected from the electrolyte by a porous membrane. A platinum gauze electrode is placed between the cathode and anode, and the circuit is completed.

Types of Analysis

Quantitative Determination of Metal Ions

The separation and quantitative determination of metal ions from combinations of metals in alloys may be performed by using constant-

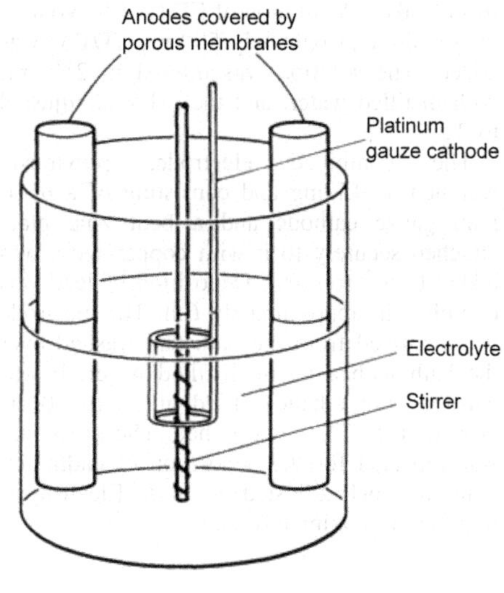

Fig. 9 Typical dual-anode cell

current electrolysis if there is sufficient difference between the decomposition potentials of the metals and other elements, such as hydrogen, being deposited on the cathode. Many elements are determined by using this method, such as Cu^{2+}, Cd^{2+}, Fe^{3+}, Sn^{2+}, and Zn^{2+}. Controlled-potential electrogravimetry may be used to determine many elements in the presence of others. Examples are copper in the presence of bismuth, cadmium, nickel, antimony, tin, or zinc and nickel in the presence of iron and zinc.

Removal of Interfering Ions

The mercury cathode is useful for removing ions that interfere with the metal of interest. Typical elements separated from those that interfere are aluminum, magnesium, titanium, and vanadium. These metals are high in the electromotive series of elements. Those much lower amalgamate with mercury and are separated before determination of those higher.

Effects of Complexation

When attempting to separate ions whose decomposition potentials are close, it is often appropriate to convert them to complexes having different potentials. A typical separation is that of bismuth and copper. When these are electrolyzed in sulfuric acid, the difference in potential is only 0.024 V. If the ions are then treated with a cyanide, the copper is reduced and complexes with cyanide:

$$Cu^+ + 3(CN)^- \rightleftharpoons Cu(CN)_3^{2-}$$

The potential of this complex is much lower than before, providing a greater difference between the potentials than previously. In this case, copper is deposited quantitatively before the bismuth separates.

Internal Electrolysis

In this technique, analysis is restricted to small samples for which separation time is brief. A typical example is the deposition of silver using $Cu/CuSO_4$ as an anode. This may be accomplished in the presence of such elements as copper, iron, nickel, and zinc, which have higher deposition potentials. Another example is the selective deposition of traces of silver, cadmium, copper, and lead using several techniques. For example, an anode of $Cu/CuSO_4$ precipitates silver; an anode of $Pb/PbSO_4$ precipitates copper and silver; an anode of $Cd/CdSO_4$ precipitates lead, copper, and silver; and an anode of $Zn/ZnSO_4$ precipitates all four metals, which can then be analyzed by difference.

Use of Potential Buffers

If chlorocuprous ions are in solution, two competing reactions exist:

$$Cu^+ + e^- \rightarrow Cu \text{ (cathode)}$$

$$Cu^+ \rightarrow Cu^{2+} + e^- \text{(anode)}$$

In the presence of sufficient hydrazine:

$$N_2H_5^+ \rightarrow N_2 + 5H^+ + 4e^- (0.17\ V)$$

$$CuCl_3^{2-} \rightarrow Cu^{2+} + 3Cl^- + e^- (0.15\,V)$$

Oxidation of hydrazine is preferred, and oxidation of Cu^+ at the anode is suppressed, allowing only reduction at the cathode to occur. The action of hydrazine, termed a potential buffer, resembles that of a buffer when stabilizing pH.

Analysis of Oxidizable Species

Analysis of oxidizable species is carried out at the anode by increasing oxidation potential in a manner similar to the method of increasing reduction potential at the cathode. The effect of using attackable anodes in internal electrolysis has been noted. The discharge of ions is usually restricted to halide, hydroxyl, and sulfide ions. In the case of other ions, the hydroxyl ion from aqueous solutions is discharged as oxygen in preference. Decomposition potential in such cases depends mainly on the metal ions present.

Oxides of a metal may be deposited on the anode. In this case, the anode, of platinum gauze, has a larger surface area than the cathode. The situation is complicated by the tendency of the oxides to form hydrates. Suitable correction factors must be applied.

Applications

Estimation of Copper in a Copper-Manganese Alloy

Approximately 1 g of the alloy, which should contain 0 to 0.5 g Cu and 0.5 to 1.0 g Mn, was weighed accurately to 0.1 mg and dissolved in 25 mL of a solution of 5 parts sulfuric acid, 3 parts nitric acid, and 7 parts water. This mixture was heated until all the oxides of nitrogen had evolved, and the reddish color of nitrogen dioxide had disappeared. The solution was cooled, diluted to 150 mL with water, and placed in a tall form beaker with platinum gauze electrodes. The cathode was weighed accurately before use.

Manganese interferes with deposition of copper unless it is reduced from the Mn^{7+} state. This can be accomplished by using a solution of ascorbic acid or of the disodium salt of ethylene diamine tetraacetic acid (EDTA). Convenient strengths are 10 g in 10 mL water for ascorbic acid and 25 g in 50 mL water for the disodium salt of EDTA. In both cases, electrolysis takes approximately 1 h, but the currents applied should be different: 3 to 4 A in the presence of disodium EDTA and 4 to 6 A in the presence of ascorbic acid. These solutions must be added dropwise from a burette to remove the color of the Mn^{7+} salt as it forms.

The cathode was then removed from the cell, washed carefully with water, dried in an oven for approximately 30 min, and weighed. It was then returned to the oven for a further short period and reweighed. This procedure was repeated until a constant weight was achieved. The experiment was repeated a sufficient number of times to provide a suitable degree of precision.

Determination of Nickel in Samples Containing Chloride Ion

A sample containing approximately 0.3 g Ni was weighed to 0.1 mg in a 250 mL beaker; 4.1 mL concentrated sulfuric acid was added. A yellow precipitate formed with the evolution of hydrochloric acid fumes. The mixture was heated to near boiling on a hot plate until the solid precipitate changed through white to pale green. At this stage, all the chloride ions had been removed. The sides of the beaker were flamed continuously with a Bunsen burner or other means. The mixture was heated for another 5 min.

Next, 65 mL distilled water and 35 mL concentrated ammonia were added, and the solution was heated to approximately 70 °C (160 °F). Using a weighed net cathode and a platinum spiral anode, electrolysis was conducted at 70 to 80 °C (160 to 175 °F) with a current of approximately 2 A. Analysis required approximately 90 min. Dimethylglyoxime was added to a drop of the solution as a check that all the nickel had been removed from the solution. The electrodes were removed while the potential was applied, and the cathode was washed with water and then ethanol. It was dried in a current of air and then weighed. Table 5 shows typical results. Cobalt may be determined similarly.

Separation of Cadmium and Lead by Internal Electrolysis

Standard solutions of cadmium and lead were prepared as the nitrates. To a solution of each nitrate, a few drops of nitric acid were added to prevent hydrolysis, and the solution was diluted to 500 mL. The cadmium and lead were standardized by complexometric titration using EDTA. A solution containing approximately 1 to 3 mg of each of the lead and cadmium ions was prepared and placed in a tall form beaker. A solution of EDTA (the volume to contain approximately 500 mg EDTA) was added. The solution was diluted to 250 mL with distilled water, and the pH was adjusted to 2.5.

The combined electrode, previously weighed to 0.1 mg and consisting of a platinum gauze cathode and a bent zinc plate attached securely to it with copper wire, was added to the solution. Deposition of lead was complete in approximately 6 h. The electrode was removed from the bath and washed over the bath with a jet of distilled water. It was washed with ethanol, dried at 70 to 80 °C (160 to 175 °F), and weighed. The procedure was repeated for the deposition of cadmium, with the pH adjusted to 4.0. Electrolysis required approximately 12 h.

Table 5 Determination of Ni^{2+} in $NiCl_2$ solution

Solution, g	Nickel, g	mmol of nickel per gram of solution
2.3625	0.2657	1.9159
2.6052	0.2929	1.9153
2.6204	0.2983	1.9133
2.8191	0.3168	1.9144
2.4832	0.2792	1.9154
2.7186	0.3056	1.9150
2.3413	0.2632	1.9151
2.4821	0.2791	1.9156
2.4626	0.2767	1.9142
2.6911	0.3024	1.9143

Source: Ref 18

Voltammetry (Ref 3)

Voltammetry is a variation of controlled-potential coulometry, where information about the analyte is obtained by measuring current as a function of a variable applied potential. The basic circuit in voltammetry (Fig. 10) is similar to that of controlled-potential coulometry with three electrodes in the solution containing analyte. The electrolysis current flows between the working electrode and counterelectrode, and a stable reference electrode of known potential acts as a sensor of the working-electrode potential. Applied voltage is varied, depending on the position of the variable resistor.

Voltammetry involves measuring the current-voltage relationships when electroactive species in solution are subject to oxidation or reduction at electrodes under carefully controlled conditions. It involves probing a small region of a solution containing, for example, metal ions, by performing small-scale electrolysis between an indicator microelectrode and a reference electrode. A reference electrode, such as the saturated calomel electrode (SCE), is by definition nonpolarizable. That is, its potential remains the same regardless of the potential difference imposed between it and the indicator electrode. The latter is described as polarizable, because it faithfully adopts any potential imposed on it relative to the reference.

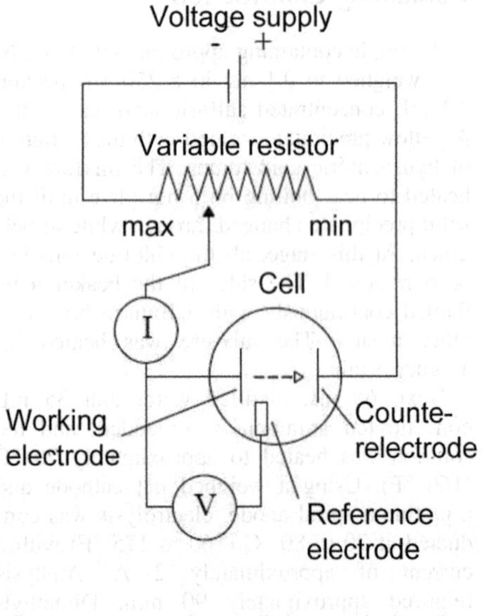

Fig. 10 Circuit of controlled-potential electrolysis cell for voltammetry

If the potential difference between indicator and reference electrode can be controlled accurately and varied uniformly, criteria which modern potentiostatic devices ensure, the corresponding currents that flow reflect the nature and concentration of oxidizable or reducible solutes in solution. Currents flow because of the exchange of electrons between the indicator electrode and electroactive solutes. The latter are frequently metal ions, and the electrode processes monitored are reductions. The indicator electrode then acts as a cathode.

Considerable care is necessary to ensure that electroreducible material reaches the indicator electrode only by natural diffusion. The other important mass-transfer processes, electrical migration and convection, are controlled rigorously; the former, an electric field effect, is related to the transport number of the metal ions and may be eliminated effectively by the presence of a large excess of a supporting, or base, electrolyte. The ionic components of this electrolyte (frequently potassium chloride) do not react with the indicator electrode at potentials at which the required species does, but the presence of the base electrolyte ensures that the transport number of the species whose analysis is required is reduced virtually to zero.

Convection effects may arise from stray vibrations and shock or even from temperature gradients within the solution. Thermostatic control and protection from any form of inadvertent stirring ensures that this interference is minimal. Under less common circumstances, controlled convection, particularly as rapidly rotated electrodes, is used to enhance current signals. For most circumstances, the current-voltage curves for metal ions in solution may be interpreted in terms of the interplay of the diffusion process, by which they arrive at the surface of the indicator electrode, and their reduction there when the applied potential has reached characteristic values.

The selection of indicator electrode material presents some problems. Prolonged accumulation of the products of reduction processes tends to alter the physical and electrochemical characteristics of solid microelectrodes. Mechanical or electrochemical removal of these depositions is not always satisfactory. That is, many solid electrodes develop an irreversible history if suitable precautions are not taken.

Electrodes

The most satisfactory microelectrode at applied potentials more negative than 0.0 V relative to the SCE is that based on mercury in the form of exactly reproducible drops issuing from the end of a capillary attached to a constant head reservoir. Voltammetry performed using the dropping mercury electrode (DME) is known as polarography.

The major limitation of using mercury as an electrode material is its restricted anodic range, which at best extends to +0.4 V versus SCE. This is not a disadvantage for metal analysis, although silver in many supporting electrolytes exhibits simply a diffusion current at 0.0 V versus SCE, the wave due to reduction of silver ion being obscured by the dissolution curve. Similar, although less serious, problems are encountered with Cu^{2+} ions in noncomplexing media.

Many electrode systems based on carbon have been devised that offer a much extended anodic range and the simplification of a stationary indicator electrode without the complications of the DME, with its continuously varying area and resultant signal. However, such electrodes do not always have the same reproducibility as those based on mercury. The hanging mercury drop electrode is based on a syringe device that combines the advantages of mercury as an electrode material with the stability of a stationary surface. After brief usage, the drop is replaced by an exactly reproduced one.

Cyclic Voltammetry

Cyclic voltammetry measures the current that develops in an electrochemical cell where the voltage is in excess of the value dictated by the Nernst equation. It is a potentiodynamic electrochemical measurement where the electrode potential is ramped linearly versus time. The applied potential is varied with time in a symmetrical sawtooth wave form (Fig. 11), and the resultant current is measured over the entire pattern of forward and reverse sweeps. Once the set potential is reached, the working-electrode potential is ramped toward the opposite direction to return to the initial potential, thus creating a cycle that can be repeated at will. The plot involves the current at the working electrode versus the applied voltage. It is generally used to study the electrochemical properties of an analyte present in a solution, often at millimolar quantities.

General uses and applications include:

- Mechanistic pathways of reactions
- Identification of compounds present in mixtures
- Identification of intermediates formed in a reaction pathway
- Determination of antioxidants in food and even skin (polyhydroxy compounds, such as flavonoids and caffeic acid esters)
- Study of electron-transfer reactions associated with host-guest complexation (oxidation of ferrocenecarboxylic acid in the presence of β-cyclodextrin)
- Presence of intermediates in redox reactions (polyaniline formation)
- Determination of diffusion coefficient of an analyte (Ref 21)
- Determination of redox mechanisms in organometallic reactions
- Determination of biochemicals in living cells (serotonin detection, dopamine uptake inhibitors and releasers, cocaine self-administration)
- Industrial studies that include polyaniline and fullerene formation, coatings of conducting polymer colloids, simultaneous registration of planar membrane conductivity and capacitance, preparation of graphene pencil graphite electrodes, in addition to determining the residual useful life of lubricants
- Mechanistic determination of redox organometallic reactions
- Study of redox activities during drug-DNA interactions against cancer
- Deciding about the electron stoichiometry of a system and the concentration of an unknown solution through generation of a calibration curve of current versus concentration

Instrumentation and Process Description

A cyclic voltammetry system involves an electrolysis cell, a potentiostat, a current-to-voltage converter, and a data-collection system. The electrolysis cell consists of a working electrode, counterelectrode, reference electrode, and electrolytic solution. Experimentally, the working-electrode potential is varied linearly with time, while the reference electrode is set at a constant potential. The counterelectrode is the electricity carrier from the signal source to the working electrode. The electrolytic solution is the source of ions to the electrodes during oxidation and reduction. A potentiostat is the device that produces a potential through a direct current power and is accurately determined. It is also used to allow small currents to be drawn into the system without voltage change. The current-to-voltage converter is needed to measure the resulting current through which the data-collection system yields the corresponding voltammogram.

Solid electrodes are used frequently, especially carbon paste, glassy carbon, or pyrolytic carbon. Material reduced and deposited in the forward part of each cycle is reoxidized in the reverse scan. The technique finds application in investigations of mechanisms, particularly identification of intermediate compounds. Short-lived species can be identified at high sweep rates. With the comparatively slow sweep rates normally used, such intermediate compounds have decomposed before their presence and response can be detected.

The potential of a working electrode (V_w) is measured against a reference electrode whose potential is known and kept constant (V_R). The resulting applied potential yields a cyclic voltammogram excitation signal that plots the applied potential versus the reference electrode one (V_W versus V_R, *y*-axis) against time (*t*, usually in seconds, *x*-axis). The same process is repeated as many times as desired (Fig. 12). The scanning starts at a greater potential (V_b) and, with elapsing time, it reaches its lowest value of potential, also known as the switching potential (V_a). At that instance, the voltage is at a value that is capable of causing the oxidation or reduction of the analyte, and the reverse scan occurs with the potential scans approaching more positive values. The process can be repeated as needed,

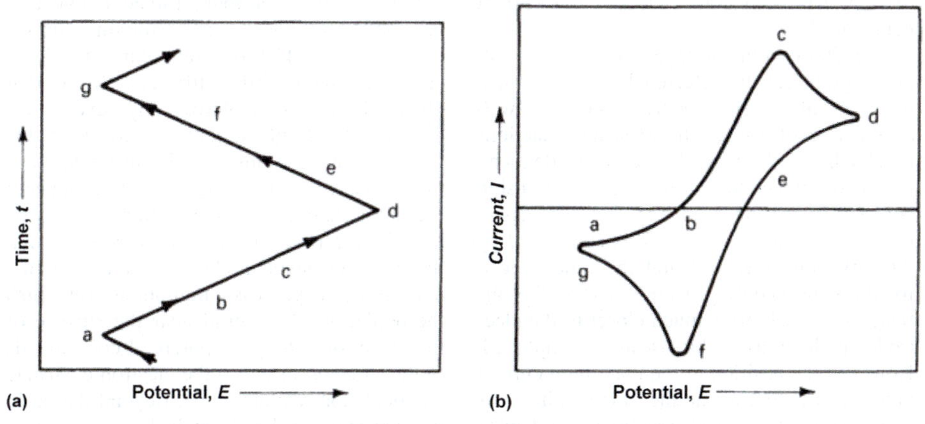

Fig. 11 Fundamentals of cyclic voltammetry. (a) Symmetrical sawtooth potential-time variation used in cyclic voltammetry. (b) Corresponding cyclic voltammogram expected for a near-reversible system. The greater the separation between the peaks for forward and reverse scans, the more irreversible the electrode reactions. Letters "a" through "g" show the stages of the cyclic variation and the corresponding positions adopted by the resultant signal.

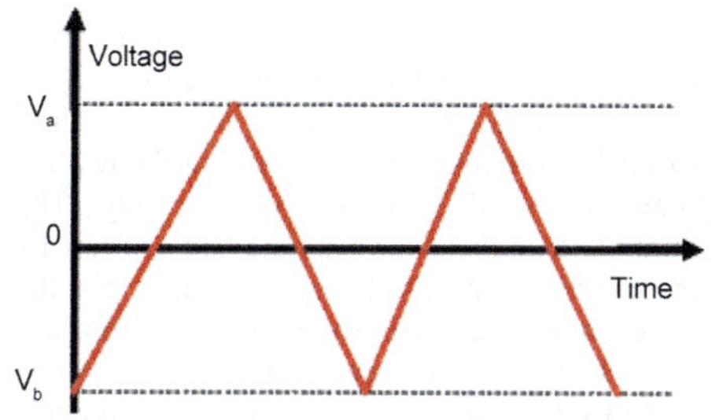

Fig. 12 Cyclic voltammetry excitation signal. V_b, greater potential; V_a, lowest potential, or switching potential

with the slope of the signal providing the value of the scan rate used.

The actual cyclic voltammogram is obtained by measuring the current at the working electrode during the scanning process. A simple, classical voltammogram of a single-electron oxidation reduction appears in Fig. 13 and generally represents the reversible equation:

$$M^+ + e^- \rightleftharpoons M$$

The reduction process occurs from the initial potential ("a") to the switching potential ("d"), resulting in the cathodic current (I_{pc}). The corresponding peak potential, known as the cathodic peak potential (E_{pc}), occurs at "c." Once the value of E_{pc} is reached, all of the substrate at the surface of the electrode has been reduced. At the switching potential ("d"), the voltage scans reversibly while registering positive values from "d" to "g." This results in the anodic current (I_{pa}), with subsequent oxidation taking place. In a similar fashion, the peak potential at "f" is called the anodic peak potential (E_{pa}) and is reached when all of the substrate at the surface of the electrode has been oxidized.

Experimentally, the solution must be stirred; otherwise, the analyte mass transport occurs only through diffusion. Fick's law, which describes diffusion, relates time, reactant concentration, and distance from the electrode to the diffusion coefficient. During the reduction process, the current increases and reaches a maximum value when all the M^+ that is exposed to the surface has been reduced to M. However, as fewer and fewer M^+ ions approach the electrode due to diffusion, the currents continuously decrease with time, and the values of I_{pa} and I_{pc} are difficult to measure.

In contrast, the stirred solution affords an up to 1 μm thick Nernst diffusion layer that lies adjacent to the electrode surface. This prevents the reacting analytes from diffusing into the bulk solution, with consequent establishing of the Nernstian equilibrium and measurement of the diffused controlled currents.

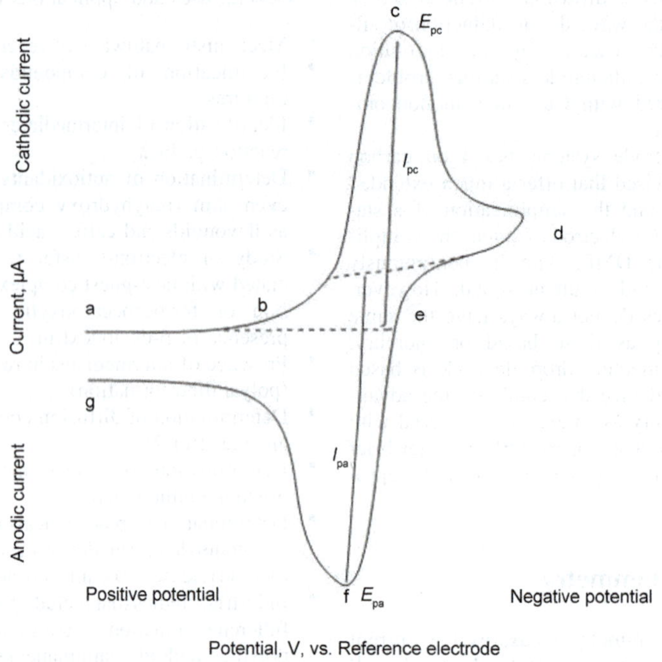

Fig. 13 Voltammogram of a single-electron oxidation reduction. See text for further explanation.

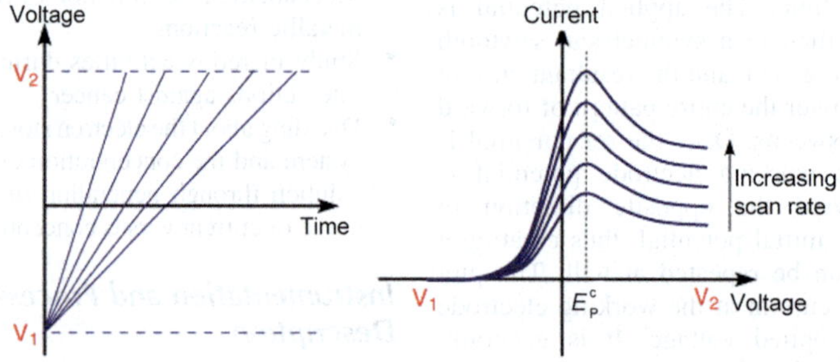

Fig. 14 Linear sweep voltammetry plot of Fe^{3+} solution under different scan rates

Linear Sweep Voltammetry

In linear sweep voltammetry, a rapidly and linearly varying potential is applied to the indicator electrode. The voltage is scanned from a lower limit to an upper one without the reverse scan employed in cyclic voltammetry. The final potential and sweep rate are chosen by the user, as long as the current range for the experiment is known. The scan starts from left to right when no current flows. Figure 14 is a plot of Fe^{3+} solution under different scan rates. As the voltage is swept toward values that are more reductive, the current starts flowing and eventually reaches a peak, at which time the reactant has been converted and the current begins to drop.

In addition to the voltage scan rate, the voltammogram can be affected by the chemical reactivity of the electroactive species as well as the rate of the electron-transfer reactions involved. In all cases, the curve is identical, but the difference lies in the direct relationship between increasing scan rate and subsequent higher total current values. This is justified by the diffusion layer size and the time needed for the scan recording. Consequently, the diffusion layer will grow much closer to the electrode as long as a fast scan is employed. However, the position of the maximum current value always occurs at the same voltage as expected for all electrode reactions that display rapid electron-transfer kinetics.

Modern instrumentation provides sweep rates to 500 mV/s. Because of the rapid removal of electroactive material from the vicinity of the electrode, forced convection is sometimes used to maintain surface concentration. The resultant current-voltage curve is then little different from that obtained by using polarography, except for the lack of current oscillations. However, different signals are observed in unstirred solutions. Peaked voltammograms are produced; the rapid change in applied potential accelerates the electrode reaction so drastically that the diffusion layer thickness increases as the mass-transfer process attempts to replenish the depletion of material near the surface of the electrode. The peak potential corresponds to the situation in which the electron exchange and depletion rate are equal. Beyond the peak, the current drops sharply as the layer of solution near the surface is deprived of reducible material. Figure 15 shows a single-sweep peaked voltammogram obtained for silver by using a carbon-wax electrode.

Direct proportionality between the magnitude of the peak current (I_p) and concentration is expressed in the following equation applicable to reversible reactions:

$$I_p = kn^{3/2}AD^{1/2}v^{1/2}C \quad \text{(Eq 13)}$$

where k is a numerical constant, A is the electrode area, D is the diffusion coefficient, v is the potential sweep rate, C is the concentration, and n represents the number of electrons transferred; although $n^{3/2}$ applies to reversible processes, it must be modified to the form $n \times (\alpha n)^{1/2}$ for irreversible cases. Alpha is the transfer coefficient.

The advantages of peaked wave forms are offset somewhat by the reduced sensitivity to irreversible reductions. Further, the position of the peak potential (E_p) is independent of sweep rate for reversible but not for irreversible systems. A considerable advantage of carbon-base electrodes is the negligible residual current obtained.

Examples of linear sweep voltammetry applications include:

- Study of irreversible reactions
- Reversible deposition of insoluble species
- Determination of direct methane production from CO_2 via a biocathode without the need of hydrogen gas
- Release of inorganic ions and DNA from an inorganic ion/DNA bilayer film in a potential gene-targeting therapy
- Determination of organic substances in food samples (e.g., isohumulones in beer by amperometric titration)

Polarography

Voltammetry performed using the dropping mercury electrode (DME) is known as polarography. Application of direct current polarography is limited at low concentrations by the large current oscillations that obscure small voltammetric signals and by the poor ratio of Faradaic to capacitance components of those signals. Modern instrumentation enables complete use of several electrochemical characteristics. The development of fast sweep rates, rapid response recorders, and the electronic sophistication to sample currents resulting from unusual potential wave forms at stable parts of a drop life have resulted in various well-defined techniques. Current-sampled polarography offers small improvements in analytical sensitivity. The current is probed once at the end of the life of each drop to obtain a stepped polarogram uncomplicated by large current oscillations.

Apart from the reproducibility of the shape and size of the electrode material, the exceptionally high overvoltage that mercury exhibits toward hydrogen is a principal advantage. Virtually all the common metal ions are reduced at potentials that are more positive than those at which hydrogen is discharged. The final current rise, which sets the practical limit to the potential range available, is due to the reduction of the cation of the supporting electrolyte.

Figure 16 illustrates current-voltage curves shown by a solution containing the species

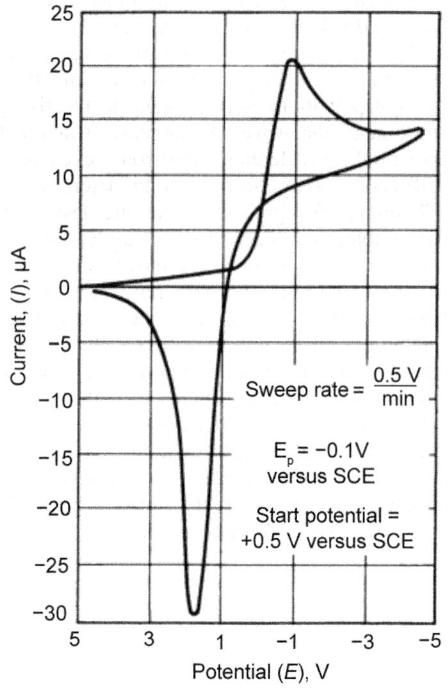

Fig. 15 Single-sweep voltammogram obtained at a carbon-wax electrode for 10^{-3} mol/L Ag^+ ion in 0.1 mol/L KNO_3 as supporting electrolyte. The reversal of the potential scan direction after the cathodic signal is fully developed produces an anodic signal whose size is enhanced relative to the first, because its origin is in material deposited and accumulated in the forward sweep. The principle, used for longer cathodic deposition times at constant potential, is the basis for stripping analysis. SCE, saturated calomel electrode

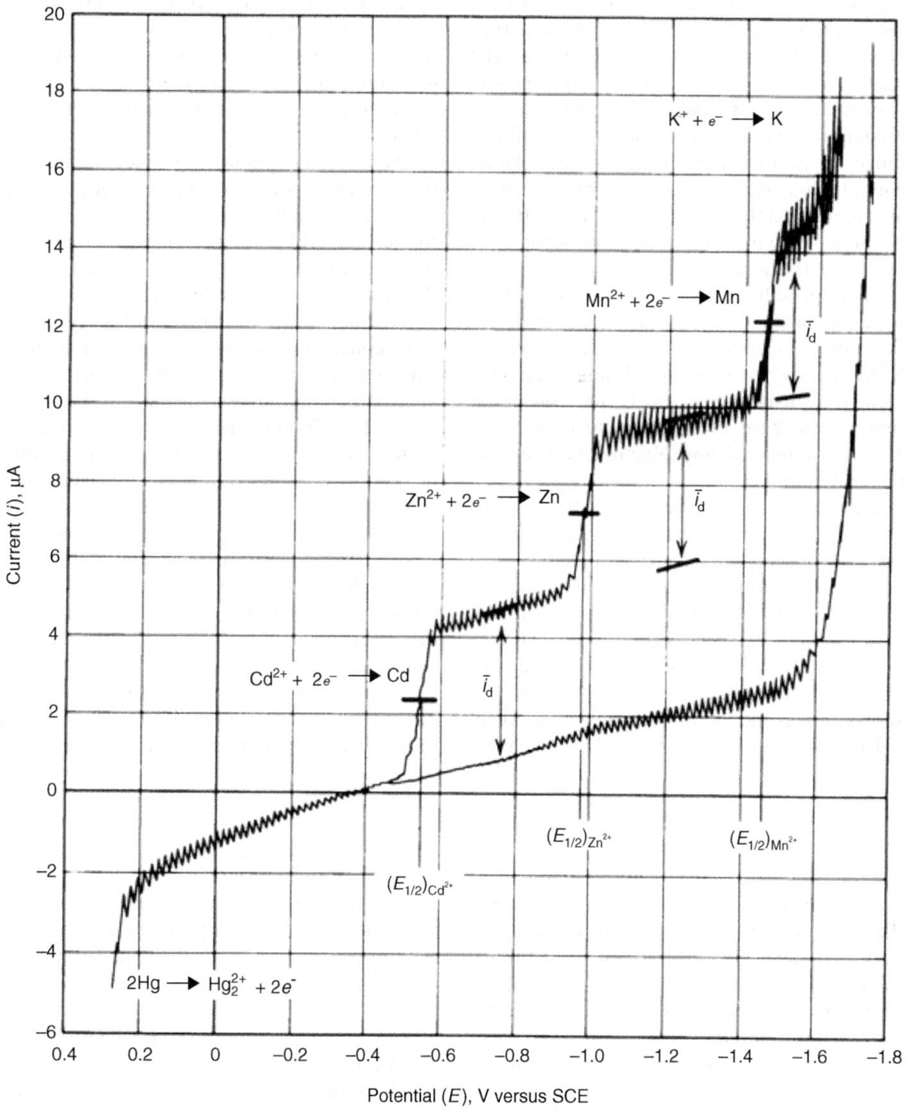

Fig. 16 Direct current polarograms of 10^{-4} mol/L Cd^{2+}, Zn^{2+}, and Mn^{2+} in 0.1 mol/L KNO_3 as supporting electrolyte. The baseline curve is that obtained with supporting electrolyte alone. SCE, saturated calomel electrode

Cd^{2+}, Zn^{2+}, and Mn^{2+}. These curves were obtained by using a recording polarograph operating on a mechanically controlled drop time of 1 s, with potassium chloride as the supporting electrolyte. Because dissolved oxygen is reduced in an elongated two-stage process whose effects influence the entire cathodic range available for analysis, this interference must be removed before measurement by purging the solution for several minutes using an oxygen-free nonreacting gas, such as nitrogen. Further, maintaining gentle flow of the gas over the surface of the working solution during measurements is usually a wise precaution. If this is not carried out, the presence of oxygen will soon become apparent again, particularly if the measurement period must be prolonged. Although a mild nuisance in probing for metals, the sensitivity of the DME to traces of oxygen is useful in determining the element.

Figure 16 shows the curve due to the supporting electrolyte alone and those due to the clearly separated reduction signals of the three cationic species; the latter are superimposed on the supporting electrolyte line. The gently sloping region between the anodic dissolution of mercury and the potassium ion reduction signals represents the available range of normal working potentials. Alkali and alkaline earth metals exhibit reduction at approximately −1.7 V versus SCE and beyond. Viewing the reduction signals of such species necessitates the use of an alkyl ammonium salt as the base electrolyte. These methods enable extending the cathodic potential range to approximately −2.3 V versus SCE. To suppress electrical migration efficiently to insignificant levels, the concentration of the base electrolyte must be 50 to 100 times that of any electroactive species under determination.

The gently sloping baseline is the residual current. This is non-Faradaic and capacitative, being associated with charging of the electrical double layer formed at each new mercury drop as it presents itself to the working solution. Faradaic signals, such as those produced during reduction of the metal ions, display the characteristic sigmoid form shown. Figure 16 shows that as the increasingly negative potential is applied, a point is reached at which this reaction occurs:

$$Cd^{2+} + 2e^- \rightarrow Cd(Hg)$$

As this begins to take place, the region of solution near the electrode surface becomes depleted of cadmium ions, and a concentration gradient is established. Additional increase of potential accelerates the aforementioned reduction process, depletes the surface concentration still further, steepens the concentration gradient, and results in an enhanced current.

A stage is reached at which the bulk concentration of Cd^{2+} can supply this species at no faster (diffusion) rate, and the concentration gradient is at its steepest, with the surface concentration effectively zero. In this condition, cadmium ions are reduced as rapidly as they arrive by natural diffusion at the electrode surface. The current can rise no further, and a plateau region is shown. No significant further current rise occurs until the reduction potential of the zinc ions is reached, and a polarographic wave for this species is observed that arises in the same way as that of Cd^{2+}. This is followed by the reduction wave for manganese and finally by a steeply rising current corresponding to the reduction of the potassium ion.

Polarography is based on a type of steady-state process in which the solution in the immediate vicinity of the drop that detaches itself from the capillary is stirred. With drop frequencies usually adopted, this stirring ensures development of subsequent drops in almost identical environments. This is shown clearly by the limiting current regions in Fig. 16; the depletion of metal ions caused by the electrochemical process taking place at one drop is replenished in time for an identical interaction with the next drop. Only the first drop, acting as an electrode in a given solution, is in an environment different from others.

The important parameters associated with a polarographic wave are the diffusion current (I_d) and the half-wave potential ($E_{1/2}$). The former, directly proportional to the concentration of the species reacting at the electrode, is the foundation for quantitative analysis. The latter, characteristic of that species regardless of its concentration, provides qualitative analysis and corresponds to the value of the applied potential at which the observed current is exactly one-half the limiting diffusion current. However, half-wave potentials must be used with some caution as a means of "fingerprinting," because their values are particularly sensitive to complexation reactions. Under appropriate circumstances, the effects of complexing may be exploited in the separation of interfering waves. This effect is discussed in the section "Applications of Polarography" in this article.

The direct proportionality between diffusion current and concentration expressed in terms of the mean diffusion current is:

$$I_d = 607nD^{1/2}m^{2/3}t^{1/6}C \qquad \text{(Eq 14)}$$

where n is the number of electrons transferred, D is the diffusion coefficient of electroactive species (cm^2/s), m is the rate of flow of mercury through the capillary (mg/s), t is the drop time in seconds, and C is the concentration of electroactive species (mmol/L).

With the units for the various parameters shown and the numerical constant of 607, the mean diffusion current is given in microamps. Equation 14 in this form provides immediate indication of the small size of the currents observed and emphasizes the minute extent of the electrochemical decomposition that takes place. This is small enough that repeated analysis does not impair the observed magnitude of limiting currents. For practical purposes, the technique is nondestructive.

Equation 14 also shows that the values of diffusion currents produced by metal ions are approximately proportional to the number of electrons exchanged in the electrode process. Thus, the wave heights for Tl^+, Cd^{2+}, and In^{3+} are, for equal concentrations, in the approximate ratio 1:2:3. The slight departure from exactness reflects relatively small variations in the diffusion coefficients.

Diffusion-controlled limiting currents are directly proportional to concentration, which may not be true of currents on the rising portion of polarographic waves. For reversible electrode reactions, diffusion is rate determining at all points on the wave; that is, the electron-transfer process is rapid. The rate constants for such rapid reactions characteristically are $\geq 2 \times 10^{-2}$ cm/s. Under such circumstances, the shape and position of the polarographic wave are:

$$E = E_{1/2} + \frac{RT}{nF} \ln \frac{\bar{I}_d - \bar{I}}{\bar{I}} \qquad \text{(Eq 15)}$$

where E and $\bar{I}$ are corresponding values of applied potential and resultant mean current on the rising portion of the wave, R is the universal gas constant, T is the absolute temperature, and F is the Faraday constant.

For irreversible electrode processes, electron transfer is slower than the rate of diffusion at potentials corresponding to the rising portion of the wave. As increase in applied potential accelerates electron exchange, diffusion finally becomes rate determining. Therefore, for processes that are less than reversible, the shapes of waves are elongated, and their positions are removed from the vicinity of their reversible redox potentials. Both of these factors may have important analytical implications for signal clarity and its accessibility on the potential scale. Electron-transfer rate constants less than 2×10^{-2} cm/s exhibit increasing degrees of irreversibility; values less than 3×10^{-5} cm/s are characteristic of totally irreversible electrode processes. Figure 17 shows a comparison of the two extremes of behavior.

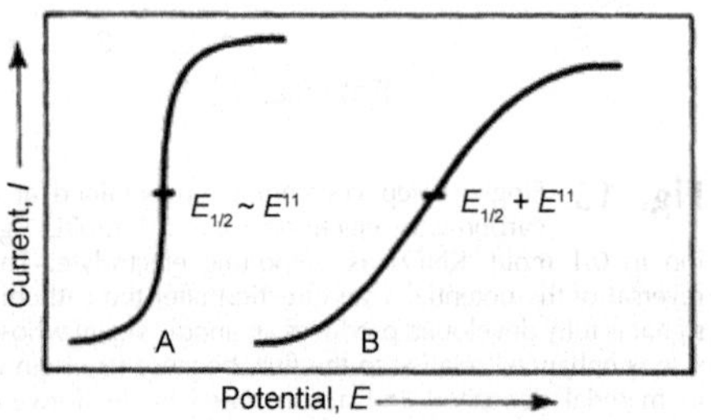

Fig. 17 Polarograms for equal concentrations of two species whose reduction involves the same number of electrons. A, reversible; B, irreversible

If unambiguous measurement of a diffusion current is possible, Eq 14 is equally valid for reversible and irreversible reductions. Unknown concentrations can be conveniently assessed by using a calibration graph or standard addition methods, which are particularly appropriate to polarographic analysis.

A serious limitation of the classical direct current technique arises at a lower concentration level in the region of 5×10^{-5} mol/L, at which the analytically significant Faradaic current decreases to and finally becomes less than the constant background capacitance current. The developments in the technique over the last 25 years have focused on improving this unsatisfactory relationship between the two types of current at low concentrations.

One advantage of polarography as an analytical tool is the possibility of multiple analyses on a single solution sample. However, signals for different species may interfere seriously with one another in the more usual supporting electrolytes, and instrumental or chemical resolution of overlapping waves is required. The former means depend on modern electronic developments; the latter require selection of a working medium that contains appropriate complexing agents showing a selective affinity for species whose waves interfere. Both methods of resolution are often used in combination. Complexation of metal ions shifts their half-wave potentials, usually to more cathodic values. For reversible reductions, the shift is a function of the thermodynamic stability of complexes formed, but for irreversible reductions it depends on the kinetics of the electrode processes.

The classical direct current circuit applies the potentials across the polarographic cell rather than across the indicator/solution interface. This demands the use of low-resistance media so that the *IR* (potential) drop through the cell is negligible. The presence of fairly high concentrations of supporting electrolyte maintains low resistance in aqueous media, although components of the cell, such as the reference electrode, may have a high resistance and distort voltammetric signals. When organic solvents or mixed aqueous/nonaqueous solvents must be used, high resistances will be encountered. Modern instruments incorporate three-electrode systems to control the potential potentiostatically at the indicator electrode/solution interface. The reference electrode, which may under these circumstances be a high-resistance type, is positioned as closely as possible to the indicator electrode and connected to the instrumentation such that no current flows through it. Even if the resistance of the solution is high to allow a significant voltage drop, an operational amplifier control loop maintains the reference electrode potential at its proper value by supplying compensating potential to a counterelectrode, usually a large platinum coil or carbon rod.

The sigmoid shape expected and observed during the polarographic reduction of many metal ions is sometimes obscured by the presence of maxima. Such phenomena (Fig. 18) derive from unusually high mass-transfer rates in some regions of the potential range in which the wave occurs. They are associated with the unstable interfacial effects that cause electroactive material to stream tangentially around the surface of the mercury drop rather than radially, as in the ideal situation. Addition of a small concentration of surface-active material eliminates such interferences. Triton-X-100 (Dow Chemical Company) is used virtually universally as the maximum suppressor. The concentration of such additives should always be kept to a minimum; too much can distort and suppress current signals. Calibration curves should be constructed using standard solutions bearing the same concentration of suppressor as the working solutions.

Pulse Polarography

The various techniques of pulse polarography derive from the principle of using a square potential waveform synchronized with the dropping electrode. After imposition of each potential change, Faradaic and capacitance currents result, but they decay over the period of constant potential that takes place before the next reversal of potential in the square-wave cycle. Although both current components decay during this period, that due to capacitance effects is reduced almost to zero so that by the end of each cycle a net Faradaic current remains. Figure 19 shows the relationship between the two currents.

Normal Pulse Polarography

In normal pulse polarography, the initial potential applied to the dropping mercury electrode is held constant until near the end of the drop life, then stepped to a new value for the remaining 60 ms of that drop (Fig. 20). Current is measured only during the last 17 ms of the pulse, that is, at the most stable part of the drop life when its area is virtually constant. Precise synchronization of voltage pulses, drop time, and current detection are crucial and may be achieved only by synchronizing the electronic parameters with mechanical detachment of drops at accurately determined intervals. Potential pulses of increasing magnitude are applied to successive drops, and the resultant plot of current flowing against applied potential constitutes the pulse polarogram. These measures improve detection limits by an order of magnitude.

Differential Pulse Polarography

Differential pulse polarography provides significant increase in detection limits; the range of trace analysis extends to 10^{-7} to 10^{-8} mol/L. A linearly increasing voltage with superimposed voltage pulses of equal magnitude is applied to the working

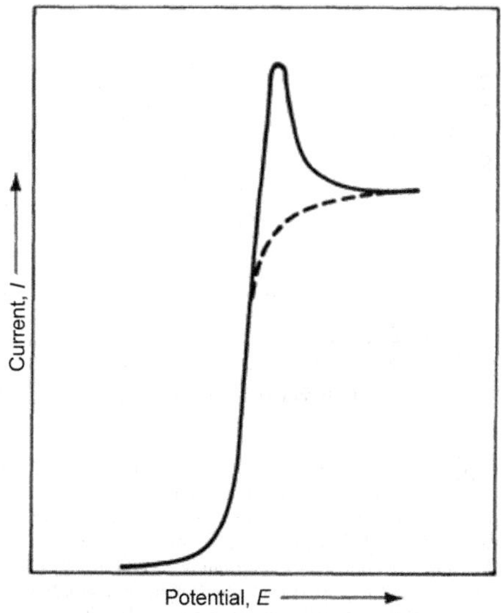

Fig. 18 The shape of a polarogram distorted by a maximum

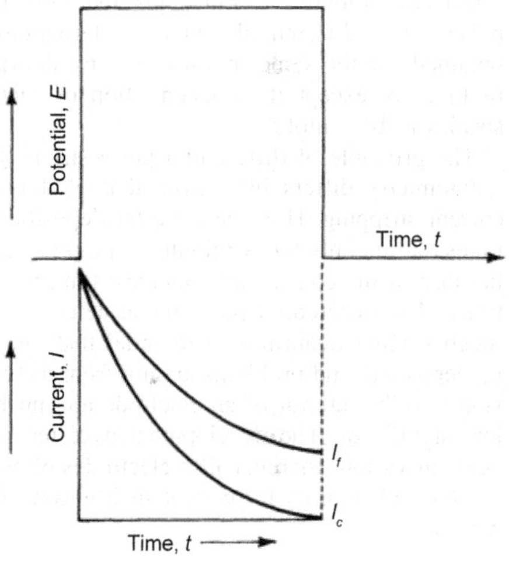

Fig. 19 Relative rates of decay of Faradaic (I_f) and capacitance (I_c) currents after imposition of voltage change in the potential square-wave profile

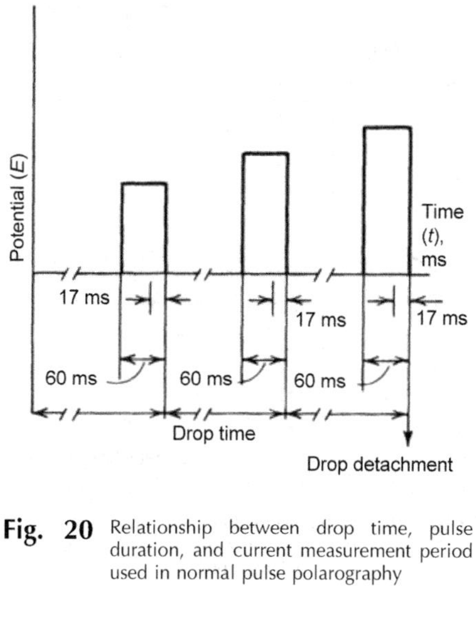

Fig. 20 Relationship between drop time, pulse duration, and current measurement period used in normal pulse polarography

electrode. Modern instrumentation provides considerable flexibility in the selection of the size of pulses, and those from 5 to 100 mV are not unusual. As with the previous technique, synchronization of applied potential and drop time is such that each pulse is imposed once during the last stages of the life of each drop. The duration of the pulse and the period during which current is measured are of the same order as those used in the normal pulse technique. However, in this case, the difference in current flowing just before application of the pulse and that flowing during the last few milliseconds of the pulse is measured. This current difference is measured instrumentally and is plotted as a function of the applied potential. Figure 21 shows the potential wave form.

The heights of the peaked polarograms obtained are directly proportional to concentration and provide improved resolution between adjacent signals compared with the direct current and normal pulse methods. The modulation amplitude may be varied to improve sensitivity. Figure 22 illustrates the normal pulse and differential pulse polarograms obtained for the same mixture of ions shown in Fig. 16, except the concentration of each species is 10^{-5} mol/L.

The principle of differential pulse stripping voltammetry differs little from that of direct current stripping. However, shorter deposition times—often only a few minutes—are possible because of the considerable increases in sensitivity. The differential pulse mode detects the much smaller quantities of material that must be deposited, and problems arising from diffusion into the interior of an electrode are much less significant. The development over recent years of various mercury film electrodes offers mechanical stability to prevent such unwanted diffusion.

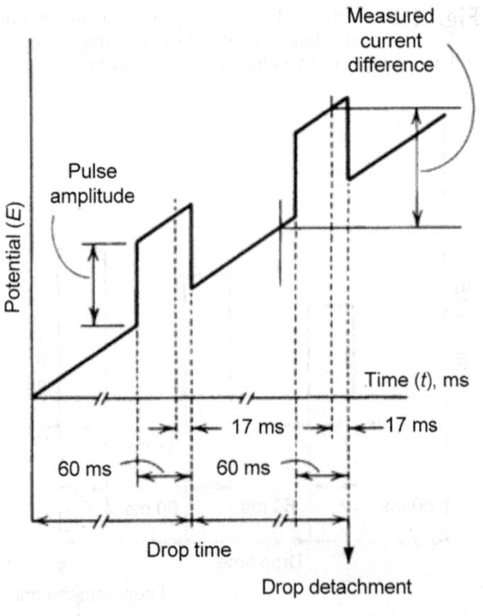

Fig. 21 Relationship between drop time, pulse duration, and current signal in differential pulse polarography

Stripping Voltammetry

Stripping voltammetry is based on the principle that a concentration of reduced metal may accumulate at an electrode by prolonged electrolysis at an appropriate potential in a solution containing such a low concentration of reducible species that its presence would be virtually undetectable by using direct voltammetry. The extent of deposition must be compatible with the production of a measurable anodic current when the deposited material is reoxidized. Application of a potential more cathodic than the half-wave potential of the metal required causes continuous reduction and accumulation as an amalgam in a hanging drop. A mercury-base electrode, which ensures occurrence of the extended reduction at a surface of constant nature and area, should be used. With solid microelectrodes, the nature of the surface changes continuously as products accumulate in the surface. Amalgam formation requires considerable time to affect seriously the surface characteristics of a drop; further, the formation of intermetallic species resulting from simultaneous deposition of two metals is decreased in significance.

In principle, determination of low concentrations—to approximately 10^{-11} mol/L—should be possible using such techniques. In practice, a concentration limit of the order of 10^{-9} mol/L is more realistic. Excessive deposition times may result in instrumental instabilities; more importantly, the last remnants of a wanted species may be removed from a solution by adsorption on the walls of the cell or its components, not by deposition. Even by using modern syringe-based systems for producing stationary mercury drops, the effective life for deposition purposes is limited because such drops cannot be maintained in a stable condition longer than 30 min. Extended deposition also encourages the undesirable diffusion of metals into the interior of the electrode. The rapid anodic polarization that follows to achieve maximum sensitivity (via the peaked anodic wave form) does not cause reoxidation, during the sweep time, of material that has penetrated significantly into the mercury.

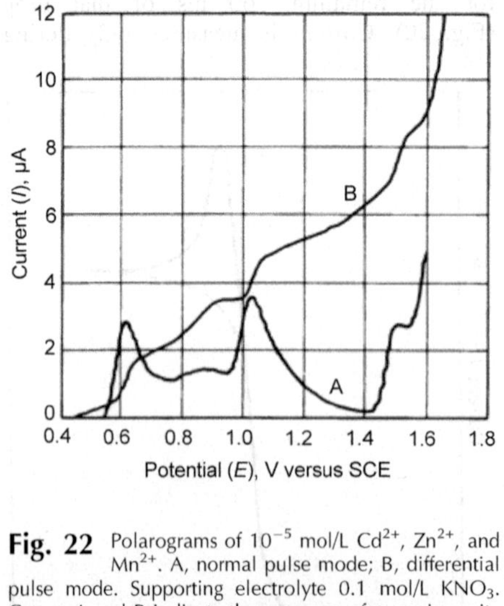

Fig. 22 Polarograms of 10^{-5} mol/L Cd^{2+}, Zn^{2+}, and Mn^{2+}. A, normal pulse mode; B, differential pulse mode. Supporting electrolyte 0.1 mol/L KNO_3. Curves A and B indicate the presence of some impurity showing a signal at approximately −0.85 V. This may originate in the supporting electrolyte and emphasizes the importance of extreme purity of such salts required in analysis at these levels. In this case, the interference does not prevent the measurement of the peak heights in curve B. SCE, saturated calomel electrode

Applications of Polarography

Determination of Trace Amounts of Nickel in Cobalt Compounds

Nickel and cobalt react similarly toward many reagents, and determination of one element in the presence of large amounts of the other is challenging. The similarity extends to the overlap of their polarographic signals in many common supporting electrolytes. However, successful separation of these signals is conducted in a medium consisting of pyridine and pyridinium chloride, the nickel step preceding that due to cobalt. Although direct current polarography is adequate for determining commercial-grade cobalt salts, which may contain 0.5% Ni, the low nickel content of analytical-grade salts, often under 0.01%, demands the additional sensitivity of the differential pulse mode.

The following details applied to the analysis of the two grades of samples of cobalt nitrate. Samples of the salts weighing 1 g were dissolved in 50 mL distilled water in 100 mL volumetric flasks. To each solution were added 2.0 mL 12 *M* hydrochloric acid, 5.0 mL pyridine, and 5.0 mL 0.2% gelatin solution. The solutions were then diluted to the mark and transferred to a clean, dry polarographic cell, then deoxygenated 5 min. They were then polarographed using direct current and differential pulse modes. Voltage scans began at −0.5 V versus SCE and were halted soon after −1.0 V versus SCE due to the onset of the cobalt step, whose magnitude would have driven the pen off the chart.

Quantification of the nickel content was accomplished by standard additions. For the higher concentrations, successive additions of 2.0 mL 0.01 *M* standard nickel solution were added to the solutions in the cell, which were then repolarographed after a few seconds of additional deoxygenation. For the much lower nickel concentrations present in solutions of the analytical-grade material, additions of 0.1 mL standard solution were found to be more appropriate.

Figures 23 and 24 illustrate experimental traces obtained for samples of both grades of the salt. Figure 18 shows that although the direct current technique clearly identifies the presence of nickel, its reduction is followed so closely by that of cobalt that the limiting currents of the required metal are less clear than desirable. Consequently, use of the differential pulse mode is advantageous because of the enhanced resolution provided by a peaked

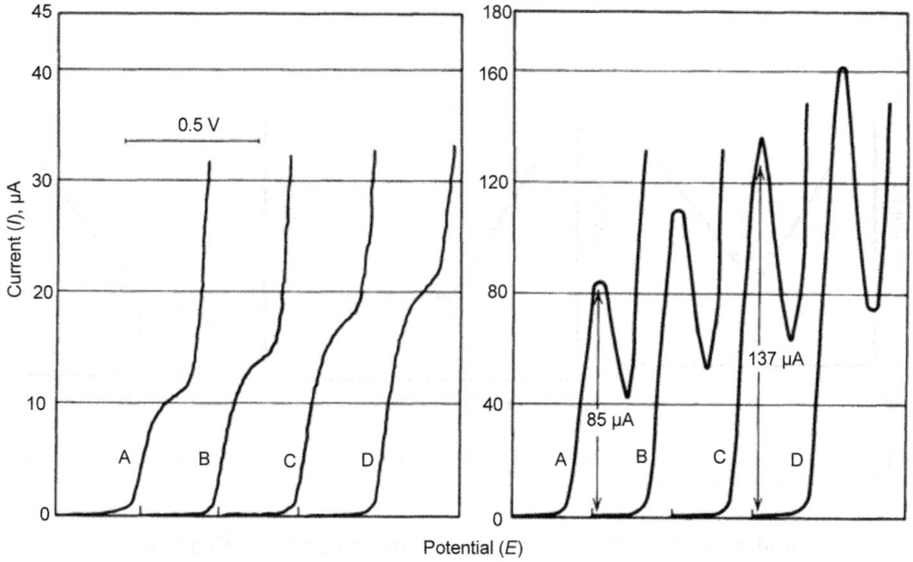

Fig. 23 Direct current and differential pulse polarograms of nickel in general-purpose cobalt nitrate. A, sample solution; B, after addition of 2 mL of 0.01 mol/L standard; C, after addition of 4 mL of standard; D, after addition of 6 mL of standard. All traces begin at −0.5 V versus saturated calomel electrode.

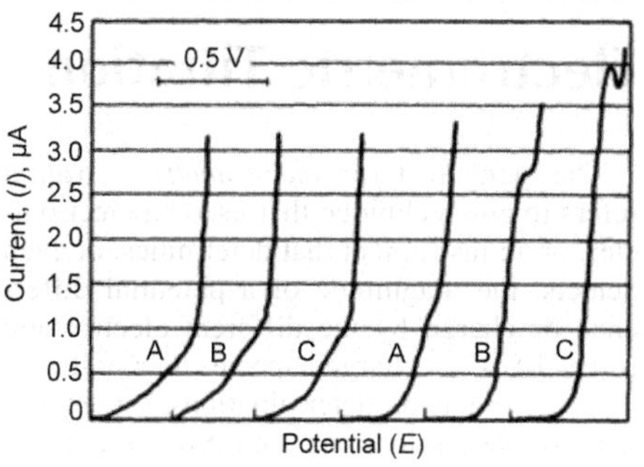

Fig. 24 Direct current and differential pulse polarograms of nickel in analytical-grade cobalt nitrate. A, sample solution; B, after addition of 0.1 mL of 0.01 mol/L standard; C, after addition of 0.2 mL of standard. All traces begin at −0.5 V versus saturated calomel electrode.

signal with consequent unambiguous measurement of its height, not the increased current sensitivity. Figure 23 illustrates that the direct current technique cannot analyze the low level of nickel present in the high-purity salt.

The nickel concentration was assessed in each case by using:

$$C_{test} = \frac{IvC_{std}}{\Delta I(V + v) + Iv} \qquad \text{(Eq 16)}$$

where I is the original signal height, ΔI is the change in signal height after standard addition, V is the original volume (100 mL), C_{test} is the concentration of test solution, and C_{std} is the concentration of standard solution.

Thus, use of Eq 16 and the peak current data for curves A and C shown in Fig. 23 yields:

$$C_{test} = \frac{85 \times 4 \times 0.01}{(52 \times 104) + (85 \times 4)} = 5.92 \times 10^{-4}\,\text{mol/L} \qquad \text{(Eq 17)}$$

Because the relative atomic mass of nickel is 58.71, this figure corresponds to a mass of nickel of 3.48×10^{-2} g/L, or 3.48×10^{-3} g/100 mL. This corresponds to the mass of nickel in 1.0263 g of cobalt salt; therefore, the proportion of nickel in the original sample is 0.34%. Similar calculations based on curves A and C of Fig. 24 produce $C_{test} = 9.04 \times 10^{-6}$ mol/L, that is, 5.31×10^{-5} g Ni/100 mL solution containing 1.0037 g of analytical-grade cobalt nitrate, or 0.0053% Ni.

Multielement Fingerprinting and Approximate Quantification in Effluent Samples

A fingerprinting method was necessary for the simultaneous detection of the presence and level of several metals in various effluent systems. This was to form the basis of a continuous monitoring setup to produce a method that could effect rapid preliminary qualitative and quantitative analysis by relatively unskilled operators. Where the presence of certain species is clear, refined subsequent analysis may be necessary—perhaps using a method other than voltammetry, such as atomic absorption.

The most appropriate supporting electrolyte was found to be a mixture of ammonium hydroxide and ammonium chloride; as a stock solution, this was prepared to a concentration of 1 *M* and used to prepare pretreated samples to a selected volume before analysis. To maintain the simplicity of the method, the reference electrode used was a mercury pool on the bottom of the cell. Major interference from iron was removed, because it was precipitated as Fe^{3+} from the medium used. The method was found to be effective for effluent samples where there was concern over the possible presence of specific metal ions. In a particular case, it was necessary to monitor for the presence of copper, lead, cadmium, nickel, and zinc.

Some pretreatment was required to destroy organic material present in the samples and to ensure that the metals were present only as noncomplexed ions before addition of the supporting electrolyte. A 100 mL sample was evaporated with 1 mL concentrate nitric acid to approximately half the initial volume. Hydrogen peroxide (6%) was added dropwise as necessary until the sample cleared. The sample was then evaporated to approximately 20 mL to ensure decomposition of excess peroxide; 1 mL 5% citric acid solution was added, and supporting electrolyte was added to a total volume of 100 mL.

Following the pretreatment to remove organic matter and iron, 10 mL working solution was deoxygenated 3 min in a cell with a mercury pool reference electrode and polarographed from −0.2 to −2.0 V relative to this. A 1 mL quantity of standard solution containing 10 mg/L of the above metals and the same supporting electrolyte was added, and the measurement repeated. Enhancement of the appropriate peaks from the first run confirmed the presence and identity of various species and approximate estimation of their concentrations by application of the standard addition formula. Appearance of a peak where there was none in the first run confirmed the absence of that species in the original sample. Figure 25 shows the results obtained for two samples.

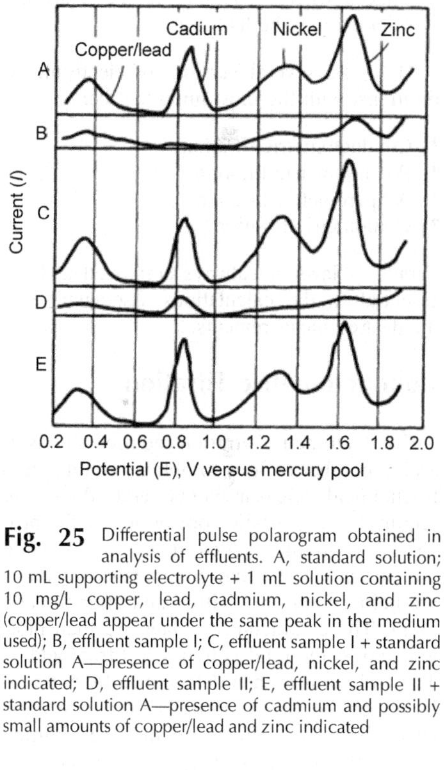

Fig. 25 Differential pulse polarogram obtained in analysis of effluents. A, standard solution; 10 mL supporting electrolyte + 1 mL solution containing 10 mg/L copper, lead, cadmium, nickel, and zinc (copper/lead appear under the same peak in the medium used); B, effluent sample I; C, effluent sample I + standard solution A—presence of copper/lead, nickel, and zinc indicated; D, effluent sample II; E, effluent sample II + standard solution A—presence of cadmium and possibly small amounts of copper/lead and zinc indicated

Electrometric Titration

The general term *electrometric titration* refers to any technique that used an electrometer, or an instrument that determines, or even detects, the magnitude of a potential difference or charge by the different electrostatic forces between charged species. General uses are high-precision determinations, volumetric analysis determinations, and process control. Examples of applications include:

- Volumetric determinations that involve change of colors and thereby no need of colored indicators
- Establishing of primary standards
- Determination of water-insoluble material by using an organic solvent
- Analysis of radioactive materials and substances where proximity to the analyzing scientist may be harmful

There are several versions of electrometric titrations, with the most important being:

- Conductometric titration
- Potentiometric titration
- Amperometric titration
- Coulometric titration

Faraday's laws are always employed and are essential in the calculations associated with the desired measurements.

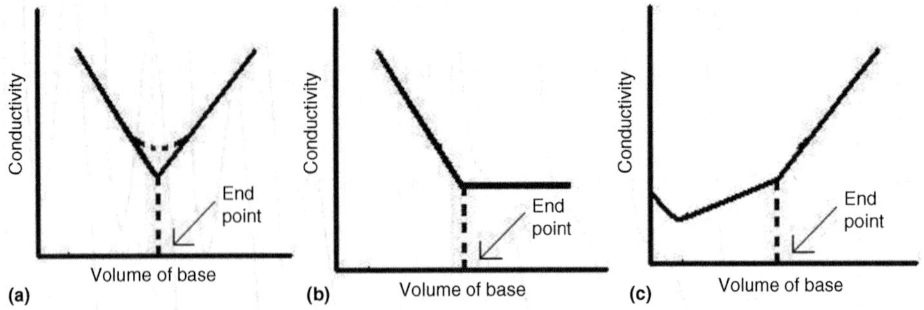

Fig. 26 Conductometric titration curves for (a) strong acid/strong base, (b) strong acid/weak base, and (c) weak acid/strong base reactants

Conductometric Titration

Conductometric titration requires a considerable number of ions to obtain reliable electrochemical measurements and does not require a thermostatic control during the process. The titrant is added incrementally while the titrant volume and conductance readings are being monitored before and after the end point. The conductance values are then plotted against the total volumes of titrant added. The end point is then determined by extrapolation of the linear portions of the titration curve branches. To overcome the dilution factor, the titrant is usually more concentrated than the solution that is being titrated. This method is used primarily in acid-base titrations. The titration curves are different based on the nature of the two reactants, with higher conductivity values being proportional to the amount of ions present. Figure 26 displays the three different combinations of acids and bases. Figure 26(a) is characteristic of a strong acid/strong base titration, with the end point having the lowest conductivity. Figure 26(b) is that of a strong acid/weak base titration, with the conductance being rather constant after the end point because the weak base does not provide as many ions. Finally, Fig. 26(c) is that of a weak acid/strong base titration, with the conductance increasing almost exponentially after the end point because more and more ions are being added. The conductometric titration of two solutions yielding an insoluble precipitate yields a graph similar to the strong acid/strong base titration (Fig. 26a). Such an example is that of sodium chloride/ silver nitrate titration.

Potentiometric Titration

The potentiometric titration process applies to all general reactions in both aqueous and nonaqueous media and is similar to a direct titration or an oxidation reduction with no need of an indicator. The developed potential changes dramatically near the end point because the potential is logarithmically related to the activity of an ion. The regular reference electrodes, such as the hydrogen calomel and silver chloride ones, are routinely used, with the indicator electrode forming an electrochemical half-cell with the ions under study. The end point is determined by reading a large change in potential, similar to the indicator procedure. The development of ion-selective electrodes has expanded the usefulness and applicability of this method.

Amperometric Titration

Amperometric titration is based on voltammetry and, as in conductometry, yields linear titration curves. The potential of the indicator electrode is fixed, which enables both the titrant and the compound to be titrated to yield a limiting current that is proportional to their concentration. A significant precaution for data acquisition is that dissolved oxygen gas must be purged before the experiment starts, because it readily undergoes electroreduction that can interfere in data acquisition. The nature of a weak organic electrolyte can provide a limiting current as large as that of a similar-concentration strong one. Organic molecules such as dimethylglyoxime can act as an active amperometric titrant. The surface area of the working electrode is of practically no significance because the magnitude of the measured current serves only as an indicator.

Coulometric Titration

Coulometric titrations use a constant-current system to determine the exact quantity of a substance in solution, with the current system serving as a titrant. The current is applied until all species become both oxidized or reduced, thus causing a significant change of the working-electrode potential and thus indicating the end point. It has great advantages in the generation of a sensitive indicator and no need for standardized solutions titrant. In addition, it has the ability to quantitatively determine unstable ions (such as Ti^{3+}, Sn^{2+}, Cr^{2+}) and is the basis for the Karl Fischer titration that determines the amount of water in a sample.

Potentiometric Membrane Electrodes (Ref 4)

Potentiometric membrane electrodes are electrochemical devices that can be used to quantify numerous ionic and nonionic species. This class of electrochemical sensors can be divided into ion-selective and gas-sensing membrane electrodes. In both cases, a selective membrane potential is related to the concentration or activity of the species of interest.

Potentiometric membrane electrode measurements require an indicating electrode and a reference electrode. The potential of the indicating electrode depends on the activity of the ion of interest, and this potential is measured with respect to the constant potential of the reference electrode by using a high-impedance potentiometer. The reference electrode is an important but frequently overlooked component of potentiometric membrane electrode measurement. Other general factors to ensure proper electrode response include temperature control, electrode storage, and sample pretreatment.

Ion-Selective Membrane Electrodes

Several types of ion-selective membrane electrodes (ISEs) have been developed in which different membrane compositions are used as the selective agent (Ref 22–24). Membranes composed of glass, solid crystalline, and polymer layers are commonly used in commercially available ISEs. Figure 27 illustrates these three basic electrode systems, as described in the previous edition of this Handbook (Ref 4). However, research on the development of ISEs remains very active (Ref 25). There also is significant practical benefit in ISE use, because analysis with an ISE does not require pretreatment of the sample, and it allows continuous monitoring in clinical, industrial, and environmental applications. Table 6 presents some examples of the applications of ISEs.

Glass Membrane Electrodes

Glass membrane electrodes are generally used to quantify monovalent cations, particularly protons (the pH electrode) and sodium ions. Figure 27(a) shows a glass membrane electrode. The ion-selective glass layer separates the sample or external solution from an internal reference solution. The composition of this internal solution remains constant throughout the operation of the electrode. In contact with this internal reference solution is a silver/silver chloride reference electrode. The several chemical potentials established throughout this system remain constant, except for the membrane potential, which is measured and related to the concentration or activity of the ion of interest. The response mechanism of glass membrane electrodes is thought to be based on selective exchange and mobility processes that occur between the glass matrix and solution ions (Ref 26). Composition of the glass membrane controls the selectivity of the electrode response (Ref 27).

Liquid or Polymer Membrane Electrodes

A similar electrode system uses a liquid or polymer membrane as the selective component (Ref 25, 28, 29). Figure 27(b) shows such an electrode system, in which a polymer membrane separates the internal and external solutions. In this case, electrode selectivity is achieved by incorporating into the polymer membrane an agent that will selectively complex with the ion of interest. During the electrode response, the complexing agent in the membrane complexes the ion of interest, creating a charge separation at the membrane/solution interface. This charge separation results in the measured membrane potential. Numerous reports are available concerning models for this type of electrode response (see also the sources in the previous edition of this Handbook, Ref 4).

Solid Crystalline Membrane Electrodes

A third ISE uses a solid crystalline membrane. Figure 27(c) shows a common arrangement for this type of electrode system. Membranes in the form of a single crystal or a pressed pellet composed of highly insoluble inorganic salts are used. For many solid crystalline membrane electrodes, the internal reference solution may be eliminated by connecting the internal reference electrode to the internal side of the solid membrane, due to the conductivity of the crystalline membrane. For the most part, the response of solid crystalline membrane electrodes is based on the solubility of the ionic species comprising the membrane.

Behavior of Ion-Selective Membrane Electrodes

The membrane potential of an ISE is described by Eq 18, which relates the measured

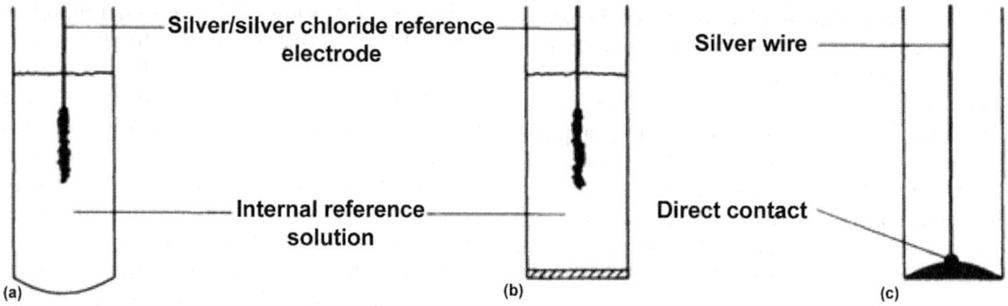

Fig. 27 Types of ion-selective membrane electrodes. (a) Glass membrane electrode. (b) Polymer membrane electrode. (c) Solid crystalline (pressed pellet or single crystal) membrane electrode

Table 6 Applications of ion-selective electrodes

Branch of activity	Typical samples	Typical analytes
Clinical analysis	Blood serum, plasma, whole blood, saliva, urine, lymph	pH, K^+, Na^+, Ca^{2+}, Li^+, Cl^-, Mg^{2+}, HCO_3^-
Agricultural industry	Soil, vegetables, fruits, milk, meat	pH, NO_3^-, K^+, NH_4^+, Ca^{2+}, Cl^-, Na^+
Industrial manufacturing	Metal plating solution, paper bleaching solutions, fertilizers	pH, Cu^{2+}, Ag, Au, NO_3^-, Ca^{2+}, K^+, Na^+, NH_4^+
Environmental monitoring	Natural, industrial, waste waters, soil, plants, human and animal tissues	pH, Pb^{2+}, Hg^{2+}, Cu^{2+}, NO_3^-, ionic and nonionic surfactants
Pharmaceutical industry	Medical drugs, liniments, mixtures	Biologically active amines, alkaloids, acids
Food industry	Juices, beverages, dough, pastry, wine	pH, Ca^{2+}, NO_3^-, CH_3COO^-
Power stations	Cooling water	pH, Na^+
Control of gases in air	Air in chemical factories	NH_3, NO_2, "acidic" gases
Control of enzymatic activity	Medical and biological liquids and tissues, pesticide-polluted soils and plants	Enzymes, substrates, enzyme inhibitors

Source: Ref 23

potential to the ionic activity of the species of interest:

$$E_{cell} = E_{const} + \frac{RT}{nF} \ln \left(\frac{a_i}{a'_i}\right) \quad \text{(Eq 18)}$$

where E_{cell} is the measured cell potential, E_{const} is a potential term that includes all cell potentials remaining constant throughout the measurement, R is the universal gas constant (8.31441 V · C/K/mol), T is the temperature of the cell in degrees Kelvin, n is the number of electrons transferred in the appropriate half reaction, F is the Faraday constant (estimated at 96 486.332 C/mol), and a'_i and a_i are the activities of the principal ion in the internal reference and sample solutions, respectively. Because the activity of the ion of interest is constant in the internal reference solution, Eq 18 can be simplified:

$$E_{cell} = \text{Constant} + 0.05916 \log a_i \quad \text{(Eq 19)}$$

Therefore, the measured cell potential is related to the logarithm of the ion activity. Electrode calibration curves are typically prepared as a plot of potential versus logarithm of standard ion activity, and sample ion activities are obtained by extrapolation from this standard curve, using a measured potential value.

Realization of the difference between ion activity and concentration is important in ISE measurements. The measured membrane potential is related to the ion activity as opposed to its actual concentration. The ion activity can be thought of as the effective or free ion concentration in solution and can be related to its concentration by:

$$a_i = \gamma_i C_i \quad \text{(Eq 20)}$$

where γ_i is the activity coefficient for species i, and C_i is the concentration. In dilute solutions, the activity coefficient goes to unity, and the activity and concentration are equivalent. At high concentrations or in solutions of high ionic strength, the activity coefficient falls below unity. The activity coefficient of an ion in a solution of known ionic strength is typically estimated by using:

$$\log \gamma_i = -Az_i I^{1/2} \quad \text{(Eq 21)}$$

where A is a solvent coefficient, z_i is the ionic charge of species i, and I is the solution ionic strength. Substituting Eq 20 into Eq 18 creates a relationship between ion concentration and measured potential. If the activity coefficient of interest remains constant between the measure of the standards and the sample, the ratio of activity coefficients will remain constant, and concentration of the free ion in solution can be determined. Therefore, concentrations can be measured by estimating the activity coefficients or by preparing the electrode calibration curve in a suitable matrix.

Equation 18 indicates that the response of an ISE depends on temperature. Many analytical applications of ISEs require temperature control. Without proper temperature control, electrode potentials drift continuously, which can complicate establishing the correct steady-state potential. Temperature changes are more problematic for gas-sensing electrodes. A temperature change of only 3 °C (5 °F) will require approximately 2 h before the electrode system can reestablish a steady-state potential. Although various factors contribute to this lengthy time period for re-equilibration, osmotic effects have been suggested as the primary cause (Ref 30). The most convenient way to control the temperature of the potentiometric membrane electrode system is to use jacketed thermostatted cells for all measurements.

Determining Electrode Selectivity

Selectivity of an ISE is extremely important and must be considered before applying such an electrode to a system. Selectivity limitations of certain ISEs are often overlooked, leading to major problems for the operator. No ISE is specific for a particular ionic species. The commonly used pH electrode, which displays one of the highest degrees of selectivity, will respond to sodium ions under certain circumstances (Ref 26). Equation 22 may be used to account for interferences:

$$E_{cell} = \text{Constant} + 0.05916 \log \left(a_i + K_{i,j}\, a_j^{(z_i/z_j)}\right) \quad \text{(Eq 22)}$$

where z_i is the ionic charge on the species of interest, and $K_{i,j}$ is the selectivity coefficient for species i relative to species j. Analysis of Eq 22 reveals that smaller values for the selectivity coefficient correspond to higher degrees of selectivity. For example, a typical selectivity coefficient for the sodium-selective glass electrode for sodium over potassium is approximately 1×10^{-5}; therefore, the electrode is approximately 10^5 times more selective for sodium than potassium. That is, such a selectivity coefficient indicates that potassium ion activity must be 5 orders of magnitude greater than sodium ion activity to produce the same electrode response.

The fixed interference and the separate solution methods may be used to measure electrode selectivity coefficients (Ref 31). It is important to know which method has been used to obtain a particular value, because the two techniques often will not agree exactly. The fixed interference method is preferred, because it characterizes electrode response under conditions that are similar to actual use. Various publications list and describe selectivity coefficients (Ref 29, 32–36).

Selectivity coefficients can change during the life of an ISE, especially for solid crystalline and polymer membrane electrodes. These changes are primarily caused by alterations in membrane composition during extended electrode use. This problem is not as severe with glass membrane electrodes. Selectivity coefficients must be considered before applying ISEs to an analysis.

Reference Electrode

The reference electrode is extremely important for ISE measurements. A poor choice of reference electrode can lead to severe interferences and electrode drift. For gas-sensing electrodes, less consideration is necessary concerning the reference electrode, because the indicator and reference elements are supplied as a single package. Figure 28 shows two typical reference electrodes that are frequently applied to ISE measurements.

Figure 28(a) illustrates a single-junction reference electrode that consists of a silver/silver chloride electrode immersed in a concentrated solution of potassium chloride. Saturated potassium chloride or 4 M potassium chloride is generally used. This internal solution contacts the sample solution through a junction that allows ions to pass from one solution to the next without permitting complete mixing of the solutions. Many types of junctions are available, but porous ceramic frits are prevalent. Ion transport across this junction provides the ion conductivity necessary to complete the electrochemical circuit. Due to ion movement, chloride and potassium ions can contaminate the sample solution. For many assays, this contamination is not important; however, when detecting potassium or chloride or when either of these ions interferes with the analyte of interest, a double-junction reference electrode must be used.

Figure 28(b) shows a double-junction reference electrode that is similar to the single-junction electrode, except an additional solution is between the sample and the internal reference solution. The composition of this middle solution is such that ions flowing into the sample will not interfere with the assay procedure. For example, potassium nitrate is generally used for chloride analyses.

Perhaps the most common reason for inaccurate results from ISE measurements stems from improper use of the reference electrode. The junction of the electrode must be open to allow free flowing of ions. These junctions

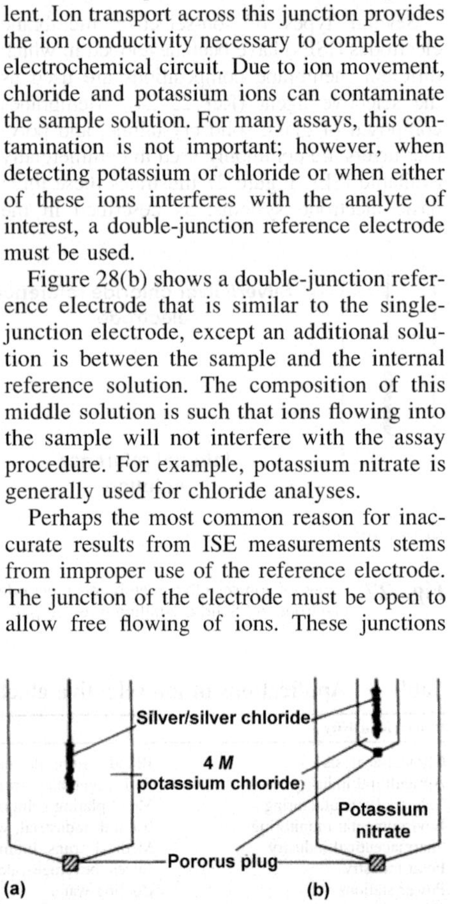

Fig. 28 Schematics of reference electrodes. (a) Single junction. (b) Double junction

often become plugged, halting ion movement and resulting in a noisy electrode system that produces inaccurate data. Therefore, suitable care and maintenance are necessary. Proper procedures for care, storage, and rejuvenation of reference electrodes have been summarized (Ref 37).

Potentiometric Gas-Sensing Electrodes

Gas-sensing membrane electrode systems are potentiometric devices in which a second membrane barrier is included with an ISE system to alter the selectivity properties of the sensor. Gas-sensing electrodes are available for ammonia, carbon dioxide, nitrogen oxide, and hydrogen disulfide (an amperometric gas sensor is available for oxygen).

Figure 29 shows a typical configuration for the ammonia gas sensor, which is a pH ion-selective electrode housed in a plastic body. Also included in this plastic body is the reference electrode, which is required to complete the electrochemical circuit. At the pH-sensing surface of the glass electrode is an electrolyte solution composed mainly of ammonium chloride. A thin gas-permeable membrane separates this thin layer of electrolyte from the external or sample solution. This membrane is generally made of a homogeneous gas-permeable polymer, such as silicon rubber, or a microporous polymer, such as microporous synthetic fluorine-containing resin.

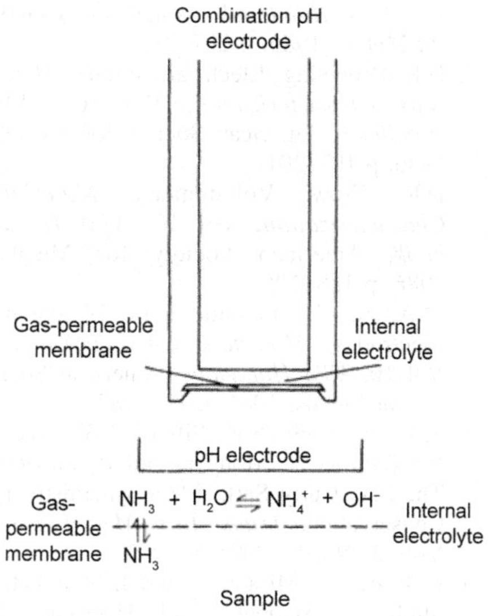

Fig. 29 Configuration of an ammonia gas-sensing membrane electrode

In either case, ammonia gas in the sample will diffuse across the gas-permeable membrane until the partial pressure of ammonia is equal on both sides. The influx of ammonia will shift the electrolyte equilibria to the right, altering the solution pH. The pH electrode senses this change in pH, and the resulting steady-state potential can be related to the sample ammonia concentration by using Eq 18. Sensors for carbon dioxide, nitrogen oxide, and hydrogen disulfide operate similarly, but the internal electrolyte is composed of a different salt.

The response of a gas sensor depends on the pH of the sample solution. For example, the ammonia sensor will not respond to ammonium ions, because only gaseous species can pass through the membrane. Therefore, the ammonia gas sensor can be used in only basic solutions. Optimal pH for this sensor is 10.5 to 12.0.

Steady-state response characteristics of gas sensors are generally favorable, with limits of detection approaching the micromolar concentration range for the ammonia sensor. A slope of 54 to 60 mV per concentration decade can be expected over a region of linearity that should extend over 2 orders of magnitude. The dynamic behavior of these electrodes can sometimes cause problems, because obtaining potentials to within 1 mV of the final steady-state potential may require up to 10 min at low analyte concentrations or as little as 30 s at the higher concentration region. In addition, 40 to 60 min may be required for the electrode to return to baseline conditions.

Gas sensors generally display a high degree of selectivity because of the gas permeability of the outer membrane. Ionic and other nonvolatile constituents in the sample, being unable to penetrate the membrane, cannot alter the inner solution, the pH of which is being monitored. In addition, the volatile substances that can pass through the polymer barrier will change the internal electrolyte pH before their presence can be sensed.

Nanometer Electrochemistry

Studies of electrochemical properties on a nanometer scale use electrodes whose size is on the order of 1 to 100 nm and are made of metals or semiconducting materials. It has created a significant impact in the development of many sensors and the study of reactions that involve extremely low concentrations. Examples of samples and analytical capabilities include:

- Single nanoparticles (gold) attached to the catalytically inert carbon surface mapped by using >3 nm radius nanoelectrodes
- Nanopores of 1 to 10 nm
- Detection of analytes in the range of 10^{-10} to 10^{-6} *M*, with lower detection limits of 1.6×10^{-11} *M* (cysteine)
- Amperometric probes have a tip radius approximately 1000 times smaller than that of a single cell.

The characterization of microscale electrode surfaces that are modified with monolayer thin films has been aided by the advancement of several modern sophisticated techniques. These include, but are not limited to, atomic force microscopy for surface morphology, x-ray photoelectron spectroscopy for elemental composition and oxidation state determination, as well as scanning electrochemical microscopy for surface description and conductivity measurements. Atomic resolution of electrochemical interfaces is aided by scanning tunneling microscopy. These techniques have succeeded in helping scientists determine the oxidation states of various materials in addition to identifying electrocatalytic "hot spots" and their impact in catalysis and analysis.

Nanometer electrochemistry is proven to play a significant role in various applications that range from energy to bioanalysis. The ability to provide measurements at the nanometer scale has opened the horizons to measurements that were otherwise extremely difficult to obtain and reproduce with accuracy. Examples of applications include:

- Measure single molecular conductance in a wide variety of environments, both polar and nonpolar, under various extreme conditions, including ultrahigh vacuum
- Applications in devices such as batteries, supercapacitors, fuel cells, sensors, electrochemical energy conversion, and electrochromic devices
- Surface nanostructure construction and nanofabrication
- Probing brain chemistry by monitoring chemical signal molecules release from single cells with spatial and temporal resolution
- Bioanalytical catalysis and electrocatalysis of carbon nanotubes, fullerenes, and semiconductor particles
- Measurement of transmembrane charge support and membrane potential to probe redox properties at the subcellular level

The designing of a large-scale homogeneous nanocup-electrode array was put together through sequential laser interference lithography. It has led to the detection of electrochemically active biomolecules and monitoring noninvasive real-time dopamine generation from human neural-stem-cell-derived midbrain neurons (Ref 38). As a result, there is hope that such nanoelectrochemistry devices can serve in stem cell transplantation, where the identification and characterization of cells is imperative before the process is administered.

Consequently, dopaminergic neurons that are generated in ex vivo conditions can quantitatively measure the dopamine quantity produced from cells. This process can serve as a hopeful stem-cell-based therapy to the eventual treatment of neurological diseases, including but not limited to Parkinson's, Alzheimer's, and Huntington's, as well as spinal cord injury.

Characterization of the intracellular redox state and high-resolution imaging of living cells were enhanced by the use of nanometer-sized amperometric probes in tandem with the scanning electrochemical microscope (SECM) to perform electrochemical studies in cultured breast cells. The tiny probe, with a radius $^1/_{1000}$th the size of a cell, penetrated and traveled through the cell with no damage to the membrane, thus proving the possibility of probing redox properties at the subcellular level (Ref 39).

Nanoelectrochemistry has also been used in probing brain chemistry by monitoring the release of chemical signal molecules from single-cell, single-vesicle, synaptic cleft, as well as the morphological analysis. An example of such a process involves the release of dopamine from a single cell through a microelectrode on a microfluidic device in conjunction with SECM technology (Ref 40).

A single-molecule conductance has been achieved in a wide variety of media, including organic liquids, water, ionic liquids, and electrolytes, through nanoelectrochemistry. The molecules are caught between a pair of facing electrodes, and the junction current response is measured of the bias voltage. The same results were obtained when either gating electrodes or a four-electrode biopotentiostatic configuration was employed. This experiment allows modern interfacial electrochemistry to monitor charge transport across single molecules as a function of electrode potential (Ref 41).

Studying the properties of single nanoscale systems overcomes the difficulties associated with complex ensemble systems. Stochastic collision nanoelectrochemistry has proven to be a convenient, fast, and recently employed single-entity electrochemical analysis method and has helped electrochemical catalysis, analysis, and even biosensing. It uses random-walk simulations and becomes significant in single-molecule experiments due to the inherent properties associated with Brownian motion (Ref 42, 43).

General uses of nanometer electrochemistry include:

- Understanding the electrocatalytic properties of metal nanoparticles on the single-particle level
- Understanding the structure-function relationship of metal nanoparticles
- Probing nanoelectrochemistry using optical microscopy

REFERENCES

1. J.E. Harrar, Controlled-Potential Coulometry, *Materials Characterization*, Vol 10, *ASM Handbook,* American Society for Metals, 1986, p 207–211
2. D.R. Browning, Electrogravimetry, *Materials Characterization*, Vol 10, *ASM Handbook,* American Society for Metals, 1986, p 197–201
3. D.R. Crow, Voltammetry, *Materials Characterization*, Vol 10, *ASM Handbook,* American Society for Metals, 1986, p 188–196
4. M.A. Arnold, Potentiometric Membrane Electrodes, *Materials Characterization*, Vol 10, *ASM Handbook,* American Society for Metals, 1986, p 181–187
5. S.A. Campbell, S.P. Gillard, I.B. Beech, W. Davies, G. Monger, and P. Lawton, The S.V. Cutty Sark: Electrochemistry in Conservation, *Trans. Inst. Met. Finish.*, Vol 83 (No. 1), 2005, p 19–26
6. E. Rocca, F. Mirambet, and J. Steinmetz, Study of Ancient Lead Materials: A Gallo-Roman Sarcophagus—Contribution of the Electrolytic Treatment to Its Restoration, *J. Mater. Sci.*, Vol 39 (No. 8), 2004, p 2767–2774
7. S.A. Pushkareva, I.P. Golovacheva, and M.V. Orlov, Technology for Restoration of Journal Roller Bearings by Compositional Iron Plating, *Vestn. Mashinostr.*, Vol 9, 1984, p 64–66
8. G. Centi and S. Perathoner, Opportunities and Prospects in the Chemical Recycling of Carbon Dioxide to Fuels, *Cataly. Today*, Vol 148 (No. 3–4), 2009, p 191–205
9. W. Plieth, Electrocrystallization—Factors Influencing Structure, *J. Solid State Electrochem.*, Vol 15 (No. 7–8), 2011, p 417–422
10. A.I. Danilov, V.N. Kudryavtsev, and Y.M. Polukarov, Theory and Practice of Electrocrystallization, *Russ. J. Electrochem.*, Vol 44 (No. 6), 2008, p 617, 618
11. J.T. Stock, The Genesis of Electrogravimetry, *Bull. Hist. Chem.*, Vol 7, 1990, p 17–19
12. V.I. Shirokova, Trends in the Development of the Potentiostatic Coulometry of Inorganic Substances, *J. Anal. Chem.* (Translation of *Zh. Anal. Khim.*), Vol 58 (No. 9), 2003, p 826–829
13. J.A. Von Fraunhofer and G.A. Pickup, Examination of Corrosion Products and Processes, Part I: Wet and Electrochemical Methods, *Anti-Corros. Methods Mater.*, Vol 14 (No. 3), 1967, p 8–11
14. E. Schmid, Potentiostatic Coulometry and Its Application to Uranium and Plutonium Determinations, *Oesterr. Chemik.-Zeit.*, Vol 67 (No. 1), 1966, p 8–16
15. L. Tang, X.-M. Liu, T.-L. Yang, and K.-M. Long, Study on Constant-Current Coulometric Titration of Potassium Dichromate by Electro-Generated Fe(II) and Its Application to Analysis of Total Uranium, *Yejin Fen.*, Vol 34 (No. 2), 2014, p 6–10
16. D.R. Browning, Ed., *Electrometric Methods*, McGraw-Hill, London, 1969, p 123, 124
17. "Coulometer," *Croatian-English Chemistry Dictionary and Glossary*, 2018, https://glossary.periodni.com
18. J.F. Owen, C.S. Patterson, and G.S. Rice, *Anal. Chem.*, Vol 55 (No. 6), 1983, p 990–992
19. G. Facsko, *Galvanotechnik*, Vol 66 (No. 5), 1975, p 391–395
20. D.D. Olm and J.T. Stock, *Mikrochim. Acta*, Vol 2 (No. 5–6), 1977, p 575–582
21. J. Moldenhauer, M. Meier, and D.W. Paul, Rapid and Direct Determination of Diffusion Coefficients Using Microelectrode Arrays, *J. Electrochem. Soc.*, Vol 163 (No. 8), p H672–H678
22. W.E. Morf, *The Principles of Ion-Selective Electrodes and of Membrane Transport*, Vol 2, Elsevier, 2012
23. K.N. Mikhelson, *Ion-Selective Electrodes*, Vol 81, *Lecture Notes in Chemistry*, Springer, Berlin, 2013
24. E. Pungor, The Theory of Ion-Selective Electrodes, *Anal. Sci.*, Vol 14 (No. 2), 1998, p 249–256
25. E. Bakker, P. Bühlmann, and E. Pretsch, Polymer Membrane Ion-Selective Electrodes—What Are the Limits? *Electroanal. Int. J. Devoted Fundament. Pract. Aspects Electroanal.*, Vol 11 (No. 13), 1999, p 915–933
26. G. Eisenman, R. Bates, G. Mattock, and S. M. Friedman, *The Glass Electrode*, Interscience, 1964
27. R.A. Durst, *Ion-Selective Electrodes*, Special Publication 314, National Bureau of Standards, 1969
28. J. Migdalski, T. Blaz, and A. Lewenstam Conducting Polymer-Based Ion-Selective Electrodes, *Anal. Chim. Acta*, Vol 322 (No. 3), April 10, 1996, p 141–149
29. E. Bakker, Selectivity of Liquid Membrane Ion-Selective Electrodes, *Electroanal.*, Vol 9 (No.1), 1997, p 7–12
30. P.L. Bailey, *Analysis with Ion-Selective Electrodes*, Heyden Press, 1980
31. A.K. Covington, *Ion-Selective Electrode Methodology*, Vol I and II, CRC Press, 1979
32. Y. Umezawa, P. Bühlmann, K. Umezawa, K. Tohda, and S. Amemiya, Potentiometric Selectivity Coefficients of Ion-Selective Electrodes, Part I: Inorganic Cations (Technical Report), *Pure Appl. Chem.*, Vol 72 (No. 10), 2000, p1851–2082
33. Y. Umezawa, Ed., *CRC Handbook of Ion-Selective Electrodes: Selectivity Coefficients*, CRC Press, Boca Raton, FL, 1990, p 3–9
34. E. Bakker, Determination of Improved Selectivity Coefficients of Polymer Membrane Ion-Selective Electrodes by Conditioning with a Discriminated Ion, *J. Electrochem. Soc.*, Vol 143 (No. 4), 1996, p L83–L85
35. T.S. Ma and S.S.M. Hassan, *Organic Analysis Using Ion-Selective Electrodes*, Vol I, *Methods,* Academic Press, 1982
36. T.S. Ma and S.S.M. Hassan, *Organic Analysis Using Ion-Selective Electrodes*, Vol II, *Applications and Experimental Procedures,* Academic Press, 1982

37. J.E. Fisher, Measurement of pH, *Am. Lab.*, Vol 16, 1984, p 54–60
38. T.H. Kim, C.-H. Yea, S.T.D. Chueng, P. T.-T. Yin, B. Conle, K. Dardir, Y. Pak, G.Y.M. Jung, J.-W. Choi, and K.B. Lee, Nanoelectrochemistry for Detection of Neurotransmitters in Dopaminergic Neuron Differentiation, *Adv. Mater.*, Vol 27 (No. 41), 2012, p 6356–6362
39. P. Sun, F.D. Laforge, T.P. Abeyweera, A. A. Rotenberg, J. Carpino, and M.V. Mirkin, Nanoelectrochemistry of Mammalian Cells, *PNAS*, Vol 105 (No. 2), 2008, p 443–448
40. J.-K. Cheng, W. Wang, W.-Z. Wu, W.-H. Huang, and Z.L. Wang, Probing Brain Chemistry—Monitoring of Chemical Signal Molecules Release from Single-Cell, Single-Vesicle, Synaptic-Cleft and Morphological Analysis with Nanoelectrochemistry, *Gaodeng Xuexiao Huaxue Xuebao*, Vol 29 (No. 12), 2008, p 2609–2617
41. R.J. Nichols and S. Higgins, Single Molecule Nanoelectrochemistry in Electrical Junctions, *Accts. of Chem. Res.*, Vol 49 (No. 11), 2016, p 2640–2648
42. P.S. Singh, E. Kaetelhoen, K. Mathwig, W. Bernrd, and S.G. Lemay, Stochasitcity in Single Molecule Nanoelectrochemistry: Origins, Consequences and Solutions, *ACS (Nano)*, Vol 6 (No. 11), 2012, p 9662–9671
43. S.J. Kwon, H. Zhou, F.R. Fan, V. Vaorobyev, B. Zhang, and A.J. Bard, Stochastic Electrochemistry with Electrocatalytic Nanoparticles at Inert Ultramicroelectrodes—Theory and Experiments, *Phys. Chem. Chem. Phys.*, Vol 13 (No. 12), 2011, p 5394–5402, DOI: 10.1039/c0cp02543g, Epub Feb 28, 2011

SELECTED REFERENCES

Selected References are provided to assist the reader in the up-to-date development and compilation of articles related to electrochemical methods.

- M. Ajmal, Electrochemical Studies on Some Metal Complexes Having Anti-Cancer Activities: A Review, *J. Coord. Chem.*, Vol 70 (No. 25), 2017, p 2551–2588
- M. Aliofkhazraei and A.S. Makhlouf, Ed., *Handbook of Nanoelectrochemistry: Electrochemical Synthesis Methods, Properties, and Characterization Techniques*, Springer Verlag, 2016
- C. Batchelor-McAuley, E. Katelhon, E.O. Barnes, R.G. Compton, E. Laborda, and A. Molina, Recent Advances in Voltammetry, *Chem. Open*, Vol 4 (No. 3), 2015, p 224–260, DOI: 10.1002/open.201500042
- N. Elgrishi, K.J. Rountree, B.D. McCarthy, E.S. Rountree, T.T. Eisenhant, and J.L. Dempsey, A Practical Beginner's Guide to Cyclic Voltammetry, *J. Chem. Educ.*, Vol 95 (No. 2), 2018, p 197–206, DOI: 10.1021/acs.jchemed.7b00361
- R. Habibur, H. Iqbal, and S.K. Manirul Haqueb, Review on Cyclic Voltammetric and Spectrophotometric Approaches for the Analysis of Drugs (Antihypertensive) Using Different Electrodes and Wavelength, *Der Pharma Chem.*, Vol 7 (No. 12), 2015, p 44–55, ISSN 0975-413X CODEN (USA): PCHHAX 44
- M.V. Mirkin and S. Amemiya, Ed., Chap. 15, Nanoelectrochemistry of Carbon, *Nanoelectrochemistry*, CRC Press, Tylor and Francis, 2015, p 293–355
- M.V. Mirkin and S. Amemiya, *Nanoelectrochemistry*, CRC Press, 2017, ISBN 9781138894662- CAT#K32906
- R. Murray, Nanoelectrochemistry: Metal Nanoparticles, Nanoelectrodes, and Nanopores, *Chem Rev.*, Vol 108 (No. 7), 2008, p 2688–2720, DOI: 10.1021/cr068077e
- M. Scampiocchio, M.S. Cosio, S.L. Mengistu, and S. Mannino, Nanoelectrochemistry: Applications Based on Electrospinning, *Agricultural and Food Electroanalysis*, A. Escarpa, M.C. Gonzalez, and M.A. Lopez, Ed., Wiley Online Library, 2015, p 357–379
- F. Scholz, Ed., *Electroanalytical Methods: Guide to Experiments and Applications*, 2nd revised and extended edition, Springer, Heidelberg, Dordrecht, London, New York, 2010, ISBN 978-3-642-02914-1, e-ISBN 978-3-642-02915-8, DOI: 10.1007/978-3-642-02915-8
- P.S. Singh, E. Goluch, H.A. Heering, and S. G. Lemay, Nanoelectrochemistry: Fundamentals and Applications in Biology and Medicine, *Applications of Electrochemistry and Nanotechnology in Biology and Medicine II*, Vol 53, Springer US, 2012, p 1–66, DOI: 10.1007/978-1-4614-2137-5_1, ISSN 0076-9924
- D.A. Skoog, F.J. Holler, and T.A. Nieman, Chap. 25, *Principles of Instrumental Analysis*, 5th ed., Brooks/Cole Publishing Co., 1997

ASM Handbook, Volume 10, *Materials Characterization*
ASM Handbook Committee
DOI 10.31399/asm.hb.v10.a0006642

Neutron Activation Analysis*

Revised by Bryan E. Tomlin, Texas A&M University

Overview

General Uses

- Nondestructive trace-element assay of essentially any material
- Ultrasensitive (as low as 10^{-12} g/g) destructive quantitative analysis
- Measurement of isotope ratios in favorable cases

Examples of Applications

- *Quality control:* For purity or composition of materials
- *Element-abundance measurements:* In geochemical and cosmochemical research
- *Resource evaluations:* Assay of surface materials, drill cores, ore samples, and so on
- *Pollution studies:* Assay of air, water, fossil fuels, and chemical wastes for toxic elements
- *Biological studies:* Determination of the retention of toxic elements in humans and laboratory animals, in some cases by in vivo measurements
- *Forensic investigations:* Trace-element assay of automobile paint, human hair, and so on

Samples

- *Form:* Virtually any solid or liquid
- *Size:* Typically 0.1 to a few grams, but the full range extends from a microgram to at least 100 kg (220 lb)
- *Preparation:* None required in many cases

Limitations

- Several elements are unobservable except through sample irradiation with high-energy neutrons or use of radiochemical (destructive) techniques.
- Access to a nuclear reactor or some other high-intensity neutron source is required.
- Turnaround time can be long (up to three weeks).
- The samples usually become somewhat radioactive, which is objectionable in some cases.
- Intense radioactivity induced in one or more of the major elements in a sample may mask the presence of some or all of the trace elements.

Estimated Analysis Time

- Counting times can range from under 1 min (for short half-life radioisotopes) to many hours (for low-intensity long-lived radioisotopes).

Capabilities of Related Techniques

- *X-ray fluorescence:* Useful for determining major and minor elements (concentrations > 0.1%), but several elements with $Z \geq 11$ can be observed at 5 to 20 ppm
- *Atomic absorption:* Can determine most elements individually. Requires dissolution of the sample. Some elements observed at levels of 10^{-9} g/mL, which usually corresponds to $\geq 10^{-7}$ g/g of the original solid sample
- *Inductively coupled plasma emission spectroscopy:* Requires dissolution of the sample. Typical detection limits are a few parts per million (ppm).
- *Inductively coupled plasma mass spectroscopy:* Can determine trace elements in the presence of high levels of major and minor elements in the sample matrix, such as in alloy and metal analysis. Requires dissolution of the sample. Detection limits for most elements are from 0.05 to 0.1 ng/mL of solution.
- *Isotope dilution mass spectrometry:* Can quantitatively measure as few as 10^6 atoms of most elements individually. Isotope ratios are used. Analysis time is at least one day per element.
- *Spark source mass spectrometry:* Simultaneous detection of most elements at levels down to approximately 1 ppm. Data are recorded on a photoplate, limiting accuracy. Sample preparation may take several hours.
- *Particle-induced x-ray emission spectroscopy:* Measures element concentrations near the surface of a sample. Several elements can be measured at a few parts per million. Access to a proton accelerator is required.

Introduction

Neutron activation analysis (NAA) is a highly sensitive and accurate method of assaying bulk materials for major, minor, and trace levels of many elements. Neutron activation analysis has found broad application in materials science and in geological, biological, archeological, health, forensic, and environmental studies (Ref 1). Much of what is known about the composition of the lunar surface was learned by using NAA. The technique is an important assay method because of its many advantages, which include:

*Revised from M.E. Bunker, M.M. Minor, and S.R. Garcia, "Neutron Activation Analysis," *Materials Characterization*, Volume 10, *ASM Handbook*, ASM International, 1986, p 233–242

- Simultaneous observation of large blocks of elements, minimizing analysis time and increasing the reliability of element ratio values
- Nondestructive analysis of virtually any material regardless of chemical form, resulting in the possibility of conducting repeated measurements on the same sample, the elimination of reagent addition (greatly reduces the possibility of sample contamination), and ease of automation of analysis
- Detection limits in the range of 10^{-6} g down to 10^{-9} g for many elements, assuming only minor interference from the radioactivity induced in the major elements of the sample (see the section "Instrumental NAA" in this article). In destructive (radiochemical) NAA, the detection limits for some elements are below 10^{-12} g.

The term *neutron activation analysis* is best reserved for describing a class of related techniques rather than a specific measurement technique. Several abbreviations are used in reference to these related techniques, and confusion can arise due to the overlapping characteristics of these designations. A few abbreviations are used more frequently than the others, and historical usage has varied, without any official standardization of terminology. The most commonly used categories of NAA are described later in this article.

Basic Principles

Neutron activation analysis is used to determine the mass or mass fraction (or volume concentration) of chemical element(s) contained in a bulk material. The NAA measurement process, like many other analytical techniques (e.g., inductively coupled plasma optical emission spectroscopy/mass spectroscopy) is based on counting discrete atoms of the analyte element(s) by using a characteristic physical property as a tag. In any such process, only a very small subset of analyte atoms is sampled, and the total number of atoms is in direct proportion to the subset of sampled atoms.

With NAA, the analyte atoms are "tagged" by means of known nuclear reactions caused by neutron bombardment. The tagging process converts a stable isotope nucleus into a radioactive nucleus. Characteristic radiation is emitted by the newly created radioactive nucleus. Both the energies and timescales of radiation emissions are unique and provide two dimensions for sorting observations. The radiation emission energies are most often discreet, with a few instances of continuum emissions being used. The half-lives of the radioactive decay range from fractions of a second to years. Electronic instrumentation is used to count the number of tagged atoms by observing and recording the individual radiation emissions.

Neutron activation analysis is a powerful assay technique primarily because neutrons, unlike charged particles (e^-) or photons (x-rays), can penetrate deeply into most bulk materials, enabling activation of the elements throughout the sample volume. In addition, the high-energy γ-ray photons (and less commonly beta particles and neutrons) emitted by the neutron-induced radioisotopes are also very penetrating and, for most materials, suffer little attenuation in escaping from the interiors of samples weighing several grams.

The NAA method most commonly used involves irradiation of a sample using low-energy neutrons in a nuclear reactor, to produce radioactive isotopes via the capture of neutrons by stable isotopes of the sample elements. Neutron irradiation is followed by observation of the γ-rays emitted from the samples by radioactive decay of the newly formed radioactive isotopes. Individual γ-rays are associated with specific elements, and the intensity of each γ-ray provides a direct measure of the element abundance.

Neutron Reactions

Slow, or thermal, neutron capture is the nuclear reaction most often used in NAA to produce radioisotopes. "Slow" denotes an energy of a few hundredths of an electron volt (eV). The capture of a slow neutron by an isotope of mass number A leads (except in the case of fission) to a product isotope of mass number A + 1, which may or may not be radioactive. In either case, immediately after its formation, the new isotope is in a highly excited state, on the order of 10^6 eV (1 MeV) above the ground state (corresponding to the neutron binding energy). De-excitation to the ground state or to a low-lying isomeric state occurs almost instantaneously by γ-ray emission. This reaction is denoted as an (n,γ) reaction, where the neutron (n) is the incoming particle and the gamma-ray photon (γ) is the outgoing particle. The γ-rays emitted are referred to as prompt neutron-capture γ-rays, and in one variation of NAA, they can be used for element assay (see the section "Prompt Gamma Activation Analysis" in this article).

The probability of a slow neutron being captured by an atomic nucleus varies dramatically throughout the periodic table. Neutron-capture probability is expressed in terms of reaction cross section (σ), which is given the unit name "barn" and has dimensions of area (1 barn = 10^{-28} m^2).

Values of σ encountered in NAA range from a fraction of a barn (<10^{-3} b) for many light-element nuclides up to 2.6×10^5 for ^{157}Gd. Some elements, such as zirconium, are not easily detectable using thermal NAA, because their neutron-capture cross sections are too low to induce sufficient radioactivity for measurements. The extremely low sigma values of light elements such as carbon, hydrogen, and oxygen make water, organic/polymeric, and biological materials effectively transparent to neutrons. This fact contributes to the matrix-independent characteristic that has been a valuable advantage of NAA over other chemical analysis techniques.

Detailed information on the approximate amount of massic activity (Bq $\cdot$ μg^{-1}) induced in each isotope of every element through irradiation for various times using slow and fast reactor neutrons and 14 MeV neutrons is cited in Ref 2. Such activity values must be used only as estimates, because of uncertainty in the effective reaction cross section, which depends on the neutron energy spectrum in which the sample is irradiated. Accurate assay measurements necessitate calibrating the detection system using standard target materials of known composition, as described subsequently.

Calibration

Neutron activation analysis is based on physical processes that are well understood and are completely described by explicit equations that can be written in terms of Système International (SI) units (Ref 3). The fundamental NAA measurement equation is stated in Eq 1. The number of counts, C, observed in a particular gamma-ray peak in a given time can be related to the mass, m_x, of the element responsible for the gamma ray by using (Ref 3):

$$m_x = \frac{\lambda \cdot C}{(1 - e^{-\lambda t_i}) \cdot e^{-\lambda t_d} \cdot (1 - e^{-\lambda t_m}) \cdot \Phi_{th} \cdot \sigma_{eff} \cdot \Gamma \cdot \varepsilon} \cdot \frac{M_a}{\theta \cdot N_A} \qquad \text{(Eq 1)}$$

where C is the net count in the gamma-ray peak (dimensionless); N_A is Avogadro's number (mol^{-1}); θ is the isotopic abundance of the target isotope (dimensionless); m_x is the mass of the analyte element (grams); M_a is the molar mass of the analyte element (g $\cdot$ mol^{-1}); Γ is the gamma-ray abundance (dimensionless); ε is the full-energy photopeak efficiency of the detector (dimensionless); Φ_{th} is the thermal neutron fluence rate for E_n < 0.55 eV (m^{-2} $\cdot$ s^{-1}); σ_{eff} is the spectrum-averaged cross section (m^2), including both thermal and epithermal resonance contributions; λ is the radioactive decay constant (s^{-1}); t_i is the duration of the irradiation (seconds); t_d is the duration of the decay interval after irradiation (seconds); and t_m is the duration of the measurement (seconds).

In practice, an absolute calibration using Eq 1 is rarely employed, primarily because accurate knowledge of the thermal neutron fluence rate, the effective cross section, and the photopeak efficiency is difficult to achieve. The lowest uncertainties and best accuracy are attained using the direct comparator method, which uses the proportionality between the analyte mass, m_x, and the net peak counts, C:

$$\left(\frac{C}{m_x}\right)_{\text{unk}} \alpha \left(\frac{C}{m_x}\right)_{\text{cal}} \quad \text{(Eq 2)}$$

where each term is used as previously defined.

A common approach is to irradiate the sample and a standard simultaneously (in the same irradiation vial, if possible) and to count the two samples later at different times using the same detector and counting geometry. The simultaneous irradiation yields identical values of t_i, Φ_{th}, and σ_{eff} for both the calibrator and the unknown sample, which allows them to be mathematically canceled in the calculation; likewise, identical counting conditions allow the photopeak efficiencies for each gamma ray to be canceled. The resulting direct comparator equation is then written:

$$m_{x(\text{unk})} = m_{x(\text{cal})} \cdot \frac{\left(\frac{C}{t_m \cdot e^{-\lambda t_d} \cdot (1-e^{-\lambda t_m})}\right)_{\text{unk}}}{\left(\frac{C}{t_m \cdot e^{-\lambda t_d} \cdot (1-e^{-\lambda t_m})}\right)_{\text{cal}}} \quad \text{(Eq 3)}$$

where $m_{x(\text{unk})}$ is the mass of analyte element in the unknown sample (grams), and $m_{x(\text{cal})}$ is the mass of analyte element in the calibration standard (grams).

Whatever calibration method is used, a calibration sample should be used that has approximately the same size and density as the sample to be analyzed. Thus, geometrical correction can be avoided, and possible corrections for gamma-ray self-absorption by the sample can be minimized.

Some of the radioisotopes observable in NAA emit many different gamma rays, each of which, in principle, could be used to determine the mass of the associated element. However, only one or two of the major gamma-ray lines are typically used, selecting those having well-resolved full-energy peaks, that is, minimal overlap with other lines in the spectrum. Tables 1 to 3 list gamma rays typically used in instrumental NAA (INAA). When more than one line from a given isotope is used, the deduced values for the element mass can be combined into a single value by appropriate statistical procedures.

The total number of counts in a given photopeak, C, can be determined by using a variety of algorithms that are available in common gamma-ray spectroscopy software packages (Ref 4). Various integration techniques are described in Ref 5. Most methods display favorable agreement for intensities extracted for intense peaks, but methods applied to weak peaks or overlapping peaks vary considerably.

In studying spectra of complex samples, there may be a few elements for which it is impossible to find an isolated γ-ray peak. Extracting net peak counts for these elements necessitates identifying all the lines in the spectrum and apportioning the area of multiplet lines to the various contributing gamma

Table 1 Typical instrumental neutron activation analysis sensitivities for elements in rock or soil samples 20 min after irradiation

Element	Radioisotope detected ($t_{1/2}$)	Indicator γ-rays, keV	Sensitivity(a), μg · g^{-1}
Sodium	^{24}Na (15 h)	1368.6, 2753.85	1000
Magnesium	^{27}Mg (9.5 min)	843.76, 1014.24	2700
Aluminum	^{28}Al (2.25 min)	1779.0	3200
Chlorine	^{38}Cl (37 min)	1642.68, 2167.5	120
Potassium	^{42}K (12.4 h)	1524.7	4500
Calcium	^{49}Ca (8.7 min)	3084.15	1500
Titanium	^{51}Ti (5.8 min)	320.08	750
Vanadium	^{52}V (3.75 min)	1434.05	8.0
Manganese	^{56}Mn (2.6 h)	846.75, 1810.66	60
Copper	^{66}Cu (5.1 min)	1039.0	350
Strontium	^{87}Sr (2.8 h)	388.52	400
Indium	^{116m}In (54.1 min)	416.9, 1097.3	0.2
Iodine	^{128}I (25 min)	442.87	40
Barium	^{139}Ba (83 min)	165.85	200
Dysprosium	^{156}Dy (2.35 h)	94.68, 361.66	0.9

Note: The superscript "m," as in ^{116m}In, represents an isomeric state. (a) Based on a nominal 4 g sample, a 4 min irradiation with 6×10^{12} cm^{-2} · s^{-1} neutron fluence rate, and an 8 min γ-ray count taken 20 min after irradiation. Spectral measurements were conducted using a 50 cm^3 germanium detector.

Table 2 Typical instrumental neutron activation analysis sensitivities for elements in rock or soil samples 5 d after irradiation

Element	Radioisotope detected ($t_{1/2}$)	Indicator γ-rays, keV	Sensitivity(a), μg · g^{-1}
Sodium	^{24}Na (15 h)	1368.6, 2753.85	300
Potassium	^{42}K (12.4 h)	1524.7	4500
Gallium	^{72}Ga (14.1 h)	834.07, 629.9	45
Arsenic	^{76}As (26.5 h)	559.09	3.0
Bromine	^{82}Br (35.4 h)	776.5, 619.1, 554.3	4.0
Molybdenum	^{99}Mo (2.76 d)	140.5	50
Antimony	^{122}Sb (2.70 d)	564.09	1.0
Barium	^{131}Ba (11.7 d)	496.2, 373.0, 216.1	800
Lanthanum	^{140}La (40.3 h)	1596.2, 487.01	7.0
Samarium	^{153}Sm (47 h)	103.18	4.0
Ytterbium	^{175}Yb (4.2 d)	396.32	1.0
Lutetium	^{177}Lu (6.7 d)	208.40	0.2
Tungsten	^{187}W (23.8 h)	685.72	5.0
Gold	^{198}Au (2.70 d)	411.79	0.04
Uranium	^{239}Np (2.35 d)(b)	106.13, 228.2, 227.6	10(c)

(a) Based on a nominal 4 g sample, a 4 min irradiation with 6×10^{12} cm^{-2} · s^{-1} neutron fluence rate, and a 0.5 h γ-ray count taken 5 d after irradiation. Spectral measurements were conducted using a 50 cm^3 germanium detector. (b) ^{239}Np is the daughter of ^{239}U (23.5 min). (c) Natural uranium can be observed with heightened sensitivity by delayed neutron activation analysis (see the section "Uranium Assay by Delayed Neutron Activation Analysis" in this article).

Table 3 Typical instrumental neutron activation analysis sensitivities for elements in rock or soil samples 21 d after irradiation

Element	Radioisotope detected($t_{1/2}$)	Indicator γ-rays, keV	Sensitivity(a), μg · g^{-1}
Scandium	^{46}Sc (83.8 d)	889.26	0.04
Chromium	^{51}Cr (28 d)	320.08	2.5
Iron	^{59}Fe (44 d)	1099.22, 1291.56	300
Cobalt	^{60}Co (5.2 yr)	1173.21, 1332.46	0.20
Zinc	^{65}Zn (244 d)	1115.52	15
Selenium	^{75}Se (120 d)	279.5, 264.5	5.0
Rubidium	^{86}Rb (18.7 d)	1076.77	20
Silver	^{110m}Ag (250 d)	657.8	5.0
Antimony	^{124}Sb (60.2 d)	602.7, 1691.0	0.50
Cesium	^{134}Cs (2.05 yr)	604.73, 795.84	0.60
Barium	^{131}Ba (11.7 d)	496.2, 373.0, 216.1	300
Cerium	^{141}Ce (32 d)	145.44	4.0
Neodymium	^{147}Nd (11.1 d)	531.0	45
Europium	^{152}Eu (13.4 yr)	1408.02, 778.87	0.09
Terbium	^{160}Tb (72 d)	879.36, 966.2	0.15
Ytterbium	^{169}Yb (30.7 d)	197.99, 307.68	0.50
Lutetium	^{177}Lu (6.7 d)	208.40	0.06
Hafnium	^{181}Hf (42 d)	482.18	0.35
Tantalum	^{182}Ta (115 d)	1221.38, 1189.02	0.45
Mercury	^{203}Hg (46.6 d)	279.2	0.90
Thorium	^{233}Th (27.4 d)	311.9	0.35

Note: The superscript "m," as in ^{110m}Ag, represents an isomeric state. (a) Based on a nominal 4 g sample, a 4 min irradiation with 6×10^{12} cm^{-2} · s^{-1} neutron fluence rate, and a 2 h γ-ray count taken 21 d after irradiation. Spectral measurements were conducted using a 50 cm^3 germanium detector.

rays by using a self-consistent least-squares calculation. Several computer codes have been written that accomplish this task (Ref 4).

Analytical Sensitivity

The analytical sensitivity—derived from the Currie limit of quantitation (Ref 6)—achievable with NAA for a given element depends on many factors: nuclide of interest half-life, sample size, neutron fluence, effective cross section, detector efficiency, and irradiation and counting times. Table 4 shows calculated order-of-magnitude sensitivities for essentially all the elements observable using INAA (see the section "Instrumental Neutron Activation Analysis" in this article), assuming that only one element in the sample has become radioactive. The footnote to Table 4 explains the experimental conditions assumed in arriving at these data. An a priori maximum acceptable standard uncertainty of 10%, corresponding to an integrated area of 100 or more counts, was chosen.

The calculated sensitivities can often be approximated experimentally, especially for elements for which the induced radioactivity is long-lived ($t_{1/2} > 5$ d). However, the radioactivity induced in other elements, particularly the major constituents, can decrease the sensitivities for the trace elements severely, sometimes by several orders of magnitude. For example, Tables 1 to 3 list typical sensitivity values for many of the trace elements often observed in rock or soil samples. The table footnotes provide the parameters of the measurements. In general, the sensitivities are not as favorable as those indicated in Table 4, primarily because the numerous radioisotopes in the irradiated sample provide a background that contributes to the peak area uncertainty. Significant improvement (as much as a factor of 10) in the indicated sensitivities for the longer-lived ($t_{1/2} > 1$ d) radioisotopes could be obtained by increasing the irradiation and counting times.

Neutron Sources

The most common source of low-energy neutrons for thermal NAA is a research reactor, where thermal and epithermal fluence rates exceeding 10^{12} cm^{-2} · s^{-1} are typically available. Other neutron sources that have been used in NAA are spontaneous fission sources, such as ^{252}Cf (Ref 7, 8), and sources based on the (α,n) reaction, such as ^{238}Pu-Be (Ref 9). The ^{252}Cf and (α,n) sources have the advantage of semiportability, making them useful for certain medical and industrial applications. However, the low fluxes available from these sources limit their value for multielement analysis. Compact deuterium-deuterium neutron generators use accelerator-driven nuclear fusion of hydrogen isotopes to produce neutron emission rates of 10^{10} s^{-1} without the need to possess fissionable radioactive source materials such as plutonium or californium (Ref 10). Such systems are available commercially with integrated plastic moderators that slow the fast fusion neutrons down to thermal energies.

Sample Handling

A primary concern with any trace-elemental analysis is avoidance of incidental contamination. Before irradiation, the sample or container should not be touched with bare hands; such handling would transfer trace levels of sodium, chlorine, and possibly other elements to the sample. Following irradiation, the optimum procedure in a nondestructive measurement is to transfer the sample from the irradiation container to an unirradiated container before recording the gamma-ray spectrum. However, such a transfer may not be possible in all cases, for example, in rapid ^{20}F measurements. An alternative is to use an irradiation container made of a material that becomes negligibly radioactive upon neutron activation and to use a very clean irradiation system that transfers a negligible amount of activity to the exterior surface of the container. An excellent material for the irradiation vial is high-density polyethylene (ethylene-butylene copolymer), which is virtually free of troublesome trace elements. For long irradiations in which high-density polyethylene can become brittle, high-purity quartz ampoules are a reliable container.

Another sample-handling technique involves the use of double containment, that is, placing the sample in a separate vial inside the irradiation container. After irradiation, the vial is removed from the container for counting. This procedure eliminates all problems associated with radioactivity of the outer container.

Automated Systems

When many similar samples are to be assayed, the use of an automated pneumatic transfer system offers considerable savings in time. In addition, a computer-controlled system provides highly accurate timing of the irradiation, delay, and counting times. Such a facility typically proceeds thus: loading of the sample into the pneumatic system, transfer to the reactor, irradiation, transfer to the counter, spectrum measurement, and transfer to a storage location. A new sample is then loaded into the system automatically. Because counting times typically exceed irradiation times, the sample throughput rate can be increased by using more than one detector, with the first sample diverted to detector No. 1, the second to detector No. 2, and so on. A four-detector system can analyze 200 samples per day for approximately 30 elements (Ref 11).

Some automated systems are designed for cyclic measurement on the same sample. This technique increases the number of counts and thus the statistical accuracy in measurements conducted on short-lived activities, such as ^{207m}Pb ($t_{1/2} = 0.8$ s) and ^{20}F. One such system is discussed in Ref 12.

Instrumental Neutron Activation Analysis

Thermal NAA is the most common NAA technique, primarily because of the relatively high-activation cross sections that many elements have for thermal neutrons. Through the instrumental (nondestructive) version of this technique, that is, with no postirradiation chemistry, as many as 40 elements can often be observed in complex materials, such as geological samples, assuming the neutron fluence rate available is at least 10^{12} cm^{-2} · s^{-1}.

Prior to the 1970s, most NAA procedures involved performing one or more postirradiation chemical separations, often to isolate individual elements or to separate all radioactive species into a few groups. At the time, the use of low-resolution radiation detection systems made chemical separations a necessity. With the appearance of higher-resolution gamma-ray detection systems in the 1970s, multielement, nondestructive analyses could be performed using strictly instrumental methods.

The full power of INAA can be realized only through the use of high-resolution solid--state gamma-ray detectors, usually high-purity

Table 4 Single-element interference-free sensitivities for instrumental neutron activation analysis

Sensitivity(a), 10^{-12} g	Elements
1–10	Mn, Rh, In, Eu, Dy
10–10^2	Ar, V, Co, I, Cs, Yb, Ir, Sm, Ho, Lu, Au
10^2–10^3	F, Na, Mg, Al, Sc, Ti, Ga, Br, Ge, As, Sr, Pd, Ag, Sb, Te, Ba, La, Nd, Er, W, Re
10^3–10^4	Cl, Cr, Ni, Cu, Zn, Se, Ru, Cd, Sn, Ce, Pr, Gd, Tb, Tm, Hf, Pt, Th, U
10^4–10^5	K, Ca, Co, Rb, Y, Mo, Ta, Os, Hg
10^5–10^6	Zr, Nb
10^6–10^7	Si, S, Fe

(a) Assumptions: Sample irradiated 1 h with a neutron fluence rate of 10^{13} cm^{-2} · s^{-1} and γ-ray counted using a 50 cm^3 germanium detector for 2 h (or one half-life if $0.25\ h < t_{1/2} < 1\ h$). If $t_{1/2} < 0.25$ h, cycle measurements conducted over a 2 h period, with timing of each cycle being $t_i = t_m = t_{1/2}$. Sensitivity was defined at a level of 10% relative uncertainty, with the source placed 2 cm (0.8 in.) from the detector face.

germanium (HPGe, or intrinsic germanium) detectors, because of the numerous gamma rays typically emitted from neutron-irradiated materials (Fig. 1). However, when the gamma-ray spectra are relatively simple, thallium-doped sodium iodide, NaI(Tl), detectors can also be used successfully. These units exhibit higher efficiency, are far less costly, have simpler associated electronics, and need not be operated at liquid-nitrogen temperatures. The NaI(Tl) detectors typically have only approximately 6% resolution at 1000 keV; large HPGe detectors have approximately 0.2% resolution at this energy. Additional information on the various γ-ray detectors is cited in Ref 4 and 13.

For INAA, optimization of detection for a particular element depends strongly on the half-life of its associated radioisotope(s), as well as complex considerations related to interfering radiation produced by other radioisotopes that may be present in the irradiated sample. For example, if equal masses of silicon and manganese were neutron irradiated for a few minutes, the induced activities, ^{31}Si ($t_{1/2}$ = 2.6 h) and ^{56}Mn ($t_{1/2}$ = 2.6 h), would differ in gamma-emission rate by a factor of approximately 5.5×10^6 (in favor of ^{56}Mn). Given these nuclear properties, INAA can easily reveal a few parts per million manganese in silicon but would not be able to detect trace levels of silicon in a manganese compound.

One consideration is that during irradiation the number of atoms, $N(t_i)$, of a given radioisotope increases almost linearly at first but eventually saturates after a few half-lives, that is, the rate of decay equals the rate of production. After an irradiation time of only two half-lives, $N(t_i)$ has already reached 75% of its saturation value; thus, it should be understood that short-lived radioisotopes generally require short irradiation periods, and long-lived radioisotopes generally require longer irradiation periods.

However, more important in NAA is that the decay rate of a radioisotope is inversely proportional to its half-life, which means that at early times after bombardment of a multielement sample, the short-lived activities dominate the gamma-ray spectrum and may impair measurements of long-lived activities.

Consequently, in nondestructive assays of complex samples, the γ-ray spectrum must be recorded at several decay times to detect the full set of observable elements. Figure 2 shows examples of the changes that can occur in the gamma-ray spectrum of a sample as a function of time. Spectral measurements conducted using postirradiation decay times of 10 s, 20 min, 4 d, and 21 d permit detection of a large fraction of the elements observable using INAA.

In addition, more than one irradiation may be necessary. For example, to observe fluorine using ^{20}F ($t_{1/2}$ = 11 s), the sample should be irradiated approximately 20 s and data recording begun within 20 s of the irradiation. Because of the rapid decay of ^{20}F, acquiring data for more than 30 s yields a diminishing return. Observing iron in the same sample, which is detected using ^{59}Fe ($t_{1/2}$ = 44.6 d), necessitates much longer irradiation and counting times. In addition, it is highly advantageous to wait at least two weeks before scanning the gamma-ray spectrum to allow the shorter-lived radioisotopes to decay, which reduces the general background upon which the ^{59}Fe gamma-ray lines must sit.

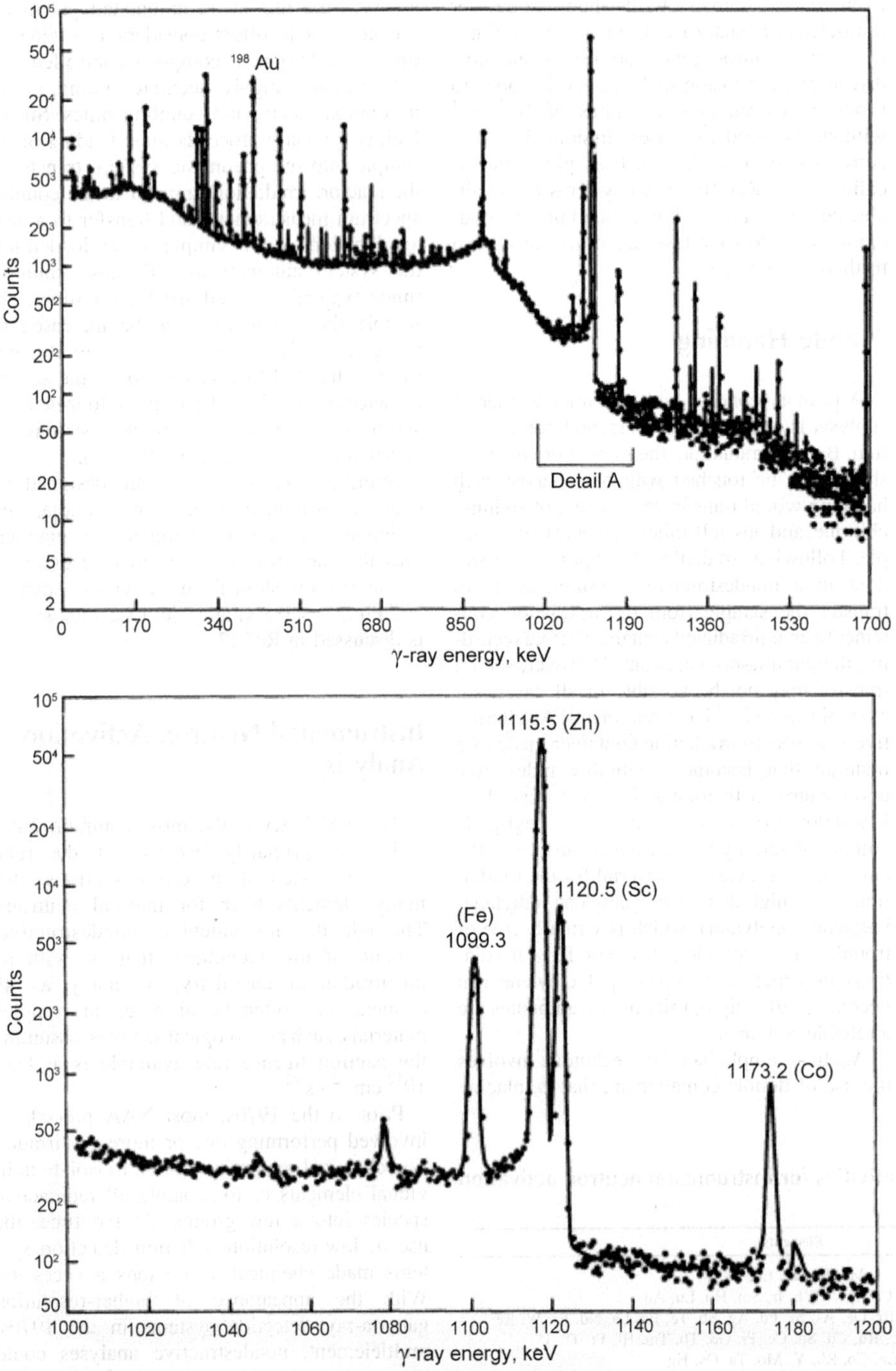

Fig. 1 Gamma-ray spectrum of a neutron-irradiated ore sample from the Jemez Mountains, New Mexico, recorded using a high-resolution detector five days after irradiation. The lower figure is an expanded view of detail A in the upper figure. The necessity of high resolution is evident from the proximity of the peaks at 1115.5 keV (^{65}Zn) and 1120.5 keV (^{46}Sc).

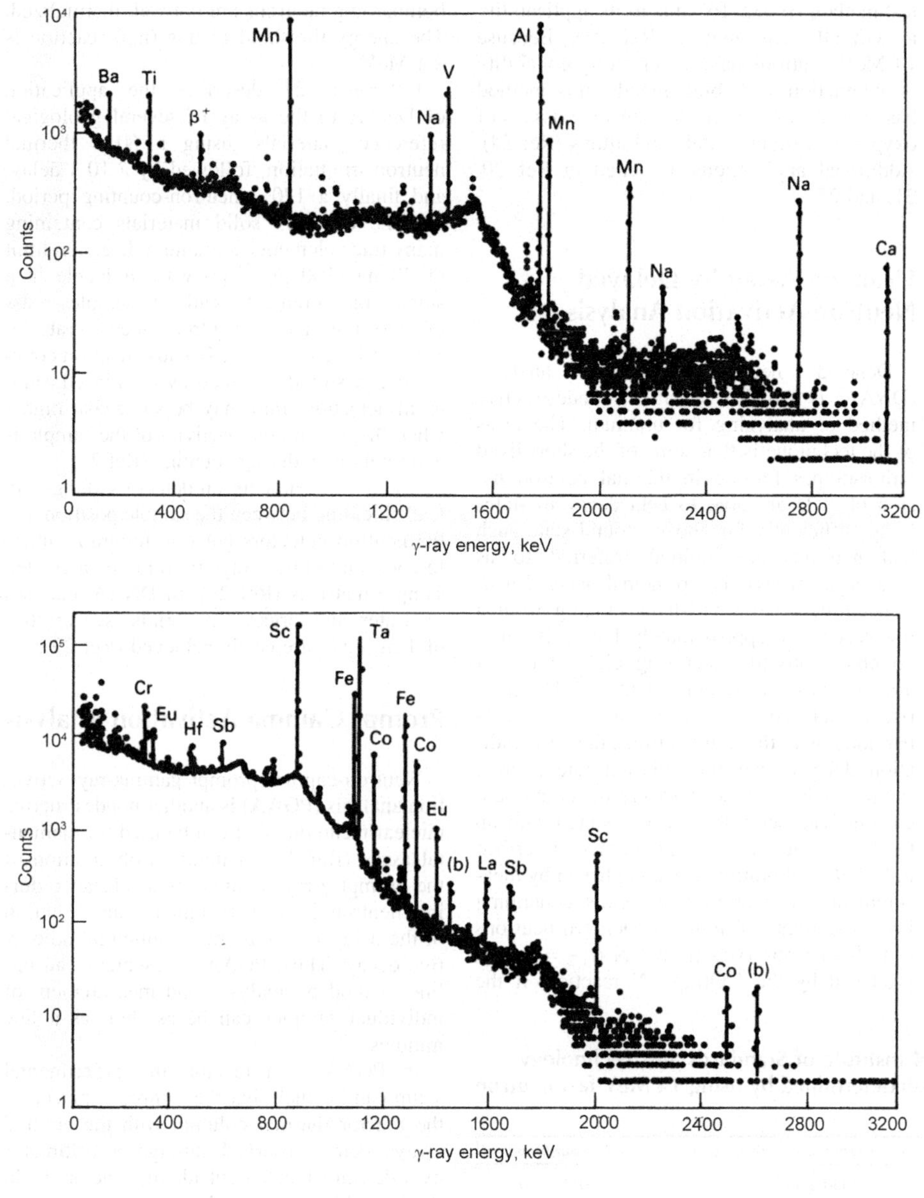

Fig. 2 Gamma-ray spectra of a neutron-irradiated National Institute of Standards and Technology standard reference material 1633a fly ash sample showing the change that occurs as a function of time. The upper spectrum was recorded in the time interval 18 to 27 min after irradiation; the lower spectrum is a 2 h count recorded after 20 d of decay. None of the peaks in the lower spectrum are visible in the upper spectrum. Peaks denoted by (b) represent background lines.

Radiochemical Neutron Activation Analysis

Certain analytes are not determined using INAA because of fundamental nuclear properties. Some elements, for example, phosphorus and thallium, upon irradiation with thermal neutrons, yield radioisotopes that are pure beta emitters or emit virtually no gamma rays. Selective observation of such a radioisotope necessitates chemical isolation of the element, meaning that the sample is chemically decomposed as part of the analysis. The beta-ray emission rate of the active isotope can then be measured, preferably using a 4π proportional counter. This counting rate can be converted to element mass using the direct comparator method of calibration.

Radiochemical neutron activation analysis (RNAA) is also useful when the fractional amount of an element is so small that the neutron-induced gamma-ray activity associated with that element cannot be detected (or is marginally detectable) above the background produced by other activities. This situation is encountered frequently in studies of lunar material, in which many of the elements of interest are present at levels of only a few parts per billion. Another example, involving observation of iridium in terrestrial rocks at the few parts per trillion level, is discussed in Example 2 in the section "Applications" in this article. Radiochemical procedures for isolating various elements (or references to such procedures) are provided in Ref 14 to 16.

Epithermal Neutron Activation Analysis

Neutrons with energies above approximately 0.5 eV and below 100 keV are referred to as epithermal, and the method that makes specific use of these neutrons for sample activation is referred to as epithermal neutron activation analysis (ENAA). At a nuclear reactor, samples can be irradiated with epithermal neutrons by shielding the sample with cadmium or boron, which have very high thermal-neutron-capture cross sections. For example, 1 mm (0.04 in.) of cadmium will absorb most of the incident neutrons of energy less than 0.5 eV. For energies above 0.5 eV, the neutron cross section for most atomic nuclei is far less than for thermal neutrons; however, for many isotopes, the curve of cross section versus energy exhibits large resonance peaks, resulting in much greater relative strength of certain induced activities after epithermal irradiation of a sample than would exist after thermal neutron irradiation (Ref 17).

Epithermal neutron irradiation of a sample results in radioisotope activities having different relative strengths than those observed after a thermal neutron bombardment (see the section "Neutron Reactions" in this article). The activity ratios often change substantially. For example, an epithermal irradiation conducted by surrounding the sample with cadmium increases the $^{99}Mo/^{24}Na$ activity ratio by a factor of approximately 40 compared with that obtained using thermal neutrons (Ref 17). If boron is used as the thermal neutron shield instead of cadmium, the relative enhancement of the $^{99}Mo/^{24}Na$ ratio is approximately 390, because boron stops more energetic neutrons than cadmium (Ref 18).

Thus, to observe molybdenum in a substance containing significant sodium, the use of epithermal neutrons to activate the sample would be highly beneficial, although the epithermal neutron flux available may be much lower than the available thermal flux. Lists of advantage factors for ENAA versus thermal INAA are available for many elements (Ref 14, 17, 19). Epithermal neutron activation analysis is advantageous for at least 20 elements, including nickel, gallium, arsenic, selenium, bromine, rubidium, strontium, zirconium, molybdenum, indium, antimony, cesium, barium, samarium, holmium, tantalum, tungsten, thorium, and uranium (Ref 19). The irradiation and counting techniques used in ENAA are essentially the same as in thermal INAA.

14 MeV Fast Neutron Activation Analysis

Fast neutron activation analysis (FNAA) is a technique that uses the unique capabilities of high-energy neutrons ($E_n > 1$ MeV). The most common method of producing high-energy neutrons is by the nuclear fusion of tritium with deuterium in an ion accelerator, which yields neutrons with energies of approximately 14 MeV (Ref 20); therefore, the method is often called 14 MeV FNAA.

Neutrons with energies above 1 MeV can induce various nuclear reactions in the target element, the most likely being (n,p), (n,α), and (n,2n). These reactions lead to different final nuclei than those produced in the (n,γ) reaction, enabling the nondestructive assay of several elements usually overlooked in thermal INAA, including oxygen, nitrogen, phosphorus, silicon, and lead. The γ-ray detection system, the timing sequence, and the calibration methods used in FNAA are essentially the same as in thermal INAA and ENAA.

Because the reaction cross sections at 14 MeV are relatively low (in many cases, less than 0.1 b), a high fluence rate of neutrons (at least 10^{10} cm^{-2} · s^{-1}) is required to produce enough activity for useful trace-element measurements. Table 5 shows an example of the quality of analytical data obtainable using a fluence rate of approximately 2×10^{11} cm^{-2} · s^{-1}. Irradiations as long as 4 h and counting times as long as 5 d were used to obtain these data.

The determination of oxygen by 14 MeV FNAA has found wide utility and has been designated as an ASTM International standard test method (E 385-16), due to its applicability to virtually any matrix (Ref 23). Because 14 MeV neutrons have a very low probability of interaction with bulk metals, this method has proven to be useful for sensitive assays of oxygen in various metals and alloys (Ref 24). Additional applications are cited in Ref 20, 21, and 25.

Uranium Assay by Delayed Neutron Activation Analysis

Delayed neutron activation analysis (DNAA) is a highly sensitive, nondestructive method of analyzing for uranium. The basis of the technique is that some of the short-lived radioisotopes formed in thermal neutron fission of ^{235}U are able to beta decay to states lying sufficiently far above ground state such that neutrons are emitted (referred to as delayed neutrons). The principal delayed neutron emitters have half-lives ranging from a few seconds to approximately 1 min. Because no other naturally occurring element except uranium (which contains 0.72% ^{235}U) gives rise to delayed neutron emission after bombardment with thermal neutrons, the postirradiation delayed neutron emission rate from a sample yields a direct measure of its uranium content. The neutrons can be counted with up to 40% efficiency by using counters described in Ref 26. Calibration is accomplished by measurements on reference materials containing known amounts of uranium. Delayed neutrons can also be observed from ^{17}N ($t_{1/2}$ = 4.2 s), produced by the ^{17}O(n,p)^{17}N reaction, if the bombarding neutrons are not well thermalized. The energy threshold of this (n,p) reaction is 8.4 MeV.

Reference 27 describes the application of DNAA to the assay of several geological reference materials, using a 60 s thermal neutron irradiation, followed by a 10 s delay, and finally a 120 s neutron-counting period. For the complex solid materials containing many trace elements, a uranium detection limit of 20 ng (200 pg ^{235}U) was attainable in a single measurement, with a sample mass of 250 mg and a neutron fluence rate of 3.4×10^{13} cm^{-2} · s^{-1}. Because neutron counters can be slightly gamma-ray sensitive, the uranium detection limit may be somewhat higher when the gamma radioactivity of the sample is extremely high during counting (Ref 27).

The use of ^{3}He neutron detectors along with lead shielding between the sample position and the neutron detectors helps to minimize interference caused by γ-rays from highly activated sample matrices (Ref 28). In DNAA analyses of water and organic materials, sensitivities of 1 ng · g^{-1} are easily achieved (Ref 29).

Table 5 Elemental mass fractions in National Institute of Standards and Technology (NIST) standard reference material 1633 fly ash determined by using 14 MeV fast neutron activation analysis

Element sought	Radioisotope detected(a) ($t_{1/2}$)	Measured element mass fraction, µg · g^{-1}	NIST value(b), µg · g^{-1}
Sodium	^{20}F (11 s), ^{22}Na (2.6 yr)	3330 ± 170	3100 ± 200
Magnesium	^{24}Na (15 h)	(2.1 ± 0.5)%	(1.6 ± 0.25)%
Aluminum	^{27}Mg (9.5 min)	(12.2 ± 0.5)%	(12.6 ± 0.5)%
Silicon	^{28}Al (2.25 min), ^{29}Al (6.5 min)	(22.4 ± 1.6)%	(22.1 ± 1.1)%
Potassium	^{38}Cl (37 min)	(1.8 ± 0.3)%	(1.69 ± 0.09)%
Calcium	^{47}Ca (4.5 d)	(4.40 ± 0.18)%	(4.6 ± 0.3)%
Scandium	^{44m}Sc (58 h)	29 ± 3	26.6 ± 1.7
Titanium	^{46}Sc (84 d), ^{48}Sc (44 h)	7600 ± 200	7300 ± 300
Chromium	^{51}Cr (28 d)	<150	128 ± 8
Iron	^{56}Mn (2.6 h)	(6.40 ± 0.15)%	(6.14 ± 0.24)%
Cobalt	^{59}Fe (44 d)	45 ± 16	40 ± 2
Nickel	^{57}Co (272 d)	106 ± 12	98 ± 6
Zinc	^{69}Zn (57 min)	<300	210 ± 9
Arsenic	^{74}As (18 d)	68 ± 12	61 ± 4
Selenium	^{75}Se (120 d)	35 ± 13	9.6 ± 0.6
Rubidium	^{84}Rb (33 d)	102 ± 5	115 ± 8
Strontium	^{87m}Sr (2.8 h)	1310 ± 60	1380 ± 100
Yttrium	^{88}Y (107 d)	150 ± 15	63 ± 7
Zirconium	^{89}Zr (78 h)	380 ± 20	300 ± 60
Molybdenum	^{99}Mo (66 h)	22.3 ± 1.6	28 ± 6
Antimony	^{120}Sb (16 min), ^{122}Sb (2.7 d)	8.3 ± 1.8	6.8 ± 0.5
Cesium	^{132}Cs (6.5 d)	10 ± 1	8.6 ± 0.7
Barium	^{135}Ba (29 h)	2250 ± 110	2650 ± 150
Cerium	^{139}Ce (138 d), ^{141}Ce (32 d)	136 ± 8	149 ± 7
Thallium	^{202}Tl (12 d)	18 ± 6	3.4 ± 0.6
Lead	^{203}Pb (52 h)	100 ± 25	72 ± 6

Note: The superscript "m," as in ^{44m}Sc, represents an isomeric state. (a) Source: Ref 21. (b) Source: Ref 22

Prompt Gamma Activation Analysis

Neutron-capture prompt gamma-ray activation analysis (PGAA) is another nondestructive nuclear technique that can be used for elemental assay (Ref 30). It involves observation of the prompt γ-rays emitted by a substance during neutron irradiation rather than detection of the delayed gamma rays emitted in radioactive decay. Thus, PGAA is essentially an online method of analysis, and measurement of individual samples can be as short as a few minutes.

In PGAA at a reactor, the experimental setup can be such that the sample is placed in the reactor thermal column, with the emitted γ-rays being observed through a collimator by a detector located outside the reactor shield (Ref 31). More commonly, a neutron beam can be extracted from the reactor and allowed to impinge on the sample, with the emergent prompt γ-rays being observed by a detector located a short distance from the sample (Ref 32). The external neutron beam configuration has several advantages, including simplified sample changing; no heating of samples, little radiation damage, and small amounts of radioactivity induced in the sample; and assay of samples too fragile, too large, or too dangerous to be placed in the reactor. Table 6 shows PGAA results obtained by using an external thermal neutron beam setup at the National Institute of Standards and Technology Center for Neutron Research.

In certain PGAA facilities, the neutron beam is passed through a moderator maintained at cryogenic temperatures (typically 20 to 30 K, or −250 to −240 °C) to reduce the average energy of the neutrons (Ref 33), producing a so-called cold neutron beam ($E_n < 0.005$ eV).

The cold neutron beam provides greater analytical sensitivities due to the higher capture cross sections for cold versus thermal neutrons.

The number of elements that can be observed in PGAA is quite broad, because the neutron capture does not need to yield a radioisotope to be observable. Most elements that can be determined by thermal INAA or ENAA are accessible by PGAA; however, PGAA can also detect a few elements, including hydrogen, boron, and carbon, that are unobservable with other types of NAA, making it a valuable complementary technique. Prompt gamma-ray activation analysis is particularly useful in analyzing many metals, alloys, and catalysts because the interfering radioactivity induced in the major matrix elements is much less than in thermal INAA (Ref 34–38).

Capture gamma rays induced by neutrons from semiportable neutron generators, such as ^{252}Cf or ^{238}Pu-Be, have proven useful in monitoring element concentrations in various industrial process streams and slurries (Ref 39). Applications to copper mining and copper mill analyses and to on-line coal analysis are discussed in Ref 40 and 41. Californium-252 neutron sources are also frequently used in borehole logging; many of the elements in the material surrounding the borehole are revealed through prompt or delayed gamma rays produced through neutron capture (Ref 42).

Applications

Example 1: Impurities in Nickel Metal

Certain experiments at Los Alamos National Laboratory require knowledge of the trace-element concentrations in high-purity nickel. Because nickel, unlike some of its impurities, becomes only slightly radioactive through neutron irradiation, NAA is an obvious choice for acquiring the assay data. The principal activity induced in nickel by slow neutrons is ^{65}Ni ($t_{1/2}$ = 2.5 h), produced though neutron capture by the rare isotope ^{64}Ni (0.9% abundance, thermal neutron cross section, σ_{th} = 1.49 b).

Figure 3 shows the γ-ray spectrum of a typical nickel sample. Although the titanium, manganese, and vanadium impurities are present at only a few parts per million, the associated γ-ray lines are easily observed in the presence of the ^{65}Ni activity. After a few days of decay, other metallic elements were detected in this sample, including cobalt (13 μg · g^{-1}), chromium (51 μg · g^{-1}), antimony (0.1 μg · g^{-1}), and tungsten (2.7 μg · g^{-1}).

Example 2: The Iridium Anomaly at the Cretaceous-Tertiary Boundary

Large iridium abundance anomalies were discovered at the Cretaceous-Tertiary boundary in marine sedimentary rocks from several world locations (Ref 43). It has been proposed that such anomalies could have been produced by the impact of a large extraterrestrial body, which caused the great extinction of plant and animal life that occurred 65 million years ago at the end of the Cretaceous Period.

Further searches for iridium anomalies at the Cretaceous-Tertiary boundary in favorable areas of the United States have been carried out using RNAA as the assay technique (Ref 44–46). Figure 4 shows one such anomaly, found at a depth of 256 m (840 ft) in the Raton Basin of New Mexico (Ref 44).

To obtain these data, 3 to 5 g samples of drill core were irradiated 24 h in a thermal neutron flux of 1.4 × 10^{11} cm^{-2} · s^{-1} to increase the activity level of ^{192}Ir ($t_{1/2}$ = 74 d). After nearly a month of decay, the platinum-group metals, including iridium, were isolated radiochemically, and the γ-ray spectrum of each residual sample was examined using a high-efficiency γ-ray detector. The lower spectrum in Fig. 5 was recorded from the chemically separated fraction of a core sample that contained 3 ng · g^{-1} iridium (near the maximum of the anomaly). All the labeled γ-rays belong to ^{192}Ir. The upper spectrum is that emitted from the gross samples, in which the strong 316.5 keV γ-ray of ^{192}Ir is barely visible. Thus, to detect a few parts per trillion of iridium in such samples, radiochemical separation is essential. The detection limit for iridium by using this technique is estimated at 0.5 ng · g^{-1}.

REFERENCES

1. S. Amiel, Ed., *Nondestructive Activation Analysis*, Elsevier, 1981
2. G. Erdtmann, *Neutron Activation Tables*, Weinheim/Verlag Chemie, 1976
3. R.R. Greenberg, P. Bode, and E.A. De Nadai Fernandes, *Spectrochim. Acta B*, Vol 66, 2011, p 193–241
4. G.R. Gilmore, *Practical Gamma-Ray Spectrometry*, John Wiley & Sons, 2008
5. M. Blaauw et al., *Nucl. Instrum. Methods Phys. Res. A*, Vol 387, 1997, p 416–432
6. L.A. Currie, *Anal. Chem.*, Vol 40, 1968, p 586–593

Table 6 Mass fractions of elements observed in National Institute of Standards and Technology standard reference material 1632a bituminous coal using prompt gamma-ray activation analysis

Element	Mass fraction, μg · g^{-1}(unless % is indicated)	Sensitivity(a), μg · g^{-1}(unless % is indicated)
Hydrogen	(3.7 ± 0.1)%	12
Boron	52.7 ± 1.8	0.05
Carbon	(71 ± 4)%	1.6%
Nitrogen	(1.27 ± 0.08)%	920
Aluminum	(3.01 ± 0.13)%	450
Silicon	(5.8 ± 0.01)%	610
Sulfur	(1.59 ± 0.02)%	180
Chlorine	784 ± 17	7.2
Potassium	(0.42 ± 0.02)%	69
Calcium	(0.24 ± 0.02)%	490
Titanium	1550 ± 40	27
Manganese	29 ± 5	49
Iron	(1.11 ± 0.06)%	16
Cadmium	0.21 ± 0.03	0.06
Neodymium	11.8 ± 0.4	7.4
Samarium	2.10 ± 0.07	0.03
Gadolinium	1.95 ± 0.03	0.05

(a) Estimate based on a 20 h count of a 1 g sample. Source: Ref 32

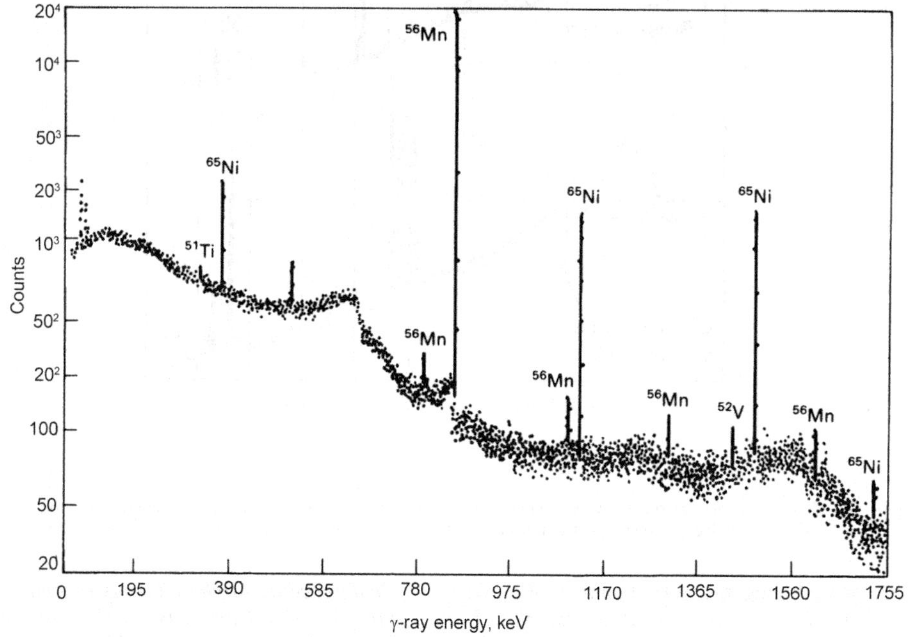

Fig. 3 Gamma-ray spectrum of a neutron-irradiated high-purity nickel sample. The spectrum, recorded in the time interval 3 to 8 min after a 20 s irradiation, shows the presence of titanium, manganese, and vanadium in the sample.

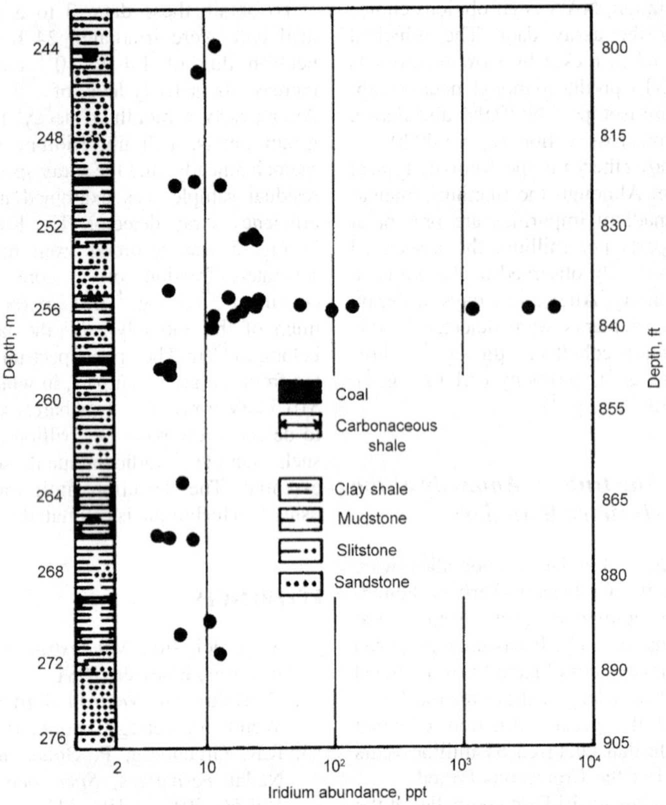

Fig. 4 Iridium mass fraction found as a function of depth in strata. The peak at approximately 256 m (840 ft) corresponds to the Cretaceous-Tertiary boundary. Source: Ref 44

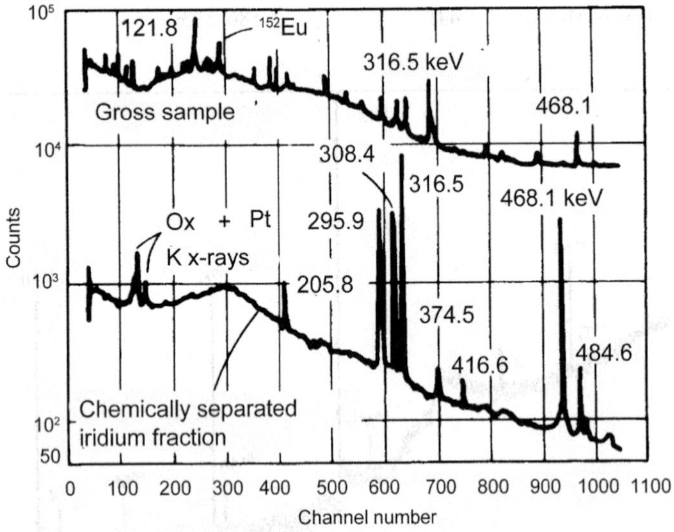

Fig. 5 Comparison of γ-ray spectra. Upper spectrum shows a neutron-irradiated rock sample that contains 3 ng · g^{-1} iridium (26 d decay). Lower spectrum shows the chemically isolated iridium fraction (^{192}Ir, $t_{1/2}$ = 74.2 d)

7. W.C. Reinig and A.G. Evans, in *Modern Trends in Activation Analysis*, Vol 2, NBS 312, National Bureau of Standards, Washington, 1969, p 953–957
8. P.E. Cummins, J. Dutton, C.J. Evans, W. D. Morgan, and A. Sivyer, in *Sixth Conference on Modern Trends in Activation Analysis (Abstracts)*, University of Toronto Press, 1981, p 242–243
9. D. Vartsky, K.J. Ellis, N.S. Chen, and S.H. Cohn, *Phys. Med. Biol.*, Vol 22, 1977, p 1085
10. J.H. Vainionpaa et al., *AIP Conf. Proc.*, Vol 1525, 2013, p 118
11. M.M. Minor, W.K. Hensley, M.M. Denton, and S.R. Garcia, *J. Radioanal. Chem.*, Vol 70, 1981, p 459, https://doi.org/10.1007/BF02516130
12. M. Wiernik and S. Amiel, *J. Radioanal. Chem.*, Vol 3, 1969, p 393, https://doi.org/10.1007/BF02513783
13. G.F. Knoll, *Radiation Detection and Measurement*, John Wiley & Sons, 2000
14. F. Girardi, in *Modern Trends in Activation Analysis,* Vol 1, NBS 312, National Bureau of Standards, Washington, 1969, p 577–616
15. A.A. Levinson, Ed., *in Proceedings of the Apollo 11 Lunar Science Conference*, Vol 2, Chemical and Isotopic Analyses, Pergamon Press, 1970
16. G.J. Lutz, R.J. Boreni, R.S. Maddock, and J. Wing, *Activation Analysis: A Bibliography through 1971,* NBS 467, National Bureau of Standards, Washington, 1972
17. E. Steinnes, in *Activation Analysis in Geochemistry and Cosmochemistry*, Universitets Forlaget, Oslo, Norway, 1971, p 113–128
18. E.S. Gladney, D.R. Perrin, J.P. Balagna, and C.L. Warner, *Anal. Chem.*, Vol 52, 1980, p 2128, https://doi.org/10.1021/ac5006 3a032
19. J.J. Rowe and E. Steinnes, *J. Radioanal. Chem.*, Vol 37, 1977, p 849, https://doi.org/10.1007/BF02519396
20. S.S. Nargolwalla and E.P. Przybylowicz, *Activation Analysis with Neutron Generators*, John Wiley & Sons, 1973
21. R.E. Williams, P.K. Hopke, and R.A. Meyer, *J. Radioanal.Chem.*, Vol 63, 1981, p 187
22. E.S. Gladney, C.E. Bums, D.R. Perrin, I. Roelands, and T.E. Gills, "1982 Compilation of Elemental Concentration Data for NBS Biological, Geological, and Environmental Standard Reference Materials," NBS 260-88, National Bureau of Standards, Washington, 1984
23. "Standard Test Method for Oxygen Content Using a 14-MeV Neutron Activation and Direct-Counting Technique," E 385-E 316, ASTM International, West Conshohocken, PA, 2016, DOI: 10.1520/E0385-16, www.astm.org
24. C.D. Fuerst and W.D. James, in *Proceedings of the 15th Int. Conf. on the Application of Accelerators in Research and Industry* (Denton, TX), 1998, p 731
25. W.D. Ehmann, *J. Radioanal. Chem.*, Vol 167, 1993, p 67–79
26. S.J. Balestrini, J.P. Balagna, and H.O. Menlove, *Nucl. Instrum. Methods*, Vol 136, 1976, p 521, https://doi.org/10.1016/0029-554X(76)90374-8
27. E.S. Gladney, D.B. Curtis, D.R. Perrin, J.W. Owens, and W.E. Goode, "Nuclear Techniques for the Chemical Analysis of Environmental Materials," Report LA-8192-MS, Los Alamos National Laboratory, Los Alamos, NM, 1980

28. S.M. Eriksson et al., *J. Radioanal. Nucl. Chem.*, Vol 298, 2013, p 1819
29. H.M. Ide, W.D. Moss, M.M. Minor, and E.E. Campbell, *Health Phys.*, Vol 37, 1979, p 405
30. R.M. Lindstrom and Z. Revay, *J. Radioanal. Nucl. Chem.*, Vol 314, 2017, p 843
31. E.T. Jurney, H.T. Motz, and S.H. Vegors, *Nucl. Phys. A*, Vol 94, 1967, p 351, https://doi.org/10.1016/0375-9474(67)90009-7
32. M.P. Failey, D.L. Anderson, W.H. Zoller, G.E. Gordon, and R.M. Lindstrom, *Anal. Chem.*, Vol 51, 1979, p 2209, https://doi.org/10.1021/ac50049a035
33. K. Ünlü and C. Ríos-Martínez, *J. Radioanal. Nucl. Chem.*, Vol 265, 2005, p 329
34. M. Heurtebise and J.A. Lubkowitz, *J. Radioanal. Chem.*, Vol 31, 1976, p 503, https://doi.org/10.1007/BF02518514
35. M. Heurtebise, H. Buenafama, and J.A. Lubkowitz, *Anal. Chem.*, Vol 48, 1976, p 1969, https://doi.org/10.1021/ac50007a040
36. H. Zwittlinger, *J. Radioanal. Chem.*, Vol 14, 1973, p 147, https://doi.org/10.1007/BF02514157
37. M.R. Najam, M. Anwar-Ul-Islan, A.F.M. Ishaq, J.A. Mirza, A.M. Khan, and I.H. Qureshi, *J. Radioanal. Chem.*, Vol 27, 1975, p115, https://doi.org/10.1007/BF02517454
38. M. Heurtebise and J.A. Lubkowitz, *J. Radioanal. Chem.*, Vol 38, 1977, p 115, https://doi.org/10.1007/BF02520188
39. D. Duffey, P.F. Wiggins, and F.E. Senftle, in *Proceedings of the American Nuclear Topical Meeting on Neutron Source Applications*, Vol 4, Savannah River, Aiken, SC, 1971, p 18–29
40. D. Duffey, J.P. Balagna, P.F. Wiggins, and A.A. El-Kady, *Anal. Chim. Acta*, Vol 79, 1975, p 149, https://doi.org/10.1016/S0003-2670(00)89427-1
41. H.R. Wilde and W. Herzog, in *Sixth Conference on Modern Trends in Activation Analysis (Abstracts)*, University of Toronto Press, 1981, p 324
42. R.M. Moxham, F.E. Senftle, and R.G. Boynton, *Econ. Geol.*, Vol 67, 1972, p 579, https://doi.org/10.2113/gsecongeo.67.5.579
43. L.W. Alvarez, W. Alvarez, F. Asaro, and H. V. Michel, *Science*, Vol 208, 1980, p 1095, https://doi.org/10.1126/science.208.4448.1095
44. C.J. Orth, J.S. Gilmore, J.D. Knight, C.L. Pillmore, R.H. Tschudy, and J.E. Fassett, *Science*, Vol 214, 1981, p 1341, https://doi.org/10.1126/science.214.4527.1341
45. J.S. Gilmore, J.D. Knight, C.J. Orth, C.L. Pillmore, and R.H. Tschudy, *Nature*, Vol 307, 1984, p 224
46. C.J. Orth, J.D. Knight, L.R. Quintana, J.S. Gilmore, and A.R. Palmer, *Science*, Vol 223, 1984, p 163

Thermal Analysis

Jeffrey A. Jansen, The Madison Group

ASM Handbook, Volume 10, *Materials Characterization*
ASM Handbook Committee
DOI 10.31399/asm.hb.v10.a0006672

Differential Scanning Calorimetry*

Adapted by Victoria Burt, ASM International

Overview

General Uses

- Characterize melting, crystallization, resin curing, loss of solvents, and other processes involving an energy change

Examples of Applications

- Characterizing relevant phase transitions that can be used to determine the best processing temperatures and maximum use temperatures
- Measuring heat capacity of pure compounds and mixtures
- Measuring heat of fusion and heat of solidification (ΔH, enthalpy change)
- Comparing quality (quality control, failure analysis, new material evaluation)
- Identifying unknown materials and determining the presence of impurities
- Evaluating formulations, blends, and effects of additives
- Determining the effects of aging
- Estimating percent crystallinity
- Determining percent purity of relatively pure organics
- Analyzing cure or crystallization kinetics
- Determining phase separation of polymer blends and copolymers
- Estimating the degree of cure; measuring residual cure
- Evaluation of eutectic point and construction of phase diagrams
- Characterizing polymorphic materials
- Evaluating thermal history of compounds
- Performing sensitive measurements of subtle, weak, or overlapping phase transitions

Samples

- *Form:* solid
- *Size:* from 0.5 to 100 mg

Limitations

- Destructive
- No direct elemental information
- Works best for samples having a surface that spreads relatively flat against the bottom of the crucible or pan
- Accurate data cannot be obtained when a decomposition or reaction event occurs within the same temperature region as the phase transition (e.g., melting)
- Mass of sample must remain constant in the pan for accurate measurement; that means no loss of sample to evaporation or sublimation during the test
- Most commercial differential scanning calorimetry instruments have a temperature range of −150 to 750 °C (−240 to 1380 °F).

Estimated Analysis Time

- Test can be performed in seconds

Capabilities of Related Techniques

- Thermogravimetric analysis provides information regarding composition and thermal stability.
- Dynamic mechanical analysis is used to assess viscoelastic properties of materials.
- Thermomechanical analysis is used to study structure by evaluating material dimensional changes.

Introduction

Differential scanning calorimetry (DSC) is the most common thermal technique for polymer characterization. Differential scanning calorimetry measures the difference in heat flow between a sample and a reference as the material is heated or cooled (Ref 1). It is easy to use, requires small samples (~20 mg), operates from –180 to 725 °C (–290 to 1335 °F), and measures heat flow associated with sample transitions as a function of temperature (or time) under controlled atmospheric conditions. Differential scanning calorimetry has been used to study thermodynamic processes (glass transition, heat capacity) and kinetic events such as cure and enthalpic relaxations associated with physical aging or stress (Ref 2). The technique is also used to evaluate thermal transitions such as melting, evaporation, crystallization, solidification, cross linking,

*Adapted from J.A. Jansen, Characterization of Plastics in Failure Analysis, *Failure Analysis and Prevention*, Volume 11, *ASM Handbook*, ASM International, 2002, p 437–459; and G. Dallas, Thermal Analysis, *Composites*, Volume 21, *ASM Handbook*, ASM International, 2001, p 973–976.

Table 1 Transitions detected by differential scanning calorimetry

Industry	Transitions	Purpose
Pharmaceuticals	Glass transition temperature (T_g)	Collapse or storage temperature, amorphous content
	Specific heat capacity (C_P)	Processing conditions
	Melting temperature (T_m)	Polymorphic forms, purity, quality control
Polymers	T_g	Indicator of material properties, quality control, effect of additives
	T_m	Polymer processing, heat history
	Exotherm	Reactions rate, curing of materials, residual care
	C_P	Energy needed to process
	Recrystallization temperature (T_c)	Recrystallization times, kinetics
Food	T_g	Storage temperature, properties
	T_m	Processing temperature

Source: Ref 4

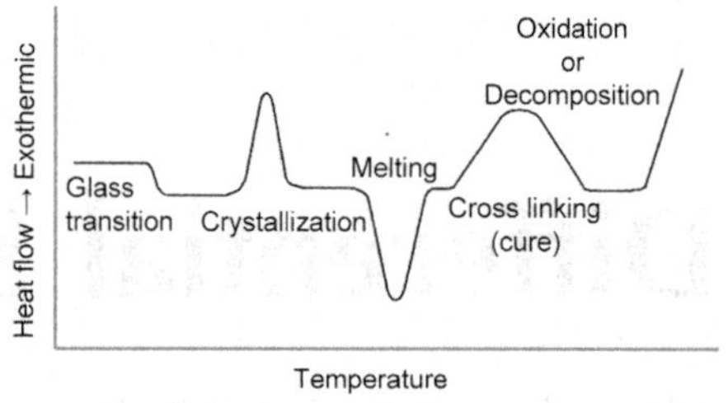

Fig. 1 Transitions of interest in a typical differential scanning calorimetry output

chemical reactions, and decomposition. Any transition in a material that involves a change in the heat content of the material can be evaluated by DSC (Ref 1), the limitation being that commercially available equipment may not be able to detect transitions within materials that are present at concentrations below 5% (Ref 3). Table 1 shows the types of transitions detected by DSC according to industry and use in that industry.

An experimental analysis related to DSC is differential thermal analysis (DTA), in which temperature differentials are measured. This information includes relative heat capacities, presence of solvents, changes in structure (that is, phase changes, such as melting of one component in a resin system), and chemical reactions. However, heat flow is not measured by the DTA method (Ref 5).

Transitions of interest in a DSC are shown in Fig 1. A typical DSC thermogram is presented in Fig. 2. The thermogram shows the heat flow in energy units or energy per mass units on the y-axis as a function of either temperature or time on the x-axis. The transitions that the sample material undergoes appear as exothermic and endothermic changes in the heat flow. Endothermic transitions require heat to proceed, while exothermic transitions give off heat.

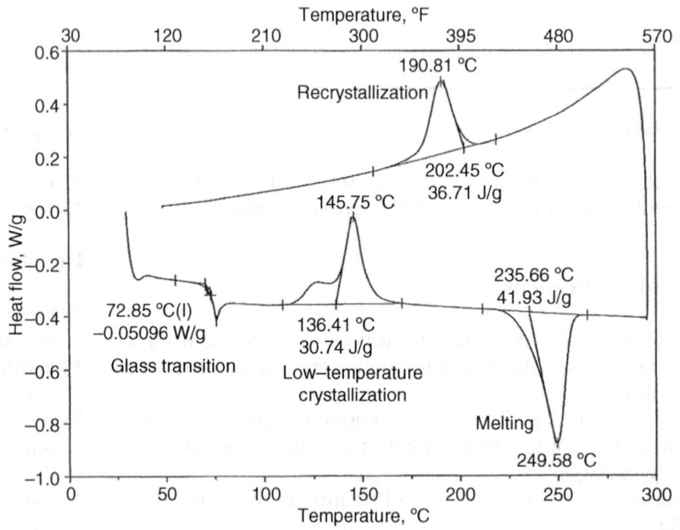

Fig. 2 Differential scanning calorimetry thermogram showing various transitions associated with polymeric materials. The (I) indicates that the numerical temperature was determined as the inflection point on the curve. Source: Ref 6

Systems and Equipment

In the DSC method, the specimen is placed within a metal pan. The pan can be open, crimped, or sealed hermetically, depending on the experiment. A reference, either in the form of an empty pan of the same type or an inert material having the same weight as the sample, is used. The most commonly used metal pan material is aluminum; however, pans made of copper and gold are used for special applications.

Differential scanning calorimetry measurements can be made in two ways: by measuring the electrical energy provided to heaters below the pans necessary to maintain the two pans at the same temperature (power compensation), or by measuring the heat flow (differential temperature) as a function of sample temperature (heat flux) (Ref 5).

Specimen size typically ranges between 1 and 10 mg, although this can vary depending on the nature of the sample and the experiment. The normal operating temperature range for DSC testing is −180 to 700 °C (−290 to 1290 °F), with a standard heating rate of 10 °C/min (18 °F/min). A dynamic purge gas is used to flush the sample chamber. Nitrogen is the most commonly used purge gas, but helium, argon, air, and oxygen can also be used for specific purposes. Often, two consecutive heating runs are performed to evaluate a sample. A controlled cooling run is performed after the initial analysis in order to eliminate the heat history of the sample. The first heating run assesses the sample in the as-molded condition, while the second run evaluates the inherent properties of the material.

Modulated DSC

Modulated DSC (MDSC), also called modulated-temperature DSC, is a high-performance version of DSC that improves the possibility for glass transition temperature (T_g) detection because it can offer a fivefold increase in sensitivity over DSC and with no loss in signal resolution (Ref 7). Figure 3 shows MDSC detection of a broad T_g in a complex glass-fiber-reinforced epoxy-aramid/polyimide composite that was unobservable by using DSC even at its highest sensitivity. Modulated DSC can also resolve the total heat-flow signal into its thermodynamic (heat capacity related) and kinetic (temperature/time related) components. This allows direct analysis of the effects of aging or process-induced stresses, which appear as an enthalpic relaxation that can mask the T_g, with the net result being a signal that looks like an endothermic melt. The total heat flow signal is shown in Fig. 4. Modulated DSC can resolve these overlapping events; the actual T_g of the epoxy composite appears in the upper curve in Fig. 4, while the relaxation is resolved into a second kinetic-related signal (nonreversing).

Specimen Preparation

To obtain good-quality measurements, sample preparation is important. Factors include choosing the appropriate crucible, good thermal contact between sample and crucible so that thermal effects are not smeared, preventing contamination of the outer surfaces of the

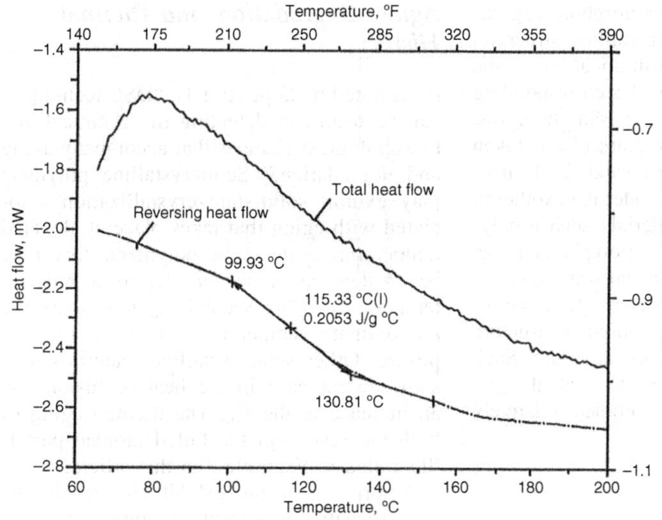

Fig. 3 High-sensitivity glass transition temperature detection using modulated differential scanning calorimetry (MDSC). Sample: glass-fiber-reinforced epoxy-aramid/polyimide; sample size: 32.9 mg; method: MDSC 2.5/60 at 1 °C/min (1.8 °F/min); crimped pan; nitrogen gas purge. The (I) indicates that the numerical temperature was determined as the inflection point on the curve. Source: Ref 8

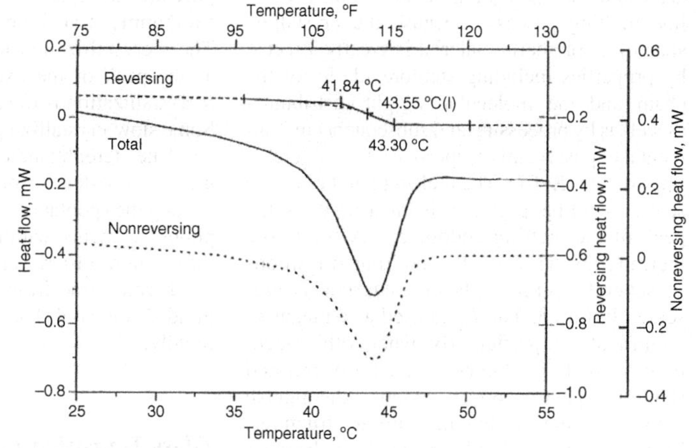

Fig. 4 Modulated differential scanning calorimetry separation of glass transition temperature (T_g) from enthalpic relaxation. Upper curve: T_g; lower curve: enthalpic relaxation. The (I) indicates that the numerical temperature was determined as the inflection point on the curve. Source: Ref 8

crucible, and the atmosphere surrounding the sample (Ref 9).

The sample must have good thermal contact with the pan so that transitions are sharp and the sample will not collapse and cause invalid peaks as it heats. If the sample is poorly encapsulated, then heat transfer will be slow, allowing buildup of thermal gradients across the sample (Ref 10).

Calibration and Accuracy

Several published methods exist that standardize test procedures and specimen dimensions for the purpose of maximizing the repeatability and reproducibility of test results. Some of the more common procedures are provided in Table 2.

Good data require adherence to applicable standards, a smooth baseline, and reasonable separation of the sample peak from any noise in the baseline (Ref 4). The baseline should be smooth, that is, no bumps or spikes, and flat, although some upward slope in temperature is expected as the heat capacity increases.

The most important DSC quantities that must be calibrated are the temperature, the heating-rate dependence of the temperature, and the heat flow (Ref 9). For many users, one-point calibrations and adjustments are sufficient. A one-point calibration uses a single reference substance and works well for a small temperature range near the melting point of the reference substance. A one-point adjustment shifts the scale in the entire temperature range by the corresponding amount. For larger temperature ranges, multipoint adjustments using two or three reference substances are recommended. Instrumentation companies provide suggestions on frequency and procedures for calibrating and adjusting DSC instruments.

Table 2 Some differential scanning calorimetry procedures and standards

Designation	Title
ASTM D 3418	"Standard Test Method for Transition Temperatures and Enthalpies of Fusion and Crystallization of Polymers by DSC"
ASTM D 3895	"Standard Test Method for Oxidative-Induction Time of Polyolefins by DSC"
ASTM D 4419	"Standard Test Method for Measurement of Transition Temperatures of Petroleum Waxes by DSC"
ASTM D 5483-05	"Standard Test Method for Oxidation Induction Time of Lubricating Greases by Pressure DSC"
ASTM D 5885-06	"Standard Test Method for Oxidative Induction Time of Polyolefin Geosynthetics by High-Pressure DSC"
ASMT D 6186-08	"Standard Test Method for Oxidation Induction Time of Lubricating Oils by Pressure DSC"
ASTM D 6604-00	"Standard Practice for Glass Transition Temperatures of Hydrocarbon Resins by DSC"
ASTM D 7426-08	"Standard Test Method for Assignment of the DSC Procedure for Determining T_g of a Polymer or an Elastomeric Compound"
ASTM E 537-12	"Standard Test Method for the Thermal Stability of Chemicals by DSC"
ASTM E 793	"Standard Test Method for Enthalpies of Fusion and Crystallization by DSC"
ASTM E 794	"Standard Test Method for Melting and Crystallization by DSC"
ASTM E 928	"Standard Test Method for Determination of Purity by DSC"
ASTM E 967-08	"Standard Test Method for Temperature Calibration of DSC and DTA Analyzers"
ASTM E 968-02	"Standard Practice for Heat Flow Calibration of DSC"
ASTM E 1269	"Standard Test Method for Determining Specific Heat Capacity by DSC"
ASTM E 1356	"Standard Test Method for Glass Transition Temperatures by DSC"
ASTM E 2160	"Standard Test Method for Heat of Thermally Reactive Materials by Differential Scanning Calorimetry"
ISO 11357-1	"Plastics—Differential Scanning Calorimetry (DSC) Part 1: General Principles"
ISO 11357-2	"Plastics—Differential Scanning Calorimetry (DSC) Part 2: Determination of Glass Transition Temperature"
ISO 11357-3	"Plastics—Differential Scanning Calorimetry (DSC) Part 3: Determination of Temperature and Enthalpy of Melting and Crystallization"
ISO 11357-4	"Plastics—Differential Scanning Calorimetry (DSC) Part 4: Determination of Specific Heat Capacity"
ISO 11357-5	"Plastics—Differential Scanning Calorimetry (DSC) Part 5: Determination of Characteristic Reaction-Curve Temperatures and Times, Enthalpy of Reaction and Degree of Conversion"
ISO 11357-6	"Plastics—Differential Scanning Calorimetry (DSC) Part 6: Determination of Oxidation Induction Time (Isothermal OIT) and Oxidation Induction Temperature (Dynamic OIT)"
ISO 11357-7	"Plastics—Differential Scanning Calorimetry (DSC) Part 7: Determination of Crystallization Kinetics"

Data Analysis and Reliability

Unlike metals, polymers have a molecular structure that includes characteristics such as molecular weight, crystallinity, and orientation, and this has a significant impact on the properties of the molded article. Additionally, plastic resins usually contain additives, such as reinforcing fillers, plasticizers, colorants, antidegradants, and process aids. It is this combination of molecular structure and complex formulation that requires specialized testing (Ref 11).

Melting Point and Crystallinity

The primary use of DSC in polymer analysis is the detection and quantification of the crystalline melting process. Because the crystalline state of a polymeric material is greatly affected by properties including stereoregularity of the chain and the molecular weight distribution as well as by processing and subsequent environmental exposure, this property is of considerable importance (Ref 1). The melting point (T_m) of a semicrystalline polymer is measured as the peak of the melting endotherm. A composite thermogram showing the melting transitions of several common plastic materials is presented in Fig. 5. The T_m is used as a means of identification, particularly when other techniques, such as Fourier transform infrared (FTIR) spectroscopy, cannot distinguish between materials having similar structures. This can be useful in identifying both the main resin and any contaminant materials. The material identification aspects of DSC are illustrated in Examples 1 to 3 in this article.

The heat of fusion represents the energy required to melt the material and is calculated as the area under the melting endotherm. The level of crystallinity is determined by comparing the actual as-molded heat of fusion with that of a 100% crystalline sample. The level of crystallinity that a material has reached during the molding process can be practically assessed by comparing the heat of fusion obtained during an initial analysis of the sample with the results generated during the second run, after slow cooling. The level of crystallinity is important, because it impacts the mechanical, physical, and chemical resistance properties of the molded article. In general, rapid or quench cooling results in a lower crystalline state. This is the result of the formation of frozen-in amorphous regions within the preferentially crystalline structure. Recrystallization, or the solidification of the polymer, is represented by the corresponding exothermic transition as the sample cools. The recrystallization temperature (T_c) is taken as the peak of the exotherm, and the heat of recrystallization is the area under the exotherm. Some slow-crystallizing materials, such as polyethylene terephthalate and polyphthalamide, undergo low-temperature crystallization, representing the spontaneous rearrangement of amorphous segments within the polymer structure into a more orderly crystalline structure. Such exothermic transitions indicate that the as-molded material had been cooled relatively rapidly.

Glass Transition in Amorphous Plastics

Polymers that do not crystallize and semicrystalline materials having a significant level of amorphous segments undergo a phase change referred to as a glass transition. The glass transition represents the reversible change from/to a viscous or rubbery condition to/from a hard and relatively brittle one (Ref 12). The glass transition is observed as a change in the heat capacity of the material. The T_g can be defined in several ways but is most often taken as the inflection point of the step transition. A composite thermogram showing the glass transitions of several common plastic materials is presented in Fig. 6. The T_g of an amorphous resin has an important impact on the mechanical properties of the molded article, because it represents softening of the material to the point that it loses load-bearing capabilities.

Aging, Degradation, and Thermal History

As noted by Sepe (Ref 1), "DSC techniques can be useful in detecting the chemical and morphological changes that accompany aging and degradation." Semicrystalline polymers may exhibit solid-state crystallization associated with aging that takes place at elevated temperatures. In some polymers, this may be evident as a second T_m at a reduced temperature. This second T_m represents the approximate temperature of the aging exposure. Other semicrystalline materials may show an increase in the heat of fusion and an increase in the T_m. The thermal aging of both the resin and the failed molded part is illustrated in Example 3 in this article.

Amorphous resins exhibit changes in the glass transition as a result of aging. In particular, physical aging, which occurs through the progression toward thermodynamic equilibrium below the T_g, produces an apparent endothermic transition on completion of the glass transition.

Degradation and other nonreversible changes to the molecular structure of semicrystalline polymers can be detected as reduced values for the T_m, T_c, or heat of fusion. Similarly, degradation in amorphous resins can be observed as a reduction in the T_g or in the magnitude of the corresponding change in heat capacity.

Further, the resistance of a polymer to oxidation can be evaluated via DSC by standard methods or experiments involving high-pressure oxygen or air exposure. Such evaluations usually measure the oxidative induction time or the temperature at which oxidation initiates under the experimental conditions. This can be used to compare two similar

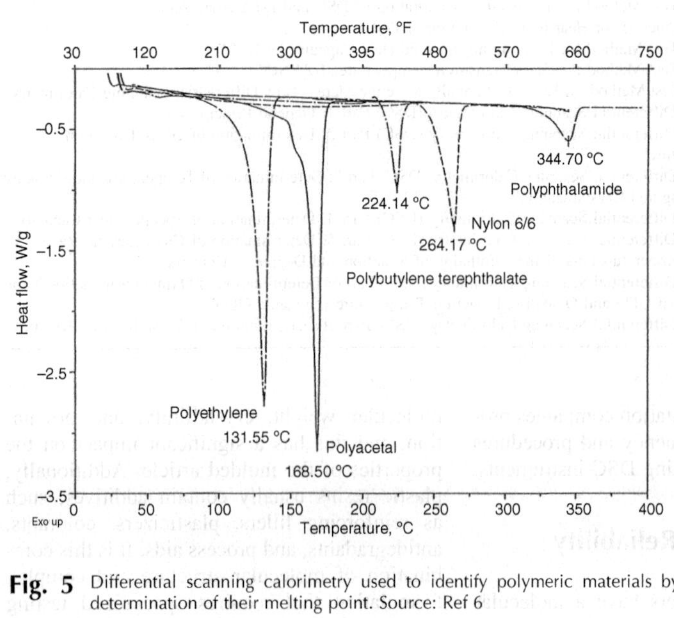

Fig. 5 Differential scanning calorimetry used to identify polymeric materials by determination of their melting point. Source: Ref 6

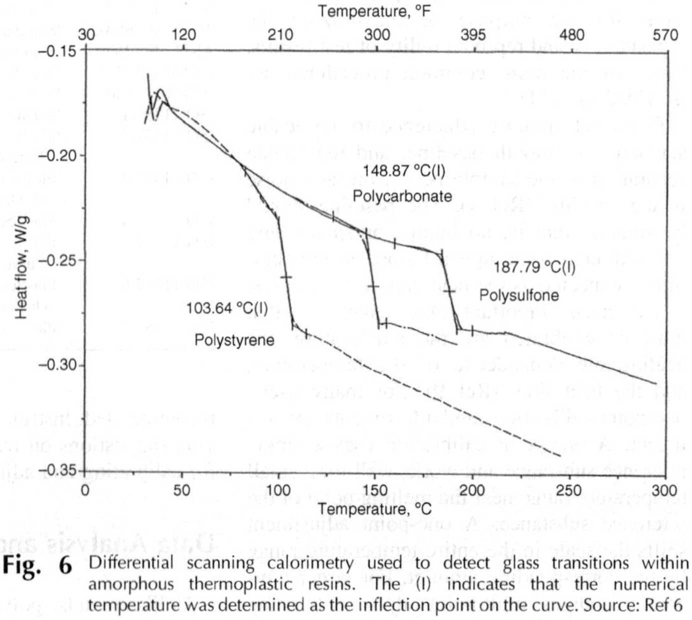

Fig. 6 Differential scanning calorimetry used to detect glass transitions within amorphous thermoplastic resins. The (I) indicates that the numerical temperature was determined as the inflection point on the curve. Source: Ref 6

materials or to determine whether a plastic resin has undergone partial oxidation.

Applications and Interpretation

Example 1: Relaxation of Nylon Wire Clips

A production lot of plastic wire clips was failing after limited service. The failures were characterized by excessive relaxation of the clips, such that the corresponding wires were no longer adequately secured in the parts. No catastrophic failures had been encountered. Parts representing an older lot, which exhibited satisfactory performance properties, were also available for reference purposes. The clips were specified to be injection molded from an impact-modified grade of nylon 6/6. However, the part drawing did not indicate a specific resin.

Tests and Results

A visual examination of the clips showed that the failed parts were off-white in color, while the control parts had a pure white appearance. An analysis of both sets of parts was performed using micro-FTIR in the attenuated total reflectance (ATR) mode. A direct comparison of the results produced a good match, with both sets of spectra exhibiting absorption bands that were characteristic of a nylon resin. The comparison, however, revealed subtle differences between the two sets of clips. The spectrum representing the reference parts showed a relatively higher level of a hydrocarbon-based impact modifier, while the results obtained on the failed parts showed the presence of an acrylic-based modifier. The differences in the spectra suggested that the two sets of clips were produced from resins having different formulations, particularly regarding the impact modifier.

The clip materials were further analyzed using DSC. The thermogram representing the reference part material, as shown in Fig. 7, exhibited an endothermic transition at 264 °C (507 °F), characteristic of the melting point of a nylon 6/6 resin. Additionally, the results contained a second melting point, of lesser magnitude, at 95 °C (203 °F). This transition was indicative of a hydrocarbon-based impact modifier, as indicated by the FTIR results. The thermogram obtained on the failed clip material also showed a melting point characteristic of a nylon 6/6 resin. However, no evidence was found to indicate a transition corresponding to the hydrocarbon-based modifier found in the control clip material.

Conclusions

It was the conclusion of the analysis that the control and failed clips had been produced from two distinctly different resins. While both materials satisfied the requirements of an impact-modified nylon 6/6 resin, differences in the impact modifiers resulted in the observed performance variation. From the results and the observed performance, it appeared that the material used to produce the failed clips had different viscoelastic properties, which produced a greater predisposition for stress relaxation.

Example 2: Embrittlement of Nylon Couplings

Molded plastic couplings used in an industrial application exhibited abnormally brittle properties, as compared to previously produced components. The couplings were specified to be molded from a custom-compounded glass-filled nylon 6/12 resin. An inspection of the molding resin used to produce the discrepant parts revealed differences in the material appearance, relative to a retained resin lot. Specifically, physical sorting resulted in two distinct sets of molding resin pellets from the lot that had generated the brittle parts. Both of these sets of pellets had a coloration that varied from that of the retained reference resin pellets. A sample of retained molding resin, which had produced parts exhibiting satisfactory performance, was available for comparative analysis.

Tests and Results

Micro-FTIR in the ATR mode was used to analyze the molding resin samples. The results obtained on the three molding resin samples were generally similar, and all of the spectra exhibited absorption bands characteristic of a nylon resin. Further analysis of the resin samples using DSC indicated that the control material results exhibited a single endothermic transition at 218 °C (424 °F), consistent with the melting point of a nylon 6/12 resin, as specified.

The DSC thermograms obtained on the two resin samples that produced brittle parts also exhibited melting point transitions associated with nylon 6/12. However, additional transitions were also apparent in the results, indicative of the presence of contaminant materials. The results obtained on one of the resin samples, as presented in Fig. 8, showed a secondary melting point at 165 °C (330 °F), indicative of polypropylene. The thermogram representing the second resin sample, as included in Fig. 9, displayed a second melting transition at 260 °C (500 °F), characteristic of a nylon 6/6 resin.

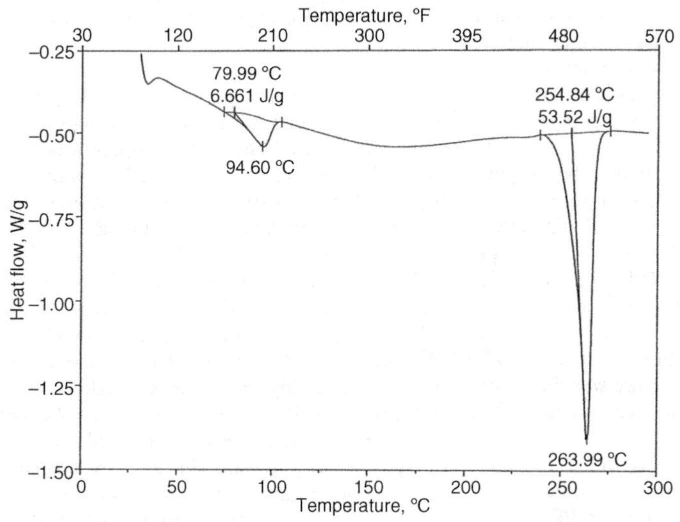

Fig. 7 Differential scanning calorimetry thermogram representing the reference clip material, exhibiting an endothermic transition characteristic of the melting of a nylon 6/6 resin. The results also showed a second melting transition attributed to a hydrocarbon-based impact modifier. Source: Ref 6

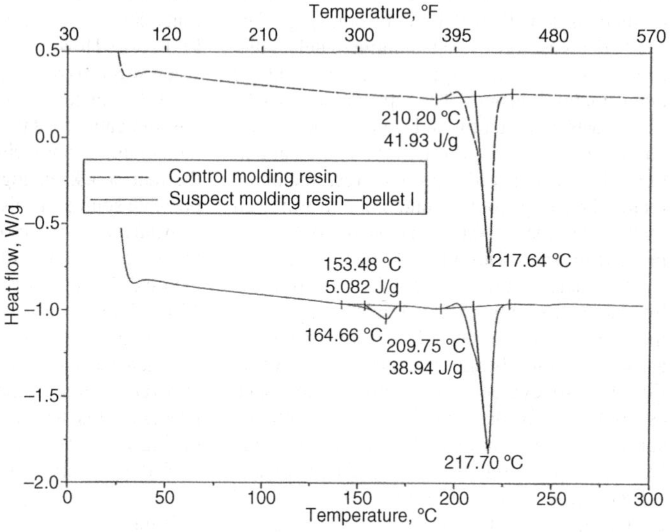

Fig. 8 Differential scanning calorimetry thermogram representing a molding resin pellet that had produced brittle parts. The thermogram shows a major melting transition associated with nylon 6/12 and a weaker transition attributed to polypropylene. Source: Ref 6

Conclusions

It was the conclusion of the analysis that the molding resin used to produce the brittle couplings contained a significant level of contamination, which compromised the mechanical properties of the molded components. Two distinct contaminants were found mixed into the molding pellets. The contaminant materials were identified as polypropylene and nylon 6/6. The source of the polypropylene was likely the purging compound used to clean the compounding extruder. The origin of the nylon 6/6 resin was unknown but may represent a previously compounded resin.

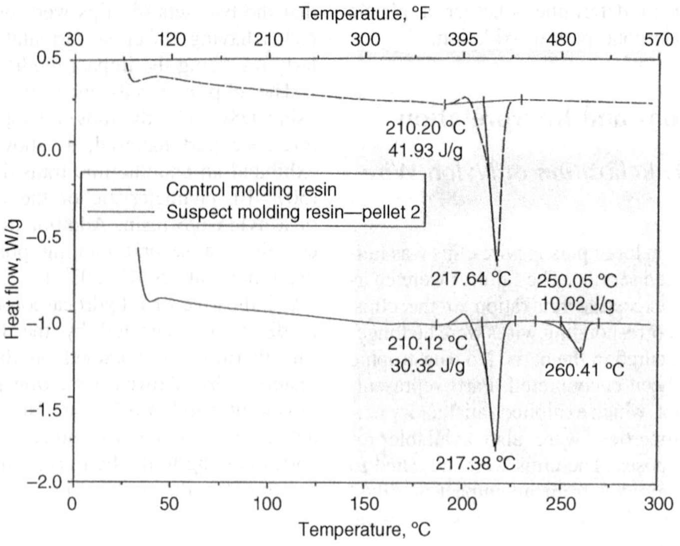

Fig. 9 Differential scanning calorimetry thermogram representing a molding resin pellet that had produced brittle parts. The thermogram shows a major melting transition associated with nylon 6/12 and a weaker transition attributed to polypropylene. Source: Ref 6

Example 3: Failure of Polyethylene Terephthalate (PET) Assemblies

Several assemblies used in a transportation application failed during an engineering testing regimen. The testing involved cyclic thermal shock, immediately after which cracking was observed on the parts. The cracking occurred within the plastic jacket, which had been injection molded from an impact-modified, 15% glass-fiber-reinforced PET resin. The plastic jacket had been molded over an underlying metal coil component. Additionally, a metal sleeve was used to house the entire assembly. Prior to molding, the resin had reportedly been dried at 135 °C (275 °F). The drying process usually lasted 6 h, but occasionally, the material was dried overnight. The thermal shock testing included exposing the parts to alternating temperatures of −40 and 180 °C (−40 and 360 °F). The failures were apparent after 100 cycles. Molding resin and nonfailed parts were also available for analysis.

Tests and Results

The failed assemblies were visually and microscopically examined. The inspection showed several different areas within the overmolded jacket that exhibited cracking. The cracked areas were located immediately adjacent to both the underlying metal coil and the outer metal housing. The appearance of the cracks was consistent with brittle fracture, without significant signs of ductility. The examination also revealed design features, including relatively sharp corners and nonuniform wall thicknesses, that appeared to have likely induced molded-in stress within the plastic jacket.

The fracture surfaces were further inspected by using scanning electron microscopy, and the examination revealed features generally associated with brittle fracture, as shown in Fig. 10. No evidence of microductility, such as stretched fibrils, was found. The fracture surface features indicated that the cracking had initiated along the outer jacket wall and subsequently extended through the wall and circumferentially around the wall. Throughout the examination, no indication of postmolding molecular degradation was found.

Micro-FTIR was performed in the ATR mode on a core specimen of the jacket material. The resulting spectrum was consistent with a thermoplastic polyester resin. Such materials, including PET and polybutylene terephthalate, cannot be distinguished spectrally, and a melting point determination is usually used to distinguish these materials. The failed jacket material and reference molding resin were analyzed using thermogravimetric analysis, and the results obtained on the two samples were generally consistent. This included relatively comparable levels of volatiles, polymer, carbon black, and glass reinforcement. Further, the results were in excellent agreement with those expected for the stated PET material.

The failed jacket and reference materials were evaluated via DSC. Analysis of the failed jacket material produced results that indicated a melting transition at 251 °C (484 °F), consistent with a PET resin. However, a second endothermic transition was also present. This transition, at 215 °C (420 °F), suggested the melting of annealed crystals, indicating that the part had been exposed to a temperature approaching 215 °C (420 °F). The thermal shock testing appeared to be the only possible source of this thermal exposure.

Analysis of the molding resin also produced results consistent with a PET resin. The results also exhibited a second melting endotherm at 174 °C (345 °F). Again, this transition was associated with melting of annealed crystals for material exposed to this temperature. The apparent source of the exposure was the drying process. This was well in excess of the stated drying temperature.

Further analysis of the assembly materials using thermomechanical analysis produced significantly different results for the PET jacket and the steel housing material. Determination of the coefficients of thermal expansion (CTEs) showed approximately an order of magnitude difference between the two mating materials.

Fig. 10 Scanning electron image showing brittle fracture features on the failed jacket crack surface. Original magnification: 20×. Source: Ref 6

An assessment of the molecular weight of the failed jacket samples as well as a nonfailed part and the molding resin samples was performed using several techniques. A combination of melt flow rate (MFR), intrinsic viscosity, and finally, gel permeation chromatography (GPC) was used, because of conflicting results. The MFR determinations showed that the drying process produced a considerable increase in the MFR of the resin, corresponding to molecular degradation in the form of chain scission. This was contrasted by the results generated by the intrinsic viscosity testing. These results showed an increase in the viscosity of the dried resin relative to the virgin resin. This increase was suggestive of an increase in molecular weight, possibly through partial cross linking. Testing of the resin samples and the molded parts via GPC produced results that reconciled the discrepancy. The GPC results showed that the drying process produced competing reactions of chain scission and cross linking. The net result was severe degradation of the dried resin, which predisposed the molding material to produce jackets having poor mechanical properties. The GPC testing showed that the molded jackets were further degraded during the injection molding process.

Conclusions

It was the conclusion of the investigation that the assemblies failed via brittle fracture associated with the exertion of stresses that exceeded the strength of the resin as-molded. The stresses were induced by the thermal cycling and the dimensional interference caused by the disparity in the CTEs of the PET jacket and the mating steel sleeve. However, several factors were significant in the failures. It was determined that the resin drying process had exposed the resin to relatively high temperatures, which caused substantial molecular degradation. The drying temperature was found to be approximately 173 °C (344 °F), well in excess of the recommendation for the PET resin. Further degradation was attributed to the molding process itself, leaving the molded jacket in a severely degraded state. This degradation limited the ability of the part to withstand the applied stresses. Additionally, the testing itself exposed the parts to temperatures above the recognized limits for PET, and this may have significantly lowered the mechanical properties of the part.

REFERENCES

1. M.P. Sepe, *Thermal Analysis of Polymers*, RAPRA Technology, Shawbury, U.K., 1997, p 3, 4, 8, 17, 19, 22, 24, 33
2. R.B. Prime, Chap. 6, *Thermal Characterization of Polymeric Materials*, Vol 2, 2nd ed., E.A. Turi, Ed., Academic Press, 1997
3. J. Scheirs, *Compositional and Failure Analysis of Polymers*, John Wiley & Sons, 2000, p 109, 138, 153, 393, 415
4. "Differential Scanning Calorimetry: A Beginner's Guide," PerkinElmer Inc., https://www.perkinelmer.com/CMSResources/Images/44-74542GDE_DSCBeginnersGuide.pdf
5. M.M. Gauthier, Thermal Analysis and Properties of Polymers, *Engineered Materials Handbook Desk Edition*, ASM International, 1995
6. J.A. Jansen, Characterization of Plastics in Failure Analysis, *Failure Analysis and Prevention*, Vol 11, *ASM Handbook*, ASM International, 2002, p 437–459
7. L.C. Thomas, "Modulated DSC Theory," Technical Publication TA 211, TA Instruments, 1993
8. G. Dallas, Thermal Analysis, *Composites*, Vol 21, *ASM Handbook*, ASM International, 2001, p 973–976
9. M. Wagner, Chap. 7, *Thermal Analysis in Practice: Fundamental Aspects*, Hanser Publishers, 2018
10. S. Gaisford, V. Kett, and P. Haines, 5.5.8 Sample Mass, *Principles of Thermal Analysis and Calorimetry*, 2nd ed., Royal Society of Chemistry, 2016
11. J.A. Jansen, Plastic Component Failure Analysis, *Adv. Mater. Process.*, May 2001, p 56, 58, 59
12. L.C. Roy Oberholtzer, General Design Considerations, *Engineering Plastics*, Vol 2, *Engineered Materials Handbook*, ASM International, 1988, p 21

ASM Handbook, Volume 10, *Materials Characterization*
ASM Handbook Committee
DOI 10.31399/asm.hb.v10.a0006673

Thermogravimetric Analysis*

Overview

General Uses

- Materials characterization
- Thermal stability of related materials
- Compositional analysis
- Corrosion studies, including oxidation

Examples of Applications

- Thermal stabilities assessment and moisture content of polymers
- Determining the quantity of filler content in polymers
- Assessment of compositional analysis of polymer blends
- Quantitative measurement of a low level of volatile matter evolved by samples

Specialty Applications

- High-resolution thermogravimetric analysis (TGA)
- Decomposition kinetics
- Evolved gas analysis when the TGA is connected to an evolved gas analyzer (e.g., TGA/mass spectrometry or TGA/Fourier transform infrared spectroscopy)
- Vacuum TGA

Samples

- 10 to 20 mg for most applications
- 50 to 100 mg for measuring volatiles
- Liquids, powders, films, solids, or crystals

Limitations

- Evolved products are identified only when the TGA is connected to an evolved gas analyzer (e.g. TGA/mass spectrometry or TGA/Fourier transform infrared spectroscopy)

Estimated Analysis Time

- Standard runs are 1 to 2 h. Times increase for specialty applications.

Introduction

Thermogravimetric analysis (TGA) is a thermal analysis technique that measures the amount and rate of change in the weight of a material as a function of temperature or time in a controlled atmosphere. The weight of the evaluated material can decrease due to volatilization or decomposition or increase because of gas absorption or chemical reaction. Thermogravimetric analysis can provide valuable information regarding the composition and thermal stability of polymeric materials. The obtained data can include the volatiles content, inorganic filler content, carbon black content, the onset of thermal decomposition, and the volatility of additives such as antioxidants (Ref 1).

The results obtained as part of TGA evaluation are presented in a thermogram. The TGA thermogram illustrates the sample weight, usually in percent of original weight, on the *y*-axis as a function of time or, more commonly, temperature on the *x*-axis. The weight-change transitions are often highlighted by plotting the corresponding derivative on an alternate *y*-axis.

Changes in mass occur when the sample loses material or reacts with the surrounding atmosphere. This creates steps on the TGA curve. There are several reasons why a sample would lose (or gain) mass, producing these steps in the curve, including (Ref 2):

- Evaporation of volatile constituents, drying, desorption and adsorption of gases, moisture and other volatile substances, loss of water or crystallization
- Oxidation of metals in air or oxygen
- Oxidative decomposition of organic substances in air or oxygen
- Thermal decomposition in an inert atmosphere with the formation of gaseous products
- Heterogeneous chemical reactions in which a starting material is taken up from the atmosphere
- Ferromagnetic materials: If the sample is measured in an inhomogeneous magnetic field, the change in magnetic attraction at the transition generates a TGA signal.

Thermogravimetric analysis instruments consist of two primary components: a microbalance and a furnace (Fig. 1). The sample is suspended from the balance while heated in conjunction with a thermal program. A ceramic or, more often, a platinum sample pan is used for the evaluation. As part of the TGA evaluation, the sample is usually heated from ambient room temperature to 1000 °C (1830 °F) in a dynamic gas purge of nitrogen, air, or a consecutive switch program. The composition of the purge gas can have a significant effect on the TGA results and, as such, must be properly controlled. The size of the sample evaluated usually ranges between 5 and 100 mg, with samples as large as 1000 mg possible. Minimal sample preparation is required for TGA experiments.

*Adapted from J.A. Jansen, Characterization of Plastics in Failure Analysis, *Failure Analysis and Prevention*, Volume 11, *ASM Handbook*, ASM International, 2002, p 437–459

Specimen Preparation

To obtain good-quality measurements, sample preparation is important. Samples should be representative of the material being analyzed. The mass of the sample should be sufficient to achieve the precision required for the test. The sample should be changed as little as possible during preparation, and it should not be contaminated by the preparation process (Ref 2). Use approximately the same sample weight during each experiment to ensure reproducibility. Many small pieces of sample are better than one large chunk, and it is better to have a large surface area exposed to the sample purge (Ref 3).

Several published methods exist that standardize test procedures and specimen preparation. Common procedures are listed in Table 1.

Calibration and Accuracy

A thermogravimetric analyzer instrument is calibrated for mass, temperature, and gas flow-rate control. The manufacturer is responsible for the gas flow-rate control during installation, but the values can be monitored by the user. Calibration can be affected by purge gas and flow rate, type of specimen pan, and heating rates (Ref 4). Prior to calibration, the TGA pan should be cleaned, using a flame torch to burn off organic residue on platinum or alumina pans, if necessary. The purge gas flow-rate setting should be set (flow rates differ depending on furnace design and internal volume) and not deviate by more than ±5 mL/min.

ASTM E 2040, "Standard Test Method for Mass Scale Calibration of TGAs," describes the calibration or performance confirmation of the mass scale of TGAs. It does not address temperature effects on mass calibration.

ASTM E 1582, "Calibration of Temperature Scale for Thermogravimetry," describes two methods by which the TGA can be calibrated for temperature: by melting point or by magnetic transition. The most common approach for a TGA would be the magnetic transition approach.

Data Analysis

Thermogravimetric analysis curves are typically classified into types according to their shapes. Figure 2 shows a schematic of the types of TGA curves (Ref 5). Curve 1 depicts no mass change over the temperature range, suggesting that the decomposition temperature is greater than the temperature range of the instrument. Curve 2 shows a large mass loss followed by a plateau. This is formed when evaporation of volatile product(s) during desorption, drying, or polymerization takes place. Curve 3 is typical of single-stage decomposition temperatures. Curve 4 shows multistage decomposition as a result of various reactions. Curve 5 is similar to curve 4 but is due to either a fast heating rate or no intermediates. Curve 6 is an atmospheric reaction, showing an increase in mass. It may be due to reactions such as surface oxidation reactions in the presence of an interacting atmosphere. Curve 7 is similar to curve 6, but the product decomposes at high temperatures, for example, the reaction of surface oxidation followed by decomposition of reaction product(s).

Thermogravimetric Analysis in Failure Analysis of Plastics

Thermogravimetric analysis is a key analytical technique used in the assessment of the composition of polymeric-based materials. The quantitative results obtained during a TGA evaluation directly complement the qualitative information produced by Fourier transform infrared spectroscopy analysis. The

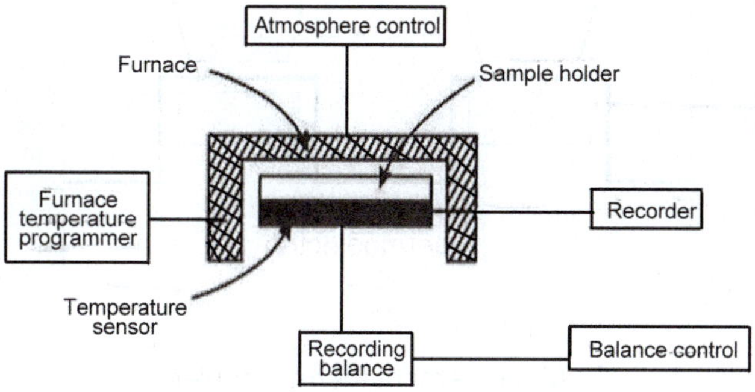

Fig. 1 Block diagram of a thermogravimetric instrument setup

Table 1 Some thermogravimetric analysis procedures and standards

Designation	Title
ISO 11358	Thermogravimetry of Polymers
ASTM E 1131	Standard Test Method for Compositional Analysis by TGA
ASTM E 1641	Standard Test Method for Decomposition Kinetics by TGA
ASTM D 7582	Standard Test Methods for Proximate Analysis of Coal and Coke by Macro Thermogravimetric Analysis
ASTM D 3850	Standard Test Method for Rapid Thermal Degradation of Solid Electrical Insulating Materials by Thermogravimetric Method (TGA)
ASTM C 1872	Standard Test Method for Thermogravimetric Analysis of Hydraulic Cement
ISO 247-2	Rubber—Determination of Ash—Part 2: Thermogravimetric Analysis (TGA)
ISO/DTS 11308	Nanotechnologies—Characterization of Carbon Nanotubes Using Thermogravimetric Analysis
ASTM E 2105	TGA/FTIR
ASTM D 6375	Standard Test Method for Evaporation Loss of Lubricating Oils by Thermogravimetric Analyzer (TGA) Noack Method

Note: Others can be found by searching "thermogravimetric," "TGA," or "thermogravimetric analysis" on the websites of ASTM International, American National Standards Institute, or International Organization for Standardization.

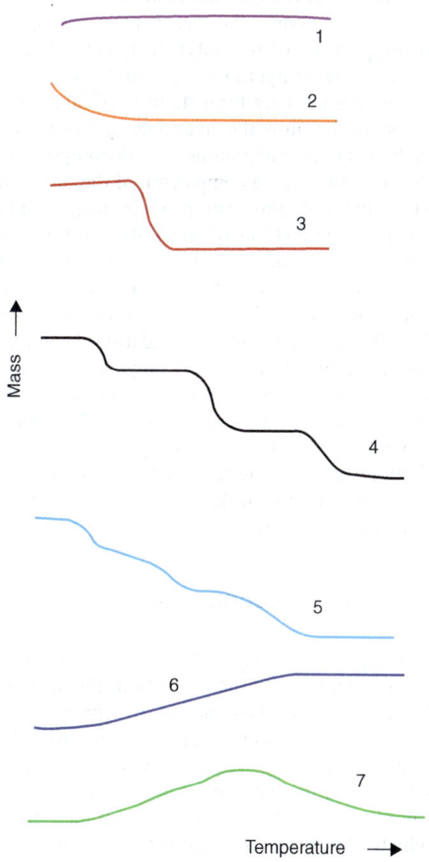

Fig. 2 Seven types of thermogravimetric analysis curves. See text for discussion. Source: Ref 5

relative loadings of various constituents within a plastic material, including polymers, plasticizers, additives, carbon black, mineral fillers, and glass reinforcement, can be assessed. The assessment of a plastic resin composition is illustrated in Fig. 3. These data are important as part of a failure analysis in order to determine if the component was produced from the correct material. The weight-loss profile of the material is evaluated, and, ideally, the TGA results obtained on the material exhibit distinct, separate weight-loss steps. These steps are measured and associated with transitions within the evaluated material. A thorough knowledge of the decomposition and chemical reactions is required to properly interpret the obtained results. In most situations, however, distinct weight-loss steps are not obtained, and, in these cases, the results are complemented by the corresponding derivative curve. Noncombustible material remaining at the conclusion of the TGA evaluation is often associated with inorganic fillers. Such residue is often further analyzed by using energy-dispersive x-ray spectroscopy to evaluate its composition. The use of TGA in characterizing plastic composition is presented in Example 1 in this article.

Thermogravimetric analysis data can also be used to compare the thermal and oxidative stability of polymeric materials. The relative stability of polymeric materials can be evaluated by assessing the onset temperature of decomposition of the polymer. Quantitatively, these onset temperatures are not useful for comparing the long-term stability of fabricated products, because the materials are generally molten at the beginning of decomposition (Ref 6). However, a comparison of the obtained TGA thermograms can provide insight into possible degradation of the failed component material. Example 2 in this article illustrates a comparison of the thermal stability of two polymeric materials, while Example 3 shows the effects of molecular degradation. Degradation experiments involving polymeric materials can also provide information regarding the kinetics of decomposition. Such studies provide information regarding the projected lifetime of the material. Such measurements, however, provide little information pertinent to a failure analysis.

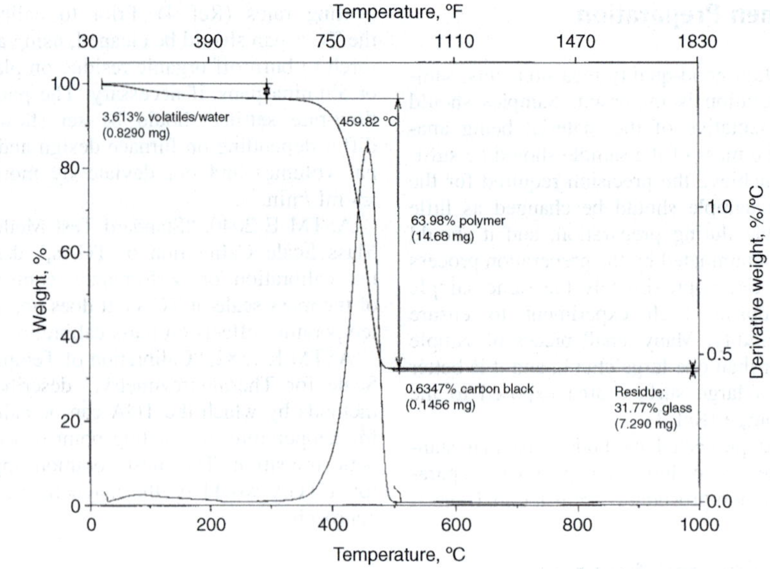

Fig. 3 Thermogravimetric analysis thermogram showing the weight-loss profile for a 30% glass-fiber-reinforced material. Courtesy of Element New Berlin

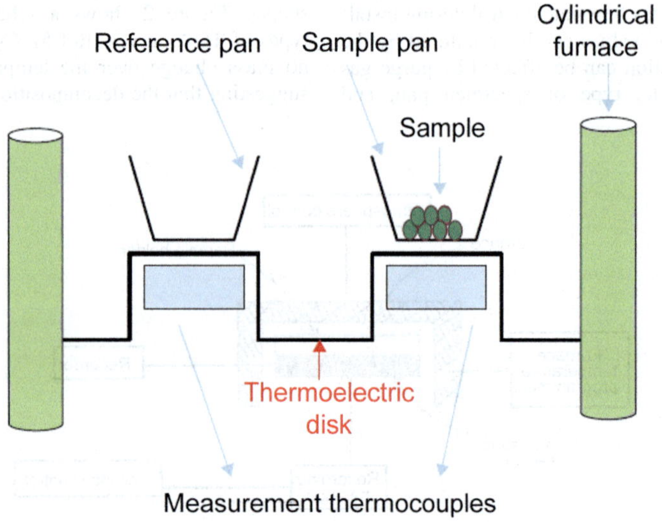

Fig. 4 Generic thermogravimetric analysis/differential scanning calorimetry setup. Source: Ref 7

Combined Techniques

Sometimes a single TGA test is not sufficient to understand the changes in the sample. For this reason, TGA and other techniques are often combined. For instance, on its own, differential scanning calorimetry (DSC) calculates the difference in heat flow between a sample and an inert reference substance, from which the glass transition temperature, the degree of crystallinity, cross-linking reactions, decomposition behavior, and more can be analyzed. Figure 4 shows the TGA-DSC setup to measure heat flux (Ref 7). The sample is placed inside a crucible, which is placed inside the measurement cell (furnace) of the DSC system, along with an empty reference pan. The pans are placed on a plate containing thermocouples, which detect the temperature difference between the two pans and convert this to a thermal energy difference (Ref 8). This setup saves the time needed to acquire two sets of data, and the technique is able to determine energy changes related to thermal decomposition reactions. For example, a chemical reaction such as curing of thermosets is a condensation reaction, and water or ammonia may be released during the process of reaction. This results in a change in mass of the sample in addition to the heat of reaction, which can be measured by DSC. Normally, the physical changes in polymer structure, such as those that accompany glass transition, crystallization, or melting, cannot be detected by TGA, but information from DSC can be useful. For example, the thermal stability of fabricated polyvinyl alcohol (PVA)/zinc acetate dihydrate nanofibers was investigated by using TGA-DSC from 25 to 600 °C (75 to 1110 °F) (Ref 9). Figure 5 shows the thermal activities of the generated nanofibers. There are three significant weight losses occurring in the thermogravimetric characteristics curve. The first weight loss is almost 22% in the range of 40 to 120 °C (105 to 250 °F). This specific weight loss is due to the loss of the residual water molecules or surface-absorbed

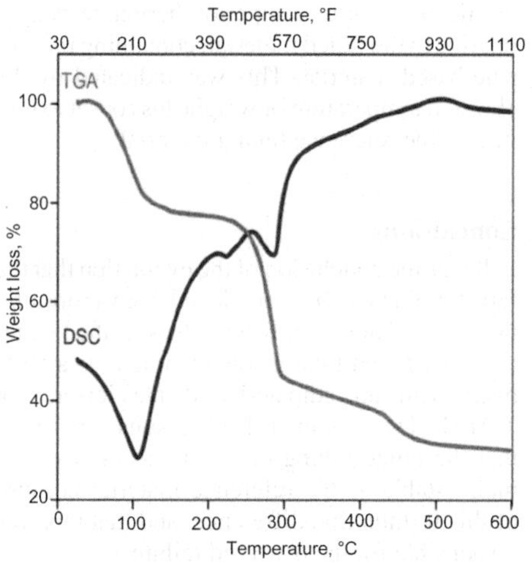

Fig. 5 Thermogravimetric analysis (TGA) and differential scanning calorimetry (DSC) graphs of polyvinyl alcohol and zinc acetate dihydrate nanofibers. Source: Ref 9

water in the precursor composite fibers. The first endothermic peak appeared at approximately 105 °C (220 °F) in the DSC curve (Fig. 5). This was detected due to the decomposition of the acetate group and the evaporation of water. It is reported that standard crystallization starts at approximately 250 °C (480 °F). The second weight loss is approximately 33% in the range of 200 to 303 °C (390 to 577 °F), caused by the loss of the volatile components, including H_2O, CO, and CO_2. The third weight loss is nearly 20% in the thermogravimetric curve, occurring from 303 to 480 °C (577 to 895 °F). This weight loss corresponds to the decomposition of the PVA. The exothermic peaks absorbed at approximately 255, 285, and 500 °C (490, 545, and 930 °F) in the DSC curve can be attributed to the vaporization of the acetate side chain and the main chain of PVA. The decomposition of the PVA/zinc becomes constant beyond 480 °C (895 °F). A thermal analysis result very clearly clarifies that no weight loss occurs after 480 °C.

Another type of thermal analysis, called evolved gas analysis (EGA), measures the amount of gaseous volatile products evolved by samples as a function of temperature (Ref 7). Two types of techniques combine EGA with TGA: Fourier transform infrared (FTIR) spectroscopy and mass spectrometry (MS). Table 2 lists the advantages of TGA-FTIR and TGA-MS techniques.

The combination of TGA with gas chromatography/mass spectrometry (GC/MS) allows gases to be transferred to the gas chromatograph, where the components can be collected (Ref 3). The sample can then be run by GC to separate the material, and the peaks can be identified by MS. The TGA-GC/MS has a good ability to detect very low levels of material in complex mixtures, making it good for quality control, safety, and product development.

Table 2 Comparison of thermogravimetric analysis/Fourier transform infrared spectroscopy (TGA-FTIR) and thermogravimetric analysis/mass spectrometry (TGA-MS)

Technique	Features	Benefits
TGA-FTIR	High chemical specificity and fast measurement	TGA-FTIR is able to characterize substances by determining their compounds and functional groups. This technique is ideal for the online measurement of substances that indicate medium-to-strong infrared absorption.
TGA-MS	High sensitivity coupled with very fast measurement	TGA-MS is only able to detect extremely small amounts of substances. This technique is ideal for the online characterization of all types of volatile compounds.

Applications and Interpretation

Example 1: Failure of Nylon Hinges

A production lot of mechanical hinges had failed during incoming quality-control testing. The hinges were used in an automotive application and had cracked during routine actuation testing. Similar parts had been through complete prototype evaluations without failure. However, a change in part supplier had taken place between the approval of the prototype parts and the receipt of the first lot of production parts. The mechanical hinges were specified to be injection molded from an impact-modified, 13% glass-fiber-reinforced nylon 6/6 resin. A resin substitution was suspected, corresponding to the supplier change. Samples representing the failed components and the original prototype parts were available for the failure investigation.

Tests and Results

A visual examination of the failed parts confirmed catastrophic cracking within the mechanical hinge in an area that would be under the highest level of stress during actuation. The failures did not show signs of macroductility, which would be apparent in the form of stress whitening and permanent deformation. The fracture surfaces of the failed parts were further inspected via scanning electron microscopy (SEM). While the presence of glass-reinforcing fibers can render a plastic resin inherently more brittle, a certain level of ductility is still expected at the 13% glass level. This ductility is often only apparent at high magnification and only between the individual glass fibers. However, the failed hinge components did not exhibit any signs of ductility even at high magnification, with the fracture surface showing only brittle features. A laboratory failure was created on one of the prototype parts by overloading the component. Examination of the fracture surface using SEM showed the normally anticipated level of ductility, as indicated by the overlapping, rose-petal morphology. The crack surfaces of both the failed part and the laboratory fracture are shown in Fig. 6.

Analysis of the failed components and the corresponding molding resin via micro-FTIR produced results characteristic of a nylon resin. The molding resin and failed parts generated

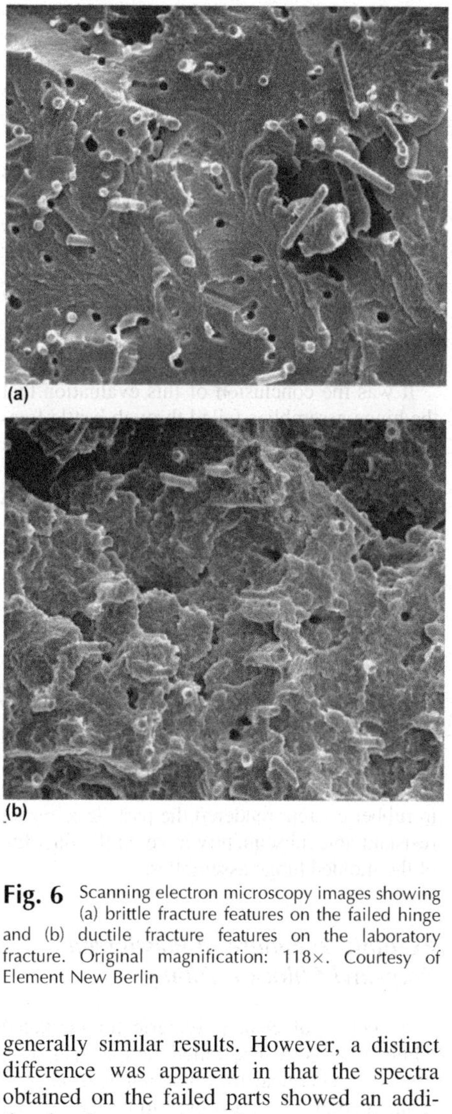

Fig. 6 Scanning electron microscopy images showing (a) brittle fracture features on the failed hinge and (b) ductile fracture features on the laboratory fracture. Original magnification: 118×. Courtesy of Element New Berlin

generally similar results. However, a distinct difference was apparent in that the spectra obtained on the failed parts showed an additional absorption band at approximately 1740 cm^{-1}, indicative of partial oxidative degradation of the resin. A spectral comparison illustrating this is presented in Fig. 7. Because the parts had not yet been in service, this degradation was thought to have occurred during the molding process.

The failed parts were further tested by using DSC. The obtained DSC results showed a melting point of 263 °C (505 °F), consistent with a nylon 6/6 resin. The molding resin was also analyzed via DSC, and a comparison of the results further indicated degradation of the molded nylon resin. This was apparent by a noted reduction in the heat of fusion in the results representing the failed parts.

The failed parts and the prototype parts were also analyzed by using conventional TGA, and both analyses produced results indicative of a nylon resin containing approximately 13% glass-fiber reinforcement. Further testing was performed using TGA in the high-resolution mode. This analysis was conducted to assess the level of impact-modifying rubber resin. The weight loss associated with the rubber was observed as a shoulder on the high-temperature side of the weight loss representing the nylon resin. This weight loss was particularly evident in the derivative curve. Because the weight losses could not be totally resolved, an absolute level of rubber could not be determined. However, a comparison of the results allowed a determination of the relative level of the impact modifiers in the two materials. This comparison showed a distinctly higher level of impact modifier in the prototype part material, relative to the failed part material.

Conclusions

It was the conclusion of this evaluation that the hinge assemblies failed through brittle fracture associated with stress overload during the actuation of the parts. The failed part material was found to be degraded, as indicated by both the FTIR and DSC analysis results. This degradation likely occurred either during the compounding of the resin or during the actual molding of the parts. A significant factor in the hinge failures is the conversion to a different grade of resin to produce the failed production parts as compared to the prototype parts. While both resins produced results characteristic of a 13% glass-fiber-reinforced, impact-modified nylon 6/6, the failed part material contained a significantly lower level of rubber. This decrease in rubber content rendered the parts less impact resistant and subsequently lowered the ductility of the molded hinge assemblies.

Example 2: Failure of Plasticized Polyvinyl Chloride Tubing

A section of clear polymeric tubing failed while in service. The failed sample had been used in a chemical transport application. The tubing had also been exposed to periods of elevated temperature as part of the operation. The tubing was specified to be a polyvinyl chloride (PVC) resin plasticized with trioctyl trimellitate (TOTM). A reference sample of the tubing, which had performed well in service, was also available for testing.

Tests and Results

The failed and reference tubing samples were analyzed by using micro-FTIR in the attenuated total reflectance mode, and the results representing the reference tubing material were consistent with the stated description: a PVC resin containing a trimellitate-based plasticizer. However, the spectrum representing the failed tubing material was noticeably different. While the obtained spectrum contained absorption bands characteristic of PVC, the results indicated that the material had been plasticized with an adipate-based material, such as dioctyl adipate. This identification is shown in Fig. 8.

To assess their relative thermal stability, the two tubing materials were analyzed via TGA. Both sets of results were consistent with those expected for plasticized PVC resins. The thermograms representing the reference and failed sample materials showed comparable plasticizer contents of 28 and 25%, respectively. The results also showed that the reference material, containing the trimellitate-based plasticizer, exhibited superior thermal resistance relative to the failed material, containing the adipate-based material. This was indicated by the elevated temperature of weight-loss onset exhibited by the reference tubing material.

Conclusions

It was the conclusion of the evaluation that the failed tubing had been produced from a formulation that did not comply with the specified material. The failed tubing was identified as a PVC resin with an adipate-based plasticizer, not TOTM. The obtained TGA results confirmed that the failed tubing material was not as thermally stable as the reference material because of this formulation difference, and that this was responsible for the observed failure.

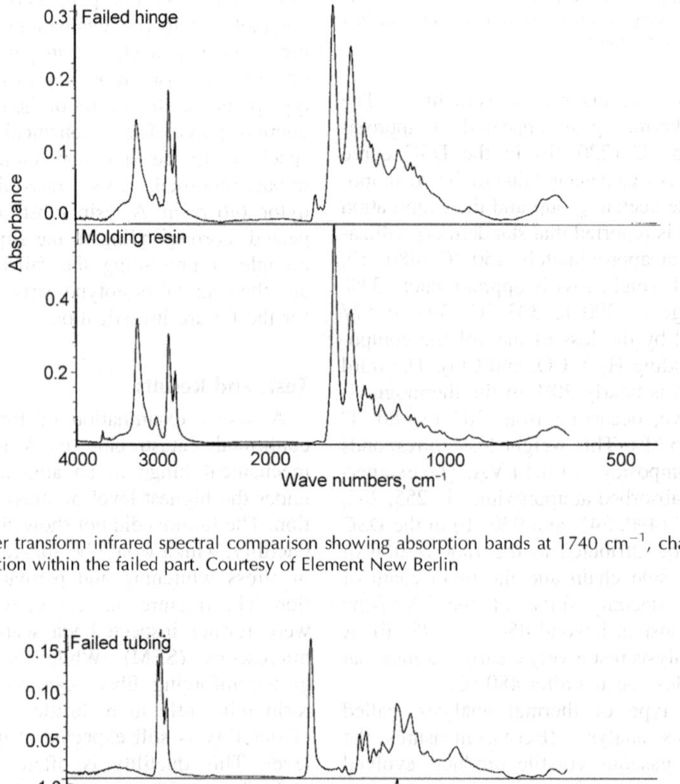

Fig. 7 Fourier transform infrared spectral comparison showing absorption bands at 1740 cm^{-1}, characteristic of oxidation within the failed part. Courtesy of Element New Berlin

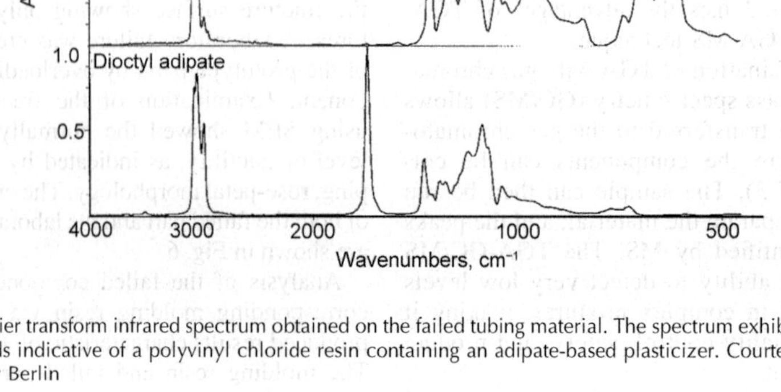

Fig. 8 Fourier transform infrared spectrum obtained on the failed tubing material. The spectrum exhibits absorption bands indicative of a polyvinyl chloride resin containing an adipate-based plasticizer. Courtesy of Element New Berlin

Example 3: Failure of a Nylon Filtration Unit

A component of a water filtration unit failed while being used in service for approximately eight months. The filter system had been installed in a commercial laboratory, where it was stated to have been used exclusively in conjunction with deionized water. The failed part had been injection molded from a 30% glass-fiber- and mineral-reinforced nylon 12 resin.

Tests and Results

A visual examination of the filter component revealed significant cracking on the inner surface. The cracking ran along the longitudinal axis of the part and exhibited an irregular pattern. The surfaces of the part presented a flaky texture, without substantial integrity, and displayed significant discoloration. The irregular crack pattern, flaky texture, and discoloration were apparent on all surfaces of the part that had been in contact with the fluid passing through the component. Several of the crack surfaces were further examined by using SEM. The fracture surface exhibited a coarse morphology, as illustrated in Fig. 9. The reinforcing glass fibers protruded unbounded from the surrounding polymeric matrix. The fracture surface also showed a network of secondary cracking. Overall, the observations made during the visual and SEM inspections were consistent with molecular degradation associated with chemical attack of the filter component material.

To allow further assessment of the failure, a mounted cross section was prepared through one of the cracks. The cross section, as presented in Fig. 10, showed a clear zone of degradation along the surface of the part that had contacted the fluid passing through the filter. The degradation zone extended into the cracks, which indicated massive chemical attack. The prepared cross section was analyzed by using energy-dispersive x-ray spectroscopy, and the results obtained on the base material showed relatively high concentrations of silicon, calcium, and aluminum, with lesser amounts of sulfur and sodium in addition to carbon and oxygen. The results were consistent with a mineral- and glass-filled nylon resin. Analysis of the surface material, which exhibited obvious degradation, showed a generally similar elemental profile. However, significant levels of silver and chlorine were also found. This was important, because aqueous solutions of metallic chlorides are known to cause cracking and degradation within nylon resins.

Analysis of the base material produced results characteristic of a glass- and mineral-filled nylon resin. However, analysis of the surface material showed additional absorption bands characteristic of substantial oxidation and hydrolysis of the nylon. A spectral comparison showing this is presented in Fig. 11. The presence of these bands is consistent with the high level of molecular degradation noted during the visual and SEM examinations.

Comparative TGA of the base material and the surface material also showed a significant difference. In particular, the results obtained on the surface material showed a lower temperature corresponding to the onset of polymer decomposition. This is illustrated in Fig. 12.

Conclusions

It was the conclusion of the evaluation that the filter component failed as a result of molecular degradation caused by the service conditions. Specifically, the part material had

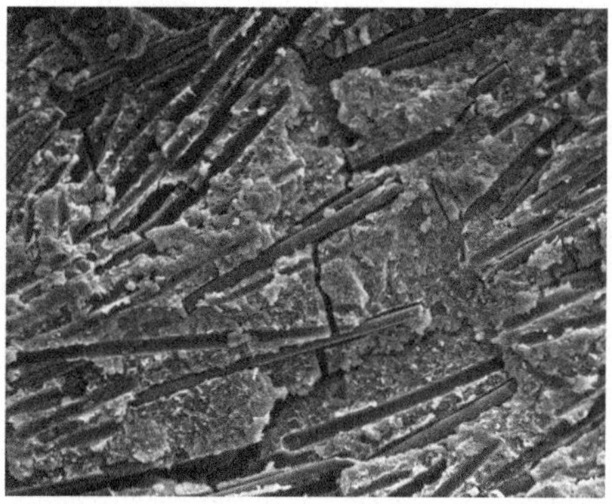

Fig. 9 Scanning electron microscopy image showing features characteristic of severe degradation of the filter material. Original magnification: 118×. Courtesy of Element New Berlin

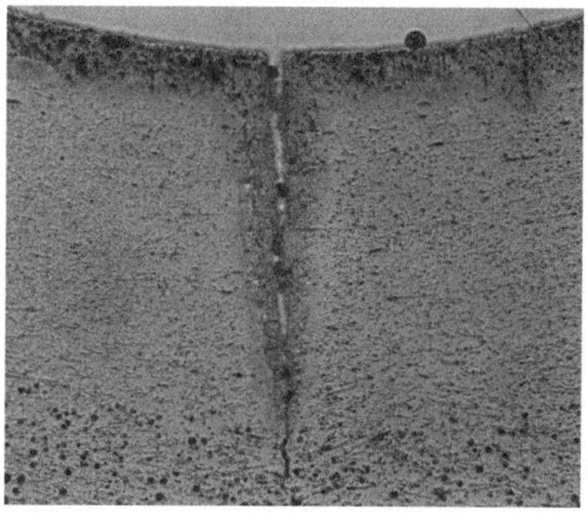

Fig. 10 Micrograph showing the cross section prepared through the filter component. Original magnification: 9×. Courtesy of Element New Berlin

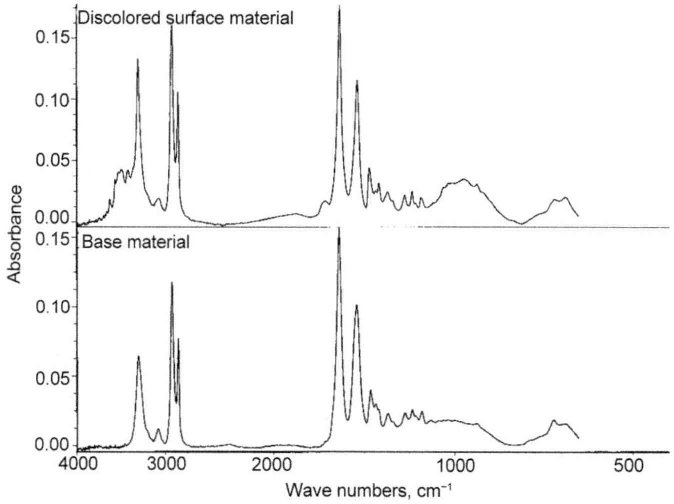

Fig. 11 Fourier transform infrared spectral comparison showing absorption bands associated with hydrolysis at 3350 cm^{-1} and oxidation at 1720 cm^{-1} in the results obtained on the discolored surface. Courtesy of Element New Berlin

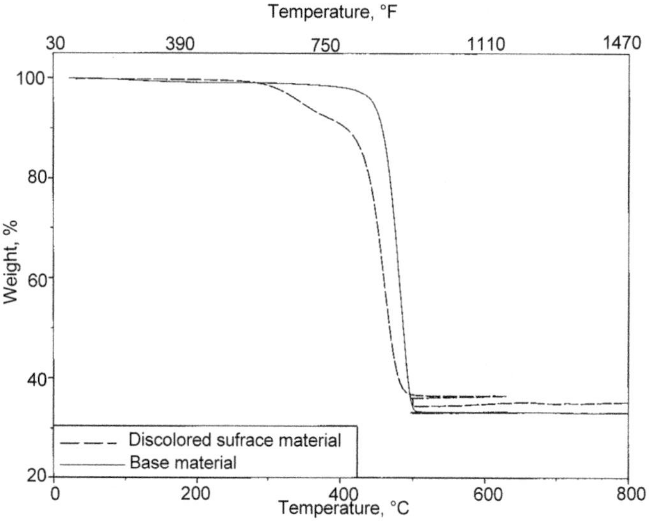

Fig. 12 Thermogravimetric analysis weight-loss profile comparison showing a reduction in thermal stability of the discolored surface material relative to the base material. Courtesy of Element New Berlin

undergone severe chemical attack, including oxidation and hydrolysis, through contact with silver chloride. The source of the silver chloride was not established, but one potential source was photographic silver recovery.

REFERENCES

1. J. Scheirs, *Compositional and Failure Analysis of Polymers*, John Wiley & Sons, 2000, p 109, 138, 153, 393, 415
2. M. Wagner, Thermogravimetric Analysis, Chap. 10, *Thermal Analysis in Practice: Fundamental Aspects*, Hanser Publishers, 2018, p 162–186
3. "Thermogravimetric Analysis: A Beginner's Guide," PerkinElmer, Inc., 2015, https://www.perkinelmer.com/lab-solutions/resources/docs/faq_beginners-guide-to-thermogravimetric-analysis_009380c_01.pdf
4. K. Mohomed, "Thermogravimetric Analysis (TGA) Theory and Applications," TA Instruments, 2016
5. P. Alagarsamy, "Characterization of Materials, Lecture 23: Thermogravimetric Analysis," Online Course, National Programme on Technology Enhanced Learning, 2016
6. M.P. Sepe, *Thermal Analysis of Polymers*, RAPRA Technology, Shawbury, U.K., 1997, p 3, 4, 8, 17, 19, 22, 24, 33
7. H.M. Ng, N.M. Saidi, F.S. Omar, R. Kasi, R.T. Subramaniam, and S.B. Baig, *Thermogravimetric Analysis of Polymers*, 2018, p 1–29, DOI:10.1002/0471440264.pst667
8. E.L. Charsley, in *Thermal Analysis: Techniques and Applications*, E.L. Charsley and S.B. Warrington, Ed., The Royal Society of Chemistry, Cambridge, U.K., 1992
9. T. Krishnasamy, A. Balamurugan, T. Venkatachalam, and E. Ranjith Kumar, Structural, Morphological and Optical Properties of ZnO Nano-Fibers, *Superlattices Microstruct.*, Vol 90, 2015, DOI:10.1016/j.spmi.2015.12.004

ASM Handbook, Volume 10, *Materials Characterization*
ASM Handbook Committee
DOI 10.31399/asm.hb.v10.a0006676

Dynamic Mechanical Analysis

Victoria Burt, ASM International

Overview

General Uses

- Young's modulus and shear modulus
- Damping characteristics and viscoelastic behavior
- Polymer structure and morphology
- Flow and relaxation behavior

Examples of Applications

- measuring the glass transition temperature
- the curing behavior of resins
- the frequency-dependent mechanical behavior of materials

Samples

- *Form:* Bars, films, fibers, coatings, pellets, rods, and cylinders
- *Size:* Depends on which measurement mode is used
- *Preparation:* Test specimens must be plane-parallel and surfaces should be smooth

Limitations

- DMA is most useful when supported by other thermal data
- Relatively long measurement time due to low heating rate
- Sample prep can take a long time
- Requires proper choice of sample geometry and holder

Capabilities of Related Techniques

- *Differential scanning calorimetry:* Indicates glass transition by a change in heat capacity
- *High-temperature X-ray diffraction:* Strain at the molecular level
- *Hot-stage microscopy:* Observation and photography of thermal dilation in real-time

Introduction

Dynamic mechanical analysis (DMA) is a powerful tool for studying the viscoelastic properties and behavior of a range of materials as a function of time, temperature, and frequency. Because polymers are viscoelastic in nature, DMA is well suited for plastic materials, but it is also used to analyze adhesives, paints and lacquers, films and fibers, composites, foodstuffs, pharmaceuticals, fats and oils, ceramics, constructional materials and metals.

DMA provides quantitative and qualitative information on:

- Damping characteristics and viscoelastic behavior
- Polymer structure and morphology
- Primary and secondary relaxation behavior
- Crystallization processes
- Influence of fillers in polymers

In DMA, the sample is subjected to a periodic stress in one of several different modes of deformation (bending, tension, shear, and compression). Modulus as a function of time or temperature is measured and provides information on phase transitions.

DMA technology is well suited for applications where maximum accuracy is required and the material has to be characterized over a wide range of stiffness and/or frequency. In addition, DMA technology is extremely versatile and can characterize materials even in liquids or at specific relative humidity levels. Table 1 (Ref 1) lists effects and properties that can be studied by DMA.

DMA is generally a more sensitive technique for detecting transitions than differential scanning calorimetry (DSC) and differential thermal analysis (DTA) (Ref 2). This is because the properties measured are the dynamic modulus and damping coefficient. Both change significantly when crystalline structures transition to the amorphous phase. The operating principle is that in these transitions, a proportionally larger change takes place in the mechanical properties of a polymer than in its specific heat.

Systems and Equipment

A generic DMA setup (Fig. 1) consists of a displacement sensor such as a linear variable differential transformer, which measures a change in voltage as a result of the instrument probe moving through a magnetic core, a temperature control system or furnace, a drive motor (a linear motor for probe loading which provides load for the applied force), a drive shaft support and guidance system to act as a guide for the force from the motor to the sample, and sample clamps in order to hold the sample being tested.

Table 1 DMA Applications Overview

Viscoelastic behavior
Relaxation behavior
Glass transition
Mechanical modulus
Damping behavior
Softening
Viscous flow
Crystallization and melting
Phase separation
Gelation
Structural change
Composition of blends
Filler activity
Material defects
Curing reactions
Crosslinking reactions
Vulcanization systems

Source: Ref 1

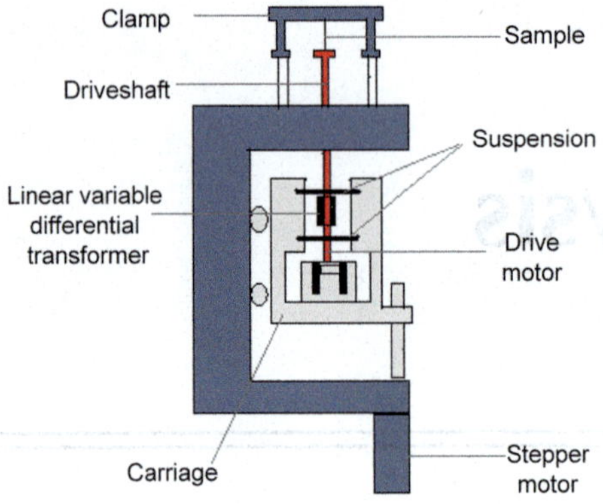

Fig. 1 Generic setup of a dynamic mechanical analyzer

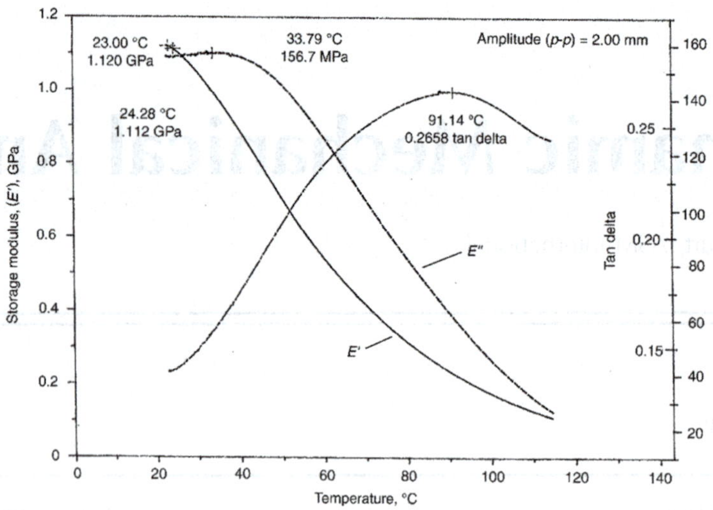

Fig. 2 Dynamic mechanical analysis thermogram showing the results obtained on a typical plastic resin. Tan delta is the ratio of the loss modulus to the storage modulus. Source: Ref 3

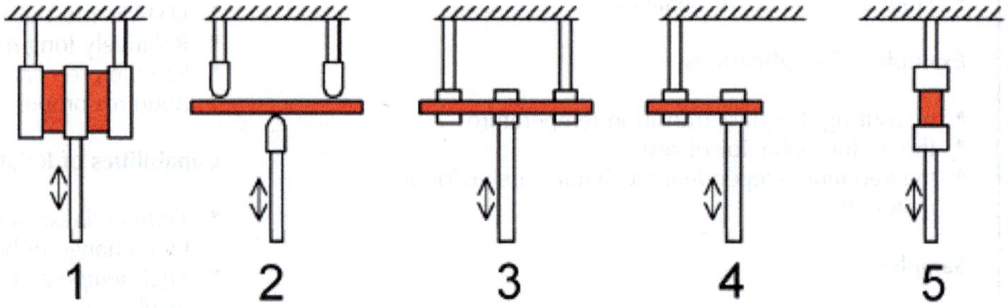

Fig. 3 The most important dynamic mechanical analysis measurement modes. 1: shear; 2: three-point bending; 3: dual cantilever; 4: single cantilever; 5: tension or compression. Adapted from Ref 1

The sample is subjected to a periodic (sinusoidal) mechanical stress, which causes it to undergo deformation with the same period. The dynamic mechanical analysis method determines elastic modulus (or storage modulus, E'), viscous modulus (or loss modulus, E''), and damping coefficient (tan Δ) as a function of temperature, frequency, or time. (Ref 2) A typical DMA thermogram is presented in Fig. 2 (Ref 3).

As mentioned earlier, DMA experiments can be performed using one of several configurations (single and dual cantilever, three-point bending, shear, compression, and tension). The mode of the analysis determines which type of modulus is evaluated (see Fig. 3) (Ref 1). The shear mode determines shear modulus and is ideal for soft samples with a modulus range of 0.1 kPa to 5 GPa. In three-point bending mode, the sample has to be pre-stressed so that it remains in contact with each support during measurement. This mode is good for samples with a modulus of 100 kPa to 1000 GPa such as fiber-reinforced polymers, metals, and ceramic materials. In dual cantilever mode the sample is clamped securely at three positions and not free to expand on warming. The modulus range is 10 kPa to 100 GPa. The single cantilever mode avoids the problem of restricted thermal expansion or contraction of the dual cantilever mode and the modulus range is 10 kPa to 100 GPa. The tension mode is ideal for films, fibers, and samples in the shape of thin rods and the modulus range is 1 kPa to 200 GPa. In compression mode sample pre-stressing is necessary to ensure the sample is always contacting the clamping plates. Modulus range is 0.1 kPa to 1 GPa. Table 2 shows details of the deformation modes used to measure the properties of different types of materials (Ref 4).

The measurement of modulus across a temperature range is referred to as temperature sweep. Dynamic mechanical analysis offers an advantage over traditional tensile or flexural testing in that the obtained modulus is continuous over the temperature range of interest. In addition, special DMAs can also be conducted to evaluate creep through the application of constant stress or stress relaxation by using a constant strain. Dynamic mechanical analysis studies can be performed from −150 to 600 °C (−240 to 1110 °F), usually employing a 2 °C/min (4 °F/min) heating rate.

Specimen Preparation

The shape of the sample will be different according to which measurement mode is used (Ref 1). In shear measurement mode, use quadratic or round test specimens with a thickness of 0.5 to 1 mm. For bending mode, use flat parallel-sided test bars with a thickness of 0.1 to about 3 mm, a width of 2-4 mm, and a length of about 90 mm, or 50 for single cantilever. In tension mode, films are cut with uniform width and a thickness of 0.005 – 0.5 mm. In compression mode, test specimens should be cube or cylinder-shaped and plane parallel.

Additional points that should also be taken into account for DMA samples (Ref 4) include:

- The geometry of the sample must be known, and be accurate. Inaccurate dimensions will lead to a large error in the modulus. Use smooth and parallel surfaces only.
- Stiffness of the sample must be at least 3 to 5 times less than the stiffness of the clamps and the instrument. If this is not the case, the modulus that is measured will be too low.
- The sample must be clamped correctly (sufficiently tightly but not overstressed), otherwise the movement of the sample in the clamp is measured and not the deformation of the sample. In this case, the modulus that is measured will be too small.
- Samples that are clamped above the glass transition temperature and then cooled to below the glass transition should be re-clamped at the start temperature.
- The thermocouple must be carefully and reproducibly positioned – temperature differences of up to 5 K between measurements can occur if the position of the thermocouple is not exactly the same.
- The thermocouple must not touch the sample or the furnace.
- Be aware of the clamping direction for non-isotropic samples.

Test specimens for DMA measurements must be plane-parallel in order to avoid measurement errors and surfaces should be

Table 2 Deformation modes used to measure the properties of different type of materials

Modes	Advantages	Comments	Applications	Modulus measured
3-point bending 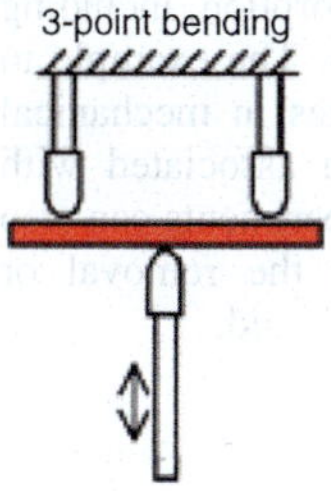	It yields very accurate modulus values for hard samples. A large range of sample dimensions is possible.	It requires on offset force. It puts high demands on sample preparation (parallel flat surfaces).	Samples with small modulus changes (composites, etc.). All glassy materials.	Young's modulus (E' and E'')
Dual cantilever 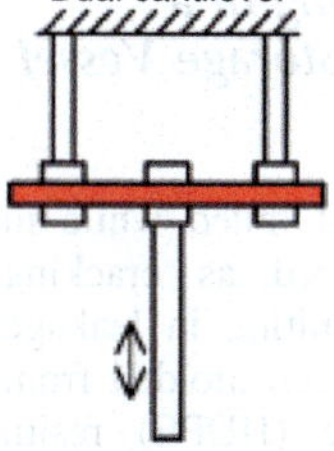	A large range of sample dimensions is possible.	The sample length is not well-defined. Thermal expansion of the sample leads to horizontal stresses in the sample holder. This results in artifacts in the measurements curves.	Samples that soften, for example during a glass transition. Thermoplastics. Composites.	Young's modulus (E' and E'')
Single cantilever 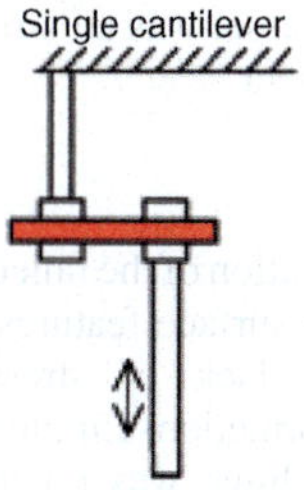				
Shear 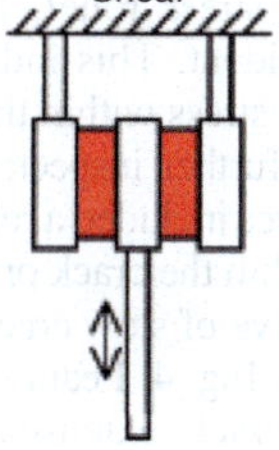	This is the only mode that yields the shear modulus. High frequencies are possible.	In general, it corresponds to the true deformation of materials in practical situations. The sample is held in place only by friction. The sample needs reclamping after cooling.	All polymers. Powders (as pressed tablets). Pastes. Viscous liquids (e.g. bitumen, waxes, oils).	Shear modulus (G' and G'')
Tension 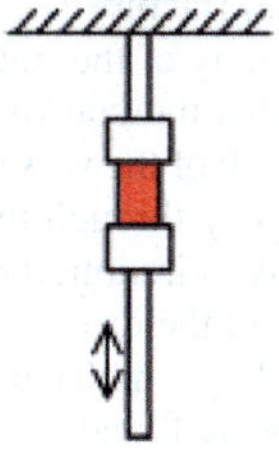	Yields the most accurate modulus values (Young's modulus). Easy to calculate the geometry factor.	Requires an offset force. High modulus values can be determined with suitable sample geometry.	Films and fibers. Thermopiastics. Elastomers are also possible.	Young's modulus (E' and E'')
Compression 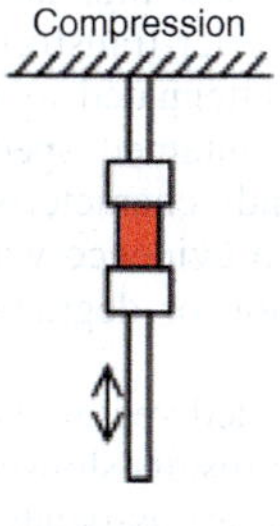	Is the only way to determine the Young's modulus of foams. It is easy to calculate the geometry factor.	Requires an offset force. Unsuitable for stiff samples.	Foams made from polymeric materials. Elastomers are also possible.	Young's modulus (E' and E'')

smooth. (Ref 4). The properties of the sample shouldn't change during preparation of the sample, in particular, plastics mustn't heat up to more than about 40 C during mechanical processing.

Calibration and Accuracy

There are several considerations when calibrating the DMA instrument. The thermocouples of the clamping assembly and the sample temperature measurement require calibration (Ref 1). The displacement sensor is adjusted using accurate gauges, and the force sensor is adjusted using a certified spring. ASTM D4065, ASTM D4440, ASTM D5279, ASTM D7028, and ASTM E1867 are a few of the standards that describe test methods and temperature calibration for DMA instruments.

Data Analysis and Reliability

When interpreting DMA curves, it's common to display the data on a logarithmic presentation. Because moduli can change by several orders of magnitude, a linear presentation doesn't adequately display measurement.

Interpreting the temperature dependence of DMA Curves

Interpreting DMA curves is discussed in detail in Ref 5 and summarized here. The storage (elastic) modulus of commonly used materials decreases with increasing temperature. The storage modulus of metals such as steel or aluminum alloys hardly changes up to temperatures of 400 C. Stepwise changes are caused by relaxation transitions (e.g. glass transition) or phase transitions (e.g. melting and crystallization). Peaks in the loss (viscous) modulus and the loss factor, correspond to the steps in the storage modulus.

Amorphous materials go through a glass transition on heating or cooling. The modulus changes by one to four decades. The same occurs when the crystallites of semicrystalline polymers melt. Such phase transitions do not exhibit the large frequency dependence of relaxation transitions. Commonly used thermoplastics such as polyvinylchloride and polystyrene have a modulus of elasticity of about 3 GPa. Their glass transition temperatures lie between room temperature and approx. 200 C. At about 100 K above the glass transition, the polymers flow and can be plastically deformed.

Elastomers such as natural rubber exhibit a glass transition below room temperature and, because of chemical cross-linking, do not flow. This low degree of cross-linking occurs during vulcanization of the originally thermoplastic rubber.

Thermosets such as epoxy resins are three-dimensionally cross-linked macro-molecules. Their glass transition region is significantly above room temperature. Due to their three-dimensional cross-linking, they do not flow when the temperature is increased. The starting materials of thermosets consist of several different components, which are usually referred to as the "resin" and the "hardener" or "curing agent." When the thermoplastic starting materials harden or cure, a three-dimensional network is produced and the glass transition temperature increases by 50 K to300 K.

Interpreting the frequency dependence of DMA curves

In comparison to temperature-dependent measurements, frequency-dependent measurements provide additional information about material properties and on molecular processes in particular. (Ref 1). Knowledge of the frequency dependence of mechanical behavior is important for practical application and optimization of materials. The frequencies at which the different processes occur correlate with characteristic volumes of the corresponding molecular regions. At higher frequencies, smaller molecular regions are observed. Ref 1 presents detailed descriptions and theoretical explanations for interpreting these frequency dependent curves.

Uses of DMA in Failure Analysis (Ref 3)

The temperature-dependent behavior of polymeric materials is one of the most important applications of DMA. In a standard temperature-sweep evaluation, the results show the storage modulus, loss modulus, and the tan delta as a function of temperature. The storage modulus indicates the ability of the material to accommodate stress over a temperature range. The loss modulus and tan delta provide data on temperatures where molecular changes produce property changes, such as the glass transition and other secondary transitions not detectable by other thermal analysis techniques. The superiority of DMA over DSC and TMA for assessing the glass transition is well documented (Ref 6). Secondary transitions of lesser magnitude are also important, because they can relate to material properties such as impact resistance. The ability of a plastic molded component to retain its properties over the service temperature range is essential and is well predicted by DMA.

Changes in the mechanical properties of plastic resins that arise from molecular degradation or aging can be evaluated via DMA. Such changes can significantly alter the ability of the plastic material to withstand service stresses. While the cause and type of degradation cannot be determined, DMA can assess the magnitude of the changes. This can provide insight into potential failure causes.

Sepe (Ref 6) notes that "DMA is sensitive to structural changes that can arise when a solid polymer absorbs a liquid material." This effect can arise from the absorption of water or organic-based solvents. Dynamic mechanical analysis experiments can assess changes in the physical properties of a plastic material that can result from such absorption, including loss of strength and stiffness. The example in this article shows the changes in mechanical properties of a plastic resin associated with chemical absorption. The experiments can also evaluate the recovery after the removal or evaporation of the absorbed liquid.

Applications

Failure Analysis—Cracking of a Polyethylene Chemical Storage Vessel (Ref 3)

A chemical storage vessel failed while in service. The failure occurred as cracking through the vessel wall, resulting in leakage of the fluid. The tank had been molded from a high-density polyethylene (HDPE) resin. The material held within the vessel was an aromatic hydrocarbon-based solvent.

Tests and Results

A stereomicroscopic examination of the failed vessel revealed brittle fracture surface features. This was indicated by the lack of stress whitening and permanent deformation. Limited ductility, in the form of stretching, was found exclusively within the final fracture zones. On cutting the vessel, significant stress relief, in the form of distortion, was evident. This indicated a high level of molded-in stress within the part. The fracture surface was further inspected using SEM. The observed features included a relatively smooth morphology within the crack origin location, which was indicative of slow crack initiation. This area is shown in Fig. 4. Features associated with more rapid crack extension, including hackle marks and river markings, were found at the midfracture and final fracture areas, as represented in Fig. 5. The entirety of the fracture surface features indicated that the cracking had initiated along the exterior wall of the vessel. The cracking extended transversely through the wall initially, and subsequently, circumferentially around the wall. Throughout the examination, no signs of postmolding molecular degradation or chemical attack were found.

The failed vessel material was first analyzed using micro-FTIR (Fourier-transform infrared spectroscopy) in the attenuated total reflection (ATR) mode. The obtained spectrum exhibited absorption bands characteristic of a polyethylene resin. No evidence was found to indicate contamination or degradation of the material.

Material excised from the failed vessel was then analyzed using DSC. The results showed a single endothermic transition associated with the melting point of the material at 133 °C (271 °F). The results were consistent with those expected for a high density polyethylene (HDPE) resin. The results also showed that the HDPE resin had a relatively

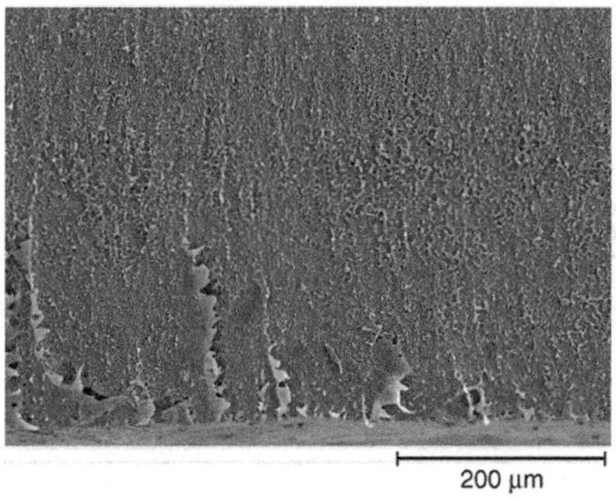

Fig. 4 Scanning electron microscopy image showing features associated with brittle fracture and slow crack growth within the crack initiation site. Original magnification: 100×

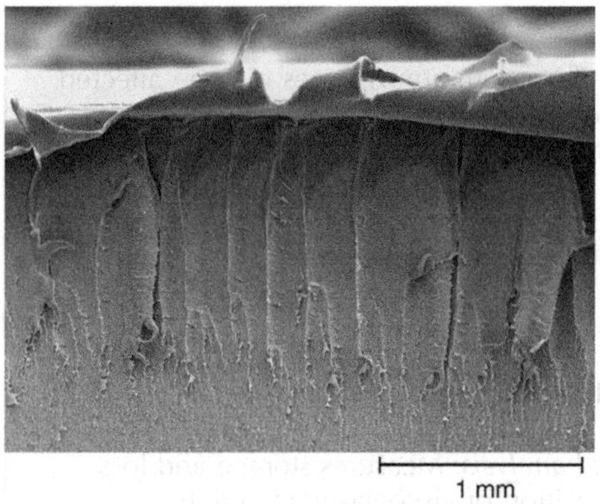

Fig. 5 Scanning electron microscopy image showing features indicative of rapid crack extension within the final fracture zone. Original magnification: 20×

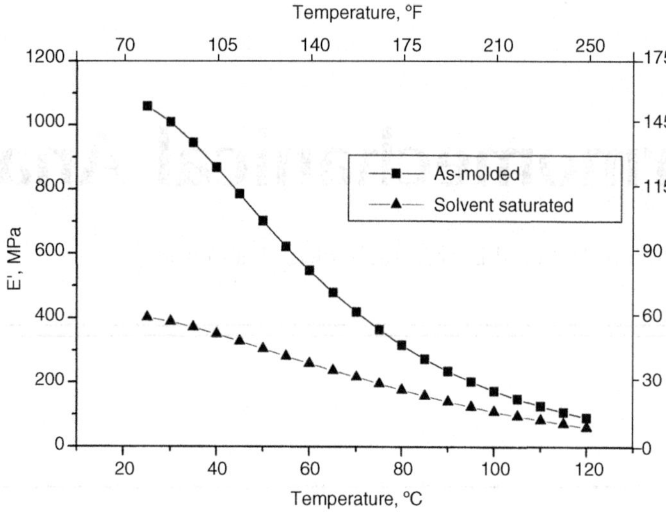

Fig. 6 Comparison of dynamic mechanical analysis results, showing a loss of more than 60% in elastic modulus (E′) as a result of solvent effects

high level of crystallinity, as indicated by the elevated heat of fusion.

Thermogravimetric analysis (TGA) was performed to further evaluate the failed vessel material. The TGA testing showed that the HDPE absorbed approximately 6.3% of its weight in the aromatic hydrocarbon-based solvent. Overall, the TGA results were consistent with those expected for a HDPE resin.

The melt flow rate of the vessel material was evaluated, and the testing produced an average result of 3.8 g/10 min. This is excellent agreement with the nominal value indicated on the material data sheet, 4.0 g/10 min. As such, it was apparent that the vessel material had not undergone molecular degradation. The specific gravity of the resin was measured. The material produced a result of 0.965. This indicated that the material had a relatively high level of crystallinity, as suggested by the DSC results.

In order to assess the effects of the hydrocarbon-based solvent on the HDPE vessel, the material was evaluated using dynamic mechanical analysis. The vessel material was analyzed in two conditions. Material samples representing the vessel material in the as-molded condition as well as material from the failed vessel were evaluated. A comparison of the DMA results showed that in the saturated, equilibrium state, the HDPE resin lost over 60% of its elastic modulus at room temperature, because of the plasticizing effects of the solvent. A comparison of the DMA results, indicating the reduction in mechanical properties, is shown in Fig. 6.

Conclusions

It was the conclusion of the investigation that the chemical storage vessel failed via a creep mechanism associated with the exertion of relatively low stresses. Given the lack of apparent ductility, the stresses responsible for the failure appear to have been below the yield strength of the material. The source of the stress was thought to be molded-in residual stresses associated with uneven shrinkage. This was suggested by the obvious distortion evident on cutting the vessel. The relatively high specific gravity and the elevated heat of fusion are indicative that the material has a high level of crystallinity. In general, increased levels of crystallinity result in higher levels of molded-in stress and the corresponding warpage. The significant reduction in the modulus of the HDPE material, which accompanied the saturation of the resin with the aromatic hydrocarbon-based solvent, substantially decreased the creep resistance of the material and accelerated the failure. The dramatic effects of the solvent had not been anticipated prior to use.

REFERENCES

1. Wagner, Matthias. (2018). *Thermal Analysis in Practice - Fundamental Aspects – Ch 12 Dynamic Mechanical Analysis.* Hanser Publishers
2. Surface Treatment of Materials for Adhesive Bonding (Second Edition) 2014, Pages 39–75
3. J.A. Jansen, "Characterization of Plastics in Failure Analysis," ASM Handbook Volume 11: Failure Analysis and Prevention, ASM International, 2002 p 437–459
4. Thermal Analysis in Practice: Tips and Hints, Volume 2, Mettler Toledo
5. Georg Widmann, Dr. Jürgen Schawe, Dr. Rudolf Riesen, Interpreting DMA curves, Part 1, UserCom 15, Mettler Toledo, 2002
6. M.P. Sepe, Thermal Analysis of Polymers, RAPRA Technology, Shawbury, U.K., 1997, p 3, 4, 8, 17, 19, 22, 24, 33

SUGGESTED REFERENCES

- M. Akay, "Aspects of Dynamic Mechanical Analysis in Polymeric Composites," Composites Science and Technology, Elsevier, 1993
- Esmaeeli, Roja & Jbr, Chiran & Aliniagerdroudbari, Haniph & Hashemi, Seyed Reza & Farhad, Siamak. (2019). Designing a New Dynamic Mechanical Analysis (DMA) System for Testing Viscoelastic Materials at High Frequencies. Modelling and Simulation in Engineering. 2019. DOI: 10.1155/2019/7026267
- K.P. Menard, Dynamic Mechanical Analysis: A Practical Introduction, 2nd edition, CRC Press, 2008
- J.D. Menczel, R.B. Prime, Thermal Analysis of Polymers: Fundamentals and Applications, Wiley, 2009
- Rieger, Jens. "The glass transition temperature Tg of polymers—comparison of the values from differential thermal analysis (DTA, DSC) and dynamic mechanical measurements (torsion pendulum)." *Polymer testing* 20, no. 2 (2001): 199-204.
- Rotter, G. and Ishida, H., 1992. Dynamic mechanical analysis of the glass transition: curve resolving applied to polymers. *Macromolecules*, *25*(8), pp.2170–2176.
- Shepard, D.D. and Twombly, B., 1996. Simultaneous dynamic mechanical analysis and dielectric analysis of polymers. *Thermochimica Acta*, *272*, pp.125–129.

ASM Handbook, Volume 10, *Materials Characterization*
ASM Handbook Committee
DOI 10.31399/asm.hb.v10.a0006674

Thermomechanical Analysis

Richard E. Chinn, National Energy Technology Laboratory

Overview

General Uses

- Thermal deformation of solids and liquids
- Phase transformations
- Sintering studies
- Characterization of viscoelastic materials
- Creep and stress relaxation
- Data for quench-cooling diagrams of hardenable alloys

Examples of Applications

- Coefficient of thermal expansion and contraction
- Sintering shrinkage and rate of densification
- Glass transitions of viscoelastic materials
- Softening and viscosity of viscoelastic materials
- Pressure-volume-temperature data of viscoelastic materials
- Storage and loss moduli of viscoelastic materials

Samples

- *Form:* Solids and liquids. Solids are usually in the shape of a right cylinder. Fibers and films in some instruments
- *Size:* Up to 25 mm (1.0 in.) long by 12 mm (0.5 in.) in diameter, in most instruments
- *Preparation:* Smooth parallel ends on cylinders

Limitations

- Larger specimens and vacuum atmospheres may be affected by low thermal conductivity.
- Slow measurement rate, especially for refractory ceramics and sintering studies

Estimated Analysis Time

- Typically 2 to 16 h

Capabilities of Related Techniques

- *Dynamic mechanical analysis:* Measures storage and loss moduli and glass transition of viscoelastic materials
- *Differential scanning calorimetry:* Indicates glass transition by a change in heat capacity
- *High-temperature stress-strain (e.g., Instron, Tinius Olsen, MTS):* Larger specimens than thermomechanical analysis; measures creep and stress relaxation
- *High-temperature x-ray diffraction:* Strain at the molecular level
- *Hot-stage microscopy:* Observation and photography of thermal dilation in real-time

Introduction

Thermomechanical analysis (TMA) is a thermal analysis technique in which the length of a specimen is precisely measured versus temperature and time as the specimen is subjected to controlled heating and cooling. The most common form of TMA is dilatometry, where a cylindrical specimen is under light compression as its coefficient of linear thermal expansion (CTE) or contraction is measured. The CTE is also known as expansivity. In addition to dilatometry, TMA includes tensile, flexural, and penetration modes of stress. Thermal expansion at the molecular level is measured by high-temperature x-ray or neutron diffractometry.

Historically, quantitative thermal expansion enabled the invention of the mercury thermometer by Fahrenheit in 1714 and the alcohol thermometer by Réaumur in 1730. The bimetallic strip was invented by Harrison circa 1759. S.P. Rockwell, of Rockwell hardness fame, invented one of the earliest dilatometers and was awarded a patent for it in 1929. P. Chevenard in France in 1930, S. Sato in Japan in 1933, H. Riepert in Germany in 1939, H.W. Dietert in the United States in 1942, and several other inventors in subsequent years received patents for their dilatometer designs. Levy and Tabeling of the DuPont Company introduced the TMA in 1969. Baldwin and Ruff of the Hewlett-Packard Company invented the optical dilatometer in 1974. Theta Industries Inc. invented the quenching dilatometer (Clusener and Kurth) in 1974, the differential dilatometer (Clusener) in 1975, and cryogenic dilatometry (Baricevac and Raffalski) in 1988. Simmonds invented the capacitance dilatometer in 2015.

An early distinction between TMA and dilatometry was that TMA was the measurement of deformation of a material under a nonoscillatory load, while dilatometry was the measurement of the dimensions of a material, as a function of temperature at a controlled heating or cooling rate. The distinction has faded with time and technology, and TMA and dilatometry are now more or less synonymous.

The trends in development of TMA, like so many other scientific instruments, include automation and programmability, removal of artifacts by software, noncontact measurement replacing the contact methods, electromagnetically applied forces and stresses replacing deadweights, and nanometer sensitivity replacing micrometer-level sensitivity.

General Principles

The thermomechanical analyzer (also TMA) in Fig. 1 consists of a specimen chamber inside a tube furnace. The right end of the chamber is a dead stop, while the left end has a pushrod that moves as the specimen dilates or shrinks. The hot end of the pushrod is pressed against the specimen, and the cold end has a ferromagnetic core inside a linear variable differential transducer (LVDT). Movement of the ferromagnetic core inside the LVDT conductive coil generates an electric signal by Faraday's law of induction that is calibrated to microscopic displacement. A thermocouple parallel to the pushrod reports the temperature of the specimen. Water cooling is often used to maintain the LVDT at a reference temperature, nominally 25 °C (75 °F), or 298 K. The system is designed so that expansion of the pushrod is nullified by expansion of the chamber.

The CTE, symbolized with α, is defined in Eq 1, where the subscript F signifies constant force. The constant force may be a spring in a horizontal TMA or a deadweight in a vertical TMA maintaining contact between the pushrod and the specimen, typically on the order of 30 cN. A TMA thermogram is normally plotted as change in length divided by the original length of the specimen, dL/L_0, as a function of temperature, T, increasing or decreasing at a constant rate, dT/dt. The CTE is the slope of this plot. Phase changes usually cause a sharp change in slope, but the CTE is often approximated as linear, hence the subscript L in Eq 1, within a single-phase temperature interval. The CTE can be thought of as a measure of thermal strain:

$$\alpha_L = \frac{1}{L_0}\left(\frac{\partial L}{\partial T}\right)_F \quad \text{(Eq 1)}$$

The CTEs of several common alloys and other engineered materials are given in Table 1. Additional CTE data can be found in the latest edition of the *CRC Handbook of Chemistry and Physics* as well as many industry trade associations, materials science textbooks, and literature from manufacturers of alloys, ceramics, and plastics. The values for copper, fused silica, sapphire, AISI 446 stainless steel, and tungsten in Table 1 demonstrate the deviation of the single-phase CTE from linearity over a broad temperature range.

Thermal expansion and contraction are actually volume phenomena, not one-dimensional. The volumetric thermal expansion is defined in Eq 2 at constant pressure, P. The instantaneous volume, V, and initial volume, V_0, are substituted for L and L_0 in Eq 1. The cross section expands or contracts with temperature change just as the length does. If the specimen is isotropic, as cubic single crystals and annealed polycrystalline metals often are, and the thermal strain is very small relative to the dimensions of the specimen, the volumetric thermal expansion can be approximated as the sum of the three linear coefficients, or thrice α_L, in Eq 3:

$$\alpha_V = \frac{1}{V_0}\left(\frac{\partial V}{\partial T}\right)_P \quad \text{(Eq 2)}$$

$$\alpha_V \approx \alpha_{Lx} + \alpha_{Ly} + \alpha_{Lz} \approx 3\alpha_L \quad \text{(Eq 3)}$$

The amplitude of vibration of atoms increases with increasing temperature, causing the bonds between neighboring atoms and the interatomic distance to increase. The increase in atomic spacing with temperature is, generally, inversely related to the bond strength. That is, the strongest solids also tend to have the lowest α values. Bond angles also change with temperature and significantly affect the CTE.

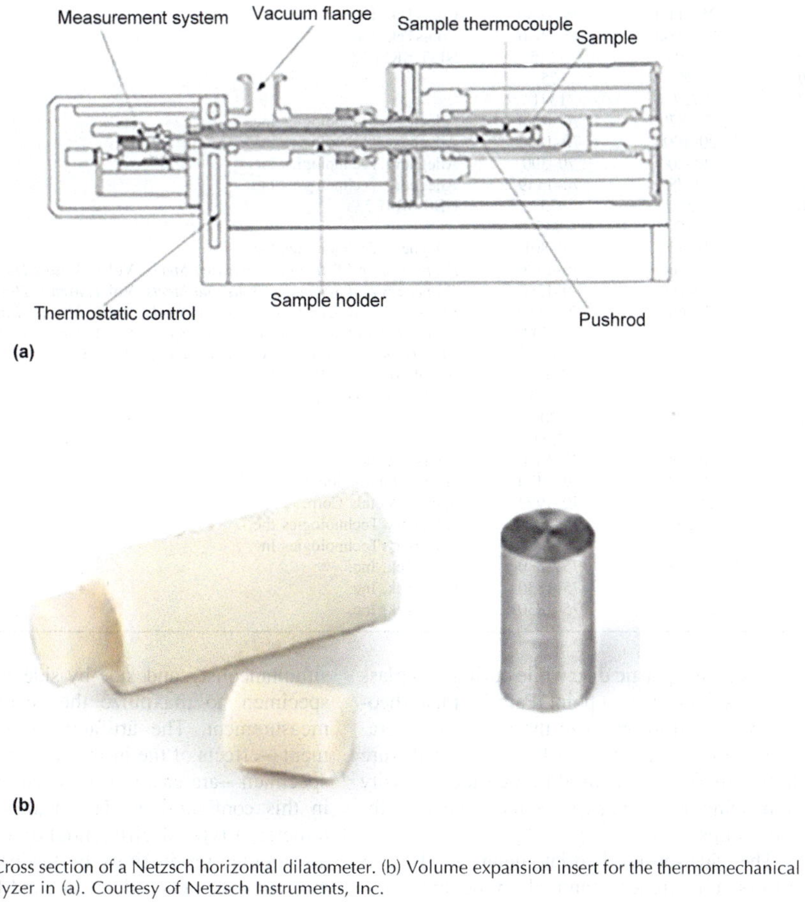

Fig. 1 (a) Cross section of a Netzsch horizontal dilatometer. (b) Volume expansion insert for the thermomechanical analyzer in (a). Courtesy of Netzsch Instruments, Inc.

Systems and Equipment

Commercial thermomechanical analyzers come in a variety of configurations, including horizontal, vertical, dual, quenching, and high pressure. Older instruments record an analog signal that may require a specific specimen length for utmost accuracy. The load on the specimen in older instruments usually comes from a deadweight. Newer instruments are entirely digital and programmable and use software to correct the imperfections of the measurement signals. The load in newer instruments is applied electromagnetically and can be programmed so as to maintain a constant stress as the cross-sectional area of the specimen changes with temperature, a specific loading rate, or an oscillatory load.

The horizontal TMA in Fig. 1(a) keeps the transducer isolated from the rising heat of the furnace and is well suited purely as a high-temperature dilatometer for metals and ceramics. The specimen requires flat, smooth, parallel faces that are normal to the expansion axis. The volume expansion accessory in Fig. 1(b) is used to measure the dilation of a powder, liquid, or an odd-shaped specimen embedded in alumina powder.

The vertical TMA in Fig. 2(a) can measure CTE similar to a horizontal TMA but has the advantage of gravity parallel to the expansion axis to do more than mere dilatometry. The standard pushrod can be replaced with the more specialized fixtures in Fig. 2(b to e) to measure in tension, softening, parallel-plate viscosity, and flexure. The vertical TMA is well suited for plastics and very small specimens. Specimens with very small cross sections, such as fibers and films, cannot support compression and therefore cannot easily be measured in a dilatometer. The CTE of fibers and films can be

Table 1 Coefficients of thermal expansion

Material	α_L, 10^{-6} K^{-1}	Temperature range		Source
		°C	°F	
Al_2O_3, 96%	8.2	25–1000	75–1830	CoorsTek, Inc.
Al_2O_3, 99.9%	8.0	25–1000	75–1830	CoorsTek, Inc.
Aluminum	25	25	75	*CRC Handbook of Chemistry and Physics,* 59th ed.
Aluminum, 3004	23.9	20–100	70–210	ASM Specialty Handbook: *Aluminum and Aluminum Alloys*
Aluminum, 356.0	21.4	20–100	70–210	ASM Specialty Handbook: *Aluminum and Aluminum Alloys*
Aluminum, 6061	23.6	20–100	70–210	ASM Specialty Handbook: *Aluminum and Aluminum Alloys*
Aluminum, 7075	23.6	20–100	70–210	ASM Specialty Handbook: *Aluminum and Aluminum Alloys*
Brass, cartridge, 70Cu-Zn	19.08	20–300	70–570	Olin Brass
Brass, red, 85Cu-Zn	17.72	20–300	70–570	Olin Brass
Brass, yellow, 66Cu-Zn	20.3	20–300	70–570	Olin Brass
Cast iron, ductile, ferritic	15.3	20–870	70–1600	*Properties and Selection: Irons and Steels,* Vol 1, *Metals Handbook,* 9th ed.
Cast iron, ductile, pearlitic	15.3	20–870	70–1600	*Properties and Selection: Irons and Steels,* Vol 1, *Metals Handbook,* 9th ed.
Cast iron, gray	13	0–500	30–930	*Properties and Selection: Irons and Steels,* Vol 1, *Metals Handbook,* 9th ed.
Cast iron, malleable	11.88	21–400	70–750	*Properties and Selection: Irons and Steels,* Vol 1, *Metals Handbook,* 9th ed.
Copper	0.23	−253	−423	NBS SRM 736
	16.59	25	75	
	20.09	527	981	
Hastelloy C-276, Ni-Cr-Mo	13.8	24–600	75–1110	Haynes International
Haynes 263, Ni-Co-Cr-Mo	18.1	25–1000	75–1830	Haynes International
Ice	51	0	32	engineeringtoolbox.com
Incoloy 800, Fe-Ni-Cr	18.0	20–800	70–1470	Special Metals Corp.
Inconel 625, Ni-Cr-Mo-Fe	16.2	21–927	70–1701	Special Metals Corp.
Invar 36, Fe-Ni	1.6	25–93	75–199	Carpenter Technology Corporation
Invar 32-5, Fe-Ni-Co	0.72	25–93	75–199	Carpenter Technology Corporation
Iron	12	25	75	*CRC Handbook of Chemistry and Physics,* 59th ed.
Kovar, Fe-Ni-Co	5.86	25–100	75–210	Carpenter Technology Corporation
Mercury	61	...	...	engineeringtoolbox.com
Monel K-500, Ni-Cu	17.3	21–800	70–1470	Special Metals Corp.
Nickel	13	25	75	*CRC Handbook of Chemistry and Physics,* 59th ed.
Nimonic 75, Ni-Cr	18.2	20–1000	70–1830	Special Metals Corp.
Nylon 11	100	...	...	*Engineering Plastics,* Vol 2, *Engineered Materials Handbook*
Platinum	9	25	75	*CRC Handbook of Chemistry and Physics,* 59th ed.
Polycarbonate	68	...	...	*Engineering Plastics,* Vol 2, *Engineered Materials Handbook*
Polyethylene terephthalate	65	...	...	*Engineering Plastics,* Vol 2, *Engineered Materials Handbook*
Sapphire (pure Al_2O_3), *c*-axis	5.38	20	70	NBS SRM 732
	11.38	1727	3141	
Si_3N_4	3	25–1000	75–1830	CoorsTek, Inc.
SiC, α	4.4	25–1000	75–1830	CoorsTek, Inc.
SiO_2, fused	−0.70	−193	−315	NIST SRM 739
	0.49	25	75	
	0.37	727	1341	
Stainless steel, 304	19.8	20–870	70–1600	Alleghery Technologies Inc.
Stainless steel, 316	19.5	20–1000	70–1830	Alleghery Technologies Inc.
Stainless steel, 420	7.0	20–200	70–390	Alleghery Technologies Inc.
Stainless steel, 430	11.9	20–787	70–1449	Alleghery Technologies Inc.
Stainless steel, 446	9.76	25	75	NBS SRM 738
	12.66	507	945	
Stainless steel, 17-4PH	11.3	21–427	70–801	Alleghery Technologies Inc.
Steel, 1020	14.8	20–700	70–1290	*Properties and Selection: Irons and Steels,* Vol 1, *Metals Handbook,* 9th ed.
Steel, 1080	14.7	20–700	70–1290	*Properties and Selection: Irons and Steels,* Vol 1, *Metals Handbook,* 9th ed.
Steel, 4140	14.5	20–600	70–1110	*Properties and Selection: Irons and Steels,* Vol 1, *Metals Handbook,* 9th ed.
Steel, maraging, 18Ni(250)	10.1	24–284	75–543	*Properties and Selection: Irons and Steels,* Vol 1, *Metals Handbook,* 9th ed.
Titanium	8.5	25	75	*CRC Handbook of Chemistry and Physics,* 59th ed.
Titanium-6% Al-4%V ATI Grade 5 Ti)	9.5	0–316	32–600	Alleghery Technologies Inc.
Tungsten	2.30	−193	−315	NBS SRM 737
	4.42	20	70	
	6.07	1527	2781	
Tungsten carbide	5.1	25–1000	75–1830	CoorsTek, Inc.
Ultimet, Co-Cr	16.1	26–800	79–1470	Haynes International
Waspaloy, Ni-Co-Cr	18.7	21–1093	70–1999	Special Metals Corp.
Zr-1.5%Sn (Zircaloy-2)	6	25	75	Alleghery Technologies Inc.
Zr-2.5%Nb (Zircadyne-705)	5.9	20–371	70–700	Alleghery Technologies Inc.
ZrO_2 (transformation-toughened zirconia-MgO)	10.1	25–1000	75–1830	CoorsTek, Inc.
ZrO_2 (tetragonal zirconia polycrystal-Y_2O_3)	10.3	25–1000	75–1830	CoorsTek, Inc.
ZrO_2 (zirconia densified with yttria), cubic	10.5	25–1000	75–1830	CoorsTek, Inc.

measured in the tension fixture in Fig. 2(b). The softening point of glass can be determined with a pushrod that has a sharp point (Fig. 2c) rather than a flat face, and the force is in the direction of the sharp point. The fixtures in Fig. 2(b and c) are applicable to horizontal dilatometers, too. The parallel-plate viscometer in Fig. 2(d) measures the time derivative of displacement dL/dt isothermally to calculate the viscosity of a plastic disc while curing or a glass disc near its softening point. Parallel-plate rheometry is used to optimize the processing parameters of thermoset cross-linking. The flexure fixture in Fig. 2(e) is used to measure viscosity in bending or tensile expansion of a beam without clamps.

The dual-pushrod dilatometer in Fig. 3 allows a reference standard to be measured simultaneously and side-by-side with the test specimen, to maximize the accuracy of the measurement. The artifacts of the measurement—effects of the instrument rather than the specimen—are easily recognized and removed in this configuration. The dual-pushrod dilatometer, a type of differential dilatometer, can also double productivity by testing two specimens at once under identical conditions.

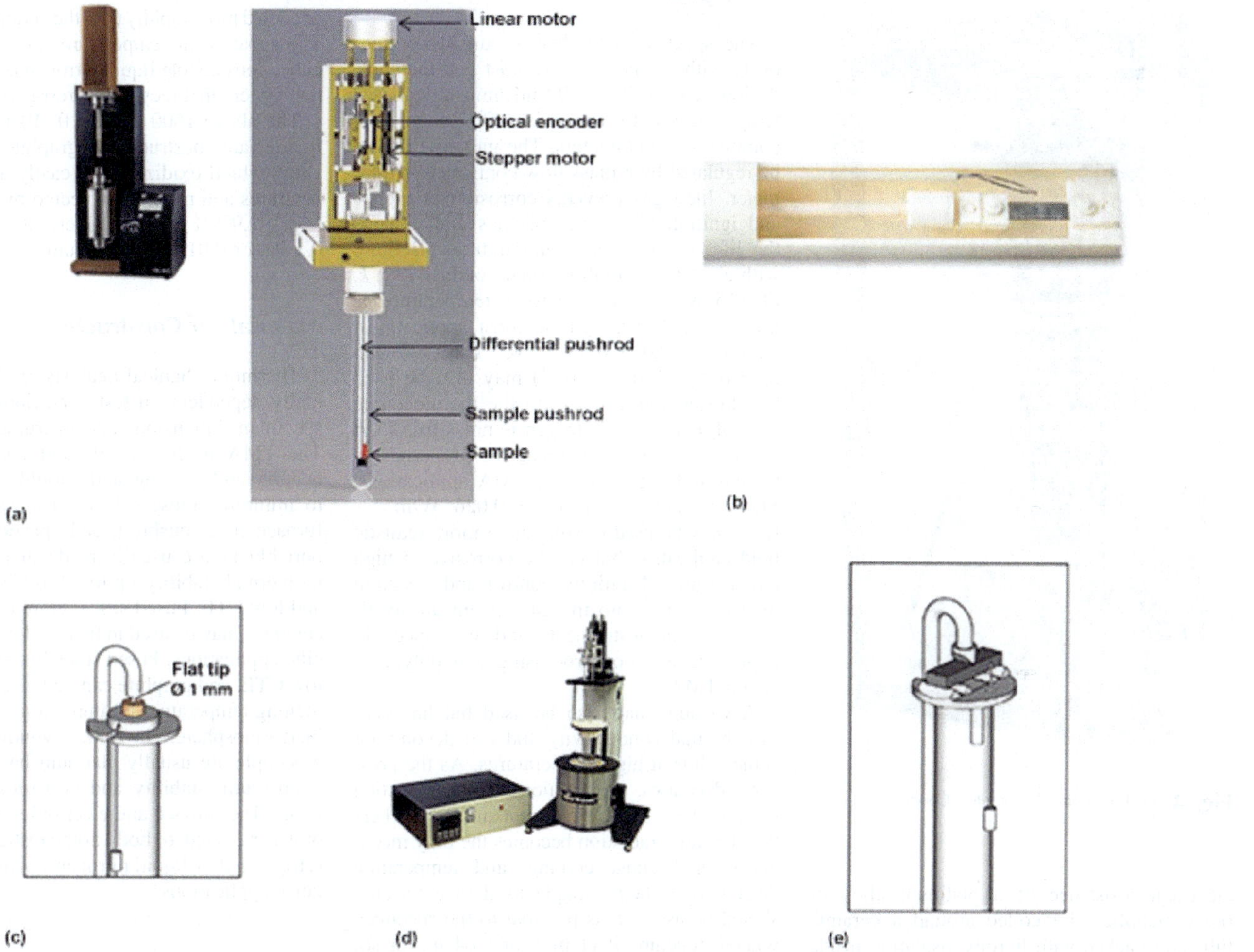

Fig. 2 (a) Vertical thermomechanical analyzer (TMA). Courtesy of TA Instruments Inc. (b) Netzsch TMA tension accessory. (c) Netzsch pushrod with sharp point for softening-temperature determination. (d) Dilatometer customized as a parallel-plate rheometer. Courtesy of Edward Orton, Jr. Ceramic Foundation. (e) Netzsch three-point-bending fixture for TMA

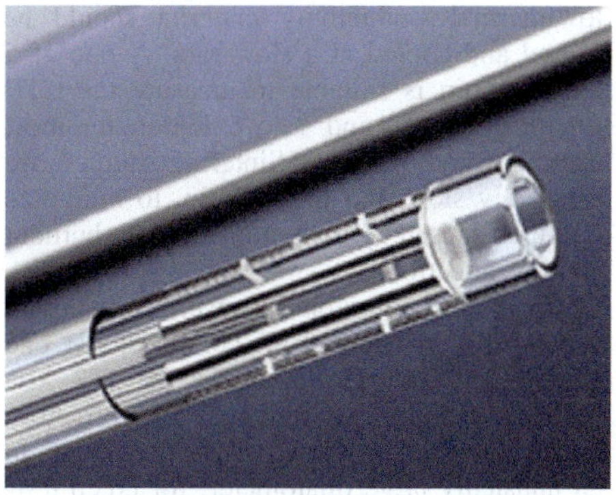

Fig. 3 Dual-pushrod dilatometer with the thermocouple between the two specimens. Courtesy of TA Instruments Inc.

An air- and water-cooled, resistance-heated TMA, such as that in Fig. 1, can heat and cool at a rate up to approximately 100 K/min. The water cools the instrument only, and the specimen is gas quenched. While suitable for most applications, these relatively slow rates are inadequate for certain studies, such as the martensitic transformation in tool steels. The quenching dilatometer can heat up to a rate of approximately 4000 K/s by induction and cool up to a rate of 2500 K/s by liquid quenching the specimen. The quenching dilatometer can provide data for continuous-cooling-transformation and isothermal time-temperature-transformation diagrams for hardenable steel heat treating and precipitation hardening of many other alloys.

The high-pressure dilatometer in Fig. 4 can be used to measure the specific volume, that is, reciprocal of density, of plastics as a function of temperature and pressure. High-pressure dilatometer data, commonly known as pressure-volume-temperature (PVT) data, are equation-of-state thermodynamic properties that characterize the volume expansion and compressibility of viscoelastic materials. The PVT data are used to model injection-molding processes. Another use for the high-pressure dilatometer is the construction of unary and nonambient phase diagrams.

Temperature Measurement

The temperature in a TMA is usually measured by a thermocouple, with the tip of the thermocouple within approximately 1 mm (0.04 in.) of the specimen. The instruments that can operate above 1600 °C (2910 °F) may be equipped with a pyrometer rather than a thermocouple. Subambient temperature may be detected by a platinum resistance thermometer in lieu of a thermocouple. The furnace has its own thermocouple embedded in the winding, independent of the sample thermocouple, to control the furnace heating and cooling rates. Most TMA furnaces are heated by the

Fig. 4 High-pressure dilatometer. Courtesy of Linseis Inc.

electrical resistance of a platinum alloy or other metallic wire coiled around a ceramic tube and packed with fibrous insulation inside a steel shell.

Other Length Measurements

Optical dilatometers use visible light in various configurations to detect specimen length and displacement. Some instruments have both an optical detector and a pushrod. Interferometric dilatometers use no-contact interferometry rather than an LVDT to measure the displacement. In Michelson interferometry, a laser beam is split into two paths, reflected by the specimen face and a reference platen, recombined, and analyzed for phase shifts in the fringe pattern. Interferometry is sensitive to displacements in nanometers. The advantage of no-contact detection is that the specimen shape is less important than it is for LVDT detection, the instrument does not impede the specimen expansion, and the faces do not have to be parallel or smooth. Parallel-plate capacitance is an alternative to the LVDT and interferometer, where a mobile plate in contact with the specimen moves relative to a fixed plate. The capacitance between the plates varies with the gap size and can be calibrated to the displacement in nanometers.

Atmosphere

The specimen chamber is purged continuously with an inert gas in most test methods. A flow rate of 50 to 100 mL/min at standard temperature and pressure is sufficient in most commercial dilatometers. The gas flow should be regulated by a mass flow controller or rotameter. Inert gas prevents corrosion of metals and ignition of volatile plastics and extends the life of the furnace and fixtures. Helium, with a relatively high thermal conductivity, k, of 0.15 $W/m^{-1} \cdot K^{-1}$ at room temperature, is the preferred purge gas in most applications. Argon (k = 0.016 $W/m^{-1} \cdot K^{-1}$) and nitrogen (k = 0.024 $W/m^{-1} \cdot K^{-1}$) may also be used but do not transmit heat to the thermocouple as well as helium. Nitrogen is not a true inert gas but has similar nonreactive properties to helium and argon in most TMA applications. Static or flowing air (k = 0.026 $W/m^{-1} \cdot K^{-1}$) may be used to simulate a more realistic field application but can be corrosive at high temperatures. Humidity control and injection of water vapor into the gas stream are available in some instruments and are especially applicable to wood-processing and polymerization TMA.

A vacuum may also be used but has very low thermal conductivity and can decompose some solids at high temperatures. As the pressure decreases, conduction and convection become less significant mechanisms of heat transfer, until radiation becomes the only mechanism of heating, cooling, and temperature detection. A thermocouple used in a vacuum should be as close as possible to the specimen without touching it (1 mm, or 0.04 in., or so) and within a line of sight of the specimen.

Reactive, corrosive, or reducing gases, including hydrogen, ammonia, and carbon monoxide, may be used in TMA to simulate an application such as reactive sintering but should always be used with great caution, typically with the instrument inside a fume hood or glove box.

Temperature Range

The temperature range of TMA spans cryogenic (approximately −150 °C, or −240 °F) to 2400 °C (4350 °F), but no single instrument configuration is valid over the entire range. The most common temperature range, room temperature up to approximately 1600 °C (2910 °F), is served by an air-cooled furnace with an alumina or silicon carbide tube, platinum resistance winding, and platinum-rhodium thermocouples, for example, type R or S. The pushrod and sample holder are also alumina or SiC, although fused silica fixtures may be used for the most sensitive measurements up to approximately 1100 °C (2010 °F). Fused silica is noted for its near-zero CTE, but it slowly crystallizes at elevated temperatures and loses its usefulness as it devitrifies.

A metal furnace, typically of steel or copper alloy, is used for cryogenic to above-ambient temperatures. The metal furnaces can be heated or cooled more rapidly than the ceramic furnaces. The subambient temperatures are achieved by either consumable liquid nitrogen or a refrigeration system that recycles its refrigerant.

The above-1600 °C (2910 °F) furnaces and fixtures are constructed of graphite or tungsten alloy, which oxidizes very easily at high temperatures and must be protected by high-purity (e.g., 99.999% or better) inert gas or high (P < 0.1 Pa, or 0.01 mbar) vacuum.

Materials of Construction

Thermomechanical analysis results are generally dependent on test conditions, including specimen dimensions and instrument settings. The TMA itself expands and contracts with heating and cooling and should be designed to minimize those effects. The sample tube, furnace tube, pushrod, end spacers, and support blocks are usually made of alumina, for its thermal stability up to ~1600 °C (2910 °F) and low CTE. Fused silica, graphite, and other ceramics may be used in lieu of alumina for specific applications. Fused silica has an extremely low CTE, and graphite can be heated above the melting temperature of alumina in a vacuum or inert atmosphere. The furnace winding and thermocouple are usually platinum base, for high-temperature stability and corrosion resistance in air. The furnace and electronics may be fan- or water-cooled, or both. Some instruments have refrigerated or liquid nitrogen cooling for cryogenic applications.

Instrument Settings

The heating and cooling rates, like the specimen size, can affect the dilatometry data. A high rate enables fast throughput in a busy laboratory but is also associated with large temperature gradients. A large specimen or low-thermal-conductivity material should be heated and cooled at a slow rate. A rate of 5 to 10 K/min is suitable to measure the CTE for most metals, and 1 to 5 K/min for most ceramics and plastics. Sintering studies are especially slow, 0.5 to 2 K/min, to minimize fractures caused by nonuniform binder burnout and densification.

The load or stress on the specimen is highly dependent on the type of TMA test. At one extreme, ASTM C 832 calls for 172 kPa (25 psi) of compression on a 38 mm refractory brick cross section, a load of 248 N (56 lbf) in an unusually large dilatometer, for creep testing. A more typical dilatometer load on a metal bar 20 mm (0.8 in.) long and 6 mm (0.24 in.) in diameter is 30 cN (0.07 lbf), for a compressive stress of ~11 kPa (1.6 psi) for CTE measurement. A polyethylene fiber of 0.5 mm (0.02 in.) diameter and 10 MPa (1.5 ksi) tensile strength would be limited to approximately 2 N (0.45 lbf) of force. The test procedures in Table 2 provide guidance for the choice of load or stress and other instrument settings.

Specimen Preparation

The size and shape of the specimen may affect the apparent CTE. The usual shape is a solid cylinder with a round or rectangular cross section. The expansion is usually measured in the direction of the longest axis. The two ends are smooth and parallel to one another and normal to the expansion axis. The as-sawed surface from a metallographic saw is generally sufficiently smooth, as long as the sawed ends are without burrs. The specimen is homogeneous, of uniform cross section, and free of flaws, cracks, notches, pores, and other discontinuities, except as specified in a particular test protocol. Most commercial dilatometers accept a specimen up to approximately 25 mm (1 in.) long and 12 mm (0.5 in.) in diameter or diagonally. Sheet metal or a small-diameter rod can be measured in a dilatometer with the aid of support blocks that fix the orientation of the specimen. A tube or other hollow shape can be measured with the aid of alumina disc spacers between the pushrod and the specimen. Platinum foil or alumina powder can be used to prevent the specimen from staining or cementing itself to the pushrod and other TMA components.

In a TMA comparison of two or more specimens, the results are most meaningful if all the specimens and standards have approximately the same dimensions, instrument settings, and preparation methods.

Calibration and Accuracy

Several published methods exist that standardize test procedures and specimen dimensions for the purpose of maximizing the repeatability and reproducibility of test results. Some of the more common procedures are provided in Table 2. The ASTM procedures are available from ASTM International (www.astm.org). The GOST, ISO, and JIS procedures are available from the Deutsches Institut für Normung e.V. (www.din.de).

Standards

The accuracy of the TMA should be verified periodically with a standard specimen, such as the no-longer-available Standard Reference Materials (SRM) 732, 736, 737, 738, and 739, from the National Institute of Standards and Technology (NIST, www.nist.gov). All five are cited in Table 1. SRM 732 is a single-crystal sapphire (pure Al_2O_3) rod, 6.4 mm (0.25 in.) in diameter, oriented along its *c*-axis. The oriented crystal axis gives SRM 732 very repeatable and reproducible CTE values. A more recent (still available in 2019) SRM is 731, a borosilicate glass, that includes subambient CTE values. Both of these standards were measured by NIST in a dilatometer equipped with a thermocouple and an interferometer to measure the temperature and displacement, respectively. In standards, the CTE is often modeled by polynomial functions of the absolute temperature, T, in the form of Eq 4, where A through D are empirical constants:

$$\alpha = A + BT + CT^2 + DT^3 \qquad \text{(Eq 4)}$$

A larger specimen in TMA has less sensitivity to microscopic flaws, such as pores and secondary phases, than a smaller specimen but is more sensitive to temperature gradients. A temperature gradient means the core of the specimen is cooler than the surface upon heating and warmer upon cooling, yielding erratic dilatometry data. The effect is often negligible in metals but very significant in nonmetallic solids, especially in low-thermal-conductivity material such as refractory ceramics.

Data Analysis and Reliability

Thermomechanical analysis data yield a number of important material properties and characteristics, including CTE, phase-change temperatures, glass transition, sintering onset and endpoint, thermal softening, viscosity, creep, and stress relaxation. The reliability of the data depends on the calibration of the instrument with an SRM, the size and condition of the specimen, thermocouple proximity to the specimen, condition of the instrument, instrument settings, and many other factors.

In any instrument, the measurement signal, ΔL, from the specimen includes a contribution from the instrument. The instrument contribution to the signal may be negligibly small in some cases but should always be verified experimentally. The contribution from the instrument or system comes from the difference between the certified value of the standard and its measured value in Eq 5. The true signal from the specimen comes from the sum of its measured value and the system value in Eq 6. To determine the true signal from the specimen, a standard of similar size and thermal conductivity is first scanned in the TMA and compared to its certified values. Second, the specimen is scanned under the same conditions as the standard, and its curve is corrected for system effects. The two corrections are made by the software of the newer instruments but must be performed manually on the data provided by analog instruments:

$$\Delta L_{\text{system}} = \Delta L_{\text{std certificate}} - \Delta L_{\text{std measured}} \qquad \text{(Eq 5)}$$

$$\Delta L_{\text{specimen}} = \Delta L_{\text{specimen measured}} + \Delta L_{\text{system}} \qquad \text{(Eq 6)}$$

Table 2 Some thermomechanical analysis procedures

Designation	Title
ASTM C 832	Standard Test Method of Measuring Thermal Expansion and Creep of Refractories under Load
ASTM D 696	Standard Test Method for Coefficient of Linear Thermal Expansion of Plastics Between −30 °C and 30 °C with a Vitreous Silica Dilatometer
ASTM E 473	Standard Terminology Relating to Thermal Analysis and Rheology
ASTM E 1363	Standard Test Method for Temperature Calibration of Thermomechanical Analyzers
ASTM E 1545	Standard Test Method for Assignment of the Glass Transition Temperature by Thermomechanical Analysis
ASTM E 1824	Standard Test Method for Assignment of a Glass Transition Temperature Using Thermomechanical Analysis: Tension Method
ASTM E 2092	Standard Test Method for Distortion Temperature in Three-Point Bending by Thermomechanical Analysis
ASTM E 2113	Standard Test Method for Length Change Calibration of Thermomechanical Analyzers
ASTM E 2206	Standard Test Method for Force Calibration of Thermomechanical Analyzers
ASTM E 228	Standard Test Method for Linear Thermal Expansion of Solid Materials with a Push-Rod Dilatometer
ASTM E 2347	Standard Test Method for Indentation Softening Temperature by Thermomechanical Analysis
ASTM E 2769	Standard Test Method for Elastic Modulus by Thermomechanical Analysis Using Three-Point Bending and Controlled Rate of Loading
ASTM E 289	Test Method for Linear Thermal Expansion of Rigid Solids with Interferometry
ASTM E 2918	Standard Test Method for Performance Validation of Thermomechanical Analyzers
ASTM E 831	Test Method for Linear Thermal Expansion of Solid Materials by Thermomechanical Analysis
BS ISO 11359-1	Plastics. Thermomechanical Analysis (TMA). General Principles
BS ISO 11359-2	Plastics. Thermomechanical Analysis (TMA). Determination of Coefficient of Linear Thermal Expansion and Glass Transition Temperature
GOST R 56723	Plastics. Thermomechanical Analysis (TMA). Part 3: Determination of Penetration Temperature
GOST R 57754	Polymer Composites. Test Method for Determination of Linear Thermal Expansion by Thermomechanical Analysis
ISO 349	Hard Coal—Audibert-Arnu Dilatometer Test
ISO 7884-8	Glass. Viscosity and Viscometric Fixed Points. Part 8: Determination of (Dilatometric) Transformation Temperature
JIS K 7196	Testing Method for Softening Temperature of Thermoplastics Film and Sheeting by Thermomechanical Analysis

The CTE comes from the slope of a corrected strain-versus-temperature plot. The plot may have more than one CTE, depending on phase changes, devitrification, or viscoelasticity. A difference in CTE between heating and cooling over the same temperature range often indicates an irreversible transformation. The slope at a point on the digital curve is the physical alpha value, calculated from two consecutive discrete data points. The slope between two nonconsecutive points is the technical alpha value. In most applications, the physical alpha curve is useful as an indicator of changes in the CTE, while the CTE itself is the technical alpha value calculated over as wide a single-phase temperature range as possible. The technical alpha value is also the mean slope between any two temperatures.

Reports of TMA data should include, in addition to the strain-versus-temperature plot, the heating and cooling rates, the time and temperature of any isothermal holds, the dimensions and composition of the specimen, the applied forces or stresses, the calibration standard, instrument identity, test procedure, and any other pertinent information.

Applications and Interpretation

Example 1: Steel

The measurement of CTE in steel and titanium is demonstrated in Fig. 5 and 6. The sharp change in slope in each case indicates the beginning of a phase change. In Fig. 5, a blunt pushrod was in contact with a 21 mm (0.83 in.) long, 6 mm (0.24 in.) diameter hypoeutectoid steel rod in a horizontal dilatometer. The atmosphere was helium flowing at 75 mL/min at standard temperature and pressure. The CTE of ferrite or α-Fe, the body-centered cubic low-temperature phase, is 14.7 $\times$ 10^{-6} K^{-1} over the range of 25 to 700 °C (75 to 1290 °F). The CTE of austenite or γ-Fe, the face-centered cubic high-temperature phase, is 22.8 $\times$ 10^{-6} K^{-1}, half again that of ferrite, from 1100 to 700 °C (2010 to 1290 °F) upon cooling. The CTE values are normally expressed in terms of 10^{-6} K^{-1}, equivalent to parts per million of strain per Kelvin, rather than standard scientific notation. The dip in the heating curve between 742 (Ac_1) and 781°C (Ac_3) (1368 and 1438 °F) indicates that ferrite shrinks slightly as it transforms into close-packed austenite; that is, the austenite is denser at the phase transformation than ferrite, even though austenite has a much higher CTE than ferrite. The phase-change hysteresis-boundary temperatures—Ac_1 and Ac_3 upon heating, and Ar_3 (614 °C, or 1137 °F) and Ar_1 (584 °C, or 1083 °F) upon cooling—can be estimated graphically from the intersections of extensions of the curve segments. The size of the hysteresis is heating- and cooling-rate dependent and is also a function of alloy content, especially carbon. At a very slow temperature rate, say ±0.1 K/min, $Ac_1 \approx Ar_1$ and $Ar_3 \approx Ac_3$. The phase transformation in both directions is rapid but not instantaneous, corresponding to the boundaries of the α + γ region of the iron-carbon phase diagram. The gradual bend in the upper part of the heating curve at ~900 °C (1650 °F) may be due to a combination of slowly transforming ferrite, slowly dissolving cementite, natural curvature, and instrument effects. (The curve is corrected for instrument expansion but not necessarily perfectly so.) The unusual density decrease between Ar_3 and Ar_1, combined with the finite solubility and diffusion rates of carbon and other elements in the two phases, enables rapidly quenched austenite to transform into the metastable martensite phase instead of ferrite under the right hardenability conditions.

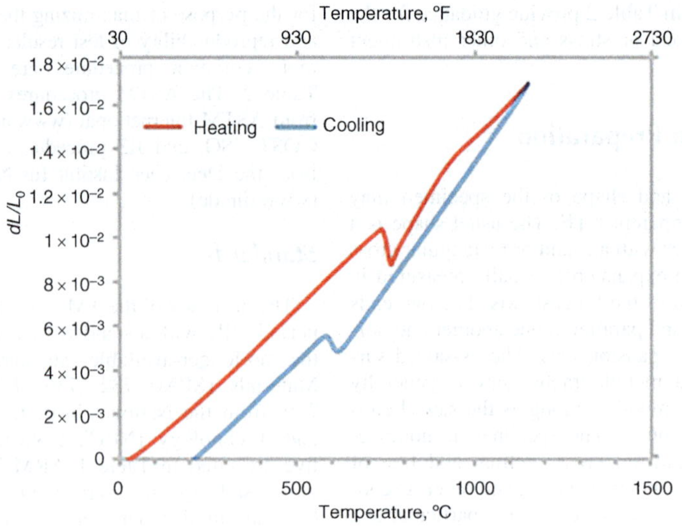

Fig. 5 Dilatometry of steel

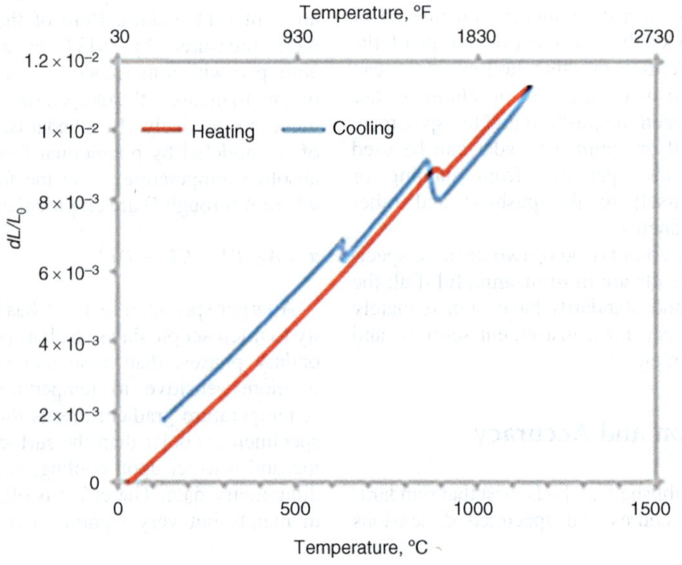

Fig. 6 Dilatometry of a titanium alloy

Example 2: Titanium

In Fig. 6, the CTE of α-Ti, the close-packed hexagonal low-temperature phase, is 10.3 $\times$ 10^{-6} K^{-1} from 25 to 850 °C (75 to 1560 °F). The CTE of β-Ti, the body-centered cubic high-temperature phase, is 10.9 $\times$ 10^{-6} K^{-1} upon heating from 950 to 1100 °C (1740 to 2010 °F) at 10.0 K/min and a little higher at 12.2 $\times$ 10^{-6} K^{-1} upon cooling at the same rate. The phase-change temperature at the jog in the heating curve, the β-transus, begins at 843 °C (1549 °F) and ends at 910 °C (1670 °F). In the cooling curve, a distinct two-phase region exists between the two transuses. The two-phase region transforms completely to α-Ti between 630 and 622 °C (1166 and 1152 °F) during cooling. The hysteresis of α- and β-transus temperatures is explained in part by the relatively low thermal conductivity of titanium and its alloys, the phase-stabilizing effects of aluminum and vanadium, and the kinetics of the phase transformation. The Ti-6%Al-4%V rod was 22 mm (0.87 in.) long and 6 mm (0.24 in.) in diameter. The atmosphere was helium flowing at 75 mL/min at standard temperature and pressure, and the horizontal dilatometer was calibrated with SRM 732.

Example 3: Glass Transition

The glass transition temperature (T_g) in amorphous and viscoelastic materials is readily apparent in a TMA plot from the abrupt increase in slope with increasing temperature in Fig. 7. Glass and most plastics are viscoelastic. Below the T_g, a viscoelastic material has predominantly elastic mechanical behavior, similar to a rigid crystalline solid, and its viscosity approaches infinity with decreasing temperature. Above the T_g, the same material has predominantly viscous mechanical behavior, similar to soft rubber initially. As the temperature increases above the T_g, the viscoelastic material becomes more like a liquid, and its elastic modulus approaches zero. The heat capacity and some other properties also change at the T_g. Thermodynamically, the glass transition is a second-order transition, unlike melting and freezing, which are first order. The temperature-rate-dependent T_g is not a unique temperature as much as a narrow range where molecules suddenly become much more mobile as the temperature increases. The volume change at the T_g is continuous, not a step increase or decrease as the phase changes in Fig. 5 and 6. The slight increase in the T_g in the second heating curve of the epoxy in Fig. 7 may be due to relaxation effects or postcuring of the polymer. The specimen was 6 mm (0.24 in.) long, and the heating rate in both curves was 2 K/min. A common practice in TMA of polymers is heating, slow cooling, and at least one reheating. The initial heating and slow cooling anneal the specimen, that is, "erase" its thermomechanical history. The CTE and T_g are determined from the subsequent heating step.

Example 4: Glass Softening

The T_g can also be represented by the glass softening temperature in Fig. 8. A sharp-ended pushrod was used to concentrate the stress on a glass rod. The abrupt change in the expansion curve corresponds to the transition from rigid to pliant. The advantage of this technique over conventional dilatometry is that it can be used on coated substrates to measure the T_g of the coating. An alternative to the sharp tip is a pushrod with a hemispherical tip, for materials of very low viscosity. The borosilicate glass in Fig. 8, approximately 15 mm (0.6 in.) long, was heated at 5 K/min. The first slope change at 551 °C (1024 °F) is the T_g.

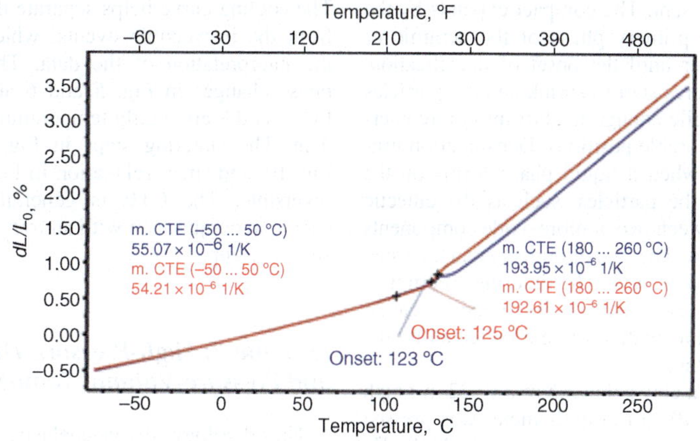

Fig. 7 Glass transition of epoxy resin, twice heated. CTE, coefficient of thermal expansion. Courtesy of Netzsch Instruments Inc.

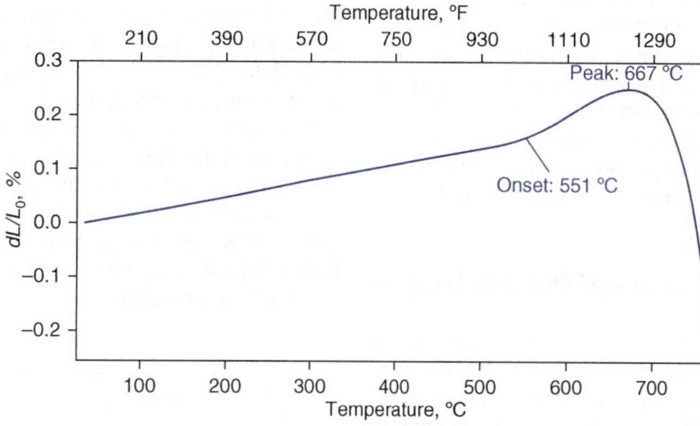

Fig. 8 Softening point of borosilicate glass. Glass transition temperature = 551 °C (1024 °F). Courtesy of Netzsch Instruments Inc.

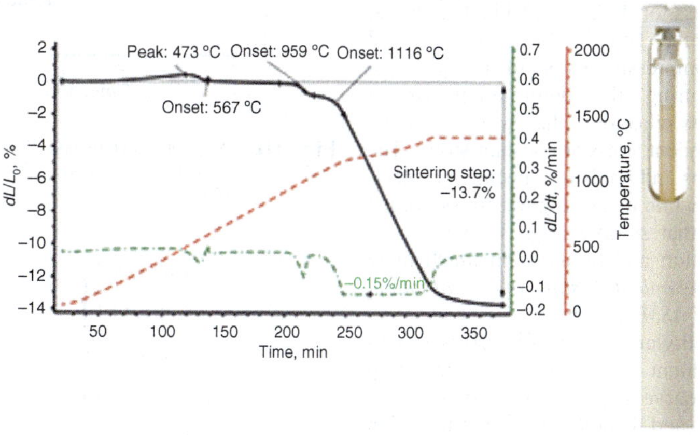

Fig. 9 Rate-controlled sintering dilatometry of clay. The expansion (dL/L_0) curve is solid, the temperature curve is dashed, and the time derivative of expansion (dL/dt) is dashed and dotted. Courtesy of Netzsch Instruments Inc.

The peak at 667 °C (1233 °F) is the dilatometric softening point.

Example 5: Sintering and Densification

The onset of densification in a sintering operation is demonstrated by the higher-temperature slope change in Fig. 9. In sintering, a very fine ceramic or metal powder is compacted in a press along with a uniformly distributed organic binder to maintain the shape. The binder becomes fluid at 100 to 200 °C (210 to 390 °F) and evaporates or oxidizes at 200 to 400 °C (390 to 750 °F). The lower end of the curve is mostly the expansion and loss of the binder, if present. The compact expands by the CTE of the primary phase of the ceramic or metal powder until the onset of densification, where the pores start to shrink and the particles become mobile enough to shift into more energetically favorable positions. Densification usually begins when a liquid phase forms on the surfaces of the particles, such as the eutectic reaction between two or more oxide components in a porcelain. The loss of porosity causes a step in the expansion curve. The sintering is complete when the curve resumes its predensification slope, but at a smaller volume indicated by the size of the step.

In Fig. 9, a clay pellet 5.5 mm (0.22 in.) high and 6.5 mm (0.26 in.) in diameter was pressed and heated at 5 K/min to 1350 °C (2460 °F). The pellet was held at 1350 °C (2460 °F) for 1 h. The rate control began at 1040 °C (1905 °F). The disturbance at ~110 min and 473 °C (883 °F) was dihydroxylation of kaolinite, and quartz transition at 567 °C (1053 °F). A second phase change occurred at 215 min, indicated by the slope change in the expansion curve and the valley in the *dL/dt* curve. The sintering began at 250 min, with a heating-rate-controlled constant expansion rate of −0.15%/min. The total linear sintering shrinkage was 13.7%.

Example 6: Creep and Stress Relaxation

The more sophisticated TMA instruments can characterize creep and stress relaxation by plotting strain or stress versus time at constant temperature. Creep is time-dependent strain at constant stress. At lower temperatures, stress and strain are proportional by the elastic modulus in most solids. At elevated temperatures, a solid can deform continuously under constant stress or lose its residual stress at constant strain. The TMA applies a fixed tensile or compressive stress to the specimen while measuring the deformation, as in Fig. 10. Stress relaxation, the inverse of creep, is time-dependent stress at constant strain. The TMA deforms and holds the specimen a specified amount while measuring the stress needed to maintain that strain in Fig. 11. Creep and stress relaxation are discussed in much more detail in *Mechanical Testing and Evaluation*, Volume 8 of *ASM Handbook*.

The polyethylene film in Fig. 10 is in tension at ambient temperature in the creep curve. The curve begins with instantaneous deformation and bends into the retardation and linear regions. The tension is removed in the recovery curve, where most, but not all, of the deformation is quickly recovered. The data can also be plotted as compliance and recoverable compliance. The same polyethylene film in Fig. 11 was stressed to a known strain. The residual stress, as relaxation modulus, decayed as a function of logarithmic time at the fixed strain.

As a rule, the cooling curve in a TMA thermogram is as important as the heating curve. The cooling curve helps separate the reversible from the irreversible events, which simplifies the interpretation of the data. The reversible phase changes in Fig. 5 and 6 and the T_g in Fig. 7 and 8 are highly temperature rate dependent. The sintering steps in Fig. 9, creep in Fig. 10, and stress relaxation in Fig. 11 are not reversible. The CTE is generally reversible except in conjunction with an irreversible event such as creep.

Example 7: High-Pressure Dilatometry and Pressure-Volume-Temperature

The rheology of viscoelastic composites, such as metal injection molding feedstock, may be simulated with the assistance of the high-pressure dilatometer. The high-pressure dilatometer plots specific volume—the reciprocal of density—versus temperature at a fixed hydrostatic pressure, collectively known as PVT data. The PVT data are thermodynamic equation-of-state properties used to calculate the Tait compressibility constants that are used in injection molding simulation software. In the PVT plot in Fig. 12, the abrupt change in slope at 345 to 370 K for each pressure corresponds to the glass transition of the feedstock binder. The heating rate was 3 K/min. The T_g is a function of pressure as well as dT/dt.

Example 8: Density versus Temperature

The density of a solid as a function of temperature is important in thermal conductivity (k) measurements. The laser-flash analyzer actually measures the thermal diffusivity (δ) and calculates k as a function of temperature from δ, the heat capacity (C_p), and density (ρ) at that same temperature in Eq 7. The heat capacity is measured by differential scanning calorimetry. The density in a region of linear thermal expansion is approximated by Eq 8, where ρ_0 is the density at T_0, usually 298 K:

$$k(T) = \delta(T) \cdot C_p(T) \cdot \rho(T) \quad \text{(Eq 7)}$$

$$\rho(T) \approx \frac{\rho_0}{1 + \alpha_V(T - T_0)} \quad \text{(Eq 8)}$$

Example 9: Densification Kinetics

Several dilatometry techniques have been applied to sintering kinetics in powder metallurgy and ceramics, to model densification as a function of time and temperature. In the master sintering curve (MSC) method, green compacts are sintered in the TMA over a range of fixed heating rates. The thermal expansion versus temperature is converted to

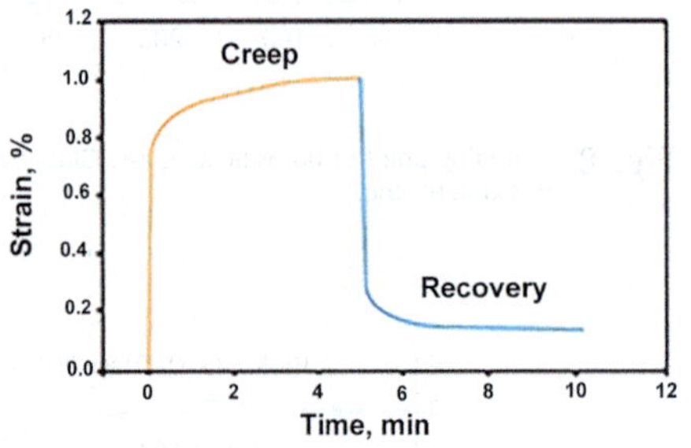

Fig. 10 Creep of polyethylene film in tension at ambient temperature. Courtesy of TA Instruments Inc.

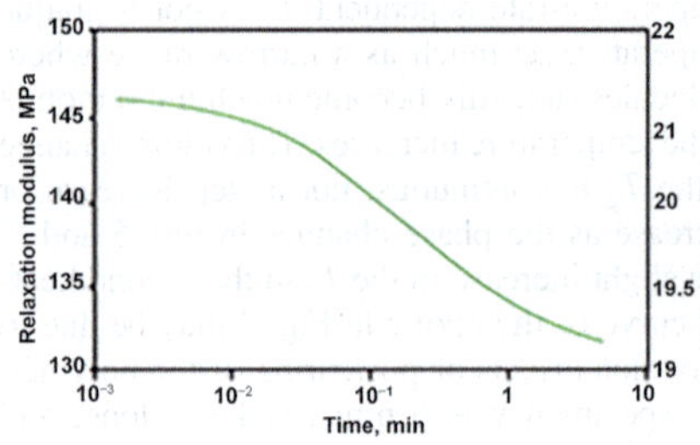

Fig. 11 Stress relaxation of polyethylene film in tension at ambient temperature. Courtesy of TA Instruments Inc.

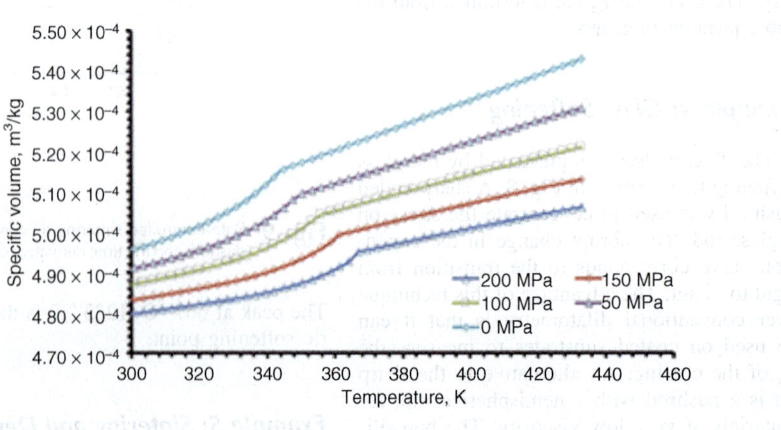

Fig. 12 Pressure-volume-temperature diagram of specific volume as a function of thermoplastic feedstock temperature and pressure

density, as in Eq 8, and plotted as a function of a sintering parameter Θ in Eq 9 that is a function of time t and $T(t)$. The activation energy, Q, in Eq 9 can also be evaluated in the MSC, where R is the ideal gas constant. The MSC plot can then be used to predict the degree of densification of the compact at a specified time and temperature:

$$\Theta(t, T(t)) = \int_0^t \frac{1}{T} e^{\left(\frac{-Q}{RT}\right)} dt \qquad \text{(Eq 9)}$$

Example 10: Materials with Unusual Thermal Expansion Behavior

A few solids have zero or even a negative CTE over a limited temperature range. Invar, a face-centered cubic alloy of 64 mass% Fe and 36% Ni, has approximately one-tenth the CTE of other steels from cryogenic temperatures up to approximately 100 °C (212 °F). The low CTE of Invar is a result of its spontaneous magnetostriction varying with temperature inversely to thermal expansion. Invar is used for wristwatch components and other precision instruments and won its inventor, C.E. Guillaume of Switzerland, the Nobel Prize in Physics in 1920. Invar has been modified with other alloying elements, such as cobalt, to reduce its CTE even further.

Fused silica, also known as fused quartz or quartz glass, has a low-density structure that accommodates thermal expansion internally, thus giving it an unusually low CTE up to approximately 1100 °C (2010 °F). Its CTE is slightly negative at cryogenic temperatures. Fused silica is nearly pure SiO_2, unlike window glass, which has minor components of CaO and Na_2O, or borosilicate glass, which contains B_2O_3. The low CTE over a wide temperature range is why fused silica is often used for TMA hot-zone components. Above ~1100 °C (2010 °F), approaching its T_g, fused silica is susceptible to creep and devitrification. Once devitrified, it has the same CTE as the mineral quartz.

Negative thermal expansion occurs in a few solids that do not have a close-packed crystal structure. These materials, such as $AlPO_4$, orthorhombic $Sc_2(MoO_4)_3$, ScF_3, cubic ZrW_2O_8, and ZrV_2O_7, contract when heated and expand when cooled. The temperature range of the negative CTE may be very narrow, such as in liquid water below 4 °C (40 °F). The mechanisms of negative CTE are changes in bond angles and symmetry, rotations of functional groups such as WO_4, and repositioning of atoms due to dielectric and magnetic ordering that oppose and exceed in magnitude the lengthening of bonds with increasing temperature. Negative CTE materials have the potential of being a component in a composite with a precisely tunable net CTE that can be matched to adjoining materials, such as dental restorations, so as to minimize interfacial stresses during heating and cooling.

Kovar, an alloy of nominally 54 mass% Fe, 29% Ni, and 17% Co, is an example of a material designed to have a specific CTE. Kovar is primarily used for glass-to-metal seal interfaces, because it has the same CTE as glass. Applications for Kovar include x-ray tubes, cathode ray tubes, and some lightbulbs.

A bimetallic strip, consisting of two dissimilar metal sheets such as steel and brass rolled or bonded together, uses differential thermal expansion to create a switch or sensor. At room temperature, the two bonded strips are the same size, and the stress at the interface is negligible. As the temperature increases, one strip tries to grow longer than the other but is constrained by the bond. The stress at the interface causes bending or curling toward the strip with the lower CTE. Applications for the bimetallic strip include the thermostats inside drip coffee makers and similar appliances, temperature-compensating mechanical clock components, and dial thermometers such as those used in cooking.

Example 11: Extended TMA Capabilities

Thermomechanical analysis should not be confused with dynamic mechanical analysis (DMA). Dynamic mechanical analysis uses low-frequency vibrations to evaluate the storage modulus (E'), loss modulus (E''), and phase angle (δ) between E' and E'' of viscoelastic solids as a function of temperature. In Eq 10, E' represents the elastic modulus, and E'' is the viscosity in the complex modulus, E^*. In a purely elastic (Hookean) solid, $\delta = 0$, and in a purely viscous (Newtonian) liquid, $\delta = 90°$. In a viscoelastic material, $0 < \delta < 90°$, and the phase angle can be derived from Eq 11. The more sophisticated TMA instruments may be capable of mimicking the DMA or providing other specialty measurements described in this section:

$$E^* = E' + iE'' \qquad \text{(Eq 10)}$$

$$\tan\delta = \frac{E''}{E'} \qquad \text{(Eq 11)}$$

Dynamic or force-modulated TMA applies a sinusoidal force and constant temperature rate to measure δ, E', and E'' from TMA data as functions of temperature, time, and stress.

Temperature-modulated TMA applies a sinusoidal and linear temperature rate and constant force to separate reversible events, such as glass transition, from irreversible events, such as stress relaxation, softening, and heat shrinkage, in one curve. The sinusoidal temperature may also be used isothermally and is considered a forcing function that elicits a sinusoidal response of expansion and contraction simultaneously. The response is deconvoluted by a Fourier transformation into reversing and nonreversing signals. The reversing signal is associated with properties dependent on dT/dt. The nonreversing signal is associated with kinetic events dependent on both temperature and time. The total expansion at any time and temperature is the sum of the two signals. The rate of total expansion is given in Eq 12, where the reversing component is represented by the first term on the right side, and the nonreversing by the second term:

$$\frac{dL}{dt} = \alpha_L \frac{dT}{dt} + f(t, T) \qquad \text{(Eq 12)}$$

Rate-controlled sintering in TMA adjusts the temperature rate so as to maintain a constant strain rate, that is, a constant shrinkage rate during the sintering of a powder compact. The strain rate is related to the first derivative of the dilation with respect to time, dL/dt. Rate-controlled sintering optimizes the sintering process by yielding more uniform and homogeneous densification, with less grain growth, than sintering at a constant temperature rate.

SELECTED REFERENCES

- R.L. Blaine, "Modulated Thermomechanical Analysis—Measuring Expansion and Contraction Simultaneously," Application Brief TA-311, TA Instruments
- Creep and Stress-Relaxation Testing, *Mechanical Testing and Evaluation*, Vol 8, *ASM Handbook,* ASM International, 2000, p 359–424, ISBN 9780871703897
- "Expansion of a Steel between Ambient and 1400 °C," Application Note AN373, Setaram
- "Gelation of Epoxy-Glass Prepreg by Parallel Plate Rheometry," Application Brief TA-126, TA Instruments
- I.F. Groves, T.J. Lever, and N.A. Hawkins, Dynamic Mechanical Analysis—A Versatile Technique for the Viscoelastic Characterisation of Materials, *Int. Labmate*, Vol XVII (No. II), Application Brief TA-070, TA Instruments
- T. Hatakeyama and F.X. Quinn, *Thermal Analysis: Fundamentals and Applications to Polymer Science*, 2nd ed., Wiley, 1999, p 125–135, ISBN 978-0471983620
- J.D. Menczel and R.B. Prime, Ed., *Thermal Analysis of Polymers: Fundamentals and Applications*, Wiley, 2009, p 319–386, ISBN 978-0471769170
- W. Miller, C.W. Smith, D.S. Mackenzie, et al., Negative Thermal Expansion: A Review, *J. Mater. Sci.*, Vol 44, 2009, p 5441, https://doi.org/10.1007/s10853-009-3692-4
- H. Palmour, Rate Controlled Sintering for Ceramics and Selected Powder Metals, *Science of Sintering*, D.P. Uskoković, H. Palmour, and R.M. Spriggs, Ed., Springer, 1989, ISBN 978-1-4899-0935-0, doi.org/10.1007/978-1-4899-0933-6_29
- E. Post, J. Blumm, L. Hagemann, and J.B. Henderson, TA for Ceramic Materials, *Netzsch*

Industrial Applications, 2nd ed., Sept 1997, p 6–9, 44–50, 54–56, 59, 65, 66, 68
- A.T. Riga and C.M. Neag, Ed., *Materials Characterization by Thermomechanical Analysis*, STP 1136, American Society for Testing and Materials, 1991
- R.F. Speyer, *Thermal Analysis of Materials*, Marcel Dekker Inc., 1994, p 165–197, ISBN 0-8247-8963-6
- H. Su and D.L. Johnson, Master Sintering Curve: A Practical Approach to Sintering, *J. Am. Ceram. Soc.*, Vol 79 (No. 12), 1996, p 3211–3217
- *"The Correlation of TMA with ASTM Modulus Data," Application Brief TA-138*, TA Instruments
- W.W. Wendlandt and P.K. Gallagher, Thermomechanical Analysis (TMA), *Thermal Characterization of Polymeric Materials*, E.A. Turi, Ed., Academic Press Inc., 1981, p 68–74, ISBN 0-12-703780-2
- B. Wunderlich, *Thermal Analysis*, Academic Press Inc., 1990, p 311–369, ISBN 0-12-765605-7
- M. Nijman, "Curve Interpretation Part 5: TMA Curves," Mettler Toledo Thermal Analysis Application No. UC421, 2016

X-Ray Analysis

ASM Handbook, Volume 10, *Materials Characterization*
ASM Handbook Committee
DOI 10.31399/asm.hb.v10.a0006645

X-Ray Spectroscopy*

Revised by S. Lampman, ASM International

Overview

Capabilities

- Detection and quantification of elements with atomic number 5 or higher. (Older energy-dispersive units with beryllium window detectors are limited to atomic number 11 or higher.)
- Detection limits for bulk determinations are normally a few parts per million (ppm) to a few tens of ppm, depending on the x-ray energy used and the sample matrix composition.
- For thin-film samples, detection limits are approximately 100 ng/cm^2.

Detection Threshold and Precision of Energy-Dispersive Detectors

- *Threshold sensitivity:* ~0.02%
- *Precision of quantitative analyses:* ~1% relative or 0.02% absolute, depending on count time

Detection Threshold and Precision of Wavelength-Dispersive Detectors

- *Threshold sensitivity:* ~0.005%
- *Precision of quantitative analyses:* ~0.2% relative or 0.005% absolute, whichever is greater

Typical Uses

- Qualitative and quantitative chemical analysis for major and minor elements in metals and alloys
- Determination of composition and thickness of thin-film deposits

Samples

- Samples may be bulk solids, powders, pressed pellets, glasses, fused disks, or liquids.
- Bulk metal samples typically are ground to produce a flat, uncontaminated surface for analysis. Typical samples have dimensions of several centimeters; however, most instruments can accommodate samples 10 cm (4.0 in.) or more in diameter.
- Powder samples are typically attached to a non-x-ray-producing substrate or are pressed into pellets.
- The technique is completely nondestructive.
- Sampling depth may range from a few micrometers to a millimeter or more, depending on the x-ray energy used and the matrix composition of the sample.

Capabilities of Related Techniques

- Inductively coupled plasma optical emission spectroscopy and atomic absorption spectrometry have better detection limits for most elements than x-ray spectrometry and are often better choices for liquid samples; elements of low atomic number can be determined by using these techniques.

Introduction

X-ray spectroscopy, or x-ray fluorescence (XRF) spectrometry, is an emission spectroscopic technique that has found wide application in elemental identification and determination. The technique depends on the emission of characteristic x-radiation, usually in the 1 to 60 keV energy range, following excitation of atomic electron energy levels by an external energy source, such as an electron beam, a charged particle beam, or an x-ray beam. In most sample matrices, laboratory x-ray spectrometry can detect elements at concentrations of less than 1 μg/g of sample (1 ppm); in a thin-film sample, it can detect total amounts of a few tenths of one microgram.

X-ray spectrometry is one of the few techniques that can be applied to solid samples of various forms. Early types of x-ray spectrometers were used in laboratories, finding application in routine analyses for production and quality control and in specialized tasks. Initially, x-ray spectrometry found wide acceptance in applications related to metallurgical and geochemical analyses. Examples of specialized applications have included the determination of sulfur and wear elements in petroleum products, the analysis of forensic samples, and the measurement of electronic and computer-related materials.

Growth in the capability and economy of microcomputer technology has continued to enhance the applications of XRF spectrometry.

* Revised from D.E. Leyden, X-Ray Spectrometry, *Materials Characterization,* Vol 10, *ASM Handbook,* ASM International, 1986, p 82–101

In particular, the development of battery-operated, field-portable x-ray analyzers has broadened the application of XRF instruments. Since the 1970s, portable XRF instruments have been used for in situ nondestructive characterization. One early application of portable XRF instruments was the in situ measurement of lead in painted surfaces in housing (Ref 1). Continuing advances in computer and detector technology have led to further advances in the capabilities and sensitivity of portable XRF instruments for applications such as soil analysis and nondestructive qualitative and quantitative analysis of inorganic elements in archaeological objects. Portable XRF instruments cannot generally provide the low detection limits attained by laboratory methods, but the sensitivity of portable XRF instruments has increased and made possible their use for measurement of lead and other metal in dusts and soils (Ref 2). For example, field XRF can easily provide detection limits for lead in soil of less than 100 ppm, well below typical regulatory levels of 300 to 1500 ppm.

Operating Principles

X-rays, discovered by Roentgen in 1895, are high-energy photons that occur from nuclear decay or when electrons in a disturbed or excited atom move to a lower energy level. The electrons in an atom have discrete (quantum) energy levels or orbitals, and if electron vacancies occur in the inner orbitals, the decay or relaxation of outer electrons to the stable ground state results in the emission of x-rays. These energy transitions of electrons within an atom lead to the emission of sharp x-ray lines characteristic of the target element and the transition involved.

H.G.J. Moseley developed the relationships between atomic structure and x-ray emission and in 1913 published the first x-ray spectra, which are the basis for modern x-ray spectrometry. Moseley recognized the potential for quantitative elemental determinations using x-ray techniques. The development of routine x-ray instrumentation, leading to the x-ray spectrometer known today (2019), took place over the following decades. Coolidge designed an x-ray tube in 1913 that is similar to those currently used. Soller achieved collimation of x-rays in 1924. Improvements in the gas x-ray detector by Geiger and Mueller in 1928 eventually led to the design of the first commercial wavelength-dispersive x-ray spectrometer by Friedman and Birks in 1948.

A simplified schematic of an XRF spectrometer is shown in Fig. 1. A beam of x-rays is produced by an x-ray tube, and the x-ray beam is directed onto the sample surface. The beam penetrates some small distance into the sample, typically 10 to 100 μm, depending on the atomic numbers of the elements in the sample. Penetration depths are greater for low-atomic-number elements. With a sufficient amount of x-ray energy, some of the electrons in the sample (of unknown composition) are excited to higher quantum states (energy levels) or removed from the atom (ionization). These "excited" atoms are quickly "relaxed" when electrons from higher energy levels fill the vacated levels. When this happens, photons are emitted whose energies are equal to the differences between the two energy levels involved; this process is called fluorescence. As the excited atoms relax, x-ray photons are emitted corresponding to the differences in the characteristic energy levels of the elements in the sample.

Spectrometers detect and measure the spectrum of x-rays emitted from the sample. The spectrum of electromagnetic radiation is generally characterized in terms of both radiation energy and radiation intensity. Due to the quantum wave-particle duality of electromagnetic radiation, a beam of x-rays can be thought of as being both a wave (defined by beam wavelength) and a particle or photon energy (where photon energy, E, is inversely related to the wavelength, λ, such that $E = hc/\lambda$, where h is Planck's constant, and c is the velocity of light. The energy of an x-ray is thus defined by the energies of the photons or the wavelength of the beam, while the intensity of x-rays is defined by the number of photons or the amplitude (height) of the wave.

When the excited atoms "relax" by electrons filling the vacated levels, the characteristic energies span a range of the electromagnetic spectrum. Electron transitions between inner shells typically produce x-rays (photons with energies in the 200 to 20,000 eV range, characterized by wavelengths of 6 to 0.06 nm). The net result is that each element has a unique set of known electron energy levels. Similarly, the set of energy differences between these electron energy levels is also unique for each element and constitutes a characteristic "fingerprint" by which each element can be identified. In XRF spectroscopy (as well as many other analytical methods), the combined electron energy level "fingerprints" of the elements present in the sample are experimentally obtained and are then compared to those of known elements. From these comparisons, it is possible to identify the elements and their compositions present in a sample.

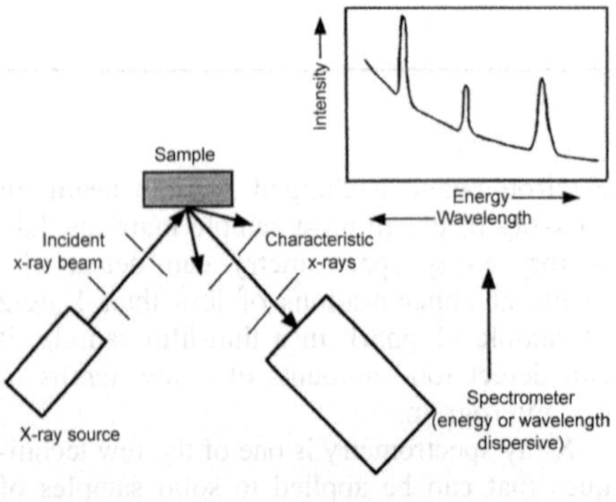

Fig. 1 Schematic of x-ray fluorescence spectrometer. X-rays emitted from the sample are analyzed to determine the characteristic energies (or wavelengths) of x-rays emitted and the intensities of the various characteristic energies.

Spectrometers are designed to measure the spectrum of x-rays in terms of either wavelength or energy. The first commercial x-ray spectrometers were wavelength-dispersive spectrometers (WDS) developed by Friedman and Birks in 1948. In WDS, emitted x-rays from the sample impinge upon crystal, which results in diffraction (constructive or destructive interference) of the beam. For a crystal of interplanar spacing d, Bragg's law gives the relationship between the x-ray wavelength, λ, and the Bragg critical angle, θ_B, at which constructive interference occurs:

$$n\lambda = 2d \sin \theta_B \quad \text{(Eq 1)}$$

where n is an integer. Depending on the interplanar spacing (d) of the analyzing crystal and the direction of the diffraction pattern (θ), the WDS detector can identify the peak intensities in the x-ray spectrum.

The spectrum of x-ray energies and intensities also can be measured by an energy-dispersive spectrometer (EDS), where the energy of individual photons is converted into electronic pulses that are processed and counted. As these energies are measured, a histogram of the numbers of photons counted corresponding to each energy is plotted. The detector in an EDS system is a lithium-doped silicon semiconductor. X-ray photons enter the semiconductor detector crystal, where numerous electron-hole pairs are created as photons expend their energy with the atoms in the detector. The energy of the photon is proportional to the number of electron-hole pairs, which are analyzed and counted by pulse-processing electronics. The detector electronics sense when each photon enters the detector and require that a photon energy be analyzed before accepting input from any additional photons. Typically, several thousand photons are analyzed per second. As these energies are measured, a histogram of the numbers of photons counted corresponding to each energy is plotted on a cathode ray tube. The result is a plot of intensity versus energy, similar to what is obtained from WDS (Fig. 1). The electronics of an EDS system are discussed in more detail in the section "Energy-Dispersive Spectrometry" in this article.

Comparison of WDS and EDS

The detectors in WDS and EDS systems operate on different principles, but both can indicate the characteristic peaks of x-rays associated with fluorescence (from electrons that move from outer atomic orbits to inner orbits of the sample). There is some background x-ray radiation with a continuum of energies from many different interactions, but the characteristic energies of electrons moving within atoms can have strong peaks of intensity. With these characteristic peaks (as a function of either wavelength or energy), it is possible to analyze the bulk elemental composition of a sample from its x-ray fluorescence.

Although x-ray spectra can be measured with both the EDS and WDS methods, they have some differences in capabilities that can complement each other. One basic difference is that the characteristic peaks of the WDS method can be resolved to sharper peak lines than the EDS methods (Fig. 2). The resolution of the WDS easily separates peaks that are poorly resolved in the EDS, such as the peaks for Zn Kα and Cu Kβ, or completely unresolved, such as those for Mn Kα and Ba L$\alpha_{2,3}$. The superior resolution of the WDS often makes it the spectrometer of choice for the heavier elements in a specimen (atomic number $Z > 20$), because they produce L- or M-family x-rays, which will often be found to interfere with the K-lines of the light elements. In light-element analysis (atomic number $Z < 10$), both the EDS and the WDS can detect x-rays from elements as low in atomic number as boron ($Z = 5$).

The primary advantage of an EDS system is speed. An EDS can collect a complete spectrum with several hundred thousand counts in approximately 1 min. The EDS provides rapid qualitative analysis of major and minor constituents, while the slower WDS provides superior resolution. The ideal analytical instrument should be equipped with both EDS and WDS. Such an arrangement provides optimum capabilities for addressing a wide variety of practical problems. The choice between EDS and WDS systems is based on the specific problems to be solved. The EDS is the proper choice for general-purpose work in which the predominant need is qualitative analysis of a wide variety of unknowns with quantitative analysis of major and minor constituents. If the predominant problems involve x-ray mapping, the detection of trace or minor constituents, or the separation of serious peak overlaps, then the WDS is the proper choice.

Qualitative analysis (determination of which elements are present) is done by comparing the energies of the x-rays emitted from the sample with the known characteristic x-ray spectra of each element (Fig. 3). Quantitative determination of the concentration of each element present is computed based on the intensities of the various characteristic x-ray energies, also shown in Fig. 3. Quantitative analyses can be most accurate by comparing the x-ray intensities from the unknown sample with their counterparts from a series of standard similar and known compositions. All modern instruments are equipped with computers to facilitate this calibration and measurement process. The use of progressively more powerful computer hardware and software has substantially decreased the need for standards with compositions tailored to specific classes of alloys. Many current analyses are done based only on pure element standards, using the computer to make composition-dependent corrections by iterative means.

X-Ray Radiation

X-rays are high-energy photons that range from hard (higher-energy) x-rays with shorter wavelengths of 0.01 to 0.1 nm (0.1 to 1 Å) to lower-energy (softer) x-rays with longer wavelengths of 0.1 to 1 nm (1 to 10 Å). If the radiation is being described as wave, then the proper meaning of the intensity of electromagnetic radiation is the energy per unit area per unit time. If the radiation is being described as particle interaction, then the intensity refers to the number of counts per unit time from the detector.

X-rays occur from various sources. One source is nuclear decay. This source of high-energy electromagnetic radiation is typically referred to as gamma rays. X-rays are generated from the disturbance of the electron orbitals of atoms, and various energy sources are used to disturb atoms that produce x-ray emission. The most common method is the bombardment of a target element with high-energy electrons. This method requires a vacuum to avoid energy loss of the electron beam. In scanning electron microscopy (SEM) and electron microprobe analysis, for example, an electron beam impinges directly on the sample in a vacuum chamber. Placing the specimen in a vacuum chamber is impractical for x-ray spectrometry, and therefore, x-rays from an x-ray tube are the most practical energy source

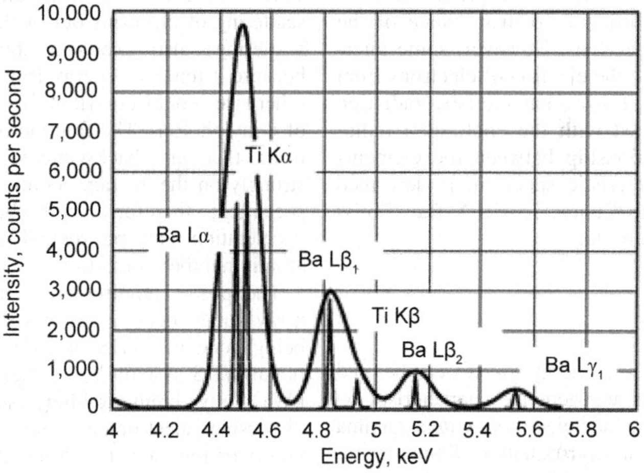

Fig. 2 Superimposed spectra of $BaTiO_3$ obtained from energy-dispersive spectrometer (EDS) and wavelength-dispersive spectrometer (WDS) systems, where the WDS spectrum is replotted on the energy scale rather than wavelength. X-ray detection in WDS systems is based on Bragg's law of constructive and destructive interference and resolves the spectrum of characteristic lines with a much sharper energy resolution than an EDS system. WDS enables resolution of nearby peaks that overlap one another on the EDS spectrum. The WDS spectrum also has less background noise. Source: Ref 3

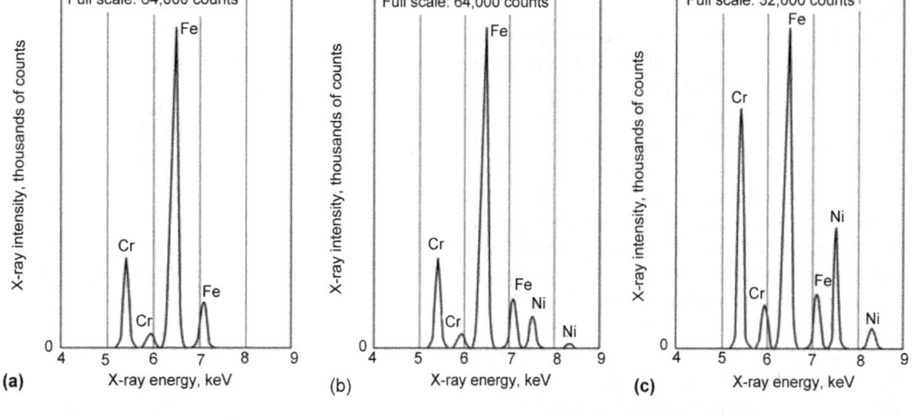

Fig. 3 X-ray fluorescence spectra of (a) Fe-16.4%Cr, (b) Fe-12.3%Cr-12.5%Ni, and (c) Fe-25.7%Cr-20.7%Ni. The iron, chromium, and nickel peaks occur at the same characteristic energies, but the intensities of the peaks increase with concentration. Courtesy of Jim Brangan, Sandia National Laboratories

to disturb atoms for XRF spectroscopy. Radioactive isotopes that emit gamma rays are another possibility for excitation of atoms to emit x-rays. However, the x-ray flux from isotopic sources that can be safety handled in a laboratory is too weak for practical use. Because gamma-ray sources usually emit only a few narrow x-ray lines, several are required to excite many elements efficiently. Another method is particle-induced x-ray emission applied to special samples, but, similar to the SEM, this also requires putting the sample in a vacuum chamber.

Electron bombardment of a target results in two types of x-ray emission from the target: a continuum of x-ray energies (bremsstrahlung radiation) as well as the characteristic radiation of the target element. Both types of x-ray emission are encountered in x-ray spectrometry. Most of the electrons impinging on a target interact with the orbital electrons of the target element in nonspecific interactions and result in little or no disturbance of the inner orbital electrons. However, some interactions result in the ejection of electrons from these orbitals. These characteristic radiation lines are emitted with the continuum radiation. The relationship between the elements and the characteristic spectrum is described in the section "Characteristic X-Ray Emissions" in this article.

X-Ray Tubes

Several different energy sources can be used to excite atoms and generate characteristic x-rays. Sources include electrons, x-rays, gamma rays, and synchrotron radiation. Sometimes a source of continuum (bremsstrahlung) x-ray radiation is used to generate specific x-radiation from an intermediate pure element sample called a secondary fluorescer. Most XRF spectrometers use x-ray tubes, which may operate with a primary target (primary mode) or a secondary target (secondary fluorescer mode). Both WDS and EDS systems use primary-mode x-ray tubes. Secondary-target EDS systems also are used.

Modern x-ray tubes are descendants of the Coolidge tube (Fig. 4). All components are in a high vacuum. A filament (typically tungsten) is heated by a filament voltage of 6 to 14 V. The heated filament thermally emits electrons. A potential of several kilovolts is applied between the filament (cathode) and the target anode, which serves as the acceleration potential for the electrons. The primary source unit consists of a very stable high-voltage generator, capable of providing a potential of typically 40 to 100 kV. The anode is usually copper, and the target surface is plated with high-purity deposits of such elements as rhodium, silver, chromium, molybdenum, or tungsten. The flux of electrons that flows between the filament and the target anode must be highly regulated and controlled.

The electrons strike the target with a maximum kinetic energy equivalent to the applied tube potential. If the kinetic energy of the electron exceeds the absorption-edge energy corresponding to the ejection of an inner orbital electron from atoms of the target material, the tube will emit x-ray lines characteristic of the target element. Interaction of the electrons in the beam with electrons of the target element will also lead to emission of a continuum. The area of the continuum and the wavelength of maximum intensity will depend on the potential, current, and anode composition.

The deceleration of the electron in the target and the probability of x-ray generation in the process is a function of the overall composition of the target; it depends mainly on the atomic number of the target components. Castaing proposed that an "atomic number correction" must be applied to take into account its effects on the x-ray emission (Ref 4). The scattering of electrons out of the target (backscattering) also exerts a significant effect, because it removes energy from the target that otherwise would contribute to the production of x-ray photons. The fraction of primary electrons that are backscattered also depends strongly on the average atomic number of the specimen; therefore, both backscatter and deceleration can be treated together in an atomic number correction.

The power requirements and selection of x-ray tubes depend on whether analysis is being done by WDS or EDS. Most conventional WDS systems use a high-power (2 to 4 kW) x-ray bremsstrahlung source. Energy-dispersive spectrometers use either a high-power or low-power (0.5 to 1.0 kW) primary source, depending on whether the spectrometer is used in the secondary or primary mode. Low-power tubes for EDS analysis may range from 9 to 100 W and are usually air cooled. Various anode materials are available, and manufacturers of x-ray spectrometers offer different tube features. Many designs have the traditional "side window" design similar to the Coolidge tube (Fig. 4), although much smaller than those used in WDS systems.

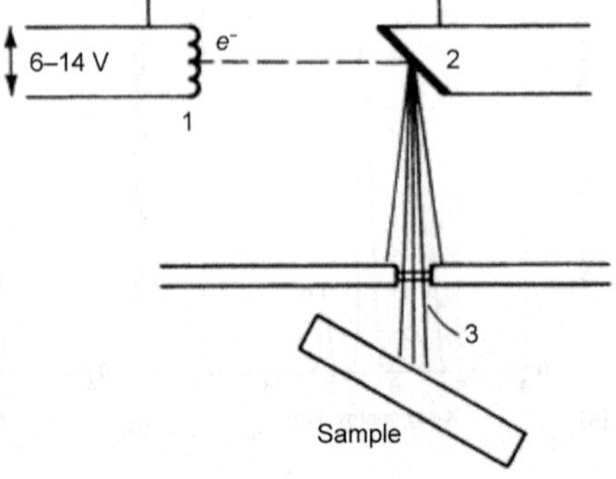

Fig. 4 Coolidge x-ray tube. 1, filament; 2, target anode; 3, beryllium window

Continuum Emission

Emission of x-rays with a smooth, continuous function of intensity relative to energy is called continuum, or bremsstrahlung, radiation. An x-ray continuum may be generated in several ways. However, the most useful is the electron beam used to bombard a target in an x-ray tube (tubes used in x-ray spectrometry are discussed subsequently). The continuum is generated as a result of the progressive deceleration of high-energy electrons impinging on a target, which is a distribution of orbital electrons of various energies. As the impinging electrons interact with the bound orbital electrons, some of their kinetic energy is converted to radiation; the amount converted depends on the binding energy of the electron involved. Therefore, a somewhat statistical probability exists as to how much energy is converted with each interaction.

This ever-present background limits the detection of weak characteristic x-ray signals, and consequently, the sensitivity (the ability to determine low concentrations of elements in the specimen) is low. The definition of limits of detection depends on the interpretation of statistical parameters to produce estimates of confidence limits. The probability of an impinging electron interacting with an orbital electron of the target element should increase with the atomic number of the element; thus, the intensity of the continuum emission should increase with the atomic number of the target element. Further, the probability of an interaction increases with the number of electrons per unit time in the beam, or flux. Therefore, the intensity of the continuum increases with electron beam current (I), expressed in milliamperes.

Moreover, the ability of the impinging electrons to interact with tightly bound electrons of the target element increases with the kinetic energy of the bombarding electrons. Because the kinetic energy of the electrons in the beam increases with acceleration potential, the integrated intensity of the continuum should increase with electron acceleration potential (V), expressed in kilovolts. Finally, the maximum energy manifested as x-ray photons equals the kinetic energy of the impinging electron, which in turn relates to acceleration potential. These concepts can be approximated quantitatively (Ref 5, 6):

$$I_\lambda = K_i Z \left(\frac{\lambda}{\lambda_{min}} - 1\right)\lambda^{-2} \quad \text{(Eq 2)}$$

$$I_{int} = (1.4 \times 10^{-9}) I Z V^2 \quad \text{(Eq 3)}$$

Other relationships have been proposed. Differentiation of an expression given by Kulenkampff (Ref 7) yields an expression that demonstrates that the energy of the maximum intensity in the continuum lies at approximately two-thirds the maximum emitted energy. The shape of the continuum predicted by Eq 2 and 3 is approximate. These functions do not include the absorption of x-rays within the target material or the absorption by materials used for windows in the x-ray tube and detectors. Therefore, some modification of the intensity distribution may occur, especially at low x-ray energies.

Characteristic X-Ray Emissions

The photoelectric effect is an x-ray absorption mechanism by which unstable states in the electron orbitals of atoms are created. Once the vacancies in the inner orbitals are formed, relaxation to the stable ground state may occur by the emission of x-rays characteristic of the excited element. Each of the transitions that may occur leads to the emission of sharp x-ray lines characteristic of the target element and the transition involved. These characteristic radiation lines also are emitted with the continuum emission.

The concept of families of x-ray lines is paramount in understanding the nature of characteristic x-ray formation and the spectral artifacts that are encountered in detecting x-rays in the EDS and WDS analysis. The energy difference between the initial and final states of the transferred electron results in characteristic x-rays with energies proportional to the differences in the energy states of atomic electrons. The relationship between the wavelength of a characteristic x-ray photon and the atomic number (Z) of the excited element first established by Moseley is (Ref 8, 9):

$$\frac{1}{\lambda} = K(Z - \sigma)^2 \quad \text{(Eq 4)}$$

where K is a constant that takes on different values for each spectral series, a shielding constant (σ) has a value of just less than unity, and the wavelength (λ) of the x-ray photon is inversely related to the energy, E, of the photon as follows:

$$\lambda\left(\text{in Angstroms, Å}\right) = \frac{12.4}{E\ (\text{in keV})} \quad \text{(Eq 5)}$$

With increasing atomic number, the electronic structure of the atom becomes more complicated, and when a vacancy is created in a particular shell, it can be filled by a transition from two or more shells or subshells, which leads to the formation of characteristic x-rays of different energies. These various transitions occur with different probability, so that the different x-ray energies associated with a given shell, for example, the K- or L-family, have different relative intensities.

The multiplicity of lines in a family can be large. For example, the L-family of a heavy element such as gold consists of approximately 35 different lines. Fortunately, not all of these will be seen due to low relative intensities for many and the limitations imposed by the resolution of the spectrometer. The information available on relative intensities is a considerable aid in qualitative analysis. For a major constituent, the identification of an entire family of lines greatly increases the confidence with which the assignment of the presence of that element can be made. By the same token, unless all members of the family of x-ray lines are identified, including those with low relative intensity, then it is very likely that a low-relative-intensity member will be misidentified as a peak of a minor constituent later in the procedure.

A second related point is that for a given beam energy from the x-ray tube, all characteristic x-rays whose critical excitation energy is exceeded may be found in the spectrum. Thus, if a K-family is identified for an element, the analyst should also locate the possible L- and M-family at lower energy in the spectrum. Certain spectral artifacts result in parasitic peaks, such as escape peaks or sum peaks, generally from the high-intensity peaks in the spectrum. These artifact peaks also must be identified when a major constituent is recognized in the spectrum, to prevent their subsequent misidentification later in the procedure. The qualitative analysis procedure thus consists of a methodical recognition and labeling process. Because of the differences in energy-dispersive and wavelength-dispersive spectra, different qualitative analysis procedures are followed for each.

As an example, Fig. 5 illustrates excitation and emission for the photoejection of a K-($1s$) electron of copper. Figure 5(a) shows a plot of mass absorption coefficient of copper versus x-ray energy from 0 to 20 keV, with K_{abs} at 8.98 keV. Figure 5(b) depicts an electronic energy level diagram for copper. Irradiation of copper with an x-ray of just greater than 8.98 keV will photoeject an electron from the K-shell. This is an ionization of the copper atom from the inner shell rather than the outer valence electrons, as is the case with chemical reactions. The energy of the $1s$ electron is shielded from the state of the valence electrons such that the absorption-edge energy and the energy of the emitted x-rays are essentially independent of the oxidation state and bonding of the atom.

In practice, the number of lines observed from a given element will depend on the atomic number of the element, the excitation conditions, and the wavelength range of the spectrometer employed. Generally, commercial spectrometers cover the K-, L-, and M-series, corresponding to transitions to the K- (inner shell), L- (second shell), and M-levels, respectively. There are many more lines in the higher series; for a detailed list of all of the reported wavelengths, the reader is referred to the work of Bearden (Ref 10). In x-ray spectrometry, most of the analytical work is carried out by using either the K- or L-series wavelengths. Most commercially available x-ray spectrometers have a range from approximately 0.2 to 20 Å (60 to 0.6 keV), which allows measurement of the K-series from fluorine (Z = 9) to lutetium (Z = 71), and for the L-series from manganese (Z = 25) to uranium (Z = 92). Other line series can occur from the M- and N-levels, but these have little use in analytical x-ray spectrometry (Ref 11).

K-Lines

Once the photoelectric effect creates a vacancy in the K-shell, the excited state relaxes by filling the vacancy with an electron from an outer orbital. Only certain transitions are allowed because of quantum mechanical rules called selection rules. Some of these are:

$$\Delta n > 0$$

$$\Delta l = \pm 1$$

$$\Delta j = \pm 1 \text{ or } 0$$

where n is the principal quantum number, l is the angular quantum number, and $j = l + s$ is the vector sum of l and s (the spin quantum number). The transitions that follow the selection rules are termed allowed (diagram) lines, those that do not are called forbidden, and those that result in atoms with two or more vacancies in inner orbitals at the time of the emission are called satellite (nondiagram) lines. The scheme of notation of x-ray spectral lines is unconventional; additional information is provided in Ref 12 and 13.

Figure 5(b) shows the transition for the K-lines of copper. These are called the K-lines because the original vacancy was created in the K-shell of copper by photoejection. These examples may be related to the general transition diagrams for all elements. The number of K-lines, and the exact one observed for an element, depends in part on the number of filled orbitals. The forbidden $K\beta_5$-line for copper is observed, because there are no $4p_{1/2,3/2}$ electrons to provide the $K\beta_2$-line of nearly the same energy that would obscure the much weaker $K\beta_5$-line.

The table in Fig. 5(b) shows some relationships among the relative intensities of the K-lines. The $K\alpha_1$- and $K\alpha_2$-lines arise from transitions from the L_{III}- ($2p_{3/2}$) and the L_{II}- ($2p_{1/2}$) levels, respectively. The former orbital contains four electrons; the latter, two electrons. The observed 2:1 intensity ratio for the $K\alpha_1$- and $K\alpha_2$-lines results from the statistical probability of the transition. Although these two lines arise from different transitions, their energies are so similar that they are rarely resolved. It is common to report only a weighted average energy for these lines:

$$E_{\bar{\alpha}} = \frac{2E^{\alpha 1} + E^{\alpha 2}}{3} \quad \text{(Eq 6)}$$

The Kβ-lines occur at an energy higher than the Kα-lines. The relative intensity of the Kα-to-Kβ-lines is a complex function of the difference between the energy levels of the states involved in the transition; therefore, relative intensity varies with atomic number. However, as an example, the Kβ-lines for elements of atomic number 24 to 30 are approximately 10 to 13% of the total Kα + Kβ intensity.

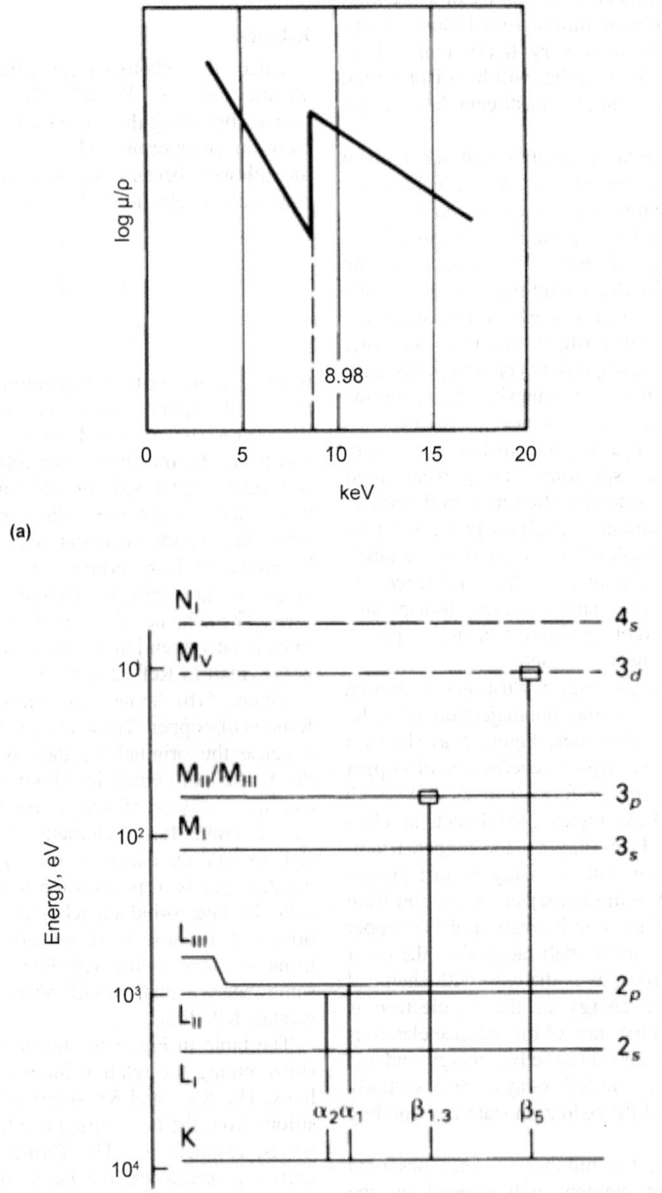

Line	Transition	Relative intensity	E, keV
$K\alpha_1$	$L_{III} \rightarrow K(2P_{3/2} \rightarrow 1S_{1/2})$	63	8.047
$K\alpha_2$	$L_{II} \rightarrow K(2P_{1/2} \rightarrow 1S_{1/2})$	32	8.027
$K\beta_1$	$M_{III} \rightarrow K(3P_{3/2} \rightarrow 1S_{1/2})$	10	8.903
$K\beta_3$	$M_{II} \rightarrow K(3P_{1/2} \rightarrow 1S_{1/2})$	10	8.973
$K\beta_5$	$M_{V} \rightarrow K(3D_{5/2} \rightarrow 1S_{1/2})$	<1 (forbidden)	8.970

Fig. 5 Transition diagram for copper. (a) Absorption curve. (b) Transitions

L-Lines

Because the practical energy range for most wavelength-dispersive x-ray spectrometers is 0 to 100 keV, and 0 to 40 keV for energy-dispersive x-ray spectrometers, the use of emission lines other than the K-lines must be considered. For a given element, L-lines are excited with lower x-ray energy than K-lines. Because there are three angular-momentum quantum numbers for the electrons in the L-shell, corresponding to the $2s_{1/2}$, $2p_{1/2}$, and $2p_{3/2}$ orbitals, respectively, there are three L-absorption edges: L_I, L_{II}, and L_{III}. To excite all L-lines, the incident x-ray photon energy must have a value greater than that corresponding to L_I. The use of L-lines is particularly valuable for elements with atomic numbers greater than approximately 45.

M-Lines

M-lines find limited application in routine x-ray spectrometry. The lines are not observed for elements with atomic numbers below approximately 57, and, when observed, the transition energies are low. The only practical use for these lines is for such elements as thorium, protactinium, and uranium. They should be used only in these cases to avoid interferences with L-lines of other elements in the sample.

Fluorescent Yield

When an outer electron moves to fill the inner-shell vacancy, not all vacancies result in the production of characteristic x-ray photons. Another competing effect is the emission of secondary (Auger) electron (Ref 14, 15). One of these events occurs for each excited atom, but not both. Therefore, secondary electron production competes with x-ray photon emission from excited atoms in a sample. The fraction of the excited atoms that emits x-rays is termed the fluorescent yield. Fluorescent yield values are several orders of magnitude less for the very low atomic numbers. This value is a property of the element and the x-ray line under consideration. For a given atomic number, the fluorescent yield for an L-line emission is always less by approximately a factor of 3 than the fluorescent yield of a corresponding K-line emission. Figure 6 shows a plot of x-ray fluorescent yield versus atomic number of the elements for the K- and L-lines.

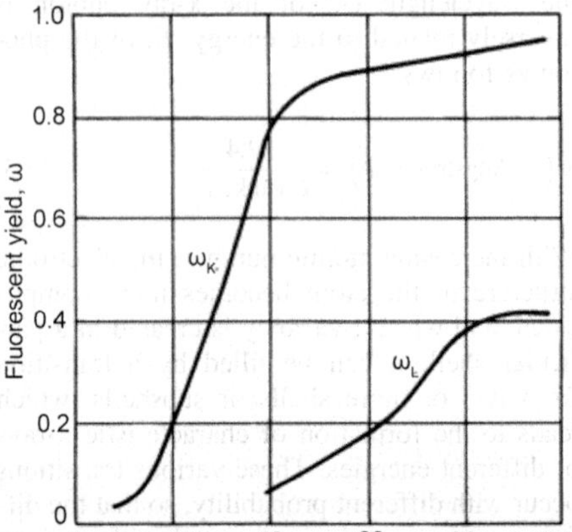

Fig. 6 Fluorescent yield versus atomic number for K- and L-lines

Low-atomic-number elements have low fluorescent yield. This, coupled with the high mass absorption coefficients that low-energy x-rays exhibit, makes the detection and determination of low-atomic-number elements challenging by x-ray spectrometry.

Interelement or Matrix Effects

For transitions in x-ray spectrometry, no emission line for a given series (K, L, M) of an element has energy equal to or greater than the absorption edge for that series. An important result is that the x-rays emitted from an element cannot photoeject electrons from the same orbital of other atoms of that element. For example, the most energetic K-line of copper is below the K-absorption edge for copper. This eliminates direct interaction of atoms of the same element in a sample.

However, a sample composed of a mixture of elements may exhibit interactions that are often called interelement or matrix effects. Generally, the relationship between intensities and concentrations for a specific analyte is not linear because of the various elements or matrix surrounding each atom of the analyte. The effects of the elements of the matrix on the analyte intensity are called interelement effects or matrix effects. When the characteristic energy of analyte line A is slightly higher than the absorption edge of a particular element B in the matrix, then the A-line is highly absorbed by element B. This has the effect of reducing the A-line intensity and increasing the proportion of emissions to the B-concentration. For example, the absorption edge of chromium is 5.99 keV, while the Kα-line for iron is 6.40 keV. As a result, the x-ray intensities per unit concentration (sensitivity) from a sample containing chromium and iron will be affected by the composition. Because the x-radiation emitted from iron will photoeject K-shell electrons from chromium, the chromium x-ray intensity will be higher than expected. The chromium absorbs some of the Kα and Kβ x-rays from iron that would otherwise typically be detected, causing a lower intensity for iron than would be anticipated.

It is the combination of both these effects, absorption and enhancement, that is called the matrix effect. The most significant matrix effects occur for the Kα-lines of the midrange elements of the periodic table and when the atomic number difference of the two considered elements is 2 (Ref 16). Such interactions of elements within a sample often require special data analysis. See ASTM E 1361, "Standard Guide for Correction of Interelement Effects in X-Ray Spectrometric Analysis."

X-Ray Absorption

X-ray photons may interact with orbital electrons of elements to be absorbed or scattered. The relationship between absorption and the atomic number of the element is important in selecting optimum operating conditions for x-ray spectrometry. Scatter of x-rays leads to background intensity in the observed spectra.

The absorption of x-ray emission within the specimen is similarly compensated by an absorption correction. The loss depends on the average exit path length of the x-ray photons and thus on the angle over which the x-ray spectrometer subtends the specimen and on the distribution in depth of the x-ray generation. This distribution, in turn, is a function of the electron beam energy and the composition of the specimen. In addition, the absorption varies strongly with the x-ray absorption coefficient of the specimen for the radiation of interest, which also is composition dependent. The correction is usually performed by means of semiempirical expressions (Ref 17).

Mass Absorption

When an x-ray beam passes through a material, the photons may interact in nonspecific ways with electrons in the orbitals of the target elements, attenuating the intensity of the x-ray beam. The interactions may lead to photoelectric ejection of electrons or scatter of the x-ray beam. In either case, the overall result is frequently described in terms of an exponential decrease in intensity with the path length of the absorbing material:

$$I_\lambda = I_0 \cdot \exp - \left[\left(\frac{\mu}{\rho}\right)\rho x\right] \qquad \text{(Eq 7)}$$

where I_λ is the intensity of a beam of wavelength λ after passing through a length x (cm) of an absorber; I_0 is the initial intensity of the beam; μ/ρ is the mass absorption coefficient of the absorber (cm^2); and ρ is the density of the absorber (g/cm^3). The mass absorption coefficient is characteristic of a given element at specified energies of x-radiation. Its value varies with the wavelength of the x-radiation and the atomic number of the target element. These relationships are discussed in the section "Mass Absorption Coefficients" in this article.

Photoelectric Effect

The photoelectric effect is the most important of the processes leading to absorption of x-rays as they pass through matter. The photoelectric effect is the ejection of electrons from the orbitals of elements in the x-ray target. This process is often the major contributor to absorption of x-rays and is the mode of excitation of the x-ray spectra emitted by elements in samples. Primarily as a result of the photoelectric process, the mass absorption coefficient decreases steadily with increasing energy of the incident x-radiation. The absorption-versus-energy curve for a given element has sharp discontinuities. These result from characteristic energies at which the photoelectric process is especially efficient. Energies at which these discontinuities occur are discussed in the section "Absorption Edges" in this article.

X-Ray Scattering

When x-ray photons impinge on a collection of atoms, the photons may interact with electrons of the target elements to result in the scatter of the x-ray photons, as illustrated in Fig. 7. Scatter of x-rays from the sample is the major source of background signal in the spectra obtained in x-ray spectrometry. The scatter of x-rays is caused mainly by outer, weakly held electrons of the elements. If the collisions are elastic, scatter occurs with no loss of energy and is known as Rayleigh scatter; if inelastic, the x-ray photon loses energy to cause the ejection of an electron, and the scatter is incoherent. The path of the x-ray photon is deflected, and the photon has an energy loss or a longer wavelength. This is Compton scatter.

Scatter affects x-ray spectrometry in two ways. First, the total amount of scattered radiation increases with atomic number because of the greater number of electrons. However, samples with low atomic number matrices exhibit a larger observed scatter because of reduced self-absorption by the sample. Second, the ratio of Compton-to-Rayleigh scatter intensity increases as the atomic number of the sample matrix decreases.

The energy loss associated with Compton scatter results in a predictable change in the wavelength of the radiation:

$$\Delta\lambda_{\text{cm}} = \left(\frac{h}{m_e c}\right)(1 - \cos\phi) \qquad \text{(Eq 8)}$$

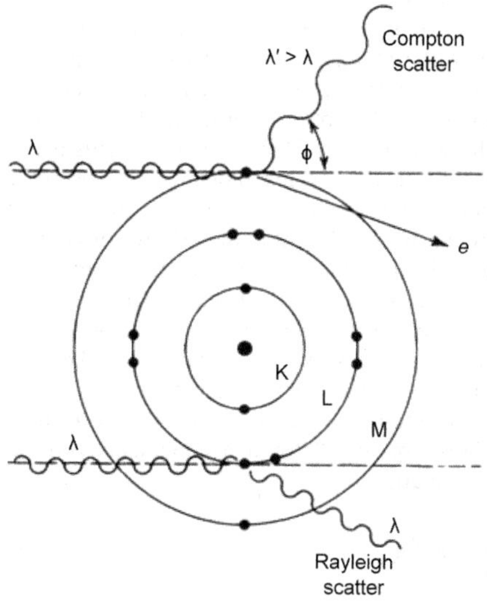

Fig. 7 Rayleigh and Compton scatter of x-rays. K, L, and M denote electron shells of principal quantum numbers 1, 2, and 3, respectively; φ is the angle between the incident and scattered rays.

where $\Delta\lambda_{cm}$ is the change in wavelength (cm), h is Planck's constant (6.6×10^{-27} erg · s), m_e is the electron mass (9.11×10^{-28} g), c is the velocity of electromagnetic radiation (3×10^{10} cm/s), and ϕ is the angle between the scattered and incident x-ray paths. Substitution of the aforementioned values into Eq 9 yields:

$$\Delta\lambda = 0.0243(1 - \cos\phi) \quad \text{(Eq 9)}$$

Because most x-ray spectrometers have a primary beam-sample-detector angle of approximately 90°, ϕ = 90° and cos ϕ = 0. Therefore, for many spectrometers:

$$\Delta\lambda = 0.024\ \text{Å} \quad \text{(Eq 10)}$$

This is known as the Compton wavelength. In energy-dispersive systems, the Compton shift may be more conveniently represented:

$$E' = \frac{E}{1 + 0.00196E(1 - \cos\phi)} \quad \text{(Eq 11)}$$

where E and E' are the x-ray energies in kiloelectron volts of the incident and scattered radiation, respectively. For a spectrometer with beam-sample-detector geometry of 90°, a Compton-scattered silver Kα-line (22.104 keV) from a silver x-ray tube will be observed at 21.186 keV. The intensity of the Compton scatter of the characteristic lines from the x-ray tube can be useful in certain corrections for matrix effects in analyses.

Mass Absorption Coefficients

Mass absorption coefficients differ for each element or substance at a given energy of x-ray and at each energy of x-ray for a given element or substance. Because of the greater probability of interaction with orbital electrons, the mass absorption coefficient increases with the atomic number of the element of the target material. At a given atomic number, the mass absorption coefficient decreases with the wavelength of the x-radiation. This is illustrated in the log-log plot of mass absorption coefficient versus wavelength for uranium given in Fig. 8, which also shows discontinuities in the relationship at certain wavelength values. These result from specific energies required for the photoelectric ejection of electrons from the various orbitals of the atom and are characteristic of the element.

A detailed analysis of data similar to those shown in Fig. 8 for many elements confirms the relationship:

$$\frac{\mu}{\rho} = KZ^4\lambda_{cm}^3 \quad \text{(Eq 12)}$$

where Z is the atomic number of the target element, λ is the wavelength of the incident x-ray, and K is the variable at each absorption edge of the target element.

Absorption Edges

Absorption edges, which are discontinuities or critical points in the plot of mass absorption versus wavelength or energy of incident x-radiation, are shown in Fig. 8. Absorption-edge energy is the exact amount that will photoeject an electron from an orbital of an element. Figure 9 shows the electron shells in an atom. The familiar K, L, and M notation is used for the shells of principal quantum numbers 1, 2, and 3, respectively. The lower the principal quantum number, the greater the energy required to eject an electron from that shell. As shown in Fig. 9, the wavelength of an x-ray that can eject an L-electron is longer (of less energy) than that required to eject an electron from the K-shell. That is, the K-absorption-edge energy (K_{abs}) is greater than the L-absorption-edge energy (L_{abs}) for a given element.

The photoelectric process leads to the unstable electronic state, which emits characteristic x-rays, as illustrated in Fig. 10. Figure 10(a) shows a plot of absorbance versus energy for radiation lower in energy than the x-ray region. In this case, photon energy is used to promote electrons from low-lying orbitals to higher ones. The transition is from a stable quantized state to an unstable quantized state. The atom, ion, or molecule that is the target defines the energy difference. The sample absorbs only photons with energy very close to this energy difference. The result is the familiar absorption peak found in visible, ultraviolet, and other forms of spectroscopy.

Figure 10(b) illustrates radiation in the x-ray energy range. The electron is ejected from a stable low-lying orbital of a given quantized energy level to the continuum of energy of an electron removed from the atom. Any excess energy in the x-ray photon is converted to kinetic energy of the ejected electron (measurement of the kinetic energy of these electrons is the basis of x-ray photoelectron spectroscopy). Therefore, instead of the absorption peak shown in Fig. 10(a), an absorption edge or jump is observed when the x-ray photon energy is sufficient to photoeject the electron. Selection of the x-ray photon energy for excitation of the elements in the sample will be based on these considerations. For example, 8.98 keV x-rays are required to photoeject the K- (1*s*) electrons from copper, but x-rays of only approximately 1.1 keV are required for the 2*s* or 2*p* electrons. For magnesium, the values are 1.3 and 0.06 keV, respectively. The energy of the absorption edge of a given orbital increases smoothly with the atomic number of the target element.

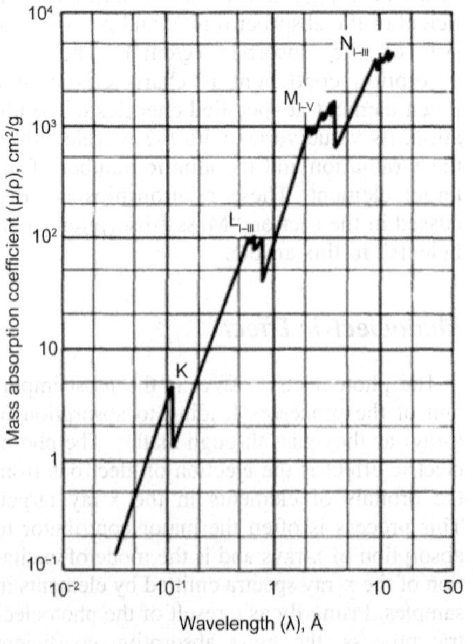

Fig. 8 X-ray absorption curve for uranium as a function of wavelength

Wavelength-Dispersive Spectrometry

As noted, the method of XRF spectrometry dates back to the early 20th century, when H.G.J. Moseley developed the relationships between atomic structure and x-ray emission. The first commercial introduction of x-ray spectrometry began in the 1950s with a type of WDS system, where the radiation emitted from a sample is collimated using a Soller collimator. The collimated beam impinges on an analyzing crystal, which diffracts the radiation to different extents according to Bragg's law, depending on the wavelength or energy of the x-radiation.

The separation of x-rays of various energies is achieved by making use of the wave nature of the photon through the phenomenon of diffraction. A WDS consists of a high-precision mechanical system for establishing the critical Bragg angle between the specimen and the

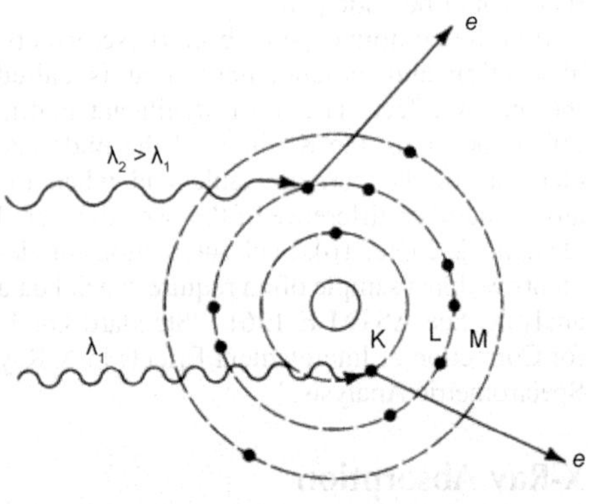

Fig. 9 Photoejection of K-electrons by higher-energy radiation and L-electrons by lower-energy radiation

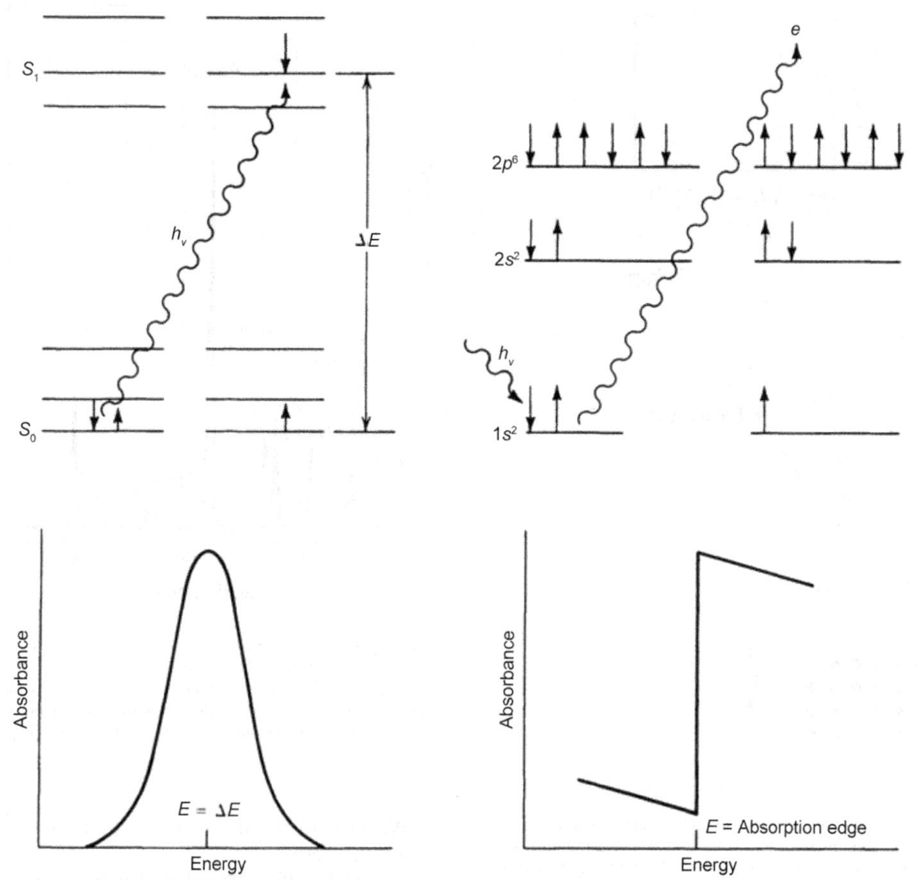

Fig. 10 Excitation of electronic energy levels. (a) Transition between two quantized energy levels. (b) Photoejection of electrons by x-radiation

diffracting crystal and between the diffracting crystal and the x-ray detector (Fig. 11). For a crystal of interplanar spacing *d*, Bragg's law (Eq 1) gives the relationship between the x-ray wavelength, λ, and the Bragg critical angle, θ_B, at which constructive interference occurs. Bragg's law (Eq 1) permits calculation of the angle θ_B at which a wavelength λ will be selected if the analyzing crystal has a lattice spacing of *d*; *d* and λ are in angstroms.

Use of a goniometer permits precise selection of the angle θ. Because of the mechanical arrangement of the goniometer, it is convenient to use 2θ rather than θ. The value of *n* can assume integer values 1, 2, 3, and so on. The resulting values of λ, λ/2, λ/3, ··· that solve Bragg's law are called first-order lines, second-order lines, and so on; any of these present in the sample will reach the detector. Because the numerical value of 2*d* is needed for Bragg's law, the 2*d* value is often tabulated for analyzing crystals. Table 1 shows some common analyzing crystals and their 2*d* spacing. Using the information in Fig. 5(b) and Table 1, the first-order Kα-line for copper is determined to be at a 2θ angle of 44.97° if a LiF(200) analyzing crystal is used. The typical resolution of a WDS peak in energy terms is 10 eV or less for Mn Kα.

An analyzing crystal should be selected that allows the desired wavelength to be detected from 20 to 150° 2θ. Simultaneous instruments normally contain several sets of analyzing crystals and detectors; one is adjusted for each desired analyte in the sample. Sequential WDS may be computer controlled for automatic determination of many elements. To cover the range of x-ray wavelengths to be measured, for example, 6.8 nm at B Kα (183 eV) to 0.092 nm at U Lα (13.4 keV), several different diffraction crystals must be used, and spectrometers often include up to four interchangeable crystals mounted on a turret. Although expensive, these instruments are efficient for routine determination of preselected elements but are not easily converted to determine elements other than the ones selected at installation. More common are sequential instruments that contain a mechanical system known as a goniometer that varies the angle among the sample, analyzing crystal, and detector. In this way, the desired wavelength of x-radiation may be selected by movement of the goniometer.

Actual detection of the x-ray in a wavelength-dispersive spectrometer is accomplished in a flow-proportional detector. The x-ray is absorbed by an argon atom in the detector, and the ejected photoelectron ionizes other atoms, producing a cascade of ejected electrons that are accelerated by a bias applied to a wire in the center of the detector. This bias is chosen so that the pulse of charge collected on the wire is proportional to the energy of the x-ray photon, which allows for the possibility of electronic discrimination of x-ray energies in addition to the physical discrimination provided by the diffraction process.

Detectors and associated electronics in WDS detect x-rays diffracted from the analyzing crystal and reject undesired signals such as higher- or lower-order diffraction by the analyzing crystal or detector noise. Two detectors are commonly positioned in tandem. The first is a gas-filled or flowing-gas proportional detector. These detectors consist of a wire insulated from a housing. Thin polymer windows on the front and back of the housing permit entry and possible exit of x-radiation. A bias potential of a few hundred volts is applied between the wire and housing.

Although many gases may be used, the typical gas is P-10, a mixture of 90% argon and 10% methane. When x-rays enter the detector, the argon is ionized to produce many Ar^+-e^- pairs. The anodic wire collects the electrons, and the electrons at the cathodic walls of the housing neutralize the Ar^+ ions. The result is a current pulse for each x-ray photon that enters the detector. The P-10-filled proportional detectors are most efficient for detecting x-ray photons of energies less than approximately 8 keV (wavelengths greater than approximately 0.15 nm). More energetic x-radiation tends to pass through the proportional detector.

A second detector often located behind the proportional counter is usually a scintillation detector. This detector consists of a thallium-doped sodium iodide crystal [NaI(Tl)], which emits a burst of blue (410 nm) light when struck by an x-ray photon. The crystal is mounted on a photomultiplier tube that detects the light pulses. The number of light photons produced is proportional to the energy of the incident x-ray photon. After electronic processing, the scintillation burst is converted into a voltage pulse proportional in amplitude to the x-ray photon energy.

These two detectors may be operated independently or simultaneously. In simultaneous operation, the detector operating potential and output gain must be adjusted so that an x-ray photon of a given energy produces the same pulse-height voltage from both detectors. Both detector types require approximately 1 μs to recover between pulses. Some counts may be lost at incident photon rates greater than approximately 30,000/s. Pulse-height discrimination of the x-ray pulses from the detector(s) rejects higher- or lower-order x-rays diffracted from the analyzing crystal.

In addition to the excellent spectral resolution of the WDS, the dead time of the associated pulse-processing system can be 1 μs or less, which allows limiting count rates of

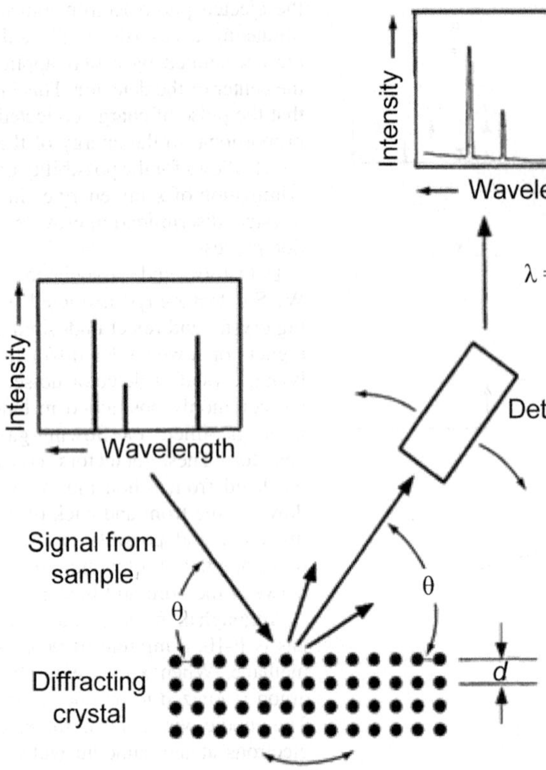

Fig. 11 Schematic of wavelength-dispersive x-ray detector. Detector can mechanically scan a range of angles to produce a plot of intensity vs. wavelength, or it can be set at specific angles corresponding to the characteristic wavelengths of elements known to be in the sample, counting the x-ray intensity at each angle.

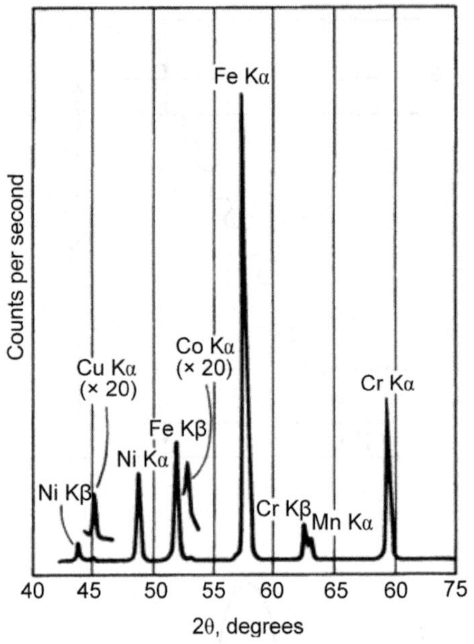

Fig. 12 Wavelength-dispersive x-ray spectrum of AISI type 347 stainless steel. Philips PW-1410 sequential x-ray spectrometer; molybdenum x-ray tube, 30 kV, 30 mA; P-10 flow-proportional detector; LiF(200) analyzing crystal; fine collimation; 100 kcps full scale

Table 1 Common analyzing crystals

Chemical name, common name(a)	Chemical formula	2d, Å
Lithium fluoride, LiF(220)	LiF	2.848
Lithium fluoride, LiF(200)	LiF	4.028
Sodium chloride, NaCl	NaCl	5.641
Germanium, Ge(111)	Ge	6.532
Pentaerythritol, PET(002)	$C(CH_2OH)_4$	8.742
Ammonium dihydrogen phosphate, ADP(101)	$NH_4H_2PO_4$	10.640

(a) Numbers in parentheses are Miller indices to show the diffracting plane.

the order of 10^5 counts per second (cps). Because only x-rays with a narrow energy range—for example, 10 eV—can reach the detector in the WDS due to the selectivity of the diffraction process, the full limiting count rate is available from the peak of interest, while other x-rays produced from the specimen usually have no effect on the count rate. The high limiting count rates that can be accommodated by the WDS detector, coupled with the extremely narrow photon energy range that actually reaches the detector, allow analysts to accumulate the necessary x-ray counts in a shorter time and to characterize trace constituents.

Quantitative applications of automated WDS are efficient, because the instrument can be programmed to go to the correct angles for desired determinations; however, qualitative applications are less efficient, because the spectrum must be scanned slowly. Figure 12 shows a wavelength-dispersive spectrum of an AISI type 347 stainless steel taken with a WDS. Approximately 30 min were required to obtain this spectrum. Examples of standards in WDS analysis include:

- ASTM C 1605, "Standard Test Methods for Chemical Analysis of Ceramic Whiteware Materials Using Wavelength-Dispersive X-Ray Fluorescence Spectrometry"
- ASTM D 6247, "Standard Test Method for Determination of Elemental Content of Polyolefins by Wavelength-Dispersive X-Ray Fluorescence Spectrometry"
- ASTM D 6443, "Standard Test Method for Determination of Calcium, Chlorine, Copper, Magnesium, Phosphorus, Sulfur, and Zinc in Unused Lubricating Oils and Additives by Wavelength-Dispersive X-Ray Fluorescence Spectrometry (Mathematical Correction Procedure)"
- ASTM D 7039, "Standard Test Method for Sulfur in Gasoline, Diesel Fuel, Jet Fuel, Kerosene, Biodiesel, Biodiesel Blends, and Gasoline-Ethanol Blends by Monochromatic Wavelength-Dispersive X-Ray Fluorescence Spectrometry"
- ASTM E 1085, "Standard Test Method for Analysis of Low-Alloy Steels by Wavelength-Dispersive X-Ray Fluorescence Spectrometry"
- ASTM E 1361, "Standard Guide for Correction of Interelement Effects in X-Ray Spectrometric Analysis"
- ASTM E 1621, "Standard Guide for Elemental Analysis by Wavelength-Dispersive X-Ray Fluorescence Spectrometry"
- ASTM E 2465-13, "Standard Test Method for Analysis of Ni-Base Alloys by Wavelength-Dispersive X-Ray Fluorescence Spectrometry"
- ASTM E 322, "Standard Test Method for Analysis of Low-Alloy Steels and Cast Irons by Wavelength-Dispersive X-Ray Fluorescence Spectrometry"
- ASTM E 572, "Standard Test Method for Analysis of Stainless and Alloy Steels by Wavelength-Dispersive X-Ray Fluorescence Spectrometry"

X-Ray Tubes for WDS

The x-ray tubes used for WDS analysis operate at 2 to 3 kW. Much of this power dissipates as heat, and provisions for water cooling of the x-ray tube are necessary. The power supplies and associated electronics for these x-ray tubes are large.

When a sample is considered and the analyte element selected, the first decision is to select the emission line. In the absence of specific interferences, the most energetic line plausible is typically used. For elements with atomic numbers less than approximately 75, this will usually be the K-line, because many WDS can operate to 100 kV potentials for the x-ray tubes. When possible, an x-ray tube is selected that emits characteristic lines at energies just above the absorption edge for the line to be used for the analyte element. When such a tube is not available, the excitation must be accomplished by use of the continuum for an available x-ray tube.

The potential of the x-ray tube should be set approximately 1.5 times the absorption-edge energy or greater. The detector(s) must be selected based on the wavelength region to be used. The proportional counter should be used for x-rays longer than approximately 0.6 nm (6 Å), the scintillation detector for wavelengths shorter than approximately 0.2 nm (2 Å), and both for the overlapping region of 0.2 to 0.6 nm (2 to 6 Å).

Energy-Dispersive Spectrometry

The basic components of an EDS consist of electronic circuitry that processes and counts electrical pulses that are created after x-rays interact with lithium-doped silicon semiconductor detectors. The detectors convert the energy of an x-ray photon into an electrical pulse with specific characteristics of amplitude and width. The circuity for pulse processing (Fig. 13) basically consists of a field-effect transistor (FET) preamplifier maintained at cryogenic temperatures (to reduce electronic noise), a main amplifier, various other signal-processing functions, and a multichannel analyzer (MCA).

As noted, EDS became possible with the development of solid-state, silicon drift x-ray detectors in the 1960s. The Si(Li) detector (Fig. 14) is a layered structure with a lithium-diffused active region that separates a *p*-type entry side from an *n*-type side. Under reversed bias of approximately 600 V, the active region acts as an insulator with an electric-field gradient throughout its volume. When an x-ray photon enters the active region of the detector, photoionization occurs with an electron-hole pair created for each 3.8 eV of photon energy. By the early 1970s, this detector was firmly established in the field of x-ray spectrometry and was applied as an x-ray detection system for SEM and x-ray spectrometry.

The semiconductor detector is cryogenically cooled and in a vacuum, which thus requires a protective window around the detector and is used as a low-density protective window. Older EDS units have a beryllium window (typically with approximately 8 μm nominal thickness), which precludes any analysis of the light elements with $Z < 10$. Thinner windows transmit x-rays more efficiently, especially at low x-ray energy, but are more susceptible to breakage. A system used for the determination of low-atomic-number elements, for which sensitivity and resolution are important, should have a thin window and small- or medium-area detector. In contrast, a system to be used in a factory for the determination of transition elements in alloys should have a thick window and larger-area detector. In the latter case, resolution usually is not a major factor. Windows also can be made of an aluminized polymer that will maintain vacuum integrity around the detector while passing x-rays of energy as low as that of carbon.

Upon entering the Si(Li) detector, an x-ray photon is converted into an electrical charge that is coupled to a FET. The FET and the electronics comprising the preamplifier produce (Fig. 13) an output proportional to the energy of the x-ray photon. Using a pulsed optical preamplifier, this output is in the form of a step signal. Because photons vary in energy and number per unit time, the output signal, due to successive photons being emitted by a multielement sample, resembles a staircase with various step heights and time spacing. When the output reaches a determined level, the detector and the FET circuitry reset to their starting level, and the process is repeated.

The preamplifier output is coupled to a pulse processor that amplifies and shapes the signal into a form acceptable for conversion to a digital format by an analog-to-digital converter (ADC). Amplification is necessary to match the analog signal to the full-scale range of the ADC. This process involves the energy calibration, which must be routinely checked. Drift in the gain and/or offset (zero) of the amplification will result in errors in the energy assigned to the x-ray photons producing the signal. Therefore, these calibrations must be as stable as possible.

The electronic pulse is processed by electronic circuitry so a photon energy can be analyzed before accepting input from any additional photons. The function of the MCA is to measure the pulse from the amplifier and increment the appropriate "channel" (memory location) in the display, where the channel location (number) is proportional to the energy of the photon. Because the detector is sensitive to all x-ray energies, the entire spectrum of interest can be measured with no changes in detector parameters. Typically, several thousand photons are analyzed per second. As these energies are measured, a histogram of the numbers of photons counted corresponding to each energy is plotted.

Early spectrometers were heavy, unwieldy units that used hard-wired MCAs that could acquire data but could do little to process it. With advances in microelectronics, however, the flexibility and capabilities of EDS have expanded to include handheld XRF instruments with low-wattage x-ray tubes for field applications in bulk analysis.

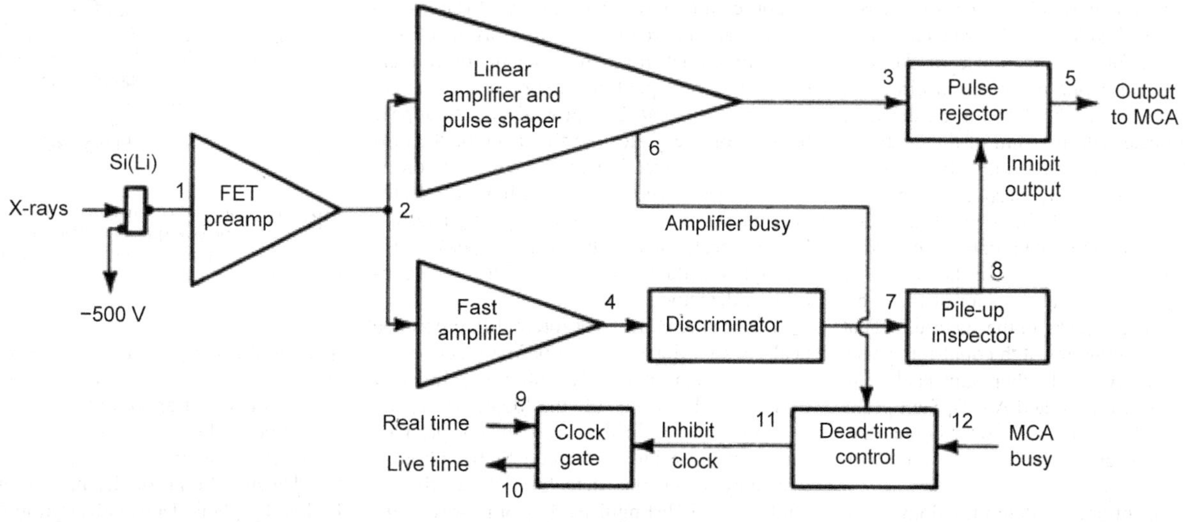

Fig. 13 Schematic diagram of a complete energy-dispersive x-ray spectrometer. Various pulse-processing functions and the multichannel analyzer (MCA) are shown. FET, field-effect transistor

Fine-beam (micro-XRF) instruments also have been developed to characterize compositional uniformity and to map compositional variations within a sample. The implementation of capillary optics for focusing x-ray beams, introduced in the 1980s, has led to instruments with the ability to obtain point spectra and elemental maps. Such instruments typically are capable of collimating the incident x-ray beam to smaller diameters (as low as ~0.1 mm, or 0.004 in.), thus enabling operator-defined adjustment of lateral spatial resolution. A second lens is placed between sample and detector to restrict the volume in the sample from which XRF radiation can reach the detector (Fig. 15). Appropriate alignment of both lenses defines an ellipsoid-like sampling volume somewhere inside the sample, from which XRF signals can be recorded. This map provides a method for characterizing chemical inhomogeneities on a spatial-resolution scale midway between the ~1 cm (0.4 in.) range of bulk XRF and the ~1 μm scale of electron probe microanalysis.

EDS Performance Factors

System performance of EDS also depends on pulse-processing circuitry and processing speed. Although it may appear that the entire spectrum is accumulated simultaneously, the EDS system must process x-ray pulses sequentially. The electronics in an EDS system have pulse pile-up rejection circuitry, to prevent measuring the energy of more than one photon (frequently called sum peaks). The pulse-inspection circuitry has a limit to its time resolution (dead time) to ensure that pulses that enter within the time resolution are in fact detected as a single pulse rather than measuring the energy of two photons.

The pulse-inspection circuitry can exclude coincident events separated by a few microseconds in time. However, there is inevitably a limitation to the time resolution of the inspection circuit, particularly for low-energy photons below 3 keV, which produce pulses just above the fundamental noise of the system. Despite pulse pile-up rejection circuitry, photons coincident in the detector are detected as a single pulse with the sum of the energies. Sum peaks are observed only from major peaks in the spectrum, particularly in the low-energy region, and the size of the sum peak depends on the input count rate. Sum peaks may be observed at twice the energy of an intense peak and/or at the sum of the energies of two intense peaks in the spectrum. That is, if a spectrum contains high count-rate peaks for constituents A and B, then sum peaks can occur for A + A, B + B, and A + B. Sum peaks decrease rapidly in intensity with count rate. Figure 16(b) is an example of spectra with sum peaks.

Precise quantitative determinations also necessitate awareness of the detector efficiency and the occurrence of artifacts. Ideally, the detector should completely collect the charge created by each photon entry and result in a response for only that energy. From 1 to 20 keV, an important region in x-ray spectrometry, silicon detectors are efficient for conversion of x-ray photon energy into charge. Nonetheless, some background counts appear because of energy loss in the detector. Incomplete charge collection in the detector contributes to background counts, although detector efficiency can be controlled by engineering. Escape peaks and false silicon peaks also are signal artifacts associated with the silicon-lithium detector.

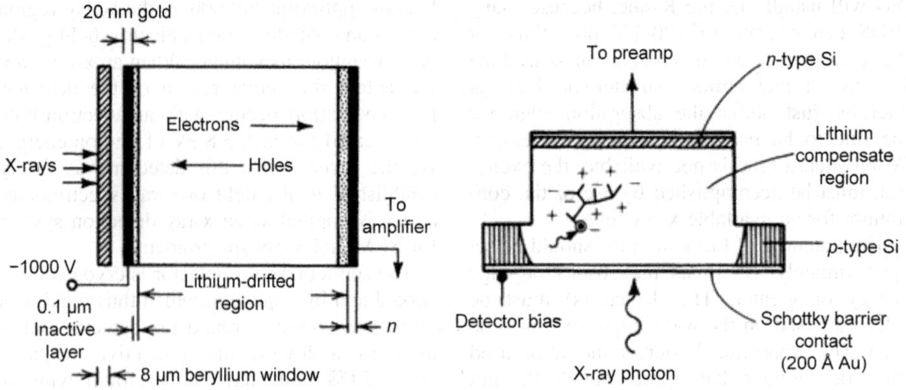

Fig. 14 Schematic diagrams of silicon-lithium semiconductor x-ray detector used in an energy-dispersive x-ray spectrometer

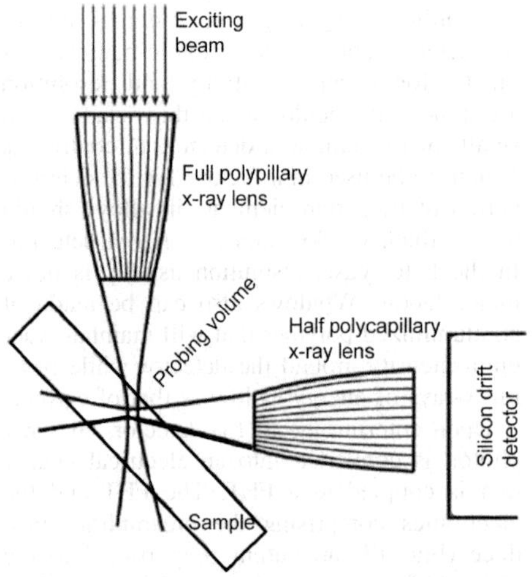

Fig. 15 Confocal lensing used in fine-beam x-ray energy-dispersive spectrometry. Adapted from Ref 18

A false silicon peak (or internal fluorescence peak) is an artifact due to a partially active silicon layer approximately 100 nm thick that exists on the front of the detector. X-ray photons captured in this layer do not produce a normal pulse but can excite Si Kα photons that may enter the detector. These Si Kα photons are indistinguishable from x-rays generated in the sample. A false silicon contribution is thus made to the spectrum, as shown in Fig. 16(a) for a manganese spectrum. In a quantitative analysis of an iron alloy, this false silicon peak appears as an apparent concentration of silicon of approximately 0.1 wt%.

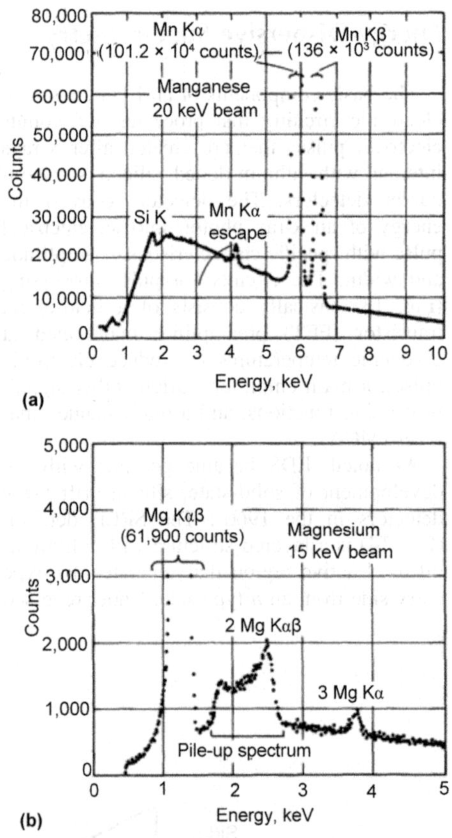

Fig. 16 Examples of sum and escape peaks in energy-dispersive spectra. (a) Escape peaks from Mn Kα and Kβ. (b) Sum peaks in a spectrum of magnesium. Source: Ref 19

Escape peaks are artifacts that arise from an imperfection in the photon-capture process in the detector. Some of the photon energy may be lost by photoelectric absorption of the incident x-ray, creating an excited silicon atom that relaxes to yield a silicon Kα x-ray. This x-ray may "escape" from the detector, resulting in an energy loss equivalent to the photon energy; for silicon Kα, this is 1.74 keV. Therefore, an escape peak 1.74 keV lower in energy than the true photon energy of the detected x-ray may be observed for intense peaks. For Si(Li) detectors, these are usually a few tenths of one percent and never more than a few percent of the intensity of the main peak. The escape peak intensity relative to the main peak is energy dependent but not count-rate dependent. The escape peak intensity decreases with increasing energy of the parent peak and is approximately 1% for a phosphorus K-peak (Fig. 16a).

Energy Resolution

Energy resolution in EDS is inherently limited by electronic noise, statistical variations in conversion of the photon energy, and the counting processes. Electronic noise is minimized by cooling the detector and the associated preamplifier with liquid nitrogen. Half of the peak width is often a result of electronic noise. Resolution of an EDS is normally expressed as the full width at half maximum (FWHM) amplitude of the manganese x-ray at 5.9 keV. The resolution will be energy dependent and somewhat count-rate dependent. Commercial spectrometers are routinely supplied with detectors that display approximately 145 eV (FWHM at 5.9 keV).

The conversion of photon energy into discrete charge carriers also is subject to statistical fluctuations. As a result of these fluctuations, the assignment of a channel in the MCA to a given pulse is distributed about a mean value. When an EDS system is operated at the best possible resolution, the peaks have an FWHM value that is approximately 2.5% of the peak energy, that is, 145 eV FWHM for a Mn Kα peak at 5890 eV. This can be compared to the natural width of the Mn Kα x-ray emission, which is approximately 2 eV. The substantial degradation of the peak width that is produced in the EDS can lead to significant problems with peak overlap (interference) in many practical analytical applications.

The contribution of the continuous bremsstrahlung background at a given x-ray energy can also be important. Because the continuum intensity increases proportionally to the atomic number, the intensity in a given energy window may rise due to changes in composition through effects on the continuum. Such intensity changes may be misinterpreted as changes in intensity of a characteristic peak of a minor constituent located at the same energy. Because the peak-to-background ratio of the EDS is at least ten times poorer than that for the WDS, the EDS is more susceptible to continuum effects. Fortunately, because the EDS constantly measures the entire spectrum, an energy window that corresponds to true bremsstrahlung can be used to normalize the signal from the characteristic window of interest for variations in the bremsstrahlung intensity with composition.

Dead Time

The pulse-inspection circuitry has a limit to its time resolution so that pulses that enter within the time resolution are in fact detected as a single pulse that is the sum of the two individual pulses. Normal operation in x-ray spectrometry is to set the time on the system clock to be used to acquire the spectrum. When a pulse is detected and processing initiated, the clock is "stopped" until the system is ready to process a new photon. The length of time the clock is off is called dead time; the time the clock is on is called live time. Their total is real time. The system monitors live time. If the spectrometer is operated with a 50% dead time, the real time is twice the live time.

The practical effect of dead time in the spectral measurement process is to set a limit to the rate at which pulses can be counted. At high count rates, the time required may become significant. The time necessary to process a pulse, the dead time, is of the order of 100 μs at the best possible system energy resolution. A dead time of 100 μs results in a count-rate limitation of 5000 cps for 50% system dead time. Higher input count rates would actually result in a lower output (recorded) count rate. This counting limitation refers to the integrated count rate over the whole x-ray spectrum, not just a single peak of interest, because there is no way to discriminate among incoming photons in advance of the photon-capture process.

The long dead time of the EDS combined with its poor peak-to-background ratio results in poor detection limits, typically approximately 0.1 wt%, while the high peak-to-background ratio and the high counting rates on the peak of interest available in the WDS can give detection limits of 0.01 wt% (100 ppm) or lower. Shorter values of the system dead time can be selected, but only at the expense of degradation of the energy resolution. For example, by choosing faster amplifier time constants, a limiting input count rate of 20,000 cps may be possible at 50% dead time, but the energy resolution would degrade to 190 eV. To make quantitative measurements of x-ray spectra, it is important to make corrections for dead time, so that the x-ray counts of all lines are recorded on the basis of constant electron dose into the sample. Thus, a low-intensity peak from a minor constituent in a sample may be effectively underreported if the system dead time arises from high-intensity peaks of major constituents. In EDS systems, the dead time is quantitatively and automatically corrected by pulse-inspection circuitry that adds on extra clock time to compensate for the dead-time losses associated with processing x-ray pulses.

X-Ray Tubes for EDS

Energy-dispersive x-ray spectrometers use x-ray tubes that operate at much lower power than tubes for WDS. Various anode materials are available, and each manufacturer of x-ray spectrometers offers special x-ray tube features. However, after many trials of tube design, most remain with the traditional side window design similar to the Coolidge tube (Fig. 4), although it is much smaller than those used in wavelength-dispersive systems. A major factor in the design of the tube and associated power supply is the stability of the tube and voltage.

An alternative to the direct x-ray tube excitation is the use of secondary-target excitation. In this mode, an x-ray tube is used to irradiate a secondary target, whose characteristic x-ray fluorescence is in turn used to excite the x-ray emission of the sample. Because of substantial efficiency loss when using a secondary target, higher-wattage x-ray tubes are required than would be needed for direct excitation.

Secondary-target excitation sometimes affords significant advantages. For example, to determine the low concentration levels of vanadium and chromium in an iron sample, these elements could be excited with an iron secondary target without excitation of the iron in the sample. With direct-tube excitation this would be difficult. Several secondary targets would be required to cover a wide range of elements. Use of secondary-target excitation has been supported as a source of monochromatic radiation for excitation. The significance of this advantage is that many of the fundamental-parameter computer programs, used to compute intensities directly from the basic x-ray equations, require monochromatic excitation radiation.

In practice, secondary-target excitation only approaches the ideal monochromatic radiation. Direct-tube excitation with appropriate primary filters performs well when compared to secondary-target techniques (Ref 20). Therefore, direct x-ray tube excitation remains the most practical for the largest number of applications of EDS. The main strength of the energy-dispersive technique lies in its simultaneous multielement analysis capabilities. Although special cases will occur in which selective excitation is desirable, this often can be accomplished with intelligent use of an appropriate x-ray tube and filter. Any fundamental design features that limit the simultaneous multielement capability will diminish the advantage of the EDS.

Because direct x-ray tube excitation is the most common method used in EDS, selection and efficient use of a single x-ray tube is important in the configuration of an EDS system. In WDS techniques, several x-ray tubes are usually available for the spectrometer. These may be changed for different applications. This is not commonly the case with EDS systems, because many WDS have few if any choices of primary filters. In WDS techniques, it is customary to attempt to excite the desired element by the characteristic emission lines of the tube anode material, but the continuum is used more efficiently in EDS. The use of EDS has been enhanced by computer control of tube current and voltage and by selection of the primary filter.

Characteristic lines emitted by an x-ray tube have much greater intensity at their maxima than the continuous radiation emitted. These lines should be used for excitation whenever possible. In addition, use of a primary filter between the x-ray tube and the sample can effectively approximate monochromatic radiation impinging on the sample from these characteristic lines. Commercial energy-dispersive x-ray systems usually offer various x-ray tube anode materials. To select the x-ray tube anode material, the applications most likely to be encountered should be considered.

The principal concern is to select an anode that has characteristic lines close to, but always higher, in energy than the absorption-edge energies to be encountered. None of the characteristic lines should create spectral interference with elements to be determined. This includes consideration of such details as the Compton scatter peak for the characteristic lines. In addition, it is difficult to perform determinations of the element of the anode material. This is especially true with samples having low concentrations of that element.

Rhodium is a favorable tube anode material for general-purpose use. The characteristic lines of this element are efficient for the excitation of elements with absorption edges to approximately 15 keV. The excitation efficiency for the K-lines of the transition elements ($Z = 22$ to 30) is low; however, the continuum can be used efficiently in this region. Rhodium also has characteristic L-lines at approximately 2.7 to 3.0 keV. These are efficient for the excitation of the K-lines of low-atomic-number elements, such as aluminum, silicon, phosphorus, and sulfur. However, in these cases, a silver anode may be preferable because of the Compton scatter radiation from the rhodium lines. The characteristic lines and the continuum from the x-ray tube may be used for excitation.

Although the elements of many samples can be excited effectively using a combination of the characteristic x-ray lines from the tube anode element and the continuum, more monochromatic radiation is sometimes desired. One such situation involves enhancing the use of fundamental-parameter computations that permit quantitative determination of elements without the need for several concentration standards.

A more frequent situation is the need to reduce the background in the spectrum energy range to be used in the analysis. Use of primary filters placed between the x-ray tube and the sample can be effective in these cases and are usually incorporated under computer control in commercial spectrometers. The object is to filter the primary radiation from the x-ray tube and selectively pass the characteristic lines of the anode element. This is accomplished using a filter made of the same element as the tube anode. Because x-rays of a given line (K, L, and so on) of an element are lower in energy than the absorption edge for that element, the photoelectric component of the mass absorption coefficient is small. Such a filter does not efficiently absorb the characteristic line emitted by the x-ray tube. The higher-energy x-rays from the continuum are efficient for the photoelectric process in the filter and are highly attenuated by absorption. X-rays of lower energy than the filter material absorption edge are absorbed more efficiently as the energy decreases.

The result is x-radiation striking the sample with an intensity that is largely determined by the characteristic lines of the tube anode and that approximates monochromatic radiation. Increasing the thickness of the filter decreases the total intensity, with further gain in the monochromatic approximation. Figure 17 shows the spectrum of a silver anode x-ray tube with and without a silver primary filter. The use of filters may be applied to the L- and K-lines. A filter with low mass absorption coefficient, such as cellulose, is required.

EDS Operation

Optimum conditions for EDS operation and analysis depend on a numbers of factors, and the simultaneous multielement capability of EDS also complicates the selection of optimum conditions. Operational factors depend on the elements being considered, the variety of samples to be encountered, and the interactive nature of instrumentation parameters (a change in one parameter may dictate adjustment of another). For example, selection of a thicker primary filter or a decrease in the tube voltage may require an increase in the tube current. Subjective factors, such as the importance of a particular element of interest in a mixture, may alter the usual guidelines to enhance the intensity of x-rays from that element. For accurate results, reference spectra for spectrum fitting must be obtained under the same conditions as those for the analyses.

The stability, linearity, and proper calibration of pulse-processing components are also important in EDS system performance. The amplifier provides gain and zero controls for the energy calibration of the spectrometer. Drift in the gain and/or offset (zero) of the amplification will result in errors in the energy assigned to the x-ray photons producing the signal. Therefore, these calibrations must be as stable as possible, and calibration must be routinely checked. The energy calibration is important for qualitative identification of the elements and for precise quantitative results when using spectrum-fitting programs. A sample with two intense peaks of roughly equal magnitude, such as a mixture of titanium ($K\alpha = 4.058$ keV) and zirconium ($K\alpha = 15.746$ keV) metal powder cast in polyester resin, makes an excellent and convenient calibration standard. Software is usually supplied to facilitate the adjustment.

The initial selection of instrument operating conditions can follow a logical sequence of decisions, but compromises in spectroscopy must be made. Exceptions may challenge the rule, and so comments offered here must be taken as guidelines. The comments also are directed to quantitative determinations. Qualitative analysis will

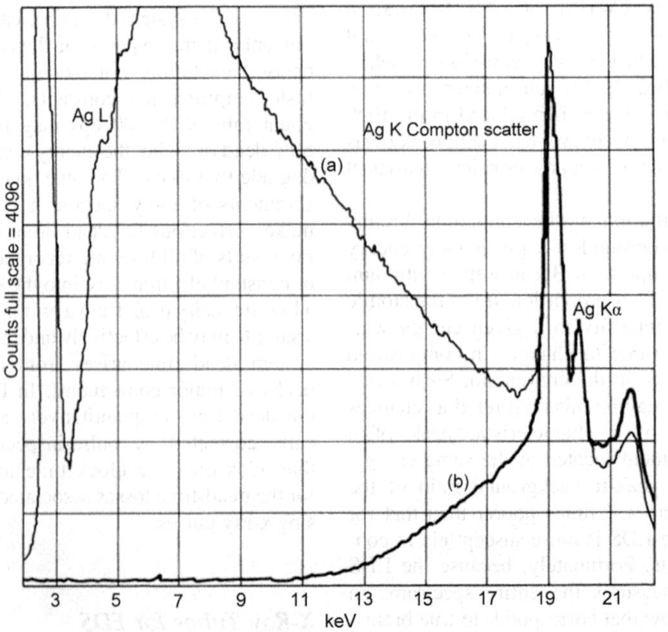

Fig. 17 Spectrum of silver x-ray tube emission. (a) Unfiltered. (b) Filtered with 0.05 mm (0.002 in.) thick silver filter

require similar procedures, usually with less stringent requirements. Experimentation is encouraged.

Once a sample is received for analysis and the elements to be determined by x-ray spectrometry are identified, the next decision is to ascertain which x-ray lines are to be used for the determinations. As a general rule, K-lines are used up to a K-absorption-edge energy a few kiloelectron volts below the characteristic line of the x-ray tube anode element. For example, operation of a rhodium x-ray tube usually necessitates using the K-lines of the elements up to approximately atomic number 40 (zirconium; K_{abs} = 18.0 keV). The continuum may be used for excitation if the voltage to the x-ray tube is set sufficiently high to place the continuum maximum at an energy higher than the absorption edge and if a background filter is used. In these cases, K-absorption-edge energies can be used up to approximately 66% of the maximum operating kilovolts of the x-ray tube. However, the observed peaks will lie on a continuum background and reduce the signal-to-noise ratio.

For a 50 kV x-ray tube, absorption edges as high as 30 keV (Z = 51, antimony; K_{abs} = 30.5 keV) may be used if the element is present in sufficient concentration. For a 30 kV rhodium or silver tube, one is restricted essentially to excitation by the characteristic tube lines. This is of no great concern unless there is a special interest in the elements between atomic numbers 41 and 50 (niobium to tin).

Elements above atomic number 50 (40 for a 30 kV system) must generally be determined using the L-lines of their x-ray spectra. To excite all L-lines, the incident x-ray photon energy must exceed the L_I absorption edge. For practical use, the energy of the L-lines must be greater than approximately 1 keV. For the L-line spectra, this requires atomic numbers greater than 30 (zinc). At such low x-ray energies, absorption of the x-rays and low fluorescent yield in the L-emission in this region require high concentration of the element to be determined and excellent sample preparation. Overlap of the K-lines of the low-atomic-number elements in this region also causes difficulty. For example, the K-lines of phosphorus overlap with the L-lines of zirconium and the M-lines of iridium at approximately 2.0 keV. These problems must be considered but are, to a large degree, solved by careful use of processing software.

Once the x-ray spectral lines are selected for determination of the elements, the next step is to decide whether all analyte elements in the sample can be determined with one instrumental setting. Although the multielement capability of EDS is useful, all elements in every sample cannot be determined with a single set of instrument parameters. Some applications require more than one condition, such as a mixture of low-atomic-number elements and transition elements. The transition elements are best determined by excitation using the K-lines of rhodium or silver and the low-atomic-number elements with the L-lines or a properly adjusted continuum using a background filter. Computer control of instrument parameters facilitates changing the conditions. Whether automatic or manual control is used, all samples should be analyzed under one set of conditions, then analyzed again using the alternate set. This is preferred over changing conditions between samples.

X-ray tube operating voltage will affect the efficiency of excitation of each element in the spectrum and the integrated x-ray photon flux from the tube. The tube current will affect the flux only. Therefore, once the operating kV has been set, the tube current typically is adjusted until the system is processing counts efficiently. System dead time should be maintained below, but near, 50%. The voltage and current settings for the x-ray tube have a surprisingly sensitive effect on the rate of information acquisition and count distribution among the respective spectral peaks for a given type of sample (Ref 21, 22).

Selection of primary tube filter thickness is important. If the filter is changed, the tube current, and sometimes the voltage, will often require resetting because the filter alters the intensity distribution of the x-rays striking the sample. When characteristic tube lines are used for excitation, the filter is usually made from the tube anode element. The intensity of the transmitted x-rays will decrease exponentially with increasing filter thickness. It is common to have two or three primary filters made from the tube anode element in the filter holder. The selection should reflect optimum count rate commensurate with reasonable current and voltage settings. Thicker filters will attenuate lower-energy radiation more effectively and reduce the excitation efficiency for the element with low absorption coefficients.

The remaining decision is the choice of atmosphere in the sample chamber. If x-rays below approximately 5 keV are to be implemented, use of a vacuum may be advantageous. Intensity may increase sufficiently to reduce significantly the counting time required to obtain an adequate number of counts. If the concentration of elements yielding these x-rays is sufficiently high, the vacuum may not be needed. Because of the extra precautions required in sample criteria and handling, a vacuum path should not be used unless significant benefit is realized. Similar reasoning applies to the helium atmosphere.

EDS Applications

XRF Analysis of Cement

Cement can be analyzed using EDS, after being dried and ground to a fine particle size and prepared as pressed pellet specimens. National Institute of Standards and Technology (NIST) cement standards (633 to 639) were pressed into pellets as-received at approximately 276 MPa (40 ksi). A rhodium x-ray tube was operated at 10 kV and 0.02 mA with no primary filter. A vacuum was used. The samples were irradiated for 200 s live time. Magnesium, aluminum, silicon, sulfur, calcium, iron, potassium, and titanium were determined in the samples. Reference files were created for spectra of these elements from their respective oxides, and intensity ratios were obtained using a fitting program. These ratios were related to the concentrations of the aforementioned elements (reported as oxides) in the NIST standards. The results are summarized in Table 2.

Table 2 shows a comparison of three methods of analysis of the data. For the empirical calculation, seven NIST standards were used to obtain the calibration curves and the empirical coefficients. NIST 634 was then analyzed again as an unknown. The data were treated using the empirical correction software. For the fundamental parameter methods, NIST 638 was used as a single standard for calibration of the instrument. The results show that for cement samples the fundamental-parameters software performs exceptionally well.

XRF Analysis of Petroleum Products

Energy-dispersive x-ray spectrometry, with its multielement capabilities, is ideally suited for many analyses of petroleum products. In the following examples, the liquid sample was placed in a 32 mm (1.25 in.) diameter sample cup with a polypropylene window. Polypropylene has better transmission characteristics for low-energy x-rays than polyester film and less interference with the low-Z elements. A silver x-ray tube was operated in the pulsed mode.

For the low-Z elements (aluminum, silicon, phosphorus, sulfur, calcium, and vanadium), the tube was operated at 8 kV, 0.4 mA, without a filter in a helium atmosphere, and the samples were irradiated 300 s live time. For the mid-Z and heavy elements (chromium, manganese, iron, nickel, copper, zinc, lead, and barium), the tube was operated at 21 kV, 0.8 mA, with a 0.05 mm (0.002 in.) silver filter and air path, and the samples were irradiated 100 s live time. The peak intensities were extracted using the XML software (Ref 24) and reference spectra of the elements required

Table 2 Cement analysis for NBS 634

	Concentration, %			
Oxide	Given	Empirical corrections	XRF-11(a)	PC-XRF(b)
MgO	3.30	3.16	3.16	3.19
AlO	5.21	5.01	5.00	5.05
SiO_2	20.73	20.45	20.10	20.35
SO_3	2.21	2.54	2.30	2.32
K_2O	0.42	0.43	0.41	0.35
CaO	62.58	62.55	63.34	64.03
FeO	2.84	2.83	2.81	2.84
TiO_2	0.29	0.29	0.29	0.30

XRF, x-ray fluorescence. (a) See Ref 23. (b) See Ref 24

for the fit. The intensity and concentration data were correlated using a linear or quadratic function.

Analysis of Conostan C-20 100 ppm standard resulted in the calculation of detection limits for elements in oil, as shown in Table 3. Energy-dispersive x-ray spectrometry can perform rapid, accurate determinations of sulfur in oil. The results obtained from a set of standards and unknowns are shown in Table 4.

Sample Preparation

The care taken to determine the best method of sample preparation for a given material and careful adherence to that method often determine the quality of results obtained. Sample preparation is the single most important step in an analysis, yet it is all too often given the least attention. In most cases, the stability and overall reproducibility of commercial x-ray instrumentation is the least significant factor affecting the precision of analytical measurements. Too often, the precision of analytical results expected from x-ray spectrometric determinations is expressed in terms of the theoretical statistics of measurement of x-ray intensities.

Sample cleaning is very important before XRF analysis. For common metals, such as iron or copper, the excitation x-rays penetrate less than 100 μm into the sample and only tens of micrometers for more dense materials, such as gold or lead. The surface thus must accurately reflect the base composition. Surface contamination, thin films, or residues can skew the analysis. In addition to preparation of the sample, precise positioning of the sample in the spectrometer is critical to quantitative determinations. Additional information is available in Ref 13.

When replicate samples are prepared and actual standard deviations measured, deviations are found to be larger than those predicted by counting statistics. If precision is poor, any one analytical result may also be poor, because it may differ substantially from the "true" value. The variety of sample types that may be analyzed using x-ray spectrometry necessitates various sample-preparation techniques.

Samples are often classified as infinitely thick or infinitely thin based on measurement of the attenuation of x-rays. Samples are considered to be infinitely thick if further increase in the thickness yields no increase in observed x-ray intensity. The critical value for infinite thickness will depend on the energy of the emitted x-radiation and the mass absorption coefficient of the sample matrix for those x-rays. For pure iron, the critical thickness is approximately 40 μm for iron x-rays.

An infinitely thin sample is defined as one in which $m(\mu/\rho) \leq 0.1$, where m is the mass per unit area (g/cm^2), and μ/ρ is the sum of the mass absorption coefficients for the incident and emitted x-radiation (Ref 25). Although infinitely thin samples afford many advantages, it is rarely feasible to prepare them from routine samples. Many samples fall between these two cases and require extreme care in preparation.

Solid Samples

Solid samples are defined as single bulk materials, as opposed to powders, filings, or turnings. Solid samples may often be machined to the shape and dimensions of the sample holder. The processing must not contaminate the sample surface to be used for analysis. In other cases, small parts and pieces must be analyzed as-received. The reproducible positioning of these samples in the spectrometer will be critical. It is often useful to fashion a wax mold of the part that will fit into the sample holder. Using the mold as a positioning aid, other identical samples may be reproducibly placed in the spectrometer. This technique is especially useful for small manufactured parts.

Samples taken from unfinished bulk material will often require surface preparation prior to quantitative analysis. Surface finishing may be performed by using a polishing wheel, steel wool, or belt grinder, with subsequent polishing using increasingly fine abrasives. Surface roughness less than 100 μm is usually sufficient for x-ray energies above approximately 5 keV, but surface roughness of less than 20 to 40 μm is required for energies down to approximately 2 keV.

Several precautions are necessary. Alloys of soft metals may smear on the surface as the sample is polished, resulting in a surface coating of the soft metal that will yield high x-ray intensities for that element and subsequently high analytical results. For matrices of low atomic number, such as papers and plastics, all samples should be infinitely thick for the most energetic x-ray used or should be the same thickness. Polishing grooves on the surface of the sample may seriously affect the measured intensity of low-energy x-rays. This can be examined by repetitive measurement of the intensity of a sample after 45 or 90° rotation. Use of a sample spinner reduces this effect. If a sample spinner is not available, the sample should be placed in the spectrometer such that the incident x-radiation is parallel to the polishing direction.

Powders and Briquets

Powder samples may be received as powders or prepared from pulverized bulk material too inhomogeneous for direct analysis. Typical bulk samples pulverized before analysis are ores, refractory materials, and freeze-dried biological tissue. Powders may be analyzed using the spectrometer, pressed into pellets or briquets, or fused with a flux, such as lithium tetraborate. The fused product may be reground and pressed or cast as a disk. For precise quantitative determinations, loose powders are rarely acceptable, especially when low-energy x-rays are used. Pressed briquets are more reliable. However, experience indicates that the best compromise is reground and pressed fusion products. This technique eliminates many problems associated with particle-size effects.

Particle-size effects result from the absorption of the incident and emitted x-rays within an individual particle. If the mass absorption coefficient of the sample matrix is high for the x-radiation used, particles even a few micrometers in diameter may significantly affect attenuation of the radiation within each particle. If the sample consists of particles of various sizes, or the particle size varies between samples, the resulting x-ray intensities may be difficult to interpret. This problem is compounded by the tendency of a material composed of a mixture of particle sizes to segregate when packed. Determination of elements using low-energy x-radiation may lead to errors from particle-size effects of as much as 50%.

If the required speed of analysis prohibits use of fusion techniques, direct determination from packed powders may be considered. The sample should be ground, if possible, to a particle size below the critical value. The grinding time required often may be ascertained by measuring the intensity from a reference sample at increasing grinding times until no further increase is observed. The lowest-energy x-ray to be used in analysis should be selected for this test. Mathematical methods of correction for particle-size effects have been developed but frequently are not useful because the particle-size distribution of the sample is required and not known.

Briquets or pressed powders yield better precision than packed powder samples and are relatively simple and economical to

Table 3 Detection limits of minor elements in oil

Elements	Detection limit, ppm
Ti	6.5
V	3.9
Cr	2.7
Mn	1.7
Fe	1.3
Ni	0.6
Cu	0.4
Zn	0.3
Pb	0.4

Table 4 Results of sulfur determination in oil

Concentration, %		
Given	Found	Relative error, %
0.010	0.007	29.5
0.141	0.156	10.4
0.240	0.249	3.7
0.719	0.691	−3.9
0.950	0.933	−1.7
1.982	2.008	1.3
4.504	4.499	0.1

prepare. In many cases, only a hydraulic press and a suitable die are needed. In the simplest case, the die diameter should be the same as the sample holder so that the pressed briquets will fit directly into the holder. The amount of pressure required to press a briquet that yields maximum intensity depends on the sample matrix, the energy of the x-ray to be used, and the initial particle size of the sample. Therefore, prior grinding of the sample to a particle size less than 100 μm is advisable.

A series of briquets should be prepared from a homogeneous powder using increasing pressure. Safety precautions must be observed, because dies may fracture. The measured intensity of the x-ray lines to be used in the analysis is plotted versus the briqueting pressure. The measured intensity should approach a fixed value, perhaps asymptotically. Pressures of 138 to 276 MPa (20 to 40 ksi) may be required. For materials that will not cohere to form stable briquets, a binding agent may be required.

Acceptable binding agents include powdered cellulose, detergent powders, starch, stearic acid, boric acid, lithium carbonate, polyvinyl alcohol, and commercial binders. Experimentation is usually required with a new type of sample. Briquets that are not mechanically stable may be improved by pressing them into the backing of prepressed binder, such as boric acid, or by the use of a die that will press a cup from a binding agent. The sample powder may then be pressed into a briquet supported by the cup. Metal cups that serve this purpose are commercially available. Improved results are often obtained if approximately 0.1 to 0.5 mm (0.004 to 0.020 in.) is removed from the surface of the briquet prior to measurement.

Fusion of Samples

Fusion of materials with a flux may be performed for several reasons. Some refractory materials cannot be dissolved, ground into fine powders, or converted into a suitable homogeneous form for x-ray spectrometric analysis. Other samples may have compositions that lead to severe interelement effects, and dilution in the flux will reduce these. The fused product, cast into a glass button, provides a stable, homogeneous sample well suited for x-ray measurements. The disadvantages of fusion techniques are the time and material costs involved as well as the dilution of the elements that can result in a reduction in x-ray intensity. However, when other methods of sample preparation fail, fusion will often provide the required results.

Low-temperature fusions may be carried out using potassium pyrosulfate. More common are the glass-forming fusions with lithium borate, lithium tetraborate, or sodium tetraborate. Flux-to-sample ratios range from 1:1 to 10:1. The lithium fluxes have lower mass absorption coefficients and therefore less effect on the intensity of the low-energy x-rays. An immense variety of flux-additive recipes are reported for various sample types. Lithium carbonate may be added to render acidic samples more soluble in the flux; lithium fluoride has the same effect on basic samples. Lithium carbonate also reduces the fusion temperature. Oxidants, such as sodium nitrate and potassium chlorate, may be added to sulfides and other mixtures to prevent loss of these elements. Several detailed fusion procedures are provided in Ref 13. Routine production of quality specimens requires considerable practice.

Filters and Ion-Exchange Resins

Various filters, ion-exchange resin beads, and ion-exchange resin-impregnated filter papers have become important sampling substrates for samples for x-ray spectrometric analysis. Filter materials may be composed of filter paper, membrane filters, glass-fiber filters, and so on. Filters are used in a variety of applications.

One widely used application is in the collection of aerosol samples from the atmosphere. Loadings of several milligrams of sample on the filter may correspond to sampling several hundred cubic meters of atmosphere. Such sampling may be performed in any environment. Many elements may be determined directly on these filters by x-ray spectrometric analysis. Particulate samples collected in this way present problems, stemming primarily from particle-size effects, which are reduced in part by the need to collect two particle-size regions using dichotomous samplers. With these units, particles are separated into those smaller and those larger than approximately 2 μm in diameter. The smaller particles tend to represent man-made materials; the larger ones are of natural origin. The smaller particles exhibit fewer particle-size effects, and x-ray spectrometric determination of even low-atomic-number elements, such as sulfur, is possible. Glass-fiber filters are often used for this purpose. The Environmental Protection Agency has established guidelines for these determinations.

Filters may also be used for nonaerosol atmospheric components, such as reactive gases. Filter materials may be impregnated with a reagent reactive to the gas that will trap it chemically. Sampling is accomplished by conveying atmospheric gases through a treated filter under carefully controlled conditions. An example is a damp filter treated with ferric ion solution used to trap hydrogen sulfide (H_2S). The excess iron can be rinsed from the filter, but the precipitated ferrous sulfide (Fe_2S_3) will remain. The sulfur can be determined directly or indirectly by measuring the iron x-radiation. The key to determining atmospheric components is the development of suitable standards. Some standards for aerosols are commercially available.

Filters can be used to determine solution components in ways parallel to those described for atmospheric components. Particulate materials may be filtered directly from solution. For example, particulate materials in environmental water samples are defined as that which is filtered using a 0.45 μm pore diameter membrane filter. Therefore, filtration of particles from water can be accomplished using such filters, and direct x-ray spectrometric analysis performed.

Application of filter sampling to dissolved elements in water is becoming more common. The principle is similar to the reactive reagent-impregnated filter application to atmospheric gases. In some cases, the filter may be impregnated with ion-exchange resins that will trap ions as the solution passes through the filter. Some varieties of these filters are commercially available.

Procedures using ion-exchange resin-impregnated filters must be carefully checked, because several passes of the solution may be required, and distribution of the ions across the paper thickness is seldom uniform. However, for solutions, a reaction may be performed prior to filtration. For example, many ions can be precipitated quantitatively from aqueous solution, even at parts per billion concentration levels. Commercially available or easily prepared reagents may be used (Ref 26). The precipitates can be collected using 0.45 μm pore-diameter membrane filters, which are then mounted between two polyester film sheets retained by ring clips on a standard plastic sample cup. Simultaneous multielement determinations are then performed using XRF.

Detection limits on the filters of as low as a few tenths of a microgram are common. If 100 g of sample solution is used, this corresponds to the detection limits of a few parts per billion in the sample. Standards are easily prepared as aqueous solutions. Standard Reference Materials (SRM) for environmental waters and industrial effluent water are available from the Environmental Protection Agency and commercial sources. The energy-dispersive x-ray spectrum of a precipitate of an SRM sample is shown in Fig. 18.

Thin-Film Samples

Thin-film samples are ideal for x-ray spectrometric analysis. The x-ray intensity of an infinitely thin sample is proportional to the mass of the element on the film, and the spectral intensities are free of interelement and mass absorption coefficient effects. However, in practice, perfect thin-film samples are rarely encountered. Powder samples of sufficiently small and homogeneous particle size may be distributed on an adhesive surface, such as cellophane tape, or placed between two drumtight layers of polyester film mounted on a sample cup.

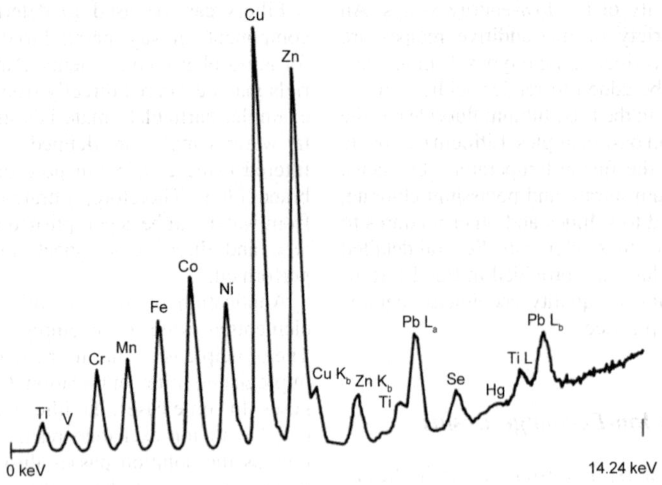

Fig. 18 Spectrum of elements in a preconcentrated standard reference material for industrial effluent water

More important thin-film types are platings and coatings on various substrates. Analysis of these sample types is increasingly important for the electronics industry. Of particular concern are measurements of film thickness and composition. Several techniques may be used, including the substrate intensity attenuation method, the coating intensity method, various intensity ratio methods, and the variable takeoff angle method. The last method is not practical in most commercial spectrometers. These techniques are discussed in Ref 13. To be infinitely thin to most x-rays used in x-ray spectrometric analyses, the specimen must be 10 to 200 μm thick.

Liquid Samples

Liquids may also be analyzed using x-ray spectrometry. The design of x-ray spectrometric instrumentation using inverted optics, in which the specimen is above the x-ray source and detector, facilitates the use of liquid samples. This convenient geometry demands caution in the preparation of liquid samples to avoid damaging the source or detector by such accidents as spills and leaking sample cups.

Quantitative standards are easily prepared for liquid samples. However, because solvents are usually composed of low-atomic-number elements, the Rayleigh and Compton scatter intensity is high, which increases background and leads to high limits of detection. These problems can be minimized by use of suitable primary tube filters, which reduce the scattered x-radiation in the analytically useful region.

Care must be taken with liquids containing suspended solids. If the suspension settles during the measurement time, the x-ray intensity of the contents of the sediment will be enhanced. The x-ray intensity from solution components or homogeneous suspension may decrease as a result of sediment absorption, which leads to erroneous results. This possibility is tested by brief, repetitive measurements, beginning immediately after a sample is prepared. Any observed increase or decrease in intensity with time indicates segregation in the sample. In these cases, an additive that stabilizes the suspension may be used, or the suspended content may be collected on a filter for analysis.

Special Sample Types

Applications of x-ray spectrometric analysis do not always provide convenient samples that can fit one of the aforementioned categories. Nondestructive analyses are occasionally required on production products that are not 32 mm (1.25 in.) diameter circles of infinite thickness. Examples include computer disks, machined parts, and long, coated strips or wire. In these cases, a sample compartment that will accommodate the sample can often be designed. With the development of the mercuric iodide detector, which can provide adequate resolution for many analyses without a liquid nitrogen dewar, special analytical systems for on-line and nondestructive analysis of large samples may become increasingly feasible.

Qualitative Analysis

Qualitative analysis (determination of which elements are present) is done by comparing the energies of the x-rays emitted from the sample with the known characteristic x-ray spectra of each element. Energy-dispersive x-ray spectrometry is a very useful method for elemental qualitative analysis. The technique can qualitatively identify elements from atomic number 11 to the end of the periodic table at levels from a few hundred nanograms in thin films to a few parts per million in bulk samples. Liquids and solids may be analyzed directly, and, in some cases, gaseous materials may be collected for analysis on filters or in chemical traps. With few exceptions, x-ray spectrometry provides no information on the oxidation state, bonding, or other chemical properties of the element.

The primary basis of the identification of elements in a sample is the energy and relative intensity of the K-, L-, or M-spectral lines. Elements can often be identified by simple use of "KLM-markers" on the display screen. Precise energy calibration of the spectrometer is required, and the position and the relative intensity of the lines must be well matched with those displayed by the markers. This is because of coincidental overlap of Kα-lines of an element of atomic number Z with the Kβ-lines of element $Z - 1$ from 3 to 9 keV.

From 1 to 5 keV, L- and M-lines of high-atomic-number elements overlap with K-lines of low-atomic-number elements. The data system will indicate the symbol for the element whose K-, L-, or M-lines are shown. Similar procedures may be used for WDS. However, the need to scan the spectrum impedes the process. Obtaining a suitable scan requires some prior knowledge of the composition to establish suitable operating conditions for the spectrometer.

X-ray spectrometry has good spectroscopic selectivity. That is, there are few circumstances in which spectral overlap of the x-ray energies cannot be adequately handled. For example, the lead L-spectrum consists of an Lα-line at 10.55 keV and Lβ-lines at 12.6 keV. The arsenic K-spectrum shows a Kα-peak at 10.53 keV and Kβ at 11.73 keV. If the lead spectrum is not intense, the Lα-peak at 14.76 keV may be missed. Using the KLM-markers, the K-spectrum of arsenic will not properly fit the L-spectrum of lead, yet in a complex mixture, these may be misinterpreted.

Two alternatives remain. Lead will show an M-line at approximately 2.3 keV, whereas arsenic L-lines would appear at approximately 1.3 keV. The lead M-lines will be weak because of low fluorescent yield. A remaining source of information is the absorption-edge energy values. If the x-ray tube voltage is decreased in steps and spectra acquired, the lead L-spectrum will disappear at its L_{III}-absorption-edge equivalent to 13.0 kV tube voltage. The arsenic K-lines will persist to 11.9 kV. Rarely will such efforts be required for qualitative analysis.

Qualitative WDS

Qualitative analysis for major constituents is relatively straightforward using WDS. The identification of minor and trace constituents, however, is considerably complicated by the multiplicity of lines that can be observed from heavy elements that form major constituents in the sample, generally exceeding that observed in EDS. A major consideration is the database to be used for analysis. The database supplied with most computer-based EDS systems is not adequate for WDS analysis. Reference 27 is a comprehensive source of x-ray energies suitable for WDS analysis.

In addition to the problem of identifying a large family of x-ray lines, WDS is complicated by the nature of the diffraction measurement process. The presence of the integer multiplier in Eq 1 results in several orders appearing in a spectrum from a single parent peak; the higher-order lines may appear on a crystal other than that of the parent peak. Another complication is the possible appearance of satellite lines that arise as a result of bonding effects.

A practical procedure for WDS qualitative analysis can be based on using information initially obtained from an EDS spectrum to identify the major constituents in the sample (Ref 28). From this information, the family of lines (K and L or L and M) of the major constituents should be identified in the set of wavelength scans made for all of the different diffraction crystals available in the spectrometer(s). Next, higher-order (n = 2, 3, 4, and so on) reflections of each of these principal lines should be located in the scans. For low-energy lines of principal constituents, satellite lines should be located.

After all lines associated with major constituents are located, then remaining peaks can be assigned to possible minor constituents. The principal source of ambiguity in qualitative analysis of minor and trace constituents is the possibility that a low-intensity peak is actually a high-order reflection of a peak that is produced by a major constituent. Thus, bookkeeping during the process of identifying the major constituents is critical to accurate peak assignments to minor and trace constituents. Starting with the shortest wavelengths (highest x-ray energies), candidate peaks should be checked against an x-ray data compilation to determine possible elements.

Following the same procedure described previously for major constituents, when a possible assignment of a minor constituent is made to a peak, this assignment should be confirmed by locating the other members of its family of lines. Because of the superior resolution of the WDS, it should be possible to obtain more than one x-ray line to identify minor constituents. For trace constituents near the limit of detectability (approximately 100 ppm), it will probably be possible to locate only one peak per element, particularly for K-lines, where the intensity of the Kβ-peak is reduced by a factor of 10 or more below that of the Kα-peak. With only one peak to identify a trace constituent, the confidence with which an identification can be made is necessarily poorer than the confidence with which a major or minor constituent can be identified.

With WDS, even under computer control, it is not practical to carry out a complete qualitative analysis at each location on the sample. However, the EDS inevitably accumulates the entire spectrum each time a measurement is made, so that the possibility exists of performing a complete qualitative analysis at all analysis locations, at least for major and minor constituents. This capability is especially valuable when unexpected changes in composition may occur.

A useful procedure when examining an unknown for the first time is to accumulate an EDS spectrum while scanning a large area on the sample. Although such a spectrum is not useful for quantitative analysis, it can serve to provide a useful starting point for determining the elements that may be encountered in the sample. The analyst should recognize that minor constituents in a minor phase may not be detected in such an overscan spectrum except at very long spectrum accumulation times.

Quantitative Analysis

Quantitative XRF analysis involves conversion of net line intensity (or counts per second) to analyte concentration. The high sensitivity and excellent precision of XRF spectrometers are suited to quantitative analysis, provided that sources of systematic error are addressed. Precision is on the order of approximately 0.1%, and the main sources of random error are the x-ray source (tube and voltage generator) and the statistics of the counting process. Formulas derived from the Poisson distribution that applies to counting statistics are frequently used to estimate overall accuracy and related parameters, such as the limit of detection. Such formulas are optimistic limits of accuracy valid only when all sources of error except counting statistics are negligible. These sources of error cannot be neglected, particularly in the use of EDS, where considerable corrections for background and line overlaps must often be applied. The representativeness of the analyzed volume must be considered both from a macroscopic and a microscopic point of view.

Systematic errors in quantitative x-ray spectrometry arise mainly from absorption- and specimen-related factors. Absorption factors include both primary and secondary absorption. Primary absorption occurs when specimens absorb photons from the primary source. Secondary absorption refers to the effect of the absorption of characteristic analyte radiation by the specimen matrix. As characteristic radiation passes out from the specimen in which it was generated, it will be absorbed by all matrix elements, by amounts relative to the mass attenuation coefficients of these elements. Qualitative analysis is therefore important to identify elements and the sources of primary and secondary absorptions.

Specimen-related factors include sample size and heterogeneity. Specimens should be large enough to cover the entire x-ray beam, because smaller specimens affect scattering. Specimens should have a homogeneous distribution of elements within the layer of material being excited by the x-ray source. The main problem with heterogeneous materials is secondary absorption of characteristic x-rays. The characteristic peak is reduced for elements that are further from the surface, while peaks are accentuated for elements closer to the surface. Specimens also should have densities similar to reference materials.

The simplest quantitative analysis is determination of a single element in a known matrix. In this instance, a simple calibration curve of analyte concentration versus line intensity is sufficient for quantitative determination. The most complex case is the analysis of all, or most, of the elements in a sample, about which little or nothing is known. This requires qualitative analysis to identify matrix elements followed by quantitative analysis by one of three general methods:

- Comparison with standard similar and known compositions
- Influence-coefficient methods
- Fundamental-parameter technique

Quantitative analyses can be most accurate by comparing the x-ray intensities from the unknown sample with their counterparts from a series of standard similar and known compositions. Both the influence-coefficient method and the fundamental-parameter technique are intensity/concentration algorithms that are used to calculate the concentration values without recourse to the use of standards (see also the section "Fundamental-Parameter Methods" in this article). These mathematical methods require computer-based calculations, and advances in computer technology have brought these correction procedures into practical use for quantitative inorganic analysis with an accuracy of a few tenths of a percent for elements from fluoride (Z = 9) to higher atomic numbers (Ref 11).

Before the widespread use of computer technology in x-ray spectrometers, sample-preparation techniques were frequently more practical for the reduction of matrix effects than application of mathematical models. Dilution of elements in a glass matrix by fusion of the sample with a flux often minimizes absorption/enhancement effects; a constant matrix is provided as well. Sample fusion is selected to provide accurate results. Although the mathematical manipulation of x-ray data is performed easily, use of proper sample-preparation techniques to minimize the need for such processing frequently improves the quality of the analysis.

The lack of linear correlation between x-ray intensity and concentration of element in a sample matrix is sometimes more a result of general variation in mass absorption coefficient than specific interelement absorption/enhancement effects. These cases are especially prevalent in minerals and ores. Considerable improvement in results will often be observed if the mass absorption coefficient for the x-ray energy (excitation and emission) used is determined for the sample and this value used to correct the intensities for each sample. To do this rigorously would be tedious.

One of the simplest methods to approximate and simplify this measurement frequently provides surprisingly favorable results. The Compton scatter intensity measured for the characteristic lines of the x-ray tube increases as the mass absorption coefficient of the sample matrix decreases (Fig. 19). However, all else being equal, the intensity of emission lines of a sample increases as the mass absorption coefficient of the sample matrix decreases.

The ratio of the x-ray emission intensity to the intensity of the Compton scatter peak from the x-ray tube characteristic lines is often more linearly correlated to the concentration of analyte elements than the intensity alone. Figure 20(a) shows a plot of intensity versus concentration, and Fig. 20(b) shows the same concentration data plotted versus the ratio of the intensity to the Compton scatter peak of the x-ray tube. This procedure, sometimes referred to as the compensation method, is used in quantitative analysis of soils, glass, ceramics, textile, and wood (Ref 29). The method uses an algorithm developed from known standards that are calibrated to the inelastic (Compton) scattering from the spectrum. The Compton peak will change in terms of energy range, shape, and intensity depending on the density and composition of the specimen.

Most manufacturers use fundamental parameters or Compton scattering or a combination of both for their quantification methods (Ref 29). Some manufacturers allow the user to create empirical calibrations. There has been recent interest in developing theoretical calibrations, such as the Monte Carlo model, for archaeological materials. This model is currently being used to perform analyses of heterogeneous materials and investigations of complex photon interactions within a sample. Basically, it is a fundamental-parameter method, but it applies statistical models of photon interaction rather than determining concentration strictly through algorithms and equations (Ref 29).

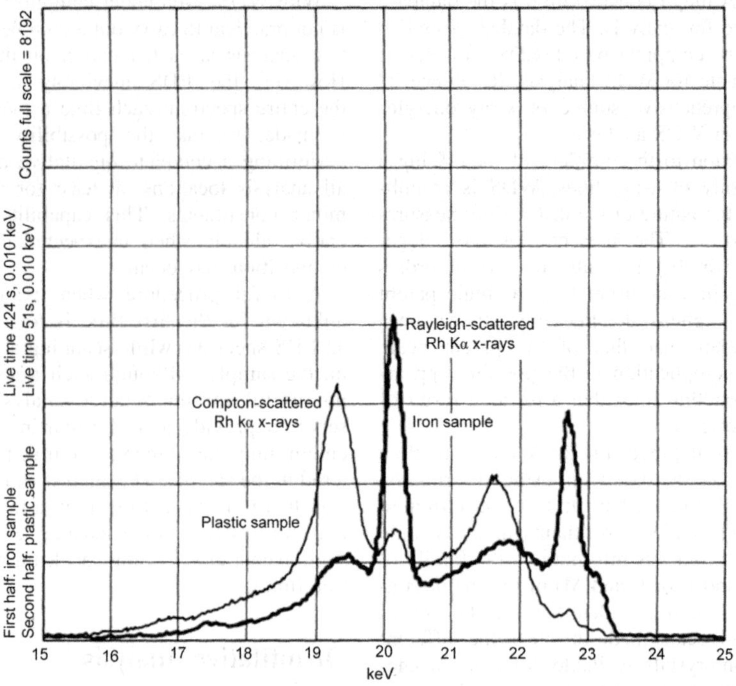

Fig. 19 Compton scatter for rhodium tube from iron and plastic

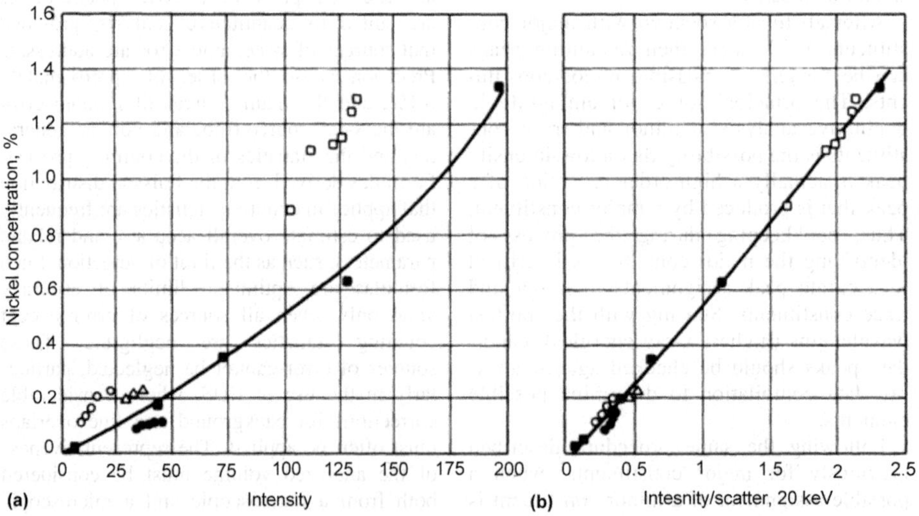

Fig. 20 Ni Kα intensity versus percentage of nickel in nickel ores. (a) Nickel intensity. (b) Nickel intensity divided by scatter at 20 KeV. Symbols represent iron concentration ranging from 10 to 65%.

Counting Statistics and Detection Limit

Quantitative applications of x-ray spectrometry are based on the relationship between the intensity of the x-rays emitted by an element in the specimen and the concentration of that element in a thick specimen or on the total amount of the element in an infinitely thin specimen. The intensities are measured using WDS by setting the goniometer at the 2θ angle for the element of interest and counting x-ray pulses for a period of time to acquire sufficient counts to satisfy the statistical requirements discussed in this section. The background is taken in a similar way by carefully selecting an angle at which only background is measured. In EDS, the intensity is normally found by fitting the spectrum to a set of computer-generated peaks or to reference spectra of single elements previously acquired. The ratio of the area of the reference spectrum required to fit the experimental spectrum is used as the signal proportional to the intensity of x-ray emission by the element of interest. Instrument suppliers offer software for the fitting process.

Intensity is normally expressed as the number of counts per second, I (cps), from the detector or the total number of counts, N, obtained in a fixed period of time, such as 100 s. Because x-ray emission is an example of a random-event process, the precision of intensity measurements can be predicted from theoretical considerations. These events follow Poisson statistics that enable calculation of the standard deviation in the number of counts:

$$\sigma = (N_t)^{1/2} \quad \text{(Eq 13)}$$

where σ is the standard deviation in the counts, and N_t is the number of counts collected in the time t. The relative standard deviation (RSD) in percent is:

$$\text{RSD} = \left[\frac{(N_t)^{1/2}}{N_t}\right] \times 100 = \frac{100}{(N_t)^{1/2}} \quad \text{(Eq 14)}$$

This result is convenient. Unlike most spectroscopic techniques, instrument precision can be adjusted to some degree by the period of time t spent acquiring N_t counts from an x-ray line of intensity I. For example, an x-ray intensity of 100 csp counted 1 s gives N_t = 100, with RSD = 10%. If the same intensity is counted 100 s, N_t = 10,000, and RSD = 1%. These statistics are for the counting process and represent only the theoretical instrumentation limit.

The electronics of most x-ray spectrometers are sufficiently stable that repetitive counting of an unmoved, stable specimen will result in a calculated standard deviation very close to the theoretical value. However, simply removing and replacing the sample will usually increase the standard deviation. The standard deviation obtained from replicate samples is even larger. In fact, the total variance (square of the standard deviation) is the sum of the variances of each step in the process of analyzing specimens from a sampling protocol:

$$V_{\text{total}} = V_{\text{inst}} + V_{\text{pos}} + V_{\text{prep}} + V_{\text{sampling}} + \ldots \quad \text{(Eq 15)}$$

where V_{total} is the total variance, and V_{inst}, V_{pos}, V_{prep}, and V_{sampling} are the variances resulting from instrumentation, specimen positioning, specimen preparation, and sampling, respectively. Therefore, the total standard deviation in an analytical procedure is normally larger than that predicted by counting statistics alone.

In most determinations, the instrumentation contributes the least of all the aforementioned components to the total standard deviation. Precision for a procedure should not be suggested from the counting alone. It is more reliable to prepare replicate samples, measure the intensities, convert these to concentrations, and calculate precision from these data.

The lower limit of detection (LLD) of an element in a sample is a parameter important to the evaluation of instrumentation and the prediction of its applicability to certain analyses. The value of reported LLD frequently misleads because of lack of uniformity in the criteria for the definition. Calculation of an LLD value for x-ray spectrometry is simple. The lowest amount or concentration of an element that can be detected by the instrumentation and procedure used must be established. The magnitude of background in the region of the peak used is important.

If a total of N_b counts are taken in the background, the standard deviation in those counts is $(N_b)^{1/2}$. Assuming Gaussian statistics, 68% of a large number of replicate measurements would give background readings of $N_b \pm (N_b)^{1/2}$ (background ± one standard deviation). Therefore, if a net number of counts in excess of the background, $N_n = N_i - N_b$, is taken that equals one standard deviation, that is, $N_n = (N_b)^{1/2}$, these counts are expected to be greater than the background for only 68% of the measurements.

If the net counts are twice the background standard deviation, this probability increases to 95%, and for three times, the standard deviation increases to greater than 99% probability. The conservative definition of detection limit is often used as the quantity or concentration that yields a net signal equal to three times the standard deviation of the background. Using this definition, the minimum detectable concentration in a sample is:

$$C_{\text{LLD}} = \frac{3(I_b)^{1/2}}{M(t)^{1/2}} \quad \text{(Eq 16)}$$

where C_{LLD} is the minimum detectable concentration, I_b is the intensity of the background (cps), M is the intensity per unit concentration of the analyte (cps/%), and t is the counting time. Such coefficients as 2 or $2 \times (2)^{1/2}$ may be used. Definitive work on the statistical basis of establishing detection limits for methods involving radiation counting is cited in Ref 30.

When comparing detection limits obtained using different spectrometers, it must be ascertained that the same method of computation is used. The detection limit for different elements depends on the instrument and excitation conditions as well as on the matrix composition of the sample. Criteria for establishing concentration levels required for quantitative determinations are also ambiguous. The criterion that the concentration must exceed three times the detection limit for quantitative measurements is often used.

Calibration Curves

Once the intensities of peaks in a spectrum have been extracted, the data must be processed to relate the intensity of the respective peaks to concentration. This is most easily accomplished by plotting the x-ray counts (or counts per second) versus the concentration of the respective analyte element in standards. Such a plot is called a working curve or calibration curve and is the most fundamental way of relating the data to concentration. The relationship between intensity and concentration in x-ray spectrometry often depends on the total sample composition rather than only the element of interest; this is a result of matrix effects. Such cases require simultaneous or iterative computations of data for many elements in the sample. Use of the computer permits trying different mathematical models for the intensity-concentration relationships.

The ideal analytical relationship is one in which the signal or intensity is linearly related to concentration:

$$I_i = a_1 C_i + a_0 \quad \text{(Eq 17)}$$

where I is the intensity, and C is the concentration of the analyte element i. The coefficients a_1 and a_0 represent the slope (sensitivity) and intercept (blank + background), respectively. Taking a set of data from standards will result in random errors and a plot of the data scatter about a line. Therefore, a "best" line must be drawn using the data. Software using techniques known as least-squares fit, or linear regression, calculates the a_1 and a_0 values for this line.

Variations in the composition of a sample may cause deviations from linearity. Measurements of the intensity of K-lines of titanium in a low-atomic-number matrix, such as a plastic material, represent a simple example. When the titanium concentration is low, the mass absorption coefficient at the energy of the titanium K-lines is essentially that of the plastic. Therefore, the titanium concentration increases with the intensity of the titanium line and the mass absorption coefficient of the sample for the titanium K-lines, resulting in a negative deviation from linearity, as shown in Fig. 21. In this example, the deviation is understood and can be corrected using mass absorption coefficients and iterative techniques. However,

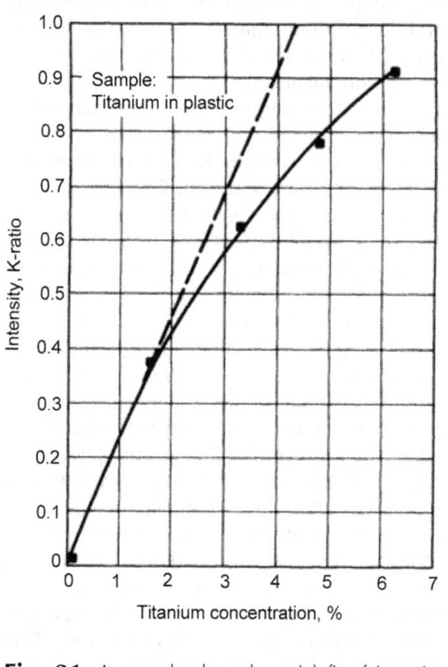

Fig. 21 A second-order polynomial fit of intensity versus concentration

often it is easier to "fit" the curve with an empirical expression rather than use fundamental information that may not be readily available. For this purpose, polynomial regressions may be used:

$$I = a_n C^n + a_{n-1} C^{n-1} + \ldots a_0 \quad \text{(Eq 18)}$$

If a high-degree polynomial, such as third or fourth order, is used, a good fit may be obtained. However, requiring polynomials higher than second order indicates the presence of several interacting processes. Figure 21 shows a second-order polynomial fit to the data. Whether a linear or polynomial regression is used, the concentration of analyte elements can be calculated from the measured intensities of lines from samples. Regression equations for several elements may be stored in the computer.

In many types of samples suitable for XRF, such as alloys, minerals and ores, and various composite materials, mass absorption coefficient changes resulting from the sample composition are affected. Interelement effects may also result. These processes are based on the physics of the absorption-emission processes. The number of simultaneous processes occurring over a wide x-ray energy range complicates attempts to solve the corresponding mathematical relationships exactly. This is especially true when excitation is performed with polychromatic radiation, such as a continuum from an x-ray tube. It is often easier and more effective to use model approximations of these processes. A variety of models have been proposed and applied to various circumstances and are described in Ref 13.

The basic processes that lead to interelement effects are absorption of x-rays emitted in a sample by another element in the sample and the resulting enhancement of the fluorescence of the latter element, thus often called absorption/enhancement effects. Consider a stainless steel sample containing chromium, iron, and nickel in substantial amounts. Nickel x-rays from the K-lines are above the K_{abs} of chromium and iron. Therefore, nickel x-rays traveling through the sample may be absorbed by chromium and iron as a result of the photoelectric process, thereby diminishing the nickel intensity. The probability of this happening will increase as the chromium and/or iron concentration increases. The nickel intensity will appear to decrease with increasing chromium and/or iron content in the sample even with constant nickel concentration. Except for the specific photoelectric absorption of nickel x-rays by iron and chromium, the three elements would have similar mass absorption coefficients. In the transition element series, it may be noticed that the K-lines of an element with an atomic number Z are above K_{abs} for the elements with atomic number $Z - 2$ and below. This observation is useful in predicting absorption/enhancement possibilities.

In addition to the loss of intensity because of absorption of x-rays emitted by an element, the absorbing element may exhibit enhanced intensity. Absorption is significant because of the photoelectric effect. In the previous example, the chromium atom with the nickel x-radiation will emit chromium x-rays. This extra mode of excitation enhances the chromium signal proportional to an increase in the nickel concentration.

The intensity of the nickel x-radiation will be proportional to the nickel concentration but will be decreased by an amount proportional to the chromium concentration:

$$I_{Ni} = a_{1,Ni} C_{Ni} + a_{0,Ni} + b_{Ni,Cr} C_{Cr} \quad \text{(Eq 19)}$$

where $b_{Ni,Cr}$ is the coefficient for the effect of chromium on the intensity of nickel x-rays and will have a negative value. The corresponding relationship for chromium is:

$$I_{Cr} = a_{1,Cr} C_{Cr} + a_{0,Cr} + b_{Cr,Ni} C_{Ni} \quad \text{(Eq 20)}$$

where $b_{Cr,Ni}$ is the coefficient for the effect of nickel on the intensity of the chromium x-rays and will have a positive value.

In applying such relationships as Eq 19 and 20, the b coefficients are included to reflect the qualitatively known effects of x-ray absorption and emission. However, the degree of these effects is not easily computed from first principles. As a result, the coefficients are determined empirically by computation from intensity data taken from standards. Assuming the data are background corrected, the a_0 terms in Eq 19 and 20 can be assumed to be negligible. Using this assumption for two elements, there are four unknowns—$a_{1,Ni}$, $a_{1,Cr}$, $b_{Ni,Cr}$, and $b_{Cr,Ni}$—requiring four equations. The data for these four equations may derive from intensities measured from four standards.

Because of experimental errors, it is preferable to use an overdetermined system in which substantially more than the minimum required data are available. Least-squares methods are used to obtain the coefficients. Obtaining reliable coefficients requires a minimum number of standards equal to the number of analyte elements plus the number of interactions to be considered.

Equations 18 and 19 are commonly written as:

$$\frac{C_i}{R_i} = 1 + a_{ij} \cdot C_j \quad \text{(Eq 21)}$$

where R_i is the measured x-ray intensity of an element in a sample relative to that of the pure element under identical conditions; C_i is the concentration of the analyte i in the sample, a_{ij} values are termed alpha coefficients; and C_j is the concentration of the element j interacting with the analyte element i. Equation 21 represents a binary sample. For multicomponent cases, the relationship is:

$$\frac{C_i}{R_i} = 1 + a_{ij} C_j + a_{ik} C_k + \ldots \quad \text{(Eq 22)}$$

Equation 22 has become known as the Lachance-Traill equation. Other modifications of the basic equations are the Lucas-Tooth and Pyne model (Eq 23) and the Rasberry-Heinrich model (Eq 24):

$$C_i = B_i + I_i \left(K_0 + \sum_{j=1}^{j=n} K_{ij} I_j \right) \quad \text{(Eq 23)}$$

$$\frac{C_i}{R_i} = 1 + \sum_{K \neq 1} A_{ik} C_k + \sum_{j \neq 1} \frac{B_{ij}}{1 + C_i} \cdot C_j \quad \text{(Eq 24)}$$

Equation 23 is based on intensities rather than concentrations and is especially useful when the concentration of interfering elements is not known in the standards. The basic empirical correction equations assume that an element pair exhibits only absorption or enhancement and treat enhancement as a "negative absorption." This is not the case with many types of samples.

Development of Eq 24 was based on analysis of Cr-Fe-Ni alloy systems. The coefficients A_{ij} are absorption coefficients; B_{ij} are enhancement coefficients. Equation 24 permits independent and simultaneous consideration of absorption and enhancement. For example, in the determination of potassium, calcium, and titanium in paint, the L-lines of barium in the paint enhance the potassium, calcium, and titanium intensity. The mass absorption coefficient of barium for the K-lines of potassium, calcium, and titanium is large. Barium must be included as an enhancer and absorber for these elements. This increases the number of standards needed to determine the coefficients.

The algorithms proposed are based on a variety of assumptions, and none appears to work for all types of samples. Alpha coefficients are not constant over large changes in composition (concentration) when polychromatic radiation is used for excitation. Careful experimentation will reveal the model that provides optimum results for the sample types, analytes, and concentration range of concern. Software is available that allows selection of various models. Many of the models are described in Ref 31.

In using empirical corrections software, once the instrument conditions to be used are established, spectra of the standards are obtained. The software provides least-squares solutions for the values of the coefficients. A minimum of one more standard than the number of coefficients to be calculated is required. More standards are required to use the full capability of Eq 24. Empirical parameter software has two parts. First, coefficients are calculated from the intensity and concentration data of the standards. Second, the coefficients are used to compute the concentration of the analyte elements in subsequent unknowns. The x-ray intensities of the unknowns must be measured under conditions identical to those used for the standards. These methods are only a best fit of a function to a set of standards and should not be used for samples whose composition falls outside the range represented by those standards.

Fundamental-Parameter Methods

Fundamental-parameter (FP) methods comprise a wide variety of applications and mathematical approaches with the level of achievable accuracy that depends on the range of the analyte elements and the analyte lines, the type of mathematical models (such as conventional FP-equations, Monte Carlo approaches, or theoretical influence coefficients), the type of specimen (bulk material, layered material, fused samples, etc.), and the instrumentation, including the geometrical setup, excitation source, and detection system. In addition, the reliability of reference materials, if required, and the values of FPs also are key elements in accurate determination of elemental concentrations from the intensity peaks of characteristic x-rays.

Mathematical models of x-ray photon interactions within a specimen are to develop used intensity/concentration algorithms that allow the calculation of the concentration values without recourse to the use of standards. These methods can provide accurate quantitative results, provided that the simplifying assumptions are understood in order to make the mathematical calculations manageable. Necessary assumptions may include sample homogeneity, sample orientation, sample shape, and physical factors such as fluorescent yield.

The basic question in quantitative XRF analysis is the limitations on accuracy if fluorescent intensities can be determined with high precision. In analyzing the accuracy of XRF spectroscopy, there are two basic methods in developing the intensity/concentration models that allow the calculation of the concentration values: the influence coefficient methods and the FP technique. Both of these methods start from the same equation that describes the production of x-rays in a specimen that is irradiated with x-rays (as described in many books, such as Ref 31 to 35):

$$P_i = q\,E_i\,C_i \int_{\lambda_0}^{\lambda_{\mathrm{abs},i}} \frac{\mu_{i,\lambda}\,I_\lambda \mathrm{d}\lambda}{\mu_{s,i} + A\,\mu_{s,\lambda_i}} \qquad \text{(Eq 25)}$$

where P_i is the fluorescence based on primary excitation only, E_i contains element-dependent FPs such as fluorescence efficiencies, q contains geometrical factors, and C_i is the weight fraction of the relevant element i. The integral in the equation describes the contribution of the exciting x-rays, I_λ, to the process of generating fluorescence. The $\mu_{i,\lambda}$ and $\mu_{s,\lambda}$ are the absorption coefficients at wavelength λ of the element i and the sample s, respectively, while A is a geometrical constant. The integration is from the shortest wavelength available, λ_0, to the relevant absorption edge of element i ($\lambda_{\mathrm{abs},i}$).

Influence Coefficient Methods

In the 1950s, Sherman (Ref 36) derived the basic equation for what is now generally known as the influence coefficient or α-method. In this approach, the integration is replaced by evaluation of the integrand at a so-called effective wavelength. The aim of the influence coefficient method is to transform nonlinear equations into a set of linear ones. A consequence is the conversion of integration over many wavelengths into evaluation at a single, effective, excitation wavelength. Furthermore, it includes the influence of all other elements j on element i through the influence coefficient ($\alpha_{i,j}$). In calculating specimen composition, the general relation between the concentration (C_i) of element i and the measured intensity (I_i) is:

$$C_i = K_i\,I_i\,M_{i,s}$$

where K_i is a calibration constant factor of analyte i, and $M_{i,s}$ is a correction factor for matrix effects of the sample s on i. The models incorporate effects of third elements by introduction of either more coefficients or by performing a calibration in various concentration ranges.

Since Sherman, many efforts have continued in the development and refinement of the influence coefficient methods (Ref 37). Many books have been written on the number of coefficients needed in multicomponent systems and on the exact ways to determine the effective excitation wavelength, which is dependent on the sample composition. Influence coefficient correction procedures can be divided into three basic types (Ref 11): fundamental, derived, and regression. Fundamental models are those that require starting with concentrations, then calculating the intensities. Derived models are those that are based on some simplification of a fundamental method but that still allow concentrations to be calculated from intensities.

Regression models are those that are semiempirical in nature and that allow the determination of influence coefficients by regression analysis of data sets obtained from standards. All regression models have essentially the same form and consist of a weight fraction term, W (or concentration, C); an intensity (or intensity ratio) term, I; an instrument-dependent term that essentially defines the sensitivity of the spectrometer for the analyte in question; and a correction term that corrects the instrument sensitivity for the effect of the matrix.

The major advantage to be gained by use of influence coefficient methods is that a wide range of concentration ranges can be covered using a relatively inexpensive computer for the calculations. However, a large number of well-analyzed standards may be required for the initial determination of the coefficients. In practice, a series of type standards is needed to do a calibration. This means that the procedure is fit for process control but not for analysis of really unknown samples. The influence coefficient methods only work when all elements in the sample have been accounted for in the calibration. Results are typically normalized to 100%, and if any elements are not represented in the calibration, they will not be recognized in the results. The method also only applies to bulk materials and is not suitable for layered samples. Nonetheless, where adequate precautions have been taken to ensure correct separation of instrument- and matrix-dependent terms, the correction constants are transportable from one spectrometer to another and, in principle, need only be determined once (Ref 11). There are strict guidelines suggested for producing reference materials for XRF, and there are many companies that specialize in their production.

Standardless Fundamental-Parameter Method

The underlying idea of an FP method is to use basic physical factors in calculating concentration values without recourse to the use of standards. This idealized case is sometimes referred to as the standardless FP method (Ref 38). There are many commercial products termed as standardless, with the understanding of employing precalibrated instruments, where data from reference materials are stored in the factory and remain valid over extended periods of time due to the exceptional long-term stability of modern instruments. In this context, a truly standardless method would be one that is based on absolute counts rather than count-rate ratios with no reference materials for calibration (Ref 38). In practice, more reliability is possible if at least one standard similar to the unknown is used for instrument calibration (Ref 39).

The development of FP methods began in the 1960s to 1970s (Ref 40–42) with overlapping developments in the use of standards and applications in the uses of influence coefficients (Ref 37). A key early part of the FP method was describing the intensity distribution from the x-ray tubes (Ref 43). It is possible to calculate the theoretical x-ray intensities provided that the FPs for each element (including fluorescence intensities, absorption coefficients, emission wavelengths, and absorption edges) are known along with sample composition, instrument parameters (e.g., tube excitation voltage, x-ray optical geometry), x-ray tube spectrum, and specimen size/configuration. The benefit is that the method uses an absolute intensity/concentration algorithm for quantitative analysis of bulk specimens as well as thin-film specimens, including multilayer thin films with a few pure element or multielement standards.

The relative intensity of an x-ray spectral line excited by monochromatic radiation can be computed for a given element, specific transition, and known spectrometer geometry:

$$I_{\mathrm{L}} = I_0\,\omega_{\mathrm{A}} g_{\mathrm{L}}\,\frac{r_{\mathrm{A}} - 1}{r_{\mathrm{A}}}\,\frac{d\Omega}{4\pi}\,\frac{C_{\mathrm{A}}\mu_{\mathrm{A}}(\lambda_{\mathrm{pri}})\csc\phi}{\mu_{\mathrm{M}}(\lambda_{\mathrm{pri}})\csc\phi + \mu_{\mathrm{M}}(\lambda_{\mathrm{L}})\csc\psi} \qquad \text{(Eq 26)}$$

Table 5 Definitions of symbols used in Eq 26

Symbol	Definition
I_L	Analyte line intensity
I_0	Intensity of the primary beam with effective wavelength λ_{pri}
λ_{pri}	Effective wavelength of the primary x-ray
λ_L	Wavelength of the measured analyte line
ω_A	Fluorescent yield of analyte A
g_L	Fractional value of the measured analyte line L in its series
r_A	Absorption-edge jump ratio of analyte A
C_A	Concentration of analyte A
$d\Omega/4\pi$	Fractional value of the fluorescent x-ray directed toward a detector
$\mu_A(\lambda_{pri})$	Mass absorption coefficient of analyte A for λ_{pri}
$\mu_M(\lambda_{pri})$	Mass absorption coefficient of the matrix for λ_{pri}
$\mu_M(\lambda_L)$	Mass absorption coefficient of the matrix for analyte line λ_L
φ	Incident angle of the primary beam
ψ	Takeoff angle of fluorescent beam

where the terms are as defined in Table 5. However, if polychromatic excitation is used, and if the sample has many elements to consider, Eq 26 becomes very complex; but such computations can be performed by computers and software. Basically, FP software contains an algorithm that solves the set of nonlinear equations describing the dependence of the intensity on the concentration and layer thickness of each element (Ref 44). The solution is of nonlinear equations—no linearization is done at any stage, as is customary for the influence coefficient methods. The spectrometer transmission is determined by using at least one standard per element that can differ, both in concentration and composition, from the unknown. The standards are measured and the ratios of the experimental intensities to the intensities calculated using the FP package are measured (Ref 45). When more than one standard is used and some elements occur in more than one standard, a linear regression can be performed.

Besides experimental errors, the main source of inaccuracy in the FP method is inaccurate values of FPs, variations in the tube spectra, and incomplete accounting of geometric factors and geometries and indirect excitation effects, particularly when light elements or L- and M-lines are involved (Ref 38). The current error-limit of absolute (truly standardless) XRF by FP methods under favorable circumstances (fully calibrated synchrotron beam lines, medium-Z elements) is 3 to 4% relative. By using relative intensities with conventional instrumentation accuracies of a few 0.1 wt%, relative can be achieved with K-analyte lines in the case of well-prepared, homogeneous specimens (Ref 38).

Applications in Quantitative Analysis

XRF Analysis of Coal

The British thermal unit and ash content of coal can be estimated from its mineral composition. X-ray spectrometry is a rapid and economical method for determining major and minor elements in coal. The coal may be dried and ground (<325 mesh), and specimens prepared as pellets. In the examples given subsequently, a chromium x-ray tube was operated in the pulsed mode at 40 kV with no primary filter. Samples were irradiated 100 s live time. A set of well-characterized standards encompassing the range of concentrations found for each element in all unknowns was used. A typical set of results is given in Table 6; the values are expressed as the percent of each element based on dry coal.

XRF Analysis of Stainless Steel

Each alloy type requires development of data-reduction procedures. The following example of the analysis of 300/400-series stainless steels illustrates the options for data reduction and standardization. The samples were cut on a lathe with no further surface polishing. A silver x-ray tube was operated in the pulsed mode at 30 kV and 0.08 mA with a 0.05 mm (0.002 in.) silver primary filter. Samples were irradiated 256 s live time, with the dead time typically at 22%.

Several approaches to standardization and data reduction were implemented. A quadratic regression fit of each element to the intensity data of the standards was used, as was Eq 24 for absorption/enhancement effects. Finally, FP software (Ref 23) was used with one standard. Table 7 lists a typical set of results for an AISI type 309 stainless steel sample for the various methods of data treatment.

Detailed evaluation of the data shows that none of the methods of data treatment is significantly better than the others. Equation 24 yields favorable results but is sensitive to the quality of the data entered in the program. The FP software outlined in Ref 23 works well and can accept more than one standard by using an adaptive regression technique. This option was not used in this example. Minimizing systematic errors in the software in Ref 23 depends on the quality of the standard selected. These analyses were performed using a system operated at 30 kV, indicating that the newer low-cost 30 kV power supplies and x-ray tubes are applicable.

XRF Analysis of Thin Films

The intensity of a characteristic x-ray line measured from a thin film depends on the thickness of the film or layer of material emitting the x-radiation. Figure 22 is a generic plot of x-ray intensity versus thickness of such a film for a single element. The plot may be characterized by three regions. In the first region of very thin film thickness, the intensity of the x-radiation increases linearly with thickness of the film. In an intermediate region, the intensity varies exponentially with the thickness. At a higher film thickness, intensity of the emitted x-radiation does not change with increased film thickness. These are the regions of infinitely thin, intermediate thickness, and infinitely thick x-ray samples.

The value of the film thickness for these regions depends on the composition of the film, geometric factors, and the energy of the x-radiation used. For chromium x-rays in a typical EDS, an infinitely thin region is that below approximately 1 μm; the infinitely thick region is that above approximately 15 μm. If the thin layer on a dissimilar substrate

Table 6 Typical results for analysis of coal

Element	Concentration, % Given	Found	Error
Na	0.070	0.153	0.083
Mg	0.100	0.176	0.076
Al	3.750	3.710	−0.040
Si	7.510	7.497	−0.013
S	0.720	0.662	−0.058
Cl	0.020	0.020	0.000
K	0.143	0.143	0.000
Ca	0.070	0.061	−0.009
Ti	0.249	0.247	−0.002
Fe	0.330	0.299	−0.031

Table 7 Results of analysis of AISI type 309 stainless steel

Element	Concentration, % Given	Quadratic	Eq 24	XRF-11(a)
Cr	23.8	23.5	23.7	23.5
Fe	59.8	59.1	56.7	58.1
Ni	13.5	14.5	16.4	15.6
Mn	1.83	1.80	2.09	1.81
Mo	0.19	0.24	0.24	0.18

(a) See Ref 23

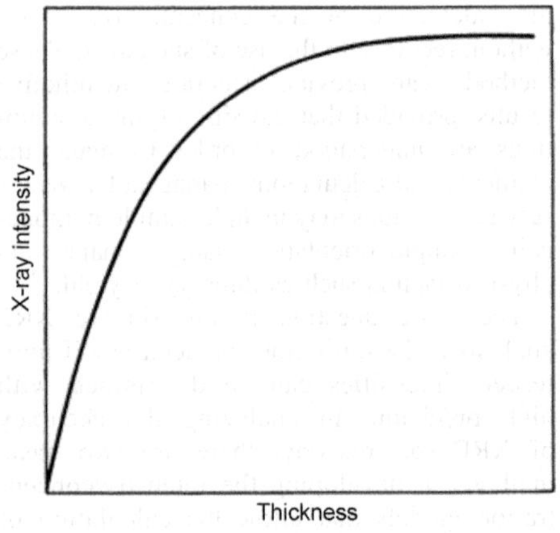

Fig. 22 Theoretical intensity versus thickness for a single element on a dissimilar substrate

Table 8 Results of Permalloy film thickness determination on ceramic

Given	Calculated
Thickness, μm	
1.61	1.60
2.59	2.58
2.11	2.13
1.75	1.75
2.35	2.34
Iron concentration, %	
19.2	19.2
19.2	19.1
18.9	19.1
19.3	19.2
19.2	19.2

is an alloy, interelement effects cause the measured intensity from the sample to deviate from the simple model. If a suitable set of standards is available, and if the measurements are applied to a small range of film thickness, an empirical, linear model can be used.

A set of standards of 20% Fe and 80% Ni (Permalloy) on a ceramic substrate was measured over a thickness range of 1.6 to 2.6 μm. A molybdenum x-ray tube was operated at 10 kV and 0.10 mA using a cellulose primary filter. The samples were irradiated 100 s live time. The software uses the intensity to compute the thickness of the alloy film and then uses this result to compute the composition of the alloy.

The results in Table 8 show that x-ray spectrometry is useful for rapid, on-line measurement to film thickness and composition in a variety of coating and plating products. Each case will require the development of data-treatment software and a set of well-characterized standards.

ACKNOWLEDGMENTS

Portions of this article were adapted from:

- K.H. Eckelmeyer, Bulk Elemental Analysis, *Metals Handbook Desk Edition,* ASM International, 1998, p 1411–1416
- K.F.J. Heinrich and D.E. Newbury, Electron Probe X-Ray Microanalysis, *Materials Characterization,* Vol 10, *ASM Handbook,* ASM International, 1986, p 516–535
- R. Jenkins, X-Ray Techniques: Overview, *Encyclopedia of Analytical Chemistry,* R.A. Meyers, Ed., John Wiley & Sons Ltd., 2000, p 13269–13288
- D.E. Leyden, X-Ray Spectrometry, *Materials Characterization,* Vol 10, *ASM Handbook,* ASM International, 1986, p 82–101

REFERENCES

1. R.M. Reece, A.J. Reed, C.S. Clark, R.C. Angoff, K.R. Casey, R.S. Challop, and E. McCabe, Elevated Blood Lead and the In Situ Analysis of Wall Paint by X-Ray Fluorescence, *Am. J. Dis. Child.*, Vol 24, 1972, p 500–502
2. S. Clark, W. Menrath, M. Chen, S. Roda, and P. Succop, Use of a Field Portable X-Ray Fluorescence Analyzer to Determine the Concentration of Lead and Other Metals in Soil Samples, *Ann. Agricul. Environ. Med.*, Vol 6 (No. 1), 1999, p 27–32
3. C. Brundle, C. Evans, and S. Wilson, *Encyclopedia of Materials Characterization*, Butterworth-Heinemann, 1992, p 128
4. R. Castaing, *Adv. Electr. Elec. Phys.*, Vol 13, 1960, p 317
5. A.H. Kramers, *Philos. Mag.*, Vol 46, 1923, p 836
6. R.T. Beatty, *Proc. R. Soc. London, Ser. A*, Vol 89, 1913, p 314
7. H. Kulenkampff, *Ann. Phys.*, Vol 69, 1923, p 548
8. H.G.J. Moseley, High Frequency Spectra of the Elements: Part I, *Philos. Mag.*, Vol 26, 1912, p 1024–1034
9. H.G.J. Moseley, High Frequency Spectra of the Elements: Part II, *Philos. Mag.*, Vol 27, 1913, p 703–713
10. J.A. Bearden, "X-Ray Wavelengths," U.S. Atomic Energy Commission Report NYO-10586, Oak Ridge, TN, 1964, p 533
11. R. Jenkins, X-Ray Techniques: Overview, *Encyclopedia of Analytical Chemistry*, R.A. Meyers, Ed., John Wiley & Sons Ltd., 2000, p 13269–13288
12. E.P. Bertin, *Introduction to X-Ray Spectrometric Analysis*, Plenum Press, 1978
13. E.P. Bertin, *Principles and Practice of X-Ray Spectrometric Analysis*, 2nd ed., Plenum Press, 1975
14. P. Auger, Compound Photoelectric Effect, *Compt. Rend.*, Vol 180, 1925, p 65
15. P. Auger, *J. Phys.*, Vol 6, 1925, p 205
16. R.M. Rosseau, Concept of the Influence Coefficient, *Rigaku J.*, Vol 18 (No. 1), 2001
17. K.F.J. Heinrich, *Electron Beam X-ray Microanalysis*, Van Nostrand Reinhold, 1981
18. K. Janssens et al., Recent Trends in Quantitative Aspects of Microscopic X-Ray Fluorescence Analysis, *Trends Anal. Chem.*, Vol 29 (No. 6), 2010
19. C.E. Fiori and D.E. Newbury, *Scan. Elec. Microsc.*, No. 1, 1978, p 401
20. R.A. Vane, *Adv. X-Ray Anal.*, Vol 26, 1983, p 369
21. W. Wegscheider, B.B. Jablonski, and D.E. Leyden, *X-Ray Spectrom.*, Vol 8, 1979, p 42
22. B.B. Jablonski, W. Wegscheider, and D.E. Leyden, *Anal. Chem.*, Vol 51, 1979, p 2359
23. J.W. Criss, *Adv. X-Ray Anal.*, Vol 23, 1980, p 93
24. Tracor X-Ray Inc., Mountain View, CA
25. J.R. Rhodes, *Am. Lab.*, Vol 5 (No. 7), 1973, p 57
26. A.T. Ellis, D.E. Leyden, W. Wegscheider, B.B. Jablonski, and W.B. Bodnar, *Anal. Chim. Acta*, Vol 142, 1982, p 73, 89
27. J.A. Bearden, "X-Ray Wavelengths and X-Ray Atomic Energy Levels," NSRDS-NBS 14, National Bureau of Standards, National Standard Reference Data Series, Washington, D.C., 1967
28. J.I. Goldstein, D.E. Newbury, P. Echlin, D.C. Joy, C. Fiori, and Ee. Lifshin, *Scanning Electron Microscopy and X-Ray Microanalysis*, Plenum Press, 1981
29. A.N. Shugar, Portable X-Ray Fluorescence and Archaeology: Limitations of the Instrument and Suggested Methods to Achieve Desired Results, *Archaeological Chemistry VIII*, Vol 1147, American Chemical Society, Washington, D.C., 2013, p 173–193
30. L.A. Currie, *Anal. Chem.*, Vol 40, 1968, p 586
31. R. Tertian and F. Claisse, *Principles of Quantitative X-Ray Fluorescence Analysis*, Heyden and Son, 1982
32. R. Jenkins, *Quantitative X-Ray Spectrometry*, CRC Press, 1995
33. B. Beckhoff, B. Kanngießer, N. Langhoff, R. Wedell, and H. Wolff, Ed., *Handbook of Practical X-Ray Fluorescence Analysis*, Springer Science and Business Media, 2007
34. E.P. Bertin, *Principles and Practice of X-Ray Spectrometric Analysis*, Springer Science and Business Media, 2012
35. R. Van Grieken and A. Markowicz, Ed., *Handbook of X-Ray Spectrometry*, CRC Press, 2001
36. J. Sherman, Theoretical Derivation of Fluorescent X-Ray Intensities from Mixtures, *Spectrochim. Acta*, Vol 7, 1955, p 283–306
37. R.M. Rousseau, The Quest for a Fundamental Algorithm in X-Ray Fluorescence Analysis and Calibration, *Open Spectrosc. J.*, Vol 3, 2009, p 31–42
38. M. Mantler and N. Kawahara, How Accurate Are Modern Fundamental Parameter Methods? *Rigaku J.*, Vol 21 (No. 2), 2004, p 17–25
39. Y. Kataoka, Standardless X-Ray Fluorescence Spectrometry: Fundamental Parameter Method Using Sensitivity Library, *Rigaku J.*, Vol 6 (No. 1), 1989
40. J.W. Criss and L.S. Birks, Calculation Methods for Fluorescent X-Ray Spectrometry—Empirical Coefficients vs. Fundamental Parameters, *Anal. Chem.*, Vol 40, 1968, p 1080–1086
41. J.W. Criss, L.S. Birks, and J.V. Gilfrich, Versatile X-Ray Analysis Program Combining Fundamental Parameters and Empirical Coefficients, *Anal. Chem.*, Vol 50 (No. 1), 1978, p 33–37
42. M. Mantler and H. Ebel, X-Ray Fluorescence Analysis without Standards, *X-Ray Spectrom.*, Vol 9 (No. 3), 1980, p 146–149
43. J.V. Gilfrich and L.S. Birks, Spectral Distribution of X-Ray Tubes for Quantitative X-Ray Fluorescence Analysis, *Anal. Chem.*, Vol 40, 1968, p 1077–1080
44. D.K.G. Boerde and P.N. Brouwer, *Adv. X-Ray Anal.*, Vol 33, 1990, p 237
45. H.A. Sprangvan, Fundamental Parameter Methods in XRF Spectroscopy, *Adv. X-Ray Anal.*, Vol 42, 2000, p 1–10

ASM Handbook, Volume 10, *Materials Characterization*
ASM Handbook Committee
DOI 10.31399/asm.hb.v10.a0006665

Extended X-Ray Absorption Fine Structure*

Overview

General Uses

- Determination of local structure (short-range order) about a given atomic center in all states of matter
- For crystals and oriented surfaces, structural information on next-nearest and further-out neighbors can also be determined.
- Identification of phases and compounds containing a specific element
- Orientation of adsorbed molecules on single-crystal surfaces
- Combined with near-edge structure, bonding and site symmetry of a constituent element in a material can be determined.

Examples of Applications

- Bond distance, coordination, and type of nearest neighbors about a given constituent atomic species in disordered systems, such as glasses, liquids, solutions, and random alloys, or complex systems, such as catalysts, biomolecules, and minerals
- In ordered systems, such as crystals and oriented surfaces, structural information on the next-nearest neighbors can also be obtained.
- Structural evolution in amorphous to crystalline transformation can be followed.
- Geometry of chemisorbed atoms or molecules on single-crystal surfaces
- In situ structure determination of active sites in catalysts
- In vivo structure determination of active sites in metalloproteins
- Combined with x-ray absorption near-edge structure, bonding and local structure of trace impurities in natural materials (for example, coal) and synthetic materials (for example, diamond) can be determined.

Samples

- *Form:* Solids, liquids, and gases. Solids should ideally be in thin foil (metals and alloys) or uniform films of fine powders (~400 mesh or finer).
- *Size:* Area—25 by 5 mm (1 by 0.2 in.) minimum. Thickness—bulk samples: 1 to 2 absorption lengths at the absorption edge of the element of interest; dilute samples: up to a few millimeters thick
- *Preparation:* Mainly required for solid samples—rolled metal foil, sputtered or evaporated thin films, and uniformly dispersed powder films. Aerobic or anaerobic environments must be used for biological materials and checks for radiation damage (loss of biological activity).

Limitations

- Nonunique results if the element of interest exists in multiple nonequivalent sites or valence states
- Structural results can depend strongly on model systems used in quantitative analysis.
- Low concentration limit is near 100 ppm or a few millimolar in favorable cases.
- For dilute and surface systems, synchrotron radiation is necessary.

Estimated Experimental Scan Time

- 30 min to 1 h for bulk samples (single scan)
- 5 to 10 h or more for dilute samples and surfaces (multiple scans)

Estimated Data-Analysis Time

- 2 to 20 h or more per spectrum

Capabilities of Related Techniques

- *X-ray or neutron diffraction:* identification of bulk crystalline phases; radial distribution functions of bulk amorphous phases
- *X-ray anomalous scattering:* characterization of bulk amorphous materials
- *Electron energy loss spectroscopy:* extended fine structure in solids; sample must withstand vacuum
- *Magic-angle spinning nuclear magnetic resonance:* for solid-state studies

*Revised from J. Wong, Extended X-Ray Absorption Fine Structure, *Materials Characterization*, Vol 10, *ASM Handbook*, ASM International, 1986, p 407–419

Introduction

X-ray absorption fine structure is a well-established technique providing reliable and useful information about the chemical and physical environment of the probe atom. It is somewhat less common than the other spectroscopic analytical methods because the energy-tunable x-ray source means it is primarily done with synchrotron radiation sources (Ref 1).

Extended x-ray absorption fine structure (EXAFS) (Ref 2) is an oscillatory modulation of the absorption coefficient on the high-energy side of an x-ray absorption edge of a given constituent atom in a material. Figure 1 shows the K-edge EXAFS of nickel metal. When an x-ray beam passes through a medium, its intensity is attenuated exponentially:

$$I = I_0 \exp(-\mu x) \quad \text{(Eq 1)}$$

where I and I_0 are the transmitted and incident intensities, respectively; μ is the linear absorption coefficient; and x is the sample thickness. In general, μ is a function of photon energy. When the x-ray energy $h\nu$ equals or exceeds the binding energy E_b of a core electron, the latter is emitted by a photoelectric process from the atom with kinetic energy E, conserving energy:

$$E = h\nu - E_b \quad \text{(Eq 2)}$$

For pure nickel, when $h\nu$ = 8332.8 eV (the binding energy of the innermost K-electron in nickel), μ increases sharply, producing a characteristic K-absorption edge (Fig. 1). On the high-energy side of the absorption edge, μx fluctuates with increasing photon energy, extending to a few hundred electron volts beyond the edge. These oscillations are understood theoretically to be a final-state electron effect resulting from the interference between the outgoing photoejected electron and that fraction of the photoejected electron that is backscattered from the neighboring atoms. The interference directly reflects the net phase shift of the backscattered electron near the central excited atom, which is largely proportional to the product of the electron momentum k and the distance traversed by the electron. The type of central absorbing atom and backscattering neighboring atoms (that is, their positions in the periodic table) also affects the interference event. Therefore, EXAFS is considered effective for probing the atomic environment of matter, particularly since the advent of intense continuous synchrotron radiation in the x-ray region.

The fine structure above x-ray absorption edges had been reported as early as 1920 (Ref 3–5), based on studies with the K-edges of magnesium-, iron-, and chromium-containing compounds and with the L-edges of elements from cesium to neodymium, respectively. Progress was slow before 1970, primarily because the physical processes associated with EXAFS were not well understood; therefore, there was no adequate theory to account for the observed spectra. In addition, the experiments were tedious before the availability of synchrotron radiation.

The early theories proposed to explain EXAFS can be classified as long-range order (LRO) and short-range order (SRO). The LRO theories require the existence of lattice periodicity characteristic of crystalline solids and assume transition to quasi-stationary states to explain the fine structure (Ref 6–8). However, the LRO theories do not adequately predict the shape of the experimental absorption curve, because the dominant matrix-element effects are neglected (Ref 9, 10). The early LRO Kronig theory (Ref 6, 7) also failed to explain EXAFS in polyatomic gases and amorphous materials.

However, ample experimental evidence supports the SRO approach (Ref 10). Extended x-ray absorption fine structure has been observed in simple gaseous molecular systems, such as $GeCl_4$ (Ref 11) and, more recently, Br_2 (Ref 12). Figure 2 shows the germanium EXAFS above its K-edge in $GeCl_4$, taken by using synchrotron radiation. The fine structure arises from backscattering of the germanium K-photoelectron by the four chlorine atoms bonded to germanium in the tetrahedral molecule. In another early study, a similarity was observed between the EXAFS spectra of a series of chromium, manganese, and cobalt crystalline compounds and those of their aqueous solution; the region of influence in an EXAFS event was concluded to be only 4 to 5 Å from the center of the atom being excited (Ref 14).

In solids, perhaps the first convincing experiments to demonstrate the SRO effects associated with EXAFS were conducted in 1962 (Ref 15). The EXAFS was measured above the germanium K-edge in glassy GeO_2 to 350 eV and compared to those of the hexagonal and tetragonal crystalline polymorphs shown in Fig. 3. The extended absorption fine structures of glassy and hexagonal GeO_2, in which germanium is fourfold coordinated by oxygen, were similar but differed notably from that of tetragonal GeO_2, in which germanium is sixfold coordinated by oxygen. These observations were reconfirmed in 1965 (Ref 2). The measurement was extended to 1100 eV beyond the K-absorption edge of germanium. Because LRO, that is, lattice periodicity, does not exist in the amorphous phase, the fine structure must be concluded to be strongly influenced by the arrangement of neighboring atoms about the germanium. An account of the history and modern practice of EXAFS since 1970 is cited in Ref 16.

EXAFS Fundamentals

Interest in EXAFS expanded in 1970 (Ref 17). A single-scattering approximation was used to show that the observed fine

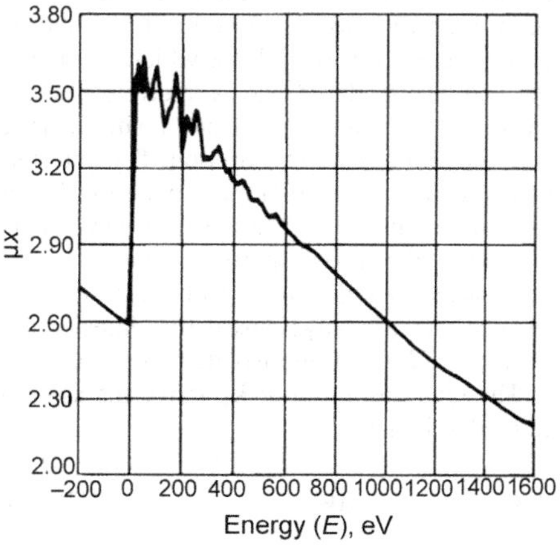

Fig. 1 Experimental extended x-ray absorption fine structure scan of nickel metal taken by using synchrotron radiation above the K-absorption edge of nickel at 8332.8 eV. The energy is labeled regarding the K-edge of nickel as zero.

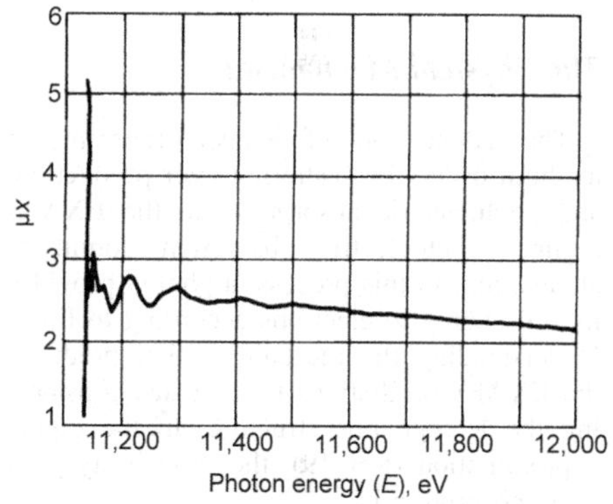

Fig. 2 Experimental extended x-ray absorption fine structure scan of germanium in $GeCl_4$ molecule taken by using synchrotron radiation above the K-absorption edge of germanium at 11,103.3 eV. Source: Ref 13

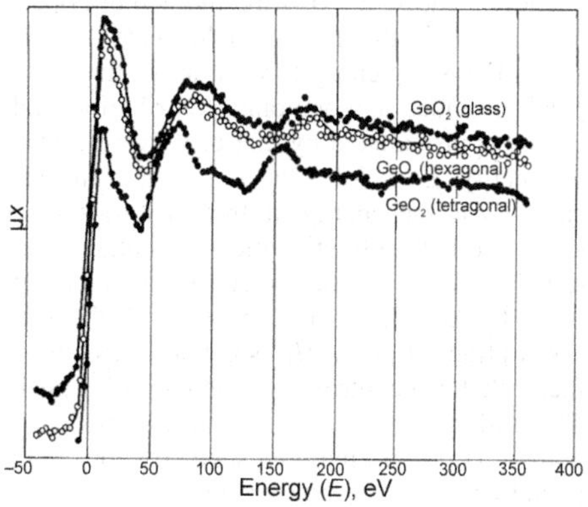

Fig. 3 Experimental K-edge extended x-ray absorption fine structure spectrum of germanium in crystalline and glassy GeO_2. The energy is labeled regarding the K-edge of germanium at 11,103.3 eV as zero. Source: Ref 15

structure oscillations can be understood in terms of interference between the outgoing photoelectron wave near the central atom and that portion of it backscattered from neighboring atoms. Furthermore, the problem can be inverted to obtain interatomic distances r_j from a Fourier analysis of EXAFS data. A Fourier transform of EXAFS data in k-space for crystalline and amorphous germanium revealed that peaks in the transforms correspond to various atomic shells (Ref 18). Analysis of the EXAFS can yield the distance, number, and type of nearest-neighbor atoms about the central atom.

Another milestone in the development of EXAFS was the availability of synchrotron radiation in the x-ray region (Ref 9). The 10^4 to 10^6 increase in intensity in tunable x-rays over a broad spectral region enables the ability to obtain EXAFS spectra with excellent signal-to-noise ratios within minutes. EXAFS has become prevalent as a structural tool for studying various materials for which such conventional techniques as x-ray diffraction and conventional electron microscopy are less useful or ineffective.

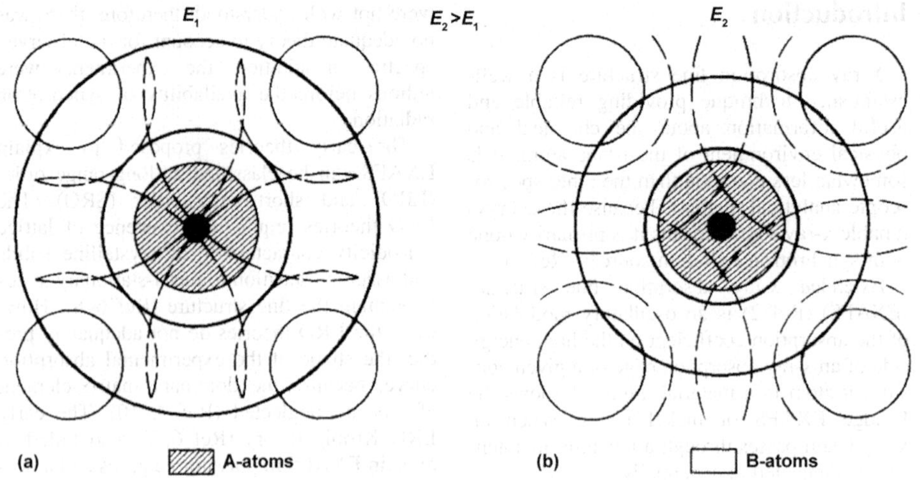

Fig. 4 Schematic representation of extended x-ray absorption fine structure event. The excited electronic state is centered about the A-atom. The solid-line circles represent the crests of the outgoing part of the electronic state. The surrounding B-atoms backscatter the outgoing part, as shown by the dashed-line circles. (a) Constructive interference. (b) Destructive interference

The Physical Mechanism

The attenuation of x-rays traversing a medium occurs by scattering, pair production, and photoelectric absorption. In the EXAFS regime, photoelectric absorption dominates attenuation. In this process, a photon provides its full energy to electrons according to Eq 2. Understanding the mechanism that produces the EXAFS oscillations necessitates considering the K-edge fine structure. In the dipole approximation (Ref 19), the probability p of x-ray absorption is:

$$p = 2\pi^2 e^2 (\omega c^2 m)^{-1} |M_{fs}|^2 \rho(E_f) \qquad \text{(Eq 3)}$$

where $M_{fs} = \langle f|\boldsymbol{\varepsilon}\cdot\mathbf{p}|s\rangle$, $|s\rangle$ is the K-shell s state, $\langle f|$ is the final unoccupied state of p symmetry, $\rho(E_f)$ is the density of states per unit energy at the energy E_f of the final state, $2\pi\omega$ is the frequency of the x-ray, $\mathbf{p}$ is the momentum operator, and $\boldsymbol{\varepsilon}$ is the electric field vector of the x-ray. For x-ray energies well above the edge, $\rho(E_f)$ provides a monotonic contribution and can be approximated by that of a free electron of energy $E = \hbar^2\mathbf{k}^2(2m\ )^{-1} + E_0$. In this case, E_0 is the energy of free electrons with $\mathbf{k} = 0$ and is the effective mean potential experienced by an excited electron. It is often termed the threshold energy. With this assumption for $\rho(E_f)$, M_{fs} is the only remaining factor that can contribute to the EXAFS signal. The initial state $|s\rangle$ is fixed and does not vary with ω. However, the final state $\langle f|$ varies with ω and produces the fine structure.

Further, the wave function $\langle f|$ is a sum of two contributions. If the atom is isolated, the excited photoelectron would be in a solely outgoing state from the central atom, as shown in Fig. 4 by the outgoing solid-line rings. In this case, M_{fs} exhibits no fine structure, and the x-ray absorption coefficient would vary monotonically with ω. This is the case for a monatomic gas such as krypton, whose spectrum beyond the K-edge at 14,326 eV follows a decay predicted by the photoelectric effect and reveals no fine structure (Fig. 5).

If the x-ray absorbing atom is surrounded by other atoms, as in a molecule such as $GeCl_4$ (Fig. 2) or in a condensed phase, whether liquid, glassy, or crystalline, the outgoing electron is scattered by the surrounding atoms producing incoming waves, as depicted by the dashed lines in Fig. 4. These incoming or backscattered waves can interfere constructively or destructively with the outgoing wave near the origin where $|s\rangle$ exists. As shown in Fig. 4(a), the amplitudes of the outgoing and backscattered waves add at the central A-atom site, leading to a maximum in x-ray absorption probability. In Fig. 4(b), the x-ray energy has been increased to E_2, leading to a shorter photoelectron wavelength for which the outgoing and backscattered waves interfere destructively at the absorbing A-atom site, with a resulting minimum in absorption. This interference produces an oscillatory variation in M_{fs} as ω is varied, changing the electron wavelength and thus the phase between the outgoing and backscattered waves. Constructive interference increases M_{fs}; destructive interference decreases M_{fs} from the isolated atom value.

The total absorption $\mu(\mathbf{k})$ above the absorption edge is then:

$$\mu(\mathbf{k}) = \mu_0(\mathbf{k})[1 + \chi(\mathbf{k})] \qquad \text{(Eq 4)}$$

where $\mu_0(\mathbf{k})$ is the smooth varying portion of $\mu(\mathbf{k})$ and corresponds physically to the absorption coefficient of the isolated atom (Ref 20). Therefore, the fine structure $\chi(\mathbf{k}) = [\mu(\mathbf{k}) - \mu_0(\mathbf{k})]/\mu_0(\mathbf{k})$ is due to interference

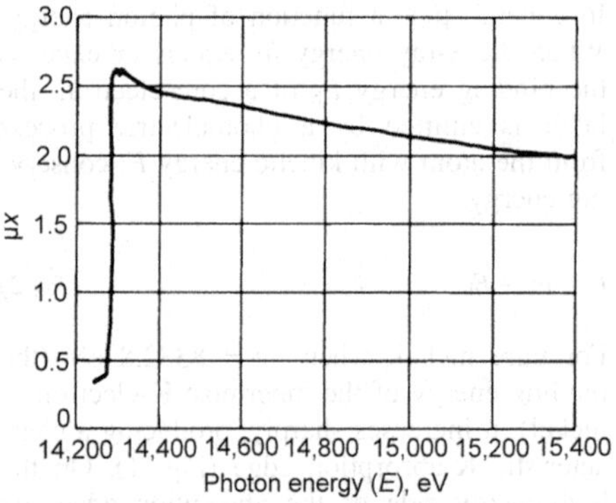

Fig. 5 K-edge absorption spectrum of krypton gas. Source: Ref 13

between backscattered and outgoing photoelectron waves in the photoabsorption-matrix element.

The Single-Scattering Approximation

The first successful working theory of EXAFS was derived based on the previously mentioned physical mechanism (Ref 17). It was subsequently modified to a more general form (Ref 10) and further refined (Ref 21, 22). For an unoriented sample, the fine structure above the K or L_I edge can be described by using:

$$\chi(\mathbf{k}) = -\frac{1}{\mathbf{k}}\sum_j \frac{N_j}{r_j^2}\exp\left(\frac{-2r_j}{\lambda}\right)\exp\left(-2\sigma_j^2\mathbf{k}^2\right) f_j(\pi,\mathbf{k})\sin\left[2\mathbf{k}r_j + \delta_j(\mathbf{k})\right] \qquad \text{(Eq 5)}$$

where $\mathbf{k} = [2m(E - E_0)/\hbar^2]^{1/2}$ is the wave vector of the ejected photoelectron of energy E, and E_0 is the inner potential or threshold energy caused by the atomic potentials and represents the threshold above which the kinetic energy must be added to determine the total energy E. The summation is over shells of atoms that are at a distance r_j from the absorbing atom and that contain N_j atoms (the coordination number). Lambda is the mean free path of the photoelectron. The second exponential containing σ_j^2 is a Debye-Waller term in which σ_j^2 is not the usual mean-square vibrational amplitude of an atom but the mean-square relative positional fluctuation of the central and backscattering atoms. The fluctuations can be static (structural disorder) or dynamic (thermal) in origin.

In this form, the resultant EXAFS is a sum of sine waves with periods $2\mathbf{k}r_j$ from each jth shell with an amplitude that represents the number of neighbors modified by an envelope due to the scattering amplitude $f_j(\pi,\mathbf{k})$, the Debye-Waller damping, and the mean-free-path damping. Besides the usual $2\mathbf{k}r_j$ that accounts for the phase difference of a free electron returning to the neighbor, additional phase shifts $\delta_j(\mathbf{k})$ are necessary to account for the potentials due to the central atom and backscatterers. The factor r_j^{-2} arises from the product of the amplitudes of the outgoing and backscattered waves, both of which decay as r_j^{-1} because of their spherical nature. For a single-crystal sample, the factor $3\cos^2\theta_j$ must be included in the summation, where θ_j is the angle the jth neighbor makes with the polarization vector of the x-ray. This factor averages to 1 for polycrystalline or amorphous materials. Conceptually, EXAFS can be considered a mode of electron diffraction in which the source of electrons is generated from within a particular atomic species participating in the absorption event.

The derivation of Eq 5 is based on two assumptions. First, because the atomic radius is small enough for the curvature of the incident wave on the neighboring atoms to be neglected, the incident wave can be approximated by using a plane wave. This is achieved mathematically by replacing the Hankel function with its asymptotic form (Ref 22, 23), which yields the factor $1/\mathbf{k}$ in Eq 5. Second, only single scattering by the neighboring atoms is included. These assumptions have been examined in detail for face-centered cubic (fcc) copper (Ref 22); electron scattering was first treated by using a spherical wave expansion to take account of the finite size of the atoms. Although large, the effects appear to make quantitative but not qualitative changes on the single-scattering description. As the size of the scattering atoms increases, significant deviations in phase and amplitude are noted between the spherical wave calculation and the asymptotic plane wave approximation (Ref 24).

Multiple-Scattering Effects

Extended x-ray absorption fine structure has been reduced to the problem of the scattering of photoelectrons by atoms analogous to low-energy electron diffraction (LEED), in which an electron beam with several hundred electron volts of energy is scattered by a crystal. Because multiple scattering is important in the interpretation of LEED data (Ref 25), this calls into question the adequacy of the single-scattering description for EXAFS.

Each multiple-scattering process can be described by an effective interference path length equal to the sum of the scattering paths (Ref 22). In **k**-space, they give rise to rapidly oscillating terms that tend to average out. Multiple scattering (band-structure and chemical bonding effects) becomes important only near the absorption edge within approximately 30 eV, the x-ray absorption near-edge structure (Ref 26). This is because at low energy the scattering becomes more isotopic, and the electron mean free path becomes long. Alternatively, if the data are Fourier transformed, the multiple-scattering contribution will appear farther out in the transformed spectrum. In particular, because the path length for multiple scattering must be larger than that for the dominant first-shell interaction, its contribution will not influence the nearest-neighbor distance that is a predominant feature in the radial structure function of such disordered systems as glasses and isolated impurities.

However, multiple scattering in EXAFS is important when an inner-shell atom shadows an outer-shell atom, as was first realized when the fourth shell in the copper EXAFS was seen to have an anomalously large amplitude and phase shift. The reason for this observation is that in an fcc lattice a nearest-neighbor atom is directly in the line of sight of the fourth-shell atom. The outgoing electron is strongly forward scattered, enhancing the electron amplitude in the fourth shell (Ref 22). This produces the focusing effect analogous to an amplifying relay system. This focusing effect has been used in developing a formalism to determine bond angle of nearly colinear systems, such as those of M-C-O in metal carbonyl complexes (Ref 27). It has also been used to "see" hydrogen atoms in EXAFS studies of metal-hydrogen systems (Ref 28).

Synchrotron Radiation as X-Ray Source for EXAFS

Synchrotron radiation is the major energy-loss mechanism of accelerated charged particles, such as electrons and positrons, traveling at relativistic velocities. The properties of synchrotron light emitted from electrons with velocities near that of the light are drastically different from classical dipole radiation (Ref 29) and constitute the importance of synchrotron radiation as an effective light source (Ref 30). Having been measured (Ref 31) and studied theoretically (Ref 32, 33), the properties can be summarized (Ref 32, 33) as follows:

- Continuous spectral distribution from the infrared to the x-ray region, which is ideal as a light source for ultraviolet and x-ray spectroscopies
- Higher intensity, permitting use of monochromators with narrow band pass
- Plane polarized, with the electric vector in the orbital plane of the circulating particles
- Extremely high collimation, which is important to lithography of submicrometer structure
- Sharply pulsed-time structure

For dilute systems, a fluorescence technique (Ref 34) has been devised to experimentally enhance the relatively weak EXAFS signal from the bulk absorption background of the matrix. This detection technique uses the principle that an inner-shell vacancy may relax by undergoing a radiative transition from a higher-energy occupied shell. The fluorescence yield is a monotonically increasing function of atomic number and is expected to be independent of excitation energy for above threshold but may vary slightly near threshold. Thus, the fluorescence intensity is a direct measure of absorption probability, the mechanism of interest in EXAFS. Figure 6(b) shows a typical fluorescence EXAFS setup. The original configuration (Ref 34) has been further modified for improved solid-angle collection, filtering of Compton and elastic scattering of the incident beam by the sample (Ref 36), and discriminative energy detection (Ref 37, 38).

Data Analysis

Experimentally, the EXAFS spectrum shown in Fig. 1 appears as low-intensity oscillations (relative to the jump at the absorption edge) superimposing on the smooth atomic absorption background that decays with increasing energy above the absorption edge. Therefore, the fine structure according to Eq 4 is $\chi(\mathbf{k}) = [\mu(\mathbf{k}) - \mu_0(\mathbf{k})]/\mu_0(\mathbf{k})$, where $\mu(\mathbf{k})$ is the total absorption measured above the edge, and $\mu_0(\mathbf{k})$ is the smooth atomic contribution. A fairly standardized procedure has been established to extract the EXAFS signal $\chi(\mathbf{k})$ from the experimental x-ray absorption spectrum (Ref 39–42). It consists of correcting for spectrometer shift, deglitching, pre-edge and post-edge background removal, edge normalization, extraction of the EXAFS signal $\chi(\mathbf{k})$, Fourier transform of $\chi(\mathbf{k})$, and inverse transform to isolate the EXAFS contribution from a selected region in real space.

Background Removal

Because the smooth absorption of an isolated atom $\mu_0(\mathbf{k})$ is not generally available experimentally, and present theoretical calculations of $\mu_0(\mathbf{k})$ are not sufficiently accurate for most EXAFS investigations (approximately 0.1%), the smooth part of $\mu(\mathbf{k})$ is assumed to represent the desired $\mu_0(\mathbf{k})$. With this assumption, the remaining oscillatory part of $\mu(\mathbf{k})$ is taken as $\Delta\mu = \mu(\mathbf{k}) - \mu_0(\mathbf{k})$ to yield $\chi(\mathbf{k}) = \Delta\mu(\mathbf{k})/\mu_0(\mathbf{k})$. The post-edge background above 30 eV in the EXAFS region can then be generated analytically by fitting $\mu(\mathbf{k})$ (including the EXAFS) with a series of cubic splines of equal segments (Ref 43). The ends of each segment are connected such that the derivatives are continuous across the ends. Three to five such splines are adequate for data extending 1000 eV above the absorption edge. When the number of segments is too small, the background is not separated well enough; when too large, the background follows the EXAFS oscillations, especially at low energy, and reduces its intensity (Ref 42).

A least-squares fit with such a spline function enables removal of low-frequency background components from $\mu(\mathbf{k})$ without affecting the higher-frequency EXAFS oscillations. Spline fitting is essentially a local fitting procedure in that the polynomial function within each interval is determined mainly by the local quality of the fit. The pre-edge background from −200 to −20 eV is obtained by a linear regression analysis of the first ten raw data points.

Figure 7 shows the result for nickel. The solid line in Fig. 7(a) is the raw experimental K-edge EXAFS scan of a 5 μm thick nickel foil taken at 90 K. The spectrum was recorded with the Stanford Position Electron Accelerator Ring operating at 2.6 GeV electron energy and beam current at approximately 30 mA. The broken line is the smooth post-edge background derived from a cubic spline fitting with five segments from 30 to 1200 eV.

Other background-removal methods have been used, such as a single polynomial fit over the whole range of the data and the sliding boxcar window fitting (Ref 44). However, in all these procedures, a single bad data point, noise, or end-point effects can introduce systematic errors.

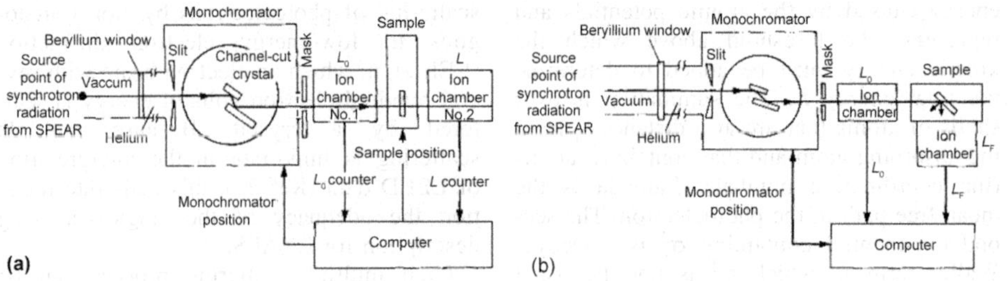

Fig. 6 Extended x-ray absorption fine structure experimental apparatus at the Stanford Synchrotron Radiation Laboratory. (a) Transmission mode of detection. (b) Fluorescence mode of detection. SPEAR, Stanford Position Electron Accelerator Ring. Source: Ref 35

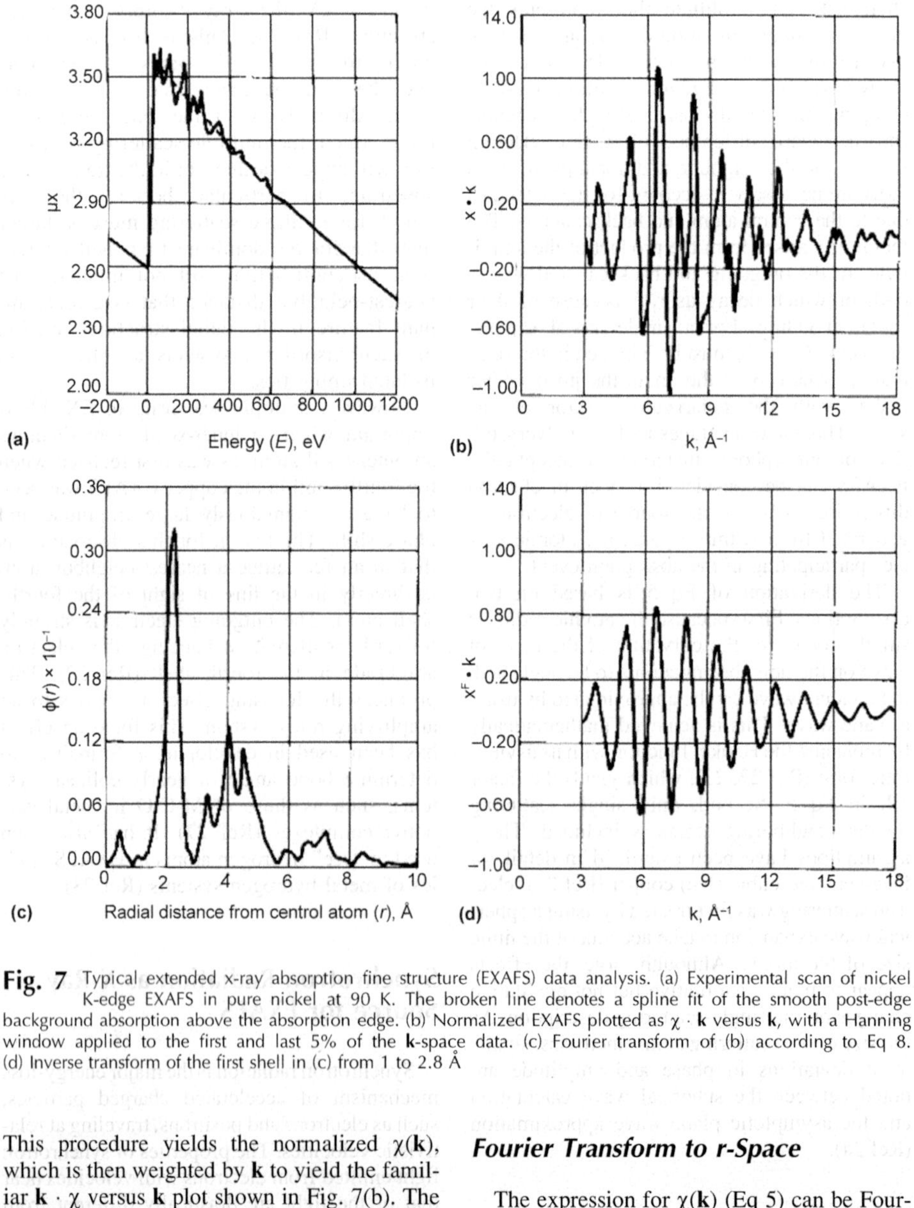

Fig. 7 Typical extended x-ray absorption fine structure (EXAFS) data analysis. (a) Experimental scan of nickel K-edge EXAFS in pure nickel at 90 K. The broken line denotes a spline fit of the smooth post-edge background absorption above the absorption edge. (b) Normalized EXAFS plotted as $\chi \cdot \mathbf{k}$ versus $\mathbf{k}$, with a Hanning window applied to the first and last 5% of the $\mathbf{k}$-space data. (c) Fourier transform of (b) according to Eq 8. (d) Inverse transform of the first shell in (c) from 1 to 2.8 Å

EXAFS Extraction

The energy scale is converted to $\mathbf{k}$-scale by using $\mathbf{k} = [0.263(E - E_0)]^{1/2}$, where E_0, the energy threshold of the absorption edge, is located experimentally by the first maximum in the derivative spectrum of the absorption curve. The EXAFS $\chi(\mathbf{k})$ at energies above approximately 30 eV is obtained by subtracting the smooth post-edge background $\mu_0(\mathbf{k})$ from the measured absorption $\mu(\mathbf{k})$ and dividing by the step jump S at the absorption edge with the McMaster correction (Ref 20), $M(\mathbf{k})$, as a function of energy:

$$\chi(\mathbf{k}) = \frac{\mu(\mathbf{k}) - \mu_0(\mathbf{k})}{S.M(\mathbf{k})} \qquad \text{(Eq 6)}$$

This procedure yields the normalized $\chi(\mathbf{k})$, which is then weighted by $\mathbf{k}$ to yield the familiar $\mathbf{k} \cdot \chi$ versus $\mathbf{k}$ plot shown in Fig. 7(b). The $\mathbf{k}$-weighting or, more generally, the $\mathbf{k}^n$-weighting is discussed subsequently in conjunction with Fourier analysis.

Fourier Transform to r-Space

The expression for $\chi(\mathbf{k})$ (Eq 5) can be Fourier transformed to yield a radial structure function $\varphi(r)$ that contains structural information on the absorbing atom. The Fourier inversion

(Ref 18), a significant step in the development of modern EXAFS, converts the technique from a qualitative to a quantitative effect (Ref 39):

$$\phi(r) = (2\pi)^{-1/2} \int \chi(\mathbf{k}) \exp(2i\mathbf{k})\, d\mathbf{k}$$
$$= \sum_{j} N_j \int \frac{dr'}{r^2 T(r-r')} \exp\left(\frac{-2\left(r-r_j'\right)^2}{\sigma_j^2}\right) \quad \text{(Eq 7)}$$

where $\phi(r)$, the Fourier transform of the EXAFS, consists of a sum of radial peaks located at r_j and determines the spatial variation of the scattering matrix. Because in practice an EXAFS spectrum is taken over a finite energy range (hence **k**-space), the Fourier transform taken is:

$$\phi(r) = (2\pi)^{-1/2} \int_{\mathbf{k}\min}^{\mathbf{k}\max} W(\mathbf{k})\mathbf{k}^n \chi(\mathbf{k}) \exp(2i\mathbf{k}r)\, d\mathbf{k} \quad \text{(Eq 8)}$$

where $\mathbf{k}_{max}$ and $\mathbf{k}_{min}$ are the maximum and minimum **k**-values of the usable experimental data, and $\mathbf{k}^n$ is a weighting function used to compensate for amplitude reduction as a function of **k** (Ref 39), especially for low-Z scatterers. Values of n = 1, 2, and 3 have been suggested (Ref 43) for backscatterers with $Z > 57$, $36 < Z < 57$, and $Z < 36$, respectively. The $\chi \cdot \mathbf{k}^1$-transform is sensitive to $\mathbf{k}_{min}$ between the origin and r_1, the first peak in $\phi(r)$ (Ref 41). The $\chi \cdot \mathbf{k}^3$-transforms can be approximated to a pseudo-charge density insensitive to $\mathbf{k}_{min}$ and E_o. The $\mathbf{k}^3$-transform weights less at low **k** and more at high **k**, at which the EXAFS effect is better approximated by the single-scattering expression given in Eq 5 but experimentally is of poorer quality because of a poorer signal-to-noise ratio.

The factor $W(\mathbf{k})$ on the right side of Eq 8 is a window function that, when multiplied by the integrand, converts the finite data set to an infinite set necessary for Fourier transform. This is accomplished by selecting functions that smoothly set the raw data points to zero at $\mathbf{k}_{min}$ and $\mathbf{k}_{max}$. An example of $W(\mathbf{k})$ is a Hanning function (Ref 45) defined in terms of **k**:

$$W(\mathbf{k}) = \frac{1}{2} \frac{(1-\cos 2\pi)(\mathbf{k}-\mathbf{k}_{min})}{(\mathbf{k}_{max}-\mathbf{k}_{min})} \quad \text{(Eq 9)}$$

It is easily seen that $W(\mathbf{k}) = 0$ at $\mathbf{k} = \mathbf{k}_{min}$ and $\mathbf{k}_{max}$. This window function is applied to the first and last 5% of the normalized nickel data discussed earlier and plotted as $\chi \cdot \mathbf{k}$ versus **k** in Fig. 7(b). The Fourier transform thus obtained is shown in Fig. 7(c). The transform is conducted relative to $\exp(2i\mathbf{k}r)$ without including the phase shift $\delta_j(\mathbf{k})$. This shifts all the peaks in $\phi(r)$ closer to the origin to $r_j - \delta'$, where δ' is some average of the first derivative of $\delta(\mathbf{k})$ relative to **k**. In Fig. 7(c), the first peak is the nearest-neighbor position in free nickel shifted to 2.24 Å. The crystallographic value from diffraction is 2.492 Å, so that δ' for j = 1 is 0. 25 Å.

In general, the effect of $\delta_j(\mathbf{k})$ on the transform can be corrected for empirically by measuring the EXAFS spectrum of a standard or model compound of known structure. As in the case of complex biomolecules, several such model compounds are used on a trial-and-error basis to deduce a model of the unknown structure. Alternatively, theoretical values of $\delta_j(\mathbf{k})$ (Ref 43) can be used in the Fourier transform, and r_j obtained directly.

Figure 7(d) shows an inverse transform of the first shell in Fig. 7(c) from 1 to 2.8 Å. This essentially isolates the EXAFS contribution due to the 12 nearest neighbors. The inverse signal in **k**-space can then be used to derive structural parameters by simulation.

Structural Information in k-Space

In addition to interatomic distances r_j, EXAFS contains other structural information, such as the coordination number N_j and type of jth atoms in the shell at r_j and their relative mean square disorder σ_j^2 about the average distance r_j. These structural parameters can be obtained by measuring the EXAFS of model compounds under identical conditions and using the transferability of phase shifts (Ref 46). The structure of the unknown is then modeled by curve-fitting procedures to arrive at a calculated EXAFS that best fits the experimental values. Modeling can be performed effectively in combination with back-Fourier transforming, especially in systems in which the coordination shells are well separated in r-space or whose $\phi(r)$ is dominated by the nearest-neighbor shell, as in amorphous materials. The shell-by-shell back-Fourier transform enables determination of a self-consistent phase shift and experimental envelope function for each jth shell. Various curve-fitting techniques for extracting structural information from EXAFS data have been prescribed (Ref 23, 41, 47–50). A more widely-used procedure is described as follows (Ref 50).

Using the single-scattering expression given in Eq 5, the observed EXAFS $\chi(\mathbf{k})$ can be described by:

$$\chi(\mathbf{k}) = \frac{-1}{\mathbf{k}} \sum_{j} A_j \sin\left[2r_j\mathbf{k} + \phi_j(\mathbf{k})\right] \quad \text{(Eq 10)}$$

having oscillatory terms with frequencies $[2r_j\mathbf{k} + \phi_j(\mathbf{k})]$ and amplitude terms A_j, given by:

$$A_j = \frac{N_j}{r_j^2} f_j(\pi, \mathbf{k}) \exp\left(\frac{-2r_j}{\lambda}\right) \exp\left(-2\sigma_j^2\mathbf{k}^2\right) \quad \text{(Eq 11)}$$

The parameters on the right side of Eq 10 and 11 can be classified as scattering parameters (phase shift, $\phi_j(\mathbf{k})$; backscattering amplitude, $f_j(\pi, \mathbf{k})$; and mean free path, λ) and structural parameters (coordination number, N_j; bond distance, r_j; and Debye-Waller factor, σ_j). The summation is over all coordination shells j participating in the EXAFS event. In a model system for which N_j and r_j are known crystallographically, EXAFS can be used to generate a set of self-consistent scattering parameters. This information can then be applied to an unknown system of similar chemical nature, for example, a glass of the same composition, to determine structural parameters.

A least-squares procedure (Ref 50) is set up to minimize the variance S:

$$S = \sum_{i}^{n} \left(\chi_i^F - \chi_i\right)^2 \quad \text{(Eq 12)}$$

where χ_i^F represents the Fourier-filtered experimental data, and χ_i is the analytical expression given in Eq 10 that describes χ_i^F for n data points. Because $\chi(\mathbf{k})$ is not a linear function of the various parameters, a Taylor series expansion is used that expresses $\chi(\mathbf{k})$ in terms of approximate parameter values, P_j, and parameter adjustments, $\Delta P_j = P_j - P'_j$. Applying the least-squares condition yields a set of simultaneous equations in terms of ΔP_j rather than P_j. The equations are solved for the adjustment, ΔP_j, and the parameters were adjusted by ΔP_j to provide a new set of estimates. The procedure was then repeated with the new estimates P'_j and so on until the new solution differed from the last by less than a desired value, usually 1%.

The phase and envelope function for the nickel-nickel pair (Ref 51) is generated by performing a self-fitting of the filtered EXAFS (Fig. 7d) of the first shell of 12 nearest neighbors in free nickel with the following fixed inputs: $N_1 = 12$, $r_1 = 2.492$ Å, $\Delta E_o = 0$, and $\sigma_1^2 = 0$. The fitting was performed in $\chi \cdot \mathbf{k}^3$ space to weigh the contribution of nickel at high **k**. The results are shown in Fig. 8; the curve denotes the filtered experimental EXAFS, and the points denote the simulated spectrum. This simulation has a standard deviation of 5% of the maximum amplitude of the experimental $\chi^F \cdot \mathbf{k}^3$ spectrum. The nickel-nickel phase parameters thus obtained can then be used as initial inputs to simulate the filtered transform arising from the nickel-nickel subshells in crystalline Ni_2B. The nickel-nickel phase shifts as derived from crystalline Ni_2B have in turn been transferred directly to determine nickel-nickel bond distances and coordination numbers of the various subshells in amorphous Ni_2B. Details of such systematic simulations are cited in Ref 51 to 53.

Near-Edge Structure

In EXAFS analysis for structural determination, the data within approximately 30 eV of the absorption edge are generally ignored because their interpretation is complicated by multiple scattering and chemical bonding effects. Phenomenologically, as the region near an x-ray absorption edge is scanned in energy, the ejected photoelectron probes the empty electronic levels of the material sequentially. The resulting x-ray absorption near-edge structure (XANES) (Ref 26) spectrum within 30 eV of threshold has been realized to be rich in chemical information and is receiving increased attention. Although the study of XANES has a long history (Ref 54), these spectra can now be measured more quickly and simply and with greater resolution due to the availability of intense and well-collimated synchrotron x-ray sources and improved experimentation.

Vanadium forms a series of oxides over a range of formal oxidation states. The crystal structures of VO, V_2O_3, V_4O_7, V_2O_4, and V_2O_5 are known. These oxides provide a useful series of materials for systematic study of the effects of valence-site symmetry and coordination geometry on the XANES spectrum of the central metal atom coordinated by the same ligand (Ref 55).

The NaCl structure of VO has regular octahedral VO_6 units. In the corundum structure of V_2O_3, V^{3+} ions are sixfold coordinated by oxygen ions at two distinct distances: 1.96 and 2.06 Å. A mixed-valence oxide, V_4O_7, consists of V^{3+} and V^{4+} ions. The structure consists of a distorted hexagonal close-packed oxygen array with vanadium atoms occupying the octahedral sites (distorted) to form rutile blocks extending indefinitely in the triclinic a-b plane. The rutile blocks are four octahedra thick along the perpendicular to this plane. There are four crystallographic nonequivalent vanadium sites with V-O distances 1.883 to 2.101 Å. The crystal structure of V_2O_4 is monoclinic and is a distorted form of rutile. The vanadium atoms are again sixfold coordinated by oxygens but are much displaced from the center of the octahedron, resulting in a short V-O bond of 1.76 Å. In V_2O_5, the vanadium is fivefold coordinated in a distorted tetragonal pyramid of oxygen. The apex oxygen distance is only 1.585 Å, but the basal V-O distances vary from 1.78 to 2.02 Å. The site symmetry of the vanadium atom decreases from O_h to VO to C_3 in V_2O_3, C_1 in V_4O_7 and V_2O_4, and C_s in V_2O_5.

The vanadium K-edge XANES spectra in these oxides (Fig. 9) exhibit a pre-edge absorption feature that grows in intensity from V_2O_3 to V_2O_5, followed by a weak shoulder on a rising absorption curve (the absorption edge) that culminates in a strong peak near approximately 20 eV. This strong peak has been assigned as the allowed transition $1s \rightarrow 4p$ (Ref 56), the lower-energy shoulder as the $1s \rightarrow 4p$ shakedown transition, and the pre-edge feature at threshold as the forbidden transition $1s \rightarrow 3d$ (Ref 56). At energies equal to and above the $1s \rightarrow 4p$ transition, absorption features may arise from transition to higher *np* states, shape resonances (Ref 57), and/or multiple scattering (Ref 58). The latter two effects are much more complicated to analyze.

Because the initial 1*s* state is a ground state, the $1s \rightarrow 3d$ transition is strictly dipole forbidden, as it is for VO, which contains regular octahedral VO_6 units having a center of inversion. When the symmetry of the ligands is lowered from O_h, the inversion center is broken, as in V_2O_3, V_4O_7, and V_2O_4 with distorted octahedral VO_6 groups and in V_2O_5 with distorted square pyramidal VO_5 groups. The pre-edge absorption becomes dipole-allowed due to a combination of stronger 3*d*-4*p* mixing and overlap of the metal 3*d* orbitals with the 2*p* orbitals of the ligand (Ref 56).

The intensity variation of the pre-edge peak across the oxide series is noteworthy. As shown in Fig. 10, the oscillator strength increases with progressive relaxation from perfect octahedral symmetry, as in VO, to distorted octahedral VO_6 groups, as in V_2O_3, V_4O_7, and V_2O_4, and to a lower coordination with a short V-O bond in a square pyramidal symmetry, as in V_2O_5. The molecular-cage effect on the oscillator strength of this transition to the 3*d* orbitals in K-edge spectra (Ref 59) appears to be operative in this case.

Closer examination of the V_2O_3 spectrum reveals a multiplet structure in the pre-edge

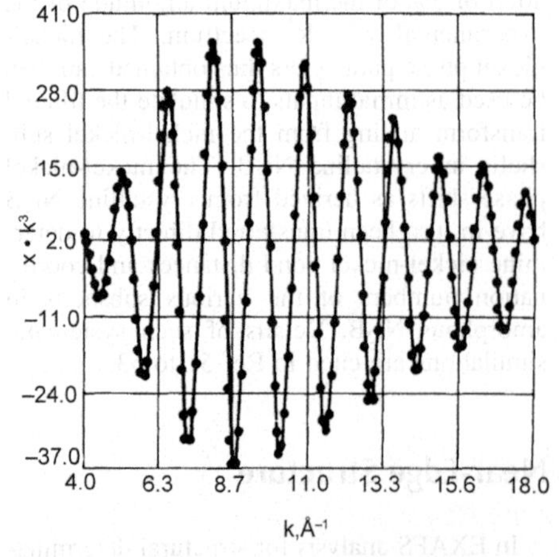

Fig. 8 Experimental (line) and simulated (points) extended x-ray absorption fine structure of the first shell of 12 neighbors from 1 to 2.8 Å about a nickel atom in face-centered cubic nickel at 90 K

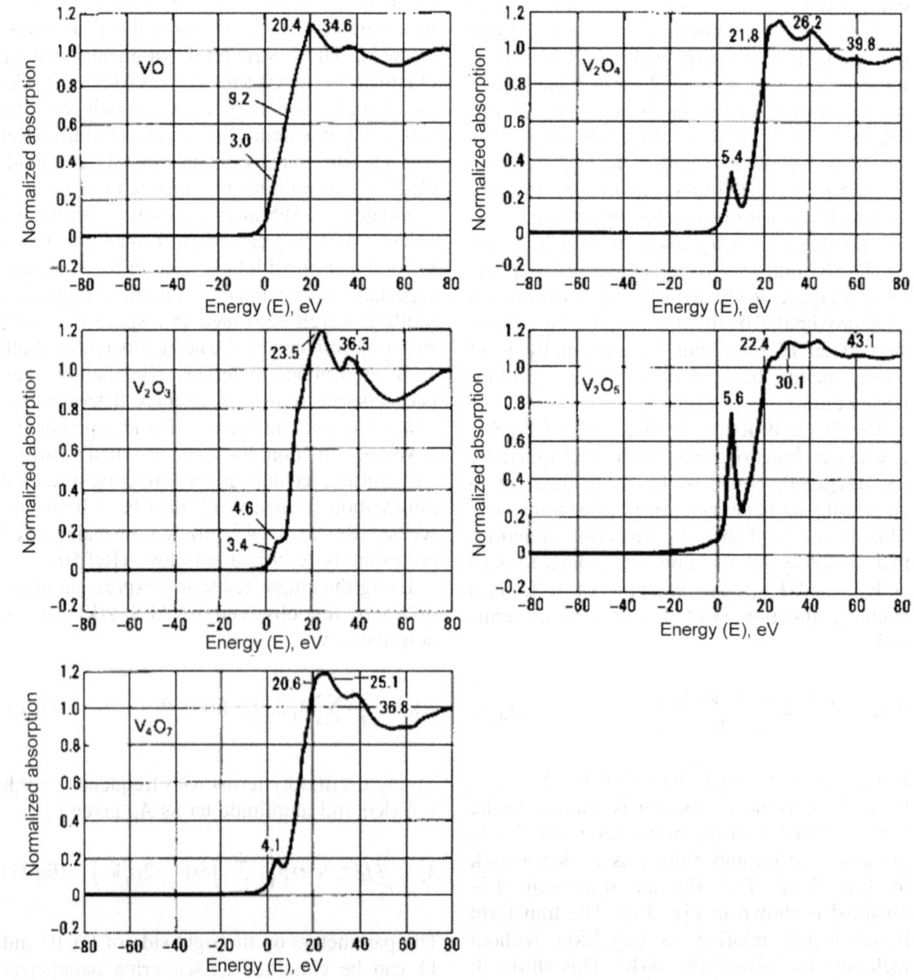

Fig. 9 The vanadium K-edge x-ray absorption near-edge structure spectra of a series of vanadium oxides. The zero of energy is taken at the K-edge of vanadium metal at 5465 eV in all cases.

peak region. The multiplet structure exhibits splitting of approximately 1.3 and 2.0 eV. The splitting in the $1s \rightarrow 3d$ transition is caused by crystal-field splitting of the ground state (Ref 58), and in the case of V_2O_3, the d levels of V^{3+} ions in C_3 sites are split into A + 2E symmetries.

The energy positions of various absorption features correlate with the oxidation state (formal valency) of vanadium in the oxides. With increases in oxidation state, the absorption threshold as defined by the position of the first peak in the derivative spectrum, the absorption edge as defined by the second peak in the derivative curve, the energy of the pre-edge peak, and the $1s \rightarrow 4p$ transition above the absorption edge shift to higher energies. The energy shifts, or chemical shifts, follow Kunzl's law (Ref 60) and vary linearly with the valence of the absorbing vanadium atoms (Fig. 10). The positive shift in the threshold energy with valence increase can be understood conceptually to be due to an increase in the attractive potential of the nucleus on the $1s$ course electron and a reduction in the repulsive core Coulomb interaction with all the other electrons in the compound.

The lines shown in Fig. 10 are least-squares-fitted lines with slopes of 1.4, 1.1, 2.5, and 3.2 eV per valence increase for the threshold, the pre-edge peak, the absorption edge, and the $1s \rightarrow 4p$ transition, respectively. The increase in slope reflects tighter binding of the inner $3d$ and $4s$ levels relative to the outermost $4p$ levels, which are more easily perturbed by valence change.

Unique Features of EXAFS

Figure 11(a) shows the EXAFS above the K-edges of iron and nickel in a body-centered cubic (bcc) iron-nickel alloy containing 80 at. % Fe. These were obtained in one experimental scan by tuning the synchrotron radiation near the K-absorption edge of iron at 7.11 keV, scanning the iron EXAFS over a 1000 V range, and continuing scanning another 1000 eV beyond the nickel K-edge to obtain its EXAFS. Iron and nickel are separated by two units in atomic number, yet their K-absorption edges are far apart in energy, so the EXAFS of iron is not overlapped by the onset of the K-absorption of nickel. Therefore, structural information extracted from analyzing each EXAFS spectrum is atom-specific in that the central atom is defined; consequently, the origin of each of the $\phi(r)$ is known. This demonstrates the atomic selectivity of EXAFS for studying multiatomic systems.

Using the data-reduction procedure described earlier in the section "Data Analysis" in this article, the normalized EXAFS for nickel is obtained and plotted as $\chi \cdot \mathbf{k}$ versus $\mathbf{k}$ (Fig. 11b). This is then Fourier transformed relative to $\exp(2i\mathbf{k}r)$ to obtain $\phi(r)$ (Fig. 11c). The Fourier transform is dominated essentially by a strong radial structure peak above 2 Å. Higher coordination shells are also visible to 5 to 6 Å. The oscillations in the low-r side of the first peak are due to termination errors of the transform and are not structural in origin. The transform shown in Fig. 7(c) for pure nickel, which is fcc and has 12 nearest neighbors, differs significantly from the transform pattern shown in Fig. 11(c) for nickel in the bcc iron-nickel alloy. The latter is characteristic of a bcc structure such as iron, which has eight nearest neighbors (Ref 44). This is also apparent in the raw spectra shown in Fig. 12 (a) in that the nickel-iron EXAFS pattern is isomorphic with that of iron and is a direct consequence of the alloying effect that results in structuring the nickel atoms in a bcc lattice. Finally, Table 1 compares EXAFS with conventional x-ray diffraction.

Applications

Because it is element-specific, x-ray absorption fine structure (XAFS) places few restrictions on the form of the sample and can be used in a variety of systems and bulk physical environments, including crystals, glasses, liquids, and heterogeneous mixtures (Ref 1). Additionally, XAFS can often be done on low-concentration elements (typically down to a few parts per million) and thus has applications in a wide range of scientific fields, including chemistry, biology, catalysis research, materials science, environmental science, and geology. General information on EXAFS is cited

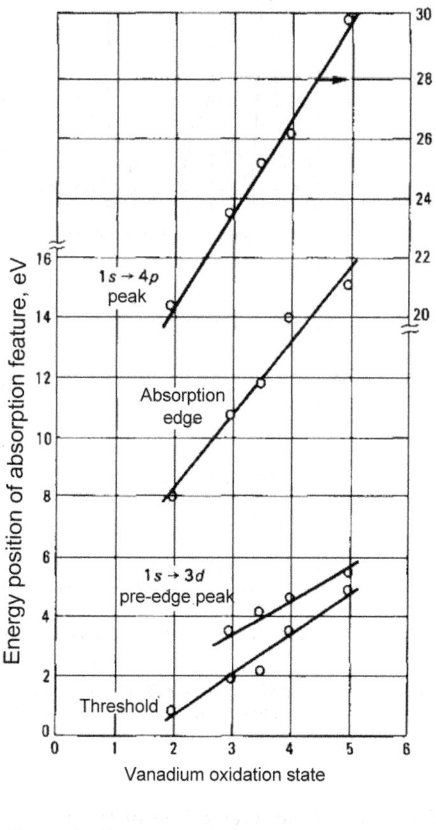

Fig. 10 Oxidation state versus energy positions of various absorption features in the vanadium K-edge x-ray absorption near-edge structure spectra of various vanadium oxides shown in Fig. 9

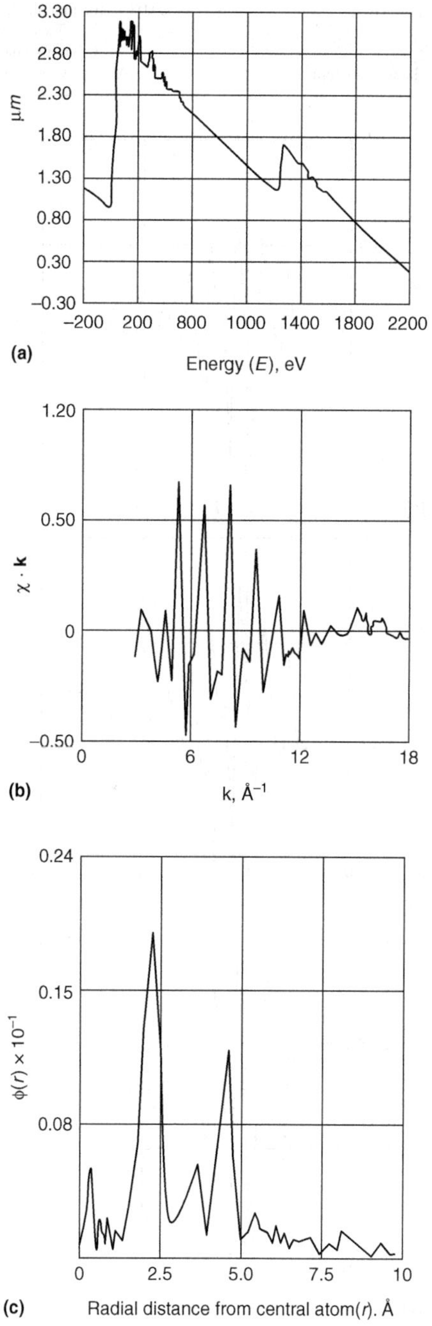

Fig. 11 Body-centered cubic (bcc) iron-nickel alloy containing 80 at.% Fe. (a) Experimental extended x-ray absorption fine structure (EXAFS) spectra above the K-edges of iron and nickel. (b) Normalized EXAFS plotted as $\chi \cdot \mathbf{k}$ versus $\mathbf{k}$ for the EXAFS. (c) Fourier transform of (b). The peaks on the low-r side of the first main peak at 2.2 Å are spurious effects of the transform and therefore nonphysical. This radial structure function for nickel in a bcc environment differs from that for pure face-centered cubic nickel shown in Fig. 7(c).

Table 1 Comparison of extended x-ray absorption fine structure (EXAFS) and x-ray diffraction

	EXAFS	X-ray diffraction
r-space range	Short, because $\chi(\mathbf{k}) \sim 1/r^2$ and $\lambda \leq 10$ Å	Long
k-space range	$0 < k < 50$ Å^{-1}	$0 \leq q < 25$ Å^{-1}
Selectivity	Atom-specific	All atoms diffract
N_j	±10%	±1% or better
r_j	±0.02 Å for first shell; ±0.1 Å for second shell	0.001
σ_j	Two-body average	One-body average
ϕ_j	Contains phase information	No phase
Material systems	All states of matter, bulk and dilute	Crystalline solids, bulk

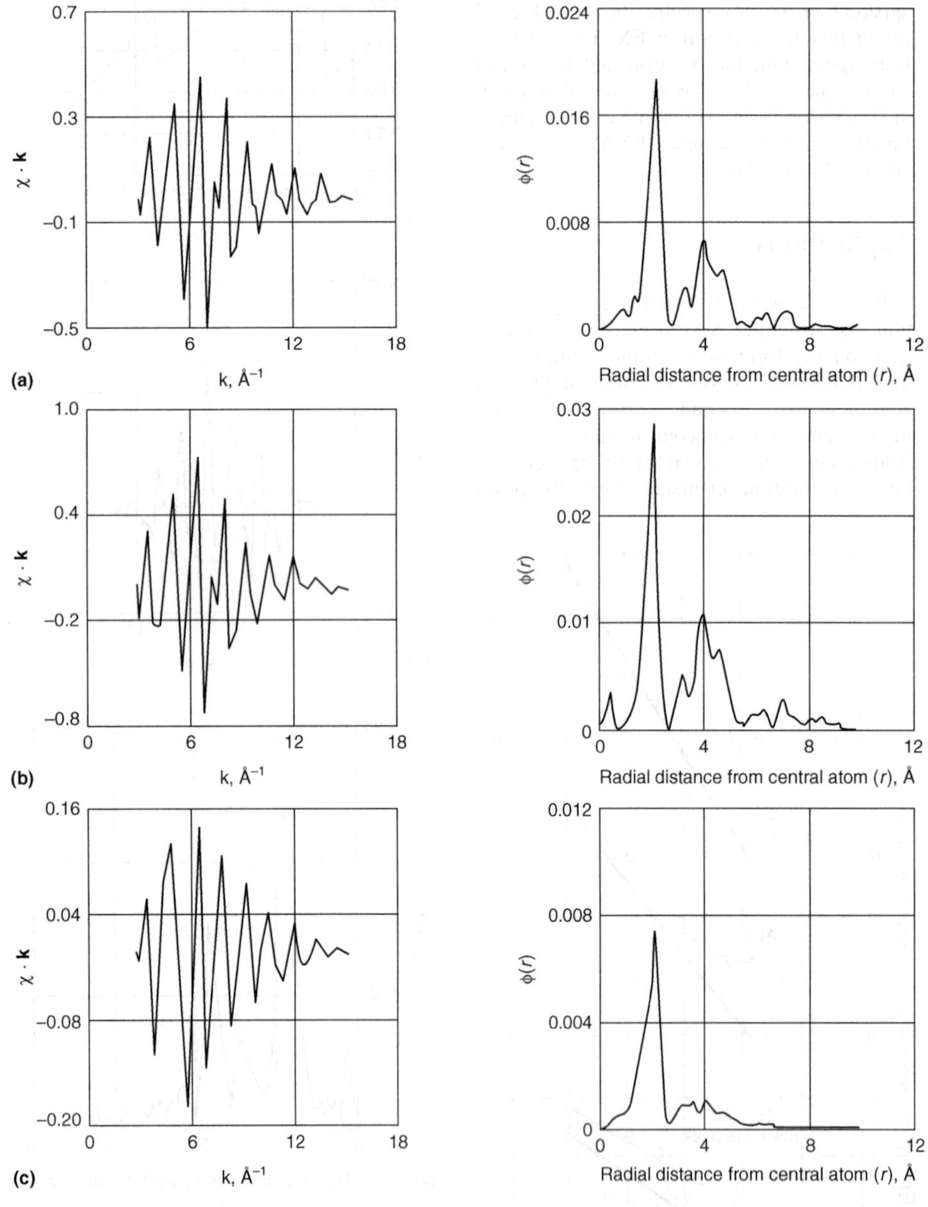

Fig. 12 Normalized nickel K-edge extended x-ray absorption fine structure plotted as $\chi \cdot \mathbf{k}$ versus $\mathbf{k}$ and corresponding Fourier transform. (a) Nickel impurity in synthetic diamond. (b) Face-centered cubic nickel. (c) Ni_3B. r is the radial distance (phase shift not included) from a central x-ray-absorbing nickel atom.

in Ref 40 and 47, with EXAFS citations as applied to biology in Ref 61 and 62; to catalysts in Ref 63; to amorphous materials in Ref 44, 64, and 65; and to geology in Ref 66.

Example 1: Characterization of Metal Impurities in Synthetic Diamond

Characterization of metal impurities incorporated in synthetic diamond crystals during growth at high temperature and high pressure illustrates the use of EXAFS as a phase-identification tool (similar to powder x-ray diffractometry) (Ref 67). In Fig. 12(a), the room-temperature nickel K-edge EXAFS from the nickel impurity in synthetic diamond is plotted as $\chi \cdot \mathbf{k}$ versus $\mathbf{k}$. The corresponding Fourier transform, shown on the right side, is plotted as $\phi(r)$ versus r, where r is the radial distance from the central atom. The corresponding results for nickel and Ni_3B are shown in Fig. 12(b) and (c). The spectrum of nickel in the synthetic diamond is directly identifiable as that of pure nickel, which yields a radial structure function (right side of Fig. 12b) characteristic of the fcc structure and consists of four resolved peaks associated with the first four coordination shells about the central atom (compare with Fig. 7 for nickel at 90 K). Like Ni_3C, Ni_3B has a cementite structure (Schoenflies symbol V_h^{16}, Hermann-Mauguin space group *Pbnm*; four molecules per unit cell) (Ref 68). There are two nonequivalent nickel sites: Ni(I) has 11 nickel neighbors at distances from 2.43 to 2.74 Å, two boron atoms at approximately 2.0 Å, and one boron at 2.30 Å; Ni(II) has 12 nickel atoms at distances from 2.50 to 2.79 Å, two boron atoms at 2.05 Å, and one boron at 2.60 Å. The nickel EXAFS in Ni_3B shown in Fig. 12(c) differs from that found in the synthetic diamond. The Fourier transform shown on the right side of Fig. 12(c) consists basically of a radial peak centered at approximately 2 Å, which is rather broad, reflective of the distribution of nickel-nickel distances in the structure. No dominant features are evident beyond 3 Å in the radial structure function. Thus, EXAFS can be used to fingerprint an unknown phase containing a selectively known constituent element.

REFERENCES

1. G.S. Henderson, D.R. Neuville, and R.T. Downs, 2.1 Introduction, *Reviews in Mineralogy and Geochemistry*, Vol 78, *Spectroscopic Methods in Mineralogy and Materials Sciences*, De Gruyter, 2015
2. F.W. Lytle, in *Physics of Non-Crystalline Solids*, J.A. Prins, Ed., North-Holland, 1965, p 12–25
3. H. Fricke, *Phys. Rev.*, Vol 16, 1920, p 202
4. G. Hertz, *Z. Phys.*, Vol 21, 1920, p 630
5. G. Hertz, *Z. Phys.*, Vol 3, 1920, p 19
6. R. De L. Kronig, *Z. Phys.*, Vol 70, 1921, p 317
7. R. De L. Kronig, *Z. Phys.*, Vol 75, 1932, p 191
8. T. Hayashi, *Sci. Rep. Tohoku Univ.*, Vol 33, 1949, p 123, 183
9. L.V. Azaroff, *Rev. Mod. Phys.*, Vol 35, 1963, p 1012
10. E.A. Stern, *Phys. Rev. B*, Vol 10, 1974, p 3027

11. J.D. Hanawalt, *Phys. Rev.*, Vol 37, 1931, p 715
12. B.M. Kincaid and P. Eisenberger, *Phys. Rev. Lett.*, Vol 34, 1975, p 1361
13. B.M. Kincaid, Ph.D. thesis, Stanford University, Stanford, CA, 1975
14. R.A. Van Ordstrand, in *Non-Crystalline Solids*, V.D. Frechette, Ed., John Wiley & Sons, 1960, p 108
15. W.F. Nelson, I. Siegel, and R.W. Wagner, *Phys. Rev.*, Vol 127, 1962, p 2025
16. F.W. Lytle, D.E. Sayers, and E.A. Stern, in *Advances in X-Ray Spectroscopy*, C. Bonnelle and C. Mande, Ed., Pergamon Press, 1982, p 267
17. D.E. Sayers, F.W. Lytle, and E.A. Stern, *Adv. X-Ray Anal.*, Vol 13, 1970, p 248
18. D.E. Sayers, E.A. Stern, and F.W. Lytle, *Phys. Rev. Lett.*, Vol 27, 1971, p 1204
19. H. Bethe and E. Salpeter, *Quantum Mechanics of One- and Two-Electron Systems*, Springer-Verlag, 1959
20. W.H. McMaster, N. Nerr del Grande, J.H. Mallett, and J.H. Hubbell, "Compilation of X-Ray Cross Sections," Report UCRL-50/74, Sec. 2, Rev. 1, Lawrence Radiation Laboratory, Berkeley, CA, 1969
21. C.A. Ashley and S. Doniach, *Phys. Rev. B*, Vol 11, 1975, p 1279
22. P.A. Lee and J.B. Pendry, *Phys. Rev. B*, Vol 11, 1975, p 2795
23. P.A. Lee and G. Beni, *Phys. Rev. B*, Vol 15, 1977, p 2862
24. R.F. Pettifer and P.W. McMillan, *Philos. Mag.*, Vol 35, 1977, p 871
25. F. Jona, *Surf. Sci.*, Vol 68, 1977, p 204
26. A. Bianconi, *Appl. Surf. Sci.*, Vol 6, 1980, p 392
27. B.K. Teo, *J. Am. Chem. Soc.*, Vol 103, 1981, p 3390
28. B. Lengeler, *Phys. Rev. Lett.*, Vol 53, 1984, p 74
29. J. Schwinger, *Phys. Rev.*, Vol 75, 1949, p 1912
30. H. Winick, in *Synchrotron Radiation Research*, H. Winick and S. Doniach, Ed., Plenum Press, 1980, p 11
31. F.R. Elder, R.V. Langmuir, and H.D. Pollock, *Phys. Rev.*, Vol 74, 1948, p 52
32. A.D. Baer, R. Gaxiola, A. Golde, F. Johnson, B. Salsburg, H. Winick, M. Baldwin, N. Dean, J. Harris, E. Hoyt, B. Humphrey, J. Jurow, R. Melen, J. Miljan, and G. Warren, *IEEE Trans. Nucl. Sci.*, NS-22, 1975, p 1794
33. S. Doniach, I. Lindau, W.R. Spicer, and H. Winick, *J. Vac. Sci. Technol.*, Vol 12, 1975, p 1123
34. J. Jaklevic, J.A. Kirby, M.P. Klein, A.S. Robertson, G.S. Brown, and P. Eisenberger, *Solid State Commun.*, Vol 23, 1977, p 679
35. S.H. Hunter, Ph.D. thesis, Stanford University, Stanford, CA, unpublished, 1977
36. F.A. Stem and S.M. Heald, *Rev. Sci. Instrum.*, Vol 50 (No. 12), 1979, p 1579
37. J.B. Hasting, P. Eisenberger, B. Lengeler, and M.C. Perlman, *Phys. Rev. Lett.*, Vol 43, 1979, p 1807
38. M. Marcus, L.S. Powers, A.R. Storm, B. M. Kincaid, and B. Chance, *Rev. Sci. Instrum.*, Vol 51, 1980, p 1023
39. F.W. Lytle, D.E. Sayers, and E.A. Stern, *Phys. Rev. B*, Vol 11, 1975, p 4825
40. P.A. Lee, P.H. Citrin, P. Eisenberger, and B.M. Kincaid, *Rev. Mod. Phys.*, Vol 53 (No. 4), 1981, p 769
41. E.A. Stern, D.E. Sayers, and F.W. Lytle, *Phys. Rev. B*, Vol 11, 1975, p 4836
42. B. Lengeler and P. Eisenberger, *Phys. Rev. B*, Vol 21, 1980, p 4507
43. B.K. Teo and P.A. Lee, *J. Am. Chem. Soc.*, Vol 101, 1979, p 2815
44. J. Wong, in *Topics in Applied Physics*, Vol 46, H.-J. Guntherodt and H. Beck, Ed., Springer-Verlag, 1981
45. C. Bingham, M.D. Godfrey, and JW. Turkey, *IEEE Trans.*, Vol Aug-15 (No. 2), 1967, p 58
46. P.H. Citrin, P. Eisenberger, and B.M. Kincaid, *Phys. Rev. Lett.*, Vol 36, 1976, p 1346
47. T.M. Hayes and J.B. Boyce, *Solid State Phys.*, Vol 37, 1982, p 173
48. B.K. Teo, P. Eisenberger, J. Reed, J.K. Barton, and S.J. Lippard, *J. Am. Chem. Soc.*, Vol 100, 1978, p 3225
49. S.P. Cramer, K.O. Hodgson, E.I. Stiefel, and W.R. Newton, *J. Am. Chem. Soc.*, Vol 100, 1978, p 2748
50. G.H. Via, J.H. Sinfelt, and F.W. Lytle, *J. Chem. Phys.*, Vol 71, 1979, p 690
51. J. Wong and H.H. Lieberman, *Phys. Rev. B*, Vol 29, 1984, p 651
52. K.J. Rao, J. Wong, and M.J. Weber, *J. Chem. Phys.*, Vol 78, 1983, p 6228
53. J.H. Sinfelt, G.H. Via, and F.W. Lytle, *J. Chem. Phys.*, Vol 72, 1980, p 4832
54. M.C. Srivastava and H.L. Nigam, *Coord. Chem. Rev.*, Vol 9, 1972, p 275
55. J. Wong, F.W. Lytle, R.P. Messmer, and D.H. Maylotte, *Phys. Rev. B*, Vol 30, 1984, p 5596
56. R.G. Shulman, Y. Yafet, P. Eisenberger, and W.E. Blumberg, *Proc. Nat. Acad. Sci.*, Vol 13, 1976, p 1384
57. J.L. Dehmer, *J. Chem. Phys.*, Vol 56, 1972, p 4496
58. P.J. Durham, J.B. Pendry, and C.H. Hodges, *Solid State Commun.*, Vol 38, 1981, p 159
59. F.W. Kutzler, C.R. Natoli, D.K. Misemer, S. Doniach, and K.O. Hodgson, *J. Chem. Phys.*, Vol 73, 1980, p 3274
60. V. Kunzl, *Collect. Trav. Cjim. Techecolovaquie*, Vol 4, 1932, p 213
61. S.P. Cramer and K.O. Hodgson, *Prog. Inorg. Chem.*, Vol 25, 1979, p 1
62. S. Doniach, P. Eisenberger, and K.O. Hodgson, in *Synchrotron Radiation Research*, H. Winich and S. Doniach, Ed., Plenum Press, 1980, p 425–458
63. F.W. Lytle, G.H. Via, and J.H. Sinfelt, in *Synchrotron Radiation Research*, H. Winich and S. Doniach, Ed., Plenum Press, 1981, p 401–424
64. S.J. Gurman, *J. Mater. Sci.*, Vol 17, 1982, p 1541
65. S.J. Gurman, in *Extended X-Ray Absorption Fine Structure*, R.W. Joyner, Ed., Plenum Press, 1985
66. K.O. Hodgson, B. Hedman, and J.E. Penner-Hahn, Ed., *Part VII: Geology and Geochemistry of EXAFS and Near-Edge Structure III*, Springer-Verlag, 1984, p 336–390
67. J. Wong, E.F. Koch, CA. Hejna, and M.F. Garbauskas, *J. Appl. Phys.*, Vol 58, 1985, p 3388
68. S. Rundquist, *Acta Chem. Scand.*, Vol 12, 1958, p 658

SELECTED REFERENCES

- C.H. Booth and F. Bridges, Improved Self-Absorption Correction for Fluorescence Measurements of Extended X-Ray Absorption Fine-Structure, *Phys. Scr. Vol. T*, Vol 115, 2005, p 202–220
- G.E. Brown, G. Calas, G.A. Waychunas, and J. Petiau, X-Ray Absorption Spectroscopy: Applications in Mineralogy and Geochemistry, *Rev. Min. Geochem.*, Vol 18, 1998, p 431–512
- G. Bunker, *Introduction to XAFS: A Practical Guide to X-Ray Absorption Fine Structure Spectroscopy,* Cambridge University Press, 2010
- S. Calvin, *XAFS for Everyone,* CRC Press, 2013
- A. Filipponi, A. Di Cicco, and C.R. Natoli, X-Ray-Absorption Spectroscopy and n-Body Distribution Functions in Condensed Matter, *Phys. Rev. B*, Vol 52, 1995, p 15122–15134
- D.C. Koningsberger and R. Prins, Ed., *X-Ray Absorption: Principles, Applications, Techniques of EXAFS, SEXAFS, and XANES,* John Wiley & Sons, 1998
- M. Newville, B. Boyanov, and D.E. Sayers, Estimation of Uncertainties in XAFS Data, *J. Synchrotron Radiation,* Vol 6, 1999, p 264–265
- M. Newville, P. Livins, Y. Yacoby, J.J. Rehr, and E.A. Stern, Near-Edge X-Ray-Absorption Fine Structure of Pb: A Comparison of Theory and Experiment, *Phys. Rev. B*, Vol 47, 1993, p 14126–14131
- J.J. Rehr and R.C. Albers, Theoretical Approaches to X-Ray Absorption Fine-Structure, *Rev. Mod. Phys.,* Vol 72 (No. 3), 2000, p 621–654

ASM Handbook, Volume 10, *Materials Characterization*
ASM Handbook Committee
DOI 10.31399/asm.hb.v10.a0006666

Particle-Induced X-Ray Emission*

Overview

General Uses

- Nondestructive multielemental analysis of thin samples, sodium through uranium, to approximately 1 ppm or 10^{-9} g/cm^2
- Nondestructive multielemental analysis of thick samples for medium and heavy elements
- Semiquantitative analysis of elements versus depth
- Elemental analyses of large and/or fragile objects through external beam proton milliprobe
- Elemental analyses using proton microprobes, spatial resolution to a few micrometers, and mass detection limits below 10^{-16} g

Examples of Applications

- Analysis of air filters for a wide range of elements
- Analysis of atmospheric aerosols by particle size for source transport, removal, and effect studies
- Analysis of powdered plant materials and geological powders for broad elemental content
- Analysis of elemental content of waters, solute, and particulate phases, including suspended particles
- Medical analysis for elemental content, including toxicology and epidemiology
- Analysis of materials for the semiconductor industry and for coating technology
- Archaeological and historical studies of books and artifacts, often using external beams
- Forensic studies

Samples

- *Form:* Thin samples (generally no more than a 10 mg/cm^2 thick solid) are analyzed in vacuum, as are stabilized powders and evaporated fluids. Thick samples can be any solid and thickness, but proton beam penetration is typically 30 mg/cm^2 or approximately 0.15 mm (0.006 in.) in a geological sample.
- *Size:* The sample area analyzed is on the order of millimeters to centimeters, except in microprobes, in which beam spot sizes approaching 1 μm are available.
- *Preparation:* None for air filters and many materials. Powders and liquids must be stabilized, dried, and generally placed on a substrate, such as plastic. Thick samples can be pelletized.

Limitations

- Access to an ion accelerator of a few mega electron volts is necessary.
- Generally, no elements below sodium are quantified.
- Elements must be present above approximately 1 ppm.
- Sample damage is more likely than with some alternate methods.
- No chemical information is generated.

Estimated Analysis Time

- 30 s to 5 min in most cases; thousands of samples can be handled in a few days.

Capabilities of Related Techniques

- *X-ray fluorescence:* With repeated analyses at different excitation energies, essentially equivalent or somewhat superior results can be obtained when sample size and mass are sufficient.
- *Neutron activation analysis:* Variable elemental sensitivity to neutron trace levels for some elements, essentially none for other elements. Neutron activation analysis is generally best for detecting the least common elements but performs the poorest on the most common elements, complementing x-ray techniques.
- *Electron microprobe:* Excellent spatial resolution (approximately 1 μm) but elemental mass sensitivity only approximately one part per thousand
- *Optical methods:* Atomic absorption or emission spectroscopy, for example, is generally applicable to elements capable of being dissolved or dispersed for introduction into a plasma.

Introduction

Particle-induced x-ray emission (PIXE) is one of several elemental analyses based on characteristic x-rays. These methods can be classified by the method of excitation and the nature of x-ray detection. The excitation source creates inner electron shell atomic vacancies that cause x-ray emission when filled by outer electrons. X-ray fluorescence (XRF) uses x-rays for this purpose; electrons are used to cause vacancies in electron microprobes and some scanning electron microscopes. Particle-induced x-ray emission uses beams of energetic ions, normally protons of

*Revised from T.A. Cahill, Particle-Induced X-Ray Emission, *Materials Characterization*, Vol 10, *ASM Handbook*, ASM International, 1986, p 102–108

a few mega electron volts, to create inner electron shell vacancies. Regarding detection, the most widely used methods involve wavelength dispersion, which is scattering from a crystal, or energy dispersion, which involves direct conversion of x-ray energy into electronic pulses in silicon or germanium diodes.

Scientists in various fields are continuously finding new uses for PIXE. From polymer testing and failure analysis to deformulation and pharmaceutical product development, PIXE is a premium nondestructive method for delivering incredibly accurate information on the elemental level.

Principles

Elemental analysis requires that some method of excitation reach the atom of interest and that the information obtained during de-excitation reach an appropriate detector. In the case of PIXE, the exciting radiation is an ion beam consisting of protons at an energy level of 2 to 5 MeV. These ions have a limited distance of penetration in a target material (Table 1A). Table 1B lists the approximate stopping distance in milligrams per square centimeter for typical targets often encountered in PIXE analysis: low-Z (atomic number) targets, such as air filters and biological samples; medium-Z targets, such as silicon chips, rocks, and pottery; and high-Z targets, such as transition metals. In addition, the thicknesses of targets necessary to meet a thin-target criterion, selected in Table 1(a) to produce a 10% loss of primary beam energy or attenuation, are also given in milligrams per square centimeter.

Thin targets, in which corrections for primary beam excitation energy and secondary x-ray absorption and refluorescence are small and easily calculable, provide highly accurate, absolute values that compare well with the best alternate nondestructive elemental techniques, such as XRF and neutron activation analysis (NAA). Thus, many PIXE research programs have worked to prepare samples, such as thin air filters and powdered biological and geological samples, that meet these energy-loss criteria. Such thin-target analysis was not normally performed before development of intrinsic germanium, lithium-doped germanium, and lithium-doped silicon energy-dispersive x-ray detectors, because wavelength-dispersive x-ray detection is highly inefficient and generally uses more massive targets with significant and difficult correction factors. Figure 1 shows a typical PIXE setup in a thin-target mode.

Once inner-shell vacancies have been created in the atoms of interest, the resulting characteristic x-rays must be able to leave the sample and reach a detector to be useful. This poses severe problems even in thin samples (Table 2). Table 2 lists the transmission of radiation through a low-Z thin sample. The first type of required correction is a loading correction that reflects the passage of x-rays through a uniform layer of deposit. The second is a particle size diameter correction associated with a spherical particle of a given diameter and composition typical in ambient aerosol studies.

The severe limitations on sample thickness for low-Z elements are shown clearly in Table 2; in practice, this limits the utility of PIXE for such elements. X-rays of elements as light as those of boron are visible by using windowless detectors, but results can never be more than qualitative for such elements. The lightest element normally reported and quantified is sodium (Z = 11), whose Kα x-ray energy is approximately 1 keV. The ability to observe and quantify every element from Z = 11 to 92 in a single analysis is the most important feature of a successful PIXE investigation. In some cases, nuclear methods such as Rutherford backscattering are used to measure elements lighter than sodium, down to and including hydrogen, allowing complete elemental analysis capable of direct comparison to total mass (determined by weighing the sample).

The ability of the incident ion to form a vacancy depends strongly on the energy of the ion. Although the process can be described in some detail by such models as the Plane Wave Born Approximation (Ref 2) and the Binary Encounter Model (Ref 3), realizing that the coulomb interaction that results in ionization depends heavily on the relative velocities of the ion and the electron yields considerable insight into vacancy formation. Thus, a very fast proton and a very slow electron behave

Table 1A Penetration of a 4 MeV proton in a low-Z matrix

Target configuration	Thickness, mg/cm^2
Thin target (10% energy loss)	5
Thick target (stopped beam)	28

Note: Effective atomic number, Z_{eff} = 9

Table 1B Transmission of secondary radiation through a low-Z sample approximately 5 mg/cm^2 thick

Secondary radiation	Energy	Criterion
Photons	3 keV	10% attenuation
Electrons	140 keV	10% energy loss
Protons	4 MeV	10% energy loss
Deuterons	5.5 MeV	10% energy loss
Tritons	6.5 MeV	10% energy loss
^{3}He	14.4 MeV	10% energy loss
^{4}He	16 MeV	10% energy loss
Neutrons	Thermal	10% attenuation, good geometry

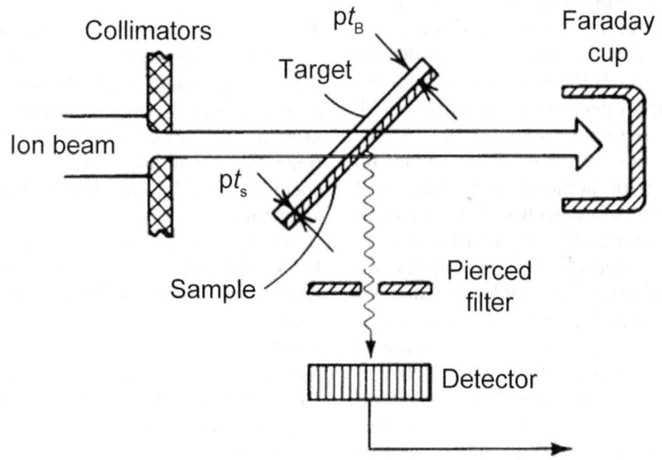

Fig. 1 Typical setup for particle-induced x-ray emission analysis. The entire apparatus is contained in a vacuum chamber. ρt_B, maximum target thickness; ρt_S, maximum sample thickness

Table 2 Typical loading and particle size diameter corrections for thin particle-induced x-ray emission samples

Element	Kα x-ray energy, keV	Maximum sample thickness(a)	Loading correction(b) for ρt_S = 300 μg/cm^2	Correction for particle size diameter 30 μm(c)	10 μm(d)	1 μm(d)
Sodium	1.041	150 μg/cm^2	1.70	4.7	2.5	1.32
Magnesium	1.255	230 μg/cm^2	1.39	4.2	2.0	1.22
Aluminum	1.487	390 μg/cm^2	1.23	2.6	1.7	1.16
Silicon	1.739	640 μg/cm^2	1.14	2.3	1.5	1.12
Phosphorus	2.014	1 mg/cm^2	1.09	2.1	1.37	1.09
Sulfur	2.307	1.5 mg/cm^2	1.06	1.9	1.26	1.07
Chlorine	2.622	2.2 mg/cm^2	1.04	1.8	1.20	1.0
Potassium	3.312	4.3 mg/cm^2	1.02	1.5	1.12	1.025
Calcium	3.690	(e)	1.02	1.4	1.09	1.02
Titanium	4.508	(e)	1.01	1.3	1.07	1.017
Iron	6.400	(e)	1.01	1.1	1.03	1.01

(a) Assuming a 30% self-attenuation correction at Z_{eff} = 9 for a target at 45° to the detector (Ref 1). (b) True value = Correction × Observed value. ρt_S, maximum sample thickness. (c) Assumes Earth crustal composition and ρ = 2.2 g/cm^3. (d) Assumes Z_{eff} = 9 and ρ = 1.5 g/cm^3. (e) Limitation set by energy loss in ion beam much like a very fast electron and a very slow proton in that the probability of ionization in either case is low.

When the velocities are approximately matched, the cross section is maximum. This occurs for K-shell x-rays with a 2 MeV proton on aluminum and a 10 MeV proton on iron—and higher for heavier elements. Another insight is that, to a good approximation, electrons bound with equal energy have approximately equal velocities. Thus, x-ray production for the Kα x-ray of arsenic (per atom) occurs with approximately the same probability as that of an Lα x-ray of lead (per atom), because the Kα of arsenic and the Lα of lead have the same energy. Obtaining the x-ray production versus mass requires factoring in the relative masses of arsenic and lead. Figure 2 illustrates x-ray production cross sections, showing examples of such approximations.

Combining these cross-sectional curves with the range/energy relationship enables calculation of yield versus depth in thick samples. One difficulty in this calculation derives from the principle that protons lose energy as they penetrate the sample. The cross sections for x-ray production generally decrease, but this varies for each element. Another difficulty is that the attenuation of x-rays exiting the sample is also a function of depth in the sample, atomic number in the element, and the transition lines (K, L, M,. . .) used for identification and quantification. If the sample is nonuniform in depth or heterogeneous in structure, yield-versus-depth calculations may not be possible. These difficulties are less troublesome for uniformly deep or homogeneous samples. A corollary of the aforementioned approximate yield-versus-depth relationship is that only a correction factor scaled by the x-ray energy, E_x, need be considered to first order, because the x-ray production and the mass attenuation of x-rays exiting the sample are similar for x-rays of the same energy (whether K, L, or M shell).

These concepts are useful in estimating x-ray production rates; however, approximations, published cross sections, and theoretical predictions are not adequate in practice to achieve the absolute accuracies (within a few percent) reported in analytical intercomparisons. This requires several gravimetric thin standards covering a range of atomic numbers. A fit to such multiple standards generally provides a better fit than any single standard alone.

Calculations of minimum detectable limits require evaluation of the characteristic x-ray production and the x-ray background (Fig. 3). The standard relationship for identifying a peak as statistically significant is:

$$N_{x,z} \geq 3.3\sqrt{N_B + N_{B,x}} \quad \text{(Eq 1)}$$

where $N_{x,z}$ is the number of x-rays, N_B is the number of background counts due to the substrate, and $N_{B,x}$ is the number of background counts due to the sample (Fig. 3). The quantity of background counts versus x-ray depends crucially on the matrix. If a low-Z matrix, such as a synthetic fluorine-containing resin air filter or a biological powder, is selected, the background is dominated by bremsstrahlung generated by "free" or loosely bound electrons. For a proton of E_P (MeV), these electrons can generate bremsstrahlung to approximately $2E_P$ (keV):

$$E_x(\text{keV}) \approx 2E_P(\text{MeV}) \quad \text{(Eq 2)}$$

The result is an 8 keV x-ray from a 4 MeV proton. This is approximately the point at which the bremsstrahlung curve flattens; it generally represents nearly optimum sensitivity of a PIXE system for a range of elements. Figure 4 illustrates theoretical and experimental background spectra for 3 MeV protons on transition metal substrates.

Approximately two-thirds of all PIXE analyses, standard or proton microprobe, operate with protons between 2 and 4 MeV, although energies as low as 0. 15 MeV or as high as 50 MeV can be used (Ref 6). In addition, alpha particles with energies as high as 30 MeV are sometimes used. Figure 5 shows the elemental sensitivity of PIXE analysis based on typical parameters discussed in this article.

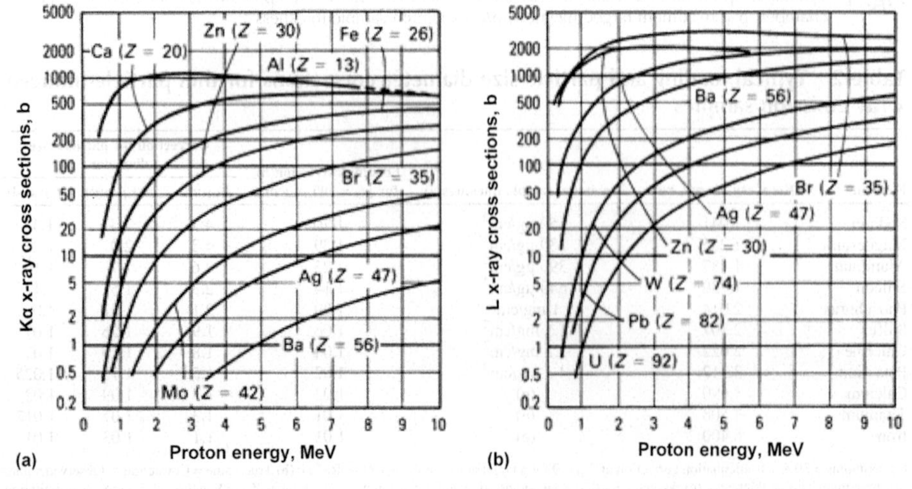

Fig. 2 (a) Kα and (b) L x-ray production cross sections versus proton energy and atomic number. Source: Ref 4

Calibration and Quality-Assurance Protocols

Calibration for most PIXE analyses is performed by using gravimetric standards thin enough to avoid serious absorption, refluorescence, or ion energy-change corrections. Because the expected x-ray yields are well known and follow a smooth dependence from element to element, intervening elements are located with little additional uncertainty. Thus, in reality, the calibration value should fall close to the yields predicted from the x-ray literature, or the presence of some error can be assumed. Once calibrated, a separate verification step is essential. This is ideally accomplished by using thin multielement glass

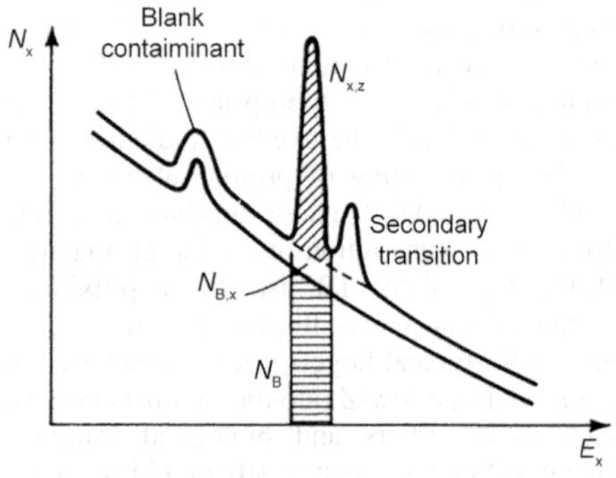

Fig. 3 Portion of an x-ray spectrum from a sample layered on a supporting substrate. Characteristic x-rays from the substrate (blank contaminant) and the sample are shown superimposed on a characteristic smooth background of x-rays due to bremsstrahlung caused by scattered electrons. The number of x-rays due to elements with atomic number Z in the sample, $N_{x,z}$, must be statistically greater than the uncertainty or fluctuations in the background due to the substrate, N_B, and the sample, $N_{B,x}$, for detection and quantification.

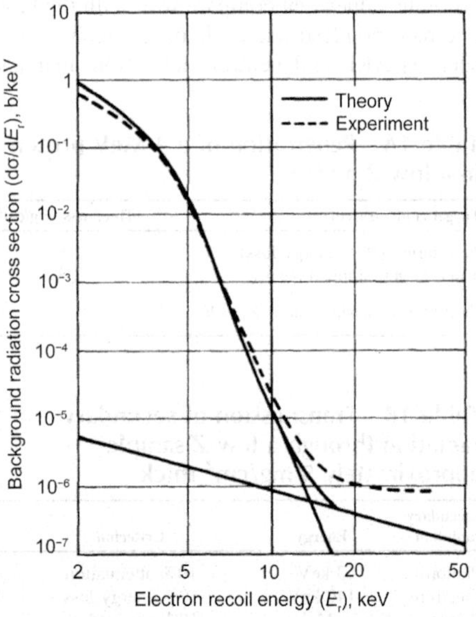

Fig. 4 Background spectra, experimental and theoretical, for 3 MeV protons on thin, clean transition metal substrates. E_r is the energy of the electron recoil caused by a 3 MeV proton on carbon. Source: Ref 5

standards (Ref 7), but other standards can also be prepared. Because PIXE is nondestructive, a previously analyzed sample can be analyzed again as a secondary standard. Finally, participation in the difficult but useful formal intercomparisons is possible. In all cases, formal protocols must be established and adhered to strictly.

Comparison of Particle-Induced X-Ray Emission with X-Ray Fluorescence

Particle-induced x-ray emission and XRF are types of x-ray analysis. Although the excitation modes differ, neither generates the massive bremsstrahlung of the direct electron beam excitation used in electron microprobes or scanning electron microscopes. Thus, both reach parts per million detection levels, as opposed to 0.1% levels in electron excitation. In most modern systems, PIXE and XRF use energy-dispersive lithium-doped silicon detectors. Although highly efficient, these detectors do not attain the energy resolution of wavelength-dispersive systems. Data-reduction codes can be essentially identical. Given these similarities, essentially equivalent detectable limits (in the parts per million range) are achievable in the same amount of time (a few minutes). Figure 6 compares the detection limits for XRF and PIXE analysis.

Particle-induced x-ray emission and XRF are also similar in that both have performed well in published interlaboratory and intermethod intercomparisons. Both methods can analyze samples as-received or with such minimal preparation as grinding or evaporating. This allows both to operate in a truly nondestructive mode that separates these methods from any other elemental analysis technique in common use. Table 3 lists the results of published interlaboratory intercomparisons using several elemental analysis procedures. Samples were prepared from solutions and placed in absorbant paper substrates, ground Canadian geological standard dusts, and ambient aerosols collected on filters. The standard deviation given in Table 3 represents the scatter of the ratio of the results to the "correct" value for many elements, ranging from aluminum to lead, and the scatter of the various reporting laboratories.

Results given in Table 3 compared reasonably well when standards were prepared from solutions. When ground rock and aerosol standards were used (such standards require the elements to be put into solution), results from atomic absorption (AA) and emission spectroscopy (ES) varied widely from correct values. Table 3 shows discrepancies as high as 47 standard errors from the correct value when using AA and ES. After this test, one major environmental agency limited its use of AA spectrometry on environmental samples to species that go into solution easily, for example, lead and cadmium, or those that x-ray or activation methods could not accommodate, such as beryllium.

Applications

Three areas in which PIXE has become a major analytical force are atmospheric physics and chemistry, external proton milliprobes and historical analysis, and PIXE microprobes. Particle-induced x-ray emission is a powerful yet nondestructive elemental analysis technique now used routinely by geologists, archaeologists, art conservators, and others to help answer questions of provenance, dating, and authenticity. Additional information on the applications of PIXE is cited in Ref 8 to 10.

Atmospheric Physics and Chemistry

One of the earliest uses of PIXE, involving analysis of atmospheric particles for elemental content (Ref 11), is also its most common application. The problem of atmospheric particles entails personal health, atmospheric visibility, and ecological effects, including ties to acid rain. Because of the complexity of the particles, existing analytical methods were inadequate. Generally, only mass was measured (by weighing), and a few chemical compounds, such as nitrates, sulfates, organic species, and lead, were studied using AA spectrophotometry. Sampling was performed using high-volume samplers that delivered a 20 by 25 cm (8 by 10 in.) sheet of fiberglass filter material for analysis. The complexity of the problem rendered such techniques as NAA and XRF ineffective.

Various elements are present in a given air sample; 22 elements occur at the 0.1% level in an average urban sample. Atmospheric variability demands timely information (sometimes as short as hours) and many analyses for statistical reliability. Particulate size is essential to understanding particulate sources, transport, transformation, effects on health and welfare, and removal mechanisms. Although PIXE has reasonable success in analysis of such air samples, the parallel development of specialized air samplers has expanded the capabilities of PIXE for analyzing small amounts of mass in a small beam size.

Simultaneous size chemistry profiles are essential to atmospheric chemistry studies, spanning questions from particulate sources through transport to effects and health and safety. However, size-collecting devices are

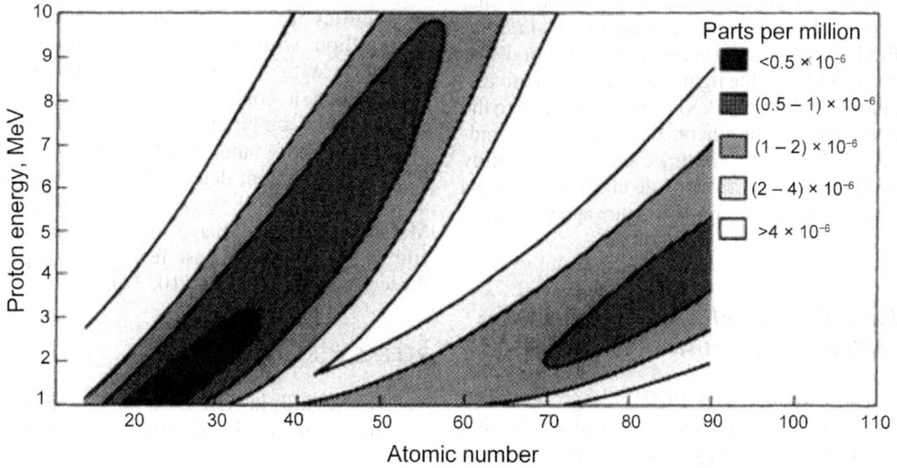

Fig. 5 Sensitivity of particle-induced x-ray emission analysis versus proton energy and atomic number based on typical parameter given in the text. Source: Ref 4

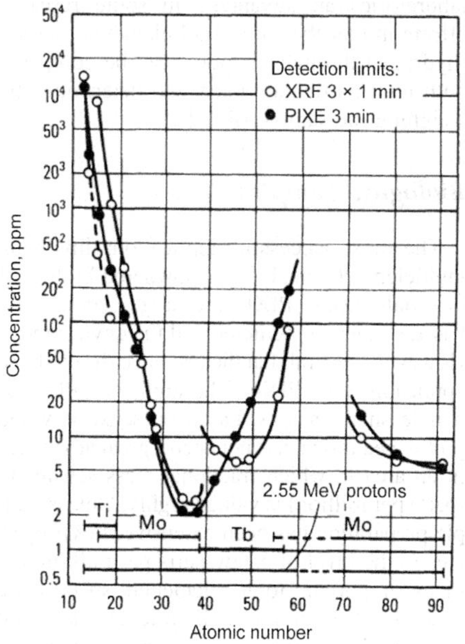

Fig. 6 Comparison of detection limits for x-ray fluorescence (XRF) and particle-induced x-ray emission (PIXE) analyses

Table 3 Results of formal interlaboratory analyses comparing particle-induced x-ray emission (PIXE) with other elemental analysis techniques

The values given represent the standard deviation (the ratio of the laboratory results to the established "correct" values).

Method	Number of groups reporting data(a)	Solution standards(b)	Ground rock standards(c)	Aerosol standards(d)	Aerosol samples(e)
PIXE	7	1.03 ± 0.16	0.99 ± 0.29	0.99 ± 0.19	0.98 ± 0.08; 1.01 ± 0.16
X-ray fluorescence (XRF)	8	0.97 ± 0.12	1.07 ± 0.20	1.03 ± 0.14	0.97 ± 0.08; 1.08 ± 0.15
Wavelength XRF	3	1.19 ± 0.34	1.12 ± 0.47	1.37 ± 0.50	...
AA, ES (f)	3	0.88 ± 0.17	0.40 ± 0.31	0.47 ± 0.29	1.04(g); 0.84(h)
ACT(i)	...	...	...	...	0.76 ± 0.15(h)

(a) Each result represents the mean and standard deviation for all laboratories using the method for all elements quoted. (b) Two samples, each including aluminum, sulfur, potassium, vanadium, chromium, manganese, iron, zinc, cadmium, and gold. (c) Two samples, each including aluminum, silicon, potassium, calcium, titanium, manganese, and iron. (d) Two samples, each including aluminum, silicon, sulfur, potassium, calcium, titanium, manganese, iron, copper, zinc, selenium, bromine, and lead. (e) Three samples or more, including up to 20 elements, of which sulfur, calcium, titanium, iron, copper, zinc, selenium, bromine, and lead are intercompared. Each result represents a single laboratory, with the result being the mean and standard deviation for each element as compared to the referees. (f) Results from atomic absorption (AA) and emission spectroscopies (ES). (g) Laboratory reported sulfur and lead results from AA only. (h) Laboratory reported sulfur only. (i) Charged particle activation analysis. Source: Ref 6

extremely limited in the amount of mass they collect. All elements in the atmosphere are important because they are part of the aerosol mass, and all results must be quantitative. These severe, simultaneous but important requirements demand the specialized capabilities of PIXE. Particle-induced x-ray emission spectroscopy "sees" more elements in a single analysis than any other method by using microgram samples delivered by particulate collectors with sizing capabilities. Devices such as a solar-powered aerosol impactor and an eight-stage rotating-drum impactor have been designed and built for use in PIXE analysis of particulate matter. Because particle accelerators are used for PIXE (a limitation to PIXE use), nuclear techniques, such as Rutherford backscattering, can be used in conjunction with PIXE to determine very light elements (down to hydrogen). This affords PIXE laboratories an advantage in some research efforts in that the sum of all elemental masses (hydrogen through uranium) can be compared with total particulate mass and determined by weighing the sample (Ref 12).

Biological Samples

The most successful applications involving multielemental analysis of powdered biological materials using PIXE are large, statistically based studies in medicine and the environment. However, several handicaps have limited the applicability of PIXE. The ability of PIXE to handle small masses, based on sensitivity and proton beam size, has not been a great advantage in an area in which fractional mass sensitivity (parts per million) is valued highly. Unless sample preparation has been extensive, PIXE operates at the level of a few parts per million, but many important toxic elemental species are desired at the parts per billion level. Elemental methods do not provide the important chemical nature of the materials. The nondestructive nature of PIXE is unimportant for such samples, and competitive methods are firmly established in the medical community. Therefore, PIXE programs in the biological sciences have met with limited success.

In other areas, the special characteristics of PIXE are making an impact. Tree ring analysis requires a small analytical area and a wide range of elements. Statistical studies in agriculture are also promising. These studies, which require low sensitivity (100 ppm) and the ability to handle large volumes of samples at low cost (a few dollars/sample), concentrate on the light, biologically important elements from sodium through zinc.

Materials Science in the Semiconductor Industry

This area was considered intensively in the first few years of the development of PIXE. An important advantage is that many of the laboratories conducting these analyses possess a particle accelerator for the Rutherford backscattering spectrometry of layered semiconductor structures. Almost all such laboratories use PIXE, but it has not become the mainstay that had been predicted. One reason is that depth information is difficult to obtain from PIXE but is essential to the semiconductor industry. In addition, the light-element (sodium to calcium) capability of PIXE is not useful due to the dominance of the silicon x-ray peak in the middle of the range. The trace element capability of PIXE finds relatively little utility in materials carefully prepared from ultrapure and well-characterized sources.

Particle-Induced X-Ray Emission Milliprobe and Historical Studies

The external-beam PIXE milliprobe exemplifies an unanticipated successful application of PIXE. Low-energy proton beams are often brought out of vacuum into air or helium atmospheres to avoid the expense of large vacuum-tight chambers. The external proton beam collimated to millimeter size and closely coupled to an x-ray detector has revealed new opportunities in analysis of large and/or fragile objects that could not be readily placed in vacuum. The high x-ray cross sections in PIXE allow use of very low beam currents, 10^{-9} A or less, solving potential damage problems. Thus, a major use of PIXE has developed for historical studies (Ref 13). Particle-induced x-ray emission has been successful because virtually no other nondestructive technique can provide parts per million analysis of a broad elemental range in minutes. A recent page-by-page analysis of a Gutenberg Bible proved the safety of the technique and revitalized an entire field of historical study (Ref 13).

The key advantage to these uses is that external-beam PIXE does not require material removal from rare and/or fragile items; therefore, no damage to these objects results when the technique is used properly. As a result, much can be learned about the early history of printing and ancient techniques in ceramics and metallurgy. Particle-induced x-ray emission also offers a useful means of detecting forgeries.

Particle-Induced X-Ray Emission and the Proton Microprobe

Particle-induced x-ray emission has become the most important analytical method used with accelerator-based proton microprobes, which are low-energy accelerators modified to provide reasonably intense beams of protons with spatial dimensions of a few micrometers in diameter. Such beams can penetrate more deeply into a matrix than the standard electron microprobes while maintaining a small diameter versus depth. The beams of protons or other light ions allow use of several nuclear and atomic analytical methods, granting capabilities absent from electron microprobes.

Because proton beams have no primary bremsstrahlung background, unlike electrons, sensitivities are significantly better—parts per million with protons versus parts per thousand with electron beams. Figure 7 shows such a comparison for a biological matrix. The combination of high fractional mass sensitivity and small beam diameter results in minimum detectable limits of mass down to the 10^{-16} g range and below (Fig. 8). Many microprobes have been constructed and used during the past few years to gain these advantages (Ref 6, 10, 14).

REFERENCES

1. T.A. Cahill, in *New Uses for Low Energy Accelerators*, J. Ziegler, Ed., Plenum Press, 1975
2. E. Merzbacker and H.W. Lewis, in *Encyclopedia of Physics*, S. Flugge, Ed., Springer Verlag, 1958

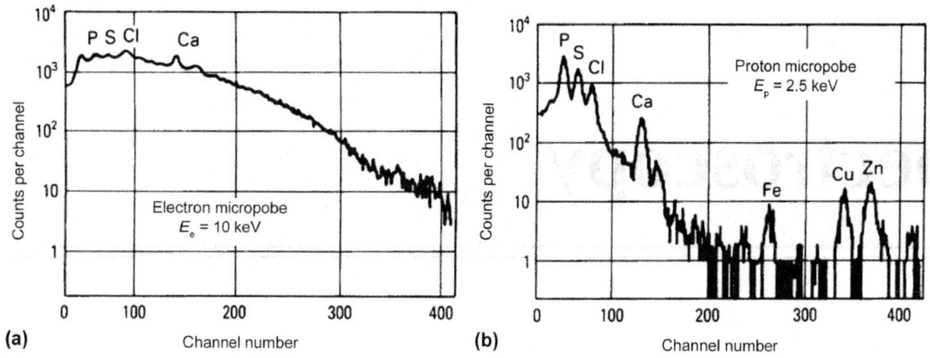

Fig. 7 Comparison of (a) electron microprobe and (b) proton microprobe analyses of a biological sample

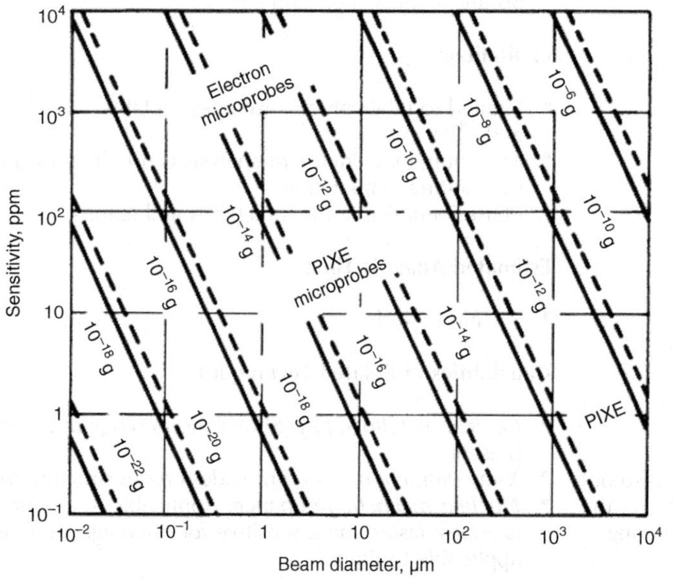

Fig. 8 Minimum detectable limits versus sensitivity and beam diameter for electron and particle-induced x-ray emission (PIXE) microprobes

3. J.D. Garcia, R.D. Fortner, and T. Kavanaugh, *Rev. Mod. Phys.*, Vol 45, 1973, p 111
4. T.B. Johansson and S.A.E. Johansson, *Nucl. Instrum. Methods*, Vol 137, 1976, p 473
5. F. Folkmann, *J. Phys. E*, Vol 8, 1975, p 429
6. T.A. Cahill, *Annu. Rev. Nucl. Part. Sci.*, Vol 30, 1980, p 211
7. *Special Reference Material Catalog 1533*, National Bureau of Standards, Gaithersburg, MD, 1984
8. K. Siegbahn, in *Proceedings of the International Conference on Particle Induced X-Ray Emission and Its Analytical Applications*, Vol 142 (No. 1, 2), 1977
9. S.A.E. Johansson, in *Proceedings of the Second International Conference on PIXE and Its Analytical Applications*, Vol 181, 1981
10. B. Martin, in *Proceedings of the Third International Conference on PIXE and Its Analytical Applications*, Vol B3, 1984
11. T.B. Johansson, R. Akselsson, and S.A.E. Johansson, *Nucl. Instrum. Methods*, Vol 84, 1970, p 141
12. T.A. Cahill, R.A. Eldred, D. Shadoon, P.J. Feeney, B. Kusko, and Y. Matsuda, *Nucl. Instrum. Methods*, Vol B3, 1984, p 191
13. T.A. Cahill, B.H. Kusko, R.A. Eldred, and R.H. Schwab, *Archeometry*, Vol 26, 1984, p 3
14. J.A. Cookson, *Nucl. Instrum. Methods*, Vol 65, 1979, p 477

SELECTED REFERENCES

- J.R. Bird, P. Duerden, and P.J. Wilson, *Nucl. Sci. Appl.*, Vol 1, 1983, p 357
- T.B. Johansson, R. Akselsson, and S.A.E. Johansson, *Adv. X-Ray Anal.*, Vol 15, 1972, p 373
- S.A.E. Johansson, J.L. Campbell, K.G. Malmqvist, and J.D. Winefordner, Ed., *Particle-Induced X-Ray Emission Spectrometry (PIXE)*, Wiley, 1995
- V. Lazic, M. Vadrucci, R. Fantoni, M. Chiari, A. Mazzinghi, and A. Gorghinian, Applications of Laser-Induced Breakdown Spectroscopy for Cultural Heritage: A Comparison with X-Ray Fluorescence and Particle Induced X-Ray Emission Techniques, *Spectrochim. Acta B, At. Spectrosc.*, Vol 149, 2018, p 1–14, ISSN: 0584-8547
- W. Maenhaut, X-Ray Fluorescence and Emission | Particle-Induced X-Ray Emission, *Encyclopedia of Analytical Science*, 3rd ed., P. Worsfold, C. Poole, A. Townshend, and M. Miró, Ed., Academic Press, 2019, p 432–442
- M.J. Owens and H.I. Shalgosky, *J. Phys. E*, Vol 7, 1974, p 593
- I.V. Popescu, A. Ene, C. Stihi, A.I. Gheboianu, G. Dima, T. Badica, and V. Ghisa, Analytical Applications of Particle-Induced X-Ray Emission (PIXE), *AIP Conf. Proc.*, Vol 899, 2007, p 538, DOI: 10.1063/1.2733279
- V. Volkovic, *Contemp. Phys.*, Vol 14, 1973, p 415

ASM Handbook, Volume 10, *Materials Characterization*
ASM Handbook Committee
DOI 10.31399/asm.hb.v10.a0006646

Mössbauer Spectroscopy*

Overview

General Uses

- Phase analysis
- Study of atomic arrangements
- Study of critical-point phenomena
- Magnetic-structure analysis
- Diffusion studies
- Surface and corrosion analysis

Examples of Applications

- Measurements of retained austenite in steel
- Analysis of corrosion products on steel
- Effects of grinding of the surface of carbon steel
- Measurement of δ-ferrite in stainless steel weld metal
- Curie and Neel point measurements

Samples

- *Form:* Solids (metals, ferrites, geological materials, and so on)
- *Size:* For transmission—powders or foils. Size varies but of the order of 50 μm thick and 50 mg of material. For scattering—films, foils, or bulk metals with an area of the order of 1 cm^2 (0.15 $in.^2$) or greater. As source—of the order of 1 to 100 mCi of radioactive material incorporated within approximately 50 μm of the sample surface

Limitations

- Limited to relatively few isotopes, notably ^{57}Fe, ^{119}Sn, ^{121}Sb, and ^{186}W
- Maximum temperature of analysis is usually only a fraction of the melting temperature
- Phase identification is sometimes ambiguous

Estimated Analysis Time

- 30 min to 48 h

Capabilities of Related Techniques

- *Optical metallography:* Shows morphology of the phases present
- *X-ray diffraction:* Faster; provides crystal structure information
- *Nuclear magnetic resonance:* Applicable to a wider range of isotopes; faster; more sensitive for nonmagnetic materials; applicable to liquids

Introduction

The Mössbauer effect (ME) is a spectroscopic method for observing nuclear γ-ray fluorescence using the recoil-free transitions of a nucleus embedded in a solid lattice. It is sometimes referred to as nuclear gamma-ray resonance. Most Mössbauer spectra are obtained by Doppler scanning, that is, by observing a count rate as a function of the relative velocity between the suitable γ-ray source and an absorber or scatterer. The spectrum then consists of a count rate versus relative velocity plot. The information in such a spectrum includes the amount of resonant absorption or scattering, line patterns characteristic of various phases or chemical species, the relative position of the spectrum (the isomer shift), and line splittings caused by nuclear hyperfine interactions. This information can then be interpreted to provide insight into the local atomic environment of those atoms responsible for the resonance. General information is provided in Ref 1 to 13.

Fundamental Principles

The Recoil-Free Fraction

The basis of the ME is the existence of a recoil-free fraction, *f*, for γ-rays emitted or absorbed by a nucleus embedded in a solid lattice. This recoil-free fraction, sometimes referred to as the Lamb-Mössbauer factor, yields the fraction of emitted or absorbed γ-rays that are unshifted by nuclear recoil. Only these unshifted γ-rays will be observed by the Mössbauer effect. The Lamb-Mössbauer factor is analogous to the Debye-Waller factor for x-ray diffraction in solids, in which the intensities of the Bragg reflection peaks decrease with increasing temperature. A temperature-dependent factor, *W*, that gives recoilless fraction is customarily defined as:

$$f = e^{-2W} \quad \text{(Eq 1)}$$

In the Debye model for the spectrum of lattice vibrations, the solid is described as an isotropic medium that undergoes vibrations over a continuous range of frequencies up to a maximum cutoff frequency, ω_{max}. A characteristic temperature, θ_D, where $k\theta_D = \hbar\omega_{max}$, is introduced. For this model, *W* is:

$$W = \frac{3E_R}{k\theta_D}\left[\frac{1}{4} + \left(\frac{T}{\theta_D}\right)^2 \int_0^{\theta_D/T} \frac{x\,dx}{e^x - 1}\right] \quad \text{(Eq 2)}$$

where *T* is the temperature, and E_R is the recoil energy of the free nucleus:

$$E_R = \frac{E_\gamma^2}{2Mc^2} \quad \text{(Eq 3)}$$

where E_γ is the energy of the γ-ray, *M* is the mass of the nucleus, and *c* is the velocity of light. Useful low- and high-temperature approximations for Eq 2 are:

$$W \cong \frac{3E_R}{4k\theta_D} \text{ for } T < \frac{\theta_D}{4} \quad \text{(Eq 4a)}$$

and

$$W \cong \frac{3E_R T}{k\theta_D^2} \text{ for } T > \frac{\theta_D}{2} \quad \text{(Eq 4b)}$$

*Revised from L.J. Swartzendruber and L.H. Bennett, Mössbauer Spectroscopy, *Materials Characterization*, Vol 10, *ASM Handbook*, ASM International, 1986, p 287–295

Table 1 lists approximate Debye temperatures for elements used in ME. When the Mössbauer emitting or absorbing atom is an isolated impurity in a metallic matrix with Debye temperature θ_D, a useful approximation for an effective Debye temperature can be obtained using:

$$\theta_{eff} = \left(\frac{M_0}{M_I}\right)^{1/2} \theta_D \qquad \text{(Eq 5)}$$

where M_0 is the atomic mass of the matrix atoms, and M_I is the atomic mass of the impurity atoms. Equation 5 neglects the effects of localized vibrational modes.

The Absorption Cross Section

The maximum cross section, σ_m, of a single nucleus for absorption of an incident Mössbauer γ-ray of energy E_γ is:

$$\sigma_m = f\sigma_0 \qquad \text{(Eq 6)}$$

where:

$$\sigma_0 = 2\pi \frac{\hbar^2 c^2}{E_\gamma^2} \frac{2I_e + 1}{2I_g + 1} \frac{1}{1+\alpha} \qquad \text{(Eq 7)}$$

where I_e is the spin of the excited nuclear state, I_g is the spin of the ground state, and α is the internal conversion coefficient. When the excited- and ground-state energy levels are split

Table 1 Some properties of Mössbauer transitions

Atomic number, Z	Atomic weight, A	Element	E_γ, keV	Isotopic abundance, %	$t_{1/2}$, ns	I_e	I_g	σ_0, 10^{-20} cm^2	W_0, mm/s	E_γ/k, K	θ_D, K
19	40	Potassium	29.6	0.0117	4.26	−3	−4	28.97	2.184	135	91
26	57	Iron	14.4125	2.19	97.81	−3/2	−1/2	256.6	0.1940	22.7	470
28	61	Nickel	67.42	1.25	5.06	−5/2	−3/2	72.12	0.8021	11.6	450
30	67	Zinc	93.32	4.11	9150	3/2	5/2	10.12	0.000320	810	327
32	73	Germanium	13.26	7.76	4000	5/2	9/2	361.2	0.005156	15.0	374
32	73	Germanium	68.75	7.76	1.86	7/2	9/2	22.88	2.139	403	374
36	83	Krypton	9.40	11.55	147	7/2	9/2	107.5	0.1980	6.63	72
44	99	Ruthenium	89.36	12.72	20.5	3/2	5/2	14.28	0.1493	503	600
44	101	Ruthenium	127.22	17.07	0.585	3/2	5/2	8.687	3.676	998	600
50	119	Tin	23.871	8.58	17.75	3/2	1/2	140.3	0.6456	29.8	200
51	121	Antimony	37.15	57.25	3.5	7/2	5/2	19.70	2.104	71.1	211
52	125	Tellurium	35.46	6.99	1.48	3/2	1/2	26.56	5.212	62.7	153
53	127	Iodine	57.60	100	1.9	7/2	5/2	21.37	2.500	163	...
54	129	Xenon	39.58	26.44	1.01	3/2	1/2	23.31	6.843	75.7	64
54	131	Xenon	80.16	21.18	0.50	1/2	3/2	7.183	6.825	306	64
55	133	Cesium	80.997	100	6.30	5/2	7/2	10.21	0.5361	307	38
59	141	Praseodymium	145.2	100	1.85	7/2	5/2	10.67	1.018	931	130
60	145	Neodymium	67.25	8.30	29.4	3/2	7/2	3.809	0.1384	194	140
60	145	Neodymium	72.50	8.30	0.72	5/2	7/2	5.916	5.240	226	140
62	147	Samarium	122.1	14.97	0.80	5/2	7/2	6.153	2.800	632	140
62	149	Samarium	22.5	13.83	7.12	5/2	7/2	7.106	1.708	21.2	140
62	152	Samarium	121.78	26.73	1.42	2	0	35.86	1.582	608	140
62	154	Samarium	81.99	22.71	3.00	2	0	30.08	1.112	272	140
63	151	Europium	21.64	47.82	9.7	7/2	5/2	11.42	1.303	19.3	140
63	153	Europium	83.3652	52.18	0.82	7/2	5/2	6.705	4.002	283	130
63	153	Europium	97.4283	52.18	0.21	5/2	5/2	17.97	13.37	387	130
63	153	Europium	103.1774	52.18	3.9	3/2	5/2	5.417	0.6798	434	130
64	154	Gadolinium	123.14	2.15	1.17	2	0	36.67	1.899	614	200
64	155	Gadolinium	60.012	14.73	0.155	5/2	3/2	9.989	29.41	145	200
64	155	Gadolinium	86.54	14.73	6.32	5/2	3/2	34.40	0.5002	301	200
64	155	Gadolinium	105.308	14.73	1.16	5/2	3/2	24.88	2.239	446	200
64	156	Gadolinium	88.967	20.47	2.22	2	0	30.42	1.385	316	200
64	157	Gadolinium	54.54	15.68	0.187	5/2	3/2	9.071	26.82	118	200
64	157	Gadolinium	64.0	15.68	460	5/2	3/2	44.79	0.009292	163	200
64	158	Gadolinium	79.51	24.87	2.46	2	0	27.88	1.399	249	200
64	160	Gadolinium	75.3	21.90	2.63	2	0	21.15	1.381	221	200
65	159	Terbium	58.0	100	13	5/2	3/2	9.827	0.3628	132	...
66	160	Dysprosium	86.788	2.29	1.98	2	0	29.42	1.592	293	210
66	161	Dysprosium	25.65	18.88	28.1	5/2	5/2	95.34	0.3795	25.2	210
66	161	Dysprosium	43.84	18.88	920	7/2	5/2	28.29	0.006782	74.4	210
66	161	Dysprosium	74.57	18.88	3.35	3/2	5/2	6.755	1.095	215	210
66	162	Dysprosium	80.7	25.53	2.25	2	0	26.09	1.507	251	210
66	164	Dysprosium	73.39	28.18	2.4	2	0	20.86	1.553	205	210
67	165	Holmium	94.70	100	0.0222	9/2	7/2	3.552	130.12	339	220
68	164	Erbium	91.5	1.56	1.73	2	0	28.10	1.728	318	220
68	166	Erbium	80.56	33.41	1.82	2	0	23.56	1.866	244	220
68	167	Erbium	79.321	22.94	0.103	9/2	7/2	7.715	33.48	235	220
68	168	Erbium	79.80	27.07	1.91	2	0	12.80	1.795	236	220
68	170	Erbium	79.3	14.88	1.92	2	0	24.31	1.797	220	220
69	169	Thulium	8.42	100	3.9	3/2	1/2	21.17	8.330	2.61	230
70	170	Ytterbium	84.262	3.03	1.60	2	0	23.93	2.029	260	120
70	171	Ytterbium	66.74	14.31	0.87	3/2	1/2	9.004	4.711	162	120
70	171	Ytterbium	75.89	14.31	1.7	5/2	1/2	13.14	2.120	210	120
70	172	Ytterbium	78.67	21.82	1.8	2	0	20.80	1.932	224	120
70	174	Ytterbium	76.5	31.84	1.76	2	0	20.69	2.032	210	120
70	176	Ytterbium	82.1	12.73	2.0	2	0	20.16	1.666	239	120
71	175	Lutetium	113.81	97.41	0.10	9/2	7/2	7.154	24.04	461	210
72	176	Hafnium	88.36	5.20	1.39	2	0	25.27	2.227	276	252
72	177	Hafnium	112.97	18.50	0.5	9/2	7/2	5.990	4.843	449	252
72	178	Hafnium	93.17	27.14	1.50	2	0	25.16	1.957	304	252

(continued)

Note: E_γ is the γ-ray energy of the Mössbauer transition, $t_{1/2}$ is the half-life of the excited Mössbauer level, I_e and I_g are the spins of the excited- and ground-state nuclear levels, σ_0 is the resonant absorption cross section. W_0 is the full width at half maximum of the unbroadened line (twice the natural width), $E\gamma/k$ is the recoil energy of the free nucleus divided by the Boltzmann constant, and θ_D is the low-temperature limit. Source: Ref 14

Table 1 (Continued)

Atomic number, Z	Atomic weight, A	Element	E_γ, keV	Isotopic abundance, %	$t_{1/2}$, ns	I_e	I_g	σ_0, 10^{-20} cm^2	W_0, mm/s	E_γ/k, K	θ_D, K
72	180	Hafnium	93.33	35.24	1.50	2	0	25.53	1.954	301	252
73	181	Tantalum	6.23	99.99	6800	9/2	7/2	167.6	0.006457	1.34	240
73	181	Tantalum	136.25	99.99	0.0406	9/2	9/2	5.968	49.45	639	240
74	180	Tungsten	103.65	0.135	1.47	2	0	25.88	1.795	372	400
74	182	Tungsten	100.102	26.41	1.37	2	0	25.17	1.995	343	400
74	183	Tungsten	46.4837	14.40	0.183	3/2	1/2	5.523	32.16	77.1	400
74	183	Tungsten	99.0788	14.40	0.692	5/2	1/2	14.95	3.990	334	400
74	184	Tungsten	111.192	30.64	1.26	2	0	26.04	1.953	419	400
74	186	Tungsten	122.5	28.41	1.01	2	0	31.35	2.211	503	400
75	187	Rhenium	134.24	62.93	0.01	7/2	5/2	5.371	203.8	600	430
76	186	Osmium	137.157	1.64	0.84	2	0	28.39	2.374	630	500
76	188	Osmium	155.03	13.3	0.695	2	0	27.96	2.539	797	500
76	189	Osmium	36.22	16.1	0.50	1/2	3/2	1.151	15.10	43.2	500
76	189	Osmium	69.59	16.1	1.64	5/2	3/2	8.419	2.397	160	500
76	189	Osmium	95.23	16.1	0.3	3/2	3/2	3.503	9.575	299	500
76	190	Osmium	186.9	26.4	0.47	2	0	33.61	3.114	1146	500
77	191	Iridium	82.398	37.3	4.02	1/2	3/2	1.540	0.8258	222	420
77	191	Iridium	129.400	37.3	0.089	5/2	3/2	5.692	23.75	546	420
77	193	Iridium	73.028	62.7	6.3	1/2	3/2	3.058	0.5946	172	420
77	194	Iridium	138.92	62.7	0.080	5/2	3/2	5.833	24.61	623	420
78	195	Platinum	98.857	33.8	0.170	3/2	1/2	6.106	16.28	312	240
78	195	Platinum	129.735	33.8	0.620	5/2	1/2	7.425	3.401	538	240
79	197	Gold	77.35	100	1.90	1/2	3/2	3.857	1.861	189	164
80	201	Mercury	32.19	13.22	<0.2	1/2	3/2	1.935	42.49	32.1	75
90	232	Thorium	49.369	100	0.345	2	0	1.667	16.06	65.5	165
92	238	Uranium	44.915	99.27	245	2	0	0.917	0.2486	52.8	200
93	237	Neptunium	59.537	(radioactive)	68.3	5/2	5/2	32.55	0.06727	93.2	75

Note: E_γ is the γ-ray energy of the Mössbauer transition, $t_{1/2}$ is the half-life of the excited Mössbaucr level, I_e and I_g are the spins of the excited- and ground-state nuclear levels, σ_0 is the resonant absorption cross section. W_0 is the full width at half maximum of the unbroadened line (twice the natural width), $E\gamma/k$ is the recoil energy of the free nucleus divided by the Boltzmann constant, and θ_D is the low-temperature limit. Source: Ref 14

by a hyperfine field, the cross section is divided proportionally among the various possible transitions. Table 1 lists values for σ_0. As the energy of the γ-ray increases, *f* is reduced (Eq 1, 2), and the absorption cross section is also reduced (Eq 7). Partly because of these two practical considerations, ME has been observed only for γ-ray energies below 200 keV.

Selection Rules and γ-Ray Polarization

The relative intensities of the individual components of lines split by a hyperfine interaction depend on the selection rules governing the nuclear transition. The probability of the emission of the γ-ray also takes on an angular dependence, depending on the angle between the magnetic field at the nucleus and the direction of propagation of the γ-ray. The selection rules for ME transitions depend on the multipolarity of the nuclear transition between ground and excited states.

The laws of conservation of angular momentum and parity determine the possible transitions between the excited and ground states of a nucleus. Defining the momentum and parities of the excited nuclear state as I_e and P_e and of the ground state as I_g and P_g (the parities can have only one of two values, 1 or −1), the angular momentum L of the emitted (or absorbed) γ-ray can take on only the values $|I_e - I_g| < L < |I_e + I_g|$. The number 2^L defines the multipolarity of the transition, that is, $L = 1$ (dipole), $L = 2$ (quadrupole), $L = 3$ (octapole), and so on.

For each multipolarity L, there are two types of nuclear transitions: electric EL and magnetic ML. If P_e and P_g are equal, EL transitions are allowed only for even Ls, and ML transitions are allowed only for odd Ls. If P_e and P_g differ, EL transitions are allowed only for odd Ls, and ML transitions are allowed only for even Ls. For example, if $I_e = 3/2$, $P_e = -1$, $I_g = 1/2$, and $P_g = -1$, only $M1$ and $E2$ transitions are possible.

For a given energy of transition, the excited-state lifetime increases rapidly with multipolarity. Most usable Mössbauer transitions have energies of approximately 10 to 100 keV and have lifetimes from 10^{-6} to 10^{-10} s. This limits most Mössbauer transitions to pure $E1$, such as ^{237}Np; nearly pure $M1$, such as for ^{57}Fe and ^{119}Sn; $M1$ with a small mixture of $E2$, such as for ^{197}Au; or pure $E2$, such as for the even-even rare earth nuclei that all have $I_g = 0$, $P_g = 1$, and $I_e = 2$, $P_e = 1$, for example, ^{178}Hf. The $E2/M1$ ratios have often been determined by internal conversion studies. For ^{197}Au, experimental values of ME line-intensity ratios correspond closely to the theoretical values for a 90% $M1$ plus 10% $E2$ transition.

In the presence of internal fields, the nuclear excited and ground state can be split into sublevels specified by a quantum number m, with $-I < m < I$. In this case, the γ-ray emissions will consist of a set of transitions between these sublevels. The relative intensities, I, of the various transitions are:

$$I = \left\{ \begin{matrix} I_f & L & I_i \\ m_f & m & -m_i \end{matrix} \right\}^2 G^L \Delta_m(\phi) \qquad \text{(Eq 8)}$$

where the Wigner 3-j coefficient (the symbol in braces) is a function of the initial (I_i) and final (I_f) nuclear spins, the initial (m_i) and final (m_f) quantum numbers for the sublevels, and the multipolarity (L) of the transition. The function $G^L_m(\varphi)$ determines the angular distribution of the radiation as a function of the angle φ between the internal field and the direction of γ-ray propagation. Table 2 lists the angular functions for the cases in which $L = 1$ and 2.

Table 2 Angular distribution functions for nuclear transitions of multipolarity $L = 1$ and $L = 2$

Δm	$L = 1$, $G^1\Delta m(\phi)$	$L = 2$, $G^2\Delta m(\phi)$
±2	...	$(1 - \cos^4\phi)$
±1	$(1 + \cos^2\phi)$	$(4\cos^4\phi - 3\cos^2\phi + 1)$
0	$\sin^2\phi$	$3\sin^2\phi\cos^2\phi$

For the transition $2 \rightleftharpoons 0$, that is, $I_i = 2 \to I_f = 0$, or $I_i = 0 \to I_f = 2$, the square of all the Wigner coefficients is 1/5, and relative emission intensities are given by $I = G^2\Delta_m(\varphi)/5$. Table 3 lists numbers proportional to the squares of the Wigner coefficients for the often encountered $3/2 \rightleftharpoons 1/2$ and $5/2 \rightleftharpoons 3/2$ cases.

A typical ME experiment uses a single-line unpolarized source. If the absorber is also unpolarized, that is, if the internal fields lie in directions random to the direction of γ-ray propagation, then all angular dependencies average to the same value, and the relative line intensities are proportional to squares of the Wigner coefficients. If the absorber is polarized, that is, if an internal magnetic field is aligned at a definite angle with the γ-ray propagation, then the angular dependencies must be included.

Table 3 Table of numbers proportional to the squared Wigner coefficients for multipolarity $L = 1$ and $L = 2$ and for $3/2 \rightleftharpoons 1/2$ and $5/2 \rightleftharpoons 3/2$ transitions

$I_1 = 3/2 \rightleftharpoons I_2 = 1/2$

		m_1			
	m_2	3/2	1/2	−1/2	−3/2
$L = 1$	1/2	3	2	1	0
	−1/2	0	1	2	3
$L = 2$	1/2	1	2	3	4
	−1/2	4	3	2	1

$I_1 = 5/2 \rightleftharpoons I_2 = 3/2$

		m_1					
	m_2	5/2	3/2	1/2	−1/2	−3/2	−5/2
$L = 1$	3/2	10	4	1	0	0	0
	1/2	0	6	6	3	0	0
	−1/2	0	0	3	6	6	0
	−3/2	0	0	0	1	4	10
$L = 2$	3/2	30	36	27	12	0	0
	1/2	40	2	6	25	32	0
	−1/2	0	32	25	6	2	40
	−3/2	0	0	12	27	36	30

Population of the Excited Nuclear Level

Observation of the ME requires a source of γ-rays from an appropriate excited nuclear level. This excited level is usually obtained from a radioactive parent that populates the excited level during its natural decay. Table 4 lists radioactive parents and their principal means of production. A notable example is ^{57}Fe; the radioactive parent most often used is ^{57}Co, which has a 270 day half-life (Fig. 1). It is usually produced by bombarding ^{56}Fe with deuterons in a cyclotron and is commercially available, usually as a chloride in solution. The solution can be dried on a rhodium foil, reduced to cobalt metal in a hydrogen atmosphere, and diffused into the foil by heating. Usable activities range from several milliCuries for transmission experiments to more than 100 mCi for scattering experiments.

The Isomer Shift

If the source and absorber differ chemically, the center of a ME spectrum will be shifted away from zero relative velocity. This isomer shift is due to the Coulomb interaction between the nuclear charge and the electronic charge (Fig. 2). The magnitude of the isomer shift, S, is:

$$S = \frac{2}{5}\frac{\pi c}{E_\gamma} Ze^2 \left[R_e^2 - R_g^2\right]\left[\rho_a(0) - \rho_s(0)\right] \qquad \text{(Eq 9)}$$

where c is the velocity of light, Z is the atomic number of the nucleus, e is the electronic charge, R_e and R_g are the radii of the excited- and ground-state nuclear levels, and $\rho_a(0)$ and $\rho_s(0)$ are the electronic densities at the nucleus in the absorber and in the source.

In ^{57}Fe, $R_e < R_g$; therefore, the isomer shift will be negative if the electronic density at

Table 4 Principal methods used for producing Mössbauer effect sources

Atomic number, Z	Atomic weight, A	Element	Radioactive parent (decay, half-life)(a)	Principal means of production	Favorable reaction energy, MeV	Convenient source hosts	Single-line absorbers
19	40	Potassium	Neutron capture	^{39}K(n, γ)	...	KF	KCl, KF
26	57	Iron	^{57}Co (EC, 270 d)	^{56}Fe(d, n)	9.5	Cr, Cu, Rh, Pd	$K_4Fe(CN)_6$
			^{57}Mn (β$^-$, 1.7 min)	^{54}Cr(α, n)	21	Cr	
28	61	Nickel	^{61}Co (β$^-$, 99 min)	^{64}Ni(p,α)	22	$NiV_{0.14}$	Ni
30	67	Zinc	^{67}Ga (EC, 78 h)	^{66}Zn(d,n)	12	ZnO	...
32	73	Germanium	^{73}As (EC, 110 d)	^{74}Ge(p,2n)	24	Ge	Ge
32	73	Germanium	Coulomb excitation	^{73}Ge(O^{4+})	25	Cr	Ge
36	83	Krypton	^{83m}Kr (IT, 1.86 h)	^{82}Kr(n,γ)	...	Kr	Kr ice
			^{83}Br (β$^-$, 2.41 h)	^{82}Br(n,γ)	...	KBr	
44	99	Ruthenium	^{99}Rh (EC, 16 d)	^{99}Ru(d,2n)	20	Ru, Rh	Ru
44	101	Ruthenium	^{101}Rh (EC, 3 yr)	^{101}Ru(d,2n)	20	Ru	Ru
50	119	Tin	^{119m}Sn (IT, 250 d)	^{118}Sn(n,γ)	...	V, $CaSnO_3$	$CaSnO_3$
51	121	Antimony	^{121m}Sn (β$^-$, 76 yr)	^{120}Sn(n,γ)	...	SnO_2, $CaSnO_3$	InSb
52	125	Tellurium	^{125m}Te (IT, 58 d)	^{124}Te(n,γ)	...	Cu, ZnTe	ZnTe
53	127	Iodine	^{127}Te (β$^-$, 9.4 h)	^{126}Te(n,γ)	...	ZnTe	Kl
			^{127m}Te (β$^-$, 109 d)	^{126}Te(n,γ)	...	ZnTe	
54	129	Xenon	^{129}I (β$^-$, 10^7 yr)	Fission product	...	$Na_3H_2IO_6$, KIO_4	Na_4XeO_6
54	131	Xenon	^{131}I (β$^-$, 8 d)	Fission product	...		Na_4XeO_6
55	133	Cesium	^{133}Ba (EC, 7.2 yr)	Fission product	...	$BaAl_4$	CsCl
59	141	Praseodymium	^{141}Ce (β$^-$, 33 d)	^{140}Ce(n, γ)	...	CeO_2	Pr_2O_3
60	145	Neodymium	^{145}Pm (EC, 17.7 yr)	^{144}Sm(n,γ)	...	Nd_2O_3	$NdCl_2$, Nd_2O_3
62	147	Samarium	^{147}Eu (EC, 22 d)	^{148}Sm(p,2n)	12	Sm_2O_3	SmB_6
62	149	Samarium	^{149}Eu (EC, 106 d)	^{150}Sm(p,2n)	12	Sm_2O_3	SmB_6
62	152	Samarium	^{152}Eu (EC, 9.3 h)	Eu(n,γ)	...	Gd_2O_3	SmB_6
62	154	Samarium	Coulomb excitation	^{154}Sm(p)	3	Sm_2O_3	...
63	151	Europium	^{151}Sm (β$^-$, 87 yr)	^{150}Sm(n,γ)	...	SmF_3, Sm_2O_3	$Cs_2NaEuCl_6$
63	153	Europium	^{153}Sm (β$^-$, 47 h)	Sm(n,γ)	...	Sm_2O_3	$EuPd_3$, EuF_3
63	153	Europium	^{153}Gd (EC, 242 d)	^{152}Gd(n,γ)	...	GdF_3, Gd_2O_3	EuF_3
63	153	Europium	^{153}Sm (β$^-$, 47 h)	Sm(n,γ)	...	$SmPd_3$, Sm_2O_3	$EuPd_3$, EuF
64	154	Gadolinium	^{154}Eu (β$^-$, 16 yr)	^{153}Eu(n,γ)	...	EuF_3	...
64	155	Gadolinium	^{155}Eu (β$^-$, 1.81 yr)	^{154}Sm(n,γ)	...	$SmPd_3$	$GdCo_2$, Cs_2NGdC
64	155	Gadolinium	^{155}Eu (β$^-$, 1.81 yr)	^{154}Sm(n,γ)	...	$SmPd_3$	$GdCo_2$, Cs_2NaG
64	155	Gadolinium	^{155}Eu (β$^-$,1.81 yr)	^{154}Sm(n,γ)	...	$SmPd_3$	$GdCo_2$, Cs_2Na
64	156	Gadolinium	^{156}Eu (β$^-$, 15 d)	^{154}Sm(n,γ)	...	SmF_3	$GdCo_2$, $Cs_2NaGdCl_6$
64	157	Gadolinium	^{157}Eu (β$^-$, 15.2 h)	^{158}Gd(γ,p)	...	CeO_2, EuF_3	$CdCo_2$, $Cs_2NaGdCl_6$
64	157	Gadolinium	^{157}Eu (β$^-$, 15.2 h)	^{158}Gd(γ,p)	...	EuF_3, CeO_2	$GdCo_2$, $Cs_2NaGdCl_6$
64	158	Gadolinium	Neutron capture	^{157}Gd(n,γ)	...	$YA1_2$	$GdCo_2$, $Cs_2NaGdCl_6$
64	160	Gadolinium	Coulomb excitation	^{160}Gd(Cl)	64	...	$GdCo_2$, $Cs_2NaGdCl_6$
65	159	Terbium	^{159}Dy (EC, 144 d)	^{158}Dy(n,γ)	...	...	...
66	160	Dysprosium	^{160}Tb (β$^-$,72.1 d)	Tb(n,γ)	...	$Tb_{0.1}Er_{0.9}H_2$	$Dy_{0.4}Sc_{0.6}H_2$
66	161	Dysprosium	^{161}Tb (β$^-$, 6.9 d)	^{160}Gd(n,γ)	...	GdF_3 (300 K)	DyF_3 (300 K)
66	161	Dysprosium	Coulomb excitation	^{161}Dy(α)	3.3	DyF_3	...
66	161	Dysprosium	^{161}Tb (β$^-$, 6.9 d)	^{160}Gd(n,γ)	...	GdF_3	...
66	162	Dysprosium	Coulomb excitation	^{162}Dy(α)	3	...	...
66	164	Dysprosium	Coulomb excitation	^{164}Dy(α)	3	...	...

(continued)

(a) The superscript m, as in ^{83m}Kr, represents the metastable state. EC, electron capture; β$^-$, β decay; IT, isomeric transition; α, α decay. Source: Ref 10

Table 4 (Continued)

Atomic number, Z	Atomic weight, A	Element	Radioactive parent (decay, half-life)(a)	Principal means of production	Favorable reaction energy, MeV	Convenient source hosts	Single-line absorbers
67	165	Holmium	^{165}Dy (β^-, 2.3 h)	^{164}Dy(n,γ)	...	...	...
68	164	Erbium	^{164}Ho (β^-,39 min)	Ho(γ,n)	50	...	...
68	166	Erbium	^{166}Ho (β^-,27 h)	Ho(n,γ)	...	$H_{0.39}Y_{0.61}H$	ErH_2
68	167	Erbium	^{167}Ho (β^-,3.1 h)	^{170}Er(p,α)	...	...	...
68	168	Erbium	^{168}Tm (EC, 86 d)	Tm(γ,n)	50	$TmAl_2$	ErH_2
68	170	Erbium	Coulomb excitation	^{166}Er ( )	3	...	ErH_2
69	169	Thulium	^{169}Er (β^-, 9.4 d)	Er(n,)	...	$ErAl_3$ (300 K)	$TmAl_2$ (300 K)
70	170	Ytterbium	^{170}Tm (β^-, 130 d)	Tm(n,)	...	$TmAl_2$, $TmB_{,2}$	$YbAl_3$
70	171	Ytterbium	^{171}Tm (β^-, 1.92 yr)	^{170}Er(n,)	...	$ErAl_3$	$YbAl_3$
70	171	Ytterbium	^{171}Lu (EC, 8.3 d)	^{169}Tm(,2n)	23	$TmAl_3$	$YbAl_3$
70	172	Ytterbium	^{172}Lu (EC, 6.7 d)	^{172}Yb(d,2n)	15	$YbAl_2$	$YbAl_3$
70	174	Ytterbium	^{174}Lu (EC, 3.6 yr)	Lu(γ,n)	50	Lu	...
70	176	Ytterbium	Coulomb excitation	^{176}Yb(α)	3	...	...
71	175	Lutetium	^{175}Yb (β^-, 101 h)	^{174}Yb(n,γ)	...	$YbAl_2$	...
72	176	Hafnium	^{176m}Lu (β^-, 3.7 h)	Lu(n,γ)	...	$LuAl_2$	$HfZn_2$
72	177	Hafnium	^{177}Lu (β^-, 6.7d)	^{176}Lu(n,γ)	...	$LuAl_2$	$HfZn_2$
72	178	Hafnium	^{178}Ta (EC, 9.4 min)	Ta(d,5n)	...	Ta	$HfZn_2$
72	180	Hafnium	^{180m}Hf (IT, 5.5 h)	^{179}Hf(n,γ)	...	...	$HfZn_2$
			^{180m}Ta (EC, 8.1 h)	Ta(γ,n)	17	Ta	...
73	181	Tantalum	^{181}W (EC, 140 d)	^{180}W(n,γ)	...	W	Ta
73	181	Tantalum	^{181} HF (β^-, 42.5 d)	^{180}Hf(n,γ)	...	$HfZn_2$	Ta
74	180	Tungsten	^{180m}Ta (β^-, 8.1 h)	Ta(γ,n)	17	Ta	W
74	182	Tungsten	^{182}Ta (β^-, 115 d)	Ta(n,γ)	...	Ta	W
74	183	Tungsten	^{183}Re (EC, 71 d)	Ta(α,2n)	...	Ta	W
			^{183}Ta (β^-, 5.1 d)	^{181}Ta(n,γ), ^{182}Ta(n,γ)	...	Ta	...
74	183	Tungsten	^{183}Re (EC, 71 d)	Ta(α,2n)	...	Ta	W
			^{183}Ta (β^-, 5.1 d)	^{181}Ta (n,γ), ^{182}Ta(n,γ)	...	Ta	...
74	184	Tungsten	^{184}Re (EC, 38 d)	^{185}Re(p,pn)	32	Re	W
74	186	Tungsten	^{186}Re (EC, 90 h)	^{185}Re(n,γ)	...	Re	W
75	187	Rhenium	^{187}W (β^-, 23.9 h)	^{186}W(n,γ)	...	W	Re
76	186	Osmium	^{186}Re (β^-, 90 h)	^{186}Re(n,γ)	...	Re	K_2OsCl_6
76	188	Osmium	^{188}Re (β^-, 16.7 h)	^{187}Re(n,γ)	...	Re	K_2OsCl_6
76	189	Osmium	^{189}Ir (EC, 13.3 d)	^{189}Os(d,2n)	13	Ir	K_2OsCl_6
76	189	Osmium	^{189}Ir (EC, 13.3 d)	^{189}Os(d,2n)	13	Ir	K_20sCl_6
76	189	Osmium	^{189}Ir (EC, 13.3 d)	^{189}Os(d,2n)	13	Ir	K_2OsCl_6
76	190	Osmium	^{190}Ir (EC, 11 d)	^{190}Os(d,2n)	13	Cu	...
77	191	Indium	^{191}Pt (EC, 3 d)	^{191}Ir(d,2n)	13	Ir	Ir
77	191	Indium	^{191}Pt (EC, 3 d)	^{191}Ir(d,2n)	13	Ir	Ir
77	193	Indium	^{193}Os (β^-, 31 h)	^{192}Os(n,γ)	...	Os, V, Pt, Nb	Ir
77	193	Indium	^{193}Os (β^-,31 h)	^{192}Os(n,γ)	...	Os	Ir
78	195	Platinum	^{193}Au (EC, 183 d)	^{195}Pt(d,2n)	13	Pt	Pt
78	195	Platinum	^{195m}Pt (IT, 4.0 d)	^{194}Pt(n,γ)	...	Pt	Pt
79	197	Gold	^{197}Pt (β^-, 18 h)	^{196}Pt(n,γ)	...	Pt	Au
80	201	Mercury	^{201}Tl (EC, 73 h)	Tl(γ,2n)	...	$T1_2O_3$	...
90	232	Thorium	Coulomb excitation	^{232}Th(α)	6	Th, ThO_2	ThO_2
92	238	Uranium	^{242}Pu (α, 3.79 $\times 10^5$ yr)	...	...	PuO_2	UO_2
93	237	Neptunium	^{237}U (β^-, 6.75 d)	^{236}U(n,γ), ^{238}U(γ,n)	...	UO_2	NpO_2
			^{237}Pu (EC, 45.6 d)	^{237}Np(d,2n)	15	...	...
			^{241}Am (α, 458 yr)	...	...	Am, Th	...

(a) The superscript m, as in ^{83m}Kr, represents the metastable state. EC, electron capture; β^-, β decay; IT, isomeric transition; α, α decay. Source: Ref 10

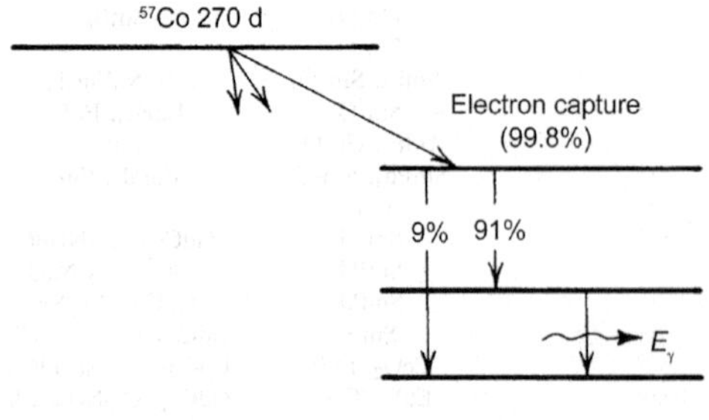

Spin (*I*)	Parity (*P*)	Energy, keV	Half-life, μs
5/2	−1	136	0.009
3/2	−1	14.4	0.098
½	−1	0	Stable

Fig. 1 Radioactive decay scheme showing the excited nuclear states of ^{57}Fe populated by decay of the radioactive precursor ^{57}Co. The ^{57}Fe Mössbauer effect uses the 14.4 keV transition between the first excited and the ground state.

the absorber nucleus exceeds that at the source nucleus. It is customary to refer to isomer shifts relative to a given absorber. For example, the isomer shift of pure iron as an absorber is arbitrarily assigned the value zero. The isomer shift for other absorbers (using the same source) is then given relative to pure iron. By convention, positive velocities refer to the source and absorber approaching each other.

Quadrupole Interaction

The energy levels of the ground and excited nuclear states can be split by the hyperfine interaction between an electric-field gradient at the nuclear site and the electric quadrupole moment, *Q*, of the nucleus. This quadrupole interaction offers an opportunity to detect variations in crystal structure, local atomic environment, lattice defects, and conduction electron states.

Unless there is perfect cubic (or tetrahedral) symmetry, it is not generally possible to avoid nuclear quadrupole effects by selecting a suitable nucleus. This is because, even if the nuclear ground-state spin has no quadrupole moment, that is, $I_g = 0$ or ½, the excited state generally does (or conversely). Thus, even when the measurement of the quadrupole energy is not the object of the experiment, it is still often necessary to understand its effect on the spectrum to extract other parameters.

The electric-field gradient at the nucleus can be expressed using a symmetric tensor consisting of the components $-V_{i,j}$ (i,j = 1 to 3):

$$V_{i,j} = \frac{\partial^2 V}{\partial x_i \, \partial x_j} \qquad \text{(Eq 10)}$$

where x_1, x_2, and x_3 are equated to spatial coordinates ($x_1 = x$, $x_2 = y$, $x_3 = z$), and V is the electric potential. A set of axes, termed the principal axes, can be selected so that only V_{xx}, V_{yy}, and V_{zz} are different from zero. Further, the Laplacian of the electric potential vanishes:

$$V_{xx} + V_{yy} + V_{zz} = 0 \qquad \text{(Eq 11)}$$

Thus, there are only two independent parameters, and they are usually defined as q, the largest of the three components of the electric-field gradient (in units of the proton charge e), and η, the asymmetry parameter:

$$eq = V_{zz} \qquad \text{(Eq 12)}$$

and

$$\eta = \frac{V_{xx} - V_{yy}}{V_{zz}} \qquad \text{(Eq 13)}$$

In addition, $|V_{zz}| \geq |V_{yy}| \geq |V_{xx}|$ are usually selected such that $0 \leq \eta \leq 1$. If the crystal-point symmetry is cubic or tetrahedral, then $V_{xx} = V_{yy} = V_{zz}$ and, by Eq 11, each component must be zero and the quadrupole interaction vanishes.

The quadrupole Hamiltonian is:

$$H_Q = \frac{e^2 qQ}{4I(2I-1)}\left[3I_z^2 - I(I+1) + \frac{1}{2}\eta(I_+^2 + I_-^2)\right] \qquad \text{(Eq 14)}$$

where $I_+ = I_x + iI_y$ and $I_- = I_x - iI_y$ are, respectively, the raising and lowering operators for the spin angular momentum (i is the imaginary number $\sqrt{-1}$). The simplest case is axial symmetry ($\eta = 0$), for which the z-axis is the axis of symmetry, and the eigenvalues are:

$$E_m = \frac{e^2 qQ}{4I(2I-1)}\left[3m^2 - I(I+1)\right] \qquad \text{(Eq 15)}$$

where m is the same quantum number as for the magnetic case.

The relative transition probabilities and their angular distribution can be obtained from Tables 2 and 3 for the case in which $\eta = 0$. The angle referred to in Table 2 is the angle between the direction of γ-ray propagation and the principal axis. For example, the spin $I = \frac{1}{2}$ ground state of ^{57}Fe has $Q = 0$ and is not affected by an electric-field gradient. The excited state, with spin $I = \frac{3}{2}$, is split into two levels, one with $m = \pm\frac{3}{2}$ and the other with $m = \pm\frac{1}{2}$. The energy separation of the two levels is $eQV_{xx}/2$. For a single crystal whose z-axis is at an angle φ to the direction of γ-ray propagation, the ME spectrum consists of two lines with an intensity ratio:

$$\frac{I\left(\frac{3}{2} \rightarrow \frac{1}{2}\right)}{I\left(\frac{1}{2} \rightarrow \frac{1}{2}\right)} = \frac{1 + \cos^2\phi}{\frac{5}{3} - \cos^2\phi} \qquad \text{(Eq 16)}$$

For a powder sample with no preferred orientation, the two transitions occur on the average with equal probability, assuming no anisotropy in the recoilless fraction. In ^{57}Fe, this results in the familiar doublet pattern.

Magnetic Interaction

The hyperfine interaction between the nuclear magnetic moment and a magnetic field at the nuclear site is useful for phase identification, for observing magnetic transitions, and for studying the local atomic environment of the resonating nucleus. This Zeeman effect is characterized by the Hamiltonian:

$$H = -\mu \cdot H_0 = -\gamma\hbar I \cdot H_0 \qquad \text{(Eq 17)}$$

The equally spaced energy levels resulting from this Hamiltonian are given by $E = -\gamma\hbar m H_0$, where $m = -I, -I+1, \cdots, I-1, I$; H_0 is the magnetic field at the nuclear site; $\hbar$ is Planck's constant over 2π; and γ is the gyromagnetic ratio of the nucleus.

Transitions and relative line intensities for the familiar and useful case of ^{57}Fe are illustrated in Fig. 3. For other cases, the relative line intensities of the transitions can be found by using Tables 2 and 3.

Experimental Arrangement

The basic components of a Mössbauer spectrometer are a source, sample, detector, and drive to move the source or absorber; these are shown in Fig. 4 (Ref 15). The source is moved toward and away from the sample, varying velocity linearly with time. The synchrotron Mössbauer oscillates the sample while leaving the source stationary. The location of the detector relative to the source and the sample defines the geometry of the

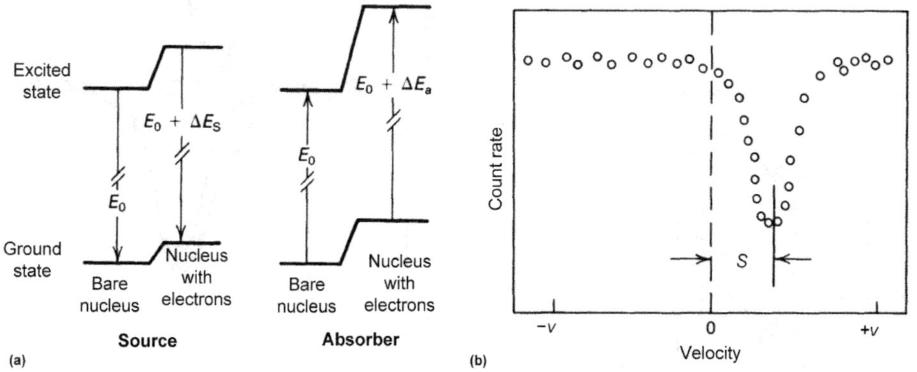

Fig. 2 Illustration of the isomer shift, *S*. (a) The ground- and excited-state energies are shifted by different amounts in the source and the absorber. The transition energy is larger in the absorber. (b) The shifted spectrum that results from such situations as (a)

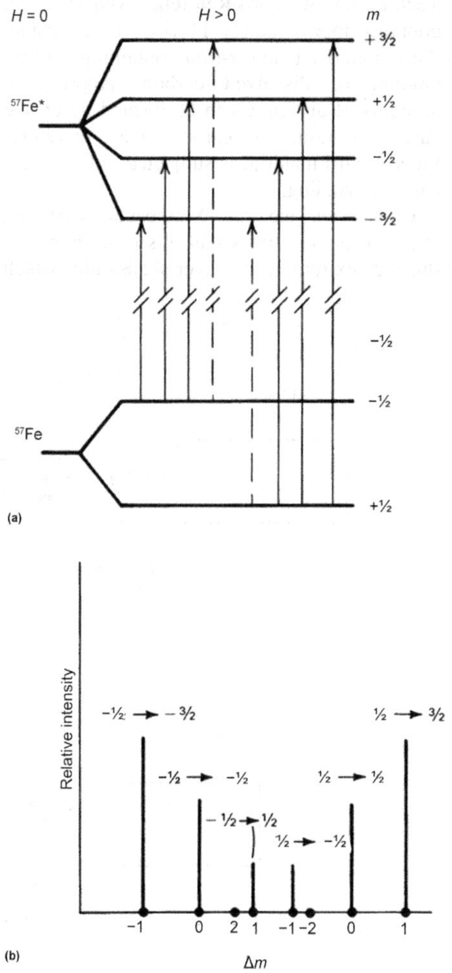

Fig. 3 (a) Transitions and (b) relative line intensities for magnetic hyperfine interactions in ^{57}Fe

experiment, and either transmission or backscatter mode is commonly used.

To obtain the required γ-rays, a radioactive precursor, or parent, is used. For example, ^{57}Co is a suitable precursor for populating the 14.4 keV level of ^{57}Fe, the most heavily studied Mössbauer isotope. Figure 1 shows the appropriate decay scheme for ^{57}Fe. Table 1 lists isotopes that can be used for ME studies, along with some of their properties. Table 4 lists radioactive parents and their principal means of production. By far the two most frequently used isotopes are ^{57}Fe and ^{119}Sn.

Applications

Example 1: Surface-Phase Detection

The ^{57}Fe Mössbauer spectrum obtained by counting conversion electrons provides information from a thin layer on the surface of a metal, the depth of the layer being limited by the range of these electrons within the metal. For iron-rich alloys, the depth being probed is of the order of 30 nm.

A sample of spheroidized iron carbide in ferrite, National Institute of Standards and Technology Standard Reference Material 493, contains approximately 14 vol% iron carbide, the remainder being α-iron containing a small amount of dissolved carbon. Appreciable amounts of austenite can be formed by coarse surface grinding of carbon steel; therefore, the standard material is subjected only to light surface grinding.

Conversion electron Mössbauer scattering (Ref 16) demonstrates that this procedure produces an extremely thin layer of austenite, which is apparent in Fig. 5. The 14.4 keV spectrum (spectrum A) probes a depth of approximately 30 μm and therefore represents the bulk sample. The inner two lines from the six-line iron spectrum and the six-line iron carbide spectrum are indicated (there is some overlap of the α-iron and Fe_3C lines). Spectrum B has an additional line identified as austenite. The austenite layer is too thin to be observed using conventional metallographic techniques, including glancing-angle x-ray diffraction, and may be too strained to be readily observed using low-energy electron diffraction.

Example 2: Phase Analysis of Hydrided TiFe

The intermetallic compound TiFe, with stoichiometric proportions of titanium and iron, has application as a hydrogen gas storage device. This compound readily absorbs and desorbs hydrogen, forming ternary $TiFeH_x$ phases. The ^{57}Fe Mössbauer transmission spectrum can be used to probe the phases formed (Ref 17).

A small amount of hydrogen dissolves in the TiFe crystal structure (Fig. 6). Spectrum A shows that for a composition of $TiFeH_{0.1}$ the ^{57}Fe spectrum has a single peak, essentially identical to that of hydrogen-free TiFe. This cubic phase, based on TiFe, is termed α. The spectrum for a composition of $TiFeH_{0.9}$, spectrum B, can be resolved into two lines: one identified with α and the second, a quadrupole doublet, associated with a noncubic β-phase. The lever rule can be used to determine the location of this composition in the two-phase (α + β) region. Spectrum C is for a $TiFeH_{1.7}$ alloy and is in the two-phase region between β and a third noncubic phase γ. A small amount (approximately 2%) of TiFe is also evident. Because the spectra for $TiFeH_x$ are not completely resolved, this method of phase analysis is not unique, but it is quick and sensitive to the phases present.

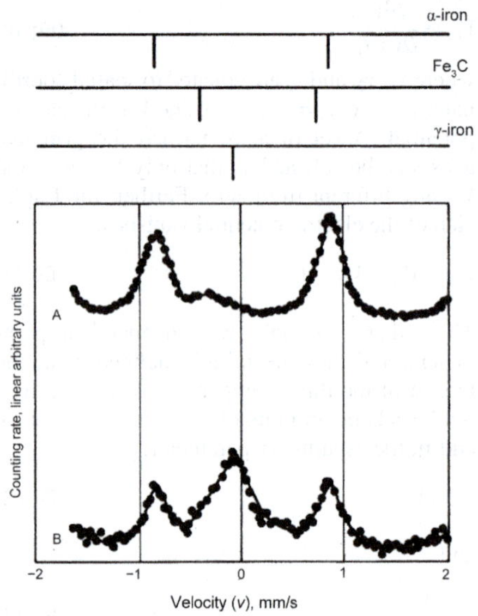

Fig. 5 Backscattered ^{57}Fe Mössbauer spectra from the lightly ground surface of an iron-iron carbide alloy (National Institute of Standards and Technology Standard Reference Material 493). Only the central region of each spectrum is shown. A, 14.4 keV γ-rays counted; B, conversion electrons counted

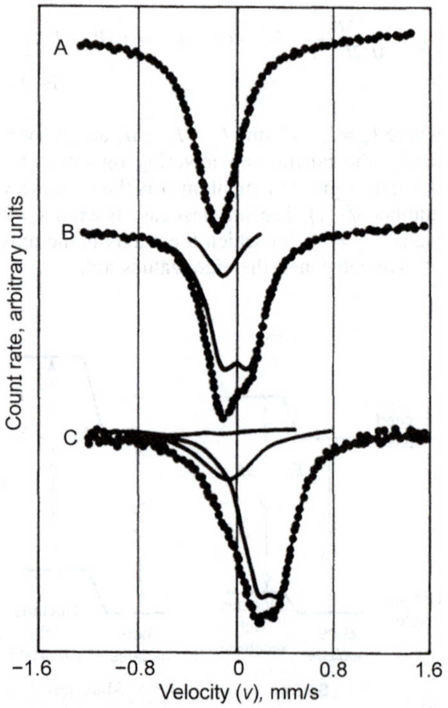

Fig. 6 ^{57}Fe Mössbauer absorption spectra for $TiFeH_x$. The lines shown through the data points are least-square fits of the separate components. A, x = 0.1; B, x = 0.9; C, x = 1.7

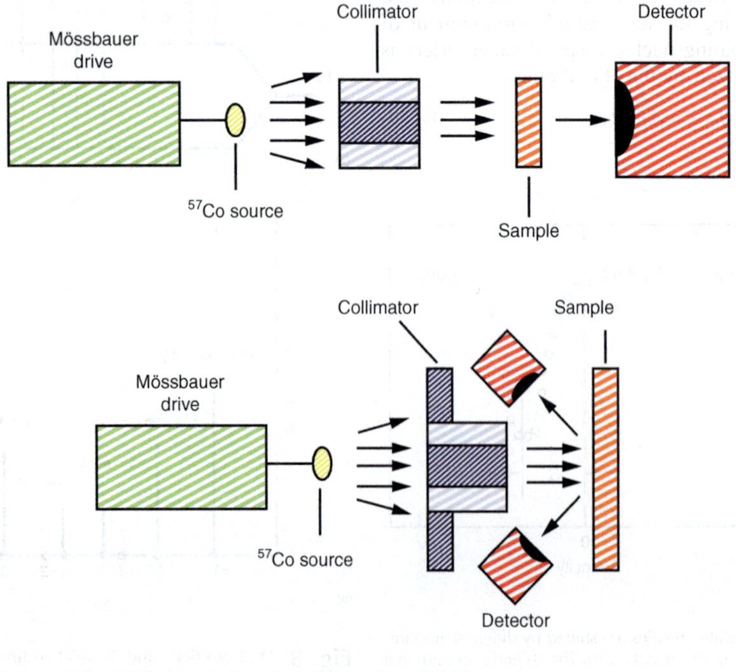

Fig. 4 Elements of a Mössbauer spectrometer. Adapted from Ref 15

Example 3: Analysis of Phases of Iron in Copper

The ME readily distinguishes four types of iron in copper-iron alloys (Ref 18). Body-centered cubic precipitates (α-iron) are characterized by a six-line magnetic hyperfine field pattern. Isolated iron atoms in solid solution, termed γ_0-iron, provide a single line centered at 0.24 mm/s. Iron atoms in solid solution with other iron atoms as near neighbors, termed γ_2-iron, yield a quadrupole split doublet centered at 0.25 mm/s with a splitting of 0.58 mm/s. Iron in face-centered cubic coherent precipitates, termed γ_1-iron, gives a single line centered at -0.10 mm/s (values given are at room temperature and are referenced to the center of a pure α-iron spectrum).

A room-temperature spectrum from a quenched sample of $Cu_{0.97}Fe_{0.03}$ is shown in Fig. 7. Two of the lines, the singlet line from γ_1-iron and one of the doublet lines from γ_2-iron, are coincident. Their relative magnitudes can be determined closely by assuming that the γ_2-iron doublet represents a powder pattern and is therefore symmetric. The transformations that occur upon cold working or annealing the sample can be readily observed from the spectra. The coherent γ_1-iron precipitates can grow quite large before they transform to α-iron.

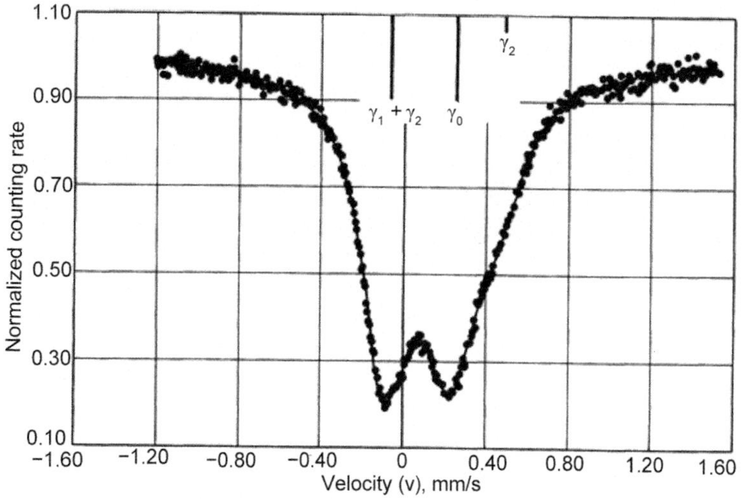

Fig. 7 Room-temperature ^{57}Fe Mössbauer absorption spectrum for a rapidly quenched $Cu_{0.97}Fe_{0.03}$ sample. The lines at the top indicate the positions of the lines that constitute the spectrum. The contribution from three types of iron (see text) is indicated. The zero of velocity represents the center of a pure iron absorption spectrum at room temperature. The source used was approximately 10 mCi of ^{57}Co in palladium.

Example 4: Extended Solubility of Iron in Aluminum

The equilibrium solid solubility of iron in aluminum is extremely small (less than 0.01 wt% Fe at 500 °C, or 930 °F). However, a large supersaturation can be obtained by rapid solidification techniques. Supersaturated solid solutions containing up to 8 wt% Fe have been reported (Ref 19). Mössbauer measurements, combined with x-ray diffraction, have been performed (Ref 20) and confirm supersaturation up to this amount for splat-quenched samples. The spectra show, however, that the number of iron-iron near neighbors is much greater than that expected for a random solid solution.

REFERENCES

1. A. Abragam, *L'effet Mössbauer*, Gordon and Breach, 1964
2. S.G. Cohen and M. Pasternak, Ed., *Perspectives in Mössbauer Spectroscopy*, Plenum Press, 1973
3. H. Frauenfelder, *The Mössbauer Effect*, Benjamin, 1962
4. U. Gonser, Ed., *Mössbauer Spectroscopy*, Springer-Verlag, 1975
5. N.N. Greenwood and T.C. Gibb, *Mössbauer Spectroscopy*, Chapman and Hall, 1972
6. L. May, Ed., *An Introduction to Mössbauer Spectroscopy*, Plenum Press, 1971
7. I.D. Weisman, L.J. Swartzendruber, and L.H. Bennett, *Techniques in Metals Research*, Vol 6, R.F. Bunshah, Ed., John Wiley & Sons, 1973, p 165–504
8. G.K. Wertheim, *Mössbauer Effect: Principles and Applications*, Academic Press, 1964
9. J.G. Stevens and G.K. Shenoy, Ed., *Mössbauer Spectroscopy and Its Applications*, American Chemical Society, Washington, 1981
10. G.K. Shenoy and F.E. Wagner, Ed., *Mössbauer Isomer Shifts*, North Holland, 1978
11. J.G. Stevens, V.E. Stevens, and W.L. Gettys, *Cumulative Index to the Mössbauer Effect Data Indexes*, Plenum Press, 1979
12. I.J. Gruverman and C.W. Seidel, Ed., *Mössbauer Effect Methodology*, Vol 10, Plenum Press, 1976
13. A. Vertes, L. Korecz, and K. Burger, *Mössbauer Spectroscopy*, Elsevier, 1979
14. J.G. Stevens and V.E. Stevens, *Mössbauer Effect Data Index*, Plenum Press, 1973
15. M.D. Dyar, "Mössbauer Spectroscopy," Department of Astronomy, Mount Holyoke College, https://serc.carleton.edu/msu_nanotech/methods/mossbauer.html, accessed March 2019
16. L.J. Swartzendruber and L.H. Bennett, *Scr. Metall.*, Vol 6, 1972, p 737
17. L.J. Swartzendruber, L.H. Bennett, and R. E. Watson, *J. Phys. F, Met. Phys.*, Vol 6, 1976, p 331
18. L.H. Bennett and L.J. Swartzendruber, *Acta Metall.*, Vol 18, 1970, p 485
19. T.R. Anatharaman and C. Suryanarayana, *J. Mater. Sci.*, Vol 6, 1971, p 1111
20. S. Nasu, U. Gonser, P.H. Shingu, and Y. Murakami, *J. Phys. F, Met. Phys.*, Vol 4, 1974, p L24

SELECTED REFERENCES

- M. Filatov, On the Calculation of Mössbauer Isomer Shift, *J. Chem. Phys.*, Vol 127 (No. 8), 2007, p 8
- G.J. Long and F. Grandjean, Ed., *Mössbauer Spectroscopy Applied to Magnetism and Materials Science*, Vol 1, Plenum, New York, 1993
- G.J. Long and F. Grandjean, Chap. 2.20, Mossbauer Spectroscopy: Introduction, *Comprehensive Coordination Chemistry II*, Elsevier, 2003
- J.L. Que, Ed., *Physical Methods in Bioinorganic Chemistry*, University Science Books, Sausalito, CA, 2000

X-Ray and Neutron Diffraction

Erik Mueller, National Transportation Safety Board

ASM Handbook, Volume 10, *Materials Characterization*
ASM Handbook Committee
DOI 10.31399/asm.hb.v10.a0006643

Introduction to Diffraction Methods

Matteo Leoni, University of Trento-Italy, ICDD, and Saudi Aramco

Introduction

Diffraction techniques are some of the most useful in the characterization of crystalline materials, such as metals, intermetallics, ceramics, minerals, polymers, plastics, or other inorganic or organic compounds. X-ray diffraction nowadays is employed to identify the phases present in samples—from raw starting materials to finished product—and to provide information on the physical state of the sample, such as grain size, texture, and crystal perfection. Most x-ray diffraction techniques are rapid and nondestructive; some instruments are portable and can be transported to the sample. The sample may be as small as an airborne dust particle or as large as an airplane wing. The sample can be characterized at ambient conditions or in a controlled environment both in static and in situ/operando conditions. This article describes the methods of x-ray diffraction analysis, the types of information that can be obtained, and its interpretation.

In general, x-ray analysis is restricted to crystalline materials, although some information may be obtained on amorphous solids and liquids using the pair distribution function method or complementary techniques (see the articles "Small-Angle X-Ray and Neutron Scattering" and "Micro X-Ray Diffraction" in this Division). Similar or complementary information often can be obtained using electron diffraction or neutron diffraction, but the sample limitations usually are more severe and the equipment considerably more elaborate and costly.

Samples are acceptable in many forms, depending on the availability of the material and the type of analysis to be performed. Single crystals from a few micrometers to a few centimeters in size or loose or consolidated aggregate of many small crystals can be used. Although the overall size of the sample may be large, the actual area of the sample examined in a given experiment rarely exceeds 1 cm^2 (0.16 $in.^2$) and may be as small as 10 μm^2.

The type of information desired may range from the question of sample crystallinity or its composition, to details of the crystal structure or the state of orientation of the crystallites, to information about residual macrostresses, to size and defects of the crystallites. Crystal structure determination is more easily performed on single crystals, as small as a few tens of micrometers in diameter. Small organic molecules and inorganic compounds also can be solved from powder diffraction data. Phase identification can be conducted on virtually all single-crystal or powder samples. Also useful are measurements of the physical state of a sample that detect differences from the ideal crystal. The evolution of the material under different conditions or environment (e.g., during mechanical testing) also can be tracked in situ or operando. Table 1 lists types of x-ray diffraction analysis, indicates specific techniques, and describes the required form of the sample.

In general, the techniques are classified as single crystal or powders/polycrystalline. A large quantity of information can be obtained from the position, intensity, and shape of the diffraction peaks. The positions contain all the information on the geometry of the crystalline lattice. The intensities are related to the types of atoms and their spatial arrangement in the unit cell of the lattice. The sharpness of the diffraction peaks is an indicator of the perfection of the single crystal or of the crystals forming the powder or the polycrystalline aggregate. Together with the shape, the breadth of the peaks is broadly related to the nanostructure, including size and shape distribution of the crystallites as well as lattice defect type and content. Diffraction experiments can be designed to measure any or all of these features.

Table 1 X-ray diffraction analysis

Type of analysis	Method	Sample
Lattice parameters and symmetry	Automatic: Indexing, space group determination from extinct reflections, refinement	Single crystal
	Indexing, space group determination from powder line extinction (difficult), refinement	Powder
Arrangement of atoms	Automatic: direct methods, solution and refinement of intensities	Single crystal
	Global optimization, direct, real space and hybrid methods, Rietveld refinement of whole pattern	Powder
Identification of compound	Identification of cell parameters	Single crystal
	Matching of *d-I* set with database	Powder
Orientation studies	Orientation of the crystal with respect to the mount	Single crystal
	Orientation of the crystallites in the specimen (texture analysis)	Powder, polycrystalline compact
Size of crystallites	Line profile analysis	Powder
Magnitude of macrostress (residual-stress analysis)	Line shifts	Polycrystalline specimen
Amount of phase	Quantitative analysis	Powder
Change of state	Special atmosphere chambers and specimen holders	Single crystal or powder
Crystal perfection	Direct imaging	Single crystal
	Line profile analysis	Powder

Theory

Nature and Generation of X-Rays

X-rays are a portion of the electromagnetic spectrum having wavelengths from 10^{-11} to 10^{-8} m (0.1 to 100 Å). In laboratory instruments, discrete wavelengths in the range of 0.5 to 2.5 Å are commonly employed. A wider range, and some possibility of selecting the working wavelength, is offered by synchrotrons and x-ray free-electron lasers (X-FELs). As electromagnetic radiation, x-rays have wave and particle properties. Derivation of all angle-dependent phenomena is based on their wave characteristics. However, electronic detectors measure photon properties, with intensity measurements usually reported as counts.

Most commercial x-ray sources use a high-energy (50 kV) electron beam directed into a cooled metal target. As the electrons are decelerated in the target, several events produce x-radiation. Most of the electron beam energy is lost in collisions that set the atoms in motion and produce heat, which must be dissipated using air (in smaller devices such as microfocus sources) or water-cooling circuits (e.g., sealed x-ray tubes and rotating anodes). Some electron energy is caught in the electric fields of the atom as the electrons decelerate and is reradiated as x-rays. Most of this radiation has energy near the excitation potential but may have any value down to zero. This bremsstrahlung or white radiation, so termed because it is polychromatic, produces a continuum of all energies between the extremes.

A small but significant portion of the electron beam collides with the electrons of the target atoms. Some target electrons are knocked out of their orbitals, leaving the target atoms in a high-energy excited state. This excited state is brief, and the stored energy is released as the electrons from other orbitals drop into the vacant orbital. These electron transitions are of distinct energy jumps, that is, they are quantized, and the radiation emitted has specific wavelengths. Therefore, x-rays exiting the target have a few strong characteristic concentrations of specific wavelengths superimposed on the white radiation. The characteristic spectrum of an x-ray tube is simple, and a few wavelengths have strong peaks (Fig. 1a).

Synchrotrons and X-FELs are large-scale installations to partly replace and highly complement the laboratory sources. The generation of the x-rays in those machines follows completely different principles, and the intensity of the beams (measured in terms of brilliance) is orders of magnitude larger than that available in the laboratory. In a synchrotron (more appropriately a storage ring), one or more bunches of electrons are circulated in a closed path at approximately the speed of light (relativistic conditions). The path includes straight and bent sections. Electromagnets are employed to locally change the trajectory of the electrons. A change in the trajectory of the electrons causes the emission of radiation over a broad spectrum along the initial trajectory. This radiation, which also includes visible radiation and x-rays, is termed synchrotron radiation or synchrotron light. The spectrum does not contain any strong characteristic peak. A larger emission intensity can be obtained by forcing the electron beam through a finely spaced periodic arrangement of alternating magnets (undulator). The electron path is oscillatory, and the synchrotron radiation emitted at each change of direction sums up. X-ray free-electron lasers are very long (kilometer-length) undulators. The x-rays emitted at every trajectory change, interfering with all previous ones; constructive interference, possibly guided by using mirrors on the ends of the device or intense laser-driving signals, leads to large amplification of the resulting beam.

In those large installations, beamlines are built to confine the emitted radiation and direct it on the experimental station. Multiple beamlines operate simultaneously on different types of experiments or analytical technique. Contrary to laboratory sources, the x-ray beam in synchrotrons and X-FELs is pulsed, because bunches of electrons are involved. The difference in brilliance between laboratory sources and large-scale facilities currently can be up to 20 orders of magnitude.

Both the white radiation and the characteristic spectrum have utility in x-ray diffraction. For most experiments, the incident beam usually consists of only a single wavelength; that is, it is monochromatic. The output beam can be monochromatized variously. One of the simplest methods is to use a thin foil of an appropriate metal. The characteristic spectrum depends on the target element, but the white radiation does not. Copper typically is used as the target because the Kα characteristic radiation is a useful wavelength, 1.5406 Å, and the target is easily cooled for high efficiency. For copper radiation, a nickel foil absorbs most of the white radiation and the other characteristic peaks (namely Kβ), transmitting essentially pure Kα radiation (Fig. 1b). As a rule of thumb, the filter for a given target material is chosen from among the elements preceding it in the periodic table (e.g., nickel for a copper target, iron for a cobalt target, vanadium for a chromium target, etc.), because these elements show characteristic absorption edges (slightly) above the characteristic emission wavelength of the target. The thin-foil filter does not provide a truly monochromatic beam. The fine structure of the characteristic radiation is unaltered, and two rather similar wavelengths are usually present (Kα_1 and Kα_2, the second one roughly half intense).

An alternate method of monochromatizing the beam, employed in both laboratories and large-scale facilities, involves the use of a single crystal as a diffractor that is set to allow diffraction of only the desired wavelength. Crystals of graphite, silicon, germanium, and quartz often are used for this purpose. A sequence of multiple crystals can be employed to narrow the bandwidth. On laboratory tubes, one of the characteristic lines (e.g., Kα_1) can be selected. Thin-film multilayers (alternating thin layers of highly different density and refracting index) also can be used as artificial monochromator crystals. They can be fabricated with a desired lattice spacing and to allow for a simultaneous change in the angular characteristics of the rays in the beam (focused versus parallel).

A further method is possible when the x-ray detector can discriminate the energies of the individual x-rays. Such a detector can be set electronically to accept only the energy corresponding to the characteristic radiation.

Some experiments use white radiation. This radiation is not characteristic of any specific target, and target materials whose characteristic spectrum is only partially excited by the 50 kV electron beam, such as tungsten, are commonly used. The small amount of characteristic radiation generated usually does not interfere with the experiment.

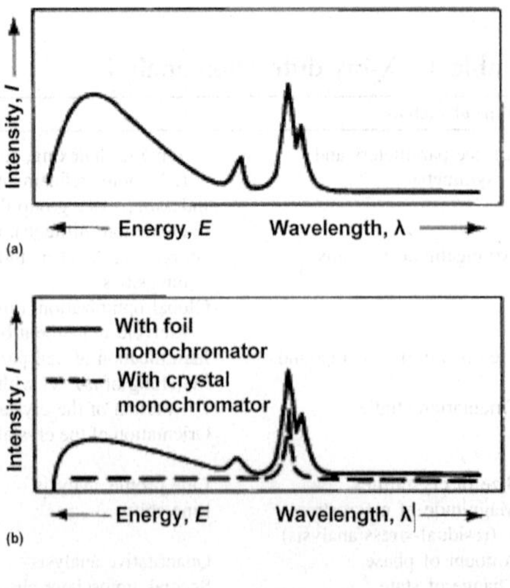

Fig. 1 Characteristic emission spectra from an x-ray sealed tube. (a) Without a filter. (b) With a thin-foil monochromator and a crystal monochromator

Detection of X-Rays

Use of x-rays in experiments necessitates detection of diffracted beams such that their intensities and positions in space are measurable.

Photographic films are the oldest types of x-ray detectors. Special photographic films, used as flat or cylindrical detectors, were used to collect many photons and beams simultaneously. The diffracted maxima appear as dark spots or lines on the negative, whose angular coordinates can be determined by its position on the film and the film geometry. The darkening of the spot is proportional to the beam intensity.

With the need for digital information, films have been entirely superseded by 0D, 1D, and 2D (pixelated) photon detectors. The 0D photon detector measures only a single beam at any given time and must be repositioned to detect a different beam. The actual intensity measurement can be direct (photon counters) or indirect. In photon counters, the charge generated in a semiconductor by each incoming x-ray photon is directly converted into an electric pulse, and the various pulses then are counted. In indirect detectors, the x-rays excite a crystal that emits radiation at a lower energy, which usually is amplified and collected using a photomultiplier. A fixed acquisition time usually is chosen based on the type of experiment and data quality required.

The position-sensitive detector (1D detector) simultaneously detects a photon and its position along a line or along a circular segment. Older gas-filled detectors consist of a wire placed to intersect the beam over a range of angles. The location on the wire that detects a photon can be identified electronically. The photon interactions are totaled for each spot on the wire, providing intensity and position information simultaneously for a range of positions. More modern 1D detectors are built as a sequence of 0D detectors and usually are created on silicon or another semiconductor, using microelectronic techniques. Each of the 0D detectors can be a photon counter, providing a signal for each discrete position, proportional to the number of photons detected. Nowadays, 2D detectors are widely employed, consisting of a bidimensional array of 0D detectors.

A wavelength-dispersive detector operates on electronic energy discrimination of the components of a polychromatic beam, resulting in the simultaneous accumulation of the intensities of many different wavelengths. A cooling system (liquid nitrogen or Peltier cooler) is employed to reduce the readout noise.

Energy-dispersive pixelated detectors exist, being able to provide a signal corresponding to the energy of each photon hitting one of the pixels.

The size of the pixels in a pixelated detector determines the resolution of the system and, together with the detector size, defines the portion of detected diffracted beams (angles). Larger (angular) areas can be covered by assembling smaller detectors into larger arrays.

Crystalline State

In a simplistic view, a crystalline material is a three-dimensional periodic arrangement of atoms in space. This arrangement is best depicted by describing a unit cell having all the fundamental properties of the crystal as a whole, that is, the basic repeating unit. This unit cell is a prismatic region of space, having typical edge dimensions of 3 to 20 Å for a large quantity of inorganic solids. Some proteins may have cell dimensions exceeding 1500 Å. Within the unit cell, the arrangement of atoms depends on the types of atoms, the nature of their bonds, and their tendency to minimize the free energy by a high degree of organization. This organization usually results in some degree of geometric symmetry and unit-cell shapes reflecting this symmetry.

The crystal as a whole consists of unit-cell building blocks packed like bricks in a wall. The resulting crystal exhibits shapes and symmetry controlled by the unit cell. Thus, a goal of x-ray diffraction is to characterize the unit cell, that is, to determine its size, shape, symmetry, and the arrangement of atoms. These characteristics can be determined from the collection of angles at which diffracted x-ray beams are detected. The arrangement of atoms and the details of symmetry require interpretation of the intensities of the diffracted beams. The former is the geometry of the unit cell; the latter, the crystal structure. Reference 1 covers the topic extensively.

Geometry of Unit Cells and Diffraction

A lattice is the arrangement of single geometric points defining the positions of each unit cell in space. The lattice is fundamental to the geometry of any diffraction experiment. The lattice reflects exactly the size and shape of the unit cells and their periodic arrangement in space. The shape of both the unit cell and the lattice is limited by the presence of symmetry. In some cases, for an easier visualization of the atomic arrangement, a conventional unit cell enclosing more than one lattice point is used.

Diffraction is a phenomenon of electromagnetic radiation scattered from a periodic arrangement of scattering centers with spacings of the same order of magnitude as the wavelength of the radiation. Interference of the scattered rays in most directions results in cancellation and absence of detectable beams; however, in a few selected directions, reinforcement of all of the scattered rays occurs, and a strong beam results. The periodicity may be in one, two, or three dimensions. Figure 2 shows a wave incident on a one-dimensional row of scattering centers. The incident beam has all the rays in phase. At each scattering center, new rays emanate in all directions. The beams will not be in phase in most directions, and cancellation will occur. In the directions indicated, all wavefronts are in phase, and addition occurs, sending out

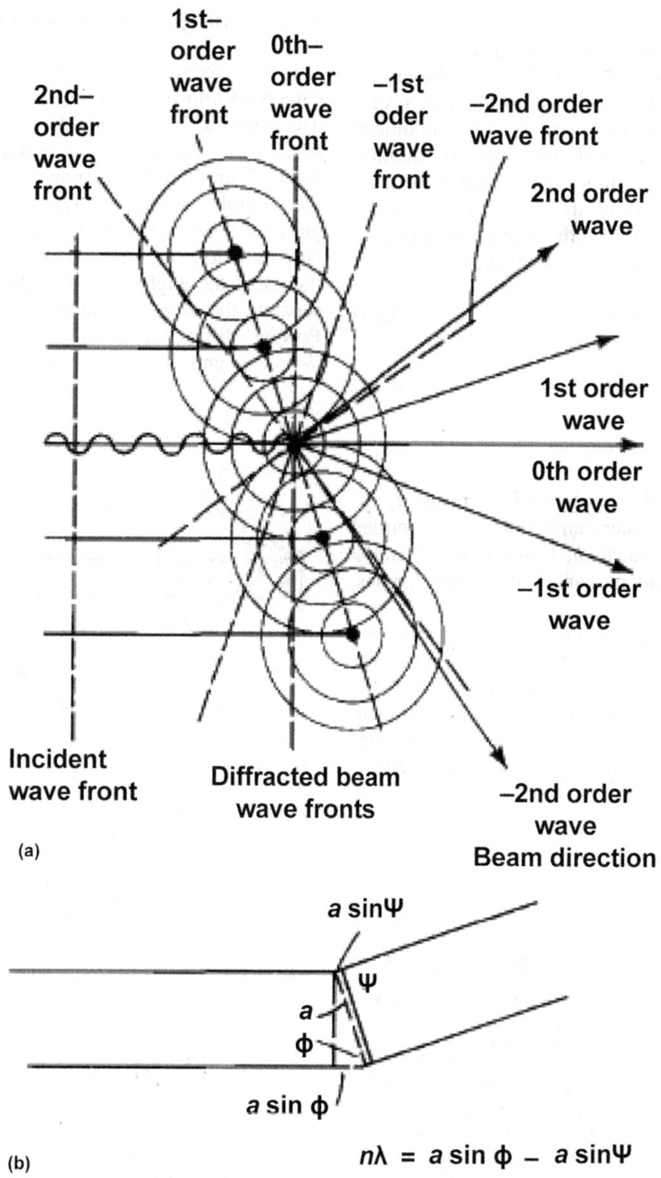

Fig. 2 Diffraction of x-rays. (a) Diffraction conditions from a row of scattering centers. Most diffracted rays interfere and cancel each other; however, in some directions, reinforcement occurs, and a strong beam results. (b) Conditions for reinforcement

strong beams. These are the diffracted beams. Similar interference phenomena can be observed in the waves emanated by a set of stones thrown in calm water.

Although the incident beam and scattering centers lie in the plane of Fig. 2, the diffracted beams are not confined to this plane. Each scattering center emits rays in all directions, and the resulting diffracted rays define a family of cones in space (Fig. 3). This type of experiment is easy to conduct as an optical analog if a (the distances between equally spaced holes in an opaque mask) is approximately 50 μm or less. The row of equally spaced holes can be placed in a light beam, and the diffraction pattern observed on a screen positioned some distance beyond the mask. The diffraction pattern is a series of lines at which the cones intersect the screen.

This one-dimensional arrangement is only part of crystal diffraction. Because the crystal is periodic in three dimensions, the lattice sites act as the scattering centers. The conditions illustrated then must be satisfied simultaneously in three dimensions. Scattering must be considered from three noncollinear rows of the lattice; all of the equations therefore must be satisfied simultaneously:

$$n_a\lambda = a(\sin\phi_a - \sin\psi_a) \qquad \text{(Eq 1)}$$

$$n_b\lambda = b(\sin\phi_b - \sin\psi_b) \qquad \text{(Eq 2)}$$

$$n_c\lambda = c(\sin\phi_c - \sin\psi_c) \qquad \text{(Eq 3)}$$

where a, b, and c are the lattice repeats; n_a, n_b, and n_c are the order numbers usually identified as the Miller indices h, k, and l; and ϕ and ψ are the incident and diffracted angles, respectively.

The three-dimensional lattice of scattering centers restricts a diffraction experiment severely. Few directions can diffract, and diffraction can occur only when the incident beam makes precisely the correct angle relative to the crystal. Equations 1 to 3 define these restrictions. It usually is easier to visualize these conditions by reducing Eq 1, 2, and 3 to that shown in Fig. 4. The lattice is considered to be planes of lattice points, and the x-ray beam acts as if it reflects off these planes. Constructive interference (i.e., the x-ray beams scattered from each layer are in phase) occurs only when the incidence angle and diffraction angle, θ, satisfy the Bragg equation:

$$\lambda = 2d\sin\theta$$

where d is the perpendicular spacing between the lattice planes. The angle θ is a function of the ϕ and ψ angles, and d is a function of a, b, c, h, k, l and the lattice angles α, β, and γ. Equations 1 to 3, being more realistic, identify diffraction as scattering and interference from the periodically arranged unit cells.

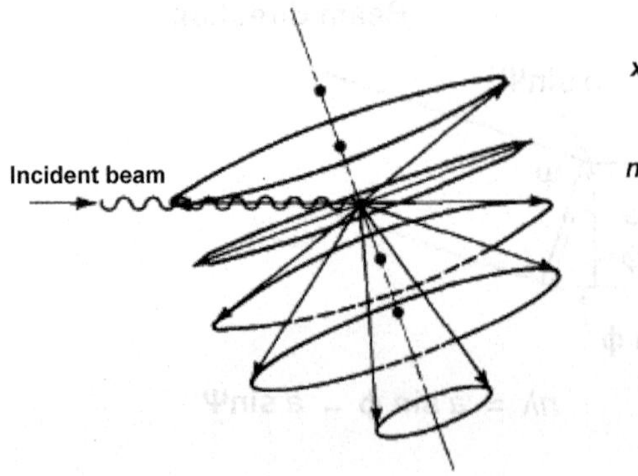

Fig. 3 Cones of diffraction

Determining Lattice Geometries

Lattice geometries are classified into seven categories known as crystal systems (Table 2). The angles α, β, and γ are the interaxial angles $b\wedge c$, $c\wedge a$, and $a\wedge b$, respectively. Also in Table 2 are the expressions for determining the possible d values for specific lattices. Single-crystal diffraction experiments allow direct determination of all the lattice constants, including the angles, because the experiments separate the three-dimensional aspect of data. Powder diffraction produces only a set of d-values, and the lattice constants must be determined by solving one of the d-spacing equations in Table 2, although it is not known a priori which equation to solve. The simultaneous presence of multiple phases in the powder (mixture) can complicate the process. Software has been developed to assist in the indexing process from powder data, but success is never guaranteed a priori.

The lattice constants of a crystal may be sufficient to identify an unknown compound when it is compared to a tabulation of the lattice data of all known compounds. Such compendia exist, for example, as the International Centre for Diffraction Data Powder Diffraction File (PDF), the National Institute of Standards and Technology (NIST) Crystal Data Database (inorganic and organic phases), the Inorganic Crystal Structure Database, the Cambridge Structure Database, and the Crystallographic Open Database. Using the PDF, identification may be possible based only on the d-spacings without knowledge of lattice constants. Accurate d-spacing or lattice data on experimental products often can indicate chemical and physical differences upon comparison with data from well-characterized related crystals or powders. Changes in lattice constants as a function of temperature and/or pressure provide fundamental thermodynamic data on the compound under study. Table 3 lists additional applications based solely on the analysis of the position of the diffraction maxima.

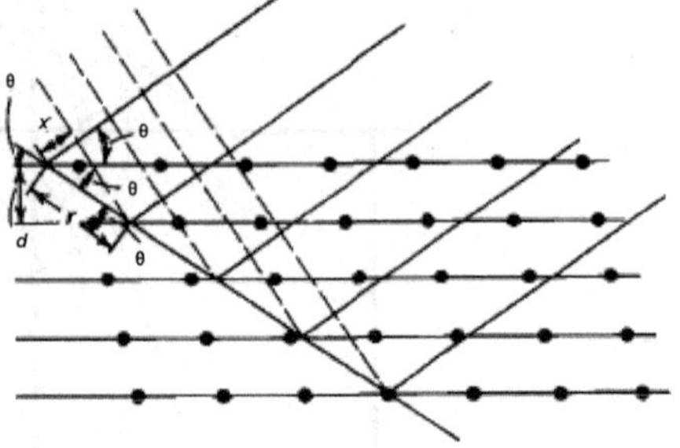

$$n\lambda = r - x$$

$$r = \frac{d}{\tan\theta} = d\,\frac{\cos\theta}{\sin\theta}$$

$$x = r\cos 2\theta = d\,\frac{\cos\theta\cos 2\theta}{\sin\theta}$$

$$n\lambda = d\left(\frac{\cos\theta}{\sin\theta}\right)(1 - \cos 2\theta)$$

$$= d\left(\frac{\cos\theta}{\sin\theta}\right)(1 - \cos^2\theta + \sin^2\theta)$$

$$= d\left(\frac{\cos\theta}{\sin\theta}\right)(2\sin^2\theta) = 2d\sin\theta$$

Fig. 4 Diffraction in a crystal lattice and the derivation of the Bragg equation

Intensities of Diffracted Beams

The geometry of the crystal lattice controls the directions of diffracted beams. The intensities of these beams depend on the types of atoms in the crystal and their arrangement in the unit cell. If a unit cell was composed of a single electron, all of the diffracted beams would have identical intensities. Real crystals consist of atoms having clouds of electrons. The scattering from all the electrons in the unit cell results in complex interference effects that enhance some beams and diminish others. These interference effects must be evaluated. In addition, the amplitude of the diffracted beam is the sum of the amplitudes of the component rays. The measured intensity is the square of this sum. Table 4 lists factors that control line intensity and the information obtainable from the intensity of diffracted x-rays.

Figure 5 shows a single atom, within which are two electrons located at some instant at points 1 and 2. Each electron scatters equally in all directions, and the scattering does not modify the phase regardless of the scattering angle. The scattering amplitude of an individual electron is given by the Thompson scattering coefficient (classical electron radius):

$$e^2/(4\pi\varepsilon_0 m_e c^2) = 10^{-7}e^2/m_e$$
$$= 2.8179403227(19)\times 10^{-15}\,\text{m}$$

Table 2 Formulas for calculating interplanar spacing d_{hkl}

Crystal System	Axial Translations	Axial Angles	d_{hkl}
Cubic	$a = b = c$	$\alpha = \beta = \gamma = 90°$	$a(h^2 + k^2 + l^2)^{-1/2}$
Tetragonal	$a = b \neq c$	$\alpha = \beta = \gamma = 90°$	$[(h^2/a^2) + (k^2/a^2) + (l^2/c^2)]^{-1/2}$
Orthorhombic	$a \neq b \neq c$	$\alpha = \beta = \gamma = 90°$	$[(h^2/a^2) + (k^2/b^2) + (l^2/c^2)]^{-1/2}$
Hexagonal	$a = b \neq c$	$\alpha = \beta = 90°,\ \gamma = 120°$	$[(4/3a^2)(h^2 + k^2 + hk) + (l^2/c^2)]^{-1/2}$
Rhombohedral	$a = b = c$	$\alpha = \beta = \gamma \neq 90°$	$a\left[\dfrac{(h^2+k^2+l^2)\sin^2\alpha + 2(hk+hl+kl)(\cos^2\alpha - \cos\alpha)}{1+2\cos^3\alpha - 3\cos^2\alpha}\right]^{-1/2}$
Monoclinic	$a \neq b \neq c$	$\alpha = \beta = 90°,\ \beta > 90°$	$\left[\dfrac{(h^2/a^2) + (l^2/c^2) - (2hl/ac)\cos\beta}{\sin^2\beta} + \dfrac{k^2}{b^2}\right]^{-1/2}$
Triclinic	$a \neq b \neq c$	$\alpha \neq \beta \neq \gamma \neq 90°$	$\left[\dfrac{\frac{h}{a}\begin{vmatrix} h/a & \cos\gamma & \cos\beta \\ k/b & 1 & \cos\alpha \\ l/c & \cos\alpha & 1 \end{vmatrix} + \frac{k}{b}\begin{vmatrix} 1 & h/a & \cos\beta \\ \cos\gamma & k/b & \cos\alpha \\ \cos\beta & l/c & 1 \end{vmatrix} + \frac{l}{c}\begin{vmatrix} 1 & \cos\gamma & h/a \\ \cos\gamma & 1 & k/b \\ \cos\beta & \cos\alpha & l/c \end{vmatrix}}{\begin{vmatrix} 1 & \cos\gamma & \cos\beta \\ \cos\gamma & 1 & \cos\alpha \\ \cos\beta & \cos\alpha & 1 \end{vmatrix}}\right]^{-1/2}$

Note: Simpler expressions involving reciprocal lattice units are cited in Ref 2. Source: Ref 3

Table 3 Position of intensity maxima (peaks, spots)

- Controlled by:
  1. Geometry of unit cell (lattice)
  2. Wavelength of x-ray beam
  3. Instrumental and sample aberrations
- Applications:
  - Accurate lattice parameters
    - Least-squares refinements/Rietveld refinements
    - Phase characterization
    - Identification (qualitative analyses)
      - Recognition by familiarity
      - *d*-value searching (database lookup)
      - Cell data (database lookup)
  - Lattice changes
    - Lattice parameters shifts
      - Composition changes
      - Thermal expansion
      - Compressibility
      - Macrostrain
      - Stacking defects (anisotropic shift)
    - Phase changes (polymorphism)

Source: Ref 4

Table 4 Line intensity

- Controlled by:
  1. Types of atoms and atomic arrangement
  2. Amount of sample that can diffract
  3. Intensity-correction factors
- Applications:
  - Identification (qualitative analyses)
    - *d-I* search/match (combined always with peak position)
    - Intensity changes with element substitution
  - Crystal structure analysis
    - Differences in fine-grained state from single-crystal state
    - Structures of materials that only occur in fine-grained state
  - Quantitative analysis
    - Phase composition
    - Order-disorder ratios
    - Percent crystallinity
      - Polymers
      - Calcine reactions
    - Chemical kinetics
  - State of polycrystalline aggregate
    - Preferred orientation
    - Texture

Source: Ref 4

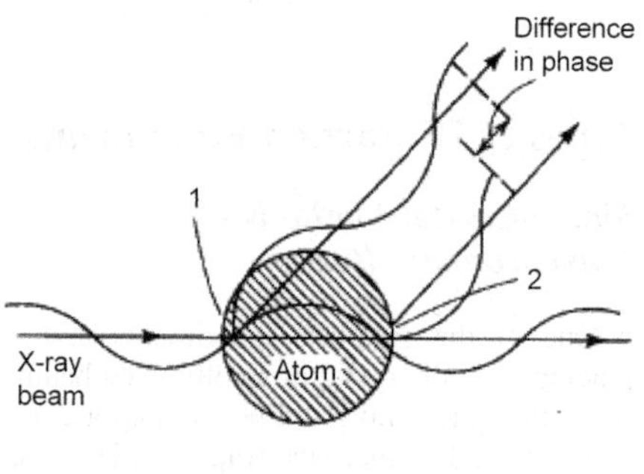

Fig. 5 Phase difference in scattering from different electrons within an atom

where e is the charge of an electron of rest mass, m_e; ε_0 is the electric constant; and c is the velocity of light (NIST Committee on Data for Science and Technology values: e = 1.6021766208(98) $\times$ 10^{-19} C; $\varepsilon_0 = 10^7/(4\pi c^2)$ F/m; m_e = 9.10938356(11) $\times$ 10^{-31} kg; c = 299,792,458 m/s). All of the rays in the incident beam are in phase, and the forward-scattered rays, that is, a scattering angle of zero, from electrons 1 and 2 remain in phase. Thus, the total amplitude of the forward-scattered beam is the sum of the contributions of all the electrons in the atomic system:

$$f_{\theta=0} = Z$$

where Z is the number of electrons.

As the scattering angle deviates from zero, the path lengths of the rays from the source to the observation point change, and the scattered rays arrive at the wave front slightly out of phase. This phase difference causes some destructive interference that diminishes the amplitude of the total beam. Figure 6 illustrates the change with angle for a typical atom. The greater the angle, the larger the interference, but because the size of atoms is approximately the same as the wavelength of the radiation, total destruction is never reached.

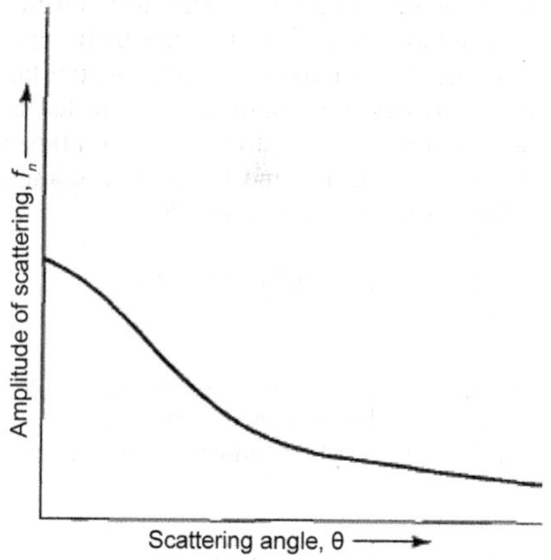

Fig. 6 Change in amplitude of the total diffracted beam as a function of the scattering angle. Higher scattering angles result in greater phase differences among the diffracted beams, decreasing the amplitude of the total diffracted beam.

The scattering factor from an atom may be measured experimentally using simple compounds of an element, or it may be calculated based on one of the electron orbital models of the atom. Tables of scattering values may be found in several sources for all the atoms and most of their ionic states. These scattering values are amplitudes normalized to the number of electrons involved at the scattering angle $\theta = 0$. Usually termed the atomic scattering factor or form factor, they are identified in intensity equations as f.

In a structure composed of a single atom in the unit cell, observed intensities decrease as the scattering angle increases proportionately to f^2. One of the forms of polonium has such a structure (simple cubic: the atoms sit only at the corners of a cubic unit cell), but most metals and other compounds have more complex structures consisting of several atoms in the unit cell. The treatment of the phase interference of two or more atoms is similar to the treatment of the effect of every electron in the atomic cloud. The phase factor depends on the relative positions of the atoms in the unit cell.

Figure 7 shows a unit cell containing two atoms. The positions of these atoms are described by their position coordinates based on the fractions of the unit-cell dimensions in the three principal directions from some origin. The choice of origin is not critical; it usually is selected on some symmetry element for convenience. Because the scattering effect of each atomic cloud has been considered in the atomic scattering factor f, each atom can be considered as a point atom concentrated at its center. The problem then reduces to determining the interference effect of the scattering from the atom centers due to their positions in space. For forward-scattered beams, $\theta = 0$, all the rays are in phase, and their amplitudes are additive. At $\theta \neq 0$, the path lengths of the scattered rays are different, and a phase difference necessarily results in some interference. The measure of this phase difference is contained in an exponential factor that expresses the amplitude shift in terms of the positions of the atoms. The combination of this phase factor and the atomic scattering factor is the structure factor, **F**:

$$\mathbf{F}(hkl) = \sum_n f_n e^{2\pi i(hx_n + ky_n + lz_n)}$$

where f_n is the atomic scattering factor for atom n; i is the complex number $\sqrt{-1}$; h, k, and l are the Miller indices of the diffracted direction; and x_n, y_n, and z_n are the position coordinates of atom n.

The amplitude of the total beam diffracted from the entire crystal is the sum of the contributions from each unit cell. In the directions of diffraction allowed by the lattice geometry, the resulting rays always are in phase and are additive. In any other direction, they interfere and totally cancel.

Several correction factors that enter into the intensity calculations are functions of the experimental technique and the nature of the sample. They correct for such components as atom vibrational motion due to temperature, beam polarization, varying diffraction times, and beam absorption in the sample. Considering these elements and the previous discussion, the intensity, I, of a diffracted beam is:

$$\begin{aligned} I(hkl) = I_0 &\times (\text{electron scattering coefficient})^2 \\ &\times (\text{experimental correction terms}) \\ &\times (\text{structure amplitude})^2 \end{aligned}$$

Types of Diffraction Experiments

Single-Crystal Methods—Polychromatic Beams

One of the simplest experiments involves placing a small crystal in a collimated beam of polychromatic x-rays and positioning a 2D flat detector on the other side (Fig. 8a). This transmission experiment, first performed by Laue in 1912, proved the wave nature of x-rays but is rarely used. The setup still is available at synchrotron and neutron sources where the emitted x-ray spectrum does not contain any characteristic line, and the diffraction pattern is collected using one or more large, pixelated 2D detectors. If the crystal is small enough to be x-ray transparent and if it is aligned with a symmetry element along the beam direction, the resulting diffraction pattern reveals the symmetry. Because the crystal is stationary, different wavelengths in the incident beam are required to satisfy the Bragg conditions from the various lattice planes that can diffract. Because neither the wavelength nor the d-spacing are known a priori, Laue patterns are difficult to analyze.

An alternate arrangement involving large crystals and placement of the detector on the source side of the crystal was used in the past for rapid crystal alignment. This is the back-reflection Laue method (Fig. 8b).

Single-Crystal Methods—Monochromatic Beams

Because a monochromatic beam can satisfy no more than one diffraction condition at a

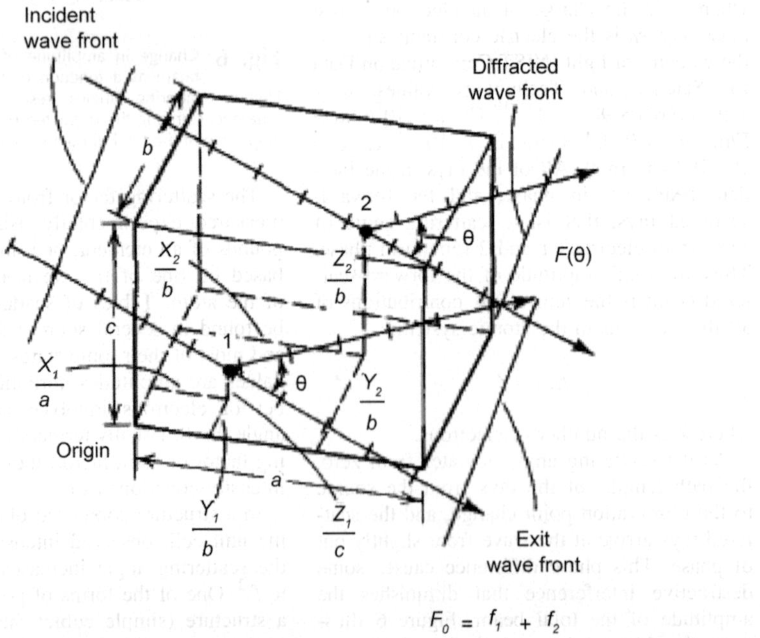

Fig. 7 Diffraction in a unit cell and the structure factor equation for the cell. The positions of atoms 1 and 2 are defined by their position coordinates based on fractions of the unit-cell dimensions in directions a, b, and c. $x_1 = X_1/a$, $y_1 = Y_1/b$, and $z_1 = Z_1/c$

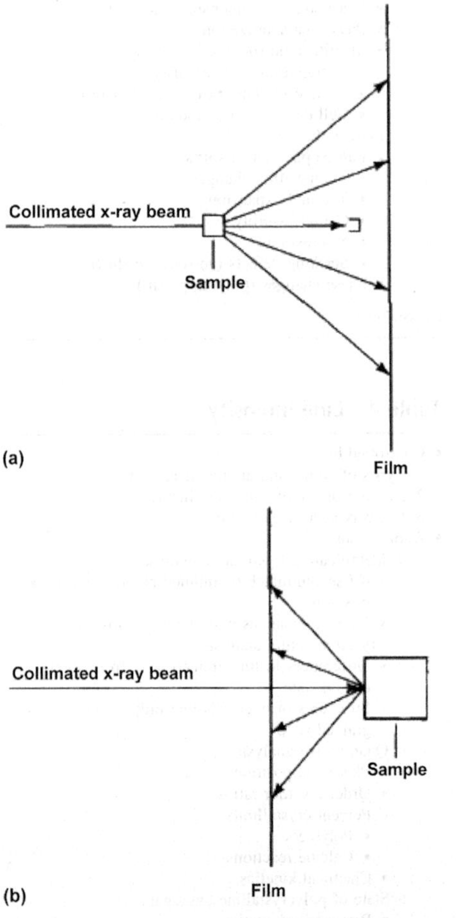

Fig. 8 Two types of single-crystal diffraction experiments. (a) Transmission arrangement. (b) Back-reflection arrangement

time, the crystal must be moved to observe the diffraction pattern.

Several moving crystal/moving film techniques using different crystal and film motions have been developed in the past to try collecting the maximum information on a single film during an experiment. The most common movement is an oscillation or rotation of the crystal about an axis perpendicular to the x-ray beam. If the crystal is aligned with a lattice row along the rotation axis, the resulting pattern of spots is easily interpreted. In the older instruments, a film usually was placed like a cylinder around the rotation axis (Fig. 9). When the film is unrolled, the spots form rows that can be easily related to the lattice geometry. Figure 10 shows an example of a rotation pattern. This rotation method was used to characterize the lattice geometry and to check the quality of a crystal. For symmetry analysis and measurement of intensities, a screen was employed to limit the beams that could reach the film, then moving the film synchronously with crystal rotation. This is the Weissenberg method, which allows unequivocal identification of every diffracted beam. Figure 11 illustrates a Weissenberg pattern of the 0th row of the same crystal used for Fig. 10. Another example is the precession method, whose pattern is shown in Fig. 12. All of these methods can be used to obtain essentially the same type of information. Interpreting the patterns of spots in space enables determining both lattice periodicity dimensions, the angles between the lattice rows, and most of the symmetry information. By measuring intensity values using the relative darkness of the spots on the films, the data necessary to determine the crystal structure also can be obtained.

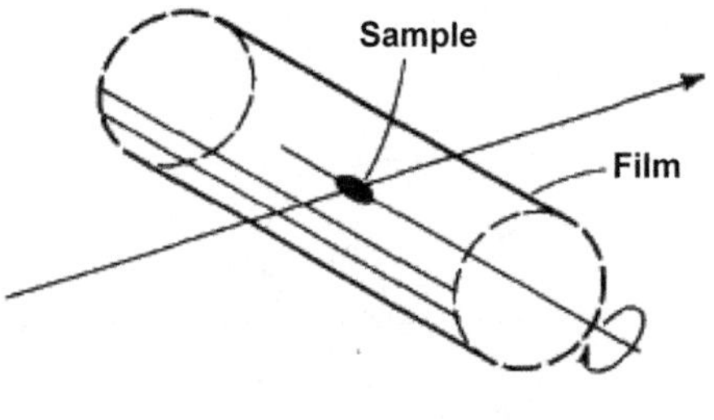

Fig. 9 Single-crystal experiment using monochromatic beams. The sample is rotated about an axis perpendicular to the beam, resulting in a pattern of spots on the cylindrical film.

Nowadays, pixelated detectors and automatic single-crystal diffractometers have superseded rotation and precession camera systems based on films. The most successful system is a diffractometer that automatically positions a crystal and uses a 2D detector to collect a large set of diffracted beams while changing the crystal position. Multiple rotation stages (goniometers) are employed to allow for any orientation of the crystal in the space around the rotation center of the system. Figure 13 shows the most common Euler and kappa geometries employed on single-crystal diffractometers. Short wavelengths (e.g., Mo or Ag Kα) are employed in laboratory machines to increase the density of intensity maxima present in a single measured image. Problems arise for proteins, with their unusually large unit cells and the possible large sensitivity to radiation. Protein crystallography beamlines, based on large, fast, and accurate detectors, are designed to perform automatic single-crystal studies on protein crystals as small as a few cubic micrometers. Nowadays, these beamlines are abundant in all synchrotron radiation facilities around the world.

Single-Crystal Topography

X-ray topography is a unique application of single-crystal analysis. It essentially is x-ray diffraction radiography in which a large area of a single crystal produces a single diffracted beam that creates a 1:1 image of the diffracting volume of the sample crystal on a 2D detector. The image reveals the presence of such crystal defects as dislocations and subgrain boundaries. The technique is effective because regions that contain defects scatter more strongly than defect-free regions. The real image produced shows the distribution of the defects in the crystal. Several experimental techniques that use this principle are considered in the article "X-Ray Topography" in this Volume.

Powder Diffraction Methods

Use of a single crystal is impossible or inappropriate for many samples. X-ray powder diffraction (XRPD) techniques use a sample composed of a large number of small crystallites either in true powder form or under the form of a polycrystalline aggregate. Uses for XRPD techniques range from phase identification, to quantitative analysis of mixtures, to determining the atomic positions, to evaluating size distribution and defect in the crystallites, to measuring strain in a weld joint or evaluating the grain elongation and

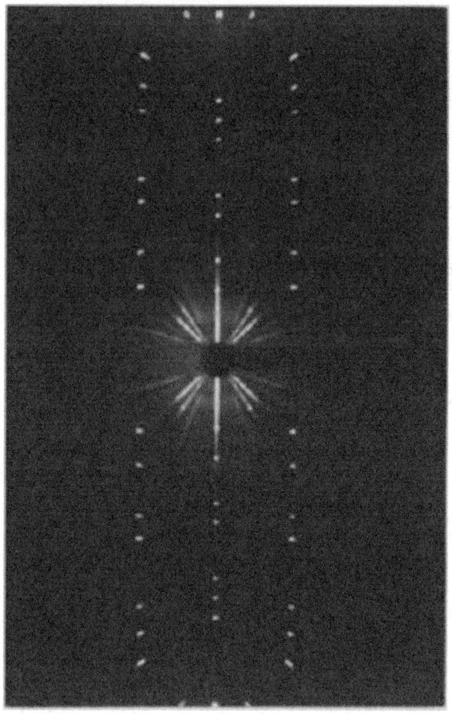

Fig. 10 Rotation pattern for an NaCl crystal. The pattern was obtained using the method shown in Fig. 9.

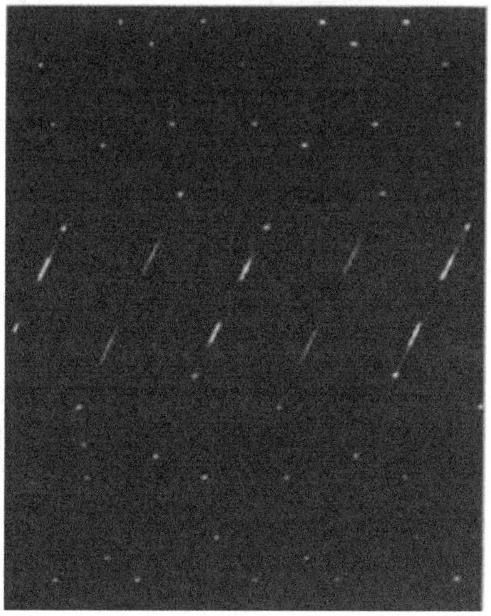

Fig. 11 Weissenberg pattern of the same crystal as in Fig. 10. In this method, the film is moved synchronously with the crystal rotation.

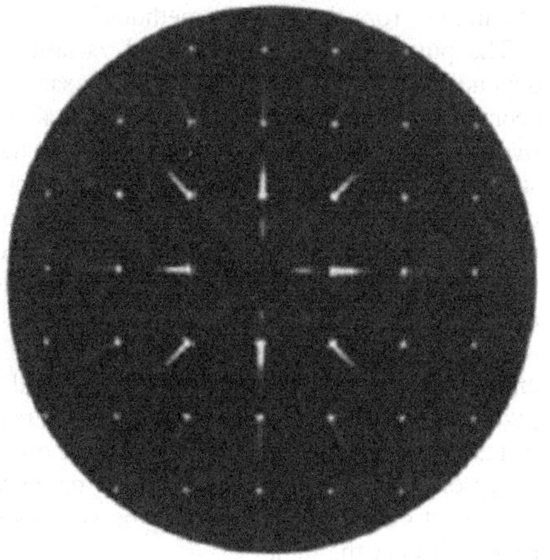

Fig. 12 Pattern for the NaCl crystal used in Fig. 10 and 11 obtained using the precession method

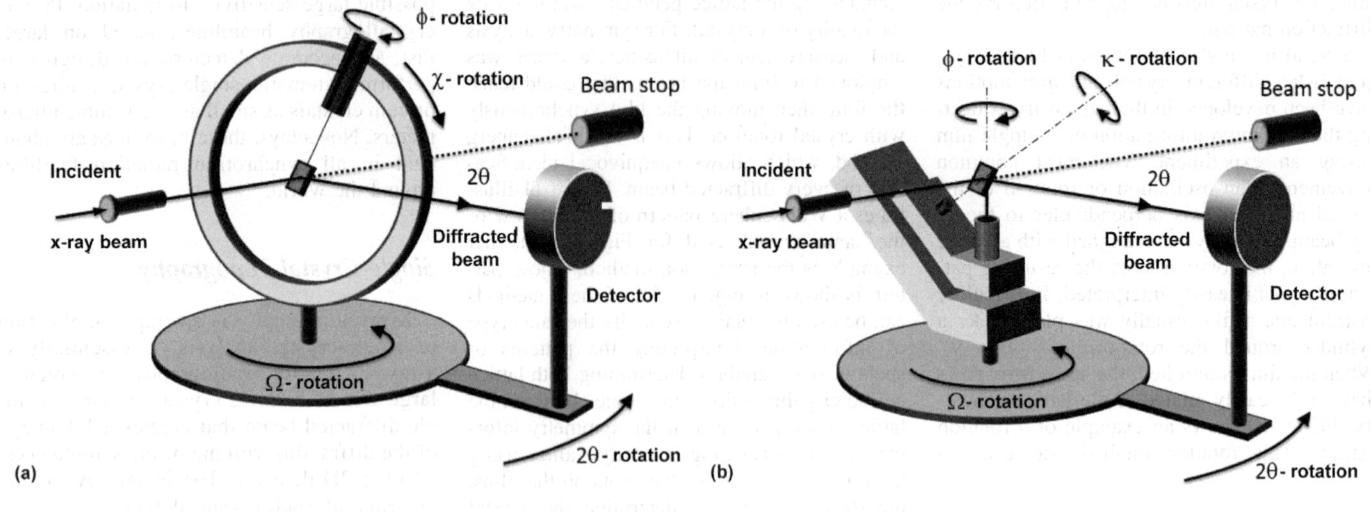

Fig. 13 Single-crystal diffractometer. (a) Euler and (b) kappa geometries

reorientation caused by rolling. These applications are considered in the article "X-Ray Powder Diffraction" in this Volume and in Ref 5 and 6.

The basic powder experiment uses a powder sample placed in a collimated monochromatic beam. The ideal sample contains a large number of crystallites to cover all possible orientations uniformly. Each crystallite behaves like a single crystal and may diffract one beam. The total collection of crystallites behaves like a single crystal that is moved into all possible orientations, diffracting all possible beams. The result is many coaxial cones of diffracted beams emanating from the sample (Fig. 14a). There is a complete cone for every possible diffracted beam from the crystal. The cone half-angles are 2θ. Because the crystallites are oriented randomly in the sample, the only measurable position parameter is 2θ, which translates into a *d*-value through the Bragg relation. No direct information on the lattice constants, the distances, or the angles is obtainable from the powder methods.

The powder method has several variations, depending on diffraction geometry, sample geometry, and method of detection (Fig. 14b–f). In the Debye-Scherrer method/geometry, the sample is a ball or rod of powder on a fiber or in a rotated capillary tube, and the detector is placed on (a fraction of) a cylinder around the sample axis. All of the diffracted beams are detected simultaneously. On modern systems, a 1D position-sensitive detector is employed. When resolution is an issue (the pixel size of a 1D detector may be too large to resolve small features in the pattern), especially on synchrotron beamlines, one or more 0D detectors are moved around the specimen axis to collect the data.

In the diffractometer method, the sample is flat while the source and 0D detector are put on two arms at the same distance from the

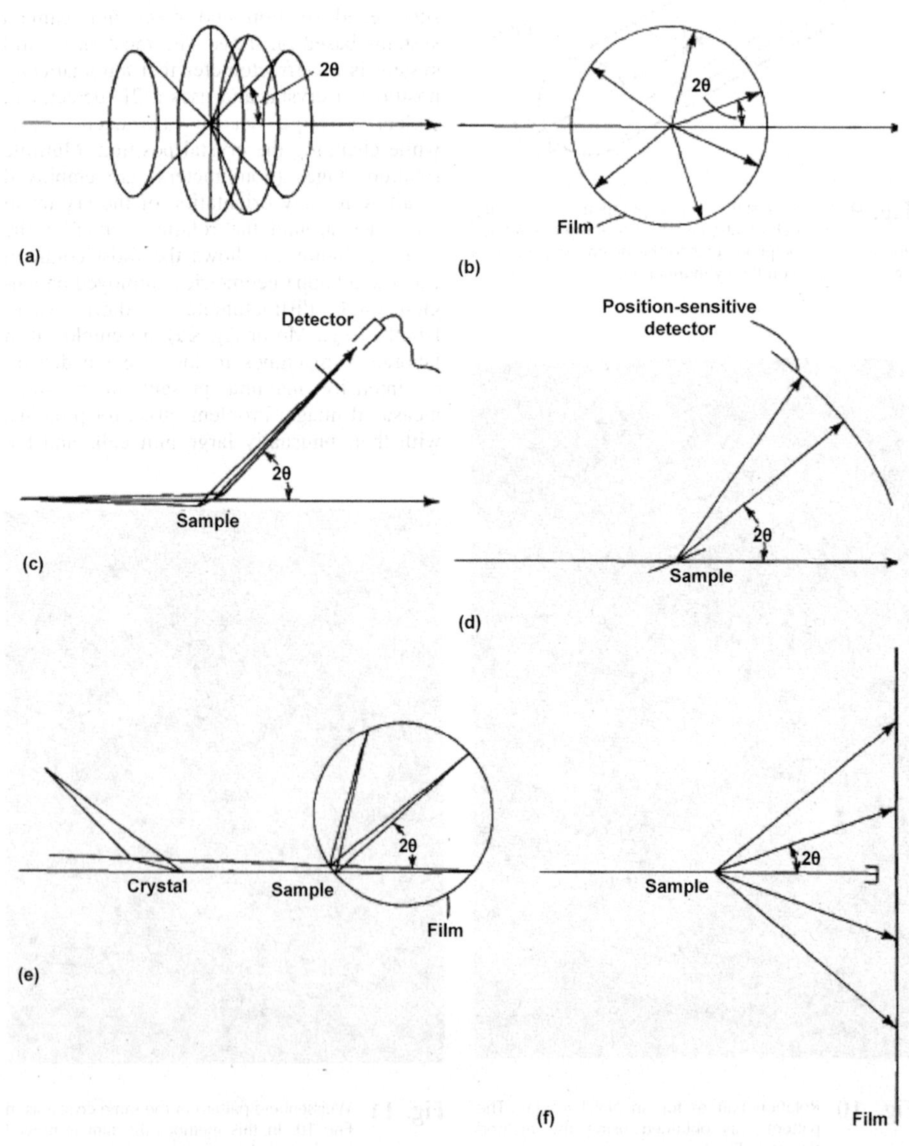

Fig. 14 Geometry of powder diffraction and several detection methods used in x-ray powder diffraction. (a) Cones of diffracted beams emanating from a powder sample. (b) Debye-Scherrer detection method. (c) Diffractometer method of detection. (d) Position-sensitive detection. (e) Guinier method of detection. (f) Flat plate texture method of detection

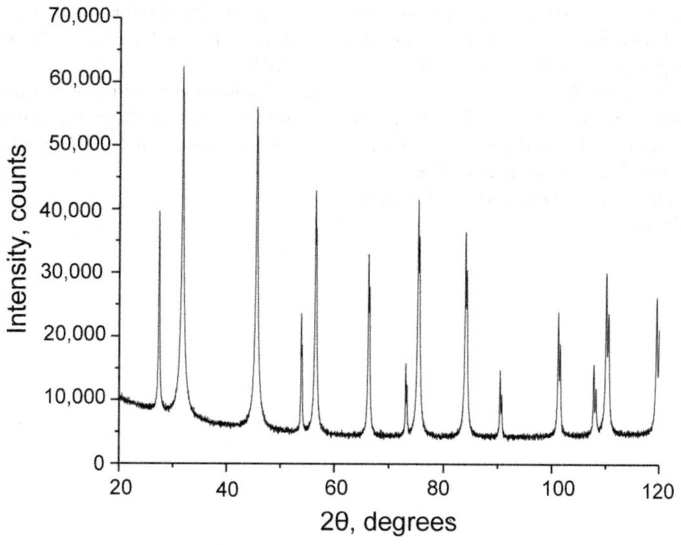

Fig. 15 Powder diffraction pattern of NaCl

specimen surface, centered on the rotation axis of the system. Specimen, tube, and detector are >moved relative to each other, and the intensity is collected as a function of the diffraction angle 2θ. In the most common Bragg-Brentano configuration, the angle between the surface of the specimen and the source and the angle between the surface of the specimen and the detector are the same (θ). Any type of specimen, including liquids, can be analyzed. Figure 15 shows a powder pattern example. Nowadays, both large, curved 1D position-sensitive detectors collecting the whole diffraction pattern simultaneously, and movable 2D (or short 1D) detectors collecting locally a few degrees in 2θ, are employed to increase data throughput and to collect larger sections of the diffraction cones.

Geometric focusing results in effective use of beam intensity. The Guinier method uses a crystal monochromator to focus the incident beam. In the old camera setup, the sample was a thin foil, and a film was the detector. In a more modern setup, a position-sensitive detector can be used.

A simple method using monochromatic radiation and a 2D detector, located as in the Laue method mentioned previously, can be used to analyze texture and to survey for grain size. If the crystallites in the sample are not oriented randomly, the diffraction cones are not complete, and arcs will be visible on the detector instead of rings. If the grain size is large, the rings appear spotty instead of smooth.

Due to its flexibility, nondestructive nature, and limited cost, XRPD is used extensively in various fields of science and technology. Sample chambers allow measurements to be made in controlled atmospheres, at temperatures from 4 to 3500 K, and at pressures from 1.3×10^{-6} Pa to a few gigapascals. Modern instrumentation achieves pattern collection and analysis in seconds or less, allowing time-resolved studies of rapid reactions.

Crystal Imperfections

Crystals are rarely perfect. They contain deformed regions and mistakes in the way the structure fits together. Local and global lattice distortions, dislocations, stacking faults, subgrain structure, and impurities disrupt the perfection of a crystal and thus affect the x-ray diffraction patterns. These imperfections affect the positions and shapes of the diffracted beams, that is, the distribution of intensity in space. This broadens the spots/peaks observed and shifts them from their ideal angular position. In a powder, the limited size of the crystallites (which can go down to the nanometer range) can further contribute to this broadening.

Table 5 lists factors that control the shapes of diffracted beams and applications of line profile analysis. Further information can be found in the article "X-Ray Powder Diffraction" in this Volume.

In single crystals, imaging of the internal defects is possible (see the article "X-Ray Topography" in this Volume). In powders, this information can be obtained indirectly by modeling the powder diffraction pattern. Spot patterns show misshapen spots and streaking (the latter resulting from the presence of stacking defects); powder patterns show only broadening, which can be symmetric or asymmetric. Symmetric broadening is caused by a small crystallite size or by the presence of localized strains (e.g., dislocations). The presence of defects such as dislocations results also in an anisotropic (i.e., direction-dependent) broadening. Asymmetric broadening may be the result of, for example, a compositional gradient. Peak shifting usually is the result of either the presence of stacking defects or a macroscopic lattice deformation (macrostress). Understanding the effects of these defects is important in allowing correct interpretation of the experimental data and in using the data to acquire information on the causes and nature of these defects. Making use of this information is elaborated in the article "X-Ray Diffraction Residual-Stress Techniques" in this Volume.

Table 5 Line profile

- Controlled by:
  1. State of crystallite perfection and sample homogeneity
  2. Instrumental aberrations
  3. Spectral distribution
- Applications:
  - Crystallite size
  - Crystallite size distributions
  - Defects (microstrain effects)
    - Lattice distortions
    - Crystal defects type and quantity
    - Antiphase boundaries
  - Inhomogeneous strain
  - Sample inhomogeneity
  - Crystallinity
    - Amorphous state

Source: Ref 4

Refinement and Rietveld Refinement

When dealing with diffraction data, the terms (structure) *refinement* and *Rietveld refinement* are often met, respectively, in the single-crystal and the powder diffraction communities. X-ray diffraction is a quantitative technique, and it is possible to link the atomic positions with the corresponding diffraction pattern, that is, to generate a synthetic pattern based on a given model. The goal of a diffraction experiment therefore is to extract some realistic starting model parameters for the material and then refine those model parameters to obtain the best agreement with the measurement. This idea was proposed in the powder diffraction community by H.M. Rietveld in the late 1960s to refine a structure model based on powder data (Ref 7). Nowadays, the Rietveld method has been extended to include the handling of most of the techniques proposed for powders in Table 1. Therefore, most of the results from diffraction analysis are based on Rietveld modeling of the whole diffraction pattern. A Rietveld-type refinement guarantees a higher accuracy of the results if one understands the limitations of some of the models employed for the generation of the pattern and considers them when reporting the analysis results.

ACKNOWLEDGMENT

Revised from D.K. Smith, Introduction to Diffraction Methods, *Materials Characterization*, Vol 10, *ASM Handbook*, ASM International, 1986, p 325–332

REFERENCES

1. C. Giacovazzo, H.L. Monaco, G. Artioli, D. Viterbo, M. Milanesio, G. Gilli, P. Gilli, G. Zanotti, G. Ferraris, and M. Catti, *Fundamentals of Crystallography*, 3rd ed., Oxford University Press, Oxford, 2011
2. *Mathematical, Physical and Chemical Tables,* Vol C, *International Tables for X-Ray Crystallography*, Wiley, 2006
3. H.P. Klug and L.E. Alexander, *X-Ray Diffraction Procedures for Polycrystalline and Amorphous Materials*, 2nd ed., John Wiley & Sons, 1974
4. D.K. Smith, C.S. Barrett, D.E. Leyden, and P.K. Predecki, Ed., *Advances in X-Ray Analysis*, Vol 24, Plenum Press, 1981
5. B.E. Warren, *X-Ray Diffraction*, Addison-Wesley, 1969
6. Powder Diffraction, Vol H, *International Tables for X-Ray Crystallography*, Wiley, 2019
7. H.M. Rietveld, A Profile Refinement Method for Nuclear and Magnetic Structures, *J. Appl. Crystallogr.*, Vol 2, 1969, p 65

ASM Handbook, Volume 10, *Materials Characterization*
ASM Handbook Committee
DOI 10.31399/asm.hb.v10.a0006680

X-Ray Powder Diffraction*

Matteo Leoni, University of Trento, ICDD, and Saudi Aramco

Overview

General Uses

- Identification of crystalline phases contained in unknown sample
- Quantitative determination of the weight fraction of crystalline phases in multiphase materials
- Fingerprinting of a materials formulation
- Characterization of solid-state phase transformations
- Lattice-parameter and lattice-type determinations
- Analysis of domain size and defects in nanocrystalline materials
- Stereographic projections (preferred orientation)

Examples of Applications

- Qualitative and quantitative analysis of crystalline phases in soils, coal ash, ceramic powders, corrosion products, and so on
- Determination of phase diagrams
- Determination of pressure- and/or temperature-induced phase transformations
- Quantitative analysis of solid solutions from lattice-parameter measurements
- Determination of anisotropic thermal expansion coefficients
- Determination of the atomic arrangement
- Evaluation of the orientation of the grains and of their residual stress
- Quantitative analysis of the nanostructure (crystallite shape and size distribution, type and quantity of lattice defects)

Samples

- *Form:* Crystalline solids (metals, ceramics, geological materials, and so on)
- *Size:* Powder samples from 1 to 10 mg are usually adequate
- *Preparation:* Sometimes none; sample may require crushing to fit into the sample holder
- *Flat metal samples:* Diffractometers can usually accommodate samples with lateral dimensions up to 5 cm (2 in.) and thicknesses up to 5 mm (0.2 in.). Ideally, the surface should be free of deformation in order to obtain sharp diffraction peaks. Chemical or electropolishing can be used to remove the vestiges of deformation in ground and/or polished samples.

Limitations

- Phase identification is easiest for crystalline materials. Poorly ordered or amorphous materials require an experimental reference pattern for identification.
- Identification requires existence of standard patterns: International Centre for Diffraction Data's "Powder Diffraction File (PDF)" (inorganic and organic phases); National Institute of Standards and Technology's "NIST Crystal Data" (inorganic and organic phases); "Inorganic Crystal Structure Database" (inorganic structure data); "Cambridge Structural Database" (organic structure data); "Crystallographic Open Database" (inorganic and organic data); "Protein Data Bank" (proteins)
- Results represent average of many grains or crystals in the sample, not an individual particle. Fine beams (down to ~100 μm diameter) can be used to characterize some individual particles.
- Crystallite size should be below ~100 nm for the result to be accurate/meaningful.

Estimated Analysis Time

- Depending on the source of x-rays, major phases can be identified in milliseconds.
- Quantitative analysis, after a procedure is set up, requires several seconds to several hours.

Capabilities of Related Techniques

- *X-ray spectrometry, inductively coupled plasma atomic emission spectroscopy, atomic absorption spectrometry, classical wet chemical analysis:* Quantitative and qualitative elemental information
- *Auger electron spectroscopy:* Elemental and structural data on small portions of the samples
- *Single-crystal x-ray diffraction:* Crystal structure using small, single crystals
- *Infrared and Raman spectroscopy:* Molecular structure and sometimes crystal structure
- *Neutron diffraction:* Similar and, in many cases, complementary information. Particularly suited for light or magnetic structures.

*Adapted from R.P. Goehner and M.C. Nichols, X-Ray Powder Diffraction, *Materials Characterization*, Vol 10, *ASM Handbook*, ASM International, 1986, p 333–343

X-RAY POWDER DIFFRACTION (XRPD) techniques are used to characterize samples in the form of loose powders, aggregates of finely divided material or polycrystalline specimens (Ref 1). These techniques cover various investigations, including qualitative and quantitative phase identification and analysis as well as determination of crystallinity, lattice parameter, structure, and micro/nanostructure. In situ and operando studies at ambient and nonambient conditions can also be routinely made.

The powder method, as it is referred to, is an ideal phase-characterization tool partly because it can routinely differentiate between phases having the same chemical composition but different crystal structures (polymorphs). Although chemical analysis can indicate that the empirical formula for a given sample is $FeTiO_3$, it cannot determine whether the sample is a mixture of two phases (FeO and one of the three polymorphic forms of TiO_2) or whether the sample is the single-phase mineral $FeTiO_3$ or ilmenite. The ability of XRPD to perform such identifications more simply, conveniently, and routinely than any other analytical method explains its importance in many industrial applications as well as its wide availability and prevalence.

In general, an XRPD characterization of a substance consists of placing a sample in a collimated monochromatic x-ray beam. For crystalline materials, diffraction takes place as described in the article "Introduction to Diffraction Methods" in this Division to produce a diffraction pattern. The diffraction pattern is recorded using a point, line, or area detector to be digitally processed.

In XRPD analysis, samples usually exist as finely divided powder (or can be reduced to powder form) or as polycrystalline aggregates. The upper limit for the grain size is on the order of micrometers, to guarantee that enough grains with different orientation are bathed by the x-ray beam. The particles in a sample comprise one or more independently diffracting regions that coherently diffract the x-ray beam. These small crystalline regions are termed coherent scattering domains or crystallites. Consolidated samples, such as ceramic bodies or as-received metal samples, will likely have crystallites small enough to be useful for powder diffraction analysis, although they may appear to have considerably larger particle sizes. This occurs because a given grain or particle may consist of several crystallites (independently diffracting regions). Although larger grain sizes can sometimes be used to advantage in XRPD, the size limitation is important because most applications of powder diffraction rely on x-ray signals from a statistical sample of crystallites.

In a simplified view, the angular position, θ, of the diffracted x-ray beam maxima depends on the spacings, *d*, between planes of atoms in a crystalline phase and on the x-ray wavelength, λ, according to the Bragg equation:

$$n\lambda = 2d\sin\theta \quad \text{(Eq 1)}$$

The intensity of the diffracted beam depends on the arrangement of the atoms on these planes.

X-ray powder diffraction techniques usually require some sample preparation. This may involve crushing the sample to fit inside a glass capillary tube or rolling it into a very thin rod shape for the Debye-Scherrer geometry, spreading it as a thin layer on a sample holder, or packing it into a sample holder of a certain size for other XRPD techniques. In some cases, samples compatible with metallographic examination can be accommodated in powder diffractometers, but some form of sample preparation will usually be necessary. Preparation will depend on the equipment available and the nature of the examination.

A diffraction pattern is usually displayed as a graph of intensity versus interplanar distance, *d*, or as a function of diffraction angle 2θ. Two dimensional plots, where the 2θ direction is radial and the tangential direction represents the angle with respect to the specimen normal, are also quite common due to the abundance of two-dimensional detectors. Original equipment manufacturers and free software exist for data reduction, conversion, and processing.

A powder pattern from a single-phase material will have maxima at positions dictated by the size and shape of its unit cell and will increase in complexity as the symmetry of the material decreases. For example, many metal patterns and those from simple compounds that tend to be mostly of cubic symmetry having small unit cell edges will produce powder patterns having fewer lines or maxima than would be expected from a compound of lower symmetry or one having a very large unit cell. A pattern of a mixture of phases in which all the individual patterns are superimposed will produce a complex experimental pattern, especially when the number of phases present in the mixture exceeds approximately three or when the phases constituting the mixture are all of very low symmetry or have very large unit cell dimensions.

Phase identification using XRPD is based on the unique pattern produced by every crystalline phase. Much as a fingerprint is unique for each person, the diffraction pattern can act as an empirical fingerprint for that phase, and qualitative identification of phases can be accomplished by pattern-recognition methods, most of which are based on heuristic algorithms. All of these methods make use of one of the available databases containing fingerprint data (peak position and intensity). The most comprehensive edited database designed for this purpose is the "Powder Diffraction File (PDF)," maintained by the International Centre for Diffraction Data (ICDD) (Ref 2). As a materials database, the PDF contains further information useful for phase identification and studies, including structural data for several inorganic and organic compounds. The other available databases do not contain peak positions but just structural data from which the peak list can, in principle, be obtained. Among the alternatives are crystal data of inorganic and organic phases from the National Institute of Standards and Technology (NIST) (Ref 3) as well as:

- The free "Crystallographic Open Database" (inorganic and organic data) at http://www.crystallography.net
- "Inorganic Crystal Structure Database" at http://www2.fiz-karlsruhe.de/icsd_home.html
- "Cambridge Structural Database" (organic structure data) at https://www.ccdc.cam.ac.uk/solutions/csd-system/components/csd/
- "Protein Data Bank" at http://www.rcsb.org/

X-Ray Powder Diffraction Instrumentation

X-ray diffraction analyses have historically been conducted using two types of equipment: either with cameras or with x-ray diffractometers, as described in other sections of this article. A pinhole camera (Fig. 1, 2) with photographic film is the oldest method of detecting x-rays. It was first used in 1912 by Laue in his discovery of the wave nature of x-rays (Ref 4), and powder diffraction patterns were first recorded on film in 1913 (Ref 5, 6).

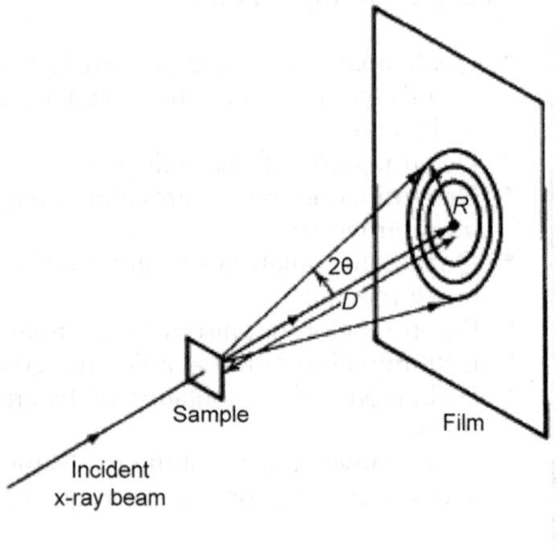

Fig. 1 Schematic of a transmission pinhole camera

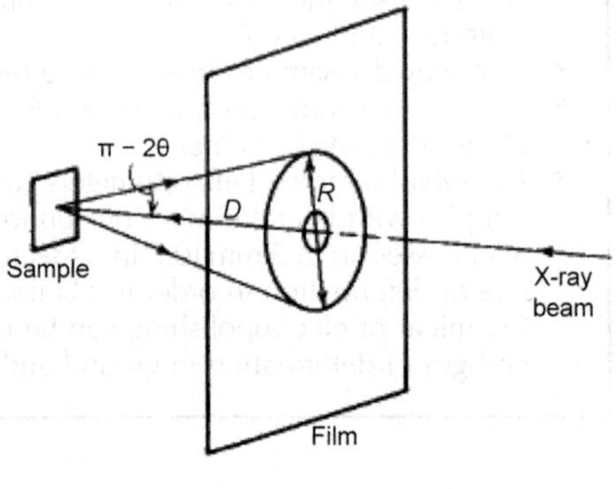

Fig. 2 Schematic of a back-reflection pinhole camera

Nowadays, films have been completely superseded by digital detectors that provide a signal proportional to the number of intercepted x-ray photons. The spatial position (along a line or within the detector area) may also be simultaneously available. Together with modern instruments, the older ones are described here for completeness: Data collected using cameras is extensive in the literature. Some of the older geometries are still employed with more modern one-dimensional (1-D) and two-dimensional (2-D) detectors or with a zero-dimensional detector in scanning mode, to acquire the data that were once impressed on a film, in a more easily manageable digital form.

The x-ray diffractometer avoids the use of film and is far better suited to automation. The sample is flat, typically either a polished metal surface or a powder adhered to a flat glass slide. The sample is exposed to the incident x-ray beam, and a counter is scanned over the desired range of θ angles (Fig. 3). The result is a plot of diffracted intensity versus diffraction angle. All modern machines are fully automated, in the sense that all motors are computer controlled and measurement routines are readily available. Fully automated analysis systems are also available for online monitoring in production plants. In recent years, position-sensitive wire detectors and solid-state charge-coupled device detectors (in essence, a high-resolution array of solid-state light sensors) have permitted all of the diffracted signals to be collected simultaneously, thus overcoming the need to mechanically scan a detector scintillation counter over the range of diffraction angles.

X-Ray Camera Techniques

The simplest instrument for diffraction studies is the pinhole camera (Fig. 1). This camera exists in four variations, depending on the placement of the detector and the use of white radiation or monochromatic radiation. Films were used in the older cameras, whereas pixelated 2-D detectors are employed in most modern versions. If white radiation is used, the discussion is usually about Laue diffraction, and the instrument can be referred to as a transmission Laue camera. The symmetry of the spot pattern can be used to determine the orientation of the single crystal relative to the x-ray beam. When monochromatic radiation is used and the sample is polycrystalline, a set of concentric rings can be detected, known as Debye rings. In this case, the *d*-spacing can be calculated by using:

$$\theta = \arctan\left(\frac{R}{D}\right) \quad \text{(Eq 2)}$$

where *R* is the radius of the Debye ring, and *D* is the detector-to-sample distance. Once θ is known, *d* can be calculated using Eq 1 as:

$$d = \frac{\lambda\sqrt{D^2 + R^2}}{2R} \quad \text{(Eq 3)}$$

In the transmission pinhole camera, the samples must be thin enough for the x-rays to penetrate. Using laboratory x-rays, metallic samples typically must be less than 0.1 mm (0.004 in.) thick. The use of synchrotron radiation or neutrons can overcome this limitation.

Figure 2 shows a back-reflection pinhole camera. If white radiation is used, the camera is usually referred to as a back-reflection Laue camera. This technique was widely used to determine the crystallographic orientation of single crystals. The back-reflection Laue technique and diffractometer methods can be used to align crystals for cutting. When monochromatic radiation is used for polycrystalline samples, the detected signal will consist of concentric rings, as in the transmission pinhole camera. These Debye rings will be from the back-reflection ($2\theta > 90°$) region.

The major disadvantage with the back-reflection pinhole camera is that the reflections observed do not include the front-reflection lines ($2\theta < 90°$), which differ more in intensity and position between different phases than the back-reflection lines. Thus, the back-reflection results do not provide a useful characterization tool for phase identification. Another disadvantage with the front- and the back-reflection pinhole techniques is that the detector-to-sample distance, *D*, usually cannot be measured with sufficient precision to provide accurate *d*-spacings. The transmission and back-reflection measurements can be easily differentiated, because the back-reflection recording has a hole in the center where the collimator went through.

The monochromatic pinhole camera was used primarily to determine the preferred orientation (texture) in a sample and to survey the sample for crystallite size and plastic deformation. Preferred orientation occurs if certain crystallographic planes have a higher occurrence in certain directions in the sample. For example, cold-drawn aluminum wire has a [111] preferred orientation along its axis and a random orientation about its circumference. This type of preferred orientation is usually referred to as a fiber texture because it also occurs in natural and artificial fibers. If the crystallite size of the sample exceeds approximately 30 μm, the Debye rings begin to appear grainy. When the crystallite size becomes less than a few thousand angstroms, the Debye rings broaden. Plastic deformation will also broaden the rings in fine-grained samples. In single-crystal or coarse-grained samples, the spots will form streaks (asterism) due to the internal microstrain in the grains. See also section "Estimation of Crystallite Size and Defects" in this article for more details.

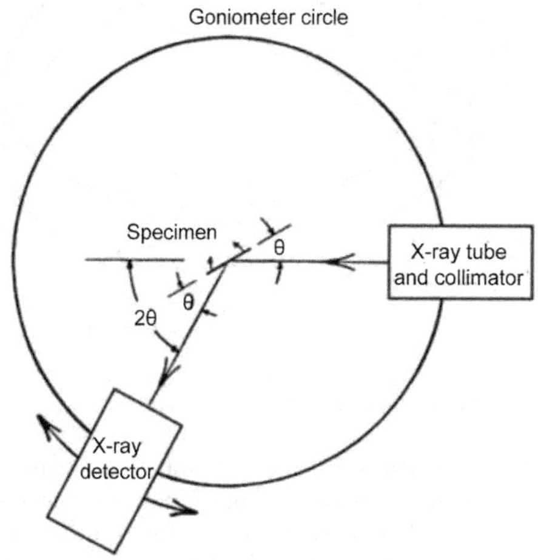

Fig. 3 Schematic of x-ray diffractometer. Typically, the x-ray tube remains stationary while the detector mechanically scans a range of θ angles. The sample also rotates with the detector such that diffraction is recorded from planes parallel to the sample surface.

Debye-Scherrer Technique

Figure 4 shows the Debye-Scherrer geometry and a typical recording. In the Debye-Scherrer technique, the sample exists as a small cylinder of finely divided powder (with a random orientation of the crystallites) that is rotated about its long axis in a monochromatic x-ray beam. This situation is usually obtained by loading the powder into a thin-walled glass capillary. Alternative capillary materials (e.g., kapton polyimide film, platinum, MgO) are available for special applications. In this random assemblage, some of the particles will be oriented to the proper angle to diffract x-rays from their (100) planes; others will have an orientation that can allow diffraction from their (110) planes, and so on. There is a continuum of orientations that allows a given set of planes to make the proper angle with the beam, producing a cone of x-rays coming from the sample for each set of planes diffracting. Figure 4(a) shows three such cones and their intersection with the film that has been placed around the sample. Besides finding applications in the laboratory (due to the ease of mount), the Debye-Scherrer geometry (cylindrical specimen in transmission) is nowadays widely used on diffractometers located at large-scale facilities (synchrotrons) for such operations as the precision determination of lattice parameters or atomic structure determination.

Gandolfi Camera Technique

The Gandolfi camera is an adaption of the Debye-Scherrer geometry that is useful for examining materials that may not exist as a random collection of finely divided powders. The principal difference from the Debye-Scherrer camera is in the motion of the sample. In the Debye-Scherrer geometry, the sample is rotated about an axis perpendicular to the x-ray beam. In the Gandolfi camera, the sample is mounted as a small sphere or assemblage of one or more grains on the end of a fiber mount. The mount is positioned such that it remains in the x-ray beam at all times, but it rotates about two axes rather than one. One of these motions rotates the sample about an axis as in the Debye-Scherrer geometry, but another axis of rotation exists at an angle with the normal Debye-Scherrer axis. Thus, the sample has two ways in which it is rotating in the x-ray beam, which presents more possible orientations to the beam. This increases the number of ways each grain or crystallite can diffract the x-ray beam. This method is excellent for examining samples that consist of just a few or even one crystal, such as may result from the extraction of grain-boundary precipitates from a metallurgical sample or the physical separation of one or more phases in a geological sample.

Guinier Geometry

The Guinier geometry combines a focusing monochromator and a focusing camera (Fig. 5). The curved crystal monochromator has a very narrow bandwidth. On laboratory instruments it can be tuned, for example, to $K\alpha_1$ radiation, thus eliminating the $K\alpha_2$ component from the experiment. Figure 6 illustrates the different arrangements for the Huber Guinier camera. The focusing monochromator increases the resolution of the instrument with respect to the Debye-Scherrer setting. Measurements conducted using a Guinier camera implementing an internal-standard technique exhibit excellent reproducibility and precision. The high-quality diffraction patterns collected at Dow Chemical using Guinier cameras were the basis for the ICDD database.

The camera is well suited to studying complex diffraction patterns of mixtures of phases and low-symmetry materials. The Guinier technique has several disadvantages. First, alignment of the monochromator requires considerable effort, at least initially. Second, the camera in any one setting cannot cover a large angle region. In the asymmetric transmission mode, the angle region is only approximately 50° in 2θ. Third, the camera requires the use of an internal standard to calculate peak positions. This is especially true in the asymmetric mode of operation. The Debye-Scherrer camera, having a geometry that can usually allow for absolute internal calibration, typically does not require an internal standard.

The Guinier geometry can still be employed on synchrotron radiation beamlines to collect high-resolution data. Working at short wavelengths with modern 1-D and 2-D detectors and with high-performance monochromators removes some of the disadvantages of the old cameras and compresses a large quantity of data at low angles, where the Guinier geometry is optimal.

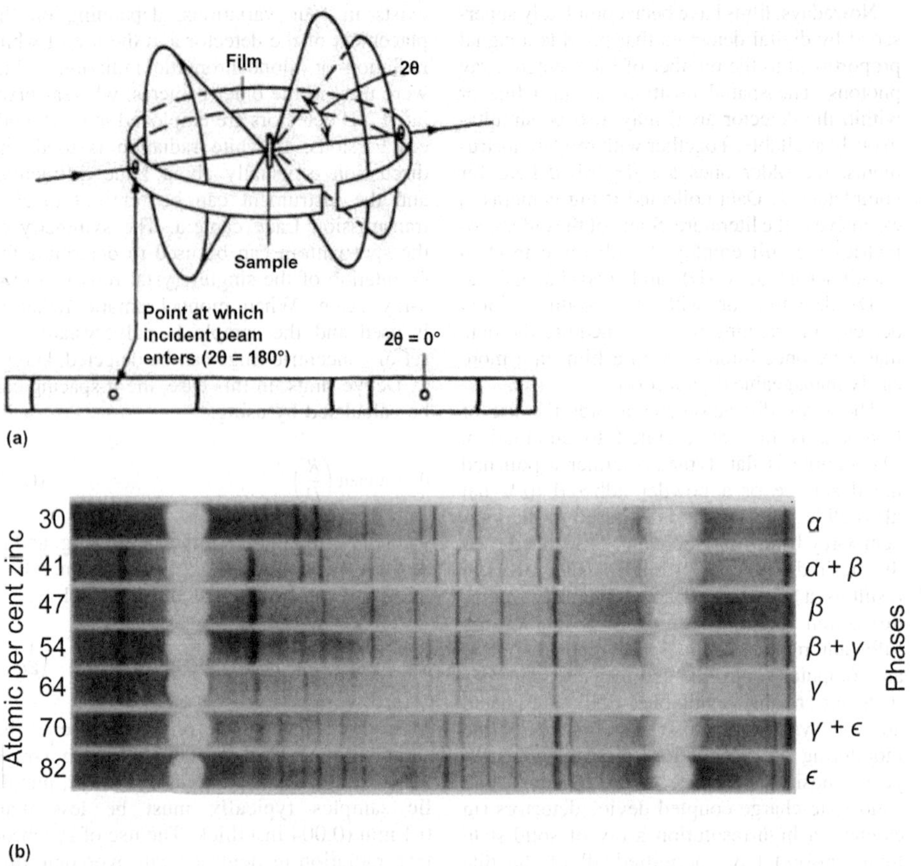

Fig. 4 Debye-Scherrer method of x-ray powder diffraction. (a) Geometric relationship of film to sample, incident beam, and diffracted beams and film when developed and laid flat. (b) Example of Debye-Scherrer films identifying phases in copper-zinc alloys of various compositions. Source: Ref 7

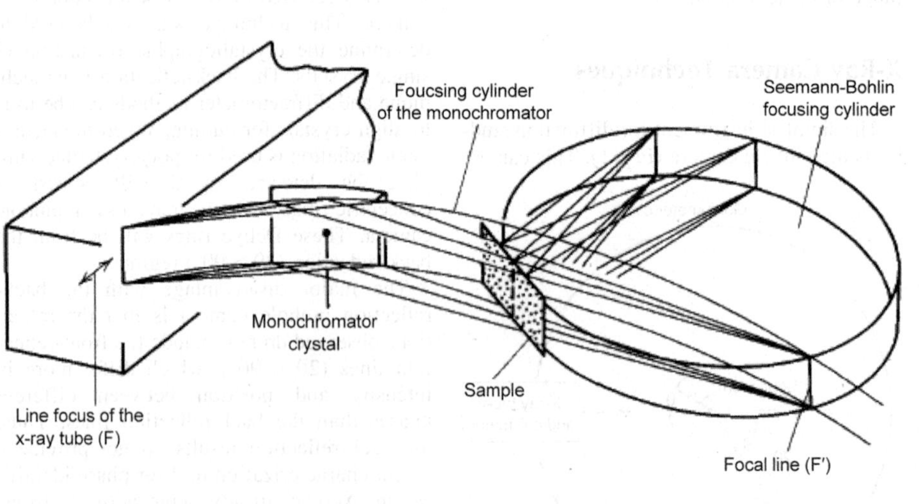

Fig. 5 Schematic of a Guinier camera in the asymmetric transmission arrangement

Microcameras

Back-reflection Laue, monochromatic pinhole, and Debye-Scherrer cameras can be equipped with very fine collimators of the order of 100 μm. These collimators are used when only a small amount of material is available or when a pattern is needed from a very small region of a sample. Diffraction patterns can be obtained with as little as 10 μg of sample. Back-reflection Laue cameras equipped with very fine collimators can often be used to obtain the orientation of small (0.5 mm, or 0.02 in.) grains in metallurgical samples. The micro-Laue camera is useful in the study of oxidation rates to determine, for example, why certain grains oxidize faster than others in the same sample or if an observed etching rate is dependent on crystallographic orientation. The microdiffractometer (described later) is a nonfilm device that was employed in the past to obtain powder patterns from very small samples and small areas on samples.

Microcameras have several disadvantages. The first is the alignment of the microcollimator of the camera to the x-ray beam, because the latter is no longer visible using a fluorescent screen, except when high-intensity sources are used. Second is the exceptionally long exposure time—one to two days using conventional tubes. Last is the positioning of the sample to the exact region of interest. For these reasons, the current applications of Laue microdiffraction are at large-scale facilities (e.g., synchrotron diffraction beamlines). A single Laue pattern contains information about the orientation of the crystal hit by the beam and about the internal stresses (macro and micro). By moving the specimen under the beam, one can effectively implement scanning Laue diffraction microscopy. The overall intensity of a Laue pattern can be employed to obtain a qualitative image of the specimen. Simultaneous chemical mapping can be obtained by employing a fluorescence detector (see Ref 8 for a review).The critical step in data analysis is still indexing of the Laue pattern. Additional information is available in the article "X-Ray Diffraction Residual-Stress Techniques" in this Division.

Glancing-Angle Camera

Several other camera geometries have been proposed in the past to overcome the limitations of the traditional Debye-Scherrer and Guinier geometries in terms of capabilities or types of specimen. Most of them employed an x-ray beam impinging the surface of the specimen at a glancing angle. The Read camera (Ref 9–11) is a Debye-Scherrer-type camera that uses a special sample holder and small collimators (Fig. 7). The sample holder allows flat samples, such as silicon wafers, to be mounted and the angle, γ, the sample makes to the entrance collimator to be set at any desired value. This camera was employed to survey deposited layers for crystallinity. If the film is polycrystalline and has preferred orientation, the observed diffraction lines have discontinuous rather than continuous arcs. The extent of these arcs indicates the degree of preferred orientation of the crystalline layer. As the texture becomes stronger, the arcs become sharper until a spot will be observed for a single-crystalline layer instead of an arc. The Read camera needed long measurement times and was rather inaccurate (difficult to set a zero, nonfocusing geometry, rather sensitive to the nanostructure).

The G.E.C. x-ray texture camera (Fig. 8) was another type of glancing-angle camera (Ref 12). The x-ray beam is incident along the axis of the cylinder. In addition, the Debye cones intercept the film such that the pattern appearing on the film is a series of straight lines.

The same type of measurement performed with those cameras is nowadays possible on stress-texture diffractometers, with the advantage of a better optical setup and a more accurate survey of the diffraction space.

Powder Diffractometers

Although film techniques have been in use since the inception of x-ray diffraction, the advent of powder diffractometers has been more recent. Powder diffractometers have been in use as a laboratory tool since the late 1950s and early 1960s with the development of x-ray detectors. The detectors are moved over different positions along the goniometer circle (Fig. 3) to obtain diffracted x-ray counts (intensity) as a function of diffraction angle (θ).

Automation of these devices began in the 1970s but became commonplace only in the 1980s. Initially, data were collected in analog form, with the results displayed and stored on chart recordings, then reduced manually. Automation of these instruments has enabled collection of digital data that can be easily processed by dedicated software. Powder diffractometers and some applications are discussed further in Ref 1 and 13 to15.

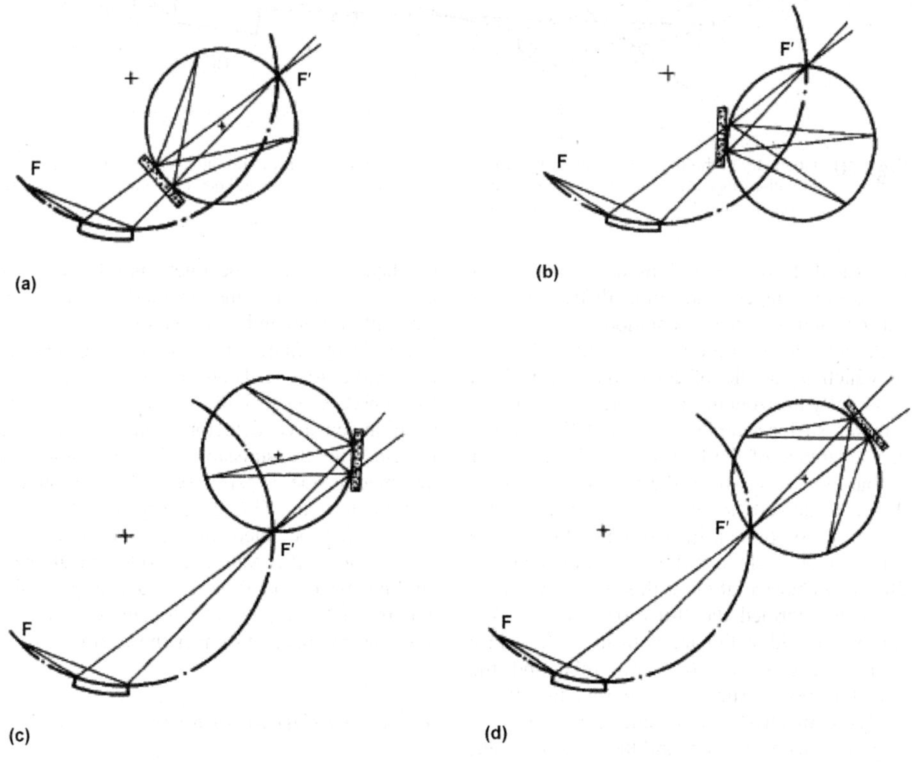

Fig. 6 Different arrangements of the Huber Guinier camera. (a) Symmetric transmission. (b) Asymmetric transmission. (c) Asymmetric back reflection. (d) Symmetric back reflection

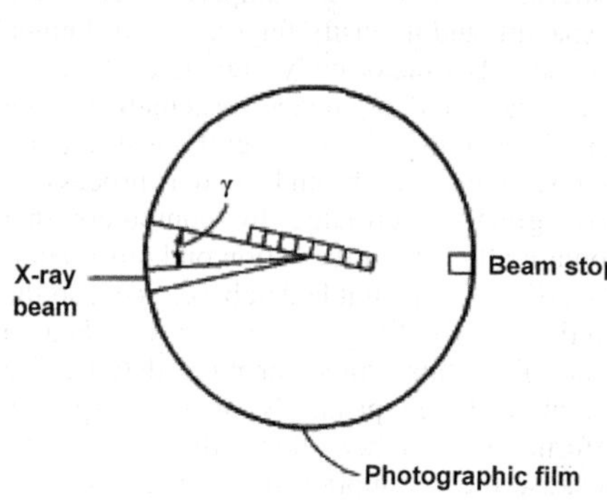

Fig. 7 Schematic of a Read camera

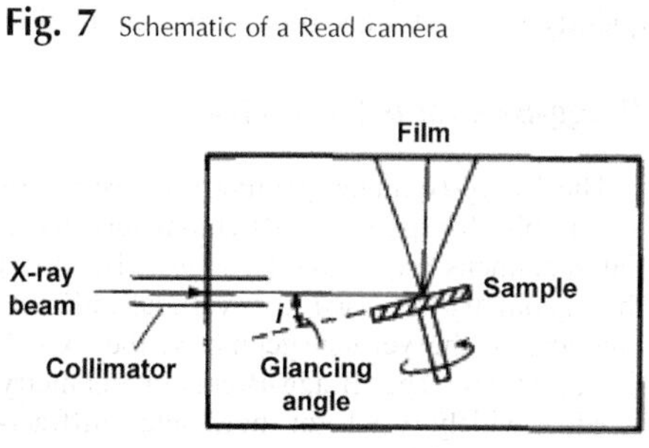

Fig. 8 Schematic of a G.E.C. x-ray texture camera

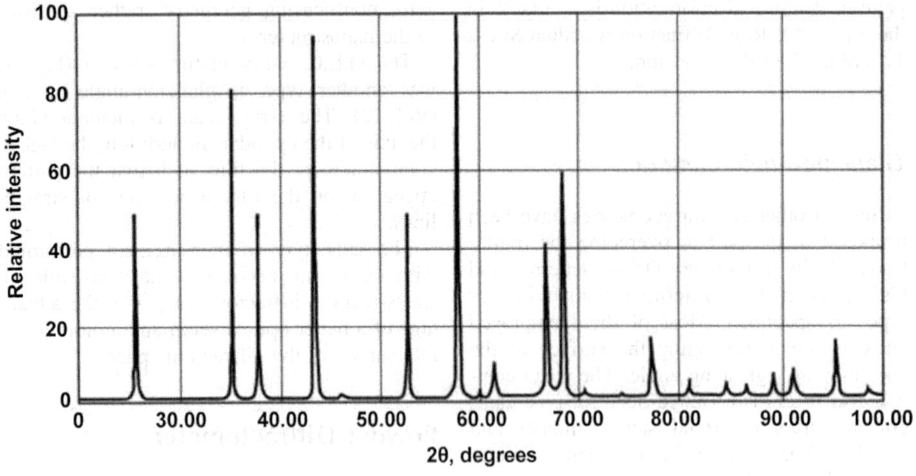

Fig. 9 X-ray powder pattern of Al_2O_3. Source: Ref 16

A diffractogram is shown in Fig. 9 for Al_2O_3. The plot of diffracted intensity versus diffraction angle exhibits a number of peaks, each corresponding to a particular set of crystallographic planes and its characteristic *d*-spacing. For the purpose of chemical analysis (identification of the compounds), the pattern can also be summarized as a table of *d*-spacings (each corresponding to a θ angle) and corresponding intensities, typically expressed as a percentage of the most intense peak, as shown in Table 1. For a sample made up of randomly oriented crystals of a given metal or compound, the relative intensities of the various diffraction peaks are predictable and reproducible within a few percent.

In solving a particular powder diffraction pattern, results are compared with the *d*-spacing and intensity fingerprints of known compounds—historically tabulated on an index card and organized systematically by the *d*-spacings of the several most intense peaks. This search and match process is now greatly facilitated by computers that contain all of the information on known compounds in an updatable database. See also the section "Qualitative Analysis (Search Match)" in this article for more details. The search is based primarily on the *d*-spacing information, rather than the intensities, because the assumption that the sample consists of randomly oriented crystals is frequently violated, thus altering the intensities.

Table 1 Identification of powder diffraction pattern from Al_2O_3 using the Hanawalt search method

The diffracting angles and intensities (area under each peak) are measured from the unknown sample (in this example from the diffractometer trace in Fig. 9). The *d*-spacings are then calculated for each diffraction peak using Bragg's law and the known wavelength of radiation used (in this case, Cu Kα, λ = 1.54178 Å).

2θ, degrees	Intensity, 0 to 100	*d*, Å	Miller indices
25.62	67	3.477	012
35.20	89	2.549	104
37.78	35	2.381	110
41.70	1	2.166	006
43.43	100	2.084	113
46.20	1	1.965	202
52.60	45	1.740	024
57.55	100	1.601	116
59.82	4	1.546	211
61.30	12	1.512	122, 018
66.60	40	1.404	214
68.28	53	1.374	300
70.40	2	1.337	125
74.30	2	1.277	208
77.15	22	1.236	1010, 119
80.80	8	1.189	220
84.50	6	1.146	223
86.50	8	1.125	312, 128
89.08	10	1.099	0210
91.00	12	1.081	0012, 134
95.34	18	1.043	226
98.50	1	1.018	042

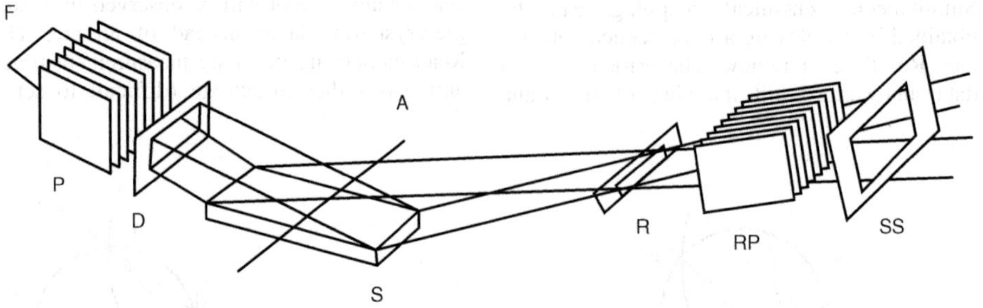

Fig. 10 Geometry of the Bragg-Brentano diffractometer. F, line source of x-rays from the anode of the x-ray tube; P, Soller slits (collimator); D, divergent slit; A, axis about which sample and detector rotate; S, sample; R, receiving slit; RP, receiving Soller slits; SS, scatter slit

Bragg-Brentano Geometry

The Bragg-Brentano geometry is used most commonly for powder diffractometers when flat specimens are available. Figure 10 shows this geometry as used for vertical diffractometers. In the vertical geometry, the axis A is horizontal. The Bragg-Brentano geometry is also widely used for horizontal diffractometers in which the apparatus shown in Fig. 10 is turned on its side such that axis A is vertical. Both orientations have advantages and disadvantages, and some diffractometers can be used in either orientation.

An advantage of the vertical unit is the way in which a sample of loose powder is held essentially horizontal; in the horizontal diffractometer, such a sample can fall out. The principal advantage of the horizontal diffractometer is that any weight from the sample is directed down on the vertical axis, allowing heavy samples to be supported. In both units, the sample rotates at an angle θ, and the detector rotates at 2θ. In another variation of this Bragg-Brentano geometry, termed the θ-θ diffractometer, the sample is held stationary with its surface in a horizontal plane as the x-ray tube and the detector move. Such a system is useful for studying materials in nonambient conditions (e.g., studies near or beyond the melting point) or to analyze liquids. Several types of sample holder and chambers, as well as optical elements and detectors are available to condition the beam and the specimen and to perform any sort of in situ and operando analyses at ambient and nonambient conditions.

The Debye-Scherrer geometry is a possible alternative to the Bragg-Brentano geometry for powders, especially when a curved 1-D position-sensitive detector or a 2-D detector is used in transmission. The specimen is mounted inside a capillary and spun on the goniometer axis. This geometry is particularly valuable for specimens that are radiation sensitive, because the pattern is collected at once and the specimen can be cooled with a cryojet and translated on its axis to always keep an undamaged specimen under the beam.

Thin-Film Diffractometers

In the Bragg-Brentano geometry, the sensitivity for the analysis of thin films is lost because the penetration of x-rays into the sample generally exceeds the thickness of the film. In conventional diffractometers, the effective depth of penetration of the x-rays varies as sin(θ)/μ. The linear mass-absorption coefficient, μ, is a function of the wavelength of the x-rays. Thus, at low Bragg angle, the penetration depth is small, and at a Bragg angle of 90°, it is at maximum. Generally, the longer the wavelength, the less penetration. Thus, a long wavelength radiation such as Cr Kα is used to increase sensitivity when thin films are studied. Synchrotron sources are ideal here, because the penetration can be finely tuned; by working near the absorption edges of the elements constituting the film, further tuning is possible.

If the surface of the sample is placed at a glancing angle of 5 to 10° to the x-ray beam, the penetration of the x-rays into the sample is decreased by an order of magnitude. This can be accomplished in modern conventional diffractometers having independent θ and 2θ stepping drives by setting the angle θ at a glancing angle and stepping 2θ over the desired range (Fig. 11). Because θ remains fixed, the x-rays no longer focus at the receiving slit, and the diffraction peaks broaden as 2θ increases. The diffractometer optics can be modified by removing the conventional slits and replacing them with a set of Soller collimators. This is usually referred to as parallel optics. The resolution of the instrument is determined by the distance between the Soller baffles and the length of the assembly. Coarse Soller slits are used for high intensity and low resolution, and fine Soller slits for high resolution and lower intensity.

The Seeman-Bohlin geometry was proposed in the past to examine polycrystalline thin films. Figure 12 illustrates the Seeman-Bohlin arrangement with a curved crystal monochromator. The principal advantage of this geometry is its ability to study thin polycrystalline films down to a few hundred angstroms in thickness. This type of diffractometer has been abandoned in favor of stress-texture diffractometers or high-resolution thin-film diffractometers, with more stringent optics (to limit the divergence of the beam) and synchrotron radiation facilities.

Microdiffractometers

The microdiffractometer enables examination of very small areas of samples. Collimators of 100, 30, and even 10 μm have been used with such an instrument. Its basic sample geometry is similar to that of a pinhole camera or a Debye-Scherrer camera (Fig. 4) in that a small beam of x-rays is impinged on what is ideally a random assemblage of very small grains, producing cones of diffracted radiation. The principal differences between the cameras and the microdiffractometers are in the nature of the detector system, which, for the microdiffractometer, consists of an annular detector that intercepts the entire cone of radiation. The motion of this annular detector toward and away from the sample allows the interception of individual cones of diffracted radiation by the annular detector at different distances from the sample. This interception of the entire cone of radiation and the small x-ray spot size contrast the microdiffractometer with the conventional diffractometer, in which the detector samples only a small section of a diffraction cone.

Figure 13 shows the essential geometry of the system, which operates in a transmission or reflection mode. This device is effective for phase characterization of small areas and has been used to determine variations in residual stress over very small areas in microdevices. As with any of the microtechniques, care must be taken to ensure that the crystallite size of the material to be examined is small enough that meaningful statistics can be collected from the crystallites in the x-ray beam.

Microdiffractometers have been almost completely replaced by synchrotron radiation beamlines and instruments with microfocus optics and 2-D detectors that can examine all the diffraction cones simultaneously.

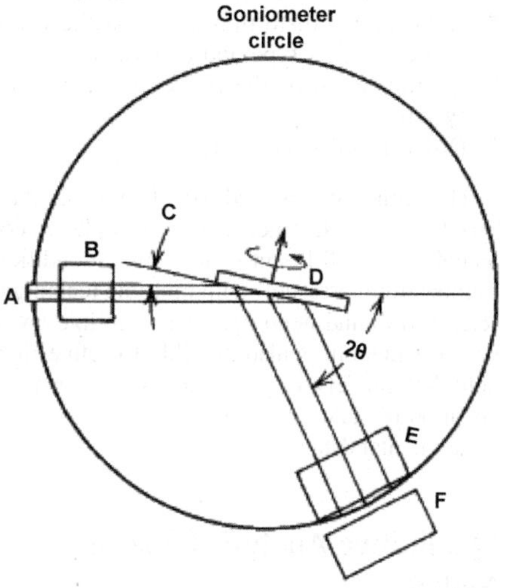

Fig. 11 Schematic of a thin-film diffractometer. A, line source of x-rays; B, axial divergence of Soller slit; C, glancing angle; D, sample; E, equatorial divergence Soller slit; F, detector

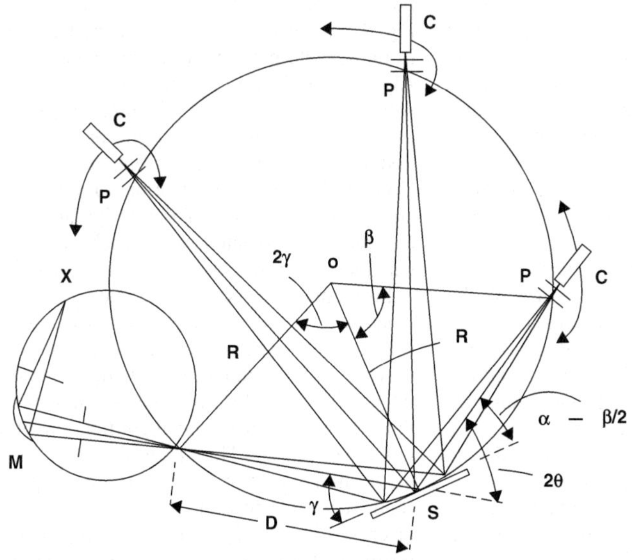

Fig. 12 Geometry of the Seeman-Bohlin arrangement. X, x-ray line source; M, curved monochromator crystal; S, sample; R, radius of focusing circle; D, focus-to-sample distance; P, receiving slits; C, counter

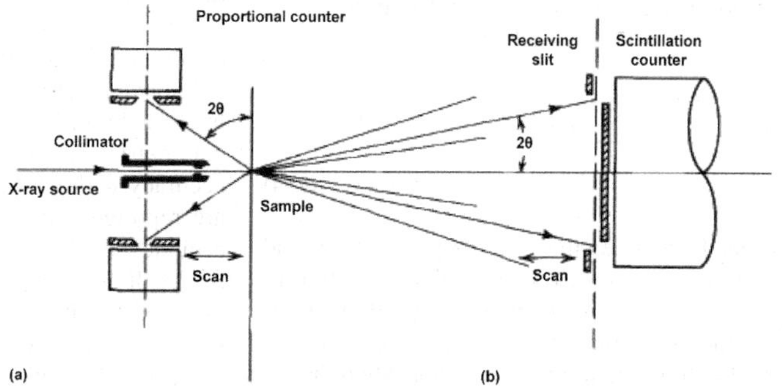

Fig. 13 Geometry of the microdiffractometer. (a) Reflection arrangement. (b) Transmission arrangement

Stress-Texture and High-Resolution Diffractometer

The traditional powder diffractometer, independent of the actual specimen geometry or optics, allows only for rotations of the specimen around the goniometer axis. To measure the residual stress or the preferred orientation or to work at glancing incidence, more degrees of freedom are necessary. The stress-texture diffractometer adds extra degrees of freedom to the conventional setup by introducing additional circles for rotation and a specimen translation stage. Four circle diffractometers are common. The optics can be tuned to deliver a quasiparallel small beam on the specimen, to minimize the aberrations related to the specimen surface and orientation with respect to both the incoming beam and the detector. This is particularly useful for the analysis of residual stresses. The extra degrees of freedom are necessary to access those grains that are not properly oriented with respect to the specimen normal, which is the only direction surveyed in a traditional setup. Additional information is available in the article "X-Ray Diffraction Residual-Stress Techniques" in this Division.

By using high-resolution optics (narrow bandwidth), accurate mapping of the diffraction peaks is possible for texture studies and to understand the epitaxy and growth of thin films. With the aid of high-performance detectors, x-ray reflectivity studies can also be performed for accurate film microstructure, thickness, and roughness analysis.

Rietveld Method of Diffraction Analysis

Most of the quantitative capabilities of diffraction rely nowadays on the use of the Rietveld method (Ref 1, 17, 18) for data analysis. In his 1969 seminal paper (Ref 17), H.M. Rietveld proposed to analyze the diffraction data by generating a pattern from materials structure and instrument characteristics and then to refine the model parameters to obtain the best match between data and model. As shown in Eq 4, the intensity, y_{ci}, at each point i in the pattern is the sum of the contributions from all the peaks plus a background term, y_{bi}:

$$y_{ci} = L \sum_{j}^{N\text{ phases}} f_j \sum_{k}^{N\text{ peaks}} S_j |F_{k,j}|^2 \phi(2\theta_i - 2\theta_{k,j}) P_{k,j} A_j + y_{bi} \quad \text{(Eq 4)}$$

where L includes the Lorentz polarization and multiplicity factors, f_j is the volume fraction of phase j, $|F_{k,j}|^2$ is the square modulus of the structure factor for the k-th peak of phase j (see the article "Introduction to Diffraction Methods" in this Division), ϕ is the peak function, $P_{k,j}$ is the correction for preferred orientation (texture), and A_j is the absorption factor for phase j. A nonlinear least-squares routine is employed to minimize the weighted sum of squares of the difference between data and model.

Initially proposed for structure (i.e., atomic position) refinement, the Rietveld method has grown as a general procedure to model the whole diffraction pattern and extract most of the information contained in peak position, intensity, and breadth/shape, including:

- Lattice, symmetry, positions of the atoms in the cell, substitution of atoms, ease of movement (vibration properties)
- Preferred orientation (texture)
- Residual macrostress
- Shape and size distribution of the coherently scattering domains (crystallites)
- Local lattice distortion and possible origin (lattice defects type and quantity)
- Quantification of the phases present in the system
- Order/disorder parameters

The simultaneous analysis of multiple datasets (surface refinement), for example as collected at different ambient conditions (temperature, pressure) with multiple techniques (e.g., x-rays and neutrons) or at multiple specimen orientation, is also possible to refine more complex models (e.g., of thermal expansion, equation of state, phase transformations) on the experimental data.

Qualitative Analysis (Search Match)

XRPD can provide general-purpose qualitative and quantitative information regarding the crystalline phases present in an unknown mixture. Other techniques, such as infrared or Raman spectroscopy, differential thermal analysis (see the article "Differential Scanning Calorimetry" in this Volume), and extended x-ray absorption fine structure yield some information on specific phases or nearest-neighbor atoms from phases in a mixture. However, these methods cannot provide general-purpose, routine identification of phases. See also the articles "Raman Spectroscopy" and "Extended X-Ray Absorption Fine Structure" in this Volume.

The identification of phases using powder techniques is based on a comparison of the unknown pattern with known ones collected in a database. In the ICDD PDF database, the data entries come from many sources, and the accuracy with which d-spacings and intensities are reported can have a large variability. The accuracy of the experimentally collected data that will be compared with the file must be assessed; limits of error within which a given spacing will be considered a match between standard and unknown must then be determined. While the resolution of the cameras was highly dependent on the geometry, modern diffractometers give an accuracy better than 0.001 Å over the whole accessible range. Matching is usually done in 2θ considering a possible variation range (that accounts for compositional variations, possible accuracy issues in the database entry, and aberrations in specimen mounting) of one measurement step (0.01 to 0.05 Å).

Manual and computer methods are the two principal ways the file can be used to identify crystalline phases from their powder patterns. The identification of phases from x-ray powder diffraction was pioneered by Hanawalt (Ref 19, 20) using a manual search technique. For the manual technique, the *Hanawalt Manual*, updated yearly by the ICDD and now discontinued, contained standard phases from the ICDD PDF file, along with their eight most intense d-spacings and intensities. For example, a set of 25 compounds with similar intense peaks is shown in Table 2, ordered by the d-spacings of their most intense peaks. This table allows tentative identification of the sample. For example, note that the eight most intense peaks for Al_2O_3 correspond to intense peaks from the sample pattern given in Fig. 9 and Table 1.

A card containing all of the information for the tentatively identified compound is compared with the diffraction information from the sample. The search manual was arranged into groups and subgroups based on pairs of lines from the standard patterns that are present as multiple entries based on different pairs of lines. For a single-phase unknown, the d-value of the strongest line on the pattern was used to determine which group is to be consulted in the manual. The d-value of the second strongest line in the unknown determined in which subgroup to search for one or more patterns matching the other intense lines on the unknown pattern. Any lines from the unknown not matching lines of the standard may indicate the presence of other phases. For example, the card for Al_2O_3 is shown in Fig. 14. Note that all of the diffraction peaks in the sample pattern can be accounted for by Al_2O_3. If unidentified lines were present, it would indicate either that the tentative identification was incorrect or that one or more additional compounds were present in the sample along with Al_2O_3. The information on the card enables identification of the Miller indices of the planes associated with each diffraction peak, shown in the last column of Table 2.

Problems can occur if the standard for the unknown pattern is not present in the file. In this case, a pattern similar to the unknown may possibly be found that has the same crystal structure as the unknown compound and thus can aid in the final identification of the unknown phase. This identification can be made more certain if the standard pattern found can be shown to have some chemical or crystal-chemical relationship with the composition of the unknown phase. A search of the file need not use any chemical information, although it is useful in making a final selection, especially if the data in the file or the data

Table 2 Example of *d*-spacings for Al_2O_3 and other compounds with similar intense peaks for tentative identification of sample in Hanawalt search method

Note that the eight most intense peaks for Al_2O_3 correspond to intense peaks from the sample pattern given in Table 1.

2.12_g	2.55_g	4.89_g	1.50_g	1.63_g	1.10_x	1.43_x	2.99_9	cF56	$Li_{0.75}Mn_{0.25}Ti_2O_4$	40–406
2.12_8	2.55_x	1.98_x	1.27_x	1.24_8	3.15_5	1.34_5	1.19_5	tP10	FeW_2B_2	21–437
2.11_x	2.55_5	2.79_3	1.37_1	1.09_1	1.98_1	1.51_1	1.16_1	cF*	$Ce_{0.78}Cu_{8.76}In_{3.88}$	43–1269
2.11_8	2.55_x	2.44_x	2.29_7	1.50_7	1.34_7	7.31_5	3.20_5	hP22	K_6MgO_4	27–410
2.10_x	2.55_7	2.61_4	1.45_4	1.29_4	1.80_3	3.88_2	2.49_2	tI10	Pd_2PrSi_2	32–721
2.10_x	2.55_x	2.43_5	1.39_2	0.85_2	3.71_1	1.50_1	1.17_1	hP8	$PmCl_3$	33–1085
2.09_8	2.55_x	6.31_8	1.68_8	3.16_6	2.79_6	1.61_6	1.56_6	···	$K_{0.72}In_{0.72}Sn_{0.28}O_2$	34–711
2.09_8	2.55_x	2.63_8	1.65_8	1.79_7	2.66_6	1.88_6	3.05_5	mC18.80	$IrB_{1.35}$	17–371
2.09_x	2.55_9	1.60_8	3.48_8	1.37_5	1.74_5	2.38_4	1.40_3	hR10	Al_2O_3	10–173
2.09_x	2.55_x	1.60_x	3.48_7	1.37_6	1.74_5	2.38_4	1.40_4	hR10	Al_2O_3	43–1484
2.08_x	2.55_8	3.22_8	1.57_6	2.00_4	1.61_4	1.75_3	2.40_2	hP6	$EuAl_2EuSi_2$	45–1237
2.08_x	2.55_8	2.16_8	1.18_x	1.17_x	2.02_8	1.16_8	2.33_6	oC20	Cr_2VC_2	19–334
2.08_8	2.55_x	2.14_4	1.23_x	1.32_8	1.17_8	1.30_6	1.64_4	oC8	HfPt	19–537
2.16_x	2.54_6	2.74_5	2.19_4	2.51_4	1.38_2	1.27_2	1.50_1	hR12	$ErFe_3$	43–1373
2.16_x	2.54_x	2.33_x	2.12_x	1.42_9	1.54_8	1.38_8	1.32_8	hP12	Cr_2Hf	15–92
2.16_x	2.54_7	1.38_2	4.14_2	1.46_2	1.27_2	2.07_1	0.93_1	cF24	Co_2Ho	29–481
2.16_x	2.54_6	1.38_2	2.07_2	1.27_2	1.46_2	4.14_1	0.93_1	cF24	$TbNi_2$	38–1472
2.14_x	2.54_8	2.35_8	1.36_7	2.16_6	0.85_6	1.86_6	0.88_5	oP8	TiB	5–700
2.12_x	2.54_x	2.33_x	2.16_x	1.42_9	1.54_8	1.38_8	1.32_8	hP12	Cr_2Hf	15–92
2.11_x	2.54_7	1.49_6	4.87_6	0.86_3	1.62_2	0.94_1	1.22_1	cF*	$AlVO_3$	25–27
2.10_x	2.54_7	2.07_6	1.16_8	1.08_8	1.36_7	1.09_7	1.07_7	cF112	$RbZn_{13}$	27–566
2.10_8	2.54_x	1.57_x	1.21_x	1.72_8	1.68_8	1.51_8	1.45_8	tP5	LuB_2C_2	27–301
2.10_x	2.54_x	1.49_7	4.86_6	1.62_4	2.97_2	2.43_2	1.40_1	cF*	$Mg_{1.5}VO_4$	19–778
2.09_9	2.54_x	3.49_8	6.39_7	3.69_5	2.76_4	2.13_4	1.38_3	hP8	$SmCl_3$	12–789
2.09_x	2.54_x	2.59_8	1.28_8	3.85_5	2.46_5	1.63_5	1.60_5	tI10	$LaPd_2P_2$	37–994

Al O
 2 3

Aluminum Oxide

Corundum, syn

Rad: CuKa1 Lambda: 1.5405 Filter: Ni d-sp:
Cutoff: Int: Diffractometer I/Icor: 1.00
Ref: Natl. Bur. Stand. (U.S.), Circ. 539, 9 3 (1960)

Sys: Rhombohedral (Hex) S.G.: R-3c (167)
a: 4.758 b: c: 12.991 A: C: 2.7303
A: B: C: Z: 6 mp: 2050 deg.
Ref: Ibid.

Dx: 3.99 Dm: 4.05 SS/FOM: F30=50(.019,32)

ea: 1.7604 nwB: 1.7686 ey: Sign: - 2V:
Ref: Dana's System of Mineralogy, 7th Ed., 1 520

Color: Blue, colorless, yellow purple to violet, green, pink to deep pigeon-blood red
Pattern made at 26 C. Sample annealed at 1400 C for four hours in an Al2O3 crucible. Spectroscopic analysis showed <0.1% K, Na, Si; <0.01% Ca, Cu, Fe, Mg, Pb; <0.001% B, Cr, Li, Mn, Ni. Also called: ruby. Also called: sapphire. Al2O3 type. Corundum group, corundum subgroup. Also called: alumina. Also called: alundum. Also called: diamonite. PSC: hR10. Mwt: 101.96. Volume[CD]: 254.70.

d A	Int.	h k l
3.479	75	0 1 2
2.552	90	1 0 4
2.379	40	1 1 0
2.165	<1	0 0 6
2.085	100	1 1 3
1.964	2	2 0 2
1.740	45	0 2 4
1.601	80	1 1 6
1.546	4	2 1 1
1.514	6	1 2 2
1.510	8	0 1 8
1.404	30	2 1 4
1.374	50	3 0 0
1.337	2	1 2 5
1.276	4	2 0 8
1.239	16	1 0 10
1.2343	8	1 1 9
1.1898	8	2 2 0
1.1600	<1	3 0 6
1.1470	6	2 2 3
1.1382	2	1 3 1
1.1255	6	3 1 2
1.1246	4	1 2 8
1.0988	8	0 2 10
1.0831	4	0 0 12

d A	Int.	h k l	d A	Int.	h k l	d A	Int.	h k l
1.0781	8	1 3 4	0.9076	14	3 2 4	0.8137	4	1 1 15
1.0426	14	2 2 6	0.9052	4	0 1 14	0.8072	11	4 0 10
1.0175	2	0 4 2	0.8991	8	4 1 0	0.7988	7	0 5 4
0.9976	12	2 1 10	0.8884	<1	2 3 5	0.7970	14	1 0 16
0.9857	<1	1 1 12	0.8804	4	4 1 3	0.7931	13	3 3 0
0.9819	4	4 0 4	0.8698	2	0 4 8			
0.9431	<1	3 2 1	0.8580	12	1 3 10			
0.9413	<1	1 2 11	0.8502	4	3 0 12			
0.9345	4	3 1 8	0.8460	4	2 0 14			
0.9178	4	2 2 9	0.8303	22	1 4 6			

Fig. 14 Example of card in Hanawalt search method for compound tentatively identified as Al_2O_3. The information on the card enables identification of the Miller indices of the planes associated with each diffraction peak, shown in the last column of Table 2.

collected for the unknown could not be obtained with the highest possible accuracy.

A mixture of several patterns can complicate identification, because it is unclear whether the strongest lines in an unknown pattern represent the same unknown phase. The probability is high that the strongest lines from a multiphase unknown do not come from the same phase. Therefore, a search of the Hanawalt file for such an unknown must consider several permutations and combinations of the strongest unknown lines. As the nature of the unknown becomes increasingly complex (for example, when more than four phases are present in a mixture), the use of manual methods for the identification can become tedious and subject to error.

Computer-based search/match programs have nowadays superseded the manual search. The major powder diffractometer manufacturers provide software packages with their automated powder diffraction systems. Some of the database vendors and third-party free and commercial software also offer similar capabilities.

In general, the computer approach to the search/match problem differs from manual methods. The manual search needs to compare lines from the unknown pattern in various permutations and combinations with the standard file; the most common computer technique involves the opposite, that is, comparing the standards with the unknown. The computer asks whether a given standard could be a subset of the unknown; that is, could the standard being compared with the unknown be one phase comprising that unknown. If so, that standard pattern is ranked according to how well it matches the unknown, and another standard is compared with the unknown until all standards have been compared. Use of chemical information to aid in this type of search sacrifices the ability to locate structurally related compounds having chemistry different from the unknown. A collection of the best matches with the unknown is usually saved and displayed at the end of a search. The program or the investigator then selects one or more of the phases, examines them in greater detail, subtracts one or more standard patterns from the unknown, and then runs the residue through the searching process again.

Although computer search/match programs have been successful, they cannot fully replace a competent analyst who must still make the critical decisions. Nevertheless, such computer programs have proven valuable and expeditious and can take advantage of the multiple fields present in the diffraction database, such as chemistry, color, hardness, unit cell edges, and density, that may help make an identification even more certain. The use of subfiles, grouping the phases in broad families (organics, inorganics, ceramics, zeolites, pigments, bioactive, etc.), is also valuable to restrict the search range and limit the number of candidate phases.

Quantitative Phase Analysis

X-ray diffraction quantitative analysis of phase mixtures has been in use since 1925, when it was implemented to determine the amount of mullite in a fired ceramic (Ref 21). X-ray diffraction is useful for bulk analysis on chemically similar phases, because it is one of the few techniques that can identify and quantify crystalline polymorphs with good reliability and accuracy (e.g., see quantitative phase analysis round robin, Ref 22 and 23, and the outcomes of the Reynolds Cup contest in quantitative analysis on minerals, Ref 24). For example, diffraction can quantify the amounts of rutile and anatase in a paint pigment mixture. These two minerals are TiO_2 polymorphs; that is, they have the same chemistry but different crystal structures. For other analyses in which the information desired is elemental or the phases are elementally distinct, x-ray fluorescence spectrometry can provide the necessary quantification more easily and often more accurately (Ref 25).

With the addition of automatic sample changers, the analysis can proceed unattended once the samples have been loaded. Instruments coupled with fast analysis software exist for online quality control (e.g., chemical or cement plants).

Except for the lattice-parameter method, x-ray diffraction quantitative phase analysis is based on the premise that each crystalline material in the sample has a unique diffraction pattern and that the intensity of the peaks in that pattern varies directly with its concentration. Many factors prevent the direct comparison of concentration with peak intensity. The basic factor is the different x-ray absorption properties of the substances in the sample (Ref 26). The pioneering work of Chung (Ref 27, 28) and the more recent extensive application of the Rietveld method have increased the reliability of and simplified the methodologies for phase analysis with and without the use of standards for an absolute quantification.

Lattice-Parameter Method

The lattice-parameter method of quantitative analysis is applicable to those compounds and metals that form continuous solid solutions, often from one pure end member to the other (Ref 29). This technique is an accurate method of determining the chemical compositions of many types of materials by measuring accurate lattice parameters, particularly if the lattice constant varies linearly between the end members. It is especially useful when x-ray spectrometry cannot be easily used, for example, when determining the amount of oxygen in AlN, or when the composition of a mullite is needed in the presence of quartz or corundum (Ref 25). Unlike typical quantitative diffraction techniques, the lattice-parameter method is used to determine the composition of a single phase, not the amounts of the different phases present. Use of the technique is limited, because lattice-parameter data versus composition are not readily available and, in some cases, are of dubious quality. Accuracy depends on the slope of the calibration curve and the precision of the lattice-parameter measurement. Several data series useful for this type of analysis can be extracted from the ICDD PDF database.

Absorption Diffraction Method

The absorption diffraction method requires the measurement of the intensity from a diffraction peak in the mixture and from a pure standard of the material:

$$\frac{I}{I_{\text{pure}}} = \frac{(\mu/\rho)}{(\mu/\rho)_m} X \qquad \text{(Eq 5)}$$

where I_{pure} is the intensity of a peak from the pure phase, I is the intensity of the same peak of the phase in a mixture, X is the weight fraction of the phase in the mixture, (μ/ρ) is the mass-absorption coefficient of the phase, and $(\mu/\rho)_m$ is the mass-absorption coefficient of the entire sample. In general, the mass-absorption coefficient of the sample and the phase under analysis must be known. If the absorption coefficient of the sample is almost the same as each of its components, the ratio of intensity to the pure standards depends on the concentration of the phase in the sample. The accuracy of this technique depends strongly on consistent sample-preparation procedures and operating conditions and on appropriate pure standards, which can be difficult to obtain.

Spiking Method

The spiking method, sometimes referred to as the method of standard additions or the doping method (Ref 30), is applicable only to powder samples. It consists of measuring the diffraction-peak intensity from the phase of interest, then adding a small amount of this phase to the sample and remeasuring the intensities. The concentration C_0 of the phase of interest is then given by:

$$C_0 = \frac{I_1 C_1}{I_2 - I_1} \qquad \text{(Eq 6)}$$

where I_1 is the intensity of a diffraction line from the sample, I_2 is the intensity of the same line after the sample has been spiked, and C_1 is the amount of phase added to spike the sample. This procedure is useful when only one phase is to be quantified. It assumes a linear change in intensity with concentration; that is, the absorption coefficient of the spiked material is the same as that of the sample. If only small amounts are spiked, this assumption is generally valid. The accuracy of this technique depends on the difference in intensity between the spiked and unspiked sample, the mixing procedure, and the sample-preparation techniques. Pure phases of the material of interest are required.

Internal Standard Method

The internal standard method is one of the most accurate and widely used procedures for quantifying phase mixtures in powder samples. The technique involves adding a fixed amount of a known standard to a sample, then measuring the intensities of the phases of interest relative to the internal standard intensities:

$$\frac{I}{I_s} = KX \qquad \text{(Eq 7)}$$

where I is the intensity of the phase to be analyzed in the sample, I_s is the intensity of the internal standard, and K is the slope of the calibration curve of the pure phase material. The National Institute for Standards and Technology has produced some standard powders for the purpose of accurate quantification. If the diffraction peaks of the sample phases and the internal standard peak are close in 2θ, the sample-preparation problems are less severe. The problem of obtaining a homogeneous mixture of the internal standard and the material to be analyzed remains (Ref 29).

External Standard Method

The external standard method (Ref 27, 28, 31, 32) is similar to the internal standard method. In this case, the intensity ratio is measured against phases already contained in the sample:

$$\frac{X_a}{X_b} = K\frac{I_a}{I_b} \qquad \text{(Eq 8)}$$

where X_a is the weight fraction of phase a, X_b is the weight fraction of phase b, I_a is the intensity from a peak of phase a, I_b is the intensity from a peak of phase b, and K is a constant ($K = (I_b/I_a)$ for a 1:1 mixture). Calibration curves of single mixtures are used to obtain the K constant. If the appropriate ratios are measured on all the phases in the material, a complete quantification can be conducted. In general, this method is as accurate as the internal standard method and depends on measured intensities and the accuracy of K. Pure standards are needed to determine K.

Reference Intensity Ratio Method

The reference intensity ratio (RIR) method, also known as the I/I_c method (Ref 33, 34), provides the ability to perform rapid semiquantitative phase analysis without pure standards. The weight fraction X_a of phase a is calculated by using:

$$X_a = \frac{(I_a/I_c)_{\text{unk}}}{(I_a/I_c)_{\text{ref}}} \qquad \text{(Eq 9)}$$

where X_a is the weight fraction of phase a in the sample, $(I_a/I_c)_{\text{unk}}$ is the intensity of the 100% peak of phase a divided by the intensity of the 100% peak of corundum in a 1:1 mixture of sample and corundum, and $(I_a/I_c)_{\text{ref}}$ is the RIR. The RIR values for several phases can be found in the ICDD PDF database. Some of them, especially the older ones, may be subject to some error, because they are calculated based on peak intensities and not on peak areas. Alternatively, they can be determined experimentally or calculated by using the Rietveld method when the structure of the compounds is known. Because only single lines are used, internal checks cannot be conducted.

Direct Comparison Method

The direct comparison method was used in the past mainly for retained austenite quantification (Ref 35, 36). It can have more general utility for several materials, particularly those for which pure phase standards are difficult to obtain. Direct comparison is based on the ratio of intensities of phases in the sample:

$$\frac{I_a}{I_b} = \frac{X_a R_a}{X_b R_b} \quad \text{(Eq 10)}$$

where I_a and I_b are the intensities of peaks from phases a and b, and X_a and X_b are weight fractions of phases a and b in the sample; R_a and R_b are constants of phase a and phase b calculated by using crystal structure information and the fundamental intensity equation. The accuracy of this technique depends on the calculation of R-values and on the accuracy of the crystal structure information.

Quantitative Phase Analysis Using the Rietveld Method

In all presented methods, determination of peak intensities and consistency among the values (when using different peaks for the analysis) are the critical issues. A higher accuracy can be obtained when the Rietveld method is employed. All the phases are first identified. The diffraction pattern for the mixture is generated from suitable structure models for the phases (available in the databases). The scale factors from the various phases, refined to obtain the best agreement from models and data, are employed for the quantification (Ref 37, 38). The density for phases and mixtures is calculated in the process based on the refined parameters. The quantification is then standardless and based on a weighted use of the scale factors:

$$X_a = \frac{S_a(ZMV)_a}{\sum_j S_j(ZMV)_j} \quad \text{(Eq 11)}$$

where Z is the number of formula units in the unit cell, M is the molecular mass of the formula unit, and V is the unit cell volume. The aggregated quantity ZMV is employed for each phase. The resulting fractions refer to the set of phases modeled in the pattern and are therefore absolute only when all phases have been identified and considered in the analysis.

Partial or No Known Crystal Structure Method

The partial or no known crystal structure (PONKCS) method has the potential of being able to quantify phase mixtures for which no standards exist and when the complete crystal structure is not known (Ref 39). The older standardless techniques (Ref 40, 41) required extensive sample handling to produce different mountings of the sample in which the weight fractions of the phases in the original mixture are significantly modified. The PONKCS method is instead based on the use of the Rietveld method and on the availability of the unknown phase in pure form. In Eq 12, the *ZMV* term multiplying the scale factor is a constant that depends on the characteristics of each single phase. The value can be calculated if the structure is known, or it can be obtained experimentally for a given unknown phase by considering the corresponding pattern as a set of peaks with fixed positions, relative shape, and relative intensities and preparing a mixture with a standard. The mixture of the unknown phase a and the standard s yields:

$$\frac{X_a}{X_s} = \frac{S_a(ZMV)_a}{S_s(ZMV)_s} \Rightarrow (ZMV)_a = \frac{X_a S_s}{X_s S_a}(ZMV)_s \quad \text{(Eq 12)}$$

where X_s is the (known) weight fraction of the standard, and $(ZMV)_s$ is the (known) constant for the standard. The procedure is valid independently of the nature of phase a, meaning that any unknown crystalline and amorphous phase can be handled. The factor can then be employed in the Rietveld method in place of the structural data for the unknown phase.

Quantification of an Amorphous Fraction

The Rietveld method can be employed for the quantification of the amorphous content in a mixture. An internal standard is added to the unknown mixture containing an amorphous component. The fractions of the crystalline phases obtained from the Rietveld analysis are then rescaled such that the quantity of the standard corresponds to the expected one. The weight content of the amorphous phase is:

$$X_{\text{amorphous}} = \frac{1}{1 - X_s}\left(1 - \frac{X_s}{X_{s,m}}\right) \quad \text{(Eq 13)}$$

where X_s is the true weight fraction of the standard in the specimen, and $X_{s,m}$ is the measured (apparent) weight fraction of the standard in the specimen, as calculated by the Rietveld method. The total fraction of the amorphous phase is obtained. Multiple amorphous phases simultaneously present in the mixture can hardly be quantified independently. A possible alternative is to employ the PONKCS method if the *ZMV* factor is known for the various amorphous phases present in the specimen.

In fortunate cases, the amorphous phase can be approximately modeled as a crystalline structure with a very small domain size, that is, very broad peaks (Ref 42). This effectively simulates a material possessing only short-range order, such as a glass.

Sensitivity, Precision, and Accuracy

Typical sensitivity and experimental limits on precision of XRPD analysis are:

- *Threshold sensitivity:* A phase or compound must typically represent 1 to 2% of the sample to be detected.
- *Precision of interplanar spacing and lattice-parameter measurements:* ~0.3% relative in routine measurements; within ~0.003% relative in experiments optimized for this purpose
- *Precision of quantitative analyses of percentage of individual phases or compounds present in samples containing multiple compounds:* 5 to 10% relative or 1 to 2% absolute, whichever is greater (presumes the use of calibrated standards)

Other systematic sources of errors that affect accuracy may include:

- Crystal-particle statistics
- Preferred orientation within the sample
- Instrumentation
- Peak broadening. Diffraction peaks can be broadened by samples having phases with very small crystallite sizes, local lattice distortions, disorder, stacking faults, dislocations, and inhomogeneous solid solutions. If integrated intensity is measured, the errors caused by these effects are minimized.
- Matrix absorption effects, which cause nonlinear change in intensity versus concentration when the matrix consists of phases having different absorption coefficients. Except when analyzing polymorphs, the calibration curves will generally not be linear.
- Compositional differences between standards and samples. Cation substitutions and solid solution occur in many types of materials. The compositional changes alter the intensity and position of the peak. Laboratory standards are often not representative of similar phases found in naturally occurring or manufactured materials.
- Primary extinction. The intensity from perfect crystals is less than that for imperfect crystals. This effect is known as primary extinction. Quartz is notorious for demonstrating primary extinction. This effect is usually related to particle size and increases with size.
- Microabsorption and particle absorption. If the absorption coefficients of the phases in

a sample are large, the highly absorbing phase will tend to mask the low-absorbing phase. The problem is alleviated by reducing the particle size.

- Sample preparation. Different sample-preparation techniques can result in intensity changes due to such factors as the packing density of powder into the holder, the size and shape of the holder, and the thickness of the sample.

Crystal-Particle Statistics

Crystal-particle statistics can be a major source of error. Favorable statistics require on the order of 6×10^8 crystallites (Ref 33). Therefore, a particle size of approximately 5 μm is necessary for a diffractometer sample. Rotating the sample in the diffractometer will bring more crystallites into the beam and provide improved intensities. One method of showing whether particle statistics are a problem is to conduct intensity measurements at different rotational placements of the sample. The intensity should remain constant as the sample is rotated. A second method is to rotate θ about its nominal value; there should be no sharp intensity changes when rotating over a range of a degree or two. A third method entails locating changes in relative intensity with different sample mountings.

Preferred Orientation within Sample

Preferred orientation within the sample can highly affect the accuracy of the quantitative phase analysis. If only one peak from each phase is being used for the analysis, preferred orientation of the peak being measured can cause widely different answers. Preferred orientation is caused in a powder sample by the morphology of the particles constituting the sample. For example, if the particles have a platelike or needlelike habit, the crystals in the powder will tend to lie flat when a sample holder is packed from the top or from the bottom. This tendency toward preferred orientation is a function of the amount, size, and shape of the other phases in the mixture and the packing pressure. If the degree of orientation can be controlled, calibration curves will yield favorable results. This approach is used to analyze clay minerals (Ref 43). Spray drying is a sample-preparation technique that can eliminate preferred orientation problems (Ref 44, 45). A suitable source of information on sample-preparation techniques is provided in Ref 46 and 47.

Instrument Calibration

The stability of the x-ray source, specimen, and detector is important in obtaining suitable-intensity data. Any changes in the slits, voltage, current, and detector settings will cause variations in the intensity. When using a modern pixelated detector, calibration is usually necessary to correct for differences in the performance of each pixel.

Intensity Measurement Procedures

Integrated intensity measurements, where possible, are preferred to peak height measurements. Poor crystallinity caused by defect structures, small crystallite size, or microstrain alters the peak height but not the integrated intensity. The background must be properly taken into account when the intensity is determined. Overlapping of peaks from one or more phases can cause large errors. This is true when deconvolution procedures are used, unless they are well controlled. The optimum procedure involves the use of peak-fitting software or some variations of the Rietveld method to correctly extract all the intensity, properly considering peak overlapping and aberrations.

Other Possible Sources of Error

Changes in barometric pressure and humidity can alter the absorption characteristics of the air and thus the intensity of the peaks in a sample. Changes in temperature can alter the cell parameter. Be warned that due to the presence of x-ray absorption edges, the relative peak intensities (in a phase and between phases) can change within a pattern when changing the measurement wavelength. Some samples can hydrate, oxidize, or carbonate when exposed to air. High humidity can also cause samples to pop out of their holders, raising the sample surface above its proper position. In some cases, grinding the samples to reduce the particle size will induce a phase transformation by causing the phases present to react with each other. In some materials, especially certain organics, the x-rays will damage the structure.

Estimation of Crystallite Size and Defects

Right after the discovery of diffraction, a simple technique was proposed by Scherrer to estimate the size of the domains in the materials responsible for the diffracted signal (Ref 48). Those domains are often referred to as crystallites and represent the primary particles in the system. Monocrystalline regions enclosed in grain boundaries or high-density dislocation networks are seen as crystallites. The grains in a powder or in a polycrystalline aggregate can be aggregates of crystallites. In a nanocrystalline material or powder, grain and crystallite are commonly the same entity.

Traditional Techniques (Integral Breadth Methods)

The presence of defects (e.g., grain boundaries, dislocations, stacking faults, antiphase boundaries) in the material results in a broadening of the diffraction peaks and in peak shape changes. The techniques developed to study this phenomenon are collectively known as line profile analysis. A qualitative average crystallite size in the *hkl* direction $\langle D \rangle_{hkl}$ can be obtained from the breadth (e.g., estimated as the full width at half maximum) of the peaks β_{hkl} evaluated in 2θ:

$$\langle D \rangle_{hkl} = \frac{K_w \quad \lambda}{\beta_{hkl} \cos \theta_{hkl}} \quad \text{(Eq 14)}$$

where K_w is a constant on the order of magnitude 1, λ is the wavelength of the x-rays, and θ_{hkl} is the diffraction angle. The contribution of the instrument to the peak breadth and shape is considered as negligible. The breadth is usually obtained by fitting the experimental data with a bell-shaped curve (e.g., Gaussian, Lorentzian, or Voigtian that represents a combination of the previous ones) plus a background. Both the peak width or the so-called integral breadth (ratio between area and maximum intensity of the peak) are used. Care should be taken to handle cases with overlapping peaks. The constant K_w is related to the actual shape of the crystallites. The results of the analysis are qualitative and can be employed for an order-of-magnitude estimate and for comparison between specimens (Ref 1). A direct comparison of the values from Eq 14 with average-sized data obtained from the transmission electron microscope (TEM) is not possible, because they represent different averages (numeric in the case of the TEM; volumetric in the case of diffraction).

The Scherrer formula does not account for the presence of defects in the material. The Williamson and Hall plot (Ref 49) can be employed to estimate simultaneously a single average size of the crystallites and a microstrain, defined as local deformation of the lattice. The microstrain should not be confused with a macrostrain; a macrostrain involves the change in the average cell parameter (peak shift), whereas the presence of microstrain, that is, a local change in the spacing between the atomic layers, affects the breadth of the peaks. Under crude approximations (Ref 1, 49), a plot of the breadth of the peaks multiplied by (cos θ)/λ as a function of the reciprocal *d*-spacing is linear with:

$$\beta \frac{\cos \theta}{\lambda} = \frac{K_w}{\langle D \rangle} + \frac{2e \sin \theta}{\lambda} = \frac{K_w}{\langle D \rangle} + \frac{e}{d} \quad \text{(Eq 15)}$$

where *e* (nonnegative term) accounts for the presence of a local strain. For a Gaussian distribution of local strains, the root mean strain (microstrain) can be calculated as $\langle \varepsilon^2 \rangle^{1/2} = e\sqrt{2/\pi}$. The intercept in the plot is inversely proportional to the size, whereas the slope gives the microstrain. Due to the various approximations and assumptions on which these formulas are based, the validity of these traditional approaches is in most cases limited to a qualitative assessment. Crystallites larger than approximately 100 nm give negligible contribution to the peak broadening, and the experimental determination of their size becomes highly inaccurate. The limit can be pushed further on synchrotron radiation beamlines where the instrumental contribution is smaller than on laboratory diffractometers, but the use of diffraction to estimate the size of very large crystallites is not advised.

Whole Pattern Methods

A whole pattern technique, like the Rietveld method, can be employed for the analysis by generating the peak profiles from physical models of the nanostructure of the material. Whole powder pattern modeling (WPPM) (Ref 1, 50) is the state-of-the-art approach and employs a mathematically complex approach where the nanostructural feature in the specimen is directly transformed into the corresponding peak profile. The contribution from different features is then combined via convolution and the model parameters refined on the experimental data. The focus of WPPM is the analysis of the size, size distribution, type, and quantity of the defects (e.g., dislocations, stacking faults, antiphase boundaries, compositional variations) in the material, so the structure aspects are not fully considered. The approach can be combined with the Rietveld idea to constrain the peak intensities to a structural model, when the latter is known.

The results of the analysis can be directly compared with the TEM observation.

Application Examples of X-Ray Powder Diffraction

X-ray powder diffraction is used to identify the crystalline phases present in a sample. Examples of the types of questions that can be answered by using XRPD include:

- Is a heat treated steel sample 100% martensite, or does it contain some retained austenite?
- What compounds are present in the corrosion product that formed when an aluminum alloy was exposed to sea spray?
- What compounds are present in the scale formed on an ingot during high-temperature forging?

Solving these types of problems by XRPD is the most common use of x-ray diffraction in metallurgy. The following are some examples.

Qualitative Analysis of a Surface Phase on Silicon

A small, oddly shaped sample was examined using x-ray analysis to determine the nature of a thin, powdery coating on its surface. The bulk sample was known to be elemental silicon. A small amount (<0.01 mg) of the deposit was obtained by carefully abrading part of the sample surface and collecting the resulting material on a glass slide. The powder on the slide was incorporated into a mount suitable for a Debye-Scherrer camera by depositing it onto the end of a dampened gelatin fiber that had been cut from the side of a medicinal capsule as a narrow wedge shape ending in a sharp point.

The sample was mounted and exposed in the camera, the film developed, and the data (Table 3) obtained by using a plastic overlay for quick reading of the *d*-spacings directly from the film. The intensities were estimated visually during determination of the *d*-spacings. The ICDD PDF database (Ref 2) for the identification of unknown phases was then accessed. Use of the file involves computer-based search/match techniques or manual methods in which the most intense *d*-spacings of the unknown pattern are used to indicate one or more phases that are then examined in greater detail to identify the unknown. An example of the cards used in the manual search is shown in Fig. 15. A computer search was made, and the phase α-SiO_2 (quartz) was found to be the only possible phase that matched the submitted pattern. No other phases were found. The entire search process was almost instantaneous.

Table 3 Diffraction data from unknown phase(s) on silicon substrate

Interplanar spacing (*d*), Å	Relative intensity, I/I_0
4.25	3
3.34	10
2.45	1
2.28	1
2.23	1
2.12	1
1.98	1
1.81	3
1.67	1
1.54	2
1.38	1
1.37	3
1.29	1

Quantitative Analysis of ZnO in Calcite

Many calcite ($CaCO_3$) and ZnO samples were analyzed over an extended period of time by using a diffractometer equipped with a copper x-ray tube. The absorption diffraction method was selected because it yields a calibration curve against which the samples can be easily compared. First, the molecular weights of calcite and ZnO were calculated. Second, the mass-absorption coefficients of calcite and ZnO were calculated from the elemental mass absorption for copper radiation found in Appendix V of Ref 29. Table 4 shows calculated values of the concentration of ZnO in calcite versus the intensity of the strongest line of ZnO in the sample, divided by the intensity of the line obtained from pure ZnO.

A graph can be produced from the data in Table 4 for use in determining the concentration of ZnO in an unknown sample. The graph does not have a linear relationship because of the difference in the absorption of copper radiation between calcite and ZnO. This technique is effective only when there is no preferred orientation and very reproducible sample-preparation techniques are used.

d	3.34	4.26	1.82	4.26	SiO_2	★
I/I_1	100	22	14	22	Silicon Oxide	Quartz, low

Rad. CuKα₁ λ 1.540598 Filter Mono. Dia.
Cut off I/I_1 Diffractometer I/I cor.
Ref. Nat. Bur. Stand. (U.S.) Monogr. 25, Sec. 18 (1981)

Sys. Hexagonal S.G. $P3_121$ (152)
a_0 4.9133(2) b_0 c_0 5.4053(4) A C 1.1001
α β γ Z 3 Dx 2.649
Ref. Ibid.

εα nωβ 1.544 εγ 1.553 Sign +
2V D 2.656 mp Color Colorless
Ref. Ibid.

Sample from the Glass Section at the National Bureau of Standards; ground single crystals of optical quality, locality unknown. Pattern at 25°C.
Silicon (a_0=5.43088Å) used as internal standard.
F_{30} = 76.6(0.0126,31). Quartz group.
To replace 5-490.

d A	I/I_1	hkl	d A	I/I_1	hkl
4.257	22	100	1.2285	1	220
3.342	100	101	1.1999	2	213
2.457	8	110	1.1978	1	221
2.282	8	102	1.1843	3	114
2.237	4	111	1.1804	3	310
2.127	6	200	1.1532	1	311
1.9792	4	201	1.1405	<1	204
1.8179	14	112	1.1143	<1	303
1.8021	<1	003	1.0813	2	312
1.6719	4	202	1.0635	<1	400
1.6591	2	103	1.0476	1	105
1.6082	<1	210	1.0438	<1	401
1.5418	9	211	1.0347	<1	214
1.4536	1	113	1.0150	1	223
1.4189	<1	300	0.9898	1	402
1.3820	6	212	.9873	1	313
1.3752	7	203	.9783	<1	304
1.3718	8	301	.9762	1	320
1.2880	2	104	.9636	<1	205
1.2558	2	302	6 reflections to 0.9089		

Fig. 15 International Centre for Diffraction Data (ICDD) POF(R) card giving x-ray powder diffraction and other associated data for the mineral quartz

Table 4 Calculated absorption coefficients and relative intensities of ZnO in calcite

Zn, wt%	Absorption coefficient	Relative intensity of zinc, I/I_{pure}
0	75.6	0.00
10	73.0	0.07
20	70.4	0.14
30	67.8	0.22
40	65.2	0.30
50	62.6	0.40
60	60.1	0.50
70	57.5	0.61
80	54.9	0.72
90	52.3	0.86
100	49.7	1.00

Qualitative Analysis of Crystallite Size in Cerium Oxide

A nanocrystalline cerium oxide specimen produced by sol-gel was measured on a laboratory machine using Cu Kα radiation from a sealed tube, to characterize the nanostructure (Ref 50, 51).

Ceria crystallizes in the face-centered cubic system, and the most intense peak is the (111), which slightly overlaps with the (200) peak. The part of the pattern is shown in Fig. 16. The peaks are broad, indicating that the size of the crystallites is likely in the nanometer range. It is not possible to observe the two characteristic wavelength components typical of the emission of the tube. Without a proper peak fitting, one can choose an average wavelength for the system. The peak position for the (111) peak in 2θ is 28.515°, and its full width at half maximum in 2θ is 1.53°. From the Scherrer formula one can estimate:

$$\frac{\langle D \rangle}{K_w} = \frac{0.15406}{1.53\pi/180 \cos(28.515/2)} \text{nm} = 5.95 \text{ nm} \quad \text{(Eq 16)}$$

Care should be posed in checking the details concerning the definition of the angles to avoid miscalculations. The size distribution obtained from a TEM analysis and from a WPPM analysis, based on the same pattern employed here, agrees on a mean size of approximately 4.3 to 4.4 nm. The differences between those numbers and the value obtained from the Scherrer equation are mainly due to:

- The instrument, which partly contributes to the profile breadth and shape
- The presence of two emission components that, even if not visible, are always present
- The presence of defects (A dislocation density of ~1.6×10^{16} m^{-2} can be estimated by using the WPPM.)
- Peculiar definition of "average size" in the Scherrer equation, which does not correspond to the mean of the size distribution

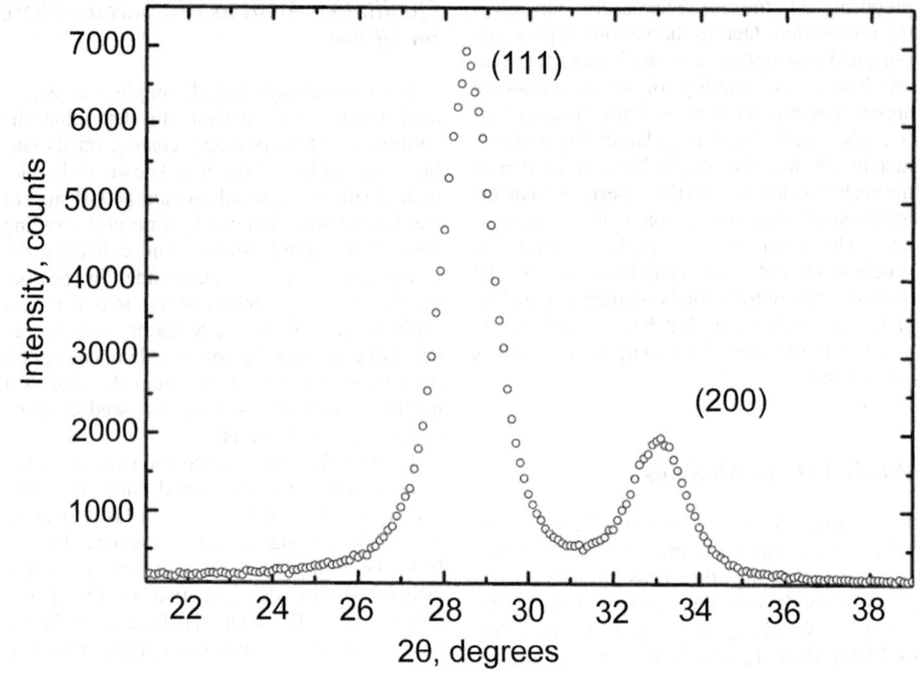

Fig. 16 Selected portion of the diffraction pattern of nanocrystalline ceria

ACKNOWLEDGMENT

Portions of this article were adapted from:

- K.H. Eckelmeyer, X-Ray Diffraction for Bulk Structural Analysis, *Metals Handbook Desk Edition*, ASM International, 1998, p 1416–1422
- R.P. Goehner and M.C. Nichols, X-Ray Powder Diffraction, *Materials Characterization*, Vol 10, *ASM Handbook*, ASM International, 1986, p 333–343

REFERENCES

1. *International Tables for X-Ray Crystallography*, Vol H: Powder Diffraction, Wiley, 2019
2. S. Kabekkodu, Ed., "PDF 2019 (Database)," International Centre for Diffraction Data, Newtown Square, PA, 2018, http://www.icdd.com
3. "NIST Crystal Data," National Institute of Standards and Technology, ftp://ftp.ncnr.nist.gov/pub/cryst/powdersuite/crystaldata.html
4. W. Friedrich, P. Knipping, and M.V. Laue, *Ann. Phys.*, Vol 411, 1912, p 971
5. W. Friedrich, *Phys. Z.*, Vol 14, 1913, p 317
6. H.B. Keene, *Nature*, Vol 91, 1913, p 607
7. C. Barrett and T. Massalski, *Structure of Metals*, McGraw-Hill, 1966
8. G. Zhou, J. Kou, Y. Li, W. Zhu, K. Chen, and N. Tamura, *Quantum Beam Sci.*, Vol 2, 2018, p 13
9. M.H. Read, *Thin Solid Films*, Vol 10, 1972, p 123–135
10. R.W. Bower, R.E. Scott, and D. Sigurd, *Solid State Elec.*, Vol 16, 1973, p 1461
11. S.S. Lau, W.K. Chu, K.N. Tu, and J.W. Mayer, *Thin Solid Films*, Vol 23, 1974, p 205–213
12. C.A. Wallace and R.C.C. Ward, *J. Appl. Crystallogr.*, Vol 8, 1975, p 255–260, 545–556
13. J.L. Amoros, M.J. Buerger, and M. Canut de Amoros, *The Laue Method*, Academic Press, 1975
14. B.D. Cullity, *Element of X-Ray Diffraction*, 2nd ed., Addison Wesley, 1978, p 175–177
15. B.E. Warren, *X-Ray Diffraction*, Addison-Wesley, 1969
16. R. Jenkins and R. Snyder, *Introduction to X-Ray Powder Diffractometry*, John Wiley, 1996
17. H.M. Rietveld, A Profile Refinement Method for Nuclear and Magnetic Structures, *J. Appl. Crystallogr.*, Vol 2, 1969, p 65
18. R.A. Young, *The Rietveld Method*, Oxford University Press, 1993
19. J.D. Hanawalt and H.W. Rinn, *Ind. Eng. Chem. Anal.*, Vol 8, 1936, p 244
20. J.D. Hanawalt, *Adv. X-Ray Anal.*, Vol 20, 1976, p 63
21. A.L. Navias, *Am. Ceram. Soc.*, Vol 8, 1925, p 296
22. I.C. Madsen, N.V.Y. Scarlett, L.M.D. Cranswick, and T. Lwin, *J. Appl. Crystallogr.*, Vol 34, 2001, p 409

23. N.V.Y. Scarlett, I.C. Madsen, L.M.D. Cranswick, T. Lwin, E. Groleau, G. Stephenson, M. Aylmore, and N. Agron-Olshina, *J. Appl. Crystallogr.*, Vol 35, 2002, p 383
24. M.D. Raven and P.G. Self, *Clays Clay Miner.*, Vol 65 (No. 2), 2017, p 122
25. M.F. Garbauskas and R.P. Goehner, *Adv. X-Ray Anal.*, Vol 25, 1982, p 283
26. L. Zwell and A.W. Danko, *Appl. Spectrosc. Rev.*, Vol 9, 1975, p 178
27. F.H. Chung, *J. Appl. Crystallogr.*, Vol 7, 1974, p 519
28. F.H. Chung, *J. Appl. Crystallogr.*, Vol 8, 1975, p 17
29. H.P. Klug and L.E. Alexander, *X-Ray Diffraction Procedures for Polycrystalline and Amorphous Materials*, John Wiley & Sons, 1974, p 562
30. S. Popovic and B. Grzeta-Plenkovic, *J. Appl. Crystallogr.*, Vol 12, 1979, p 205
31. L.E. Copeland and R.H. Bragg, *Anal. Chem.*, Vol 30 (No. 2), 1958, p 196
32. R.P. Goehner, *Adv. X-Ray Anal.*, Vol 25, 1982, p 309
33. C.R. Hubbard and D.K. Smith, *Adv. X-Ray Anal.*, Vol 20, 1977, p 27
34. F.H. Chung, *Adv. X-Ray Anal.*, Vol 17, 1974, p 106
35. C.F. Jatczak, J.A. Larson, and S.W. Shin, *Retained Austenite and Its Measurements by X-Ray Diffraction*, SP-453, Society of Automotive Engineers, Warrendale, PA, 1980
36. D.K. Smith and M.C. Nichols, Report SAND81-8226, Sandia National Laboratories, Livermore, CA, 1981
37. D.L. Bish and S.A. Howard, *J. Appl. Crystallogr.*, Vol 21, 1988, p 86
38. R.J. Hill and C.J. Howard, *J. Appl. Crystallogr.*, Vol 20, 1987, p 467
39. N.V.Y. Scarlett and I.C. Madsen, *Powder Diffr.*, Vol 21 (No. 4), 2006, p 278
40. L.S. Zevin, *J. Appl. Crystallog.*, Vol 10, 1977, p 147
41. J. Fiala, *Anal. Chem.*, Vol 52, 1980, p 1300
42. A. Le Bail, *J. Non-Cryst. Solids*, Vol 183, 1995, p 39
43. H.F. Shaw, *Clay Min.*, Vol 9, 1972, p 349
44. S.T. Smith, R.L. Snyder, and W.E. Browell, *Adv. X-Ray Anal.*, Vol 22, 1979, p 77, 181
45. L.D. Calvert, A.F. Sirianni, and G.J. Gainsford, *Adv. X-Ray Anal.*, Vol 26, 1983, p 105
46. D.K. Smith and C.S. Barrett, *Adv. X-Ray Anal.*, Vol 22, 1979, p 1
47. V.E. Buhrke, R. Jenkins, and D.K. Smith, *A Practical Guide for the Preparation of Specimens for X-Ray Fluorescence and X-Ray Diffraction Analysis*, Wiley-VCH, 1998
48. P. Scherrer, *Nachr Ges Wiss Goettingen*, Math-Phys Kl, 1918, p 98
49. G.K. Williamson and W.H. Hall, *Acta Metall.*, Vol 1 (No. 1), 1953, p 22
50. M. Leoni, Whole Powder Pattern Modelling Microstructure Determination from Powder Diffraction Data, International Tables for *X-Ray Crystallography*, Vol H: Powder Diffraction, Wiley, 2019
51. M. Leoni, R. Di Maggio, S. Polizzi, and P. Scardi, *J. Am. Ceram. Soc.*, Vol 87, 2004, p 87

SELECTED REFERENCES

- L.E. Alexander, *X-Ray Diffraction Methods in Polymer Science*, Wiley-Interscience, 1980
- L.V. Azaroff and M.J. Buerger, *The Powder Method in X-Ray Crystallography*, McGraw-Hill, 1958
- C.S. Barrett and T.B. Massalski, *Structure of Metals*, Pergamon Press, 1980
- S. Block and C.R. Hubbard, Ed., *Accuracy in Powder Diffraction*, NBS 567, National Bureau of Standards, Washington, 1980
- G.W. Brindley and G. Brown, *Crystal Structures of Clay Minerals and Their X-Ray Identification*, Mineralogical Society of America, Washington, 1980
- B.D. Cullity, *Elements of X-Ray Diffraction*, Addison Wesley, 1978
- C. Giacovazzo, H.L. Monaco, G. Artioli, D. Viterbo, M. Milanesio, G. Gilli, P. Gilli, G. Zanotti, G. Ferraris, and M. Catti, *Fundamentals of Crystallography*, 3rd ed., Oxford University Press, Oxford, 2011
- A. Guinier, *X-Ray Diffraction in Crystals, Imperfect Crystals, and Amorphous Bodies*, W.H. Freeman and Company, 1963
- *International Tables for X-Ray Crystallography, Vol H: Powder Diffraction*, Wiley, 2019
- R.W. James, *The Optical Principles of the Diffraction of X-Rays*, Ox Bow Press, 1982
- H.P. Klug and L.E. Alexander, *X-Ray Diffraction Procedures for Polycrystalline and Amorphous Materials*, John Wiley & Sons, 1974
- H.S. Peiser, H.P. Rooksby, and A.J. Wilson, Ed., *X-Ray Diffraction by Polycrystalline Materials*, Institute of Physics, London, 1955, p 278–297
- E. Preuss, B. Krahl-Urban, and R. Butz, *Laue Atlas*, John Wiley & Sons, 1974
- A. Taylor, *X-Ray Metallography* John Wiley & Sons, 1961
- B.E. Warren, *X-Ray Diffraction*, Addison-Wesley, 1969
- A.J.C. Wilson, *X-Ray Optics*, 2nd ed., Methuen, 1962

ASM Handbook, Volume 10, *Materials Characterization*
ASM Handbook Committee
DOI 10.31399/asm.hb.v10.a0006631

Single-Crystal X-Ray Diffraction

Vladislav V. Gurzhiy, St. Petersburg State University

Overview

General Uses

- Unit cell parameter identification
- Space group determination
- Location of atomic sites within the unit cell (i.e., determination of the crystal structure), and hence determination of atom coordination numbers and geometries, interatomic bond lengths and angles, the structural architecture motif and topology of the structural units, molecular structure, and absolute configurations

Examples of Applications

- Identification of previously studied crystalline phases by comparison of unit cell parameters
- Characterization of new crystalline compounds
- For rather simple synthetic compounds, the analysis of chemical composition
- Interpretation of physical and chemical properties and even growth conditions of crystalline compounds in terms of interatomic interactions

Samples

- *Form*: Single crystal or its fragment, preferably well shaped and, if translucent, without visible flaws; it is preferable to evaluate the quality of the crystal under a polarizing microscope using crossed Nikol prisms or Polaroids to avoid twinned crystals
- *Size*: 0.05 to 0.2 mm, depending on the generation of the diffractometer and diffraction ability of the crystal
- *Preparation*: Generally, suitable crystals can be grown using various crystallization techniques, but it is preferable to control the stationarity of the system to avoid an atomic or molecular disorder in the structure. Air-stable crystals are usually mounted on glass fibers or cryoloops using any type of glue or inert pitch; air-sensitive crystals can be sealed inside capillaries or fully covered by glue or pitch
- *Data collection*: It is preferable to cool the sample to reduce the atomic thermal displacement parameters (especially required for atoms with low atomic numbers, for instance, in organic molecules), if the crystal does not undergo phase transition

Limitations

- Sample must be a single crystal. Samples that appear to be a single crystal may consist of several domains with different orientations or may not even be crystalline at all. The crystal selected for data collection may not be representative of the bulk sample
- For unit cell identification, literature data or a structural database must be available
- Success in the determination of the crystal structure can be impeded by ambiguities in the assignment of the crystal class and/or space group, the presence of a sub- or superlattice, twinning effects, disorder or partial occupancy of particular atomic sites, inaccurate data collection or reduction

Estimated Analysis Time

- Two stages should be kept in mind: data collection and structure determination. In analysis with the perfect crystal and using a modern diffractometer, each of the stages can last about 30 min. The smaller the crystal, the longer the data collection time. A simple check of unit cell parameters usually takes several minutes

Capabilities of Related Techniques

- *Single-crystal neutron diffraction* can be used to determine the atomic structure of crystals. Advantages include higher sensitivity to atoms with low atomic numbers (especially hydrogen), ability to distinguish isotopes, deep penetration, and information about magnetic structure. Disadvantages include the requirement for larger crystals (about 1 mm^3), longer data collection times, and a nuclear reactor
- *Electron diffraction* is similar to x-ray and neutron diffraction; it is usually used in electron microscopy as well as electron backscatter diffraction. It is usually used for symmetry determination and raw determination of unit cell parameters; however, recent techniques can determine the structures from these data. Advantages include a nanometer size of single crystals. Disadvantages include the requirement for the sample to be electron transparent, and the possibility of incident electrons damaging the sample
- *Rietveld refinement of x-ray or neutron powder diffraction data* can be used to determine crystal structure. Advantages include use of a crystalline powder rather than a single crystal. Disadvantages include the requirement of a starting atomic model
- *Structure determination using powder diffraction data* is probably the youngest approach for the determination of atomic sites within the unit cell of a crystalline compound. Advantages include use of a crystalline powder and no need for a starting atomic model. Disadvantages include the need for high-precision diffraction data and relatively large computing power, as well as a limited number of specialists in the technique

Introduction

The main application of single-crystal x-ray diffraction (XRD) is to characterize the crystal structure, including determining symmetry, unit cell parameters, atomic coordinates and thermal displacement parameters, bond lengths and angles between the atoms, and structural motive (or topology). During the structure description process it is possible to refine such crystallographic characteristics as order or disorder of cations (anions), site occupancy factors, and possible sub- or superstructures. The results of crystal structure analysis can become the basis for broader conclusions. Thus, determination of crystal structure is quite essential to understanding the formation conditions of crystalline substances. For instance, compounds of the same chemical composition can exist in various polymorph modifications, which directly correlates with the thermodynamic parameters of their genesis. Another example is nondestructive determination of the crystal phase using the values of the unit cell parameters, which is very useful for the jewelry industry.

Historical Review

Modern x-ray analysis has become a powerful tool for studying the structure of substances, revealing many interesting facts and allowing a new look at a number of natural phenomena. The total number of solved structures to date exceeds 1.3 million (more than 100,000 structures are deposited in the Cambridge Structural Database, a repository for small-molecule organic and metal-organic crystal structures; and about 200,000 structures are deposited in the Inorganic Crystal Structure Database), and annually more than 60,000 novel natural and synthetic compounds are structurally characterized. Due to improved techniques, automated equipment, and computing facilities, it became possible to determine the structures of very complex crystals, such as proteins, which contain hundreds or even thousands of atoms. Pressures exceeding millions of times greater than atmospheric is attained within special x-ray chambers, which makes it possible to model and study the state of matter in the deep shells of the Earth.

Structural analysis makes probably the most significant contribution to the mineralogy and materials sciences, where XRD methods play a key role in studying the composition and structure of substances, expanding scientific concepts of the taxonomy, forms, and concentration of chemical elements in the geosphere as well as isomorphism, polymorphism, and many other crystal chemical phenomena in natural and synthetic compounds.

X-rays were first systematically studied by the German physicist Wilhelm Conrad Röntgen at the University of Würzburg in 1895. Within the first weeks, Röntgen made some very important discoveries. First of all, the absorption of x-rays depends on the atomic numbers of the material through which they are passed. For instance, platinum or lead absorbs the rays more strongly than silver or copper, and absorption by aluminum is quite insignificant. He also raised a question that was solved only after 17 years: are x-rays particles or waves?

Conceptions of the structure of gases and liquids were much more complete than those of crystals within materials science at that time. Max von Laue, during a 1912 discussion in Munich with another well-known German physicist, Paul Peter Ewald, suggested that crystals can diffract x-rays. Röntgen's students Paul Knipping and Walter Friedrich were involved in the testing of Laue's idea, and after a series of unsuccessful experiments, the first diffraction pattern was obtained from the crystal of blue vitriol (copper sulfate pentahydrate).

The final evidence of the electromagnetic nature of x-rays was presented by British scientists William Henry Bragg and his son William Lawrence Bragg. It was suggested that x-rays are reflected within the crystal from translucent mirror planes formed by atoms. Registered reflections from lamellar crystals of mica confirmed this theory. If the system of parallel planes in a mica crystal is inclined to a beam, then waves reflected from neighbor planes strengthen each other only for a certain orientation of the planes.

For these discoveries, Nobel Prizes in Physics were awarded to Laue in 1914 for "his discovery of the diffraction of X-rays by crystals" and the Braggs in 1915 for "their services in the analysis of crystal structure by means of X-rays."

One of the most famous Russian crystallographers, M.A. Poray-Koshits, called the middle of the 20th century "the era of romantic x-ray analysis: crystal structure investigation was a very fascinating and challenging task, similar to the solution of chess puzzles." Today, this situation has significantly changed. For the vast majority of compounds, both organic and inorganic, crystal structure analysis has become more of a standard task, largely due to the wide availability of modern high-precision instruments and high-performance computers. However, it would be erroneous to state that work in the field of XRD analysis does not still require special preparation now. So, before we go on to consider the modern techniques of crystal structure investigation, first we discuss the fundamentals of crystallography and the symmetry of crystalline substances, as well as some aspects and formulas of XRD theory (Ref 1–13).

Crystal Symmetry

The crystalline state of a substance is characterized by the three-dimensional periodicity of atomic and molecular arrangement. This feature underlies the diffraction of x-rays passed through the crystal and hence is the basis of the XRD structural analysis of crystals. The periodic repeatability of identical atomic groups (in other words, the translational symmetry in their arrangement) is an obligatory property of any crystal. But atoms in a crystal can be connected not only by translations but also by other symmetry operations, whose presence also affects the diffraction pattern and therefore must be taken into account during crystal structure determination.

Within symmetry theory, a lattice is a group of translations, where nodes are mathematical points, not material points (atoms). To specify the crystal lattice in the general case (Fig. 1a), it is necessary to set three vector parameters (a, b, and c) and the angles between their directions (α, β, and γ). These six parameters are called *lattice* or *unit cell parameters*, and the figure constructed on them is a parallelepiped of repeatability. If the x, y, and z axes are chosen in accordance with the rules accepted

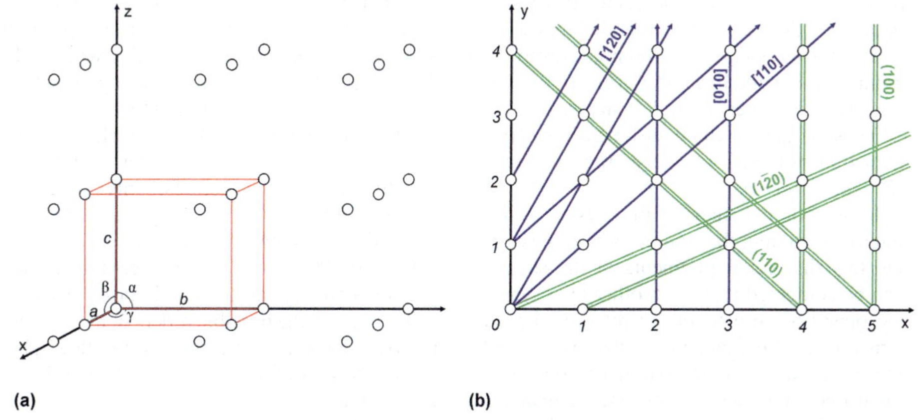

Fig. 1 (a) General case of a unit cell selection within the system of crystallographic axes. (b) Indexation of the nodal rows (blue arrows) and nets (green double lines). The graph shows reflections for the [010], [120], and [110] directions.

in crystallography, then such a box is called the *unit cell* of the crystal.

In descriptions of a crystal lattice, one of its nodes is selected as the origin of coordinates, and all other nodes are numbered in order along the coordinate axes (Fig. 1b). The coordinate of each node is determined by three integers, *x*, *y*, and *z*, called indexes. It is possible in a lattice to set a number of nodal rows and nets of different orientation. The [*hkl*] index is assigned to the series of parallel node rows (index of the closest to the origin node, through which the row intersecting the origin is passing). The slope of the nodal nets is described by the (*hkl*) index, indicating the number of parts into which the edge of the unit cell is divided.

It is worth noting that further in the text the Hermann–Mauguin notation (sometimes called international notation) is used to represent the symmetry elements, and point and space groups. These symbols and classification, compared with the Schoenflies notation, are recommended by the International Union of Crystallographers and preferred because they provide a better understanding of the directions of the symmetry axes and translational symmetry element arrangement.

Bravais Lattices

A lattice where the nodes are located only at the corners of a unit cell (Fig. 2) is called *primitive* (*P*). In the presence of translations that link the vertexes of cells with equivalent nodes inside or on the faces of unit cells, the lattice is considered to be centered. There are three types of centered unit cells: *body-centered* (*I*), in which an additional node is located in the middle of the unit cell's volume; *base-centered* (*A*, *B*, or *C*—centering of the *bc*, *ac*, or *ab* faces, respectively), in which additional nodes appear in the middle of any two opposite faces of the unit cell; and *face-centered* (*F*), in which additional nodes are located in the middle of all six faces. In the trigonal system, a double-centered lattice is possible, the primitive parallelepiped for which has the shape of a rhombohedron; such a unit cell is called *rhombohedral* (*R*). Taking into account seven crystal systems, there are 14 possible lattices in three-dimensional space, named after French physicist Auguste Bravais (*Bravais lattices*).

In the external shape of the crystal, as in its internal structure, the symmetry is expressed in the fact that equal fragments of the structure can be combined with each other by a symmetry operation—reflection or rotation (i.e., the structure or the fragment of the crystal looks the same after some symmetry operation). Such operations and corresponding geometric images (symmetry elements) are taken as a basis for the description of symmetry groups (Fig. 3).

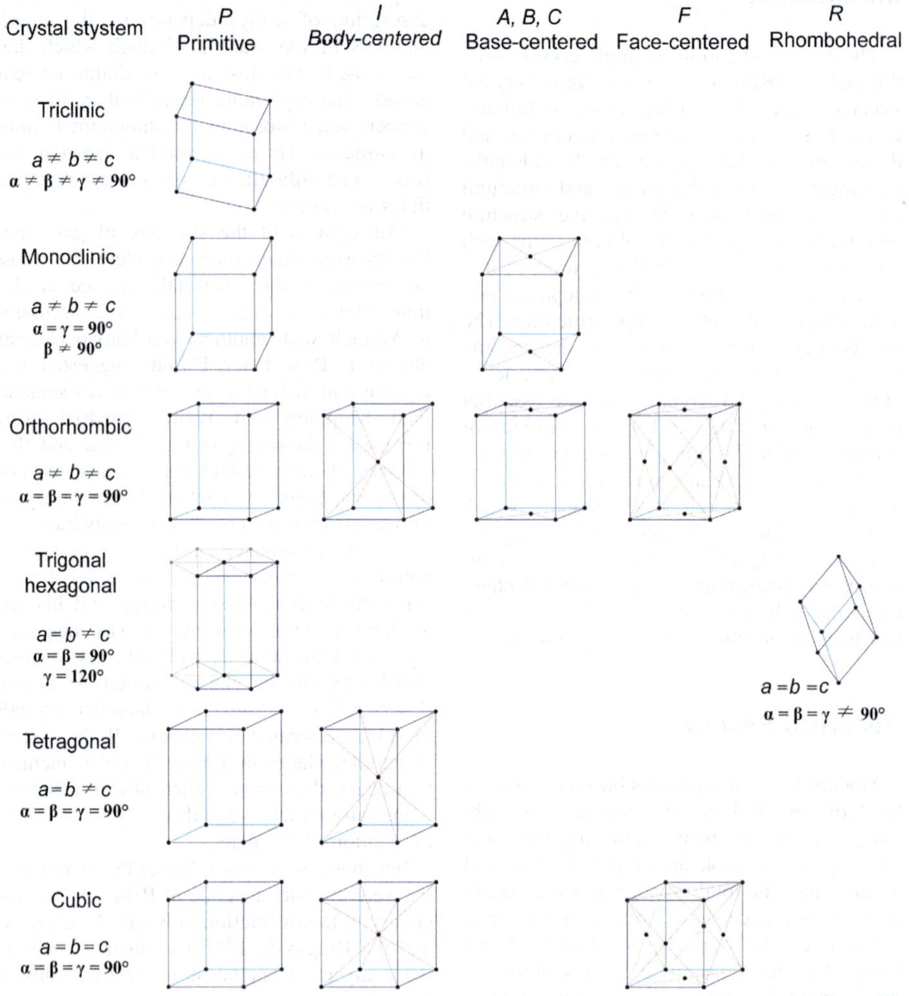

Fig. 2 The 14 conventional Bravais lattices distributed among the seven crystal systems

Symmetry Elements

The first symmetry operation is *rotation*, and the corresponding symmetry element is called a *rotation axis*. A rotation axis is a straight line where, when rotating around it at a certain angle, a figure coincides with itself. The order of the axis is given by *n*, where $n = 360°/\alpha$, and α is a minimal angle required to reach a position indistinguishable from the starting point. In crystal lattices and consequently in crystal structures, only 1-, 2- 3-, 4-, and 6-fold rotation axes can occur. Accordingly, the elementary rotation angles are 360°, 180°, 120°, 90°, and 60°. The three-dimensional periodicity of crystal structure imposes stringent restrictions on the possible orders of the symmetry axes. Thus in conventional crystallography, symmetry axes of orders of 5 and higher than 6 are forbidden.

A further symmetry operation is *reflection*, and the corresponding symmetry element is called a *mirror plane* (with a symbol *m*). Any point on one side of the mirror plane is matched by the generation of an equivalent point on the other side from the plane at the same distance along the line normal to it.

The symmetry operation called *inversion* relates pairs of points that are equidistant and are arranged on the opposite sides of the central point (called an *inversion center* or the center of symmetry).

The combinational symmetry of rotation and inversion is called *rotoinversion* and is present when all points remain unchanged after rotation around the *n*-fold axis and inversion with respect to the point arranged on the same axis. Its symmetry elements are *inversion axes* or rotoinversion axes, whose written symbol is $\overline{n}$ ("-*n*" is less correct but more common in recent scientific publications; read as "minus *n*"). There are only five possible inversion axes, corresponding to the rotation axes: $\overline{1}, \overline{2}, \overline{3}, \overline{4}$ and $\overline{6}$. The rotoinversion operation $\overline{1}$ is identical to the inversion center. Operation $\overline{2}$ is identical to the mirror plane and represents a direction perpendicular to *m*.

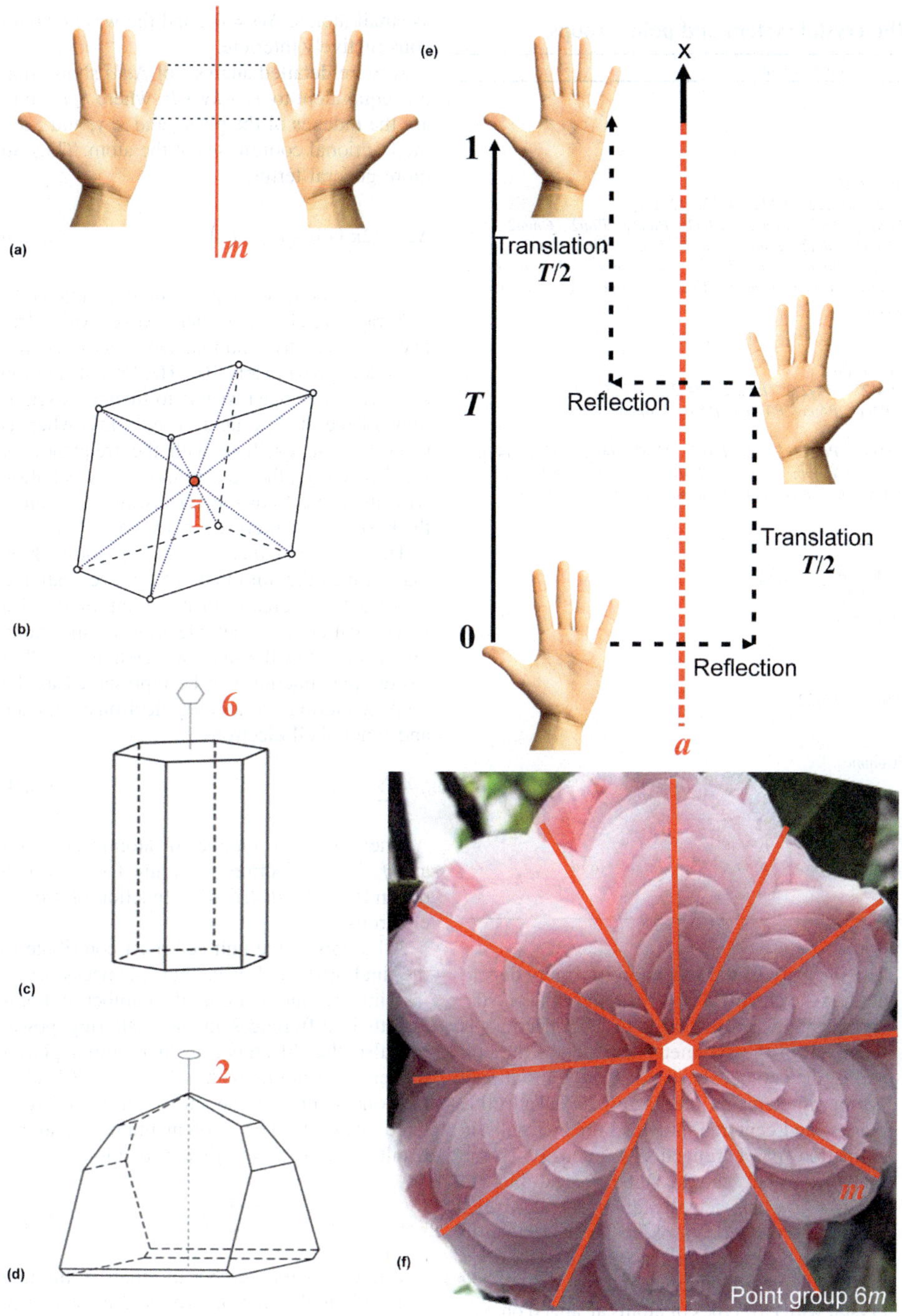

Fig. 3 Illustration of symmetry operations and elements: (a) reflection and mirror plane; (b) a general lattice, showing inversion at ½, ½, ½; (c) 6-fold axis in the beryl crystal model, point group 6/*mmm*; (d) 2-fold axis in the struvite crystal model, point group *mm*2; (e) glide reflection and glide plane of *a* (*b*, *c*) type; (f) hexagonal symmetry of the *camellia Japonica* flower

Rototranslational symmetry, with its symmetry element called a *screw axis*, combines two operations: rotation around the axis of an order of *n* and displacement along the axis by the translational component *t*. Screw axes are labeled as n_m, so $t = m/n$ is a portion of the parallel lattice vector. For instance, 2_1 is a 180° (2-fold) rotation followed by a translation of ½ of the lattice vector; 4_1 is a 90° (4-fold) rotation followed by a translation of ¼ of the lattice vector. The possible screw axes are 2_1, 3_1, 3_2, 4_1, 4_2, 4_3, 6_1, 6_2, 6_3, 6_4, and 6_5. It should be noted that such axes as 3_1 and 3_2, or 6_1 and 6_5, are enantiomorphic axes. Their resulting symmetry operations are identical and differ only in the direction of rotation; that is, 3_1 is a 120° (3-fold) clockwise rotation followed by a translation of ⅓ of the lattice vector, and 3_2 is a 120° (3-fold) clockwise rotation followed by a translation of ⅔, or it is equal to a 120° (3-fold) counterclockwise rotation followed by a translation of ⅓.

The combinational symmetry of reflection and translation is *glide reflection*, and the corresponding symmetry element is called a *glide plane*. There are three types of glide planes. The first is denoted by *a*, *b*, or *c*, depending on which axis the glide is along; translation occurs along ½ of the lattice vector of this face (translation of ½ along the only crystallographic direction). The second is the *n* plane, which refers to a glide along ½ of a face diagonal (½ diagonally along two crystallographic directions). The last is the *d* plane, which is along ¼ of either a face or space diagonal of the unit cell. The *d* plane is often called a diamond glide plane since it is present in the structure of diamond.

Point and Space Groups

Combination of the symmetry elements that do not have any translation operations results in 32 *point groups*, which are the set of symmetry operations that leave at least one point fixed while another fragment of the crystal is moved until it reaches a position indistinguishable from the starting position. These 32 point groups (Table 1) are the same as the 32 types of morphological (external) crystalline forms derived in 1830 by Johann Friedrich Christian Hessel from a consideration of observed shapes of crystals.

Addition of translational symmetry elements such as the lattice type, screw axes, and glide planes to the set of 32 point groups results in a total of 230 *space groups* describing all possible crystal symmetries (Table 1). Space groups are often called crystallographic or Fedorov groups, after the Russian crystallographer Evgraf Fedorov, who in 1891 was the first to publish a correct list of space groups.

X-Ray Diffraction

X-rays are a form of electromagnetic radiation. The great utility of x-rays for determining the structure and composition of materials is due to the differential absorption of x-rays by materials of different density, composition, and homogeneity. The higher the atomic number of a material, the more x-rays are absorbed. X-rays cause photochemical reactions and luminescence, and undergo scattering, reflection, interference, and diffraction. The range of wavelengths of x-rays is placed between the ultraviolet region and γ-rays, but the most useful for crystallography purposes ranges between 0.4 and 2.5 Å (0.04–0.25 nm), which is comparable with interatomic distances in crystals and why a periodic crystal structure can be used as a diffraction lattice. In addition, the wavelength depends on the anode material of the x-ray source (for instance, $\lambda_{Mo} = 0.71$ Å and $\lambda_{Cu} = 1.54$ Å).

Table 1 The 230 space groups arranged by the crystal system and point groups

Crystal system	Point group	Space groups
Triclinic	1	$P1$
	$\bar{1}$	$P\bar{1}$
Monoclinic	2	$P2\quad P2_1\quad C2$
	m	$Pm\quad Pc\quad Cm\quad Cc$
	$2/m$	$P2/m\quad P2_1/m\quad C2/m\quad P2/c\quad P2_1/c\quad C2/c$
Orthorhombic	222	$P222\quad P222_1\quad P2_12_12\quad P2_12_12_1\quad C222_1\quad C222\quad F222\quad I222\quad I2_12_12_1$
	$mm2$	$Pmm2\quad Pmc2_1\quad Pcc2\quad Pma2\quad Pca2_1\quad Pnc2\quad Pmn2_1\quad Pba2\quad Pna2_1\quad Pnn2\quad Cmm2\quad Cmc2_1\quad Ccc2\quad Amm2\quad Abm2\quad Ama2\quad Aba2\quad Fmm2\quad Fdd2\quad Imm2\quad Iba2\quad Ima2$
	mmm	$Pmmm\quad Pnnn\quad Pccm\quad Pban\quad Pmma\quad Pnna\quad Pmna\quad Pcca\quad Pbam\quad Pccn\quad Pbbcm\quad Pnnm\quad Pmmn\quad Pbcn\quad Pbca\quad Pnma\quad Cmcm\quad Cmca\quad Cmmm\quad Cccm\quad Cmma\quad Ccca\quad Fmmm\quad Fddd\quad Immm\quad Ibam\quad Ibca\quad Imma$
Tetragonal	4	$P4\quad P4_1\quad P4_2\quad P4_3\quad I4\quad I4_1$
	$\bar{4}$	$P\bar{4}\quad I\bar{4}$
	$4/m$	$P4/m\quad P4_2/m\quad P4/n\quad P4_2/n\quad I4/m\quad I4_1/a$
	422	$P422\quad P42_12\quad P4_122\quad P4_12_12\quad P4_222\quad P4_22_12\quad P4_322\quad P4_32_12\quad I422\quad I4_122$
	$4mm$	$P4mm\quad P4bm\quad P4_2cm\quad P4_2nm\quad P4cc\quad P4nc\quad P4_2mc\quad P4_2bc\quad I4mm\quad I4cm\quad I4_1md\quad I4_1cd$
	$\bar{4}2m$	$P\bar{4}2m\quad P\bar{4}2c\quad P\bar{4}2_1m\quad P\bar{4}2_1c\quad P\bar{4}m2\quad P\bar{4}c2\quad P\bar{4}b2\quad P\bar{4}n2\quad I\bar{4}m2\quad I\bar{4}c2\quad I\bar{4}2m\quad I\bar{4}2d$
	$4/mmm$	$P4/mmm\quad P4/mcc\quad P4/nbm\quad P4/nnc\quad P4/mbm\quad P4/mnc\quad P4/nmm\quad P4/ncc\quad P4_2/mmc\quad P4_2/mcm\quad P4_2/nbc\quad P4_2/nnm\quad P4_2/mbc\quad P4_2/mnm\quad P4_2/nmc\quad P4_2/ncm\quad I4/mmm\quad I4/mcm\quad I4_1/amd\quad I4_1/acd$
Trigonal	3	$P3\quad P3_1\quad P3_2\quad R3$
	$\bar{3}$	$P\bar{3}\quad R\bar{3}$
	32	$P312\quad P321\quad P3_112\quad P3_121\quad P3_212\quad P3_221\quad R32$
	$3m$	$P3m1\quad P31m\quad P3c1\quad P31c\quad R3m\quad R3c$
	$\bar{3}m$	$P\bar{3}1m\quad P\bar{3}1c\quad P\bar{3}m1\quad P\bar{3}c1\quad R\bar{3}m\quad R\bar{3}c$
Hexagonal	6	$P6\quad P6_1\quad P6_5\quad P6_2\quad P6_4\quad P6_3$
	$\bar{6}$	$P\bar{6}$
	$6/m$	$P6/m\quad P6_3/m$
	622	$P622\quad P6_122\quad P6_522\quad P6_222\quad P6_422\quad P6_322$
	$6mm$	$P6mm\quad P6cc\quad P6_3cm\quad P6_3mc$
	$\bar{6}m2$	$P\bar{6}m2\quad P\bar{6}c2\quad P\bar{6}2m\quad P\bar{6}2c$
	$6/mmm$	$P6/mmm\quad P6/mcc\quad P6_3/mcm\quad P6_3/mmc$
Cubic	23	$P23\quad F23\quad I23\quad P2_13\quad I2_13$
	$m\bar{3}$	$Pm\bar{3}\quad Pn\bar{3}\quad Fm\bar{3}\quad Fd\bar{3}\quad Im\bar{3}\quad Pa\bar{3}\quad Ia\bar{3}$
	432	$P432\quad P4_232\quad F432\quad F4_132\quad I432\quad P4_332\quad P4_132\quad I4_132$
	$\bar{4}3m$	$P\bar{4}3m\quad F\bar{4}3m\quad I\bar{4}3m\quad P\bar{4}3n\quad F\bar{4}3c\quad I\bar{4}3d$
	$m\bar{3}m$	$Pm\bar{3}m\quad Pn\bar{3}n\quad Pm\bar{3}n\quad Pn\bar{3}m\quad Fm\bar{3}m\quad Fm\bar{3}c\quad Fd\bar{3}m\quad Fd\bar{3}c\quad Im\bar{3}m\quad Ia\bar{3}d$

Source: Ref 1, 9, 12

The diffraction of x-rays by crystal can be described as a reflection of rays from the set of lattice planes. A parallel, monochromatic beam of x-rays (i.e., characterized by a single wavelength λ) fall on a set of lattice planes of (hkl) indexes separated one from another by the spacing of d and making an angle of incidence θ with them. The diffraction condition can be derived from the scheme illustrated in Fig. 4(a). An interference maximum is observed when both scattered waves remain in phase, so the difference between their path lengths should be equal to an integer multiple n of the wavelength. The resulting equation is called Bragg's law, or the Wulff-Bragg condition, named after Sir William Lawrence Bragg and Russian crystallographer George Wulff, who separately derived this equation in 1913:

$$n\lambda = 2d sin\theta \qquad \text{(Eq 1)}$$

where d is the spacing between the lattice planes, θ is the scattering angle, λ is the wavelength of the incident wave, and n is the order of interference.

Diffraction spots are usually called reflections, because these spots are left by the ray reflected from the plane. Each XRD experiment gives a number of spots with two characteristics: the angle of diffraction and intensity of the spot. The intensity of the reflection is largely dependent on the chemical composition of the crystal, and the angular distribution of the reflections is the result of the symmetry and size of the unit cell of the crystal. Presence of the translational symmetry elements (i.e., screw axes and glide planes) and centered Bravais lattices results in extinction of a reflection's intensity: some reflections that should appear according to Bragg's law are actually absent. The regular absence of the reflections makes it possible to determine the particular space group of the crystalline compound. Figure 4(b) shows a section of crystal with two types of atoms arranged on parallel planes with Δd spacing. Phases of the x-rays reflected from the series of large- and small-atom planes considered separately are matches, so they should constructively interfere. However, considering the diffraction system together, one can find that the diffracted waves from a large-atom plane lag behind those from small atoms by the value of $\Delta\varphi$:

$$\Delta\varphi = 2\pi\,\frac{\Delta d}{d} \qquad \text{(Eq 2)}$$

If the plane of large atoms is arranged exactly in the middle of the planes of small atoms, $\Delta\varphi$ = 180°, so the waves would destructively interfere. If the large atoms are arranged in the same plane as small atoms, $\Delta\varphi$ = 0°, and the waves would constructively interfere.

A more detailed analysis of $\Delta d/d$ shows that it is equivalent to $hx + ky + lz$, where h, k, and l are the indexes of the plane, and x, y, and z are the fractional coordinates of the atom. Thus, in more general terms:

$$\Delta\varphi_j = 2\pi\left(hx_j + ky_j + lz_j\right) \qquad \text{(Eq 3)}$$

where j refers to the jth atom in the unit cell.

X-rays are electromagnetic waves with a frequency of electric and magnetic vector oscillations at approximately 10^{18} Hz. Protons are too massive—they react weakly to such fast oscillations of the electric field of the x-ray, whereas electrons can oscillate with the frequency of incident x-rays, thereby emitting x-ray radiation with the same frequency. Thus, the electron is the basic unit scattering x-ray waves.

The *atomic scattering factor* f_j shows how many times the amplitude of a wave scattered by an atom is greater than the amplitude of a wave scattered by an electron in the same direction and of the same wavelength. In other words, the function can be represented as the total scattering of atomic electrons: valence and inner shell electrons.

$$f_j = \sum\nolimits_{j=1}^{z} f_e \qquad \text{(Eq 4)}$$

where Z is the number of atomic electrons and f_e is the electron scattering factor, which depends on the distribution function of atomic electrons.

All atoms in the unit cell must contribute to the final diffracted wave, but it is necessary to take into account not only the number of atoms and their differences in the scattering power but also the difference in their initial phases (different locations within the unit cell lead to noncoincidence of the initial phases of scattered waves). The aforementioned features result in the *structure factor* equation:

$$F_{hkl} = |F_{hkl}|e^{i\varphi_{hkl}} = \sum\nolimits_{j=1}^{N} f_j\, e^{i2\pi\left(hx_j+ky_j+lz_j\right)} \qquad \text{(Eq 5)}$$

where N is the number of atoms in the unit cell. This is the first of two basic equations of structural analysis; it expresses the dependence of the amplitude and the phase of any diffraction beam on the coordinates of atoms in the structure and allows one to solve the inverse task facing the researcher: to determine the intensity of diffracted rays from the known atomic coordinates. But physically it is more correct to consider the electron density at each point. According to this approach, the summing in Eq (5) should be replaced by integration over the unit cell, and the atomic scattering factor by the amplitude of scattering from the electron density in an infinitesimal volume dV:

$$F_{hkl} = \int_V \rho(x,\, y,\, z)e^{i2\pi\left(hx_j+ky_j+lz_j\right)}dV \qquad \text{(Eq 6)}$$

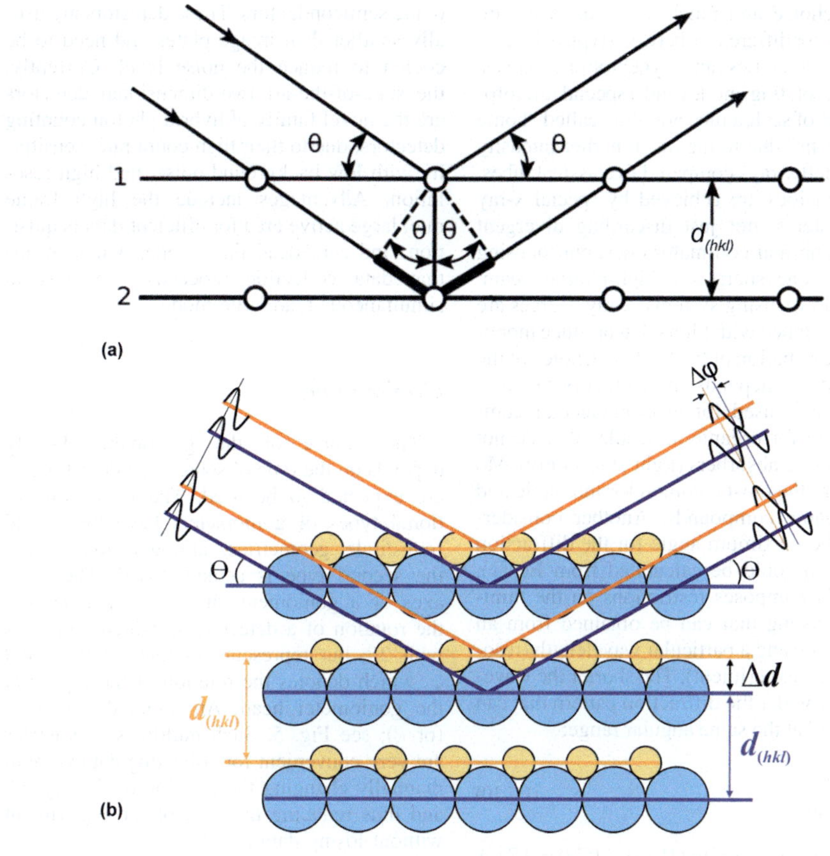

Fig. 4 Diffraction of x-rays from a series of lattice planes: (a) illustration of Bragg's law; (b) extinction of reflected x-rays due to destructive interference

Since the distribution of the electron density in the crystal is periodic and is repeated in each cell, Eq (6) becomes a triple Fourier series:

$$\rho(xyz) = \frac{1}{V_0}\sum\nolimits_{-\infty}^{\infty} h \sum\nolimits_{-\infty}^{\infty} k \sum\nolimits_{-\infty}^{\infty} l\, F_{hkl}\, e^{-i2\pi(hx+ky+lz)} \qquad \text{(Eq 7)}$$

This is the second basic equation of structural analysis. It expresses the dependence of the electron density at some point of the unit cell on the set of structural amplitudes of the rays diffracted by the crystal. If the structural amplitudes of all the reflections are known, then one can find the value of the electron density at any point of the unit cell, and hence the positions of the density maxima—centers of the electron clouds (i.e., atoms). The structure factor is directly related to the intensity of the diffracted wave:

$$I_{hkl} = \left|F_{hkl}^2\right| \cdot LP \cdot A \qquad \text{(Eq 8)}$$

where LP is a combined Lorentz and polarization factor, which depends on the scattering angle θ, and A is an absorption correction factor, which depends on the size, shape, and chemical composition of the crystal.

Phase Problem

The main problem of x-ray structural analysis is the *phase problem.* As seen from Eq (5), the structure factor is composed of an amplitude $|F_{hkl}|$ and a phase, but only amplitudes can be obtained from the experimental diffraction pattern. Without phase information, the electron densities cannot be determined. A general solution has not been found; however, there are several methods that can be successfully applied to circumvent the problem. First attempts were made in a so-called "trial-and-error method," in which the solution was simply to guess the primary structural model. If the proposed structure is correct, the intensities calculated from the model should match those from the experiment.

The next step was made by Arthur Lindo Patterson in 1935. He suggested the first "direct" method (known as the *heavy-atom method* or the *Patterson function*) of structure solution based on diffraction data analysis and not requiring a structural model. The idea is that one atom having many electrons predominates in determining the phases of structure factors. Thus, if the heavy atom can be placed correctly within the unit cell, its phases will be close to the phases of the structure factors. An electron density map based on the heavy-atom phases and the observed structure amplitudes reveals the locations of the lighter atoms.

Another way to solve the phase problem is to use a statistical approach that does not require the presence of heavy atoms. It is known as the *direct methods* and was introduced by Jerome Karle and Herbert Hauptmann, who shared the Nobel Prize in Chemistry in 1985. The idea is based on the fact that the phases of reflections are not independent of each other. Phase information can be obtained by comparing the intensities of related reflections. For instance, if three reflections with indexes $h_1k_1l_1$, $h_2k_2l_2$, and $h_3k_3l_3$ are subject to the equations $h_1 + h_2 + h_3 = 0$, $k_1 + k_2 + k_3 = 0$, and $l_1 + l_2 + l_3 = 0$, then it is most likely that $\varphi(h_1k_1l_1) + \varphi(h_2k_2l_2) + \varphi(h_3k_3l_3) = 0°$. If two of the phases can be assigned, the phase of the third reflection can be calculated. The calculation of an electron density map using only the intense reflections reveals a significant portion of atoms in the structure, which in turn can be used to phase the structure amplitudes, and to construct further, more detailed maps of electron density.

The most recent method for the solution of the phase problem was introduced in the charge flipping algorithm, whose unique feature is the ability to solve structures in arbitrary dimensions. The structure solution proceeds in iterative cycles, initialized by assigning random phases to the experimental structure amplitudes. Then electron density is calculated and modified so that grid points with values below some positive threshold are multiplied by -1. New temporary structure factors are calculated, their phases are combined with the experimental amplitudes, and the dataset enters the next cycle of iteration. An important feature is that the algorithm does not use the symmetry of the structure during density reconstruction ($P1$ symmetry is used); a real space group is assigned afterward.

Validation of Structural Model

It is noteworthy that regardless of the chosen method, only more or less the initial structural model will be obtained, which then should be refined until the closest agreement with the experimental data is achieved. The most common indicator of the agreement between the calculated and observed structure amplitudes is known as the *residual index* or *agreement index*, or simply the *R-factor*:

$$R = \sum \frac{\left|\left|F_{hkl}^{calc}\right| - \left|F_{hkl}^{obs}\right|\right|}{\left|F_{hkl}^{obs}\right|} \qquad \text{(Eq 9)}$$

Favorable matches yield low values of R, with a range between 0.01 and 0.04 (1 to 4%) indicating an excellent structural model and a range between 0.04 and 0.08 (4 to 8%) indicating rather good refinement. Higher values of the R-factor might cause questions from peer reviewers, so it is advisable to explain the reasons to prevent negative comments. But sometimes, crystallographers, especially mineralogists, are faced with the task of investigating very complex mineral structures, for which a value of about 0.15 would be acceptable. The decisive factor here is a reasonable crystal chemical explanation of the structural model. The *R*-factor itself is not a criterion of structure validity; it is of great importance that the structural model does not contradict scientific data.

Experimental Procedure

Each structural determination usually starts with a single crystal selection. A suitable crystal is generally characterized by well-formed faces and uniform optical properties when observed under a polarizing microscope using crossed Nicol prisms or Polaroids. In practice, crystals are often grown under less than ideal conditions and shape, providing only few recognizable faces due to twinning, intergrowth, or other types of defects. Additionally, not all crystals can be examined using the transmitted light due to the opacity of the samples (ore minerals or metal phases). Therefore, the most reliable test for any single crystal is to check its diffraction pattern. Observed diffraction spots should be discrete and more or less rounded, and their indexes should be in good agreement with the known unit cell parameters.

After a suitable crystal has been selected under the microscope, it should be mounted on top of the glass fiber or special nylon loop. A crystal can be fixed by any type of glue (e.g., cyanoacrylate, epoxy resin) that will not react with the crystal. If the sample will be cooled, it is possible to use inert viscous liquids (e.g., motor or Vaseline oil), which will not crystallize at low temperatures.

Then the crystalline sample should be placed for data collection into a single-crystal x-ray diffractometer. Today there are several world-leading manufacturers of such equipment, which have distinct advantages and disadvantages, but fundamentally, all the devices operate according to the same principle, which has not significantly changed since the works of Laue and Braggs. The scheme can be represented as follows: source of x-rays → goniometer (responsible for crystal rotation) → detector of scattered x-rays. Each of these components has quite noticeably evolved over the past century.

X-Ray Sources and Choice of Radiation Type

There are two conventional types of x-ray sources used in laboratories: sealed tube and rotating anode. The anode material in both cases is limited to several metals, among which the most common are Mo and Cu, while others (e.g., Co, Ni, Ga, Ag) are very rare. Limited choice and much lower intensity are the two major differences between typical laboratory x-ray facilities and synchrotron sources. However, rotating anodes and a special microfocused type of sealed tube are often called "home synchrotrons" due to the much higher intensity of generated x-rays compared to classical tubes. High intensities are achieved by special x-ray optics systems, not just discarding divergent rays, as a normal collimator does, but focusing them into one sharp and high-photon beam. Along with focusing systems, x-ray sources are usually equipped with filters that produce monochromatic radiation of the Kα line. Choice of the anode material depends on several crystal parameters. Cu is used for macromolecular compounds and for organic molecules that do not contain strong absorbers (light atoms only). Mo sources are the most common for inorganic and metal-organic compounds. Another consideration is the maximum angle for the diffraction data, which could be calculated from Eq (1). Bragg's law imposes restrictions on the minimum *d* spacing that can be obtained from an experiment using a particular wavelength (resolution of the experiment). The shorter the wavelength, the wider the diffraction pattern that can be obtained at the same angular range:

$$d_{min} = \frac{\lambda}{2sin\theta} \quad \text{(Eq 10)}$$

Thus, according to Eq (10), $d_{min}(Cu) = 0.77$ Å and $d_{min}(Mo) = 0.35$ Å. Here one should also bear in mind that the higher unit cell parameters are, the closer to each other the diffraction spots are. So, for higher accuracy of calculations, reflections need to be separated one from another, which could be done by choosing a larger wavelength. Thus, it makes sense even for inorganic or metal-organic compounds with high unit cell periodicity (greater than 30 Å) to perform measurement using Cu radiation.

X-Ray Detectors

There are three types of detectors: point (scintillation photon counter), linear, and area. The highest precision can be obtained using a point detector, but to reduce the time of data collection, area detectors are preferable for single-crystal diffractometers, among which several varieties exist. An *image plate* is a modern reusable analog of a photographic plate. X-rays produce a latent image on a plate that is then excited by helium–neon laser stimulation. Generated light is irradiated from the areas that were hit by the x-ray photons. This phenomenon is known as photostimulated luminescence. Positions of these spots are analyzed by a scanner; afterward, the surface of the plate is reset by intense light. Another variety is a semiconductor detector, based on charge-coupled device (CCD) technology. It is a set of independent detector elements (up to 4096 × 4096), each composed of metal-oxide semiconductors. These detectors are usually smaller than image plates and need to be cooled to reduce the noise level. Currently, the state-of-the-art two-dimensional detectors are the novel family of hybrid photon counting detectors, due to their high count rate, sensitivity with low background noise, and high resolution. Advantages include the high frame rate, large active area for efficient data acquisition, and zero dead time mode, which means that data collection proceeds very fast in simultaneous read/write mode.

Goniometers

The choice of the goniometer largely depends on the type of source and detector that are expected to be used. The most conventional types of goniometers have three- and four-circle geometries, although today only the second type is actually used. The main axes of a goniometer are 2θ, which denotes the rotation of a detector; ω, which coincides with 2θ, but represents sample rotation; and φ, which denotes the rotation of the crystal at the goniometer head. An optional axis is χ (or κ); see Fig. 5. Such multiaxis geometries are very convenient for collecting data without manually changing the position of the crystal, and thus reducing the time of the experiment without losing data quality.

Experimental Data Reduction

Each manufacturer of x-ray diffractometers uses its own software, but the overall scheme of the reduction procedure is more or less the same for each program. Usually an experiment starts by measuring several patterns to evaluate the quality of the crystal. Each pattern is a diffraction data file (called a frame, Fig. 6), which is obtained by rotating the crystal around one of the axes (usually φ or ω) for a set time (for instance, 1° at ω for 15 s). Then a small series of scans (10 to 20 consecutive frames) are performed to analyze the symmetry of the crystal and raw unit cell parameters: the higher the symmetry, the lower the number of frames that will be needed, and the shorter the experiment is. Afterward, the program analyzes the collected data and suggests a strategy for the full diffraction experiment (multiple sets of sequentially measured frames). Otherwise, the whole diffraction sphere can be collected, which, although valuable for low-symmetry crystals (triclinic and monoclinic), might be too redundant for high-symmetry crystals (tetragonal or cubic). After the end of the experiment, unit cell parameters are refined using the entire set of the diffraction data, and a file in *.*hkl* format, which contains structure amplitudes, is produced as the result of the data integration procedure and introduction of several corrections. The absorption correction is one of those.

Absorption reduces the intensity of an x-ray beam passing through the crystal by an amount that depends on the size and chemical composition of the crystal. The linear absorption coefficient μ can be calculated for the given wavelength from the mass absorption coefficients μ_m of the atoms filling the unit cell, but no structural information is required. Thus,

$$\mu = \rho \sum_i g_i \mu_m^i \quad \text{(Eq 11)}$$

where g_i is the mass fraction of the ith element present within the cell, μ_m^i is its mass absorption coefficient, and ρ is the crystal density. On the one hand, the larger the crystal, the stronger the intensity of the diffraction spots is, but on the other hand, a diffracted beam of low intensity could be simply blocked by the crystal's absorption. Previously crystallographers tried to select a crystal with the absorption-to-size ratio $\mu{\cdot}r = 1$ ($mm^{-1}{\cdot}mm$). But today very sensitive detectors are used in instruments, allowing crystallographers to collect data from tiny crystals, for which an absorption correction could be introduced analytically, according to the shape of the crystal, or even ignored (which, of course, is not optimal).

After all corrections are introduced, the space group for the compound is defined by analyzing the data array via a statistical approach. Only after that, the process of crystal structure evaluation starts.

The first step of a structure description, called *structure solution*, is to determine the initial structural model, which entails finding positions of some portion of atoms. Using modern crystallographic software, this stage lasts from several seconds to usually not more than half a minute. From the basics of the aforementioned diffraction theory, this is the most important step upon which the ultimate success of structural description largely depends. However, today it takes even less time, and what is now called "structural analysis" begins after this stage. In order to obtain an appropriate initial model, it is desirable to indicate the correct chemical

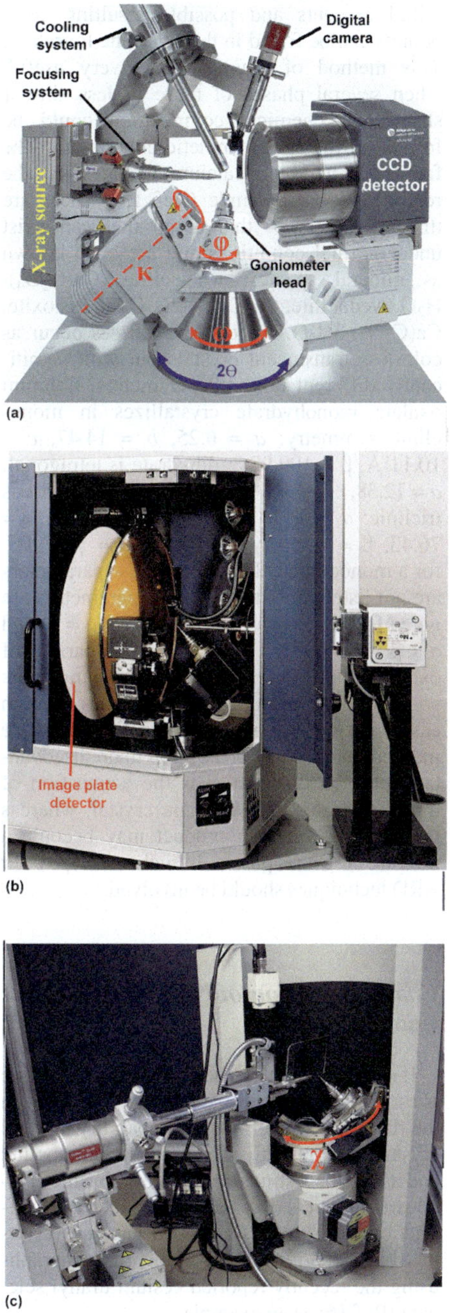

Fig. 5 Single-crystal diffractometers: (a) combination of four-circle kappa goniometer and charge-coupled device detector (image of Rigaku Oxford Diffraction instrument; source: Ref 14); (b) combination of three-circle goniometer and image plate detector (image of STOE IPDS II instrument); (c) combination of three-circle χ goniometer and curved image plate detector (image of Rigaku R-Axis Rapid II instrument)

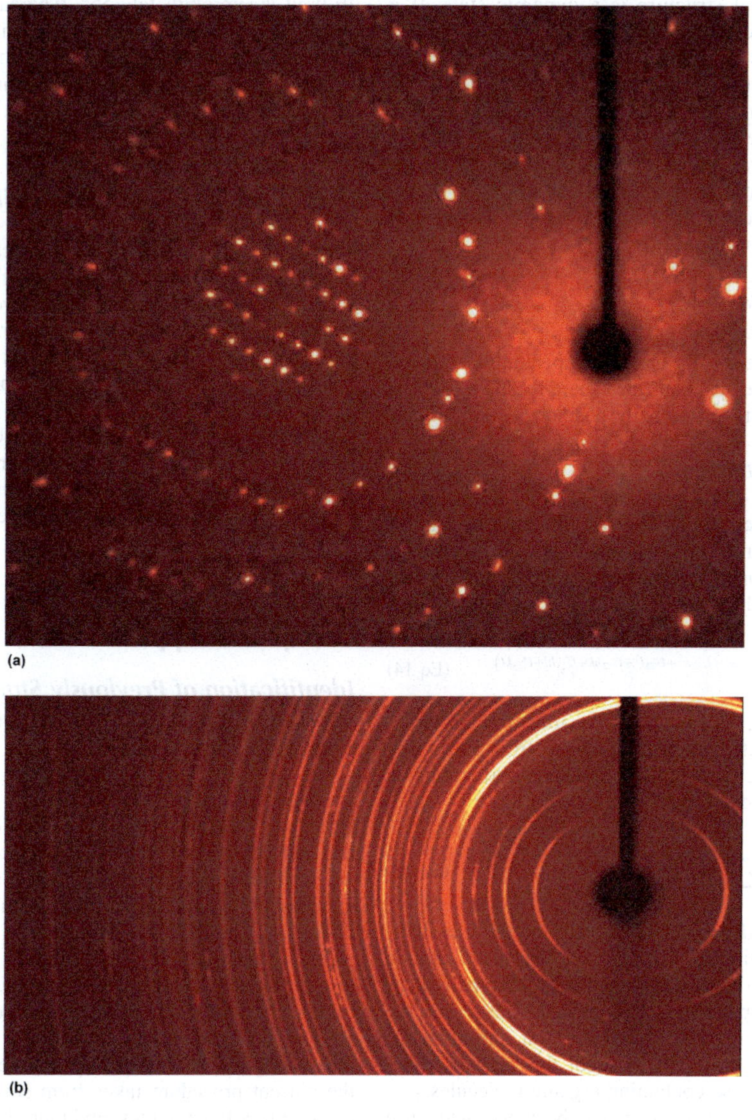

Fig. 6 Diffraction patterns obtained from (a) single-crystal and (b) polycrystalline samples and recorded using a charge-coupled device detector

composition of the compound before solving the structure, so that the respective atomic factors can be properly assigned; however, experienced crystallographers could guess the chemical composition, although it could take more time.

The next step, called *structure refinement*, is to "polish" the structural model by considering a number of effects, such as thermal displacement of atoms, the degree of perfection of a crystal, and site occupancy factors. Structure refinement is a cyclic process. First, an initial model (which can be partial) is defined. Theoretical intensities are calculated on its basis, they are compared with experimental ones, and subsequently a new model is defined. Then some changes are made to this model, and a new set of intensities is calculated. This process continues until a final reasonable crystal chemical model of the structure is reached and an acceptable value of the R-factor is obtained.

Crystal structure is a dynamic state of a substance, because all atoms oscillate within the crystal lattice. Atomic *thermal vibrations* spread the electron density of atoms averaged over time, which introduces correction into the atomic scattering factor. First, the structural model usually operates only with mean atomic positions and thermal vibrations described in an isotropic (spherical) approximation:

$$\tau = e^{-Bsin^{2\theta}/\lambda^2} \quad \text{(Eq 12)}$$

where B is the atomic temperature factor, and U^2 is a dispersion (the square mean shift of the atom with respect to the position of equilibrium):

$$B = 8\pi^2 U^2 \quad \text{(Eq 13)}$$

In fact, atoms do not vibrate equally in all directions; their vibrations are considered an anisotropic approximation and are described by an ellipsoid:

$$\tau = e^{-(U_{11}h^2+U_{22}k^2+U_{33}l^2+U_{12}hk+U_{13}hl+U_{23}kl)} \quad \text{(Eq 14)}$$

Thus, the complete equation for the structure factor could be obtained from Eq (5) by introducing temperature factor corrections as a coefficient f_j:

$$F_{hkl} = \sum_{j=1}^{N} \tau_j f_j e^{i2\pi(hx_j+ky_j+lz_j)} \quad \text{(Eq 15)}$$

In order to define the orientation of the thermal ellipsoid centered on the mean atomic position with respect to crystallographic axes, six U_{ij} parameters are needed. The lighter the atom, the higher the vibration amplitude is, so it is preferable to cool the crystals, especially those containing organic molecules.

In summary, it should be mentioned once again that the R-factor is a useful but not the only indicator of the structure refinement precision. When monitoring its values, one should not forget about precision, the rationality of the structural model (e.g., interatomic distances and angles, site occupancy factors), and the chemical logic. To obtain an appropriate model, the diffraction dataset should be sufficient. One refined parameter should account for five to ten registered reflections. Taking into account that one atom in a model has nine or ten refined parameters (three fractional coordinates, six anisotropic thermal parameters, and optional site occupancy), a simple structure of ten atoms has about 100 refined parameters (the aforementioned atomic and common ones, such as a scale factor), so the dataset should consist of at least 500 (or better yet 1000) reflections. Next, the redundancy and completeness of the dataset are addressed. Far-angle reflections affect the precision of interatomic parameters, so there is an upper limit for each wavelength. For instance, with Mo $K\alpha$ radiation, the required upper 2θ limit is approximately 55°, whereas for Cu $K\alpha$, it is approximately 140°. The diffraction dataset should also be intense; the mean signal-to-noise ratio (I/σ_I) is recommended to be not less than 10.

Crystallographic Information File

After the structure is solved, a special *Crystallographic Information File* (*.cif) should be generated and carefully checked or revised if needed. These files contain comprehensive information about an array of diffraction data, crystal structure, and the origin and properties of the crystal. The *.cif format and its structure were developed by the International Union of Crystallography with the goal of creating a standardized dataset that most modern crystallographic software packages could recognize, process, and even create.

Examples of Applications

Identification of Previously Studied Crystalline Phases by Comparison of Unit Cell Parameters

Diagnostics of a material by the unit cell parameters has become one of the most rapid and effective methods of crystalline phase determination. For instance, during a synthetic experiment, a number of crystals suitable for diffraction studies are formed. To analyze whether the reaction is successful, one can select a single crystal, place it into the diffractometer, and collect the starting dataset. Using modern diffractometers and keeping in mind the average size and quality of the crystal, the current procedure takes from several seconds up to a few minutes, and there is no need to use the whole amount of the reaction product. Afterward, the data reduction program reports rather accurate values of the unit cell parameters for the selected single crystal. Then, these parameters should be compared with the values of already known crystalline compounds. The most comprehensive structural data for the comparison are contained within special databases (see Databases in Single-Crystal XRD Analysis). However, if one is working in a well-known synthetic system, crystallographic information about the initial reagents and possible resulting compounds can be found in the scientific literature. This method of diagnostics is very useful, when several phases of more or less similar shape and chemical composition could be formed in the same synthetic experiment. The family of hydrated calcium oxalates can be regarded as an example (Ref 15). There are three members of this group that can exist under normal conditions and are even known as mineral species: whewellite, $Ca(C_2O_4)\cdot H_2O$, weddellite, $Ca(C_2O_4)\cdot 2H_2O$, and caoxite, $Ca(C_2O_4)\cdot 3H_2O$. All of these phases occur as colorless translucent crystals but with significantly different unit cell parameters. Calcium oxalate monohydrate crystallizes in monoclinic symmetry: a = 6.25, b = 14.47, c = 10.11 Å, β = 109.98 °; dihydrate is tetragonal: a = 12.38, c = 7.37 Å; and the trihydrate phase is triclinic: a = 6.11, b = 7.16, c = 8.44 Å, α = 76.43, β = 70.19, γ = 70.91 ° (α = γ = 90 ° for a monoclinic cell and thus these parameters are not listed; only two variable parameters are given for the tetragonal lattice since $a = b$, and α = β = γ = 90 °). Therefore, by comparison of experimentally determined unit cell parameters with those known from previous studies, one can easily assign the obtained synthetic crystalline material to the exact phase. The main disadvantage of such diagnostics is the acquisition of information for only a single crystal, whereas the resulting reaction product may become a mixture of several phases. In such a case, powder XRD techniques should be involved.

Characterization of New Crystalline Compounds

If it is not possible to determine the crystalline phase during unit cell parameter verification in the databases, there is high probability that the crystal under investigation is a novel compound and it has a new structure. To determine the crystal structure, it is necessary to conduct a full diffraction experiment, some details and the results of which are given in the following paragraphs using the recently reported cesium uranyl selenate (Ref 16) as an example.

Yellow-green lamina crystals of $Cs_2[(UO_2)_2(SeO_4)_3]$ were obtained during evaporation at room temperature from an aqueous solution of $(UO_2)(NO_3)_2\cdot 6H_2O$, $CsNO_3$, and H_2SeO_4. Diffraction data were collected using monochromated Mo Kα radiation at 150 K with frame widths of 0.5° in ω and φ, and an exposure time of 50 s per each frame. Data were integrated and corrected for background, Lorentz, and polarization effects. Absorption correction was applied using an empirical spherical model. The unit cell parameters ($P\bar{4}2_1m$, a = 9.8560 (16), c = 8.1587(14) Å) were refined by the least-squares technique on the basis of 5428 reflections with 2θ in the range of 4.13 to 55.00°. The structure was solved by direct methods and refined to R_1 = 0.020 (wR_2 = 0.043) for 918 unique reflections with $I \geq 2\sigma(I)$. The final model included coordinates (Table 2) and anisotropic displacement parameters for all atoms.

The crystal structure of $Cs_2[(UO_2)_2(SO_4)_3]$ contains one crystallographically nonequivalent U atom with two short $U^{6+}{\equiv}O^{2-}$ bonds (1.756(8)–1.774(10) Å) forming an approximately linear UO_2^{2+} uranyl ion (*Ur*). *Ur* is coordinated by five O atoms (U1–O_{eq} = 2.332 (12)–2.413(9) Å) that belong to selenate groups and that are arranged in the equatorial plane of the UrO_5 pentagonal bipyramid. Two symmetrically nonequivalent Se^{6+} sites are tetrahedrally coordinated by four O^{2-} atoms each. The $(Se1O_4)^{2-}$ group is three-connected, sharing three vertexes with three adjacent U polyhedra. Unshared vertexes of the tetrahedra are oriented either *up* or *down* relative to the plane of the layer. The $(Se2O_4)^{2-}$ group exhibits four-connected bridging, having all four O atoms common with U polyhedra. The crystal structure of $Cs_2[(UO_2)_2(SeO_4)_3]$ is based upon the infinite $[(UO_2)_2(SeO_4)_3]^{2-}$ layers (Fig. 7) arranged parallel to (001) that are formed by linkage of the UO_7 pentagonal bipyramids and SO_4 tetrahedra via common vertexes (O atoms). These layers are corrugated due to the presence of four-connected tetrahedra in an "egg tray" manner with an approximately 6 Å diameter of the "tray's cell."

Analysis of Chemical Composition

The aforementioned compound was obtained during phase formation studies in a mixed sulfate-selenate U-bearing aqueous system, and except for the pure sulfate and selenate crystals, seven compounds with variable content of S and Se were also synthesized (Ref 16). In order to control the chemical composition of crystals obtained during the structural studies, energy-dispersive x-ray spectra were measured using a scanning electron microscope, and discrepancies in the values were not higher than the first percentile. Moreover, chemical analysis based on single-crystal XRD data can show much more detailed information, such as the distribution of atoms between various sites. When more than one atom occupies the same crystallographic site, its scattering ability (site-scattering factor) is equal to the total electron density, which is proportional to the contribution of each atom. The distinctive feature of $Cs_2[(UO_2)_2(TO_4)_3]$ (T = S, Se) structures is the selective distribution of Se^{6+} and S^{6+} cations between the two tetrahedral sites. Thus, the site-scattering factor for the position fully occupied by S atom should be close to 16 *electrons per formula unit* (epfu), and that for Se atom should be close to 34 epfu. It appeared that for the crystal of $Cs_2[(UO_2)_2(SeO_4)_{1.16}(SO_4)_{1.84}]$ composition, the *T1* site (site-scattering factor = 25.3 epfu) is half occupied by Se and S atoms (Se:S = 0.50:0.50), whereas the *T2* site (site-scattering factor = 18.9 epfu) is predominately occupied by S atoms with a significantly lower amount of Se (Se:S = 0.16:0.84). The closer to each other in the periodic table are the atoms, the more difficult it is to evaluate their contribution, because the total electron density is close to their individual values. In such complicated cases, collection of high-redundancy diffraction data could help (full sphere even for high-symmetry crystals with high-intensity diffraction spots even at the far 2θ range).

Table 2 Atomic coordinates and isotropic displacement parameters, Å², for $Cs_2[(UO_2)_2(SeO_4)_3]$

Atom (a)	x(b)	y	z	B_{iso}
U1	0.80273(4)	0.30273(4)	0.35074(7)	0.01319(13)
Se1	0.83188(9)	0.66812(9)	0.2535(2)	0.0147(3)
Se2	0.5	0.5	0.5	0.0118(4)
O1	0.7213(7)	0.2213(7)	0.1837(12)	0.018(2)
O2	0.8825(8)	0.3825(8)	0.5175(14)	0.023(3)
O3	0.8009(14)	0.5128(9)	0.1988(11)	0.056(3)
O4	0.5855(10)	0.4067(11)	0.3801(11)	0.050(2)
O5	0.7331(8)	0.7669(8)	0.1623(16)	0.050(2)
O6	0.8199(8)	0.6801(8)	0.4549(15)	0.050(2)
Cs1	0.5	0	0.1537(2)	0.0195(3)
Cs2	0.5	0.5	0	0.0237(3)

(a) Only crystallographically nonequivalent atoms are listed, which is half of the atoms in the empirical formula and a fourth part of the unit cell content. The rest of the atoms are generated by symmetry operations. (b) Coordinates are given as a fraction of the translation, ranging from 0 to 1; estimated standard deviations are given in parentheses for the variable parameters and are omitted for the fixed (for instance, a Se2 atom occupies a special position on the fourfold inversion axis and thus its coordinates cannot deviate).

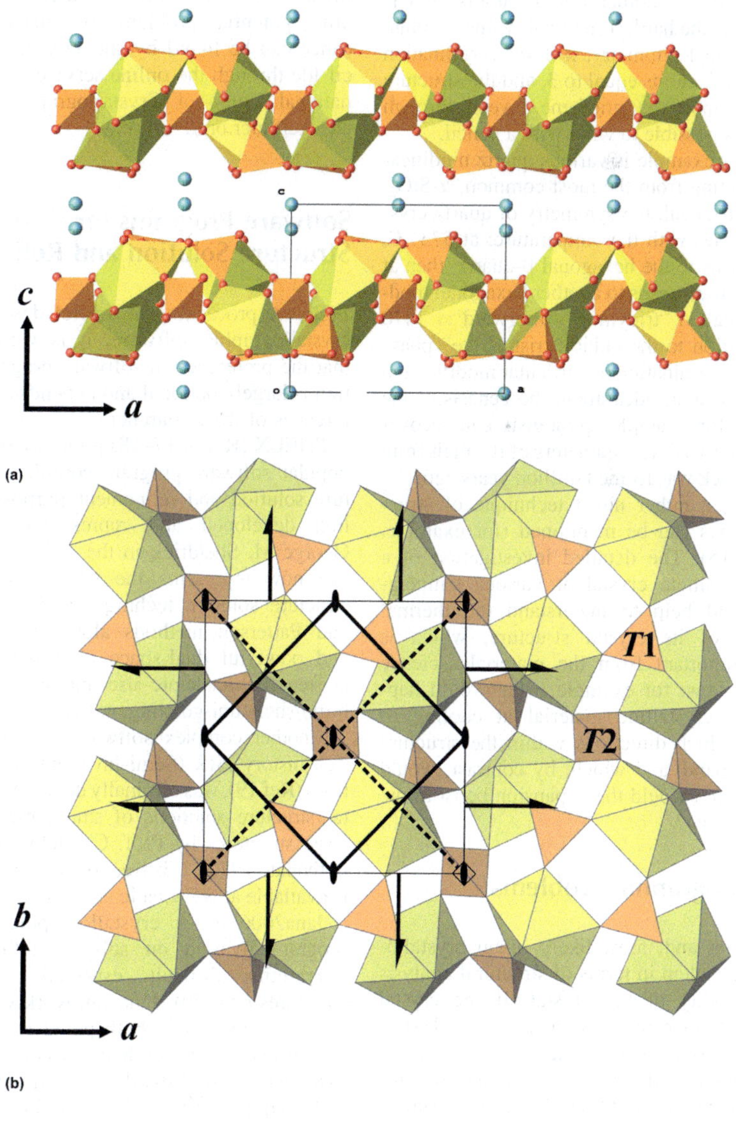

Fig. 7 Polyhedral representation of the $Cs_2[(UO_2)_2(SeO_4)_3]$ crystal structure: (a) view along the layers, and (b) projection of the uranyl-selenate layer, with symmetry elements overlaid. Legend: U polyhedra = yellow, SeO_4 tetrahedra = orange, Cs atoms = cyan, O atoms = red

Interpretation of Physical and Chemical Properties, and Growth Conditions

Of course, any properties of crystalline compounds (such as luminescence characteristics, magnetic susceptibility, piezo- or pyroelectric properties) can be evaluated without structural studies. On the other hand, only precise determination of a material's crystal structure can provide an answer as to *why* this compound has such properties. There are numerous well-known examples, especially among the polymorph modifications (which have identical chemical composition but different crystal structures), that demonstrate how the specific atomic arrangement governs the manifestation of physical and chemical properties. For instance, consider diamond and graphite (two polymorph modifications of C): diamond has a cubic framework structure with all C atoms tetrahedrally coordinated by another four C atoms, and is regarded as the hardest material; while graphite has hexagonal symmetry, with the coordination number of C atoms equal to 3, and the structure is based upon the graphene layers, through which it is possible to write with a pencil.

Another example is various quartz modifications. Starting from the most common, α-SiO_2, which is trigonal, the symmetry of quartz crystals increases with the temperature: at 573 °C, it transforms to the hexagonal β-quartz; then at 870 °C, it transforms to another hexagonal modification called "tridymite"; afterward at 1470 °C, transition to the cubic "cristobalite" phase occurs. Determination of particular modification could provide an idea about the genesis of the material; for example, geologists can recover the thermodynamic parameters of the melt from which a rock was formed million years ago.

Finally, a rather novel technique of structural studies can be mentioned (for example, see Ref 15). The detailed investigation of a particular single crystal at various temperatures could help to understand the thermal behavior of its crystal structure, which is highly important from the materials science point of view: for example, what would happen with crystalline material at cooling or heating; which directions within the structure would expand, and which, by contrast, would contract; and would the expansion be isotropic or anisotropic?

Crystallographic Problems

The first and, most likely, main crystallographic problem in terms of structural analysis is the quality of the crystal. If the crystal quality is poor (e.g., of a very small size, defective, not a single crystal), it is quite difficult to obtain diffraction data suitable for further structure description. Once a crystal is selected, it might be short lived in air. To keep materials stable, a type of capillary or immediate freezing in liquid nitrogen might be used, followed by a measurement at low temperatures. Instability of the structure or movability of some structural parts (for instance, terminal *tert*-butyl groups, partially evaporated solvent molecules, unevenly distributed ions or water molecules within any type of structural voids) can result in structural disorder and uncertainty of the assignment of its positions. Determination of the crystal symmetry can be ambiguous. For instance, correct space group determination sometimes can be made only after full cycles of solution and refinement followed by precise analysis and re-revision of the structural model (quite often such a problem relates to the presence or absence of a center of symmetry). Another example is an erroneously defined crystal system: orthorhombic instead of monoclinic with a β angle of 90° within experimental error. Sub- or superlattices are features necessitating a detailed look at the weak reflections on the diffraction pattern. Most of the errors mentioned here and some other potential problems of various significance can be found by checking the resulting cif file through the online service of the International Union of Crystallography at http://checkcif.iucr.org/.

Software Programs for Crystal Structure Solution and Refinement

Before proceeding to a brief description of crystallographic software, it is worth noting that the preference in software program selection is largely optional and depends on the preferences of the researcher.

SHELX (Ref 8, 17–18) probably is the most popular software program complex for structure solution and refinement purposes. Being first developed and represented by Prof. George M. Sheldrick in the 1970s, this package now represents the whole spectrum of structure solution techniques: classical direct and Patterson methods along with a novel and powerful dual-space algorithm. SHELX is free for academic use and is available at http://shelx.uni-goettingen.de/.

Another complex software program with a rich history, SIR (Semi-Invariants Representation) (Ref 19), was originally developed for crystal structure solutions of small, medium, and large structures by Prof. Carmelo Giacovazzo in the late 1970s. It is free for academic use and is available at www.ba.ic.cnr.it/softwareic/sir/.

Jana2006 is a crystallographic software program focused on solving, refining, and interpreting difficult, especially modulated structures (Ref 20). Jana can work with multiphase structures (for both powder and single-crystal data) and merohedric twins as well as twins with partial overlap of diffraction spots and commensurate and composite structures. It is free for academic use and is available at http://jana.fzu.cz/. Jana can also be used to solve structures from powder diffraction data; however, one should keep in mind more specialized software programs (for instance, see Ref 21).

SUPERFLIP is a software program for application of the charge-flipping algorithm to crystal structure solutions from diffraction data (Ref 22). The program is freeware and available for download at http://superflip.fzu.cz/.

WinGX provides a user-friendly graphical user interface (GUI) for some of the best and most popular publicly available crystallographic software programs (such as SHELX and SIR) (Ref 23). WinGX is provided free of charge for academic, scientific, and educational users by its author Prof. Louis J. Farrugia at www.chem.gla.ac.uk/~louis/software/wingx/.

Olex2 is a simple to use but very powerful software program package, providing everything needed to solve, refine, and finalize a small-molecule crystal structure using an intuitive GUI (Ref 24). The software has its own solution and refinement programs but also is fully compatible with popular ones, such as SHELX and SUPERFLIP. It is free to use and is available at www.olexsys.org/.

The software program PLATON was developed by Prof. Anthony L. Spek and is also known as a "multipurpose crystallographic tool" (Ref 25–26). For instance, it helps to find the higher symmetry of a structure or the twinning law. Another very useful option is a SQUEEZE routine, which treats the disordered molecules as a diffuse contribution to the overall scattering without specified atom positions. The International Union of Crystallography online check-cif service is operated by PLATON algorithms. The program is available for free download at www.cryst.chem.uu.nl/spek/platon/.

Visualization of Crystal Structures

Some software programs dealing with structure refinement have their own GUIs, which sometimes are very powerful, as in the case of Olex2. Other programs provide only graphical support and analysis of structural data without model refinement mechanisms. Visualization programs can be divided into groups for the convenience of working with structures of inorganic and organic compounds, although all of them can process any of the structure data files. The most common and popular programs are Atoms (www.shapesoftware.com), CrystalMaker (www.crystalmaker.com); Diamond (www.crystalimpact.com/diamond/), Mercury (supported by the Cambridge Crystallographic Data Centre and available at www.ccdc.cam.ac.uk/solutions/csd-system/components/mercury/), and VESTA (http://jp-minerals.org/vesta/en/).

Some examples of structure visualization in different representations are provided.

Figure 8(a) shows disordering of a nitro moiety due to a reverse at the chiral C6 center in a molecule of 3-parachlorophenyl-6-methyl-6-nitro-1-carbamoylhexahydrothieno[2,3-d] pyrazole-4,4-dioxide, resulting in the presence of two diastereomers within the crystal structure (Ref 27). C, N, O, Cl, and S atoms are gray, blue, red, green, and orange, respectively; thermal ellipsoids are shown at the 50% probability level; and the picture was taken in the Olex2 program.

Figure 7 provides a polyhedral representation of the $Cs_2[(UO_2)_2(SeO_4)_3]$ crystal structure (Ref 15); the pictures were taken in the Olex2 (Fig. 7a) and Atoms (Fig. 7b) programs.

A combination of capped-stick and ball-and-stick representations of the $[(\mu 3\text{-}Bu^{t}NAu_3)_2(PPh_2C_6H_4PPh_2)_3]^{2+}$ molecule is depicted in Fig. 8(b) (Ref 28); the picture was taken in the Mercury program.

Figure 8(c) and (d) shows an $[Cu_3(SeO_4)_2(SeO_3OH)_2(H_2O)_{12}]$ octahedral-tetrahedral heptamer in ball-and-stick and polyhedral representations, respectively (Ref 29); the pictures were taken in the CrystalMaker program.

Databases in Single-Crystal XRD Analysis

The number of crystal structures is steadily growing. Most structures are being reported in the scientific literature, and also can be found within several databases that are available both online and by using special software that can be installed on personal computers.

The Cambridge Structural Database (CSD) was established in 1965, and is the world's repository for small-molecule organic and metal-organic crystal structures. Structural information can be analyzed in a program window or by downloading crystallographic information files in the conventional *.cif format. The database available at www.ccdc.cam.ac.uk is a commercial product; however, a particular structure file can be obtained free of charge. In

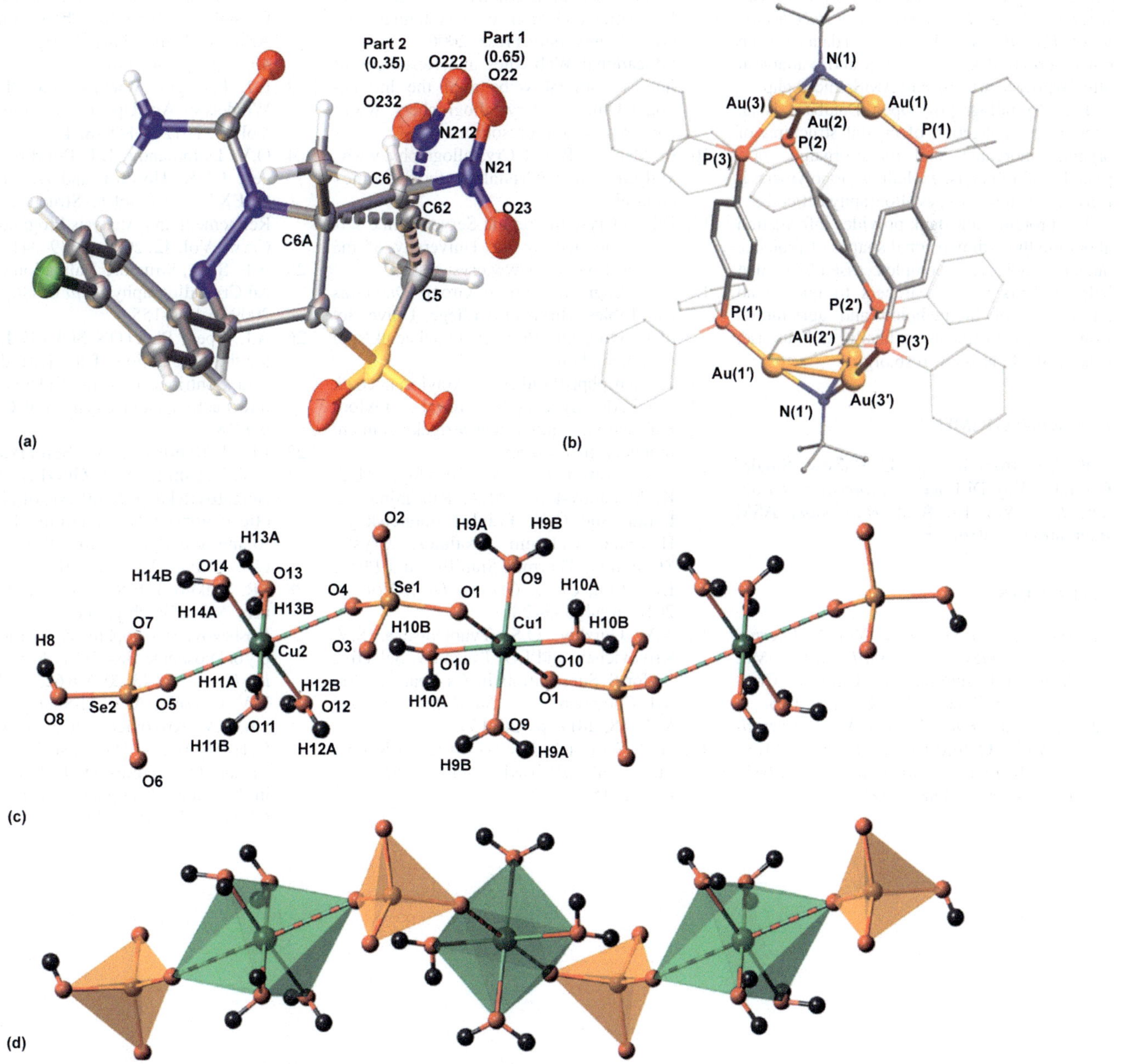

Fig. 8 Examples of structure visualization in different representations: (a) disordering of nitro moiety; (b) combination of capped-stick and ball-and-sticks representations; (c) octahedral-tetrahedral heptamer shown in ball-and-stick representation; (d) octahedral–tetrahedral heptamer shown in polyhedral representation

addition, the CSD offers a selection of products and services free of charge and covering a wide range of crystallographic tools, from data collection, validation, and visualization to teaching, research, and analysis.

The Inorganic Crystal Structure Database contains a list of known inorganic crystal structures (pure elements, minerals, metals, and intermetallic compounds) published since 1913, including their symmetry and unit cell parameters, atomic coordinates, thermal displacement parameters, mineral names, bibliographic data, and synthesis conditions. The database is available at www2.fiz-karlsruhe.de/icsd_home.html.

The American Mineralogist Crystal Structure Database contains every mineral-related structure published in the journals *American Mineralogist*, *The Canadian Mineralogist*, *European Journal of Mineralogy*, and *Physics and Chemistry of Minerals*, as well as selected datasets from other journals. It is free of charge and available at http://rruff.geo.arizona.edu/AMS/amcsd.php.

The Crystallography Open Database is an open-access collection of crystal structures of organic, inorganic, and metal-organic compounds and minerals, excluding biopolymers; it is available at www.crystallography.net/cod/.

The Protein Data Bank provides information about the three-dimensional shapes of proteins, nucleic acids, and complex assemblies that help students and researchers to understand all aspects of biomedicine and agriculture, from protein synthesis to health and disease; it is available at www.rcsb.org.

ACKNOWLEDGMENT

Revised from Richard L. Harlow, Single-Crystal X-Ray Diffraction, *Materials Characterization*, Vol 10, *ASM Handbook*, ASM International, 1986.

REFERENCES

1. *Space-Group Symmetry*, Vol A, *International Tables for Crystallography*, M.I. Aroyo, Ed., International Union of Crystallography/John Wiley & Sons, 2016
2. A. Authier, *Early Days of X-Ray Crystallography*, Oxford University Press, 2013
3. M.J. Buerger, *Crystal-Structure Analysis*, John Wiley & Sons, 1960
4. M.J. Buerger, *Elementary Crystallography*, John Wiley & Sons, 1963
5. M.J. Buerger, *Vector Space and Its Application in Crystal-Structure Investigation*, John Wiley & Sons, 1959
6. *Crystal Structure Analysis: Principles and Practice*, No. 13, *IUCr Texts on Crystallography*, 2nd ed., W. Clegg, Ed., International Union of Crystallography/Oxford University Press, 2009
7. *Fundamentals of Crystallography*, No. 15, *IUCr Texts on Crystallography*, 3rd ed., C. Giacovazzo, Ed., International Union of Crystallography/Oxford University Press, 2011
8. *Crystal Structure Refinement: A Crystallographer's Guide to SHELXL*, No. 8, *IUCr Texts on Crystallography*, P. Muller, Ed., International Union of Crystallography/Oxford University Press, 2006
9. Educational Web Sites and Resources of Interest, official website of the International Union of Crystallographers, www.iucr.org/education/resources
10. M. Martínez-Ripoll, Crystallography, www.xtal.iqfr.csic.es/Cristalografia/welcome-en.html
11. Bilbao Crystallographic Server online service supported by the University of the Basque Country, www.cryst.ehu.es
12. Crystallographic Space Group Diagrams and Tables, Birkbeck College, University of London, http://img.chem.ucl.ac.uk/sgp/mainmenu.htm
13. xForum, phpBB Ltd., www.xrayforum.co.uk
14. "XtaLAB Synergy-S," Rigaku Oxford Diffraction, https://www.rigaku.com/en/products/smc/synergy
15. A.R. Izatulina, V.V. Gurzhiy, M.G. Krzhizhanovskaya, M.A. Kuz'mina, M. Leoni, and O.V. Frank-Kamenetskaya, Hydrated Calcium Oxalates: Crystal Structures, Thermal Stability and Phase Evolution, *Cryst. Growth Des.*, Vol 18, 2018, p 5465–5478
16. V.V. Gurzhiy, O.S. Tyumentseva, S.V. Krivovichev, and I.G. Tananaev, Selective Se-for-S Substitution in Cs-Bearing Uranyl Compounds, *J. Solid State Chem.*, Vol 248, 2017, p 126–133
17. G.M. Sheldrick, Crystal Structure Refinement with SHELXL, *Acta Cryst.*, Vol C71, 2015, p 3–8
18. G.M. Sheldrick, SHELXT - Integrated Space-Group and Crystal-Structure Determination, *Acta Cryst.*, Vol A71, 2015, p 3–8
19. M.C. Burla, R. Caliandro, B. Carrozzini, G.L. Cascarano, C. Cuocci, C. Giacovazzo, M. Mallamo, A. Mazzone, and G. Polidori, Crystal Structure Determination and Refinement via SIR2014, *J. Appl. Cryst.*, Vol 48, 2015, p 306–309
20. V. Petricek, M. Dusek, and L. Palatinus, Crystallographic Computing System JANA2006: General Features, *Z. Kristallogr.*, Vol 229, 2014, p 345–352
21. R. Cerny, V. Favre-Nicolin, J. Rohlicek, and M. Husak, FOX, Current State and Possibilities, *Crystals*, Vol. 7, 2017, p 322
22. L. Palatinus and G. Chapuis, SUPERFLIP – A Computer Program for the Solution of Crystal Structures by Charge Flipping in Arbitrary Dimensions, *J. Appl. Cryst.*, Vol 40, 2007, p 786–790
23. L.J. Farrugia, WinGX and ORTEP for Windows: An Update, *J. Appl. Cryst.*, Vol 45, 2012, p 849–854
24. O.V. Dolomanov, L.J. Bourhis, R.J. Gildea, J.A.K. Howard, and H. Puschmann, OLEX2: A Complete Structure Solution, Refinement and Analysis Program, *J. Appl. Cryst.*, Vol. 42, 2009, p 339–341
25. A.L. Spek, Structure Validation in Chemical Crystallography, *Acta Cryst.*, Vol D65, 2009, p 148–155
26. A.L. Spek, PLATON SQUEEZE: A Tool for the Calculation of the Disordered Solvent Contribution to the Calculated Structure Factors, *Acta Cryst.*, Vol C71, 2015, p 9–18
27. I.E. Efremova, A.V. Serebryannikova, L.V. Lapshina, V.V. Gurzhiy, and V.M. Berestovitskaya, Synthesis of New Bicyclic Compounds Containing Fused Sulfolane and Pyrazolidine Rings, *Russ. J. Gen. Chem.*, Vol 86, 2016, p 481–488
28. J.R. Shakirova, E.V. Grachova, A.J. Karttunen, V.V. Gurzhiy, S.P. Tunik, and I.O. Koshevoy, Metallophilicity-Assisted Assembly of Phosphine-Based Cage Molecules, *Dalton Trans.*, Vol 43, 2014, p 6236–6243
29. V.V. Gurzhiy, A.A. Al-Shuray, S.N. Britvin, S.V. Krivovichev, $Cu_3(SeO_4)_2(SeO_3OH)_2(H_2O)_{16}$ – The First Example of a Linear Octahedral-Tetrahedral Heptamer in Inorganic Compounds, *Eur. J. Inorg. Chem.*, Vol 2015, 2015, p 5311–5313

ASM Handbook, Volume 10, *Materials Characterization*
ASM Handbook Committee
DOI 10.31399/asm.hb.v10.a0006656

Micro X-Ray Diffraction

Thomas N. Blanton, International Centre for Diffraction Data
Gregory Schmidt, Thermo Fisher Scientific

Overview

General Uses

- Confirmation of crystallinity
- Phase identification of crystalline phases and some amorphous phases (qualitative analysis)
- Quantitative phase analysis
- Texture (orientation) analysis
- Stress analysis

Examples of Applications

- Phase analysis of small quantities (single crystals, cluster of crystals, powders) of inorganics, including ceramics and metals, organics, polymers, etc.
- Identification of pigments in cultural heritage items
- Orientation of films and fibers
- Residual stress in metal articles
- Changes in crystallite size in thermally treated articles
- Alignment of single crystals

Samples

- *Form:* fine powder, single crystal, crystal cluster, film, wire, small region on large object
- *Size:* micrograms to milligrams of powder; single crystal or crystal cluster 1 mm down to 50 μm; possible to analyze samples < 10 μm when comprised of high-*Z* elements; film or wire tens to hundreds of micrometers in thickness, depending on elements present
- *Preparation:* load into a glass or polyimide film capillary; mount on the end of a glass capillary; film and wire as-received

Limitations

- Phase identification requires commercial (i.e., International Centre for Diffraction Data Powder Diffraction File) or user database
- Amorphous phases difficult to characterize
- Samples comprised of low-*Z* elements are poor x-ray scatterers
- Requires careful handling of small quantities of sample
- Samples exhibiting significant x-ray fluorescence contribute to excessive background
- Analyzed sample may not be representative of bulk sample

Estimated Analysis Time

- Data-collection time can range from a few minutes to hours, depending on the amount of sample, the composition of the sample, and the analysis method employed.
- Data-analysis time for phase identification of major phases is typically less than an hour.
- Analysis of multiphase samples, residual-stress measurements, texture studies, etc. on can require a few hours.

Capabilities of Related Techniques

- *Synchrotron radiation:* beamline high-flux x-ray source coupled to a microdiffractometer
- *X-ray fluorescence:* elemental composition
- *Electron microscope with x-ray detector:* elemental composition
- *Electron microscope:* selected-area electron diffraction or electron backscatter diffraction phase identification
- *Single-crystal x-ray diffraction:* crystal structure determination
- *Microinfrared and micro-Raman spectroscopy:* molecular structure and phase identification
- *Polarized optical microscopy:* orientation of polymer films

Introduction

X-ray diffraction (XRD) has been a fundamental tool in the field of crystallography as applied to practically every industrial sector for over a century. Conventionally, XRD was constrained to bulk analysis on crystalline materials that could be ground to a sufficient particle size, to not bias the particular system used. Historically, several different types of diffractometers/detection methods have been used, including the pinhole camera/Laue camera, Debye-Scherrer/Gandolfi camera, Guinier camera, back-reflection Laue, and various others. While initial diffractometers tended to rely on film, a transition was made (with the advent of digital systems) to simple point counters in the early 1970s. These systems tended to operate in either a θ-2θ, or Guinier multicircle, geometry. Most recently, in approximately the last 15 years, conventional diffractometers have been mostly standardized to use a Bragg-Brentano geometry operating in a θ-θ mode, although occasionally θ-2θ still can be found. Exceptions also exist in the case of asymmetric systems that rely on a fixed goniometer equipped with a curved position-sensitive (CPS) detector or image plate area detector. These modern systems employ digital detectors with capabilities spanning 0-D (scintillation counter), 1-D (silicon strip and CPS), and 2-D (area detectors), with the latter two being the most prevalent.

Micro x-ray diffraction (μXRD) is a tool for the examination of materials in situ. Samples are either part of a rock, slab, or polished thin section, or they are encased in a medium, such as epoxy, a capillary, etc. This allows for the sampling of specimens ranging from single crystals to fine-grained mixtures. Unlike conventional XRD, which analyzes a bulk sample, μXRD analysis is applied at the microscopic scale. This allows for the correlation of phase and structural data to other microscopic and microanalytical techniques, such as scanning electron microscopy/energy-dispersive spectroscopy, optical microscopy, or Raman/Fourier transform infrared spectroscopy, on a point-by-point analysis. Microdiffraction also allows for analysis of materials in capillaries as well as small-volume samples/crystal clusters mounted on a glass fiber.

Principles

The μXRD analysis geometry is similar to that of conventional powder diffraction techniques. Generally operating in a reflection geometry, the Bragg equation (Ref 1) can be satisfied for μXRD for a specific point on a sample instead of the average over the entire bulk:

$$n\lambda = 2d \sin \theta \quad \text{(Eq 1)}$$

This equation was shown to be applicable to polycrystalline powders as well as single crystals. A comparison of transmission Laue diffraction patterns for single-crystal versus 5 μm powder α-Al_2O_3 (corundum) is shown in Fig. 1.

The single-crystal diffraction pattern is comprised of spots representing specific lattice planes. The position of the spots is dependent on the orientation and symmetry of the single crystal (in this case a (0001) single-crystal wafer). The powder diffraction pattern is comprised of rings also representing lattice planes and indicating there is a random orientation of the crystallites in the x-ray beam. In this case of a random oriented powder, each crystallite contributes a spot pattern, and the summation of spot patterns results in diffraction rings (Debye rings).

The small spot size of the μXRD beam allows for analyses in cross section, transition, in situ, etc. without the need to alter the sample by grinding. Thus, objects can be analyzed without sampling, by collimating the incident x-ray beam and aligning it on an area of interest. Using a conventional laboratory x-ray source, a target spot size diameter as low as 30 μm can be achieved while still maintaining a realistic counting timeframe (Ref 2). Microdiffraction systems employ a highly collimated beam, in accordance with a laser-guided camera, *XYZ* stage, capillary/glass fiber or Eulerian cradle, and, most commonly, either a position-sensitive or 2-D area detector (Fig. 2).

Phase identification from a μXRD diffraction pattern can be carried out once the diffraction data have been converted to *d*-spacings. For many years, the process of phase identification relied on the use of search manuals based on the Hanawalt technique (Ref 4). The technique still has application today (2019), although modified and carried out by using computer-based search-match methods (Ref 5). All computer methods make use of a diffraction pattern reference database, i.e., the International Centre for Diffraction Data Powder Diffraction File (PDF) (Ref 6). For μXRD data collected using a 2-D detector, the observed diffraction data are integrated and converted to an *X-Y* plot of 2θ versus intensity. The resulting diffraction peaks are rank-ordered based on intensity, with the most intense first and the least intense last. Search-match methods then compare this peak list with reference data to find a phase match.

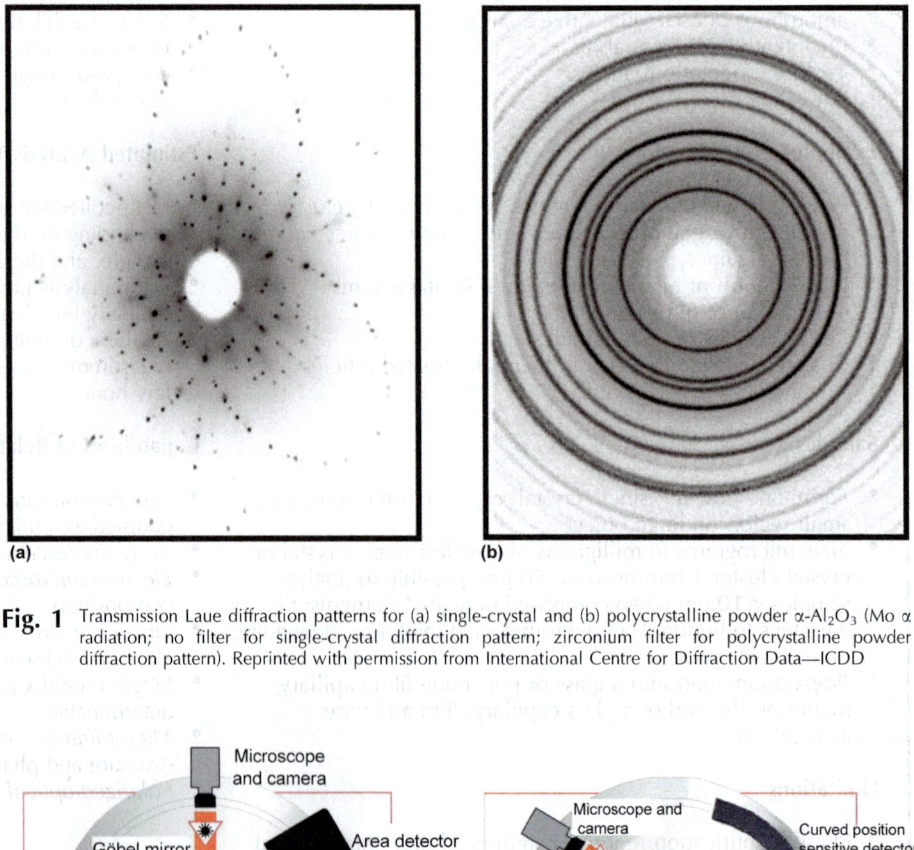

Fig. 1 Transmission Laue diffraction patterns for (a) single-crystal and (b) polycrystalline powder α-Al_2O_3 (Mo α radiation; no filter for single-crystal diffraction pattern; zirconium filter for polycrystalline powder diffraction pattern). Reprinted with permission from International Centre for Diffraction Data—ICDD

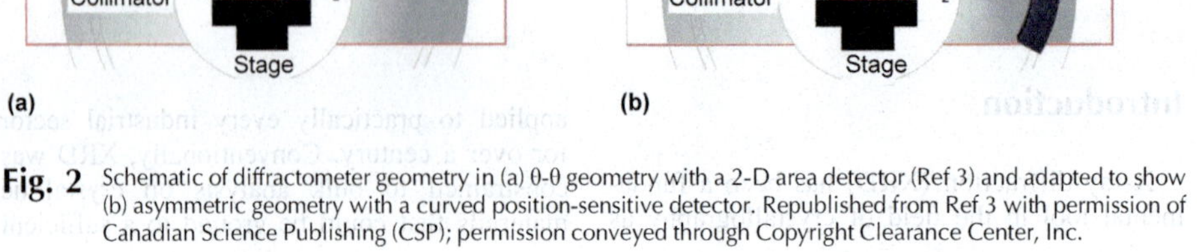

Fig. 2 Schematic of diffractometer geometry in (a) θ-θ geometry with a 2-D area detector (Ref 3) and adapted to show (b) asymmetric geometry with a curved position-sensitive detector. Republished from Ref 3 with permission of Canadian Science Publishing (CSP); permission conveyed through Copyright Clearance Center, Inc.

Systems and Equipment

X-rays produced from common sources have both bremsstrahlung and characteristic radiation. Methods to isolate the useful characteristic radiation with a reduced amount of unwanted radiation include the use of a Kβ filter (for example, nickel foil used with a copper x-ray tube), a primary beam monochromator, or focusing optics or mirrors. Regardless of the method used, there will be a loss in intensity of the desired characteristic radiation.

Large beamline facilities that have synchrotron radiation capabilities can provide a near-monochromatic-energy x-ray incident beam, with a flux several orders of magnitude more intense than a laboratory x-ray source. While these sources are the brightest and most configurable for performing μXRD, this section focuses on the laboratory-scale instrumentation, because synchrotron physics is significantly different and less accessible for routine analysis. Laboratory systems are constantly evolving to handle more complex tasks that used to be performed only with a synchrotron.

For μXRD analysis, there have been many advances in technology, from hardware to software, with the principal components of a μXRD system being an x-ray tube coupled with focusing optics, a laser-guided microscope camera system for picking the analysis spot(s), a stage compatible with the sample type, and a detector, usually 1-D or 2-D, although scintillation detectors still are uncommonly used.

Micro x-ray diffraction as a technique saw a decline in use as diffractometers grew in popularity and users moved away from using film as a detector. A renaissance in μXRD occurred with the development of the linear position-sensitive detector (PSD), a line detector capable of collecting a selected 2θ range faster than a conventional point detector (Ref 7). As PSD technology advanced, curved PSDs with a 2θ range of up to 120° provided a fast electronic detector capable of being used in a Debye-Scherrer geometry instrument, a potential but costly replacement for film. More recently, another line detector that has been developed is a silicon strip detector comprised of diodes on a silicon strip. These detectors offer high speed for data collection, high spatial resolution, high count rate before saturation, and high dynamic range.

A further improvement in μXRD analysis was the development of improved 2-D detectors. One replacement for photographic film is the image plate (IP) storage phosphor (Ref 8), which uses a phosphor (i.e., BaBrF doped with europium) that, when irradiated by x-rays, stores the x-ray signal until it is read by a laser scanner. Advantages of the IP detector include:

- It is an order of magnitude faster than photographic film.
- It can be coated on a polymer support such that the IP can be directly placed in any instrument where x-ray film is used.
- The dynamic range is 6 orders of magnitude compared to 2 for photographic film.

Other 2-D detectors that have come into use for μXRD applications include the multiwire proportional counter PSD, charge-coupled device (CCD), hybrid photon counting, pixel, and strip detectors (Ref 7). It is with the availability of 2-D detectors that the full power of μXRD can be realized.

X-Ray Tubes

Conventional

Conventional x-ray tubes have been employed for many years. The tube housing is comprised of either glass or ceramic. The vacuum-sealed interior consists of a tungsten filament cathode and a target anode, most commonly copper, cobalt, molybdenum, chromium, or tungsten. When energized, electrons are generated from the cathode and impact the anode, and the resulting interaction produces x-rays of a wavelength dependent on the target material (Table 1). Beryllium windows at the base of the tube allow the x-rays to exit the tube into the optic to focus on the sample.

Generally, these tubes have two beam shapes: a line and spot focus. These usually are offset 90° from each other on the tube. For μXRD experiments, the spot focus is ideal, because the beam exiting the tube is already the ideal shape. Using a line focus source results in the eventual waste of beam, because the majority of the beam exiting the window will have to be cut off by the optic to generate the final spot on the sample.

Conventional tubes are powered by a high-power (3 to 3.5 kW) generator. Under normal running conditions, the majority of energy is converted to heat and must be removed by means of an external circulating water chiller so as not to damage the x-ray tube.

Microsource

More recently, technological advances have led to the creation of microsource tubes. These tubes function similarly to a conventional x-ray tube but require only a fraction of the power requirements to achieve a sufficiently bright beam. The generator required to produce x-rays usually is 10 to 50 W, requiring significantly less infrastructure to support the instrumentation.

X-rays exiting the tube are passed through focusing mirrors to create either a line or a spot profile before passing through any secondary optics. This generates an effectively bright source using minimal power. Paired with multilayer optics, these tubes can achieve intensities (W/mm^2) similar to rotating anodes. Currently only copper, cobalt, and molybdenum sources are commercially available.

Optics

When x-rays exit the window of the tube, the beam paths are divergent. Beam optics are employed to either constrain or converge and focus the individual beam paths to reduce the effects of anomalous dispersion as the incident beam impinges on the sample.

Göbel Mirror

Parabolic-graded multilayers (Göbel mirrors) serve as monochromators. The mirror is wavelength dependent and works by fulfilling Bragg's law across the entire length of the mirror (Fig. 3). This modifies the divergent x-ray

Table 1 Common x-ray sources and properties

Anode	Wavelength $K\alpha_1$, Å	Applications	Fluorescence from sample
Mo	0.70930	Heavy absorbing samples, high penetration depth	Y, Sr, Rb
Cu	1.54056	Standard powder analysis, high-resolution x-ray diffraction	Co, Fe, Mn
Co	1.78897	High-iron samples, stress and texture analyses	Mn, Cr, V
Cr	2.2897	Large lattice constants, stress and texture analyses	Ti, Sc, Ca
W	Continuous	Laue diffraction	...

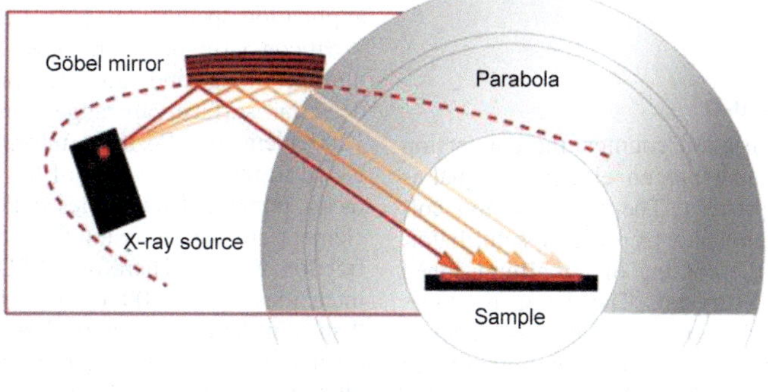

Fig. 3 Schematic of a parabolic Göbel mirror focusing emitted x-rays from the source. Source: Ref 9

beam emerging from a point or line focus source in such a way that it can optimally be applied with Debye-Scherrer, Bragg-Brentano, or parallel beam geometries. The parabolic curve and gradient of the mirror allows for the conversion of a divergent beam into a parallel beam. The use of a parallel beam in μXRD, high-resolution XRD, reflectometry, and grazing-incidence diffractometry results in significant intensity improvements (Ref 10, 11).

Mono- and Polycapillary

Capillary optics are used to guide and shape x-ray radiation. Such systems are either monocapillaries (such as cylindrical, ellipsoidal, or paraboloidal) or polycapillaries (consisting of a monolithic system of many hollow capillary channels), as shown in Fig. 4. Their function is based on the effect of total external reflection of x-rays from the internal smooth surfaces of the capillary channels. The basic material for x-ray capillaries is glass. The low roughness of reflective glass surfaces leads to a low portion of diffuse scattered x-rays, so an efficient transport of x-radiation through glass capillaries can be realized. Due to these properties, capillary optics are able to guide x-rays from source to sample in such a way that a high-intensity beam is concentrated in an accurately defined area on the sample surface. Polycapillary optics are especially useful in point-by-point or raster-style analysis μXRD.

Pinhole Collimator

The cheapest and most simplistic focusing attachment is the pinhole collimator. The collimator is a snub-nosed metal cylinder that attaches to the end of the beam tube/slit assembly. It works by allowing only the x-rays that pass through the aperture opening to impinge on the sample. There is no intensity gain, as in the case of the capillary optics. For increased intensity, the collimator can be placed on the end of a mirror such as the Göbel mirror. The pinhole collimator assures an exact spot size. There are several aperture size options, and the low cost of the accessory makes it a viable option for many laboratories to be able to vary their beam on target with just the change of the collimator.

Detectors

Scintillation—0-D

In the scintillation counter, the conversion of x-ray photons into an electrical signal is a two-stage process. The x-ray photon collides with a phosphor screen, or scintillator, which forms the coating of a thallium-doped sodium iodide crystal. The latter produces photons in the blue region of the visible spectrum. These visible photons subsequently are converted to voltage pulses by means of a photomultiplier tube attached directly behind

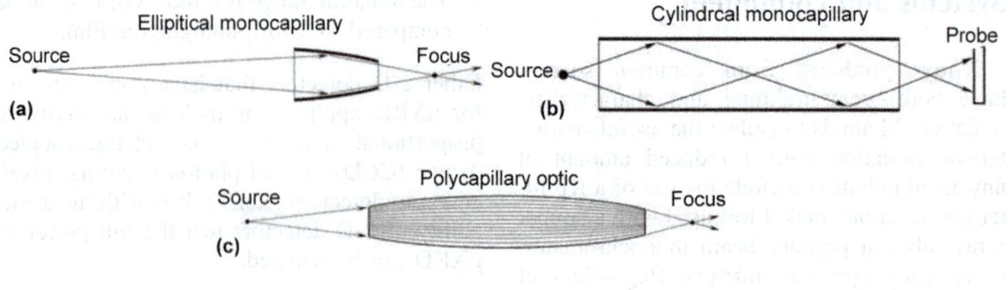

Fig. 4 Schematics of (a) single-bounce elliptical monocapillary, (b) multibounce cylindrical monocapillary, and (c) polycapillary optic

the scintillator. The number of electrons ejected by the photocathode is proportional to the number of visible photons that strike it; that in turn is proportional to the energy of the original x-ray photon. Due to a large number of losses, the energy resolution of the detector is poor, and as such, it cannot be used to resolve x-ray photons due to Kα and Kβ radiation. However, it has a high quantum efficiency and a low dead time, making it the ideal detector for the point-intensity measurements required for step-scanning diffractometers.

While these detectors have significantly decreased in usage over the last decade, they still have some relevant applications in analysis types such as rocking curves, which can be a very useful application for μXRD. The high spatial resolution of the detector results in a high degree of confidence in peak shape, with any asymmetry due to the sample. The most significant negative factor for a scintillation detector is data-collection time. Due to the nature of the detector, it is not uncommon for full scan times to be on the order of an hour for one sample.

Silicon Strip—1-D

The silicon strip detector (SSD) initially was developed as a beam tracker for high-energy physics, and it has evolved for applications requiring an x-ray detector (Fig. 5). Its unique characteristics, such as single-photon counting capability, high spatial resolution, good energy resolution, and room-temperature operation, are ideal for use with an in-house XRD system. However, the performance of an SSD is strongly reliant on its read-out electronics, especially a read-out application-specific integrated circuit.

Silicon strip detectors have a layout with multiple long and thin silicon strips. The detection surface can range from 256 to over 1200 elements. Position information is collected in the 2θ direction of the detection surfaces. Integrated intensities collected by each element are output as intensity at specific diffraction angles. Therefore, SSDs can collect intensities up to 100 times (or more) faster than scintillation detectors (Ref 12). The newest iterations of SSDs are referred to as hybrid photon counters, which are capable of operating in 0-D, 1-D, or 2-D modes.

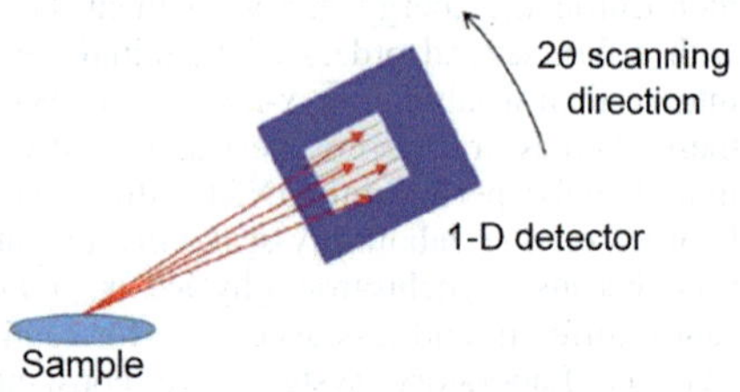

Fig. 5 Schematic of a silicon strip detector showing individual strips totaling ~5° 2θ swath. Adapted from Ref 12

Curved Position Sensitive—1-D

The CPS detectors are gas-filled (ethane/argon) curved chambers consisting of a solid blade anode and segmented cathode that act as a proportional counter. When an x-ray photon becomes incident on the detector anode, an electrical charge develops on the cathode at a position that is spatially coincident with the incident photon. This electrical charge is measured in terms of its intensity and position by way of a delay line. This signal is digitized and then stored in an electronic card. The desired signal-to-noise ratio is achieved by multiscan accumulation. In practice, an excellent diffraction pattern can be recorded in just a few seconds at a high spatial resolution.

The CPS detectors are essentially spatial detectors used to directly intersect the diffracted, reflected, or transmitted x-ray beam (Fig. 6). The length (or radius) of the detector determines its spatial (or angular) position about the x-ray beam (or 2θ), and the distance to the sample also determines resolution.

Area—2-D

Over the last decade, many advances have been made in the field of area detector technology. Current commercially available detectors tend to incorporate one of three major technologies: MIKROGAP, hybrid pixel detectors, and electronic IP (Fig. 7). Other 2-D technologies still exist, such as CCDs and complementary metal oxide semiconductors (CMOS), but they are at a disadvantage due to their limitations for μXRD analysis.

In a MIKROGAP detector, a sealed vessel contains the grid, anode, delay lines, and

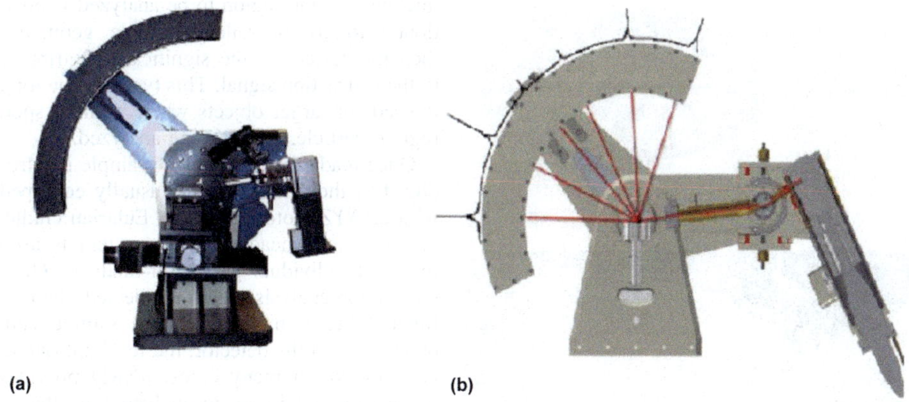

Fig. 6 (a) Micro x-ray diffraction system equipped with curved position-sensitive (CPS) detector and (b) representation of CPS simultaneous analysis

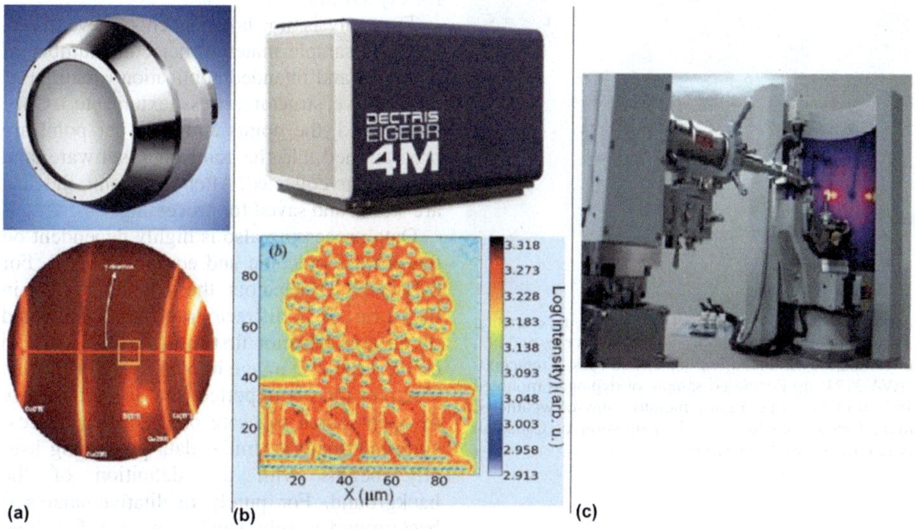

Fig. 7 (a) MIKROGAP detector with sample output. (b) Hybrid photon detector with example output (reprinted from Ref 13 with permission of Wiley. Copyright © International Union of Crystallography). (c) Micro x-ray diffraction Rigaku image plate system with simulated diffraction reflections visible

$Xe\text{-}CO_2$ gas mixture. Incoming x-ray photons produce primary electrons by gas ionization. These electrons drift toward the anode under a high electric field, and the number of electrons is multiplied in the amplification gap. This signal amplification is by far larger than any solid-state detector provides; thus, very weak signals can be detected. The location of each x-ray photon is determined by readout strips through the X and Y delay lines. The key feature of the MIKROGAP technology is the presence of a resistive anode to allow a thin amplification gap, which greatly improves the local count rate.

Hybrid pixel detectors operate by directly converting x-rays into an electronic signal, unlike other types of x-ray detectors that rely on intermittent steps to capture and convert x-rays into an electronic signal. The CCD and CMOS active pixel detectors, for instance, must convert x-rays to visible light first. Scattering of light by the phosphor screen smears out the signal and thus decreases the spatial resolution. Fiberglass optics then transduce the light on the chip, which causes further loss and distortion of the signal. In a hybrid pixel detector, every pixel is comprised of two components: a sensor pixel and a readout pixel. X-ray photons are directly converted into an electric charge in the sensor pixel, which the readout pixel processes and counts. Each hybrid pixel essentially is an independent x-ray detector, which results in low point spread, high sensitivity, and speed.

The versatile IP detector can perform a wide variety of XRD applications. Its unique, curved large-area detector subtends a 2θ range of 204° at a single detector setting for maximum reciprocal space coverage. The large curved active area is advantageous because a massive solid angle of data is collected in a single exposure. While the wide dynamic range eliminates worrying about detector saturation, the curved-plate geometry of the IP detector reduces oblique-incidence x-ray absorption effects seen with flat detectors of any kind.

Procedure

The samples that can be analyzed by using μXRD are diverse in type and size. Sometimes the sample is micrometers in size; other times it is a large object with a small region of interest. An excellent reference for XRD (and x-ray fluorescence) sample preparation can be found in Ref 14. Materials can be prepared for analysis by using such methods as:

- Encasing small samples in an epoxy or similar binder
- Polishing a flat surface of metal alloys
- As-received powders on substrates
- Directly on the original material in cases such as pigment analysis

In other fields, such as pharmaceuticals, a well plate of different compounds can be introduced to analyze many small-volume samples in one experimental set (Fig. 8).

In preparing a powder specimen for μXRD analysis, it is important to remember that with a small x-ray beam irradiating the specimen, a statistically representative sampling may not occur. There generally are two approaches for μXRD analysis of powder: loaded into a thin-wall glass capillary or on the outside or tip of a glass fiber. Glass capillaries (Charles Supper Co.) with an inside diameter of 100 to 300 μm are recommended for loading powders. A larger diameter may be too thick for transmission of the x-ray beam, particularly for high-Z elements.

Conventional XRD specimen preparation works best when using a powder with an ideal particle size of <10 μm; however, it is not often achieved. Still, particles less than 45 μm in size (400 mesh) are best for good particle statistics. The problem for μXRD is that often only small quantities of a sample are available, and any attempt to grind the sample may result in a loss of some or all of the sample. When loading particles in a capillary or coating onto a glass fiber, it is important to remember that if the beam size irradiating the sample is 300 μm, very few 100 μm particles are going to be in the incident beam. The μXRD analysis of large particles will result in a diffraction ring that is spotty, while small particles will produce what appears to be a continuous ring (Fig. 9).

For sheet samples such as polymer films or thin metal foils, samples can be mounted for transmission analysis (beam through the film or foil) or along the sample edge in a grazing-incidence alignment that is similar to a reflection-mode diffractometer measurement. For transmission work, the thickness, density, and mass attenuation coefficient of the

Fig. 8 Examples of samples mounted for micro x-ray diffraction. (a) Powdered sample of synthetic hydroxyapatite. (b) Polished thin section of metamorphic rock. (c) Cut slab of Southampton pallasite (length: 13 cm, or 5 in.). (d) Epoxy mount of garnet single crystals from Roberts Victor Kimberlite, South Africa. (e) Whole rock, Cretaceous lobster. (f) Whole rock with cut surface, Martian meteorite NWA 3171. (g) Powdered sample of dypingite mounted on pedestal. (h) Cut slab of concrete (width: 3 cm, or 1.2 in.). (i) Optical microscope monitor view of weathered copper leaf (width: 4 cm, or 1.6 in.) exhibiting patina. Republished from Ref 3 with permission of Canadian Science Publishing (CSP); permission conveyed through Copyright Clearance Center, Inc.

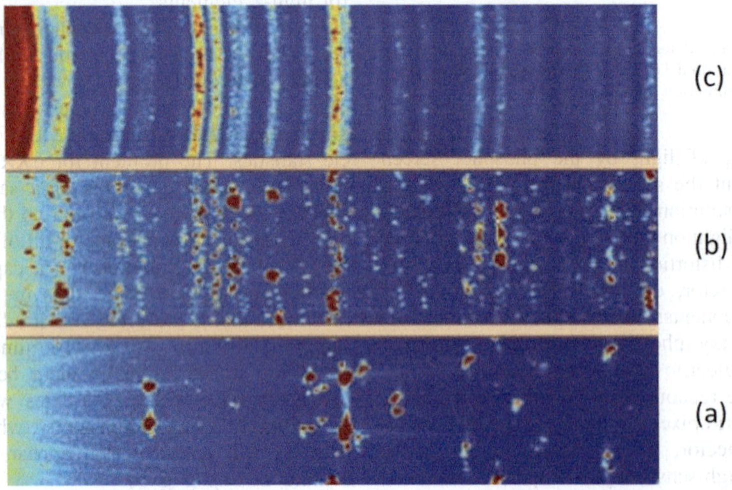

Fig. 9 X-ray diffraction data collected for (a) coarse, (b) medium, and (c) fine particle specimens (Cu Kα radiation, reflection-mode geometry diffractometer, and pixel detector). Source: Ref 15.

specimen will determine if the x-ray beam can penetrate the sample. For low-*Z*-element samples, the density and mass attenuation coefficient generally are low and can be several hundred micrometers in thickness, whereas high-*Z*-element samples with a higher density and mass attenuation coefficient will need to be a few tens of micrometers or less in thickness. If the region to be analyzed is positioned totally in reflection-mode geometry, then the surface is the significant contributor to the diffraction signal. This type of mounting is used for larger objects where a small spot, region, particle, etc. is being analyzed.

Once ready for analysis, the sample is introduced to the diffractometer, usually equipped with an *XYZ* motorized stage, Eularian cradle, or goniometer head. A video system is used to target individual points for analysis. Once the sample analysis region is selected, the collimated beam impinges on the sample and, depending on the detector, the resultant output is recorded. In many cases, μXRD provides much more flexibility in analysis type than a conventional powder XRD system. Single or multiple points can be analyzed, a sample can be rastered (collecting a line or square region on a sample), texture maps can be performed for crystallinity, etc.

The analysis time is dependent on several factors: the sample material, detector, number of analyses, and intended application (qualitative, quantitative, structure, stress/texture, etc.). Once determined, the points and time per point are programmed into the controlling software, and the data are collected. After each point, the data are stored and saved for processing.

Data processing also is highly dependent on the desired outcome and equipment used. For 0-D and 1-D detectors, the data already are in the form of a diffractogram. Data collected on an area detector first must be integrated to convert the 2-D image into an *X-Y* plot format. The plot then is imported into any proprietary or open-source software programs for processing and analysis. Initial data processing usually begins with the definition of the background. For purely qualitative analyses, background is subtracted. However, for other forms of analysis, the background is left as-is to be factored into the final statistical analysis. A peak search then is employed to mark the location in either 2θ or *d*-spacing of peaks in the plot. This usually is an automated process, with the software using several adjustable limits to identify peaks. The identified peaks then are searched against a database, most commonly the International Centre for Diffraction Data PDF (Ref 6).

More detailed analyses, such as quantitative phase analysis (QPA) (with or without crystallite size and strain), texture, and structural analysis, require more in-depth processing. While several methods exist to perform QPAs, with the power and ease of computing today (2019), most methods rely at least in part on the Rietveld method (Ref 16). This method compares an experimentally collected dataset to a modeled plot by using the Rietveld equation:

$$I_i^{\text{calc}} = S_F \sum_{j=1}^{N \text{ phases}} \frac{f_j}{V_j^2} \sum_{k=1}^{N \text{ peaks}} L_k \left|F_{k,j}\right|^2 S_j(2\theta_i - 2\theta_{k,j}) P_{k,j} A_j + bkg_i \quad \text{(Eq 2)}$$

where I_i^{calc} are the calculated intensities at each point i, which are then compared to the diffracted intensities described at each point i, where $S_F \sum_{j=1}^{N \text{ phases}} \frac{f_j}{V_j^2}$ is the scale factor for each phase, L_k is the Lorentz polarization factor, $|F_{k,j}|^2$ is the structure factor for each phase, $S_j(2\theta_i - 2\theta_{k,j})$ is the profile shape function, $P_{k,j}$ is the texture (or preferred orientation), A_j is the absorption factor, and bkg_i is the background at each point.

Stress and texture analyses generally are performed on a Eularian cradle goniometer with either a 2-D or CPS detector, where the sample can be analyzed in several different orientations. These scans then are processed and combined to produce an overall plot or series of pole figures (Fig. 10). A typical experiment can last for as little as ~30 min to several hours, depending on sample quality and the detector used. Like Rietveld analysis, several programs, both proprietary and open source, exist to perform such analyses, and it is up to the user to determine which software best suits their need.

Calibration and Accuracy

There are three primary considerations for calibrating a detector for μXRD. Two deal with the detector itself: uniform response and spatial correction (Ref 7). These parameters are instrument and detector dependent, requiring procedures provided by the manufacturer. They are mentioned here as a caution that if the response is not uniform across a detector, a correction is required or else diffraction intensities will not be registered correctly. If the spatial correction is not applied, diffraction ring positions will be in error, resulting in incorrect 2θ (and d-spacing) results.

The third calibration that is required for a microdiffractometer is determination of the sample-to-detector distance. For some instruments, the sample and detector positions are fixed, and calibration is established during instrument installation. In those instruments in which the sample position can be changed or the detector position can be moved closer to (more 2θ range coverage, less resolution) or farther away from (less 2θ range coverage, more resolution) the sample, distance calibration is required each time the detector is moved. Perhaps the single largest source of error in μXRD measurements is improper specification of the detector position. Many instruments have a ruler on the detector arm that is used for defining the sample-to-detector distance, but these rulers are an approximation.

The proper way to perform the detector position calibration is to run a standard material with well-defined d-spacings, corresponding to 2θ values that can be used to calibrate the sample-to-detector distance. One source of certified standards is the National Institute of Standards and Technology, which has available Standard Reference Material powder XRD standards that can be used for calibrating a microdiffractometer, including α-Al_2O_3, silicon, and LaB_6.

Data Analysis and Reliability

Figure 11 shows a diffraction pattern collected from a powder sample, packed in a glass capillary, by using a micro x-ray diffractometer. After data collection, diffraction pattern integration to 2θ versus intensity, determination of peak d-spacings (d-spacing because different radiations will change the 2θ of diffraction, but d-spacing is fixed for a defined crystalline phase) and intensities, and a computer search-match program, Sieve + (Ref 17), were used to identify the sample as α-Al_2O_3.

When the sample is comprised of more than one phase, identification of all phases can be more difficult. Critical to successful phase analysis is a well-aligned (and calibrated) instrument, properly prepared sample, quality data (i.e., good counting statistics), and paying attention to the peak search results to ensure accurate peak positions and intensities. Because a small beam size is used for μXRD data collection, count rates at each pixel of the detector typically are low.

However, if the signal-to-noise ratio is good, the low count rate may not be a problem. One area of concern when collecting μXRD data is sample fluorescence. The configuration of a microdiffractometer does not permit a diffracted beam monochromator, allowing sample fluorescence to be directed to the detector and increasing background signal. For example, a specimen containing iron that is irradiated with Cu Kα radiation will have a significant noise component in the diffraction pattern. The signal will be strong, but the signal-to-noise ratio will be low.

Errors in diffraction ring positions due to instrument misalignment, improper calibration, and detector resolution can affect accuracy in the determination of d-spacings. With careful sample preparation and proper data collection and peak searching, deviation in derived d-spacings can be limited to:

Interplanar d-spacing, Å	Deviation in d-spacing, Å
5	0.04
4	0.02
3	0.01
2	0.004
1.5	0.002
1.0	0.0007

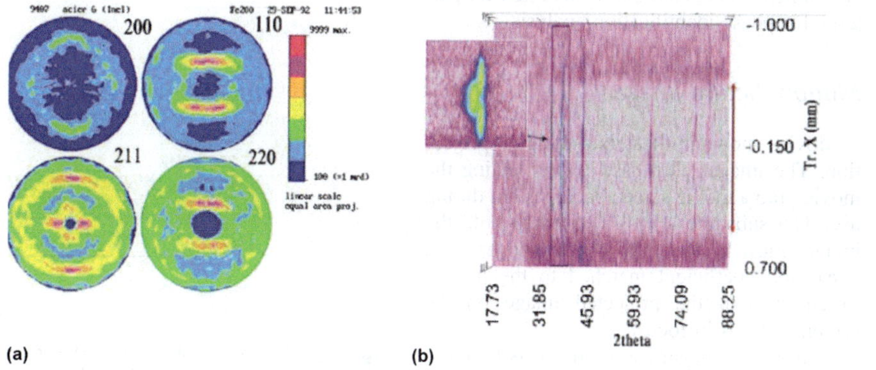

Fig. 10 (a) Texture analysis on stainless steel foil. (b) Stress analysis along edge of machined aluminum plate

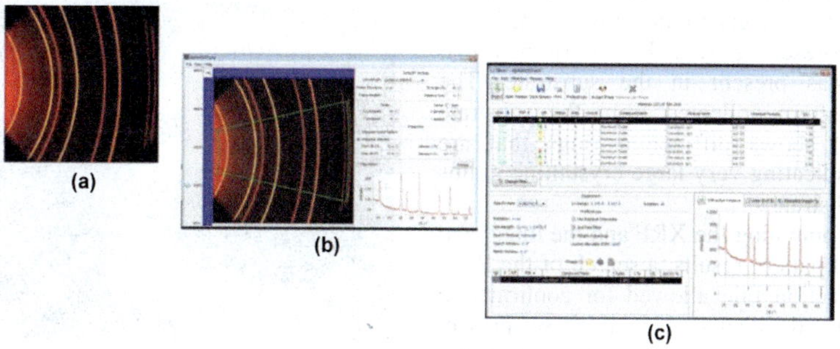

Fig. 11 (a) Micro x-ray diffraction pattern (b) converted to a 2θ-versus-intensity plot, (c) followed by peak searching and phase identification, α-Al_2O_3 (Cu Kα radiation; nickel filter; sample-to-detector distance: 10 cm, or 4.0 in.). Reprinted with permission from International Centre for Diffraction Data—ICDD

The most intense diffraction data generally occur at lower 2θ (larger *d*-spacing), the region where the largest error in *d*-spacing typically is observed.

Applications and Interpretation

Micro-XRD data from a motion picture imager, cathode ray tube cathode assemblies, plastic gears, and polymer films were collected by using a microdiffractometer comprised of a Rigaku copper rotating anode x-ray source and a Bruker goniometer consisting of cross-coupled mirrors aligned for Cu Kα radiation, *XYZ* sample stage, video alignment camera, and a general area-detector diffraction system (GADDS) two-dimensional (2-D) multiwire detector (Ref 18). The operating power for the x-ray source was 30 kV and 60 mA. The final x-ray beam size was controlled with collimators and ranged from 50 to 300 μm in diameter, depending on the sample size. Data-collection times ranged from 60 s to 10 min, depending on the material composition, sample size, and sample crystallinity. Samples were evaluated as-received and were mounted on an appropriate sample holder for data collection. Data analysis was performed by using software from Bruker (GADDS and EVA) or Materials Data, Inc. (JADE). Some 2-D diffraction patterns were converted to one-dimensional (1-D) 2θ-versus-intensity patterns for phase-identification analysis.

Motion Picture Imager

Movies shown in theaters use digital projection. The imager chip used for projecting the movie onto a movie screen becomes hot during use. The substrate (Fig. 12) used to hold the imager must be thermally stable and have an expansion coefficient matched to the imager to ensure that the projected image on the screen remains in focus.

A group of imager assemblies was found to produce unacceptable image quality. The cause was attributed to either the imager chips or the substrates provided by an external supplier. A request was made to determine the phase composition of these substrates without exposing the imager to an x-ray beam. Micro x-ray fluorescence (μXRF) determined that silicon was present in the substrate. The μXRD pattern collected from the substrate (Fig. 13) shows diffraction rings that are spotty, indicating very large crystalline grains in the substrate.

After combining the XRF and the integrated 1-D XRD pattern results, a search of the Powder Diffraction File allowed for confirmation of silicon and silicon carbide as phases present in the substrate. With this information, it was determined that the composition of the imager substrate was not correct, resulting in a thermal expansion coefficient that was not matched with the imager. This mismatch explained the defocusing of the projected image. Because of the small size of the x-ray beam during the collection of μXRD data, the imager devices had not been exposed to a direct x-ray beam. The imagers were preserved and could be removed from the defective substrates to be used on other imager assemblies.

Fig. 12 Backside of a motion picture imager assembly. The asterisk marks the location of the micro x-ray diffraction analysis. The rectangular box marks the location of the imager on the frontside of the substrate. Reprinted with permission from International Centre for Diffraction Data—ICDD

Cathode Ray Tube Cathode Assemblies

Digital print services use high-intensity light sources for printing images. These light sources use cathode ray tubes (CRTs) containing cathode assemblies (Fig. 14) that are expected to be in service for greater than 20,000 h. Early failure (<2000 h) of some CRTs prompted an investigation to determine the cause. The supplier of these cathodes indicated that the electron-generating portion was comprised of tungsten and the surrounding cup was comprised of tantalum, and that nothing had changed in the manufacturing process of the CRTs.

Micro-XRD patterns (Fig. 15) were collected in the tungsten region for a new cathode, a cathode operated for 26,000 h and still working, and a cathode that failed after 1000 h of operation.

Phase identification determined that two phases were present in the tungsten region of the new cathode: tungsten and osmium (confirmed by μXRF). The supplier did not indicate

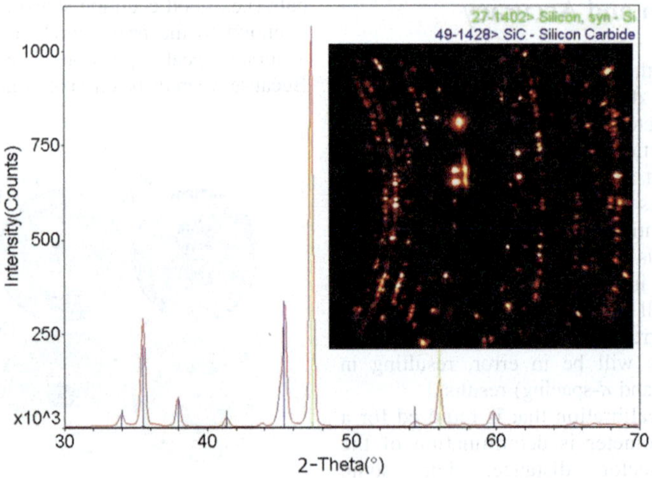

Fig. 13 Micro x-ray diffraction (XRD) data from the backside of the imager substrate. Inset is the original two-dimensional (2-D) pattern. The *X-Y* plot is a result of integration of the 2-D pattern and conversion to a traditional one-dimensional XRD pattern. Reprinted with permission from International Centre for Diffraction Data—ICDD

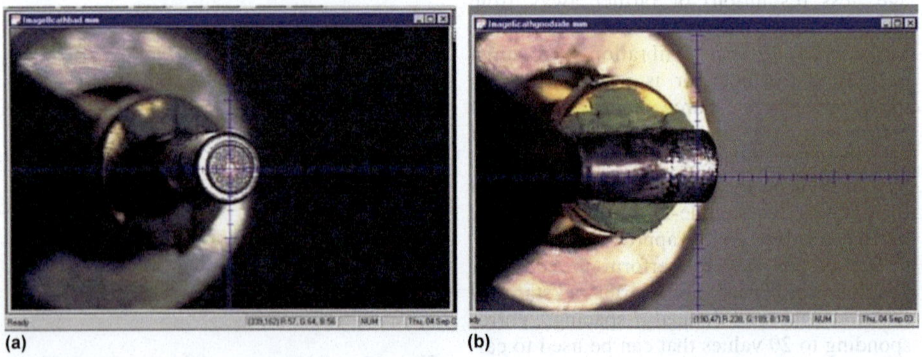

Fig. 14 Cathode ray tube cathode. (a) Tungsten region. (b) Tantalum cup

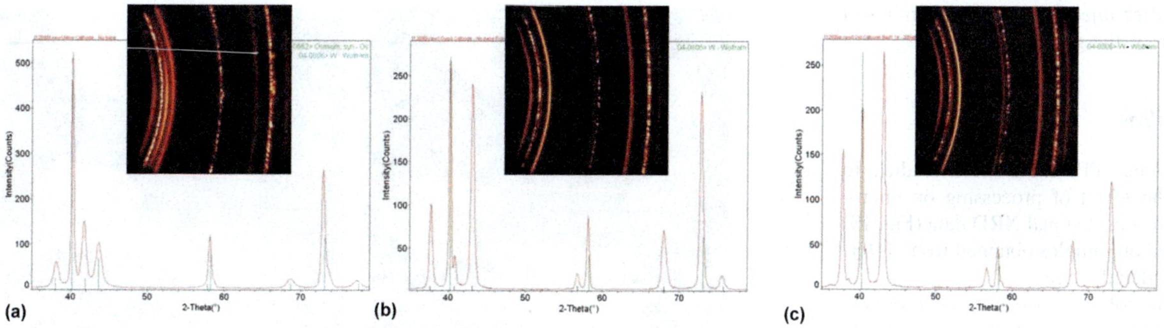

Fig. 15 Micro x-ray diffraction two-dimensional (inset) and corresponding one-dimensional diffraction patterns for (a) new cathode, (b) working cathode after 26,000 h of operation, and (c) failed cathode after 1000 h of operation, collected in the respective tungsten regions

the presence of osmium. The absence of complete knowledge of the composition of components often is a problem when dealing with external suppliers. Note that in the 2-D μXRD pattern in Fig. 15(a), the diffraction rings due to tungsten are spotty (large grain), whereas the rings due to osmium are continuous (small grain). This microstructure observation would not be possible when using a conventional point detector for XRD data collection. For the used cathodes, both showed the presence of two phases. Tungsten is present at reduced intensity, and osmium is absent. In both of these samples, the additional phase could be indexed based on a hexagonal unit cell with lattice constants of $a = 2.755$ Å and $c = 4.416$ Å. A match for any tungsten or osmium phases or alloys with this unit cell was made for an $Os_{(1-x)}W_x$ phase with a magnesium structure type, using Pearson's Handbook (Ref 19). At this point, there was no obvious answer to explain why one cathode was working after 26,000 h and the other failed after 1000 h.

The next step was to analyze the tantalum cup region. Analysis of the new and 26,000 h cathodes revealed the presence of tantalum metal. No other phases were detected. When the 1000 h failed cathode was analyzed, several additional diffraction rings in addition to tantalum metal were observed (Fig. 16).

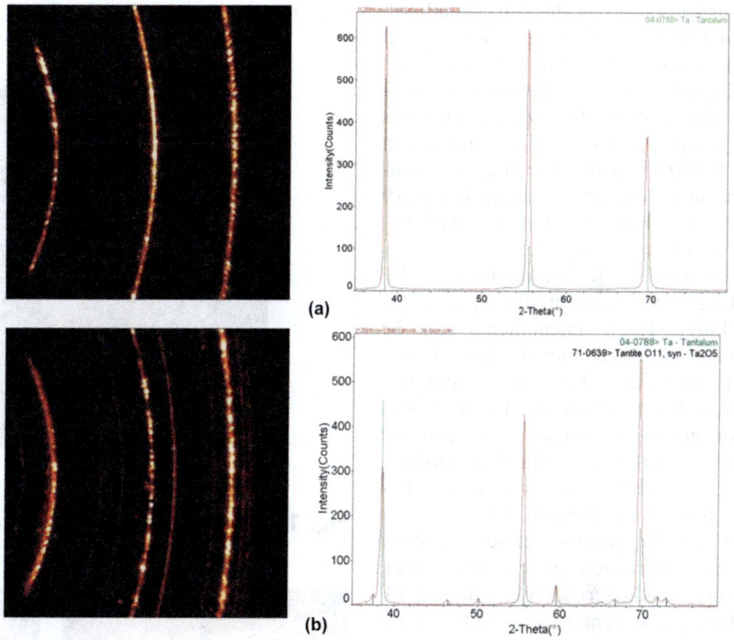

Fig. 16 Micro x-ray diffraction two-dimensional and corresponding one-dimensional diffraction patterns for (a) working cathode after 26,000 h of operation and (b) failed cathode after 1000 h of operation, both collected in the respective tantalum cup regions

Phase identification for the failed cathode revealed that Ta_2O_5 was present. Tantalum is an excellent gettering element for oxygen, and the presence of Ta_2O_5 explains why the cathode failed prematurely. With these findings, the manufacturer of the CRTs concluded that a new type of seal being used (although the original discussion indicated no manufacturing changes) was not working properly, and oxygen was being allowed to enter the CRT. Replacement of the defective seals solved the problem, and service lifetimes returned to the expected 20,000+ h.

Plastic Gears

Gears can be made of plastic to help reduce weight and cost. Typically, these gears are made by using an injection molding process, followed by a cooling period before the finished gear is removed from the mold. At one site manufacturing plastic gears, the gear teeth were found to be distorted, rendering the gear unusable. Figure 17 shows a visual comparison of a good versus a bad gear.

Micro-Fourier transform infrared spectroscopy confirmed that the polymer in both samples was polymethylene oxide (PMO). An XRD analysis was requested to determine if there were crystallinity differences between the two samples. The use of a conventional diffractometer would not allow characterization in only the region of the gear teeth, because the x-ray beam would be as large as the entire gear. Micro-XRD was well suited for this analysis because the x-ray beam could be collimated to irradiate a single tooth. Two-dimensional diffraction patterns for both samples are shown in Fig. 18.

Both diffraction patterns in Fig. 18 show the (100) and (200) PMO diffraction rings. In Fig. 18(a), the rings were uniform in intensity, indicating a random orientation of crystallites. In Fig. 18(b), the rings were not uniform in intensity, indicating that a texture (preferred orientation) of the crystallites was present. The preferred orientation was an indication that the bad gear tooth was pulled during processing at a temperature below the melting point of PMO. A review of the manufacturing procedure revealed that a change in the process recently had been instituted. The hold time between injection molding and release from the mold was shortened, with the intention of increasing the number of gears produced in a production cycle. With the reduced hold time, the gear teeth were sticking to the mold and being stretched when removed during the release step. A return to the original hold time for cooling after injection molding eliminated the distortion of the gear teeth.

Polymer Films

Polypropylene (PP) films were studied to understand the effect of processing on microstructure. Two-dimensional XRD data (Fig. 19) were collected on samples obtained from different stages of manufacture.

Melted PP polymer was extruded onto a rotating chilled stainless steel drum, quickly quenching the initial film. The 2-D diffraction pattern in Fig. 19(a) shows broad Debye rings, indicating the film was an amorphous or meso phase. Upon heating the film at 120 °C (248 °F) for 10 s (heat set), the film crystallized, as shown in Fig. 19(b), where the 2-D diffraction pattern showed narrow Debye rings consistent with a randomly oriented polycrystalline sample. After the heat-set step, the PP film was stretched in one direction 6*x* its original length along the extruder machine direction. The heat-set/stretch sample diffraction pattern in Fig. 19 (c) was now dominated by short arcs, with evidence of low-intensity rings.

The arcs were an indication that the PP lattice planes had preferentially aligned due to stretching. These phenomena can be referred to as preferred orientations of lattice planes. However, the alignment was not perfect, because the arcs indicate that there was some distribution of orientation, as opposed to a single crystal that would show diffraction spots. The other indicator that this sample did not have perfect lattice plane alignment was the presence of low-intensity rings in the diffraction pattern (Fig. 19c), due to some residual randomly oriented PP lattice planes. The benefit of a 2-D detector for orientation studies was demonstrated in this example, because a point or line detector would not have the detection area to observe the complete diffraction pattern shown in Fig. 19(c).

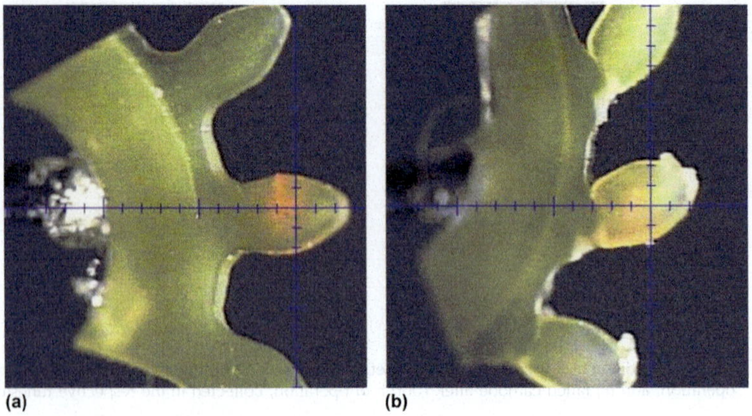

Fig. 17 Sections of (a) good plastic gear with undistorted teeth and (b) bad plastic gear with distorted teeth

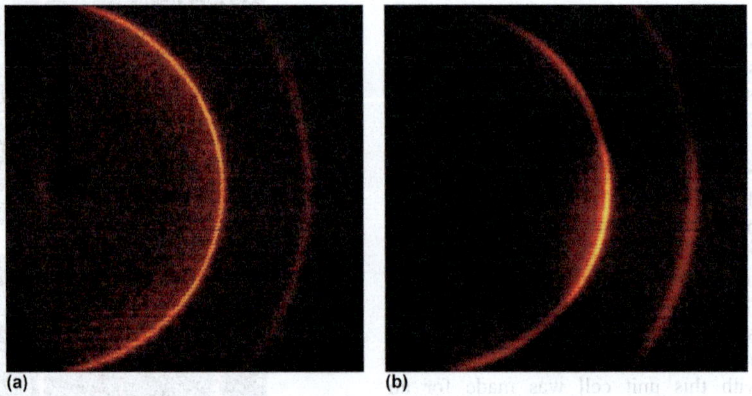

Fig. 18 Micro x-ray diffraction two-dimensional patterns for (a) good gear tooth and (b) bad gear tooth

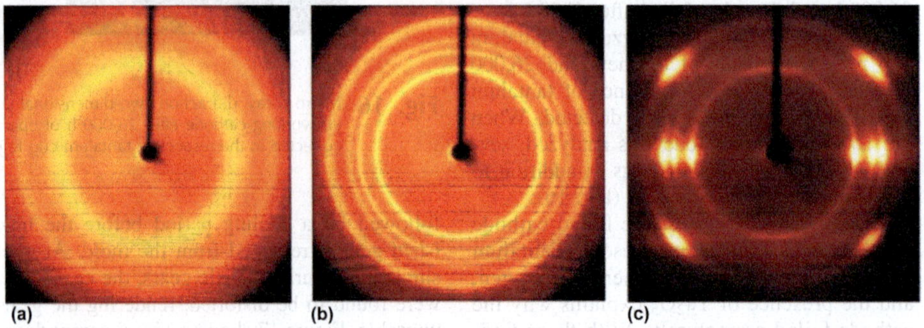

Fig. 19 Two-dimensional diffraction patterns of polypropylene films (a) as-cast, (b) heat set at 120 °C (248 °F) for 10 s, and (c) heat set at 120 °C for 10 s and then uniaxially stretched 6*x* its original length

Qualitative Phase Identification

Flemming (Ref 3) used μXRD mineralogical mapping to reveal the distribution of phases in an inhomogeneous sample, a specimen from the Toluca (Estado de México, Mexico) iron meteorite. Figure 20 shows the reproduced Widmanstätten pattern, caused by the exsolution of kamacite from taenite upon cooling. Kamacite forms in a fixed geometrical relation to taenite. The etched surface of the Toluca meteorite, exhibiting Widmanstätten pattern, is shown in Fig. 20(a and c). In Fig. 20(b), diffraction patterns were collected at 1312 spots located on a 15.5 by 20 mm (0.6 by 0.8 in.) grid, with 0.5 mm (0.02 in.) spacing between spots, using a 500 μm beam.

The 2-D image from one location on the surface is shown in Fig. 20(b). Unit-cell dimensions of kamacite differ from taenite, which make their *d*-spacings distinguishable as distinct arcs with different 2θ values. Integrating the peak intensities along the kamacite and taenite arcs gives the respective kamacite and taenite maps (Fig. 20d, e), nearly reproducing the Widmanstätten pattern.

Quantitative Analysis

In a study on endontic rotary tips, Alapati (Ref 20) showed the ability to analyze multiple targeted points on a sample to determine the amount of austenite in the metal. This technique has proved fundamental in recent years in the areas of failure analysis, alloy control, lamellar cross sections, corrosion analysis, semiconductors, and manufacturing. Through the analysis of different tips and batches, the study was able to show a variation in the amount of austenite at each level (Fig. 21).

Rietveld Structure Refinement

Schmidt (Ref 21) collected μXRD data by using a Thermo Scientific ARL Equinox 5000 equipped with a Gandolfi stage on a single crystal of acetaminophen. Intensities were extracted from a 1 h dataset and imported into GSAS-II (Ref 22), where the ab initio structure solution was performed by applying a Pawley refinement followed by charge-flipping methods to map peaks of the electron density over a range of 10 to 70° 2θ (Fig. 22). The final parameters of the solution and Rietveld refinement resulted in $R_w = 8.128$ with a unit cell of $a = 12.816(9)$, $b = 9.3254(8)$, $c = 7.0629(29)$, $\beta = 115.654(9)$, and a volume of 760.93(22) with a space group of P21/*a*.

Stress/Texture

Gloaguen et al. (Ref 23) analyzed the texture and residual stress by using μXRD on zirconium alloy cladding tubes after an industrial mechanical process. The last rolling pass was completely characterized by μXRD in the transition area between the raw tube and the finished one (Fig. 23). A quantitative interpretation of the influence of intergranular stresses on the development of analyzed stress was performed. The study highlighted the usefulness of a polycrystal model for exploring the active deformation modes in hexagonal close-packed alloys. Reasonable agreement between model predictions and experimental measurements was observed for both residual stresses and evolved deformation textures. The stresses measured by XRD differed significantly from one lattice plane family to another. These results were explained by the presence of first- and second-order stresses of plastic origin. The choice of prismatic slip as the main deformation mode explained the experimental texture and stress values for the six studied planes.

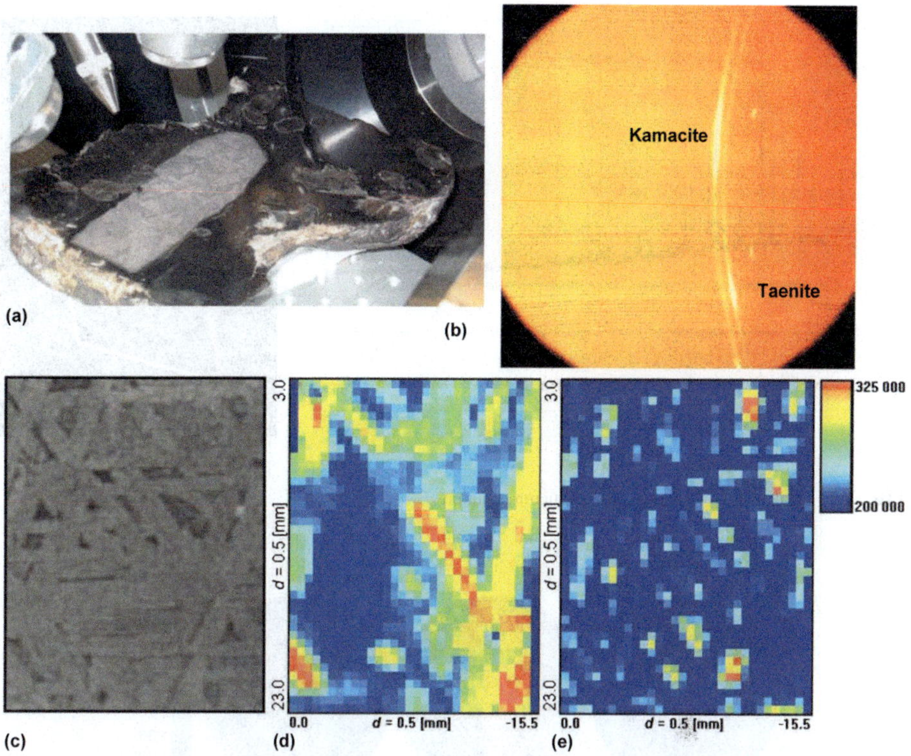

Fig. 20 Micro x-ray diffraction (μXRD) maps of kamacite and taenite showing the Widmanstätten pattern in a Toluca iron meteorite. (a) Toluca iron meteorite on the diffractometer. Etched portion of the cut surface shows the Widmanstätten pattern. (b) Two-dimensional (2-D) image collected at a single location on the Toluca meteorite showing polycrystalline Debye rings with a strong preferred orientation. (c) Photograph of the sample area mapped in 2-D mineral maps. (d and e) 2-D kamacite and taenite maps, respectively, of area shown in (c). μXRD images were collected at 0.5 mm (0.02 in.) intervals.

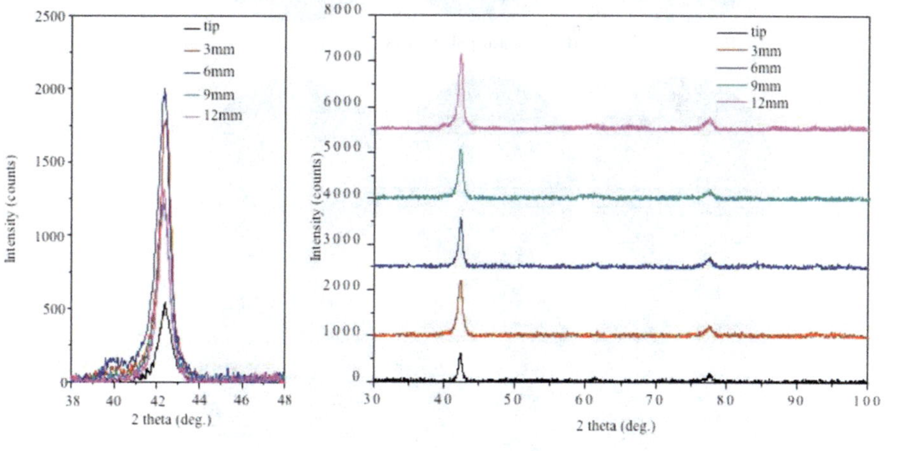

Fig. 21 Micro x-ray diffraction patterns for an as-received K3 NiTi rotary instrument showing the variation in intensity at various distances from the tip for the main austenite peak. Source: Ref 19

REFERENCES

1. W.H. Bragg and W.L. Bragg, *X-Rays and Crystal Structure*, G. Bell and Sons, Ltd., London, England, 1915
2. D. Lau, D. Hay, and W. Wright, Micro X-Ray Diffraction for Painting and Pigment Analysis, *AICCM Bull.*, Vol 30, 2007, p 38–43
3. R. Flemming, Micro X-Ray Diffraction (μXRD): A Versatile Technique for Characterization of Earth and Planetary Materials, *Can. J. Earth Sci.*, Vol 44, 2007, p 1333–1346, https://doi.org/10.1139/e07-020
4. J.D. Hanawalt, H.W. Rinn, and L. Frevel, Chemical Analysis by X-Ray Diffraction—Classification and Use of X-Ray Diffraction Patterns, *Ind. Eng. Chem. Anal.*, Vol 10, 1938, p 457–512
5. T.G. Fawcett, S.N. Kabekkodu, J.R. Blanton, and T.N. Blanton, Chemical Analysis by Diffraction: The Powder Diffraction File, *Powder Diffr.*, Vol 32 (No. 2), 2017, p 63–71
6. S. Kabekkodu, Ed., "ICDD PDF-4+ 2020 (Database)," International Centre for Diffraction Data, Newtown Square, PA, 2019
7. B.B. He, *Two-Dimensional X-Ray Diffraction*, 2nd ed., John Wiley & Sons, 2018, p 100, 119–136, 151–190
8. T.N. Blanton, X-Ray Diffraction Orientation Studies Using Two-Dimensional Detectors, *Adv. X-Ray Anal.*, Vol 37, 1994, p 367–373

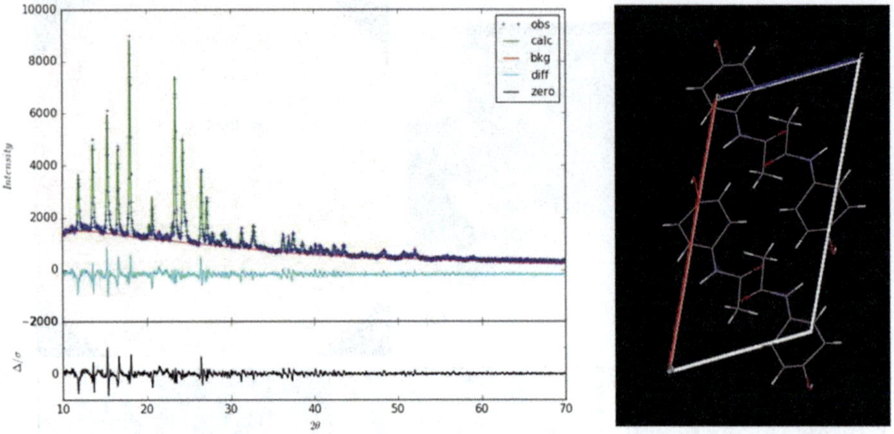

Fig. 22 Final GSAS-II fit and resulting structure solution

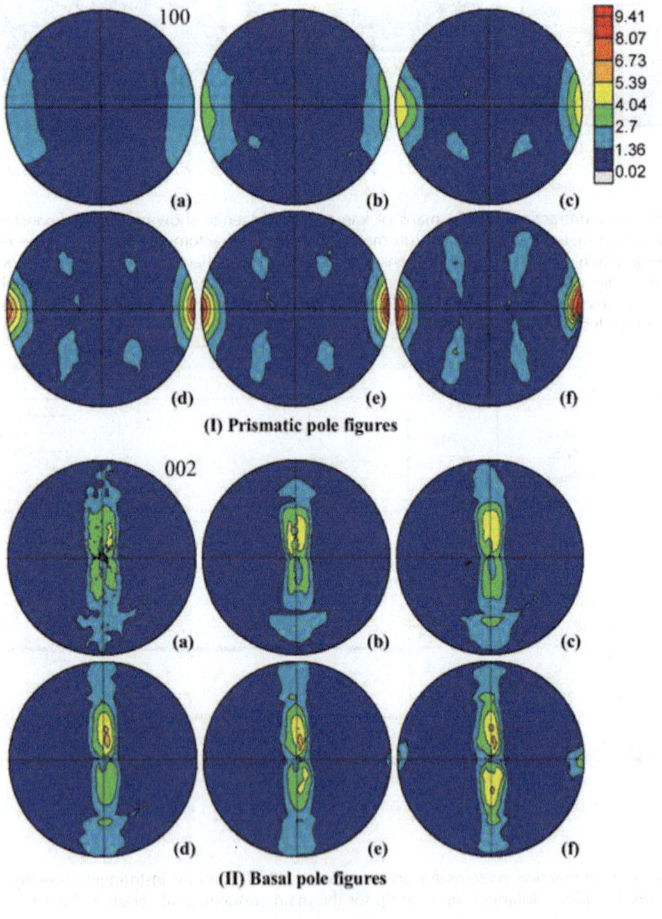

Fig. 23 (I) Experimental prismatic and (II) basal pole figures, recalculated using the WIMV method, at (a) 0% (initial texture), (b) 21%, (c) 51%, (d) 124%, (e) 159%, and (f) 195% total macroscopic strain. Reprinted from Ref 23 with permission of Wiley. Copyright © International Union of Crystallography

9. "Göbel Mirrors for Parallel Beam," CCP14, Crystallography Laboratory University of Nijmegen, http://www.ccp14.ac.uk/ccp/web-mirrors/dirdif/xtal/documents/equipment/d8/accessories/goebel.html
10. C. Michaelsen, P. Ricardo, D. Anders, M. Schuster, J. Schilling, and H. Gobel, Improved Graded Multilayer Mirrors for XRD Applications, *Adv. X-Ray Anal.*, Vol 42, 2000, p 308–320
11. H. Mai, Göbel X-Ray Mirrors, *Mater. World*, Vol 7 (No. 10), 1999, p 616–618
12. Y. Namatame, T. Kuzumaki, Y. Shiramata, and K. Nagao, Use of Multi-Dimensional Measurement in Powder X-Ray Diffraction, *Rigaku J.*, Vol 34 (No. 1), 2018, p 9–13
13. G.A. Chahine, M.-I. Richard, R.A. Homs-Regojo, T.N. Tran-Caliste, D. Carbone, V.L.R. Jaques, R. Grifone, P. Boesecke, J. Katzer, I. Costina, H. Djazouli, T. Schroeder, and T.U. Schülli, Imaging of Strain and Lattice Orientation by Quick Scanning X-Ray Microscopy Combined with Three-Dimensional Reciprocal Space Mapping, *J. Appl. Crystallogr.*, Vol 47, 2014, p 762–769, doi:10.1107/S1600576714004506
14. V.E. Buhrke, R. Jenkins, and D.K. Smith, *A Practical Guide for the Preparation of Specimens for X-Ray Fluorescence and X-Ray Diffraction Analysis*, Wiley-VCH, 1997
15. Private Communication
16. H.M. Rietveld, *J. Appl. Crystallogr.*, Vol 2, 1969, p 65–71
17. "ICDD Sieve+ Release 2019 (Computer Software)," International Centre for Diffraction Data, Newtown Square, PA, 2018
18. T. Blanton, Applications of X-Ray Microdiffraction in the Imaging Industry, *Powder Diffr.*, Vol 21 (No. 2), 2006, p 91–96
19. P. Villars and L.D. Calvert, *Pearson's Handbook of Crystallographic Data for Intermetallic Phases*, Vol 2, American Society for Metals, 1987
20. S. Alapati, "An Investigation of Phase Transformation Mechanisms for Nickel-Titanium Rotary Endontic Instruments," Doctoral dissertation, The Ohio State University, 2006
21. G. Schmidt, The Gandolfi Stage: A Novel Approach for the Analysis of Single Crystals and Small Volume Samples, *J. Appl. Crystallogr.*, 2019, in press
22. B.H. Toby and R.B. Von Dreele, GSAS-II: The Genesis of a Modern Open-Source All-Purpose Crystallography Software Package, *J. Appl. Crystallogr.*, Vol 46 (No. 2), 2013, p 544–549
23. D. Gloaguen, J. Fajoui, E. Girard, and R. Guillén, X-Ray Measurement of Residual Stresses and Texture Development during a Rolling Sequence of Zirconium Alloy Cladding Tubes: Influence of Plastic Anisotropy on the Mechanical Behaviour, *J. Appl. Crystallogr.*, Vol 43, 2010, p 890–899, https://doi.org/10.1107/S0021889810019989

SELECTED REFERENCES

- T. Blanton, X-Ray Film as a Two-Dimensional Detector for X-Ray Diffraction Analysis, *Powder Diffr.*, Vol 18 (No. 2), 2003, p 91–98
- B.D. Cullity, *Elements of X-Ray Diffraction*, Addison-Wesley, 1978

- P. Debye and P. Scherrer, *Interferenzen an regellos orientierten Teilchen im Röntgenlicht. I.*, Nachrichten von der Gesellschaft der Wissenschaften zu Göttingen, Math. Phys. Kl, 1916, p 1–15
- R.E. Dinnebier and S.J.L. Billinge, *Powder Diffraction Theory and Practice*, RSC Publishing, 2008
- R.T. Downs and M. Hall-Wallace, *The American Mineralogist Crystal Structure Database*, *Am. Mineralog.,* Vol 88, 2003, p 247–250
- H. Friedman, Method and Means for Measuring X-Ray Diffraction Patterns, U.S. Patent 2,386,785, 1945
- W. Friedrich, P. Knipping, and M. Laue, *Interferenz-Erscheinungen bei Röntgenstrahlen*, Bayerische Akad. d. Wissenshaften zu Munchen, Sitzungsberichte, math.-phys. Kl, 1912, p 303–322
- R.P. Goehner and M.C. Nichols, X-Ray Powder Diffraction, *Materials Characterization*, Vol 10, *ASM Handbook*, R.E. Whan, Ed., American Society for Metals, 1986, p 333–343
- S. Gražulis, D. Chateigner, R.T. Downs, A.T. Yokochi, M. Quiros, L. Lutterotti, E. Manakova, J. Butkus, P. Moeck, and A. LeBail, Crystallography Open Database—An Open-Access Collection of Crystal Structures, *J. Appl. Crystallogr.*, Vol 42, 2009, p 726–729
- S. Gražulis, A. Daškevič, A. Merkys, D. Chateigner, L. Lutterotti, M. Quirós, N.R. Serebryanaya, P. Moeck, R.T. Downs, and A. LeBail, Crystallography Open Database (COD): An Open-Access Collection of Crystal Structures and Platform for World-Wide Collaboration, *Nucleic Acids Res.*, Vol 40, 2012, p D420–D427
- S. Gražulis, A. Merkys, A. Vaitkus, and M. Okulič-Kazarinas, Computing Stoichiometric Molecular Composition from Crystal Structures, *J. Appl. Crystallogr.*, Vol 48, 2015, p 85–91
- B.B. He, *Two-Dimensional X-Ray Diffraction*, 2nd ed., John Wiley & Sons, 2018
- A.W. Hull, A New Method of X-Ray Crystal Analysis, *Phys. Rev.*, Vol 10 (No. 6), 1917, p 661–696
- R. Jenkins and R.L. Snyder, *X-Ray Powder Diffractometry*, John Wiley & Sons, 1996
- A. Merkys, A. Vaitkus, J. Butkus M. Okulič-Kazarinas, V. Kairys, and S. Gražulis, COD::CIF::Parser: An Error-Correcting CIF Parser for the Perl Language, *J. Appl. Crystallogr.*, Vol 49, 2016
- W. Parish and S.G. Gordon, Precise Angular Control of Quartz-Cutting with X-Rays, *Am. Mineralog.*, Vol 30 (No. 5–6), 1945, p 326–346
- V.K. Percharsky and P.Y. Zavalij, *Fundamentals of Powder Diffraction and Structural Characterization of Materials*, 2nd ed., Springer, 2009

ASM Handbook, Volume 10, *Materials Characterization*
ASM Handbook Committee
DOI 10.31399/asm.hb.v10.a0006632

X-Ray Diffraction Residual-Stress Techniques

Paul S. Prevéy and Douglas J. Hornbach, Lambda Research, Inc.

Overview

General Uses

- Determination of residual stresses from lattice strain in crystalline materials

Macrostress measurement

- Determination of subsurface residual-stress distributions from machining, shot peening, carburizing, welding, etc.
- Nondestructive surface residual-stress measurement for quality control
- Measurement of residual stresses supporting fatigue or stress-corrosion failure analyses
- Mapping of residual stress distributions from welding, heat treating, or forming

Microstress measurement

- Determination of the percent cold work with stress
- Measurement of hardness in steels in thin layers
- Assessment of thermal and mechanical stress stability

Examples of Applications

- Depth and magnitude of the compressive layer and hardness produced by carburizing steels
- Uniformity of the surface compressive residual stresses produced by shot peening in complex geometries
- Surface residual stresses and hardness on the raceway of ball and roller bearings
- Alteration of residual stress and cold work distributions by stress-relieving heat treatment or forming
- Mapping of residual stresses parallel and perpendicular to a weld fusion line with depth and distance
- Direction of maximum residual stress and percent cold work gradient caused by machining

Estimated Analysis Time

- 1 min to 1 h per measurement depending on the diffracted x-ray intensity and technique used. Typically, 1 h per measurement for subsurface work, including material removal and sample repositioning

Samples

- *Form:* Polycrystalline solids, metallic or ceramic, moderate to fine grained
- *Size:* Various, with limitations dictated by the type of apparatus, the stress field to be examined, and x-ray optics
- *Preparation:* Generally, none. Large samples and inaccessible areas may require sectioning with prior strain gaging to record the resulting stress relaxation. Careful handling or protective coatings may be required to preserve surface stresses.

Limitations

- Delicate apparatus generally limited to a laboratory or to quality-assurance testing.
- Only a shallow (<0.01 mm, or 0.0005 in.) surface layer is measured nondestructively; requiring electrolytic polishing to remove layers for subsurface measurement
- Samples must be polycrystalline, of reasonably fine grain size, and not severely textured.
- Precise positioning and orientation of the sample and instrument is required.
- Access to the measurement location is required for the incident and diffracted x-ray beams.

Capabilities of Related Techniques

- Linear-elastic diffraction techniques that calculate stress from total elastic strain include: *Neutron diffraction*
- Mechanical relaxation methods include: *General Dissection Methods, Center-Hole Drilling, Ring-Core. Deep-Hole-Drilling, Contour Method, Slitting Method*
- Nonlinear elastic methods using other stress-dependent properties include: *Ultrasonic methods (speed of sound), and Barkhausen magnetic methods (magnetic hysteresis)*

Introduction

Residual stresses generally are caused by nonuniform thermal and/or mechanical plastic deformation, as in forming, machining, grinding, shot peening, welding, quenching, or virtually any thermal-mechanical process that leaves a distribution of elastic strains. Phase transformations that produce nonuniform volume changes in a part, as in carburizing or case hardening of steel or austenite transformations in service, also generate residual stresses, generally compressive, in the expanding hardened layer. Although nonuniform plastic strain or phase changes produce residual stresses, once the part is in equilibrium, the residual stresses produced are entirely elastic. The prior complex thermal-mechanical history and plastic strain details of casting, forging, forming, heat treatment, or machining need not be known. Only the elastic strains remaining in the free part at thermal equilibrium contribute to the residual-stress distribution in the part.

Although the term *residual-stress measurement* is widely used, stress is an extrinsic property that is not directly measurable. All methods of stress determination require measurement of some intrinsic properties, such as strain and elastic constants, or force and area, from which the associated stress is calculated. X-ray and neutron diffraction methods calculate the residual stress from the elastic strain measured in the crystal lattice without altering the part. Mechanical methods calculate the residual stress from the change in strain relaxed (or dimensional change) caused by sectioning, slitting, drilling, or trepanning the part. The nonlinear elastic methods, including Barkhausen noise, eddy current, and ultrasonic, rely on higher-order effects of stress on the magnetic, electrical, or acoustical properties of the part.

If the part is not externally constrained, then the residual-stress distribution must be in equilibrium. Equilibrium requires that the integral of the forces and the moments acting on any entire plane through the body must both sum to zero. It is not necessary that the residual stress at any one point that happens to be measured must show equilibrium with depth into the part. For example, if a point in the center of a plate is heated to incandescence and cools, then the center will be in residual tension entirely through the thickness, surrounded by equilibrating residual compression. A bend formed in a tube will be in residual tension through the wall that was deformed in compression, and in residual compression through the side deformed in tension. Cross-roll straightened bar stock can have spiral patterns of residual tension and compression. The complex residual-stress distribution in a weld depends on the geometry, constraints, order of fusion, temperature distribution, and cooling. There really is no limit to the complexity of residual-stress distributions that can be formed, but they must always be in elastic equilibrium.

X-ray diffraction (XRD) residual-stress methods calculate the residual stress present from the strains measured in the crystal lattice of the grains in the sample. Because the residual stresses are entirely elastic, the entire stress present is measured nondestructively without altering the sample. To determine the stress in one direction on the sample surface, the strain in the crystal lattice must be measured for at least two precisely known orientations relative to the sample surface. Therefore, XRD residual-stress measurement is applicable to materials that are crystalline, relatively fine grained, and produce diffraction for any orientation of the sample surface. Samples may be metallic or ceramic, provided a diffraction peak of suitable intensity and free of interference from neighboring peaks can be produced in the high-back-reflection region with the radiations available. X-ray diffraction residual-stress measurement is unique in that macroscopic and microscopic residual stresses can be determined nondestructively, but only in a very thin layer, nominally 0.01 mm (0.0005 in.) deep.

Macroscopic Stresses

Macroscopic stresses, or macrostresses, which extend over distances that are large relative to the grain size of the material, are of general interest to engineers. These are the stresses of primary interest in component design, finite-element modeling, and fatigue or stress-corrosion failure analysis. Macrostresses are homogeneous, in the sense that they extend uniformly over the various metallurgical features of grains, grain boundaries, and precipitates. Macrostresses are second-order tensor quantities, with shear and normal stress magnitudes both varying in three directions at a single point in a body. The macrostress for a given location and direction is determined by measuring the strain in that direction at a single point. When macrostresses are determined in at least three known directions, and a condition of plane stress is assumed at the free surface, the three stresses can be combined using Mohr's circle for stress to determine the maximum and minimum residual stresses, the maximum shear stress, and their orientation relative to a reference direction. Macrostresses strain many crystals uniformly in the surface. This uniform elongation or compression of the crystal lattice shifts the angular position of the diffraction peak selected for residual-stress measurement. The lattice strain is calculated from the small angular shift in the position of the diffraction peak. The residual stress in the surface is calculated from the strain measured in crystals oriented at two or more angles to the surface.

Microscopic Stresses

Microscopic stresses, or microstresses, are inhomogeneous, varying over minute distances in both magnitude and orientation. Microstresses exist between different phase particles, regions of different crystallographic orientation, and the subgrain regions between dislocation tangles. The literature contains various attempts to classify microstresses as different types and to relate them to the phase-selective XRD measurement of lattice strain and residual stresses. Interphase stresses result from different states of stress in the phases present in the sample matrix. Examples are stresses developed between the martensite and austenite phases after heat treatment of steel, or between precipitates and the matrix phase of an alloy. Because XRD stress measurement is inherently phase selective, any stress calculated is for the phase producing the diffraction peak measured. This makes XRD methods unique in being able to measure the different phase stresses separately. However, it must be kept in mind when interpreting the results obtained on multiphase materials that the stress of only one phase may be known. Other than the interphase stresses, microstresses cannot be measured directly in the minute individually stressed regions.

Diffraction-Peak Breadth

Diffraction-peak breadth is of more practical use as a measure of the magnitude of microstresses present in the material. The aggregate effect on the range of lattice strain then can be treated as a scalar property of the material in the diffracting volume. The percent of cold work or hardness are scalar properties without direction that result from dislocations and imperfections in the crystal lattice causing the microstresses. Diffraction line broadening is associated with strains in the regions between dislocation tangles within the crystal lattice that traverse distances much less than the dimensions of the crystals. Broadening of the diffraction peaks used for macroscopic residual-stress measurement arises from variation in the local lattice spacing of the coherent diffracting domains or crystallites, the perfectly crystalline material between dislocations, and from the range of strains in these regions. Compressively stressed crystallites contribute to the high-angle side of the peak, and tensile regions to the low-angle side. Measurable broadening of the diffraction peak from the related reduction in crystallite size begins as the domains become smaller than a few hundred atomic dimensions, insufficient to ensure the constructive and destructive interference of Bragg's law. A scalar measure of the range of microstresses can be determined from the diffraction-peak breadth measured in conjunction with macroscopic residual-stress measurement. The peak breadth can be related empirically to material properties useful in materials engineering, including cold work level, yield strength, and hardness.

Principles of X-Ray Diffraction Stress Measurement

In XRD residual-stress measurement, the strain in the crystals making up the sample itself provides the strain gage needed to calculate the residual stresses present. The method uses linear elastic theory to calculate stress from strain, but on an atomic scale rather than the millimeter scale of electrical resistance strain gages. Therefore, samples must be crystalline, fairly fine grained, and not too highly oriented so that well-defined diffraction peaks are produced by many crystals at any angle to the surface. Fortunately, most high-strength forgings and machined or shot-peened surfaces meet these requirements. Coarse-grained materials, such as castings, may not be suitable.

The selective nature of XRD allows the spacing of a specific set of crystal lattice planes with the Miller indexes (*hkl*) oriented at a precise angle to the sample surface to be measured with the accuracy needed to determine the strain in the surface. Figure 1 shows the diffraction of a monochromatic beam of x-rays at a high diffraction angle, 2θ, from the surface of a stressed sample for two orientations of the sample relative to the x-ray beam. The angle ψ, defining the orientation of the sample surface, is the angle between the normal of the surface and the incident and diffracted beam bisector, which is also the angle between the normal to the diffracting lattice planes and the sample surface.

Diffraction occurs at an angle 2θ, defined by Bragg's law:

$$n\lambda = 2d \sin \theta$$

where n is an integer denoting the order of diffraction ($n = 1$, normally), λ is the x-ray wavelength, d is the lattice spacing of crystal planes, and θ is the diffraction angle. For the monochromatic x-rays produced by the metallic target of an x-ray tube, the wavelength (in nanometers) is known to 1 part in 10^5. Any change in the lattice spacing, d, results in a corresponding shift in the diffraction angle, 2θ. The angular shift caused by residual stresses in the position of a typical diffraction peak with a half-width of several degrees is generally less than a degree, requiring high accuracy in sample placement, peak location, and angular measurement.

Figure 1(a) shows the sample in the $\psi = 0$ orientation. The only crystals that satisfy Bragg's law and diffract are parallel to the surface. The presence of a tensile stress in the sample results in a Poisson's ratio contraction, reducing the lattice spacing and slightly increasing the diffraction angle, 2θ, of the diffracting crystals. If the sample is then rotated through some known angle ψ (Fig. 1b), the tensile stress present in the surface now increases the lattice spacing of the crystals diffracting at that orientation relative to the stress-free state, decreasing 2θ. Measuring the change in the angular position of the diffraction peak for at least two orientations of the sample defined by the angle ψ enables calculation of the stress present in the sample surface in the direction defined by the plane of diffraction, which contains the incident and diffracted x-ray beams. To measure the stress in different directions at the same point, the sample is rotated through an angle ϕ about its surface normal to align the direction of interest with the diffraction plane.

Because only the elastic strain changes the mean lattice spacing, only elastic strains are measured using XRD for the determination of macrostresses. When the elastic limit is exceeded, further strain results in dislocation motion, disruption of the crystal lattice, and the formation of microstresses, but no additional increase in macroscopic stress. Although residual stresses result from nonuniform plastic deformation, all residual macrostresses remaining after deformation are necessarily elastic. X-ray diffraction determines the total elastic strain and, therefore, the total residual stress present in the diffracting volume of material, a very thin surface layer, without altering the sample.

The residual stress determined using XRD is the arithmetic average stress in a volume of material defined by the irradiated area, which may vary from square centimeters to square millimeters, and the shallow depth of penetration of the x-ray beam. The linear absorption coefficient for the sample material and radiation used governs the depth of penetration, which can vary considerably. However, in iron-, nickel-, and aluminum-base alloys, 50% of the radiation is diffracted from a layer approximately 0.005 mm (0.0002 in.) deep for the radiations generally used for stress measurement. This shallow depth of penetration allows high-resolution determination of residual-stress distributions as functions of depth, nominally 10 to 100 times the resolution of mechanical or neutron diffraction methods. A condition of plane stress exists in the thin diffracting surface layer, so that no normal or shear stresses are acting out of the free surface. Therefore, the stresses of interest in the plane of the surface can be determined without the need for reference to an unstressed lattice spacing standard, as shown in the following derivation.

Although in principle virtually any interplanar lattice spacing may be used to measure strain in the crystal lattice, the wavelengths available from commercial x-ray tubes limit the choice to a few possible planes. The choice of radiation and diffraction peak selected determine the precision of the strain measurement. The higher the diffraction angle, 2θ, the greater the precision of the strain calculated from the measured angular shift in diffraction-peak position. Practical techniques generally require diffraction angles, 2θ, greater than 120°.

Table 1 lists some recommended diffraction techniques for various alloys. The relative sensitivity is shown by the value of K_{45}, the magnitude of the stress necessary to cause a 1° shift in the diffraction-peak position for a 45° ψ tilt. As K_{45} increases, sensitivity decreases.

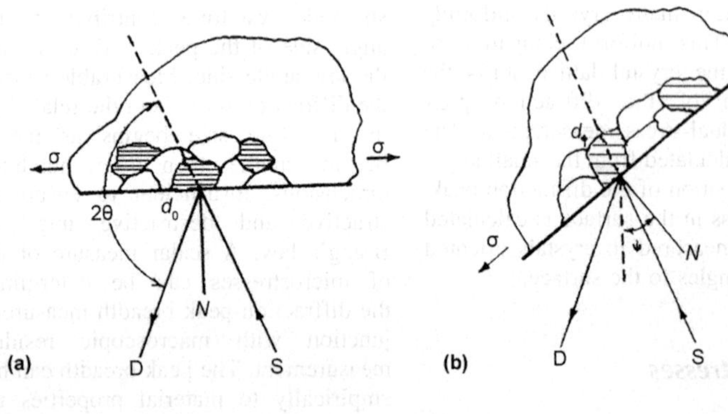

Fig. 1 Principles of x-ray diffraction residual-stress measurement. D, x-ray detector; S, x-ray source; N, normal to the surface. (a) Ψ = 0: Poisson's ratio contraction of lattice spacing. (b) Ψ > 0: Tensile extension of lattice planes by stress σ

Plane-Stress Elastic Model

The radiations suitable for XRD stress measurement are very "soft," with low energies (typically 5 to 8 keV) and wavelengths comparable to the lattice spacing to be measured. The diffraction peaks then occur at high Bragg angles, providing maximum lattice strain resolution. The soft x-ray penetration is very shallow and attenuated exponentially. Nearly all of the diffracted radiation comes from a layer only approximately 0.025 mm (0.001 in.) thick in most materials of engineering interest, confining the measurement to the very near surface of the sample. Electropolishing is used to remove successive layers, exposing new surfaces for subsurface measurement without introducing any plastic deformation that would alter the residual stresses present. Corrections for the exponential attenuation of the radiation penetrating the subsurface stress distribution and for relaxation of the stresses present due to layer removal are then necessary, as discussed subsequently. Because the surface is

Table 1 Recommended diffraction techniques, x-ray elastic constants, and bulk values for various ferrous and nonferrous alloys

Alloy	Radiation	Lattice plane, (*hkl*)	Diffraction angle (2θ), degrees	Elastic constants(a) $E/(1+\nu)$, GPa (10^6 psi)		Bulk error, %	K_{45}(b)		Linear absorption coefficient (μ)	
				(*hkl*)	Bulk		MPa	ksi	cm^{-1}	$in.^{-1}$
Aluminum-base alloys										
2014-T6	Cr Kα	(311)	139.0	59.4 ± 0.76 (8.62 ± 0.11)	54.5 (7.9)	−8.3	387	56.2	442	1124
2024-T351	Cr Kα	(311)	139.3	53.8 ± 0.55 (7.81 ± 0.08)	54.5 (7.9)	+1.1	348	50.5	435	1105
7075-T6	Cr Kα	(311)	139.0	60.9 ± 0.48 (8.83 ± 0.07)	53.8 (7.8)	−11.4	397	57.6	...	...
7050-T6	Cr Kα	(311)	139.0	57.1 ± 0.41 (8.28 ± 0.06)	53.8 (7.8)	−5.8	372	54.0	443	1126
Iron-base alloys										
Incoloy 800	Cu Kα	(420)	147.2	148.2 ± 2.8 (21.5 ± 0.4)	147.5 (21.4)	−0.4	758	110.0	1656	4205
304L	Cu Kα	(420)	147.0	157.2 ± 2.8 (22.8 ± 0.4)	151.0 (21.9)	−3.9	814	118.0	2096	5321
316	Cu Kα	(420)	146.5	132.4 ± 1.4 (19.2 ± 0.2)	153.8 (22.3)	+16.0	696	101.0	2066	5245
Invar	Cu Kα	(420)	147.0	108.2 ± 4.1 (15.7 ± 0.6)	112.4 (16.3)	+3.8	560	81.2	1706	4330
410 (22 HRC)	Cr Kα	(211)	155.1	176.5 ± 0.7 (25.6 ± 0.1)	155.8 (22.6)	−11.7	680	98.6	840	2129
410 (42 HRC)	Cr Kα	(211)	155.1	173.1 ± 1.4 (25.1 ± 0.2)	155.8 (22.6)	−9.9	667	96.7	840	2129
1050 (56 HRC)	Cr Kα	(211)	156.0	184.1 ± 2.1 (26.7 ± 0.3)	148.2 (21.5)	−19.4	683	99.0	885	2244
4340 (50 HRC)	Cr Kα	(211)	156.0	168.9 ± 2.8 (24.5 ± 0.4)	156.5 (22.7)	−7.3	627	90.9	909	2307
6260	Cr Kα	(211)	155.5	169.6 ± 2.8 (24.6 ± 0.4)	158.9 (23.0)	−6.5	643	93.2	894	2271
9310	Cr Kα	(211)	155.5	172.4 ± 2.8 (25.0 ± 0.4)	160.0 (23.2)	−7.2	653	94.7	894	2271
52100	Cr Kα	(211)	156.0	173.7 ± 2.1 (25.2 ± 0.3)	153.8 (22.3)	−11.5	645	93.5	714	1807
M50 (62 HRC)	Cr Kα	(211)	154.0	179.3 ± 2.1 (26.0 ± 0.3)	157.9 (22.9)	−11.9	724	105.0	1000	2490
17-4PH	Cr Kα	(211)	155.0	180.0 ± 0.7 (26.1 ± 0.1)	158.9 (23.0)	−11.9	696	101.0	888	2254
Nickel-base alloys										
Inconel 600	Cu Kα	(420)	150.8	159.3 ± 0.7 (23.1 ± 0.1)	165.5 (24.0)	+3.9	724	105.0	896	2275
Inconel 718	Cu Kα	(420)	145.0	140.0 ± 2.1 (20.3 ± 0.3)	156.5 (22.7)	−8.9	772	112.0	1232	3127
Inconel X-750	Cu Kα	(420)	151.0	160.6 ± 1.4 (23.3 ± 0.2)	160.6 (24.0)	+3.0	724	105.0	813	2062
Incoloy 901	Cu Kα	(420)	146.0	134.4 ± 3.4 (19.5 ± 0.5)	158.6 (23.0)	+17.9	717	104.0	1408	3569
René 95	Cu Kα	(420)	146.7	168.9 ± 0.7 (24.5 ± 0.1)	164.1 (23.8)	−2.8	882	128.0	935	2370
Titanium-base alloys										
Commercially pure Ti	Cu Kα	(21.3)	139.5	90.3 ± 1.4 (13.1 ± 0.2)	84.8 (12.3)	−6.1	581	84.3	917	2320
Ti-6Al-4V	Cu Kα	(21.3)	141.7	84.1 ± 0.7 (12.2 ± 0.1)	84.8 (12.3)	+0.8	509	73.9	867	2203
Ti-6Al-2Sn-4Zr-2Mo	Cu Kα	(21.3)	141.5	102.0 ± 1.4 (14.8 ± 0.2)	86.2 (12.5)	−15.5	622	90.2	866	2200

(a) Constants determined from four-point bending tests. (b) K_{45} is the magnitude of the stress necessary to cause an apparent shift in diffraction-peak position of 1° for a 45° angle tilt.

not disturbed in order for the lattice strain to be measured, XRD stress measurement is the only method that can measure the actual surface stress nondestructively.

The diffracting surface layer is so thin that a condition of plane stress can be assumed to exist at this free surface. That is, a stress distribution described by principal stresses σ_1 and σ_2 exists in the plane of the surface, and no stress, normal or shear, is acting perpendicular to the unrestrained free surface, $\sigma_3 = 0$. A strain component ε_3 does exist normal to the surface as a result of the Poisson's ratio contractions caused by the two principal stresses shown in Fig. 2.

The strain, $\varepsilon_{\phi\psi}$, in the direction defined by the angles ϕ and ψ is:

$$\varepsilon_{\phi\psi} = \left[\frac{1+\nu}{E}(\sigma_1\alpha_1^2 + \sigma_2\alpha_2^2)\right] - \left[\frac{\nu}{E}(\sigma_1 + \sigma_2)\right] \quad \text{(Eq 1)}$$

where E is the modulus of elasticity, ν is the Poisson's ratio, and α_1 and α_2 are the angle cosines of the strain vector:

$$\alpha_1 = \cos\phi \sin\psi$$
$$\alpha_2 = \sin\phi \sin\psi \quad \text{(Eq 2)}$$

Substituting for the angle cosines in Eq 1 and simplifying enables expressing the strain in terms of the orientation angles:

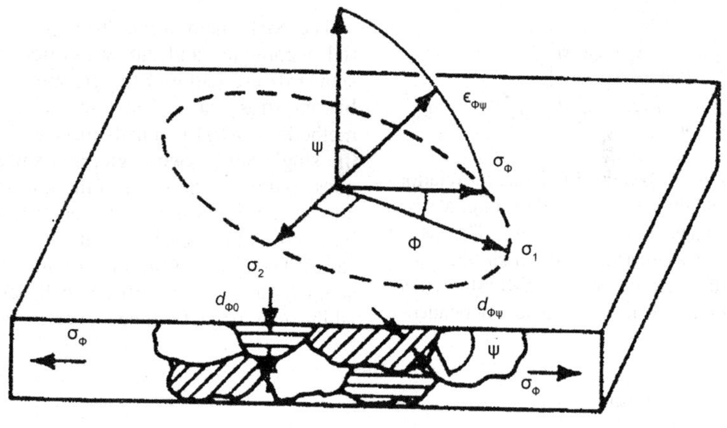

Fig. 2 Plane-stress elastic model

$$\varepsilon_{\phi\psi} = \left[\frac{1+\nu}{E}(\sigma_1\cos^2\phi + \sigma_2\sin^2\phi)\sin^2\psi\right] - \left[\frac{\nu}{E}(\sigma_1 + \sigma_2)\right] \quad \text{(Eq 3)}$$

If the angle ψ is taken to be 90°, the strain vector lies in the plane of the surface, and the surface stress component, σ_ϕ, is:

$$\sigma_\phi = (\sigma_1\cos^2\phi) + (\sigma_2\sin^2\phi) \quad \text{(Eq 4)}$$

Substituting Eq 4 into Eq 3 yields the strain in the sample surface at an angle ϕ from the principal stress σ_1:

$$\varepsilon_{\phi\psi} = \left[\frac{1+\nu}{E}\sigma_\phi\sin^2\psi\right] - \left[\frac{\nu}{E}(\sigma_1 + \sigma_2)\right] \quad \text{(Eq 5)}$$

Equation 5 relates the surface stress, σ_ϕ, in any direction defined by the angle ψ to the strain, ε, in the direction (ϕ,ψ) and the principal stresses in the surface. Note that Eq 5 describes any elastic plane-stress condition at a surface regardless of how the stresses may be measured.

The XRD stress measurement is now introduced through the use of the strain measured in the crystal lattice. If $d_{\phi\psi}$ is the spacing between the lattice planes measured in the direction defined by ϕ and ψ, the strain can be expressed in terms of changes in the linear dimensions of the crystal lattice:

$$\varepsilon_{\phi\psi} = \frac{\Delta d}{d_0} = \frac{d_{\phi\psi} - d_0}{d_0}$$

where d_0 is the stress-free lattice spacing. Substitution into Eq 5 yields:

$$\frac{d_{\phi\psi} - d_0}{d_0} = \left[\left(\frac{1+\nu}{E}\right)_{(hkl)} \sigma_\phi \sin^2\psi\right] - \left[\left(\frac{\nu}{E}\right)_{(hkl)} (\sigma_1 + \sigma_2)\right] \quad \text{(Eq 6)}$$

where the elastic constants $((1 + \nu)/E)_{(hkl)}$ and $(\nu/E)_{(hkl)}$ are not the bulk values determined in a tensile test but the values for the crystallographic direction normal to the lattice planes in which the strain is measured as specified by the Miller indexes (*hkl*). Because of elastic anisotropy, the elastic constants in the (*hkl*) direction often vary significantly from the bulk mechanical values, which, for a randomly oriented material, are an average over all possible directions in the crystal lattice.

The lattice spacing for any orientation then is:

$$d_{\phi\psi} = \left[\left(\frac{1+\nu}{E}\right)_{(hkl)} \sigma_\phi d_0 \sin^2\psi\right] - \left[\left(\frac{\nu}{E}\right)_{(hkl)} d_0(\sigma_1 + \sigma_2)\right] + d_0 \quad \text{(Eq 7)}$$

Equation 7 describes the fundamental relationship between lattice spacing and the biaxial stresses in the surface of the sample. The lattice spacing $d_{\phi\psi}$ is a linear function of $\sin^2\psi$, a critically important requirement for XRD stress measurement. Figure 3 shows the actual dependence of $d(311)$ for ψ ranging from 0 to 45° for shot-peened 5056-O aluminum with a surface stress of −148 MPa (−21.5 ksi), to which a straight line has been fitted by least-squares regression.

The intercept of the plot at $\sin^2\psi = 0$:

$$d_{\phi 0} = d_0 - \left(\frac{\nu}{E}\right)_{(hkl)} d_0(\sigma_1 + \sigma_2) = d_0\left[1 - \left(\frac{\nu}{E}\right)_{(hkl)}(\sigma_1 + \sigma_2)\right] \quad \text{(Eq 8)}$$

which equals the unstressed lattice spacing, d_0, minus the Poisson's ratio contraction caused by the sum of the principal stresses. The slope of the plot is:

$$\frac{\partial d_{\phi\psi}}{\partial \sin^2\psi} = \left(\frac{1+\nu}{E}\right)_{(hkl)} \sigma_\phi d_0$$

which can be solved for the stress σ_ϕ:

$$\sigma_\phi = \left(\frac{E}{1+\nu}\right)_{(hkl)} \frac{1}{d_0}\left(\frac{\partial d_{\phi\psi}}{\partial \sin^2\psi}\right) \quad \text{(Eq 9)}$$

The x-ray elastic constants can be determined empirically, but the unstressed lattice spacing, d_0, is generally unknown and may depend on local composition. However, because $E \gg (\sigma_1 + \sigma_2)$, the value of $d_{\phi 0}$ from Eq 8 differs from d_0 by not more than ±1%, and σ_ϕ may be approximated to this accuracy using:

$$\sigma_\phi = \left(\frac{E}{1+\nu}\right)_{(hkl)} \frac{1}{d_{\phi 0}}\left(\frac{\partial d_{\phi\psi}}{\partial \sin^2\psi}\right) \quad \text{(Eq 10)}$$

The XRD method then becomes a differential technique, and no stress-free reference standards are required to determine d_0 for the biaxial stress case. The three most common methods of XRD residual-stress measurement, the single-angle, two-angle, and $\sin^2\psi$ techniques, assume plane stress at the sample surface and are based on the fundamental relationship between lattice spacing and stress given in Eq 7. The residual stresses of interest in the sample surface and with depth by electropolishing can be determined accurately even if the nominal lattice spacing varies with alloying, carburizing, or cold work. The XRD method is a differential technique requiring only the measurement of lattice spacing at two or more angles to the surface.

In contrast, neutron diffraction or XRD techniques using high-energy synchrotron radiation that penetrates deep into the surface cannot assume that plane stress exists in the diffracting volume. To calculate the residual stresses from the lattice strains, the full stress tensor must be solved and the unstressed lattice spacing must be independently known, which may be impractical in inhomogeneous materials, such as a case-hardened steel.

If the lattice spacing is determined not to be a linear function of $\sin^2\psi$, then the XRD method should not be attempted. Nonlinear d versus $\sin^2\psi$ dependence can be due to shear stresses acting out of the surface or extreme preferred orientation, but this is very rarely seen in practice. Nonlinearity is most commonly caused by instrument misalignment, poor x-ray optics, grain sizes too coarse to produce well-defined peaks, poor diffraction-peak-locating algorithms, severe preferred orientation causing variation in the elastic constants with ψ, and/or nonuniform stress in the irradiated area varying with ψ. In such cases, simply fitting a straight line by linear regression to the data will not produce a valid result.

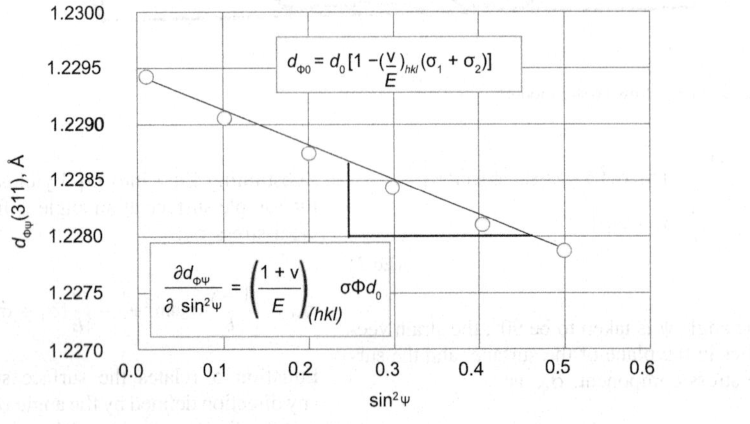

Fig. 3 $d(311)$ versus $\sin^2\Psi$ plot for shot-peened 5056-O aluminum having a surface stress of −148 MPa (−21.5 ksi)

Single-Angle Technique

The single-angle technique, or single-exposure technique, derives its name from early photographic methods that require a single exposure of the film (Ref 1, 2). Position-sensitive detectors have replaced x-ray film. The method is generally considered less sensitive than the two-angle or $\sin^2\psi$ techniques, primarily because the possible range of ψ is limited by the diffraction angle 2θ, but it has the advantage of not requiring any instrumental movements. Figure 4 shows the basic geometry of the method.

A collimated beam of x-rays is inclined at a known angle, β, from the sample surface normal. X-rays diffract from the sample, forming a cone of diffracted radiation originating at point 0. The diffracted x-rays are recorded using film or position-sensitive detectors placed on either side of the incident beam. The presence of a stress in the sample surface varies the lattice spacing slightly between the diffracting crystals shown at points 1 and 2 in Fig. 4, resulting in slightly different diffraction angles on either side of the x-ray beam. If S_1 and S_2 are the arc lengths along the surface of the film or detectors at a radius R from the sample surface, the stress is:

$$\sigma_\phi = \left(\frac{E}{1+\nu}\right)_{(hkl)} \frac{S_1 - S_2}{2R}\left(\frac{\cot\theta}{\sin^2\psi_1 - \sin^2\psi_2}\right) \quad \text{(Eq 11)}$$

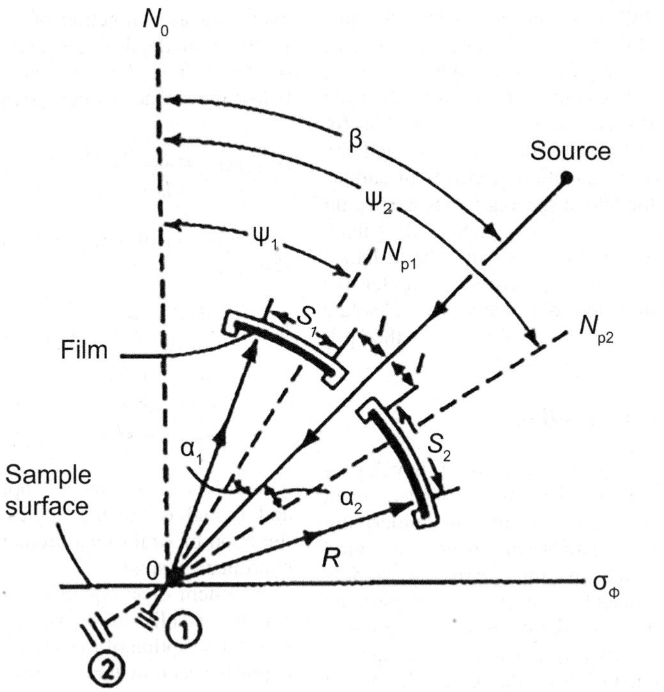

Fig. 4 Basic geometry of the single-angle technique for x-ray diffraction residual-stress measurement. β, angle of inclination of the instrument; 0, point at which a cone of diffracted radiation originates; 1 and 2, points of the diffracting crystals; S_1 and S_2, the arc lengths along the surface of the film; *R*, radius; ψ_1 and ψ_2, angles that are related to the Bragg diffraction angles; N_p, normal to the lattice planes; N_0 normal to the surface. Source: Ref 2

The angles ψ_1 and ψ_2 are related to the Bragg diffraction angles, θ_1 and θ_2, and the angle of inclination of the instrument, β, by:

$$\psi_1 = \beta + \theta_1 - \frac{\pi}{2}$$

and

$$\psi_2 = \beta - \theta_2 + \frac{\pi}{2}$$

The precision of the method is limited by the fact that increasing the diffraction angle 2θ to achieve precision in the determination of lattice spacing reduces the possible range of $\sin^2\psi$, lessening sensitivity. The single-angle technique with position-sensitive detectors is being used for high-speed measurement in quality-control and automated layer-removal applications.

Cosine Alpha Technique

The cosine alpha method is instrumentally similar to the single-angle technique and was first developed in Japan, initially to measure the stress in extremely small irradiated areas for the purpose of mapping stresses around the tip of cracks in fatigue crack growth samples (Ref 3). The technique uses data from the entire Debye ring collected with a two-dimensional area detector positioned where the cone of diffraction intersects the plane of the two detectors shown in Fig. 4. The peak positions at four quadrants of the Debye ring are measured, and the stress is calculated in two directions on the surface. Although commercial instruments are offered, there is only limited literature describing the means of calculating the stresses and the potential experimental errors. Sensitivity of the stresses reported to the precise position of the incident beam position and the angle ψ are noted, but no quantitative assessments or comparisons to other techniques are currently available. Except in limited cases, the cosine alpha method does not appear to offer a significant advantage over the conventional XRD methods described subsequently, if a sufficient diffracted beam intensity is available.

Two-Angle Technique

Equation 7 and Fig. 3 show that if the lattice spacing, $d_{\phi\psi}$, is a linear function of $\sin^2\psi$, the stress can be determined by measuring the lattice spacing for any two ψ angles. In this respect, the two-angle technique is similar to the single-angle, but ψ values can be selected to provide optimal sensitivity or to avoid interference with the x-ray beam. The technique has been thoroughly investigated by SAE International and is widely accepted (Ref 1). Selecting ψ angles to provide as large a range of $\sin^2\psi$ as possible within the limitations imposed by the diffraction angle 2θ and the sample geometry maximizes sensitivity of the method. Lattice spacing is determined precisely at two extreme values of ψ, typically 0 and 45°, and the stress is calculated using Eq 10. Because only the slope of *d* versus $\sin^2\psi$ is required to calculate the stress, it is easily shown that the accuracy of measurements using just the two extreme ends of the $\sin^2\psi$ range is comparable to least-squares fitting to numerous points, provided $d_{\phi\psi}$ is a linear function of $\sin^2\psi$.

$Sin^2\psi$ Technique

The $\sin^2\psi$ technique (Ref 1) is identical to the two-angle technique, except lattice spacing is determined for multiple ψ tilts, a straight line is fitted by least-squares regression (as shown for the shot-peened aluminum sample in Fig. 3), and the stress is calculated from the slope of the best-fit line using Eq 10. The method is a standard procedure that is widely used and described in SAE H784 and European specification BS EN 15305:2008. It requires measurement time in proportion to the number of ψ tilts used. Positive ψ tilts are recommended due to the increased experimental error associated with negative values. The method provides no significant improvement in accuracy over the two-angle technique using the two ψ tilts at the extreme ends of a linear *d* versus $\sin^2\psi$ range. However, it does allow nonlinearity to be detected, and it is recommended when initially investigating measurement of samples that may have large grain size.

The primary benefit of the $\sin^2\psi$ technique, considering the additional time required for data collection, is in establishing the linearity of *d* as a function of $\sin^2\psi$ to demonstrate that XRD residual-stress measurement is possible on the sample of interest. As noted, XRD measurements should not be attempted if the dependence is not linear. Simply fitting a line to nonlinear data, which unfortunately is commonly observed, will not produce a valid stress value.

Marion-Cohen Technique

The Marion-Cohen technique characterizes the dependence of lattice spacing on stress in highly textured materials (Ref 4). The method assumes a biaxial stress field with an additional dependence of the lattice spacing on a texture distribution function, *f*(ψ), a measure of the (*hkl*) pole density calculated from the diffracted intensity over the range of ψ tilts used for stress measurement. The model assumes a lattice spacing dependence of:

$$d_{\phi 0} = \left(\frac{1+\nu}{E}\right)_{(hkl)} \sigma_\phi d_0 \sin^2\psi + (d_{\max} - d_B) f(\psi) + d_B \quad \text{(Eq 12)}$$

where d_{max} and d_B are the maximum and minimum lattice spacings in the range investigated. The method requires simultaneous determination of the preferred orientation, or texture, in the sample to determine $f(\psi)$ along with lattice spacing and is solved by multiple linear regression over the functions $f(\psi)$ and $d_{\psi\phi}$ as functions of $\sin^2\psi$ to determine σ_ϕ, d_{max}, and d_B.

The assumption that the lattice spacing and preferred orientation present at the time of measurement resulted entirely from the same origin limits practical application of the method. Residual stresses produced by shot peening, grinding, or machining in most materials of practical interest yield virtually identical results when measured by the Marion-Cohen, two-angle, and $\sin^2\psi$ methods (Ref 5).

Full-Tensor Determination

An expression for the lattice spacing can be formulated as a function of ϕ and ψ, assuming that a triaxial rather than plane-stress state may exist in the layers penetrated by the x-rays below the free surface. Shear and normal stresses may then exist in all directions, as in neutron diffraction from the sample interior. Triaxial stresses in the surface layers penetrated by the x-ray beam is a possible explanation for nonlinear dependence of the lattice spacing on $\sin^2\psi$ reported in severely ground steel or shot peening at steep angles. Nonlinearities in the form of elliptical curvature of the d versus $\sin^2\psi$ plots resulting in "ψ splitting" are attributable to shear stresses σ_{13} and σ_{23}, acting normal to the surface, where σ_{33} is the stress normal to the surface. All three must be zero at the free surface in plane stress. Using ψ splitting results in different values of the lattice spacing for positive and negative ψ tilts, and potential error in stress calculation if linearity is assumed.

In principle, the full-tensor method (Ref 6, 7) can determine the near-surface stresses without assuming plane stress at the free surface. However, extensive data collection is required, generally exceeding that acceptable for routine testing. True ψ splitting caused by out-of-plane shear stresses will cause an elliptical separation of positive and negative ψ data. Subsurface stress gradients in plane stress will also cause nonlinear d versus $\sin^2\psi$ plots, but the curvature is the same for positive and negative ψ tilts. Before the full-tensor methods can be applied, the raw data must first be corrected for penetration of the x-ray beam into the subsurface stress gradient. Unfortunately, sample and instrumental misalignment, x-ray beam divergence, stress gradients along the surface or with depth, and peak location errors can produce similar nonlinearity, especially if negative ψ tilts are used. Using only positive ψ tilts and comparing data for $\phi = 0$ and 180° sample orientations eliminates the instrumental alignment error contributions, allowing true ψ splitting to be detected.

Unlike the plane-stress methods, determination of the full stress tensor requires absolute knowledge of the unstressed lattice spacing, d_0, at the accuracy required for strain measurement (1 part in 10^5) to calculate the stress tensor from the measured strains. In many cases, such as for plastically deformed surfaces generated by machining or composition gradients in carburized steels, the lattice spacing varies as a result of deformation or heat treating, precluding independent determination of the unstressed lattice spacing with sufficient precision. The extensive data collection and dependence on absolute knowledge of d_0 limit the full-tensor method primarily to research applications.

X-Ray Integral Method

Studies have been directed at developing nondestructive residual-stress measurement XRD methods of recovering the underlying residual-stress distribution from measured nonlinear lattice spacing versus $\sin^2\psi$ data (Ref 7). Attempts have been made to estimate both high-stress gradients and shear components acting normal to the surface through the depth of penetration of the x-ray beam. All such methods assume some functional form to describe the subsurface strain (or stress) distribution and seek to find the form of that function which best describes the observed attenuation-weighted integral of lattice spacing with depth. The true strain or stress subsurface distribution (z-profile) is calculated from the measured weighted integral of the lattice spacing with depth (τ-profile).

A method capable of recovering a generalized approximation of the stress function has been described by Wern and Suominen (Ref 8). A method known as the x-ray integral method (RIM) has been developed that is capable of recovering a generalized approximation of the stress function (Ref 8). This method nondestructively determines the full triaxial state of stress within the depth of the x-ray penetration, allowing for both a full stress tensor and variation in all of the stress components with depth. Published results show that the necessary equilibrium condition ($\sigma_{33} = 0$ at the surface) is achieved in the preliminary tests, even though this condition is not required by the method of solution. The method also does not depend on lattice spacing measurements at extremely small grazing angles, minimizing defocusing errors in peak location, error due to surface roughness, and the difficulties of the LaPlace transform solution method.

The RIM method allows calculation of residual stresses from strain distributions measured as a function of depth below the surface. The method is based on approximating the unknown z-profile of strain, $\varepsilon(z)$, shown in Eq 13, by using Fourier trigonometric series expansion. No prior knowledge of the residual-stress distributions is required; the stress distribution is not forced to follow a linear pattern. Standard XRD equipment can be used to collect the data.

The average measured strain profile can be expressed as a function of τ where D is the information depth defined by the penetration of the diffracted x-rays, and z is the depth below the surface of the specimen:

$$\langle \varepsilon_{\phi\psi} \rangle(\tau) = \frac{\int_0^D e^{-\frac{z}{\tau}} \varepsilon(z) dz}{\int_0^D e^{-\frac{z}{\tau}} dz} \qquad \text{(Eq 13)}$$

The equation of x-ray strain determination is shown as:

$$\begin{aligned}\langle \varepsilon \rangle_{\phi\psi} = \frac{\langle d_{\phi\psi} \rangle - d_0}{d_0} = A(\tau)\big[(\varepsilon_{11}(z)\cos^2\phi \\ + \varepsilon_{12}(z)\sin 2\phi + \varepsilon_{22}(z)\sin^2\phi)\sin^2\psi \\ + (\varepsilon_{13}(z)\cos\phi + \varepsilon_{23}(z)\sin\phi)\sin 2\psi \\ + \varepsilon_{23}(z)\cos^2\psi\big]\end{aligned} \qquad \text{(Eq 14)}$$

where $A(\tau)$ is the integral operator in Eq 13, and ψ and ϕ are the angles that define the direction of strain measurement in the sample coordinate system.

A system of m equations with n unknowns can be established by substituting a Fourier series description of the strain distribution with depth for each of the six strain profiles shown in Eq 14. Direct methods of solving for the unknown coefficients in the system of equations in general fail due to the nearly singular condition of the matrix. A technique known as the method of conjugate gradients can be used to determine the coefficients of a poorly conditioned set of equations.

Basic Procedure

Sample Preparation

If the geometry of the sample does not interfere with the incident or diffracted x-ray beams, sample preparation is generally minimal. Preparation of the sample surface depends on the nature of the residual stresses to be determined. If the stresses of interest are produced by such surface treatments as machining, grinding, or shot peening, the residual-stress distribution is usually limited to less than 0.5 mm (0.02 in.) of the sample surface. Therefore, the sample surface must be carefully protected from secondary abrasion, corrosion, or etching. Samples should be oiled to prevent corrosion and packed to protect the surface during handling. Secondary abrasive treatment, such as wire brushing or sand blasting, radically alters the surface residual stresses, generally producing a shallow, highly compressive layer replacing the original residual-stress distribution.

If the stresses of interest are those produced by carburizing or heat treatment, it may be advisable to electropolish the surface of the sample, which may have undergone finish grinding or sand blasting after heat treatment. Electropolishing eliminates the shallow, highly stressed surface layer, exposing the subsurface stresses before measurement, without introducing any deformation and altering residual stresses.

To measure the inside surface of tubing, in bolt holes, between gear teeth, and other restrictive geometries, the sample must be sectioned to provide clearance for the incident and diffracted x-ray beams. Unless prior experience with the sample under investigation indicates that no significant stress relaxation occurs upon sectioning, electrical resistance strain-gage rosettes should be applied to the measurement area to record the strain relaxation that occurs during sectioning. Unless the geometry of the sample clearly defines the minimum and maximum directions of stress relaxation, a full rectangular (three-gage) strain-gage rosette should be used to calculate the true stress relaxation in the direction of interest from the measured strain relaxation.

Following XRD residual-stress measurements, the total stress before sectioning can be calculated by subtracting algebraically the sectioning stress relaxation from the XRD results. If only near-surface layers are examined on a massive sample, a constant relaxation correction can be applied to all depths examined. If a significant volume of material is removed, as in determination of the stress distribution through the carburized case of a thin bearing race, a more accurate representation of sectioning relaxation can be achieved by applying strain-gage rosettes to both the inner and outer surfaces and by assuming a linear relaxation of stress through the sample thickness.

Sample Positioning

Improper positioning of the sample and instrument is the most common source of error. Because the diffraction angles must be determined to accuracies of approximately ±0.01°, the sample must be positioned in the x-ray beam at the true center of rotation of the ψ and 2θ axes. The angle ψ must be essentially constant throughout the irradiated area. Therefore, extremely precise positioning of the sample to accuracies of approximately 0.025 mm (0.001 in.) is critical. Further, the size of the irradiated area must be limited to an essentially flat region on the sample surface in order for ψ to be constant. Stress measurements on small-diameter samples or small-radius fillets, thread roots, and fine-pitched gears are subject to large errors if the x-ray beam is not confined to an essentially flat region at a known ψ tilt on the curved surface. If the irradiated area is allowed to span a curved surface, ψ will not be constant during determination of lattice spacing. These restrictions imposed by the sample geometry may prohibit XRD residual-stress measurement in many areas of primary concern, such as the roots of threads.

Irradiated Area and Measurement Time

The residual stress determined by XRD is the arithmetic average stress in the volume defined by the dimensions of the x-ray beam and the depth of penetration. Consideration must be given to an appropriate beam size for the nature of the stress to be investigated. If average stresses over significant areas are of interest, the maximum beam size allowed by the geometry of the sample would be an appropriate choice. If local variations in residual stress, such as those produced by individual passes of a grinding wheel, are of interest, a smaller irradiated area with a geometry appropriate for the investigation should be selected. Practical dimensions of the irradiated area may range from circular zones 1 mm (0.040 in.) in diameter to a range of rectangular geometries from approximately 0.5 to 13 mm (0.020 to 0.5 in.). The maximum irradiated area generally feasible is approximately 13 by 8 mm (0.5 by 0.3 in.).

As the irradiated area is increased, the diffracted beam intensity increases, and the data collection time necessary to achieve adequate precision for residual-stress measurement diminishes. The precision with which the diffracted intensity can be determined varies as the inverse of the square root of the number of x-rays collected. To determine the intensity to an accuracy of 1% at a single point on the diffraction peak, 10^4 x-rays must be counted, regardless of the time required. With diffracted intensities typically available on a fixed slit diffractometer system, this may require collection times of approximately 30 s for each point on the diffraction peak. If seven data points are collected on each diffraction peak for a two-angle technique, total measurement time may be 10 to 15 min. Reducing the irradiated area sufficiently to decrease the diffracted intensity by an order of magnitude increases the data-collection time proportionally for the same precision in measurement. If high background from sample fluorescence in the x-ray beam is not a problem, position-sensitive detectors can be used to collect data simultaneously at numerous points across the diffraction peak, with some sacrifice in angular precision, reducing data-collection time by an order of magnitude.

Diffraction-Peak Location

To achieve the lattice strain resolution and accuracy required for XRD residual-stress measurement, diffraction peaks with widths of several degrees must be located with a precision on the order of 0.01°. A variety of mathematical methods have been developed to locate diffraction peaks, with varying degrees of success. The calculated diffraction angle positon can shift as the shape of the diffraction peak changes as ψ is changed. As material is removed by electropolishing for subsurface measurement, the shape of the diffraction peak changes as the hardness in case-hardened steels or cold working of machined or shot-peened surfaces varies with depth. Reported nonlinear d versus $\sin^2\psi$ behavior is more often caused by instrument alignment or inaccurate diffraction-peak location than stresses acting out of a free surface. Errors in locating the diffraction peak are a primary source of experimental error in XRD stress measurement.

The transition metal targets of the x-ray tubes used for stress measurement produce a continuous spectrum of white radiation and three monochromatic high-intensity lines. The three lines are the $K\alpha_1$, $K\alpha_2$, and K_β characteristic radiations, with wavelengths known to high precision. The $K\alpha_1$ and $K\alpha_2$ lines differ too little in wavelength and energy to allow separation of the diffraction peaks produced. The highest-intensity $K\alpha_1$ line is nominally twice the intensity of the $K\alpha_2$ line, making it the preferred wavelength for residual-stress measurement. The higher-energy K_β line has a significantly shorter wavelength and can generally be separated from the $K\alpha$ lines by filtration, the use of detectors with high-energy resolution, or crystal diffracted-beam monochromators. The K_β line intensity is typically one-fifth that of the $K\alpha_1$ line and is generally too weak for practical XRD residual-stress measurement on plastically deformed surfaces.

Because the $K\alpha$ doublet is generally used for residual-stress measurement, the diffraction peaks produced consist of a superimposed pair of peaks, as shown in Fig. 5 for four cases, indicating the various degrees of broadening that may be encountered. The variable blending of the $K\alpha$ doublet typical of an annealed sample is indicated by curve *A*, and a fully hardened or cold-worked sample by curve *D*. Because the accuracy of XRD residual-stress measurement depends on the precision with which the diffraction peak can be located, the method used to locate broadened doublet peaks is of primary importance.

Precise determination of the position of the diffraction peak at each ψ tilt begins with collection of raw intensity data at several points on the peak. The diffracted intensity (x-rays counted per unit time) or inverse intensity (time for a fixed number of x-rays to be counted) is determined to a precision exceeding 1% at several fixed diffraction angles, 2θ, spanning the diffraction peak. Depending on the method to be used for peak location, 3 to 15 individual data points and 2 background points are measured using standard diffractometer techniques. If data are collected using a position-sensitive detector, the diffracted intensity can be determined at dozens of data points spanning the diffraction peak. Sharp diffraction peaks, such as those shown in curve *A* in Fig. 5, may be located using intensity data of lower precision than that required for broad peaks, as in curve *D*. The number of x-rays to be collected, and therefore the time required for stress measurement to a fixed precision, increases as the diffraction peaks broaden.

Before determining a diffraction-peak position, the raw measured intensities must be corrected for Lorentz polarization and absorption. The background intensity is subtracted generally assuming a linear variation beneath the diffraction peak. Various numerical methods are available to calculate the position of the diffraction peak. The simplest method, incorporated in early automated diffraction equipment, locates 2θ positions on either side of the peak at which the intensity is equal and assume the peak position to be at the midpoint. A straight line can be fitted to the opposing sides of the diffraction peak and the point of intersection of the two lines taken as a peak position (Ref 9). Early SAE International literature recommends calculating the vertex of the parabola defined by three points confined to the top 15% of the peak (Ref 10). A significant improvement in precision can be achieved, approaching the 0.01° resolution of most diffractometers, by collecting 5 to 15 data points in the top 15% of the peak and fitting a parabola by least-squares regression before calculation of the peak vertex.

If the intensity is measured at many points ranging across the entire Kα doublet, the peak position can be calculated as the centroid of the area above the background or by autocorrelation. Both of these area-integration methods are independent of the peak shape but are quite sensitive to the precision with which the tails of the diffraction peak can be determined and therefore to the accuracy of the background correction.

All of the aforementioned methods are effective, regression-fit parabola being superior, if applied to a single symmetrical diffraction-peak profile, such as the separated $K\alpha_1$ peak shown in curve *A* in Fig. 5, or the broad fully combined doublet shown in curve *D*. All can lead to significant error in the event of partial separation of the doublet, as shown in curves *B* or *C* of Fig. 5. Partial separation commonly results from defocusing as the sample is tilted through a range of ψ angles. If residual stresses are measured as a function of depth, diffraction peaks can vary from breadths similar to curve *D* at the cold-worked or -hardened surface through a continuous range of blending to complete separation beneath the surface, as shown in curve *A*. All the techniques of peak location discussed can lead to significant error in stress measurement as the degree of doublet separation varies.

The Rachinger correction (Ref 11) can be applied to separate the Kα doublet before fitting a parabola to the $K\alpha_1$ peak, but the precision of the correction diminishes on the $K\alpha_2$ side of the combined profile and is generally inadequate for precise residual-stress measurement. Fitting Pearson VII distribution functions (Cauchy to Gaussian bell-shaped curves) separately to the $K\alpha_1$ and $K\alpha_2$ diffraction peaks, assuming a Kα doublet separation based on the difference in wavelengths, provides a method of peak location that overcomes most of the problems outlined previously (Ref 12, 13).

Figures 6 and 7 show the effect of the peak-location method on the results obtained. Figure 6 illustrates comparison of the same data reduced using Pearson VII distribution functions and a five-point least-squares parabolic fit for ground Ti-6Al-4V using the (21-13) planes for residual-stress measurement. Apparent nonlinearities in *d* versus $\sin^2\psi$ for the parabola fit are due to inaccurate diffraction-peak location in the presence of partial blending of the Kα doublet. Figure 7 shows the difference in stress measurement by the two methods of peak location applied to the identical data for the entire stress profile. The errors for the Pearson VII distribution function fit are smaller than the plotting symbols at all depths. Notice that even the sign of the residual stress calculated is affected by the peak-location method.

Microstress Determination and Line Broadening

Diffraction-peak broadening is caused by imperfections in the crystal. Plastic deformation creates and drives dislocations through the crystalline grains, breaking them into smaller crystallite or coherent diffracting domain regions of perfect crystalline stacking order between the dislocation tangles. Phase transformations, notably the martensitic and austenitic phases in steels, produce comparable effects. As the dislocation density increases, these diffracting regions become smaller. As the coherent diffracting regions become smaller than approximately 1000 atomic layers, the lattice periodicity can no longer support perfect scattered x-ray interference, and the peaks become broader. In addition

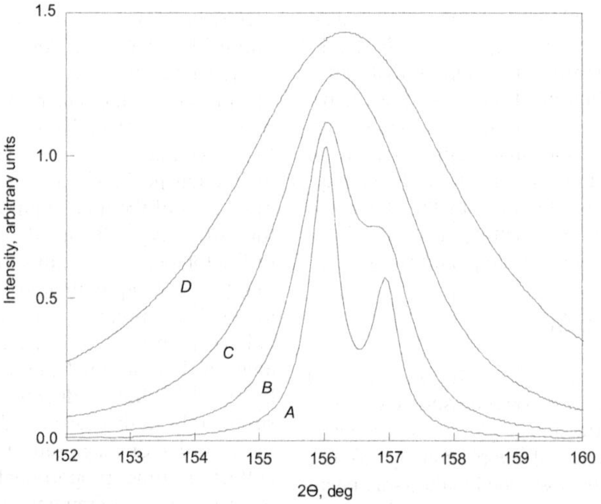

Fig. 5 Range of Kα doublet blending for a simulated steel (211) Cr Kα peak at 156.0°. *A*, fully annealed; *B* and *C*, intermediate hardness; *D*, fully hardened

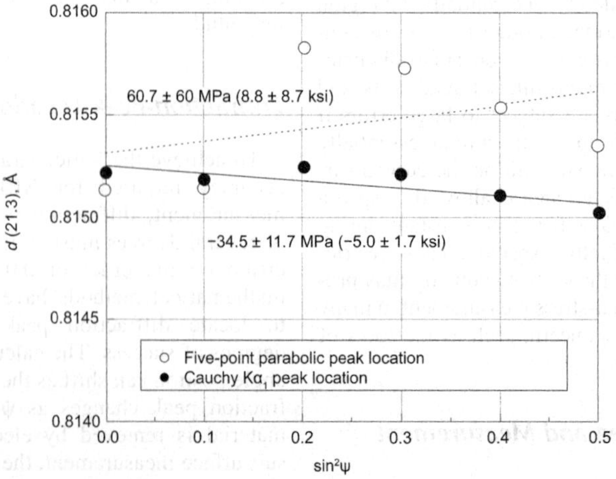

Fig. 6 Comparison of *d* (21.3) versus $\sin^2\psi$ data taken 0.18 mm (0.0069 in.) below the surface of a ground Ti-6Al-4V sample using two diffraction-peak location methods

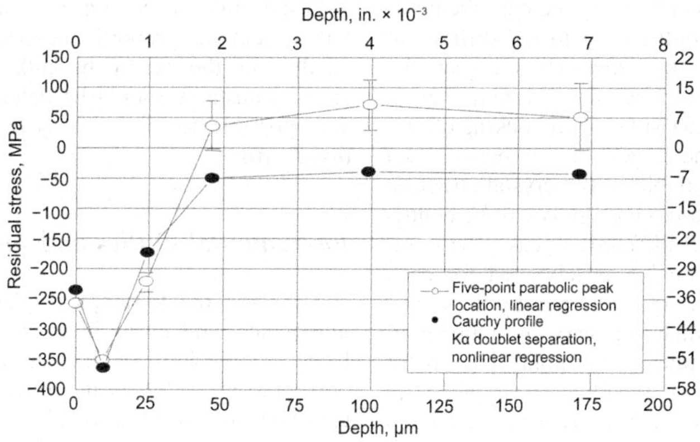

Fig. 7 Comparison of residual-stress patterns derived by using Cauchy and parabolic peak location for a ground Ti-6Al-4V sample using a six-angle $\sin^2\psi$ technique. Errors in stress measurement by two methods of diffraction-peak location are shown.

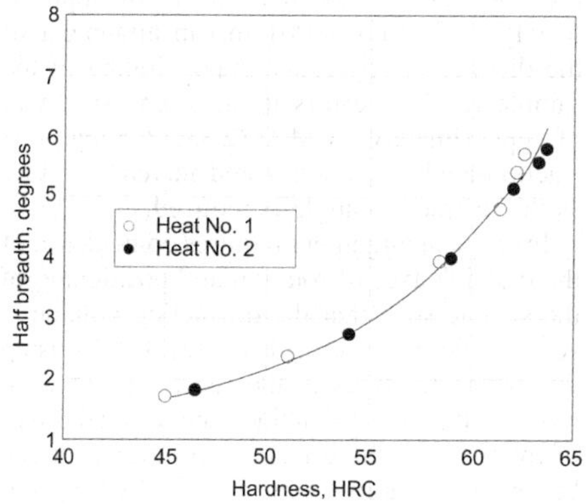

Fig. 8 Diffraction-peak breadth at half height for the (211) peak for M50 high-speed tool steel as a function of Rockwell hardness

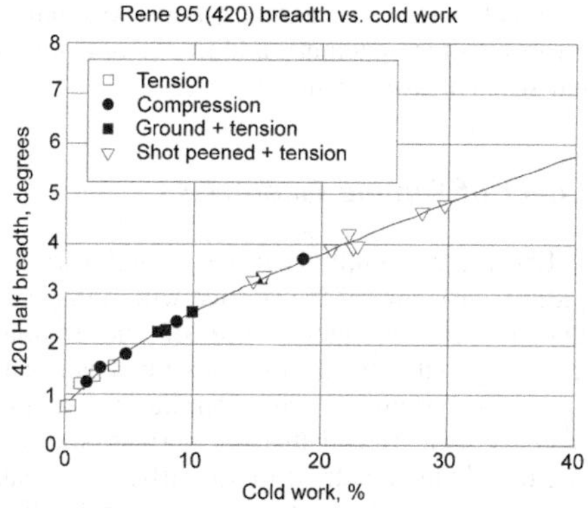

Fig. 9 Diffraction-peak breadth at half height for the (420) peak for René 95 as a function of cold working percentage

to reduced crystallite size, the difference in lattice strain between the individual elastically deformed crystallites causes further broadening by contributing diffracted intensity to opposing sides of the diffraction peak. Tensile-strained regions diffract to the low side, and compressive regions to the high-angle side. The individual crystallite microstresses cannot be measured individually, but the range of microstress can be assessed in terms of the diffraction-peak broadening.

The contributions to diffraction-peak broadening can be separated into components due to strain in the crystal lattice and size. First, the broadening that is of instrumental origin must be separated from that due to crystallite lattice strain and size using Fourier analysis of the diffraction-peak profile and data collection sufficient to define the precise shape of the entire diffraction peak. Analysis of the Fourier series terms then allows separation of the components of broadening attributable to lattice strain from that caused by reduction in the crystallite size. However, this method requires extensive data collection and is very dependent on the precision with which the tails of the diffraction peak can be separated from the background intensity.

For most routine analyses of microstresses associated with cold working or heat treatment, separation of the strain and size components is not necessary, and much simpler determinations of diffraction-peak breadth are adequate. The diffraction-peak breadth can be quantified either as the integral breadth (total area under the peak divided by diffraction-peak height) or the width at half the height of the diffraction peak. The width of the diffraction peak can be calculated directly from integrated data points or from the width of the function fitted to the diffraction-peak profile during macrostress measurement. Microstresses and macrostresses can then be determined simultaneously from the peak breadth and position.

Figures 8 and 9 show empirical relationships established between diffraction-peak breadth at half height for the (211) peak for M50 high-speed tool steel as a function of hardness and for the (420) peak breadth as a function of percent cold work for René 95, respectively. These empirical curves can be used to calculate the hardness or percent cold work in conjunction with macroscopic measurement. For the percent cold work curve, samples are heat treated and then deformed in tension, compression, or combined means to produce a series of coupons with various known amounts of cold work. Note that because cold work is defined in terms of the true plastic strain, the peak width is independent of the mode of deformation, whether in tension or compression, and is cumulative for repeated deformation by different processes (see Ref 13 for discussion.) Because the initial heat treatment may alter significantly the initial peak breadth before cold work, the coupons must receive the same heat treatment as the samples to be measured before inducing known amounts of cold work.

Sample Fluorescence

Sample fluorescence complicates the selection of the radiation to be used for residual-stress measurement. Just as ultraviolet light causes some minerals to fluoresce in the optical spectrum, x-rays of higher energy than the emission lines of the irradiated sample can cause the sample to fluoresce, emitting lower-energy x-rays that produce a high background intensity. The radiation producing strong high-2θ diffraction peaks giving the highest precision in strain measurement may cause fluorescence of the elements present in the sample. The use of Cu Kα radiation for residual-stress measurement in alloys containing iron, chromium, or titanium can result in fluorescent background intensities emitted by the sample that are as or more intense than the diffracted radiation, greatly reducing the signal-to-noise ratio.

Failure to eliminate fluorescence can severely degrade the accuracy with which the diffraction peak can be located, significantly increasing random experimental error. Sample fluorescence may be reduced sufficiently, with some loss of intensity, using incident and/or diffracted beam filters. Diffracted beam crystal or graphite monochromators, or high-energy-resolution solid-state Si(Li) detectors used on standard laboratory diffractometers, give superior peak resolution with minimal loss of intensity. Portable instruments generally use position-sensitive detectors (PSDs) for residual-stress measurement that are of the gas-filled proportional counter, fluorescence screen, or diode array types. Gas-filled proportional detectors can provide moderate energy resolution using single-channel analyzers to count only x-rays in the selected energy range. Fluorescence screen and diode array PSDs do not detect x-rays individually but rather the integrated optical intensity or charge collected, respectively, so energy resolution is not possible.

Instrumental and Positioning Errors

The principal sources of error in XRD residual-stress measurement are related to the high precision with which the diffraction-peak position must be located. Errors of approximately 0.025 mm (0.001 in.) in alignment of the diffraction apparatus or positioning of the sample result in errors in stress measurement of approximately 14 MPa (2 ksi) for high-diffraction-angle techniques and increase rapidly as the diffraction angle is reduced.

Instrument alignment requires coincidence of the θ and ψ axes of rotation and positioning of the sample such that the diffracting volume is centered on these coincident axes. If a focusing diffractometer is used, the receiving slit must move along a true radial line centered on the axes of rotation. All these features of alignment can be checked readily using a stress-free powder sample (Ref 14). If the diffraction apparatus is properly aligned for residual-stress measurement, a loosely compacted powder sample producing diffraction at approximately the Bragg angle to be used for residual-stress measurement should indicate not more than ±14 MPa (±2 ksi) apparent stress. Alignment and positioning errors result in systematic additive error in residual-stress measurement.

Effect of Sample Geometry

Excessive sample surface roughness or pitting, curvature of the surface within the irradiated area, or interference of the sample geometry with the diffracted x-ray beam can result in systematic errors similar to sample displacement. Recall that the derivation relating the strain and stress to the diffraction-peak position assumed the plane-stress model with a flat diffracting surface oriented at the known angles ψ and ϕ. The incident beam must be cropped to ensure that these conditions are met. Errors in ψ will produce nonlinear *d* versus $\sin^2\psi$ dependence and errors in measurement proportional to the error in the slope of those data. Improper ψ setting, as in displacing the irradiated area on a curved radius, can completely change the sign of the results.

Effect of Sample Crystallinity

Coarse grain size, often encountered in cast materials, can lessen the number of crystals contributing to the diffraction peak such that the peaks become asymmetrical, resulting in random error in diffraction-peak location and residual-stress measurement. It is not the grains seen in a micrograph but the perfectly crystalline coherent diffracting domains or crystallites between the dislocation tangles that diffract to produce the diffraction peak. Even coarse-grained samples deformed by machining, grinding, shot peening, or forming may produce suitable diffraction peaks, allowing near-surface measurement in the deformed surface layers.

Rocking of coarse-grained samples, or alternately, the portable diffractometer, about the ψ-axis through a range of a few degrees during measurement can effectively increase the number of crystals contributing to the diffraction peak. Residual-stress measurement can be made on coarse-grained samples with a grain size as large as ASTM No. 1 by rocking during measurement. The larger the number of (*hkl*) lattice planes available in the crystal structure of the sample, the more crystals will be contributing to the (*hkl*) diffraction peak. The number of planes available is tabulated for each possible (*hkl*) and crystal system by the multiplicity factor, which can vary by a factor of 8 for the lattice planes commonly chosen for strain measurement. Residual stress generally cannot be measured reliably using XRD in samples with coarser grain sizes.

X-Ray Elastic Constants

A major source of potential systematic proportional error arises in determination of the x-ray elastic constants $E/(1 + \nu)_{(hkl)}$. The residual stress calculated from the lattice strain is proportional to the value of the x-ray elastic constants, which may differ by as much as 40% from the bulk value due to elastic anisotropy for each crystallographic direction in the crystal. To account for substitutional alloying and multiphase effects, the x-ray elastic constant should be determined empirically by loading a sample of the material to known stress levels and measuring the change in the (*hkl*) lattice spacing as a function of applied stress and ψ tilt (Ref 15, 16). The x-ray elastic constant can then be calculated from the slope of a line fitted by least-squares regression through the plot of the change in lattice spacing for the ψ tilt used as a function of applied stress. If empirical determination is not possible, x-ray elastic constants can be estimated with lower accuracy from single-crystal data.

Figure 10 shows data obtained for determination of the x-ray elastic constants in Inconel 718. With instrumented samples placed in four-point bending, the x-ray elastic constant can typically be determined to an accuracy of ±1%. Table 1 lists elastic constants determined in four-point bending for various alloys along with the bulk elastic constants and the potential systematic proportional error that could result from the use of the bulk values. X-ray elastic constants should be determined whenever possible to minimize systematic proportional error.

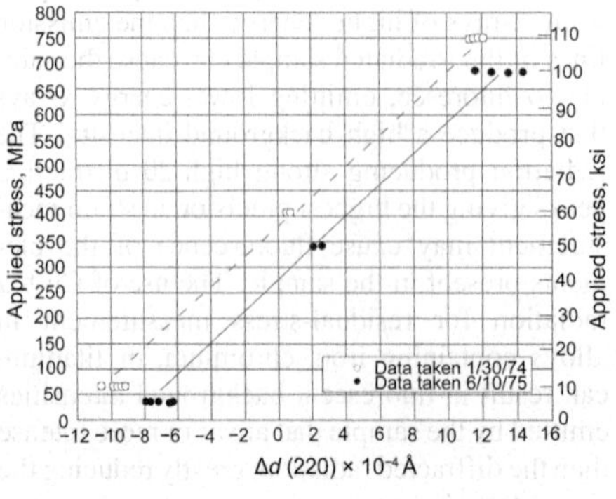

Fig. 10 X-ray elastic constant determination for Inconel 718, (220) planes. $\Delta\psi = 45°$, $d_0 = 1.1272$ Å

Instrumental Optics and Alignment

The errors due to sample displacement, beam divergence, and errors in ψ setting are worse for a negative ψ tilt and increase in inverse proportion to the radius of the diffractometer. The use of negative ψ tilts always increases the errors in peak location because of the asymmetric spreading of the diverging incident x-ray beam on the sample surface. Small portable PSD instruments are prone to report nonlinear *d* versus $\sin^2\psi$ results when using both positive and negative ψ tilts. A triaxial stress state at the free surface is frequently proposed as the cause of nonlinearity rather than the limitations of the small radius, instrument alignment, and/or peak-detection methods. The problems of negative ψ tilts can be eliminated when investigating ψ-splitting by measuring the lattice spacing as functions of $\sin^2\psi$ using only positive ψ tilts and comparing results obtained with the sample rotated about the surface normal to ϕ = 0 and ϕ = 180° orientations. In this way, the lattice strain of grains that would have contributed to both the positive and negative ψ orientations is measured, but the effect of negative ψ tilts on alignment and irradiated area is eliminated.

Subsurface Measurement and Required Corrections

Measuring residual-stress distributions as functions of depth into the sample surface with high-depth resolution is one of the most important uses of the XRD method. The shallow penetration that gives high-depth resolution necessitates electropolishing layers of material to expose the subsurface layers. Electropolishing is preferred for layer removal because no residual stresses are induced, and, if properly performed, preferential etching of the grain boundaries does not occur. Any mechanical method of removal, regardless of how fine the abrasive or machining method, deforms the surface and induces residual stresses, altering severely the state of stress present in the sample. Such methods must be avoided. Thick layers can be removed using a combined machining or grinding procedure, followed by electropolishing to remove at least 0.2 mm (0.008 in.) of material to eliminate the machining or grinding residual stresses.

Subsurface Stress Gradients

The x-ray beam penetrates only to shallow depths (50% to less than approximately

0.005 mm, or 0.0002 in.) beneath the exposed surface. Recall that the derivation of the stress dependence on strain assumed a uniform stress with depth throughout the diffracting volume. However, the residual-stress distributions produced by many processes of interest, including machining and grinding, may vary significantly with depth within the diffracting volume. The incident x-ray beam is attenuated exponentially both as it passes into and again as the diffracted beam comes back out of the sample. Therefore, stress measurements conducted in the presence of such a subsurface stress gradient yield an exponentially weighted average of the stress at the exposed surface and in the layers below. Fortunately, it is possible to unfold this exponential weighting.

The intensity of the radiation penetrating to a depth x is exponentially attenuated:

$$I(x) = I_0 e^{-\mu x}$$

where I_0 is the initial intensity, μ is the linear absorption coefficient, and e is the base of the natural logarithms. If the linear absorption coefficient is known, this exponential weighting can be unfolded, provided that measurements have been conducted at a sufficient number of closely spaced depths to define the stress gradient adequately. Correction for penetration of the radiation into the subsurface stress gradient requires calculating the derivative of the lattice spacing at each ψ tilt as a function of depth. The linear absorption coefficient is calculated from the chemical composition, mass absorption coefficients for the elemental constituents of the alloy, density of the alloy, and radiation used. Failure to correct for penetration of the radiation into the stress gradient can lead to errors as large as 345 MPa (50 ksi) and even change the sign of the stress calculated.

Figure 11 shows an example of the effect of the correction on the residual-stress profile produced in ground 4340 steel. Errors due to the subsurface stress gradient are generally maximal at the surface of the sample and become minimal beneath the highly deformed surface layer. Nondestructive surface residual-stress measurements cannot be corrected for the presence of a subsurface stress gradient and may be subject to significant error on machined, ground, or even shot-peened surfaces due to the presence of a subsurface stress gradient. If nondestructive surface measurements are being considered for quality-control testing, determination of the subsurface stress distributions with proper correction for any stress gradient must be undertaken to fully characterize the stress field and establish any corrections that may be appropriate for the surface measurements.

Stress Relaxation Caused by Layer Removal

In contrast to all mechanical methods, the XRD method of residual-stress measurement measures the strain in each layer before it is removed, rather than measuring the strain relaxation after material is removed. However, some relaxation of residual stress does occur in the surface of each layer exposed by electropolishing during XRD subsurface measurement. The potential error caused by the relaxation of the stresses present increases with depth and the magnitude of the residual stresses present and can be quite large, even altering the sign of the stress as the depth increases, as seen in Fig. 12.

If the sample geometry and nature of the residual-stress distribution conform to the simple geometries of flat plates or cylindrical bodies, closed-form solutions are available to correct the results obtained on the surfaces exposed by electropolishing for removal of the stressed layers above (Ref 17). These corrections involve integration over the residual stress measured in the layers removed from the exposed layer back to the original surface to calculate the amount of relaxation that occurred to reach each depth. The corrections are analogous to the calculations made in mechanical layer-removal methods of residual-stress measurement but are only applied as a correction to the stress measured in each exposed layer.

Finite-element-based relaxation corrections have been developed for complex geometries, such as gear teeth or airfoils, but these are sample geometry specific. Often, the simpler closed-form solutions are sufficiently accurate. The accuracy of the stress-relaxation corrections depend on the depth resolution with which the stress distribution is measured in order to adequately define the uncorrected residual-stress distribution to be integrated.

Often, a sample must be sectioned to expose the measurement location of interest and/or allow access for the incident and diffracted x-ray beams. Correction for layer removal can be combined with correction for relaxation during sectioning of a sample measured with electrical resistance strain gages to determine the total state of residual stress before sectioning of the sample. This allows the full state of stress on the inside of a pipe or bolt hole, disk bore, thread root, and similar parts to be determined.

The magnitude of the layer-removal stress-relaxation correction, which depends on the stress in the layers removed and the moment of inertia for the sample geometry, increases with the total strain energy released. For massive samples from which only thin layers have been removed or for any sample geometry in which no significant stresses are present, correction will be insignificant. However, the correction can be large for some combinations of stress distribution and geometry, such as the longitudinal residual-stress distribution with complete removal of the carburized case on a steel shaft shown in Fig. 12.

Figure 12 also shows that failure to perform the corrections for stress relaxation from layer removal will produce subsurface stress distributions that are not in equilibrium. Because the tension in the core of the shaft exists only because equilibrium is imposed by the compressive case, the residual stress remaining and measured in each exposed layer will gradually diminish to zero. The equilibrating tension and equilibrium in the residual-stress distribution is only seen if the corrections for stress relaxation are applied.

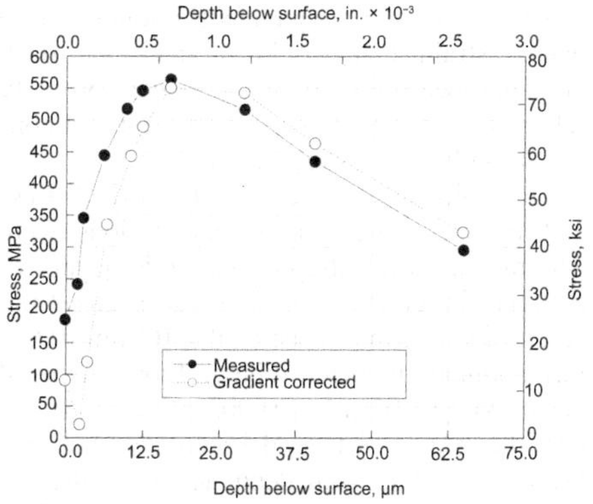

Fig. 11 Effect of stress gradient correction on the measurement of near-surface stresses for ground 4340 steel (50 HRC)

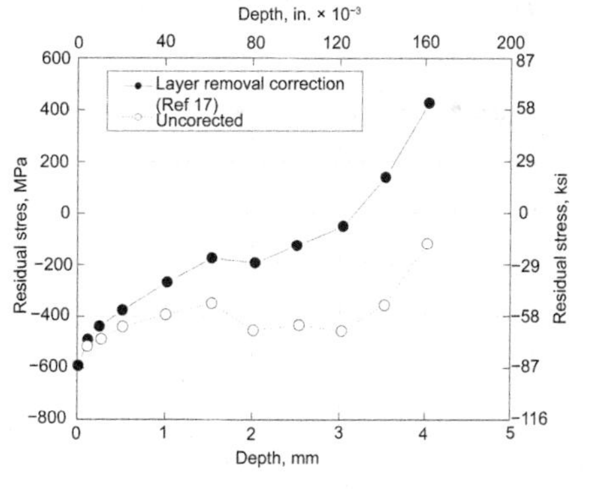

Fig. 12 Longitudinal residual-stress distribution with and without correction for removal of the carburized case from a 16 mm (5/8 in.) diameter 1070 steel shaft

Examples of Applications

The following examples of applications of XRD residual-stress measurement are typical of industrial metallurgical, process development, and failure analysis investigations undertaken at Lambda Research. Diffraction measurements were made on horizontal laboratory Bragg-Brentano diffractometers designed for stress measurement and instrumented with a lithium-doped silicon solid-state detector for suppression of sample fluorescence. The $\sin^2\psi$ or two-angle methods are used, after verification of linear d versus $\sin^2\psi$ dependence confirming a biaxial stress state at the free surface. The angular position of the diffraction peak was located by either fitting of a parabola by regression to the top 15% of the very broad blended peaks on hardened steels in the earlier work, or fitting Pearson VII functions to separate the $K\alpha_1$ and $K\alpha_2$ doublet and using the position and width of the $K\alpha_1$ peak. Results were corrected for Lorentz polarization and absorption and background intensity. Subsurface results were corrected for penetration of the radiation into the subsurface stress gradient and for sectioning and layer-removal stress relaxation, as appropriate in accordance with SAE HS-784 (Ref 1).

The elastic constants used to calculate macroscopic stress from strain in the crystal lattice were obtained empirically by loading an instrumented beam of the alloy under investigation in four-point bending, with the surface stress calibrated and monitored with electrical resistance strain gages, in accordance with ASTM E 1426 (Ref 16). The samples were positioned to the center of the diffractometer using a mechanical gage capable of repeat positioning precision of ±0.05 mm (±0.002 in.). The alignment of the diffractometers was established and checked using alloy or base metal powder incapable of supporting macroscopic residual stress, in accordance with ASTM E 915 (Ref 14).

Example 1: Subsurface Residual-Stress and Hardness Distributions in an Induction-Hardened Steel Shaft

The longitudinal residual-stress and hardness distributions through the case produced by induction hardening of a 1070 carbon steel shaft were investigated to verify a modification of the induction-hardening procedure. The sample consisted of a nominally 205 mm (8 in.) long shaft of complex geometry. A 16 mm (0.625 in.) diameter induction-hardened bearing surface was the region of interest.

The sample was first sectioned to approximately 100 mm (4 in.) in length to facilitate positioning on the diffractometer. Because the sample was cut a distance of several diameters from the area of interest, no attempt was made to monitor sectioning stress relaxation, which was assumed to be negligible. X-ray diffraction macroscopic residual-stress measurements were performed using the two-angle Cr $K\alpha$ (211) technique in the longitudinal direction as a function of depth to approximately 4 mm (0.16 in.) beneath the original surface, fully removing the hardened case by electropolishing. Complete cylindrical shells were removed, conforming to the Moore and Evans closed-form solution to correct cylindrical geometries for layer-removal stress relaxation (Ref 17). Simultaneous determinations of the breadth of the Pearson VII diffraction-peak profile fitted to the $K\alpha_1$ peak were used to calculate the hardness of the material by using an empirical relationship previously established for 1070 steel, similar to that shown in Fig. 8.

Figure 12 shows the longitudinal residual-stress distribution before and after correction for penetration of the radiation into the stress gradient, essentially negligible for the gradual stress gradient produced by induction hardening, and for stress relaxation due to layer removal. The stress-relaxation correction begins as zero at the surface, where no material has been removed, and increases to 550 MPa (80 ksi) as the compressive case material is removed at the maximum depth. The fully corrected results show surface compression of approximately −550 MPa (−80 ksi) diminishing initially in a near-exponential fashion, then more gradually beyond depths of approximately 1.5 mm (0.060 in.). The stress distribution crosses into tension at a nominal depth of 3 mm (0.125 in.) and rises to relatively high tension in the core of the shaft, approaching 517 MPa (75 ksi) at the maximum depth of 4 mm (0.160 in.) examined. Note that the interior is only in residual tension because the compressive case is formed by induction hardening, so that the shaft is in equilibrium. Without correcting for stress relaxation due to layer removal, the raw data would just show a gradual reduction in compression until the shaft is electropolished away. The interior tension would not be revealed, and the shaft would not appear to be in equilibrium.

Figure 13 illustrates the hardness distribution calculated from the breadth of the (211) diffraction-peak profile fitted using a Pearson VII distribution function to separate the $K\alpha$ doublet. The hardness was found to be extremely uniform, varying between 59 and 60 HRC to a depth of 3 mm (0.120 in.). At approximately the depth at which the longitudinal residual-stress distribution goes into tension, the hardness begins to diminish linearly, dropping to approximately 35 HRC at the maximum depth examined in the core of the shaft. The combination of residual stress and material property data derived from line broadening, HRC hardness in this case, can be very useful in assessing the properties of a part in failure analysis, including yield strength, ductility, and fatigue performance.

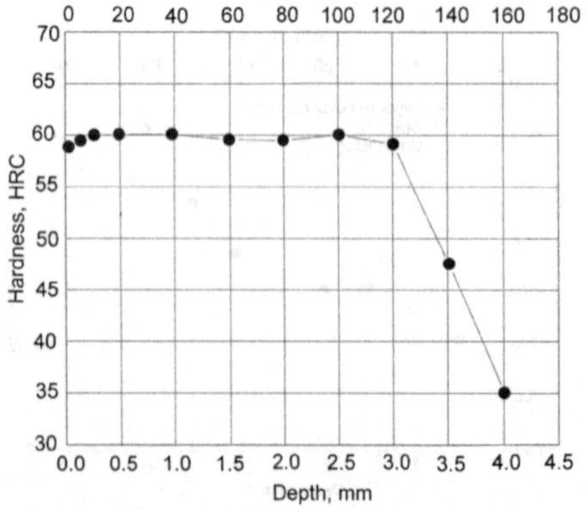

Fig. 13 Rockwell C scale hardness distribution in an induction-hardened 1070 carbon steel shaft with residual-stress distribution shown in Fig. 12

Example 2: Residual Stress and Percent Cold Work Distribution in Belt-Polished and Formed Inconel 600 Tubing

Inconel 600 tubing of the type used for steam generators subject to potential stress-corrosion cracking is fabricated by cross roll straightening and belt polishing of the outer diameter surface. Belt polishing is a cold abrasive process that removes material by chip forming on a fine scale and induces residual stress and cold work distributions in the surface layers. The plastic deformation of the face-centered cubic alloy during the abrasion of the surface creates a yield strength gradient with depth that influences the state of residual stress present in the tubing when it is formed into U-bends.

A single sample of mill-annealed and belt-polished straight tubing was investigated to determine the longitudinal subsurface residual stress and percent plastic strain distribution as functions of depth produced by belt polishing. X-ray diffraction macrostress and line broadening measurements were performed using a Cu $K\alpha$ (420) two-angle technique. The $K\alpha_1$ diffraction peak was separated from the doublet by fitting Pearson VII diffraction-peak profiles to the doublet. The x-ray elastic constant required had been determined previously by loading a strain-gage-instrumented sample of the alloy in four-point bending. The variation in the (420) diffraction-peak width with plastic deformation was established by annealing and then drawing samples of tubing to true plastic strain levels in excess of 20%, generating an empirical relationship similar to that shown in Fig. 9. The measure of plastic strain was taken to be the equivalent amount of true plastic strain that would produce the peak breadth measured. It is a scalar property referred to as percent cold work to avoid confusion with the plastic strain tensor.

The subsurface longitudinal residual-stress and percent plastic strain distributions were determined by electropolishing thin layers of material in complete cylindrical shells from around the circumference of the 16 mm (0.625 in.) nominal diameter tubing. Layer removal began with 0.005 mm (0.0002 in.) thick layers near the sample surface, the increment between layers increasing with depth to nominally 0.4 mm (0.017 in.) beneath the original surface. Corrections were applied for penetration of the radiation to the stress gradient and for stress relaxation in the layers exposed by material removal.

The subsurface longitudinal residual-stress and percent cold work distributions are shown in Fig. 14. The residual-stress distribution shows a pronounced gradient from approximately −35 MPa (−5 ksi) at the surface to a maximum compressive value of approximately −150 MPa (−20 ksi) at a nominal depth of 0.05 mm (0.002 in.). With increasing depth, the stress distribution rises back into tension at approximately 0.13 mm (0.005 in.), with a low-magnitude equilibrating tensile maximum of nominally 55 MPa (8 ksi) at greater depths. The cold work distribution shows a slight hook near the surface of the sample, with a maximum of 19% at a nominal depth of 5 m (0.0002 in.). With increasing depth, the cold work distribution decreases nearly exponentially to negligible values beyond approximately 0.13 mm (0.005 in.) beneath the belt-polished surface.

A 63 mm (2.5 in.) radius U-bend manufactured from Inconel 600 tubing was strain gaged at the apex and sectioned to remove approximately a 50 mm (2 in.) arc length. This portion of the U-bend was mounted in a special fixture providing precision orientation around the circumference of the tubing to an accuracy of 0.1°. X-ray diffraction residual-macrostress measurements were made on the existing surface as a function of angle around the circumference of the tubing.

The longitudinal surface residual-stress distribution around the bent tubing is shown in Fig. 15. The stress is plotted as a function of the quantity (1 + cos θ) to expand the central portion of the plot, at which the sharp transition occurs between maximum compression and tension. The position around the circumference of the tubing ranges from the outside of the bend at the origin, around the flank to the neutral axis at (1 + cos θ) = 1, and around to the inside of the bend at 2.0. The results shown as open circles indicate the longitudinal residual stress around one side of the tubing; closed circles indicate comparable measurements made on the opposing side.

The x-ray beam was limited to a height of 0.5 mm (0.020 in.) and a width of 2.5 mm (0.1 in.) along the axis of the tubing. The small beam size was necessary to optimize spatial resolution in the presence of the pronounced stress gradient occurring on the flank of the tubing. The compressive stresses produced around the outside of the bend exceed −550 MPa (−80 ksi) in a material with a nominal annealed yield strength of 240 MPa (35 ksi). The presence of these high stresses after forming result from cold working at the tubing induced during belt polishing. Cold working of Inconel 600 to 20% increases yield strength to approximately 690 MPa (100 ksi). Cold-worked surface layers in components subjected to subsequent forming frequently result in complex residual-stress distributions having magnitudes often exceeding the yield strength of the undeformed material.

Example 3: Local Variations in Residual Stress Produced by Surface Grinding

The high spatial resolution of XRD residual-stress measurement was applied to determine the longitudinal surface and subsurface residual-stress variation near grinder burns produced by traverse grinding of a sample of 4340 steel with a hardness of 50 HRC. Three samples were initially investigated: two were ground abusively with a dull wheel and loss of coolant to produce grinder burns, and one was ground gently using a sharp, newly dressed wheel and adequate coolant. X-ray diffraction residual-stress measurements were performed initially on only the surfaces of the three samples using a Cr Kα (211) two-angle technique. The diffraction-peak positions were located using a five-point parabolic regression procedure, assuming the Kα doublet to be completely blended into a single symmetrical peak for all measurements performed in the hardened steel. The irradiated area was 0.5 by 6.4 mm (0.020 by 0.250 in.), with the long axis aligned in the grinding direction. Measurements were conducted using the narrow irradiated area as a function of distance across the surface of each sample. A single measurement, using a 12.5 by 6.4 mm (0.5 by 0.250 in.) irradiated area spanning nearly the entire region covered by the series of measurements made with the smaller irradiated zone, was then performed on each sample.

Figure 16 shows the results of the surface measurements. The individual measurements made using the 0.5 mm (0.02 in.) wide irradiated area are shown as circles. The single result obtained using the large 13 mm (0.5 in.) wide beam is plotted as a dashed line. The bounds on the line indicate the approximate extent of the large irradiated area. The gently ground sample C was found to be uniformly in compression, with surface stresses ranging from approximately −400 to −520 MPa (−60 to −75 ksi) at all points examined. The measurement made with the large irradiated area equals the arithmetic average over the region, as expected for the combined diffracting volume.

The abusively ground sample A was found to be entirely in tension; the values range from 275 to 825 MPa (40 to 120 ksi) across the width of the sample. Abusively ground sample B shows regions of compression and tension, with visible grinder burn revealed as dark stripes associated with the tensile peaks occurring above approximately 275 MPa (40 ksi) near the center of the sample. The results for the large irradiated area provide nominally the arithmetic

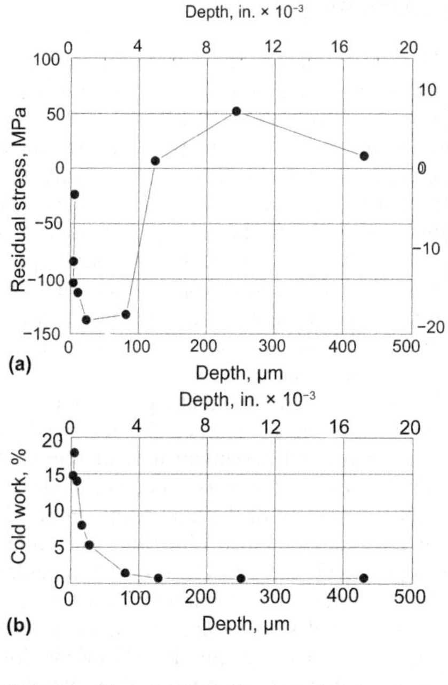

Fig. 14 (a) Longitudinal residual-stress and (b) percent cold work distributions in belt-polished Inconel 600 tubing

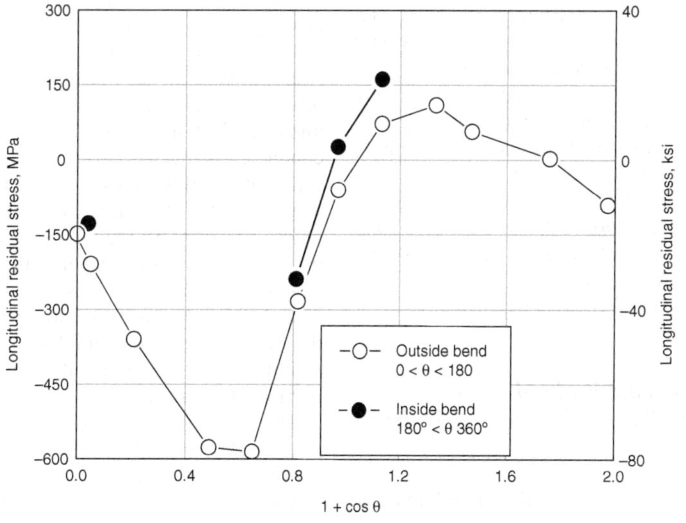

Fig. 15 Longitudinal residual stress as a function of the quantity (1 + cos θ) for a 63 mm (2.5 in.) Inconel 600 tubing U-bend

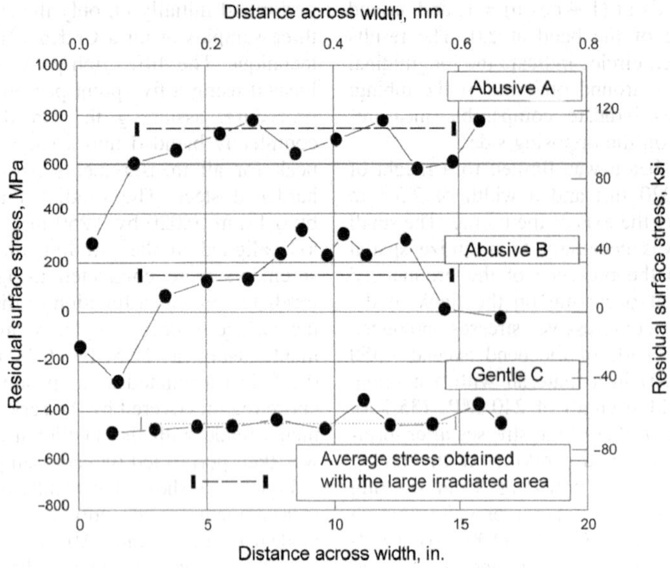

Fig. 16 Variations in longitudinal surface residual stress produced by surface grinding 4340 alloy steel (50 HRC) samples

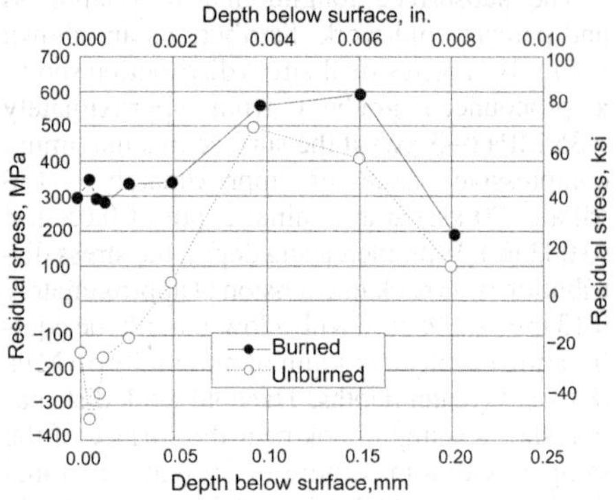

Fig. 17 Subsurface residual-stress profiles produced in burned and unburned regions of abusively ground 4340 steel (50 HRC)

average of the small-area results for both of the abrasively ground samples.

The subsurface residual-stress distribution was then determined at the points of maximum compression and maximum tension on the abusively ground sample B using the 0.5 mm (0.020 in.) irradiated area. The sample was electropolished completely across the width as measurements were conducted at the two locations of interest. The subsurface results shown in Fig. 17 indicate compressive stresses near the edge of the unburned sample at the point of maximum surface compression that extend to a nominal depth of 0.05 mm (0.002 in.) and then rise into tension approaching 500 MPa (70 ksi) at greater depths. The burned region shows entirely tensile stresses ranging from approximately 275 to 345 MPa (40 to 50 ksi) to a depth of 0.05 mm (0.002 in) and then rises into tension of approximately 600 MPa (90 ksi) further below the surface.

The residual stresses produced by many grinding and machining operations can vary significantly over local distances, particularly if there is significant heat input caused by loss of coolant or friction from dull tooling. As seen in Fig. 17, a nondestructive surface measurement of residual stress may not reveal subsurface tensile residual stresses that could severely degrade fatigue performance.

Example 4: Longitudinal Residual-Stress Distribution in Welded Railroad Rail

Continuously welded railroad rail may be subject to high tensile or compressive applied stresses resulting from seasonal thermal contraction and expansion in the field as well as cyclic loading of the rolling cars. Rail fatigue failure at the welded joints is a primary cause of derailments. Residual stresses in the flash-butt-welded joints of the continuously welded rail add a residual mean stress that can contribute to fatigue failures initiating near the welds. The head of modern rail is also often hardened to minimize wear, particularly for rail installed at curves where the wheel may slide on the top of the rail.

To determine the longitudinal residual stresses in the hardened head of welded rail in the vicinity of the weld, a nominally 200 mm (8 in.) portion of rail containing the weld was band sawed from a section of continuous rail after butt welding. Sectioning stress relaxation was assumed to be negligible.

The surface of the rail head was prepared by electropolishing to a nominal depth of 0.25 mm (0.010 in.) to remove any surface residual stresses that may have originated from sources other than welding. X-ray diffraction longitudinal residual-stress measurements were then conducted using the two-angle technique at a series of positions across the centerline of the weld, which was located by etching with nital before electropolishing. A Cr Kα (211) technique was used, locating the diffraction peak by using a parabolic regression procedure. The rail head was induction hardened, and the Kα doublet was completely blended and symmetrical throughout the hardened head portion of the rail.

The longitudinal residual-stress distribution is shown in Fig. 18. The longitudinal residual-stress distribution in the head of the rail is entirely compressive near the weld, revealing an asymmetrical oscillating pattern of residual compression different from what would have been predicted by analytical solution for a uniformly fused and cooled simple butt joint. The results of repeat measurements confirmed the nature of the stress distribution.

The analytical methods for predicting the residual stresses produced by welding generally predict a symmetrical residual-stress distribution around the weld fusion line; however, the actual stress distributions revealed by measurement are often substantially more complex than those predicted. The complexity may be due to deformation of the hot weld and heat-affected zones during the cooling stages of the mechanized field welding process.

Example 5: Determination of the Magnitude and Direction of the Principal Residual Stresses Produced by Machining

The direction of the maximum principal residual stress, that is, the most tensile or least compressive, is often assumed to occur in the cutting or grinding direction during most machining operations. This is frequently the case, but the maximum stress often occurs at significant angles to the cutting direction. Furthermore, the residual-stress distributions produced by many cutting operations, such as turning, may be highly eccentric, producing a high tensile maximum stress and a high compressive minimum stress.

The residual-stress field at a point, assuming a condition of plane stress, can be described by the minimum and maximum normal principal residual stresses, the maximum shear stress, and the orientation of the maximum stress relative to some reference direction. The minimum principal stress is always perpendicular to the maximum. The maximum and minimum normal residual stresses are shown as σ_1 and σ_2 in Fig. 2. The magnitude and orientation of the principal stresses relative to a reference direction can be calculated along with the maximum shear stress using Mohr's circle for

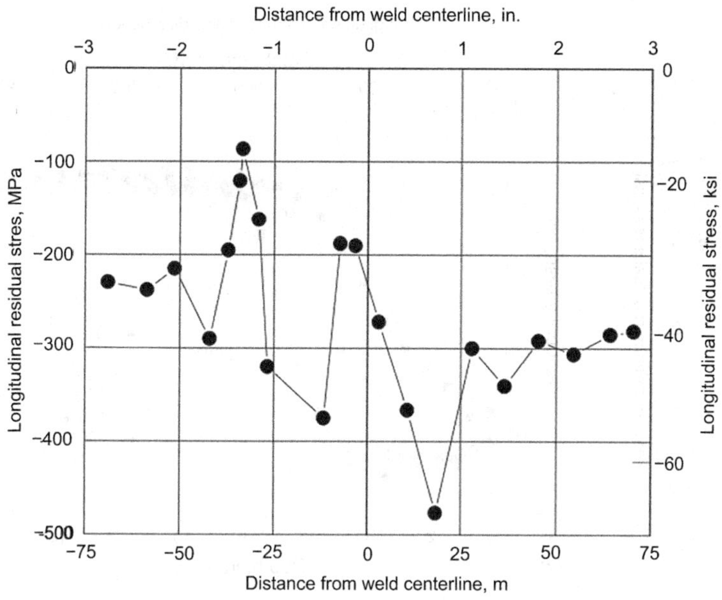

Fig. 18 Longitudinal residual-stress distribution across a flash-butt-welded induction-hardened railroad rail head

stress. Solution requires determining the stress σ_ϕ for three different values of ϕ.

To investigate the minimum and maximum normal residual stresses and their orientation produced by turning an Inconel 718 cylinder, XRD residual-stress measurements were performed in the longitudinal, 45°, and circumferential directions at the surface and at subsurface layers to a nominal depth of 0.1 mm (0.004 in.). Subsurface depths were exposed by electropolishing complete cylindrical shells around the cylinder. The cylinder was nominally 19 mm (0.75 in.) in diameter and uniformly turned along a length of several inches. The irradiated area was limited to a nominal height of 1 mm (0.05 in.) around the circumference by 2.5 mm (0.10 in.) along the length. Measurements were conducted using a Cu Kα (420) two-angle technique, separating the Kα_1 peak from the doublet by using a Pearson VII peak profile.

The measurements performed independently in the three directions were combined using Mohr's circle for stress at each depth to calculate the minimum and maximum normal residual stresses and their orientation. The orientation was defined by the angle ϕ, taken to be a positive angle counterclockwise from the longitudinal axis of the cylinder. Figure 19 shows the maximum and minimum principal residual-stress depth profiles and their orientation relative to the longitudinal direction. The maximum stresses are tensile at the surface, in excess of 140 MPa (20 ksi), dropping rapidly into compression at a nominal depth of 0.005 mm (0.0002 in.). The maximum stress returns into tension at depths exceeding 0.025 mm (0.001 in.) and remains in slight tension to the greatest depth of 0.1 mm (0.004 in.) examined. The minimum residual stress is in compression in excess of −480 MPa (−70 ksi) at the turned surface and diminishes rapidly in magnitude with depth to less than −138 MPa (−20 ksi) at a depth of 0.013 mm (0.0005 in.). The minimum stress remains slightly compressive and crosses into tension only at the maximum depth examined. The orientation of the maximum stresses is almost exactly in the circumferential direction (90° from the longitudinal), the cutting direction, for the first two depths examined. For depths of 0.013 mm (0.0005 in.) to the maximum depth of 0.1 mm (0.004 in.), the maximum stress is rotated to within approximately 10 degrees of the longitudinal direction.

The results appear to indicate that stresses within approximately 0.013 mm (0.0005 in.) of the sample surface are dominated by chip formation during machining, which resulted in a maximum stress direction essentially parallel to the cutting action. At greater depths, the stress distribution may be influenced by residual stresses due to prior forging, heat treatment, or straightening.

Example 6: Optimizing Residual Stress and Cold Work in Shot Peening

Shot peening is by far the most widely used surface-enhancement process to improve fatigue performance. Shot peening is controlled by selection of a shot material and size, and the peening intensity is measured by deflection of an Almen strip to produce a required coverage. Coverage is generally at least "100%," meaning that essentially all of the surface is impacted and dimpled by shot. Often, more than 100% is used, meaning the surface is impacted repeatedly in an attempt to ensure that every point on the surface is impacted.

Excessive coverage can damage the surface, reduce ductility, increase costs, and reduce production rates without further fatigue performance benefit. The Almen strip serves the intended purpose as a measure of the peening apparatus operation, but it is manufactured from 1070 steel and heat treated so that it neither work hardens nor softens. The strip does not reveal the actual residual stress or cold work developed in the shot-peened component. Subsurface XRD residual-stress and cold work measurements made on the shot-peened component can be used to determine the minimum coverage needed to achieve the depth and magnitude possible for a given peening process.

The subsurface residual-stress and cold work distributions produced in 4340 steel, 50 HRC, by shot peening with 3 to 200% coverage with cut wire 14 shot at 9A intensity are shown in Fig. 20. The large number of measurements shown were obtained with an automated electropolishing apparatus developed at Lambda Research mounted on an automated diffractometer. The full depth and magnitude of the beneficial compressive layer is achieved with minimal cold work with as little as 20% coverage. The regions between the peening dimples are in compression, as has been confirmed by finite-element modeling and fatigue testing. Further coverage only produces more cold work but gives no further residual compression or fatigue benefit. Because of the random nature of shot impacts, full coverage is approached exponentially, with 80% coverage achieved in only 20% of the peening time. Therefore, production rates can be increased fivefold by peening to 80% rather than 100% coverage. Optimizing peening processes using XRD stress and cold work measurements can improve production rates and reduce costs without sacrificing fatigue performance.

Example 7: Prediction of Yield Strength Gradient and Residual-Stress Inversion from Plastic Deformation

Fatigue-critical areas of components subject to low-cycle fatigue (LCF), such as compressor and turbine disks, are frequently shot peened to introduce a layer of compression to improve fatigue performance. Disk bores and dovetail slots will yield, usually a fraction of a percent, with each cycle at the high LCF service loads. If the yield strength of the surface material has been altered by cold working during shot peening, the benefit of the compressive layer from shot peening may be lost or even inverted into tension when the softer, less compressive subsurface material yields.

The surface layers of a metallic component are plastically deformed and cold worked by machining, grinding, shot peening, or other

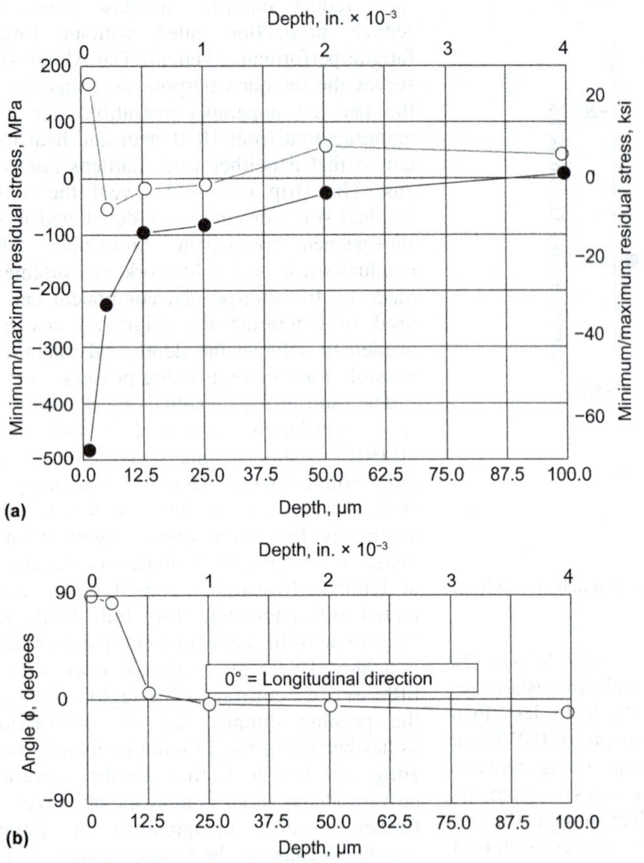

Fig. 19 (a) Minimum and maximum principal residual-stress profiles and (b) their orientation relative to the longitudinal direction in a turned Inconel 718 cylinder

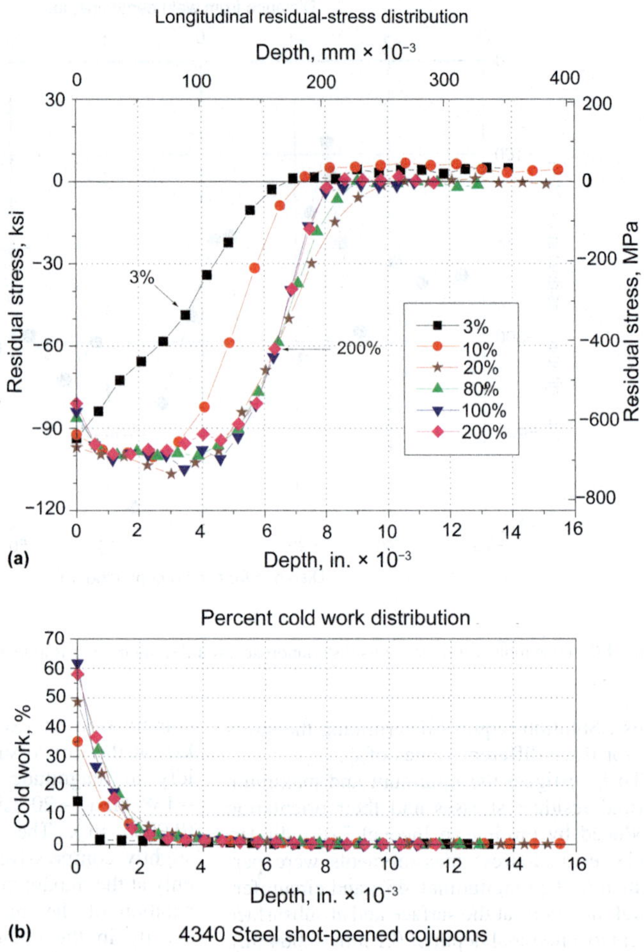

Fig. 20 (a) Subsurface residual-stress and (b) cold work distributions in 4340 steel (50 HRC) shot peened at 3 to 200% coverage

mechanical processing. As noted previously, the dislocation density and lattice strain range increase from cold working. In a work-hardening alloy, the yield strength increases in relation to the degree of cold working induced at each depth. The amount of cold work can be measured as a function of depth during electropolishing for subsurface XRD residual-stress measurement. Defining the measure of cold work as the equivalent amount of true plastic strain required to produce the line broadening measured, a true stress-strain curve extending to the plastic strain levels produced by the processing can be used to determine the change in yield strength at each depth.

The effect of 2% tensile plastic deformation on the beneficial compressive layer produced by shot peening of a nickel-base disk superalloy tensile sample was measured and compared to the original compressive profile before deformation. The yield strength at each depth was estimated from the available true stress-strain curve. Finite-element analysis was then used, with the yield strength gradient applied at each depth to predict the change in the residual-stress distribution due to 2% plastic deformation.

The residual-stress distributions as-shot-peened and after 2% plastic elongation are shown in Fig. 21. The highly compressive shot-peened layer was inverted from over −1000 MPa (−145 ksi) to tension approaching +400 MPa (60 ksi) in a single half-cycle. The cold work measured by XRD line broadening and the yield strength gradient estimated from the true stress-strain curve are as functions of depth in the bottom of Fig. 21. The finite-element analysis (FEA) prediction with the yield strength gradient was in remarkably good agreement with the measured residual stress after 2% elongation. The variation in yield strength was found to dominate the FEA prediction, and the existing compressive residual-stress distribution had little influence.

Example 8: Mapping Residual-Stress Distributions from Welding

Welding can develop complex residual-stress distributions that may include local areas of high residual tension that leave the welded assembly subject to fatigue or stress-corrosion failure. The residual stresses developed upon cooling are determined by the temperature distribution and properties of the weld metal, heat-affected zones, and parent metal as functions of time and the physical constraints imposed on the structure. The order of welding determines both the temperature distributions over time and the constraints as different portions of the weld fuse and contract upon cooling. The combination of variables can be very difficult to predict. X-ray diffraction residual-stress measurements made using automated sample or instrument positioning can be programmed to perform a series of measurements in a grid pattern. Contour maps of the complex stress distributions can then be created.

A contour map of the residual stress parallel to the *x*-axis of a T-weld is shown in Fig. 22. The weldment was made from the three pieces of plate shown. Welds were made in the directions shown, with the first weld attaching the two 30 by 40 mm (1.2 by 1.6 in.) plates to

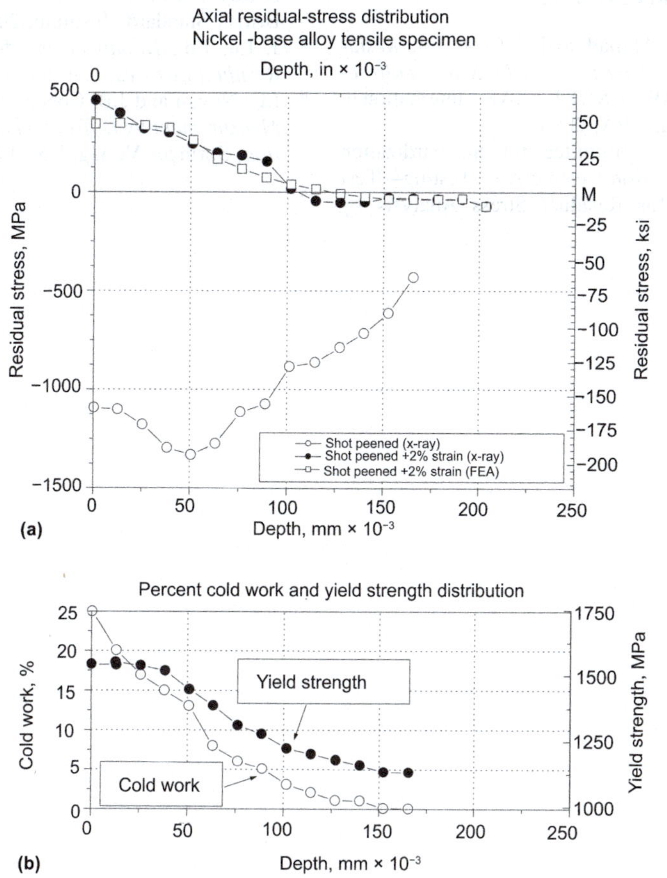

Fig. 21 (a) Subsurface residual-stress, (b) cold work, and yield strength distributions show inversion to tension with 2% extension and finite-element analysis (FEA) confirmation

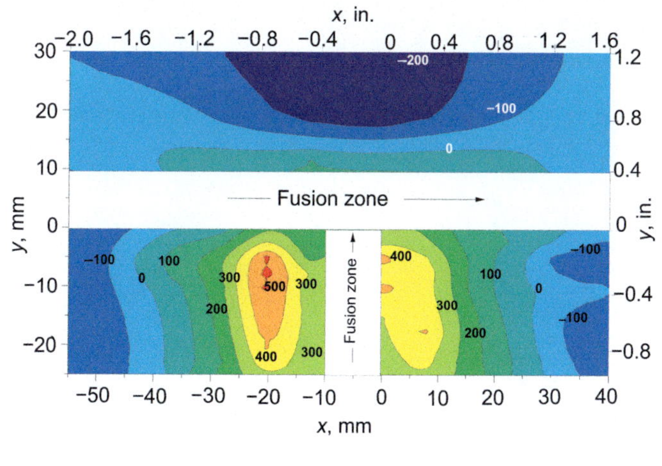

Fig. 22 Contour map of the residual-stress distribution produced in a T-weld of three steel plates

the upper plate, welding from left to right. The final 30 mm (1.2 in.) vertical weld joined the two smaller plates.

Residual-stress measurements were made in the direction parallel to the *x*-axis using an automated *x*-*y* positioning stage programed to perform measurements at 5 mm (0.2 in.) increments. Measurements were not made in the fusion zone of the weld because of the coarse grain size and irregular surface topography.

The contour plot reveals that the highest stress occurs on either side of the final vertical weld in both of the smaller plates. However, the maximum stress, over 500 MPa (70 ksi), occurs to the left of the final weld in the first small plate that was constrained by the initial horizontal weld. The small plate and the portion of the weld on the left had more time to cool before the final vertical weld was made. Yield strength of the material would then be higher as the final weld cooled and contracted, allowing higher residual stresses to be developed on the left side of the weld. The upper plate is seen to be drawn into equilibrating compression by the tensile zone across the lower vertical weld.

ACKNOWLEDGMENT

Revised from P.S. Prevey, X-Ray Diffraction Residual Stress Techniques, *Materials Characterization*, Vol 10, *ASM Handbook*, ASM International, 1986, P 380–392.

REFERENCES

1. *Residual Stress Measurement by X-Ray Diffraction*, SAE HS-784, Society of Automotive Engineers, Warrendale, PA, 2003, p 27–29, 61
2. B.D. Cullity, *Elements of X-Ray Diffraction*, 2nd ed., Addison-Wesley, 1978, p 470
3. K. Hirastsuka et al., *Adv. X-Ray Anal.*, Vol 46, 2003, p 61
4. R.H. Marion and J.B. Cohen, *Adv. X-Ray Anal.*, Vol 18, 1975, p 466
5. P.S. Prevéy, *Adv. X-Ray Anal.*, Vol 19, 1976, p 709
6. H. Dölle and J.B. Cohen, *Metall. Trans. A*, Vol 11, 1980, p 159
7. B. Eigenmann, B. Scholtes, and E. Macherauch, Determination of Residual Stresses in Ceramics and Ceramic-Metallic Composites by X-Ray Diffraction Methods, *Mater. Sci. Eng. A*, Vol 118, 1989, p 1–17
8. H. Wern and L. Suominen, Self-Consistent Evaluation of Non-Uniform Stress Profiles and X-Ray Elastic Constants from X-Ray Diffraction Experiments, *Adv. X-Ray Anal.*, 1996, p 39
9. A.L. Christenson and E.S. Rowland, *Trans. ASM*, Vol 45, 1953, p 638
10. D.P. Koistinen and R.E. Marburger, *Trans. ASM*, Vol 51, 1959, p 537
11. W.A. Rachinger, *J. Sci. Instr.*, Vol 25, 1948, p 254
12. S.K. Gupta and B.D. Cullity, *Adv. X-Ray Anal.*, Vol 23, 1980, p 333
13. P.S. Prevéy, *Adv. X-Ray Anal.*, Vol 29, Plenum Press, 1986, p 103
14. "Standard Method for Verifying the Alignment of X-Ray Diffraction Instrumentation for Residual Stress Measurement," E 915, *Annual Book of ASTM Standards*, Vol 03.01, American Society for Testing and Materials, Philadelphia, 1984, p 809–812

15. P.S. Prevéy, *Adv. X-Ray Anal.*, Vol 20, 1977, p 345
16. "Standard Test Method for Determining the X-Ray Elastic Constants for Use in the Measurement of Residual Stress Using X-Ray Diffraction Techniques," E 1426-14, ASTM International, West Conshohocken, PA, 2014
17. M.G. Moore and W.P. Evans, *Trans. SAE*, Vol 66, 1958, p 340

SELECTED REFERENCES

- P. Prevéy Ahmad, and C. Ruud, Ed., *Residual Stress Measurement by X-Ray Diffraction,* SAE HS-784, SAE International, Warrendale, PA, 2003
- European Committee for Standardization (CEN), "Non-Destructive Testing—Test Method for Residual Stress Analysis by X-Ray Diffraction," BS EN 15305:20008, British Standards Institute, 2009
- J. Lu, Ed., *Handbook of Measurement of Residual Stresses*, Fairmont Press, 1996
- I.C. Noyan and J.B. Cohen, *Residual Stress Measurement by Diffraction and Interpretation*, Springer-Verlag, NY, 1987

ASM Handbook, Volume 10, *Materials Characterization*
ASM Handbook Committee
DOI 10.31399/asm.hb.v10.a0006644

X-Ray Topography

Balaji Raghothamachar and Michael Dudley, Stony Brook University

Overview

General Uses

- Imaging of crystallographic defects such as dislocations, point defect clusters, precipitates, inclusions, and stacking faults in near-perfect crystals
- Nondestructive characterization of twins and polytypes, lattice distortion, and strain fields due to defects and defect accumulations such as low-angle grain boundaries and growth-sector boundaries in imperfect single crystals and polycrystalline aggregates
- Measurement of crystal defect densities as well as subgrain sizes and shapes
- Quantitative analysis of the Burgers vector and line direction of dislocations, tilt angles across subgrain boundaries, interfacial defects and strains, domain structures, and other substructural entities

Examples of Applications

- Study of crystal growth, recrystallization, and phase transformations, focusing on crystal perfection and attendant defects
- Characterization of deformation processes and fracture behavior
- Correlation between crystal defects and electronic properties in solid-state device materials, including device failure analysis
- Use of synchrotron radiation topography for the study of dynamic processes, such as magnetic domain motion, in situ transformations (solidification, polymerization, recrystallization), radiation damage, electronic device failure, and high-temperature deformation

Samples

- *Form:* For defect imaging (transmission or reflection case), flat, relatively perfect ($<10^6$ dislocations/cm^2) single crystals with uniform thickness
- *Material:* Evaluation of lattice distortions, texture, substructure, and surface relief in monocrystals, polycrystalline aggregates, ceramic or metal alloys, or composites
- *Size:* Typically 1 by 1 cm (0.4 by 0.4 in.) (as small as 0.1 mm, or 0.004 in., to 15 cm, or 6 in., diameter or larger wafers). Thicknesses from a few micrometers to several millimeters thick; thin films 100 nm and thicker
- *Preparation:* Desirable to remove surface damage due to cutting, abrading, and so on, from virgin material by chemical or electrolytic polishing

Limitations

- Sample must be crystalline.
- Relatively defect-free crystals are required for defect-imaging techniques.
- Thickness of single-crystal or polycrystalline samples that can be studied in transmission arrangement is limited by intensity and wavelength of incident radiation used as well as absorption by the sample.
- Direct images are actual size. Further magnification must be obtained optically; that is, grain size of photographic plate emulsion or pixel size of electronic detector must be small enough to allow substantial enlargement.

Estimated Analysis Time

- Several minutes to hours exposure time for conventional photographic (plate or film) methods, in addition to developing/enlarging time
- Milliseconds to several seconds using synchrotron radiation and/or electronic or electrooptical imaging systems

Capabilities of Related Techniques

- *Optical microscopy:* Characterization of grain size and shape, subgrains, phase morphology, and slip traces using suitable etchants; estimation of low dislocation densities and determination of slip systems by etch-pit techniques
- *Scanning electron microscopy:* Observation of irregular surfaces, surface relief, and various features induced by deformation, such as slip bands and rumpling; examination of fracture surfaces to evaluate crack initiation and propagation
- *Electron channeling:* Qualitative evaluation of crystal perfection over shallow surface layer of crystals with high-symmetry orientations
- *Transmission electron microscopy:* Imaging of line and planar defects and estimation of defect densities; substructural and morphological characterization of thin foils prepared from bulk sample or replicas taken from the surface
- *Neutron diffraction and topography:* Study of very thick or heavy metals in transmission arrangement and of magnetic domain structure

Introduction—History and Development Trends

X-ray topography is the general term for a family of x-ray diffraction imaging techniques capable of providing information on the nature and distribution of imperfections, such as dislocations, inclusions/precipitates, stacking faults, growth-sector boundaries, twins, and low-angle grain boundaries, as well as other lattice distortions in crystalline materials of a wide range of chemical compositions and physical properties, such as semiconductors, oxides, metals, and organic materials. The complete name *x-ray diffraction topography* is a little more informative, indicating that the technique is concerned with the topography of the internal diffracting planes, that is, local changes in spacing and rotations of these planes, rather than with external surface topography. This technique usually is nondestructive and suitable for imaging single crystals of large cross section with thickness ranging from hundreds of micrometers to several millimeters. Different imperfections are identified through interpretation of contrast using well-established kinematical and dynamical theories of x-ray diffraction. The capability of in situ characterization during crystal growth, heat treatment, stress application, device operation, and so on allows for the study of the generation, interaction, and propagation of defects, making it a versatile technique to study many materials processes.

The first topographic image of a single crystal was recorded as early as 1931 by Berg (Ref 1) and later by Barrett (Ref 2), developing the technique commonly known as Berg-Barrett reflection topography, which is used to study large-sized crystals. Subsequently, the x-ray topography technique was developed to image and characterize defects using characteristic radiation (Ref 3, 4) as well as white radiation (Ref 5). The high-resolution double-crystal technique (Ref 6) used a monochromator, making it highly sensitive to lattice misorientations (lattice tilts and lattice parameter changes down to 10^{-8}). The real potential of x-ray topography was realized in 1958 when Lang (Ref 7) developed the projection topography system to image individual dislocations in a silicon crystal in transmission. This marked a milestone in the field of x-ray topography and was partly driven by the availability of high-quality semiconductor crystals such as silicon and germanium, due to improved crystal growth technology. Based on the Lang technique, Schwuttke developed the scanning oscillator technique (Ref 8) that allowed recording transmission topographs of large-sized wafers containing appreciable amount of elastic and/or frozen-in strain. Tuomi et al. (Ref 9) performed the first x-ray topography experiments using synchrotron radiation on silicon samples. While conventional radiation-based x-ray topography techniques continue to be used, the availability of numerous synchrotron radiation facilities providing intense, low-divergent x-ray beams of wide spectral range has allowed the extensive development of synchrotron x-ray topography techniques.

General Principles

In x-ray diffraction topography, a collimated area-filling ribbon of x-rays is incident on the single-crystal sample at a set Bragg angle, and the corresponding area-filling diffracted beam is projected onto a high-resolution x-ray film or detector (Fig. 1a). The two-dimensional diffraction spot thus obtained constitutes an x-ray topograph (Fig. 1b), and it precisely displays the variation of the diffracted intensity as a function of position, depending on the local diffracting power as well as the prevailing overall diffraction conditions. Local diffracting power is affected by the distorted regions surrounding an imperfection, leading to differences in intensities between these regions and the surrounding more-perfect regions. This intensity variation gives rise to contrast, and different defect types can be characterized from the specific contrast produced by the way they distort the local crystal lattice and thereby the local diffracting power. The absence of magnification enables the correlation of the relative position of the image of a defect with its location inside the crystal. Identification and characterization of defects is achieved by detailed interpretation of the variations in contrast obtained under different diffraction conditions by applying the kinematical and dynamical theories of x-ray diffraction, as explained briefly in "Appendix A: Kinematical and Dynamical Theories of X-Ray Diffraction" in this article and also addressed in detail elsewhere (Ref 10–12).

Contrast Mechanisms on X-Ray Topographs

Orientation contrast results from inhomogeneous intensity distributions arising from the overlap and/or separation of diffracted x-rays with varying directions. This contrast is observed in crystals containing regions of different orientations, such as grains, subgrains, and twins. Misorientations caused by dilations as well as rotations of the lattice also can lead to orientation contrast.

Orientation Contrast from Subgrain Boundaries and Twins

The occurrence of orientation contrast depends on the nature of the x-rays as well as the nature of the misorientation, as illustrated in Fig. 2. If either the incident beam divergence or the range of wavelengths available in the incident x-ray spectrum is smaller than the misorientation between the blocks projected onto the incidence plane (Fig. 2a), then only one of the blocks can diffract at a given time. However, if either the incident beam divergence or the range of wavelengths available in the incident spectrum allows each block to diffract independently (Fig. 2b, c), then the diffracted beams emanating from the individual blocks will travel in slightly different directions and will give rise to image overlap if they converge or separation if they diverge. A white-beam x-ray topograph (Fig. 2d), recorded from a ZnO crystal that shows the overlap and separation of the subgrain images, demonstrates orientation contrast.

Orientation Contrast from Dislocations and Inclusions

The distorted regions around dislocations and inclusions diffract x-rays kinematically according to their local lattice orientation.

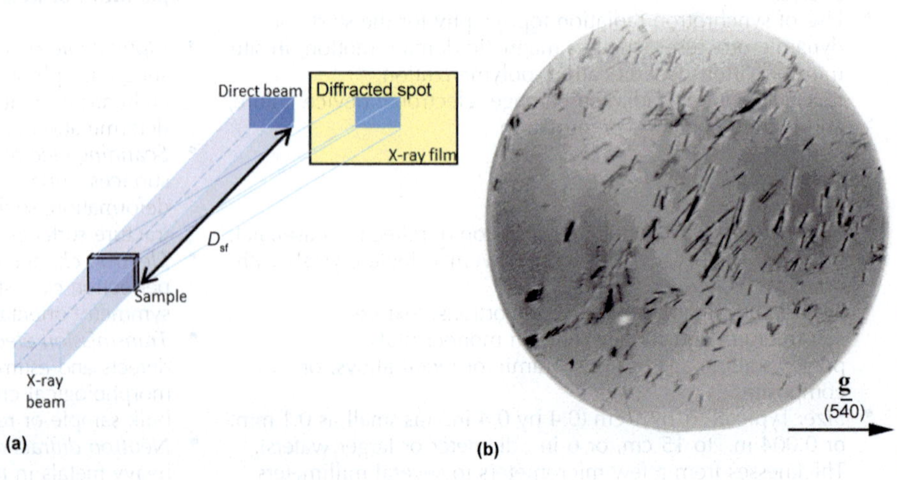

Fig. 1 (a) Schematic of x-ray topography technique. (b) Typical transmission x-ray topograph from a quartz wafer showing images of dislocations

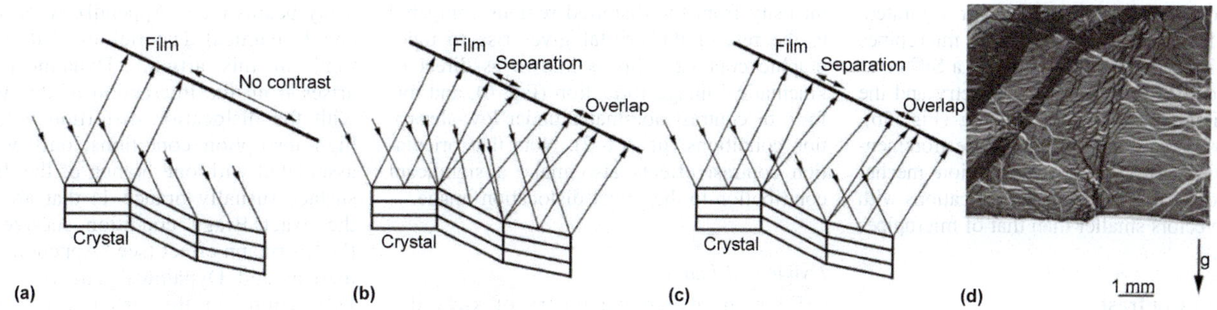

Fig. 2 Orientation contrast arising from misoriented regions. (a) Monochromatic radiation (beam divergence < misorientation). (b) Monochromatic radiation (beam divergence > misorientation). (c) Continuous radiation. (d) Reflection topograph from a ZnO single crystal. The white bands correspond to separation between images of adjacent subgrains, while the dark bands correspond to image overlap.

The overlap and separation of these inhomogeneously diffracted x-rays with continuously varying directions is projected onto the recording plate to produce a defect image due to orientation contrast (Fig. 3).

Ray-Tracing Simulations

A ray-tracing simulation based on the orientation contrast mechanism provides an excellent tool to qualitatively and quantitatively interpret the dislocation images on x-ray topographs. It has been successfully used in back-reflection x-ray topography to clarify the screw character of the micropipes in SiC (Ref 13). It also has been used to reveal the dislocation sense of screw dislocations (Ref 14), the Burgers vectors of threading-edge dislocations (Ref 15), the core structure of Shockley partial dislocations (Ref 16), and the sign of Frank partial dislocations (Ref 17). In ray-tracing simulations (Fig. 4), the mosaic region around the dislocation (or other defect) is divided into a large number of cubic diffraction units, with their local misorientations coinciding with the long-range displacement field of the dislocation. The plane normal $\boldsymbol{n}(x, y, z)$ after distortion due to the strain field associated with the dislocation varies from the original plane normal $\boldsymbol{n}^{\mathbf{0}}(x, y, z)$ before distortion, and it is calculated for each constant area. The contrast on the simulated image is determined by the superimposition (or separation) of beams diffracted from the individual small areas on the specimen surface. Modified ray-tracing simulations also can take into account diffraction from areas beneath the crystal surface, allowing for the effects of photoelectric absorption.

In ray-tracing simulation, the key point is to obtain the plane normal after distortion due to the strain fields associated with the defects. The plane normal $\overline{\boldsymbol{n}}(x, y, z)$ after distortion is given by:

$$\overline{\boldsymbol{n}}(x, y, z) = \overline{\boldsymbol{n}}^{\mathbf{0}}(x, y, z) - \nabla\left[\overline{\boldsymbol{n}}^{\mathbf{0}}(x, y, z) \cdot \overline{\boldsymbol{u}}(x, y, z)\right] \quad \text{(Eq 1)}$$

where $\overline{\boldsymbol{n}}^{\mathbf{0}}(x, y, z)$ is the plane normal before distortion, and $\overline{\boldsymbol{u}}(x, y, z)$ is the displacement field of the dislocation or other defect. When x-rays are incident (with incidence wave vector $\boldsymbol{s}_0$) on the crystal surface, local distortion on the reflecting planes will result in the variation of diffracted beam directions, $\boldsymbol{s}_g$. The diffracted beam direction is given by $\boldsymbol{s}_0 \times \overline{\boldsymbol{n}} = -\overline{\boldsymbol{n}} \times \boldsymbol{s}_g$. Diffracted beams with slightly different directions will be collected on the recording plane, and contrast is formed as the

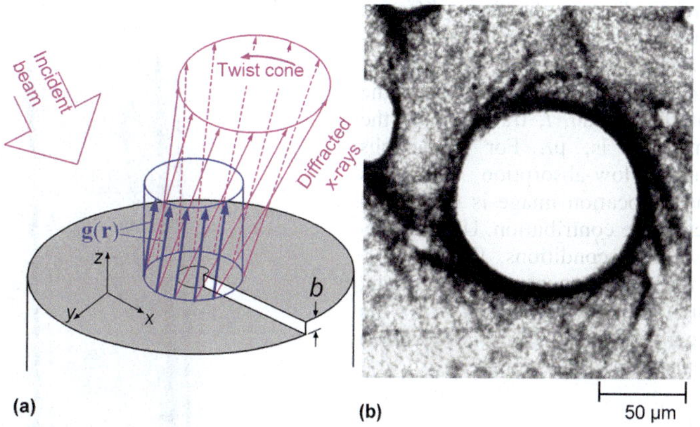

Fig. 3 (a) Schematic showing the overlap and separation of inhomogeneously diffracted x-rays from a screw dislocation. (b) Back-reflection x-ray topograph of a (0001) 4H-SiC wafer showing the resultant circular image of a superscrew dislocation (g = 00016)

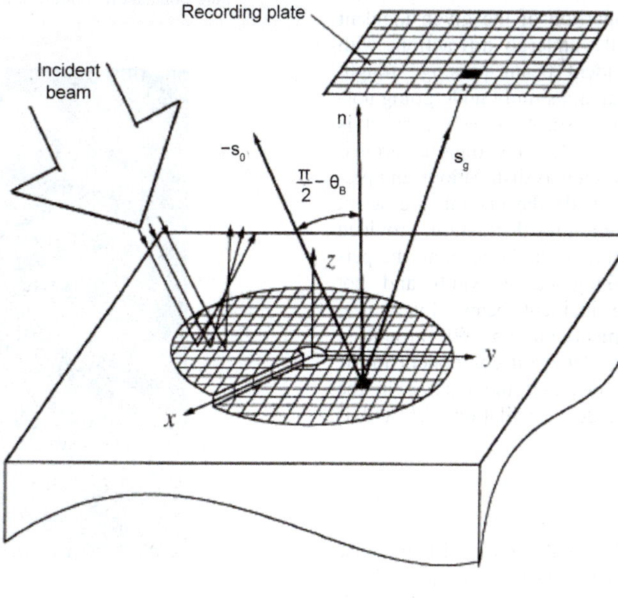

Fig. 4 Schematic showing the principle of ray-tracing simulation

diffracted beams are overlapped or separated. Figure 5(a) shows images of micropipes (screw dislocations with $\boldsymbol{b} > 2c$) in a SiC crystal recorded in transmission geometry and the image simulated using ray tracing (Fig. 5b). Under low-absorption conditions, the aforementioned orientation contrast formation mechanism also applies to ordinary dislocations with Burgers vectors smaller than that of micropipes (Ref 18).

Extinction Contrast

Extinction contrast arises when the scattering power around the defects differs from that of the rest of the crystal. Topographic contrast of dislocations (as well as other defects) consists of direct, dynamic, and intermediary images corresponding to the three different types of extinction contrast. Absorption conditions usually are defined by the product of the linear absorption coefficient, μ, and the thickness of the crystal, t, traversed by the x-ray beam, that is, μt. For topographs recorded under low-absorption conditions ($\mu t < 1$), the dislocation image is dominated by the direct image contribution. Under intermediate-absorption conditions, that is, $5 > \mu t > 1$, all three components can contribute, while for high-absorption cases ($\mu t > 6$), the dynamical contribution (in this case known as the Borrmann image) dominates.

Direct Image

The direct dislocation image is formed when the angular divergence or wavelength bandwidth of the incident beam is larger than the angular or wavelength acceptance of the perfect crystal (Ref 19). Under this condition, only a small proportion of the given incident beam will actually undergo diffraction, with most of the incident beam passing straight through the crystal and simply undergoing normal photoelectric absorption. However, it is possible that the deformed regions around structural defects, such as dislocations and precipitates present inside the crystal, are set at the correct orientation for diffraction, provided that their misorientation is larger than the perfect crystal rocking curve width and not greater than the incident beam divergence. The effective misorientation, $\delta\theta$, around a defect is the sum of the tilt component in the incidence plane, $\delta\varphi$, and the change in the Bragg angle, θ_B, due to dilation, δd, and is given by:

$$\delta\theta = -\tan\theta_B \,{}^{\delta d}/_{d} \pm \delta\varphi \qquad \text{(Eq 2)}$$

Therefore, the distorted region will give rise to a new diffracted beam. Further, if the distorted region is small in size, then this region will diffract kinematically and will not suffer the effective enhanced absorption associated with extinction effects to which the diffracted beams from the perfect regions of the crystal are subjected. The enhanced diffracted intensity from the distorted regions compared to the rest of the crystal gives rise to topographic contrast. This is known as direct or kinematical image formation (Fig. 6), and this form of contrast dominates under low-absorption conditions ($\mu t < 1$–2). Note that orientation contrast effects also make a significant contribution to the direct dislocation image.

Dynamical Image

From the dynamical theory of x-ray diffraction, the wavefield propagating in the crystal is represented by a tie point on the dispersion surface and comprises two waves corresponding to the incident and diffracted x-ray beams (see "Appendix A: Kinematical and Dynamical Theories of X-Ray Diffraction" in this article). Dynamical contrast arises from the interaction of this wavefield with the dislocation distortion field. Under high-absorption conditions, only wavefields associated with one branch of the dispersion surface (usually branch 1) that are close to the exact Bragg condition survive, due to the Borrmann effect (see "Appendix A: Kinematical and Dynamical Theories of X-Ray Diffraction" in this article). For a perfect crystal, the incident boundary conditions determine the position of the tie point on the dispersion surface. However, for an imperfect crystal, the local lattice distortion

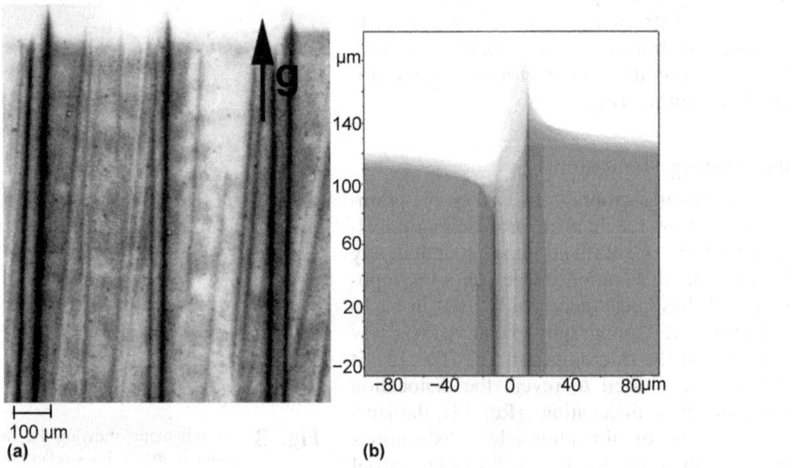

Fig. 5 (a) Synchrotron transmission topograph of superscrew dislocations in 6H-SiC (**g** = 0006). (b) Simulation of pure orientation contrast of a $5c$ superscrew dislocation

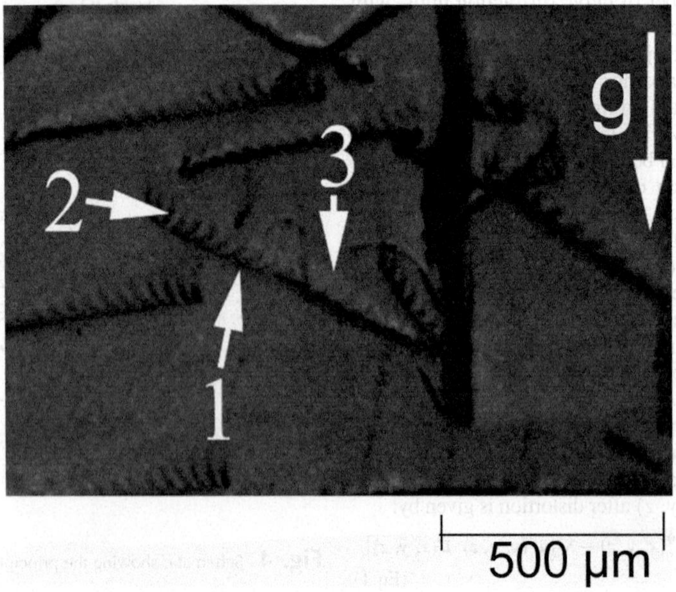

Fig. 6 Synchrotron white-beam x-ray topography transmission topograph ($\mathbf{g} = 10\bar{1}0$, $\lambda = 0.75$ Å) recorded from an AlN single crystal under high-absorption conditions ($\mu t = 8$) showing the direct (1), intermediary (3), and dynamical (2) images of a dislocation

can modify the position of the tie point of a wavefield as it passes through the crystal.

This can occur by two mechanisms, depending on the nature of the distortion field. Tie point migration along the dispersion surface occurs when the wavefield encounters a shallow misorientation gradient upon passing through the crystal (e.g., regions away from the dislocation line). The variation of the misorientation should be less than the rocking curve width over an extinction distance, ξ_g. Under these conditions, as a wavefield approaches the long-range distortion field of a dislocation, rays will bend in opposite directions on either side of the core, potentially producing opposite contrast. Net contrast is observed, because any deviation of the wavefields from the direction of propagation corresponding to the perfect crystal region forces them to experience enhanced absorption, thereby producing a loss of intensity. Under these conditions, the dislocation image will appear white and diffuse. In cases where the dislocation is close to the exit surface of the crystal, the lattice curvature above and below the defect is asymmetric, due to the requirements of surface relaxation. This means that the ray bending experienced above the defect is no longer compensated by that experienced below, resulting in opposite contrast observed from regions on either side of the defect. On the other hand, when the wavefield encounters a sharp misorientation gradient upon passing through the crystal (e.g., regions close to the dislocation core), the strain field would completely destroy the conditions for propagation and force the wavefield to decouple into its component waves. When these component waves reach the perfect crystal on the other side of the defect, they will excite new wavefields. These newly created wavefields will have tie points that are distributed across the dispersion surface, and because only those wavefields with tie points close to the exact Bragg condition survive, they will be heavily attenuated, leading to a loss of intensity from the region surrounding the dislocation. Such images, known as Borrmann images, appear white on a dark background. An example is shown in Fig. 6, which is a detail from a white-beam x-ray topograph recorded from an AlN single crystal.

Intermediary Image

The intermediary image arises from interference effects at the exit surface between the new wavefields created below the defect and the undeviated original wavefield propagating in the perfect regions of the crystal. Usually, these images appear as a beadlike contrast along direct dislocation images on projection topographs. Under moderate-absorption conditions in which the defect (e.g., dislocation line) is inclined to the surface, the intermediary image forms a fan within the intersections of the exit and entrance surface of the dislocation and has an oscillatory contrast with depth periodicity of an extinction distance ξ_g. Again, this is illustrated in Fig. 6.

Systems and Equipment

X-ray topographic techniques can be broadly classified into two types: conventional x-ray topographic techniques, based on using laboratory x-ray sources using characteristic radiation (copper, molybdenum, tungsten, etc.), and synchrotron x-ray topographic techniques, based on using synchrotron continuous (white-beam) radiation.

Section X-Ray Topography

A section x-ray topograph (Fig. 7c) is produced when a narrow x-ray beam approximately 10 μm wide is used (Fig. 7a) and images a restricted volume of the crystal as defined by the Borrmann fan (see "Appendix A: Kinematical and Dynamical Theories of X-Ray Diffraction" in this article). The incident beam is diffracted at different depths, resulting in a topographic image representing a defect profile across the thickness of the crystal. The two sets of wavefields produced within the Borrmann fan are successively in and out of phase at the base of the fan, resulting in Pendellösung fringes on the corresponding section topograph from perfect crystals (Fig. 7b). The presence of defects and strain distort the fringe pattern. At high levels of strain, the fringes are no longer visible. The section topograph is the fundamental x-ray topograph, with the standard projection topograph considered essentially an overlap of a series of section topographs in which the depth information is lost.

Conventional X-Ray Topography Techniques

Berg-Barrett Topography

The Berg-Barrett topography experimental setup (Fig. 8a) uses an extended x-ray source, and the crystal is aligned so that the diffracting conditions are satisfied for the characteristic K_α lines from a set of Bragg planes (Ref 1, 2). The crystal is cut such that the incident beam makes a small angle to the specimen surface, and the diffracted beam emerges almost normal to the specimen surface. The photographic

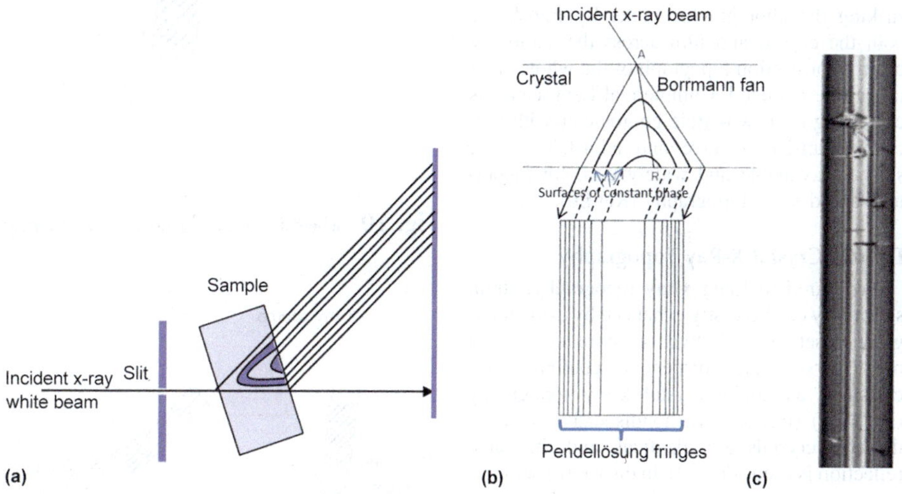

Fig. 7 (a) Schematic of section x-ray topography. (b) Wavefields in Borrmann fan leading to formation of Pendellösung fringes. (c) Section x-ray topograph from a diamond crystal showing Pendellösung fringes that are distorted by dislocations

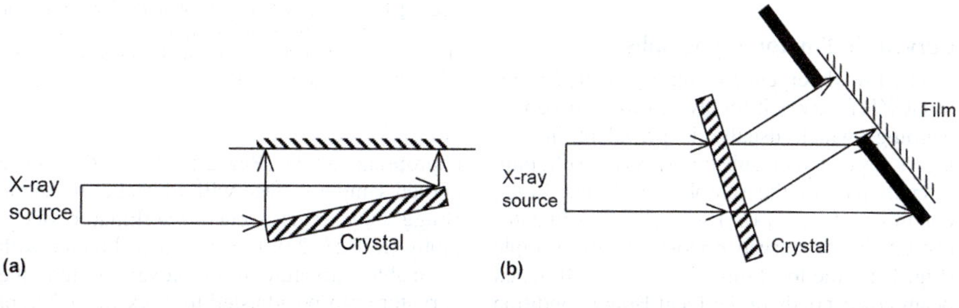

Fig. 8 Schematic diagrams of (a) reflection and (b) transmission Berg-Barrett techniques of topography

film can be placed very close to the specimen surface (as low as 1 mm, or 0.04 in.). The use of reflection geometry permits imaging up to only the x-ray penetration depth below the surface of the crystal, as determined by the extinction length or absorption distance. This allows for the study of high-dislocation-density materials (~10^6 cm^{-2}). Limitations of this technique include image doubling due to the K_α doublet as well as significant loss of spatial resolution with increasing specimen-film distance. This technique often is used for initial assessment of crystals of new materials (Ref 20). The transmission Berg-Barrett method (Fig. 8b) (also known as Barth-Hosemann geometry) is similar to the Lang technique except that it suffers from high background of scattered radiation that limits its use for studying dynamical images.

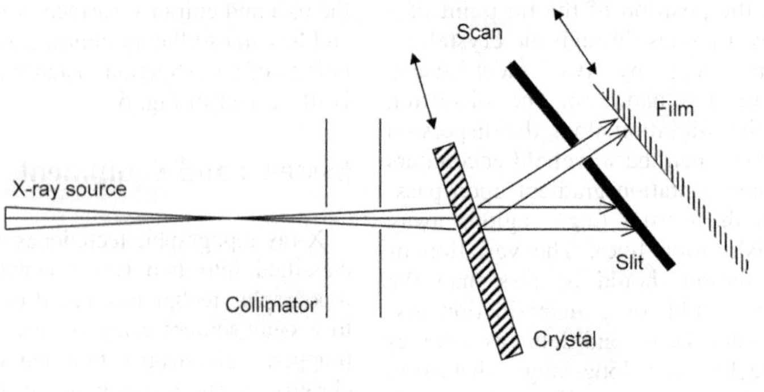

Fig. 9 Schematic diagram of the Lang projection technique. The topograph image due to $K\alpha_1$ alone is recorded on the film. The secondary slit blocks the direct beam while allowing the diffracted beam to pass through. The crystal and the film are translated synchronously, and the whole image of the crystal is recorded.

Lang X-Ray Topography

This is the most widely used laboratory technique and is based on transmission geometry (Ref 7, 21). Figure 9 shows the basic experimental setup, in which the x-ray source is collimated to allow diffraction from one K_α line (usually $K_{\alpha 1}$). The diffracting planes typically are nearly perpendicular to the crystal surface, and the diffracted beam passes through a secondary slit that blocks the direct beam before striking the photographic plate. By translating both the crystal and film across the stationary beam, a projection topograph of the whole crystal can be obtained. Commercial Lang cameras consisting of a two-circle goniometer with precision translation stage and adjustable beam slits are available and widely used for laboratory-based x-ray topography (Ref 20).

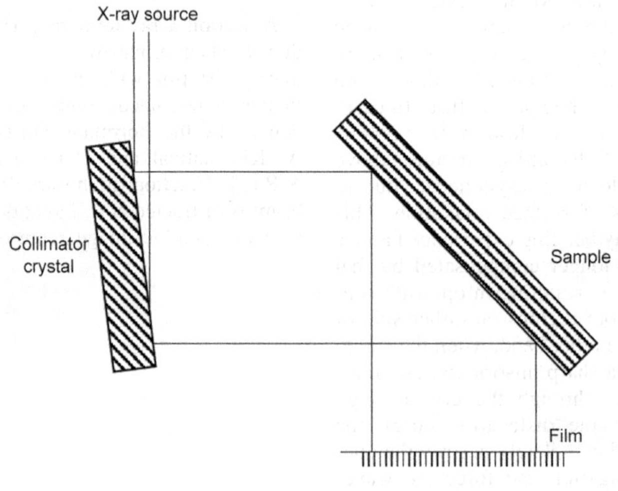

Fig. 10 Schematic diagram of double-crystal topography in nondispersive setting

Double-Crystal X-Ray Topography

Compared to Lang x-ray topography, strain sensitivity can be vastly enhanced by using topographic setups with two or more crystals in nondispersive and dispersive configurations, classified as double-crystal x-ray topography (DCXRT) (Ref 22). In nondispersive settings, the two crystals are identical, and the same reflection is used (Fig. 10); in dispersive settings, the two crystals may be different, and different reflections can be used. In either case, the first crystal collimates the beam to an approximate plane wave and also may serve to expand the beam. Strain sensitivity as high as 10^{-6} to 10^{-7} can be achieved.

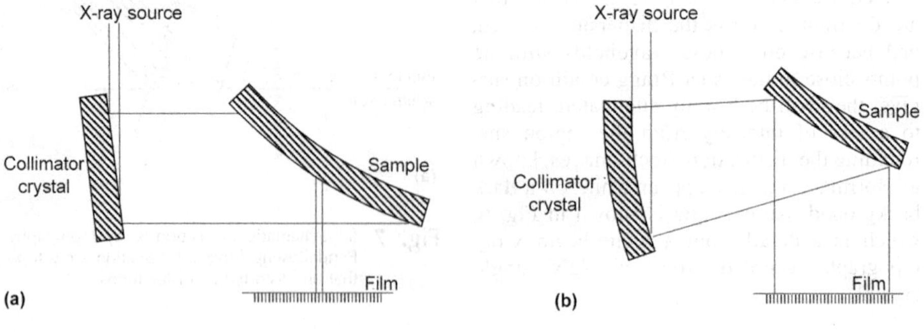

Fig. 11 Schematic diagrams of (a) double-crystal topography with a curved sample in which only a small part of the sample fulfills the Bragg condition, and (b) curved collimator topography in which the collimator curvature is adjusted to compensate for sample curvature, resulting in the whole homogeneously bent sample fulfilling the Bragg condition

Curved Collimator Topography

The low divergence of the collimated beam in DCXRT makes it highly sensitive to sample warping, which usually is prevalent in epilayers deposited on substrates. As a result, only a small part of the sample fulfills the Bragg condition (Fig. 11a). By using a collimator crystal that is curved opposite to the sample (Fig. 11b), the local direction of the collimated beam corresponds to the local Bragg condition at the sample, if the collimator is set to the appropriate angle (Ref 22). Thus, the entire curved sample surface will diffract at the same Bragg angle. This is curved collimator topography (Ref 23, 24), in which a collimator with a tunable curvature is employed so that the curvature can be adjusted to work for different homogeneously bent samples.

Synchrotron-Radiation-Based X-Ray Topography Techniques

The advent of dedicated sources of synchrotron radiation has enabled the development of synchrotron x-ray topography techniques that are endowed with numerous advantages over conventional radiation techniques and derive from the high brightness, energy tunability, and natural collimation of synchrotron radiation.

Synchrotron White-Beam X-Ray Topography

The natural collimation (typically $\approx 2 \times 10^{-4}$ radians in the vertical plane, coupled with an acceptance angle of typically a few milliradians in the horizontal plane) of the synchrotron beam allows the use of very long beamlines (~25 to 50 m, or 85 to 165 ft) to maximize the area of the beam delivered at the sample without incurring significant losses in total intensity originally available at the tangent point (Ref 9, 25). The large beam area delivered at the sample location allows studies to be carried out on relatively large-scale single crystals, and crystals as large as 150 mm (6 in.) or even 300 mm (12 in.) in diameter can be imaged by using precision translation stages similar to those used in the Lang technique (although exposure times are much shorter) (Ref 26).

If a single crystal is oriented in the beam and the diffraction pattern is recorded on a photographic detector (i.e., x-ray film), each diffraction spot constitutes a map of the diffracting power from a particular set of planes as a function of position in the crystal with excellent point-to-point resolution (typically of the order of less than 1 μm). In other words, each diffraction spot will be an x-ray topograph (Fig. 12a, b). The high intensity over the wide spectral range of the radiation drastically reduces the exposure times necessary to record a topograph, from the order of hours on conventional systems to seconds on a synchrotron, and a multiplicity of images is recorded simultaneously (Fig. 12c). The multiplicity of images also enables extensive characterization of strain fields present in the crystal. For example, instantaneous dislocation Burgers vector analysis can become possible by comparison of dislocation images obtained on several different Laue spots. Similarly, lattice rotation in the specimen can be characterized by analysis of the asterism (the spreading of a Laue spot) observed on several different Laue spots.

Synchrotron white-beam x-ray topography can be used to image the surfaces of the as-grown boules or large-sized crystal plates in reflection geometry. It can reveal the overall distribution of defects and distortion around the cylindrical surface of these crystals. Investigating as-grown boules enables observation of the true microstructures and striations developed during growth, and it can substantially reduce the time and process costs in cutting and polishing. Topographs could be recorded covering the entire length of the boule in strips by using the synchrotron beam (Fig. 13).

Synchrotron x-ray topography in the reflection geometry also can be used to examine substrate/epilayer systems that have devices fabricated on them. The features that make up the device topology typically provide contrast on x-ray topographs. The contrast usually originates from the strain experienced by the crystal at the edges of growth mesas, or metallization layers, although some absorption contrast also may superimpose on this. Topographs recorded from such structures provide an image of the defect microstructure that may also be superimposed on the backdrop of the device topology. Direct comparisons can be drawn between the performance of specific devices and the distribution of defects within their active regions. This has made it possible to determine the influence of threading screw dislocations (closed and hollow core) on device performance (Ref 27, 28). Back-reflection geometry is particularly useful here, because it gives a clear image of the distribution of screw dislocations on the background device topology that is imaged with sufficient clarity to unambiguously identify the device. An example of a back-reflection image recorded from a crystal with devices fabricated on it is shown in Fig. 14.

The wide spectrum available in the synchrotron beam allows the following crystals to be imaged in a single exposure:

- Uniform range of lattice orientation (e.g., uniformly bent by a small amount)
- Several regions of distinctly different orientations (e.g., containing subgrains, grains, or twinned regions)
- Regions of different lattice parameters (e.g., containing more than one phase or polytype)

Analysis of Laue spot shape or Laue spot asterism (the deviation from the shape expected from an undistorted crystal) enables quantitative analysis of lattice rotation. For those crystals that contain several regions of distinctly different orientations, so-called orientation contrast becomes evident, whereby the two neighboring regions of crystal, separated by the boundary (twin boundary or grain boundary, for example), each give rise to diffracted beams that travel in different directions in space, leading to image shifts on the detector. Analysis of the directions of these diffracted beams from the measured image

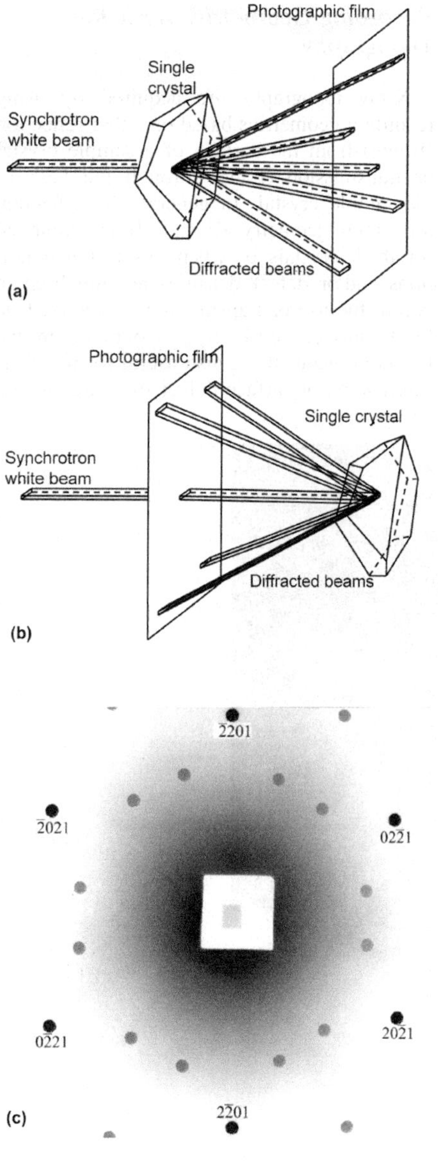

Fig. 12 Schematic diagrams of the white-beam diffraction pattern recorded in (a) transmission geometry and (b) back-reflection geometry. (c) Actual transmission x-ray diffraction pattern recorded from an (0001) AlN single crystal on an SR-45 20.3 by 25.4 cm (8 by10 in.) x-ray film

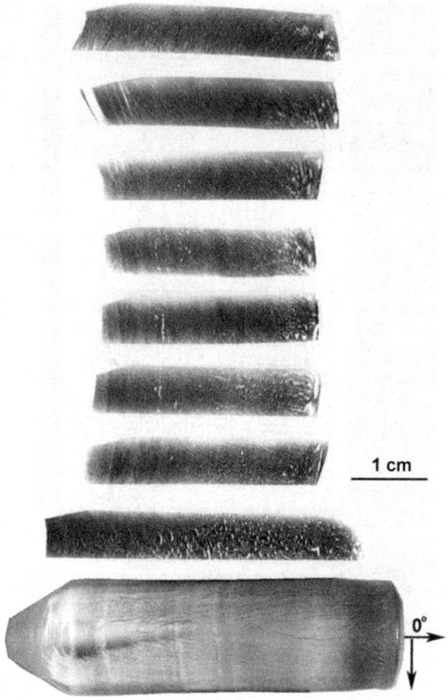

Fig. 13 Optical picture of an as-grown langasite boule along with a series of reflection topographs recorded from the surface of the boule showing the microstructure on the surface

shifts can enable the orientation relationships between the two regions of a crystal to be established, leading to, for example, determination of twin relationship. An example of a diffraction pattern recorded from a nominally (111) CdZnTe single crystal containing twins is shown in Fig. 15. Detailed analysis of the orientation relationships between the segments of the various diffraction spots enables the twin operation to be defined as a 180° rotation about [111].

Synchrotron Monochromatic-Beam X-Ray Topography

When the synchrotron white beam is passed through a monochromator, an x-ray topograph is obtained when the crystal is set to the Bragg angle for a specific set of lattice planes for the selected x-ray energy (Ref 29). Images from different atomic planes are acquired by orienting the sample to satisfy the Bragg condition for those planes and orienting the detector to the new scattering angle ($2\theta_B$) to record the image. With monochromatic radiation, only one topograph is recorded at a time, but the experimenter controls the energy or wavelength of the x-ray beam, the x-ray collimation, the energy or wavelength spread of the x-ray beam, and the size of the incident beam on the sample crystal. Monochromators used at synchrotron radiation facilities can be either single-crystal or multiple-crystal designs, which can condition the x-ray beam to achieve optimal spatial and angular resolutions. X-ray topographs (Fig. 16) recorded in the transmission and grazing incidence geometry from a SiC single crystal show different types of dislocations. The presence of misorientations greater than a few arc seconds leads to the situation in which only part of the crystal fulfills the diffracting condition at a given time (i.e., Bragg contours are produced). These contours delineate those regions of the crystal that are in the diffracting condition from those that are not. Such contours can be used to obtain quantitative information on the distribution of lattice tilt and lattice strain across the wafer.

Recording Geometries for X-Ray Topography

X-ray topographs are acquired by using recording geometries based on either reflection (Bragg) from the surface of a sample crystal or transmission (Laue) through the bulk of the sample crystal. In general, the reflection topograph geometry (Fig. 17d) is employed for thick crystals or when absorption conditions and/or defect densities are too high to permit the use of transmission geometry. Due to its surface sensitivity, topography in the reflection geometry is also useful for the characterization of surface defect structures within semiconductor heterostructures and epitaxial thin films. In all reflection-type geometries, defect information can be obtained from the volume defined by the effective area of the incident beam on the crystal and the penetration depth of the x-ray beam. The penetration depth of x-rays is determined either by the kinematical penetration depth (in imperfect crystals) or by the dynamical penetration depth (in highly perfect single crystals). The kinematical penetration depth (t_p^k) can be determined simply by geometrical relations between the incident and diffracted beams and the sample surface and is given by:

$$t_p^k = \frac{\mu_0(\lambda)}{\csc \Phi_0 + \csc \Phi_H} \qquad \text{(Eq 3)}$$

where $\mu_0(\lambda)$ is the linear absorption coefficient, and Φ_0 and Φ_H are the angles of the incident and diffracted beams with respect to the surface, respectively. The dynamical penetration depth (t_p^d) is defined to be equal to half the extinction distance, ξ_g. In transmission topography (Fig. 17a), all of the defects within the crystal volume are recorded, provided that the absorption is low enough to permit sufficient transmission through the crystal. Because the x-rays pass through the entire thickness of the sample, this technique is used to characterize overall bulk defect content of a crystal, such as dislocation networks and inclusions. The back-reflection technique (Fig. 17b),

Fig. 14 Back-reflection white-beam x-ray topograph recorded from a 6H-SiC single crystal with thyristors fabricated on it. The small white spots distributed over the image are 1c and larger screw dislocations. The location of these dislocations with respect to the device topology can be clearly discerned, enabling the influence of the defects on device performance to be determined. The large white feature corresponds to damage inflicted by a probe.

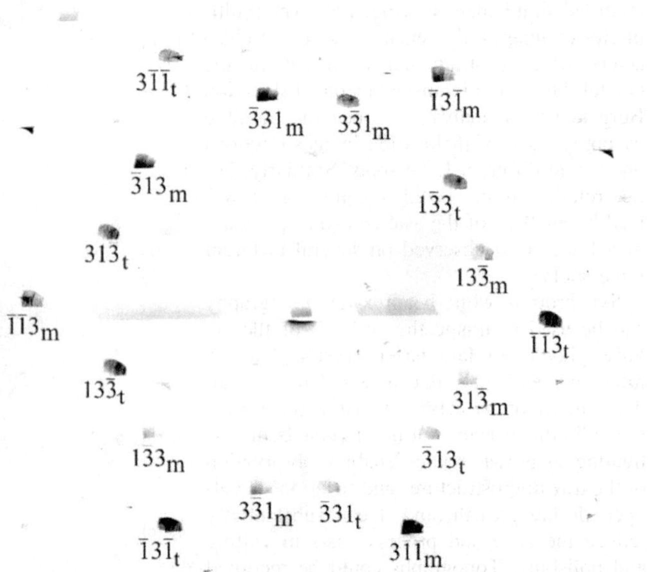

Fig. 15 Transmission x-ray diffraction pattern recorded from a CdZnTe crystal showing the presence of a twin. Detailed analysis of the orientation relationships between the segments of the various diffraction spots enables the twin operation to be defined as a 180° rotation about [111].

commonly used for orienting single crystals, also can be used to record x-ray topographs of crystals containing specific defects, such as superscrew dislocations (micropipes) in SiC (Ref 30). The grazing Bragg-Laue (Ref 31) and the grazing incidence reflection (Fig. 17c) geometries allow precise tuning of the penetration depth for depth-profiling studies of epitaxial thin films.

Techniques Based on X-Ray Topography

X-Ray Reticulography

X-ray reticulography is x-ray topography imaging with a fine-scale x-ray-absorbing mesh placed between the diffracting specimen and the detector (x-ray film) or in the path of the incident beam (Fig. 18) (Ref 32–37). The precision mesh (typically made of a heavy metal such as gold or tungsten that strongly absorbs x-rays) splits the diffracted beam into an array of microbeams. Orientation differences between the crystal elements reflecting individual microbeams are measured by tracking the angular dispersal of microbeams as the mesh-to-plate distance is changed. A reticulograph then gives directly a true map of misorientation vectors over the area of the specimen imaged. This method is suitable for mapping and quantifying long-range deformation that does not result in dislocation/grain boundary formation. Using x-ray reticulography, a stress-mapping analysis by ray-tracing technique (Ref 38) has been developed to map the stress tensor in single-crystal materials. The principle is based on the fact that there exists a relationship between the stress state in a crystal and the local lattice plane orientation. This relationship can be exploited to determine the full stress tensor as a function of position inside the crystal. This technique essentially is the inverse of ray-tracing simulation.

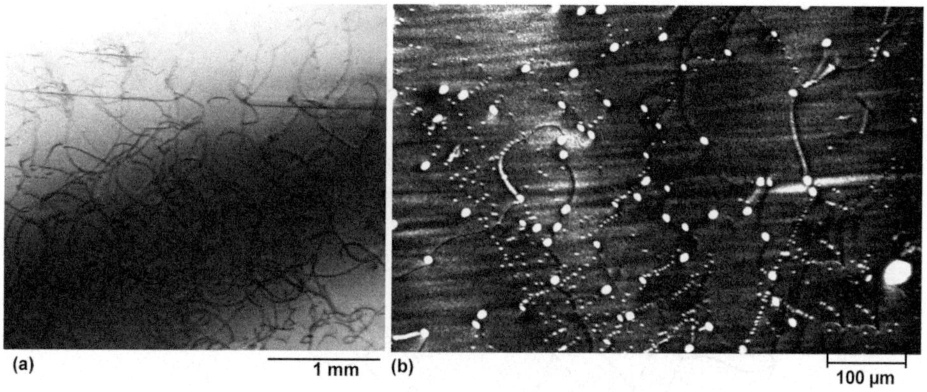

Fig. 16 (a) Transmission and (b) grazing incidence x-ray topographs recorded from a SiC crystal by using a monochromatic synchrotron beam

X-Ray Topographic Contour Mapping

When a single crystal is illuminated by monochromatic x-ray radiation of certain divergence, only a limited region will diffract (Ref 39–46). This is due to the existence of lattice deformation (effective misorientation) that deviates from the rest of the crystal from perfect Bragg condition by $\Delta\omega$, and thus, only a small region is accepted for diffraction. With a single exposure, a so-called equimisorientation contour can be obtained on the recording plate, as shown in Fig. 19. Therefore, a contour map can be generated by rocking the crystal through the perfect Bragg condition with small steps of angular rotation and taking exposure at each step. According to Bragg's law, for an arbitrary location in the crystal referenced to perfect lattice, the deviation from perfect Bragg condition locally is due to the convoluted effect of lattice dilation/compression ($\partial u_i/\partial x_i$) and lattice shear/rotation ($\partial u_i/\partial x_k$). This can be expressed in Eq 4:

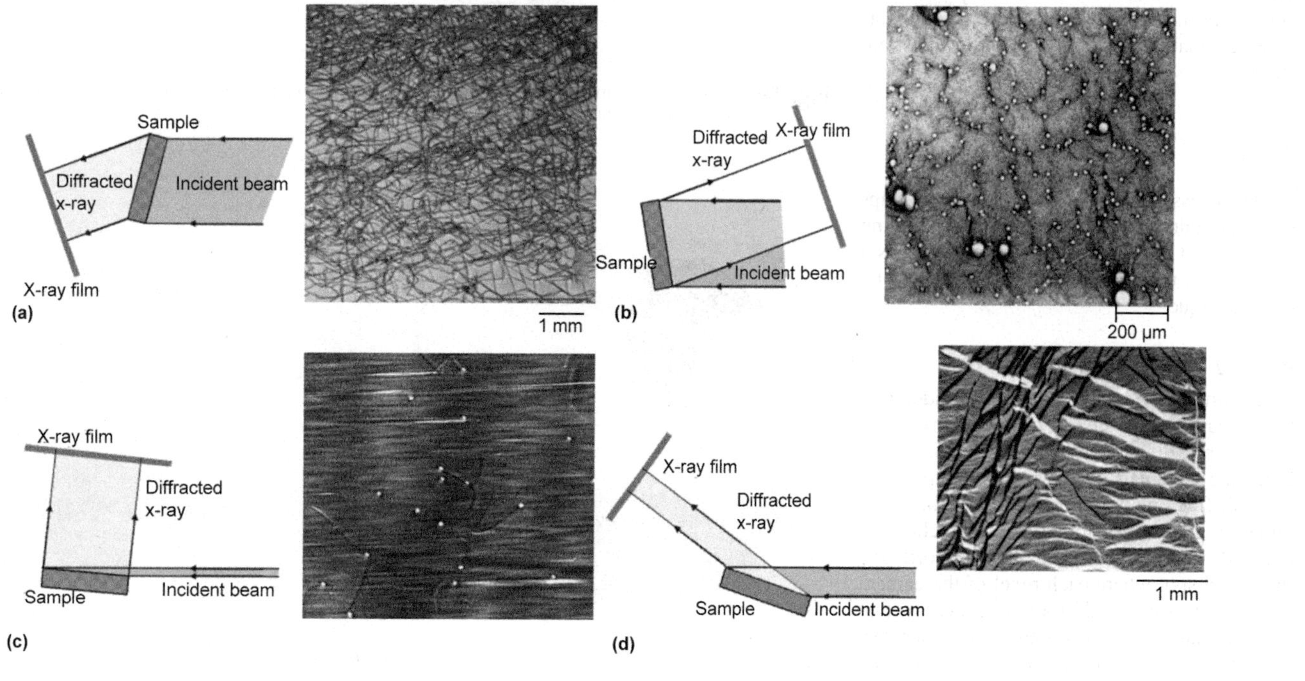

Fig. 17 Schematic of recording geometries and example of an x-ray topograph recorded in each geometry. (a) Transmission: image from a 4H-SiC wafer showing basal plane dislocations. (b) Back reflection: image from a 6H-SiC wafer showing threading screw dislocations. (c) Grazing incidence: image from a 4H-SiC wafer showing threading screw dislocations. (d) Reflection: image from a ZnO wafer showing low-angle grain boundaries

$$\omega^{x_i}(x_j) = \frac{\partial u_i}{\partial x_i}\tan\theta_B + \frac{\partial u_i}{\partial x_k} \quad \text{(Eq 4)}$$

where $\omega^{x_i}(x_j)$ denotes the angular deviation from perfect Bragg condition, and, if a reference point is selected in the crystal to be free of strain, $\omega^{x_i}(x_j)$ equals the amount of angular movement one must apply to shift the equimisorientation contour from the reference point to the location of interest by rotating the diffracting plane x_i about the x_j axis. θ_B is the perfect Bragg angle. To deconvolute these two strain components, an additional contour map is needed with diffraction taking place on the same lattice plane but with opposite diffraction vector ($-\vec{\mathbf{g}}$). This can be done practically by first rotating the crystal about the diffraction plane normal 180° and recording another contour map in the same manner. In the $-\vec{\mathbf{g}}$ contour map, the local deviation from perfect Bragg's condition is given by:

$$\omega^{-x_i}(x_j) = \frac{\partial u_i}{\partial x_i}\tan\theta_B - \frac{\partial u_i}{\partial x_k} \quad \text{(Eq 5)}$$

where the shear/rotation component $\partial u_i/\partial x_k$ makes an opposite contribution to the deviation in the $-\vec{\mathbf{g}}$ map and thus has a negative sign in front of it. Simple addition and subtraction of these two equations yield:

$$\frac{\partial u_i}{\partial x_i} = \frac{\omega^{x_i}(x_j) + \omega^{-x_i}(x_j)}{2\tan\theta_B} \quad \text{(Eq 6)}$$

$$\frac{\partial u_i}{\partial x_k} = \frac{\omega^{x_i}(x_j) - \omega^{-x_i}(x_j)}{2} \quad \text{(Eq 7)}$$

The values of these two strain components for any location in the crystal thus can be quantitatively estimated. The complete set of the strain tensor:

$$\varepsilon_{ij} = \frac{1}{2}(\partial u_i/\partial x_i + \partial u_i/\partial x_j) \quad \text{(Eq 8)}$$

can be accessed from various $\pm\vec{\mathbf{g}}$ contour maps that are obtained by rotating different diffracting planes about different axes. Figure 20 shows a set of strain maps for a SiC wafer obtained by this technique.

Rocking Curve Imaging

Rocking curve imaging can be considered to be a quantitative version of monochromatic-beam x-ray diffraction topography that combines the advantages of x-ray topography and x-ray diffractometry (Ref 48–50). In this technique, a two-dimensional detector (charged-coupled device, or CCD, camera) records the diffracted spot where each pixel of the camera records its own "local" rocking curve, so that several images (or maps) can be reconstructed by extracting data from these local rocking curves. In particular, maps of three parameters of interest for each rocking curve can be obtained:

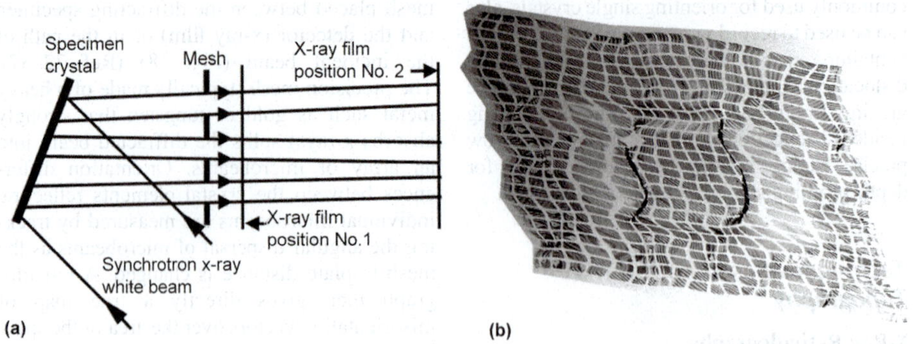

Fig. 18 (a) Schematic for synchrotron white-beam x-ray reticulography measurements. (b) X-ray reticulograph from a packaged silicon wafer

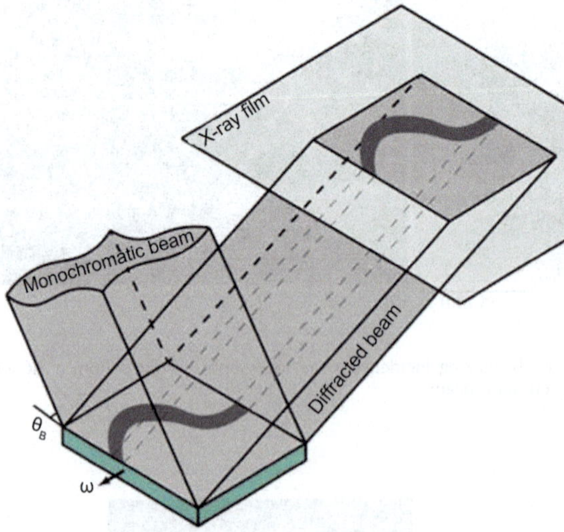

Fig. 19 Equimisorientation contour obtained when the crystal is illuminated by monochromatic radiation in which only a limited region diffracts

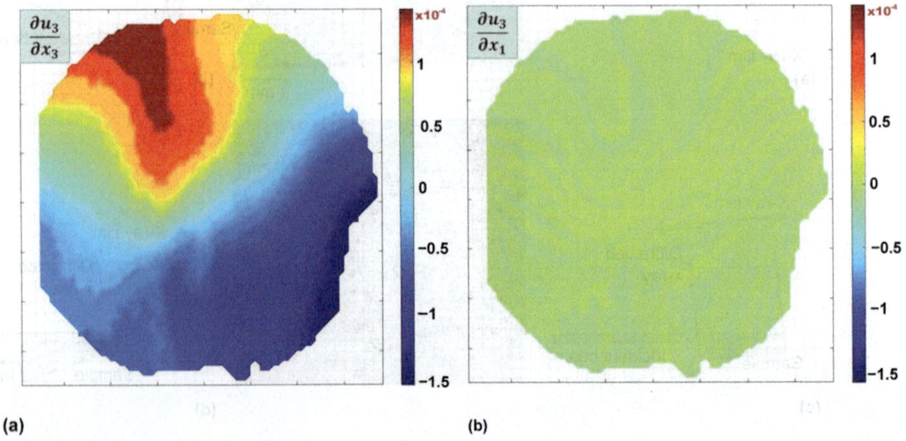

Fig. 20 Strain maps of different strain components. (a) $\partial u_3/\partial x_3$ and (b) $\partial u_3/\partial x_1$ for a SiC wafer obtained by x-ray topographic contour mapping. Source: Ref 47

- Angular peak position, which indicates macroscopic curvature and local lattice tilts
- Peak intensity
- Peak full width at half maximum, which indicates the local crystalline perfection

These maps provide a visual impression of the defect distribution and quantitative measures of lattice homogeneity (Fig. 21).

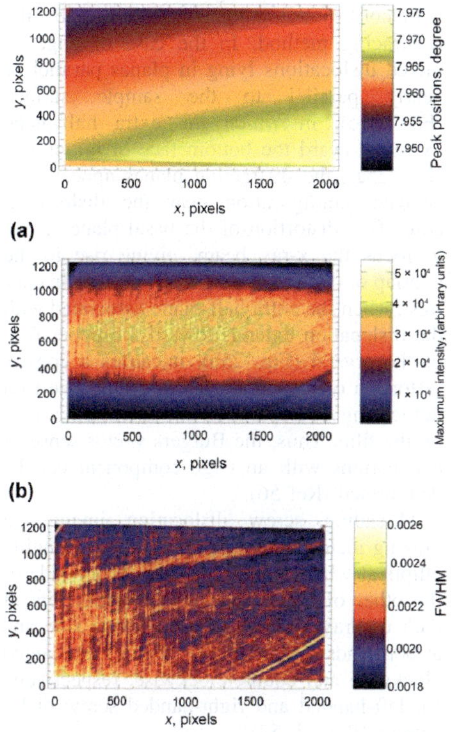

Fig. 21 Visualization of the doping concentration variations across a SiC substrate with maps of (a) angular peak position, (b) peak intensity, and (c) full width at half maximum (FWHM) obtained by rocking curve imaging

Detectors for X-Ray Topography

Photographic films continue to be the primary detectors of choice for recording x-ray topographs, although large-area CCD detectors can be used when high resolution is not paramount. Holographic films (Slavich VRP-M) are capable of recording x-ray topographs with submicrometer resolution but require long exposure times even for the intense synchrotron beam. Nuclear emulsions (Ilford plates) that have a grain size below 1 μm also are suitable to record high-resolution x-ray topographs, but they are expensive and require special handling and developing procedures. Single-side-coated high-resolution x-ray films (Agfa D3-SC, Fuji IX20, etc.) that have a grain size of the order of 1 μm usually are adequate for recording most x-ray topographs. These have sufficient contrast and resolution and can be developed fast enough to allow real-time observation and feedback for recording multiple topographs. For in situ imaging of dynamic processes, high-resolution x-ray camera systems (~1 μm) in which CCD cameras are coupled with optimized scintillators are used, but the field of view is limited to the order of 1 cm^2 (0.16 in.^{2}). By using conventional sealed-tube or microfocus x-ray sources in combination with high-resolution CCD cameras, digital x-ray topography systems have been developed, such as the BedeScan (Ref 51) and XRTmicron (Ref 52), that are capable of rapidly imaging wafers as large as 300 mm (12 in.) and conducting automated dislocation analysis.

Specimen Preparation

Sample preparation is critical for x-ray topography to obtain the highest-quality images. X-rays are highly sensitive to surface conditions, and it is important to eliminate any surface features that can introduce unwanted contrast on the images (Fig. 22). Generally, the specimens must have smooth, flat surfaces free from scratches, artifacts, and subsurface polishing damage. Imaging of as-grown surfaces of crystal boules (Fig. 23a) by synchrotron white-beam x-ray topography (SWBXT) does not require any specific surface preparation because these are generally strain-free, and any surface preparation actually may remove essential information. For reflection geometry, the surface to be imaged is

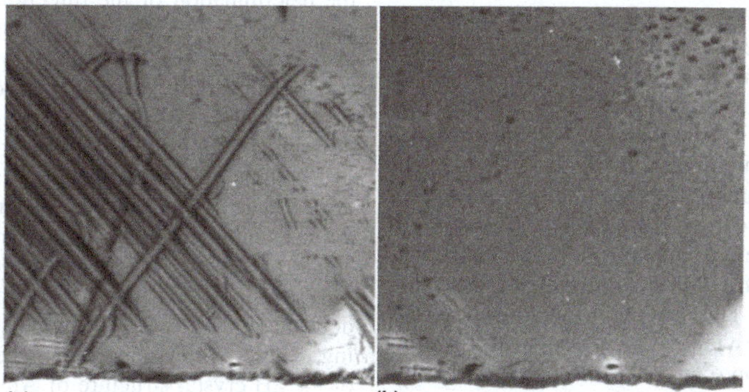

Fig. 22 Transmission x-ray topographs from an AlN wafer (a) showing contrast from scratches due to poor polishing and (b) after repolishing to remove scratches, showing dislocation-free region

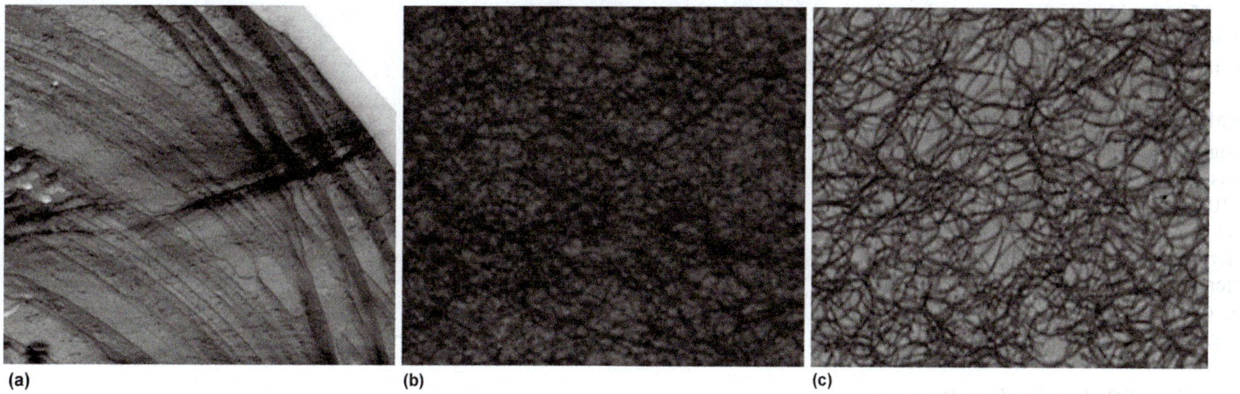

Fig. 23 (a) Reflection x-ray topograph from as-grown surface of a sapphire boule showing growth steps and low-angle grain boundaries. (b) Transmission x-ray topograph from a sapphire wafer polished on one side (other side ground) showing poor contrast from dislocations. (c) Transmission x-ray topograph from a sapphire wafer polished on both sides showing sharp contrast from dislocations

mechanically or chemomechanically polished to a smooth finish, followed by mild chemical etching to remove subsurface damage layers. For transmission geometry, both surfaces (x-ray entrant and exit) of the specimen must be polished to a smooth finish. Often, wafers are polished on one side for thin-film deposition, while the backside is ground. In that case, the polished side must be mounted to be on the exit side of the x-rays facing the x-ray film, to minimize contrast from ground surface. The overall image quality still will be degraded compared to double-sided polished specimens (Fig. 23b, c).

Calibration and Accuracy

On an x-ray topograph, the contrast from individual defects will be clearly discernible only if the spatial resolution is adequate. There is no magnification involved in x-ray topography, and spatial resolution is controlled solely by geometrical factors. The effective resolution can be approximately written as:

$$R = \frac{SD}{C} \quad \text{(Eq 9)}$$

where R is the effective resolution, S is the maximum source dimension in the direction perpendicular to the plane of incidence, D is the specimen-film distance, and C is the source-specimen distance. Because the spatial resolution is proportional to the distance between the crystal and detector, this distance should be minimized. However, the optimum spatial resolution of the photographic detectors typically used in topography is limited to approximately 1 μm by photoelectron tracking between adjacent grains in the emulsion of the film. The specimen-film distance usually is set to yield a calculated spatial resolution that approximately coincides with this. Based on this resolution of approximately 1 μm, defect visibility and characterization by x-ray topography is limited to samples with defect densities below 10^6/cm^2. This is because topographic dislocation images can be anywhere from approximately 5 to approximately 15 μm wide, so that greater densities would lead to image overlap and therefore loss of information.

Applications and Interpretation

Topography, both synchrotron and conventional, is well suited for analysis of low densities (<10^6 cm^{-2}) of dislocations in crystals.

Basic dislocation analysis can be used for the determination of dislocation line direction, the determination of Burgers vector direction, and the determination of Burgers vector sense and magnitude.

Determination of Dislocation Line Direction

For dislocations created by slip processes, knowledge of the line direction as well as detailed information on the Burgers vector of the dislocations is required to fully assign the active slip system. Knowledge of both line direction and Burgers vector of crystal-growth-induced dislocations also is important in understanding their origin and for developing strategies to reduce the density of such dislocations. The projected directions of direct images of growth dislocations also have been used very successfully to compare with line energy calculations designed to determine why particular line directions are preferred by such dislocations in crystals (Ref 53). The line direction of a dislocation can be obtained by analyzing its direction of projection on two or more topographs recorded with different reciprocal lattice vectors. The use of analytical geometry enables the line direction to be determined either directly from the measured direction of projection (Ref 54) or indirectly by comparing calculated projected directions of expected dislocation line directions for the material of interest with the measured projection directions (Ref 55).

Determination of Burgers Vector Direction

For sufficiently low dislocation densities, standard Burgers vector analysis, which enables the determination of the direction of the Burgers vector, is readily carried out in the low-absorption regime, using the $\mathbf{g} \cdot \mathbf{b} = 0$ criterion for the invisibility of screw dislocations, and the combination of $\mathbf{g} \cdot \mathbf{b} = 0$ and $\mathbf{g} \cdot \mathbf{b} \times \mathbf{l} = 0$ criterion for the invisibility of edge or mixed dislocations (where **b** is the dislocation Burgers vector, and **l** is the dislocation line direction). An example of dislocation analysis in a single crystal of AlN is presented in Fig. 24. These dislocations likely were formed by the deformation process during postgrowth cooling. These dislocations lie in the basal plane of the 2H crystal structure of AlN and are visible in Fig. 24(a) ($\mathbf{g} = 1\bar{1}01$) and invisible in the $\bar{1}011$ (Fig. 24b) and $10\bar{1}0$ (Fig. 24c) reflections, although weak contrasts are observed. Application of the $\mathbf{g} \cdot \mathbf{b} = 0$ criterion to possible $\frac{1}{3}\langle 11\bar{2}0\rangle$ Burgers vectors that lie in the basal plane shows that the dislocations have Burgers vector along $[1\bar{2}10]$.

Determination of Burgers Vector Sense and Magnitude

The Burgers vector sense of dislocations can be determined by comparing the back-reflection or grazing incidence synchrotron x-ray topographic images of the dislocations with corresponding images simulated by the ray-tracing method. In the case of edge or mixed dislocations lying in planes parallel or nearly parallel to the sample surface (Fig. 25c), in which the extra half-plane extends toward the bottom face of the crystal (Fig. 25a), the diffracting planes appear as a concave configuration near the dislocation core. This distortion of the basal planes tends to focus the x-ray beam, giving rise to the overlap of diffracted x-rays on the film. In contrast, when the extra half-plane associated with the dislocation extends toward the upper face of the crystal (Fig. 25b), it causes a convex distortion of the basal planes, giving rise to a defocusing or separation of diffracted x-rays on the film. Thus, the Burgers vector sense of dislocations with an edge component can be determined (Ref 56).

Threading screw dislocation images in grazing incidence geometry appear as roughly elliptical white features, canted to one side or the other of the **g** vector, with perimeters of dark contrast that thicken along one side and at both ends (Fig. 26). The features are canted clockwise and counterclockwise, respectively, for left-handed and right-handed screw dislocations (Ref 14, 57).

Similarly, ray-tracing simulation images of threading-edge dislocations in grazing incidence geometry can be used to determine Burgers vector sense (Ref 15). Figure 27 shows the topographic images of six types of

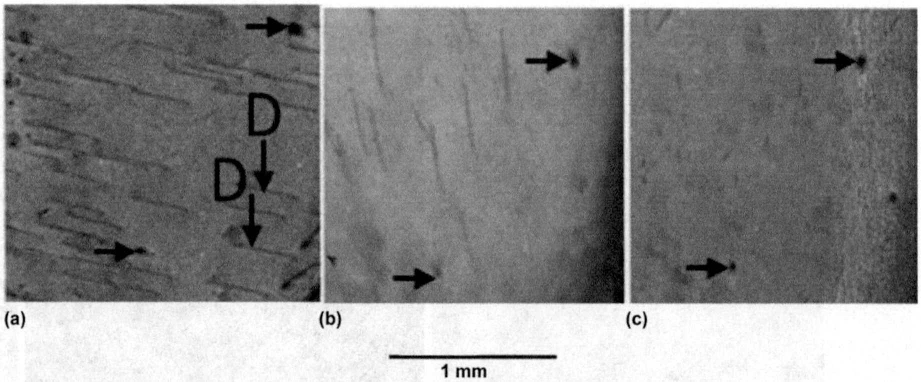

Fig. 24 Synchrotron white-beam topographs recorded in transmission geometry from an AlN single crystal containing slip dislocations. (a) $\mathbf{g} = 1\bar{1}01$. (b) $\mathbf{g} = \bar{1}011$. (c) $\mathbf{g} = 10\bar{1}0$. Note the disappearance of dislocation segments, indicated by the arrows in (a) through (c). The Burgers vector of these dislocations is determined to be along $[1\bar{2}10]$.

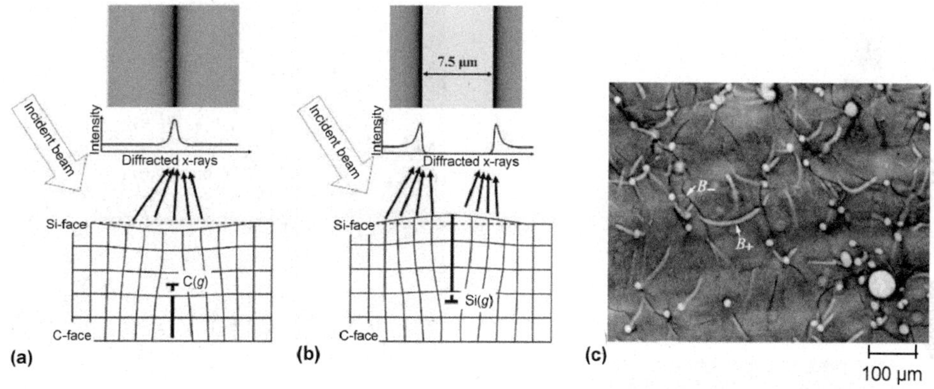

Fig. 25 (a) Focusing and (b) defocusing of x-ray beams by the concave and convex basal planes, caused by the displacement field associated with the edge component of the partial dislocations, giving rise to the dislocation image of a narrow dark line or white stripe. (c) X-ray topograph showing dark and white images of dislocations depending on the Burgers vector sense of the edge component of dislocations

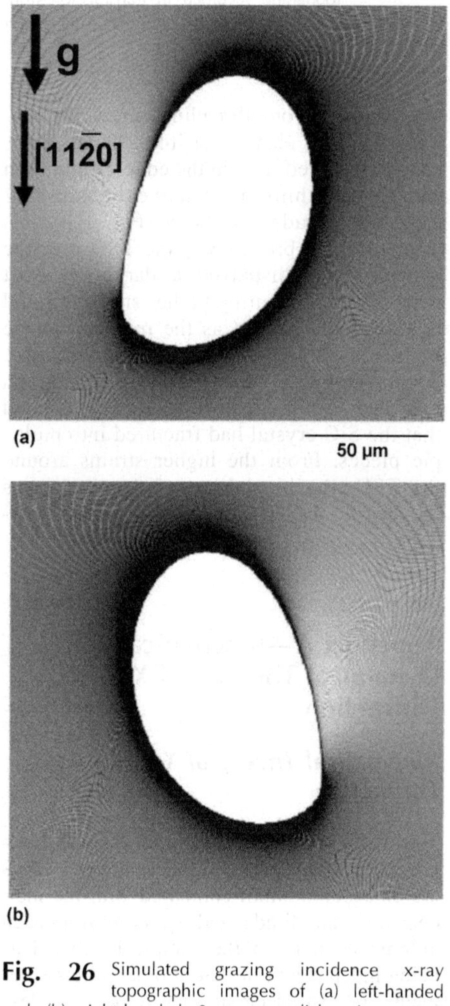

Fig. 26 Simulated grazing incidence x-ray topographic images of (a) left-handed and (b) right-handed 8*c* screw dislocations at a specimen-film distance of 35 cm (14 in.). 11$\bar{2}$0 reflections are simulated. Both images appear as roughly white ellipses canted (a) clockwise or (b) counterclockwise from the vertical configuration.

threading-edge dislocations observed in 4H-SiC crystals, along with their schematics according to the extra atomic half-planes associated with them. They appear as two dark arcs canted to one side or the other of the **g** vector, and these two dark arcs are either shifted vertically (Fig. 27a, b) or are separated by an area of white contrast (Fig. 27c–f).

Contrast from Inclusions

While individual point defects are not visible on x-ray topographs, when such defects cluster to form a precipitate or inclusion, contrast can be observed. Under low-absorption conditions, direct or kinematical images of precipitates are formed on x-ray topographs, and these typically consist of two dark half-circles separated by a line of no contrast perpendicular to the projection of the diffracted vector. This is simply due to the fact that distortions parallel to a given set of atomic planes are not discernible. An example is shown in Fig. 28, which is a transmission SWBXT image recorded from an AlN crystal. Under higher-absorption conditions, dynamical contrast can be observed. When the precipitate is close to the x-ray exit surface, opposite contrast on either side of the defect can usually be observed. This contrast usually will reverse with reversal of the sign of the reflection vector. The contrast is produced by tie point migration in the region above the defect. Because the defect is close to the exit surface, the reflecting planes rotate sharply to meet the surface at the preferred angle, and thus the curvature becomes too large for the Eikonal theory to handle. Consequently, the contrast developed above the defect is "frozen-in." The black-white contrast not only reverses with the sense of the reflection vector but also with the sense of the strain in the lattice. This can be used to determine the nature of the precipitate. If the contrast is enhanced on the side of positive **g**, then the lattice is under tension; if reduced, it is under compression. This empirical rule was first determined by Meieran and Blech (Ref 58).

Stacking-Fault Analysis

The contrast from stacking faults in x-ray topography arises from the phase shift experienced by the x-ray wavefields as they cross the fault plane. This phase shift has been computed to be equal to $\delta = (-2\pi\, \mathbf{g} \cdot \mathbf{R})$, where **g** is the active reciprocal lattice vector for the reflection, and **R** is the fault vector. Contrast is expected to disappear when $\delta = 0$ (corresponding to $\mathbf{g} \cdot \mathbf{R}$ = integer). To determine the fault vectors of stacking faults in 4H-SiC crystals (faults A and B in Fig. 29), multiple reflections are recorded and analyzed based on expected contrast based on calculated $\mathbf{g} \cdot \mathbf{R}$ values. Fault contrast is expected to be very weak when $\delta = \pm\pi/6$ (corresponding to $\mathbf{g} \cdot \mathbf{R} = \pm 1/12, \pm 11/12$) and weak but visible for $\delta = \pm\pi/3$ (corresponding to $\mathbf{g} \cdot \mathbf{R} = \pm 1/6, \pm 5/6$). Contrast should be well marked for $\delta = \pm\pi$ and $\pm 2\pi/3$ (corresponding to $\mathbf{g} \cdot \mathbf{R} = \pm 1/2, \pm 1/3$, and $\pm 2/3$). Detailed analysis of the fault contrast on different reflections shows that these observations are consistent with a fault vector, $\mathbf{R}_B$, of $c/2$ for faults B and $c/2$ plus a Shockley displacement for faults A, that is, $\mathbf{R}_A$ = 1/6 <20$\bar{2}$3> (Ref 59). The $\mathbf{g} \cdot \mathbf{R}$ values for stacking faults A and B are:

R \ g·R \ g	(01$\bar{1}$0)	(0$\bar{1}$11)	($\bar{2}$110)
$\mathbf{R}_A$ = 1/6 <20$\bar{2}$3>	1/3	1/6	−1
$\mathbf{R}_B$ = 1/2[0001]	0	1/2	0

In Situ X-Ray Topography

The high intensities, wide spectral range, and good signal-to-noise ratio associated with SWBXT open up the possibility of direct imaging of crystals under in situ or in operando conditions. The excellent geometrical resolution capability also relaxes the requirement of having small specimen-film distances to achieve good resolution. Thus, crystals can be surrounded with elaborate environmental chambers, necessitating considerable increases in specimen-detector distances without significant loss of resolution. Truly dynamic, quasi-real-time studies of crystals subjected to some type of external stimulus (such as applied fields, applied stress, heating, cooling, etc.) can be carried out.

In Situ Observation of Operation of Frank-Read Sources

White-beam x-ray diffraction topography experiments on 4H-SiC samples were undertaken in transmission mode in a double-ellipsoidal mirror furnace at temperatures up to 1600 °C (2910 °F) with heating/cooling rates between 10 and 100 °C/min (18 and 180 °F/min) (Ref 60). The x-ray topographs were recorded at regular intervals using a high-resolution camera/scintillator. The evolution of double-ended Frank-Read sources of dislocations is shown in Fig. 30(e–h). This deduced how such sources are created in SiC crystals. The

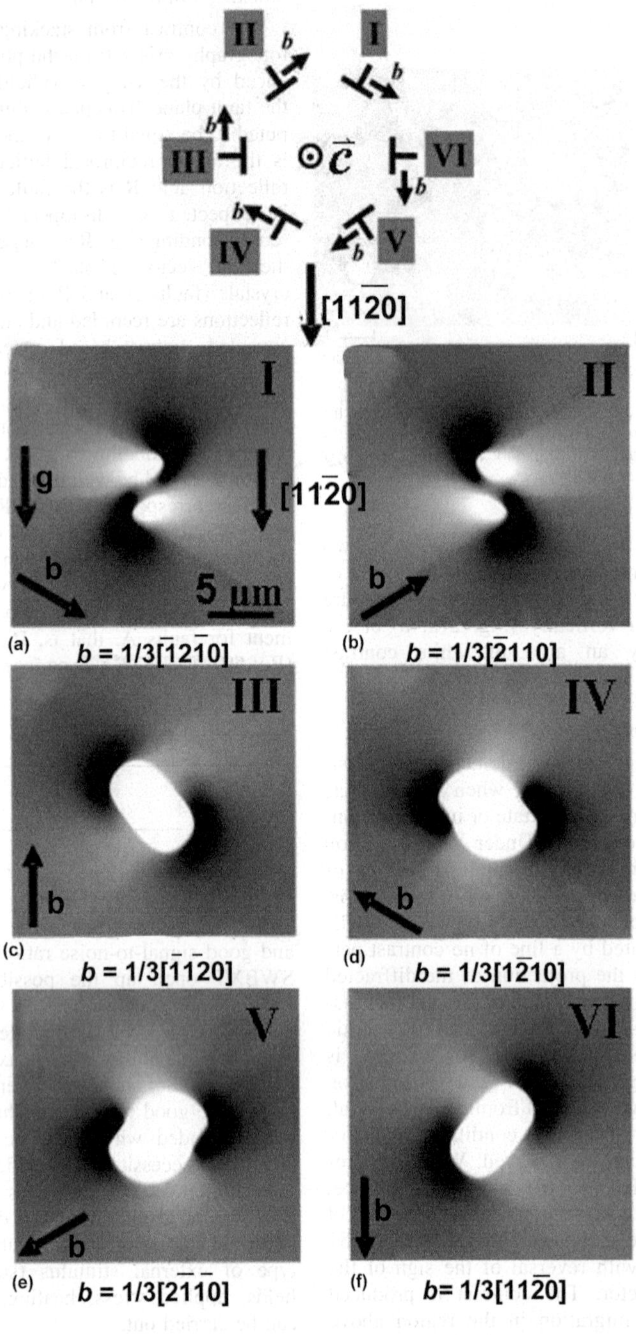

Fig. 27 (a)–(f) Simulated $11\bar{2}0$ grazing incidence x-ray topography images of threading-edge dislocations (TEDs) with six different Burgers vectors. Top: The six types of TEDs are illustrated according to the position of the extra atomic half-planes associated with them.

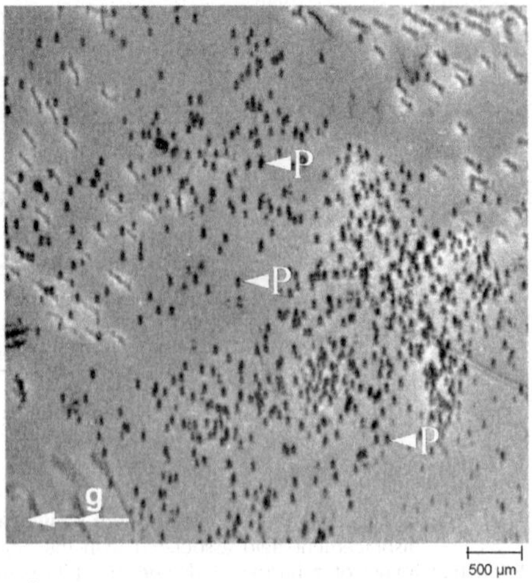

Fig. 28 Detail from a synchrotron white-beam topograph recorded in transmission from an AlN single crystal showing precipitate contrast (P) under low-absorption conditions

mechanism involves deflection of threading-edge dislocations onto the basal plane to form basal plane dislocations, followed by redeflection back to the threading direction due to competing step flows of macrosteps and screw dislocations. The threading segments act as pinning points, and the basal plane dislocation segment in between can glide to operate as a double-ended Frank-Read source, as shown schematically in Fig. 30(a–d) and as recorded by in situ x-ray topography in Fig. 30(e–h).

In Operando Study of Device Failure

A commercial 600 V/23 A-rated SiC Schottky barrier diode was subject to large switching reverse bias while back-reflection x-ray topographs were recorded periodically (Ref 61). During the experiment, reverse voltage was increased from 0 to 900 V at an interval of 100 V until final device breakdown. An x-ray topograph (Fig. 31a) from the diode recorded at 700 V is dominated by contrast from the two contact leads on the SiC crystal, but contrast from threading screw dislocations (TSDs) also is visible. Three TSDs can be observed: two in the edge-termination area and the third in the active die area near the contact leads. At 900 V (Fig. 31b), just prior to diode breakdown, the TSD near the contact leads displayed a darker contrast than before, indicating higher strains around this particular defect as the reverse voltage is increased. At 900 V, the diode broke down, and the back-reflection topograph recorded after breakdown (Fig. 31c) revealed that the SiC crystal had fractured into multiple pieces. From the higher strains around the TSD, the breakdown of the diode was induced by the TSD in a region that is favorable to current filamentation.

Appendix A—Kinematical and Dynamical Theories of X-Ray Diffraction

Kinematical Theory of X-Ray Diffraction

In the kinematical theory of x-ray diffraction, the amplitudes of the scattered waves are considered small compared with the incident wave amplitude, and scattering from each volume element in the sample is treated as being independent of that of other volume elements. For small crystals, of dimensions less than a micrometer in diameter, and in heavily deformed crystals in which the dislocations act to divide the crystal into a mosaic structure of independently diffracting cells, the kinematical theory may be employed satisfactorily to predict diffracted intensities. However, for large single crystals that are also highly perfect, the kinematical theory breaks down, and

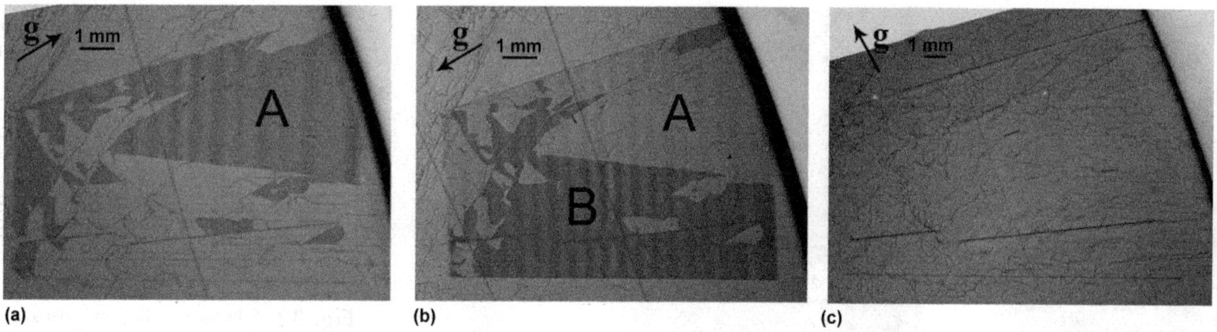

Fig. 29 Synchrotron white-beam x-ray topography transmission images recorded from a region near the edge of a 76 mm (3.0 in.) wafer cut with 4° offcut toward [$11\bar{2}0$]. (a) $01\bar{1}0$ reflection showing stacking-fault contrast from fault A only. (b) $0\bar{1}11$ reflection showing strong fault contrast from fault B and weak fault contrast from fault A. (c) $\bar{2}110$ reflection showing absence of all fault contrast

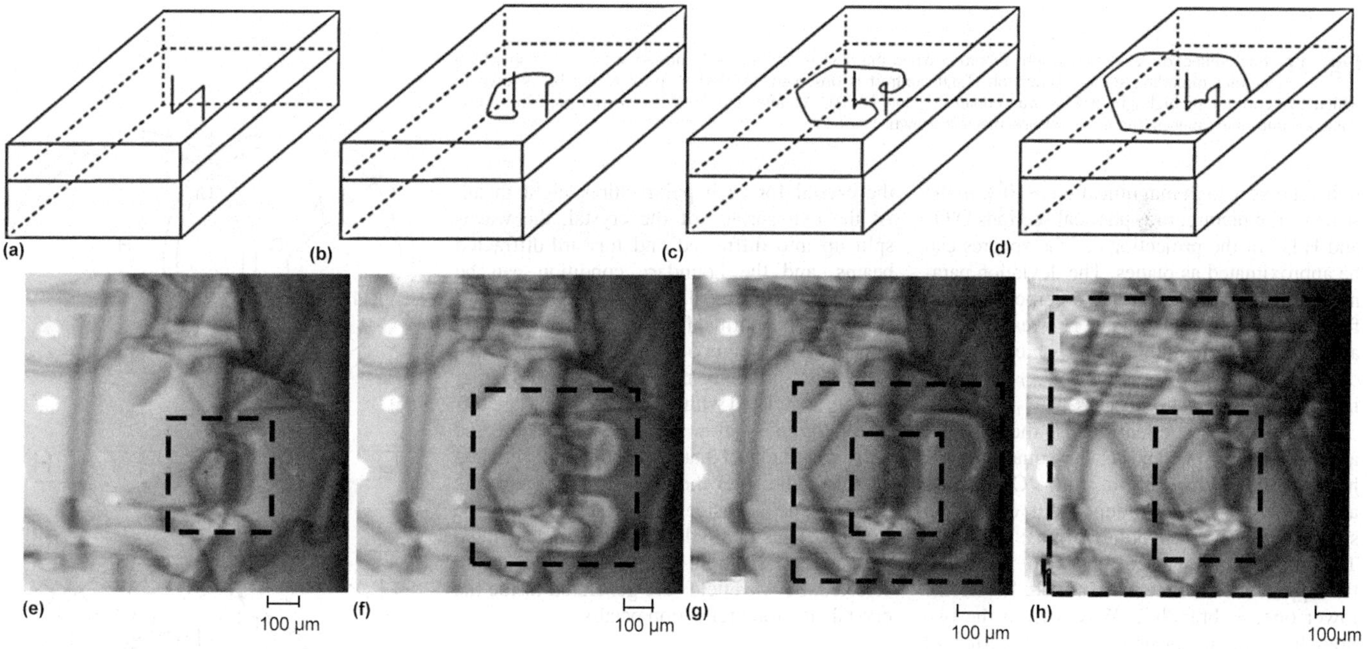

Fig. 30 (a) Schematic of a basal plane dislocation (BPD) segment pinned by threading-edge dislocations (TEDs) formed by deflection of a TED onto the basal plane by a macrostep, followed by redeflection back into the threading direction through the encounter between the macrostep and the TSD spiral advanced in the opposite direction during physical vapor transport growth. (b)–(d) Glide of pinned BPD segment leading to activation of double-ended Frank-Read sources under elevated temperature after growth. (e)–(h) Series of recorded $11\bar{2}0$ transmission x-ray topography images showing the operation of a double-ended Frank-Read source during high-temperature treatment of the 4H-SiC sample. The gliding dislocation segment spirals around the pinned points, resulting in a closed dislocation loop (shown in g), and the regenerated dislocation segment continues to glide to form a second loop (shown in h).

the volume elements can no longer be treated as independent of one another. From wave theory, x-rays diffracted once from an atomic plane experience a phase change of 90°. When these waves become scattered again by the backside of the diffracting planes, they propagate in the same direction as the incident beam but are 180° out of phase. This gives rise to an attenuation of the incident intensity due to destructive interference between the primary incident and the secondary scattered beams, which, in turn, leads to a reduction in the total diffracted beam intensity. This is the so-called primary extinction effect, shown schematically in Fig. 32.

Dynamical Theory of X-Ray Diffraction

The dynamical theory of x-ray diffraction considers the total wavefield inside a crystal while diffraction is taking place as a single entity. The fundamental problem is finding solutions to Maxwell's equations in a periodic medium (i.e., the crystal) matched to solutions that are plane waves (the incident $\vec{k}_0$ and diffracted $\vec{k}_h$ x-ray beams). These solutions must reflect the periodicity of the crystal, and such functions are known as Bloch or lattice functions. The wave equations satisfying Maxwell's equations can be represented geometrically by a construction known as the dispersion surface, illustrated in Fig. 33 (Ref 10). In the kinematical condition, the center of the Ewald sphere is at the Laue point "L," which is at the intersection of spheres of radius k about the origin "O" and reciprocal lattice point "H" in the dispersion plane. In the dynamical condition, the wave vector inside the crystal is corrected for the mean refractive index, resulting in a shorter wave vector, and therefore, the loci of these wave vectors are represented by spheres of diameter $k(1 + \chi/2)$ about "O" and "H." This results in a shift of the intersection of these spheres and the dispersion plane from the Laue point "L" to the Lorentz point "L_0" (Fig. 33a). Figure 33(b) shows the region around the Laue

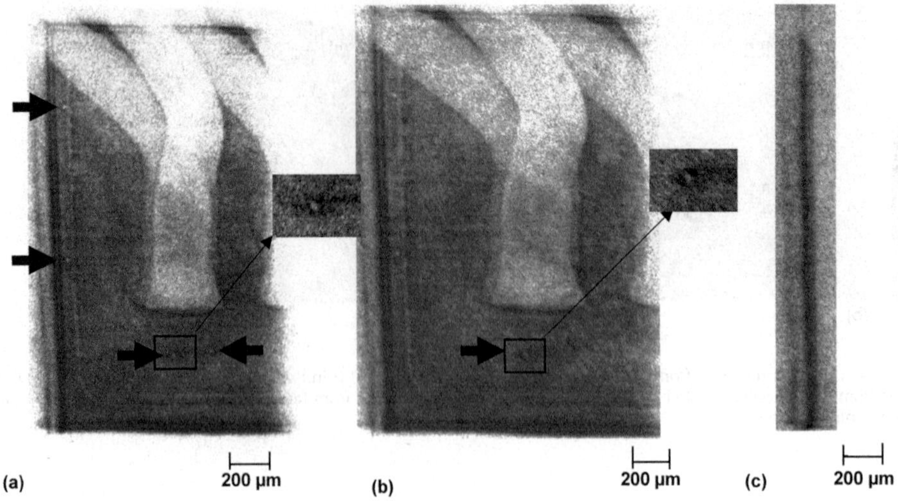

Fig. 31 Back-reflection x-ray topographs recorded when the device was reversely biased at (a) 700 V, in which regular threading screw dislocation (TSD) contrast is observed; (b) 900 V, prior to the breakdown in which the contrast of highlighted TSD was enhanced; and (c) 900 V, after the breakdown in which the diode fractured into many pieces (the image of one piece is shown)

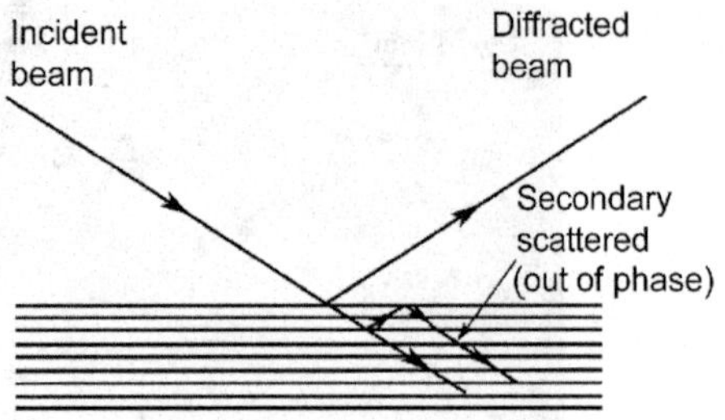

Fig. 32 Schematic diagram demonstrating the phenomenon of primary extinction in a perfect crystal. The diffracted beam (90° out of phase) is scattered by the backside of the diffracting planes to produce a secondary scattered beam that is 180° out of phase with the incident beam, resulting in the attenuation of the intensity of the incident beam, which, in turn, reduces the total diffracted beam intensity.

point at a very high magnification (~10^6). In the scale of the picture, the spherical sections O′O″ and H′H″ of the projections of the spheres can be approximated as planes. The deviation parameters α_0 and α_h are measured perpendicularly from the planes O′O″ and H′H″, respectively, and denote the tie point "A" at which the tails of wave vectors $\vec{K}_0$ from "O" and $\vec{K}_h$ from "H" intersect and diffraction occurs.

The dispersion equation is the loci of all such tie points and is an equation of a hyperboloid of revolution, with "OH" as its axis. There are two independent dispersion surfaces for the two polarizations: σ and π. The dispersion surface has two branches: the upper one, denoted as branch 1, and the lower one, as branch 2. Waves from the two branches are in antiphase. The direction of energy flow is described by the Poynting vector, parallel to $\vec{E} \times \vec{H}$, and it has been shown that this is perpendicular to the dispersion surface at the tie point (Ref 62, 63).

The boundary conditions at the crystal surface require that the tangential components of both $\vec{E}$ and $\vec{H}$ of the wave vectors should be continuous across the surface. These waves must be matched in amplitude at the crystal surface and in phase velocity parallel to the surface. The wave vectors inside the crystal differ from that outside only by a vector normal to the crystal surface, that is, $\vec{K}_0 - \vec{k}_0 = \delta\vec{\mathbf{n}}$, where $\vec{\mathbf{n}}$ is a unit vector normal to the surface, and δ is a scalar variable. A line normal drawn from the tip of the incident wave vector $\vec{k}_0$ intersects the dispersion surface at the excited tie point and determines the tail of the wave vector $\vec{K}_0$. In the Laue case (Fig. 34a), there are two points excited: one on each branch, labeled "A" and "B." From each tie point, wave vectors directed toward "O" and "H" can be generated. There are thus four wave vectors generated in the crystal for each polarization, eight in all. At the exit surface of the crystal, the waves split up into diffracted and forward-diffracted beams, and the boundary condition can be similarly determined. In the Bragg case (Fig. 34b), the normal from the surface intersects either two tie points on the same branch of the dispersion surface or none at all. The Poynting vectors associated with the two tie points are different; the energy flow from one point is directed into the crystal, but that from the other is directed outward. The latter therefore does not generate any wavefields inside the crystal and can be ignored. Thus, a single wavefield is generated for each polarization. When no tie points are selected, no wavefields are generated inside the crystal, and total reflection occurs.

Borrmann Effect

The amplitudes of the wavefields are Bloch functions and are modulated with the periodicity corresponding to the Bragg planes. The maxima and minima of the standing wavefield occur either at or halfway between the atomic planes (Fig. 35). The wavefield with intensity maxima at the atomic planes (branch 2) will suffer greater photoelectric absorption because the electron density is at its maximum at the atomic planes, whereas the branch 1 wavefield with intensity maxima between the atomic planes suffers minimum absorption. This effect is known as anomalous transmission, or the Borrmann effect, and was discovered by Borrmann in calcite (Ref 64, 65). The presence of the Borrmann effect is indicative of crystal perfection.

Pendellösung Effect

An incident plane wave excites two tie points on the dispersion surface and generates two Bloch wavefields. The wave vectors

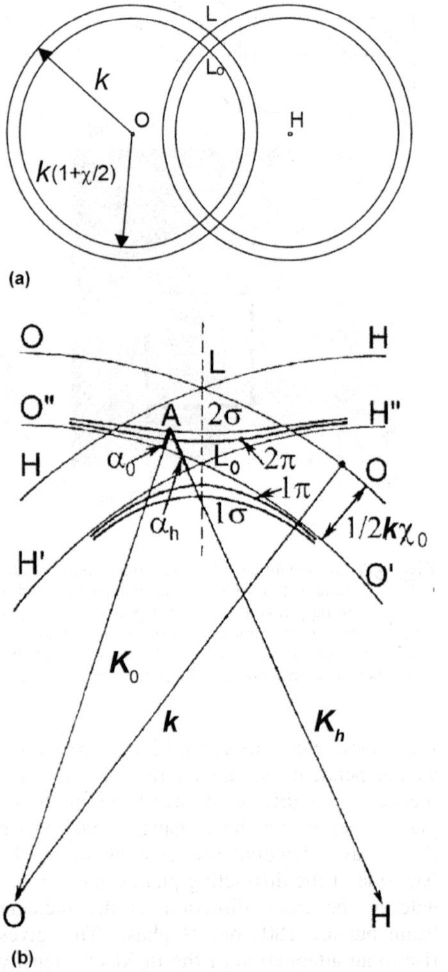

Fig. 33 Construction of the dispersion surface. (a) Spheres of radius k and $k(1 + \chi/2)$ about the origin "O" and reciprocal lattice point "H" in reciprocal space showing the position of the Laue point "L" and Lorentz point "L_0." (b) Dispersion surface for σ and π polarization states

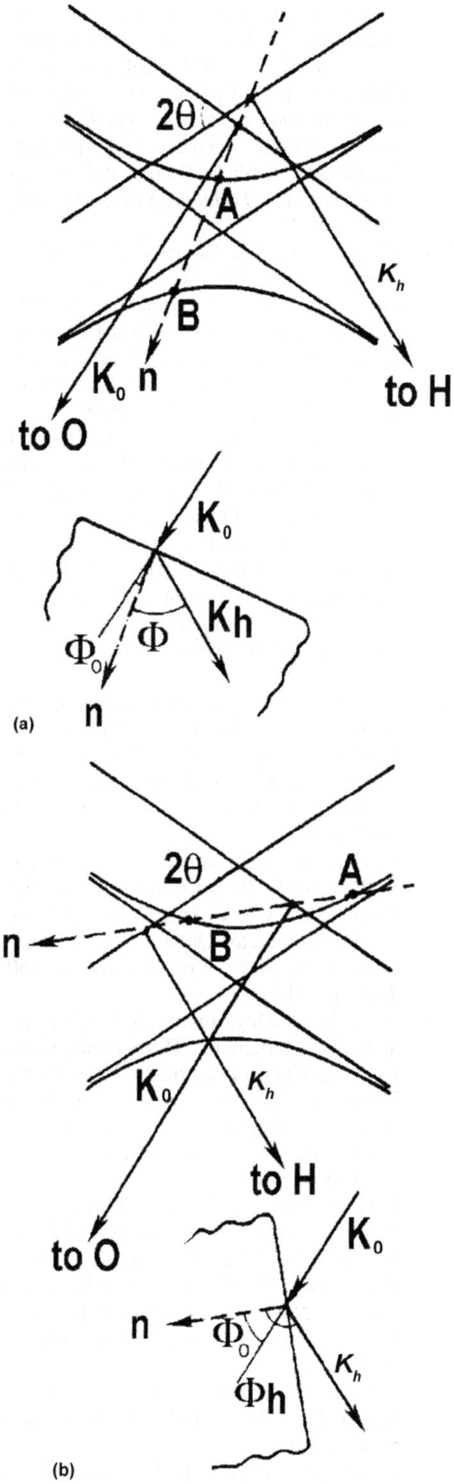

Fig. 34 Dispersion surface construction showing the tie points excited by an incident wave in (a) the Laue geometry, where one tie point in each branch ("A" on branch 1 and "B" on branch 2) is excited, and (b) the Bragg geometry, where two tie points ("A" and "B") on the same branch are excited

associated with these tie points differ. The difference in wave vector leads to a difference in the propagation velocity, and interference effects can occur between the Bloch

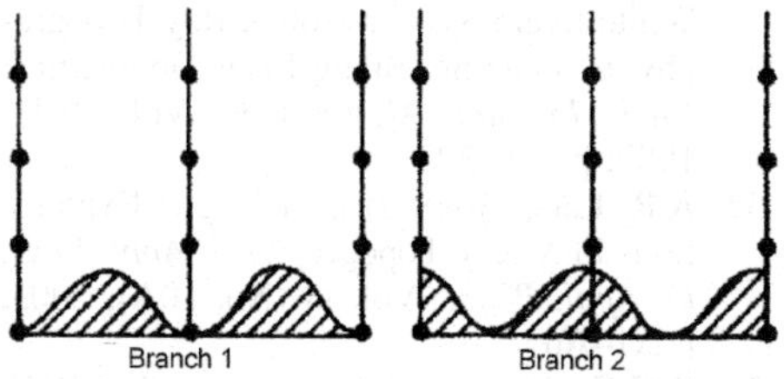

Fig. 35 Standing wavefields with a period corresponding to the spacing between the Bragg planes produced at the exact Bragg condition. Branch 1 waves, which have a minimum intensity at the atomic positions, suffer minimal absorption, while branch 2 waves, which have a maximum intensity at the atomic positions, are strongly absorbed because of maximum electron density at the atomic planes.

waves. This gives rise to the production of beats, a phenomenon referred to as the Pendellösung effect. The period of the beats is given by the extinction distance, ξ_g, which is the reciprocal of the dispersion surface diameter, d_h, for the case of exact fulfillment of the Bragg condition. The aforementioned results can be extended to cover asymmetric reflections by the introduction of the terms γ_0 and γ_h, the cosines of the angles between the surface normal and the incident and diffracted beams, respectively, in the appropriate places.

The previous analysis applies to an incident plane wave. However, in practice, x-ray sources have a significant angular divergence because the entire dispersion surface is excited simultaneously. This gives rise to energy flow within the Borrmann fan bounded by the incident (*AB*) and diffracted (*AC*) beam directions (Fig. 36). Defects at any point within the Borrmann fan may contribute to the change of diffracted intensity at the exit surface of the crystal and thus to image contrast.

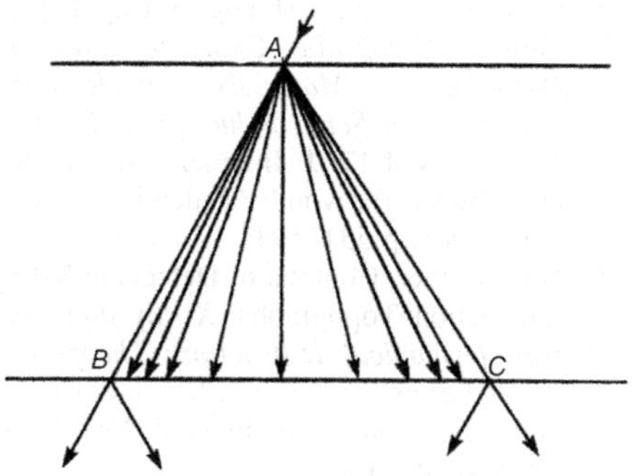

Fig. 36 Borrmann fan bounded by the incident (*AB*) and diffracted (*AC*) beams showing the distribution of energy for an incident spherical wave that excites all tie points along the dispersion surface

ACKNOWLEDGMENT

Revised from Robert N. Pangborn, X-Ray Topography, *Materials Characterization*, Vol 10, *ASM Handbook*, ASM International, 1986.

REFERENCES

1. V.W. Berg, Uber eine Rontgenographische Methode zur Untersuchung von Gitterstorungen an Kristallen, *Naturwissenschaften*, Vol 19, 1931, p 391–396
2. C.S. Barrett, A New Microscopy and Its Potentialities, *Trans. Am. Inst. Min. Metall. Eng.*, Vol 161, 1945, p 15–65
3. W. Wooster and W.A. Wooster, X-Ray Topographs, *Nature*, Vol 155, 1945, p 786–787
4. J.B. Newkirk, Method for the Detection of Dislocations in Silicon by X-Ray Extinction Contrast, *Phys. Rev.*, Vol 110, 1958, p 1465–1466
5. L.G. Schulz, Method of Using a Fine-Focus X-Ray Tube for Examining the Surfaces of Single Crystals, *J. Met.*, Vol 6, 1954, p 1082–1083
6. W.L. Bond and J. Andrus, Structural Imperfections in Quartz Crystals, *Am. Mineral.*, Vol 37, 1952, p 622–632
7. A.R. Lang, Direct Observation of Individual Dislocations by X-Ray Diffraction, *J. Appl. Phys.*, Vol 29, 1958, p 597–598
8. G.H. Schwuttke, New X-Ray Diffraction Microscopy Technique for Study of Imperfections in Semiconductor Crystals, *J. Appl. Phys.*, Vol 36, 1965, p 2712–2714
9. T. Tuomi, K. Naukkarinen, and P. Rabe, Use of Synchrotron Radiation in X-Ray-Diffraction Topography, *Phys. Status Solidi (a)*, Vol 25, 1974, p 93–106
10. B.W. Batterman and H. Cole, Dynamical Diffraction of X-Rays by Perfect Crystals, *Rev. Mod. Phys.*, Vol 36, 1964, p 681–717
11. B.K. Tanner, *X-Ray Diffraction Topography*, Pergamon Press, Oxford, 1976
12. D.K. Bowen and B.K. Tanner, *High Resolution X-Ray Diffractometry and Topography*, Taylor & Francis, London, 1998
13. X.R. Huang, M. Dudley, W.M. Vetter, W. Huang, S. Wang, and C.H. Carter, Jr., *Appl. Phys. Lett.*, Vol 74, 1999, p 353
14. Y. Chen and M. Dudley, Direct Determination of Dislocation Sense of Closed-Core Threading Screw Dislocations Using Synchrotron White Beam X-Ray Topography in 4H Silicon Carbide, *Appl. Phys. Lett.*, Vol 91, 2007, p 141918
15. I. Kamata, M. Nagano, H. Tsuchida, Y. Chen, and M. Dudley, High-Resolution Topography Analysis on Threading Edge Dislocations in 4H-SiC Epilayers, *Mater. Sci. Forum*, Vol 600–603, 2009, p 305–308
16. Y. Chen, M. Dudley, K.X. Liu, J.D. Caldwell, and R.E. Stahlbush, Synchrotron X-Ray Topographic Studies of Recombination Activated Shockley Partial Dislocations in 4H-SiC Epitaxial Layers, *Mater. Sci. Forum*, Vol 600–603, 2009, p 357–360
17. M. Dudley, Y. Chen, and X.R. Huang, Aspects of Dislocation Behavior in SiC, *Mater. Sci. Forum*, Vol 600–603, 2009, p 261–266

18. M. Dudley and X. Huang, X-Ray Topography, *Microprobe Characterization of Optoelectronic Materials (Optoelectronic Properties of Semiconductors and Superlattices)*, Vol 17, J. Jimenez, Ed., Gordon and Breach/Harwood Academic, Amsterdam, 2003, p 531–594
19. B.K. Tanner, Contrast of Defects in X-Ray Diffraction Topographs, *X-Ray and Neutron Dynamical Diffraction: Theory and Applications*, A. Authier, S. Lagomarsino, and B.K. Tanner, Ed., Plenum, New York, 1996, p 147–166
20. D.K. Bowen and B.K. Tanner, *High Resolution X-Ray Diffractometry and Topography*, Taylor & Francis, London, 1998, p 174
21. A.R. Lang, Point-by-Point X-Ray Diffraction Studies of Imperfections in Melt-Grown Crystals, *Acta Crystallogr.*, Vol 10, 1957, p 839
22. U. Bonse, *Z. Phys.*, Vol 153, 1958, p 278
23. R. Kohler, High Resolution X-Ray Topography, *Appl. Phys. A*, Vol 58, 1994, p 149–157
24. B. Jenichen, R. Kohler, and W. Mohling, Double Crystal Topography Compensating for the Strain in Processed Samples, *Phys. Status Solidi (a)*, Vol 89, 1985, p 79–87
25. J. Miltat, White Beam Synchrotron Radiation, *Characterization of Crystal Growth Defects by X-Ray Methods*, B.K. Tanner and D.K. Bowen, Ed., Plenum Press, New York, 1980, p 401–420
26. S. Stoupin, B. Raghothamachar, M. Dudley, Z. Liu, E. Trakhtenberg, K. Lang, K. Goetze, J. Sullivan, and A. Macrander, Projection X-Ray Topography System at 1-BM X-Ray Optics Test Beamline at the Advanced Photon Source, *AIP Conf. Proc.*, Vol 1741, 2016, p 050018-1 to 050018-4, doi: 10.1063/1.4952938
27. P.G. Neudeck, Electrical Impact of SiC Structural Defects on High Electric Field Devices, *Mater. Sci. Forum*, Vol 338–342, 2000, p 1161–1166
28. P.G. Neudeck, W. Huang, and M. Dudley, Breakdown Degradation Associated with Elementary Screw Dislocations in 4H-SiC P+N Junction Rectifiers, *Symp. E: Power Semiconductor Materials and Devices, Mater. Res. Soc. Symp. Proc.*, Vol 483, S.J. Pearton et al., Ed., Materials Research Society, Warrendale, PA, 1998, p 285–294
29. D.R. Black and G.G. Long, "X-Ray Topography," Special Publication 0960-10, National Institute of Standards and Technology, 2004
30. M. Dudley, S. Wang, W. Huang, C.H. Carter, Jr., V.F. Tsvetkov, and C. Fazi, White Beam Synchrotron Topographic Studies of Defects in 6H-SiC Single Crystals, *J. Phys. D: Appl. Phys.*, Vol 28, 1995, p A63–A68
31. M. Dudley, J. Wu, and G.-D. Yao, Determination of Penetration Depths and Analysis of Strains in Single Crystals by White Beam Synchrotron X-Ray Topography in Grazing Bragg-Laue Geometries, *Nucl. Instrum. Methods B*, Vol 40/41, 1989, p 388–392
32. A.R. Lang, Some Bristol-Prague Explorations in X-Ray Topography, *J. Appl. Phys. D, Appl. Phys.*, Vol 38 (No. 10A), 2005, p A1–A6
33. R.G.C. Arridge, A.R. Lang, and A.P.W. Makepeace, Elastic Deformation in a Crystal Plate where Lattice-Parameter Mismatch is Present between Adjacent Growth Sectors, Part II: Measurement of Lattice Tilts by Synchrotron X-Ray Reticulography, *Proc. R. Soc. (London) A, Math. Phys. Eng. Sci.*, Vol 458, 2002, p 2623–2643
34. R.G.C. Arridge, A.R. Lang, and A.P.W. Makepeace, Elastic Deformation in a Crystal Plate where Lattice-Parameter Mismatch is Present between Adjacent Growth Sectors, Part I: Anisotropic Elasticity Theory, with Application to Lattice-Parameter Measurements, *Proc. R. Soc. (London) A, Math. Phys. Eng. Sci.*, Vol 458, 2002, p 2485–2521
35. A.R. Lang and A.P.W. Makepeace, Synchrotron X-Ray Reticulographic Measurement of Lattice Deformations Associated with Energetic Ion Implantation in Diamond, *J. Appl. Crystallogr.*, Vol 32 (No. 6), 1999, p 1119–1126
36. A.R. Lang and A.P.W. Makepeace, Synchrotron X-Ray Reticulography: Principles and Applications, *J. Phys. D: Appl. Phys.*, Vol 32 (No. 10A), 1999, p A97–A103
37. A.R. Lang and A.P.W. Makepeace, Synchrotron X-Ray Reticulography: A Versatile New Technique for Mapping Misorientations in Single Crystals, *Microsc. Semicond. Mater. (IOP Conf. Ser.)*, Vol 157, 1997, p 457–460
38. B. Raghothamachar, V. Sarkar, V. Noveski, M. Dudley, and S. Sharan, A Novel X-Ray Diffraction-Based Technique for Complete Stress State Mapping of Packaged Silicon Dies, *MRS Symp. Proc.*, Vol 1158E, 2009, p 1158-F01-07
39. S. Kikuta, K. Kohra, and Y. Sugita, Measurements on Local Variations in Spacing and Orientation of the Lattice Plane of Silicon Single Crystals by X-Ray Double-Crystal Topography, *Jpn. J. Appl. Phys.*, Vol 5 (No. 11), 1966, p 1047
40. S.J. Barnett, B.K. Tanner, and G.T. Brown, Investigation of the Homogeneity and Defect Structure in Semi-Insulating LEC GaAs Single Crystals by Synchrotron Radiation Double Crystal X-Ray Topography, *Mater. Res. Soc. Symp. Proc.*, Vol 41, 1985, p 83
41. C. Ferrari, D. Korytar, and J. Kumar, Study of Residual Strains in Wafer Crystals by Means of Lattice Tilt Mapping, *Il Nuovo Cimento D*, Vol 19 (No. 2–4), 1997, p 165
42. S.R. Stock, H. Chen, and H.K. Birnbaum, The Measurement of Strain Fields by X-Ray Topographic Contour Mapping, *Philos. Mag. A*, Vol 53 (No. 1), 1986, p 73
43. D.J. Larson, Jr., R.P. Silberstein, D. DiMarzio, F.C. Carlson, D. Gillies, G. Long, M. Dudley, and J. Wu, Compositional, Strain Contour and Property Mapping of CdZnTe Boules and Wafers, *Semicond. Sci. Technol.*, Vol 8 (No. 6S), 1993, p 911
44. M. Jackson, M.S. Goorsky, A. Noori, S. Hayashi, R. Sandhu, B. Poust, P. Chang-Chien, A. Gutierrez-Aitken, and R. Tsai, Determination of Stress Distribution in III-V Single Crystal Layers for Heterogeneous Integration Applications, *Phys. Status Solidi (a)*, Vol 204 (No. 8), 2007, p 2675
45. A.T. Macrander, S. Krasnicki, Y. Zhong, J. Maj, and Y. Chu, Strain Mapping with Parts-per-Million Resolution in Synthetic Type-Ib Diamond Plates, *Appl. Phys. Lett.*, Vol 87, 2005, p 194113
46. G. Yang, R. Jones, F. Klein, K. Finkelstein, and K. Livingston, Rocking Curve Imaging for Diamond Radiator Crystal Selection, *Diam. Relat. Mater.*, Vol 19, 2010, p 719
47. J. Guo, Y. Yang, B. Raghothamachar, M. Dudley, and S. Stoupin, Mapping of Lattice Strain in 4H-SiC Crystals by Synchrotron Double-Crystal X-Ray Topography, *J. Electron. Mater.*, Vol 47 (No. 2), 2017, p 903
48. D. Lubbert, T. Baumbach, J. Hartwig, E. Boller, and E. Pernot, μm-Resolved High Resolution X-Ray Diffraction Imaging for Semiconductor Quality Control, *Nucl. Instrum. Methods Phys. Res. B*, Vol 160, 2000, p 521–527
49. A. Philip, J. Meyssonnier, R.T. Kluender, and J. Baruchel, Three-Dimensional Rocking Curve Imaging to Measure the Effective Distortion in the Neighbourhood of a Defect within a Crystal: An Ice Example, *J. Appl. Cryst.*, Vol 46, 2013, p 842–848
50. R.T. Kluender, A. Philip, J. Meyssonnier, and J. Baruchel, *Phys. Status Solidi (a)*, Vol 208 (No. 11), 2011, p 2505–2510
51. D.K. Bowen, M. Wormington, and P. Feichtinger, A Novel Digital X-Ray Topography System, *J. Phys. D: Appl. Phys.*, Vol 36, 2003, p A17–A23
52. "X-Ray Topography Imaging System," Rigaku Corp., https://rigaku.com/en/products/xrm/xrtmicron
53. H. Klapper, Defects in Non-Metal Crystals, *Characterization of Crystal Growth Defects by X-Ray Methods*, B.K. Tanner and D.K. Bowen, Ed., Plenum Press, New York, 1980, p 133–160
54. D. Yuan and M. Dudley, Dislocation Line Direction Determination in Pyrene Single Crystals, *Mol. Cryst. Liq. Cryst.*, Vol 211, 1992, p 51–58
55. J. Miltat and M. Dudley, Projective Properties of Laue Topographs, *J. Appl. Crystallogr.*, Vol 13, 1980, p 555–562

56. X.R. Huang, D.R. Black, A.T. Macrander, J. Maj, Y. Chen, and M. Dudley, High-Geometrical-Resolution Imaging of Dislocations in SiC Using Monochromatic Synchrotron Topography, *Appl. Phys. Lett.*, Vol 91, 2007, p 231903
57. Y. Chen, G. Dhanaraj, M. Dudley, E.K. Sanchez, and M.F. MacMillan, Sense Determination of Micropipes via Grazing-Incidence Synchrotron White Beam X-Ray Topography in 4H Silicon Carbide, *Appl. Phys. Lett.*, Vol 91, 2007, p 071917
58. E.S. Meieran and I.A. Blech, X-Ray Extinction Contrast Topography of Silicon Strained by Thin Surface Films, *J. Appl. Phys.*, Vol 36, 1965, p 3162
59. M. Dudley, F. Wu, H. Wang, S. Byrappa, B. Raghothamachar, G. Choi, S. Sun, E.K. Sanchez, D. Hansen, R. Drachev, S.G. Mueller, and M.J. Loboda, Stacking Faults Created by the Combined Deflection of Threading Dislocations of Burgers Vector *c* and *c+a* during the Physical Vapor Transport Growth of 4H-SiC, *Appl. Phys. Lett.*, Vol 98, 2011, p 232110
60. Y. Yang, J. Guo, B. Raghothamachar, M. Dudley, S. Weit, A.N. Danilewsky, P.J. McNally, and B.R. Tanner, In Situ Synchrotron X-Ray Topography Observation of Double-Ended Frank-Read Sources in PVT-Grown 4H-SiC Wafers, *Mater. Sci. Forum,* Vol 924, 2018, p 172–175
61. K. Shenai, A. Christou, M. Dudley, B. Raghothamachar, and R. Singh, Crystal Defects in Wide Bandgap Semiconductors, *ECS Trans.,* Vol 61 (No. 4), 2014, p 283–293
62. N. Kato, The Flow of X-Rays and Materials Waves in Ideally Perfect Single Crystals, *Acta Crystallogr.,* Vol 11, 1958, p 885–887
63. N. Kato, The Energy Flow of X-Rays in an Ideally Perfect Crystal: Comparison between Theory and Experiments, *Acta Crystallogr.,* Vol 13, 1960, p 349–356
64. G. Borrmann, The Extinction Diagram of Quartz, *Physik. Z.,* Vol 42, 1941, p 157–162
65. G. Borrmann, Absorption of Röntgen Rays in the Case of Interference, *Physik. Z.,* Vol 127, 1950, p 297–323

ASM Handbook, Volume 10, *Materials Characterization*
ASM Handbook Committee
DOI 10.31399/asm.hb.v10.a0006654

Synchrotron X-Ray Diffraction Applications

Wenqian Xu, Saul H. Lapidus, Andrey Y. Yakovenko, Youngchang Kim, Olaf J. Borkiewicz, and Kamila M. Wiaderek, Argonne National Laboratory
Yu-Sheng Chen, The University of Chicago
Tiffany L. Kinnibrugh, DeNovX

Overview

General Uses

- Determination of crystal structures
- Identification and/or quantification of phases in a multiphase sample (powder diffraction)
- Charge-density analysis (single-crystal diffraction)
- Characterization of crystallite size, strain, and texture in polycrystalline materials (powder diffraction)
- Characterization of structural changes as a function of temperature, pressure, and other environmental parameters

Examples of Applications

- Determination of structures of proteins, macromolecule assemblies, protein complexes with nucleic acids and small-molecule ligands, and studying structural basis of mechanisms of molecular recognition and regulation in atomic details
- Determination of crystal structure of an inorganic solid solution, resolving the percentage of substituting ions at a crystallographic site
- Quantitative analysis of lattice contraction or expansion due to an external force or other physical or chemical processes of interest

Samples

- *Form:* Single crystal, polycrystalline solid, powder, nanomaterial, amorphous solid
- *Size:* 1 to 300 μm edge length for single crystals; 1 to 10 mm^3 volume for powder sample

Limitations

- Available at only approximately 50 synchrotron facilities around the world

Estimated Analysis Time

- *Acquisition:* 3 to 30 min for single-crystal x-ray diffraction (XRD); 10 to 60 min for high-resolution powder XRD. For powder XRD with area detector, data-acquisition time varies from a few minutes down to the subsecond level.
- *Analysis:* From a few hours to several weeks, depending on the nature of the sample and the target information

Capabilities of Related Techniques

- *Neutron diffraction:* Highly complementary; comparably more sensitive to low-*Z* atoms, for example, hydrogen and oxygen. Sensitive to magnetic structure
- *Transmission electron microscopy:* Imaging; qualitative analysis of lattice features and particle size
- *Cryogenic electron microscopy:* Determining structures of biomolecules in solution at cryogenic temperatures; does not require crystallization
- *X-ray absorption spectroscopy:* Probing local atomic arrangement surrounding atoms of a specific element.

Introduction

X-ray diffraction (XRD) is a powerful technique for characterizing structures of materials to the atomic resolution level. A major advantage of synchrotron XRD over its laboratory counterpart is the super brightness of the synchrotron radiation source, which is many orders of magnitude higher than that of a contemporary laboratory x-ray source (Ref 1–3). The brightness of an x-ray source is not only related to the total flux of photons it emits but is also dependent on the angular divergence of the radiation. Inside a laboratory x-ray tube, x-rays are generated in all directions from the metal anode, and only a small fraction come through the beryllium window of the tube. This fraction of x-rays is too divergent to be directly used for any diffraction measurements. Laboratory diffractometers have presample slits or optics to trim and collimate the radiation, which significantly reduces the flux of photons on the sample. Synchrotron x-rays, on the other hand, are highly collimated in nature, due to the relativistic effects, and have an angular divergence of less than a few milliradians. This small divergence leads to much less photon loss in the beam-manipulation process and a relatively larger portion of the beam delivered to the sample. The high brightness of synchrotron x-rays results in excellent counting statistics on the detector within a short period of time, enabling fast collection of high signal-to-noise data inconceivable on a laboratory instrument. This really paves the way for high-throughput measurement of large numbers of samples and also for in situ experiments where samples are continuously monitored by diffraction to capture structural changes as a function of time or other stimuli, for example, temperature, pressure, magnetic field, gas or liquid dopants, and so on. In addition, the low divergence makes the beam easy to focus to a small spot size of a few micrometers with more concentrated photon density. This in turn reduces the minimum sample volume required for adequate data quality and increases the spatial resolution in examination of complex or inhomogeneous materials. The microfocusing is particularly helpful for solving structures of weak scattering microcrystals that are difficult or impossible to be synthesized to a larger size (Ref 4, 5). Figure 1 compares diffraction images of such a microcrystal collected with both a laboratory source and a synchrotron source.

Other than the brightness, another advantage of synchrotron radiation (SR) is its wide and continuous spectrum of x-ray photon energies, which allows easy energy turning for a particular experiment, for example, selecting an energy to avoid x-ray fluorescence or absorption edges by some elements contained in a sample, measuring at several energies across an absorption edge to use the anomalous scattering effect, and so on. Such flexibility does not exist on a laboratory diffractometer, where the working photon energy is fixed to the characteristic Kα radiation of the installed metal anode. In fact, due to the heat load issue, only a few metal anodes are commercially available, including copper, molybdenum, silver, chromium, iron, and cobalt. In addition, the wide energy range of SR extends to well beyond 100 kiloelectron volts (keV), equal to a wavelength of less than 0.1 Å. The high-energy portion (>30 keV) of the synchrotron x-ray is invaluable for research on dense materials such as metals and ceramics, which require probing below the surface of the sample (Ref 6), and for obtaining ultra-high-resolution protein crystal structure (Ref 7). No laboratory source can produce such high-energy x-rays.

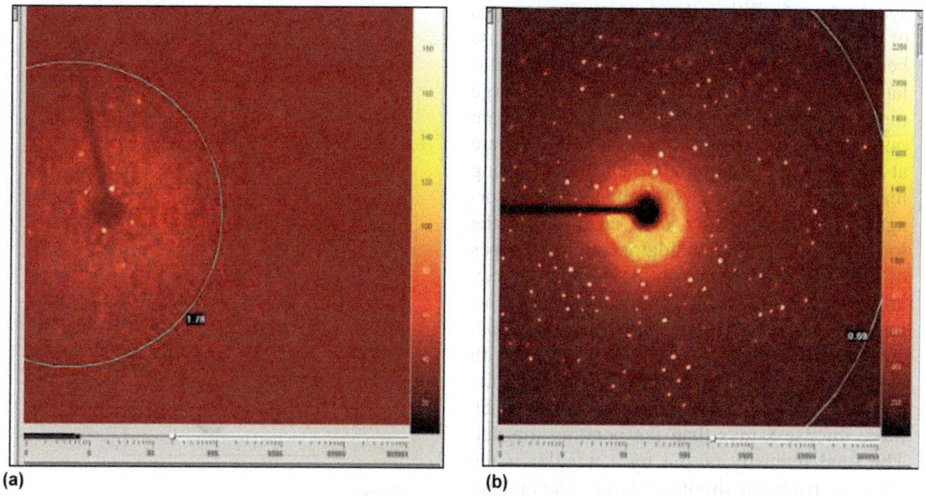

Fig. 1 The same crystal measured by using (a) a laboratory-based molybdenum source (0.71073 Å) with an exposure time of 120 s and (b) the ChemMatCARS beamline with a wavelength of 0.41328 Å and an exposure time of 0.2 s

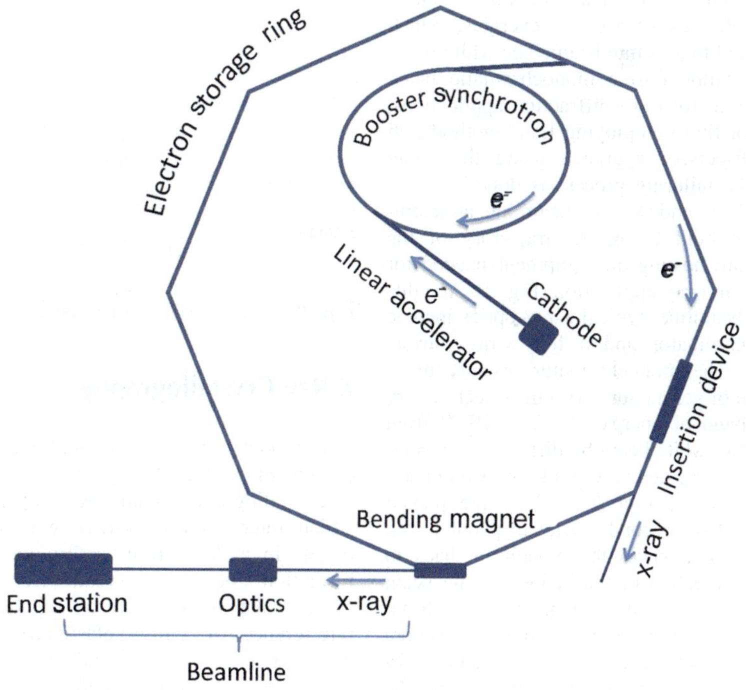

Fig. 2 Schematic diagram of a synchrotron light source

X-Ray Generation and Monochromation

A synchrotron light source is a combination of linear and circular accelerators where electrons produced from a cathode are accelerated by high-voltage alternating electric fields that eventually travel at close to the speed of light (>99.99% light speed) in a closed loop, called an electron storage ring (Fig. 2). Along the storage ring are multiple magnetic and electronic fields for navigating electrons in the orbit while maintaining their speed. Electromagnetic radiation is emitted when electrons are forced to travel in a curved path in the magnetic fields, and the radiation is linearly polarized in the orbit (horizontal) plane. The majority of the emitted radiation falls into the energy range of x-rays, spanning from a few hundred electron volts (eV) to over several hundred kiloelectron volts. At third-generation synchrotron facilities, the three common types of electromagnets for producing x-rays are bending magnet, multipole wiggler, and undulator (Fig. 3). Bending magnets are large dipole magnets that bend the electron beam only once, so that the outgoing electrons are at an angle to the incoming electrons. Bending magnets are necessary components of the storage ring, because they keep electrons in the loop. Wigglers and undulators are two types of the so-called insertion devices, and they are placed in the straight sections of the storage ring between bending magnets. They both consist of an array of dipole magnets with alternating magnetic fields that deflect the electron beam multiple times, producing much more intense x-rays than the bending magnet. The total intensity output by a wiggler is the intensity from one dipole scaled by the number of dipoles (N) in the wiggler. For an undulator, x-rays emitted from each dipole interfere with rays emitted from other dipoles; as a result, the total intensity is scaled by the square of the dipole number (N^2), making the undulator an even brighter source than the wiggler.

X-rays emitted from a synchrotron source contain photons of various energies, which are referred to as white beam. The white beam must be tailored to a monochromatic beam (monobeam) for most diffraction applications, except for those employing Laue methods, an energy-dispersive approach using the white beam. The tailoring process is done by x-ray optics, filters, and slits at a beamline, an instrumentation built along the trajectory of the x-ray beam hosting all equipment needed for a particular x-ray application (Fig. 2). At a diffraction beamline, typical x-ray optics include a monochromator and a few x-ray mirrors (Fig. 4). The monochromator, usually made of silicon or germanium crystals, selects a very narrow band of energy ($\Delta E/E \approx 10^{-4}$) from the incident white beam by diffraction. A common type is the double-crystal monochromator, in which the first crystal selects the photon energy and the second crystal in parallel with the first one bounces the monobeam back to the same direction of the incident white beam with a vertical offset so that the white beam does not contaminate the monobeam. Before the monochromator, a vertical collimating mirror often is used as the first optic facing the source. With its surface slightly curved along the beam direction, the mirror reflects the initial white beam at a grazing angle of incidence to a vertically parallel white beam. Collimating the white beam boosts the flux of photons selected by the monochromator. Additional mirrors are placed after the monochromator to focus the monobeam in both the horizontal and vertical directions. These mirrors, along with beam-cutting slits placed at various locations along the beam path, allow one to adjust the size and shape of the beam footprint on the sample. For general XRD applications, the beam spot size usually is set to several micrometers to a few hundred micrometers.

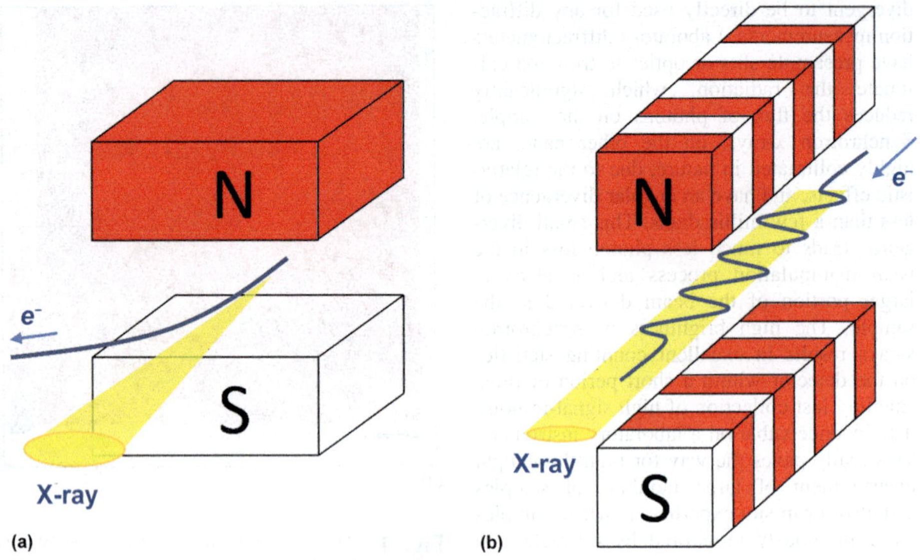

Fig. 3 Schematics of (a) a bending magnet and (b) an undulator

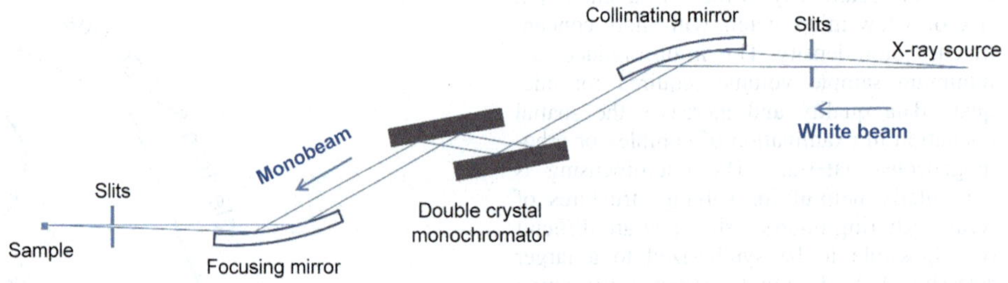

Fig. 4 Example layout of x-ray optics at a diffraction beamline

X-Ray Crystallography

The power of x-ray crystallography was evident more than 100 years ago. The core of crystallography is an XRD experiment in which interaction of x-rays with a crystal results in a diffraction pattern carrying information about the crystal structure. The diffraction pattern is a result of constructive interference of x-rays elastically scattered by electrons in the crystal. Each electron scatters a tiny portion of the incident x-ray to all directions in space. Scattered waves from all the electrons superimpose in space and in most directions, resulting in zero-summed amplitude due to destructive interference because of the periodic arrangement of atoms in the crystal. The other directions where there are non-zero-net amplitudes are scattered in space and are seen as spots on an area x-ray detector (Fig. 1). The spots are called reflections or peaks, and their occurrence and directions in space are according to Bragg's law, which is detailed in the next paragraph. The peaks have different intensities, reflecting the electron distribution (how atoms are arranged) in the unit cell. Structure determination by XRD relies on the directional and intensity information of the reflections to reconstruct the crystal structure.

Bragg's law was derived by William Lawrence Bragg in 1912 to explain when diffraction would take place (Ref 8). It refers to Eq 1:

$$n\lambda = 2d\sin\theta \qquad \text{(Eq 1)}$$

where n is a positive integer, λ is the x-ray wavelength, d is the intraplanar distance of a set of crystal planes, and θ is the angle between the incident or diffracted x-ray beam and the corresponding crystal planes (Fig. 5). Crystal planes are a set of abstract parallel planes that are equally spaced going through the whole crystal. Atoms can be related to the planes, either in them or between them. Figure 5 shows a simple scenario where all atoms are in the planes. X-rays scattered by the atoms in the same plane have the same path length in the specified incoming and outgoing directions and therefore are in constructive interference. The x-rays scattered by the atoms in the next plane travel in a longer path, and the extra distance is equal to $2d\sin\theta$. For scattered x-rays from both planes to constructively interfere, the extra path length should be equal to the wavelength or its multiples, which is exactly the condition specified in Eq 1. Because the planes are evenly spaced, the scattered x-rays from all of the planes will have constructive interference at the same θ angles, at which point x-rays appear to be "reflected" by the planes. This is why diffraction peaks also are called reflections. The concept of crystal planes and the analogy of diffraction to reflection, which is also proposed by W.L. Bragg, is a very intuitive way of understanding diffraction.

Crystal planes are identified by Miller indices to indicate their orientations relative to the crystal lattice. A Miller index is a set of three integer numbers, *h*, *k*, and *l*, which are the reciprocals of the fractional intercepts of the plane on the three crystallographic axes. For a crystal lattice with three axis lengths of *a*, *b*, and *c*, the (*hkl*) crystal planes cross the three axes at distances of $m \cdot a/h$, $m \cdot b/k$, and $m \cdot c/l$, respectively, where *m* represents all integers. Figure 6 shows an example of the (232) planes intersecting the lattice axes. If there are zero numbers among *h*, *k*, and *l*, the planes are parallel to the corresponding axes. For example, (012) planes are parallel to the *a* axis, and (200) planes are parallel to the *bc* plane. Other than indicating orientation, the Miller index also relates interplanar spacing of crystal planes, also known as *d*-spacing, to the lengths and angles of lattice axes, or lattice parameters. Therefore, lattice parameters can be retrieved from a series of *d*-spacings with known Miller indices. Experimentally, *d*-spacings are calculated from the θ angles of the observed reflections by using Eq 1 rewritten as:

$$d_{hkl} = \frac{\lambda}{2\sin\theta} \quad \text{(Eq 2)}$$

Note that *n* in Eq 1 is now set to unity, which means a set of (*hkl*) planes is associated with only one reflection. Higher-order ($n \geq 2$) reflections can be viewed as being reflected from higher-order crystal planes, (*nh nk nl*), which are in the same orientation as the (*hkl*) planes but with closer *d*-spacings. While *d*-spacings can be readily obtained from XRD data, the Miller indices are not immediately known. However, given enough numbers of observed reflections, the Miller indices can be assigned in the same process of obtaining lattice parameters. The process is known as indexing, which usually is the first step involved in solving unknown structures.

Bragg's law predicts where to find reflections but provides no information on reflection intensities. The intensities are reflective of how electrons are distributed in the unit cell. In other words, atom types, positions, and vibrations affect reflection intensities. It is important to keep in mind that the intensity of a single reflection is contributed by all of the

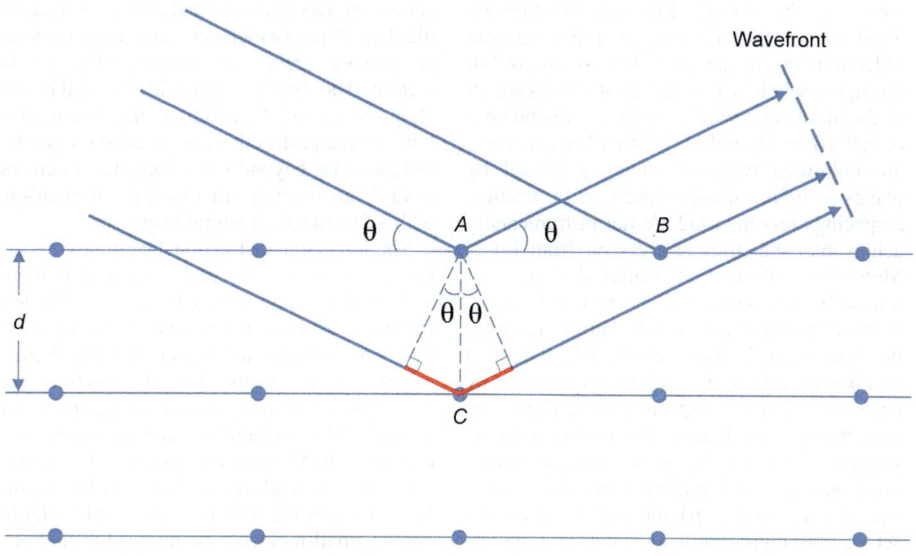

Fig. 5 Illustration of Bragg's law. The path in red marks the extra length the x-rays go through when scattered by the second plane of atoms.

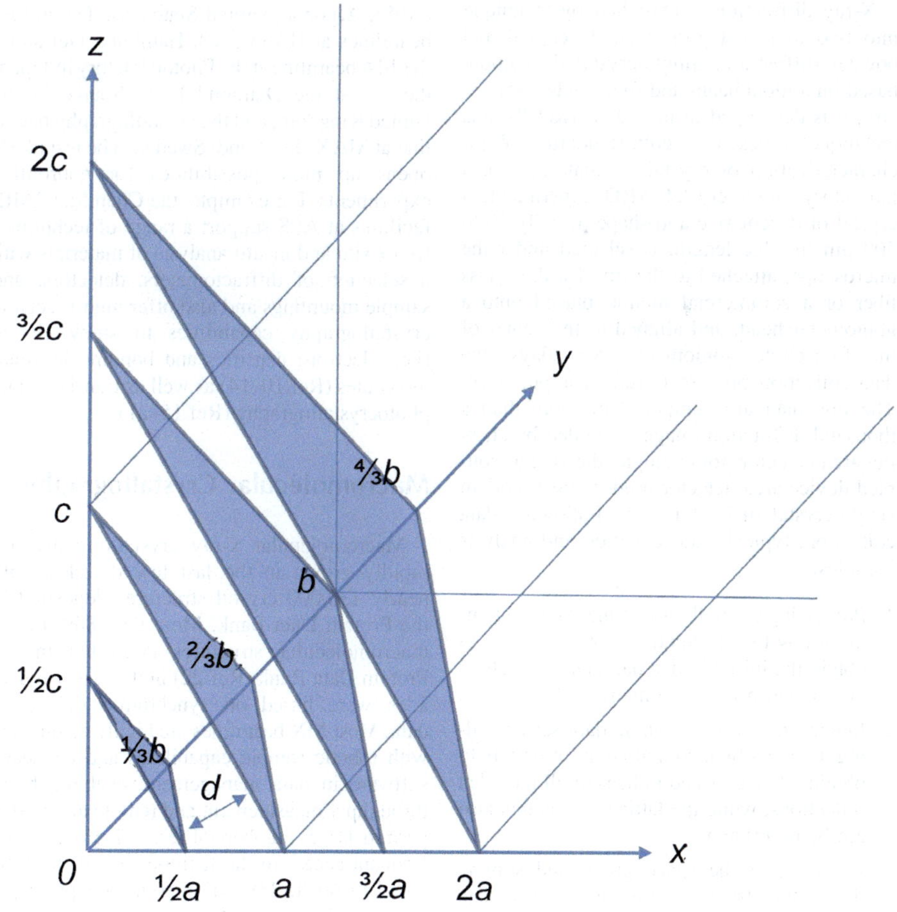

Fig. 6 Illustration of the (232) crystal planes and their relationship to the crystal lattice

atoms of the crystal. Structure solution by XRD is not through one or a few specific reflections to locate one or two atoms but through a whole set of reflections to locate all of the atoms as a whole. In theory, the number of reflections should be infinite, because there are unlimited ways of slicing a crystal by planes of infinitesimal *d*-spacings. In practice, *d*-spacings less than $\lambda/2$ are not instrumentally achievable, which is easily seen from Eq 2. Moreover, reflection intensities decrease significantly with smaller *d*-spacings and usually become undetectable much before reaching the limit of $\lambda/2$. As a result, only a limited number of reflections can be experimentally observed. Small *d*-spacing reflections are important for resolving fine details of a crystal structure, for example, precise atom positions, bond lengths, and angles. Crystallographers use the minimum *d*-spacing covered in a dataset, d_{min}, to represent data resolution. Data resolution depends on sample crystallinity as well as the instrument. A synchrotron diffraction beamline can provide data of much higher resolution than a laboratory diffractometer (Fig. 1).

Single-Crystal Diffraction

X-ray diffraction is classified by technique into two main categories: single crystal and powder diffraction. Single-crystal diffraction, based on a monobeam and four-circle goniometry, was developed in the 1960s (Ref 9); this technique serves as a gold-standard tool for characterization of crystalline materials. In a laboratory single-crystal XRD experiment, a crystal of decent size and shape, usually 30 to 300 μm in edge length, is selected under the microscope, attached to the tip of a thin glass fiber or a commercial mount, placed onto a goniometer head, and aligned to the center of the four-circle goniometer. Nowadays, the data-collection process is fully computerized. The raw data are composed of more than a thousand diffraction images recorded by an x-ray area detector, for example, the charge-coupled device area detector used in most modern single-crystal diffractometers. Following data collection, typical data reduction and analysis includes:

1. Processing a small part of the data containing a few tens to hundreds of reflections to obtain the initial lattice parameters, a lattice group, and orientation matrix
2. Integrating over the whole data set according to the obtained lattice information to obtain intensities and indices of all recorded reflections, while the lattice parameters also are being refined
3. Determining the space group and solving the crystal structure with the reflection data

The same sample handling and data processing applies to routine experiments on a single-crystal beamline, except that the beamline allows crystals of much smaller sizes, for example, 1 to 10 μm (which generally are considered as powder grains or microcrystals), to be measured because they often diffract well in synchrotron x-rays. In addition, the synchrotron light source pushes this classic crystallographic technique far beyond solving crystal structures, toward uncovering microscopic mechanisms with a functional structural response.

Single-crystal diffraction beamlines benefit two major areas of research: structural biology, and chemistry and materials sciences. Modern structural biology relies heavily on synchrotron single-crystal diffraction and has been a major science thrust for construction and development of many macromolecular crystallography (MX) beamlines and even new synchrotron light sources around the world. There are currently more than 130 MX beamlines worldwide. On the other side, single-crystal small-molecule beamlines for chemical crystallography also are available, although fewer than the MX beamlines. Examples are the ID11 and ID13 beamlines at the European Synchrotron Radiation Facility (ESRF) in Grenoble, France; the DuPont-Northwestern-Dow Collaborative Access Team beamline and the ChemMatCARS Collaborative Access Team beamline at the Advanced Photon Source (APS), Argonne, United States; the D3 and F1 beamlines at HASYLAB, Hamburg, Germany; the 14A beamline at the Photon Factory in Japan; the I19 at the Diamond Light Source in the United Kingdom; and the crystallography beamline at MAX II in Lund, Sweden. The use of SR opens up more possibilities for nonroutine experiments. For example, the ChemMatCARS facilities at APS support a range of techniques for ex situ and in situ analysis of materials with a selection of diffractometers, detectors, and sample mountings and also offer high-precision crystallography capabilities to study charge (i.e., electron) densities and bonding in small molecules (Ref 10–14) as well as transient-state photocrystallography (Ref 15–21).

Macromolecular Crystallography

Macromolecular x-ray crystallography has rapidly grown in the last few decades, with nearly 130,000 crystal structures deposited in the Protein Data Bank. More than 80% of the macromolecular structures deposited in the Protein Data Bank (Ref 22) in the last two decades were based on synchrotron diffraction data. Most MX beamlines are highly automated, with robotic sample capabilities and advanced software in data management, enabling high-throughput measurement and remote access that permits larger numbers of research groups more frequent access to the facilities in smaller time chunks (Ref 4, 23). The high-throughput capability is also due to the development of low-noise, fast-readout area detectors, such as PILATUS and EIGER, in which the read-out time is less than a millisecond.

Despite the easy beamline accessibility and increasing number of crystal structures solved each year, MX experiments still suffer from two major bottlenecks: the availability of large, well-diffracting crystals, and radiation damage. Production of well-diffracting large crystals is the most fundamental problem in MX. Quite often, microcrystals of less than a few micrometers in size are produced, and they are not suitable for regular-sized x-ray beams, for example, 50 to 200 μm. Many MX beamlines have microfocusing capabilities that can focus x-ray beams to near 1 μm, enabling the microcrystals to be successfully measured (Ref 24–26). The beam size usually is matched to the size of the crystal to minimize background scattering from air and to maximize the signal-to-noise ratio. For the radiation damage, one solution is cryocrystallography, that is, flash-freezing a crystal to cryogenic temperatures (typically approximately 100 K) that stabilize its structure long enough for a complete dataset collection. Cryocrystallography is the current standard method for data collection (Ref 27–33). However, cryogenic cooling has its own limitations because it is likely to increase mosaicity of the crystal, which makes indexing large unit cells more difficult. Also, cryocooled crystals may present different protein conformations from when they are at ambient temperature. In addition, not all crystals can survive long enough even under cryogenic conditions.

Another way of mitigating the radiation damage is serial crystallography, in which multiple crystals of the same sample are measured to build a complete dataset (Ref 34–36). In serial crystallography, the lifetime of a single crystal under radiation is no longer a concern, and the data usually are taken at room temperature, close to conditions at which the proteins function. Also, serial crystallography does not require individual crystals to be rotated with a goniometer, and this allows more freedom in how crystals are mounted and fed into the x-ray beam, including arrays of crystals on fixed sample holders (Ref 37–43) or hundreds to thousands of micro- or nanocrystals introduced to the synchrotron beam via a jet stream of liquid or gas (Ref 44–46). The latter also is called serial millisecond crystallography, which has been revived in part by the success of serial femtosecond crystallography at x-ray free-electron laser (XFEL) beamlines (Ref 47). Serial millisecond crystallography uses a high-viscosity medium to carry microcrystals, so that the crystals can retain their orientations during collection of a single diffraction image, which takes a few milliseconds.

In macromolecular crystallography, experimental phasing remains a major challenge for solving structures. The phase problem is a classic problem in x-ray crystallography. To obtain the correct structure model, one needs not only the reflection intensities, which are recorded by XRD detectors, but also the relative phases of the reflections, which are not and unable to be measured by the detectors. Solving the phase problem usually is easy for small-molecule structures, because the relatively small number of reflections in a dataset allow phases to be calculated by direct methods with current computation capability, but not for macromolecules. Molecular replacement, developed in the 1960s, used a macromolecular model, which was a previously solved structure and homologous to the target unknown structure, to estimate the initial phases for the diffraction data of the target. Molecular replacement is still the predominant method for phasing.

When a homologous model is unavailable, diffraction phases can be estimated with the help of purposely-placed heavy atoms in the protein structures. Multiple isomorphous replacement uses multiple heavy-atom isomorphous crystals to determine phases for the native form; it used to be a common phasing method when options of laboratory x-ray wavelengths were limited. With the ability to tune the x-ray wavelength at synchrotron beamlines, single-wavelength anomalous diffraction (SAD) and multiple-wavelength anomalous dispersion (MAD) methods became widely used. Both methods take advantage of the anomalous scattering of the heavy atoms nearing the absorption edges. The MAD method provides more phasing information than SAD but also requires longer data-collection time and thus longer x-ray exposure on the crystal, because multiple wavelengths are used. With the current advancement of beamline software and hardware, SAD now is a more efficient and popular method. For both methods, selenium anomalous scattering is the most commonly used, because it is relatively easy to incorporate selenomethionine (SeMet) to many proteins without altering their structures. In some occasions where SeMet protein or SeMet protein crystals cannot be produced, native SAD phasing still is possible by using sulfur atoms in Cys and Met amino acids, or phosphorus atoms in nucleic acids, or other atoms native in the protein structures, such as zinc, manganese, nickel, iron, and chlorine. Because some of these elements are neither similar to nor as good anomalous scatterers as selenium, and their absorption edges are at lower energies, the x-ray beam must be tuned much lower to get close to the absorption edges to maximize these anomalous signals. Typical wavelengths at MX beamlines are between 0.9 and 3.3 Å. The first native SAD experiment on the sulfur edge was performed at the Beamline I23 of the Diamond Light Source, using a wavelength of 4.96 Å to phase the diffraction data of thaumatin (Ref 48).

Powder Diffraction

Powder x-ray diffraction (PXRD) is a useful bulk probe for characterization of materials. Samples for PXRD generally are in powder form but also can be unground polycrystalline solids. Because x-rays are diffracted by millions of randomly orientated crystallites, the diffracted beams have the shape of coaxial cones that have their axes along the incident beam and the semivertical angles equal to the diffraction angles, 2θ (Fig. 7). If projected to an area detector, the cones will be shown as concentric rings, which are called diffraction rings or Debye-Scherrer rings. The PXRD data usually are presented as diffracted intensities as a function of 2θ, *d*-spacing, or the scattering vector magnitude (Q), which is related to θ by Eq 3:

$$Q = \frac{4\pi \sin\theta}{\lambda} \qquad \text{(Eq 3)}$$

where Q is in units of inverse angstrom (Å^{-1}) or inverse nanometer (nm^{-1}). Q is independent to the x-ray wavelength and more convenient to use when comparing data taken with different x-ray sources. Similar to d_{min}, the maximum Q value reached by PXRD data, Q_{max}, is also often used to indicate the data resolution. In the powder diffraction community, the term *resolution* also is used to describe how sharp the peaks are in the 2θ domain, as indicated by the full width at half maximum (FWHM) of the peaks, or $\Delta d/d$, which is the Δd value corresponding to the FWHM divided by the *d*-spacing of the peak. For clarity, the term *angular resolution* is used in this article to refer to the resolution in the 2θ domain.

The PXRD technique can leverage a number of useful characteristics of SR to vastly improve capabilities compared to laboratory measurements, specifically the increased photon flux, variable energy, and inherent beam collimation. Collimation and increased flux increase the ability of transmission geometry measures, which further allow for capillary measurements and a wide variety of sample environments. Variable energy allows for selection of an energy that reduces sample absorption and fluorescence. The combination of these differences allows for synchrotron PXRD measurements to vastly improve upon laboratory diffraction measurements in resolution, angular resolution, time scale, and unique sample environments.

There are three common experimental setups for PXRD measurements based on the type of detector used, each with their own advantages and disadvantages. These types of detectors can be delineated by the dimensionality of the detector: 0-D (point), 1-D (strip), and 2-D (area). The trade-off between these types of detectors is between angular resolution and temporal resolution.

With a point-detector setup, the point detector is scanned through the diffraction rings, and effectively, a small strip of the diffraction pattern is measured (Fig. 7). Commonly, an analyzer crystal is put before the detector. This crystal is aligned so that only x-rays coming from the sample satisfy the Bragg condition. This has the effect of increasing the angular resolution and reducing the background, because no air scattering or fluorescence from the sample can reach the detector. Additionally, these point detectors are photon-counting scintillation detectors, recording very precise intensities of the diffraction peaks. Typical setups of this type have $\Delta d/d$ on the order of 10^{-4}, which allows for detection of subtle phase transitions and detailed structural

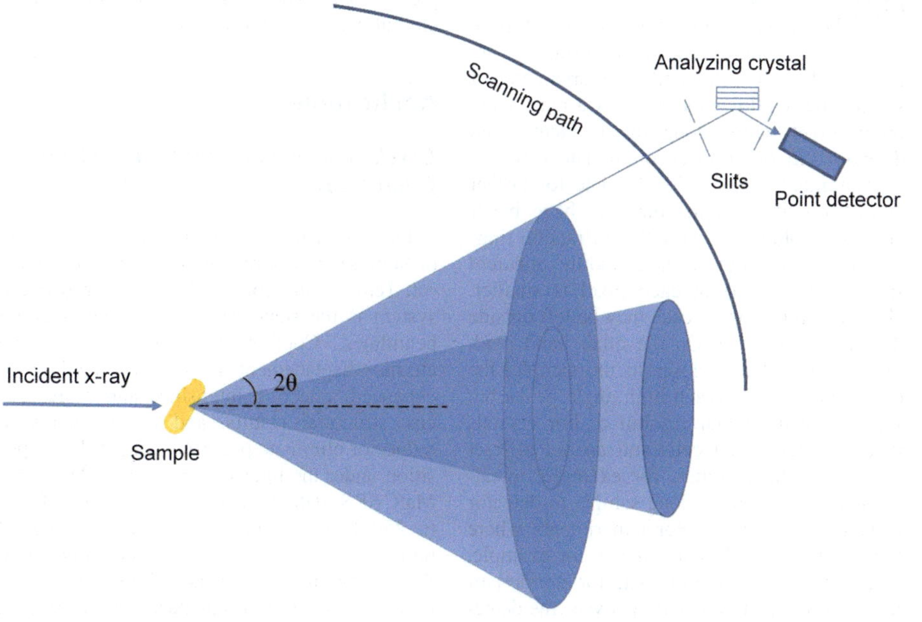

Fig. 7 Illustration of the diffraction cones in powder x-ray diffraction and the geometry of a point-detector setup

studies. All world high-resolution PXRD beamlines (e.g., 11-BM at APS, I11 at Diamond Light Source, and ID22 at ESRF) are based on the point-detector setup. The tradeoff for this increased resolution is longer scan times, generally on the order of an hour, and the necessity to spin the sample, because any granularity will present itself strongly in the diffraction pattern.

An area detector measures complete diffraction rings, which allows for quicker measurements with a time scale of less than a second to a minute but trades this ability for reduced angular resolution ($\Delta d/d$ on the order of 10^{-3}) and higher backgrounds from air scattering and fluorescence. Additionally, spinning of the sample is not needed as often because the measurement of the entire ring assists with good powder averaging. In addition, the 2-D data provide a direct way of evaluating sample texture or preferred orientation through uneven intensity distribution in the diffraction rings at different azimuth angles. The area-detector setup is particularly suited to in situ diffraction experiments to record rapid structural changes, total scattering, and high-throughput measurements.

A common intermediate choice between these two detector setups is a strip detector. This detector is placed to measure a cut through the diffraction rings, similar to the scanning of the point detector. However, due to the extent of the length of this detector, it usually provides enough 2θ coverage without scanning. This allows for quicker data collection, on the order of seconds to minutes, and generally has an angular resolution between that of an area-detector and a point-detector setup. For example, Beamline I11 at Diamond Light Source has a Mythen-2 silicon strip detector covering 90° of the scattering angle and able to measure at a rate of one powder pattern per second.

Unlike the point- or strip-detector setup, in which the sample-to-detector distances are fixed and well calibrated, the area-detector setup usually allows the detector a reasonably large travel range along the incident beam direction, in order to accommodate different measurement needs. With the detector farther away from the sample, the data have better angular resolution because the diffraction rings are more spread out in space, and the apparent size (angular size) of each pixel is smaller. However, high angle peaks may be left outside the detecting range. On the other hand, with the detector placed closer to the sample, the collected data can reach high angle peaks but with reduced angular resolution. For crystals that have large unit cells and do not diffract strongly at high angles, for example, metal-organic frameworks, a long sample-to-detector distance is preferred. For experiments where the Q range must be maximized, for example, collecting x-ray scattering data for atomic pair distribution function (PDF) analysis, the detector is put close to the sample, usually between 100 and 400 mm (4 and 16 in.).

The PDF analysis is a method of processing and analyzing scattering data, and it is particularly useful for characterizing systems that lack long-range periodicity, such as defect-rich crystalline compounds, nanoscale and amorphous solids, as well as liquids, gels, alloys, and melts. The atomic PDF describes the distribution of atom pairs as a function of interatomic distance and can be experimentally obtained by Fourier transform of the normalized structure factor derived from x-ray powder data in which the raw intensities are corrected for background, inelastic scattering, and other contributions. The intensities taken into the PDF analysis include both Bragg and diffuse scattering, the latter referring to the x-ray intensities distributed outside the Bragg peak positions. Diffuse scattering arises from aperiodic structural features, such as structural defects, surface modifications to crystallites, aperiodic arrangement of atoms in amorphous solids, and so on. It is the inclusion of diffuse scattering in the PDF analysis that enables this technique to possibly resolve those structural features and differentiates it from other whole-pattern-fitting methods that focus on the Bragg reflections, for example, Rietveld refinement. The PXRD data suitable for PDF analysis must have an extremely high signal-to-noise ratio and cover a broad range of Q, that is, $Q_{max} > 20$ Å^{-1}, in order to achieve adequate resolution in the real space after Fourier transform. For that reason, high-energy x-rays at 50 keV or above ($\lambda < 0.25$ Å) often are used in combination with large area detectors for collecting PDF-quality data. Thanks to the high brightness of high-energy x-rays available at synchrotrons and the advancement of area detectors, rapid measurement of PDF-quality data is becoming routinely available at many high-energy PXRD beamlines. Examples of PDF analysis are not given in this article but can be found in Ref 49 and 50.

Applications

Crystals at Nonambient and Extreme Conditions

Phase transitions due to thermal expansion or compressibility are often of research interest. Temperature control through a cryostream system is the standard setup at single-crystal beamlines. Cooling a crystal reduces the atomic displacement parameters, leading to enhanced reflection intensities and more precise structural results, and it also helps to reduce or eliminate potential sample decomposition under the intense x-ray beam. At ChemMatCARS, data collection in the range of 100 to 250 K is considered routine, and special requests to 10 K also can be accommodated. Sample heating usually is a deliberate operation. At Daresbury, it has been used in studies of microporous materials, for monitoring the decomposition and removal of template molecules during calcination processes (Ref 51), and for investigation of phase transitions between normal crystalline and liquid crystalline forms (Ref 52).

Synchrotron radiation particularly benefits high-pressure studies. A method of generating controlled high pressures on the order of gigapascals (GPa) is through a diamond anvil cell (DAC). Due to mechanical restrictions, DAC allows only a limited angular access for the incident and diffracted x-ray beams, plus the cell materials significantly absorb x-rays. Both of these problems are much reduced when shorter-wavelength x-rays are used, giving lower absorption and a compression of the diffraction pattern to smaller Bragg angles. Here, both the high photon intensity and the wavelength selectability of SR are important. Due to the high beam intensity, smaller crystals can be used, reducing the likelihood of crystal damage at high pressure (Ref 53, 54).

Photocrystallography

Photocrystallography explores the metastable molecular state induced by light. With carefully synchronized laser pulses, motor control, and x-ray measurement, the time-resolved photocrystallography visualizes light-induced reactions in four dimensions (x, y, z, and time). Figure 8 shows the structural changes of trans-[FeII(abpt)$_2$(NCS)$_2$] (abpt = 4-amino-3,5-bis (pyridine-2-yl)-1,2,4-triazole) from a low-spin ground state to a high-spin state.

High-Resolution Charge-Density Study

Charge-density (CD) studies examine more than a basic geometrical structure, and they seek to reveal and model the valence electron density in a material. This provides information on intermolecular interaction energy, electrostatic potential of molecules, atomic charge, H-bonds, bonding contributions, and nonbonding features such as lone pairs of electrons (Fig. 9). To refine the much larger number of parameters needed to describe the nonspherical distribution of electrons in an atom produced by its valence electrons, and to provide an effective decoupling of this from the nonspherical dynamic distribution due to atomic displacements (thermal motion), the temperature of the sample must be as low as possible, and the data of significant intensity are required to be a much higher resolution (lower d-spacings, equivalent to higher sin θ/λ in the Bragg equation) than for standard structure determinations, unless complementary neutron diffraction data also are available.

With conventional laboratory x-ray equipment and Mo-Kα radiation, this involves measuring data to a considerably higher maximum 2θ angle, which can be time-consuming

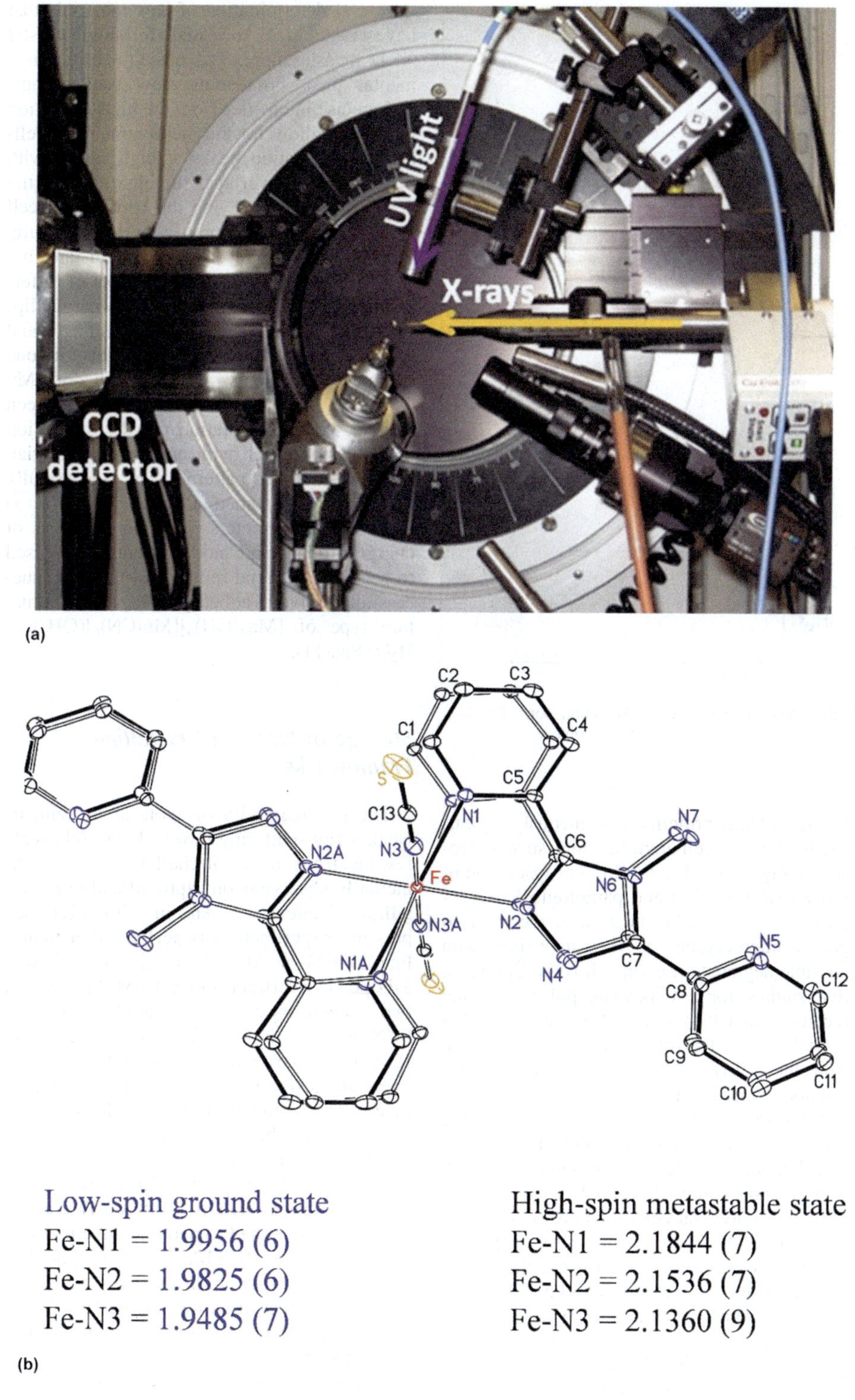

Fig. 8 (a) Photocrystallography experiment setup at the ChemMatCARS beamline. CCD, charge-coupled device; UV, ultraviolet. (b) Superimposed molecular diagram of trans-$[Fe(abpt)_2(NCS)_2]$ at low-spin and high-spin states

because high-angle reflections are relatively weak. An experiment can take many days with a modern area detector and weeks or months with a four-circle diffractometer. The Ag-Kα radiation, with a shorter wavelength, can be used to compress the diffraction pattern to lower angles, but it is intrinsically much weaker in intensity. Synchrotron radiation brings considerable advantages here through the use of high-intensity short-wavelength radiation. Thereby, the diffraction pattern is compressed to lower angles, so that more reflections are measured at once on an area detector, greatly speeding up the process, and intensities still can be high. The shorter wavelength also means reduced systematic errors from absorption and extinction effects, which are further reduced because smaller crystals can be used with the high intensity of SR. A high-quality full set of data for a charge-density study, with high redundancy of symmetry-equivalent data, can be achieved within 1 to 2 days. A number of CD studies have been performed at various synchrotron facilities (Ref 55–61).

Resonant Diffraction and Diffraction Anomalous Fine Structure

The resonant diffraction method uses anomalous dispersion effects by tuning the wavelength relative to absorption edges of elements present in the sample. In macromolecular crystallography, the main use of this is to obtain information to help to solve the phase problem in determining large structures, as detailed in the previous MX section. In chemical crystallography, the method is used to produce significant contrast in the scattering factors of atoms having similar electron density, thereby enabling a clear differentiation between these atoms in a structure (Ref 62). Examples include distinguishing zinc and gallium in microporous materials, where these atoms may be ordered or disordered (Ref 63), and similar problems in minerals and alloys, for example, aluminum versus silicon, and zinc versus copper. In some cases, it even is possible to distinguish atoms of the same element in different oxidation states, a technique known as valence-difference contrast (Ref 62, 64–66). Moreover, x-ray resonant diffraction combined with spectroscopic analysis is a strong tool to identify a selected element at a given crystallographic site.

Structural Dynamic at Controlled Atmosphere

Crystals for single-crystal diffraction measurements can be enclosed in environmental control cells (ECCs) for in situ observation of the structure in response to a change of environment. One such ECC is shown in Fig. 10. The cell has a quartz capillary enclosing the single crystal supported on a thin glass fiber (Ref 67). The cell base is connected to tubing through which gas or liquid can be injected into the capillary. A vacuum also can be generated if the capillary is sealed at the top. The cell was used to study the hydration and dehydration behavior of $[Co(5\text{-}NH_2\text{-}bdc)(bpy)_{0.5}(H_2O)] \cdot 2H_2O$ (5-NH_2-bdc = 5-aminoisophthalate; bpy = 4,40-bipyridine), in which the two free water molecules and one coordinated water molecule could be removed at dry nitrogen flow without destructing the crystal (Fig. 10).

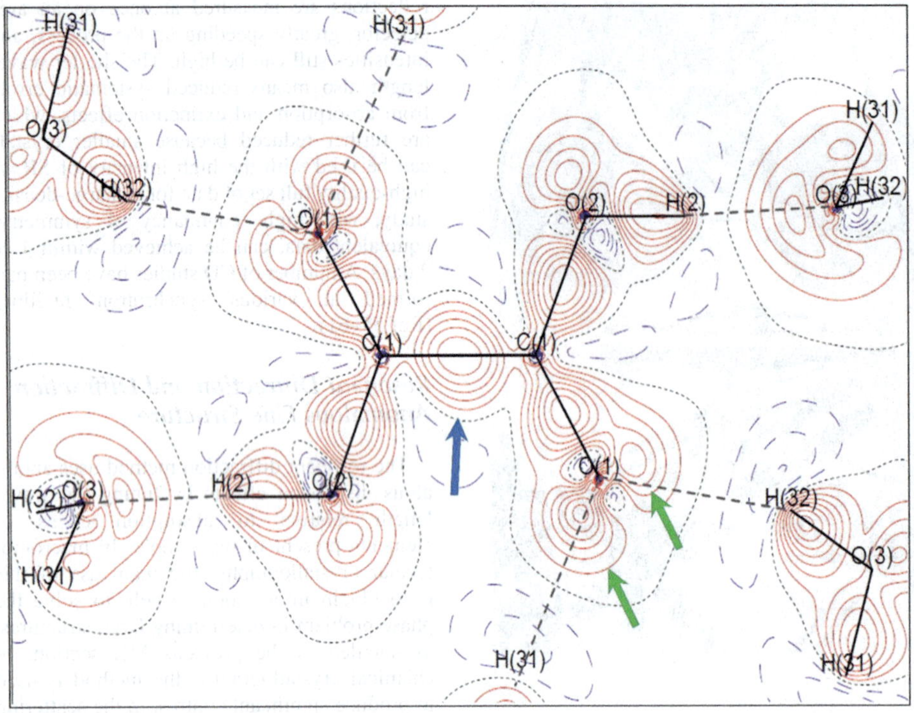

Fig. 9 Electron density map of oxalic acid based on high-resolution single-crystal diffraction data. The C–C bonding electrons (blue arrow) and the oxygen lone pair (green arrows) are clearly resolved.

In Situ and In Cellulo MX Studies

Protein crystal growth involves systematically screening as many crystallization conditions as possible, including different pHs, precipitants, salts, additives, and ligands; hence, the number of combinations can be well over several hundred. The growth usually is carried out with commercially available screens in a 96-well format. The crystals produced in the wells then are checked for diffraction. Traditionally, a crystal is picked out from each well, cryofreezed, and measured in the x-ray beam. It takes tremendous labor and time to check hundreds of crystals. More recently, in situ approaches without crystal harvesting have been developed at several light sources, for example, Swiss Light Source (Ref 68), VMXi at the Diamond Light Source, the Structural Biology Center and APS, and MX3D at the Australian Synchrotron Source. Different sample platforms were designed and tested, including microcapillaries (Ref 69), microfluidic devices (Ref 70–72), chips (Ref 73), micromeshes (Ref 74), and regular crystallization plates (Ref 68, 75). Because the platforms restrict the range of rotation and the crystals degrade under x-rays at ambient temperatures, this technique requires scanning several crystals to collect a complete dataset. On the positive side, likely crystal damage from the harvesting and freezing processes is avoided.

In cellulo MX is similar to the in situ approach in that no crystal harvesting is required. Many proteins spontaneously crystallize in complex cell environments, such as protein storage in seeds (Ref 76), encapsulation of viruses (Ref 77, 78) and pathological overexpression (Ref 79). Purification of the in-vivo-grown microcrystals is time-intensive, with possible degradation upon cell lysis. Comparative studies for a cypovirus polyhedrin and recombinant CPV1 polyhedrin suggested that crystals are adversely affected by removal from the cell (Ref 80, 81). For in cellulo diffraction experiments, crystal-containing cells first are separated from empty cells by using flow cytometry and then stained to facilitate visualization. The cells are mounted on mesh grid supports and flash-cooled with liquid nitrogen. Individual cells then are sequentially exposed to the x-ray beam. Several protein structures have been studied using the in cellulo approach at synchrotron and XFEL beamlines, including a natural insecticidal toxin Cry3A (Ref 82), the alcohol oxidase from yeast, and the human PAK4 in complex with Inka1, a potent endogenous kinase inhibitor (Ref 83–85).

Ab Initio Structure Determination from High-Resolution Powder Diffraction Data

Synchrotron high-resolution powder diffraction data are preferred or required for ab initio structural determination of new materials that have crystal sizes too small for single-crystal XRD measurement. The close-to-zero instrumental peak broadening, low backgrounds, and precise intensities from the high-resolution beamlines allow for the measurement of well-defined diffraction peaks at high angles with minimal peak overlaps, therefore giving the best chance of retrieving the correct unit cell and peak intensities for solving the structure. The ab initio analyses in general are not straightforward and require a number of different possible techniques, such as charge flipping (Ref 86), simulated annealing, and Rietveld refinements of trial models. In one example (Ref 87), the reaction of Mn $(O_2CMe)_2$ and NaCN formed a light-green powder. A high-resolution powder diffraction scan did not match with any known material. The diffraction pattern was successfully indexed and assigned a space group ($R\ \bar{3}$) and lattice parameters. This combination of charge flipping and model building was used to produce an initial model that then was successfully Rietveld refined, yielding a novel structure type of $[Mn_4(OH)_4][Mn(CN)_6](OH_2)_6 \cdot H_2O$ (Fig. 11).

Storage of Neon in Crystalline Frameworks

The low reactivity of neon, along with its weak x-ray-scattering power due to relatively few electrons, makes it challenging to experimentally observe neon captured within a crystalline framework. An in situ high-gas-pressure experiment was performed at Beamline 17-BM at APS (λ = 0.72768 Å) using 2-D powder diffraction (Ref 88). Samples of two metal-organic frameworks (MOFs), PCN-200 and NiMOF-74, were loaded into Kapton capillaries of 1.0 mm (0.04 in.) diameter and packed by glass wool. The capillaries then were mounted into a gas-loading cell connected to the gas control system of the beamline. Samples first were activated at elevated temperatures in a helium atmosphere and then cryocooled to 100 K. The atmosphere was switched to neon, and the pressure was increased to 10 MPa (100 bar). The data collected during the entire process were used to elucidate the structures of the two MOFs with neon molecules captured within the frameworks (Fig. 12). This study marked the first observation of interactions between neon and a transition metal.

Pressure-Induced Phase Transition

The magnetic molecular framework material $Co(dca)_2$ (dca = dicyanamide or $N(CN)^{2-}$) was examined for its pressure-dependent structural properties (Ref 89). An orthorhombic ($P\ mnn$) to monoclinic ($P\ 2_1/n$) phase transformation in

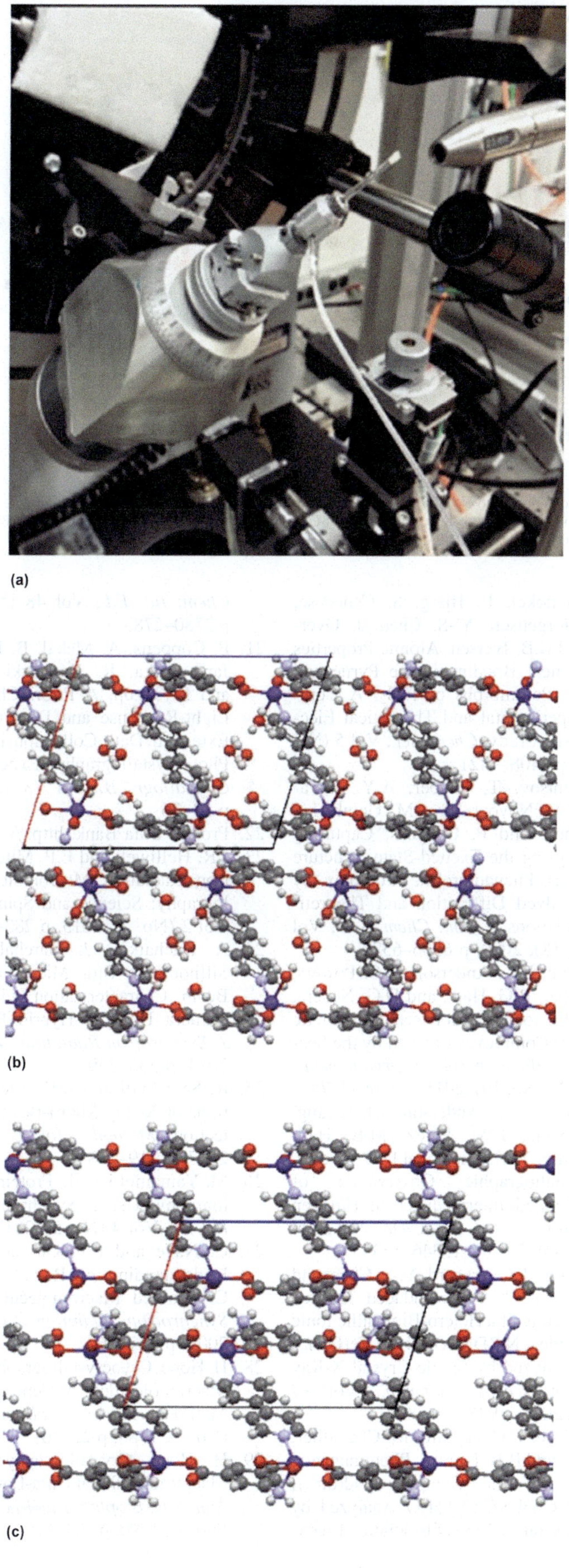

Fig. 10 (a) Photograph of an environmental control cell (ECC) used at the ChemMatCARS beamline. (b) Structure of $[Co(5\text{-}NH_2\text{-} bdc)(bpy)_{0.5}(H_2O)] \cdot 2H_2O$ before and (c) after dehydration; studied with the ECC

this material occurs at 1.1 GPa (0.16×10^6 psi) (Fig. 13). Structural determination of the high-pressure phase, namely γ-$Co(dca)_2$, showed that the rutilelike topology of the pristine material was retained, and the lower symmetry was due to a progression of volume-reducing structural distortions.

ACKNOWLEDGMENTS

Macromolecular x-ray crystallography research reported in this publication was supported by the National Institute of General Medical Sciences of the National Institutes of Health under Award Number R44GM108158. The National Science Foundation's ChemMatCARS Sector 15 is principally supported by the Divisions of Chemistry (CHE) and Materials Research (DMR), National Science Foundation, under grant number NSF/CHE-1834750. Use of the Advanced Photon Source, an Office of Science User Facility operated for the U.S. Department of Energy (DOE) Office of Science by Argonne National Laboratory, was supported by the U.S. DOE under Contract No. DE-AC02-06CH11357.

REFERENCES

1. J. Als-Nielsen and D. McMorrow, *Elements of Modern X-Ray Physics*, 2nd ed., Wiley, 2011, p 29–60
2. P. Willmott, *An Introduction to Synchrotron Radiation: Techniques and Applications*, Wiley, 2011, p 1–13
3. G.E. Ice, Synchrotron Diffraction: Capabilities, Instrumentation and Examples, *Modern Diffraction Methods*, Wiley, 2013, p 439–444
4. R.L. Owen, J. Juanhuix, and M. Fuchs, Current Advances in Synchrotron Radiation Instrumentation for MX Experiments, *Arch. Biochem. Biophys.*, Vol 602, July 2016, p 21–31
5. C. Riekel, M. Burghammer, and G. Schertler, Protein Crystallography Microdiffraction, *Curr. Opin. Struct. Biol.*, Vol 15 (No. 5), Oct 2005, p 556–562
6. Y. Ren, High-Energy Synchrotron X-Ray Diffraction and Its Application to In Situ Structural Phase-Transition Studies in Complex Sample Environments, *J. Min., Met., Mater. Soc.*, Vol 64 (No. 1), Jan 2012, p 140–149
7. G. Rosenbaum, S.L. Ginell, and J.C. Chen, Energy Optimization of a Regular Macromolecular Crystallography Beamline for Ultra-High-Resolution Crystallography, *J. Synchrotron Radiation*, Vol 22 (No. 1), Jan 2015, p 172–174
8. W.L. Bragg, The Diffraction of Short Electromagnetic Waves by a Crystal, *Proc. Cambridge Philos. Soc.*, Vol 17, 1913, p 43–57
9. W.R. Busing and H.A. Levy, Angle Calculations for 3- and 4-Circle X-Ray and

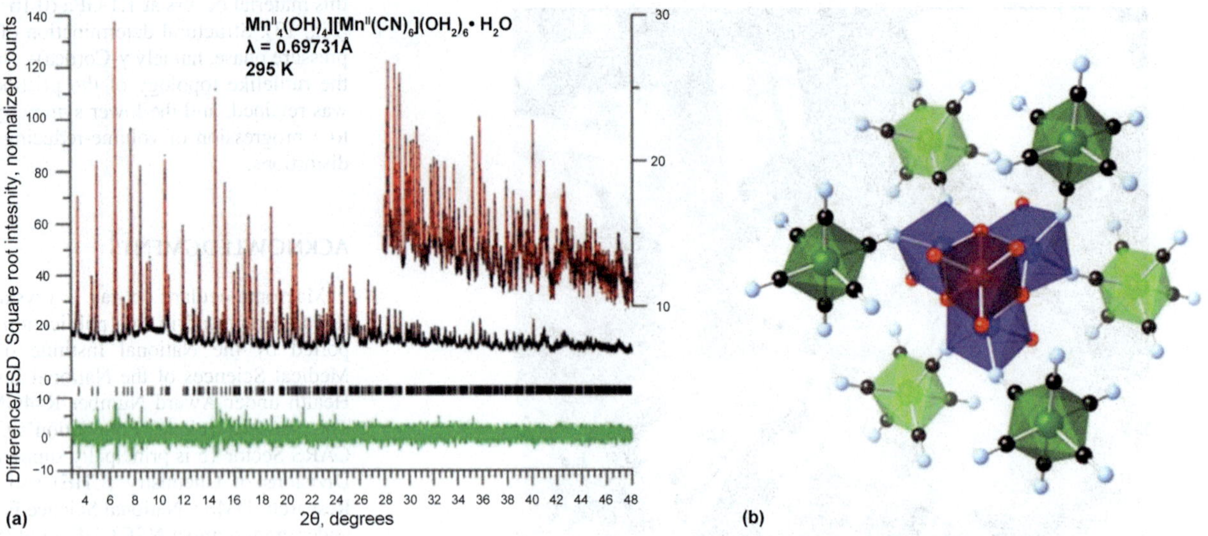

Fig. 11 (a) High-resolution synchrotron powder diffraction data (black dots) and Rietveld fit (red line) of the data. The lower (green) trace is the difference (measured minus calculated), normalized to statistical uncertainty of the raw counts. ESD, electrostatic discharge. (b) Drawing of the solved structure. Source: Ref 87. Reprinted with permission from Wiley. https://onlinelibrary.wiley.com/doi/abs/10.1002/chem.201804935

Neutron Diffractometers, *Acta Crystallogr.*, Vol 22 (No. 4), 1967, p 457–464

10. Y.-C. Chuang, C.-F. Sheu, G.-H. Lee, Y.-S. Chen, and Y. Wang, Charge Density Studies of 3D Metal (Ni/Cu) Complexes with a Non-Innocent Ligand, *Acta Crystallogr. B*, Vol 73 (No. 4), 2017, p 634–642
11. H.F. Clausen, M.R.V. Jørgensen, S. Cenedese, M.S. Schmøkel, M. Christensen, Y.-S. Chen, G. Koutsantonis, J. Overgaard, M.A. Spackman, and B.B. Iversen, Host Perturbation in a β-Hydroquinone Clathrate Studied by Combined X-Ray/Neutron Charge-Density Analysis: Implications for Molecular Inclusion in Supramolecular Entities, *Chem. Eur. J.*, Vol 20 (No. 26), 2014, p 8089–8098
12. H.F. Clausen, Y.-S. Chen, D. Jayatilaka, J. Overgaard, G.A. Koutsantonis, M.A. Spackman, and B.B. Iversen, Intermolecular Interactions and Electrostatic Properties of the β-Hydroquinone Apohost: Implications for Supramolecular Chemistry, *J. Phys. Chem. A*, Vol 115 (No. 45), 2011, p 12962–12972
13. S. Grabowsky, M. Weber, D. Jayatilaka, Y.-S. Chen, M.T. Grabowski, R. Brehme, M. Hesse, T. Schirmeister, and P. Luger, Reactivity Differences between α,β-Unsaturated Carbonyls and Hydrazones Investigated by Experimental and Theoretical Electron Density and Electron Localizability Analyses, *J. Phys. Chem. A*, Vol 115 (No. 45), 2011, p 12715–12732
14. H.F. Clausen, J. Overgaard, Y.S. Chen, and B.B. Iversen, Synchrotron X-Ray Charge Density Study of Coordination Polymer Co3(C8H4O4)4(C4H12N)2(C5H11NO)3 at 16 K, *J. Am. Chem. Soc.*, Vol 130 (No. 25), 2008, p 7988–7996,
15. M.S. Schmøkel, L. Bjerg, S. Cenedese, M.R.V. Jørgensen, Y.-S. Chen, J. Overgaard, and B.B. Iversen, Atomic Properties and Chemical Bonding in the Pyrite and Marcasite Polymorphs of FeS_2: A Combined Experimental and Theoretical Electron Density Study, *Chem. Sci.*, Vol 5 (No. 4), 2014, p 1408–1421
16. I.I. Vorontsov, T. Graber, A.Y. Kovalevsky, I.V. Novozhilova, M. Gembicky, Y.-S. Chen, and P. Coppens, Capturing and Analyzing the Excited-State Structure of a Cu(I) Phenanthroline Complex by Time-Resolved Diffraction and Theoretical Calculations, *J. Am. Chem. Soc.*, Vol 131 (No. 18), 2009, p 6566–6573
17. S.J. Hwang, B.L. Anderson, D.C. Powers, A.G. Maher, R.G. Hadt, and D.G. Nocera, Halogen Photoelimination from Monomeric Nickel(III) Complexes Enabled by the Secondary Coordination Sphere, *Organometallics*, Vol 34 (No. 19), 2015, p 4766–4774
18. D.C. Powers, B.L. Anderson, S.J. Hwang, T.M. Powers, L.M. Pérez, M.B. Hall, S.-L. Zheng, Y.-S. Chen, and D.G. Nocera, Photocrystallographic Observation of Halide-Bridged Intermediates in Halogen Photoeliminations, *J. Am. Chem. Soc.*, Vol 136 (No. 43), 2014, p 15346–15355
19. H. Svendsen, J. Overgaard, Y.-S. Chen, and B.B. Iversen, A Photo-Induced Excited State Structure of a Hetero-Bimetallic Ionic Pair Complex, $Nd(DMA)_4(H_2O)_4Fe(CN)_6 \cdot 3H_2O$, Analyzed by Single Crystal X-Ray Diffraction, *Chem. Commun.*, Vol 47 (No. 33), 2011, p 9486–9488
20. H. Svendsen, J. Overgaard, M. Chevallier, E. Collet, and B.B. Iversen, Photomagnetic Switching of the Complex $[Nd(dmf)_4(H_2O)_3(\mu\text{-}CN)Fe(CN)_5]\cdot H_2O$ Analyzed by Single-Crystal X-Ray Diffraction, *Angew. Chem. Int. Ed.*, Vol 48 (No. 15), 2009, p 2780–2783
21. P. Coppens, A. Makal, B. Fournier, K.N. Jarzembska, R. Kaminski, K. Basuroy, and E. Trzop, A Priori Checking of the Light-Response and Data Quality before Extended Data Collection in Pump-Probe Photocrystallography Experiments, *Acta Crystallogr. B*, Vol 73 (No. 1), 2017, p 23–26
22. Protein Data Bank, http://www.rcsb.org
23. J.R. Helliwell and E.P. Mitchell, Synchrotron Radiation Macromolecular Crystallography: Science and Spin-Offs, *IUCrJ.*, Vol 2 (No. 2), 2015, p 283–291
24. A. Kachatkou, J. Marchal, and R. van Silfhout, In Situ Micro-Focused X-Ray Beam Characterization with a Lensless Camera Using a Hybrid Pixel Detector, *J. Synchrotron Radiation*, Vol 21 (No. 2), 2014, p 333–339
25. R. Sanishvili and R.F. Fischetti, Applications of X-Ray Micro-Beam for Data Collection, *Methods Mol. Biol.*, Vol 1607, 2017, p 219–238
26. M. Yamamoto et al., Protein Microcrystallography Using Synchrotron Radiation, *IUCrJ.*, Vol 4 (No. 5), 2017, p 529–539
27. C. Nave and E.F. Garman, Towards an Understanding of Radiation Damage in Cryocooled Macromolecular Crystals, *J. Synchrotron Radiation*, Vol 12 (No. 3), 2005, p 257–260
28. H. Hope, Cryocrystallography of Biological Macromolecules: A Generally Applicable Method, *Acta Crystallogr. B*, Vol 44 (No. 1), 1988, p 22–26
29. H. Hope, Crystallography of Biological Macromolecules at Ultra-Low Temperature, *Ann. Rev. Biophys. Biophys. Chem.*, Vol 19 (No. 1), 1990, p 107–126

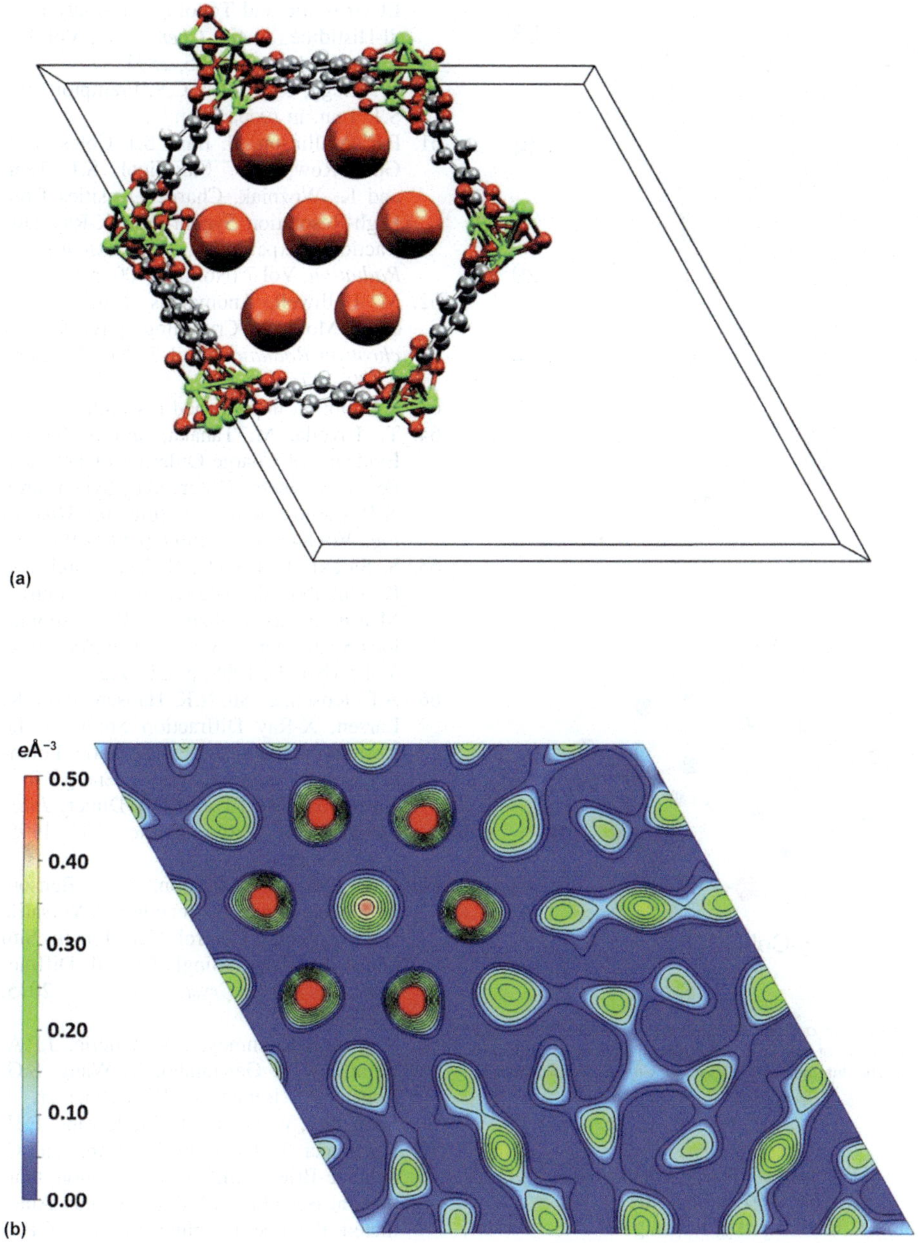

Fig. 12 (a) Structure model of neon atoms inside the pores of NiMOF-74 and (b) Fourier difference map constructed based on the in situ powder x-ray diffraction data at 100 K and 10 MPa (100 bar) of neon gas pressure. Source: Ref 88. Reproduced with permission from The Royal Society of Chemistry

30. B.W. Low et al., Studies of Insulin Crystals at Low Temperatures: Effects on Mosaic Character and Radiation Sensitivity, *Proc. Natl. Acad. Sci. U.S. Am.*, Vol 56 (No. 6), 1966, p 1746–1750
31. T.-Y. Teng, Mounting of Crystals for Macromolecular Crystallography in a Free-Standing Thin Film, *J. Appl. Crystallogr.*, Vol 23 (No. 5), 1990, p 387–391
32. P. Macchi, Cryo-Crystallography: Diffraction at Low Temperature and More, *Topics in Current Chemistry*, K. Rissanen, Ed., Springer, Berlin, Heidelberg, 2011, p 33–67
33. J.W. Pflugrath, Macromolecular Cryocrystallography—Methods for Cooling and Mounting Protein Crystals at Cryogenic Temperatures, *Methods*, Vol 34 (No. 3), 2004, p 415–423
34. T. Weinert et al., Serial Millisecond Crystallography for Routine Room-Temperature Structure Determination at Synchrotrons, *Nat. Commun.*, Vol 8 (No. 1), 2017, p 542
35. J.M. Martin-Garcia et al., Serial Millisecond Crystallography of Membrane and Soluble Protein Microcrystals Using Synchrotron Radiation, *IUCrJ.*, Vol 4 (No. 4), 2017, p 439–454
36. J. Standfuss and J. Spence, Serial Crystallography at Synchrotrons and X-Ray Lasers, *IUCrJ.*, Vol 4, 2017, p 100–101
37. C.-Y. Huang et al., In Meso In Situ Serial X-Ray Crystallography of Soluble and Membrane Proteins, *Acta Crystallogr. D*, Vol 71 (No. 6), 2015, p 1238–1256
38. U. Zander et al., MeshAndCollect: An Automated Multi-Crystal Data-Collection Workflow for Synchrotron Macromolecular Crystallography Beamlines, *Acta Crystallogr. D*, Vol 71 (No. 11), 2015, p 2328–2343
39. R.B. Doak et al., Crystallography on a Chip—Without the Chip: Sheet-on-Sheet Sandwich, *Acta Crystallogr. D*, Vol 74 (No. 10), 2018, p 1000–1007
40. K. Hasegawa et al., Development of a Dose-Limiting Data Collection Strategy for Serial Synchrotron Rotation Crystallography, *J. Synchrotron Radiation*, Vol 24 (No. 1), 2017, p 29–41
41. N. Coquelle et al., Raster-Scanning Serial Protein Crystallography Using Micro- and Nano-Focused Synchrotron Beams, *Acta Crystallograph. D*, Vol 71 (No. 5), 2015, p 1184–1196
42. C. Gati et al., Serial Crystallography on In Vivo Grown Microcrystals Using Synchrotron Radiation, *IUCrJ.*, Vol 1 (No. 2), 2014, p 87–94
43. A.Y. Lyubimov et al., Capture and X-Ray Diffraction Studies of Protein Microcrystals in a Microfluidic Trap Array, *Acta Crystallogr. D*, Vol 71 (No. 4), 2015, p 928–940
44. S. Botha et al., Room-Temperature Serial Crystallography at Synchrotron X-Ray Sources Using Slowly Flowing Free-Standing High-Viscosity Microstreams, *Acta Crystallogr. D*, Vol 71 (No. 2), 2015, p 387–397
45. F. Stellato et al., Room-Temperature Macromolecular Serial Crystallography Using Synchrotron Radiation, *IUCrJ.*, Vol 1 (No. 4), 2014, p 204–212
46. P. Nogly et al., Lipidic Cubic Phase Serial Millisecond Crystallography Using Synchrotron Radiation, *IUCrJ.*, Vol 2 (No. 2), 2015, p 168–176
47. U. Weierstall et al., Lipidic Cubic Phase Injector Facilitates Membrane Protein Serial Femtosecond Crystallography, *Nat. Commun.*, Vol 5, 2014, p 3309
48. O. Aurelius, R. Duman, K. El Omari, V. Mykhaylyk, and A. Wagner, Long-Wavelength Macromolecular Crystallography—First Successful Native SAD Experiment Close to the Sulfur Edge, *Nucl. Instrum. Methods Phys. Res. B*, Vol 411, 2017, p 12–16
49. T. Egami and S.J.L. Billinge, Underneath the Bragg Peaks—Structural Analysis of Complex Materials, *Pergamon Materials Series*, Vol 7, Pergamon, 2003
50. K.W. Chapman and P.J. Chupas, Pair Distribution Function Analysis of High-

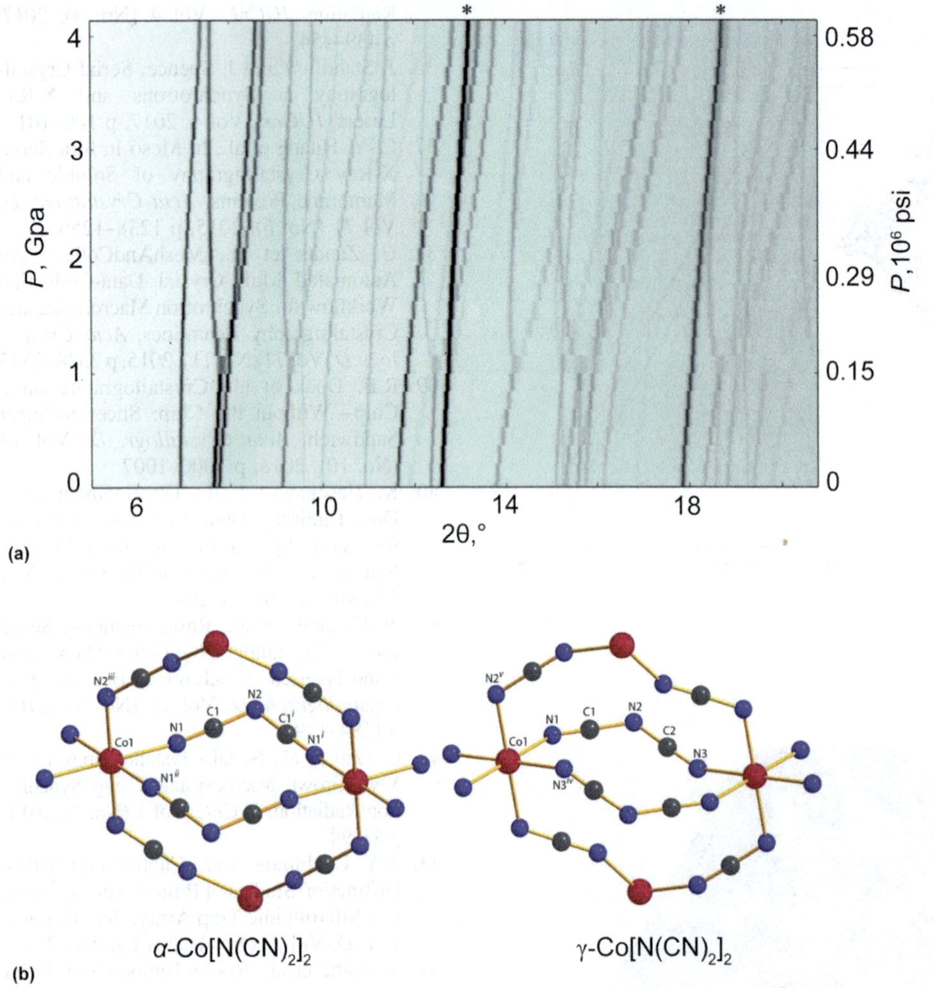

Fig. 13 (a) Contour plot of in situ powder x-ray diffraction patterns of $Co(dca)_2$ upon compression, displaying a phase transition at 1.1 GPa (0.16×10^6 psi). (b) View of metal-ligand connectivity before and after transition. Source: Ref 89. Reprinted with permission from The International Union of Crystallography (IUCr). https://doi.org/10.1107/S2052520615005867

Energy X-Ray Scattering Data, *In-Situ Characterization of Heterogeneous Catalysts*, J.A. Rodriguez, J.C. Hanson, and P.J. Chupas, Ed., Wiley, 2013, p 147–168

51. G. Muncaster, G. Sankar, C.R.A. Catlow, J.M. Thomas, R.G. Bell, P.A. Wright, S. Coles, S.J. Teat, W. Clegg, and W. Reeve, An In Situ Microcrystal X-Ray Diffraction Study of the Synthetic Aluminophosphate Zeotypes DAF-1 and CoAPSO-44, *Chem. Mater.*, Vol 11 (No. 1), 1999, p 158–163
52. M. Helliwell, unpublished research
53. D.R. Allan, M.I. McMahon, and R.J. Nelmes, unpublished research
54. H. Kitahara Eba, N. Ishizawa, F. Marumo, and Y. Noda, Synchrotron X-Ray Study of the Monoclinic High-Pressure Structure of AgGaS_{2}, *Phys. Rev. B*, Vol 61 (No. 5), 2000, p 3310–3316
55. H. Graafsma, S.O. Svensson, and Å. Kvick, An X-Ray Charge-Density Feasibility Study at 56 keV of Magnesium Formate Dihydrate Using a CCD Area Detector, *J. Appl. Crystallogr.*, Vol 30 (No. 6), 1997, p 957–962
56. T. Koritsánszky, R. Flaig, D. Zobel, H.-G. Krane, W. Morgenroth, and P. Luger, Accurate Experimental Electronic Properties of DL-Proline Monohydrate Obtained within 1 Day, *Science*, Vol 279 (No. 5349), 1998, p 356–358
57. P. Coppens, Charge-Density Analysis at the Turn of the Century, *Acta Crystallogr. A*, Vol 54 (No. 6-1), 1998, p 779–788
58. R. Flaig, T. Koritsánszky, J. Janczak, H.-G. Krane, W. Morgenroth, and P. Luger, Fast Experiments for Charge-Density Determination: Topological Analysis and Electrostatic Potential of the Amino Acids L-Asn, dl-Glu, dl-Ser, and L-Thr, *Angew. Chem. Int. Ed.*, Vol 38 (No. 10), 1999, p 1397–1400
59. P. Coppens, Y. Abramov, M. Carducci, B. Korjov, I. Novozhilova, C. Alhambra, and M.R. Pressprich, Experimental Charge Densities and Intermolecular Interactions: Electrostatic and Topological Analysis of dl-Histidine, *J. Am. Chem. Soc.*, Vol 121 (No. 11), 1999, p 2585–2593
60. W. Clegg, S.J. Coles, C.S. Frampton, and S.J. Teat, in preparation
61. P.R. Mallinson, G. Barr, S.J. Coles, T.N. Guru Row, D.D. MacNicol, S.J. Teat, and K. Wozniak, Charge Densities from High-Resolution Synchrotron X-Ray Diffraction Experiments, *J. Synchrotron Radiation*, Vol 7 (No. 3), 2000, p 160–166
62. M. Helliwell, Anomalous Scattering for Small-Molecule Crystallography, *J. Synchrotron Radiation*, Vol 7 (No. 3), 2000, p 139–147
63. R.H. Jones, unpublished research
64. T. Toyoda, M. Tanaka, and S. Sasaki, Evidence of Charge Ordering of Fe^{2+} and Fe^{3+} in Magnetite Observed by Synchrotron X-Ray Anomalous Scattering, *Am. Mineralog.*, Vol 84 (No. 3), 1999, p 294–298
65. S. Sasaki, T. Toyoda, K. Yamawaki, and K. Ohkubo, Valence-Difference Contrast Measurements Utilizing X-Ray Anomalous Scattering, *J. Synchrotron Radiation*, Vol 5 (No. 3), 1998, p 920–922
66. A.F. Jensen, Z. Su, N.K. Hansen, and F.K. Larsen, X-Ray Diffraction Study of the Correlation between Electrostatic Potential and K-Absorption Edge Energy in a Bis (μ-oxo)Mn(III)-Mn(IV) Dimer, *Inorganic Chem.*, Vol 34 (No. 16), 1995, p 4244–4252
67. J.M. Cox, I.M. Walton, C.A. Benson, Y.-S. Chen, and J.B. Benedict, A Versatile Environmental Control Cell for In Situ Guest Exchange Single-Crystal Diffraction, *J. Appl. Cryst.*, Vol 48, 2015, p 578–581
68. R. Bingel-Erlenmeyer, V. Olieric, J.P.A. Grimshaw, J. Gabadinho, X. Wang, S.G. Ebner, A. Isenegger, R. Schneider, J. Schneider, W. Glettig, C. Pradervand, E.H. Panepucci, T. Tomizaki, M. Wang, and C. Schulze-Briese, SLS Crystallization Platform at Beamline X06DA—A Fully Automated Pipeline Enabling In Situ X-Ray Diffraction Screening, *Cryst. Growth Des.*, Vol 11 (No. 4), 2011, p 916–923
69. M.K. Yadav et al., In Situ Data Collection and Structure Refinement from Microcapillary Protein Crystallization, *J. Appl. Crystallogr.*, Vol 38 (No. 6), 2005, p 900–905
70. M. Heymann et al., Room-Temperature Serial Crystallography Using a Kinetically Optimized Microfluidic Device for Protein Crystallization and On-Chip X-Ray Diffraction, *IUCrJ.*, Vol 1 (No. 5), 2014, p 349–360
71. M. Maeki et al., X-Ray Diffraction of Protein Crystal Grown in a Nano-Liter Scale Droplet in a Microchannel and Evaluation of Its Applicability, *Anal. Sci.*, Vol 28 (No. 1), 2012, p 65
72. S.L. Perry et al., A Microfluidic Approach for Protein Structure Determination at Room

Temperature via On-Chip Anomalous Diffraction, *Lab Chip*, Vol 13 (No. 16), 2013, p 3183–3187
73. G. Kisselman et al., X-CHIP: An Integrated Platform for High-Throughput Protein Crystallization and On-the-Chip X-Ray Diffraction Data Collection, *Acta Crystallogr. D*, Vol 67 (No. 6), 2011, p 533–539
74. X. Yin et al., Hitting the Target: Fragment Screening with Acoustic In Situ Co-Crystallization of Proteins plus Fragment Libraries on Pin-Mounted Data-Collection Micromeshes, *Acta Crystallogr. D*, Vol 70 (No. 5), 2014, p 1177–1189
75. D. Axford et al., In Situ Macromolecular Crystallography Using Microbeams, *Acta Crystallogr. D*, Vol 68 (No. 5), 2012, p 592–600
76. P.M. Colman, E. Suzuki, and A. Van Donkelaar, The Structure of Cucurbitin: Subunit Symmetry and Organization In Situ, *Eur. J. Biochem.*, Vol 103 (No. 3), 1980, p 585–588
77. H.M.E. Duyvesteyn et al., Towards In Cellulo Virus Crystallography, *Sci. Rep.*, Vol 8 (No. 1), 2018, p 3771
78. K. Anduleit et al., Crystal Lattice as Biological Phenotype for Insect Viruses, *Protein Sci.*, Vol 14 (No. 10), 2005, p 2741–2743
79. T. Wang et al., Overexpression of the Human ZNF300 Gene Enhances Growth and Metastasis of Cancer Cells through Activating NF-kB Pathway, *J. Cell. Molec. Med.*, Vol 16 (No. 5), 2012, p 1134–1145
80. M. Boudes et al., A Pipeline for Structure Determination of In Vivo-Grown Crystals Using In Cellulo Diffraction, *Acta Crystallogr. D*, Vol 72 (No. 4), 2016, p 576–585
81. D. Axford et al., In Cellulo Structure Determination of a Novel Cypovirus Polyhedrin, *Acta Crystallogr. D*, Vol 70 (No. 5), 2014, p 1435–1441
82. M.R. Sawaya et al., Protein Crystal Structure Obtained at 2.9 Å Resolution from Injecting Bacterial Cells into an X-Ray Free-Electron Laser Beam, *Proc. Natl. Acad. Sci. U.S. Am.*, Vol 111 (No. 35), Sept 2, 2014, p 12769-12774
83. M. Duszenko et al., In Vivo Protein Crystallization in Combination with Highly Brilliant Radiation Sources Offers Novel Opportunities for the Structural Analysis of Post-Translationally Modified Eukaryotic Proteins, *Acta Crystallogr. F*, Vol 71 (No. 8), 2015, p 929–937
84. A.J. Jakobi et al., In Cellulo Serial Crystallography of Alcohol Oxidase Crystals inside Yeast Cells, *IUCrJ.*, Vol 3 (No. 2), 2016, p 88–95
85. Y. Baskaran et al., An In Cellulo-Derived Structure of PAK4 in Complex with Its Inhibitor Inka1, *Nat. Commun.*, Vol 6, 2015, p 8681
86. L. Palatinus, Ab Initio Determination of Incommensurately Modulated Structures by Charge Flipping in Superspace, *Acta Crystallogr. A*, Vol 60, 2004, p 604–610
87. S.H. Lapidus et al., Anomalous Stoichiometry and Antiferromagnetic Ordering for the Extended Hydroxymanganese (II) Cubes/Hexacyanometalate-Based 3D-Structured $[Mn^{II}{}_4(OH)_4][Mn^{II}(CN)_6](OH_2)_6 \cdot H_2O$, *Chem. Eur. J.*, 2019
88. P.A. Wood, A.A. Sarjeant, A.A. Yakovenko, S.C. Ward, and C.R. Groom, Capturing Neon—The First Experimental Structure of Neon Trapped Within a Metal-Organic Environment, *Chem. Commun.*, Vol 52, 2016, p 10048–10051
89. A.A. Yakovenko, K.W. Chapman, and G.J. Halder, Pressure-Induced Structural Phase Transformation in Cobalt(II) Dicyanamide, *Acta Crystallogr. B*, Vol 71, 2015, p 252

ASM Handbook, Volume 10, *Materials Characterization*
ASM Handbook Committee
DOI 10.31399/asm.hb.v10.a0006667

Neutron Diffraction

António M. dos Santos, Melanie Kirkham, and Christina Hoffmann,
Oak Ridge National Laboratory

Overview

General Uses

- Nuclear and magnetic structure solution and refinement
- Pair distribution function measurements in amorphous and crystalline materials
- Studies on texture and residual-stress mapping in engineering materials

Examples of Applications

- Location of light atoms, hydrogen storage, battery materials (lithium)
- Magnetic structure determination
- Texture measurements on bulk engineering parts
- Structure solution and refinement of proteins and other biological molecules

Limitations

- Only available at large facilities
- May require deuteration
- Natural isotopes of some elements (e.g., gadolinium, cadmium, lithium, boron) are strong absorbers.
- Limited spatial and time resolution
- Samples may become radiologically activated.

Samples

- Loose or compacted powders
- Single crystals, bulk polycrystalline samples
- Preferably, but not necessarily, single phase

Estimated Data-Collection Time

- *Single powder pattern:* As fast as a few minutes for crystalline powders, comprising neutron-"friendly" species and simple crystal structures. Approximately 8 to 24 h for parametric studies
- *Full single-crystal datasets (comprising multiple orientations):* One day per temperature

Capabilities of Related Techniques

- X-ray diffraction probes similar information and can be performed in-house.
- Use of synchrotron sources allows picogram samples and millisecond data-collection times.
- Modern techniques include diffraction tomography and anomalous scattering.
- Structural information, including determination, can be obtained through electron diffraction.

Background

The existence of the neutron as a fundamental atomic particle was predicted by Rutherford in 1920 (Ref 1). Subsequent experimental efforts by James Chadwick to confirm its existence were successful in 1932 (Ref 2). Chadwick's experiments developed in parallel with the efforts of Bothe and Becker (1930) (Ref 3), which observed the interaction of a "highly penetrating radiation" with matter after alpha-particle bombardment of beryllium, as well as with Irène and Frédérick Joliot-Curie's (1932) (Ref 4) work, which found that the said radiation dislodged protons from hydrogenous materials.

Neutron diffraction was first demonstrated by von Halban and Preiswerk (1936) (Ref 5) on iron powder and by Mitchell and Powers (1936) (Ref 6) on single-crystalline magnesium oxide. In 1937, Schwinger (Ref 7) showed that neutrons have a spin, quickly followed in 1939 by Halpern and Johnson's magnetic scattering theory (Ref 8) and the measurement of the neutron magnetic moment by Alvarez and Block in 1940 (Ref 9). This groundwork on the nature of the neutron as a free particle was essential for the development (in 1942) of the Chicago Pile-1, (CP-1), the first fission reactor for demonstration purposes, as this reaction was sustained by the natural release and absorption of neutrons. In CP-1, the fast neutrons resulting from fission were slowed down (or moderated) using graphite. A major breakthrough was achieved shortly after by sucessfully maintaining a sustained nuclear reaction in another graphite reactor, named X-10, located at what became the Oak Ridge National Laboratory in East Tennessee. This was followed by a heavy-water-moderated reactor CP-3 at the original location of the Argonne National Laboratory, in Cook County, Illinois. The main mission of the reactors was plutonium production; however, within months of starting operation, W.H. Zinn, at Argonne, and Ernest O. Wollan, at Oak Ridge, pursued fundamental neutron experiments to determine the scattering amplitudes of elements and diffraction from single crystals, respectively. In 1994, the Nobel prize in physics was awarded jointly to Clifford Shull, who had joined Wollan in Oak Ridge right after the end of the war and later went on to Massachusetts Institute of Technology, for the development of elastic neutron scattering (diffraction) and to Bertram Brockhouse, a researcher at the Chalk River reactor in Canada and later at McMaster University, for inelastic neutron scattering (spectroscopy). Neutron scattering had already been instrumental in two earlier Nobel awards: to Linus Pauling on the nature of the chemical bond in 1954 (chemistry) and Louis Néel for work on ferromagnetism and antiferromagnetism in 1970 (physics).

Since then, as more intense sources, improved detector technology, neutron optical devices, and added computer power became available, neutron diffraction, as well as other associated techniques (spectroscopy, radiography, etc.), have experienced remarkable advances.

In terms of neutron production, the intrinsic power limitations of reactor sources, arising from the technical challenge of cooling a high-power density source, were finally overcome when neutron beams were produced via spallation from heavy, stable metals rather than from fissionable uranium fuel in reactors (Fig. 1).

Nonetheless, both fission and spallation remain quite energy intensive, and thus, usable neutron beams for materials science are only available in large-scale research facilities that house a nuclear reactor or a spallation neutron source. While this requirement has arguably been the biggest hurdle to the growth of neutron sciences, the importance—and uniqueness—of neutron diffraction is demonstrated by the construction of high-flux-spallation neutron facilities and reactors as well as technical upgrades and expansions to existing facilities across the world. Improved access has significantly grown the user community and moved the technique from esoteric measurements that once required several cubic centimeter samples measured over days or weeks to time-resolved measurements on samples of a few hundred milligrams or less.

General Principles of the Neutron

The neutron, along with the protons are the fundamental constituents of the atomic nucleus; orbiting electrons balance the charge, and a combination of all three form every atom. In fact, light hydrogen, ^{1}H, is the only element without neutrons as the name suggests, neutrons do not carry electric charge but have mass, $m_n = 1.67495 \times 10^{-27}$ kg. Like electrons, neutrons have a magnetic moment, $\mu_n = -1.913043(1)\ \mu_N$ (where μ_N is the nuclear magneton), with a corresponding spin of ½. Free neutrons can occur from radioactive decay of unstable isotopes and are themselves unstable, with a half-life of approximately 14.7 min. The high flux required for materials research purposes is currently produced via fission (in nuclear reactors) or spallation. These neutron sources are always used in conjunction with a moderator, whose purpose is to slow down neutrons to energies more suitable for scattering experiments (see the section "Neutron Generation" in this article).

Despite its mass, which is considerably larger than electrons, neutrons share in the wave/particle duality with electrons and photons. As such, a beam of neutrons with the appropriate wavelength can diffract when interacting with matter. Directly correlated with this de Broglie wavelength (λ, in Å) is a neutron energy (E, in meV) and a temperature (T, in Kelvin), associated with the corresponding Maxwell-Boltzmann distribution (because moderated neutron beams can be approximated as ideal gases). Such characteristics can be inferred from knowledge of the neutron velocity (υ, in m/s) obtained from measurements of its travel time (time of flight, or TOF) over a known distance (L, in meters):

$$E = \frac{1}{2} m_n \upsilon_n{}^2 = \frac{h^2}{2\, m_n\, \lambda^2} = \frac{m_n\, L^2}{2\, TOF^2} \sim k_B T \qquad \text{(Eq 1)}$$

Or, rearranging to:

$$E(\text{meV}) = \frac{81.81}{\lambda^2} \text{ and } \upsilon(\text{km/s}) = \frac{3.956}{\lambda} \qquad \text{(Eq 2)}$$

with the neutron mass (m_n, in kg), where k_B is Boltzmann's constant, and h is Planck's constant. For example, neutrons with a wavelength of 1.8 Å have a corresponding energy of 25.3 meV, a temperature of 293.6 K, and are said to be thermal and well suited for diffraction experiments, because their wavelength matches the interatomic distances found in solids.

Neutron-Scattering Theory

Similar to x-rays or other diffraction techniques, neutron diffraction requires a beam with the appropriate wavelength and can be described according to Bragg's law. This relates the incident radiation wavelength (λ) with the interplanar spacing between adjacent atomic planes (d) and the scattering angle (2θ), as expressed in Eq 3:

$$\lambda = 2 d_{hkl} \sin\theta \qquad \text{(Eq 3)}$$

For the simplest case of monochromatic radiation, a diffraction pattern is simply the angular distribution of scattered intensity from which information about the interplanar spacings may be derived.

These measured Bragg intensities are directly related to the crystal structure and chemical composition of the material via the structure factors (F_{hkl}) by:

$$I_{hkl} = k\, A\, L\, |F_{hkl}|^2 \qquad \text{(Eq 4)}$$

where k is a scale factor that combines the incident flux and detector efficiency, A is the absorption, and L is the Lorentz factor, which is, in the monochromatic case:

$$\text{L} = \frac{\lambda^3}{\sin 2\theta} \qquad \text{(Eq 5)}$$

and for polychromatic beams:

$$\text{L} = \frac{\lambda^4}{\sin^2\theta} \qquad \text{(Eq 6)}$$

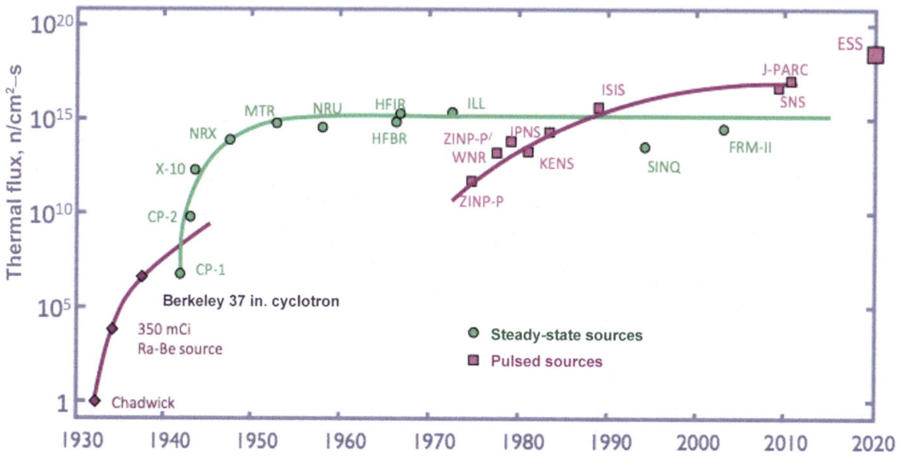

Fig. 1 Historical evolution of the available thermal neutron flux. For pulsed sources, peak flux is reported. Reactor sources output has remained largely constant since the 1960s, while major advances have been based on spallation. Early sources are shown as red diamonds, reactor sources as green circles, and spallation sources as red squares. Note that SINQ is a spallation-based steady-state source. The value for the European Spallation Source (ESS), indicated by the larger red square, is an estimate, because it is not yet operating at the time of publication. This figure was compiled first in Ref 10 and updated in Ref 11 (as shown).

The structure factor (F_{hkl}) that contains information on the chemical makeup of the unit cell is:

$$F_{hkl} = \sum_j b_j e^{2\pi i(hx_j + ky_j + lz_j)} \quad \text{(Eq 7)}$$

The summation is over j atoms, with the isotope-specific neutron-scattering length b_j in the unit cell.

The nuclear density (ρ_{hkl}) is then obtained through a Fourier transform, which relates diffraction, reciprocal, or Q-space to real (or atomic) space by:

$$\rho_{xyz} = \frac{1}{V} \sum_{hkl} F_{hkl} e^{-2\pi(hx+ky+lz)} \quad \text{(Eq 8)}$$

By collecting a sufficiently complete dataset of individual reflections, one can determine the location of all atoms in the unit cell. This formalism is almost identical to the one used for x-ray diffraction but replaces the isotopic neutron-scattering lengths by the atomic x-ray scattering factors.

Despite sharing the scattering mechanism, the interaction of neutrons with matter is fundamentally different than that of x-rays as photons strongly interact with the electron shell of the atom. Because all electrons interact with radiation, independent of which atoms they are part of, the scattering for x-rays is related only to the electronic density, being proportional to Z^2, where Z is the atomic number.

The interaction of neutrons with matter is far more complex, because they scatter via a weak interaction with the nucleus. In this process, they can be absorbed, scattered coherently, or scattered incoherently, depending on the combined effects of nuclear isotopes and spin. Indeed, the interaction characteristics are strongly dependent on the makeup of the nucleus, which varies for each element and isotope and can be in or out of phase (resulting in positive or negative scattering lengths). One important consequence of this is that the scattering cross section, which is the scattering probability for a particular species, varies for each isotope in a fashion uncorrelated with the atomic number, Z (Fig. 2). Diffraction from materials only involves elastic scattering (a change of wave vector direction without energy gain or loss). Inelastic scattering, where there is transfer of energy between the beam and the sample, can be very informative as well and is studied with neutron spectrometers, but is outside the scope of this article.

Finally, for the purposes of neutron scattering, only the bound coherent cross section is relevant, because it pertains to regimes where atoms are bound in a condensed phase with a binding energy much higher than the one of the incident neutrons; thus, any bond breaking or recoil effects from the scattering process are negligible. For very high-incident-energy neutrons, an unbound coherent cross section must be used, but this is seldom, if ever, necessary in the context of neutron diffraction. A more in-depth description of the fundamental interactions of the neutron with matter can be found in Berk's review (Ref 12).

Another important distinction between x-rays and neutrons is that, with respect to other dimensions present in the material (bond lengths) and the wavelength of the radiation, the nucleus presents itself as a "point scatterer," and consequently, neutron scattering is not affected by an angular-dependent attenuation, known as form factor, which, for x-rays, results in weaker intensities at higher momentum transfers (shorter d-spacings). The only remaining major contribution to the angular, or Q-dependent, intensity attenuation, is due to thermal vibrations of atoms, expressed in the Debye-Waller factor: Since the vibrations are random and in a much faster time scale than the measurements, they increase the apparent size of the scattering center. This phenomena is observed in both neutron and x-ray scattering.

One important feature that is often explored in neutron diffraction is that scattering intensity from "light" elements, such as oxygen (bound coherent scattering length = b_{coh} = 5.81 fm), is comparable to much heavier elements, such as uranium (b_{coh} = 8.42 fm), while scattering contributions of neighboring elements iron (b_{coh} = 9.54 fm) and manganese (b_{coh} = −3.73 fm) can be significantly different and allow structural details of each species to be clearly distinguished (Ref 13).

Finally, a notable property of neutrons is that, due to their intrinsic magnetic dipole moment, their polarization can be affected by magnetic fields. This interaction is observed in coherent scattering from the magnetic field arising from unpaired electrons in ordered magnetic structures. The scattering amplitudes from magnetic and nuclear scattering are roughly equivalent. However, magnetic scattering has a vectorial character; therefore, for each atom where the ordered magnetic spin resides, up to three orientational terms must be refined to obtain the total magnetic moment on that site. However, the use of magnetic symmetry can reduce significantly the number of degrees of freedom to refine. Unlike nuclear scattering, here the size of the scattering unit (the valence electrons) is comparable to the incident neutron wavelength, and a magnetic form factor is observed. This is actually more pronounced than the nuclear (nonmagnetic) form factor for x-rays, because it originates primarily from more outwardly, valence electrons. Table 1 summarizes the main contrasts between neutron and x-ray diffraction.

Neutron Generation

Neutrons are either produced in a nuclear reactor from natural or enriched fission materials or extracted by means of high-energy protons impinging on a neutron-rich element, a process known as spallation. Nuclear reactors maintain a self-sustained nuclear chain reaction by partial recapture of neutrons from spontaneous nuclear fission in close proximity. The production of neutrons is thus (almost always) continuous.

The generation of neutrons via spallation is done through bombarding a heavy metal target of, for example, tungsten or mercury, with a high-energy proton beam (several hundreds of mega-electron volt to giga-electron volt). The proton beam typically reaches the target in carefully shaped and timed discrete bunches or pulses. Each impact starts a cascade of

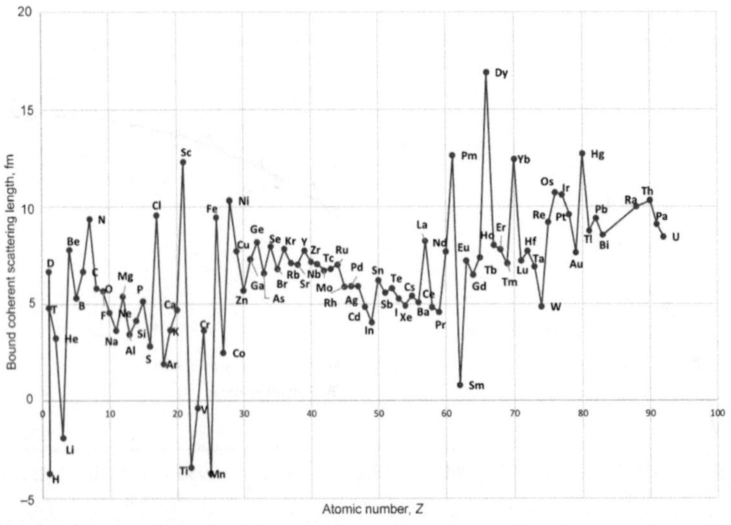

Fig. 2 Graphical display of the neutron bound coherent scattering length for the naturally occurring elemental abundances. Not captured in this plot is the fact that ^{3}He, ^{10}B, and some rare earth isotopes from the elements cadmium, samarium, gadolinium, and dysprosium are highly absorbing, effectively precluding their use in neutron diffraction. Also, ^{1}H and isotopes from neodymium, samarium, and gadolinium contribute incoherent scattering, observed experimentally as an increase in background.

nuclear events that release neutrons. This impact from the proton into the target is designated t_0 ("t-zero") and corresponds to the start time for the TOF determination.

In both reactor sources and spallation, the neutrons produced have too-high kinetic energy (fast neutrons) for scattering and must therefore be attenuated in moderators. These are reservoirs containing an appropriate low-Z material with weak absorption cross section (e.g., hydrogen, water, methane, graphite, or other) and are kept at a prescribed temperature. The purpose of these devices is to gradually thermalize the average energy of the neutron. Emanating from the moderator is a neutron beam with an approximately Maxwell-Boltzmann distribution of epithermal and thermal neutrons (~0.4 to 3 Å, or ~10 to 500 meV) and a peak wavelength ideally at approximately 1.2 Å (for diffraction). The "temperature" of the moderator is strongly related to its wavelength distribution. Specialized "hot" or "cold" moderators (also called sources) may be employed to provide high- (fast) or low- (slow) energy neutrons, respectively.

Types of Incident Radiation—Monochromatic, Polychromatic, and Pulsed

A reactor source emits a continuous (except the Dubna reactor in the Russian Federation, which operates in pulsed mode) and approximately constant flux of neutrons, moderated to thermal and epithermal energies. A conventional arrangement for diffraction is to monochromatize the neutron beam by placing an appropriately oriented, large single crystal, called a monochromator, placed in the full beam cross section in an orientation such that one strong Bragg peak is scattered to the sample position. The signal diffracted from the sample is collected as a function of the angle, thus generating an angle-resolved diffraction pattern. This technique is generally employed in both powder and single-crystal diffraction instruments at reactor sources.

A special case is a polychromatic or Laue diffractometer at a reactor source, in which the sample is illuminated from the moderator, using a wide incident wavelength band. The single-crystal sample diffracts all incident neutron wavelengths into a suitable wide-angle detector. Conventional Laue diffraction does not resolve overlapping, higher-order reflections, and it is generally only used for compounds with large unit cells, for example, proteins.

Spallation neutron sources, on the other side, are almost always pulsed and operate in the TOF mode, wherein the travel time between generation (t_0) and detection, termed TOF, is recorded for every neutron. Given a known flight path, the velocity (and therefore energy and wavelength) of each neutron may be calculated from the TOF. A polychromatic beam of neutrons reaches the sample, and the whole range of thermal and epithermal neutrons can be used for diffraction. Each neutron event is characterized by its TOF and position in the detector, and this is used to calculate the diffraction pattern, which combines energy and angle-resolved diffraction. One exception in the spallation context is the Paul Scherrer Institute's spallation source in Switzerland, which operates in continuous mode (and consequently at a much lower peak power than other state-of-the-art spallation sources).

As mentioned earlier, the neutron emission from the moderator has a wavelength-dependent spectral shape. As the beam is transported to the sample, this distribution can be affected by neutron guides, apertures, and sample environment. It is critical to know this distribution at the sample position, with the instrument and sample environment set in a configuration as close as possible to the actual measurement. The most common approach is to collect data from a strong incoherent scatterer, such as vanadium. The data from such a measurement yield the incident spectrum convoluted with the wavelength-dependent detector efficiency. As described subsequently, such information is essential for further data analysis.

Single-Crystal Diffraction

Single-crystal neutron diffraction (SCND) instruments are designed to measure the Bragg intensities (I_{hkl}) from single crystals. Each reflection is observed as separated in reciprocal space and thus mostly free from the overlapping observed in powder diffraction.

To measure one complete dataset of single-crystal diffraction, it is required to collect several crystal orientations with respect to the incident beam, while keeping the center of mass of the crystal as much as possible at the center of the instrument. The Ewald sphere is a graphical representation of this process, relating Bragg's law to reciprocal space (Fig. 3). The monochromatic case shows one sphere radius representing the incident wavelength. When a lattice point intersects the sphere, it fulfills the diffraction condition; meaning it is, it is observed. Sequential rotations of the reciprocal space grid (or sample) through the surface of the sphere are therefore required to sample a majority of the reflections. For Laue and TOF Laue, a range of radii representing the longest and shortest incident wavelength are present, and the volume between contains the Bragg peaks measured concurrently; therefore, a larger section of reciprocal space can be measured efficiently. An example of a steady-state single-crystal neutron diffractometer is shown in Fig. 4.

A main purpose of single-crystal neutron diffraction is structure analysis for precise and accurate unit cell, structure symmetry, and atomic coordinates with their uncertainties and concentrations. Furthermore, the material response to externally applied conditions, such as magnetic field, temperature, and specific gas environment, can be probed following structure changes or the emergence of a magnetic order. Effects that are directional in a structure can be resolved with SCND because each sample

Table 1 Comparison of neutron and x-ray diffraction characteristics

Characteristic	Neutron	X-rays
Atomic number dependent	✗	✓
Wide range of scattering strengths	✗	✓
Form factor attenuation	✗	✓
Sensitive to hydrogen	✓	✗
Neighboring element contrast	✓	✗
Penetrating (better sample averaging)	✓	✗
Excellent time and spatial resolution	✗	✓
Sensitive to magnetic order	✓	✗

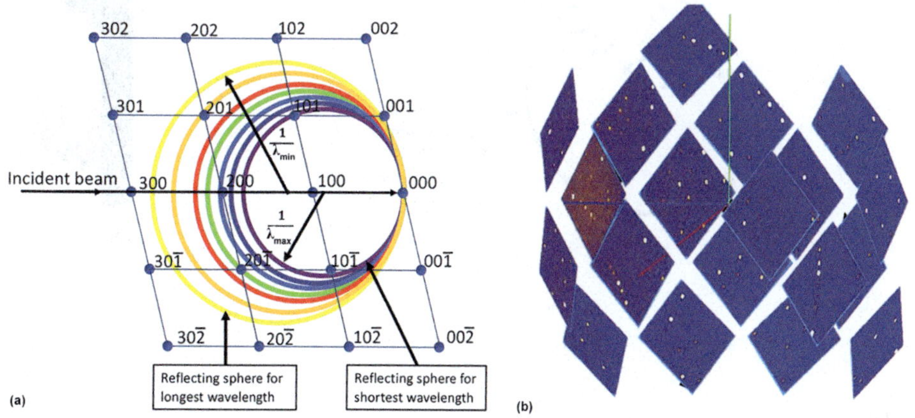

Fig. 3 (a) The Ewald construction describes Bragg's law in reciprocal space. For a polychromatic beam, the Ewald sphere of each wavelength can be described as contracting through the reciprocal space. (b) Raw data from one orientation as detected from a single crystal in time-of-flight mode. Data collected at the TOPAZ diffractometer at the Spallation Neutron Source. The large number of reflections observed is a result of the extensive simultaneous sampling of reciprocal space in both angle and wavelength. Note the spatial resolution required for peak integration. (See the section "Detectors" in this article.)

direction is measured independently. This allows the measurement of effects arising from stacking faults, defects, and anisotropic diffusion that are not easily disentangled in radially averaging methods where these manifest as superstructures or diffuse scattering.

Powder Diffraction

Powder diffraction instruments are designed to measure the scattering intensity from polycrystalline samples, whether composed of individual particles, as in powders, or a single solid piece, for example, metals or ceramics. It follows the same basic principles as single-crystal diffraction, except that powder diffraction consists of the superposition of thousands of individual single-crystal diffraction patterns from each individual grain. The orientational information is thus lost as everything is collapsed into one dimension; therefore, one single measurement suffices. However, powder diffraction has proven invaluable when, as is often the case, single-crystal samples of sufficient size and quality are not available. Additionally, powder diffraction is employed in solid-state chemistry and engineering for phase quantification of multiphase samples and for operando measurements. Special care must be taken, because the accuracy of the method can be severely compromised by preferred orientation of the sample, texture, or other artifacts. Schematic representations of reactor and time-of-flight powder diffractometers are shown in Fig. 5.

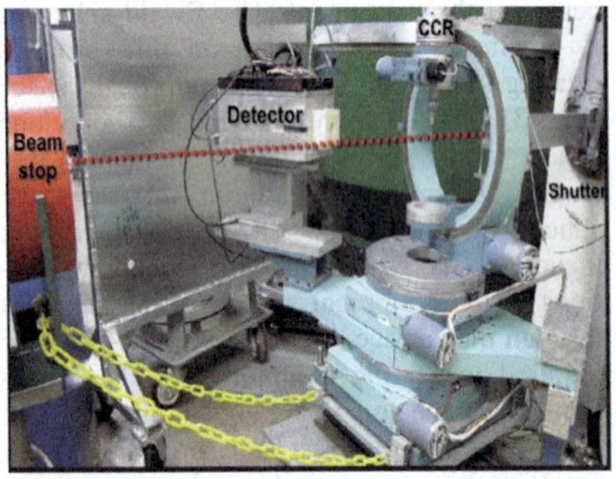

Fig. 4 The High-Flux Isotope Reactor four-circle diffractometer with a Eulerian cradle with ϕ (sample axis rotation) and χ (large rotation) circles on a vertical ω axis, all sitting on a θ moving the detector, typical of a monochromatic single-crystal diffractometer. CCR, closed-cycle refrigerator. Source: Ref 14

Pair Distribution Function Analysis

Pair distribution function (PDF) measurements are, in many ways, similar to powder diffraction measurements. The main difference is as follows. For data collected for "standard" powder diffraction, the attention is focused on obtaining accurate relative intensities of the Bragg peaks; for PDF measurements, the aim is to collect quantitative data across the widest momentum transfer ($Q = 2\pi/d$) range possible. Once all the appropriate corrections are applied, a Fourier transform is applied to the data to convert from reciprocal space into real space. From the resulting radial distribution function, G(r), direct measurement of the bond lengths and coordination environments (e.g., number of nearest neighbors and respective bonds) can be extracted for each of the components of the system. One big advantage of PDF analysis is that it can be model-free and can therefore be applied to all types of samples: extended and nanocrystalline solids as well as amorphous samples and melts, provided that the chemical composition and density are known.

Neutron-scattering techniques, in particular TOF, are arguably the gold standard for PDF measurements. This is because, due to the vanishing form factor, the data quality persists through rather high Q (low d-spacing), with the observed decay being solely due to the thermal parameters. This wide d- (or Q-) range of high-quality data ensures that the Fourier transform can be relatively free from artifacts that hamper the interpretation of the low-r data, where the information regarding nearest neighbors and coordination environment is more critical. Furthermore, the data are additive, in the sense that the measured data

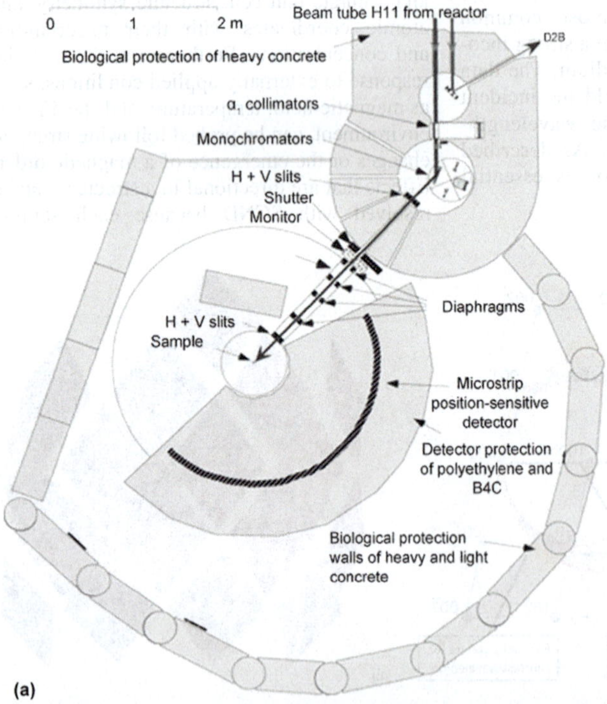

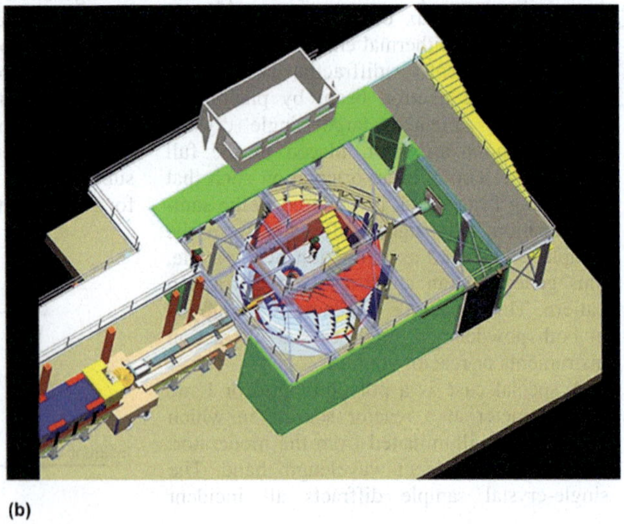

Fig. 5 Schematic representations of state-of-the-art neutron diffractometers. (a) Reactor-based instrument, D20 at the Institut Laue-Langevin in France, allowing access to multiple incident wavelengths through gaps in the biological shielding of heavy concrete. These are placed to match the takeoff angle from a number of available monochromators. Reprinted with permission from T.C. Hansen et al., *Meas. Sci. Technol.*, Vol 19, 2008, p 034001, https://doi.org/10.1088/0957-0233/19/3/034001. (b) Idealized layout of POWGEN at Oak Ridge National Laboratory's Spallation Neutron Source, with a tear-shaped detector bank optimized for resolution-matching data collection, thus allowing single dataset analysis, and opposed to the traditional joint refinement of multiple detector banks.

correspond to the superposition of all the atomic pairs, called partial structure factors. For example, in SiO_2, there would be three partials: Si-O, Si-Si, and O-O. Because the relative contribution of the partials is proportional to the cross section of the element to the incident radiation, neutron PDF measurements are ideally suitable for compounds comprising low-Z elements, especially when in the presence of heavier ones, as well as being more sensitive to bond lengths between metals and light anions (deuterium, carbon, oxygen, nitrogen, fluorine). Note that the separation of total radial distribution function in its constitutive partials is nontrivial; one such example is described in the section "Local Structure and Short-Range Order" in this article.

Time of flight surpasses both laboratory x-rays and reactor-based neutron sources because of the availability of epithermal neutrons with short wavelengths. Even for a silver x-ray tube (λ = 0.56 Å), the highest momentum transfer available is $Q = 22.4$ Å^{-1}, while for $\lambda = 0.3$ Å, a readily available wavelength at appropriate moderators in spallation sources, the range goes up to $Q = 41.9$ Å^{-1}. To have a sense of the improvement, the resolution, in real space, of a PDF measurement is related to $2\pi/Q_{max}$. Therefore, for the latter, one can resolve a feature (in real space) of 0.15 Å, while for the former, only features down to 0.35 Å are visible.

Relationship between Detector Space and Reciprocal Space

Diffraction data must be converted from the units in which they are collected, for example, scattering angle and wavelength, into physically meaningful units, for example, d-spacing. In the case of monochromatic diffractometers as typically found at reactors, the diffraction profile is obtained by collapsing any out-of-plane contributions into a scattering angle plot, which can then be readily converted into d-spacing via Bragg's law, in the same way as for x-ray diffraction. In the case of TOF diffraction, data are collected in two dimensions: scattering angle and TOF. The neutron wavelength can be extracted from TOF via the de Broglie relationship:

$$\lambda = \frac{h}{m_n v} \text{with } v = L/TOF \qquad \text{(Eq 9)}$$

where λ is the neutron wavelength, h is Planck's constant, and v is the velocity, as determined from the TOF and the neutron path length, L. This, combined with Bragg's law, suffices to convert every detected neutron into a corresponding d.

Although such formalism would indicate a linear relationship between d and t, instrument calibration and sample-specific corrections may need to be applied to account for wavelength-dependent sample absorption (a) as well as a zero offset (z). The equation to convert from TOF to d-spacing becomes:

$$TOF = \frac{2m_n}{h} L \sin\theta \times d + a \times d^2 + z \qquad \text{(Eq 10)}$$

In the case of single-crystal diffraction, this equation suffices. However, in the case of powder diffraction, because the angular information that is required for conversion to d-spacing is superfluous for analysis, a strategy to aggregate essential information, called time focusing, is applied. This consists of applying Eq 5 to all the pixels, followed by adding all the data in d-spacing. Because the incident wavelength and scattering angle are needed to apply scattering corrections, most refinement software packages require the time-focused data to be converted back into TOF units calculated on a single detector at a representative scattering angle that approximates the experimental conditions. However, work has begun on avoiding the need for time focusing (Ref 15).

Systems and Equipment

Every neutron diffraction instrument is comprised of four main components:

- Neutron source, either constant wavelength or TOF
- Properly conditioned neutron beam in terms of size, divergence, and spectral distribution
- Sample stage to place and orient the sample in the beam as well as expose it to the desired conditions
- Detector or detector array

In this section, a brief introduction is given to these instrument components.

Instrument Components

Neutron Beam Conditioning

One of the main challenges present in instrument design is to deliver as high a neutron flux as possible to the sample, keeping the beam divergence at a suitable level for the application. However, due to its neutral charge and consequent weak interaction with matter, neutron transport, steering, and focusing are challenging.

In the case of monochromatic radiation, the beam can be effectively steered by using Bragg diffraction from an appropriately aligned crystal. Increased flux can also be achieved by highly mosaic crystals, bent crystals, or stacks of carefully positioned crystals.

Time-of-flight beam manipulation presents a different challenge, because the neutron optical elements must operate across the entire wavelength band of interest. Here, the transport mechanism is based on external reflection on nickel or, even more effective, on supermirror guides, which are devices that consist of alternating thin layers of nickel and titanium. These guides can effectively reflect neutrons with wavelength as low as 0.5 Å; however, the critical angle for total reflection is shallow, which requires relatively large distances for beam steering and focusing. The figure of merit for neutron guides is m, which relates the critical angle, φ_c, with the incident wavelength, λ (Ref 16):

$$\varphi_{c,m} = 1.73 \frac{\text{mrad}}{\text{Å}} \lambda \cdot m = \frac{0.1^\circ}{\text{Å}} \lambda \cdot m \qquad \text{(Eq 11)}$$

A value of $m = 1$ indicates that the surface will reflect neutrons with 1 Å wavelength with a critical angle of 0.1°, the critical angle of reflection for pure nickel; $m = 2$ will reflect the same wavelength but with twice the critical angle. An example of the focusing effect of such a supermirror guide is represented in Fig. 6.

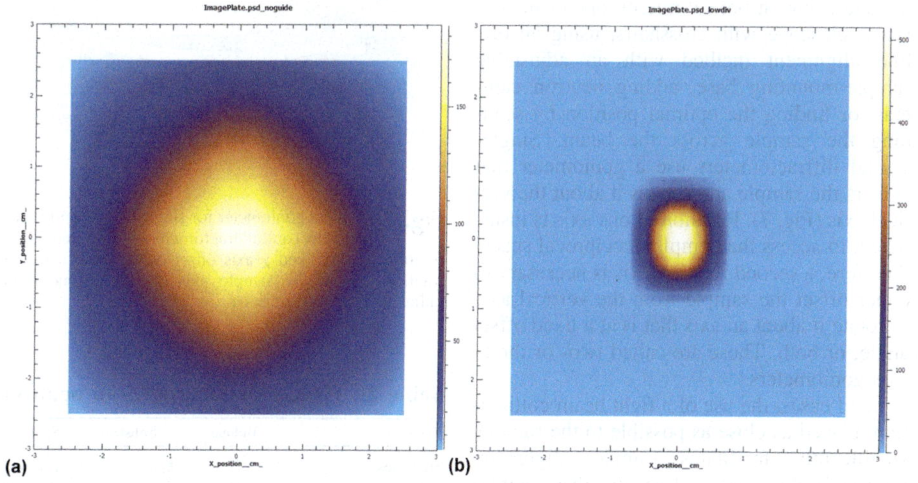

Fig. 6 Monte Carlo simulations of a focusing guide performance. (a) Unfocused beam at the sample position. (b) Incident beam through a 2.4 m (7.9 ft) focusing square guide with logarithmic spiral curvature. Note that the color scale at right is much higher (500 versus 200 counts).

It is noteworthy that supermirror guide technology is used to great effect in reactor sources as well. For example, the Wombat diffractometer at the Australian Centre for Neutron Scattering has a competitive neutron flux on the sample when compared with much higher-power reactor sources, due to a state-of-the-art neutron guide system (Ref 17).

With neutron scattering, as in all diffraction experiments, there is a trade-off between intensity and resolution. While there are scientific cases where high flux is a requirement, such as measuring small samples or in high-pressure research, there are many other instances where resolution cannot be sacrificed. This can more easily be achieved by building long instruments, because large distances from the source increase time resolution and reduce background. It is fortunate that such geometry is ideally coupled to the shallow critical angles characteristic of neutron optics components. As a result, longer diffraction beamlines, especially those based on spallation sources, will typically have better resolution, at the cost of weaker flux on the sample.

Sample Positioning

A critical requirement for high-quality data collection is the reproducible positioning of the sample. While traditional neutron instruments involved wide beams illuminating large samples, which resulted in relatively poor resolution (when compared with x-rays), modern beamlines can deliver submillimeter beam dimensions to equally microscopic samples. However, this makes sample alignment more challenging. Furthermore, as neutron optics deliver a tighter focused beam, a small deviation from the ideal sample position can deeply affect the incident illuminating spectrum (as can be seen in Fig. 6b). Strategies to align powder and single-crystal samples in the beam include using an optical microscope or magnifying telescope with crosshairs, using an offline alignment method with an adjustable sample-mounting base, adding neutron cameras, or finding the optimal position by scanning the sample across the beam. Single-crystal diffractometers use a goniometer that centers the sample and rotates it about the vertical axis (Fig. 7). Rotation of one axis is insufficient to access the complete reciprocal space; therefore, a second rotation axis is necessary to either offset the sample from the vertical axis or rotate it about an axis that is at a fixed offset angle, or both. These are called two- or three-axis goniometers.

In all cases, the use of a tight beam collimation, placed as close as possible to the sample, significantly facilitates sample alignment, reduces background, and ensures that complementary measurements (empty instrument, incident spectra, etc.) are as representative as possible of the measured sample.

Detectors

Neutron detectors are typically an integral part of a beamline, in the sense that they are very often custom built for it, and the detector choices have important implications for many other aspects. Because they are bulky, expensive, wired with complex electronics, and sometimes inside a vacuum chamber, they are not meant to be removed from the instrument. Consequently, detector type, solid angle coverage, and placement are defined at the design stage for (almost) the lifetime of the beamline.

Key parameters of interest when deciding the detection strategy for an application include size (i.e., solid angle coverage and sample-to-detector distance), detection efficiency at the wavelength(s) of interest, spatial resolution, count rate (maximum flux supported), time resolution (although this is highly dependent on instrument design parameters, such as flight path), and gamma discrimination (the ability to distinguish between neutrons and gamma rays). For the purposes of neutron diffraction, detectors measure events, that is, neutron counts, but are not sensitive to neutron energies. In reactor instruments, this is not an issue, because only the position is recorded; however, if energy resolution is a requirement, as in TOF measurements, this is achieved via the use of fast electronics (tens of nanosecond resolution) to record the neutron travel time.

As mentioned earlier, neutrons have almost no interaction with electrons; hence, the detection process for neutrons necessarily involves a reaction with the nucleus, resulting in a photon or charged particle. In most materials, however, the neutron absorption cross section, which determines the probability of neutron capture by the nucleus, is relatively low. There are only a select number of isotopes with sufficiently high absorption cross sections to make useful neutron-detector materials. The most commonly used isotopes are ^{3}He (5333 barn), ^{6}Li (940 barn), ^{10}B (3835 barn), and ^{nat}Gd (29,400 barn).

^{3}He (sometimes written helium-3) gas-filled, position-sensitive, proportional counters are the most widely used, but other configurations, including multiwire proportional chambers, are also common. Lithium-6 and boron-10 are most often employed in solid form, either incorporated in scintillator materials or deposited as thin convertor layers. Figure 8 shows a scintillator-based detector module, called an Anger camera. It has submillimeter spatial resolution and is deployed both in constant-wavelength and TOF diffractometers. In each case, the neutron-nucleus reaction products generate a signal charge, either directly (as in proportional detectors) or indirectly (as in scintillators), where an initial photon is then converted via photomultipliers and is subsequently measured by signal-processing electronics. Some of the key characteristics for the main types of neutron detectors in use are listed in Table 2.

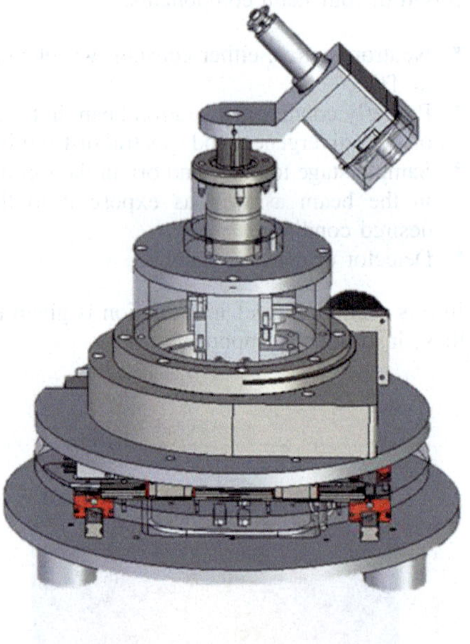

Fig. 7 Typical goniometer for single-crystal and time-of-flight Laue diffractometers with a vertical ω rotation axis, a fixed χ axis offset of 45°, and a full ϕ rotation around the sample axis. It can be mounted either vertically up or down.

Fig. 8 Individual Anger camera module with scintillator 15 by 15 cm (6 by 6 in.) active area. Optics package and photomultipliers are at the very front. To optimize speed, a significant portion of the signal processing is made on the boards at the back.

Table 2 Typical characteristics of neutron detectors for diffraction

Type	Isotope	Substrate	Spatial resolution	Notes
He tubes	^{3}He	Gas	Linear, poor	Poor time resolution but good gamma discrimination
Multiwire	^{3}He	Gas	Area, good	Same as tubes; difficult to scale up active area
Anger camera	^{6}Li	Glass (GS10)	Very good	Fast but moderate gamma discrimination
Wavelength shifting	^{6}Li	LiF, ZnS(As)	Medium	Moderate count rate and good gamma discrimination
Converter foils	Gd, ^{6}Li, ^{10}B	Gas chamber	Excellent	Very fast response but poor gamma discrimination

Sample Environment

Sample environment equipment allows neutron diffraction data to be collected in situ on samples while subjected to controlled conditions. This includes temperature, pressure, and magnetic field (Ref 18). With their high transmission through most materials, neutron beams offer exciting flexibility in sample environment design and materials choices. In that sense, this section presents a cursory survey of the standard sample environments that are common at user facilities. Nevertheless, custom-made systems for individual experiments or areas of research are not uncommon.

Low Temperature

Measurements under cryogenic conditions are one of the most sought after in neutron diffraction, due to improved data quality (a result of reduced thermal motion of the atoms) and because many of the phenomena that match well to the unique characterization capabilities of neutron scattering are found at low temperatures, such as the ordered state of magnetic materials. To this effect, several different approaches to cool samples may be employed. Helium-bath (or wet) cryostats use a controlled flow of cold gas from evaporating liquid helium or nitrogen cryogens to cool samples down to minimum temperatures of 4.2 K (the boiling temperature of liquid helium at ambient pressure) or even down to 1.5 K by reducing the pressure in the sample chamber.

Measurements below this temperature require specialized inserts placed inside conventional cryostats. Helium-3 inserts can reach temperatures of approximately 300 mK, while dilution refrigerators can reach temperatures of approximately 30 mK. Due to the complexities of the equipment, the use of these inserts requires extensive setup time. Changing a sample often takes many hours, which must be considered when planning a neutron diffraction experiment. Another important aspect to consider is that the thermal conductivity of the sample is negligible at these temperatures; therefore, great care must be taken to accurately determine the measurement temperature, especially in powder samples.

Closed-cycle refrigerators (CCRs) use compressors with helium gas as the working fluid. They can have a minimum operating temperature of approximately 5 K but, unlike wet cryostats, do not consume liquid helium, a rather expensive consumable. Another advantage is that some CCRs are especially designed to allow heating above room temperature (up to approximately 800 K). Another option for intermediate temperatures is a cryostream setup, which evaporates liquid cryogens and directs a laminar stream of the cold gas, for example, helium, nitrogen, or argon, to envelop the sample without need of any additional shielding. The gas may also be heated to reach up to approximately 500 K. Two important advantages of cryostream systems are low background, due to the absence of heat shields and vacuum shrouds, and optical access to the sample, allowing for precise alignment. However, with the nozzle diameter size being only a few millimeters, this option is limited to small samples. Without careful gas stream alignment, this open system can be prone to ice accumulation from atmospheric moisture.

Furnaces

Measurements above room temperature are often desired, for example, when dealing with phase transitions, chemical reactions, and melts. Vacuum furnaces operate through resistive heating of a metal foil heating element, most typically vanadium or niobium. Vanadium is chosen because it has extremely low coherent scattering, contributing only low-intensity sharp Bragg peaks in the diffraction pattern; it can operate to approximately 1200 °C (2190 °F) in inert gases. Niobium can reach higher temperatures, approximately 1600 °C (2910 °F), but can result in significant niobium Bragg peaks in the diffraction pattern. As the name suggests, vacuum furnaces must be kept in a vacuum to prevent oxidation of the heating elements and to minimize heat leaks to the outside. This environment effectively becomes slightly reducing for the sample and may preclude its rapid cooling. For heating in air or inert gas environments, atmosphere furnaces with ceramic heating elements ($MoSi_2$) and insulation can reach temperatures up to 1500 °C (2730 °F). However, insulation can result in higher background and attenuated sample signal. Lamp furnaces, also known as mirror furnaces, use halogen lamps with reflectors to direct heat onto a sample. They generally have lower background (Ref 19), but the heated zone tends to be smaller. Induction heaters may be used with electrically conductive samples, often in the context of engineering diffraction. The maximum temperatures attainable depend on the resistivity of the sample; steels, including stainless, and nickel alloys are the most suitable, while highly conductive metals, such as copper and aluminum, have a lower operating range.

The highest temperatures, up to approximately 3000 °C (5430 °F), may be reached by laser heating. One important parameter for this option is how well the sample couples to the available excitation wavelength. Due to the fact that few container materials can withstand such temperatures and because of the small focal spot of the laser, this technique requires intense, narrow neutron beams and, whenever possible, is used in conjunction with sample levitators. Aerodynamic levitators (Ref 20), where the sample sits on a cushion of flowing gas, are more common, because they are compatible with all types of samples. Metal samples may also be placed in an electrostatic levitator, as long as the surface tension of the liquid is sufficient to hold its shape. For either type of levitator, samples are limited to beads of approximately 3 mm (0.12 in.) diameter. Sample levitation allows for containerless measurements, which are often important for experiments where exact background subtraction is critical, such as PDF analysis of weak scatterers or when the sample will be melted.

When dealing with elevated temperatures, several factors must be considered. Particular care must be taken to ensure that the sample and container are compatible over the full temperature range of the study, keeping in mind that dissimilar metals in contact can often alloy into eutectics with lower melting temperatures. Additionally, overpressurizing sample containers at high temperatures due to outgassing must be prevented. Finally, the possibility of a significant temperature gradient must be considered, because the measured temperature in the furnace may not be the same as that of the sample. This offset must be determined, typically by measuring a standard sample with either a well-known thermal expansion or a phase transition at a known temperature.

High Pressure

Pressure is a vital parameter to control, because it allows the tuning of the material density without having to resort to chemical manipulation, which can add disorder and modify the electronic ground state of the parent compound. The high penetration of neutrons allows for the use of bulky cells and equipment required for high pressure (Ref 21–23). Time-of-flight diffraction, being an energy-resolved technique, is particularly well suited for extreme pressure, because it is still possible to collect a reasonable interatomic spacing range of the diffraction pattern, even when the incident and scattered beams are confined to small apertures.

Gas pressure cells can reach moderate pressures, up to roughly 0.7 GPa (7 kbar) if made from aluminum (desirable because it is more neutron transparent) or up to 1 GPa (10 kbar) if made from steel. These devices can pressurize samples of several cubic centimeters. The pressure is continuously monitored and may be easily adjusted online. One important aspect that must be considered is the pressure-temperature phase diagram of the pressurization gas in order to prevent condensation or blockages arising from freezing it, which may pose a serious safety hazard. Clamp cells can reach higher pressures, up to 2 to 3 GPa (20 to 30 kbar), but the pressure must be manually adjusted offline, and the sample size is smaller, on the order of 200 mm^3 (0.012 $in.^3$). The highest pressures are reached via the use of opposed anvil devices. The most ubiquitous of this type of press is the Paris-Edinburgh style (Ref 24), typically fitted with toroidal anvils made from cubic boron nitride or polycrystalline diamond. These cells can pressurize a sample of approximately 80 mm^3 (0.005 $in.^3$) up to 25 GPa (250 kbar), and they have been successfully cooled to 2 K and heated to 2000 K. For even higher pressures, gem cells, a type that has been widely used in the context of Raman spectroscopy and synchrotron radiation, have now been expanded to neutron scattering. Here, the anvils are made from single-crystal hard materials, ideally single-crystal diamond (Fig. 9b), but, for lower pressures, sapphire (Al_2O_3) or moissanite (SiC) can be used. Due to the minute sample size (<1 mm^3, or 6×10^{-5} $in.^3$), measurement times can take multiple hours; therefore, samples should ideally be comprised of strong scatterers with simple crystal structures. Measurements using single-crystal diamonds have reached a record 94 GPa (940 kbar) (Ref 25), and measurements above 40 GPa (400 kbar) are now routine (Ref 26). These cells have been regularly cooled to less than 10 K, and ongoing efforts aim to couple them to resistive and laser heating, although these efforts are relatively new.

One noteworthy development is the deployment of an Atsuhime, a multianvil press, at the PLANET beamline, the high-pressure diffractometer at the Materials and Life Science Experimental Facility in Japan (Fig. 9a).

Magnetic Field

An applied magnetic field is a parameter of interest for many materials, because it couples directly with the magnetic (and sometimes nuclear) structure, an area where neutron scattering is ideally suited. Fields up to approximately 14 T may be generated with commonly available equipment. Most magnets have a vertical-field, split-coil design. Horizontal-field magnets are also available, and some specialized magnets allow an arbitrary field direction but at a reduced maximum field. Because these magnets are based on superconductors, the coils must remain cold, which can limit the sample temperature to below 325 K. Magnets with large-bore diameters can accept dilution or ^{3}He inserts to reach ultralow temperatures. Persistent-mode magnets are designed to hold a constant field without requiring active electrical current. Nonpersistent-mode magnets must continuously apply electrical current but can change the magnetic field strength more quickly. One important consideration for magnetic measurements is to understand the forces that may be generated on the sample; the sample must be securely attached to prevent movement under field. Additionally, powder samples must be fixed in some way, often by pressing into a solid pellet, to prevent individual grains from aligning with the magnetic field. One drawback of magnets is that they require a complex equipment surrounding the sample position, which can reduce the detector field of vision and can result in parasitic scattering in the diffraction pattern if the diffracted beam is not well collimated.

Specialized Sample Environment

A variety of other types of sample environment equipment are available, including gas flow (Ref 27–31) and applied electric fields (Ref 32). In situ mechanical testing is often of interest to industrial researchers. Loads up to approximately 100 kN (22,500 lbf) may be generated in tension or compression (Ref 33). For these measurements, the strain rate may be limited by the minimum time needed to collect sufficient quality data, depending on the sample and what is being interrogated. In recent years, there has been much interest in diffraction experiments on batteries during charging and discharging (Ref 34). More specialized operando experiments are facilitated by open-table instrument designs, which allow large custom equipment to be placed in the sample position.

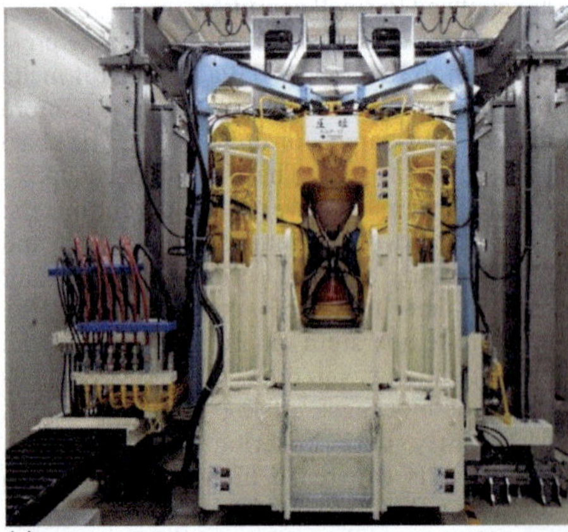

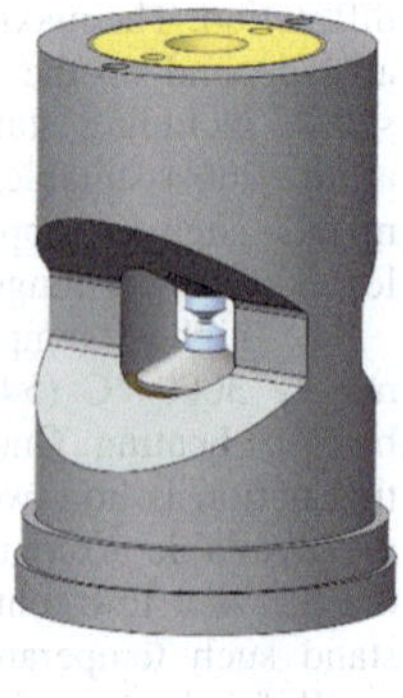

Fig. 9 (a) Atsuhime multianvil press, deployed at the PLANET beamline, Materials and Life Science Experimental Facility (MLF), Japan Proton Accelerator Research Complex. It comprises six independent axes (500 tons per axis) capable of reaching 10 GPa (100 kbar) and 2000 K on large (~1 cm^3, or 0.06 $in.^3$) samples. (b) Neutron diamond anvil cell developed at the Spallation Neutrons and Pressure beamline at the Spallation Neutron Source. Allows in situ pressure changes and can regularly reach 40 GPa (400 kbar) and 5 K. The diameter of the top surface is 50 mm (2 in.), and the diamond culets are 2 mm (0.08 in.). A typical sample size is ~600 μm in diameter and ~150 μm tall. Courtesy of T. Hattori, MLF Facility, Japan

Sample Preparation

For single-crystal neutron diffraction experiments, typical sample sizes are between 0.1 and 100 mm^3 (6×10^{-6} and 6×10^{-3} $in.^3$), depending on the specific problem, material, scattering strength, and instrument. Prior to exposure to neutrons, a candidate single-crystal or single-domain material, free from twinning, inclusions, or similar defects, is selected by close examination with a polarized optical microscope. The selected crystal is affixed to a thin pin or plate of aluminum for mounting into the diffractometer, typically with either a wire wrap for larger crystals or glue for smaller crystals (Ref 35). The crystallographic orientation of the sample within the instrument frame (UB matrix) is determined by using a small number of reflections. Once the selected crystal is deemed to be of suitable quality, all crystallographic-relevant directions are measured by reorienting the sample successively. Knowledge of the UB matrix allows the tracking of specific directions of interest for phase transitions or magnetic and structure form-factor measurements to be tracked. The crystal orientation, or face indexing, may also be determined by a laboratory single-crystal x-ray diffractometer before a neutron experiment.

For powder neutron diffraction, sample sizes depend significantly on the instrument as well as on sample factors such as purity, crystallinity, structure (e.g., unit cell size and crystal symmetry), and neutron-scattering properties (neutron cross section, attenuation, etc.). Researchers are encouraged to use available resources to estimate sample and measurement time requirements (e.g., the National Institute of Standards and Technology, or NIST, cross-section database, Ref 36; or the Institut Laue-Langevin "blue book," Ref 37). For instruments designed for high intensity, typical samples are approximately 100 to 200 mg, and the minimum sample size is approximately 10 mg. For instruments designed for high resolution, typical samples are approximately 1 g, and the minimum sample size is approximately 200 to 300 mg. Sample containers for powder are typically vanadium cans (Ref 38). Vanadium has a low coherent scattering length (0.0184 barns) and therefore has weak diffraction signal. Sample containers are typically backfilled with helium gas before sealing. The exclusion of air is necessary for cryogenic experiments, which can easily reach temperatures below the freezing point of most gases, and the relatively high thermal conductivity of helium improves the thermal equilibrium of the sample. If the sample amount is limited or if the empty container must be exactly subtracted, pure quartz (boron-free SiO_2 glass) or polyimide (Kapton, DuPont) capillaries are sometimes used.

Special care must be taken with highly absorbing samples. For experiment-planning purposes, the absorption of a sample may be estimated with online calculators (Ref 39, 40). To address this issue, one can reduce the absorption of the sample by diluting it with a well-known material of low neutron absorption. Alternatively, it is possible to reduce the neutron path length through the sample by using either a sample can with a small diameter if the beam is appropriately focused, or an annular can with a thin ring-shaped cross section to match the illuminated sample to the beam.

Calibration and Accuracy

Single Crystal

Single-crystal neutron diffraction instrumentation requires calibration for sample centering, detector positioning, detector efficiency, and incident neutron flux. The latter two are especially critical for TOF measurements. A standard sample, for example, silicon or sodium chloride, is used to calibrate the instrument geometry, while an isotropic scattering sample is used for the detector efficiency and incident neutron flux characterization. This is typically done in advance by the instrument team. The experiment will only start once the instrument is characterized, with the sample precisely placed in the predetermined location in the instrument (see the section "Sample Positioning" in this article). The accuracy of the data is dependent on reliable sample centering, because the sample must move in a eucentric motion between exposures. Typically, the resolution to a specific minimum d (d_{min}) value determines the accuracy of the refinement. For the relative atomic positions, the d_{min} must reach beyond the shortest bond distance, which is typically a carbon-hydrogen bond of approximately 0.9 Å. Data collected to a resolution of $d_{min} > 0.4$ Å can resolve bond density and nuclear density distributions.

Powder

The first step in calibrating a neutron powder diffractometer is to determine the detector positions and the incident wavelength range, which is accomplished by collecting data on a strongly diffracting standard with certified (or at least well-known) lattice parameters, for example, diamond or silicon, and then matching the observed peak positions measured at each point on each detector with expected values. The next step is to determine the instrumental peak shape parameters by using data collected on a standard having a known sample-dependent line broadening, such as NIST SRM 660b ($La^{11}B_6$). Peak shape from reactor sources tend to be Gaussian, because the incident beam originates from a single-crystal monochromator, which is easily well fitted. Modeling the peak shape is one of the most challenging aspects of TOF neutron diffraction, because it is strongly affected by the dimensions of the moderator from where the neutrons emanate. This results in a TOF-dependent exponential decay on the leading and trailing edges of each peak.

In addition to the instrument parameters, several corrections are typically applied to the experimental data. The data are normalized by data collected on an isotropic incoherent scatterer, typically vanadium, to correct for variations in detector efficiency and, in the case of TOF, the incident spectrum. Also, data collected on an empty sample container are subtracted from the experimental data to remove contributions from the can and sample environment equipment. Note that when measuring highly absorbing samples, direct subtraction of the empty can measurement may overcorrect, leading to small negative peaks, which must be appropriately treated because refinement software cannot handle negative intensities (e.g., postprocessing subtracted or excluded).

Instrument staff should be able to supply instrument parameter files formatted for major Rietveld refinement packages. These account for instrument dimensions and resolution. Ideally, users should refine only sample-dependent contributions to the measured pattern.

Data Analysis and Reliability

Single Crystal

Single-crystal neutron data are generally used for refining structural parameters, the most prominent being the atomic positions (x, y, z), with associated isotropic or anisotropic displacement parameters and occupancies. This requires a resolution of the data below the shortest bonding distances of 0.9 Å. Single-crystal neutron diffraction is a premier probe to localize and quantify structural hydrogen next to metals or in high-Z compounds. A section of reciprocal space, representative of the material symmetry, contains all the information needed to refine structure parameters, making it feasible to measure complex arrangements in a relatively short time and with high accuracy. Single-crystal neutron diffraction is reliable and repeatable, because the number of independently measured observables (the individual reflections) is much greater than the unknowns (the atomic coordinates). For large structural units and macromolecular materials, the resolution is often not sufficient to refine an atomic- or molecular-level model with only 2 Å resolution. Data analysis for inorganic compounds or small organic molecules can be done by using various commercially available or free software packages, such as SHELX (Ref 41), FULLPROF (Ref 42), JANA2006 (Ref 43), and OLEX2 (Ref 44). For biological compounds, the refinement software in use for these materials is PHENIX (Ref 45), which refines the placements of molecular clusters such as amino acids as whole units rather than as individual atoms. These systems are so complex that special packages, such as the Crystallographic Object-Oriented Toolkit (COOT) (Ref 46), are used for visualization purposes.

Powder

The primary technique used for analyzing powder diffraction data is Rietveld refinement, a well-established technique (Ref 47, 48). Rietveld refinement uses a least-squares minimization to match the observed data with a structural model of the sample and the instrumental parameters. The great innovation of this approach is that the calculated data not only include the intensity expected for each Bragg peak but also envelope that information in a peak shape model and background functions. The available variables are sequentially refined to optimize a model that includes structural parameters (e.g., lattice parameters and atomic coordinates), crystallite size, microstrain, orientation texture, and other microstructural parameters. Neutron diffraction is particularly sensitive to atomic parameters: x, y, z positions, fractional occupancies, and displacement parameters (isotropic or anisotropic). Despite the usefulness of computer programs to help automate least-squares refinement, they should not be treated as a "black box;" careful attention must be given to areas where the calculated and experimental patterns do not agree. Instrumental parameters are provided by the instrument staff and should only be changed with care. As with any fitting technique, the minimum number of parameters necessary to obtain a good fit should be refined. When available, co-refining neutron data with good-quality x-ray data may improve the sensitivity and reliability of the results, beyond what is possible with either approach alone. However, care must be taken to properly weight the neutron and x-ray data.

In situations where the primary concern is to determine lattice parameters rather than refine atomic parameters, Pawley or LeBail refinements may provide more accurate results, particularly when sample factors, such as texture, affect the peak intensities. These refinement techniques allow the peak intensities to freely adjust while constraining the peak positions according to the known cell metrics and space group. To determine full structural models, that is, structure solution, the most common technique is single-crystal x-ray diffraction, because it requires only small crystals, can be done in-house, and is free from peak overlap, thus yielding more accurate structure factor extraction. However, when single crystals are not available, structures may also be determined from powder diffraction data with more advanced analysis techniques (Ref 49, 50). Again, the combined treatment of neutron and x-ray diffraction is preferred.

Application Examples

Magnetic Structures

Traditional magnetization measurements are suitable to track magnetic features (i.e., measure transition temperatures, suggest the magnetic order type, and determine saturation magnetic moments), neutron scattering is the sole technique that can elucidate the magnetic structure, determining the orientation and magnitude of each magnetically ordered spin. The magnetic scattering from an ordered structure appears as Bragg scattering (i.e., peaks or reflections). Care must be taken when dealing with ferromagnetic structures because this extra scattering overlaps with nuclear (i.e., nonmagnetic) peaks, which may be difficult to resolve, especially in the case of weak magnetism. To solve this problem, one can use polarized neutrons, in which differential spin-up/spin-down measurements will remove the nuclear contribution but not the magnetically ordered component. These measurements can be complemented with a second set of measurements above the Curie temperature, where no ordering should be present. An illustrative example of the power of this technique is Fe_4O_5. Due to its ability to stabilize in a variety of oxidation states, iron can form complex oxides, where it can be present in several oxidation states and local environments. Many of these phases are recovered from high-pressure conditions and are therefore difficult to grow in large amounts. This oxide not only exhibits a complex crystal structure, with three iron sites, but also shows magnetic reorientation with temperature, as well as a dimer-trimer structural ordering of the iron octahedra, shown in Fig. 10. The magnetic scattering contribution arising from each iron site and the structural signatures associated with these transitions were determined by using neutron diffraction (Ref 51).

Local Structure and Short-Range Order

Pair distribution function (PDF) analysis gives information about all the pairs of atoms within a sample, starting from the nearest neighbors and progressing outward. This function can be obtained directly from a Fourier transform of carefully collected total scattering measurements. The result is a histogram containing the sum of all the partial structure factors, where the partial structure factor is the contribution to the data from a specific pair of atoms in the structure. While traditional diffraction measurements require a crystalline sample, PDF analysis can be performed on amorphous materials such as glasses and liquids. One advantage of neutron diffraction is that it can make use of isotopically enriched samples. These may have quite different neutron-scattering lengths, despite being chemically identical. The difference in datasets collected on isotopically different but otherwise similar samples allows the extraction of the radial distribution of atom pairs containing the isotopes in question, resulting in the experimental separation of element-specific partial radial distribution functions. The work of Ahn et al. (Ref 52) illustrates this quite effectively by investigating a series of bulk metallic glass samples with different isotopes of nickel and neodymium (Fig. 11). Here, the local environment of each component of the complex, aluminum-rich bulk metallic glass, $Al_{87}Ni_7Nd_6$, was studied by collecting a set of

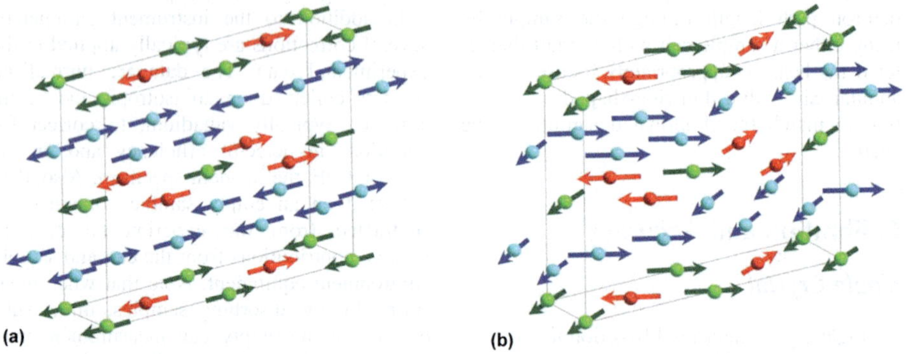

Fig. 10 Complex magnetic structures of Fe_4O_5. (a) Collinear magnetic structure measured at 150 K. (b) Canted magnetic structure develops upon cooling at 85 K (here solved at 10 K), which gives rise to the observed increase in susceptibility. Source: Ref 51

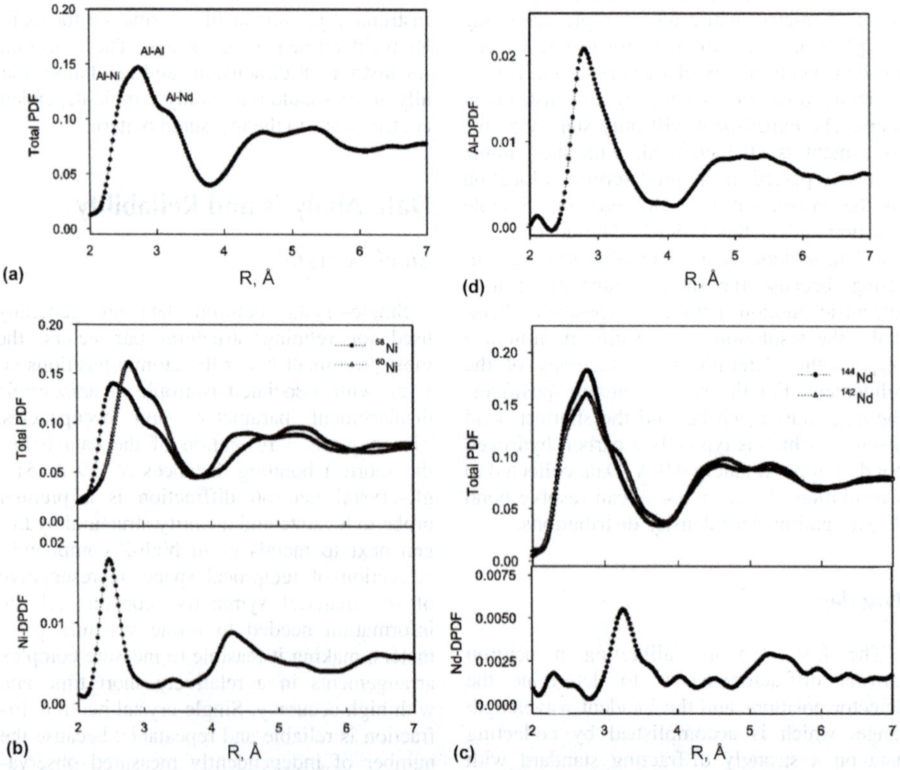

Fig. 11 Series of experimental and differential radial distribution functions of $Al_{87}Ni_7Nd_6$. (a) Total pair distribution function (PDF) with natural composition. (b) Measurements with different nickel isotopes (above) and nickel differential pair distribution function (DPDF) (below). (c) Measurements with different neodymium isotopes (above) and neodymium DPDF (below). (d) Aluminum-only partial radial distribution function (corresponding to aluminum-aluminum bonding structure) extracted from nickel and neodymium DPDFs. Source: Ref 52

differential pair distribution functions. The existence of sufficiently contrasting isotopes within the composition was critical for this work: ^{58}Ni b = 14.4 fm, ^{60}Ni b = 2.8 fm, ^{142}Nd b = 7.7 fm, and ^{144}Nd b = 2.8 fm.

High-Pressure Research

The in situ study of samples under static applied pressure can provide insights from mimicking natural conditions, such as those observed in the Earth's interior; from operational conditions, such as for structural or armor applications; and from providing experimental data to test the understanding of physical or chemical phenomena, such as lattice instabilities or superconductivity.

The sensitivity of neutrons to structure and magnetism and their penetrating power make coupling neutrons with pressure particularly interesting. It is true that, due to the higher flux and smaller x-ray beams available at synchrotron sources, one can study samples subjected to much higher pressures, in the hundreds of gigapascals. Notwithstanding this, the importance of coupling pressure and neutron scattering is attested by the fact that all operating spallation sources (as of 2019) have a dedicated high-pressure diffraction beamline. An example that well illustrates this came about with a notable study in doped iron arsenide superconductors (the AFe_2As_2 family, where "A" is alkaline earth). The observation of this new type of superconductivity (SC) spurred a great deal of research, especially regarding the role of structure and magnetism in the superconducting state. The case of its barium analog ($BaFe_2As_2$) was especially interesting, because SC could be observed in both doped systems or under application of pressure but not in the parent compound under ambient pressure. Questions regarding if both pressure and doping activated similar SC mechanisms were finally addressed through a set of neutron powder diffraction measurements up to 6 GPa, or 60 kbar and low temperature (down to 4 K). A depiction of the phase diagram explored in this study is shown in Fig. 12. These eventually showed that indeed pressure acted in the structure much in the same way as doping (Ref 53).

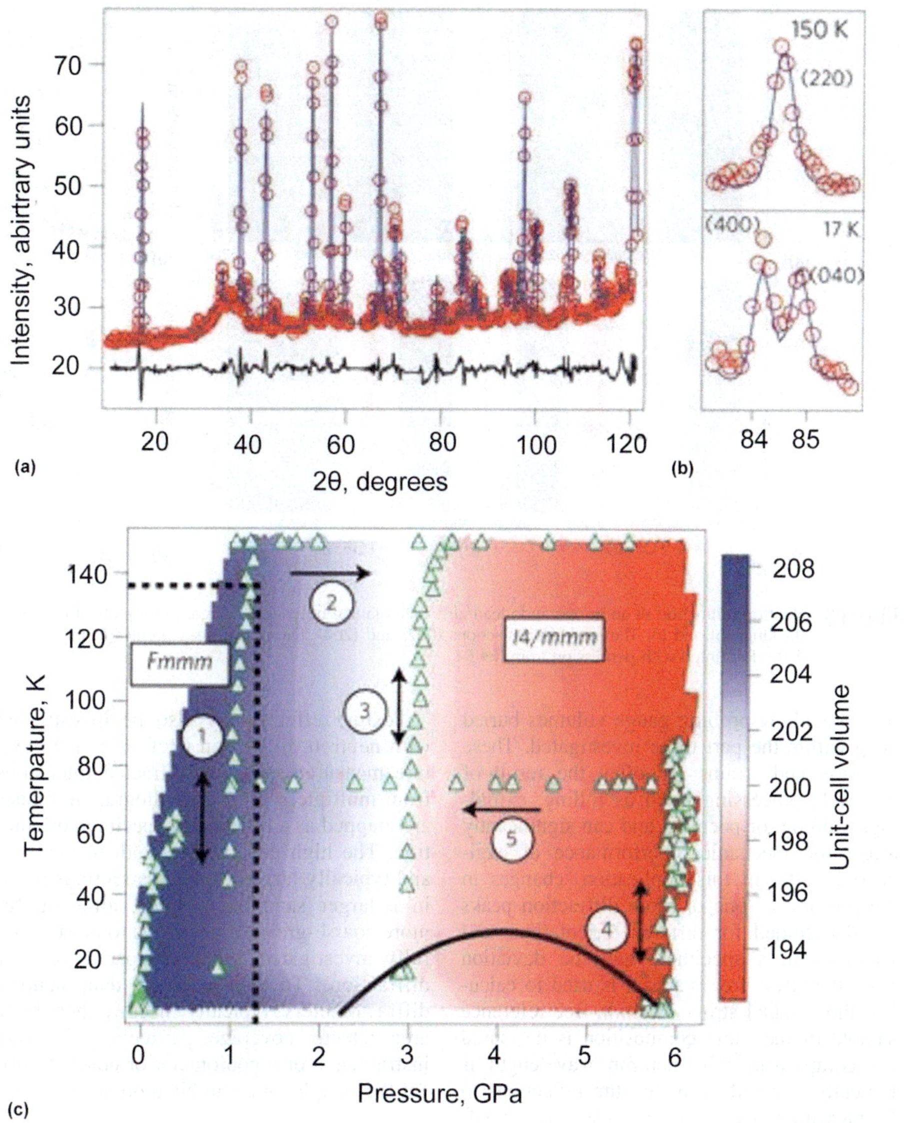

Fig. 12 In situ pressure and temperature measurements on $BaFe_2As_2$. (a) Diffraction pattern of $BaFe_2As_2$ collected at 1 GPa (10 kbar) and 17 K. (b) Peak splitting characteristic of the tetragonal-to-orthorhombic distortion observed in the magnetically ordered phase. (c) Pressure and temperature phase diagram mapping the presence of structural distortion. White triangles indicate data collection. Reprinted with permission from Ref 53

Stroboscopic Measurements

Even in the most modern neutron facilities, data-collection time remains in the range of minutes or longer, rendering some transient states inaccessible for neutron diffraction. The recent ability to operate in event-mode data collection (meaning that the individual clock time of each detected neutron is recorded) enables after-the-fact filtering of the data against not only time but also any other logged meta-data parameter. One way this new strategy has been used to great effect is by making stroboscopic measurements, accessing states that, by their nature or by the thermodynamic conditions in which they exist, would not be present long enough to make a measurement with sufficient statistics. The only requirement is that the sample reverts unchanged to the "normal" state, with no hysteresis, so that the cycling process truly probes the same transition. To make a stroboscopic measurement, the sample is repeatedly cycled across the desired transition while diffraction data are collected. The data are then filtered against a tracked applied condition (e.g., temperature or electric field), and the data from the region of interest are then accumulated until sufficient statistics are acquired. Figure 13 shows such an example, whereby a lead-zirconate-titanate-base multilayer-type piezoelectric actuator was cycled with a square wave (0.5 Hz), and the voltage was monitored along with a continuous neutron measurement over 12 h. Sorting of neutron events against the voltage allowed the tracking of strain of the layers in the parallel and perpendicular directions to the stacking (and electric field) (Ref 54). This work greatly benefited from neutron diffraction sensitivity to the oxygen position in the vicinity of the much heavier lead.

Residual-Stress/Strain Mapping and Texture Analysis

Neutron diffraction can also be a powerful technique for investigating residual stresses and internal strains (Ref 55, 56), because

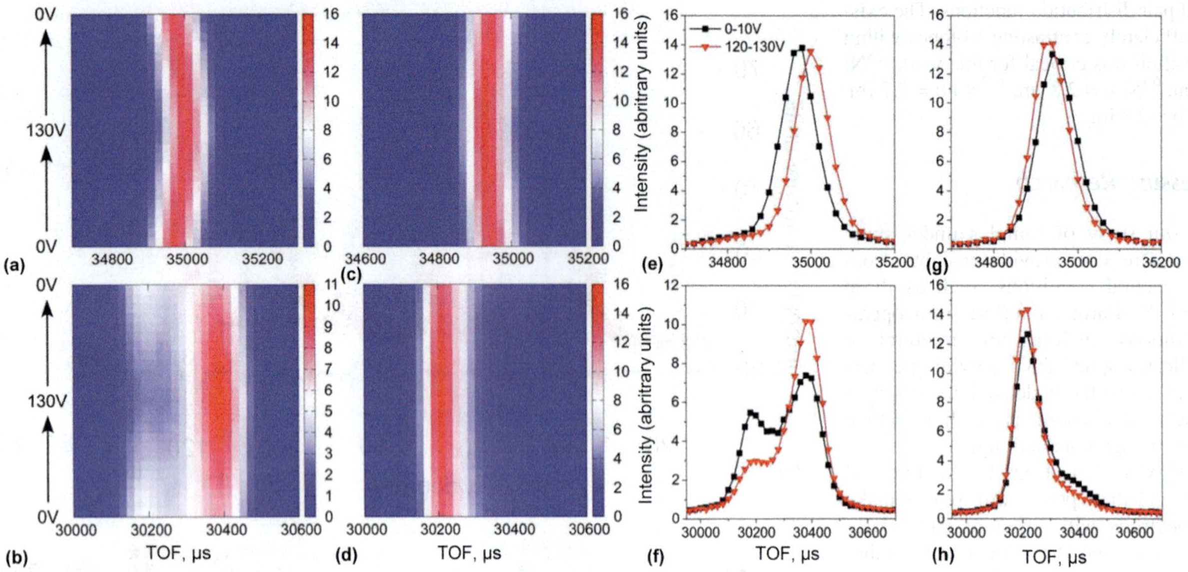

Fig. 13 Neutron diffraction of an in operando lead zirconate titanate multilayer device. Color plots of the strain across the (111) reflection (a) parallel and (c) normal to the stacking. (b) and (d) The same for reflections (002) and (200). Diffraction peak profiles in the same directions as (a) to (d) are shown in (e) to (h). TOF, time of flight. Reprinted with permission from Ref 54

neutrons allow probing gauge volumes buried deep within the part to be investigated. These stressess and strains are often the result of materials processing, such as rolling, extruding, welding, or peening, and can significantly affect the mechanical performance of engineered parts. In this application, changes in the position of one or more diffraction peaks are determined for different orientations and locations in a specimen, and the deviation from the stress-free material is used to calculate the residual stress. A strain-free reference sample of the same composition is measured for comparison. The neutron wavelength is typically selected such that the reflections to be measured scatter at an angle close to 90° 2θ. Time-of-flight techniques, being energy dispersive, allow a wider *d*-range to be accessed from a single detector, which is centered at this scattering angle.

The gauge volume to be measured is defined by slits and collimators on both the incident and diffracted neutron beams. Centering the detector at 90° 2θ simplifies the treatment of the gauge volume, which becomes more rectangular. The size of the gauge volume must be selected to optimize the required spatial resolution with a statistically significant assembly of grains and adequate Bragg peak intensity. The specimen is then scanned across the beam to populate a map of the strains within. Recently, neutron residual-stress mapping has found increased application in the development of additive manufacturing techniques. For example, residual stresses have been mapped in 718 stainless steel pieces manufactured by selective laser sintering by using different printing sequences to help determine the best technique to avoid warping during additive manufacturing of rocket engine components (Fig. 14) (Ref 57).

Texture effects may also be investigated with neutron diffraction (Ref 58–61). In texture measurements, the diffracted intensities from multiple sets of crystallographic planes are mapped as a function of specimen orientation. The high penetration depth of neutrons and typically large beam cross sections result in a larger sampling volume, allowing for more coarse-grained specimens to be successfully investigated (when compared to x-ray diffraction). To collect texture data, neutron diffractometers typically need either large area detector coverage, particularly for TOF instruments, or a goniometer or cradle to orient the sample, or a combination of these two.

Future Developments

This article aims to provide a brief introduction to neutron diffraction as well as its state-of-the-art capabilities. The opportunities afforded by this technique keep spurring a great deal of interest, and, with the currently available measurement speeds and the small sample requirements, it is only expected that the scientific community engaged in neutron scattering will continue to grow. To conclude, it is appropriate to look forward to developments that, while not yet fully mature, can provide game-changing capabilities in the next 10 to 15 years in this field. New facilities are being designed with higher-brightness moderators; in particular, one-dimensional para-hydrogen moderators promise a flux in the cold neutron range approximately three to five times brighter than current sources (Ref 62, 63). These smaller sources will allow new optics that will deliver increased flux to smaller samples. These ultrabright sources will allow new collection modes, such as monitoring in situ chemical reactions and catalytical processes or accessing transient states via pump-probe-type experiments. The spatial resolution is also increasing, enabling diffraction tomography with submillimeter resolution.

The prospect of long-pulse facilities may enable experiment-dependent resolution and flux selection as well as better defined peak shapes. Faster beamlines and new sample environments are also bound to provide new opportunities. Dynamic nuclear polarization may eliminate the need for deuteration even in biological molecules and other hydrogen-rich compounds. The combination of diamond anvil pressure cells, presented in the section "High Pressure" in this article, with the new sources may make 100 GPa (1 Mbar) pressures accessible to neutron diffraction. Finally, the coupling of neutron measurements with the now seemingly ubiquitous computational resources will enable the maturing of the "instrument on a chip" approach, whereby the data from a particular sample will be combined with the user's understanding of the science and the neutron instrument performance, allowing real-time model refinement against the measured data.

Finally, and farther out in the future, there are exciting developments in neutron production from compact sources based on electrostatic accelerators that could potentially be deployed in smaller institutions at a fraction of the cost of the existing large facilities. There are already several pilot projects (e.g., Neutrons Obtained Via Accelerator for Education and Research Activities, or NOVA ERA, at the Jülich Centre for Neutron Science in Germany; and SONATE, an accelerator-driven

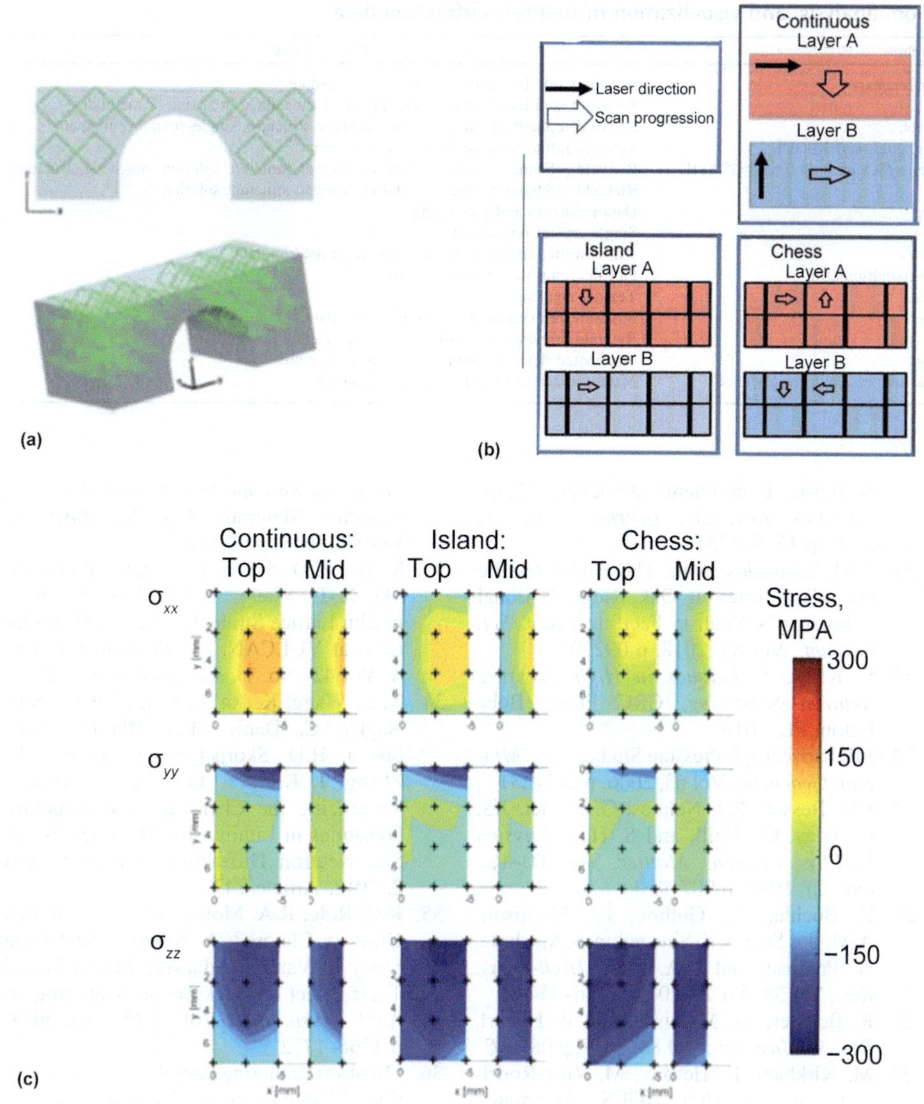

Fig. 14 Residual-stress map of additively manufactured arches with different printing sequences: continuous printing, island printing, and chess printing. (a) Test piece geometry and gauge volume mapping. The piece outer dimensions are 20 mm (0.8 in.) in length × 10 mm (0.4 in.) in diameter × 8 mm (0.3 in.) in height. (b) Printing sequence diagram. (c) Volumetric residual-stress results for the three printing sequences. Source: Ref 57

neutron source at Saclay in France), and it is hoped that such technology will revolutionize the availability of neutron techniques for the study of materials.

Resources

Table 3 lists user facility resources, while Table 4 lists software for reduction, analysis, and visualization of neutron diffraction data.

ACKNOWLEDGMENTS

The authors wish to thank B.C. Chakoumakos, E.A. Payzant, and A.D. Stoica for a very thorough review of the manuscript, and K. Berry, M. Everett, M. Frost, F. Gallmeier, A. Huq, G. Lynn, G. Martin, B. Mills, R. Riedel, L. Robertson, and B. Winn for helpful discussions. This manuscript used resources at Oak Ridge National Laboratory's Spallation Neutron Source, a DOE Office of Science User Facility, managed by the Oak Ridge National Laboratory.

An earlier chapter on this topic was published by W.B. Yelon, F.K. Ross, and A.D. Krawitz, Neutron Diffraction, *Materials Characterization*, Volume 10, *ASM Handbook*, American Society for Metals, 1986, p 420–426.

REFERENCES

1. E. Rutherford, Bakerian Lecture: Nuclear Constitution of Atoms, *Proc. R. Soc. (London) A*, Vol 97 (No. 686), 1920, p 374–400, https://doi.org/10.1098/rspa.1920.0040
2. J. Chadwick, The Existence of a Neutron, *Proc. R. Soc. (London) A*, Vol 136, 1932, p 692–708
3. W. Bothe and H. Becker, Künstliche Erregung von Kern-γ-Strahlen, *Z. Phys.*, Vol 66 (No. 5–6), 1930, p 289–306
4. I. Joliot-Curie and F. Joliot-Curie, Émission de Protons de Grande Vitesse par les Substances Hydrogénées sous l'influence des Rayons ? Très Pénétrants, *C.R. Acad. Sci.*, Vol 194, 1932, p 273
5. H. von Halban and P. Preiswerk, Preuve Experimental de la Diffraction des Neutrons, *C.R. Acad. Sci.*, Vol 203, 1936, p 73–45

Table 3 Operating and planned user facilities providing neutron diffraction instruments

Name	Location	Website	Notes
Australian Centre for Neutron Scattering (ANSTO)	Sydney, Australia	https://neutron.ansto.gov.au	Reactor
China Spallation Neutron Source	Dongguan, Guangdong, China	http://csns.ihep.ac.cn/english/index.htm	Spallation
European Spallation Source (ESS)	Lund, Sweden	https://europeanspallationsource.se/	Under construction; spallation
High Flux Isotope Reactor	Oak Ridge, Tennessee, U.S.	https://neutrons.ornl.gov/	Reactor
IBR-2	Dubna, Russia	http://flnp.jinr.ru/en/facilities/ibr-2	Pulsed reactor
Institut Laue-Langevin	Grenoble, France	https://www.ill.eu/	Reactor
ISIS	Abingdon, U.K.	https://www.isis.stfc.ac.uk	Spallation (two target stations)
MLF (J-PARC)	Tokai, Japan	http://j-parc.jp/MatLife/en/	Spallation
MLZ (FRM-II)	Garching, Germany	https://mlz-garching.de/englisch	Reactor
NIST Center for Neutron Research	Gaithersburg, Maryland, U.S.	https://www.nist.gov/ncnr	Reactor
RA-10 (LAHN)	Buenos Aires, Argentina	http://www.lahn.cnea.gov.ar/	Under construction; reactor
SINQ (Paul Scherrer Institut)	Villigen, Switzerland	https://www.psi.ch/sinq/	Spallation (not pulsed)
Spallation Neutron Source	Oak Ridge, Tennessee, U.S.	https://neutrons.ornl.gov/	Spallation

Table 4 Commonly used software for reduction, analysis, and visualization of neutron diffraction data

Program	Ref	Website	Primary focus
EXPO2014	64	http://www.ba.ic.cnr.it/softwareic/expo/	Structure solution from powder, direct methods
FOX	65	https://fox.vincefn.net/	Structure solution from powder, simulated annealing, "real space" methods
FULLPROF	42	https://www.ill.eu/sites/fullprof/	Rietveld refinement—nuclear and magnetic structure; single-crystal refinement
GSAS/EXPGUI	66, 67	https://subversion.xray.aps.anl.gov/trac/EXPGUI	Rietveld refinement; single-crystal refinement
GSAS-II	68	https://www.aps.anl.gov/Science/Scientific-Software/GSASII	Rietveld refinement; single-crystal refinement; structure solution; magnetic structures
JANA2006	43	http://jana.fzu.cz/	Rietveld refinement; single-crystal refinement; structure solution
Mantid	69	http://www.mantidproject.org	Data reduction and processing
OLEX2	44	http://www.olexsys.org/	Single-crystal refinement
PDFgetN	70	http://pdfgetn.sourceforge.net/	Pair distribution function extraction from neutron data
PDFgui	71	https://www.diffpy.org/products/pdfgui.html	Pair distribution function analysis
MAUD	72	http://maud.radiographema.eu/	Texture analysis
SHELX	41	http://shelx.uni-goettingen.de/	Structure refinement from single-crystal data
TOPAS	73	http://www.topas-academic.net/	Rietveld refinement—fundamental parameters a possibility
PHENIX	45	https://www.phenix-online.org	Automated determination of molecular structures
COOT	46	https://www2.mrc-lmb.cam.ac.uk/personal/pemsley/coot/	Macromolecular model building; validation

6. D.P. Mitchell and P.N. Powers, Bragg Reflection of Slow Neutrons, *Phys. Rev.*, Vol 50, 1936, p 486–487
7. J.S. Schwinger, On the Spin of the Neutron, *Phys. Rev.*, Vol 52, 1937, p 1250
8. O. Halpern and M.H. Johnson, *Phys. Rev.*, Vol 55, 1939, p 898–923
9. L.W. Alvarez and F. Bloch, A Quantitative Determination of the Neutron Moment in Absolute Nuclear Magnetons, *Phys. Rev.*, Vol 57, 1940, p 111–122
10. K. Sköld and D.L. Price, Ed., *Neutron Scattering*, Academic Press, 1986
11. K.H. Andersen and C.J. Carlile, *J. Phys.: Conf. Ser.*, 2016, p 746 012030
12. N.F. Berk, *J. Res. Natl. Inst. Stand. Technol.*, Vol 98 (No. 1), Jan–Feb 1993, p 15–30
13. V.F. Sears, Neutron Scattering Lengths and Cross Sections, *Neutron News*, Vol 3 (No. 3), 1992, p 29–37, https://doi.org/10.1080/10448639208218770
14. B.C. Chakoumakos, H. Cao, F. Ye, A.D. Stoica, M. Popovici, M. Sundaram, W. Zhou, J.S. Hicks, G.W. Lynn, and R.A. Riedel, Four-Circle Single-Crystal Neutron Diffractometer at the High Flux Isotope Reactor, *J. Appl. Crystallogr.*, Vol 44, 2011, p 655–658, https://doi.org/10.1107/S0021889811012301
15. P. Jacobs, A. Houben, W. Schweika, A.L. Tchougréeff, and R. Dronskowski, *J. Appl. Crystallogr.*, Vol 48, 2015, p 1627–1636
16. C.P. Cooper-Jensen, A. Vorobiev, E. Klinkby, V. Kapaklis, H. Wilkens, D. Rats, B. Hjörvarsson, O. Kirstein, and P.M. Bentley, *J. Phys.: Conf. Ser.*, Vol 528, 2014, p 012005
17. G.J. McIntyre and P.J. Holden, *J. Phys.: Conf. Ser.*, Vol 746, 2016, p 012001
18. I.F. Bailey, A Review of Sample Environments in Neutron Scattering, *Z. Kristallogr.*, Vol 218, 2003, p 84–95
19. G. Lorenz, R.B. Neder, J. Marxreiter, F. Frey, and J. Schneider, A Mirror Furnace for Neutron Diffraction up to 2300 K, *J. Appl. Crystallogr.*, Vol 26, 1993, p 632–635
20. C. Landron et al., Aerodynamic Laser-Heated Contactless Furnace for Neutron Scattering Experiments at Elevated Temperatures, *Rev. Sci. Instrum.*, Vol 71, 2000, p 1745–1751
21. A.M. Santosdos et al., The High Pressure Gas Capabilities at Oak Ridge National Laboratory's Neutron Facilities, *Rev. Sci. Instrum.*, Vol 89, 2018, p 092907
22. S. Klotz, *Techniques in High Pressure Neutron Scattering*, CRC Press, Boca Raton, FL, 2013
23. J.B. Parise, High Pressure Studies, *Rev. Mineral. Geochem.*, Vol 63, 2006, p 205–231
24. J.M. Besson, R.J. Nelmes, G. Hamel, J.S. Loveday, G. Weill, and S. Hull, *Physica B: Phys. Condens. Mattter*, Vol 180–181 (No. 2), 1992, p 907–910
25. R. Boehler, M. Guthrie, J.J. Molaison, A.M. dos Santos, S. Sinogeikin, S. Machida, N. Pradhan, and C.A. Tulk, *High-Press. Res.*, Vol 33 (No. 3), 2013, p 546–554
26. R. Boehler, J.J. Molaison, and B. Haberl, *Rev. Sci. Instrum.*, Vol 88, 2017, p 083905
27. M. Kirkham, L. Heroux, M. Ruiz-Rodriguez, and A. Huq, AGES: Automated Gas Environment System for In Situ Neutron Powder Diffraction, *Rev. Sci. Instrum.*, Vol 89, 2018, p 092904
28. D. Olds et al., A High Temperature Gas Flow Environment for Neutron Total Scattering Studies of Complex Materials, *Rev. Sci. Instrum.*, Vol 89, 2018, p 092906
29. R. Haynes et al., New High Temperature Gas Flow Cell Developed at ISIS, *J. Phys.: Conf. Ser.*, Vol 251, Nov 2010, p 012090
30. J.F.C. Turner, R. Done, J. Dreyer, W.I.F. David, and C.R.A. Catlow, On Apparatus for Studying Catalysts and Catalytic Processes Using Neutron Scattering, *Rev. Sci. Instrum.*, Vol 70 (No. 5), May 1999, p 2325–2330
31. M.G. Kibble, A.J. Ramirez-Cuesta, C.M. Goodway, B.E. Evans, and O. Kirichek, Hydrogen Gas Sample Environment for TOSCA, *J. Phys.: Conf. Ser.*, Vol 554, Nov 2014, p 012006
32. T.-M. Usher et al., Time-of-Flight Neutron Total Scattering with Applied Electric Fields: Ex Situ and In Situ Studies of Ferroelectric Materials, *Rev. Sci. Instrum.*, Vol 89, 2018, p 092905
33. K. An, H.D. Skorpenske, A.D. Stoica, D. Ma, X.-L. Wang, and E. Cakmak, First In Situ Lattice Strain Measurements under Load at VULCAN, *Metall. Mater. Trans. A*, Vol 42 (No. 1), Jan 2011, p 95–99
34. X.-L. Wang, K. An, L. Cai, Z. Feng, S.E. Nagler, C. Daniel, K.J. Rhodes, A.D. Stoica, H.D. Skorpenske, C. Liang, W. Zhang, J. Kim, Y. Qi, and S.J. Harris, Visualizing the Chemistry and Structure Dynamics in Lithium-Ion Batteries by In-Situ Neutron Diffraction, *Sci. Rep.*, Vol 2, 2012, Article 747
35. K.C. Rule, R.A. Mole, and D. Yu, Which Glue to Choose? A Neutron Scattering Study of Various Adhesive Materials and Their Effect on Background Scattering, *J. Appl. Crystallogr.*, Vol 51 (No. 6), 2018, p 1766–1772
36. "Neutron Scattering Lengths and Cross Sections," NIST Center for Neutron Research, 2018, https://www.ncnr.nist.gov/resources/n-lengths/
37. A.-J. Dianoux and G. Lander, Ed., *Neutron Data Booklet*, 2nd ed., OCP Science, Philadelphia, PA, 2003
38. M. Potter, H. Fritzsche, D.H. Ryan, and L.M.D. Cranswick, Low-Background Single-Crystal Silicon Sample Holders for Neutron Powder Diffraction, *J. Appl. Crystallogr.*, Vol 40 (No. 3), 2007, p 489–495
39. "Compute Neutron Attenuation and Activation," NIST Center for Neutron Research, 2018, https://www.ncnr.nist.gov/instruments/bt1/neutron.html
40. "Neutron Calculator V2," neutroncalc, FRM II, 2018, https://webapps.frm2.tum.de/intranet/neutroncalc/
41. G.M. Sheldrick, A Short History of SHELX, *Acta Crystallogr. A*, Vol 64, 2008, p 112–122
42. J. Rodriguez-Carvajal, *Phys. B*, Vol 192, 1993, p 55
43. V. Petricek, M. Dusek, and L. Palatinus, Crystallographic Computing System, JANA2006:

General Features, *Z. Kristallogr.*, Vol 229 (No. 5), 2014, p 345–352
44. O.V. Dolomanov, L.J. Bourhis, R.J. Gildea, J.A.K. Howard, and H. Puschmann, OLEX2: A Complete Structure Solution, Refinement and Analysis Program, *J. Appl. Crystallogr.*, Vol 42, 2009, p 339–341
45. P.D. Adams, P.V. Afonine, G. Bunkoczi, V.B. Chen, I.W. Davis, N. Echols, J.J. Headd, L.W. Hung, G.J. Kapral, R.W. Grosse-Kunstleve, A.J. McCoy, M.W. Moriarty, R. Oeffner, R.J. Read, D.C. Richardson, J.S. Richardson, T.C. Terwilliger, and P.H. Zwart, PHENIX: A Comprehensive Python-Based System for Macromolecular Structure Solution, *Acta Crystallogr. D*, Vol 66, 2010, p 213–221
46. P. Emsley, B. Lohkamp, W.G. Scott, and K. Cowtan, Features and Development of COOT, *Acta Crystallogr. D*, Vol 66, 2010, p 486–501
47. L.B. McCusker, R.B. Von Dreele, D.E. Cox, D. Louër, and P. Scardi, Rietveld Refinement Guidelines, *J. Appl. Crystallogr.*, Vol 32 (No. 1), Feb 1999, p 36–50
48. B.H. Toby, R Factors in Rietveld Analysis: How Good Is Good Enough? *Powder Diffr.*, Vol 21 (No. 1), March 2006, p 67–70
49. W.I.F. David, K. Shankland, L.B. McCusker, and C. Barlocher, Ed., *Structure Determination from Powder Diffraction Data*, International Union of Crystallography, 2006
50. W.I.F. David and K. Shankland, Structure Determination from Powder Diffraction Data, *Acta Crystallogr. A*, Vol 64 (No. 1), Jan 2008, p 52–64
51. S.V. Ovsyannikov, M. Bykov, E. Bykova, D.P. Kozlenko, A.A. Tsirlin, A.E. Karkin, V.V. Shchennikov, S.E. Kichanov, H. Gou, A.M. Abakumov, R. Egoavil, J. Verbeeck, C. McCammon, V. Dyadkin, D. Chernyshov, S. van Smaalen, and L.S. Dubrovinsky, *Nature Chem.*, Vol 8, May 2016, p 501
52. K. Ahn, D. Louca, S.J. Poon, and G.J. Shiflet, *Phys. Rev. B*, Vol 70 (No. 22), 2004, p 224103
53. S.A.J. Kimber, A. Kreyssig, Y.-Z. Zhang, H.O. Jeschke, R. Valentí, F. Yokaichiya, E. Colombier, J. Yan, T.C. Hansen, T. Chatterji, R.J. McQueeney, P.C. Canfield, A.I. Goldman, and D.N. Argyriou, *Nature Mater.*, Vol 8, June 2009, p 471
54. T. Kawasaki, Y. Inamura, T. Ito, T. Nakatani, S. Harjo, W. Gonga, and K. Aizawa, *J. Appl. Crystallogr.*, Vol 51, 2018, p 630–634
55. M.T. Hutchings, P.J. Withers, T.M. Holden, and T. Lorentzen, *Introduction to the Characterization of Residual Stress by Neutron Diffraction*, CRC Press, New York, 2004
56. V. Davydov, *Neutron Diffraction Analysis of Internal Stresses and Microstructure in Single and Two Phase Metals and Alloys*, Lambert Academic Publishing, Germany, 2013
57. S. Bagg, L.M. Sochalski-Kolbus, and J. Bunn, "The Effect of Laser Scan Strategy on Distortion and Residual Stresses of Arches Made with Selective Laser Melting," NASA Technical Reports Server, NASA, 2016, p 6
58. P. Xu et al., Progress in Bulk Texture Measurement Using Neutron Diffraction, *Proc. Second Int. Symp. Sci. J-PARC*, Vol 8, 2015, p 031022
59. M.-J. Li et al., The Neutron Texture Diffractometer at the China Advanced Research Reactor, *Chin. Phys. C*, Vol 40 (No. 3), 2016, p 036002
60. H.-R. Wenk, L. Lutterotti, and S.C. Vogel, Rietveld Texture Analysis from TOF Neutron Diffraction Data, *Powder Diffr.*, Vol 25 (No. 3), 2010, p 283–296
61. D. Ma, A.D. Stoica, Z. Wang, and A.M. Beese, Crystallographic Texture in an Additively Manufactured Nickel-Base Superalloy, *Mater. Sci. Eng. A*, Vol 684, Jan 27, 2017, p 47–53
62. F. Mezei, L. Zanini, A. Takibayev, K. Batkov, E. Klinkby, E. Pitcher, and T. Schönfeldt, *J. Neutron Res.*, Vol 17 (No. 2), 2014, p 101–105
63. K.-H. Andersen, M. Bertelsen, L. Zanini, E.B. Klinkby, T. Schönfeldt, P.M. Bentleya, and J. Saround, *J. Appl. Crystallogr.*, Vol 51, 2018, p 264–281
64. A. Altomare, C. Cuocci, C. Giacovazzo, A. Moliterni, R. Rizzi, N. Corriero, and A. Falcicchio, EXPO13: A Kit of Tools for Phasing Crystal Structures from Powder Data, *J. Appl. Crystallogr.*, Vol 46, 2013, p1231–1235
65. V. Favre-Nicolin and R. Černý, FOX, Free Objects for Crystallography: A Modular Approach to Ab Initio Structure Determination from Powder Diffraction, *J. Appl. Crystallogr.*, Vol 35, 2002, p 734–743
66. A.C. Larson and R.B. Von Dreele, "General Structure Analysis System (GSAS)," Los Alamos National Laboratory Report LAUR 86-748, 2000
67. B.H. Toby, EXPGUI, A Graphical User Interface for GSAS, *J. Appl. Crystallogr.*, Vol 34, 2001, p 210–213
68. B.H. Toby and R.B. Von Dreele, GSAS-II: The Genesis of a Modern Open-Source All-Purpose Crystallography Software Package, *J. Appl. Crystallogr.*, Vol 46 (No. 2), 2013, p 544–549
69. O. Arnold et al., Mantid—Data Analysis and Visualization Package for Neutron Scattering and μSR Experiments, *Nucl. Instrum. Methods Phys. Res. A*, Vol 764, 2014, p 156–166
70. P.F. Peterson, M. Gutmann, T. Proffen, and S.J.L. Billinge, PDFgetN: A User-Friendly Program to Extract the Total Scattering Structure Function and the Pair Distribution Function from Neutron Powder Diffraction Data, *J. Appl. Crystallogr.*, Vol 33, 2000, p 1192
71. C.L. Farrow, P. Juhás, J.W. Liu, D. Bryndin, E.S. Božin, J. Bloch, T. Proffen, and S.J.L. Billinge, PDFfit2 and PDFgui: Computer Programs for Studying Nanostructure in Crystals, *J. Phys.: Condens. Matter*, Vol 19, 2007, p 335219
72. L. Lutterotti, M. Bortolotti, G. Ischia, I. Lonardelli, and H.-R. Wenk, Rietveld Texture Analysis from Diffraction Images, *Z. Kristallogr. Suppl.*, Vol 26, 2007, p 125–130
73. A.A. Coelho, TOPAS and TOPAS-Academic: An Optimization Program Integrating Computer Algebra and Crystallographic Objects Written in C++, *J. Appl. Crystallogr.*, Vol 51 (No. 1), 2018, p 210–218

SELECTED REFERENCES

- D.N. Argyriou, A.J. Allen, M. Arai, K.W. Herwig, F. Meilleur, K. Nakajima, and D.A. Neumann, Ed., Advanced Neutron Scattering Instrumentation Special Issue, *J. Appl. Crystallogr.*, Vol 51, June 2018
- T. Chatterji, Ed., *Neutron Scattering from Magnetic Materials*, Elsevier Science, 2006,
- A. Furrer, J. Mesot, and T. Strässle, *Neutron Scattering in Condensed Matter Physics (Series on Neutron Techniques and Applications)*, World Scientific Publishing Co. Inc., 2009
- C. Giacovazzo, Ed., *Fundamentals of Crystallography*, International Union of Crystallography, Oxford Science Publication, ISBN-10: 9780199573660
- E.H. Kisi and C.J. Howard, *Applications of Neutron Powder Diffraction*, Oxford University Press, 2008
- S. Klotz, *Techniques in High Pressure Neutron Scattering*, 1st ed., CRC Press, Dec 4, 2012
- L. Liang and R. Rinaldi, *Neutron Applications in Earth, Energy and Environmental Sciences (Neutron Scattering Applications and Techniques)*, Springer, 2009
- S.W. Lovesey, *Theory of Neutron Scattering from Condensed Matter*, Clarendon Press, Oxford Science Publications, ISBN: 9780198520290
- *Neutron Scattering in Biology: Techniques and Applications*, Springer Verlag
- Powder *Diffraction*, Vol H, *International Tables for Crystallography*, International Union of Crystallography
- D.L. Price and F. Fernandez-Alonso, Chap. 1, An Introduction to Neutron Scattering, *Neutron Scattering—Fundamentals (Experimental Methods in the Physical Sciences)*, Vol 44, Academic Press, Nov 22, 2013
- Special Topic on Advances in Modern Neutron Diffraction at Oak Ridge National Laboratory, *Rev. Sci. Instrum.*, Vol 89, 2018
- G.L. Squires, *Introduction to the Theory of Thermal Neutron Scattering*, Cambridge University Press, Online ISBN: 9781139107808

- J.B. Suck, in *Methods in the Determination of Partial Structure Factors of Disordered Matter by Neutron and Anomalous X-Ray Diffraction, Proceedings of the ILL/ESRF Workshop*, Sept 10–11, 1992 (Grenoble, France), World Scientific Publishing Co., 1993
- H.-R. Wenk, Ed., *Neutron Scattering in Earth Sciences (Reviews in Mineralogy and Geochemistry)*, Vol 63, Mineralogical Society of America, 2006
- H.-R. Wenk, Ed., *Neutron Scattering in Earth Sciences (Reviews in Mineralogy and Geochemistry)*, de Gruyter, April 9, 2018
- B.T.M. Willis and C.J. Carlile, *Experimental Neutron Scattering*, Oxford University Press, ISBN: 9780199673773
- C.C. Wilson, *Single Crystal Neutron Diffraction from Molecular Materials*, World Scientific Publishing Co. Inc., 2000

Light Optical Metallography

George F. Vander Voort, Vander Voort Consulting L.L.C.

ASM Handbook, Volume 10, *Materials Characterization*
ASM Handbook Committee
DOI 10.31399/asm.hb.v10.a0006684

Light Optical Metallography

George F. Vander Voort, Vander Voort Consulting L.L.C.

Overview

General Uses

- Imaging of microstructural phases and constituents on as-polished and polished and etched specimen surfaces at magnifications up to ~1500×
- Characterization of grain structures and measurement of grain size plus the amount, size, distributions, and spacing of phases and constituents

Examples of Applications

- Observation of microstructural phases and constituents using bright-field illumination, dark-field illumination, polarized light, and Nomarski differential interference contrast to effectively reveal the microstructure
- Comparison of images of the same specimen using two or more of these illumination techniques
- Illustrative examples of etchants used to properly reveal the microstructures

Samples

Form: Metals and alloys, polymers, ceramics, composites, and minerals
Size: Dimensions usually range from <1 to 25–30 mm^2, although larger specimens can be prepared and examined
Preparation: Typically, manufactured parts and mill products are sectioned to provide a specimen with a polished surface area of ≤30 mm^2, which is mounted in the appropriate polymeric resin for the task and then ground and polished to yield a flat, scratch-free surface free of preparation-induced damage that would make examination problematic. Samples are examined as-polished first to detect nonmetallic inclusions, nitrides, and cracks and voids and then etched, if required, to examine the microstructure

Limitations

- Resolution is limited to ~1 μm
- Depth of field decreases with increasing magnification and the numerical aperture of the objective. Computer-controlled methods are now available to capture a number of images focused at different depths; the in-focus portions of each image are merged to create an overall in-focus image as a function of depth
- Light optical microscopes do not provide chemical and crystallographic information about the observed microstructural elements. Color tint etching reveals if grains are randomly oriented or if a preferred orientation exists, but cannot define the nature of the preferred orientation

Estimated Preparation Time

- Specimen preparation can be automated so that multiple specimen holders, with typically six specimens per holder, can be prepared in 20 to 30 min for most specimens, ignoring the time to mount specimens (which can vary from ~5 min/specimen to 10 to 18 h for low-viscosity epoxy resins). Certain metals and alloys, such as face-centered cubic metals, could require a final polish using a vibratory polisher, which can add 20 to 60 min per specimen (multiple specimens can be prepared simultaneously). Manual preparation can be performed and takes 20 to 30 min per specimen

Capabilities of Related Techniques

- Scanning electron microscopy (SEM): Provides the ability to examine microstructural features at magnifications well above the limit of the light microscope and with greater resolution, enabling resolution of features not possible with the light microscope, such as fine interlamellar spacing in pearlite. Scanning electron microscopy has a much greater depth of field than light microscopy, enabling it to provide fully in-focus images of rough surfaces such as those of metal fractures. Scanning electron microscopes can be equipped with devices for chemical analysis, energy dispersive spectroscopy (EDS) and wavelength dispersive spectroscopy (WDS). Both can provide maps of elemental distributions relative to the microstructure and provide qualitative and semiquantitative chemical analysis of selected locations. Historically, WDS has been more effective for analyzing low-atomic-number elements than EDS. At magnifications <1500×, microstructural images obtained with SEM are inferior to those captured with light microscopy, and natural colors and colors produced by etchants cannot be detected with SEM
- Electron microprobe analysis: Provides the highest-quality quantitative chemical analysis using WDS detectors for nearly all chemical elements. Electron microprobe analysis also makes excellent maps of elements in a field being examined
- Transmission electron microscopy: Provides the highest-resolution images of fine microstructural details over the greatest magnification range. Specimens can be prepared in several ways: replicas of the structure of a polished and

(continued)

etched specimen, extraction replicas with fine second-phase particles removed from the matrix (which makes chemical analysis easier because the elements in the matrix around the particles are not detected), and thin sections viewed in transmission. Phases can be identified by chemical analysis and by x-ray diffraction. Crystallographic details can also be collected and analyzed to obtain the orientation of phases

The reflected light microscope is the most commonly used tool to study the microstructure of metals, composites, ceramics, minerals, and polymers. It has long been recognized that the microstructure has a profound influence on many metal and alloy properties. Mechanical properties (e.g., strength, toughness, ductility) are influenced much more than physical properties (many are insensitive to microstructure). The structure of metals and alloys can be viewed at a wide range of levels: macrostructure, microstructure, and ultramicrostructure. Macrostructure can be seen with the human eye and using stereo-zoom microscopes, generally up to a maximum of 20 to 100× magnification. Light microscopes usually cover the range from 50 to 1000×, but can reach 1500× with the use of oil-immersion lenses. Above 1000×, structure is examined using electron metallographic techniques, such as scanning and transmission electron microscopies. In recent years, many published technical articles contain SEM images, usually secondary electron images, at magnifications well under 1000×. These images are markedly inferior to what can be achieved using the light microscope, as shown in this article. For the study of the microstructure of metals and alloys, light microscopy is employed in the reflected-light mode using either bright-field illumination, dark-field illumination, polarized light illumination, oblique illumination (not common today), and differential interference contrast, generally by the Nomarski technique. Bright-field illumination is usually the starting point for examination of microstructures, which should start before etching to detect nonmetallic inclusions, nitrides, carbides, and possibly voids and cracks.

Introduction

The reflected light microscope is the most commonly used tool to study the microstructure of metals, composites, ceramics, minerals, and polymers. It has long been recognized that the microstructure has a profound influence on many metal and alloy properties. Mechanical properties (e.g., strength, toughness, ductility) are influenced much more than physical properties (many are insensitive to microstructure). The structure of metals and alloys can be viewed at a wide range of levels: macrostructure, microstructure, and ultramicrostructure. Macrostructure can be seen with the human eye and using stereo-zoom microscopes, generally up to a maximum of 20 to 100× magnification. Light microscopes usually cover the range from 50 to 1000×, but can reach 1500× with the use of oil-immersion lenses. Above 1000×, structure is examined using electron metallographic techniques, such as scanning and transmission electron microscopies. In recent years, many published technical articles contain scanning electron microscope (SEM) images, usually secondary electron images, at magnifications well under 1000×. These images are markedly inferior to what can be achieved using the light microscope, as shown in this article. For the study of the microstructure of metals and alloys, light microscopy is employed in the reflected-light mode using either bright-field illumination, dark-field illumination, polarized light illumination, oblique illumination (not common today), and differential interference contrast, generally by the Nomarski technique. Bright-field illumination is usually the starting point for examination of microstructures, which should start before etching to detect nonmetallic inclusions, nitrides, carbides, and possibly voids and cracks.

In the study of microstructure, the metallographer identifies what phases and constituents are present; their relative amounts; and their size, spacing, and arrangement. The microstructure is established based on the chemical composition of the alloy and the processing steps. A small specimen is cut from a larger mass (for example, a casting, forging, rolled bar, plate, sheet, or wire) for evaluation. The product must be sampled with consideration of where and how the plane of polish should be located to assess the variations possible in the product. Sectioning must be performed with care as this step introduces the greatest amount of damage in the preparation sequence. Every step must be performed properly because the removal rate of the next step may not be adequate to remove the damage from the previous step. You cannot interpret what you cannot see. Details of specimen preparation are not covered in this article as they are covered in great depth for all commercially used alloy systems, together with the use of the light microscope, interpretation of microstructure, and color metallography (Ref 1–18). This article concentrates on how to reveal microstructure properly to enable properly identifying the constituents and, if needed, measuring the amount, size, and spacing of constituents. The best tool to achieve this is the light optical microscope.

First and foremost, the specimen must be polished to a very high quality, free from any damage introduced by sectioning, grinding, and polishing, or the true structure will not be revealed and interpretation will be inaccurate. Specimens are generally viewed in the as-polished condition using bright-field illumination initially to observe constituents that have a natural color or reflectivity difference from the bulk metal. This procedure is commonly used to examine graphite and nonmetallic inclusions and other small particles that might be present, such as nitrides, carbonitrides, and borides. Some other small precipitates that have essentially the same reflectivity as the metal can also be observed if they have a much different hardness and polishing rate than the surrounding metal. They will either stand above or below the matrix phase and can be easily observed, particularly if differential interference contrast illumination (DIC) is used. However, bright-field illumination is the starting point for examination of a polished specimen and after etching, and is the most commonly used examination mode in metallography. Polarized light is widely used to examine the structure of as-polished metals with a noncubic crystal structure. However, it is still useful to examine the as-polished surface in bright field to detect inclusions and other small phases harder or softer than the matrix, as well as cracks and voids, which usually are not visible using polarized light to reveal the matrix structure. Dark-field and Nomarski DIC illumination methods usually must be tried to see if they provide useful details that cannot be seen using bright field. It is difficult to predict when these other illumination methods can help without trying them.

Examination of Microstructures Using Different Illumination Methods

Bright-Field Illumination

Figure 1 shows a schematic of the light path in an upright reflected light microscope. The path is basically the same as that in an inverted light microscope, except the polished face of the specimen is pointed downward rather than upward. Figure 2 shows how the numerical aperture of the objective lens establishes the maximum resolution of the microstructure as a function of the color of the light. Historically, it was common to insert a green filter

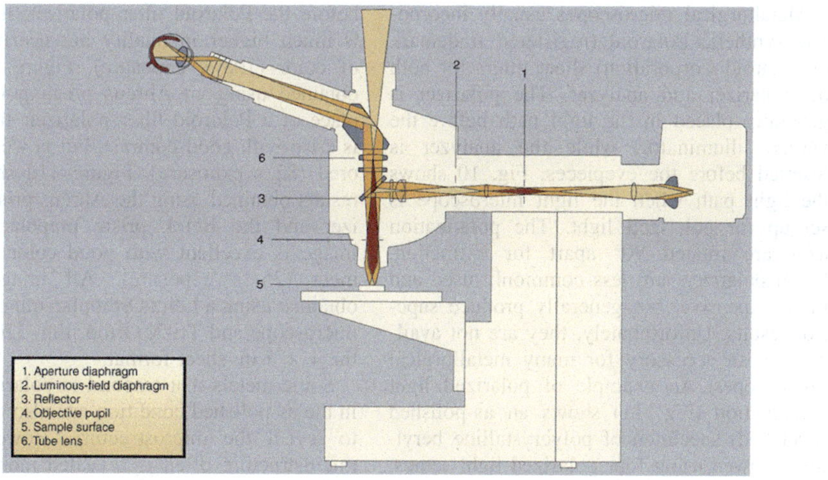

Fig. 1 Schematic of the light path in an upright light microscope. Courtesy of the Carl Zeiss Co.

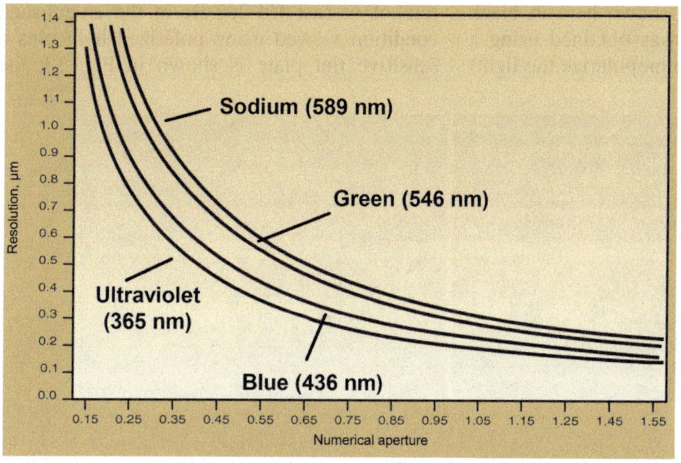

Fig. 2 Relationship between the numerical aperture of the objective lens and the color of the light on the resolution of a light microscope. Courtesy of the Carl Zeiss Co.

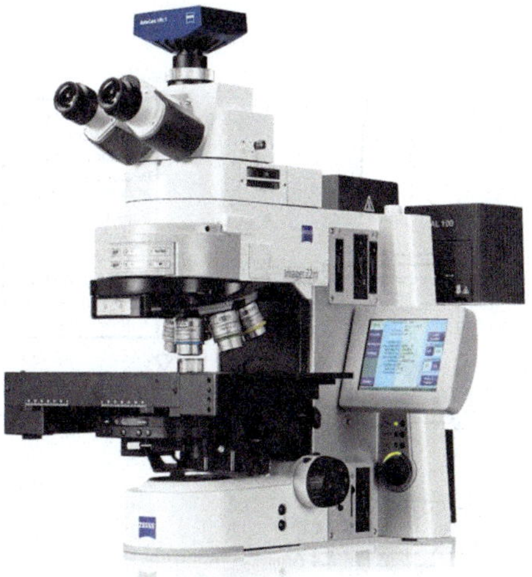

Fig. 4 Axio Imager.Z2m upright light microscope. Courtesy of Zeiss Microscopy

in the light path to control the aperture rating, and orthochromatic film was used to record the image. However, the end the age of film before the 21st century began ended the use of green filters. Figure 3 illustrates the effect of numerical aperture of the objective lens on depth of focus; depth of focus decreases with increasing resolution. The trend is the same as objective magnification is increased. Figure 4 shows an example of a current light microscope. A schematic of the light path in bright-field illumination is shown in Fig. 5.

An example of revealing grain boundaries in an interstitial-free (IF) sheet steel using Marshall's reagent plus a small amount of hydrofluoric acid is shown in Fig. 6. The commonly used nital etchant does not reveal the structure of an IF steel. Only a few portions of grain boundaries are missing. Figure 7 depicts delta ferrite in a billet of 17-4 PH martensitic stainless steel revealed selectively by etching with Murakami's reagent at ~100 °C (~210 °F). The uniform coloring of delta ferrite is ideal for image analysis phase detection and measurement. The microstructure of annealed SAE 4140 alloy steel consisting mainly of coarse pearlite and proeutectoid

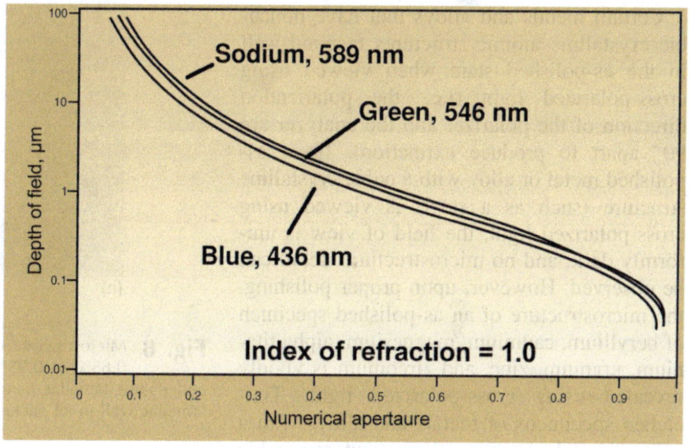

Fig. 3 Relationship between the numerical aperture and the color of light on the depth of focus of a light microscope. Courtesy of the Carl Zeiss Co.

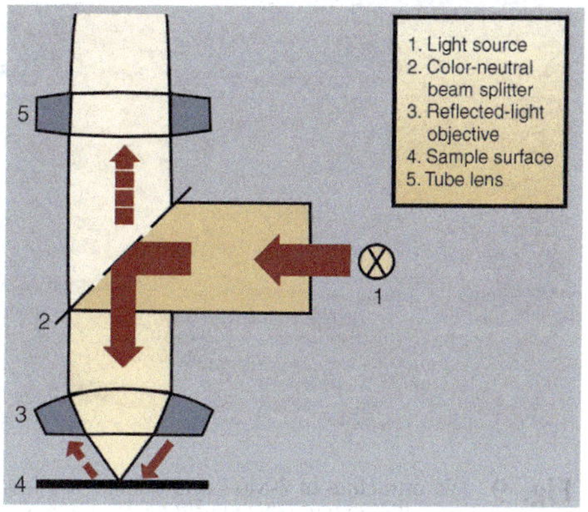

Fig. 5 Schematic of the light path through a light microscope in bright-field illumination. Courtesy of the Carl Zeiss Co.

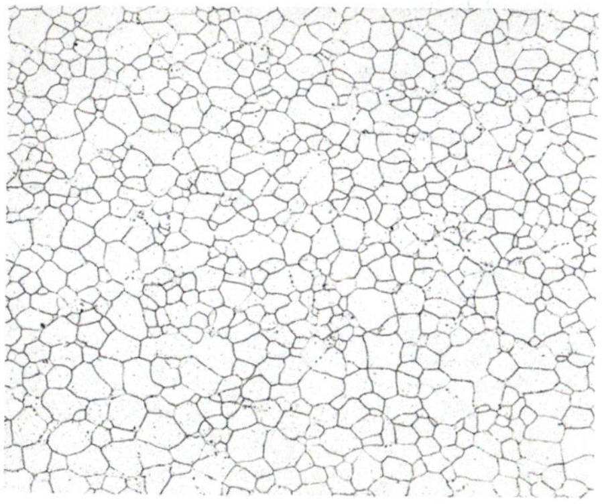

Fig. 6 Ferrite grain boundaries in interstitial-free sheet steel etched using Marshall's reagent plus ~1% hydrofluoric acid; nital has no effect on interstitial-free steels. Original magnification: 200×

ferrite etched with 4% picral and 2% nital is shown in Fig. 8. Picral uniformly dissolves the ferrite, but nital is orientation sensitive and dissolves ferrite at rates depending on the crystallographic orientation of ferrite grains. The lines in Fig. 8 indicate pearlite colonies where cementite is barely visible, and the areas could be misidentified as ferrite. Cementite lamellae in the specimen etched with picral are more uniformly revealed, because picral etches ferrite at the same rate regardless of crystal orientation of the colonies. Picral is a far more effective etchant for as-rolled, normalized, and annealed carbon and low-alloy steels. Picral can be used to reveal such structures in many tool steels, but Vilella's reagent must be used for grades with higher chromium content. Figure 9 shows ferrite in Carpenter Technology Corporation's 7-Mo Plus duplex stainless steel revealed by a slight modification of Beraha's BI tint etch. Color is vividly revealed in bright field, and the image is perfect for image analysis measurements of the microstructure.

Polarized Light

Certain metals and alloys that have noncubic crystalline atomic structures respond well in the as-polished state when viewed using cross-polarized light (i.e., the polarization direction of the polarizer and the analyzer are 90° apart to produce extinction). If an as-polished metal or alloy with a cubic crystalline structure (such as a steel) is viewed using cross-polarized light, the field of view is uniformly dark, and no microstructural detail can be observed. However, upon proper polishing, the microstructure of an as-polished specimen of beryllium, cadmium, magnesium, alpha titanium, uranium, zinc, and zirconium is vividly revealed using cross-polarized light. Tint-etched specimens of metals and alloys with a cubic crystal structure (body-centered cubic and face-centered cubic) can be viewed in polarized light (with or without the sensitive tint filter) to enhance coloration if it is weak in bright field.

Metallurgical microscopes usually incorporate synthetic Polaroid (registered trademark of Polaroid Corporation) sheet filters for both the polarizer and analyzer. The polarizer is generally placed in the light path before the vertical illuminator, while the analyzer is inserted before the eyepieces; Fig. 10 shows the light path when the light microscope is set up for polarized light. The polarization axes are rotated 90° apart for extinction. Prism polarizers are less commonly used and more expensive, but generally produce superior results. Unfortunately, they are not available as an accessory for many metallurgical microscopes. An example of polarized light examination (Fig. 11) shows an as-polished (unetched) specimen of polycrystalline beryllium viewed using four polarized light setups. Figure 11a shows the structure viewed using Polaroid filters (in the crossed position) for both the polarizer and analyzer and photographed with Tri-X Ortho 4 × 5 in. sheet film (32 s exposure). The image is flat and devoid of color (all images are shown here in black and white). Figure 11b was obtained using a Berek prism polarizer (to prepolarize the light) before the Polaroid filter polarizer; the image is much higher in quality but nearly devoid of color (108 s exposure). Figure 11c was obtained using an Ahrens prism polarizer in place of a Polaroid filter polarizer; the image is crisp with good contrast, but is weakly colored (82 s exposure). Figure 11d shows the results obtained using the Ahrens prism polarizer and the Berek prism prepolarizer; the image is excellent with good color development (25 s exposure). All images were obtained using a Leitz Orthoplan mineralogical microscope and Tri-X Ortho film, 320 ISO, in the 4 × 5 in. sheet format.

Some metals that respond to polarized light in the as-polished condition can also be etched to reveal the microstructure. However, the microstructure often is revealed more clearly using cross-polarized light examination of an as-polished, unetched specimen. Figure 12 shows the twinned microstructure of as-cast, pure orthorhombic bismuth in the as-polished condition in polarized light. The microstructure of as-cast Cd-20%Bi in the as-polished condition viewed using polarized light plus a sensitive tint plate is shown in Fig. 13; the

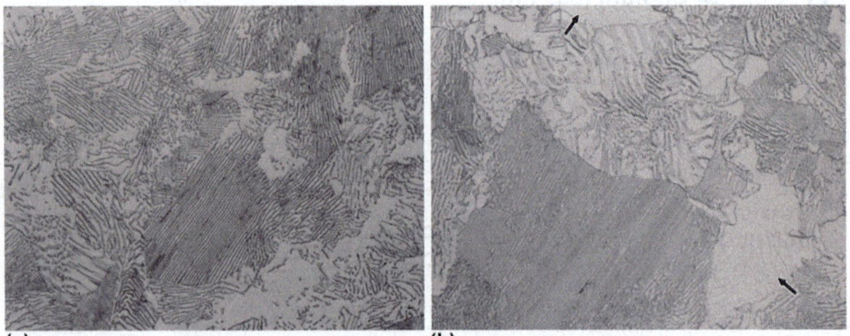

(a) (b)

Fig. 8 Microstructure of annealed (870 °C, or 1600 °F, for 1 h, slow cooled) SAE 4140 alloy steel (Fe-0.4%C-0.85%Mn-0.95%Cr-0.2%Mo) etched using 4% picral (a) and 2% nital (b), revealing proeutectoid ferrite and coarse lamellar pearlite. Nital, because it is sensitive to crystal orientation, does not reveal the cementite lamellae well in all packets (arrows point to examples). Original magnification: 1000×

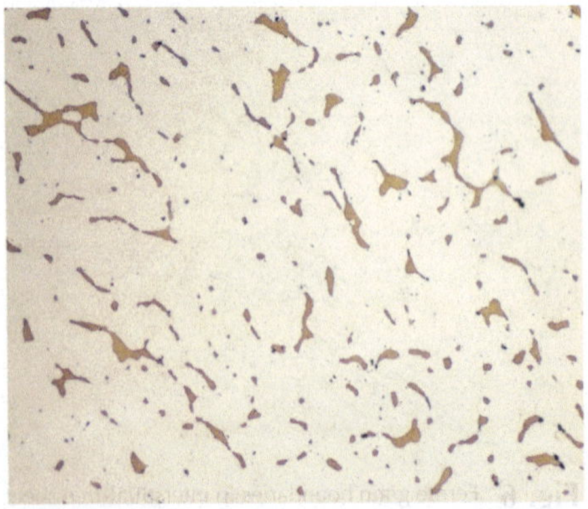

Fig. 7 Delta ferrite in a billet of 17-4PH revealed by etching using Murakami's reagent at 100 °C (210 °F). Magnification: 200×

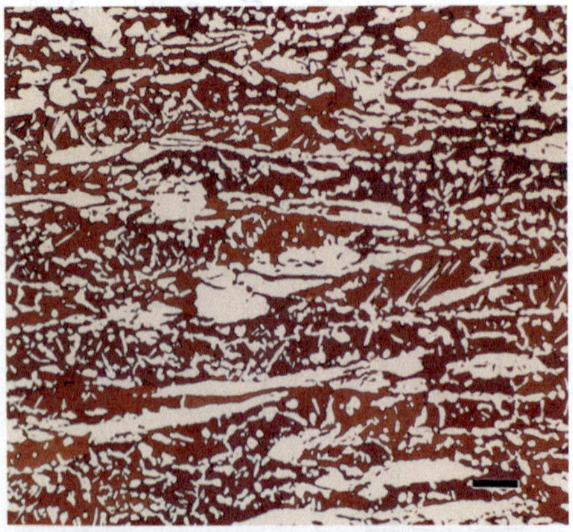

Fig. 9 Microstructure of 7-Mo PLUS duplex stainless steel (Fe-<0.03C-<2%Mn-27.5%Cr-4.85%Ni-1.75%Mo-0.25%N) etched using Beraha's reagent (15 mL HCl, 85 mL water, 1 g $K_2S_2O_5$) and viewed using bright-field illumination. Ferrite is colored and austenite is unaffected. The magnification bar is 50 μm long.

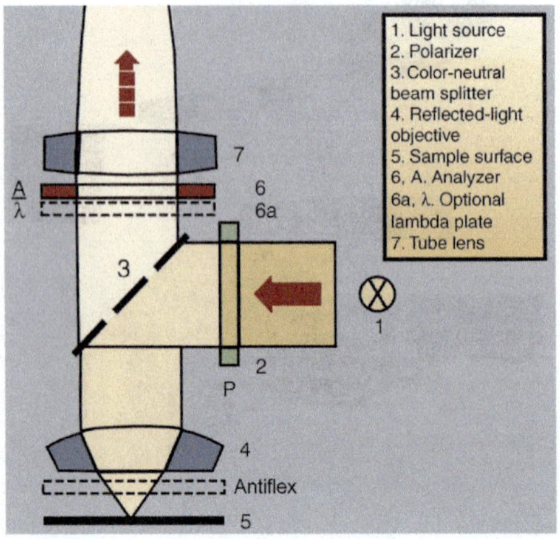

Fig. 10 Schematic of the light path through a light microscope in polarized light. (The lambda plate, No. 6a, is another term for a sensitive tint plate, an optional accessory.) Courtesy of the Carl Zeiss Co.

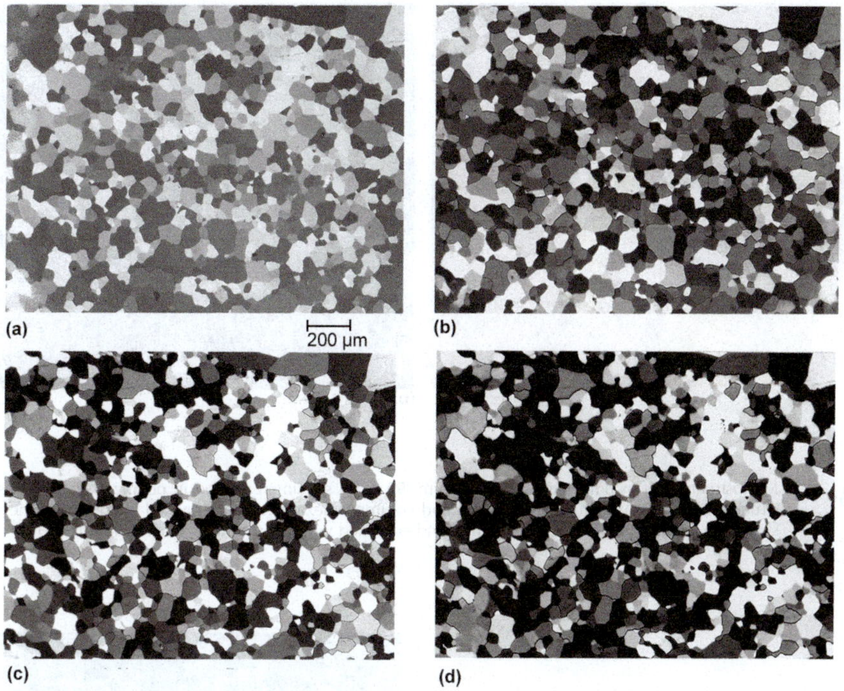

Fig. 11 Influence of polarized light setup on image quality using (a) Polaroid filters for polarizer and analyzer, (b) Berek prism prepolarizer, (c) Ahrens prism polarizer and Polaroid filter analyzer, and (d) Ahrens prism polarizer and Polaroid filter analyzer, with Berek prism prepolarizer added

Fig. 12 As-cast pure orthorhombic bismuth in the as-polished condition viewed using polarized light shows mechanical twins. Magnification: 50×

Fig. 14 Martensite formed in heat treated (heated to 900 °C, or 1650 °F, held 1 h, water quenched) wrought eutectoid aluminum bronze (Cu-11.8%Al). Specimen unetched and viewed using cross-polarized light. Original magnification: 200×

variation in colors of Cd-rich dendrites is due to variations in crystallographic orientation. The Cd-Bi eutectic in the interdendritic regions is too fine to see at 50× magnification. The martensitic structure of a heat treated wrought aluminum bronze binary eutectoid alloy held at 900 °C (1650 °F) for 1 h and water quenched is shown in Fig. 14. The structure is shown in polarized light and the unetched condition, revealing three colors of the martensite plates. Figure 15 illustrates the grain structure of wrought, pure hexagonal close-packed hafnium in the as-polished condition revealed using polarized light and a sensitive tint plate; fine mechanical twins appear along the upper edge, probably from clamping pressure during sectioning. The microstructure of wrought, pure magnesium in the as-polished condition viewed using a Foster prism for the polarized light plus a sensitive tint filter is shown in Fig. 16; mechanical twins with different orientations appear on the opposite sides of the grain boundary. Figure 17 depicts the grain structure of pure hexagonal close-packed ruthenium in the as-cast and wrought conditions. Both are as-polished; there is a mix of coarse equiaxed and columnar grains plus a few small shrinkage cavities in the as-cast specimen at 50×and a much finer equiaxed grain structure in the wrought specimen at 100×. Figure 18 shows the microstructure of as-cast, pure tin in the as-polished condition, revealing a very coarse columnar grain structure and mechanical twins at 50× in polarized light.

Fig. 13 Microstructure of Cd-20%Bi in the as-cast condition, unetched, viewed using polarized light (slightly off the crossed position) plus sensitive tint, revealing cadmium dendrites of various orientations. The interdendritic constituent is a eutectic of cadmium and bismuth, but it is too fine to resolve at this magnification.

Polarized Light versus Bright Field

Figure 19 shows two unusual clusters of nodular graphite with apparent intergrowth between nodules in an as-polished specimen of austempered ductile iron viewed with polarized light plus a sensitive tint filter. Figure 20 depicts graphite nodules in a pearlitic ductile iron specimen etched using 2% nital and viewed with bright-field illumination and polarized light plus a sensitive tint filter, which produces a superior image of the microstructure, particularly of the growth pattern in the nodules. Pearlite responds to polarized light even without etching due to its fine interlamellar structure. The microstructure of a Zn-0.1%Ti-0.1%Cu alloy hot-rolled to 6 mm (0.250 in.) thick is shown in Fig. 21.

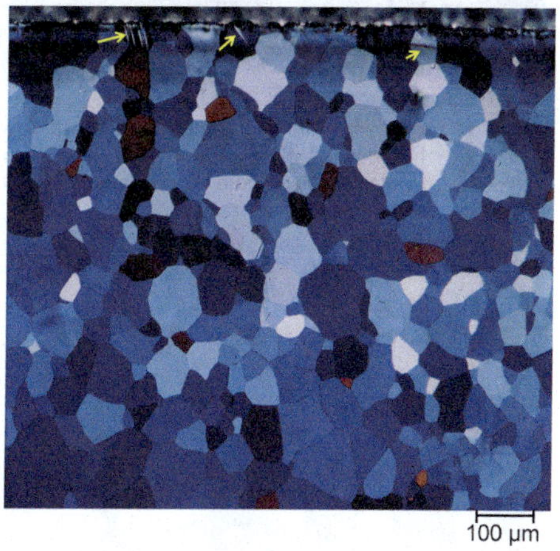

Fig. 15 Microstructure of as-polished wrought, pure hexagonal close-packed hafnium specimen viewed in polarized light plus sensitive tint reveals an equiaxed alpha grain structure and a few mechanical twins at the surface (yellow arrows).

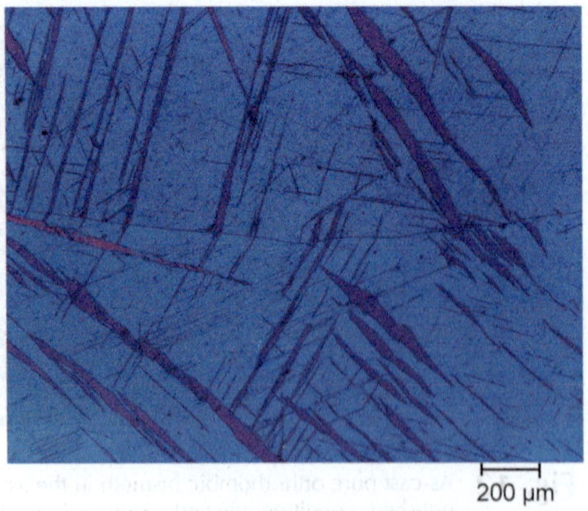

Fig. 16 Microstructure of as-polished wrought, pure hexagonal close-packed magnesium specimen viewed in cross-polarized light using a microscope with a Foster prism and a sensitive tint plate. Note the change in direction of the mechanical twins at the grain boundary.

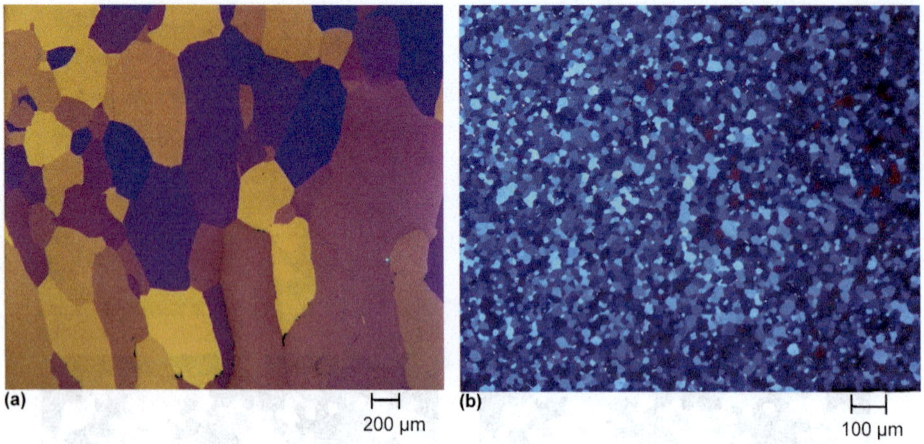

Fig. 17 Microstructure of as-polished pure ruthenium (99.95%) specimen viewed in polarized light: (a) as-cast structure contains a mixture of equiaxed and columnar hexagonal close-packed grains and some small shrinkage cavities (black); (b) fine-grained and equiaxed wrought structure. Magnification: 100×

Fig. 18 Microstructure of as-polished, as-cast pure hexagonal close-packed tin specimen viewed using polarized light (note the mechanical twins)

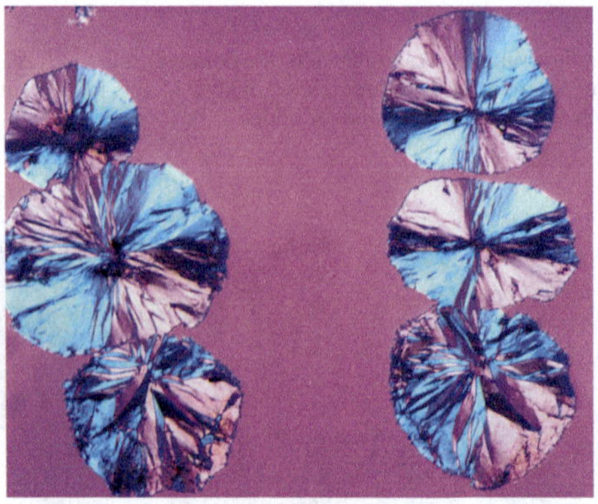

Fig. 19 Two clusters of spheroidal graphite nodules in as-polished austempered ductile iron viewed using cross-polarized light plus sensitive tint. Original magnification: 500×

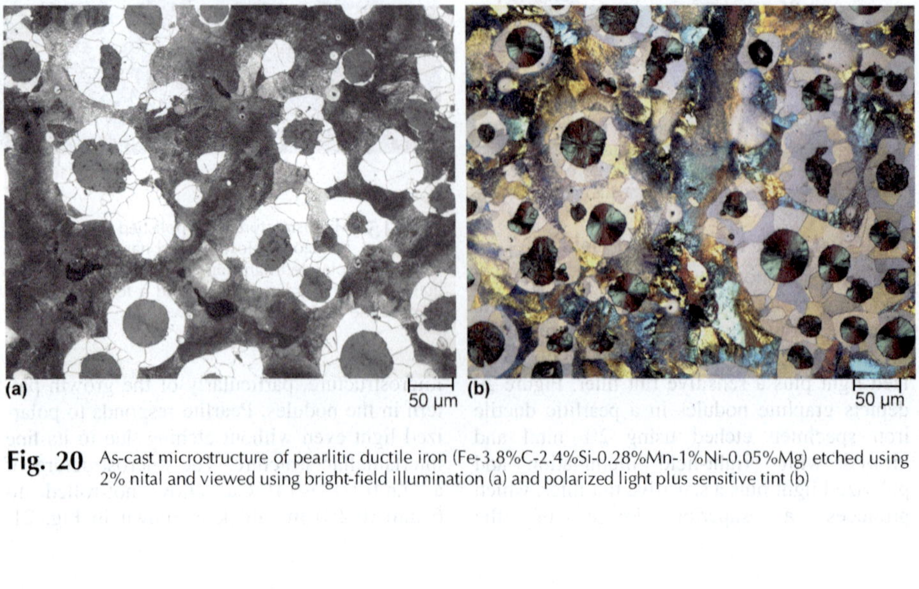

Fig. 20 As-cast microstructure of pearlitic ductile iron (Fe-3.8%C-2.4%Si-0.28%Mn-1%Ni-0.05%Mg) etched using 2% nital and viewed using bright-field illumination (a) and polarized light plus sensitive tint (b)

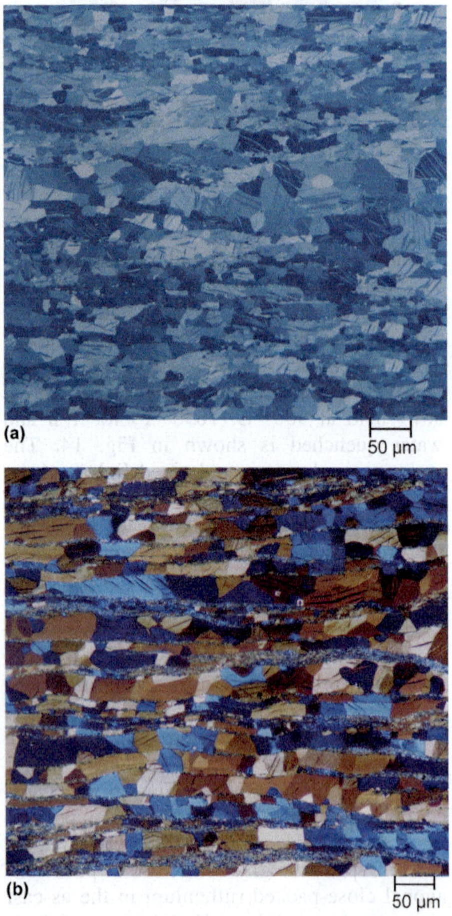

Fig. 21 Microstructure of Zn-0.1%Ti-0.1%-Cu hot-rolled to 6 mm (0.250 in.) thickness: (a) view in the as-polished condition using polarized light reveals elongated hexagonal close-packed grains containing mechanical twins and some fine precipitates present in the grain boundaries, but not clearly; (b) etching using Palmerton reagent and observation using polarized light plus sensitive tint reveals precipitates more clearly

In the as-polished condition, viewing using polarized light weakly reveals the grain structure, with some small twins visible. Etching using Palmerton reagent and viewing with polarized light plus a sensitive tint plate vividly reveals the structure, including remnants of the cold-worked, two-phase constituent running longitudinally, which were not visible without etching. Figure 22 shows the microstructure of hot-rolled, annealed, and cold-drawn Monel 400 tint-etched using Beraha's selenic acid reagent. Bright-field examination weakly reveals the grain structure with several nonmetallic inclusions (indicator lines). Polarized light with the sensitive tint filter vividly reveals the grain structure, but not the inclusions (inclusions should be observed unetched in bright field for highest detectability). The microstructure of annealed wrought aluminum brass after etching using potassium dichromate (Fig. 23) reveals the grain structure and annealing twins in black and white in a grain-contrast mode. When an etchant produces a grain-contrast etch in bright field, it can be viewed in color by switching to polarized light, and adding the sensitive tint filter usually enhances the coloration. A reversal of results is shown in Figs. 24 and 25. An AISI 8620

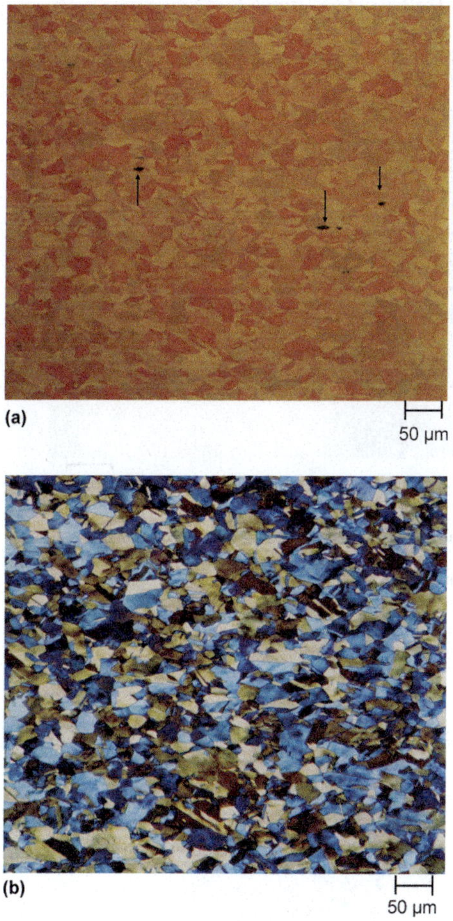

Fig. 22 Microstructure of hot-worked, annealed, and cold-drawn Monel 400 (Ni-32%Cu-<0.3%C-<2%Mn-<0.5%Si) revealed using Beraha's selenic acid etch for copper (longitudinal axis is horizontal) and viewed using bright field (a) and Polarized light plus sensitive tint (b). Monel alloys, especially wrought alloys (as-cast alloys are easier), are very difficult to color-etch. Bright field reveals a weak image as the interference film produced is thin (inclusions are visible, indicated by arrows). Polarized light often dramatically enhances image quality (the sensitive tint filter enhances coloration). Note the deformed twinned face-centered cubic alpha grain structure.

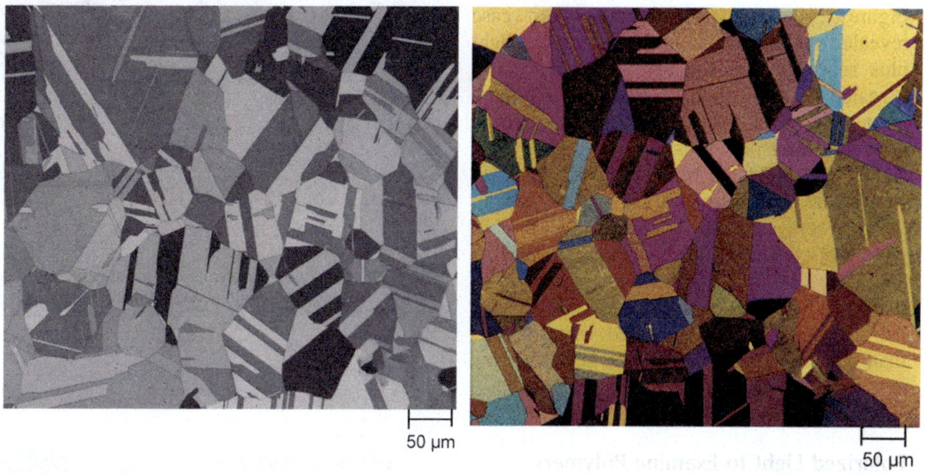

Fig. 23 Equiaxed alpha grains containing annealing twins in annealed (at 750°C, or 1380 °F) wrought aluminum brass (Cu-22%Zn-2%Al) revealed by using potassium-dichromate etch, a grain-contrast etchant that responds well to polarized light

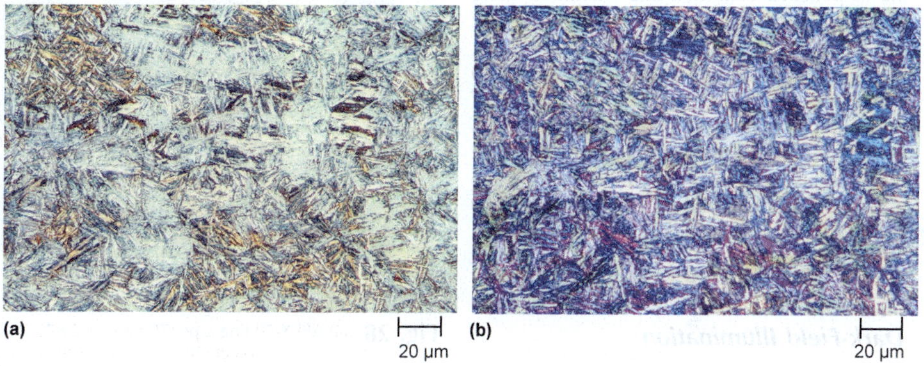

Fig. 24 Microstructure of lower bainite case of carburized (0.95% C potential) 8620 alloy steel etched using 10% sodium metabisulfite and viewed using bright field (a) and polarized light plus sensitive tint (b). The steel was carburized at 955 °C (1750 °F), quenched into a 50-50 mix of sodium nitrite and potassium nitrate at 250 °C (480 °F) and held for 120 min, air-cooled, and tempered at 250 °C for 240 min to an aim case hardness of 52–60 HRC. A lower bainite case performs better under low-cycle-fatigue conditions.

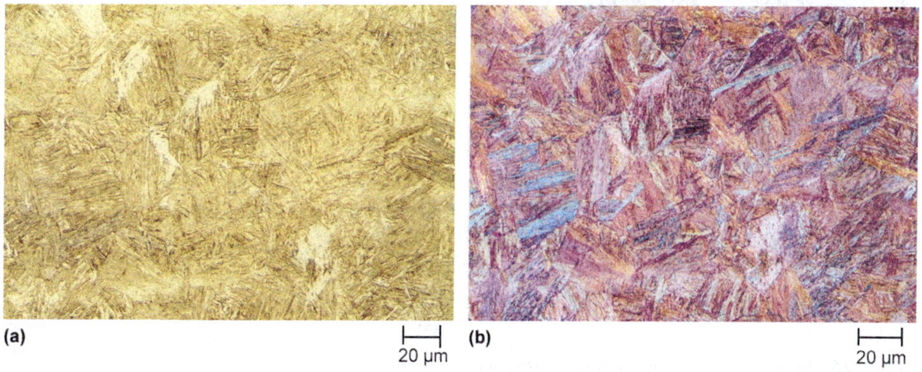

Fig. 25 Lath martensitic core structure with some ferritic areas of carburized (0.95% C potential) 8620 alloy steel revealed using 10% sodium metabisulfite etchant and viewed using bright field (a) and polarized light plus sensitive tint (b). The lath martensite is seen more clearly in polarized light.

alloy gear was carburized and heat treated to form lower bainite in the case (for higher resistance to failure under low-cycle fatigue conditions) and lath martensite in the core. The sample was etched using 10% sodium metabisulfite, an effective tint etchant for alloy steels. Figure 24 shows the lower bainite case revealed in bright field and in polarized light plus sensitive tint; the case is more clearly revealed in bright field. In Figure 25, the lath martensite core is more vividly revealed in polarized light plus sensitive tint and is more weakly revealed in bright field. Figure 26 shows the microstructure of Invar powder plasma sprayed onto an Invar (Fe-36%Ni) plate tint-etched using Klemm's III reagent and viewed with polarized light plus a sensitive tint plate; the grain structures of the weld deposit and plate are revealed more vividly than can be achieved in bright field or using any standard black and white reagent.

Polarized Light to Examine Polymers

Polarized light is useful to study polymer microstructures. Figure 27 shows the microstructure of a magnesium diboride (MgB_2) fiber-reinforced polymer-matrix composite in the as-polished condition and examined using polarized light plus a sensitive tint filter at 100 and 200× magnification, vividly revealing the MgB_2 fibers. Figure 28 shows an API X-70 alloy line pipe coated with three layers of polymers for salt water corrosion protection as well as the pipe microstructure and the three polymer layers at 50×. Figure 29 shows the pipe microstructure and polymer layers at 100 and 200× magnification (only two layers can be seen at 200×).

Dark-Field Illumination

Dark-field illumination, while available for many years, is not commonly used by metallographers. However, it does provide value. Dark

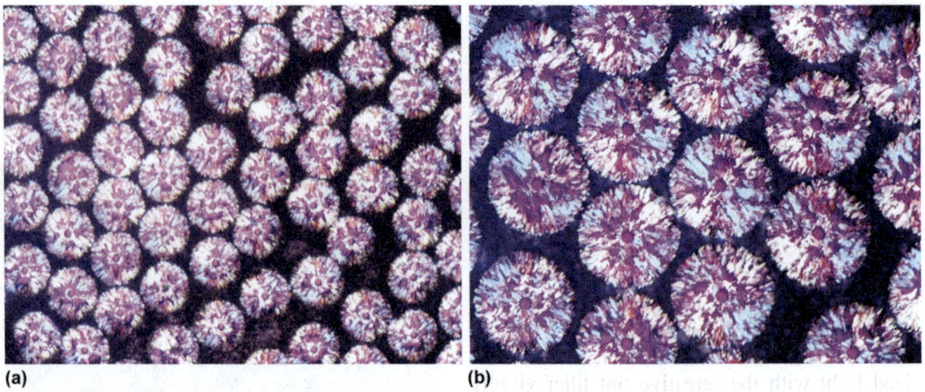

Fig. 27 As-polished polymer composite containing MgB_2 reinforcement fibers viewed using polarized light and a sensitive tint filter. Magnification: 100× (a) and 200× (b)

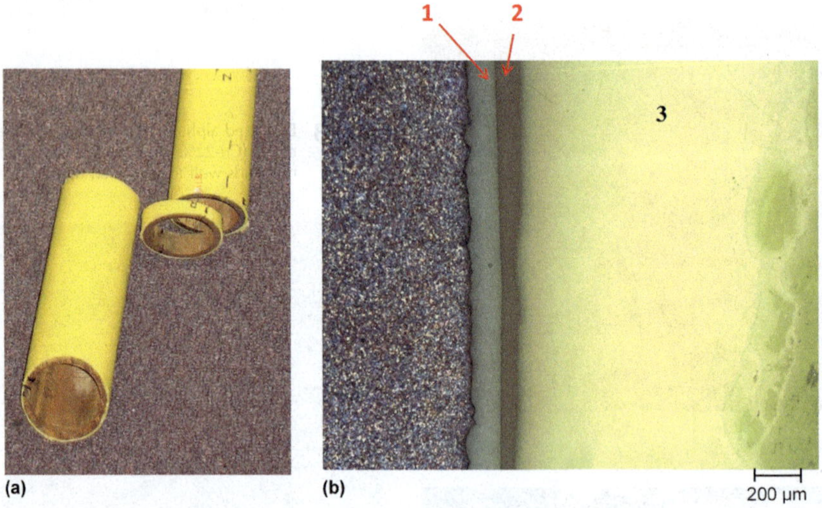

Fig. 28 (a) API X-70 line pipe (75 mm, or 2.875 in., diameter) coated with three layers of polymers for salt water corrosion protection; (b) line pipe steel on left side and polymer layers (Nos. 1, 2, and 3) on right side etched using Klemm's I reagent and viewed using polarized light

Fig. 26 Interface between Invar powder plasma and an Invar (Fe-36%Ni) plate onto which it was sprayed revealed using Klemm's III etchant and viewed using polarized light plus sensitive tint. Magnification: 100×

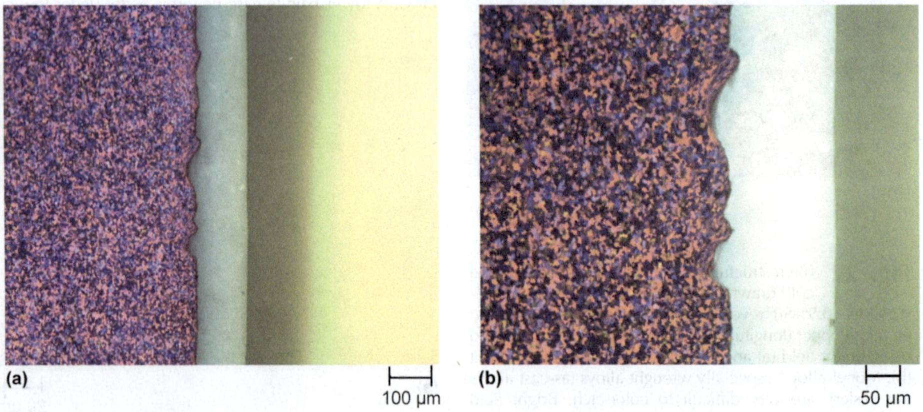

Fig. 29 Etched (Klemm's reagent) polymer-coated API X-70 line pipe viewed using polarized light: (a) pipe grain structure and polymer layers at 100× magnification; (b) grain structure and inner two polymer layers at 200× magnification

field captures light that was scattered and could not be seen in bright field, making it much like an inverse image of bright field. The light path in a light microscope set up for dark-field illumination is shown in Fig. 30. Figure 31 shows a powder metallurgy sample of René 95 nickel-base superalloy after hot isostatic pressing that contains coarse carbides and coarse gamma prime (too coarse to improve strength) in an austenite matrix after etching using glyceregia. Bright-field and dark-field images at 1000× reveal carbides in the grain boundaries and also the coarse gamma prime, but the dark-field image is more vivid. A dark-field image of cuprous oxides in hot-extruded, cold-drawn tough-pitch copper is shown in Fig. 32. In bright field, cuprous oxide and cuprous sulfide look similar in color and shape, but in dark field, cuprous oxide is revealed in bright red while cuprous sulfide is not visible. Cracks in the cuprous oxide are due to cold drawing. Figure 33 shows the microstructure of a walnut tree specimen where the image is perpendicular to the trunk axis. Cells and pores are visible in dark field, but not in bright field.

Nomarski Differential Interference Contrast

Although incident bright-field illumination is by far the most common illumination mode used by metallographers (except for those who work exclusively with metals such as beryllium, alpha titanium, tin, zinc, uranium, cadmium, and zirconium, where polarized light is used regularly), DIC illumination is seeing increased use. The Nomarski system is most commonly used for DIC. Figure 34 shows the light path in a light microscope when it is set up for Nomarski DIC. DIC brings out height differences that might not be observable using bright field. Previously, height differences were revealed using oblique illumination. However, the method yields uneven illumination across the field and degrades resolution, and it is rarely available today because DIC has become popular. Nomarski DIC reveals features not visible in bright field by improving image contrast, not by increasing resolution.

Figure 35 reveals comet tails after manual polishing of a powder metallurgy specimen of AISI 434 stainless steel (with a nickel addition) that did not undergo hot isostatic pressing, hot rolling, or forging after manual polishing. Voids due to lack of consolidation can be seen in bright field and with DIC. Nomarski DIC also reveals many small beads of water on the surface that were not visible in bright field. A continuously cast specimen of aluminum alloy 3004 etched using Keller's reagent is shown in Fig. 36. Bright field reveals only intermetallic particles and some shrinkage cavities, which can also be seen in the as-polished condition, whereas DIC also reveals the dendritic structure, although the

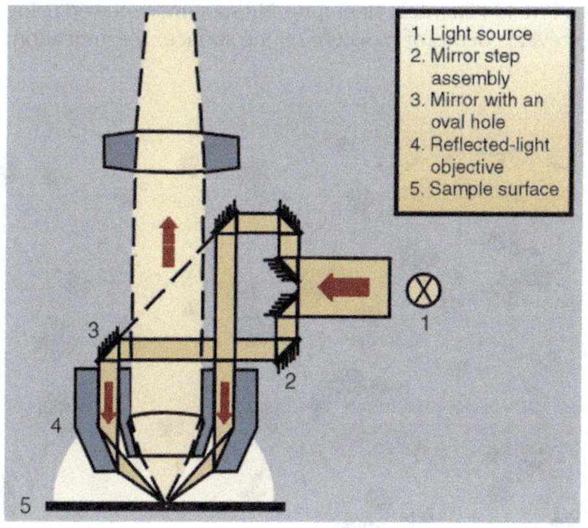

Fig. 30 Schematic of the light path through a light microscope in dark-field illumination, which detects the scattered light. Courtesy of the Carl Zeiss Co.

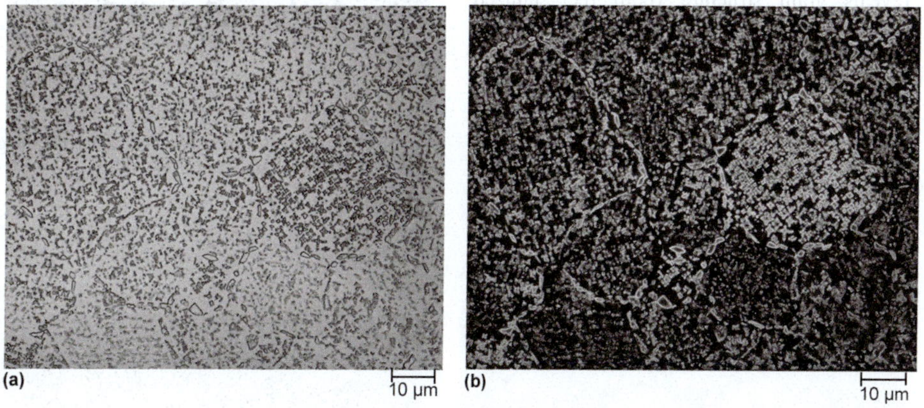

Fig. 31 The microstructure of powder metallurgy René 95 alloy (Ni-0.15C-8Co-14Cr-3.5Mo-3.5W-2.5Ti-3.5Al-3.5Nb) that underwent hot isostatic pressing, was etched using glyceregia, and was viewed using bright field (a) and dark field (b) reveals grain boundary carbides and coarse gamma prime in an austenitic matrix.

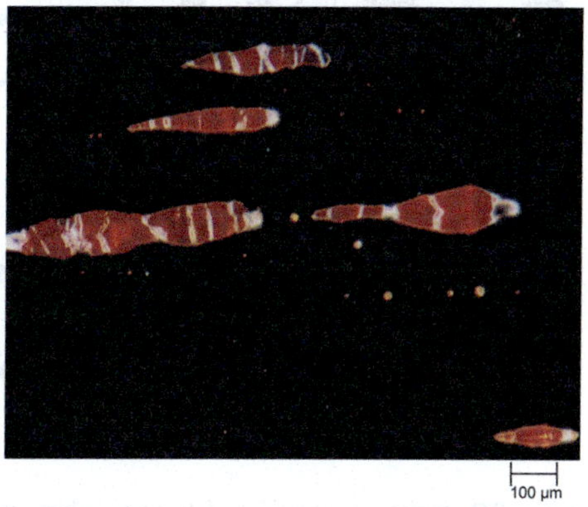

Fig. 32 Cuprous oxide in hot-extruded, cold-drawn tough-pitch arsenical copper viewed in dark field reveals the classic ruby-red color.

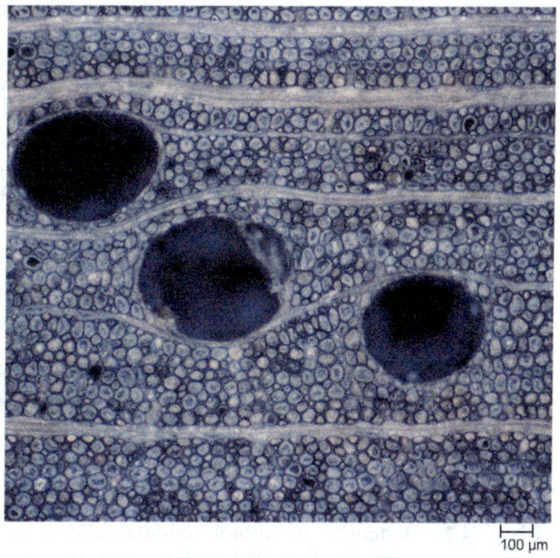

Fig. 33 The microstructure of walnut (plane perpendicular to the trunk axis) viewed using dark-field illumination reveals the cells and pores.

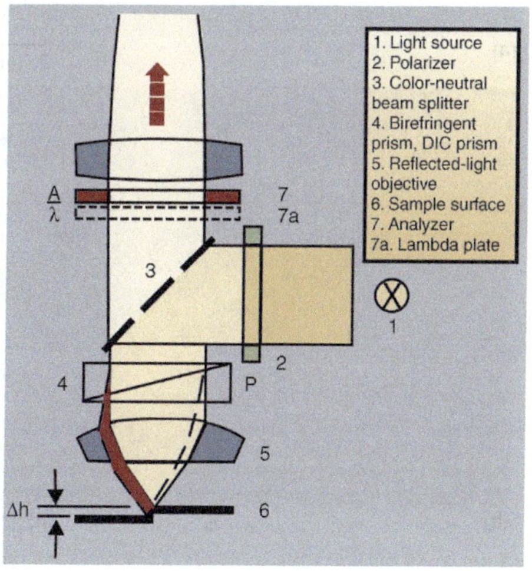

Fig. 34 Schematic of the light path through a light microscope in Nomarski DIC, which uses polarized light (the lambda plate, No. 7a, is another term for a sensitive tint plate, an optional accessory). Courtesy of the Carl Zeiss Co.

intermetallic structures are hard to see. In Fig. 37, martensite in a nitinol shape memory alloy (SMA) after etching using equal parts nitric acid, acetic acid, and hydrofluoric acid is revealed more clearly using DIC than bright field.

The grain structure of temper carbon in malleable cast iron etched using nital is more clearly revealed using DIC than bright field (Fig. 38). Figure 39 shows the grain structure of 14-karat gold etched using equal parts 10% NaCN and 30% concentration hydrogen peroxide; DIC reveals the grain structure much more effectively than bright field. Figure 40 shows the microstructure of Elgiloy (a face-centered cubic cobalt-base alloy), that was hot-rolled and solution-annealed (1090 °C, or 1995 °F), resulting in partial recrystallization. The specimen was swab-etched using 15 mL of HCl, 10 mL of acetic acid, and 10 mL of nitric acid. Nomarski DIC reveals grain structure (both recrystallized and nonrecrystallized) and longitudinal segregation more clearly than bright-field illumination. Figure 41 reveals alpha prime martensite in a binary Ti-3%Cr alloy after heating to 1040 °C (1900 °F), holding for 10 min, and water quenching. The specimen was etched using Kroll's reagent, and Nomarski DIC revealed the microstructure more effectively than bright field. The dendritic microstructure of a sand-cast Cu-4%Sn binary alloy tint-etched using Beraha's PbS reagent for copper-base alloys is shown in Fig. 42. Nomarski DIC reveals the dendritic structure more clearly than bright field.

Figure 43 shows low-carbon martensite formed at the polished surface of Carpenter Technology Corporation's Temperature Compensator "30" alloy, type 2, after refrigeration in liquid nitrogen. Unstable austenite was transformed to low-carbon martensite, and shear deformation at the previously polished surface produces a roughness that reveals the

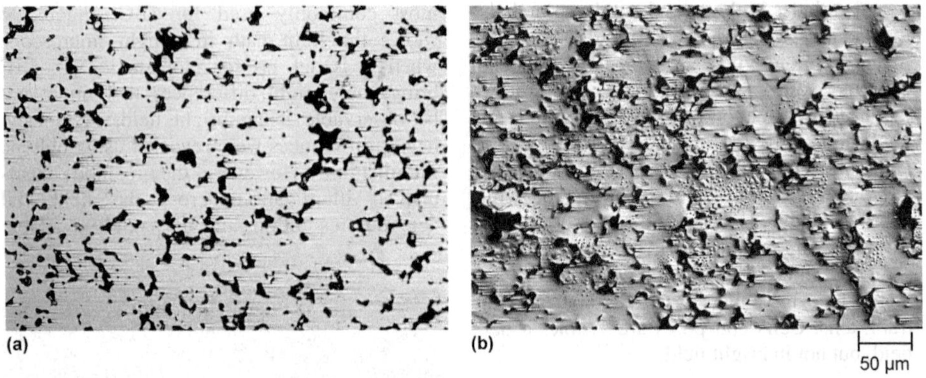

Fig. 35 Example of comet tails after manual polishing of a PM 434 stainless steel (plus Ni) specimen viewed using bright field (a) and Nomarski DIC (b), which also reveals beads of moisture on the surface. Magnification: 200×

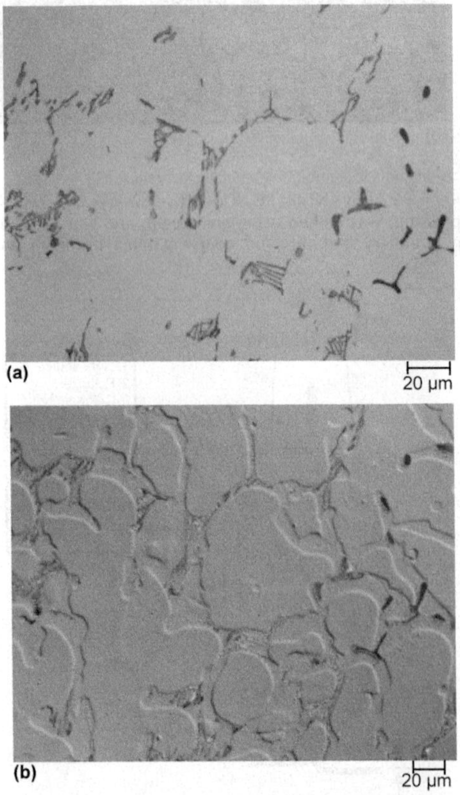

Fig. 36 Continuously cast aluminum alloy 3004 (Al-1.25%Mn-1.05%Mg) etched using Keller's reagent and viewed using bright field (a) and DIC (b). Dendrites are visible using DIC but not in bright field, while intermetallic particles between dendrites are easier to see in bright field.

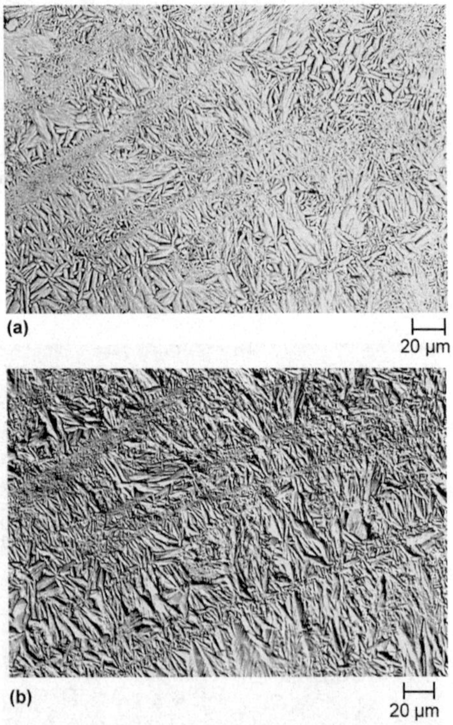

Fig. 37 Martensite in nitinol (Ni-50at.%Ti) shape memory alloy revealed by etching using equal parts HNO_3, acetic acid, and hydrofluoric acid, and viewed using bright field (a) and Nomarski DIC (b). Nomarski DIC reveals more detail compared with bright field.

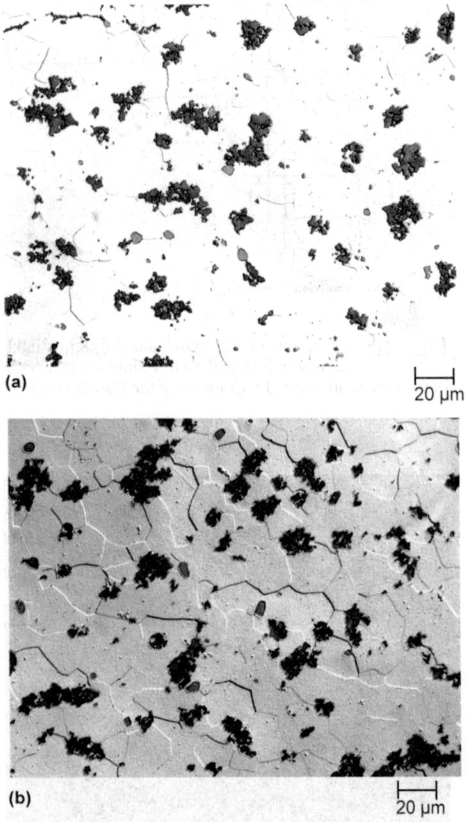

Fig. 38 Surface topography of temper carbon in malleable cast iron etched using nital and viewed using bright field (a) and Nomarski DIC (b). DIC reveals the surface topography more effectively than bright field.

martensitic structure formed by refrigeration using DIC without etching. This is more useful than polishing the specimen after refrigeration and etching it to reveal the newly formed martensite. Figure 44 shows a somewhat similar result obtained using a Cu-26% Zn-5%Al SMA polished and thermally cycled through the SMA process. β_1 martensite is shown at different locations using polarized light and Nomarski DIC. Both reveal the structure without etching. Figure 45 shows the microstructure of Spangold SMA, developed by Mintek, South Africa (Ref 19), for use in jewelry applications. The specimen was hot-mounted to determine if the thermal cycle for hot mounting would produce β_1 martensite. After polishing, it was heated to 100 °C (210 °F), held for 2 min, and water quenched, which produced more β_1 martensite than hot mounting. The sample surface in bright field and in Nomarski DIC and the "criss-crossed" regions are where β_1 martensite formed in one orientation during hot mounting and another orientation when the SMA cycle was used after the specimen was polished.

Nomarski Differential Interference Contrast Used to Examine Composites, Polymers, and Ceramics

Figures 46 and 47 show the microstructure of a Ti-6Al-4V metal-matrix composite reinforced by fibers of SiC grown around tungsten (transverse views). Figure 46 shows bright-field views at 100 and 200×, while Fig. 47 shows bright-field and Nomarski DIC images at 1000×. The structure is clearly visible with both illumination methods, but DIC reveals the grain structure of the Ti-6A-4V matrix in

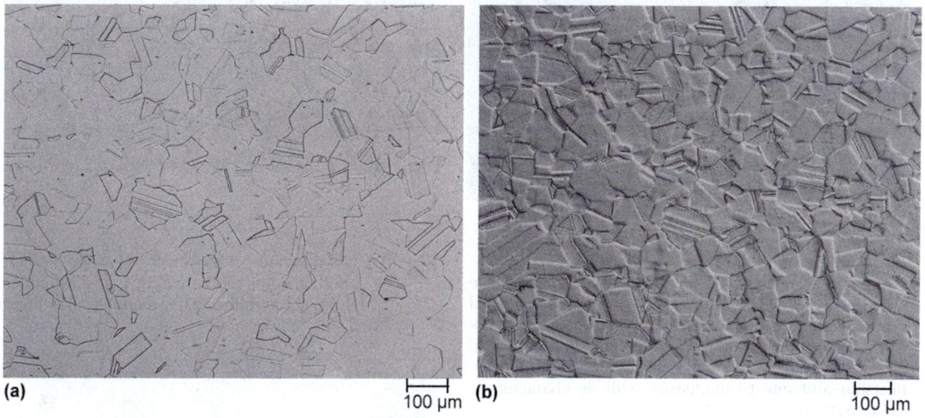

Fig. 39 Attack-polished 14-karat gold etched using equal parts 10% NaCN and H_2O_2 (30% concentration) and viewed using bright field (a) and Nomarski DIC (b). DIC reveals the grain structure much more effectively than bright field.

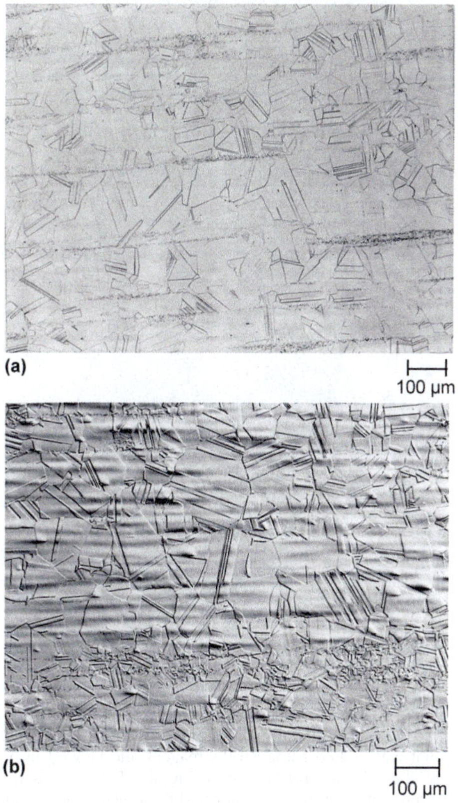

Fig. 40 Microstructure showing partial recrystallization produced by hot rolling and solution annealing (1090 °C, or 1995 °F, for 2 h, water quenched) Elgiloy (Co-20%Cr-5%Fe-5%Ni-%Mo-%Mn-0.05%B-0.15%C). The specimen was etched using 15 mL of HCl, 10 mL of acetic acid, and 10 mL of HNO_3 and viewed using bright field (a) and Nomarski DIC (b); DIC reveals the microstructure more effectively than bright field.

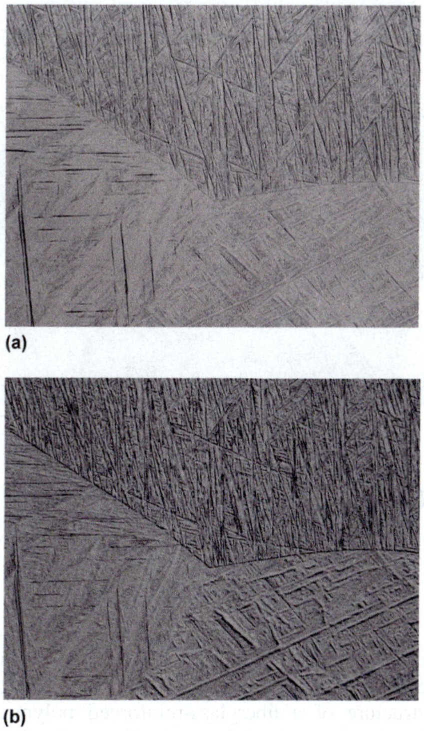

Fig. 41 Alpha prime martensite in microstructure of Ti-3%Cr heated to 1040 °C, or 1900 °F, held 15 min, and water quenched. The specimen was etched using Kroll's reagent and viewed using bright field (a) and Nomarski DIC (b); DIC reveals the microstructure more clearly than bright field (note the prior beta grain boundary). Magnification bars are 20 μm long.

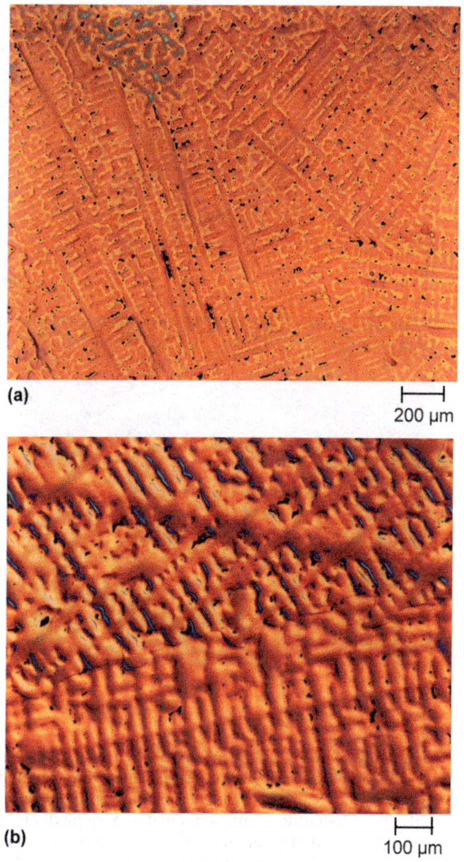

Fig. 42 Microstructure of sand-cast Cu-4%Sn (63 HV hardness) etched using Beraha's PbS tint etch and viewed using bright field (a) and Nomarski DIC (b). DIC reveals the dendritic cast structure more clearly than bright field.

Fig. 43 Nomarski DIC reveals low-carbon martensite formed on the as-polished surface of Carpenter Technology Corp.'s Temperature Compensator "30" alloy, type 2 (Fe-0.12%C-0.6%Mn-0.25%Si-30%Ni) after refrigeration in liquid nitrogen, which converted any unstable austenite to martensite with its characteristic shear reaction.

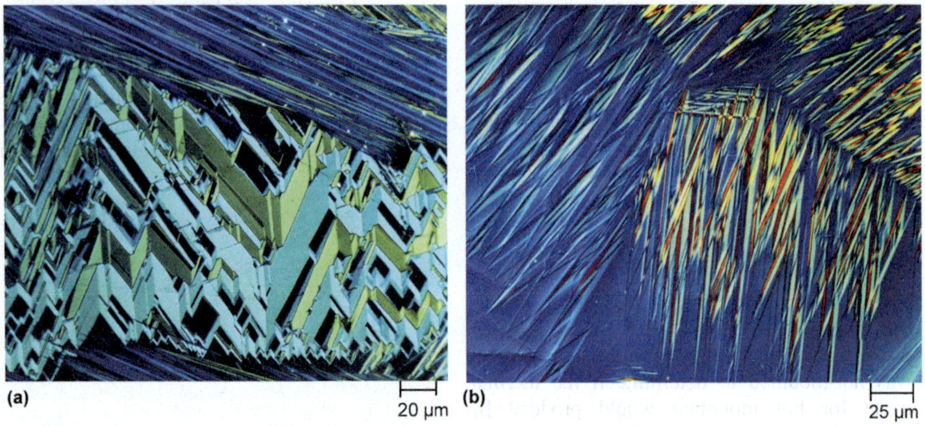

Fig. 44 As-polished Cu-26%Zn-5%Al shape memory alloy viewed using polarized light (a) and Nomarski DIC (b). Both imaging modes vividly reveal β_1 martensite formed in face-centered cubic alpha phase by cycling the alloy through the shape memory alloy thermal cycle.

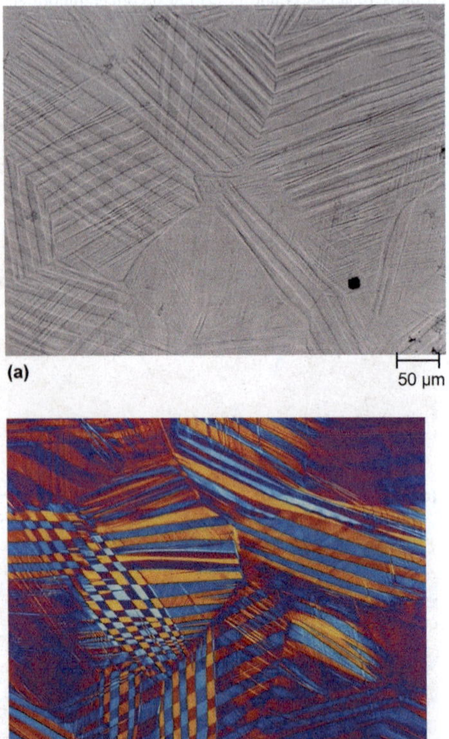

Fig. 45 Microstructure of as-polished Spangold (Au-19%Cu-5%Al), a jewelry alloy developed in South Africa, viewed in bright field (a) and Nomarski DIC (b). Martensite formation creates a rippled effect ("spangles") on the surface. The specimen was hot-mounted, polished, and given the shape memory alloy heat treatment (held at 100 °C, or 210 °F, for 2 min and water quenched), forming martensite, which produces shear at the free surface. The crisscrossed pattern is produced by forming some martensite during the hot mounting cycle, polishing, and then forming new martensite using the shape memory alloy cycle.

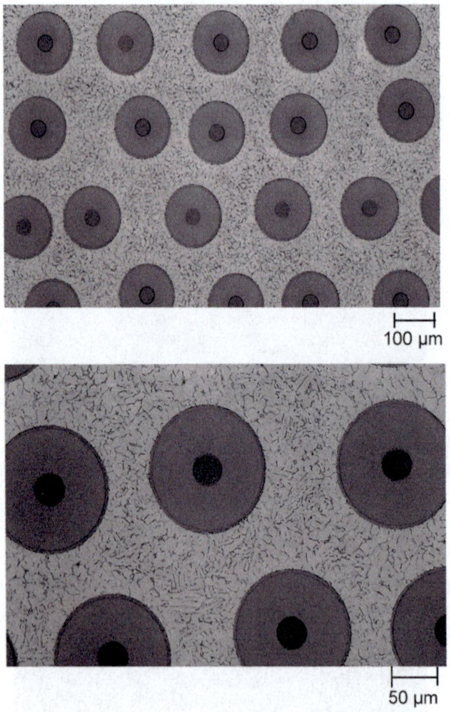

Fig. 46 Microstructure of SiC fiber-reinforced Ti-6Al-4V metal-matrix composite (transverse view) that was etched using Kroll's reagent and viewed using bright field

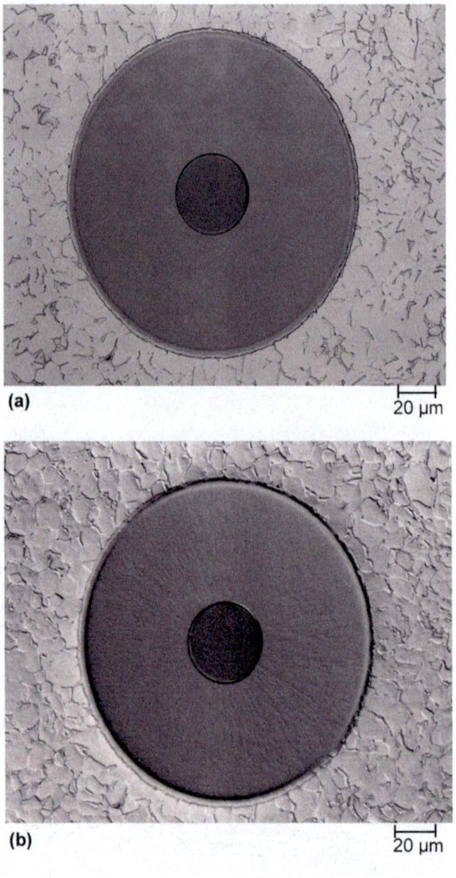

Fig. 47 Microstructure of SiC fiber-reinforced Ti-6Al-4V metal-matrix composite (transverse view) that was etched using Kroll's reagent and viewed using bright field (a) and Nomarski DIC (b). DIC reveals the grain pattern in the SiC fiber and the metal-matrix grain structure more clearly than bright field.

more detail and the growth pattern of the SiC more clearly. Figure 48 shows the microstructure of a fiberglass-reinforced polymer-matrix composite I-beam in Nomarski DIC. The fiberglass is clearly visible, but could not be seen in bright field. Figure 49 shows the structure of high-density polyethylene containing (unknown) filler particles (gray) in what appears to be a longitudinally oriented plane. Figure 50 shows an as-polished PZT-Ag/Pd ceramic multilayer actuator, with the PZT being (Pb, Zr, Ti)0_3, visualized using bright field and Nomarski DIC. The white lines are Ag/Pd. The structure is visible in both illumination modes, but DIC reveals the texture of the PZT ceramic.

Comparison of Imaging Modes

The microstructure of full-annealed cartridge brass (Cu-30%Zn) tint-etched using Klemm's I reagent is shown in Fig. 51. The specimen was somewhat over-polished in an attempt to remove all possible residual surface damage from preparation, resulting in some surface relief. Height differences are visible in the bright-field, oblique, and Nomarski DIC imaging modes. Figure 52 shows the same specimen as that in Fig. 14: Cu-11.8% Al heated to 900 °C (1650 °F), held for 1 h, and water–quenched, forming martensite. Four imaging modes (bright field, dark field, polarized light, and Nomarski DIC in black and white) were used on an as-polished specimen for Fig. 52. The martensite can be seen in bright field, although rather faintly.

Figure 53 shows the microstructure of the Ni-base superalloy Waspaloy in the solution-annealed and double-aged condition (~42 HRC), after swab etching with a fresh mixture of 15 mL of HCl, 10 mL of acetic acid, and 5 mL nitric acid, and viewed using bright field, dark field, and Nomarski DIC. Grain and twin boundaries can be seen using all three imaging modes, but dark field reveals twin boundaries most effectively. DIC reveals the grain structure well, but the image would not be the best to use when trying to measure grain size. A similar example, alloy 330 (Fe-0.05%C-36%Ni-19%Cr) electrolytically etched with aqueous 10% oxalic acid at 6 V dc for 10 s, is shown in Fig. 54. Grain structure is viewed using bright field, dark field, and Nomarski DIC. All imaging modes reveal grain structure, but dark field has the greatest contrast between grain and twin boundaries and the austenitic matrix for performing grain size measurement. DIC vividly reveals alloy segregation.

Figure 55 shows the as-cast eutectic microstructure of Cu-8.4%P consisting of alpha copper and Cu_3P. This is not a commercial alloy. The structure looks similar in nature to coarse lamellar pearlite in steels. It was etched using KBI reagent (100 mL water, 3 g ammonium persulfate, and 1 mL ammonium hydroxide). The microstructure is shown using bright field, dark field, and Nomarski DIC at 1000×. In dark field, all copper-phosphide lamellae appear to be outlined with bright light, and the copper lamellae between the Cu_3P lamellae are black. This very strong contrast makes the structure easy to see and would make it easy to perform an interlamellar spacing measurement. The DIC image shows that the etchant dissolved the alpha copper phase, leaving the Cu_3P lamellae in relief above the matrix.

Figure 56 shows an as-polished specimen of titanium diboride viewed using bright field, polarized light, and Nomarski DIC. Each imaging mode provides different information, with polarized light showing the most. These examples illustrates that illumination modes besides bright field should be tried to learn the most about the microstructure of a specimen. Also, all etchants are not equal, and nital, used probably over 90% of the time by metallographers studying iron and steel specimens, is frequently not the optimal reagent for viewing the structure.

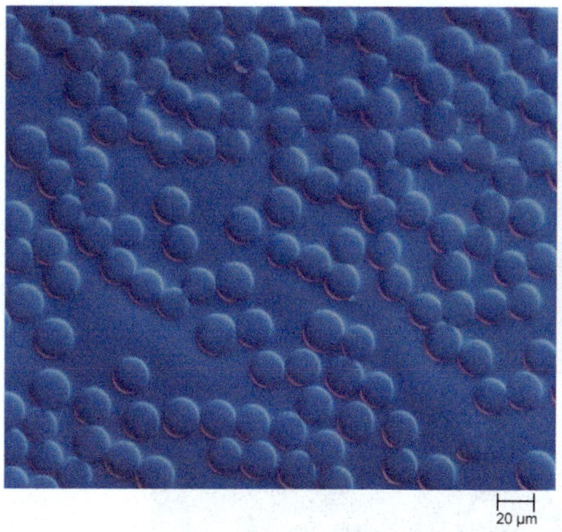

Fig. 48 Microstructure of fiberglass-reinforced polymer-matrix composite I-beam viewed using Nomarski DIC. Nothing is visible in bright field.

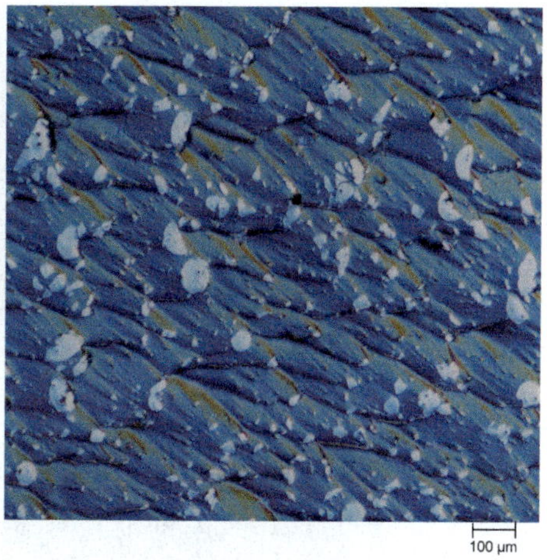

Fig. 49 Microstructure of as-polished high-density polyethylene containing a filler revealed using Nomarski DIC

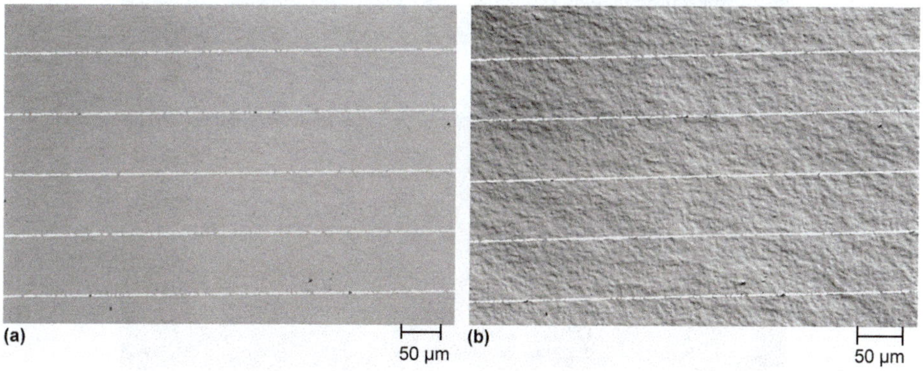

Fig. 50 As-polished microstructure of PZT [(Pb, Zr, Ti)O_3]-Ag/Pd ceramic multilayer actuator viewed using bright field (a) and Nomarski DIC (b). DIC reveals details more effectively.

Light Optical Images versus Scanning Electron Microscopy Images of Microstructure

As mentioned previously, many images of microstructures at 1000× and lower are being published in metallurgical journals using secondary electron and backscattered electron scanning electron microscopy (SEM) images rather than light microscope images. This is often poor practice, because microstructural details in SEM images at 1000× and lower are usually markedly inferior to light optical images at that magnification. The forte of SEM microstructural imaging is at magnifications higher 1000×, and SEM does not satisfactorily reveal fracture surface details at magnification lower 1000×. Optimal methods for using SEM to image microstructures can be found in Ref 20.

Figure 57 shows the as-rolled ferrite-pearlite microstructure of properly and improperly prepared AISI 10B21 carbon steel, both etched using 2% nital and observed under a light microscope at 1000×. Backscattered electron images of the same specimen, properly and poorly prepared, obtained using SEM at 1000× are shown in Fig. 58. Virtually no details can be seen in either image, and secondary electron images at 1000× are even poorer. Scanning electron microscopy backscattered electron images at 5000 and 10,000× of the same specimen, properly and poorly prepared, are shown in Fig. 59. Higher magnification reveals more details about the pearlitic structure. Figure 60 shows a properly

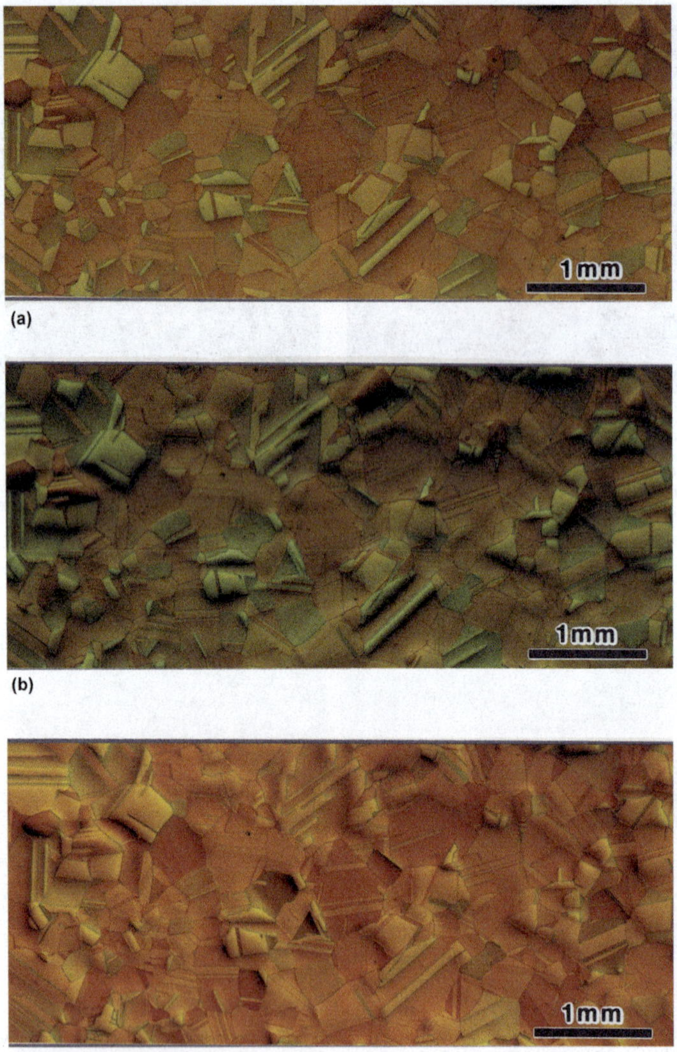

Fig. 51 Microstructure of full-annealed Cu-30%Zn etched using Klemm's I reagent and viewed using (a) bright field, (b) oblique illumination, and (c) Nomarski DIC. Excessive relief was produced during final polishing. Magnification: 15×

prepared specimen of as-rolled 9254 alloy steel etched using 4% picral and observed at 1000× using a light microscope. The interlamellar spacing is too fine to see at 1000×. Scanning electron microscopy backscattered electron images of the properly prepared specimen at 5000 and 10,000× are also shown, and the interlamellar spacing can be seen more clearly as the magnification increases. This is an example of the proper use of SEM to examine microstructures.

Summary

To examine specimens and correctly interpret their microstructure, properly prepared specimens must first be examined in the as-polished condition to detect nonmetallic inclusions, nitrides, cracks, and voids using a light microscope before the specimen is etched to reveal the matrix microstructure. To see the microstructure of noncubic-crystal-structure metals and alloys, polarized light should be used on properly prepared, unetched specimens. In some cases, the structure of these metals can be seen in the etched condition, but not always as well as with polarized light using an as-polished specimen. Etched specimens, depending on the etch chosen, can be viewed in bright field, dark field, polarized light, and Nomarski DIC, and different details are sometimes observed in each illumination mode. In some cases, bright field is not the best illumination mode to use. Other modes must be tried to determine if a mode other than bright field will produce better images. It is often difficult to predict beforehand that another illumination mode will work better.

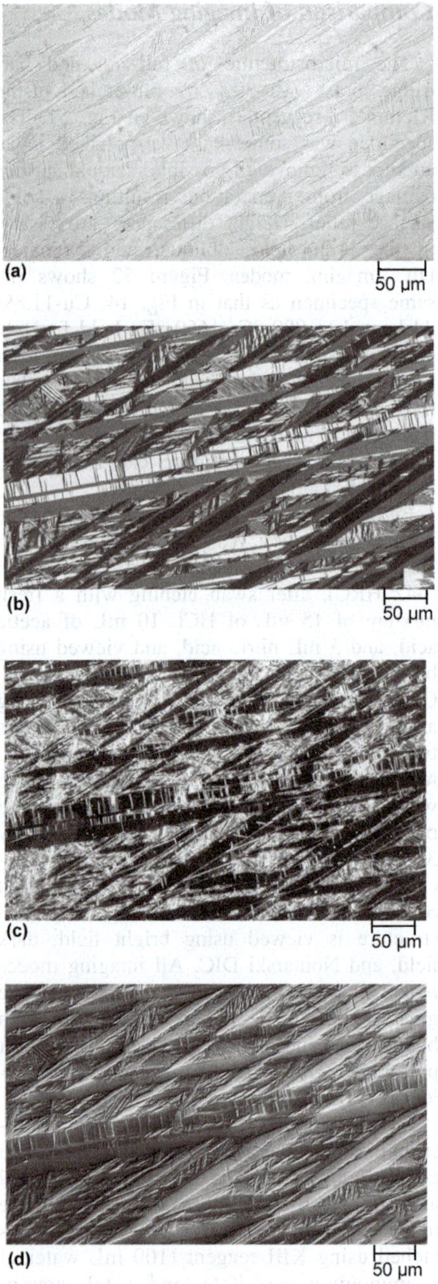

Fig. 52 As-polished Cu-11.8%Al specimen (same as shown in Fig. 14, but in black and white) viewed using (a) bright field, (b) polarized light, (c) dark field, and (d) Nomarski DIC; bright field only weakly reveals martensite. Magnification: 200×

REFERENCES

1. G.F. Vander Voort, Ed., *Metallography and Microstructures*, Volume 9, *ASM Handbook*, ASM International, 2004
2. G.L. Kehl, *The Principles of Metallographic Laboratory Practice*, 3rd ed., McGraw-Hill Book Co., 1949
3. J.H. Richardson, *Optical Microscopy for the Materials Sciences*, Marcel Dekker, 1971

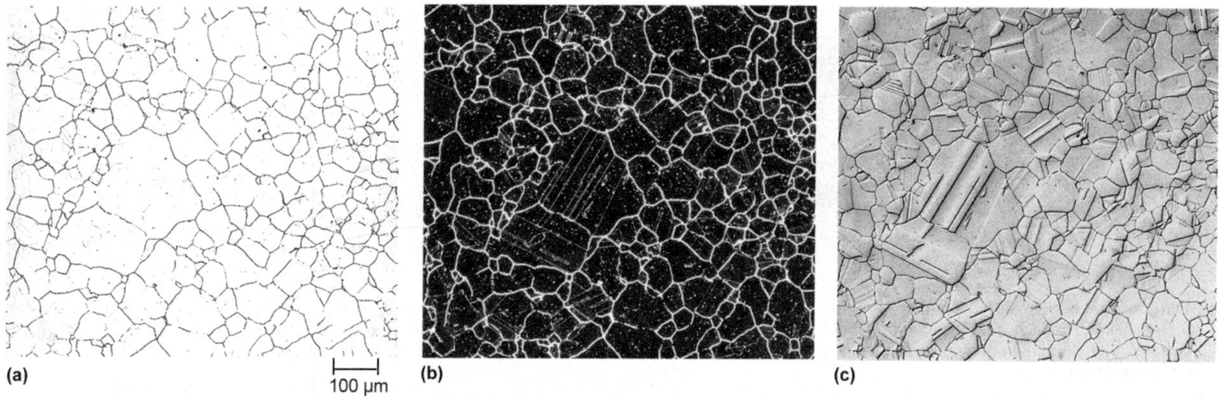

Fig. 53 Solution-annealed and double-aged Waspaloy etched using 15 mL of HCl, 10 mL of acetic acid, and 5 mL of HNO_3 and viewed using (a) bright field, (b) dark field, and (c) Nomarski DIC. All three imaging modes reveal grain and twin boundaries, but dark field reveals twin boundaries most clearly. DIC reveals the grain structure well, but the image would not be suitable for measuring grain size. Magnification: 100×

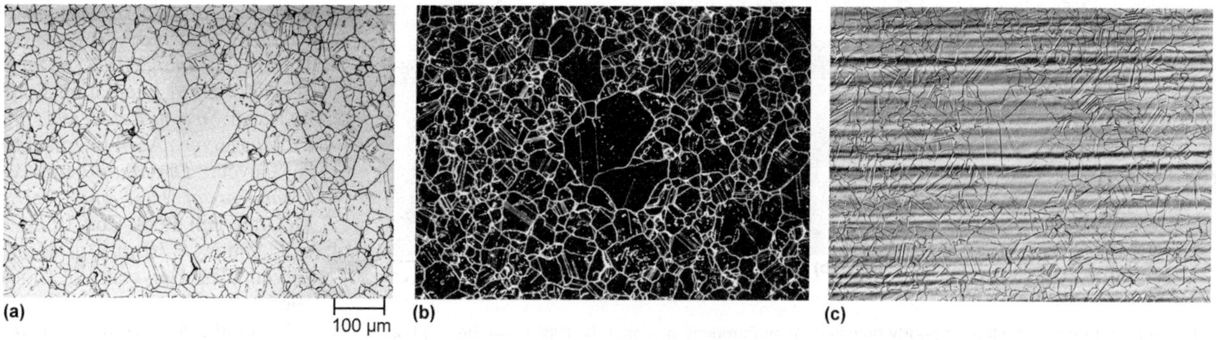

Fig. 54 Microstructure of as-rolled alloy 330 solution annealed at 1080 °C (1975 °F), electrolytically etched using 10% oxalic acid at 6 V dc for 10 s and viewed using (a) bright field, (b) dark field, and (c) Nomarski DIC. All imaging modes reveal grain structure, but dark field has the greatest contrast between grain and twin boundaries and the austenitic matrix for performing grain size measurement. DIC vividly reveals alloy segregation.

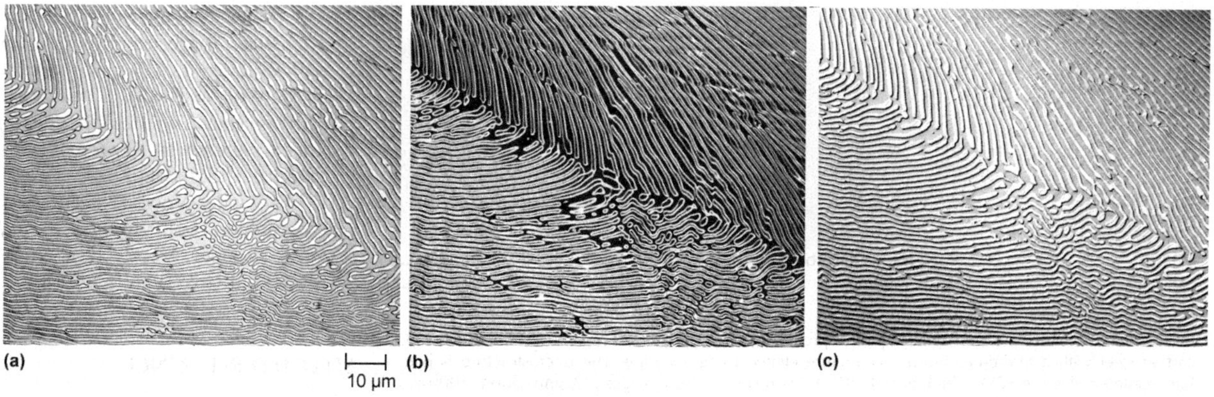

Fig. 55 Microstructure of as-cast Cu-8.4%P eutectic structure of alpha copper and Cu_3P etched using KBI reagent (100 mL water, 3 g ammonium persulfate, and 1 mL ammonium hydroxide) and viewed using (a) bright field, (b) dark field, and (c) Nomarski DIC. In dark field, all copper-phosphide lamellae appear to be outlined with bright light, and the copper lamellae between the Cu_3P lamellae are black, contrast that makes the structure easy to see and enabling easy interlamellar spacing measurement. DIC shows that the etchant dissolved the alpha copper phase, leaving the Cu_3P lamellae in relief above the matrix. Magnification: 1000×

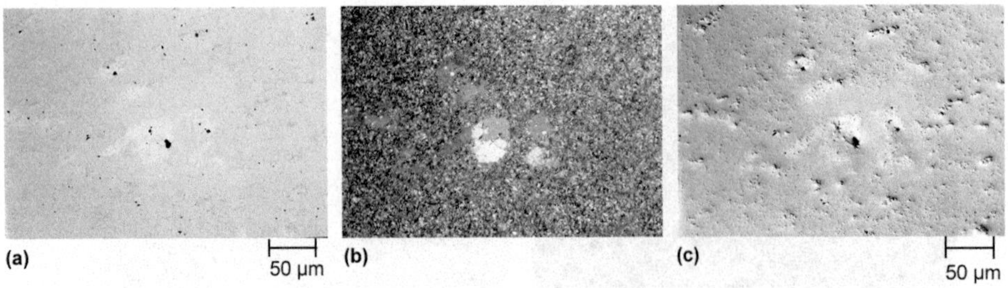

Fig. 56 Microstructure of as-polished titanium diboride viewed using (a) bright field, (b) polarized light, and (c) Nomarski DIC. Each imaging mode provides different information, with polarized light showing the most. This illustrates that illumination modes besides bright field should be tried to learn the most about the microstructure of a specimen.

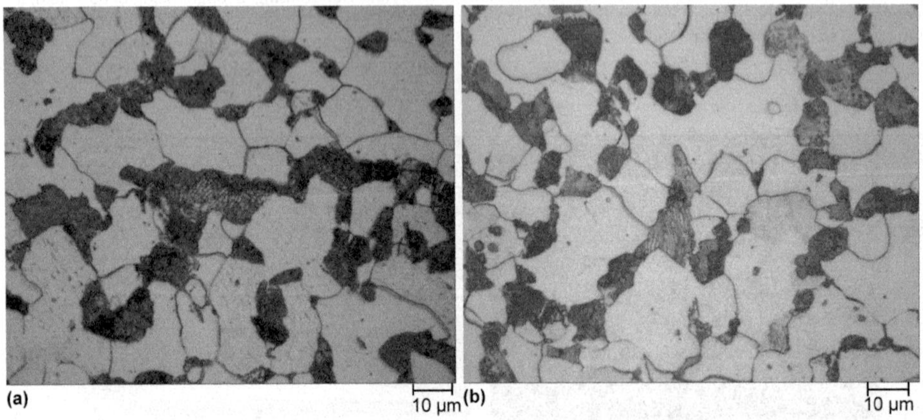

Fig. 57 Ferrite-pearlite microstructure of poorly prepared (a) and properly prepared (b) 10B21 as-rolled carbon-steel specimen etched using 2% nital and viewed using bright field. Magnification: 1000×

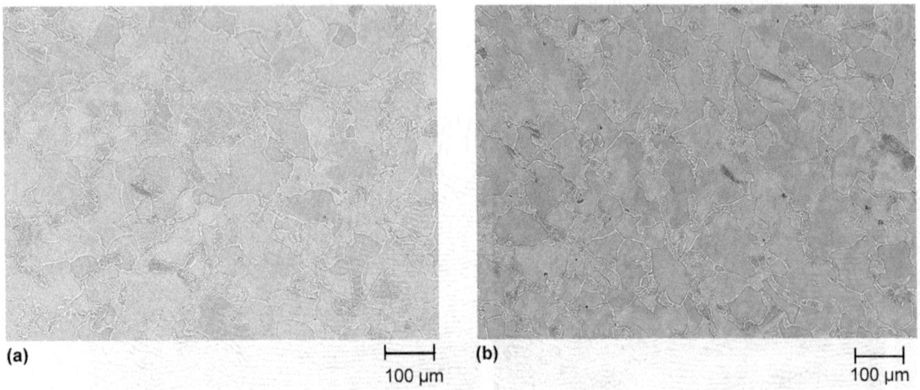

Fig. 58 Backscattered electron SEM images of poorly prepared (a) and properly prepared (b) as-rolled 10B21 carbon steel with a ferrite-pearlite microstructure etched using 2% nital. The microstructure is barely visible with backscattered electron SEM (which is better than secondary electron images). Magnification: 1000×

4. L.E. Samuels, *Light Microscopy of Carbon Steels*, 2nd ed., ASM International, 1999
5. G.F. Vander Voort, *Metallography: Principles and Practice*, McGraw-Hill Book Co., 1984; ASM International, 1999
6. J.L. McCall and P.M. French, *Interpretive Techniques for Microstructural Analysis*, Plenum Press, 1977
7. J.L. McCall and P.M. French, Ed., *Metallography as a Quality Control Tool*, Plenum Press, 1980
8. G.F. Vander Voort, Ed., *Applied Metallography*, Van Nostrand Reinhold Co., 1986
9. G.F. Vander Voort, F.J. Warmuth, S.M. Purdy, and A. Szirmae, Ed., *Metallography: Past, Present and Future, 75th Anniversary Volume*, ASTM STP 1165, ASTM International, 1993
10. B.L. Bramitt and A.O. Benscoter, *Metallographer's Guide: Practices and Procedures for Irons and Steels*, ASM International, 2002
11. K. Geels, *Metallographic and Materialographic Specimen Preparation, Light Microscopy, Image Analysis and Hardness Testing*, ASTM International, 2007
12. A. Tomer, *Structure of Metals Through Optical Microscopy*, ASM International, 1991
13. A. Sauveur, *The Metallography and Heat Treatment of Iron and Steel*, 4th ed., McGraw-Hill Book Co., 1935.
14. C.S. Smith, *A History of Metallography*, 2nd ed., The University of Chicago Press, 1965
15. C.S. Smith, *A Search for Structure*, The MIT Press, 1981
16. G.L. Turner, *The Great Age of the Microscope, A Collection of the Royal Microscopical Society Through 150 Years*, Adam Hilger, 1989
17. Symposium on Metallography in Color, ASTM, STP 86, Philadelphia, 1949
18. E. Beraha and B. Sphigler, *Color Metallography*, American Society for Metals, 1977
19. I.M. Wolf and M.B. Cortie, *The Aesthetic Enhancement or Modifications of Articles or Components Made of Non-Ferrous Metals*, U.S. Patent 5, 503, 691, 2 Apr 1996
20. G.F. Vander Voort, The SEM as a Metallographic Tool, *Applied Metallography*, G.F. Vander Voort, Ed., Van Nostrand Reinhold Co., 1986, p 139–170

SELECTED REFERENCE

- M.R. Louthan, Optical Metallography, *Materials Characterization*, Vol 10, *ASM Handbook*, ASM International, 1986

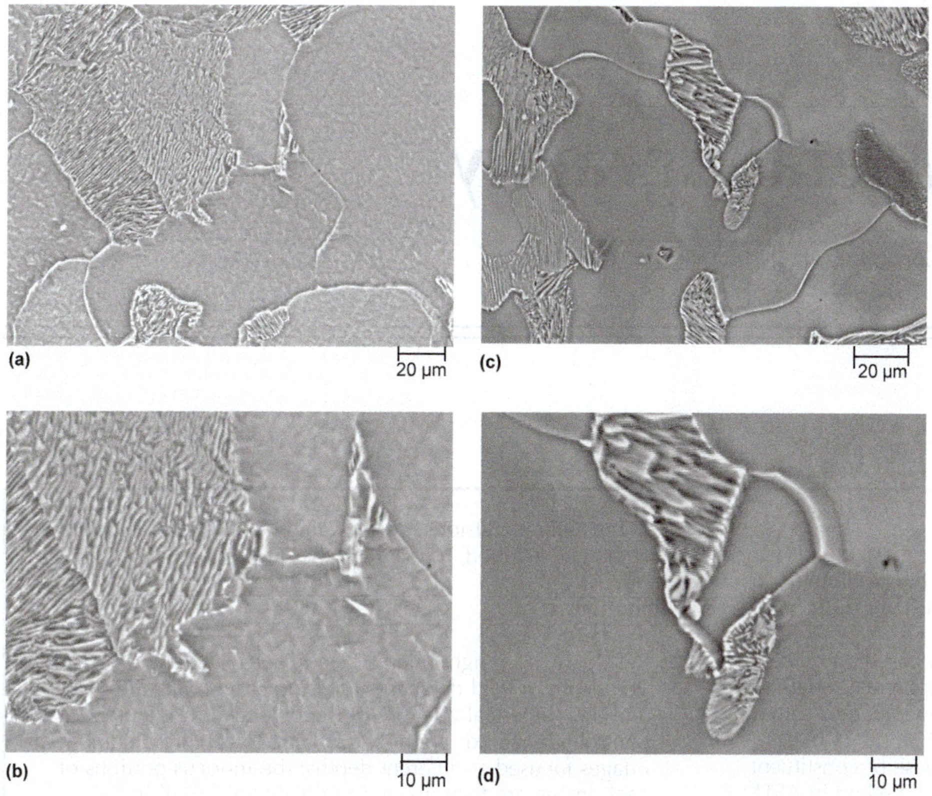

Fig. 59 Higher-magnification backscattered electron SEM images of (a, b) poorly prepared and (c, d) properly prepared as-rolled 10B21 carbon steel with a ferrite-pearlite microstructure etched using 2% nital. Magnification: (a) and (c), 5000×; (b) and (d), 10,000×. Higher magnification improves results compared with those in Fig. 58.

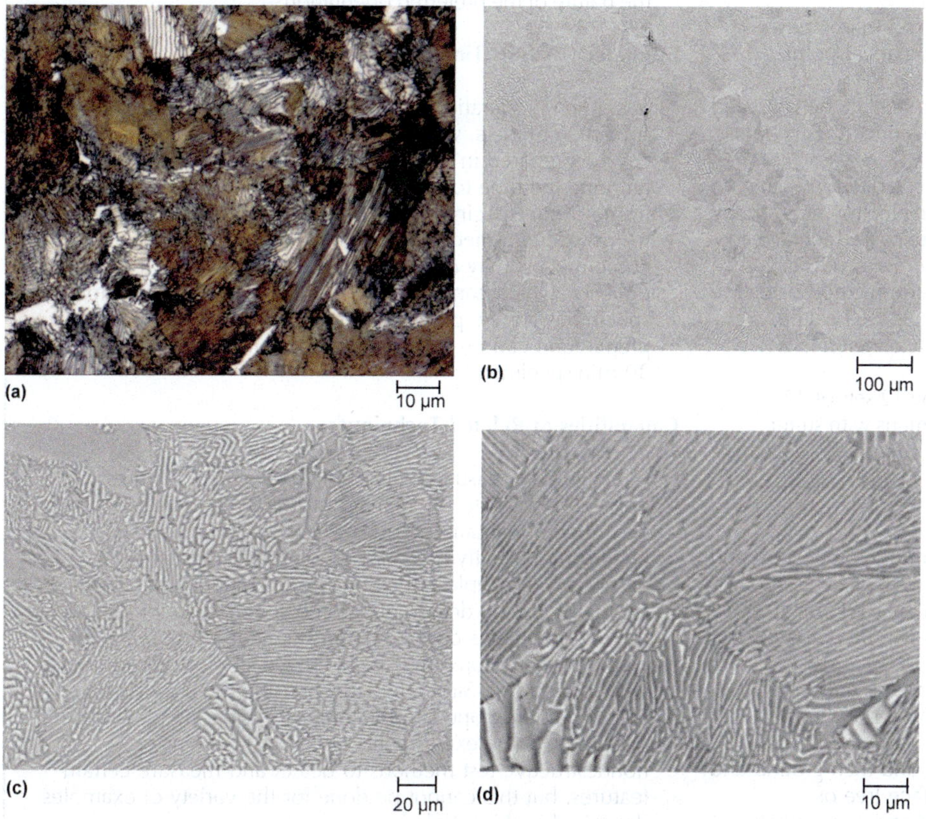

Fig. 60 Comparison of (a) bright-field light optical microscopy image (magnification: 1000×) and (b–d) backscattered electron SEM images (magnification: 1000, 5000, and 10,000×, respectively) of properly prepared as-rolled 9254 alloy steel with very fine pearlite. Increasing magnification increasingly reveals the interlamellar spacings.

ASM Handbook, Volume 10, *Materials Characterization*
ASM Handbook Committee
DOI 10.31399/asm.hb.v10.a0006682

Quantitative Metallography

George F. Vander Voort, Vander Voort Consulting, L.L.C

Overview

General Uses

- Measurement of grain size using either the planimetric or intercept method manually, based on ASTM E 112, or by image analysis based on ASTM E 1382
- Measurement of nonmetallic inclusions based on manual chart ratings using the image analysis approach (originally in ASTM E 1122) in ASTM E 45 or by image analysis using the stereological approach in ASTM E 1245
- Measurement of the amount of each phase or constituent present using the manual point-counting method in ASTM E 562 or the image analysis method in ASTM E 1245

Examples of Applications

- Measurement of electroless nickel plating using standard metrology thickness measurements
- Measurement of three different images by 33 people to determine the percent of each phase by manual point counting
- Grain size and grain elongation of ferritic steel in the annealed condition and after cold deformation to reduce the thickness 12, 30, and 70%
- Interlamellar spacing of pearlite using directed test lines and randomly oriented test lines. Transmission electron microscopy thin foils were tilted to reveal the true spacings, which agreed very closely with one-half the mean random spacing.
- Grain size measurement by the planimetric method using the Jeffries circle and the Saltykov rectangle as well as the intercept method using a single-phase specimen and a specimen with two constituents
- Nonmetallic inclusion measurement using 12 sets of 90 measurements (1080 total) on nine specimens with sulfur contents from 0.020 to 0.34% using 16, 32, and 80× objectives

Samples

- *Form:* Metals and alloys, polymers, ceramics, composites, and minerals
- *Size:* Dimensions usually range from <1 to 100 mm^2 (0.0015 to 0.15 in.^{2}), although larger specimens can be prepared and examined.
- *Preparation:* Normally, manufactured parts or mill products are sectioned to provide a specimen with a polished surface area of ≤100 mm^2 (0.15 in.^{2}), which is mounted in the appropriate polymeric resin for the task and then ground and polished to yield a flat, scratch-free surface free of preparation-induced damage that would make examination problematic. They are examined as-polished first to detect nonmetallic inclusions, nitrides, and cracks or voids and then etched, if required, to examine the microstructure.

Limitations

- Resolution with light microscope images is limited to ~1 μm.
- The depth of field decreases with increasing magnification and the numerical aperture of the objective. Computer-controlled methods are now available to capture a number of images focused at different depths; the in-focus portions of each image are then merged to create an overall in-focus image as a function of depth.
- Light optical microscopes do not provide chemical or crystallographic information about the observed microstructural elements. Color tint etching will reveal if the grains are randomly oriented or if a preferred orientation exists, but it cannot define the nature of the preferred orientation.

Estimated Analysis Time

- Specimen preparation can be automated so that multiple specimen holders, with typically six specimens per holder, can be prepared in 20 to 30 min for most specimens, ignoring the time to mount specimens (which can vary from ~5 min/specimen to 10 to 18 h for low-viscosity epoxy resins). Certain metals and alloys, such as face-centered cubic metals, may require a final polish using a vibratory polisher, which can add 20 to 60 min/specimen (multiple specimens can be prepared simultaneously). Manual preparation can be performed and would take 20 to 30 min/specimen.

Capabilities of Related Techniques

- Quantitative measurements of microstructural features can only be made with the manual or image analysis techniques described. Qualitative "guesstimates" of microstructural features are usually made based on chart images, as described, or simple visual guesses, as mentioned regarding evaluation of the degree of carbide spheroidization. Some of the measurements described can be made using scanning electron microscopy images, as long as the features of interest can be detected. Similarly, images using the transmission electron microscope can be analyzed using the methods discussed. Some experiments are being made using nondestructive test methods to detect and measure certain features, but this cannot be done for the variety of examples described in this article.

Introduction

Many tasks performed by metallographers are done simply by visual examination of specimens and qualitative assessments of the structural features being examined. A number of ASTM E-4 international "round-robins" to assess the precision and bias associated with different test methods have shown that qualitative assessments of microstructural features are imprecise and often rate structures incorrectly. This has been demonstrated for grain size measurements, inclusion ratings, and decarburization measurements. By comparison, actual quantitative measurements used in such studies reveal far better precision in assessing microstructural features. Chart-type ratings showing grain size and inclusion ratings are fraught with problems. In this article, the focus is placed on the more commonly used measurements made by metallographers, particularly in relation to production manufacturing work and failure analysis studies. Measurement of the interlamellar spacing of cementite in pearlite is an exception, because this cannot be done qualitatively. However, for the evaluation of properly isothermally transformed pearlite that will be cold drawn to produce wire, it is a necessary procedure to ensure that formability and the resulting properties are optimum.

The theory of quantitative microstructural measurements can be traced to 1950 when Sarkis A. Saltykov (1905–1983) from Armenia published the first edition of his book *Stereometric Metallography*. The second edition was published in 1958 and was translated into English in 1961, Czechoslovakian in 1962, and German in 1974. The third edition was published in 1970, and the fourth edition, for students, was published in 1976 (Ref 1–4). Earlier attempts to quantify microstructure were made, but the understanding of how microstructure was formed, how it could be altered by processing, and how it affected mechanical properties was too primitive for the science to be developed. Efforts to measure grain size can be found prior to 1945, but the procedures were not well developed. Starting in the early 1960s, the technical development of measurement procedures began to reach a level useful for research and commercial purposes. The development of image analysis equipment began in the late 1960s, but computer systems were not technically advanced at that time. The history of quantitative metallography is not long, but its use has become very important.

Although the fundamental relationships for stereology, the foundation of quantitative metallography, were known for some time, implementation of these concepts was limited because they could only be performed manually due to the tremendous effort required. Further, while humans are good at pattern recognition, as in the identification of complex structures, they are less successful at repetitive counting. Many years ago, George Moore (Ref 5) and members of ASTM International Committee E-4 on Metallography conducted a simple counting experiment. Approximately 400 people were asked to count the number of times the letter "e" appeared in a paragraph without physically marking or striking out the letters as they counted. The correct answer was obtained by only 3.8% of the people. Results were not Gaussian, however, as only 4.3% had higher values while 92% had some much lower values. The standard deviation was 12.28. This experiment revealed a basic problem with manual ratings. In this case, the subject was one very familiar to the raters, yet only 3.8% obtained the correct count. What degree of counting accuracy can be expected when the subject is less familiar, such as for microstructural features? Image analyzers, on the other hand, are quite good at counting but not as competent at recognizing features of interest. To enhance feature detection, it is necessary to use etchants that are selective as to which constituent or phase they color or darken and which they do not. In general, the commonly used etchants for qualitative evaluation of the microstructure are often inadequate for image analysis work. Fortunately, there has been tremendous progress in the development of powerful computers and user-friendly image analyzers since their initial development in the late 1960s and early 1970s.

Chart methods for rating microstructures have been used for many years to evaluate microstructures, chiefly for conformance to specifications. True quantitative procedures are replacing chart methods for such purposes, and they are being used in quality control and research studies. Examples of the applications of stereological measurements have been reviewed by Underwood (Ref 6, 7). In 1967, chart ratings and visual examinations were the main approach to evaluate microstructures for grain size and inclusion content. For example, mill metallographers used a light microscope to evaluate how well spheroidized the carbides in annealed tool steel structures were, determining, for example, 90% spheroidized (many raters would never say 100%), or 60% spheroidized and 40% lamellar tending to spheroidize. Or, without referring to the chart, they would determine that the grain size was, for example, 100% 6 to 8 or 70% 8 and 30% 3 to 5, if it appeared to be bimodal in appearance.

Two types of measurements of microstructures are commonly made. The first group includes measurements of depths, that is, depth of decarburization, depth of surface hardening, or thickness of coatings or platings. These measurements are made at a specific location (the surface) and can be subject to considerable natural thickness variations along the surface of the product. To obtain reproducible data, these surface conditions must be measured at a number of positions on a given specimen, and on several specimens, if the material being sampled is rather large. Sampling is a very important aspect in obtaining reliable measurements. Standard metrology methods are used, and these can be automated. Metrology methods are also used for individual feature analysis, for example, particle size, elongation, and shape measurements. Figure 1 illustrates metrology used to measure the depth of electroless nickel plated on 4150 alloy steel as a function of plating time. Measurements were made at 1000× with up to 21 locations per specimen.

The second group of measurements belongs to the field referred to as stereology. This is the body of measurements that describe the relationships between measurements made on the two-dimensional plane-of-polish to predict the nature of the three-dimensional microstructural features being measured. To facilitate communications, the International Society for Stereology proposed a standard system of notation, as shown in Table 1 (Ref 6–8). This table lists the most commonly used notations.

These measurements can be made manually with the aid of templates outlining a fixed field area, systems of straight or curved lines of known length, or a number of systematically spaced points. The simple counting measurements P_P, P_L, N_L, P_A, and N_A are most important and are easily made. These measurements are useful by themselves and can be used to derive other important relationships. These measurements can also be made by using automatic image analyzers.

In this article, the basic rules of stereology are described with emphasis on how these

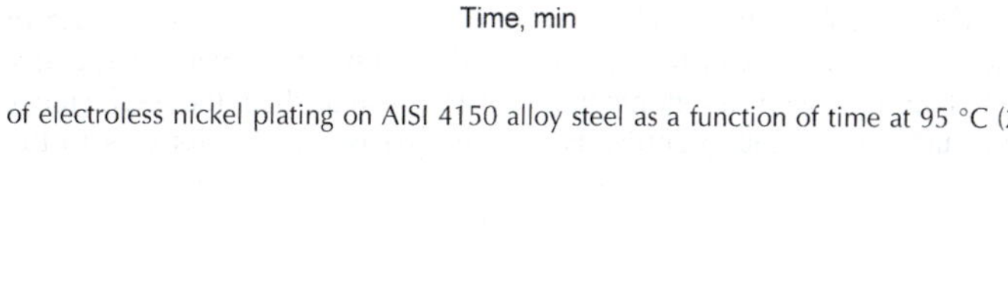

Fig. 1 Depth of electroless nickel plating on AISI 4150 alloy steel as a function of time at 95 °C (205 °F)

Table 1 International Society for Stereology nomenclature for stereological measurements

Symbol	Dimensions (arbitrarily shown in millimeters)	Definition
P	...	Number of point elements, or test points
P_P	...	Point fraction. Number of points (in areal features) per test point
P_L	mm^{-1}	Number of point intersections per unit length of test line
P_A	mm^{-2}	Number of points per unit test area
P_V	mm^{-3}	Number of points per unit test volume
L	mm	Length of lineal elements, or test line length
L_L	mm/mm	Lineal fraction. Length of lineal intercepts per unit length of test line
L_A	mm/mm^2	Length of lineal elements per unit test area
L_V	mm/mm^3	Length of lineal elements per unit test volume
A	mm^2	Planar area of intercepted features, or test area
S	mm^2	Surface or interface area (not necessarily planar)
A_A	mm^2/mm^2	Area fraction. Area of intercepted features per unit test area
S_V	mm^2/mm^3	Surface area per unit test volume
V	mm^3	Volume of three-dimensional features, or test volume
V_V	mm^3/mm^3	Volume fraction. Volume of features per unit test volume
N	...	Number of features (as opposed to points)
N_L	mm^{-1}	Number of interceptions of features per unit length of test line
N_A	mm^{-2}	Number of interceptions of features per unit test area
N_V	mm^{-3}	Number of features per unit test volume
$\overline{L}$	mm	Average lineal intercept, L_L/N_L
$\overline{A}$	mm^2	Average areal intercept, A_A/N_A
$\overline{S}$	mm^2	Average surface area, S_V/N_V
$\overline{V}$	mm^3	Average volume, V_V/N_V

Source: Ref 6–8

procedures are applied manually, which should be understood before using image analysis for the measurements. Image analysis users must understand these principles before attempting to perform them. When developing a new measurement routine, it is good practice to compare image analysis data with data developed manually, because it is easy to make a mistake in setting up a measurement routine.

Sampling

Sampling material is important and must be done properly so that the measurement results are representative of the material. Ideally, random sampling is best, but it can rarely be performed, except for small parts such as fasteners where a specific number of fasteners can be drawn from a production lot at random. However, with a large forging or casting, it is generally impossible to select specimens at random from the product mass, so test specimens are generally taken from excess material added to the bulk for such purposes. For a casting, it may be possible to trepan sections at locations that will undergo machining at some later step. The practice of casting a small keel block for test specimens can produce results markedly different than would be obtained from the actual casting if there are significant differences in the keel block versus casting size and their solidification and cooling rates. After the specimens are obtained, there is still a sampling problem, particularly in wrought material. Microstructural measurements made on a plane parallel to the deformation axis will often be quite different from those taken on a plane perpendicular to the deformation axis, especially for features such as nonmetallic inclusions. In such cases, the practice is to compare results on similarly oriented planes. Few people in industry have the time to measure the microstructural feature of interest on the three primary planes in a flat product, such as plate or sheet, so that the true three-dimensional nature of the structure can be determined, except, perhaps, in research studies. In general, when only one plane is evaluated, the longitudinal plane is most commonly chosen for inclusion and grain size measurements, and a transverse plane is chosen for measurements of the depth of decarburization, carburization, nitriding, coatings, and platings, because they will vary more going around the periphery than along the edge of a longitudinal plane.

The sampling plan must also specify the number of specimens to be tested and the number of fields per specimen. In practice, the number of specimens chosen is a compromise between minimizing testing cost and the desire to perform adequate testing to characterize the lot statistically. Excessive testing is rare. Inadequate sampling is more likely, due to physical constraints with some components and a desire to control testing costs. In the case of inclusion ratings, the testing plan was established years ago by the chart method in ASTM E 45, "Standard Methods for Determining the Inclusion Content of Steel." The procedure was to sample billets at locations representing the top and bottom or top, middle, and bottom of the first, middle, and last ingots on a heat. The plane-of-polish is longitudinal (parallel to the hot working axis) at the midthickness location. This yields six or nine specimens. The area examined is 160 mm^2 (0.250 $in.^2$) per specimen, that is, 960 or 1440 mm^2 (1.5 or 2.25 $in.^2$) for six or nine specimens, respectively. This small area established the inclusion content and is the basis for decisions as to the marketability of a heat of steel that could weigh from (typically) 50 to 300 tons (45,000 to 270,000 kg). For bottom-poured heats, there is no first, middle, and last ingot, and continuous casting eliminates ingots. So, alternative sampling plans are needed. Image analysis inclusion testing using ASTM E 1245, "Standard Practice for Determining the Inclusion or Second-Phase Constituent Content of Metals by Automatic Image Analysis," characterization was markedly improved by the use of at least 18 or 27 specimens per heat where the subsurface, midradius, and center locations at each billet location (top and bottom or top, middle, and bottom of the first, middle, and last top-poured ingots, or three ingots at random from a bottom-poured heat) were evaluated. ASTM E 1245 produces mean and standard deviation values for each measurement. Therefore, valid statistical analysis techniques can be used to compare the inclusion content of one heat versus another or one melting practice versus another and determine if the differences are significant and at what level of significance. ASTM E 45 chart ratings of inclusions are worst-field ratings, and a valid quantitative statistical comparison of the inclusion content in one heat versus another, or one melting practice versus another, cannot be performed.

Specimen Preparation

The measurement part of the task is quite straightforward, and much of the difficulty encountered is in preparing the specimens properly so that the true structure can be observed. Measurement of inclusions is conducted on as-polished specimens, because etching brings out other extraneous details that will obscure the detection of the inclusions. Measurement of graphite in cast iron is also performed on as-polished specimens. However, it is possible that shrinkage cavities, which are often present in castings, can cause interference with image analysis detection of the graphite because they have overlapping gray scales. When the specimen must be etched to see the constituent of interest, it is best to etch the specimen so that only the constituent of interest is revealed or everything but the constituent of interest is revealed. Selective etchants are best.

Specimen preparation is easier than ever with the introduction of automation in sample-preparation equipment. The biggest source of preparation-induced damage is sectioning. This step must be done carefully using the least damaging approach, especially when the sectioning is performed to obtain the plane to be polished. Every step in the preparation sequence must be done properly, because the subsequent steps usually have lower removal rates and cannot correct excessive damage from the prior step. Specimens prepared with automated devices exhibit better flatness than manually prepared specimens. Mounting of specimens is a necessity when details at the surface are being assessed, because the edges of an unmounted specimen are always rounded. Edge-retention results of different mounting resins do vary, and the best resins for edge retention should be used. If the edges are not being examined and measurements are only performed away from the edge, then the choice of the best resin for edge retention is not critical. The preparation sequence must establish the true structure, free of any artifacts. Automated equipment can produce a much greater number of properly prepared specimens per day than the best manual operator.

Quantitative Microstructural Measurements

Common quantitative microstructural measurements include the volume fraction, number per unit area, intersections and intercepts per unit length, grain size, and inclusion content.

Volume Fraction, V_V

The amount of a second phase or constituent in a two-phase alloy can have a significant influence on the material properties and behavior (Ref 9). Consequently, determination of the amount of the second phase is an important measurement. The amount of a second phase is defined as the volume of the second phase per unit volume, or volume fraction, V_V. There is no simple, direct experimental technique for measuring this value. Three practical experimental approaches for estimating the volume fraction have been developed using microscopy methods: the area fraction, the lineal fraction, and the point fraction.

The volume fraction was first estimated by areal analysis by Delesse, a French geologist, in 1848 (Ref 10). He showed that the area fraction was an unbiased estimate of the volume fraction. Several procedures have been used on real structures. One is to trace the second phase or constituent with a planimeter and determine the area of each particle. These areas are summed and divided by the field area to obtain the area fraction, A_A. Another approach is to weigh a photograph and then cut out the second-phase particles and weigh them. The two weights are used to calculate the area fraction, because the weight fraction of the micrograph should be equivalent to the area fraction. Both techniques are only realistically possible with large second phases or constituents. A third approach is the so-called occupied-squares method. A clear plastic grid containing 500 small square boxes is superimposed over a micrograph or live image. Then, the operator counts the number of grid boxes that are completely filled, ¾ filled, ½ filled, and ¼ filled by the second phase or constituent. These data are used to calculate the area covered by the second phase and then is divided by the image area to obtain the area fraction. All three methods give a reasonably precise measurement of the area fraction of one field, but they require an enormous amount of effort per field. However, it is well recognized that the field-to-field variability in volume fraction has a larger influence on the precision of the volume fraction estimate than the error in rating a specific field, regardless of the procedure used. Therefore, it is not wise to spend a great deal of effort in obtaining a very precise measurement on one field or only a few fields.

Delesse also stated that the volume fraction could be determined using a lineal analysis approach, but he did not define such a method. This was done in 1898 by Rosiwal, a German geologist, who demonstrated that a sum of the lengths of line segments within the phase of interest divided by the total length, L_L, provided a valid estimate of the volume fraction with less effort than areal analysis (Ref 11). In approximately 1930, several workers in different fields and countries (e.g., Thomson, Ref 12; Glagolev, Ref 13; and Chalkley, Ref 14) showed that the percentage of points on a test grid lying in the phase of interest was equal to the volumetric percentage, that is, $P_P = V_V$.

These studies have shown that the third method, point counting, is a more efficient method than lineal analysis; that is, it gives the best precision with the least effort (Ref 8). Point counting is described in ASTM E 562, "Standard Test Method for Determining Volume Fraction by Systematic Point Count," and ISO 9042, "Steels: Manual Point Counting Method for Statistically Estimating the Volume Fraction of a Constituent with a Point Grid," and has been widely used to estimate volume fractions of microstructural constituents. To perform the test, a clear plastic grid with a number of systematically spaced points (usually crosses are used, where the "point" is the intersection of the arms), typically 9, 16, 25, 49, 64, or 100, is placed on a micrograph, on a projection screen, or inserted as an eyepiece reticle. The number of points lying on the phase or constituent of interest is counted and divided by the total number of grid points. Points lying on a boundary are counted as half-points. This procedure is repeated on a number of fields selected without bias, that is, without looking at the image.

The point fraction, P_P, is given by:

$$P_P = P_\alpha / P_T \quad \text{(Eq 1)}$$

where P_α is the number of grid points lying inside the feature of interest, α, plus one-half the number of grid points lying on particle boundaries, and P_T is the total number of grid points. Studies have shown that the point fraction is equivalent to the lineal fraction, L_L, and the area fraction, A_A, and all three are unbiased estimates of the volume fraction, V_V, of the second-phase particles:

$$P_P = L_L = A_A = V_V \quad \text{(Eq 2)}$$

Point counting is much faster than lineal or areal analysis and is the preferred manual method. Point counting is always performed on the minor phase, that is, where $V_V \leq 0.5$. The amount of the major (matrix) phase can be determined by difference.

The fields measured should be selected at locations over the entire polished surface selected randomly, that is, not confined to a small portion of the specimen surface. The field measurements should be averaged, and the standard deviation can be used to assess the relative accuracy of the measurement, as described in ASTM E 562.

In general, the number of points on the grid should be increased as the volume fraction of the feature of interest decreases. One study (Ref 8) suggested that the optimum number of grid test points is $3/V_V$. Therefore, for volume fractions of 0.5 (50%) and 0.01 (1%), the optimum numbers of grid points are 6 and 300, respectively. If the structure is heterogeneous, measurement precision is improved by using a lower-point-density grid and increasing the number of fields measured. The field-to-field variability in the volume fraction has a greater influence on the measurement than the precision in measuring a specific field. Thus, it is better to assess a greater number of fields using a lower-point-density grid than a small number of fields using a high-point-density grid, where the total number of points is constant. The adage "do more less well" in manual measurement refers to this problem.

To illustrate the point-counting procedure, Fig. 2 shows a 1000× image of a synthetic microstructure consisting of fifteen 17 μm diameter circular yellow "particles" within a field area of 25,900 μm^2 (0.026 mm^2). The total area of the 15 circular particles is 3405 μm^2, which is an area fraction of 0.1315 (i.e., 13.15%). This areal measurement is a very accurate estimate of the volume

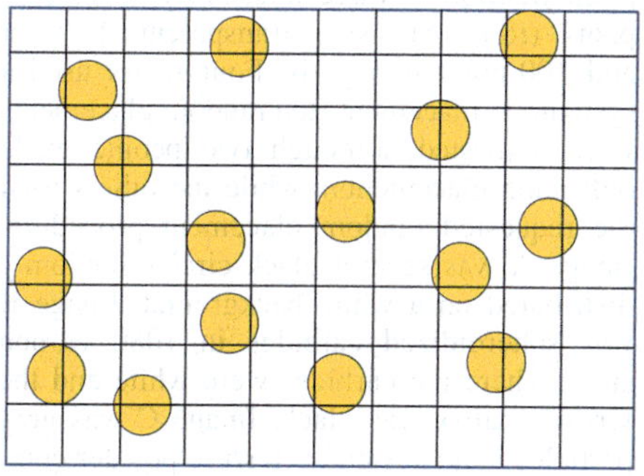

Fig. 2 Synthetic microstructure of 15 yellow particles, each 17 μm in diameter, within a test field measuring 25,900 μm^2 at 1000×. There are nine vertical and seven horizontal grid lines with a total true length of 2645 μm, and there are 63 grid intersections.

fraction for such a geometrically simple microstructure and is considered to be the true value. To demonstrate the use of point counting to estimate the volume fraction, a grid pattern was drawn over this field, producing 63 intersection points, 10 of which are inside the circles. Thus, P_P is 10/63 or 0.1587 (i.e., 15.87%), which agrees reasonably well with the calculated area fraction (2.72% greater). For a real microstructure, the time required to point count one field is far less than to do an areal analysis on that field. In practice, several fields would be point counted, and the average value would be a good estimate of the volume fraction in a small fraction of the time required to do areal analysis on an adequate number of fields. A manual area fraction measurement can only be done easily when the feature of interest is large in size and of simple shape. Point counting is the simplest and most efficient technique to use to assess the volume fraction. The area fraction, A_A, and the point fraction, P_P, are unbiased estimates of the volume fraction, V_V, as long as the sectioning plane intersects the structural features at random.

The lineal fraction, L_L, can also be determined for the synthetic microstructure shown in Fig. 2. The measured length of the horizontal and vertical line segments within the circular particles was 371.9 μm. The total test line length was 2645 μm. Consequently, the lineal fraction is 371.9/2645 or 0.1406 or 14.06%. This is a slightly higher estimate (approximately 0.91% greater) than obtained by areal analysis but better than the point-count estimate. If several fields were measured, the average would be a good estimate of the volume fraction. Lineal analysis becomes more tedious as the structural features become smaller, but for coarse structures, it is simple to perform. Lineal analysis is commonly performed when a structural gradient must be measured, that is, a change in second-phase concentration as a function of distance from a surface or an interface.

ASTM Committee E-4 on Metallography conducted a round-robin measuring the point fraction of three images by 33 participants (Ref 15) using transparent 25-point and 100-point test grids. Four raters used a systematic placement (ten random placements were requested, although two people made only nine placements), while the others used the requested random placement procedure. Image A was several black circles randomly distributed on a white background. Image B was spheroidized carbides in 1080 carbon steel, where the carbides were white and the ferritic matrix was black. Image C was gray particles in an austenitic-ferritic powder compact. Several of the 33 raters made consistent errors, especially when using the 25-point grid. The P_P data by the 33 raters are shown in Fig. 3(a–c) for images A, B, and C, and Table 2 summarizes the data for all 33 raters per point-count grid density and per image.

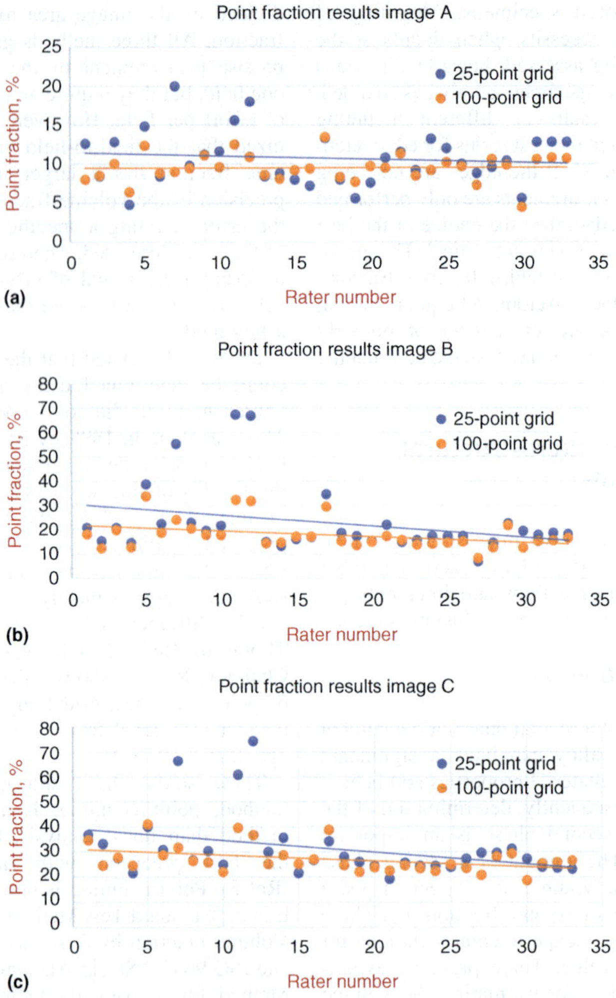

Fig. 3 (a–c) Results of the ASTM International round-robin to point count the volume fraction of second-phase particles using 25-point and 100-point transparent test grids by 33 different people. Results with the 25-point test grid were poorer than with the 100-point test grid.

Table 2 Summary of ASTM International point-counting round-robin experiment

	25-point test grid			100-point test grid		
Image	P_P, %	Standard deviation	95% confidence limit	P_P, %	Standard deviation	95% confidence limit
A	10.74	3.398	±1.214	9.51	1.759	±0.635
B	22.79	14.216	±5.132	17.76	6.127	±2.212
C	31.31	12.522	±4.520	27.10	5.471	±1.975

Source: Ref 15

The 100-point-count grid produced better results. The experiment showed that people do not always read the instructions and follow the guidelines when doing measurements, a not uncommon problem.

Number per Unit Area, N_A

The count of the number of particles within a given measurement area, N_A, is a useful microstructural parameter and is used in other calculations. Referring to Fig. 2, there are 15 particles in the measurement area (25,900 μm^2, or 0.0259 mm^2). Thus, the number of particles per unit area, N_A, is 0.000579 per μm^2, or 579 per mm^2. The average cross-sectional area of the particles is calculated by dividing the volume fraction, V_V, by N_A:

$$\bar{A} = V_V / N_A \quad \text{(Eq 3)}$$

This yields an average area, $\bar{A}$, of 227.08 μm^2, which agrees well with the calculated area of a

17 µm diameter circular particle of 227 µm². The example illustrates the calculation of the average area of particles in a two-phase microstructure using stereological field measurements rather than individual particle measurements.

Intersections and Interceptions per Unit Length, P_L and N_L

Counting of the number of intersections of a line of known length with particle boundaries or grain boundaries, P_L, or the number of interceptions of particles or grains by a line of known length, N_L, provides useful microstructural parameters. For space-filling grain structures (single phase), $P_L = N_L$, while for two-phase structures, $P_L = 2N_L$ (this may differ by one count in actual cases).

Grain-Structure Measurements

For single-phase grain structures, it is usually easier to count the grain-boundary intersections with a line of known length, especially for circular test lines. This is the basis of the Heyn-Hilliard-Abrams intercept grain size procedure described in ASTM E 112, "Standard Test Methods for Determining Average Grain Size." For most work, a circular test grid composed of three concentric circles with a total circumferential line length of 500 mm (20 in.) is recommended. Grain size is determined by measuring the mean lineal intercept length, l:

$$l = 1/P_L = 1/N_L \qquad \text{(Eq 4)}$$

This equation must be modified, as described later, for two-phase structures. The value l is used to calculate the ASTM International grain size number using an empirical equation. Grain size measurements are discussed later in more detail.

The P_L measurements are used to define the surface area per unit volume, S_V, or the length per unit area, L_A, of grain boundaries:

$$S_V = 2P_L \qquad \text{(Eq 5)}$$

and

$$L_A = (\pi/2)\,(P_L) \qquad \text{(Eq 6)}$$

For single-phase structures, P_L and N_L are equal, and either measurement can be used. For two-phase structures, it is best to measure P_L to determine the phase-boundary surface area per unit volume, or phase-boundary length per unit area.

Partially Oriented Structures

Many deformation processes, particularly those that are not followed by recrystallization, produce partially oriented microstructures. It is not uncommon to see partially oriented grain structures in metals that have been cold deformed. The presence and definition of the orientation (Ref 7) can only be detected and defined by examination of specimens on the three principal planes. Certain microstructures exhibit a high degree of preferred directionality on the plane of polish or within the sample volume. A structure is completely oriented if all of its elements are parallel. Partially oriented systems are those with features having both random and oriented elements. Once the orientation axis is defined, then a plane parallel to the orientation axis can be used for measurements. The P_L measurements are used to assess the degree of orientation of lines or surfaces.

Several approaches are used to assess the degree of orientation of a microstructure. For single-phase grain structures, a simple procedure is to make P_L measurements parallel and perpendicular to the deformation axis on a longitudinally oriented specimen, because the orientation axis is usually the longitudinal direction. The degree of grain elongation is the ratio of perpendicular to parallel P_L values, that is, $P_{L\perp}/P_{L\parallel}$. Another useful procedure is to calculate the degree of orientation, Ω, using the P_L values:

$$\Omega = \frac{P_{L\perp} - P_{L\parallel}}{P_{L\perp} + 0.571\,P_{L\parallel}} \qquad \text{(Eq 7)}$$

To illustrate these measurements, a section of low-carbon steel sheet was cold rolled to reductions in thickness of 10, 30, and 70% (Fig. 4a–c). The $P_{L\perp}$ and $P_{L\parallel}$ measurements were made using a grid with parallel straight test lines on a longitudinal section from each of four specimens (one specimen for each of the three reductions, plus one specimen of as-received material). The results are given in Table 3. As shown, cold working produces an increased orientation of the grains in the longitudinal direction.

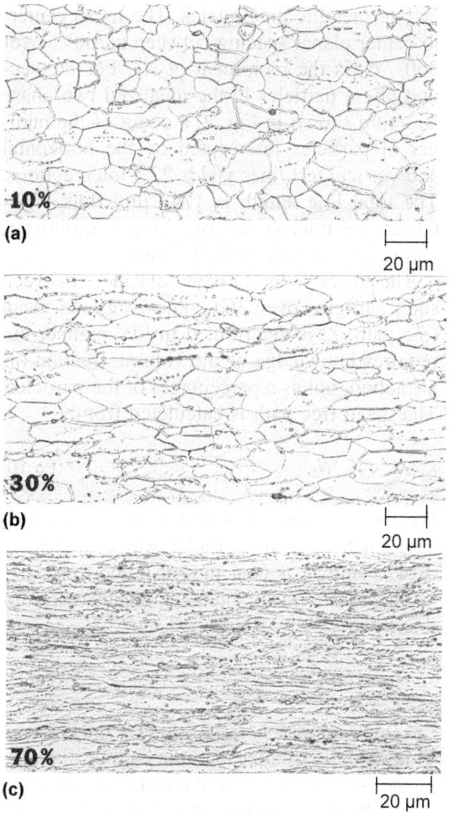

Fig. 4 Microstructure of a cold rolled low-carbon sheet steel reduced (a) 10, (b) 30, and (c) 70%. Etched with 2% nital

Table 3 Grain orientation for four specimens of low-carbon sheet steel

Sample	$P_{L\perp}$	$P_{L\parallel}$	$P_{L\perp}/P_{L\parallel}$	Ω, %
Annealed	114.06	98.86	1.15	8.9
10% cold reduction	126.04	75.97	1.66	29.6
30% cold reduction	167.71	60.6	2.77	52.9
70% cold reduction	349.4	34.58	10.1	85.3

Note: P_L is the number of grain-boundary intersections per millimeter.

Spacing

The spacing between second-phase particles or constituents is a structure-sensitive parameter influencing strength, toughness, and ductile fracture behavior. The N_L measurements are more easily used than the P_L measurements in the study of the spacing of two-phase structures. The most common spacing measurement is interlamellar spacing of eutectoid (such as pearlite) or eutectic structures (Ref 16, 17). The true interlamellar spacing, σ_t, is difficult to measure, but the mean random spacing, σ_r, is readily assessable and is directly related to the mean true spacing:

$$\sigma_t = \sigma_r/2 \qquad \text{(Eq 8)}$$

The mean random spacing is determined by placing a test grid consisting of one or more concentric circles on the pearlite lamellae in an unbiased manner. The number of interceptions of the carbide lamellae with the test line(s) is counted and divided by the true length of the test line to obtain N_L. The reciprocal of N_L is the mean random spacing:

$$\sigma_r = 1/N_L \qquad \text{(Eq 9)}$$

The mean true spacing, σ_t, is ½σ_r. To make accurate measurements, the lamellae must be clearly resolved; hence, the use of transmission electron microscope (TEM) replicas is quite common. If scanning electron microscope images are used, the etched specimen surface must be perpendicular to the beam; otherwise, the magnification will vary across the image at an unknown rate.

The N_L measurements are also used to measure the interparticle spacing in a two-phase alloy, such as the spacing between carbides or intermetallic precipitates. The mean center-to-center planar spacing between particles over 360°, σ, is the reciprocal of N_L. For the second-phase particles in the idealized two-phase structure shown in Fig. 2, a count of the number of particles intercepted by the horizontal and vertical test lines yields 22.5 interceptions. The total line length is 1743 µm; thus, N_L = 0.0129 per micrometer, or 12.9 per millimeter, and σ is 77.5 µm, or 0.0775 mm.

The mean edge-to-edge distance between such particles over 360°, known as the mean free path, λ, is determined in like manner but requires knowledge of the volume fraction (as a fraction, not as a percentage) of the particles. The mean free path is calculated from:

$$\lambda = 1 - V_V/N_L \quad \text{(Eq 10)}$$

For the structure illustrated in Fig. 2, the volume fraction of the particles was estimated as 0.147. Thus, λ is 66.1 µm, or 0.066 mm.

The mean lineal intercept distance, l_α, for these particles is determined by:

$$l_\alpha = \sigma - \lambda \quad \text{(Eq 11)}$$

For this example, l_α is 11.4 µm, or 0.0114 mm. This value is smaller than the caliper diameter of the particles because the test lines intercept the particles at random, not only at the maximum dimension. The calculated mean lineal intercept length for a circle with a 15 µm diameter is 11.78 µm. Stereological field measurements can be used to determine a characteristic dimension of individual features without performing individual particle measurements.

Measurement of the interlamellar spacing of pearlite has been performed mainly in research studies which show that refining the interlamellar spacing improves both the strength and the toughness and ductility of pearlitic steels, particularly near-eutectoid carbon steels widely used in cold-drawn wires to make cables. Several approaches have been used to measure interlamellar spacings. The most commonly used is to draw straight test lines perpendicular to the regions in a colony with parallel cementite lamellae, a so-called directed spacing measurement. A less common approach is the use of a circle placed over a pearlite colony where all lamellae are resolvable, as shown in Fig. 5. This should be performed at the lowest possible magnification to obtain adequate resolution of the lamellae, good area coverage, and, with a number of measurements, good data. Table 4 gives interlamellar spacing data for lead-patented rods of near-eutectoid carbon content using the circular test grid. The interceptions of the test line with lamellae are counted and a random spacing is determined. The mean random spacing was predicted by Saltykov (Ref 2) to be twice the mean true spacing. The mean true spacing can be measured using TEM thin foils where the specimen is tilted so that the width of the cementite lamellae is as thin as possible and parallel to the electron beam. A large

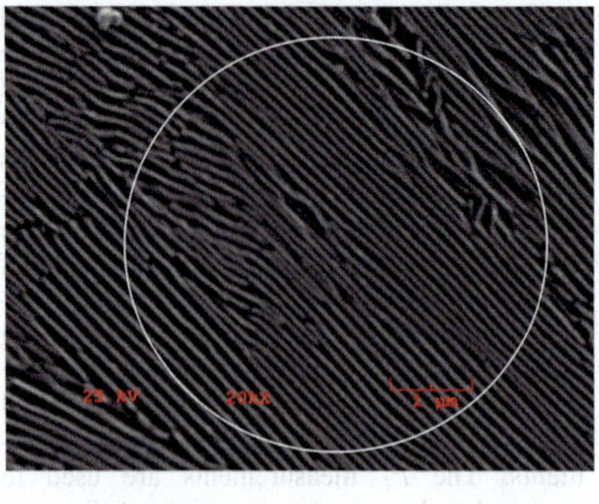

Fig. 5 Scanning electron microscope secondary electron image of pearlite revealed by etching with 4% picral and a test circle superimposed to make a measurement of the mean random spacing of the lamellae

number of such colonies must be measured to obtain good statistical data. Pearlite that forms isothermally exhibits a much narrower range of true spacings than pearlite that forms over a range of temperatures, as in an as-rolled specimen or a normalized specimen.

Figure 6 shows the microstructure of as-rolled 1040 carbon steel consisting of ferrite and pearlite. Etching with 4% picral is preferred to etching with 2% nital, because picral uniformly dissolves ferrite, while nital dissolves ferrite as a function of crystallographic orientation. Figure 7 shows a TEM thin foil where the cementite lamellae are parallel to the electron beam, and it is possible to see the true thickness of the cementite lamellae. Figure 8 shows a TEM replica of the pearlite etched with 4% picral. This image can be evaluated with the circular test grid, as shown in Fig. 5, to obtain the mean random spacing, which is twice the mean true spacing.

Measurement with the more commonly used directed test lines perpendicular to the lamellae was performed using TEM replicas of the polished and etched specimen, as shown in Fig. 8, where the lamellae are at variable angles to the specimen surface. The tests yielded a total of 183 measurements (three per pearlite colony). The mean value was 333.4 ± 24.8 nm (95% confidence limit) with a relative accuracy of 7.4%. Figure 9 shows the distribution plot of these measurements. The minimum spacing measured was 123.7 nm, while the maximum spacing measured was 1130.2 nm. The mean random spacing was determined by placing test circles of known size over pearlite colonies where the lamellae were completely resolvable at the chosen magnification using the same TEM replicas of the etched specimen surface, such as shown in Fig. 8. The process consisted of placing 92 test circles on 92 such TEM replica images and counting the cementite lamellae intercepts. Figure 10 shows a plot of the distribution of the random spacing measurements. The mean random spacing was 507.5 ± 51.6 nm (95% confidence limit) with a relative accuracy of 10.17%. Dividing the mean random spacing

Table 4 Summary of interlamellar spacing measurements of lead-patented rods

	10 mm (0.4 in.) diameter, 0.87% C rod		8 mm (0.3 in.) diameter, 0.79% C rod	
Parameter	σ_r, nm	σ_t, nm	σ_r, nm	σ_t, nm
Number of measurements	10	10	12	12
Mean	181.47	90.73	184.89	92.45
Standard deviation	40.9	20.45	22.43	11.22
95% confidence limit	±30.84	±15.42	±14.89	±7.45
Relative accuracy, %	17.0	17.0	12.1	12.1

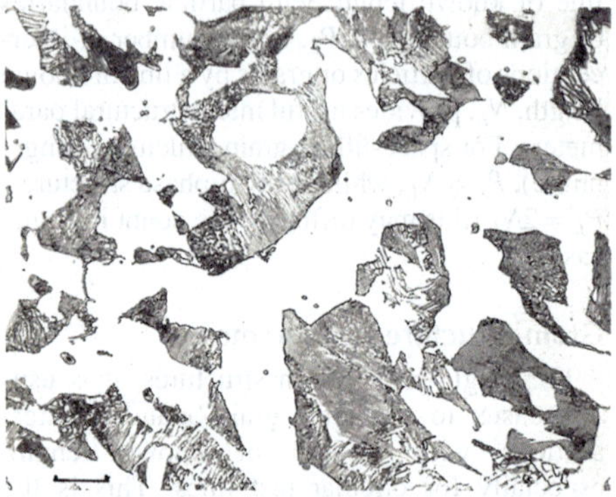

Fig. 6 Light optical micrograph of the ferrite-pearlite microstructure revealed using 4% picral in as-rolled (continuously cooled) AISI 1040 carbon steel. Original magnification: 500×

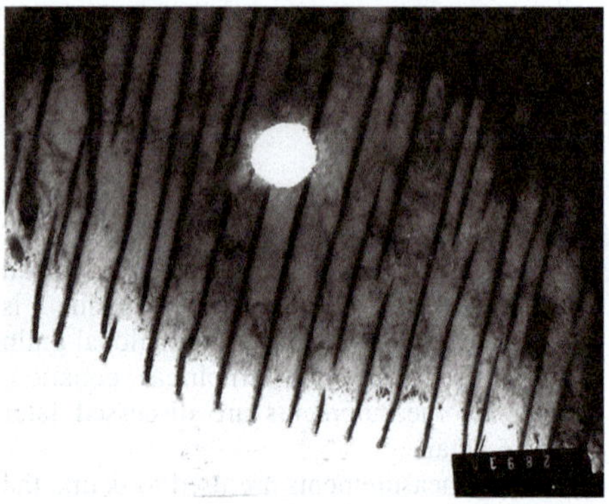

Fig. 7 Transmission electron microscope thin foil rotated under the beam until the cementite lamellae were parallel to the electron beam to measure the true spacing. Original magnification: 22,000×

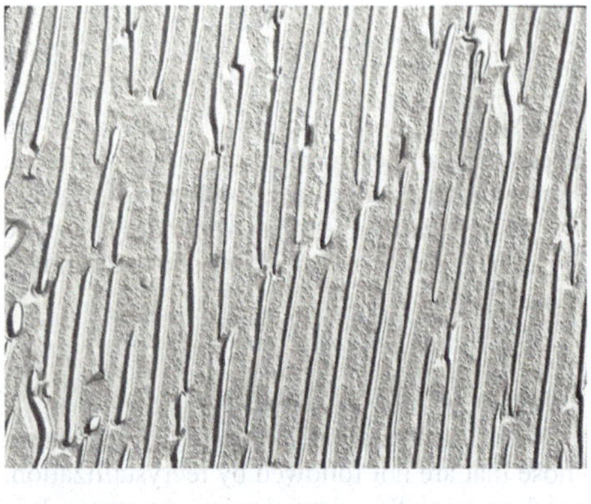

Fig. 8 Transmission electron microscope replica of a lightly etched (with 4% picral) specimen. Original magnification: 20,000×

by 2 (Eq 8) yields a mean true spacing of 253.8 nm. The minimum mean random spacing measured was 201.4 nm, and the maximum mean random spacing measured was 1566.6 nm, considerably higher than using the directed test lines.

Another measurement technique using the TEM thin foil, as shown in Fig. 7, consisted of making a number of measurements of the true interlamellar spacing by choosing a colony and then tilting the foil under the beam until the true spacing was visible, photographing the colony, and then repeating the measurements numerous times. Three measurements were made of the true spacing, producing a total of 218 measurements. The mean true spacing using this method was 254.5 ± 9.4 nm (3.7% relative accuracy), which is in good agreement with the previously mentioned measurement of the mean true spacing (253.8) obtained by dividing the mean random spacing by 2 (Eq 8). For the data using the tilted TEM thin foils, the minimum true spacing measured was 100 nm, and the maximum true spacing measured was 449.9 nm—the narrowest range of the three methods. Figure 11 shows the distribution of the true interlamellar spacing of the pearlite using the tilted TEM thin foils.

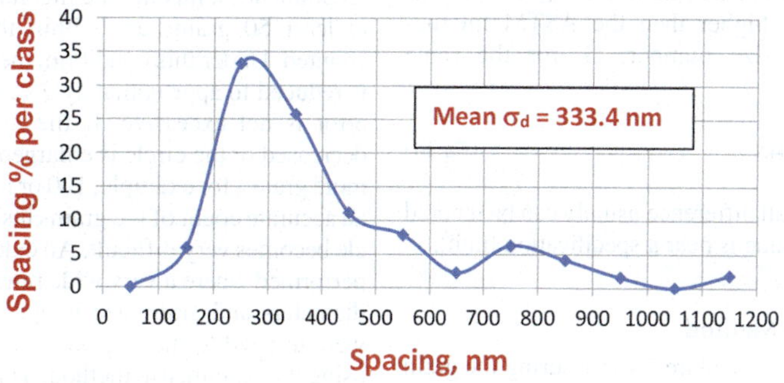

Fig. 9 Plot of the distribution of interlamellar spacing of cementite in as-rolled 1040 carbon steel using directed test lines

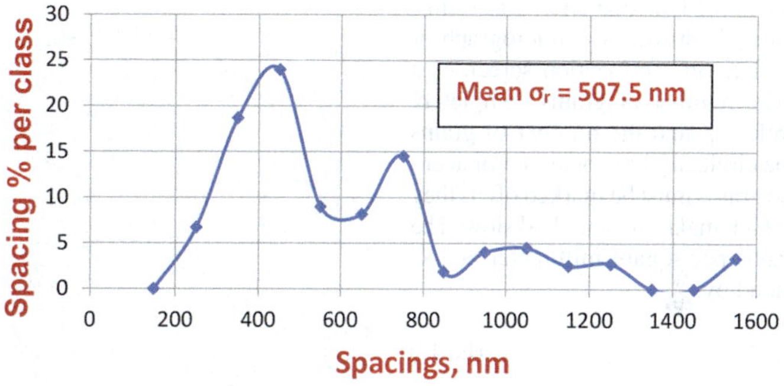

Fig. 10 Plot of the distribution of interlamellar spacing of cementite in as-rolled 1040 carbon steel using random circular test lines

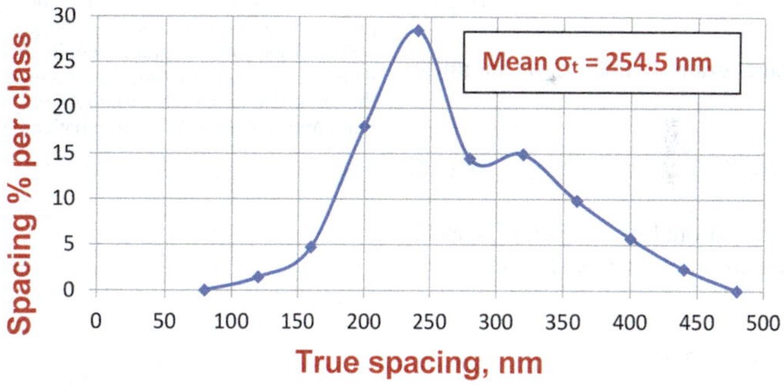

Fig. 11 Plot of the distribution of interlamellar spacing of cementite in as-rolled 1040 carbon steel using tilted transmission electron microscope foils, revealing the true spacing

Grain Size

The most common quantitative microstructural measurement is the grain size of metals, alloys, and ceramic materials. Numerous procedures have been developed to estimate grain size; these procedures are summarized in detail in ASTM E 112 (manual methods) and E 1382, "Standard Test Methods for Determining Average Grain Size Using Semiautomatic and Automatic Image Analysis," and described in Ref 18 to 33. Several types of grain sizes can be measured: ferrite grain size, austenite grain size, and prior-austenite grain size. Each type presents particular problems associated with revealing these boundaries so that an accurate rating can be obtained. While this relates specifically to steels, ferrite grain boundaries are identical (geometrically) to grain boundaries in any alloy that does not exhibit annealing twins, while austenite grains are identical (geometrically) to grain boundaries in any alloy that exhibits annealing twins. Therefore, charts depicting ferrite grains in steel can be used to rate grain size in metals such as aluminum, chromium, and titanium, while charts depicting austenite grains can be used to rate grain size in metals such as copper, brass, cobalt, and nickel. Qualitative chart ratings (Ref 18) are the simplest way to rate grain size. An ASTM E-4 round-robin on grain size measurements, mentioned in Appendix X1 of ASTM E 112, found that chart ratings of grain size were typically 1 G value lower in number (coarser rating) than planimetric and intercept measurements of the same images by the same raters (Ref 25). No bias was seen between the planimetric or intercept measurements.

Parameters used to measure grain size include:

- Average grain area, A
- Number of grains per unit area, N_A
- Average intercept length, l
- Number of grains intercepted by a line of known length, N_L
- Number of grain-boundary intersections by a line of known length, P_L

These parameters have been related to the ASTM International grain size number, G.

The ASTM International grain size scale was established using the Imperial system of units, but the use of metric units (more common worldwide) does not introduce difficulty. The ASTM International grain size equation is:

$$n = 2^{G-1} \qquad \text{(Eq 12)}$$

where n is the number of grains per square inch at 100×. Multiplying n by 15.5 gives grains per square millimeter at 1×, N_A. The metric grain size number, G_M, which is used by the International Organization for Standardization and many other countries, is based on the number of grains per square millimeter at 1×, m, using the formula:

$$m = 8(2^{G_M}) \qquad \text{(Eq 13)}$$

The metric grain size number, G_M, is approximately 4.5% higher than the ASTM International grain size number, G, for the same structure, or:

$$G = G_M - 0.045 \quad \text{(Eq 14)}$$

This very small difference usually can be ignored (unless the value is near a specification limit).

Planimetric Method

The oldest procedure for measuring the grain size of metals is the planimetric method introduced by Zay Jeffries (Ref 19), a founding member of ASTM Committee E-4 on Metallography, in 1916 based on earlier work by his Ph.D. advisor, Albert Sauveur. A circle of known size (generally 79.8 mm, or 3.14 in., diameter; 5000 mm^2, or 7.75 in.^{2}, area) is drawn on a micrograph or used as a template on a projection screen or a photograph. The number of grains completely within the circle, n_1, and the number of grains intersecting the circle, n_2, are counted. For accurate counts, the grains must be marked off as they are counted, which makes this method slow. The number of grains per square millimeter at 1×, N_A, is determined by:

$$N_A = f(n_1 + n_2/2) \quad \text{(Eq 15)}$$

where f is the magnification squared divided by 5000 (the circle area). The average grain area, A, in square millimeters, is:

$$A = 1/N_A \quad \text{(Eq 16)}$$

and the average grain diameter, d, in millimeters, is:

$$d = (A)^{1/2} = 1/N_A{}^{1/2} \quad \text{(Eq 17)}$$

The ASTM International grain size, G, can be found by using the tables in ASTM E 112 or by the equation:

$$G = [3.322\,(\log N_A) - 2.95] \quad \text{(Eq 18)}$$

Figure 12 illustrates the planimetric method. Expressing grain size in terms of d is being discouraged by ASTM Committee E-4 on Metallography because the calculation implies that grain cross sections are square in shape, which they are not.

In theory, a straight test line will, on average, bisect grains intercepting a straight line. If the test line is curved, however, bias can be introduced (Ref 2, 29, 30). Saltykov (Ref 2) suggested that such bias could occur, but he did not do any tests to evaluate this potential problem. Two other studies (Ref 29, 30) claimed that if such bias occurred, it was not evaluated properly. Theoretically, bias should decrease as the number of grains within the circle increases. If only a few grains are within the circle, the error could be large, for example, a 10% error if only ten grains are within the circle. ASTM E 112 recommends adjusting the magnification so that at least 50 grains are within the field to be counted. Under this condition, the possible error is reduced to approximately 2%. This degree of error is not excessive. If the magnification is decreased or the circle is enlarged to encompass more grains, for example, 100 or more, obtaining an accurate count of the grains inside the test circle becomes very difficult. An extensive test was performed where a very wide range of test circle diameters and grain size image magnifications were utilized to make grain size measurements using the planimetric method. This test revealed that bias did not result. At very small numbers of grains within the circle and intercepting the circle (Eq 15), data scatter increased, but no bias was observed (Ref 31). In this experiment using

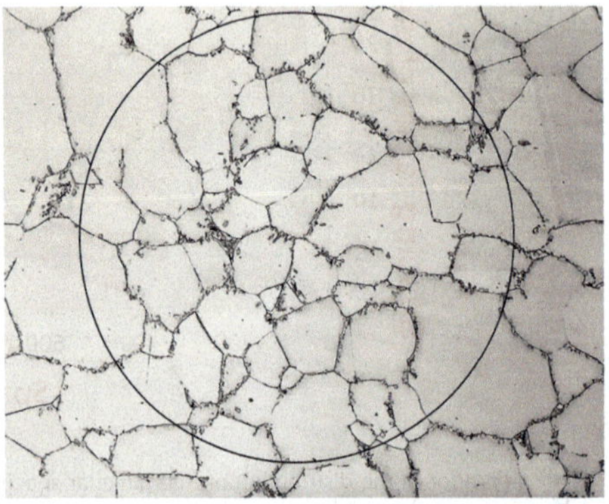

Fig. 12 Microstructure of an austenitic manganese steel, solution annealed and aged to precipitate a pearlitic phase on the grain boundaries (at 100×). There are 43 grains within the circle (n_1), and there are 25 grains intersecting the circle (n_2). The area of the test circle is 0.5 mm^2 (0.0008 in.^{2}) at 1×. N_A is 111 grains/mm^2, and the ASTM International grain size is 3.8.

the planimetric method with test circles, data scatter was observed when the value of (n_1 + 0.5n_2) was below ~30. For the Saltykov rectangle, data scatter was observed when (n_1 + 0.5n_2) was below ~11. For the triple-point-count method, data scatter was observed when (n_1 + 0.5n_2) was below ~25. Bias was not observed in any of these methods (and not with the intercept method, either).

A simple alternative to the data-scatter problem exists and is amenable to image analysis. If the test pattern is a square or rectangle, rather than a circle, data scatter occurs with low counts per placement, and the extensive test results (Ref 31) revealed the approach exhibited the least data scatter at the lowest grain count. However, counting grains intersecting the test line, n_2, is slightly different using a square or rectangular grid. In this method, grains intercept the four corners of the square or rectangle. Statistically, the portions intercepting the four corners would be in parts of four such contiguous test patterns. Thus, when counting n_2, the grains intercepting the four corners are not counted but are weighted as one intercepted grain. A count of all other grains intercepting the test square or rectangle (of known size) is weighted as half inside. Equation 15 is modified as follows:

$$N_A = f(n_1 + (n_2/2) + 1)$$

where n_1 is still the number of grains completely within the test figure (square or rectangle), n_2 is the number of grains intercepting the sides of the square or rectangle but not the four corners, 1 is for the four grains that intercept the four corners, and f is the magnification divided by the area of the square or rectangle grid. Figure 13 demonstrates such a counting procedure.

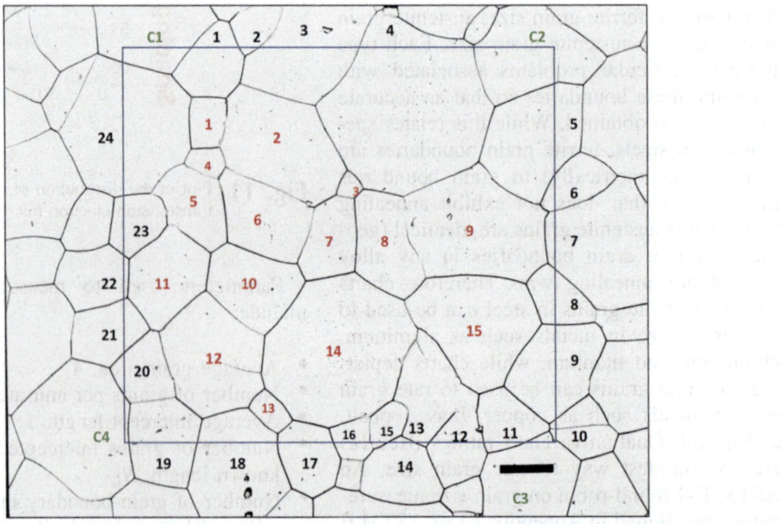

Fig. 13 Grain structure of Monit, a ferritic stainless steel at 400× magnification (the magnification bar is 25 μm). There are 15 grains completely inside the 102 × 114 mm (4 × 4.5 in.) rectangle and 24 grains that intersect the rectangle, ignoring the four grains at the corners (coded C1 to C4), which adds 1 to Eq 15. The area of the rectangle is 0.07258 mm^2 (0.00011 in.^{2}) at 1×. N_A is 385.7785 mm^2 (0.60 in.^{2}), and the ASTM International grain size is 5.64.

Intercept Method

The intercept method, developed by Emil Heyn (Ref 20) in 1903, is faster than the planimetric method because the micrograph or template does not require marking to obtain an accurate count. ASTM E 112 recommends use of a template (Ref 22) consisting of three concentric circles with a total circumferential line length of 500 mm (20 in.) (template available from ASTM International). The template is placed over the grain structure without bias, and the number of grain-boundary intersections, P, or the number of grains intercepted, N, is counted. Dividing P or N by the true line length, L, gives P_L or N_L, which are identical for a single-phase grain structure. It is usually easier to count grain-boundary intersections for single-phase structures. If a grain boundary is tangent to the line, it is counted as ½ of an intersection. If a triple-point line junction is intersected, it is counted as 1½ or 2. The latter is preferred because the small diameter of the inner circle could introduce a slight bias to the measurement, which is offset by weighing a triple-point test line intersection as two hits.

The mean lineal intercept length, l, determined as shown in Eq 4, is a measure of ASTM International grain size. This length is smaller than the maximum grain diameter because the test lines do not intersect each grain at their maximum breadth. The ASTM International grain size, G, can be determined by using the tables in ASTM E 112 or can be calculated from:

$$G = [-6.644\,(\log l) - 3.288] \qquad \text{(Eq 19)}$$

where l is in millimeters. Figure 14 illustrates the intercept method for a single-phase alloy.

Nonequiaxed Grains

Nonequiaxed grain structures require measurements on the three principal planes, that is, the longitudinal, planar, and transverse planes. (In practice, measurements on any two of the three are adequate.) For such structures, the intercept method is preferred, but the test grid should consist of a number of straight, parallel test lines of known length, rather than circles, oriented as described subsequently. Because the ends of the straight lines generally end within grains, these interceptions are counted as half-hits. Three mutually perpendicular orientations are evaluated using grain-interception counts:

- $N_{L\parallel}$: parallel to the grain elongation, longitudinal or planar surface
- $N_{L\perp}$: perpendicular to the grain elongation (through-thickness direction), longitudinal or transverse surface
- N_{LP}: perpendicular to the grain elongation (across width), planar or transverse surface

The average N_L value is obtained from the cube root of the product of the three directional N_L values. Grain size number, G, is determined from the tables in ASTM E 112 or by using Eq 19 (l is the reciprocal of N_L, Eq 4).

Two-Phase Grain Structures

The grain size of a particular phase in a two-phase structure requires determination of the volume fraction of the phase of interest, for example, by point counting. The minor phase (second phase) is point counted, and the volume fraction of the major phase (matrix phase) is determined by difference.

Next, a circular test grid is applied to the microstructure without bias, and the number of grains of the phase of interest intercepted by the test line, N_α, is counted. The mean lineal intercept length of the alpha grains, l_α, is determined by:

$$l_\alpha = (V_V)(L/M)/N_\alpha \qquad \text{(Eq 20)}$$

where V_V is a fraction (not a percent), L is the line length, and M is the magnification. The ASTM International grain size number can be determined from the tables in ASTM E 112 or by using Eq 19. The method is illustrated in Fig. 15. Again, a circular test grid could introduce some data scatter in counting (as described previously for the planimetric method) if the number of interceptions per grid placement is quite low. This can be eliminated by using a square or rectangular test grid and counting as described earlier. (The grains intercepted by the four corners are counted as 1, while all of the grains intercepting the sides are weighted as one each.)

Inclusion Content

Assessment of inclusion types and amounts (Ref 34–50) is commonly performed on high-quality steels. Production evaluations use comparison chart methods such as those described in ASTM E 45, "Standard Test Methods for Determining the Inclusion Content of Steel," SAE J422a, "Microscopic Determination of Inclusions in Steels," ISO 4967, "Steels—Determination of Content of Non-Metallic Inclusions—Micrographic Method Using Standard Diagrams," or the German standard SEP 1570 (DIN 50602), "Metallographic Examination: Microscopic Examination of Special Steels Using Standard Diagrams to Assess the Content of Non-Metallic Inclusions." In these chart methods, the inclusion images are defined by type and graded by severity (amount, width, and length). Either qualitative procedures (worst rating of each type observed) or quantitative procedures (all fields in a given area rated) are used. Only the Japanese standard JIS-G-0555, "Microscopic Testing Method for Non-Metallic Inclusions in Steel," uses actual volume fraction measurements for the rating of inclusion content (although the statistical significance of the data is questionable due to the limited number of counts required). Qualitative chart ratings of inclusions using methods such as ASTM E 45 yield poor evaluations of inclusions by type and by severity in interlaboratory round-robins conducted by a number of mill metallographers doing such work regularly (Ref 50). Raters often misidentified A- and C-type inclusions, and the severity ratings for the same specimens often covered the full severity range.

Manual measurement of the volume fraction of inclusions requires substantial effort to obtain acceptable measurement accuracy due to the rather low volume fractions usually encountered (Ref 39, 42). When the volume fraction is below 0.02 (2%), which is the case for inclusions, even in free-machining steels, acceptable relative accuracies (Eq 22) cannot

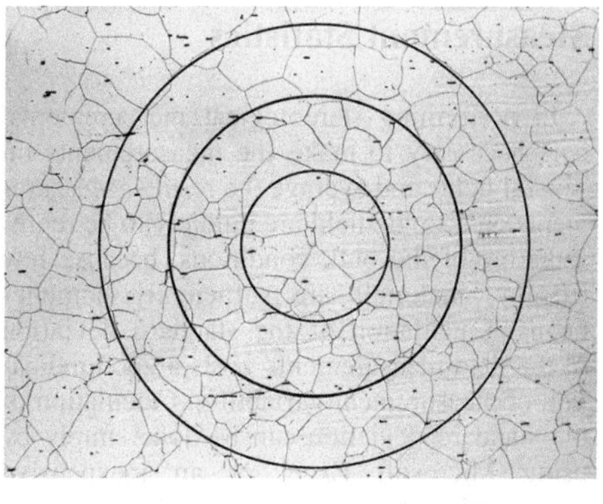

Fig. 14 100× micrograph of 304 stainless steel etched electrolytically with 60% HNO_3 (0.6 V direct current, 120 s, platinum cathode) to suppress etching of the twin boundaries. The three circles have a total circumference of 500 mm (20 in.). A count of the grain-boundary intersections yielded 75 (P = 75). The mean lineal intercept length is 0.067 mm (0.003 in.), and the ASTM International grain size number, G, is 4.5.

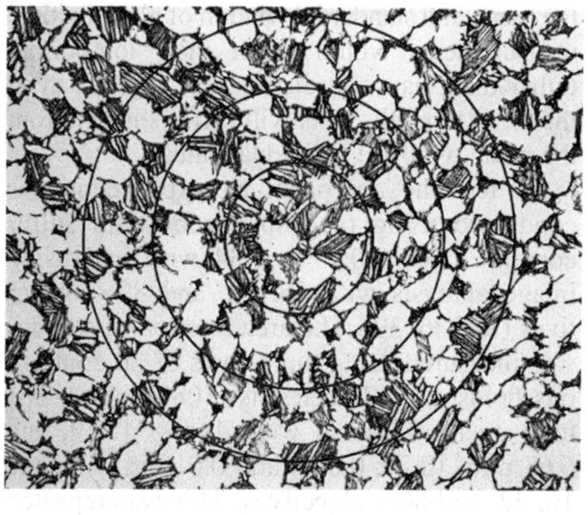

Fig. 15 500× micrograph of Ti-6242 alpha/beta forged and alpha/beta annealed, then etched with Kroll's reagent. The circumference of the three circles is 500 mm (20 in.). Point counting revealed an alpha-phase volume fraction of 0.485 (48.5%). The three circles intercepted 76 alpha grains. The mean lineal intercept length of the α-Ti phase is 0.00638 mm (0.00025 in.), and the ASTM International grain size, G, is 11.3.

be obtained using manual point counting without a large amount of counting time (Ref 39). The use of image analyzers has overcome this problem. Image analyzers separate the oxide and sulfide inclusions on the basis of their gray-level differences. By using automated stage movement and autofocusing, enough field measurements can be made in a relatively short time to obtain reasonable statistical precision. Image analysis is also used to measure the length of inclusions and to determine stringer lengths.

Two image-analysis-based standards have been developed: ASTM E 1122, "Standard Practice for Obtaining JK Inclusion Ratings Using Automatic Image Analysis" (Ref 41), and ASTM E 1245, "Standard Practice for Determining the Inclusion or Second-Phase Constituent Content of Metals by Automatic Image Analysis" (Ref 42, 44, 47, 48). ASTM E 1122 produces JK ratings by image analysis that overcome most of the weaknesses of manual JK ratings (in 2006, ASTM E 1122 information was merged into ASTM E 45). ASTM E 1245 is a stereological approach defining, for oxides and sulfides, the volume fraction (V_V), number per unit area (N_A), average length, average area, and the mean free path (mean edge-to-edge spacing). These data are easily stored in a database and are statistical. This enables developing means and standard deviations to compare data from different tests to determine if the differences between the measurements are valid at a particular confidence level.

At an ASTM E-4 meeting, George A. Moore of the former National Bureau of Standards (now the National Institute of Standards and Technology) recommended measuring the area fraction using eight to twelve sets of 30 to 100 sequential measurements of the area fraction on each specimen. The mean value of the area fraction would be calculated based on the eight to twelve set means. This would enable obtaining a valid arithmetic standard deviation. Calculating the mean and standard deviation of all individual measurements would result in a log-normal rather than Gaussian distribution. The mean of the eight to twelve sets will differ slightly from the mean of all the individual values.

To test the approach, an experiment was conducted using nine specimens with sulfur contents from 0.020 to 0.34 wt% and using image analysis with 16, 32, and 80× objectives (Ref 26, 39). The area fraction of sulfides was measured using three rows of 30 contiguous fields and 12 such sets (1080 fields). The total areas examined were 165.8, 42.69, and 6.63 mm^2 (0.26, 0.067, and 0.010 in.^{2}) for the 16, 32, and 80× objectives. This was repeated using manual point counting at 500× and using a 100-point grid and 10 sets of 10 contiguous fields (1 h/specimen). Lineal analysis used a Hurlbut counter at 1000× and using eight to ten linear measurements on each specimen, with 15 min duration for each run. Results are shown in Fig. 16 for the manual point-count method, Fig. 17 for the lineal analysis data, and Fig. 18 for the image analysis data. As would be expected, the results were best, and took the least time and effort, for the image analysis experiment.

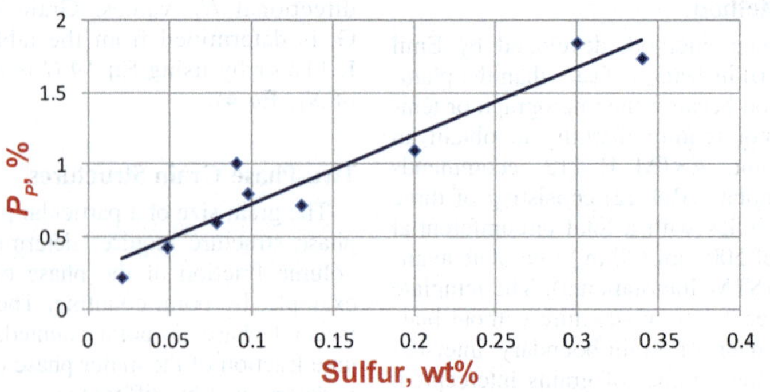

Fig. 16 Manganese sulfides in each specimen were manually point counted with a 100-point grid using 10 sets of 10 contiguous fields (~1 h/specimen) to obtain the data shown. The correlation between the wt% S and the manual point fraction was quite good, although the time required was impractical. P_P = 4.7278 (%S) + 0.2561. Correlation coefficient, r = 0.9555

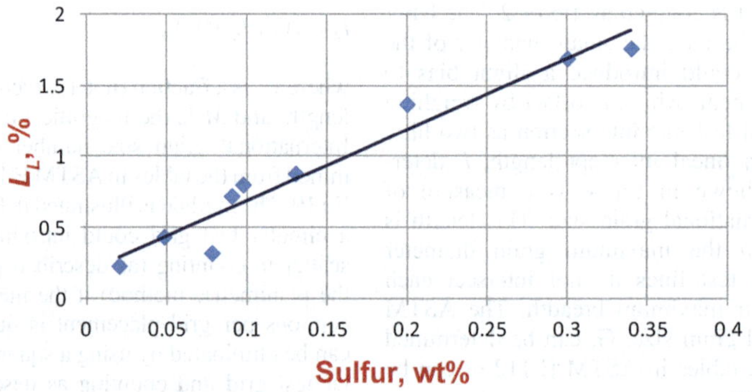

Fig. 17 Lineal analysis was performed using a Hurlbut counter at 1000× with 8 to 10 linear measurements on each specimen, each lasting 15 min. The correlation between the wt% S and the lineal fraction, L_L, was slightly better than the point fraction data, but the amount of effort was more than double per specimen. L_L = 4.9808 (%S) + 0.1904. Correlation coefficient, r = 0.972

Measurement Statistics

In performing stereological measurements, it is necessary to make the measurements on several fields and average the results. Measurements on a single field are unlikely to be representative of the bulk conditions, because few (if any) materials are sufficiently homogeneous. Calculation of the standard deviation of the field measurements gives a good indication of measurement variability. Calculation of the standard deviation can be done simply by using Microsoft Excel or an inexpensive pocket calculator.

A further refinement of statistical analysis is calculation of the 95% confidence limit (CL) based on the standard deviation, s, of the field measurements. The 95% CL is calculated from:

$$95\%\,\text{CL} = ts/(n-1)^{1/2} \quad \text{(Eq 21)}$$

where t is the student's t-value that varies with n, the number of measurements. Many users standardize on a single value of t, 2, for calculations regardless of n. Table 5 lists the student's t-values for calculating the 95% CL as a function of the number of measurements, n, and the degrees of freedom, $n - 1$. The measurement value is expressed as the average, X, plus and minus the 95% CL value. This means that if the test was conducted 100 times, the average values would be between plus and minus the average, X, in 95 of the measurements. Next, one can calculate the relative accuracy, %RA, of the measurement by:

$$\%\text{RA} = (95\%\,\text{CL}) \cdot 100/X \quad \text{(Eq 22)}$$

Usually, a 10% relative accuracy is adequate. DeHoff (Ref 51) developed a simple formula to determine how many fields, N, must be measured to obtain a specific desired degree of relative accuracy at the 95% CL:

$$N = \left(\frac{200}{\%\text{RA}} \cdot \frac{s}{X}\right)^2 \quad \text{(Eq 23)}$$

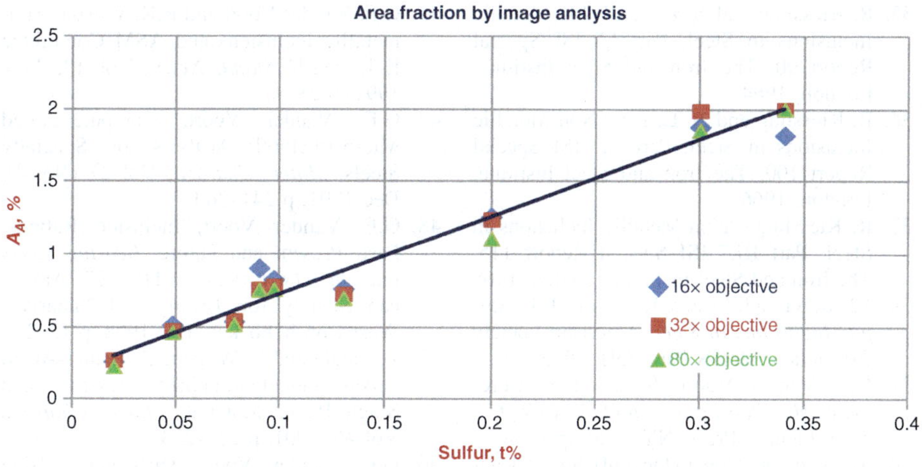

Fig. 18 Image analysis measurements were performed with three objectives using three rows of 30 contiguous fields and a total of 12 sets per specimen (1080 fields/specimen). The area evaluated was 165.8, 42.69, and 6.63 mm^2 (0.26, 0.067, and 0.010 in.^{2}) for the 16, 32, and 80× objectives, respectively. Image analysis gave the best correlation and was fastest. For the 16, 32, and 80× objective data, r-values were 0.9764, 0.9884, and 0.9878. A_A = 5.3285 (%S) + 0.1909. Correlation coefficient, r = 0.9866 (all data)

Table 5 Student's *t*-values for a 95% confidence limit, where *n* is the number of measurements, and *n* – 1 is the degrees of freedom

n – 1	*t*-values	*n* – 1	*t*-values
1	12.706	18	2.101
2	4.303	19	2.093
3	3.182	20	2.086
4	2.776	21	2.080
5	2.571	22	2.074
6	2.447	23	2.069
7	2.365	24	2.064
8	2.306	25	2.060
9	2.262	26	2.056
10	2.228	27	2.052
11	2.201	28	2.048
12	2.179	28	2.045
13	2.160	30	2.042
14	2.145	40	2.021
15	2.131	60	2.000
16	2.120	120	1.980
17	2.110	∞	1.960

Image Analysis

The measurements described in this brief review, and other measurements not discussed, can be made by use of automatic image analyzers. These devices rely primarily on the gray level of the image on the display monitor to detect the desired features. In some instances, complex image editing can be used to aid separation. Use of image analysis to perform these measurements is discussed elsewhere in this Volume and in Ref 52 to 58.

Summary

This article reviewed many commonly used stereological counting measurements and the relationships based on these parameters. The measurements described are easy to learn and use. They enable metallographers to discuss microstructures in a more quantitative manner, and they reveal relationships between the structure and properties of the material. The measurements also enable determination of whether one batch of metal products is better, worse, or the same quality as another, or if one manufacturing process produced better quality, equal quality, or poorer quality than another. Qualitative analysis work using chart methods cannot meet the statistical quality needed to make such evaluations unless the results are vastly different, which is possible but not likely.

REFERENCES

1. S.A. Saltykov, *Introduction to Stereometric Metallography,* Ed. SSR, Yerevan, Armenia, 1950
2. S.A. Saltykov, *Stereometric Metallography*, 2nd ed., Metallurgy of Steel and Alloys, Moscow, 1958
3. S.A. Saltykov, *Stereometric Metallography*, 3rd ed., Metallurgy, Moscow, 1970
4. S.A. Saltykov, *Textbook on Stereometric Metallography*, Metallurgy, Moscow, 1976
5. G.A. Moore, Is Quantitative Metallography Quantitative? *Application of Modern Metallographic Techniques*, STP 480, American Society for Testing and Materials, Philadelphia, PA, 1970, p 3–48
6. E.E. Underwood, Applications of Quantitative Metallography, *Metallography, Structures and Phase Diagrams*, Vol 8, *Metals Handbook*, 8th ed., American Society for Metals, Metals Park, OH, 1973, p 37–47
7. E.E. Underwood, *Quantitative Stereology*, Addison-Wesley, Reading, MA, 1970
8. J.E. Hilliard and J.W. Cahn, An Evaluation of Procedures in Quantitative Metallography for Volume-Fraction Analysis, *Trans. AIME*, Vol 221, April 1961, p 344–352
9. R.H. Greaves and H. Wrighton, *Practical Microscopical Metallography*, Chapman and Hall, London, 1st ed., 1924; 2nd ed., 1933; 3rd ed., 1939
10. A. Delesse, Procédé Méchanique pour Déterminer la Composition des Roches, *Ann. Mines*, Vol 13 (No. IV), 1848, p 379–388
11. A. Rosiwal, About Geometric Rock Analysis. A Simple Way to Numerically Determine the Quantitative Ratio of the Mineral Constituents of Mixed Rocks, *Negotiations of K.K. Geological Reichsanstalt*, Vol 5/6, 1898, p 143–175 (in German)
12. E. Thompson, Quantitative Microscopic Analysis, *J. Geol.*, Vol 38, 1930, p 193–222
13. A.A. Glagolev, Quantitative Analysis with the Microscope by the "Point" Method, *Eng. Min. J.*, Vol 135, 1934, p 399
14. H.W. Chalkley, Method for Quantitative Morphological Analysis of Tissues, *J. Natl. Cancer Inst.*, Vol 4, 1943, p 47–53
15. H. Abrams, Practical Applications of Quantitative Metallography, *Stereology and Quantitative Metallography*, STP 504, American Society for Testing and Materials, Philadelphia, PA, 1972, p 138–182
16. G.F. Vander Voort and A. Roósz, Measurement of the Interlamellar Spacing of Pearlite, *Metallography*, Vol 17, Feb 1984, p 1–17
17. G.F. Vander Voort, The Measurement of the Interlamellar Spacing of Pearlite, *Pract. Metallogr.*, Vol 52, Aug 2015, p 419–436
18. F.C. Hull, A New Method for Making Rapid and Accurate Estimates of Grain Size, *Trans. AIME*, Vol 172, 1947, p 439–451
19. Z. Jeffries, A.H. Kline, and E.B. Zimmer, The Determination of the Average Grain Size in Metals, *Trans. AIME*, Vol 54, 1916, p 594–607
20. E. Heyn, Short Reports from the Metallurgical Laboratory of the Royal Mechanical and Testing Institute of Charlottenborg, *Metallographist*, Vol 5, 1903, p 37–64
21. J. Hilliard, Estimating Grain Size by the Intercept Method, *Met. Prog.*, Vol 85, May 1964, p 99–101
22. H. Abrams, Grain Size Measurements by the Intercept Method, *Metallography*, Vol 4, 1971, p 59–78
23. G.F. Vander Voort, Grain Size Measurement, *Practical Applications of Quantitative Metallography*, STP 839, American Society for Testing and Materials, Philadelphia, PA, 1984, p 85–131
24. G.F. Vander Voort, Examination of Some Grain Size Measurement Problems,

Metallography: Past, Present and Future, STP 1165, American Society for Testing and Materials, Philadelphia, PA, 1993, p 266–294
25. G.F. Vander Voort, Precision and Reproducibility of Quantitative Measurements, *Quantitative Microscopy and Image Analysis*, ASM International, Metals Park, OH, 1994, p 21–34
26. G.F. Vander Voort, *Metallography: Principles and Practice*, McGraw-Hill Book Co., NY, 1984; ASM International, Materials Park, OH, 1999
27. G.F. Vander Voort, Grain Size Measurements Using Circular or Rectangular Test Grids, *Pract. Metallogr.*, Vol 50 (No. 1), Jan 2013, p 17–31
28. G.F. Vander Voort, Measuring the Grain Size of Specimens with Non-Equiaxed Grains, *Pract. Metallogr.*, Vol 50 (No. 4), April 2013, p 239–251
29. H. Li, Improvement on the Formulas of the Sum of Grains on a Circle Area and Triple Point Count Method, *MC95: Proceedings of the International Metallography Conference*, May 10–12, 1995 (Colmar, France), ASM International, Materials Park, OH, p 157–160
30. H. Li and X. Zhou, Bias in the Planimetric Procedure According to ASTM E 112 and Correction Method, *J. ASTM Int.*, Vol 8 (No. 5), 2011, p 1–6
31. G.F. Vander Voort, Is there Possible Bias in ASTM E 112 Planimetric Grain Size Measurements? *Mater. Perform. Charact.*, Vol 2 (No. 1), 2013, p 194–205
32. G.F. Vander Voort, Grain Size Measurements by the Triple Point Count Method, *Pract. Metallogr.*, Vol 51 (No. 3), March 2014, p 201–207
33. G.F. Vander Voort, Measurement of Grain Shape Uniformity, *Pract. Metallogr.*, Vol 51, May 2014, p 367–374
34. L.F. McCombs and M. Schrero, *Bibliography of Non-Metallic Inclusions in Iron and Steel*, Mining and Metallurgical Advisory Boards, Pittsburgh, PA, 1935
35. R. Kiessling and N. Lange, "Non-Metallic Inclusions in Steel, Part I," ISI Special Report 90, The Iron and Steel Institute, London, 1964
36. R. Kiessling and N. Lange, "Non-Metallic Inclusions in Steel, Part II," ISI Special Report 100, The Iron and Steel Institute, London, 1966
37. R. Kiessling, "Non-Metallic Inclusions in Steel, Part III," ISI Special Report 115, The Iron and Steel Institute, London, 1968
38. J.J. deBarbadillo and E. Snape, Ed., *Sulfide Inclusions in Steel*, American Society for Metals, Metals Park, OH, 1975
39. G.F. Vander Voort, Inclusion Measurement, *Metallography as a Quality Control Tool*, Plenum Press, NY, 1980, p 1–88
40. S. Johansson, How to Quantify Steel Cleanliness, *Swedish Symposium on Non-Metallic Inclusions in Steel*, Uddeholms AB, Hagfors, Sweden, April 27–29, 1981, p 221–233
41. G.F. Vander Voort and J.F. Golden, Automating the JK Inclusion Analysis, *Microstructural Science*, Vol 10, Elsevier North-Holland, NY, 1982, p 277–290
42. G.F. Vander Voort, Measurement of Extremely Low Inclusion Contents by Image Analysis, *Effect of Steel Manufacturing Processes on the Quality of Bearing Steels*, STP 987, American Society for Testing and Materials, Philadelphia, PA, 1988, p 226–249
43. D.W. Hetzner and B.A. Pint, Sulfur Content, Inclusion Chemistry, and Inclusion Size Distribution in Calcium-Treated 4140 Steel, *Inclusions and Their Influence on Materials Behavior*, ASM International, Metals Park, OH, 1988, p 35–48
44. G.F. Vander Voort, Characterization of Inclusions in a Laboratory Heat of AISI 303 Stainless Steel, *Inclusions and Their Influence on Materials Behavior*, ASM International, Metals Park, OH, 1988, p 49–64
45. D.B. Rayaprolu and D. Jaffrey, Sources of Errors in Particle Size Distribution Parameters, *Inclusions and Their Influence on Materials Behavior*, ASM International, Metals Park, OH, 1988, p 109–121
46. G.F. Vander Voort and R.K. Wilson, Nonmetallic Inclusions and ASM Committee E-4, *ASTM Stand. News*, Vol 19, May 1991, p 28–37
47. G.F. Vander Voort, Computer-Aided Microstructural Analysis of Specialty Steels, *Mater. Charact.*, Vol 27 (No. 4), Dec 1991, p 241–260
48. G.F. Vander Voort, Inclusion Ratings: Past, Present and Future, *Bearing Steels into the 21st Century*, STP 1327, American Society for Testing and Materials, West Conshohocken, PA, 1998, p 13–26
49. Z. Latała and L. Wojnar, Computer-Aided versus Manual Grain Size Assessment in a Single-Phase Material, *Mater. Charact.*, Vol 46, 2001, p 227–233
50. G.F. Vander Voort, Difficulties Using Standard Chart Methods for Rating Nonmetallic Inclusions, *Ind. Heat.*, Vol 84 (No. 1), Jan 2016, p 34–37
51. R.T. DeHoff, Quantitative Metallography, *Techniques of Metals Research*, Vol II (Part 1), Interscience, NY, 1968, p 221–253
52. *Image Analysis: Principles & Practice*, Joyce-Loebl Inc., England, 1985
53. H.E. Exner and H.P. Hougardy, Ed., *Quantitative Image Analysis of Microstructures*, DGM Informationsgesellschaft, Germany, 1988
54. G.F. Vander Voort, Ed., *MiCon 90: Advances in Video Technology for Microstructural Control*, STP 1094, ASTM, Philadelphia, PA, 1991
55. K.J. Kurzydłowski and B. Ralph, *The Quantitative Description of the Microstructure of Materials*, CRC Press, Boca Raton, FL, 1995
56. L. Wojnar, *Image Analysis: Applications in Materials Engineering*, CRC Press, Boca Raton, FL, 1999
57. J. Ohser and F. Mücklich, *Statistical Analysis of Microstructures in Materials Science*, John Wiley & Sons, Ltd., Chichester, 2000
58. *Practical Guide to Image Analysis*, ASM International, Materials Park, OH, 2000

Microscopy and Microanalysis

Ryan Deacon, United Technologies Research Center
Neal Magdefrau, Electron Microscopy Innovative Technologies

ASM Handbook, Volume 10, *Materials Characterization*
ASM Handbook Committee
DOI 10.31399/asm.hb.v10.a0006668

Scanning Electron Microscopy

Yoosuf N. Picard, Carnegie Mellon University

Overview

General Uses

- Imaging of surface features from 10 to 100,000× magnification. Resolution is typically 1 to 10 nm, although <1 nm is possible.
- Depending on available detectors, the microscope enables differentiation of polycrystalline grains for unetched samples, microstructural components with and without surface etching, topographical features, crystallographic structure and material phase, extended defects, and local variations in material properties.

Examples of Applications

- Examination of metallographically prepared samples at magnifications well above the capabilities of optical microscopy
- Examination of fracture surfaces and deeply etched surfaces requiring depth of field well beyond that possible with the optical microscope
- Evaluation of phase and crystallographic orientation for individual grains, precipitates, microstructural constituents, and nanoscale particles
- Identification of the chemistry for micrometer and submicrometer features, such as inclusions, precipitate phases, and wear debris
- Mapping of elemental composition across micrometer to millimeter length scales
- Examination of semiconductor devices for failure analysis, critical length determination, and design verification

Samples

- *Form:* Any low-vapor-pressure (0.1 Pa, or $\leq 10^{-3}$ torr), conductive, solid material including particles, thin films, and bulk samples. Nonconductive solids can be imaged using a conductive coating. Uncoated nonconductive solids can be imaged using low-beam energy or using a variable-pressure/environmental scanning electron microscope (SEM). Wet and fixed biological specimens can also be imaged using variable-pressure/environmental SEM.
- *Size:* Limited by the SEM specimen chamber size and stage travel. Generally, samples as large as ~20 cm (8 in.) can be placed in the microscope, but regions on such samples that can be examined without repositioning are limited to ~5 cm (2 in.).
- *Preparation:* Standard metallographic polishing and etching techniques are adequate for electrically conductive materials. Nonconducting materials are generally coated with a thin layer of carbon or metal. Samples must be electrically grounded to the holder, and fine samples, such as powders, can be dispersed on an electrically conducting adhesive. Samples must be free of high-vapor-pressure liquids, such as water, organic cleaning solutions, and remnant oil-based films, unless imaged inside a variable-pressure/environmental SEM.

Limitations

- Image quality on relatively flat samples, such as metallographically polished and etched samples, is generally inferior to the optical microscope below 300 to 400× magnification.
- Point resolution, although much better than the optical microscope, is inferior to the transmission electron microscope and the scanning transmission electron microscope.

Capabilities of Related Techniques

- *X-ray diffraction:* provides bulk crystallographic information
- *Optical microscopy:* faster, less expensive, and provides superior image quality on relatively flat samples at less than 300 to 400× magnification
- *Transmission electron microscopy:* provides similar information with superior resolution but is more expensive and requires thinned specimens

Introduction

A scanning electron microscope (SEM) is a type of instrument that magnifies and images sample surfaces through controlled rastering of a highly focused electron beam across the area of interest. A variety of signals are produced, particularly backscattered and secondary electrons, as the electron beam interacts with the sample surface; these signals provide local topographic and compositional information regarding the specimen. The SEM was invented in 1937 (Ref 1) and was first commercialized in 1965 (Ref 2). There have been continual improvements in SEM resolution, dependability, ease of operation, and reduction in instrument size. Scanning electron microscopes are regularly used in materials research, forensics, failure analysis, geological studies, biological imaging, metallurgy, nanomaterials development, microelectronics, and fractography. Scanning electron microscopes are a common instrument in most materials characterization laboratories and are increasingly used for immediate, on-site metrology and quality control at the manufacturing floor.

The primary use of an SEM is to produce high-resolution images at magnifications unattainable by optical microscopy, while also providing direct topographic and compositional information. Magnifications up to 100,000× or more are possible by modern SEMs. Scanning electron microscopes produce nanoscale-resolution imaging and mapping because the electron beam is focused to an ~1 to 10 nm sized probe. However, issues consider regarding electron probe size and depth sensitivity are the influence of accelerating voltage, beam current, sample composition, and the specific signal being detected. Signals are most commonly secondary electrons, backscattered electrons, and x-rays. Other possible signals that can be generated and detected in the SEM include visible light (cathodoluminescence), beam induced current, and Auger electrons.

This article provides detailed information on the instrumentation and principles of the SEM, including:

- Description of the primary components of a conventional SEM instrument
- Advantages and disadvantages of the SEM compared with other common microscopy and microanalysis techniques
- Critical issues regarding sample preparation
- Details on the physical principles regarding electron beam-sample interaction
- Mechanisms for many types of image contrast
- Details of SEM-based techniques
- Specialized SEM instruments
- Example applications using various SEM modes

The Microscope

Figure 1 illustrates the basic components of a typical SEM. The components can be categorized as the electron-optics column, the specimen chamber, the support system, and the control and imaging system. The electron beam is generated at an electron gun and accelerated toward the sample housed inside a specimen chamber, typically below the electron-optics column. Electromagnetic lenses below the electron gun focus the electron beam to a small probe at the sample surface. Scanning coils deflect the electron probe across the sample surface, and detectors housed either in the specimen chamber or in the electron-optics column collect and report resultant signal intensity as a function of beam position. The SEM operator uses a computer to control the electron-optics, detectors, and a motorized stage for sample positioning. During SEM operation, a support system provides cooling water for the electro-magnetic lenses and maintains a low pressure within the electron-optics column and the specimen chamber using vacuum pump.

Electron-Optics Column

The electron-optics column produces a narrowly divergent beam of electrons along the centerline of the column and steers the beam onto the sample surface. The major components of the electron-optics column are the electron gun, condenser lens, objective lens, and scanning coils.

Electron Gun

The electron gun contains the source of electrons and components that accelerate electrons to high energies. Two example guns are shown in Fig. 2. The electron gun produces a beam of electrons and focuses them to a spot of diameter, d_0, and divergence half-angle, α_0. The electron source is typically a tungsten hairpin, tungsten needle (Fig. 2), or a sharp-tipped crystal composed of a rare earth metal hexaboride (LaB_6 or CeB_6). Electron sources are generally classified as thermionic sources or field-emission sources. Thermionic sources use heat to energize and release electrons from the source material; field-emission sources primarily rely on an electrostatic field for electron beam generation.

Thermionic emission sources heat a tungsten filament or La/CeB_6 crystal to temperatures of ~2500 or 1600 °C (4500 or 2900 °F), respectively. Current applied to the thermionic source resistively heats the material and energizes electrons from the sharpest radius-of-curvature (ROC) point on the source. The thermal energy imparted to the electrons at the surface must be sufficient to overcome the work function of the source material surface. Most metals reach or exceed their melting temperature before the electrons can overcome the work function. Tungsten has the highest melting temperature, lowest vapor pressure, lowest thermal expansion coefficient, and a very high tensile strength. Thus, tungsten is an ideal metal for an electron source. Tungsten filaments are bent into a "hairpin" shape with an ROC of ~100 μm (Ref 4). The LaB_6 is preferable to tungsten, because it has a lower work function and higher electron emissivity, thus requiring a lower operating temperature while emitting more electrons. The CeB_6 is less susceptible to contamination than LaB_6 and thus is seeing increased use. Stable hexaboride crystals have 90° cone tips with ~15 μm ROC.

Field-emission sources use a high electric field in the vicinity of the source. Field enhancement at the sharpest point in the source tunnels electrons off the source surface while also greatly localizing electron emission. Therefore, field-emission sources produce

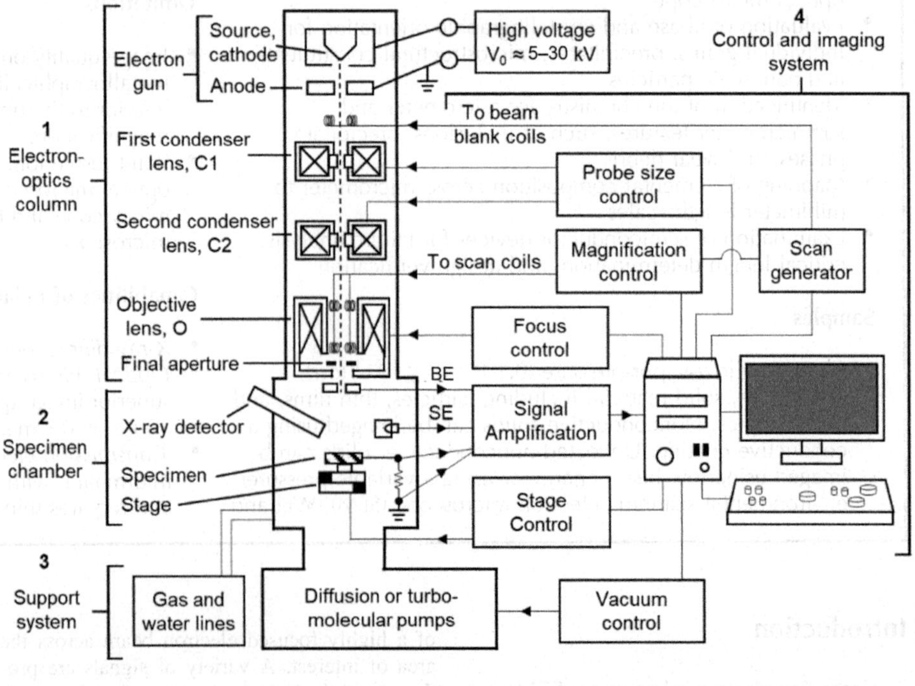

Fig. 1 Schematic of the basic components of a conventional scanning electron microscope. BE, backscattered electron; SE, secondary electron

much higher electron current densities (1000 to 10,000 A/cm^2) than thermionic sources (1 to 10 A/cm^2). Field-emission sources, or emitters, are typically etched tungsten single-crystal needles with nanometer-sized ROCs. A ZrO coating on sharp tungsten needles (tens of nanometers ROC) creates a local Schottky barrier within the electric field, which lowers the work function of the tungsten (100) facet, thus requiring a much lower operating temperature (1500 °C, or 2700 °F) than a thermionic tungsten source. The ZrO-tungsten sources are called Schottky field-emission sources (Fig. 2b). Single-crystal (310) tungsten can be etched to achieve even sharper needle tips (few nanometers ROC), so the local electric field alone is enough to tunnel electrons off the tungsten (310) facet surface; these sources are called cold field-emission sources, since they operate at room temperature.

Within the electron gun, electron extraction from the electron source requires a voltage potential between the source (cathode) and another electrode in proximity (anode). The anode accelerates the electrons to a specific kinetic energy and directs the initial electron beam trajectory down the electron-optics column. The cathode is negatively biased, and the accelerating anode is positively biased. For thermionic sources, the cathode is considered "hot," and another electrode with an aperture is positioned near the tungsten filament or hexaboride crystal. The aperture electrode, the Wehnelt cylinder, is negatively biased to a few hundred volts and functions as an electrostatic lens, which helps control electron extraction from the thermionic source. Farther away, the accelerating anode is strongly biased positive. For field-emission guns (FEGs), the appreciable voltage gradient present between the anode and the cathode provides a sufficiently strong electrostatic field, so the much sharper tip of the field-emission sources produces adequate field enhancement for electron extraction. A negatively biased electrode, the suppressor, helps localize the electron-extraction region at the source, while a positively biased extracting anode farther away pulls the electrons toward the accelerating anode.

The most important parameter distinguishing these four types of electron guns is the brightness, β:

$$\beta = \frac{I}{\left(\frac{\pi d_0^2}{4}\right)(\pi \alpha_0^2)} \quad \text{(Eq 1)}$$

where β equals the amount of current, I, focused on an area, $\pi d_0^2/4$, entering and exiting this area through a solid angle, $\pi\alpha_0^2$ (Ref 5). Increasing β improves the overall performance of the SEM. The value of β is primarily a function of the electron source material and the type of electron gun. Cold FEG sources provide the highest β values but require ultrahigh-vacuum conditions, because the room-temperature emitter is more susceptible to contamination. Contaminants that attach to the cold FEG source are removed by flashing the emitter with bursts of current. Cold FEG sources greatly increase SEM instrumentation costs but are highly stable and long-lasting sources. Schottky FEG sources produce lower β values than cold FEG sources, but they are also less expensive and require less stringent vacuum demands. Thermionic sources have much lower β values than FEG sources but are also much cheaper. They do require occasional source replacement, particularly the less expensive tungsten filaments as opposed to the more expensive but higher-brightness hexaboride crystals. Table 1 compares important properties for the four types of electron sources.

Lenses

Lenses within electron microscopes use electromagnetism to focus electron beams. Electromagnetic lenses (Fig. 3a) consist of copper wire windings inside an iron fixture, which is carefully machined to specific dimensions with pole pieces designed to localize magnetic fields. Current passed through the windings of copper wire magnetizes the iron and produces a magnetic field, which is radially symmetric about the lens axis. As an electron moves through the magnetic field, it experiences a radial force inward, which is proportional to the Lorentz force, $v \times \mathbf{B}$, where v is the electron velocity, and $\mathbf{B}$ is the magnetic flux density. The function of the pole pieces is to produce progressively higher density magnetic flux lines farther away from the lens axis. Electrons traveling parallel, but at different distances to the lens axis, experience different inward Lorentz forces, so that they converge at the focal plane. The resulting lensing action is comparable to the function of an optical lens, in which a ray parallel to the axis of the lens is bent toward the lens axis. Eventually, the ray meets the lens axis at a distance below the lens principal plane that corresponds to the focal length, f, of the lens (Fig. 3b).

For an optical lens, f is determined by the lens surface curvature and refractive index of the lens medium; the focal length is fixed. The focal length of an electromagnetic lens depends on two factors: the acceleration voltage (which determines the electron velocity, v) and the amount of current passing through the copper winding (which determines the flux density, $\mathbf{B}$). Therefore, the operator can control and tune the electromagnetic lens focal length by adjusting the currents supplied to them; an

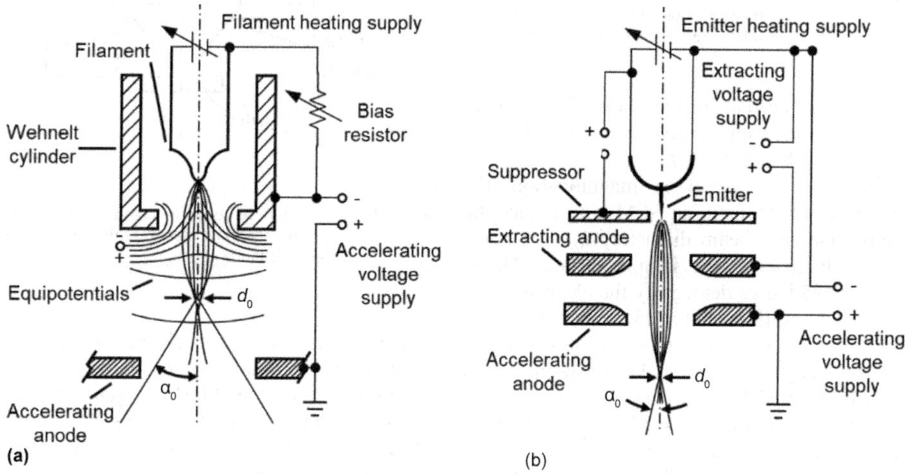

Fig. 2 Schematics of conventional (a) thermionic tungsten hairpin filament gun and (b) Schottky field-emission gun. Adapted from Ref 3

Table 1 Comparison of electron source properties

Property	Thermionic tungsten filament	Thermionic LaB_6/CeB_6	Schottky field-emission ZrO/W (100)	Cold field-emission tungsten (310)
Brightness (β), A/cm^2 · sr	10^6	10^7	5×10^8	10^9
Crossover or virtual source diameter (d_0), nm	$>10^4$	$>10^3$	15–25	3–5
Work function, eV	4.5	2.6/2.4	2.8	4.3
Energy spread, eV	1–3	>1	0.3–1	0.2–0.3
Operating temperature, °C (°F)	2500 (4500)	1600 (2900)	1500 (2700)	25 (75)
Typical service lifetime, h	40–100	1000/1500	>2000	>2000
Operating vacuum, Pa (torr)	10^{-4} (7.5×10^{-7})	10^{-6} (7.5×10^{-9})	10^{-8} to 10^{-9} (7.5×10^{-11} to 7.5×10^{-12})	10^{-9} to 10^{-11} (7.5×10^{-12} to 7.5×10^{-14})

Source: Ref 6

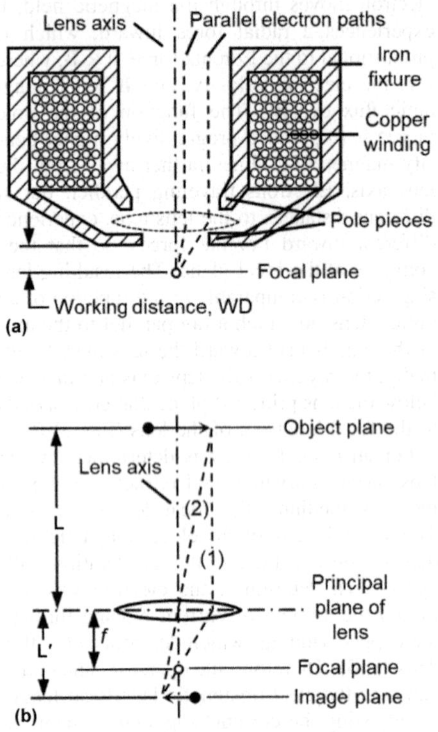

Fig. 3 (a) Schematic cross section of conventional electromagnetic objective lens. (b) Ray diagram of a standard lens. Magnification, $M = L'/L$

increase in current increases the overall radial force experienced by all electrons within the beam and thus reduces f.

Rather than achieving magnification, the purposes of lenses in an SEM is to reduce the initial electron beam diameter, d_0, to a much small diameter at the sample surface. Therefore, SEM lenses demagnify the electron beam diameter, as illustrated in Fig. 3(b). The arrow in the object plane is reproduced upside down in the image plane, and the arrow-tip image can be located by following rays (1) and (2). The magnification is $M = L'/L$; as focal length f is reduced, the value of L' is reduced, reducing magnification, M. Therefore, if the length of the arrow is taken as the electron beam diameter, d_0, produced by the gun, then the beam diameter, d_1, after passing through the first condenser lens is $d_1 = M_1 d_0$, where M_1 is the magnification of the first condenser lens.

Figure 4 shows the coupling of two condenser lenses and one objective lens; the object plane for a given lens is the image plane from the lens above it. The net result is that the diameter of the electron beam at the sample surface, d_3, is:

$$d_3 = d_0 \cdot M_1 \cdot M_2 \cdot M_3 \qquad \text{(Eq 2)}$$

where M_1, M_2, and M_3 are the demagnification factors for each lens. Spot size, or actual beam size, on the sample surface, d_s, is somewhat larger than d_3 due to lens aberrations. Considering a typical value of d_0 = 15 μm for a thermionic tungsten filament gun, and the value of d_3 may reach 15 nm, this corresponds to $M_1 \cdot M_2 \cdot M_3 = 1000$. Note that the 1000× demagnification has no direct relationship to the actual magnification achieved by the SEM using scanning coils, as described later. Also, apertures located near each lens serve to limit the angular spread of the electron beam as it progresses along the electron-optics column axis. Apertures are metal discs with machined holes of various diameters ranging from 5 to 1000 μm. The user can often select aperture sizes to adjust beam convergence angle, depth of field, and beam current (number of electrons) arriving at the specimen.

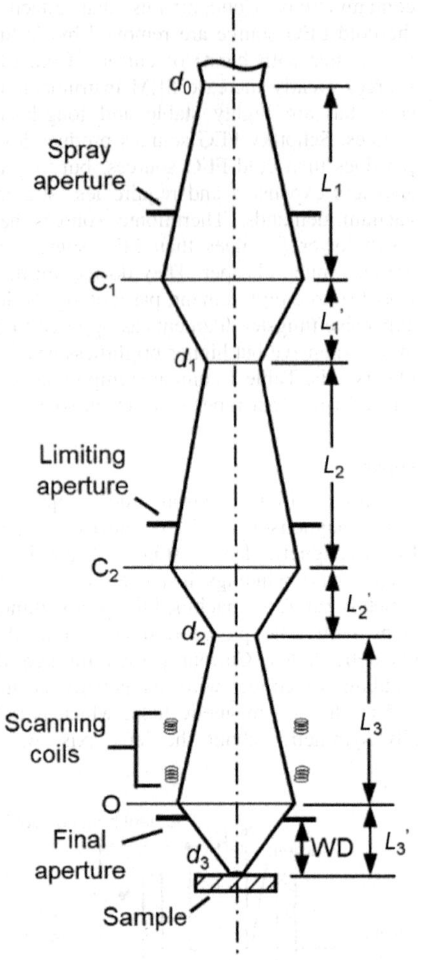

Fig. 4 Ray diagram of beam demagnification by three lenses in a scanning electron microscope. WD, working distance

The sample position relative to the objective lens focal plane, L_3' in Fig. 4, can also be independently adjusted by the user via stage positioning inside the specimen chamber. The distance of the focal plane below the bottom of the objective lens, and/or final aperture, is termed the working distance (WD). The center axis of the electro-optics column is typically considered the z-axis direction with respect to sample orientation. If the sample surface position is changed vertically along the z-axis, the objective lens current must be adjusted to have L_3' fall on the sample surface. The objective lens current adjustment is the focusing procedure performed by the operator to obtain the sharpest image possible. Because f_3 and therefore L_3' are functions of current applied to the objective lens, a unique value of WD corresponds to objective lens current; WD is often displayed on the control system. Therefore, once the operator ensures the sample is at focus, the associated WD value can often serve as a useful z-axis value for the stage position.

Stronger lens focusing produces a smaller beam diameter. Therefore, image resolution improves as WD is decreased. In the past, special high-resolution SEMs used an immersion lens that greatly reduced WD by positioning the sample inside the objective lens. Modern SEMs might incorporate a snorkel lens, which extends the electromagnetic field below the objective lens pole piece, effectively immersing the sample within the focusing lens field. These modes greatly reduce f_3 and further reduce d_3 to enhance resolution.

As mentioned previously, the actual beam spot size at the sample, d_s, is larger than d_3 due to lens aberrations, including spherical aberration, chromatic aberration, and astigmatism. Of these, only astigmatism is regularly corrected in a typical SEM and primarily for the objective lens, since it is the probe-forming lens of the SEM. Astigmatism arises from imperfections in the machined soft iron pole piece of the electromagnetic lenses, which diminish the radial symmetry of the electromagnetic field produced by the lens. This leads to a larger beam diameter at the lens focal plane. To improve the field radial symmetry, an array of independently controlled stigmator coils are arranged around the inside circumference of a lens. Each coil produces additional magnetic field lines, enabling the user to adjust and balance the overall electromagnetic lens action of the entire assembly. Figure 5 illustrates the effect of astigmatism on SEM images of gold nanoparticles. When the sample is slightly out of focus, features exhibit a stretched appearance because of the elongated beam diameter overlapping adjacent image pixels. The directionality of the stretching is orthogonal when going from an underfocus to overfocus condition or vice versa. Adjusting the stigmators to ensure the beam diameter is circular in underfocus and overfocus conditions serves to reduce the beam size when at focus. For this reason, Fig. 5(d) exhibits sharper image quality than Fig. 5(a), even though both images are technically in focus. Astigmatism correction for the objective lens is a common task during SEM operation, particularly for higher magnifications where it becomes crucial to obtain the smallest possible spot size, d_s.

It may be necessary to increase the number of electrons arriving at the sample, even at the expense of producing a larger spot size. More electrons impinging the sample lead to greater overall signal, which is particularly desirable for x-ray analysis and backscattered electron imaging. The condenser lens immediately above the objective lens enables controlling both spot size and number of arriving electrons, as shown in Fig. 6. The L_2' and d_2 decrease with increasing condenser lens strength. For the same objective lens strength, a smaller d_2 beam diameter produces a smaller d_3 beam diameter at the sample surface. However, as the cross-over point for the condenser lens moves farther away from the objective lens and objective aperture, fewer electrons can pass through the objective aperture. Therefore, any reduction in the spot size accomplished by increasing condenser lens strength also results in fewer electrons arriving at the sample surface. Sacrificing beam current for smaller spot size is a standard trade-off in the SEM.

Scan Coils

The main function of scan coils is scanning the focused electron beam across the sample surface to produce an image. Two sets of scan coils located in the bore of the objective lens cage shown in Fig. 1 perform the scanning function. The coils, further detailed in Fig. 7, deflect the beam to scan over a square area of size $r \times r$ on the sample surface. The scanned area is generally termed the raster. Although the beam is shown in Fig. 7 as a line, it is diverging as it passes through the scan coils. However, because the divergence half-angle is in the milliradian range, representation as a line is reasonable. For simplicity, Fig. 7 depicts the scanning process at approximately midway during the formation of a single frame.

Scanning electron microscopes achieve magnification using this double-deflection system, with the beam deflected by the Lorentz force produced by low-impedance coil windings driven by a low-voltage power supply. The scan generator produces a voltage across each coil pair. The upper scan coil pair, X_1, produces a magnetic field at time 1, which provides a Lorentz force that deflects the beam to the right through angle θ_{max}. The lower scan coil pairs, X_2, deflect the beam back to the left through angle $2\theta_{max}$ so it strikes the sample at the left edge of the raster. The X_1 - X_2 voltage signal decreases stepwise with time, as shown at the upper left of Fig. 7. For each short time interval, termed dwell time, the beam sits at a single position on the sample surface. From time 1 to time 5, the beam scans along the line of length r on the sample surface.

At time 5, the $X_1 - X_2$ scan voltage "flies back" quickly to 1′, causing the beam to return rapidly to the left side of the raster.

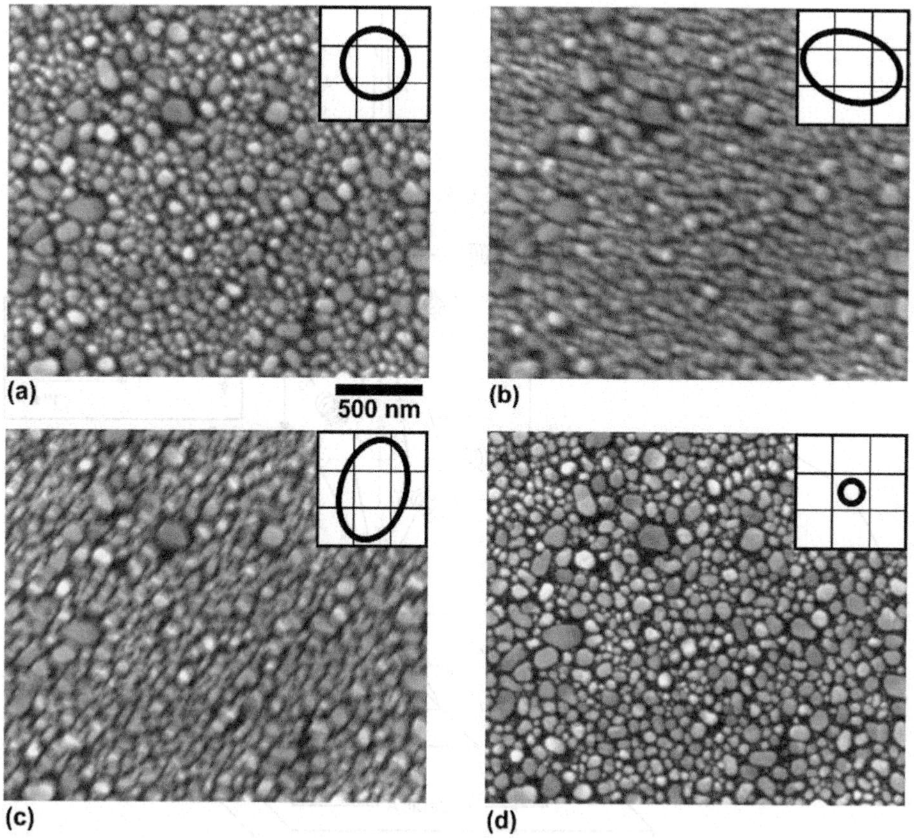

Fig. 5 Scanning electron microscope images of gold nanoparticles obtained before astigmatism correction. (a) At focus. (b) Underfocus. (c) Overfocus. (d) At focus after astigmatism correction. Inset depicts electron beam spot size relative to pixel size.

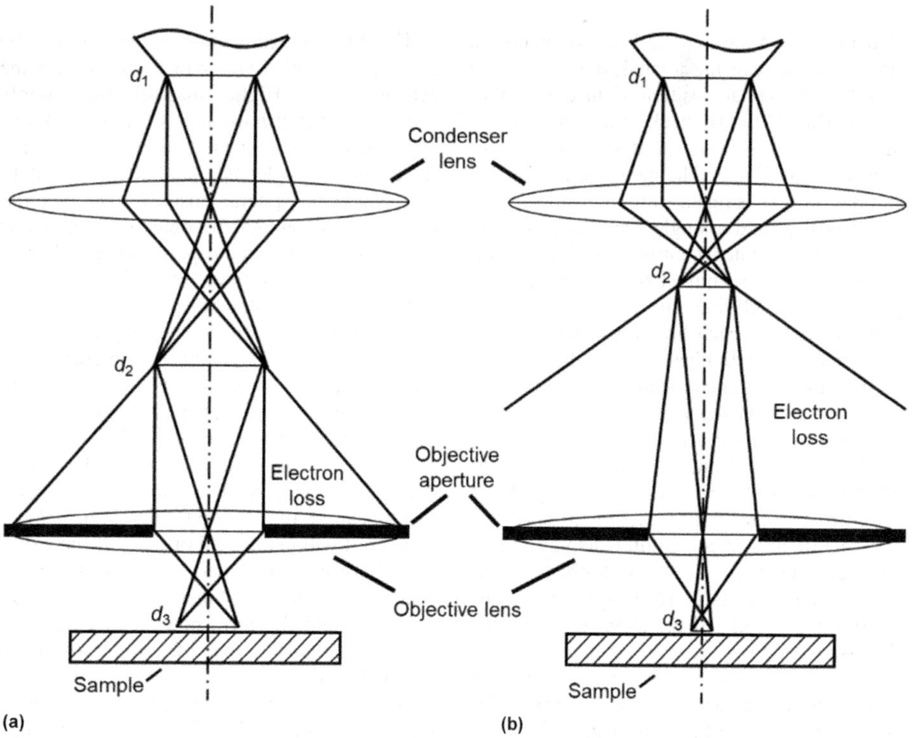

Fig. 6 Illustration of (a) lower condenser lens strength producing a larger beam diameter, d_3, of sample surface with more electrons. vs (b) increased condenser lens strength producing a smaller beam diameter, d_3, at sample surface but with fewer electrons. Adapted from Ref 4

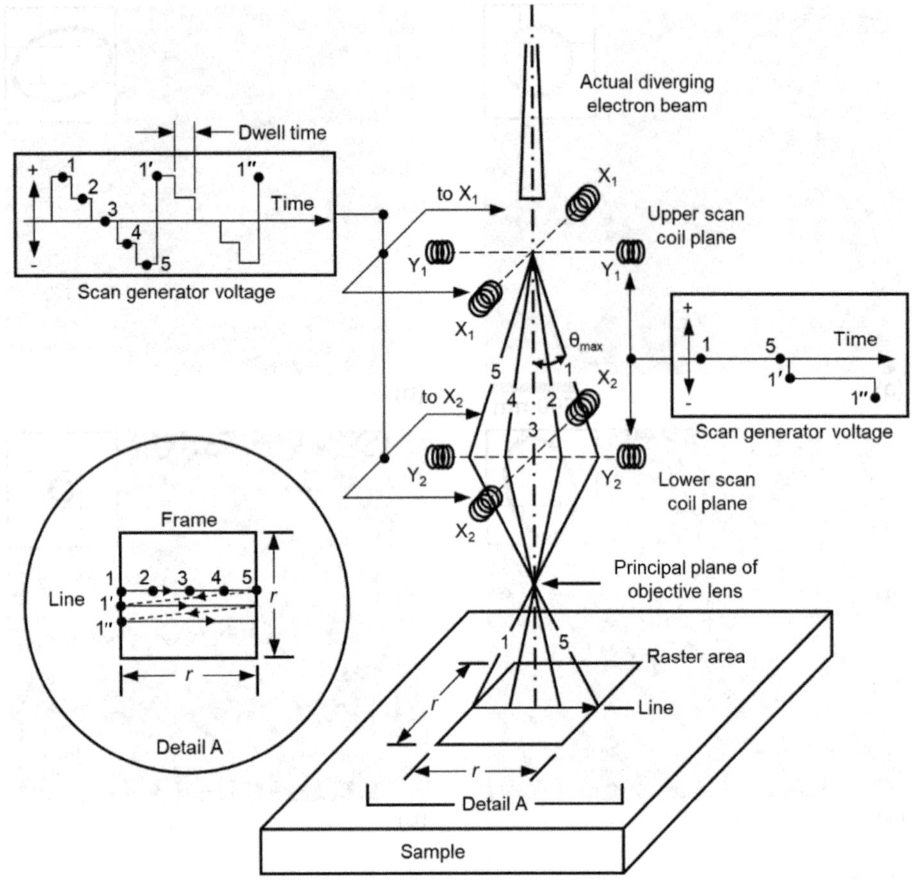

Fig. 7 Diagram of double-deflection system showing progressive line scanning to produce a frame of size $r \times r$

Concurrently, Y_1 and Y_2 coil pairs now have a small voltage, causing a small deflection so a new line begins at position 1′ in detail A of Fig. 7. During the fly-back time from 5 to 1′, the beam moves along the dashed line shown in detail A in Fig. 7. The next line is scanned, and the process is repeated until a full raster is accomplished and a single frame is produced. The scan generator controls the frame size, number of lines within each frame, number of positions on each line, and the dwell time for each position on each line. Each position corresponds to an individual pixel in the resultant digital image. The pixel value in terms of grayscale depends on the relative signal intensity collected during the dwell time interval.

The magnification, *M*, achieved by the SEM is simply the ratio of the displayed size to the raster size, *r*. For example, consider a 50 cm (20 in.) screen digital display that is 44.3 cm wide by 24.9 cm high (17.5 by 10 in.) and projects in 16:9 aspect ratio with 1600 × 900 pixels. If the raster size matches the aspect ratio, but over a much smaller region corresponding to 32.0 μm by 18.0 μm of the sample surface, then the magnification is $M = 44.3$ cm/0.0032 cm or $M = 24.9$ cm/0.0018 cm, which corresponds to a magnification of ~13,800×.

The SEM magnification is determined by the user-controlled size *r* of the raster via the scanning coils. Higher magnification simply requires reducing *r* through smaller variations in the scan generator voltage, thus decreasing the deflection angles θ_{max} and $2\theta_{max}$. The double-deflection system ensures the electron beam passes through the objective lens plane consistently at the same point on the electro-optics axis. Therefore, the objective lens current can be maintained, and the sample generally remains in focus as the magnification is altered. The SEMs typically enable controlling and tuning the directionality of the raster, enabling scan rotation that effectively rotates the apparent image on the digital screen. Sometimes reduced areas within the frame can be selected, often so microscope parameters can be tuned based on a single feature of interest within the field of view. If the working distance is increased with no change to scanning conditions, then the raster size increases and *M* correspondingly decreases; the lowest available magnification depends on the WD used.

Because the objective lens current, is a known function of WD for a focused sample, magnification is reported based, in part, on this lens current; SEM instruments show the magnification directly on the digital display. More importantly, a scale bar is also provided. The scale bar provides a direct correlation between the pixel units of the digital image to the actual physical size of the raster and associated surface features being imaged in the SEM. Commercially available specimens with patterned features of well-established physical sizes are used to calibrate any SEM. The calibration step is crucial to ensure the accuracy of both the displayed magnification and associated scale bar.

Specimen Chamber

The specimen chamber of an SEM contains the sample and the primary detectors for imaging and analysis. Specimen chambers range in size from small enclosures used for tabletop SEMs housing centimeter-sized samples, to medium-range sizes typical for research-grade SEMs capable of handling up to ~20 cm (8 in.) samples, to large enclosures designed to handle specimens up to meters in size. The chambers are vacuum capable enclosures usually constructed of nonmagnetic stainless steel. The inside chamber walls must be maintained dry and free of residue to facilitate obtaining the necessary low operating pressures during imaging and analysis, while also minimizing sample surface contamination. Plasma cleaners can be attached to SEM chambers to enable removal of residual contamination during operation. Specimen chambers can be vented to atmosphere and opened for sample loading. For SEMs that regularly operate at high-vacuum conditions $<1.33 \times 10^{-5}$ Pa ($<10^{-8}$ torr), a load-lock apparatus is attached to the specimen chamber. This enables the operator to load the sample into a secondary chamber that is pumped to low pressures before sample insertion to the main specimen chamber, thus avoiding the need to vent the entire chamber to atmosphere.

Stage

When the sample is loaded into the SEM specimen chamber, it sits on a motorized, computer-controlled stage. Some SEMs with manual micrometer control and no motorization are still in use, but they are becoming increasingly obsolete. The SEM operator positions the sample in the electron beam path and selects regions of interest for imaging and/or analysis. Simple SEM systems may only have a two-axis stage, which enables *x*- and *y*-axis control of the sample position. More advanced SEMs typically have a five-axis stage for *x*-, *y*-, and *z*-axis positioning plus sample rotation and tilt control. As mentioned previously, the *z*-axis is conventionally defined parallel to the center axis of the electro-optics column. The *y*-axis is often oriented relative to the direction of the secondary electron detector. In an analytical SEM, the *y*-axis may be oriented with respect to the energy-dispersive spectrometer.

Detectors

Standard detectors are mounted on the SEM chamber, as shown in Fig. 8, and are designed to detect secondary electrons (SEs) and/or backscattered electrons (BEs), which is discussed later in this article. Most standard SEM detectors are scintillation and solid-state detectors. Historically, the most used detector has been the Everhart-Thornley detector (ETD) (Ref 7), which consists of a scintillator coupled to a photomultiplier tube with a Faraday cage typically biased by a positive voltage (50 to 300 V) to attract and enhance the signal of SEs. The ETDs are often mounted a few centimeters away from the sample surface and employ the Faraday cage to enhance SE capture. Alternatively, scintillation crystal-based detectors, such as the Robinson detector (Ref 8), are positioned directly above (within a few millimeters of) the sample surface, usually just below the pole piece. These scintillation detectors can have larger collection areas and provide high BE collection yields.

Another common method for BE detection in this geometry is using semiconductor-based solid-state detectors, which are often segmented so the BE signal to different regions of the detector can be isolated and mixed. The resultant BE images from such detectors can be added, subtracted, and/or mixed to enhance compositional contrast, topographic contrast, and quantitative surface information for three-dimensional reconstruction. Pole-piece-positioned scintillation and solid-state detectors have an orifice in the center to enable the incoming electron beam arrival to the sample surface. Many pole-piece-positioned detectors are mounted on a retractable device so they can be inserted or removed by the operator without venting the SEM chamber and manually positioning the detector.

A newer development for SEM detectors is positioning scintillation and solid-state detectors below the sample or above the sample inside the electron-optics system, as shown in Fig. 8. To detect electrons emanating from the underside of a specimen, it is necessary to thin specimens to ~100 nm thickness and use higher acceleration voltages (>20 kV) so samples are somewhat transparent to the incoming electron beam. This scanning transmission electron microscopy (STEM) mode and associated STEM detectors can be used to detect BEs (dark field) and transmitted electron beam signal (bright field). The STEM imaging provides enhanced imaging resolution and a variety of contrast mechanisms, which provides complex information (Ref 9, 10).

In-lens and in-column detectors selectively detect SEs and BEs that have a high take-off angle from the sample surface. An internal direct current bias inside the electron-optics system facilitates SE collection, while in-lens BE detectors rely on line-of-sight positioning of the sample for optimized detection. In-lens detectors are particularly useful when imaging at low accelerating voltages and small WDs. Detector efficiencies, ε, relate the performance of various electron detectors (Ref 11). In-column SE detectors yield ε values of ~0.6, far superior to conventional ETDs with ~0.1 to 0.05 efficiencies (Ref 12).

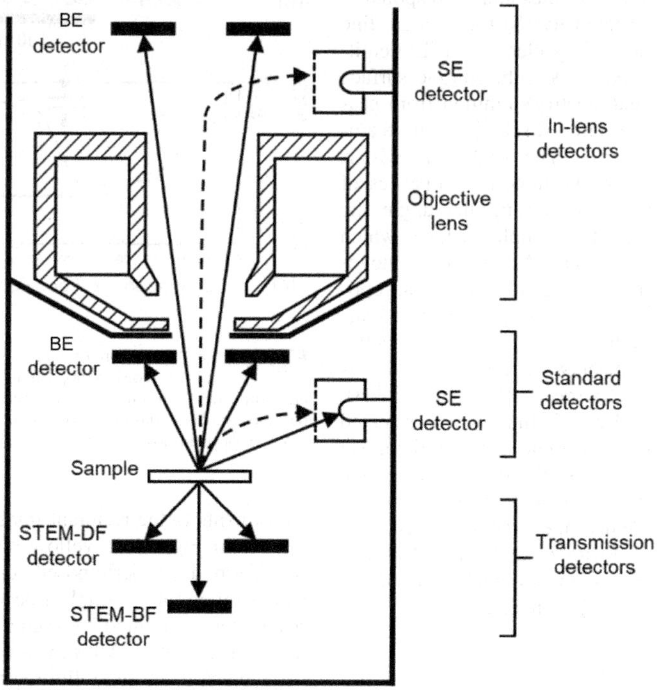

Fig. 8 Schematic of various detector positions and associated electron trajectories for secondary electrons (SE; dotted arrows) and backscattered electrons (BE; solid arrows). STEM, scanning transmission electron microscopy; DF, dark field; BF, bright field

Support System

The required support system varies depending on the comparative image resolution and functional capabilities of the SEM. At a minimum, an SEM requires electrical power to supply the voltage and current required for the electron beam. Water lines may be required to cool the electromagnetic lenses. Air or dry nitrogen is often used as a purging gas during the venting process. Higher-resolution SEM instruments require an air table to dampen external vibrations.

An SEM also requires that both the electron-optics column and specimen chamber operate under high vacuum conditions to ensure the electron beam is not scattered by gas molecules before arriving at the sample surface. Therefore, the sample and mounting materials used to hold the sample in place must not have a high vapor pressure. A valve isolates the electron-optics column from the specimen chamber so the electron-optics column remains under vacuum during sample loading. This arrangement enables the electron gun to remain activated during specimen exchange, hastening turnaround time for changing specimens. For SEMs that require more stringent specimen chamber vacuum conditions, an additional sample-loading chamber enables inserting a sample into the specimen chamber without venting the entire chamber to atmosphere.

Different vacuum pumps are used to achieve different levels of vacuum for SEMs. Mechanical rotary pumps are typically used as a first-stage pump for initial evacuation of SEM specimen chambers down to 10^{-2} Pa (7.5×10^{-4} torr). Oil-based diffusion pumps and cleaner turbomolecular pumps can further reduce pressures down to 10^{-5} to 10^{-7} Pa (7.5×10^{-7} to 7.5×10^{-9} torr), respectively. Within the electron-optics column, ion getter pumps attain low pressures in the 10^{-8} Pa (7.5×10^{-10} torr) range.

Control and Imaging System

Control of the modern microscope involves a computer-user interface that enables electron beam manipulation, image acquisition, sample positioning, detector selection, and vacuum operation. The user can define the probe based on the acceleration voltage and spot size. A scan generator enables the user to select magnification, dwell time per pixel, and number of pixels per image; a frame grabber enables digital image capture and storage on the computer. Quality of the images is a function of both the number of pixels to provide sharpness and the contrast. Optimizing focus through control of the objective lens current and/or specimen z-position further enhances image sharpness. Contrast can be enhanced by several factors, depending on the conditions of the electron probe, the type of signal detected, and the detector gain settings used.

Image Formation

Image formation is based on detected variations in the collected signal as a function of probe position as the electron beam is scanned across the sample surface. Figure 9(a) depicts a single line scan from positions 1 to 5 across a topographical bump on a surface. Possible trajectories of SE paths are shown for beam positions 2 and 4, and the SE paths can curve to the ETD due to the positively biased Faraday cage. However, many electrons from position 2 may never reach the SE detector due to their initial escape trajectory. Consequently, the SE signal is much higher at position 4 due to its direct line of sight to the detector. Figure 9(b) illustrates the variation in SE signal intensity as the beam is scanned across the bump, demonstrating how SE signal variation can be correlated to a topographical feature.

If multiple line scans are conducted, a two-dimensional representation of detected SE signal intensity can be presented as an array of pixels, each with a corresponding grayscale value. During SEM operation, the scan generator that drives the scan coils of the electron column coordinates with both the detector and the digital display readout. Therefore, as shown in Fig. 9, there is a synchronous positioning of the beam within the raster on the sample surface and the pixel position defined in the digital display.

To prepare a picture, the grayscale intensity of the individual pixels is modulated proportionally to the magnitude of the signal from the detector. When the beam is at position 2 on the sample, the signal is low; therefore, the pixel corresponding to point 2′ on the digital display is dark. Similarly, the pixel at position 4′ is bright, because position 4 faces the detector and provides a larger signal. Therefore, as the beam scans the full raster, a full frame is produced on the digital display; the side of the bump facing the SE detector appears bright and the opposite side dark.

An image can be prepared with the signal from any of the detectors shown in Fig. 8. The topographic contrast in the image when using the SE signal is like the optical case in which the eye is above the sample and a light shines on the sample surface from the direction of the detector position. The side of the bump facing the light source would appear bright, and the other side would be dark, as illustrated in Fig. 9(c). Therefore, assuming the source of contrast is topographic, observed shadowing effects in SE or BE images from the SEM are comparable to the effects of typical light-source images obtained by optical microscopy.

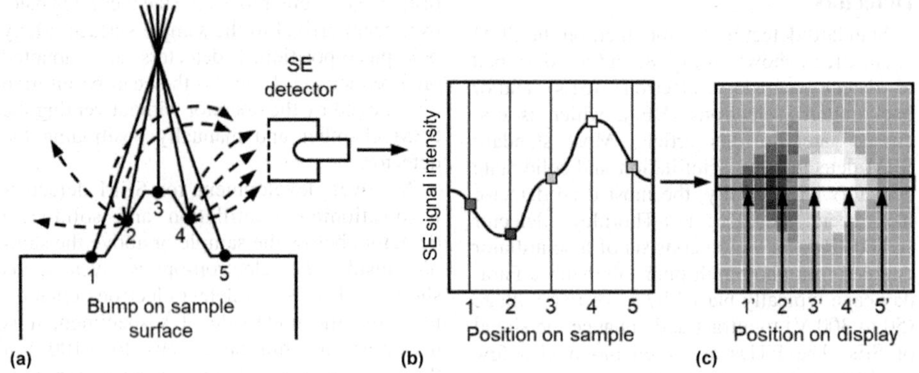

Fig. 9 Illustrations depicting (a) a line scan of the electron beam across a bump, (b) the synchronous secondary electron (SE) signal intensity, and (c) pixel grayscales in the resultant digital image

Contrast

Contrast is the quantitative differences in grayscale values for different pixels. Percent contrast can be determined as:

$$C(\%) = (S_2 - S_1)/(S_2 + S_1) \times 100 \quad \text{(Eq 3)}$$

where S_2 and S_1 are signal intensities at two arbitrary raster positions. Information is contrast that corresponds to physical characteristics of the specimen surface. Therefore, extracting information from SEM micrographs requires evaluation and interpretation of signal differences (ΔS) attributed to features of interest as opposed to random signal fluctuations, called noise (N).

Figure 10 presents an example SEM micrograph of gold nanoparticles and corresponding grayscale signal intensity (S) for a single line of 48 pixels. Secondary electrons (SEs) emitted when the beam strikes the sample surface are weakly bound electrons emitted from near the surface (see later discussion on beam-sample interactions). The yield of SEs is increased either by the material properties of the surface or by decreasing the angle the beam makes with the sample surface, which is termed the tilt angle. As the beam moves from points 3 to 4 in Fig. 10, there is an optimally large change in material properties (carbon to gold) and beam surface angle geometry. Consequently, ΔS between points 3 and 4 is large and the contrast, $\Delta S/S$, is highest. Between points 4 and 5, the beam is scanned across the top of a single gold particle, yet some S variation is detected despite the smooth surface. The fluctuation in S is attributed to noise. When the signal change, ΔS, exceeds N by a factor of at least 5 (Ref 13), the signal change is conventionally considered meaningful in the SEM image.

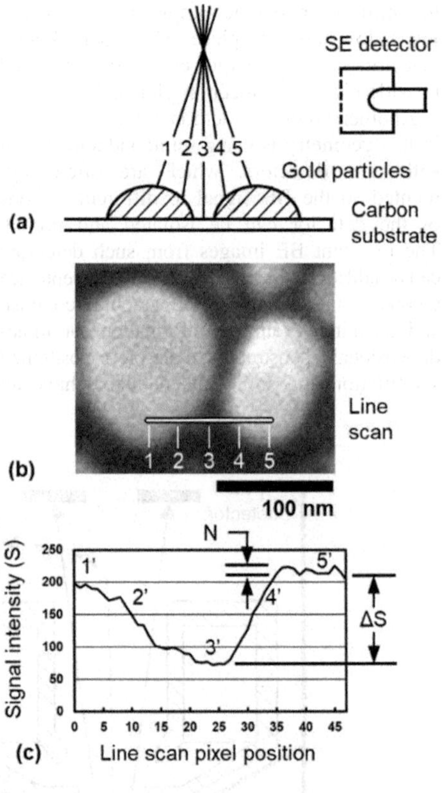

Fig. 10 (a) Illustration of electron beam scanning across gold nanoparticles. SE, secondary electron. (b) Scanning electron micrograph of gold particles. (c) Secondary electron signal intensity, S, as a function of pixel position from line scan denoted in (b)

Brightness

Brightness and contrast are image settings that the user can tune through control of the detector signal gain. Figure 11 demonstrates the effect of brightness and contrast gain adjustments to the resultant image. For all five images in Fig. 11, a yellow plot termed the waveform is superimposed on the image, which represents a single line of pixel grayscale intensity plotted between 0 and 255, similar to Fig. 10(c). All waveforms in Fig. 11 correspond to a centrally located horizontal line of pixels. Figure 11(c) presents an optimized brightness and contrast setting. All pixel intensities fall within the 0 to 255 range. As such, the profile of the waveform spans all

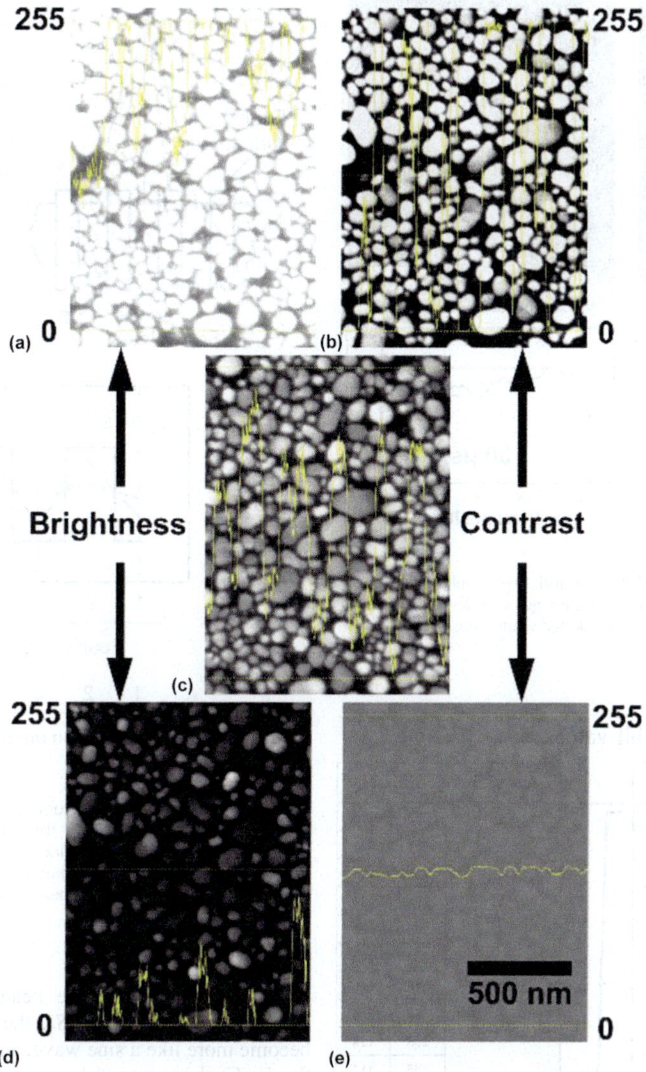

Fig. 11 Five scanning electron micrographs of gold nanoparticles recorded at various brightness and contrast settings, with superimposed waveform (yellow plot). (a) Excessive brightness leads to loss of information. (b) Excessive contrast results in saturation for both 255 and 0 grayscale values. (c) Optimized brightness and contrast settings result in all pixel intensities falling within the 0 to 255 range. (d) Too-low brightness produces a saturation where much of the pixel intensity falls below the 0 grayscale value. (e) Excessively low contrast setting produces an image where features become difficult to resolve

grayscale values without exceeding the top dotted line (255 grayscale value) or bottom dotted line (0 grayscale value). Also, the waveform exhibits seven individual peaks that correspond to seven individual gold nanoparticles spanned by the central horizontal line of pixels (central dotted line).

Increasing the brightness gain settings increases the grayscale values for all pixels. However, excessive brightness increase can lead to loss of information (Fig. 11a). If much of the waveform exceeds the 255 grayscale value, then numerous adjacent pixels are solid white and yield no contrast or useful information. This loss of information is termed saturation. Lowering brightness too far also produces a saturation, where much of the pixel intensity falls below the 0 grayscale value (Fig. 11d).

While brightness adjustments aim to position the waveform between the 255 and 0 grayscale values, contrast adjustments should aim to maximize the waveform amplitude while also avoid exceeding the 255 and 0 values. Figure 11(b) demonstrates when the contrast is too high. Saturation occurs for both 255 and 0 grayscale values. Reducing contrast decreases the waveform amplitude size. While this can ensure that saturation is avoided, an excessively low contrast setting produces an image where features become difficult to resolve (Fig. 11e).

Dwell Time

Dwell time, as mentioned previously in Fig. 7, is the amount of time the electron beam spends on each point in the raster. Dwell times typically vary from milliseconds down to nanoseconds. Setting a short dwell time is advantageous for fast scan rates approaching "live" imaging of the specimen. Each frame may only take a fraction of a second to acquire, so images are rapidly updated. This is useful when navigating to different positions on the specimen, adjusting focus, or correcting astigmatism. Longer dwell times enable more signal collection for each pixel position, producing images with less noise, as illustrated in Fig. 12. Improving signal-to-noise (S/N) ratio helps to better define contrast and associated information regarding the specimen, such as feature width and surface topography. Using longer dwell times could mean a single frame might require a few seconds up to a few minutes to acquire. Slight drift of the beam and/or the specimen during such long acquisition times could produce a smeared image. Control software often provides drift correction combined with frame integration to mitigate this problem.

Digital images

Digital images acquired by SEM are typically stored in the tagged image file format. Scanning electron microscopes generally enable the user to define the number of pixels that make up the image and the range of possible grayscale values each pixel could be assigned. An image with more pixels provides more spatial information but at the expense of requiring longer time to acquire a single frame. Digital grayscale images can be recorded at 8 or 16 bits for 256 or 65,536 grayscale values, respectively. The human eye can resolve ~16 different shades of gray, which corresponds to ~16 units of grayscale intensity differences for 8-bit values, as shown in Table 2. Recording digital images in 24-bit format retains color associated with any graphics or annotations added to the SEM micrograph (data bar, measurements, and notes). The amount of memory required for any digital image file is a function of the total number of pixels acquired and the grayscale intensity range used. Table 2 provides example colors and bit values that are standard for digital images.

Advantages of the SEM

Two major advantages of the SEM over the optical microscope as a tool for examining surfaces are improvements in resolution and depth of field. Information on the determination of resolution and depth of field in the optical microscope is provided in the article "Light Microscopy" in *Metallography and Microstructures*, Volume 9, *Handbook*, ASM 2004. Because SEMs are also often compared with transmission electron microscopes (TEMs), this section also details various types of SEM samples and mounting strategies to highlight potential advantages over the TEM.

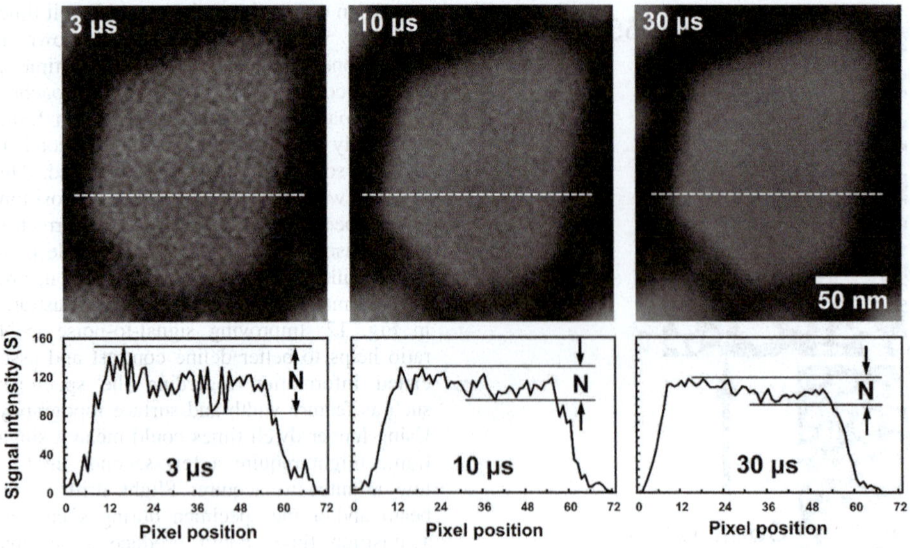

Fig. 12 Scanning electron micrographs of a single gold nanoparticle recorded at 3, 10, and 30 μs dwell times. Longer dwell times enable more signal collection for each pixel position, producing images with less noise. Signal intensity plots correspond to the horizontal row of 72 pixels denoted by the dotted white line.

Table 2 Example colors and grayscales for 24-, 8-, and 16-bit values

Color		24-bit values		
White		255	255	255
Gray		128	128	128
Black		0	0	0
Red		255	0	0
Green		0	255	0
Blue		0	0	255
Yellow		255	255	0
Cyan		0	255	255
Magenta		255	0	255
Orange		255	192	0
Purple		112	48	160
Pink		255	153	204

Grayscale	8-bit values
	255
	240
	224
	208
	192
	176
	160
	144
	128
	112
	96
	80
	64
	48
	32
	16
	0

Grayscale	16-bit values	
	255	255
	255	240
	255	224
	255	208
	255	192
	255	176
	255	160
	255	144
	255	128
	255	112
	255	96
	255	80
	255	64
	255	48
	255	32
	255	16
	255	0

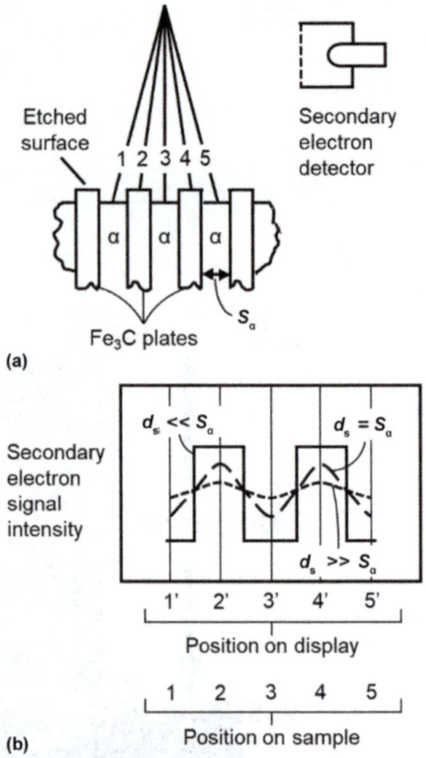

Fig. 13 (a) Line scan across an etched pearlite sample surface. (b) Effect of beam diameter, d_s, on secondary electron signal intensity. A beam with d_s much smaller than the lamella spacing, S_α, results in lamella plates appearing in sharp contrast on the digital display.

Resolution

Resolution can be defined in terms of a point resolution and an image resolution. Point resolution is the crucial parameter that defines SEM performance and its ability to resolve features down to a minimum threshold size. Point resolution for the SEM is directly determined by spot size, which is the electron beam diameter at the sample surface plane, d_s. However, it should not be automatically assumed that the electron beam/probe size sets the point resolution. Point resolution is also influenced by the type of signal used for image formation, discussed later in this article. Additional information on the resolution limits of the SEM can be found in Ref 13 and 14.

Figure 13 illustrates the relationship between d_s and feature size for resolving topographical details of interest. Pearlite consists of alternating plates of Fe_3C in a matrix of α-iron. When etched using nital or picral, the Fe_3C plates stand in relief, as shown in Fig. 13(a). Because the Fe_3C plates are in relief, the SE yield from them is very high. Consider a scanning electron beam having a d_s much smaller than the Fe_3C spacing, S_α. As the beam scans from positions 1 to 5, the scanning electron signal intensity exhibits the sharp square wave form shown in Fig. 13(b), and the Fe_3C plates appear in sharp contrast on the digital display. If the beam diameter is increased such that $d_s = S_\alpha$, the signal would become more like a sine wave, and the edges of the Fe_3C plates would become very fuzzy on the digital display. Finally, for $d_s > S_\alpha$, the Fe_3C plates would no longer be distinguishable on the display. In general, to resolve a feature of size S_α, a beam diameter less than S_α is required.

A line scan across a feature produces signal intensity variations, ΔS, visible as various grayscales in the digital display (Fig. 10). A generally accepted rule for the ability of the eye to discern a feature is that ΔS > 5N (see Fig. 10 for an explanation of S, ΔS, and N). Based on this criterion, the signal current, i_s, required to see a feature that produces a signal jump of ΔS is (Ref 15):

$$i_s = i_B \varepsilon > \frac{(4 \times 10^{-18} \text{coulombs})\, n_{\text{pixels}}}{\left(\frac{\Delta S}{S}\right)^2 t_f} \quad \text{(Eq 4)}$$

where i_B is the beam current in the beam diameter, d_s; ε is the detector efficiency in terms of signal collection per incident electron; n_{pixels} is the number of pixels in a digital image; and t_f is the time to acquire the image, that is, the frame time. As the spot size is made smaller, it contains less current. The current in the beam spot is a function of the brightness of the electron gun, β; the beam voltage, V; and the aberrations of the electron lens.

Spherical aberration is generally considered to be the limiting factor for SEM resolution, and it is characterized by a constant, C_s, where the larger the value of C_s, the greater the aberration. For conditions of optimum aperture size selection, beam current varies with beam diameter on the sample (Ref 14):

$$d_s = \left[\frac{16 i_B}{3\pi^2 \beta}\right]^{3/8} C_s^{1/4} \quad \text{(Eq 5)}$$

The value of d_s is larger than the d_3 value given by Eq 2 due to lens aberrations. Equations 4 and 5 illustrate the important factors in resolution limits with an SEM. Assuming a single megapixel (10^6) digital image, resolving a feature with ΔS/S greater than a certain minimum value requires a certain critical minimum beam current in picoamps, $(i_B)_c$:

$$(i_B)_c = \frac{4}{\varepsilon \left[\frac{\Delta S}{S(\min)}\right]^2 t_f} \quad \text{(Eq 6)}$$

By increasing the collector efficiency, ε, and by using longer frame times, t_f, the value of $(i_B)_c$ can be reduced. Equation 5 shows that reductions of d_s are limited, because decreasing d_s decreases i_B, and eventually it will fall below $(i_B)_c$. For the crucial value of $(i_B)_c$, Eq 5 shows that smaller beam diameters are achieved by increasing β and decreasing the spherical aberration coefficient, C_s. Maximum resolution requires a minimum d_s, which requires maximum values of β, ε, and t_f and minimum C_s values. Field-emission sources have higher brightness than thermionic sources and thus are preferable for high-resolution imaging. In general, C_s decreases as the sample is moved closer below the objective lens; that is, as the WD decreases.

More advanced commercial SEMs provide specialized lens configurations to obtain ultrahigh resolution (UHR). One type of lens is an immersion lens, where the sample is positioned inside a gap between the upper and lower parts of an objective lens pole piece. Another lens type, the snorkel lens, extends the magnetic field below the lens and into the sample. In either case, the lenses greatly reduce the WD and reduce C_s significantly, but they require the use of in-lens detectors. Immersion lens instruments provide improved resolution, and current UHR-SEMs have point resolution <0.5 nm.

Any point resolution value reported for a given SEM depends on many factors, including whether the electron source is thermionic versus a field-emission source, what type of signal is detected, and what lens configuration is used. Typical resolution values claimed by manufacturers for general types of SEMs are given in Table 3 and are compared with those of optical microscopy (OM). The SEM offers a distinct advantage over OM for high-resolution imaging. Table 3 also lists the maximum useful magnification ranges; at magnifications above approximately twice these values, features become fuzzy and unresolvable.

Another way to compare SEM and OM resolution is to assess micrographs of fine pearlite, as shown in Fig. 14. The optical micrograph (Fig. 14a) of a pearlitic gray iron specimen was taken at the upper magnification limit of an optical microscope using optimum resolution conditions (Ref 16). Pearlite lamellar spacing of 1 to 2 μm can be defined, but finer spacing is unresolvable. Figures 14(b) and (c) show micrographs for a hypereutectoid steel acquired using an SEM at higher magnifications. The pearlite lamellae are much finer in the hypereutectoid steel. Nevertheless, the SEM can readily resolve the pearlite plate spacing (~200 nm) at a magnification 20 times higher than OM.

Image resolution is often mistakenly used in place of point resolution. Image resolution refers to the quality of the digital image itself in resolving fine-scale features. Image resolution is influenced by the digital image characteristics, such as contrast gain settings, grayscale range, and number of pixels used to resolve features. As such, image resolution pertains to the quality of the digital image instead of the actual point resolution, probe size, or performance of the SEM.

Depth of Field

At a distance below and above the focal plane, the beam diameter becomes large enough to overlap adjacent pixels. Depth of field refers to the distance above and below the focal plane where features are still resolvable, and the image is still considered in focus. Calculating depth of field requires consideration of how the primary beam diameter broadens from a diameter at focus, d_s, to a diameter large enough to overlap an adjacent pixel. Therefore, depth of field depends on the magnification used as well as the beam convergence angle, α. A useful criterion is to presume a typical observer can discern image blur once two adjacent pixels completely overlap (Ref 17). Thus, depth of field (DOF) is given as:

$$\mathrm{DOF(mm)} = \frac{2p}{\alpha \times M} \quad \text{(Eq 7)}$$

where M is the magnification, α is the convergence angle of the beam striking the sample, and p is the physical pixel size in millimeters. For example, a standard 21.5 in., 1080p digital display has a p value of 0.248 mm. Table 4 presents calculated values of depth of field based on Eq 7 for beam convergence angles, α, of 2 and 30 milliradians. At higher magnifications, the pixel size approaches the focused beam diameter, and features quickly become blurry and out of focus when displaced even a small distance from the focal plane. Therefore, higher magnification leads to a smaller depth of field.

Table 4 also indicates that reducing beam convergence angle, α, increases depth of focus. The value of α is determined by the

Table 3 Typical resolution and magnification ranges for general types of scanning electron microscopes and for an optical microscope

Microscope	Minimum resolution range, nm	Maximum useful magnification
General-purpose scanning electron microscope	5–10	20,000–50,000×
General-purpose field-emission scanning electron microscope	1.0–2.5	250,000–400,000×
Ultrahigh-resolution field-emission scanning electron microscope	0.3–1.0	400,000–1,000,000×
Optical microscope	100–200	1,000–2,000×

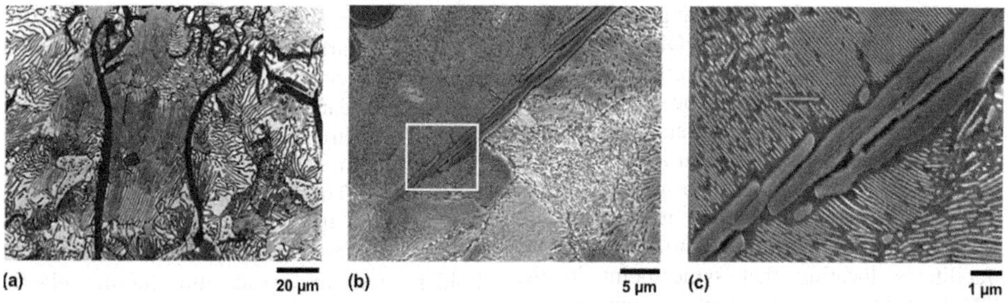

Fig. 14 (a) Optical micrograph of pearlitic gray iron taken using a high-quality optical microscope with an oil immersion lens and green filter; 4% picral etch. Fine pearlite lamellar spacing is unresolvable. Source: Ref 16. (b) Scanning electron micrograph of pearlitic hypereutectoid steel taken using a conventional scanning electron microscope, secondary electron detection, and 15 keV beam; 4% nital etch. Courtesy of M. Hecht, Carnegie Mellon University. (c) The center region denoted by white rectangle in image (b) with magnification increased by a factor of 4. Scanning electron microscopy resolves spacing (~200 nm) at a magnification 20 times higher than optical microscopy.

Table 4 Calculated values for depth of field based on Eq 7 for a digital scanning electron micrograph

		Depth of field, μm	
Magnification	Image width, μm	α = 2 mrad	α = 30 mrad
10×	47,600	24,800	1,653
50×	9,520	4,960	331
100×	4,760	2,480	165
500×	952	496	33
1,000×	476	248	17
10,000×	48	25	1.7
100,000×	4.8	2.5	0.17

Note: Micrograph encompasses entire standard 21.5 in., 1080p digital display, with a 19 in. wide horizontal display composed of 1920 pixels.

WD and the diameter of the objective aperture, d_a:

$$\alpha \approx \frac{d_a/2}{WD} \quad \text{(Eq 8)}$$

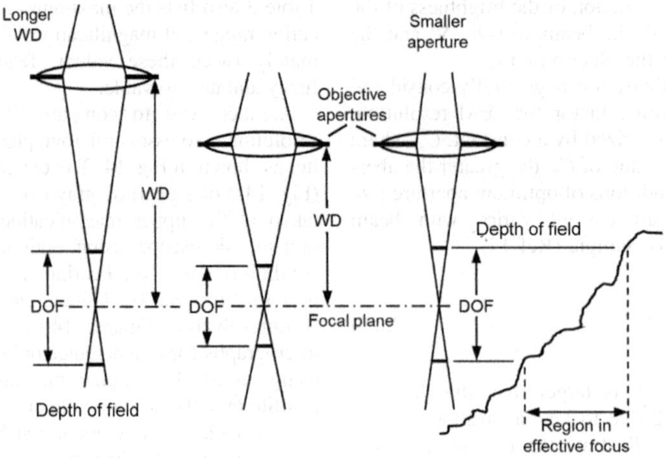

Fig. 15 Effect of (left) longer working distance (WD) and (right) smaller objective aperture size on depth of field (DOF) and associated size of region in effective focus

Figure 15 illustrates the effect of increased WD and reduced objective aperture size in extending depth of focus. From Eq 7 and 8, for a given magnification, the user can increase the spatial region that is in effective focus for a highly tilted or irregular surface using a longer WD and/or smaller d_a.

Depth of field of an SEM is very good compared with an OM. The difference in the two microscopes is illustrated in Fig. 16, which compares images of 1 to 10 μm-sized tin spheres obtained by OM using a 50× objective lens and by SEM. The OM image is focused to the smallest (~1 μm) spheres and exhibits significant blurring for the larger-sized spheres. However, spheres of all sizes are in focus in the SEM image. In general, the SEM depth of field exceeds that of the OM by a factor of approximately 300. Therefore, SEMs have found widespread use for examination of fracture surfaces and deeply etched samples.

The SEM is generally inferior to the OM for routine examination of samples prepared using standard metallographic techniques when examined at magnifications less than 300 to 400×. Standard metallographic techniques can produce surface relief (especially after etching) suitable for contrast in the SEM at these lower magnification ranges. However, color variations detectable by OM can provide subtle and useful material information not obtainable by SEM. Due to the relatively limited contrast of the SEM on metallographically polished and etched samples at low magnifications, it is generally useful to use microhardness indentations as fiducial marks. Placing such marks near a point examined in the OM facilitates locating that same point in the SEM for examination at higher magnification.

Samples

Scanning electron microscopy is amenable to a broad array of samples, but there are some limitations. Conventionally, samples should be electrically grounded to the holder and SEM stage. As such, solid specimens that are electrically conductive require no sample preparation. Nonconductive solids can be imaged if coated with a conductive metallic or graphitic carbon thin film that grounds the imaged surface. If there are concerns with surface coatings, uncoated nonconductive solids can be imaged using low-electron-beam energy (Ref 18) or using a variable-pressure or environmental SEM (Ref 15), which can remove surface charge in the absence of electrical grounding. Another approach is the use of flood guns, an add-on capability for SEMs. Flood guns direct charged particles at the sample surface, thereby removing surface charge for nonconductive specimens. Wet and fixed biological specimens can also be imaged using a variable-pressure or environmental SEM in this manner. New SEM stages have been developed to allow imaging of liquid specimens (Ref 19).

Because specimens are placed under vacuum inside the SEM, samples should have a low vapor pressure (≤0.01 Pa, or 10^{-3} torr) and be free of any high-vapor-pressure liquids, such as water, organic cleaning solutions, and remnant oil-based films (Ref 20). Unwanted carbon buildup at the imaged region often occurs. This buildup occurs due to electron-beam-stimulated surface reactions with hydrocarbon contamination originally trapped within sample surface pores and surface cracks. Therefore, it is important to dry the samples thoroughly, to use gloves to avoid skin contact with the sample and holder, and to avoid introducing plastic mounting materials into the specimen chamber, which can have high vapor pressures and trap hydrocarbon films at the mount-sample interface. Add-on accessories to the SEM chamber can generate a plasma to help capture and remove hydrocarbon contamination during SEM analysis (Ref 21).

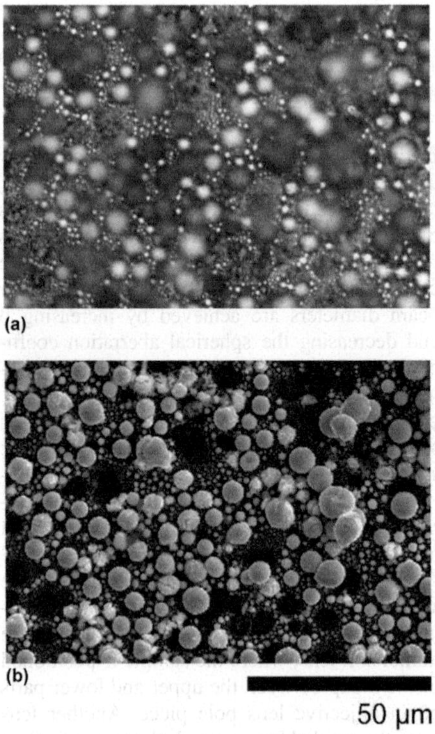

Fig. 16 Micrographs of tin spheres recorded by (a) optical microscope (OM) and (b) scanning electron microscope (SEM). Focusing the OM on the smallest spheres results in blurring of larger spheres, while spheres of all sizes are in focus with the SEM due to better depth of field.

The SEM offers an important advantage over the TEM with regard to sample size and preparation issues. Samples as large as 20 cm (~8 in.) can be placed in the SEM, although regions on such samples that can be examined without repositioning are limited to ~5 cm (2 in.). Nevertheless, this is an exceptional advantage over the TEM, where samples can only be as large as 3 mm (0.125 in.) and areas for analysis must be transparent to the electron beam. Electron transparency requires that TEM specimens are thinned down to <200 nm at any region of interest. Because the SEM is primarily a surface analysis technique, specimen thinning is not necessary. However, thinned specimens amenable for TEM analysis can also be mounted, imaged, and analyzed in the SEM using transmission-mode detectors described previously (Fig. 8). Transmission-mode SEM (Ref 9) offers a cost advantage for those interested in nanoscale imaging using Z-contrast and diffraction-contrast modes instead of using the TEM. Also, the use of transmission Kikuchi diffraction (Ref 22) offers an exceptional capacity for nanoscale crystallographic orientation and phase analysis in the SEM instead of using precession-diffraction techniques in the TEM (Ref 23).

The SEM sample preparation varies depending on the material to be analyzed and the type of SEM analysis to be performed. Consideration must be given to the electrical conductivity of the specimen and if a conductive coating should be used. For bulk samples, consideration should also be given regarding the desired configuration of the surface in relation to the beam and detectors, and the desired state of that surface (as-is, polished, etched). Samples are conventionally attached to a metal support, such as a brass cylinder or aluminum pin stub. These supports are often designed for an SEM manufacturer. Pin stubs may be flat or could provide a particular pretilt angle (45°, 90°). Some supports hold the samples in place using metal pins or screws. Otherwise, bulk materials are conventionally affixed to flat pin stubs using a vacuum-compatible, conductive adhesive, such as graphitic carbon or copper adhesive tape. Carbon (graphitic flakes) paste and colloidal silver paint are also common adhesives, but they must be cured for some time in air or in an oven (~80 °C, or 175 °F). Carbon tape and conductive paints can later be removed using solvents such as acetone. The support stub has a pin that fits inside the SEM stage. The stub is held in place either by a spring-mounted connection or by tightening a set-screw. Good mechanical and electrical contact between the support stub and the SEM stage surface ensures minimal vibration and charging effects during SEM analysis. For more details, an excellent reference on SEM sample preparation can be found in Ref 24.

Electron Beam Interactions with Samples

An image of the scanned surface region can be acquired using any signal generated by the electron beam. In a typical SEM, it is possible to use as many as three or more different signals to generate complimentary images detailing the topography and composition of a sample. To understand the potential utility of scanning electron microscopy, it is necessary to understand some elementary ideas regarding the nature of signal generation when high-energy electrons strike a sample surface.

Types of Signals

When an electron beam strikes a solid surface, a variety of signals can be generated, as illustrated in Fig. 17. Backscattered electrons (BEs) are those electrons from the primary electron beam that have scattered back out of the specimen surface. Deceleration of the high-energy electron beam by the specimen produces x-rays and other electron signals. Electrons bound to atoms within the specimen can be energized by the primary beam. Secondary electrons (SEs) are electrons initially within the specimen that have absorbed sufficient energy from the primary electron beam to escape the specimen surface. Electron transitions within an atom that result from the interactions associated with SE production will also produce x-rays of specific energies characteristic to the elemental identify of that atom. The electron energy transitions resulting from the primary beam can also produce Auger electrons; semiconductors also produce light through a process called cathodoluminescence. Electrons from the primary beam that do not escape as BEs are absorbed electrons, which can produce an electric current. If the specimen is sufficiently thin, some electrons from the primary beam could pass through the sample to become transmitted electrons.

Interaction Volume and Escape Depth

Most of the signals previously mentioned emanate from different depths within the specimen. Figure 18 depicts a typical interaction volume where the primary electron beam produces various signals within the specimen surface. As high-energy electrons penetrate the surface, they undergo elastic and inelastic collisions, which lead to the various signals mentioned previously. The collisions scatter the primary beam electrons mostly in forward directions but also laterally within the sample. This gives rise to the characteristic "teardrop" shape of the interaction volume in Fig. 18. Many signal types are generated throughout the entire interaction volume. However, each signal type varies considerably in its ability to escape the specimen surface. The escape depth is the extent within the specimen surface that a signal can exit the sample and be detected.

Electron signal escape depths are a direct function of the type and energy as well as the chemical composition of the sample. Electromagnetic radiation, such as x-rays and visible light, travels through solids much easier than electrons. Therefore, x-rays can be detected even when generated from the bottom of the

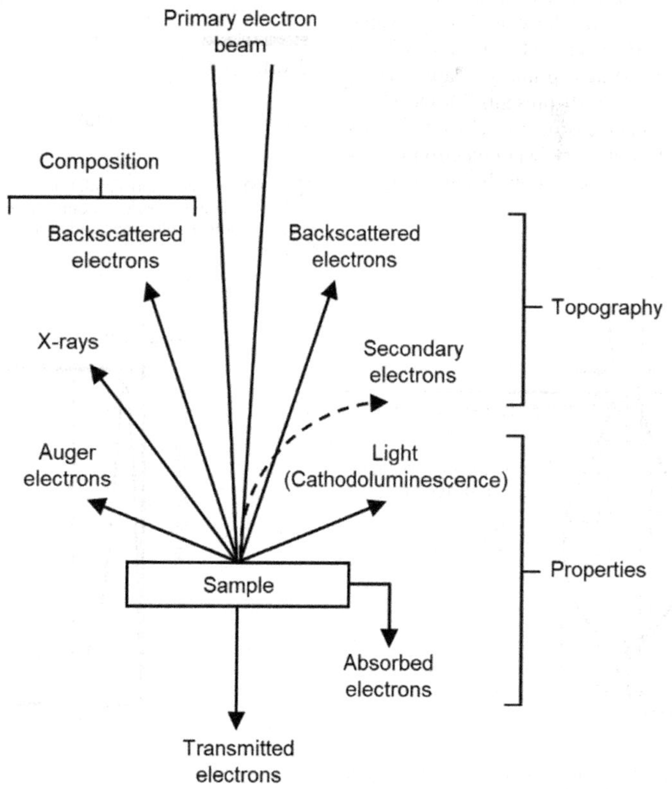

Fig. 17 Various signals generated when an electron beam interacts with a specimen

interaction volume, where electrons from the primary beam have penetrated deep into the specimen (Fig. 1). Therefore, x-rays have an associated information depth. Secondary electrons are also generated at these depths but are readsorbed by the specimen well before escaping the surface. Only SEs generated within ~5 to 50 nm of the surface can escape and reach a SE detector placed above the sample. Metals absorb SEs more readily than insulators, so SE escape depths are generally a factor of 10 smaller for metals than insulators. Backscattered electrons have higher energy than SEs and thus can escape from greater depths than SEs. Escape depths for BEs and information depth for x-rays change based on the primary beam energy, but SE escape depths remain constant.

Depth of penetration by the primary electron beam is a function of electron beam energy and the chemical composition of the specimen. A useful expression for determining the maximum range, R (nm), of primary electron beam penetration into a specimen is the Kanaya-Okayama formula (Ref 25):

$$R = \frac{27.6AE_0^{(5/3)}}{\rho Z^{(8/9)}} \quad \text{(Eq 9)}$$

where E_0 is the primary electron beam energy (keV), A is the atomic weight (g/mol), ρ is the density (g/cm^3), and Z is the atomic number of the specimen. From this expression, the depth of the beam-sample interaction volume increases with higher SEM acceleration voltage. For example, the calculated R value for iron using Eq 9 is 0.51, 1.6, and 3.2 μm at 10, 20, and 30 keV primary beam energy, respectively. Figure 19 presents Monte Carlo simulations, generated using CASINO v3.3 (Ref 26), of 100 primary beam electron trajectories into iron at different beam energies. Monte Carlo simulations help to visualize the degree of depth penetration and lateral travel the primary electrons undergo inside the specimen.

It is crucial to consider this interaction volume when evaluating the SEM as a probe into the chemical composition and structure of a specimen. Spot size, d_s, is an important parameter regarding the physical size of the electron beam diameter as it enters the specimen. However, the lateral size and sampling depth of the actual electron probe is influenced by the interaction volume and specific signal to be detected. Figure 18 shows that the same probe produces x-rays from a much larger volume and wider lateral probe size, x_X, of the sample than BEs and SEs, corresponding to x_B and x_S, respectively. Therefore, the actual point resolution of the SEM is a direct function of acceleration voltage and detected signal, in addition to spot size. This raises important concerns regarding how to interpret images generated simultaneously inside an SEM by one signal type versus another; discrepancies between an x-ray map and SE/BE image obtained in the SEM can be readily attributed to different signal escape information depths within the interaction volume. This point is discussed later in the section "Sampling Volume for Various Signals" in this article.

Electrons

Electron signal emitted by the sample can be partitioned into three types based on energy distribution, as shown in Fig. 20: secondary, backscattered, and Auger. Secondary electrons are <50 eV. Backscattered electrons range from energies equal to the primary beam energy, E_0, down to ~50 eV. Auger electrons can exceed the number of detected BEs at specific energies characteristic of the elements within the specimen. Auger electron energies fall in the 50 to 2000 eV range. Energies of SEs and Auger electrons are fixed, but BEs

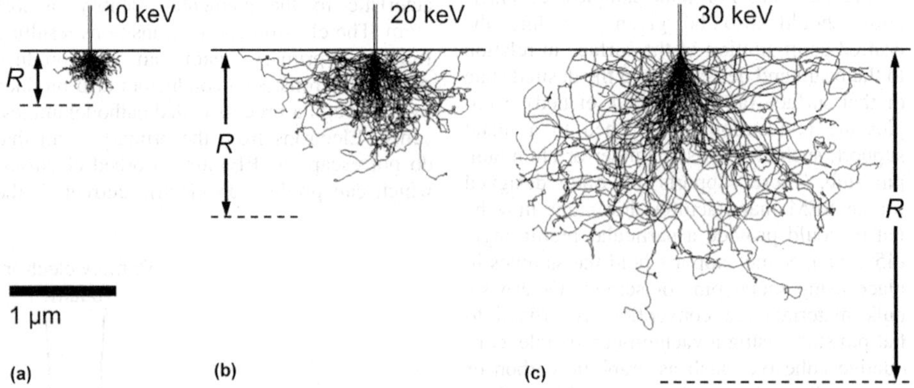

Fig. 19 Monte Carlo simulations of 100 primary beam electron trajectories in iron for (a) 10, (b) 20, and (c) 30 keV beam energies. Monte Carlo simulations help to visualize the degree of depth penetration, R, and lateral travel the primary electrons undergo inside the specimen. Reprinted/adapted from Ref 27 with permission of Springer Nature

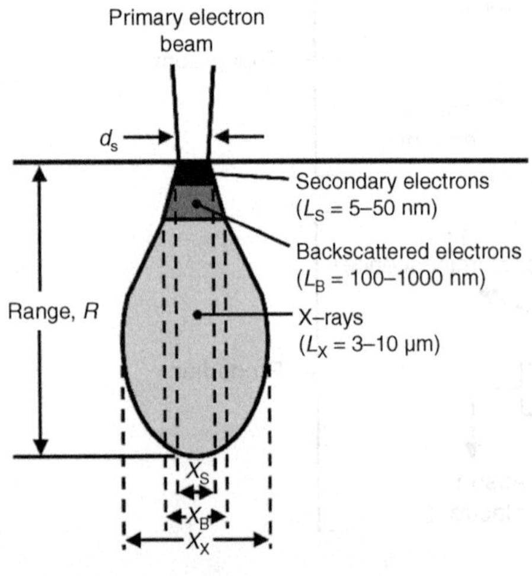

Fig. 18 Illustration of interaction volume of depth, R, with corresponding escape depths and information depths, L, and lateral probe size, x, for different signal types

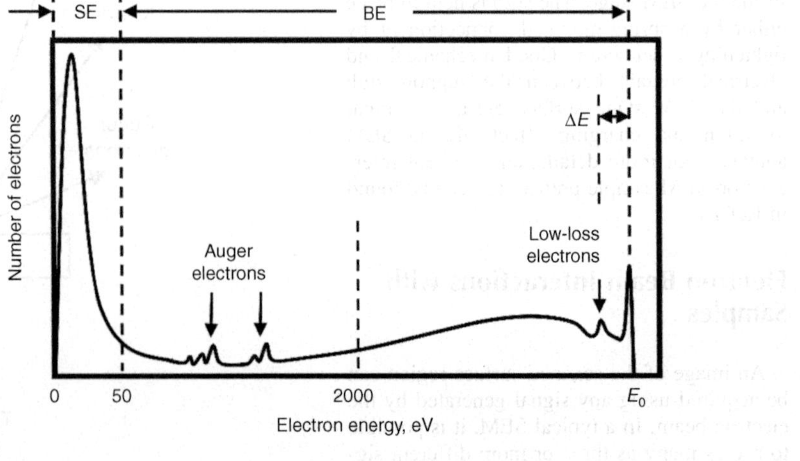

Fig. 20 Energy distribution of emitted electrons produced by a primary electron beam energy, E_0. SE, secondary electrons; BE, backscattered electrons

shift their energy values as the primary beam energy changes. The BEs and SEs can be further partitioned into three types, so all possible electron signals include:

- Backscattered electrons
  a. Type 1: elastically scattered
  b. Type 2: inelastically scattered
  c. Type 3: plasmon and interband transition low loss
- Secondary electrons
  a. Type 1: inside primary beam
  b. Type 2: outside primary beam
  c. Type 3: other nonspecimen surfaces
- Auger electrons

Backscattered Electrons

When a primary beam electron strikes the sample, it can take one of many possible pathways through the sample. A number of these trajectories terminate inside the sample, particularly those trajectories that penetrate deep into the bulk, but some trajectories curve their way back to the surface (Fig. 19). Primary electrons that escape the surface by such trajectories are termed backscattered, and there are three general processes by which this can occur. Figure 21 illustrates primary electron (PE) trajectories into a sample surface. The first PE trajectory, (A), illustrates a PE that, upon entry into the specimen, immediately scatters elastically out of the specimen and emerges with the same energy as the beam energy, E_0; this type of BE is called BE(1). A second PE trajectory, (B), propagates a longer distance within the specimen surface before escaping, losing some energy through inelastic collisions along the way. This type of BE is called BE(2) and can have any energy value below E_0. The PE trajectory (C) illustrates a PE that has penetrated so deep into the specimen that it is unable to escape and becomes absorbed. The BE(3) is a primary beam electron scattered by interactions that produce either a plasmon oscillation of the electrons in the sample material or a transition of specimen electrons between different energy bands. The PE trajectory (D) illustrates a PE that produces a plasmon and immediately escapes the specimen without losing additional energy. The energy to excite plasmons or interband excitations have fixed values denoted as ΔE in Fig. 20; therefore, BEs that emerge after one of these interactions have an energy lower than E_0 by the amount ΔE. Two types of plasmon oscillations occur in which surface electrons or bulk electron oscillations are excited. The interband transitions also require a specific ΔE. The ΔEs for all these small peaks differ from element to element, and sometimes they differ according to whether the element is present as a pure element or an oxide, hydride, nitride, and so on. The BE (3)s are often termed low-loss electrons (LLEs) because they lose a small but specific amount of energy. The LLE energy is readily measured in the TEM through electron energy loss spectrometers (EELS) to provide a variety of useful information on chemical identity, chemical bonding, and plasmonics (Ref 28). However, SEMs conventionally use BE detectors that measure the number of collected BEs and not the energy of individual BEs. The incorporation of EELS in the SEM is under development.

The BE yield is quantified by a coefficient, η, that relates the ratio of number of BEs produced per number of primary beam electrons. Two factors that influence η are the atomic number of the specimen, Z, and the tilt angle, θ, of the primary beam with respect to the sample surface normal direction. Therefore, detected variations in η give rise to BE image contrast sensitive to specimen composition and/or surface topography. Figure 22 plots calculated BE yields, η, as a function of θ for different elements. For a 0° tilt angle, values of η range from 0.05 to 0.55 from low to high atomic number. At high sample tilt angles, η values increase to a range of 0.6 to 0.8.

Figure 23 illustrates why BE yield increases with both higher atomic number and higher tilt angle. As shown in Fig. 23 (a), for a given primary beam energy, the penetration range decreases with higher atomic number. Therefore, a greater portion of the interaction volume is closer to the surface and within the BE escape depth, L_{BE}, which leads to a higher probability for primary beam electrons to escape the surface. A similar effect occurs with increasing tilt angle, illustrated in Fig. 23(d, e). The interaction volume is brought closer to the surface, enhancing the probability for primary beam electrons to escape the surface as BEs.

The angular distribution of BEs is important to consider. Figure 23(b) depicts the angular distribution of BE yield as a function of take-off angle, Φ, for a nontilted specimen ($\theta = 0°$). The shaded circular area represents the total number of BEs generated, so the size of the circle can be related to η (Ref 29). The angular distribution is a sine function of Φ as defined here, so the number of BEs generated can be related to a maximum value when $\Phi = 90°$. For example, the number of BEs emitted along trajectory 1 in Fig. 23(b) is $\eta_n \sin(\Phi_1)$, where η_n is the fraction of total BE yield, η, emanating parallel to the sample normal direction, **n**. As the take-off angle decreases, the number of BEs emitted at that trajectory decreases. Therefore, vector 1 is a larger value than vector 2 in Fig. 23(b). This demonstrates why BE detectors are conventionally positioned directly above the sample. Preferential collection of high-take-off-angle trajectories maximizes BE collection for nontilted samples.

As the sample is tilted, the angular distribution becomes an ellipse, with the maximum BE yield located at an angle, $\Phi = 90° - \theta$

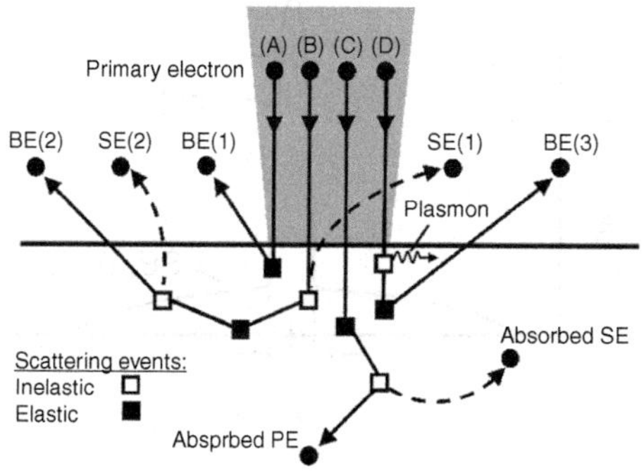

Fig. 21 Illustration of various types of backscattered electrons (BE) and secondary electrons (SE) generated by primary electrons (PE)

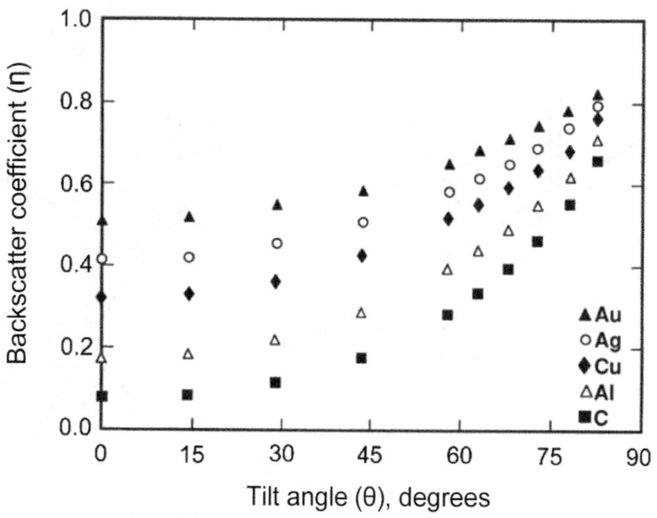

Fig. 22 Backscattered electron yield as a function of tilt angle, θ, calculated for different elements using Monte Carlo simulations. Source: Ref 27

(Ref 29). Thus, for a sample tilt of 60°, the take-off angle with the highest BE emission is $\Phi = 30°$, as illustrated by trajectory 4 in Fig. 23(c). Note that Φ_4 is similar in magnitude to Φ_2, but the BE yield is substantially higher for Φ_4 due to the 60° sample tilt. As such, angles Φ_3 and Φ_1 are also similar in magnitude, but the corresponding BE yield is much lower for Φ_3. Therefore, a BE detector positioned at the electron source collects a very small number of BEs from a highly inclined surface. This effect gives rise to topographic contrast due to BE signal-collection variations.

Lower-atomic-number elements show a more dramatic increase in BE yield as a function of sample tilt angle. It is instructive to consider values of η for copper and aluminum in relation to Fig. 22 and 23. According to Fig. 22, η_{Cu} is ~0.31 and η_{Al} is ~0.18 at $\theta = 0°$. The differences in η are reflected in the areas of the shaded circles depicting angular BE distribution in Fig. 23(a) for the high-Z (copper = 29) and low-Z (aluminum = 13) cases. According to Fig. 22, η_{Al} increases to ~0.35 at a sample tilt angle of $\theta = 60°$. This corresponds to the depicted shaded ellipse of angular BE emission, η_{LZ}, in Fig. 23(e), which has an area comparable to the η_{HZ} shaded circle in Fig. 23(a). It should be noted that as a larger portion of the interaction volume reaches the BE escape depth, as shown in Fig. 23(e), the spatial region of BE emission increases. This effect makes the point resolution worse when BE imaging highly tilted samples. Note that the take-off angle and the symbol Φ are defined differently in this article compared with Ref 27 and 29.

Secondary Electrons

Secondary electrons are electrons that originally existed in the specimen and have been energized by the primary beam with enough excess kinetic energy (<50 eV) to escape the surface. Secondary electrons are predominantly in the 5 to 10 eV range (Fig. 20). Because of their lower energy, SE escape depths are 1% of BE signal depths for 10 to 30 keV incident beam energies (Ref 13). Secondary electron escape depth is also independent of the primary beam energy. Image resolution is generally superior when detecting more localized SEs versus more spatially extensive BEs.

Like BEs, SEs are distinguished by three types, but the distinction is determined based on the proximity of their generation in relation to the primary beam position rather than by energy. Figure 24 shows the main sources of SEs using an etched pearlite sample as an example. Type 1 SEs are generated within the area of the primary beam, while type 2 SEs emanate from the surrounding area. Additional type 2 SEs are generated by primary electrons that emerge through the protruding cementite phase and strike the surrounding sample surfaces, such as the etched ferrite. Type 3 SEs are generated when BEs strike surfaces elsewhere in the SEM chamber, such as the bottom plate of the objective lens. Thus, SE(1)s and SE(2)s contribute signal from the sample, while SE(3)s contribute noise and are undesirable.

Even for a flat specimen, the effect of SE (2) signal on the point resolution can be significant. Figure 25 illustrates the spatial distribution of SE signal in relation to the primary beam spot size, d_s, at the specimen surface. As primary beam electrons propagate into the specimen, the lateral spread of primary electrons within the interaction

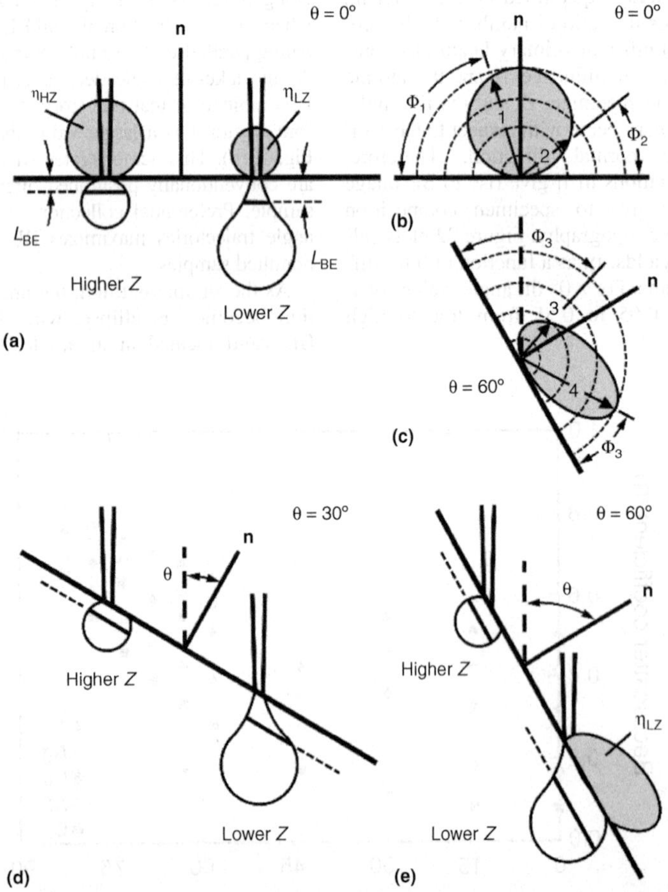

Fig. 23 (a) Illustration of primary beam interaction with 0° sample tilt (θ) for higher and lower atomic number (Z) and the resulting interaction volumes, backscattered electron (BE) escape depths (L_{BE}) and BE yields (η). Angular distribution of BEs as a function of take-off angle (Φ) for (b) 0° and (c) 60° sample tilt. Illustration of primary beam interaction with (d) 30° and (e) 60° sample tilt for higher and lower Z and resulting interaction volumes

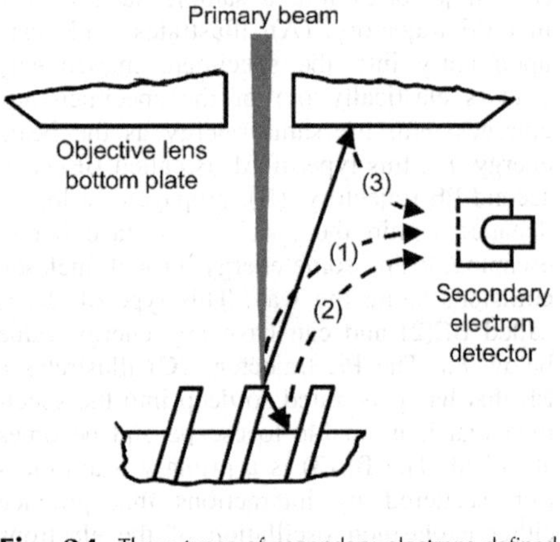

Fig. 24 Three types of secondary electrons defined by proximity to the primary beam

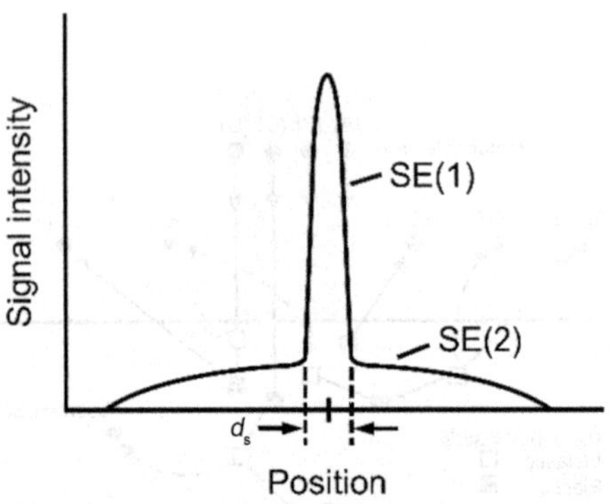

Fig. 25 Illustration of secondary electron (SE) spatial distribution in relation to primary beam diameter at specimen surface, d_s, for both SE(1) and SE(2) signals

volume produces SE signal outside the spot size diameter.

Secondary electron yield is quantified by a coefficient, δ, which relates the ratio of number of SEs produced per number of primary beam electrons. Similar to η, δ is influenced by the tilt angle, θ, of the primary beam with respect to the sample surface normal. Unlike η, δ can change significantly depending on the primary beam energy; SE yield increases substantially below 10 keV, because a significant portion of the interaction volume is brought within the SE escape depth, ~5 nm for metals and ~50 nm for insulators (Ref 27). The escape depth is larger for insulators, because there are fewer free electrons that can interact and absorb the <50 eV SEs as they propagate to the free surface. Chemical composition has very little effect on δ, so SE imaging is generally used for topographic contrast and not compositional contrast.

Auger Electrons

Auger electrons are generated as a result of incoming electrons knocking out intershell electrons (K, L, and M, depending on atomic number) from atoms near the surface. The knock-out events occur within the interaction volume (Fig. 18). After a K electron is knocked out, the surface atom emits either a characteristic x-ray or an Auger electron, as illustrated in Fig. 26. The probability for Auger emission exceeds that for x-ray emission at low atomic number. This is one of the reasons why Auger analysis has some advantages for light-element analysis. As with the characteristic x-ray emission, the energy of the Auger electron is different for each element; therefore, analysis of Auger energies yields information on chemical identity. In addition, as with LLEs, Auger energy levels sometimes shift when an atom becomes oxidized, nitrided, and so on; therefore, information on the chemical state of the surface atoms may sometimes be obtained from Auger analysis.

A highly specialized SEM-type instrument operates at chamber pressure levels in the 10^{-7} Pa (7.5×10^{-9} torr) range to detect and analyze these low-energy Auger electrons. Scanning Auger electron spectroscopy (AES) requires cleaning surfaces by ion sputtering inside the chamber or by fracturing the sample in the chamber and examining the fractured surface. Auger electrons are collected from sample depths of 0.5 to 3 nm below the surface, depending on their energy. Auger electron energies are relatively low, and only those electrons near the sample surface can escape without suffering additional energy loss. Therefore, the Auger signal is very surface sensitive. This is particularly important when considering metal samples, which are typically covered with a thin oxide layer. The energy-loss electrons can be used as a signal in the AES instrument; the depth of its sample volume is a function of the beam energy contrary to the Auger signal. By using beam energies from 70 to 100 eV, surface sensitivities of less than half that of the Auger signal can be achieved, that is, down to 0.25 nm, or approximately one monolayer. More details on AES can be found in the article "Auger Electron Spectroscopy" in this Volume.

X-Rays

X-rays are emitted by a specimen under high-energy electron bombardment. Bremsstrahlung x-rays are "braking x-rays" produced as the primary beam electrons are decelerated by the specimen. These x-rays are a continuum in energy distribution and form a background signal in x-ray spectroscopy. Characteristic x-rays are another type of x-ray produced by the primary beam and in higher numbers than bremsstrahlung x-rays, but only at specific energy values. Both x-ray types are illustrated in Fig. 27.

Similar to Auger electrons, characteristic x-rays are generated as a result of incoming electrons knocking out intershell electrons (K, L, and M, depending on atomic number) for atoms within the specimen. After an intershell electron is knocked out, a higher-energy electron from an outershell moves to replace that electron. This transition requires the emission of a photon that corresponds to the difference in outershell and innershell energy. Every element has specific energy values, ranging from hundreds to tens of thousands of electron volts, for its different electron shells. Therefore, the emitted photon is an x-ray with energy characteristic of the element. Figure 27 depicts three characteristic x-rays emitted by iron: $K_\alpha = 6.4$ eV, $K_\beta = 7.1$ eV, and $L_\alpha = 0.7$ eV. The α and β x-rays correspond to single- and two-level energy shell transitions, respectively. Thus, K_α is a single-level transition to the K-shell,

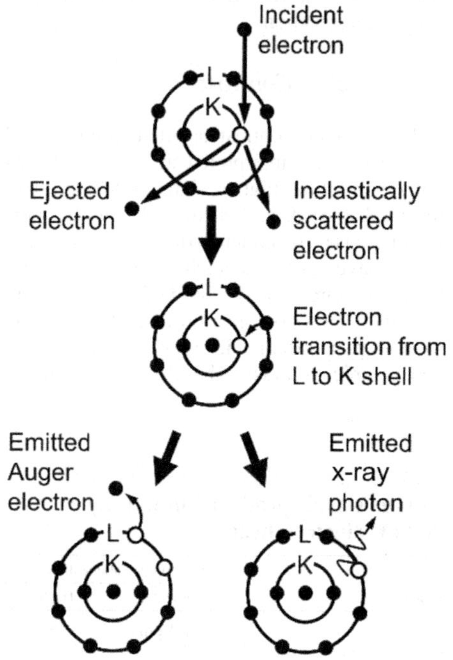

Fig. 26 Generation of Auger electron and x-ray photon by inner shell ionization of an atom and subsequent electron transitions

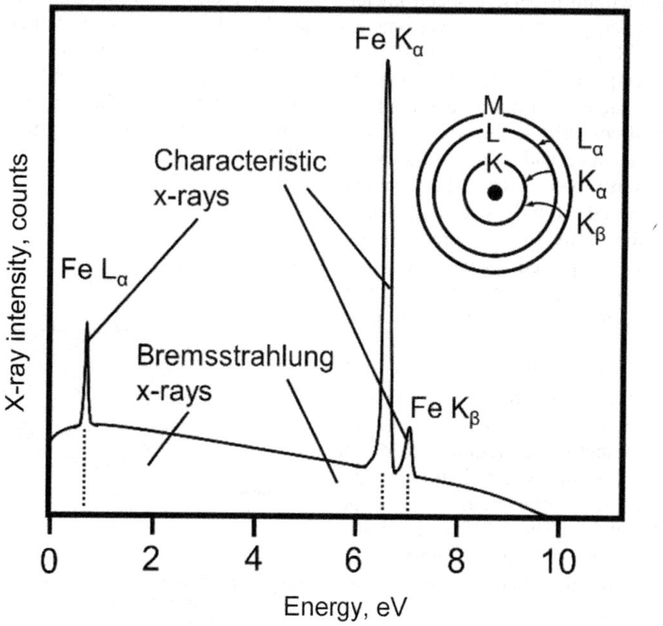

Fig. 27 Illustration of Bremsstrahlung x-rays and characteristic x-rays produced by a pure iron specimen. Energy transitions between M, L, and K shells are diagrammed for different characteristic x-rays.

L_α is a single-level transition to the L-shell, and K_β is a two-level transition to the K-shell. Note that the sum of the energies for the iron K_α and L_α peaks equals the energy of the K_β. Each element has specific characteristic x-ray energies that can be used to identify elements and quantify chemical composition inside the SEM. The larger the atomic number, Z, the higher the x-ray energies will be and the more transitions are possible to produce various characteristic x-rays.

The addition of an x-ray detector enables determination of the energy of emitted characteristic x-rays. Because each element in the periodic table has a different set of characteristic energies, the x-ray analyzer enables SEM determination and quantification of the elements present in the sample, from point to point on the sample surface. After obtaining a scanned image of the surface with the SE or BE detector, the user can position the electron probe to a specific pixel position on the static image. The scan coils deflect the probe to the user-defined position, such as a particle or region of interest. An x-ray spectrum can be collected from that point. Additionally, the user can define multiple analysis points, often equidistant, as defined by a line, rectangle, or other arbitrary shape. A single x-ray spectrum can be integrated from all points, or individual x-ray spectra can be recorded for each individual sample position. Furthermore, based on elements of interest, the measured intensity of specific x-ray energies can be plotted as a function of position to produce elemental distribution maps. A more detailed discussion of x-ray analysis can be found in the article "Electron Probe X-Ray Microanalysis" in this Volume.

Two approaches for x-ray analysis in the SEM are wavelength-dispersive spectrometers (WDSs) and energy-dispersive spectrometers (EDSs). Most SEMs with x-ray analysis capabilities are equipped with EDS detectors. X-ray detectors, such as lithium-doped silicon detectors or silicon-drift detectors, absorb x-rays, one at a time, and convert the x-ray energy to electron-hole pairs. The resultant charge is measured, stored as a single count assigned on the basis of measured energy, and the process is repeated for the next x-ray. The result is an x-ray spectrum plotted as number of x-rays, or counts, as a function of energy.

The EDS detectors detect elements with atomic number, Z, above boron ($Z = 5$) with 0.1 to 1.0 at.% sensitivity; more specialized "windowless" detectors are required to detect beryllium and lithium, while hydrogen and helium produce no characteristic x-rays and thus are undetectable. The WDS method also enables detection of all elements with $Z = 5$ and above, but with improved sensitivity (analysis down to approximately 0.01%), enabling superior quantitative analysis and resolving x-ray peaks that may overlap in EDS-produced spectra. However, WDS detectors are large, slow, and not compatible with commercial TEMs. Specialized SEMs can be designed to accommodate multiple WDS detectors; the instruments are commonly referred to as electron microprobe analyzers. A detailed discussion of WDS can be found in the article "Electron Probe X-Ray Microanalysis" in this Volume.

Sampling Volume for Various Signals

To correctly interpret the physical significance of various signals used in scanning electron microscopy, the volume below the surface from which the signal is originating must be known. Monte Carlo simulations (Fig. 28) depict 2000 electron trajectories for a 20 keV primary beam in iron. The BEs correspond to the red trajectories and help to provide an estimate of the BE signal volume. The BEs emerge from depths extending up to ~0.5 μm from the surface, according to Fig. 28. The x-ray signal can be presumed to emanate from a volume that corresponds to the entire interaction volume, as denoted by the blue trajectories. Therefore, although the electron spot size, d_s, may be 1 to 10 nm, x-rays are generated over a volume exceeding ~1 μm^3. Table 5 lists some estimates of sampling volume in iron for various signals generated by a 20 keV electron beam.

In general, some considerations must be made when quantitatively analyzing size and composition for particles with diameters less than approximately 1 to 2 μm (Ref 30). It is important to note that the x-ray sampling volume and shape vary with the electron beam energy and the sample atomic number. In general, higher voltages, lower density, and lower-atomic-number elements produce larger volumes, which tend to "balloon out" below the beam. Thin foil samples are used to reduce BE and x-ray generation volume. Sampling diameters <5 nm can be achieved simply by using thin foil samples (Ref 23).

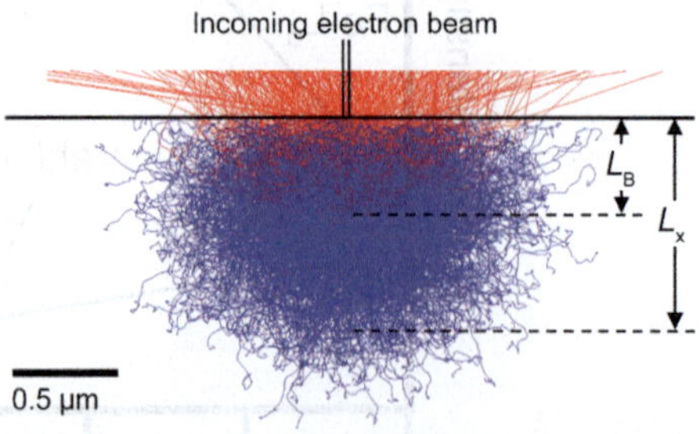

Fig. 28 Monte Carlo simulations of 2000 electron trajectories for a 20 keV primary beam in iron; red denotes trajectories that produce backscattered electrons, and blue denotes those that do not. Backscattered electrons emerge from depths extending up to ~0.5 μm from the surface.

Image Contrast

The primary types of image contrast in the SEM are topographic and compositional. Topography refers to the physical shape and arrangement of surface features. Composition refers to the identity, density, and distribution of chemical elements present in the specimen. The SEM image results from variation in signal intensity, S, as the beam moves from point to point on the sample surface. Information on composition and topography of the sample requires interpretation of contrast, $\Delta S/S$, between different points or areas in the image. Variations in the detected signal from two points on a surface can arise from two physically different mechanisms: emission (the number of electrons emitted from the surface) and collection (the number of electrons reaching the detector).

Consider the SE signal shown in Fig. 9(a). When the beam is at positions 2 and 4, the number of SEs emitted from the surface is roughly the same, because the beam/surface tilt angle is the same. However, because point 4 faces the detector and 2 does not, the SE signal is much higher at point 4. Consequently, in this example, the contrast between points 2 and 4 results entirely from differences in the number of electrons collected. When two points have similar line of sight to the detector, as is the case for positions 3 and 4 shown in Fig. 9 (a), the contrast results mainly from a variation in SE emission due to differences in the beam-surface tilt angle. Therefore, interpreting contrast requires careful consideration of both the effects of variations in electron emission and collection as a function of local topography and/or composition.

Topographic Contrast

Topographic contrast is often intuitive and can be directly interpreted as prominent features exhibiting higher signal intensity than recessed, buried, and shadowed features. Different phase constituents of a microstructure often have different etch rates. Therefore, microstructures can be studied, defined, and identified by a combination of metallographic sample preparation and SEM imaging using topographic contrast. The SEM micrographs

Table 5 Estimation of the volume of various signals produced in iron by a 20 keV electron beam

	Approximate volume dimensions	
Signal	**Diameter**	**Depth**
X-ray	~1.3 μm	~1 μm
Backscattered electrons (BE)		
Elastic (BE 1)	~d_s	...
Inelastic (BE 2)	~0.8 μm	~0.5 μm
Low loss (BE 3)	~d_s	...
Secondary electrons	~1.2 d_s	~10 nm

are usually recorded with the beam at normal incidence to the surface. Variations in detected signal attributed to topographic features are a function of variations in both collected and emitted electrons.

Collection Variation due to Topography

The SE/BE detector position and collection angle both influence topographic contrast, because surface features influence local collection efficiency for both SEs and BEs emitted from various points on the sample surface. Figure 29(a) depicts the primary beam at two positions relative to a topographic surface bump. At position (1), any emitted SE has clear line of sight to the SE detector (consider an Everhart-Thornley detector), because the surface is high and prominent. At position (2), some SEs can feel the positive +300 V potential at the detector and swing a wide arc to fly over the bump. However, many other SEs with an initial trajectory already toward the detector are reabsorbed by the bump. Therefore, position (1) has higher SE signal intensity than position (2) due to collection variations. Note that from the detector perspective, position (2) lies inside a shadowed region.

The same shadowing effect is also true for BEs. Many BEs emanating from the sample at position (1) in Fig. 29(b) have clear line of sight to a BE detector positioned similar to the SE detector. The collection variation due to topography is perhaps more pronounced for BEs than SEs if considering similar detector geometries. Higher-energy BEs do not change trajectory in the presence of a +300 V bias like the <50 eV SEs. Therefore, virtually none of the BEs emitted at position (2) can reach the detector in Fig. 29(b).

The resulting shadowing effect can provide intuitive assessment of surface topography. Raised areas of the surface inclined toward the detector are bright, and recessed and shadowed regions are dark. The resultant image appears as if the surface was illuminated by a light source located at the detector position. The degree of shadowing can be enhanced by lowering the collection angle. Everhart-Thornley detectors are conventionally mounted above, but to the side of, the specimen. Solid-state detectors (SSDs) positioned directly above the sample are sufficiently large enough in size to collect electrons at both low and high take-off angles. Many SSDs compartmentalize circular collection areas into segments such as quadrants and annular rings. This enables the user to select the electron signal from particular take-off angle ranges. In-lens detectors can only collect high-take-off-angle electrons, so shadowing effects and associated topographic contrast from collection variations are low.

Emission Variations due to Topography

The relative angle between the incoming primary electron beam and sample surface has a strong influence on the resultant emission of both SEs and BEs. As mentioned previously, it is standard in the SEM for the primary beam to be normal, or 90°, to the sample surface plane. Decreasing the incident angle of the primary beam serves to lift more of the interaction volume closer to the surface, as previously shown in Fig. 23 in the case of sample tilting.

Figure 30(a) depicts a similar situation for a topographic feature, where a larger fraction of the interaction volume is closer to the surface at position (1) than position (2). For the same SE escape depth in both situations, more SEs are emitted at position (1) than for position (2). Therefore, detected SE signal is higher at position (1) than for position (2) due to emission differences. Note that position (2) in Fig. 30 is located far enough away from the bump that the collected SE signal is comparable to position (1) in Fig. 29 (the top of the bump). Thus, Fig. 30(a) illustrates topographic contrast purely due to emission differences instead of collection differences. The same topographic contrast result also occurs for the case of BE emission, although the corresponding BE escape depth is larger.

Edge effect is a phenomenon where electron emission is significant as the primary beam approaches edges nearly parallel to the incident beam direction; edge effect is more pronounced at higher accelerating voltages. Comparing Fig. 30(a) and (b) illustrates the edge effect. Figure 30(b) depicts a higher-energy primary beam interacting with the same

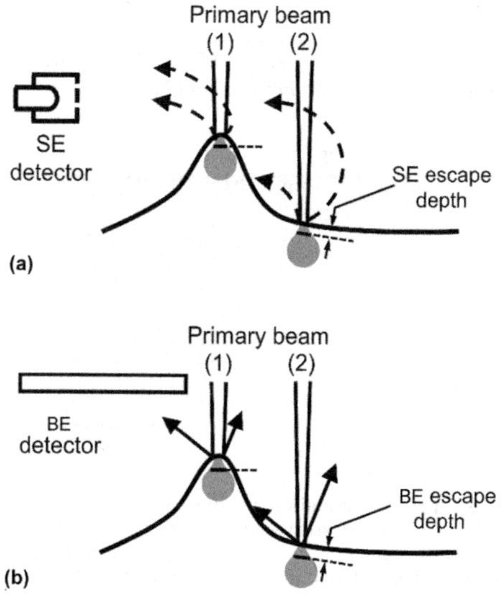

Fig. 29 Illustration of (a) secondary electron (SE) and (b) backscattered electron (BE) collection variations due to line of sight that produces topographic contrast during scanning electron microscope imaging. Position (1) produces more signal intensity than position (2).

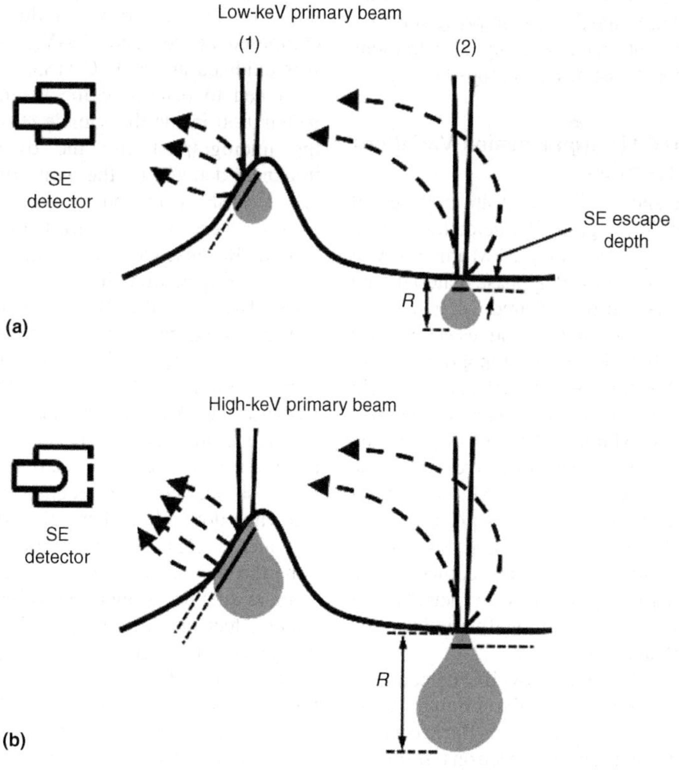

Fig. 30 Illustration of secondary electron (SE) emission variations for (a) low- and (b) high-kiloelectron volt (keV) primary beam energy. Topographic contrast is produced due to sample tilt effects on SE emission variations, such that position (1) produces more signal intensity than position (2). This effect is more pronounced at higher primary beam energies.

bump feature as Fig. 30(a). The interaction volume and corresponding penetration depth range, R, is larger for the higher-energy primary beam. However, the SE yield does not change significantly at position (2), because the interaction volume is extended away from the flat surface. However, increasing the interaction volume at an inclined surface extends the amount of interaction volume that falls within the SE escape depth. Therefore, increasing the energy of the primary beam (higher acceleration voltage) produces higher SE emission at position (1). In addition to the intuitive shadowing effect described previously, SEM imaging at higher (>15 kV) acceleration voltages produces a significant glow at steep edges due to the edge effect. The edge effect is less pronounced for BEs, because the average take-off trajectory from steep edges is often too low to reach the BE detector positioned somewhere above the specimen.

Compositional Contrast

Backscattered electron detection is advantageous for producing SEM images with compositional contrast. Compositional contrast exhibits relatively brighter and darker regions based on the local average atomic number and material density. Composition contrast is indirectly possible by SE detection even though SE emission is generally insensitive to atomic number. Enhancing overall SE collection yield improves the total signal, S, and thus can improve sensitivity to small but detectable signal variations induced by local changes in composition.

Backscattered Electron Emission Variations due to Composition

While SE and BE detection are both useful for topographic imaging, BE detection is also advantageous for producing SEM images with compositional contrast. Polyphase materials and complex alloys can be prepared using standard metallographic polishing techniques yet produce no surface relief. Therefore, topographic contrast by SEM imaging is minimal, and no insight on the microstructure can be obtained. Compositional contrast exhibits relatively brighter and darker regions based on the local average atomic number and material density. Microstructures composed of various constituent phases can be analyzed and identified based on the spatial variations in grayscale attributed to compositional contrast. Backscattered electron detection is more sensitive to compositional variations of a flat surface than SE detection.

Compositional contrast by BE detection is explained by considering the primary beam interaction with the specimen. Backscattered electrons are high-energy electrons from the primary beam that are scattered back out from the specimen surface. It is the individual atomic nuclei within the specimen that serve to scatter primary beam electrons. The scattering is a strong function of the physical size of the atomic nuclei, which is defined by the element and associated atomic number, Z. With higher Z, more primary beam electrons are scattered closer to the surface. Therefore, the overall probability of BEs escaping the interaction volume is increased, and higher BE emission is produced with larger Z.

Figure 31(a) illustrates composition contrast by BE imaging. Position (1) corresponds to a higher-Z region, which produces a smaller interaction volume than the lower-Z region corresponding to position (2). In Eq 9, R decreases with increasing Z and ρ. Because R is smaller in the higher-Z region, more of the interaction volume falls within the BE escape depth. Therefore, for the same primary beam electron energy, position (1) emits more BEs than position (2). Because BE detection relies on line of sight, a large-area BE detector position above the specimen surface can collect a large fraction of BEs produced at both positions. The result is clear contrast due to BE emission differences between the high-Z region (high BE signal intensity) and low-Z region (low BE signal intensity). Figure 31(b) presents measured SE and BE yield as a function of atomic number. The SE yield increases moderately with higher atomic number. However, BE yield increases significantly with higher atomic number.

Figure 31(c) illustrates the enhanced compositional contrast provided by BE imaging over SE imaging. The iron sample in Fig. 31(c) contains 3 wt% Y, forming a dual-phase alloy composed of iron and $Fe_{17}Y_2$. The specimen was oxidized at 1000 °C (1830 °F) and cross sectioned to determine the extent of oxygen penetration inside the sample (from the top in the micrographs) and the oxidized region thickness denoted by the white line in Fig. 31 (c). The SE image on the left in Fig. 31(c) exhibits very little contrast throughout the sample. Because the $Fe_{17}Y_2$ phase has a higher average Z than iron, it also has a higher BE yield. Therefore, the BE image on the right in Fig. 31(c) exhibits clear composition contrast between the $Fe_{17}Y_2$ (bright regions) and iron (dark regions) below the oxidation zone. Because $Fe_{17}Y_2$ preferentially oxidizes, the average Z differences between the two phases are reduced inside the oxidation zone, so the compositional contrast becomes weaker above the white dotted line. Thus, the BE image also provides a clear indication of the oxidation zone through compositional contrast.

Increasing the acceleration voltage has only a small effect on overall BE yield. The higher-energy primary beam produces larger R, deeper penetration, and larger interaction volumes. In the higher-Z region, a slight lateral extension of the interaction volume leads to slightly more area within the BE escape depth where BE emission can occur for position (1). So, a slight BE emission increase is possible with higher acceleration voltage for high-Z regions. However, the lower-Z region does not scatter the primary beam as efficiently as the higher-Z region. Therefore, much of the interaction volume progresses deeper into the low-Z region with corresponding increase in the BE escape depth, because the primary beam electrons have more energy. The result is a negligible change in overall BE emission with increased acceleration voltage for low-Z regions.

Figure 32 presents SEM micrographs of an as-polished (not etched) multiphase alloy acquired using a segmented, annular pole-piece BE detector. The insets in Fig. 32 depict which segment of the BE detector is used. Two phases, denoted as areas (1) and (2), exhibit different signal intensities due to compositional contrast in Fig. 32(a) and (b) when using either half-segment of the BE detector, A or B. However, the compositional contrast becomes stronger in Fig. 32(c) when both half-segments are used together, A + B. When segment B signal is subtracted from segment A, the resultant

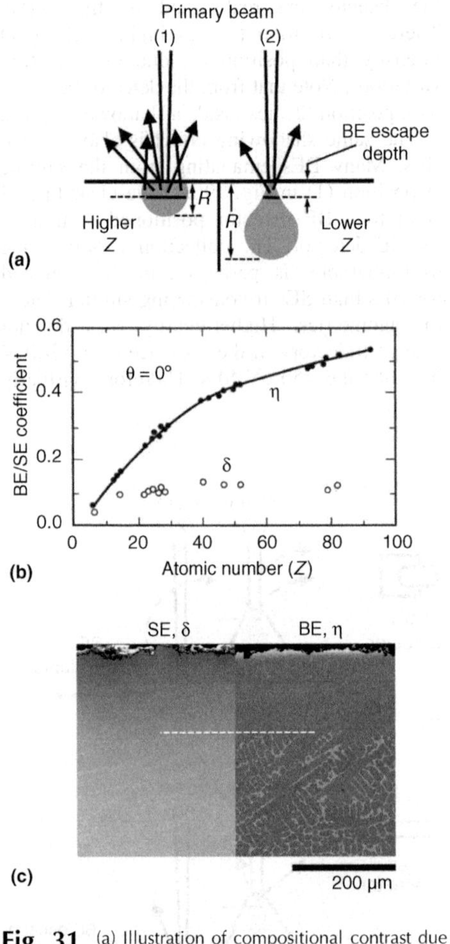

Fig. 31 (a) Illustration of compositional contrast due to backscattered electron (BE) emission variations from a flat surface with regions of higher and lower average atomic number, Z. (b) Plot of measured secondary electron (SE) and BE yield as a function of atomic number. Source: Ref 27. (b) Reprinted/adapted from Ref 27 with permission of Springer Nature. Copyright 2003. (c) Cross-sectional scanning electron micrograph of an oxidized dual-phase iron-yttrium alloy recorded by (left) SE and (right) BE detection. Courtesy of A. Weiss and B. Webler, Carnegie Mellon University

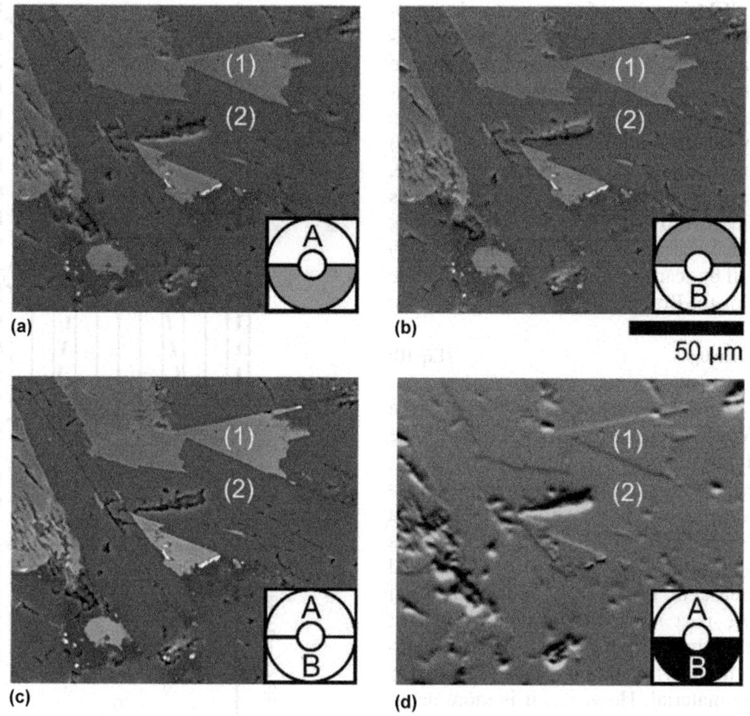

Fig. 32 Backscattered electron (BE) micrographs of a polished multiphase alloy surface recorded using a segmented, annular BE detector mounted to the scanning electron microscope pole piece. Insets indicate BE signal contribution from the detector segments: (a) A, (b) B, (c) A + B, and (d) A − B. Courtesy of Thermo Fisher/FEI

image shows no compositional contrast between the two phases. An elongated gouge near area (2) in the center of the image provides a useful topographic, fiducial feature. In Fig. 32(a), the lower edge of the gouge is bright due to higher BE emission, and the upper edge is dark due to lower BE collection when using the A segment. The topographic contrast is reversed for the case of the B segment in Fig. 32(b). Summing A and B segment signals together serves to reduce topographic contrast, as shown in Fig. 32(c), while subtracting B signal from A serves to enhance topographic contrast, as shown in Fig. 32(d). Topographic contrast in Fig. 32(d) is sufficiently strong enough to indicate that area (1) is raised in relation to area (2). This shows that signals from segmented BE detectors can be used to preferentially boost or suppress either contrast mechanism.

Secondary Electron Emission Variations due to Composition

Composition contrast is indirectly possible by SE detection even though SE emission is generally insensitive to atomic number, as was shown in Fig. 31(b). Secondary electron emission is a strong function of electrical properties of the material surface, so SE contrast between metals and insulators can be significant (Ref 31). Even variations in the doping level of semiconductors can be resolved by SE emission variations (Ref 32). Enhancing overall SE collection yield improves total signal, S, and thus can improve sensitivity to small but detectable signal variations induced by local changes in composition. Reducing acceleration voltage decreases the range of beam penetration, *R*. Consequently, more primary beam electrons undergo scattering closer to the near surface, improving overall SE emission. Thus, lowering the acceleration voltage for a very flat sample produces higher S, lower N, and potentially enough ΔS, which could be attributed to composition or general material variations.

Another way to enhance SE signal sensitivity is by optimizing the SE detector position. Like BE emission, SE emission increases at higher take-off angles. In-lens detectors are optimally positioned to detect these high-take-off SEs. Shadowing due to line-of-sight effects is mitigated by the detector geometry, so any competing contrast from specimen topographic effects is relatively low. It is possible to optimize the collection of SEs by using a combination of low (<10 kV) acceleration voltage, smaller WD, and in-lens/in-column detectors. This approach helps to enhance point resolution, because high-take-off SEs are predominantly SE(1) type. Although this combination of approaches also improves SEM sensitivity to possible local SE emission variations due to composition, great care must be exercised in attributing SE signal variations to composition (Ref 31). Many other physical characteristics of the sample surface can influence SE emission, as discussed in the next section.

Other Contrast Mechanisms

In addition to topography and composition, several other contrast mechanisms are directly related to SE and BE emission variations. Contrast mechanisms described are readily accessible in conventional SEMs using typical SE and BE detectors. This section describes SEM contrast due to a variety of material surface properties, including surface charge, surface potentials, magnetic properties, and crystal structure/orientation.

Charge Contrast

During SEM imaging, the sample is bombarded by negatively charged electrons. If unable to dissipate, electrons accumulate and produce electrostatic charge of the surface. Negative charge buildup lowers the primary beam landing energies and enhances SE emission. Charge buildup can be stochastic and fluctuate the SE signal during beam raster across insulating surfaces. The resulting charge contrast is generally unwanted in SEM images and can be problematic when imaging nonconductive samples. Generally, nonconductive samples are coated with a thin layer of an electrically conducting material to avoid charging effects. However, charge contrast can arise even in nominally conductive samples, such as a steel containing ceramic inclusions. Despite being flat and having lower average atomic number, an insulating particle, such as an alumina (Al_2O_3) inclusion in steel, exhibits higher signal due to charge contrast than the surrounding iron matrix. Charge contrast must be considered when SEM imaging composite materials containing nonconductive components. Charge contrast can be mitigated by using a higher acceleration voltage, lower beam current, and faster scan speed. Backscattered electron imaging is less susceptible to charge contrast than SE imaging.

Voltage Contrast

Electrically isolated or floating structures at the specimen surface can produce different surface potentials. As mentioned previously, the presence of a surface potential alters the landing energy of the primary beam in the SEM. At low SEM acceleration voltages (<5 kV), any variation in the landing energy of the primary beam makes detectable changes in the SE yield. Therefore, local SE emission variations, or voltage contrast, can be used to image and even measure surface potentials for various electrical structures and devices. In-depth studies of electrical device behavior can be performed by applying an external bias to device elements and spatially measuring surface potentials via voltage contrast SEM. Another approach to obtain voltage contrast in the SEM is by measuring SE energy distribution using an energy analyzer. Applied voltage to a specimen induces a systematic, measurable shift of the entire SE energy spectrum. More details regarding voltage contrast in the SEM can be found in Ref 33.

Magnetic Contrast

It is possible to obtain images of magnetic domains in the SEM. Ferromagnetic materials are composed of small subgrain-sized regions termed domains. In every domain, the magnetic moment of each electron has a common direction along a certain crystallographic axis, which is often termed the easy axis. In certain crystals, such as cobalt, there is only one easy axis; such crystals are termed uniaxial crystals. In the crystals, the magnetic moment at a surface often has a component normal to the surface, which means that a small magnetic flux, **B**, will "leak" out of the surface. Thus, an SE ejected from the surface of a uniaxial crystal with velocity v experiences a Lorentz force $v \times \mathbf{B}$. Because the local external flux changes sign over each domain, the change in the Lorentz force causes the SE detector signal to change when the beam moves over each domain; thus, images of domains can be obtained using the SE signal. This contrast is known as type I magnetic contrast.

In most ferromagnetic crystals, there is more than one easy axis; for example, in α-iron they are [100], [010], and [001]. In these crystals, closure domains form at the surface that have their moment lying along the easy axis most closely parallel to the surface, which greatly reduces the magnetic flux leakage outside the surface. Domains are revealed by type II magnetic contrast in these crystals. Inside the metal, there is an abrupt change in the magnetization direction at a domain boundary. This means that the direction of **B** switches from domain to domain inside the metal. Therefore, after entering the sample, a primary electron experiences a Lorentz force in different directions from domain to domain, which results in changes in the backscattering yield as the primary beam sweeps across a domain boundary. Therefore, the BE detector is used to detect type II magnetic contrast. More details regarding magnetic contrast can be found in Ref 29 and 34. Detailed information on techniques used to observe domain structures can be found in the article "Microstructure and Domain Imaging of Magnetic Materials," *Metallography and Microstructures*, Volume 9, *ASM Handbook*, 2004.

Channeling Contrast

Channeling contrast relates to emission variations due to the local crystal structure and orientation of the specimen. Figure 33 illustrates the mechanism for electron channeling contrast. Consider the case of low-magnification (<100×) SEM imaging of a monoatomic, topographically flat, nominally strain-free single-crystal sample with a crystallographic surface plane oriented normal with respect to the incoming primary electron beam. Because of the cubic single-crystal orientation, at least one set of (*hkl*) planes lies vertically with respect to the surface, as illustrated in Fig. 33(a). As the beam scans from positions 1 to 4, it scans over angular range, γ, while also warning an angle θ with this (*hkl*) plane set. If θ_B scans over angular range, γ, while also varying the Bragg angle for diffraction, then:

$$n\lambda = 2d_{hkl} \sin \theta_B \qquad \text{(Eq 10)}$$

where λ is the wavelength of the electron beam. Wavelength varies inversely with the acceleration voltage.

Figure 33(b) illustrates BE emission variations due to electron channeling plotted as a function of angle θ. As the primary beam scans across θ_B at points 2 and 3, a strong drop and then increase in BE yield results. The term *channeling* is a slight misnomer, because it implies that the primary beam electrons channel down the (*hkl*) planes and penetrate deeper within the material. However, it is more accurate to regard the electrons as having a stronger interaction with the vertical planes when $\theta < |\theta_B|$; this produces elevated BE yields (Ref 35). Therefore, when the beam lies between positions 2 and 3, the BE signal is higher over a spatial angular width, *B*, with strong dips in signal intensity just outside positions 2 and 3.

Figure 33(c) presents an experimental BE image obtained at low magnification of a single-crystal aluminum specimen with the (111) crystallographic surface oriented normal to the incoming 30 keV electron beam. The black cross denotes θ = 0°, where the electron beam is exactly parallel to the [111] direction. Assuming the (111) surface is exactly normal to the electron-optic axis, this point also corresponds to γ = 0°. Thus, this position is considered the center of the electron channeling pattern. Three pairs of channeling lines are clearly observable, and each line corresponds to a specific set of aluminum {220} planes. The white dotted line in Fig. 33(c) corresponds to the BE intensity profile illustrated in Fig. 33 (b). Using Eq 10, θ_B is ~0.7° for the aluminum {220} family of planes. Therefore, the channeling bandwidth, *B*, for each pair of channeling lines corresponds to the $2\theta_B$ value, or 1.4°. This illustrates that the electron channeling pattern (ECP) provides angular information regarding the primary beam trajectory with respect to crystallographic planes. Any location along a single channeling line represents satisfaction of the Bragg condition for that specific set of planes. For example, points 2′, 2a′, and 2b′ in Fig. 33(c) represent positions where the primary beam is at an angle θ_B with respect to the (−220) planes.

The crystallographic orientation of the single-crystal region beneath the beam can be evaluated from the ECP. Decreasing image magnification increases the range of scan angle, γ, as well as the image width, *W*. The result is a broader crystallographic view of the single-crystal sample. Considering typical SEM acceleration voltages of 2 to 30 kV, the nonrelativistic wavelength can vary from 0.007 to 0.027 nm. Because the values are an order of magnitude smaller than typical crystallographic plane spacings, d_{hkl}, the Bragg angles from Eq 10 range in values from ~0.3 to 5°. Therefore, only very low-magnification (<100×) imaging can swing the beam over an angular range, γ, large enough to span $2\theta_B$

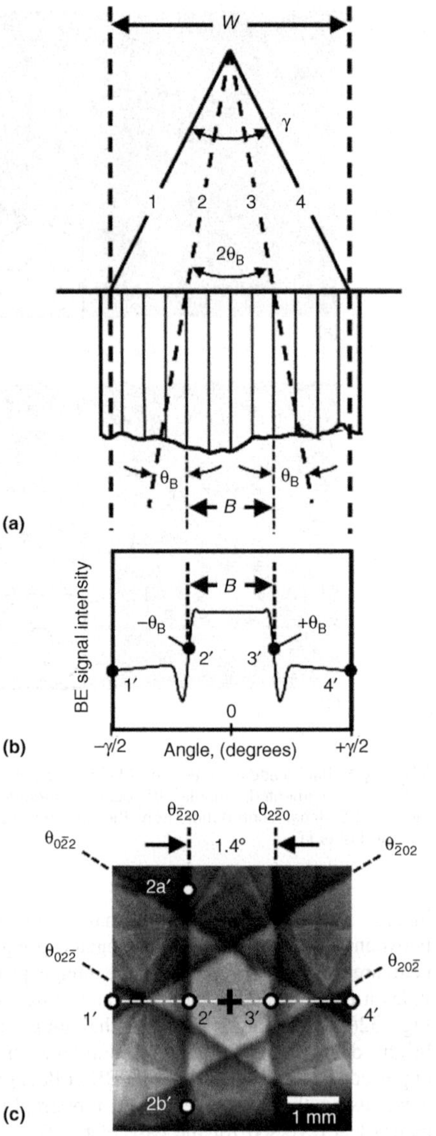

Fig. 33 Electron channeling contrast. (a) Illustration of low-magnification, underfocused imaging of a single-crystal sample. (b) Resultant backscattered electron (BE) signal variation due to electron channeling. (c) Experimental low-magnification BE image of an aluminum (111) single crystal exhibiting channeling contrast that results in an electron channeling pattern. Vertical {220} planes with respect to the aluminum (111) surface are identified. Courtesy of J. Kamaladasa, Carnegie Mellon University

for any set of crystallographic planes. Electron channeling patterns are essentially the same as Kikuchi patterns obtained from thin foil samples in transmission electron microscopes. Just as Kikuchi patterns enable evaluation of crystallographic orientation more precisely than is possible with the electron diffraction pattern, the ECP enables more precise orientation evaluation than an x-ray Laue pattern. In general, ECPs for phase and orientation determination have not seen development as compared to the more established and commercialized electron backscatter diffraction approach, described in the section "Special Techniques" in this article.

Channeling contrast is surface sensitive, because signal contribution is dominated by BEs with little or no energy loss. Therefore, crystallographic information is carried by electrons scattered from depths below the surface of only 10 to 100 nm, so sample preparation is important. For metals, the worked surface layer produced by standard metallographic specimen-preparation techniques degrades and, in some cases, eliminates ECPs due to the variation in d_{hkl} produced by residual strains. With more brittle materials, such as intermetallic compounds and semiconductors such as germanium and silicon, residual strains are less of a problem. Because the effect that produces channeling contrast involves BEs with minimal energy loss, channeling patterns and channeling contrast are better observed using dedicated BE detectors, where maximizing overall BE collection efficiency and increasing signal gain is critical.

Figure 34 illustrates channeling contrast as a method to observe grain size and twinning for a polished, highly flat polycrystalline copper sample. Grain boundaries would not be visible in the optical microscope and may only vaguely be detected by the SE imagining in the SEM. The contrast shown in Fig. 34 is due to the different crystallographic orientations of the various grains, which result in grayscale variations across the BE image. As the sample is tilted or rotated, the BE emission for each grain changes. For the largest grain located at the bottom-center in the images, a 1° stage tilt is enough to cause inversion of the dark/light contrast compared with microtwins within the grain. Examples 3 and 6 in the section "Applications" in this article show BE micrographs exhibiting orientation contrast by electron channeling.

Material extended defects, such as dislocations and stacking faults, induce localized atomic structure variations due to elastic strain, which results in subtle orientation changes detectable by electron channeling contrast only when the beam is oriented near a diffraction condition. Used as an imaging technique, electron channeling contrast enables SEM-based defect imaging and identification. The electron channeling contrast imaging (ECCI) method relies on controlled orientation of the incoming electron beam to a specific Bragg angle, θ_B.

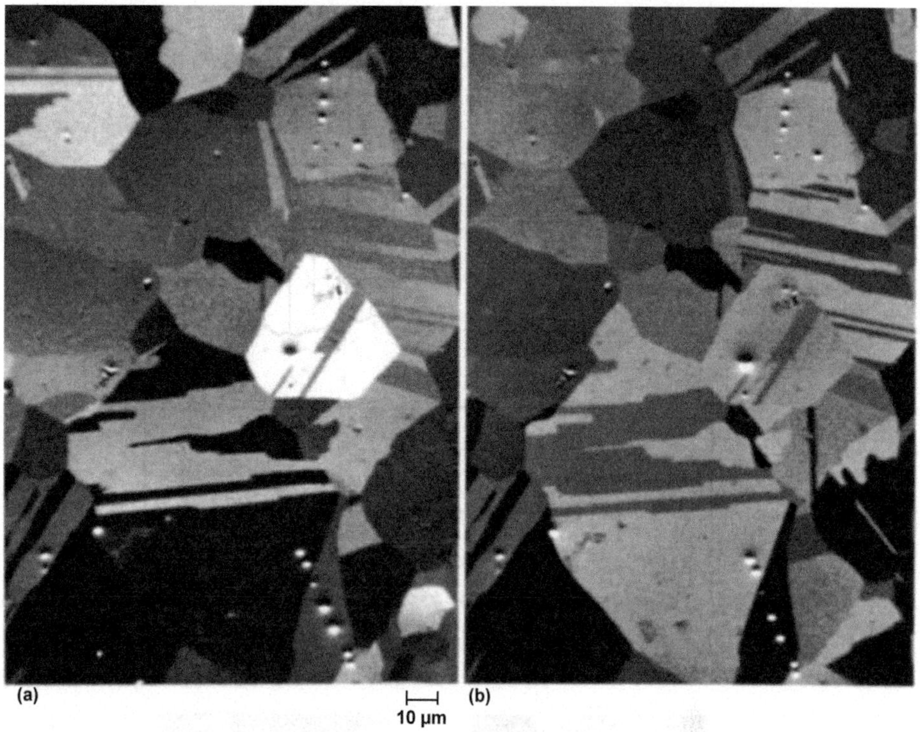

Fig. 34 Backscattered electron micrographs of electrolytically polished polycrystalline copper exhibiting orientation contrast due to electron channeling effects. A 1° tilt produces significant changes in the grain contrast between (a) and (b). Reprinted/adapted from Ref 14 with permission of Springer Nature. Copyright 1998

Any local lattice strain that bends the specific set of crystallographic planes corresponding to θ_B yields BE contrast, as illustrated in Fig. 35.

Figure 35(a) illustrates high-magnification SEM imaging of a single-crystal surface containing a single near-surface edge dislocation. The dislocation produces elastic strain that varies the local crystallographic plane orientations with respect to the incoming electron beam. An angle, $\Delta\theta_S$, denotes this orientation variation induced by the defect. The beam is scanned across this defect at an initial trajectory that satisfies the Bragg condition for these planes: $+\theta_B$ at points 1 and 5, and $-\theta_B$ at points 6 and 10. The resulting BE yield is presented in Fig. 35(b) as a function of angular deviation from the Bragg angle, θ_B. Points 2 and 9 in Fig. 35(a) correspond to conditions where the crystallographic planes are oriented more parallel to the incoming electron beam. This leads to a higher BE yield, as illustrated for points 2′ and 9′ in Fig. 35(b). Likewise, there is a drop in BE yield for points 4 and 7 where crystallographic planes are bent away from the Bragg condition toward a direction less parallel to the incoming electron beam trajectory.

Figure 35(c) presents BE images of near-surface dislocations in a $SrTiO_3$ (100) single-crystal sample. The dark-light vertical line feature is a near-surface dislocation parallel to the surface; spot features denote surface-penetrating dislocations inclined 45° to the (100) surface. The observed channeling contrast of the dark-light vertical line correlates to the BE intensity profiles illustrated in Fig. 35(b). It should be noted that the change in sample tilt/orientation is only ~1° to go from $+\theta_B$ to $-\theta_B$, yet the features in Fig. 35(c) exhibit a complete reversal in BE intensities. Thus, features cannot be attributed to topographic contrast or composition contrast. The directionality of the BE intensity fluctuation depends on whether the incoming trajectory is initially at $+\theta_B$ or $-\theta_B$ for a specific set of crystallographic planes.

The implication regarding channeling contrast is that the user must select which set of planes will yield channeling contrast. This is accomplished by obtaining an ECP and adjusting the sample orientation so the channeling line is centered on the optic axis. This ensures that the primary electron beam is nominally at an angle corresponding to θ_B. The diffraction vector, **g**, defines the direction normal to the planes selected for channeling contrast. A series of ECCI micrographs can be recorded at different channeling conditions for dislocations of interest. Based on the contrast behavior as a function of **g**, dislocations can be identified by their Burgers vector either through comparisons to dynamical diffraction simulations (Ref 36) or through the use of the invisibility criterion (Ref 37).

Because it is necessary to obtain crystal orientation information from a localized region,

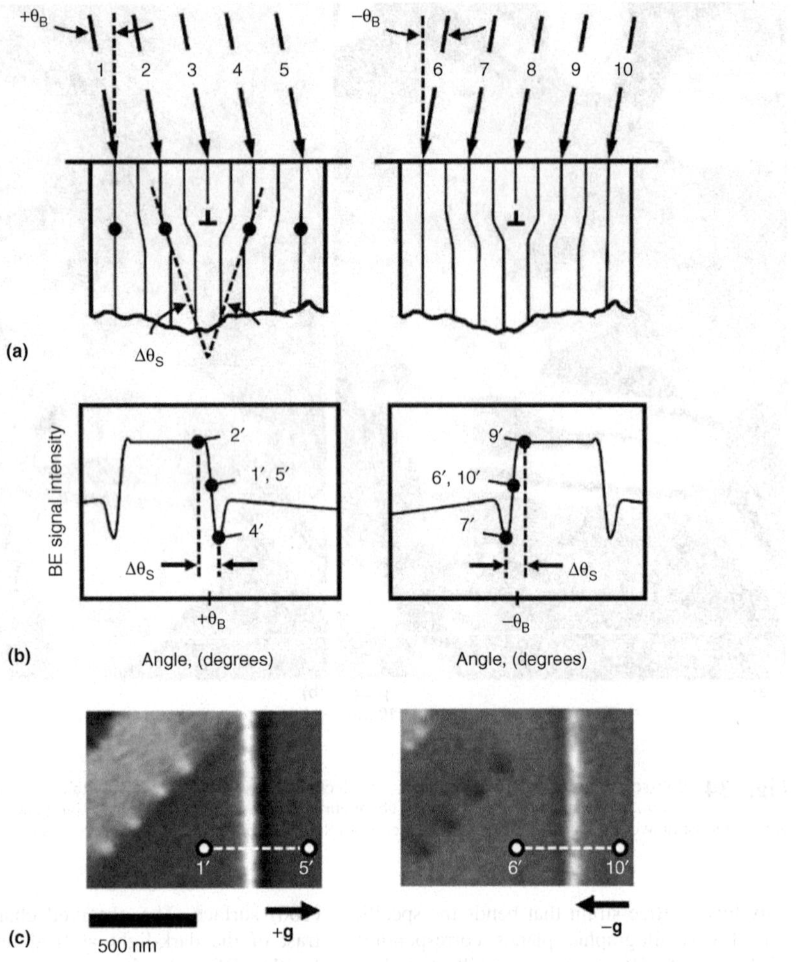

Fig. 35 (a) Illustration of scanning electron microscope imaging of a single-crystal surface containing a single near-surface edge dislocation producing local orientation variation, $\Delta\theta_S$. (b) Resultant backscattered electron (BE) yield as a function of beam position. (c) Experimental electron channeling contrast imaging micrographs of near-surface dislocations in a $SrTiO_3$ (100) single-crystal surface imaged under two different tilt conditions, corresponding to trajectory angles $+\theta_B$ and $-\theta_B$, or **+g** and **-g**. Experimental micrographs courtesy of J. Kamaladasa, Carnegie Mellon University

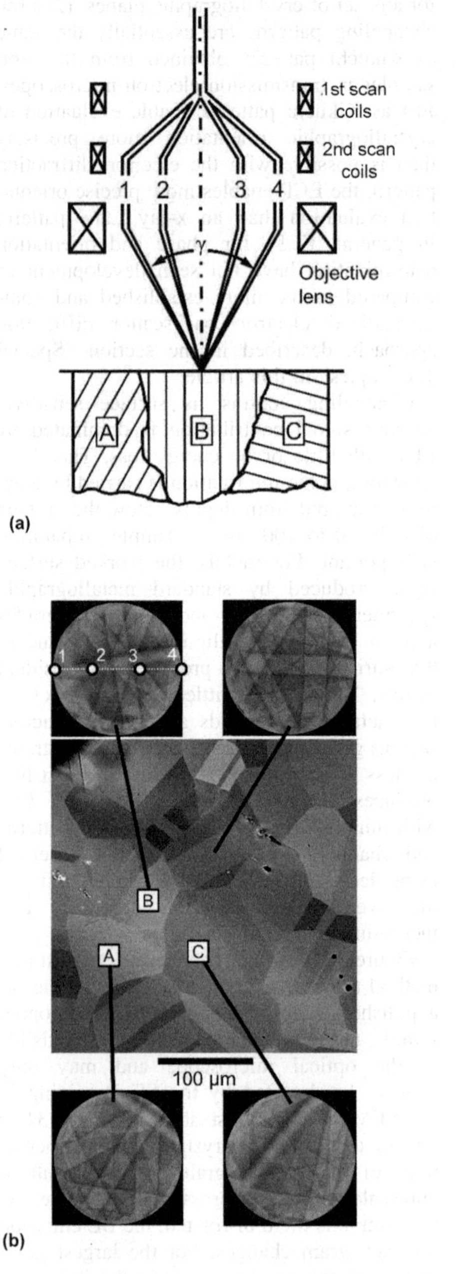

Fig. 36 (a) Illustration of beam rocking to produce a selected-area channeling pattern (SACP) from a single grain (grains denoted by A, B, and C). (b) Backscattered electron micrograph of a polished steel sample with four example SACPs acquired from individual grains. Courtesy of S. Singh and M. De Graef, Carnegie Mellon University

the SEM must acquire an ECP for a specific location. Therefore, the SEM must swing the incoming electron beam across a wide angle, γ, but consistently sample a selected area of the specimen surface. An example electron-optics arrangement is shown in Fig. 36(a), which can accomplish the selected-area channeling pattern (SACP) acquisition using a conventional SEM. Both sets of scan coils coordinate with the objective lens to maintain a consistent rocking point at the sample surface. Rocking angles of ~5 to 15° are possible for selected areas from 20 μm down to <1 μm in size (Ref 38). Figure 36(b) shows an example BE micrograph of a steel sample where individual grains exhibit orientation contrast due to channeling. The SACPs acquired from four different grains highlight the orientation differences from grain to grain. Some commercial SEMs are designed with a beam-rocking or SACP mode to facilitate the use of channeling contrast.

Special Techniques

Special techniques include electron backscatter diffraction capable of determining local crystal structure to map microstructure; electron beam induced current, which is useful for semiconductor device characterization, defect imaging, and failure analysis; cathodoluminescence, where detected photons have energies characteristic of the local sample electrical and optical properties; in situ and in operando methods for imaging and analysis of specimens under specific conditions to observe dynamic phenomena, such as phase transitions, surface chemical reactions, and microstructural transformations; low-voltage SEM analysis performed using a low acceleration voltage that leads to improved detail of fine surface features when detecting SEs or BEs; and commercial TEM detectors inside the SEM chamber to facilitate transmission-mode imaging for enhanced image resolution sufficient to resolve atomic structure and produce diffraction contrast images.

Electron Backscatter Diffraction

Electron backscatter diffraction (EBSD) is an SEM-based approach capable of determining the local crystal structure, including phase and orientation, to map microstructure. Electron backscatter diffraction is used to produce a variety of maps that provide tremendous insight on the crystallographic properties of materials, including elastic strain, plastic deformation, grain size, grain boundaries, geometrically necessary dislocations, and texture. While predominantly applied for investigating metal microstructures and geological samples, EBSD is being increasingly used for semiconductors and nanomaterials.

Electron backscatter diffraction, illustrated in Fig. 37, uses a two-dimensional scintillator detector and a charge-coupled device camera to image the angular spread in BE intensity emanating from a single point. Because the electron beam undergoes both inelastic and elastic scattering within the BE escape depth of the specimen, the projected BE intensity on an area detector produces a Kikuchi pattern. The pattern contains bands, each of which represents a set of crystallographic planes, similar to the channeling bands described in the earlier section "Channeling Contrast" in this article. Band configurations can be used to identify both the crystallographic phase and the orientation of the local crystal structure. Point-by-point acquisition of multiple Kikuchi patterns produces a comprehensive map of crystallographic structure for in-depth analysis of material microstructure.

Typically, the sample is tilted ~70° with respect to the incoming electron beam; the EBSD camera is brought within a few centimeters of the inclined sample surface. This geometry is advantageous, because the forescattering electron yield is substantially higher than for a backscattered geometry at 0° sample tilt, as illustrated in Fig. 23 and demonstrated in Fig. 22. Gain control and background subtraction are used to improve the Kikuchi pattern quality. Conventionally, a Hough transform is applied to an acquired Kikuchi pattern so individual bands can be identified. The user proposes possible crystal structures, and the software compares the configuration of the detected bands to a look-up table of interplanar angles for the phases of interest. A best match is determined, and the corresponding orientation is assigned to each pixel position. The orientation is assigned based on three Euler angles, which define the necessary rotation of the local crystal structure to the sample reference orientation. As such, each individual point is assigned both a phase and orientation to produce a map of crystal structure. Adjacent pixel positions with identical phase and similar orientations (within some user-defined threshold misorientation angle) can be assigned to a single grain. Larger changes in orientation for adjacent pixels define the location of grain boundaries. See Examples 2 and 3 in the section "Applications" in this article for cases using EBSD analysis. More details on the EBSD method can be found in the article "Crystallographic Analysis by Electron Backscatter Diffraction in the Scanning Electron Microscope" in this Volume.

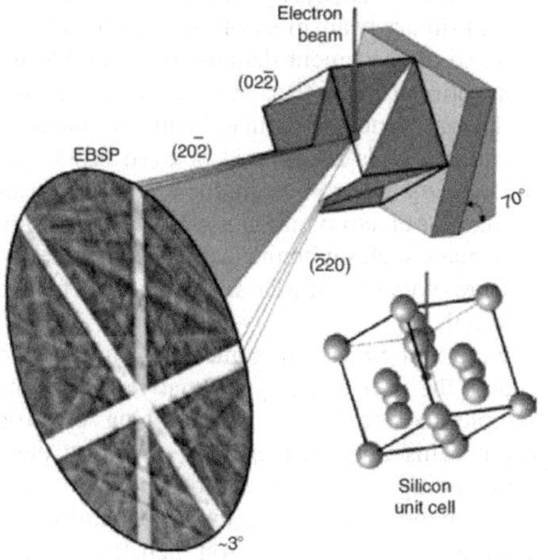

Fig. 37 Illustration of electron backscatter diffraction method where a sample is tilted 70° and the band configurations within an acquired electron backscatter pattern (EBSP) indicate the local crystal orientation. Reprinted/adapted from Ref 39 with permission of Springer Nature. Copyright 2006

Electron Beam Induced Current

This technique is useful for semiconductor device characterization, defect imaging, and failure analysis. Electron beam induced current (EBIC) works by measuring the current between two electrodes in contact with the semiconductor sample. The current is induced by the electron beam as it is scanned across the contact-semiconductor interface and used to modulate grayscale intensity when producing a digital image. When the electron beam strikes the sample surface, electron-hole pairs are created within some interaction volume. If the semiconductor contains an internal electric field near these charge carriers due to, for example, a depletion layer at a *p-n* junction, as shown in Fig. 38, the charge carriers create a current between the two attached electrodes.

To study crystals that do not contain *p-n* junctions, a surface Schottky barrier is fabricated on the crystal surface by deposition of a thin metallic layer. The thickness of material beneath the junction depleted of mobile carriers is a function of the applied reverse bias voltage. Crystal defects and composition gradients within this zone that affect the charge dissipation current can be imaged by EBIC. The current used in the EBIC technique is different from the current flow to ground used for specimen current detectors.

It is possible to obtain images of localized crystal defects such as dislocations and stacking faults. In addition, crystal growth defects, such as dopant inhomogeneities, can also be imaged. Factors that control image formation are reviewed in Ref 40. Many electrical parameters of semiconductor devices can be evaluated by using quantitative EBIC techniques, for example, doping level, diffusion length and lifetime of minority carriers, mobility and surface recombination velocity, surface potential, electric field, multiplication factor, and ionization coefficients. In addition, the EBIC technique is used for failure analysis and device diagnostics in such ways as locating a defective device in an array, locating a defect in a device or junction, or locating an irregularity in a junction. A review of EBIC is provided in Ref 41.

Cathodoluminescence

Cathodoluminescence (CL) is the emission of light by a solid under electron beam bombardment. A sufficiently high-energy electron beam excites electrons within the material to elevated energy states that emit photons as they decay back down to their original energy (Ref 42). The photons sit in the visible, near-infrared, and ultraviolet regions of the spectrum with photon energies ranging from 0.5 to 6 eV. Furthermore, the photon energies are

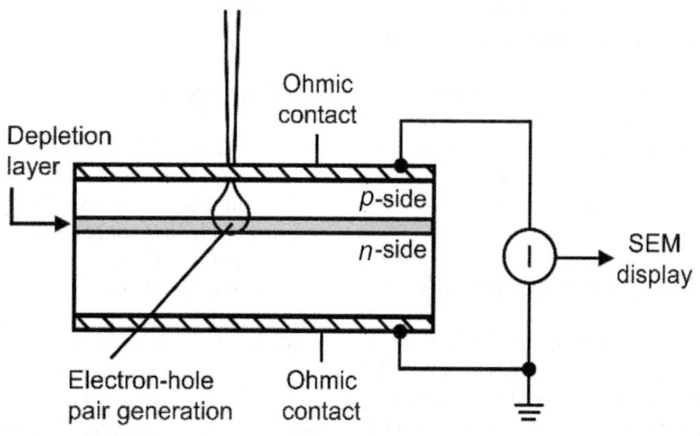

Fig. 38 Illustration of electron beam induced current setup for a *p-n* junction. SEM, scanning electron microscope

characteristic of the local sample electrical and optical properties. To use CL in the SEM, the light given off by the sample under electron beam irradiation must be collected using a special photon analyzer inside the specimen chamber. This photon analyzer/collector converts the emitted light intensity to an electrical signal. The detected light intensity can be modulated to produce a digital image or measured as a function of photon energy for spectral analysis.

Cathodoluminescence can be performed in the SEM using different modes and detector configurations. By implementing a photomultiplier tube or solid-state diode (SSD), the overall light signal, regardless of color, can be measured as a function of electron beam position to produce two-dimensional panchromatic images. An optical filter or exit slit enables the user to select a specific wavelength or photon energy of interest, thus resulting in monochromatic images. A CL spectrometer disperses all emitted light from a single position, enabling the recording of light intensity as a function of wavelength for one specific pixel. A series of CL spectra recorded at multiple pixel positions produces a three-dimensional, hyperspectral data volume, which can be further analyzed postacquisition. See Example 4 in the section "Applications" in this article for an example monochromatic CL image.

Cathodoluminescence is applicable for any material with an energy gap between the valence band and the conduction band, which applies to semiconductors and insulators. In the case of semiconductors, electron-hole pairs are generated by the energetic electron beam, and the radiative recombination of those electron-hole pairs produces a light signal indicative of the semiconductor bandgap and other optoelectronic properties. The physical basis of CL in semiconductors is comparable to EBIC, being governed by generation, motion, and recombination of excess charge carriers. Therefore, with semiconductors, CL is broadly applied for measuring electronic bandgap, imaging dislocations, and evaluating minority carrier diffusion length and life (Ref 43). It is often used as a complementary tool with EBIC for semiconductor materials and devices. Ceramics, minerals, and gemstones also yield useful information when analyzed by CL (Ref 44). Trace elements present at concentrations too low for detection by EDS or WDS can often be detected and even mapped by CL. Minerals can be dated for geochronology based on CL characteristics (Ref 45).

In Situ and In Operando Methods

For SEM, *in situ* and *in operando* refer to the imaging and analysis of specimens under specific conditions. It may be desirable to perform SEM analysis of specimens at a specific temperature, under a certain strain condition, under applied voltage, or within a liquid/gas environment. In this way, dynamic phenomena, such as phase transitions, surface chemical reactions, and microstructural transformations, can be observed.

Commercial in situ stages are available to apply one or more combinations of external loads. Hot stages are designed to elevate the specimen temperature to as high as 2500 °C (4530 °F) (Ref 46). Water cooling and heat shielding could be required, depending on the target temperature. Cold stages can bring specimens down to cryogenic temperatures, usually through Peltier cooling, liquid nitrogen, or liquid helium. Tensile and compressive stress testing is possible using commercial mechanical stages (Ref 47). These systems typically use a force transducer to monitor and measure applied force and displacement for producing stress-strain measurements. Other mechanical stages provide movable tips and appendages to controllably manipulate features on the specimen surface with submicrometer control. The nanomanipulators can also be used to apply voltage or current to a specific location in a controlled manner. Alternatively, stages can be equipped with prepatterned electrical leads so the specimen can be placed in electrical contact with an external voltage/current source for in situ biasing studies. See Example 6 in the section "Applications" in this article for an example study using a hot stage inside an SEM.

Specialized stages are produced with environmental cells that enable the flow of liquids or gases inside the SEM (Ref 19). The cells are encased in a thin membrane capable of retaining the internal pressure of the cell. The membrane also enables transmission of the primary electron beam to the sample as well as signals to the SEM detectors. The user has external control of the interior cell pressure or fluid flow rate during SEM analysis. Noise reduction is crucial under these circumstances because of extensive unwanted scattering due to the cell membrane and the fluid surrounding the specimen. In cases where actual operating conditions (temperature, applied voltage, gas environment) for a specimen are replicated inside the SEM, the analysis is referred to as in operando.

Low-Voltage SEM

Low-voltage SEM generally refers to analysis performed using an acceleration voltage lower than 5 kV (Ref 48). The primary effect of using such low-energy electrons is significant reduction in the interaction volume, which leads to improved detail of fine surface features when detecting SEs and BEs. Tilt influence on SE yield decreases with lower voltage, leading to reduced edge effect and improved definition of surface features. For beam energies less than 1 keV, BE escape depths approach the same scale as SE escape depths, leading to BE image resolution comparable to SE imaging (Ref 49). Other advantages of low-voltage SEM include the ability to image nonconductive specimens, analyze beam-sensitive materials, and access other contrast mechanisms linked to material properties such as electrical conductivity. More details regarding low-voltage SEM can be found in Ref 48 and 49.

Nonconductive specimens present a challenge for conventional SEM due to unwanted charge buildup while the electron beam impacts the specimen. At high voltages, the number of electrons accelerated into the target specimen exceeds the number of BEs and SEs produced. Therefore, excess negative charge accumulates at the site of imaging unless the charge is conducted to ground, either by the specimen itself or with the help of a conductive surface coating. In the absence of a pathway to ground, the buildup of negative charge serves to decelerate incoming electrons and reduce beam penetration to the sample. Lowering incident electron energy has the effect of increasing overall BE and SE yield until a balance is reached and charge is conserved. The critical energy value, E_2, is an important parameter for low-voltage SEM (Ref 18). The E_2 value is material specific (ranging from 0.5 to 3.0 keV), and represents the desired electron beam energy that ensures charge balance and thus optimal imaging in the SEM for nonconductive specimens. The E_2 values for a variety of glasses, ceramics, and polymers are provided in Ref 18 and 48.

Low-voltage SEM also provides an advantage in reducing the possibility of electron beam induced damage to the specimen, including electrostatic charging, atomic displacement, and specimen heating. Excess charge accumulation can produce electrostatic forces strong enough to modify features on the sample surface. Therefore, attaining charge balance during imaging is a first step in minimizing surface alterations. Damage by individual atom displacement is statistically unlikely to occur when using primary beam electron energies below 100 keV, so displacement damage is negligible for conventional SEM (Ref 50). However, atomic/molecular structural damage can also occur if chemical bonds are altered or even broken by the incident electron beam due to inelastic scattering. For electron energies <30 keV, this form of damage scales with the incident beam energy, so low-voltage SEM is best for sensitive materials.

Note that SEMs are sometimes equipped to enable the user to apply a negative bias at the sample to decelerate the incoming electron beam. This approach enables obtaining electron landing energies down to the electron volt range without sacrificing image resolution. At such low energies, electron penetration into the sample may only extend 1 nm or less. Therefore, contrast is strongly influenced by surface chemistry and surface electrical/magnetic properties. Interpretation of image contrast at such low landing energies is still being explored (Ref 49).

Transmission-Mode SEM

Historically, the primary distinctions between SEM and TEM were the energy range of the primary electron beam and the detector geometry. The SEM uses 5 to 30 keV beam energies compared with the 80 to 300 keV range for conventional TEM. The higher beam energy for TEM also means a preference for positioning detectors below the specimen, because beam-sample interactions predominantly result in very low-angle scattering. Commercial TEM detectors (Fig. 8) were introduced inside the SEM chamber to facilitate transmission-mode imaging. Advantages of conducting SEM in transmission mode include enhanced image resolution sufficient to resolve atomic structure (Ref 51) and producing diffraction contrast images (Ref 9). See Example 5 in the section "Applications" in this article for further demonstration of transmission-mode SEM.

Scanning Electron Beam Instruments

Historically, the SEM was an advanced instrument requiring significant training and extensive maintenance. Conventional SEMs required mounting the electron-optics column and sample chamber on an air table with a standard desktop personal computer adjacent to the microscope for user control. These instruments had to be housed in a dust-free environment, preferably on the ground/basement level within a building. Users had to manually control many aspects of the instrument during imaging. Specimens had to be vacuum compatible or risk significant degradation to the contamination-sensitive electron-optical components.

Developments in SEM instrumentation during the 1990s and 2000s produced compact SEMs, which are smaller and more robust instruments, including some that are portable for remote analysis. More powerful computers, automated stages, and automated SEM control enabled significant advances in high-throughput analysis. Focused ion beams were combined with SEM to produce dual-beam instruments. Specially designed SEMs, called variable-pressure/environmental SEMs, are equipped with differential pumping so the specimen chamber can operate under higher background pressure environments, a capability that accommodates high-vapor-pressure specimens inside the SEM, such as biological samples, polymers, and liquids.

Compact SEMs

During the 2010s, commercial SEMs were sufficiently scaled down in size so they can sit on a simple table. "Tabletop" instruments are largely self-contained and are relatively easy to relocate. Often, the instruments are limited to x and y control with limited sample tilt capability. Sample sizes can be limited due to smaller specimen chamber sizes. Compact SEMs generally use thermionic sources and are limited to magnifications no higher than ~50,000×. Nevertheless, the systems are designed for ease of use, require less maintenance, and are more user friendly for new users. Most compact SEMs connect to vacuum pumps displaced a few feet away. Many systems integrate a silicon-drift detector for EDS in addition to SE and BE detection using an annulus, pole-piece-mounted SSD, or scintillator detector.

Variable-Pressure/Environmental SEM

Variable-pressure/environmental SEMs (VP-ESEMs) are a specific class of SEM instruments that enable the user to flow gas to the specimen surface inside the SEM specimen chamber. Gas incorporation at the imaging site provides many advantages. Primary beam electrons ionize gas molecules, producing positively charged ions, which can balance negative charge accumulation during SEM imaging of an insulating sample surface. Thus, VP-ESEMs facilitate imaging of nonconducting specimen surfaces without the use of a conductive coating. The VP-ESEMs can also introduce water vapor to the specimen surface, suppressing evaporation for moist or liquid specimens inside the SEM chamber. User control of gas type and local pressure enables SEM imaging of surface reactions occurring at the specimen surface, including oxidation, reduction, hydration, evaporation, corrosion, phase transformations, and so on.

The primary distinction for a VP-ESEM is differential pumping, the main objective of which is to keep as much of the primary electron beam pathway at extremely low pressures (<1.33 × 10^{-4} Pa, or <10^{-7} torr) while also maintaining a much higher pressure (100 to 1000 Pa, or 0.75 to 7.5 torr) immediately at the specimen surface. To accomplish this, the VP-ESEM uses multiple pressure-limiting apertures (PLAs) to separate different pumping zones between the electron source and the specimen surface, as illustrated in Fig. 39. Each pumping zone is evacuated by its own pump. Ion getter pumps maintain the 1.33 × 10^{-6} Pa (~10^{-9} torr) pressure within the electron-optics column (zone 1), which constitutes most of the primary electron beam path to the sample. The next pumping zone (zone 2) includes the interior of the objective lens and a small distance between PLA 1 and PLA 2. This zone is maintained at a slightly higher pressure (0.133 to 0.0133 Pa, or 10^{-4} to 10^{-6} torr) by a turbomolecular pump. Finally, the electron beam passes through PLA 2 and into the high-pressure (100 to 1000 Pa, or 0.75 to 7.5 torr) zone 3 inside the SEM specimen chamber. An external source provides the desired background gas (water vapor, oxygen, argon, nitrogen, etc.) through an inlet attached to the specimen chamber.

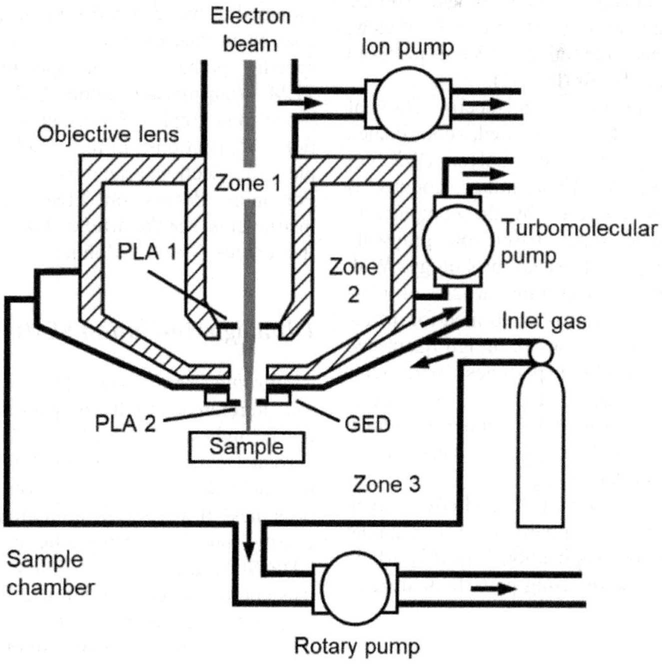

Fig. 39 Schematic illustration of a variable-pressure scanning electron microscope system with three distinct differentially pumped pressure zones: zone 1 (1.33 × 10^{-4} to 1.33 × 10^{-6} Pa, or 10^{-7} to 10^{-9} torr), zone 2 (0.133 to 0.0133 Pa, or 10^{-4} to 10^{-6} torr), and zone 3 (100 to 1000 Pa, or 0.75 to 7.5 torr). Each zone is separated by a pressure-limiting aperture (PLA). GED, gaseous electron detector

The primary electron beam path inside zone 3 must be minimized, because the gas molecules significantly increase primary electron scattering prior to their arrival at the specimen surface, surface; thus, the smallest WD must be used. Specialized gaseous electron detectors (GEDs) can be mounted at the bottom of the objective lens pole piece to extend zone 2 closer to the sample surface. The bottom surface of a GED consists of silicon-drift detectors or scintillator detectors for high-take-off-angle SE and BE collection. Alternatively, a narrow conical component with a PLA at the nose can be mounted at the bottom of the objective lens pole piece. The conical component ensures extension of zone 2 right to the sample surface, as well as clear line of sight for x-ray spectroscopic analysis during VP-ESEM operation. See Example 6 in the section "Applications" in this article for an example experiment using a VP-ESEM. An excellent resource covering VP-ESEMs can be found in Ref 15.

High-Throughput Instruments

Developments in SEM design enabled automated acquisition of images and spectral analysis over numerous locations of a specimen surface. A multibeam SEM separates a single, high-current primary electron beam into dozens of individual lower-current primary beams. The primary beams are scanned in parallel across different points of the specimen surface. Individual detectors for each primary beam collect SEs that emanate from each primary beam location. In this manner, dozens of SEM images can be acquired in parallel over ~100 μm field of view, with each image containing sufficient pixels to resolve <10 nm details (Ref 52).

A particle analyzer is a specific class of SEM designed for automated, coordinated imaging and spectroscopic analysis (Ref 53). After recording an SE/BE image, contrast is interpreted and analyzed by the computer to identify particle types based on grayscale values and/or particle size/morphology. With input from the user regarding particles of interest, point EDS analysis is performed for different particle classifications to assign quantified composition to different particles. Next, automated SE/BE imaging and EDS analysis is performed over additional specimen locations, producing large-scale maps and highly statistically significant details regarding the morphologies and compositions for hundreds or even thousands of individual particles. See Example 1 in the section "Applications" in this article for an example study using particle analyzers.

Dual-Beam Instruments

Dual-beam instruments combine SEM with focused ion beam (FIB), another scanning beam instrument (Ref 54). Focused ion beams generate and accelerate beams of ionized atoms (typically gallium or xenon) to high energies (2 to 30 keV). The ion beams can be electrostatically focused and scanned across a sample surface in a manner similar to scanned electron beams. The FIBs are capable of controllable deposition of materials by a process called ion-beam-assisted chemical vapor deposition (CVD). The process uses the ion beam to stimulate CVD reactions at the surface while a reactive precursor gas is injected at the sample site using a nozzle. Additionally, FIBs direct high-energy ions to controllably remove, or sputter, material away from the sample. Because the 2 to 30 keV ion beam can be focused to <10 nm spot sizes, FIB sputtering is similar to atomic sandblasting or nanoscale milling.

Enabling site-specific, cross-sectional SEM is a major advantage of a dual-beam instrument. The instruments enable deposition of a protective layer over a region of interest, nanoscale precision sputtering of unwanted material adjacent to the region of interest, and SEM imaging/analysis of the newly revealed surface. Dual-beam instruments have automated control of the ion beam, electron beam, and stage for three-dimensional analysis. Surfaces can be ion milled, imaged, and ion milled again to form a new surface for subsequent imaging. The process is repeated hundreds or thousands of times, so numerous SEM images can be obtained and combined to reconstruct the three-dimensional physical characteristics of the material. Dual-beam instruments are commonly equipped with nanomanipulators so that thin specimens can be removed and mounted to predesigned sample holders. Thus, dual-beam instruments are often used for site-specific preparation of specimens for the TEM, transmission-mode SEM, and atom probe tomography. See Example 6 in the section "Applications" in this article for an example cross-sectional analysis by FIB milling. For more details on FIBs and dual-beam instruments, see the article "Focused Ion Beam Instruments" in this Volume.

Ultrahigh-Resolution SEMs

Ultrahigh-resolution (UHR) is a general term used in reference to SEMs specifically designed for the highest point resolution obtainable using current technology. The UHR-SEMs typically use cold field-emission guns to obtain maximum brightness, β, as well as minimum primary beam energy spread, or chromatic aberration. The UHR-SEMs use either immersion lens or snorkel lens configurations to minimize WD and minimize spherical aberration, C_S. The UHR-SEM instruments use extensive vibration damping and electromagnetic field shielding, yet they still require stringent ambient conditions with minimal external vibrations and electromagnetic fields. Sample size can be limited, especially for cases where an immersion lens is used. Nevertheless, the SEM immersion lens configuration is similar to TEM, so transmission-mode SEM can be readily implemented in certain UHR-SEMs; scanning TEM detectors are often available in UHR-SEMs for transmission-mode imaging (Ref 55). As mentioned in Table 3, UHR-SEMs can achieve point resolution down to 0.3 nm.

Applications

This section presents examples in which SEM instruments are used for particle analysis, EBSD analysis, EDS and contrast by electron channeling, monochromatic CL imaging and ECCI, transmission-mode SEM, and in situ investigation using an environmental SEM.

Example 1: EDS for Steel Inclusion Analysis

Molten steel contains dissolved oxygen, which can form unwanted porosity during solidification. Aluminum added to molten steel captures the oxygen, typically forming spinel inclusions, which are solid solutions of $MgAl_2O_4$ and Al_2O_3. High-melting-temperature inclusions present a problem where the ceramic particles can agglomerate and degrade crucial components of the steel casting process. Calcium treatment of the molten steel serves to partially or completely liquefy the spinel inclusions. This study used particle analyzers to investigate the morphology and composition of spinel inclusions at various stages of the calcium treatment process. Particle analyzers, combining automated SEM with EDS, are well established for analyzing steel inclusions (Ref 53).

Two SEMs operating at 20 kV acceleration voltage were used to study inclusions in cross-sectioned steel samples from laboratory tests and industrial casting operations. Figure 40 (a) presents SE images of the spinel inclusions at various stages of calcium treatment. Inclusion sizes ranged from 1 to 2.5 μm. The calcium addition initially converted faceted Mg-Al-O spinel inclusions to a transient Al_2O_3-CaS combination, or a rarely observed spinel-CaS combination. After 4 to 5 min of calcium treatment, spinel inclusions were either partially or fully liquefied, with the liquid phase being lower MgO-Al_2O_3 mixtures. The results indicate that calcium modifies the spinel inclusions through a transient calcium sulfide formation process, with magnesium eventually dissolving into the steel. Figure 40(b) further illustrates the phenomenon through analysis of hundreds of inclusions for three industrial samples: ladle 1 (no calcium), ladle 2 (immediate calcium treatment), and ladle 3 (5 min after calcium treatment). Initial inclusion compositions for ladle 1 are either Al_2O_3 or $MgAl_2O_4$-Al_2O_3. Inclusion compositions in ladles 2 and 3

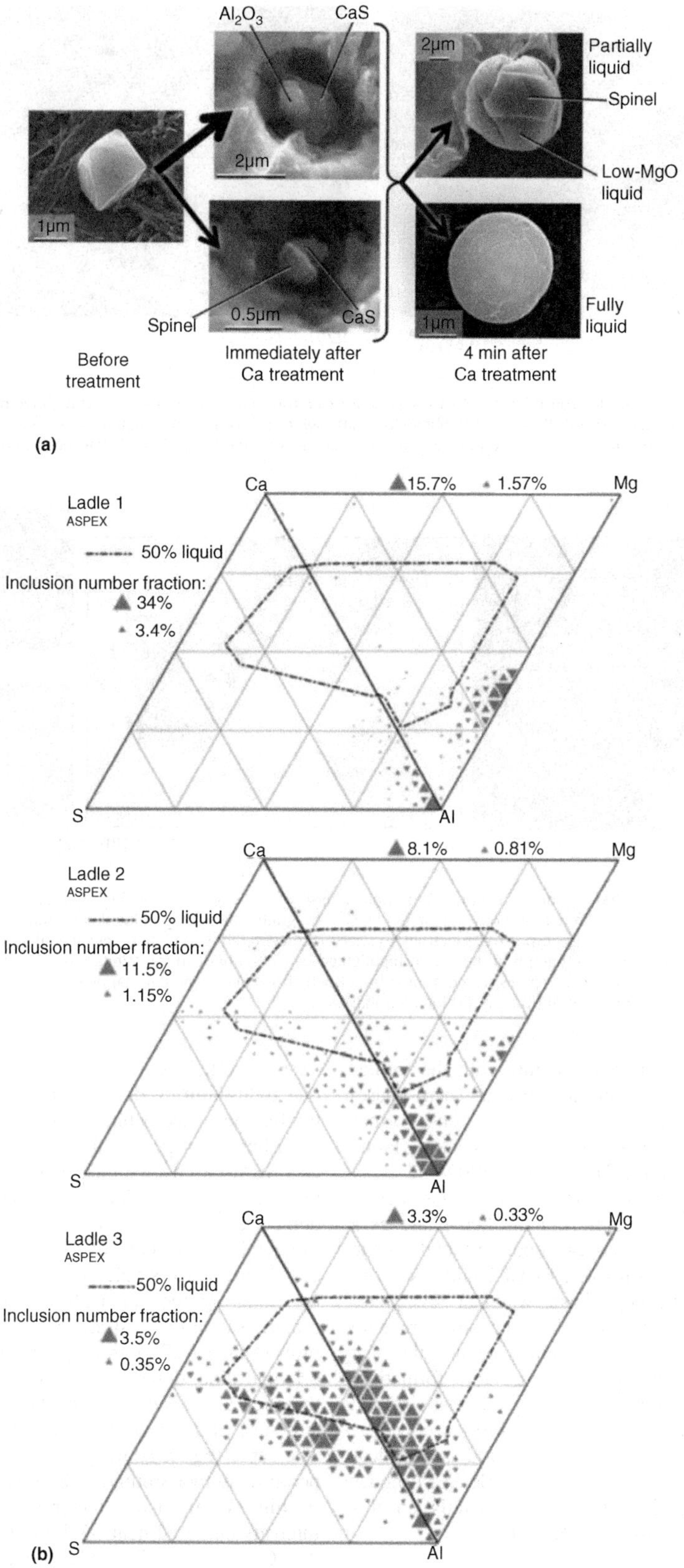

Fig. 40 (a) Secondary electron images of spinel inclusions at various stages of calcium treatment in molten steel. (b) Inclusion compositions for industrial samples before calcium treatment (ladle 1), immediately after calcium treatment (ladle 2), and 5 min after calcium treatment (ladle 3). Source: Ref 56

exhibit depletion of magnesium with calcium treatment. The formation of CaS occurs immediately upon calcium incorporation and then decreases over time as the inclusions exhibit predominantly calcium aluminate compositions after 5 min of calcium treatment. An important implication from this study was that while calcium treatment demonstrated liquefication of spinel inclusions, magnesium dissolved back into the steel and could serve to form new, unwanted spinel inclusions in the molten steel. The particle analyzer SEMs in this study enabled morphological and compositional analysis of hundreds of micrometer-scale particles to obtain these insights. More details on this study can be found in Ref 56.

Example 2: EBSD and Fractography for Additive Manufacturing

This example highlights the use of EBSD and fractography to correlate fracture behavior and tensile strength to the microstructure of Ti-6Al-4V specimens produced by additive manufacturing (AM). Ti-6Al-4V is an alloy with a high strength-to-weight ratio and good corrosion resistance and is widely used in aerospace and biomedical applications. This dual-phase alloy is composed of a hexagonal close-packed α phase and a body-centered cubic β phase. Depending on the cooling rate from the higher-temperature β-phase region (above 995 °C, or 1820 °F) through the $\alpha + \beta$ region, α-phase microconstituents can form globular morphologies or Widmanstätten platelets. Sufficiently fast cooling can produce martensitic phases, including a hexagonal α' structure. The mechanical properties can vary significantly based on the thermal history and resultant α-β microstructure.

Selective laser melting (SLM) is a type of AM process that uses a scanned laser to melt controllably fed powder in a layer-by-layer process. Local microstructure can be tailored by controlling laser conditions and feed powder. A study was conducted to explore the mechanical properties of SLM-produced Ti-6Al-4V samples fabricated under various build orientations. Starting powder was plasma-atomized Ti-6Al-4V consisting entirely of the α' phase. During SLM, melting and resolidification of the powders determined the β grain size and texture as well as the eventual α-phase formation during cooling.

Inverse pole figure (IPF) maps obtained by the EBSD method show both the orientations of the hexagonal α phase and a reconstructed IPF map for the prior-β-phase grains. Electron backscatter diffraction was performed for specimens produced at different build orientations by SLM to evaluate both α- and β-phase grain size and texture. For all SLM build orientations, prior-β phase exhibited a dominant (100) solidification texture.

Fractography was performed using SE imaging on fracture surfaces of specimens tensile stressed to failure, as shown in Fig. 41. At higher magnifications, terracelike features were observed, denoted by the black arrows in Fig. 41(b). The terrace size and aspect ratio were consistent with prior-β grains and seemed to indicate crack propagation through prior-β grains. Additionally, it was found that fracture surface profiles differ significantly based on build profile primarily due to prior-β grain-boundary orientation with respect to the external axial loading direction during tensile testing. Ultimately, AM-produced Ti-6Al-4V specimens appeared to be dominated by intergranular-mode fracture along α grain boundaries and along prior-β grain boundaries. In this study, the coordination between EBSD and SE imaging in the SEM was critical in elucidating the mechanical behavior of these AM-produced specimens. More details of this study can be found in Ref 57.

Fig. 41 Scanning electron micrographs by secondary electron detection of a fractured additive-manufacturing-produced Ti-6Al-4V specimen obtained at (a) lower and (b) higher magnifications. Black arrows in (b) denote terraces speculated to occur at prior-β grain boundaries. Reprinted from Ref 57 with permission of Elsevier

Example 3: EDS and Electron Channeling for High-Entropy Alloys

This example demonstrates coordinated SEM analysis using EBSD, EDS, and electron channeling to investigate the microstructure of a high-entropy alloy (HEA). The HEAs are alloys formed by mixing multiple elemental components, resulting in a crystal structure consisting of elements occupying random locations in the lattice. Interest in HEAs is increasing due to their unusual structure-property relationships and potential superior performance over conventional alloys.

A study was conducted to investigate the addition of interstitial carbon to an HEA. Figure 42 presents SEM analysis of an $Fe_{49.5}Mn_{30}Co_{10}Cr_{10}C_{0.5}$ (at.%) HEA after cold rolling and recrystallization annealing. Electron backscatter diffraction, shown in Fig. 42 (a), confirmed the HEA was primarily face-centered cubic (fcc) with an ~4 μm average grain size and that twins were present. The EBSD phase map with grain boundaries was correlated to a BE image (Fig. 42b), showing grain orientation contrast via the electron channeling mechanism. The white circles in Fig. 42 (b) denote dark particles confirmed by EDS in Fig. 42(c) to be iron-deficient and chromium-rich. Coordinated analysis by atom probe tomography confirmed increased carbon content for these chromium-rich particles. Results indicate that the presence of carbon provided an avenue for forming nanoscale chromium carbides within the HEA microstructure.

Room-temperature mechanical testing indicated improved tensile strength compared with HEAs with similar compositions that did not contain carbon. The mechanical property improvement was tied to the existence of multiple deformation mechanisms for the new alloy composition, including interstitial solid-solution strengthening, microcomposite effects due to a dual-phase structure with the formation of a hexagonal close-packed (hcp) phase, nanotwinning-induced plasticity, and formation of extended defects such as stacking faults and dislocations. The use of EBSD and x-ray diffraction could confirm the dual-phase nature of this HEA, quantifying the hcp phase as a function of processing conditions and tensile deformation. However, ECCI could also confirm that the hcp phase, when at low phase fractions, existed as a lamellar structure with associated stacking faults within the fcc matrix. Twinning, stacking fault formation, and dislocation formation were also imaged by ECCI and correlated to deformation behavior. All of these insights required the application and coordination of multiple SEM modes (EBSD, ECCI, EDS). More details on this study can be found in Ref 58.

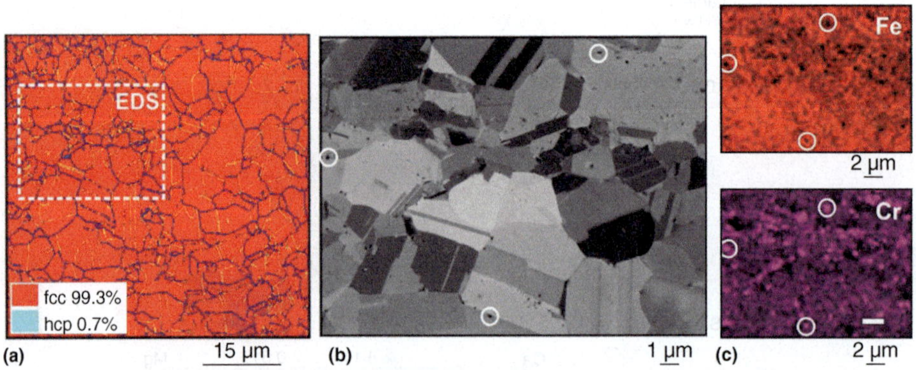

Fig. 42 Coordinated scanning electron microscope analysis of an Fe-Mn-Co-Cr-C high-energy alloy. (a) Electron-backscatter-diffraction-produced phase map with grain boundaries (blue lines) and twin boundaries (yellow lines). EDS, energy-dispersive spectrometry; fcc, face-centered cubic; hcp, hexagonal close-packed. (b) Backscattered electron micrograph of the region highlighted by dotted lines in (a). White circles in (b) denote iron-depleted and chromium-rich regions, as confirmed by (c), which is EDS elemental mapping. Reprinted from Ref 58 under a Creative Commons Attribution 4.0 International License

Example 4: CL and ECCI for Wide-Bandgap Semiconductors

Reducing extended defects in GaN, AlN, and InN is an important way to improve the quantum efficiency of green, blue, and ultraviolet light-emitting diodes. Directly confirming the luminescence behavior of extended defects is an important goal in this field. Coordinating two SEM-based methods, ECCI and CL, provides a direct correlation between individual defects and the local luminescence behavior. Figure 43(a) presents ECCI images of a silicon-doped GaN single-crystal film deposited by metal-organic vapor-phase epitaxy onto a *c*-plane sapphire substrate. The dark/light features in the ECCI micrograph denote vertically oriented line defects, dislocations that have threaded along the *c*-axis of the GaN wurtzite structure and penetrated the (0002) surface. The same region is analyzed using room-temperature monochromatic CL imaging, shown in Fig. 43(b), using integrated light signal intensity centered near ~3.4 eV. Dark spots in the CL map exhibit one-to-one correlation with the dislocation positions determined by ECCI. The coordination of two SEM techniques, ECCI and CL, succeeds in confirming the nonradiative recombination behavior associated with these dislocations. More details regarding this study can be found in Ref 59.

Example 5: Transmission-Mode SEM for Nanoparticles

Nanoparticles range in size from 1 to 100 nm and exhibit a host of unique properties, depending on composition, shape, morphology, and surface structure. Different SEM detectors are used to determine various nanoparticle characteristics. Figure 44 presents SEM micrographs of gold and TiO_2 nanoparticles on a lacey carbon substrate using 20 kV acceleration voltage. Figure 44(c) is an SE image recorded using an in-lens detector. Both isolated nanoparticles and agglomerates are visible, but there is no obvious delineation between the gold and TiO_2 nanoparticles. A modular aperture system (Ref 9) enabled angular selectivity for a commercial SSD positioned below the sample. The result was transmission SEM images produced by collecting electrons over a specific half-angle range, 24 to 48 milliradians in Fig. 44(a) and 496 to 547 milliradians in Fig. 44(b). The larger atomic number of gold can appreciably scatter more electrons in the larger-collection half-angle than the TiO_2 nanoparticles. Thus, the bright features in Fig. 44(b) can be identified as individual gold nanoparticles. For the smaller-collection half-angle image in Fig. 44(a), gold nanoparticles exhibit a more complex contrast (Ref 10), while the TiO_2 nanoparticles, denoted by white arrows, yield uniform bright intensity. Higher-magnification micrographs (not shown here) highlight the superior atomic number (*Z*) contrast for a transmission-mode SEM compared with conventional BE imaging. This study highlights potential advantages in using a transmission mode while also demonstrating the capability of SEM in resolving different nanoparticles. More details regarding this study can be found in Ref 9.

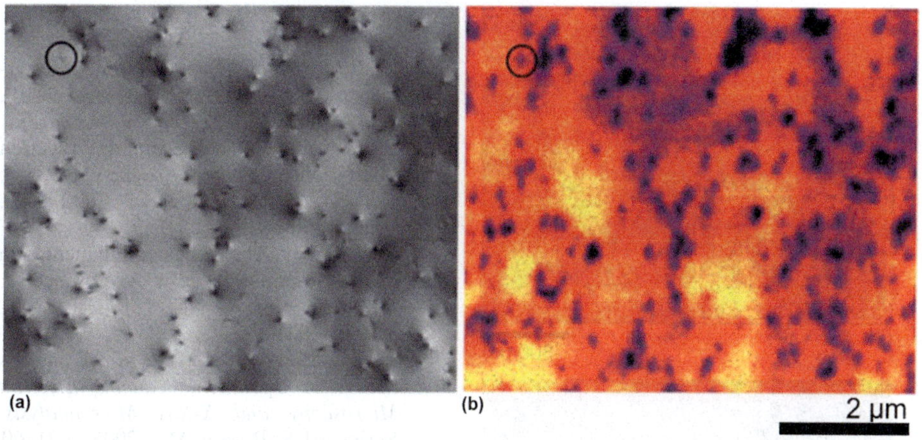

Fig. 43 Coordinated scanning electron microscope analysis of threading dislocations in a GaN (0002) single crystal using (a) electron channeling contrast imaging and (b) cathodoluminescence. Reprinted from Ref 59 with permission of Cambridge University Press

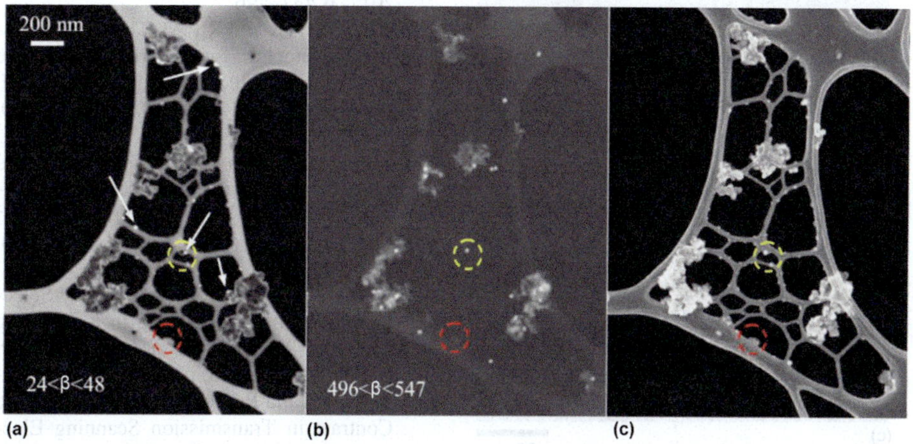

Fig. 44 Scanning electron micrographs of gold and TiO_2 nanoparticles on a lacey carbon substrate acquired at (a) low-angle and (b) high-angle transmission modes using a commercial solid-state detector, and (c) a conventional secondary electron image. Reprinted from Ref 9 with permission of Elsevier

Example 6: Environmental SEM, FIB, and In Situ Heating for Iron Oxidation

Oxidation of metals, such as iron, is investigated in situ by using an environmental SEM (ESEM) in conjunction with a heating stage. An advantage of this approach is the observation of oxidation as a function of local metal grain orientation. Figure 45(a) presents an SEM image of an iron sample prior to oxidation experiments. Individual grains were delineated by their orientation based on the channeling contrast visible due to forescattered electron (FSE) detection. The SSDs used for FSE detection were mounted on an EBSD system, so the sample was tilted to 70° for FSE imaging. During in situ heating, the sample was at 0° tilt and was imaged using a special heat- and light-insensitive SE detector. A commercial ESEM facilitated in situ exposure to water vapor, dry air (<6 ppm H_2O), and wet air (1% H_2O) during heating.

After 1 h of exposure to water vapor at 500 °C (930 °F) inside the ESEM, different oxide scales were observed for different iron grains, as shown in Fig. 45(b, c). Different oxide thicknesses were determined by using an FIB to cross section two adjacent iron grains. Figure 45(d) shows the thicker and thinner oxide scales revealed by FIB milling. Cross-sectional imaging using the FIB with SE detection provided ion channeling contrast of the iron oxide grain structure. These images (not shown here) revealed that the oxide layers consisted of three layers when ESEM reactions used dry and wet air: a porous magnetite layer, a fully dense magnetite layer, and a thin, fine-grained hematite layer. Oxidation experiments using water vapor showed only a duplex magnetite layer structure. The duplex structure suggested an initial magnetite microstructure lattice matched to the underlying iron grain; the magnetite then transitioned to a densified, columnar magnetite microstructure. The columnar grains were larger for the water vapor oxidation case than for the wet/dry air oxidation case. This study demonstrates how an ESEM can monitor a process such as oxidation while also coordinating analysis using other SEM-related methods, such as FSE imaging and FIB milling, to correlate oxidation behavior to the local microstructure and oxidation conditions. Additional details of this study can be found in Ref 60.

ACKNOWLEDGMENT

Revised from John D. Verhoeven, Scanning Electron Microscopy, *Materials Characterization*, Vol 10, *ASM Handbook*, ASM International, 1986.

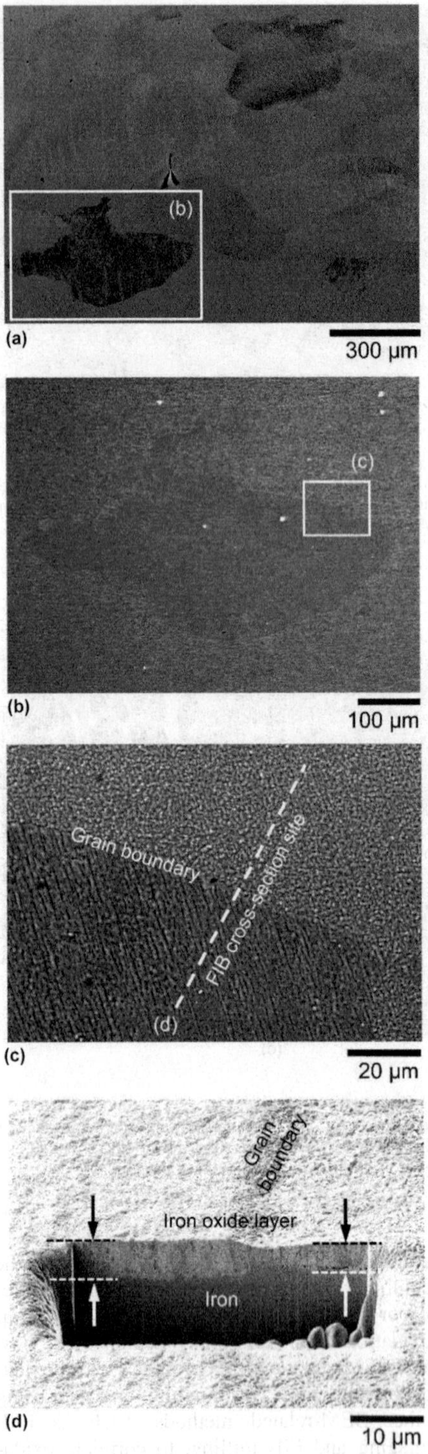

Fig. 45 (a) Forescattered electron micrograph of iron surface before oxidation. (b, c) Environmental scanning electron microscope/secondary electron micrographs during oxidation. (d) Secondary electron micrograph of focused ion beam (FIB)-milled cross section showing oxide scale thickness differences between adjacent grains. Adapted from Ref 60. Courtesy of T. Jonsson, Chalmers University of Technology

REFERENCES

1. M. Ardennevon, Das Elektronen-Rastermikroskop, *Z. Phys.*, Vol 109 (No. 9–10), Sept 1938, p 553–572
2. C.W. Oatley, The Scanning Electron Microscope, *Science Progress (1933–)*, Vol 54, Science Reviews 2000 Ltd., 1966, p 483–495
3. C.W. Oatley, The Tungsten Filament Gun in the Scanning Electron Microscope, *J. Phys. E.*, Vol 8 (No. 12), Dec 1975, p 1037–1041
4. J.I. Goldstein et al., The SEM and Its Modes of Operation, *Scanning Electron Microscopy and X-Ray Microanalysis*, Springer US, Boston, MA, 2003, p 21–60
5. J. D. Verhoeven and E.D. Gibson, Evaluation of a LaB_6 Cathode Electron Gun, *J. Phys. E.*, Vol 9 (No. 1), Jan 1976, p 65–69
6. A.E. Vladár and M.T. Postek, The Scanning Electron Microscope, *Handbook of Charged Particle Optics*, CRC Press, 2017, p 437–496
7. T.E. Everhart and R.F.M. Thornley, Wide-Band Detector for Micro-Microampere Low-Energy Electron Currents, *J. Sci. Instrum.*, Vol 37 (No. 7), July 1960, p 246–248
8. V.N.E. Robinson, The Construction and Uses of an Efficient Backscattered Electron Detector for Scanning Electron Microscopy, *J. Phys. E.*, Vol 7 (No. 8), Aug 1974, p 650–652
9. J. Holm and R.R. Keller, Angularly-Selective Transmission Imaging in a Scanning Electron Microscope, *Ultramicroscopy*, Vol 167, Aug 2016, p 43–56
10. T. Woehl and R. Keller, Dark-Field Image Contrast in Transmission Scanning ElectronMicroscopy: Effects of Substrate Thickness and Detector Collection Angle, *Ultramicroscopy*, Vol 171, Dec 2016, p 166–176
11. D.C. Joy, C.S. Joy, and R.D. Bunn, Measuring the Performance of Scanning Electron Microscope Detectors, *Scanning*, Vol 18 (No. 8), Dec 1996, p 533–538
12. B.J. Griffin, A Comparison of Conventional Everhart-Thornley Style and In-Lens SecondaryElectron Detectors—A Further Variable in Scanning Electron Microscopy, *Scanning*, Vol 33 (No. 3), May 2011, p 162–173
13. J.I. Goldstein et al., *Scanning Electron Microscopy and X-Ray Microanalysis*, Springer US, Boston, MA, 2003
14. L. Reimer, *Electron Optics of a Scanning Electron Microscope*, 1998, p 13–56
15. D.J. Stokes, *Principles and Practice of Variable Pressure/Environmental Scanning Electron Microscopy (VP-ESEM)*, John Wiley & Sons, Ltd., Chichester, U. K., 2008
16. B.L. Bramfitt and A.O. Benscoter, *Metallographer's Guide : Practice and Procedures for Irons and Steels*, ASM International, 2001
17. J.I. Goldstein et al., Image Formation and Interpretation, *Scanning Electron Microscopy and X-Ray Microanalysis*, Springer US, Boston, MA, 2003, p 99–193
18. D.C. Joy, Control of Charging in Low-Voltage SEM, *Scanning*, Vol 11 (No. 1), Jan 1989, p 1–4
19. A. Kolmakov, Membrane-Based Environmental Cells for SEM in Liquids, *Liquid Cell Electron Microscopy*, Cambridge University Press, 2016, p 78–105
20. M.T. Postek and A.E. Vladár, Does Your SEM Really Tell the Truth?—How Would You Know? Part 1, *Scanning*, Vol 35 (No. 6), Nov 2013, p 355–361
21. T.C. Isabell, P.E. Fischione, C. O'Keefe, M.U. Guruz, and V.P. Dravid, Plasma Cleaning and Its Applications for Electron Microscopy, *Microsc. Microanal.*, Vol 5 (No. 2), March 1999, p 126–135
22. P.W. Trimby, Orientation Mapping of Nanostructured Materials Using Transmission Kikuchi Diffraction in the Scanning Electron Microscope, *Ultramicroscopy*, Vol 120, Sept 2012, p 16–24
23. E.F. Rauch and M. Véron, Automated Crystal Orientation and Phase Mapping in TEM, *Mater. Charact.*, Vol 98, Dec 2014, p 1–9
24. P. Echlin, *Handbook of Sample Preparation for Scanning Electron Microscopy and X-Ray Microanalysis*, Springer US, Boston, MA, 2009
25. K. Kanaya and S. Okayama, Penetration and Energy-Loss Theory of Electrons in Solid Targets, *J. Phys. D. Appl. Phys.*, Vol 5 (No. 1), Jan 1972, p 308
26. H. Demers et al., Three-Dimensional Electron Microscopy Simulation with the CASINO Monte Carlo Software, *Scanning*, Vol 33 (No. 3), May 2011, p 135–146
27. J.I. Goldstein et al., Electron Beam-Specimen Interactions, *Scanning Electron Microscopy and X-Ray Microanalysis*, Springer US, Boston, MA, 2003, p 61–98
28. R.F. Egerton, *Electron Energy-Loss Spectroscopy in the Electron Microscope*, Springer, 2011
29. L. Reimer and M. Riepenhausen, Detector Strategy for Secondary and Backscattered Electrons Using Multiple Detector Systems, *Scanning*, Vol 7 (No. 5), Jan 1985, p 221–238
30. D. Tang, M.E. Ferreira, and P.C. Pistorius, Automated Inclusion Microanalysis in Steel by Computer-Based Scanning Electron Microscopy: Accelerating Voltage, Backscattered Electron Image Quality, and Analysis Time, *Microsc. Microanal.*, Vol 23 (No. 6), Dec 2017, p 1082–1090
31. J. Cazaux, On Some Contrast Reversals in SEM: Application to Metal/Insulator Systems, *Ultramicroscopy*, Vol 108 (No. 12), Nov 2008, p 1645–1652

32. S. Chung et al., Secondary Electron Dopant Contrast Imaging of Compound Semiconductor Junctions, *J. Appl. Phys.*, Vol 110 (No. 1), 2011, p 014902
33. V.G. Dyukov, S.A. Nepijko, and G. Schönhense, Voltage Contrast Modes in a Scanning Electron Microscope and Their Application, *Adv. Imag. Electron Phys.*, Vol 196, Jan 2016, p 165–246
34. J.I. Goldstein et al., Special Topics in Scanning Electron Microscopy, *Scanning Electron Microscopy and X-Ray Microanalysis*, Springer US, Boston, MA, 2003, p 195–270
35. D.C. Joy, D.E. Newbury, and D.L. Davidson, Electron Channeling Patterns in the Scanning Electron Microscope, *J. Appl. Phys.*, Vol 53, 1982,
36. Y.N. Picard, M. Liu, J. Lammatao, R.J. Kamaladasa, and M. De Graef, Theory of Dynamical Electron Channeling Contrast Images of Near-Surface Crystal Defects, *Ultramicroscopy*, Vol 146, 2014, p 71–78
37. S. Zaefferer and N.-N. Elhami, Theory and Application of Electron Channelling Contrast Imaging under Controlled Diffraction Conditions, *Acta Mater.*, Vol 75, Aug 2014, p 20–50
38. J. Guyon, H. Mansour, N. Gey, M.A. Crimp, S. Chalal, and N. Maloufi, Sub-Micron Resolution Selected Area Electron Channeling Patterns, *Ultramicroscopy*, Vol 149, Feb 2015, p 34–44
39. T. Maitland and S. Sitzman, Backscattering Detector and EBSD in Nanomaterials Characterization, *Scanning Microscopy for Nanotechnology*, Springer, New York, NY, 2006, p 41–75
40. J.I. Hanoka and R.O. Bell, Electron-Beam-Induced Currents in Semiconductors, *Ann. Rev. Mater. Sci.*, Vol 11, 1981, p 353–380
41. P.M. Haney, H.P. Yoon, P. Koirala, R.W. Collins, and N.B. Zhitenev, Electron Beam Induced Current in the High Injection Regime, *Nanotechnology*, Vol 26 (No. 29), July 2015, p 295401
42. F.J. García de Abajo, Optical Excitations in Electron Microscopy, *Rev. Mod. Phys.*, Vol 82 (No. 1), Feb 2010, p 209–275
43. P.R. Edwards and R.W. Martin, Cathodoluminescence Nano-Characterization of Semiconductors, *Semicond. Sci. Technol.*, Vol 26 (No. 6), June 2011, p 064005
44. J. Götze, Application of Cathodoluminescence Microscopy and Spectroscopy in Geosciences, *Microsc. Microanal.*, Vol 18 (No. 6), Dec 2012, p 1270–1284
45. G. Koschek, Origin and Significance of the SEM Cathodoluminescence from Zircon, *J. Microsc.*, Vol 171 (No. 3), Sept 1993, p 223–232
46. R. Podor, G.I.N. Bouala, J. Ravaux, J. Lautru, and N. Clavier, Working with the ESEM at High Temperature, *Mater. Charact.*, Vol 151, May 2019, p 15–26
47. C. Jiang, H. Lu, H. Zhang, Y. Shen, and Y. Lu, Recent Advances on In Situ SEM Mechanical and Electrical Characterization of Low-Dimensional Nanomaterials, *Scanning*, Vol 2017, Oct 2017, p 1–11
48. D.C. Joy and C.S. Joy, Low Voltage Scanning Electron Microscopy, *Micron*, Vol 27 (No. 3–4), June 1996, p 247–263
49. N. Brodusch, H. Demers, and R. Gauvin, Low Voltage SEM, *Field Emission Scanning Electron Microscopy*, Springer, 2018, p 37–46
50. R.F. Egerton, P. Li, and M. Malac, Radiation Damage in the TEM and SEM, *Micron*, Vol 35 (No. 6), Aug 2004, p 399–409
51. M. Konno, T. Ogashiwa, T. Sunaoshi, Y. Orai, and M. Sato, Lattice Imaging at an Accelerating Voltage of 30 kV Using an In-Lens Type Cold Field-Emission Scanning Electron Microscope, *Ultramicroscopy*, Vol 145, Oct 2014, p 28–35
52. A.L. Eberle, S. Mikula, R. Schalek, J. Lichtman, M.L.K. Tate, and D. Zeidler, High-Resolution, High-Throughput Imaging with a Multibeam Scanning Electron Microscope, *J. Microsc.*, Vol 259 (No. 2), Aug 2015, p 114–120
53. M. Reischl, G. Frank, C. Martinez, S. Aigner, B. Lederhaas, and S. Michelic, Automated SEM/EDX Particle Analysis to Determinate Non-Metallic Inclusions in Steel Samples: Round Robin Tests Aiming at Studying the Comparability of Results from Different Measurement Systems, *Pract. Metallogr.*, Vol 48 (No. 12), Dec 2011, p 643–659
54. L.A. Giannuzzi and F.A. Stevie, Ed., *Introduction to Focused Ion Beams*, Springer US, Boston, MA, 2005
55. C. Sun, E. Müller, M. Meffert, and D. Gerthsen, On the Progress of Scanning Transmission Electron Microscopy (STEM) Imaging in a Scanning Electron Microscope, *Microsc. Microanal.*, Vol 24 (No. 2), April 2018, p 99–106
56. N. Verma, P.C. Pistorius, R.J. Fruehan, M. S. Potter, H.G. Oltmann, and E.B. Pretorius, Calcium Modification of Spinel Inclusions in Aluminum-Killed Steel: Reaction Steps, *Metall. Mater. Trans. B*, Vol 43 (No. 4), Aug 2012, p 830–840
57. M. Simonelli, Y.Y. Tse, and C. Tuck, Effect of the Build Orientation on the Mechanical Properties and Fracture Modes of SLM Ti-6Al-4V, *Mater. Sci. Eng. A*, Vol 616, Oct 2014, p 1–11
58. Z. Li, C.C. Tasan, H. Springer, B. Gault, and D. Raabe, Interstitial Atoms Enable Joint Twinning and Transformation Induced Plasticity in Strong and Ductile High-Entropy Alloys, *Sci. Rep.*, Vol 7 (No. 1), Dec 2017, p 40704
59. G. Naresh-Kumar et al., Coincident Electron Channeling and Cathodoluminescence Studies of Threading Dislocations in GaN, *Microsc. Microanal.*, Vol 20 (No. 1), Feb 2014, p 55–60
60. T. Jonsson et al., An ESEM In Situ Investigation of the Influence of H_2O on Iron Oxidation at 500 °C, *Corros. Sci.*, Vol 51 (No. 9), Sept 2009, p 1914–1924

ASM Handbook, Volume 10, *Materials Characterization*
ASM Handbook Committee
DOI 10.31399/asm.hb.v10.a0006660

Crystallographic Analysis by Electron Backscatter Diffraction in the Scanning Electron Microscope

Joseph R. Michael, Sandia National Laboratories

Overview

General Uses

- Texture measurement
- Orientation relationship determination
- Phase identification
- Plastic and elastic strain measurement

Examples of Applications

- Texture of deformation-processed materials
- Annealing textures
- Interphase orientation relationship determination

Sample Requirements

- The surface must be free from artifacts introduced by sample preparation.
- Sample preparation may include electropolishing, standard metallographic practice, ion milling, and focused ion beam preparation.

Limitations

- Heavily deformed materials may be difficult.
- Minimum step size of bulk electron backscatter diffraction (EBSD) is approximately 50 nm; minimum step size of transmission Kikuchi diffraction is approximately 2 nm.
- Some nanomaterials may not produce useable patterns.
- Reasonably high beam currents may cause damage to beam-sensitive samples.
- Angular resolution of EBSD is approximately 0.1 to 0.25°.

Analysis Time

- Analysis time varies depending on hardware and software.
- Fastest systems can now acquire EBSD patterns at an effective rate of 3000 patterns per second.
- Some materials may not be able to be run at high speed due to EBSD pattern quality.

Related Techniques

- X-ray diffraction, selected-area electron channeling patterns, transmission electron microscopy (TEM) electron diffraction, precession electron diffraction in the TEM

AUTOMATED ELECTRON BACKSCATTER DIFFRACTION (EBSD) is a technique that allows the crystallography of a sample to be determined in a suitably equipped scanning electron microscope (SEM). In brief, a prepared specimen that is flat and free from damage at the surface is scanned in a suitably equipped SEM. When the electron beam is left stationary at one location on the specimen surface with a 20° incidence angle between the beam and the surface, the backscattered electrons diffract, giving rise to sets of cones of high and low intensity. To image the pattern, a fluorescent screen is placed near the specimen. The fluorescent screen converts the incident electrons to photons, which in turn are converted to a digital image with a charge-coupled device (CCD) camera. The collected EBSD patterns are then compared or indexed electronically, where a consistent set of crystallographic directions and planes are assigned to the EBSD pattern with respect to a given reference or candidate crystal structure. If agreement can be found between the expected patterns from the candidate phase and the experimental pattern, then the pattern is considered indexed. Commercial vendors have various techniques for indexing the EBSD patterns; the Hough-based indexing method has been used for many years and has proven effective. More recently, the introduction of dictionary-based indexing can improve the accuracy of EBSD in determining crystal orientation.

In a typical polycrystalline material with grain size between 2 and 200 μm, EBSD can map a large number of grains over a range of length scales spanning nanometers to centimeters. Not only is the average texture of the material available directly, but local information is also available, such as grain size, grain shape, and the crystallographic character of the grain and phase boundaries. Especially when combined with other SEM-based techniques, such as energy-dispersive and wavelength-dispersive analysis of composition, phase identification is also possible.

The use of EBSD has now become a "must have" SEM accessory for any laboratory attempting to do serious characterization of crystalline materials. The EBSD technique has proven to be very useful in the measurement of crystallographic textures, orientation relationships between phases, and both plastic and elastic strains. It has made the SEM a more complete tool for microstructural characterization by providing the linkage between microstructure and crystallography. When EBSD is combined with other microstructural characterization methods, such as imaging and elemental analysis, it provides a more complete material description and allows for a more complete understanding of materials performance.

Crystallography is important in materials because, in many cases, the final properties of the material that make it useful are controlled or at least influenced by the crystallography of the sample. The EBSD technique is capable of providing three complementary descriptions of the crystallography of a sample. The most common description is generally referred to as texture, and this is a description of how the unit cells in the sample are aligned with some external frame of reference. For example, in metal rolling it is important to understand the crystallographic directions that are or are not aligned with the rolling direction of the metal sheet. An example of this is shown in Fig. 1, where an inverse pole figure map and the pole figures obtained from an EBSD map are shown from a grain-oriented electrical steel. Another important use of EBSD is the determination of crystallographic orientation relationships across interfaces. In this case, the interface can be a grain boundary where the crystal phase is the same on either side, or it can be more complex, where the interface is an interphase interface with phases of differing crystallography on either side of the interface.

The EBSD technique is also used for the discrimination of phases within a microstructure. Phase discrimination through EBSD is based on the comparison of the collected patterns with a list of known candidate crystal structures, and thus, identification is performed through the specific crystallography. Figure 2 is an example of EBSD mapping of a multiphase Fe-Al-Ni alloy that was twin roll cast and then annealed (Ref 1). Figure 2(a) is an inverse pole figure (IPF) map with respect to the sample surface normal. The orientation data determined through EBSD and displayed here as an IPF map can allow the orientation relationships between all the phases to be determined. Figure 2(b) is a phase map where the three phases present are colored differently, and the phases have been determined not from their elemental chemistry but from their different crystallography. A further use of EBSD is that of strain determination, where the acquired pattern is compared to a reference pattern and the distortions in the collected pattern are used to determine the strain that could cause the observed distortions.

Electron backscatter diffraction has grown from an interesting laboratory development to a capability that should be considered as

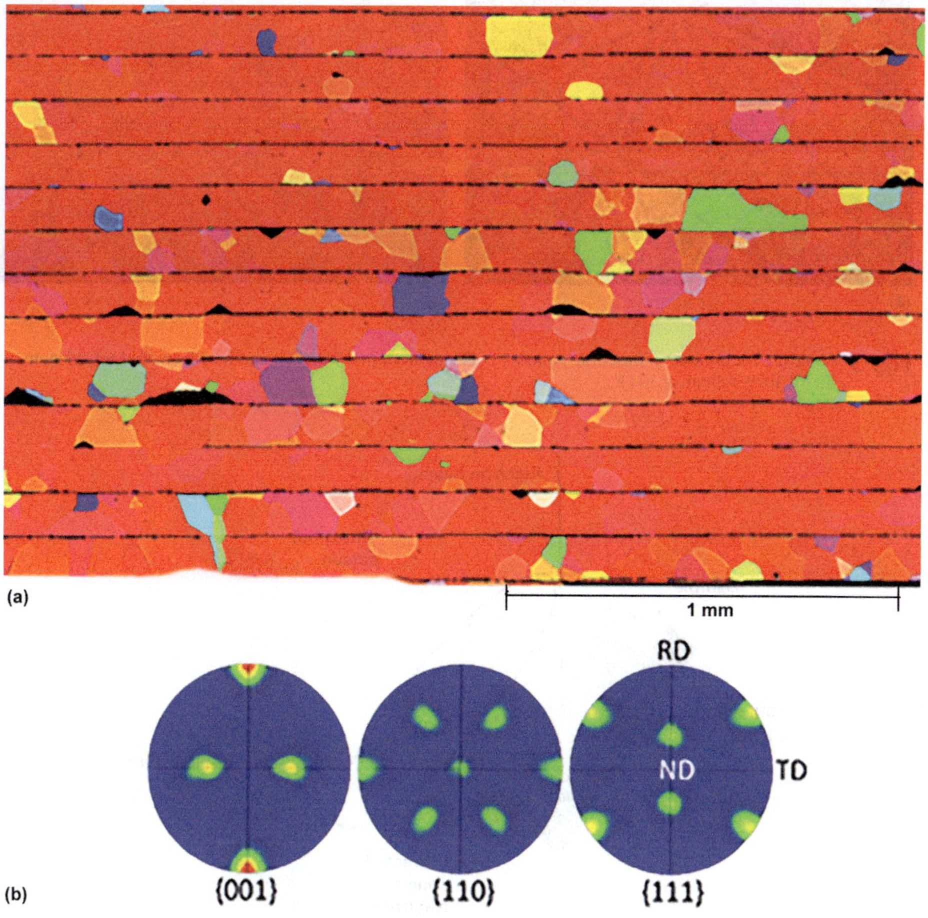

Fig. 1 Inverse pole figure orientation map and pole figures from a grain-oriented electrical steel transformer. (a) Inverse pole figure orientation map with respect to the rolling direction. (b) {001}, {110}, and {111} pole figures extracted from the electron backscatter diffraction map data. ND, normal direction; RD, rolling direction; TD, transverse direction

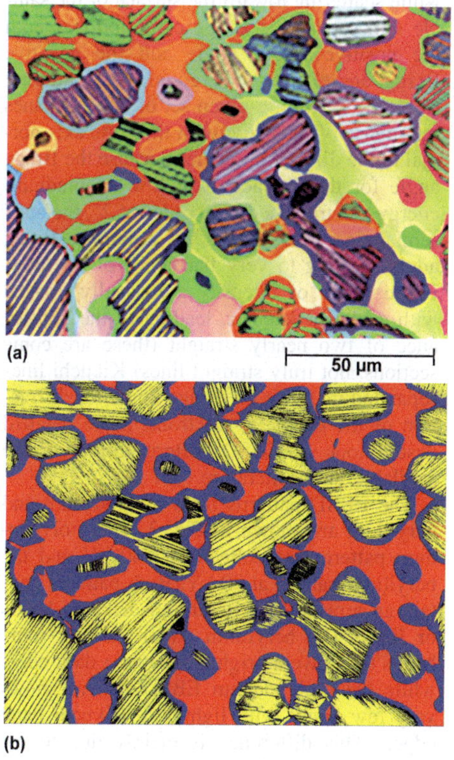

Fig. 2 Electron backscatter diffraction maps obtained from a twin roll cast Fe-Al-Ni alloy that was annealed at 1000 °C (1830 °F). (a) Inverse pole figure map with respect to the surface normal. (b) Phase discrimination map with NiAl shown in blue, Ni_3Al in red, and NiAl martensite in yellow, determined by the crystallography of the phases present

an essential component of SEM analysis of crystalline materials. Due to speed increases and SEM stability improvements, it is now possible to use EBSD to provide orientation measurements over large areas of a sample. The main emphasis of this article is on backscatter diffraction in an SEM, but this article also describes transmission Kikuchi diffraction (TKD). The TKD technique is similar, although a thin sample is required rather than a bulk sample, and the same type of information can be obtained at a much higher spatial resolution. Development of the TKD technique now allows similar measurements to be made on the scale of a few nanometers.

Origin of EBSD Patterns

The EBSD patterns are obtained by illuminating a sample with a stationary electron beam. The sample may be highly tilted for conventional bulk EBSD, or, if TKD is being conducted, the thin sample is mounted so that the electron beam is normal or up to 30° from normal from the thin sample surface. Kikuchi patterns arise from the elastic scattering of previously inelastically scattered electrons. In the case of traditional EBSD, the primary-beam electrons are subject to undergoing backscattering. Some of the backscattered electrons while scattering toward the surface of the sample will satisfy the Bragg condition (for both the positive Bragg angle and the negative Bragg angle) and undergo elastic scattering (diffraction) from a given set of crystal planes into cones of intensity, with the cone axis normal to the diffraction plane. Because the Bragg angle for the diffraction of low- to medium-energy electrons is on the order of 2°, the apex semiangle of the cone is 90° minus the Bragg angle, leading to very flat cones of intensity. These cones of intensity are usually imaged with a planar detector, leading to the appearance of two nearly straight (these are conic sections, not truly straight lines) Kikuchi lines separated by twice the Bragg angle for diffraction of the energetic electrons. More detailed descriptions of EBSD pattern formation are available in the literature (Ref 2–5).

The EBSD patterns from molybdenum and beryllium are shown in Fig. 3. The molybdenum pattern was collected at 40 kV and the beryllium pattern at 30 kV. Both patterns contain an array of nearly straight Kikuchi bands or lines. The beryllium pattern shows some bands that have a bright edge and a dark edge, while the molybdenum EBSD pattern shows very few bands with light/dark Kikuchi band edges. This difference is mainly due to the larger amount of electron scattering that occurs in the higher-atomic-number material. It is important to be aware of the difference in the EBSD patterns caused by variations in atomic number, because some light-element patterns can be more difficult to index due to the light/dark Kikuchi band edges. In the images shown in Fig. 3, there are places where many of the Kikuchi bands intersect. These are called zone axes and represent crystallographic directions within the unit cell. The angles between the Kikuchi bands and the angles between the zone axes are specific to a given crystal structure and given lattice parameters. This property is useful when the individual EBSD patterns are indexed. It is also important to note that the patterns are fixed with respect to the unit cell. Thus, if the unit cell is tilted or undergoes a change in orientation, the EBSD pattern will move with the unit cell. This allows the Kikuchi pattern to be used to accurately determine the orientation of grains within a given sample. In addition, there are other features in the EBSD patterns that can help determine information about the crystal structure. The circular and parabolic-appearing features in the molybdenum EBSD pattern in Fig. 3(a) are the result of diffraction from higher-order Laue zones. The diameter of the rings has been shown to be a sensitive measure of the lattice spacing along the specific zone axis (Ref 6).

Collection of EBSD Patterns

Acquisition of EBSD patterns in the SEM is not difficult. Many early studies used film to collect the EBSD patterns (Ref 7, 8). All commercial EBSD systems now use a fluorescent screen placed inside the SEM sample chamber that is imaged by an external camera separated from the vacuum of the microscope. Figure 4 is a schematic of the typical EBSD experiment configuration in the SEM. The sample is most commonly set to 70° with respect to the horizontal. This exact angle is not critical to the

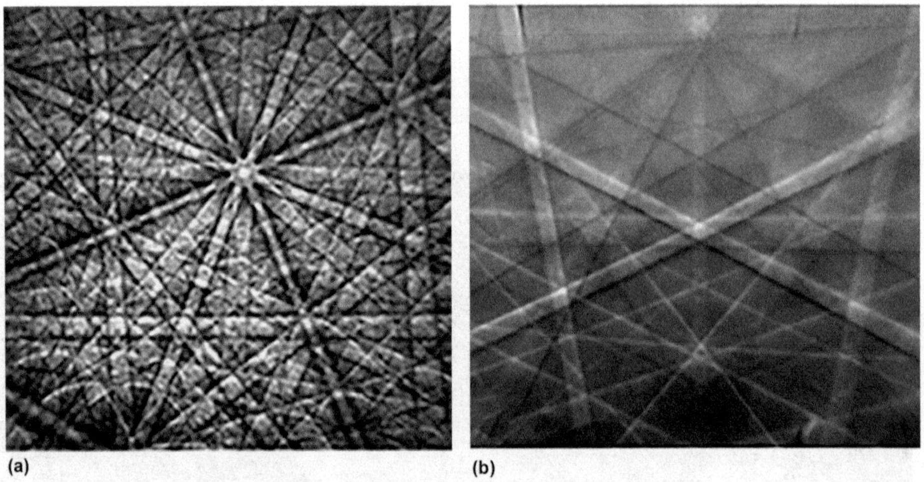

Fig. 3 Electron backscatter diffraction patterns collected from molybdenum and beryllium. (a) Molybdenum at 40 kV. (b) Beryllium at 30 kV

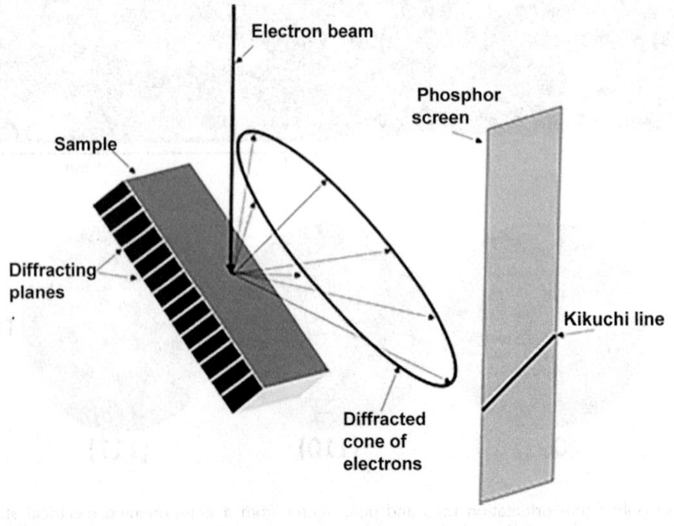

Fig. 4 Schematic of the experimental arrangement for a typical electron backscatter diffraction measurement. Note that the sample is highly titled with respect to the electron beam.

collection of quality EBSD patterns (Ref 9, 10). However, for subsequent calculations of orientations, this angle must be known. Several types of cameras have been used for EBSD. The cameras have evolved from television-rate low-light-level cameras, to CCD-based imaging cameras, to the next generation of complementary metal oxide semiconductor (CMOS)-based imagers. The CMOS-based imagers are now the most common type of camera and provide both speed and sensitivity. A view of an EBSD camera from inside an SEM chamber is shown in Fig. 5, where the arrangement of the sample, the pole piece of the SEM, and the EBSD camera can be seen. Users of EBSD should note that there is limited space available to move the sample, and much care should be taken to avoid damaging the sample, the EBSD camera, or the microscope pole piece. Imaging with the standard electron detectors in the SEM is possible, but due to the tilted sample and the EBSD camera, these signals may be compromised. To allow quality images of the sample to be readily obtained, most modern cameras offer forescattered electron detectors located at the lower edge of the fluorescent screen. There may be other detectors located on the top edge or sides of the fluorescent screen for the collection of more typical backscattered electron images. Images with interesting and informative channeling or grain contrast can be obtained from the forescattered detectors (Ref 11–13).

The EBSD camera or detector must be mounted on a precision slide (usually motorized) to allow the phosphor screen to be moved into position when EBSD experiments are performed. It is often advantageous to place the EBSD camera close to the sample, because closer approach distances allow more of the EBSD pattern in angular space to be imaged, and this can favorably influence the system ability to index EBSD patterns (Ref 14). It is also beneficial to use energy-dispersive spectrometry combined with EBSD for better phase discrimination when the phases have similar crystal structures (Ref 15). The slide must be able to accurately and reproducibly place the screen so that calibration of the camera is more easily accomplished. The movement of the camera also allows oddly shaped or large samples to be accommodated. Many of the modern EBSD cameras also allow the camera to be moved or tilted vertically within the sample chamber without opening the sample chamber. This is extremely useful because the vertical position of the detector can be fine-tuned to maximize the quality of the collected EBSD patterns.

It is possible to collect EBSD patterns on any SEM equipped with an EBSD camera. To collect the highest-quality EBSD maps, one needs an SEM that can generate a reasonably high probe current while producing a small spot, a stage that allows the sample to be placed in the optimal location with respect to the EBSD camera, and a stage that is stable over time frames of hours so that sample drift does not produce a distorted view of the microstructure. With modern EBSD systems on modern microscopes, the use of SEM beam energies from 5 kV up to the maximum beam energy of the SEM (usually 30 kV) is possible.

The stability of the SEM and the ability to produce a finely focused, high-current electron probe on the sample is critical to obtaining quality EBSD measurements. Typically, thermionic sources (such as tungsten or LaB_6 sources) can produce high beam currents but at the expense of beam size. Besides the larger probe size produced by the thermionic sources, the other main disadvantage is that these sources tend to have limited lifetimes. This can be an important consideration if large-area maps are of interest. Field emission sources, specifically Schottky-type emitters, are ideal for EBSD measurements. These sources can be operated to produce high beam currents (>20 nA) into a very small focused spot. The advantages and disadvantages of these sources are fully discussed elsewhere in this Volume and in other texts (Ref 10).

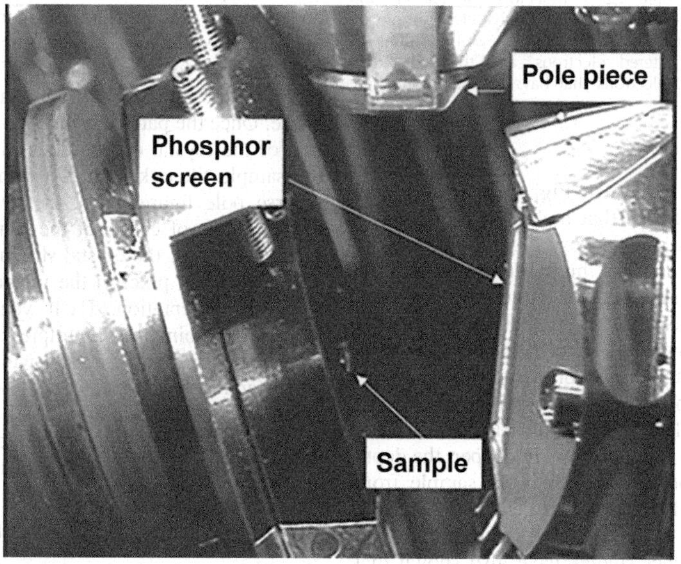

Fig. 5 View of the electron backscatter diffraction (EBSD) camera in the scanning electron microscope (SEM) sample chamber for the experimental configuration used for conventional EBSD measurements. The EBSD phosphor screen, the sample, and the SEM pole piece are shown.

During EBSD measurements, the sample is generally tilted to a relatively high angle of approximately 70°. The sample stage must exhibit minimal drift at these high tilts so that distortion-free maps can be collected. Positioning of the sample for EBSD is greatly simplified if the stage has a motion that is parallel to the optic axis of the SEM. It is helpful if the SEM is equipped with both tilt correction and what is called dynamic focus. Tilt correction is a simple correction for the foreshortening of the image that occurs when a highly tilted sample is observed. This is important if quantitative microstructural measurements (such as grain size and shape) are to be extracted from the EBSD maps. Dynamic focus is quite useful for EBSD due to the high tilt of the sample that is required. Dynamic focus adjusts the strength of the final lens so that the electron probe remains in focus over the entire surface of the tilted sample. This is most important when large-area EBSD maps are acquired, and it avoids the problems encountered when trying to image the extremes of the tilted sample.

Electron backscatter diffraction is useful for a range of samples that include electrical conductors, semiconductors, and insulators. In all cases, a suitable ground path must be established to avoid charging of the sample and the resulting sample drift or image distortion. Charging of the sample is a result of the energetic electron beam depositing charge into the sample that cannot be conducted to ground. The charged region of the sample will distort the electron beam or cause what appears to be sample drift to occur. Insulators and metallographically mounted samples are most often coated with a thin conductive layer to prevent charging of the sample. The mounting material for metallographic samples may also charge from the interaction with scattered electrons. Even though a metal sample in the mount may be conductive, efforts must be made to remove charge from the insulating mounting material. Thin conductive coating of metals or carbon can be used to remove the charge, but care must be taken to keep the coatings very thin so as not to distort or obscure the EBSD patterns. One method that has proven to be quite useful in mitigating charge in the sample is to introduce a low pressure of gas into the sample chamber. The SEM manufacturers have called this capability VP for variable pressure, LV for low vacuum, and other names, but they all work approximately the same. The introduction of the gas allows the surface charge buildup to be reduced to acceptable levels. Some trial and error may be needed to set the vacuum levels so that the charging is mitigated and quality EBSD patterns can be collected (Ref 16).

EBSD Spatial Resolution

The utility of any microanalytical technique depends on the minimum volume or the spatial resolution from which a signal can be generated. In many applications of EBSD, the highest spatial resolution is not always required for mapping the crystallographic texture of a sample. There are applications requiring high spatial resolution, including mapping of fine-grained materials and deformed microstructures.

Electron backscatter diffraction is a diffraction technique and thus relies on a small spread in the wavelength of the diffracted electrons. If a large spread exists in the exiting electron wavelength or energy, there would be no distinct diffraction lines visible in the EBSD pattern due to diffraction. It has been shown that the Kikuchi lines are in nearly the exact correct positions as described by Bragg diffraction for the energy that the SEM is operated. Thus, it can be concluded that the electrons leaving the sample surface and contributing to the EBSD pattern are of nearly the same energy. There are some electrons that have lost larger amounts of energy due to scattering in the sample. These electrons contribute to the background observed in EBSD patterns and not the features associated with diffraction (Ref 17).

The high sample tilt used during EBSD enables high-quality EBSD patterns to be acquired. The best EBSD patterns are generally obtained at higher tilt angles consistent with ease of imaging the highly tilted sample. The high sample tilt results in many of the electrons not losing much energy while exiting the sample (Ref 10, 18). The EBSD pattern is formed by those electrons that have lost very little energy. These electrons have been shown to exit the sample very close to the initial beam position, resulting in spatial resolutions in typical transition metals to be better than 0.1 μm for beam voltages up to 20 kV (Ref 10). Monte Carlo electron trajectory simulations of this effect are shown in Fig. 6. Figure 6(a) shows the results of Monte Carlo electron trajectory simulations of the energy of the backscattered electrons from a nickel sample where the primary beam energy was 20 kV for a sample at normal incidence and tilted to 70°. The important feature to note is that the energy distribution for the highly tilted sample is peaked near the initial beam energy, leading to the conclusion that the electrons that contribute to the EBSD pattern generally have lost little energy. The simulations shown in Fig. 6(b) demonstrate that the electrons that have lost little energy arise from a region very close to the initial beam position. The high tilt of the sample results in an elongation of the electron beam down the sample perpendicular to the tilt axis, as shown in Fig. 6(b). Due to the high sample tilt, the resolution parallel and perpendicular to the tilt axes is different.

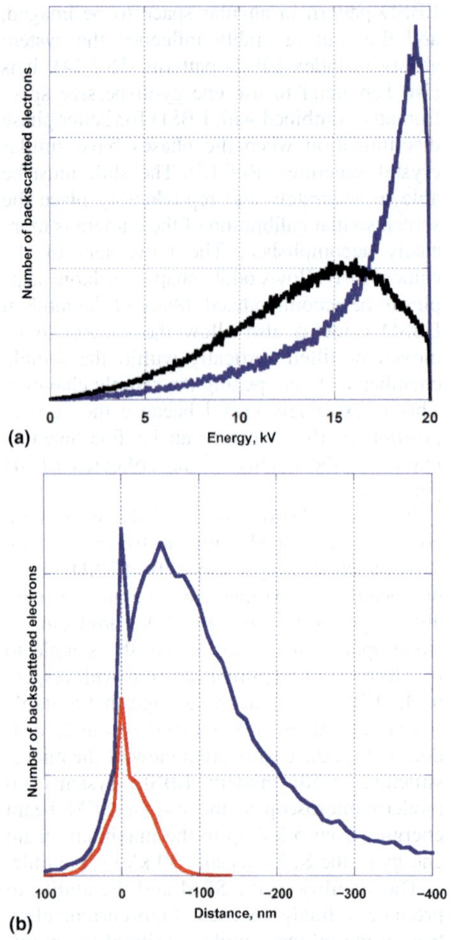

Fig. 6 Monte Carlo electron trajectory simulations for 20 kV primary beam energy. (a) Simulations of the backscattered electron energy distributions for normal (black curve) and tilted (blue curve) incidence of the electron beam. Note that the energy distribution for the tilted sample is highly peaked near the beam energy. (b) Monte Carle trajectory simulations for backscattered electrons from a highly tilted sample of nickel. The blue curve represents the exit positions of all the backscattered electrons, and the red curve indicates those electrons that have lost no more than 200 V.

The depth resolution of EBSD has been shown to be quite shallow. In copper, the depth resolution or the depth into the sample from which the EBSD pattern is generated has been shown to be 46 nm at 10 kV and 38 nm at 5 kV (Ref 19). Other studies have also shown that the depth resolution of EBSD is quite shallow and on the order of tens of nanometers (Ref 20, 21).

EBSD System Operation

All of the modern orientation mapping systems from any of the commercial sources work in a very similar manner using the same steps. The object of orientation mapping is to collect EBSD patterns and then convert the patterns to an orientation with respect to the sample or some external frame of reference. Orientation mapping steps include pattern collection, extraction of the Kikuchi band positions, indexing the pattern by comparing to a known crystal description, and determining the orientation of the crystal with respect to some external frame of reference. Careful alignment of the sample with the external frame of reference (tilt axis of the SEM stage, for example) is important because this will allow preferred orientations to be more easily observed and compared between samples. Once an EBSD pattern has been collected, the software uses an algorithm to determine the positions of approximately 12 Kikuchi lines in the pattern. This is most commonly done with a technique called the Hough or Radon transform. This is a complex subject, and there are many good descriptions in the literature of how the Hough transform works (Ref 22). The Hough transform is used to detect straight lines in digital images. Thus, the Hough transform efficiently parameterizes the location of straight lines or bands made up of pixels. The Hough transform is used to locate approximately 12 lines in the EBSD pattern. At least three are required to unambiguously determine the crystal structure and the orientation. The system then compares the angles between the Kikuchi bands in the experimental pattern with a table of the ideal angles between planes, derived from the candidate crystal structures input by the analyst. If agreement can be found between the expected patterns from the candidate phase and the experimental pattern, then the pattern is considered indexed. Each vendor uses its own metric for the quality of the match between the experiment and the candidate phase. Once the pattern is indexed, the orientation of the crystal can be determined because the sample tilt is known. To create the color inverse pole figure maps, the system uses a look-up table of colors of the primary stereographic zone for the crystal structure to assign a color to each pixel in the image that represents the orientation. Each vendor has its own way of doing this. Examples of color keys or legends for a cubic crystal with m3m symmetry and a hexagonal crystal with 6/mmm symmetry are shown in Fig. 7. Other color legends have been suggested and have been successfully applied to EBSD data (Ref 23).

Performing an EBSD Experiment

Although the actual details of how a modern EBSD system produces indexed EBSD patterns are not identical, the general steps

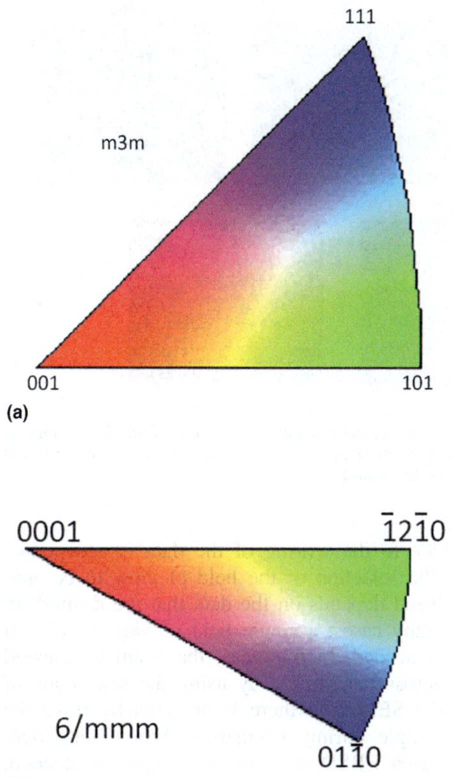

Fig. 7 Stereographic triangle color keys for cubic and hexagonal crystal structures. (a) Cubic m3m symmetry. (b) Hexagonal 6/mmm symmetry

are very similar. The sample and the EBSD camera must be positioned so that the area of interest on the sample can be imaged. The SEM beam parameters and the EBSD camera acquisition parameters must then be chosen to provide EBSD patterns that are of usable quality. The geometrical parameters of the EBSD system are then calibrated, if necessary. It is highly recommended that the calibration be checked periodically. Now the actual experiment can be conducted where the electron beam is stepped across the sample surface. At each pixel, an EBSD pattern is collected and indexed (*indexed* implies that a consistent set of crystallographic directions and planes are assigned to the EBSD pattern with respect to a given candidate phase structure within a desired level of agreement); thus, the crystal orientation can be calculated. Commercial vendors have various techniques for indexing the EBSD patterns and measures of what constitutes correct indexing of the pattern.

As a first step, the operator positions the EBSD detector and sample in the SEM to image the sample area of interest. Television-rate cameras (sometimes called chamberscopes), which provide views inside the SEM chamber during sample movement and positioning and EBSD camera insertion, are very helpful. They are almost mandatory to avoid expensive collisions between the sample and the various parts of the SEM and the EBSD camera. The specific type of SEM and EBSD system may limit the maximum sample size that can be accommodated and analyzed. It is important for the analyst to understand the frames of reference used by the EBSD system with respect to the sample orientation (Ref 24). It is most common to associate directions in the sample image with the stage tilt axis and the stage movements.

Electron backscatter diffraction experiment success requires selection of the correct phases for indexing. Electron backscatter diffraction is used more for phase discrimination in polyphase materials rather than direct phase identification. This means that it is up to the user to make sure that the proper crystallographic descriptions of the phases present in the sample are included during the setup of the system. If the correct phases or correct crystal structures have not been selected for the analysis, the user will have little success achieving superior-quality orientation mapping. The analyst should also be aware of the shortcomings of the common currently used indexing routines. Many issues occur with cubic crystals in that one can use any face-centered cubic metal phase description to index other face-centered cubic phases; thus, the current indexing methods cannot discriminate the various face-centered cubic metals. This is a result of the indexing methods making use of the angles between diffracting planes in the sample, and all face-centered cubic elements and compounds will index successfully. There are other examples of this in other cubic crystal structures.

Quality EBSD patterns are critical to obtaining good orientation information and phase discrimination. Many EBSD cameras have the capability to vary the number of pixels used to collect EBSD patterns. Higher numbers of pixels can produce very detailed EBSD patterns. However, for most EBSD experiments, one does not need the highest-quality pattern. Thus, the EBSD pattern is binned (adjacent pixels are combined) in 2 × 2, 4 × 4, or even 8 × 8 groupings. The advantage of binning is that the time required to acquire each pattern is greatly reduced. Binning can also reduce noise within the EBSD pattern and can allow the SEM to be operated at a lower beam-current setting and thus at a smaller probe size. The disadvantage to binning is a small loss in orientation accuracy. During the measurement of strain using EBSD, it is often necessary to use the highest pixel resolution possible. Once the degree of binning is selected, the user must decide the dwell time or exposure time per pixel. Exposure times of a fraction of a millisecond to a couple milliseconds are typical values, and these depend on the sample and the electron optical conditions selected (beam current and beam voltage). Many of the commercial vendors now have automated routines that will set the optimal exposure for a given set of conditions. Longer exposure times will result in the acquisition taking longer.

The user has many choices over both the SEM and the EBSD camera operating parameters. The overall contrast in EBSD patterns is generally quite low. The low contrast is a result of the collection of not only the electrons that have diffracted from the sample but also the unintended but unavoidable collection of the electrons that have undergone larger energy-loss interactions. To compensate for the low contrast, there are methods for removing the background. Some systems now use a software-generated background to normalize and reduce the background intensity and thereby increase the EBSD pattern contrast. This method is very effective but can impact the speed of pattern acquisition. Another approach is to use the actual sample to generate the background image for normalization. This background is collected by scanning the electron beam over many differently oriented grains in a polycrystalline sample. If the grain size is large or the sample is a single crystal, the dynamic background may be the best choice. This background image is then used to normalize the raw EBSD pattern to increase the contrast (Ref 25, 26).

Selection of the camera setting and the processing parameters in the software are important in obtaining fast indexing rates while maintaining a high hit rate. (Hit rate is the fraction of patterns indexed in a scan compared with the total number of pixels scanned.) Figure 8 demonstrates the relationship between the exposure dwell time and the number of patterns processed per second for both the static background and the dynamic background corrections. Note that the static background correction (maximum speed of ~3000 patterns processed per second) allows the system to run at a much higher processing rate than does the dynamic method (maximum speed of ~2000 patterns processed per second). Thus, although the dynamic background correction is a simpler method, there are sacrifices made in the total speed of the system. Also displayed in Fig. 8 is the hit rate for each of the dwell settings. Note that for a well-prepared sample with proper microscope settings, there is only a minor loss of hit rate at the highest speeds. However, the user should realize that the phases present in the sample may require longer dwell times to obtain quality patterns. As the speed of acquisition is increased (dwell time per pixel decreased), the EBSD pattern noise increases. Figure 9 shows the pattern quality that can be achieved at a range of dwell times by using the dynamic background method. It should be noted that even though the patterns collected at the shortest dwell times are noisy, modern indexing algorithms can still index these patterns.

The EBSD camera calibration is extremely important to achieving a quality EBSD

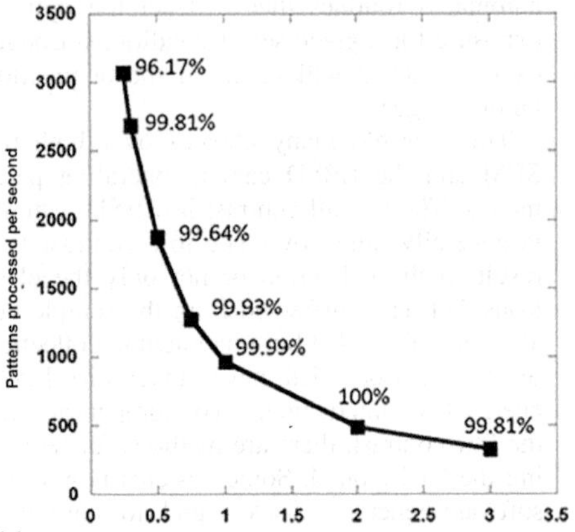

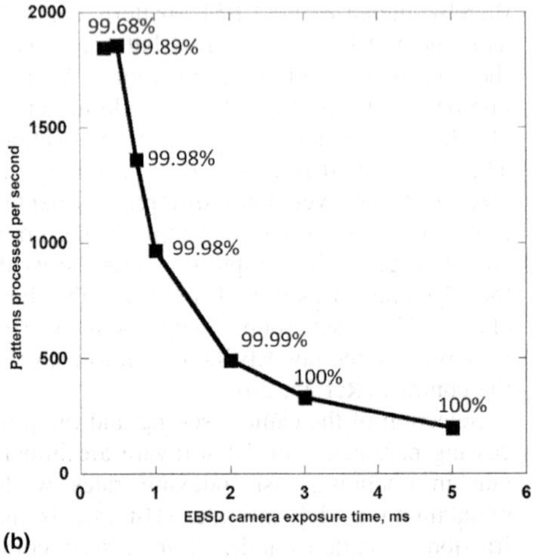

Fig. 8 Camera acquisition parameters impact the speed of an electron backscatter diffraction (EBSD) measurement. Processing speed of EBSD patterns processed per second from a polished tantalum sample at 20 kV and 10 nA of beam current for a modern complementary metal oxide semiconductor EBSD camera. The numbers next to the curves are the fraction in percent of the patterns indexed for each speed. (a) Processed patterns per second using static background correction. (b) Processed patterns per second using dynamic background correction

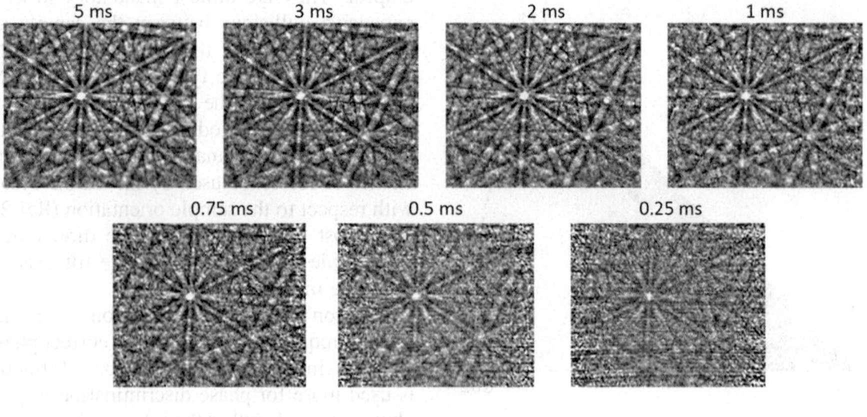

Fig. 9 Electron backscatter diffraction patterns collected by using the dynamic background mode for a range of dwell times. Note the decrease in the pattern quality with decreasing dwell time. Patterns were collected from a tantalum sample by using a beam voltage of 20 kV and 10 nA.

experimental result. Calibration has become quite simple since the vendors have implemented automated routines for this. Previous accurate camera calibrations were mostly achieved through manual means (Ref 26). The automated routines require that EBSD patterns be collected from a known phase in the sample. The software then determines the distance from the sample to the pattern center and the location of the pattern center by minimizing the error in the fitting of the indexed pattern to the calculated pattern for that phase. Each vendor has a slightly different method for doing this. Some rely on an average calibration from multiple patterns (Ref 3, 27).

If the user has properly set up the EBSD system, it is prudent to run at the optimum speed for the experiment being conducted. The goal of the experiment will dictate the parameters used for a given experiment. Maximum speed consistent with obtaining indexable patterns is probably adequate for orientation measurements. Experiments where misorientations or phase interfaces are to be studied should be conducted at slower speeds so that better-defined EBSD patterns can be obtained. The operator should collect and index a few randomly selected patterns over the area of interest on the sample. If successful indexing is achieved, the operator must select the pixel step size. The step size should be selected to allow the microstructural details of interest to be sampled appropriately by the EBSD map. Thus, it is helpful to have a step size that is smaller than the smallest features of interest. It has been suggested that the step size be one-tenth the feature size (ten pixels across the smallest feature to be studied) (Ref 28, 29). Figure 10 demonstrates the effect of step size or beam spacing on the resulting inverse pole figure maps. The microstructure is much clearer with the smaller step sizes of 0.25 or 0.5 μm. A step size that is too large may not properly sample the microstructure, and the resulting data may not be representative of the sample. A smaller step size requires more pixels to cover the same area, and therefore, more time is required for the map to be completed.

It is useful to collect a small test map from a representative area of the sample to make sure a large fraction of pixels is indexed. This is also an opportunity to decide if a phase has been missed and should be added to the list of potential phases for indexing. After the analyst is satisfied that the test maps display the desired results, then the remaining decision is the size of the area analyzed and how this will be performed. An alternative to this approach is to store the EBSD patterns or a representation of the EBSD patterns at every pixel or for only the pixels that cannot be indexed. Many of the modern EBSD systems can then reanalyze the data with new candidate phases or crystal structures to fill in the missing data. This is a good approach if sufficient storage space is available and the decrease in speed due to the writing of the data is acceptable. The selection of the field of view to be analyzed depends on the data that are desired. In many cases, a single field of view is all that is needed. In this case, the beam is scanned across the sample by using the scan coils of the SEM, and there is no need to move the sample during acquisition. Where data over larger EBSD areas of the sample are needed, a different approach is required. Most modern SEMs have a limit to the lowest magnification or the largest field of view that can be imaged. The SEM should not be operated at the largest possible field of view (lowest magnification) for reasons related to image distortions that are often inherent to the SEM (Ref 30). A field of view of 1 mm (0.04 in.) or smaller will reduce this distortion to an acceptable level. If this area is smaller than the desired map size, a hybrid or combined strategy of beam scanning and stage scanning is typically employed. Combined stage and beam scanning can map very large sample areas, mainly limited by the time required for acquisition of patterns over the large area. In a combined strategy, beam scanning maps each subregion, and the stage of the SEM is used to move the sample to the next field of view for beam scanning. This is continued until the desired area of the sample has been covered. The resulting maps must be stitched together to allow the microstructure to be visualized.

Sample Preparation for EBSD

As in any experiment, sample preparation is critical to the success or usefulness of an EBSD experiment. The main requirement for EBSD is that the prepared surface be representative of the starting material. As was discussed earlier in this article, the depth resolution or the depth from which the patterns are generated in the sample is quite shallow. This means that the surface prepared for EBSD should be free from damage associated with any sectioning and

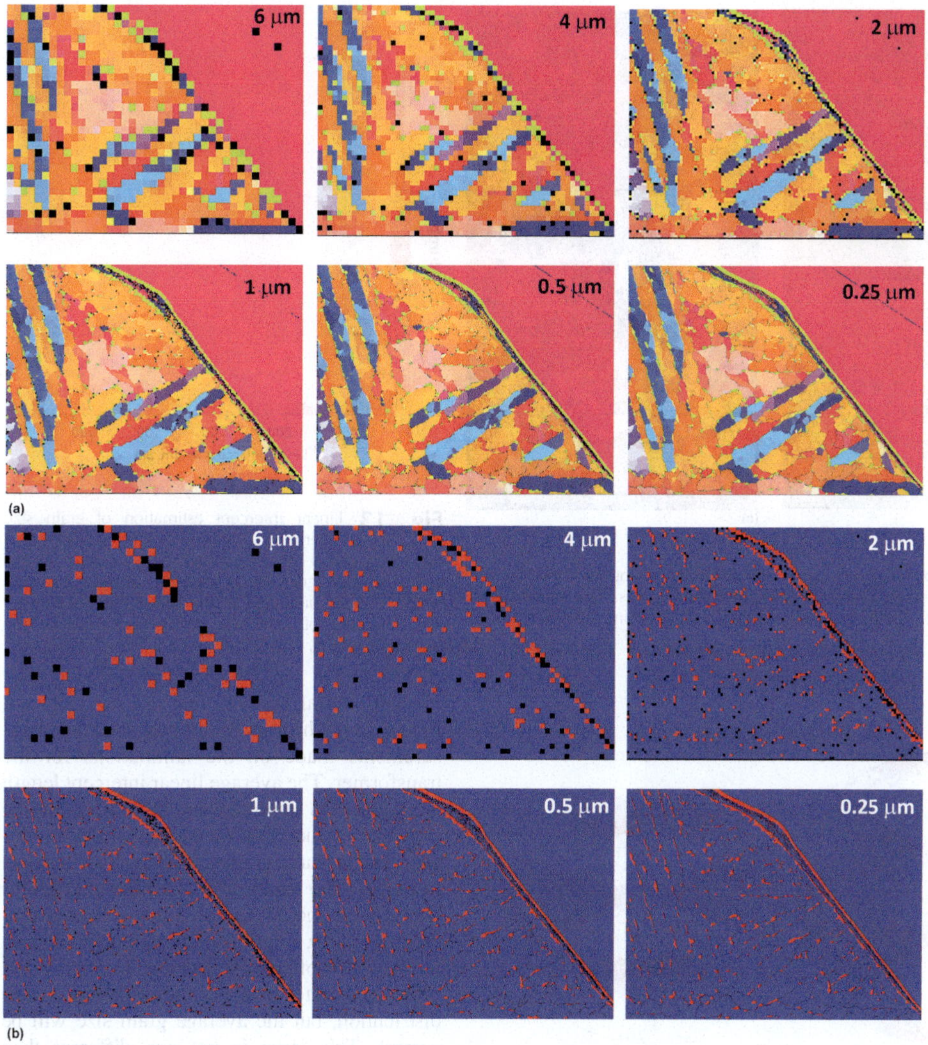

Fig. 10 Comparison of maps obtained from a two-phase region of austenite and ferrite. The step size was varied from 6 to 0.25 μm. Finer step sizes produce better sampling of the fine structure of this sample at the expense of speed, because more smaller pixels are required to sample the same area. The 6 μm step-size map required 1 s to collect, while the 0.25 μm step-size map required 12 min and 24 s. (a) Inverse pole figure maps. (b) Phase maps with austenite shown in red and ferrite in blue

polishing steps. Due to the limited depth, any damage that is induced by the sample-preparation technique will either obscure or distort the patterns and may make them unsuitable for indexing. There are a limited set of samples that may not require any additional sample preparation prior to performing EBSD. Examples of this are as-grown films on surfaces and some naturally or laboratory-grown single crystals. Electron backscatter diffraction has been performed to determine the growth directions of individual tin whiskers with no sample preparation (Ref 31). However, most samples will require some sample preparation.

In most cases, standard careful metallographic sectioning and polishing techniques should produce adequate samples for EBSD. The standard preparation steps leading up to the final polishing step include sectioning, mounting, and grinding; these should be conducted in a manner to minimize extensive subsurface preparation-induced damage. The final polishing step is critically important to obtaining a surface that will yield quality EBSD patterns. Final polishing of the sample may include colloidal suspensions performed manually or with a vibratory polisher (Ref 32–36). Electropolishing and ion beam polishing have also been used successfully for the final polishing step for EBSD. Chemical etching of the sample is sometimes helpful but should be applied with caution, because the introduction of surface topography or oxide films can greatly reduce the pattern quality (Ref 10, 37). Low-energy polishing using a focused ion beam tool or a low-energy argon ion polishing system has been shown to be very effective in developing damage-free surfaces suitable for EBSD for some materials (Ref 38).

It is entirely acceptable to apply conductive coatings to samples that present imaging difficulties due to a lack of electrical conductivity. This is known as sample charging, due to the energetic SEM beam interacting with the sample. This also applies to samples that are mounted in epoxy or other insulating mounting media, due to the potential for charging of the mount at a location remote from the analysis area. The analyst should always use the minimum coating thickness needed to reduce charging, because thicker conductive layers tend to reduce EBSD pattern visibility (Ref 39, 40). Great success in reducing the effects of sample charging has been achieved by using poor vacuum modes when available. The lowest pressure possible that mitigates sample charging should be used. Too high of a chamber pressure will result in the blurring of the EBSD patterns, due to electron scattering in the gas from the sample to the EBSD camera (Ref 41).

Applications of EBSD to Bulk Samples

Understanding Texture and Grain Size in Iron-Cobalt Transformer Material

The performance of a laminated electrical transformer depends on the material used in its construction. The crystallographic texture of the laminated core of the transformer affects the transformer efficiency. In this case, a customer was interested in determining the texture of the iron-cobalt body-centered cubic (bcc) alloy that had been used to make an older custom transformer that needed refurbishing. A cross section of the transformer core was metallographically prepared, including vibratory polishing as a last step. This produced a sample that was suitable for EBSD. During mounting, the major directions associated with processing must be understood so that the resulting EBSD data are more easily interpreted. The sample was prepared with the transverse direction (TD) of rolling normal to the prepared surface. The sample was mounted in the SEM with the rolling direction (RD) parallel to the sample tilt axis; the TD was parallel to the sample normal, and the RD was normal to the TD and the normal direction (ND), placing it in the plane of the sample. The SEM was operated at 20 kV and 10 nA of beam current to obtain the EBSD map. A large-area map with a 2 μm step size was obtained with combined beam and stage motion. Inverse pole figure maps with respect to the RD, TD, and ND are shown in Fig. 11. Note that a complete view of the sample requires all three (or at least two of these, because the third can be inferred) properly color-coded IPF maps, because each map can only show one direction (Ref 23).

Discreet and contoured <001>, <110>, and <111> pole figures from the IPF maps shown in Fig. 11 are shown in Fig. 12(a) and (b). The pole figures are complicated but show the expected α- and γ-fibers expected from

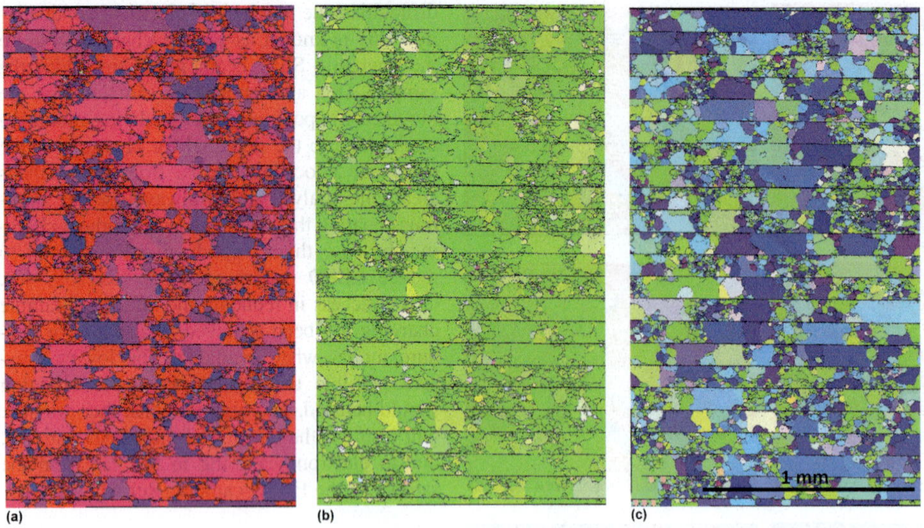

Fig. 11 Electron backscatter diffraction (EBSD) inverse pole figure maps with respect to the rolling direction, normal direction, and tranverse direction acquired from iron-cobalt transformer laminations. The sample was prepared by using standard metallographic techniques, and the EBSD maps were obtained at 20 kV and a 2 μm step size.

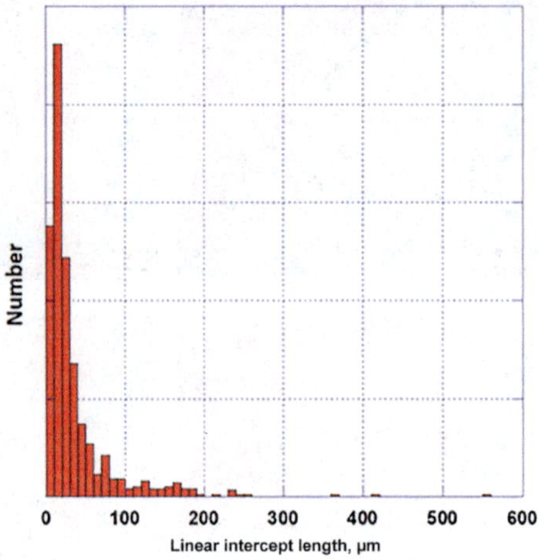

Fig. 13 Linear intercept estimation of grain size determined from electron backscatter diffraction maps of iron-cobalt alloys, using a 10° definition of a grain boundary. The average intercept length is 35 μm. Intercept length data should not be used as a measure of grain-size distributions, due to the sectioning effects.

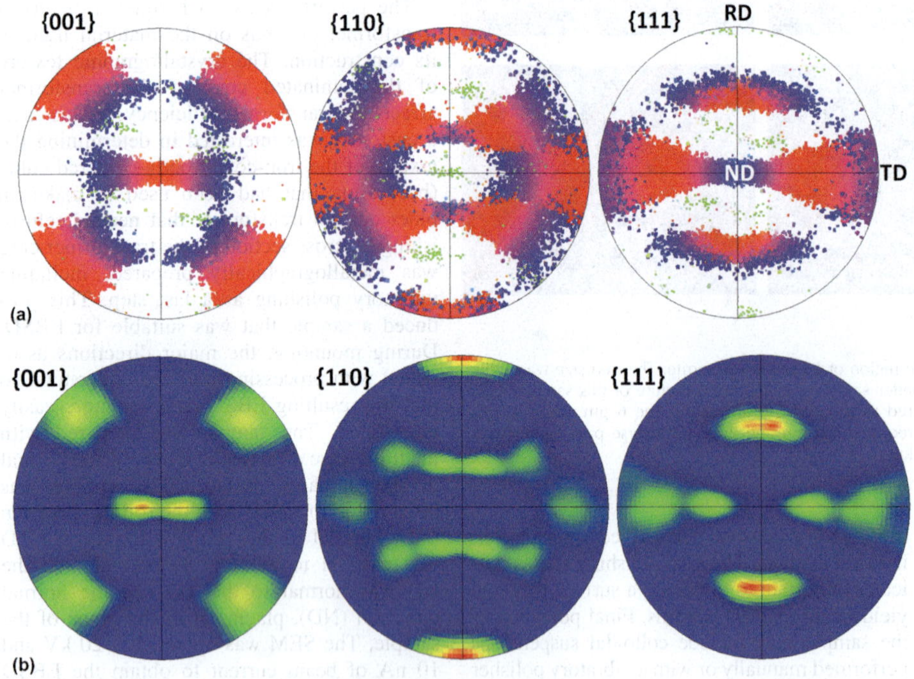

Fig. 12 Pole figures from iron-cobalt transformer laminations. (a) Discreet pole figure. RD, rolling direction; ND, normal direction; TD, transverse direction. (b) Contoured pole figure. Both show the presence of alpha and gamma fiber textures common in rolled and annealed body-centered cubic metals.

rolled bcc metals (Ref 42). It is also possible to provide measures of microstructural aspects from EBSD maps and data. One interesting measure is grain size. Grain size has been more traditionally measured from etched metallographic samples, with the assumption that the applied etch delineates high-angle grain boundaries. Once an EBSD map has been obtained, the grain size for a given quantitative grain-boundary description can be determined (Ref 43, 44). ASTM E 2627 provides guidelines for average grain-size measurement by using EBSD (Ref 45). The EBSD maps are analyzed and grain boundaries are determined by user-selected misorientations (usually greater than 10°) between pixels. It is also possible to ignore twins or other microstructural features that are not relevant. Figure 13 is a histogram of linear intercept measurements made on the laminations for the transformer. The average linear intercept length is 35 μm. It is important to understand that the step size of the EBSD measurement can influence the measured grain size. Steps must be taken to remove the smallest of the measured grains (Ref 46). It should be noted that this measurement is from a single section with no correction for the effect of cross sectioning. Thus, this data must be corrected to determine the true grain-size distribution, but the average grain size will be correct. This issue is not any different than grain-size measurement made from a single image from any imaging technique. Corrections for this are available in the literature, although none are accurate for typical grain shapes (Ref 47, 48).

Understanding Austenite/Ferrite Phase Relationships in Iron Meteorites

Iron meteorites have been studied extensively because the phases (mostly ferrite or kamacite with smaller amounts of austenite or taenite), orientation relationships, and elemental distributions can be used to learn about the origins of the universe (Ref 49, 50). Iron meteorites are also quite interesting from the aspect of phase transformations, because these meteorites have been slowly cooled over many millions of years, resulting in near-equilibrium phase distributions as well as well-formed orientation relationships between the transformation products of austenite and ferrite (Ref 51, 52). Understanding this orientation relationship is important for many metallurgical products and processes, such as welding and additive manufacturing. Specific meteorites were prepared for study of the regions that consisted of fine-grained austenite and ferrite, also called plessite (Ref 51, 53–56). A variety of EBSD maps obtained from the Tawallah Valley meteorite are shown in Fig. 14. The band contrast map in Fig. 14(a) is a useful way to show the general sample microstructure. All commercial EBSD systems offer some display that is associated with the quality of the EBSD patterns obtained at each pixel, and these maps are useful as a general indicator of the microstructure of the analyzed area. Figure 14(b) is the ferrite IPF map with respect to the horizontal direction of the sample. Figure 14(c) and (d) are austenite IPF maps with respect to the horizontal direction (IPF *X*) and the vertical direction (IPF *Y*). The color legend used is shown in Fig. 7(a) for both the austenite and the ferrite, because both phases are cubic with m3m symmetry. Note that the ferrite IPF map shows that there are multiple orientations present. The austenite IPF maps shown in Fig. 14(c) and (d) indicate that most of the austenite is present as a single orientation, with some small regions that are of a different orientation.

It is well known that ferrite that forms from austenite or austenite that transforms from ferrite has well-defined orientation relationships. These have been described in the literature and are called either Nishiyama-Wassermann (NW) or Kurdjumov-Sachs (KS) (Ref 57–60). The KS relationship is $\{110\}_{\text{ferrite}} \parallel \{111\}_{\text{austenite}}$ and $\langle 111\rangle_{\text{ferrite}} \parallel \langle 110\rangle_{\text{austenite}}$. The NW relationship has the same relationship between planes as the KS, but the directions are described by $\langle 001\rangle_{\text{ferrite}} \parallel \langle 0\text{-}11\rangle_{\text{austenite}}$. The KS orientation relationship has the close-packed planes and close-packed direction in the austenite and ferrite aligned, while the NW orientation relationship has the close-packed planes aligned. The NW and KS relationships differ by approximately 6°. Pole figures calculated from the data used to produce the IPF maps in Fig. 14 for $\{111\}_{\text{austenite}}$ and $\{110\}_{\text{ferrite}}$ are shown in Fig. 15. The austenite <111> pole figure in Fig. 15(a) indicates that the austenite is nearly all of the same orientation, with some very small regions of austenite that exhibit other orientations. The ferrite <110> pole figure in Fig. 15(b) demonstrates the multiple orientations of ferrite that have formed consistent with the NW and KS orientation relationships during the cooling of the large parent body in space. One can compare the two pole figures and see that the majority of the <111> austenite poles are overlapped by the ferrite <110> poles, as expected by the KS and NW orientation relationship.

Figure 16 shows the idealized pole figures that are expected from the austenite-to-ferrite transformation if the orientation relationship that develops is a mix of NW and KS. Figure 16(a) is the <111> pole figure for the parent austenite, and Fig. 16(b) is the expected orientations of the daughter ferrite that forms from the austenite. A comparison of Fig. 15 and 16 demonstrates excellent agreement between the experimental measurement and the known NW and KS orientation relationships. The interesting observation is that the austenite in the meteorite has the same orientation across the area shown in Fig. 14. This leads to the conclusion that the austenite is a retained phase from the austenitic high-temperature parent phase and leads to a modified cooling and phase-transformation sequence (Ref 53, 54).

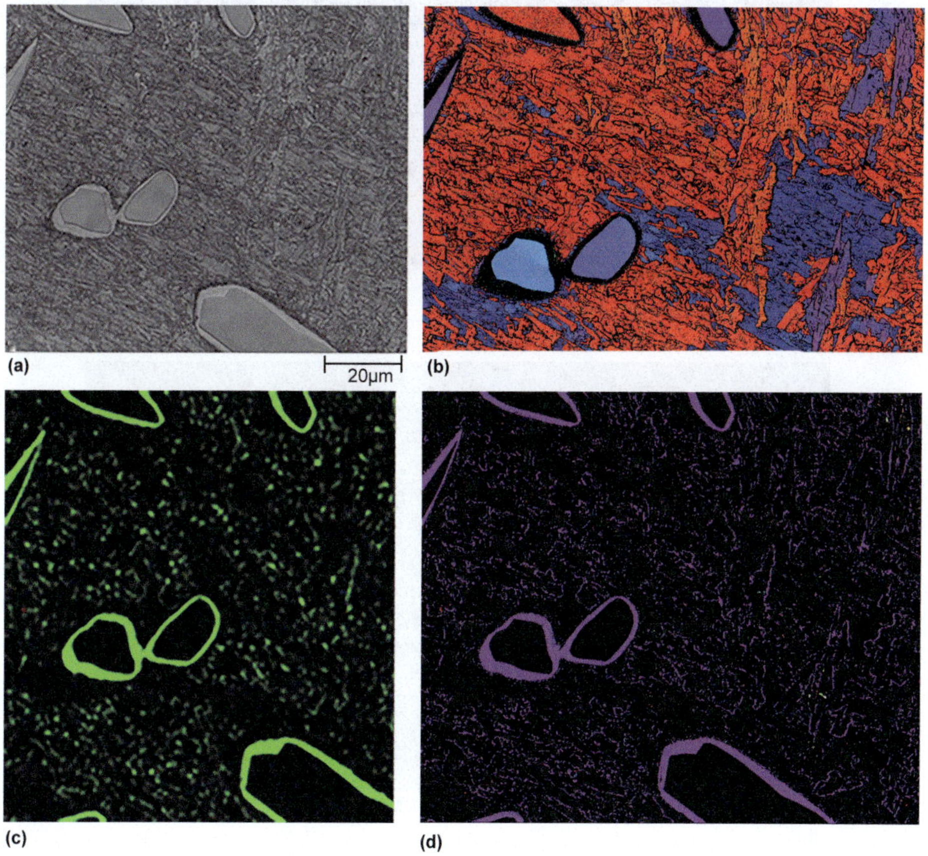

Fig. 14 Electron backscatter diffraction (EBSD) maps acquired from the Tawallah Valley meteorite. (a) Band contrast image produced by measuring the pattern contrast in the EBSD pattern from each pixel. (b) Ferrite inverse pole figure (IPF) horizontal direction. (c) Austenite IPF horizontal direction. (d) Austenite IPF vertical direction

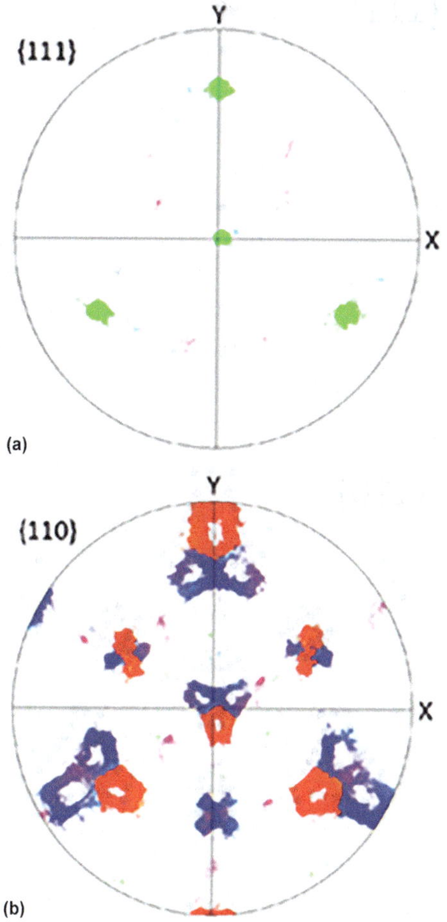

Fig. 15 Pole figures produced from the electron backscatter diffraction maps of Tawallah Valley. (a) Austenite or face-centered cubic {111} pole figure. (b) Ferrite or body-centered cubic {110} pole figure

Electron backscatter diffraction and x-ray diffraction studies of a variety of iron meteorites have shown that the austenite present has the same orientation and the ferrite is the transformation product that forms with the NW or KS relationship to the austenite (Ref 61). To demonstrate the application of large-area mapping with EBSD, maps were obtained from the Gibeon meteorite over an area of 21.9 by 90.4 mm (0.86 by 3.6 in.) and were stitched from 2170 individual fields, requiring 9 h. The pattern-quality map from the large area of Gibeon is shown in Fig. 17. This shows that the large-scale map captures an appropriate area when compared to the length scales of the microstructure of this sample. Figure 18 is an IPF map with respect to the *X*-direction, along with an inset that shows some of the detail in the map. Note that the austenite has the same orientation throughout the area imaged. A small-area inset is shown to demonstrate the detail available and the consistent orientation of the austenite. The best way to show the orientations of the ferrite and austenite is by pole figures, as shown in Fig. 19. Here, the <111> and <110> pole figures are shown for the area mapped in Fig. 18. Clearly, the austenite is nearly all the same orientation, while the ferrite has orientations that would be expected from the KS and NW orientation relationships. Thus, it can be inferred that the Widmanstätten ferrite has formed from a parent large single-crystal austenite grain during slow cooling of the large parent meteorite body (Ref 61, 62). These conclusions are similar to those drawn from the smaller maps shown in Fig. 14.

Fig. 17 Large-area electron backscatter diffraction pattern-quality map of the Gibeon meteorite. The area covers 21.9 by 90.4 mm (0.86 by 3.6 in.) and is 2170 individual fields stitched. The map was obtained at a 2 μm step size at 1894 pixels per second for a total acquisition time of 9.5 h.

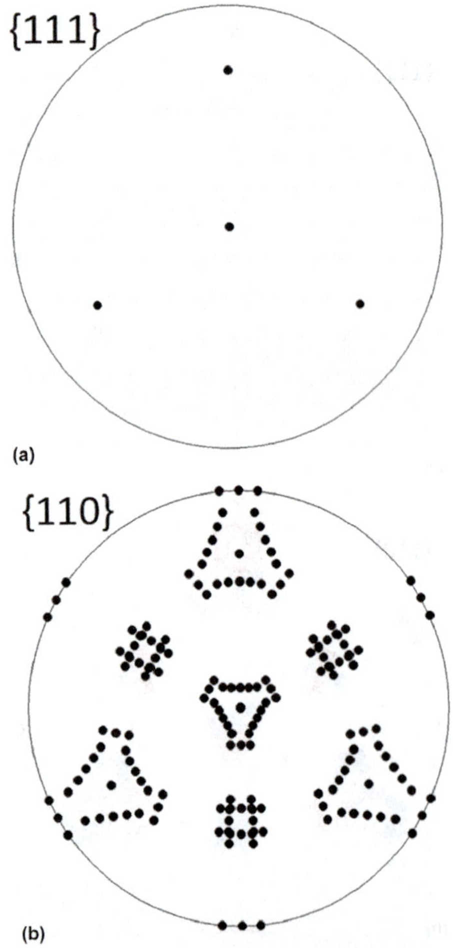

Fig. 16 Calculated pole figures for the combined Nishiyama-Wassermann and Kurdjumov-Sachs orientation relationships. (a) Face-centered cubic {111} pole figure. (b) Ferrite body-centered cubic {110} pole figure

(a)

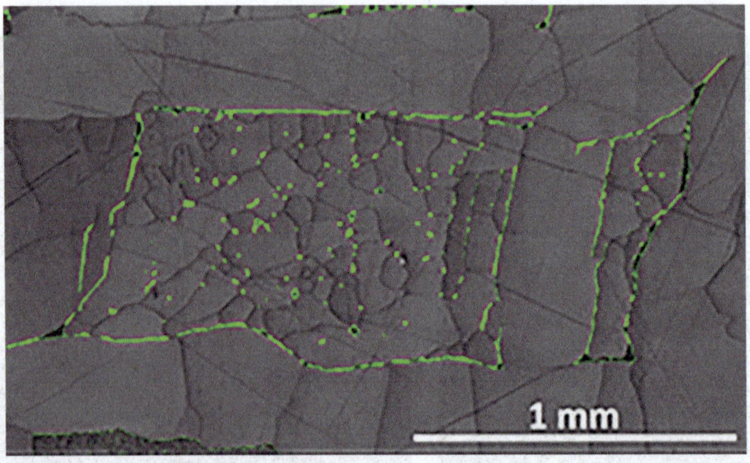

(b)

Fig. 18 Inverse pole figure (IPF) X-map of the area shown in Fig. 17. (a) IPF map with respect to the horizontal direction. (b) Composite map from the region outlined in white in (a) of the ferrite band contrast map and the IPF map for austenite. Note the high-quality map that can be extracted from the large-area data.

Specialized Applications

Transmission Kikuchi Diffraction

Orientation mapping is most commonly applied to polished bulk samples and is performed in the conventional manner, as shown in Fig. 4 and 5. The interaction volume limits the spatial resolution attainable with this method. Transmission Kikuchi diffraction (TKD) was developed to allow extremely high-resolution orientation maps to be obtained by using the same hardware and software used for conventional mapping (Ref 63–65). This technique has also been referred to as transmission EBSD. Transmission Kikuchi diffraction differs from conventional EBSD in two ways. First, the sample must be made thin so that the electron beam can penetrate through the sample. Secondly, the camera is placed below or on the exit side of the sample opposite the side that is imaged by the SEM. There are two arrangements of the camera and sample that are currently used, and these are shown in Fig. 20. Figure 20(a) is an example of TKD with a conventional vertical phosphor screen, and Fig. 20(b) is an example with a horizontal phosphor screen, or what is termed on-axis. The on-axis approach is the more recent arrangement and requires modification of the camera or the fitment of an appropriate phosphor screen to a standard EBSD camera (Ref 66, 67). The advantages of the conventional arrangement for TKD are that no additional hardware is required and that the existing EBSD hardware and software can be used. The disadvantage to the conventional arrangement is that the resulting diffraction patterns may be significantly distorted by the camera arrangement, causing difficulty in indexing. This distortion is due to the gnomonic projection distortion and can be reduced by tilting the sample away from the EBSD phosphor screen by up to 30°. Most modern EBSD systems can deal with gnomonically distorted patterns. The on-axis arrangement does not suffer from the elongation of the diffraction patterns, and, because of the intense forward-scattered nature of the electrons, shorter dwell times are needed to produce patterns that are suitable for automated indexing (Ref 68). Alternatively, one can apply a lower beam current to allow a smaller electron probe to be used to improve spatial resolution of the measurement (Ref 67).

Figure 21 is a comparison of TKD patterns from a thin gold sample collected with the conventional arrangement in Fig. 21(a) and the on-axis arrangement of the phosphor screen in Fig. 21(b). The pattern collected in the conventional arrangement in Fig. 21(a) demonstrates the gnomonic elongation of the pattern that results from the geometric location of the sample with respect to the phosphor screen location. Note that the on-axis pattern in Fig. 21(b) has a very intense bright spot in the middle of the pattern that corresponds to the transmitted beam of electrons. This spot is ignored by the indexing software but cannot be allowed to be too big, or fewer Kikuchi bands will be detected. In either case, indexing software can effectively determine the Kikuchi band positions and index the patterns.

There are some significant difficulties that the analyst should keep in mind when conducting a TKD study. The sample thickness is critical to the success of the analysis. Given that the maximum accelerating voltage in most modern SEM tools is 30 kV, there is a maximum useful thickness for TKD that depends on the sample atomic number and density. Good samples for TKD have been obtained by using focused ion beam (FIB) techniques or electropolishing; both techniques are well known to produce thin samples for transmission electron microscopy (TEM). As in bulk sample EBSD, it is important to keep track of the sample alignment for the experiment, if required. This may be relatively easy for FIB-produced samples, because these types of samples can be made with excellent alignment of the thin section of the sample with a known physical direction. Electropolished samples present difficulties, because there is often no way to control exactly where the thin region of the sample is located. It is prudent for the analyst to test sample-orientation measurements with known samples to understand that the system hardware and software produce orientations and how these orientations are related to the sample axes of interest.

The patterns obtained from TKD are not a through-thickness measurement of the sample, as in TEM imaging and analysis. In TKD, the patterns are formed in the last few tens of nanometers of the sample exit surface. Because of this exit surface sensitivity, the microstructural observation made in TEM may be difficult to correlate with the TKD measurements (Ref 68).

As was shown in Fig. 10 for EBSD from bulk samples, selection of step size for TKD is equally important. Sample details cannot be resolved if the step size is too large.

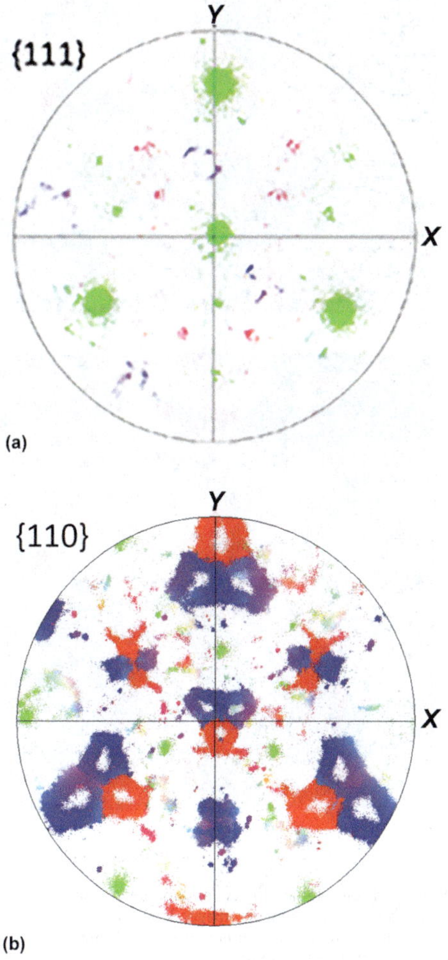

Fig. 19 Pole figures produced from electron backscatter diffraction maps of the large area of the Gibeon meteorite, demonstrating the Kurdjumov-Sachs and Nishiyama-Wassermann orientation relationships present in the large area of the sample. (a) Austenite or face-centered cubic {111} pole figure. (b) Ferrite or body-centered cubic {110} pole figure

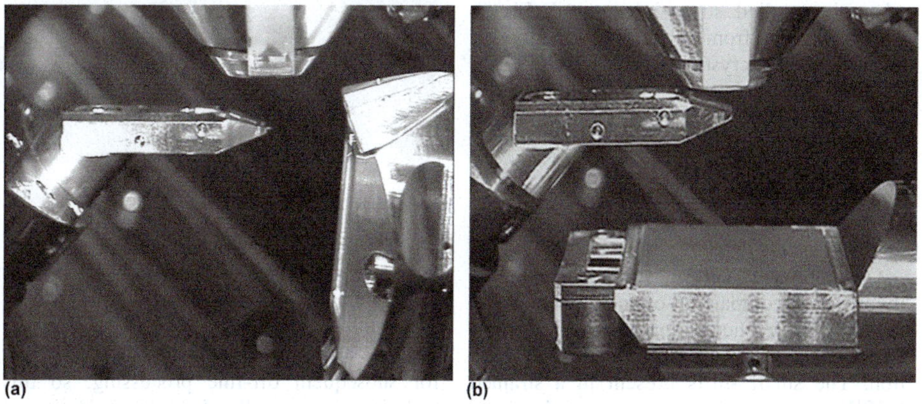

Fig. 20 Detector and sample positioning for transmission Kikuchi diffraction. (a) Conventional arrangement with vertical or conventional phosphor screen location. (b) On-axis orientation with horizontal phosphor screen

Further, too small a step will yield greater detail but at the expense of added time required to collect the TKD map. This is demonstrated in Fig. 22, where maps have been collected at 30 kV from a thin gold sample at 25 and 10 nm step sizes by using an on-axis phosphor screen. Depending on the information needed, the coarser step size may be adequate; however, the 10 nm step size shows that fine-scale microstructural details can be resolved in this gold sample.

An application of TKD to wear is shown in Fig. 23, where a thin section of a wear scar was prepared by FIB techniques. The wear scar was formed on a single crystal of {110} nickel, and sliding was in the <1-10> direction. In this example, a vertical phosphor screen geometry was used, as shown in Fig. 21(a). A dark-field 30 kV transmitted electron image of the sample is shown in Fig. 23(a). Figure 23 (b) is a pattern-quality map of the analyzed region of the sample, where the patterns were obtained by using a 3 nm step size. Figure 23 (c) shows an IPF map with respect to the wear direction (horizontal direction) of the wear scar. Transmission Kikuchi diffraction orientation mapping allows the microstructural and textural changes that occur during wear to be visualized.

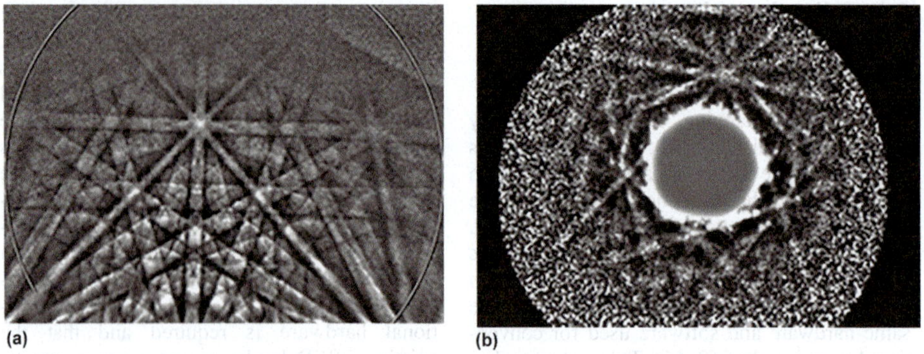

Fig. 21 Transmission Kikuchi diffraction patterns from gold at 30 kV. (a) Conventional positioning. (b) On-axis positioning

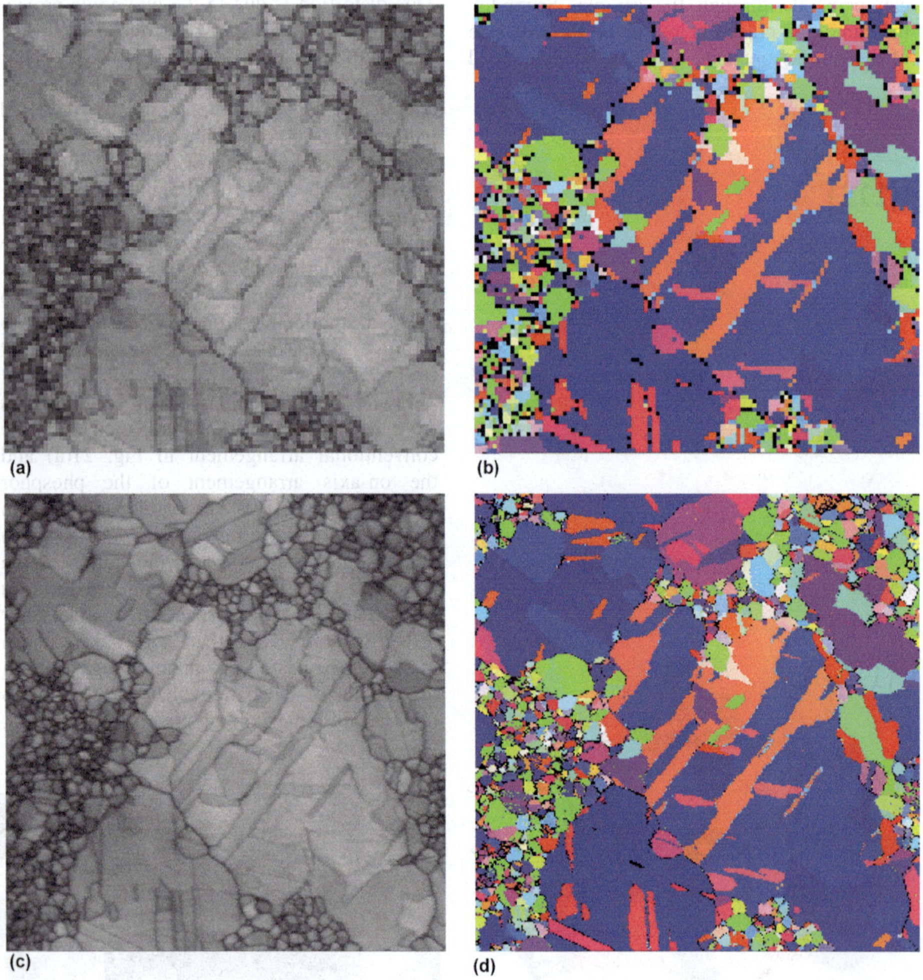

Fig. 22 Transmission Kikuchi diffraction maps of a thin gold sample at 30 kV obtained by using on-axis arrangement with 25 and 10 nm step sizes. Note the improvement in image detail with the smaller step size. (a) Band contrast map, 25 nm step size. (b) Inverse pole figure vertical direction map, 25 nm step size. (c) Band contrast map, 10 nm step size. (d) Inverse pole figure vertical direction map, 10 nm step size

Strain Analysis

There has been increasing interest in the use of EBSD to qualitatively and quantitatively measure strain in microstructures (Ref 69). There are EBSD-based methods that can be used for elastic strain and others for plastic strain. Some of these methods rely on analysis of conventional EBSD data, while other methods require EBSD patterns to be stored for comparison against ideal or patterns obtained from unstrained data. Plastic or elastic strain produces different observable changes in the EBSD patterns.

Elastic strain results when there is a linear relationship between the applied stress and the resulting strain or elongation of the sample (Ref 70). Elastic strains are fully recovered when the applied stress is removed. Elastic strains resulting from applied stress cause distortions of the crystal lattice, which causes changes in the observed EBSD pattern. The lattice distortion may cause small shifts in some of the zone axes, along with changes in the width of the Kikuchi bands. Also, hydrostatic stresses may only cause a change in the bandwidths without any relative movement of the zone axes.

One approach to elastic strain measurement involves the comparison of each pattern to an unstrained reference pattern. This comparison is typically done by means of cross correlation. The small shifts present in a strained EBSD pattern can be determined and used to calculate a strain tensor at every point in the EBSD map. Application of this method requires that every EBSD pattern be stored for subsequent off-line processing, so this technique is generally slow. The selection of a representative unstrained pattern for comparison is important to the success of the method. Much work is being done in this area, and some methods use calculated patterns, while some require actual patterns acquired from the sample (Ref 71–75).

Plastic strain occurs when the distortions in the crystal lattice due to an applied stress are

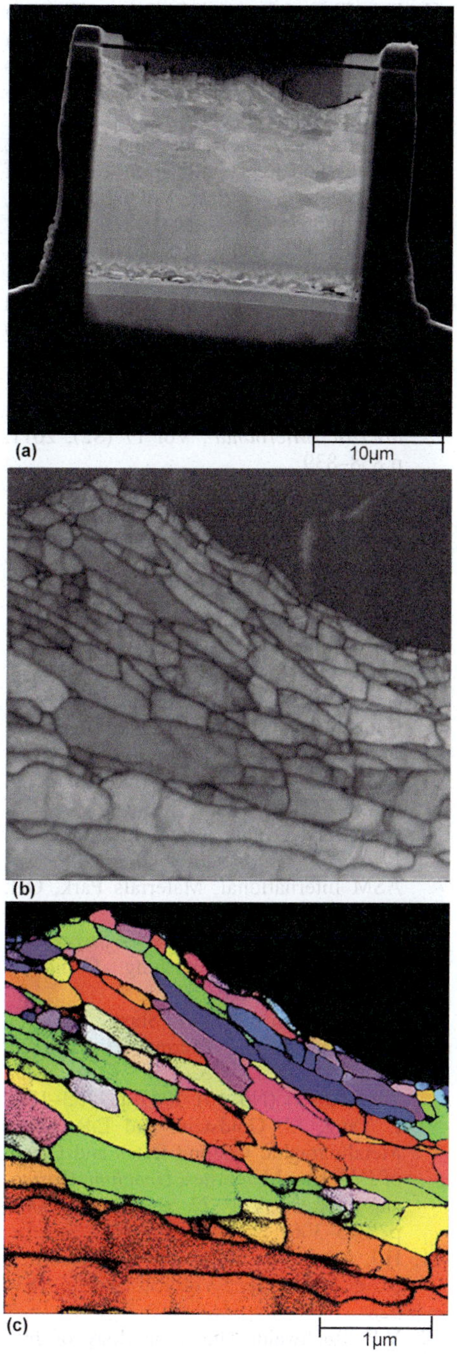

Fig. 23 Transmission Kikuchi diffraction maps of a worn nickel surface obtained at 30 kV by using the conventional arrangement and a 3 nm step size. (a) Scanning transmission electron microscope image of the thin focused-ion-beam-prepared sample. (b) Band contrast image of a small region of the sample shown in (a). (c) Orientation map with respect to the horizontal direction or the sliding direction of wear

accommodated by the formation of dislocations and other crystalline defects. The defects have been shown to degrade the quality of the EBSD patterns. The degradation of the EBSD pattern quality is reflected in the band contrast or image quality maps that are obtained from the sample. This approach can qualitatively show where the dislocation and defect density is high, but there are many other factors, such as surface contamination, poor sample preparation, surface topography, and others, that can reduce the EBSD pattern quality as well. As plastic strain is accumulated in the lattice, the residual strain will result in lattice rotations that can be measured from the EBSD orientation measurements. These orientations can be used to compare to other orientations within the map to determine the amount of plastic strain that is present in the microstructure. This is usually done with kernel operations, where a 3×3 kernel (larger may be used, depending on the microstructure and parameters needed) is used as a mask on the data. The orientation difference between the nine pixels is then plotted as an image to indicate something about the plastic strain state within the sample. This method has been shown to be a useful qualitative tool for assessing the relative amounts of plastic strain within a microstructure (Ref 69, 76, 77). These methods suffer because they are not generally considered quantitative. There have been many of these kernels developed, and each application is dependent on the information that is desired. For example, one can compare the average misorientation within a grain to determine the plastic strain within a sample. This is sometimes referred to as the grain-averaged misorientation.

New EBSD Indexing Methods

The traditional use of the Hough transform and pattern-indexing procedures has proven to be quite useful, but these techniques are somewhat limited in the angular accuracy of the orientation measurements. A new method is called the dictionary method, where EBSD patterns are calculated for all of the possible distinct orientations for the given crystal structure, followed by a comparison of the experimental patterns with the calculated patterns to determine the sample orientation (Ref 78, 79). This method has been shown to provide orientation accuracies that are better than the traditional Hough-based approach, but at the expense of speed and complexity (Ref 80). The pattern-matching or dictionary approach has been shown to be useful where the EBSD patterns are not of high quality, allowing much better indexing in cases of poor pattern quality. Further developments in algorithms and computer hardware are necessary to provide needed analysis speed increases (Ref 78).

DISCLAIMER

Sandia National Laboratories is a multimission laboratory managed and operated by National Technology & Engineering Solutions of Sandia, LLC, a wholly owned subsidiary of Honeywell International Inc., for the U.S. Department of Energy's National Nuclear Security Administration under contract DE-NA0003525. This article describes objective technical results and analysis. Any subjective views or opinions that might be expressed in the article do not necessarily represent the views of the U.S. Department of Energy or the United States Government.

REFERENCES

1. B.L. Bramfitt and J.R. Michael, AEM Microanalysis of Phase Equilibria in Ni_3Al Intermetallic Alloys Containing Iron, *Materials Problem Solving with the Transmission Electron Microscope*, Materials Research Society, 1985, p 201–208
2. A. Winkelmann, Dynamical Simulation of Electron Backscatter Patterns, *Electron Backscatter Diffraction in Materials Science*, Springer U.S., New York, 2009
3. V. Randle, *Microtexture Determination and Its Applications*, 2nd ed., Maney, London, 2013
4. L. Reimer, *Scanning Electron Microscopy: Physics of Image Formation and Microanalysis*, Springer-Verlag, New York, 1985
5. J.M. Cowley, *Diffraction Physics*, North-Holland, New York, 1990
6. J.R. Michael and J.A. Eades, Use of Reciprocal Lattice Layer Spacing in Electron Backscatter Diffraction Pattern Analysis, *Ultramicroscopy*, Vol 81 (No. 2), 2000, p 67–81
7. D.J. Dingley, K. Baba-Kishi, and V. Randle, *Atlas of Electron Backscatter Diffraction Patterns*, Institute of Physics Publishing, Bristol, 1995
8. M.N. Alam, M. Blackman, and D.W. Pashley, High-Angle Kikuchi Patterns, *Proc. R. Soc. (London)*, Vol 221, 1954, p 224–242
9. R.A. Schwarzer, Automated Crystal Lattice Orientation Mapping Using a Computer-Controlled SEM, *Micron*, Vol 28 (No. 3), 1997, p 249–265
10. J.I. Goldstein, D.E. Newbury, J.R. Michael, N.W. Ritchie, J.H.J. Scott, and D.C. Joy, *Scanning Electron Microscopy and X-Ray Microanalysis*, Springer, New York, 2018
11. D.J. Prior, P.W. Trimby, U.D. Weber, and D.J. Dingley, Orientation Contrast Imaging of Microstructures in Rocks Using Forescatter Detectors in the Scanning Electron Microscope, *Mineralog. Mag.*, Vol 60 (No. 6), 1996, p 859–869
12. A.P. Boyle, D.J. Prior, M.H. Banham, and N.E. Timms, Plastic Deformation of Metamorphic Pyrite: New Evidence from Electron-Backscatter Diffraction and Forescatter Orientation-Contrast Imaging, *Miner. Depos.*, Vol 34 (No. 1), 1998, p 71–81

13. S.I. Wright, M.M. Nowell, R. de Kloe, P. Camus, and T. Rampton, Electron Imaging with an EBSD Detector, *Ultramicroscopy*, Vol 148, 2015, p 132–145
14. M.M. Nowell and S.I. Wright, Orientation Effects on Indexing of Electron Backscatter Diffraction Patterns, *Ultramicroscopy*, Vol 103 (No. 1), 2005, p 41–58
15. M.M. Nowell and S.I. Wright, Phase Differentiation via Combined EBSD and XEDS, *J. Microsc.*, Vol 213 (No. 3), 2004, p 296–305
16. B.S. El-Dasher and S.G. Torres, Electron Backscatter Diffraction in Low Vacuum Conditions, *Electron Backscatter Diffraction in Materials Science*, A.J. Schwartz, M. Kumar, B.L. Adams, and D.P. Field, Ed., Kluwer Academic/Plenum Publishers, New York, 2009
17. A. Deal, T. Hooghan, and A. Eades, Energy-Filtered Electron Backscatter Diffraction, *Ultramicroscopy*, Vol 108 (No. 2), 2000, p 116–125
18. J.R. Michael, Phase Identification Using Electron Backscatter Diffraction in the Scanning Electron Microscope, *Electron Backscatter Diffraction in Materials Science,* M. Kumar, B.L. Adams, and A.J. Schwartz, Ed., Kluwer Academic/Plenum Publishers, New York, 2000, p 75–89
19. D. Chen, J.C. Kuo, and W.T. Wu, Effect of Microscopic Parameters on EBSD Spatial Resolution, *Ultramicroscopy*, Vol 111 (No. 9–10), 2011, p 1488–1494
20. S. Zaefferer, On the Formation Mechanisms, Spatial Resolution and Intensity of Backscatter Kikuchi Patterns, *Ultramicroscopy*, Vol 10 (No. 2–3), 2007, p 254–266
21. T.C. Isabell and V.P. Dravid, Resolution and Sensitivity of Electron Backscattered Diffraction in a Cold Field Emission Gun SEM, *Ultramicroscopy*, Vol 67 (No. 1–4), 2007, p 59–68
22. V.F. Leavers, *Shape Detection in Computer Vision Using the Hough Transform*, Springer, London, 1992
23. G. Nolze and R. Hielscher, Orientations—Perfectly Colored, *J. Appl. Crystallogr.*, Vol 49 (No. 5), 2016, p 1786–1802
24. T.B. Britton, J. Jiang, Y. Guo, A. Vilalta-Clemente, D. Wallis, L.N. Hansen, A. Winkelmann, and A.J. Wilkinson, Tutorial: Crystal Orientations and EBSD—Or Which Way Is Up? *Mater. Charact.*, Vol 117, 2016, p 113–126
25. J.R. Michael and R.P. Gohner, Crystallographic Phase Identification in the Scanning Electron Microscope: Backscattered Electron Kikuchi Patterns, *51st Annual Meeting of the Microscopy Society of America* (Cincinnati, OH), San Francisco Press,1993, p 772–773
26. R.P. Goehner and J.R. Michael, Phase Identification in a Scanning Electron Microscope Using Backscattered Electron Kikuchi Patterns, *J. Res. Natl. Inst. Stand. Technol.*, Vol 101 (No. 3), 1996, p 301–308
27. V. Randle and O. Engler, *Introduction to Texture Analysis: Macrotexture, Microtexture and Orientation Mapping*, Gordon and Breach Science Publishers, Amersterdam, 2000
28. V. Randle, Electron Backscatter Diffraction: Strategies for Reliable Data Acquisition and Processing, *Mater. Charact.*, Vol 60 (No. 9), 2009, p 913–922
29. T.J. Ruggles, T.M. Rampton, A. Khosravani, and D.T. Fullwood, The Effect of Length Scale on the Determination of Geometrically Necessary Dislocations via EBSD Continuum Dislocation Microscopy, *Ultramicroscopy*, Vol 164, 2016, p 1–10
30. G. Nolze, Image Distortions in SEM and Their Influences on EBSD Measurements, *Ultramicroscopy*, Vol 107 (No. 2–3), 2007, p 172–183
31. J.R. Michael, B.B. McKenzie, and D.F. Susan, Application of Electron Backscatter Diffraction for Crystallographic Characterization of Tin Whiskers, *Microsc. Microanal.*, Vol 18 (No. 4), 2012, p 876–884
32. G.F. Vander Voort, *Metallography: Principles and Practice*, McGraw-Hill Book Company, New York, 1984
33. G.F. Vander Voort, *Metallography: Principles and Practice*, ASM International, Materials Park, OH, 1999
34. G.F. Vander Voort, Metallographic Specimen Preparation for Electron Backscattered Diffraction, *Metall. Ital.*, Vol 11, 2009, p 71–79
35. G.F. Vander Voort, Metallographic Specimen Preparation for Electron Backscattered Diffraction: Part I, *Prakt. Metallogr.*, Vol 48 (No. 9), 2011, p 454–473
36. G.F. Vander Voort, Metallographic Specimen Preparation for Electron Backscattered Diffraction: Part II, *Prakt. Metallogr.*, Vol 48 (No. 10), 2011, p 527–543
37. P. Echlin, *Handbook of Sample Preparation for Scanning Electron Microscopy and X-Ray Microanalysis*, Springer, New York, 2011
38. J.R. Michael and L. Giannuzzi, Improved EBSD Sample Preparation via Low Energy Ga^+ FIB Ion Milling, *Microsc. Microanal.*, Vol 13 (S02), 2007, p 926–927
39. J.K. Farrer, J.R. Michael, and C.B. Carter, EBSD of Ceramic Materials, *Electron Backscatter Diffraction in Materials Science,* M. Kumar, B.L. Adams, and A.J. Schwarz, Ed., Kluwer Academic/Plenum Publishers, New York, 2000, p 299–318
40. J.I. Goldstein, D.E. Newbury, D.C. Joy, C.E. Lyman, P. Echlin, E. Lifshin, L.C. Sawyer, and J.R. Michael, *Scanning Electron Microscopy and X-Ray Microanalysis*, 3rd ed., Kluwer Academic/Plenum Publishers, New York, 2003
41. B.S. El-Dasher and S.G. Torres, Electron Backscatter Diffraction in Low Vacuum Conditions, *Electron Backscatter Diffraction in Materials Science*, Springer, Boston, 2009, p 339–344
42. M. Hölscher, D. Raabe, and K. Lücke, Rolling and Recrystallization Textures of bcc Steels, *Steel Res.*, Vol 62 (No. 12), 1991, p 567–575
43. N. Gao, S.C. Wang, H.S. Ubhi, and M.J. Starink, A Comparison of Grain Size Determination by Light Microscopy and EBSD Analysis, *J. Mater. Sci.*, Vol 40 (No. 18), 2005, p 4971–4974
44. J. Friel, S. Wright, and S. Sitzman, ASTM Grain Size by EBSD—A New Standard, *Microsc. Microanal.*, Vol 17 (S2), 2011, p 838–839
45. "Standard Practice for Determining Average Grain Size Using Electron Backscatter Diffraction (EBSD) in Fully Recrystallized Polycrystalline Materials," *ASTM Annual Book of Standards,* ASTM International, West Conshohocken, PA
46. K.P. Mingard, B. Roebuck, E.G. Bennett, M.G. Gee, H. Nordenstrom, G. Sweetman, and P. Chan, Comparison of EBSD and Conventional Methods of Grain Size Measurement of Hard Metals, *Int. J. Refract. Met. Hard Met.*, Vol 27 (No. 2), 2009, p 213–223
47. L. Wojnar, Analysis and Interpretation, *Practical Guide to Image Analysis*, ASM International, Materials Park, OH, 2000, p 145–202
48. E.E. Underwood, *Quantitative Stereology*, Addison-Wesley Publishing Company, Menlo Park, CA, 1970
49. J.I. Goldstein and J.M. Short, The Iron Meteorites, Their Thermal History and Parent Bodies, *Geochim. Cosmochim. Acta*, Vol 31 (No. 10), 1967, p 1733–1770
50. J. Yang, J.I. Goldstein, and E.R. Scott, Metallographic Cooling Rates and Origin of IVA Iron Meteorites, *Geochim. Cosmochim. Acta*, Vol 72 (No. 12), 2008, p 3043–3061
51. H.J. Axon, The Metallurgy of Meteorites, *Prog. Mater. Sci.*, Vol 13, 1968, p 183–228
52. V.F. Buchwald, The Mineralogy of Iron Meteorites, *Philos. Trans. R. Soc. (London) A*, Vol 286 (No. 1336), 1977, p 453–491
53. G. Nolze and V. Geist, A New Method for the Investigation of Orientation Relationships in Meteoritic Plessite, *Cryst. Res. Technol.*, Vol 39 (No. 4), 2004, p 343–352
54. J.I. Goldstein and J.R. Michael, The Formation of Plessite in Meteoritic Metal, *Meteorit. Planet. Sci.*, Vol 41 (No. 4), 2006, p 553–570
55. B. Hutchinson and J. Hagström, Austenite Decomposition Structures in the Gibeon Meteorite, *Metall. Mater. Trans. A*, Vol 37 (No. 6), 2006, p 1811–1818

56. J. Yang, J.I. Goldstein, E.R.D. Scott, J.R. Michael, P.G. Kotula, T. Pham, and T.J. McCoy, Thermal and Impact Histories of Reheated Group IVA, IVB, and Ungrouped Iron Meteorites and Their Parent Asteroids, *Meteorit. Planet. Sci.*, Vol 46 (No. 9), 2011, p 1227–1252
57. Z. Nishiyama, X-Ray Investigation of the Mechanism of the Transformation from Face Centered Cubic Lattice to Body Centered Cubic, *Sci. Rep. Tohoku Univ.*, Vol 23, 1934, p 637–664
58. G. Wassermann and K. Mitt, About the Mechanism of α-γ Transformation of the Iron, *Wilh. Inst. Wisenforsche*, Vol 17, 1935, p 149–155
59. G. Kurdjumov and G. Sachs, Over the Mechanism of Steel Hardening, *Z. Phys.*, Vol 64, 1930, p 325–343
60. G. Nolze, Determination of Orientation Relationships between fcc/bcc Lattices by Use of Pole Figure, *Proceedings of Channel Users Meeting* (Ribe, Denmark), 2004, p 37–45
61. H.J. Bunge, W. Weiss, H. Klein, L. Wcislak, U. Garbe, and J.R. Schneider, Orientation Relationship of Widmannstätten Plates in an Iron Meteorite Measured with High-Energy Synchrotron Radiation, *J. Appl. Crystallogr.*, Vol 36 (No. 1), 2003, p 137–140
62. G. Nolze, V. Geist, R.S. Neumann, and M. Buchheim, Investigation of Orientation Relationships by EBSD and EDS on the Example of the Watson Iron Meteorite, *Cryst. Res. Technol.: J. Exp. Ind. Crystallogr.*, Vol 40 (No. 8), 2005, p 791–804
63. R.H. Geiss, R.R. Keller, and D.T. Read, Transmission Electron Diffraction from Nanoparticles, Nanowires and Thin Films in an SEM with Conventional EBSD Equipment, *Microsc. Microanal.*, Vol 16 (S2), 2010, p 1742–1743
64. R.R. Keller and R.H. Geiss, Transmission EBSD from 10 nm Domains in a Scanning Electron Microscope, *J. Microsc.*, Vol 245 (No. 3), 2012, p 245–251
65. P.W. Trimby, Orientation Mapping of Nanostructured Materials Using Transmission Kikuchi Diffraction in the Scanning Electron Microscope, *Ultramicroscopy*, Vol 120, 2012, p 16–24
66. J.J. Fundenberger, E. Bouzy, D. Goran, J. Guyon, H. Yuan, and A. Morawiec, Orientation Mapping by Transmission-SEM with an On-Axis Detector, *Ultramicroscopy*, Vol 161, 2016, p 17–22
67. H. Yuan, E. Brodu, C. Chen, E. Bouzy, J.J. Fundenberger, and L.S. Toth, On-Axis versus Off-Axis Transmission Kikuchi Diffraction Technique: Application to the Characterisation of Severe Plastic Deformation-Induced Ultrafine-Grained Microstructures, *J. Microsc.*, Vol 267 (No. 1), 2017, p 70–80
68. J.D. Sugar, J.T. McKeown, D.C. Bufford, and J.R. Michael, Crystallographic Orientation Image Mapping with Multiple Detector Configurations at 30–300 kV, *Microsc. Microanal.*, Vol 23 (S1), 2017, p 534–535
69. S.I. Wright, M.M. Nowell, and D.P. Field, A Review of Strain Analysis Using Electron Backscatter Diffraction, *Microsc. Microanal.*, Vol 17 (No. 3), 2011, p 316–329
70. R.W. Herztberg, *Deformation and Fracture Mechanics of Engineering Materials*, John Wiley and Sons, New York, 1976
71. A.J. Wilkinson, Measurement of Elastic Strains and Small Lattice Rotations Using Electron Back Scatter Diffraction, *Ultramicroscopy*, Vol 62 (No. 4), 1996, p 237–247
72. A.J. Wilkinson, G. Meaden, and D.J. Dingley, High-Resolution Elastic Strain Measurement from Electron Backscatter Diffraction Patterns: New Levels of Sensitivity, *Ultramicroscopy*, Vol 106 (No. 4–5), 2006, p 307–313
73. A.J. Wilkinson and T.B. Britton, Strains, Planes, and EBSD in Materials Science, *Mater. Today*, Vol 15 (No. 9), 2012, p 366–376
74. D. Fullwood, M. Vaudin, C. Daniels, T. Ruggles, and S. Wright, Validation of Kinematically Simulated Pattern HR-EBSD for Measuring Absolute Strains and Lattice Tetragonality, *Mater. Charact.*, Vol 107, 2015, p 270–277
75. S. Villert, C. Maurice, C. Wyon, and R. Fortunier, Accuracy Assessment of Elastic Strain Measurement by EBSD, *J. Microsc.*, Vol 233 (No. 2), 2009, p 290–301
76. L.N. Brewer, D.P. Field, and C.C. Merriman, Mapping and Assessing Plastic Deformation Using EBSD, *Electron Backscatter Diffraction in Materials Science*, Springer, Boston, 2009, p 251–262
77. J.F. Bingert, V. Livescu, and E.K. Cerreta, *Characterization of Shear Localization and Shock Damage with EBSD*, Springer, Boston, 2009
78. Y.H. Chen, S.U. Park, D. Wei, G. Newstadt, M.A. Jackson, J.P. Simmons, M. De Graef, and A.O. Hero, A Dictionary Approach to Electron Backscatter Diffraction Indexing, *Microsc. Microanal.*, Vol 21 (No. 3), 2015, p 739–752
79. S.I. Wright, M.M. Nowell, S.P. Lindeman, P.P. Camus, M. De Graef, and M.A. Jackson, Introduction and Comparison of New EBSD Post-Processing Methodologies, *Ultramicroscopy*, Vol 159, 2015, p 81–94
80. F. Ram, S. Wright, S. Singh, and M. De Graef, Error Analysis of the Crystal Orientations Obtained by the Dictionary Approach to EBSD Indexing, *Ultramicroscopy*, Vol 181, 2017, p 17–26

ASM Handbook, Volume 10, *Materials Characterization*
ASM Handbook Committee
DOI 10.31399/asm.hb.v10.a0006681

Transmission Electron Microscopy

Masashi Watanabe, Lehigh University

Overview

General Uses

- Imaging of microstructural features at 1,000 to 2,000,000× in conventional transmission electron microscopy (TEM) instruments and to 20,000,000× in aberration-corrected scanning transmission electron microscopy (STEM) instruments, respectively. Microstructural detail resolution can be as good as ~0.1 nm.
- Lattice imaging of crystals with interplanar spacings ~0.12 nm in conventional TEM instruments and ~0.05 nm in aberration-corrected TEM/STEM instruments, respectively
- Crystal structure and orientation determination of microstructural features as small as 2 to 3 nm
- Qualitative elemental and quantitative composition analysis of microstructural features as small as 1.0 nm in conventional TEM instruments and 0.1 nm in aberration-corrected TEM/STEM instruments, respectively

Specimens

- *Form:* Solids (metals, ceramics, minerals, polymers, biological, etc.)
- *Size:* A typical TEM specimen fits into a slot of a specimen holder in a disk shape with a diameter of 3 mm (0.12 in.) and a thickness of ~100 to 150 μm. For TEM characterization, electron-transparent thin regions are required within the disk. Conventional imaging and x-ray analysis typically require 50 to 200 nm thick specimens, whereas 10 to 100 nm or less is essential for atomic-resolution imaging and electron energy loss spectrometry (EELS) analysis.
- *Preparation:* Conventionally, bulk specimens must be sectioned and electropolished or ion milled to produce regions that permit transmission of the electron beam. If materials are relatively soft, for example, polymers and biomaterials, ultramicrotomy with a diamond knife can be used to slice electron-transparent thin sections. In addition, the focused ion beam (FIB) can fabricate thin specimens from metals/alloys, ceramics, polymers, and even biomaterials. Electron-transparent regions typically are less than 100 nm thick. Nanoparticles can be dispersed on a thin carbon (including graphene) or silicon nitride substrate.

Limitations

- *Specimen preparation:* Specimen preparation is tedious. However, the FIB significantly eases difficulties in specimen preparation.
- *Electron diffraction:* Minimum size of region analyzed is approximately 1 nm in diameter. Crystal structure identification is limited to phases or compounds tabulated in powder diffraction files (over 800,000 phases or compounds). Orientation relationship between two coexisting phases can be determined. Determination of full space and point groups is possible only by using specialized convergent-beam electron diffraction techniques. Precession electron diffraction can offer more accurate determination of crystal structures, including crystal symmetry, measurement of lattice parameters, and refinement of atomic structures without any prior knowledge.
- *Elemental analysis:* Typical minimum size of region analyzed is as low as 1 nm in diameter if a TEM instrument with a field-emission gun is used. Threshold analytical sensitivity is approximately 0.1 to 1 wt% in both x-ray analysis and EELS analysis in terms of minimum mass fraction, which indicates the lowest composition detectable. Conversely, the analytical sensitivity in terms of the minimum detectable mass, which corresponds to detectable number of atoms in the analyzed volume, is as good as a single atom in both x-ray analysis and EELS analysis in aberration-corrected STEM instruments. Accuracy of quantification should be 1 to 10% (relative) if appropriate quantification is applied. By x-ray analysis quantification is possible only for elements with atomic number ≥5 (boron). By EELS analysis, quantification can be applied to lithium (atomic number 3) or higher-atomic-number elements.

Capabilities of Related Techniques

- *X-ray diffraction:* Gives bulk crystallographic information
- *Optical metallography:* Faster; lower-magnification (up to approximately 2000×) overview of sample microstructure.
- *Scanning electron microscopy/electron probe x-ray microanalysis:* Faster; lower magnification than TEM (up to approximately 20,000×), with image resolution of approximately 5 to 10 nm in bulk samples. Both qualitative and quantitative chemical analysis can be performed by x-ray analysis.
- *Atom probe tomography:* Higher analytical sensitivity with single-atom detectability as well as atomic-level spatial resolution.

Introduction

Microstructure observation is an essential approach for materials characterization, which is primarily performed by using light optical microscopy and then scanning electron microscopy (SEM). Transmission electron microscopy (TEM) is used to perform further detailed microstructure characterization. The TEM approach is unique among materials characterization techniques in that it enables essentially simultaneous examination of microstructural features through imaging from lower magnifications to atomic resolution and the acquisition of chemical and crystallographic information from small (submicrometer down to atomic level) regions of the thin specimen. Figure 1 shows modern TEM instruments from selected manufacturers. Some new instruments are fully covered with a box to isolate them from the environment and achieve better stability. This article discusses fundamentals of the technique, especially for solving materials problems. The article is intended to be as practical as possible, with less emphasis on associated mathematical theories. Background information is provided to help understand basic operations and principles, including instrumentation, physics of signal generation and detection, image formation, electron diffraction, and spectrometry techniques with data analysis.

A major benefit of TEM is its superior resolution over optical microscopy and SEM. The intensity distribution of light or electrons from a single point should ideally be a single point. However, if light or electrons pass through a round hole or an aperture, the trajectories no longer converge to a single point but are slightly enlarged as concentric rings with a very intense circle at the center, as shown in Fig. 2. This enlargement (blurring) is caused by diffraction at the aperture, and its distribution size depends on the aperture size against the focal length (i.e., the angle, α) and the wavelength (λ) of light or electrons, which is called the diffraction-limited intensity distribution. The center circle is called the Airy disk, and its diameter is expressed as:

$$d = \frac{1.22\lambda}{\sin\alpha} \cong \frac{1.22\lambda}{\alpha} \qquad \text{(Eq 1)}$$

The diffraction limit is reduced if a larger aperture (i.e., larger angle) is used.

Resolution is defined as the minimum distance to distinguish two point objects; resolution can be determined based on the diffraction-limited intensity distributions of the two point objects. Figure 3 compares the diffraction-limited intensity distributions from the two point objects: completely separated (resolved) (Fig. 3a), intermediate (Fig. 3b), and not separated (unresolved) (Fig. 3c). When the two objects are located very far away, the intensity distributions are completely separated, and the two objects are resolved. Conversely, when the two objects are very close, the intensity distributions are superimposed, and thus, the two objects are unresolved. When the two objects are separated by a distance of 0.61 λ/α (normalized by the wavelength, λ, and aperture angle, α), the local minimum between the two objects in the intensity distribution appears as 75% of the peak intensity (Fig. 3b). This distance is defined as the Rayleigh criterion, which is generally accepted as the definition of the image-resolution limit. Therefore, the resolution of microscopy to distinguish two point objects is given as:

$$R = \frac{0.61\lambda}{\alpha} \qquad \text{(Eq 2)}$$

Resolution is directly related to wavelength. Table 1 summarizes wavelengths of different types of radiation and different energy values of electrons. For example, when λ = 500 nm (visible green light), R can be as good as 3 μm with α = 0.1 rad in Eq 2, which could be the limit of resolution in optical microscopy. Conversely, for an electron microscope operated at 200 kV (λ = 0.00251 nm) with α = 0.01 rad, R approaches 0.15 nm. To obtain higher resolution, a shorter-wavelength radiation should be used. This is the one of the major reasons why electrons are used instead of light and is a motivation for the development of TEM instruments, which typically operate at accelerating voltages greater than 100 kV.

The primary motivation to use electrons rather than visible light is for improvement of resolution. Achievement of the best possible resolution is the backbone of TEM

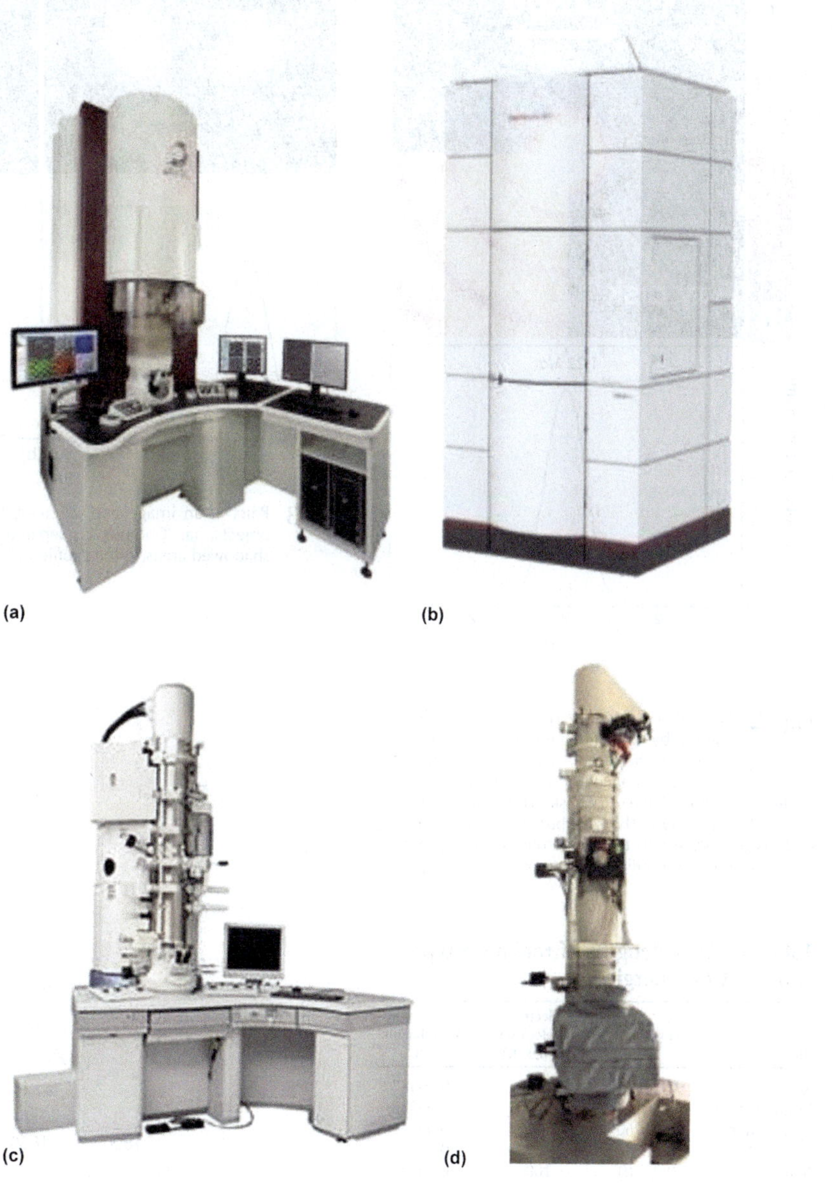

Fig. 1 Various transmission electron microscopy instruments. (a) JEOL JEM-ARM300F Grand ARM. Courtesy of JEOL. (b) Thermo/Fisher Themis Z. Source: Thermo/Fisher. (c) Hitachi HF-3300. Courtesy of Hitachi. (d) Nion UltraSTEM 200. Courtesy of O. Krivanek, Nion

instrumentation development. Figure 4 shows advances in microscopy resolution over time. When the first TEM was developed, the resolution was ~100 μm, which is equivalent to that achieved in optical microscopes. In the 1980s, high-voltage (1 MV) TEM instruments were developed to improve the resolution, and ~1.6 Å was achieved due to the short wavelength (87 pm) at 1 MV. The milestone of 1 Å resolution was achieved in the late 1990s using an intermediate-voltage (300 kV) TEM instrument associated with computer-based processing, such as the exit-wave reconstruction.

Further breakthrough was achieved in the early 2000s by developing improved optics, that is, aberration correctors. The resolution is now as good as 0.5 Å in aberration-corrected TEM instruments (Ref 3), as shown in Fig. 4.

Transmission Electron Microscopy Specimens: Dimensions

Much higher-energy electrons are used in TEM compared with SEM, typically 60 to 300 keV in modern instruments. Even with such high energies, most of the incident electrons must be transmitted through a specimen, so the specimen should be very thin. The ideal thickness for TEM characterization depends on the characterization type. For example, conventional imaging and x-ray analysis typically require 50 to 200 nm thick specimens, whereas 10 to 100 nm or less is essential for atomic-resolution imaging and electron energy loss analysis at 200 keV.

A typical TEM specimen fits into a slot of a disk-shaped specimen holder 3 mm (0.12 in.) in diameter and is ~100 to 150 μm thick, as shown in Fig. 5(a). This thickness is not electron transparent, but it is useful from a practical standpoint to handle using tweezers. For a self-supported specimen, there should be a thinner part (up to 200 nm) near the center

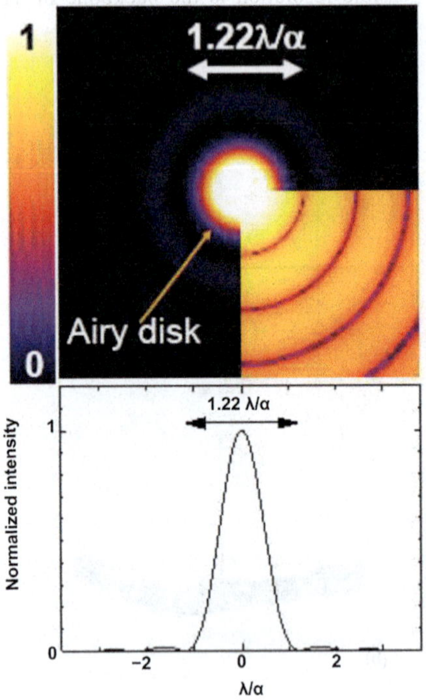

Fig. 2 Normalized intensity distribution of blurring caused by circular aperture diffraction with its image. The distribution due to the aperture diffraction appears as concentric rings, with the intense center circle called the Airy disk. The diameter of the Airy disk is 1.22 λ/α. The distribution image in the log scale is also shown in the right bottom corner, and an extracted intensity profile is shown below the image.

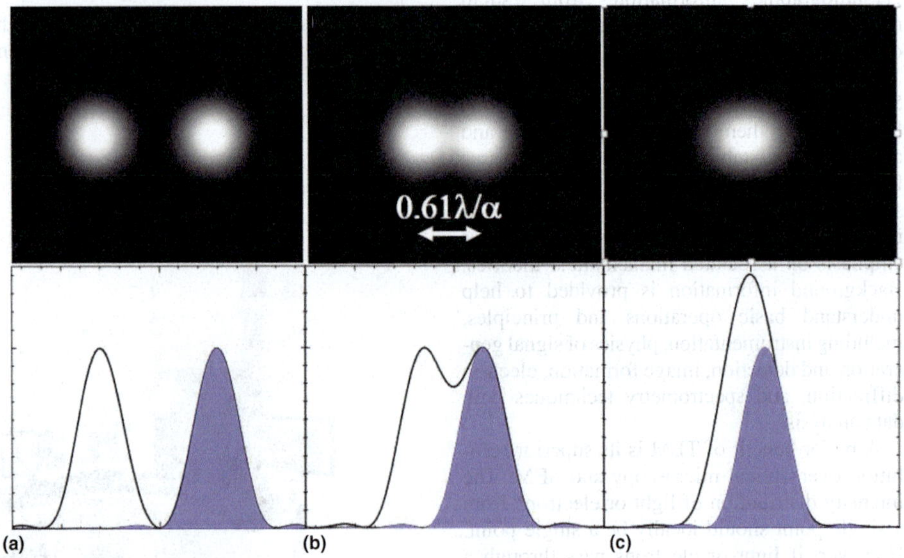

Fig. 3 Pairs of an image and extracted-line profile of diffraction-limited intensity distributions from two point objects. (a) Completely separated (resolved). (b) Intermediate. (c) Not separated (unresolved). The shadowed areas in the profile indicate the intensity distribution from the single point object.

Table 1 Wavelengths of radiation types and electron energies

Radiation	Wavelength, nm	Electron accelerating voltage, kV	Wavelength, nm
Radio wave	10^6–10^{12}	60	0.00487
Infrared ray	10^3–10^5	100	0.00387
Visible light	390–760	200	0.00251
Ultraviolet ray	13–390	400	0.00164
X-ray	0.05–10	1000	0.00087
γ-ray	0.005–0.1	···	···

Source: Ref 1

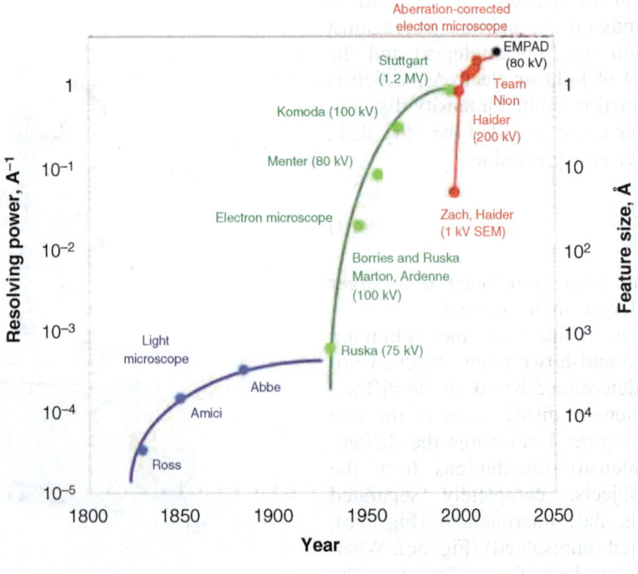

Fig. 4 Advances in microscopy resolution. EMPAD, **electron microscope pixel array detector.** Reprinted from Ref 2 with permission of Springer Nature. Copyright 2009

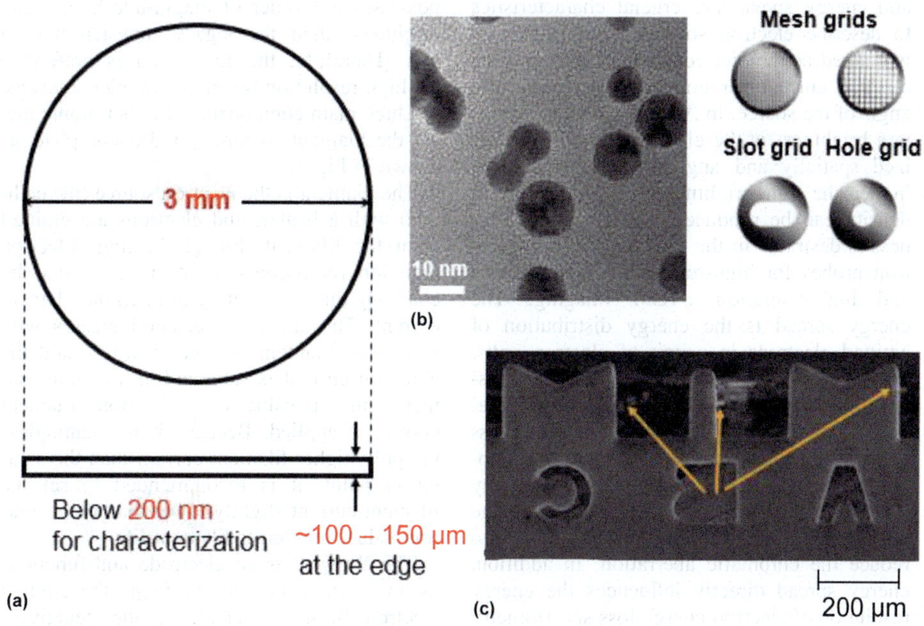

Fig. 5 (a) Typical transmission electron microscopy (TEM) specimen shape fits into a TEM specimen holder. (b) Gold nanoparticles supported by amorphous carbon film on a grid (left), and various TEM specimen grids (right). Source: Electron Microscopy Sciences. (c) Three rectangular TEM specimens (indicated by arrows) prepared by focused ion beam and mounted on a half-grid

of the specimen. If specimens cannot be supported by themselves due to their sizes and shapes (e.g., nanoparticles and nanorods), TEM grids with support films made of carbon and silicon nitrides are required (Fig. 5b). Grids are typically 3 mm (0.12 in.) in diameter and ~100 µm thick. A focused ion beam (FIB) instrument can also be used for specimen preparation. In this case, electron-transparent coupons 20 by 10 μm^2 are fabricated using the FIB and are mounted on a half-cut grid, as shown in Fig. 5(c). More details about the FIB can be found elsewhere in this Volume.

Operation Modes in TEM Instruments: TEM and STEM Modes

Two main operation modes used in TEM instruments, TEM mode and scanning TEM (STEM) mode, are discussed here prior to introducing the details of TEM instrumentation and applications.

Transmission electron microscopy mode is the more traditional mode in TEM operation, in which a parallel stationary beam is illuminated on a specimen. This mode is useful for conventional TEM imaging, such as bright-field (BF) and dark-field (DF) TEM imaging, high-resolution TEM imaging, and conventional electron diffraction analysis, including selected-area electron diffraction. This mode is also used for energy-filtering TEM analysis through an electron energy filter.

A focused electron beam (referred to as a converged electron probe) is used in the STEM mode. Similar to SEM, the focused probe is dynamically scanned over a thin specimen in this mode. Images are formed from signals either transmitted through the thin specimen or reflected on the specimen surface. Thus, BF- and DF-STEM images, secondary electron images, and backscattered electron images can be recorded in the STEM mode. In addition, this mode is very useful for chemical analysis via x-ray signals and/ or via energy loss electrons generated from the specimen.

More details of operations, including imaging and analysis, are described later in this article. In this section, TEM instruments and interactions of incident electrons with a thin specimen are introduced, and electron diffraction and imaging are explained. Spectrometry-basis analysis via x-rays and energy loss electrons is also covered.

Further detailed information on instrumentation and applications can also be found in the literature (Ref 4–9).

Instrumentation and Hardware

In general, it is still possible to use a relatively complicated instrument such as a TEM without detailed knowledge of its hardware components. Modern electron microscopes consist of a set of complex components, including an electron source, electromagnetic lenses, recording devices, and so on, which must be optimized for certain applications. It is difficult to understand all hardware components in the instruments and how they work together properly. In addition, computer-control schemes have been developed for these instruments, and thus, an ordinary standard operation of electron microscopes has already been performed as a routine in the modern instruments. Therefore, users can focus on the materials characterization task itself rather than on the complex operations of the electron microscope. However, materials scientists and engineers are often challenged to improve existing characterization schemes and to develop new characterization methods; both would be impossible without knowledge of the hardware and an understanding of the complex components in electron microscopes. Therefore, fundamental hardware components used in modern TEM instruments are briefly summarized here for practical operation purposes.

Vacuum Requirements

All electron microscopes require a high vacuum, which is one of the reasons for the high cost of the instruments. Due to the pressure difference between the inside and outside of a microscope column, careful handling is required, especially in loading specimens and in exchanging photographic plates (in relatively old instruments). Two major reasons to maintain higher vacuum in the microscope column are to ensure that electrons can travel a longer distance without scattering (defined as the electron mean-free path) and to reduce sources of contamination not only on thin-film specimens but also within the microscope column.

The mass of a stationary electron is ~2000 times lighter than the lightest hydrogen atom. Because of the large mass difference, an electron is easily scattered away by an atom not only in a solid form but also in liquid or even in gas. Thus, the electron mean-free path is strongly dependent on the density of scattering objects that exist in the forward direction of electrons, that is, the vacuum level in electron microscopes. Higher vacuum conditions (i.e., lower-density status of gas atoms and molecules) are preferable to achieve longer mean-free paths. Therefore, all TEMs require higher vacuum in the microscope columns for electrons to travel from the electron source through a thin-film specimen to the recording devices and detectors. A typical vacuum level near the specimens in the microscope column is $\sim 10^{-5}$ Pa (7.5×10^{-8} torr) or better (i.e., lower values in pascals). Near the electron source, the vacuum level is 1 or 2 orders better than that near the specimens. A cold field-emission gun (FEG) source requires ultrahigh vacuum (UHV) ($\sim 10^{-9}$ Pa, or 7.5×10^{-12} torr) to prevent gas adsorption to the gun tip, which significantly degrades the electron emission. This is why a cold FEG tip must be heated up in a few hours to remove the adsorbed gas molecules from the tip.

These higher-vacuum conditions are generally achieved by using a combination of rotary pumps for lower-vacuum levels with diffusion pumps and oil-free turbo pumps. Initially, rotary pumps are used to achieve intermediate-vacuum levels, then diffusion pumps or turbo pumps are used with rotary pumps for backing. Because diffusion pumps use oil to achieve high vacuum, which could be a major source of contamination in the microscope column and/or on the specimen surface, oil-free turbo pumps are preferable in modern instruments. To achieve UHV levels required for a cold FEG or for a special-purpose UHV-TEM, additional ion-based pumps and sublimation pumps are used rather than diffusion and turbo pumps. For the cold FEG source especially, the gun area is typically separated from lower-column parts by several differential pumping apertures to compensate for the difference in vacuum between the gun and lower column.

When specimens or legacy recording media, such as films and imaging plates, are exchanged, the vacuum must be broken. The specimen chamber and film chamber in the TEM should be isolated during the exchange. However, careless mistakes in the exchange process could lead to crashing the entire vacuum system in the microscope, which ends up shutting down the system. Therefore, users must pay great attention during exchange operations.

Electron Sources

All electron microscopes (including both SEM and TEM instruments) require reliable illumination sources, that is, electron sources that provide stable amounts of electrons. An extra amount of energy is required (defined as the work function) to obtain electrons from the sources. Two main ways to overcome the work function for electron extraction from the source are thermal heating for electron emission and applying electrical fields. The former is the thermionic gun, and the latter is the FEG. More details on these electron sources are provided later.

It is worth mentioning some of the important characteristics of electron guns as illumination sources prior to describing the source types. Source stability and life as well as the price are important parameters, from a practical point of view. In addition, gun brightness and energy spread are crucial characteristics to describe electron sources. Gun brightness is defined as the electron current density (i.e., electron current per unit area) per unit solid angle of the source, in A/m^2/sr. In other words, gun brightness is the electron current normalized spatially and angularly. Therefore, the higher the gun brightness, the more current density can be produced. Higher gun brightness is desirable in the formation of fine electron probes for high-resolution microanalysis and high-resolution STEM imaging. The energy spread is the energy distribution of emitted electrons in a unit of electron volts. Similar to visible light and x-rays, energy distribution is related to the "color" of illumination. Narrower energy distributions mean less colorful, that is, less chromatic toward monochromatic. Electron sources with low energy spread are preferable to achieve higher image resolution in both TEM and STEM modes to reduce the chromatic aberration. In addition, energy spread directly influences the energy resolution of electron energy loss spectrometry (EELS), which is used for chemical analysis of the TEM specimen. Again, the lower the energy spread, the higher the energy resolution achieved, which can provide more detailed features in EELS spectra.

Some features of typical electron sources are summarized in Table 2.

Thermionic Sources

Thermionic electron sources have traditionally been used in all SEMs and TEMs. They are less expensive and relatively robust and thus are still used in some modern instruments. Electrons are emitted from the thermionic source by heating to overcome the work function. In other words, electron emission occurs due to the thermal activation process, which results in higher energy spreads during the operation.

A tungsten hairpin gun is common and is still used in entry-level instruments, because of the relatively low price and robustness of the source. Lanthanum hexaboride (LaB_6) has become a more popular thermionic source in modern TEM instruments. Compared with the tungsten hairpin source, LaB_6 has a lower work function, which requires lower heating temperature, and thus, the energy spread becomes lower. In addition, the LaB_6 source possesses a 1 order of magnitude higher gun brightness than the legacy tungsten hairpin gun. Therefore, the LaB_6 gun is preferable for high-resolution imaging and microanalysis.

Three main components of a thermionic gun are the filament, Wehnelt, and anode plate, as shown in Fig. 6.

The filament is the electron source (the cathode) with a heater, and electrons are emitted from the filament through heating. Electron emission (emission current) is controlled by changing filament heating through the filament current. The emission current increases with increasing filament current (heating), and the emission current is saturated at a certain filament current value even if more filament current is applied. Because it is meaningless to apply higher filament current than the saturation point, it is recommended to set the filament current slightly lower than the saturation value, for longer filament life.

The Wehnelt is an electrode and functions as an electrostatic lens to focus the emitted electron beam by changing the negatively charged Wehnelt bias. When the Wehnelt bias is too low or not applied at all, emitted electrons never focus but instead diverge from the source, which enlarges the virtual source size and thus degrades the image resolution. If the Wehnelt bias is too high, emitted electrons are completely repulsed at the Wehnelt, and they never escape from the gun part. In appropriate Wehnelt bias conditions, emitted electrons are focused between the Wehnelt and the anode plate, and then the first crossover of the electron beam is formed. The size of the gun crossover (called the virtual source size) is smaller than the actual source size, which improves image resolution. Therefore, it is important to set up the optimum Wehnelt bias to form the first crossover of the electron beam for appropriate microscope operation.

Table 2 Features of electron sources used in transmission electron microscopy

	Type of electron source			
Features	Tungsten hairpin	LaB_6	Schottky field-emission gun	Cold field-emission gun
Work function, eV	4.5	2.4	3.0	4.5
Operating temperature, K	2700	1700	1700	300
Cross-over size, nm	$>10^5$	10^4	15	3
Brightness, A/m^2/sr	10^{10}	5×10^{11}	5×10^{12}	10^{13}
Energy spread, eV	3	1.5	0.7	0.3
Required vacuum, Pa (torr)	$1.33\ (10^{-2})$	$1.33 \times 10^{-2}\ (10^{-4})$	$1.33 \times 10^{-4}\ (10^{-6})$	$1.33 \times 10^{-7}\ (10^{-9})$
Life, h	100	1000	>5000	>5000

Adapted from Ref 4

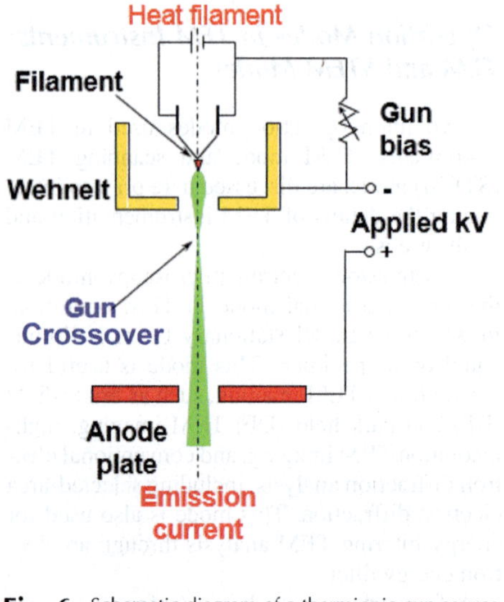

Fig. 6 Schematic diagram of a thermionic gun source for transmission electron microscopy

Field-Emission Gun Sources

The FEG consists of the filament and two anodes, as shown in Fig. 7. Compared with thermionic gun tips, FEG tips are much finer, so source sizes, including the virtual source sizes at the first crossover, are reduced as well, which improves image resolution. At the first anode, a positive charge is applied up to several KeV against the tip, which is used for extracting electrons from the sharp tip of the filament. Thus, the charge at the first anode is called the extraction voltage. These electrons extracted by the first anode are accelerated to high voltages (e.g., 200 kV) by the second anode. The FEG source components, including the filament and two anodes, function as an electrostatic lens. In modern FEG instruments, an additional electromagnetic lens is equipped at the gun to control the sources more effectively, which is called a gun lens.

The FEG sources that produce electrons purely by applying electrical field, as explained previously, are called cold FEG (no cooling, just at room temperature). Compared with thermionic sources, the energy distributions of electrons extracted from the cold FEG sources are much narrower (lower energy spread), because there is no activation process by heating. Therefore, cold FEG sources are used in electron microscopes for high-resolution imaging and microanalysis. However, the major disadvantage of nonheating is that cold FEG sources require UHV environments (~10^{-9} Pa, or 7.5×10^{-12} torr). This is to prevent residual gas molecules in the gun from adsorbing on the filament tip surface and disturbing the field emission of electrons, which would degrade the effective brightness of the cold FEG source. Even though cold FEG sources are operated in UHV, the field-emission process is gradually degraded over time. Therefore, cold FEG sources are required to heat up (called "flashing") within a few hours prior to and/or during an operation.

Other types of FEG sources with thermal assistance include thermal FEG and Schottky FEG. The tips of these thermally assisted FEGs are coated with ZrO_2 to reduce the work function for electron emission. The energy spread and the brightness of the Schottky FEG are slightly degraded compared with those of the cold FEG. However, instrument operation with the Schottky source is much easier than cold FEG instruments. Because electrons must be extracted all the time even though an instrument is not in use for stable electron emission, daily start-up protocols are not necessary. The TEM operator can simply open a valve between the gun and microscope column as long as the system vacuum is sufficient enough for an operation. Consequently, tip flashing is no longer required for the Schottky FEG source. Overall performance of the Schottky FEG source is very high.

Electron Lenses

Paths of moving charged particles such as electrons can be modified by applying electrical fields and magnetic fields. Therefore, moving electrons accelerated after the emission process can be controlled as long as the fields are created. The fields function as lenses for electrons, essentially equivalent to glass lenses for visible light. The electrical field is created by an electrostatic lens, and the magnetic field is created by an electromagnetic lens in electron microscopes. Because aberrations of electrostatic lenses are rather high, electromagnetic lenses are generally used in electron microscopes.

In TEMs, multiple lenses are combined, as shown in Fig. 8. Lenses before a thin specimen (termed condenser lenses) control illumination. The most important objective lens is located at the specimen position, and there are several lenses after the specimen, which are intermediate lenses and projector lenses. Prior to describing roles and functions of TEM lenses, it is worth reviewing general principles of optical lenses and their properties.

Optical Lens Properties

Figure 9(a) shows the relationship between a point object and its image through an optical round lens. In an ideal optical lens, the point object is imaged as a point. Two important planes are the object plane, where the object is located, and the image plane, where the image of the object is formed. The object and its image should have one-to-one relation, and the object and image planes are optically equivalent. The three lines drawn from the object to the image represent paths of light rays; one path is parallel to the optical axis. An image with ray trajectories similar to those in Fig. 9 is called a ray diagram.

Figure 9(b) shows another ray diagram in which an object with an asymmetrical shape (an arrow in this example) located at the object plane is imaged off axis with 180° rotation at the image plane. Three rays are drawn from the center, tip, and end of the arrow. Parallel rays (lines of the same color in the figure) are focused at one plane between the lens and image plane. The third plane, in addition to the object and image planes, is defined as the back focal plane. In TEMs, the back focal plane is very important, because scattered electrons from the specimen are focused at the back focal plane. So, information on electron scattering (an angular distribution of the specimen) is formed at the back focal plane, whereas a spatial distribution of the object is expressed at the image plane. In other words, the spatial and angular distributions of the specimen are formed at the image and back focal planes, respectively. However, information contained in both planes is essentially the same, just different expressions.

If distances between the object plane and the lens, the lens and the image plane, and the lens

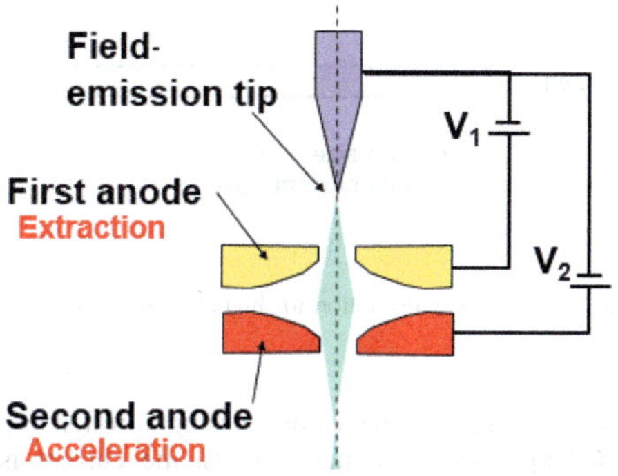

Fig. 7 Schematic diagram of a field-emission gun source for transmission electron microscopy

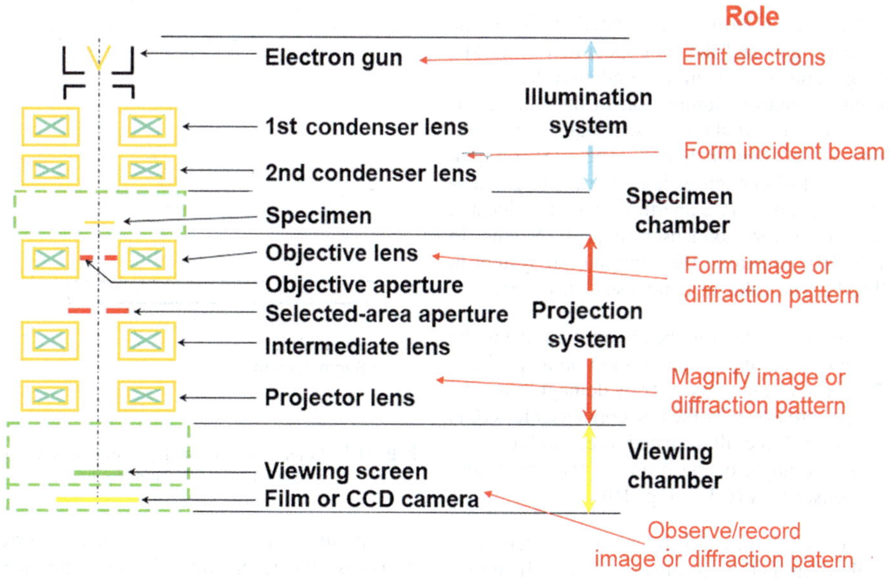

Fig. 8 Summary of configurations of an electron gun and lenses in transmission electron microscopy. CCD, charge-coupled device

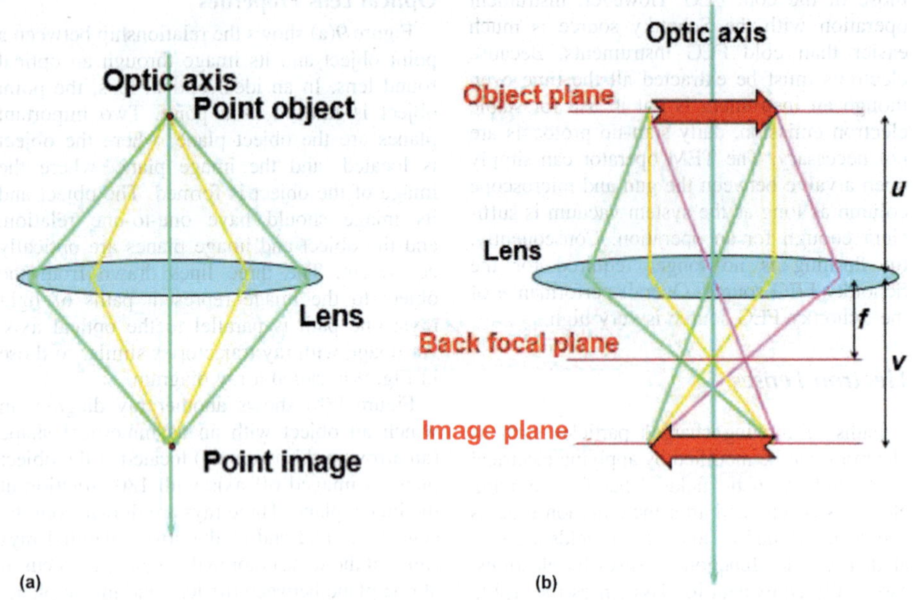

Fig. 9 Ray diagrams of (a) a single object through a lens and (b) an arrow object through a lens

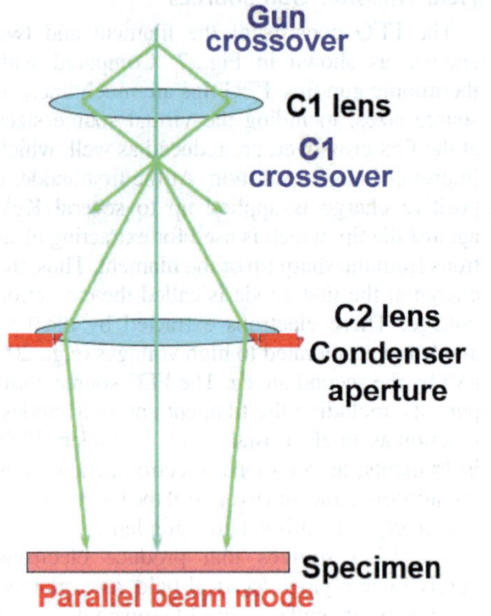

Fig. 10 Schematic diagram of a condenser lens system in a transmission electron microscopy instrument including two condenser lenses and a condenser aperture

and the back focal plane are denoted as u, v, and f, respectively, then the relationship among the three distances is given as:

$$\frac{1}{u}+\frac{1}{v}=\frac{1}{f} \quad \text{(Eq 3)}$$

This equation is known as Newton's lens equation, and f is defined as the focal length. In addition, the magnification of the lens, M, is given as:

$$M=\frac{v}{u} \quad \text{(Eq 4)}$$

Condenser Lenses

Condenser lenses are located between the electron gun and the thin specimen in TEMs. In general, two or more condenser lenses are used to control illumination to the thin specimen. Figure 10 shows a schematic configuration of condenser lenses. The electron gun and a set of condenser lenses, including a condenser aperture (C2 aperture), which is located at the lower condenser lens (C2 lens in Fig. 10), constitute the illumination system in TEMs. The roles of condenser lenses are:

- Form an electron beam emitted from the gun to an appropriate region on a specimen
- Form a beam crossover by demagnifying the gun crossover by the first condenser lens (C1)
- Control the illuminated area and convergence angle of the beam by the lower condenser lens (C2 in Fig. 10)

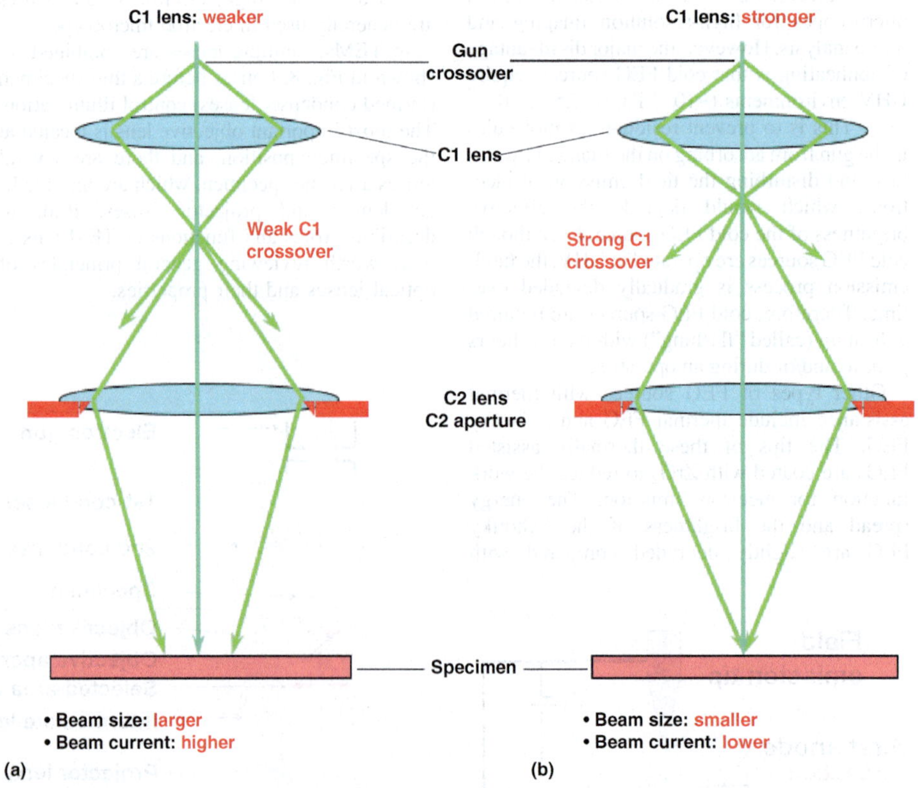

Fig. 11 Comparison of different illumination settings. (a) Weak C1 lens excitation for higher current mode. (b) Strong C1 lens excitation for finer probe formation

Figure 11 summarizes how the C1 lens controls illumination and beam current. If the C1 lens is weakly excited, then the crossover is formed near the C2 lens and the aperture, which allows more electrons to pass through the aperture. As a result, the beam current becomes higher, and the relative beam size is enlarged. Conversely, if the C1 lens is set to a stronger excitation condition, the crossover shifts upward toward the C1 lens side, and the amount of electrons passing through the C2 aperture is reduced. Thus, the current is reduced, which results in the formation of a finer beam. The total amount of electrons under illumination can also be controlled by selecting a different size of C2 aperture. In addition, the C2 aperture size does influence the convergent angle of illumination.

Objective Lens

The objective lens is the most important lens in any microscope, including a TEM instrument. For superior image resolution, it is crucial to use the best possible objective lens in the TEM instrument. Figure 12 shows the relationship between a thin specimen and an objective lens in a typical TEM system. The diagram is essentially equivalent to the optical lens diagram shown in Fig. 9(b). The specimen position is the object plane, and the image plane is underneath the objective lens. This is the first magnified image of the specimen, and thus, it is called the first-image plane. In addition, parallel rays diffracted from the specimen via scattering are focused at the back focal plane. So, electron diffraction patterns form at the back focal plane as a result of scattering. The image of the specimen exhibits a spatial distribution of electron-specimen interactions as well as the diffraction pattern of the specimen, which contains information about the scattering angle, that is, angular distribution. Information contained in both the image and the diffraction pattern is the same but simply expressed differently.

In TEM instruments, the object lens is located below the thin specimen (the forward-scattering direction), as shown in Fig. 12. In modern TEM instruments, there is an additional prefield above the specimen created by an additional lens, which is called a condenser objective (CO) lens, also known as a condenser minilens or the top part of a twin objective lens setting. Sometimes, the CO lens is referred to as a C3 lens. This additional prefield is used to form much higher convergence angles for finer probe formations. Finer probes are essential for STEM imaging and high-resolution microanalysis.

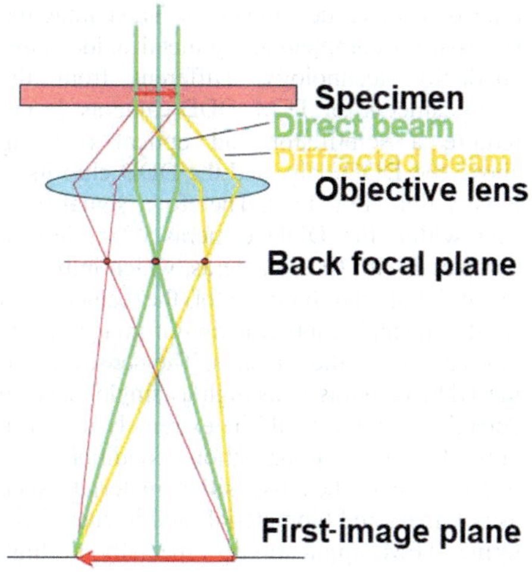

Fig. 12 Relationship between a thin specimen and an objective lens in a transmission electron microscopy instrument

Postspecimen Lenses

Two types of postspecimen lenses are an intermediate lens and a projector lens. As shown in Fig. 13, the primary role of postspecimen lenses is to magnify the first image formed by the object lens to reach desired magnification. The final magnification is given as a product of magnifications at individual lenses from the object lens through the projector lens.

In modern TEM instruments, there are multiple intermediate lenses, so image rotations in magnifying images are seamlessly compensated. More importantly, intermediate lenses switch the image mode to the diffraction mode, or vice versa. Figure 14 compares the difference between image and diffraction modes. If the intermediate lens in the system is strongly excited, an image is formed at the screen, that is, the image mode. By contrast, if the intermediate lens is weaker, a diffraction pattern appears on the screen. Switching between the two modes is carried out by a button selection and is frequently done for BF and/or DF imaging.

Electron Detectors

A human's eyes are able to detect visible light with wavelengths between 380 and 740 nm. High-energy electrons used in TEM instruments have much shorter wavelengths (3.70 to 1.97 pm, or 0.00370 to 0.00197 nm) for 100 and 300 keV electrons with the relativistic correction, respectively. Human eyes cannot see electrons. Therefore, devices are required to convert high-energy electrons to a form visible to the human eye.

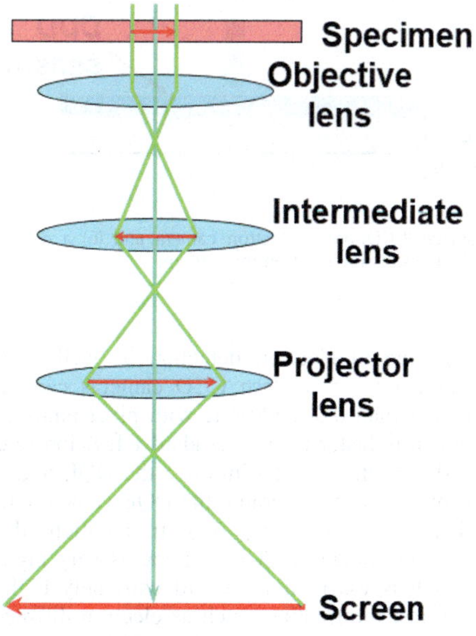

Fig. 13 Summary of postspecimen lenses in a transmission electron microscopy instrument

Traditionally, photographic film has been used in TEM instruments, similar to that used with conventional cameras. The top of the film contains silver halide grains with sizes of a few micrometers; the halide grains transform to silver when electrons interact with them. The transformation to silver creates contrast in the film. The film is very sensitive to electrons and has a high detector quantum efficiency (DQE), which is the ratio of input/output signal-to-noise ratios, and indicates how much noise is induced by the detector; it should be <1. Recording area sizes are typically high (100 by 100 mm^2, or 0.15 by 0.15 $in.^2$). These are still significant advantages compared with newer recording devices discussed later. However, dynamic ranges (displayable contrast levels) of films are limited. The number of films that can be loaded into a cartridge is significantly limited (at most, 50 films per cartridge), which limits TEM operations (the end of films is the end of operation!), and it is very time-consuming to develop negatives and photos in a dark room. A significant disadvantage of using film in TEM is that the system vacuum must be broken when the film cartridge is exchanged. Furthermore, films themselves, with large surface areas, are the worst sources of contamination. Photographic film is still used in some older instruments. Because of the disadvantages, film has become nearly obsolete, and digital recording media are the mainstream for electron detection. This media conversion occurred much earlier in scanning instruments (STEM and SEM). Initially, photographic film was used to record images in STEM and SEM instruments, but this option is now completely obsolete.

Digital recording devices used in modern TEM and STEM instruments are introduced in the following subsections. Development of recording devices has been significant.

Charge-Coupled Device Cameras

Charge-coupled device (CCD) cameras (similar to thosc in digital cameras and phones) are the most popular recording devices used in TEM instruments. The CCD chips are based on metal-insulator-silicon structures and consist of several megapixels, for example, 4000 × 4000 = 16 million pixels. Charge generated by light is stored at individual cells, which are electrically isolated from others in the CCD camera. Because the stored charge is proportional to the intensity of illumination, digital images can be formed when the charge is detected at the cell arrays in the CCD camera. Typical CCD cameras used for electron detection in TEM have a yttrium-aluminum-garnet (YAG) single crystal or a phosphor powder scintillator on the top, coupled with the CCD chip through fiber or lens optics (Ref 10), as shown schematically in Fig. 15(a).

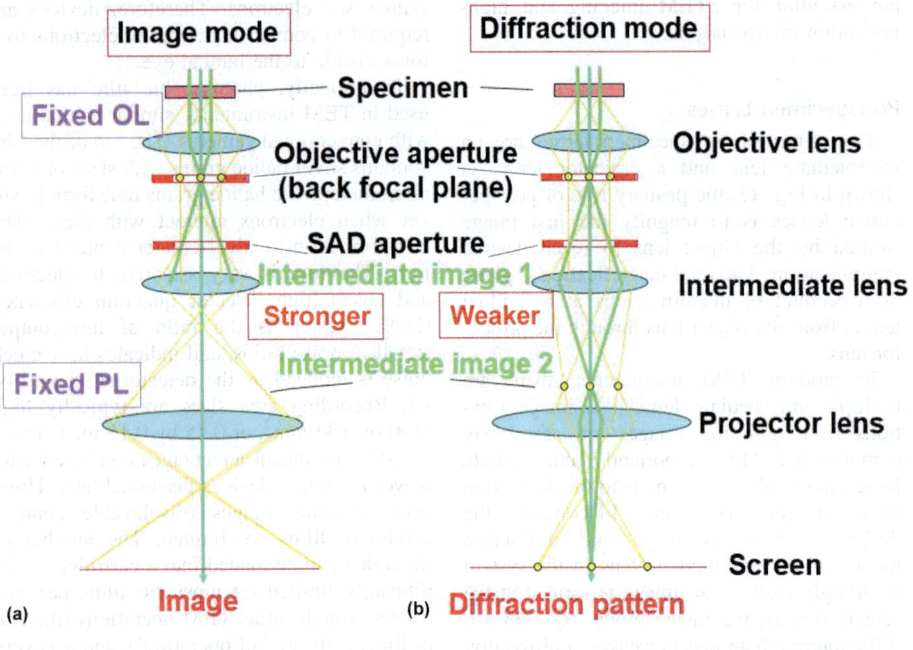

Fig. 14 Comparison of postspecimen lens settings for (a) imaging and (b) diffraction modes. OL, objective lens; SAD, selected-area diffraction; PL, projector lens

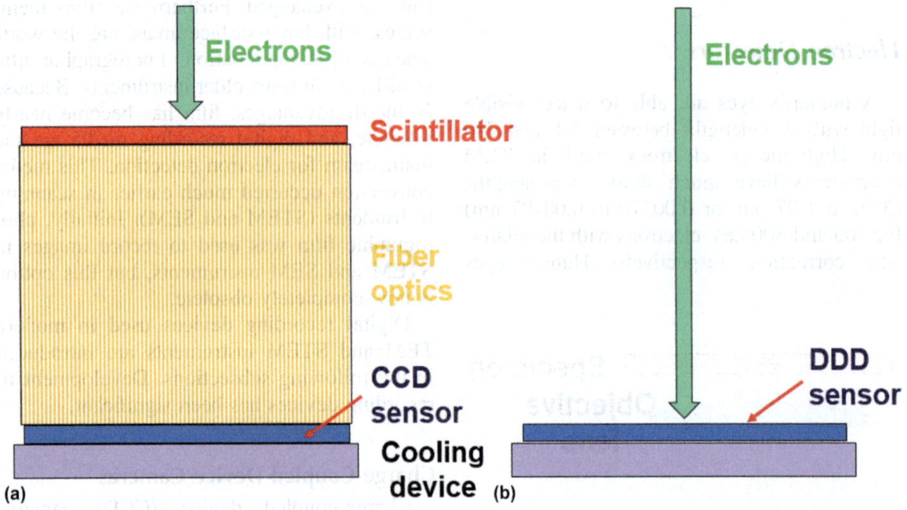

Fig. 15 Schematic diagrams of (a) a charge-coupled device (CCD)-based electron camera and (b) a direct-detection device (DDD)-based electron camera for transmission electron microscopy

The scintillator at the top of the camera first converts the primary electrons to photons, which transfer through the optical coupling to reach the CCD chip. Both the scintillator and optical coupling are essential to prevent the CCD chip from damage due to the illumination of high-energy electrons. However, both the primary electrons and the emitted photons experience scattering in the top layers, which leads to blurring of the image.

In contrast to conventional films, images are instantly available in digital form using the CCD camera. The CCD cameras are more sensitive to electron detection as well. The frame rate of modern CCD cameras can be higher than 0.001 s/frame with pixel binning, which is faster than a standard television rate (~0.033 s/frame). To improve the DQE, CCD cameras must be cooled; the DQE value could then reach ~0.5 (Ref 11). In addition, the dynamic range of CCD cameras is very high, which is essential to record extremely high-contrast information, such as electron diffraction patterns. In the modern CCD systems, the dynamic range is ~14 bits, which means that each pixel in the CCD camera can store up to 14 bits (2^{14} = ~16,000 counts). Practically, target intensities of such CCD cameras should be half of this value (~8,000 counts) to record sufficient signals but not to saturate the CCD camera accidentally.

Images and diffraction patterns recorded by CCD cameras require two corrections: gain normalization and dark-current correction. The CCD chip consists of arrayed cells, which would have slightly different response functions against photons emitted by electrons in TEM instruments. Gain normalization corrects the difference in response among the individual CCD cells. For gain normalization, a reference image containing response information of the individual CCD cells must be multiplied to an image or a diffraction pattern recorded in the CCD camera. Figure 16(a) shows a reference image obtained by averaging multiple images acquired without any specimen under homogeneous illumination. In a large CCD camera for TEM, multiple CCD chips may be combined, for example, 2 × 2 (= 4) CCD chips as one camera, to enhance readout speed. In this case, in addition to the gain normalization of individual CCD cells, the response of each CCD chip must also be normalized.

In a CCD camera, charge would be stored in the chip even without electron illumination; this is called the dark current. A typical dark-current image obtained in a TEM instrument is shown in Fig. 16(b). The dark-current image should ideally be acquired prior to every image acquisition, because the charge is different in individual acquisition conditions and is dependent on a CCD temperature. For dark-current correction, the acquired dark-current image must be subtracted from an image or a diffraction pattern.

Direct-Detection Device Cameras

A direct-detection device (DDD), a new type of camera developed for TEM imaging, is based on complementary metal oxide semiconductor technology. Different from the CCD cameras for TEM, DDD cameras do not require a scintillator and optical coupling above the DDD chip, and the DDD chip itself is very thin (Fig. 15b). Therefore, signal blurring within the DDD cameras is minimized compared with CCD cameras, which improves the point-spread function of the camera and results in significant reduction of image distortion caused by the cameras. The sensitivity of the DDD cameras is as high as single-electron detection, which would be extremely useful in very low-dose illumination conditions for observation of beam-sensitive materials such as polymers and biomaterials and in energy-filtering TEM applications, especially at high energy loss ranges.

In addition, DDD cameras can be operated at extremely high frame rates (e.g., over 50 frames per second without pixel binning), which enables recording dynamic experiments, such as in situ heating, electrical, and deformation experiments, including rapid phase transformations and defect motion. The DDD cameras are also extremely popular for detailed structure determinations of viruses and related particles in combination with cryo-TEM. The DDD cameras are still new to TEM, so further advancements in development and new applications can be expected. More information on the DDD cameras and their applications can be found in the literature (Ref 10–12).

Solid-State Semiconductor Detectors

Semiconductor detectors are frequently used for detection of backscattered electron (BE) signals in SEM. A similar BE detector can also be added for STEM imaging. In addition, semiconductor detectors are primarily used for detecting forward-scattered electrons through thin specimens in the STEM mode. By placing the semiconductor detector on the optic axis in the forward-scattering side (i.e., postspecimen side), BF signals can be detected to form a BF-STEM image. If the detector is in the off-optic axis, electrons that are not scattered toward the optic axis (deflected electrons) can be collected. These are DF signals, and a DF-STEM image can be formed by collecting them in the STEM mode.

The semiconductor detector consists of a *p-n* junction. Electron-hole pairs are generated when an electron hits the detector. Subsequently, both generated electrons and holes move in opposite directions in the detector due to the *p-n* junction, which can be measured as an electrical current externally. Semiconductor detectors convert detected electrons into a measurable current. Because the mobility of electrons and holes in silicon-base semiconductors is very high, a high signal response is achieved in semiconductor detectors. The detectors are very efficient and relatively inexpensive to fabricate. More importantly, their flexibility enables making different forms, such as an annular (or ring) shape, which is used for annular DF-STEM imaging. It is also possible to make several individual segments within a detector. The semiconductor-basis segmented detector opened a new imaging approach, the so-called differential phase contrast imaging in STEM (Ref 13). The concept of detector segmentation has also led to the recent development of pixelated detectors for STEM (Ref 14).

Scintillator-Photomultiplier Detectors

Scintillator-photomultiplier (PM) detectors, commonly used for secondary electron (SE) imaging in SEM, can also be used for SE detection in STEM. When an SE hits the scintillator (usually made of cerium-doped YAG or doped glass), visible light is emitted at the scintillator. The emitted visible light is amplified in a PM part attached to the scintillator via a light pipe. The SE signals are not very popular for imaging thin specimens in STEM. However, SE signals are useful to observe surface features in the detector side of the specimen, because generated SEs emit only from the surface, even in a thin specimen, due to their lower energies. In addition, it is possible to record SE images of single atoms and to produce atomic-resolution SE images in aberration-corrected STEMs (Ref 15). The SE signals are useful for characterizing thin specimens in STEM.

Electron-Specimen Interaction

Figure 17 shows interactions between electrons and a thin specimen in TEM instruments. A majority of incident electrons (defined as primary electrons) accelerated at typically 100 to 300 kV or higher should transmit through the thin specimen. These electrons are called forward-scattered electrons. Some (termed direct electrons) still remain aligned with the optic axis even after scattering in the specimen, while others in the off-optic axis are deflected (termed indirect electrons), whose directions are modified by scattering in the specimen. Imaging (discussed later) by selecting direct and deflected electrons is BF and DF imaging, respectively. Selection of these different types of electrons can be done at the back focal plane of the objective lens. In other words, different types of images can be formed by selecting electrons at different scattering angles, that is, imaging based on angular selection.

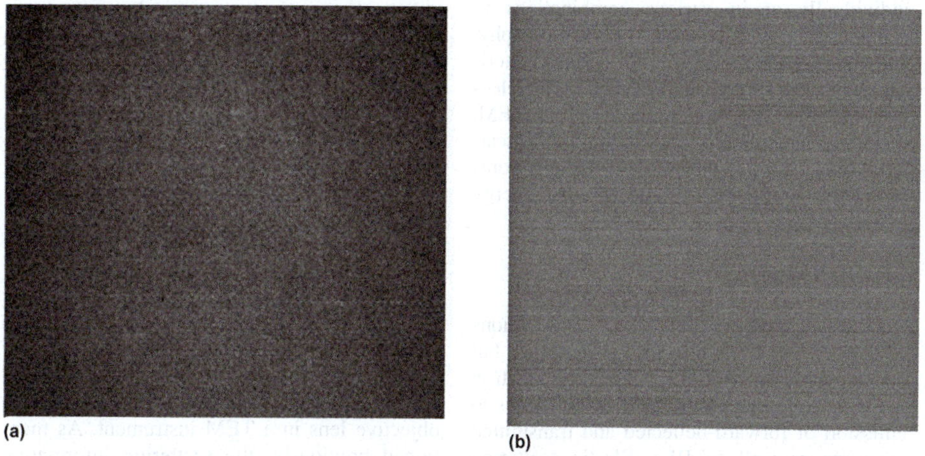

Fig. 16 (a) Reference image showing channel-to-channel gain difference in a charge-coupled device (CCD) chip. (b) Typical dark-current image of a CCD camera

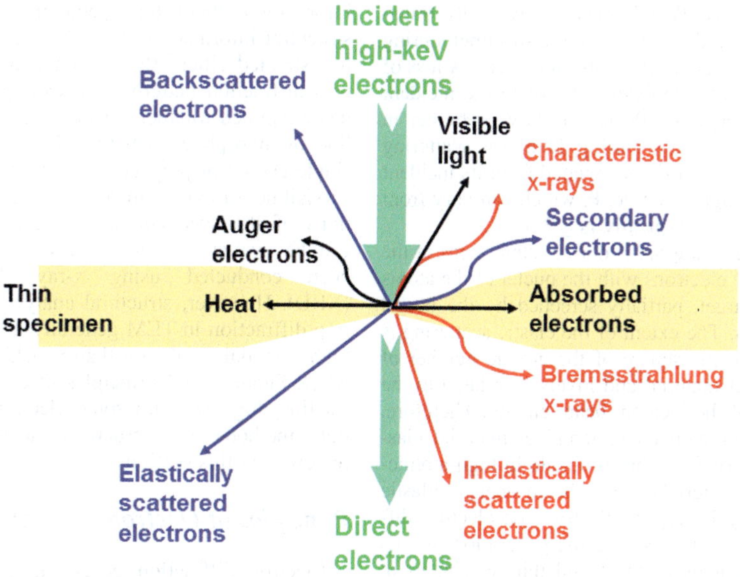

Fig. 17 Schematic summary of interactions between an electron beam and a thin specimen in transmission electron microscopy

Some electrons can be scattered at very high angles (>90°), which is the incident beam side of the specimen. They are defined as the BEs, which are very common signals used for imaging in SEMs. However, in the TEM, electrons that impinge on the very thin specimen have much higher energies compared with SEM; therefore, the fraction of generated BEs is rather lower. Even in the STEM mode, it is not very common to use BEs for imaging. Other electrons can also be absorbed in the thin specimen, which could turn to heat or be used to produce electron-hole pairs if the specimen is a semiconductor material such as silicon and germanium. Other important signals generated from the thin specimen after interaction with the primary electrons include energy loss electrons, characteristic x-rays, and so on, as shown in Fig. 17. These signals can be used individually or in various combinations to characterize microstructure, crystallography, and compositional variations in various materials. A number of signals produced by the electron-specimen interaction are used in TEM, including unscattered and inelastically scattered transmitted electrons, deflected electrons, SEs, BEs, emitted x-ray spectra, and energy loss electrons.

Elastic Scattering

Elastic scattering events affect the directions of primary electrons (electron trajectories) but do not significantly affect primary electron velocities and kinetic energies, which leads to emission of forward-deflected and transmitted electrons as well as BEs. Elastic scattering changes the direction component of electron velocity, v, but not the magnitude. Because the velocity is unchanged, the kinetic energy of the electron, $E = 1/(2m_e v^2)$, where m_e is the electron mass, is also essentially unchanged. Less than 1 eV of energy is typically transferred from the electron to the specimen during an individual elastic scattering event. A loss of 1 eV is not significant relative to the incident beam energy of 100 to 300 keV or higher in TEM instruments. During the elastic scattering event, the electron is deviated from its incident path through an angle, θ, which can vary from 0 to 180° but typically is $\leq$2 to 3°.

Elastic scattering is due to interactions of the energetic electrons with the nuclei of the atoms in the target, partially screened by the bound electrons. The extent of the elastic scattering is related to the square of the atomic number of the target material and inversely related to the square of the incident beam energy. Therefore, for a specific thickness of a given material, elastic scattering is more probable in high-atomic-number materials at low beam energy. Elastic scattering is responsible for both electron diffraction (thin foils) and the generation of BEs (bulk samples in SEM and thin specimens in TEM). Coherent elastic scattering of electrons in the forward direction produces the form of electron diffraction commonly encountered in TEM. The origin and use of electron diffraction is discussed later in the section "Electron Diffraction" in this article. In addition, elastic-scattered electrons are the major signal sources in high-angle annular dark-field STEM imaging as well.

Inelastic Scattering

Inelastic scattering is the second category of scattering by which the energy of primary electrons is transferred to the target atoms, and thus, the kinetic energy of the incident electrons is reduced. Conversely, electron trajectories after inelastic scattering are changed very little. Inelastic scattering events generate SEs; Auger electrons; characteristic and continuum x-rays; long-wavelength electromagnetic radiation in the visible, ultraviolet, and infrared regions; electron-hole pairs; lattice vibrations (phonons); and electron oscillations (plasmons), as shown in Fig. 17. Therefore, most of the signals related to elemental and compositional information and material properties are the result of inelastic scattering events. Several inelastic scattering processes are possible in TEM.

Electron Diffraction

A TEM instrument is used primarily for imaging the microstructure of a material. In addition, information about the scattering of incident electrons through a thin specimen can be observed at a back focal plane of the objective lens in a TEM instrument. As mentioned previously, the scattering information is essentially the same as an image from the same field of view of the thin specimen, but in a different form. Information formed at the back focal plane is called an electron diffraction pattern, which contains structural information about the specimen. Therefore, structural information about the specimen can be extracted efficiently by analyzing the electron diffraction patterns. For example, it can be determined whether the specimen is crystalline or amorphous, whether the specimen is single crystal or polycrystal, and what type of crystalline structure. In addition, the orientations of the specimen and defect structures can be evaluated. These analyses have also been conducted using x-ray diffraction (XRD). However, structural analysis by electron diffraction in TEM generally offers much higher spatial resolution than is achievable by XRD. Fundamental principles of electron diffraction, together with some electron diffraction methods for structural analysis, are reviewed in this section.

Principles of Electron Diffraction

Electron diffraction occurs as a result of electron-scattering events in the specimen. Electron scattering is revisited from different points of view for electron diffraction.

Consider electron scattering from an isolated atom. Two ways that an incident electron can be scattered by an isolated atom are through interaction with inner-shell electrons and by the nucleus. Electron scattering through interactions with inner shell electrons, i.e. the electron cloud, is due to Coulombic repulsion, which results in lower scattering angles. In general, most electron-electron interactions result in inelastic scattering. Scattering caused by the nucleus is called the electron-nucleus interaction. Because the nucleus is positively charged, these types of interactions are also based on Coulombic forces but are attractive in nature. Consequently, electron scattering by the nucleus tends to be at higher angles compared with electron-electron scattering. Typical electron diffraction through a thin specimen in the TEM is due to the nucleus-electron interaction. Therefore, only elastic scattering is taken into account in this section.

Figure 18(a) shows a schematic diagram of electron scattering from an isolated atom. The angle is defined as low if the electron is scattered to the forward direction, nearly parallel to the incident electron direction. The electron-scattering event with the isolated single atom can be described by the atomic scattering factor, $f(\theta)$, which is given as:

$$f(\theta) = \frac{[1+E/(mc^2)]}{8\pi^2 a_0}\left[\frac{\lambda}{\sin(\theta/2)}\right]^2 (Z - f_x) \qquad \text{(Eq 5)}$$

where a_0 and Z are the Bohr radius and the atomic number of the atom, respectively; c and m are the velocity of light and the mass of the electron, respectively; E and λ are the incident electron energy and the electron wavelength, respectively; and θ is the scattering angle. In addition, f_x is the scattering factor for x-rays, which is well known and can be found in Ref 16. The atomic scattering factor has the unit of distance. Figure 18(b) compares $f(\theta)$ values of several elements, plotted as a function of the scattering angle, θ. The value of $f(\theta)$ is higher at low values of θ, so low-angle scattering (less than 1 to 2°) occurs with much higher frequencies than higher-angle scattering. In addition, heavier atoms scatter an electron more strongly than lighter atoms at any angle. According to Eq 5, electron scattering is reduced at higher accelerating voltages. Note that $f(\theta)$ essentially exhibits the amplitude of the electron wave scattered from an atom. Therefore, $|f(\theta)|^2$ is proportional to the scattering intensity.

The electron-scattering model from repeated atomic arrangements (i.e. crystalline structure) can be derived based on the concept of electron scattering from the isolated atom. Scattering from a crystal is dramatically influenced by the periodicity of the crystal. Electron scattering from a crystalline structure is schematically shown in Fig. 19. By taking into consideration the periodicity of atomic arrangements in the crystal, a new term, called the

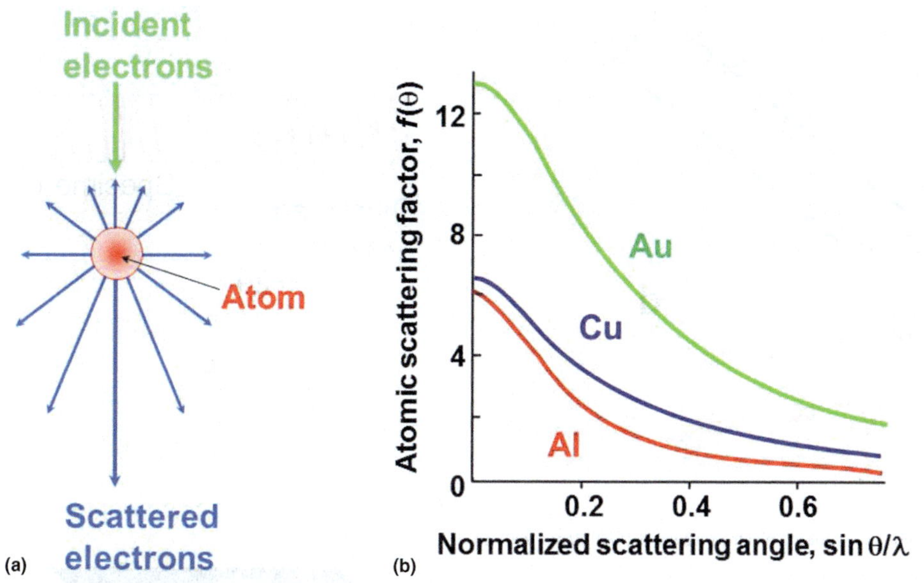

Fig. 18 (a) Schematic illustration of electron scattering from an isolated atom. (b) Atomic scattering factors of aluminum, copper, and silver as a function of the scattering angle normalized by the electron wavelength

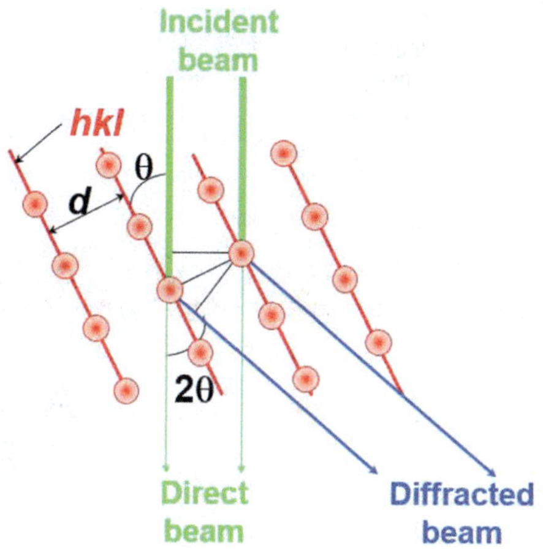

Fig. 19 Illustration of relationship between direct and diffracted beams at an (*hkl*) plane of a crystalline structure

structure factor, $F(\theta)$, can be defined as a sum of the atomic scattering factors from all the atom positions in the unit cell:

$$F(\theta) = \sum_j f_j \exp\left[2\pi i\left(hx_j + ky_j + lz_j\right)\right] \qquad \text{(Eq 6)}$$

where x_j, y_j, and z_j are atom positions in the unit cell, and h, k, and l are Miller indices of an atomic plane where electrons are scattered, as shown in Fig. 19. Similar to the atomic scattering factor $f(\theta)$, the structure factor is the amplitude of the electron wave scattered by a unit cell of crystal structure. In other words, the structure factor describes contributions of the entire unit cell to the diffraction intensity of a particular spot. In fact, $|F(\theta)|^2$ is proportional to the intensity of a particular reflection (hkl).

Figure 20 schematically compares the structure factor of crystalline and amorphous materials as a function of the scattering angle. In comparison to the atomic scattering factors shown in Fig. 18(b), the structure factor from the crystal has a discrete distribution, which indicates that the intensity of diffracted beams takes strong maxima at specific angles where more electrons are strongly scattered. This is due to the periodicity of fixed atomic positions. Conversely, $F(\theta)$ of amorphous materials (Fig. 20b) is rather similar to the atomic scattering factor shown in Fig. 18(b), except for a few local maxima at certain angles. The local maxima correspond to the first- and second-nearest neighbor spacings of short-range ordering in amorphous materials.

Coherent Bragg Diffraction

The directions in which electrons are elastically scattered from a crystal are not random but depend on the geometric arrangement of the atomic planes. Scattering from specific atomic planes is coherent; that is, the electrons scatter in specific directions determined by the crystallographic nature (orientation between the given atomic planes and the incident beam) of the specimen. Electron scattering in the specimen creates many electron beams traveling at specific angles relative to a single incident beam.

Because electrons also have properties of waves, it is reasonable to treat the scattered electrons as the wave propagation in a crystalline specimen. Waves reinforce one another when they are in phase (constructive interference) and cancel one another if they are out of phase (destructive interference), which is well known as the wave theory. The reinforcement of the electron waves in the crystalline specimen can be explained by Bragg's law. Figure 21 shows a configuration of incident and diffracted beams at a certain crystalline plane with the spacing of d (interplanar spacing). Electron waves reflected from different crystalline planes should be in phase for reinforcement. To reinforce those waves, the path difference between different waves must be equal to an integral number of the electron wavelength, λ. The path difference is AB + BC (Fig. 21), which is expressed as $2d \sin \theta$ geometrically. Therefore, Bragg's law can be derived as:

$$n\lambda = 2d \sin\theta \qquad \text{(Eq 7)}$$

Because λ is extremely short in TEM, $\theta \ll 1$. If only the first-order diffraction is taken into account, $n = 1$. Then, Eq 7 can be simplified as:

$$\lambda = 2d\theta \qquad \text{(Eq 8)}$$

Based on the geometry in a TEM instrument, Bragg's law can be further modified.

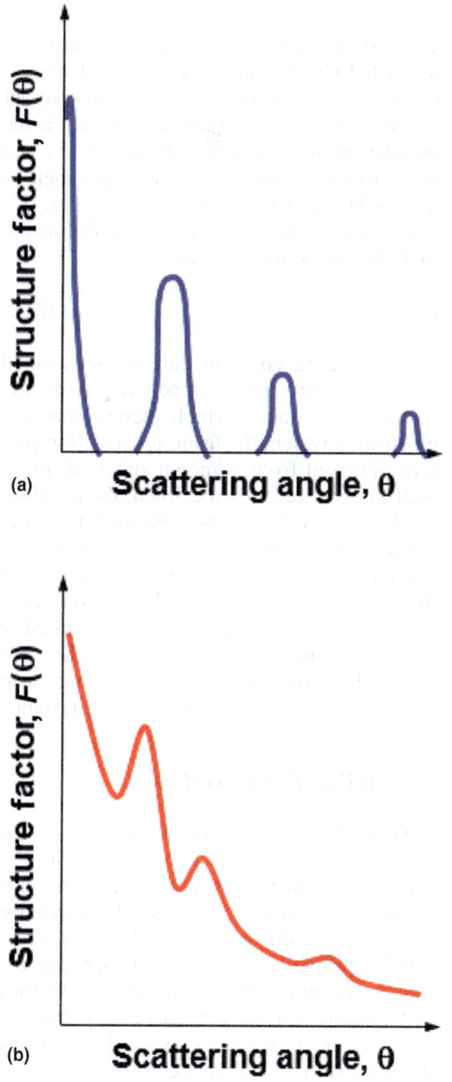

Fig. 20 Structure factors, $F(\theta)$, of (a) crystalline and (b) amorphous materials, schematically plotted as a function of θ

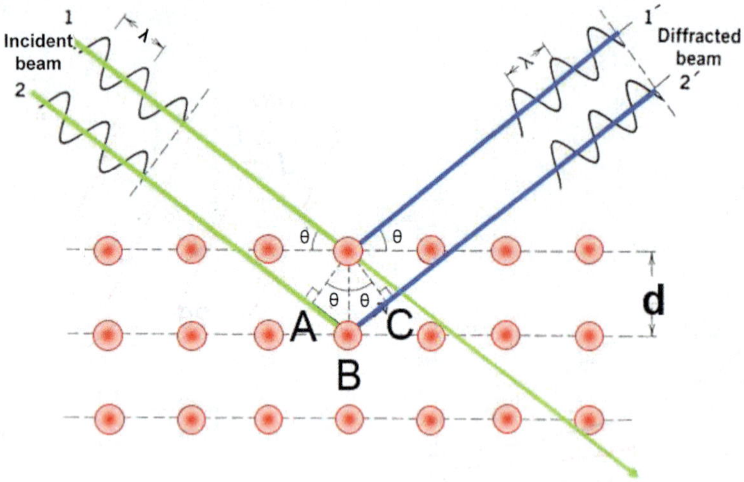

Fig. 21 Illustration of the Bragg condition between incident and scattered electron waves at {*hkl*} planes with an interplanar spacing, *d*

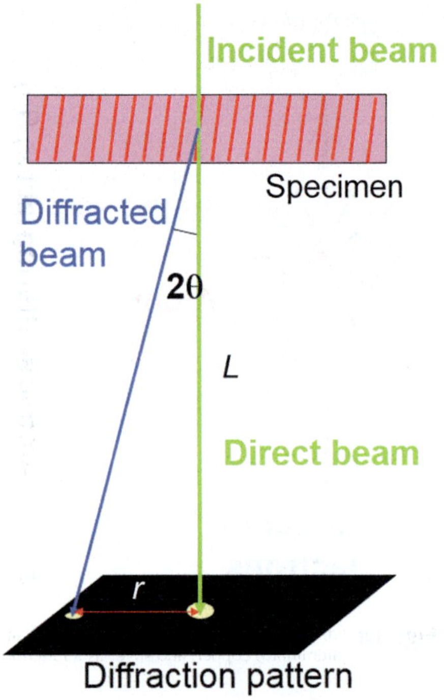

Fig. 22 Geometrical configurations of a camera length, *L*, and a distance between direct and diffracted spots, *r*, at the recording plane

The distance from the transmitted beam to the diffracted beam, r, at the recording plane of a CCD camera is related to the camera constant of the microscope. The camera constant is the product of the electron wavelength, λ, and the effective distance from the specimen to the CCD camera, L, as shown in Fig. 22. Therefore, Bragg's law for electron diffraction in TEM can be rewritten as:

$$rd = \lambda L \qquad \text{(Eq 9)}$$

The camera constant can be determined for a specific microscope at a specific operating voltage (which determines the electron wavelength) from a diffraction pattern obtained from a known standard. For a well-known standard specimen, the d values are known, and r can be measured from the diffraction pattern. Therefore, λL can be calculated. If the determined λL values from the d and r values are not constant, then either the d spacings or the measured r values are incorrect. Furthermore, the geometrical L value can also be calibrated once the λL values are confirmed to be constant.

Diffraction from Crystals

Because electrons have both particle and wave properties, diffraction of electrons by crystals is mathematically identical to the diffraction of x-rays by a crystal. The difference between x-ray diffraction and electron diffraction lies primarily in the wavelength of the diffracting radiation. Due to the wave properties of the electron, Eq 6 indicates modification of the wave due to scattering, which is defined as the phase shift. Therefore, a complex expression is used in the structure factor equation to describe the wave-phase shift. If the crystal unit cell possesses a center of symmetry, the imaginary parts of the expression (actually, the sine terms after rearrangement using Euler's formula) are canceled. Then, the structure factor can be simplified by only using the real parts, which are expressed by cosine terms as:

$$F(\theta) = \sum_j f_j \cos\left[2\pi\left(hx_j + ky_j + lz_j\right)\right] \qquad \text{(Eq 10)}$$

The intensity of diffraction spots is proportional to $|F(\theta)|^2$. So, it is worth exploring when (for which scattering angles) the reflection is enhanced. From Eq 10, it is understood that only certain atomic planes will diffract the electron beam. Atomic planes that will diffract are dependent on the crystal structure. Therefore, characteristics of the structure factors are determined mathematically for selected crystals in this section.

Simple Cubic Structure

A unit cell of a simple cubic (sc) structure is shown in Fig. 23(a). The structure is categorized as a primitive cubic P in the Bravais lattice description and as Pm$\bar{3}m$ (No. 221) in the space group. In the structure, ⅛ atom is located at each corner, so the total number of atoms in the cell is 1. Thus, the atomic position in this structure is represented as (000). The structure factor of an sc crystal can be described as:

$$F(\theta) = f\cos[2\pi(h \times 0 + k \times 0 + l \times 0)] = f\cos(0) = f \qquad \text{(Eq 11)}$$

This indicates that diffracted intensities appear at every crystallographic (*hkl*) plane in the sc crystal. If N is defined as $N = h^2 + k^2 + l^2$, the Miller indices of planes where incident electrons are reflected in the sc crystal are summarized in ascending order of N in Table 3. These reflection indices appear as points in the reciprocal lattice, where the electron diffraction patterns are formed. The reciprocal lattice of the sc crystal is shown in Fig. 23(b) with those reflection points. If the reciprocal lattice is projected from a certain direction, then an electron diffraction pattern from the crystal can ideally be constructed. Figure 23(c) summarizes electron diffraction patterns simulated for some of the major directions, including [001], [011], and [111].

Body-Centered Cubic Structure

A unit cell of the body-centered cubic (bcc) structure is shown in Fig. 24(a). Because the structure can be constructed from the primitive cell with a body-centering symmetry operation, it can be described as I in the Bravais lattice. Its space group is then Im$\bar{3}m$ (No. 229). There are two atoms in the unit cell, at the locations of (000) and (1/2 1/2 1/2). The structure factor of a bcc crystal is given as:

$$F(\theta) = f + f\cos[\pi(h+k+l)]$$
$$= f[1 + \cos\{\pi(h+k+l)\}] \qquad \text{(Eq 12)}$$

If $h + k + l$ is an even value, the cosine term becomes 1. So, the structure factor is enhanced as $2f$. Conversely, if the $h + k + l$ term is odd, then $F(\theta)$ becomes 0. In the latter case, scattered waves are no longer enhanced by one another, and the reflection of the particular *hkl* planes never appears. The *hkl* planes that can be enhanced are also summarized in Table 3. Based on the reflection planes in the bcc crystal, the reciprocal lattice

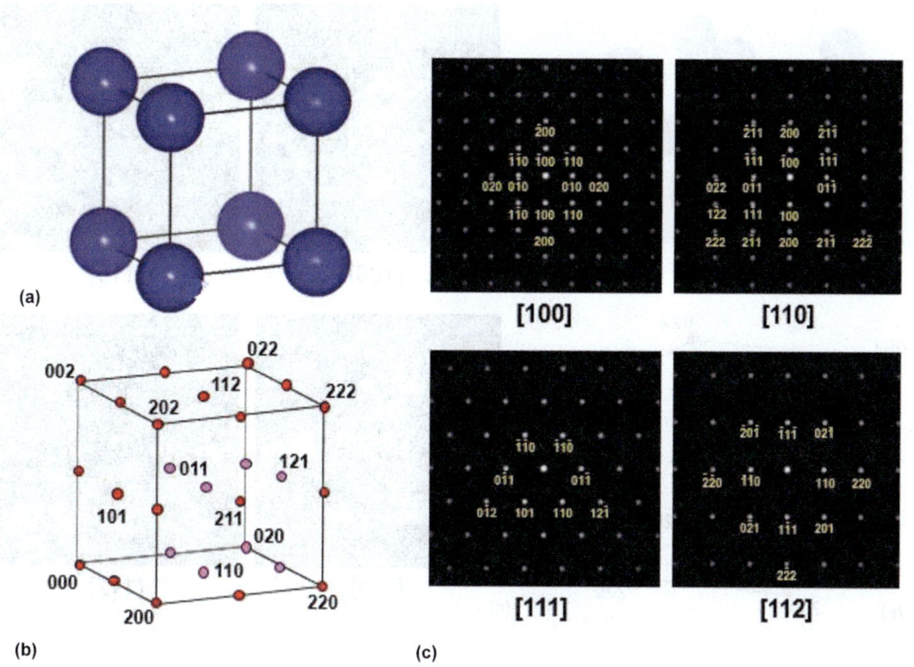

Fig. 23 (a) Real and (b) reciprocal lattices of a simple cubic structure. (c) Simulated electron diffraction patterns for a simple cubic structure in the [001], [011], [111], and [112] directions

Table 3 Miller indices of planes where incident electrons are reflected in simple cubic, body-centered cubic, and face-centered cubic structures

		Crystal structure		
$N = h^2 + k^2 + l^2$	$\{h\ k\ l\}$	Simple cubic	Body-centered cubic	Face-centered cubic
1	100	√	...	...
2	110	√	√	...
3	111	√	...	√
4	200	√	√	√
5	210	√	...	...
6	211	√	√	...
7	...	...	...	...
8	220	√	√	√
9	300, 221	√	...	...
10	310	√	√	...
11	311	√	...	√
12	222	√	√	√
13	320	√	...	...
14	321	√	√	...
15	...	...	...	...
16	400	√	√	√
17	410	√	...	...
18	411, 330	√	...	...
19	331	√	...	√
20	420	√	√	√
21	421	√	...	...
22	332	√	...	...
23	...	...	...	...
24	422	√	√	√
25	500	√	...	...

can also be constructed with the reflection plane information as points (Fig. 24b). The reciprocal lattice of the bcc crystal with the reflection points looks like a face-centered cubic crystal structure. Some electron diffraction patterns simulated for the major directions are summarized in Fig. 24(c).

Face-Centered Cubic Structure

A face-centered cubic (fcc) unit cell is shown in Fig. 25(a). This structure is constructed from the primitive cell with a face-centering symmetry operation, which is described as F-centering in the Bravais lattice. The space group of the fcc structure is $Fm\overline{3}m$ (No. 225), and the structure contains four atoms total, positioned at (000), (1/2 1/2 0), (1/2 0 1/2), and (0 1/2 1/2). Therefore, the structure factor of the fcc lattice is given as:

$$F(\theta) = f[1 + \cos\{\pi(h+k)\} + \cos\{\pi(h+l)\}] + \cos\{\pi(k+l)\} \quad \text{(Eq 13)}$$

When all h, k, and l values become either even or odd, then all cosine terms in Eq 13 become 1, and thus, the structure factor is $4f$. Otherwise, the structure factor is no longer enhanced. The reflection planes of the fcc structure are also summarized in Table 3, and its reciprocal lattice is shown in Fig. 25(b). The reciprocal lattice of the fcc crystal looks like a bcc crystal lattice. Similar to structures explained previously, simulated electron diffraction patterns for the major directions of the fcc structure are shown in Fig. 25(c).

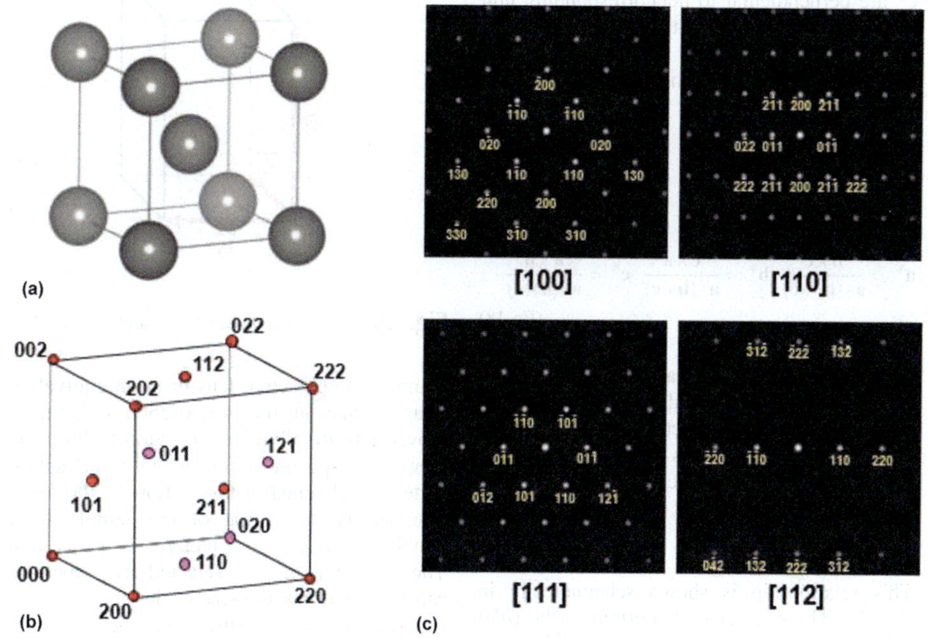

Fig. 24 (a) Real and (b) reciprocal lattices of a body-centered cubic structure. (c) Simulated electron diffraction patterns for a body-centered cubic structure in the [001], [011], [111], and [112] directions

Properties of Reciprocal Lattices

As addressed in the previous section, it is useful to construct reciprocal lattices to understand electron diffraction patterns from crystalline materials. In the reciprocal lattices, the reflected planes are represented as spots, which are the diffraction spots in the diffraction patterns. There are relationships between real and reciprocal lattices. Properties of reciprocal lattices are summarized here.

Reciprocal lattices are useful in the interpretation of diffraction patterns. In other words, each of the points in the reciprocal lattice represents a set of reflecting planes in the crystal and has the same Miller indices as the corresponding plane. Therefore, in the reciprocal lattice, sets of parallel (hkl) atomic planes with the interplanar distance d_{hkl} are represented by a single point located at a distance of $1/d_{hkl}$ from the origin (i.e., the direct beam position, 000). In the original real lattice, the lattice vector can be expressed as:

$$\mathbf{r} = n_1\mathbf{a} + n_2\mathbf{b} + n_3\mathbf{c} \quad \text{(Eq 14)}$$

where **a**, **b**, and **c** are unit vectors of the original lattice, and n_1, n_2, and n_3 are integer values. The lattice vector in the corresponding reciprocal lattice is then given as:

$$\mathbf{r}^* = m_1\mathbf{a}^* + m_2\mathbf{b}^* + m_3\mathbf{c}^* \quad \text{(Eq 15)}$$

where **a***, **b***, and **c*** are unit vectors of the reciprocal lattice, and m_1, m_2, and m_3 are integer values. The relationship between the real and reciprocal lattices is schematically shown in Fig. 26. Because each edge length of the reciprocal lattice has a reciprocal relationship with that of the original lattice, the following relationship can be satisfied:

$$\mathbf{a}\cdot\mathbf{a}^* = \mathbf{b}\cdot\mathbf{b}^* = \mathbf{c}\cdot\mathbf{c}^* = 1 \quad \text{(Eq 16)}$$

In addition, the reciprocal vectors **a***, **b***, and **c*** are perpendicular to noncorresponding unit vectors in the original lattice, respectively, for example, $\mathbf{a}^* \perp \mathbf{b}$ and $\mathbf{a}^* \perp \mathbf{c}$. Therefore, the following relationship can be held:

$$\mathbf{a}^*\cdot\mathbf{b} = \mathbf{a}^*\cdot\mathbf{c} = \mathbf{b}^*\cdot\mathbf{c} = \mathbf{b}^*\cdot\mathbf{a} = \mathbf{c}^*\cdot\mathbf{a} = \mathbf{c}^*\cdot\mathbf{b} = 0 \quad \text{(Eq 17)}$$

and

$$\mathbf{a}^* = \frac{\mathbf{b}\times\mathbf{c}}{\mathbf{a}\cdot(\mathbf{b}\times\mathbf{c})} \quad \mathbf{b}^* = \frac{\mathbf{c}\times\mathbf{a}}{\mathbf{a}\cdot(\mathbf{b}\times\mathbf{c})} \quad \mathbf{c}^* = \frac{\mathbf{a}\times\mathbf{b}}{\mathbf{a}\cdot(\mathbf{b}\times\mathbf{c})} \quad \text{(Eq 18)}$$

It should be noted that $\mathbf{a}\cdot(\mathbf{b}\times\mathbf{c})$ represents the volume of the original lattice.

Furthermore, the reflection vector in a diffraction pattern indicates an (hkl) plane can also be expressed as **g** vector by the reciprocal unit vectors as:

$$\mathbf{g}_{hkl} = h\mathbf{a}^* + k\mathbf{b}^* + l\mathbf{c}^* \quad \text{(Eq 19)}$$

This relationship is shown schematically in Fig. 27. The $\mathbf{g}_{hkl}$ vector is normal to the (hkl) plane, and the magnitude of $\mathbf{g}_{hkl}$ is $1/d_{hkl}$. The reciprocal space is useful when electron diffraction is taken into account.

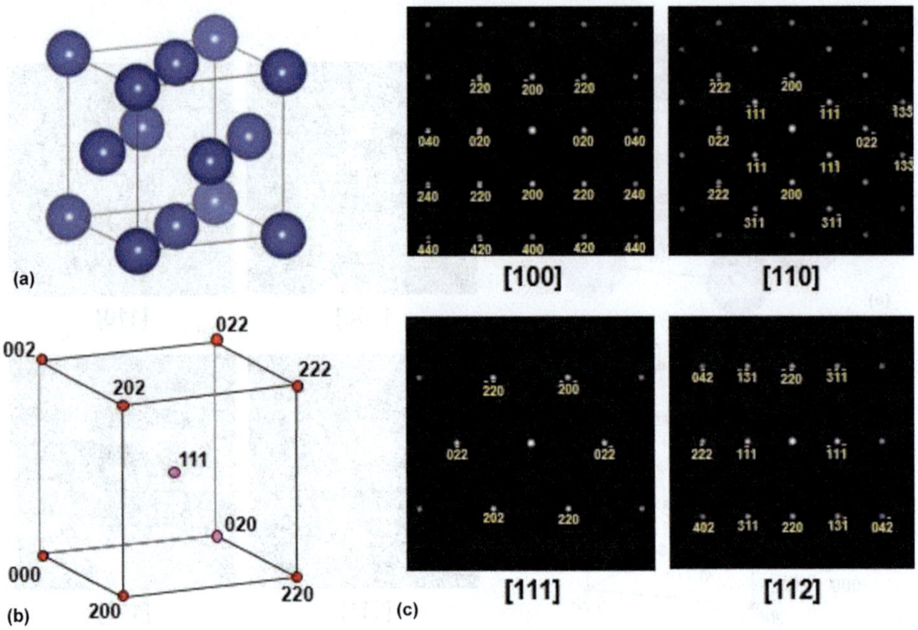

Fig. 25 (a) Real and (b) reciprocal lattices of a face-centered cubic structure. (c) Simulated electron diffraction patterns for a face-centered cubic structure in the [001], [011], [111], and [112] directions

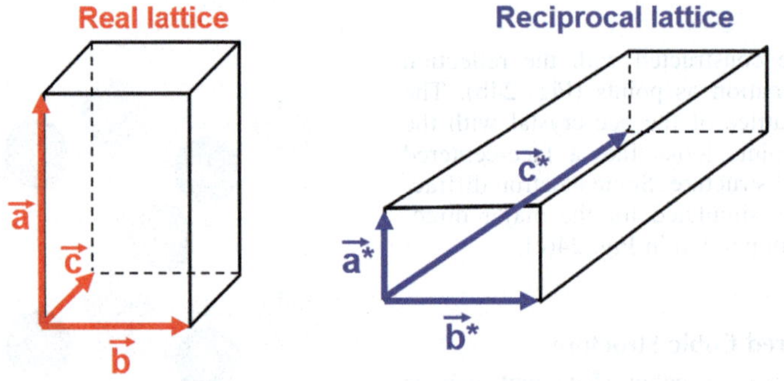

Fig. 26 Relationship between real and reciprocal lattices

Selected-Area Electron Diffraction

The standard method for generating diffraction patterns using conventional TEM is by selected-area electron diffraction (SAED). In this method, an aperture (appropriately termed the selected-area aperture) located at the image plane of the objective lens or at an equivalent image plane in the microscope column is moved into the illuminated region to limit the area of the specimen from which a diffraction pattern is obtained. The position of the aperture selects the region of the sample from which the diffraction pattern is generated. The diameter of the selected-area aperture can be varied, but the size of the aperture projected to the image plane limits the minimum area of the specimen from which an SAED pattern can be generated.

Any of three types of electron diffraction patterns can be generated in the SAED approach:

- Ring patterns from fine-grained polycrystalline materials, in which diffraction occurs simultaneously from many grains with different orientations relative to the incident beam
- Spot patterns, in which diffraction occurs from a single-crystal region of the specimen
- Kikuchi lines, in which diffraction occurs from a single-crystal region of the specimen, which is sufficiently thick that the diffracting electrons have undergone simultaneous elastic and inelastic scattering, known as Kikuchi scattering

Ring Patterns

Ring patterns arise when many randomly oriented fine grains contribute to the observed diffraction pattern. Ring pattern analysis, similar to powder XRD analysis, is used to identify unknowns and to characterize the crystallography of a new crystalline material. Ring patterns are commonly used to assist identification of fine precipitates and other secondary phases in a matrix. Ring patterns can also be generated from a fine polycrystalline matrix, such as nanostructures produced by certain chemical vapor deposition processes and other fabrication processes. In

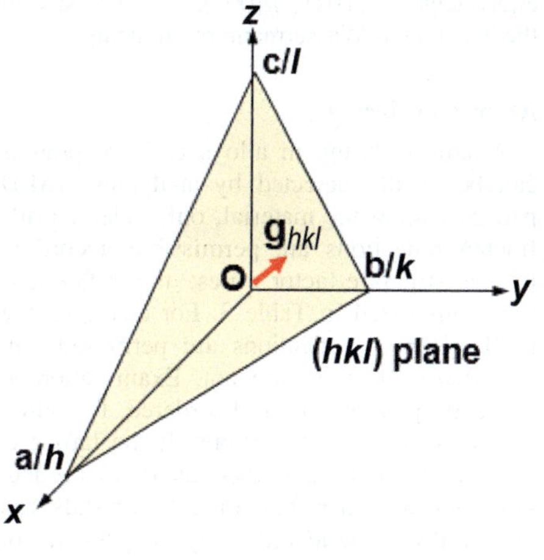

Fig. 27 Definition of the reflection vector g_{hkl} at a crystallographic plane (*hkl*)

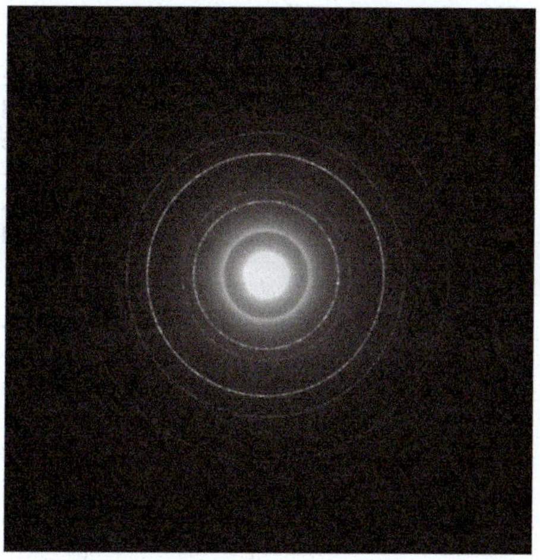

Fig. 28 Ring pattern obtained from a thin polycrystalline aluminum. Courtesy of E. Musterman and J.P. Cline

Fig. 29 Simple spot pattern from a single-crystal region of a [011]-projected thin stainless steel specimen. Courtesy of J.P. Cline

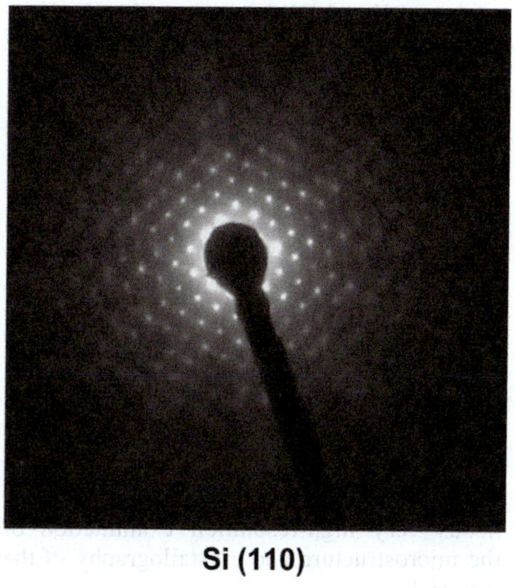

Fig. 30 Typical Kikuchi pattern of silicon in the [011] projection

these cases, the crystal structure and lattice parameters of the material producing the ring pattern can be determined.

In its applications and mechanics of use, an electron diffraction ring pattern is analogous to a Debye-Scherrer XRD pattern. Figure 28 shows a ring pattern generated from a thin specimen of polycrystalline aluminum. The ratio of ring diameters proves that the crystal has an fcc structure. From Table 3, the rings correspond to (111), (200), (220), (311), (222), (400), and so on. The interplanar spacings are related to the ring radii by the camera constant of the microscope. The camera constant is the product of the electron wavelength (λ) and the camera length (distance from the specimen to the recording plane, L), and the λL value should be constant under a given operating condition. Therefore, by measuring the radii, r, of the given (*hkl*) diffraction rings from the ring patterns, the (*hkl*) interplanar d-spacings can be determined, and the lattice parameter(s) can then be calculated from standard crystallographic relationships. Conversely, by measuring the ring radii of a known standard crystal specimen, the L value can also be calibrated.

Alternatively, a common technique for identifying an unknown from its ring pattern involves measuring the d-spacings of the various rings and comparing the results to the spacings of suspected candidates listed in the powder diffraction file data (Ref 17).

Spot Patterns

Spot patterns arise when the incident electron beam irradiates a portion of a single crystal within the specimen. The spot pattern is qualitatively analogous to the single-crystal XRD patterns obtained using the Laue technique. Figure 29 shows a simple spot pattern from a single-crystal region of a thin stainless steel specimen. The sample is tilted so a simple crystallographic direction, [011], which is aligned parallel to the incident beam. The transmitted spot appears at the center, and the various (*hkl*) reflections appear at specific distances and angles from the transmitted spot. The transmitted spot corresponds to the point at which the transmitted beam intersects the recording device, such as a CCD camera; each diffracted spot corresponds to the point at which a given diffracted beam intersects the recording device. The crystal structure and orientation of the single-crystal region of the specimen being irradiated can be identified from its spot pattern. These patterns can be indexed, and the (*hkl*) plane that produced each diffraction spot can also be determined. Similar to the ring pattern analysis described previously, the pattern is solved using the modified form of Bragg's law. Simple crystal structures and orientations can often be identified by comparing the unknown spot pattern to standard tabulations.

Kikuchi Patterns

Kikuchi patterns can be observed in a single-crystal SAED pattern taken from a thick region of the thin specimen. If the specimen is sufficiently thick, some electrons will scatter multiple times. Some of these scattering events are inelastic and some are elastic. Some inelastically scattered electrons are subsequently elastically scattered by Bragg diffraction from the various (*hkl*) atomic planes. Two cones of diffracted electrons are generated from each set of (*hkl*) planes, and they intersect the recording device. The arc of the cones as they intersect the recording device appears as lines termed Kikuchi lines. One cone of radiation is more intense than the general background and produces a bright line on the pattern; the bright Kikuchi line is the excess line. The second cone of radiation is less intense than the inelastic background and produces a dark line; the dark Kikuchi line is the defect line. Each pair of parallel bright and dark Kikuchi lines has a characteristic crystallographic spacing, which is equivalent to twice the scattering angle, 2θ. Figure 30 shows a typical Kikuchi pattern of silicon. The symmetry apparent in the pattern indicates that the pattern was taken with the electron beam exactly parallel to the [110] zone axis. Kikuchi patterns provide significant information about the crystallography of the specimen. The geometry of Kikuchi patterns is much more precise than spot patterns, enabling very accurate determination of crystal orientation. Kikuchi lines can also be observed in convergent-beam electron diffraction patterns described later and in electron backscatter diffraction (EBSD) patterns obtained from a bulk sample in an SEM instrument. More details on the EBSD method can be found in the article

"Crystallographic Analysis by Electron Backscatter Diffraction in the Scanning Electron Microscope" in this Volume.

Kikuchi lines are also useful for navigating in the reciprocal space. The distance between a pair of parallel bright and dark Kikuchi lines is 2θ, that is, the magnitude of $\mathbf{g}_{hkl}$ (= $1/d_{hkl}$). The narrower the Kikuchi line spacing, the larger the interplanar spacing. Therefore, the major zone axes can be reached by tilting a specimen along a narrower Kikuchi line pair, which appear in the reciprocal space, that is, in the diffraction mode.

The Kikuchi pattern is also invaluable in establishing the proper diffracting conditions for obtaining high-contrast images. Very near the exact Bragg condition for a set of (*hkl*) planes, the excess (*hkl*) Kikuchi line passes through the (*hkl*) spot, and the defect (*hkl*) Kikuchi line passes through the (000) transmitted spot. The relative positioning of diffraction spots and Kikuchi lines occurs under dynamical conditions for which high-contrast BF and DF images can be obtained. Under kinematical conditions, far from the exact Bragg condition, excess and defect Kikuchi lines do not pass near the (*hkl*) and (000) spots, respectively (Fig. 31). Low-contrast images are obtained under kinematical conditions.

Analysis of ring, spot, and Kikuchi patterns reveals a great deal about the nature of the specimen. The patterns can be indexed to determine the crystal structure and lattice parameters of an unknown material. Advantages of such determinations using the analytical electron microscope relative to XRD include the use of much smaller volumes of material, the use of small volume fractions of a second phase in a matrix, and simultaneous very high-resolution examination of the microstructure and crystallography of the material.

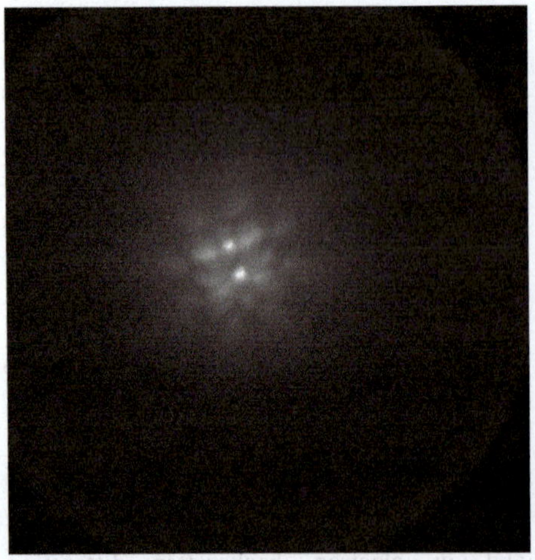

Fig. 31 Kikuchi pattern of thin stainless steel specimen recorded away from any zone axis. Courtesy of J.P. Cline

Fine Structure Effects

In addition to crystal structure and lattice parameters, other useful information related to the solid-state nature of materials can be determined by analyzing the geometry and orientation of fine structure diffraction effects that appear in SAED patterns (ring, spot, and Kikuchi), including:

- Orientation relationships between coexisting phases
- Dislocation Burgers vectors
- Defect habit planes
- Characterization of order/disorder in the crystal
- Characterization of coherent precipitation
- Characterization of spinodal decomposition

A few examples are discussed here.

Orientation Relationships

Even at a very fine microstructural scale, it is possible to determine the orientation relationship between two coexisting phases from an examination of the relative orientation of their spot SAED patterns. Orientation relationships are typically expressed as a pair (one in each phase) of parallel directions in a pair of parallel planes. A common example is the Nishiyama-Wassermann relationship between fcc and bcc materials: $(111)_{fcc}//(110)_{bcc}$, $[\bar{1}01]_{fcc}//[001]_{bcc}$, and $[\bar{1}2\bar{1}]_{fcc}//[\bar{1}10]_{bcc}$ (Ref 18, 19). These relationships are best illustrated in the spot pattern from a bcc-Fe/fcc-Ni interface shown in Fig. 32, in which patterns from both bcc and fcc phases are superimposed (Ref 20). The diffraction pattern from the fcc phase is [112]-projected, and that from the bcc phase is [110]-projected. As shown in Fig. 32, the 200 spot of bcc is parallel to the 220 spot of fcc, which is equivalent to $[\bar{1}01]_{fcc}//[001]_{bcc}$, expressed in the Nishiyama-Wassermann relationship.

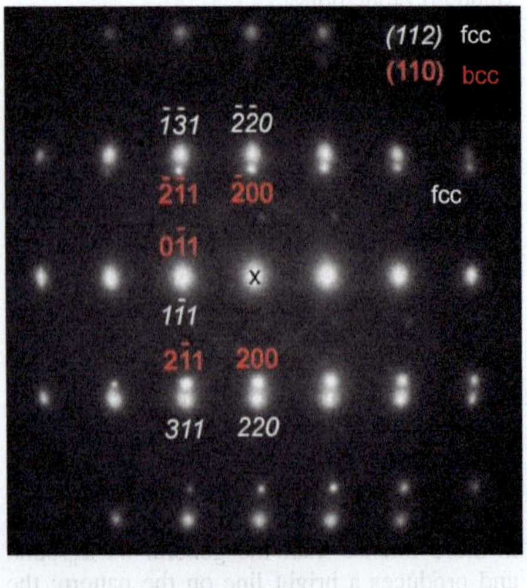

Fig. 32 Selected-area electron diffraction pattern from a body-centered cubic (bcc)-Fe/face-centered cubic (fcc)-Ni interface showing the Nishiyama-Wassermann orientation relationship. Courtesy of K. Lorcharoensery

Atomic Ordering

Atomic ordering in alloys and compounds can be readily detected by analyzing SAED patterns. In an fcc material, only selected diffracted reflections are permissible according to the structure-factor rules; the reflections are summarized in Table 3. For example, the (020) (N = 4) reflections are permitted, but the (010) reflections are not. Examination of the spot pattern for a disordered fcc alloy uncovers no (010) reflections. If the lattice is ordered, the (010) and other superlattice reflections appear. Therefore, ordered crystals can almost always be identified by the presence of superlattice spots in the spot pattern. Figure 33 shows the image, spot diffraction pattern, and indexed schematic diffraction pattern from an iron-base superalloy sample containing an ordered fcc-based precipitate, γ-Ni_3(Ti,Al), in a disordered fcc austenitic matrix. The presence of ordered precipitates can be identified unambiguously by imaging the structure with one of the suspected superlattice diffraction spots. If ordered, the phase appears in the image created by using the diffracted beam. The image verification procedure is illustrated in Fig. 34.

Strain-Induced Defects

The lattice strain associated with crystal defects or precipitation can lead to streaking of some spots in the diffraction pattern. One use of elastic strain-induced streaking is the study of coherent precipitation (Ref 21, 22). Figure 35 shows a diffraction pattern in which elastic strains due to coherent precipitation have streaked the diffraction spots. Although streaking is helpful in identifying the presence of coherent precipitation, it is difficult to quantify precipitation through diffraction pattern analysis alone. Streaking of diffraction spots has several other causes, including precipitate shape effects, stacking faults, twins, and surface films.

Satellite Spots

Features sometimes observed in diffraction patterns are satellite spots. Satellites are small additional diffraction spots occurring near the major diffraction spots. Composition modulations, such as those from spinodal decomposition, are one source of satellite spots. Satellite spots lie in the same crystallographic direction as the composition fluctuation and are due to a variation in the lattice parameter and the nature of the elastic scattering of the electrons. The spacing of satellite spots is related to the wavelength of the compositional modulation. The elastic strain induced by the composition modulation can also induce streaking in diffraction patterns. Satellite spots also form due to regular arrays of dislocations, twins, antiphase domain boundaries, and magnetic domains.

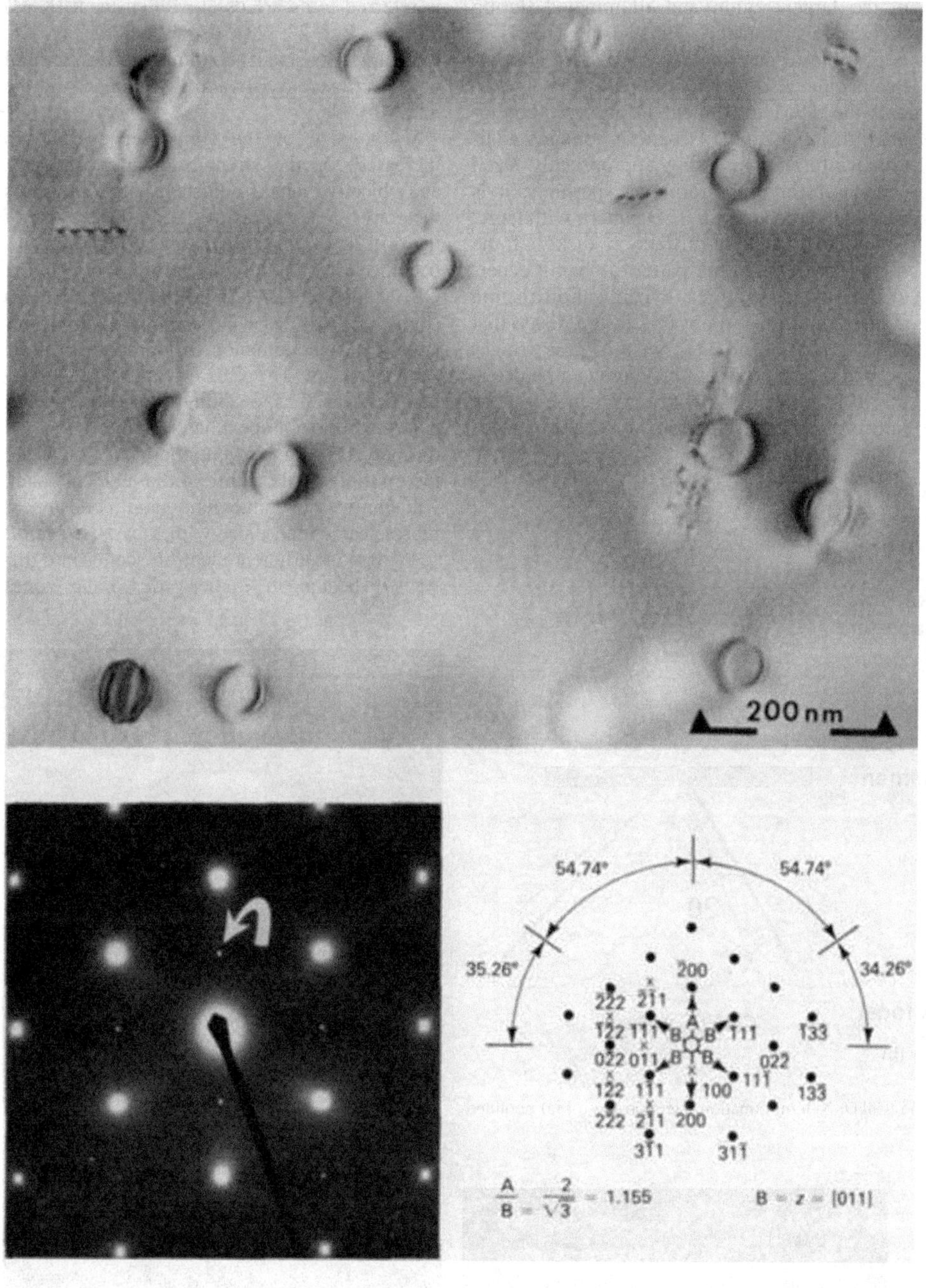

Fig. 33 Bright-field image and diffraction pattern from a face-centered cubic (fcc) matrix (austenite) containing precipitates having the ordered fcc ($L1_2$) superlattice, $Ni_3(Ti,Al)$. The fcc spot pattern is indexed in the schematic key, and the superlattice reflections are indicated by an arrow in the diffraction pattern. The beam direction, B, and the zone axis, *z*, are parallel. Courtesy of T.J. Headley

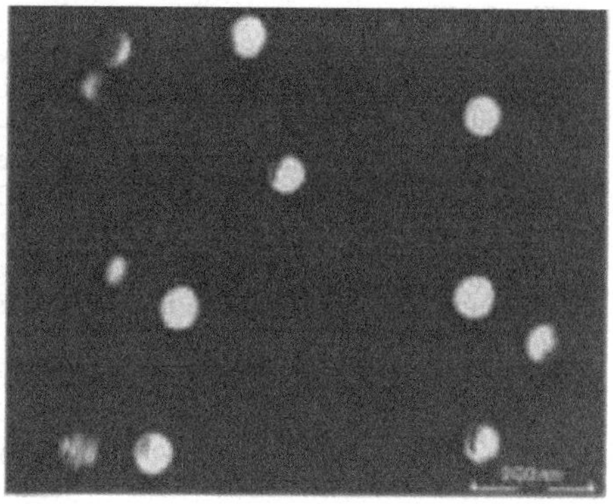

Fig. 34 Image of ordered $Ni_3(Ti,Al)$ precipitates. Image formed with the superlattice spot indicated by the arrow in Fig. 33. Courtesy of T.J. Headley

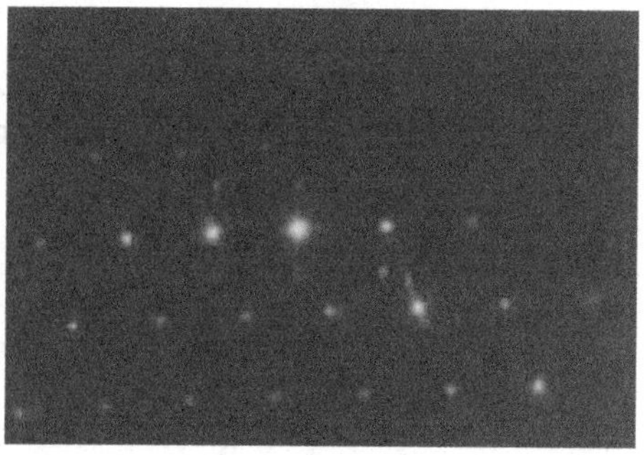

Fig. 35 Diffraction pattern from a specimen of shock-loaded Nitronic 40. The streaking is due to precipitation of a coherent second phase and deformation-induced twinning. The shape of the coherent precipitates and the elastic strain they induce caused streaking of the precipitate diffraction spots. Courtesy of C.R. Hills

Convergent Beam Electron Diffraction

The advent of the modern TEM instrument with STEM capability and/or the ability to form STEM-like fine probes in the TEM mode enable examination of the crystallographic nature of materials on a very small spatial scale using micro/nanodiffraction. Generally, micro/nanodiffraction is defined as any diffraction technique that enables collection of diffraction data from specimen volumes ≤0.5 μm in diameter. The most common micro/nanodiffraction technique in use is convergent beam electron diffraction (CBED).

In the standard SAED approach using the conventional TEM mode, the specimen is illuminated by essentially parallel rays of electrons, producing a standard SAED pattern with sharply defined diffraction spots. In the STEM mode, a diffraction pattern is also produced by the scanning fine probe. The diffraction pattern produced by a fine STEM probe is a stationary diffraction pattern. To generate a diffraction pattern from a specific small volume of the specimen area, the scanning probe can be stopped and placed at the position of interest. However, because the electron probe is now highly convergent, the diffraction pattern is a series of disks rather than sharp spots, and disk diameters are proportional to the beam convergence angle. The disk pattern formation under a convergent incident beam is shown in Fig. 36. Because the convergent beam contains various tilt-angle components within the illumination, the components off from the optic axis appear at corresponding angular position within the disk (Fig. 36b), which is why the disk pattern is formed by the convergent beam. Diffraction patterns produced by convergent beams are termed CBED patterns. Figure 37 shows a comparison between conventional SAED and CBED patterns generated from the same specimen ([111]-projected silicon). The CBED patterns that have spot arrangements analogous to conventional SAED patterns are termed zero-order Laue zone CBED (ZOLZ-CBED) patterns. The procedure for indexing ZOLZ-CBED patterns is the same as that used to index conventional SAED patterns.

The CBED patterns are used in a number of ways to extract quantitative crystallographic information from the specimen. Perhaps the most routine application of CBED is crystallographic analysis of small volumes, such as fine precipitates, in thin foil specimens. The CBED pattern is formed from only the electron-scattering volume around the fine beam probe. The electron-scattering volume in thin foils is typically <50 nm in diameter. Therefore, CBED patterns can be generated easily from these small volumes. This type of CBED (ZOLZ-CBED) pattern does not inherently contain more information than a conventional SAED pattern, but it may be more convenient to perform the desired analysis using CBED because extraneous spots generated elsewhere in the larger region and illuminated during SAED are eliminated. Thus, using CBED, it is possible to determine the structure and lattice parameter of small regions.

If the TEM instrument is properly configured, the ZOLZ-CBED pattern contains additional information. Using appropriately sized condenser (or probe-forming) apertures, it is possible to produce a CBED pattern with large, yet nonoverlapping, diffraction disks. If the region from which the pattern is being generated is sufficiently thick, dynamical diffraction contrast information appears in the disks; that is, it is possible to observe thickness fringes in each disk. Specimen thickness can be determined from the disks with the accuracy of the lattice parameter. This form of CBED pattern is referred to as a Kossel-Mollenstaedt (K-M) pattern (Ref 23, 24), as shown in Fig. 38. Kossel-Mollenstaedt conditions are required to form discrete BF and DF images in the STEM mode.

If the size of the condenser aperture is increased, or the strength of the condenser and objective lens is adjusted, which increases the convergence angle of the probe, the size of the diffraction disks increases. When the disks become large enough to overlap, the resulting CBED pattern is termed a Kossel pattern (Fig. 39). This term is taken from the analogous XRD phenomenon that produces Kossel x-ray patterns. The CBED patterns also contain Kikuchi lines, enabling precise orientation determinations of specimen volumes too small to analyze using conventional SAED. Due to the multiple-angle nature of the incident beam, Kikuchi lines in convergent beam mode appear more clearly than those with a parallel beam.

If there is sufficient electron scatter into diffracted beams in higher-order Laue zones

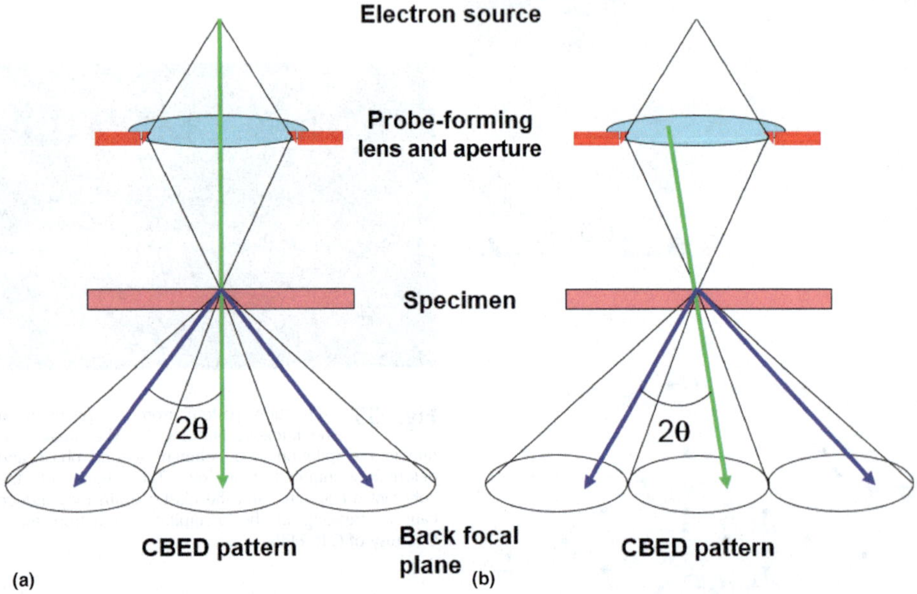

Fig. 36 Concept of convergent beam electron diffraction (CBED) pattern formation. Electron rays of (a) nontilted and (b) tilted incident beams

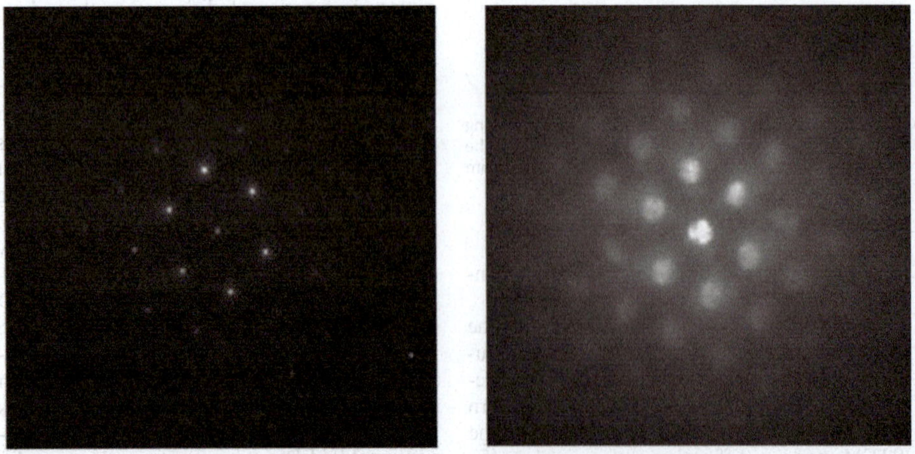

Fig. 37 Comparison of (a) selected-area electron diffraction and (b) convergent beam electron diffraction patterns from a [011]-projected stainless steel specimen. Courtesy of J.P. Cline

Fig. 38 Kossel-Mollenstaedt pattern from [011]-projected silicon

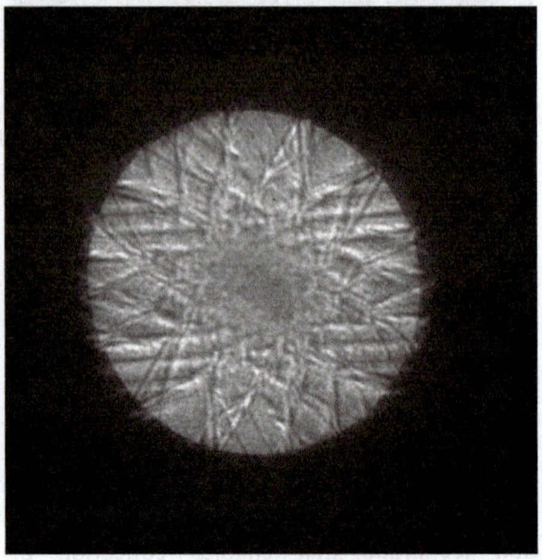

Fig. 39 Kossel pattern from [011]-projected silicon

(HOLZ) (that is, $n = 2, 3$, etc. in the Bragg equation), three-dimensional crystallographic information can be determined from a single CBED pattern. Higher-order zones appear as rings around the ZOLZ-CBED pattern. The first-order Laue zone (FOLZ) corresponds to $n = 1$, the second-order Laue zone (SOLZ) to $n = 2$, and so on. Such a pattern is often referred to as a HOLZ pattern (Fig. 40). Usually, only the FOLZ is present in the CBED pattern. The HOLZ pattern shown in Fig. 40 contains only the FOLZ ring and is typical of many published HOLZ patterns.

Higher-order Laue zone patterns are obtained by operating under K-M conditions. Observation of the HOLZ rings requires a wide-angle view (typically greater than ±10 to 12°) of the diffraction pattern. Diffraction disks within the HOLZ rings and the 000 transmitted disk can also contain HOLZ lines, which are essentially extensions of the Kikuchi lines formed in the ZOLZ pattern. Analysis of the HOLZ lines enables the determination of lattice parameters as accurate as 5×10^{-4} nm. The HOLZ lines are very sensitive to the local lattice distortions. Therefore, HOLZ lines can be applied to measure local strain difference in the crystal, such as strain fields around coherent precipitates and in the vicinity of epitaxial interfaces. Complete analysis of HOLZ rings, HOLZ lines, and the intensity distribution in the ZOLZ disks in the HOLZ pattern enables complete determination of the full point and crystal symmetry and thus the space group of the crystal. Additional information on CBED patterns, HOLZ patterns, and crystal structure determination can be found in Ref 4, 7, and 9.

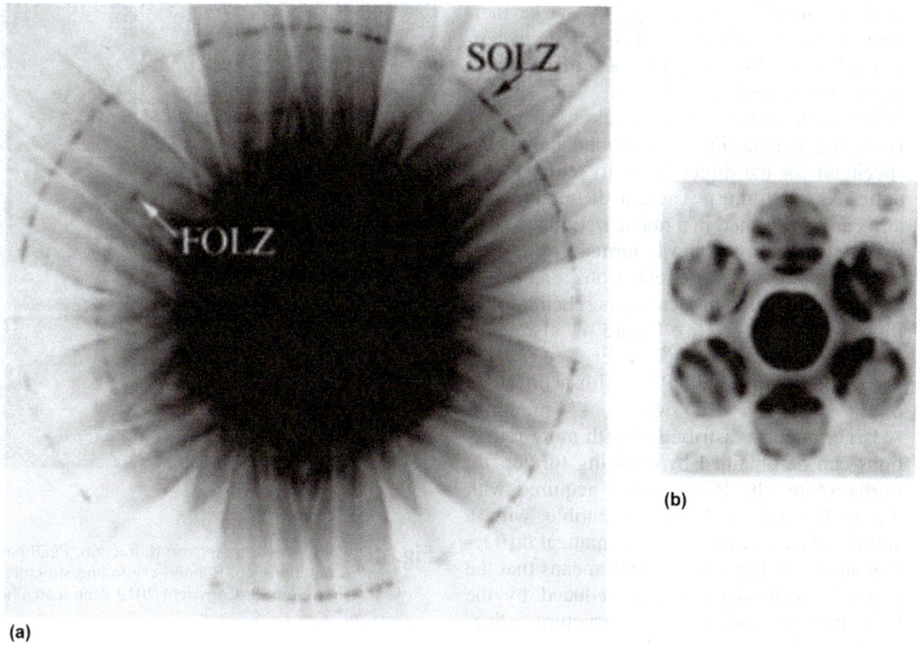

Fig. 40 Convergent beam electron diffraction (CBED) patterns from [111]-projected $BaBiO_3$. (a) Complete CBED pattern including higher-order Laue zone rings (both first-order Laue zone, or FOLZ, and second-order Laue zone, or SOLZ). (b) Zero-order Laue zone disks. Reprinted with permission from Ref 25. Copyright 2001 by the American Physical Society

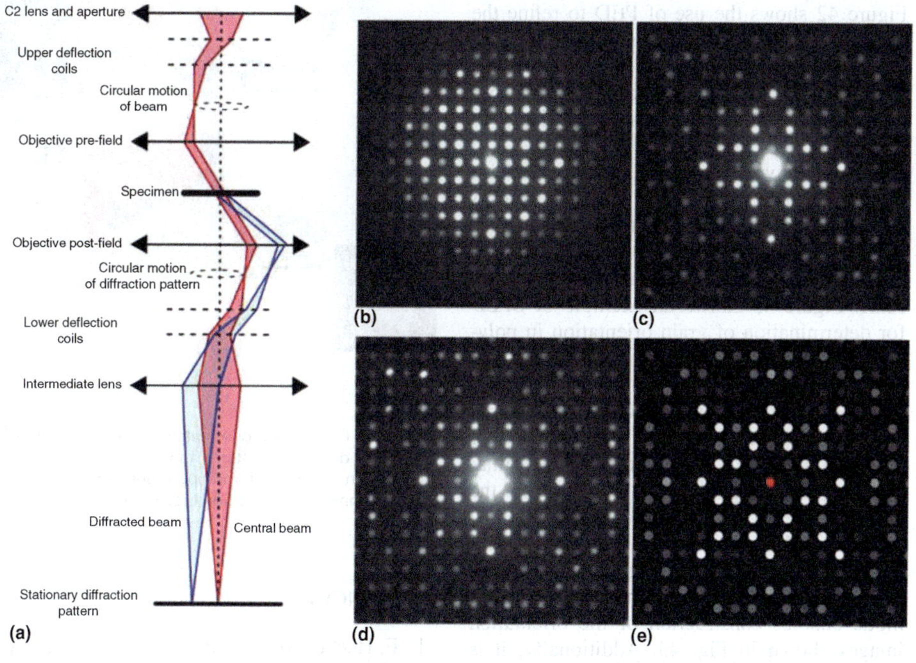

Fig. 41 (a) Schematic diagram for the precession electron diffraction (PED) method, illustrating tilting and detilting of the beam before and after the specimen. (b) Conventional selected-area electron diffraction pattern from $Er_2Ge_2O_7$. (c, d) PED patterns with tilt angles of 20 and 47 mrad, respectively. (e) Simulated diffraction pattern based on kinematical diffraction. Source: Ref 27. Reproduced with permission of the International Union of Crystallography (IUCr)

Precession Electron Diffraction

As described previously, electron diffraction techniques are used to identify crystalline structures at the nanometer scale in TEM. Through electron diffraction analysis, it is possible not only to determine atomic positions but also to refine the charge-bonding densities around atoms if detailed analysis is applied to electron diffraction data. However, there are a few issues in the electron diffraction approaches by TEM. In TEM, only forward-scattered electrons are used to form diffraction patterns. Therefore, the angular range of diffraction patterns is rather limited in TEM compared with other diffraction-based methods. For example, the radius of the HOLZ ring shown in Fig. 40 is ~50 mrad at 200 keV, so the typical angular range recordable in TEM can be as large as ~100 mrad, which is only ~6°. Conversely, the recordable angular range in the SEM-EBSD approach is more than 50°. Due to the limited angular range in TEM, structure analysis can be degraded. In addition, detailed structure determination, which requires evaluation of not only diffraction spot positions but also spot intensities, would be problematic due to complex dynamic scattering of electrons in a thin specimen.

To overcome these issues in detailed structure determination using the TEM-based electron diffraction approach, the precession electron diffraction (PED) method was developed (Ref 26). In the PED method, the incident electron beam is tilted and rotated along the tilt angle (termed beam rocking). After the incident beam is tilted, a diffraction pattern formed at the objective back focal plane is also tilted correspondingly, because any angular information appears at the back focal plane. However, countertilting is applied below the specimen for the direct beam to return to the optic axis (detilting). Then, a diffraction pattern can be obtained in a manner similar to that by a fixed electron beam in terms of reflection positions, but with more reflections. Precession electron diffraction processing is schematically illustrated in Fig. 41(a). Figure 41 also compares diffraction patterns of $Er_2Ge_2O_7$ by a conventional approach: PED with a tilt angle of 20 mrad and PED with a 47 mrad tilt. A larger angular distribution with more reflections can be obtained by applying further tilt. Furthermore, the PED pattern acquired with the higher tilt angle is compatible with a simulated pattern based on kinematical diffraction shown in Fig. 41(e), which means that the dynamic scattering effect is reduced by the PED method, and thus, the structure refinement can also be improved.

There are basically two application types of the PED method. Primary applications are the determination of crystal structures, including crystal symmetry, measurement of lattice parameters, and refinement of atomic structures in crystals without any prior knowledge. Figure 42 shows the use of PED to refine the crystal structure of beam-sensitive $LiBH_4$. From various $LiBH_4$ particles, PED patterns (Fig. 42a) were acquired in very low-dose conditions. The structure was determined as shown in Fig. 42(b).

The other type of application is to perform PED in the STEM mode, which is defined as scanning PED (SPED). This technique enables orientation imaging of multiple grains in a thin specimen and strain mapping at nanometer scale. Figure 43 shows an example of SPED for determination of grain orientation in polycrystalline alumina. Phase and grain orientation were determined by indexing PED patterns acquired from individual pixels and comparing them with patterns from neighbor pixels. Accuracy of the orientation determination has been significantly improved in the SPED approach (Fig. 43b) compared with a series of SAED patterns acquired in the STEM mode without precession (Fig. 43a). The series of PED pattern acquisition in the STEM mode enables construction of the orientation images shown in Fig. 43. Additionally, it is possible to extract virtual bright-field (VBF) and dark-field (VDF) images by selecting specific reflections (direct for VBF and deflected for VDF, respectively) from the series of PED pattern. Both PED and SPED methods offer high-quality data for crystal structure determination and detailed microstructure characterization compared with conventional electron diffraction approaches. More extensive review articles on the PED-based approaches can be found elsewhere (Ref 30).

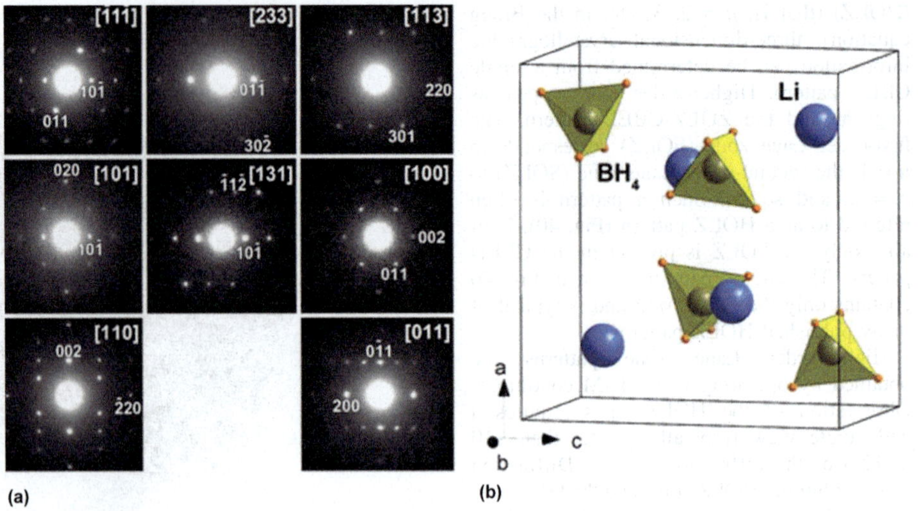

Fig. 42 (a) Precession electron diffraction (PED) patterns from various $LiBH_4$ particles acquired in very low-dose conditions. (b) Refined crystalline structure of $LiBH_4$ using the PED patterns. Reprinted with permission from Ref 28. Copyright 2012 American Chemical Society

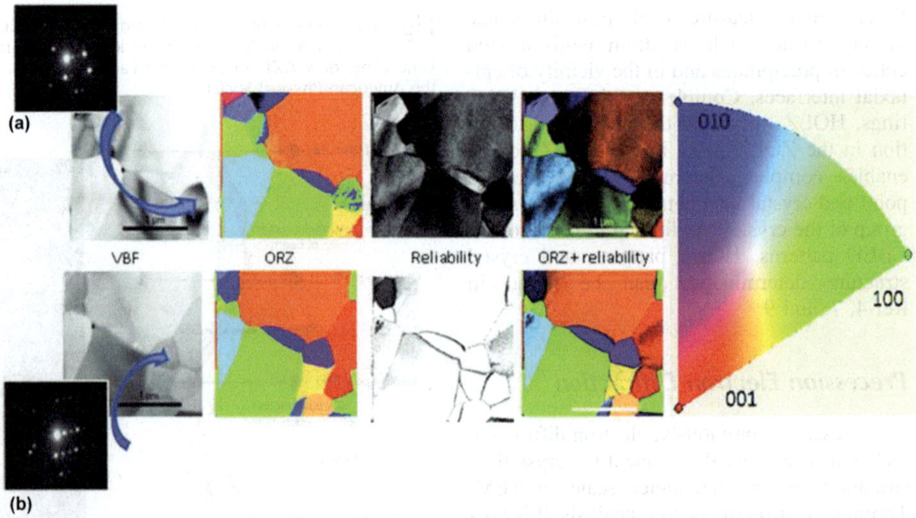

Fig. 43 Comparison of orientation zone (ORZ) distribution mapping based on (a) a series of selected-area electron diffraction patterns and (b) a series precession electron diffraction patterns, both obtained in the scanning transmission electron microscopy mode. VBF, virtual bright-field. Source: Ref 29. Reprinted under the terms of the Creative Commons Attribution License (CC BY)

REFERENCES

1. E. Hecht, *Optics*, 3rd ed., Addison-Wesley, 1988
2. D.A. Muller, Structure and Bonding at the Atomic Scale by Scanning Transmission Electron Microscopy, *Nature Mater.*, Vol 8 (No. 4), 2009
3. C. Kisielowski, B. Freitag, M. Bischoff, H. van Lin, S. Lazar, G. Knippels, P. Tiemeijer, M. van der Stam, S. von Harrach, M. Stekelenburg, M. Haider, S. Uhlemann, H. Müller, P. Hartel, B. Kabius, D. Miller, I. Petrov, E.A. Olson, T. Donchev, E.A. Kenik, A. Lupini, J. Bentley, S. Pennycook, I.M. Anderson, A.M. Minor, A.K. Schmid, T. Duden, V. Radmilovic, Q. Ramasse, M. Watanabe, R. Erni, E.A. Stach, P. Denes, and U. Dahmen, Detection of Single Atoms and Buried Defects in Three Dimensions by Aberration-Corrected Electron Microscopy with 0.5 Å Information Limit, *Microsc. Microanal.*, Vol 14, 2008, p 469–477
4. D.B. Williams and C.B. Carter, *Transmission Electron Microscopy: A Textbook for Materials Science*, 2nd ed., Springer, 2009

5. L. Reimer and H. Kohl, *Transmission Electron Microscopy: Physics of Image Formation*, 5th ed., Springer, 2008
6. J. Pennycook and P.J. Nellist, Ed., *Scanning Transmission Electron Microscopy: Imaging and Analysis*, Springer, 2011
7. C.B. Carter and D.B. Williams, Ed., *Transmission Electron Microscopy: Diffraction, Imaging, and Spectrometry*, Springer, 2016
8. R.F. Egerton, *Physical Principles of Electron Microscopy: An Introduction to TEM, SEM, and AEM*, 2nd ed., Springer, 2016
9. J.M. Zuo and J.C.H. Spence, *Advanced Transmission Electron Microscopy: Imaging and Diffraction in Nanoscience*, Springer, 2016
10. O.L. Krivanek and P.E. Mooney, Applications of Slow-Scan CCD Cameras in Transmission Electron-Microscopy, *Ultramicroscopy*, Vol 49, 1993, p 95–108
11. R.R. Meyer and A.I. Kirkland, Characterization of the Signal and Noise Transfer of CCD Cameras for Electron Detection, *Microsc., Res. Tech.*, Vol 49, 2000, p 269–280
12. G. McMullan, A.R. Faruqi, and R. Henderson, Direct Electron Detectors, *Meth. Enzymol.*, Vol 579, 2016, p 1–17
13. N. Shibata, S.D. Findlay, Y. Kohno, H. Sawada, Y. Kondo, and Y. Ikuhara, Differential Phase-Contrast Microscopy at Atomic Resolution, *Nature Phys.*, Vol 8, 2012, p 611–615
14. H. Ryll, M. Simson, R. Hartmann, P. Holl, M. Huth, S. Ihle, Y. Kondo, P. Kotula, A. Liebel, and K. Müller-Caspary, A *pn*CCD-Based, Fast Direct Single Electron Imaging Camera for TEM and STEM, *J. Instrum.*, Vol 11, 2016, p 4006
15. Y. Zhu, H. Inada, K. Nakamura, and J. Wall, Imaging Single Atoms Using Secondary Electrons with an Aberration-Corrected Electron Microscope, *Nature Mater.*, Vol 8, 2009, p 808–812
16. P.A. Doyle and P.S. Turner, Relativistic Hartree-Fock X-Ray and Electron Scattering Factors, *Acta Crystallogr. A*, Vol 24, 1968, p 390–397
17. PDF, International Center for Diffraction Data, Newtown Square, PA
18. Z. Nishiyama, X-Ray Investigation of the Mechanism of the Transformation from Face-Centered-Cubic Lattice to Body-Centered Cubic, *Sci. Rep. Res. Inst. Tohoku Univ.*, Vol 23, 1934, p 637–664
19. G. Wassermann, Influence of the α-γ Transformation of an Irreversible Ni Steel onto Crystal Orientation and Tensile Strength, *Arch. Eisenhuttenwes.*, Vol 16, 1933, p 647
20. K.W. Andrew, D.J. Dyson, and S.R. Keown, *Interpretation of Electron Diffraction Patterns*, Plenum Press, 1967
21. A. Guinier, Heterogeneities in Solid Solutions, *Phys. Status Solidi*, Vol 9, 1959, p 293–398
22. L.E. Tanner, Diffraction Contrast from Elastic Shear Strains due to Coherent Phases, *Philos. Mag.*, Vol 14, 1966, p 111–130
23. P. Goodman, A Practical Method of Three-Dimensional Space-Group Analysis Using Convergent-Beam Electron Diffraction, *Acta Crystallogr. A*, Vol 31, 1975, p 804–810
24. J.W. Steeds, Microanalysis by Convergent Beam Electron Diffraction, *Quantitative Microanalysis with High Spatial Resolution*, G.W. Lorimer, M.H. Jacobs, and P. Doig, Ed., The Metals Society, London, 1981, p 210–216
25. T. Hashimoto, K. Tsuda, J. Shiono, J. Mizusaki, and M. Tanaka, Determination of the Crystal System and Space Group of $BaBiO_3$ by Convergent-Beam Electron Diffraction and X-Ray Diffraction Using Synchrotron Radiation, *Phys. Rev. B*, Vol 64, 2001
26. R. Vincent and P.A. Midgley, Double Conical Beam-Rocking System for Measurement of Integrated Electron Diffraction Intensities, *Ultramicroscopy*, Vol 53, 1994, p 271–282
27. P.A. Midgley and A.S. Eggeman, Precession Electron Diffraction—A Topical Review, *Inter. Union Crystall. J.*, Vol 2 (Part 1), 2015, p 126–136
28. J. Hadermann, A.M. Abakumov, S. Van Rompaey, T. Perkisas, Y. Filinchuk, and G. Van Tendeloo, Crystal Structure of a Lightweight Borohydride from Submicrometer Crystallites by Precession Electron Diffraction, *Chem. Mater.*, Vol 24, 2012, p 3401–3405
29. D. Viladot, M. Véron, M. Gemmi, F. Peiró, J. Portillo, S. Estradé, J. Mendoza, N. Llorca-Isern, and S. Nicolopoulos, Orientation and Phase Mapping in the Transmission Electron Microscope Using Precession-Assisted Diffraction Spot Recognition: State-of-the-Art Results, *J. Microsc.*, Vol 252, 2013, p 23–34
30. A.S. Eggeman and P.A. Midgley, Chap. 1, Precession Electron Diffraction, *Adv. Imag. Electron Phys.*, Vol 170, 2012, p 1–63

ASM Handbook, Volume 10, *Materials Characterization*
ASM Handbook Committee
DOI 10.31399/asm.hb.v10.a0006638

Electron Probe X-Ray Microanalysis

Dale E. Newbury and Nicholas W. M. Ritchie, National Institute of Standards and Technology

Overview

General Uses

- Qualitative and quantitative elemental microanalysis of solids for all elements except hydrogen and helium
- Applicable to major (concentration $C > 10$ wt%), minor (1 wt% $\leq C \leq 10$ wt%), and trace constituents ($C < 1$ wt%, to limits of detection of 0.01 wt% with WDS and 0.05 wt% with EDS)
- Accuracy demonstrated with both WDS and EDS using the standards-based intensity ratio/matrix corrections analysis protocol: 95% of analyses fall within ±5% relative of the correct value for major and minor constituents
- Standards can be pure elements (e.g., Fe, Cu, Ni, Cr) or simple compounds (e.g., CaF_2, GaP, FeS_2)
- Lateral and depth spatial resolution of the order of 1 μm with submicrometer resolution possible (beam energy, specific element, and matrix dependent)
- Elemental compositional area mapping, lateral dimensions up to centimeters with spatial resolution to 1 μm or less.
- Used on SEM, enabling correlation of elemental composition with structure (via backscattered electron and secondary electron imaging) and crystallography (via electron backscatter diffraction)

Examples of Applications

- Elemental analysis of individual phases at the microstructural level in multiphase samples, for example, analysis of individual inclusions in steels and other alloys
- Analysis of compositional gradients at boundaries
- Determination of compositional homogeneity at the micrometer scale in single-phase materials
- Compositional mapping of heterogeneous specimens to produce images of elemental location and concentration correlated with SEM structural & crystallographic images

Samples

- *Form:* A bulk solid sample materialographically polished to a mirror finish (< 50 nm rms roughness) is ideal for optimum analytical accuracy. Other forms include rough surfaces, individual particles, and films on substrates, but these nonideal shapes require special analytical procedures
- Insulating materials, including those embedded in a conducting matrix (e.g., inclusions in steels), must be coated with a conductive film, typically carbon 5 to 10 nm thick, that is then connected to electrical ground
- *Size:* Usually 25 mm dia × 10 mm thick, but can be larger depending on the configuration of the instrument specimen stage

Limitations

- Detects elements with atomic number ≥ 3 (Li), but sensitivity is limited for lithium and beryllium and other low-atomic-number elements depending on the matrix
- Accurate analysis for elements with atomic number ≥ 5 (B)
- Lateral and depth spatial resolution approximately 1 μm, limited by electron scattering in the specimen. Operation at low beam energy, $E_0 \leq 5$ keV, can improve spatial resolution by an order of magnitude or more at the expense of more challenging x-ray spectrometry
- Quantitative analysis is limited to flat, polished, unetched specimens. Geometric effects encountered with rough surfaces, individual particles, and films on substrates degrade the accuracy, in some situations severely
- Standardless analysis software for EDS produces degraded accuracy: 95% of analyses fall within ±25% relative of the correct value for ideal flat samples
- Elemental mapping requires quantitative spectrum processing to remove artifacts that arise from spectrum background and peak interference, especially for minor and trace constituents

Estimated Analysis Time

- Individual analyses: 10 to 100 s accumulation time for accurate analysis of major and minor constituents; longer times may be necessary for trace constituents
- Elemental mapping: 100 s to 30 min or more depending on map size, concentration level visibility desired

Capabilities of Related Techniques

- *Analytical electron microscopy:* Extends spatial resolution of EPMA down to 10 nm or less and provides high-resolution imaging and electron diffraction for crystallographic information at this spatial scale
- *Secondary ion mass spectrometry:* Provides coverage of all elements and distinguishes isotopes at part per million sensitivity and with nanometer depth resolution but destructively and with larger uncertainties in quantitative analysis
- *Auger microprobe analyzer:* Provides elemental analysis with surface sensitivity (approximately 1 to 5 nm deep) of the sample at the same lateral resolution as EPMA at the same E_0 and has special sensitivity to low-atomic-number elements
- *Raman and infrared microprobe:* Provide molecular (compound) analysis at micrometer spatial level

Electron-excited x-ray microanalysis is a technique that enables spatially resolved elemental composition measurement at the micrometer to submicrometer lateral and depth resolution (Ref 1). With the exception of hydrogen and helium, which do not produce characteristic x-rays, all elements of the periodic table can be measured. Generally referred to as *electron probe microanalysis* (EPMA), the technique is typically performed in a scanning electron microscope equipped with an energy dispersive x-ray spectrometer. Scanning electron microscopy with energy dispersive spectrometry (EDS) is capable of quantitative measurements of constituents at the major (concentration $C >$ 10 wt%), minor (1 wt% $\leq C \leq$ 10 wt%), and trace (0.05 wt% $< C <$ 1 wt%) levels. In tests on known materials, demonstrated accuracy is generally within ±5% relative in 95% of analyses for major and minor constituents. Low-atomic-number elements (e.g., F, O, N, C, and B) can be measured with useful accuracy as major constituents in fluorides, oxides, nitrides, carbides, and borides. Quantitative compositional mapping can be performed on x-ray spectrum image databases, resulting in elemental images in which the gray or color scale is related to elemental concentration.

Physical Basis of Electron-Excited X-Ray Microanalysis

The x-ray microanalyst needs to be aware of the principles of x-ray generation, x-ray propagation through the target, and spectral measurement to make optimal use of the technique (Ref 1).

Origin of Characteristic X-Rays

An energetic electron with kinetic energy in the kilo-electron volt range that exceeds the binding energy of an inner shell atomic electron can scatter and eject it, leaving the atom in an ionized, energetic state with a vacancy in the atomic shell, as shown schematically in Fig. 1. The excited, ionized atom subsequently will lower its energy state by filling that vacancy through transitions of electrons between the atomic shells. Because the shells have precisely defined energy levels, the difference in energy when an electron moves between the shells is also precisely defined and can be released from the atom in either of two ways: (1) as shown in the left branch in Fig. 1, the energy can be transferred to another outer shell electron that is ejected, leaving the atom doubly ionized (the Auger effect); or (2) as shown in the right branch in Fig. 1, the difference in energy can be expressed as a photon of sharply defined electromagnetic energy, a "characteristic x-ray" because it is specific to the particular atom species emitting the photon. All elements of the periodic table except hydrogen and helium have electrons that occupy two or more shells necessary to produce characteristic x-rays. The lowest-atomic-number elements (Li to Ne) have electrons occupying only the K- and L-shells; consequently, only one characteristic x-ray energy is possible (e.g., C K-$L_{2,3}$ at 0.282 keV), with a very narrow energy width, approximately 1 eV, as shown in Fig. 1. As the atomic number increases and the shell structure becomes more complex with electrons filling outer shells, with sufficiently high electron beam energy, additional characteristic x-rays can be produced, for example, for the Fe K-shell, the Fe K-family consisting of Fe K-$L_{2,3}$ and Fe K-$M_{2,3}$, and for the Fe L-shell, the Fe L-family (L_3-M_1, L_2-M_1,L_3-$M_{4,5}$, L_2-M_4, L_1-$M_{2,3}$, L_2-$M_{1,3}$, etc.), as shown in Fig. 2. For higher-atomic-number elements, the K-shell ionization energy eventually becomes so high that it exceeds the beam energy available (typically a maximum of 30 keV) in electron-beam instruments such as the scanning electron microscope, and the L-family and M-family must be used instead for microanalysis. For example, the characteristic x-rays available for Au are the Au L family (L_3-$M_{4,5}$, L_2-M_4, L_1-$M_{2,3}$, L_2-$M_{1,3}$, etc.) and the Au M-family ($M_{4,5}$-$N_{2,3}$, M_5-$N_{6,7}$, M_4-N_6, M_3-N_5, etc.), as shown in Fig. 3. The sharply defined x-ray energies and the multiplicity of characteristic x-rays form the basis for performing qualitative analysis in which the atomic species present are identified.

Origin of Continuum (Bremsstrahlung) X-Rays

A second process involving deceleration of the energetic electron in the coulombic field of the atoms, shown schematically in Fig. 4, operates concurrently with inner shell ionization and generates x-ray photons at all energies from the spectrometer detection threshold up to the incident beam energy E_0 to form a continuous x-ray spectrum (continuum). As shown in Fig. 4 (a spectrum of carbon modeled to show the characteristic and continuum x-rays produced within the specimen before propagation to the spectrometer), this x-ray continuum forms a background under the characteristic x-ray peaks that impacts the correct measurement of the characteristic x-ray intensity and sets the eventual statistical limit for those measurements when the elemental concentration decreases.

Intensity of Characteristic and Continuum X-Rays

The incident electron kinetic energy, E_0, must exceed the shell electron binding energy, E_c, for ionization to occur. Characteristic x-ray production begins when the "overvoltage" $U_0 = E_0/E_c$ for a particular shell exceeds unity:

$$U_0 = E_0/E_c > 1 \quad \text{(Eq 1)}$$

and the generated intensity (as opposed to the emitted intensity) of characteristic x-rays, I_{ch}, increases with increasing U_0 as:

$$I_{ch} \sim i_b(U_0 - 1)^n \quad \text{(Eq 2)}$$

where i_b is the beam current, and n has a value in the range 1.5–2, depending on the atom species and shell. To produce adequate

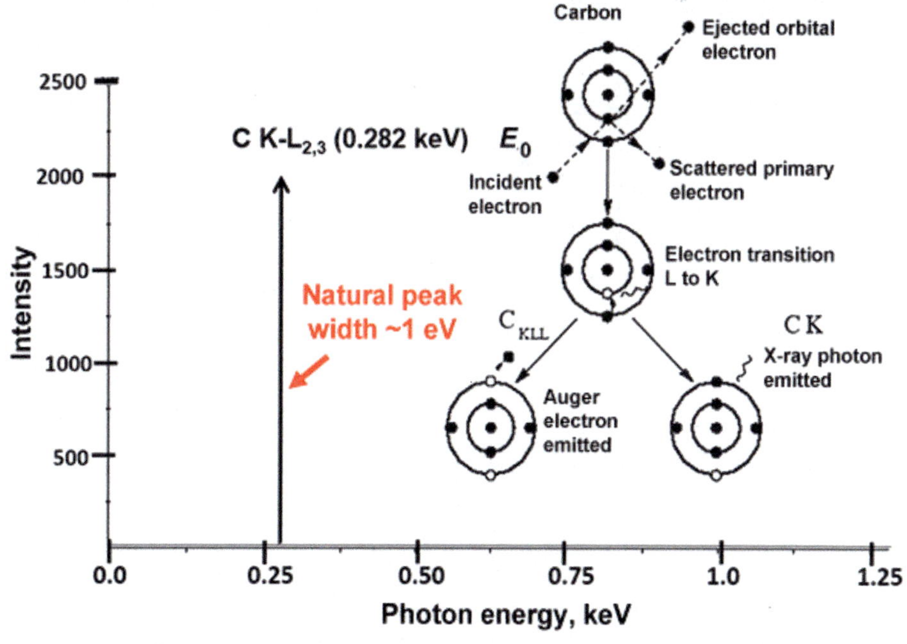

Fig. 1 Electron-induced inner shell ionization tree with ideal characteristic x-ray spectrum of carbon

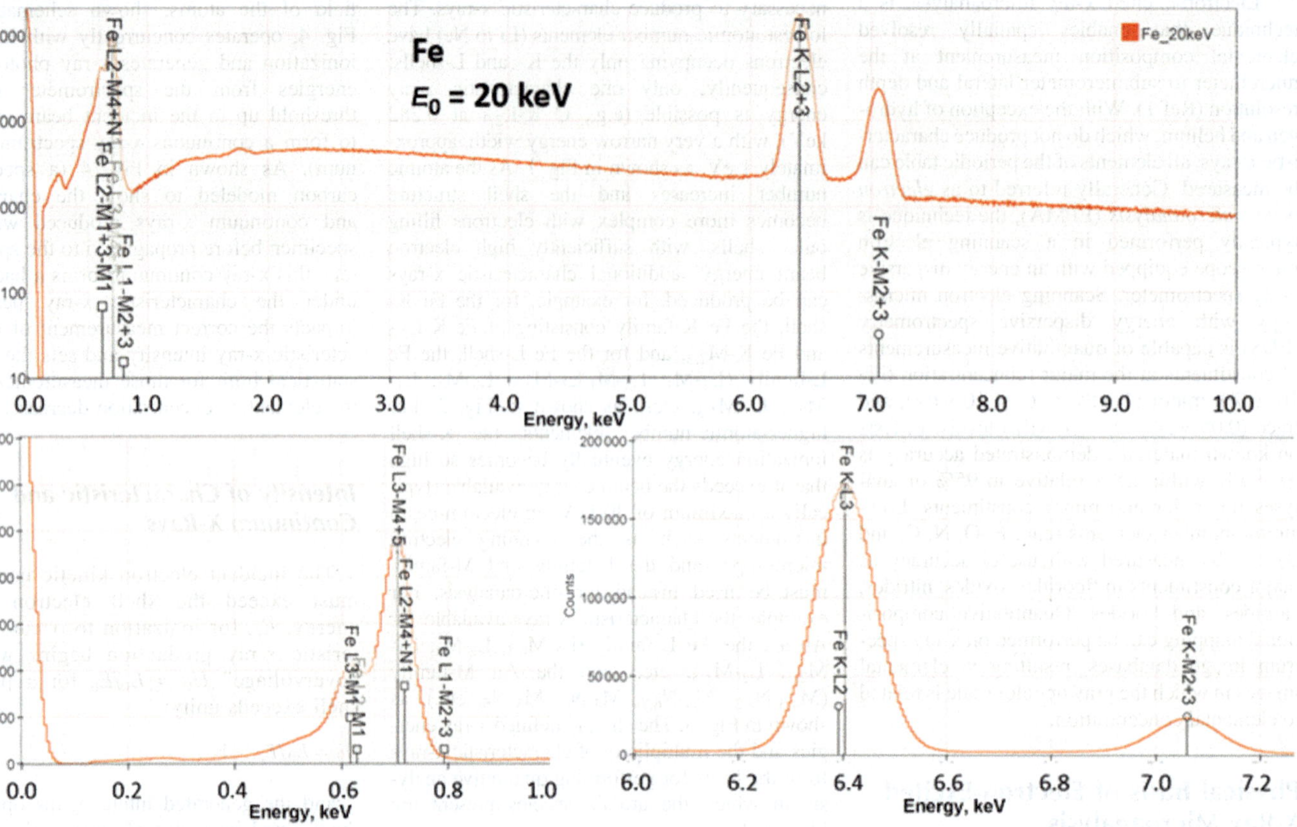

Fig. 2 Measured EDS x-ray spectrum of iron at E_0 = 20 keV: upper, 0–10 keV with logarithmic intensity scale; lower, Fe L-family and Fe K-family, linear intensity scale

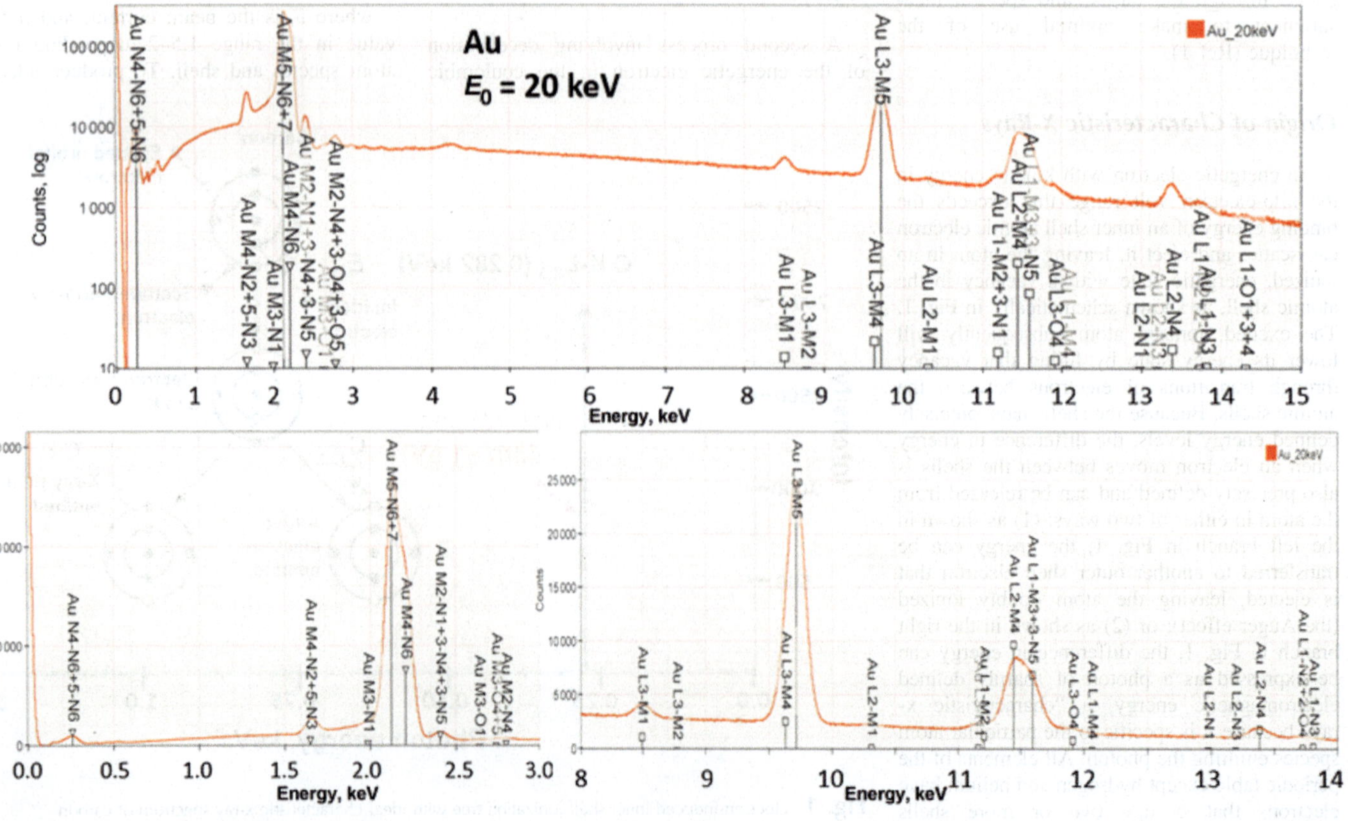

Fig. 3 Measured EDS x-ray spectrum of gold at E_0 = 20 keV: upper, 0–15 keV with logarithmic intensity scale; lower left, Au N-family and M-family; lower right, Au L-family, linear intensity scale

characteristic x-ray intensity for compositional measurements, a useful strategy is to choose an incident beam energy such that $U_0 > 2$ for the highest value of E_c represented by the elements in the specimen.

Intensity of the X-Ray Continuum

The generated intensity of the x-ray continuum, I_{cm}, at a particular photon energy, E_ν, depends on the overvoltage and on the average atomic number, $\dot{Z}$ (atomic number of each element present scaled by the mass concentration, C_m), of the target:

$$I_{cm} \sim i_b \dot{Z}[(E_0 - E_\nu)/E_\nu] \quad \text{(Eq 3a)}$$

For the continuum intensity generated at the energy equivalent to a particular elemental absorption energy, setting $E_\nu = E_c$ transforms Eq 3a to:

$$I_{cm} \sim i_b \dot{Z}(U_0 - 1) \quad \text{(Eq 3b)}$$

Generated Peak-to-Background Ratio

The peak-to-background ratio, $P/B = I_{ch}/I_{cm}$, is given by the ratio of Eq 2 and 3b:

$$P/B = I_{ch}/I_{cm} = i_b (U_0 - 1)^n / \{i_b \dot{Z} (U_0 - 1)\} = (U_0 - 1)^{n-1}/Z \quad \text{(Eq 4)}$$

Equation 4 indicates that to achieve a useful sensitivity for analysis, the value of U_0 should be greater than 2. Lower values of U_0 approaching unity can be used, but with an increasingly large penalty in elemental sensitivity.

Range of Characteristic and Continuum X-Ray Production

Upon entering the specimen, energetic beam electrons scatter elastically, which changes the direction of travel, and inelastically through a variety of mechanisms including x-ray generation, which causes progressive loss of kinetic energy that is transferred to the specimen atoms. The incremental changes in the travel path due to elastic scattering have a random component that rapidly degrades the highly directional incident beam. The cumulative effect of elastic scattering causes some beam electron trajectories to reach the specimen surface, where they emerge as backscattered electrons with energy less than the incident energy due to inelastic scattering. For those beam electron trajectories that remain in the specimen, the cumulative effects of inelastic scattering eventually rob the energetic electron of all of its energy so that it is absorbed by the specimen. The combined effects of elastic and inelastic scattering constrain the beam electrons that travel within the specimen, creating the interaction volume, shown in Fig. 5, as calculated by means of a Monte Carlo electron trajectory simulation. The interaction volume defines the region within which beam electrons generate characteristic and continuum x-rays. The limiting depth of the interaction volume (for normal beam incidence, i.e., a specimen tilt of 0°) can be estimated through a *range equation* as:

$$R_{KO}(\text{nm}) = [(27.6\,A)/(Z^{0.89}\rho)]E_0^{1.67} \quad \text{(Eq 5a)}$$

where A is the atomic weight (g/mol), ρ is the density (g/cm^3), and E_0 is in kilo-electron volt units. Characteristic x-rays for a particular atomic binding energy, E_c, can be generated as long as the beam electron energy $E > E_c$. The limiting range of characteristic x-ray production for that binding energy can be estimated by modifying Eq 5a for the x-ray production threshold at E_c:

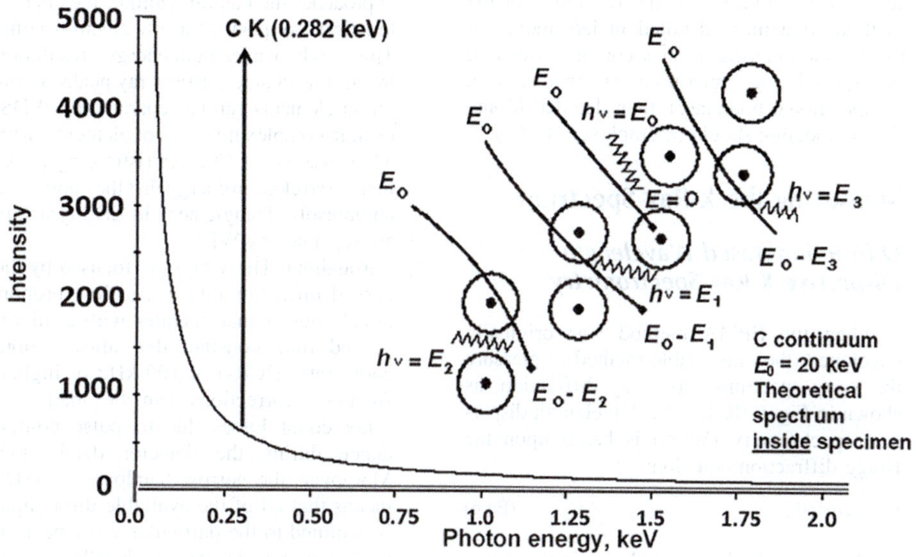

Fig. 4 Mechanism for x-ray bremsstrahlung (continuum) emission with simulated characteristic and continuum x-ray spectrum as generated for carbon

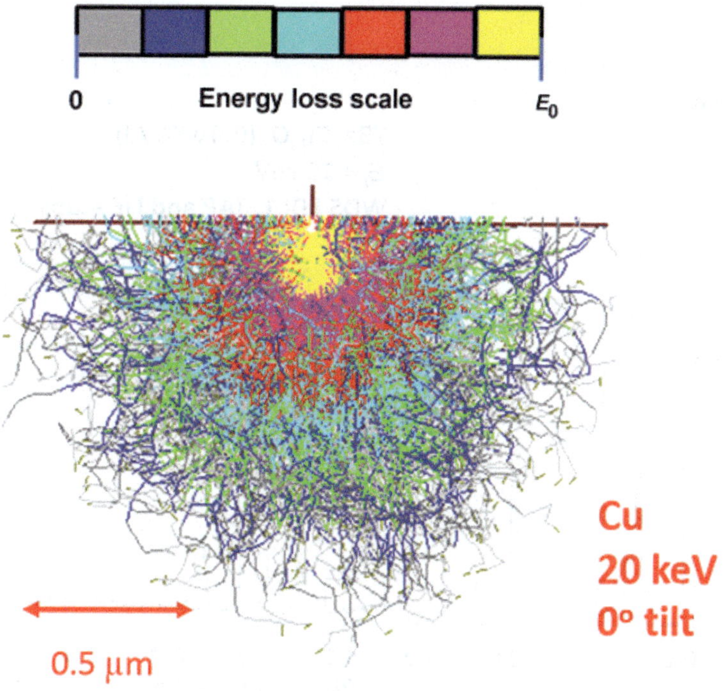

Fig. 5 Monte Carlo simulation of interaction volume for copper at E_0 = 20 keV

$$R_{KO,x\text{-}ray}(nm) = [(27.6\,A)/(Z^{0.89}\rho)]\,[E_0^{1.67} - E_c^{1.67}] \quad \text{(Eq 5b)}$$

The range values given by Eq 5a and 5b should be considered imprecise because a single length metric is used to attempt to describe the complex, three-dimensional interaction volume within which is a highly nonuniform distribution of energy deposition and x-ray generation, as shown in the Monte Carlo simulation in Fig. 5. Nevertheless, Eq 5b is useful for estimating the spatial resolution of the method. If a more detailed understanding of the interaction volume in a complex situation is required (e.g., measurements that cross a compositional boundary), then detailed Monte Carlo modeling should be employed (Ref 1).

Measuring the X-Ray Spectrum

Diffraction-Based Wavelength Dispersive X-Ray Spectrometry

When the EPMA method was originally developed, the only viable method to measure the x-ray spectrum was x-ray diffraction, as shown in Fig. 6 (Ref 1, 2). Wavelength dispersive spectrometry (WDS) is based upon the Bragg diffraction equation:

$$n\lambda = 2d\sin\Theta_B \quad \text{(Eq 6)}$$

where n is the integer order, λ is the wavelength of the x-rays, d is the spacing of the planes of the diffractor element, and Θ_B is the Bragg angle for constructive interference of the diffracted waves.

Wavelength Dispersive Spectrometry Advantages

(1) Resolution: The diffraction process is sharply peaked over a narrow wavelength (energy) range so that the width of the diffraction peak for some diffractors approaches the natural width of the characteristic x-ray peak of a few electron volts. This high wavelength/energy resolution means that characteristic x-ray peaks of different elements can be separated by WDS, even in complex mixtures of elements, with a few exceptions. Characteristic x-ray peaks that are so close in energy that they appear as an unresolved single peak in EDS typically are separated by WDS.

(2) Throughput: The x-rays are focused by the curved diffractor into a gas flow proportional counter that operates with a microsecond time constant that allows input count rates (ICRs) of 100 kHz or higher. Accurate corrections can be made for x-ray count losses due to pulse coincidence during the detector dead time. Moreover, the narrow bandpass of WDS means that all of the available throughput is confined to the particular x-ray peak or background region to which WDS is set. The rest of the x-ray spectrum generated by the beam is physically excluded by the diffraction process from reaching the WDS detector and thus does not contribute to the dead time, which is a particular advantage when trace constituents are to be measured. If the specimen can withstand the necessary increased beam current, the count rate for the trace constituent can be increased without a dead time penalty from the more abundant constituents that produce higher x-ray intensities.

(3) Superior trace sensitivity: The combination of the resolution and the throughput of WDS enables higher trace sensitivity; that is, lower limits of detection can be achieved in comparison with EDS. Depending on the exact circumstances of the specimen composition, limits of detection lower than 0.01 wt% (100 ppm) can be reached.

Wavelength Dispersive Spectrometry Limitations

(1) Narrow wavelength (energy) bandpass: The narrow instantaneous wavelength/energy bandpass of WDS means that in order to view the extended x-ray spectrum to determine which characteristic peaks are present for qualitative analysis, mechanical scanning through a range of angles must be performed. Moreover, any given diffractor spans a finite range of diffraction angles and therefore x-ray wavelengths (energy).

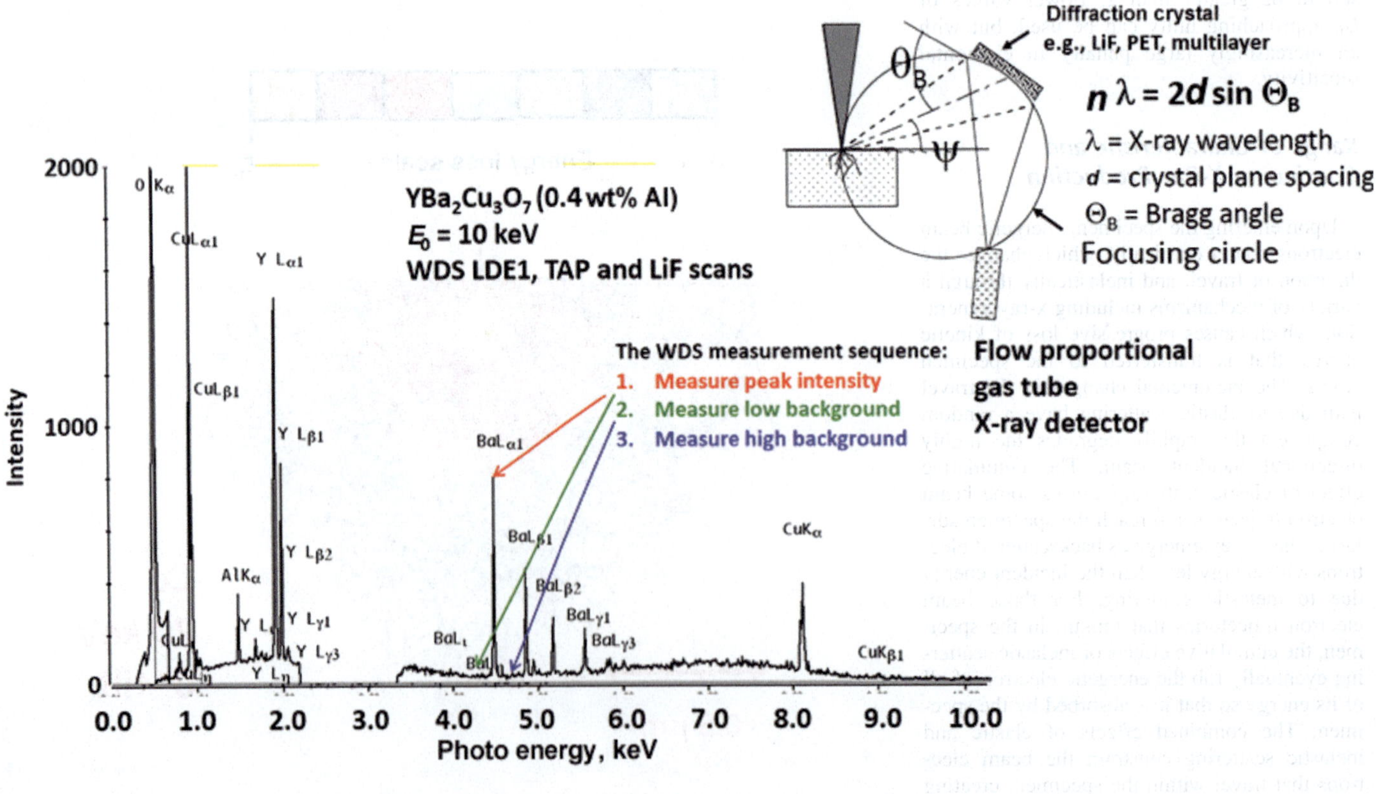

Fig. 6 Principle of the wavelength dispersive x-ray spectrometer (WDS) with the spectrum of $YBa_2Cu_3O_7$-Al measured with scans of three diffractors (LDE1, TAP, and LiF) to measure all of the characteristic peaks. LDE, layered dispersive element; lithium fluoride, LiF; PET, pentaerythritol; thallium acid phthalate, TAP

Measurement of the x-ray energy range from 100 eV to 10 keV needed for comprehensive qualitative analysis requires at least four diffractors, which normally are accommodated on a turret. In the example for $YBa_2Cu_3O_7$ shown in Fig. 6, three diffractors were needed to scan through all of the elements. Thus, the need for WDS scanning makes qualitative analysis a time-intensive process that is not typically performed at every location measured on a specimen during an analytical campaign.

(2) Complexity from multiple reflections: Because of the integer order of reflection, n, in Eq 5, a given wavelength λ can diffract at multiple values of θ, including appearing on more than one diffractor during wavelength scanning. Multiple orders of diffraction for a given wavelength complicate the identification of peaks during qualitative analysis, especially when higher-order diffraction peaks from a major constituent can be mistaken for low-relative-intensity peaks from possible minor or trace constituents.

(3) Efficiency: The area and distance of the diffractors from the electron-excited source results in a modest solid angle of collection, which when combined with the inherent inefficiency of the diffraction process leads to a requirement for high electron beam current. Typically 100 nA or higher is required to excite a sufficiently high x-ray flux to accomplish measurements with adequate counting statistics within a practical time.

(4) Single element mapping: For x-ray mapping, only one element can be mapped with each wavelength dispersive spectrometer during electron beam scanning of an area, often requiring multiple area scans if only one wavelength spectrometer is available or if a very complex specimen is to be mapped where the number of elements of interest exceeds the number of wavelength spectrometers available.

Semiconductor Energy Dispersive X-Ray Spectrometry

Semiconductor EDS operates on the principle of photoelectric absorption of an x-ray photon in a silicon crystal, as shown in Fig. 7 (Ref 1). The ejected energetic photoelectron subsequently scatters inelastically within the silicon and promotes electrons from the filled valence band of the semiconductor into the empty conduction band, leaving behind positively charged "holes." The promoted electrons are free to move under an applied potential to the positive electrode while the holes drift in the opposite direction to the negative electrode. The quantity of electrons collected at the positively charged electrode constitutes a charge that is proportional to the energy of the photon.

Energy Dispersive Spectrometry Advantages

(1) Complete spectrum measurement: EDS can measure the x-ray spectrum from a threshold as low as 40 eV, enabling measurement of the Li K-L photon at 54 eV, to photon energies of 30 keV or higher. Thus, for an incident beam energy of 30 keV or lower, typically the upper limit of the beam energy available with the scanning electron microscope, the entire x-ray spectrum can be recorded at every electron beam location, thus capturing the information needed for a comprehensive qualitative analysis with every measurement. Unexpected variations in the composition of a specimen at any measured location, such as the localized occurrence of an elevated level of a bulk trace constituent, thus can be recognized during the qualitative analysis of each measured spectrum.

(2) Large solid angle of collection: The development of the *silicon drift detector* (SDD) EDS design with electronic (Peltier) cooling has eliminated the need for the liquid nitrogen cooling of the earlier *lithium-drifted silicon* [Si(Li)-EDS] design. This

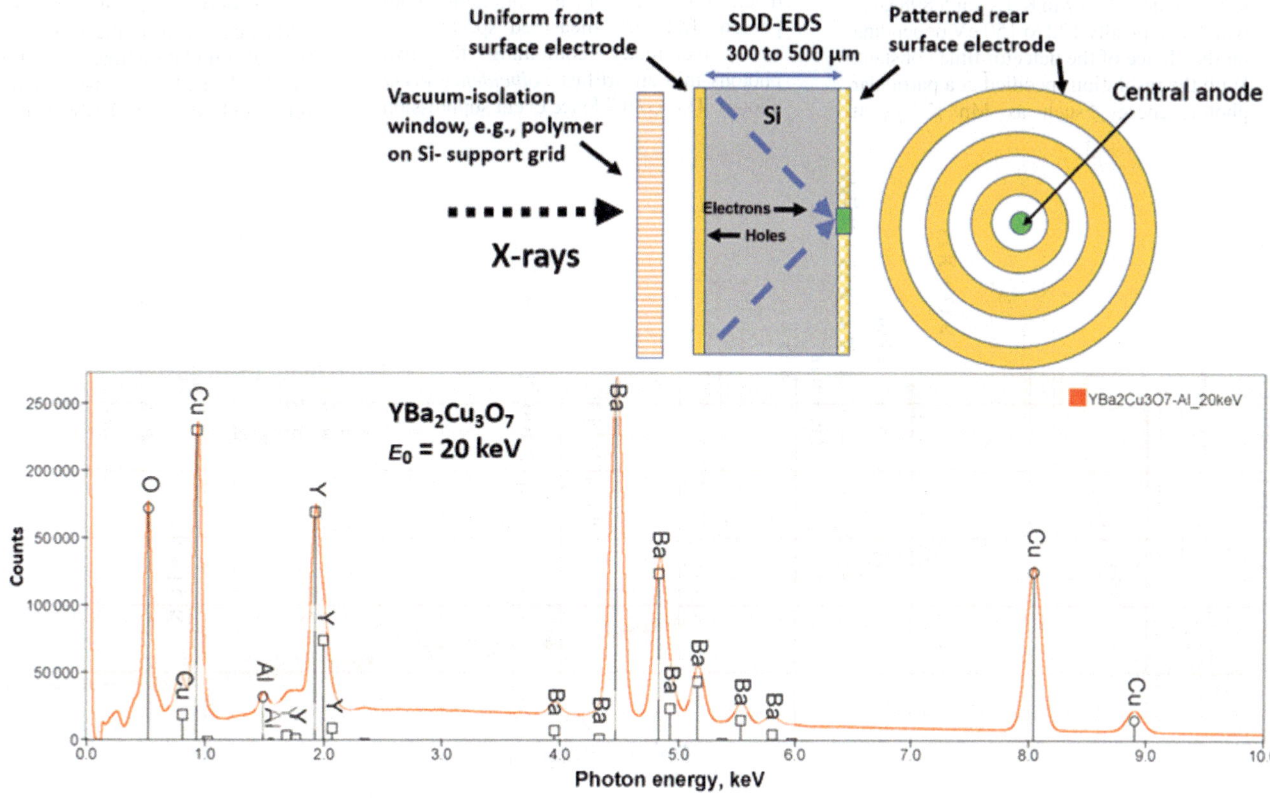

Fig. 7 Principle of the semiconductor silicon drift detector energy dispersive x-ray spectrometer (SDD-EDS) with the spectrum of $YBa_2Cu_3O_7$-0.4wt%Al

simpler cooling scheme has enabled the development of large solid angle detectors that can be placed in close proximity to the specimen; for example, a detector with an active area of 30 mm^2 can be placed at a distance s = 30 mm from the electron-excited x-ray source, giving a solid angle $\Omega = A/s^2$ = 0.033 sr. Moreover, multiple arrays of SDD-EDS detectors can be accommodated, giving a total collection angle of 0.1 sr or higher for a four-detector array. Such a large solid angle of collection enables the use of electron beam currents of 1 nA or lower to achieve useful x-ray count rates.

(3) Simultaneous mapping of all elements: For x-ray mapping, the EDS collection of the complete x-ray spectrum at each beam location means that positional information has been captured on all possible elements (excepting H and He) in the specimen. X-ray maps for all of these elements can be recovered from the collected data structure.

Energy Dispersive Spectrometry Disadvantages

(1) Energy resolution: The resolution of EDS is inherently limited by the statistics of electron-hole production. The resolution, arbitrarily defined as the full peak width at half the peak intensity (FWHM), is dependent on the energy of the photon measured. The industry-standard figure of merit is the resolution achieved for Mn K-$L_{2,3}$ at 5.898 keV, which is typically 122 to 150 eV depending on the choice of the detector time constant. With the resolution specified at a particular photon energy, such as Mn K-$L_{2,3}$ at 5.898 keV, the resolution at any other energy can be estimated as:

$$\text{Resolution (FWHM)} = \left[2.5\,(E - E_{\text{ref}}) + FWHM^2{}_{\text{ref}}\right]^{0.5} \quad \text{(Eq 7)}$$

where E, E_{ref}, and $FWHM$ are specified in electron volts (Ref 1).

(2) Peak interference situations: The EDS resolution is approximately a factor of 5 to 20 poorer than that of WDS for a given photon energy, leading to much more frequent situations of peak interference when complex compositions are analyzed by EDS.

(3) Detector dead time and artifacts: When an EDS spectrum is observed during active accumulation, it appears that all photon energies are measured simultaneously. However, the EDS detector is in fact capable of processing only one photon at a time. The maximum throughput is determined by the EDS processing time (detector time constant), a parameter that the user typically chooses with the following trade-off between throughput and resolution: A short time constant allows a higher throughput at the expense of poorer resolution. The time period during which the detector system is processing a photon and unavailable to process a second photon is referred to as *dead time*. If a second photon arrives at the detector while the first photon is being processed, a software inspection function deletes this measurement, eliminating both photons from the measured spectrum to avoid incorrectly combining the two photons into an artifact *coincidence event* photon. This coincidence event suppression function has an inevitable timing resolution limit, so that at high count rates when two photons are more likely to arrive too close in time to be distinguished, some coincidence events are not excluded, resulting in the appearance artifact *coincidence (sum) peaks* in the spectrum. Figure 8 shows an example of the ingrowth of coincidence peaks as a function of EDS system dead time. System dead time typically is reported by the EDS operational software, and a prudent analyst seeks to choose measurement conditions that restrain the dead time to 30% or lower in order to minimize coincidence peaks in the spectrum.

(4) Throughput limitations: The progressive loss of coincidence events as the ICR increases is manifested as a decrease in the output count rate (OCR) relative to the ICR, as shown in Fig. 9. The OCR eventually rises to a maximum rate, and then decreases with further increases in the ICR. At a sufficiently high ICR, the OCR drops toward zero, creating a situation of *paralyzable dead time*. For a constant beam current, the dead time varies from one material to another because of two factors: (1) differences in characteristic x-ray generation rates that arise from different overvoltages for different elements, and (2) the continuum intensity rate, which scales with the average atomic number. To establish the basis for quantitative measurements, the EDS software corrects for dead time by noting the lengths of detector "busy" periods and then adding extra time to the user-specified spectrum accumulation time (clock time or "live time"). Thus, for a user-selected beam current and desired EDS live time, all

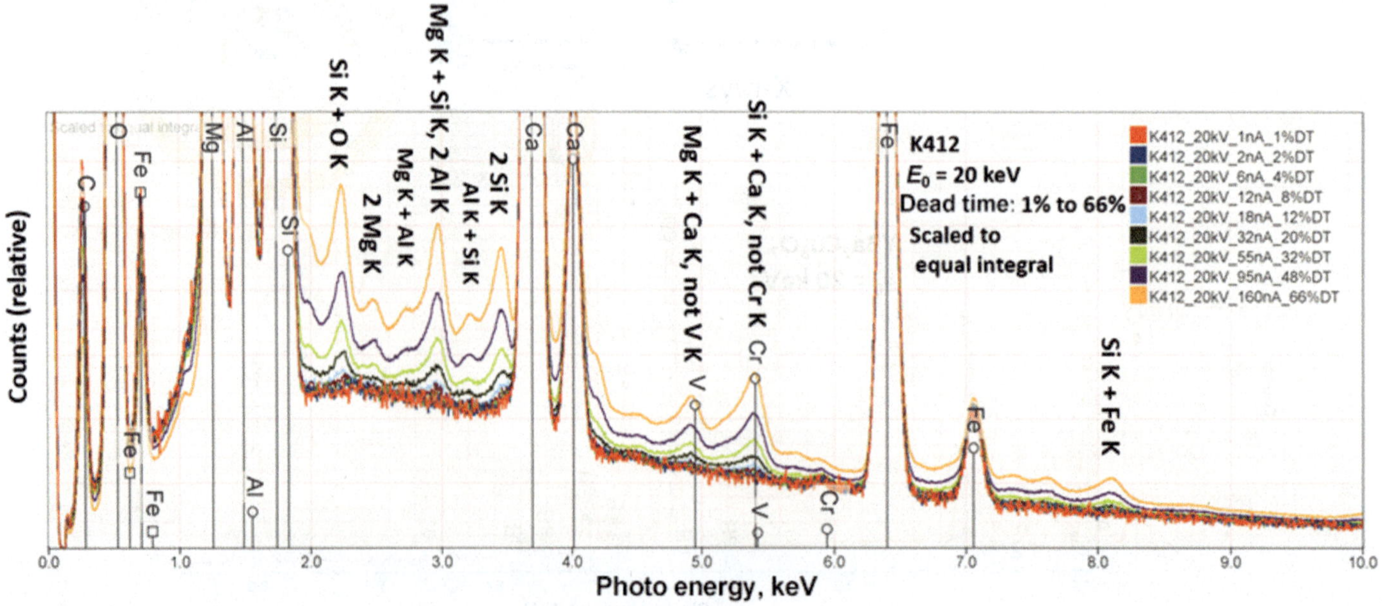

Fig. 8 EDS spectra at increasing dead time showing ingrowth of sum peaks as measured for NIST Standard Reference Material 470 (glass K412); E_0 = 15 keV. Spectra overlaid on the basis of the normalized spectrum integral from 0.1 to 15 keV

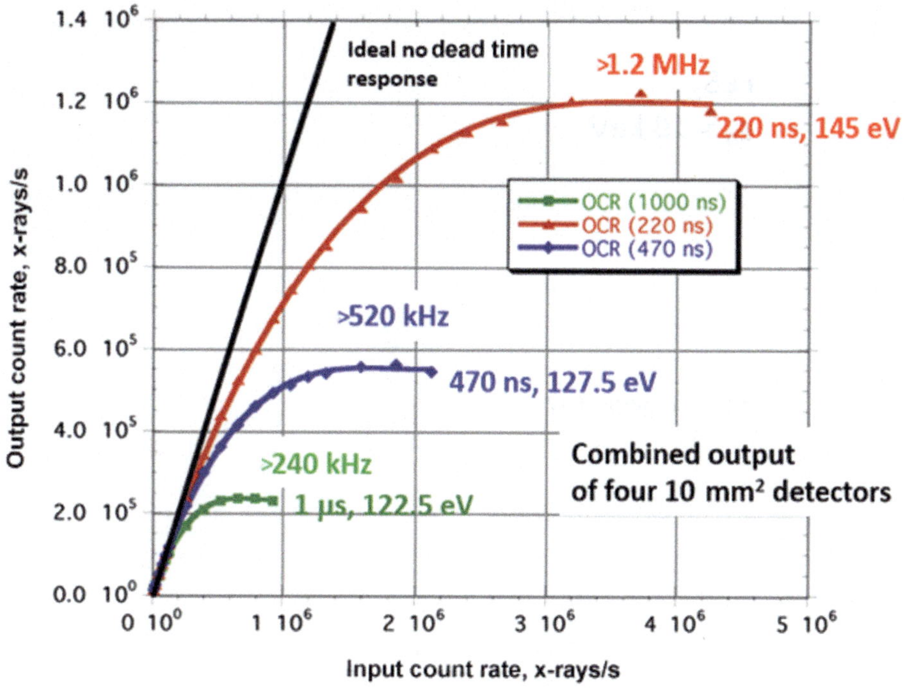

Fig. 9 Output count rate versus input count rate for three time constants (resolution at Mn K-$L_{2,3}$ is noted for each time constant); ganged output of four 10 mm^2 SDD-EDS detectors

measurements are made with a known equivalent electron dose, regardless of the dead time, which is critical for accurate standards-based quantitative analysis.

Energy Dispersive Spectrometry Replaces Wavelength Dispersive Spectrometry for Electron Probe Microanalysis

The development of SDD-EDS has greatly advanced the performance of EDS over that of the previous technology, Si(Li)-EDS (Ref 1). SDD-EDS has increased throughput by a factor of 50 or more for the same energy resolution while improving the stability of the spectrum in both peak position (energy calibration) and peak width (resolution) over the full count rate range. With this performance, it has been demonstrated that SDD-EDS can solve problems involving complex spectra with severe peak overlaps for constituents at the major (concentration $C > 10$ wt%, 0.1 mass concentration), minor (1 wt% $\leq C \leq$ 10 wt%, 0.01 to 0.1 mass concentration), and trace levels (0.05 wt% $< C <$ 1 wt%, 0.0005 or 500 ppm to 0.01 mass concentration) (Ref 1, 3, 4). While WDS remains the spectrometry choice for detecting trace levels below C = 0.05 wt% (0.0005 mass concentration or 500 ppm), this specialized use is not considered further in this article, which focuses on the utilization of EDS for qualitative and quantitative analysis to solve materials characterization problems.

Qualitative Analysis

The assignment of elemental identities to the characteristic peaks recognized in a spectrum is an automated function performed by vendor software, *automatic peak ID*. While automatic peak ID generally gives a correct result for major constituents, the possibility exists for misidentification of an element, even at the level of a major constituent, and the likelihood of misidentification increases significantly for minor and trace constituents (Ref 5, 6). The prudent analyst considers the automatic peak ID result as a useful starting point and always performs manual peak identification to confirm or correct the result. Thus, an important part of learning EDS analysis is to become familiar with the physics that governs characteristic x-ray emission. Key concepts for performing qualitative analysis include:

(1) As expressed in Eq 1, in order for characteristic x-ray emission to occur for a particular shell of an element, the overvoltage $U_0 = E_0/E_c$ must exceed unity for that shell's binding energy, E_c.
(2) As noted earlier, x-ray families have two or more characteristic x-ray energies for elements with atomic numbers greater than 10. With EDS spectral resolution, the presence of two peaks becomes noticeable for phosphorous (K-$L_{2,3}$ = 2.013 keV and K-$M_{2,3}$ = 2.139 keV) as an asymmetry on the high-energy side of the major peak, and more obvious for sulfur (K-$L_{2,3}$ = 2.307 keV and K-$M_{2,3}$ = 2.464 keV), as shown in Fig. 10.
(3) Depending on the primary beam excitation energy, E_0, two or more families of x-rays (K-shell, L-shell, M-shell, and N-shell) may be excited for intermediate- and high-atomic-number elements, such as the Fe K-shell and Fe L-shell families, as shown in Fig. 3, and the Ta L-shell and M-shell families, as shown in Fig. 4. With the low-photon-energy performance of EDS, L-shell x-rays can be readily detected for titanium, as shown in Fig. 11, and some EDS systems can detect the Ca L-family.
(4) EDS artifacts include sum (coincidence) peaks, described earlier, and silicon escape peaks, in which the charge deposition process in the detector is robbed of the energy of a Si K-$L_{2,3}$ x-ray that is emitted after the initial photon capture by photoionization of a silicon atom and that occasionally escapes the detector, producing an artifact peak located 1.74 keV below the parent peak. Both of these artifact peaks occur at low relative intensity compared with the high-intensity characteristic x-ray peaks of the major constituents from which they arise. The artifact peaks must be properly recognized and assigned to avoid being misinterpreted as the peaks of minor- or trace-level constituents.

Manual peak identification makes use of the extensive database of characteristic x-ray energies embedded in all vendor software to generate *KLM* lines that mark the positions of the characteristic peaks and give the general intensity ratios of peaks within a family (Ref 1). The KLM markers also should mark the escape and possible sum peaks. Performing due diligence, the analyst should use these tools to review the elemental identifications proposed by the automatic peak ID software. Note that the relatively poor EDS resolution impacts qualitative analysis through the frequent occurrence of peak overlaps between elements whose characteristic peaks have similar energies for example, the region around 2.3 keV contains peaks for the S K-family (K-$L_{2,3}$ = 2.307 keV and K-$M_{2,3}$ = 2.464 keV), the Mo L-family (L_3-$M_{4,5}$ = 2.293 keV, L_2-M_4 = 2.395 keV), and the Pb M-family ($M_{4,5}$-$N_{2,3}$, M_5-$N_{6,7}$, M_4-N_6, M_3-N_5, etc.). As a result of the possibility of peak interferences, especially between the high-intensity peaks of major constituents and the low-intensity peaks of minor and trace constituents, rigorous qualitative analysis entails an iterative approach involving peak fitting and subtraction of the fitted intensities of the assigned elemental peaks to reveal lower-intensity peaks of minor and trace constituents hidden by the high-intensity peaks, as described in the section "Analysis when Severe Peak Overlap Occurs" in this article.

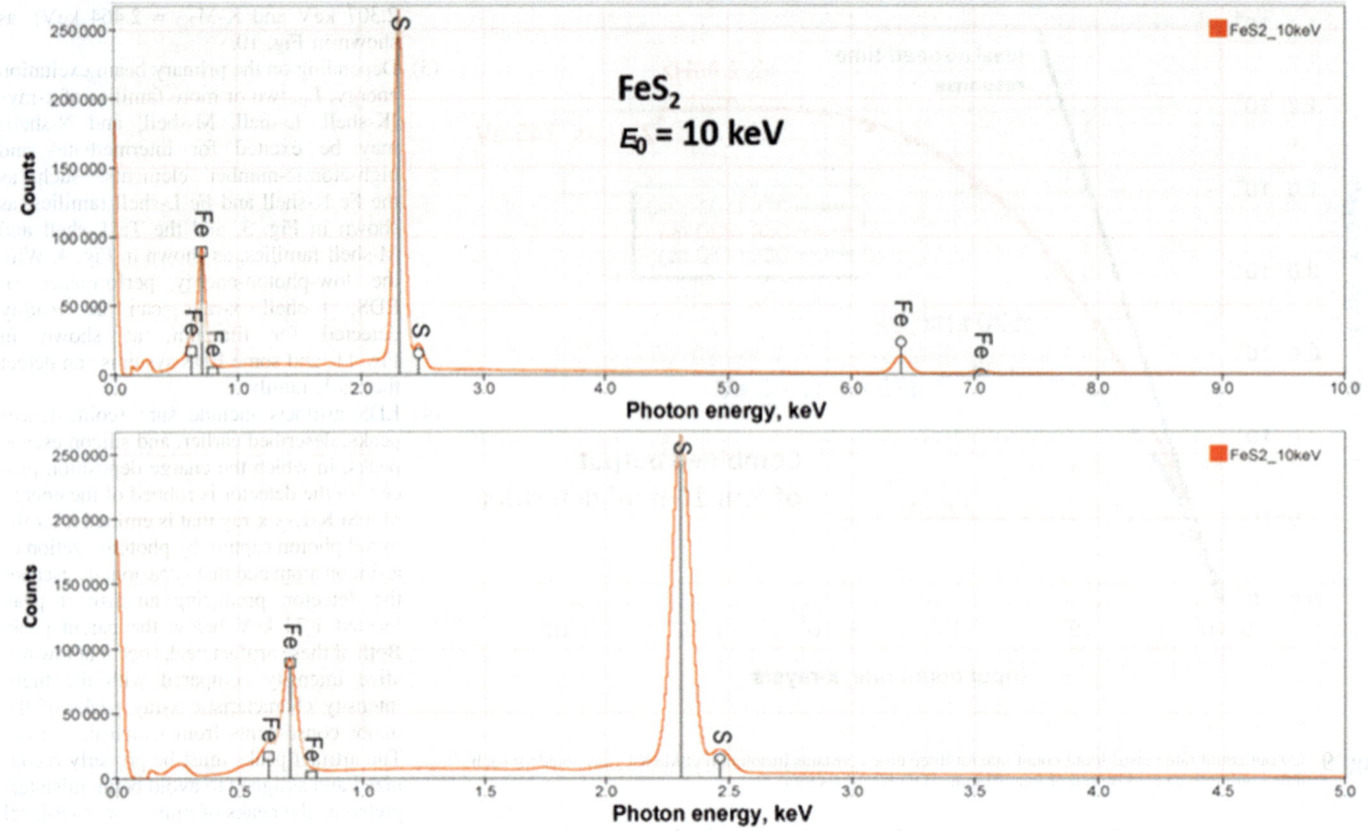

Fig. 10 EDS spectrum of FeS$_2$ at E_0 = 10 keV

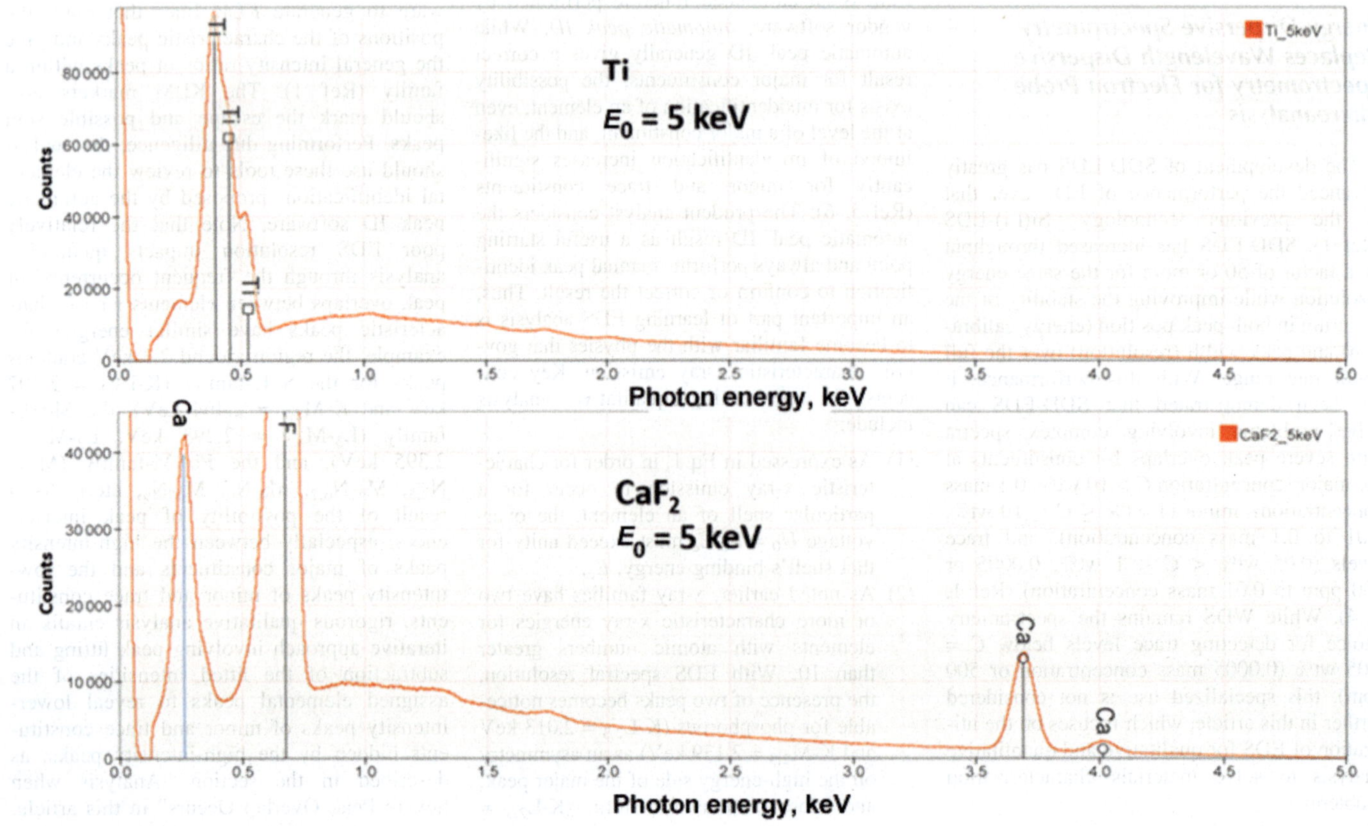

Fig. 11 EDS spectra for titanium and calcium showing the L-family peaks; E_0 = 5 keV

Quantitative Analysis

The basis for quantitative analysis by electron-excited x-ray spectrometry is the *k*-ratio/matrix corrections protocol proposed by the inventor of the EPMA technique, Raymond Castaing (Ref 2). The *k*-ratio involves the measurement of the intensity, I_X of the characteristic peak of an element in the unknown relative to that same element in a standard of known composition under identical conditions of incident beam energy, known electron dose, and spectrometer efficiency:

$$k = I_{X,\text{unknown}}/I_{X,\text{standard}} \quad \text{(Eq 8)}$$

Suitable standards for the measurement of the *k*-ratio can be pure elements. For those elements whose properties are incompatible with the vacuum of an electron beam instrument, such as gases, a compound of known stoichiometry that is homogeneous on a submicrometer scale can be used. An example of a standard for fluorine is CaF_2. Chemically active pure elements that are difficult to prepare, such as gallium and phosphorus, can be analyzed with a compound such as GaP. Compounds also can be used for those elements that are unstable under electron beam bombardment, such as sulfur, examples of standards for which include CuS and FeS_2. It is important to note that this simple standards suite of pure elements and stoichiometric compounds enables the analyst to solve the composition of nearly any unknown. Multielement standards can be used when available and are advantageous because using a similar composition to that of the unknown reduces the magnitude of the matrix corrections, described in the section "Matrix Correction Factors" in this article. However, the strict requirement of homogeneity on the scale of the electron beam excitation volume, which can be submicrometer, eliminates many multielement mixtures from consideration as standards due to the heterogeneity resulting from the formation of two or more phases on the microstructural level.

Measurement of the intensities for the *k*-ratio by EDS requires separation of the characteristic x-ray intensity from the continuum background and from the peaks of other elements that interfere, which is accomplished by multiple linear least squares peak fitting (Ref 1). Peak fitting makes use of a library of *peak references*, which is a channel-by-channel complete span of a single peak, such as Cu K-$L_{2,3}$, or of a group of closely spaced, unresolved peaks within a family, such as the Fe L-family (L_3-M_1, L_2-M_1,L_3-$M_{4,5}$, L_2-M_4, L_1-$M_{2,3}$, L_2-$M_{1,3}$). References may be constructed from ideal Gaussian descriptions of the peak or group of peaks to create the peak fitting library, or they may be experimentally measured on the EDS system that actually is used for analysis. Peak references that are measured locally on the EDS system that is used for analysis have the advantage of automatically including instrumental deviations from the ideal peak shape. Deviations from the ideal peak shape include incomplete charge collection that distorts the low-energy side of a peak and the nonlinearity seen in the calibration of low-energy photon peaks, typically below 300 eV, which causes deviation from the ideal photon energy of the characteristic peaks in this range. Note that some EDS systems apply corrections for this nonlinear energy behavior.

Matrix Correction Factors

When a suite of *k*-ratios is measured for a complex composition, it is found that for each element *i* in the set of *k*-ratios:

$$C_i \approx k_i \quad \text{(Eq 9)}$$

where C_i is the mass concentration for each element. The relationship between the *k*-ratio and the mass concentration is not an equality because of the action of *matrix effects*, that is, interelement influences on the generation and propagation of the characteristic x-rays. Matrix effects are classified broadly into three categories (Ref 1):

1. *Atomic number effect* (*Z*): Beam electrons entering the target suffer directional change through elastic scattering, the cumulative effects of which lead to backscattering. Backscattering is a strong function of the atomic number of the target and thus of the composition of the target. The escape of these beam electrons through backscattering represents a loss of a fraction of the total possible characteristic x-ray generation. Beam electrons also undergo inelastic scattering, which causes them to lose energy in the target. The loss of energy represents a loss of ionization power, as expressed in Eq 2. The rate of energy loss also depends on the atomic number of the target and thus is composition dependent.
2. *Absorption effect* (*A*): Characteristic x-rays are generated over a range of depth into the target, from a fraction of a micrometer to several micrometers, depending on the incident beam energy and the target composition, as given by Eq 5b. To reach the x-ray spectrometer, the x-rays must propagate through the target, where they are subject to loss through photoelectric absorption. (Note that with electron excitation, the characteristic x-rays are produced over a relatively shallow range. The distances that the x-rays must travel to escape are sufficiently short that inelastic scattering of the x-rays, which would cause a loss in energy resulting in deviation from the original characteristic value, is not a significant factor.) X-ray absorption follows an exponential behavior:

$$I/I_0 = \exp[-(\mu/\rho)\rho s] \quad \text{(Eq 10)}$$

where I_0 is the initial intensity, and I is the intensity that emerges after the x-ray passes through a distance s of the material with density ρ and a mass absorption coefficient (μ/ρ) for that photon energy. The mass absorption coefficient for a material depends on all of the elements present and thus is composition dependent.

3. *Fluorescence effect* (*F*, *c*): A consequence of photoelectric absorption is that the absorbing atom is ionized in a core shell, and its energy is subsequently lowered through transitions of electrons between the outer and inner shells, with the emission of a characteristic photon for a fraction of these ionizations. Thus, the fluorescence effect due to absorption of characteristic x-rays (*F*) creates characteristic x-rays of the absorbing atoms in addition to those created by beam electron-induced inner shell ionization. There also is a contribution to fluorescence emission due to the electron-induced x-ray continuum, which creates photons at all energies from that of the binding energy of a particular atom, E_c, up to the incident beam energy, E_0, leading to a continuum fluorescence effect (*c*). Both *F* and *c* depend on the atoms present and thus upon the composition.

The exact relation between the measured *k*-ratio and the concentration of each element *i* is given by:

$$C_i/C_{i,\text{standard}} = k_i/(Z_iA_iF_ic_i) \quad \text{(Eq 11)}$$

The composition affects the value of each of the matrix corrections, Z_i, A_i, F_i, and c_i, which are calculated relative to the known composition of the standard. However, the composition of the sample being measured at the beam location is unknown; therefore, to initiate a quantitative calculation, an initial estimate of the sample composition is needed. This initial estimate is made by normalizing the suite of measured *k*-ratios for all constituents:

Initial estimate of sample composition:

$$C_{i,n} = k_i/\sum k_i \quad \text{(Eq 12)}$$

This initial suite of estimated concentrations is used to calculate the first set of matrix corrections for the sample relative to the standards for all elements. Using Eq 11, a new set of calculated *k*-ratios is determined and is compared to the set of measured *k*-ratios. This procedure is followed iteratively until the estimated *k*-ratios from the suite of calculated concentrations converge to the measured *k*-ratios, generally within a few iterations. Table 1 gives an example of the analysis of an Al-Ti-Nb-W alloy (National Institute of Standards and Technology [NIST] Standard Reference Material [SRM] 2061) with the NIST DTSA-II EDS

Table 1 Information reported in a DTSA-II analysis for Standard Reference Material 2061, Al-Ti-Nb-W alloy

Element	Standard (element)	Peak(s)	*k*-ratio	Counts (std)	Counts (unk)	*Z*	*A*	*F*	*C* (mass frac)	Combined uncertainty	Raw total (mass frac)	*C* (norm mass)	*C* (atom)
Al	Al (element)	K-family	0.2019 ± 0.0001	±7.2 × 10^{-5}	±0.0001	1.097 ± 3 × 10^{-5}	0.614 ± 0.0014	1.003	0.2990	±0.0014	1.0065	0.2971 ± 0.0014	0.4599 ± 0.0021
Ti	Ti (element)	K-family	0.5147 ± 0.0003	±0.0002	±0.0003	0.985 ± 8 × 10^{-6}	0.940 ± 0.0003	1.001	0.5557	±0.0005	...	0.5521 ± 0.0005	0.4819 ± 0.0004
Nb	Nb (element)	Nb L_3-M_5	0.0796 ± 0.0002	±4 × 10^{-5}	±0.0002	0.890 ± 7 × 10^{-6}	0.827 ± 0.0006	1.006	0.1083	±0.0007	...	0.1076 ± 0.0007	0.0484 ± 0.0003
W	W65	W L_3M_5	0.0319 ± 0.0005	±5 × 10^{-5}	±0.0006	0.725 ± 1.6 × 10^{-7}	1.015 ± 3.3 × 10^{-5}	0.997	0.0435	±0.0006	...	0.0432 ± 0.0006	0.0098 ± 0.0001

std, standard; unk, unknown; *Z*, atomic number effect; *A*, absorption effect; *F*, fluorescence effect; *C*, concentration; mass frac, mass fraction; norm mass, normalized mass

analysis software showing the measured *k*-ratios; the *Z*, *A*, and *F* factors; and the calculated concentrations (Ref 7, 8).

Measurement Uncertainties

Measurement uncertainties in the *k*-ratio/matrix corrections protocol arise from several sources:

(1) The counting statistics in the characteristic peak and the continuum background for both the unknown and the corresponding standard used for that characteristic x-ray family
(2) Uncertainty in the composition of the standard, which can be minimized by using a pure element or stoichiometric compound of exact composition. The presence of a thin native oxide on most pure elements represents a finite deviation from the pure value. The impact of the native oxide can be minimized by operating at higher incident beam energy, $E_0 \geq 15$ keV, but the effect of the native oxide (or contamination layers) can be significant for low-beam-energy analysis (e.g., $E_0 \leq 5$ keV).
(3) The matrix corrections have inherent uncertainty that arises from limitations on the accuracy of the ionization depth distribution function that forms the basis for calculating the corrections, as well as uncertainty in the physical parameters that are needed, especially the mass absorption coefficients. The uncertainties are dependent on composition and, in the case of the absorption correction, the photon energy of the x-ray used for the measurement. Generally, if the incident beam energy is high enough to excite two x-ray families for an element (e.g., K and L or L and M), by choosing the higher energy x-ray family, the absorption correction can be minimized and the uncertainty in the correction reduced.

The NIST DTSA-II software provides the major components of the uncertainty budget: the statistics of the measurement of the known and standard, and estimates of the uncertainty in the atomic number (*Z*) correction and in the absorption (*A*) correction based on the uncertainties in the physical parameters needed to calculate *Z* and *A* (Ref 7, 8, 9). The components of the uncertainty budget for the analysis of the Ti-Al-Nb-W alloy are given in Table 1. For each element, the uncertainty contributions from the counting statistics of the unknown and standard are listed as well as the uncertainties in the atomic number and absorption factors. The uncertainties then are combined in quadrature to give the overall uncertainty budget for the measurement of each element.

The Raw Analytical Total

The rigorous *k*-ratio/matrix corrections protocol is performed with careful attention to measuring the unknown and standard at known dose (beam current × detector live time). A special advantage of this procedure is the significance of the raw analytical total, that is, the sum of the unnormalized mass concentrations for all measured elements and any elements, such as oxygen, that are calculated on the basis of assumed stoichiometry. Because each element in the unknown is measured independently against an appropriate standard, the sum of the mass concentrations (raw analytical total) rarely equals exactly unity (100 wt%) because of the combined uncertainties over the measurement suite. In Table 1 for the analysis of the Al-Ti-Nb-W alloy, the raw analytical total is 1.0065 (100.65 wt%). When all constituents are determined, including oxygen (or another element) by the method of assumed stoichiometry, the analytical total typically should fall between 0.99 and 1.01 (99 to 101 wt%). Deviations outside this range should be carefully considered and resolved, if possible. The deviation could be due to variations in the measurement conditions, such as uncorrected drift in the beam current (dose), or there could be differences in the conductive coating of the unknown and standard or the quality of the specimen surface, including local topography (pits, ridges), or unexpected surface layers due to contamination or oxide formation. However, for analytical totals that fall significantly below unity, there is a strong possibility that there are elements present in the specimen but not recognized and included in the quantitative analysis. Thus, a low total should prompt the analyst to carefully inspect the EDS spectrum for the x-ray peaks of these missing elements.

Accuracy of the Standards-Based *k*-Ratio/Matrix Corrections Protocol with EDS

The standards-based *k*-ratio/matrix corrections protocol can be tested by analyzing as unknowns certain multielement materials whose composition is known by independent means, such as bulk elemental analysis techniques, and whose microhomogeneity has been verified by extensive measurements of the electron-excited x-ray spectrum at fixed beam locations selected randomly and systematically over a spatial grid. A number of such standards that are qualified for electron beam x-ray microanalysis are available from the various national measurement institutes such as NIST (Ref 7). Other materials that can serve as suitable test unknowns include stoichiometric compounds (i.e., materials with exact formulae) whose microhomogeneity also has been confirmed.

A parameter that serves as a measure of accuracy when testing materials of known composition is the relative deviation from expected value (RDEV), expressed as:

$$RDEV(\%) = [(Measured\ Value - True\ Value)/True\ Value] * 100\% \quad \text{(Eq13)}$$

By this definition, a positive RDEV indicates an overestimate of the true value, while a negative RDEV indicates an underestimate.

Examples of EDS analysis of the suite of NIST SRM (metal alloys and glasses) specifically created for microanalysis are presented in Tables 2–10. These analyses were performed using the NIST DTSA-II software platform, which uses multiple linear least squares fitting to determine the intensity values need for the *k*-ratios. As part of the procedure to determine the characteristic x-ray intensities, DTSA-II constructs the peak fitting residual spectrum, which shows the spectrum after the fitted intensities have been removed, as shown in Fig. 12 for the analysis of SRM 2061 (Al-Ti-Nb-W alloy). The matrix correction calculations in DTSA-II are based upon the

Table 2 Analysis of NIST Standard Reference Material (SRM) 478, cartridge brass

Element	SRM value	DTSA-II	Combined uncertainties	RDEV %	σ_{Rel}, % (9 rep)
Cu	0.7285	0.7301	±0.0009	0.22	0.16
Zn	0.2710	0.2699	±0.0007	−0.41	0.43

E_0 = 20 keV; pure element standards; K-shell x-rays; values in normalized mass concentration; combined uncertainties estimated from NIST DTSA-II; RDEV%, relative deviation from expected value expressed as percentage; σ_{Rel}, relative standard deviation; rep, replicates

Table 3 Analysis of NIST Standard Reference Material (SRM) 479, stainless steel

Element	SRM value	DTSA-II	Combined uncertainties	RDEV %	σ_{Rel}, % (7 rep)
Cr	0.183	0.1846	±0.0002	0.87	0.12
Fe	0.710	0.7106	±0.0015	0.08	0.08
Ni	0.107	0.1049	±0.0007	−2.0	0.55

E_0 = 20 keV; pure element standards; K-shell x-rays; values in normalized mass concentration; combined uncertainties estimated from NIST DTSA-II; RDEV%, relative deviation from expected value expressed as percentage; σ_{Rel}, relative standard deviation; rep, replicates

Table 4 Analysis of NIST Standard Reference Material (SRM) 480, molybdenum-tungsten alloy

Element	SRM value	DTSA-II	Combined uncertainties	RDEV %	σ_{Rel}, % (7 rep)
Mo	0.2150	0.2167	±0.0039	0.81	0.72
W	0.7850	0.7833	±0.0017	−0.22	0.25

E_0 = 20 keV; pure element standards; L-shell x-rays; values in normalized mass concentration; combined uncertainties estimated from NIST DTSA-II; RDEV%, relative deviation from expected value expressed as percentage; σ_{Rel}, relative standard deviation; rep, replicates

Table 5 Analysis of NIST Standard Reference Material (SRM) 481, gold-silver alloys

Element	SRM value	DTSA-II	Combined uncertainties	RDEV %	σ_{Rel}, % (5 rep)
Ag	0.7758	0.7698	±0.0039	−0.77	0.17
Au	0.2243	0.2302	±0.0020	2.5	0.57
Ag	0.5993	0.5972	±0.0049	−0.36	0.06
Au	0.4003	0.4028	±0.0025	0.63	0.09
Ag	0.3992	0.3922	±0.0046	−2.0	0.07
Au	0.6005	0.6078	±0.0023	1.2	0.05
Ag	0.1996	0.1919	±0.0029	−3.9	0.30
Au	0.8005	0.8081	±0.0015	0.95	0.07

E_0 = 20 keV; pure element standards; Ag L and Au M; values in normalized mass concentration; combined uncertainties estimated from NIST DTSA-II; RDEV%, relative deviation from expected value expressed as percentage; σ_{Rel}, relative standard deviation; rep, replicates

Pouchou and Pichoir φ(ρz) ionization depth distribution model (Ref 10). Inspecting the results in Tables 2–10 reveals that all of the RDEV values for these 50 individual elemental analyses are less than ±4% relative, and the majority (84%, 45 out of 50) are less than ±2% relative, with 56% (28 out of 50) less than ±1% relative.

Table 6 Analysis of NIST Standard Reference Material (SRM) 482, gold-copper alloys

Element	SRM value	DTSA-II	Combined uncertainties	RDEV%	σ_{Rel}, % (5 rep)
Cu	0.1983	0.1970	±0.0005	−0.57	0.17
Au	0.8015	0.8030	±0.0027	0.19	0.04
Cu	0.3964	0.3945	±0.0006	−0.48	0.47
Au	0.6036	0.6055	±0.0026	0.31	0.31
Cu	0.5992	0.5989	±0.0008	−0.05	0.33
Au	0.4019	0.4011	±0.0023	0.03	0.49
Cu	0.7985	0.8021	±0.0009	0.46	0.11
Au	0.2012	0.1979	±0.0017	−1.7	0.46

E_0 = 20 keV; pure element standards; Ag L and Cu K; values in normalized mass concentration; combined uncertainties estimated from NIST DTSA-II; RDEV%, relative deviation from expected value expressed as percentage; σ_{Rel}, relative standard deviation; rep, replicates

Table 7 Analysis of NIST Standard Reference Material (SRM) 483, Fe-3Si transformer steel alloy

Element	SRM value	DTSA-II	Combined uncertainties	RDEV %	σ_{Rel}, % (9 rep)
Si	0.0322	0.0313	±0.0001	−2.8	0.24
Fe	Not certified	0.9687	±0.0019	NA	0.02

E_0 = 20 keV; pure element standards; L-shell x-rays; values in normalized mass concentration; combined uncertainties estimated from NIST DTSA-II; RDEV%, relative deviation from expected value expressed as percentage; NA, not applicable; σ_{Rel}, relative standard deviation; rep, replicates

Table 8 Analysis of NIST Standard Reference Material (SRM) 2061, Al-Ti-Nb-W alloy

Element	SRM value	DTSA-II	Combined uncertainties	RDEV %	σ_{Rel}, % (11 rep)
Al	0.3031	0.2976	±0.0014	−1.8	0.14
Ti	0.5392	0.5515	±0.0005	2.3	0.17
Nb	0.1078	0.1074	±0.0007	−0.40	0.41
W	0.0438	0.0435	±0.0006	−0.60	2.5

E_0 = 20 keV; pure element standards; W L; values in normalized mass concentration; combined uncertainties estimated from NIST DTSA-II; RDEV%, relative deviation from expected value expressed as percentage; σ_{Rel}, relative standard deviation; rep, replicates

Table 9A Analysis of Standard Reference Material (SRM) 470, K411 glass

Element	SRM value	DTSA-II	Combined uncertainties	RDEV %	σ_{Rel}, % (9 rep)
O	0.4276	0.4280 (stoi)	±0.0004	1	0.04
Mg	0.0885	0.0881	±0.0002	−0.4	0.13
Si	0.2538	0.2565	±0.0002	1.1	0.09
Ca	0.1106	0.1117	±0.0002	0.96	0.19
Fe	0.1121	0.1156	±0.0007	3.1	0.27

E_0 = 10 keV; oxide standards except for Fe; O by assumed stoichiometry; K-shell x-rays; values in mass concentration; combined uncertainties estimated from NIST DTSA-II; RDEV%, relative deviation from expected value expressed as percentage; σ_{Rel}, relative standard deviation; rep, replicates

In general, extensive testing of the accuracy of the standards-based *k*-ratio/matrix corrections procedure with EDS spectrometry reveals that 95% of analyses fall within ±5% of the correct value for major and minor constituents (Ref 1).

Table 9B Analysis of Standard Reference Material (SRM) 470, K411 glass

Element	SRM value	DTSA-II	Combined uncertainties	RDEV %	σ_{Rel}, % (9 rep)
O	0.4276	0.4372 (MgO)	±0.0046	3.2	0.10
Mg	0.0885	0.0870	±0.0002	−1.6	0.17
Si	0.2538	0.2525	±0.0002	−0.12	0.07
Ca	0.1106	0.1097	±0.0002	−0.85	0.22
Fe	0.1121	0.1135	±0.0007	1.2	0.32

E_0 = 10 keV; oxide standards except for Fe; O directly analyzed (MgO); K-shell; values in mass concentration; combined uncertainties estimated from NIST DTSA-II; RDEV%, relative deviation from expected value expressed as percentage; σ_{Rel}, relative standard deviation; rep, replicates

Table 10A Analysis of Standard Reference Material (SRM) 470, K412 glass

Element	SRM value	DTSA-II	Combined uncertainties	RDEV %	σ_{Rel}, % (9 rep)
O	0.4276	0.4303 (stoi)	±0.0003	0.64	0.04
Mg	0.1166	0.1160	±0.0002	−0.48	0.23
Al	0.0491	0.0485	±0.0001	−1.2	0.14
Si	0.2120	0.2136	±0.0002	0.75	0.09
Ca	0.1090	0.1121	±0.0002	2.8	0.46
Fe	0.0774	0.0794	±0.0006	2.6	0.60

E_0 = 10 keV; oxide standards except for Fe; O by assumed stoichiometry; K-shell; values in normalized mass concentration; combined uncertainties estimated from NIST DTSA-II; RDEV%, relative deviation from expected value expressed as percentage; σ_{Rel}, relative standard deviation; rep, replicates

Table 10B Analysis of Standard Reference Material (SRM) 470, K412 glass

Element	SRM value	DTSA-II	Combined uncertainties	RDEV %	σ_{Rel}, % (9 rep)
O	0.4276	0.4347 (MgO)	±0.0043	1.7	0.58
Mg	0.1161	0.1153	±0.0002	−1.1	0.14
Al	0.0491	0.0482	±0.0001	−1.9	0.47
Si	0.2120	0.2120	±0.0002	0	0.49
Ca	0.1090	0.1111	±0.0002	2.0	0.90
Fe	0.0774	0.0787	±0.0006	1.7	0.79

E_0 = 10 keV; oxide standards except for Fe; O by assumed stoichiometry or directly analyzed (MgO); K-shell; values in normalized mass concentration; combined uncertainties estimated from NIST DTSA-II; RDEV%, relative deviation from expected value expressed as percentage; σ_{Rel}, relative standard deviation; rep, replicates

Analysis when Severe Peak Overlap Occurs

All of the examples presented in Tables 2–10 involve EDS spectra in which there are no significant overlaps between elements for the characteristic peaks used for analysis. For these analyses, the only test of the peak fitting step of the analytical procedure is the separation of the characteristic x-ray intensity from the continuum background intensity. The quality of the peak fitting result is illustrated by the peak fitting residual spectrum, which shows the continuum intensity

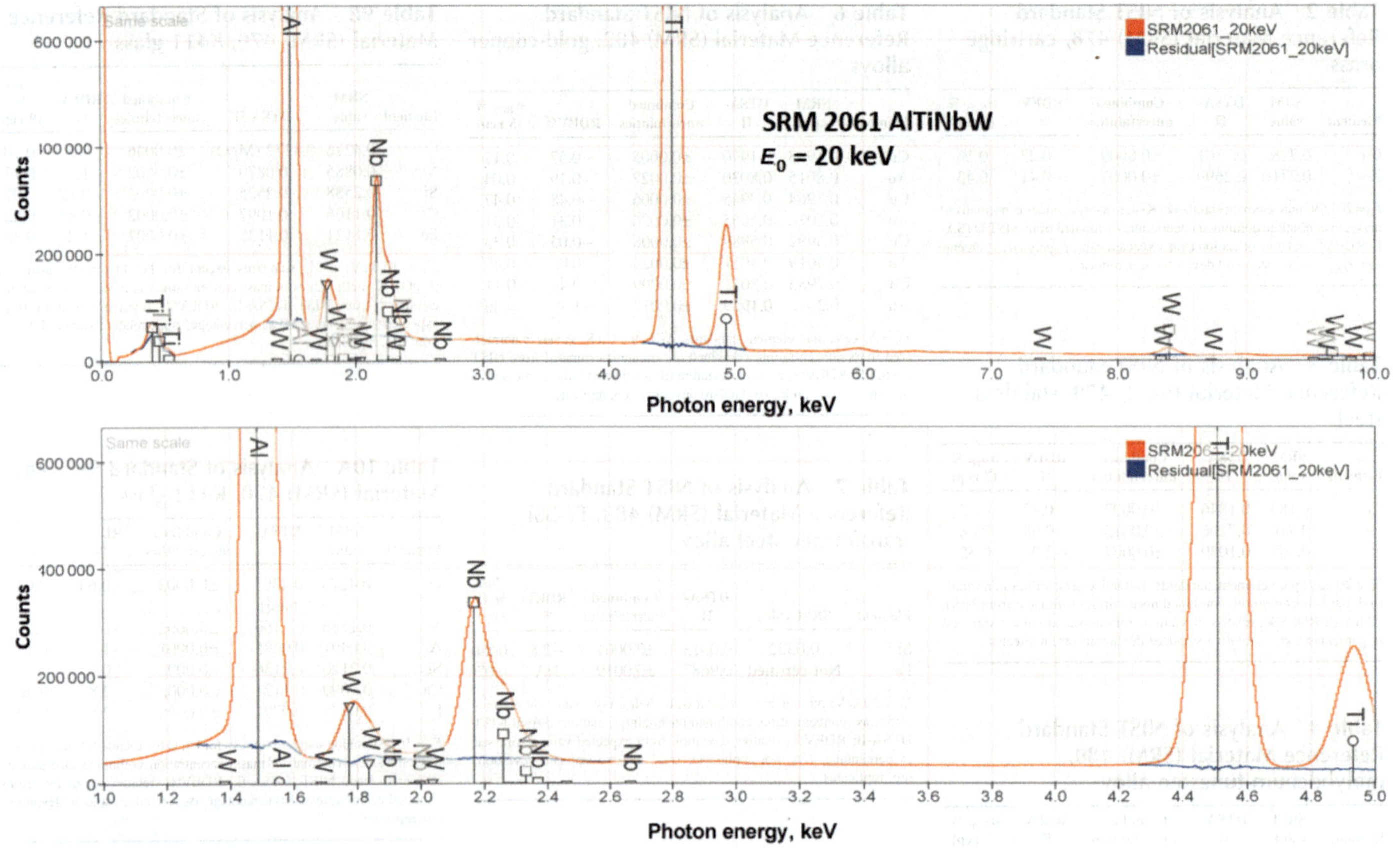

Fig. 12 EDS spectrum for NIST Standard Reference Material (SRM) 2061 (Al-Ti-Nb-W alloy) at E_0 = 20 keV with peak fitting residual spectrum

remaining after fitting for the characteristic x-ray peaks in Fig. 12. When materials that have significant peak overlaps are analyzed, peak fitting and the construction of the fitting residual spectrum have the added value of revealing unexpected constituents. This is illustrated in Fig. 13(a) for major constituents in the analysis of PbS, where the separation of the principal characteristic x-ray peaks is 39 eV (S K-$L_{2,3}$ at 2.307 keV and Pb M_5-$N_{5,7}$ at 2.346 keV) at an energy where the EDS resolution is approximately 87 eV (for a detector resolution of 129 eV at 5.895 keV corresponding to Mn K-$L_{2,3}$), which constitutes a significant interference situation. If the analyst was unaware of the interference and peak fitting is performed for only the Pb M-family, the fitting residual reveals the presence of the S K-$L_{2,3}$ peak, as shown in Fig. 13(b). Similarly, if the peak fitting is performed for only the S K-family, the Pb M-family is revealed in the residuals (Fig. 13c). When fitting is performed for both sulfur and lead, the nearly featureless residual spectrum reveals the continuum background beneath the interfering peaks (Fig. 13d). The quantitative results from this analysis are presented in Table 11, which show RDEV values lower than 1.1% relative for both constituents.

A severe interference situation is encountered when analyzing MoS_2, as shown in Fig. 14, where the separation of the principal characteristic x-ray peaks is 14 eV (Mo L_3-$M_{4,5}$ at 2.293 keV and S K-$L_{2,3}$ at 2.307 keV) at an energy where the EDS resolution is approximately 87 eV (for a detector resolution of 129 eV at 5.895 keV corresponding to Mn K-$L_{2,3}$). The peak fitting residual shows the continuum background under the characteristic peaks. The quantitative results from this analysis are presented in Table 12, which shows RDEV values below 1% relative for both constituents.

The EDS spectra of PbS and MoS_2 involve mutual interference by major constituents. Peak fitting of EDS analysis with high counts can deal with problems involving severe peak interference between a major constituent and a minor or trace constituent. Tables 13 and 14 illustrate this for the interference between the Ti K-family and the Ba L-family, for which the separation of the major peaks, Ti K-$L_{2,3}$ (4.508 keV) and Ba L_3-$M_{4,5}$ (4.467 keV), is 41 eV at an energy where the EDS resolution is approximately 115 eV (for a detector resolution of 129 eV at 5.895 keV corresponding to Mn K-$L_{2,3}$). The results given in Table 12 for the quantitative analysis of barium titanate ($BaTiO_3$), in which both titanium and barium are major constituents, show RDEV values less than 1% relative for both elements. NIST glass K2496 contains barium as a major constituent (0.4299 mass concentration) and titanium as a minor constituent (0.0180 mass concentration), giving a ratio Ba/Ti = 23.9. Despite this large elemental ratio and the severe interference, peak fitting accurately recovers the peak intensities, and the concentrations calculated give RDEV values of −2.5% for titanium and 1.7% for barium with the fitting residual shown in Fig. 15(a). Moreover, if the analyst was unaware of the presence of the minor titanium constituent, whose K-family peaks are hidden under the much higher-intensity Ba L-family peaks, inspection of the peak fitting residual spectrum when only the Ba L-family is included in the fitting procedure reveals the presence of the low-relative-intensity Ti K-family peaks, as shown in Fig. 15(b).

Low-Atomic-Number Elements

Accurate analysis of low-atomic-number elements (Li–F) with electron-excited x-ray microanalysis has been problematic for several reasons: (1) The low energy of the photons of these elements results in high

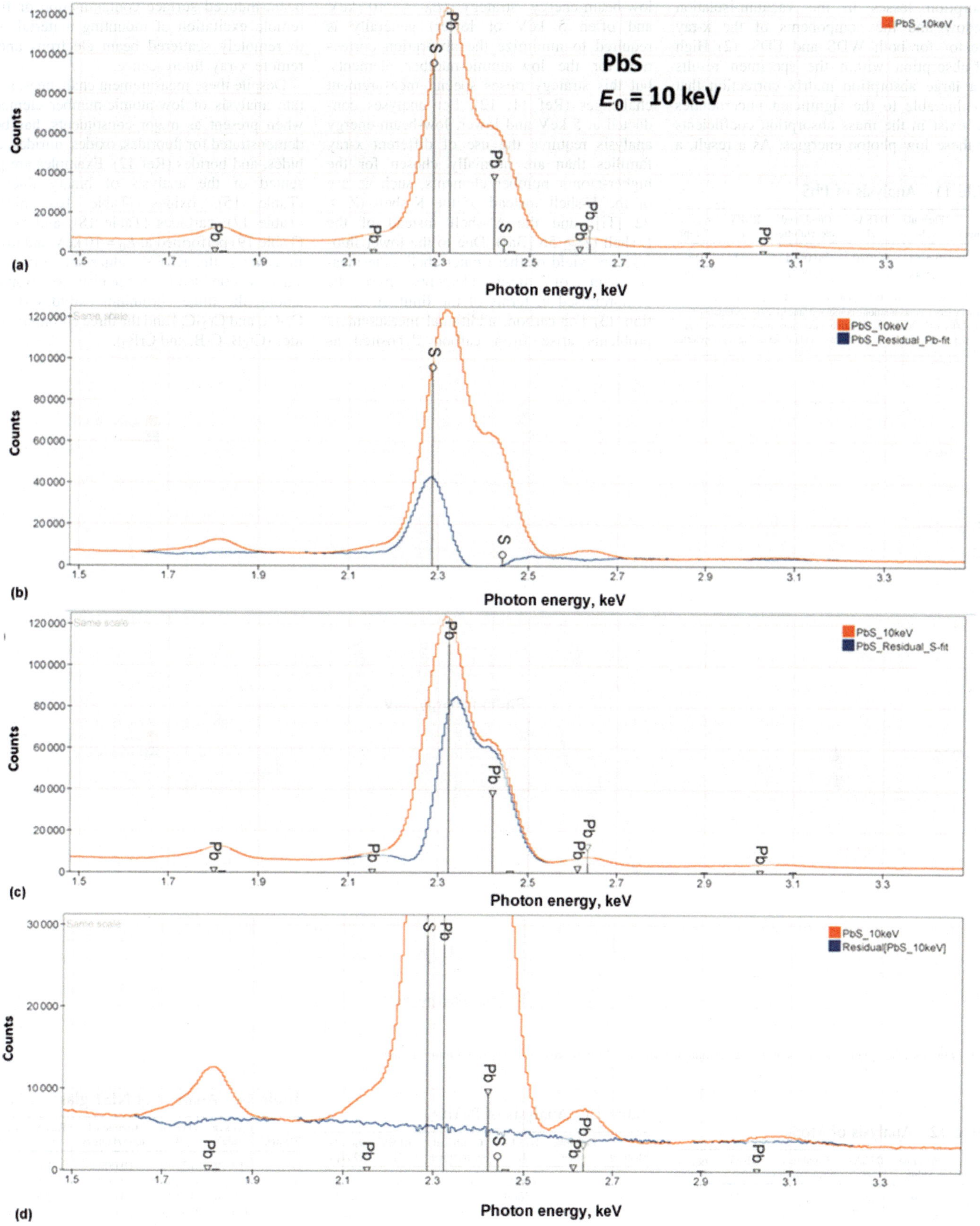

Fig. 13 (a) EDS spectrum for PbS at E_0 = 10 keV; (b) peak fitting residual spectrum after fitting only for the Pb M-family; (c) peak fitting residual spectrum after fitting only for the S K-family; (d) peak fitting residual spectrum after fitting for both the S K-family and the Pb M-family

absorption losses in the vacuum-isolation window and the components of the x-ray detector for both WDS and EDS. (2) High self-absorption within the specimen results in a large absorption matrix correction that is vulnerable to the significant uncertainties that exist in the mass absorption coefficients for these low photon energies. As a result, a low-beam-energy strategy ($E_0 \leq 10$ keV and often 5 keV or lower) generally is required to minimize the absorption correction for the low-atomic-number elements, but this strategy raises special measurement challenges (Ref 11, 12). For analyses conducted at 5 keV and lower, low-beam-energy analysis requires the use of different x-ray families than are normally chosen for the higher-atomic-number elements, such as use of the L-shell instead of the K-shell ($Z \geq 22$ [Ti]) and the M-shell instead of the L-shell ($Z \geq 56$ [Ba]). Due to the lower fluorescence yield of these outer shells, the measurement of these elements may be compromised in terms of the limit of detection. (3) For carbon, additional measurement problems arise from carbon deposited as beam-induced surface contamination or from remote excitation of mounting materials due to remotely scattered beam electrons and/or remote x-ray fluorescence.

Despite these measurement challenges, accurate analysis of low-atomic-number elements, when present as major constituents, has been demonstrated for fluorides, oxides, nitrides, carbides, and borides (Ref 12). Examples are presented of the analysis of binary fluorides (Table 15), oxides (Table 16), nitrides (Table 17), carbides (Table 18), and borides (Table 19) performed at $E_0 = 10$ keV and lower. In general, the RDEV values are sufficiently small to confidently distinguish, for example, among the three chromium carbides (Cr_3C_2, Cr_7C_3, and $Cr_{23}C_6$) and the three chromium borides (Cr_2B, CrB, and CrB_2).

Table 11 Analysis of PbS

Element	Formula value	DTSA-II	Combined uncertainties	RDEV %	σ_{Rel}, % (7 rep)
S	0.5000	0.4970	±0.0021	−0.6	0.06
Pb	0.5000	0.5030	±0.0018	0.6	0.06

$E_0 = 10$ keV; standards: PbSe and CdS; Pb M-family and S K-family; values in atom concentration; combined uncertainties estimated from NIST DTSA-II; RDEV%, relative deviation from expected value expressed as percentage; σ_{Rel}, relative standard deviation; rep, replicates

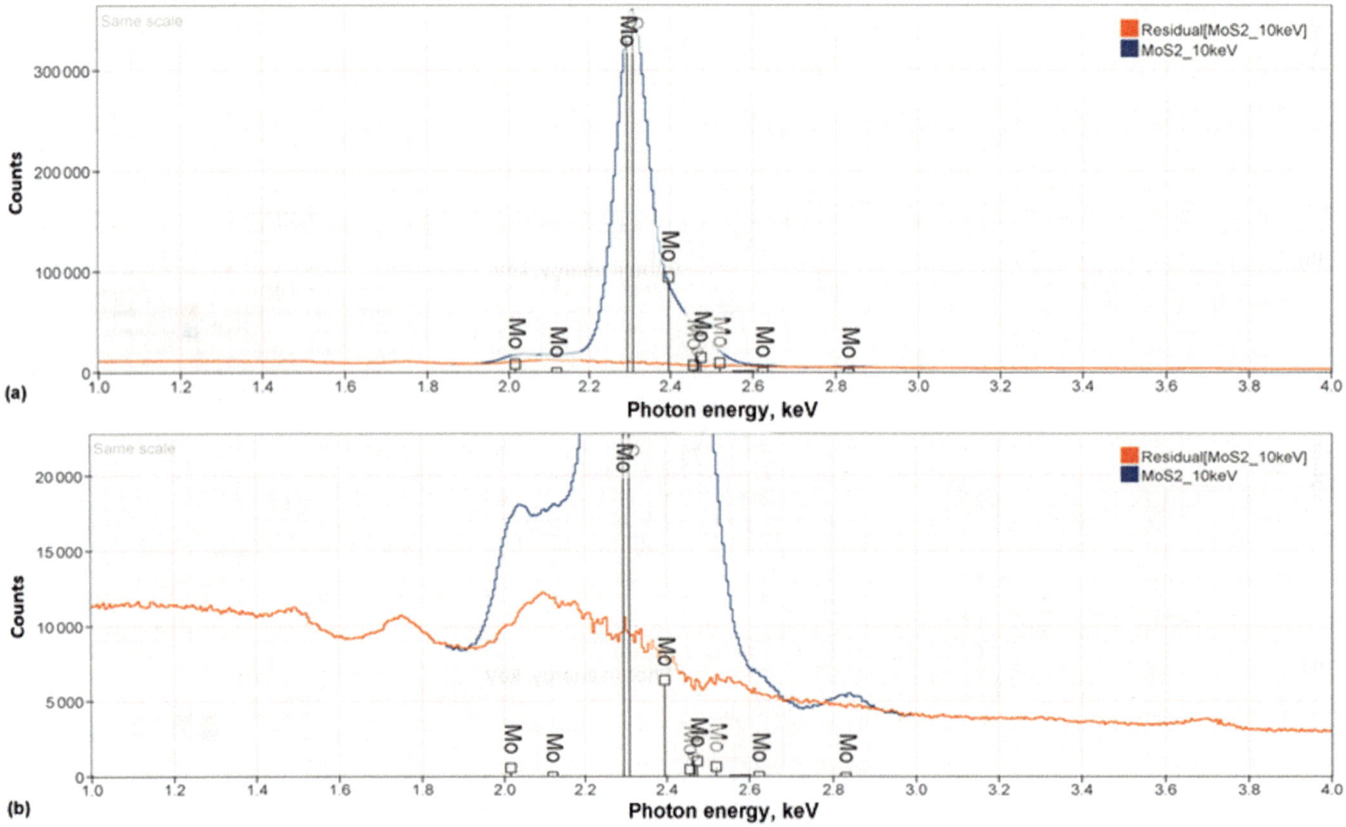

Fig. 14 (a) EDS spectrum and peak fitting residual for MoS_2 at $E_0 = 10$ keV; (b) expanded intensity axis

Table 12 Analysis of MoS_2

Element	Formula value	DTSA-II	Combined uncertainties	RDEV %	σ_{Rel}, % (11 rep)
S	0.6667	0.6630	±0.0014	−0.56	0.30
Mo	0.3333	0.3370	±0.0008	1.1	0.59

$E_0 = 10$ keV; standards: CuS and Mo; S K-family and Mo L-family; values in atom concentration; combined uncertainties estimated from NIST DTSA-II; RDEV%, relative deviation from expected value expressed as percentage; σ_{Rel}, relative standard deviation; rep, replicates

Table 13 Analysis of $BaTiO_3$

Element	Formula value	DTSA-II	Combined uncertainties	RDEV %	σ_{Rel}, % (9 rep)
O	0.6000	0.6005	±0.0011	0.08	0.02
Ti	0.2000	0.2009	±0.0005	0.46	0.14
Ba	0.2000	0.1986	±0.0005	−0.69	0.22

$E_0 = 10$ keV; standards: Ti; $BaSi_2O_5$ (sanbornite) for Si, Ba; O by stoichiometry; values in atom concentration; combined uncertainties estimated from NIST DTSA-II; RDEV%, relative deviation from expected value expressed as percentage; σ_{Rel}, relative standard deviation; rep, replicates

Table 14 Analysis of NIST glass K2496

Element	As-syn value	DTSA-II	Combined uncertainties	RDEV %	σ_{Rel}, % (9 rep)
O	0.3230	0.3197	±0.0004	−1.0	0.09
Si	0.2291	0.2256	±0.0002	−1.5	0.13
Ti	0.0180	0.0175	±0.0004	−2.5	1.8
Ba	0.4299	0.4373	±0.0013	1.7	0.16

$E_0 = 10$ keV; standards: Ti; $BaSi_2O_5$ (sanbornite) for Si, Ba; O by stoichiometry; values in mass concentration; the reference glass composition is the as-synthesized value; RDEV%, relative deviation from expected value expressed as percentage; σ_{Rel}, relative standard deviation; rep, replicates

Iterative Qualitative-Quantitative Analysis for Complex Compositions

To analyze the complex multielement compositions frequently encountered in materials science applications, an analysis strategy is needed for the frequently encountered situation in which high-intensity peaks of major constituents mask the low-relative-intensity peaks of minor and trace constituents. A useful approach for complex specimens is iterative qualitative-quantitative analysis. The spectrum is recorded to contain an adequate number of counts, which depends on the concentration levels sought. In the first analysis round, qualitative analysis is performed to assign elements to the recognizable peaks (using the automatic peak identification software tool but always confirmed by manual inspection of the proposed elemental assignments). After the suite of elements is identified, standards-based quantitative analysis is performed with peak fitting to construct the residual spectrum that remains after removal of the peak intensities for those identified elements. The raw analytical total is inspected as a possible indicator of missing constituents. For the second analysis round, this residual spectrum is inspected to identify any newly emerged peaks previously hidden under the major peaks. Qualitative analysis then is applied to assign elements to these residual spectrum peaks, and the second round of quantitative analysis is performed with the newly extended suite of recognized elements. This qualitative-quantitative analysis sequence is repeated until the residual spectrum no longer reveals new peaks and the raw analytical total approaches 100 wt% (unity mass concentration). An example of the iterative qualitative-quantitative analysis procedure is presented in Fig. 16 for the analysis of an inclusion in NIST SRM 168, in which the first fitting residual reveals the presence of minor tungsten and the second fitting residual reveals trace molybdenum. The sequence of quantitative analysis results is presented in Table 20.

Standardless Analysis

Virtually all vendor EDS software includes *standardless analysis* as an option. In standardless analysis, the intensity of the standard required for the denominator in Eq 8 is supplied either from first principles calculation of x-ray generation, emission, and detection or, more commonly, from a library of vendor-measured standard intensities recorded over a wide range of incident beam energy (Ref 1). The user only needs to provide the incident beam energy for the analysis. The library intensity for a particular x-ray peak is adjusted for the efficiency of the local EDS by a channel-by-channel comparison of a locally measured spectrum (e.g., Cu) to the library reference spectrum for the same element. If a beam energy is used that is not represented in the library, the physical equations of x-ray generation are used to adjust the library value to the required beam energy. Because the electron dose is not available, the results of standardless analysis must be normalized to 100 wt% to place the calculated concentrations on a sensible basis. Thus, the raw analytical total is not available as a possible indication of missing constituents.

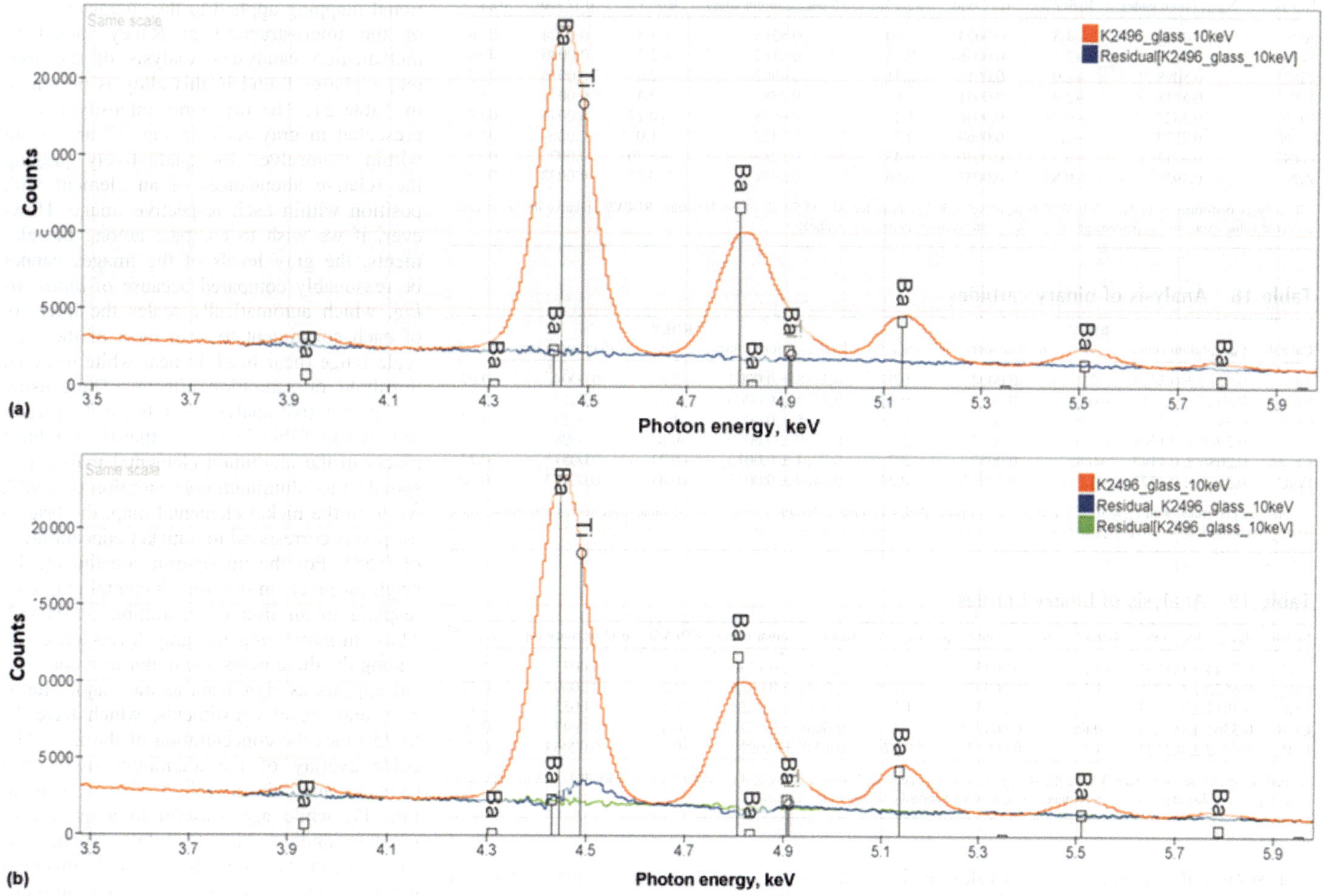

Fig. 15 (a) EDS spectrum and peak fitting residual for glass K2496 at E_0 = 10 keV; (b) peak fitting residual spectrum shown when titanium is not included in the fit (blue)

Table 15 Analysis of binary fluorides

Fluoride	F_{mean} (atom conc)	RDEV%	σ (7 rep)	$σ_{Rel}$, %	$Metal_{mean}$ (atom conc)	RDEV%	σ (7 rep)	$σ_{Rel}$, %
NaF	0.5143	+2.9	0.00069	0.13	0.4857	−2.9	0.00069	0.14
CaF_2	0.6686	+0.29	0.00074	0.11	0.3314	−0.57	0.00074	0.22
SrF_2	0.6611	−0.83	0.00021	0.032	0.3389	+1.7	0.00021	0.062
BaF_2	0.6527	−2.1	0.00152	0.23	0.3473	+4.2	0.00152	0.44
LaF_3	0.7600	+1.3	0.00283	0.37	0.2400	−4.0	0.00283	1.2

All analyses performed with E_0 = 10 keV and the K-$L_{2,3}$ peak (Na and Ca) or the L_3-$M_{4,5}$ peak (Sr, Ba, and La); RDEV%, relative deviation from expected value expressed as percentage; $σ_{Rel}$, relative standard deviation; rep, replicates

Table 16 Analysis of binary oxides

Oxide	O_{mean} (atom conc)	RDEV%	σ (7 rep)	$σ_{Rel}$, %	$Metal_{mean}$ (atom conc)	RDEV%	σ (7 rep)	$σ_{Rel}$, %
MgO	0.4966	−0.68	0.00064	0.13	0.5034	+0.67	0.00064	0.13
Al_2O_3	0.5905	−1.6	0.00039	0.07	0.4095	+2.4	0.00039	0.10
SiO	0.4989	−0.20	0.00033	0.07	0.5011	+0.20	0.00033	0.07
SiO_2	0.6568	−1.5	0.00030	0.05	0.3432	+3.0	0.00030	0.09
TiO_2	0.6702	+0.54	0.0011	0.16	0.3297	−1.1	0.0011	0.32
Cr_2O_3	0.5962	−0.64	0.0117	2.0	0.4038	+0.95	0.0117	2.9
Fe_2O_3	0.5988	−0.20	0.00068	0.11	0.4012	+0.30	0.00068	0.17
Cu_2O	0.3292	−1.2	0.00069	0.21	0.6708	+0.62	0.00069	0.10
CuO	0.5065	+1.30	0.0022	0.44	0.4935	−1.3	0.0022	0.45
ZnO	0.4905	−1.90	0.0058	1.2	0.5095	+1.10	0.0058	1.10
Y_2O_3	0.5998	−0.04	0.00071	0.12	0.4002	+0.05	0.00071	0.18

All analyses performed with E_0 = 10 keV and the K-$L_{2,3}$ peak, except for Zn L_3-$M_{4,5}$ and Y L_3-$M_{4,5}$; RDEV%, relative deviation from expected value expressed as percentage; $σ_{Rel}$, relative standard deviation; rep, replicates

Table 17 Analysis of binary nitrides

Nitride	N_{mean} (atom conc)	RDEV%	σ (7 rep)	$σ_{Rel}$, %	$Metal_{mean}$ (atom conc)	RDEV%	σ (7 rep)	$σ_{Rel}$, %
AlN	0.4784	−4.3	0.0024	0.50	0.5216	4.3	0.0024	0.46
Si_3N_4	0.5857	+2.5	0.0036	0.61	0.4143	−3.3	0.0036	0.86
TiN	0.5098	+2.0	0.0016	0.31	0.4902	−2.0	0.0016	0.32
VN	0.5118	+2.4	0.0051	1.0	0.4882	−2.4	0.0051	1.0
Cr_2N	0.3322	−0.33	0.0046	1.4	0.6678	+0.17	0.0046	0.69
Fe_3N	0.2573	+2.9	0.0069	2.7	0.7427	−1.0	0.0069	0.93
GaN	0.4717	−5.7	0.0020	9.43	0.5283	+5.70	0.0020	0.38
ZrN	0.4959	−0.82	0.0033	0.66	0.5041	0.82	0.0033	0.65

All analyses performed with E_0 = 5 keV (ZrN at 10 keV); K-$L_{2,3}$ peak for Al and Si; all others L-family; RDEV%, relative deviation from expected value expressed as percentage; $σ_{Rel}$, relative standard deviation; rep, replicates

Table 18 Analysis of binary carbides

Carbide	C_{mean} (atom conc)	RDEV %	σ (5 or more rep)	$σ_{Rel}$, %	$Metal_{mean}$ (atom conc)	RDEV %	σ (5 or more rep)	$σ_{Rel}$, %
SiC	0.5112 ± 0.3925	+2.2	0.0004	0.07	0.4888 ± 0.0002	−2.2	0.0004	0.07
VC	0.4963 ± 0.2517	−0.73	0.0017	0.33	0.5037 ± 0.0006	+0.73	0.0017	0.33
Cr_3C_2	0.3925 ± 0.2127	−1.9	0.0029	0.73	0.6075 ± 0.0008	+1.3	0.0029	0.47
Cr_7C_3	0.2960 ± 0.1608	−2.0	0.0097	3.3	0.7059 ± 0.0009	+0.84	0.0098	1.4
$Cr_{23}C_6$	0.2086 ± 0.1143	+0.82	0.0057	2.7	0.7914 ± 0.0010	−0.21	0.0057	0.72
Fe_3C	0.2531 ± 0.1572	+1.2	0.00073	0.29	0.7469 ± 0.0015	−0.41	0.00073	0.10

all analyses performed with E_0 = 10 keV and the K-$L_{2,3}$ peak; RDEV%, relative deviation from expected value expressed as percentage; $σ_{Rel}$, relative standard deviation; rep, replicates

Table 19 Analysis of binary borides

Boride	B_{mean} (atom conc)	RDEV%	σ (5 or more rep)	$σ_{Rel}$, %	$Metal_{mean}$ (atom conc)	RDEV%	σ (5 or more rep)	$σ_{Rel}$, %
B_4C	0.7744 ± 0.1104	−3.2	0.0033	0.43	0.2256 ± 0.1240	13	0.0033	1.5
CrB_2	0.6552 ± 0.5276	−1.7	0.00092	0.14	0.3448 ± 0.0034	3.5	0.00092	0.27
CrB	0.4913 ± 0.4563	−1.7	0.0055	1.1	0.5087 ± 0.0028	1.7	0.0055	1.1
Cr_2B	0.3362 ± 0.3283	0.86	0.0047	1.4	0.6638 ± 0.0026	−0.42	0.0047	0.70
LaB_6	0.8702 ± 0.2183	1.5	0.00063	0.07	0.1298 ± 0.0012	−9	0.00063	0.49

all analyses performed with E_0 = 5 keV; B K-$L_{2,3}$ peak, C K-$L_{2,3}$ peak, Cr L-family peaks, and La M-family peaks; RDEV%, relative deviation from expected value expressed as percentage; $σ_{Rel}$, relative standard deviation; rep, replicates

Testing of the accuracy of standardless analysis procedures, as implemented in vendor software, using known materials has shown that approximately 95% of analyses fall within ±25% relative of the correct value, which is a factor of 5 poorer performance than standards-based analysis with local measurement of the standards (Ref 1, 13, 14).

Elemental/Compositional Mapping

Mapping elemental distributions with micrometer to submicrometer lateral resolution is an important operational mode of electron-excited x-ray microanalysis (Ref 1). The high throughput that is now possible with SDD-EDS makes it possible to perform elemental mapping in the x-ray spectrum image mode of data collection, whereby a complete EDS spectrum is collected at each pixel of an *x-y* image scan. Having the complete EDS spectrum available at every pixel enables full quantitative analysis at every pixel, including peak fitting for overlap and background correction that enables implementation of the standards-based *k*-ratio/matrix correction protocol. Thus, the raw x-ray intensity values at each pixel used for simple elemental mapping can be replaced with rigorously calculated compositional values. The gray or color scale that is assigned to the pixel thus is related directly to concentration and not mere x-ray intensity. X-ray intensity maps are subject to the strong effects of overvoltage, absorption, EDS efficiency, and other factors that make map interpretation problematic. Figures 17 and 18 present an example of elemental and compositional mapping applied to the characterization of the microstructure of Raney nickel (a methanation catalyst). Analysis of the four major phases found in this alloy is presented in Table 21. The raw x-ray intensity images presented in gray scale in Fig. 17 are useful within themselves for qualitatively judging the relative abundances of an element with position within each respective image. However, if we wish to compare among the elements, the gray levels of the images cannot be reasonably compared because of *autoscaling*, which automatically scales the intensity of each constituent to span most of the gray scale range (near black to near white to avoid threshold or saturation effects). Thus, using the quantitative analysis results for the phases reported in Table 21 reveals that the brightest pixels in the aluminum elemental map correspond to an aluminum concentration of 0.987, while in the nickel elemental map, the brightest pixels correspond to a nickel concentration of 0.585. For the minor iron constituent, the brightest pixels in the iron elemental map correspond to an iron concentration of 0.0437. Thus, in comparing the gray levels observed among the three maps, the minor iron constituent appears as significant as the major aluminum and nickel constituents, which have 15 to 25 times the concentration of the iron. The color overlay of the aluminum (red), iron (green), and nickel (blue) also shown in Fig. 17, while again useful in a qualitative sense for understanding the relative spatial distributions of the elements, is equally misleading in terms of judging the relative amounts at each location. Colors (with variables such as hue and saturation) are even more difficult

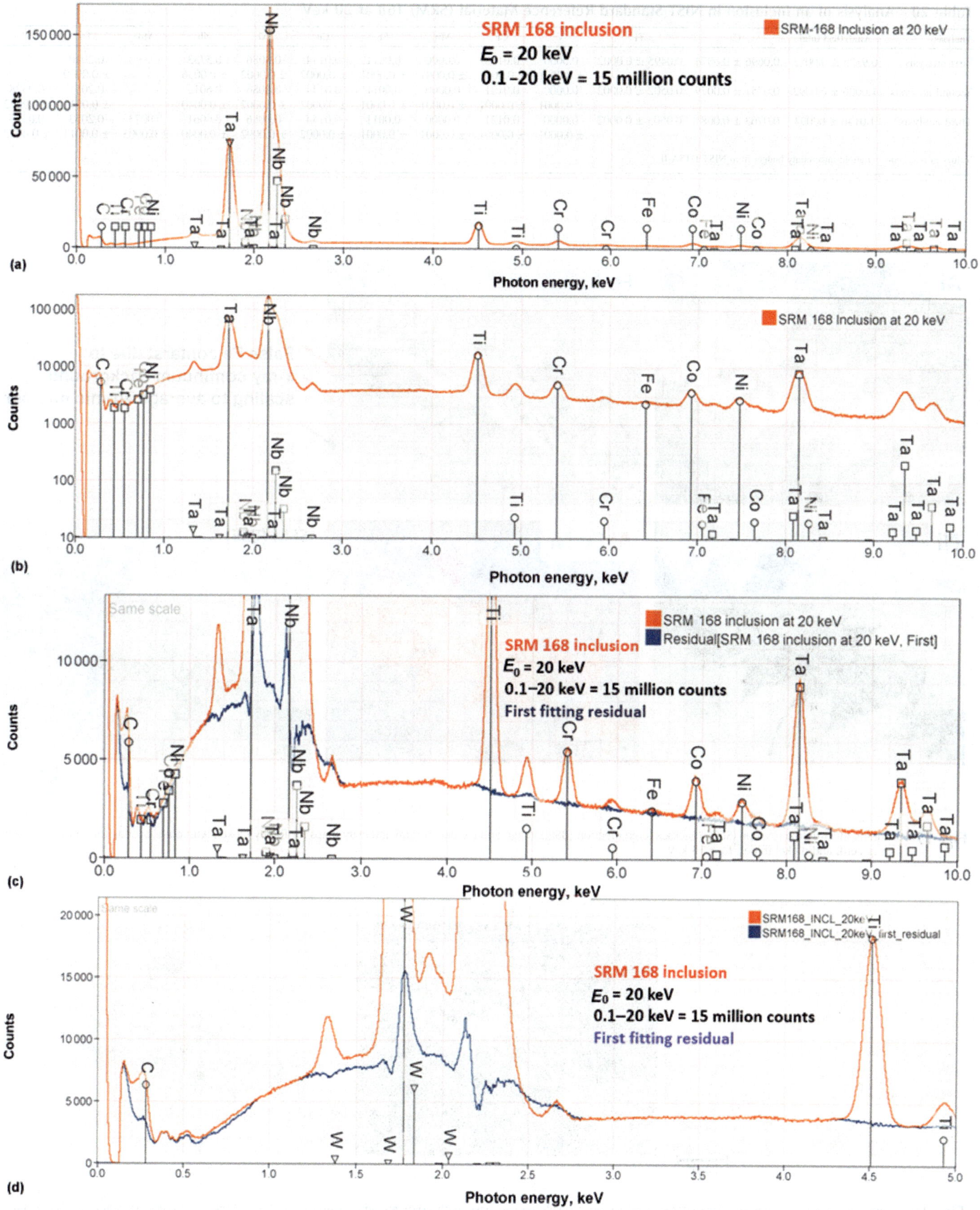

Fig. 16 (a) EDS spectrum for an inclusion in NIST Standard Reference Material (SRM) 168; (b) logarithmic display showing minor constituents; (c) original spectrum and peak fitting residual 0–10 keV; (d) expanded region 0–5 keV: peak fitting residual spectrum revealing the W M-family

Table 20 Analysis of an inclusion in NIST Standard Reference Material (SRM) 168 at 20 keV

Analysis	Analytical total	C	Ti	V	Cr	Mn	Fe	Co	Ni	Nb	Mo	Ta	W
First analysis	0.9325 ± 0.0083	0.0696 ± 0.0074	0.0495 ± 0.0002	0.0000 ± 0.0001	0.0120 ± 0.0001	0.0000 ± 0.0001	0.0011 ± 0.0001	0.0144 ± 0.0002	0.0086 ± 0.0002	0.5733 ± 0.0036	...	0.2040 ± 0.0009	...
Second analysis	1.0006 ± 0.0092	0.0751 ± 0.0079	0.0502 ± 0.0002	0.0000 ± 0.0001	0.0121 ± 0.0001	0.0000 ± 0.0001	0.0011 ± 0.0001	0.0144 ± 0.0002	0.0086 ± 0.0002	0.6012 ± 0.0040	...	0.2051 ± 0.0011	0.0328 ± 0.0022
Third analysis	1.0146 ± 00094	0.0763 ± 0.0081	0.0503 ± 0.0002	0.0000 ± 0.0001	0.0121 ± 0.0001	0.0000 ± 0.0001	0.0011 ± 0.0001	0.0144 ± 0.0002	0.0086 ± 0.0002	0.6061 ± 0.0040	0.0073 ± 0.0005	0.2053 ± 0.0011	0.0329 ± 0.0022

Values in mass concentration; uncertainty budget from NIST DTSA-II

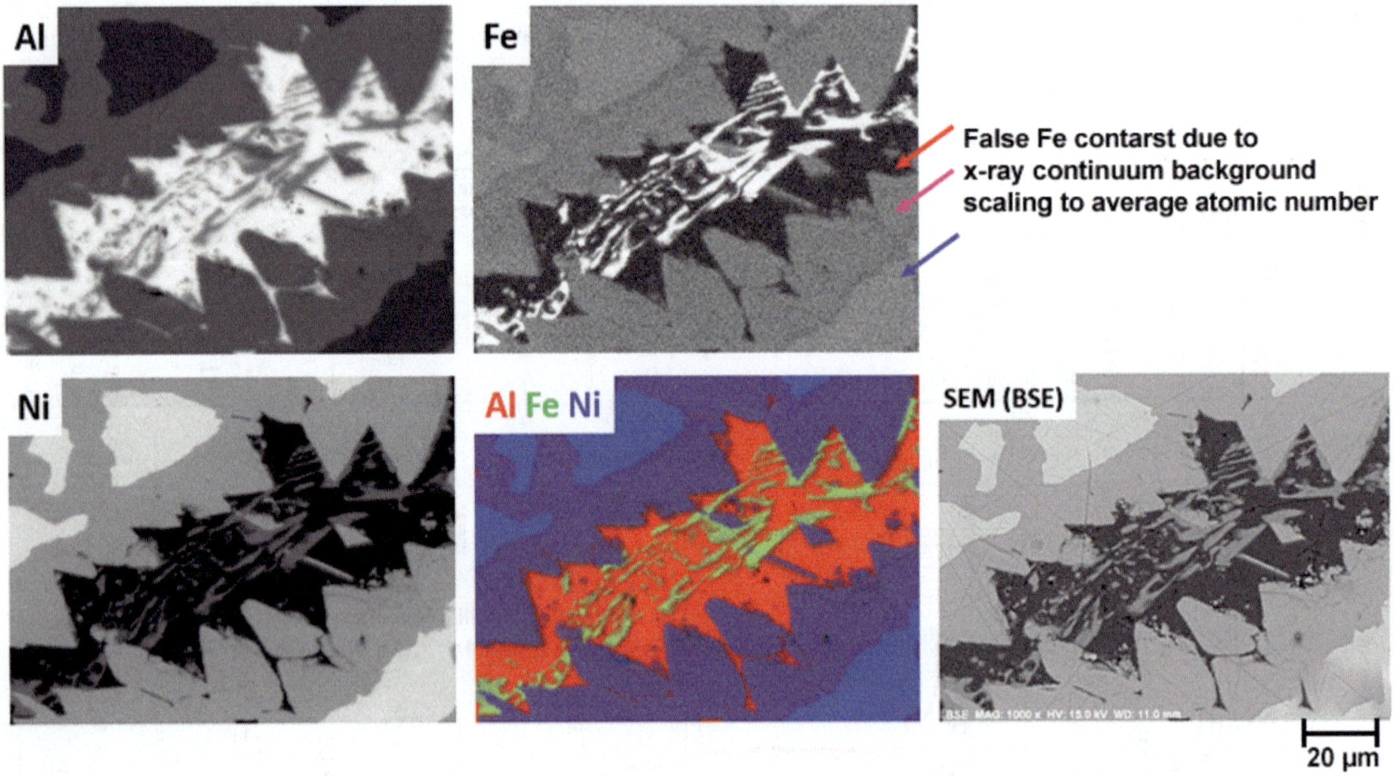

Fig. 17 Scanning electron microscope (SEM) (backscattered electron [BSE]) image and elemental x-ray intensity maps for Raney nickel, with color overlay for aluminum (red), iron (green), and nickel (blue); E_0 = 15 keV

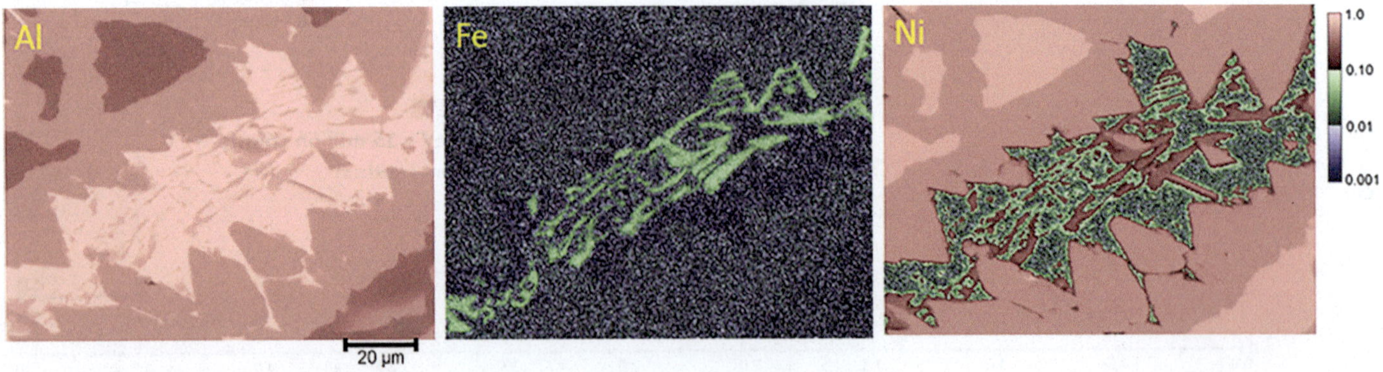

Fig. 18 Same area as Fig. 17, but after quantitative analysis on a pixel-by-pixel basis with color encoding using logarithmic three-band color encoding, enabling direct comparison of concentration levels among the three constituents. Note the elimination of false contrast in the iron image.

Table 21 Analysis of phases in Raney nickel

Phase	Al	σ (7 rep)	σ_{Rel}, %	Fe	σ (7 rep)	σ_{Rel}, %	Ni	σ (7 rep)	σ_{Rel}, %
Al-rich	0.9867 ± 0.0005	0.0015	0.16	0.0002 ± 0.0001	0.0001	50	0.0132 ± 0.0003	0.0016	12
Fe-rich	0.6866 ± 0.0039	0.0152	2.2	0.0427 ± 0.0003	0.0063	15	0.2707 ± 0.0008	0.011	4
Ni-int	0.5754 ± 0.0047	0.0019	0.3	0.0027 ± 0.0002	0.0002	8	0.4219 ± 0.0009	0.0021	0.5
Ni-rich	0.4143 ± 0.0043	0.0031	0.8	0.0008 ± 0.0002	0.00007	9	0.5849 ± 0.0011	0.0031	0.5

E_0 = 15 keV; Al, Fe, and Ni standards; values in mass concentration

than gray levels for an observer to interpret in making any quantitative judgments.

Moreover, there are artifacts that arise because of the x-ray physics. Thus, in a raw intensity map, the presence of both characteristic and continuum (bremsstrahlung) x-rays within the energy band selected to span the peak means that as the concentration of an element falls, the x-ray continuum, which scales with the overall average atomic number, becomes increasingly significant in the image. In the iron image in Fig. 17, while the iron-rich phase is correctly portrayed, there appears to be a significant level of iron in the intermediate-nickel and high-nickel regions, but the iron analyses of these regions reported in Table 16 show that, in reality, compared with the iron-rich phase, the iron level is reduced by a factor of 16 in the intermediate-nickel phase and a factor of 53 in the high-nickel phase. Thus, the apparent iron contrast seen in the nickel-rich phases relative to the aluminum-rich phase level is an artifact of the x-ray continuum background under the iron characteristic x-ray peak that is not corrected in the raw x-ray intensity elemental maps.

By using the complete spectrum collected at each pixel in the x-ray spectrum image, the composition at each pixel can be determined using NIST DTSA-II with proper corrections made for the background and any peak interference, enabling replacement of the raw x-ray intensity scale with a concentration scale. Figure 18 shows the same area as Fig. 17, but with quantified compositional maps. In these compositional maps, the concentration scale has been encoded following the logarithmic three-band scheme (Ref 15). Major constituents (0.101 to 1, or 10.1 to 100 wt%) are shown from deep red to pink; minor constituents (0.01 to 0.1, or 1 to 10 wt%) are shown from deep green to green pastel; and trace constituents (0.001 to 0.0099, or 0.1 to 0.99 wt%) are shown from deep blue to blue pastel (Ref 14). This type of color encoding enables the analyst to quickly assess the relative amounts of each constituent. This scale is not limited to three constituents, but can be applied to as many constituents as are measured. Note also that the false contrast in the iron map in Fig. 18 has been greatly reduced in the compositional map.

Summary

Electron-excited x-ray microanalysis with EDS following the standards-based matrix corrections protocol can achieve a level of analytical accuracy that can solve elemental analysis problems in materials science for constituents present at major (concentration $C > 10$ wt%), minor (1 wt% $\leq C \leq$ 10 wt%), and trace (0.05 wt% $< C <$ 1 wt%) levels.

(1) Accurate qualitative analysis for the identification of the elements requires the analyst to perform a careful review of each spectrum to confirm and correct the results of the automatic peak identification software procedure typically embedded in vendor software.

(2) Accurate quantitative analysis is achieved by following the standards-based intensity-ratio method with matrix corrections. Extensive testing of known homogeneous materials indicates that standards-based results fall within ±5% of the reference values for 95% of analyses. In comparison, vendor-supplied standardless analysis procedures are likely to produce results for which approximately 95% of analyses fall within ±25% of the correct value, a factor of 5 poorer performance compared with the standards-based procedure.

(3) This level of accuracy for standards-based analysis with EDS can be achieved even when severe peak overlap occurs between constituents with a large concentration ratio, 20:1 or higher.

(4) By constructing the peak fitting residual spectrum as part of the EDS quantitative analysis procedure, the peaks of minor or trace elements that were hidden under the higher-intensity peaks of major constituents can be revealed, enabling discovery of unexpected constituents.

(5) Iterative qualitative and quantitative analysis with review of the raw analytical total and the peak fitting residual spectrum after each round enables comprehensive characterization of complex compositions.

(6) Low-atomic-number elements, including fluorine, oxygen, nitrogen, carbon, and boron, can be accurately measured by operating at reduced beam energy, $E_0 \leq 10$ keV, to minimize x-ray absorption effects.

(7) EDS enables comprehensive elemental mapping for major and minor constituents. By collecting a complete EDS spectrum at each picture element, quantitative mapping can be performed. Artifacts in the gray scale presentation of elemental maps can be minimized by using the logarithmic three-band color scheme to encode quantitative compositional results.

ACKNOWLEDGMENT

Article revised from Kurt F.J. Heinrich and Dale E. Newbury, Electron Probe X-Ray Microanalysis, *Materials Characterization*, Vol 10, *ASM Handbook*, ASM International, 1986

REFERENCES

1. J.I. Goldstein, D.E. Newbury, J.R. Michael, N.W.M. Ritchie, J.H.J. Scott, and D.C. Joy, *Scanning Electron Microscopy and X-Ray Microanalysis*, 4th ed., Springer, 2018
2. R. Castaing, "Application of Electron Probes to Local Chemical and Crystallographic Analysis," Ph.D. thesis, University of Paris, 1951
3. D.E. Newbury and N.W.M. Ritchie, Review: Performing Elemental Microanalysis with High Accuracy and High Precision by Scanning Electron Microscopy/Silicon Drift Detector Energy Dispersive X-Ray Spectrometry (SEM/SDD-EDS), *J. Mater. Sci.*, Vol 50, 2015, p 493–518
4. N.W.M. Ritchie, D.E. Newbury, and J.M. Davis, EDS Measurements at WDS Precision and Accuracy Using a Silicon Drift Detector, *Microsc. Microanal.*, Vol 18, 2012, p 892–904
5. D.E. Newbury, Misidentification of Major Constituents by Automatic Qualitative Energy Dispersive X-Ray Microanalysis: A Problem that Threatens the Credibility of the Analytical Community, *Microsc. Microanal.*, Vol 11, 2005, p 545–561
6. D.E. Newbury, Mistakes Encountered during Automatic Peak Identification of Minor and Trace Constituents in Electron-Excited Energy Dispersive X-Ray Microanalysis, *Scanning*, Vol 31, 2009, p 91–101
7. National Institute of Standards and Technology Standard Reference Materials, 2018, information available at https://www.nist.gov/srm
8. N.W.M. Ritchie, *NIST DTSA-II*, a software engine for electron-excited energy dispersive X-ray spectrometry, 2018, available free at www.nist.gov (search "DTSA-II")
9. N.W.M. Ritchie and D.E. Newbury, Uncertainty Estimates for Electron Probe X-Ray Microanalysis Measurements, *Anal. Chem.*, Vol 84, 2012, p 9956–9962
10. J.L. Pouchou and F. Pichoir, Quantitative Analysis of Homogeneous or Stratified Microvolumes Applying the Model "PAP" in Electron Probe Quantitation, K.F.J.

Heinrich and D.E. Newbury, Ed., Plenum, p 31–75
11. D.E. Newbury and N.W.M. Ritchie, Electron-Excited X-Ray Microanalysis at Low Beam Energy: Almost Always an Adventure!, *Microsc. Microanal.*, Vol 22 (4), 2016, p 735–753
12. D.E. Newbury and N.W.M. Ritchie, Quantitative Electron-Excited X-Ray Microanalysis of Borides, Carbides, Nitrides, Oxides, and Fluorides with Scanning Electron Microscopy/Silicon Drift Detector Energy-Dispersive Spectrometry (SEM/SDD-EDS) and NIST DTSA-II, *Microsc. Microanal.*, Vol 21, 2015, p 1327–1340
13. D.E. Newbury, C.R. Swyt, and R.L. Myklebust, 'Standardless' Quantitative Electron Probe Microanalysis with Energy-Dispersive X-Ray Spectrometry: Is It Worth the Risk?, *Anal. Chem.*, Vol 67, 1995, p 1866–1871
14. N.W.M. Ritchie and D.E. Newbury, Standardless Analysis – Better But Still Risky, *Microsc. Microanal.*, Vol 20 (Suppl 3), 2014, p 696–697
15. D.E. Newbury and D.S. Bright, Logarithmic 3-Band Color Encoding: Robust Method for Display and Comparison of Compositional Maps in Electron Probe X-Ray Microanalysis, *Microsc. Microanal.*, Vol 5, 1999, p 333–343

ASM Handbook, Volume 10, *Materials Characterization*
ASM Handbook Committee
DOI 10.31399/asm.hb.v10.a0006677

Focused Ion Beam Instruments

Nabil Bassim, McMaster University
John Notte, Carl Zeiss SMT, Inc.

Overview

General Uses

- Focused ion beam (FIB) instruments can be used for imaging, modification, or analysis of samples. Lateral resolution is limited by the ion beam type from ~1 to 50 nm. While gallium is the most commonly available ion beam, other instruments can produce beams with other ion species.

Examples and Applications

- The FIB instrument can be used for preparing lamella-shaped or needle-shaped portions of a sample to be fabricated, extracted, and then analyzed in the transmission electron microscope or atom probe.
- The FIB instrument can be used to expose subsurface layers and cross sections for subsequent imaging and analysis by other techniques. When repeated in an alternating fashion, this can provide three-dimensional reconstructed information.
- The FIB instrument can provide images of surfaces with high lateral resolution and useful contrast mechanisms that often are different from scanning electron microscopy images.
- The FIB Instrument can be used for fine-scale nanofabrication by several methods, both additive and subtractive, at the 5 nm and larger length scales.
- The FIB instrument can be used for implanting ions or functionalizing a material with precision on the order of 10 nm, and with well-controlled dose.

Samples

- *Form:* Solid samples (e.g., metals, semiconductors, insulators, or organics)
- *Size:* Overall sample size can be 50 by 50 mm (2 by 2 in., typical) or can be larger or smaller, depending on the instrument. The area to be processed can be 100 nm to 1 mm, depending on the desired application.
- *Preparation:* Samples must be prepared for vacuum. In some cases, insulating samples may require conductive coating. When coupled with a cryogenic stage, liquid/hydrated samples also may be observed.

Limitations

- When ion beams are used for a given application, there sometimes are unintended consequences that must be considered (residual gallium implanted, surface sputtering, or subsurface damage).
- For heavier ions especially, imaging at high magnification causes appreciable sample damage.

Estimated Analysis Time

- Depending on the application, this typically ranges from 30 min to 8 h. Some tasks requiring three-dimensional datasets can take considerably longer but are amenable to automation.

Capabilities of Related Techniques

- The *FIB* techniques for site-specific sample preparation are vital for both transmission electron microscopy and atom probe techniques.
- *Scanning electron microscopy:* Often is a complimentary imaging technique with lower damage rates and the ability to generate characteristic x-rays.
- The sample interaction of FIB is related to ion-scattering spectroscopies such as *low-energy ion-scattering spectroscopy, Rutherford backscattering spectroscopy,* and *medium-energy ion-scattering spectroscopy.*
- *Secondary ion mass spectroscopy* analysis is a closely related technique that combines a high-performance spectrometer with ion beams of different species.
- Broad ion beams often can take the place of FIBs for applications in which lateral precision is not required or where precision can be attained with the use of apertures and masks.

Introduction

Focused ion beam (FIB) instruments can be thought of as the tools that can help humans to see, manipulate, and analyze matter at the smallest length scales. At the most fundamental level, the virtue of ion beams arises from the unique sample interactions that are distinctly different from electron beam instruments, scanned probes, or optical instruments. In most conventional uses, the FIB is used as a nanoscale "sand blaster" to remove materials from the sample by a subtractive process. In other use cases, the ion beam of the FIB can add material to the sample or otherwise modify the sample by other processes. Increasingly, there are applications in which the ion beam is used to provide high-resolution images of the sample.

This article covers only those ion beam instruments capable of achieving probe sizes smaller than approximately 50 nm, even if they sometimes are operated otherwise. This criterion for the term *focused* in the FIB is somewhat arbitrary because the capabilities have changed historically. Our present-day applications, however, demand precision, acuity, and fidelity that help to define our current usage of the term *focused*.

The population of FIB instrumentation has grown steadily, and it now represents a substantial fraction of the wider "charged particle beam" family that includes scanning electron microscopes (SEMs) and transmission electron microscopes (TEMs). The FIB instrument is an increasingly essential tool in the academic research setting, the analytical electron microscopy laboratory, a failure analysis workspace, and in evaluating process control in a manufacturing environment. Within the FIB family the different varieties of instrument and different beam species serve very diverse purposes, some being better suited for specific imaging, analysis, or nanofabrication tasks. A researcher who is equipped with a broad understanding of these instruments has the best chance of making the best use of their FIB instrument and arriving at the best possible results.

This article is intended to provide the reader with a good understanding of the underlying science, technology, and the most common applications of FIB instruments. Organizationally, it begins with a survey of the various types of FIB instruments and their configurations. It discusses the essential components and explains their function only to the extent that it helps the operator obtain the desired results. Specifically, the article introduces the ion optical column and explains how its components shape and steer the ion beam to the desired target locations. Also provided is a thorough review of the available ion sources, because this component, more than anything else, dictates the overall instrument performance. The article then reviews the many diverse accessories and options that can be included with the instrument and that enable it to realize its full potential across all of the varied applications. Following this, a detailed analysis is provided of the physical processes associated with the ion beam interacting with the sample. An understanding of these processes is vital for attaining the desired effects and minimizing the artifacts associated with any FIB application. Finally, a complete survey of the most prominent FIB applications is presented. For each application, the article discusses the application goals and explains the methods by which the FIB instrument is used to achieve these goals; example results are provided. Limitations and challenges associated with each application are also presented.

FIB Instrument Configurations

The simplest FIB instrument is the stand-alone system composed of a single FIB column mounted to a vacuum vessel equipped with the necessary supporting hardware to make it work. Such instruments can be used for ion beam lithography, sample modification, or a variety of other tasks, which are discussed later. However, the more common configuration today (2019) is the FIB-SEM (also known as dual beam or cross beam). These instruments are a combination of an FIB and an SEM in which the ion beam and the electron beam are nominally coincident, that is, aimed at the same location, which is nominally on the sample surface. These instruments commonly are used for precision tasks in which the electron beam is essential for imaging with negligible damage, while the ion beam can remove material from the tens of nanometers to hundreds of micrometers scale. The FIB-SEM instruments are versatile combination tools, together providing a means of seeing and manipulating matter at the smallest length scales on a single platform. Both of these configurations (Fig. 1) are commercially available from several scientific instrument manufacturers. There also are very specialized configurations for narrow applications, such as neural mapping in biology (Ref 1), and within the semiconductor industry (Ref 2), such as metrology, mask repair, and circuit edit. Some instruments have the FIB at a 90° angle relative to the SEM. In some cases, the FIB is vertical and the SEM is inclined. There even are some configurations in which the FIB is inverted, aiming up, with the sample above it.

Essential Components of an FIB Instrument

Any of these FIB instruments will come equipped with the standard set of requisite components to make the instrument fully functional. This section provides a detailed look at essential components, such as the vacuum system, sample stage, airlock, and vibration isolation, before delving into the FIB column and the varieties of FIB ion sources.

Vacuum System

The vacuum system provides the vacuum levels that are necessary for the ion source to function and for the ion beam to propagate to the sample without excessive scattering. Typical vacuum levels for the ion source part of the ion column vary depending on source type, with ultrahigh vacuum required for gas field ion sources, high vacuum for liquid metal ion sources, and medium vacuum for plasma sources. It is typical for the sample chamber to be 1×10^6 torr or better (lower pressure), but some applications will have more demanding vacuum levels for the benefit of sample cleanliness. A typical FIB system will have integrated turbo pumps, roughing pumps, gages, and valves. All of these are controlled by the necessary electronics enclosed in the plinth or a nearby rack and with semiautomatic control via firmware, software, and a user interface. It is not uncommon to have a camera and lighting integrated into the vacuum chamber to help the operator to visualize the sample motions and avoid collisions between the sample and the FIB column. Systems often will have provisions for periodically cleaning the chamber, sample stage, and sample holders to maintain good vacuum levels. This could include an internally integrated plasma cleaner, internal baking lights, or external chamber heaters. Most systems will have an airlock for sample exchange. The benefit of the airlock is that the sample can be introduced or removed from the sample chamber without the time required to vent and re-evacuate the

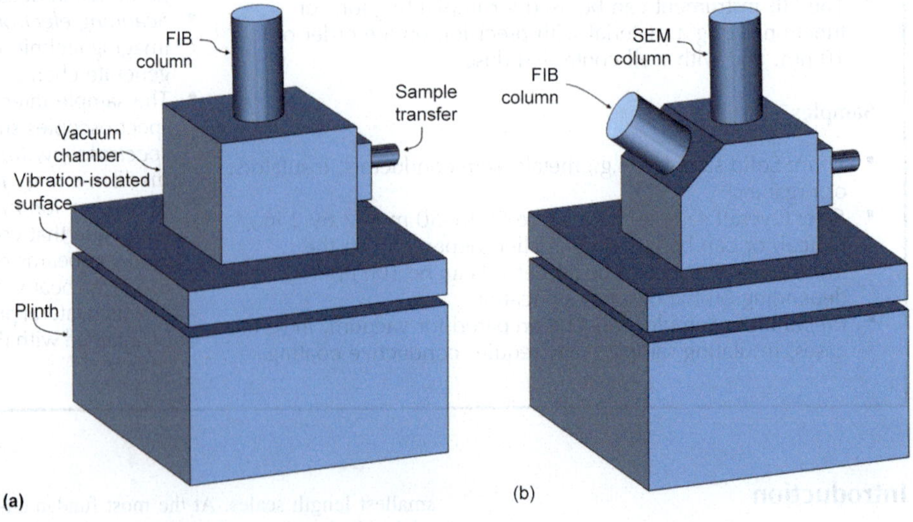

Fig. 1 (a) Stand-alone or single-column focused ion beam (FIB) instrument. (b) The more common configuration, with both FIB and scanning electron microscope (SEM) on a single platform

sample chamber every time, although this can impose sample size limits. Not shown in Fig. 1 are the additional vacuum ports, or flanges, onto which one can attach important accessories that increase the scope and capabilities of the FIB microscope. These can be placed at a variety of positions relative to the FIB or FIB-SEM and are discussed later.

Sample Stage

The sample is attached to a multiaxis stage for controlling the position of the sample relative to the beam. In most cases, the stage consists of a stack of axes such as *X*, *Y*, and *Z* for translation and *T* and *R* for tilt and rotation. The stacking order will vary from one model of instrument to another, as will the range of motions available. An example of a particular stage concept is show in Fig. 2. Often the stage is described as having eucentric tilt if the projected tilt axis would intersect with the beam of the FIB. A sample that is positioned at this height through *Z* adjustment is said to be at eucentric height. Under these circumstances, the feature of interest will appear relatively motionless as the tilt was changed. In a typical FIB-SEM system, the SEM also is aligned to have the same eucentric point in space, making both ion and electron beams coincident at the eucentric point. In modern systems, the stage motion is controlled with high precision and with accurate readbacks through a convenient user interface. Stages often use a combination of mechanical gears and piezoelectric motors to accurately position features of interest under the beam. There are some specialized FIB systems for lithography that include laser encoders to allow for ~1 nm positioning accuracy. The sample size usually dictates the range of motion of the stage. Some FIB systems allow for 300 mm (12 in.) semiconductor wafers (and thus require a vacuum chamber approximately twice this size). Some systems will have a touch alarm that abruptly stops the stage motion if it detects the first sign of contact between a sample and another component of the system. Most stages have the uppermost level electrically insulated to allow for a sample bias to be applied. Systems with an airlock have a detachable sample holder with a repeatable mounting scheme that can be engaged via a remote actuator.

Ion Optical Column

The task of shaping and steering the ion beam is the responsibility of the ion optical column. Despite the various commercial providers, ion sources, and applications, all ion optical columns are remarkably similar. A basic description of a typical column and the components that comprise it are described herein. It must be recognized that different systems will be configured somewhat differently and will be operated differently. Distinctly different from the SEM, the FIB favors the use of electric fields over magnetic fields for purposes of beam shaping and control. This arises from the fact that ions, with their larger mass, are less easily controlled by modest-strength magnetic fields that impart a significant force on the electrons.

A generic ion optical column is shown in cross section in Fig. 3 to guide the subsequent discussion. Only the essential components are shown, and these are simplified for pedagogical purposes. The full length of a typical FIB column often is approximately half a meter (1.5 ft), and the ion beam diameter (shown in yellow) may be ~1 mm (0.04 in.) at its widest. Starting at the top, there usually is some provision to mechanically align the ion gun relative to the column so that the initial beam is directed along the optical axis. The ion beam then enters an electrostatic condenser lens (Fig. 4) that serves to limit the otherwise divergent beam. Most commonly, this first lens simply limits the divergence, but in some configurations, the lens can cause the beam to become fully collimated or even to converge toward and pass through a crossover point so it diverges again. An aperture (Fig. 5) serves to select just the core or central portion of the emitted ion beam to limit its diameter and, correspondingly, the current available in the remaining beam. For example, there may be

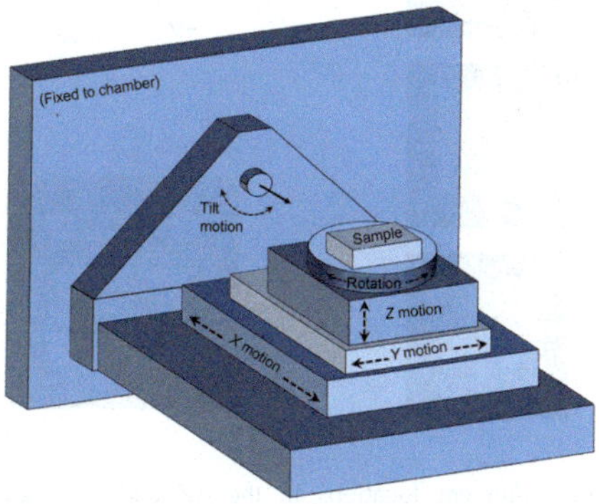

Fig. 2 Typical sample stage with motion axes labeled

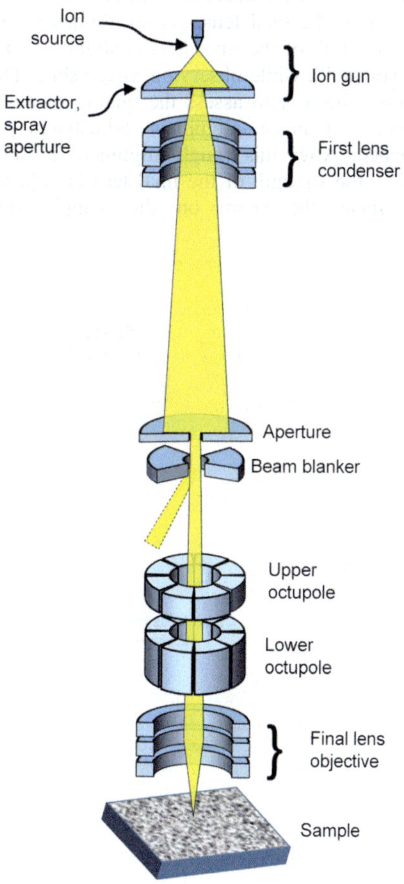

Fig. 3 Basic components of a typical ion optical column

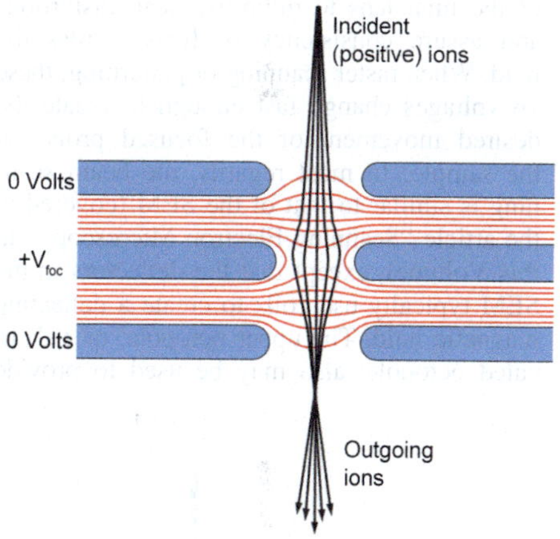

Fig. 4 Schematic of an electrostatic lens, with electrodes in blue and equipotential lines in red. This example shows a decelerating Einzel lens, since the ion energy is the same before and after and because the energy within the lens is reduced. Courtesy of A. Lombardi

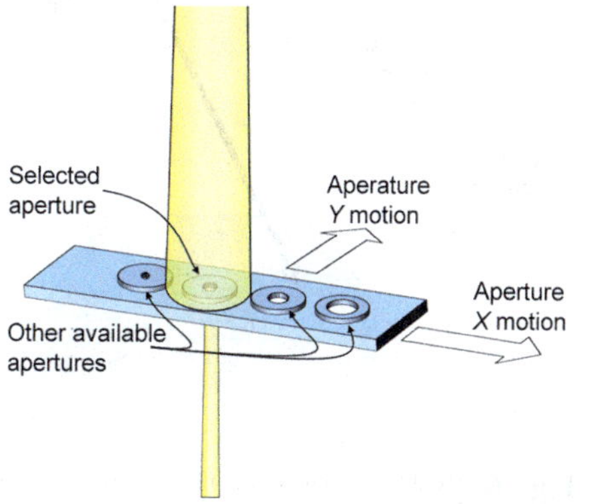

Fig. 5 Aperture blade typical of what is integrated in most focused ion beam columns

200 nA of beam current above the aperture but only 10 pA below the aperture in a conventional gallium FIB. One can change the final beam current by selecting a different aperture. Most systems will come equipped with a beam blanker that provides a method of occasionally deflecting the beam off axis by means of a strong electric field. This provides the ability to "turn off" the ion beam so it does not reach the sample. Time to blank and time to unblank are critical characteristics of a beam blanker. The blanked beam often is directed to a beam dump or Faraday cup with an attached picoammeter so the probe current can be measured whenever the beam is blanked.

Most FIB columns will be equipped with a double-deflection scanning system comprised of two octopoles. Appropriate voltages applied to these 16 combined plates or poles push the beam one direction and then the opposite direction to cause the beam to land at the desired location on the sample while passing through a virtual pivot point in the final lens (Fig. 6). This pivot point must be in the center of the final lens to minimize scan distortions and assure consistency of focus across the field. When raster scanning or patterning, these 16 voltages change fast enough to create the desired movement of the focused probe on the sample. In most regards, the beam scanning is similar to that of the SEM (covered in the article "Scanning Electron Microscopy" in this Volume), except that the deflectors in the SEM typically use coils to create a deflecting magnetic field. The upper octopole, or a dedicated octopole, also may be used to provide stigmation correction (Fig. 7) to assure that the ion beam converges in a symmetric fashion to its final focused probe. The final lens or objective lens is an adjustable-strength electrostatic lens similar to the condenser lens shown in Fig. 3 and allows the operator to adjust the height of the focused probe to be at the surface of the sample. Not shown in Fig. 3 is a conventional Everhart-Thornley detector, which is positioned outside the column just above the sample. More advanced detector options are discussed in later sections.

Column Alignment

All of the earlier-mentioned ion optical components in Fig. 3 will perform optimally when the beam is directed so it passes through the central axis of each component. Starting at the top, the ion source often is mechanically aligned so it is aimed through the center of the first condenser lens. Depending on the FIB instrument, the beam may be electrostatically deflected to aim the beam toward the condenser lens. For ion sources with larger emission angles, the aperture can be used to effectively select the portion of the beam that will reach the center of the final lens. Thus, fine motion of the aperture causes the transmitted portion of the beam to be aimed toward the center of the final lens. As an alignment aid, the strength of the final lens is slightly varied ("wobbled") while observing image shift. This allows the user to assess the quality of alignment and make alignment adjustments as needed. After this rough alignment is complete, the strength of the final lens is adjusted to focus the beam on the sample. The stigmation then can be optimized to correct for any asymmetries in the column. Lastly, the final lens strength is again optimized.

While this description is typical, there are many variations on different systems (Ref 3). For example, some systems include additional spray apertures to control beam diameter, and additional lenses to allow the beam diameter to be controlled independently of the delivered probe current. Other ion columns have additional deflectors for alignment to direct the beam more precisely from one optical component to the next.

For FIB systems in which a range of ion species is expected (such as the liquid metal alloy ion source, discussed later), a Wien filter (also known as an "*E* cross *B*" filter or velocity filter) is incorporated into the column. The Wien filter applies both transverse electrostatic (*E*) and magnetic fields (*B*) to a limited portion of the beam path (Fig. 8). The resulting forces are perfectly balanced for ions traveling at the chosen velocity, v_z, allowing them to continue undeflected. Ions with lower mass (or higher mass) will travel faster (or slower) for the same fixed energy and will be deflected off the optical axis, to be stopped by the aperture plate. In effect, the Wien filter serves as a mass filter and allows only one component of the beam to pass directly through the column.

Basic Scanning Operation

With the components of the FIB column now introduced, the basic operations can be described. Raster imaging is a routine and fundamental operation of the FIB and essential for

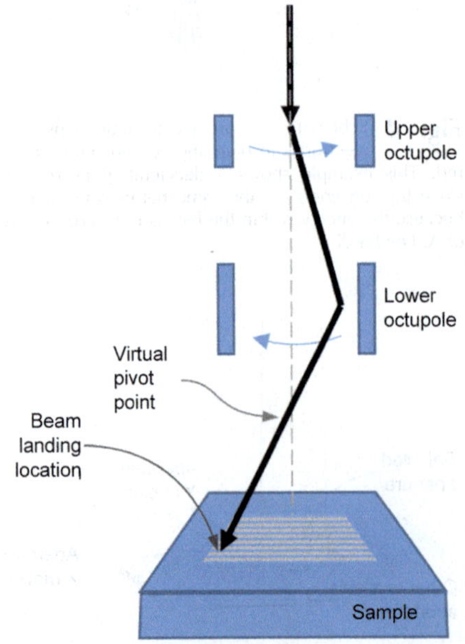

Fig. 6 Double-deflection scanning causes the ion beam to be pushed in one direction and then the opposite direction in order to pass through a virtual pivot point in the center of the final lens and land at the intended location on the sample.

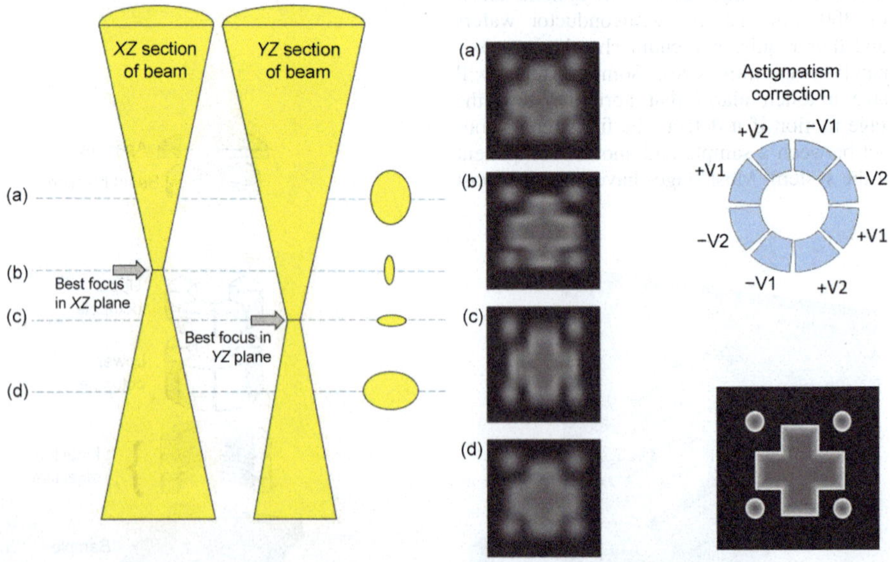

Fig. 7 Stigmatic beam showing that the best focus is at different locations in the *XZ* and *YZ* planes. Correspondingly, the probe shape changes as the focus is adjusted, producing images that are blurrier in one direction and sharper in another. Corresponding beam profiles and resulting images are labeled as (a), (b), (c), and (d). The octupole with appropriate voltages V1 and V2 allows for the corrections to yield the optimal image.

image formation. During raster imaging, the voltages on the scanning deflectors are precisely controlled over time to cause the beam to land at the desired locations on the sample, each of which corresponds to a pixel in the image. The beam is conventionally moved in a raster pattern, starting in the upper left corner, with the beam moving across each row before being advanced to the next row and repeated. During the retrace back to the left edge (i.e., the flyback), the beam is blanked to minimize sample damage. During raster scanning, the beam will remain resident at a single location (i.e., pixel) for a dwell time, T_{pixel}, which can be adjusted by the operator from hundreds of nanoseconds to milliseconds. The size of the rastered region of the sample is the field of view and can be controlled by the operator to be under a micrometer to over a millimeter. The number of pixels in the rastered region may consist of, for example, 1024 locations per row and 800 rows in the full rectangular region, which would produce a 1024 × 800 pixel image. For conventional imaging, the gray level of each pixel is assigned an 8 or 16 bit value that is proportional to the number of secondary electrons produced while the beam is at the corresponding location. Usually, image acquisition can be configured to involve averaging of successive line scans or successive frames to reduce the overall noise content in the image.

While raster scanning consists of rectangular regions with consistent pixel dwells per site, patterning offers much more flexibility. *Patterning* is the term used when the trajectory of the beam on the sample is scripted as a sequence of locations on the sample, each with a prescribed dwell time and provisions for blanking the beam as needed. Patterning can include circular shapes, spaced lines, L-shapes, and more elaborate shapes. Such patterning capabilities are essential for nanofabrication and many other applications discussed later in this article. Patterning strategies distinguish different ways to achieve the same beam dose on the sample. For example, it is possible to produce the same dosage in a rectangular region with a traditional raster scan, as described earlier, or with a spiral scan that proceeds clockwise (for example) from the outer perimeter and moving toward the center. For some applications, these beam strategies can be useful in mitigating some of the undesired artifacts (discussed later).

Ion Sources

The ion source, more than any other component, determines the characteristics of the ion beam and, correspondingly, the utility of the FIB instrument into which it is integrated. A great variety of ion sources exists (Ref 4), largely because any one technology can only offer a limited range of performance characteristics. The key characteristics include the available ion species (e.g., Ga^+, Ar^+, O^- or He^+), the available current (measured, for example, in pA, nA, or μA), the extraction energy (e.g., in keV), and the energy spread (e.g., in eV). Also important are the size of the ion emission region (e.g., in nm^2) and the solid angle of the emitted beam (e.g., in steradians), the product of which is called the etendue. A more frequently used metric for ion sources is the brightness (conventional units of amps cm^{-2} steradian^{-1}), which represents the emitted current divided by the etendue. Together, high brightness and low energy spread (i.e., monochromaticity) are the two most important factors in determining the ultimate focused probe size. Tondare (Ref 5) provides a review of this topic and identifies some of the most promising ion source technologies for further development. In the following sections, the most prevalent and most promising ion source technologies are introduced and their principles of operation are explained. Their strengths and limitations also are detailed.

Liquid Metal Ion Source

By far, the most prevalent ion source is the liquid metal ion source (LMIS) operated with gallium. In fact, unless otherwise stated, the term *FIB* without any other qualifier almost always refers to the LMIS running with gallium as the chosen liquid metal. This technology grew rapidly from research laboratories in the 1970s (Ref 6, 7) and established itself as a workhorse for specific applications, such as TEM lamella preparation (Ref 8–10), semiconductor circuit edit (Ref 11) and failure analysis (Ref 2), mask repair for lithography applications, and various other material-removal tasks. At present, >99% of the install base and >90% of annual sales are FIB instruments with the gallium LMIS. Gallium is the most popular choice for LMIS because of the economy of scale but also because of its low vapor pressure, low melting point, and high fraction of singly charged ions.

The typical LMIS (Ref 12) is constructed from a precisely shaped pointy needle (usually made from tungsten) that is wetted with a liquid metal such as gallium and is connected to a reservoir supply of the same metal (Fig. 9). The source requires some provisions for occasional heating but otherwise operates near room temperature. For nongallium LMIS sources, additional heating may be needed. The pointed end is electrically biased positive with respect to the nearby extractor electrode, so as to produce a strong electric field, in the presence of which the liquid metal is drawn into an equilibrium shape known as a Taylor cone. At the apex of the cone, where the

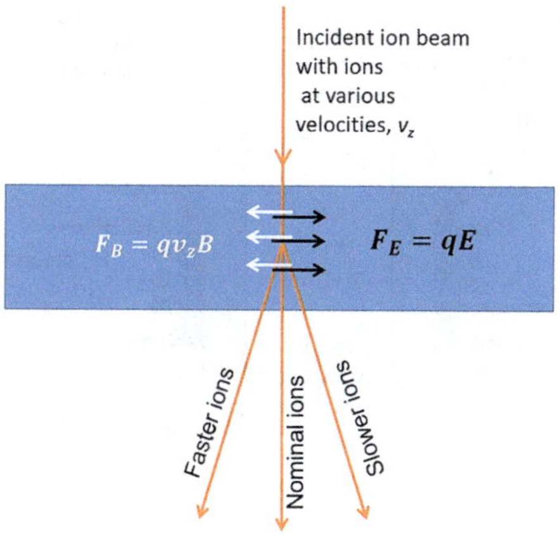

Fig. 8 Wien filter with electric and magnetic fields providing opposite forces to perform mass selection

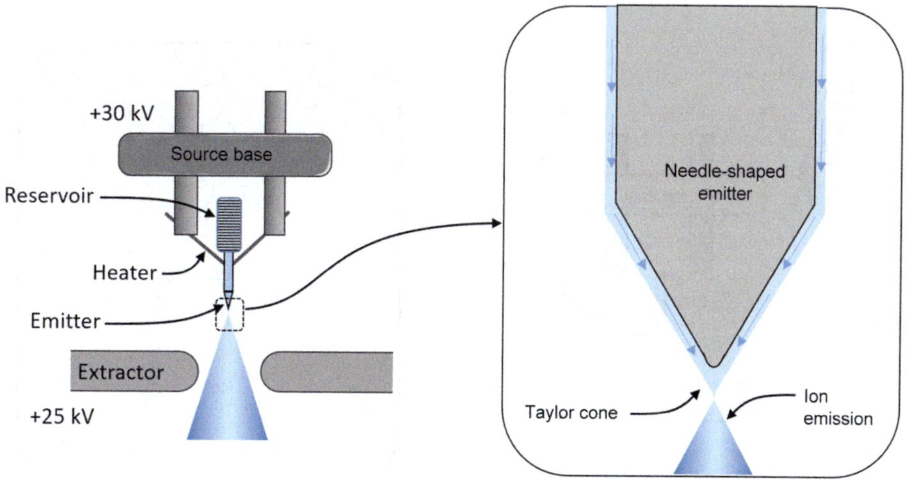

Fig. 9 Schematic of liquid metal ion source showing gallium forming the Taylor cone, from which the gallium ions are emitted

electric field is greatest, individual atoms can be ionized and pulled from the surface at a steady rate. Provided that the electric field is neither too strong nor too weak, this process can be continued, with the lost gallium being continually replenished from the reservoir until the latter is fully depleted, usually approximately 2000 h of operation time. Some level of routine and predictable maintenance calls for the gallium to be heated from time to time. Otherwise, the ion source can provide relatively high currents for thousands of hours before it needs replacement. Other species of LMIS, such as lithium, aluminum, tin, and mercury, have been demonstrated but are not routinely used or commercially established.

A simple variant of the LMIS is the liquid metal alloy ion source (LMAIS), which is comprehensively detailed in a review article (Ref 13). The LMAIS operates similar to the LMIS (Fig. 9), except the liquid metal is composed of two or more materials that form a low-melting-temperature alloy, usually at a eutectic composition. There is a small collection of such alloys that offer sufficient thermal and hydrodynamic characteristics to make them viable for LMAIS. Examples of commercially available alloy ion sources are Au-Ge-Si and Ga-Bi-Li. Because such an ion source will emit a mixture of ion species, the Wien filter described earlier is an essential component to select the desired ion species, although this necessarily reduces the available probe current. However, sources do not have to be monoelemental. There has also been some preliminary exploration of ionic liquid ion sources that use the LMIS concept to produce molecular fragments such as $(EMI\text{-}Im)_n\ EMI^+$ and $(EMI\text{-}Im)_n\ Im^-$ from a 1-ethyl-3-methylimidazolium bis(triflouromethylsulfonyl)amide (EMI-Im) source. Interestingly, ionic liquid sources can produce both positive and negative ions depending on the bias of the extractor, because they essentially are complex molten salts (Ref 14, 15).

Gas Field Ion Source

The gas field ion source (GFIS) is shown as a simplified schematic in Fig. 10. This ion source relies on the physics of field ionization, wherein a neutral gas atom can be ionized in the presence of a sufficiently high electric field. In practice, the high electric field is established adjacent to a few atoms at the apex of an atomically precise pyramid that is formed at the apex of a pointed needle. The needle is biased positively relative to a nearby extractor and must be maintained at cryogenic temperatures. The admission of neutral gases, such as helium or neon, into this region allows the ionization to happen at a very controllable rate, producing an emitted current that typically is less than 100 pA. One of the advantages of this technology is the very high brightness, because the beam is emitted from a single atomic site and is relatively well collimated (Ref 17). The high brightness makes it possible to form focused probes smaller than can be achieved with the gallium FIB. Probe sizes as small as 0.25 nm have been reported for helium, but only for very low currents (<0.2 pA). The available ion species are fairly limited, but helium and neon are routinely used, and there are demonstrations of the viability of hydrogen, N_2, as well as other noble gases (Ref 18).

Plasma Ion Source

The plasma ion source was one of the original ion sources (Ref 19) that did not enjoy early success, while the gallium FIBs grew in popularity. The interest in these sources has renewed in the last decade as the technology has matured to address needs not easily met with the gallium FIB (Ref 20). The present-day plasma FIB offers alternative beam species (e.g., xenon, O_2^+, O^-) and high currents (as high as several microamps). This continuous ionization of the supplied gas is achieved by the action of radio-frequency fields, either inductively coupled or capacitively coupled. The ions then are pulled from the plasma through an aperture by an applied electric field from an adjacent extraction electrode (Fig. 11). The large extracted currents make this a very compelling choice when large volumes of materials must be sputtered away at high rates. The ultimate focused probe size is limited by the brightness and the energy spread but has advantages over the duoplasmatron ion sources when it comes to well-focused oxygen beams. A few of the commercial providers of charged particle beam systems have now added plasma FIBs to their available menu of FIB instruments, making these available in the lab, with xenon as the most commercially available plasma source.

Cold Atom Ion Source

The most recently introduced ion source technology relies on lasers to produce high-brightness ion beams from the ionization of "cold atoms." This technology, shown schematically in Fig. 12, takes advantage of the laser-based magneto-optical trap concept to continuously collect atoms (Ref 21). Further lasers act to provide compression and then cooling of the atoms to temperatures as low as 30 μK. In one scheme, two intersecting laser beams are used for finally ionizing the atoms through a two-step excitation. In other schemes, the atoms are excited to Rydberg states and then field ionized (Ref 23). The resulting ion beam is characterized by both high brightness and low energy spread, making it suitable for focusing to a very small probe size. A broad variety of atomic species can be ionized in this manner

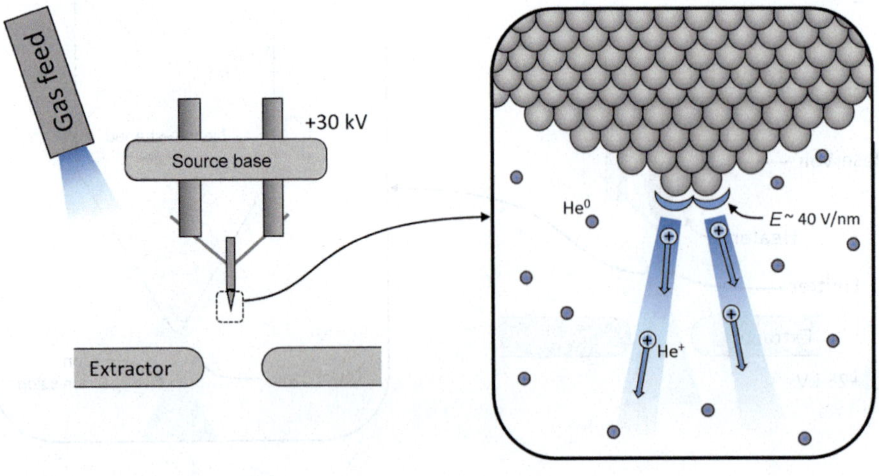

Fig. 10 Schematic diagram of gas field ion source. Adapted from Ref 16

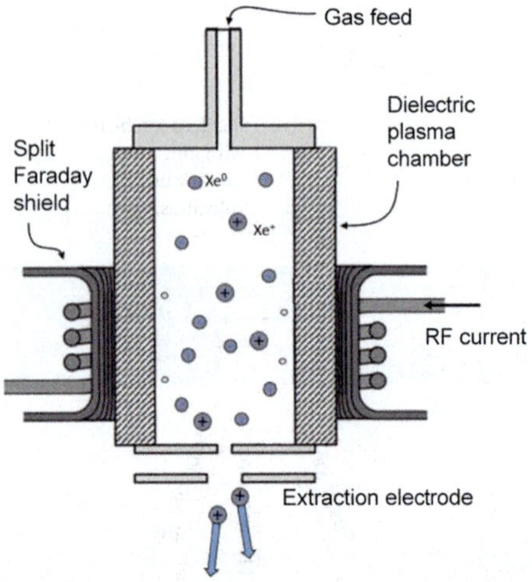

Fig. 11 Schematic diagram of plasma ion source. RF, radio frequency; MOT, magneto-optical trap. Adapted from Ref 16. Reprinted with permission from Cambridge University Press

by proper selection of lasers tuned to their atomic transitions.

Ion Source Summary

The ion sources described herein are uniquely suitable for typical FIB applications. Table 1 summarizes these ion source technologies and their characteristics. There also are ongoing efforts to develop new ion sources to expand the application space addressed by FIB instrumentation. Figure 13 provides convincing evidence that ions of many elements have been or can be made available by one of the earlier listed ion source technologies. Going beyond the elemental ion species, there are polyatomic ion sources under active development. While they do not offer extremely high brightness and hence the small probe size necessary for FIB, there is increasing usage of cluster ion sources (Ref 26, 27) such as C_{60}, $(H_2O)_{5000}$, Ar_{1500}, and $(CO_2)_{1831}$.

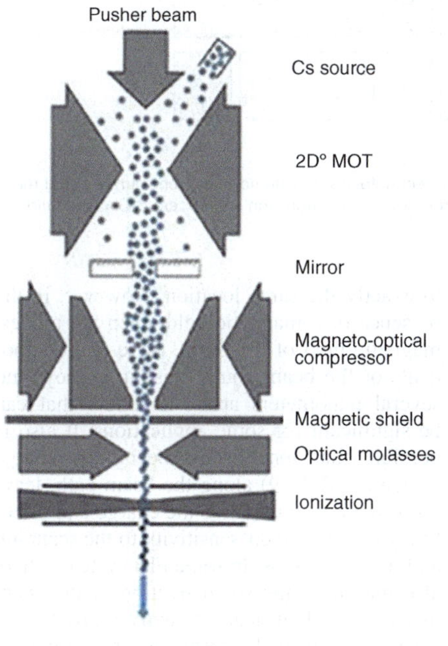

Fig. 12 Schematic diagram of laser-based cold ion source. Adapted from Ref 22

Probe Size

The ultimate size of the focused probe is one of the key metrics that determines the fidelity of a given ion beam for imaging or precision milling applications. As alluded to earlier, probe sizes are limited in part by the brightness of the ion source. Other factors that determine the probe size are the chromatic and spherical aberration of the final lens, and these factors tend to dominate for larger beam currents. Because ion beams are much more massive than electrons, diffraction is almost never a consideration in determining the focused probe size. The ultimate probe size can be as small as 3 nm for the commercially available gallium FIBs when operated at low probe currents and fully optimized. Some of the new FIBs based on cold atom sources and GFISs have reported probe sizes under 1 nm (Ref 21, 28). The earlier-mentioned four contributors to focused probe size are shown graphically in Fig. 14, in which they are plotted against the beam final convergence angle, α. Depending on some details of the column and ion source, larger α generally permits larger available probe current, but at the expense of a larger probe size. While a small probe size is important, under many circumstances the physics of the beam sample interaction size is the factor that limits precision and resolution of the instrument.

Reducing the working distance from the final lens to the sample is another consideration to help keep the focused probe size as small as possible. However, this comes with a trade-off, because it may restrict the sample size or may impact the collection of secondary electrons by the off-axis Everhart-Thornley detector.

In practice, the probe size is easily measured by imaging an atomically sharp edge, such as a fractured piece of crystalline silicon, and assessing the ion-induced secondary electron signal as it transitions from dark to bright (Fig. 15a). Such methods can reveal the current density distribution of the beam, $J(r)$, which is predominantly described by a Gaussian profile (Fig. 15b) characterized by a width parameter, σ:

$$J(r) = I\frac{1}{\pi\sigma^2}e^{-\left(\frac{r}{\sigma}\right)^2} \quad \text{(Eq 1)}$$

Here, I is the total probe current, and the parameter σ can be estimated from the transition width between the 25 and 75% levels, which, in this simple model, would be 0.954 σ. For such an ideal Gaussian, the beam diameter that contains 50% of the beam current, d_{50}, is 1.665 σ. However, a Gaussian profile is an ideal situation, while in reality there are other contributions that often are worth considering.

Beam Profile Nonidealities

There are several additional beam profile considerations for applications in which the beam profile is critical. The first effect is described as beam tails or a halo, that is, the portion of the beam that is not well described by a simple Gaussian. The existence of beam tails originates from nonideal characteristics of the ion source or column and therefore is different for different instruments. The following equation shows the current density consisting of a narrow central Gaussian and beam tails approximated by a secondary Gaussian and an exponential function:

$$J(r) = I_1\frac{1}{\pi\sigma_1^2}e^{-\left(\frac{r}{\sigma_1}\right)^2} + I_2\frac{1}{\pi\sigma_2^2}e^{-\left(\frac{r}{\sigma_2}\right)^2} + I_3\frac{1}{2\pi\sigma_3^2}e^{-\left(\frac{r}{\sigma_3}\right)} \quad \text{(Eq 2)}$$

In this hypothetical case, $I_1 = 5$ pA, $\sigma_1 = 1.5$ nm, $I_2 = 5$ pA, $\sigma_2 = 10$ nm, $I_3 = 5$ pA, and $\sigma_3 = 20$ nm. Figure 16 shows the corresponding beam profile in a semilog plot. The effect of these beam tails is merely to provide a hazy background for FIB imaging. However, for precision milling, lithography, or analysis, the beam tails can have a more significant effect. The beam tails can be controlled somewhat by employing defocus, but this often degrades the more centralized Gaussian at the same time. In some applications, such as TEM lamella preparation, the low current densities in the beam tails are relied on for a gentle final polishing.

The beam profile also can be distorted in a nonsymmetric "comma" shape, named for its elongated cometlike shape. This usually is associated with the beam traveling far off axis, such as when the alignment is not optimized or when scanning or panning the beam to larger angles. For imaging applications with a large field of view, the effect often is hidden, provided the distorted beam shape is not much larger than the pixel size. However, for nonimaging applications, especially patterning (called direct-write lithography), there often is evidence that the beam size is stretched and elongated, especially at the corners of large scan fields or when panning to regions away from the nominal optical axis.

Table 1 Various ion source technologies used in conventional and promising new focused ion beam instruments

Ion source technology	Prevalence	Beam current range	Energy spread, eV	Brightness, A m^{-2} Sr^{-1} V^{-1}	Species available
Liquid metal ion source (Ga)	95% de facto standard	1 pA to 100 nA	~5	1×10^6	Ga
Liquid metal alloy ion source	~2% flexible alternative	Varied	Varied	Varied	Si, Au, In, Li, Bi, Ge
Plasma	<5% new high-current option	<3 μA	~7–10	1×10^4	Xe, O_2, O
Gas field ion source	<2% precision	<10 pA; <1 pA typical	<1	1×10^9	He, Ne, H, N_2
Cold atom	Coming soon	0.5 pA to 1 nA	0.3	2×10^7	Li, Cs

Source: Ref 13, 16, 24

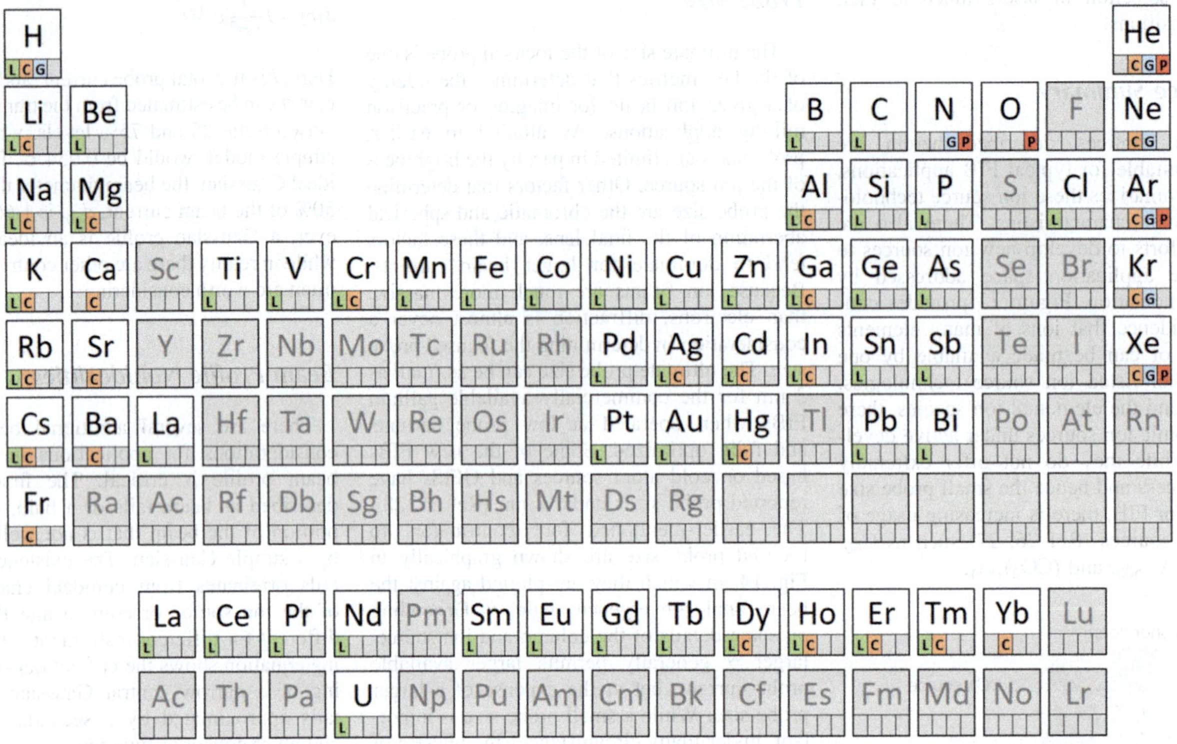

Fig. 13 Various species of elemental ion beams that have been or could be demonstrated using one of the described technologies. L = liquid metal ion source, liquid metal alloy ion source, or ionic liquid ion source; C = cold atom ion source (e.g., magneto-optical trap ion source, low-temperature ion source, etc.); G = gas field ion source; P = plasma ion source. Adapted from Ref 13, 21, 25

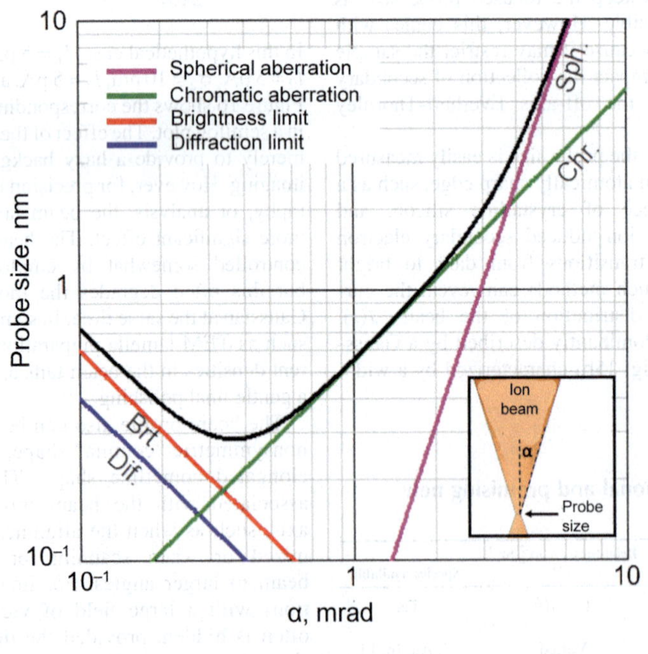

Fig. 14 Probe size versus convergence angle and the relative impact of the four contributing effects

A different artifact can be seen if the beam is found to contain populations of ions with a different mass or a different charge state. For example, a gallium LMIS may have the natural abundance of gallium isotopes (~60% mass ~69 amu, and ~40% mass ~71 amu), and under electrostatic fields both of these would follow exactly the same trajectories and be focused to exactly the same location. However, in the presence of a magnetic field, even the modest magnetic field of the Earth, these two components of the beam could be split so they land several nanometers apart, an effect that can be significant for some applications. It also is possible that ions in the beam can charge exchange (Ref 29) along the beam path, leaving some ions with twice the charge and effectively doubling their sensitivity to the scanning and focusing fields. In some FIB systems, there also may be some small fraction of the beam that is neutral (instead of singly ionized), and these are especially pernicious because they cannot be scanned, focused, or blanked. Fortunately, because they are unfocused, they represent a low number of ions per area (areal dose) and are seldom a problem if the column isolation valve is closed when the FIB system is idle.

Transit Time

Transit time of the ions through the column sometimes is an important point of consideration for high-speed applications. Due to the larger mass of ions compared to electrons, the transit time of the ion from source to sample is considerably longer than the journey of an electron in the conventional SEM. For example, it would take a 30 keV gallium ion approximately 347 ns to transit 100 mm (4 in.), which may be the distance from the blanker, through the deflectors, to the sample. High-speed systems include provisions that adjust the timing of the control signals to the various optical components to compensate for this effect. However, transit time effects cannot be completely corrected for and may be revealed during high-speed operation (Ref 30) as undesired patterning artifacts (e.g., for pixel dwell times of 500 ns or faster).

Beam Placement

Apart from probe size and shape, image quality and patterning fidelity also depend on beam placement accuracy. This is a measure of the landing location of the beam relative to the intended location. In principle, this can be less than 0.1 nm if it was determined only by the precision of the deflection electronics and of the pattern generator or scan generator that control them. Other factors to consider are vibration and drift, both factors that introduce relative motion between the source and the sample. Vibration and drift can arise from some nanometer-scale flexure in the mechanical connections between the source to the sample. The sample stage often has five degrees of freedom and therefore is vulnerable to resonant vibrations excited by ambient energy from acoustics, temperature changes, or floor vibrations. Hence, it is best practice to have these instruments installed on the ground floor or in basements in rooms that are designed for low acoustic energy and high thermal stability. High-performance systems sometimes can be influenced by alternating current magnetic fields from nearby equipment, but this usually can be corrected for with a commercial magnetic field cancellation system. The concept of beam placement accuracy inherently refers to a target location relative to another landmark feature on the sample.

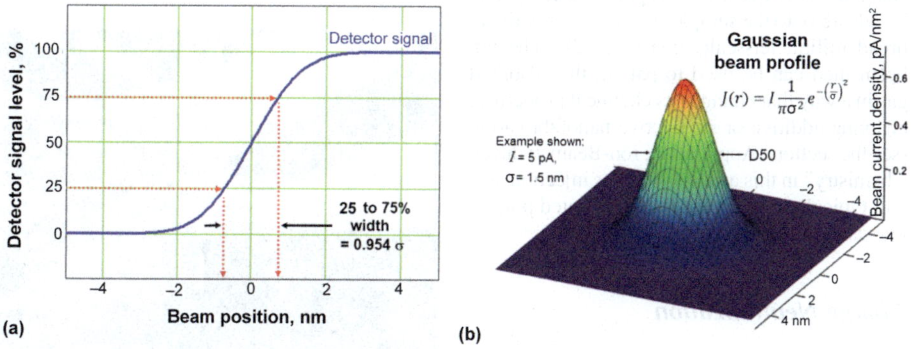

Fig. 15 (a) Edge profile that may be generated from scanning a beam over an atomically sharp edge. (b) Symmetric Gaussian beam profile that may be inferred from such an edge profile

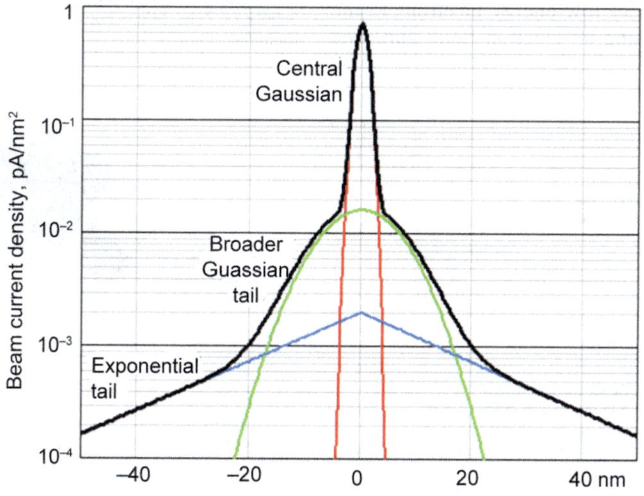

Fig. 16 Hypothetical beam profile (black) composed of the sum of a central Gaussian (red), a superimposed broader Gaussian tail (green), and an exponential tail (blue), shown on a logarithmic scale

Accessories and Options

This section provides a survey of the many specialized accessories and options that have been developed to extend the basic FIB system for higher performance or more specialized tasks. The same commercial vendors that provide the FIB instrument make their own limited set of accessories and options available. However, a larger pool of third-party vendors also makes available different accessories or develops custom components to help broaden the available applications addressed by the FIB. Because the applications are discussed in subsequent sections, the description herein is limited to brief overviews.

Special Sample Stages

There are several specialized sample stages beyond the standard option (e.g., Fig. 2) that can enable greater capabilities in the FIB instrument. These include stages with larger ranges of travel (e.g., 300 mm, or 12 in., in *X* and *Y*) or additional degrees of freedom. One option is called a rocking stage, which provides a method of mounting the sample on top of the original stage with the provisions for rocking the sample by several degrees on an axis that is perpendicular to the FIB column and perpendicular to the original tilt axis, as show in Fig. 17. Such a rocking stage provides a means of reducing curtaining artifacts during cross-sectional milling (see the section "Application: Exposing Cross Sections" in this article). Another stage accessory is the flip stage, which allows the sample to be tilted by a larger angle than the basic stage could otherwise achieve. This is used mostly for the purpose of preparing thin lamella with the FIB and then transmission imaging (scanning transmission electron microscopy) in the same FIB-SEM instrument. There also are specialized stages that offer long-range nanometer accuracy in *X*,*Y* positioning through the use of laser

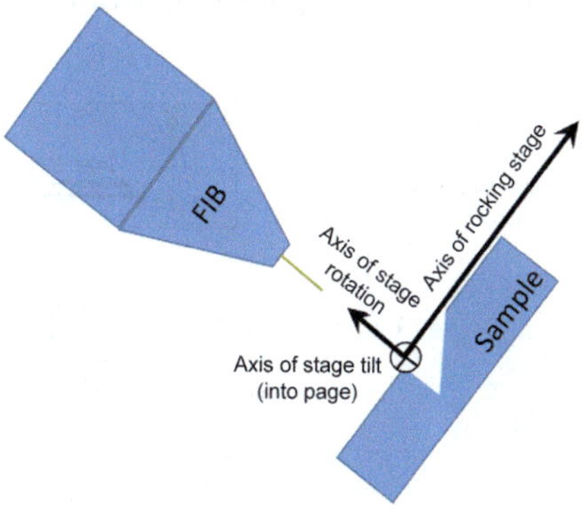

Fig. 17 Geometry of the rocking stage that often is mounted on top of the standard sample stage. FIB, focused ion beam

interferometers. Such precision is essential for lithographic applications when multiple write fields must be matched to create a larger-scale pattern (see the section "Application: Direct-Write Lithography with Resist" in this article).

In Situ Control

The various options for many forms of in situ FIB experiments are quickly gaining popularity. These include heating stages that allow the instrument to work on samples undergoing processes at elevated temperatures (as high as 1000 °C, or 1830 °F). This is especially helpful when the FIB is used to expose or prepare a sample that then can be thermally processed while the SEM observes the exposed region. Another option is an in situ mechanical actuator, so that a sample is deformed and observed during deformation. Focused ion beams may be used to mill fiducial patterns that help to monitor real-time strain in the sample during loading. There also are cooled stages that allow the sample to be maintained at cryogenic temperatures (see the section "Application: Cryogenic FIB" in this article). This is an invaluable tool for biologists who may need to use the FIB to process fully hydrated samples. For many cases, the cold stage also requires a special sample-transfer mechanism that maintains the sample in a cold state. Other in situ options allow the user to transfer in millimeter-sized environmental cells (Fig. 18) encapsulated by a thin membrane within which a gas or liquid sample can be contained. In these cases, the ion beam can pass through a small region with zero or minimal corruption of the chamber vacuum (Ref 31).

Manipulator and Contact Probes

A selection of in situ probes is increasingly available to achieve useful functions beyond the capabilities of the traditional sample stage. Such probes can lift out small subsamples that have been cut away and detached from the larger sample by using the FIB. Such probes can change the orientation of the subsample or can facilitate the transfer of the subsample to a different substrate. This technique is commonly used to prepare specimens for TEM or atom probe instruments and is described in later sections. Also, a set of such probes can offer electrical contact at precise locations on the sample, so that voltages and currents can be applied or measured when prototyping devices (Fig. 19; also see the section "Application: Semiconductors" in this article). The probe and the actuators are either mounted directly to an access port on the chamber or to some portion of the stage of the system.

Gas Injection System

This accessory delivers a gas-phase precursor chemistry through a small nozzle that is directed at the sample surface under the ion beam. The chemistry usually evolves from a liquid or solid supply reservoir that is heated to produce the correct vapor pressure of a precursor gas. When enabled, the delivered gas supply will temporarily adsorb onto the sample surface on a relatively broad millimeter-scale area (Fig. 20). The ion beam then can be used to pattern the adsorbed gas in a way that changes its chemical properties, enabling additive or subtractive nanofabrication (see the section "Application: Ion-Beam-Induced Chemistry" in this article). The gas injection system typically is mounted to a designated port on the vacuum chamber.

Charge Neutralization

The imaging, patterning, or analysis of insulating samples is facilitated through the introduction of a charge compensation system. Because the FIB typically will cause a positive charge imbalance at the sample surface, good success is sometimes obtained by introducing very low-energy electrons to the vicinity by using an electron flood gun. The low-energy flood electrons (and the secondary electrons they generate) are readily attracted to the positive surface charge until it is fully neutralized. Other charge neutralization schemes have used intense light levels to generate photoelectrons to achieve the same purpose. Conductive overcoatings, such as carbon or platinum, also can be introduced on the insulating sample, but these run the risk of obscuring the finer details of the sample. Flood gun conditions often must be adjusted manually to obtain optimal results.

Pattern Generators

A standard FIB will always come with some software suitable for controlling the scan electronics, thereby controlling where the beam hits the sample and synchronously coordinating the beam blanking and detector readback. This "native" patterning provides basic capabilities suitable for simple tasks. However, for advanced patterning, many FIB operators

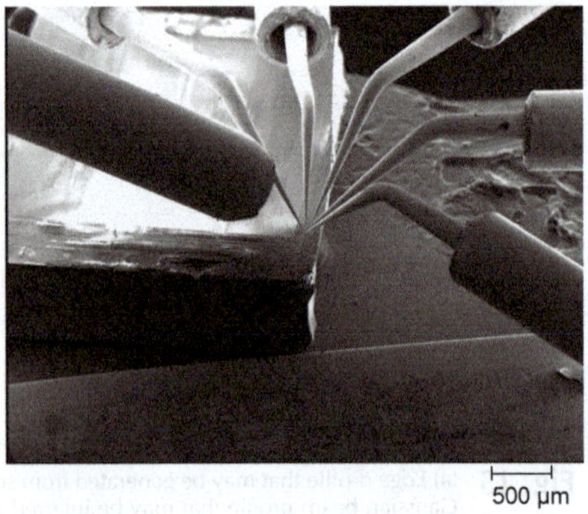

Fig. 19 Probes making electrical contact with a sample. Also note the gas injection system nozzle for gas delivery to the same area. Courtesy of H. Schulz of Zeiss Microscopy GmbH and A. Rummel of Kleindiek, GmbH

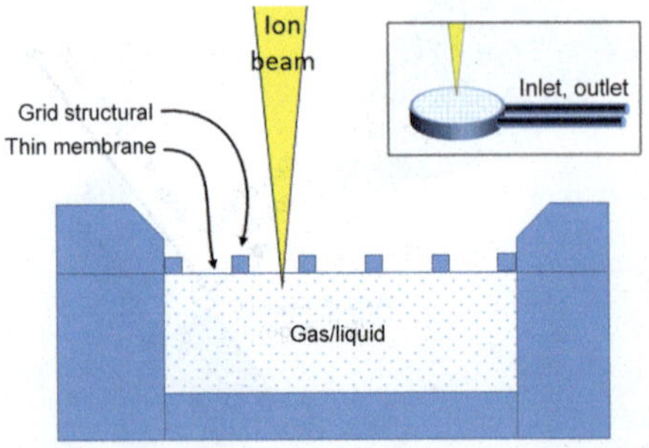

Fig. 18 Environmental cell allows focused ion beam experiments with liquids or high gas pressures

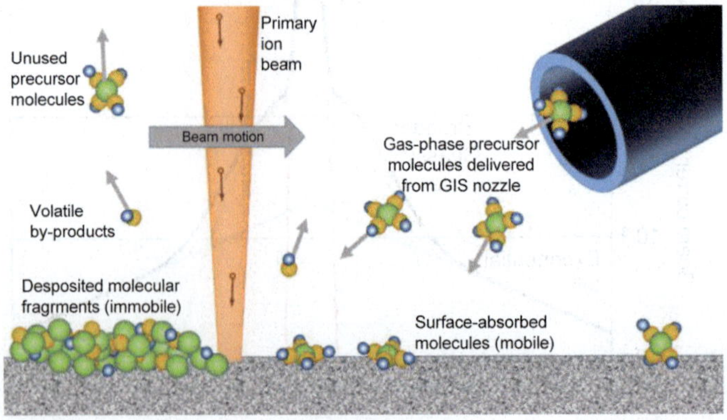

Fig. 20 Diagram of gas injection system (GIS) nozzle and the beam dissociation process. The molecules remain mobile while on the surface before either being dissociated and immobilized or desorbing from the surface.

prefer to use a third-party patterning system that may have greater capabilities, be easier to use, or perhaps is simply more familiar. The most advanced patterning systems will allow for complex shapes and provide many options for controlling the dose distribution and dosing strategy. Other patterning engines allow for navigational aids such as a graphic overlay or a "blueprint" to aid in finding subsurface features, which is especially valuable in semiconductor applications.

Sample Cleaning

Most samples introduced to the FIB or SEM will have a layer of hydrocarbons that is the result of air exposure, storage conditions, sample handling, imperfect vacuum systems, or contaminated sample stages. For heavier ion beam species, the contaminants are quickly sputtered away, but for light ion beam species or for the SEM, the layers of hydrocarbons will present a challenge. First, ion beams tend to be more efficient at cracking and immobilizing these hydrocarbons so as to grow a deposit (Ref 32). Second, because the FIB is such a surface-sensitive instrument, the effect of these hydrocarbon deposits is readily seen. Therefore, the imaging FIB demands high levels of sample cleanliness and, consequently, a routine practice of cleaning the sample, sample holder, and vacuum vessel. Many FIB and FIB-SEM systems are equipped with a commercial plasma cleaner system (Ref 33). The action of heat, ozone, oxygen radicals, and ultraviolet light can help to break bonds and desorb the surface-bound hydrocarbons.

Advanced Imaging Detectors

When an FIB is used for imaging, the gray level of each pixel is assigned a value in proportion to some detected signal. Most FIB instruments will come equipped with a basic Everhart-Thornley detector for the collection of secondary electrons. This is the conventional means by which images are generated in the FIB. Because FIBs are inherently damaging to the sample, it is important to gain the most information before an interesting feature is sputtered away. Therefore, some additional detectors have been developed to provide additional information, such as a novel contrast or an improved signal level. A microchannel plate detector has been used to generate images based on relative quantity of the primary ions that have backscattered from the sample (Ref 34). There are commercially available secondary electrons secondary ions detectors that collect either secondary electrons or positive sputtered ions, the latter revealing contrast related to chemical information. Some researchers have developed a detector (Ref 35) for transmitted ions that, in some systems, can be fitted under the sample. Such detectors can indicate when a beam first penetrates through a thin film or can reveal interesting contrast as ions pass through the channeling directions of crystalline materials. Also, there are third-party detectors designed to detect photons (Ref 36) that sometimes are generated from the sample atoms as the ion beam rasters over the sample. Finally, there are some applications in which the net current received by the sample is used for image generation.

Advanced Analytical Detectors

There also are accessories available that can ascertain further information about the composition of the sample, specifically, the mass of the individual atoms at a particular location. One such analytical detector scheme relies on measurements of the angle and energy of backscattered primary ions. The backscattered energy can be ascertained with either a solid-state detector (Ref 37) or a time-of-flight detector (Ref 38). More elaborate detectors rely on sputtered atoms that are ejected with a positive or negative charged state. These secondary ions can be mass analyzed by standard principles adopted from the well-established commercial secondary ion mass spectrometry (SIMS) instruments. Some commercial FIB instruments have optional SIMS detectors. One type extracts these secondary ions and passes them through a magnetic field, which causes the different masses to follow different trajectories and hit detectors positioned appropriately for each mass (Ref 39). Another type of third-party SIMS detector can ascertain the mass of the sputtered ions by first accelerating them to a known energy and then measuring the time it takes them to traverse a known distance from the sample to a high-speed detector (Ref 40). Examples of analytical options for the FIB are discussed in the section "Focused Ion Beam Applications" in this article.

Scanning Electron Microscope

Lastly, the most popular accessory for the FIB is its long-time partner and accomplice, the SEM. This combination is sometimes termed FIB-SEM, cross beam, or dual beam, and it is the most prevalent configuration from all of the equipment vendors. They are frequently paired because of their complementary capabilities: The FIB excels at removing materials to expose an underlying surface, and the SEM can provide the high-resolution images of the newly exposed surfaces. When configured with the SEM, there are many SEM accessories that become relevant (e.g., secondary electron and backscattered electron detectors, energy-dispersive spectroscopy, wavelength-dispersive spectroscopy, electron backscatter diffraction, etc.), most of which are detailed elsewhere in this Volume, including the article "Crystallographic Analysis by Electron Backscatter Diffraction in the Scanning Electron Microscope."

Beam-Sample Interactions

The great virtue of ion beams, and their major distinction from electron beams, is their unique and varied sample interactions. The beam-sample interaction is the basis for the many useful FIB applications, and an understanding of these interactions helps the user to anticipate any potential adverse effects. The physics of how an individual ion interacts with an individual atom is fairly well understood; however, experimental conditions (exactly where the ion will land relative to the location of the sample atoms) prevents full predictability of the fate of an individual ion. Nonetheless, the underlying physics can be modeled by using computer simulations. After averaging the outcomes of many thousands of simulations of incident ions, useful statistical results can be attained that closely match the experimental outcomes. Such Monte Carlo simulations (Ref 41) are invaluable for planning any experimental work with FIB instruments. The freely available Stopping and Range of Ions in Matter/Transport of Ions in Matter (SRIM/TRIM) (Ref 42, 43) package is most widely used and usually provides good predictions for the interaction of a given ion beam with a given sample. All of the simulation results presented throughout this section were generated with the SRIM/TRIM software. Some of the assumptions and limitations of the SRIM package include the following: The sample must be modeled as a stack of planar slabs; sample damage effects, such as sputtering or redeposition, do not accumulate; and the sample is treated as amorphous with no opportunity for channeling effects.

There are several software packages of varying sophistication, ease of use, and specialization. They are mentioned here in brief so the reader can investigate as needed: SDTrimSP (Ref 44), CrystalTRIM (Ref 45), SIMNRA (Ref 46), IoniSE and EnvizION (Ref 47), LAMMPS (Ref 48), TRIDYN (Ref 49), TRI3-DYN (Ref 50), and COMSOL (Ref 51). All of these simulation engines require some basic experimental details, such as the ion species, energy, charge state, and its angle relative to the sample local surface normal. Also necessary are details about the composition of the sample and its geometrical shape. With these parameters known, the Monte Carlo simulations can be performed to provide useful statistical predictions for the ion beam interaction with the sample. Any FIB operator is encouraged to run simulations in advance of choosing an ion beam and energy for a given sample, so they know what to expect.

Collision Cascade

The term *collision cascade* refers to the process of the ion passing into the sample and eventually dissipating all of its energy until it is effectively stopped, coming to thermal equilibrium with the sample. For simplicity, the collision cascade of one ion is considered in Fig. 21, with important features identified. The path of the ion is a series of nearly straight line segments, deflected where the ion scatters from a nucleus. Some of these collisions bring about secondary cascades, where the recoiled atom goes in a different direction before displacing several other atoms and coming to rest. Notice how one of these secondary cascades approaches the surface some distance away and causes a sputtered atom. In the subsequent figures, collision cascades produced from SRIM simulations show many overlapping collision cascades. Note that it is quite reasonable to assume the ions do arrive one at a time. For example, when using a 100 pA beam current, the average interval between ion arrivals is approximately 1.6 ns, and the collision cascade of an ion is comparatively brief, being resolved in just a few picoseconds or less. To understand the collision cascade and significant outcomes, the interactions of the incident ion with both the electrons comprising the sample and the nuclei comprising the sample are considered separately. The distinction is quite natural because the physics of scattering is characterized by the mass ratio.

Stopping Force

As the incident ion enters the sample, it interacts with both the electrons and the nuclei that comprise atoms of the sample. Through these encounters there is a general energy transfer from the ion to the particles it interacts with until full thermalization is achieved. The total stopping force, $(dE/dS)_t$, is usually expressed in eV/nm and is the sum of both the nuclear and electronic stopping forces:

$$\left(\frac{dE}{dS}\right)_t = \left(\frac{dE}{dS}\right)_n + \left(\frac{dE}{dS}\right)_e \qquad \text{(Eq 3)}$$

Here, dS represents the path length, which is not generally aligned to any axis. For a silicon sample, Fig. 22 shows the total stopping force versus energy for six commonly used ion species distributed in the periodic table. The total stopping force will vary for different samples, but the general trends hold: Stopping force is higher for more massive ions, and the stopping force diminishes as the energy is reduced. The total stopping force dictates one of the most fundamental measures of the interaction: the penetration depth of the ion.

Figure 23 shows the SRIM-calculated collision cascades for five different ion species normally incident on a silicon sample. For each case, the ion has the same incident energy, so the effect of ion mass and the consequential stopping force is clearly evident. Most apparent is the longitudinal depth and lateral spread of the collision cascade. The inset table displays this information as well as the sputter yield and the vacancies produced per ion. The heavier species are clearly better suited for applications such as milling, where sputtering is valued, whereas the lighter ion species are instead suited for imaging. The numeric values in Fig. 23 are based on statistics of 100,000 incident ions, but only 30 ion trajectories are drawn for clarity.

The effect of changing beam energy for a single ion species is seen in Fig. 24. As expected from the total stopping force, lower-energy ion beams will have a shallower and narrower penetration. The sputter yield will change as well, but this trend is not consistent for different ion species, as is evident later.

The effect of changing the sample for a given ion beam is shown in Fig. 25. The variations are significant here, and it is hard to draw upon any firm conclusion except that high-atomic-number samples and high-density samples generally produce shallower penetrations. The FIB user is strongly encouraged to perform their own SRIM simulations for their intended beam species, beam energy, and sample composition.

Implantation

The fact that incident ions will come to rest within the sample is sometimes regarded as an unavoidable nuisance of the FIB. For the gallium FIB, for example, the implanted gallium can produce optical staining (i.e., locally changing the index of refraction) or can alter the electrical properties. Light, inert ions such as helium can diffuse out of most noncrystalline materials at room temperature, but for more rigid lattices, implanted ions can introduce stress, dislocations, swelling, and subsurface bubbles (Ref 52). It should be recognized that implantation can be an asset: Historically, the FIB was used to introduce essential dopants to enable some of the earliest semiconductor devices. (Broad ion beam implantation still is the primary method for doping semiconductors.) Adjustable beam energy and species gives good control over the depth and width of the implantation zone.

Electronic Stopping Force

It is helpful to consider the electronic portion of the total stopping force, because this determines how an ion interacts with the electrons of the sample. Because an incident ion will be much more massive than the electrons it interacts with, the trajectory of the ion will be deflected only slightly by these interactions. In aggregate, the repeated scattering with the

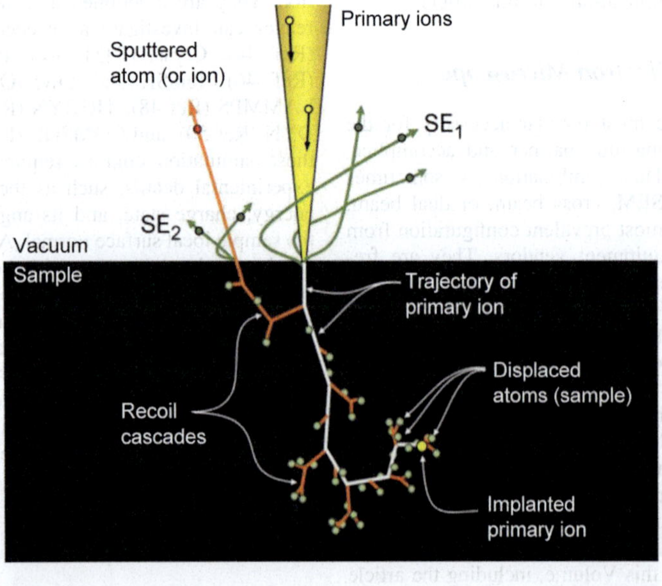

Fig. 21 A single representative ion trajectory as it interacts with the sample. Key interactions include recoils, displaced atoms, sputtered atoms, and secondary electrons (SE) type 1 and 2. Not shown is a backscattered primary ion.

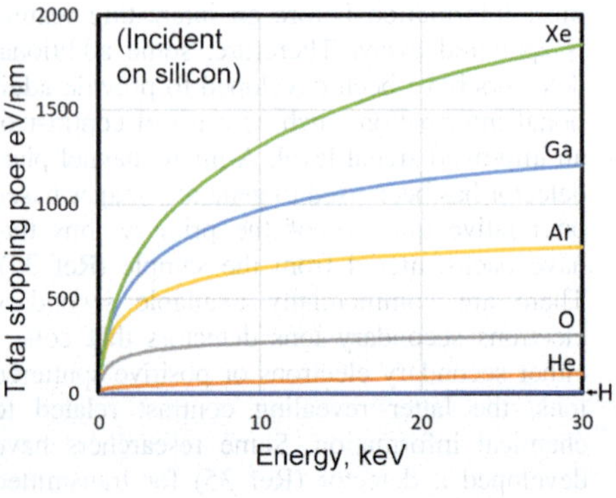

Fig. 22 Total stopping force of various ions into a silicon sample

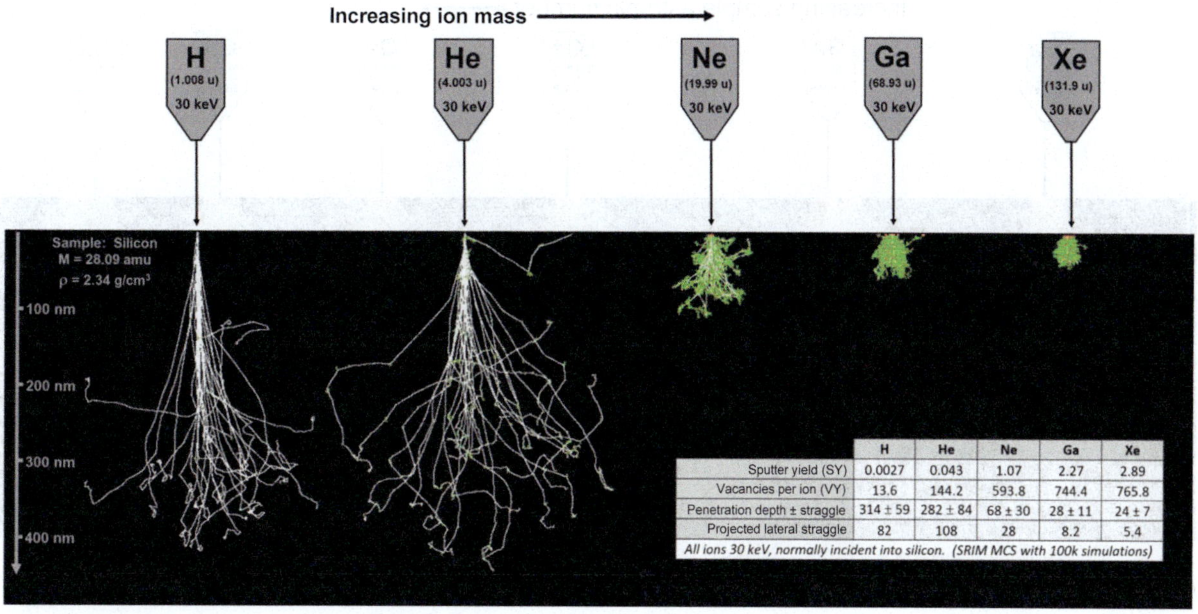

	H	He	Ne	Ga	Xe
Sputter yield (SY)	0.0027	0.043	1.07	2.27	2.89
Vacancies per ion (VY)	13.6	144.2	593.8	744.4	765.8
Penetration depth ± straggle	314 ± 59	282 ± 84	68 ± 30	28 ± 11	24 ± 7
Projected lateral straggle	82	108	28	8.2	5.4
All ions 30 keV, normally incident into silicon. (SRIM MCS with 100k simulations)					

Fig. 23 Collision cascades for five different ion species of the same energy incident on a silicon substrate. The choice of ion species plays a fundamental role in determining the sample interactions.

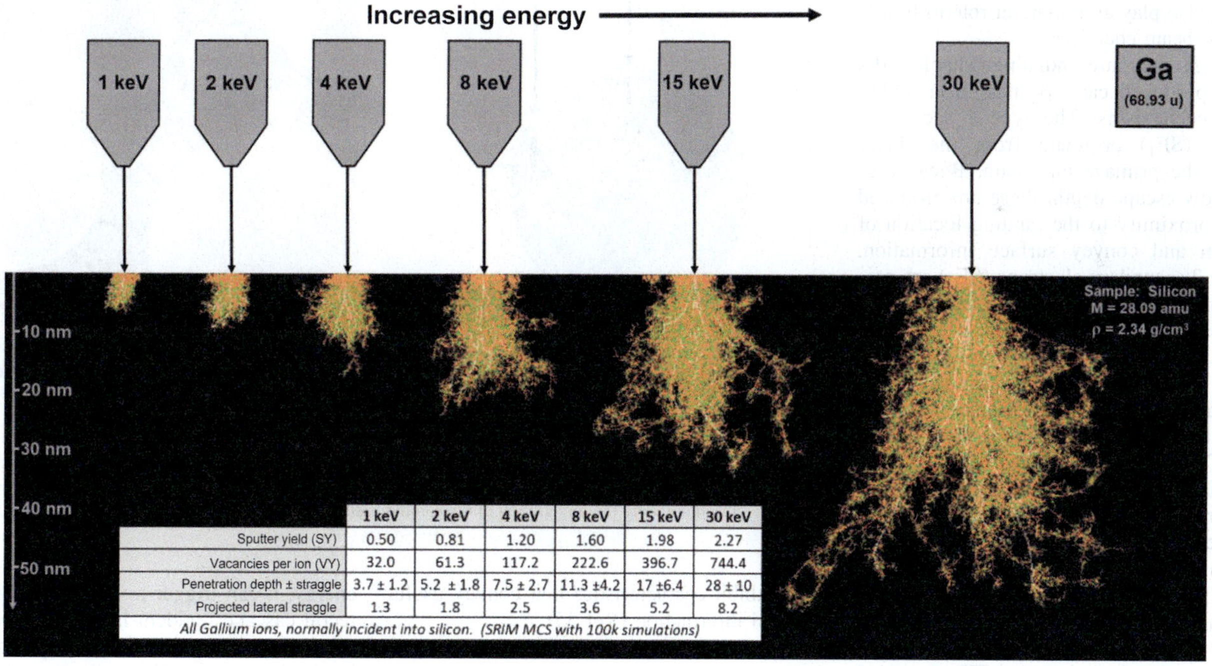

	1 keV	2 keV	4 keV	8 keV	15 keV	30 keV
Sputter yield (SY)	0.50	0.81	1.20	1.60	1.98	2.27
Vacancies per ion (VY)	32.0	61.3	117.2	222.6	396.7	744.4
Penetration depth ± straggle	3.7 ± 1.2	5.2 ± 1.8	7.5 ± 2.7	11.3 ±4.2	17 ±6.4	28 ± 10
Projected lateral straggle	1.3	1.8	2.5	3.6	5.2	8.2
All Gallium ions, normally incident into silicon. (SRIM MCS with 100k simulations)						

Fig. 24 Collision cascades for five different energies for the same incident ion species and same sample. The energy of the ion beam is also of great importance in determining the sample interactions.

low-mass electrons does, however, produce a loss of energy, $(dE/dS)_e$. This energy transfer is sometimes referred to as inelastic, but of course, the energy lost by the ion is gained by electrons in the sample. The energy gained by a single electron typically is several electron volts or less. When electrons are excited near the surface, they can escape into the vacuum and become secondary electrons (SEs) that can be collected and are the basis for generating SE images with ion beams. The characteristic escape depth, λ, for SEs varies for different samples but typically is only a few nanometers (Ref 53). For this reason, the SE images tend to provide surface-specific information. The average number of SEs produced per incident ion is known as the SE yield (Y_{SE}). It varies somewhat for different beam energies, different ion species, and different samples. However, there always is a dramatic increase of Y_{SE} for glancing angles, because the ion spends much of its trajectory near the surface where electrons can more easily escape. This gives rise to the classic

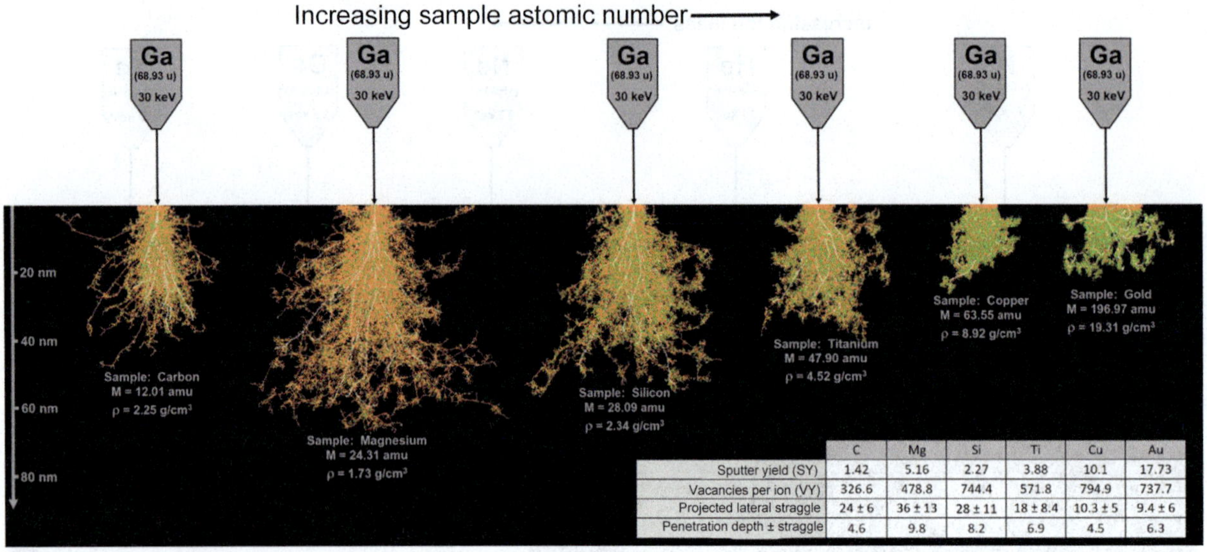

	C	Mg	Si	Ti	Cu	Au
Sputter yield (SY)	1.42	5.16	2.27	3.88	10.1	17.73
Vacancies per ion (VY)	326.6	478.8	744.4	571.8	794.9	737.7
Projected lateral straggle	24 ± 6	36 ± 13	28 ± 11	18 ± 8.4	10.3 ± 5	9.4 ± 6
Penetration depth ± straggle	4.6	9.8	8.2	6.9	4.5	6.3

Fig. 25 Collision cascades for six different samples with the same incident beam. The trends for different samples are less intuitive.

bright-edge effect, which provides the eye with valuable topographic cues that make the images easy to interpret (Fig. 26). The excited electrons also play an important role in lithography and beam chemistry.

Although they are indistinguishable, SEs that are produced can originate from either of two mechanisms. The type 1 secondary electrons (SE_1) originate from the direct effect of the primary ion beam. Because of the shallow escape depth, these are produced in local proximity to the landing location of the beam and convey surface information. The type 2 secondary electrons (SE_2) are produced through some follow-on mechanism in the collision cascade. This may occur when energetic particles (such as a backscattered incident ion or recoiled secondary ion) approach the surface some distance from the primary beam after scattering up from deeper below. Thus, the SE_2 will necessarily convey nonlocal information and subsurface information. For a given energy, lighter ion beams penetrate deeper and produce fewer recoil cascades near the surface and hence produce a higher proportion of SE_1 over SE_2, therefore producing high-fidelity, surface-specific images.

It also is worth noting that FIB-generated SEs have a lower energy distribution (Ref 54) compared to electron-beam-induced SEs. This arises from the simple physics of binary collisions, which conserve energy and momentum. The greatest energy transfer occurs in the case of equal masses (i.e., electrons scattering from other electrons), so SEM-induced SEs can have much higher energies. This has important consequences for voltage contrast effects, which are discussed in the applications section of this article. There even is some evidence that the energy spectra of the SEs in the FIB

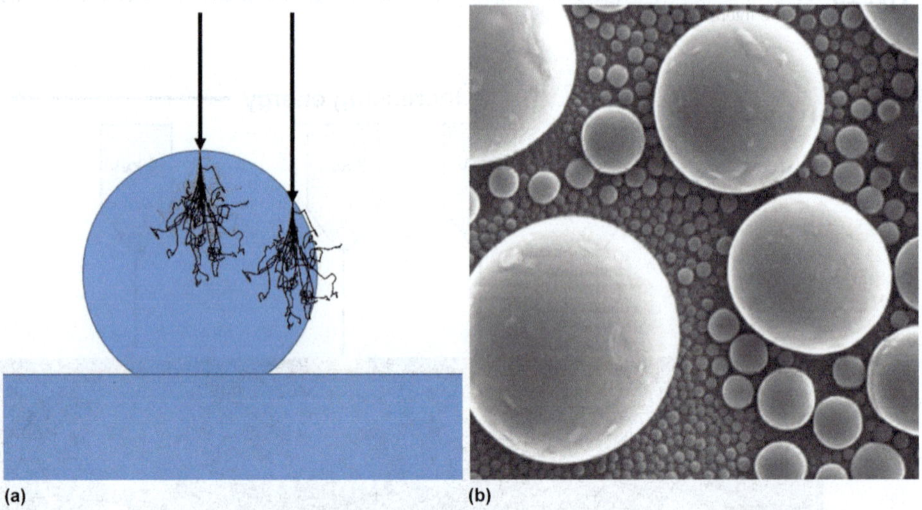

Fig. 26 (a) Secondary electron yield increases dramatically for glancing angles because an appreciable fraction of the collision cascade is near the surface. (b) The resulting focused ion beam image shows these edges as bright, making the topography of the sample easily recognized.

may exhibit characteristics that convey useful material information (Ref 55).

Nuclear Stopping Force

As with electronic interactions, the incident ion will lose energy at an average rate, $(dE/dS)_n$, through scattering with the nuclei of the sample. Because of the comparable masses, each collision is capable of causing substantial deflections of the incident ion and also substantial energy transfers. If the recoiling sample atom receives more than a few electron volts of energy, it can be displaced from its original site and leave a vacancy in its place. If the sample atoms receive considerably more energy, it can trigger a secondary collision or recoil cascade (evident in Fig. 21), where the recoiled atom continues colliding and producing further effects. This collection of cascades eventually comes to an end when the energetic particles have fully shared their energy and they approach thermal energies. The collected possible outcomes from these nuclear interactions include sputtering, backscattering, vacancies, and displacements. Depending on the application, any one of these can be the desired effect or an undesired effect. The experienced FIB user will know how to choose the ion beam species and energy to optimize the desired effect while minimizing any undesired effects.

With the fundamental FIB mechanisms ascribed to either nuclear or electronic interactions, it is worth considering the relative measure of these two phenomena. Figure 27 is a complement to Fig. 22, because it shows the ratio of electronic to nuclear stopping forces as a function of energy for the same six ion beam species incident on a silicon sample. For the heavier ions (e.g., argon and above), nuclear scattering clearly dominates, and consequently, the primary observed effects are sputtering, displacements, and so on. For the lighter ions (e.g., neon and below), electronic stopping forces plays a significant role. Figure 27 shows that, for some cases, the ion may enter the sample with a high energy, in which electronic effects are dominant, but as it penetrates deeper and loses energy, nuclear effects become dominant. Thus, there are circumstances in which an ion beam can produce very little damage at the surface but can still introduce subsurface damage.

Sputtering

Sputtering usually is a strongly desired feature of FIBs and is exploited in several popular applications described in the applications section of this article. *Sputter yield* is the term given to the number of sample atoms sputtered from the surface per incident ion. This is a dimensionless number that is largely influenced by the $(dE/dS)_n$ near the surface. The sputter mechanism usually involves a secondary cascade directed toward the surface. As seen in Fig. 28, the heavier ion species produce more sputtering at the typical energy range of 30 keV, but this observation is not consistently upheld at lower energies. Neon, for example, can sputter with a higher yield than gallium or xenon at lower energies. It also is generally true that lower-atomic-number (*Z*) sample materials tend to have a lower sputter yield. It also should be noted that generally the sputtered atoms are mostly in a neutral state, with a very small fraction charged either positive or negative. The ionization fraction can be augmented with deliberate additions of cesium or oxygen to the sample surface for critical applications such as secondary ion imaging or SIMS.

It should be recognized that the sputtered atoms will not be ejected from the surface at the precise location where the ion beam was incident. The sputtered atoms originate from the collision cascade, so its characteristic width at the surface gives some expectation for the distribution of the sputtered atoms. Note that this distinction was not significant until focused probe sizes were improved and became smaller than the collision cascade. The sputter distribution is important for applications such as precision milling and for applications that entail SIMS analysis with high lateral resolution. Figure 29 shows a representative lateral distribution of the sputtered atoms as they leave the surface for the case of an infinitesimal incident neon beam.

It also is worth noting that sputter yield can be greatly enhanced when the ion beam strikes the sample with a glancing angle, as shown in Fig. 30. This technique can be exploited in certain milling applications to remove materials at a faster rate. This effect arises simply because the collision cascade is closer to the surface, as shown in Fig. 26, making it more probable for energetic recoils to leave the surface. The effect is similar to the edge-enhanced SE yield.

Sputter Yield versus Sputter Rate

When considering material removal, the sputter rate (SR; e.g., units of μm^3/s) often is more relevant than the previously introduced sputter yield (SY; e.g., atoms sputtered per incident ion), but they are closely related:

$$SR = SY\left(\frac{I}{q}\right)\left(\frac{M}{\rho}\right) \quad \text{(Eq 4)}$$

Here, *M* is the average atomic weight, ρ is the density of the material, *I* is the current in the beam, and *q* is the charge of the ions (e.g., in Coulombs). As a computational example, consider a gallium beam milling tungsten at 30 keV and with a probe current of 100 pA. The sputter yield is easily calculated from the SRIM Monte Carlo simulator to be 7.6, and the atomic mass and density of tungsten are taken to be 183.84 u and 19.3 g/cm^3, respectively:

$$SR = 7.6\left(\frac{100\ \text{pA}}{1.6\times10^{-19}\text{C}}\right)\left(\frac{183.84\times1.66\times10^{-24}\text{g}}{19.3\frac{\text{g}}{\text{cm}^3}}\right) \quad \text{(Eq 5)}$$

$$SR = 0.075\ \mu\text{m}^3/\text{s}$$

So, for these materials and beam currents, 1 μm^3 could be removed in under 14 s, assuming redeposition is not a factor. Another practical way to use this result is to divide it by the exposed area to provide a milling depth rate (e.g., nm/s). For the same example, if the current is applied over an area of 4 by 20 μm, then

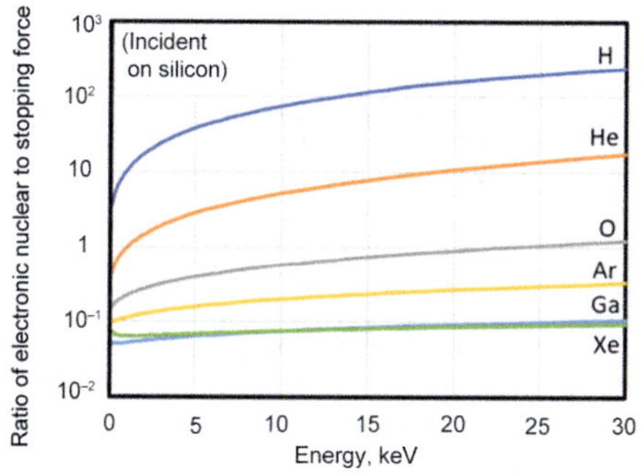

Fig. 27 Ratio of electronic stopping force to nuclear stopping force as predicted by the Stopping and Range of Ions in Matter simulation. The vertical scale is logarithmic to reveal the vastly different interaction regimes between the lighter ions and the heavier ions.

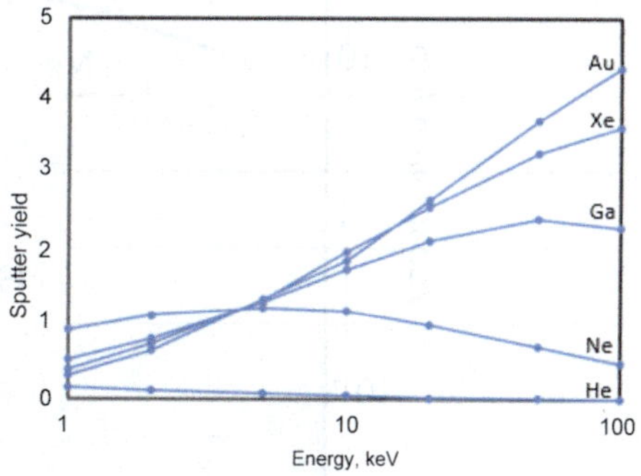

Fig. 28 Sputter yield versus energy for ion beams of Au, Xe, Ga, Ne, and He incident on silicon. Note the different energy trends for atoms heavier and lighter than the atoms of the sample.

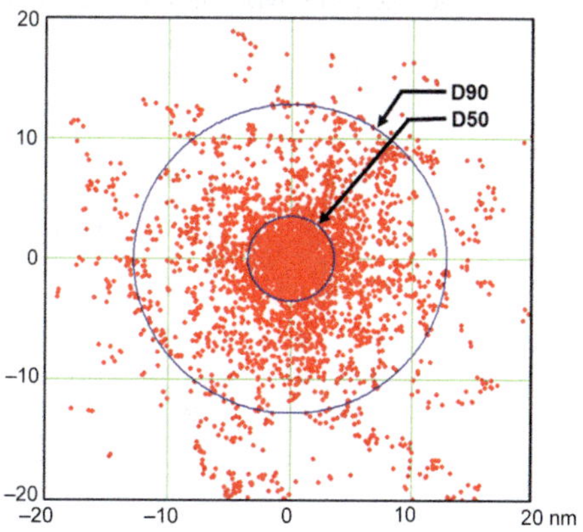

Fig. 29 Lateral distribution of sputtered silicon atoms leaving the surface from a 10 keV incident neon beam. D50 and D90 represent the diameters that contain 50 and 90% of the sputtered atoms. Shown here are just 5000 sputtered atoms, but the statistical measures are based on 130,000 sputtered atoms.

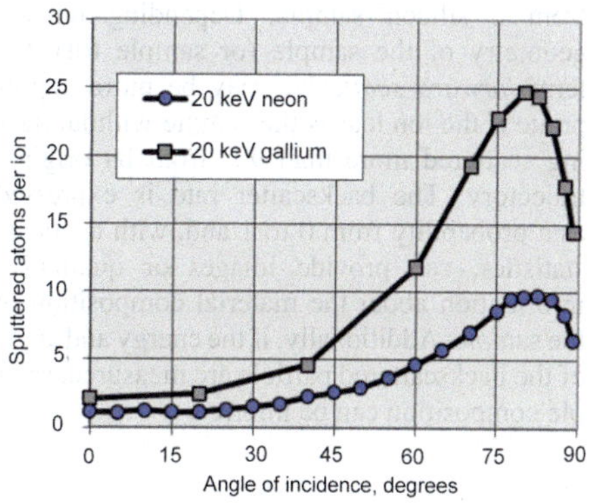

Fig. 30 Surface sputtering of silicon. Sputter rates increase for angles of approximately 80 to 85°, because the collision cascade is proximal to the surface.

the surface in this region will be sputtered away and recede at a rate of approximately dZ/dt = 0.94 nm/s.

Vacancies and Displaced Atoms

Along the collision cascade, there usually are many nuclear interactions in which sufficient energy is transferred to the sample atoms so that they are knocked from their original locations. The original location remains vacant, and the knocked atom eventually comes to rest at a new location as an interstitial. For a typical gallium beam in silicon, the number of displacements and vacancies can be several hundred per incident ion. The distribution versus depth is determined primarily by the mass of the ion and the energy, as evident in Fig. 31. Note that the more massive ions introduce the most vacancies, and these are closer to the surface. The lighter ions cause fewer total vacancies, and these are distributed deeper under the surface. The tables inset into Fig. 23 to 25 also show the total number of vacancies for the selected ion beams. For crystalline materials, a sufficient density of defects leads to partial or full amorphization of the original crystal lattice. For applications such as TEM or atom probe sample preparation, this effect must be minimized because the amorphous zone interferes with diffraction methods or mixes the local chemistry in the sample. The introduction of interstitials and vacancies is sometimes an adverse effect, but there are some applications discussed later that rely on these processes.

Backscattering

Backscattering occurs when the incident ion incurs one or more large-angle scatterings and re-emerges from the sample. Not surprisingly, this is most common for light ions incident on a sample with more massive atoms. For example, lithium ions on a tungsten sample could have a 20% backscatter yield, but gallium is much less likely to be backscattered from a silicon sample. Depending on the geometry of the sample (or sample tilt), the term *forward scattering* may be more appropriate if the ion leaves the sample without having scattered more than 90° from its original trajectory. The backscatter rate is expressed as a probability from 0 to 1 and, with adequate statistics, can provide images or qualitative information about the material composition of the sample. Additionally, if the energy and angle of the backscattered particle are measured, sample composition can be inferred.

Channeling

For many materials, the atoms are arranged into regions of repeating crystalline structure commonly known as grains. Such grains can be macroscopic or just several nanometers in length. The incident ion beam may be traveling along one of the low-index directions, such as hkl = (−1,0,0), as shown in the left half of Fig. 32. In this case, the foremost nucleus effectively shields a whole string of nuclei below it. Hence, the chance of nuclear scattering and $(dE/dS)_n$ is relatively low. The string of nuclei acts as an electrostatic conduit or channel in which the incident ion will be guided deeply with a low chance of high-angle scattering. In contrast, when a grain is not oriented along a low-index direction, as shown in the right half of Fig. 32, then each nucleus is a scattering opportunity, and ultimately, the incident ion will have a close encounter before it travels very deep. In this case, scattering is much more likely, as is the chance of lattice damage near the surface, so beam penetration will be much less deep. Channeling effects such as these are not considered by SRIM, but the FIB effect of channeling is often seen in many materials.

Ion Neutralization

For simplicity, throughout this article the particles in the primary beam are referred to as ions, even when they enter the sample. However, in reality the incident ion often is neutralized as it approaches and enters the sample surface. In effect, the electric field of the positive ion often is strong enough to "grab" an electron at the surface of the sample even before it enters the sample and continues its trajectory as a neutral atom, possibly being reionized and reneutralized again and again as it penetrates into the sample. The charge state of the backscattered and transmitted ions (atoms) has been measured and gives some confirmation of this charge-exchange process (Ref 56).

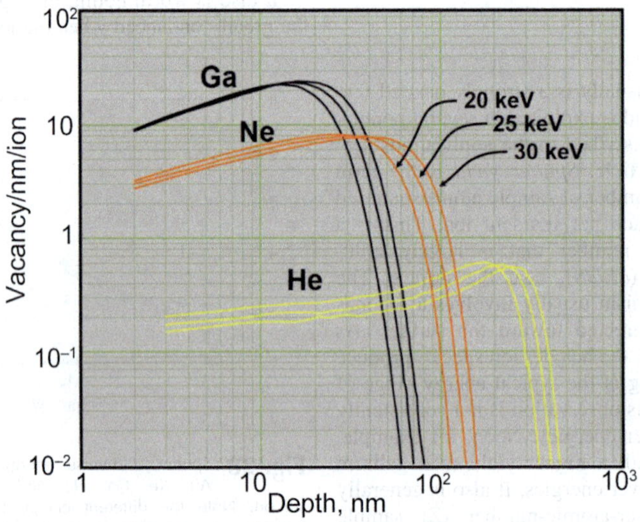

Fig. 31 Distribution of vacancies under the surface for representative ion beams. Three energies are shown for each species to convey the energy dependence.

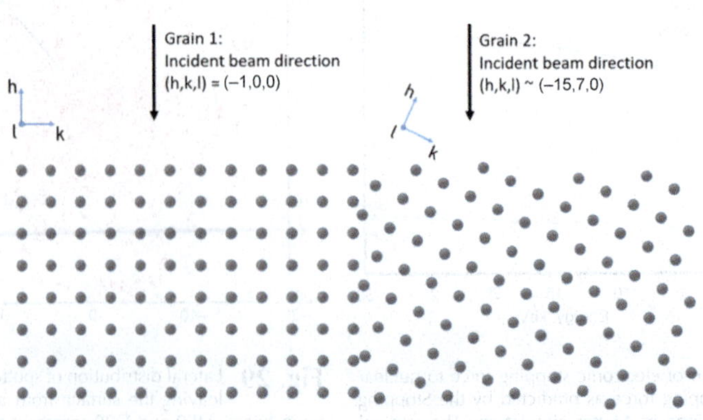

Fig. 32 Different crystal grains may present their atoms with good alignment to the incident beam (left) or with bad alignment (right).

Sample Charging

The process described earlier of an incident ion acquiring an electron from the surface leaves the surface positively charged. Similarly, the ejection of one or more secondary electrons from the surface leaves the sample surface positively charged, whereas the implanted ion (atom) generally results in very little subsurface charging, because it usually is neutral below the surface. Thus, FIB instruments tend to induce charging only at the surface, and this is necessarily positive, as shown on the left side of Fig. 33. If there are appreciable recoils, as is the case for heavier ions, then there can be additional surface charging from the SE emissions they induce. For samples that have some level of electrical conductivity, any charge is quickly dissipated. However, for insulating materials (e.g., polymers, ceramics, glass, etc.), the surface charge could accumulate. For these cases, the surface charge often can be neutralized with the benefit of a low-energy electron flood gun. In contrast, for the SEM (right side of Fig. 33), there is surface charging from the ejection of SEs that is inherently positive, and there will be subsurface charging that is inherently negative, producing a net charge that can be either positive or negative. The ejection of backscattered electrons produces additional positive surface charging through the ejection of SE_2. Under certain SEM conditions there may be no net charging (Ref 57, 58), but this is just to say that both positive and negative charging effects are present equally. Image distortion and sample charging still are possible artifacts when imaging insulators with electrons. For the SEM, an electron flood gun is not able to solve the charging imbalance arising from imaging insulating materials. Similarly, a variable-pressure SEM cannot address the subsurface charging from the SEM. See the article "Scanning Electron Microscopy" in this Volume for more details.

Ion Beam Dose

To many FIB users, experienced or otherwise, it is remarkable that ion beams can be used for such diverse and seemingly incompatible applications that may include lithography, imaging, milling, and patterning. Part of the flexibility of an FIB comes from the ability to control the areal dose. Areal dose, D_2, often is expressed in units of ions/cm^2 and can be varied over several orders of magnitude, for example, from 10^{10} to 10^{19} ions/cm^2. In different fields of research, alternative units for D_2 include nC/μm^2 or nC/nm^2. Regardless of the preferred units of measure, controlling D_2 is critical for a given FIB application and for limiting the undesired effects. Other common terms for D_2 include fluence or dose. Other related terms are D_1 or line dose, which is the amount of ions or charge delivered per length (e.g., ions/nm), and D_0 or point dose, which is the total number of ions or charge delivered to a point location (e.g., measured as a number of ions or in Coulombs). Note that both D_0 and D_1 can be converted to an areal dose by dividing by the probe area or probe diameter, respectively.

As an example calculation of D_2, consider the process of acquiring an FIB image. As a rule of thumb, a basic image requires approximately 25 ions delivered to each pixel to generate enough SEs to achieve an acceptable signal-to-noise ratio. If the chosen field of view (FOV) is 100 × 100 μm and the image size is 1024 × 1024 pixels, this would imply an areal dosage of D_2 = 25 ions × 1024 × 1024/(100 × 100 μm) = 2.6 × 10^{11} ions/cm^2. The total ion dose would be 2.6 × 10^7 ions or 4.2 picoCoulombs, which would take just 84 ms for a 50 pA ion beam. Note that increasing the magnification by a decade to a 10 μm FOV would increase the dosage, D_2, by 2 decades.

Dose Rate

In most circumstances, D_2 is of prime consideration for the FIB operator because most effects (both the preferred and the undesired) are proportional to it. However, some effects are sensitive to the rate at which the given dose is applied. As an example, consider rastering an ion beam over a 1 μm^2 region with a 0.1 pA beam for 16 s, or with a 100 pA beam for 0.016 s. Both achieve the same areal dose, D_2 = 1 × 10^{15} ions/cm^2, but the areal dose rates, D_2', (e.g., units of ions cm^{-2} s^{-1}) differ by 3 orders of magnitude. The situations in which dose rate plays a role are those in which there are simultaneously competing rate effects. This can happen, for example, with beam chemistry, charge dissipation, or where thermal effects are important.

Patterning and Milling Strategies

To take advantage of the higher sputter rate at glancing angles (Fig. 30), it is common practice to mill away materials with a chosen speed so that the beam is continuously advancing on an edge to give the higher sputter rate. When milling a trench by this technique, however, much of the sputtered materials may be redeposited, and the "redep" must ultimately be cleaned up by a repeated milling. The alternative strategy (Fig. 34) calls for many repeated fast scans, which minimize redeposition but at the expense of a lower sputter yield. Some of the more sophisticated patterning engines allow for a specified dose to be applied with different strategies to produce the best possible effects. Other fill strategies include serpentine scanning, double-serpentine scanning, and spiral scanning.

Thermal Effects

When a beam impinges on a sample, it delivers an average power that is the product of its beam voltage (e.g., 30 kV) and the beam current (e.g., 100 pA), which, for this example, amounts to 3 μW. The vast majority of this power, generally more than 95%, stays in the sample. A small fraction of the energy leaves the sample by way of ejected SEs, sputtered atoms, ion-induced luminescence, and possibly backscattering. The balance of the power is transferred to the sample via electronic excitations and lattice vibrations (i.e., phonons for crystalline materials). Following Melngailis (Ref 59) with certain reasonable assumptions, an analytic solution to the heat equation can be found. The incident beam is characterized by a voltage, V, a

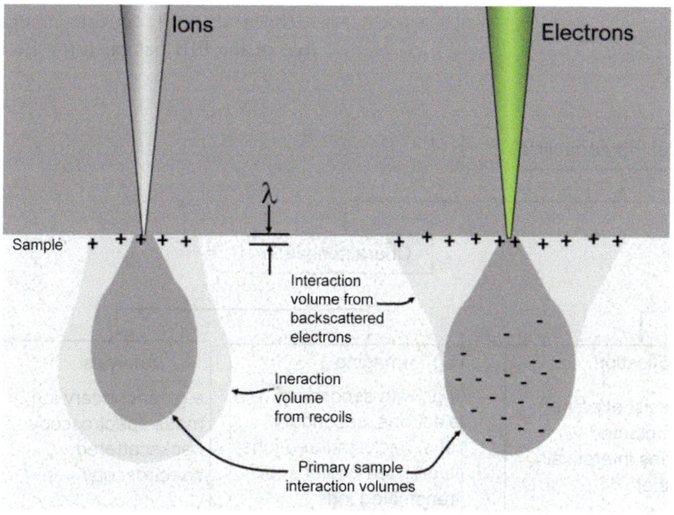

Fig. 33 Charging mechanisms for the focused ion beam (left) and scanning electron microscope (right). λ = characteristic escape depth for secondary electrons

current density, J, and is fully contained within a radius, r_o. The geometry is for a uniform semiinfinite slab of material with thermal conductivity, k, and the beam normally incident upon it. The temperature rise, ΔT, at the sample surface at the circumference of the beam then is given by:

$$\Delta T = \frac{JV}{2k} r_o \quad \text{(Eq 6)}$$

The calculated result for selected materials (Ref 60) indicates modest temperature rises. For conductive materials, the tendency is for a very low temperature rise, which usually can be neglected (e.g., <5 °C, or 9 °F). For materials with low thermal conductivity, low melting point, or both (i.e., polymers), much larger temperature rises are expected (e.g., >100 °C, or 180 °F) and can lead to loss of mechanical stability or melting. Before any generalizations are made, however, some special considerations should be pointed out. First, for special geometries, such as cross sections, lamella, thin films, or pillars, higher temperatures may be expected because of the reduced pathways for heat conduction. Some of these special geometries are treated by using Green's function methods by Ishitani (Ref 61). Second, it must be recognized that the treatment by both Melngailis and Ishitani considers the heat to be deposited at the surface only and within the radius of the beam, when in fact the heat is delivered into a large subsurface volume determined by the size of the collision cascade. This correction yields much lower temperatures, especially for lower-mass ions with their larger and deeper interaction volume. Finally, for lighter ion species, an appreciable fraction of the ion beam energy is transferred to the electrons (Fig. 27), and electrons are known to have a relatively long mean free path (Ref 62) before they dissipate their energy and are thermalized. Accounting for this would tend to further lower any temperature rise.

Finally, it sometimes is important to recognize that the ion beam is not a continuous supply of energy to the sample, because of the corpuscular nature of the ion beam. Instead, the energy is delivered in a series of pulses, as ions one at a time. For a 30 keV singly charged ion beam with 100 pA, the energy delivered per ion is 4.8×10^{-15} J, with the ions arriving sporadically with an average interval of 1.6 ns. Melngailis points out that the effects from individual ions perhaps should not be considered as temperature effects. The FIB operator who suspects they are seeing symptoms of a thermal effect is advised to apply the same dosage with a 10 times lower dose rate. The single-particle effects (sputtering, amorphization, etc.) will be the same, but any temperature rise would be 10 times less.

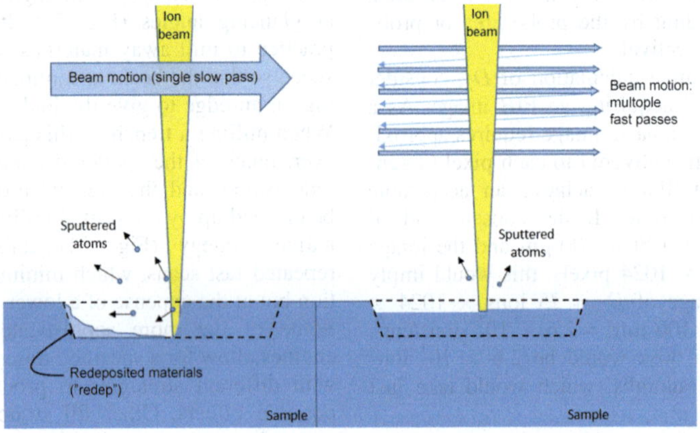

Fig. 34 Scanning strategies for milling include slow scan versus repeated fast scans.

Focused Ion Beam Applications

There are a great variety of applications enabled by the available FIB instruments, a selection of which is described in this section. The term *applications* is used here to describe the FIB end usages, whether they be for routine purposes or for cutting-edge scientific research. The applications are distinguished from the basic capabilities, that is, the fundamental ways focused ions beams can be exploited, regardless of the end purpose. A tree chart of FIB capabilities is provided in Fig. 35. The broad range of these capabilities is a direct consequence of the remarkably diverse selection of beam species, beam energies, and the accessories available in the modern FIB. All of the FIB applications described in this section rely on one or more of these capabilities. For example, the most common FIB application for TEM lamella preparation relies on the following selected subset of capabilities: imaging, deposition, and sputtering. Conversely, there are some applications that require single capabilities, such as FIB imaging. A survey of some of the more common and more interesting applications follows.

Application: Milling for Sample Preparation

Since its early development, the primary role of the FIB has been for the bombardment

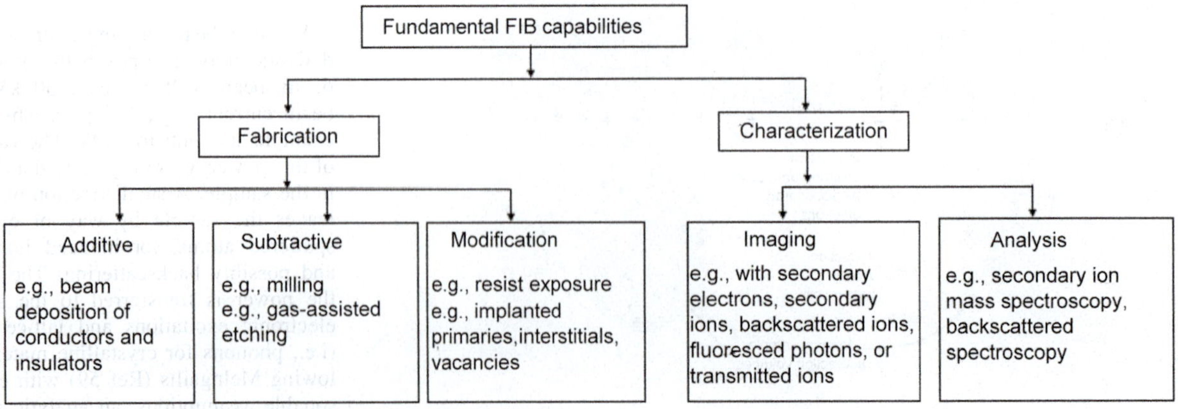

Fig. 35 The various focused ion beam (FIB) applications are built on one or more of these capabilities.

of samples with the expressed purpose of inducing the sputtering process, that is, causing sample atoms to be ejected through momentum transfer, as described in the section "Beam-Sample Interactions" in this article. Alternately termed as milling or erosion, this remains the predominant use for FIB instruments. Several applications that rely principally on material removal are first described.

The sputtering process is very useful for cleaning surfaces that may have been oxidized, hydrocarbon coated, or otherwise covered in order to expose the underlying material of interest. Typically, this is done with the incident beam orthogonal to the sample (Fig. 36). Sometimes this process can lead to nonuniform surfaces, because some materials will sputter faster than others. Techniques to generate more uniform milling include using multiple angles of incidence (i.e., rocking) or etching gases.

Note that the milling of amorphous materials such as polymers and glasses tends to proceed at a very uniform rate, which is well predicted by SRIM. However, for crystalline materials, the milling rate can vary considerably from the predictions, from one crystalline grain to another of the same material. The effect of nonuniform milling rate is easily observed during the milling of polycrystalline materials, in which a milled region will exhibit an uneven milled surface or a nonsmooth floor. Grains that have low-index directions (e.g., (100) or (110)) oriented parallel to the ion beam tend to sputter away at a much slower rate. As explained in a previous section, ions entering a grain at such angles are effectively captured and experience lower stopping force, both nuclear and electronic. This effect is commonly observed (Ref 63) and can be minimized by changing the tilt by 4° from the channeling direction or by the introduction of etching gases (described later).

Application: Exposing Cross Sections

A routine FIB application is the cross sectioning of materials to reveal the subsurface layers or defects under the surface. The process sometimes is called block-face imaging. This is conventionally done in a combined FIB-SEM instrument, so the adjacent SEM can image the newly exposed surfaces. To provide the optimal imaging perspective for the SEM, a wedge of material is removed with the FIB using a ramped dosage, as show in Fig. 37(a). An example of the SEM image of such a cross-sectional face is shown in Fig. 37(b). Redeposition of the previously sputtered material on the cross-sectional face can be corrected for by using a final low-current "cleaning cut" of the cross-sectional face. The newly exposed features now are available for SEM imaging in an FIB-SEM. Such cross sections also can be used for image-based metrology, energy-dispersive spectroscopy (EDS) mapping, or electron backscatter diffraction (EBSD) analysis (described in the article "Crystallographic Analysis by Electron Backscatter Diffraction in the Scanning Electron Microscope" in this Volume). The plasma FIB, with its high beam currents, is particularly good at exposing large cross-sectional areas in a relatively short time.

It is routinely observed that the sputter rate will be different for the various components of composite materials or materials with porosity. This certainly can lead to a protective effect or "shadowing" of the materials located physically below it. Figure 38 demonstrates the effect of curtaining, because the hard-to-sputter layers serve to protect the material below from the full impact of the ion beam. Curtaining can be mitigated by applying a uniform protective coating at the top surface. The protective coating may be added from a gas injection system precursor gas (usually tungsten, platinum, or carbon) or by a conductive polymer (Ref 64) or potentially ionic liquid (Ref 65). Alternatively, curtaining can be reduced appreciably by milling the same planar surface from a variety of angles, which is

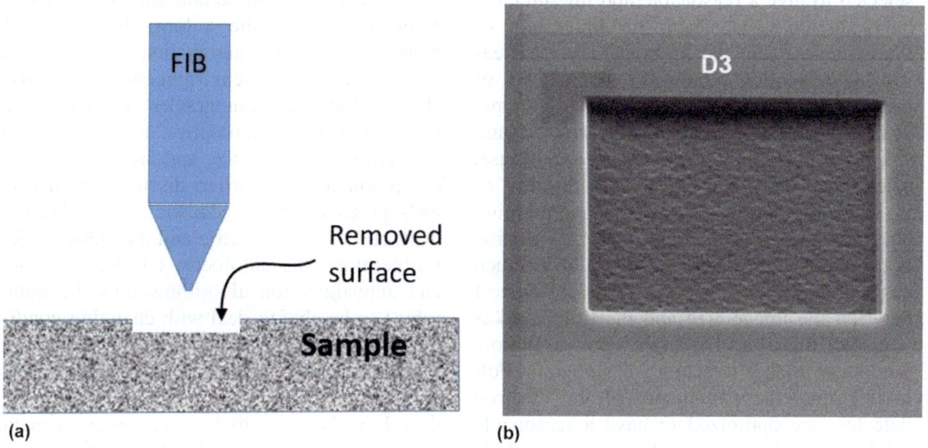

Fig. 36 (a) Typical focused ion beam (FIB)/sample orientation for top-down delayering. (b) FIB image of newly exposed surface

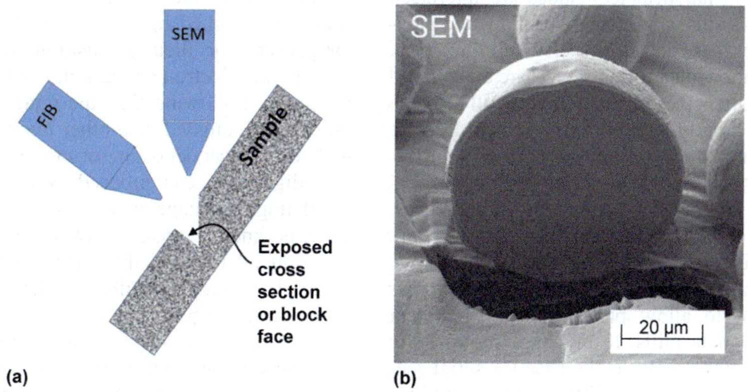

Fig. 37 (a) Typical focused ion beam (FIB) and scanning electron microscope (SEM) geometry for producing a cross section. (b) Cross section of a spherical particle from an iron powder. SEM image nicely shows the surface newly exposed by FIB milling. Courtesy of Tescan Orsay Holding, a.s.

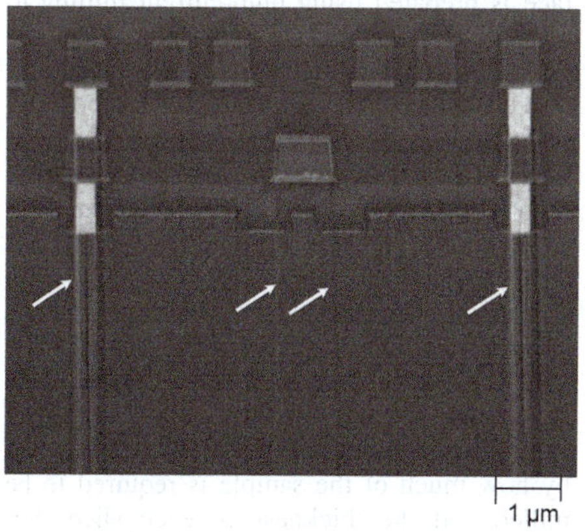

Fig. 38 Curtaining effects demonstrated in a cross section of a semiconductor device that was cross sectioned by a gallium beam and imaged with a scanning electron microscope. The bright structures are tungsten interconnects, and these are largely responsible for the curtaining effects seen below them. Courtesy of H. Schulz, Zeiss Microscopy GmbH

conveniently achieved with a rocking stage (Ref 66).

By recognizing the curtains from the background, image-processing algorithms that take advantage of their Fourier frequencies can be used to remove the curtains from the images. Commercial and open-source software is available for image processing the curtains out of the images (Ref 67, 68). While this produces cleaner images, if the curtains are severe, they may impart surface roughness or hide authentic features with similar appearances.

Application: Serial Sectioning and Three-Dimensional Tomography

Once the original cross section is prepared, the FIB can be used to further advance the cut face (i.e., shave off another slice from the originally exposed cross section) so as to expose successively deeper planes for subsequent imaging. When repeated again and again, the successive images can be used to reconstruct three-dimensional (3-D) data sets or tomograms. Sometimes called serial block-face imaging or serial sectioning, most of the major instrument providers offer automated workflows that can be customized by the operator.

A typical procedure requires the following steps. First, a region of interest (ROI) must be identified, where a serial-sectioning tomography experiment will be performed and the position adjusted to the eucentric point. A protective layer usually is deposited with a gas injection system (GIS) to put a tungsten, carbon, or platinum coating on the top surface to protect the sample surface during FIB imaging. Then, a fiducial mark is patterned into the top coating using the FIB to precisely position the beam for the next cut, thereby avoiding thermal and other drift issues. Next, a block face is prepared using high-current milling to a depth greater than the desired depth of the cross section. Additionally, trenches adjacent to the block face are prepared as a depository for redeposited materials. At this point, the sample is ready for an automated serial-sectioning acquisition. Figure 39 shows such a sectioning in the middle of tomogram acquisition.

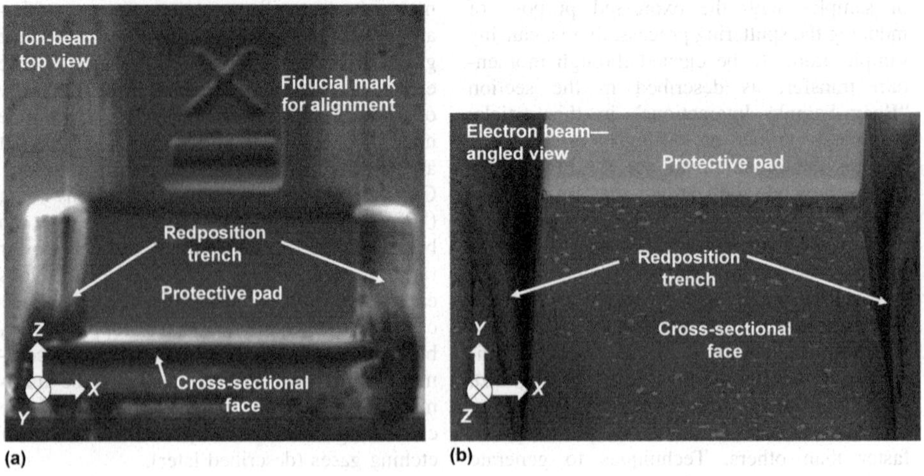

Fig. 39 (a) Ion beam and (b) electron beam images of a plasma focused ion beam serial-sectioning tomography acquisition on an aluminum alloy showing key geometric features, including the cross-sectional face, redeposition trench, fiducial marks, and protective pad on the top surface. In (b), a tilted *X-Y* face is imaged, and the *Z*-direction is orthogonal to that face. The cross-sectional face has a 120 μm width. Courtesy of H. Yuan, Canadian Centre for Electron Microscopy

The ion beam and electron beam parameters are selected for the experiment. Usually the ion beam parameters are selected to optimize a relevant viewing width (x-parameter) and depth (y-parameter). The y-distance is dictated by how much of the sample is required to be milled and the thickness of each slice. For example, to mill a z-depth of 10 μm with 50 nm slice thickness, it would require $10{,}000 \div 50 = 200$ slices (and images). With 10 nm slices, it requires 5 times as many slices (i.e., 5 times as long and 5 times more data).

The electron beam parameters are selected as well. First, the voltage of the beam is selected to give a reasonable and interpretable backscattered yield. Then, the dimensions of the image are selected in order to have a reasonable resolution relative to the features of importance within the serial-sectioned volume. This typically relies on beam current, beam size, and pixel dimensions. For ease of use, often the pixel size is selected so that the x-, y-, and z-dimensions are all equal, presenting equal-sized voxels. Because the z-dimension is the slice depth, x- and y- resolution is much easier to adjust. Again, there is a tradeoff between the image size and the time it takes to gather the data. The higher the resolution, the longer it takes to acquire a tomogram. Following this, the parameters of the electron detectors are optimized to have a reasonable brightness and contrast. Other options include selection of bit depth for the image. Finally, other features, such as beam shifts, autofocus, and autostigmation algorithms, often are selected for the acquisition.

If other analytical techniques are interlaced within this slice and image workflow, the parameters for these detectors (i.e., dwell time, peak identification, data formats, etc.) are all selected as well. Often, these measurements will be made with third-party accessories and will require some communication between the FIB instrument software and the accessory software.

After setup, the acquisition commences, with a certain ROI and slice depth being milled. The images are all referenced relative to the milled fiducial mark and usually cross correlated to allow precise beam placement for subsequent mills. If any stage motions are necessary (as usually is the case for serial sectioning in single-beam FIBs or in tomograms employing EBSD as a detector), the fiducial marks are very important for realigning the stage and the electron and ion beams.

It is very important to note that these experiments can take quite a long time (tens of hours) and rely on automation software and instrumental and environmental stability. High-performance tomography machines have stable electron and ion optics, are well aligned for optimal performance, and are placed in an environment (see earlier discussion) that is well protected from acoustic noise, thermal drift, and external electric and magnetic fields. Furthermore, the autofocus, cross-correlation, and autostigmation algorithms must be quite robust to be able to deal with changing conditions in the sample.

After a tomogram has been acquired, there now are more steps required to deal with the data. Usually, the output of such an experiment is several hundreds or thousands of images. This data now must be processed. There are commercial as well as open-source software packages available for dealing with these data. First, the images must be aligned relative to the fiducials in the image. Then, the images often are stretched in the vertical direction to compensate for their acquisition at a tilted angle by the electron column; this is termed tilt correction. Following this, the images usually are cropped, aligned, and other imaging filters applied, such as denoising or curtain removal. The alignments can proceed either by cross correlating an image to a previous image in the experimental image stack or by aligning to a fiducial mark in the image that did not change (i.e., was not milled by the ion beam) during image acquisition.

The images now are ready for quantification and visualization. Within this 3-D image it is possible to determine the volume fractions of multiphase materials, usually with simple techniques such as thresholding. Other, more complex image-processing techniques, such as edge detection, segmentation, and surface rendering, are out of the scope of this article. Typical quantitative information that can be obtained from serial sectioning includes identifying a percolation path, relative phase fractions, porosity, embedded particle size analysis, and (when coupled with EBSD or EDS) chemical and crystalline orientation relationships within the material. Figure 40 is an image of a plasma FIB serial-sectioned tomogram of ultrahigh-performance concrete, with the thresholded visualization of the internal porosity and cracking based on the backscattered contrast.

Serial sectioning has grown in popularity recently. Pioneering work has demonstrated that a gallium beam can achieve a slice resolution as small as 2 nm (which is competitive with TEM tomography). On the other end of the spectrum, the use of xenon plasma FIB has enabled very large serial sections with an *x-z* ROI dimension of up to 500 μm on a side. Furthermore, the xenon beam leaves a lateral surface damage layer that is small enough to enable high-quality EBSD patterns for serial sectioning interlaced with EBSD at technologically relevant length scales important in mesoscale applications (Ref 1).

Another enabling technology made possible by FIB serial sectioning is the development of the correlative microscopy workflow. In this case, a feature of interest can be tagged or identified from a lower-resolution, nondestructive technique, such as magnetic resonance imaging or micro-computed tomography x-ray imaging. The FIB then can be used to locate the identical spot and use the ion beam to mill to that location. From there, a serial-sectioning experiment can be set up to perform high-resolution imaging, followed by a deliberate experimental step to prepare a TEM lamella or atom probe tomography needle for even higher-resolution analysis. This has been performed in such disparate fields as biological imaging (Ref 69) and mechanical testing, among others. At this writing, correlative microscopy and the targeted search for critical features is a rapidly advancing field, and many of the microscope vendors offer a fully integrated correlative workflow with matching software to enable data correlation over a variety of length scales and types of data (*x*,*y*,*z*, composition, etc.). Researchers are developing algorithms to cross correlate regions of interest in 3-D to glean new insights about the materials they are characterizing. Because FIB milling features and imaging resolution stand at an intermediate length scale between macro-mesoscale measurements and truly nanoscale measurements, FIB is a critical technique in correlative workflows.

Application: Sample Preparation for Micromechanical Testing

As the trends toward miniaturization of devices continue, engineers find bulk material properties are no longer accurate representations for very small samples of these same materials. Performance and reliability do not necessarily persist as devices become smaller. Therefore, testing of the very smallest samples is needed to understand the material properties at this scale. For metallic samples, one of the critical characteristics is the number of grains present in the sample. Only when this number is very large is bulk behavior realized. Preparing micro- and nanoscale samples with the FIB gives the researcher specific advantages. First, the channeling contrast makes the grain structure plainly visible, so samples with specifically chosen grains can be selected, and grain-boundary effects can be isolated. Also, samples typically experience less mechanical damage when they are cut with the FIB instrument as opposed to mechanical cutting, grinding, or polishing to create the same sample. While there are some potential disadvantages (curtaining artifacts, amorphization layer, gallium-implantation effects), the skilled FIB operator can confine these effects to within approximately 10 nm of the cut. In situ mechanical testing includes nanoindent testing, compression testing, and tensile testing. Figure 41 shows a nickel film milled into a T-shape or "half dog bone" geometry suited for tensile testing. Because of the high currents available in the plasma FIB, this structure took just 20 min to fabricate. The structure subsequently was mounted to an in situ tensile testing jig for stress-strain characterization to the point of failure. Bending failures also can be examined closely by creating appropriate structures by FIB milling and testing them in the same FIB-SEM instrument with an internal nanoindenter. Figure 42 shows a single-crystal copper specimen that has been gallium milled to create a cantilever for bend testing.

Application: Sample Preparation for TEM

Transmission electron microscope sample preparation is the most widely used application of the FIB. This application derives its value from the ease of navigation afforded by the electron beam or ion beam imaging to identify an ROI, from which a sample may be extracted for higher-resolution imaging using the TEM. The development of FIB-based TEM sample preparation has displaced many of the traditional TEM sample-preparation techniques, such as electropolishing, dimple grinding, or broad ion beam milling, because of its site-specificity, repeatability, and reliability when performed by a skilled technician. Samples prepared using the FIB ideally should be under 100 nm in thickness, with samples thinner than 30 nm required for atomic-scale resolution and

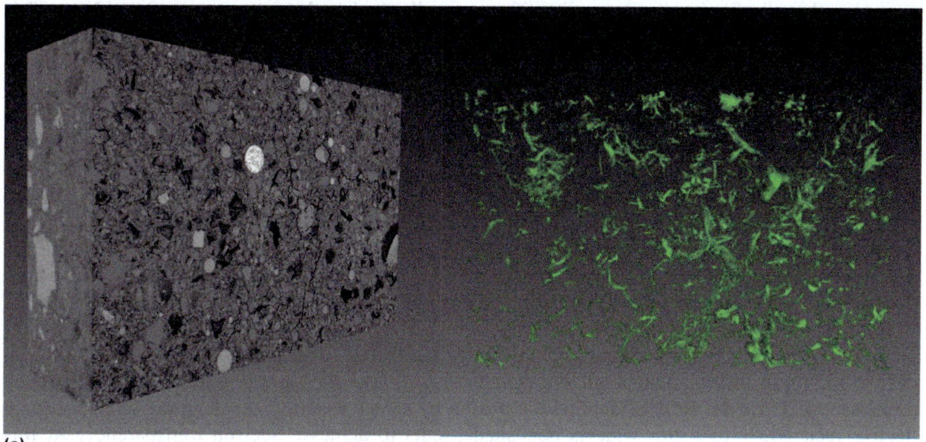

(a) (b)

Fig. 40 (a) Three-dimensional rendering of serial-sectioned volume of ultrahigh-performance concrete from a perspective angle by plasma focused ion beam. (b) Internal porosity and crack distribution obtained using image analysis software. Sample volume is 150 × 100 × 38 μm. Courtesy of K. Wille, University of Connecticut

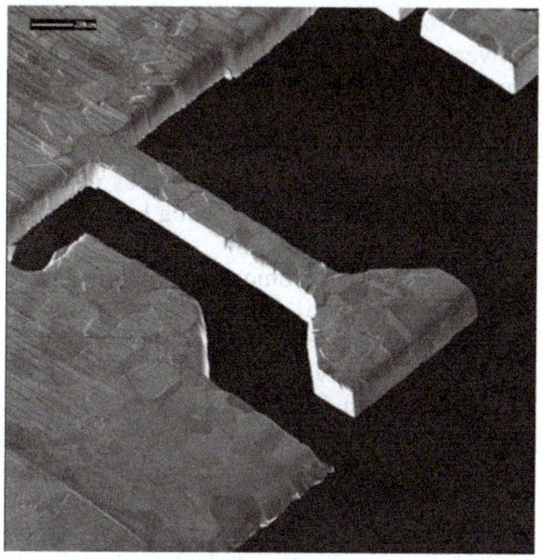

Fig. 41 T-shaped tensile strength sample manufactured with the high currents available in the xenon plasma focused ion beam. Courtesy of N. Smith; copyrighted by Oregon Physics and used with permission

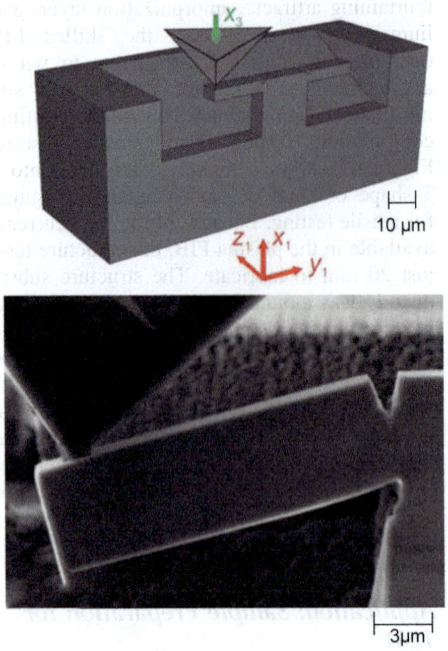

Fig. 42 Single-crystalline copper sample gallium focused ion beam milled to create cantilevers for testing. Reprinted with permission from Ref 70

high-resolution electron energy loss spectroscopy analysis. A typical TEM sample prepared with the FIB is 5 to 10 µm wide (*X*) by 5 to 10 µm deep (*Y*).

There are three main methods for preparing TEM samples using the FIB. These are the H-bar method, the ex situ liftout method, and the in situ liftout method. The H-bar geometry is depicted in Fig. 43. A small piece of the sample is diced and thinned mechanically and mounted to a 3 mm (0.12 in.) outside diameter half-grid. The sample then is thinned with box cuts, followed by cleaning cuts in each side of the ROI until the membrane between each side is electron transparent. This method is most convenient if specialized accessories such as nanomanipulator probes are not available or if a single-beam FIB is used. It has significant disadvantages related to a limited viewing area if the sample is tilted in the TEM, as well as a long milling time to hollow out a large-enough viewing area.

The next two techniques involve using a micromanipulator to pluck a sample from an FIB-prepared ROI and remove it to a grid for further TEM analysis. The primary difference between ex situ liftout and in situ liftout is where the sample is removed. In ex situ liftout, the sample is thinned to electron transparency inside the FIB and then removed from the FIB instrument. It is transferred to a TEM grid by using a needle that "plucks" the sample from its surroundings. Typically, an ex situ system requires a glass needle, a hydraulic apparatus for moving the needle, a moving stage, and an optical microscope (Ref 71).

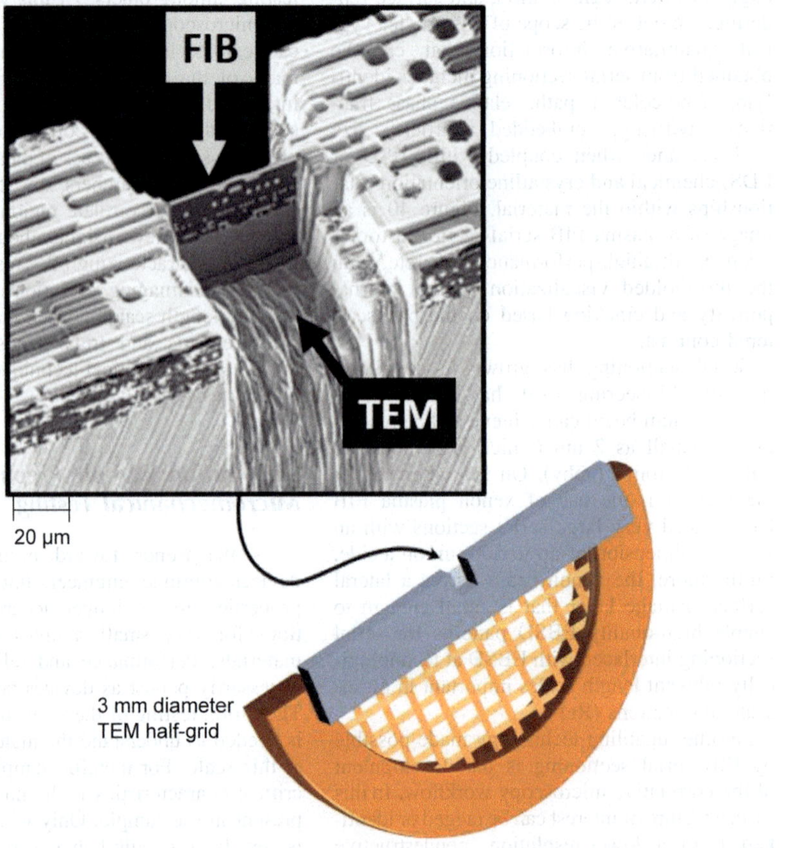

Fig. 43 H-bar geometry for transmission electron microscope (TEM) sample preparation. FIB, focused ion beam. Inset image courtesy of Fibics Incorporated, Ottawa, Canada

The sample then is transferred to a 3 mm (0.12 in.) TEM grid, where it remains attached by using some sort of glue or electrostatic forces. In the in situ case, the sample is thinned to a medium thickness and then welded to an in situ micromanipulator by using a GIS to attach it to the TEM grid.

Figure 44 shows the procedure required to prepare a lamella for in situ or ex situ liftout. In steps 1 and 2, a sample is prepared in a similar manner for both ex situ and in situ liftout. The operator uses the SEM or FIB in an imaging condition to identify an ROI. The ROI may be identified as a specific defect, an interface, or a chemical heterogeneity (as identified by EDS). This also may be a feature identified by another cross-correlated technique, such as micro-computed tomography or fluorescence microscopy, with a map to the feature guided by specific features and navigation markings (see the section "Application: Serial Sectioning and Three-Dimensional Tomography" in this article for more information about correlative microscopy). Once the ROI is identified, a protective rectangle 1 to 2 µm wide is deposited onto the sample, usually the length of the desired sample (5 to 15 µm) plus 2 µm on each side for in situ liftout using GIS precursors. This step is meant to protect the top surface from further ion milling and redeposition during subsequent ion milling steps. Next, a trench is cut adjacent to the sample, usually in a ramp or stairstep orientation, to arrive at a sloping geometry to the bottom of the intended lamella. The depth of the cut usually is set to a depth and ramp length of 1.5 to 2:1 in length; that is, if the TEM sample is intended to be 6 µm deep, typically the length of the ramp is 9 to 12 µm. This allows the SEM to image the bottom of the sample. Next, another ramp or rectangular box is cut on the opposite side of the sample, to provide access to the sample from the opposite side and to free it from the bulk. At this point, the sample still is attached at its base and sides. The sample is tilted to 0° and a box is milled under the lamella, as well as two cuts on the side of the sample, leaving a tab that keeps the sample attached at the top of the lamella to hold it in place.

The ex situ and in situ liftout techniques diverge at this point. To finish the ex situ workflow, the sample is thinned in this trench by using a series of cleaning cross-sectional cuts to achieve electron transparency. At this point, the FIB is used to cut free the lamella at the tabs, and it is allowed to rest within the trench. Thereafter, the sample is removed from the FIB. At an ex situ station, a microscope

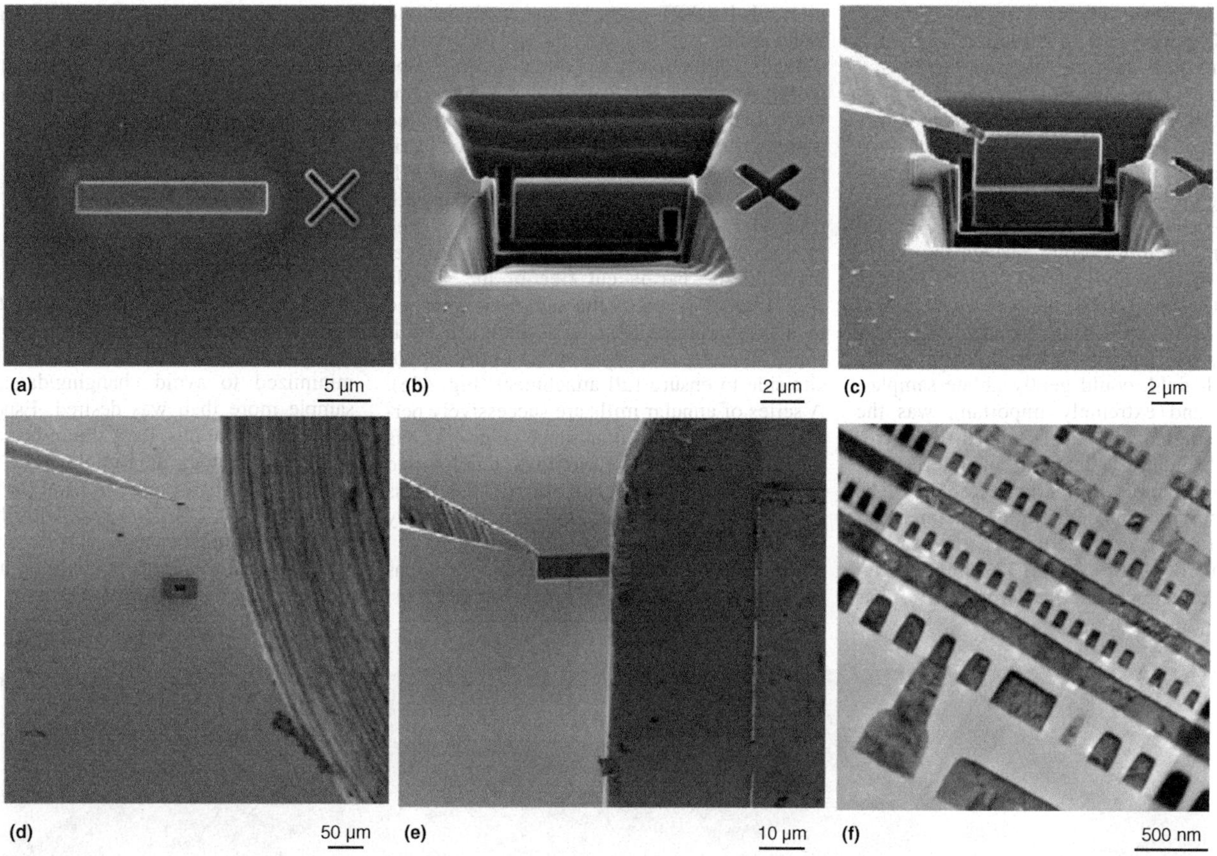

Fig. 44 The transmission electron microscope (TEM) lamella is created by a series of steps. (a) Platinum overcoating to protect the lamella and minimize curtaining. Also evident is the fiducial mark used to designate the feature of interest. (b) Ramp-profiled regions milled out in front of and behind the lamella. The side and undercuts also are visible. (c) Micromanipulator probe being attached after the final side cut is made. (d) Lamella attached to the manipulator probe now removed from the sample. The gas injection system nozzle is in the foreground. (e) Lamella now attached to a transferrable TEM grid and thinned to electron transparency. (f) Final TEM image of the lamella. Courtesy of Carl Zeiss

and pneumatically controlled manipulator arm are used to bring a needle to the sample surface and to attach the thinned specimen to the needle using van der Waals forces (Ref 72). Following this, the sample is transferred to a TEM grid. (There are several grid designs that work well for ex situ liftout.)

The in situ workflow continues in Fig. 44, with all of the remaining steps done in the microscope. Following the undercut and side cuts, the sample then is welded to a micromanipulator arm using GIS platinum, carbon, or tungsten, and using the SEM and FIB beam to align all of the operations at the eucentric point. After that, the sample is cut free from the substrate while still attached to the manipulator probe, and the probe is retracted to a safe position. The stage is translated to position a previously loaded TEM grid under the beam. The lamella sample then is repositioned to the side or top of the TEM grid and welded in place using a glue or GIS precursor on the grid. Once welded, the sample is detached from the grid by cutting the tip of the micromanipulator, using the ion beam. The sample then is thinned to electron transparency in a series of cross-sectional cuts from each side, gradually stepping down the beam current and eventually the voltage to minimize the amount of amorphization damage to the sample. Note that all of the thinning steps are performed in a geometry where the lamella cut face is parallel to the beam to prevent full penetration of the incident ion. Typical Monte Carlo methods model the sputter damage by examining an 89° incidence and looking at the lateral straggle of the ions as a measure of the interaction volume.

Relative advantages of the H-bar technique include the lack of need for pricey accessories. Disadvantages include a lack of tilt availability before sidewall occlusion in the TEM and a relatively long milling time, as well as polishing and sample-preparation time prior to H-bar preparation. Ex situ workflows are rapid and most amenable to automation, because preparation of the lamella can be repeated by using image recognition software that guides the beam. Ex situ preparation often puts TEM specimens on carbon support films, at which point they cannot be thinned further. In situ workflows have the advantage of final thinning at low voltages, as well as full transfer, which is suitable for one-of-a-kind samples. They have the disadvantage of being time-consuming and prone to error with novice FIB operators.

Application: Sample Preparation for the Atom Probe

An important subset of sample-preparation techniques enabled by the FIB is the preparation of atom probe tomography (APT) samples. Atom probe tomography is a technique that developed in the 1960s (Ref 73) in which a high electric field is applied to the apex of a very sharp needle in an ultrahigh-vacuum environment, causing progressive field evaporation of the atoms at the tip. The ionized atoms travel a prescribed distance to a position-sensitive detector. The time of flight provides sufficient information to determine the mass of the individual atoms. This allows the atom-by-atom reconstruction of an object and allows sensitivities on a spatial scale in the parts per million. See the recent reviews of the atom probe technique in Ref 74.

Using this technique, one can examine many localized phenomena, such as grain-boundary segregation (Ref 75), nanoscale precipitation

(Ref 76), dopant concentration (Ref 77), electronic device dopant distribution (Ref 77), and many other subjects. Original attempts at sample preparation involved the use of electropolishing to create a sharp needle from metallic samples, but this method was limited to metals and often was random in terms of obtaining any site-specific information.

Atom probe tomography expanded its size and scope of applications through the development of two innovations. The addition of a pulsed laser (Ref 78) enabled development of the APT application for nonconducting samples, because the laser, in combination with high field, could gently ablate samples. Second, and extremely important, was the use of single- and dual-beam FIB microscopes to identify features of interest for FIB-based sample extraction (Ref 79). The site-specificity of the FIB, plus its amenability for tracking or targeting specific features (through EBSD [Ref 80], EDS, simple imaging, or other correlative workflows), has allowed the APT research and analysis field to flourish.

Figure 45 shows a typical atom probe workflow for an FIB liftout of a needle sample. a) The area of interest is identified, and a protective layer (typically tungsten or platinum) is deposited (Fig. 45a). A series of FIB cuts are made to create a cantilever bar with triangular cross section (Fig. 45b). The micromanipulator is attached to the cantilever bar, and the bar is cut free of the sample (Fig 45c). A piece of the sample is attached to a presharpened post (Fig. 45d). An additional bonding coating is made on the opposite side to ensure full attachment (Fig. 45e). A series of annular mills are successively performed to create a pillar shape (Fig. 45f). As the sample thins, one employs a defocused beam to sharpen the needle for APT readiness (Fig. 45g). At this point, the sample is ready for APT characterization.

The APT-based sample preparation also has recently been adapted for cryogenic applications, and more and more correlative techniques to target specific areas have been developed, including transmission Kikuchi diffraction (Ref 81) and correlative workflows to FIB-APT preparation (Ref 82).

Application: FIB Imaging

Even the earliest FIB instruments had basic imaging capabilities so the operator could "see" the sample and thus precisely target the region to be sputtered away. While imaging is regarded as useful, it was something to be minimized to avoid changing/damaging the sample more than was desired. Especially at high magnifications, the gallium beam induces sputtering, making the sample erode from one image to the next, especially along topographic edges (Ref 83, 84). However, with the advent of lighter ion beams, such as hydrogen, helium, and lithium, there has been a growing appreciation that ion beams can be used for

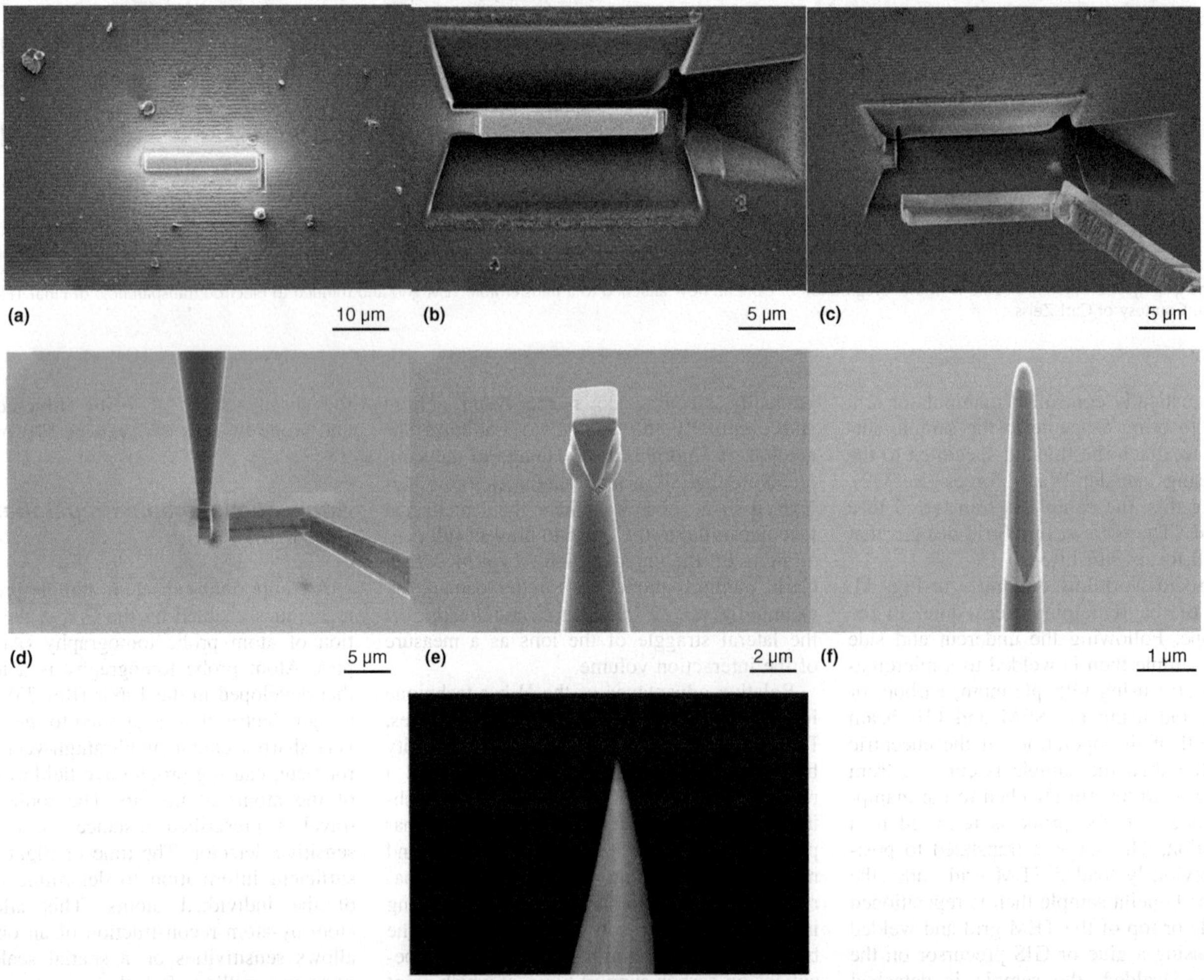

Fig. 45 Atom probe tomography sample-preparation workflow using the focused ion beam. See text for description. Courtesy of B. Langelier, Canadian Centre for Electron Microscopy

high-magnification imaging, with several unique advantages highlighted as follows.

High Resolution

It is surprising to many SEM enthusiasts that an ion beam can be focused to a probe size comparable to, or smaller than, that of an SEM. After all, it sometimes is reasoned that an ion must be larger than an electron. In reality, however, the size of the particle is not a limiting factor in either case. Diffraction is the factor that limits the probe size for most SEMs, and the much more massive ion beams do not suffer appreciably in this regard. Further, the gas field ion sources (GFISs) and laser-cooled atom sources have significantly higher brightness than any other charged-particle sources; therefore, the brightness limit is less significant for these beams. The relatively low energy spread of these same ion sources also helps to reduce chromatic aberrations, a factor described earlier that limits probe sizes for most SEMs (especially at low accelerating voltages) and the gallium liquid metal ion source FIB.

The high-resolution imaging of light ion beams also is partly attributed to the nature of the beam-sample interactions. For the lighter-ion-species FIBs, the beam spread is quite narrow near the surface, and the collision cascade is mostly contained under the surface (Fig. 46). Consequently, the produced SEs are primarily type 1 (see the section "Electronic Stopping Force" in this article) and provide surface-specific information from the region immediately surrounding the primary beam (Ref 85, 86). In the SEM, the incident electrons spread rapidly under the surface as they frequently scatter from identical mass electrons of the sample. The occasional scattering from nuclei generates backscattered electrons (BEs), which can produce SEs some distance from the primary beam. Such type 2 secondary electrons (SE_2) therefore convey nonlocal and subsurface information, in effect blurring the image. For SEMs, the interaction volume can be reduced by using lower beam energies, but the contribution of SE_2 still will be dominant. Furthermore, BEs that re-emit into the vacuum chamber hit chamber sidewalls and generate even more useless electrons (SE_3, etc.). The heavier ion species (gallium and xenon, for example) can have collision cascades near the surface that can generate SE_2 some distance from the primary beam. Figure 47 shows a comparison of an SEM image and a helium FIB image of aluminum posts. The SEM image shows a superimposed haze due to the contribution of SE_2 information and the nonlocal and subsurface information they convey.

At present, the resolution of the light ion beams is limited in part by sample damage. The highest-resolution images usually are obtained with a thin, unsupported edge so that subsurface nuclear effects (displacements, implantation, and swelling) are minimized. The development of lighter ion species (e.g., hydrogen) or higher beam energies (e.g., >40 keV) would lower the surface-damage rates and afford significant resolution improvements.

Surface Sensitivity

It already has been observed that light ion beams provide surface-specific information due to the predominance of type 1 SEs. Figure 48 shows an example of this surface-sensitive imaging of a monolayer of graphene suspended on GaN pillars. It was imaged with a 30 keV helium ion beam with a subnanometer probe size. The single atomic layer of carbon atoms generates sufficient SE signal to produce an image, due to the high electronic stopping force. Because the nuclear stopping force is relatively low at the surface, very little damage is produced. Other examples include the imaging of self-assembled monolayers and thin surface coatings (Ref 87).

Channeling Contrast

Another FIB contrast mechanism is channeling, which is the modulation of the SE yield for samples comprised of crystalline grains that are differently oriented with respect to the ion beam. This effect often is seen when looking at polycrystalline metal samples (Fig. 49) and provides a clear indication of the extent and boundary of each grain (Ref 88). The effect is most

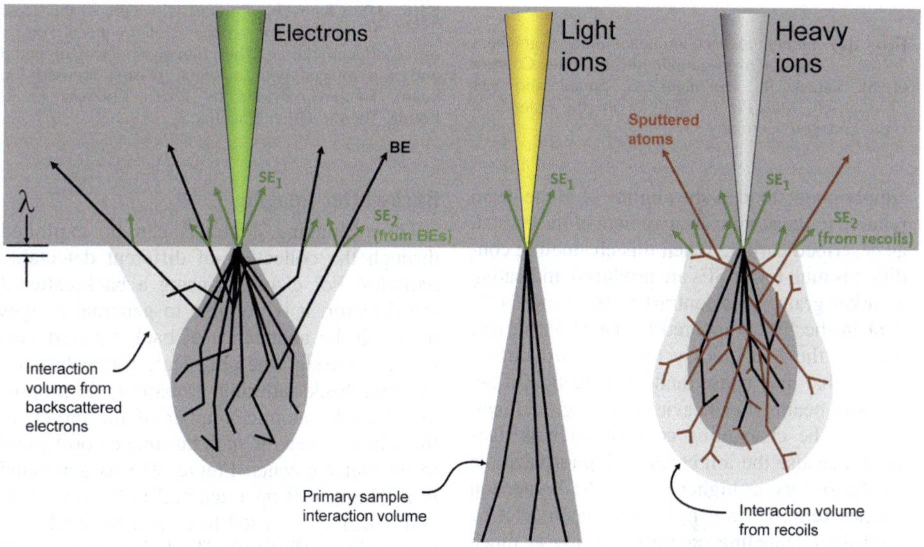

Fig. 46 Diagram comparing the origination of SE_1 and SE_2 for incident electrons, light ions, and heavy ions. The scanning electron microscope signal includes a large fraction of SE_2 from the backscattered electrons (BEs). The heavy ions generate a large SE_2 signal from the recoil cascades near the surface. The signal from light ions is dominated by SE_1, which provides information proximal to the beam location. The parameter λ represents the secondary electron (SE) escape depth.

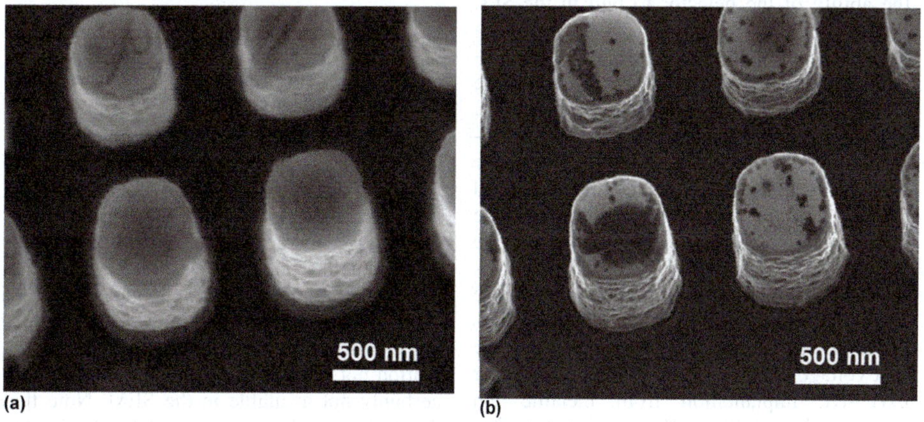

Fig. 47 Aluminum posts on silicon substrate images by (a) scanning electron microscope (SEM) and (b) helium focused ion beam (FIB). The helium image offers surface-specific information, while the SEM-sample interaction produces a mix of surface and deeper information.

Fig. 48 Helium focused ion beam image of graphene supported by gallium nitride pillars. Courtesy of M. Latzel, M. Heilmann, G. Sarau, and S.H. Christiansen, Max Planck Institute for the Science of Light, Erlangen, Germany

Fig. 49 Channeling contrast is evident in this cross section of copper interconnects on a semiconductor device. The cross section was prepared and also imaged with a xenon plasma focused ion beam. Sample provided by IMEC. Courtesy of R. Young, Thermo Fisher Scientific

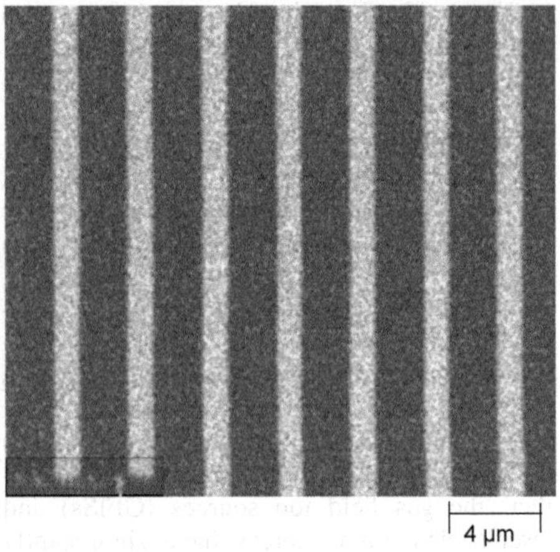

Fig. 50 Image generated from backscattered ions/atoms provides strong atomic number contrast

simply understood as the alignment of the beam relative to the atomic arrangement of the crystal, as described earlier. When this channeling condition is met, fewer SEs are produced, indicating a darker grain. This contrast is much more evident in the FIB compared to the SEM, in part because the FIB quickly removes any surface oxides that may cause random deflection of the incident beams. For heavier ions (e.g., gallium, xenon), the channeling contrast often is transient, because the ion beam will amorphize the crystal surface at higher doses. The lighter ion species (e.g., hydrogen, helium, neon) can be used for channeling contrast imaging at much higher magnifications because the ratio of electronic to nuclear stopping force is much higher.

Voltage Contrast

This contrast arises because of the presence of varying voltage on the sample, affecting the ability of the detector to pull in the SEs. A region of the sample that has a positive voltage on it will look darker because the Everhart-Thornley detector will be less effective in drawing away the SEs from this location. The voltage can be induced through the charging effect from the incident beam (passive voltage contrast) or deliberately applied through a contacting probe (active voltage contrast). For passive voltage contrast, the effect reveals capacitance and resistance properties of a sample. It can be used to interrogate and understand electrical components such as batteries, capacitors, or complex circuits. Note that voltage contrast effects can be muted after excessive implantation from metallic ion beams such as gallium. Voltage contrast often is used in circuit edit applications (Ref 89), some of which is described in the section "Circuit Edit and Debug" in this article.

Backscatter Imaging

Other imaging contrasts can be exploited through the collection of different detectable particles. For example, using a backscattered ion detector, it is possible to generate images in which the abundance of backscattered ions is used to assign the gray scale for each pixel. Because backscattering probability is proportional to the atomic number of the target, Z, the whiter pixels in such an image correspond to heavier elements. Figure 50 was generated in a lithium FIB operated at 2 keV, and a special detector was used to collect the backscattered lithium (Ref 90). The lighter regions are the underlying silicon substrate, while the darker regions are patterned low-*Z* resist on top of the silicon. In this example, the low beam energy assures that backscattering occurs near the surface. Because the backscatter rate can be relatively low for some ion species, high areal dosages are required to produce such an image, introducing some level of damage to the sample.

Secondary Ion Imaging

Some FIBs are equipped with a special detector that collects secondary ions (SIs), that is, the low-energy atoms that are sputtered from the sample and leave with some positive or negative charge state. The SI yield is relatively small (e.g., $Y_{SI} \sim 10^{-2}$ to 10^{-5}) and depends very much on the chemical nature of the sputtered atom and the matrix of surrounding atoms. The SI image often provides information not readily seen in the SE images and certainly not available in the SEM. Note that this technique does not absolutely identify the SIs, because there is no spectrometer involved. Figure 51 shows a comparison of a single region that was imaged using SEs (Fig. 51a) or SIs (Fig. 51b). Notice how the foreign particle is plainly evident due to its high SI yield. Some systems are equipped with a secondary electrons secondary ions detector capable of being used to detect SEs or SIs. Because the SI yield normally is so low, considerable areal dosages are required to generate such images, often causing appreciable material loss and limiting the resolution of this technique.

Imaging of Insulating Materials

The imaging of insulators such as polymers, ceramics, glass, and minerals is a challenge for the SEM. The e-beam tends to implant its negative electrons some distance under the surface, while leaving the top surface positively charged through SE emission. Charging can damage materials through dielectric breakdown, or cause bright and dark streaks in the image, or even cause image motion. Some SEM users will resort to using a variable-pressure SEM or coating their sample with conductive layers of carbon or platinum. However, these remedies will limit the ultimate resolution, obfuscating details smaller than 10 nm. For the FIB, however, the situation is much improved because there is only positive charging, and this is limited to the topmost few nanometers (Fig. 33). The introduction of an electron flood gun allows the imaging to be interleaved with the flood gun, so charge accumulation can be neutralized. The flooding dose is adjustable, and the interleaving can happen at the frame scan rate or line scan rate. For example, hydroxyapatite minerals on a polymethyl methacrylate (PMMA) polymer substrate are difficult to image because both are electrical insulators. High-voltage and low-voltage SEMs are prone to charging effects and radiolysis that cause damage to the polymer (Fig. 52a). When imaged in the helium FIB instrument (Fig. 52b), the fine details of the minerals and the polymer are easily seen, with little evidence of charging or damage.

Long Depth of Focus

Imaging with long depth of field means that features are sharp in both the foreground and background. This is useful for imaging 3-D structures and also for nanofabrication. Optical microscopy ultimately suffers from the wave nature of light and cannot generate long-depth-of-field images at high magnification. The SEMs have considerable advantage versus light, but FIBs generally provide the best available performance regarding depth of focus. Ultimately, the FIB advantage is rooted in the mass of the particles, which reduces diffraction effects and encourages the operation at small convergence angles, α (Fig. 14). The convergence angle plays a critical role in determining how the beam diameter (and blurriness) grows for features not at the optimal focus. Formally, the depth of focus is the range of height variations (above or below the ideal focus), which results in a doubling of the resolution. For many FIB instruments, the convergence angle may be approximately 1 milliradian, compared to many high-performance SEMs, which may be over 10 milliradians. Correspondingly, the depth of focus of the FIB can be 5 times longer than the SEM under comparable magnifications. Figure 53 shows an image of the 3-D pillar shapes and reveals the high depth of field available in helium FIB images.

Application: Precision Milling

The previous sections on sample preparation introduced applications in which the goal is to reveal some feature or characteristic inherent in the sample, such as exposing a cross-sectional face or preparing a lamella. This section is distinctly different in that it covers applications in which the FIB is used for precision sputtering for nanofabrication purposes, that is, to remove material in order to fabricate features of interest that were otherwise not present. In contrast to the previous sections, relatively small amounts of materials are removed; hence, the lower sputter rates of light ion beams are of greater value.

An example of precision milling is the fabrication of plasmonic devices. Plasmonic devices exhibit resonances through the collective motion of electrons under the influence of external electric fields. The resonance frequency and the modes of oscillation depend largely on the precise geometry of the metal structure. If properly tuned, such devices provide a novel way to couple electromagnetic waves to devices. Typically, the device is fabricated from a thin metal film (e.g., 10 to 100 nm thick) that is covering a substrate such as sapphire or quartz or perhaps a freestanding foil that has been transferred to a partly supported grid. The ion beam is used to mill away material and leave the plasmonic structure intact. An example of a plasmonic device is show in Fig. 54. Because of the reduced sputtering rate and the absence of gallium residue, the helium beam is commonly used for such applications, although neon beams and other noble gas ion beams are being explored.

Another type of device that is commonly fabricated by ion beams is pores, precisely cut holes through an otherwise contiguous film. The pores can be generated in simple films (e.g., silicon nitride or graphene) or as functional pores in which the electrodes are integrated into the layer stacks. Such fabricated pores subsequently are used for wet

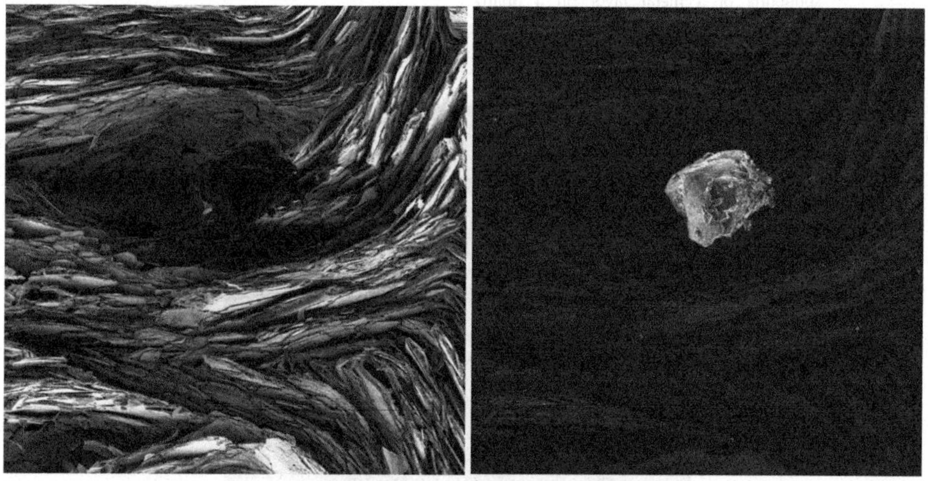

(a) (b)

Fig. 51 Cesium focused ion beam images of pencil lead taken with a 20 μm field of view with a foreign particle at the center. The secondary electron image (a) offers topographic information, while the positive secondary ion image (b) provides striking contrast arising from the compositional information.

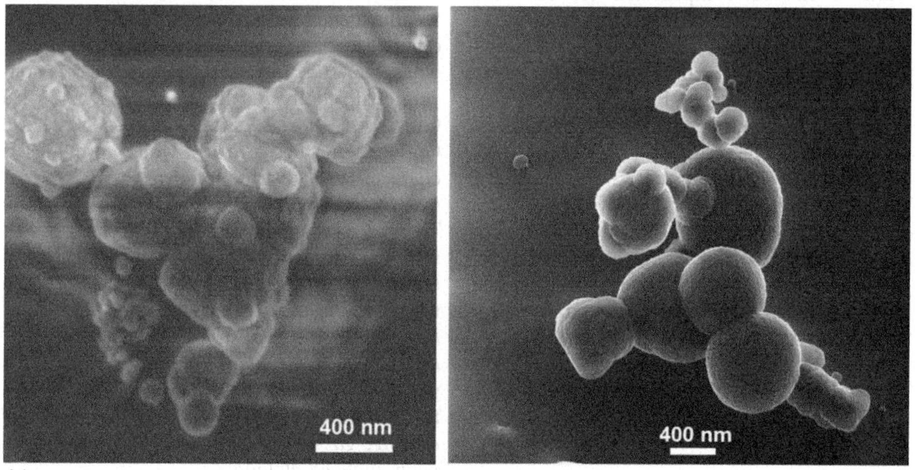

(a) (b)

Fig. 52 Hydroxyapatite minerals on a polymer substrate imaged with the (a) scanning electron microscope and (b) helium focused ion beam

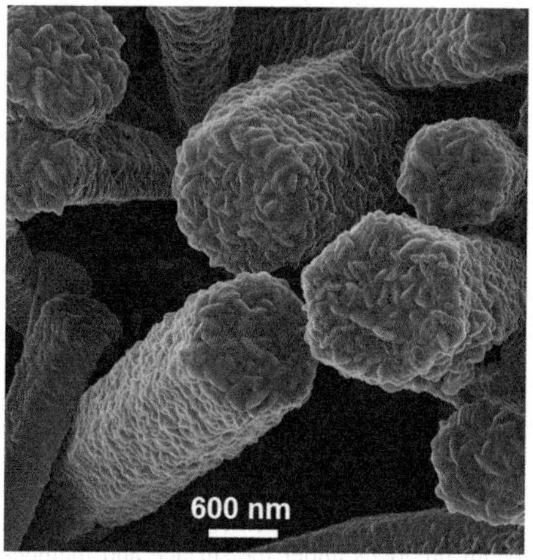

Fig. 53 Palladium nanorods grown on a sacrificial template of ZnO nanowires. The depth of focus is evident in the high resolution in both the foreground and the background.

chemical processes, such as for molecular sensors or sieves (Ref 92).

Application: Direct-Write Lithography with Resist

Lithography is a very general term referring to almost any process in which information is written to a substrate. The qualifier term *direct write* usually implies the writing device is moved on the substrate to scribe the desired pattern. The alternative, indirect write, indicates that a mask or stencil contains the pattern information, which is projected and transferred to the substrate. The term *resist* refers to a thin film that coats the substrate and is designed to be sensitive to the writing process via exposure by either photons or charged particles, which are more relevant in this application. For electron and ion beams, commonly used resists include PMMA (Ref 93), hydrogen silsesquioxane (Ref 94), and there is active research concerning several exploratory resists (Ref 95–97). With a positive-tone resist, the exposure achieves cross linking, and this exposed portion will not be dissolved in the following development process. (Examples of such techniques are found in Fig. 55 and 56.) If it is a negative-tone resist, then the exposure breaks the chemical bonds, and this area will be removed during the development process. Often, the patterning and development of a resist is just the first step, which is followed by a transfer of the patterned resist to other materials through additive and/or subtractive processes (e.g., deposition and etching). Industrial-scale lithography usually is done with photons, even extreme ultraviolet photons, and is the basis of modern semiconductor manufacturing (Ref 100). The FIB instruments offer unique advantages (Ref 101) for direct-write lithography because of their high beam-placement accuracy, small focused probe size, and unique sample interaction. Their principal limitation is the speed with which they can produce patterns; that is, they operate in a serial, not parallel, fashion. Hence, they are used for exploratory work or for prototyping novel devices.

Note that there are dedicated lithography instruments using ion and electron beams that are so specialized they often are not considered SEMs or FIBs. These instruments may have laser stages, higher beam energies (e.g., 50 to 100 keV), and the ability to deflect the beam to larger "write fields" without suffering from excessive aberrations. For smaller scale and proof-of-principle demonstrations, the multipurpose FIB and SEM are widely used. The electron beam is more commonly used for lithography but does have some disadvantages. Because the collision cascade of the electron beam broadens under the surface and includes a large fraction of backscatter events, the resist will be partly exposed in the area surrounding the incident beam location. This so-called proximity effect partly exposes the resist some micrometers away from the desired region; while it can be compensated for to some extent, it limits the overall utility of e-beams for this application (Ref 102). Ion beams, however, have some advantages for direct-write lithography with resist. First, the ion beam has a low backscatter rate, and the collision cascade is much narrower near the surface, especially for light ions (Fig. 46). Both of these greatly reduce the proximity effect compared to electron beams. Second, ion beams interact very strongly with the electrons in the sample and hence require a much lower dosage to reach the threshold needed for exposure (Ref 103).

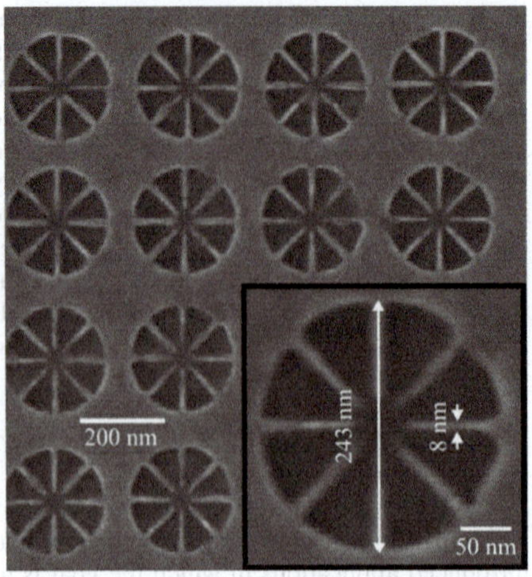

Fig. 54 Nanoplasmonic devices fabricated by direct sputtering of a metal layer on a quartz substrate. Adapted and reprinted with permission from Ref 91. ©2019 IEEE

Fig. 55 Gears created using a helium beam writing of palladium mercaptide resist. Source: Ref 98. Unpublished results, used with permission

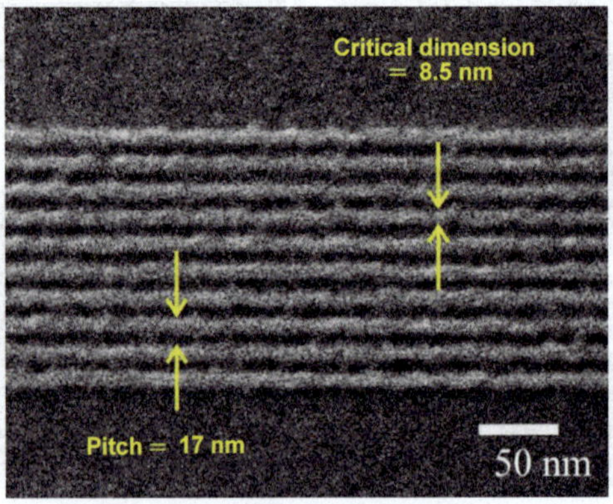

Fig. 56 Resist based on C60 fullerenes. Adapted and reprinted from Ref 99. Copyright 2016, with permission from Elsevier

Application: Ion-Beam-Induced Chemistry

Ion-beam-induced chemistry allows for versatile micro- and nanoscale fabrication using both additive and subtractive processes (Ref 104). The process usually involves a gas-phase precursor that is delivered broadly to the sample substrate, where the ion beam locally induces some change to dissociate or otherwise functionalize the precursor molecules (Ref 105, 106). The precursor gas is conventionally delivered through the nozzle of a GIS (see the section "Accessories and Options" in this article), which provides a characteristic distribution size of approximately 1 mm (0. 04 in.) on the sample surface. The precision of this process is limited by the sample interaction size of the ion beam (approximately tens of nanometers). When the goal is to deposit materials, the terms *ion-beam-assisted deposition* or *focused-ion-beam-induced deposition* are used. When the goal is to enhance the removal of materials, the terms could be *ion-beam-assisted chemical etching*, *gas-assisted etching*, or *focused-ion-beam-induced etching*.

For deposition, the interaction of the ion beam leaves the precursor molecules nonvolatile and relatively immobile, causing an accumulation of material where the ion beam is positioned. A commonly used precursor for deposition is tungsten hexacarbonyl, $W(CO)_6$. The action of the ion beam is to remove one or more of the carbonyl groups, leaving the remaining molecule nonvolatile. Because of the metal composition, the deposited material is conductive, although the residual carbon and oxygen limit the conductivity to be considerably less than that of bulk tungsten. It also is known that the depositions can be contaminated by some amount of the primary beam species. Other commonly used precursors for the deposition of conductors are platinum, carbon, or cobalt. In Fig. 57, a 30 nm wide cobalt line was deposited over four pre-existing metal contacts so the resistivity could be measured. The cobalt lines were deposited with a helium ion beam and resulted in line widths as small as 10 nm and resistivities as low as 60 μΩ-cm (Ref 107).

Deposition of insulators is likewise possible with the appropriate gas-phase precursor, such as pentamethyl cyclopentasiloxane. Optimal insulating qualities can be achieved if the incident ion species is nonmetallic, thus avoiding gallium contamination. The silicon-oxide-rich pads in Fig. 58 were deposited with the helium FIB (Ref 108). Their dark appearance is evidence of their insulating qualities, and a current-voltage measurement indicated a resistivity on the order of 10^{13} Ω-cm.

Models (Ref 109) have been developed to help understand the deposition mechanisms and have been used successfully to help design sophisticated 3-D shapes suitable for focused-ion-beam-induced deposition (FIBID) (Ref 110). Figure 59 is an example of such a scaffolding fabricated with the use of a focused helium beam and a GIS equipped with a platinum precursor. The versatility of FIBID makes this an effective 3-D printer at the nanoscale.

Some limited work has been done with liquid-phase precursors. This offers the advantage of high precursor density but introduces the challenge of maintaining a liquid in a vacuum environment. Recently, ion beams have been used to induce deposition with commercially available in situ wet cells, in which a membrane separates the liquid from the vacuum environment (Ref 31). Of course, the membrane must be thin enough to allow the beam to penetrate but thick enough to prevent rupturing and damage to the FIB system.

For material removal, focused-ion-beam-induced etching offers some advantages over simple sputtering without an etchant gas. Similar to the deposition process, the precursor gases, such as XeF_2 and Cl_2, are delivered by the GIS and are temporarily adsorbed onto the surface. The ion beam causes dissociation into reactive components, where they can play beneficial roles. First, the reactive components can attack the sample and remove materials as volatile by-products. Additionally, the reactive components can help to volatize the sputtered atoms, allowing the redeposition effects to be reduced (Ref 112, 113). It has been reported in the literature (Ref 114) that material-removal rates can be enhanced by 10 to 100 times compared to sputtering without etch enhancement. Also, the presence of an etchant gas reduces channeling artifacts. Finally, specific etching gases offer selectivity of one material over the other, giving the instrument operator greater ability to stop milling once a specific material at a certain depth is reached (also called endpointing).

For beam chemistry applications, the ion beam flux (ions/cm^2/s) at a given location must be carefully considered to correspond to the arrival-rate precursor molecules. In deposition applications, if the ion beam flux is too high, sputtering may dominate, effectively reducing any desired deposition effect. Likewise, for gas-assisted etching applications, the best benefit can be obtained with a specific ratio of precursor arrival rate and ion arrival rate. Most FIB instruments will have a strict upper limit of gas load that can be delivered without causing adverse effects, so ion beam flux often is the easiest parameter to adjust. Many patterning systems will provide special patterning strategies, or refresh times, to allow for best beam chemistry results.

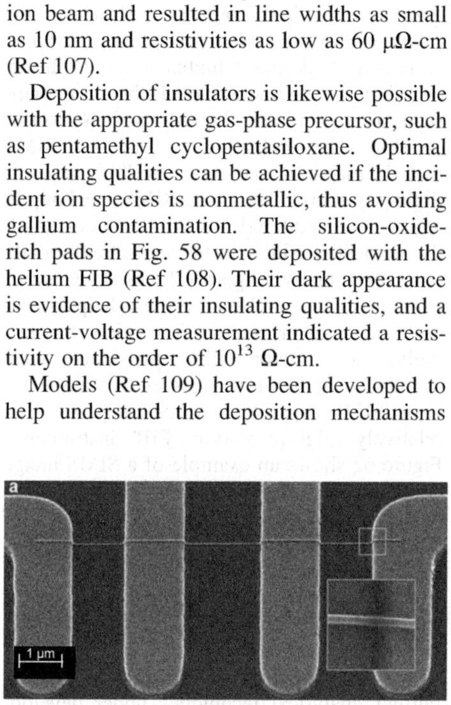

Fig. 57 Fine cobalt line deposited across metal pads with a demonstration of good step coverage. Adapted from Ref 107. Reprinted with permission from Springer Nature

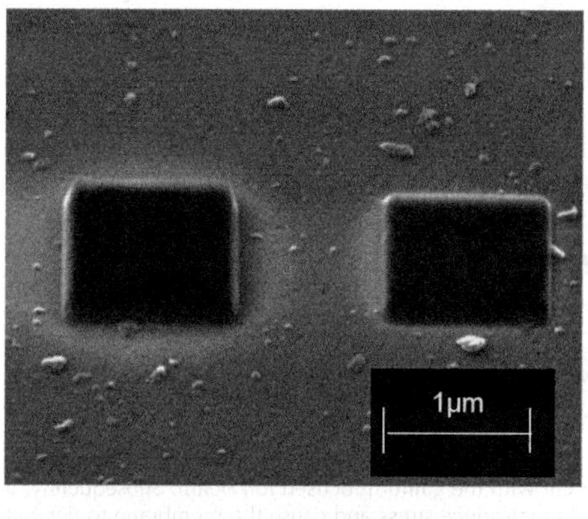

Fig. 58 Focused-ion-beam-deposited insulating silicon oxide pads. Courtesy of Carl Zeiss

Application: Ion Beam Functionalization

One of the remaining techniques for nanofabrication with ion beams includes ion beam functionalization, sometimes described as defect engineering or ion beam modification of materials. This process entails the modification of the substrate in a manner that is distinctly different from beam chemistry, resist exposure, and precision sputtering. Generally, the ion beam is used with precision to introduce defects, such as vacancies, interstitials, or implanted primary ions, into the substrate. The accuracy of the ion beam placement and the easy control of the areal dosage over several orders of magnitude give tremendous flexibility to the FIB operator. In one instance, researchers used a helium beam to introduce disorder or amorphization in a precise region so as to alter the thermal conductivity of silicon (Ref 115). In other cases, an alloy of

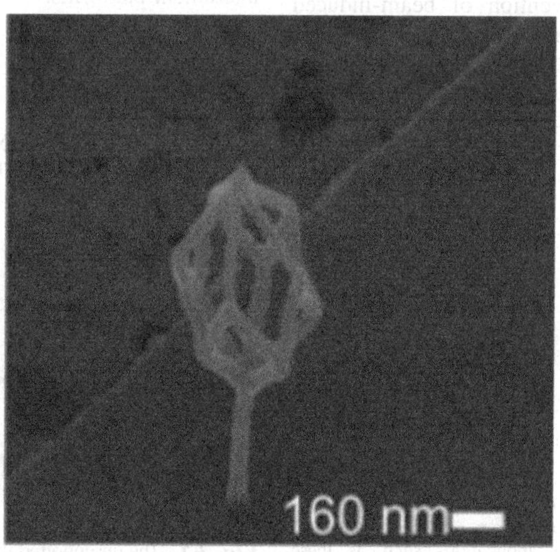

Fig. 59 Three-dimensional helium-focused-ion-beam-grown structure using the platinum precursor $Me_3PtCpMe$. Source: Ref 111. Reprinted with permission from Cambridge University Press

$Fe_{60}Al_{40}$ was subjected to ion beam irradiation to introduce ferromagnetism in patterned domains as small as 50 nm (Ref 116). Superconductive materials such as Y-Ba-Cu-O (Ref 117) and NbN (Ref 118) have been patterned with ion beams to interrupt their conductivity, thereby making Josephson junctions, which can be used for superconducting circuits or magnetic field sensors (Fig. 60). Both graphene and transition metal dichalcogenides have been similarly patterned using such ion beams. Such devices offer tunable performance and may be used for advanced computing and memory devices. At much higher dosages, ions can be implanted into membranes to introduce stress, causing a membrane to flex under the influence, as shown in Fig. 61.

Application: Cryogenic FIB

The development of cryogenic stages for electron microscopy (Ref 120) has enabled new insights in the field of biology, soft materials, and liquid-solid interfaces. This work has been developing in parallel (see the section "Application: Sample Preparation for TEM" in this article) for many years but usually for TEM applications or for surface-based SEM of structures. The ability of the FIB to produce 3-D images through serial milling and imaging has spurred the further development of cryogenic devices and transfer tools aimed at biological (Ref 121), geological (Ref 122), and materials (Ref 123) applications. General characterization at liquid nitrogen temperatures has many benefits. These include the reduction of radiolysis damage in electron- and ion-sample interactions (Ref 124), an improvement in mechanical rigidity in soft samples (Ref 60), the vitrification of liquid-based systems so that their spatial state can be preserved (Ref 125), and the prevention of beam-induced phase transformations (Ref 126).

A typical cryogenic FIB-SEM setup involves the plumbing of liquid nitrogen feed lines or a Peltier-cooled thermoelectric stage to cool the sample stage. Depending on the application, there also is the need for additional equipment to introduce a cryogenically frozen sample into the FIB-SEM. Usually, this freezing process is done ex situ, where the sample is plunge-frozen in liquid ethane and transferred to the load lock of the system. Additional advancements include an in-load lock gold sputtering system to coat nonconductive newly frozen samples and a cryogenic temperature shuttle from the load lock to the sample stage.

Some limitations of cryo-FIB operation include a lack of degrees of freedom to tilt and rotate the stages, because the cryogenic plumbing is relatively fixed. Another limitation is the preclusion from use of many of the precursor gases for GIS deposition and etching, because at cryogenic temperatures, their diffusion on the surface is lowered and they are not easy to remove from the vacuum, resulting in repeated dissociation and deposition over the sample surface as the beam rasters on the surface. Efforts to mitigate this effect and still produce a protective coating have been developed (Ref 127), but care must be taken in sample manipulation at low temperatures.

In recent years, an interesting subset of cryo-FIB applications has been developed: cryo-FIB liftout sample preparation for TEM and APT. First proposed and demonstrated by Marko for biological applications (Ref 128), this technique has been used for structural biology (Ref 129), battery applications (Ref 130), and metallurgical applications (Ref 131). This technique involves preparing samples at cryogenic temperatures and then manipulating in situ through a cryogenic transfer method in situ into an APT or TEM system. Further information on the details of these manipulations are available in the literature, and, as of this writing (2019), developments are being made by microscope vendors to make this level of control more turnkey across instrument platforms.

Application: Analysis Techniques

Ions have a long history of use for analytical purposes, including elemental analysis (Ref 132), film-thickness measurement (Ref 133), and even providing insight into atomic arrangement (Ref 134). In most cases, however, the ions are not well focused, so they do not provide lateral information at a resolution of 50 nm or better. Some of these techniques have been introduced recently to a growing family of FIB instrumentation.

Secondary ion mass spectrometry (SIMS) spans a variety of techniques by which the ion beam sputters away atoms and molecules of the sample that are subsequently mass analyzed to determine their composition. The lateral resolution is determined by the smaller of either the focused probe size or the collision cascade, which can be as small as 15 nm (Ref 135). The mass resolution can provide elemental identification with the ability to recognize specific isotopes, allowing the researcher to trace chemical processes. Some systems provide much higher mass resolution, enabling the distinction of isobaric overlaps such as CO (mass 28.010) and N_2 (mass 28.013). The technologies used for SIMS include magnetic sectors as well as time-of-flight systems. Because SIMS techniques require the sputtered atom to be nonneutral, the dedicated SIMS instruments traditionally have used relatively poorly focused cesium or oxygen ion beams, because these help to promote negative or positive ionization. Increasingly, the major FIB providers have introduced optional accessories that allow for basic SIMS capabilities to be integrated into relatively high-resolution FIB instruments. Figure 62 shows an example of a SIMS image generated by using a highly focused neon ion beam with a magnetic spectrometer (Ref 136). Other accessories also provide for time-of-flight detectors for elemental identification (Ref 137). Other FIB analytical efforts have used the angle and energy of backscattered ions to infer the mass of the scattering nuclei or the thickness of an overcoating (Ref 138). Further analytical techniques under development include ion-induced luminescence (Ref 139) and ion-blocking patterns (Ref 140) to assess crystalline information.

It is also important to note that aside from the aforementioned techniques, in which the ion beam is fundamental to the analysis, there are many cases where FIBs play a supporting

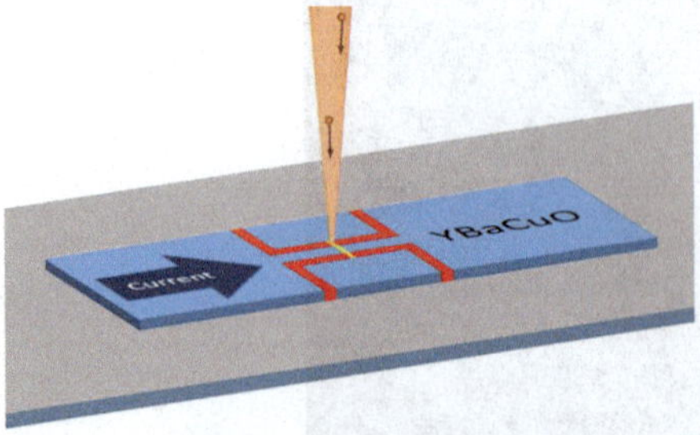

Fig. 60 Thin film of Y-Ba-Cu-O patterned by a focused ion beam (FIB) to make a Josephson junction. In applications such as these, sputtering by the FIB is not essential; the ion beam need only disrupt the otherwise critical atomic structure through displacements.

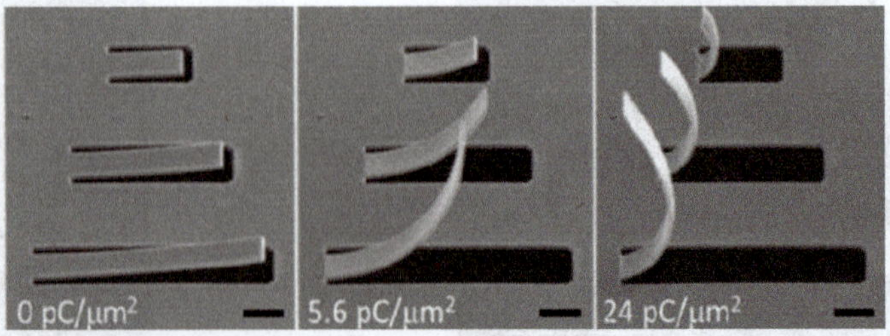

Fig. 61 The unsupported 80 nm thick gold film was first cut with the gallium focused ion beam. Subsequently, a much lower areal dosage (indicated) was applied to introduce stress and cause the membrane to flex out of the original plane. Adapted from Ref 119. Reprinted under the terms of the Creative Commons Attribution Non-Commercial license for open-access articles

role. For example, ion beams are commonly used in FIB-SEM platforms to isolate a feature of interest for subsequent analysis by EDS, wavelength-dispersive spectroscopy, EBSD, electron energy loss spectroscopy, or Auger electron spectroscopy.

Application: Semiconductors

Since the 1980s, the semiconductor industry has played an important role in the development and maturation of FIB technologies. At the very onset, FIB instruments were used for ion implantation and photomask repair. Throughout the following years, the FIB has been one of the tools of choice to help this industry chart its remarkable march to miniaturization (Ref 141). In the present day (2019), voltage contrast imaging (already discussed) of the FIB is routinely used for identifying and localizing flaws in the manufacturing process that give rise to undesired electrical short circuits or open circuits. The FIB-SEM is used for cross-sectional imaging (already discussed) during the defect-review process to detect and classify defects that may be under the surface or are otherwise unrecognizable by optical techniques. The previously discussed technique of 3-D tomography is used for metrology, and lamella preparation is used for routine process control for semiconductor manufacturing. Two additional FIB-based semiconductor applications are discussed herein.

Mask Repair

Each of the many layers of a semiconductor device is patterned with one or more lithographic photomasks. The photomask consists of a metal overlayer, such as chromium or $MoSi_2$ on a transparent quartz substrate. The set of photomasks represent the template for mass production and are a significant financial investment. Defects come in the form of missing metal or excess metal. The ability to repair such defects represents a significant advantage over the alternative of fabricating a new mask in terms of lost production time. Gallium beams are well established (Ref 142) for making these repairs through sputtering away of excess metal or ion-beam-induced deposition of missing metal (both variants of applications already discussed). However, more advanced mask sets, such as those used for extreme ultraviolet (EUV) lithography, are intolerant of the unavoidable residual gallium that is implanted into the otherwise transparent substrate. To avoid these artifacts, FIBs based on the GFIS source have been introduced to use nanometer-scale hydrogen and nitrogen beams for the express purpose of mask repair. Figure 63 shows an EUV mask before and after the repair of a defect in which excessive metal caused the desired pattern to be obscured.

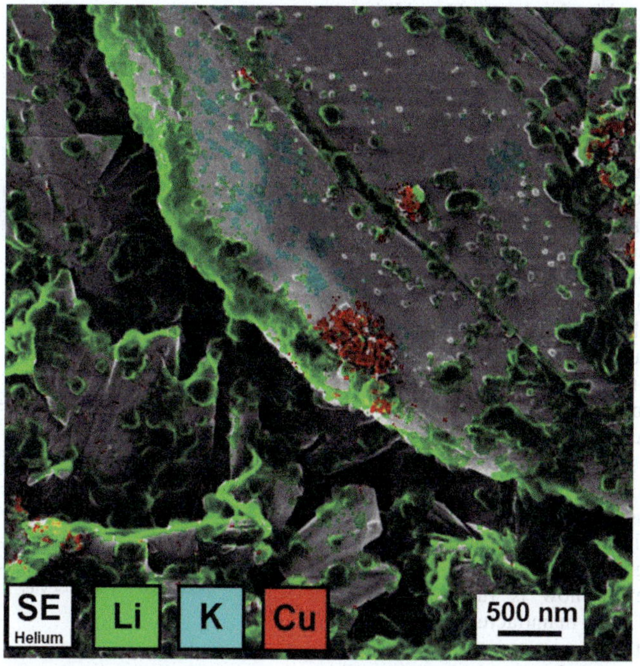

Fig. 62 Neon secondary ion mass spectrometry image of the anode of a commercial lithium ion battery. The underlying gray-scale image is based on helium-generated secondary electrons (SEs); the colorization is based on the detected elements, as shown. Courtesy of B. Lewis, Carl Zeiss

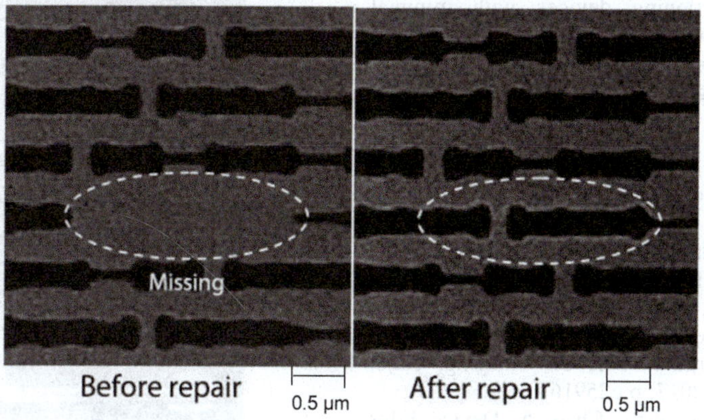

Fig. 63 $MoSi_2$ metal layers repaired by and imaged with an advanced gas field ion source focused ion beam instrument dedicated to mask repair. Adapted from Ref 143. Reprinted with permission from SPIE and from F. Aramaki and T. Kozakai

Circuit Edit and Debug

Modern processor chips contain several billion transistors, and the various levels of copper interconnects can total several miles (Ref 144). While sophisticated design tools help to make these chips function with high predictability, there occasionally are design flaws or manufacturing errors that prevent a given chip from functioning as expected. The FIB provides an opportunity to isolate and understand the root cause of the fault (Ref 11). The FIB also provides an opportunity to validate a proposed rewiring scheme without the investment of time and money in a new mask set. These demanding applications often push the limits of modern FIB capabilities and explain why semiconductor manufacturers have promoted the development of FIB technology throughout the decades.

Before reaching the FIB, the chip to be analyzed and edited will have most of the bulk silicon removed by mechanical or laser techniques to make the circuitry accessible to the FIB through a relatively thin layer of silicon. The FIB system usually will be equipped with an infrared microscope able to see the metal lines through the silicon, and a navigational overlay of the design drawings to help the operator find the suspected region. Some systems allow the chip to be electrically connected so it can be actively operated in situ to provide additional diagnostic abilities. As the FIB is milling through various layers, fluctuations in the yield of SEs provide an indication of the different materials penetrated through. Thus, monitoring this signals an indication of the milled depth and provides a means of predicting the endpoint of the milling. Figure 64 shows an example of that signal as the neon FIB milled through various layers. The voltage contrast of the FIB, described previously, helps to identify the unexpected short or open circuits in the now-exposed metal lines. When the fault is properly identified, FIB sputtering, with or without gas-assisted etching, is used to make precise cuts to interrupt failed conductive pathways. Ion-beam-assisted deposition (IBAD) often is used to refill the void with a suitable dielectric material. New conductive lines can be laid down by using IBAD with metal bearing precursors to reroute signals to functioning devices. If the failed device now is functioning adequately, the chip manufacturer can confidently invest in a corresponding design change for mass production through a modified process or photomask. Neon beams have been found to offer high-precision machining while making high-aspect-ratio cuts (27:1) through a 380 nm membrane in close proximity to functioning devices with minimal adverse effects (Ref 146). A demonstration of these circuit edit capabilities is demonstrated in Fig. 65, in which subsurface wiring has been cut and rerouted to a new location.

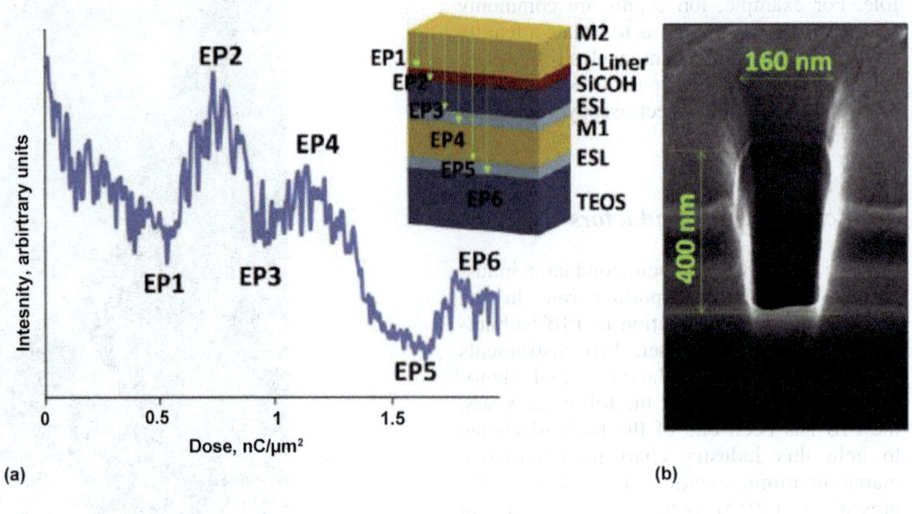

Fig. 64 (a) Depth-profiling signature serves as a depth gage while milling through layers. (b) The milled hole was later cross sectioned to better reveal the different layers. Adapted and reprinted from Ref 145 with permission from Elsevier

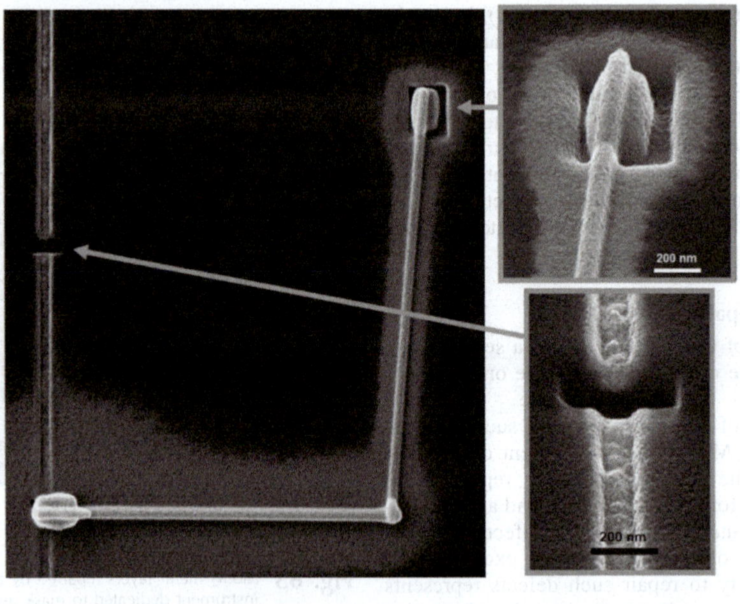

Fig. 65 Most basic circuit edit with focused ion beam (FIB) cuts and FIB-deposited metal lines to connect one layer to the next

REFERENCES

1. C.S. Xu, K.J. Hayworth, Z. Lu, P. Grob, A.M. Hassan, J.G. García-Cerdán, K.K. Niyogi, E. Nogales, R.J. Weinberg, and H.F. Hess, Enhanced FIB-SEM Systems for Large-Volume 3D Imaging, *eLife*, Vol 6, 2017, p e25916
2. K.N. Hooghan, Chap. 5, Device Edits and Modifications, *Introduction to Focused Ion Beams*, L.A. Giannuzzi and F.A. Stevie, Ed., Springer, 2005
3. J. Orloff, M. Utlaut, and L. Swanson, Chap. 3 and 5 in *High Resolution Focused Ion Beams: FIB and Its Applications*, Kluwer Academic, 2003
4. B. Wolf, Ed., *Handbook of Ion Sources*, CRC Press, 1995
5. V.N. Tondare, Quest for a High Brightness, Monochromatic Gas Field Ion Source, *J. Vac. Sci. Technol. A*, Vol 23 (No. 6), 2005, p 1498–1508
6. V.E. Krohn and G.R. Ringo, Ion Source of High Brightness Using Liquid Metal, *Appl. Phys. Lett.*, Vol 27 (No. 9), 1975, p 479–481
7. R. Clampitt, K.L. Airken, and D.K. Jeffries, Abstract: Intense Field-Emission Ion Source of Liquid Metals, *J. Vac. Sci. Technol.*, Vol 12, 1975, p 1208
8. R. Anderson and F.A. Stevie, Chap. 9, Practical Aspects of FIB TEM Specimen Preparation, *Introduction to Focused Ion Beams*, L.A. Giannuzzi and F.A. Stevie, Ed., Springer, 2005
9. L.A. Giannuzzi, B.W. Kempshall, S.M. Schwarz, J.K. Lomness, B.I. Prenitzer, and F.A. Stevie, Chap. 10, FIB Lift-Out Specimen Preparation Techniques, *Introduction to Focused Ion Beams*, L.A. Giannuzzi and F.A. Stevie, Ed., Springer, 2005
10. T. Kamino, T. Yaguchi, T. Hashimoto, T. Ohnishi, and K. Umemura, Chap. 11, A FIB Micro-Sampling Technique and a Site Specific TEM Specimen Preparation Method, *Introduction to Focused Ion Beams*, L.A. Giannuzzi and F.A. Stevie, Ed., Springer, 2005
11. B. Holdford, Chap. 6, The Uses of Dual Beam FIB in Microelectronic Failure Analysis, *Introduction to Focused Ion Beams*, L.A. Giannuzzi and F.A. Stevie, Ed., Springer, 2005
12. J. Orloff, M. Utlaut, and L. Swanson, Chap. 2 in *High Resolution Focused Ion Beams: FIB and Its Applications*, Kluwer Academic, 2003
13. L. Bischoff, P. Mazarov, L. Bruchhaus, and J. Gierak, Liquid Metal Alloy Ion Sources—An Alternative for Focused Ion

Beam Technology, *Appl. Phys. Rev.*, Vol 3, 2016, p 021101
14. P.C. Lozano, Energy Properties of an EMI-IM Ionic Liquid Ion Source, *J. Phys. D, Appl. Phys.*, Vol 39 (No. 1), 2006, p 126–134
15. C. Perez-Martinez, S. Guilet, J. Gierak, and P. Lozano, Ionic Liquid Ion Sources as a Unique and Versatile Option in FIB Applications, *Microelectron. Eng.*, Vol 88 (No. 8), Aug 2011, p 2088–2091
16. N. Smith, J. Notte, and A. Steele, Advances in Source Technology for Focused Ion Beams, *MRS Bull.*, Vol 39, April 2014, p 329–335
17. R. Hill, J. Notte, and L. Scipioni, Scanning Helium Ion Microscopy, *Adv. Imag. Electron Phys.*, Vol 170, 2012, p 65–148
18. H. Shichi, S. Matsubara, and T. Hashizume, Characteristics Comparison of Neon, Argon, and Krypton Ion Emissions from Gas Field Ionization Sources with a Single-Atom Tip, *Microsc. Microanal.*, Vol 25 (No. 1), 2019, p 1–10
19. A.R. Hill, Uses of Fine Focused Ion Beams with High Current Density, *Nature*, Vol 218, April 1968, p 202, 203
20. N.S. Smith, P.P. Tesch, N.P. Martin, and R.W. Boswell, New Ion Probe for Next Generation FIB, SIMS, and Nano-Ion Implantation, *Microsc. Today*, Sept 2009, p 18–22
21. J.J. McClelland, A.V. Steele, B. Knuffman, K.A. Twedt, A. Schwarzkopf, and T.M. Wilson, Bright Focused Ion Beam Sources Based on Laser-Cooled Atoms, *Appl. Phys. Rev.*, Vol 3, 2016, p 011302
22. B. Knuffman, A.V. Steele, and J.J. McClelland, Cold Atomic Beam Ion Source for Focused Ion Beam Applications, *J. Appl. Phys.*, Vol 114, 2013, p 044303
23. L. Kime, A. Fioretti, Y. Bruneau, N. Porfido, F. Fuso, M. Viteau, G. Khalili, N. Šantić, A. Gloter, B. Rasser, P. Sudraud, P. Pillet, and D. Comparat, High-Flux Monochromatic Ion and Electron Beams Based on Laser-Cooled Atoms, *Phys. Rev. A*, Vol 88 (No. 3), 2013, p 033424
24. N.S. Smith, W.P. Skoczylas, S.M. Kellogg, D.E. Kinion, and P.P. Tesch, High Brightness Inductively Coupled Plasma Source for High Current Focused Ion Beam Applications, *J. Vac. Sci. Technol. B*, Vol 24 (No. 6), 2006, p 2902–2906
25. L. Bruchhaus, P. Mazarov, L. Bischoff, J. Gierak, A.D. Wieck, and H. Hövel, Comparison of Technologies for Nano Device Prototyping with a Special Focus on Ion Beams: A Review, *Appl. Phys. Rev.*, Vol 4, 2017, p 011302
26. O.F. Hagena, Cluster Ion Sources (Invited), *Rev. Sci. Instrum.*, Vol 63 (No. 4), 1992, p 2374–2379
27. I. Yamada, J. Matsuo, Z. Insepov, D. Takeuchi, M. Akizuki, and N. Toyoda, Surface Processing by Gas Cluster Ion Beams at the Atomic (Molecular) Level, *J. Vac. Sci. Technol. A*, Vol 14 (No. 3), May/June 1996, p 781–785
28. R. Hill, J. Notte, and B. Ward, The ALIS He Ion Source and Its Application to High Resolution Microscopy, *Phys. Proced.*, Vol 1, 2008, p 135–141
29. C.F. Barnett and P.M. Stier, Charge Exchange Cross Sections for Helium Ions in Gases, *Phys. Rev.*, Vol 100 (No. 2), Jan 1958, p 385–390
30. C.M. Scheffler, R.H. Livengood, H.E. Prakasam, M.W. Phaneuf, and K. Lagarec, Patterning in an Imperfect World: Limitations of Focused Ion Beam Systems and Their Effects on Advanced Applications at the 14 nm Process Node, *Proc. 42nd Int. Symp. Test. Failure Anal.*, Nov 2016
31. A.V. Ievlev, J. Jakowski, M.J. Burch, V. Iberi, H. Hysmith, D.C. Joy, B.G. Sumpter, A. Belianinov, R.R. Unocic, and O.S. Ovchinnikova, Building with Ions: Towards Direct Write of Platinum Nanostructures Using In Situ Liquid Cell Helium Ion Microscopy, *Nanoscale*, Vol 9, 2017, p 12949–12956
32. S.A. Boden, Chap. 6, Introduction to Imaging Techniques in the HIM, *Helium Ion Microscopy*, G. Hlawacek and A. Gölzhäuser, Ed., Springer International, 2016
33. E.A. Kenik, H.M. Meyer, R. Vane, and V. Carlino, Effects of Evactron Plasma Cleaning on X-Ray Detector Windows, *Microsc. Microanal.*, Vol 11 (S02), 2005, p 1368, 1369
34. S. Kostinski and N. Yao, Rutherford Backscattering Oscillation in Scanning Helium-Ion Microscopy, *J. Appl. Phys.*, Vol 109, 2011, p 064311
35. J. Wang, S.H.Y. Huang, C. Herrmann, S. A. Scott, F. Schiettekatte, and K.L. Kavanagh, Focused Helium Ion Channeling through Si Nanomembranes, *J. Vac. Sci. Technol. B*, Vol 36 (No. 2), 2018, p 021203
36. J.R. Huddle, P.G. Grant, A.R. Ludington, and R.L. Foster, Ion Beam-Induced Luminescence, *Nucl. Instrum. Methods Phys. Res. B*, Vol 261 (No. 1–2), 2007, p 475, 476
37. S. Sijbrandij, J. Notte, L. Scipioni, C. Huynh, and C. Sanford, Analysis and Metrology with a Focused Helium Ion Beam, *J. Vac. Sci. Technol. B*, Vol 28 (No. 1), Jan/Feb 2010, p 73–77
38. V. Manichev, "NanoFabrication and Characterization of Advanced Materials and Devices," Doctoral dissertation, Rutgers University, 2018
39. S. Sijbrandij, A. Lombardi, A. Sireuil, F. Khanom, B. Lewis, C. Guillermier, D. Runt, and J. Notte, NanoFab SIMS: High Spatial Resolution Imaging and Analysis Using Inert-Gas Ion Beams, *Microsc. Today*, May 2019
40. N. Klingner, R. Heller, G. Hlawacek, J.V. Borany, J. Notte, J. Huang, and S. Facsko, Nanometer Scale Elemental Analysis in the Helium Ion Microscope Using Time of Flight Spectrometry, *Ultramicroscopy*, Vol 162, 2016, p 91–97
41. W. Eckstein, *Computer Simulation of Ion-Solid Interactions*, Springer, Berlin, 1991
42. J. Ziegler, J.P. Biersack, and M.D. Ziegler, "*SRIM*—The Stopping and Range of Ions in Solids," SRIM Co., Chester, 2008, www.srim.org
43. J.F. Ziegler, M.D. Ziegler, and J.P. Biersack, SRIM—The Stopping and Range of Ions in Matter, *Nucl. Instrum. Methods Phys. Res. B*, Vol 268 (No. 11–12), 2010, p 1818–1823
44. H. Hofsäss, K. Zhang, and A. Mutzke, Simulation of Ion Beam Sputtering with SDTrimSP, TRIDYN and SRIM, *Appl. Surf. Sci.*, Vol 310, 2014, p 134–141
45. M. Posselt, Crystal-TRIM and Its Application to Investigations on Channeling Effects during Ion Implantation, *Radiat. Eff. Defects Solids*, Vol 130 (No. 1), 1994, p 87–119
46. M. Mayer, "SIMNRA User's Guide," Report IPP 9/113, Max Planck Institut für Plasmaphysik, Garching, Germany, 1997
47. K.T. Mahady, S. Tan, Y. Greenzweig, A. Raveh, and P.D. Rack, Simulating Advanced Focused Ion Beam Nanomachining: A Quantitative Comparison of Simulation and Experimental Results, *Nanotechnology*, Vol 29 (No. 49), 2018, p 495301
48. S. Plimpton, Fast Parallel Algorithms for Short-Range Molecular Dynamics, *J. Comp. Phys.*, Vol 117 (No. 1), 1995, p 1–19
49. W. Möller and W. Eckstein, Tridyn—A TRIM Simulation Code Including Dynamic Composition Changes, *Nucl. Instrum. Methods Phys. Res. B*, Vol 2 (No. 1–3), 1984, p 814–818
50. W. Möller, TRI3DYN—Collisional Computer Simulation of the Dynamic Evolution of 3-Dimensional Nanostructures under Ion Irradiation, *Nucl. Instrum. Methods Phys. Res. B*, Vol 322, 2014, p 23–33
51. "Particle Tracing Module User Guide," 2016, https://www.comsol.com/release/5.2/particle-tracing-module, accessed Jan 6, 2019
52. R. Livengood, S. Tan, Y. Greenzweig, J. Notte, and S. McVey, Subsurface Damage from Helium Ions as a Function of Dose, Beam Energy, and Dose Rate, *J. Vac. Sci. Technol. B*, Vol 27 (No. 6), Nov/Dec 2009, p 3244–3249
53. R. Ramachandra, B. Griffin, and D. Joy, A Model of Secondary Electron Imaging in the Helium Ion Scanning Microscope, *Ultramicroscopy*, Vol 109, 2009, p 748–757
54. Y.V. Petrov, O.F. Vyvenko, and A.S. Bondarenko, Scanning Helium Ion Microscope Distribution of Secondary Electrons and Ion Channeling, *J. Surf. Invest., X-Ray, Synch. Neutron Tech.*, Vol 4 (No. 5), 2010, p 792–795

55. N. Stehling, R. Masters, Y. Zhou, R. O'Connell, C. Holland, H. Zhang, and C. Rodenburg, New Perspectives on Nano-Engineering by Secondary Electron Spectroscopy in the Helium Ion and Scanning Electron Microscope, *MRS Commun.*, Vol 8 (No. 2), 2018, p 226–240
56. T.M. Buck, G.H. Wheatley, and L.C. Feldman, Charge Stages of 25–150 keV H and ^{4}He Backscattered from Solid Surfaces, *Surf. Sci.*, Vol 35, 1973, p 345–361
57. L. Reimer, *Scanning Electron Microscopy*, 2nd ed., Springer, 1998, p 121
58. J.I. Goldstein, D.E. Newbury, J.R. Michael, N.W.M. Ritchie, J.H.J. Scott, and D.C. Joy, *Scanning Electron Microscopy and X-Ray Microanalysis*, 4th ed., Springer, 2018, p 139
59. J. Melngailis, Critical Review: Focused Ion Beam Technology and Applications, *J. Vac. Sci. Technol. B*, Vol 5 (No. 2), March/April 1987, p 469–495
60. N.D. Bassim, B.T. De Gregorio, A.L.D. Kilcoyne, K. Scott, T. Chou, S. Wirick, G. Cody, and R.M. Stroud, Minimizing Damage during FIB Sample Preparation of Soft Materials, *J. Microsc.*, Vol 245 (No. 3), 2012, p 288–301
61. T. Ishitani and H. Kaga, Calculation of the Local Temperature Rise in Focused Ion Beam Sample Preparation, *J. Electron. Microsc.*, Vol 44, 1995, p 331–336
62. B. Ziaja, R. London, and J. Hajdu, Ionization by Impact Electrons in Solids: Electron Mean Free Path Fitted over a Wide Energy Range, *J. Appl. Phys.*, Vol 99, 2006, p 033514
63. B.W. Kempshall, S.M. Schwarz, B.I. Prenitzer, L.A. Giannuzzi, R.B. Irwin, and F.A. Stevie, Ion Channeling Effects on the Focused Ion Beam Milling of Cu, *J. Vac. Sci. Technol. B*, Vol 19 (No. 3), 2001, p 749–754
64. J.A. Taillon, V. Ray, and L.G. Salamanca-Riba, Teaching an Old Material New Tricks: Easy and Inexpensive Focused Ion Beam (FIB) Sample Protection Using Conductive Polymers, *Microsc. Microanal.*, Vol 23 (No. 4), 2017, p 872–877
65. N. Brodusch, H. Demers, and R. Gauvin, Ionic Liquid Used for Charge Compensation for High-Resolution Imaging and Analysis in the FE-SEM, *Microsc. Microanal.*, Vol 20 (S3), 2014, p 38–39
66. A. Denisyuk, T. Hrnčíř, J.V. Oboňa, Sharang M. Petrenec, and J. Michalička, Mitigating Curtaining Artifacts during Ga FIB TEM Lamella Preparation of a 14 nm FinFET Device, *Microsc. Microanal.*, Vol 23 (No. 3), June 2017, p 484–490
67. C. Schankula, C. Anand, and N. Bassim, Multi-Angle Plasma Focused Ion Beam (FIB) Curtaining Artifact Correction Using a Fourier-Based Linear Optimization Model, *Microsc. Microanal.*, Vol 24 (No. 6), 2018, p 657–666
68. J. Schwartz, Y. Jiang, Y. Wang, A. Aiello, P. Bhattacharya, H. Yuan, Z. Mi, N. Bassim, and R. Hovden, Removing Stripes, Scratches, and Curtaining with Nonrecoverable Compressed Sensing, *Microsc. Microanal.*, Vol 25 (No. 3), 2019, p 705–710, doi: 10.1017/S1431927619000254
69. K. Narayan, C.M. Danielson, K. Lagarec, B. Lowekamp, P. Coffman, A. Laquerre, M. Phaneuf, T. Hope, and S. Subramaniam, Multi-Resolution Correlative Focused Ion Beam Scanning Electron Microscopy: Applications to Cell Biology, *J. Struct. Bio.*, Vol 185 (No. 3), 2014, p 278–284
70. D. Kiener, C. Motz, G. Dehm, R. Pippan, and R. Pippan, Overview on Established and Novel FIB-Based Miniaturized Mechanical Testing Using In Situ SEM, *Int. J. Mater. Res.* (formerly *Z. Metallkd.*), Vol 100 (No. 8), Aug 2009, p 1074–1087
71. L.A. Giannuzzi, XpressLO for Fast and Versatile FIB Specimen Preparation, *Microsc. Microanal.*, Vol 18 (S2), 2012, p 632, 633
72. L.A. Giannuzzi, Z. Yu, D. Yin, M.P. Harmer, Q. Xu, N.S. Smith, L. Chan, J. Hiller, D. Hess, and T. Clark, Theory and New Applications of Ex Situ Lift Out, *Microsc. Microanal.*, Vol 21, 2015, p 1034–1048
73. E.W. Müller, J.A. Panitz, and S.B. McLane, The Atom-Probe Field Ion Microscope, *Rev. Sci. Instrum.*, Vol 39 (No. 1), 1968, p 83–86
74. B. Gault, M.P. Moody, J.M Cairney, and S.P. Ringer, *Atom Probe Microscopy*, Springer, New York, 2012
75. M. Herbig, D. Raabe, Y.J. Li, P. Choi, S. Zaefferer, and S. Goto, Atomic-Scale Quantification of Grain Boundary Segregation in Nanocrystalline Material, *Phys. Rev. Lett.*, Vol 112 (No. 12), 2014, p 126103
76. Z.B. Jiao, J.H. Luan, M.K. Miller, and C. T. Liu, Precipitation Mechanism and Mechanical Properties of an Ultra-High Strength Steel Hardened by Nanoscale NiAl and Cu Particles, *Acta Mater.*, Vol 97, 2015, p 58–67
77. A.K. Kambham, J. Mody, M. Gilbert, S. Koelling, and W. Vandervorst, Atom-Probe for FinFET Dopant Characterization, *Ultramicroscopy*, Vol 111 (No. 6), 2011, p 535–539
78. B. Gault, F. Vurpillot, A. Vella, M. Gilbert, A. Menand, D. Blavette, and B. Deconihout, Design of a Femtosecond Laser Assisted Tomographic Atom Probe, *Rev. Sci. Instrum.*, Vol 77 (No. 4), 2006, p 043705
79. M. Miller, K. Russell, K. Thompson, R. Alvis, and D. Larson, Review of Atom Probe FIB-Based Specimen Preparation Methods, *Microsc. Microanal.*, Vol 13 (No. 6), 2007, p 428–436
80. M. Herbig, M. Kuzmina, C. Haase, R.K. W. Marceau, I. Gutierrez-Urrutia, D. Haley, D.A. Molodov, P. Choi, and D. Raabe, Grain Boundary Segregation in Fe-Mn-C Twinning-Induced Plasticity Steels Studied by Correlative Electron Backscatter Diffraction and Atom Probe Tomography, *Acta Mater.*, Vol 83, 2015, p 37–47
81. K. Babinsky, R. De Kloe, H. Clemens, and S. Primig, A Novel Approach for Site-Specific Atom Probe Specimen Preparation by Focused Ion Beam and Transmission Electron Backscatter Diffraction, *Ultramicroscopy*, Vol 144, 2014, p 9–18
82. R.P. Kolli, Atom Probe Tomography: A Review of Correlative Analysis of Interfaces and Precipitates in Metals and Alloys, *JOM*, Vol 70 (No. 9), 2018, p 1725–1735
83. V. Castaldo, "High Resolution Scanning Ion Microscopy," Ph.D. thesis, Technical University of Delft, 2011
84. J. Orloff, L.W. Swanson, and M. Utlaut, Fundamental Limits to Imaging Resolution for Focused Ion Beams, *J. Vac. Sci. Technol. B*, Vol 14 (No. 6), Nov/Dec 1996, p 3759–3763
85. T. Ishitani, T. Yamanaka, K. Inai, and K. Ohya, Secondary Electron Emission in Scanning Ga Ion, He Ion and Electron Microscopes, *Vacuum*, Vol 84, 2010
86. S. Sijbrandij, J. Notte, C. Sanford, and R. Hill, Analysis of Subsurface Beam Spread and Its Impact on the Image Resolution of the Helium Ion Microscope, *J. Vac. Sci. Technol. B*, Vol 28 (No. 6), Nov/Dec 2010
87. A. Beyer, H. Vieker, R. Klett, H.M. zu Theenhausen, P. Angelova, and A. Gölzhaäuser, Imaging of Carbon Nanomembranes with Helium Ion Microscopy, *Beilstein J. Nanotechnol.*, Vol 6, 2015, p 1712–1720
88. R.E. Franklin, E.C.G. Kirk, J.R.A. Cleaver, and H. Ahmed, Channeling Ion Image Contrast and Sputtering in Gold Specimens Observed in a High-Resolution Scanning Ion Microscope, *J. Mater. Sci. Lett.*, Vol 7, 1988, p 39–41
89. D. Xia, S. McVey, C. Huynh, and W. Kuehn, Defect Localization and Nanofabrication for Conductive Structures with Voltage Contrast in Helium Ion Microscopy, *ACS Appl. Mater. Interfaces*, Vol 11, 2019, p 5509–5516
90. A. Twedt, L. Chen, and J.J. McClelland, Scanning Ion Microscopy with Low Energy Lithium Ions, *Ultramicroscopy*, Vol 142, July 2014, p 24–31
91. M. Semple, E. Baladi, and A.K. Iyer, Optical Metasurface Based on Subwavelength Nanoplasmonic Metamaterial Lined Apertures, *IEEE J. Select. Topics*

Quant. Electron., Vol 25 (No. 3), May/June 2019, p 1–8
92. D. Xia, C. Huynh, S. McVey, A. Kobler, L. Stern, Z. Yuan, and X.S. Ling, Rapid Fabrication of Solid-State Nanopores with High Reproducibility over a Large Area Using a Helium Ion Microscope, *Nanoscale*, Vol 10, 2018, p 5198–5204
93. W. Hu, K. Sarveswaren, M. Lieberman, and G.H. Bernstein, Sub-10 nm Electron Beam Lithography Using Cold Development of Poly(Methylmethacrylate), *J. Vac. Sci. Technol. B*, Vol 22 (No. 4), July/Aug 2004, p 1711–1716
94. D. Winston, B.M. Cord, B. Ming, D.C. Bell, W.F. DiNatale, L.A. Stern, A.E. Vladar, M.T. Postek, M.K. Mondol, J.K. W. Yang, and K. Berggren, Scanning Helium Ion Beam Lithography with Hydrogen Silsesquioxane Resist, *J. Vac. Sci. Technol. B*, Vol 27 (No. 6), Nov/Dec 2009, p 2702–2706
95. F. Luo, V. Manichev, M. Li, G. Mitchson, B. Yakshinskiy, T. Gustafsson, D. Johnson, and E. Garfunkel, Helium Ion Beam Lithography (HIBL) Using HafSOx as the Resist, *Proc. SPIE Adv. Lithogr. Conf.*, Vol 9779, 2016, p 977928
96. Z. Xianghui, H. Vieker, A. Beyer, and A. Gölzhäuser, Fabrication of Carbon Nanomembranes by Helium Ion Beam Lithography, *Beilstein J. Nanotechnol.*, Vol 5 (No. 1), 2014, p 188–194
97. V. Manichev, F. Yu, D. Hutchison, M. Nyman, T. Gustafsson, L. Feldman, and E. Garfunkel, Novel Sn-Based Photoresist for High Aspect Ratio Patterning, *Proc. SPIE Adv. Lithogr. Conf.*, Vol 10586, 2018, p 105860K-1058608
98. V. Viswanathan, H. Hao, M.S.M. Saifullah, R. Ganesan, S. Lim, H. Hussain, and D. Pickard, National University of Singapore and Institute of Material Research and Engineering, Singapore, unpublished results,
99. X. Shi, P. Prewett, E. Huq, D.M. Bagnall, A.P.G. Robinson, and S.A. Boden, Helium Ion Beam Lithography on Fullerene Molecular Resists for Sub-10 nm Patterning, *Microelectron. Eng.*, Vol 155, 2016, p 74–78
100. B. Wu and A. Kumar, Extreme Ultraviolet Lithography: A Review, *J. Vac. Sci. Technol. B*, Vol 25 (No. 6), Nov 2007, p 1743–1761
101. J. Melngailis, Focused Ion Beam Lithography, *Nucl. Instrum. Methods Phys. Res. B*, Vol 80–81 (Part 2), 1993, p 1271–1280
102. S.J. Wind, P.D. Greber, and H. Rothuizen, Accuracy and Efficiency in Electron Beam Proximity Effect Correction, *J. Vac. Sci. Technol. B*, Vol 16 (No. 6), 1998, p 3262–3268
103. N. Kalhor, W. Mulckhuyse, P. Alkemade, and D. Maas, Impact of Pixel-Dose Optimization on Pattern Fidelity for Helium Ion Beam Lithography on EUV Resist, *Proc. SPIE Adv. Lithogr. Conf.*, Vol 9425, 2015, p 942513
104. I. Utke, P. Hoffmann, and J. Melngailis, Gas-Assisted Focused Electron Beam and Ion Beam Processing and Fabrication, *J. Vac. Sci. Technol. B*, Vol 26 (No. 4), July/Aug 2008, p 1197–1276
105. C.-S. Kim, S.-H. Ahn, and D.-Y. Jang, Review: Developments in Micro/Nanoscale Fabrication by Focused Ion Beams, *Vacuum*, Vol 86 (No. 8), 2012, p 1014–1035
106. C. Kang, C. Chandler, and M. Weschler, Chap. 3, Gas Assisted Ion Beam Etching and Deposition, *Focused Ion Beam Systems*, N. Yao, Ed., Cambridge University Press, 2007
107. H. Wu, L.A. Stern, D. Xia, D. Ferranti, B. Thompson, K.K. Klein, C.M. Gonzalez, and P.D. Rack, Focused Helium Ion Beam Deposited Low Resistivity Cobalt Metal Lines with 10 nm Resolution: Implications for Advanced Circuit Editing, *J. Mater. Sci.: Mater. Electron.*, Vol 25, 2014, p 587–595
108. H. Wu, L.A. Stern, and D. Ferranti, "Insulator Deposition Induced by Gas Field Ion Source (GFIS) Column: Ultrahigh Resistivity and High Resolution with Zeiss Orion NanoFab," https://www.selectscience.net/application-notes/insulator-deposition-induced-by-gas-field-ion-source-(gfis)-column-ultrahigh-resistivity-and-high-resolution-with-zeiss-orion-nanofab/?artID=35730, accessed Feb 7, 2019
109. I. Utke, Chap. 6, FEB and FIB Continuum Models for One Adsorbate Species, *Nanofabrication Using Focused Ion and Electron Beams*, I. Utke, S. Moshkalev, and P. Russell, Ed., Oxford University Press, New York, NY, 2012
110. S. Matsui, Three-Dimensional Nanostructure Fabrication by Focused-Ion Beam Chemical Vapor Deposition, *Proc. 12th Int. Conf. Solid-State Sensors, Actuators, Microsyst.*, Vol 1, 2003, p 179–181
111. A. Belianinov, M.J. Burch, S. Kim, S. Tan, G. Hlawacek, and O.S. Ovchinnikova, Noble Gas Ion Beams in Materials Science for Future Applications and Devices, *MRS Bull.*, Vol 42, 201, p 660–666
112. S. Tan, R.H. Livengood, Y. Greenzweig, Y. Drezner, R. Hallstein, and C. Scheffler, Characterization of Ion Beam Current Distribution Influences on Nanomachining, *Conf. Proc. 38th Int. Symp. Test. Failure Anal.*, 2012, p 436–439
113. S. Tan and R. Livengood, Chap. 19, Application of GFIS in Semiconductors, *Helium Ion Microscopy*, G. Hlawacek and A. Gölzhäuser, Ed., Springer International, 2016
114. D. Xia, J. Notte, L. Stern, and B. Goetze, Enhancement of XeF_2-Assisted Gallium Ion Beam Etching of Silicon Layer and Endpoint Detection from Backside in Circuit Editing, *J. Vac. Sci. Technol. B*, Vol 33 (No. 6), Nov/Dec 2015, p 06F501
115. Y. Zhao, D. Liu, J. Chen, L. Zhu, A. Belianinov, O.S. Ovchinnikova, R.R. Unocic, M.J. Burch, S. Kim, H. Hao, D. S. Pickard, B. Li, and J.T.L. Thong, Engineering the Thermal Conductivity along an Individual Silicon Nanowire by Selective Helium Ion Irradiation, *Nature Commun.*, Vol 8, 2017, p 15919
116. R. Bali, S. Wintz, F. Meutzner, R. Hübner, R. Boucher, A.A. Ünal, S. Valencia, A. Neudert, K. Potzger, J. Bauch, F. Kronast, S. Facsko, J. Lindner, and J. Fassbender, Printing Nearly-Discrete Magnetic Patterns Using Chemical Disorder Induced Ferromagnetism, *Nano Lett.*, Vol 14, 2014, p 435–441
117. S.A. Cybart, E.Y. Cho, T.J. Wong, B.H. Wehlin, M.K. Ma, C. Huynh, and R.C. Dynes, Nano Josephson Superconducting Tunnel Junctions in Y-Ba-Cu-O Direct-Patterned with a Focused Helium Ion Beam, *Nature Nanotechnol.*, Vol 10 (No. 7), 2015, p 598–602
118. J. Burnett, J. Sagar, P.A. Warburton, and J.C. Fenton, Embedding NbN Nanowires into Quantum Circuits with a Neon Focused Ion Beam, *IEEE Trans. Appl. Superconduct.*, Vol 26 (No. 3), April 2016
119. Z. Liu, H. Du, J. Li, L. Lu, Z.-Y. Li, and N.X. Fang, Nano-Kirigami with Giant Optical Chirality, *Sci. Adv.*, Vol 4 (No. 7), 2018
120. P. Echlin, *Low-Temperature Microscopy and Analysis*, Springer Science and Business Media, 2013
121. A. Rigort and J.M. Plitzko, Cryo-Focused-Ion-Beam Applications in Structural Biology, *Arch. Biochem. Biophys.*, Vol 581, 2015, p 122–130
122. G. Desbois, J. Urai, C. Burkhardt, M. Drury, M. Hayles, and B. Humbel, Cryogenic Vitrification and 3D Serial Sectioning Using High Resolution Cryo-FIB SEM Technology for Brine-Filled Grain Boundaries in Halite: First Results, *Geofluids*, Vol 8 (No. 1), 2008, p 60–72
123. N. Antoniou, A. Graham, C. Hartfield, and G. Amador, Failure Analysis of Electronic Material Using Cryogenic FIB-SEM, *Conf. Proc. 38th Int. Symp. Test. Failure Anal.* (Phoenix, AZ), 2012, p 399–405
124. A. Meents, S. Gutmann, A. Wagner, and C. Schulze-Briese, Origin and Temperature Dependence of Radiation Damage in Biological Samples at Cryogenic Temperatures, *Proc. Natl. Acad. Sci.*, Vol 107 (No. 3), 2010, p 1094–1099
125. M.J. Zachman, Z. Tu, S. Choudhury, L. A. Archer, and L.F. Kourkoutis, Cryo-STEM Mapping of Solid-Liquid Interfaces and Dendrites in Lithium-Metal Batteries, *Nature*, Vol 560 (No. 7718), 2018, p 345

126. K.E. Knipling, D.J. Rowenhorst, R.W. Fonda, and G. Spanos, Effects of Focused Ion Beam Milling on Austenite Stability in Ferrous Alloys, *Mater. Charact.*, Vol 61 (No. 1), 2010, p 1–6
127. M. Hayles, D. Stokes, D. Phifer, and K.C. Findlay, A Technique for Improved Focused Ion Beam Milling of Cryo-Prepared Life Science Specimens, *J. Microsc.*, Vol 226 (No. 3), 2007, p 263–269
128. M. Marko, C. Hsieh, P. Schalek, J. Frank, and C. Mannella, Focused-Ion-Beam Thinning of Frozen-Hydrated Biological Specimens for Cryo-Electron Microscopy, *Nature Meth.*, Vol 4 (No. 3), 2007, p 215
129. M. Schaffer, J. Mahamid, B. Engel, T.T. Laugks, W. Baumeister, and J. Plitzko, Optimized Cryo-Focused Ion Beam Sample Preparation Aimed at In Situ Structural Studies of Membrane Proteins, *J. Struct. Bio.*, Vol 197 (No. 2), 2017, p 73–82
130. M. Zachman, E. Asenath-Smith, L. Estroff, and L. Kourkoutis, Revealing the Internal Structure and Local Chemistry of Nanocrystals Grown in Hydrogel with Cryo-FIB Lift-Out and Cryo-STEM, *Microsc. Microanal.*, Vol 21, 2015, p 2291
131. Y. Chang, W. Lu, J. Guénolé, L. Stephenson, A. Szczpaniak, P. Kontis, A. Ackerman, F. Dear, I. Mouton, X. Zhong, and S. Zhang, Ti and Its Alloys as Examples of Cryogenic Focused Ion Beam Milling of Environmentally-Sensitive Materials, *Nature Commun.*, Vol 10 (No. 1), 2019, p 942
132. A. Benninghoven, F.G. Rudenauer, and H.W. Werner, *Secondary Ion Mass Spectrometry*, Wiley Interscience, New York, NY, 1987
133. L.C. Feldman and J.W. Mayer, *Fundamentals of Surface and Thin Film Analysis*, P.T.R. Prentice-Hall, Upper Saddle River, NJ, 1986
134. J.W. Rabalais, *Principles and Applications of Ion Scattering Spectroscopy*, Wiley Interscience, Hoboken, NJ, 2003
135. B. Lewis, F. Khanom, and J. Notte, NanoFab with SIMS—Recent Results from the BAM-L200 Analytical Standard and Semiconductor Samples, *Microsc. Microanal.*, Vol 24 (S1), 2018, p 850–851
136. P. Gratia, G. Grancini, J.N. Audinot, X. Jeanbourquin, E. Mosconi, I. Zimmermann, D. Dowsett, Y. Lee, M. Grätzel, F. De Angelis, K. Sivula, T. Wirtz, and M.K. Nazeeruddin, Intrinsic Halide Segregation at Nanometer Scale Determines the High Efficiency of Mixed Cation/Mixed Halide Perovskite Solar Cells, *J. Am. Chem. Soc.*, Vol 138, 2016, p 15821–15824
137. N. Klingner, R. Heller, G. Hlawacek, S. Facsko, and J. Boranyvon, Time-of-Flight Secondary Ion Mass Spectrometry in the Helium Ion Microscope, *Ultramicroscopy*, Vol 198, March 2019, p 10–17
138. S. Sijbrandij, B. Thompson, J. Notte, B. W. Ward, and N.P. Economou, Elemental Analysis with the Helium Ion Microscope, *J. Vac. Sci. Technol. B*, Vol 26 (No. 6), 2008, p 2103–2106
139. S.A. Boden, T.M.W. Franklin, L. Scipioni, D.M. Bagnall, and H.N. Rutt, Ionoluminescence in the Helium Ion Microscope, *Microsc. Microanal.*, Vol 18 (No. 6), 2012, p 1253–1262
140. R.A. Schwarzer, Spatial Resolution in ACOM—What Will Come after EBSD? *Microsc. Today*, Jan 2008
141. G. Moore, Cramming More Components onto Integrated Circuits, *Electronics*, Vol 38 (No. 8), April 1965
142. P.D. Prewett, A.W. Eastwood, G.S. Turner, and J.G. Watson, Gallium Staining in FIB Repair of Photomasks, *Microelectron. Eng.*, Vol 21, 1993, p 191–196
143. F. Aramaki, T. Kozakai, O. Matsuda, O. Takaoka, Y. Sugiyama, H. Oba, K. Aita, and A. Yasaka, Photomask Repair Technology by Using Gas Field Ion Source, *Bacus News*, Vol 29 (No. 4), 2013
144. L. Zhao, All about Interconnects, *Semicond. Eng.*, Dec 2017
145. H. Wu, D. Ferranti, and L. Stern, Precise Nanofabrication with Multiple Ion Beams for Advanced Circuit Edit, *Microelectron. Reliab.*, Vol 54, 2014, p 1779–1784
146. S. Tan, R. Hallstein, R.H. Livengood, and W. Ali, GFIS in Semiconductor Applications, *Microsc. Microanal.*, Vol 22 (S3), 2016, p 156, 157

Surface Analysis

Dehua Yang, Ebatco

ASM Handbook, Volume 10, *Materials Characterization*
ASM Handbook Committee
DOI 10.31399/asm.hb.v10.a0006675

Introduction to Surface Analysis

Dehua Yang, Ebatco

THIS DIVISION, "Surface Analysis," is a new addition in *Materials Characterization*, Volume 10 of *ASM Handbook* (2019). This division includes newly developed surface-analysis techniques, such as scanning probe and atomic force microscopy. This division focuses on the analysis of surface layers that are less than 100 nm. There is a broad spectrum of techniques that fit into this range of definition. Some techniques not covered in this division include those rarely used or used mostly for academic and research purposes. More commonly used and practical techniques (not already covered in Division 2, "Spectroscopy," or Division 9, "Microscopy and Microanalysis," in this Volume) are covered in this division.

Some methods/techniques that have not been included in this division include:

- Glow discharge optical emission spectroscopy, a surface-sensitive technique (covered in Division 2, "Spectroscopy," in this Volume)
- Electron energy loss spectroscopy (covered in Division 9, "Microscopy and Microanalysis," in this Volume)
- Ion-scattering spectrometry (covered in Division 3, "Mass and Ion Spectrometry," in this Volume)
- Contact angle, a surface-sensitive technique for single-layer atom detection and surface free-energy analysis
- White light interferometry, good for surface roughness analysis and three-dimensional contour determination of the surface
- Atom probe tomography
- Ellipsometry, mostly used for thin-film thickness measurement

The techniques covered in this division are based on probing methods using direct probe contact, electron, ion, photon, thermal, or x-ray interaction between the analytical instrument and the material surfaces. They fit well in the scope of this Volume because the methods provide information about composition, structure, and defects. Further, they are focused on determining and revealing characteristics of surface layers of materials that are less than 100 nm. They also are suited for characterization of other surface properties, such as electrical, magnetic, mechanical, and thermal properties of the surface layers.

The articles included in this division have been contributed by materials characterization and surface-analysis specialists in their particular fields. The articles represent the most widely used, surface-sensitive, practically valuable, advanced, and cutting-edge surface-analysis methods. These techniques are capable of providing elemental composition, chemical state, and other important properties of the outermost atomic layers of metals, semiconductors, ceramics, organic materials, and biomaterials in bulk, thin-film, and coating, powder, or particulate format. To aid readers in achieving speedy selection of these surface-analytical methods for a specific application, Table 1 provides a quick reference summary of the analytical methods presented in this division.

Table 1 Quick reference summary of surface-analysis methods

Analysis method	Analysis probe	Detection signal	Analysis information	Lateral resolution	Depth resolution	Typical applications
Atomic force microscopy (AFM)	Coated or noncoated cantilever probes made of various materials	Laser light	Images of surface topographical features or other near-field and far-field interactions between probe and sample surface	0.2–10 nm	10–80 pm	Biological molecules, biomaterials, cells, crystallography, electrochemistry, polymer chemistry, thin-film studies, nanomaterials, nanotechnology, failure analysis, process development, process control, surface metrology
Auger electron spectroscopy (AES)	Electrons	Auger electrons	Elemental composition analysis for all elements except H and He	20 nm	0.5–5 nm	Adhesion, catalysis, corrosion, oxidation, surface chemical reaction, surface contamination,wear, depth profile of each element with ion gun sputtering
Low-energy electron diffraction (LEED)	Electrons	Diffraction electrons	Surface crystallography and microstructure	10 μm	0.4–2 nm	Adsorption, catalysis, chemical reactions, crystallography, crystal structure in epitaxial growth, film-growth kinetics, grain size and boundary, microstructure, reconstruction, segregation
Scanning probe microscopy (SPM)	Probes made of various materials	Laser light, electrical current, or other probe-sample interactions	Three-dimensional image with atomic resolution of surface properties such as height, electron tunneling current, electrostatic force, magnetic force, etc.	0.2–10 nm	10–80 pm	Broad usage in research and development of nanomaterials, applications of nanotechnology, and micromanufacturing that involves understanding, characterization, and manipulating surfaces at atomic or nanometer scale
Secondary ion mass spectroscopy (SIMS)	Ion beam	Secondary ions	Elemental, isotopic, or molecular composition of the surface through detection of the species with different mass-to-charge ratios	50 nm–10 μm	0.5–10 nm	Concentration depth profiling, identification of inorganic or organic surface layers, isotopic abundances in geological and lunar samples, trace elements detection including hydrogen

(continued)

Table 1 (Continued)

Analysis method	Analysis probe	Detection signal	Analysis information	Lateral resolution	Depth resolution	Typical applications
Thermal desorption spectroscopy (TDS)	Thermal energy	Desorbed atoms and/or molecules	Desorption rate of desorbing gases from surfaces as a function of temperature	Not applicable	Not applicable	Adsorption, desorption, and reaction of adsorbed atoms and molecules on surfaces; catalysis; personal exposure to toxic chemicals; indoor air-quality monitoring; identification of volatiles in soil and water; analysis of environmental pollutants; quantification of hydrogen in metals (especially steels); corrosion mechanisms; electrochemistry; tribology
X-ray photoelectron spectroscopy (XPS)	X-ray	Photoelectrons	Chemical state and composition	20–500 μm for monochromatic x-ray; 10–30 mm for nonmonochromatic x-ray	10 nm	Elemental analysis of all elements with an atomic number of lithium and higher, chemical state identification of surface elements, composition and chemical state depth profiles with ion gun sputtering, determination of oxidation states of metal atoms in metal compounds, identification of surface contaminations, measurement of surface film thickness, identification and degradation of polymers

ASM Handbook, Volume 10, *Materials Characterization*
ASM Handbook Committee
DOI 10.31399/asm.hb.v10.a0006657

Auger Electron Spectroscopy

A. Joshi, Advanced Technology Center, Lockheed Martin Space
D.F. Paul, Physical Electronics USA

Overview

General Uses

- Compositional analysis of the 0.5 to 5 nm region near the surface for all elements except hydrogen and helium.
- Chemical-state analysis is possible in selected cases.
- Depth-compositional profiling and thin-film analysis
- High lateral-resolution surface analysis in areas ≥20 nm

Examples of Applications

- Analysis of surface contamination of materials to investigate its role in properties such as adhesion, wear, corrosion, secondary electron emission, and catalysis
- Identification of chemical-reaction products, for example, in oxidation and corrosion
- In-depth compositional evaluation of surface films, coatings, and thin films used for various metallurgical surface modifications and microelectronics applications
- Evaluation of surface and buried defects in microelectronics
- Analysis of grain-boundary chemistry to evaluate the role of precipitation and impurity segregation on mechanical properties, corrosion, and stress-corrosion cracking

Samples

- *Form:* Solids (metals, ceramics, and organic materials with relatively low vapor pressures). High-vapor-pressure materials can be handled by sample cooling. Many liquid samples can be handled by sample cooling or by applying the sample as a thin film onto a conductive substrate.
- *Size:* Individual powder particles as small as 20 nm in diameter can be analyzed. Maximum sample size depends on the instrument; typical is 6 cm (2.4 in.) in diameter by 1.5 cm (0.6 in.) in height.
- *Surface topography:* Flat surfaces are preferable, but rough surfaces such as fracture surfaces can be analyzed.
- *Preparation:* Frequently not necessary. Samples must be free of fingerprints, oils, and other high-vapor-pressure materials.

Limitations

- Cannot detect hydrogen and helium
- The accuracy of quantitative analysis is limited to ±30% of the element present when calculated using published elemental sensitivity factors (Ref 1–5). Better quantification is possible by using standards that closely resemble the sample.
- Electron beam damage can severely limit the useful analysis of organic and biological materials and some ceramic materials.
- Electron beam charging can limit analysis when examining highly insulating materials. Ion beam neutralization and other approaches are available to minimize charging.
- Quantitative detection sensitivity for most elements is in the 0.1 to 1.0 at.% range.
- Ultrahigh vacuum is required.

Estimated Analysis Time

- Usually 5 to 30 min for a complete survey spectrum, longer times needed when using smaller beam sizes and current. Selected peak analyses for studying chemical effects, elemental imaging, and depth profiling generally take longer times.

Capabilities of Related Techniques

- *X-ray photoelectron spectroscopy (XPS):* The most nondestructive among the surface analytical methods, providing compositional and chemical-state information with better accuracy compared with AES. Depth-profiling capability with ion beam sputtering.
- *Ion scattering spectroscopy:* Provides top atomic-layer surface composition for all elements heavier than primary ion, specificity of surface atomic bonding in selected cases, depth-profiling information with ion beam sputtering
- *Secondary ion mass spectroscopy (SIMS):* High elemental-detection sensitivity from parts per million to parts per billion levels; surface compositional information; molecular bonding information leading to chemical compound identification in organic materials
- *Scanning electron microscopy with x-ray analysis using energy-dispersive spectroscopy or wavelength-dispersive spectroscopy:* Analysis to 1 μm depth in conventional operation; can be quantitative

Background and Principles

Auger electron spectroscopy (AES), x-ray photoelectron spectroscopy, secondary ion mass spectroscopy, and low-energy ion-scattering spectroscopy, discussed in articles in this Volume, are among the most widely used surface-sensitive analytical techniques capable of providing elemental composition of the outermost atomic layers of a solid. These techniques are used to investigate the surface chemistry and interactions of solid surfaces of metals, semiconductors, ceramics, organic materials, and biomaterials. The techniques use electrons, x-rays, and ions as the probing sources, and the surface chemical information is derived from analysis of electrons and ions emitted from the surface.

Exposing a sample to high-energy electrons emits a variety of electrons, including secondary electrons, backscattered electrons, and Auger electrons. Auger electron spectroscopy involves precise measurement of the number of emitted Auger electrons as a function of kinetic energy. Auger electrons are characteristic of the element from which they are emitted and are useful in qualitative and semiquantitative surface analysis. Auger electrons were discovered in 1925 (Ref 6), and the utility of the technique for surface analysis was demonstrated in 1968 (Ref 7). Numerous advances in experimental methods (Ref 8, 9), spectral interpretations, data-manipulation techniques, and various application fields made AES an effective surface analytical technique. Detailed descriptions of the principles and applications of AES are available in Ref 10 to 23.

Auger electrons are produced when incident radiation (photons, electrons, ions, or neutral atoms) interacts with an atom, and the radiation energy exceeds that necessary to remove an inner-shell electron (K, L, M, etc.) from the atom. The interaction, or scattering process, leaves the atom in an excited state with a core hole, or missing inner-shell electron. Excited atoms are unstable, and de-excitation occurs immediately, resulting in the emission of an x-ray or a low-energy electron termed an Auger electron (Fig. 1). Figure 1(a) illustrates the process of an atom in the initial state excited by an electron or other incident particle, resulting in the emission of an x-ray with a characteristic energy of $E_K - E_{L2}$. Alternatively, an Auger electron may be emitted with a probable transition shown in Fig. 1(b), which leaves the atom in a doubly ionized final state. The electron is identified as a KL_2L_3 Auger electron, or simply as a KLL electron, representative of its shells of origination.

For the first approximation, ignoring the binding energy changes due to ionization, the kinetic energy of this Auger electron E_{ke} is given by:

$$E_{ke} \approx E_K - E_{L2} - E_{L3} - \Phi \qquad \text{(Eq 1)}$$

where Φ is the work function of the sample, and E_{ke} is the kinetic energy of the electron escaping the sample surface. In practice, when measurements are conducted using an electron spectrometer in electrical contact with the sample, the work function of the analyzer is more significant than that of the sample (Ref 24). The typical work function of an analyzer is in the 3 to 5 eV range and rarely changes. Thus, the kinetic energy of an Auger electron is characteristic of the three electron binding levels of the atom from which it originates. Generation of an Auger electron requires participation of at least three electrons, which excludes hydrogen and helium from being detected by AES.

Auger Emissions and Light-Element Sensitivity

An excited atom with a core-level (electron) hole can decay to a lower-energy state in several ways, of which the Auger and x-ray emission (fluorescence) processes are the most probable. Auger de-excitation is a common mode for orbitals involving low energies; x-ray emission is equally probable or dominant for strongly bound orbitals. The competitive yields for Auger and fluorescence are shown in Fig. 2. All elements (except hydrogen and helium) produce high Auger yields, making AES highly sensitive for light-element detection.

Auger Emission Probabilities and Qualitative Analysis

The example shown in Fig. 1 describes emission of a specific Auger electron via a KL_2L_3 transition. Because most atoms have several electron shells and subshells, emission of various other electrons becomes probable. The series of transitions from various levels, often represented by the series KLL, LMM, MNN, and so on, are often the dominant transitions. For example, Fig. 3 shows the energies at which principal Auger electrons appear for each element and their relative intensities. Other Auger transitions involving the valence band, such as KVV, LVV, and so on, are also possible as valance band electrons fill the core holes.

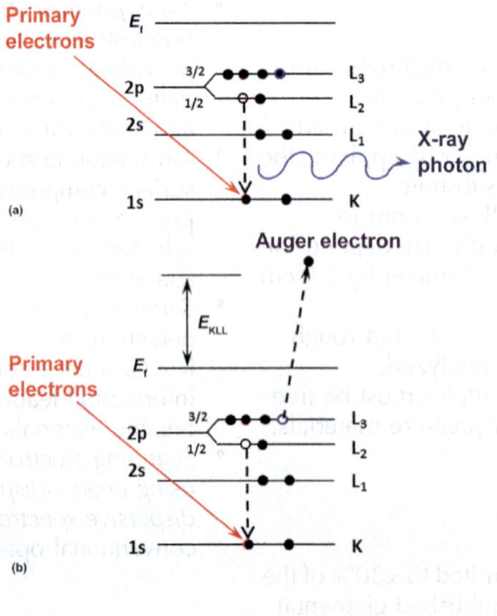

Fig. 1 Schematic of energy-level diagrams showing (a) x-ray photon emission and (b) Auger electron emission. Courtesy of Physical Electronics, USA

Mode of Primary Excitation

Primary excitation can be achieved by various energetic particles, such as x-rays, electrons, and ions, resulting in a final state culminating in Auger electron emission. Electron beam excitation is the most common because of its ability to finely focus to a small spot, enabling small-area analysis and imaging using secondary, backscattered, and Auger electrons. In x-ray photoelectron spectroscopy (XPS), bombardment of the sample surface with x-rays results in photoelectron and Auger electron emission. Thus, an XPS spectrum contains both sets of peaks representing the sample surface and provides additional chemical-state information.

Electron Energy Distribution

Figure 4(a) shows the intensity of electrons, expressed as $EN(E)$, as a function of energy obtained from a nickel specimen. Auger electron peaks are readily visible in the spectrum, but they are relatively small because only ~0.1% of the total current is typically contained in Auger peaks. Low-energy peaks, such as the 64 eV nickel peak, lie on a large slope of true secondary electrons, and high-energy peaks reside on a high background of backscattered electrons. Taking the derivative, $dN(E)/dE$, of the spectrum (Fig. 4b) resolves the high background issue and has become the common practice of displaying AES spectra. The peak positions in the $dN(E)/dE$ spectrum are commonly used in identifying elements, and it must be noted that their positions are slightly different in the $EN(E)$ spectrum. The Auger peak-to-peak heights are measured from the most positive to the most negative excursion in the dN/dE spectra and are

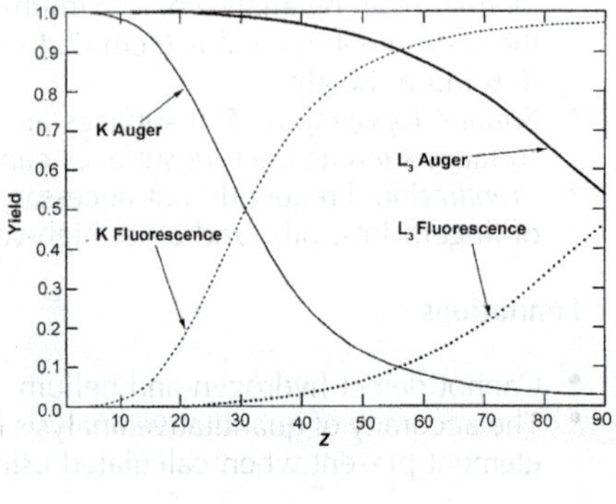

Fig. 2 Efficiencies of x-ray emission (fluorescence) compared to Auger yields. Source: Ref 25, 26

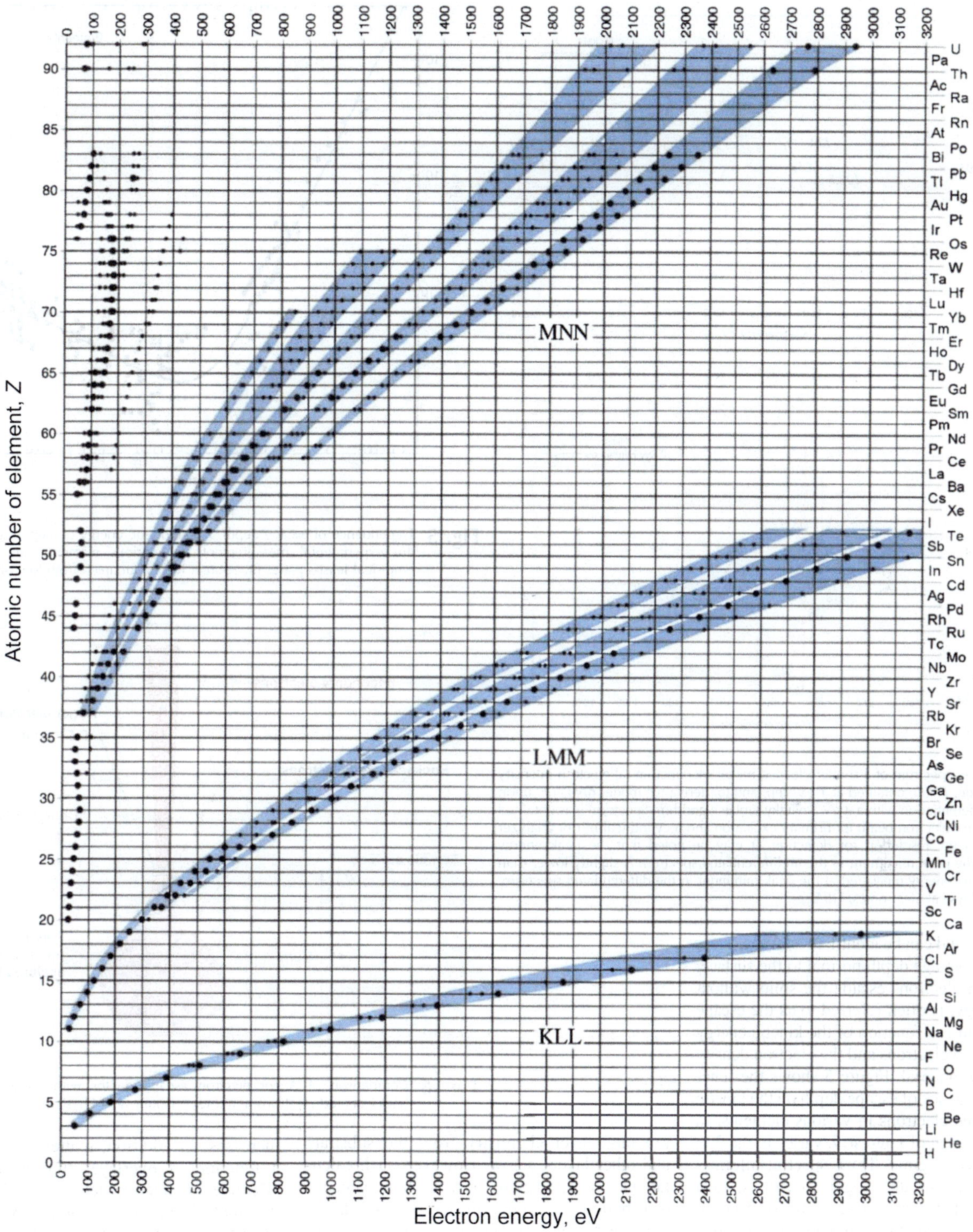

Fig. 3 Chart of principal Auger electron energies. Data points indicate the electron energies of the principal Auger peaks from each element. Larger data points represent the most intense peaks for each element. Source: Ref 1. Courtesy of Physical Electronics, USA

proportional to the number of atoms emitting the Auger electrons. These measurements are commonly used to quantify Auger spectra. The peak areas under the $N(E)$ curve are also proportional to the number of atoms emitting the Auger electrons, but they are not widely used due to difficulties in determining the end points in a rapidly varying background, particularly at low energies.

Surface Sensitivity

The most useful kinetic energy range of Auger electrons for analytical purposes is from 20 to 2500 eV, which corresponds to electrons with high-scattering cross sections in solids. As these low-energy electrons emanate from the solid, they undergo inelastic scattering events and lose their characteristic energy. Only those electrons from the near-surface region can escape without losing a significant portion of

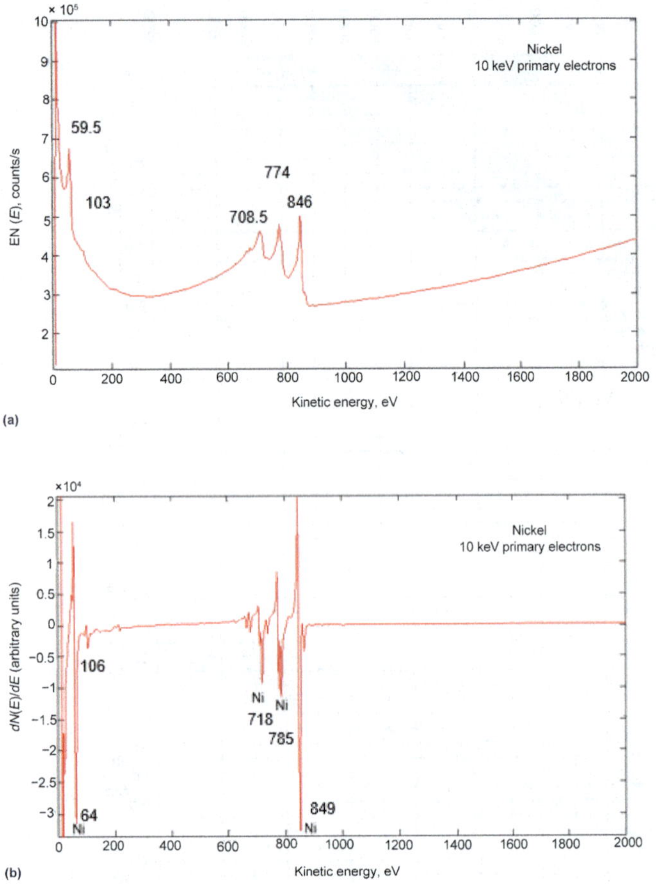

Fig. 4 Auger spectrum of a sputter-cleaned nickel specimen in (a) electron-intensity mode, *EN*(*E*), obtained at 1 eV/step, and (b) derivative mode, *dN*(*E*)/*dE*, upon numerical differentiation with a 4 eV Savitzky-Golay algorithm. Spectrum was obtained by using 10 keV electron beam for primary excitation and a cylindrical-mirror analyzer with 0.6% energy resolution for detection. It may be noted that the peak energy measured at the peak maximum in the undifferentiated spectrum is slightly lower than that of the peak energy measured at the peak minimum in the differentiated spectrum. Courtesy of Physical Electronics, USA

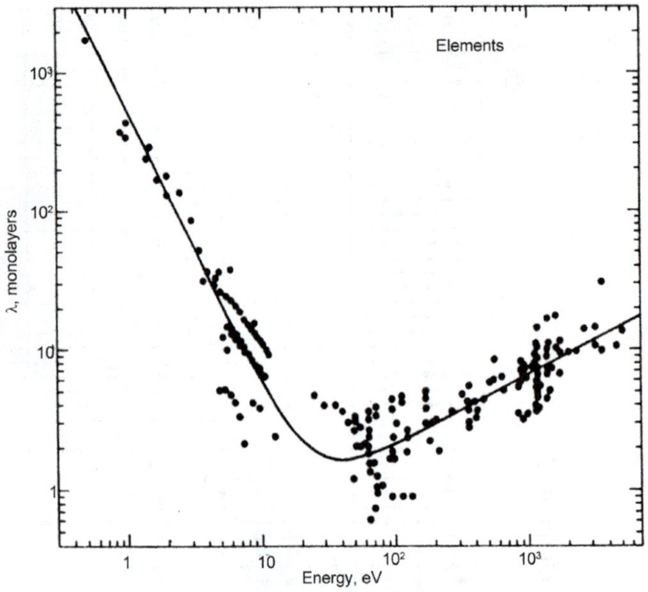

Fig. 5 Dependence of escape depth, λ, on kinetic energy of electrons. Data points are experimental measurements for various elements; full curve is an empirical least-squares fit. Source: Ref 27 (adapted from Ref 14)

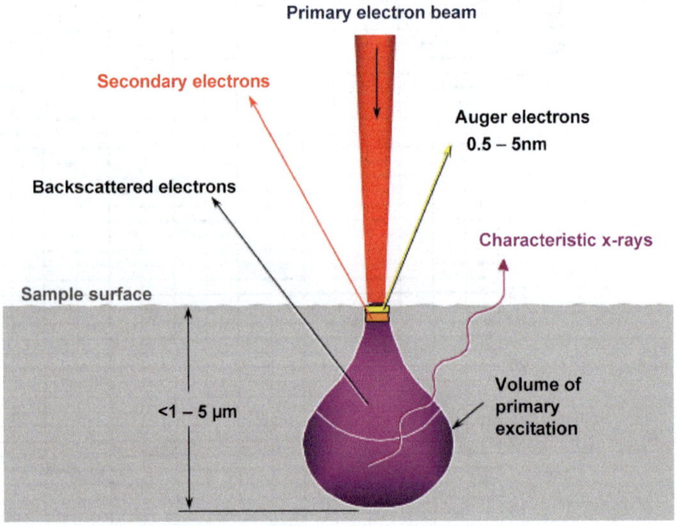

Fig. 6 Comparison of Auger electron escape depths with emission depths of backscattered electrons and x-rays. Courtesy of Physical Electronics, USA

their energy and can be identified as Auger electrons. This average depth normal to the surface, from which electrons escape the solid without losing energy, is often referred to as the escape depth, λ, and is a function of the kinetic energy of the emitted electrons and the electron density of the solid material. Figure 5 shows the functional dependence of escape depth on the kinetic energy of the electrons in various elements. In the energy range of interest, λ varies in the 2 to 10 monolayer regime. Thus, spectral information contained in the Auger spectra is, to a greater extent, representative of the top 5 nm of the surface, making the technique surface sensitive. Note that escape depth is independent of the primary electron energy used for excitation.

Auger electrons undergo exponential decay as a function of distance, *d*, from the point of origin, following the expression exp ($-d/\lambda$). Nearly 63.2% of the signal seen at the surface is from the depth λ, and 95% of the signal is from 3λ. Thus, a typical spectrum has 95% of the Auger electrons detected from a depth of no more than 10 monolayers. This exponential dependence can also be used to distinguish top monolayer chemistry from the substrate and other layers below. However, such calculations are applicable only to homogeneous distribution of atoms in the surface layer, which is rare in many practical situations. Tougaard (Ref 28) and Seah (Ref 29) discuss several approaches to quantify surface layers and other forms of surface inhomogeneity.

While inelastic scattering of electrons is the primary mechanism leading to loss in energy, there can be a significant fraction of Auger electrons (or photoelectrons, in case of XPS) that undergoes elastic electron collisions. Elastic scattering effects are significant at large angles of emission (greater than 60° to the surface normal), where the attenuation length increases. In addition, scatter effects are also a function of overlayer film thickness. The effective attenuation lengths (EALs) can be different from the theoretically calculated inelastic mean free paths (IMFPs) and become important in quantification, in thickness estimates of a surface layer, and in measurements involving a thin marker layer close to the surface (Ref 30, 31). The National Institute of Standards and Technology (NIST) developed EAL databases to correct for the IMFP values (Ref 32).

The escape depth of Auger electrons is small compared with that of characteristic x-rays used in microprobe analysis. The x-ray analysis volume (Fig. 6) is typically ~1 μm^3

and can complement AES information derived from the top 5 nm.

Chemical Effects in Auger Electron Spectroscopy

Changes in the chemical environment of atoms in the surface region of a solid can affect the measured Auger spectrum. For example, the kinetic energy at which an Auger transition occurs, Auger peak shape, and/or the loss structure associated with Auger transitions can change. Precise measurements to determine energy shifts and peak shape changes are useful in identifying the chemical states of surface atoms.

Energy shifts are expected when there is a charge transfer from one atom to another. In ionic bonding, the net electron transfer causes core-level electrons of electronegative elements to shift to lower binding energies and those of electropositive elements to shift to higher binding energies. The result is often a shift of several electron volts in the Auger peaks compared with their zero-valence states.

Auger line shape changes occur due to changes in bonding, particularly when one or two valence electrons are involved in the process. Although line shape could be related to the energy distribution of electrons in the valence band, this relationship is not simple, because some materials exhibit quasi-atomic Auger spectra rather than bandlike spectra. Experimentally observed changes in Auger line shapes are useful in identifying chemical states of elements at surfaces, for example, in carbon, sulfur, nitrogen, and oxygen. Figure 7 illustrates changes in carbon KLL and tin MNN line shapes and peak positions. Auger line shape and peak shift information for selected peaks is available in various handbooks (Ref 1–5) and NIST Standard Reference Databases 20 and 100.

Changes in the composition of elements in metal alloys would not be expected to produce measurable changes in the Auger energies (for core levels) of the elements. However, submonolayer quantities of oxygen adsorbed on clean metal surfaces can produce measurable changes in the metal Auger peaks, with the shift increasing with oxygen coverage. For most metals, such shifts are typically ≤1 eV. In the case of CO adsorption on W(112) surface, peak shifts of as little as 0.5 eV in the low-energy tungsten peak have been observed (Ref 33). When bulk sulfides, carbides, or oxides form at the surface, the shifts usually exceed 1 eV, for example, 6 eV for tantalum in tantalum oxide (Ta_2O_5). Observed shifts typically increase with differences in electronegativities, but the oxidation number and relaxation effects also affect the magnitude of a shift.

Auger peak widths are often larger than their corresponding core-level binding energy peak widths in XPS, which makes it more difficult to resolve mixed chemical states in AES compared with XPS. However, it is possible to take the advantage in cases where the size of the Auger chemical shift is larger than the photoelectron chemical shift. Figure 8 shows Auger chemical shifts observed in alumina compared with metallic aluminum. Large shifts of the order 17 to 18 eV are evident in the low-energy (68 eV) LVV transition and the high-energy (1396 eV) KLL transition. Line shape changes can also occur when only core levels are involved in the Auger process, due to changes in the electron energy-loss mechanism; for example, aluminum KLL Auger electrons suffer strong plasmon losses escaping from the pure metal but not the oxide.

Recognizing that the chemical shifts in Auger peaks are sometimes larger than their corresponding core-level photoelectron lines, a modified Auger parameter, α', is an additional way in XPS to evaluate the chemical state of elements at a surface during XPS analysis. The modified parameter is expressed as:

$$\alpha' = E_{ke}(\text{Auger}) + E_b(\text{photoelectron}) \qquad \text{(Eq 2)}$$

where E_{ke} is the kinetic energy, and E_b is the binding energy. Wagner (Ref 35) provided a comprehensive survey of α' and Auger and photoelectron energies for a number of elements and compounds. Chemical-state plots of Auger kinetic energy as a function of photoelectron binding energy for a variety of compounds, often referred to as Wagner

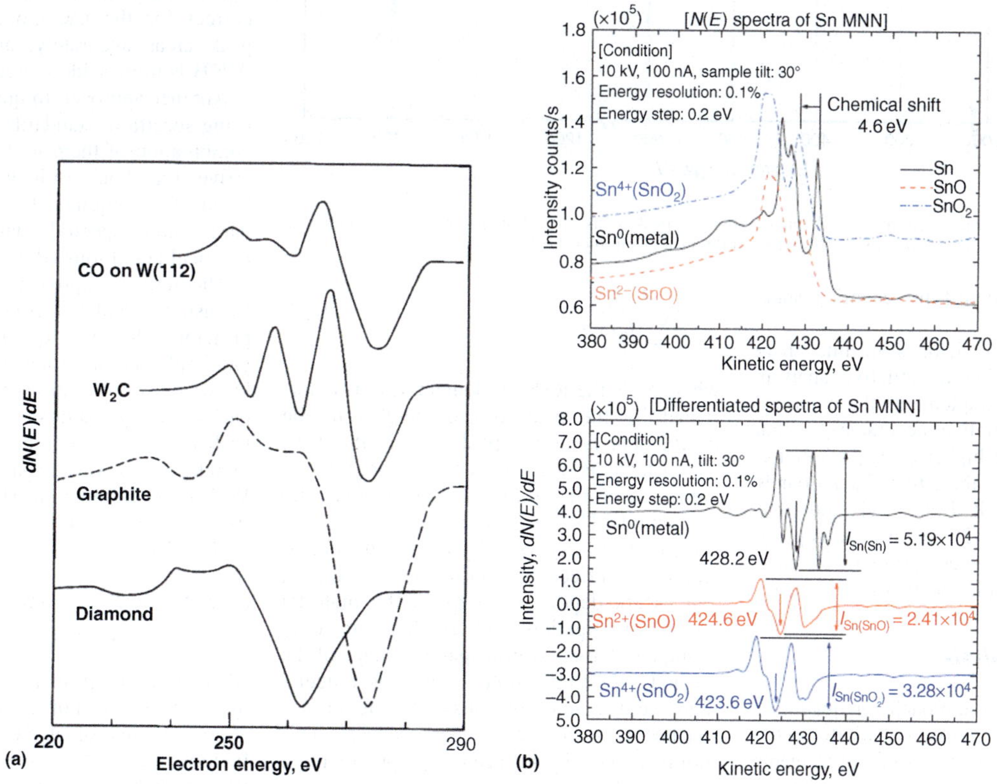

Fig. 7 (a) Carbon KLL Auger spectra from various forms of carbon: CO on W(112), tungsten carbide (W_2C), graphite, and diamond. Diamond peak position may have a drift due to sample charging. Source: Ref 33. (b) Tin MNN Auger spectra in the forms of Sn, SnO, and SnO_2. Source: Ref 34

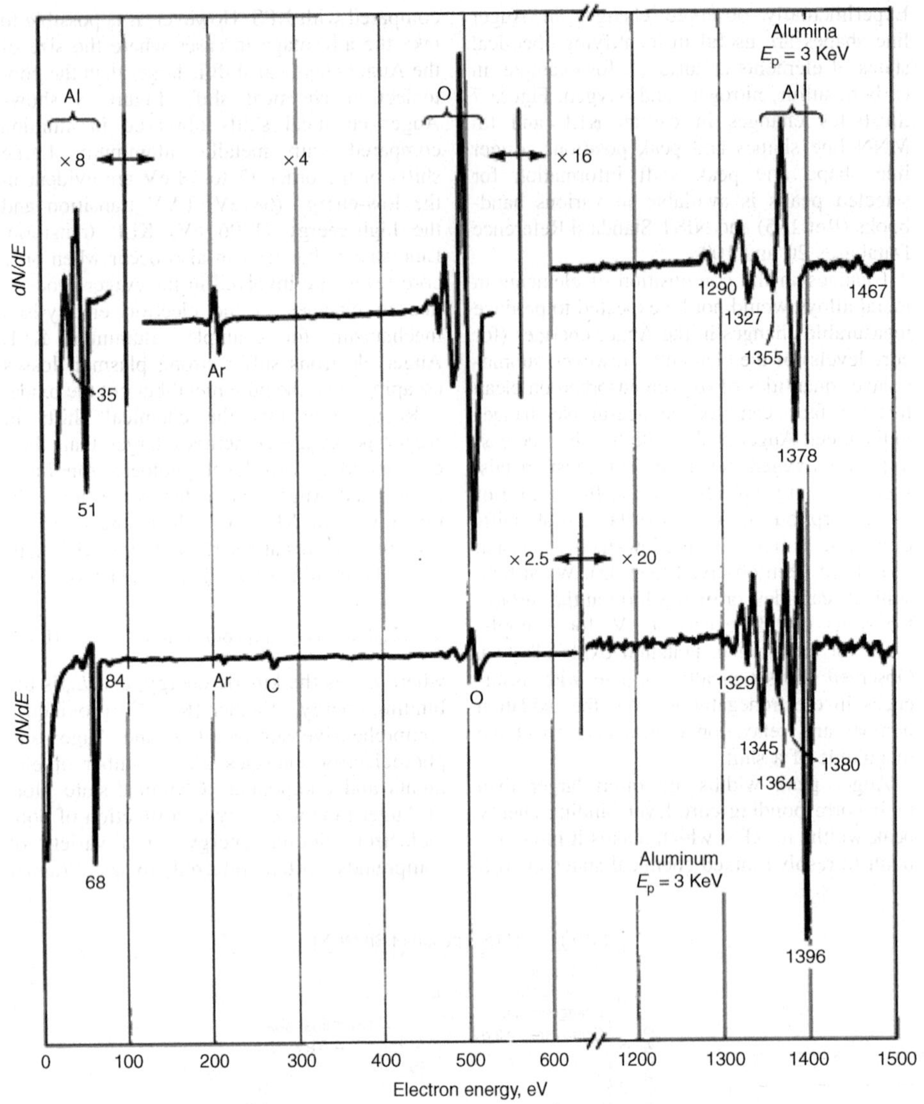

Fig. 8 Auger spectra from alumina and aluminum showing peak shifts and plasmon-loss peak structures in elemental aluminum. Source: Ref 1. Courtesy of Physical Electronics, USA

plots, are useful in identifying chemical states (Ref 36–38).

Energy shifts and changes in line shape can significantly affect quantitative analysis. Methods used to implement these effects and improve quantitative accuracies include obtaining the data at higher energy resolution, curve fitting, and subtracting backgrounds. Linear least-squares fit (Ref 39–41) and target factor analysis (Ref 42) are used to separate components of an element present in a spectrum, in a depth profile, or in Auger images.

Quantitative Analysis

The most common approach for quantification involves using relative sensitivity factors (RSFs) derived from Auger spectra of elements. It is based on measuring relative Auger intensities, I_x, of element x and calculating the atomic concentrations, C_x, using the expression:

$$C_x = \frac{(I_x/S_x)}{\sum_{\infty}(I_\alpha/S_\alpha)} \quad \text{(Eq 3)}$$

where S_x is the RSF of element x. I_x is determined by the peak-to-peak heights in the *dN*/*dE* data or as the peak area in the *N*(*E*) data. Elemental sensitivity factors are relative values derived from pure elements and compounds. Use of these matrix-independent sensitivity factors typically ignores the chemical-state effects, backscattering factor, and escape-depth variations in the material under investigation and is therefore only semiquantitative. A principal advantage of the method is the elimination of standards. Calculations are also somewhat insensitive to surface roughness, because all Auger peaks are uniformly affected by surface topography to a first approximation.

The Auger electron intensity of a given transition is proportional to the Auger peak-to-peak height (APPH) in the *dN*/*dE* data and to the area under the curve, as well as peak height corrected to the background intensity in the *N*(*E*) data. Auger data for standards are available in *dN*/*dE* mode (Ref 1, 4, 5) and in both *N*(*E*) and *dN*/*dE* modes (Ref 4). An example of RSFs derived from *dN*/*dE* data using a cylindrical-mirror analyzer is shown in Fig. 9. Sensitivities vary widely among the elements. These and similar RSFs are applicable only to data obtained by using analyzers with similar transmission function and detection schemes. Accuracies of concentration values obtained depend on the nature of the material, the accuracies of Auger intensity measurements, and S_x used in the calculation. Materials with little or no strong binding-state effects produce accurate quantitative determinations, typically within ±10% of actual concentrations. In ionic compounds, large errors (±200%) are not inconceivable. In such instances and when more favorable accuracies are required, a more effective approach would involve use of composition standards close to the material under investigation.

Auger intensities derived from *N*(*E*) data (peak areas) provide a more accurate quantitative analysis. Peak areas inherently contain complete Auger emission currents and are not influenced by most chemical-state effects. However, the background level is high in the electron-excited Auger data, and it varies rapidly with energy, particularly in the low-energy range where secondary electron contribution is significant. This makes it difficult to correct for the background and to measure peak areas accurately, and for this reason, APPH is more widely used in quantification.

Another approach to quantification involves using spectra of standards and making a linear combination of them in desired proportions to arrive at the final result close to that being analyzed. This requires the availability of standards and spectral data obtained under identical conditions (Ref 43).

Theoretical approaches have been made by using calculated Auger yields from first principles (Ref 15, 16, 20, 29, 43, 44). Auger yield calculations from first principles require knowledge of ionization cross sections, electron backscattering factors, and Auger transition probabilities for the material and transitions of interest, as well as analyzer characteristics. With some assumptions (Ref 10), the detected Auger current, I_α, for a WXY Auger transition is given by:

$$I_\alpha(\mathrm{WXY}) = I_p T N_\alpha x_\alpha \gamma_\alpha(\mathrm{WXY}) R \sigma_\alpha(E_{p'} E_W) \lambda (1 + R_B) \quad \text{(Eq 4)}$$

where I_p is the primary electron current; T is the analyzer response function; N_α is the atomic density of element α; x_α is the atom fraction; γ_α(WXY) is the WXY Auger transition probability factor; E_W is the binding energy of the core-level W with respect to the Fermi surface; $\sigma_\alpha(E_{p'}, E_W)$ is the ionization

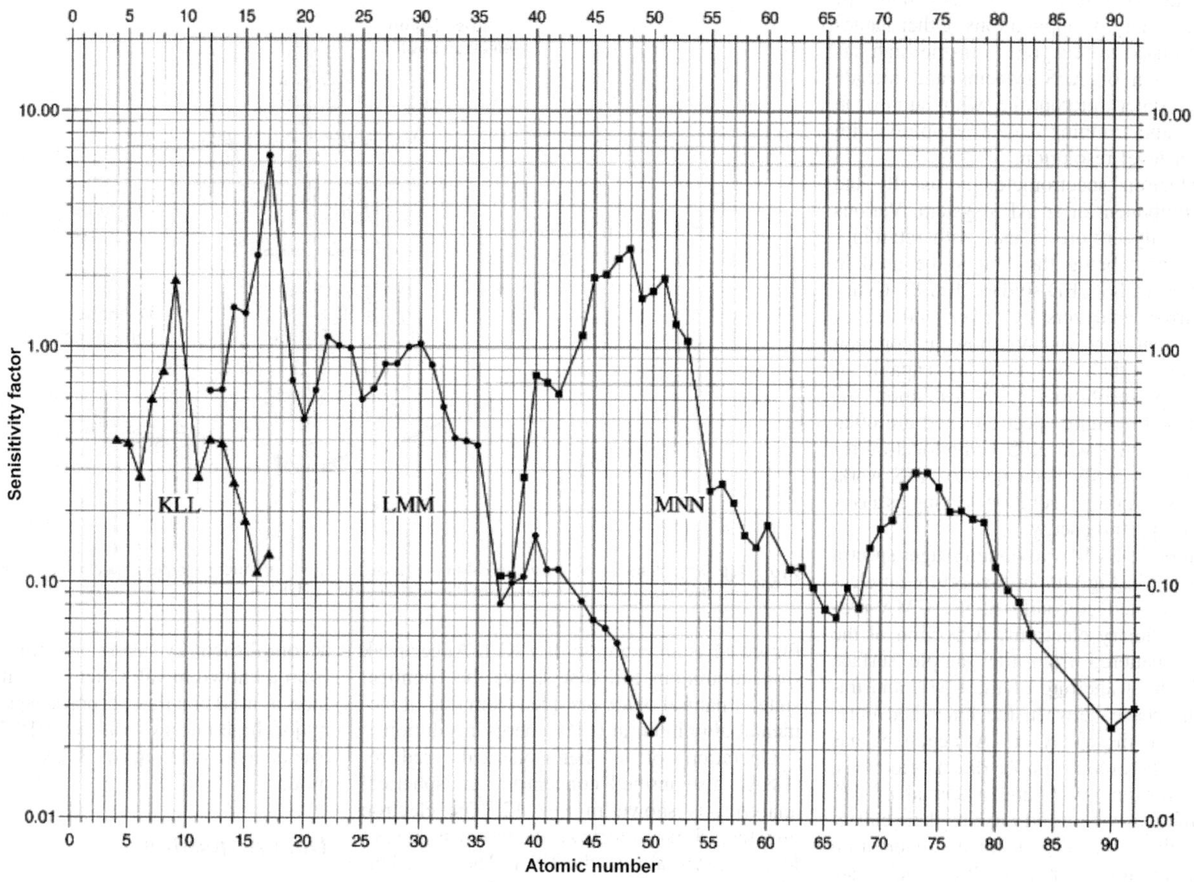

Fig. 9 Relative sensitivity factors derived from Auger peak-to-peak height data using a cylindrical-mirror analyzer with 10 keV primary beam. For relative sensitivities, the 10 keV copper LMM is set to unity. Source: Ref 5. Courtesy of Physical Electronics, USA

cross section of the core-level W, which is a function of the primary electron energy, E_p; λ is the escape depth of Auger electrons in the material; R is the surface-roughness factor; and R_B is the electron backscattering factor.

For a given Auger transition and primary energy, σ decreases smoothly with the atomic number of the element. The cross section goes through a maximum as a function of E_p. The maximum in E_p is observed commonly from four to eight times E_W and indicates the importance of selecting an appropriate value of E_p for a given experiment. With better understanding of backscattering and escape-depth contributions and some assumptions, average matrix RSFs have been calculated and show good agreement with experimental values for the major Auger transitions above 180 eV (Ref 29, 44). The results below 180 eV are less reliable, likely due to high background contribution of secondary electrons.

Detailed quantitative analysis of outermost monolayers and fractional layers as well as in-depth analysis using sputtering can be found in the literature (Ref 14, 28, 29). Powell and Seah (Ref 45) reviewed various factors affecting accuracy, precision, and uncertainty in quantitative analyses of AES and XPS.

Experimental Methods

Instrumentation typically used in AES includes an electron gun for primary electron excitation of the sample, an electron spectrometer for energy analysis of the Auger electrons, a secondary electron detector for secondary electron imaging, a stage for sample manipulation, and an ion gun for sputter removal of atoms from the sample surface. Stages are also equipped with sample rotation capability, such as Zalar rotation (Ref 46). All components are housed in an ultrahigh-vacuum system capable of vacuum in the 10^{-8} to 10^{-10} torr. Lower pressures are necessary to maintain surface cleanliness over time, which may be needed for studies involving surface reactions, deposition and sputter cleaning, and specimen fracture. The vacuum system may also be equipped with special-purpose auxiliary equipment such as a fracture attachment for in situ fracture studies, an evaporation unit for thin-film deposition, and a hot/cold stage to conduct elevated-temperature studies and to maintain a low vapor pressure for the sample. Systems are also equipped with sample-preparation chambers for special treatments prior to placing the sample into the analytical chamber. Some systems combine other techniques, such as XPS, energy-dispersive x-ray spectroscopy (EDS), secondary ion mass spectroscopy, low-energy ion-scattering spectroscopy, low-energy electron diffraction (LEED), and reflection high-energy electron diffraction, to obtain complementary information. Among the most common combination systems with AES are XPS and EDS because of the commonality of the detection system for XPS and the commonality of the source for EDS evaluations. Modern systems use sophisticated electronics to operate the various components and computers for data acquisition and manipulation.

Primary electron excitation is often accomplished by using 1 to 30 keV electrons. Earlier-generation systems used electron guns based on thermionic emission with beams focused to <1 μm in diameter. Second-generation instruments used lanthanum hexaboride (LaB_6) filaments capable of beam sizes to <100 nm. Current state-of-the-art instruments use field-emission sources generating spot sizes as small as 3 nm for secondary electron imaging and tens of nanometers for analytical Auger spectroscopy. This ability of electron beams to generate Auger signal from very small areas remains the major advantage of AES in a variety of applications. Other names have been used in recognition of the high lateral resolution of AES systems and include scanning Auger microprobe with submicron resolution and scanning Auger nanoprobe with nanometer-level resolution.

The electron spectrometer is usually the central component of an AES system. Various types of analyzers are in use, with retarding-field analyzers being the first to be used in conjunction with LEED studies (Ref 8). Cylindrical-mirror analyzers (CMAs) and hemispherical analyzers and are now in common use. Hemispherical analyzers are capable of high energy resolution, with resolution ΔE fixed across the entire spectrum, and are often used in conjunction with XPS. Cylindrical-mirror analyzers are mostly dedicated to AES work. They provide energy resolution varying as a percent of the spectral energy.

Operation of a CMA to obtain Auger spectra is illustrated in Fig. 10. In this setup, an electron gun located coaxially to the CMA provides the primary electrons. A portion of the electrons scattered from the sample surface enters the inlet aperture of the CMA and traverses the space between the inner and outer cylinders of the analyzer. A negative bias applied to the outer cylinder directs electrons, with specific energy E and a spread of ΔE, toward the CMA axis and enables their collection at the coaxially located channeltron electron multiplier or a multichannel plate electron multiplier. Multichannel plates in this configuration enable the anode to be divided into an appropriate number of channels (eight, in this illustration) and enables simultaneous detection of electrons in multiple energy channels. A spectrum, $EN(E)$ as a function of E, is generated by sweeping the bias applied to the outer cylinder. The energy resolution of the analyzer, $\Delta E/E$, is variable and is typically in the 0.1 to 1% range. Cylindrical-mirror analyzers have a large acceptance angle (42 ± 6° for the Physical Electronics CMA) integrated around the full 360° azimuthal angle and provide for high transmission. The wide angular acceptance enables signal reception simultaneously from a variety of angles. Coaxial location of the electron gun also minimizes shadowing of rough surfaces, because the signal generated is picked by the analyzer from a variety of angles. This is an advantage when examining samples having high surface roughness, such as fracture surfaces, and when sampling an area located in a trench. Some CMAs have coaxial electron guns, while others operate with electron sources located externally.

Because the analyzer energy resolution varies as $\Delta E/E$, the resolution is low at higher energies. Typically, this is not an issue because the high-energy peaks often have high natural peak widths. Situations with sharp peaks, peak shifts, and peak overlaps are often resolved using the higher-resolution mode of operation.

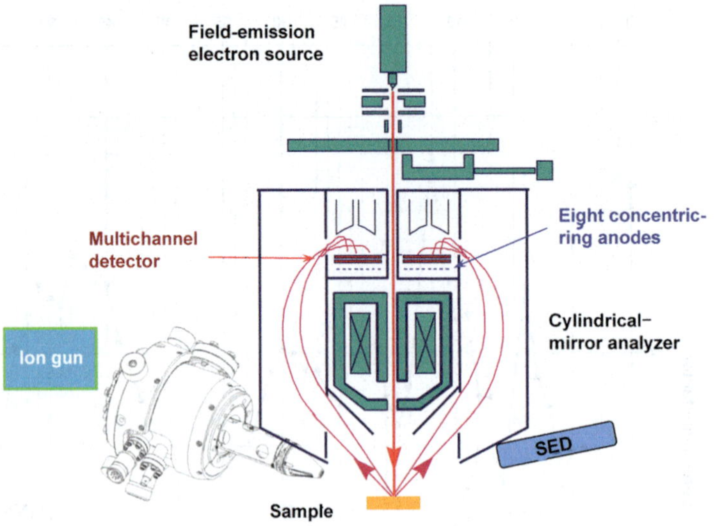

Fig. 10 Key elements of a cylindrical-mirror analyzer used for Auger electron spectroscopy. SED, secondary electron detector. Courtesy of Physical Electronics, USA

Figure 11 is an example of a KLL Auger peak from aluminum foil that shows the chemical state of natural oxide resolved from that of elemental aluminum at 0.1% energy resolution.

Operation of a hemispherical analyzer (HSA), also termed concentric hemispherical analyzer, is shown in Fig. 12. The electron gun for primary electron excitation is always located externally to the HSA. The input lens of the HSA receives a portion of the electrons scattered from the sample surface, retards them to a desired low energy, and focuses them onto the entrance aperture of the HSA. The applied voltage between concentric hemispheres serves as the energy filter and enables filtered electrons to be collected at the multichannel detector. For typical AES operation, the input lens and the energy differential at hemispheres is swept over the energy range of interest, while maintaining a constant retard ratio of input electron energy to the analyzer pass energy. The spectrum is generated as $EN(E)$, and the energy resolution $\Delta E/E$ remains constant across the entire spectrum, similar to CMA operation.

The HSA can also be operated with a fixed differential voltage, or pass energy, at the hemispheres, while the input lens is swept over a range of energies of spectral interest. This mode of operation is conducted by using an electron source for AES or an x-ray source for XPS investigations. The spectrum is generated as $N(E)$, and the resolution ΔE remains fixed across the entire spectrum as determined by the pass energy, which is also variable.

Energy calibration of a CMA is accomplished simply by focusing a known energy of primary electrons, for example, 3 keV, onto a sample, adjusting the sample position for maximum point transmission, and calibrating the reflected beam to the same energy. Calibration of either CMA or HSA is also accomplished by using spectral information from a known metal. Often a clean copper surface is used for this purpose, because it has a major high-energy peak at 918.7 eV and a low-energy peak at 63.3 eV in the $EN(E)$ mode.

Data Acquisition

Modern Auger systems acquire data in digital form either by direct pulse counting from the channeltron output or by voltage-to-frequency conversion. Both methods enable display of the electron energy distribution in $EN(E)$ mode or in dN/dE format upon digital differentiation (Fig. 4).

The electron beams used in AES offer many benefits in examining a given surface. Foremost, it generates secondary and backscattered electron images that help in locating the area of interest. Auger spectra can be obtained from large areas by electron beam defocusing and by beam rastering, or from small areas as defined approximately by the size of the finely focused electron beam. Auger images and line scans are used to define surface chemical inhomogeneities uniquely. Auger imaging is performed in a manner similar to x-ray mapping in a scanning electron microscope equipped with EDS. A selected area of a sample surface is scanned by the electron beam as the AES elemental signal generated by the electron beam and detected by the analyzer is used to generate an intensity-modulated image. Figure 13 is an example of a defect on a silicon wafer generated during etching. Auger spectrum obtained from the central region of the defect shows it has silicon, carbon, aluminum, fluorine, and oxygen along with small amounts of nitrogen and sulfur. Auger images of carbon, aluminum, and silicon are combined in Fig. 13(b) to show the distribution of these

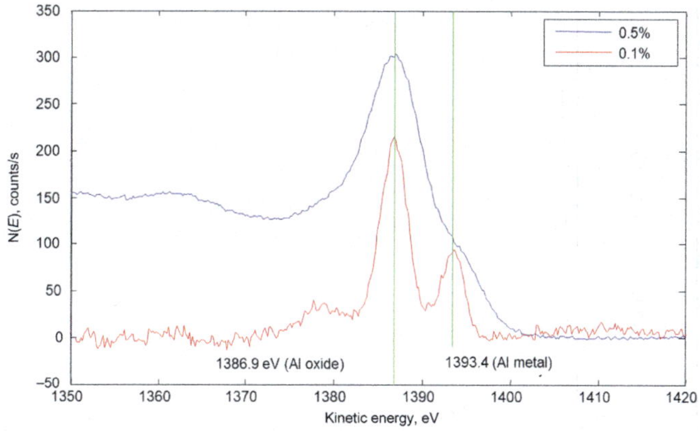

Fig. 11 KLL Auger spectrum of aluminum foil with native oxide obtained by using cylindrical-mirror analyzer resolutions at 0.5% (blue) and 0.1% (red). Courtesy of Physical Electronics, USA

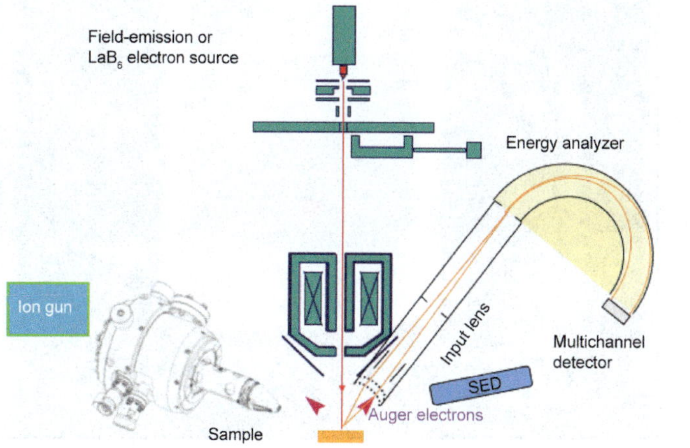

Fig. 12 Setup of a typical Auger instrument using a hemispherical analyzer. SED, secondary electron detector. Courtesy of Physical Electronics, USA

elements on the surface. Aluminum is located only in the central region of the defect, with carbon and fluorine present at distinguishably higher levels over the majority of the defect area. This example shows the capabilities in obtaining AES spectra from small areas, and secondary electron and Auger images at submicrometer resolution.

Auger composition-depth profiling using ion beam sputtering is widely used to expose the subsurface region of the sample, despite the sputtering artifacts discussed later; the method is used in applications relating to deposited thin films, coatings, and various surface-interface phenomena, such as segregation, corrosion, and surface reactions. An example of an AES depth profile obtained from a via contact of a semiconductor device is shown in Fig. 14. The via contact is approximately 10 μm in size. The composition of the aluminum metallization over silicon as a function of sputtering time was obtained from a small area of ~1 μm. Compucentric Zalar rotation was used to minimize surface cone formation and enhance depth resolution. The measured APPHs were converted to atomic concentrations by using appropriate RSFs. The profile data show a thin layer of aluminum oxide present on top of the aluminum metallization.

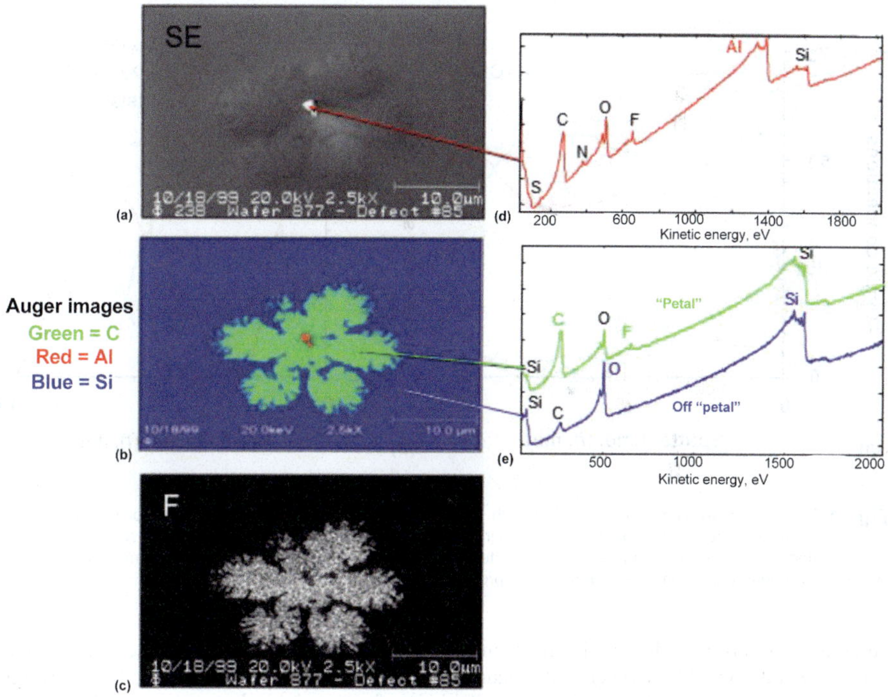

Fig. 13 Etch defect on a silicon wafer. (a) Secondary electron (SE) image. (b) Compilation of Auger images of aluminum (red), carbon (green), and silicon (blue). (c) Fluorine Auger image. Auger spectra are from (d) the central region of the defect and (e) the carbon-rich region of the defect and the silicon-rich surrounding region. These images and spectra are obtained by using Physical Electronics Scanning Auger Nanoprobe. Courtesy of Physical Electronics, USA

Data Processing

Many advancements in data processing are available with most commercial instrumentation. These include automatic peak identification, a variety of smoothing and differentiation routines, background subtraction, deconvolution to remove energy-loss peaks, curve fitting, atomic concentration calculations of depth-profile data, linear least-squares (LLS) fit, and target factor analysis. The following example illustrates the need to use LLS fitting to extract meaningful data, which were not readily observable in the as-collected depth-profile data.

Linear Least-Squares Fit

In addition to obtaining elemental composition from APPH analysis, the shape and position of an Auger peak can provide more quantitative understanding of the chemical nature of an element on the surface, particularly when combined with an LLS methodology (Ref 40, 47–52). Compared with photoelectrons measured in XPS, experimentally observed shifts in Auger peak position are often significantly greater (>10 eV) and, with modern instrumentation, offer insight on the chemical speciation of surface elements on the nanoscale. When combined with sputter depth profiling, the use of LLS analysis of Auger peak shapes in place of peak-to-peak heights can substantially improve discrimination of layers composed of different chemical species. Figure 15 illustrates the increased clarity provided regarding chemical composition by applying multivariate analysis. At the conclusion of the depth profile, LLS analysis determined the sample to

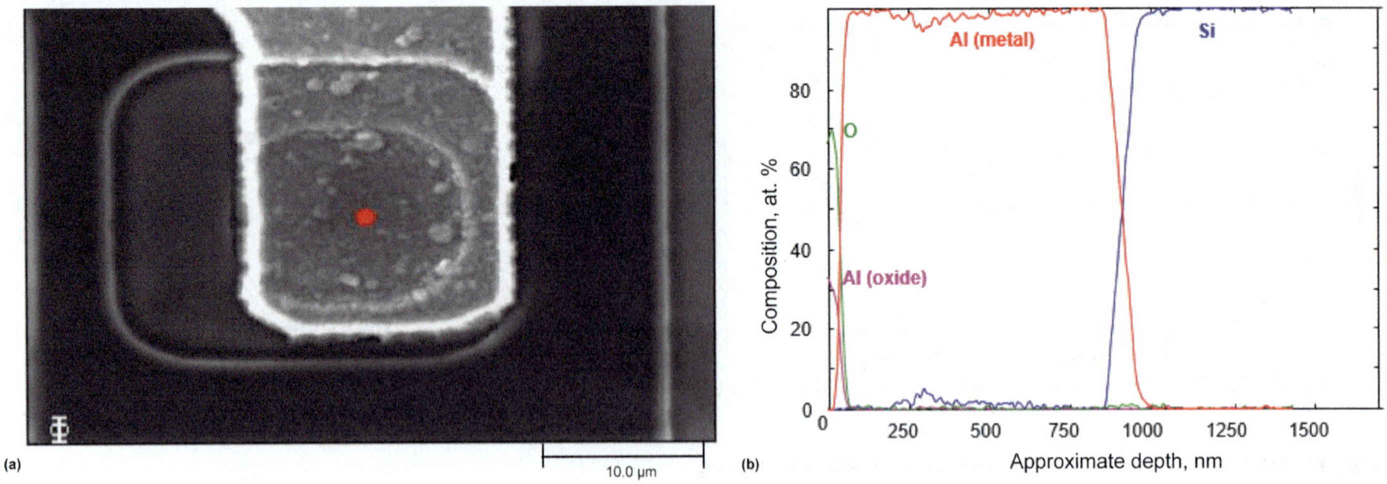

Fig. 14 (a) Secondary electron image of a 10 μm via contact. (b) Auger electron spectroscopy depth profile of aluminum metallization obtained with compucentric Zalar rotation. Courtesy of Physical Electronics, USA

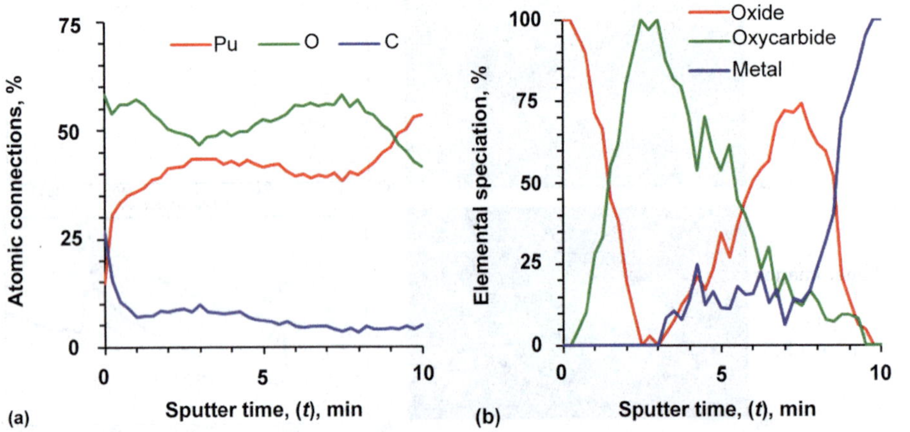

Fig. 15 Variation in sample composition with sputter time as calculated from (a) Auger peak-to-peak height measurements and (b) the same experimental data worked up using linear least-squares analysis of differentiated Auger spectra acquired during sputter depth profiling. Sputter rate for plutonium metal under the given experimental conditions is ~15 nm/min. Reproduced with permission from Ref 51

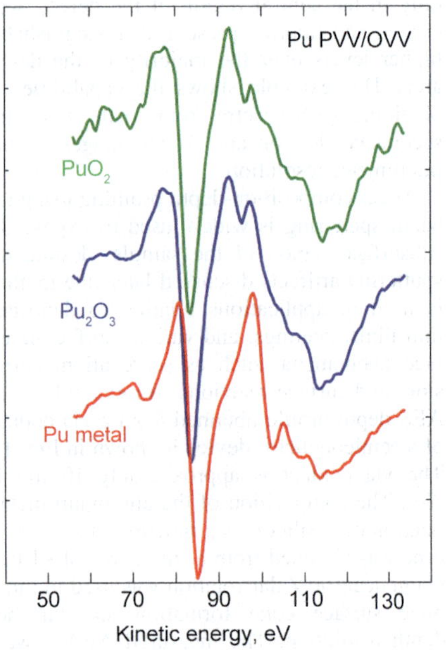

Fig. 16 Differentiated Auger spectra taken from reference materials for the plutonium P_1VV and $O_{45}VV$ Auger transition for plutonium in the dioxide, sesquioxide, and metallic chemical states used for initial linear least-squares (LLS) analysis. Note the significant variation in shape but small shifts in peak position as a function of chemical species. These reference spectra were obtained under the same instrument settings as the experimental spectra, and the entire $dN(E)/dE$ range as shown is used in LLS fits. Reproduced with permission from Ref 51

be almost entirely metallic, while APPH measurements indicated 40% oxygen still remained on the surface. In cases where only small shifts in peak energy together with little variation in peak shape are observed between species, the changes may be too subtle to enable reliable identification by line-shape analysis alone. However, even in the absence of significant variation in peak energy, but having variations in peak shapes, as shown in Fig. 16, LLS methodology still enables separation of different chemical species (Ref 51).

Linear least-squares analysis is accomplished by combining concentrations or weighting factors, C, of multiple representative spectra, R, in such a way as to minimize the difference between the acquired sample spectra, D, while fitting the basic expression $D = R \cdot C$. The process is repeated for the complete set of recorded spectra. The primary difference between LLS fitting and other commonly used fitting analyses (e.g., target factor analysis) is that the former is used to solve for the weighting factors given the experimentally acquired spectra and representative spectra for each previously identified component, while the latter relies on experimental data to solve for the weighting factors, representative spectra, and number of components (Ref 52).

Experimental Limitations

Elemental Detection Sensitivity

Limitations of the technique include its insensitivity to hydrogen and helium and its relatively low detection sensitivity for all elements compared with bulk analysis methods. Because the Auger process involves three electrons in a given atom, hydrogen and helium cannot be detected. However, hydrogen effects on the valence-electron distribution of selected metals have been obtained by using AES. The AES detection sensitivity for most elements is from 0.1 to 1.0 at.%. Sampling volume and associated number of atoms examined using AES are often small compared with x-ray and other bulk analysis techniques, which limit the sensitivity of the technique. Use of high currents and longer time-signal averaging often improves detection level, but overall detection is limited by time and shot noise associated with the background current upon which the Auger peaks are superimposed.

Electron Beam Artifacts

Another limitation involves the primary electron beam used in Auger electron excitation. Electrons interact with matter more readily than x-rays and can cause electron beam artifacts. Electron beams can change the surface by promoting migration of atoms into or out of the area of analysis, which includes surface migration, reactions, and desorption. Poor thermal conductivity of the sample also causes localized heating and associated artifacts. Such effects include decomposition of surface materials (prominent with organic, biological, and selected inorganic compounds) and polymerization. Many oxides, although relatively strong compounds, can be reduced to a lower oxidation state. These artifacts are generally total flux dependent, and the effects are minimized by using low primary current densities (associated with some sacrifice of detection sensitivity), low primary beam voltages, and electron beam rastering of large areas.

Sample Charging

An electron beam impinging on an insulating or an ungrounded sample can cause surface charge to build up and result in peak shifts, spectral distortions, arcing, and occasionally alter the surface. It is not always a major issue, because even relatively highly resistive materials often provide sufficient conductivity for typical AES analyses (Ref 53). Most common glasses and other insulating compounds are sufficiently impure and permit reasonable analysis using AES. The high surface sensitivity of the technique generally precludes metal coating of the analysis surface, a common practice in scanning electron microscopy (SEM). Using low-energy positive ions to neutralize surface charge is a logical approach and has become common with the advent of commercially available ion sources. Typically, a low-energy argon ion beam under 70 eV is used to preclude sputtering. An example of Auger images obtained from a submicrometer particulate mix of quartz and diamond by using ion beam neutralization is shown in Fig. 17. Other common methods to minimize sample-charging effects include the use of low-electron-beam current densities, optimizing voltages (typically 1 to 3 keV), and glancing angles of incidence, which aid in increased secondary electron emission to balance the surface charge buildup. A high beam voltage sometimes can help penetrate an insulating surface film and aid in charge dissipation. Another practice involves masking the surface with a conductive metal grid or depositing conductive masks on the sample surface, which function as a local sink to the electron charge. Baer et al. (Ref 54) reviewed common issues with electron beams and methods to minimize charging. When using electron and ion beams, it is important to be aware of their potential artifacts occurring at surfaces.

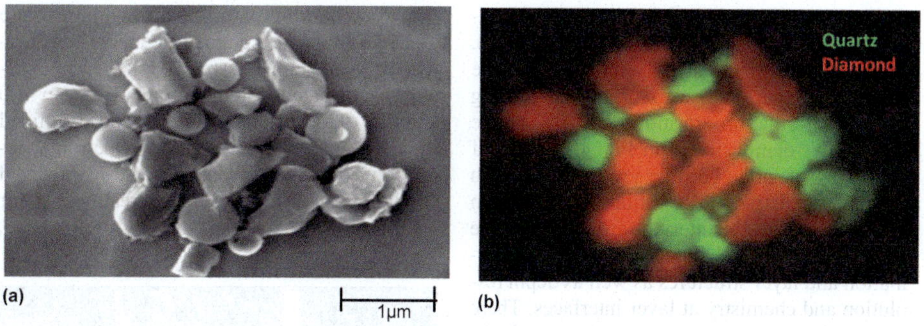

Fig. 17 (a) Secondary electron image and (b) Auger images of quartz (green) and diamond (red) obtained from a submicrometer particle mix of quartz and diamond. These data were collected by using ion neutralization while acquiring the images. Courtesy of Physical Electronics, USA

Due to these limitations, most common AES applications involve good conductors, such as metals, alloys, and semiconductors. Highly insulating oxides, other insulators, and organic materials have been examined but often with limited success.

Spectral Peak Overlap

Spectral peak overlap is a problem in a few situations in AES. It occurs when one of the elements is present in a small concentration, and its primary peaks are overlapped by peaks of a major constituent in the sample. Often the effect is significant degradation of sensitivity. For example, titanium and nitrogen, iron and manganese, and sodium and zinc are frequently encountered combinations in which peak overlap is of concern. The problem is most severe when one of the elements has only one peak, such as nitrogen. In most cases, one or both elements have several peaks, and the analysis can be performed using one of the nonoverlapping peaks. Peak overlaps are often resolved by acquiring spectra at higher energy resolution. Overlaps are also resolved in many instances by deconvolution of peaks in the $N(E)$ mode and by spectral subtraction. Spectral subtraction of data in the derivative mode is also used to determine small energy shifts resulting from chemical-state changes (Ref 55, 56). Linear least-square fitting and target factor analysis (Ref 41, 42) are also used to separate spectral components and derive quantitative information in the form of depth profiles and maps.

High-Vapor-Pressure Samples

Another limitation for AES analysis is with high-vapor-pressure samples. A sample is vacuum compatible if its vapor pressure is low enough not to degrade or ruin the vacuum in the spectrometer, which typically is less than 10^{-8} torr. If the sample outgasses at a high rate, surface chemistry can change upon introduction into the vacuum chamber, and the vacuum can degrade. Sample size often can be decreased to a workable level to minimize outgassing, as long as the surface chemistry change resulting from outgassing is not a concern. In some situations, such as with fluid samples, a very thin film of the fluid can be applied onto a metal substrate, and the spectral information can be obtained. Another approach is to use an appropriate cold stage to cool the sample to liquid nitrogen temperature, thereby reducing outgassing. This procedure helps to maintain the appropriate vacuum in the spectrometer and the original surface chemistry of the sample.

Sputtering Artifacts

Sputtering is the most common method used for depth profiling and surface cleaning and is associated with numerous artifacts, including ion beam mixing, preferential sputtering, surface diffusion, amorphization, chemical degradation, and surface roughening (Ref 17, 57, 58). Ion beam mixing arises from the knock-on and implantation processes associated with energetic impingement of ions, and it can influence surface composition and sputter rates in the profiles. Preferential sputtering is a result of differences in sputter rates of the elements and compounds involved, often leading to surface enrichment of an element with the lower yield.

Chemical degradation is readily observed in many materials; some compounds depart quickly from stoichiometry upon sputtering, and metal oxides undergo bond breaking, lose some oxygen preferentially, and exhibit a lower oxidation state. Surface roughening almost always occurs in sputtering, but lower ion beam energies minimize it. Surface roughening becomes very pronounced if cone formation occurs and markedly affects depth resolution in profiles.

An example of cone formation while sputtering a via contact is shown in Fig. 18(a). Cone orientation is typically aligned with the direction of the ion beam. Surface roughening and cone formation are minimized by sample rotation, as demonstrated in Fig. 18(b). All of these artifacts often become severe in the depth profile of relatively thick films (>1 μm). In multilayer films, sputtering artifacts become very pronounced and affect compositional information and layer structures as well as depth resolution and chemistry at layer interfaces. There are a few alternatives to sputtering (Ref 17), which are discussed in the section "Thick Films and Coatings" in this article.

Applications

Auger electron spectroscopy is used in many basic, materials science, and industrial applications. It has widespread applications in the areas of metals, semiconductors, ceramics, composites, catalysts, nanomaterials, electronics, and magnetic materials. Understanding of surface and interface chemistry led to the development of new materials and processes and resolved issues related to surface contamination, segregation, adhesion, catalysis, oxidation and other reactions, thin films, fracture, creep, corrosion, and stress-corrosion cracking. The following examples highlight some capabilities of the technique.

Thin Films and Interfacial Reactions

Thin films are important in microelectronics, especially in integrated circuits with layered structures containing metallization, diffusion barriers, and insulating layers. Thin films are also used to understand the early stages of reactions in bulk materials, illustrated in the following applications that involve light elements and refractory materials.

Light Elements

High detection sensitivity coupled with surface sensitivity makes AES an ideal technique in studies involving light elements such as lithium, beryllium, carbon, and their alloys. Beryllium has a high elastic modulus and low density, making it attractive for use in many aerospace structural applications. For example, beryllium mirrors are produced by physical vapor deposition onto a polished metal substrate, a process in which a problem of adhesion in localized areas was encountered (Ref 59). A release agent, such as carbon, deposited onto the substrate prior to beryllium deposition often minimizes adhesion issues. However, interfacial reactions can occur at high temperatures, limiting the effectiveness of the release agent. Reactions between beryllium and carbon were performed to determine the temperature limits of its application.

Carbon-beryllium diffusion couples were prepared by sputter depositing amorphous carbon on beryllium substrates. They were vacuum heat treated at temperatures between 400 and 800 °C (750 and 1470 °F) and examined by using AES and XPS. Auger electron spectroscopy depth-profile data in the as-received condition and upon heat treatments for 30 min at 450, 550, and 600 °C (840, 1020, and 1110 °F) are shown in Fig 19. Results indicate that reaction between carbon and beryllium begins at or below 450 °C and is complete at 600 °C. The thin oxide layer at the beryllium surface appears not to limit beryllium and carbon reactions, and beryllium appears to diffuse outward to react with carbon. Quantification with XPS and beryllium 1s peak positions are consistent with Be_2C

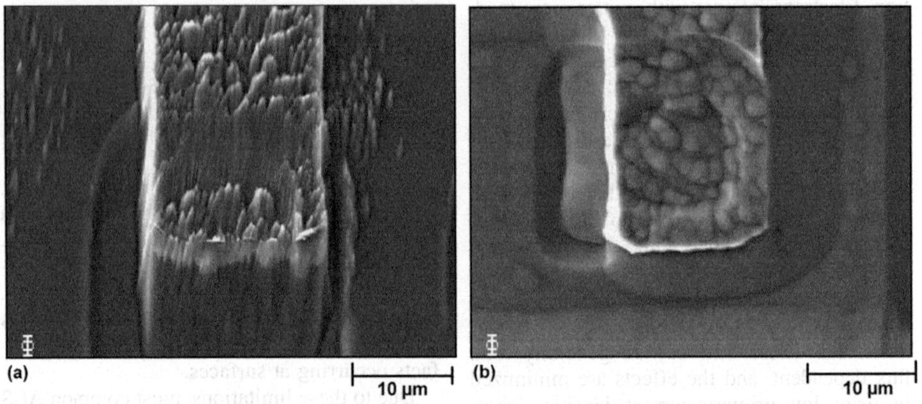

Fig. 18 Via contacts similar to that seen in Fig. 14. Cone formation at a surface during depth profiling (a) without sample rotation and (b) with Zalar sample rotation. Courtesy of Physical Electronics, USA

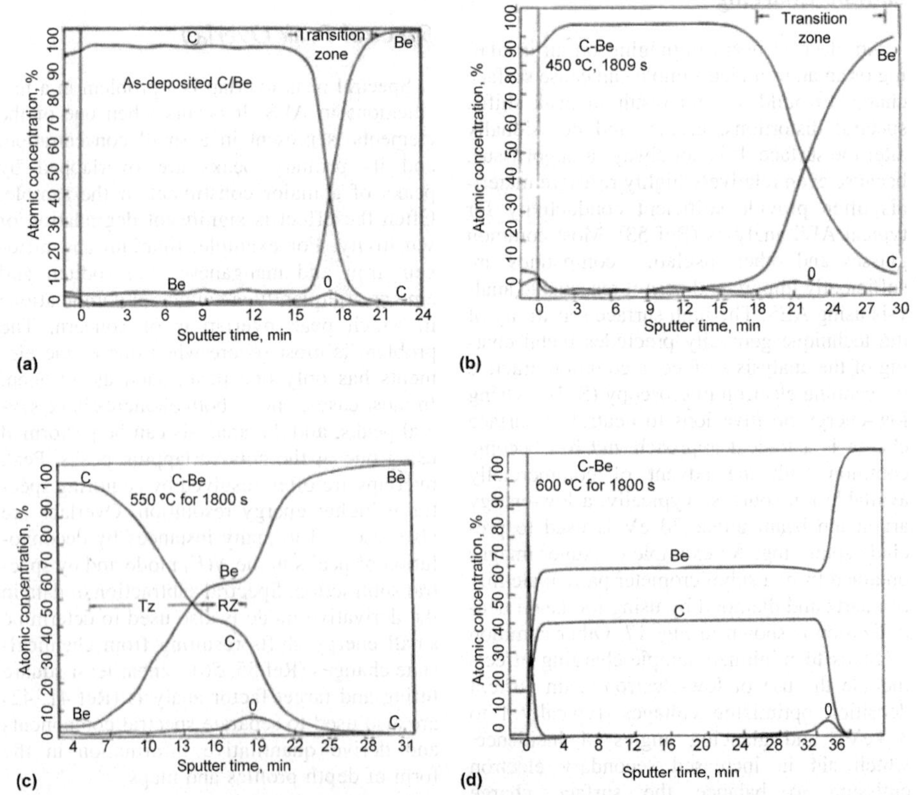

Fig. 19 Auger electron spectroscopy depth profiles of carbon-beryllium reactions. (a) As-deposited 200 nm carbon on beryllium substrate and after vacuum heating at (b) 450 °C (840 °F) for 30 min, (c) 550 °C (1020 °F) for 30 min, and (d) 600 °C (1110 °F) for 30 min. TZ, transition zone including interdiffusion; RZ, reaction zone where compound formation was observed. Source: Ref 59

formation at high temperatures, although Auger profiles indicated a Be:C atomic ratio of ~3:2 rather than 2:1. Beryllium Auger peak positions and shape changes are shown in Fig. 20, and XPS and AES peak positions are summarized in Table 1.

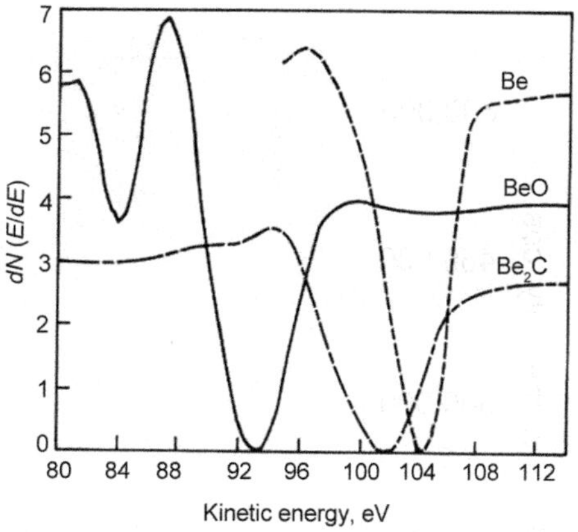

Fig. 20 Beryllium KLL Auger peak shape and position in beryllium, BeO, and Be_2C chemical states. Source: Ref 59

Table 1 Beryllium and carbon peak positions in x-ray photoelectron spectroscopy and Auger electron spectroscopy

Peak	Photoelectrons (1s binding energy, eV)	Auger electrons (KLL kinetic energy, eV)
Be in Be	111.5 ± 0.1	104.3 ± 0.4
Be in BeO (surface oxide)	114.1 ± 0.1	93.0 ± 0.4
Be in Be_2C	112.4 ± 0.1	101.0 ± 0.4
C in Be_2C	282.1 ± 0.1	...
C (hydrocarbon reference used)	284.60	...

Source: Ref 59

High-Temperature Materials

Metal-ceramic composites offer high potential in structural applications operating at high temperatures (Ref 60). Metals such as titanium, niobium, and nickel; superalloys; and intermetallic compounds such as nickel aluminides and titanium aluminides are among the strong candidates for matrices. Because of their low density, good mechanical strength, and thermal stability at elevated temperatures, SiC and Al_2O_3 are two common reinforcement candidates. Formation of interfacial reaction layers as thin as 0.1 μm can be detrimental to composite strength and ductility and, in some instances, to corrosion behavior, which suggests the importance of understanding the early stages of interfacial reactions between the matrix and reinforcement. Studies on thin-film couples using surface analytical techniques offer relatively quick answers to the nature and degree of interactions that can occur at high temperatures.

Interactions between candidate materials including niobium, nickel, and cobalt (Ref 61, 62), stainless steels (Ref 62, 63), and Ti_3Al (Ref 64) with SiC, Al_2O_3, and Si_3N_4 have been reported. Both bulk diffusion and thin-film couples have been studied to understand interdiffusion and second-phase formation.

In this example, reactions of niobium (the lightest of the refractory metals) with SiC were investigated using thin-film diffusion couples prepared by depositing niobium films (~1 μm thick) on SiC substrates (Ref 61). Interfacial reactions occurring between 800 and 1200 °C (1470 and 2190 °F) were studied after vacuum annealing the couples, followed by Auger depth profiling. The profiles in Fig. 21 show that some interdiffusion of niobium and carbon occurs near the niobium/SiC interface upon annealing at 800 °C for 4 h. The slight increase in oxygen and carbon levels at the film surface is likely due to pickup from the annealing environment. Niobium reacted extensively with SiC at temperatures above 900 °C (1650 °F), forming reaction products that included Nb_2C, NbC, and Nb_xSi_yC. Compositions at various depths and chemical-state information in the profiles (Fig. 22) helped to understand the layered structures. Detailed transmission electron microscopy evaluations (Ref 65) of the sample annealed at 1100 °C (2010 °F) for 4 h aided in identifying specific phases and the overall layered structure as $Nb\text{-}Nb_2C\text{-}Nb_5Si_4C\text{-}NbC\text{-}SiC$.

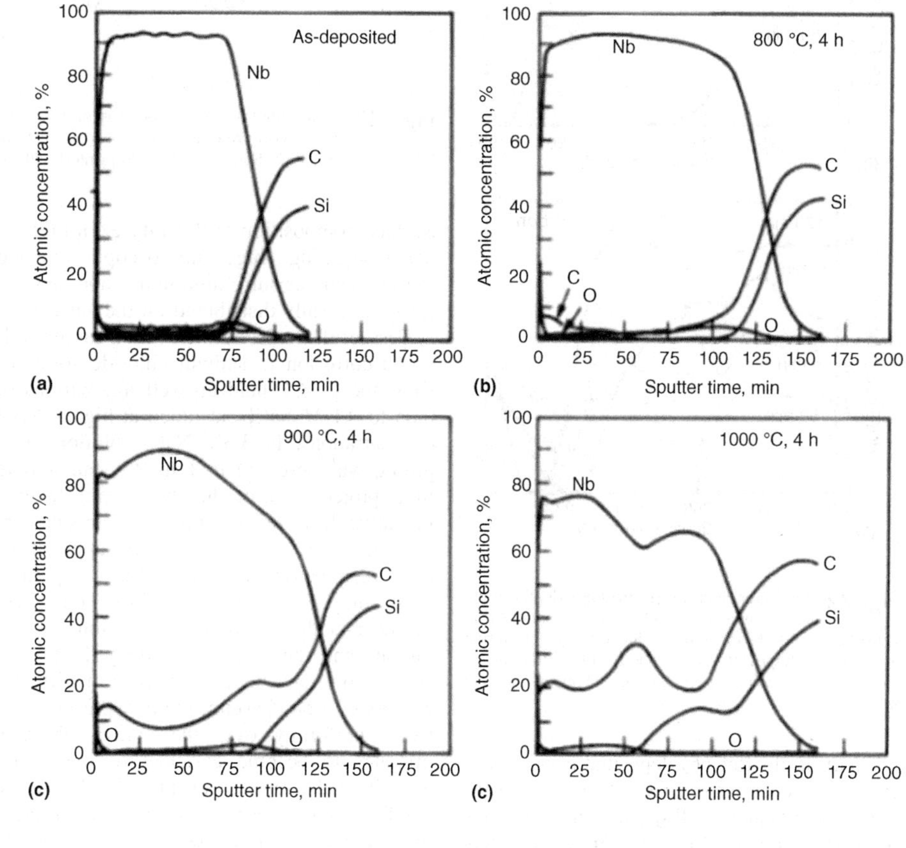

Fig. 21 Auger electron spectroscopy depth profiles from Nb-SiC couples in the as-deposited condition and upon vacuum annealing at various temperatures and times. Source: Ref 61

Corrosion and Protective Coatings

Conversion coatings are often used on steel, aluminum, zinc, and other metals and alloys to enhance corrosion resistance and improve paint adhesion. The multistep chemical process typically involves removal of surface residues and natural oxides, followed by chemical treatments to develop a protective coating. Chromate conversion coatings have been most widely used for aluminum alloys. However, hexavalent chromium present in some coatings is considered an environmental pollutant and an occupational health and safety hazard. Research and development of alternative coatings with hexavalent-free chromium, such as cerium-base conversion coatings, is ongoing. Coatings are typically formed from solutions involving cerium chlorides and nitrates and consist of insoluble cerium oxides and hydroxides. Application of AES and XPS together with other methods has aided in developing these protective coatings (Ref 66–68).

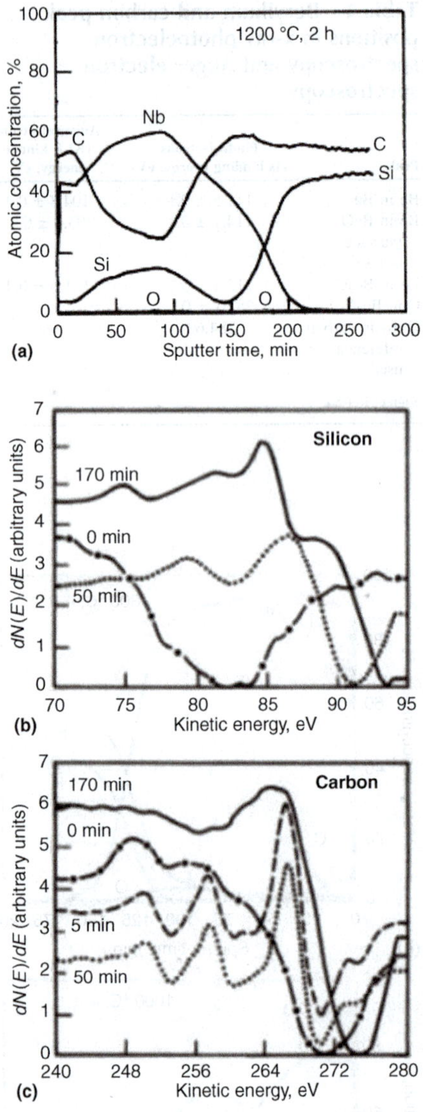

Fig. 22 (a) Auger electron spectroscopy depth-profile measurement of Nb-SiC sample annealed at 1200 °C (2190 °F) for 2 h. (b) Silicon and (c) carbon Auger peak shapes at various sputter times (depths) in the profile, showing chemical-state changes. Source: Ref 61

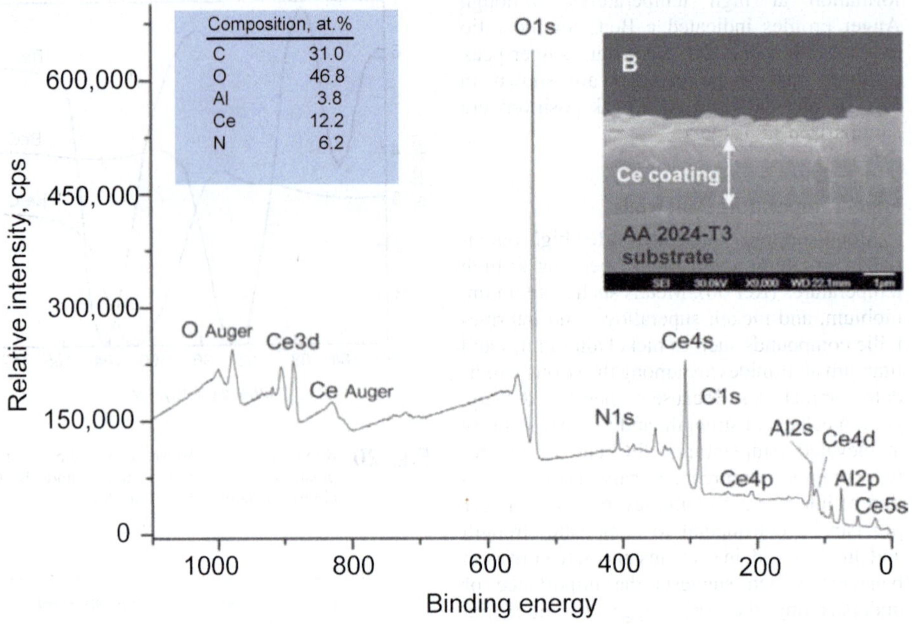

Fig. 23 X-ray photoelectron spectroscopy spectrum with compositional estimates and scanning electron microscopy micrograph of a cross section of a cerium-base conversion coating over aluminum alloy 2024-T3. Source: Ref 66. Reprinted with permission from Elsevier

Uhart et al. (Ref 66) deposited cerium-base conversion coatings (Ce CC) at room temperature by immersing an alloy panel in an aqueous solution containing $Ce(NO_3)_3{,}6H_2O$ at different cerium concentrations (0.01, 0.05, 0.1, and 0.5 *M*). Aluminum alloy 2024-T3 was examined by using XPS before and after surface preparation and after the application of Ce CC. An example of a coating formed in 0.1 *M* solution is shown in Fig. 23. The XPS spectra and high-resolution scans indicated that the coating is cerium-rich oxide containing Ce +III and Ce +IV chemical states. Cross-sectional evaluations using SEM indicated that the ~2 μm thick coating is relatively uniform. A few surface defects, including cracks and voids, are visible. Surface composition uniformity examined by using scanning Auger microscopy revealed the elements cerium, aluminum, and oxygen fairly uniformly distributed on the surface.

Test results to determine coating susceptibility to corrosion in sodium chloride solutions show the Ce CC held up well in solutions of 0.05 to 0.1 *M* NaCl but failed at higher NaCl concentrations. In 3.5% NaCl solution tests, pitting was observed in 1 h. Pits can initiate in a protective film due to several factors, including localized chemical and mechanical damage, nonuniformity of film application, and precipitation of intermetallic compounds in the alloy. Scanning Auger evaluations aided in understanding surface compositional distributions and the nature of pitting. Figure 24 shows how the chemistry varied in different regions of a pit. Cerium, initially present relatively uniformly over the entire substrate, migrates to become enriched in the center and the halo regions of the pit. Thus, the anodic regions of a pit tend to receive protection from cerium-hydrated oxides, potentially enabling self-passivation.

Thick Films and Coatings

Thick films and coatings can be depth profiled similar to thin films, but limitations emerge due to ion beam mixing and surface roughening at large depths, which leads to a loss in depth resolution. While these artifacts can be minimized by using sample rotation, the time required to profile thick films could be prohibitive. A suggested approach to circumvent this problem entails chemical etching of the top layer(s). For example, a thick top coat on a thin-film layer at an interface can potentially be etched to a depth leaving only a thin top coat, thereby reducing the total sputter time and enhancing the depth resolution in the profile.

The high lateral resolution offered by AES coupled with high surface sensitivity enables analyzing thick films, coatings, and layered structures in cross-sectional mode. Small, near-grazing cross sections produced by using focused ion beam and other ion cross-sectioning/polishing methods enable geometric magnification of structures previously not possible. Ion beam methods also add another dimension by generating clean surfaces available for AES analysis.

Scheithauer (Ref 69–71) describes another method to generate low-angle cross sections inside an Auger chamber by using the existing ion gun. Placing a mask above the sample and argon ion sputtering at a low angle enables generating a shallow bevel, which at any location on the bevel can be Auger depth profiled in conjunction with sputtering. A schematic of the sputtering setup with mask to generate a geometric magnification across the layered structure is shown in Fig. 25.

A fracture across a coating or layered structure does not necessarily provide any geometric magnification, but useful information can still be obtained by direct analysis. The superlattice structure of GaAs-AlAs thin films highlights the potential of this approach (Ref 72). The detailed superlattice structure grown by molecular beam epitaxy consists of a GaAs cap layer/[GaAs (50 nm)/AlAs

(50 nm) × 5]/GaAs//[GaAs (20 nm)/AlAs (20 nm) × 5]/GaAs//[GaAs (10 nm)/AlAs (10 nm) × 5]. The sample was cleaved and examined using AES. Figure 26 shows the secondary electron image of the cleaved section, with the grouped 50, 20, and 10 nm lattices identified. Figure 27 shows the secondary electron image of the 10 nm structure and a gallium Auger image. The 10 nm GaAs (yellow regions) and the AlAs (red regions) are readily distinguishable, and the spacings are measurable. Line scans of gallium, arsenic, and aluminum obtained from the 50, 20, and 10 nm lattices are shown in Fig. 28 and illustrate its chemical analysis potential. Examination of bevel samples, with the benefit of geometric magnification coupled with small beam size and surface sensitivity of AES, offers many possibilities to examine not only thick films and coatings but also other common and uncommon materials and structures.

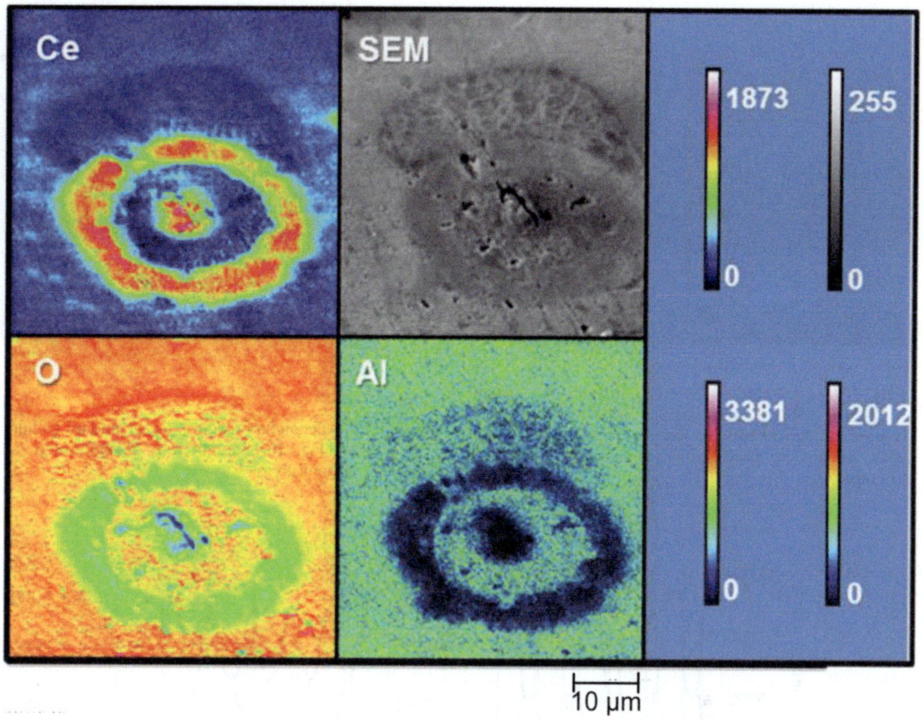

Fig. 24 Scanning Auger images of cerium, oxygen, and aluminum showing elemental distributions, and scanning electron microscopy (SEM) image of the corresponding region. The relative Auger intensity of each element (color scales on the right) is proportional to its atomic concentration. Source: Ref 66. Reprinted with permission from Elsevier

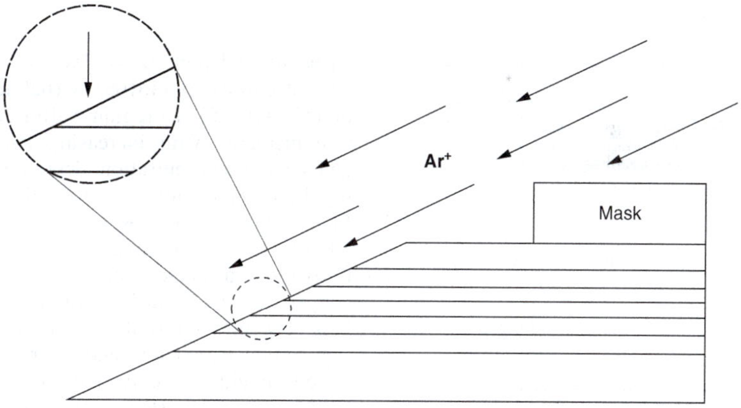

Fig. 25 Schematic of sputtering setup with mask to generate a bevel. Adapted from Ref 69

Surface Segregation

Segregation of alloying elements and impurities to the surface from the bulk alters the surface chemistry of any material and can influence its physical and chemical properties, including adhesion, surface reactivity such as in oxidation and corrosion, catalysis, and sintering behavior. Often, surface segregation of an element may not reach its equilibrium state in reasonable time, due to its low bulk diffusion rates, interaction with other constituents that are segregated, surface contamination and modification, and site competition (Ref 17, 73). Knowledge of the state of segregation is desired to better understand chemistry-property relationships. Studying surface segregation can also be a guide to segregation at grain boundaries, which are not as easy to access as surfaces but markedly influence many properties. Grain boundaries and free surfaces are thermodynamically similar, and segregation relationships between the two have been explored (Ref 23, 74–77).

Auger spectroscopy is frequently used to study surface segregation due its high surface sensitivity to detect submonolayer levels of segregation as well as its high rate of data acquisition, which is needed to determine segregation kinetics (Ref 78–80). Auger electron spectroscopy enables making these measurements at a sufficiently low temperature range, which is often not accessible when using conventional methods of diffusivity measurements.

Segregation of an element to the surface is primarily controlled by its rate of diffusion from the bulk material, and attempts have been made successfully to determine diffusion coefficients from early-stage kinetic data, where equilibrium processes are not limiting the segregation. In this example, copper and sulfur surface segregation in a Ni-18.7at.%Cu alloy with 0.0007 at.% S was examined (Ref 80). Initially, the alloy was mechanically polished to a surface roughness of ≤0.05 μm and annealed at 961 K (690 °C, or 1270 °F) for 24 h to restore the initial condition and homogenize the bulk concentrations. To remove surface contaminants such as carbon, oxygen, and nitrogen, the alloy surface was cleaned by sputtering, followed by repeated heating to 727 K (455 °C, or 850 °F) and sputter cleaning the surface at room temperature. Figure 29 shows Auger spectra after the 24 h anneal and after sputtering in preparation for heating and segregation measurements.

Surface compositional measurements were made using in situ linear programmed heating in the temperature range of 423 to 1121 K (150 to 850 °C, or 300 to 1560 °F) at a constant rate of 0.02 K/s (5.5 °C/s, or 10 °F/s). Copper segregation to the surface started to occur at 560 K (285 °C, or 550 °F) and reached the maximum surface coverage values of 41.4 at.% at 896 K (620 °C, or 1150 °F). At higher temperatures, sulfur segregation displaced copper from the surface until it reached a maximum surface coverage value of 17.1 at. %, as shown in Fig. 30. Quantification of the segregated material is done by using a monolayer segregation method (Ref 81), which assumes that copper and sulfur segregation

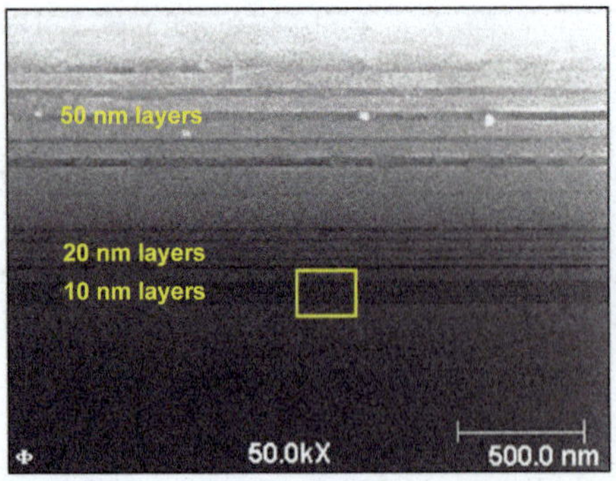

Fig. 26 Secondary electron image of cleaved superlattice sample showing the three layered structures. Source: Ref 72. Reprinted (adapted) with permission from Japan Society for Analytical Chemistry

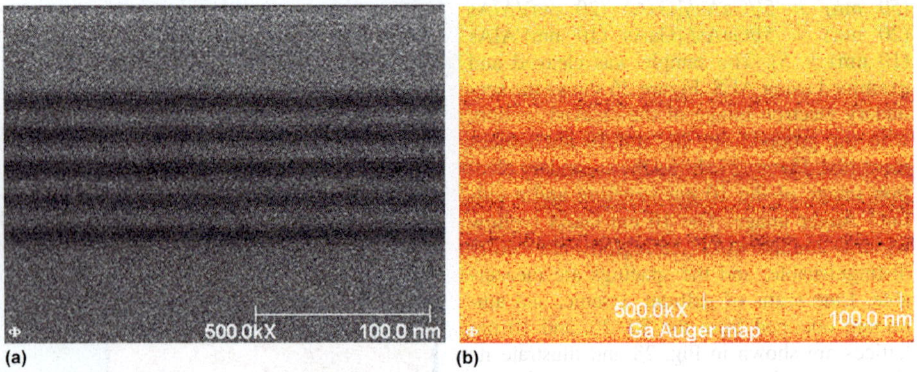

Fig. 27 (a) Secondary electron image and (b) gallium Auger image collected from the 10 nm layer region indicated in Fig. 26. Source: Ref 72. Reprinted (adapted) with permission from Japan Society for Analytical Chemistry

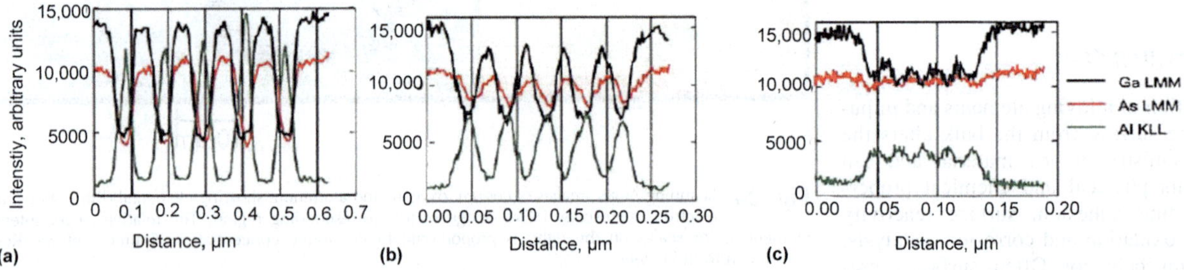

Fig. 28 Auger line scan traces of gallium, arsenic, and aluminum across the lattice structures of (a) 50 nm, (b) 20 nm, and (c) 10 nm lattices shown in Fig. 26. Source: Ref 72

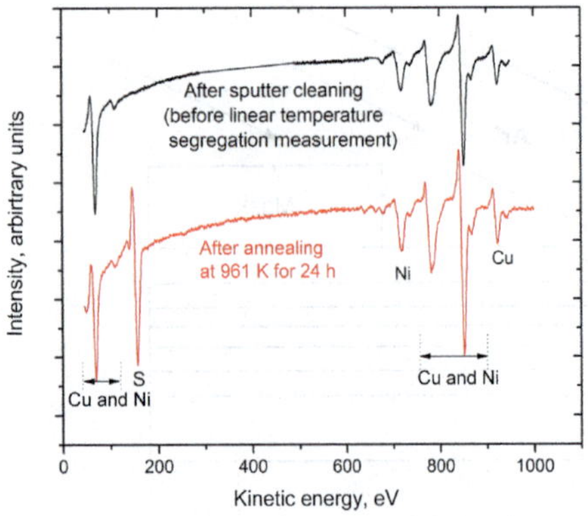

Fig. 29 Auger spectrum of a ternary Ni-Cu(S) alloy annealed at 961 K (690 °C, or 1275 °F) for 24 h (red) and after sputter cleaning with Ar^+ ion (prior to segregation measurements; black). Source: Ref 80

was restricted to the monolayer on the surface, and that copper and sulfur concentrations below the surface layer were equal to the bulk concentration.

In deriving the segregation parameters in the kinetic region of 560 to 896 K (285 to 620 °C, or 550 to 1150 °F), the best fit was obtained by using Fick's diffusion expression, which enabled calculating D_0 and Q values of the Arrhenius relation $D = D_0 \exp(-Q/RT)$ (Fig. 30). In the 896 to1065 K (620 to 790 °C, or 1150 to 1455 °F) region, sulfur concentration increased with increasing temperature, and copper concentration decreased due to repulsive interactions between these atoms. In the 1065 to 1121 K (790 to 850 °C, or 1455 to 1560 °F) region, the surface concentration of sulfur reached a maximum, and copper was at 17.1 at.%, a value close to the bulk concentration. In this equilibrium region, segregation parameters and interaction parameters could be extracted by using the Guttmann model (Ref 82). The modified Darken model is a unified model covering the entire temperature range. It assumes the driving force in the segregation system is the minimization of the total energy of the crystal instead of the concentration gradient, as assumed by Fick's diffusion theory. Therefore, segregated atoms move from a position with a higher chemical potential (the bulk) to a position with a lower chemical potential (the surface), and, as a result, the total energy of the crystal will be lower. The fit with this model is shown in Fig. 31, which enabled determining segregation parameters for both copper and sulfur in the Ni-Cu(S) ternary alloy, namely the segregation energy (ΔG^0), the interaction energies (U), and the diffusion parameters (pre-exponential factor, D_0, and activation energy, Q). Determined values are shown in Table 2.

Grain-Boundary Chemistry

Grain boundaries in metals and alloys are high-energy regions and often exhibit a chemistry different from that of the bulk material. Second-phase precipitation and solute and impurity segregation occur at the boundaries and influence a number of material properties, including strength, some forms of hydrogen embrittlement, intergranular corrosion, and stress-corrosion cracking. Bulk composition and thermal and thermomechanical treatments also exert influence on the grain-boundary chemistry and thereby properties. Auger electron spectroscopy is the primary tool in grain-boundary investigations to reveal the identity, quantity, thickness, and lateral distribution of segregated species. Examples include studies of embrittlement caused by segregation of antimony, phosphorus, sulfur, and tin in steels (Ref 83–91), bismuth in copper (Ref 92, 93), phosphorus in tungsten (Ref 94), phosphorus in nickel (Ref 95), sulfur in nickel (Ref 96), boron and sulfur in Ni_3Al (Ref 76, 97), corrosion in steels (Ref 98) and nickel alloys (Ref 77, 96), and stress-corrosion cracking (Ref 99–102). The thickness of segregated

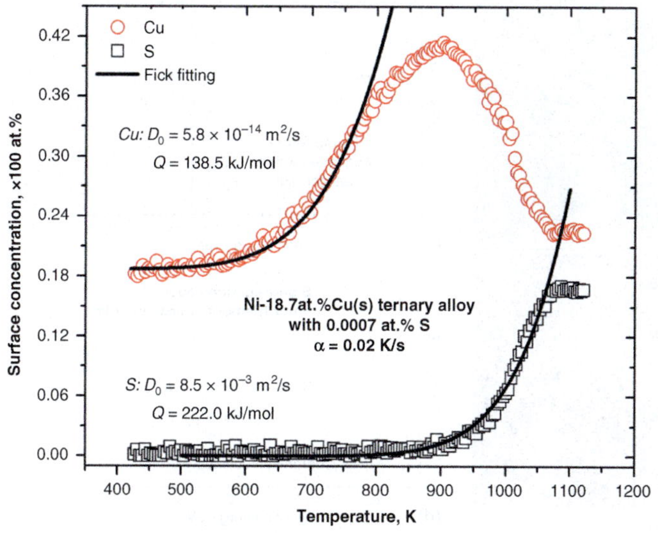

Fig. 30 Surface segregation of copper and sulfur as a function of the temperature for the ternary Ni-18.7at.% Cu(S) alloy. The solid lines are the best fits with the Fick's diffusion expression, which yields the diffusion parameters (pre-exponential factor, D_0, and activation energy, Q) for both copper and sulfur in the Ni-Cu(S) ternary alloy. Source: Ref 80

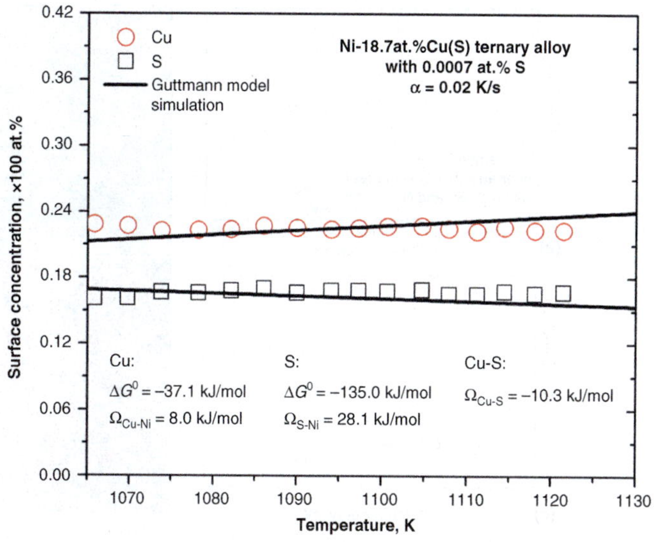

Fig. 31 Surface segregation equilibrium region of copper and sulfur as a function of the temperature for the ternary Ni-18.7at.%Cu(S). The solid lines are the best simulation fits with the Guttmann model, which yields the segregation parameters (segregation energy, ΔG^0, and interaction energy, Ω) for both copper and sulfur in the ternary Ni-Cu(S) alloy. Source: Ref 80

layers in several steels determined by argon ion sputtering has a characteristic depth of 0.45 nm (Ref 103), which amounts to no more than two atomic layers. Grain-boundary segregation studies using AES are a topic of great interest and have been widely reviewed (Ref 10, 14, 17, 19, 23, 104–106).

In contrast to common second-phase precipitation, segregation often occurs in the monolayer and submonolayer regime, making it difficult for evaluation by using common metallographic and SEM-EDS methods. Evaluation of fracture surfaces by using AES is ideally suited for these investigations, because it offers high surface sensitivity coupled with high lateral resolution. Modern instruments have spatial resolution exceeding 5 nm, similar to that of scanning electron microscopes, which enables distinguishing grain boundaries from cleavage and ductile rupture features and identifying second-phase precipitates.

Methodology used to determine grain-boundary segregation involves fracturing samples in the vacuum chamber of the spectrometer or in an adjoining vacuum chamber and examining the grain-boundary regions of the freshly exposed surface. The in situ fractured surface is chemically very active and adsorbs active gases quickly. Even in a good vacuum (better than 1 × 10^{-9} torr), notable adsorption of oxygen and carbon occurs from residual gases in the vacuum chamber, thereby limiting analysis time, often to no more than a couple of hours. The most common approach to create a fracture surface is impact fracture or a combination of sample cooling and impact fracture. For many metals and alloys that exhibit a ductile-to-brittle transition temperature (DBTT), sample cooling below the DBTT is effective. However, many other metals and alloys will not fracture along grain boundaries. For such materials, it may be possible to expose sufficient grain boundaries by hydrogen charging the sample and/or using an in situ slow strain-rate fracture apparatus.

Embrittlement of low-alloy steels is a problem area that has been investigated extensively by using AES, which shows that embrittlement is caused by grain-boundary segregation of certain metalloid impurity elements, such as sulfur, phosphorus, antimony, and tin. High-spatial-resolution analysis enables examination of the composition of grain boundaries in detail. A secondary electron image of the fracture surface obtained from a low-alloy steel of composition 3.5 wt% Ni, 1.6 wt% Cr, 0.39 wt% C, 50 ppm P, 620 ppm Sb, and balance iron is shown in Fig. 32, which shows nearly 100% grain-boundary fracture, and much of the texture seen at grain-boundary facets is due to the presence of carbides. Auger mapping for various elements and Auger spectra obtained from selected regions are shown Fig. 33. The most notable observation is that some boundaries exhibit grain-boundary enrichment of antimony and nickel, while others do not. The various regions are identified as ferrite-ferrite boundaries with antimony and nickel enrichment (Fig. 33a), ferrite-ferrite boundaries with no enrichment (Fig. 33b), chromium-rich carbides (Fig. 33c), and chromium sulfides (Fig. 33d). Ferrite-ferrite boundaries with antimony and nickel enrichment are believed to be

Table 2 Segregation, diffusion, and interaction parameters using modified Darken model simulation

	Segregation species	
Segregation parameter	**Copper**	**Sulfur**
Activation energy (Q), kJ/mol	145.2	224.0
Pre-exponential factor (D_0), m^2/s	8.6 × 10^{-14}	9.2 × 10^{-3}
Segregation energy (ΔG^0_{Cu}), kJ/mol	−36.1	...
Segregation energy (ΔG^0_S), kJ/mol	...	−136.0
Interaction energy (Ω_{CuNi}), kJ/mol	7.6	...
Interaction energy (Ω_{SNi}), kJ/mol	...	28.1
Interaction energy (Ω_{CuS}), kJ/mol	...	...

Source: Ref 80

Fig. 32 Secondary electron image collected from the surface of a low-alloy steel fractured in situ within the Auger chamber showing nearly 100% intergranular fracture. Courtesy of Physical Electronics, USA

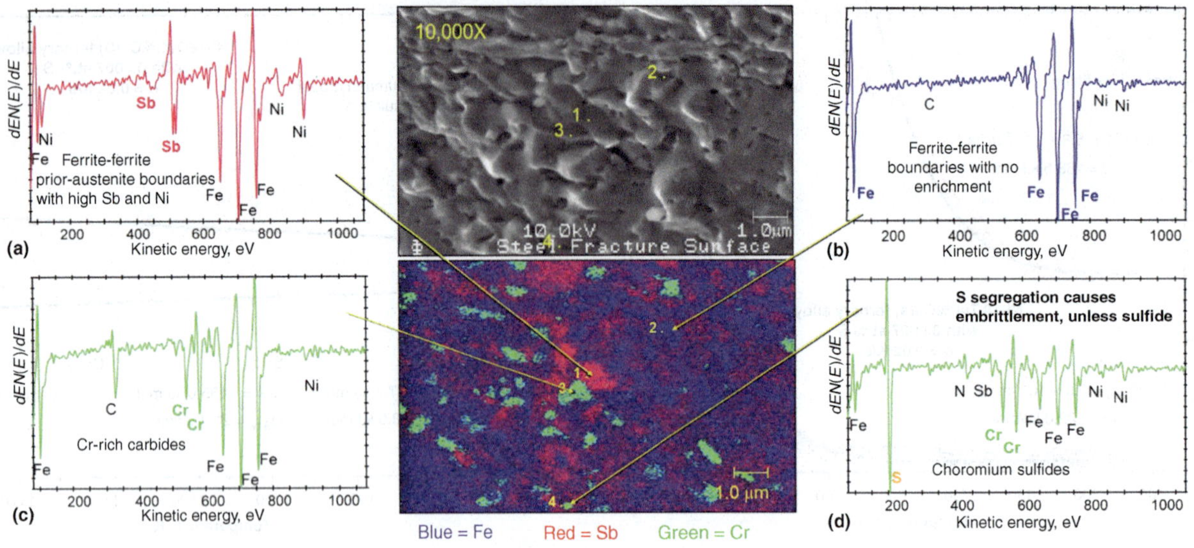

Fig. 33 Auger electron spectroscopy analysis of an in situ fractured low-alloy steel showing (a) ferrite-ferrite boundaries with antimony and nickel enrichment, (b) ferrite-ferrite boundaries with no enrichment of antimony and nickel, (c) chromium-rich carbides, and (d) chromium sulfides. Courtesy of Physical Electronics, USA

prior-austenite grain boundaries, which had little opportunity to lower their energy upon quenching from the austenite phase, while others with no enrichment are believed to be ferrite-ferrite subboundaries that nucleated and grew in the tempering range and had an opportunity to remain at lower energy (Ref 89). Grain boundaries with high energy are expected to draw segregation and thereby lower their energy, while low-energy boundaries have little or no driving force to accumulate segregated constituents.

The second example illustrated here is phosphorus segregation to grain boundaries in a NiCrMoV steel (Ref 107). It emphasizes an understanding of segregation at specific mating locations of ferrite-ferrite grain boundaries and ferrite/carbide interfaces. The steel has a composition of 3.43 wt% Ni, 1.7 wt% Cr, 0.61 wt% Mo, 0.1 wt% V, 0.25 wt% C, 0.02 wt% P, 0.02 wt% Si, 0.32 wt% Mn, 0.02 wt% Sn, 0.005 wt% S, and balance iron. It was air cooled from an austenitizing temperature of 985 °C (1810 °F), tempered to a hardness level of 20 or 30 HRC, and aged at 480 °C (895 °F) for 2400 h to induce grain-boundary segregation. Samples for Auger analysis were prepared with a 2 by 2 mm (0.08 by 0.08 in.) cross section and had a notch to facilitate in situ fracture at the desired location. A small nickel strip was spot welded to the sides behind the notch so that, upon fracture, the detached segment can be bent back and the two fracture surfaces positioned next to each other for AES evaluation.

Fracture was conducted in situ at or near liquid nitrogen temperatures, with fracture surfaces examined immediately after the fracture. Surfaces were nearly 100% intergranular, as shown in Fig. 34. An Auger spectrum obtained by using a 10 keV primary electron beam showed phosphorus along with all other major

Fig. 34 Secondary electron image of a 20 HRC NiCrMoV steel fracture surface obtained in the Auger electron spectroscopy instrument. The steel was aged at 480 °C (895 °F) for 2400 h. Source: Ref 107

elements of the alloy (Fig. 35). The composition determined by using elemental sensitivity factors (Ref 1) yields 7.4 at.% P at this surface. As seen later, the mating surfaces have almost equal amounts of phosphorus, thus amounting to an estimated 14.8 at.% on average at these boundaries. This represents an enrichment ratio of approximately 370 compared with the bulk concentration of 0.04 at.%. Similar enrichments are seen in numerous works for phosphorus and antimony grain-boundary segregation in steels.

Composition depth profiles indicate phosphorus is segregated near the top monolayer. Because AES analysis comprises an estimated depth of approximately seven monolayers, corresponding to the escape depth of emitted Auger electrons, and phosphorus is present only in the top layer, the actual quantity of

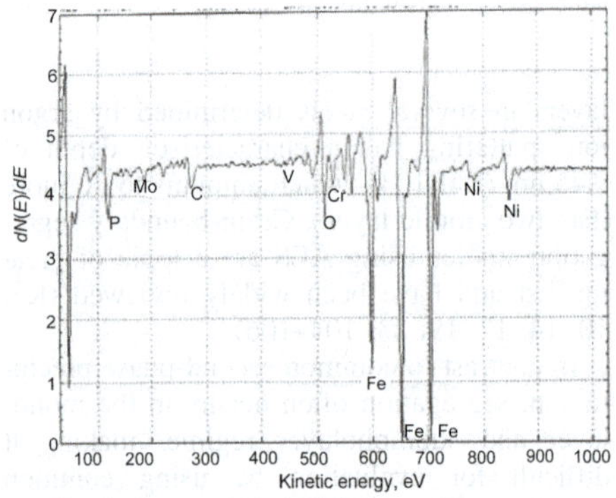

Fig. 35 Auger spectrum of 20 HRC NiCrMoV steel fracture surface obtained from the entire grain-boundary region shown in Fig. 34. Source: Ref 107

phosphorus at grain boundaries is higher than the measured value. Seah's approach (Ref 29) to quantify thin overlayers suggests the actual amount at grain boundaries of many of the segregation constituents is likely more than four to ten times that measured on one side of the fracture surface.

In this study, AES spectra were obtained from a number of mating grain facets, similar to those shown in Fig. 36. Auger peak-to-peak heights of the phosphorus peak at 120 eV and the iron peak at 703 eV were measured, and phosphorus-iron peak height ratios (PHRs) were calculated for various boundaries. Results show the grain-boundary compositions are very similar at their mating grain-boundary facets, indicating that phosphorus segregation splits to both sides during fracture.

A closer examination of the fracture surface (Fig. 37) shows several carbide precipitates present at the ferrite grain boundaries. Several of the carbide precipitates and their mating

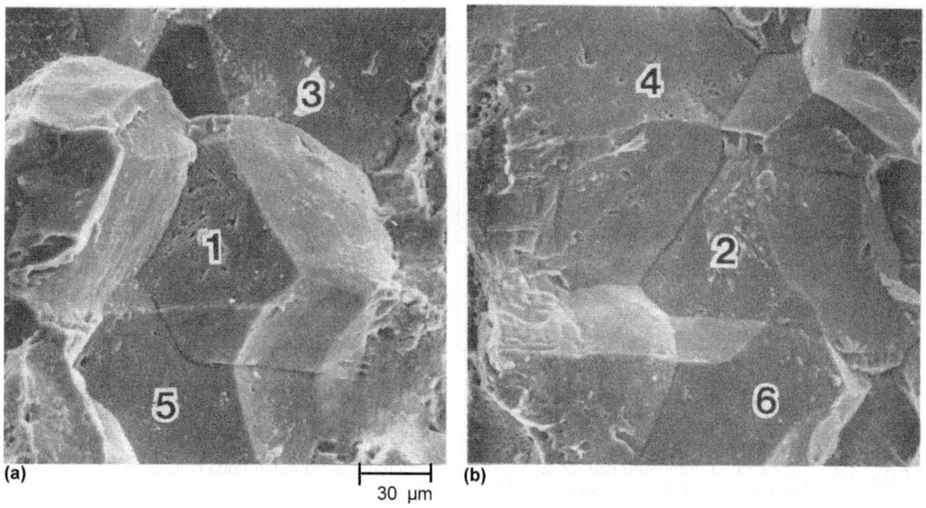

Fig. 36 Secondary electron images of 30 HRC NiCrMoV steel identifying the mating grains (1 and 2, 3 and 4, 5 and 6) from which Auger spectra were obtained to determine the extent of phosphorus segregation. Source: Ref 107

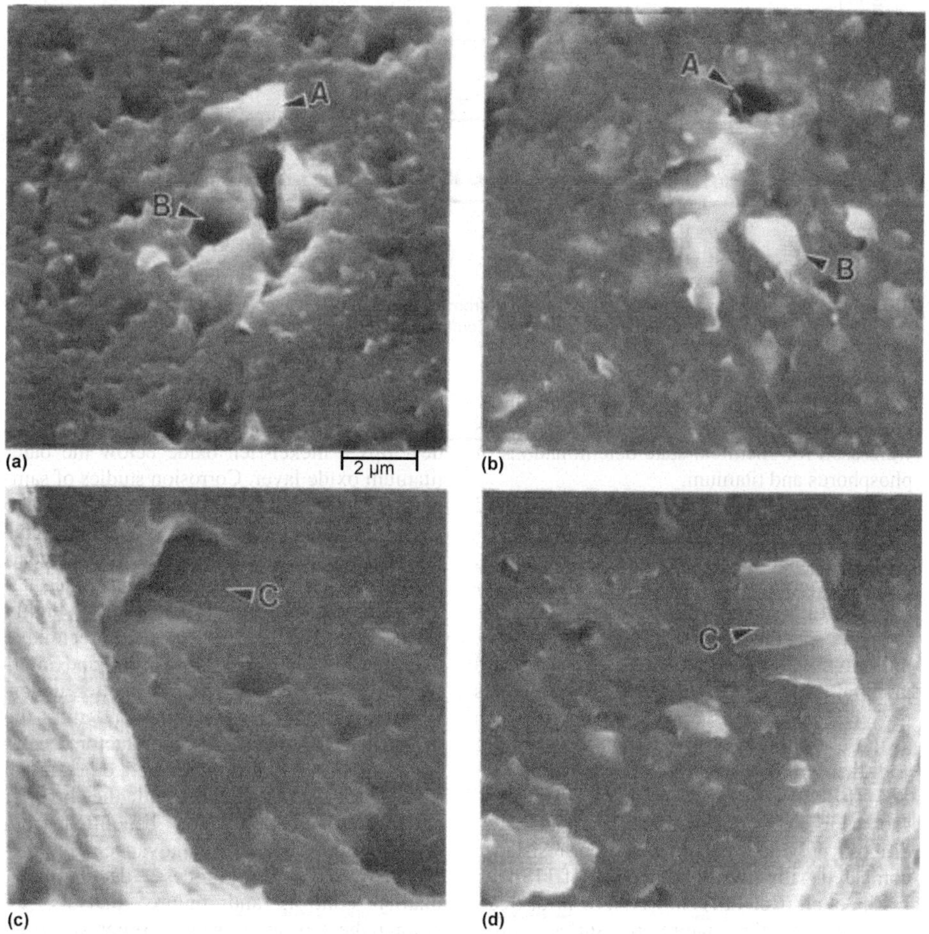

Fig. 37 Phosphorus segregation was investigated from mating areas of carbide precipitates on grain-boundary facets of 30 HRC NiCrMoV steel. Source: Ref 107

locations on the ferrite facets were selected and evaluated by using AES. Results show that phosphorus-iron PHRs at these carbide/ferrite interfaces are slightly lower than those seen at the ferrite/ferrite interfaces. Changes in the carbide/ferrite interface during tempering and aging heat treatments could account for this result. As carbides grow, they form a lower-energy interface with the ferrite matrix, creating fewer sites for segregation of impurity elements and less driving force for segregation. Similar reorientation of the ferrite-ferrite boundaries is limited by pinning from carbides, and therefore, they remain as higher-energy boundaries and accommodate a greater concentration of phosphorus.

This example illustrates how the high lateral resolution coupled with the high surface sensitivity of AES made it possible to determine the chemistry of mating interfaces.

Electronic Materials

Auger electron spectroscopy is widely used in the electronics industry at all levels of semiconductor processing, for example, in studies of discrete devices, hybrids, and printed wiring boards as well as in failure analysis, such as in wire-bonding failures, surface contamination, corrosion, defect identification, and metallization (Ref 11, 13,108–110). The high surface sensitivity of AES, coupled with secondary electron imaging and small-area analysis capabilities, makes it the choice for use in applications including wire-bonding issues, defect identification in semiconductor processing, and metallization for integrated circuits.

Wire-bonding issues are a common problem in the electronics industry, because even low levels of impurities, often arising from the process, can lead to poor bonding. For example, in a case where a suspect chip (Fig. 38) was analyzed, AES showed high levels of carbon, fluorine, and oxygen at the aluminum pad surface. Fluorine appeared to be high at its surface compared with a pad in the chip that did not exhibit bond failures (Fig. 39). Depth profiles conducted on both pads (Fig. 40) led to a better understanding of the problem. The pad from the contaminated chip had ~160 nm thick aluminum oxide and fluorine compared with that of a nominal chip pad that revealed approximately 4 nm. Knowing the nature of the contamination enabled identifying the source of the contamination.

Defect identification is crucial in semiconductor processing, and demands continue to grow as feature size shrinks to meet smaller, faster device requirements. The example in Fig. 41 shows a partially buried defect before and after it was cross sectioned by using a focused ion beam. The AES spectra were obtained from various locations on the section, showing varying levels of silicon, oxygen, and tungsten. Auger images were obtained for

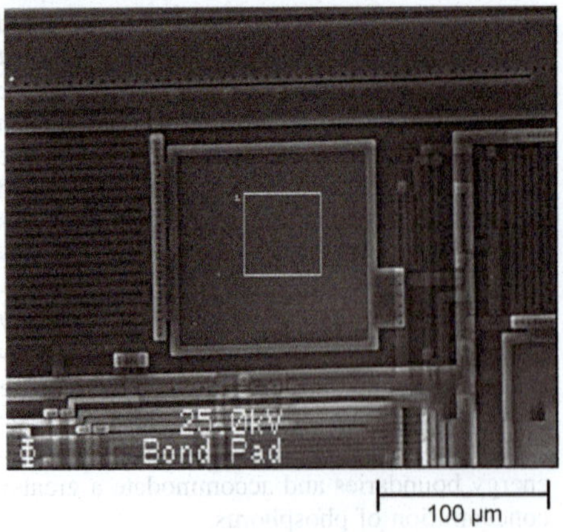

Fig. 38 Secondary electron image of a bond pad from a chip suspected of contamination and exhibiting bond failures. These secondary electron images did not appear much different from the nominal pads. Courtesy of Physical Electronics, USA

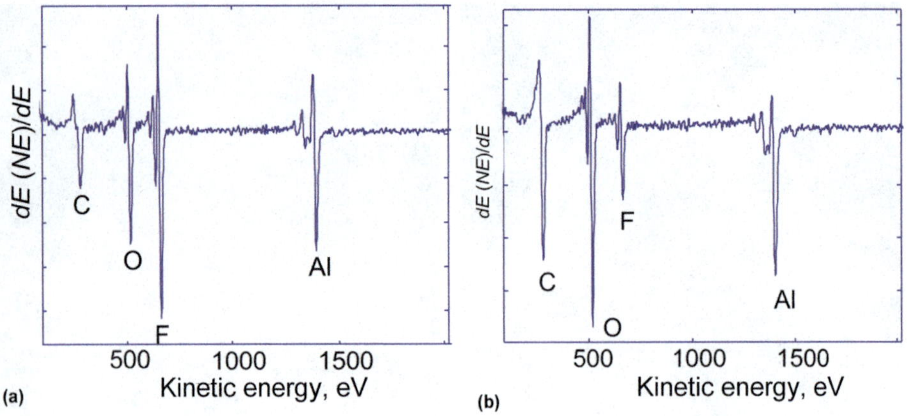

Fig. 39 Auger spectra from bond pads of (a) a contaminated chip that exhibited bond failures and (b) a nominal chip. Courtesy of Physical Electronics, USA

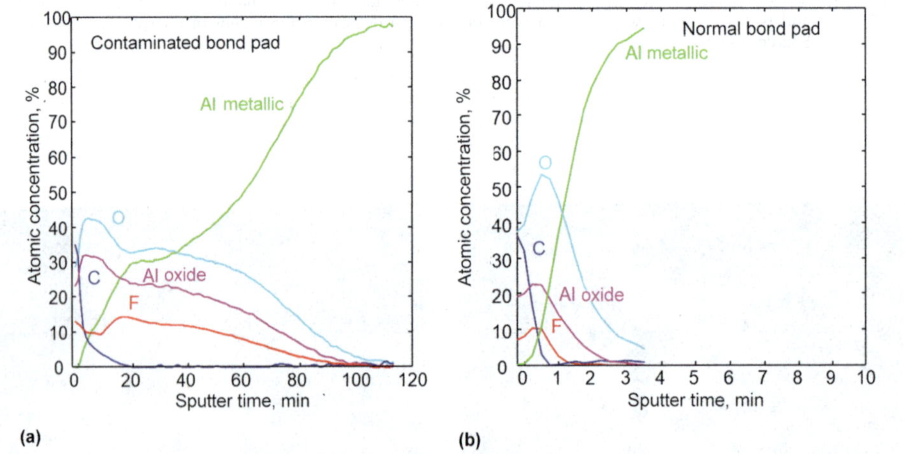

Fig. 40 Auger electron spectroscopy depth profiles from bond pads of (a) a contaminated chip and (b) a nominal chip. Estimated sputter rate is ~2 nm/min. Courtesy of Physical Electronics, USA

tungsten and for silicon in both elemental and oxide forms. The composite Auger image in Fig. 42 illustrates the compositional variations and points to a SiO_2 particle as the central defect, which became covered with elemental silicon during polysilicon deposition.

Metallization for integrated circuits is a key area that benefits from AES, particularly in testing the accuracy and compositions of metallization layers and in developing diffusion barriers. Many pure metals and alloys, including chromium, molybdenum, and titanium-tungsten, are not sufficiently good diffusion barriers and must have impurities such as oxygen or nitrogen to become effective. Figure 43 illustrates the benefit of AES depth profiling on a molybdenum diffusion barrier. Figure 43(a) shows a layered structure of Au-Mo-Si in the as-deposited condition. Upon heating to 450 °C (840 °F) for 30 min, gold diffuses through molybdenum and alloys with silicon (Fig. 43b), indicating that molybdenum is not a good diffusion barrier to gold. Nitrogen introduced during molybdenum deposition forms a thin nitride layer within molybdenum (Fig. 43c), restricting diffusion of gold and thus making molybdenum an effective diffusion barrier.

Biomaterials

Auger spectroscopy is used to evaluate the surface condition of metallic implants, aid in the development of protective coatings, and help in understanding the interaction of implant surfaces with their biological hosts. Stainless steels, titanium, and alloys of cobalt-chromium and nickel-titanium are among the chief implant materials used in a variety of applications, and they often require surface treatments for protection against bodily corrosion. Ong and Lucas (Ref 112) used AES to study titanium and hydroxyapatite coatings to determine oxide thickness and distribution of phosphorus in the coating. Auger peak positions and peak shapes were used for chemical-state determinations of phosphorus and titanium.

Nickel-titanium alloys are commonly used for stents because of their superelastic properties, shape memory behavior, and radiopacity, but they require surface coatings (typically oxides) for corrosion protection. Coatings require a specific chemistry, optimal thickness, uniform coverage, and the ability to withstand a certain degree of deformation, properties governed by prior processing parameters such as those used in chemical and electropolishing and thermal oxidative treatments (Ref 113, 114). In corrosion tests, thinner oxide layers, such as in electropolished and passivated conditions, provided better corrosion resistance compared with other treatments (Ref 113).

Zhu et al. (Ref 114) showed how prior treatment affects the nature of surface films (Fig. 44). In the electropolished condition, nitinol has a thin titanium oxide layer that offers protection from corrosion (Fig. 44a). Oxidation at 400 °C (750 °F) for 3 min shows oxide growth, and after heating for 30 min, it develops a nickel-rich oxide below the outer titanium oxide layer. Corrosion studies of samples oxidized in the 400 to 1000 °C (750 to 1830 °F) temperature range show that incorporation of nickel-rich phases in thicker coatings is deleterious for corrosion resistance.

Auger electron spectroscopy is a highly surface-sensitive microanalytical technique used to study surfaces, thin films, and interfaces. Its application led to a better understanding of material behavior and the properties of metals, semiconductors, catalysts, ceramics, composites, and other novel materials. Semiconductor and the broader electronics industries have benefitted from the technique, because its advancements met industry demands. Process development, quality control, and failure analysis are among the key areas where Auger spectroscopy continues to see application. It is also finding application in areas such as biointerfaces and nanotechnology.

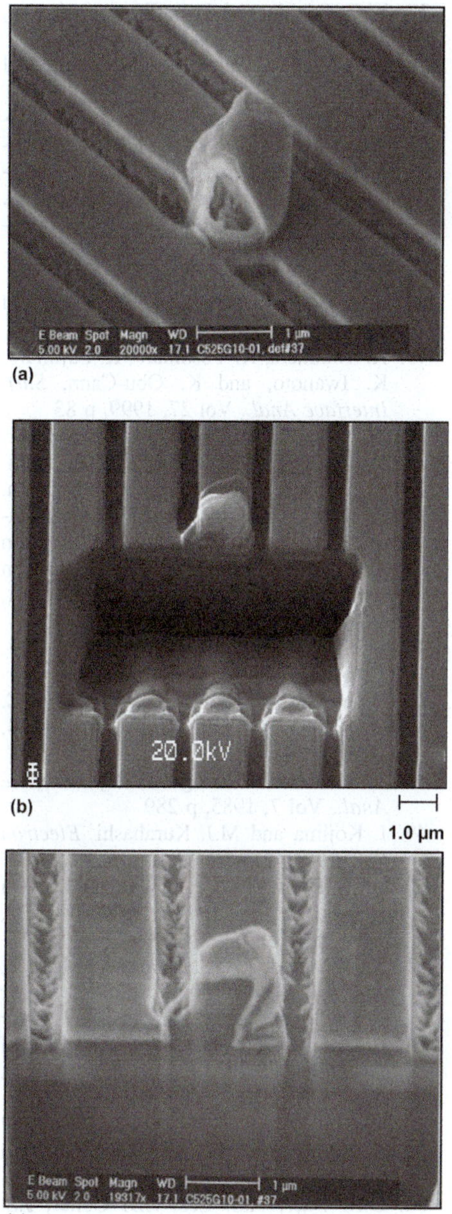

Fig. 41 Secondary electron images of ~1 µm sized defect (a) before ex situ focused ion beam cross sectioning and (b) and (c) after cross sectioning. Courtesy of Physical Electronics, USA

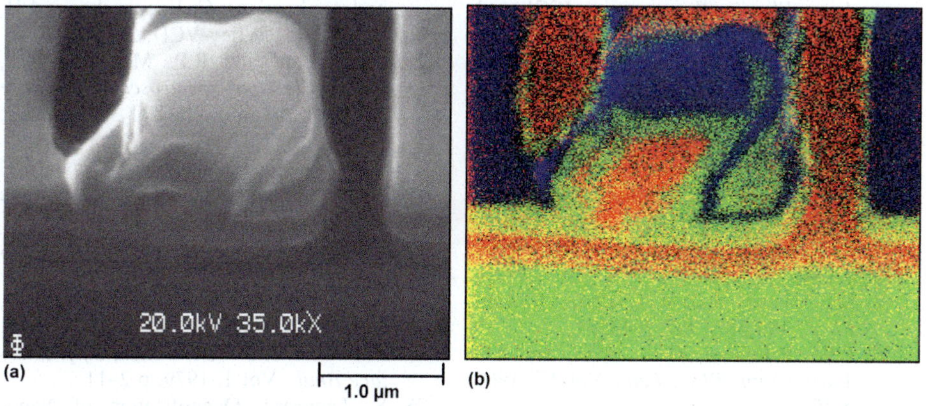

Fig. 42 Focused ion beam cross section of the defect. (a) Secondary electron image. (b) Color Auger images of silicon in the oxide form (red), silicon in the elemental form (green), and tungsten (blue). Courtesy of Physical Electronics, USA

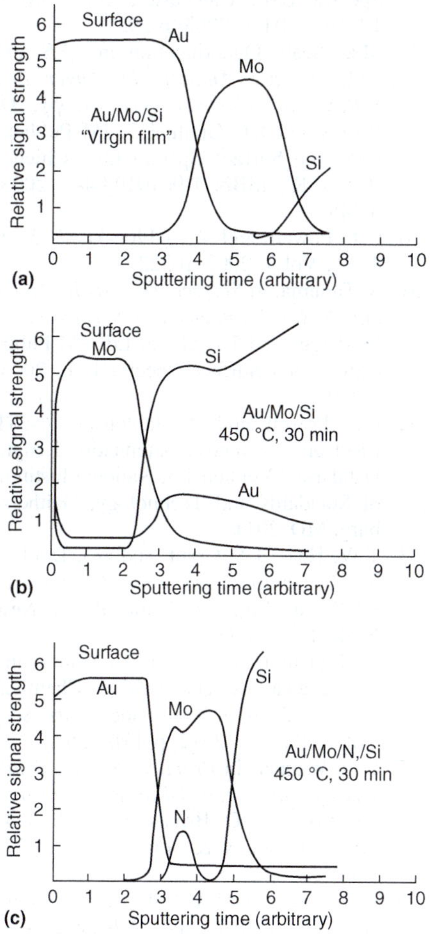

Fig. 43 Auger electron spectroscopy depth profiles. (a) As-deposited 150 nm molybdenum and 150 nm gold films on silicon. (b) Same films after 450 °C (840 °F) anneal for 30 min. (c) Nitrided molybdenum and gold films after similar heat treatments. Source: Ref 111

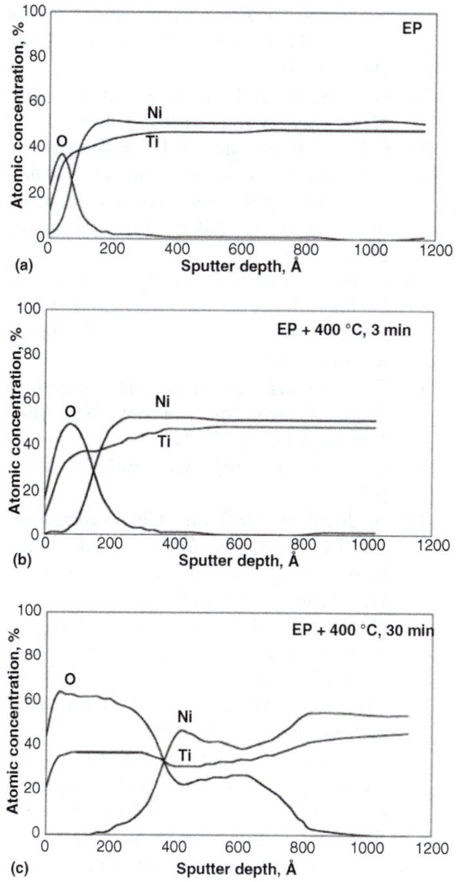

Fig. 44 Auger electron spectroscopy depth profiles of NiTi wire in (a) as-electropolished (EP) condition and after air oxidation at 400 °C (750 °F) for (b) 3 min and (c) 30 min. Note the presence of a nickel-rich region below the titanium oxide surface layer. Source: Ref 114

ACKNOWLEDGMENT

Revised from A. Joshi, Auger Electron Spectroscopy, *Materials Characterization*, Vol 10, *ASM Handbook*, ASM International, 1986.

REFERENCES

1. L.E. Davis, N.C. MacDonald, P.W. Palmberg, G.E. Riach, and R.E. Weber, *Handbook of Auger Electron Spectroscopy*, Perkin-Elmer Corporation, Physical Electronics Division, Eden Prairie, MN, 1976
2. G.E. McGuire, *Auger Electron Spectroscopy Reference Manual*, Plenum, New York, 1979
3. Y. Shiokawa, T. Isida, and Y. Hayashi, *Auger Electron Spectra Catalogue: A Data Collection of Elements*, Anelva, Tokyo, Japan, 1979
4. T. Sekine, Y. Nagasawa, M. Kudoh, Y. Sakai, A.S. Parkes, J.D. Geller, A. Mogami, and K. Hirata, *Handbook of*

Auger Electron Spectroscopy, JEOL, Ltd. Tokyo, 1982
5. K.D. Childs, B.A. Carlson, L.A. Lavanier, J.F. Moulder, D.F. Paul, W.F. Stickle, and D.G. Watson, *Handbook of Auger Electron Spectroscopy*, Physical Electronics Inc., Eden Prairie, MN, 1995
6. P. Auger, *J. Phys. Radium*, Vol 6, 1925, p 205
7. L.A. Harris, *J. Appl. Phys.*, Vol 39, 1968, p 1419
8. R.E. Weber and W.T. Peria, *J. Appl. Phys.*, Vol 38, 1967, p 4355
9. P.W. Palmberg, G.K. Bohn, and J.C. Tracy, *Appl. Phys. Lett.*, Vol 15, 1969, p 254
10. A. Joshi, L.E. Davis, and P.W. Palmberg, in *Methods of Surface Analysis,* A.W. Czanderna, Ed., Elsevier, 1975, p 160–222
11. P.H. Holloway, Fundamentals and Applications of Auger Electron Spectroscopy, *Advances in Electronics and Electron Physics 54*, Academic Press, New York, 1980, p 241
12. D.F. Stein and A. Joshi, *Annu. Rev. Mater. Sci.*, Vol 11, 1981, p 485
13. G.E. McGuire and P.H. Holloway, in *Electron Spectroscopy: Theory, Techniques and Applications*, Vol 4, C.R. Brundle and A.D. Baker, Ed., Academic Press, 1981
14. D. Briggs and M.P. Seah, Ed., *Practical Surface Analysis by Auger and X-Ray Photoelectron Spectroscopy*, John Wiley & Sons, 1983
15. C.C. Chang, Advances in Analytical Auger Electron Spectroscopy, *MRS Bull.*, Vol 12, 1987, p 70–74
16. C.C. Chang, *Surf. Sci.*, Vol 25, 1971, p 53
17. A. Joshi, in *Materials Characterization*, Vol 10, *Metals Handbook*, 9th ed., American Society for Metals, Metals Park, OH, 1986, p 550–567
18. D. Briggs and J.C. Riviere, in *Practical Surface Analysis by Auger and X-Ray Photoelectron Spectroscopy*, Vol 1, D. Briggs and M.P. Seah, Ed., John Wiley & Sons, New York, 1990
19. S. Hofmann, *Auger and X-Ray Photoelectron Spectroscopy in Materials Science: A User-Oriented Guide*, Springer Science and Business Media, 2012
20. D. Briggs and J.T. Grant, Ed., *Surface Analysis by Auger and X-Ray Photoelectron Spectroscopy*, IM Publications and Surface Spectra Ltd., Chichester, U.K., ISBN 1-901019-04-7, 2003
21. J.M. Lannon, Jr. and C.D. Stinespring, Auger Electron Spectroscopy in Analysis of Surfaces, *Encyclopedia of Analytical Chemistry*, John Wiley & Sons, Ltd., 2006
22. H. Bubert, J.C. Riviere, and W.S.M. Werner, Auger Electron Spectroscopy, *Surface and Thin Film Analysis: A Compendium of Principles, Instrumentation and Applications*, G. Friedbacher and H. Bubert, Ed., Wiley-VCH, 2011, p 43
23. C.L. Briant and R.P. Messemer, Ed., *Auger Electron Spectroscopy, Treatise on Materials Science*, Vol 30, Elsevier, 2012
24. R.S. SwingleII and W.M. Riggs, *CRC Crit. Rev. Anal. Chem.*, Vol 5 (No. 3), 1975, p 267
25. M.O. Krause, *J. Phys. Chem. Ref. Data*, Vol 8, 1979, p 307
26. A. Kotani and S. Shin, *Rev. Mod. Phys.*, Vol 73, 2001, p 203–246
27. M.P. Seah and W.A. Dench, *Surf. Interface Anal.*, Vol 1, 1979, p 2–11
28. S. Tougaard, Quantification of Nano-Structure by Electron Spectroscopy, *Surface Analysis by Auger and X-Ray Photoelectron Spectroscopy*, D. Briggs and J.T. Grant, Ed., IM Publications and Surface Spectra Ltd., Chichester, U.K., ISBN 1-901019-04-7, 2003, p 295
29. M.P. Seah, Quantification in AES and XPS, *Surface Analysis by Auger and X-Ray Photoelectron Spectroscopy*, D. Briggs and J.T. Grant, Ed., IM Publications and Surface Spectra Ltd., Chichester, U.K., ISBN 1-901019-04-7, 2003, p 345
30. C.J. Powell and A. Jablonski, *J. Surf. Anal.*, Vol 9, 2002, p 322
31. S. Tanuma, in *Surface Analysis by Auger and X-Ray Photoelectron Spectroscopy*, D. Briggs and J.T. Grant, Ed., IM Publications and Surface Spectra Ltd., 2003, p 259
32. C.J. Powell and A. Jablonski, "NIST Electron Effective-Attenuation-Length Database," Version 1.3, National Institute of Standards and Technology, Gaithersburg, MD, 2011
33. T.W. Haas, J.T. Grant, and G.J. Dooley, in *Adsorption-Desorption Phenomena*, F. Ricca, Ed., Academic Press, New York, 1972, p 359
34. K. Tsutsumi, A. Tanaka, M. Shima, and T. Tazawa, A Quantitative Chemical State Analysis of Tin Oxides, *Surf. Sci.*, Vol 33 (No. 8), 2012, p 431–436
35. C.D. Wagner, in *Practical Surface Analysis by Auger and X-Ray Photoelectron Spectroscopy*, D. Briggs and M.P. Seah, Ed., John Wiley & Sons, 1983, p 477–509
36. C.D. Wagner and A. Joshi, Auger Parameter, Its Utility and Advantages—A Review, *J. Electron Spectrosc. Relat. Phenom.*, Vol 47, 1988, p 283–313
37. "NIST X-Ray Photoelectron Spectroscopy (XPS) Database," National Institute of Standards and Technology, 2012, https://srdata.nist.gov/xps/intro.aspx
38. J.F. Moulder, W.F. Stickle, P.E. Sobol, and K.D. Bomben, in *Handbook of X-Ray Photoelectron Spectroscopy*, J. Chastain and R.C. King, Jr., Ed., Physical Electronics, USA, 1995
39. W.F. Stickle and D.G. Watson, *J. Vac. Sci. Technol. A*, Vol 10, 1992, p 2806
40. D.G. Watson, *Surf. Interface Anal.*, Vol 15, 1990, p 516
41. W.F. Stickle, Chap. 14, The Use of Chemometrics in AES and XPS Data Treatment, *Surface Analysis by Auger and X-Ray Photoelectron Spectroscopy*, D. Briggs and J.T. Grant, Ed., IM Publications and Surface Spectra Ltd., Chichester, U.K., 2003, p 377
42. S.W. Gaarenstroom, *Appl. Surf. Sci.*, Vol 26, 1986, p 561
43. H. Tokutaka, K. Yoshihara, K. Fujimura, K. Iwanoto, and K. Obu-Cann, *Surf. Interface Anal.*, Vol 27, 1999, p 83
44. M.P. Seah, I.S. Gilmore, and S.J. Spenser, *Surf. Interface Anal.*, Vol 31, 2001, p 778
45. C.J. Powell and M.P. Seah, Precision, Accuracy, and Uncertainty in Quantitative Surface Analyses by Auger Electron Spectroscopy and X-Ray Photoelectron Spectroscopy, *J. Vac. Sci. Technol. A*, Vol 8, 1990, p 735
46. A. Zalar, *Thin Solid Films*, Vol 124, 1985, p 223
47. D.G. Watson, W.F. Stickle, and A.C. Diebold, *Thin Solid Films*, Vol 193/194, 1990, p 305
48. T. Sekine and A. Mogami, *Surf. Interface Anal.*, Vol 7, 1985, p 289
49. I. Kojima and M.J. Kurahashi, *Electron Spectrosc.*, Vol 46, 1988, p 185
50. W.D. Roos, G.N. van Wyk, and J. Du Plessis, *Surf. Interface Anal.*, Vol 22, 1994, p 65
51. S.B. Donald, J.A. Stanford, W. McLean, and A.J. Nelson, Application of Linear Least Squares to the Analysis of Auger Electron Spectroscopy Depth Profiles of Plutonium Oxides, *J. Vac. Sci. Technol. A*, Vol 36, 2018, p 03E104
52. W.F. Stickle, *Anal. Sci.*, Vol 7, 1991, p 357
53. S. Hofmann, *J. Electron Spectrosc.*, Vol 59, 1992, p 15–32
54. D.R. Baer, A.S. Lea, J.D. Geller, J.S. Hammond, L. Kover, C.J. Powell, M.P. Seah, M. Suzuki, J.F. Watts, and J. Wolstenholme, *J. Electron Spectrosc. Relat. Phenom.*, doi: 10.1016/j.elspec.2009.03.021, 2009
55. M.P. Hooker and J.T. Grant, The Use of Auger Electron Spectroscopy to Characterize the Adsorption of CO on Transition Metals, *Surf. Sci.*, Vol 62, 1977, p 21
56. J.T. Grant, M.P. Hooker, and T.W. Haas, Spectrum Subtraction Techniques in Auger Electron Spectroscopy, *Surf. Sci.*, Vol 51, 1975, p 318
57. G.K. Wehner, The Aspects of Sputtering in Surface Analysis Methods, *Methods of Surface Analysis*, A.W. Czanderna, Ed., Elsevier, 1975, p 35
58. D.R. Baer, D.J. Gasper, M.H. Englehard, and A.S. Lea, in *Surface Analysis by Auger and X-Ray Photoelectron Spectroscopy*, D. Briggs and J.T. Grant, Ed.,

IM Publications and Surface Spectra Ltd., Chichester, U.K., ISBN 1-901019-04-7, 2003, p 211
59. T.G. Nieh, J. Wadsworth, and A. Joshi, Interfacial Reaction between Carbon and Beryllium, *Scr. Metall.*, Vol 20, 1986, p 87–92
60. R.Y. Lin, R.J. Arsenault, G.P. Martins, and S.G. Fishman, Ed., *Interfaces in Metal/Ceramics Composites*, TMS, Warrendale, PA, 1990
61. T.C. Chou, A. Joshi, and J. Wadsworth, High Temperature Interfacial Reactions of SiC with Metals, *J. Vac. Sci. Technol. A*, Vol 9 (No. 3), 1991, p 1525–1534
62. T.C. Chou, A. Joshi, and J. Wadsworth, Solid State Reactions of SiC with Co, Ni, and Pt, *J. Mater. Res.*, Vol 6 (No. 4), 1991, p 796–809
63. T.C. Chou and A. Joshi, Selectivity of SiC/Stainless Steel Solid State Reactions and Discontinuous Decomposition of SiC, *J. Am. Ceram. Soc.*, Vol 76, 1991, p 1364–1372
64. T.C. Chou and A. Joshi, Solid State Interfacial Reactions of Ti_3Al with Si_3N_4 and SiC, *J. Mater. Res.*, Vol 7 (No. 5), 1992, p 1253–1265
65. D.L. Yaney and A. Joshi, Reaction between Niobium and Silicon Carbide at 1373 K, *J. Mater. Res.*, Vol 5 (No. 10), 1990, p 2197–2208
66. A. Uhart, J.B. Ledeuil, D. Gonbeau, J.C. Dupin, J.P. Bonino, F. Ansart, and J. Esteban, An Auger and XPS Survey of Cerium Active Corrosion Protection for AA2024-T3 Aluminum Alloy, *Appl. Surf. Sci.*, Vol 390, Elsevier, 2016, p 751–759
67. A.E. Hughes, J.D. Gorman, P.R. Miller, B.A. Sexton, P.J.K. Paterson, and R.J. Taylor, Development of Cerium-Based Conversion Coatings on 2024-T3 Al Alloy after Rare-Earth Desmutting, *Surf. Interface Anal.*, Vol 36, 2004, p 290–303
68. S.K. Toh, A.E. Hughes, D.G. McCulloch, J. Du Plessis, and A. Stonham, Characterization of Non-Cr-Based Deoxidizers on Al Alloy 7475-T7651, *Surf. Interface Anal.*, Vol 36, 2004, p 1523–1532
69. U. Scheithauer, *Appl. Surf. Sci.*, Vol 179, 2001, p 20–24
70. U. Scheithauer, *Appl. Surf. Sci.*, Vol 255 (No. 22), 2009, p 9062–9065
71. U. Scheithauer, *J. Electron Spectrosc. Relat. Phenom.*, 10.1016/j.elspec.2015.04.012, April 2015
72. M. Suzuki, N. Urushihara, N. Sanada, D.F. Paul, S. Bryan, and J.S. Hammond, *Anal. Sci.*, Vol 5, 2010, p 203–208
73. C. Lea and M.P. Seah, Site Competition in Surface Segregation, *Surf. Sci.*, Vol 52, 1975, p 272
74. M.P. Seah and C. Lea, Surface Segregation and Its Relation to Grain Boundary Segregation, *Philos. Mag.*, Vol 31, 1975, p 627
75. C.L. White, C.T. Liu, and R.A. Padgett, Jr., Free Surface Segregation in Boron Doped Ni_3Al, *Acta Metall.*, Vol 36 (No. 8), 1988, p 2229
76. C.L. White and A. Choudhary, *MRS Proc.*, Vol 81, 1986, p 427
77. H. Chaung, J.B. Lumsden, and R.W. Staehle, Effect of Segregated Sulfur on the Stress Corrosion Susceptibility of Nickel, *Metall. Trans. A*, Vol 10, 1979, p 1853
78. A. Joshi and M. Strongin, Surface Segregation of Oxygen in Nb-O and Ta-O Alloys, *Scr. Metall.*, Vol 8, 1974, p 413
79. E.C. Viljoen and J. Du Plessis, Auger/LEED Linear Heating Study of Sn and S Bulk-to-Surface Diffusion in a Cu (111)(Sn, S) Single Crystal, *Surf. Interface Anal.*, Vol 23, 1995, p 110
80. X.L. Yan, J.Y. Wang, H.C. Swart, and J.J. Terblans, *J. Alloy. Compd.*, Vol 768, 2018, p 875–882
81. J. Du Plessis, *Solid State Phenomena, Part B: Surface Segregation*, Sci-Tech Publications, Brookfield, MA, 1990
82. M. Guttmann, Equilibrium Segregation in a Ternary Solution: A Model for Temper Embrittlement, *Surf. Sci.*, Vol 53, 1975, p 213–227
83. D.F. Stein, A. Joshi, and R.P. LaForce, Studies Utilizing Auger Electron Emission Spectroscopy on Temper Embrittlement in Low Alloy Steels, *Trans. ASM Q.*, Vol 62, 1969, p 776
84. H.L. Marcus and P.W. Palmberg, *Trans. Met. Soc. AIME*, Vol 245, 1969, p 1664
85. A. Joshi and D.F. Stein, Temper Embrittlement of Low Alloy Steels, *Temper Embrittlement of Alloy Steels*, Special Tech. Pub. 499, American Society for Testing and Materials, 1972, p 59–89
86. E.D. Hondros and M.P. Seah, *Int. Met. Rev.*, Vol 222, 1977, p 262
87. C.J. McMahon, Jr. and L. Marchut, *J. Vac. Sci. Technol.*, Vol 15, 1978, p 450
88. R.A. Mulford, C.L. Briant, and R.G. Rowe, *Scanning Electron Microsc.*, Vol 1, 1980, p 487
89. A. Joshi, Segregation at Selective Grain Boundaries and Its Role in Temper Embrittlement of Low Alloy Steels, *Scr. Metall.*, Vol 9, 1975, p 251–260
90. R. Viswanathan and A. Joshi, Effect of Microstructure on the Temper Embrittlement of Cr-Mo-V Steels, *Metall. Trans. A*, Vol 6, 1975, p 2289–2297
91. M.P. Seah and E.D. Hondros, *Proc. R. Soc. (London) A*, Vol 335, 1973, p 191–212
92. A. Joshi and D.F. Stein, An Auger Spectroscopic Analysis of Bismuth Segregated to Grain Boundaries in Copper, *J. Inst. Met.*, Vol 99, London, 1971, p 178–181
93. W.C. Johnson, A. Joshi, and D.F. Stein, The Effect of Fe Additions on the Embrittlement of Cu-Bi Alloys, *Metall. Trans. A*, Vol 7, 1976, p 949–951
94. A. Joshi and D.F. Stein, Intergranular Brittleness Studies in Tungsten Using Auger Spectroscopy, *Metall. Trans.*, Vol 1, 1970, p 2453
95. M. Guttmann, P. Dumoulin, N. Tan-Tai, and P. Fontaine, An Auger Electron Spectroscopic Study of Phosphorus Segregation in the Grain Boundaries of Nickel Base Alloy 600, *Corrosion*, Vol 37 (No. 7), 1981, p 416–425
96. S.M. Bruemmer, R.H. Jones, M.T. Thomas, and D.R. Baer, *Metall. Trans. A*, Vol 14, 1983, p 223
97. C.L. White, R.A. Padgett, C.T. Liu, and S.M. Yalisove, Surface and Grain Boundary Segregation in Relation to Intergranular Fracture: Boron and Sulfur in Ni_3Al, *Scr. Metall.*, Vol 18, 1984, p 1417
98. A. Joshi and D.F. Stein, Chemistry of Grain Boundaries and Its Relation to Intergranular Corrosion of Austenitic Stainless Steels, *Corrosion*, Vol 28 (No. 9), 1972, p 321–330
99. A. Joshi, Influence of Density and Distribution of Intergranular Sulfides on the Sulfide Stress Cracking Properties of High Strength Steels, *Corrosion*, Vol 34 (No. 2), 1978, p 47
100. M.B. Hintz, L.A. Heldt, and D.A. Koss, in *Embrittlement by the Localized Crack Environment*, R.P. Gangloff, Ed., American Institute of Mining, Metallurgical, and Petroleum Engineers, Warrendale, PA, 1984, p 229
101. C.R. Shastry, M. Levy, and A. Joshi, Effect of Solution Treatment Temperature on Stress Corrosion Susceptibility of 7075 Aluminum Alloy, *Corros. Sci.*, Vol 21 (No. 9), 1981, p 673–688
102. A. Joshi, C.R. Shastry, and M. Levy, Effect of Heat Treatment on Solute Concentrations at Grain Boundaries in 7075 Aluminum Alloy, *Metall. Trans. A*, Vol 12, 1981, p 1081–1088
103. P.W. Palmberg and H.L Marcus, *Trans. ASM*, Vol 62, 1969, p 1016
104. W.C. Johnson and J.M. Blakely, Ed., *Interfacial Segregation*, American Society for Metals, 1979
105. D. Raabe, M. Herbig, S. Sandlobes, Y. Li, D. Tytko, M. Kuzmina, D. Ponge, and P.P. Choi, Grain Boundary Segregation Engineering in Metallic Alloys: A Pathway to Design of Interfaces, *Curr. Opin. Solid State Mater. Sci.*, Vol 18, 2014, p 253–261
106. M.P. Seah, Interface Adsorption, Embrittlement and Fracture, *Surf. Sci.*, Vol 53, 1975, p 168–212
107. J. Wittig and A. Joshi, High Resolution Auger Study of Phosphorous Grain Boundary Segregation in NiCrMoV Steels, *Metall. Trans. A*, Vol 21, 1990, p 2817–2821
108. D. Skinner, The Role of Auger Electron Spectroscopy in the Semiconductor Industry, *Microsc. Microanal. Microstruct.*, Vol 6, 1995, p 321–343

109. G.B. Larrabee, Materials Characterization for VLSI, *VLSI Electronics Microstructure Science*, Vol 2, N.G. Einspruch, Ed., Academic Press, London, 1981, p 37–65
110. S. Thomas and A. Joshi, Failure Analysis of Microelectronic Materials Using Surface Analysis Techniques, *Materials and Processes—Continuing Innovations*, Vol 28, SAMPE Series, 1983, p 752–764
111. R.S. Nowicki, *Solid State Technology*, Vol 23, Technical Publishing, A Company of Dun and Bradstreet, 1980
112. J.L. Ong and L.C. Lucas, *Biomaterials*, Vol 19, 1998, p 455–464
113. C. Trépanier, M. Tabrizian, L.H. Yahia, L. Bilodeau, and D.L. Piron, *J. Biomed. Mater. Res.*, Vol 43, 1998, p 433–440
114. L. Zhu, C. Trépanier, A.R. Pelton, and J. Fino, in *Medical Device Materials: Proceedings from the Materials and Processes for Medical Devices Conference*, S. Shrivastava, Ed., ASM International, 2004, p 156–161

ASM Handbook, Volume 10, *Materials Characterization*
ASM Handbook Committee
DOI 10.31399/asm.hb.v10.a0006655

Low-Energy Electron Diffraction

Revised by Amirali Zangiabadi, Columbia University

Overview

General Uses

- Surface crystallography and microstructure
- Surface phase identification (adsorption, segregation, reconstruction)
- Analysis of surface dynamic processes (growth kinetics, thermal vibration)
- Determination of surface atom positions to 0.01 nm

Examples of Applications

- Reconstruction of semiconductor, metal, and alloy surfaces
- Analysis of chemical reactions at surfaces (chemisorbed layers)
- Influence of surface structure on catalytic processes
- Evolution of crystal structure in epitaxial growth
- Grain size determination in thin oriented films

Samples

- *Form:* Solids (metals and semiconductors; insulators in special cases). Single crystals or oriented films. Polycrystalline samples with large grain size can be analyzed under special circumstances.
- *Size:* 1 mm^2 to 25 cm^2
- *Preparation:* Samples must be polished carefully to expose the appropriate surface orientation. Surface contaminants must be removed by annealing in a vacuum, annealing in a low-pressure ($\leq 10^{-4}$ Pa) oxidizing or reducing atmosphere to clean the surface chemically, or ion beam etching and annealing in situ. Some samples can be cleaved on appropriate crystallographic planes in situ; in this case, no further preparation is required.

Limitations

- Samples must be at least slightly electrically conductive. Electrical charging of nonconductors is a problem.
- Ultrahigh vacuum is required.
- Determination of sizes of ordered regions (grains, islands, terraces, and so on) is limited by instrumental parameters to sizes less than approximately 500 nm.
- Surface preparation is extensive and can be difficult.
- High electron flux may induce surface reconstruction and accelerate surface contamination.

Estimated Analysis Time

- 10 min to 3 months, depending on information desired and the initial condition of sample

Capabilities of Related Techniques

- *Reflection high-energy electron diffraction:* High-energy (~10 keV) analog of low-energy electron diffraction. Provides basically the same information as low-energy electron diffraction. However, some measurements—for example, atom positions—are more difficult to attain than with low-energy electron diffraction; others—for example, three-dimensional crystal growth on surfaces—are facilitated. The other example is monitoring the layer-by-layer deposition by investigating the fluctuations in reflection intensities.
- *Glancing-angle x-ray diffraction:* More limited than low-energy electron diffraction, unless synchrotron radiation is used. Interpretation of atom positions is simpler. The structure of internal interfaces can be determined in special cases. Flat surfaces and large sample areas are required.
- *Transmission electron microscopy, scanning transmission electron microscopy:* Many transmission electron or scanning transmission electron microscopes allow reflection diffraction and also the electron backscattered diffraction technique, which is discussed in the article "Crystallographic Analysis by Electron Backscatter Diffraction in the Scanning Electron Microscope" in this Volume. These techniques are analogous to reflection high-energy electron diffraction, but the use of higher-energy electron beams (of the order of 10 to 30 keV) requires shallow angles of incidence for surface sensitivity.

Introduction

Low-energy electron diffraction (LEED) is a technique for investigating the crystallography of surfaces and overlayers adsorbed on surfaces. It is the surface analog of x-ray diffraction, which is sensitive to bulk crystallography. Low-energy electron diffraction generally is performed with electrons from 30 to several hundred electron volts. The limited penetration of electrons in this energy range provides information from a very top surface.

The first experimental observation of electron diffraction by a crystal lattice was performed by Davisson and Germer in 1927. After an accidental explosion of a bottle containing a nickel target at high temperature in their lab, the surface of the nickel formed a thin layer of recrystallized oxide. By repeating the electron bombardment experiment on the same target, this time they observed different but consistent electron reflection angles. By measuring the new angles, they discovered the electrons were diffracting from the nickel crystal planes according to Bragg's law. This observation was the first experimental proof of wave-particle duality and was crucial in the development of quantum mechanics and the Schrödinger equation. Based on this accidental discovery, Davisson won the 1937 Nobel Prize in Physics.

Low-energy electron diffraction measurements are conducted by using a monoenergetic beam of electrons that impinges on the crystal surface. Diffraction of electrons occurs because of the periodic arrangement of atoms in the surface. This periodic arrangement can be conceptualized as parallel rows of atoms analogous to grating lines in a diffraction grating. Thus, the diffraction in LEED occurs from rows, unlike x-ray diffraction, which can be considered as occurring from planes. The diffracted beams emanate from the crystal surface in directions satisfying interference conditions from these rows of atoms. The diffraction pattern and the intensity distribution in the diffracted beams can provide information on the positions of atoms in the surface and on the existence of various crystallographic defects in the periodic arrangement of surface atoms. In its most elementary form, LEED can be used to test for the existence of overlayer phases having a two-dimensional crystal structure different from the surface on which they are adsorbed and to test whether a surface phase is ordered or disordered.

Although LEED is the best known and most widely used surface crystallographic technique, other diffraction techniques can provide information on the surface structure. Reflection high-energy electron diffraction is applied widely to the epitaxial growth of reflection high-energy electron diffraction (RHEED) films. X-ray diffraction also can be used for surface crystallography under appropriate conditions.

Surface Crystallography Vocabulary

If an imaginary plane is drawn somewhere through a perfectly periodic three-dimensional crystal, and the two halves of the crystal are separated along this plane, ideal surfaces are formed. If the imaginary plane corresponds to an (*hkl*) plane in the bulk crystal, the surface is defined as an (*hkl*) surface, using the usual Miller indices. Bonds between atoms must be broken to create a surface. The coordination number defines the number of nearest neighbors of any atom in the crystal structure. The necessity of breaking bonds to create a surface implies a coordination number for surface atoms that is lower than that of atoms in the bulk. Surfaces can be singular or vicinal (Fig. 1). A singular surface is one in which only atoms in the outermost layer have broken nearest-neighbor bonds (reduced coordination number). It corresponds to a low Miller index plane. For example, (111) and (100) surfaces of face-centered cubic (fcc) crystals and the (110) surface of body-centered cubic crystals are singular. Most surfaces are body-centered cubic (bcc); that is, some atoms in deeper layers also have broken nearest-neighbor bonds. Such surfaces can be pictured as rougher and more open, with lower atomic density as well as a reduced average coordination number per unit surface area than singular surfaces. The average coordination number also is generally a measure of surface reactivity. Vicinal surfaces have a higher surface energy and are less stable than singular surfaces because of the larger number of broken bonds per unit area.

The periodic arrangement of atoms in the surface can be viewed as a two-dimensional lattice; that is, every point in this arrangement can be reached by a translation vector. The smallest translation vectors define the unit mesh, the two-dimensional analog of the unit cell. Primitive unit meshes contain one lattice point per mesh; nonprimitive contain more than one lattice point per mesh. Figure 2(a) shows the five two-dimensional Bravais nets. The unit mesh vectors are conventionally defined, as shown, with the angle between the unit vectors ≥90°, **a** denoting the shorter unit vector, and **b** (aligned horizontally) the longer one. A two-dimensional lattice also may have a basis. The lattice is defined by those points that can be reached by a translation vector. The basis is the conformation of atoms around each of these points.

The arrangement of lattice points in a two-dimensional lattice can be visualized as sets of parallel rows. The orientation of these rows can be defined by two-dimensional Miller indices (*hk*) (Fig. 2b). Interrow distances can then be expressed in terms of two-dimensional Miller indices, just as they are for three-dimensional crystals. For example, for square or rectangular primitive unit meshes, the interrow distance d_{hk} is:

$$d_{hk} = \left[\frac{h^2}{\mathbf{a}^2} + \frac{k^2}{\mathbf{b}^2}\right]^{-1/2} \quad \text{(Eq 1)}$$

where **a** and **b** are the unit mesh vectors.

In the discussion so far, an ideal termination of the bulk crystal has been assumed at the surface; that is, the positions of atoms in the surface have been assumed to be the same as

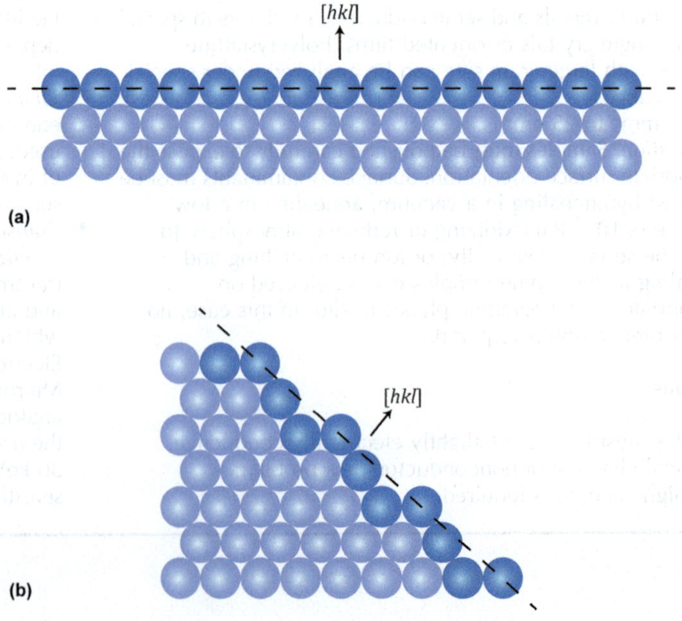

Fig. 1 (a) Singular and (b) vicinal surfaces of a cubic lattice. The solid circles represent atoms having missing nearest-neighbor bonds. The dashed line defines the surface orientation (*hkl*) with surface normal [*hkl*].

those they would have in the bulk before the surface was created. This need not be true. Reconstruction, a rearrangement of atoms in the surface and near-surface layers, occurs frequently. It is caused by an attempt of the surface to lower its free energy by eliminating broken bonds. The atomic layers participating in this reconstruction then can be considered a different phase having a different periodic arrangement of atoms. This region is sometimes referred to as a selvedge.

Even when no reconstruction occurs, adsorption of a foreign species onto the substrate surface from the ambient gas, from a deliberately created beam of atoms, or as a consequence of segregation of an impurity out of the bulk creates a surface phase termed an overlayer. An overlayer can be a fraction of a single atomic layer or several atomic layers. Thicker overlayers are referred to as thin films. The overlayer phase has its own crystal structure, which may or may not be related in a simple manner to that of the substrate surface.

An overlayer may be commensurate or incommensurate with the substrate. Commensurate overlayers have unit meshes that are related by simple rational numbers to those of the substrates on which they are adsorbed. This is not true for incommensurate layers. The unit mesh of overlayers is defined as a multiple of that unit mesh of the substrate surface that would be produced by the ideal termination of the bulk lattice. Thus, (100) and (111) surfaces of an fcc crystal have square and parallelogram-shaped primitive unit meshes, respectively (Fig. 2a), both of which would be considered (1×1) unit meshes. Commensurate overlayer unit meshes then are $p(m \times n)$ or $c(m \times n)$, where p and c refer to primitive and centered overlayer unit meshes relative to the primitive substrate unit mesh, and m and n are constants. An example of a complete unit mesh description of an overlayer on a particular substrate is W(110) $p(2 \times 1)$-O, which describes an oxygen overlayer adsorbed on the (110) surface of tungsten having a unit mesh twice that of the primitive W(110) unit mesh in the **a** direction and the same as the W(110) unit mesh in the **b** direction.

The overlayer unit mesh sometimes is rotated relative to that of the substrate (Fig. 3). An example of such a notation is Ni(111) $(\sqrt{3} \times \sqrt{3})$ R30°–O. The symmetry of the substrate often is such that it permits formation of energetically equivalent structures rotated relative to each other by some specific amount. Domains having such symmetry-related structure are called rotational antiphase domains and are degenerate; that is, they have the same free energy, and there is no preference for formation of one over the other. Overlayers with $(m \times n)$ meshes generally also form $(n \times m)$ meshes; for example, $p(2 \times 1) \rightarrow p(1 \times 2)$ and $c(2 \times 4) \rightarrow c(4 \times 2)$ (Fig. 3d).

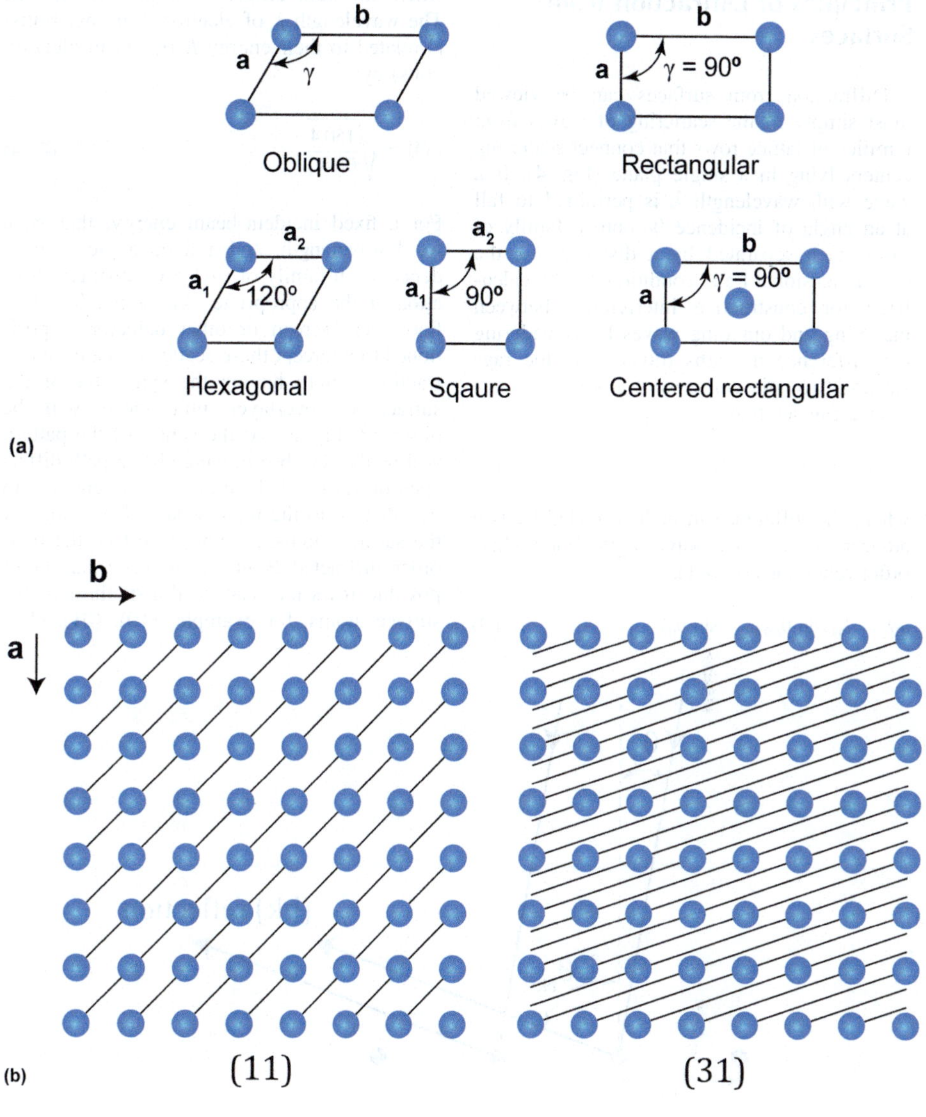

Fig. 2 Unit meshes and two-dimensional Miller indices. (a) The five two-dimensional Bravais nets. (b) Examples of families of rows with Miller indices referenced to the unit mesh vectors. (11) and (31) families of rows are shown.

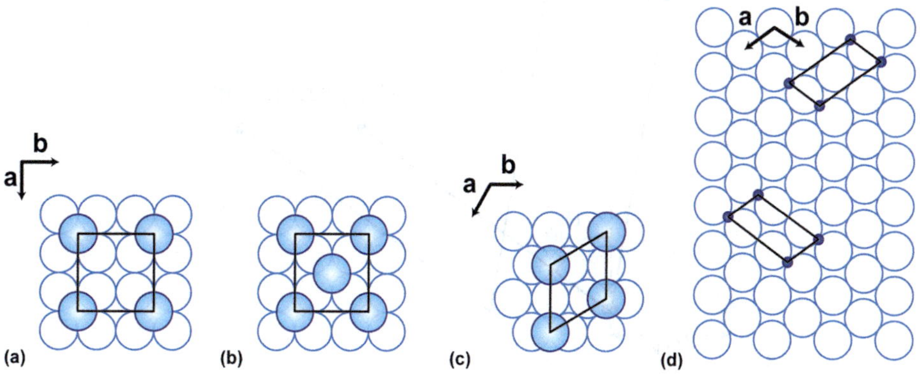

Fig. 3 Examples of overlayer structures with appropriate notation. (a) Face-centered cubic (fcc) (100)$p(2 \times 2)$. (b) fcc (100)$c(2 \times 2)$. (c) fcc (111)$p(\sqrt{3} \times \sqrt{3})$R30°. (d) Body-centered cubic (bcc) (110)$p(2 \times 1)$ and bcc (110)$p(1 \times 2)$. The two orientations for the unit mesh in (d) are both possible because of the symmetry of the substrate; they have the same free energy and are termed degenerate.

Principles of Diffraction from Surfaces

Diffraction from surfaces can be viewed most simply as the scattering of waves from families of lattice rows that connect scattering centers lying in a single plane (Fig. 4). If a wave with wavelength λ is permitted to fall at an angle of incidence θ_0 onto a family of rows, each separated by a distance d_{hk}, the two-dimensional Laue condition can be calculated for constructive interference between incoming and outgoing waves by considering the difference in paths traveled by the rays striking two adjacent rows of atoms.

One can see that:

$$\lambda = d_{hk}(\sin\theta_{hk} - \sin\theta_0) \quad \text{(Eq 2)}$$

where the difference in paths traveled by two adjacent rays is one wavelength. For higher-order reflections ($m > 1$):

$$m\lambda = d_{hk}(\sin\theta_{mh,mk} - \sin\theta_0) \quad \text{(Eq 3)}$$

where the path difference is m wavelengths. The wavelength λ of electrons (in angstroms) is related to their energy E (given in electron volts) by:

$$\lambda(\text{Å}) \cong \sqrt{\frac{150.4}{E(\text{eV})}} \quad \text{(Eq 4)}$$

For a fixed incident-beam energy, that is, a fixed wavelength, and a fixed angle of incidence, each family of (hk) rows diffracts radiation at the appropriate exit angle, θ_{hk}. If a fluorescent screen or other detector is positioned to intercept these scattered beams, a diffraction pattern having the symmetry of the surface or overlayer unit mesh will be observed (Fig. 5). At the center of the pattern will be the (00) beam, which has a path difference of zero and therefore is not sensitive to d_{hk}; that is, to the relative lateral positions of the surface atoms. Around it will be the first-order diffracted beams from each family of possible rows that can be drawn through the surface atoms, for example, (10), (01), (11), (21), and so on. In addition, there will be higher-order reflections at larger angles. For example, the (20) and (30) reflections will fall on the extension of a line connecting the (00) and (10) beams; the (00), (11), (22), (33), . . . beams will be colinear; and so forth.

If the energy of the incident beam or the angle of incidence is changed, the diffraction angles θ_{hk} adjust to continue to satisfy the Laue conditions. For example, if the energy is increased, the entire pattern will appear to shrink around the (00) beam because θ_{hk} becomes smaller.

The diffraction conditions can be depicted most easily by using a reciprocal lattice and Ewald construction (Fig. 6). The Ewald sphere provides a schematic description of the conservation of energy; that is, because diffraction involves elastic scattering, the incident and exiting beams have the same energy. Its radius is inversely proportional to the wavelength λ. The reciprocal lattice for a single layer of atoms consisting of families of rows is a set of rods (hk) normal to the crystal surface with spacing $2\pi/d_{hk}$. The intersection of the Ewald sphere and the reciprocal-lattice rods is a graphical solution of the Laue equation and therefore yields the diffraction pattern. As the energy of the incident beam or the angle of incidence is varied, the radius of the Ewald sphere or its orientation relative to the rods changes, consequently also changing the points of intersection with the rods. The directions of the outgoing vectors define the directions of the diffracted beams.

The discussion of diffraction so far has made no reference to the size of the two-dimensional grating. It has been assumed that the grating is infinite. As in three dimensions, finite sizes of the ordered regions on the surface (finite-size gratings) broaden the diffraction features. This can be confirmed by calculating the amplitude, $A(\mathbf{S})$, scattered from a grating consisting of N rows spaced a distance d_{hk} apart:

$$A(\mathbf{S}) = \sum_{n=1}^{N} f_n e^{i\mathbf{S}.\mathbf{r}_n} \quad \text{(Eq 5)}$$

where $\mathbf{S}$ is the momentum transfer vector $\mathbf{S} = \mathbf{k} - \mathbf{k}_o$ and $|\mathbf{S}| = 2\pi/\lambda - 2\pi/\lambda_o$, f_n is the atomic scattering factor of the nth atom, and $\mathbf{r}_n$ is the position of the nth atom. The exponent represents the phase factor in summing up waves scattered from different atoms at positions $\mathbf{r}_n$. The scattered intensity I_{hk} from the (hk) family of rows is:

$$I_{hk} = AA^* = \left[\frac{\sin\left(\frac{1}{2}N\mathbf{S}_{hk}d_{hk}\right)}{\sin\left(\frac{1}{2}\mathbf{S}_{hk}d_{hk}\right)}\right]^2 \quad \text{(Eq 6)}$$

where A^* is the complex conjugate of A. For $N = \infty$, I_{hk} is a δ function, but for finite N, it is a function having a width proportional to $1/Nd_{hk}$. Thus, the smaller N is, the broader the diffraction peak. A measurement of the angular

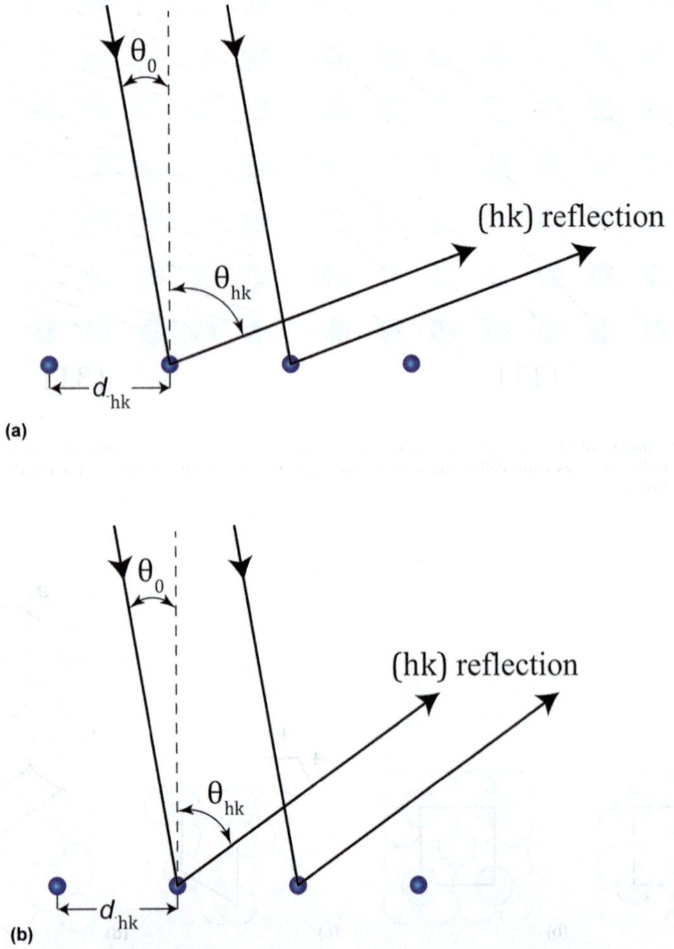

Fig. 4 Diffraction from a family of rows spaced d_{hk} apart and its dependence on wavelength. Each solid circle represents a row of atoms into the plane of the paper. Rays with wavelength λ fall on this family of rows at an angle θ_0. Interference maxima occur at angles θ_{hk} that satisfy Eq 2. Only the first-order reflection is shown. (a) Longer wavelength. (b) Shorter wavelength

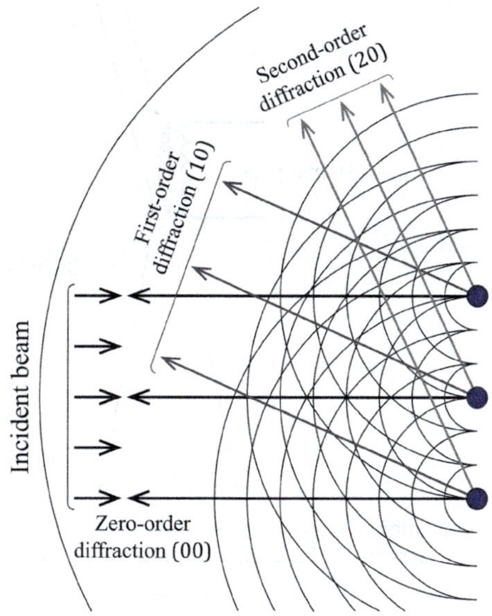

Fig. 5 Interference pattern created when regularly spaced atoms scatter a plane wave incident on them. A spherical wave emanates from each atom; diffracted beams form at the directions of constructive interference between these waves. The mirror reflection [(00) beam] and the first- and second-order diffracted beams are shown.

profile of the diffracted beams provides information on two-dimensional particle (island) size effects and other finite-size effects in the surface or overlayer. In terms of the unit mesh vectors **a** and **b:**

$$I(\mathbf{S}) = \left[\frac{\sin\left(\frac{1}{2}\mathbf{S}.N_1\mathbf{a}\right)}{\sin\left(\frac{1}{2}\mathbf{S}.\mathbf{a}\right)}\ \frac{\sin\left(\frac{1}{2}\mathbf{S}.N_2\mathbf{b}\right)}{\sin\left(\frac{1}{2}\mathbf{S}.\mathbf{b}\right)}\right]^2 \qquad \text{(Eq 7)}$$

where $N_1\mathbf{a}$ and $N_2\mathbf{b}$ are the average dimensions of the islands or domains of the surface layer. The width of the angular profile is correspondingly $1/N_1\mathbf{a}$ and $1/N_2\mathbf{b}$ in the **a** and **b** directions.

If an overlayer with a superlattice exists, additional diffracted beams appear that correspond to the periodicity of the superlattice. A larger unit mesh yields diffracted beams that are nearer to each other (Eq 2) (Fig. 7).

Diffraction Measurements

Figure 8 is a schematic of a typical LEED system. The electron gun can produce a monoenergetic beam having energies from approximately 10 to 1000 eV, with beam sizes typically a fraction of a millimeter but ranging in newer instruments to submicrometer sizes (additional information on electron guns is provided in the article "Transmission Electron Microscopy" in this Volume). The goniometer generally allows two sample motions: a rocking about an axis in the plane of the crystal surface and a rotation about an axis normal to the crystal surface. In addition, the goniometer generally permits heating and sometimes cooling of the sample as well as temperature measurement. Most LEED systems have a detector consisting of concentric grids and a fluorescent screen (Fig. 8). The grids filter inelastically scattered electrons that also emanate from the crystal surface. In addition, a Faraday cup detector for accurate measurement of beam currents and beam profiles can be used. The Faraday cup is biased to exclude inelastically scattered electrons. A reflection high-energy electron diffraction system is similar, except the beam energy is higher, and the diffraction geometry is arranged to provide grazing incidence and exit (Fig. 9).

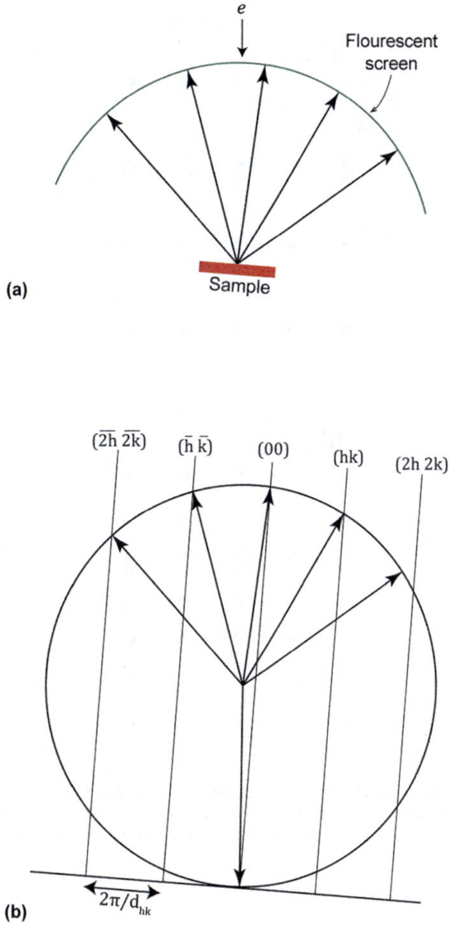

Fig. 6 Reciprocal lattice and Ewald construction corresponding to low-energy electron diffraction and comparison to real-space picture. (a) Real-space schematic diagram of diffraction from a surface. The electron beam is incident on the sample along the direction given by e^-. The five diffracted beams represent the ($\bar{2h}$, $\bar{2k}$), ($\bar{h}$ $\bar{k}$), (00), (***hk***), and (***2h***, ***2k***) beams from a family of rows (***hk***) with spacing d_{hk}. (b) The corresponding cut through the reciprocal lattice and Ewald construction. The reciprocal-lattice rods are normal to the crystal surface, given by the nearly horizontal line. They represent the same set of beams as in (a). The Ewald sphere always intersects the origin of the reciprocal lattice at the point of incidence of the incoming ray. Directions of the diffracted beams depend on the radius of the Ewald sphere ($2\pi/\lambda$). Similarly, as the orientation of the crystal surface is changed in relation to the incident beam, the reciprocal lattice rotates about its origin.

The surface sensitivity in LEED is provided by the limited mean free path for inelastic scattering of slow electrons. This mean free path is the distance traveled by an electron in the solid before it collides inelastically, loses energy, and thus becomes lost for diffraction. The mean free path in the energy range of LEED is from 0.4 to 2 nm. Because a layer spacing is of the order of 0.3 nm, these slow electrons probe only a few atomic layers. Even though in the section "Principles of Diffraction from Surfaces," it was assumed that only one layer contributed, the basic principles still apply. In RHEED, the same effect is obtained by grazing incidence, maintaining the beam near the surface in this manner. Figure 10 shows the dependence of the inelastic mean free path versus the kinetic energy of electrons traveling in a solid.

The simplest diffraction measurement is the determination of the surface or overlayer unit mesh size and shape. This can be performed by inspection of the diffraction pattern at any energy of the incident beam. The determination is simplest if the electron beam is incident normal to the surface, because the symmetry of the pattern preserved. Diffracted beams are indexed with the (*hk*) notation of the families of rows from which they are diffracted. Thus, the (10) beam is scattered from (10) rows in the direction perpendicular to these rows. The distance of the (*hk*) reflection from the mirror reflection, the (00) beam, is inversely related to the (*hk*) interrow distance (Fig. 7). Figure 11 shows diffraction patterns obtained from GaAs (110) cleaved in an ultrahigh vacuum (UHV).

In the analysis of a diffraction pattern, it first must be determined which diffracted beams belong to the substrate pattern and which are due to the superperiodicity of the overlayer. This can be accomplished experimentally by observing diffraction from the substrate surface without the overlayer adsorbed (Fig. 7a), then adsorbing the overlayer (Fig. 7b). Alternatively, the diffraction pattern of the ideal surface can be calculated from knowing the crystal structure of the bulk sample and the orientation at which it was cut and assuming that the surface is the ideal termination of the bulk lattice.

For example, the unit mesh of an fcc (110) surface appears rectangular (Fig. 7a). The diffraction pattern from a rectangle can be calculated to be a rectangle rotated by 90° relative to the real lattice (Fig. 7b). The unit distances, $\mathbf{a}^*$, $\mathbf{b}^*$, in the diffraction pattern are related to the real-space unit mesh by $\mathbf{a}^* = 2\pi/\mathbf{a}$ and $\mathbf{b}^* = 2\pi/\mathbf{b}$. Once the substrate diffraction pattern is known, the unit mesh for the overlayer can be determined by constructing reciprocal-space unit distance $\mathbf{a}^*_{\text{overlayer}}$ and $\mathbf{b}^*_{\text{overlayer}}$. If the patterns are aligned relative to the sample orientation, then:

$$\mathbf{a}^*_{\text{overlayer}} = \frac{1}{m}\mathbf{a}^*_{\text{substrate}} \qquad \text{(Eq 8)}$$

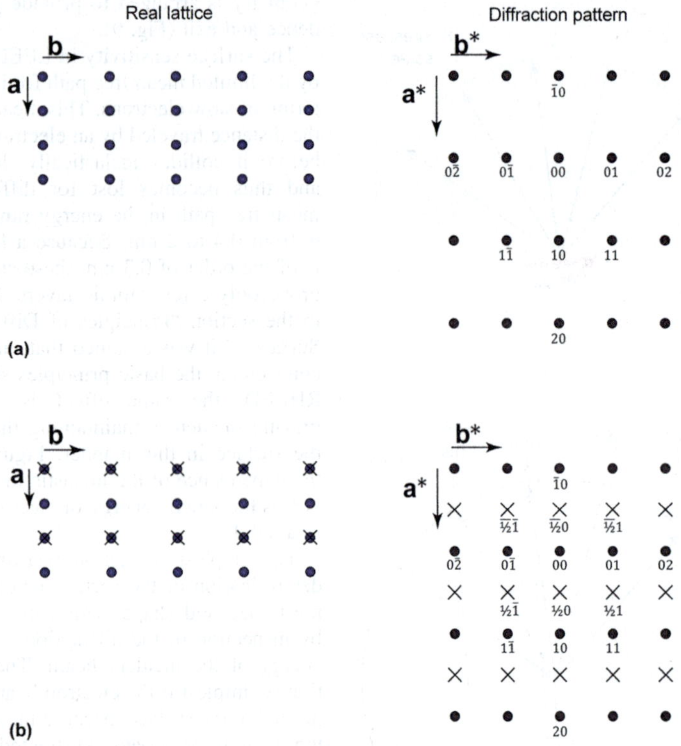

Fig. 7 Diffraction pattern from a superlattice. (a) Rectangular substrate lattice and corresponding diffraction pattern showing fundamental reflections. (b) Substrate plus $p(2 \times 1)$ overlayer and corresponding diffraction pattern showing fundamental and superlattice reflections. The overlayer atoms (x's) are arbitrarily shown to sit on top of the substrate atoms.

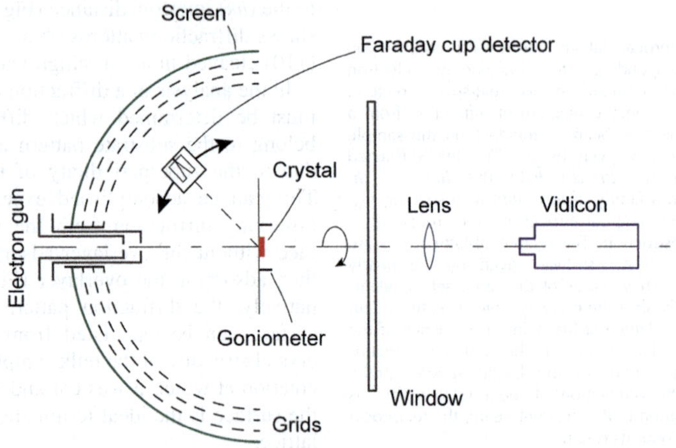

Fig. 8 Low-energy electron diffractometer. The Vidicon camera can be interfaced with a computer to record the diffraction pattern displayed on the fluorescent screen. The goniometer can be tilted in all three directions.

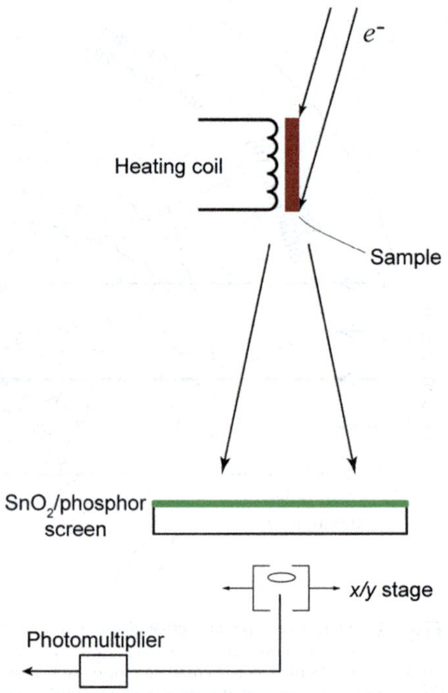

Fig. 9 Reflection high-energy electron diffraction system that uses a fluorescent screen to display and a photomultiplier to record the diffraction pattern. A heater is shown behind the sample. The incident beam (top) strikes the surface at grazing angles. (Arrows represent electrons.)

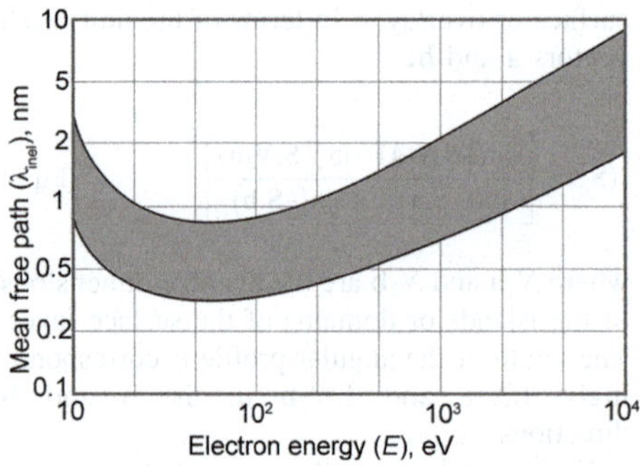

Fig. 10 Mean free path for inelastic scattering of electrons as a function of kinetic energy of electrons travelling in a solid. Electrons in the low-energy electron diffraction energy range travel only of the order of 0.4 to 2 nm in the crystal before losing energy and thus becoming lost for diffraction. Surface sensitivity is a consequence of this behavior.

$$\mathbf{b}^*_{\text{overlayer}} = \frac{1}{n}\mathbf{b}^*_{\text{substrate}} \quad \text{(Eq 9)}$$

From the relationship between reciprocal and real-space vectors, the unit mesh of the overlayer is (m, n) times that of the substrate. In the general case, the patterns may not be aligned:

$$\mathbf{a}^*_{\text{overlayer}} = \frac{1}{m_1}\mathbf{a}^*_{\text{substrate}} + \frac{1}{m_2}\mathbf{b}^*_{\text{substrate}} \quad \text{(Eq 10)}$$

$$\mathbf{b}^*_{\text{overlayer}} = \frac{1}{n_1}\mathbf{a}^*_{\text{substrate}} + \frac{1}{n_2}\mathbf{b}^*_{\text{substrate}} \quad \text{(Eq 11)}$$

The unit mesh of the overlayer can be determined directly by solving Eq 10 and 11 in terms of the relationships $\mathbf{a}^* = 2\pi/\mathbf{a}$ and $\mathbf{b}^* = 2\pi/\mathbf{b}$.

The diffraction pattern determines only the size and shape of the unit mesh. The positions of overlayer atoms relative to substrate atoms, that is, the arrangement of atoms within the unit mesh, cannot be determined from visual inspection of the diffraction pattern but must be obtained from an analysis of the intensities of the diffracted beams as a function of diffraction parameters, particularly the energy and the angles at which the diffracted beams emerge relative to the surface normal.

Figure 12 shows three fcc (100)$c(2 \times 2)$ overlayer structures. All yield the same diffraction pattern, but the intensities of the beams differ. An analogy is provided by the missing reflections in x-ray diffraction from particular three-dimensional structures, in which the intensity of certain beams is zero. In two dimensions, the intensity of some beams also may be zero, but different positions of overlayer atoms generally make some beams weaker than others at particular diffraction conditions. This can be visualized as an interference between the substrate atoms and the overlayer atoms that differs for different positions of overlayer atoms. A calculation of the geometric structure factor for each of the overlayer unit meshes will demonstrate these differences in diffracted-beam intensities.

To determine lateral and vertical positions of overlayer atoms relative to those of the substrate atoms, the intensity (I) in a diffracted beam is plotted as a function of electron kinetic energy (V); this plot is known as a LEED I-V curve. These intensity-versus-energy curves then are compared to multiple scattering calculations of the geometric structure factor. The existence of an intensity contribution in diffracted beams due to the multiple scattering of electrons off surface rows and between substrate and overlayer makes this determination more complicated than this discussion suggests. Figure 13 shows a measured intensity-versus-energy profile.

If the overlayer or surface structure is not perfect, that is, does not have long-range periodicity, the diffracted beams show a change in shape. A common occurrence in LEED or RHEED is the existence of streaks, large spots, and additional diffuse intensity in the pattern, indicating the presence of structural disorder. A measurement of the angular profile of diffracted beams provides information on disorder and finite-size effects in surfaces analogous to x-ray diffraction measurements of three-dimensional particle sizes or strains. The measurement of angular profiles as a function of diffraction parameters, such as energy, angles, and beam order, makes it possible to distinguish among various defects, such as finite island sizes, strain, crystal mosaic, monatomic and multiatomic steps at the surface, and regular or irregular domain or antiphase boundaries. To measure a profile, a detector with a narrow slit or a point aperture must be used to minimize the influence of instrumental broadening. Figure 14 shows an example of a two-dimensional angular profile (intensity versus θ_x and θ_y) using a surface-sensitive diffractometer.

A recent development in the LEED technique is spot profile analysis (SPA). SPA-LEED is one of the most powerful methods to analyze the morphology of surfaces. Deviations from simple structures caused by point defects, steps, strain, roughness, facets, or dislocations play an important role on surfaces, and they influence the adsorption properties, electronic structure, and growth of adsorbate layers. SPA-LEED has been used extensively to analyze semiconductor surfaces, including studies on the growth modes of epitaxial layers on semiconductor surfaces, defects in silicon surfaces, oxides or metals on silicon surfaces, and domain size distribution in heteroepitaxial growth on semiconductor surfaces.

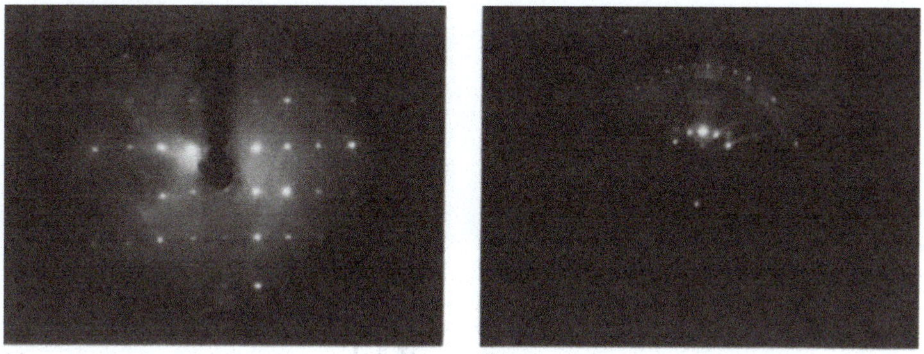

Fig. 11 Diffraction patterns from GaAs (110). (a) Low-energy electron diffraction pattern near normal incidence, E_p = 100 eV. (b) Reflection high-energy electron diffraction pattern, incident beam along [100] azimuth at grazing angle of incidence of 4.5°, E_p = 10 keV. The lowest arc of spots corresponds to the ($\bar{2}$0), ($\bar{1}$0), (00), (10), and (20) beams.

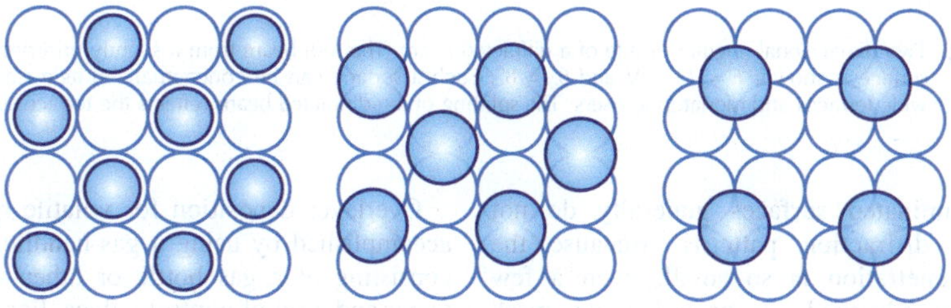

Fig. 12 Possible placements of overlayer atoms on a face-centered cubic (100) substrate that produce a $c(2 \times 2)$ overlayer unit mesh. The diffraction pattern from these structures differs in the intensities of the diffracted beams but not in the symmetry of the pattern.

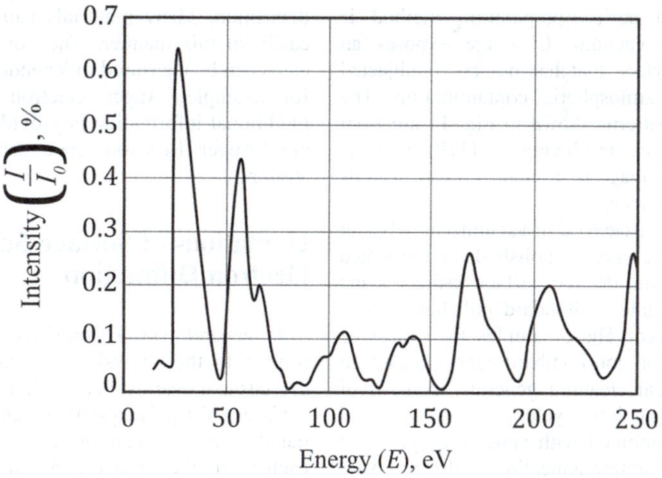

Fig. 13 Measured intensity-versus-energy curve for the (00) beam from W(110) at an incident angle θ_o = 7°

Sample Preparation

All surface-sensitive diffraction must occur in a vacuum system, and, in most cases, this must be a UHV system. A high vacuum is necessary because surface contamination rates from ambient gases are rapid. A useful criterion is that formation of one monolayer of contamination, such as hydrogen, oxygen, or carbon monoxide, on the surface requires of the order of 1 s at 1×10^{-4} Pa. Even a fraction of a monolayer is sufficient to prevent meaningful surface crystallography measurements. Thus, except for the most unreactive surfaces, a vacuum exceeding 1×10^{-8} Pa is essential.

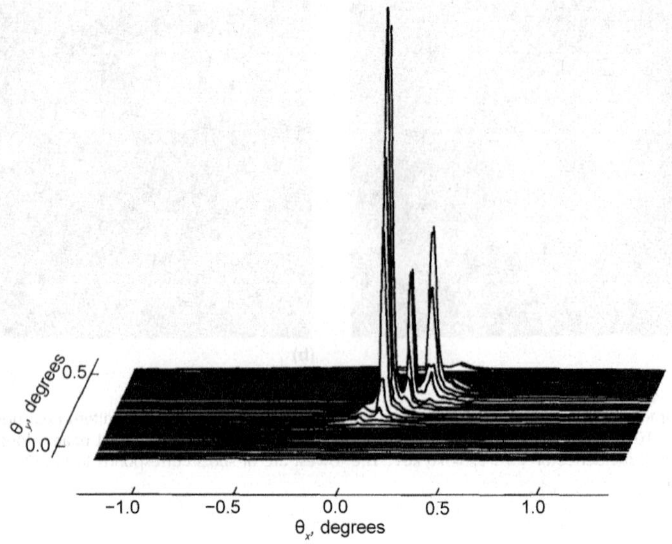

Fig. 14 Two-dimensional angular profile of a diffracted beam. The (00) beam from a slightly misoriented silicon surface is shown; E = 235 eV and θ_0 = 8°. Such a surface can be conceptualized as a staircase with wide terraces and monatomic risers. The splitting of the diffracted beam reflects the terrace size.

Contaminated surfaces generally do not provide diffraction patterns. Because the beam penetration is so small, even a few layers of adsorbed gas or oxide can mask the underlying pattern. In some cases, the oxide (or other surface phase) is crystalline and exhibits its own, generally very complicated, diffraction pattern. Therefore, the surface must be cleaned.

The ideal surface-preparation method is cleavage in vacuum. Cleavage exposes an internal interface that has not been subjected to ambient atmospheric contamination. The diffraction patterns shown in Fig. 11 are from cleaved GaAs, producing a (110) surface. However, cleavage is limited to a few crystals and a few surfaces.

Samples not cleaved in vacuum must be cut from a single crystal, polished, and oriented carefully to the desired surface using a Laue diffraction camera. Standard polishing procedures are used. These samples then must be chemically or ion-bombardment cleaned in situ. Chemical cleaning generally consists of oxidation-reduction cycles in oxygen and hydrogen combined with annealing cycles. In a few cases, simple annealing cycles in a vacuum produce clean surfaces. The most prevalent approach is a combination of inert-gas ion bombardment (sputter etching) of the surface with subsequent annealing to reorder the lattice. Ion bombardment cleans the surface effectively but leaves it so disordered that no diffraction pattern generally is observable. Subsequent annealing restores most of the order, but the reordering is never complete. Sputter etching is not advisable in some applications, notably where structural defects influence such properties as thin-film growth or chemical interactions at surfaces.

Overlayer deposition for volatile species is accomplished by using a gas-handling system consisting of a gas bottle or other source of vapor and a set of control valves. For nonvolatile species, an evaporation source is used that consists of a container for the source material inside an apertured can, a heater, and a cooled shroud that acts as a cryopump to prevent an increase of the vacuum system pressure during deposition. Many materials can be deposited easily in this manner. The coverage in both cases can be measured independently by using, for example, Auger electron spectroscopy (additional information is provided in the article "Auger Electron Spectroscopy" in this Volume).

Limitations of Surface-Sensitive Electron Diffraction

In general, surface-sensitive diffraction is limited to the analysis of surfaces of single crystals and overlayers as well as films on such surfaces. If a polycrystalline sample is illuminated by using a beam of low-energy electrons, each crystallite surface exposed will create its own diffraction pattern, all of which will be superimposed on the fluorescent screen. If more than a few orientations are illuminated by the beam, the pattern becomes too complicated to analyze. However, non-single-crystal films having preferred crystalline orientation in the surface plane can be analyzed. For example, a silver thin film, grown on mica, that has a (111) orientation and twin boundaries as its major defect appears in the LEED pattern as a (111) surface of a single crystal of silver, except the spots are broader, and their behavior with diffraction parameters reveals the existence of twin boundaries. If the film had (111) orientation but with a random azimuthal orientation (random rotations around the surface normal), diffraction still would be possible, but the spots would turn into rings.

Any two-dimensional disorder is observable. Analysis of some three-dimensional disorder—for example, the random orientation of surface normals in a polycrystalline sample—is limited essentially by the instrument, particularly the size of the incident beam. As this size is reduced to typical grain sizes ($\leq$100 nm), surface-sensitive diffraction can be used to investigate the surface crystallography of polycrystalline samples by analyzing individual grains of a polished sample sequentially.

Similarly, analysis of the surfaces of liquids and amorphous solids is not meaningful. In this case, however, the problem is fundamental and not simply instrumental. Diffraction is useful whenever there is a distinct phase relationship between scattering units. The greater the degree of this phase relationship, the better defined are the diffraction features. Three-dimensional crystals give diffraction spots. If disorder is introduced, the spots broaden and weaken. Two-dimensional structures give diffraction rods. Disorder in the plane broadens the rods and, in the case of complete in-plane randomness, converts them into cylinders observed as diffraction rings. Disorder out of the plane, as in an amorphous solid or a fine-grained polycrystalline solid with random orientations, moves these rods over all space, creating only diffuse intensity and no pattern.

Because electrons are charged, only materials having reasonable conductivity can be investigated. Insulators and ceramics pose difficulties because a charge accumulates on the sample that eventually prevents the incident beam from striking the surface. The usual approach of coating insulators to make them conductive is ineffective because the coating, although only a few atomic layers thick, prevents diffraction from the substrate surface. Most insulators can be investigated with special techniques, such as providing surface conduction paths near the analysis area or using very thin samples mounted on a conducting plate.

Applications

Example 1: Wettability of Graphite

The wettability of a material often is an important consideration. In some applications, a wettable surface is desired, for example, the flow of ink in a fountain pen; in others, a nonwettable surface is needed, such as polymer film or foil. Surface wettability can be determined by contact angle measurements.

One possible influence on surface wettability is roughness on an atomic scale. Such roughness is too small to observe in optical or scanning electron microscopes. Using these techniques, an atomically rough surface may appear featureless, yet it may have a considerably different wettability than one that is atomically smooth. Graphite is a useful material with which to conduct roughness-wettability correlations, because it is chemically fairly inert. Thus, it is possible to separate chemical changes that may influence wettability from structural ones. Graphite also illustrates the capabilities of LEED for analysis of materials that are not single crystals.

A high-purity commercially available pyrolytic graphite called oriented graphite was used. This material has the basal [(0001)] plane oriented parallel to the surface. The sample was cleaved in air and mounted in a LEED vacuum chamber.

After degassing the chamber, Auger electron spectroscopy was used to determine that the sample surface was clean. The vacuum chamber also has capabilities for ion beam etching of the surface and for contact angle measurements in situ. The experiment consisted of ion beam etching the surface to make it atomically rough, observing the changes in the diffraction pattern, and correlating the changes in contact angle to the LEED results.

Figure 15 shows the LEED results. The freshly cleaved sample (Fig. 15a) exhibits a hexagonal diffraction pattern that reflects the symmetry of the graphite unit mesh. It also shows other weaker spots, which indicate that the arrangement of atoms on the surface is not perfect but that some in-plane rotational randomness exists. Therefore, the graphite sample is oriented in terms of having only (0001) planes in the surface, but crystallites can be rotated relative to each other.

Figures 15(b) to (d) show LEED patterns from the graphite sample, with increasing surface crystallographic damage caused by ion beam etching. Two effects are evident. At first, the sample becomes more rotationally random; that is, planes with various rotations become evident but all still with a (0001) orientation in the plane. Analysis of the intensity of the various diffraction spots can determine the percentage of crystallites having a given orientation. In Fig. 15(c), the sample begins to approach the randomness that would be evident in a two-dimensional liquid, in which a bright ring would be observed.

Figure 15(d) shows a new effect. The diffraction spots begin to broaden. This is evidence of a decrease in the size of the individual graphite grains. A measurement of the increase in the width of the reflection relative to the width of reflection in the initial surface (Fig. 15a) can be interpreted in terms of the decrease in size of the crystallites. The crystallites shown in Fig. 15(a) are approximately 10 nm in diameter; those shown in Fig. 15(d) are approximately half that. In this experiment, a direct correlation between wettability and surface roughness was observed: the rougher the surface, the greater its wettability.

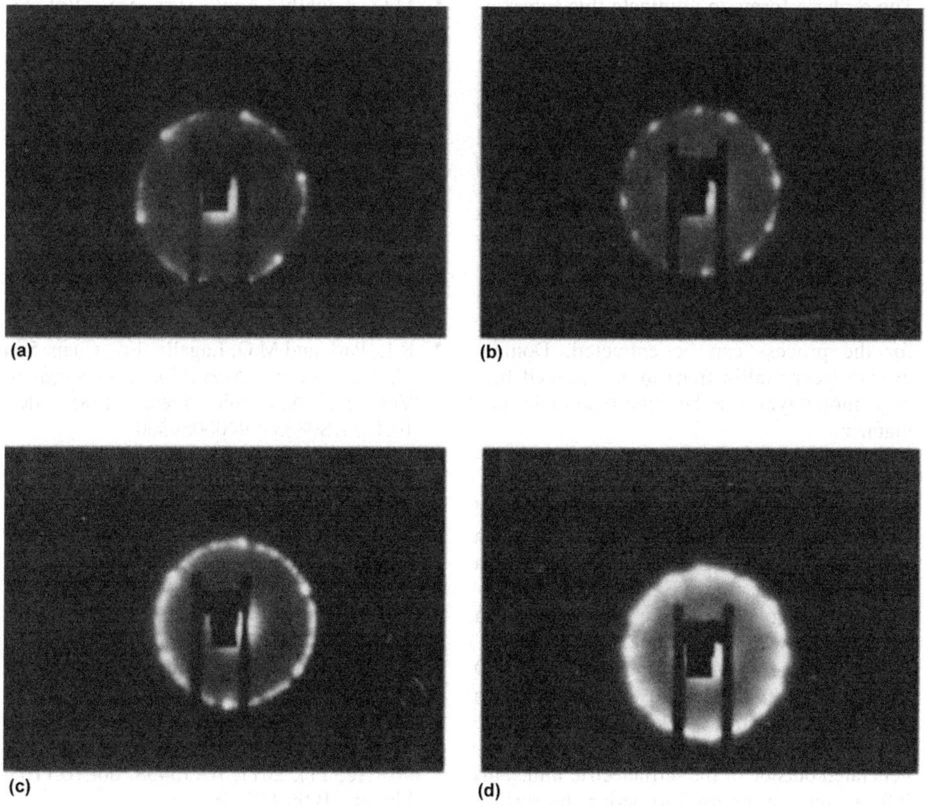

Fig. 15 Diffraction patterns from oriented pyrolytic graphite. (a) Freshly cleaved sample. (b to d) Increasing surface damage caused by ion beam etching

Example 2: Grain Size in a Silver Film Grown on Mica

Vacuum evaporation is a common technique for preparing thin-film coatings that have application in various technologies.

In vacuum evaporation, a crucible containing the source material is heated to temperatures high enough to create a significant evaporation rate. The surfaces to be coated then are placed into the evaporant stream, and the source atoms adhere to this surface. Because of the directionality of the flux and the generally low temperatures at the substrate, a columnar microstructure with a preferred orientation frequently forms. In many cases, this preference for some orientation can be beneficial.

Two aspects of the microstructure are significant: the preferred orientation and the size of the columnar grains. The preferred orientation of a film can be ascertained using a standard x-ray diffractometer by comparing a θ-2θ scan to one from a powder of the same material. This type of measurement also provides the grain size in the vertical direction (the average height of the columns) but does not easily provide information about the average diameter of the columns.

Figure 16 illustrates the desired parameters. With LEED, the preferred orientation and the grain size in the surface plane (but not in the vertical direction) can be measured. Low-energy electron diffraction (or more commonly RHEED) also can be combined with the evaporator to study the microstructure of the growing film in real time.

In this instance, a 100 nm thick silver film was grown on a clean mica substrate using an electron beam evaporation source. The film was grown at a rate of 10 nm/min. The sample was then inserted into a LEED vacuum

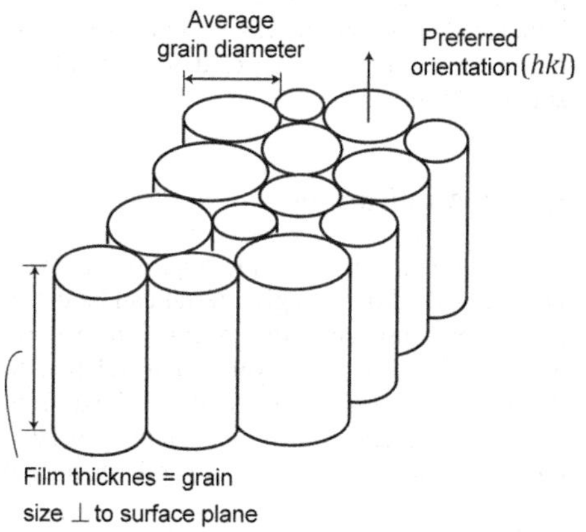

Fig. 16 Schematic diagram of a thin film with columnar growth morphology

chamber, and surface contaminants were removed by heating the sample to approximately 500 °C (930 °F)in a partial pressure of 1×10^{-3} Pa of oxygen. This removed all contaminants except for a trace of carbon.

A LEED diffraction pattern of the surface then was formed by illuminating the film surface with an electron beam of approximately 100 eV at normal incidence. If a preferred orientation is present, a diffraction pattern corresponding to the symmetry of the exposed surface plane will be observed. It was found that the silver film had a very strong preferential (111) orientation; that is, the growth direction was [111]. It was possible to determine that the average misorientation from (111) was less than 0.25°. However, extensive twinning was found in the crystal.

Twinning in this type of crystal structure is caused by stacking faults. The normal sequence of planes in an fcc (111) stacking is ABCABC. That is, every third plane is alike. A stacking fault occurs if the stacking is ABCABABC, a common occurrence in the growth of such structures. The result is crystal structures that have a 60° rotational relationship to each other. These are termed twins. Twins can be identified in the LEED pattern because they make the diffraction pattern sixfold symmetric instead of threefold symmetric, as would be expected for an fcc (111) surface.

The average grain size in the plane of the surface, that is, the average column diameter, of the oriented silver film was determined by analysis of the diffracted beam shape in the LEED pattern. The width of diffracted beams is inversely related to the average column diameter. By measuring the beam shape at various diffraction parameters, the average in-plane grain size was determined to be 7.5 ± 1.5 nm. Thus, the film consisted of oriented (111) columns of approximately 7.5 nm diameter that were rotationally very well aligned (unlike example 1) but contained a significant concentration of twins.

Any possible strain in the surface plane also was investigated by examining the diffraction profiles of higher-order reflections. The results indicate that no strain was present in the surface plane; this was expected in light of the small column size.

Other Applications

The unit mesh size and shape of an overlayer adsorbed on a single crystal can be determined by observing the diffraction pattern. The positions of overlayer atoms relative to the substrate atoms can be determined from model calculations to fit intensity-versus-energy measurements of several diffracted beams.

The density of steps on a single-crystal surface can be determined by varying the diffraction conditions such that the amplitudes of electrons scattered from terraces separated by a step interfere destructively. The angular profile of the diffracted beams then reflects the average terrace size, and from this the step density is readily obtained. For the example shown in Fig. 14, the average terrace size is 30 nm, with a distribution of terrace sizes of ±10 nm about this average value.

Order-disorder and order-order transitions in two-dimensional overlayers can be studied in the same manner that x-ray diffraction is used for three-dimensional phase transformations. The intensity of a superlattice reflection is monitored as a function of the temperature at constant coverage or as a function of the coverage at constant temperature. When a phase boundary is crossed, the superlattice beam intensity vanishes. The transition type (first or second order) and the critical exponents can be determined from the mechanics of this process. Order-disorder phenomena in surface phases of binary alloys and in reconstructed surfaces also can be investigated in this manner. The enthalpies of interaction of overlayer atoms can be modeled from the transition temperatures.

The kinetics of overlayer growth at the submonolayer level and above can be investigated by following the development of reflections belonging to the overlayer structure. Nucleation and growth of an ordered phase, α, proceeds out of a two-phase coexistence region, $\alpha + \beta$, in which the component of interest (A) has a supersaturation. The α phase forms to eliminate this supersaturation. The existence of the overlayer phase can be ascertained from the existence of diffracted beams corresponding to it. The size distribution of ordered two-dimensional islands (analogous to a three-dimensional precipitate of α in β) can be determined from the angular profiles of the relevant diffracted beams. As the islands grow, the profiles become narrower. The kinetics of growth can be investigated by measuring the average island size as a function of time for a given temperature. Activation energies for the process can be extracted. Domain growth (recrystallization) in an oriented film or a monolayer can be investigated in this manner.

The thermodynamics of surface or grain-boundary segregation of a binary or ternary trace element in a solid can be investigated by monitoring the absence or presence of a surface phase ascribable to this element as a function of temperature. Free energies for segregation and solution can be established from the behavior with temperature.

In another example, the position of LEED spots can be used to study the polarization state of ferroelectric single-crystal films. The spot position changes due to electric fields generated outside of the ferroelectric film. This information can be used to derive the surface potential energy of a ferroelectric film and to calculate the depth where the mobile charge carriers compensating the depolarization field are located.

ACKNOWLEDGMENT

This article was revised from M.G. Lagally, Low-Energy Electron Diffraction, *Materials Characterization*, Vol 10, *ASM Handbook*, ASM International, 1986, p 536–545, whose preparation was supported by the Office of Naval Research.

SELECTED REFERENCES

- C. Davisson and L.H. Germer, *Phys. Rev.*, Vol 30, Dec 1, 1927, p 705, doi:10.1103/PhysRev.30.705
- G. Ertl and J. Küppers, Chap. 9 and 10 in *Low-Energy Electrons and Surface Chemistry*, Verlag Chemie, 1974
- P.J. Estrup, in *Modern Diffraction and Imaging Techniques in Materials Science*, S. Amelinckx, R. Gevers, G. Remaut, and J. Van Landuyt, Ed., North-Holland, 1970, doi:10.1016/B978-0-444-85129-1.50007-3
- P.J. Estrup and E.G. McRae, *Surf. Sci.*, Vol 25, 1971, p 1, doi: 10.1016/0039-6028(71)90209-3
- N. Gheorghe, G. Lungu, R. Costescu, and C. Teodorescu, *Phys. Status Solidi (b)*, Vol 248, 2011, p 1919–1924, doi:10.1002/pssb.201147220
- M. Henzler, *Top. Curr. Phys.*, Vol 4, 1977, p 117, doi: 10.1007/978-3-642-81099-2_4
- M. Henzler, *Appl. Surf. Sci.*, Vol 11/12, 1982, p 450, doi: 10.1016/0378-5963(82)90092-7
- M.G. Lagally, *Appl. Surf. Sci.*, Vol 13, 1982, p 260
- M.G. Lagally, *Springer Ser. Chem. Phys.*, Vol 20, 1982, p 281
- J.J. Lander, *Prog. Solid State Chem.*, Vol 2, 1965, p 26, doi: 10.1016/0079-6786(65)90004-X
- J.W. May, *Adv. Catal.*, Vol 21, 1970, p 152, doi: 10.1016/S0360-0564(08)60565-9
- W. Moritz, Chap. 5.2 LEED, *Physics of Solid Surfaces,* Vol 45A, SpringerMaterials, Springer-Verlag Berlin Heidelberg, 2015, doi:10.1007/978-3-662-47736-6_49
- R.L. Park and M.G. Lagally, Ed., Chap. 5 in *Methods of Experimental Physics—Surfaces*, Vol 22, Academic Press, 1985, doi: 10.1557/S0883769400069360
- J.B. Pendry, *Low-Energy Electron Diffraction,* Academic Press, 1990, doi: 10.1007/978-1-4684-8777-0_7
- M.P. Seah and W.A. Dench, *Surf. Interface Anal.*, Vol 1 (No. 1), 1979, p 2, doi: 10.1002/sia.740010103
- G.A. Somorjai, *Surf. Sci.*, Vol 34, 1973, p 156, doi: 10.1016/0039-6028(73)90196-9
- G.A. Somorjai and H.H. Farrell, *Adv. Chem. Phys.*, Vol 20, 1972, p 215, doi: 10.1002/9780470143681.ch5
- C.M. Teodorescu et al., *Phys. Rev. B,* Vol 96 (No. 11), 2017, p 115438, doi:10.1103/PhysRevB.96.115438
- M.B. Webb and M.G. Lagally, *Solid State Phys.*, Vol 28, 1973, p 301, doi: 10.1016/S0081-1947(08)60205-2

ASM Handbook, Volume 10, *Materials Characterization*
ASM Handbook Committee
DOI 10.31399/asm.hb.v10.a0006633

Introduction to Scanning Probe Microscopy

Bharat Bhushan, The Ohio State University

Overview

General Uses

- Scanning tunneling microscope (STMs) and atomic force microscope (AFMs) are used for characterization on micro- to atomic scales with high resolution for studies of inorganic, organic, biological, and hybrid materials.
- They can be used to image liquids such as liquid crystals and lubricant molecules.
- They can be used in any environment, including ambient air, high temperature, and low temperature (liquid helium temperature).

Examples of Applications

- STMs and AFMs are commonly used for imaging.
- In addition to imaging, a family of SPMs have been developed for lateral force microscopy, scanning electrostatic force microscopy, scanning force acoustic microscopy, scanning magnetic microscopy, scanning near-field optical microscopy, scanning thermal microscopy, scanning electrochemical microscopy, scanning Kelvin probe microscopy, scanning chemical potential microscopy, scanning ion-conductance microscopy, and scanning capacitance microscopy.

Samples

- Inorganic, organic, biological, and hybrid samples

Limitations

- Small scan area
- Requires advanced training for operation

Estimated Analysis Time

- One to a few hours for samples

Capabilities of Related Techniques

- Scanning electron microscopes (SEMs) and transmission electron microscopes (TEMs) are commonly used for imaging down to atomic scales. However, these require sample preparations and have limitations as to the type of sample and the operating environment.

SINCE THE INTRODUCTION of the scanning tunneling microscope (STM) in 1981 and the atomic force microscope (AFM) in 1985, many variations of probe-based microscopies, referred to as scanning probe microscopes (SPMs), have been developed. While the pure imaging capabilities of SPM techniques are dominated by the application of these methods at their early development stages, the physics of probe-sample interactions and the quantitative analyses of tribological, electronic, magnetic, biological, and chemical surfaces have now become of increasing interest. This article introduces various STM and AFM designs, various operating modes, and various probes (tips).

Introduction

The STM developed by Dr. Gerd Binnig and his colleagues in 1981 at the IBM Zurich Research Laboratory, Rueschlikon, Switzerland, was the first instrument capable of directly obtaining three-dimensional (3-D) images of solid surfaces with atomic resolution (Ref 1). Binnig and Rohrer received a Nobel Prize in Physics in 1986 for their discovery. The STMs can be used only to study surfaces that are electrically conductive to some degree. Based on their design of the STM, Binnig et al. in 1985 developed an AFM to measure ultrasmall forces (less than 1 μN) present between the AFM tip surface and the sample surface (Ref 2, 3). The AFMs can be used for measurement of all engineering surfaces that may be either electrically conductive or insulating. The AFM has become a popular surface profiler for topographic and normal force measurements on the micro- to nanoscale (Ref 4, 5).

The AFMs have been modified to measure both normal and lateral forces and are called lateral force microscopes (LFMs) or friction force microscopes (FFMs) (Ref 6–13). The FFMs have been further modified to measure lateral forces in two orthogonal directions (Ref 14–18). A number of researchers have continued to improve the AFM/FFM designs and used them to measure adhesion and friction of solid and liquid surfaces on micro- and nanoscales (Ref 11–13, 19–35). The AFMs have been used for scratching, wear, and measurements of elastic/plastic mechanical properties (such as indentation hardness and modulus of elasticity) (Ref 11–13, 23, 25, 28–34, 36–47).

The AFMs have been used for manipulation of individual atoms of xenon (Ref 48), molecules (Ref 49), silicon surfaces (Ref 50), and polymer surfaces (Ref 51). The STMs have been used for formation of nanofeatures by localized heating or by inducing chemical reactions under the STM tip (Ref 52–54) and nanomachining (Ref 55). The AFMs have been used for nanofabrication (Ref 12, 38, 56–58) and nanomachining (Ref 59).

The STMs and AFMs are used at extreme magnifications ranging from 10^3 to 10^9x in x-, y-, and z-directions for imaging macro to atomic dimensions with high-resolution information and for spectroscopy. These instruments can be used in any environment, such as ambient air (Ref 2, 60), various gases (Ref 19), liquid (Ref 61–63), vacuum (Ref 1, 64), low temperatures (lower than approximately 100 K, or -173 °C) (Ref 65–69), and high temperatures (Ref 70, 71). Imaging in liquid allows the study of live biological samples and also eliminates water capillary forces present in ambient air present at the tip-sample interface. Low-temperature (liquid helium temperatures) imaging is useful for the study of biological and organic materials and the study of low-temperature phenomena such as superconductivity or charge-density waves. Low-temperature operation also is advantageous for high-sensitivity force mapping due to the reduction in thermal vibration. The STMs and AFMs also have been used to image liquids such as liquid crystals and lubricant molecules on graphite surfaces (Ref 72–75).

The pure imaging capabilities of variations of probe-based microscope techniques, referred to as scanning probe microscopy, dominated the application of these methods at their early development stages. However, the physics and chemistry of probe-sample interactions and the quantitative analyses of tribological, electronic, magnetic, biological, and chemical surfaces now have increased in interest. Nanoscale science and technology are strongly driven by SPMs, which allow investigation and manipulation of surfaces down to the atomic scale. With growing understanding of the underlying interaction mechanisms, SPMs have found application in many fields outside basic research fields. In addition, various derivatives of all these methods have been developed for special applications, some of them targeting far beyond microscopy.

A family of instruments based on STMs and AFMs, SPMs has been developed for various applications of scientific and industrial interest. These include STM, AFM, FFM (or LFM), scanning electrostatic force microscopy (SEFM) (Ref 76, 77), scanning force acoustic microscopy (SFAM, or atomic force acoustic microscopy) (Ref 23, 24, 44, 47, 78–80), scanning magnetic microscopy (SMM, or magnetic force microscopy) (Ref 81–85), scanning near-field optical microscopy (Ref 86–89), scanning thermal microscopy (Ref 90–92), scanning electrochemical microscopy (Ref 93), scanning Kelvin probe microscopy (Ref 94–98), scanning chemical potential microscopy (Ref 91), scanning ion-conductance microscopy (Ref 99, 100), and scanning capacitance microscopy (Ref 94, 101–103). The subfamily of instruments that measure forces (e.g., AFM, FFM, SEFM, SFAM, and SMM) also are referred to as scanning force microscopes. Although these instruments offer atomic resolution and are ideal for basic research, they most often are used for cutting-edge industrial applications that do not require atomic resolution.

Commercial production of SPMs started with STM in 1987 and AFM in 1989 by Digital Instruments, Inc. (now Bruker Instruments, Inc.). Numbers of these instruments are equally divided into the United States, Japan, and Europe, with industry/university and government lab splits of approximately 50/50, 70/30, and 30/70, respectively. It is clear that research and industrial applications of SPMs are rapidly expanding. For various examples, see Ref 13 and 104 to 112.

This article presents an overview of STM and AFM, with various operating modes and various probes (tips) used in these instruments. Then, details on AFM instrumentation and analyses are given.

Scanning Tunneling Microscope

The principle of electron tunneling was proposed by Giaever (Ref 113). He proposed that if an electrical potential difference is applied to two metals separated by a thin insulating film, a current will flow across the barrier because of the ability of electrons to penetrate it. To be able to measure this tunneling current, the two metals must be spaced no more than 10 nm apart. Binnig et al. (Ref 1) introduced vacuum tunneling combined with lateral scanning. The vacuum provides the ideal barrier for tunneling. The lateral scanning allows surface imaging with exquisite resolution—lateral less than 1 nm and vertical less than 0.1 nm—sufficient to define the position of single atoms. The very high vertical resolution is obtained because the tunneling current varies exponentially with the distance between the metal tip and the scanned surface acting as two electrodes. Typically, tunneling current decreases by a factor of 2 as the separation is increased by 0.2 nm. Very high lateral resolution depends on the sharp tips. Binnig et al. (Ref 1) overcame two key obstacles for damping external vibrations and for moving the tunneling probe in close proximity to the sample. Their instrument is called the scanning tunneling microscope. Today's STMs can be used in the ambient environment for atomic-scale imaging of surfaces. Excellent reviews on this subject are presented in Ref 114 to 123.

The principle of STM is straightforward. A sharp metal tip (one electrode of the tunnel junction) is brought close enough (0.3 to 1 nm) to the surface to be investigated (second electrode) so that, at a convenient operating voltage (10 mV to 1 V), the tunneling current varies from 0.2 to 10 nA, which is measurable. The tip is scanned over a surface at a distance of 0.3 to 1 nm, while the tunneling current between it and the surface is sensed. The STM can be operated in either the constant-current mode or the constant-height mode (Fig. 1). The left column of Fig. 1 shows the basic constant-current mode of operation. A feedback network changes the height of the tip z to keep the current constant. The displacement of the tip given by the voltage

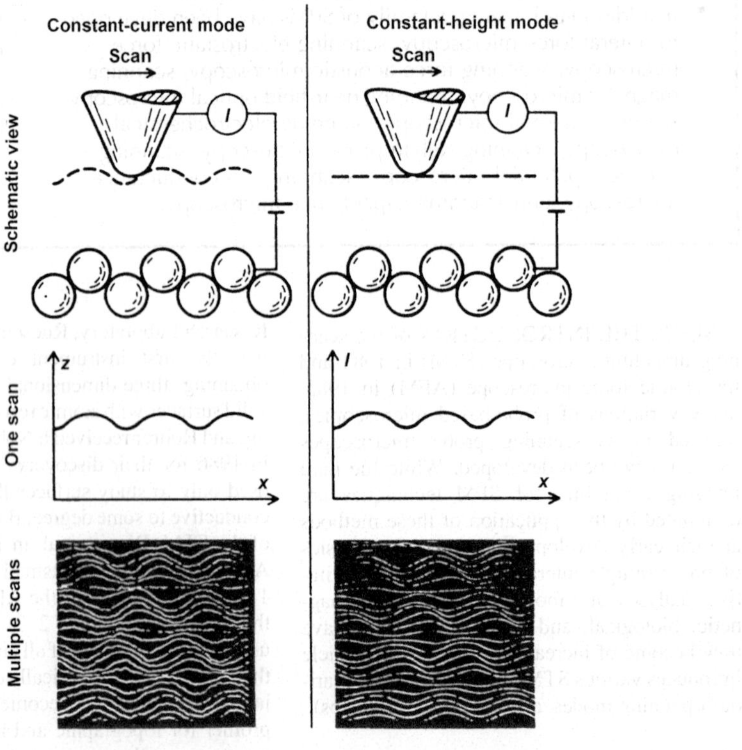

Fig. 1 A scanning tunneling microscope can be operated in either the constant-current or the constant-height mode. The images are of graphite in air. Reprinted from Ref 114, with the permission of AIP Publishing

applied to the piezoelectric drives then yields a topographic map of the surface. Alternatively, in the constant-height mode, a metal tip can be scanned across a surface at nearly constant height and constant voltage while the current is monitored, as shown in the right column of Fig. 1. In this case, the feedback network responds only rapidly enough to keep the average current constant. The constant-current mode is used generally for atomic-scale images. This mode is not practical for rough surfaces. A 3-D picture [$z(x, y)$] of a surface consists of multiple scans [$z(x)$] displayed laterally from each other in the y-direction. It should be noted that if different atomic species are present in a sample, the different atomic species within a sample may produce different tunneling currents for a given bias voltage. Thus, the height data may not be a direct representation of the topography of the surface of the sample.

Binnig et al.'s Design

Figure 2 shows a schematic of one of Binnig and Rohrer's designs for operation in ultrahigh vacuum (Ref 1, 124). The metal tip was fixed to rectangular piezodrives P_x, P_y, and P_z made out of commercial piezoceramic material for scanning. The sample is mounted on either a superconducting magnetic levitation or a two-stage spring system to achieve a stability of the gap width of approximately 0.02 nm. The tunnel current, J_T, is a sensitive function of the gap width, d, that is:

$$J_T \propto V_T \exp\left(-A\phi^{1/2}d\right)$$

where V_T is the bias voltage, ϕ is the average barrier height (work function), and the constant $A = 1.025\ \text{eV}^{-1/2}\ \text{Å}^{-1}$. With a work function of a few electron volts, J_T changes by an order of magnitude for every angstrom change of d. If the current is kept constant to within, for example, 2%, then the gap d remains constant to within 1 pm. For operation in the constant-current mode, the control unit (CU) applies a voltage V_z to the piezo P_z such that J_T remains constant when scanning the tip with P_y and P_x over the surface. At a constant work function ϕ, $V_z(V_x, V_y)$ yields the roughness of the surface $z(x, y)$ directly, as illustrated at a surface step at A. Smearing the step, δ (lateral resolution) is on the order of $(R)^{1/2}$, where R is the radius of the curvature of the tip. Thus, a lateral resolution of approximately 2 nm requires tip radii on the order of 10 nm. A 1 mm diameter solid rod ground at one end at roughly 90° yields overall tip radii of only a few hundred nanometers, but with closest protrusion of rather sharp microtips on the relatively dull end yields a lateral resolution of approximately 2 nm. In situ sharpening of the tips by gently touching the surface brings the resolution down to the 1 nm range by applying high fields (on the order of 10^8 V/cm). For example, with applications of high field for half an hour, resolutions considerably below 1 nm could be reached. Most experiments were done with tungsten wires either ground or etched to a radius typically in the range of 0.1 to 10 μm. In some cases, in situ processing of the tips was done for further reduction of tip radii.

Commercial STMs

There are a number of commercial STMs available on the market. In 1987, Digital Instruments, Inc. (now Bruker Instruments, Inc.), located in Santa Barbara, California, introduced the first commercial STM, the Nanoscope I. In a later-model Nanoscope IV STM for operation in ambient air, the sample is held in position while a piezoelectric crystal in the form of a cylindrical tube (referred to as a PZT tube scanner) scans the sharp metallic probe over the surface in a raster pattern while sensing and outputting the tunneling current to the control station (Fig. 3). The digital signal processor (DSP) calculates the desired separation of the tip from the sample by sensing the tunneling current flowing between the sample and the tip. The bias voltage applied between the sample and the tip encourages the tunneling current to flow. The DSP completes the digital feedback loop by outputting the desired voltage to the piezoelectric tube. The STM operates in both the constant-height and constant-current modes, depending on a parameter selection in the control panel. In the constant-current mode, the feedback gains are set high, the tunneling tip closely tracks the sample surface, and the variation in the tip height required to maintain constant tunneling current is measured by the change in the voltage applied to the piezo tube. In the constant-height mode, the feedback gains are set low, the tip remains at a nearly constant height as it sweeps over the sample surface, and the tunneling current is imaged.

Physically, the Nanoscope STM consists of three main parts: the head, which houses the piezoelectric tube scanner for 3-D motion of the tip and the preamplifier circuit (field-effect transistor input amplifier) mounted on top of the head for the tunneling current; the base on which the sample is mounted; and the base support, which supports the base and head (Ref 12). The base accommodates samples up to 10 by 20 mm (0.4 by 0.8 in.) and 10 mm (0.4 in.) in thickness. Scan sizes available for the STM are 0.7 (for atomic resolution), 12, 75, and 125 μm square.

The scanning head controls the 3-D motion of the tip. The removable head consists of a piezo tube scanner, approximately 12.7 mm (0.50 in.) in diameter, mounted into an invar shell used to minimize vertical thermal drifts because of good thermal match between the piezo tube and the invar. The piezo tube has separate electrodes for X, Y, and Z that are driven by separate drive circuits. The electrode configuration (Fig. 3) provides x- and y-motions that are perpendicular to each other, minimizes horizontal and vertical coupling, and provides good sensitivity. The vertical motion of the tube is controlled by the Z-electrode, which is driven by the feedback loop. The x- and y-scanning motions are each controlled by two electrodes that are driven by voltages of the same magnitudes but opposite signs. These electrodes are called $-Y$, $-X$, $+Y$, and $+X$. Applying complimentary voltages allows a short, stiff tube to provide a good scan range without large voltages. The motion of the tip due to external vibrations is proportional to the square of the ratio of vibration frequency to the resonant frequency of the tube. Therefore, to minimize tip vibrations, the resonant frequencies of the tube are high (approximately 60 kHz) in the vertical direction and approximately 40 kHz in the horizontal direction. The tip holder is a stainless steel

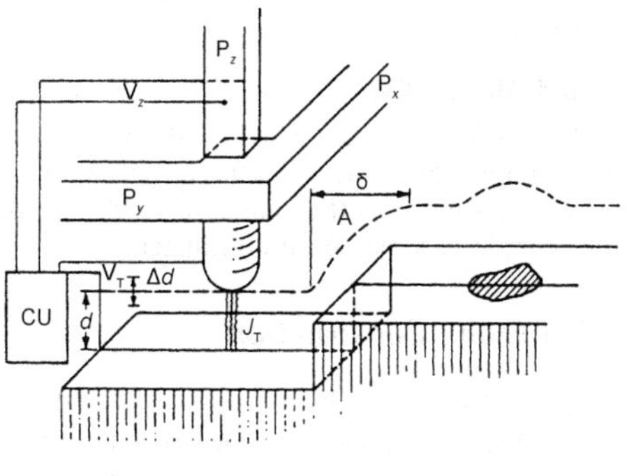

Fig. 2 Principle of operation of the scanning tunneling microscope made by Binnig and Rohrer. CU, control unit. Reprinted from Ref 124, with permission from Elsevier

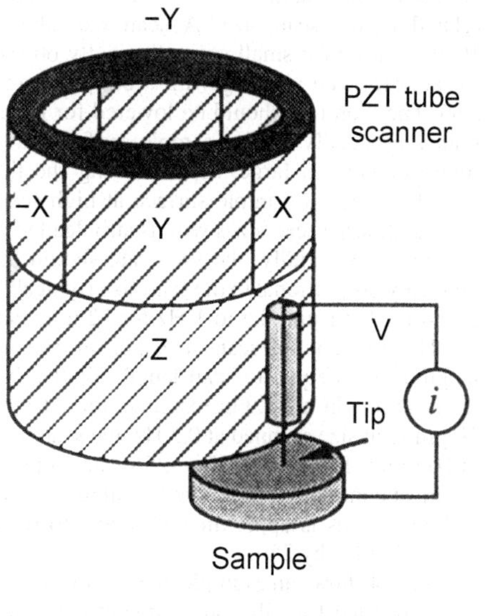

Fig. 3 Principle of operation of a commercial scanning tunneling microscope. A sharp tip attached to a piezoelectric tube (PZT) scanner is scanned on a sample.

tube with a 300 μm inner diameter for 250 μm diameter tips, mounted in ceramic to keep the mass on the end of the tube low. The tip is mounted either on the front edge of the tube (to keep mounting mass low and resonant frequency high) (Fig. 3) or on the center of the tube for large-range scanners, namely, 75 and 125 μm (to preserve the symmetry of the scanning). This commercial STM accepts any tip with a 250 μm diameter shaft. The piezo tube requires *X-Y* calibration, which is carried out by imaging an appropriate calibration standard. Cleaved graphite is used for the small-scan-length head, while two-dimensional grids (a gold-plated ruling) can be used for longer-range heads.

The invar base holds the sample in position, supports the head, and provides coarse *X-Y* motion for the sample. A spring-steel sample clip with two thumb screws holds the sample in place. An *x-y* translation stage built into the base allows the sample to be repositioned under the tip. Three precision screws arranged in a triangular pattern support the head and provide coarse and fine adjustment of the tip height. The base support consists of the base support ring and the motor housing. The stepper motor enclosed in the motor housing allows the tip to be engaged and withdrawn from the surface automatically.

Samples to be imaged with STM must be conductive enough to allow a few nanoamperes of current to flow from the bias voltage source to the area to be scanned. In many cases, nonconductive samples can be coated with a thin layer of a conductive material to facilitate imaging. The bias voltage and the tunneling current depend on the sample. Usually they are set at a standard value for engagement and fine-tuned to enhance the quality of the image. The scan size depends on the sample and the features of interest. A maximum scan rate of 122 Hz can be used. The maximum scan rate is usually related to the scan size. A scan rate above 10 Hz is used for small scans (typically 60 Hz for atomic-scale imaging with a 0.7 μm scanner). The scan rate should be lowered for large scans, especially if the sample surfaces are rough or contain large steps. Moving the tip quickly along the sample surface at high scan rates with large scan sizes will usually lead to a tip crash. Essentially, the scan rate should be inversely proportional to the scan size (typically 2 to 4 Hz for 1 μm, 0.5 to 1 Hz for 12 μm, and 0.2 Hz for 125 μm scan sizes). Scan rate in length/time is equal to scan length divided by the scan rate in hertz. For example, for a 10 by 10 μm scan size scanned at 0.5 Hz, the scan rate is 10 μm/s. Typically, 256 × 256 data formats are most commonly used. The lateral resolution at larger scans is approximately equal to scan length divided by 256.

Figure 4 shows an example of STM images of an evaporated C_{60} film on gold-coated freshly cleaved mica taken at room temperature and ambient pressure (Ref 125). Images with atomic resolution at two scan sizes are obtained. The following sections describe STM designs that are available for special applications.

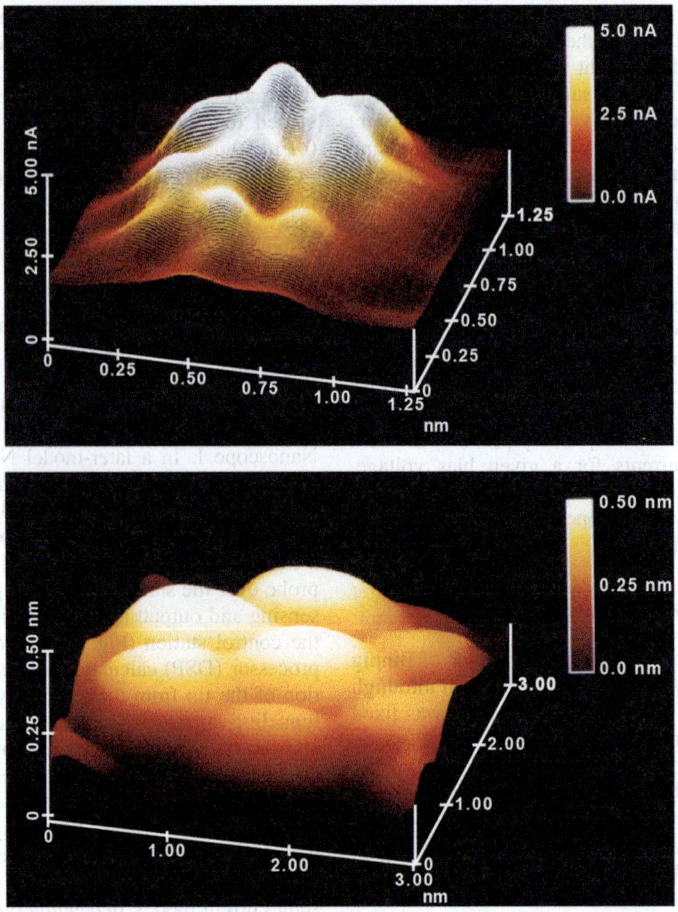

Fig. 4 Scanning tunneling microscope images of evaporated C_{60} film on gold-coated freshly cleaved mica using a mechanically sheared platinum-iridium (80/20) tip in constant-height mode. Source: Ref 125.

Electrochemical STM

Electrochemical STM is used to perform and monitor the electrochemical reactions inside the STM. It includes a microscope base with an integral potentiostat, a short head with a 0.7 μm scan range and a differential preamp, and the software required to operate the potentiostat and display the result of electrochemical reaction.

Stand-Alone STM

Stand-alone STMs, which rest directly on the sample, are able to scan large samples. They are similar to the standard STM, except the sample base has been eliminated.

STM Probe Construction

The STM probe should have a cantilever integrated with a sharp metal tip with a low aspect ratio (tip length/tip shank) to minimize flexural vibrations. Ideally, the tip should be atomically sharp, but in practice, most tip-preparation methods produce a tip with a rather ragged profile that consists of several asperities, with the one closest to the surface responsible for tunneling. The STM cantilevers with sharp tips are typically fabricated from metal wires of tungsten, platinum-iridium, or gold and sharpened by grinding, cutting with a wire cutter or razor blade, field emission/evaporator, ion milling, fracture, or electrochemical polishing/etching (Ref 126, 127). The two most commonly used tips are made from either a platinum-iridium (80/20) alloy or tungsten wire. Iridium is used to provide stiffness. The platinum-iridium tips are generally mechanically formed and are readily available. The tungsten tips are etched from tungsten wire with an electrochemical process, for example, by using 1 molar KOH solution with a platinum electrode in an electrochemical cell at approximately 30 V. In general, platinum-iridium tips provide better atomic resolution than tungsten tips, probably due to the lower reactivity of platinum. However, tungsten tips are more uniformly shaped and may perform better on samples with steeply sloped features. The tungsten wire diameter used for the cantilever is typically 250 μm, with the radius of curvature ranging from

20 to 100 nm and a cone angle ranging from 10 to 60° (Fig. 5). The wire can be bent in an L-shape, if required, for use in the instrument. For calculations of normal spring constant and natural frequency of round cantilevers, see Ref 115.

Platinum alloy and tungsten tips are very sharp and have high resolution but are fragile and sometimes break when contacting a surface. Diamond tips have been used by Kaneko and Oguchi (Ref 128). The diamond tip made conductive by boron ion implantation is found to be chip resistant.

Atomic Force Microscope

Like the STM, the AFM relies on a scanning technique to produce very high-resolution, 3-D images of sample surfaces. The AFM measures ultrasmall forces (less than 1 nN) present between the AFM tip surface and a sample surface. These small forces are measured by measuring the motion of a very flexible cantilever beam having an ultrasmall mass. While STM requires that the surface to be measured be electrically conductive, AFM is capable of investigating surfaces of both conductors and insulators on an atomic scale if suitable techniques for measurement of cantilever motion are used. In the operation of high-resolution AFM, the sample is generally scanned, rather than the tip as in STM, because AFM measures the relative displacement between the cantilever surface and reference surface, and any cantilever movement would add vibrations. For measurements of large samples, AFMs are available where the tip is scanned and the sample is stationary. As long as the AFM is operated in the so-called contact mode, little if any vibration is introduced.

The AFM combines the principles of the STM and the stylus profiler (Fig. 6). In an AFM, the force between the sample and tip is detected rather than the tunneling current to sense the proximity of the tip to the sample. The AFM can be used either in static or dynamic mode. In the static mode, also referred to as repulsive mode or contact mode (Ref 2), a sharp tip at the end of a cantilever is brought in contact with a sample surface. During initial contact, the atoms at the end of the tip experience a very weak repulsive force due to electronic orbital overlap with the atoms in the sample surface. The force acting on the tip causes a cantilever deflection that is measured by tunneling, capacitive, or optical detectors. The deflection can be measured to within 0.02 nm, so for a typical cantilever spring constant of 10 N/m (0.7 lbf/ft), a force as low as 0.2 nN (corresponding normal pressure ~200 MPa, or 29 ksi, for a Si_3N_4 tip with a radius of approximately 50 nm against single-crystal silicon) can be detected. (To put these numbers in perspective, individual atoms and a human hair are typically a fraction of a nanometer and approximately 75 μm in diameter, respectively, and an eyelash and a drop of water have a mass of approximately 100 nN and 10 μN, respectively.)

In the dynamic mode of operation for the AFM, also referred to as attractive force imaging or noncontact imaging mode, the tip is brought in close proximity to (within a few nanometers), and not in contact with, the sample (Ref 129, 130). The cantilever is deliberately vibrated in either amplitude modulation mode (Ref 76) or frequency modulation mode (Ref 87, 115, 131, 132). Very weak van der Waals attractive forces are present at the tip-sample interface. In this technique, the normal pressure exerted at the interface is close to zero (desirable to avoid any surface deformation). In the two modes, surface topography is measured by laterally scanning the sample under the tip while simultaneously measuring the separation-dependent force or force gradient (derivative) between the tip and the surface (Fig. 6).

In the contact (static) mode, the interaction force between tip and sample is measured by measuring the cantilever deflection. In the noncontact (or dynamic) mode, the force gradient is obtained by vibrating the cantilever and measuring the shift of resonant frequency of the cantilever. To obtain topographic information, the interaction force is either recorded directly or used as a control parameter for a feedback circuit that maintains the force or force derivative at a constant value. With an AFM operated in contact mode, topographic images with a vertical resolution of less than 0.1 nm (as low as 0.01 nm) and a lateral resolution of approximately 0.2 nm have been obtained (Ref 3, 61, 133–137). With a 0.01 nm displacement sensitivity, 10 nN to 1 pN forces are measurable. These forces are comparable to the forces associated with chemical bonding, for example, 0.1 pN for an ionic bond and 10 pN for a hydrogen bond (Ref 2). For further reading, see Ref 115 to 117, 121, 123, and 138 to 142.

Lateral forces being applied at the tip during scanning in contact mode affect roughness measurements (Ref 143). To minimize the effects of friction and other lateral forces in the topography measurements in contact-mode AFMs and to measure topography of soft surfaces, AFMs can be operated in the so-called tapping mode or amplitude modulation mode (Ref 37, 144).

The STM is ideal for atomic-scale imaging. To obtain atomic resolution with AFM, the spring constant of the cantilever should be weaker than the equivalent spring between atoms. For example, the vibration frequencies, ω, of atoms bound in a molecule or in a crystalline solid are typically 10^{13} Hz or higher. Combining this with the mass of the atoms, m, on the order of 10^{-25} kg, gives interatomic spring constants k, given by $\omega^2 m$, on the order of 10 N/m (0.7 lbf/ft) (Ref 138). (For comparison, the spring constant of a piece of household aluminum foil that is 4 mm, or 0.16 in., long and 1 mm, or 0.04 in., wide is approximately 1 N/m, or 0.07 lbf/ft.)

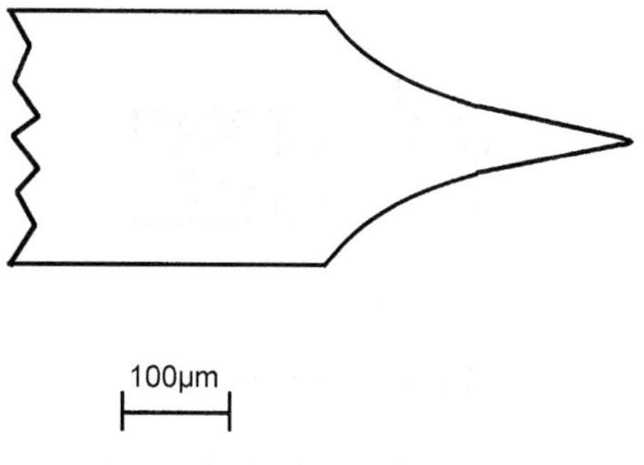

Fig. 5 Schematic of a typical tungsten cantilever with a sharp tip produced by electrochemical etching

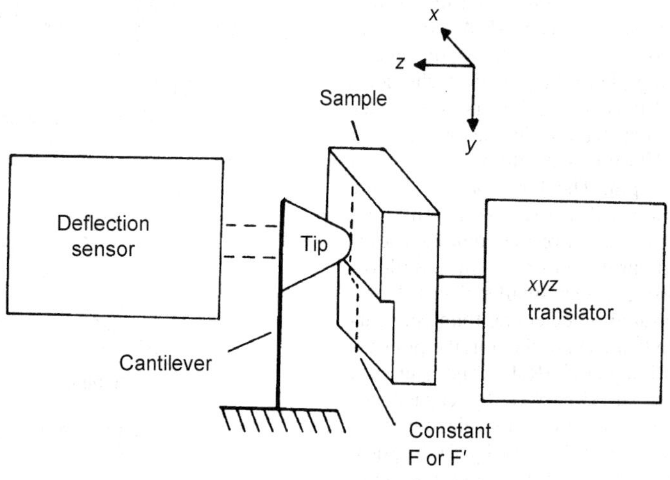

Fig. 6 Principle of operation of the atomic force microscope. A sample mounted on a piezoelectric scanner is scanned against a short tip, and the cantilever deflection is measured, mostly using a laser-deflection technique. Force (contact mode) or force gradient (noncontact mode) is measured during scanning.

Therefore, a cantilever beam with a spring constant of approximately 1 N/m or lower is desirable. Tips must be as sharp as possible. Tips with a radius ranging from 20 to 50 nm are commonly available.

Atomic resolution cannot be achieved with these tips at the normal load in the nanonewton range. Atomic structures at these loads have been obtained from lattice imaging or by imaging of the crystal periodicity. Reported data show either perfectly ordered periodic atomic structures or defects on a larger lateral scale but no well-defined, laterally resolved, atomic-scale defects such as those seen in images routinely obtained with STM. Interatomic forces with one or several atoms in contact are 20 to 40 or 50 to 100 pN, respectively. Thus, atomic resolution with AFM is only possible with a sharp tip on a flexible cantilever at a net repulsive force of 100 pN or lower (Ref 145). Upon increasing the force from 10 pN, Ohnesorge and Binnig (Ref 145) observed that monoatomic steplines were slowly wiped away and a perfectly ordered structure was left. This observation explains why mostly defect-free atomic resolution has been observed with AFM. Note that for atomic-resolution measurements, the cantilever should not be too soft, to avoid jumps. Further note that measurements in the noncontact imaging mode may be desirable for imaging with atomic resolution.

The key component in AFM is the sensor for measuring the force on the tip due to its interaction with the sample. A cantilever (with a sharp tip) with extremely low spring constants is required for high vertical and lateral resolutions at small forces (0.1 nN or lower), but at the same time, a high resonant frequency (approximately 10 to 100 kHz) is necessary to minimize the sensitivity to vibration noise from the building in which the AFM is located (close to 100 Hz). This requires a spring with an extremely low vertical spring constant (typically 0.05 to 1 N/m, or 0.003 to 0.07 lbf/ft) as well as low mass (on the order of 1 ng). Today (2019), the most advanced AFM cantilevers are microfabricated from silicon or silicon nitride by using photolithographic techniques. (For further details on cantilevers, see a later section). Typical lateral dimensions are on the order of 100 μm, with the thicknesses on the order of 1 μm. The force on the tip due to its interaction with the sample is sensed by detecting the deflection of the compliant lever with a known spring constant. This cantilever deflection (displacement smaller than 0.1 nm) has been measured by detecting tunneling current similar to that used in STM in the pioneering work of Binnig et al. (Ref 2) and later used by Giessibl et al. (Ref 67), by capacitance detection (Ref 146, 147), by piezoresistive detection (Ref 148, 149), and by four optical techniques, namely, by optical interferometry (Ref 6, 7, 150, 151) and with the use of optical fibers (Ref 68, 82), by optical polarization detection (Ref 83, 152), by laser diode feedback (Ref 153), and by optical (laser) beam deflection (Ref 8, 9, 64, 134, 135).

Schematics of the four more commonly used detection systems are shown in Fig. 7. The tunneling method originally used by Binnig et al. (Ref 2) in the first version of AFM uses a second tip to monitor the deflection of the cantilever with its force-sensing tip. Tunneling is rather sensitive to contaminants, and the interaction between the tunneling tip and the rear side of the cantilever can become comparable to the interaction between the tip and sample. Tunneling is rarely used and was mentioned earlier for historical purposes. Giessibl et al. (Ref 67) used it for a low-temperature AFM/STM design. In contrast to tunneling, other deflection sensors are far away from the cantilever, at distances of micrometers to tens of millimeters. The optical techniques are believed to be more sensitive, reliable, and easily implemented detection methods than others (Ref 115, 141). The optical beam deflection method has the largest working distance, is insensitive to distance changes, and is capable of measuring angular changes (friction forces). Therefore, it is the most commonly used in the commercial SPMs.

Almost all SPMs use piezo translators to scan the sample or, alternatively, to scan the tip. An electric field applied across a piezoelectric material causes a change in the crystal structure, with expansion in some directions and contraction in others. A net change in volume also occurs (Ref 154). The first STM used a piezo tripod for scanning (Ref 1). The piezo tripod is one way to generate 3-D movement of a tip attached to its center. However, the tripod must be fairly large (~50 mm, or 2 in.) to obtain a suitable range. Its size and asymmetric shape make it susceptible to thermal drift. Tube scanners are widely used in AFMs (Ref 155). These provide ample scanning range within a small size. Control electronics systems for AFMs can use either analog or digital feedback. Digital feedback circuits are better suited for ultralow-noise operation.

Images from AFMs must be processed. An ideal AFM is a noise-free device that images a sample with perfect tips of known shape and has perfect linear scanning piezo. In reality, scanning devices are affected by distortions for which corrections must be made. The distortions can be linear and nonlinear. Linear distortions mainly result from imperfections in the machining of the piezo translators, causing crosstalk between the Z-piezo to the X- and Y-piezos, and vice versa. Nonlinear distortions mainly result because of the presence of a hysteresis loop in piezoelectric ceramics. These may also result if the scan frequency approaches the upper frequency limit of the X- and Y-drive amplifiers or the upper frequency limit of the feedback loop (z-component). In addition, electronic noise may be present in the system. The noise is removed by digital filtering in the real space (Ref 156) or in the spatial frequency domain (Fourier space) (Ref 157).

Processed data consists of many tens of thousands of points per plane (or data set). The output of the first STM and AFM images were recorded on an X-Y chart recorder, with z-value plotted against the tip position in the fast scan direction. Chart recorders have slow response, so computers are used for display of the data. The data are displayed as wire mesh display or gray-scale display (with at least 64 shades of gray).

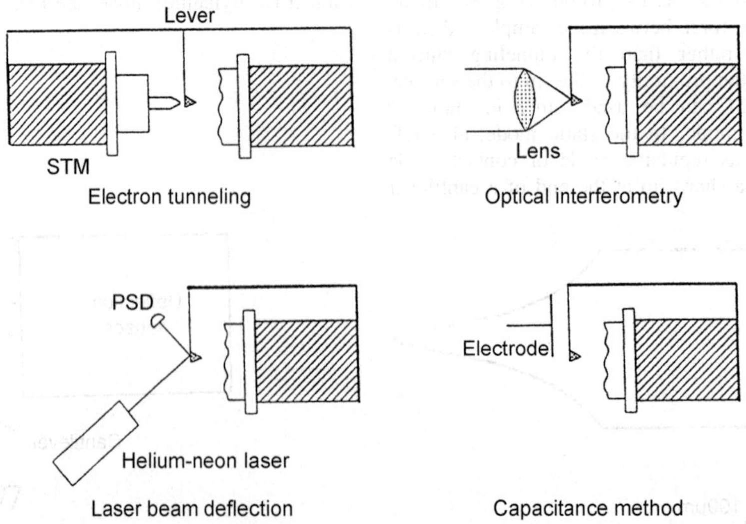

Fig. 7 Schematics of the four detection systems for measurement of cantilever deflection. In each setup, the sample mounted on a piezoelectric body is shown on the right, the cantilever in the middle, and the corresponding deflection sensor on the left. STM, scanning tunneling microscope; PSD, position-sensitive detector. Reprinted from Ref 141, with permission from Elsevier

Binnig et al.'s Design

In the first AFM design developed by Binnig et al. (Ref 2), AFM images were obtained by measurement of the force on a sharp tip created by the proximity to the surface of the sample mounted on a 3-D piezoelectric scanner. Tunneling current between STM tip and the backside of the cantilever beam with attached tip was measured to obtain the normal force. This force was kept at a constant level with a feedback mechanism. The STM tip was also mounted on a piezoelectric element to maintain the tunneling current at a constant level.

Commercial AFM

A review of early designs of AFMs is presented by Bhushan (Ref 12). There are a large number of commercial AFMs available on the market. Major manufacturers of AFMs for use in the ambient environment are Bruker Instruments, Inc. (formerly Digital Instruments, Inc. and later Veeco Instruments, Inc.), Santa Barbara, California; Agilent Technologies, Santa Clara, California; Park Systems Corp., Suwon, Korea; Asylum Research, Santa Clara, California; NT-MDT Spectrum Instruments, Moscow, Russia; Nanosurf AF, Liestal, Switzerland; Seiko Instruments, Japan; and Olympus, Japan. A major manufacturer for AFM/STMs for use in ultrahigh-vacuum and low-temperature environments is Scienta Omicron GMBH, Taunusstein, Germany.

This section describes two typical commercial AFMs—small-sample and large-sample AFMs—for operation in the contact mode, produced by Bruker Instruments, Inc., Santa Barbara, California, with scanning lengths ranging from approximately 0.7 μm (for atomic resolution) to approximately 125 μm (Ref 10, 134, 137, 158). The original design of these AFMs comes from Meyer and Amer (Ref 64). Basically, the AFM scans the sample in a raster pattern while outputting the cantilever deflection error signal to the control station. The cantilever deflection (or the force) is measured using a laser deflection technique (Fig. 8). The DSP in the workstation controls the *z*-position of the piezo based on the cantilever deflection error signal. The AFM operates in both the constant-height and constant-force modes. The DSP always adjusts the height of the sample under the tip based on the cantilever deflection error signal, but if the feedback gains are low, the piezo remains at a nearly constant height, and the cantilever deflection data are collected. With the high gains, the piezo height changes to keep the cantilever deflection nearly constant (therefore, the force is constant), and the change in piezo height is collected by the system.

To further describe the operation principles of the small-sample commercial AFM shown in Fig. 8(a), the sample, generally no larger than 10 by 10 mm (0.4 by 0.4 in.), is mounted on a piezoelectric tube (PZT) scanner that consists of separate electrodes to scan precisely the sample in the *x-y* plane in a raster pattern and to move the sample in the vertical (*z*) direction. A sharp tip at the free end of a flexible cantilever is brought in contact with the sample. Features on the sample surface cause the cantilever to deflect in the vertical and lateral directions as the sample moves under the tip. A laser beam from a diode laser (5 mW maximum peak output at 670 nm) is directed by a prism onto the back of a cantilever near its free end, tilted downward at approximately 10° with respect to the horizontal plane. The reflected beam from the vertex of the cantilever is directed through a mirror onto a quad photodetector (split photodetector with four quadrants) (commonly called a position-sensitive detector, produced by Silicon Detector Corp., Camarillo, California). The differential signal from the top and bottom photodiodes provides the AFM signal, which is a sensitive measure of the cantilever vertical deflection. Topographic features of the sample cause the tip to deflect in the vertical direction as the sample is scanned under the tip. This tip deflection will change the direction of the reflected laser beam, changing the intensity difference between the top and bottom sets of photodetectors (AFM signal). In the AFM operating mode called the height mode, for topographic imaging or for any other operation in which the applied normal force is to be kept a constant, a feedback circuit is used to modulate the voltage applied to the PZT scanner to adjust the height of the PZT. This ensures that the cantilever vertical deflection (given by the intensity difference between the top and bottom detector), and consequently normal load, will remain constant during scanning. The PZT height variation is thus a direct measure of the surface roughness of the sample.

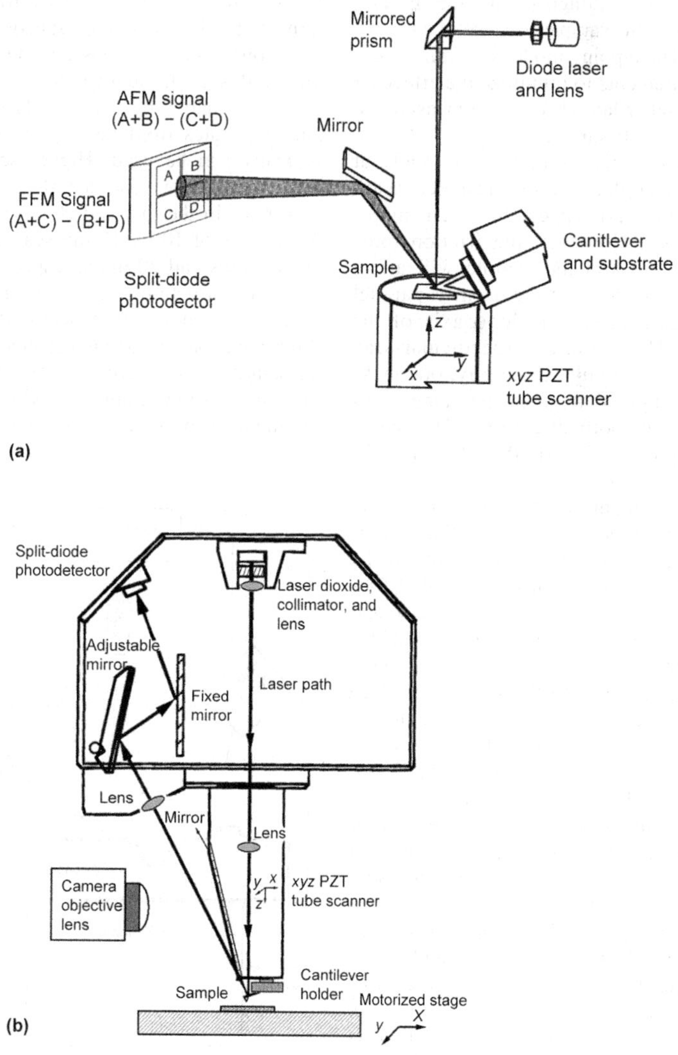

Fig. 8 Principles of operation of (a) a commercial small-sample atomic force microscope (AFM)/friction force microscope (FFM) and (b) a large-sample AFM/FFM. PZT, piezoelectric tube

In a large-sample AFM, both force sensors using the optical deflection method and scanning unit are mounted on the microscope head (Fig. 8b). Because of vibrations added by the cantilever movement, lateral resolution of this design is somewhat poorer than the design in Fig. 8(a), in which the sample is scanned instead of cantilever beamed. The advantage of the large-sample AFM is that large samples can be measured readily.

Most AFMs can be used for topography measurements in the so-called tapping mode (intermittent contact mode), also referred to as dynamic force microscopy in the amplitude mode, mentioned earlier. In the tapping mode, during scanning over the surface, the cantilever/tip assembly is sinusoidally vibrated by a piezo mounted above it, and the oscillating tip slightly taps the surface at the resonant frequency of the cantilever (70 to 400 kHz) with a constant (20 to100 nm) oscillating amplitude introduced in the vertical direction, with a feedback loop keeping the average normal force constant (Fig. 9). The oscillating amplitude is kept large enough so that the tip does not get stuck to the sample because of adhesive attractions. The tapping mode is used in topography measurements to minimize the effects of friction and other lateral forces to measure the topography of soft surfaces.

Topographic measurements are made at any scanning angle. At a first instance, scanning angle may not appear to be an important parameter. However, the friction force between the tip and the sample will affect the topographic measurements in a parallel scan (scanning along the long axis of the cantilever). Therefore, a perpendicular scan may be more desirable. Generally, one picks a scanning angle that gives the same topographic data in both directions. This angle may be slightly different than that for the perpendicular scan.

For measurement of friction force being applied at the tip surface during sliding, left-hand and right-hand sets of quadrants of the photodetector are used. In the so-called friction mode, the sample is scanned back and forth in a direction orthogonal to the long axis of the cantilever beam. A friction force between the sample and the tip will produce a twisting of the cantilever. As a result, the laser beam will be reflected out of the plane defined by the incident beam and the beam reflected vertically from an untwisted cantilever. This produces an intensity difference of the laser beam received in the left-hand and right-hand sets of quadrants of the photodetector. The intensity difference between the two sets of detectors (FFM signal) is directly related to the degree of twisting and hence to the magnitude of the friction force. This method provides 3-D maps of friction force. One problem associated with this method is that any misalignment between the laser beam and the photodetector axis introduces error in the measurement. However, by following the procedures developed by Ruan and Bhushan (Ref 158) in which the average FFM signal for the sample scanned in two opposite directions is subtracted from the friction profiles of each of the two scans, the misalignment effect is eliminated. By following the friction force calibration procedures developed by Ruan and Bhushan (Ref 158), voltages corresponding to friction forces can be converted to force units. The coefficient of friction is obtained from the slope of friction force data measured as a function of normal loads typically ranging from 10 to 150 nN. This approach eliminates any contributions due to adhesive forces (Ref 38). For calculation of the coefficient of friction based on a single-point measurement, friction force should be divided by the sum of applied normal load and intrinsic adhesive force. Furthermore, note that for a single asperity contact, the coefficient of friction is not independent of load. For further details, refer to a later section.

The tip is scanned in such a way that its trajectory on the sample forms a triangular pattern (Fig. 10). Scanning speeds in the fast and slow scan directions depend on the scan area and scan frequency. Scan sizes ranging from less than 1 by 1 nm to 125 by 125 μm and scan rates from less than 0.5 to 122 Hz typically can be used. Higher scan rates are used for smaller scan lengths. For example, scan rates in the fast and slow scan directions for an area of 10 by 10 μm scanned at 0.5 Hz are 10 μm/s and 20 nm/s, respectively.

The construction of a small-sample AFM is now described in more detail. It consists of three main parts: the optical head that senses the cantilever deflection, a PZT tube scanner that controls the scanning motion of the sample mounted on its one end, and the base that supports the scanner and head and includes circuits for the deflection signal (Fig. 11a). The AFM connects directly to a control system. The optical head consists of a laser diode stage, photodiode stage preamp board, cantilever mount and its holding arm, and deflection beam reflecting mirror (Fig. 11b). The laser diode stage is a tilt stage used to adjust the position of the laser beam relative to the cantilever. It consists of the laser diode, collimator, focusing lens, baseplate, and the *X* and *Y* laser diode positioners. The positioners are used to place the laser spot on the end of the cantilever. The photodiode stage is an adjustable stage used to position the photodiode elements relative to the reflected laser beam. It consists of the split photodiode, the base plate, and the photodiode positioners. The deflection beam reflecting mirror is mounted on the upper left in the interior of the head, which reflects the deflected beam toward the photodiode. The cantilever mount is a metal (for operation in air) or glass (for operation in water) block that holds the cantilever firmly at the proper angle (Fig. 11d). Next, the tube scanner consists of an invar cylinder holding a single tube made of piezoelectric crystal that provides the necessary 3-D motion to the sample. Mounted on top of the tube is a magnetic cap on which the steel sample puck is placed. The tube is rigidly held at one end, with the sample mounted on the other end of the tube. The scanner also contains three fine-pitched screws that form the mount for the optical head. The optical head rests on the tips of the screws, which are used to adjust the position of the head relative to the sample. The scanner fits into the scanner support ring mounted on the base of the microscope (Fig. 11c). The stepper motor is controlled manually with the switch on the upper surface of the base and automatically by the computer during the tip-engage and tip-withdraw processes.

The scan sizes available for these instruments are 0.7, 12, and 125 μm. The scan rate must be decreased as the scan size is increased. A maximum scan rate of 122 Hz can be used. Scan rates of approximately 60 Hz should be

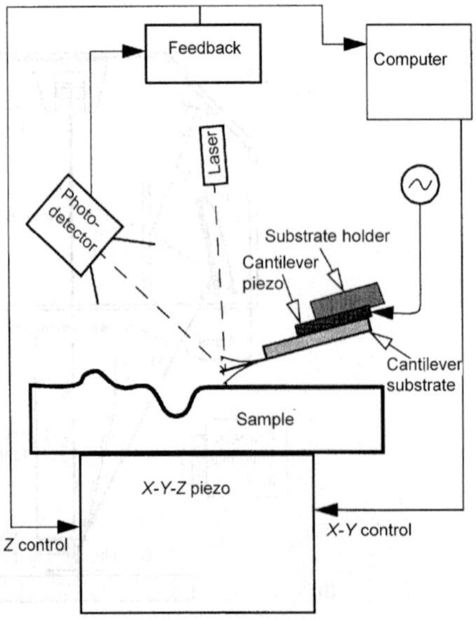

Fig. 9 Schematic of tapping mode used for surface roughness measurements

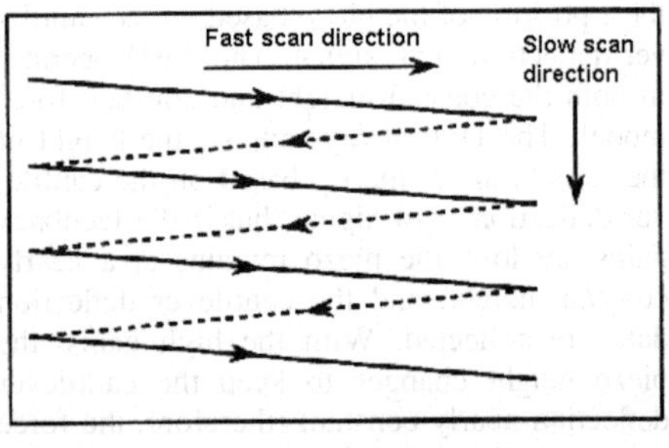

Fig. 10 Schematic of triangular pattern trajectory of the atomic force microscope tip as the sample is scanned in two dimensions. During imaging, data are recorded only during scans along the solid scan lines.

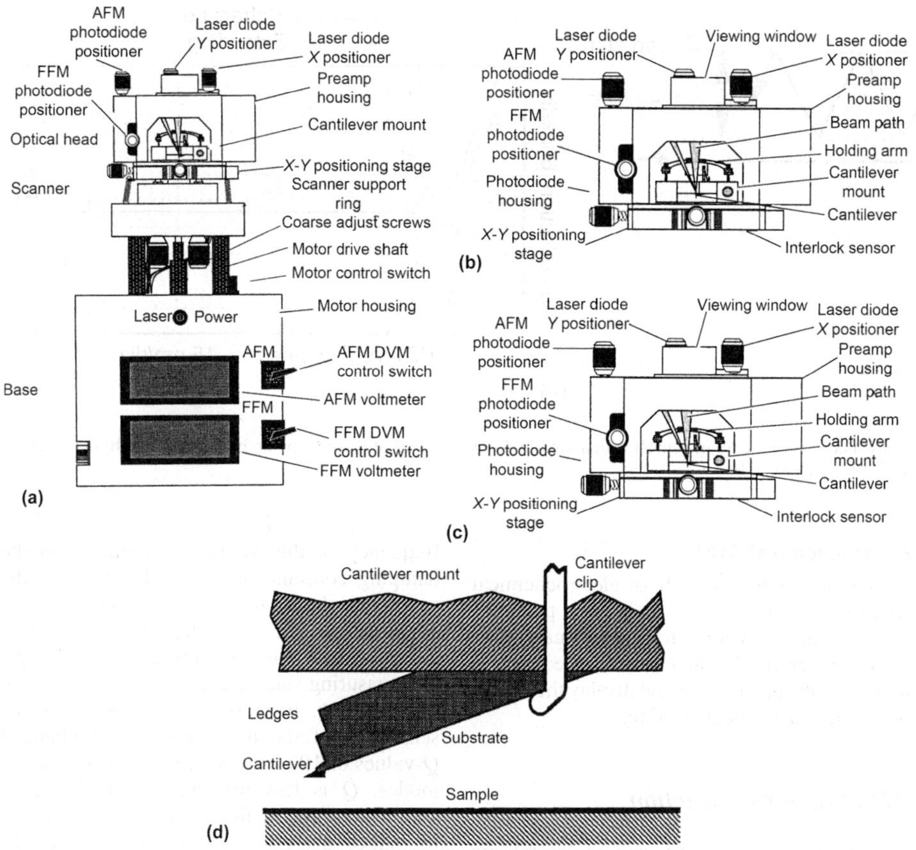

Fig. 11 Schematics of a commercial multimode atomic force microscope (AFM)/friction force microscope (FFM) made by Bruker Instruments, Inc. (a) Front view. (b) Optical head. (c) Base. (d) Cantilever substrate mounted on cantilever mount (not to scale). DVM, digital voltmeter

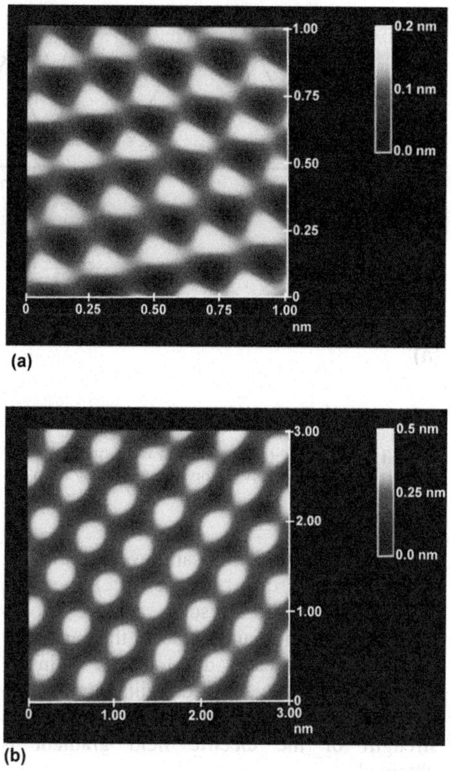

Fig. 12 Typical atomic force microscopy images of (a) freshly cleaved, highly oriented pyrolytic graphite and (b) mica surfaces taken using a square pyramidal Si_3N_4 tip

used for small scan lengths (0.7 μm). Scan rates of 0.5 to 2.5 Hz should be used for large scans on samples with tall features. High scan rates help reduce drift, but they can only be used on flat samples with small scan sizes. Scan rate, or scanning speed in length/time in the fast scan direction, is equal to twice the scan length times the scan rate in hertz, and in the slow direction, it is equal to the scan length times the scan rate in hertz divided by the number of data points in the transverse direction. For example, for 10 μm by 10 μm scan size scanned at 0.5 Hz, the scan rates in the fast and slow scan directions are 10 μm/s and 20 nm/s, respectively. Normally 256 × 256 data points are taken for each image. The lateral resolution at larger scans is approximately equal to the scan length divided by 256. The piezo tube requires *x-y* calibration, which is carried out by imaging an appropriate calibration standard. Cleaved graphite is used for small scan heads, while two-dimensional grids (a gold-plated ruling) can be used for longer-range heads.

Examples of AFM images of freshly cleaved, highly oriented pyrolytic graphite and mica surfaces are shown in Fig. 12 (Ref 61, 133, 137). Images with near-atomic resolution are obtained.

Force-calibration mode is used to study the interaction between the cantilever and the sample surface. In the force-calibration mode, the *X* and *Y* voltages applied to the piezo tube are held at zero, and a sawtooth voltage is applied to the *Z* electrode of the piezo tube (Fig. 13a). The force measurement starts with the sample far away and the cantilever in its rest position. As a result of the applied voltage, the sample is moved up and down relative to the stationary cantilever tip. As the piezo moves the sample up and down, the cantilever deflection signal from the photodiode is monitored. The force-distance curve, a plot of the cantilever tip deflection signal as a function of the voltage applied to the piezo tube, is obtained. Figure 13(b) is a typical force-distance curve showing the various features of the curve. The arrow heads reveal the direction of piezo travel. As the piezo extends, it approaches the tip, which is at this point in free air and hence shows no deflection. This is indicated by the flat portion of the curve. As the tip approaches the sample within a few nanometers (point *A*), an attractive force exists between the atoms of the tip surface and the atoms of the sample surface. The tip is pulled toward the sample, and contact occurs at point *B* on the graph. From this point on, the tip is in contact with the surface, and as the piezo further extends, the tip becomes further deflected. This is represented by the sloped portion of the curve. As the piezo retracts, the tip goes beyond the zero deflection (flat) line into the adhesive regime because of attractive forces (van der Waals forces and long-range meniscus forces). At point *C* in the graph, the tip snaps free of the adhesive forces and is in free air again. The horizontal distance between points *B* and *C* along the retrace line gives the distance moved by the tip in the adhesive regime. This distance multiplied by the stiffness of the cantilever gives the adhesive force. Incidentally, the horizontal shift between the loading and unloading curves results from the hysteresis in the PZT tube (Ref 12).

Multimode Capabilities

The multimode AFM can be used for topography measurements in the contact mode and amplitude or tapping mode, as described earlier. It also can be used for measurements of lateral (friction) force, electric force gradients, and magnetic force gradients.

The multimode AFM using a grounded conducting tip can measure electric field gradients by oscillating the tip near its resonant frequency. When the lever encounters a force gradient from the electric field, the effective

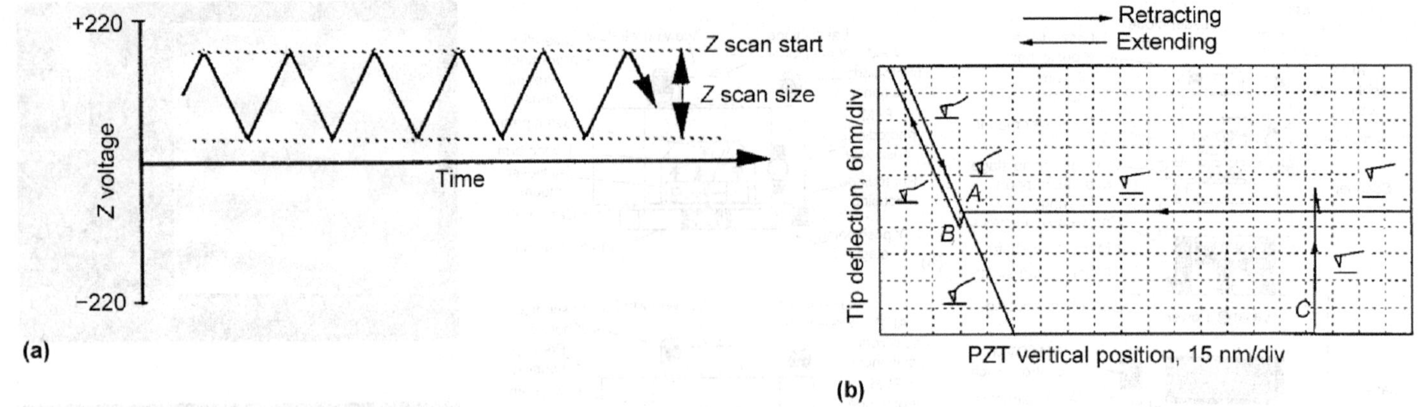

Fig. 13 (a) Force-calibration *Z* waveform. (b) Typical force-distance curve for a tip in contact with a sample. Contact occurs at point *B*; the tip breaks free of adhesive forces at point *C* as the sample moves away from the tip. PZT, piezoelectric tube

spring constant of the cantilever is altered, changing its resonant frequency. Depending on which side of the resonance curve is chosen, the oscillation amplitude of the cantilever increases or decreases due to the shift in the resonant frequency. By recording the amplitude of the cantilever, an image revealing the strength of the electric field gradient is obtained.

In a magnetic force microscope (MFM) used with a magnetically coated tip, static cantilever deflection is detected that occurs when a magnetic field exerts a force on the tip, and the MFM images of magnetic materials can be produced. The MFM sensitivity can be enhanced by oscillating the cantilever near its resonant frequency. When the tip encounters a magnetic force gradient, the effective spring constant, and hence the resonant frequency, is shifted. By driving the cantilever above or below the resonant frequency, the oscillation amplitude varies as the resonance shifts. An image of magnetic field gradients is obtained by recording the oscillation amplitude as the tip is scanned over the sample.

Topographic information is separated from the electric field gradients and magnetic field images by using a so-called lift mode. Measurements in lift mode are taken in two passes over each scan line. On the first pass, topographical information is recorded in the standard tapping mode where the oscillating cantilever lightly taps the surface. On the second pass, the tip is lifted to a user-selected separation (typically 20 to 200 nm) between the tip and local surface topography. By using the stored topographical data instead of the standard feedback, the separation remains constant without sensing the surface. At this height, cantilever amplitudes are sensitive to electric field force gradients or relatively weak but long-range magnetic forces without being influenced by topographic features. Two-pass measurements are taken for every scan line, producing separate topographic and magnetic force images.

Electrochemical AFM

This option allows study of electrochemical reactions in the AFM. It includes a potentiostat, a fluid cell with a transparent cantilever holder and electrodes, and the software required to operate the potentiostat and display the results of the electrochemical reaction.

AFM Probe Construction

Various probes (cantilevers and tips) are used for AFM studies. The cantilever stylus used in the AFM should meet the following criteria (Ref 159):

- Low normal spring constant (stiffness)
- High resonant frequency
- High quality factor of the cantilever, *Q*
- High lateral spring constant (stiffness)
- Short cantilever length
- Incorporation of components (such as a mirror) for deflection sensing
- Sharp protruding tip

To register a measurable deflection with small forces, the cantilever must flex with a relative low force (on the order of a few nanonewtons), requiring vertical spring constants of 10^{-2} to 10^{2} N/m (0.0007 to 7.0 lbf/ft) for atomic resolution in the contact-profiling mode. The data rate or imaging rate in the AFM is limited by the mechanical resonant frequency of the cantilever. To achieve a large imaging bandwidth, the AFM cantilever should have resonant frequency greater than approximately10 kHz (preferably 30 to 100 kHz) in order to make the cantilever the least sensitive part of the system. Fast imaging rates are not just a matter of convenience, because the effects of thermal drifts are more pronounced with slow scanning speeds. The combined requirements of a low spring constant and a high resonant frequency are met by reducing the mass of the cantilever. The quality factor Q $(= \omega_R/(c/m)$, where ω_R is the resonant frequency of the damped oscillator, c is the damping constant, and m is the mass of the oscillator) should have a high value for some applications. For example, resonance curve detection is a sensitive modulation technique for measuring small force gradients in noncontact imaging. Increasing the Q increases the sensitivity of the measurements. Mechanical Q-values of 100 to 1000 are typical. In contact modes, Q is less important. A high lateral spring constant in the cantilever is desirable to reduce the effect of lateral forces in the AFM, because frictional forces can cause appreciable lateral bending of the cantilever. Lateral bending results in error in the topography measurements. For friction measurements, cantilevers with less lateral rigidity are preferred. A sharp protruding tip must be formed at the end of the cantilever to provide a well-defined interaction with the sample over a small area. The tip radius should be much smaller than the radii of corrugations in the sample for these to be measured accurately. The lateral spring constant depends critically on the tip length. Additionally, the tip should be centered at the free end.

In the past, cantilevers have been cut by hand from thin metal foils or formed from fine wires. Tips for these cantilevers were prepared by attaching diamond fragments to the ends of the cantilevers by hand or, in the case of wire cantilevers, electrochemically etching the wire to a sharp point. Several cantilever geometries for wire cantilevers have been used. The simplest geometry is the L-shaped cantilever, usually made by bending a wire at a 90° angle. Other geometries include single-V and double-V geometries with a sharp tip attached at the apex of the V, and the double-X configuration with a sharp tip attached at the intersection (Ref 36, 160). These cantilevers can be constructed with high vertical spring constants. For example, a double-cross cantilever with an effective spring constant of 250 N/m (17 lbf/ft) was used by Burnham and Colton (Ref 36). The small size and low mass needed in the

AFM make hand fabrication of the cantilever a difficult process, with poor reproducibility. Conventional microfabrication techniques are ideal for constructing planar thin-film structures that have submicrometer lateral dimensions. Triangular (V-shaped) cantilevers have improved (higher) lateral spring constants in comparison to rectangular cantilevers. In terms of spring constants, triangular cantilevers are approximately equivalent to two rectangular cantilevers in parallel (Ref 159). Although the macroscopic radius of a photolithographically patterned corner is seldom much less than approximately 50 nm, microscopic asperities on the etched surface provide tips with near-atomic dimensions.

Silicon nitride cantilevers with integrated tips are less expensive than those made of other materials. They are very rugged and well suited to imaging in almost all environments. They are especially compatible to organic and biological materials. Microfabricated silicon nitride triangular beams with integrated square pyramidal tips made of plasma-enhanced chemical vapor deposition are the most commonly used (Ref 159). Figure 14(a) shows a schematic example of four cantilever beams, each with different sizes and spring constants, arranged in pairs, attached to a borosilicate glass substrate (Pyrex), and marketed by various manufacturers, including Bruker Instruments, Inc. Each pair of triangular cantilevers measures 115 and 193 μm from the substrate to the apex, with base widths of 122 and 205 μm, respectively. These cantilevers are commercially available with wide (top) and narrow (bottom) legs, all in the same thickness of 0.6 μm. Calculated spring constant and measured natural frequencies for each of the configurations are listed in Table 1. The most commonly used cantilever beam is the 115 μm long, wide-legged cantilever (vertical spring constant = 0.58 N/m, or 0.04 lbf/ft). Cantilevers with smaller spring constants should be used on softer samples. Pyramidal tips are highly symmetric, with ends having a radius of approximately 20 to 50 nm. The tip side walls have a slope of 35°, and the length of the edges

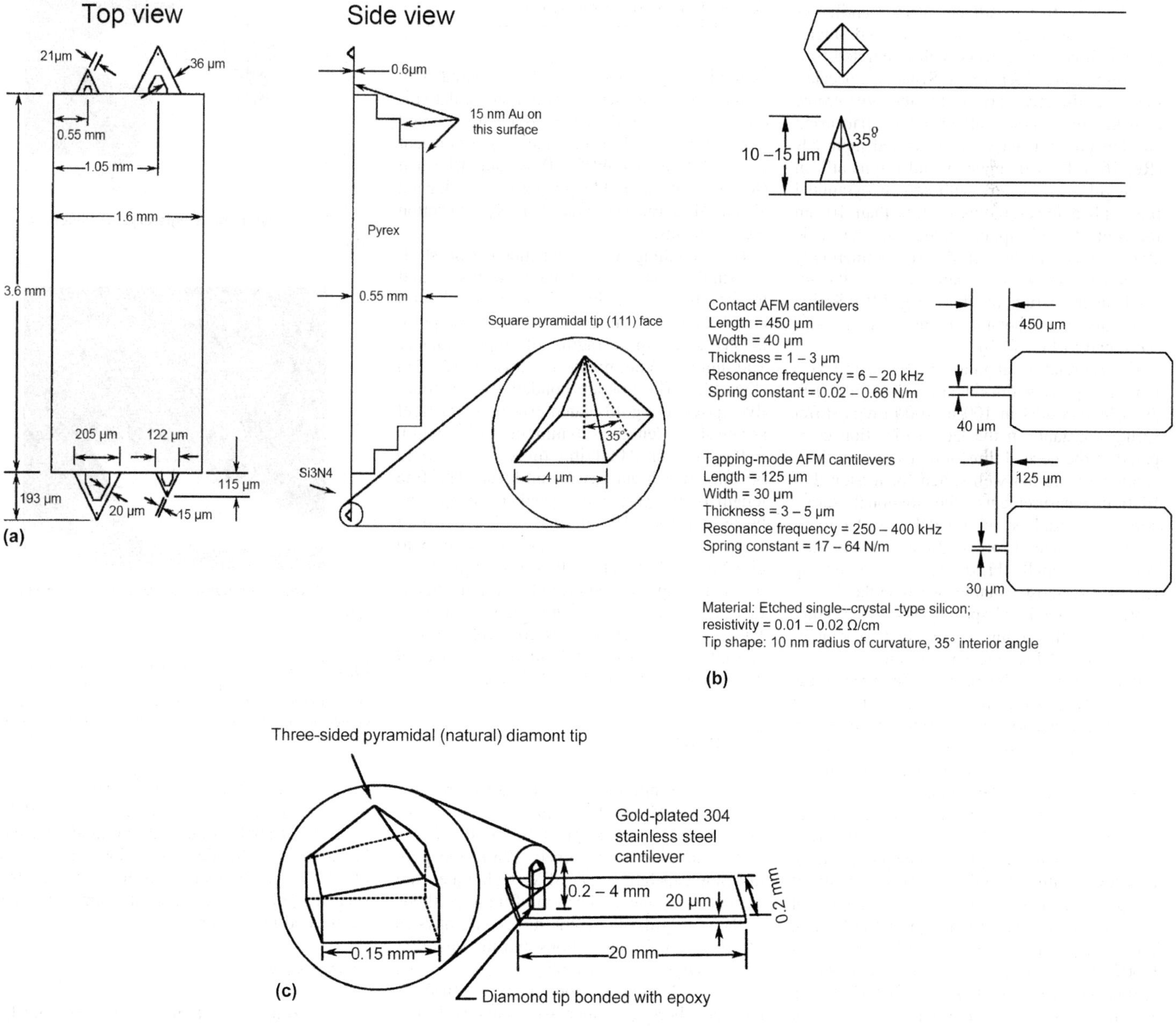

Fig. 14 Schematics of (a) triangular cantilever beam with square pyramidal tips made of plasma-enhanced chemical vapor deposition Si_3N_4, (b) rectangular cantilever beams with square pyramidal tips made of etched single-crystal silicon, and (c) rectangular cantilever stainless steel beam with three-sided pyramidal natural diamond tip. AFM, atomic force microscope

Table 1 Measured vertical spring constants and natural frequencies of triangular (V-shaped) cantilevers made of plasma-enhanced chemical vapor deposition Si_3N_4

Cantilever dimension	Spring constant (k_z) N/m	Spring constant (k_z) lbf/ft	Natural frequency (ω_0), kHz
115 µm long, narrow leg	0.38	0.026	40
115 µm long, wide leg	0.58	0.040	40
193 µm long, narrow leg	0.06	0.004	13–22
193 µm long, wide leg	0.12	0.008	13–22

Data provided by Bruker Instruments, Inc.

Table 2 Vertical (k_z), lateral (k_y), and torsional (k_{yT}) spring constants of rectangular cantilevers made of silicon (IBM Corp.) and plasma-enhanced chemical vapor deposition Si_3N_4

Dimensions/stiffness	Silicon cantilever	Si_3N_4 cantilever
Length (L), µm	100	100
Width (b), µm	10	20
Thickness (h), µm	1	0.6
Tip length (ℓ), µm	5	3
k_z, N/m (lbf/ft)	0.4 (0.03)	0.15 (0.01)
k_y, N/m (lbf/ft)	40 (2.7)	175 (12)
k_{yT}, N/m (lbf/ft)	120 (8.2)	116 (8.0)
ω_0, kHz	~90	~65

Note: $k_z = Ebh^3/4L^3$, $k_y = Eb^3h/4\ell^3$, $k_{yT} = Gbh^3/3L\ell^2$, and $\omega_o = [k_z/(m_c + 0.24\ bhL\rho)]^{1/2}$, where E is Young's modulus, G is the modulus of rigidity [$= E/2(1 + \nu)$, where ν is Poisson's ratio], ρ is the mass density of the cantilever, and m_c is the concentrated mass of the tip (~4 ng) (Ref 115). For silicon, E = 130 GPa (19 × 10^6 psi), ρg = 2300 kg/m^3, and ν = 0.3. For Si_3N_4, E = 150 GPa (22 × 10^6 psi), ρg = 3100 kg/m^3, and ν = 0.3. Data provided by Bruker Instruments, Inc.

of the tip at the cantilever base is approximately 4 µm.

An alternative to silicon nitride cantilevers with integrated tips is microfabricated single-crystal silicon cantilevers with integrated tips. Silicon tips are sharper than Si_3N_4 tips because they are directly formed by the anisotropic etch in single-crystal silicon rather than using an etch pit as a mask for deposited materials (Ref 161). Etched single-crystal n-type silicon rectangular cantilevers with square pyramidal tips with a lower radius of less than 10 nm for contact- and tapping-mode (tapping-mode etched-silicon probe) AFMs are commercially available from various manufacturers, including Bruker Instruments, Inc. (Fig. 14b). Spring constants and resonant frequencies are also presented in Fig. 14(b).

Commercial triangular Si_3N_4 cantilevers have a typical width-thickness ratio of 10 to 30, which results in 100 to 1000 times stiffer spring constants in the lateral direction compared to the normal direction. Therefore, these cantilevers are not well suited for torsion. For friction measurements, the torsional spring constant should be minimized in order to be sensitive to lateral forces. Rather long cantilevers with small thicknesses and large tip lengths are most suitable. Rectangular beams have lower torsional spring constants in comparison to the triangular (V-shaped) cantilevers. Table 2 lists the spring constants (with full length of the beam used) in three directions of typical rectangular beams. It is noted that lateral and torsional spring constants are approximately 2 orders of magnitude larger than the normal spring constants. A cantilever beam required for the tapping mode is quite stiff and may not be sensitive enough for friction measurements. Meyer et al. (Ref 162) used a specially designed rectangular silicon cantilever with a length of 200 µm, width of 21 µm, thickness of 0.4 µm, tip length of 12.5 µm, and shear modulus of 50 GPa (7 × 10^6 psi), giving a normal spring constant of 0.007 N/m (0.0005 lbf/ft) and a torsional spring constant of 0.72 N/m (0.05 lbf/ft), which gives a lateral force sensitivity of 10 pN and an angle of resolution of 10^{-7} rad. With this particular geometry, sensitivity to lateral forces could be improved by approximately a factor of 100 compared with commercial V-shaped Si_3N_4 or rectangular silicon or Si_3N_4 cantilevers used by Meyer and Amer (Ref 9) with a torsional spring constant of ~100 N/m (6.9 lbf/ft). Ruan and Bhushan (Ref 10, 158) used 115 µm long, wide-legged V-shaped cantilevers made of Si_3N_4 for friction measurements.

For scratching, wear, and indentation studies, single-crystal natural diamond tips ground to the shape of a three-sided pyramid with an apex angle of either 60 or 80° whose point is sharpened to a radius of approximately 100 nm are commonly used (Ref 12, 38) (Fig. 14c). The tips are bonded with conductive epoxy to a gold-plated 304 stainless steel spring sheet (length = 20 mm, or 0.8 in.; width = 0.2 mm, or 0.008 in.; thickness = 20 to 60 µm), which acts as a cantilever. The free length of the spring is varied to change the beam stiffness. The normal spring constant of the beam ranges from approximately 5 to 600 N/m (0.35 to 41 lbf/ft) for a 20 µm thick beam. The tips are produced by various manufacturers, including R-DEC Co., Tsukuba, Japan. Scanning electron micrographs of silicon nitride, single-crystal silicon, and natural diamond tips are shown in Fig. 15.

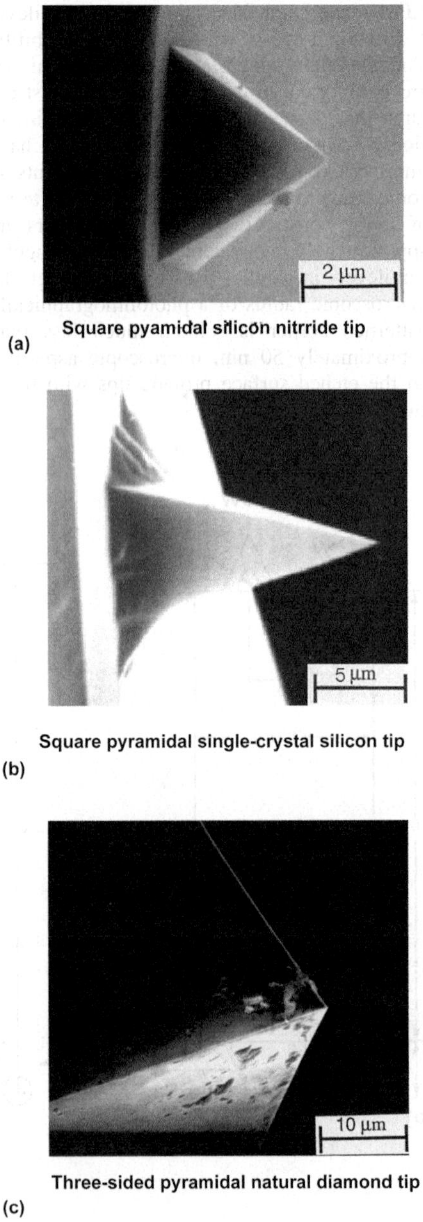

Fig. 15 Scanning electron micrographs of (a) square pyramidal plasma-enhanced chemical vapor deposition Si_3N_4 tip, (b) square pyramidal etched single-crystal silicon tip, and (c) three-sided pyramidal natural diamond tip

Summary

Since the introduction of the STM in 1981 and the AFM in 1985, many variations of probe-based microscopies, referred to as SPMs, have been developed. While the pure imaging capabilities of SPM techniques are dominated by the application of these methods at their early development stages, the physics of probe-sample interactions and the quantitative analyses of tribological, electronic, magnetic, biological, and chemical surfaces have now become of increasing interest. Nanoscale science and technology are strongly driven by SPMs, which allow investigation and manipulation of surfaces down to the atomic scale. With growing understanding of the underlying interaction mechanisms, SPMs have found application in many fields outside basic research. In addition, various derivatives of all these methods have been developed for special applications, some of them targeting far beyond microscopy.

REFERENCES

1. G. Binnig, H. Rohrer, C. Gerber, and E. Weibel, Surface Studies by Scanning Tunneling Microscopy, *Phys. Rev. Lett.*, Vol 49, 1982, p 57–61

2. G. Binnig, C.F. Quate, and C. Gerber, Atomic Force Microscope, *Phys. Rev. Lett.*, Vol 56, 1986, p 930–933
3. G. Binnig, C. Gerber, E. Stoll, T.R. Albrecht, and C.F. Quate, Atomic Resolution with Atomic Force Microscope, *Europhys. Lett.*, Vol 3, 1987, p 1281–1286
4. B. Bhushan, *Nanotribology and Nanomechanics: An Introduction,* 4th ed., Springer International, Cham, Switzerland, 2017
5. B. Bhushan, *Springer Handbook of Nanotechnology,* 4th ed., Springer International, Cham, Switzerland, 2017
6. C.M. Mate, G.M. McClelland, R. Erlandsson, and S. Chiang, Atomic-Scale Friction of a Tungsten Tip on a Graphite Surface, *Phys. Rev. Lett.,* Vol 59, 1987, p 1942–1945
7. R. Erlandsson, G.M. McClelland, C.M. Mate, and S. Chiang, Atomic Force Microscopy Using Optical Interferometry, *J. Vac. Sci. Technol. A,* Vol 6, 1988, p 266–270
8. O. Marti, J. Colchero, and J. Mlynek, Combined Scanning Force and Friction Microscopy of Mica, *Nanotechnology,* Vol 1, 1990, p 141–144
9. G. Meyer and N.M. Amer, Simultaneous Measurement of Lateral and Normal Forces with an Optical-Beam-Deflection Atomic Force Microscope, *Appl. Phys. Lett.*, Vol 57, 1990, p 2089–2091
10. B. Bhushan and J. Ruan, Atomic-Scale Friction Measurements Using Friction Force Microscopy: Part II—Application to Magnetic Media, *ASME J. Tribol.*, Vol 116, 1994, p 389–396
11. B. Bhushan, J.N. Israelachvili, and U. Landman, Nanotribology: Friction, *Wear, and Lubrication at the Atomic Scale, Nature*, Vol 374, 1995, p 607–616
12. B. Bhushan, *Handbook of Micro/Nanotribology,* 2nd ed., CRC Press, Boca Raton, FL, 1999
13. B. Bhushan, *Nanotribology and Nanomechanics I and II,* 3rd ed., Springer-Verlag, Heidelberg, Germany, 2011
14. S. Fujisawa, M. Ohta, T. Konishi, Y. Sugawara, and S. Morita, Difference between the Forces Measured by an Optical Lever Deflection and by an Optical Interferometer in an Atomic Force Microscope, *Rev. Sci. Instrum.,* Vol 65, 1994, p 644–647
15. S. Fujisawa, E. Kishi, Y. Sugawara, and S. Morita, Fluctuation in 2-Dimensional Stick-Slip Phenomenon Observed with 2-Dimensional Frictional Force Microscope, *Jpn. J. Appl. Phys. 1,* Vol 33, 1994, p 3752–3755
16. S. Grafstrom, J. Ackermann, T. Hagen, R. Neumann, and O. Probst, Analysis of Lateral Force Effects on the Topography in Scanning Force Microscopy, *J. Vac. Sci. Technol. B,* Vol 12, 1994, p 1559–1564
17. R.M. Overney, H. Takano, M. Fujihira, W. Paulus, and H. Ringsdorf, Anisotropy in Friction and Molecular Stick-Slip Motion, *Phys. Rev. Lett.,* Vol 72, 1994, p 3546–3549
18. R.J. Warmack, X.Y. Zheng, T. Thundat, and D.P. Allison, Friction Effects in the Deflection of Atomic Force Microscope Cantilevers, *Rev. Sci. Instrum.,* Vol 65, 1994, p 394–399
19. N.A. Burnham, D.D. Domiguez, R.L. Mowery, and R.J. Colton, Probing the Surface Forces of Monolayer Films with an Atomic Force Microscope, *Phys. Rev. Lett.,* Vol 64, 1990, p1931–1934
20. N.A. Burham, R.J. Colton, and H.M. Pollock, Interpretation Issues in Force Microscopy, *J. Vac. Sci. Technol. A,* Vol 9, 1991, p 2548–2556
21. C.D. Frisbie, L.F. Rozsnyai, A. Noy, M.S. Wrighton, and C.M. Lieber, Functional Group Imaging by Chemical Force Microscopy, *Science,* Vol 265, 1994, p 2071–2074
22. V.N. Koinkar and B. Bhushan, Microtribological Studies of Unlubricated and Lubricated Surfaces Using Atomic Force/Friction Force Microscopy, *J. Vac. Sci. Technol. A,* Vol 14, 1996, p 2378–2391
23. V. Scherer, B. Bhushan, U. Rabe, and W. Arnold, Local Elasticity and Lubrication Measurements Using Atomic Force and Friction Force Microscopy at Ultrasonic Frequencies, *IEEE Trans. Mag.,* Vol 33, 1997, p 4077–4079
24. V. Scherer, W. Arnold, and B. Bhushan, Lateral Force Microscopy Using Acoustic Friction Force Microscopy, *Surf. Interface Anal.,* Vol 27, 1999, p 578–587
25. B. Bhushan and S. Sundararajan, Micro/Nanoscale Friction and Wear Mechanisms of Thin Films Using Atomic Force and Friction Force Microscopy, *Acta Mater.,* Vol 46, 1998, p 3793–3804
26. U. Krotil, T. Stifter, H. Waschipky, K. Weishaupt, S. Hild, and O. Marti, Pulse Force Mode: A New Method for the Investigation of Surface Properties, *Surf. Interface Anal.,* Vol 27, 1999, p 336–340
27. B. Bhushan and C. Dandavate, Thin-Film Friction and Adhesion Studies Using Atomic Force Microscopy, *J. Appl. Phys.,* Vol 87, 2000, p 1201–1210
28. B. Bhushan, *Micro/Nanotribology and Its Applications,* Vol E330, Kluwer Academic Publishers, Dordrecht, Netherlands, 1997
29. B. Bhushan, *Principles of Tribology,* Vol 1, *Modern Tribology Handbook,* CRC Press, Boca Raton, FL, 2001
30. B. Bhushan, Fundamentals of Tribology and Bridging the Gap between the Macro- and Micro/Nanoscales, *NATO Science Series II,* Vol 10, Kluwer Academic Publishers, Dordrecht, Netherlands, 2001
31. B. Bhushan, Nanotribology and Nanomechanics, *Wear,* Vol 259, 2005, p1507–1531
32. B. Bhushan, Nanotribology, Nanomechanics and Nanomaterials Characterization, *Philos. Trans. R. Soc. A,* Vol 366, 2008, p 1351–1381
33. B. Bhushan, *Principles and Applications of Tribology,* 2nd ed., Wiley, New York, 2013
34. B. Bhushan, *Introduction to Tribology,* 2nd ed., Wiley, New York, 2013
35. M. Reinstaedtler, U. Rabe, V. Scherer, U. Hartmann, A. Goldade, B. Bhushan, and W. Arnold, On the Nanoscale Measurement of Friction Using Atomic Force Microscope Cantilever Torsional Resonances, *Appl. Phys. Lett.,* Vol 82, 2003, p 2604–2606
36. N.A. Burnham and R.J. Colton, Measuring the Nanomechanical Properties and Surface Forces of Materials Using an Atomic Force Microscope, *J. Vac. Sci. Technol. A,* Vol 7, 1989, p 2906–2913
37. P. Maivald, H.J. Butt, S.A.C. Gould, C.B. Prater, B. Drake, J.A. Gurley, V.B. Elings, and P.K. Hansma, Using Force Modulation to Image Surface Elasticities with the Atomic Force Microscope, *Nanotechnology,* Vol 2, 1991, p 103–106
38. B. Bhushan, V.N. Koinkar, and J. Ruan, Microtribology of Magnetic Media, *Proc. Inst. Mech. Eng. J, J. Eng. Tribol.,* Vol 208, 1994, p 17–29
39. B. Bhushan, A.V. Kulkarni, W. Bonin, and J.T. Wyrobek, Nano/Picoindentation Measurements Using Capacitive Transducer in Atomic Force Microscopy, *Philos. Mag. A,* Vol 74, 1996, p 1117–1128
40. B. Bhushan and V.N. Koinkar, Nanoindentation Hardness Measurements Using Atomic Force Microscopy, *Appl. Phys. Lett.,* Vol 75, 1994, p 5741–5746
41. D. DeVecchio and B. Bhushan, Localized Surface Elasticity Measurements Using an Atomic Force Microscope, *Rev. Sci. Instrum.,* Vol 68, 1997, p 4498–4505
42. B. Bhushan, Wear and Mechanical Characterisation on Micro-to-Picoscales Using AFM, *Int. Mater. Rev.,* Vol 44, 1999, p 105–117
43. B. Bhushan, Nano- to Microscale Wear and Mechanical Characterization Studies Using Scanning Probe Microscopy, *Wear,* Vol 251, 2001, p 1105–1123
44. S. Amelio, A.V. Goldade, U. Rabe, V. Scherer, B. Bhushan, and W. Arnold, Measurements of Mechanical Properties of Ultra-Thin Diamond-Like Carbon Coatings Using Atomic Force Acoustic Microscopy, *Thin Solid Films,* Vol 392, 2001, p 75–84
45. B. Bhushan and J. Qi, Phase Contrast Imaging of Nanocomposites and Molecularly-Thick Lubricant Films in Magnetic

Media, *Nanotechnology,* Vol 14, 2003, p 886–895
46. T. Kasai, B. Bhushan, L. Huang, and C. Su, Topography and Phase Imaging Using the Torsional Resonance Mode, *Nanotechnology,* Vol 15, 2004, p 731–742
47. M. Reinstaedtler, T. Kasai, U. Rabe, B. Bhushan, and W. Arnold, Imaging and Measurement of Elasticity and Friction Using the TR Mode, *J. Phys. D: Appl. Phys.,* Vol 38, 2005, p R269–R282
48. D.M. Eigler and E.K. Schweizer, Positioning Single Atoms with a Scanning Tunnelling Microscope, *Nature,* Vol 344, 1990, p 524–528
49. A.L. Weisenhorn, J.E. MacDougall, J.A. C. Gould, S.D. Cox, W.S. Wise, J. Massie, P. Maivald, V.B. Elings, G.D. Stucky, and P.K. Hansma, Imaging and Manipulating of Molecules on a Zeolite Surface with an Atomic Force Microscope, *Science,* Vol 247, 1990, p 1330–1333
50. I.W. Lyo and P. Avouris, Field-Induced Nanometer-to-Atomic-Scale Manipulation of Silicon Surfaces with the STM, *Science,* Vol 253, 1991, p 173–176
51. O.M. Leung and M.C. Goh, Orientation Ordering of Polymers by Atomic Force Microscope Tip-Surface Interactions, *Science,* Vol 225, 1992, p 64–66
52. D.W. Abraham, H.J. Mamin, E. Ganz, and J. Clark, Surface Modification with the Scanning Tunneling Microscope, *IBM J. Res. Develop.,* Vol 30, 1986, p 492–499
53. R.M. Silver, E.E. Ehrichs, and A.L. Lozannede, Direct Writing of Submicron Metallic Features with a Scanning Tunnelling Microscope, *Appl. Phys. Lett.,* Vol 51, 1987, p 247–249
54. A. Kobayashi, F. Grey, R.S. Williams, and M. Ano, Formation of Nanometer-Scale Grooves in Silicon with a Scanning Tunneling Microscope, *Science,* Vol 259, 1993, p 1724–1726
55. B. Parkinson, Layer-by-Layer Nanometer Scale Etching of Two-Dimensional Substrates Using the Scanning Tunneling Microscopy, *J. Am. Chem. Soc.,* Vol 112, 1990, p 7498–7502
56. A. Majumdar, P.I. Oden, J.P. Carrejo, L. A. Nagahara, J.J. Graham, and J. Alexander, Nanometer-Scale Lithography Using the Atomic Force Microscope, *Appl. Phys. Lett.,* Vol 61, 1992, p 2293–2295
57. B. Bhushan, Micro/Nanotribology and Its Applications to Magnetic Storage Devices and MEMS, *Tribol. Int.,* Vol 28, 1995, p 85–96
58. L. Tsau, D. Wang, and K.L. Wang, Nanometer Scale Patterning of Silicon (100) Surface by an Atomic Force Microscope Operating in Air, *Appl. Phys. Lett.,* Vol 64, 1994, p 2133–2135
59. E. Delawski and B.A. Parkinson, Layer-by-Layer Etching of Two-Dimensional Metal Chalcogenides with the Atomic Force Microscope, *J. Am. Chem. Soc.,* Vol 114, 1992, p 1661–1667
60. B. Bhushan and G.S. Blackman, Atomic Force Microscopy of Magnetic Rigid Disks and Sliders and Its Applications to Tribology, *ASME J. Tribol.,* Vol 113, 1991, p 452–458
61. O. Marti, B. Drake, and P.K. Hansma, Atomic Force Microscopy of Liquid-Covered Surfaces: Atomic Resolution Images, *Appl. Phys. Lett.,* Vol 51, 1987, p 484–486
62. B. Drake, C.B. Prater, A.L. Weisenhorn, S.A.C. Gould, T.R. Albrecht, C.F. Quate, D.S. Cannell, H.G. Hansma, and P.K. Hansma, Imaging Crystals, Polymers and Processes in Water with the Atomic Force Microscope, *Science,* Vol 243, 1989, p 1586–1589
63. M. Binggeli, R. Christoph, H.E. Hintermann, J. Colchero, and O. Marti, Friction Force Measurements on Potential Controlled Graphite in an Electrolytic Environment, *Nanotechnology,* Vol 4, 1993, p 59–63
64. G. Meyer and N.M. Amer, Novel Optical Approach to Atomic Force Microscopy, *Appl. Phys. Lett.,* Vol 53, 1988, p 1045–1047
65. J.H. Coombs and J.B. Pethica, Properties of Vacuum Tunneling Currents: Anomalous Barrier Heights, *IBM J. Res. Dev.,* Vol 30, 1986, p 455–459
66. M.D. Kirk, T. Albrecht, and C.F. Quate, Low-Temperature Atomic Force Microscopy, *Rev. Sci. Instrum.,* Vol 59, 1988, p 833–835
67. F.J. Giessibl, C. Gerber, and G. Binnig, A Low-Temperature Atomic Force/Scanning Tunneling Microscope for Ultrahigh Vacuum, *J. Vac. Sci. Technol. B,* Vol 9, 1991, p 984–988
68. T.R. Albrecht, P. Grutter, D. Rugar, and D.P.E. Smith, Low Temperature Force Microscope with All-Fiber Interferometer, *Ultramicroscopy,* Vol 42–44, 1992, p 1638–1646
69. H.J. Hug, A. Moser, T. Jung, O. Fritz, A. Wadas, I. Parashikor, and H.J. Guntherodt, Low Temperature Magnetic Force Microscopy, *Rev. Sci. Instrum.,* Vol 64, 1993, p 2920–2925
70. C. Basire and D.A. Ivanov, Evolution of the Lamellar Structure during Crystallization of a Semicrystalline-Amorphous Polymer Blend: Time-Resolved Hot-Stage SPM Study, *Phys. Rev. Lett.,* Vol 85, 2000, p 5587–5590
71. H. Liu and B. Bhushan, Investigation of Nanotribological Properties of Self-Assembled Monolayers with Alkyl and Biphenyl Spacer Chains, *Ultramicroscopy,* Vol 91, 2002, p 185–202
72. J. Foster and J. Frommer, Imaging of Liquid Crystal Using a Tunneling Microscope, *Nature,* Vol 333, 1988, p 542–547
73. D. Smith, H. Horber, C. Gerber, and G. Binnig, Smectic Liquid Crystal Monolayers on Graphite Observed by Scanning Tunneling Microscopy, *Science,* Vol 245, 1989, p 43–45
74. D. Smith, J. Horber, G. Binnig, and H. Nejoh, Structure, Registry and Imaging Mechanism of Alkylcyanobiphenyl Molecules by Tunnelling Microscopy, *Nature,* Vol 344, 1990, p 641–644
75. Y. Andoh, S. Oguchi, R. Kaneko, and T. Miyamoto, Evaluation of Very Thin Lubricant Films, *J. Phys. D: Appl. Phys.,* Vol 25, 1992, p A71–A75
76. Y. Martin, C.C. Williams, and H.K. Wickramasinghe, Atomic Force Microscope—Force Mapping and Profiling on a Sub 100-A Scale, *J. Appl. Phys.,* Vol 61, 1987, p 4723–4729
77. J.E. Stern, B.D. Terris, H.J. Mamin, and D. Rugar, Deposition and Imaging of Localized Charge on Insulator Surfaces Using a Force Microscope, *Appl. Phys. Lett.,* Vol 53, 1988, p 2717–2719
78. K. Yamanaka, H. Ogisco, and O. Kolosov, Ultrasonic Force Microscopy for Nanometer Resolution Subsurface Imaging, *Appl. Phys. Lett.,* Vol 64, 1994, p 178–180
79. K. Yamanaka and E. Tomita, Lateral Force Modulation Atomic Force Microscope for Selective Imaging of Friction Forces, *Jpn. J. Appl. Phys.,* Vol 34, 1995, p 2879–2882
80. U. Rabe, K. Janser, and W. Arnold, Vibrations of Free and Surface-Coupled Atomic Force Microscope: Theory and Experiment, *Rev. Sci. Instrum.,* Vol 67, 1996, p 3281–3293
81. Y. Martin and H.K. Wickramasinghe, Magnetic Imaging by Force Microscopy with 1000 A Resolution, *Appl. Phys. Lett.,* Vol 50, 1987, p 1455–1457
82. D. Rugar, H.J. Mamin, and P. Guethner, Improved Fiber-Optical Interferometer for Atomic Force Microscopy, *Appl. Phys. Lett.,* Vol 55, 1989, p 2588–2590
83. C. Schoenenberger and S.F. Alvarado, Understanding Magnetic Force Microscopy, *Z. Phys. B.,* Vol 80, 1990, p 373–383
84. U. Hartmann, Magnetic Force Microscopy, *Annu. Rev. Mater. Sci.,* Vol 29, 1999, p 53–87
85. A. Avila and B. Bhushan, Electrical Measurement Techniques in Atomic Force Microscopy, (Invited), *Crit. Rev. Solid State Mater. Sci.,* Vol 35, 2010, p 38–51
86. D.W. Pohl, W. Denk, and M. Lanz, Optical Stethoscopy-Image Recording with Resolution Lambda/20, *Appl. Phys. Lett.,* Vol 44, 1984, p 651–653
87. E. Betzig, J.K. Troutman, T.D. Harris, J. S. Weiner, and R.L. Kostelak, Breaking the Diffraction Barrier—Optical Microscopy on a Nanometric Scale, *Science,* Vol 251, 1991, p 1468–1470

88. E. Betzig, P.L. Finn, and J.S. Weiner, Combined Shear Force and Near-Field Scanning Optical Microscopy, *Appl. Phys. Lett.*, Vol 60, 1992, p 2484
89. P.F. Barbara, D.M. Adams, and D.B. O'Connor, Characterization of Organic Thin Film Materials with Near-Field Scanning Optical Microscopy (NSOM), *Annu. Rev. Mater. Sci.*, Vol 29, 1999, p 433–469
90. C.C. Williams and H.K. Wickramasinghe, Scanning Thermal Profiler, *Appl. Phys. Lett.*, Vol 49, 1986, p 1587–1589
91. C.C. Williams and H.K. Wickramasinghe, Microscopy of Chemical-Potential Variations on an Atomic Scale, *Nature*, Vol 344, 1990, p 317–319
92. A. Majumdar, Scanning Thermal Microscopy, *Annu. Rev. Mater. Sci.*, Vol 29, 1999, p 505–585
93. O.E. Husser, D.H. Craston, and A.J. Bard, Scanning Electrochemical Microscopy—High Resolution Deposition and Etching of Materials, *J. Electrochem. Soc.*, Vol 136, 1989, p 3222–3229
94. Y. Martin, D.W. Abraham, and H.K. Wickramasinghe, High-Resolution Capacitance Measurement and Potentiometry by Force Microscopy, *Appl. Phys. Lett.*, Vol 52, 1988, p 1103–1105
95. M. Nonnenmacher, M.P. O'Boyle, and H.K. Wickramasinghe, Kelvin Probe Force Microscopy, *Appl. Phys. Lett.*, Vol 58, 1991, p 2921–2923
96. J.M.R. Weaver and D.W. Abraham, High Resolution Atomic Force Microscopy Potentiometry, *J. Vac. Sci. Technol. B*, Vol 9, 1991, p 1559–1561
97. D. DeVecchio and B. Bhushan, Use of a Nanoscale Kelvin Probe for Detecting Wear Precursors, *Rev. Sci. Instrum.*, Vol 69, 1998, p 3618–3624
98. B. Bhushan and A.V. Goldade, Measurements and Analysis of Surface Potential Change during Wear of Single-Crystal Silicon (100) at Ultralow Loads Using Kelvin Probe Microscopy, *Appl. Surf. Sci.*, Vol 157, 2000, p 373–381
99. P.K. Hansma, B. Drake, O. Marti, S.A.C. Gould, and C.B. Prater, The Scanning Ion-Conductance Microscope, *Science*, Vol 243, 1989, p 641–643
100. C.B. Prater, P.K. Hansma, M. Tortonese, and C.F. Quate, Improved Scanning Ion-Conductance Microscope Using Microfabricated Probes, *Rev. Sci. Instrum.*, Vol 62, 1991, p 2634–2638
101. J. Matey and J. Blanc, Scanning Capacitance Microscopy, *J. Appl. Phys.*, Vol 57, 1985, p 1437–1444
102. C.C. Williams, Two-Dimensional Dopant Profiling by Scanning Capacitance Microscopy, *Annu. Rev. Mater. Sci.*, Vol 29, 1999, p 471–504
103. D.T. Lee, J.P. Pelz, and B. Bhushan, Instrumentation for Direct, Low Frequency Scanning Capacitance Microscopy, and Analysis of Position Dependent Stray Capacitance, *Rev. Sci. Instrum.*, Vol 73, 2002, p 3523–3533
104. B. Bhushan, H. Fuchs, and S. Hosaka, *Applied Scanning Probe Methods*, Springer, Heidelberg, Germany, 2004
105. B. Bhushan and H. Fuchs, *Applied Scanning Probe Methods II—Scanning Probe Microscopy Techniques, III—Characterization, and IV—Industrial Applications*, Springer, Heidelberg, Germany, 2006
106. B. Bhushan, H. Fuchs, and S. Kawata, *Applied Scanning Probe Methods V—Scanning Probe Microscopy Techniques*, Springer, Heidelberg, Germany, 2007
107. B. Bhushan and S. Kawata, *Applied Scanning Probe Methods VI—Characterization*, Springer, Heidelberg, Germany, 2007
108. B. Bhushan and H. Fuchs, *Applied Scanning Probe Methods VII—Biomimetics and Industrial Applications*, Springer, Heidelberg, Germany, 2007
109. B. Bhushan, H. Fuchs, and M. Tomitori, *Applied Scanning Probe Methods VIII—Scanning Probe Microscopy Techniques*; *IX—Characterization*; and *X—Biomimetics and Industrial Applications*, Springer, Heidelberg, Germany, 2008
110. B. Bhushan and H. Fuchs, *Applied Scanning Probe Methods XI—Scanning Probe Microscopy Techniques*; *XII—Characterization;* and *XIII—Biomimetics and Industrial Applications*, Springer, Heidelberg, Germany, 2009
111. H. Fuchs and B. Bhushan, *Biosystems—Investigated by Scanning Probe Microscopy*, Springer, Heidelberg, Germany, 2010
112. B. Bhushan, *Scanning Probe Microscopy in Nanoscience and Nanotechnology*, Vol 1–3, Springer, Heidelberg, Germany, 2010–2013
113. I. Giaever, Energy Gap in Superconductors Measured by Electron Tunneling, *Phys. Rev. Lett.*, Vol 5, 1960, p 147–148
114. P.K. Hansma and J. Tersoff, Scanning Tunneling Microscopy, *J. Appl. Phys.*, Vol 61, 1987, p R1–R23
115. D. Sarid and V. Elings, Review of Scanning Force Microscopy, *J. Vac. Sci. Technol. B*, Vol 9, 1991, p 431–437
116. U. Durig, O. Zuger, and A. Stalder, Interaction Force Detection in Scanning Probe Microscopy: Methods and Applications, *J. Appl. Phys.*, Vol 72, 1992, p 1778–1797
117. J. Frommer, Scanning Tunneling Microscopy and Atomic Force Microscopy in Organic Chemistry, *Angew. Chem. Int. Ed. Engl.*, Vol 31, 1992, p 1298–1328
118. H.J. Guntherodt and R. Wiesendanger, Ed., *Scanning Tunneling Microscopy I: General Principles and Applications to Clean and Adsorbate-Covered Surfaces*, Springer-Verlag, Berlin, 1992
119. R. Wiesendanger and H.J. Guntherodt, Ed., *Scanning Tunneling Microscopy, II: Further Applications and Related Scanning Techniques*, Springer-Verlag, Berlin, 1992
120. D.A. Bonnell, Ed., *Scanning Tunneling Microscopy and Spectroscopy —Theory, Techniques, and Applications*, VCH, New York, 1993
121. O. Marti and M. Amrein, Ed., *STM and SFM in Biology*, Academic Press, San Diego, CA, 1993
122. J.A. Stroscio and W.J. Kaiser, Ed., *Scanning Tunneling Microscopy*, Academic Press, Boston, MA, 1993
123. H.J. Guntherodt, D. Anselmetti, and E. Meyer, Ed., *Forces in Scanning Probe Methods*, Vol E286, Kluwer Academic Publishers, Dordrecht, Netherlands, 1995
124. G. Binnig and H. Rohrer, Scanning Tunnelling Microscopy, *Surf. Sci.*, Vol 126, 1983, p 236–244
125. B. Bhushan, J. Ruan, and B.K. Gupta, A Scanning Tunnelling Microscopy Study of Fullerene Films, *J. Phys. D: Appl. Phys.*, Vol 26, 1993, p 1319–1322
126. R.L. Nicolaides, W.E. Yong, W.F. Packard, H.A. Zhou, et al., Scanning Tunneling Microscope Tip Structures, *J. Vac. Sci. Technol. A*, Vol 6, 1988, p 445–447
127. J.P. Ibe, P.P. Bey, S.L. Brandon, R.A. Brizzolara, N.A. Burnham, D.P. DiLella, K.P. Lee, C.R.K. Marrian, and R.J. Colton, On the Electrochemical Etching of Tips for Scanning Tunneling Microscopy, *J. Vac. Sci. Technol. A*, Vol 8, 1990, p 3570–3575
128. R. Kaneko and S. Oguchi, Ion-Implanted Diamond Tip for a Scanning Tunneling Microscope, *Jpn. J. Appl. Phys.*, Vol 28, 1990, p 1854–1855
129. F.J. Giessibl, Y. Sugawara, S. Morita, H. Hosoi, K. Sueoka, K. Mukasa, A. Sasahara, and H. Onishi, Noncontact Atomic Force Microscopy and Related Topics, *Nanotribology and Nanomechanics I*, 3rd ed.,B. Bhushan, Ed., *Springer-Verlag*, Heidelberg, Germany, 2011, p 195–237
130. A. Schirmeisen, B. Anczykowski, H. Hoelscher, and F. Fuchs, Dynamic Modes of Atomic Force Microscopy, *Nanotribology and Nanomechanics I*, 3rd ed.,B. Bhushan, Ed., *Springer-Verlag*, Heidelberg, Germany, 2011, p 307–353
131. F.J. Giessibl, Atomic Resolution of the Silicon (111)-(7×7) Surface by Atomic Force Microscopy, *Science*, Vol 267, 1995, p 68–71
132. B. Anczykowski, D. Krueger, K.L. Babcock, and H. Fuchs, Basic Properties of Dynamic Force Spectroscopy with the Scanning Force Microscope in Experiment and Simulation, *Ultramicroscopy*, Vol 66, 1996, p 251–259
133. T.R. Albrecht and C.F. Quate, Atomic Resolution Imaging of a Nonconductor by Atomic Force Microscopy, *J. Appl. Phys.*, Vol 62, 1987, p 2599–2602

134. S. Alexander, L. Hellemans, O. Marti, J. Schneir, V. Elings, and P.K. Hansma, An Atomic-Resolution Atomic-Force Microscope Implemented Using an Optical Lever, *J. Appl. Phys.*, Vol 65, 1989, p 164–167
135. G. Meyer and N.M. Amer, Optical-Beam-Deflection Atomic Force Microscopy: The NaCl (001) Surface, *Appl. Phys. Lett.*, Vol 56, 1990, p 2100–2101
136. A.L. Weisenhorn, M. Egger, F. Ohnesorge, S.A.C. Gould, S.P. Heyn, H.G. Hansma, R.L. Sinsheimer, H.E. Gaub, and P.K. Hansma, Molecular Resolution Images of Langmuir-Blodgett Films and DNA by Atomic Force Microscopy, *Langmuir*, Vol 7, 1991, p 8–12
137. J. Ruan and B. Bhushan, Atomic-Scale and Microscale Friction of Graphite and Diamond Using Friction Force Microscopy, *J. Appl. Phys.*, Vol 76, 1994, p 5022–5035
138. D. Rugar and P.K. Hansma, Atomic Force Microscopy, *Phys. Today*, Vol 43, 1990, p 23–30
139. D. Sarid, *Scanning Force Microscopy*, Oxford University Press, New York, 1991
140. G. Binnig, Force Microscopy, *Ultramicroscopy*, Vol 42–44, 1992, p 7–15
141. E. Meyer, Atomic Force Microscopy, *Surf. Sci.*, Vol 41, 1992, p 3–49
142. H.K. Wickramasinghe, Progress in Scanning Probe Microscopy, *Acta Mater.*, Vol 48, 2000, p 347–358
143. A.J. Boefden, The Influence of Lateral Forces in Scanning Force Microscopy, *Rev. Sci. Instrum.*, Vol 62, 1991, p 88–92
144. M. Radmacher, R.W. Tillman, M. Fritz, and H.E. Gaub, From Molecules to Cells: Imaging Soft Samples with the Atomic Force Microscope, *Science*, Vol 257, 1992, p 1900–1905
145. F. Ohnesorge and G. Binnig, True Atomic Resolution by Atomic Force Microscopy through Repulsive and Attractive Forces, *Science*, Vol 260, 1993, p 1451–1456
146. G. Neubauer, S.R. Coben, G.M. McClelland, D. Horne, and C.M. Mate, Force Microscopy with a Bidirectional Capacitance Sensor, *Rev. Sci. Instrum.*, Vol 61, 1990, p 2296–2308
147. T. Goddenhenrich, H. Lemke, U. Hartmann, and C. Heiden, Force Microscope with Capacitive Displacement Detection, *J. Vac. Sci. Technol. A*, Vol 8, 1990, p 383–387
148. U. Stahl, C.W. Yuan, A.L. Delozanne, and M. Tortonese, Atomic Force Microscope Using Piezoresistive Cantilevers and Combined with a Scanning Electron Microscope, *Appl. Phys. Lett.*, Vol 65, 1994, p 2878–2880
149. R. Kassing and E. Oesterschulze, Sensors for Scanning Probe Microscopy, *Micro/Nanotribology and Its Applications*,B. Bhushan, Ed., Kluwer Academic Publishers, Dordrecht, Netherlands, 1997, p 35–54
150. C.M. Mate, Atomic-Force-Microscope Study of Polymer Lubricants on Silicon Surfaces, *Phys. Rev. Lett.*, Vol 68, 1992, p 3323–3326
151. S.P. Jarvis, A. Oral, T.P. Weihs, and J.B. Pethica, A Novel Force Microscope and Point Contact Probe, *Rev. Sci. Instrum.*, Vol 64, 1993, p 3515–3520
152. C. Schoenenberger and S.F. Alvarado, A Differential Interferometer for Force Microscopy, *Rev. Sci. Instrum.*, Vol 60, 1989, p 3131–3135
153. D. Sarid, D. Iams, V. Weissenberger, and L.S. Bell, Compact Scanning-Force Microscope Using Laser Diode, *Opt. Lett.*, Vol 13, 1988, p 1057–1059
154. N.W. Ashcroft and N.D. Mermin, *Solid State Physics*, Holt Reinhart and Winston, New York, 1976
155. G. Binnig and D.P.E. Smith, Single-Tube Three-Dimensional Scanner for Scanning Tunneling Microscopy, *Rev. Sci. Instrum.*, Vol 57, 1986, p 1688
156. S.I. Park and C.F. Quate, Digital Filtering of STM Images, *J. Appl. Phys.*, Vol 62, 1987, p 312
157. J.W. Cooley and J.W. Tukey, An Algorithm for Machine Calculation of Complex Fourier Series, *Mat. Comput.*, Vol 19, 1965, p 297
158. J. Ruan and B. Bhushan, Atomic-Scale Friction Measurements Using Friction Force Microscopy: Part I—General Principles and New Measurement Techniques, *ASME J. Tribol.*, Vol 116, 1994, p 378–388
159. T.R. Albrecht, S. Akamine, T.E. Carver, and C.F. Quate, Microfabrication of Cantilever Styli for the Atomic Force Microscope, *J. Vac. Sci. Technol. A*, Vol 8, 1990, p 3386–3396
160. O. Marti, S. Gould, and P.K. Hansma, Control Electronics for Atomic Force Microscopy, *Rev. Sci. Instrum.*, Vol 59, 1988, p 836–839
161. O. Wolter, T. Bayer, and J. Greschner, Micromachined Silicon Sensors for Scanning Force Microscopy, *J. Vac. Sci. Technol. B*, Vol 9, 1991, p 1353–1357
162. E. Meyer, R. Overney, R. Luthi, D. Brodbeck, et al., Friction Force Microscopy of Mixed Langmuir-Blodgett Films, *Thin Solid Films*, Vol 220, 1992, p 132–137

ASM Handbook, Volume 10, *Materials Characterization*
ASM Handbook Committee
DOI 10.31399/asm.hb.v10.a0006658

Atomic Force Microscopy

Paul West, AFMWorkshop

Overview

General Uses

- Measuring three-dimensional topography images of samples in ambient air, liquids, and vacuums
- Measuring physical properties associated with surfaces, such as magnetic fields, electric forces, thermal conductivity, and temperature
- Scanning samples with surface features that are less than 10 μm in the vertical axis and less than 50 μm in the horizontal axis
- Only microscope technique that can give nanometer-resolution images of samples submerged in a liquid

Examples of Applications

- Any sample surface with a surface texture at the nanometer scale, including metals, ceramics, polymers, and biomaterials
- Metrology of structured materials, including step heights, and horizontal dimensions
- Applications where images of surface physical properties at the nanometer scale are needed

Samples

- Sample surfaces must be free of contamination.
- General laboratory atomic force microscopes (AFMs) accept samples that are 25 by 25 by 13 mm (1 by 1 by ½ in.).
- Specially designed AFMs can accept any size samples, including wafers, fabricated parts, and multiple samples.
- Two classes of samples are measured: bulk materials and materials deposited on a surface.
- Materials deposited on a surface must be chemically bound to the surface.

Limitations

- Atomic force microscopy does not have high depth of field and is well suited for relatively flat samples having structures less than 10 μm high.
- Atomic force microscopes are sensitive to vibrations; resolution can be degraded by structural and acoustic vibrations in the localized environment.
- An optical microscope typically is required to identify the region for scanning.
- Scan rates are such that it is not possible to visualize large areas.

Estimated Analysis Time

- Sample preparation for scanning surfaces of bulk materials is minimal and takes less than a few minutes.
- Sample preparation for materials deposited on a surface can take a few minutes or a few hours, depending on the type of sample being imaged.
- A typical scan of a 10 by 10 μm area with high resolution would take between 5 and 10 min.
- Analysis and display of images takes a few minutes.

Related Techniques

- The atomic force microscope is one type of microscope included in the scanning probe microscope family. Scanning probe microscopes include *near-field scanning optical microscopes* and *scanning tunneling microscopes.*
- *Scanning electron microscopy and atomic force microscopy:* Give complementary types of images. Scanning electron microscopy has a high depth of field, and atomic force microscopy has a low depth of field.
- An AFM has several modes that can measure surface physical properties and change the structure and chemistry of a surface.

Introduction

There are three general types of microscopes: electron microscopes, including scanning electron microscopes (SEMs) and transmission electron microscopes (TEMs), which use focused electron beams to generate images of surfaces; optical microscopes, which use focused light to generate images of surfaces; and the newest type of scanning probe microscopes, developed in the last few decades of the 20th century, which mechanically measure surface images by monitoring the motion of a probe as it is scanned across a surface (Fig. 1). These are known as atomic force microscopes (AFMs). There are three characteristics of an AFM that differentiate it from other types of microscopes:

- *Three-dimensional (3-D) images:* AFM images have magnification in the *x*-, *y*-,

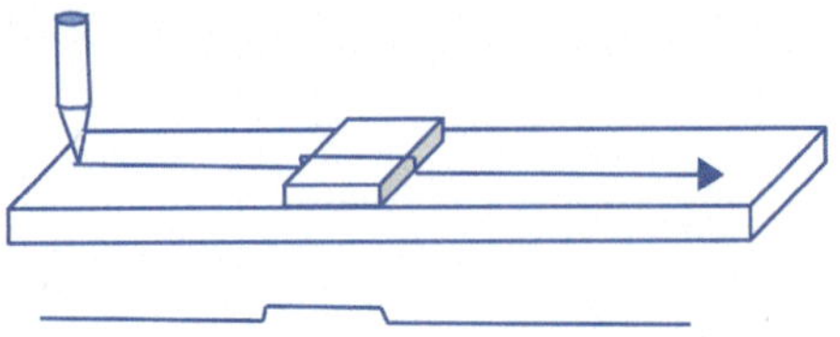

Fig. 1 As the probe moves from left to right across a surface feature, the motion of the probe is monitored, and a profile is created. An image of the surface is created by making successive profiles.

and z-axis. Magnifications in the z-axis can be as high as 1 billion and in the horizontal axis as high as 1 million.

- *Vacuums, air, and liquids:* An AFM can measure images with nanometer resolution in vacuums, ambient air, and liquids. It should be noted that the AFM is the only microscope that can do so in ambient air and liquids.
- *Physical properties of surfaces:* In an AFM, the probe is capable of measuring almost any physical property associated with a surface. This includes temperature, potential, fields, and hardness.

This article focuses on laboratory AFMs used in ambient air and liquid environments. It does not include discussion of AFMs used in ultra-high-vacuum environments or specialized AFMs used for manufacturing in the semiconductor and data-storage industries. In addition, there are several types of force sensors that are used in AFMs. These include crystals, microelectromechanical systems (MEMS) self-actuating cantilevers, and the light lever. Because 99% of force sensors used in AFMs are light levers, this article discusses light lever AFMs.

History

The origin of the AFM can be found in surface profilers first developed by precision engineers in the early 1900s (Ref 1). The instrument measured the motion of a probe at the end of a cantilever with light reflected from the cantilever end. This allowed surface features as small as 25 nm to be measured. This type of instrumentation, called a profilometer, was developed and used for many applications during the 20th century.

The horizontal resolution limitation of the profiler is established by the utilized probe size. Furthermore, the size of the probe depends on the loading force (pressure) of the probe on the surface. The probe will break under high loading pressures. Typically, the loading forces in a profiler are measured in micronewtons, and 1 μm diameter probes are used. The vertical resolution of the profiler is established by the mechanical rigidity of the instrument structure, that is, the ability to control the motion of the probe relative to the surface.

The loading force of a probe on the surface is greatly reduced by adding a feedback control system that uses a force sensor to measure the probe surface forces in combination with a precision actuator to accurately control the motion between the probe and the surface. The first of this type of instrumentation was disclosed by precision engineers (Ref 2) at the National Institute of Standards and Technology.

In the mid-1980s, Calvin Quate of Stanford University and Gerd Binnig of IBM invented the AFM (Ref 3). Their force sensor could measure forces as low as 10 pN, so they could use probes as small as a few nanometers in diameter. Piezoelectric ceramics were used to control the motion of the probe relative to the sample, and a vibration-isolation system similar to that used in the scanning tunneling microscope was used. In the initial AFM instrument, vertical and horizontal resolution was a few nanometers.

With the introduction of commercial AFMs in the 1990s, many new applications were realized. By far the largest application was for research scientists and engineers measuring nanometer surface features on a vast array of materials. Biologists began using AFMs for measuring cells, biological molecules, and biomaterials. Finally, high-technology industries used AFM for metrology applications.

There are several driving forces that expanded the use of AFMs. For measuring images of nanoscale features, an AFM is relatively inexpensive compared to other techniques, such as the SEM and TEM. The AFMs can measure high-resolution images in ambient air and liquids. Finally, the AFM can measure surface physical properties such as magnetic forces, stiffness, and nanomechanical properties.

Development Trends

The AFM instrumentation is still in its infancy, and AFM development continues to offer new and exciting types of instrumentation. There currently are two development trends occurring: innovation for instruments used by AFM experts, and those made to create products for non-AFM experts.

Instrumentation for AFM Specialists

Instrument innovators at universities throughout the world continue to create new capabilities for their AFMs. This includes new types of probes, higher-speed scanning, and new application areas. Much of the innovation then is passed on to commercial AFM vendors, who provide state-of-the-art instrumentation to their customers. Typically, this instrumentation costs from $100,000 to 400,000. Capabilities developed in the past few years for AFM specialists include high-speed scanning, combined optical and AFM for measuring surface spectral properties, and advanced scanning modes. Customers for this type of instrumentation are able to advance their research programs by having access to the latest commercial developments. These researchers typically are specialists in AFMs, and their projects demonstrate new capabilities for AFMs.

High-Speed Scanning

The AFMs have a relatively slow scan rate, with a typical image taking 5 to 10 min to generate. The primary limiting factor that controls AFM scan speeds is the bandwidth of the feedback control loop, which includes the force sensor and the piezoelectric ceramic that moves the sample and/or probe relative to each other. Additionally, the piezoelectric scanner that moves the sample and/or probe in the horizontal xy-axis must be capable of rapid motions. Finally, the data system that is used for high-speed scanning must have very fast capabilities. By optimizing these components, it is possible to obtain rates up to 30 frames per second with an AFM. The benefits of faster scan rates include monitoring dynamic processes at surfaces and rapid scanning when a large number of samples must be scanned.

Measuring Chemical Composition at the Nanoscale Level

There is great interest in being able to investigate structural chemical compositions at the nanoscale level. Combining the AFM with optical spectroscopic techniques allows for the possibility of making nanoscale chemical analysis. Such measurements are made by exposing a sample to a light source while scanning with the AFM probe. When a sample absorbs light at a specific frequency, the motion of the AFM probe is modified. Examples of this instrumentation are tip-enhanced Raman spectroscopy and nanoinfrared microscopy (Fig. 2).

Four-Dimensional AFM Imaging

Typically in an AFM, the probe tracks the surface during a scan. Another dimension that can be examined by the AFM is force interactions of the probe in the vertical direction above the surface, called a force/distance curve. It is possible to measure a force/distance curve at every image pixel point while scanning a surface. This technique has several advantages for scanning soft samples. It reduces horizontal forces between the probe and the sample while scanning, so that the probe does not tear the sample surface. A map of the force interaction above the surface is measured, giving insight into the mechanical stability of the surface.

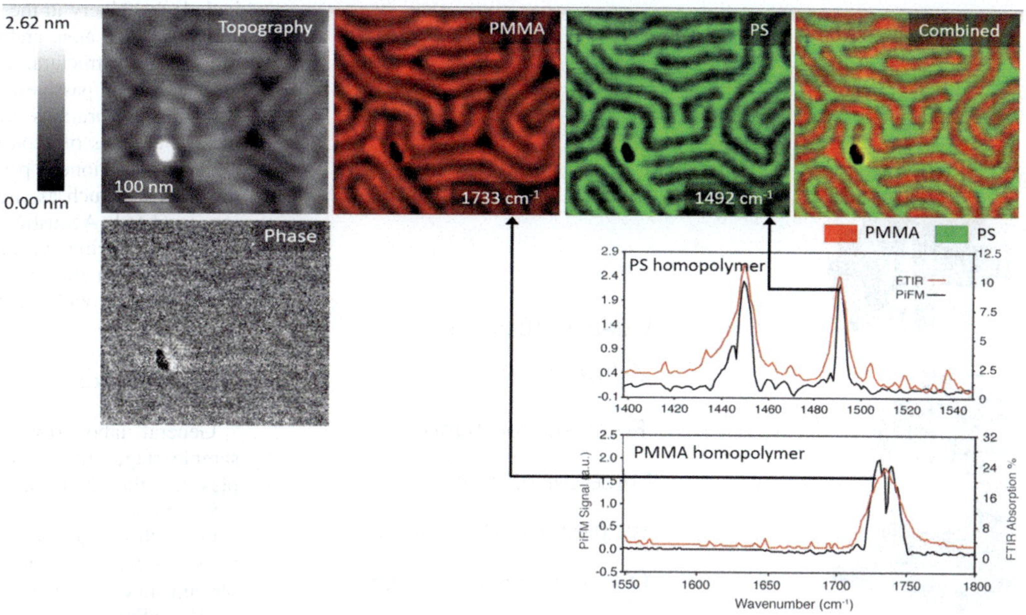

Fig. 2 Images of polystyrene (PS)/polymethyl methacrylate (PMMA) block copolymer, a sample used to gage the performance of nanoinfrared techniques. Photo-induced force microscopy (PiFM) spectra correlate with the Fourier transform infrared (FTIR) spectra quite nicely. By tuning the excitation laser to a wavenumber that excites one or the other molecules, PiFM can generate topographical and spectral images concurrently for each molecular type (PS or PMMA). The combined image shows the lamella structure expected for the block copolymer. The defect that is seen in the topography is neither PS nor PMMA but a foreign object.

Instrumentation for Non-AFM Specialists

There are a growing number of scientists and engineers that are not AFM experts but want access to the AFM as a tool for supporting their research and product development. Historically, nonspecialists would take their samples to laboratories that have advanced instrumentation operated by AFM experts. These applications usually support a research project, quality control, or product development. Instrumentation for non-AFM specialists costs $20,000 to 100,000 and has the following characteristics:

- *Reduced features:* The AFMs designed for specialists with an in-depth understanding of AFM theory have extremely complicated software with numerous options. By reducing the number of options and settings in the control software, the AFM is considerably easier to use by nonspecialists.
- *Robust:* New designs for probe holders and scanners make the AFM more robust and can withstand operation by customers with minimal training.
- *Automated scan parameters:* Through auto-optimization routines, especially for the Z feedback control, AFMs do not require theoretical training for operation.

As the AFM moves into the mainstream as a tool for routine sample analysis in research and industrial environments, there is a growing need for trained AFM operators. Thus, another use for nonspecialist AFMs is for educational purposes in institutions ranging from high schools to graduate-level universities.

General Principles

Atomic Force Microscope Instrumentation

Figure 3 illustrates the components typically found in an AFM stage. There are large-scale translators with ranges typically greater than 10 mm (0.4 in.). They include a *Z*-motion translator for moving the probe vertically above the sample, and an *XY* translator for moving the sample in large motions relative to the probe.

All AFMs have an *xyz* scanner for moving the probe and/or sample relative to each other. These scanners have a range of typically less than 100 μm in the *xy*-axis and less than 20 μm in the *z*-axis. The scanner mechanism typically is created from piezoelectric ceramics. There are several types of scanners, including tube, tripod, and flexure scanners. Additionally, some AFMs scan the probe in the *xyz*-axis, while other AFMs scan the sample in the *xyz*-axis. Alternatively, some designs move the probe in the *z*-axis and scan the sample in the *xy*-axis.

A force sensor is used for measuring the forces between a probe and a surface. The force sensor has a lower measurement range of a few piconewtons and an upper range of a few micronewtons. The most common type of force sensor is the light lever (Fig. 4) (Ref 4). In the light lever, the motion of the probe at the end of a cantilever is measured by reflecting a laser light source off of the cantilever into a photodetector. Although the light lever is the most common type of force sensor used in AFMs, it should be noted that force sensors can be made from crystals and microfabricated structures.

Because the scan range of an AFM is less than 100 μm, it is helpful to have an optical microscope for locating features and for helping to bring the probe down to the surface for scanning. When the force sensor is a light lever, the optical microscope is helpful for getting the laser focused on the cantilever.

Electronics for the AFM include drivers for the *Z* translator mechanism and the *XY* sample translator, if it is automated. Circuits are required for driving the *XY* scanner. A feedback electronic circuit (discussed in the section "Z Feedback Control" in this article) is required to measure the output of the force sensor and control the motion for the *Z* piezoelectric ceramic. The circuits required for the force sensor depend on the AFM. For the light lever, a power supply for the laser and photodetector amplifier circuits is required.

A computer with specialized software is used to control all of the AFM stage functions. Common to all AFM software platforms is a module for optimizing the force sensor; a module for bringing the probe down to the surface, commonly called probe approach; and finally, a module for scanning the sample. The scanning module includes scan size, pixel resolution, and feedback control parameters.

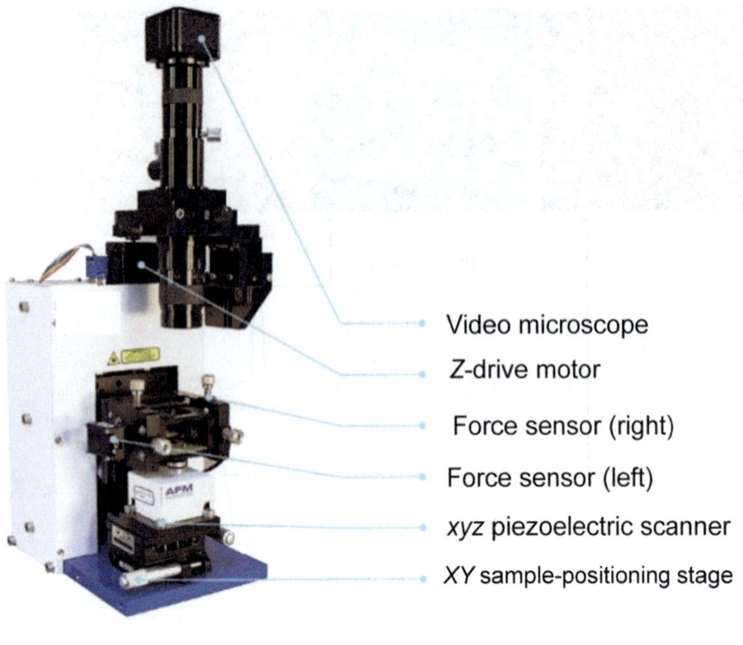

Fig. 3 Components of an atomic force microscope (AFM) stage. A mechanical structure supports both the force sensor and the *xyz* piezoelectric scanner. The vertical resolution of an AFM is primarily established by the rigidity of the mechanical structure.

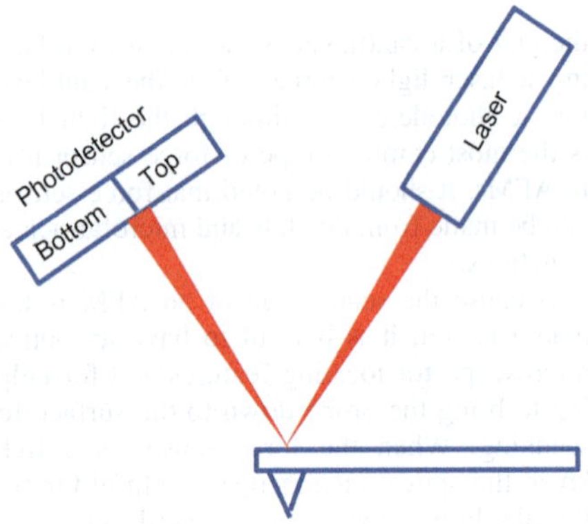

Fig. 4 In the light lever force sensor, a laser is reflected off the backside of a cantilever into a photodetector having top and bottom sections. If the cantilever bends, the amount of light reflected into the top and bottom detector changes.

Atomic Force Microscope Resolution

An AFM has three dimensions of resolution that must be considered: vertical, horizontal, and force resolution. The contributing factors for each of the three dimensions are different.

Electronic noise in circuits that drive the stage components contribute to resolution. With modern electronics, it is possible to design circuits with a sensitivity of 1 in 100,000 at a given bandwidth. Scan ranges in an AFM typically are selected so that the resolution required is balanced with the dynamic range. For example, if the maximum range of the piezoelectric ceramic used for controlling the *z*-axis is 10 μm (10,000 nm), the electronic noise will be 0.1 nm.

Vertical or height resolution depends on the ability to stabilize the probe above the surface. This requires building the AFM stage with a rigid structure and placing the stage in a low-vibration environment. It should be noted that the larger the stage, the more susceptible it is to external vibrations. In a well-constructed AFM that is operated in a good vibration environment, it is possible to obtain vertical noise floors of 10 to 80 pm.

The greatest contribution to the horizontal resolution is the diameter or shape of the probe used in the AFM. Probes manufactured with MEMS technology typically have diameters in the 10 nm range. However, through specialized processes, probes with 1 or 2 nm also are available. Specialized probes can be fabricated by using nanostructures, such as carbon nanotubes for the probe tip.

Vibration Control

As mentioned in the section "Atomic Force Microscope Resolution" in this article, an AFM is capable of measuring images with a vertical noise floor down to 20 pm. This is possible by building the AFM stage with a structure with a rigid mechanical loop. This contains all of the components in the AFM stage between the probe and the sample. Even AFMs with very rigid stage structures require some sort of vibration isolation. There are two sources of vibrations that cause problems: structural and acoustic.

Structural vibrations cause the floor where the AFM is located to move up and down in a vertical motion. Sources of this type of vibration include machinery in the same building, people walking in the area, and freeways and airports in the area. Structural vibrations are reduced with active and passive vibration tables.

Acoustic vibrations are transmitted through the air. Sources of acoustic vibrations include fans, air conditioners, people talking, and local disturbances, such as sirens and airplanes flying overhead. Acoustic isolation cabinets are used for reducing unwanted vibrations transmitted through the air and are made from a heavy material with foam on the inside.

Sample Stages

General laboratory AFMs typically have sample stages that accommodate small samples, less than 25 by 25 by 13 mm (1 by 1 by 0.5 in.). For most analytical and research applications, this is adequate. However, there are many special situations in which larger samples must be investigated. These include stages for the following applications:

- *Semiconductor wafers:* There are several applications in the semiconductor industry that require AFM scanning on structures located on a semiconductor wafer. These can be as large as 30 cm (12 in.) in diameter. The AFM stages designed for this size wafer typically use a vacuum chuck to hold the wafer onto the stage (Fig. 5).
- *Life sciences:* Biologists use inverted optical microscopes for studying cell structures. This is because inverted optical microscopes have high magnifications and can be combined with other types of microscopy, such as fluorescence microscopy. For these applications, the AFM is placed directly on the stage of the inverted optical microscope.
- *Specialized experiments:* There is interest in understanding changes in materials at the nanoscale when materials are under high stress or magnetic fields. This is possible with AFMs that scan the probe in the *xyz*-axis and can be placed on the specialized stages.

Steps to Measure an AFM Image

Sample Preparation

There are two types of samples that are routinely scanned with AFMs, and sample-preparation considerations are different for the two types. The first is the imaging of bulk material surfaces, while the second is studying structures placed on the surface of a substrate.

When imaging the surface of materials, features on the surface of the material must be less than the *Z*-range of the piezoelectric ceramic. The largest features typically imaged with an AFM are less than 10 μm. Additionally, the surface of the bulk material must be clean and free of surface contamination.

Imaging structures placed on a substrate, such as biological cells, nanotubes, DNA, nanoparticles, and graphene, requires selection of an optimal substrate, which must not have surface roughness greater than the structures being imaged. Also, there must be an affinity for the structures to the substrate that is greater than the affinity for the structures to the probe. Structures that are not tightly held by the substrate may be pushed around by the probe while scanning.

Putting the Sample and Probe in Microscopes

The procedure for inserting the probe in the AFM depends on the manufacturer's design of the probe holder. In some designs, the probe is per-mounted and can easily be picked up without special tools. In cases where the cantilever is not per-mounted, high-quality tweezers are required. Most importantly, when placed in the AFM, the probe must be returned to the same location each time.

Samples typically are mounted on a small metal disk that is magnetically held in the microscope stage. Conductive double-sticky material typically is used to fasten the sample to the sample disk. It is essential that the sample be rigidly mounted in the microscope stage. There can be a substantial reduction in resolution if the sample vibrates when mounted in the stage.

Probe Approach

Once the sample and probe are mounted in the AFM, the probe must be positioned so that it can interact with the surface forces and scanning can commence. This is commonly called probe approach. The most common type of probe approach is called the woodpecker technique.

In the woodpecker technique, the *Z* motor that moves the probe relative to the sample is used in concert with the *Z* piezoelectric ceramic that holds the sample, or the probe. The sequence used in the woodpecker method is:

1. Extend the *Z* ceramics.
2. Stop the tip approach if the tip interacts with the surface.
3. If the tip does not interact with the surface, retract the *Z* ceramics.
4. Stop the motor a smaller distance than the extension of the *Z* piezoelectric.
5. Go to Step 1.

Scanning the Sample

Once the probe is at the specimen surface, scanning can commence. There are several parameters that can be changed when scanning a sample. These are scan rate, number of lines, scan size, and scan location. These parameters are selected based on the type of sample being scanned. Typical scan rates are 0.5 Hz with 256 lines, which take approximately 5 min.

Retracting the Probe

Once the scan is complete, the probe must be retracted from the sample. This typically is done by retracting the *Z* sample and then activating the *Z* motor to move the probe a safe distance from the surface of the specimen.

Probes/Cantilevers

The most common type of force sensor used in an AFM is the light lever. This type of force sensor uses a cantilever with a probe on its end. There is a chip that supports the cantilever. Figure 6 illustrates the relationship between the probe, cantilever, and chip. Atomic force microscope operators have adopted the convention of calling the probe/cantilever/chip the "probe."

The AFM probes are created with MEMS manufacturing techniques. There are several variables that are used when creating the probes:

- *Material:* Most cantilevers/probes are made from either silicon or silicon nitride.
- *Cantilever shape:* Cantilevers are either rectangular or pyramidal.
- *Probe shape:* Probes can be pyramidal, with three or four sides.
- *Top coating:* Typically there is no coating, gold, or aluminum.
- *Bottom coating:* Bottom coating usually is metals, such as potassium, iron, or diamond.

The dimensions of an AFM cantilever can vary depending on the type of cantilever and the application. Typically, the length is 50 to 300 μm, the width is 20 to 50 μm, and the thickness is 1 to 20 μm. Based on the cantilever dimensions and its material, the force constant of the cantilever can be estimated. Force constants for AFM cantilevers are 0.01 to 100 N/m (0.0007 to 7 lbf/ft).

There are several types of AFM probes available for different scanning modes and applications. Several manufacturers have websites and catalogs that help to select the appropriate probe for an application.

Atomic Force Microscopy Image Processing

Once measured, all AFM images require some image processing, and some manufacturers offer integrated image-processing packages. Additionally, there are several image-processing packages available from independent sources, both open source and commercial. There are three basic functions of AFM image-processing software: processing, display, and analysis.

The AFM images are a 3-D array of numbers; that is, for each *xy*-location on the surface, there is a *z*-height value. Image-processing steps alter the values of the numbers in the 3-D array of numbers. Examples of process steps are leveling, filtering, and deglitching. Each of these primary image-processing steps has several subsets. For example, for leveling there is three-point, plane, polynomial, and line leveling.

There are several methods for viewing AFM images. These different display methods do not change the numbers in the array but change

Fig. 5 This atomic force microscope is designed for scanning a 20 cm (8 in.) wafer. Applications include dimensional measurements, failure analysis, and surface textures.

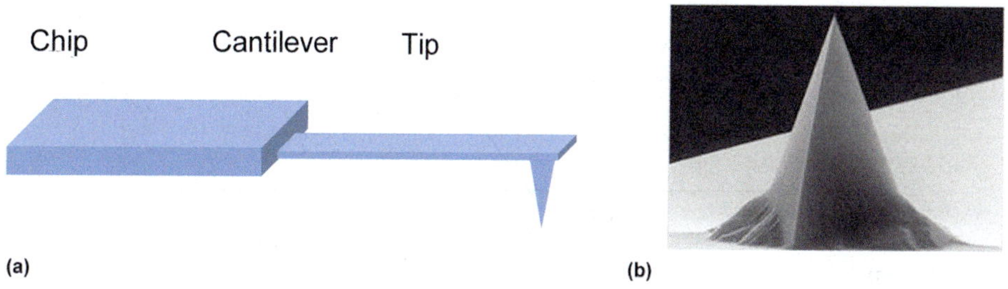

Fig. 6 (a) There are three parts to a probe used in a light lever atomic force microscope (AFM). They are the chip, cantilever, and tip. (b) Scanning electron microscopy image of a microfabricated tip used in an AFM

how the numbers are viewed. Examples include 3-D rendering, two-dimensional view, histogram adjustment, color scale, and light shading. Changing the display method is important because it helps to visualize specific features in the surface of a sample. The analysis of an AFM image generates a numerical value representative of a specific characteristic of the surface of the sample. There are several types of analysis, including line profile, surface texture, fractal, and particle. Figure 7 illustrates the capabilities of image-processing software.

Z Feedback Control

Understanding Z feedback control in the AFM is essential for optimal use of an AFM. The Z feedback control takes a signal from the light lever force sensor and the signal to control the distance between the probe and sample. Typically, a general-purpose interface bus electronic controller is used in the AFM for controlling the motion of the Z ceramic (Fig. 8).

As discussed in the section "Topography Modes" in this article, there are a few types of topography-imaging modes, the most common of which are vibrating and nonvibrating. In nonvibrating mode, the top-bottom signal is one that is directed to the differential amplifier. In vibrating mode, the output from an amplitude demodulator is input into the differential amplifier.

Other Considerations

When operating an AFM, there are a few issues that must be considered when preparing samples, scanning, and interpreting data. These issues often are not discussed but should be considered by operators. The first is the effects of surface contamination on the AFM operation, while the second is unwanted image artifacts.

Surface Contamination

Every surface that is maintained in ambient air has a small layer of contamination (Fig. 9). The contamination layer is composed of water and hydrocarbons, and the thickness can vary depending on the relative humidity of the environment and the amount of other ambient contaminants. It can range from less than a nanometer to a hundred nanometers. In the case of a fingerprint on the surface, the contamination layer is approximately 1000 nm.

When an AFM is operated in ambient air, the probe must penetrate the contamination layer before it can be scanned across the surface. The amount of force required to penetrate the contamination layer depends on the thickness and composition of the layer. Unfortunately, it is not possible to characterize contamination layers until scanning is commenced.

The greatest problem caused by the contamination layer is associated with probe approach. As the probe approaches the surface, it can start to interact with the contamination layer and stop. If the contamination layer is sufficiently thick, capillary forces between the probe and surface will pull the cantilever toward the surface. This is called false feedback. When scanning in a false feedback

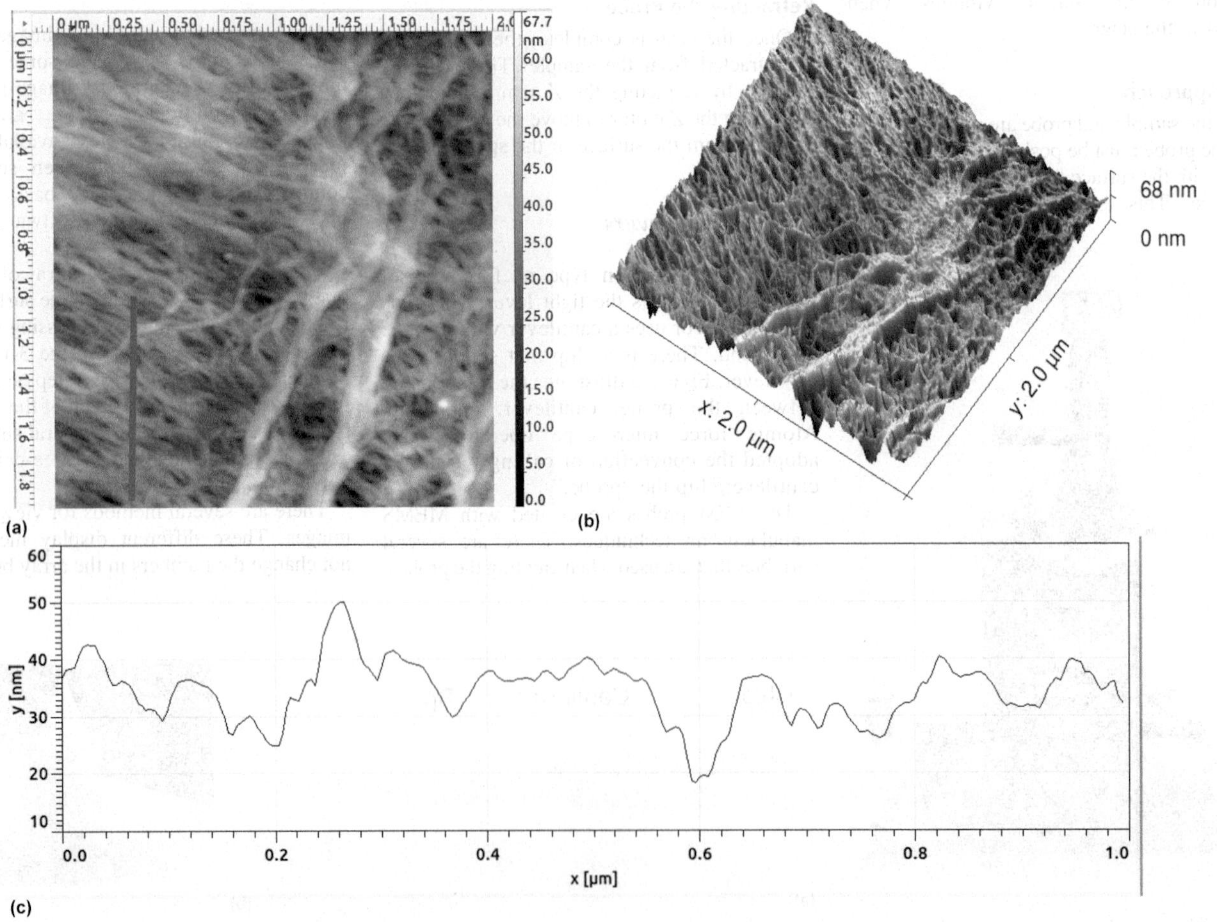

Fig. 7 A 2 × 2 µm vibrating-mode image of a biaxially oriented polypropylene polymer fiber. (a) Two-dimensional color-scale presentation. (b) Three-dimensional light-shaded representation. (c) Line profile showing vertical dimensions of the polymer surface. Created with Gwyddion software

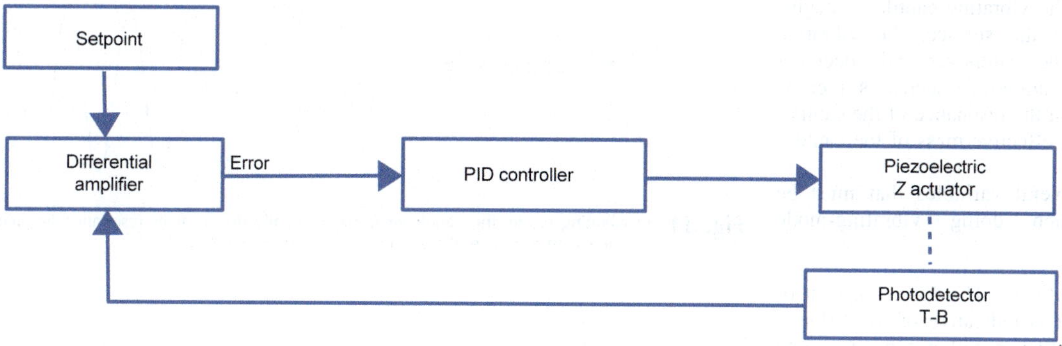

Fig. 8 Diagram illustrating the feedback loop used in nonvibrating-mode atomic force microscopy. A photodetector measures the motion of the cantilever, so when the cantilever bends, the top-bottom (T-B) signal changes. The signal is passed to a differential amplifier and compared to a setpoint voltage. An error signal is sent to a proportional, integral, derivative (PID) controller, and the output of the PID controller activates the *Z* actuator that controls the motion of the sample.

condition, the probe does not interact with the hard forces of the surface, and image resolution is very poor.

Image Artifacts

As with every other type of microscope, there are potential unwanted image artifacts that can occur in AFM images. It is important to identify these unwanted image artifacts when they occur. The most common types of artifacts are related to probes and image processing.

By far, the most common image artifacts occur when a probe tip is broken before or during a scan. Silicon probes are very susceptible to chipping, which creates an artifact. The artifact is easily identified because the features in the image appear the same and in a similar shape. It should be noted that this type of artifact most often is seen in images where the features of interest are similar in size to the probe used for scanning.

The second most common type of image artifact is created after the image is measured. A common type of image leveling is line-by-line leveling. That is, each scan line in the image is numerically fitted to a line, and the slope of the line is subtracted to create a new line, which is displayed. This type of leveling creates unwanted bands in the image and can be eliminated by using the appropriate leveling methodology.

Scan Direction

In the AFM there are two distinctive scan directions that can give differing physical interactions between the probe and the surface. In the first scan direction, the cantilever axis is parallel to the scan axis. In the second scan direction, the probe is scanned perpendicular to the primary axis of the cantilever. However, by using software, it is possible to generate scans with any scan axis from 0 to 360°. In cases in which the scan axis is not parallel or perpendicular to the primary axis of the probe, the probe-sample interaction is a hybrid of the two primary axes.

Atomic Force Microscopy Imaging Modes

An AFM is an extremely versatile instrument and can be operated in a number of different modes. Each mode can give important information about surface structure and/or physical properties. The most common modes used with an AFM are topography modes. However, as probe quality and probe technologies improve, the other modes are becoming more commonly used. This section includes a comprehensive list of AFM scanning modes. There is a brief description of how each mode works and what types of applications can be made with each mode.

Topography Modes

As initially invented in the mid-1980s, the AFM was a relatively simple technique for measuring high-resolution images of surfaces. Soon after its invention, it was realized that there were several modes of operation that could be employed for measuring surface topography.

Contact Mode

Contact-mode AFM was the first AFM mode developed. Hooke's law gives the relationship between the force required and the deflection of a cantilever (Fig. 10):

$$F = -kX$$

Thus, if a light lever force sensor can measure a deflection of 100 nm, and the force constant of a cantilever is 0.01 nm, then it is possible to measure a force of 0.05 nN. In contact-mode AFM, the deflection of a cantilever is fixed during a scan. From the deflection, it is possible to calculate the force on the surface.

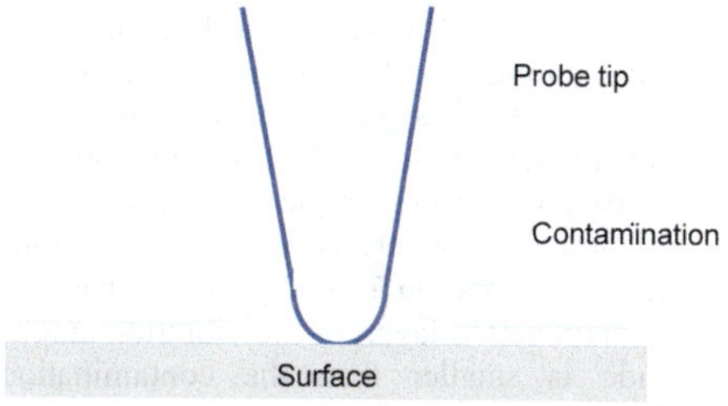

Fig. 9 Surface contamination wets the surface of the probe tip, and capillary forces pull the contamination up the probe. The surface contamination can be as small as a monolayer and as thick as 100 nm.

Fig. 10 The force required to bend the cantilever distance *d* is given by Hooke's law ($F = -kd$). In contact-mode atomic force microscopy, the deflection of the cantilever is measured by a laser reflected off the top of the cantilever.

If the probe and sample are not electrically connected, it is possible to have different charges on the surface and on the sample. If there are differing charges, the cantilever can bend from electrostatic forces derived from the differing charges. This bending can make it difficult to successfully complete a probe approach in an AFM. To increase the tip-sample interaction in the microscope, the setpoint is increased.

Vibrating Mode

Vibrating mode is the most common scanning mode used in an AFM. There are several names created for vibrating mode that vendors use for their products, such as tapping mode, intermittent-contact mode, noncontact mode, near-contact mode, and high/low-amplitude resonance.

In vibrating mode, the piezoelectric ceramic is used to drive at its resonance frequency

(Ref 5). When the vibrating cantilever begins to interact with the surface, the vibration amplitude of the cantilever will decrease (Fig. 11). As the amplitude decreases, there is a slight change in the resonance of the cantilever, because the effective mass of the cantilever increases.

There are several variables that must be considered when doing vibrating-mode scanning:

- *Cantilever stiffness:* The force constant of the cantilever is indicative of its stiffness. A stiffer cantilever will not be drawn to the surface by capillary forces as easily as a cantilever with a low constant. Also, stiffer cantilevers are not as easily bent by electrostatic forces.
- *Free-air vibration amplitude:* Before probe approach is started, the cantilever is vibrated, which is called the free-air vibration amplitude. This can vary from a few nanometers to a hundred nanometers. The magnitude of the vibration will greatly impact the probe-sample interaction while scanning. There are two cases: the amplitude of vibration is much larger than the contamination layer, or the vibration amplitude is smaller than the contamination layer. If the amplitude is larger than the contamination, as the probe vibrates it will enter and exit the contamination layer on each oscillation. This makes the AFM more stable but can result in breaking the probe. If the amplitude is smaller than the contamination, the AFM can be unstable, but higher-resolution images are measurable.
- *Decrease in vibration amplitude when scanning:* As the probe interacts with the surface, its amplitude decreases. This is expressed in percentage of free amplitude vibration. For example, scanning at 80% means that there is a 20% decrease in vibration amplitude. This is set by the setpoint.

To increase the tip-sample interaction, the setpoint is decreased.

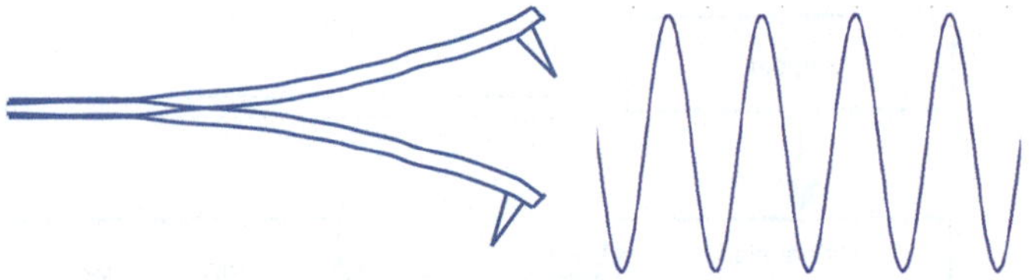

Fig. 11 In vibrating mode, the cantilever is mechanically driven at its resonance frequency. When the probe interacts with a surface, the vibration amplitude is reduced.

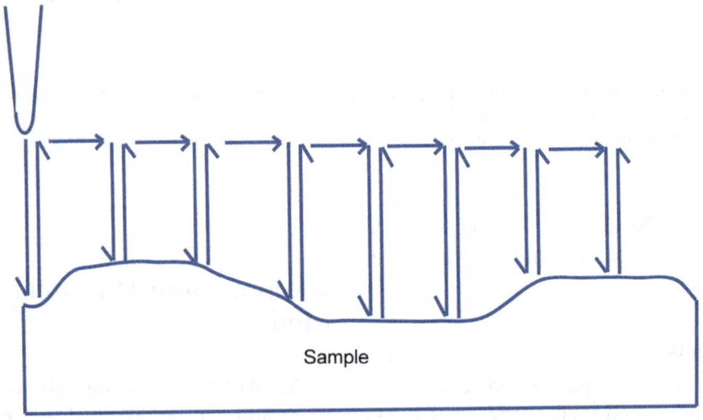

Fig. 12 A four-dimensional image is measured by moving the probe from a known distance above the surface down to the surface. The direction of the probe is reversed when a pre-established force is measured. A force-distance curve is acquired each time the probe is dropped to the surface.

Four-Dimensional Mode

Four-dimensional scanning mode is the latest topography mode; it has its origins in a technique called pulsed-for-imaging. In pulsed-for-imaging, the probe is raised and lowered at nonresonant frequency as it is scanned across a surface (Fig. 12). The direction of the probe is reversed by a specific force. A force-distance curve is created at each pixel point in the image. From the force-distance curve, several surface parameters can be calculated. The technique is similar to that developed by Hartoonian (Ref 6).

As discussed in the section "Four-Dimensional AFM Imaging" in this article, four-dimensional scanning is very helpful for scanning soft materials. One of the drawbacks of this technique is that it generates data arrays that are extremely large. For example, if a 256 × 256 image is measured with 1000 datapoints per pixel, there will be 65,000,000 individual values in the image file. It can take long time periods to process an image of this size with a personal computer.

Electrical Modes

Using a conductive probe in an AFM facilitates making a number of different types of nanoscale electrical measurements. The probe/cantilever can be fabricated from an electrically conductive material, or the bottom side of the cantilever can be coated with metal.

Conductive AFM

Conductive AFM (C-AFM) images are made while scanning in contact mode and monitoring the current flow between the probe and surface. A potential source and electrometer are connected between the probe and the sample of interest. The current then is measured while scanning, and a plot is made of the conductivity at each image pixel point. Potentials can run between 0 and 5 V and the currents between 1 pA and 1 μA. With C-AFM instrumentation, current/voltage (*I*/*V*) curves are measurable. The *I*/*V* curves are measured by holding the probe at a fixed position and ramping the voltage and monitoring the current.

It should be noted that the quality of C-AFM measurements depends on the quality and conductivity of the probe used. Additionally, surface contamination can greatly alter the quality of the measurements. In cases where a coated probe is use, results can vary if the probe coating wears off.

Electric Force Microscopy

Surfaces with differing areas of electric potential will have electric fields that emanate from the surface. These electric fields can be mapped with electric force microscopy (EFM). Most often, EFM images are measured in vibrating mode. A two-pass imaging technique is used in which the probe is scanned over the surface to measure the surface topography. The surface topography is stored in a computer, and the image is rescanned with the probe raised a fixed distance above the surface. This can be done one line at a time or for the entire image.

Scanning Kelvin Probe Microscopy

Scanning Kelvin probe microscopy (SKPM) directly measures a surface potential image. This technique relies on the fact that two vibrating plates, in close proximity, will have zero force between each other when they are at the same potential. The SKPM images are made in a vibrating topography mode. There are two drive frequencies in the instrument: one drives the tickler piezo, which physically drives the cantilever at resonance; the second drive frequency applies an oscillating voltage to the probe. A feedback loop then is used to null out the potential between the probe and surface when scanning.

Scanning Microwave Impedance Microscopy

In *scanning* microwave impedance microscopy (sMIM), the probe is used as a waveguide to probe samples with electromagnetic radiation in the millimeter wave range (30 to 39 GHz). The two primary advantages of sMIM are that subsurface images can be made, and there is no need for a conductive path.

Scanning Capacitance

Scanning capacitance microscopy is a contact-mode scanning technique. Capacitance sensor electronics measure the capacitance between the probe and the surface while scanning. The capacitance image displays the capacitance of a surface as a function of *XY* coordinates.

Magnetic Force Microscopy

Magnetic force microscopy (MFM) measures the magnetic fields that emanate from a surface with magnetic domains. The MFM images are measured in vibrating mode. A two-pass imaging technique is used in which the probe is scanned over the surface to measure the surface topography. The surface topography is stored in a computer, and the image is rescanned with the probe raised a fixed distance above the surface. This can be done one line at a time or for the entire image (Fig. 13). Probes used for MFM typically are standard silicon probes that are coated with a magnetic metal.

Liquid Scanning

An AFM is the only type of microscope that provides nanometer resolution on surfaces submerged in liquids. There are several advantages for scanning with an AFM in liquids:

- Contaminants are dissolved at the surface.
- Biomaterials can be studied in physiological-relevant environments.
- Corrosion and dissolution are observed in real-time.

There are three methods for imaging samples in liquids: the microcell, the open cell, and the closed cell.

Microcell

A microcell is the simplest technique for imaging samples submerged in liquids. A small amount of liquid is placed directly on the surface of the sample, and the probe then is submerged in the microdroplet. The advantage of this technique is its simplicity. However, the primary disadvantage is that, because only a small amount of liquid is used, it can rapidly evaporate, often before a scan is completed.

Open Cell

In an open AFM liquid cell, a small window is placed directly above the cantilever. Liquid is placed in a small vessel, with a level that is above the glass window. Enough liquid is used so that evaporation during scanning is not a problem. There are two factors that must be considered when using an open liquid cell.

In an AFM, an optical microscope typically is used for positioning a laser light onto the cantilever. Because a small amount of light is refracted at the interfaces between glass and air and between glass and liquid, there can be several "ghost" spots in the optical microscope image. The optical path of the laser light in an AFM is designed for use in ambient air. In an open liquid cell, as the light passes through the glass window and then the liquid, the direction of the light changes. This is because the refractive indexes of glass and liquid are slightly different than air.

Closed Cell

In a closed liquid cell, the entire sample and probe are enclosed in a small volume that has an inlet and outlet port. Because the cell is completely sealed, evaporation is not a problem. The only drawback of a closed cell is that there is a seal between the scanning component (sample and/or probe) and the nonscanning component (probe and/or sample). This seal makes it difficult to move the sample relative to the probe in the *xy*-axis more than a few hundred micrometers.

Electrochemical Cell

For in situ electrochemical experiments, a reference and counterelectrode can be used in either an open or closed cell. Typically, the sample is the working electrode in this type of apparatus. It is possible to directly measure surface electrochemical reactions with an AFM electrochemical cell.

Force-Distance Curves

A force-distance (F-D) curve is a graph of the motion of a cantilever/probe as it is moved toward and away from the surface of a sample. Figure 14 is an example of an F-D curve measured in ambient air. A curve, called the inbound curve, indicates when the probe approaches the surface, and another curve, called the outbound curve, indicates when the probe is moving away from the surface. There are four distinctive regions in an F-D curve:

- When the cantilever is not interacting with the surface
- In the inbound curve, when the probe first interacts with the contamination layer and capillary forces pull the cantilever inward
- As the probe interacts with the hard forces of the surface
- In the outbound curve, when the probe sticks to the surface and is about to release

There are a number of applications for the F-D curve:

- It can indicate the amount of surface contamination.
- The stiffness of a sample surface can be measured.
- In cases where the cantilever force constant is matched with the surface hardness, nanoindentations can be made.
- The gain of a light lever is determined from the F-D curve.

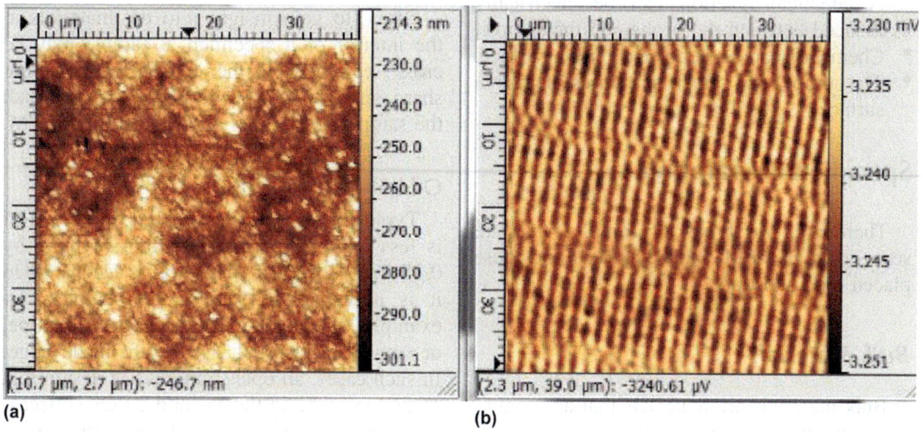

Fig. 13 Magnetic force microscopy images of magnetic recording media. (a) Topography image measured during the first pass of the probe interacting with the hard forces of the surface. (b) Magnetic field image measured during the second pass of the probe

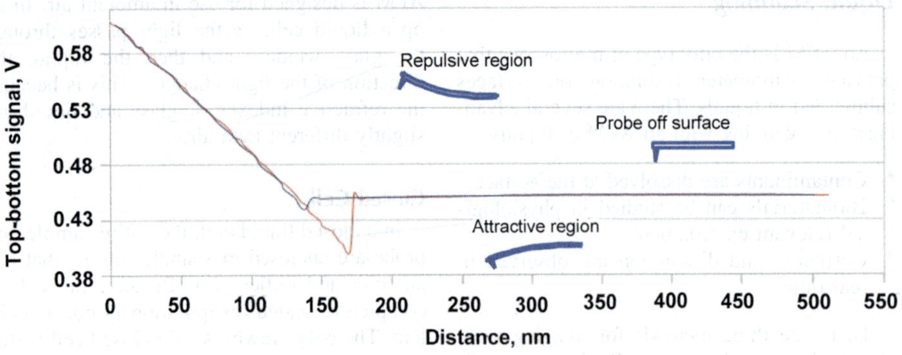

Fig. 14 In this force-distance curve, the blue line is the inbound curve, and the red line is the outbound curve. There are three separate regions: probe off the surface; attractive region, where the probe is bent downward; and repulsive region, where the probe is bent upward.

Lateral Force Microscopy

The light lever in an AFM typically has four quadrants. They are the upper left (UL), upper right (UR), lower left (LL), and lower right (LR).

For measuring a surface topography, the up-and-down (vertical) motion of the cantilever is calculated by UL + UR − (LL + LR). When the scan direction is perpendicular to the cantilever axis, it is possible to monitor the rotational motion of the cantilever by UL + LL − (UR + LR).

While scanning with the AFM, the horizontal motion of the cantilever can be a measure of the friction between the probe and the sample. By measuring the lateral forces in the forward and reverse scanning directions, a friction loop may be established.

Phase-Mode Microscopy

Phase mode is a vibrating-mode technique that gives contrast between regions of different hardness and/or adhesion on the sample surface. In vibrating-mode AFM, the Z feedback loop is used to maintain a fixed vibration amplitude while scanning. If the probe then moves over a region of differing hardness, there will be a change in the signal phase. In a phase-mode image, the changes in phase are displayed as a function of the x- and y-axes. There are many applications for phase-mode imaging on polymer samples.

Thermal Modes

A small thermal-sensing device can be placed at the end of a probe and used to monitor the thermal properties of a sample at the nanoscale. A topography and thermal map of the surface is created from this technique. There are two types of thermal sensors: thermocouples and resistive sensors. Both sensors can measure temperature at the nanoscale; however, resistive sensors are capable of measuring thermal-conductivity maps of a surface. A second application for the resistive probe is microthermal analysis. In this technique, the resistive probe is used to heat a very small region of a sample and simultaneously measure the response to the heat generated by the probe. This monitors surface-phase transitions.

Optical AFM

Combining an AFM with optical methods creates instrumentation capable of measuring nanoscale spectroscopic and thus chemical maps (see the section "Measuring Chemical Composition at the Nanoscale Level" in this article).

Atomic Force Microscope Lithography

The probe in an AFM can modify the surface of a specimen. There are several options for AFM lithography:

- The probe can mechanically alter a surface.
- A chemical reaction can be electrically induced between the probe and the sample.
- Chemicals can be deposited on the surface.
- The probe can push structures across a surface.

Specimen Preparation

There are two types of samples that are scanned in an AFM: bulk materials and those placed on a substrate.

Bulk Materials

Bulk materials are materials that are contiguous and thus comprised of a single object. This could be a polymer, ceramic, semiconductor, or processed samples. In general, these types of samples are straightforward to image with an AFM. Simply put the sample into the microscope and scan it. The following are a few guidelines.

Maximum and Minimum Dimensions of Features

Atomic force microscopes do not have great depth of resolution when compared to electron and optical microscopes. The height of features that may be scanned is limited by the range of the Z piezoelectric ceramic that moves the probe and/or sample in the z-axis, and the height of the probe at the end of a cantilever. In commercial AFMs, the range of the z-axis typically is less than 20 μm; thus, sample features must be less than 20 μm high. Often, because the probe is almost never completely perpendicular to the sample, some of the z-range will be required for the tilt of the sample.

The size of features that can be imaged also depends on the shape and length of the probe. There are a number of different shapes and probes available with differing diameters, angles, and lengths. It is critical to ensure that the probe selected for an application is appropriate.

The minimum size of features that can be examined is set by the Z noise floor in the vertical axis; in the horizontal axis, the minimum size is set by the electromechanical noise of the piezo scanners and by the geometry of the probe. Typically, the z-axis noise is >50 pm, and the xy-axis limit is 1 nm.

Cleaning the Samples

Specimens with excessive surface contamination are difficult to image with an AFM. If a sample does have excessive surface contamination, it must be cleaned with pure isopropyl alcohol or other cleaning reagent. The cleaning reagent must be pure and not leave any residual chemicals behind.

Cross Sections

Often, to gain insight into the materials at the interior of a specimen, a sample may be cross sectioned. If the cutting blade is not sharp enough, it may be necessary to polish the sample before it is scanned with the AFM.

Odd-Shaped Samples

Typically, a sample scanned in an AFM is less than 25 by 25 by 6 mm (1 by 1 by 0.25 in.). However, there are occasions when it is necessary to scan larger samples, for example, sheets of glass, sections of a pipe, or small objects such as a knife blade edge. In such cases, an open-frame AFM design that scans the probe in the xy- and z-axes is necessary. In an open-frame design, the probe extends below the structure holding the light lever and can be placed or configured for imaging odd-shaped samples.

Materials Deposited on Substrates

Common sample types scanned in an AFM include graphene, nanotubes, nanoparticles, cells, and biological molecules. In such cases, the material must be directly attached to a surface for imaging. It is essential that the material be firmly bound to a surface, or it can be pushed around by the probe. It is not possible to list all of the methods for affixing samples on an AFM. In the following two sections, approaches to imaging this type of sample are briefly discussed.

Deposit on a Substrate

The primary requirements for substrates used in AFM imaging are:

- They must be smooth, with features that are much smaller than the objects being examined (Fig. 15).
- The surface must be very clean.
- If possible, the surface should be chemically activated to hold the objects.

Two common substrate materials used for AFM imaging are mica and highly ordered pyrolytic graphite (HOPG). Mica is a layered material, so to obtain a clean surface, a razor is used to peel away a top surface. After cleaning, the mica surface can be activated chemically. In the case of HOPG, the surface is cleaned with adhesive tape. Press the tape against the HOPG and peel it away.

Another consideration is establishing the best density for surface materials. If the density is too high, the materials will stick together, while if the density is too low, there may not be objects in the field of view of the microscope.

Embed or Affix to a Substrate

For scanning larger objects in the AFM that are not bulk materials, it is sometimes possible to embed them in clay or even affix them to a sticky substrate. As an example, salt crystals can be affixed to a piece of tape for scanning, and large powders can be pushed into a clay sample.

Calibration and Accuracy

Dimensional Measurements

An AFM inherently gives the 3-D measurements of a surface structure. The accuracy and precision of AFM measurements depend on two factors: the scanner and the probe geometry.

Scanners

Piezoelectric ceramics typically are used to generate motion in AFM scanners. Piezoelectric scanners have two undesirable properties: hysteresis and creep. Correcting for these properties can be made with hardware and software.

The motion of the piezoelectric scanner can be monitored with an external sensor such as a strain gage or a capacitance sensor. In such cases, the motion of the scanner can be corrected with a signal from the external sensor. This is commonly called linearization. If the exact same measurement is being made repetitively, software can be used to correct AFM images for metrological measurements. The target image is corrected with an image of a standard sample.

External sensors for correcting ceramic problems are advantageous because they work with almost any scan range. However, external sensors have inherent noise and limit the resolution of images. Software correction has the advantage that it does not introduce unwanted noise into the images; however, as mentioned previously, it is useful only for per-established scan ranges.

Probes

The MEMS processing technology is used for fabricating most AFM probes. Thus, the errors in silicon processing contribute to errors in probe geometries. Typically, probes offered by commercial vendors have a maximum specified diameter. For example, the probe diameter can be less than 10 nm. If probes are required for a measurement with a specific tolerance, there are a few options. One of them is to measure the probe geometry with an SEM. The second is to use a probe that is sharpened with an ion mill, electrochemical process, or by mounting a nanotube on the end of the probe.

Standards and References

There are several references available from commercial vendors (Fig. 16). Typically, there is no error in the dimensional size of the features on the references. The references have a Z step-height range of 2 to 100 nm and a horizontal-dimension range of 2 to 50 μm. It should be noted that the errors in step-height references below 2 nm are very large.

Force Measurements—Cantilevers

Forces between the probe and the sample in an AFM are established by the force constant of the cantilever. The force constant of the cantilever depends on the material, length, width, and thickness of the cantilever. There is great uncertainty in the thickness of the probe, and the force constant depends on the inverse of the thickness to the third power. Thus, the uncertainty of cantilever force constants is 50 to 100%.

Light Lever Gain

The relationship between cantilever movement and the output of the photodetector is called the light lever gain. This is different for each probe and depends on the location of the laser spot on the cantilever when measurements are made.

To measure the light lever gain, a hard sample is used, such as a piece of silicon. An F-D curve is measured, and the light lever gain is

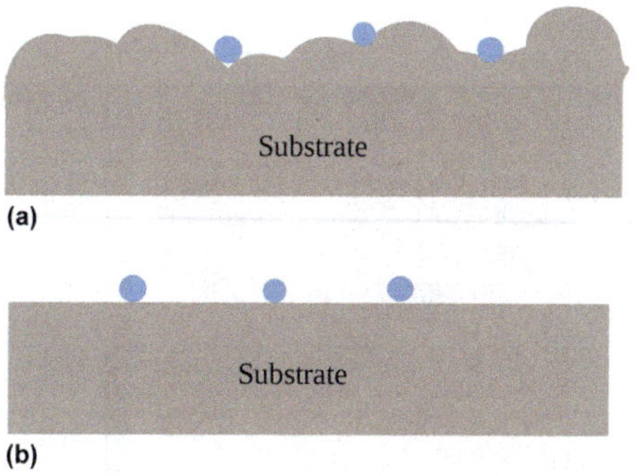

Fig. 15 When imaging structures (blue circles) deposited on a substrate, the surface texture of the substrate must be lower than the size of the feature being imaged. (a) The particles could not be distinguished from the substrate. (b) The surface texture of the substrate is much lower than the size of the features of interest.

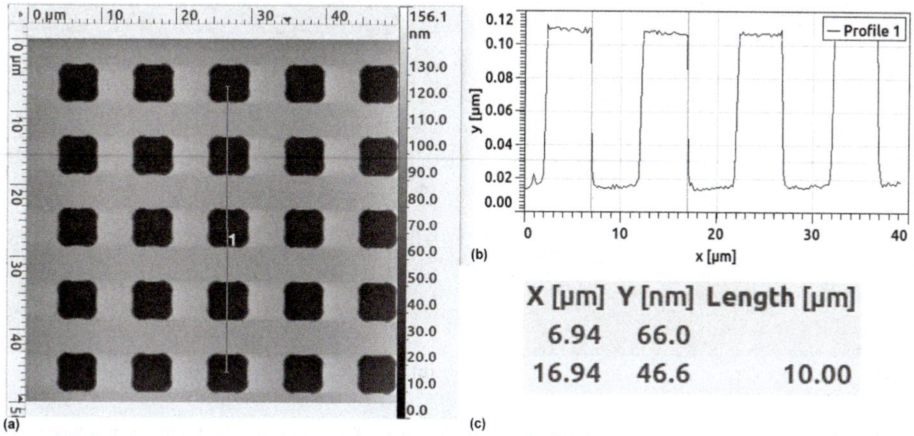

Fig. 16 (a) Vibrating-mode atomic force microscopy image of a reference sample. (b) Line profile showing dimensions of the features. (c) The pitch is 10 μm, and the height of the features is 100 nm.

calculated from the curve where the probe interacts with the hard surfaces. The slope of this line is the difference in voltage divided by the distance, or the light lever gain.

Cantilever Force Constant

Because it is not possible to accurately predict cantilever force constants, they must be measured and can vary from cantilever to cantilever in a single manufactured batch. The three methods used to calibrate cantilevers are the thermal method, the Sader method, and the vibrometer.

Data Analysis and Reliability

All AFM images require image processing. Often, the image processing is done so that it is possible to visualize features on a surface. If the end goal is visualization, there are a number of image-processing steps that may be applied to represent surface features. Image-enhancement techniques—such as filtering, deglitching, and display methods, such as light shading and 3-D display perspectives—help with visualization.

However, if the AFM is being used for metrological measurements, there is a different standard. The image-processing steps undertaken should not compromise the data. For example, excessive filtering can change the shapes of objects in an image. A detailed image-processing protocol should be followed for making metrological measurements with an AFM.

Special care should be taken if the surface features being studied are on the length scale of the probe being used. As an example, a 10 nm diameter probe should be used to image 20 nm particles. This is because a chipped probe can cause a change in probe geometry and thus impact the image. It is recommended to practice with a known sample of similar dimensions to assure that the image is being operated correctly before scanning unknown samples.

Nanostructures

The geometry of the probe must be considered when making dimensional measurements on a nanostructure with an AFM. This is especially true when nanometer-sized objects are being imaged. As the probe moves across a nanoparticle, the image is broadened by the width of the probe. In measuring the sizes of nanoparticles, the horizontal dimensions are typically larger than the actual particle; however, the height of the particle measured from the AFM will give the correct dimension.

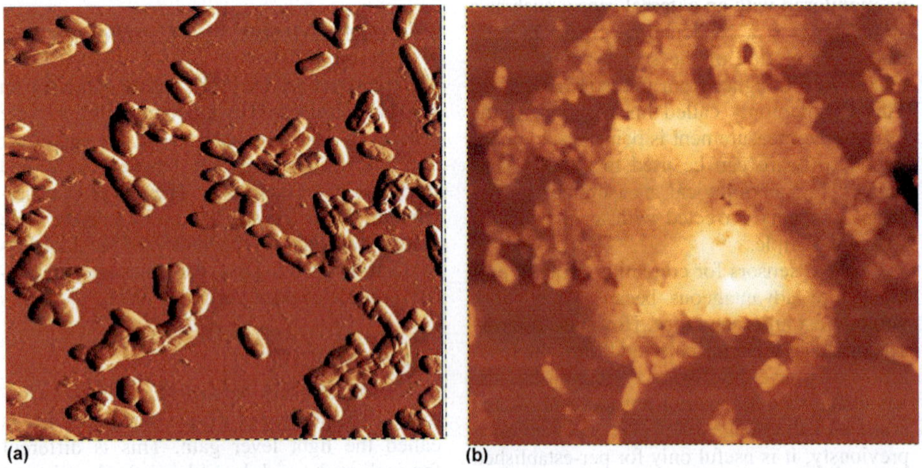

Fig. 17 Vibrating-mode images of cells. (a) Bacteria cells. (b) Epithelial cells. Both images are 50 × 50 μm. An advantage of the atomic force microscope is that these images are measurable in liquids.

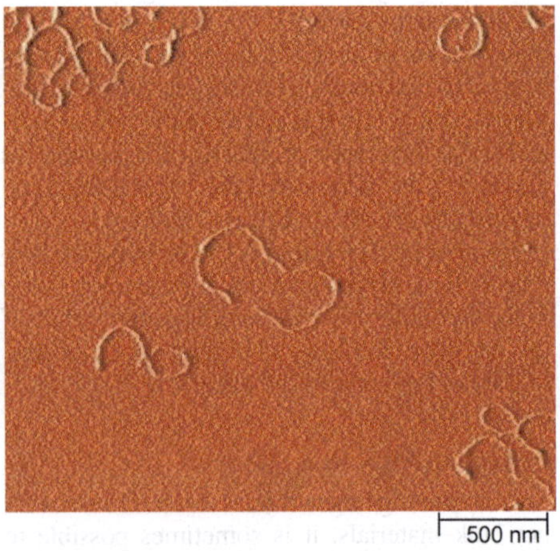

Fig. 18 Vibrating-mode atomic force microscopy image of DNA deposited on mica

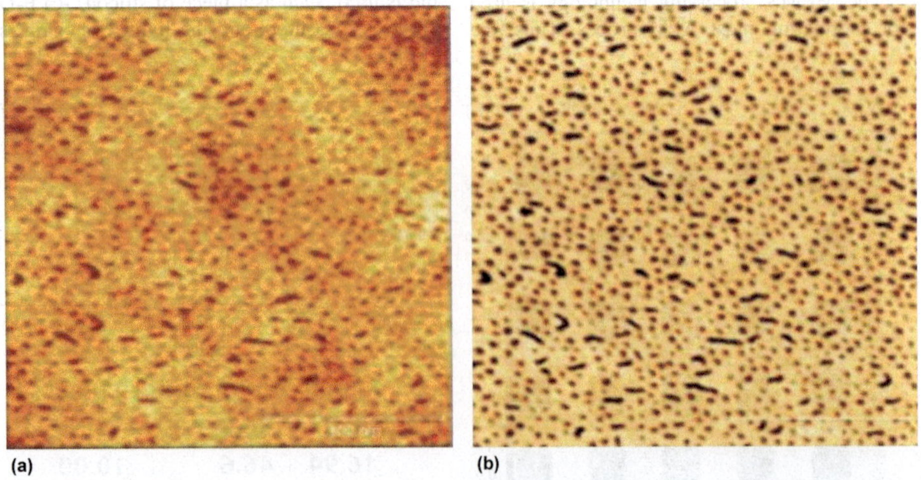

Fig. 19 (a) 1 × 1 μm phase image of styrene-ethylene/butylene- styrene polymer. (b) Corresponding topography image. Although the existence of the two domains is observed in (b), the different blocks are more clearly differentiated in (a), enabling a clearer understanding of the morphology, dispersion, and size of the various block domains.

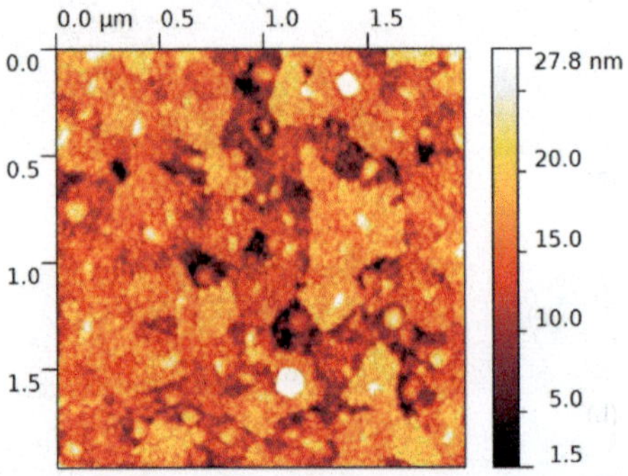

Fig. 20 A 2 × 2 μm atomic force microscopy image of an indium/tin oxide thin film deposited on a glass substrate. Nanometer-sized features are directly observed in this image. It should be noted that this material in optically transparent.

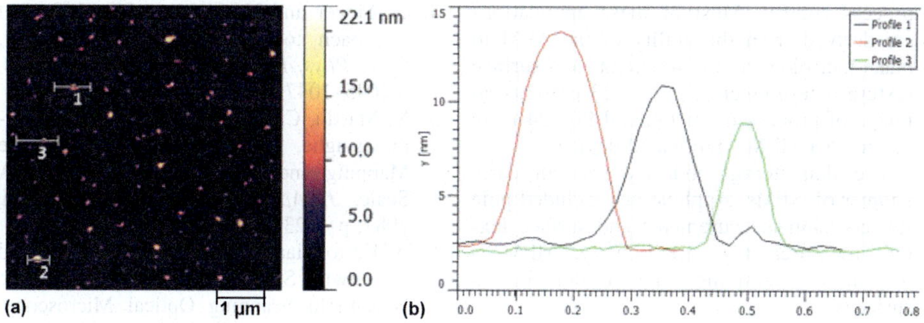

Fig. 21 (a) Image of a sample of quantum dots (nanoparticles). (b) Line profiles show three different particle sizes.

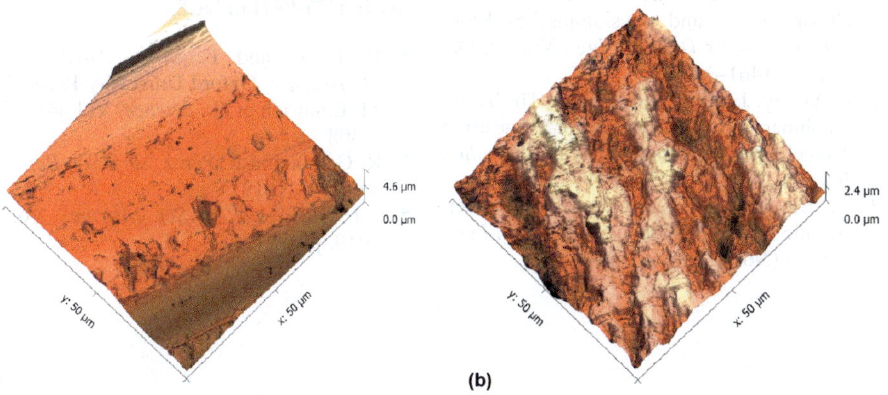

Fig. 22 Atomic force microscopy images of commercially available aluminum foil. (a) Shiny side of foil. (b) Dull side of foil. Each image is 50 × 50 μm and was measured in nonvibrating mode.

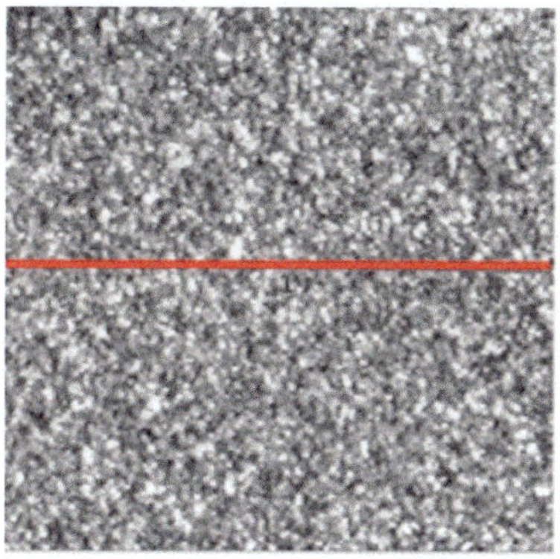

(a)

Parameters	
Average value:	15.43 nm
Minimum:	7.12 nm
Maximum:	29.85 nm
Median:	15.39 nm
Ra (Sa):	1.69 nm
Rms (Sq):	2.11 nm
Rms (grain-wise):	2.11 nm
Skew:	0.135
Kurtosis:	0.05
Surface area:	100.166 μm^2
Projected area:	100.000 μm^2
Inclination θ:	0.0 deg
Inclination φ:	-71.8 deg

(b)

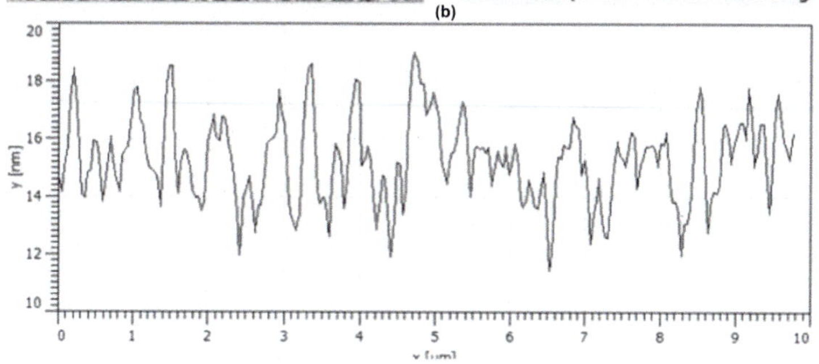

(c)

Fig. 23 (a) 40 × 40 μm atomic force microscopy image of a processed silicon sample coated with silicon oxide. (b) Surface roughness parameters for the processed silicon image. (c) Line profile of silicon image, designated by red line

Fig. 24 Atomic force microscopy image of a gear fabricated with microelectromechanical systems technology. The image is 50 × 50 μm and was measured in vibrating mode.

Device Structures

Atomic force microscopes can be helpful for measuring the dimensions of fabricated devices, such as gratings, semiconductor devices, and storage media. Horizontal dimensions, such as pitch, are readily measured with an AFM. Vertical measurements of devices can be measured. However, the same protocol should be used for repetitive measurements. If angular measurements are required, it is essential to calibrate the AFM for the specific measurement with a standard sample and a specific type of probe.

Applications

There are demonstrated applications for AFMs in every field of science, technology, and engineering. To date, many applications are made by specialists in research laboratories. It is expected that as more students are educated in the operation and application of AFMs, these instruments will become commonplace. One of the compelling reasons for this prediction is that AFMs can be manufactured and sold for a fraction of the price of other types of microscopes with extreme resolution. Due to the vast number of AFM applications reported in the literature, it is not possible to make a comprehensive review. However, the following is a partial list of applications.

Life Sciences

Life sciences applications are separated into three types of samples: cells, biological molecules, and biomaterials. There are three types of measurements: dynamic high-speed scanning, topography, and indentation/adhesion. Furthermore, measurements are made in either air or physiological conditions. Figure 17 presents images of cells, and Fig. 18 is an image of DNA.

Physical Sciences

Applications for AFM in the physical sciences include polymer chemistry, electrochemistry, crystallography, nondenominational, and thin-film studies. Figure 19 presents topography and phase images of a polymer, and Fig. 20 is an AFM image of an indium/tin oxide thin film.

Nanotechnology

With the advent of nanotechnology, a number of structures were discovered and readily examined with the AFM. These include nanotubes, nanofibers, nanoparticles, and graphene. Figure 21 is an AFM image of quantum dots, which are a type of nanoparticle.

High Technology—Industrial

Traditional industries, such as the food, paper, and construction industries, are increasingly using the AFM for product development and quality control. Figure 22 presents AFM images of commercially available aluminum foil.

There are a number of demonstrated applications in the semiconductor industry, including failure analysis, process development, and process control. Most of these applications are derived from the ability of the AFM to make metrological measurements and surface texture measurements. Figure 23 presents an image of processed silicon, and Fig. 24 is an image of a MEMS-fabricated device.

The data-storage industry was an early adopter of AFMs. Applications included pole tip recession measurements and surface texture measurements. Additionally, MFM is helpful for examining magnetic domains on surfaces.

REFERENCES

1. G. Schmaltz, Über Glätte und Ebenheit als physikalisches und physiologisches Problem, *Z. Vereins Deutsch. Ing.*, Vol 12, Oct 1929, p 1461–1467
2. R. Young, J. Ward, and F. Scire, The Topographiner: An Instrument for Measuring Tunneling Microtopography, *Rev. Sci. Instrum.*, Vol 43 (No. 7), 1972, p 999
3. G. Binnig, C.F. Quate, and C. Gerber, Atomic Force Microscope, *Phys. Rev. Lett.*, Vol 56 (No. 9), 1986, p 930
4. G. Meyer and N.M. Amer, Novel Optical Approach to Atomic Force Microscopy, *Appl. Phys. Lett.*, Vol 53 (No. 12), 1988, p 1045–1047
5. Y. Martin, C.C. Williams, and H.K. Wickramasinghe, Atomic Force Microscope Mapping and Profiling on a Sub 100-A Scale, *J. Appl. Phys.*, Vol 61 (No. 9), 1987, p 4723
6. A. Hartoonian, E. Betzig, M. Isaacson, and A. Lewis, Super-Resolution Fluorescence Near-Field Scanning Optical Microscopy, *Appl. Phys. Lett.*, Vol 49 (No. 11), Sept 15, 1986, p 674

SELECTED REFERENCES

- P. Eaton and P. West, *Atomic Force Microscopy*, Oxford University Press, 2010
- M. Emmrich et al., *Science*, Vol 348, 2015, p 308
- R. García and R. Pérez, *Surf. Sci. Rep.*, Vol 47, 2002, p 197
- F.J. Giessibi, *Rev. Sci. Instrum.*, Vol 90, 2019, p 011101

ASM Handbook, Volume 10, *Materials Characterization*
ASM Handbook Committee
DOI 10.31399/asm.hb.v10.a0006683

Secondary Ion Mass Spectroscopy*

Overview

General Uses

- Surface compositional analysis with approximately 2 to 10 nm depth resolution
- Elemental in-depth concentration profiling
- Trace element analysis at the parts-per-billion to parts-per-million range
- Isotope abundances
- Hydrogen analysis
- Spatial distribution of elemental species
- Spatial distribution of organic compounds

Examples of Applications

- Identification of inorganic or organic surface layers on metals, glasses, ceramics, thin films, or powders
- In-depth composition profiles of oxide surface layers, corrosion films, leached layers, and diffusion profiles
- In-depth concentration profiles of low-level dopants (≤1000 ppm) diffused or ion implanted in semiconductor materials
- Hydrogen concentration and in-depth profiles in embrittled metal alloys, vapor-deposited thin films, hydrated glasses, and minerals
- Quantitative analysis of trace elements in solids
- Isotopic abundances in geological and lunar samples
- Tracer studies (for example, diffusion and oxidation) using isotope-enriched source materials
- Phase distribution in geologic minerals, multiphase ceramics, and metals
- Second-phase distribution due to grain-boundary segregation, internal oxidation, or precipitation

Samples

- *Form:* Crystalline or noncrystalline solids, solids with modified surfaces, or substrates with deposited thin films or coatings; flat, smooth surfaces are desired; powders must be pressed into a soft metal foil (for example, indium) or compacted into a pellet
- *Size:* Variable but typically 1 cm by 1 cm by 1 mm (0.4 by 0.4 by 0.04 in.)
- *Preparation:* None for surface or in-depth analysis; polishing for microstructural or trace element analysis. Samples may be cross sectioned with a diamond microtome and analyzed, thus not limiting the technique to the surface. Samples may be frozen for analysis.

Limitations

- Analysis is destructive, although time-of-flight secondary ion mass spectroscopy (static) consumes so little of the surface that it is often considered nondestructive.
- Qualitative and quantitative analyses are complicated by wide variation in detection sensitivity from element to element and from sample matrix to sample matrix.
- The quality of the analysis (precision, accuracy, sensitivity, and so on) is a strong function of the instrument design and the operating parameters for each analysis.

Estimated Analysis Time

- One to a few hours per sample

Capabilities of Related Techniques

- *Auger electron spectroscopy:* Qualitative and quantitative elemental surface and in-depth analysis is straightforward, but the detection sensitivity is limited to >1000 ppm. Microchemical analysis with spatial resolution to <100 nm
- *Rutherford backscattering spectroscopy:* Nondestructive elemental in-depth profiling; quantitative determination of film thickness and stoichiometry
- *Electron microprobe analysis:* Quantitative elemental analysis and imaging with depth resolution ≥1 μm

Introduction

In secondary ion mass spectroscopy (SIMS), an energetic beam of focused ions is directed at the sample surface in a high- or ultrahigh-vacuum (UHV) environment. The transfer of momentum from the impinging primary ions to the sample surface causes sputtering of surface atoms and molecules. Some sputtered species are ejected with positive and negative charges, termed secondary ions, which are mass analyzed using a double-focusing mass spectrometer or an energy-filtered quadrupole mass spectrometer. The principles of SIMS are shown in in Fig. 1.

*Revised from C.G. Pantano, Secondary Ion Mass Spectroscopy, *Materials Characterization*, Vol 10, *ASM Handbook*, ASM International, 1986, p 610–627

This method is used to acquire information about the surface, near surface, and bulk composition of the sample, depending on instrumental parameters. Using a relatively low rate of sputtering enables recording a complete mass spectrum to provide a surface analysis of the outermost 5 nm of the sample, termed static SIMS. Although a useful mode of operation, it is not yet a routine analytical technique. Alternatively, the intensity of one or more of the peaks in the mass spectrum can be continuously recorded at a higher sputtering rate to provide an in-depth concentration profile of the near-surface region. At very high sputtering rates, trace element and impurity analysis in the bulk is possible. In addition, a secondary ion image of the surface can be generated to provide a spatially resolved analysis of the surface, near surface, and bulk of the solid. This article focuses on the principles and applications of high-sputter-rate dynamic SIMS for depth profiling and bulk impurity analysis.

Secondary ion mass spectroscopy can characterize solid samples with high spatial and in-depth resolution, which is usually achieved with high detection sensitivity due to the inherent sensitivity of mass spectroscopy. Quantitative capabilities depend on the nature of the specimen, instrument design, and particular instrument parameters and methods used in the analysis. The interpretation of SIMS spectra and depth profiles can be difficult; thus, SIMS is not yet applicable for chemical analysis of unknowns in the true sense of the word. Secondary ion mass spectroscopy has unique capabilities to answer specific questions about specimens whose stoichiometry and matrix structure are already characterized.

Sputtering

Bombardment of a solid surface with a flux of energetic particles causes ejection of atomic species. The process, termed sputtering (Ref 1–3), causes erosion or etching of the solid. Incident projectiles are often ions, because this facilitates production of an intense flux of energetic particles, which can be focused into a directed beam. Therefore, the techniques are referred to as ion sputtering and ion beam sputtering. However, in principle, sputtering (and secondary ion emission) also occurs under neutral beam bombardment. Secondary ion mass spectroscopy is typically based on ion beam sputtering of the sample surface, although there are new approaches to SIMS based on fast atom bombardment (Ref 4, 5).

At incident ion energies from 1 to 20 keV, the most important interaction with the solid surface is momentum transfer from the incident ion to the target atoms. This occurs because the primary ion penetrates the solid surface, travels some distance (termed the mean free path), and collides with a target atom. Figure 2 shows that the collision displaces the target atom from its lattice site, where it collides with a neighboring atom, which, in turn, collides with its neighbor. The succession of binary collisions, termed a collision cascade, continues until the energy transfer is insufficient to displace target atoms from their lattice positions.

Sputtering, or ejection, of target atoms and molecules occurs because much of the momentum transfer is redirected toward the surface by the recoil of the target atoms within the collision cascade. Because the lifetime of the collision cascade produced by a single primary ion is much smaller than the frequency of primary ion impingements (even at the highest primary ion beam current densities), the process is an isolated, albeit dynamic, event. Ejection of target atoms due to a single binary collision between the primary ion and a surface atom occurs infrequently.

The primary ion undergoes a continuous energy loss due to both momentum transfer and electronic excitation of target atoms. Thus, the primary ion is eventually implanted tens to hundreds of angstroms below the surface. Thus, ion bombardment of a solid surface leads not only to sputtering but also to electronic excitation, ion implantation, and lattice damage. Figure 3 summarizes the effects of ion bombardment and indicates that the sputtered species may be monatomic species or molecular clusters. The effects of ion implantation and electronic excitation on the charge of the sputtered species are discussed in the section "Secondary Ion Emission" in this article.

Sputtering yield, S, is the average number of atoms sputtered per incident primary ion. The number depends on the target material and on the nature, energy, and angle of incidence of the primary ion beam. Sputtering yield is directly proportional to the stopping power of the target (because it determines the extent of momentum transfer near the surface), and it is inversely proportional to the binding energy of surface atoms. Therefore, sputtering yield also depends on the crystallographic orientation of the material being sputtered. Sputter yield is proportional to the square root of the incident ion energy, but there is not universal agreement with experiments in the energy range from 1 to 20 keV. In most SIMS experiments, Cs^+, O_2^+, O^-, Ga^+, and Ar^+ primary ion species are used in the energy range from 2 to 20 keV at angles of incidence between 45° and 90°. Under these conditions, the sputtering

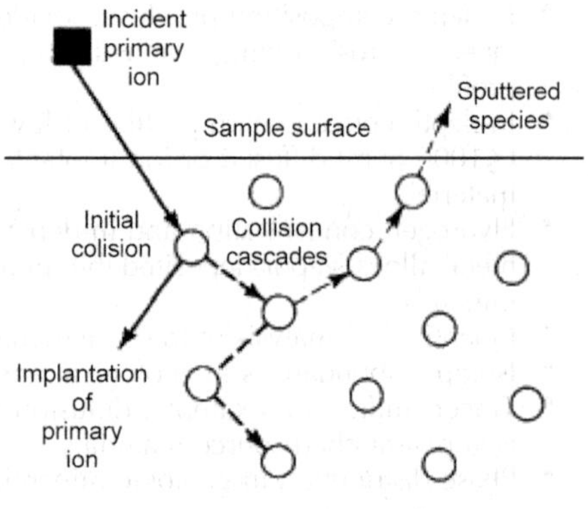

Fig. 2 Physical effects of primary ion bombardment: implantation and sputtering

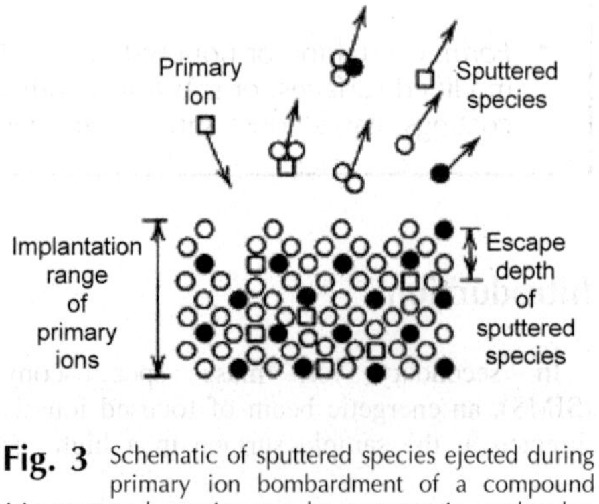

Fig. 3 Schematic of sputtered species ejected during primary ion bombardment of a compound i_xj_y; sputtered species can be monatomic, molecular, and/or incorporate implanted primary ions. i = ○, j = ●

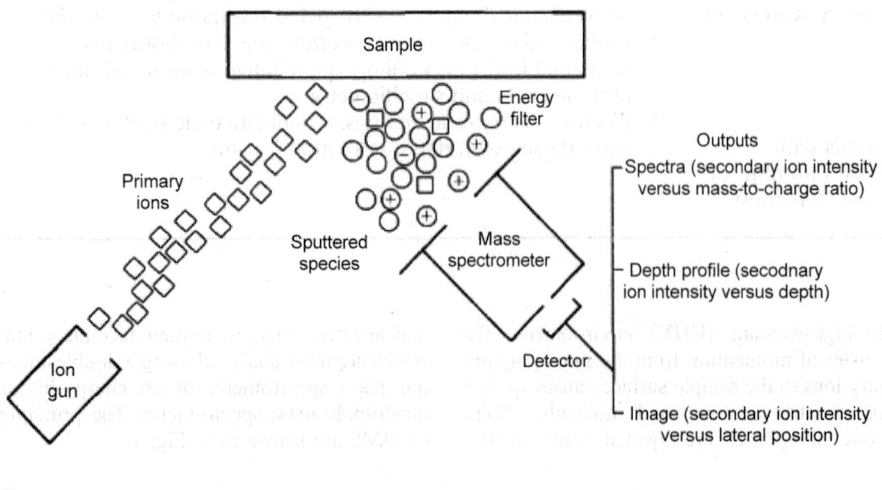

Fig. 1 Schematic representation of the principles of secondary ion mass spectroscopy

yields for most materials range from 1 to 20. Information on the effects of the primary ion beam and the target material on sputtering yields is provided in Ref 6.

Selective sputtering, or preferential sputtering, can occur in multicomponent, multiphase, and polycrystalline materials. Thus, it is possible for the composition of alloy surfaces to become modified during sputtering where the species with the lowest sputtering yield become enriched in the outermost monolayer, while the species with the highest yield are depleted. In the case of multiphase materials, phases with the higher yield are preferentially etched, which alters the phase composition at the surface and introduces microtopography and roughness. For polycrystalline materials, variation in sputter yield with crystallographic orientation can also lead to the generation of surface roughness during sputtering. These effects can influence the quality and interpretation of a SIMS analysis.

Secondary Ion Emission

The species ejected from a surface during sputtering can be monatomic, polyatomic, and multicomponent, and each of these can be positively charged (+), negatively charged (−), multiple charged (+p, −p), or neutral (0). Thus, the sputtering yield, S, for a compound of composition i_xj_y can be expanded to:

$$S = \{(i^+) + (j^+) + (i^-) + (j^-) + (i_2{}^+) + (j_2{}^+) + (i_2{}^-) + (j_2{}^-) + (i^{2+}) + (j^{2+}) \ldots (i_n{}^{\pm p}) + (j_n{}^{\pm p}) + (i_2j^+) + (i_2j^-) \ldots + (i_nj_m{}^{\pm p}) + (i^0) + (j^0) + (ij^0) \ldots + (i_n{}^0)(j_n{}^0)(i_nj_m{}^0)\} \div I_p \quad \text{(Eq 1)}$$

where ( ) represent the fluxes of secondary ions and neutrals, and I_p represents the primary ion (or neutral) beam flux.

Whereas the sputter yield, S, defines the total number of species ejected per incident ion, the secondary ion yield, $S_{in}{}^{\pm p}$, defines the number of positive ions i^+, i^{+2}, $i_2{}^+$, $i_n{}^{+p}$, or negative ions, i^-, i^{-2}, $i_2{}^-$, $i_n{}^{-p}$, and in the case of multicomponent systems, molecular ions $i_nj_m{}^{+p}$ ejected per incident primary ion. It is related to the sputter yield, S, through a factor termed the ionization probability $\gamma_{in}{}^{+p}$; for example, in the case of material i_xj_y, one can define the secondary ion yield for i^+ as:

$$S_i{}^+ = \gamma_i{}^+ S(i_xj_y) \quad \text{(Eq 2)}$$

where $\gamma_i{}^+$ is the number of i^+ ions ejected divided by the total number of species ejected that contain i, and $S(i_xj_y)$ is the sputter yield for material i_xj_y.

Secondary ion yield is an important parameter, because it determines the relative intensities of the various SIMS signals. Secondary ion yield depends on the same factors as the sputter yield, but it also depends on the ionization probability. Most models of secondary ion emission theory, particularly ionization probability, emphasize the importance of chemical and electronic effects (see the Selected References in this article). Accordingly, the presence of reactive species at the surface is believed to modify the work function, the electronic structure, and the chemical bonding, and all of these can influence the probability that a sputtered species will be ejected in a neutral or charged state.

It is necessary to recognize that the secondary ion emission phenomenon is sensitive to the presence of electropositive or electronegative species at the sample surface. Thus, sputtering of surfaces in the presence of oxygen enhances the production of many positive metal ions, because the electronegative oxygen increases the probability that the sputtered atom or molecule can release an electron. Oxygen can be a constituent of the sample (e.g., a metal oxide), it can adsorb onto a metallic or semiconductor surface during the sputtering (e.g., from an oxygen gas jet or a high residual oxygen pressure in the vacuum environment), or it can be implanted in the surface during sputtering (e.g., by using an oxygen primary ion beam). In all cases, the production of positive ions is enhanced relative to that observed in the absence of the oxygen (or any other electronegative species); Table 1 shows the effect of oxygen on the absolute secondary ion yields of various metals. Similarly, sputtering of surfaces in the presence of electropositive species (e.g., the use of Cs^+ primary ion beams) leads to an enhanced production of negative secondary ions.

There is an alternative to this chemical secondary ion emission mechanism. In the case of very clean surfaces under inert ion bombardment in a UHV environment, intrinsic secondary ion emission may be observed. Under these conditions, production of secondary ions is greatly reduced (Table 1, clean surfaces). However, this situation is seldom achieved in typical SIMS analyses because of the presence of reactive species in the specimen or vacuum environment, nor is it desirable, due to the need to enhance secondary ion production for detection sensitivity.

Table 1 Effect of oxygen on positive secondary ion yields in metals

Element (M)	Sputtering yield ($S_M{}^+$), clean surface	Sputtering yield ($S_M{}^+$), oxygen-covered surface
Magnesium	0.01	0.9
Aluminum	0.007	0.7
Titanium	0.0013	0.4
Vanadium	0.001	0.3
Chromium	0.0012	1.2
Manganese	0.0006	0.3
Iron	0.0015	0.35
Nickel	0.0006	0.045
Copper	0.0003	0.007
Strontium	0.0002	0.16
Niobium	0.0006	0.05
Molybdenum	0.00065	0.4
Barium	0.0002	0.03
Tantalum	0.00007	0.02
Tungsten	0.00009	0.035
Silicon	0.0084	0.58
Germanium	0.0044	0.02

Source: Ref 7

Secondary ion yield, which determines the measured SIMS signal, is a very sensitive function of the chemical and electronic properties of the surface under bombardment. Thus, it is dependent on the element, its matrix, and the bombarding species being implanted in the surface during the analysis. It is also influenced by residual gas pressure and composition during the analysis, because adsorbates can modify the chemical and electronic state of the surface monolayer.

In the case of inert primary beam bombardment, for example, Ar^+ on aluminum versus aluminum oxide, the positive metal ion yield is 3 to 4 orders of magnitude higher in metal oxides than in their pure metal counterparts. The ion yield dependence is due to the ionization probability, $\gamma_{Al}{}^+$, which is approximately 100 times greater for Al_2O_3 than for aluminum metal, not to the sputtering yield, which is approximately 2 times greater for the metal than for the metal oxide.

Similarly, Ar^+ bombardment of a pure aluminum metal sample produces a larger Al^+ signal in a dirty vacuum or in the presence of an intentional oxygen leak than in a nonreactive UHV environment. Therefore, most modern approaches to SIMS analysis—at least when quantitative elemental analysis is of interest—use reactive primary ion beams rather than inert ion beams; an oxygen beam ($O_2{}^+$ or O^-) or a cesium (Cs^+) beam is typically used. Thus, the surface is always saturated with a reactive species (due to the primary ion implantation) that enhances the ion yield and makes the elemental analysis less sensitive to matrix effects and/or to the residual vacuum environment during analysis.

Crystallographic orientation further compounds matrix dependence of the secondary ion yield. This is due primarily to the difference in electronic properties (e.g., work function or band structure) from one crystal face to another and to the difference in adsorptivity or implantation range from one face to another (and much less so due to variation in sputtering yield). In the case of polycrystalline and/or multiphase materials, emission intensity can vary considerably from one grain to another, which can be an important source of contrast in secondary ion emission imaging of polycrystalline materials.

Regarding the energy and angular distribution of the ejected species, secondary ions are ejected with a wide distribution of energies. The distribution is usually peaked in the range from 1 to 10 eV, but the form of the distribution varies depending on the identity, mass, and charge of the specific secondary ion. In general, the monatomic species (for example, i^+ or j^+) have the widest distribution, often extending to 300 eV under typical conditions;

the molecular species (such as i_2^+ or i_2j^+) cut off at lower energies. The energy distribution of the ejected secondary ions is relevant to the design of the SIMS instrument (because it must be energy filtered before mass analysis) and to the mode of operation (because i^+ can often be resolved, for example, from i_2^+ or j_2^+ on the basis of energy).

Instrumentation

The quality and applicability of SIMS analysis depends strongly on the details of the instrumentation. Secondary ion mass spectroscopy instruments are categorized according to versatility, mass resolution, primary beam characteristics, and, imaging capability.

Simplest Instrument

This instrument is sometimes called an add-on, macro-, or broad-beam instrument, intended primarily for surface analysis and qualitative depth profiling and less so for quantitative elemental analysis, microanalysis, and imaging. It is seldom a dedicated unit but rather a set of components often used in conjunction with an Auger or x-ray photoelectron spectrometer. Figure 1 shows a simple SIMS system. The instrument uses a standard electron-impact inert ion gun, which is often the same sputter gun used for sample cleaning or Auger sputter profiling. The mass spectrometer is usually a quadrupole type, but to select a portion of the wide energy distribution of sputtered secondary ions, it requires an energy prefilter. Other than the addition of the energy prefilter, the quadrupole mass spectrometer and the ion detector are very standard.

Dedicated Instrument

This instrument often incorporates a more intense, finely focused primary ion beam suitable for probe imaging. Because it is unnecessary to provide the flexibility for performing analyses by Auger electron spectroscopy (AES) or x-ray photoelectron spectroscopy (XPS) in this type of instrument, sample orientation and the ion-collection system can also be optimized. These instruments, often called ion microprobes, are designed primarily for quantitative in-depth profiling and microanalysis and less so for surface analysis. Figure 4 shows a typical ion microprobe that uses a duoplasmatron ion source to generate inert, as well as reactive, primary ion beams. The instrument shown in Fig. 4 uses an energy-filtered quadrupole mass spectrometer, but in principle, a magnetic analyzer could also be used.

Direct-Imaging Instrument

Usually called an ion microscope, the direct-imaging SIMS instrument creates a secondary beam with the ejected secondary ions. The secondary beam is mass analyzed in a double-focusing electrostatic/magnetic sector; transmitted secondary ions can be counted or used to create a direct image of the sample surface on a microchannel plate. Figure 5 shows a schematic diagram of a direct-imaging ion microscope in which the primary ion beam and secondary ion beam are clearly distinguished. The instrument is designed primarily for microstructural analysis and imaging, quantitative in-depth profiling, and trace element analysis. It is not suitable for true surface analysis, and due to the precise ion-collection optics, the ability to treat or manipulate the specimen in situ is essentially lost.

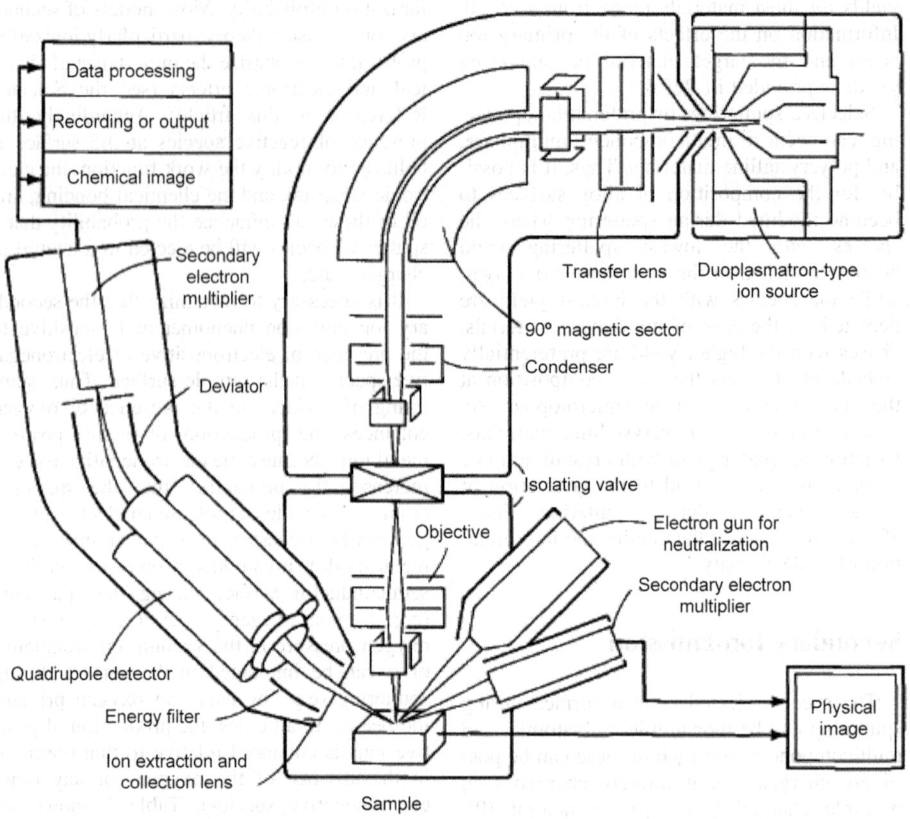

Fig. 4 Schematic of the layout and components used in an ion microprobe. Courtesy of ISA Inc., Riber Division

System Components

The design and operation of the various instrumental components is not discussed in detail. More detailed information is available in the literature (Ref 8–10). However, several instrument components and features whose specific design and/or availability can influence the quality of the analysis are summarized as follows.

The quality of the vacuum environment during SIMS analysis is important. Any variations in the vacuum pressure can influence the secondary ion intensity due to the effects of reactive species on ionization probability and, to a lesser extent, on sputtering rate. Thus, the day-to-day reproducibility and the quality of the depth profile, which must be acquired under constant vacuum pressure, are affected. The pressure and composition of the vacuum environment influence the background levels and therefore the detection limits for some species. This is usually due to their adsorption onto the surface and their subsequent ejection as secondary ions. The background species of most concern are hydrogen, oxygen, and water, because they are readily adsorbed and quite prevalent in most vacuum systems. Ideally, pressure of the order of 10^{-9} to 10^{-10} torr should be maintained at the specimen surface, especially for hydrogen (Ref 11), oxygen, and carbon analysis.

In the case of the primary ion beam, several features are of interest. First, the ability to generate reactive primary ions, for example, O_2^+, O^-, and Cs^+, requires a duoplasmatron or liquid metal source; more conventional electron-impact ion guns typically used in surface analysis are limited to inert species (He^+, Ne^+, and Ar^+). The negative oxygen species is of great benefit for the analysis of insulators. Second, beam diameter and maximum current density characteristics vary with the gun type and design. Simple ion guns usually provide beam diameters (spot sizes) of the order of 25 to 250 μm; dedicated ion microprobes and ion microscopes produce spot sizes as small as 1 to 2 μm. Third, background levels that can influence detection limits are controlled to some degree by the ion source. This refers not only to residual gases but also to metallic species, which can become incorporated in the beam, implanted in the sample, and ultimately appear in the analysis of the sample.

The ability to pump the ion source differentially (relative to the sample chamber) can greatly alleviate the background due to residual gases and other contaminants produced in the ion source. However, metallic and other impurities that become ionized and thus incorporated in the primary ion beam are most effectively eliminated using a primary beam mass filter, essentially a mass spectrometer that filters primary ions before they strike the specimens. A primary beam mass filter also rejects any neutral species in the beam. The use of a primary beam mass filter is a necessity for trace element and low-level dopant analyses. Finally, rastering of the primary ion beam over the specimen surface is required in SIMS depth profiling and secondary ion imaging.

Secondary ion-collection optics vary among commercial and laboratory-built SIMS systems. Ideally, secondary ions are extracted electrostatically from the center of the crater (over the maximum solid angle) instead of simply collecting that portion of the ejected flux of secondary ions that intersects the entrance aperture of the analyzer. Many dedicated SIMS instruments include an extraction lens over the sample, which limits the sample geometry that the instrument can tolerate and the capability for a multitechnique apparatus. Extracted secondary ions are energy analyzed to select the optimum portion of the energy distribution for subsequent mass filtering, to preclude severely degrading mass resolution. This is a necessity for quadrupole spectrometers and single-focusing magnetic sectors, whereas an electrostatic energy filter is an inherent component of higher-resolution double-focusing magnetic mass spectrometers. Various electrostatic energy filters are available for use with quadrupoles, including grids, apertures, Bessel boxes, and spherical or hemispherical plates (Ref 12); the collection efficiency, selectivity, resolution, and transmission of these filters vary and influence the overall sensitivity of the SIMS instrument.

Instrumental features required for secondary ion imaging are presented in the section "Ion Imaging" in this article, while the differences between quadrupole and high-resolution magnetic mass filters are described in the next section of this article. Data acquisition for spectral and depth-profiling analyses simply requires an ion detector and an automated control for scanning the mass spectrum or, in the case of depth profiling, for sequential acquisition of selected peak intensities in the spectrum. The ion detector should be electronically gated in all instruments except the ion microscope, in which apertures can be used to eliminate crater-edge effects. Instrumental features for computerized data acquisition, data manipulation, and output vary considerably.

Secondary Ion Mass Spectra

Figure 6 shows the positive SIMS spectrum for a very high-purity silicon sample (Ref 13). The bar graph was obtained in an ion microscope using an O_2^+ primary ion beam and a double-focusing mass spectrometer. The spectrum shows the presence of silicon isotopes $^{28}Si^+$, $^{29}Si^+$, and $^{30}Si^+$ and thus demonstrates the isotopic sensitivity of SIMS. Also shown are the polyatomic species Si_2^+ and Si_3^+ and the molecular species SiO^+, SiO_2^+, Si_2O^+, $Si_2O_2^+$, Si_3O^+, $Si_3O_2^+$, $Si_3O_3^+$, and so on. In this case, the oxide species are not representative of the silicon sample but are due to the use of the oxygen primary beam. The oxygen ions are implanted in the surface, and these charged molecular secondary ion clusters are subsequently sputter ejected. In addition, the $^{16}O^+$ and $3_2O_2^+$ secondary ions are also generated.

The use of an oxygen primary ion beam has enhanced positive secondary ion yields; therefore, the measured secondary ion signals are quite strong. If the specimen was analyzed using an inert primary ion beam (for example, argon), the positive secondary ion signal intensities would be 100 to 1000 times less intense, and the presence of the aluminum impurity ($^{27}Al^+$) probably would not have

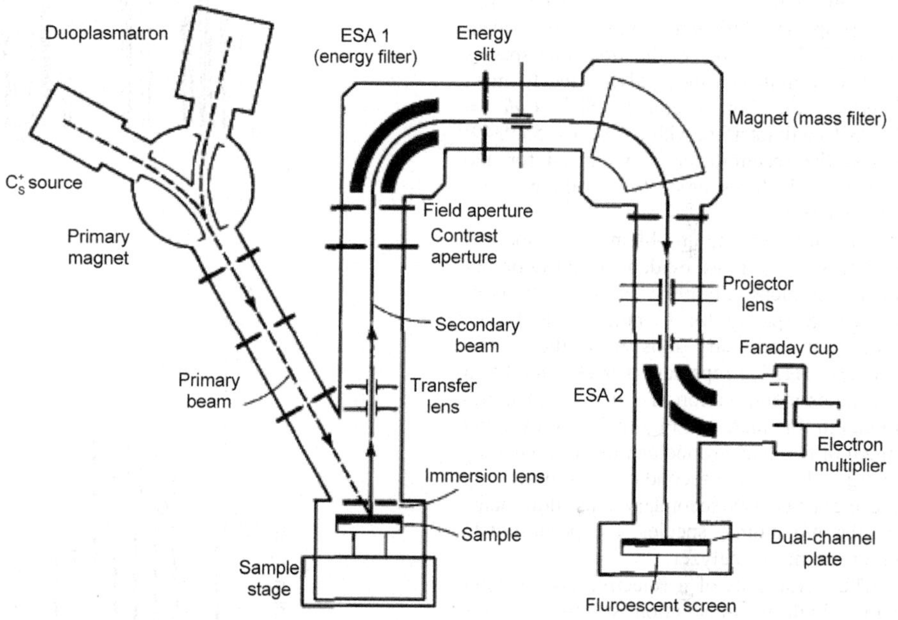

Fig. 5 Schematic of the layout and components used in an ion microscope. Courtesy of Surface Science Western

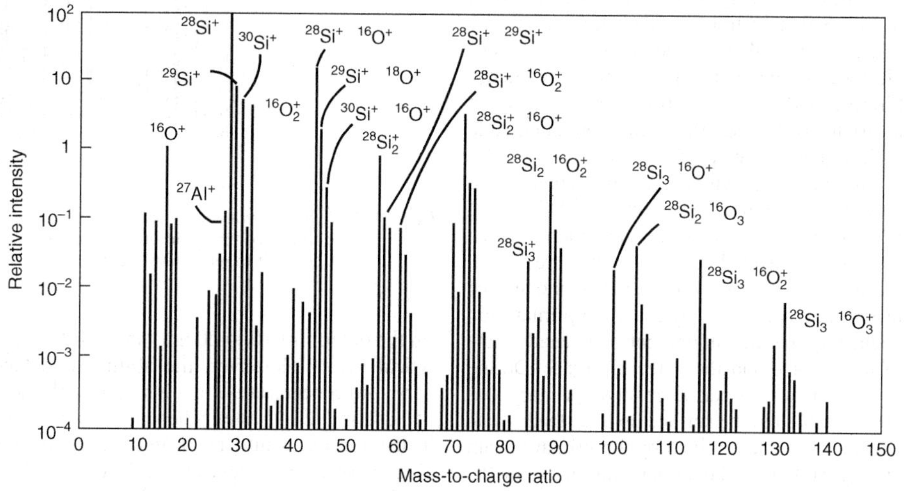

Fig. 6 Positive secondary ion mass spectroscopy spectra (in the form of a bar graph) for high-purity silicon under oxygen bombardment in an ion microscope. Source: Ref 13

been detected. In some instances, it could be considered a disadvantage to use an oxygen primary beam to enhance the secondary ion yields, because it precludes an oxygen analysis in the specimen. However, it is common to use an ^{18}O primary ion beam when ^{16}O analyses are of interest or to use a Cs^+ beam (if available), because it enhances the secondary ion yield for electronegative species such as oxygen.

Other peaks in the spectrum in Fig. 6 that have not been identified are due primarily to the molecular species that form between the silicon sample and implanted primary ions as well as between the Si_n, O_n, and Si_nO_m species and ^1H, ^{12}C, and ^{14}N in the residual vacuum. The SIMS spectrum is very complex due to this multitude of molecular secondary-ion-matrix species. Although it may not be evident in Fig. 6, the presence of the molecular species further complicates the analysis due to mass interferences. For example, Si_2^+ (m/e = 55.9539) will interfere with Fe^+ (m/e = 55.9349) unless the spectrometer is optimized for and capable of high resolution (m/e is the mass-to-charge ratio).

The interference problem and spectral complexity are more evident in multicomponent specimens. Figure 7 shows portions of the SIMS spectra for a stainless steel. The molecular secondary ions from the matrix species (Fig. 7a) represent a background that precludes identification of minor or trace elements in this material. Figure 7(b) shows that much of this background can be eliminated by energy-filtering the secondary ions; here, only the higher-energy secondary ions, dominated by the monatomic species, are permitted to enter the mass analyzer.

The advantages of a selective energy filter and a high-mass-resolution spectrometer are further demonstrated in Fig. 8. The CuTi specimen produced a doublet at mass number 63 that could be resolved to show the presence of both Cu^+ (m/e = 62.9296) and TiO^+ (m/e = 62.9467) secondary ions. The spectra shown in Fig. 8 demonstrate the detection sensitivity and resolution that can be achieved in a double-focusing magnetic spectrometer by energy filtering and high-resolution mass filtering; it is 10 to 100 times the resolution obtainable with a quadrupole spectrometer.

The spectra shown in Fig. 6 to 8 represent the bulk composition of these samples. This is due in part to the nature of the specimens, which were not subjected to any surface treatment, but it is also a result of the analysis mode. The use of an oxygen (or any reactive) primary ion beam requires the attainment of a steady state in the surface and the detected signal. During this transient time period, reactive primary ions are implanted in the surface up to a steady-state concentration determined by the primary beam energy and flux, the target material, and the total sputtering rate. Ion yields change during this period; therefore, the SIMS spectra generated during the first few minutes of sputtering (in the outermost 10 to 20 nm) are of little analytical value. Moreover, the ability to detect very low concentration levels requires the use of high sputtering rates when data acquisition in the surface would be difficult even in the absence of the transient effect. Thus, the data in Fig. 6 to 8 represent trace element analyses in the near surface or bulk, not true surface analyses.

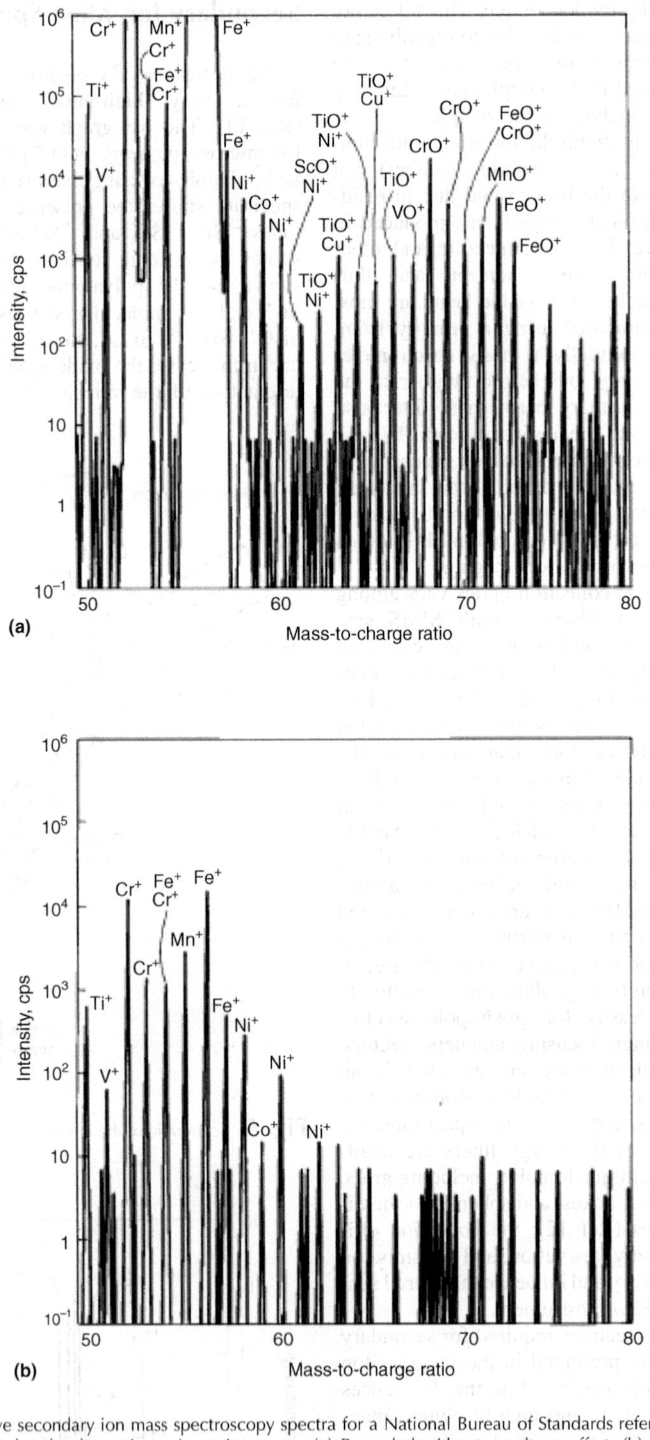

Fig. 7 Positive secondary ion mass spectroscopy spectra for a National Bureau of Standards reference steel under oxygen bombardment in an ion microscope. (a) Recorded without a voltage offset. (b) Recorded with a voltage offset to reject low-energy molecular secondary ions

Figure 9 presents the positive SIMS spectrum for a silicon substrate upon which an ~100 nm thick organometallic silicate film was deposited. In contrast to the spectra shown in Fig. 6 to 8, this spectrum was obtained in an ion microprobe using an argon primary ion beam and an energy-filtered quadrupole mass spectrometer. Due to the oxide nature of the film, the ion yields are more than enough to produce measurable intensities, even though an inert argon primary ion beam is used. Because the spectrum is presented on a linear scale, much of the background due to the molecular secondary ions is eliminated.

The spectrum indicates the presence of hydrogen and hydroxylated molecular oxide species, but the quadrupole analyzer has insufficient resolution to separate the interferences. Although some of these molecular species could be due to hydrogen background in the vacuum system, they are primarily indicative of the hydrated nature of the film. There also appear to be boron, carbon, sodium, and potassium impurities in the film. These impurities were undetected in a corresponding Auger analysis, further exemplifying the high sensitivity of SIMS. The spectrum demonstrates the use of SIMS for a thin-film analysis, in this case using an inert primary ion beam. In the next section in this article, a depth profile of this film demonstrates the effects of the matrix change (at the oxide film/semiconductor substrate interface). In-depth analysis of oxide and other nonmetallic films and coatings on metallic and semiconducting substrates is a common objective of SIMS analyses in metallurgy and materials science.

Due to the high detection sensitivity in SIMS, almost any constituent can be found in the SIMS spectrum. The analyst must determine whether or not the constituents are inherent to the sample or to the instrument background and what concentration levels the peak intensity represents. Thus, even qualitative analysis with SIMS is a nontrivial matter, because the presence of certain species in the spectra can be independent of the sample and/or can be at levels that are negligible relative to the problem.

Depth Profiles

Most analytical applications of SIMS, except static SIMS, do not emphasize true surface compositional analysis. Rather, in-depth profiling (from 20 to 2000 nm), bulk impurity analysis, and imaging of microstructural features are more common applications of SIMS. Of these, quantitative depth profiling with high detection sensitivity and depth resolution is the strength of SIMS. In a depth-profiling analysis, one or more of the secondary ion signals are monitored as a function of sputtering time (or depth) into the surface or bulk of the specimen or through an adherent thin film or coating.

The profiles shown in Fig. 10 illustrate several important points about matrix effects and the use of reactive primary ion beams in thin-film profiling. The profiles were obtained for the film whose spectrum is shown in Fig. 9. The interface between the silicate film and silicon substrate is evident. This is due to the change in secondary ion yield between the oxide thin film and the silicon substrate. Although the substrate is pure silicon, the $^{28}Si^{+}$ signal falls by nearly 2 orders of magnitude between the film and substrate. In contrast, although the effects are fundamentally related, the $^{28}Si^{2+}$ signal increases by a factor of 10 upon sputtering through to the substrate. These are matrix effects on secondary ion yields (or, more specifically, the ionization probability), not indications of any concentration or chemical structure change at the interface. If an oxygen primary ion was used in this analysis, the changes in secondary ion signal for $^{28}Si^{+}$ and $^{28}S^{2+}$ would not occur (at least under ideal conditions), although a transient would be observed in the signals at the interface (during which time the oxygen primary ion implantation would reach a steady state). Clearly, the use of an inert primary

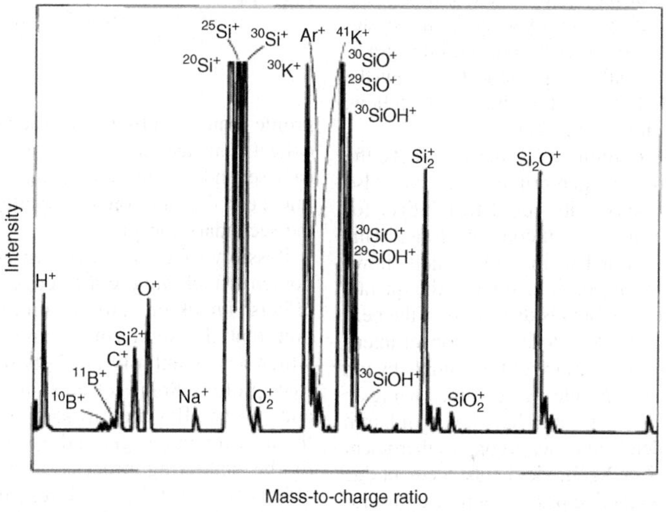

Fig. 9 Positive secondary ion mass spectroscopy spectra for an organometallic silicate film deposited on a silicon substrate acquired using a scanning ion microprobe under inert argon bombardment

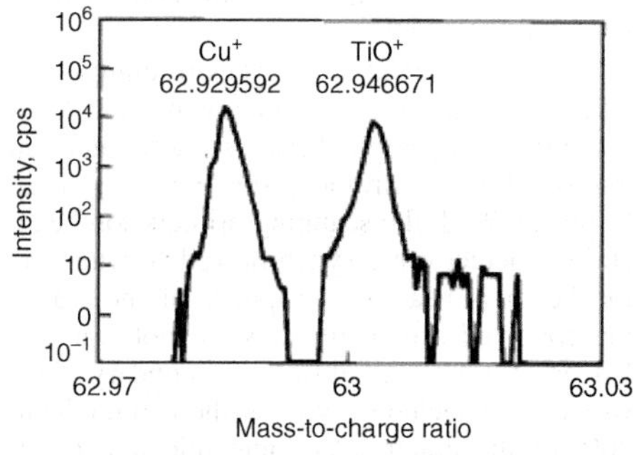

Fig. 8 High-resolution mass scan over the region of *m/e* = 63 for a CuTi specimen acquired using an ion microscope

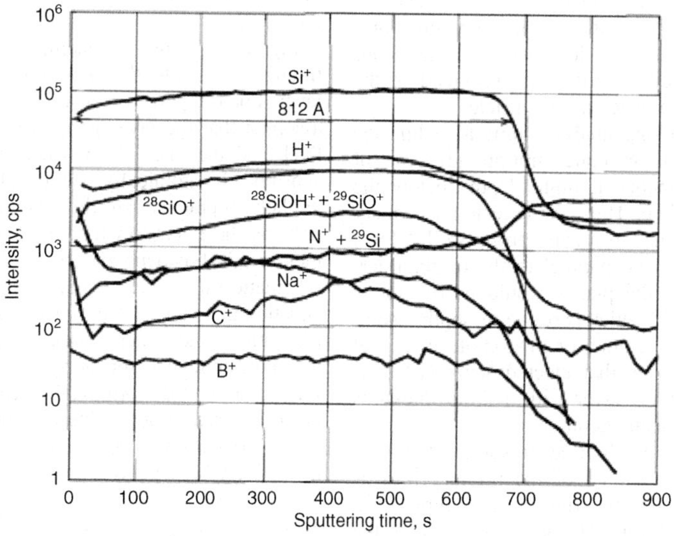

Fig. 10 Depth profiles for the organometallic silicate film shown in Fig. 9. Note the matrix effects at the film-silicon substrate interface due to the use of an inert argon ion beam. Acquired using raster gating in a scanning ion microprobe

beam enhanced location of the film-substrate interface.

Advantages from using an oxygen primary ion beam for a depth-profiling analysis of this type have little to do with the silicon species as such but rather concern the interpretation of the other signals. For example, the $^{1}H^{+}$, $^{12}C^{+}$, $^{23}Na^{+}$, and $^{11}B^{+}$ signals also drop at the interface. However, the extent to which the signal drop is due to the matrix effect versus any real change in concentration is unknown. Due to the large variation in ionization between the oxide film and semiconducting substrate, the concentrations of hydrogen, carbon, sodium, and boron could increase, even though their signals decrease. The use of a reactive primary ion beam would be of great benefit in this situation. In this case, oxygen implantation would provide a constant matrix from which secondary emission could occur throughout the depth-profiling analysis. Therefore, any changes in these signals would be due to real concentration gradients. Thus, it would be possible to follow the B^{+}, Na^{+}, and H^{+} across the interface to follow, for example, diffusion effects.

In addition to minimizing matrix effects, the use of a reactive oxygen primary ion beam also enhances the sensitivity (or detectability) for positive ions in the silicon substrate. This effect is exemplified in Fig. 11, which shows a depth profile for boron in silicon (the quantitative analysis of which is discussed in the next section of this article). Boron is a dopant intentionally put into the surface of this sample by ion implantation. The profile was obtained using an oxygen primary ion beam. These levels of boron could not be detected under argon bombardment (nor with AES or XPS); yield enhancement due to the use of a reactive primary ion beam is crucial to this analysis. In addition, the $^{30}S^{2+}$ matrix signal did not vary over the profile; that is, the observed variation in the boron signal is due to a real change in concentration.

Factors that determine or influence the depth and interface resolution of the measured profiles are discussed. More detailed discussions of SIMS depth profiling are provided in the Selected References in this article.

Depth-profiling analysis aims to define the concentration or concentration gradient of selected elements at finite depths below the original surface. Thus, the detected secondary ion signal must originate from well-defined planes below and parallel to the original surface. Ideally, the planes would be of atomic dimensions (in which case the in-depth resolution would be atomic or monomolecular), but Fig. 3 illustrates that ejected secondary ions originate from a layer with a finite thickness termed the escape depth.

The escape depth varies to some extent with the energy and mass of the primary ions as well as with the material, but it is typically 3 to 5 atomic layers. However, in practice, the theoretical limit on depth resolution is not achieved. At the very least, each composition or concentration data point defined in the depth profile requires sufficient erosion of material to collect a measurable and statistically significant secondary ion signal; this effective layer thickness depends on the sputtering rate and the secondary ion yield.

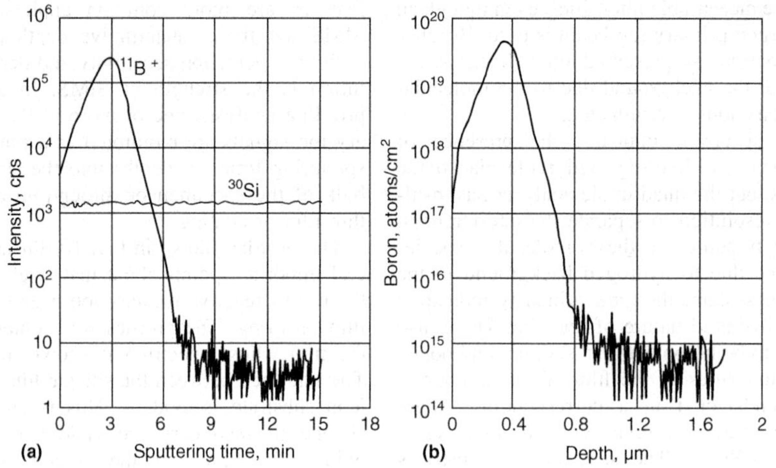

Fig. 11 (a) Raw $^{11}B^{+}$ and $^{30}Si^{2+}$ secondary ion signals versus sputtering time for a boron-implanted silicon substrate. Acquired using oxygen beam bombardment in an ion microscope. (b) Boron profile after quantitative analysis of the sputtering rate and secondary ion intensity

Possibly of greater concern are other phenomena, including crater-edge effects, halo effects, knock-on, atomic mixing, diffusion, preferential sputtering, and roughening, which are essentially artifacts that distort the true profile. Some effects are instrumental and can be alleviated to some extent through the instrument design, and others are intrinsic to the sputtering process. Altogether, these effects can limit the depth resolution in a typical SIMS depth-profiling analysis from the theoretical limit of approximately 1 nm to as much as 5 to 10 nm. Thus, special attention should be paid to the following effects when the highest possible depth resolution is required.

Figure 12 shows a schematic cross section through a sample that has a distinct surface layer. For the purposes of this discussion, it is assumed that the layer is approximately 1 μm thick and the interface between the layer and substrate is atomically sharp. (The latter is unlikely in practice, but the assumption enables defining the extent to which the SIMS measurement itself broadens the true interface.) The schematic sample design represents a common situation that can arise in practice by oxidation, nitridation, and other surface treatment of a metal or semiconductor, or by deposition of thin films and coatings. Although a SIMS analysis could be used to determine the composition or compositional uniformity of the layer, it would more likely be applied to determine:

- The impurities present in the layer, and their concentration and depth distribution
- The thickness of the film and the width of the interface
- The diffusion profile for selected surface or interfacial reactants
- Whether segregation or compound formation occurred at the interface

Effects described subsequently can influence any depth-profiling analysis, regardless of the presence or absence of a distinct surface layer, but consideration of a surface film of finite thickness with a subsurface interface best illustrates the points.

Figure 12(b) illustrates a situation that can be most detrimental to depth resolution. The crater-edge effect is the result of nonuniform erosion of the specimen caused by nonuniform distribution of ions in the primary beam. The detection system simultaneously filters and counts secondary ions over a range of depths in the sample, and depth resolution is severely degraded. The practical consequence of this in the measured depth profile is a gross broadening of interfaces, diffusion profiles, and surface layer thicknesses. The most common remedy is to produce a flat-bottomed crater (Fig. 12c) by rastering a focused ion beam in a square pattern over the specimen. In this case, most of the secondary ions originate from a well-defined depth, but there are still contributions from the side walls.

There are several approaches to eliminating side-wall contributions, which can limit depth resolution. In simple SIMS systems and ion microprobes, electronic raster gating is used. In this method, the scanning primary ion beam and the detector are synchronized so when the ion beam is near the periphery of the crater, the collected secondary ions are not counted. Under optimal conditions, secondary ions would be detected only from the central 10 to 20% of the crater; electronic rejection of 80 to 90% of the secondary ions reduces the signal and thus the detection limits. However, this is required to acquire meaningful depth

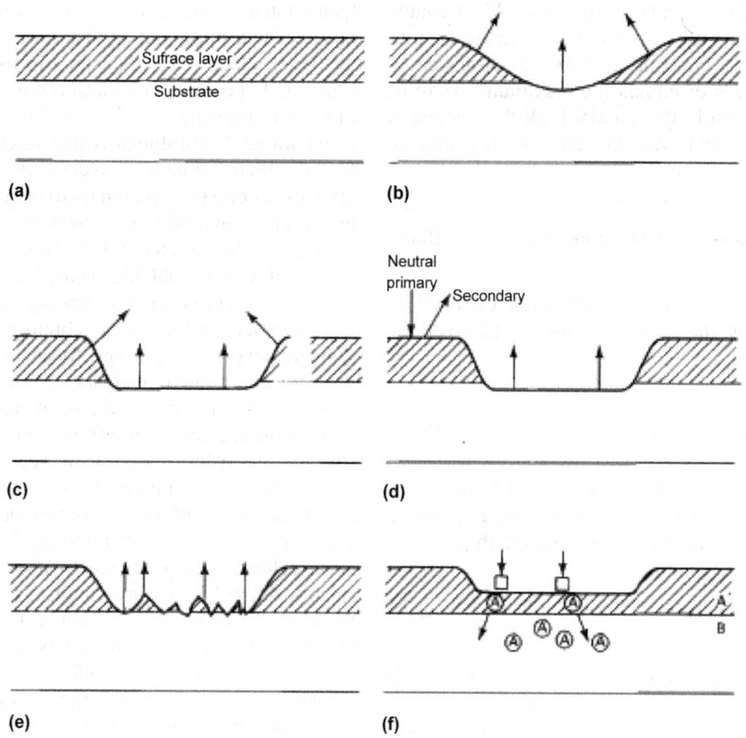

Fig. 12 Schematic representations of artifacts that can distort a secondary ion mass spectroscopy depth-profiling analysis

profiles. In the ion microscope, the direct-imaging optics preclude the need for electronic gating, because the extraction lens and field aperture combine to gate off the side-wall contributions mechanically.

The mechanical aperture used in the ion microscope has an additional advantage over electronic raster gating because it can reject secondary ions produced by neutral primary species bombarding the specimen. That is, the primary ion beam contains a neutral component of energetic species (unless the primary beam is electrostatically deflected before striking the specimen) that does not respond to the raster plates. Thus, as shown in Fig. 12(d), secondary ion ejection and detection from the original surface (and side walls) regardless of the position of the primary ion beam are possible. The analysis shows species that may be present only near the original surface even when erosion has proceeded to great depths; in this situation, it is possible to mistakenly believe that species present only at the original surface or in the overlayer have migrated or diffused deep into the specimen. A useful remedy for this (when direct-imaging ion microscopy is not available or is ineffective) is to coat the specimen with a thin film of gold or other material that is not expected to be found in the specimen during analysis.

Figure 12(e) describes another effect that must be recognized in the interpretation of depth profiles. Ion beam erosion of even atomically smooth, single-crystal materials can generate roughness, but in polycrystalline and multiphase materials, roughening can be especially severe due to differential sputtering rates (from grain to grain or phase to phase). This also broadens the appearance of interface or diffusion profiles. The scale of roughness increases with sputtering time and is usually assumed to be approximately 10% of the etch depth plus the original roughness. For example, after sputtering a smooth single crystal to a depth of 1 μm, the scale of the roughness is approximately 100 nm, and even an atomically sharp interface encountered at that depth appears to be approximately 100 nm in width.

Figure 12(f) depicts a phenomenon termed atomic mixing or knock-on, where primary ions displace atoms (e.g., across an interface) and thus distort their original distribution. This effect can be minimized to some extent by using primary ions with lower energy and higher mass. Another related source of distortion is beam-induced diffusion or migration of solute species due to heating, radiation damage, chemical interaction with implanted primary ions, and electrostatic charging of the surface. In some cases, these effects cannot be avoided, and caution is advised.

One other source of distortion in a depth profile occurs only in multicomponent materials due to a phenomenon termed preferential sputtering. This is a transient effect that occurs at the beginning of a depth profile (or at internal interfaces) due to differences in sputtering yield among the various constituents.

Preferential sputtering effect in, for example, alloy *AB* changes the relative concentration of *AB* in the sputtered surface region, which reduces the concentration of the high-yield species and enhances the concentration of the low-yield species. Thus, after the transient is complete, the steady-state sputtered surface composition yields a relative flux of secondary ions *A* and *B* that is proportional to their true concentration ratio in the material. That is, in a sample where $S_A \neq S_B$, the relative fluxes of *A* and *B* do not represent their true concentration ratio until preferential sputtering has occurred.

In a SIMS analysis, the relative secondary ion signals can change even in the absence of a true concentration gradient in the material. The transient persists only over a short range (of the order of 1 to 10 nm) and thus is another reason to ignore the first portion of the depth profile. Nonetheless, it can be a severe limitation for the surface analysis of metal alloys or for the analysis of interfaces between alloy films.

An important parameter required for the interpretation of depth profiles is the sputtering rate. The rates typically range from 0.5 to 50 nm/min and are fundamentally related to the incident ion current density at the sample surface and the sputter yield of the material for the type and energy of the primary ions being used. In practice, the sputter rate is determined for a given set of operating conditions and the material of interest by measuring the depth of the sputtered crater, which can be measured using optical interferometry or profilometry. When appropriate, sputter rate can also be obtained by depth profiling films of known thicknesses, that is, thin films whose thicknesses have been measured independently using optical techniques or Rutherford backscattering spectrometry. In this case, it must be assumed that the sputtering rate of the film is the same as that of the specimen. Therefore, the use of oxide or metallic films to determine sputtering rates of bulk materials, even if the film and the bulk are of identical composition, is not recommended.

In general, it is not necessary to measure the sputter rate for each analysis of the same material if the primary ion beam current can be accurately measured or reproduced. In some cases, it is possible to translate the sputtering rate measured for one material into a rate appropriate for another material using tabulated values of the sputter yield. However, the sputtering rate can vary through a depth profile where large concentration changes occur; that is, it is an assumption to apply a constant sputtering rate to a depth profile if the concentration or composition changes are severe.

Quantitative Analysis

The relationship between the secondary ion current, I_i^+ (that is, the positive secondary ion count rate for a monoisotopic element i), and the concentration of i in the specimen is:

$$I_i^+ = I_p \cdot S \cdot \gamma_i^+ \cdot C_i \cdot \eta \qquad \text{(Eq 3)}$$

where I_p is the primary ion beam current, S is the sputter yield, γ_i^+ is the ionization efficiency for i^+, C_i is the atomic fraction of i, and η is an instrumental factor that characterizes the collection, transmission, and detection efficiency of the instrument, that is, the ratio of ions i^+ emitted to ions i^+ detected. The problem in quantitative SIMS analysis is that the measured signal I_i^+ depends not only on the concentration of i in the specimen but also on the specimen matrix (S, γ_i^+) and the electronic properties of the surface γ_i^+.

Secondary ion mass spectroscopy—at least in a practical sense—is limited to the quantitative analysis of the impurities, dopants, and minor elements detected in a depth-profile or bulk impurity analysis; SIMS is not used extensively for quantitative analysis of the surface or bulk stoichiometry of unknowns.

A quantitative SIMS analysis requires independent knowledge of the sputtering rate (see the previous section in this article) and a calibration of the secondary ion signal using standards whose matrix and surface electronic properties match those of the specimen. The use of reactive primary ion beams facilitates reproduction of the surface electronic properties between the specimens and the standards. The methods for preparation and calibration of standards with matrices identical to the specimens vary, but ion implantation of known doses of the element of interest into the matrix of interest is the most convenient (Ref 14) and is described subsequently.

The use of ion-implanted standards for quantitative SIMS analysis assumes a linear relationship between the SIMS signal and the elemental concentration. That is, the product of I_p, S, γ_i^+, and η (Eq 3) is assumed to be a constant for the specimen and the standard (at least during analysis). This is true only at low concentrations (that is, dopants and impurities in the sub-ppm to 1000 ppm range); at higher concentrations, electronic and chemical properties of the matrix can become a function of the concentration. The basic concept behind the use of ion-implant standards is that the implant fluence F determines the total number of atoms implanted per unit area; F (atoms or ions/cm^2) can be measured and controlled to within approximately 5%. Thus, the secondary ion signal, when integrated over the entire profile measured for the implant standard (Fig. 11a), can be related to the total number of implanted atoms contained in the specimen:

$$\int_0^z I_i^+(z)dz = K\int_0^z C_i(z)(dz) \qquad \text{(Eq 4)}$$

where z is the depth of the analysis, and K is a constant that accounts for the terms in Eq 3; that is, K is the calibration factor used to convert the secondary ion signal (I_i^+ in counts per second) measured over the range dz into an average concentration (C_i in atoms/cm^3). If z exceeds or equals the maximum extent of implantation in the standard, all the implanted atoms per unit area will have been analyzed, and:

$$\int_0^z C_i(z)(dz) = F(\text{atoms or ions/cm}^2) \qquad \text{(Eq 5)}$$

Thus, the calibration factor can be obtained by integrating the secondary ion signal over the depth z:

$$K = \frac{\int_0^z I_i^+(z)dz}{F} \qquad \text{(Eq 6)}$$

Because the secondary ion signal is measured over a time interval dt, an independent measure of the sputtering rate, $\dot{z}$, is required; then:

$$K = \frac{\dot{z}}{F}\int_0^t I_i^+(t)dt \qquad \text{(Eq 7)}$$

where t is the total time over which the profile is integrated.

The calibration constant K can now be used to convert the measured secondary ion signal into absolute concentration in atoms per cubic centimeter:

$$C_i = \frac{1}{K} I_i^+ \qquad \text{(Eq 8)}$$

The secondary ion signal is usually measured in counts per second (averaged over a time interval dt); therefore, the units of K are essentially (atoms/cm^3)/cps. This constant is valid only for element i in a matrix that is identical to that of the standard. Moreover, the concentration C_i in the unknown must be in the dilute range, and the instrumental parameters used during analysis of the unknown must be the same as those used to obtain K from the unknown. Ideally, the standard is analyzed together with the unknowns. However, when the specimens and instruments are well characterized and controlled, or it is impractical to examine the implanted standard repeatedly, a normalized calibration constant, K_n, can be determined. It is assumed that the ratio of the terms in Eq 3 for the element $i(I_i^+)$ and some matrix species $M(I_M^+)$ is constant, that is:

$$K_n = \frac{\dot{z}}{F}\int_0^t \frac{I_i^+(t)}{I_M^+(t)}dt \qquad \text{(Eq 9)}$$

and then:

$$C_i = \frac{1}{K_n} \cdot \frac{I_i^+(z)}{I_M^+(z)} \qquad \text{(Eq 10)}$$

Perhaps the most obvious use of the calibration factor K is for quantitative analysis of the ion-implantation profile itself (Fig. 11a). Quantitative analysis of ion-implantation profiles is an important application of SIMS in the microelectronics industry. In this case, implanted specimens (assuming the implanted fluence is known) are self-standardizing; that is, the integrated profile provides the calibration factor, which can be used to convert the instantaneous secondary ion signals to concentration units. The conversion is shown in Fig. 11(b) for the raw data plotted in Fig. 11(a). However, the use of K or K_n (obtained using the implanted standard) to convert the secondary ion signals from another sample is more common.

Secondary ion mass spectroscopy is often used for quantitative analysis when concentrations from 0.1 to 10 at.% are of interest. In this case, it is necessary to verify the existence of a linear relationship between the secondary ion signal and concentration. Thus, it is common to prepare a set of standards that bracket the concentration range of interest. Calibration curves shown in Fig. 13 and 14 illustrate these relationships for a glass matrix (where the analysis of tin is of interest) and a steel matrix (where the analysis of nickel is of interest). Normalization of the secondary ion signal to a matrix species facilitates the definition of a practical calibration constant.

Time-of-Flight Secondary Ion Mass Spectrometry

Time-of-flight secondary ion mass spectrometry (TOF-SIMS) is an analytical technique that uses a high-energy, primary ion beam to probe the surface of a solid material. The instrument is typically operated in the

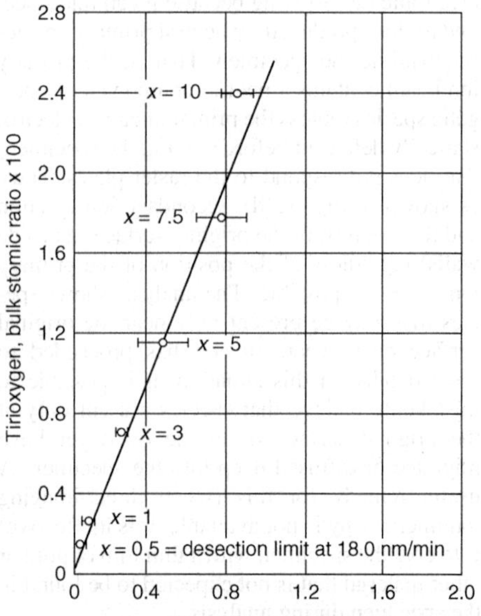

Fig. 13 Relationship between $^{120}Sn^+$ secondary ion signal (normalized to the $^{16}O^+$ signal) and relative tin content of various tin-oxide-doped silicate glasses

"static" mode to obtain elemental and molecular-chemical information from both organics and inorganics. In this mode of operation, sample integrity and chemistry are preserved by applying extremely low primary ion doses (less than 1×10^{12} ions/cm^2) during the entire experiment. This ensures that roughly less than 0.1% of the surface atoms or molecules are ever struck and damaged by the primary ion beam. Time-of-flight secondary ion mass spectrometry is the most surface sensitive of the surface analytical techniques, with a depth of analysis of only approximately 2 nm. The technique has extremely good detection sensitivities, with detection limits for most elements in the parts-per-thousand to parts-per-million range. By using a finely focused ion beam (gallium, indium, or gold), it is also possible to record the lateral distribution of chemical species across a surface with micrometer to submicrometer spatial resolution. Cluster ion beams (e.g., Bi_3, C_{60}, Ar_{5000}) have become available that yield more signal for organic materials and even allow for depth profiling of organics. Because secondary ion intensities vary dramatically from element to element and are highly dependent on the matrix from which they are sputtered, quantification with TOF-SIMS can be extremely difficult. Thus, in most cases, the technique is used more for qualitative purposes than for quantitative analyses. However, in select cases where appropriate standards are available (e.g., metals on silicon), accurate quantification can be performed (Ref 16).

When elemental depth profiles are required, most TOF-SIMS systems can also be operated as a "dynamic" SIMS instrument. In the dynamic mode, very high primary ion doses are used to obtain in-depth information. However, in this mode of operation, very little chemical specificity can be gleaned, due to the destructive nature of the primary ion beam.

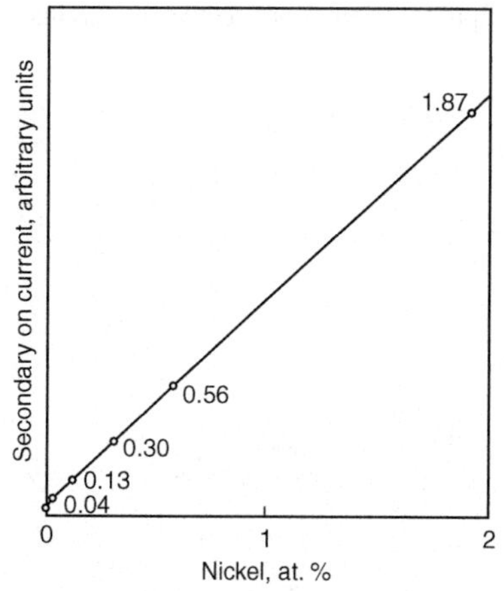

Fig. 14 Relationship between Ni^+ secondary ion signal and nickel content of National Bureau of Standards reference steels 661, 662, 663, 664, and 665. This calibration curve can be used for quantitative analysis of nickel in comparable low-alloy steels using an ion microprobe. Source: Ref 15

Therefore, the following discussion only focuses on the static mode of operation.

In a TOF-SIMS experiment, the desorption/ejection of secondary ions from the surface of a material is initiated by a short pulse (~1 ns) of primary ions that impinges on the surface at high angles of incidence. The momentum transfer from the primary beam to the solid initiates a collision cascade within the solid, much like a microscopic billiard game. A portion of this momentum is redirected back toward the surface, resulting in the ejection of atomic and molecular ions. Greater than 90% of these secondary ions originate from the outermost one to two layers of the solid, thus defining TOF-SIMS as an extremely surface-sensitive technique. By applying a potential between the sample surface and the mass analyzer, the desorbed secondary ions are extracted into a TOF mass spectrometer, where their masses are separated in flight time, with very high accuracy, based on their mass-to-charge ratio (*m*/*z*). The resulting mass spectrum, typically in the 1 to 2000 Da (dalton) range, enables unambiguous identification of chemical moieties, which can often be correlated to the original surface structure.

An example of the type of chemical/molecular information that can be obtained from secondary ion mass spectra is shown in the positive ion spectra of polyethylene terephthalate (PET), shown in Fig. 15. The spectra show a variety of large molecular cluster ions that are identified as fragments of the PET polymer chain. This provides information on the long-range molecular makeup of the sample, as well as providing a unique fingerprint for this material.

On insulating materials, the loss of secondary electrons during ion bombardment can lead to a positive charge buildup and loss of signal from the sample surface. Therefore, auxiliary electron sources are often used to supply electrons to the sample surface to help neutralize the excess charge.

Because TOF-SIMS can easily handle insulating, as well as conducting materials, the types of samples analyzed by TOF are often similar to those analyzed using XPS. However, because of its much smaller analytical probe size, TOF-SIMS systems can analyze features

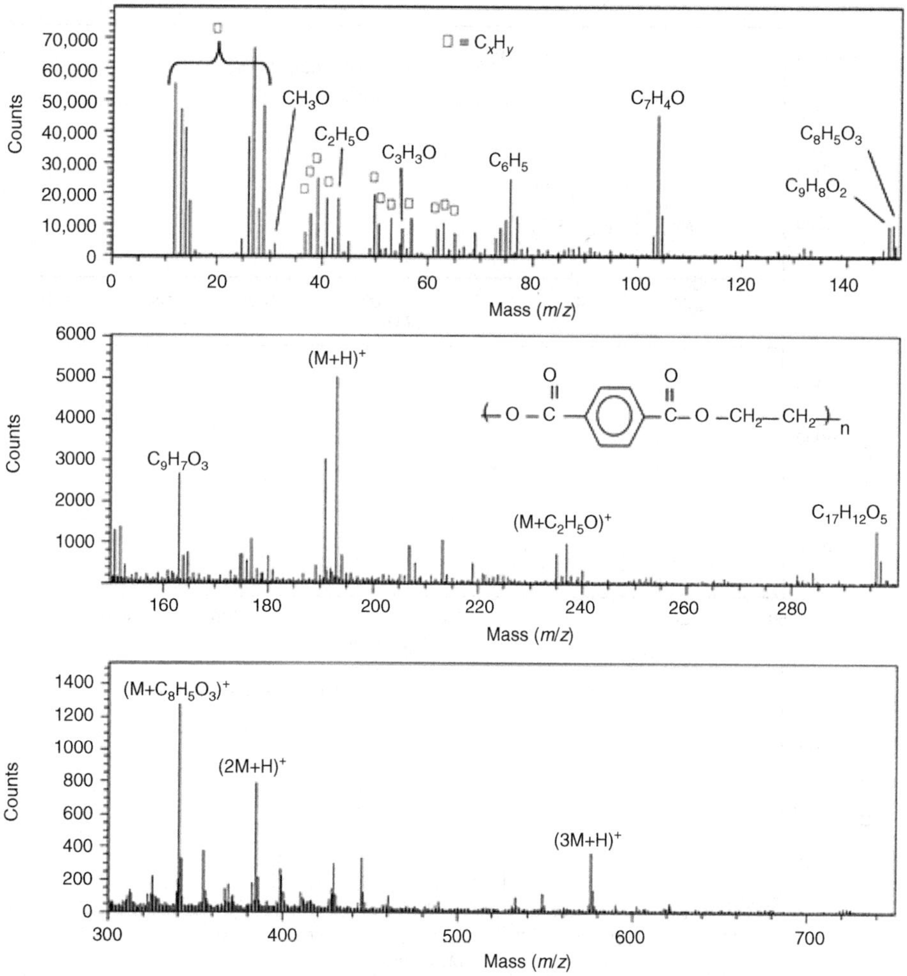

Fig. 15 Time-of-flight secondary ion mass spectroscopy mass spectra of polyethylene terephthalate

as small as a few micrometers in width and are commonly used for mapping distributions of elements and molecules across surfaces.

Application areas for TOF-SIMS include organic and inorganic contaminant identification, cleaning studies, surface segregation and modifications, corrosion, discolorations, drug distribution, diffusion studies, and chemical characterization. Two types of secondary ion data are simultaneously obtained in a TOF-SIMS experiment:

- Total-area mass spectrum
- Total secondary ion image

As the focused primary ion beam is digitally rastered across the sample surface, complete mass spectra of ion intensities versus mass-to-charge ratio are obtained at every pixel (256 × 256) within the raster. The summation of these spectra produces a total-area mass spectrum (Fig. 16). From the same acquisition, a total ion image is also obtained, which shows the lateral distribution of secondary ion signals from across the area of analysis (Fig. 17). High intensities (bright areas) on the image indicate that more secondary ion signal is present at those points than at lower-intensity (darker) points. The variation in secondary ion intensity can be due to topographical effects or from differences in chemical composition. The total ion image is used to selectively analyze the mass spectra (chemistries) from areas that show differing amounts of brightness. Conversely, the total ion mass spectrum can be used to select specific elements or molecules for display in secondary ion maps that show the relative localized abundance of these species. Because both positive and negative secondary ions are created during ion bombardment, two separate experiments must be performed for complete characterization of a given sample.

Additional information can be found in Ref 17 and 18.

Ion Imaging

Two methods used to acquire secondary ion images are the ion microprobe and the ion microscope (Fig. 18). In the case of the ion microprobe (Fig. 4, 18a), the incident primary ion beam is focused to a small spot and rastered over the sample surface. The analysis is carried out point by point, and the image is constructed by synchronizing the cathode ray tube (CRT) and the detector with the primary ion beam. That is, a particular secondary ion (for example, $^{27}Al^{+}$, FeO^{+}, etc.) is selected at the mass spectrometer, and its intensity variation from point to point is displayed on the CRT. This approach is analogous to that used for x-ray imaging in the electron microprobe. The resolution is determined by the diameter of the primary ion beam, typically 2 to 5 μm.

In contrast, the ion microscope is a direct-imaging system in which the secondary ions are simultaneously collected over the entire imaged area (Fig. 5, 18b). A strong electrostatic field between the specimen and the immersion lens preserves the spatial distribution of the emitted secondary ions. The secondary ion beam is then mass analyzed (only a double-focusing magnetic mass spectrometer is applicable for direct imaging), and the spatial distribution of the secondary ions is transmitted through the mass spectrometer. The transmitted secondary ions (for example, $^{27}Al^{+}$, FeO^{+}) strike a microchannel plate, where the image is formed. The direct-imaging ion microscope can be compared to the transmission electron microscope or emission microscope. The resolution is determined by the system optics, for which the size of the imaged area on the specimen (the field aperture) and the contrast aperture are major factors. However, the resolution is typically on the order of 0.5 to 1 μm.

The interpretation of secondary ion images must be done carefully. Because secondary ion intensity depends not only on the concentration of the imaged species but also on the chemical and electronic properties of the matrix, image contrast from point to point can have many origins. For example, in polycrystalline materials, the differences in crystallographic orientation between grains can

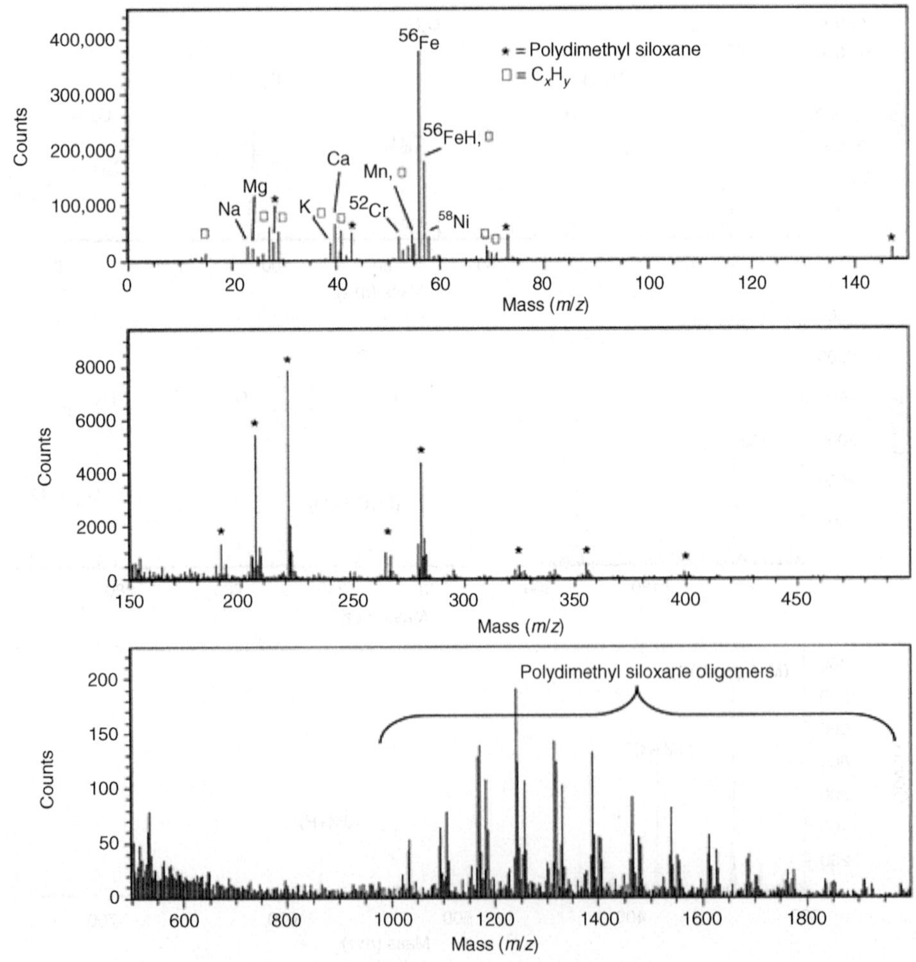

Fig. 16 Time-of-flight secondary ion mass spectrometry positive ion spectra of stainless steel surface

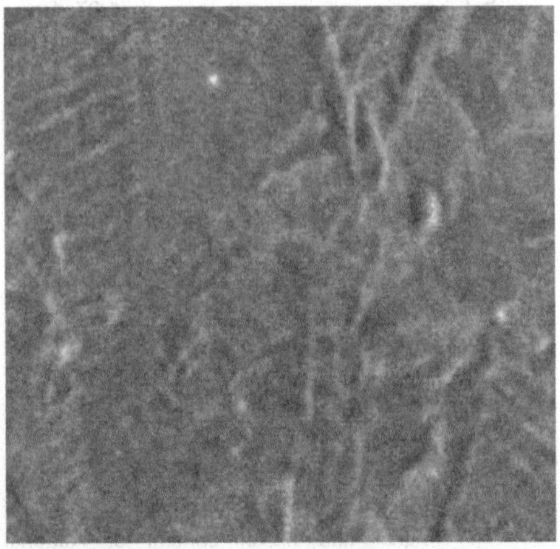

Fig. 17 Total positive ion image of stainless steel surface (50 × 50 μm) revealed by time-of-flight secondary ion mass spectrometry

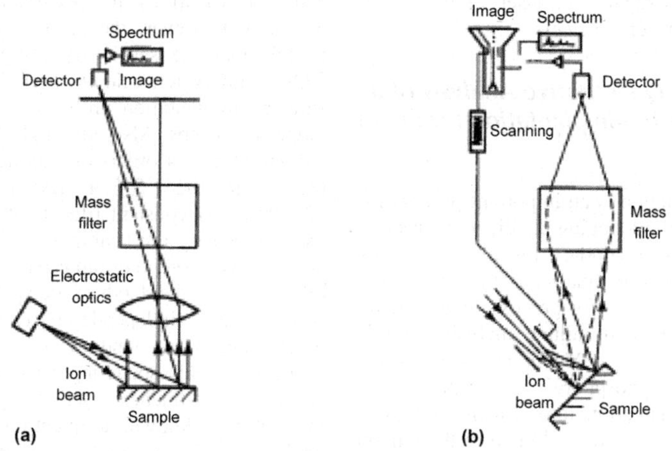

Fig. 18 Schematic diagrams of secondary ion imaging (a) in a direct-imaging ion microscope and (b) in a scanning ion microprobe

influence the adsorption of oxygen from the residual vacuum, the implantation of oxygen from the primary beam, and the sputtering rate. Any one of these can lead to contrast in the secondary ion image for one grain relative to the other, even though no concentration difference exists. The sensitivity to subtle differences in secondary ion yield between various microstructural features can be the major source of image contrast and is the basis of the ion emission microscope. The microscope is used primarily for microstructural imaging and not necessarily for the microchemical analysis of interest in SIMS.

Nonmetallic Samples

The analysis of insulators (glasses, ceramics, geological materials, and polymers) presents charging problems in SIMS, requiring special procedures and additional effort (Ref 19, 20). The charging problem associated with these materials is due to the high incident primary ion current relative to the much lower emitted secondary ion current; secondary electron emission during ion bombardment further enhances the charge imbalance. All of this leads to a net positive charge at the surface (if positive primary ions are used), and the associated surface potential influences the energy distribution of the emitted secondary ions. In general, the surface potential is unstable and thereby precludes the acquisition of meaningful spectra or depth profiles.

The optimum method for analysis of insulators depends on the specifics of the instrument and the materials. Common approaches include:

- Deposition of metal coatings on the sample that are sputtered through to provide a path for local charge neutralization at the periphery of the ion beam crater
- Use of metal grids or metal plates with apertures, which are placed over the sample for local charge neutralization
- Auxiliary charge neutralization with an electron beam or flood gun
- Use of a negative oxygen primary beam (O^-) that leads to a finite but stable surface potential after a few moments of bombardment
- Bombardment with a neutral beam (Ref 21)

Another problem associated with the analysis of insulators, especially thin insulating films and coatings, is the migration of mobile ions due to the electric field generated by the ion-bombardment-induced surface potential. Migration can severely distort the in-depth profile of, for example, sodium, potassium, chlorine, fluorine, and so on in oxide materials and films. The proper charge neutralization greatly alleviates this effect, but its presence should always be considered.

Analysis of organic, biological, and other materials with volatile constituents also requires special approaches. It is common to freeze the specimen with liquid nitrogen using a cold stage. Mass spectra of organic and biological materials are exceedingly complex. The use of SIMS to analyze these materials should be considered is cases such as for trace metal detection in biological specimens, organic polymers, and films.

Detection Limits

One of the unique characteristics of SIMS is its high detection sensitivity relative to other surface microanalytical techniques. Nonetheless, it is important to recognize that the detection limit varies with the element, sample matrix, and operating procedures of the instrument (primary ion, sputter rate, instrument transmission, and background). For example, the detection limit in metallic, semiconducting, and other nonoxide matrices is maximized only when reactive oxygen or cesium beams are used. When high spatial resolution, high depth resolution, or high mass resolution is also required, there is some loss in detection sensitivity.

One of the concerns is the rate of material consumption, because this ultimately determines the total secondary ion count rate. For example, if high spatial resolution is required in a microanalysis problem, a reduction in the beam diameter is necessary. This is often accompanied by a decrease in beam current and therefore a decrease in the rate of material consumption, even though the sputtering rate may increase. Similarly, a loss in sensitivity occurs when the sputtering rate is reduced at a constant spot size to enhance the resolution of in-depth profiles.

Another reason for a less-than-optimum detection sensitivity concerns the collection and transmission of the secondary ions. Any electronic or mechanical gating of the sputtered crater, such as that required for high-resolution in-depth profiling, reduces the detected signal. Similarly, the need for high mass resolution reduces ion transmission in the mass spectrometer and results in a loss in detection sensitivity. In addition, any background due to residual gas or primary ion beam contamination further limits detection sensitivity.

The SIMS detection limits are maximized in a bulk impurity analysis in which the highest beam current can be used, because neither depth nor spatial resolution are of concern; this assumes that low mass resolution can be tolerated. Under these conditions, the detection limits for most elements can be expected to be in the parts-per-billion to parts-per-million range, assuming an oxide matrix and/or a reactive-ion primary beam.

Applications

A variety of examples of SIMS applications are presented in the following Examples 1 to 5; some others can be found in Ref 22 to 28.

Example 1: Surface Composition Effects during Laser Treatment of AISI Stainless Steel

Laser welding of an alloy containing one or more volatile components results in selective vaporization of the more volatile components during welding, which can lead to inadequate control of the weld composition and poor mechanical properties of the fabricated product. Loss of alloying elements and the eventual properties of the weld zone are influenced by fluid flow in the weld pool, heat transfer, and the thermodynamics and kinetics of the vaporization of various components from the molten pool.

Secondary ion mass spectroscopy was used to examine the surface region of an AISI 202 stainless steel before and after various treatments with a CO_2 laser. Of particular interest were the effects on hydrogen and oxygen, because these species can profoundly influence mechanical and chemical properties. Figure 19 shows the oxygen and hydrogen depth profiles for selected samples analyzed using an ion microscope. A $^{133}Cs^+$ primary ion beam was used to enhance sensitivity to the $^{16}O^-$ and $^1H^-$ secondary ions.

The initial untreated surface is oxidized to a depth of the order of 10 μm; the surface also shows hydrogen penetration to approximately 5 μm. After laser surface treatment, oxygen and hydrogen contents of the irradiated surfaces are reduced. However, careful inspection reveals that the extent of selective vaporization of oxygen and hydrogen is reduced with increasing power levels. These observations were correlated with independent calculations and measurements of the depth of penetration of the laser treatment, liquid pool temperature, and the corresponding change in concentration profile for nickel, chromium, and manganese alloying elements. Although not shown, the latter are also measured using SIMS, but in that case, a $^{32}O_2^+$ primary ion beam is used to enhance the sensitivity and to attenuate the matrix effects for analysis of the Ni^+, Cr^+, and Mn^+ secondary ions. It was found that the observed changes in concentration of alloying elements are due to differences in their solubility in the ferrite and other nonequilibrium phases, which form after laser irradiation of the original austenitic steel.

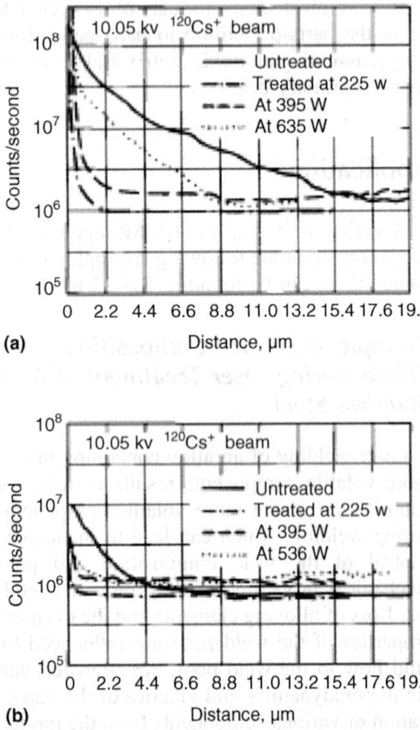

Fig. 19 Negative secondary ion mass spectroscopy depth profiles of (a) oxygen and (b) hydrogen as a function of the laser irradiation of an AISI 202 steel acquired using $^{133}Cs^+$ primary ion bombardment in an ion microscope

Example 2: Quantitative Analysis of a Phosphorus Ion-Implantation Profile in Silicon

Ion implantation is an important process for low-level, shallow doping of silicon microelectronic devices. Secondary ion mass spectroscopy is uniquely qualified to verify the implantation profiles for process development, quality control, and failure analysis. In this example, the use of SIMS is demonstrated for the analysis of phosphorus (an *n*-type dopant) after ion implantation into silicon. However, due to the presence of hydrogen in the silicon matrix and in the residual vacuum, a spectral interference occurs at $m/e \approx 31$ between P^+ (m/e = 30.9738), $^{30}SiH^+$ (m/e = 30.9816), $^{29}SiH_2^+$ (m/e = 30.9921), and $^{28}Si_3^+$ (m/e = 31.0004). Thus, a high-resolution mass spectrometer is necessary to separate the desired analyte species (P^+) from the molecular matrix ion interference (SiH^+, SiH_2^+, and SiH_3^+).

Figure 20 shows a high-resolution SIMS spectrum for the region m/e = 28 to 31, where the separation of the isotopic silicon and silicon-hydride species is demonstrated. It would not be possible to verify unequivocally the presence or concentration of phosphorus in this semiconductor without the high-resolution capability. This level of resolution cannot be achieved using a quadrupole, where resolution is typically 1 amu (atomic mass unit). The mass resolution of this double-focusing magnetic sector is 10 to 100 times better; therefore, species whose mass-to-charge ratios differ by fractions of 1 amu can be separated.

In this case, because the instrument was optimized for mass resolution, a loss in the detection sensitivity for phosphorus must be tolerated. However, the profile in Fig. 21(a) reveals that the sensitivity using a reactive $^{32}O_2^+$ primary ion beam is sufficient to measure the implantation profile over a dynamic range of 10^4 cps. Also, although the $^{31}P^+$ signal varied over 4 orders of magnitude through the profile, the $^{28}S^{2+}$ matrix signal (m/e = 13.9885) was constant (due to the use of an oxygen primary ion beam). After the profiling analysis was complete (approximately 10 min of sputtering), the depth of the sputtered crater was measured independently using a profilometer and was found to be approximately 0.11 μm to yield a sputtering rate of 110 nm/min. Because the fluence used to prepare this implant was known, a quantitative analysis could be performed without an external standard. That is, the quantified profile shown in Fig. 21(b) was obtained using Eq 7 and 8. The background concentration represents less than 1 ppm P.

Example 3: Quantitative Impurity Analysis in Low-Pressure Chemical Vapor Deposition Thin Films

Low-pressure chemical vapor deposition is used to deposit thin films. In this example, the relationship between the reactive and residual gases in the chemical vapor deposition reactor is of interest, as are the composition and hydrogen impurity content of silicon-oxynitride films. Figure 22 presents two quantified profiles for SiO_xN_y films deposited on silicon using different process parameters. In both cases, films were deposited at 910 °C (1670 °F) with a total gas pressure of 0.4 torr, and the silicon source was $Si_2H_2Cl_2$ gas. The film shown in Fig. 22(a) used an ammonia (NH_3) to nitrous oxide (N_2O) ratio of 3; the

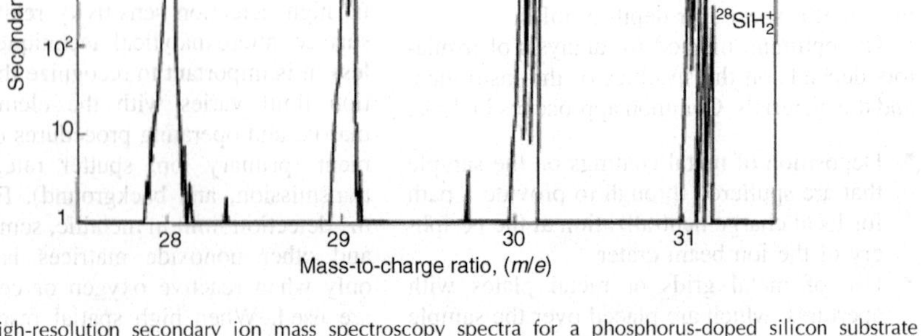

Fig. 20 High-resolution secondary ion mass spectroscopy spectra for a phosphorus-doped silicon substrate acquired using $^{32}O_2^+$ primary ion bombardment in an ion microscope

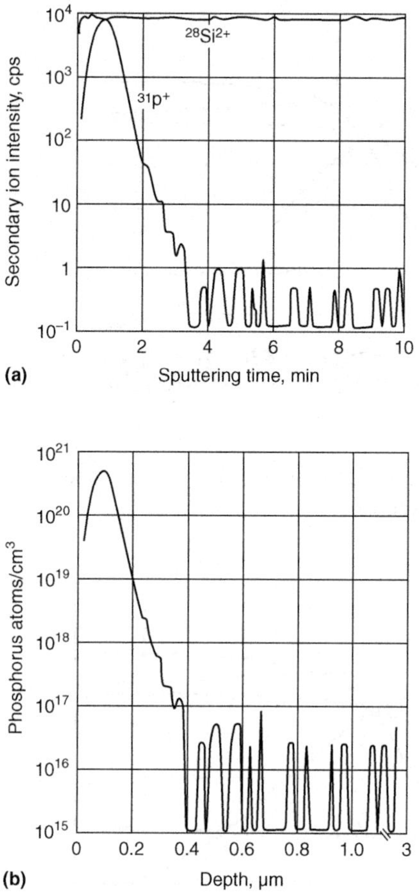

Fig. 21 Phosphorus depth profiles for an ion-implanted silicon substrate. (a) Before quantitative analysis of the positive secondary ion mass spectroscopy data. (b) After quantitative analysis. Acquired using $^{32}O_2^+$ bombardment in an ion microscope. Acquired using $^{33}Cs^+$ beam bombardment in an ion microscope

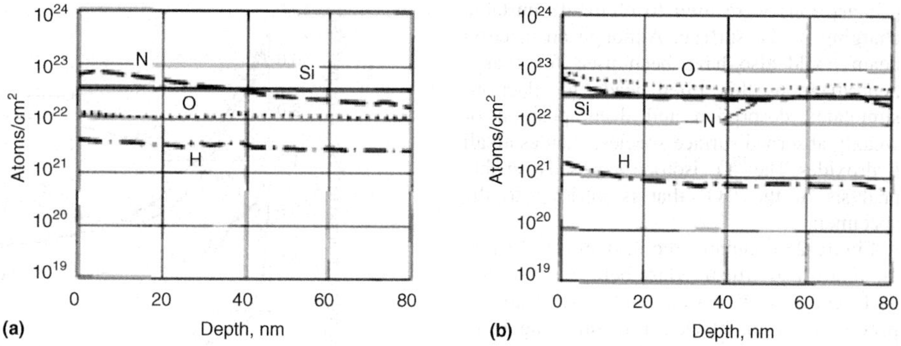

Fig. 22 Negative secondary ion mass spectroscopy depth profiles for low-pressure chemical vapor deposition SiO_xN_y thin films on silicon. (a) NH_3:N_2O = 3 during deposition. (b) NH_3:N_2O = 0.33 during deposition. Acquired using $^{133}Cs^+$ beam bombardment in an ion microscope

film in Fig. 22(b) was deposited using an NH_3:N_2O ratio of 1:3. The average silicon/nitrogen ratio of the films was not significantly affected by the NH_3:N_2O ratio, but the relative concentrations of oxygen and hydrogen were influenced. The analysis of many SiO_xN_y films processed under different conditions has revealed that NH_3 is a more efficient source of nitrogen than is N_2O and that the NH_3 activity determines the hydrogen level in the film.

The quantitative analyses represented in Fig. 22(a) and (b) were performed using selected silicon-oxynitride thin films whose absolute hydrogen, nitrogen, and oxygen concentrations were independently calibrated using Rutherford backscattering spectrometry and nuclear reaction analyses. These standards were used to calibrate the secondary ion signals for the other specimens. The sputtering rate was determined by independently measuring the film thickness using ellipsometry, then observing the time necessary to sputter through the film to the SiO_xN_y/Si interface.

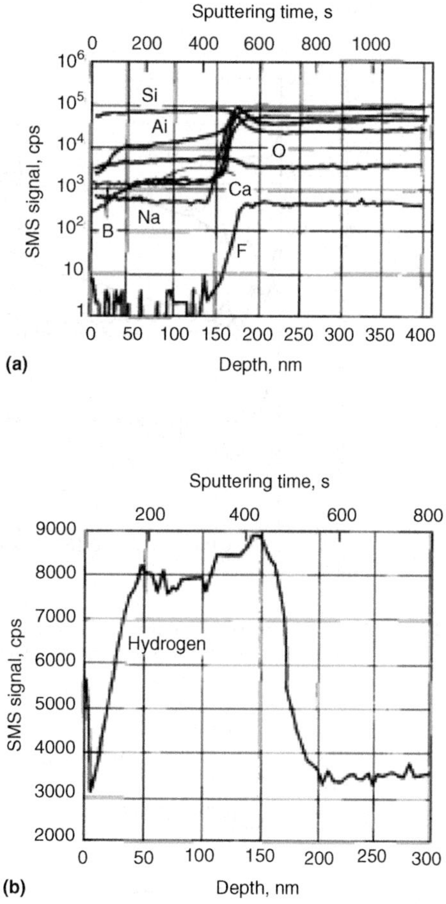

Fig. 23 Positive secondary ion mass spectroscopy (SIMS) depth profiles. (a) Various constituents. (b) Hydrogen in a calcium-boroaluminosilicate glass ribbon after acid etching 16 h in H_2SO_4. Acquired using Ar^+ primary ion bombardment in a scanning ion microprobe and an electron beam for charge neutralization

Example 4: Glass Surface Layer Analyses

The use of SIMS for depth profiling of three (unrelated) glass specimens after surface treatment or corrosion is described. Figure 23 shows the profiles for a multicomponent calcium-boroaluminosilicate glass ribbon after an acid-etching treatment. The presence of a surface film approximately 200 nm thick is evident after 16 h in concentrated sulfuric acid. The surface layer forms due to preferential leaching of the acid-soluble aluminum, boron, calcium, and sodium oxides and fluorides. The insoluble silica remains intact on the glass surface as a hydrated surface layer; the corresponding hydrogen profile is shown in Fig. 23(b). In this application, the kinetics and chemical mechanisms of this leaching process could be followed without the need for a quantitative compositional analysis. The profiles were obtained in an ion microprobe using an inert argon ion beam and raster gating; the charge neutralization was accomplished using an electron beam.

Figure 24(a) shows profiles for an alkali-lead-silicate glass (commonly termed crystal) that exhibited a haze due to weathering in a humid atmosphere. The haze is essentially a surface film that can be easily removed by rinsing in cold water. Thus, the profiles shown in Fig. 24(b) represent a piece of the same glass specimen shown in Fig. 24(a) after rinsing in deionized water. Secondary ion signals for the various elements in this alkali-lead-silicate glass were normalized to the $^{29}Si^{4+}$ signal to facilitate comparison of the profiles for the hazed and cleaned surfaces.

The haze is rich in sodium and potassium (probably hydroxides), which were leached to the surface from depths on the order of 200 to 300 nm; that is, regions depleted of sodium and potassium are apparent in the subsurface. The corresponding penetration of hydrogen to depths on the order of 200 to 300 nm indicates that water in the ambient atmosphere is probably responsible for the corrosion reaction. The glass surface, even after the cleaning, is permanently hydrated and depleted of sodium and potassium. Thus, subsequent exposure to a moist atmosphere requires a greater length of time to regenerate visible haze.

The profiles were obtained in an ion microscope using an $^{18}O^-$ primary ion beam; negative primary ions were used to eliminate unstable charging of the surface. Although an electron beam could also have been used for charge neutralization, this often leads to electron-stimulated desorption and decomposition of weakly absorbed surface species, such as alkali hydroxides. The $^{18}O^-$ isotope was used to enable analysis of the $^{16}O^+$ that is intrinsic to the specimen.

Figure 25 compares the in-depth profiles of an alkali-lead-silicate glass before and after a hydrogen reduction treatment. The treatment produces a thin, black semiconducting layer at the glass surface, which is crucial to the fabrication of microchannel plates and other charged-particle detectors. The kinetics of the hydrogen diffusion and reduction reaction are easily followed using the $^1H^+$ secondary ion profile. Note also that sodium has been depleted to some extent during the hydrogen reduction (probably due to vaporization); this alkali depletion can influence the work function and secondary electron yield of the devices.

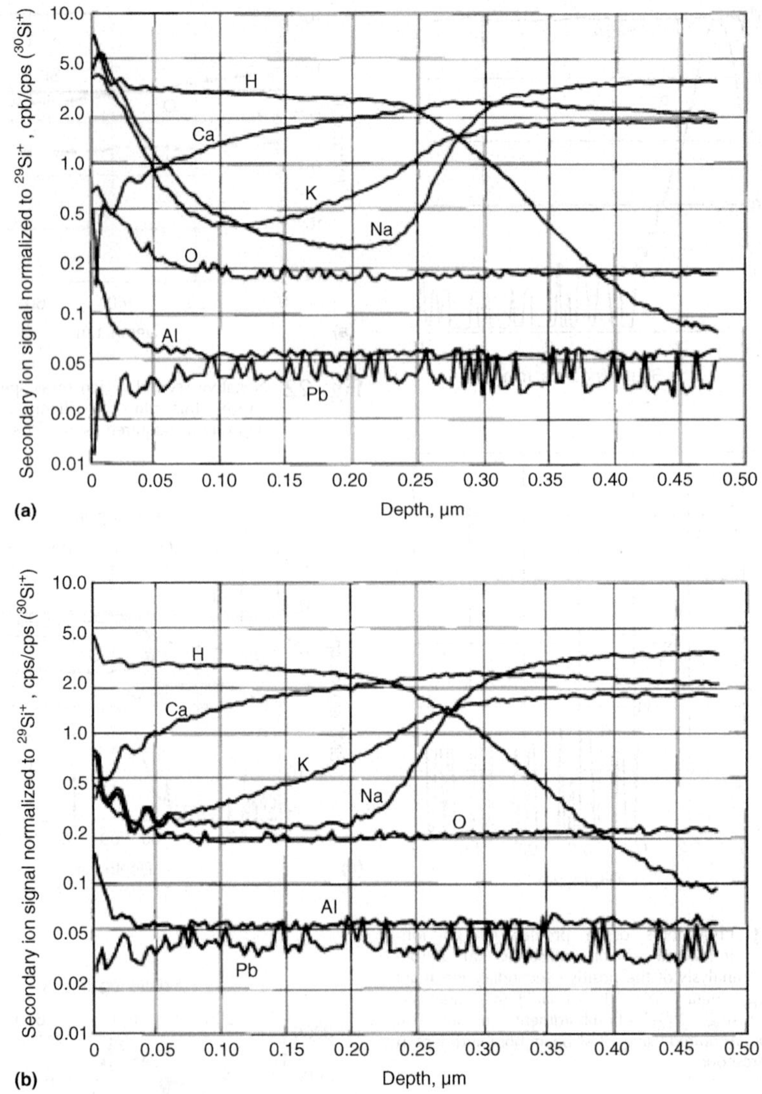

Fig. 24 Positive secondary ion mass spectroscopy depth profiles for alkali-lead-silicate crystal glass. (a) Hazed surface. (b) Cleaned surface. Acquired using $^{18}O^-$ primary beam bombardment in an ion microscope

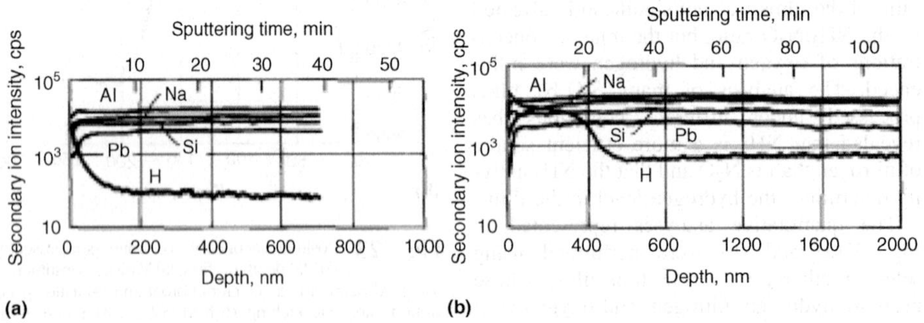

Fig. 25 Positive secondary ion mass spectroscopy depth profiles for a lead-silicate glass. (a) Before and (b) after hydrogen reduction to produce a semiconducting surface layer. Acquired using $^{32}O_2^+$ primary beam bombardment and electron beam charge neutralization in an ion microscope

Example 5: Determining Isotopes of Oxygen in an Explosive Actuator

Secondary ion mass spectroscopy is unique among surface analytical techniques for its ability to detect hydrogen and to distinguish among the various isotopes of an element. The latter capability is useful when studying the oxidation process, in that ^{18}O can be used to differentiate between the oxidation that occurs during the experiment and any oxide (from ^{16}O) that may have been present before the experiment. An example of the use of SIMS to help determine the cause of enhanced output of an explosive actuator is discussed.

The explosive actuator of interest consists of TiH_x powder mixed with $KClO_4$ powder. The powders are ignited by an electrically heated bridgewire. After extended periods of storage, the actuators experienced outputs of explosive energy that were considerably greater than their nominal value. One of the possible explanations for this occurrence was that the natural TiO_2 layer that exists on the surface of the TiH_x powder was reduced during shelf life, allowing a faster reaction with the $KClO_4$. Due to the high free energy for the oxidation of titanium, it was thought to be impossible to open the actuators and transport them to a surface analysis apparatus to locate reduced titanium. Instead, it was decided to open the actuators in an ^{18}O environment so the elemental titanium, if present, would react with the ^{18}O, forming $Ti^{18}O_2$. Secondary ion mass spectroscopy would then be used to locate the $Ti^{18}O_2$.

Five sample materials were analyzed. Two samples (TiH_x and $TiH_x/KClO_4$) were used to obtain baseline data in terms of which molecular ion species were present. The other three samples (all $TiH_x/KClO_4$) were exposed to ^{18}O. The positive SIMS spectra in the mass-to-charge region of 45 to 70 amu were recorded for each of the aforementioned samples. The region contains the lines of the elemental titanium isotopes and the various molecular ions associated with titanium, ^{16}O, ^{18}O, and hydrogen.

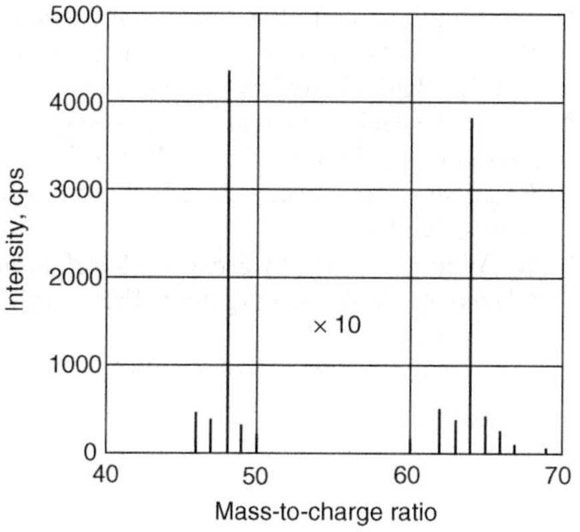

Fig. 26 Spectra acquired from a $TiH_x/KClO_4$ sample exposed to ^{18}O

Figure 26 shows an example of the spectra obtained from an ^{18}O-exposed sample. The peaks due to the various molecular ions are indicated. Because there is considerable peak overlap among the possible molecular ion species, a linear least-squares analysis was used to separate the ^{18}O contribution. In this method, standard spectra for each elemental and molecular ion are linearly combined to obtain the best least-squares fit to the experimental data. The result of the fit provides the contribution of each component to the experimental spectra. To determine the absolute concentration of each component, a sensitivity factor for each component must be known. These were not known for the current case, and absolute quantification was not attempted. However, because molecular ions due to $Ti^{16}O$ were easily observed, it was assumed that ions due to $Ti^{18}O$, if present, would be observed with the same sensitivity.

Standard spectra for each molecular species used in the least-squares fitting were calculated from tables of isotopic abundances. Standard spectra were calculated for titanium, $Ti^{16}O$, $Ti^{18}O$, TiH, TiH_2, $Ti^{16}OH$, $Ti^{18}OH$, $Ti^{16}OH_2$, and $Ti^{18}OH_2$. Table 2 shows the results from one of the samples. In all three cases analyzed, the ^{18}O contribution was zero within experimental error. These results indicate that for the samples used in these experiments, the surface $Ti^{16}O_2$ layer was not reduced and thus was not the source of the enhanced output of the explosive actuator.

ACKNOWLEDGMENT

Example 5 was provided by G.C. Nelson, Sandia National Laboratories.

Table 2 Secondary ion mass spectroscopy analysis of an explosive actuator

Component	Intensity (I)
Ti	5659
TiH	79.5
TiH_2	−3
$Ti^{16}O$	512
$Ti^{16}OH$	14
$Ti^{16}OH_2$ + $Ti^{18}O$	−0.6 ± 0.6
$Ti^{18}OH$	1.1 ± 1.8
$Ti^{18}OH_2$	−0.2 ± 0.3

REFERENCES

1. G. Garter and J.S. Colligon, *Ion Bombardment of Solids*, Heinemann, 1968
2. M. Kamisky, *Atomic and Ionic Impact Phenomena on Metal Surfaces*, Springer-Verlag, 1965
3. O. Auciello and R. Kelly, Ed., *Ion Bombardment Modification of Surfaces*, Elsevier, 1984
4. M. Barber, R.S. Bordoli, R.D. Sedgewick, and A.N. Tyler, Fast Atom Bombardment of Solids (F.A.B.): A New Ion Source for Mass Spectrometry, *J. Chem. Soc., Chem. Commun.*, Vol 7, 1981, p 325–327
5. J.H. Gross, Fast Atom Bombardment, *Mass Spectrometry*, Springer, Berlin, Heidelberg, 2004
6. G.K. Wehner, in *Methods of Surface Analysis*, A.W. Czanderna, Ed., Elsevier, 1975, p 5–38
7. A. Benninghoven, *Crit. Rev. Solid State Sci.*, Vol 6, 1976, p 291
8. K. Wittmack, in *Ion Beam Surface Layer Analysis*, Vol 2, O. Meyer et al., Ed., Plenum Press, 1976, p 649–658
9. C.W. Magee, W.L. Harrington, and R.E. Honig, *Rev. Sci. Instrum.*, Vol 49, 1978, p 477
10. J.M. Ruberol, M. Lepareur, B. Autier, and J.M. Gourgout, in *Eighth International Conference on X-Ray Optics and Microanalysis*, D.R. Beaman, Ed., Pendell Publishing, 1979, p 322–328
11. C.W. Magee and E.M. Botnick, *J. Vac. Sci. Technol.*, Vol 19 (No. 1), 1981, p 47
12. W.L. Fite and M.W. Siegel, "*Energy Filters for Secondary Ion Mass Spectrometry*," Extranuclear Laboratories, Pittsburgh, PA, 1977
13. R.J. Blattner and C.A. Evans, Jr., *Scan. Elec. Microsc.*, Vol IV, 1980, p 55
14. D.P. Leta and G.H. Morrison, *Anal. Chem.*, Vol 52 (No. 3), 1980, p 514
15. D.E. Newbury et al., in *Surface Analysis Techniques for Metallurgical Applications*, STP 596, American Society for Testing and Materials, Philadelphia, PA, 1976, p 101–113
16. M.A. Douglas and P.J. Chen, Quantitative Trace Metal Analysis of Silicon Surfaces by TOF-SIMS, *Surf. Interface Anal.*, Vol 26, 1998, p 984–994
17. P. Bertrand and W. Lu-Tao, Time-of-Flight Secondary Ion Mass Spectrometry (ToF-SIMS), *Microbeam and Nanobeam Analysis, Mikrochim. Acta Suppl.*, Vol 13, D. Benoit, J. F. Bresse, L. Van't dack, H. Werner, and J. Wernisch, Ed., Springer, Vienna, 1996
18. S. Fearn, *An Introduction to Time of Flight Mass Spectrometry (ToF-SIMS) and Its Application to Materials Science*, Morgan & Claypool Publishers, San Rafael, CA, 2015
19. H.W. Werner and A.E. Morgan, *J. Appl. Phys.*, Vol 47, 1976, p 1232
20. K. Wittmack, *J. Appl. Phys.*, Vol 50, 1979, p 493
21. D.J. Surman and J.C. Vickerman, *Appl. Surf. Sci.*, Vol 9, 1981, p 108
22. B.W. Schueler, Microscope Imaging by Time-of-Flight Secondary Ion Mass Spectrometry, *Microsc. Microanal. Microstruct.*, Vol 3, 1992, p 119–139
23. K. Takai, J. Seki, and Y. Homma, Observation of Trapping Sites of Hydrogen and Deuterium in High-Strength Steels by Using Secondary Ion Mass Spectrometry, *Mater. Trans., JIM*, Vol 36 (No. 9), 1995, p 1134–1139
24. H. Francois-Saint-Cyr, Secondary Ion Mass Spectrometry Characterization of the Diffusion Properties of 17 Elements Implanted into Silicon, *J. Vac. Sci. Technol. B*, Vol 19 (No. 5), Sept/Oct 2001, p 1769–1774
25. S.G. Boxer, M.L. Kraft, and P.K. Weber, "Advances in Imaging Secondary Ion Mass Spectrometry for Biological Samples," LLNL-JRNL-408214, Lawrence Livermore National Laboratory, Oct 2008
26. N. Saintier et al., Analyses of Hydrogen Distribution around Fatigue Crack on Type 304 Stainless Steel Using Secondary Ion Mass Spectrometry, *Int. J. Hydrog. Energy*, Vol 36 (No. 14), 2011, p 8630–8640
27. N.N. Nikitenkov, O.V. Vilkhivskaya, A.N. Nikitenkov, and V.S. Sypchenko, Fundamentals of the Layer-by-Layer Chemical Analysis of Heterogeneous Samples Using Secondary Ion Energy-Mass Spectrometry, *23rd Conf. on Application of Accelerators in Research and Industry (CAARI)*, Elsevier B.V., 2014
28. T. Wirtz et al., High-Resolution, High-Sensitivity Elemental Imaging by Secondary Ion Mass Spectrometry: From Traditional 2D and 3D Imaging to Correlative Microscopy, *Nanotechnology*, Vol 26, 2015

SELECTED REFERENCES

- A. Benninghoven et al., Ed., *Secondary Ion Mass Spectrometry—SIMS II*, Springer-Verlag, 1979
- G. Blaise, in *Materials Characterization Using Ion Beams*, J.P. Thomas and A. Cachard, Ed., Plenum Press, 1978, p 143–238
- J.W. Colbrun and E. Kay, *CRC Crit. Rev. Solid State Sci.*, Vol 4 (No. 4), 1974, p 561
- R.J. Colton, *J. Vac. Sci. Technol.*, Vol 18 (No. 3), 1981, p 737
- E. De Pauw, A. Agnello, and F. Derwa, Liquid Matrices for Liquid Secondary Ion Mass Spectrometry-Fast Atom

Bombardment: An Update, *Mass Spectrom. Rev.*, Vol 10 (No. 4), 1991, p 283–301
- K.F.J. Heinrich and D.E. Newbury, Ed., *Secondary Ion Mass Spectrometry,* NBS 427, National Bureau of Standards, Gaithersburg, MD, Oct 1975
- H. Liebl, *Anal. Chem.*, Vol 46 (No. 1), 1974, p 22A
- M. Linford, An Introduction to Time-of-Flight Secondary Ion Mass Spectrometry (ToF-SIMS), *Vac. Technol. Coat.*, Feb 2013
- J.A. McHugh, in *Methods of Surface Analysis*, A.W. Czanderna, Ed., Elsevier, 1975, p 223–278
- D.S. McPhail, Applications of Secondary Ion Mass Spectrometry (SIMS) in Materials Science, *J. Mater. Sci.*, Vol 41 (No. 3), Feb 2006, p 873–903
- G.H. Morrison and G. Slodzian, *Anal. Chem.*, Vol 47 (No. 11), 1975, p 932A
- N. Shimizu and S.R. Hart, in *Annual Review of Earth and Planetary Sciences*, Vol 10, Annual Reviews, 1981
- P. van der Heide, *Secondary Ion Mass Spectrometry: An Introduction to Principles and Practices*, John Wiley & Sons Inc., New Jersey, 1962
- N. Winograd, in *Progress in Solid State Chemistry*, Vol 13, Pergamon Press, 1982, p 285–375
- E. Zinner, *Scanning*, Vol 3, 1980, p 57

ASM Handbook, Volume 10, *Materials Characterization*
ASM Handbook Committee
DOI 10.31399/asm.hb.v10.a0006639

X-Ray Photoelectron Spectroscopy*

Binayak Panda, NASA Marshall Space Flight Center

Overview

General Uses

- Elemental analysis of surfaces of all elements with the atomic number of lithium or higher
- Analysis of the first few atomic layers on samples
- Analysis of thin surface layers that cannot be analyzed by other techniques
- Chemical state identification of surface elements
- In-depth composition profiles for elemental distribution as well as oxidation states in samples
- Surface analysis of samples when destructive effects of electron or ion beam techniques must be avoided

Examples of Applications

- Determination of oxidation states of metal atoms in metal compounds
- Identification of surface contaminants
- Measurement of surface film thickness
- Identification of polymers and evaluation of any degradation

Samples

- *Form:* Flat solids with low vapor pressure are preferred
- *Size:* Depends on machine—up to 102 mm (4 in.) diameter
- *Preparation:* Must be free of fingerprints, oils, or surface contamination

Limitations

- Data collection is slow compared with other surface analysis techniques, but analysis time can be decreased substantially when high resolution or chemical state identification is not needed
- Lower lateral resolution in imaging and analysis compared with electron microscopes
- Surface sensitive, comparable to other surface analysis techniques
- Charging effects may be a problem with insulating samples. Some instruments are equipped with charge compensation devices.

Estimated Analysis Time

- Requires few hours of vacuum pump-down before analysis
- Qualitative analysis can be performed in 5 to 10 min
- Quantitative analysis requires one hour to several hours, depending on information desired

Capabilities of Related Techniques

- *Auger electron spectroscopy:* Compositional analysis of surfaces. Faster, with better lateral resolution than x-ray photoelectron spectroscopy. Has depth-profiling capabilities. Electron beam can be very damaging; bonding and other chemical state information are not easily interpreted. Very thin surface layers cannot be analyzed.
- *Low-energy ion-scattering spectroscopy:* Sensitive to the top atomic layer of the surface and has profiling capabilities. Quantitative analysis requires use of standards; no chemical state information; poor mass resolution for high-*Z* elements
- *Ultraviolet photoelectron spectroscopy:* Ultraviolet photoelectron spectroscopy operates with the same principles as x-ray photoelectron spectroscopy, but the radiation in this case is ultraviolet light with a narrow band of wavelengths. The core-level electrons are not accessible, but do provide useful information at the surface level (valence band level). Analysis depth is estimated to be approximately 20 to 30 Å. Used for information on surface species, reaction products, catalysis, and chemisorption.

Introduction

Study of the interaction between the photon and the electron began in the early part of the 19th century, emanating from the photoelectric effect described by Albert Einstein in 1905 (Ref 1) and the redistribution of kinetic energy resulting from the interaction of x-rays and solids reported in 1925 (Ref 2). The spectrum resolutions obtained at that time were not sufficient to observe distinct peaks in spectra for materials. Thus, these phenomena hardly attracted any attention for many years following these discoveries.

Modern x-ray photoelectron spectroscopy (XPS) was made possible by the extensive and significant contribution from Kai Siegbahn and others (Ref 3, 4) of Uppsala University. Siegbahn developed and employed a high-resolution electron spectrometer that revealed electron peaks in a spectrum emerging from the interaction of x-rays and solids. Eventually, Siegbahn received the Nobel Prize in 1981 for his contributions to XPS. Around 1958, shifts

*This article was revised from J.B. Lumsden, X-Ray Photoelectron Spectroscopy, *Materials Characterization*, Volume 10, *ASM Handbook*, ASM International, 1986, p 568–580.

in elemental peaks were realized in compounds when the same elements were bound to different elements. This discovery resulted in identification of the chemical state in various chemicals as well as the oxidation states of atoms in compounds. Because of these useful physical effects, the Uppsala group called XPS electron spectroscopy for chemical analysis (ESCA). XPS or ESCA not only identifies elements but also the compound the elements form from their chemical shifts.

Compared with other microanalytical techniques such as energy dispersive and wavelength dispersive techniques, XPS analyzes only a few atomic layers present on the surface. This was discovered early in 1966 (Ref 5). While this presents value to the analytical technique, it also means that adsorbed superficial gases and contaminations on a sample can be introduced into the analytical chamber. This necessitates that the surface be cleaned and the underlying material, the material of interest, be exposed to a clean environment so that it can be analyzed. The cleaning is accomplished by a scanning ion gun within the analytical chamber of the instrument. The ion gun uses an argon gas and is commonly attached in most modern machines. Reliable and efficient vacuum systems employed in modern machines do not allow adsorbed layers to rebuild after the surface is cleaned.

Development of efficient and reliable vacuum pumps was another important step in the commercialization of XPS machines. Vacuum levels greater than 10^{-7} torr (1.33×10^{-7} mbar) are essential to increase the mean free path of electrons released from the sample surface. Thus, modern machines are equipped with high-capacity ion, turbo, or cryogenic pumps in their analytical chambers.

Today, XPS has advanced from applied physics laboratories to industry for use in quality control as well as analysis of contaminants and has taken a dominant role in microanalysis. Its uniqueness arises from the fact that it is considered nondestructive compared with other common microanalytical techniques that use electron and ion excitation sources. Polymers and plastics can be analyzed because the binding energies of saturated and unsaturated bonds in atoms can be separated. Extremely thin layers, including materials with layered structures, can be analyzed. The technique, although it did not advance for many years, eventually opened a new window for research and applications in industry due to its ability to separate and measure the chemical shifts in bound elements.

Principles

Fig. 1 (Ref 6) illustrates the electronic transitions involved in an XPS process. It shows an energetic x-ray beam impinging on the surface. Due to the high energy of the photons, they eject one or more core electrons. The ejected electrons are collected by the spectrometer and eventually detected by a multichannel analyzer. The process of this interaction conserves total energy for each interaction, and from the voltages needed for the retardation of the ejected electron, the energy of the electron is known. This associated energy leads to the identification of the element and the orbit from which the electron is released. Mathematically, if the kinetic energy KE is the core electron immediately after it was ejected, then, according to the law of conservation of energy,

$$KE = h\nu - BE + \alpha \quad \text{(Eq 1)}$$

where $h\nu$ is the energy of the exciting radiation, BE is the binding energy of the emitted electron in the solid, and α is the spectrometer work function. The value of α depends on the machine design, which varies from model to model.

Reference 7 explains the standard calibration used to eliminate the work function in surface science instruments and current updated software.

Often the vacancy created by the released electron from the photon interaction is filled by an electron from a higher energy level. The extra energy is released by photon emission or by ejection of another electron with discrete energy known as the Auger electron. The spectrometer also collects the Auger electron and displays it along with the other core-level electrons. Auger emission is the more probable transition for low-atomic-number elements.

From Eq 1 it is clear that the position of the XPS lines (binding energy of peaks) depends on the energy of the x-ray photon and hence

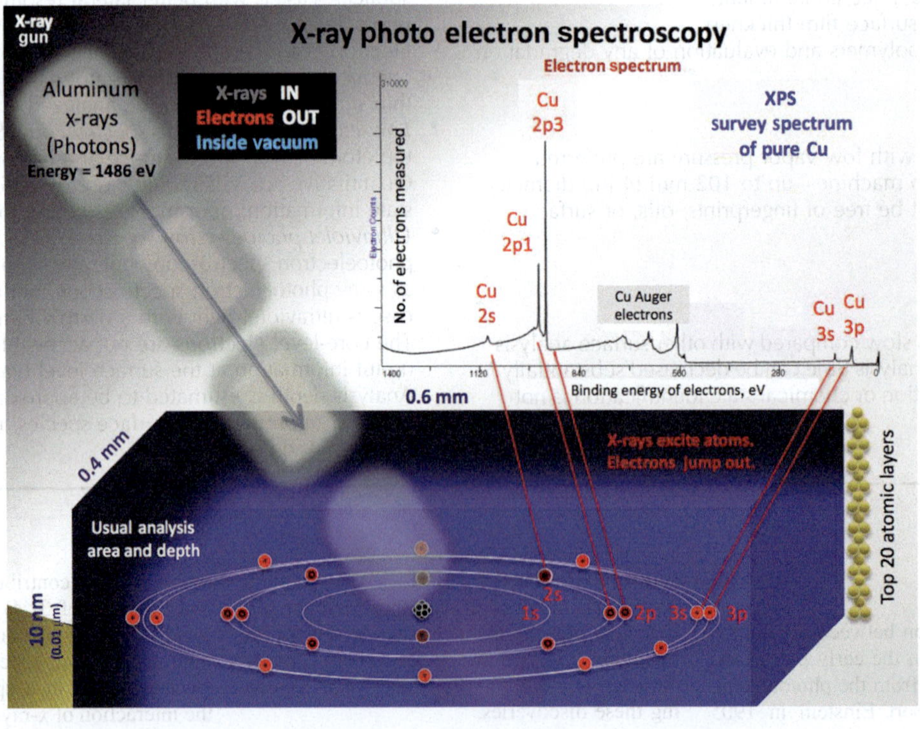

Fig. 1 The electronic transitions involved in an XPS process are shown. Figure shows aluminum monochromatic x-rays moving from left to a copper sample. Electrons ejected from the shells are knocked off and collected by the XPS spectrometer to produce a spectrum (top right). The peak correspondences are illustrated (Ref 6).

the target material of the x-ray source. The most common target materials are aluminum and magnesium. Existing databases include line positions for both aluminum and magnesium x-ray targets because the positions vary depending on the wavelengths of the exciting beams. The Auger line positions, however, do not change, as the Auger electron energies are liberated from the electron exchange process within an atom rather than by the impinging photon. When two or more elements combine to form a compound, the binding energies of their electrons change, which is reflected in shifts in their corresponding line positions in the spectrum. Aluminum and magnesium x-ray targets are selected in most XPS machines due to the strong x-ray beam intensity of these targets and the availability of inexpensive quartz crystals that can effectively filter Al Kα x-rays.

Spectra generated by XPS instruments are mostly similar to that shown in Fig. 1 (Ref 6) for Cu. Figure 1 shows both photoelectron and Auger lines, but there are also several other types of peaks found in XPS spectra largely due to aberrations and the complexity of the interactions that the emitted photoelectron experiences.

X-ray satellite peaks are found when the x-ray is not quite monochromatic. Small peaks appear toward the lower binding energies of XPS peaks. Figure 2 (Ref 8) shows a small number of photons with higher energy than the main exciting x-rays. Ghost lines appear due to impurities in the anode material. The x-rays generated from the impurity atoms can be eliminated only by using a crystal monochromator. If absorption filters are used, additional lines will be present in the spectra. Shake-up lines appear because of the retardation effect of the ions that are created by the ejection of a photoelectron. This retardation effect generates small humps toward the higher binding energy side of a peak. The $\pi \rightarrow \pi^*$ shake-up for the C 1s (see the section "Nomenclature" in this article) line is a common example. Figure 3 (Ref 8) shows shake-up lines in copper compounds.

Photoelectrons may react with the surface atoms and lose their energy. Such electrons, when detected by the spectrometer, appear as a hump at about 20 to 25 eV above the binding energy of a peak. They are called energy loss lines or plasmon lines. Some cases appear in a periodic fashion in 20 to 25 eV intervals of diminishing intensity. Figure 4 (Ref 8) shows plasmon lines in aluminum.

An XPS spectrum very close to the Fermi level (0 to 20 eV), known as the valence band spectrum, is often utilized to distinguish between conductors and insulators.

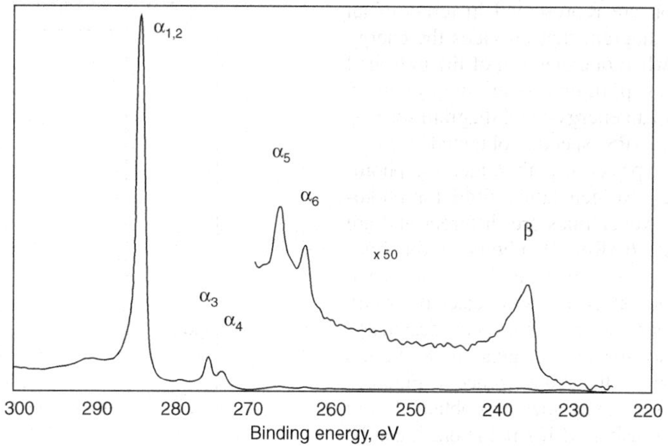

Fig. 2 Magnesium x-ray satellite peaks in C 1s spectrum of graphite. Binding energy displacements are α3 at 8.4 eV, α4 at 10.1, α5 at 17.6, α6 at 20.6, and β at 48.7 (Ref 8).

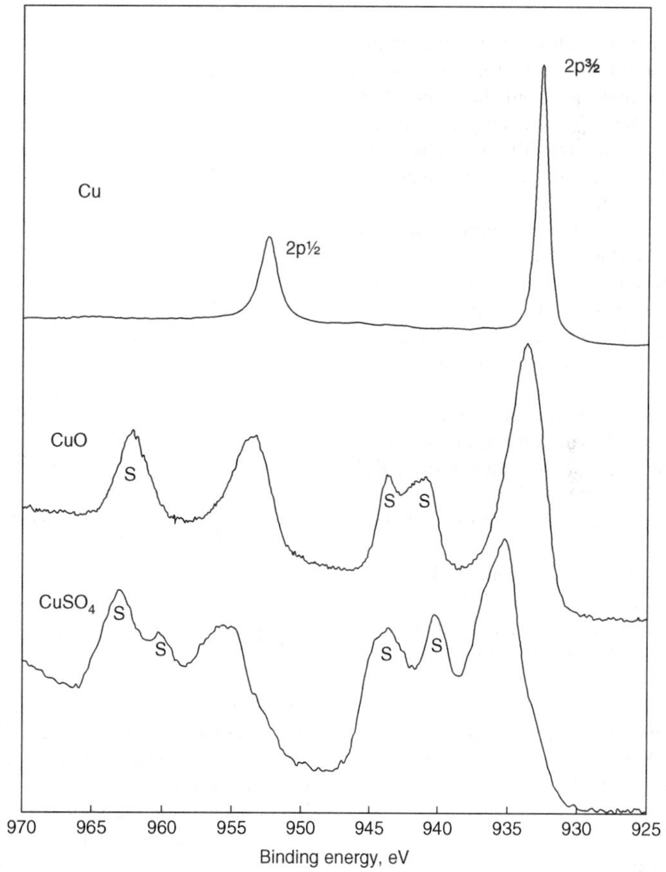

Fig. 3 Shake-up lines (2p) as seen for some copper compounds (Ref 8)

Nomenclature

The binding energy of an electron is the energy required to free the electron from its orbit. This is also known as the ionization energy and is commonly expressed in electron volts (eV). XPS data from a sample are often plotted with binding energy in electron volts on the abscissa and intensity of the peaks in the number of counts on the ordinate. Such plots are presented with electron volts increasing from right to left. Sometimes, the ejected electron energies are presented in the form of kinetic energies with electron volts increasing from left to right. Most presentations shown in this chapter are binding energy spectra.

The binding energy may be regarded as the energy difference between the initial and final states of an atom after an electron is ejected. The energy levels of an atom involved in photoemission are represented in terms of an energy-level diagram that provides the energy of the atom when one electron of the indicated energy level, s, p, d, or f, is missing. Figure 5 (Ref 8) shows an energy-level diagram for uranium metal. XPS spectra obtained from a machine show peaks of both Auger and photoelectron lines. Nomenclature used for photoelectron and Auger lines are different and are shown in Fig. 6 (Ref 8). Photoelectron lines are designated as s, p, d, and f levels (from the shell the electron was ejected from), whereas Auger electron lines are designated using the x-ray spectral designations K, L, and M, depending on the Auger process. Figure 7 (Ref 8) shows a survey spectrum obtained from uranium metal using Al Kα radiation. It can be compared with the energy levels of uranium shown in Fig. 5 (Ref 8).

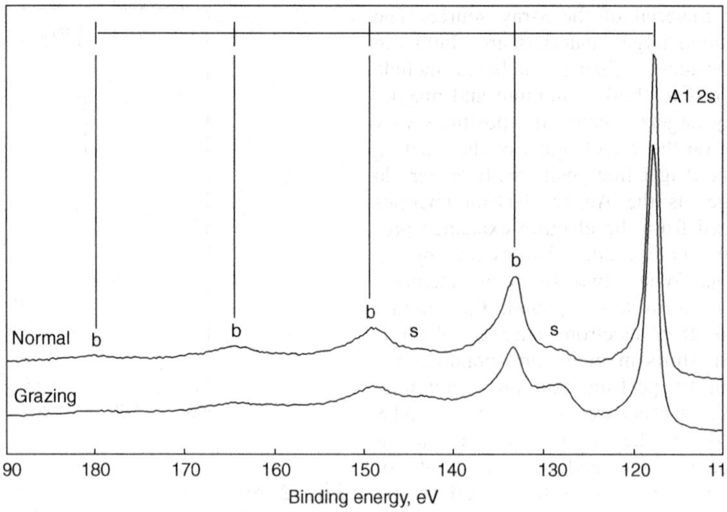

Fig. 4 Surface and bulk plasmon lines associated with Al 2s at normal and grazing takeoff angles (Ref 8)

Systems and Equipment

XPS instruments analyze the first few atomic layers of the sample. Usually, these layers also contain contaminations from the atmosphere and the environment that the sample is exposed to prior to the sample reaching the laboratory. While care is taken not to contaminate the sample any further in the laboratory, it is essential to remove the unwanted contamination from the surface of the sample to reach the base material inside. An ion gun is used for this purpose and is an essential part of a modern machine. There are several types of ion guns available, but an argon ion gun is most common.

The main part of the machine consists of an x-ray generator unit with a filtration system for the generation of monochromatic x-rays. Reference 9 provides an excellent design review of monochromatic x-ray generation, filtration, and detection in XPS systems.

Vacuum systems are inherent to the design of the instrument. Vacuum of the order of 10^{-10} torr (1.33×10^{-10} mbar) can produce excellent results, though 10^{-7} torr is generally adequate. The vacuum system is an integral part of the instrument.

Charge neutralizers are an equally important part of the XPS system when nonmetallic materials are analyzed.

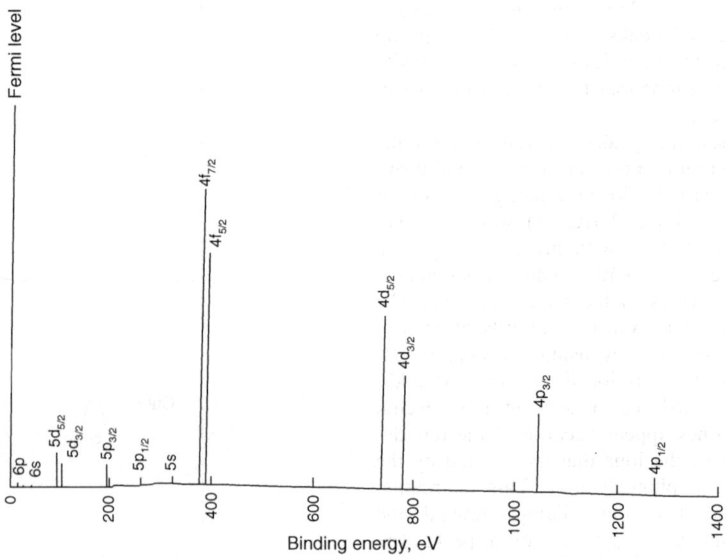

Fig. 5 Relative binding energies in electron volts (x-axis) for a uranium atom (Ref 8)

X-Ray Source

As mentioned previously, XPS literature focuses on sources that have either Al or Mg anodes for the generation of x-rays. Early on, thin metal foil was used to filter the x-ray continuum and produce a broad x-ray beam with the energy band of Al or Mg Kα. However, the x-ray beam could not be focused, and ghost lines as well as background radiation created problems of sensitivity and confusion. The first commercial instrument developed to address these issues was by Hewlett Packard; it used a quartz crystal monochromator. Kai Siegbahn was a consultant to Hewlett Packard. The instrument provided three advantages: a focused x-ray beam, very high energy resolution dependent on the spot size of the electron beam on the x-ray anode, and elimination of continuum background radiation. The early design had four monochromator crystals focused to the same point, which collected x-rays from a single anode. The electron source was not tightly focused.

The next commercial improvement was made by Surface Science Instruments (SSI), a spin-off of Hewlett Packard. This provided a focused electron gun to create x-rays from a small, variable spot on the anode. Thus, the x-ray energy band width could be varied from 0.25 to 0.82 eV. The large band width was limited by the natural width of the Al Kα line.

A narrow and often monochromatic intense beam is sought for XPS analysis. Such a beam is accomplished by filtering the white radiation generated from the source. For Al Kα radiation, a bent quartz crystal is used. The filtration reduces the intensity; consequently, to generate more intense beams in some advanced machines, seven such crystals are employed, arranged with one at the center and six others around the center one. This configuration increases the intensity 7-fold compared with a single crystal. Each crystal subtending the same solid angle with the source reflects only a narrow wavelength (Al Kα) by the quartz crystal atomic layers. These crystals are not only bent to focus the x-ray beam but also maintained at a constant temperature for constant lattice spacing. This ensures stability of the x-ray beam and intensity.

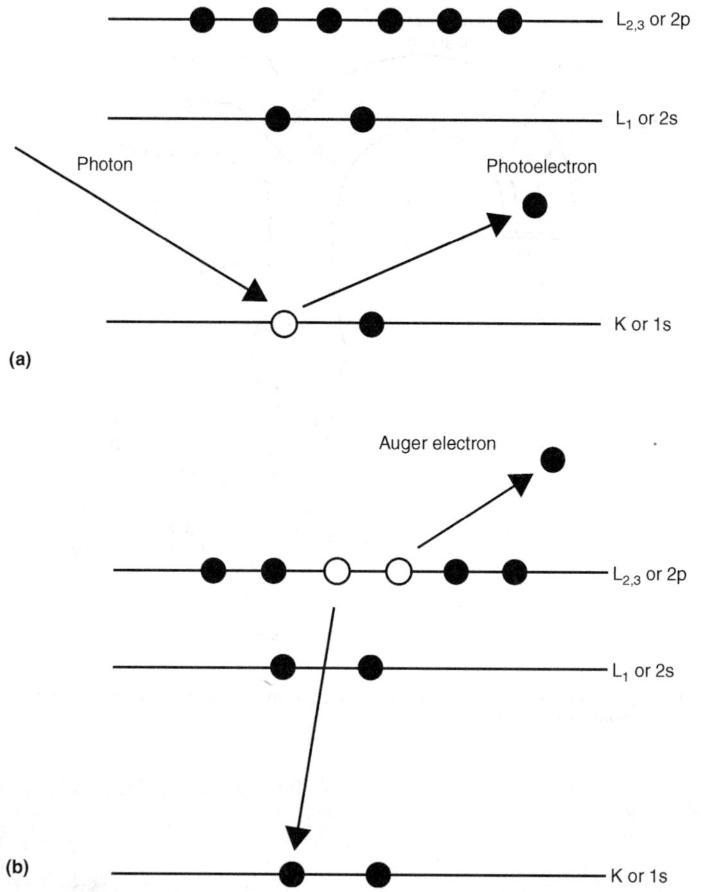

Fig. 6 XPS emission process for a model atom. An incoming photon causes the ejection of a photoelectron (a). The relaxation process results in an emission process of an Auger electron (b) $KL_{23}L_{23}$. The final arrangement results in a two-electron vacancy at the $L_{2,3}$ or 2p level (Ref 8).

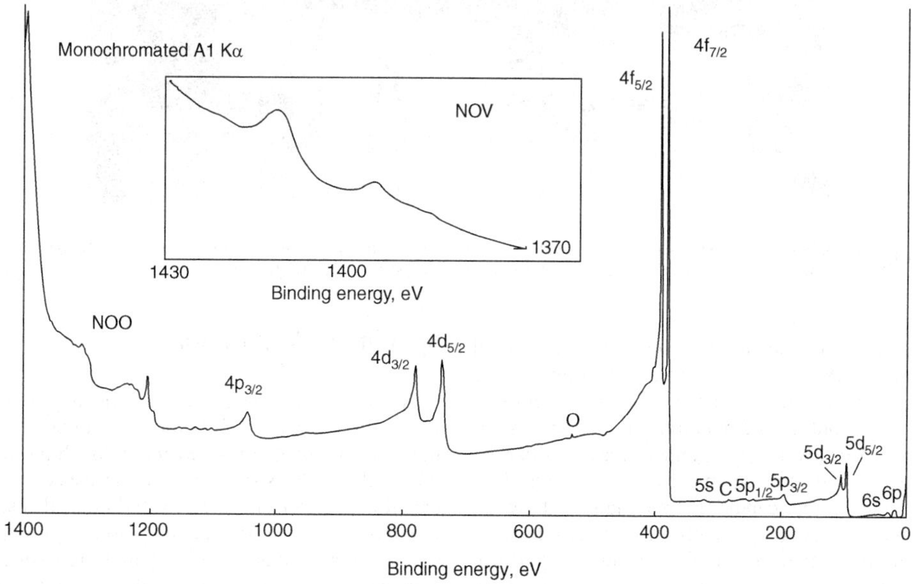

Fig. 7 Photoelectron spectra from uranium obtained using aluminum monochromatic radiation (Ref 8). NOV and NOO denote Auger series: NOV represents an initial vacancy in the N shell and a final vacancy in the O shell, with the final vacancy being in a valence level, whereas NOO denotes an initial vacancy in the N shell and a final double vacancy in the O shell.

Analyzers

Electrons emerging from the sample are collected by field lenses. Some designs may use a strong magnet below the sample to focus these electrons to the analyzer. Generally, the analyzer is either a hemispherical analyzer or a cylindrical mirror analyzer. Figure 8 (Ref 10) shows a schematic of a hemispherical analyzer. In a hemispherical or spherical sector analyzer, the two concentric charged hemispherical lens segments are charged separately such that the electrons coming from the same electronic shell are focused to a point on the detector generating the signal for a spectral line. A detector is generally a channel plate or an electron multiplier.

Modern machines employ a multichannel plate with a computer for digital counting. Photoelectrons striking the multichannel plate create pulses, which are eventually counted as a function of time. Sometimes the scans are repeated when the whole scanning process is also repeated. The number of photoelectrons gathered at the channel plate constitutes the height of a peak of a particular binding energy, and all the binding energies of all photoelectrons are covered by varying voltages of the spherical analyzers.

Vacuum Systems

The capacities of vacuum systems are such that the vacuum levels at the sample chamber and the analyzer are maintained very low to allow a long mean free path for the photoelectrons that are generated. Vacuum levels of the order of 10^{-8} or 10^{-9} torr (1.33×10^{-8} or 1.33×10^{-9} mbar) are obtained using ion, cryogenic, or turbo pumps of adequate capacity attached to the analytical chamber. These pumps are most efficient at high vacuum levels and therefore are backed by a conventional mechanical rotary or dry pump that generates a vacuum level of approximately 10^{-3} torr (1.33×10^{-3} mbar). In most instruments, there is also a sample preparation chamber prior to the analytical chamber for introducing the sample to air and expelling volatiles. The vacuum level in this chamber is maintained at a lower level (approximately 10^{-3} torr). Samples with holders are introduced to the analytical chamber after they spend enough time in the sample preparation chamber to remove most volatiles.

Charge Neutralizers

When photoelectrons leave the surface of a sample, a positive hole is created. In conductive samples, the electron hole is neutralized by electrons coming from the ground immediately (provided there is a good connection to the ground). For nonconducting samples, the positive hole remains on the surface and attracts the ejected photoelectron, reducing its kinetic energy. Due to the reduction in kinetic energy, these electrons erroneously appear at a higher binding energy level on the spectrum and broaden the peak. A charge neutralizer attempts to compensate for this effect by simply spraying low-energy electrons onto the sample surface. Thus, a charge neutralizer is nothing but an electron gun. It may appear simple, but the x-ray spot has varying intensity that requires more electrons at the center of the spot for compensation, which is not easily done. Modern XPS machines have complex charge neutralizers to counteract this effect as much as possible.

Ion Guns

Argon ion guns are extensively used for XPS analysis to remove material from or to clean the surface of the samples being analyzed. Recently, cluster atom ion guns have been gaining popularity for use in polymeric materials because the cluster ions prevent damage to the surface bonds often encountered when argon ion guns are used.

An argon ion gun generates ions when argon gas is passed over heated wires carrying electricity. While a small percentage of the flowing gas is ionized, these ions are manipulated and focused using a series of electrostatic lenses to generate a scanning ion beam. The ions, due to their heavy mass compared with an electron in an electron microscope, are difficult to manipulate and require strong electrostatic force to focus. In any case, these charged ions strike the sample surface like bowling balls and remove surface layers. The scanned area is generally larger than the area being analyzed.

Samples with an uneven surface are cleaned unevenly by an ion gun, because the gun is normally at an angle to the sample surface, with hills creating a shadowing effect. To counteract this effect, the samples are rotated around the point being analyzed. This is accomplished by placing the sample at the center of the analytical stage. Some instruments have arrangements (called eucentric stages) where the sample and the stage rotate around the point of interest rather than around the center of the stage.

Accessories

Because XPS involves an electron spectrometer, other analytical excitation sources such as an electron source for Auger electron spectroscopy and an ultraviolet photon source for ultraviolet photoelectron spectroscopy are often included as attachments to the main XPS instrument. The Kratos AXIS Ultra, shown in Fig. 9, has attachments such as a fracture stage and reaction cell, where fresh surfaces in metals and alloys can be studied in the vacuum after they are fractured inside the instrument or after a chemical reaction is carried out. In the reaction cell chamber, studies in catalysis and chemical reactions can be performed after chemical reactions take place between the candidate materials and gases, and the reaction products are transferred within the instrument under high vacuum to perform chemical analysis, without influence from reactions with the atmosphere. A fracture stage attachment is used when the grain boundaries in alloys are suspected to be responsible for low fracture test values and need investigation.

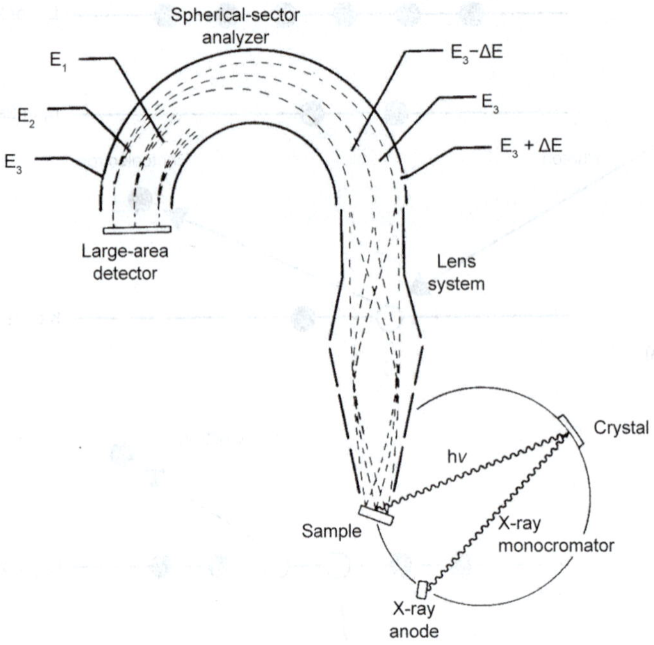

Fig. 8 Schematic presentation of a spherical-sector analyzer with a monochromatized x-ray source, reprinted with permission from Elsevier (Ref 10)

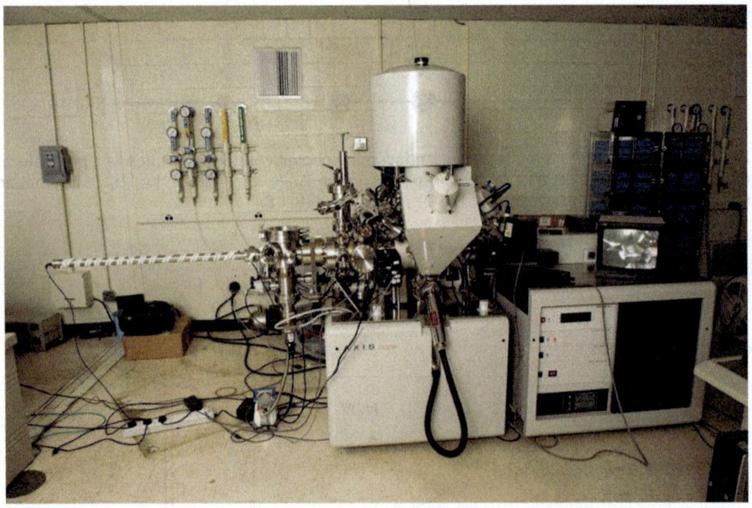

Fig. 9 XPS machine (Kratos AXIS Ultra) with a fracture stage, ultraviolet photon source, Auger attachment, and reaction cell (at the back, not shown). Courtesy of Marshall Space Flight Center

Specimen Preparation

As mentioned, XPS analyzes the first few layers on a specimen. It is therefore important not to contaminate these layers by touching or mishandling the specimen. This is especially true for specimens targeted for failure analysis. Oftentimes it becomes necessary to use gloves that are manufactured with no mold-releasing agents (generally silicones, as they migrate to the analysis sites and can give false indications for the presence of silicones).

Any sample exposed to air shows adsorbed molecules of H_2O or CO_2. These adsorbed molecules must be removed to analyze the material underneath. Ion guns are employed for this purpose. Ion guns are mounted at an angle to the sample surface, and therefore, nonuniform cleaning occurs on an uneven surface. A flat surface or sample rotation is required to avoid this.

The specimen should be placed flat (as opposed to at a small angle) with respect to the x-ray beam. If the sample plane interacts at a low angle with the x-ray beam, surface areas are analyzed more than the inside materials. This behavior can be exploited when a film or thin coating on the specimen must be characterized. This is called angle-resolved spectroscopy (AR-XPS), where the specimen is tilted at different angles and analysis is performed, providing more information on the surface or the coating.

Powder Specimens

Powder samples are mounted and analyzed using adhesive tapes. Because most tapes use organic volatiles, tapes compatible with high vacuum in the analytical chamber should be used. Usually, the powder does not cover the tape surface completely; carbon or silicon peaks on the tapes appear on the spectrum when the x-ray falls on the uncovered areas.

Another technique that is used effectively to analyze powder specimens is to mount the powder in soft metals such as indium. In this case, the powder is placed on an indium sheet and the sheet is folded over to retain the powder inside. Generally, the indium piece that is used as a sheet is cut from a small indium ingot with a sharp knife, which creates two fresh surfaces, and the powder is placed on one of the surfaces (on one side only). The folded indium sheet is then flattened further by rolling or hammering. This process not only embeds the powder into the metal sheet; the particles are also squeezed to produce good electrical contact. The two folded sides are then separated to expose the powder, and the embedded powder is analyzed. In this case, the indium lines are seen in the spectrum obtained along with the spectrum of the powder.

The powder can also be briquetted to generate a flat surface for analysis using a die and punch. While this process generates a solid specimen for analysis, there is always a small amount of material transfer from the die and punch. Care should be taken that the specimen does not transform to a compound under high pressure and adiabatic temperature rise when the die and the punch are squeezed.

Specimens required for XPS analysis are small and may be cut using a wire saw or diamond wheel. However, these surfaces are contaminated with cutting fluid or the cutting blade materials. The analyst must know the material and chemical information of the fluids and the saw materials.

Sample Charging

Sample charging is inevitable with insulators. For semiconductors, it depends on the incoming x-ray intensity. Conductive metals can also accumulate charge if the electron holes are not filled quickly by an electron from the instrument body. Most instruments incorporate spring clips or flat springs in their stage designs to make the contact as high as possible. While the charging effect distorts the spectrum completely in insulators, it can shift the spectrum slightly in alloys. It is important, therefore, that the specimen be loaded into the analytical chamber such that an adequate conduction path for electrons exists.

Mounted metallographic specimens cannot be analyzed due to lack of conductivity and the outgassing of the embedding matrix. To alleviate this problem, the mounting material must be a small amount of gold (or any other known coating) that can be applied to provide conductivity. The mounts should spend enough time in the preanalytical (or sample preparation) chamber prior to their insertion into the analytical chamber in order to remove volatiles.

Charge compensation in modern instruments is performed by supplying electrons to the location where the x-ray beam falls. However, for some instruments, the intensity is not uniform across the x-ray spot. This poses problems because some areas of the spot cannot be compensated. Some instruments have patented charge neutralizers with a combination of a low-energy electron gun and a mesh-screen device, as shown in Fig. 10. The device incorporates a very thin stretched nickel screen on top of the specimen holder, and the specimen is placed about 0.5 to 1.2 mm (0.02 to 0.05 in.) below the screen. The flood gun emits a low-energy electron onto the sample through the screen (Ref 11).

Calibration and Accuracy

Most machines are calibrated using lines for noble metals such as gold, silver, and copper. Except for gold, pure elements react with the atmosphere to form oxides or sulfides on the surface that need to be cleaned by sputtering. Thus, the calibration material is common but must be clean.

Most metals exposed to the atmosphere are covered with carbon known as "adventitious carbon." This is saturated carbon and is used as the reference for the entire XPS binding energy spectrum from 0 to 1200 eV. The position for the C 1s line is 284.6 eV (some consider it 284.4 or even 285 eV). The span of binding energy is calibrated with respect to the binding energy for Cu 3s (122.4 eV) and Cu 2p3/2 (932.5 eV).

Older systems such as those made by SSI check monthly for the Au 4f7/2 line to be at 83.98 eV and the split of 4f lines for Au to be at 3.68 eV. For the SSI X-Probe, the accuracy of the machine is checked by ensuring that the Au 4f line separation is at 3.68 eV ± 0.02 eV and the location of the Au 4f7/2 line is at 83.98 ± 0.1 eV. The width of the peak at half the height (FWHM) of the peak should be 0.95 eV. In addition, the integrated intensity of the peak should be at least 60,000 counts for a standard four-by-four matrix of spot size and pass energy that provides specifications on resolution and total counts under the Au

Fig. 10 A charge neutralization cage for insulators is installed on an SSX-100 machine. The nickel screen is not shown, but is installed on top. The spring around grabs and keeps it in place; the sample is placed inside the cage. The whole fixture is mounted at the bottom and slides to the analyzer stage.

4f7/2 line. Modern instruments that incorporate intense monochromatic x-rays such as the instrument shown in Fig. 9 are expected to yield 10 times more counts for the same lines. Kratos uses the Ag 3d5/2 peak at 368.3 eV for their intensity calibrations and system health check.

XPS machines make several adjustments to generate the desired spectra. Pass energy regulation allows control of passing electrons of a narrow energy band reaching the channel plate. A low-pass energy means that a narrower bandwidth of electrons is collected, which results in a higher-resolution spectrum. To obtain a high-resolution spectrum with sufficient counts, an intense x-ray beam is essential; however, machines with low-intensity beams accomplish high resolution by increasing the analysis time. Figure 11 shows the Si 2p line split obtained with low-pass energy using the Kratos AXIS Ultra spectrometer with high resolution.

Accuracy

The spectrum obtained from XPS has two types of information on two axes: the binding energy in electron volts and the intensity of the photoelectron line. The binding energy gives information on the binding state of an atom, such as whether oxygen is present as an oxide, dioxide, or sulfate. These states are estimated from the line shift. Such chemical shifts are often small and, when accurate, provide useful information. It is estimated that a well-performed analysis has an accuracy of about ±0.2 eV of the line position reported in the literature.

The intensity, or the area under the curve that constitutes a peak, depends on a number of factors including sample charging, interference from other peaks, sample thickness, and sample inclination to the beam. Peak integration in a resolution is estimated to be approximately 10%. The section "Data Analysis and Reliability" in this article elaborates on some aspects of this accuracy in quantitative analysis.

Data Analysis and Reliability

Quantitative surface analysis by XPS requires several operations that effect the level of accuracy, including specimen preparation, adjustment to machine parameters, and data manipulation by the operator. The final result in terms of uncertainty in measurements is the total amount of uncertainty introduced by all these operations.

The most useful aspect of XPS is the shift of binding energy lines due to chemical interactions between atoms. These shifts and the presence of other reacting elements establishes the nature of compounds that are present on the surface of the sample. Shifts in spectral lines are then compared with the shifts available in the literature and other databases and handbooks (Ref 8, 11). Fortunately, there is not a lot of variation in the measurements of line positions. In addition to variations in measurements in databases, variations in line positions can be caused by sample charging. Sample charging can shift lines to new positions. To evaluate such shifts, positions of lines such as adsorbed gases or adventitious carbon on analysis surface can be noted on a sample experiencing charging effects. Because the positions of these lines are documented and often referenced for uncharged samples, the shift can be calculated by simple addition or subtraction. Several databases are available (Ref 8, 11, 12) but should be used with caution as they may include data from older instruments, perhaps obtained using nonmonochromatic radiation. It is advisable to consult the original publication to determine how the original experiments were carried out. In any case, there may be a variation of about 0.2 eV between two experiments so far as the line positioning is concerned.

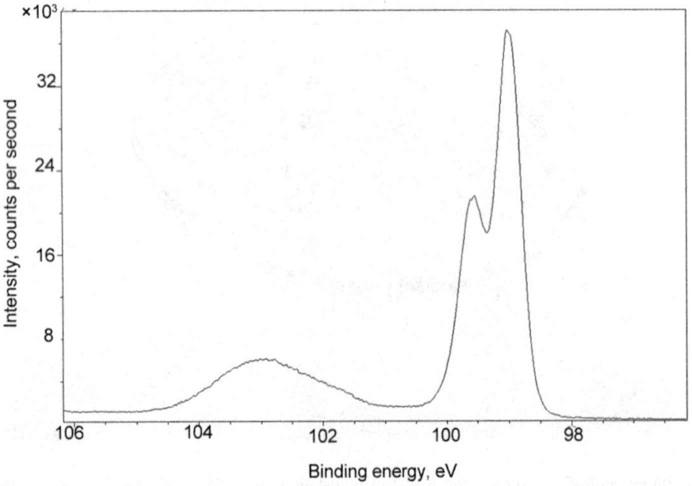

Fig. 11 Splitting of Si 2p line captured by a high-resolution spectrometer. Courtesy of Kratos Analytical

The intensity of the photoelectron lines is an aspect that reduces confidence in the quantitative analytical results of XPS. This is because in the XPS process, it is assumed that the number of electrons collected and detected for a given transition is proportional to the number on the surface that is being analyzed. What is not correct in this assumption is that all the ejected electrons are accounted for. Electrons with higher or lower energies than the ones forming the main photoelectron peak due to acceleration or retardation by overcompensation or undercompensation of the specimen charge compensation system tend to shift away from detection by the spectrometer. They broaden the peak and make the peak asymmetric. This phenomenon affects spectra for insulating samples, where the peaks are wider than those obtained for conducting specimens. In general, however, the peaks are broadened to the high binding side, and separate lines (shake-up lines in polymers) may even be created. To separate the background from the peaks, one of two popular methods is incorporated into commercial software. The background can be separated from the peak by joining the bottommost points of the peak with a straight line or a Shirley curve (Ref 13). The Shirley curve shape is determined by an iterative process where the background intensity at a point is proportional to the total peak area above the background and to higher energy.

Figure 12 provides a comparison of simple background subtractions, linear and quadratic, with a curved Shirley background. Some have claimed background corrections by Tougaard (Ref 12, 14) to be superior to Shirley corrections. The background subtraction is specific to the software used and may differ among XPS instrument manufacturers.

Once the background has been subtracted, the intensity of the peak remains, and it is processed using software, which eventually leads to a composition table. When performing these calculations, relative sensitivity factors (RSFs) are used to scale the measured peak areas so that they are related to the amount of material present on the sample. RSF calculations date back to Scofield's report (Ref 15) from the early 1970s and are modified to the instrument design. For example, for Service Physics SSX-100 software, an adjusted sensitivity factor (SF) is modified such that it is equal to SF_o multiplied by $\{(\text{Al K}\alpha - \text{B.E.})/(\text{Al K}\alpha - \text{C 1s})\}^x$, where B.E. is the binding energy of an electron from Scofield's table (Ref 14), Al Kα = 1486.6 eV, C 1s is the binding energy for C 1s, and x has a default value of 0.7. Some SF values of Fe taken with C 1s as 1.0 are:

Line	SF	Line position
C 1s	1.0	284.6 eV
Fe 2p3	10.82	710.0 eV
Fe 2p (doublets included)	16.42	710.0 eV
Fe 3p	1.669	56 eV

It is easy to see from this table why the quantification for Fe analysis could be erroneous when a single peak versus a double peak is taken. Software from Casa XPS (Ref 12) lists three RSF values for doublets depending on which one of the doublets (or both together) gains consideration for calculation. For SSX-100 software, the C 1s line has an SF of 1.0. Machines by Physical Electronics or Kratos take F 1s as reference (value taken as 1.0). Figure 13 shows the sensitivity factors for elements (at top right-hand corners for each element) for an XPS machine by Physical Electronics where the x-ray source is at 90° to the axis of the analyzer (Ref 8).

For metal and alloy analysis, polished surfaces of standard alloys can be used and peak areas evaluated for the known and unknown samples. With background appropriately subtracted, the peak area ratios between known and unknown specimens can yield reliable results.

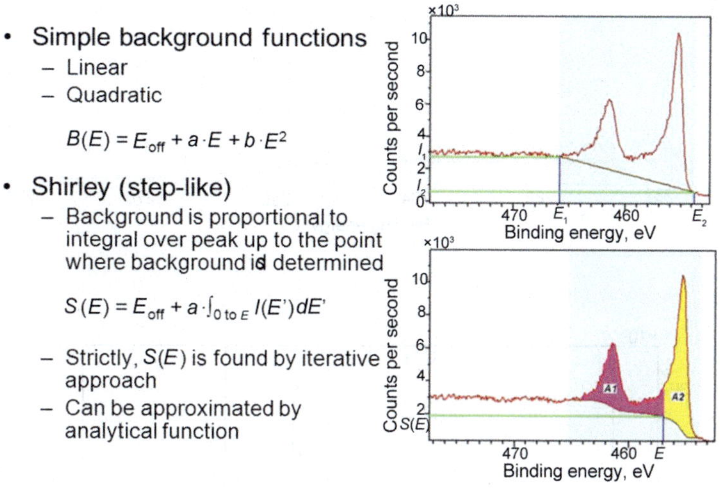
• Simple background functions
– Linear
– Quadratic
B(E) = E_off + a·E + b·E²
• Shirley (step-like)
– Background is proportional to integral over peak up to the point where background id determined
S(E) = E_off + a·∫0 to E I(E')dE'
– Strictly, S(E) is found by iterative approach
– Can be approximated by analytical function
×10³
Counts per second
470
460
Binding energy, eV
A1
A2

Fig. 12 Various background corrections applied to XPS peaks

Data Analysis

An XPS spectrum is displayed as a plot between the intensity (number of electrons of a definite binding energy) on the ordinate and the binding energy on the abscissa. Sometimes the binding energy is replaced by the electron kinetic energy. Binding energy plots increase from right to left, whereas kinetic energy plots increase from left to right. Data analysis starts with the identification of lines that are easily found in the XPS spectrum. The first step in line identification is to look for lines such as C 1s (known as adventitious carbon originating from the environment) and O1s lines. C and O are invariably present on a surface that is unsputtered and unclean. They may be present even

Key: Atomic number 9, PHI sensitivity factor for designated photoelectron transition 1.0, Element symbol F

1 H																	2 He
3 0.025 Li	4 0.074 Be											5 0.159 B	6 0.296 C	7 0.477 N	8 0.711 O	9 1.0 F	10 1.340 Ne
11 1.685 Na	12 0.252 Mg											13 0.193 Al	14 0.283 Si	15 0.412 P	16 0.570 S	17 0.770 Cl	18 1.011 Ar
19 1.30 K	20 1.634 Ca	21 1.678 Sc	22 1.798 Ti	23 1.912 V	24 2.201 Cr	25 2.42 Mn	26 2.686 Fe	27 3.255 Co	28 3.653 Ni	29 4.798 Cu	30 3.354 Zn	31 3.341 Ga	32 3.100 Ge	33 0.570 As	34 0.722 Se	35 0.895 Br	36 1.096 Kr
37 1.316 Rb	38 1.578 Sr	39 1.867 Y	40 2.216 Zr	41 2.517 Nb	42 2.867 Mo	43 3.266 Tc	44 3.696 Ru	45 4.179 Rh	46 4.643 Pa	47 5.198 Ag	48 3.444 Cd	49 3.777 In	50 4.095 Sn	51 4.473 Sb	52 4.925 Te	53 5.337 I	54 5.702 Xe
55 6.032 Cs	56 6.361 Ba	57 7.708 La	72 2.221 Hf	73 2.589 Ta	74 2.959 W	75 3.327 Re	76 3.747 Os	77 4.217 Ir	78 4.674 Pt	79 5.240 Au	80 5.797 Hg	81 6.447 Tl	82 6.968 Pb	83 7.632 Bi	84 Po	85 At	86 Rn
87 Fr	88 Ra	89 Ac															

58 7.399 Ce	59 6.358 Pr	60 4.697 Nd	61 3.754 Pm	62 2.907 Sm	63 2.210 Eu	64 2.207 Gd	65 2.201 Tb	66 2.198 Dy	67 2.189 Ho	68 2.184 Er	69 2.172 Tm	70 2.169 Yb	71 2.156 Lu
90 7.498 Th	91 Pa	92 8.476 U	93 Np	94 Pu	95 Am	96 Cm	97 Bk	98 Cf	99 Es	100 Fm	101 Md	102 No	103 Lr

Fig. 13 Sensitivity factors along with other physical properties of elements

after light sputtering. After sputtering by argon ions, Ar lines around 241 eV are expected to be seen and may be used as a reference as well. It is important that both C 1s and O1s appear as thin and sharp lines; otherwise, sample charging is suspected. C 1s and O 1s line widths can be compared with the instrument standards and deviations observed. In some instruments, charge neutralizer parameters can be adjusted such that the FWHM is minimized and the positions for these lines are noted. Even after such manipulations, the C 1s and O 1s lines may be found to be wider than the instrument specifies, which can indicate that other peaks are hiding inside. Figure 14 shows a comparison of scans obtained by two machines for a common reference polymer, polyethylene terephthalate (PET) (Ref 11, 16). This comparison illustrates that even if all carbon peaks are resolved clearly, earlier instruments provide wider peaks and in some cases may not be able to resolve peaks to the satisfaction of the operator.

Once the C 1s or O 1s line is identified and its position is determined accurately at 284.6 eV (for C 1s), other intense lines should be identified. Subsequently, smaller and perhaps wider lines such as shake-up, satellite, and Auger lines associated with the main intense lines are identified. Following this, low-intensity lines are generally identified. The most intense lines of an unknown element present in small quantities can show up as a small peak. Finally, spin doublets for p, d, and f lines are identified. They should have the correct separation and height or intensity ratios. Multiple splitting and shake-up lines can also be utilized for chemical identification (Ref 8).

An important advantage of XPS analysis is its ability to measure small shifts due to chemical interactions in compounds. These shifts are small but measurable. These shifts are based on experimental data because theoretical calculations are not satisfactory. Handbooks and reference data (Ref 8, 11) on shifts of main elemental lines have been compiled and provided for operators' reference.

Curve Fitting

Often several peaks are superimposed, resulting in skewed or wide envelopes for peaks. Most commercial XPS software provides a peak-fitting algorithm along with peak identification software. Each of these programs has options for line positions for one or more peaks inside the large peak envelope. Peak widths and different types of background subtractions can also be adjusted so that the peaks best fit the large peak envelope. The fitting is performed by the software, and the extent of the curve fitting is also calculated. The instrument operator only has to check the possibility of the existence of such peaks. See Fig. 15 (Ref 16) for the fitting inside of the C 1s line for nylon 6,6 and compare it with the binding of carbon atoms on the molecular chain.

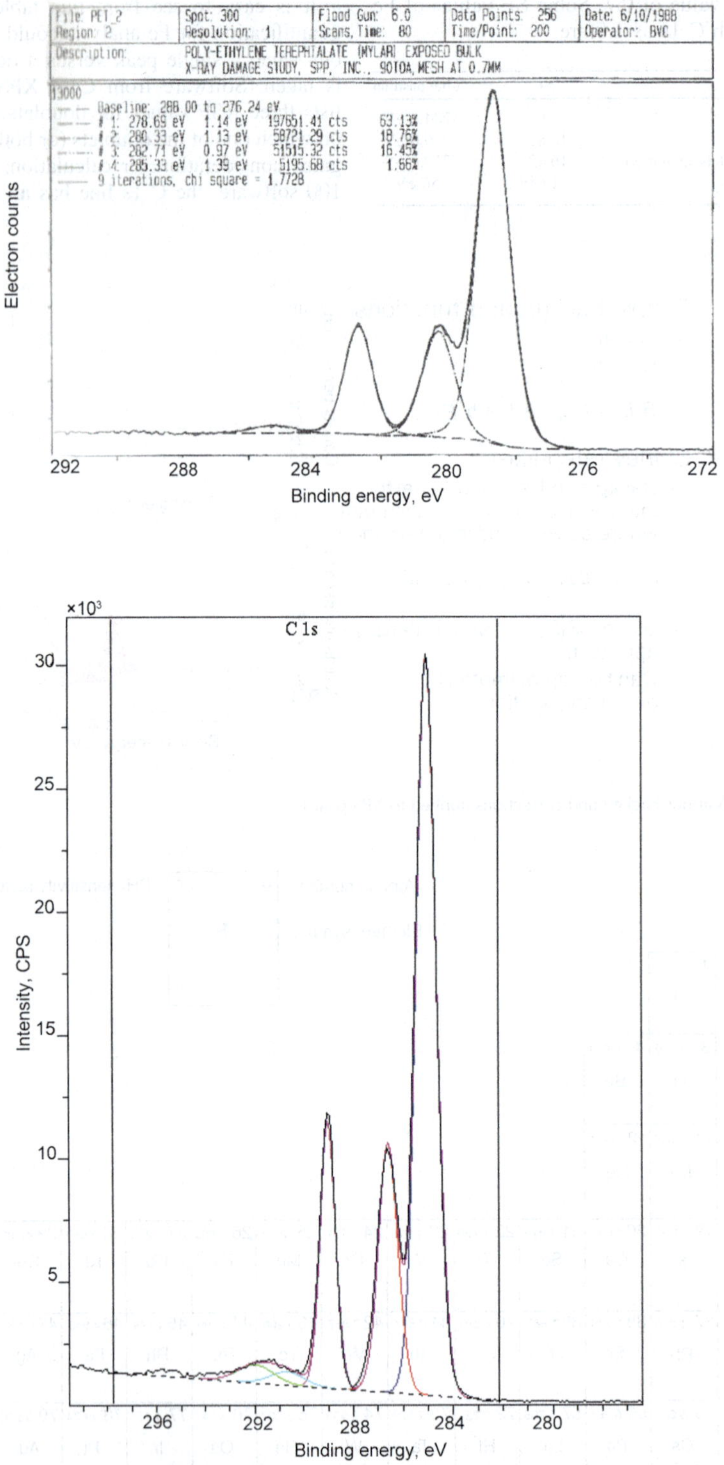

Peak	Position Binding energy (eV)	Full width at half maximum (eV)	Raw area (CPS)	Atomic concentration percentage
C 1s CC,CH	285.077	0.896	29875.7	58.96
C 1s C-O	286.626	0.896	10062.6	19.86
C 1s ester	289.053	0.690	8364.6	16.50
C 1s sat	291.988	1.712	2369.7	4.68

Fig. 14 Comparison of scans (C 1s area) of polyethylene terephthalate in two XPS machines: top, SSX-100; bottom, Kratos AXIS Ultra. Notice the peak widths. SSX-100, 300 spot, 25 eV pass energy, 0.45 eV full width at half maximum; Kratos, 150 spot, 10 eV pass energy, 0.25 eV full width at half maximum (Ref 11, 16). Courtesy of Wiley (Ref 16)

Valence Band Spectra

Valance band spectra, in the range of −5 eV to approximately 50 eV, have been found to be useful in several cases where the core level shift is subtle. These spectra have also been used as fingerprints of materials rather than identification of a specific molecular orbital. Figure 15 also shows a valence band spectrum for nylon 6,6 (Ref 16).

Applications and Interpretation

From the very onset of the development of commercial XPS systems with associated monochromators and ion guns, XPS has emerged as a very useful tool in all kinds of analysis. Due to its ability to analyze monolayers of surface materials and detect elements with an atomic number higher than that of Li, as well as the presence of elements registered in the form of sharp peaks, initial researchers used the technique extensively to understand catalysis. Because surface contamination can be removed by argon sputtering, and metals and alloys are conductive, metallurgical analyses can easily be performed and have taken center stage for the analysis of coatings and plating, study of diffusional processes, and development of materials for adhesion and metallic bonding, as well as applications in microelectronics. XPS is perhaps most advantageous in the analysis of polymers because the bonds in polymers can be resolved, particularly for failure analysis, evaluation of production issues, and overall characterization. With advancements in x-ray sources, cluster guns, and imaging capabilities, XPS is useful for investigating almost all surface problems. The following paragraphs present examples (brief summary of reports) of how XPS data can be used to solve problems with existing surface interactions, or to investigate new materials.

General applications of XPS analysis include (1) identification of unknown substances or material characterization, (2) measurement of layer thickness in layered engineering materials, (3) quality control of products with thin coatings, (4) contamination analysis, and (5) general failure analysis where failure is investigated using XPS as a tool. What follows are examples of cases where XPS has been successfully employed in resolving or analyzing problems that cannot be resolved by other analytical

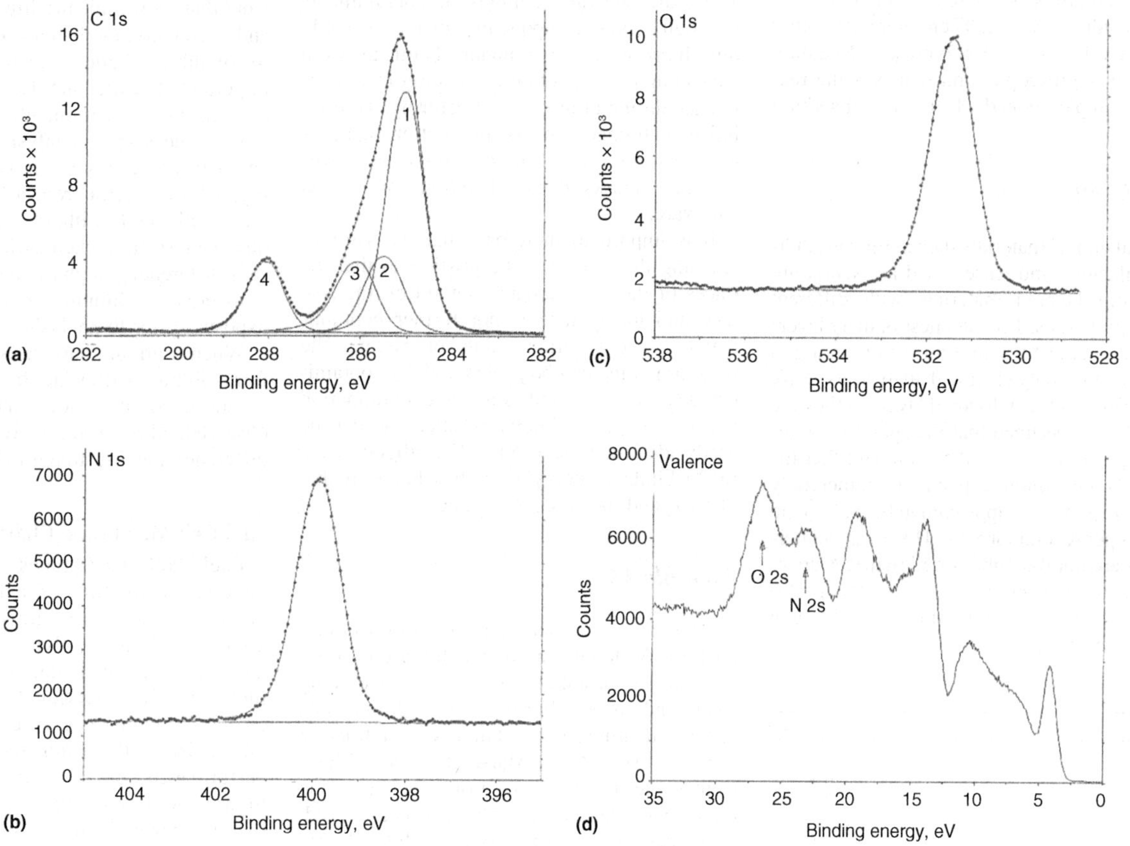

Nylon 6,6 (N66)

	C 1s				O 1s	N 1s
	1	2	3	4		
BE (eV)	285.00	285.46	286.02	288.02	531.37	399.81
FWHM (eV)	0.95	0.99	1.03	0.95	1.18	1.17
Area (%)	51	17	17	15		
A	0.07	0.06	0.16	0.11	0.13	0.10
m	0.85	0.90	0.92	0.83	0.76	0.83

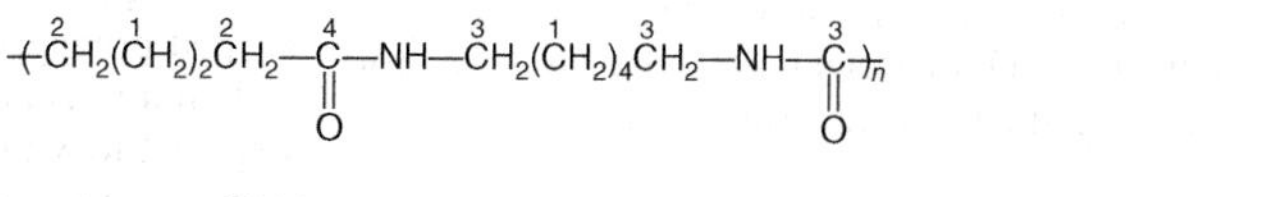

Source: Aldrich
Casting solution: 1% (w/w) in HFIP
Degradation index: 5, O/C
VB acquisition: 2
Charginq reference: lowest *BE* component

(e)

Fig. 15 Spectra from nylon 6,6. (a) Curve fitting of the C 1s line. (b) O 1s spectrum. (c) N 1s spectrum. (d) Valence band spectrum. (e) The formula and the position of various carbon peaks fitted to C 1s. Courtesy of Wiley (Ref 16) instruments such as electron microscopes or electron microprobe analyzers.

Example #1

Rhenium is an expensive but ductile high-temperature material. A rocket engine (thruster) with rhenium liner was tested, and at the end of testing there was a clear liquid found inside the engine. The samples were brought to the NASA Marshall Space Flight Center laboratory using a cotton swab. To analyze the residue, the cotton swab was rubbed on aluminum foil, and the residue, though nearly invisible to the naked eye, was transferred to the aluminum foil. The sample spent 1 h in the preparation chamber and then was analyzed using an SSI SSX-100 XPS machine. Several rhenium peaks were identified, indicating the residue as ReO_3. There were also other rhenium oxides (ReO_2, ReO) present. Rhenium oxides are hygroscopic, and that was the reason the sample looked clear after absorbing moisture.

Example #2

Several optical materials such as mirrors, radiation deflectors, and lenses used by NASA are made from layered materials with different physical properties. The thickness of these layers can be measured by sputtering using an argon ion gun, followed by chemical analysis to verify the chemistry of the individual layers as they are removed. It is assumed that the sputtering rates of various substrates are the same and that the gun at its maximum capacity (commercially available) removes approximately 1.0 Å in thickness per second for SiO_2. However, sputtering rates are not the same for all matrices under the same ion gun settings. The sputtering rates of some substrates (relative rates at 4 kV gun voltage) are (Ref 8):

Target	Sputtering rate
Ta_2O_5	1.00
Si	0.90
SiO_2	0.85
Pt	2.20
Cr	1.4
Al	0.95
Au	4.10

Therefore, care must be taken and sputtering rates for each material need to be calibrated for accurate thickness determination.

Example #3

Carbon fibers are coated with materials that aid in the adhesion of fibers to the matrix. Chemistry and the amount of the coating material can be evaluated using XPS because the fibers are very fine. To perform this analysis, a bunch of fibers are pressed and clamped to create a flat surface that is covered by fibers. The surface, with its attachment to the bundle, is introduced into the analytical chamber, and the coating can then be analyzed.

Example #4

Contamination on surfaces is a major problem when two surfaces are not able to bond, leading to exfoliation, corrosion, or bulging of paint on a painted surface. This calls for a surface analysis and a possible remedy of the failure. Contamination on one of the inside layers on a multilayered material is challenging because removal of layers by sputtering is a very slow process and may not be uniform. There are several examples of contaminants in engineering components, including body oils, lubricants, and remnant chemicals from a cleaning process, often originating from negligence during or after manufacturing. To minimize such contaminants, production facilities use clean rooms, and the operators use gloves and face masks during handling of critical hardware.

It is important, therefore, that the samples reaching the laboratory be pristine so that the traces of the contaminants remain on the surface and the surface is not further contaminated. For exfoliating paints or plating, the paint or plating can be peeled and the contaminated surface analyzed. Silicone contamination is a major cause of nonadherence and can be easily identified using XPS. For silicones, the Si 2p binding energy peak lies between that of SiO_2 and the silicone element.

Example #5

XPS can be used as a general tool for failure analysis. Materials can be identified on a fracture surface, despite the fracture surface being rough and uneven. For example, chloride salts can be identified on a stainless-steel fracture surface. At NASA Marshall Space Flight Center, the fracture surface of a rocket engine component made from niobium was analyzed. XPS analysis of the surface indicated that it reacted to fluoride-containing lubricants, resulting in compounds such as NbO_2 F. It is believed that the lubricant reacted with the surface to produce this compound and possibly helped propagate the crack.

Failure analysis of polymeric materials exposed to x-ray and other forms of radiation can be accomplished by taking advantage of loss of peak ratios from databases (Ref 16). Peak ratio losses as a function of time are shown in Fig. 16. This indicates that the polymers are degraded by exposure to x-ray during analysis, and their ratios would indicate the length of exposure that could be related to failures.

X-Ray Photoelectron Spectroscopy Machine Manufacturers

Currently, there are three main manufacturers of XPS machines: Physical Electronics Inc. (PHI), Thermo Fisher Scientific Inc., and Kratos Analytical Ltd. All three manufacturers offer advanced applications, continuously developing capabilities, and XPS applications for new materials.

Physical Electronics (Ref 17)

Lithium Battery Electrode Analysis

XPS analysis can be effectively used to study the surface composition of lithium battery electrodes as it relates to their use. This evaluation is critical to understanding mechanisms that may limit the life of the electrodes, and eventually, the battery. Due to the reactivity of lithium battery electrode surfaces, it is important that the samples are transported in a controlled environment, such as a glove box, to the surface analysis instrument under an inert cover gas or in vacuum. Shown in Fig. 17 are spectra from a lithium anode surface with and without air exposure. The observed surface chemistries show a significant difference, emphasizing how important it is to protect a lithium anode surface from air exposure prior to analysis.

When a lithium surface is exposed to air, it forms lithium carbonate, but when the surface is not exposed, oxides and hydroxides are observed. XPS surface analysis can show these differences, as illustrated in Fig. 17.

Fuel Cell Membrane Characterization

Fuel cell membranes are multilayered materials consisting of a permeation membrane (fluorocarbon) with two thin layers of noble metal in a polymer matrix acting as the anode and cathode, respectively, of the fuel cell. Cross-sectional chemical analysis of these membranes is useful for studying mechanisms that influence the life and performance of a fuel cell over time. Figure 18 shows XPS chemical state distribution images from an aged fuel cell membrane that indicate degradation of the fuel cell by fluoride products in the anode and cathode regions.

Contamination Identification on Polymer Surfaces

XPS has the capability to characterize contaminants on a polymer surface through resolving the emerging electron spectrum from the surface. In this example, a scanning x-ray beam created a secondary electron image of an optically transparent PET sample. The analysis revealed the presence of unexpected localized surface contaminants

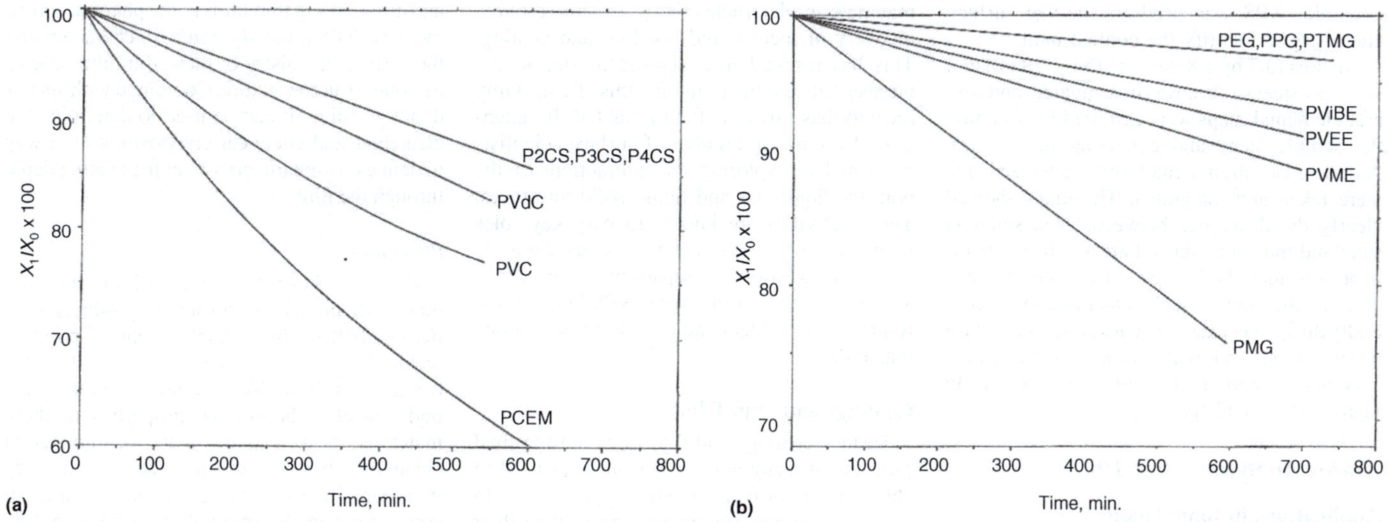

Fig. 16 Comparison of various polymers as they are damaged due to exposure to x-rays. (a) Plots of $(X_t/X_0) \times 100$ versus t for chlorine-containing polymers, showing loss of chlorine intensity. (b) Plots of $(X_t/X_0) \times 100$ versus t for aliphatic ether polymers, showing loss of oxygen peaks. X = O/C atom ratio. Courtesy of Wiley (Ref 16)

on the polymer surface. A micro-area XPS spectrum obtained with a 20 μm diameter x-ray beam identified the contaminant as a fluorocarbon (see Fig. 19).

Thermo Fisher Scientific (Ref 18)

X-Ray Photoelectron Spectroscopy Evaluation of Wear-Resistant Coatings

Lubricating oils contain additives that play an important role in forming friction-modifying layers on metal surfaces. Calcium sulfonate additives in lubricating oil, as well as zinc dialkyldiphosphonate (ZDDP), are used to deposit wear-resistant layers on steel surfaces and to reduce friction. It has been observed that ZDDP forms a protective, glassy phosphonate coating on surfaces under tribological loads. In one study, various ratios of calcium sulfonate in oil and ZDDP were used, and three steels were evaluated in these mixtures.

The three tribological samples tested in this evaluation were $GOOD_{NEW}$, BAD_{NEW}, and $GOOD_{OLD}$. Although full details of all three samples were unavilable, the $GOOD_{OLD}$ sample was in the tribology test rig for a long time and showed good tribological properties; the $GOOD_{NEW}$ sample behaved well under the tribology test rig, whereas BAD_{NEW} did show unusual tribological parameters in the test. The surfaces of all three samples were analyzed using a monochromatic XPS instrument. Some of the results are shown in Fig. 20.

Two of the samples, $GOOD_{OLD}$ and $GOOD_{NEW}$, showed good tribological properties, and XPS showed the presence of calcium carbonate tracks (identified as red peaks) on these samples. The BAD_{NEW} sample, which showed poorer tribilogical properties, had virtually no calcium carbonate on its surface.

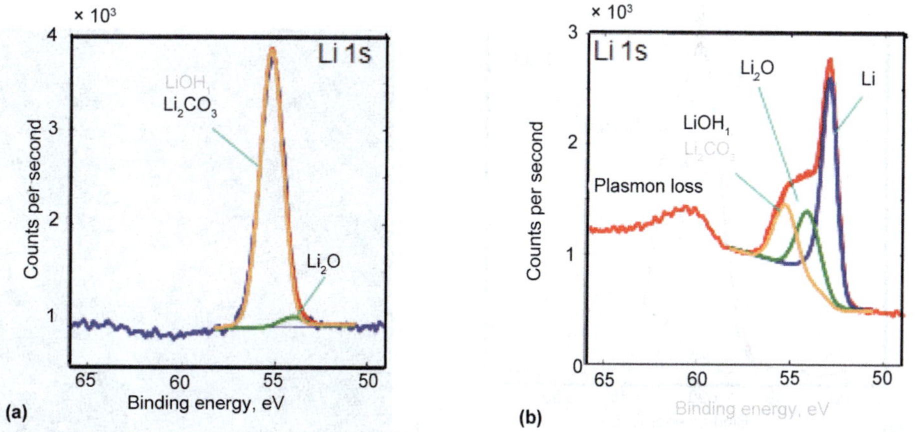

(a)

Binding energy (eV)	Bond	Area percentage
52.88	Li	0
54.06	Li_2O	3
55.26	Li_2CO_3	97

(b)

Binding energy (eV)	Bond	Area percentage
52.88	Li	55
54.06	Li_2O	25
55.26	Li_2OH	19

Fig. 17 Lithium battery electrode analysis using XPS illustrates differences between conditions (a) with and (b) without air exposure (Ref 17)

Good tribological properties are attributed to the presence of calcium carbonate. XPS analysis also showed that the correct weight ratio of ZDDP to calcium detergent results in the formation of calcium carbonate during tribological loads. Calcium carbonate, however, did not form on the BAD_{NEW} sample, indicating an unfavorable ratio of ZDDP to detergent in the oil formulation.

X-Ray Photoelectron Spectroscopy Analysis of a Surface Contamination on a Steel Sample

A contaminant-free surface is very important to guarantee that materials such as steels achieve their desired performance such as appearance and the ability to provide the expected surface finish. Surface contamination of steel can also result in problems such as adhesion failure and contact bonding problems when components are used in manufacturing. Surface contaminants often result in an "unsatisfactory" appearance for many steel finished products such as ovens and other domestic appliances. In addition, surface contaminants are the sources of cross-contamination, corrosion, and electrical contact problems. Many of these surface contamination issues are not obvious in appearance during or after production of manufactured goods using steel parts. In this example, XPS was used to analyze surface chemistry to identify the contaminant.

An area (3.7 by 4.8 mm, or .15 by .19 in.) of a stainless-steel surface was investigated, and several elemental maps were acquired by scanning the sample stage and collecting the spectra. Atomic concentration maps of the analyzed area were taken and integrated. The maps showed clearly the difference between clean stainless steel and the contaminated areas. The contaminant was identified to be an organic residue. The identification of such contamination is generally difficult because the contaminant may be a mixture of several materials and the instrument may not have the high resolution necessary to identify the chemistry.

Kratos Analytical (Ref 19)

Applications in Ionic Liquid Characterization

Ionic liquids have become an area of increasing popularity in academic research. They have been found to be candidate materials in electrochemistry, in situ resource recovery of metals, and bonding and coating. This has resulted in a significant rise in the number of publications in this field. Only recently has this new field attracted the attention of a growing number of surface scientists interested in exploring the interactions at the both the liquid-gas and liquid-solid interfaces. These activities are known to play key roles in processes such as heterogeneous catalysis and gas distillation and separation. The surface analysis of ionic liquid using XPS helps in the fundamental understanding of these unique materials.

Coatings and Thin Films

Surface coatings and thin films are the final products in many applications and are used to enhance or provide required properties to bulk engineering components targeted to their applications. Thin films can range from tens of angstroms to several microns in thickness and find application in diverse areas such as optical antireflective coatings, architectural glazing, and drug-eluting thin films in the pharmaceutical industry. XPS is ideally suited for characterizing the surface chemistry of these thin films due to its nondestructive nature. Combined with sputter depth profiling, it can be used to determine the elemental and chemical composition as a way to define coating integrity over the coating depth through the film.

Polymers

Polymers have been utilized in replacing metals in numerous consumer products over the last half century. Applications of polymer materials range from food packaging to electronics and from biomaterials to automotive body panels. The surface properties of these materials, as well as unwanted contamination changing these properties, are often very important in determining the performance of components made from polymers for a specific application. X-ray photoelectron spectroscopy is ideally suited for the surface characterization of various polymeric materials because it can provide quantitative chemical state information from the upper 10 nm of the material.

Commercial polymer materials are insulating, and therefore effective charge compensation is important in generating high-resolution spectra with monochromatic x-rays. The charge neutralization of modern spectrometers is demonstrated by the performance on PET films, where the peak width of the component peaks corresponding to the ester group is narrow with a sensitivity defined by the maximum heights of the hydrocarbon peaks.

An interesting aspect of polymer characterization is the use of cluster ion guns.

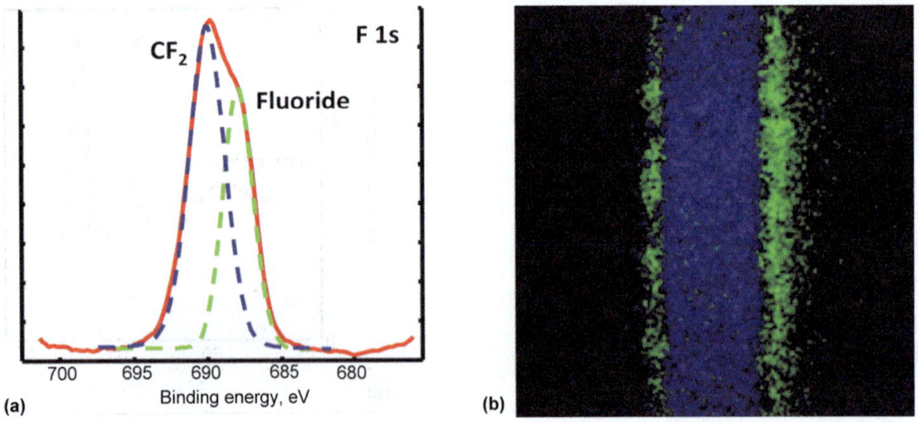

Fig. 18 (a) F 1s spectrum region corresponding to (b) the chemical state image. The chemical state image shows the presence of fluorocarbon chemistry at the center (blue) and the degradation products, fluoride chemistry, in the anode and cathode regions (green) (Ref 17)

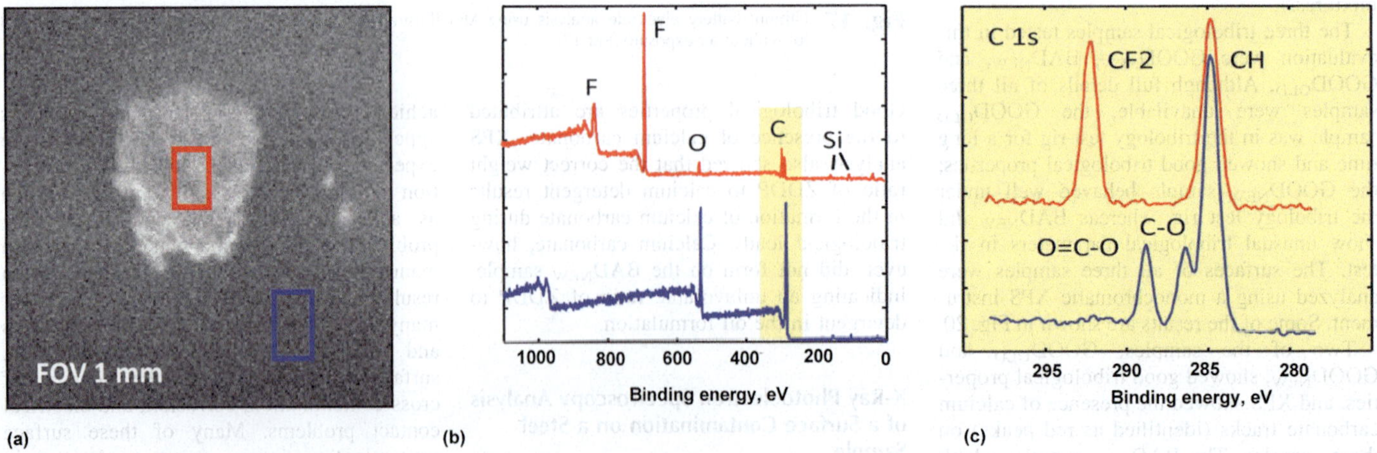

Fig. 19 Contaminant analysis on a polyethylene terephthalate sample identified as a fluorocarbon. (a) Scanning electron microscope image. (b) XPS scans from a contaminant (red) and the polyethylene terephthalate (blue). (c) Detailed C 1s region (Ref 17)

REFERENCES

1. A. Einstein, Über einen die Erzeugung und Verwandlung des Lichtes betreffenden heuristischen Gesichtspunkt (Concerning an Heuristic Point of View toward the

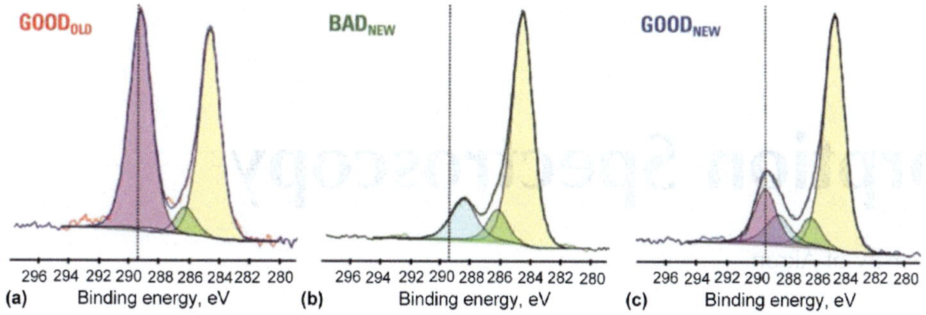

Fig. 20 Spectrum obtained around C 1s on three samples. (a) GOOD$_{OLD}$ shows a significant amount of carbonate compared with (b) and (c); (b) BAD$_{NEW}$ shows the least amount of carbonate (Ref 18).

Emission and Transformation of Light), *Ann. Phys. (Berlin)*, Vol 17, 1905, p 132

2. H. Robinson, X-Ray Terms and Intensities, *Phil. Mag.*, Vol 50, 1925, p 241
3. K. Siegbahn, *ESCA: Atomic, Molecular and Solid State Structure Studied by Means of Electron Spectroscopy*, Almqvist and Wiksells, 1967
4. K. Siegbahn, D. Hammond, H. Fellner-Feldegg, and E.F. Barnett, Electron Spectroscopy with Monochromatized X-Rays, *Science*, Vol 176, 1972, p 245 10.1126/science.176.4032.245
5. K. Larsson, C. Nordling, K. Siegbahn, and E. Stenhagen, Photoelectron Spectroscopy of Fatty Acid Monolayers, *Acta Chem. Scand.*, Vol 20, 1966, p 2880 10.3891/acta.chem.scand.20-2880
6. B. Vincent Crist, X-Ray Photoelectron Spectroscopy, Wikipedia, https://en.wikipedia.org/wiki/X-ray_photoelectron_spectroscopy
7. D. Briggs and M.P. Seah, Ed., *Practical Surface Analysis, Auger and X-ray Photoelectron Spectroscopy*, Vol 1, Wiley, 1996, p 534–536
8. J.F. Moulder, W.F. Stickle, P.E. Sobol, and K.D. Bomben, *Handbook of X-Ray Photoelectron Spectroscopy*, Perkin-Elmer Corp., 1992
9. R.L. Chaney, Recent Developments in Spatially Resolved ESCA, *Surf. Interface Anal.*, Vol 10, 1987, p 36 10.1002/sia.740100108
10. W.M. Riggs and M.J. Parker, Surface Analysis by X-Ray Photoelectron Spectroscopy, *Methods of Surface Analysis*, Vol 1, A.W. Czanderna, Ed., Elsevier Scientific Publishing Co., 1975, p 131
11. B.V. Crist, *Handbook of Monochromatic XPS Spectra*, John Wiley & Sons, 2000, p xv
12. Casa XPS databases, http://www.casaxps.com/
13. D.A. Shirley, High-Resolution X-Ray Photoemission Spectrum of the Valence Bands of Gold, *Phy. Rev. B*, Vol 5, 1972, p 4709 10.1103/PhysRevB.5.4709
14. M.P. Seah, Background Subtraction, Part I: General Behavior of Tougaard-Style Background in AES and XPS, *Surf. Sci.*, Vol 420, 1999, p 285 10.1016/S0039-6028(98)00852-8
15. J.H. Scofield, Hartree-Slater Subshell Photoionization Cross-Sections at 1254 and 1487 eV, *J. Electron Spectros. Relat. Phenomena*, Vol 8, 1976, p 129 10.1016/0368-2048(76)80015-1
16. G. Beamson and D. Briggs, *High Resolution XPS of Organic Polymers: The Scienta ESCA300 Database*, John Wiley & Sons, 1992, p 174
17. Physical Electronics Inc. website, www.phi.com
18. Thermo Fisher Scientific Inc. website, www.thermofisher.com
19. Kratos Analytical Ltd. website, www.kratos.com

ASM Handbook, Volume 10, *Materials Characterization*
ASM Handbook Committee
DOI 10.31399/asm.hb.v10.a0006661

Thermal Desorption Spectroscopy

Gary L. Doll and Paul J. Shiller, The University of Akron

Overview

General Uses

- Experimental method that measures the desorption rate of desorbing gases from surfaces as a function of temperature
- Used in the study of adsorption, desorption, and reaction of adsorbed atoms and molecules on surfaces

Examples of Applications

- Extensively used as a tool to study fundamental reactions of atoms and molecules on well-defined surfaces, especially in the area of catalysis
- Used in environmental applications such as monitoring personal exposure to toxic chemicals, indoor air-quality monitoring, identification of volatiles in soil and water, analysis of environmental pollutants, quantification of hydrogen in metals (especially steels), corrosion mechanisms, electrochemistry, and tribology

Samples

- Specimens include the substrates (solids and powders) and, in most cases, the adsorbing gas or gases.

Limitations

- Factors that must be considered when performing experiments include reactor leaks and gas impurities, residual water vapor in the reactor, system pressure and measurement, gas flow rates, adsorption and desorption temperatures, reactor size and amount of exposed substrate, temperature measurement and programmer, exit-gas temperature, pressure drop at and location of the detector, and cracking in and calibration of the mass spectrometer.

Estimated Analysis Time

- Measurement time depends on initial temperature required for adsorption, temperature ramp rate, and final temperature after complete desorption.
- Analysis time depends on the complexity of the desorbed species measured by the mass spectrometer.

Capabilities of Related Techniques

- *Thermogravimetric analysis:* A method of thermal analysis in which the mass of a sample is measured over time as the temperature changes. This measurement can provide similar information as thermal desorption spectroscopy but not to the level of accuracy. Thermal desorption spectroscopy can provide information that something happened but may not be able to provide information about what happened. Physical phenomena, such as phase transitions and absorption and desorption, as well as chemical phenomena, such as chemisorption, thermal decomposition, and solid-gas reactions (e.g., oxidation or reduction), can be interpreted from the temperature-dependent mass change of a sample.

Thermal desorption spectroscopy (TDS) measures the desorption rate of desorbing gases from a surface as a function of temperature. It is used to study adsorption, desorption, and reaction of adsorbed atoms and molecules on surfaces. The method is also referred to as temperature-programmed desorption (TPD) when the sample temperature is increased linearly as a function of time, *t*, in a programmed way. A TPD technique was first described by Apker (Ref 1) in 1948; he used a flash-filament approach to study the nature of adsorbed gases on tungsten. Since then, the use of TDS and TPD in the measurement of gas desorption from surfaces has been mentioned extensively in the literature.

Prior to the work of Apker, Taylor and Langmuir (Ref 2) used polycrystalline tungsten filaments as substrates in a flash-filament method to determine the relative surface coverage of cesium chemisorbed on tungsten. An adsorbed species can be bound to a surface by van der Waals forces (physisorption) and by covalent bonding (chemisorption). At a characteristic temperature, adsorbates begin to desorb from the surface via an activated process. Knowledge of the nature of the desorption process helps in understanding the nature of elementary chemical processes in the layer, the energetics of bonding and the specification of the chemical nature of the bound species, and the nature and magnitude of interactional effects between adsorbate species (Ref 3).

Ehrlich (Ref 4) pioneered the so-called flash-desorption method as a means for studying the adsorption of gases such as N_2, CO, and xenon on polycrystalline surfaces. Ehrlich exposed a clean polycrystalline filament to a flowing gas environment while rapidly flashing the filament at a high rate of heating (~1000 K/s). The initial coverage of the adsorbate could be varied by varying the initial time and pressure of adsorption. Redhead (Ref 5) used much slower rates of temperature increases with time, which led to an increase in resolution of the method and to the discovery of additional binding states unresolved by rapid flash desorption.

Delchar and Ehrlich (Ref 6) performed the first TPD studies using single-crystal substrates, studying N_2 adsorption on tungsten crystals having various surface planes. The use of single-crystal substrates has become a standard procedure in many surface science experiments.

Early studies of TPD were reviewed in 1972 by Cvetanovic and Amenomiya (Ref 7). In 1975, King (Ref 8) reviewed thermal desorption from metal surfaces, while Falconer and Schwarz (Ref 9) reviewed TPD applications in catalysis research in 1983, and Yates (Ref 3) provided a comprehensive treatment on the thermal desorption of adsorbed species on polycrystalline and metal catalysts in 1985. Ishii and Kyotani (Ref 10) and Ogura and Fukutani (Ref 11) described recent advancements in TDS experimental procedures.

General Principles

In a TPD experiment, a substrate is mounted on a heater in a vacuum chamber, preferably a chamber capable of achieving an ultrahigh vacuum level. After suitable pretreatment or a reduction process has been performed to obtain the desired substrate surface, the substrate is dosed with an adsorbing gas, usually through an injection of the adsorbate into an inert carrier gas upstream of the reactor. As the substrate is heated, adsorbed gases desorb, decompose, or react. The desorbed gas or gases are monitored downstream, usually by a mass spectrometer. With increased temperature, the desorption rate of a particular surface species increases, eventually goes through a maximum, and drops back to zero as the surface is depleted of adsorbate. The spectra usually have more than one maximum or peak (Ref 7), and the shape and position of the peak maxima (peak temperatures) are related to the desorption process and therefore provide information on how the gas is adsorbed on the substrate. An example of a TPD spectrum showing the desorption of CO and CO_2 from an activated carbon substrate is shown in Fig. 1.

Temperature-programmed desorption experiments typically use a linear temperature ramp given by:

$$T(t) = T_0 + \beta t \quad \text{(Eq 1)}$$

where t is time, T_0 is the starting temperature, and β is the heating rate (dT/dt). A TPD experiment doses a cold substrate to adsorb (condense) the atoms or molecules of interest by injecting the material into the vacuum chamber. After the substrate is dosed, the injection is stopped and the pressure is reduced to baseline pressures. Under these conditions, there is no longer adsorption, and desorption occurs as the temperature rises. The rate of desorption can be written as:

$$-\frac{d\Theta}{dt} = k_{-1}\Theta^n \quad \text{(Eq 2)}$$

where Θ is the surface coverage, k_{-1} is the desorption rate constant, and n is the rate order. From Eq 1 and 2, the desorption rate as a function of sample temperature can be expressed as:

$$-\frac{d\Theta}{dT} = \frac{k_{-1}}{\beta}\Theta^n \quad \text{(Eq 3)}$$

Desorption follows an Arrhenius behavior, so:

$$k_{-1} = Ae^{\frac{-E_d}{RT}} \quad \text{(Eq 4)}$$

where A is the pre-exponential factor for first-order desorption, E_d is the activation energy, and R is the ideal gas constant. Combining the Arrhenius equation with Eq 3 gives the Polanyi-Wigner equation (Ref 12):

$$-\frac{d\Theta}{dT} = \frac{A}{\beta}\Theta^n e^{\frac{-E_d}{RT}} \quad \text{(Eq 5)}$$

This is the starting point for analyses of TPD data. Whereas n provides insight into the mechanism of adsorption and desorption, E_d and A determine the desorption rate of each molecule.

Differentiating Eq 5 with respect to temperature (assuming a first-order equation; i.e., $n = 1$) and setting it equal to zero gives the temperature of peak desorption (T_p):

$$\frac{E_d}{RT_p^2} = \frac{A}{\beta}e^{\frac{-E_d}{RT_p}} \quad \text{(Eq 6)}$$

Equation 6 shows that T_p is independent of θ for a first-order reaction, so E_d can be found directly from a measurement of T_p, assuming E_d is independent of surface coverage. This is the method outlined by Redhead to analyze the thermal desorption of gases (Ref 13). Carter (Ref 14) was independently performing similar work at approximately the same time. Both researchers reached the same conclusions and submitted papers, with Carter's work published one issue later. The activation energy, E_d, is then directly related to the peak temperature. The surface coverage and desorption rate modeled from the Polanyi-Wigner equation for first order with an activation energy E_d = 30 kcal/mol from 200 to 400 K (−75 to 125 °C) is shown in Fig. 2. The relationship between E_d and T_p is nearly linear, but it depends on β. Figure 3 shows a plot of Eq 6 for various values of β, which assumes that $A = 10^{18}$/s. The activation energy, E_d, can be determined without a predetermined

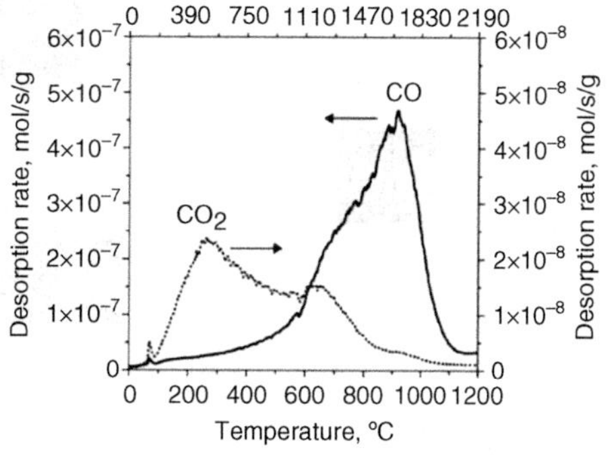

Fig. 1 Temperature-programmed desorption spectrum of the desorption of CO and CO_2 from an activated carbon substrate. Source: Ref 10

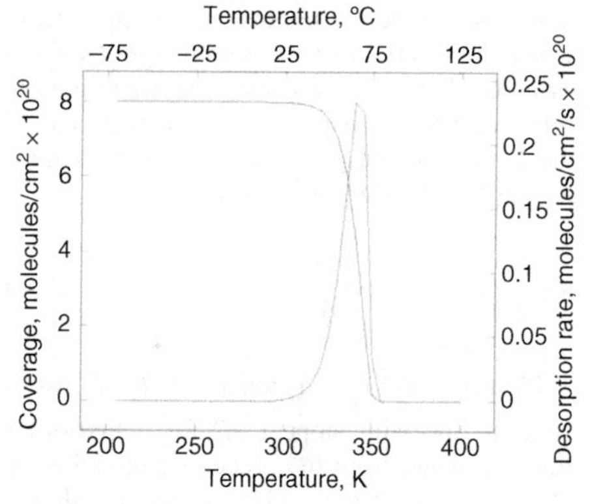

Fig. 2 Surface coverage and desorption rate modeled from the Polanyi-Wigner equation (Eq 5) for first-order desorption with an activation energy (E_d) of 30 kcal/mol from 200 to 400 K (−75 to 125 °C)

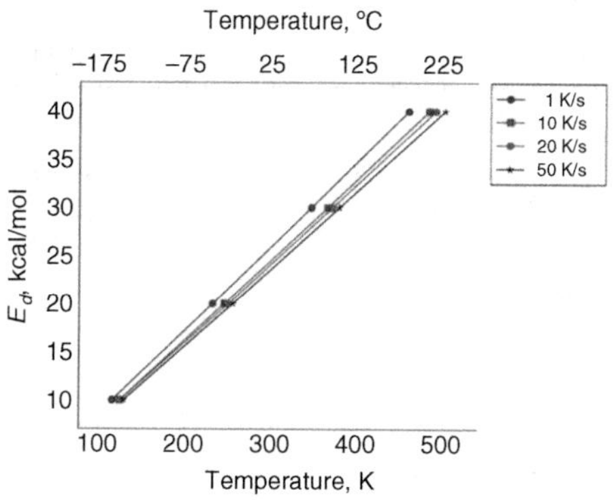

Fig. 3 Activation energy, E_d, as a function of peak desorption temperature for a first-order reaction and a linear temperature sweep for various heating-rate values for the pre-exponential factor for first-order desorption, $A = 10^{18}$/s

assumption of A by varying β and plotting ln (T_p) versus $\ln(\beta)$:

$$\frac{d(\ln\beta)}{d(\ln T_p)} = \frac{E_d}{RT_p} + 2 \quad \text{(Eq 7)}$$

Redhead (Ref 13) also worked out the relationship for a second-order desorption that includes the initial surface coverage:

$$\frac{E_d}{RT_p^2} = \frac{A_2}{\beta}\Theta_0 e^{\frac{-E_d}{RT_p}} \quad \text{(Eq 8)}$$

where A_2 is the second-order pre-exponential factor.

If the temperature of the peak decreases with increasing coverage, the reaction may be either second order with a fixed activation energy or first order with an activation energy dependent on coverage. These two cases can be distinguished by plotting $\ln(\Theta_0 T_p^2)$ versus T_p^{-1}. A second-order reaction with fixed activation energy yields a straight line.

Temperature-programmed desorption experiments are carried out under vacuum, and the total or partial pressure of a species can be measured by using a mass spectrometer. Peak desorption under a temperature ramp occurs when $dp/dT = 0$. Lord and Kittelberger (Ref 15) assumed that under fast pumping, there is very little resorption, so pressure is proportional to the desorption rate ($p \propto -d\Theta/dT$). Differentiating Eq 5 and setting $\frac{d^2\Theta}{dT^2} = 0$ yields:

$$\ln\left(\frac{T_p^2}{\beta}\right) = \frac{E_d}{RT_p} + \ln\left(\frac{E_d}{AR}\right) \quad \text{(Eq 9)}$$

Plotting $\ln\left(T_p^2/\beta\right)$ versus $1/T_p$ forms a straight line with slope E_d/R for a first-order reaction. Data from the preceding model were plotted and analyzed, and the results are shown in Fig. 4 for calculated values of E_d from the plot of $\ln(T^2/\beta)$ versus $1/T$:

E_d model	E_d from plot
10	8.09
20	18.53
30	29.47
40	40.30

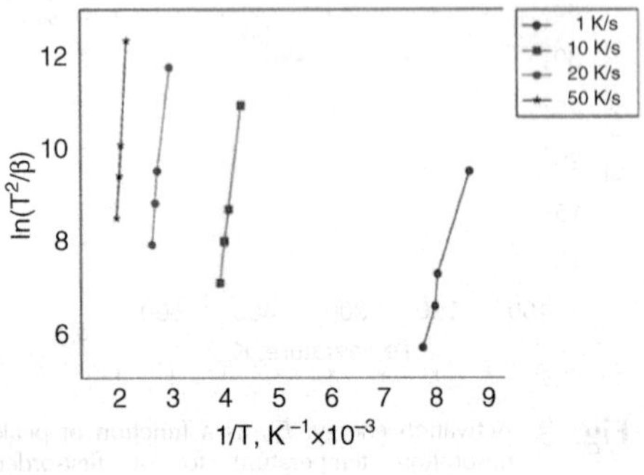

Fig. 4 Plot of calculated data for $\ln(T^2/\beta)$ for different values of β. Slopes are equal to E_d/R.

Thermal desorption spectroscopy can be also used to determine the surface coverage of adsorbates after calibration of a mass spectrometer by measuring the TDS spectrum from a standard sample with a known coverage or by measuring the ion current of the gas at a certain pressure (Ref 11). Because E_d and A often depend on Θ due to interactions between adsorbates, the interactions can be investigated by measuring TDS spectra with different initial coverages. The number of adsorption sites or adsorption phases can be determined from the number of desorption peaks in a spectrum. Surface reactions can be studied by recording TDS spectra with different masses at the same time. When TDS is applied to a system where surface reactions are involved, the method is sometimes called temperature-programmed reaction spectroscopy.

Systems and Equipment

Figure 5 illustrates an example of a TPD experiment using an apparatus consisting of a vacuum chamber and pumping system capable of achieving base pressures of approximately10^{-6} Pa (7.5×10^{-9} torr). Substrates are introduced to the chamber through a load-lock mechanism (not shown) and placed onto a holder that can be heated to 1300 K (1025 °C) and cooled to approximately 170 K (−105 °C). After reaching base pressure, the substrate is heated and a gas is introduced through the gas inlet to remove contaminants from the surface of the substrate. An Auger electron spectroscopy (AES) probe is used to establish the effectiveness of contaminant removal. After the substrate is effectively pretreated, it is cooled to cryogenic temperatures, and the adsorbing gas is introduced through the gas inlet and adsorbs onto the substrate. The typical unit of gas exposure is the Langmuir (1 L = 1.33×10^{-4} Pa-s, or 10^{-6} torr-s). After obtaining the desired dosage, the temperature of the substrate is linearly ramped, and mass spectra are collected using a quadrupole mass spectrometer as a function of temperature. The substrate temperature is increased until no more gas is desorbed from the substrate.

Figure 6 shows an example of a well-executed TPD experiment by Collins et al. (Ref 16), which was performed in a system similar to that shown in Fig. 5. The researchers explored the desorption of CO (in partial pressure) from a platinum substrate as a function of dosage (L). After inserting the 99.99% Pt substrate into the chamber, it was pretreated by heating to 1120 K (845 °C) in 1.33×10^{-4} Pa (10^{-6} torr) of O_2 for 5 h. The treatment produced a preferentially (111) oriented surface. After pretreating, the chamber was evacuated to less than 4×10^{-7} Pa (3×10^{-9} torr), and the sample was cooled to 295 K (20 °C). According to AES measurements, the procedure removed the two major contaminants (carbon and calcium) from the surface. A research-grade (99.9%) CO gas was introduced to the substrate through a variable leak valve at a pressure of 7×10^{-7} Pa (5×10^{-9} torr) for periods of time, producing the doses shown in the figure.

Whereas few TPD experiments are as sophisticated as that shown in Fig. 5, all TPD experiments share some common elements: a vacuum system, an ability to ramp up the temperature of a substrate, and a tool to detect the desorbing species as a function of temperature. Temperature-programmed desorption experiments designed to probe surface reactions also use a mechanism of cleaning and dosing the substrate surface in the vacuum chamber.

Several techniques are used to detect gas concentrations as a function of substrate temperature. The first primitive TPD studies used thermal conductivity (Ref 7) and flame detectors (Ref 17). Later, gas chromatographs were used to study the composition, especially when more than one desorbed gas was present (Ref 18). With mass spectrometer detectors, the concentration of a desorbed gas can be measured continuously as a function of temperature and time, and multiple peaks can be

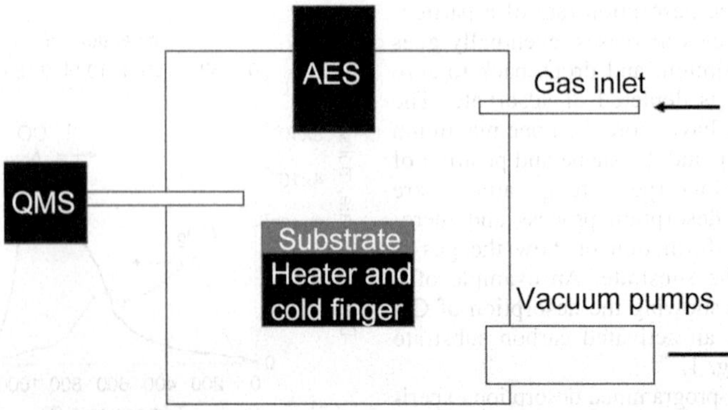

Fig. 5 Schematic of a temperature-programmed desorption system consisting of a vacuum chamber and pumping system capable of achieving base pressures of approximately 10^{-6} Pa (7.5×10^{-9} torr). QMS, quadrupole mass spectrometer; AES, Auger electron spectrometer

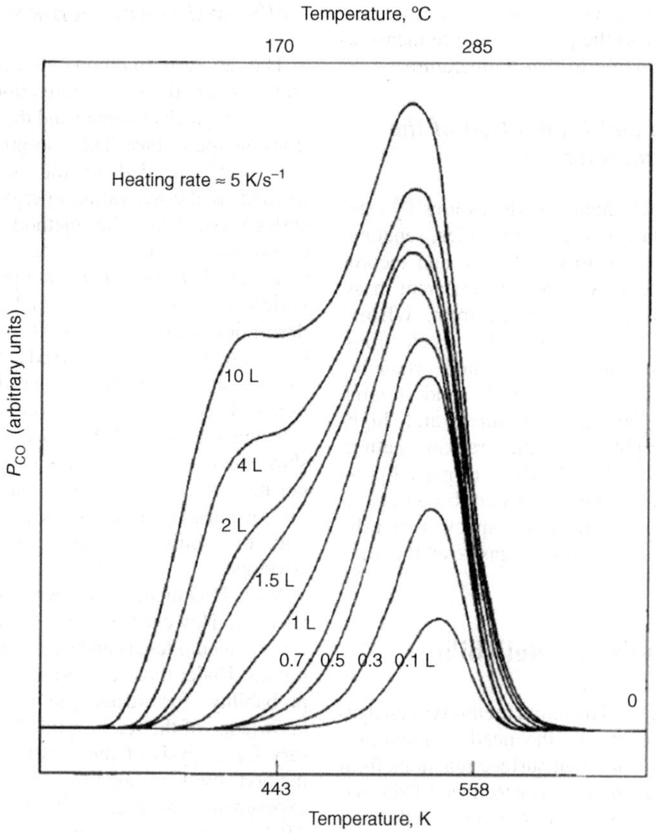

Fig. 6 Thermal desorption of CO for various exposures (in Langmuir, L) of CO at 295 K (20 °C). Source: Ref 16

simultaneously observed. Therefore, most TPD experiments performed today (2019) use mass spectrometry detectors.

Specimen Preparation

Temperature-programmed desorption specimens include the substrates and, in most cases, the adsorbing gas or gases. Because most TPD experiments are performed in reactors capable of attaining ultrahigh vacuum levels, substrates are pretreated within the chamber to expose the desirable surface for study. A good example of a pretreatment process for platinum is described in the illustrative TPD experiment contained in Ref 16. The pretreatment process depends on the material and the desirable initial state of the substrate. For example, in their study of the desorption of CO and CO_2 from activated carbon powder (Fig. 1), Ishii et al. (Ref 10) placed a carbon sample into a sample holder in ambient atmosphere and then evacuated the system to 2×10^{-5} Pa (1.5×10^{-7} torr). No additional adsorbing gases were introduced to the system, and the TPD experiment was performed by ramping the temperature from 300 to 2100 K (25 to 1825 °C).

In TPD experiments by Kang et al. (Ref 19) on adsorption of NH_3 on zirconium, prior to performing the experiments, they sputter etched zirconium disks with argon ions (2 keV, 2 μA/cm^2) prior to annealing at 840 K (565 °C) for 2 min. Their TPD system was equipped with a reverse-view low-energy electron diffraction tool used to demonstrate that the substrate pretreatment produced the desired (1 × 1) reconstructed surface on the zirconium (0001) crystal face.

In another example of a specific specimen-preparation process, Ifergane et al. (Ref 20) used TPD to study hydrogen trapping in precipitation-hardened martensitic stainless steel. Specimens consisted of 50 mm (2 in.) long by 12 mm (0.5 in.) wide rectangular pieces ground to a thickness between 0.240 and 0.260 mm (0.0095 and 0.01 in.) and polished to a surface roughness of approximately 1 μm. The specimens were rinsed in a detergent, washed in water, ultrasonically cleaned in acetone, dried using high-pressure air, washed in ethanol, and dried under high-pressure room-temperature air. Hydrogen charging was performed in 0.05 *M* NaOH solution at a negative current density of 9 mA/cm^2 for 14 days at room temperature. After charging, specimens were rinsed in distilled water, dried by pressurized air, and immersed within 30 s into liquid nitrogen for up to 5 h before inserting the charged specimen into the TPD reactor.

Calibration and Accuracy

The factors that must be considered when performing TPD experiments include: reactor leaks and gas impurities, water vapor in the reactor, system pressure and measurment, gas flow rates, adsorption and desorption temperature, reactor size and specimen surface area, temperature measurement and programmer, exit-gas temperature, pressure and location of the detector, and cracking in and calibration of the mass spectormeter.

Reactor Leaks and Gas Impurities

Small amounts of impurities in the carrier gas or leaks in the vacuum chamber can have a significant effect on TPD results. For example, ultrahigh-purity argon gas typically contains approximately 5 ppm of O_2. Depending on the flow rate of argon and the temperature and surface area of the substrate, a considerable amount of the surface area of the substrate could be poisoned prior to performing the actual experiment. Small leaks in the vacuum system have the same effect.

Water Vapor in the Reactor

At a chamber pressure of 10^{-4} Pa (7.5×10^{-7} torr), a surface with 10^{15} atoms/cm^2 can be completely covered with a monolayer of water in approximately 2.1 s. Conversely, the same surface would require almost 6 h to be covered with a monolayer of water at a chamber pressure of 10^{-8} Pa (7.5×10^{-10} torr). The flux of water vapor onto a surface (Φ) in molecules/cm^2-s is approximately 8.28×10^{21} $(P\sqrt{T})$, where the pressure (P) is in torr and the temperature (T) is in Kelvin. Therefore, TPD experiments that are sensitive to moisture contamination of the substrate surface should be performed at low pressures and elevated temperatures.

System Pressure and Measurement

Temperature-programmed desorption experiments performed at low system pressure have advantages over high-pressure experiments, including greater detection sensitivity, less high-purity gas used, and less load placed on the pumping system. Additionally, readsorption of desorbed gases is reduced at low system pressures. For example, the theoretical flux of readsorption of desorbed O_2 at 10^{-7} Pa (7.5×10^{-9} torr) is 1000 times less than that at 10^{-4} Pa (7.5×10^{-7} torr). Adsorption of a gas of interest onto a substrate is typically accomplished in a TPD experiment by injecting the gas into an inert carrier gas (such as argon) and introducing the mixture into the reactor through a leak valve. To accurately calculate the surface coverage (for example, in Langmuir) during the adsorption process, the measurement of the pressure rise in the system must be exact. Calibration of pressure gages is essential to perform these types of TPD experiments.

Gas Flow Rates

During the adsorption process of a TPD experiment, it is important to know and accurately control the flow rate of the gas mixture into the reactor. This is usually accomplished by means of mass flow meters and controllers. Calibration of these instruments is usually performed by the manufacturer.

Adsorption and Desorption Temperature

Temperature-programmed desorption experiments similar to that performed in Ref 16 strongly rely on knowing the accurate temperature of the substrate during the pretreatment, adsorption, and desorption processes. For that reason, thermocouples are usually affixed directly to solid substrate specimens or embedded in powder specimens. The objective is to minimize the temperature difference between the substrate and the thermocouple.

Reactor Size and Specimen Surface Area

Reactor size should be designed to minimize temperature gradients. For example, if pressure measurements are being performed large distances from the heated specimens, temperature differences between the locations could create pressure differences, which can negatively impact the accuracy of the TPD experiment. The size of the substrate can also affect the accuracy of a TPD experiment. This is especially true for experiments evaluating desorption from powders and other high-surface-area materials, where pressure gradients can promote an increase in readsorption.

Temperature Measurement and Programmer

Temperature-programmed desorption experiments are usually performed with a linear rise in substrate temperature across the entire temperature range of interest. Accuracy in the rate of temperature rise (dT/dt) of the substrate can be crucial to TPD experiments designed to explore desorption kinetics.

Exit-Gas Temperature

The temperature of desorbed gas at the detector affects apparent gas concentration through the ideal gas law. If the exit-gas temperature is significantly different from the gas near the substrate, a baseline shift can occur in mass spectrometric measurements of the partial pressure.

Pressure and Location of the Detector

Transit time between the substrate and the detector should be as short as possible to limit inaccuracies between the sample temperature measurement and the partial pressure measurement corresponding to that temperature.

Cracking in and Calibration of the Mass Spectrometer

A mass spectrometer produces ions by electron bombardment of gas molecules, separates the ions according to their mass, and records the relative intensity of ions of different mass, thereby producing a mass spectrum. When a gas such as CO is bombarded with electrons, ions of CO^+, C^+, and O^+ are created, producing a spectrum with peaks at 28, 12, and 16 amu, respectively. For gas mixtures and high-molecular-weight gases, the cracking pattern can be very complex. Analysis of partial pressures of gases can be performed through careful calibration of the mass spectrometer by admitting pure calibration gases at the time of the analysis.

Data Analysis and Reliability

There are several different methods of analyzing TPD data, due to the need to maximize knowledge gained about surface reactions from a limited amount of data. The Redhead TPD analytical method described previously in the section "General Principles" in this article is useful when there is a single TPD curve and an estimate of the activation energy is desired. The method is easy to apply, but an assumption of both the pre-exponential factor of the rate equation and the rate order must be made. Other analytical methods include heating-rate variation, isothermal desorption rate with and without readsorption, peak width analysis, shape index analysis, and skewness parameter analysis.

Heating-Rate Variation Method

Redhead suggested that testing under different temperature ramps can eliminate the need for assuming a pre-exponential factor (Ref 5), which is referred to as the heating-rate variation method. In this method, a number of TPD experiments are carried out by using a fixed starting surface coverage and a variable heating rate, β. Peak temperatures (T_p) and peak amplitudes (N_p) are found for each temperature-ramp experiment. A plot of $\ln(\beta/T_p^2)$ versus $1/T_p$ yields a straight line with a slope equal to $-E_d/R$. A plot of $\ln(N_p)$ versus $1/T_p$ also yields a straight line with slope equal to $-E_d/R$. Calculations of E_d using peak temperature and peak amplitude are independent and can be used to establish consistency of results. This method can be used to analyze more complicated TPD data. Falconer and Schwartz (Ref 9) suggest that heating rates must be varied by a factor of 10, while others suggest the heating rate must be varied over 2 orders of magnitude.

Isothermal Desorption-Rate Methods

The analysis methods described thus far make assumptions that activation energy does not change with coverage and there is no resorption onto the surface. The resorption limitation is not strictly needed in the heating-variation method. If the activation energy changes with surface coverage, the method of isothermal desorption rates can be used. In this method, a series of TPD experiments are performed using various heating rates and initial coverages. The desorption rates, N, are plotted versus coverages, θ, as $\ln(N)$ versus $\ln(\theta)$ at fixed temperature. The slope of these plots yields the order of the reaction, n. Then, a plot of $\ln(N)$ versus $1/T$ at a fixed θ is constructed, and the slope of that relationship equals $-E_d/R$. The temperature ramps do not need to be linear ramps, but knowledge of the coverage of the surface is necessary together with the ability to change and measure the coverage.

Most TPD methods are performed under high vacuum. However, some experiments require lower vacuum levels and even atmospheric conditions. Under those conditions, there is a strong probability for readsorption to occur, and knowledge of the readsorption kinetics is necessary for analysis of these TPD experiments. A method used to analyze high-pressure TPD experiments involves performing a series of TPD experiments using various temperature ramps and initial coverages, with the coverage going up to saturation. The order of the adsorption rate, ρ, is determined first by following a procedure such as that outlined by Falconer and Schwarz (Ref 9). Then, desorption rates and coverages at a number of temperatures are determined. This is an alteration of the desorption-rate isotherms method, but it still requires two steps for the analysis. A desorption-rate order is assumed (e.g., n = 1 or 2), and $\ln(N)$ is plotted versus $\ln(\theta/(1 - \theta)^{\rho/n}$ at fixed temperatures. The plots should yield lines that have a constant slope equal to the desorption-rate order. A second plot of $\ln(N)$ versus $1/T$ at fixed θ gives a straight line with a slope of $-\Delta H/R$. Unlike the other methods, this method yields the heat of adsorption, ΔH, rather than the desorption activation energy.

Peak Width Analysis Method

Not only can peak positions be used to analyze TPD data, but curve widths can also be used, provided the TPD data have usable curve shapes. Only one TPD experiment in which the temperature is varied is needed for peak width analysis, but the order of the reaction must be known. Measurements needed are peak temperature, full width at half maximum of the peak ($W_{1/2}$), and the full width at three-quarters maximum ($W_{3/4}$) (Fig. 7).

Chan and Weinberg (Ref 21, 22) derived equations for relating activation energy to peak temperature and peak widths, which are valid over a wider range than others. The equations to find E_d are:

First order:

$$\frac{E_d}{RT_p} = -1 + \left[1 + \frac{5.832\ T_p^2}{W_{1/2}^2}\right]^{1/2} \quad \text{(Eq 10a)}$$

$$\frac{E_d}{RT_p} = -1 + \left[1 + \frac{2.353\ T_p^2}{W_{3/4}^2}\right]^{1/2} \quad \text{(Eq 10b)}$$

Second order:

$$\frac{E_d}{RT_p} = 2\left[-1 + \left(1 + \frac{3.117\ T_p^2}{W_{1/2}^2}\right)^{1/2}\right] \quad \text{(Eq 11a)}$$

$$\frac{E_d}{RT_p} = 2\left[-1 + \left(1 + \frac{1.209\ T_p^2}{W_{3/4}^2}\right)^{1/2}\right] \quad \text{(Eq 11b)}$$

Shape Index Analysis Method

Shape index analysis of desorption peaks can reveal more information, but the procedure is difficult and limited. The method is useful in determining the rate order and presence of resorption and requires only one TPD experiment, but the desorption must be a single peak. Ibok and Ollis (Ref 23) described a peak shape index analysis method to interpret TPD data for first- and second-order desorption. The method calculates a shape index factor, S, from the ratio of the slopes of a TPD peak $(d^2\Theta/dT^2)$ at its inflection points (T_1 and T_2), as shown in Fig. 8, which can then be used to determine n and the effect of readsorption on the data. Criado et al. (Ref 24) pointed out that while the values of S at full coverage are similar for first-order desorption and second-order desorption with readsorption, S is independent of coverage for the two cases (first and second order) of no readsorption. This is important, because readsorption makes it difficult to use the variation in T_p with θ_0 to distinguish first- and second-order desorption.

Skewness Parameter (X_j) Analysis Method

Temperature-programmed desorption peaks show that first-order desorption curves are skewed, while second-order desorption curves are not (at least not to as great an extent). Chan et al. (Ref 21, 22) developed a skewness parameter analysis based on the symmetry of the TPD desorption curve. The method requires only one TPD experiment and is used to determine the rate order (n) of the desorption. Peak temperature is measured together with the lower temperature (T') and upper temperature (T'') corresponding to peak half height ($W_{1/2}$) and three-quarter height ($W_{3/4}$) according to Fig. 7 and according to the equation:

$$W_j = 200\left[\frac{(T_p - T')}{(100 - X_j)}\right] \quad \text{(Eq 12)}$$

where W_j is the full width of desorption peak at half maximum (j = ½) and three-quarters maximum (j = ¾) for first-order desorption; W_j corresponds to $W_{1/2}$ when $X_{1/2} = -15$, and to $W_{3/4}$ when $X_{3/4} = -10$. For second-order desorption, $X_{1/2} = 5$ and $X_{3/4} = 3$. Chen calculated a desorption peak symmetry factor, U_g, in a similar fashion, deriving a simpler equation (Ref 25):

$$U_g = \frac{T'' - T_p}{T'' - T'} \quad \text{(Eq 13)}$$

where the desorption is first order if $U_g = 0.42$ and second order if $U_g = 0.52$.

Applications and Interpretations

Temperature-programmed desorption is a well-established tool for studying fundamental reactions of atoms and molecules on well-defined surfaces, especially in the area of catalysis. Temperature-programmed desorption is also used in environmental applications such as monitoring personal exposure to toxic chemicals (Ref 26), indoor air-quality monitoring (Ref 27), and the identification of volatiles in soil and water (Ref 28, 29). Ho et al. (Ref 30) reviewed studies where TPD was used for the analysis of environmental pollutants. Temperature-programmed desorption is also used to quantify hydrogen in metals, especially steels (Ref 31), and to investigate larger molecules on less well-defined, more real-world surfaces with applications in corrosion (Ref 32), electrochemistry, and tribology (Ref 33). A few studies where TPD was used to elucidate surface reactions that impact the tribological performances of materials are discussed here.

MoS_2 on Molybdenum

Pierce et al. (Ref 34) used TPD to investigate the thermal decomposition of a molybdenum disulfide (MoS_2) film on a molybdenum substrate. Molybdenum disulfide is a dichalcogenide and is frequently used as a solid lubricant. The TPD experiments were performed on ~230 nm thick MoS_2 films that were burnished onto molybdenum foils. The TPD system had a base pressure of approximately 10^{-7} Pa (7.5×10^{-9} torr) and held three samples. Each sample was heated, and the evolved vapor was analyzed by using a mass spectrometer in a line-of-sight configuration. Measured data consisted of peak identification that yielded rate order, desorption energy, and pre-exponential factor, but the method of analysis was not given in the report. Three peaks were observed and identified to be from the desorption of SO_2, S_2, and CS_2. A summary of analyzed data is given in Table 1.

The authors assigned a low-temperature peak to SO_2, which was inferred to have stemmed from a reaction of residual water vapor in the chamber with the MoS_2. They suggested that the reaction may be important in the degradation process of the MoS_2 when used as a solid lubricant. It was also speculated that the high-temperature sulfur peak (S_2) comes from the decomposition of MoS_2 and thereby sets an upper temperature limit for solid lubricant applications. The CS_2 peak

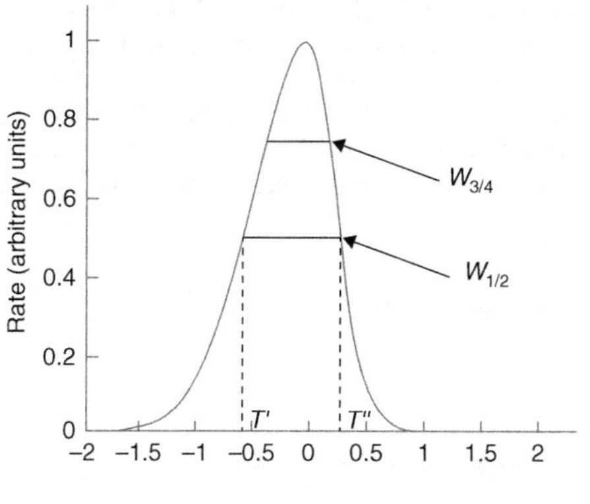

Fig. 7 Plot showing definitions of $W_{1/2}$ (full width at half maximum of the peak) and $W_{3/4}$ (full width at three-quarters maximum) for use in peak width analysis

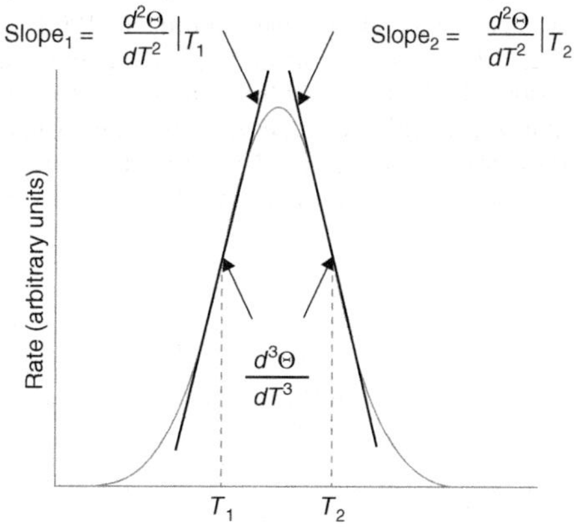

Fig. 8 Example of a plot for shape index analysis

Table 1 Data from temperature-programmed desorption analysis of thermal decomposition of MoS_2 film on a molybdenum substrate

Mass, amu	Species	Temperature of desorption peak maximum (T_p), K	Temperature of desorption peak maximum (T_p), °C	Order	Activation energy (E_d), J/mol	Pre-exponential factor for first-order Arrhenius equation, s^{-1}
64	SO_2	700	425	2	80	10^{-10}
64	S_2	1465	1190	1	339	5×10^{11}
76	CS_2	1465	1190	1	335	2×10^{11}

was assumed to be due to a reaction between the MoS_2 and carbon-containing impurities.

β-Hydride Elimination in Vapor-Phase Lubricants

Ren and Gellman (Ref 35) and Sung and Gellman (Ref 36) investigated the surface phase chemistry of vapor-phase phosphorus-containing lubricants on nickel and copper surfaces and iron surfaces, respectively. Aryl phosphates, such as tricresyl phosphate, are lubricants used in high-temperature gas turbine engines. At approximately the same time, Holbert et al. (Ref 37) investigated a similar system of trimethyl phosphite on Fe(110). Trimethyl phosphite is an analog in the study of the lubricant additive zinc dialkyl dithiophosphate (ZDDP). Both groups of researchers found CO and H_2 desorbing from the surfaces at approximately 400 K (125 °C). The reaction products were tied to the formation of a surface methoxy ($O\text{-}CH_3$) species. When the phosphorus compounds were adsorbed onto the surfaces, additional species of CH_2O and CHO were found to desorb from the nickel, copper, and iron surfaces. All investigations saw a lack of phosphorus in the desorbed molecules. It was also found that the amount of CO desorbed was related to the phosphorus coverage, with the greatest desorption on clean surfaces. Their analysis showed a differentiation in the reaction mechanisms between the alkyl and aryl phosphorus species. The alkyl phosphorus compounds decomposed via a β-hydride elimination mechanism. Aryl compounds do not have a β-hydrogen constituent, so decomposition was accomplished through a different route. After total desorption, in the case of the alkyl phosphorus compound, only phosphorus remained on the surfaces, while in the case of the aryl compound, carbon and phosphorus remained.

Analyses of the TPD data in these studies were challenging because the surface compounds decomposed or reacted before the reaction products could desorb from the surface. Activation energies of the reactions or decompositions were greater than the activation energies of desorption; that is, the β-hydride elimination reaction energy is greater than the activation energy of alkoxy desorption. The activation energy of the β-hydride elimination has been reported to be less than 20 kcal/mol (~84 kJ/mol) (Ref 38, 39), whereas a simple Redhead analysis of TPD data with a desorption peak at 400 K (125 °C) yields an activation energy of approximately 25 kcal/mol (~105 kJ/mol).

Tribochemistry

Temperature-programmed desorption was used in studies at the University of Wisconsin-Madison to investigate the chemistry that occurs between lubricants and surfaces in tribological contacts (Ref 40–42). The studies investigated phosphorus-containing compounds to understand the reactions of ZDDP with oxidized iron surfaces, which are more like actual tribological surfaces than bare iron surfaces. The system used had a base pressure of ~3 × 10^{-8} Pa (2 × 10^{-10} torr), and a mass spectrometer was used to determine the atomic masses of the desorbed material. Iron surfaces were cleaned using argon ion sputtering and annealed to 1000 K (725 °C) several times to remove carbon, sulfur, and nitrogen contaminants. Oxide layers were generated by the adsorption of O_2 for 1 h at a pressure of ~7 × 10^{-5} Pa (5 × 10^{-7} torr) and a temperature of 800 K (525 °C). The oxygenation state of the iron surfaces gaged by using AES was Fe_3O_4. After oxidation, surfaces were exposed to tributyl phosphite. A complex desorption spectrum was produced by the oxidized surface, and the analysis was more difficult to perform than the analysis of a desorption spectrum from a bare iron surface. That is, the desorption spectrum contained species created by reactions between the tributyl phosphite and the oxidized surface. The chemistry of tributyl phosphite on Fe_3O_4 was interpreted according to the scheme shown in Fig. 9 (Ref 41). Following adsorption of tributyl phosphite at ~300 K (25 °C), phosphorus-oxygen bond scission resulted in the adsorption of C_4H_9O and PO_x moieties. Some of the C_4H_9O reacted with hydrogen derived either from a 1-butyl group or from tributyl phosphite decomposition to form C_4H_9OH that desorbed at ~450 K (175 °C). The remainder of the C_4H_9O was stable up to ~630 K (355 °C), whereupon it decomposed to desorb in the form of C_3H_7CHO and C_4H_9OH.

This scheme was refined in subsequent investigations and provided insight into the tribologically driven surface reactions of phosphate lubricants with steel alloys (Ref 42). Specifically, the authors proposed that in tribological contact, tributyl phosphite decomposes on an oxidized steel surface to butoxy and adsorbed phosphorus-oxygen. This state is stable for temperatures up to ~600 K (325 °C), the temperature at which antiwear additives become active. In highly loaded contacts, local surface temperatures can approach 1000 K (725 °C), and at these high temperatures, adsorbed tributyl phosphite fragments desorb, oxidize, and release butanol, butanal, and water. Further reactions release CO and water, reducing the surface and removing the oxide layer. After the oxide layer is removed, surface chemistry involving clean metal surfaces dominates, and phosphorus and carbon from the lubricant additives can diffuse into the surfaces.

Alky-Polysulfide on Iron and the Catalytic Activity of Nickel

Seeley (Ref 43) conducted TPD investigations of surface-reaction mechanisms of liquid lubricant additives (mainly ZDDP) with iron and steel alloys by using the sulfur analog di-tert-octyl polysulfide (DTOPS). The goals were to separate thermal from tribological mechanisms, understand the catalytic activity of steel chemical compositions, and compare that to the catalytic activity of surfaces containing nickel. The vacuum system used had a base pressure of ~10^{-3} Pa (10^{-5} torr) and was equipped with a mass spectrometer for desorbed species detection.

In the first experiment, an iron surface with a native oxide (Fe_xO_y) was dosed with DTOPS ($S_n[C_8H_{16}]_2$) under atmospheric conditions. The chamber was evacuated to its base pressure, the iron substrate cooled to ~200 K (−75 °C), and then the temperature was increased at a heating rate (β) of approximately 10 K/min (10 °C/min). The TPD spectra showed that $C_8H_{16}O$ desorbed at ~400 K (125 °C) (E_d ~ 105 kJ/mol), with no identifiable desorbed sulfur compounds. Based on these observations, a mechanism was proposed that carbon-sulfur bond scission in the DTOPS occurred and produced C_8H_{16} fragments that subsequently reacted with the iron oxide to form $C_8H_{16}O$. Oxide reduction by the C_8H_{16} fragments then enabled adsorbed sulfur atoms to bond with the reduced iron oxide surface:

$$\{S_n[C_8H_{16}]_2\}_{ad} + \{Fe_xO_y\}_s \xrightarrow{400\ K} \{S_nFe_3O_{(y-2)}\}_s + 2[C_8H_{16}O]_g \quad \text{(Eq14)}$$

where the subscripts "ad," "s," and "g" refer to adsorbed, surface, and gas, respectively.

Subsequent experiments showed that $C_8H_{16}O$ desorption occurred at ~480 K (205 °C) on AISI 52100 steel and at ~437 K (165 °C) on nickel-containing steel surfaces. A Redhead analysis indicates that the desorption energy of DTOPS on nickel-containing steel surfaces is approximately 13 kJ/mol lower than on AISI 52100, which contains no nickel.

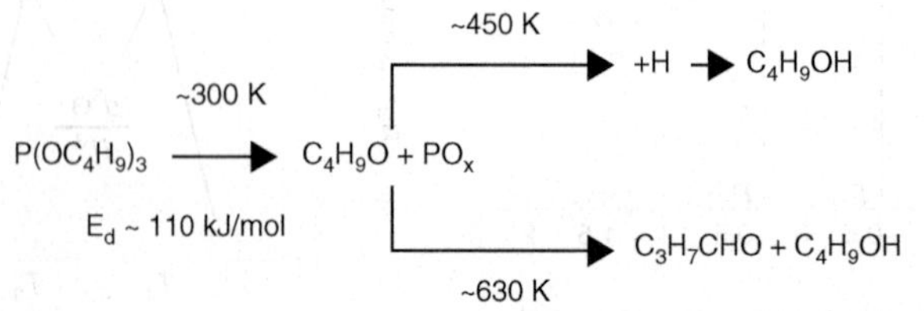

Fig. 9 Desorption scheme of tributyl phosphite on Fe_3O_4 as interpreted by Gao et al. Source: Ref 41. Reprinted (adapted) with permission from Ref 41. Copyright (2004) American Chemical Society

That is, the presence of nickel on steel surfaces enhances catalytic activity with sulfur-containing compounds. The tribological impact of a reduced desorption energy is that antiwear sulfur compounds form at lower temperatures and generate thicker tribofilms on nickel-containing steel alloys than on steel alloys without the catalytic enhancement of nickel.

List of Symbols

Symbol	Description	Units
M	Molecular weight	amu
P	Pressure	Pa (torr)
R	Gas constant (8.314×10^6)	$cm^3 \cdot Pa/K \cdot mol$
T	Absolute temperature	K
k_i	Rate constants	...
T_0	Starting point of temperature ramp	K
β	Heating rate $\left(\frac{dT}{dt}\right)$	K/s
θ	Surface coverage	...
n	Rate order	...
E_d	Activation energy	J
A	Pre-exponential factor for first-order Arrhenius equation	...
A_2	Pre-exponential factor for second-order Arrhenius equation	...
T_p	Temperature of desorption peak maximum	K
N_p	Peak amplitude at T_p	...
N	Desorption rate	...
ρ	Adsorption rate order	...
ΔH	Heat of absorption	J
W_j	Full width of desorption peak at half maximum ($j = ½$) and three-quarters maximum ($j = ¾$)	K
S	Shape index factor	...
T_i	Inflection points of desorption peak at low ($i = 1$) and high ($i = 2$) temperatures	K
T'	Lower temperature of the $W_{1/2}$ or $W_{3/4}$ peak widths	K
T''	Upper temperature of the $W_{1/2}$ or $W_{3/4}$ peak widths	K
X_j	Skewness parameter for $j = ½$ and $j = ¾$	...
U_g	Desorption peak symmetry factor	...

REFERENCES

1. L.R. Apker, Surface Phenomena Useful in Vacuum Technique, *Ind. Eng. Chem.*, Vol 40 (No. 5), 1948, p 846–847
2. J.B. Taylor and I. Langmuir, The Evaporation of Atoms, Ions and Electrons from Caesium Films on Tungsten, *Phys. Rev.*, Vol 44 (No. 6), 1933, p 423
3. J.T. Yates, Jr., The Thermal Desorption of Adsorbed Species, *Methods in Experimental Physics*, Vol 22, Academic Press, 1985, p 425–464
4. G. Ehrlich, Modern Methods in Surface Kinetics: Flash Desorption, Field Emission Microscopy, and Ultrahigh Vacuum Techniques, *Advances in Catalysis*, Vol 14, Academic Press, 1963, p 255–427
5. P.A. Redhead, Chemisorption on Polycrystalline Tungsten, Part 1: Carbon Monoxide, *Trans. Faraday Soc.*, Vol 57, 1961, p 641–656
6. T.A. Delchar and G. Ehrlich, Chemisorption on Single-Crystal Planes: Nitrogen on Tungsten, *J. Chem. Phys.*, Vol 42 (No. 8), 1965, p 2686–2702
7. R.J. Cvetanović and Y. Amenomiya, Application of a Temperature-Programmed Desorption Technique to Catalyst Studies, *Advances in Catalysis*, Vol 17, Academic Press, 1967, p 103–149
8. D.A. King, Thermal Desorption from Metal Surfaces: A Review, *Surf. Sci.*, Vol 47 (No. 1), 1975, p 384–402
9. J.L. Falconer and J.A. Schwarz, Temperature-Programmed Desorption and Reaction: Applications to Supported Catalysts, *Catal. Rev. Sci. Eng.*, Vol 25 (No. 2), 1983, p 141–227
10. T. Ishii and T. Kyotani, Temperature Programmed Desorption, *Materials Science and Engineering of Carbon: Characterization*, M. Inagaki and F. Kang, Ed., Elsevier, Amsterdam, 2016, p 287–305
11. S. Ogura and K. Fukutani, Thermal Desorption Spectroscopy, *Compendium of Surface and Interface Analysis*, Springer, Singapore, 2018, p 719–724
12. M. Polanyi and E. Wigner, Über die Interferenz von Eigenschwingungen als Ursache von Energieschwankungen und Chemischer Umsetzungen, *Z. Phys. Chem.*, Vol 139 (No. 1), 1928, p 439–452
13. P.A. Redhead, Thermal Desorption of Gases, *Vacuum*, Vol 12 (No. 4), 1962, p 203–211
14. G. Carter, Thermal Resolution of Desorption Energy Spectra, *Vacuum*, Vol 12 (No. 5), 1962, p 245–254
15. F.M. Lord and J.S. Kittelberger, On the Determination of Activation Energies in Thermal Desorption Experiments, *Surf. Sci.*, Vol 43 (No. 1), 1974, p 173–182
16. D.M. Collins, J.B. Lee, and W.E. Spicer, A Photoemission and Thermal Desorption Study of Carbon Monoxide and Oxygen Adsorbed on Platinum, *Surf. Sci.*, Vol 55 (No. 2), 1976, p 389–402
17. L.D. Krenzke, G.W. Keulks, A.V. Sklyarov, A.A. Firsova, M.Y. Kutirev, L. Y. Margolis, and O.V. Krylov, The Study of Selective Oxidation of Propylene on Complex Oxides by Means of Temperature-Programmed Desorption, *J. Catal.*, Vol 52 (No. 3), 1978, p 418–424
18. K. Fujimoto, M. Kameyama, and T. Kunugi, Hydrogenation of Adsorbed Carbon Monoxide on Supported Platinum Group Metals, *J. Catal.*, Vol 61 (No. 1), 1980, p 7–14
19. Y.C. Kang, M.M. Milovancev, D.A. Clauss, M.A. Lange, and R.D. Ramsier, Ultra-High Vacuum Investigation of the Surface Chemistry of Zirconium, *J. Nucl. Mater.*, Vol 281 (No. 1), 2000, p 57–64
20. S. Ifergane, R.B. David, E. Sabatani, B. Carmeli, O. Beeri, and N. Eliaz, Hydrogen Diffusivity and Trapping in Custom 465 Stainless Steel, *J. Electrochem. Soc.*, Vol 165 (No. 3), 2018, p C107–C115
21. C.M. Chan, R. Aris, and W.H. Weinberg, An Analysis of Thermal Desorption Mass Spectra, Part I, *Applic. Surf. Sci.*, Vol 1 (No. 3), 1978, p 360–376
22. C.M. Chan and W.H. Weinberg, An Analysis of Thermal Desorption Mass Spectra, Part II, *Applic. Surf. Sci.*, Vol 1 (No. 3), 1978, p 377–387
23. E.E. Ibok and D.F. Ollis, Temperature-Programmed Desorption from Porous Catalysts: Shape Index Analysis, *J. Catal.*, Vol 66 (No. 2), 1980, p 391–400
24. J.M. Criado, P. Malet, G. Munuera, and V. Rives-Arnau, Study of the "Shape Index" in the Analysis of Temperature-Programmed Desorption Curves, *Thermochim. Acta*, Vol 38 (No. 1), 1980, p 37–45
25. R. Chen, On the Analysis of Thermal Desorption Curves, *Surf. Sci.*, Vol 43, 1974, p 657–661
26. M. Kołtowski and P. Oleszczuk, Toxicity of Biochars after Polycyclic Aromatic Hydrocarbons Removal by Thermal Treatment, *Ecolog. Eng.*, Vol 75, 2015, p 79–85
27. C.H. Wu, M.N. Lin, C.T. Feng, K.L. Yang, Y.S. Lo, and J.G. Lo, Measurement of Toxic Volatile Organic Compounds in Indoor Air of Semiconductor Foundries Using Multisorbent Adsorption/Thermal Desorption Coupled with Gas Chromatography-Mass Spectrometry, *J. Chromatogr. A*, Vol 996 (No. 1–2), 2003, p 225–231
28. T. Sato, T. Todoroki, K. Shimoda, A. Terada, and M. Hosomi, Behavior of PCDDs/PCDFs in Remediation of PCBs-Contaminated Sediments by Thermal Desorption, *Chemosphere*, Vol 80 (No. 2), 2010, p 184–189
29. M. Bonnard, S. Devin, C. Leyval, J.L. Morel, and P. Vasseur, The Influence of Thermal Desorption on Genotoxicity of Multipolluted Soil, *Ecotoxicol. Environ. Safety*, Vol 73 (No. 5), 2010, p 955–960
30. S.S.H. Ho, J.C. Chow, J.Z. Yu, J.G. Watson, J.J. Cao, and Y. Huang, Application of Thermal Desorption Mass Spectrometry for the Analysis of Environmental Pollutants, *Chromatographic Analysis of the Environment: Mass Spectrometry Based Approaches*, CRC Press, 2017, p 79–101
31. D. Rudomilova, T. Prošek, and G. Luckeneder, Techniques for Investigation of Hydrogen Embrittlement of Advanced High Strength Steels, *Corros. Rev.*, Vol 36 (No. 5), 2018, p 413–434
32. J. Sayers, S. Ortner, K. Li, and S. Lozano-Perez, Effect of pH on Hydrogen Pick-Up and Corrosion in Zircaloy-4, *Proceedings of the 18th International Conference on Environmental Degradation of Materials in Nuclear Power Systems-Water Reactors*, Springer, 2019, p 1169–1180
33. A.D. Richardson, M.-H. Evans, L. Wang, M. Ingram, Z. Rowland, G. Llanos, and

R.J.K. Wood, The Effect of Over-Based Calcium Sulfonate Detergent Additives on White Etching Crack (WEC) Formation in Rolling Contact Fatigue Tested 100Cr6 Steel, *Tribol. Int.*, Vol 133, 2019, p 246–262

34. D.E. Pierce, R.P. Burns, H.M. Dauplaise, and L.J. Mizerka, Thermal Desorption Spectroscopy of Sputtered MoS_x Films, *Tribol. Trans.*, Vol 34 (No. 2), 1991, p 205–214
35. D. Ren and A.J. Gellman, Initial Steps in the Surface Chemistry of Vapor Phase Lubrication by Organophosphorus Compounds, *Tribol. Lett.*, Vol 6 (No. 3–4), 1999, p 191–194
36. D. Sung and A.J. Gellman, The Surface Chemistry of Alkyl and Arylphosphate Vapor Phase Lubricants on Fe Foil, *Tribol. Int.*, Vol 35 (No. 9), 2002, p 579–590
37. A.W. Holbert, J.D. Batteas, A. Wong-Foy, T.S. Rufael, and C.M. Friend, Passivation of Fe (110) via Phosphorus Deposition: The Reactions of Trimethylphosphite, *Surf. Sci.*, Vol 401 (No. 3), 1998, p L437–L443
38. B. Alcaide, P. Almendros, T.M. Campodel, and I. Fernandez, Fascinating Reactivity in Gold Catalysis: Synthesis of Oxetenes through Rare 4-Exo-Dig Allene Cyclization and Infrequent β-Hydride Elimination, *Chem. Commun.*, Vol 47 (No. 32), 2011, p 9054–9056
39. M. Castineira Reis, C.S. López, E. Kraka, D. Cremer, and O.N. Faza, Rational Design in Catalysis: A Mechanistic Study of β-Hydride Eliminations in Gold (I) and Gold (III) Complexes Based on Features of the Reaction Valley, *Inorganic Chem.*, Vol 55 (No. 17), 2016, p 8636–8645
40. O. Furlong, F. Gao, P. Kotvis, and W.T. Tysoe, Understanding the Tribological Chemistry of Chlorine-, Sulfur- and Phosphorus-Containing Additives, *Tribol. Int.*, Vol 40 (No. 5), 2007, p 699–708
41. F. Gao, O. Furlong, P.V. Kotvis, and W.T. Tysoe, Reaction of Tributyl Phosphite with Oxidized Iron: Surface and Tribological Chemistry, *Langmuir*, Vol 20 (No. 18), 2004, p 7557–7568
42. F. Gao, P.V. Kotvis, D. Stacchiola, and W. T. Tysoe, Reaction of Tributyl Phosphate with Oxidized Iron: Surface Chemistry and Tribological Significance, *Tribol. Lett.*, Vol 18 (No. 3), 2005, p 377–384
43. M.A. Seeley, "Interactions of Additives on Surfaces via Temperature Programmed Desorption," Master's dissertation, University of Akron, 2017

Reference Information

ASM Handbook, Volume 10, *Materials Characterization*
ASM Handbook Committee
DOI 10.31399/asm.hb.v10.a0006748

Glossary of Terms

A

aberration. In microscopy, any error that results in image degradation. Such errors may be chromatic, spherical, astigmatic, comatic, distortion, or curvature of field and can result from design, execution, or both.

absorbance (*A*). The logarithm to the base 10 of the reciprocal of the transmittance. The preferred term for photography is optical density.

absorption (of electromagnetic radiation). A decrease in the intensity of the beam (light, x-rays, electrons, and so on) when passing through matter. In many cases, specific wavelengths or energies are preferentially absorbed, forming the basis of absorption spectroscopy.

absorption contrast. In transmission electron microscopy, image contrast caused by differences in absorption within a sample due to regions of different mass density and thickness.

absorption edge. The wavelength or energy corresponding to a discontinuity in the plot of absorption coefficient versus wavelength for a specific medium.

absorption spectroscopy. The branch of spectroscopy treating the theory, interpretation, and application of spectra originating in the absorption of electromagnetic radiation by atoms, ions, radicals, and molecules.

absorptivity. A measure of radiant energy from an incident beam as it traverses an absorbing medium, equal to the absorbance of the medium divided by the product of the concentration of the substance and the sample path length. Also known as absorption coefficient.

accelerating voltage. In various electron beam instruments and x-ray generators, the difference in potential between the filament (cathode) and the anode, causing acceleration of the electrons by 2 to 30 keV. See also *depth of penetration* and *resolution*.

accuracy. The degree of agreement of a measured value with the true or correct value for the quantity being measured.

achromatic. Refers to an optical element that transmits light without dispersing it into its component wavelengths. See also *achromatic lens* and *apochromatic lens*.

achromatic lens. A lens that is corrected for chromatic aberration so that its tendency to refract light differently as a function of wavelength is minimized. See also *achromatic* and *apochromatic lens*.

ac noncapacitive arc. A high-voltage electrical discharge used in spectrochemical analysis to vaporize the sample material. See also *dc intermittent noncapacitive arc*.

activation analysis. A method of chemical analysis based on the detection of characteristic radionuclides following nuclear bombardment. See also *neutron activation analysis*.

adsorption chromatography. Chromatography based on differing degrees of adsorption of sample compounds onto a polar stationary phase. See also *liquid-solid chromatography*.

aliquot. A representative sample of a larger quantity.

amorphous solid. A rigid material whose structure lacks crystalline periodicity; that is, the pattern of its constituent atoms or molecules does not repeat periodically in three dimensions. See also *metallic glass*.

amperometry. Chemical analysis by methods that involve measurements of electric currents.

amperostatic coulometry. Coulometry in which the current is kept constant via an amperostat or galvanostat. Also known as coulometric titration.

analog-to-digital converter (ADC). A device that converts a continuously variable electrical signal into discrete signals suitable for analysis by a digital computer.

analysis. The ascertainment of the identity or concentration, or both, of the constituents or components of a sample. See also *determination*.

analyte. In any analysis, the substance (element, ion, compound, and so on) being identified or determined.

analytical chemistry. The science of chemical characterization and measurement. Qualitative analysis is concerned with the description of chemical composition in terms of elements, compounds, or structural units; quantitative analysis is concerned with the precise measurement of amount. A variety of physical measurements are used, including methods based on spectroscopic, electrochemical, radiochemical, chromatographic, and nuclear principles.

analytical curve. The graphical representation of a relation between (1) the intensity of the response to measurement (for example, emission, absorbance, and conductivity) and (2) the concentration or mass of the substance being measured. The curve is generated by measuring the responses for standards of known concentration. Also termed standard curve or working curve.

analytical electron microscopy (AEM). The technique of materials analysis in the transmission electron microscope equipped to detect and quantify many different signals from the specimen. The technique usually involves a combination of imaging, chemical analysis, and crystallographic analysis by diffraction at high spatial resolution.

analytical gap. The region between two electrodes in which the sample is excited in the sources used for emission spectroscopy and spark source mass spectrometry.

analytical line. In spectroscopy, the particular spectral line of an element used in the identification or determination of the concentration of that element.

analytical wavelength. In spectroscopy, the particular wavelength used for the identification or determination of the concentration of an element or compound.

analyzer. An optical device, capable of producing plane polarized light, used for detecting the state of polarization.

angle of incidence. The angle between an incident radiant beam and a perpendicular to the interface between two media.

angstrom (Å). A unit of length equal to 10^{-10} m.

anion. An ion that is negatively charged and moves toward the positive pole (anode) during electrolysis. See also *cation* and *ion*.

anti-Stokes Raman line. A Raman line that has a frequency higher than that of the incident monochromatic radiation.

aplanatic. Corrected for spherical aberration and coma.

apochromatic lens. A lens whose secondary chromatic aberrations have been substantially reduced. See also *achromatic*.

apparent density (of solids and liquids). The mass in air of a unit of volume of a material at a specified temperature.

aromatic. In organic chemistry, pertaining to or characterized by the presence of at least one benzene ring.

aspect ratio. The length-to-width ratio of a microstructural feature on the plane of polish.

assay. Determination of how much of a sample is the material indicated by the name. For example, for an assay of $FeSO_4$, the analyst would determine both iron and SO_4^{2-} in the sample.

astigmatism. A defect in an optical focusing system in which the beam profile does not maintain cylindrical symmetry due to rays propagating in two perpendicular planes arriving at different foci.

atom. The smallest particle of an element that retains the characteristic properties and behavior of the element. See also *atomic structure*, *isotope*, and *nuclear structure*.

atomic mass unit (amu). See *unified atomic mass unit* (*u*).

atomic number (Z). The number of elementary positive charges (protons) contained within the nucleus of an atom. For an electrically neutral atom, the number of planetary electrons is also given by the atomic number. Atoms with the same Z (isotopes) may contain different numbers of neutrons. Also known as nuclear charge. See also *isotope* and *proton*.

atomic number contrast. See *atomic number imaging*.

atomic number imaging. In scanning electron microscopy, a technique in which image contrast is influenced by atomic number. (Higher-atomic-number areas appear brighter, while lower-atomic-number areas appear darker.) Usually obtained by detecting backscattered electron signal during imaging. See also *backscattered electron*.

atomic structure. The arrangement of the parts of an atom, which consists of a positively charged nucleus surrounded by a cloud of electrons arranged in orbits that can be described in terms of quantum mechanics.

atomic weight. A number assigned to each chemical element that specifies the average mass of its atoms. Because an element may consist of two or more isotopes, each having atoms with well-defined but differing masses, the atomic weight of each element is the average of the masses of its naturally occurring isotopes weighted by the relative proportions of those isotopes.

atomization. The subdivision of a compound into individual atoms using heat or chemical reactions. This is a necessary step in atomic spectroscopy. See also *atomizer* and *nebulizer*.

atomizer. A device that atomizes a sample, for example, a burner, plasma, or hydride reaction chamber. See also *atomization* and *nebulizer*.

atom probe. An instrument for measuring the mass of a single atom or molecule on a metal surface; it consists of a field ion microscope with a hole in its screen opening into a mass spectrometer; atoms are removed from the specimen by pulsed field evaporation, travel through the hole, and are detected in the mass spectrometer. See also *field ion microscopy*.

attenuation. Reduction in the amplitude of a signal, including electric currents and light beams.

Auger chemical shift. The displacement in energy of an Auger electron peak for an element due to a change in chemical bonding relative to a specified element or compound.

Auger electron. An electron emitted from an atom with a vacancy in an inner shell. Auger electrons have a characteristic energy detected as peaks in the energy spectra of the secondary electrons generated.

Auger electron spectroscopy (AES). A technique for chemical analysis of surface layers that identifies the atoms present in a layer by measuring the characteristic energies of their Auger electrons.

Auger electron yield. The probability that an atom with a vacancy in a particular inner shell will relax by an Auger process.

Auger map. A two-dimensional image of the specimen surface showing the location of emission of Auger electrons from a particular element. A map is normally produced by rastering the incident electron beam over the specimen surface and simultaneously recording the Auger signal strength for a particular transition as a function of position.

Auger matrix effects. Effects that cause changes in the shape of an Auger electron energy distribution or in the Auger signal strength for an element due to the physical environment of the emitting atom and not due to bonding with other elements or changes in concentration.

Auger process. The radiationless relaxation of an atom, involving a vacancy in an inner electron shell. An electron (known as an Auger electron) is emitted.

Auger transition designations. Transitions are designated by the electron shells involved. The first letter designates the shell containing the initial vacancy; the last two letters designate the shells containing electron vacancies created by Auger emission (for example, KLL and LMN).

austenite grain size. The mean grain size that exists in a face-centered cubic alloy.

automated image analysis. See *image analysis*.

automatic image analysis. The detection and quantitative evaluation of a microstructural image in which the features are detected and measured with, or without, operator influence.

Avogadro's number. The number of molecules (6.02×10^{23}) in a gram-molecular weight of any substance. See also *gram-molecular weight* and *mole*.

B

background. Any noise in the signal due to instabilities in the system or to environmental interferences. See also *signal-to-noise ratio*.

backscattered electron. An information signal arising from elastic (electron-nucleus) collisions, wherein the incident electron rebounds from the specimen with no energy loss or some energy loss. The backscattered electron yield is strongly dependent on atomic number, primary beam energy, and take-off angle. Images formed by backscattered electron detection reveal compositional and topographic information about the specimen. See also *atomic number imaging*.

backscattered ion. An ion that has entered a sample and, through one or more collisions, has scattered out of the sample. The backscattering process is unlikely when the mass of the incident ion is larger than the mass of the atoms that comprise the sample. Backscattered ion abundance provides atomic number information. The energy and angle of backscattered ions can be used to ascertain composition of the sample.

banded microstructure. Microstructure having alternate bands parallel to the deformation axis resulting from elongation of segregated regions with intermediate regions of lower alloy content.

barn. A unit of area equal to 10^{-24} cm^2 used in specifying nuclear cross sections. See also *nuclear cross section*.

base-line technique. A method for measurement of absorption peaks for quantitative analysis of chemical compounds in which a base line is drawn tangent to the spectrum background; the distance from the base line to the absorption peak is the absorbance due to the sample under study.

basic NMR frequency. The frequency, measured in hertz, of the oscillating magnetic field applied to induce transitions between nuclear magnetic energy levels. See also *magnetic resonance*.

beam tail. The larger, usually non-Gaussian, component of a charged particle beam current density distribution on the sample surface.

Beer's law. A relationship in which the optical absorbance of a homogeneous sample containing an absorbing substance is directly proportional to the concentration of the absorbing substance. See also *absorptivity*.

beryllium window. A very thin (~7.5 μm thick), relatively x-ray-transparent window separating the x-ray detector from the vacuum chamber, which serves to protect the detector from damage.

bias. A systematic error inherent in a method (such as temperature effects and extraction inefficiencies) or caused by some artifact or idiosyncrasy of the measurement system (such as blanks, contamination, mechanical losses, and calibration errors). Bias may be both positive and negative, and several types can exist concurrently, so that the net bias is all that can be evaluated except under certain conditions. See also *blank*.

bimodal grain size distribution. A condition in which the distribution of grain areas or intercept lengths converted to ASTM International grain size numbers, based on the area percent or length percent in accordance with ASTM G class, exhibits two distinct peaks.

birefringent crystal. A crystalline substance that is anisotropic with respect to the velocity of light.

blank. The measured value obtained when a specified component of a sample is not present during the measurement. In such a case, the measured value/signal from the component is believed to be due to artifacts and thus should be deducted from a measured value to give a net value due to the component contained in a sample. The blank measurement must be made so that the correction process is valid.

blind sample. A sample submitted for analysis whose composition is known to the submitter but unknown to the analyst, used to test the efficiency of a measurement process.

blocking pattern. Angular distribution of ions or atoms as the scatter from the atoms of a sample. Usually, this pattern is observed on a specialized detector after an ion beam impinges on the sample in question.

Boltzmann distribution. A function giving the probability that a molecule of a gas in thermal equilibrium will have generalized position and momentum coordinates within a given infinitesimal range of values, assuming that the molecules obey classical mechanics.

bonded-phase chromatography (BPC). Liquid chromatography with a surface-reacted, that is, chemically bonded, organic stationary phase. See also *normal-phase chromatography* and *reversed-phase chromatography*.

Bragg equation. See *Bragg's law*.

Bragg's law. A statement of the conditions under which a crystal will diffract electromagnetic radiation. Bragg's law reads $n\lambda = 2d \sin \theta$, where n is the order of reflection, λ is the wavelength of x-rays, d is the distance between lattice planes, and θ is the Bragg angle, or the angular distance between the incident beam and the lattice planes considered.

bremsstrahlung radiation. See *continuum radiation* and *white radiation*.

bright-edge effect. The effect seen in scanning electron microscopy or focused ion beam images in which topographic edges are evident due to their increased brightness. An effect attributed to the generation and escape of additional secondary electrons as the primary beam remains relatively close to the surface.

brightness. A measure of the emissive properties of an electron or ion source with units of amps cm^{-2} sr^{-1}. High brightness is essential for focusing a charged particle beam to a small focused probe size.

buffer. A substance which by its addition or presence tends to minimize the physical and chemical effects of one or more of the substances in a mixture. Properties often buffered include pH, oxidation potential, and flame or plasma temperatures.

bulk sample. See *gross sample*.

bulk sampling. Obtaining a portion of a material that is representative of the entire lot.

buret. An instrument used to deliver variable and accurately known volumes of a liquid during titration or volumetric analysis. Burets are usually made from uniform-bore glass tubing in capacities of 5 to 100 mL, the most common being 50 mL. See also *titration* and *volumetric analysis*.

C

calomel electrode. An electrode widely used as a reference electrode of known potential in electrometric measurements of acidity and alkalinity, corrosion studies, voltammetry, and measurement of the potentials of other electrodes. See also *electrode potential*, *reference electrode*, and *saturated calomel electrode*.

carrier. In emission spectrochemical analysis, a material added to a sample to facilitate its controlled vaporization into the analytical gap. See also *analytical gap*.

cathode-ray tube (CRT). An electronic tube that permits the visual display of electronic signals.

cathodoluminescence. Emission of visible light by a specimen impacted by an energetic electron beam.

cation. A positively charged atom or group of atoms, or a radical that moves to the negative pole (cathode) during electrolysis. See also *anion*.

channeling. The process in which an electron beam or ion beam is well aligned to the low index directions of a crystal and therefore passes with minimal scattering.

channeling contrast. Contrast variations seen in secondary electron images generated by scanning electron microscopy or focused ion beam imaging in which one crystalline grain appears brighter or darker due to the channeling process. Channeling contrast commonly results from variation in backscattered electron yields for scanning electron microscopy and variation in secondary electron yields for focused ion beam imaging.

channeling pattern. A pattern of lines observed in an electron microscope when scanning an electron beam across a range of angles relative to a single grain or single-crystal sample. The pattern provides information on the structure and orientation of the crystal.

characteristic electron energy loss phenomena. The inelastic scattering of electrons in solids that produces a discrete energy loss determined by the characteristics of the material. The most probable form is due to excitation of valence electrons.

characteristic radiation. Electromagnetic radiation of a particular set of wavelengths, produced by and characteristic of a particular element whenever its excitation potential is exceeded. Electromagnetic radiation is emitted as a result of electron transitions between the various energy levels (electron shells) of atoms; the spectrum consists of lines whose wavelengths depend only on the element concerned and the energy levels involved.

charging. In scanning electron microscopy and focused ion beam applications, the progressive accumulation of positive, negative, or both signs of charge of an insulating sample under the influence of the incident beam.

chelate. A coordination compound in which a heterocyclic ring is formed by a metal bound to two atoms of the associated ligand. See also *complexation*, *coordination compound*, and *ligand*.

chemical adsorption. See *chemisorption*.

chemical bonding. The joining together of atoms to form molecules. See also *molecule*.

chemical shift. A shift in line position of an element in an x-ray photoelectron spectroscopy spectrum due to interactions of the element with other elements, such as a compound formation.

chemisorption. The binding of an adsorbate to the surface of a solid by forces whose energy levels approximate those of a chemical bond. Contrast with *physisorption*.

chromatic aberration. A failure of a lens or lens system to focus radiations of different wavelengths to the same point.

chromatogram. The visual display of the progress of a separation achieved by chromatography. A chromatogram shows the response of a chromatographic detector as a function of time.

chromatography. A separation method based on the distribution of sample compounds between the stationary phase and a mobile phase. See also *gas chromatography*, *ion chromatography*, and *liquid chromatography*.

cluster ion guns. Ion guns that emit a cluster of atomic ions instead of single atomic ions.

coefficient of linear thermal expansion (CTE or CLTE). The thermal strain of a material, from the slope of a thermomechanical analysis or dilatometer plot of dL/L_0 versus T. See also *expansivity*.

cold atom ion source. An ion source in which the atoms are first trapped and cooled well below ambient temperatures with laser techniques before being ionized.

cold finger. A liquid-nitrogen-cooled cold trap used to reduce contamination levels in vacuum chambers.

collimate. To make parallel to a certain line or direction.

collimation. The operation of controlling a beam of radiation so that its rays are as nearly parallel as possible.

color center. A point lattice defect that produces optical absorption bands in an otherwise transparent crystal.

coma. A lens aberration occurring in that part of the image field that is some distance from the principal axis of the system. It results from different magnification in the various lens zones. Extra-axial object points appear as short, conelike images with the brighter small head toward the center of the field (positive coma) or away from the center (negative coma).

compensating eyepieces. Those designed for use with objectives such as apochromats to correct chromatic aberration.

complexation. The formation of complex chemical species by the coordination of groups of atoms termed ligands to a central ion, commonly a metal ion. Generally, the ligand coordinates by providing a pair of electrons that forms an ionic or covalent bond to the central ion. See also *chelate*, *coordination compound*, and *ligand*.

composite. A sample composed of two or more increments.

compositional depth profile. The atomic concentration measured as a function of the perpendicular distance from the surface.

Compton scattering. The elastic scattering of photons by electrons. Contrast with *Rayleigh scattering*.

concave grating. A diffraction grating on a concave mirror surface. See also *diffraction grating* and *plane grating*.

concentration. The mass of a substance contained in a unit volume of sample, for example, grams per liter.

condenser, condenser lens. A term applied to lenses or mirrors designed to collect, control, and concentrate radiation in an illumination system.

conductance. The property of a circuit or its materials of composition that allows the transmission of a current when suitable potential difference exists; the reciprocal of resistance. The usual unit is the siemens. See also *siemens*.

confidence interval. That range of values, calculated from estimates of the mean and standard deviation, which is expected to include the population mean with a stated level of confidence. Confidence intervals in the same context also can be calculated for standard deviations, lines, slopes, and points.

continuum radiation. The noncharacteristic rays emitted upon irradiation of a specimen and caused by deceleration of the incident electrons by interaction with the electrons and nuclei of the specimen. See also *bremsstrahlung radiation* and *white radiation*.

controlled-potential coulometry. Measurement of the number of coulombs required for an electrochemical reaction occurring under conditions where the working electrode potential is precisely controlled.

convergent-beam electron diffraction (CBED). A technique of impinging a highly convergent electron beam on a crystal to produce a diffraction pattern composed of disks of intensity. In addition to *d*-spacing and crystal orientation information, the technique can provide information on crystallographic point or space group symmetry.

coordination compound. A compound with a central atom or ion bound to a group of ions or molecules surrounding it. Also known as coordination complex. See also *chelate*, *complexation*, and *ligand*.

coordination number. In coordination compounds, the number of atoms bonded to a central atom or ion. In a space lattice, the number of nearest neighbors of a specific atom or ion.

coulometric titration. Coulometry in which the current is kept constant via an amperostat or galvanostat. Also known as amperostatic coulometry.

coulometry. An electrochemical technique in which the total number of coulombs consumed in an electrolysis is used to determine the amount of substance electrolyzed.

counterelectrode. In emission spectroscopy, the electrode that is used opposite to the self-electrode or supporting electrode and that is not composed of the sample to be analyzed. In voltammetry, the current between the working electrode and counterelectrodes is measured. See also *self-electrode* and *supporting electrode*.

covalent bond. A bond in which two atoms share a pair of electrons. Contrast with *ionic bond*.

critical micelle concentration. The concentration of a micelle at which the rate of increase of electrical conductance with increase in concentration levels off or proceeds at a much slower rate. See also *micelle*.

cryopump. Also known as a cryogenic pump, a type of high-vacuum pump that relies on the condensation of gas molecules and atoms on internal surfaces of the pump, which are maintained at extremely low temperatures.

Curie point. See *Curie temperature*.

Curie temperature. The temperature marking the transition between ferromagnetism and paramagnetism, or between the ferroelectric phase and the paraelectric phase. Also known as Curie point. See also *ferromagnetism* and *paramagnetism*.

cyclodextrin. Cyclic degradation products of starch that contain six, seven, or eight glucose residues and have the shape of large ring molecules.

D

dc intermittent noncapacitive arc. A low-voltage electrical discharge used in spectrochemical analysis to vaporize the sample material. Each current pulse has the same polarity as the previous one and lasts for less than 0.1 s. See also *ac noncapacitive arc*.

dc plasma excitation. See *plasma-jet excitation*.

dead time. The total time during which the spectrometer is processing information and is unavailable to accept input data.

decay constant (λ). The constant in the radioactive decay law $dN = -\lambda N dt$, where N is the number of radioactive nuclei present at time t. The decay constant is related to half-life $t_{1/2}$ by the expression $t_{1/2} = \ln 2/\lambda$. See also *half-life*.

density (of gases). The mass of a unit volume of gas at a stated temperature and pressure.

density (of solids and liquids). The mass of a unit volume of a material at a specified temperature.

depth of field. The depth or thickness of the object space that is simultaneously in acceptable focus.

depth of penetration. In various analytical techniques, the distance the probing radiation penetrates beneath the surface of a sample. See also *excitation volume*. Contrast with *escape depth*.

detection limit. In an analytical method, the lowest mass or concentration of an analyte that can be measured.

determination. The ascertainment of the quantity or concentration of a specific substance in a sample. See also *analysis*.

deuteron. The nucleus of the atom of heavy hydrogen, deuterium. The deuteron is composed of a proton and a neutron; it is the simplest multinucleon nucleus. Deuterons are used as projectiles in many nuclear bombardment experiments. See also *neutron* and *proton*.

dewar flask. A vessel having double walls, the space between being evacuated to prevent the transfer of heat and the surfaces facing the vacuum being heat reflective; used to hold liquid gases and to study low-temperature phenomena.

diaphragm. A fixed or adjustable aperture in an optical system. Diaphragms are used to intercept scattered light, to limit field angles, or to limit image-forming bundles or rays.

differential scanning calorimetry (DSC). A thermoanalytical technique in which the difference in the amount of heat required to increase the temperature of a sample and reference is measured as a function of temperature.

diffraction contrast. In electron microscopy, contrast produced by intensity differences in Bragg-diffracted beams from a crystalline material. These differences are caused by regions of varying crystal orientation.

diffraction grating. A series of a large number of narrow, close, equally spaced, diffracting slits or grooves capable of dispersing light into its spectrum. See also *concave grating* and *reflection grating*. Compare with *transmission grating*.

diffraction pattern. The spatial arrangement and relative intensities of diffracted beams.

diffuse transmittance. The transmittance value obtained when the measured radiant energy has experienced appreciable scattering in passing from the source to the receiver. See also *transmittance*.

dilatometry. A thermal analysis technique in which the length or volume of a specimen is precisely measured versus temperature and time as the specimen is subjected to controlled heating and cooling. See also *thermomechanical analysis*.

dilution factor. When diluting a sample, the ratio of the final volume or mass after dilution to the volume or mass of the sample before dilution.

dimer. A condensation compound formed from two monomers or molecules.

dimerization. The formation of a dimer. See also *dimer*.

direct injection burner. A burner used in flame emission and atomic absorption spectroscopy in which the fuel and oxidizing gases emerge from separate ports and are mixed in the flame itself. One of the gases, usually the oxidant, is used for nebulizing the sample at the tip of the burner.

disordered structure. The crystal structure of a solid solution in which the atoms of different elements are randomly distributed relative to the available lattice sites. Contrast with *ordered structure*.

Donnan exclusion. The mechanism by which an ion-exchange resin can be made to act like a semipermeable membrane between an interstitial liquid and a liquid occluded inside the resin particles. Highly ionized molecules are excluded from the resin particles by electrostatic forces; weakly ionized or nonionized molecules may pass through the membrane.

Doppler effect. The change in the observed frequency of an acoustic or electromagnetic wave due to the relative motion of source and observer. See also *Doppler shift*.

Doppler shift. The amount of change in the observed frequency of a wave due to the Doppler effect, usually expressed in hertz. See also *Doppler effect*.

dosimeter. An instrument that measures the total dose of nuclear radiation received in a given period.

dot map. See *x-ray map*.

doublets. Double peaks in an x-ray photoelectron spectroscopy spectrum appearing as twin peaks formed due to spin-orbit splitting.

dropping mercury electrode. An electrode formed by a sequence of mercury continuously dropping from a reservoir through a capillary tube (internal diameter: 0.03 to 0.05 mm, or 0.0012 to 0.0020 in.) into a solution being analyzed by *polarography*.

duoplasmatron. A type of ion source in which a plasma created by an arc discharge is confined and compressed by a nonuniform magnetic field.

duplex microstructure. A microstructure consisting of two different phases or constituents.

duplicate measurement. A second measurement made on the same (or identical) sample of material to assist in the evaluation of measurement variance.

duplicate sample. A second sample randomly selected from a population to assist in the evaluation of sample variance.

dynamic mechanical analysis. A technique used to study and characterize materials. It is most useful for studying the viscoelastic behavior of polymers. A sinusoidal stress is applied and the strain in the material is measured, allowing one to determine the complex modulus.

E

elastic constants. The factors of proportionality that relate elastic displacement of a material to applied forces. See also *modulus of elasticity* and *Poisson's ratio*.

elastic scattering. Collisions between particles that are completely described by conservation of energy and momentum. Contrast with *inelastic scattering*.

electric dipole. The result of a distribution of bound charges, that is, separated charges that are bound to their centers of equilibrium by an elastic force; equal numbers of positive and negative charges must be present in an uncharged medium.

electric dipole moment. A quantity characteristic of a distribution of bound charges equal to the vector sum over the charges of the product of the charge and the position vector of the charge.

electric dipole transition. A transition of an atom, molecule, or nucleus from one energy state to another, which results from the interaction of electromagnetic radiation with the dipole moment of the molecule, atom, or nucleus.

electric field effect. See *Stark effect*.

electrode. In emission spectroscopy, either of two terminals between which an electrical discharge occurs. See also *counterelectrode*, *self-electrode*, and *supporting electrode*.

electrode potential. Voltage existing between an electrode and the solution or electrolyte in which it is immersed. Electrode potentials are referred to a standard electrode, such as the hydrogen electrode.

electrogravimetry. An electroanalytical technique in which an analyte solution is electrolyzed and the electrochemical reduction causes the analyte to be deposited on the cathode and then is weighed.

electrolysis. A method by which chemical reactions are carried out by passage of electric current through a solution of an electrolyte or through a molten salt. See also *electrolyte*.

electrolyte. A chemical compound or mixture of compounds which when molten or in solution will conduct an electric current.

electromagnetic lens. An electromagnet designed to produce a suitably shaped magnetic field for the focusing and deflection of electrons or other charged particles in electron-optical instrumentation.

electromagnetic radiation. Energy propagated at the speed of light by an electromagnetic field. The electromagnetic spectrum includes the following approximate wavelength regions:

Region	Wavelength, Å (metric)
Gamma-ray	0.005 to 1.40 (0.0005 to 0.14 nm)
X-ray	0.1 to 100 (0.01 to 10 nm)
Far-ultraviolet	100 to 2000 (10 to 200 nm)
Near-ultraviolet	2000 to 3800 (200 to 380 nm)
Visible	3800 to 7800 (380 to 780 nm)
Near-infrared	7800 to 30,000 (0.78 to 3 μm)
Middle-infrared	3×10^4 to 3×10^5 (3 to 30 μm)
Far-infrared	3×10^5 to 3×10^6 (30 to 300 μm)
Microwave	3×10^6 to 1×10^{10} (0.3 mm to 1 m)

electrometric titration. A family of techniques in which the location of the endpoint of a titration involves the measurement of, or observation of changes in, some electrical quantity. Examples of such quantities include potential, current, conductance, frequency, and phase.

electron. A negatively charged particle that resides in specific orbits around the nucleus of an atom. The electron is the lightest known particle that possesses an electric charge. Its rest mass is approximately 9.1×10^{-28} g, approximately $\frac{1}{1836}$ of the mass of the proton or neutron, which are, respectively, the positively charged and neutral constituents of the atomic nucleus. See also *neutron* and *proton*.

electron backscatter diffraction (EBSD). A scanning electron microscopy technique in which a finely focused beam of electrons is scattered from the sample to produce a Kikuchi pattern that can be used to identify the local crystallographic phase and orientation. See also *Kikuchi lines* and *Kikuchi pattern*.

electron beam induced current (EBIC). A scanning electron microscopy technique in which a finely focused beam of electrons induces a measurable current signal that can be used to spatially probe local electrical properties of semiconductors and devices.

electron diffraction. The phenomenon or the technique of producing diffraction patterns through the incidence of electrons upon crystalline matter.

electron energy loss spectroscopy (EELS). A spectrographic technique in the electron microscope that analyzes the energy distribution of the electrons transmitted through the specimen. The energy loss spectrum is characteristic of the chemical composition of the region being sampled.

electron multiplier phototube. See *photomultiplier tube*.

electron probe x-ray microanalysis (EPMA). A technique in analytical chemistry in which a finely focused beam of electrons is used to excite an x-ray spectrum characteristic of the elements in a small region of the sample.

electron scattering. Any change in the direction of propagation or kinetic energy of an electron as a result of a collision.

electronic stopping power. The progressive energy loss of a high-energy ion as it enters a sample and interacts with the electrons therein.

emission (of electromagnetic radiation). The creation of radiant energy in matter, resulting in a corresponding decrease in the energy of the emitting system.

emission lines. Spectral lines resulting from emission of electromagnetic radiation by atoms, ions, or molecules during changes from excited states to states of lower energy.

emission spectrometer. An instrument that measures percent concentrations of elements in samples of metals and other materials; when the sample is vaporized by an electric spark or arc, the characteristic wavelengths of light emitted by each element are measured with a diffraction grating and an array of photodetectors or photographic plates.

emission spectroscopy. The branch of spectroscopy treating the theory, interpretation, and application of spectra originating in the emission of electromagnetic radiation by atoms, ions, radicals, and molecules.

emission spectrum. An electromagnetic spectrum produced when radiation from any emitting source, excited by any of various forms of energy, is dispersed.

emulsion calibration curve. The plot of a function of the relative transmittance of the photographic emulsion versus a function of the exposure. The calibration curve is used in spectrographic analysis to calculate the relative intensity of a radiant source from the density of a photographically recorded image.

energy-dispersive spectrometry (EDS). The measurement of the spectrum of x-ray energies and intensities, in which the energy of individual photons is converted into electronic pulses that are processed and counted. As these energies are measured, a histogram of the numbers of photons counted corresponding to each energy is plotted. Compare with *wavelength-dispersive spectrometry*.

energy-dispersive spectroscopy (EDS). A method of x-ray analysis that discriminates by energy levels the characteristic x-rays emitted from the sample. Compare with *wavelength-dispersive spectroscopy*.

energy levels. Specific energies that electrons in an atom can have when occupying specific orbitals.

epitaxy. Oriented growth of a crystalline substance on a crystalline substrate with a fixed orientation between the two crystal lattices.

equiaxed grain(s). The grain structure of a specimen in which all of the grains exhibit similar dimensions in all measurement directions.

error. The difference between the true or expected value and the measured value of a quantity or parameter.

escape depth. A distance perpendicular to the surface from which data signals can escape before being absorbed within the specimen. Contrast with *depth of penetration* and *information depth*.

escape peak. An artifact observed in x-ray analysis; manifested as a peak at energy 1.74 keV (the silicon Kα peak) less than the major line detected. Escape peaks can be avoided by increasing the accelerating voltage.

eucentric stage. A sample stage so designed that a sample at any point on the stage can be made to rotate around the analysis point as the axis of rotation.

Euler angles. Three angular parameters that specify the orientation of a body with respect to reference axes.

excitation index. The ratio of the intensities of two selected spectral lines of an element having widely different excitation energies. This ratio serves to indicate the level of excitation energy in the source.

excitation potential (x-ray). The applied potential on an x-ray tube required to produce characteristic radiation from the target.

excitation volume. The volume within the sample in which data signals originate.

expansivity. The thermal strain of a material, from the slope of a thermomechanical analysis or dilatometer plot of dL/L_0 versus T. See also *coefficient of linear thermal expansion*.

exposure. The product of the intensity of a radiant source and the time of irradiation.

exposure index. The relative transmittance or optical density of a selected spectral line, this value serving to indicate the degree of blackening of the photographic emulsion.

extended x-ray absorption fine structure (EXAFS). The weak oscillatory structure extending for several hundred electron volts away from an absorption edge. The oscillations occur because the electromagnetic wave produced by the ionization of the absorbing atom for some energy E has a wavelength $\lambda = 1.225/(E - E_k)^{1/2}$ nm, where E_k is the energy of the absorption edge. For example, a loss of 100 eV above an edge corresponds to a wavelength of 0.12 nm, which is of the order of atomic spacing. Consequently, the wave can be diffracted from neighboring atoms and return to interfere with the outgoing wave. An analysis of EXAFS data reveals important information about atomic arrangements and bonding. Either synchrotron x-radiation or the electron beam in the analytical transmission electron microscope can be used as the excitation source. See also *analytical electron microscopy* and *synchrotron radiation*.

eyepiece. The lens system used in an optical instrument for magnification of the image formed by the objective.

F

Faraday's law. The amount of any substance deposited or liberated during electrolysis is proportional to the quantity of electric charge passed and to the equivalent weight of the substance.

far-infrared radiation. Infrared radiation in the wavelength range of 30 to 300 μm (3×10^5 to 3×10^6 Å).

feature-specific measurements (in image analysis). Individual measurements of all detected features in the image field of view.

ferrite grain size. The mean grain size of body-centered cubic ferrite in a low-carbon steel alloy.

ferromagnetic resonance. Magnetic resonance of a ferromagnetic material. See also *ferromagnetism* and *magnetic resonance*.

ferromagnetism. A property exhibited by certain metals, alloys, and compounds of the transition (iron group), rare earth, and actinide elements in which, below a certain temperature termed the Curie temperature, the atomic magnetic moments tend to line up in a common direction. Ferromagnetism is characterized by the strong attraction of one magnetized body for another. See also *Curie temperature*. Compare with *paramagnetism*.

FIB-SEM. An instrument that includes both a focused ion beam (FIB) and a scanning electron microscope (SEM) usually directed at the same point where the sample can be located. Also known as cross-beam or dual-beam.

field-emission microscopy. An image-forming analytical technique in which a strong electrostatic field causes emission of electrons from a sharply rounded point or from a specimen that has been placed on that point. The electrons are accelerated to a phosphorescent screen, or photographic film, producing a visible picture of the variation of emission over the specimen surface.

field ionization. The ionization of gaseous atoms and molecules by an intense electric field, often at the surface of a solid.

field ion microscopy. An analytical technique in which atoms are ionized by an electric field near a sharp specimen tip; the field then forces the ions to a fluorescent screen, which shows an enlarged image of the tip, and individual atoms are made visible. See also *atom probe*.

field measurements (in image analysis). An aggregate measurement of each phase or constituent in the image field of view.

field of view (FOV). A term in microscopy used to describe the size of the imaged region, for example, in micrometers or nanometers.

filter. A semitransparent optical element capable of absorbing unwanted electromagnetic radiation and transmitting the remainder. A neutral density filter attenuates relatively uniformly from the ultraviolet to the infrared, but in many applications highly wavelength-selective filters are used. See also *neutral filter*.

Firmi level. The highest energy state occupied by an electron at absolute zero temperature.

fluorescence. A type of photoluminescence in which the time interval between the absorption and re-emission of light is very short. Contrast with *phosphorescence*.

fluorimetry. See *fluorometric analysis*.

fluorometric analysis. A method of chemical analysis that measures the fluorescence intensity of the analyte or a reaction product of the analyte and a chemical reagent.

focused ion beam (FIB). An instrument used to generate a focused, nanometer-scale ion beam for material removal, material deposition, or imaging of samples. Many varieties of ion beams and instruments exist. The most common variety generates gallium ion beams with ~5 nm focused probe size.

Fourier transform infrared (FTIR) spectrometry. A form of infrared spectrometry in which data are obtained as an interferogram, which is then Fourier transformed to obtain an amplitude-versus-wavenumber (or wavelength) spectrum.

Frank-Condon principle. The principle which states that the transition from one energy state to another is so rapid that the nuclei of the atoms involved can be considered stationary during the transition.

free radical. Any molecule or atom that possesses one unpaired electron. In chemical notation, a free radical is symbolized by a single dot (to denote the odd electron) to the right of the chemical symbol.

frequency. The number of cycles per unit time. The recommended unit is the hertz, Hz, which is equal to one cycle per second.

full width at half maximum (FWHM). A measure of resolution of a spectrum or chromatogram determined by measuring the peak width of a spectral or chromatographic peak at half its maximum height.

functional group. A chemical radical or structure that has characteristic properties; examples are hydroxyl and carboxyl groups.

G

gallium FIB. The most common type of focused ion beam (FIB) instrument. Often used for material removal by the sputtering process.

gamma ray. A high-energy photon, especially as emitted by a nucleus in a transition between two energy levels.

gamma-ray spectrometry. See *gamma-ray spectroscopy*.

gamma-ray spectroscopy. Determination of the energy distribution of γ-rays emitted by a nucleus. Also known as gamma-ray spectrometry.

gas chromatography. A separation method involving passage of a gaseous mobile phase through a column containing a stationary adsorbent phase; used principally as a quantitative analytical technique for volatile compounds. See also *chromatography*, *ion chromatography*, and *liquid chromatography*.

gas constant. The constant of proportionality appearing in the equation of state of an ideal gas, equal to the pressure of the gas multiplied by its molar volume divided by its temperature. Also known as universal gas constant.

gas field ion source (GFIS). Ion source that relies on the ionization of neutral gas atoms at specific atomic sites near the apex of a sharpened needle that is positively biased.

gas injection system (GIS). A common accessory for the scanning electron microscope or focused ion beam that allows a precursor gas to be delivered to the region of the sample where the beam is incident. The GIS allows for nanometer-scale deposition or etching processes in conjunction with the charged particle beam.

gas mass spectrometry. An analytical technique that provides quantitative analysis of gas mixtures through the complete range of elemental and molecular gases.

gel-permeation chromatography (GPC). See *size-exclusion chromatography*.

glass transition temperature (T_g). The narrow temperature range in which the mechanical behavior of a viscoelastic solid changes abruptly and reversibly from rigid solid (predominantly elastic) to soft rubberlike (predominantly viscous).

goniometer. An instrument devised for measuring the angle through which a specimen is rotated or for orienting a sample (for example, a single crystal) in a specific way.

gradient elution. A technique for improving the efficiency of separations achieved by liquid chromatography. It refers to a stepwise or continuous change with time in the mobile phase composition.

grain(s). Individual crystallites with the same atomic configuration throughout a polycrystalline metal; the grains may or may not contain twinned regions or subgrains.

grain boundary. The interface separating adjacent grains where the crystal lattice orientation in one grain is substantially different from that of the adjacent grains.

grain-boundary intersection count (P). Determination of the number of times a test line (or a line of pixels) cuts across, or is tangent to, grain boundaries. (Tangent intersections are counted as one intersection; triple-point intersections are counted as 1.5 intersections.)

grain intercept count (N). Determination of the number of times a test line (or a line of pixels) cuts through individual grains on the plane of polish. (Tangent hits are counted as 0.5 intercepts; test lines that end within a grain are counted as half an intercept.)

grain size. The physical dimensions of the grains in a polycrystalline metal exclusive of regions with annealing twins or subgrain boundaries, if present.

gram-equivalent weight. The mass in grams of a reactant that contains or reacts with Avogadro's number of hydrogen atoms. See also *Avogadro's number*.

gram-molecular weight. The mass of a compound in grams equal to its molecular weight.

gross sample. One or more increments of material taken from a larger quantity (lot) of material for assay or record purposes. Also termed bulk sample or lot sample. See also *increment* and *lot*.

H

half-life ($t_{1/2}$) . The time required for one-half of an initial (large) number of atoms of a radioactive isotope to decay. Half-life is related to the decay constant λ by the expression $t_{1/2} = \ln 2/\lambda$. See also *decay constant*.

Hall effect. The development of a transverse electric field in a current-carrying conductor placed in a magnetic field.

helium ion microscope (HIM). A class of focused ion beam instrument that uses a helium beam for imaging or modifying a sample.

helium ion microscopy (HIM). The technique of using a helium ion microscope for imaging or modifying a sample. The helium ion microscope is well established for imaging of an insulating sample and for precision sputtering, for example.

heterogeneity. The degree of nonuniformity of composition or properties. Contrast with *homogeneity*.

high-temperature combustion. An analytical technique for determining the concentrations of carbon and sulfur in samples. The sample is burned in a graphite crucible in the presence of oxygen, which causes carbon and sulfur to leave the sample as carbon dioxide and sulfur dioxide. These gases are then detected by infrared or thermal conductive means.

homogeneity. The degree of uniformity of composition or properties. Contrast with *heterogeneity*.

homologous pairs. Spectral lines for different elements that respond in the same way to changes in excitation conditions. One line can be used as an interval standard line for the other.

Hooke's law. A generalization applicable to all solid material, which states that stress is directly proportional to strain and is expressed as:

$$\frac{\text{Stress}}{\text{Strain}} = \frac{\sigma}{\varepsilon} = \text{Constant} = E$$

where E is the modulus of elasticity or Young's modulus. The constant relationship between stress and strain applies only below the proportional limit. See also *modulus of elasticity*.

I

image analysis. Measurement of the size, shape, and distributional parameters of microstructural features by electronic scanning methods, usually automatic or semiautomatic. Image analysis data output can provide individual measurements on each separate feature (feature specific) or field totals for each measured parameter (field specific).

image contrast. A measure of the degree of detectable difference in intensity within an image.

image processing (in image analysis). Computer modifications of a digitized image on a pixel-by-pixel basis to better reveal certain features in the microstructure to enhance measurement precision.

increment. An individual portion of material collected by a single operation of a sampling device from parts of a lot separated in time or space. Increments may be tested individually or combined (composited) and tested as a unit.

index of refraction. See *refractive index*.

individuals. Conceivable constituent parts of a population. See also *population*.

inductively coupled plasma (ICP). An argon plasma excitation source for atomic emission spectroscopy or mass spectroscopy. It is operated at atmospheric pressure and sustained by inductive coupling to a radio-frequency electromagnetic field. See also *radio frequency*.

inelastic scattering. Any collision or interaction that changes the energy of an incident particle. Contrast with *elastic scattering*.

inert gas fusion. An analytical technique for determining the concentrations of oxygen, hydrogen, and nitrogen in a sample. The sample is melted in a graphite crucible in an inert gas atmosphere; individual component concentrations are detected by infrared or thermal conductive methods.

information depth. A measure of the sampling depth for the detected signal in Auger electron spectroscopy and x-ray photoelectron spectroscopy. Contrast with *escape depth*.

infrared radiation. Electromagnetic radiation in the wavelength range of 0.78 to 300 μm (7800 to 3×10^6 Å). See also *electromagnetic radiation*, *far-infrared radiation*, *middle-infrared radiation*, and *near-infrared radiation*.

infrared spectrometer. A device used to measure the amplitude of electromagnetic radiation of wavelengths between visible light and microwaves.

infrared spectroscopy. The study of the interaction of material systems with electromagnetic radiation in the infrared region of the spectrum. The technique is useful for determining the molecular structure of organic and inorganic compounds by identifying the rotational and vibrational energy levels associated with the various molecules.

infrared spectrum. (1) The range of wavelengths of infrared radiation. (2) A display or graph of the intensity of infrared radiation emitted or absorbed by a material as a function of wavelength or some related parameter.

in operando. A particular mode of in situ materials characterization performed under specific conditions comparable to when the specimen is in operation. See also *in situ*.

in situ. A mode of materials characterization performed under application of one or more external loads to the specimen.

instrument response time. The time required for an indicating or detecting device to attain a defined percentage of its steady-state value following an abrupt change in the quantity being measured.

intensity ratio. The ratio of two (relative) intensities.

intercept count. The number of particles or grains of a specific phase or constituent of interest that have been crossed by lines in a test grid, or by lines of pixels.

interference of waves. The process whereby two or more waves of the same frequency or wavelength combine to form a wave whose amplitude is the sum of the amplitudes of the interfering waves.

interferometer. An instrument in which the light from a source is split into two or more beams, which are subsequently reunited and interfere after traveling over different paths.

internal standard. In spectroscopy, a material present in or added to samples that serves as an intensity reference for measurements; used to compensate for variations in sample excitation and photographic processing in emission spectroscopy.

internal standard line. In spectroscopy, a spectral line of an internal standard, with which the radiant energy of an analytical line is compared.

intersection count. The number of boundaries between the matrix phase and the phase or constituent of interest that are crossed by the lines in a test grid, or by lines of pixels.

intersystem crossing. A transition between electronic states that differ in total spin quantum number.

ion. An atom, or group of atoms, which by loss or gain of one or more electrons has acquired an electric charge. If the ion is formed from an atom of hydrogen or an atom of a metal, it is usually positively charged; if the ion is formed from an atom of a nonmetal or from a group of atoms, it is usually negatively charged. The number of electronic charges carried by an ion is termed its electrovalence. The charges are denoted by superscripts that give their sign and number; for example, a sodium ion, which carries one positive charge, is denoted by Na^+; a sulfate ion, which carries two negative charges, by SO_4^{2-}. See also *atomic structure* and *chemical bonding*.

ion beam. A stream of ionized atoms accelerated to a characteristic average energy, and an energy spread. The ion beam is characterized by its brightness, which in part determines its ability to be focused to a small probe size.

ion chromatography. An area of high-performance liquid chromatography that uses ion-exchange resins to separate various species of ions in solution and elute them to a suitable detector for analysis. See also *chromatography*, *gas chromatography*, and *liquid chromatography*.

ion exchange. A reversible chemical reaction between a solid (ion exchanger) and a fluid (usually an aqueous solution) by means of which ions may be interchanged.

ion-exchange chromatography (IEC). Liquid chromatography with a stationary phase that possesses charged functional groups. This technique is applicable to the separation of ionic (charged) compounds. See also *ion chromatography*.

ion-exchange resin. A synthetic resin containing active groups (usually sulfonic, carboxylic, phenol, or substituted amino groups) that give the resin the property of combining with or exchanging ions between the resin and a solution.

ionic bond. A type of chemical bonding in which one or more electrons are transferred completely from one atom to another, thus converting the neutral atoms into electrically charged ions. These ions are approximately spherical and attract each other because of their opposite charges. Contrast with *covalent bond*.

ionic charge. The positive or negative charge of an ion.

ionization. A process by which an atom or a molecule is positively or negatively charged by acquiring or losing electron(s).

ion neutralization. The generic term for a class of charge-exchange processes in which an ion is neutralized by passage through a gas or by interaction with a material surface.

ion-pair chromatography (IPC). Liquid chromatography with a mobile phase containing an ion that combines with sample ions, creating neutral ion pairs. The ion pairs are typically separated using bonded-phase chromatography. See also *bonded-phase chromatography*.

ion pump. A type of high-vacuum pump that operates on the principle of ionizing the gas moles to be pumped out and driving them to a solid electrode to be captured.

ion-scattering spectrometry. A technique to elucidate composition and structure of the outermost atomic layers of a solid material, in which principally monoenergetic, singly charged, low-energy (less than 10 keV) probe ions are scattered from the surface and are subsequently detected and recorded as a function of the energy.

ion-scattering spectrum. A plot of scattered ion intensity as a function of the ratio of the scattered ion energy to the incident ion energy.

ion species. Type and charge of an ion. If an isotope is used, it should be specified.

irradiance (of a receiver). The radiant power per unit area incident on a receiver. See also *exposure*.

isobar. One of two or more atoms that have a common mass number, *A*, but differ in atomic number, *Z*. Thus, although isobars possess approximately equal masses, they differ in chemical properties; they are atoms of different elements. See also *nuclear structure*.

isocratic elution. In liquid chromatography, the use of a mobile phase whose composition is unchanged throughout the course of the separation process.

isotone. One of two or more atoms that display a constant difference $A - Z$ between their mass number, *A*, and their atomic number, *Z*. Thus, despite differences in the total number of nuclear constituents, the numbers of neutrons in the nuclei of isotones are the same. See also *nuclear structure*.

isotope. One of two or more atoms having an identical number of protons (*Z*) in the nucleus but a different number of neutrons (*N*). Isotopes differ in mass but chemically are the same element. See also *nuclear structure*.

J

Johnson noise. See *thermal noise*.

Joule-Thomson effect. A change in temperature in a gas undergoing Joule-Thomson expansion. See also *Joule-Thomson expansion*.

Joule-Thomson expansion. The adiabatic, irreversible expansion of a gas flowing through a porous plug or partially open valve. See also *Joule-Thomson effect*.

K

***k*-factor.** The ratio between the unknown and standard x-ray intensities used in quantitative analyses.

Kikuchi lines. Light and dark lines within an electron-diffraction pattern caused by diffraction of diffusely scattered electrons within the crystal; the pattern provides structural information on the crystal.

Kikuchi pattern. A pattern of Kikuchi lines obtained in an electron microscope that can be used to determine crystallographic orientation. See also *Kikuchi lines* and *electron backscatter diffraction*.

kinetic energy. The energy that a body possesses because of its motion; in classical mechanics, equal to one-half of the body's mass times the square of its speed.

klystron. An evacuated electron-beam tube in which an initial velocity modulation imparted to electrons in the beam subsequently results in density modulation of the beam. This device is used as an amplifier or oscillator in the microwave region.

K-radiation. Characteristic x-rays produced by an atom or ion when a vacancy in the K shell is filled by an electron from another shell.

K-series. The set of characteristic x-ray wavelengths making up K-radiation for the various elements.

K shell. The innermost shell of electrons surrounding the atomic nucleus, having electrons characterized by the principal quantum number 1.

L

laboratory sample. A sample, intended for testing or analysis, prepared from a gross sample or otherwise obtained; the laboratory sample must retain the composition of the gross sample. Reduction in particle size is often necessary in the course of reducing the quantity.

lamella. A thin slice (coupon) of a larger sample that is suitable for extraction and subsequent transmission imaging. Commonly, the lamella is less than 100 nm thick and is prepared by using a focused ion beam instrument.

Larmor frequency. The classical frequency at which a charged body precesses in a uniform magnetic field. $\omega_L = -eB/2mc$, where *e* is the electron charge, *B* is the magnetic field intensity, *m* is mass, and *c* is the velocity of light. See also *Larmor period*.

Larmor period. The inverse of the Larmor frequency. See also *Larmor frequency*.

lens. A transparent optical element, so constructed that it serves to change the degree of convergence or divergence of the transmitted rays.

ligand. The molecule, ion, or group bound to the central atom in a chelate or a coordination compound. See also *chelate* and *coordination compound*.

light. Radiant energy in a spectral range visible to the normal human eye (~380 to 780 nm, or 3800 to 7800 Å). See also *electromagnetic radiation*.

linear dispersion. In spectroscopy, the derivative $dx/d\lambda$, where *x* is the distance along the spectrum, and λ is the wavelength. Linear dispersion is usually expressed as mm/Å.

line pair. In spectroscopy, an analytical line and the internal standard line with which it is compared. See also *internal standard line*.

liquid chromatography. A separation method based on the distribution of sample compounds between a stationary phase and a liquid mobile phase. See also *chromatography*, *gas chromatography*, and *ion chromatography*.

liquid-liquid chromatography (LLC). Liquid chromatography with a stationary phase composed of a liquid dispersed onto an inert supporting material. Also termed liquid-partition chromatography.

liquid metal alloy ion source (LMAIS). As with the liquid metal ion source but in which the selected metal is an alloy.

liquid metal ion source (LMIS). Technology used for the production of bright ion beams by the ionization of a liquid metal that is drawn into a Taylor cone. This is the ion source commonly used in most gallium focused ion beam instruments.

liquid-partition chromatography (LPC). See *liquid-liquid chromatography*.

liquid-solid chromatography (LSC). Liquid chromatography with silica or alumina as the stationary phase. See also *adsorption chromatography*.

lithography. The technique of transcribing information onto a medium or substrate. Varieties include resist-based lithography, including optical lithography with masks, or direct-write lithography with electron or ion beams. There are also resistless techniques of modifying a sample directly.

lot. A quantity of bulk material of similar composition whose properties are under study.

lot sample. See *gross sample*.

low-energy electron diffraction. A technique for studying the atomic structure of single-crystal surfaces, in which electrons of uniform energy in the approximate range of 5 to 500 eV are scattered from a surface. Those scattered electrons that have lost no energy are selected and accelerated to a fluorescent screen, where the diffraction pattern from the surface is observed.

L-radiation. Characteristic x-rays produced by an atom or ion when a vacancy in the L shell is filled by an electron from another shell.

L-series. The set of characteristic x-ray wavelengths making up L-radiation for the various elements.

L shell. The second shell of electrons surrounding the nucleus of an atom, having electrons with the principal quantum number 2.

M

macroscopic stress. Residual stress in a material in a distance comparable to the gage length of strain-measurement devices (as opposed to stresses within very small, specific regions, such as individual grains). See also *microscopic stress*.

magnetic contrast. In electron microscopy, contrast that arises from the interaction of the electrons in the beam with the magnetic fields of individual magnetic domains in ferromagnetic materials.

magnetic resonance. A phenomenon in which the magnetic spin systems of certain atoms absorb electromagnetic energy at specific (resonant) natural frequencies of the system.

magnetometer. An instrument for measuring the magnitude and sometimes also the direction of a magnetic field, such as the Earth's magnetic field. See also *torque-coil magnetometer*.

magneton. A unit of magnetic moment used for atomic, molecular, or nuclear magnets. The Bohr magneton (μ_B), which has the value of the classical magnetic moment of the electron, can theoretically be calculated as:

$$\mu_B = \mu_0 = \frac{e\hbar}{2mc} = 9.2741 \times 10^{-2}\ \text{erg/G} = 9.2741 \times 10^{-24}\ J/T$$

where e and m are the electronic charge and mass, respectively; $\hbar$ is Planck's constant divided by 2π; and c is the velocity of light. See also *Planck's constant*.

magnetostriction. Changes in dimensions of a body resulting from application of a magnetic field.

magnification. A ratio of the size of an image to its corresponding object. This is usually determined by linear measurement.

mass absorption coefficient. The linear absorption coefficient divided by the density of the medium.

mass spectrometry. An analytical technique for identification of chemical structures, analysis of mixtures, and quantitative elemental analysis, based on application of the mass spectrometer.

mass spectrum. A record, graph, or table that shows the relative number of ions of various masses that are produced when a given substance is processed in a mass spectrometer.

matrix. The principal element or elements in a sample.

matrix isolation. A technique for maintaining molecules at low temperature for spectroscopic study; this method is particularly well suited for preserving reactive species in a solid, inert environment.

metallic glass. An alloy having an amorphous or glassy structure. See also *amorphous solid.*

metallography. The study of the structure of metals and alloys by various methods, especially by optical and electron microscopy.

micelle. A submicroscopic unit of structure built up from ions or polymeric molecules.

microanalysis. The analysis of samples smaller than 1 mg.

micrograph. A graphic reproduction of an object as formed by a microscope or equivalent optical instrument.

microscope. An instrument capable of producing a magnified image of a small object.

microscopic stress. Residual stress in a material within a distance comparable to the grain size. See also *macroscopic stress.*

microscopy. The science of the interpretive use and applications of microscopes.

microstructure. The structure of a properly prepared specimen as revealed by examination with the light optical microscope, the scanning electron microscope, and the transmission electron microscope.

microwave radiation. Electromagnetic radiation in the wavelength range of 0.3 mm to 1 m (3×10^6 to 1×10^{10} Å). See also *electromagnetic radiation.*

middle-infrared radiation. Infrared radiation in the wavelength range of 3 to 30 μm (3×10^4 to 3×10^5 Å). See also *infrared radiation* and *electromagnetic radiation.*

mobile phase. In chromatography, the gas or liquid that flows through the chromatographic column. A sample compound in the mobile phase moves through the column and is separated from compounds residing in the stationary phase. See also *stationary phase*.

modulus of elasticity (*E*). The measure of rigidity or stiffness of a metal; the ratio of stress, below the proportional limit, to the corresponding strain. In terms of the stress-strain diagram, the modulus of elasticity is the slope of the stress-strain curve in the range of linear proportionality of stress to strain. Also known as Young's modulus.

moiety. A portion of a molecule, generally complex, having a characteristic chemical property.

moiré pattern. A pattern developed from interference or light blocking when gratings, screens, or regularly spaced patterns are superimposed on one another.

molality. The number of gram-molecular weights of a compound dissolved in 1 L of solvent. See also *gram-molecular weight.* Compare with *molarity* and *normality.*

molarity. The number of gram-molecular weights of a compound dissolved in 1 L of solution. See also *gram-molecular weight.* Compare with *molality* and *normality.*

mole. An amount of substance of a system that contains as many elementary units (6.02×10^{23}) as there are atoms of carbon in 0.012 kg of the pure nuclide ^{12}C; the elementary unit must be specified and may be an atom, molecule, ion, electron, photon, or even a specified group of such units. See also *Avogadro's number.*

molecular fluorescence spectroscopy. An analytical technique that measures the fluorescence emission characteristic of a molecular, as opposed to an atomic, species. The emission results from electronic transitions between molecular states and can be used to detect and/or measure trace amounts of molecular species.

molecular spectrum. The spectrum of electromagnetic radiation emitted or absorbed by a collection of molecules as a function of frequency, wave number, or some related quantity.

molecular structure. The manner in which electrons and nuclei interact to form a molecule, as elucidated by quantum mechanics and the study of molecular spectra.

molecular weight. The sum of the atomic weights of all the atoms in a molecule. Atomic weights (and therefore molecular weights) are relative weights arbitrarily referred to an assigned atomic weight of exactly 12.0000 for the most abundant isotope of carbon, ^{12}C. See also *atomic weight.*

molecule. A molecule may be thought of either as a structure built of atoms bound together by chemical forces or as a structure in which two or more positively charged nuclei are maintained in some definite geometrical configuration by attractive forces from the surrounding cloud of electrons. Besides chemically stable molecules, short-lived molecular fragments termed free radicals can be observed under special circumstances. See also *chemical bonding, free radical*, and *molecular structure.*

monochromatic. Consisting of electromagnetic radiation having a single wavelength or an extremely small range of wavelengths, or particles having a single energy or an extremely small range of energies.

monochromator. A device for isolating monochromatic radiation from a beam of polychromatic radiation. See also *polychromator*.

monomer. A simple molecule capable of combining with a number of like or unlike molecules to form a polymer; a repeating structure unit within a polymer.

Monte Carlo techniques. Calculation of the trajectory of incident electrons within a given matrix and the pathway of the x-rays generated during interaction.

morphology. The shape characteristics of the structure of metals, alloys, ceramics, polymers, and composites, and the form and orientation of specific phases or constituents.

Mössbauer effect. The process in which γ-radiation is emitted or absorbed by nuclei in solid matter without imparting recoil energy to the nucleus and without Doppler broadening of the γ-ray energy.

Mössbauer spectroscopy. An analytical technique that measures recoilless absorption of γ-rays that have been emitted from a radioactive source as a function of the relative velocity between the absorber and the source.

Mössbauer spectrum. A plot of the relative absorption of γ-rays versus the relative velocity between an absorber and a source of γ-rays.

M shell. The third layer of electrons surrounding the nucleus of an atom, having electrons characterized by the principal quantum number 3.

multichannel analyzer (MCA). An instrument that splits an input signal into a number of channels with respect to a particular parameter of the input. See also *energy-dispersive spectroscopy.*

multiple scattering event. A collision process that may be described as a sequence of binary scattering events that may or may not be elastic.

multiplier phototube. See *photomultiplier tube.*

N

nanofabrication. The process of deliberate and controlled fabrication of devices or structures at the nanometer scale. It comprises a variety of techniques such as sputtering, lithography, beam chemistry, and so on.

near-infrared radiation. Infrared radiation in the wavelength range of 0.78 to 3 μm (7800 to 30,000 Å). See also *electromagnetic radiation* and *infrared radiation*.

nebulizer. A device for converting a sample solution into a gas-liquid aerosol for atomic absorption, emission, and fluorescence analysis. This may be combined with a burner to form a nebulizer burner. See also *atomization* and *atomizer*.

Neel point. See *Neel temperature*.

Neel temperature. The temperature below which spins in an antiferromagnetic material are ordered antiparallel so that there is zero net magnetic moment. Also known as Neel point.

neutral filter. A filter that attenuates the radiant power reaching the detector by the same factor at all wavelengths within a prescribed wavelength region.

neutron. An elementary particle that has approximately the same mass as the proton but no electric charge. Rest mass is 1.67495×10^{-27} kg. An unbound (extranuclear) neutron is unstable and β-decays with a half-life of 10.6 min.

neutron absorber. A material in which a significant number of neutrons entering combine with nuclei and are not re-emitted.

neutron absorption. A process in which the collision of a neutron with a nucleus results in the absorption of the neutron into the nucleus with the emission of one or more prompt γ-rays: in certain cases, emission of α-particles, protons, or other neutrons or fission of the nucleus results. Also known as neutron capture.

neutron activation analysis. Activation analysis in which the specimen is bombarded with neutrons; identification is made by measuring the resulting radioisotopes. See also *activation analysis*.

neutron capture. See *neutron absorption*.

neutron cross section. A measure of the probability that an interaction of a given kind will take place between a nucleus and an incident neutron; it is an area such that the number of interactions that occur in a sample exposed to a beam of neutrons is equal to the product of the cross section, the number of nuclei per unit volume in the sample, the thickness of the sample, and the number of neutrons in the beam that would enter the sample if their velocities were perpendicular to it. The usual unit is the barn (10^{-24} cm^2). See also *barn*.

neutron detector. Any device that detects passing neutrons, for example, by observing the charged particles or γ-rays released in nuclear reactions induced by the neutrons or by observing the recoil of charged particles caused by collisions with neutrons.

neutron diffraction. The phenomenon associated with the interference processes that occur when neutrons are scattered by the atoms within solids, liquids, and gases.

neutron flux. The number of neutrons passing through an area in a unit of time. Also called neutron fluence rate.

neutron spectrometry. See *neutron spectroscopy*.

neutron spectroscopy. Determination of the energy distribution of neutrons. Scintillation detectors, proportional counters, activation foils, and proton recoil are used.

neutron spectrum. The distribution by energy of neutrons impinging on a surface, which can be measured by neutron spectroscopy techniques or sometimes from knowledge of the neutron source. See also *dosimeter*.

Nicol prism. A prism made by cementing together with Canada balsam two pieces of a diagonally cut calcite crystal. In such a prism the ordinary ray is totally reflected at the calcite/cement interface, while the orthogonally polarized extraordinary ray is transmitted. The prism can thus be used to polarize light or analyze the polarization of light.

noise. Any undesired signal that tends to interfere with the normal reception or processing of a desired signal.

normality. A measure of the number of gram-equivalent weights of a compound per liter of solution. Compare with *molarity*.

normal-phase chromatography (NPC). This refers to liquid-solid chromatography or to bonded-phase chromatography with a polar stationary phase and a nonpolar mobile phase. See also *bonded-phase chromatography* and *liquid-solid chromatography*.

N shell. The fourth layer of electrons surrounding the nucleus of an atom, having electrons with the principal quantum number 4.

nuclear charge. See *atomic number*.

nuclear cross section (σ). The probability that a nuclear reaction will occur between a nucleus and a particle, expressed in units of area (usually barns). See also *barn*.

nuclear magnetic resonance (NMR). A phenomenon exhibited by a large number of atomic nuclei that is based on the existence of nuclear magnetic moments associated with quantized nuclear spins. These nuclear moments, when placed in a magnetic field, give rise to distinct nuclear Zeeman energy levels between which spectroscopic transitions can be induced by radio-frequency radiation. Plots of these transition frequencies, termed spectra, furnish important information about molecular structure and sample composition. See also *Zeeman effect*.

nuclear stopping power. The progressive energy reduction of a high-energy ion as it penetrates into the sample and interacts with the nuclei therein. Scattering, sputtering, vacancies, and recoil cascades are the consequences.

nuclear structure. The atomic nucleus at the center of the atom, containing more than 99.975% of the total mass of the atom. Its average density is approximately 3×10^{11} kg/cm^3; its diameter is approximately 10^{-12} cm and thus is much smaller than the diameter of the atom, which is approximately 10^{-8} cm. The nucleus is composed of protons and neutrons. The number of protons is denoted by Z, the number of neutrons by N. The total number of protons and neutrons in a nucleus is termed the mass number and is denoted by $A = N + Z$. See also *atomic structure*, *electron*, *neutron*, *proton*, *isobar*, *isotone*, and *isotope*.

nuclide. A species of atom distinguished by the constitution of its nucleus. Nuclear constitution is specified by the number of protons (Z) and the number of neutrons (N), or by atomic number and mass number.

numerical aperture (of a lens). The product of the lowest index of refraction of the object space multiplied and the sine of the half-angle from the object to the edges of the lens.

O

objective. The primary magnifying system of a microscope. A system, generally of lenses, less frequently of mirrors, forming a real, inverted, and magnified image of the object.

ocular. See *eyepiece*.

optical axis. The line formed by the coinciding principal axes of a series of optical elements comprising an optical system. It is the line passing through the centers of curvature of the optical surfaces.

optical emission spectroscopy. Pertaining to emission spectroscopy in the near-ultraviolet, visible, or near-infrared wavelength regions of the electromagnetic spectrum.

ordered structure. The crystal structure of a solid solution in which the atoms of different elements seek preferred lattice positions. Contrast with *disordered structure*.

outlier. In a set of data, a value so far removed from other values in the distribution that it is probably not a bona fide measurement. There are statistical methods for classifying a data point as an outlier and removing it from a data set.

P

paramagnetism. A property exhibited by substances that, when placed in a magnetic field, are magnetized parallel to the field to an extent proportional to the field (except at very low temperatures or in extremely large magnetic fields). Compare with *ferromagnetism*.

particle accelerator. A device that raises the velocities of charged atomic or subatomic particles to high values.

particle-induced x-ray emission. A method of trace elemental analysis in which a beam of ions (usually protons) is directed at a thin foil on which the sample to be analyzed has been deposited; the energy spectrum of the resulting x-rays is measured. See also *particle accelerator* and *proton*.

parts per billion. A measure of proportion by weight, equivalent to one unit weight of a material per billion (10^9) unit weights of compound. One part per billion is equivalent to 1 ng/g.

parts per million. A measure of proportion by weight, equivalent to one unit weight of a material per million (10^6) unit weights of compound. One part per million is equivalent to 1 mg/kg.

peak overlap. Formation of a single peak when two closely spaced x-ray peaks cannot be resolved; the energy (or wavelength) of the peak is some average of the characteristic energies (or wavelengths) of the original two peaks. See also *full width at half maximum*.

phase contrast. Contrast in high-resolution transmission electron microscope images arising from interference effects between the transmitted beam and one or more diffracted beams.

phosphorescence. A type of photoluminescence in which the time period between the absorption and re-emission of light is relatively long (of the order of 10^{-4} to 10 s or longer). See also *photoluminescence*. Contrast with *fluorescence*.

photoelectric effect. An x-ray absorption mechanism by which unstable states in the electron orbitals of atoms are created. Once the vacancies in the inner orbitals are formed, relaxation to the stable ground state may occur by the emission of x-rays characteristic of the excited element. Each of the transitions that may occur leads to the emission of sharp x-ray lines characteristic of the target element and the transition involved.

photoelectric electron-multiplier tube. See *photomultiplier tube*.

photoelectron. An electron emitted from an atom by the interaction with a photon.

photoluminescence. Re-emission of light absorbed by an atom or molecule. The light is emitted in random directions. There are two types of photoluminescence: fluorescence and phosphorescence. See also *fluorescence* and *phosphorescence*.

photometer. A device so designed that it measures the ratio of the radiant power of two electromagnetic beams.

photomultiplier tube. A device in which incident electromagnetic radiation creates electrons by the photoelectric effect. These electrons are accelerated by a series of electrodes called dynodes, with secondary emission adding electrons to the stream at each dynode. Also known as multiplier phototube, electron multiplier phototube, and photoelectric electron-multiplier tube.

photon. A particle representation of the electromagnetic field. The energy of the photon equals $h\nu$, where ν is the frequency of the light in hertz, and h is Planck's constant. See also *Planck's constant*.

physical adsorption. See *physisorption*.

physisorption. The binding of an adsorbate to the surface of a solid by forces whose energy levels approximate those of condensation. Contrast with *chemisorption*.

pi bonding. Covalent bonding in which the atomic orbitals overlap along a plane perpendicular to the sigma bond(s) joining the nuclei of two or more atoms. See also *sigma bonding*.

pi electron. An electron that participates in pi bonding. See also *pi bonding*.

pipet. A tube, usually made of glass or plastic, used almost exclusively to deliver accurately known volumes of liquids or solutions during titration or volumetric analysis.

Planck's constant (h). A fundamental physical constant, the elementary quantum of action; the ratio of the energy of a photon to its frequency, it is equal to $6.62620 \pm 0.00005 \times 10^{-34}$ J · s.

plane grating. An optical component used to disperse light into its component wavelengths by diffraction off a series of finely spaced, equidistant ridges. A plane grating has a flat substrate. See also *concave grating*, *diffraction grating*, *reflection grating*, and *transmission grating*.

plane strain. The condition in which one of the principal strains is zero.

plane stress. The condition in which one of the principal stresses is zero.

plasma. A gas of sufficient energy so that a large fraction of the species present is ionized and thus conducts electricity. Plasmas may be generated by the passage of a current between electrodes, by induction, or by a combination of these methods.

plasma-jet excitation. The use of a high-temperature plasma jet to excite an element in a sample, for example, for atomic emission spectroscopy. Also known as dc plasma excitation.

plasmon. A quantum of a collective longitudinal wave in the electron gas of a solid.

plasmon lines. Lines that appear in some x-ray photoelectron spectroscopy spectra due to the photoelectron and the conductive oscillating electrons.

point count. The total number of test points in a test grid that fall within the microstructural feature of interest, or the feature boundary (these are counted as one-half a point count).

point fraction. The ratio, often expressed as a percentage, of the number of grid points of the phase or constituent of interest on the two-dimensional image of an opaque specimen divided by the number of grid points; this is averaged over n measurement fields to obtain an unbiased estimate of the volume fraction of the phase or constituent of interest.

Poisson's ratio (ν). The absolute value of the ratio of transverse (lateral) strain to the corresponding axial strain resulting from uniformly distributed axial stress below the proportional limit of the material.

polarizing element. A general term for a device for producing or analyzing plane-polarized light. It may be a Nicol prism, some other form of calcite prism, a reflecting surface, or a polarizing filter.

polarography. *Voltammetry* performed using the dropping mercury electrode (DME). An electroanalytical technique in which the current between a DME and a counterelectrode (both of which are immersed in electrolyte) is measured as a function of the potential difference between the DME and a reference electrode.

pole figure. A stereoscopic projection of a polycrystalline aggregate showing the distribution of poles, or plane normals, of a specific crystalline plane, using specimen axes as reference axes. Pole figures are used to characterize preferred orientation in polycrystalline materials.

polychromator. A spectrometer that has many (typically 20 to 50) detectors for simultaneously measuring light from many spectral lines. Polychromators are commonly used in atomic emission spectroscopy. See also *monochromator*.

polycrystalline. The characteristic of an aggregate composed of more than one crystal, and usually a vast number of crystals.

polymorph. In crystallography, one crystal form of a polymorphic material. See also *polymorphism*.

polymorphism. In crystallography, the property of a chemical substance whereby it crystallizes into two or more forms having different crystallographic structures, such as diamond versus graphite or face-centered cubic iron versus body-centered cubic iron.

population. In statistics, a generic term denoting any finite or infinite collection of individual samples or data points in the broadest concept; an aggregate determined by some property that distinguishes samples that do and do not belong.

positive eyepiece. An eyepiece in which the real image of the object is formed below the lower lens elements of the eyepiece.

potentiometric membrane electrodes. Electrochemical sensing devices that can be used to quantify cationic and anionic substances and gaseous species in aqueous solutions. These devices are also used for analytical titrations. See also *titration*.

potentiostatic coulometry. See *controlled-potential coulometry*.

Pourbaix (potential-pH) diagram. A plot of the redox potential of a corroding system versus the pH of the system, compiled using thermodynamic data and the Nernst equation. The diagram shows regions within which the metal itself or some of its compounds are stable.

prearc (or prespark) period. In emission spectroscopy, the time interval after the initiation of an arc (or spark) discharge during which the emitted radiation is not recorded for analytical purposes.

precision. The reproducibility of measurements within a set, that is, the scatter or dispersion of a set of data points about its central axis. Generally expressed as standard deviation or relative standard deviation.

premix burner. A burner used in flame emission and atomic absorption spectroscopy in which the fuel gas is mixed with the oxidizing gas before reaching the combustion zone.

primary beam. The beam of particles such as electrons or ions that are directed toward a sample. The term distinguishes this from the secondary particles that may be generated as the primary beam interacts with the sample.

primary x-ray. The emergent beam from the x-ray source.

prism. A transparent optical element whose entrance and exit apertures are polished plane faces. Using refraction and/or internal reflection, prisms are used to change the direction of propagation of monochromatic light and to disperse polychromatic light into its component wavelengths.

probe ion. An ionic species intentionally produced by an ion source and directed onto the specimen surface at a known incident angle and a known energy.

proportional limit. The greatest stress a material is capable of developing without a deviation from straight-line proportionality between stress and strain. See also *Hooke's law*.

proton. A particle that is the positively charged constituent of ordinary matter. Its charge is identical in magnitude but of opposite sign to that of the electron. Its mass equals 1.62726 $\times 10^{-24}$ g. See also *electron* and *neutron*.

Q

qualitative analysis. An analysis in which some or all of the components of a sample are identified. Contrast with *quantitative analysis*.

quantitative analysis. A measurement in which the amount of one or more components of a sample is determined. Contrast with *qualitative analysis*.

quantitative metallography. See *image analysis*.

quantum mechanics. The modern theory of matter, of electromagnetic radiation, and of the interaction between matter and radiation; also, the mechanics of phenomena to which this theory may be applied. Quantum mechanics, also termed wave mechanics, generalizes and supersedes the older classical mechanics and Maxwell's electromagnetic theory.

quantum number. One of the quantities, usually discrete with integer or half-integer values, needed to characterize a quantum state of a physical system.

quantum theory. See *quantum mechanics*.

R

radial distribution function analysis. Diffraction method that gives the distribution of interatomic distances present in a sample along with information concerning the frequency with which the particular distances occur.

radiant energy. Energy transmitted as electromagnetic radiation.

radiant intensity. The radiant power or flux emitted per unit solid angle expressed, for example, in watts per steradian.

radiant power (flux). The energy emitted by a source or transported in a beam per unit time, expressed, for example, in ergs per second or watts.

radical. See *free radical*.

radioactivity. (1) The property of some atomic nuclei to spontaneously decay (lose energy). Usual mechanisms are emission of α, β, or other particles and splitting (fissioning). Gamma rays are frequently, but not always, given off in the process. (2) A particular component from a radioactive source, such as β radioactivity. See also *isotope* and *radioisotope*.

radioanalysis. An analytical technique, such as neutron activation analysis, that makes use of the radioactivity of an element or isotope. See also *isotope* and *radioisotope*.

radio frequency. A frequency at which coherent electromagnetic radiation of energy is useful for communication purposes; roughly the range from 10 kHz to 100 GHz.

radio-frequency spectrometer. An instrument that measures the intensity of radiation emitted or absorbed by atoms or molecules as a function of frequency at frequencies from 10 kHz to 100 GHz.

radio-frequency spectroscopy. The branch of spectroscopy concerned with the measurement of the intervals between atomic or molecular energy levels that are separated by frequencies from approximately 10^5 to 10^9 Hz as compared to the frequencies that separate optical energy levels of approximately 6×10^{14} Hz.

radioisotope. An isotope that is radioactive. Also known as a radionuclide. See also *isotope* and *radioactivity*.

radionuclide. See *radioisotope*.

Raman line (band). A line (band) that is part of a Raman spectrum and corresponds to a characteristic vibrational frequency of the molecule being probed.

Raman shift. The displacement in wave number of a Raman line (band) from the wave number of the incident monochromatic beam. Raman shifts are usually expressed in units of cm^{-1}. They correspond to differences between molecular vibrational, rotational, or electronic energy levels.

Raman spectroscopy. Analysis of the intensity of Raman scattering of monochromatic light as a function of frequency of the scattered light.

Raman spectrum. The spectrum of the modified frequencies resulting from inelastic scattering when matter is irradiated by a monochromatic beam of radiant energy. Raman spectra normally consist of lines (bands) at frequencies higher and lower than that of the incident monochromatic beam.

raster. A rectangular pattern of parallel scanning lines used to direct an energetic beam for imaging or mapping.

Rayleigh scattering. Scattering of electromagnetic radiation by independent particles that are smaller than the wavelength of radiation. Contrast with *Compton scattering*.

reagent. A substance, chemical, or solution used in the laboratory to detect, measure, or react with other substances, chemicals, or solutions. See also *reagent chemicals*.

reagent chemicals. High-purity chemicals used for analytical reactions, for testing of new reactions where the effects of impurities are unknown, and for chemical work where impurities must either be absent or at a known concentration.

reciprocal lattice. A lattice of points, each representing a set of planes in the crystal lattice, such that a vector from the origin of the reciprocal lattice to any point is normal to the crystal planes represented by that point and has a length that is the reciprocal of the plane spacing.

reciprocal linear dispersion. The derivative $d\lambda/dx$, where λ is the wavelength and x is the distance along the spectrum. The reciprocal linear dispersion usually is expressed in Å/mm.

red shift. A systematic displacement toward the longer wavelengths of the spectrum.

reduction. In sampling, the process of preparing one or more subsamples from a sample. Reduction may also refer to decreasing the valence of an ion or atom by the addition of electrons.

reference electrode. A nonpolarizable electrode with a known and highly reproducible potential used for potentiometric and voltammetric analyses. See also *calomel electrode*.

reference material. A material of definite composition that closely resembles in chemical and physical nature the material with which an analyst expects to deal; used for calibration or standardization.

reflectance. The ratio of the radiant power or flux reflected by a medium to the radiant power or flux incident on it; generally expressed as a percentage.

reflection grating. A grating that employs reflection off a series of fine, equidistant ridges, rather than transmission through a pattern of slots, to diffract light into its component wavelengths. The gratings used in optical instrumentation are almost exclusively reflection gratings. See also *concave grating*, *diffraction grating*, *plane grating*, and *transmission grating*.

reflux. Heating a substance at the boiling temperature and returning the condensed vapors to the vessel to be reheated.

refractive index. The ratio of the phase velocity of monochromatic light in a vacuum to that in a specified medium. Refractive index is generally a function of wavelength and temperature. Also known as index of refraction.

relative sensitivity factor (RSF). When quantifying peaks in x-ray photoelectron spectroscopy spectra, relative sensitivity factors are used to scale the measured peak areas so that variations in the peak areas are representative of the amount of material in the sample surface.

relative standard deviation (RSD). The standard deviation expressed as a percentage of the mean value:

$$\mathrm{RSD} = 100\left(\frac{S}{X}\right)\frac{d^2}{n-1}$$

where S is the standard deviation, d is the difference between individual results and the average, n is the number of individual results, and X is the average of individual results. Also known as coefficient of variation.

relative transmittance. The ratio of the transmittance of the object in question to that of a reference object. For a spectral line on a photographic emulsion, it is the ratio of the transmittance of the photographic image of the spectral line to the transmittance of a clear portion of the photographic emulsion. Relative transmittance may be total, specular, or diffuse. See also *transmittance*.

residual stress. Stresses that remain within a body as the result of plastic deformation. These stresses can be measured by using x-ray diffraction residual-stress techniques.

resolution. The fineness of detail revealed by an optical device. Resolution is usually specified as the minimum distance by which two lines in the object must be separated before they can be revealed as separate lines in the image.

retention time (t_R). In chromatography, the amount of time a sample compound spends in the chromatographic column.

reversed-phase chromatography (RPC). Bonded-phase chromatography with a nonpolar stationary phase and a polar mobile phase. See also *bonded-phase chromatography*.

rocking curve. A method for determining the degree of imperfection in a crystal by using monochromatic, collimated x-rays reflecting off a "perfect" crystal to probe a second test crystal. A rocking curve is obtained by monitoring the x-ray intensity diffracted by the test crystal as it is slowly rocked, or rotated, through the Bragg angle for the reflecting planes.

Rutherford backscattering spectrometry (or spectroscopy). The analytical technique in which high-energy (MeV) helium ions are scattered from a sample to assess its composition.

Rutherford scattering. A general term for the classical elastic scattering of energetic ions by the nuclei of a target material.

S

sample. A portion of a material intended to be representative of the whole. Also known as specimen.

sampling. The obtaining of a portion of a material that is adequate for making the required tests or analyses, and that is representative of that portion of the material from which it is taken.

satellite peaks. A group of peaks appearing in an x-ray photoelectron spectroscopy spectrum at lower binding energies of a dominant peak originating from unfiltered radiation from an x-ray target.

saturated bond. Organic materials containing saturated bonds have a single bond between the carbon atoms in their chemical formula. Unsaturated bonds are double or triple bonds between carbon atoms.

saturated calomel electrode. A reference electrode composed of mercury, mercurous chloride (calomel), and a saturated aqueous chloride solution.

scanning Auger microscopy (SAM). An analytical technique that measures the lateral distribution of elements on the surface of a material by recording the intensity of their Auger electrons versus the position of the electron beam.

scanning electron microscopy (SEM). An analytical technique in which an image is formed on a cathode-ray tube or digital screen whose raster is synchronized with the raster of a point beam of electrons scanned over an area of the sample surface. The brightness of the image at any point is proportional to the scattering by or secondary emission from the point on the sample being struck by the electron beam. See also *raster*.

scanning ion gun. An ion gun capable of emitting a fine ion beam with the ability to scan over a sample surface, attached to an analytical instrument.

scanning transmission electron microscopy (STEM). An analytical technique in which an image is formed on a cathode-ray tube or digital screen whose raster is synchronized with the raster of a point beam of electrons scanned over an area of the sample. The brightness of the image at any point is proportional to the number of electrons that are transmitted through the sample at the point where it is struck by the beam. See also *raster*.

scattering (of radiant energy). The deviations in the direction of propagation of radiant energy.

Schöniger combustion. A method of decomposition of organic materials by combusting them in a sealed flask that contains a solution suitable for absorbing the combustion products. The flask is swept with oxygen before ignition.

secondary electron. A low-energy electron (0 to 50 eV) emitted from a surface that is struck by particles with higher energies. The type 1 secondary electrons originate in the immediate region in which the particle struck the surface. The type 2 secondary electrons are the result of a two-step process involving a backscattered particle or a recoiling atom.

secondary ion. An ion other than the probe ion that originates from and leaves the specimen surface as a result of bombardment with a beam of primary or probe ions.

secondary ion mass spectroscopy (SIMS). An analytical technique that measures the masses of ions emitted from the surface of a material when exposed to a beam of incident ions. The incident ions are usually monoenergetic and are all of the same species, for example, 5 keV Ne^+ ions. See also *secondary ion*.

secondary x-rays. The x-rays emitted by a specimen following excitation by a primary x-ray beam or an electron beam.

segment. In sampling, a specifically demarked portion of a lot, either actual or hypothetical.

selected-area channeling pattern (SACP). A pattern of channeling lines obtained by scanning an electron beam about a pivot point located at the sample surface. See also *channeling pattern* and *channeling contrast*.

selected-area diffraction (SAD). Electron diffraction from a portion of a sample selected by inserting an aperture into the magnification portion of the lens system of a transmission electron microscope. Areas as small as 0.5 μm in diameter can be examined in this way.

selected-area diffraction pattern (SADP). An electron diffraction pattern obtained from a restricted area of a sample. The sharp spots in the pattern correspond closely to points in the reciprocal lattice of the material being studied. Usually such patterns are taken from a single crystal or a small number of crystals.

selectivity. The ability of a method or instrument to respond to a desired substance or constituent and not to others.

self-absorption. In optical emission spectroscopy, reabsorption of a photon by the same species that emitted it. For example, light emitted by sodium atoms in the center of a flame may be reabsorbed by different sodium atoms near the outer portions of the flame.

self-electrode. An electrode fabricated from the sample material and analyzed by emission spectroscopy.

self-reversal. In optical emission spectroscopy, the extreme case of self-absorption. See also *self-absorption*.

sensitivity. The capability of a method or instrument to discriminate between samples having different concentrations or amounts of the analyte.

shake-up peaks. In a photoemission process there is a finite probability that an ion after photoionization will be left in an excited state. This reduces the kinetic energy of the emitted electron that contributes to the shake-up peaks in the x-ray photoelectron spectroscopy spectrum.

siemens (S). A unit of electrical conductivity. One siemens of conductance per cubic meter with a potential of 1 V allows the passage of 1 A/m^2. See also *conductance*.

sigma bonding. Covalent bonding between atoms in which *s* orbitals or hybrid orbitals between *s* and *p* electrons overlap in cylindrical symmetry along the axis joining the nuclei of the atoms. See also *pi bonding*.

signal-to-noise ratio. The ratio of the amplitude of a desired signal at any time to the amplitude of noise signals at the same time. See also *noise*.

single crystal. A material that is completely composed of a regular array of atoms.

size-exclusion chromatography (SEC). Liquid chromatography method that separates molecules on the basis of their physical size. This technique is most often used in the analysis of polymers. Also termed gel-permeation chromatography.

solute. The substance dissolved in a solvent.

solution. In chemistry, a homogeneous dispersion of two or more types of molecular or ionic species. Solutions may be composed of any combination of liquids, solids, or gases, but they always consist of a single phase.

solvent. The part of a solution present in the largest amount, or the compound that is normally liquid in the pure state (as for solutions of solids or gases in a liquid).

spark. A series of electrical discharges, each of which is oscillatory and has a comparatively high maximum instantaneous current resulting from the breakdown of the analytical gap or the auxiliary gap, or both, by electrical energy stored at high voltage in capacitors. Each discharge is self-initiated and is extinguished when the voltage across the gap, or gaps, is no longer sufficient to maintain it.

spark source mass spectrometry. An analytical technique in which a high-voltage spark in a vacuum is used to produce positive ions of a conductive sample material. The ions are injected into a mass spectrometer, and the resulting spectrum is recorded on a photographic plate or measured using an electronic detector. The position of a particular mass spectral signal determines the element and isotope, and the intensity of the signal is proportional to the concentration.

specific gravity (of gases). The ratio of the density of a gas to the density of dry air at the same temperature and pressure.

specific gravity (of solids and liquids). The ratio of the mass of a unit volume of a material to the mass of the same volume of gas-free distilled water at the same temperature.

specimen. See *sample*.

spectral background. In spectroscopy, a signal obtained when no analyte is being introduced into the instrument, or a signal from a species other than that of the analyte.

spectral distribution curve. The curve showing the absolute or relative radiant power emitted or absorbed by a substance as a function of wavelength, frequency, or any other directly related variable.

spectral line. A wavelength of light with a narrow energy distribution or an image of a slit formed in the focal plane of a spectrometer or photographic plate that has a narrow energy distribution approximately equal to that formed by monochromatic radiation.

spectral order. The number of the intensity maxima of a given line from the directly transmitted or specularly reflected light from a diffraction grating.

spectrochemical (spectrographic, spectrometric, spectroscopic) analysis. The determination of the chemical elements or compounds in a sample qualitatively, semiquantitatively, or quantitatively by measurements of the wavelengths and intensities of spectral lines produced by suitable excitation procedures and dispersed by a suitable optical device.

spectrogram. A photographic or graphic record of a spectrum.

spectrograph. An optical instrument with an entrance slit and dispersing device that uses photography to record a spectral range. The radiant power passing through the optical system is integrated over time, and the quantity recorded is a function of radiant energy.

spectrometer. An instrument with an entrance slit, a dispersing device, and one or more exit slits, with which measurements are made at selected wavelengths within the spectral range, or by scanning over the range. The quantity detected is a function of the radiant power.

spectrophotometer. A spectrometer that measures the ratio (or a function of the ratio) of the intensity of two different wavelengths of light. These two beams may be separated in terms of time or space, or both.

spectroscope. An instrument that disperses radiation into a spectrum for visual observation.

spectroscopy. The branch of physical science treating the theory, measurement, and interpretation of spectra.

spectrum. The ordered arrangement of electromagnetic radiation according to wavelength, wave number, or frequency.

specular transmittance. The transmittance value obtained when the measured radiant energy in emission spectroscopy has passed from the source to the receiver without appreciable scattering.

spherical aberration. A lens defect in which image-forming rays passing through the outer zones of the lens focus at a distance from the principal plane different from that of the rays passing through the center of the lens.

spin glass. One of a wide variety of materials that contain interacting atomic magnetic moments and also possess some form of disorder, in which the temperature variation of the magnetic susceptibility undergoes an abrupt change in slope at a temperature generally referred to as the freezing temperature.

spin wave. A sinusoidal variation, propagating through a crystal lattice, of that angular momentum associated with magnetism (mostly spin angular momentum of the electrons). See also *spin glass*.

spot size. A parameter of an electron microscope that relates the beam size at focus as well as the beam current.

sputtering. A process in which atoms from a sample are ejected after the sample is subjected to the collision from an incident high-energy ion. This process is often the desired outcome for analytical techniques or for sample modification (for example, milling).

standard addition. A method in which small increments of a substance under measurement are added to a sample under test to establish a response function or, by extrapolation, to determine the amount of a constituent originally present in the sample.

standard electrode potential. The reversible or equilibrium potential of an electrode in an environment where reactants and products are at unit activity.

standardization. In analytical chemistry, the assignment of a compositional value to one standard on the basis of another standard.

standard reference material. A reference material, the composition or properties of which are certified by a recognized standardizing agency or group.

Stark effect. A shift in the energy of spectral lines due to an electrical field that is either externally applied or is an internal field caused by the presence of neighboring ions or atoms in a gas, solid, or liquid.

stationary phase. In chromatography, a particulate material packed into the column or a coating on the inner walls of the column. A sample compound in the stationary phase is separated from compounds moving through the column as a result of being in the mobile phase. See also *mobile phase.*

stereology. The mathematical procedures used to develop three-dimensional parameters of a microstructure observed on a two-dimensional plane of polish; extrapolation of two-dimensional data to three-dimensional space.

Stokes Raman line. A Raman line that has a frequency lower than that of the incident monochromatic beam.

stopping and range of ions in matter (SRIM). A software package that provides Monte Carlo simulation of ions interacting with matter and basic statistical outcomes of the collision cascade.

stopping power. A characteristic of an electron or ion beam that describes the rate at which energy is lost from an ion as it penetrates into a sample. Generally measured in eV/nm, the stopping power is composed of both nuclear stopping and electronic stopping components.

strata. In sampling, segments of a lot that may vary with respect to the property under study.

stress-strain diagram. A graph in which corresponding values of stress and strain are plotted against each other. Values of stress are usually plotted vertically (ordinate, or *y*-axis) and values of strain horizontally (abscissa, or *x*-axis). Also known as stress-strain curve.

structure factor. A mathematically formulated term that relates the positions and identities of atoms in a crystalline material to the intensities of x-ray or electron beams diffracted from particular crystallographic planes.

subsample. A portion taken from a sample. A laboratory sample may be a subsample of a gross sample; similarly, a test portion may be a subsample of a laboratory sample.

sum peak. An artifact encountered during pulse pileup where two x-rays simultaneously entering the detector are counted as one x-ray, the energy of which is equal to the sum of both x-rays.

superlattice. See *ordered structure.*

supporting electrode. An electrode, other than a self-electrode, on which the sample is supported during spectrochemical analysis.

synchrotron. A device for accelerating charged particles by directing them along a roughly circular path in a magnetic guide field. As the particles pass through accelerating cavities placed along their orbit, their kinetic energy is increased repetitively, multiplying their initial energy by factors of hundreds or thousands. See also *synchrotron radiation.*

synchrotron radiation. Electromagnetic radiation emitted by charged particles in circular motion at relativistic energies.

T

target. That part of an x-ray tube which the electrons strike and from which x-rays are emitted.

texture. A preferential alignment of the crystalline lattice of the various grains in a polycrystalline aggregate.

thermal noise. The electrical noise produced in a resistor by thermally generated currents. These currents average zero but produce electrical power having a nonzero average, which can affect instrument response. Also known as Johnson noise.

thermogravimetric analysis. Also known as thermal gravimetric analysis, it is a method of thermal analysis in which the mass of a sample is measured over time as the temperature changes.

thermomechanical analysis (TMA). A thermal analysis technique in which the length or volume of a specimen is precisely measured versus temperature and time as the specimen is subjected to controlled heating and cooling. See also *dilatometry.*

threshold setting (in image analysis). Isolation of a range of gray levels exhibited by a phase or constituent in the field of view selected to permit measurements of the field or constituent of interest.

tilt. In electron microscopy, the angle of the specimen relative to the axis of the electron beam; at zero tilt the specimen is perpendicular to the beam axis.

titration. A method of determining the composition of a sample by adding known volumes of a solution of known concentration until a given reaction (color change, precipitation, or conductivity change) is produced. See also *volumetric analysis.*

torque-coil magnetometer. A magnetometer that depends for its operation on the torque developed by a known current in a coil that can turn in the field to be measured. See also *magnetometer.*

torr. A unit for measuring pressure levels, often for vacuums. One atmosphere pressure is equal to 760 torr.

total transmittance. The ratio of the radiant energy leaving one side of a region between two parallel planes to the radiant energy entering from the opposite side.

transmission electron microscopy (TEM). An analytical technique in which an image is formed on a cathode-ray tube whose raster is synchronized with the raster of an electron beam over an area of the sample surface. Image contrast is formed by the scattering of electrons out of the beam.

transmission grating. A transparent diffraction grating through which light is transmitted. See also *concave grating, diffraction grating, plane grating,* and *reflection grating.*

transmittance. The ratio of the light intensity transmitted by a material to the light intensity incident upon it. In emission spectrochemical analysis, the transmittance of a developed photographic emulsion, including its film or glass supporting base, is measured by a microphotometer. In absorption spectroscopy, the material is the sample. See also *diffuse transmittance, relative transmittance, specular transmittance,* and *total transmittance.*

triggered capacitor discharge. A high-voltage electrical discharge used in emission spectroscopy for vaporization and excitation of a sample material. The energy for the discharge is obtained from capacitors that are charged from an ac or dc electrical supply. Each discharge may be either oscillatory, critically damped, or overdamped. It is initiated by separate means and is extinguished when the voltage across the analytical gap falls to a value that no longer is sufficient to maintain it.

triton. The nucleus of tritium (^3H), the triton is the only known radioactive nuclide belonging to hydrogen and β-decays to ^3He with a half-life of 12.4 years.

turbo pump. Also called turbo-molecular pump, a vacuum pump used to generate high vacuum. It works on the principle that the gas molecules can be given momentum in a desired direction by contact with moving solid surfaces of rapidly spinning pump blades. The gas molecules are directed toward the pump exhaust.

U

ultrahigh vacuum (UHV). Vacuum regime in which the pressure is measured in the 10^{-9} torr range or below.

ultraviolet. Pertaining to the region of the electromagnetic spectrum from approximately 10 to 380 nm. The term ultraviolet without further qualification usually refers to the region from 200 to 380 nm.

ultraviolet radiation. Electromagnetic radiation in the wavelength range of 10 to 380 nm. See also *electromagnetic radiation.*

ultraviolet/visible (UV/VIS) absorption spectroscopy. An analytical technique that measures the wavelength-dependent attenuation of ultraviolet, visible, and near-infrared light by an atomic or molecular species; used in the detection, identification, and quantification of numerous atomic and molecular species.

uncertainty. The range of values within which the true value is estimated to lie. It is a best estimate of possible inaccuracy due to both random and systematic error.

unified atomic mass unit (*u*). An arbitrarily defined unit expressing the masses of individual atoms. One atomic mass unit is defined as exactly $\frac{1}{12}$ of the mass of an atom of the nuclide ^{12}C, the predominant isotope of carbon. This replaced the atomic mass unit (amu). Also known as a Dalton (Da). See also *atomic weight.*

unit cell. A parallelepiped element of crystal structure, containing a certain number of atoms, the repetition of which through space will build up the complete crystal.

universal gas constant. See *gas constant.*

uranyl. The chemical name designating the UO_2^{2+} group and compounds containing this group.

V

vacancy. A location in a sample, usually a crystalline sample, that was formerly occupied by an atom. Focused ion beam instruments are known to create many vacancies in a sample for each incident ion.

valence band spectrum. Spectrum obtained in an x-ray photoelectron spectroscopy analysis in the range of zero binding energy.

velocity filter. A combination of electrostatic and magnetostatic fields that are orthogonal to each other and also to the incident charged particle beam. Under nominal operation the velocity filter allows only the particles with the ideal velocity to pass undeflected. See also *Wien filter*.

viscoelasticity. The mechanical deformation behavior of solids that are amorphous, such as glass, or semicrystalline, such as most plastics, that is characterized by both elastic (Hookean) and viscous (Newtonian) models.

visible. Pertaining to radiant energy in the electromagnetic spectral range visible to the normal human eye (~380 to 780 nm).

visible radiation. Electromagnetic radiation in the spectral range visible to the human eye (~380 to 780 nm).

voltage contrast. In scanning electron microscopy, additional contrast in an image arising from increased emission of secondary electrons from negatively biased regions of a sample. This type of contrast is often used to advantage in the examination of microelectronic devices.

voltammetry. An electrochemical technique in which the current between working (indicator) electrodes and counterelectrodes immersed in an electrolyte is measured as a function of the potential difference between the indicator electrode and a reference electrode, for example, polarography.

volumetric analysis. Quantitative analysis of solutions of known volume but unknown strength by adding reagents of known concentration until a reaction endpoint (color change or precipitation) is reached; the most common technique is by titration.

W

wavelength (λ). The distance, measured along the line of propagation of a wave, between two points that are in phase on adjacent waves. The customary units are angstroms, micrometers, and nanometers.

wavelength-dispersive spectrometry (WDS). The quantitative analyses (down to trace element levels) using the characteristic x-rays generated by individual elements. Wavelength-dispersive spectrometry can also be used to create element x-ray compositional maps over a broader area by means of rastering the beam. Together, these capabilities provide fundamental quantitative compositional information for a wide variety of solid materials. Compare with *energy-dispersive spectrometry*.

wavelength-dispersive spectroscopy (WDS). A method of x-ray analysis that employs a crystal spectrometer to discriminate characteristic x-ray wavelengths. Compare with *energy-dispersive spectroscopy*.

wave number. The number of waves per unit length. Commonly used in infrared and Raman spectroscopy, the wave number is expressed as the reciprocal of the wavelength. The usual unit of wave number is the reciprocal centimeter (cm^{-1}).

white radiation. See *bremsstrahlung radiation* and *continuum radiation*.

Wien filter. A combination of electrostatic and magnetostatic fields that are orthogonal to each other and also to the incident charged particle beam. Under nominal operation the Wien filter allows only the particles with the ideal velocity to pass undeflected. See also *velocity filter*.

working distance. The distance between the focal plane position and the physical front surface of the corresponding lens.

X

xenon focused ion beam (XeFIB). An instrument that uses a plasma ion source for the generation of a focused ion beam. Typical applications include the removal of materials by sputtering.

x-ray diffraction (XRD). An analytical technique in which measurements are made of the angles at which x-rays are preferentially scattered from a sample (as well as of the intensities scattered at various angles) in order to deduce information on the crystalline nature of the sample—its crystal structure, orientations, and so on.

x-ray diffraction residual-stress techniques. Diffraction method in which the strain in the crystal lattice is measured, and the residual stress producing the strain is calculated assuming a linear elastic distortion of the crystal lattice. See also *macroscopic stress*, *microscopic stress*, and *residual stress*.

x-ray emission spectroscopy. Pertaining to emission spectroscopy in the x-ray wavelength region of the electromagnetic spectrum.

x-ray fluorescence. Emission by a substance of its characteristic x-ray line spectrum on exposure to x-rays.

x-ray fluorescence spectrometry. An emission spectroscopic technique that has found wide application in elemental identification and determination. The technique depends on the emission of characteristic x-radiation, usually in the 1 to 60 keV energy range, following excitation of atomic electron energy levels by an external energy source, such as an electron beam, a charged particle beam, or an x-ray beam.

x-ray map. An intensity map (usually corresponding to an image) in which the intensity in any area is proportional to the concentration of a specific element in that area.

x-ray photoelectron spectroscopy (XPS). An analytical technique that measures the energy spectra of electrons emitted from the surface of a material when exposed to monochromatic x-rays.

x-ray spectrograph. A photographic instrument for x-ray emission analysis. If the instrument for x-ray emission analysis does not employ photography, it is better described as an x-ray spectrometer.

x-ray spectrometry. Measurement of wavelengths of x-rays by observing their diffraction by crystals of known lattice spacing.

Y

Young's modulus. See *modulus of elasticity*.

Z

ZAF corrections. A quantitative x-ray program that corrects for atomic number (*Z*), absorption (*A*), and fluorescence (F) effects in a matrix.

Zeeman effect. A splitting of a degenerate electron energy level into states of slightly different energies in the presence of an external magnetic field. This effect is useful for background correction in atomic absorption spectrometers.

SELECTED REFERENCES

- *Compilation of ASTM Standard Definitions,* 5th ed., American Society for Testing and Materials, Philadelphia, PA, 1982
- *Concise Encyclopedia of Science and Technology,* McGraw-Hill, 1984
- *Dictionary of Scientific and Technical Terms,* 2nd ed., McGraw-Hill, 1978
- "Standard Definitions of Terms and Symbols Relating to Emission Spectroscopy," E 135, *Annual Book of ASTM Standards,* Vol 03.06, American Society for Testing and Materials, Philadelphia, PA, 1984, p 74–80
- "Standard Definitions of Terms Relating to Metallography," E 7, *Annual Book of ASTM Standards,* Vol 03.03, American Society for Testing and Materials, Philadelphia, PA, 1984, p 12–48
- "Standard Definitions of Terms Relating to Microscopy," E 175, *Annual Book of ASTM*

Standards, Vol 14.01, American Society for Testing and Materials, Philadelphia, PA, 1984, p 215–219
- "Standard Definitions of Terms Relating to Surface Analysis," E 673, *Annual Book of ASTM Standards,* Vol 03.06, American Society for Testing and Materials, Philadelphia, PA, 1984, p 305–309
- "Standard Definitions of Terms, Symbols, Conventions and References Relating to High-Resolution Nuclear Magnetic Resonance (NMR) Spectroscopy," E 386, *Annual Book of ASTM Standards,* Vol 14.01, American Society for Testing and Materials, Philadelphia, PA, 1984, p 453–463

Index

Z